Developmental Mathematics

Fifth Edition

Marvin L. Bittinger

Indiana University—Purdue University
at Indianapolis

Judith A. Beecher

Indiana University—Purdue University
at Indianapolis

ADDISON-WESLEY

An imprint of Addison Wesley Longman, Inc.

Reading, Massachusetts • Menlo Park, California • New York • Harlow, England
Don Mills, Ontario • Sydney • Mexico City • Madrid • Amsterdam

Publisher	Jason A. Jordan
Assistant Editor	Suzanne Alley
Managing Editor	Ron Hampton
Production Supervisor	Kathleen A. Manley
Design Direction	Susan Raymond
Text Designer	Rebecca Lloyd Lemna
Editorial and Production Services	Martha Morong/Quadrata, Inc.
Illustrators	Scientific Illustrators, Gayle Hayes, and Maria Sas
Compositor	The Beacon Group
Cover Designer	Jeannet Leendertse
Cover Photographs	© Photodisc, 1998 (background) Bruce Anderson
Prepress Buyer	Caroline Fell
Manufacturing Buyer	Evelyn Beaton

PHOTO CREDITS

1, Pascal Rondeau/Tony Stone Images **4,** Copyright 1997, USA TODAY. Reprinted with permission **51,** Pascal Rondeau/Tony Stone Images **83,** Greg Probst/Tony Stone Images **118,** AP/Wide World Photos **124,** Jeff Greenberg/Photo Researchers **137,** Mark Richards/PhotoEdit **180,** Gary Gay/The Image Bank **185,** Mark Richards/PhotoEdit **196,** Focus on Sports **203,** Brian Spurlock **213,** Nicholas Devore III/Photographers Aspen **214,** AP/Wide World Photos **215,** Kenneth Chen/Envision **243,** Reuters/Jeff Christensen/Archive Photos **248,** Brian Spurlock **267, 277, 296,** Steven Needham/Envision **301,** Scott Goodman/Envision **305,** Tom Stratman **308,** Andrea Mohin/NYT Pictures **319, 333,** Eric Neurath/Stock Boston **360,** © 1998 Al Satterwhite **395,** John W. Banagan/The Image Bank **421,** AP/Wide World Photos **426,** John W. Banagan/The Image Bank **465, 493, 495, 502,** AP/Wide World Photos **505,** A. Lichtenstein/The Image Works **526,** Tom Tracy/FPG International **533,** Lee Snider/The Image Works **535,** © 1996 Churchill & Klehr **551,** Lee Snider/The Image Works **567,** AP/Wide World Photos **574,** Tom Stratman **585,** AP/Wide World Photos **607,** Bob Daemmrich/The Image Works **618,** AP/Wide World Photos **647,** Brian Spurlock **665,** Judy Gelles/Stock Boston **718,** Judy Gelles/Stock Boston **727, 775,** John Riley/Tony Stone Images **777,** Ford Motor Company **778,** AP/Wide World Photos **779,** © 1990 Glenn Randall **799, 808,** Bob Daemmrich/The Image Works **835,** © 1998 Churchill & Klehr **859,** M. Rangell/The Image Works **865,** Churchill & Klehr **867,** John Banagan/The Image Bank **870,** © 1998 Churchill & Klehr **881, 884,** Terje Rakke/The Image Bank **887,** Eric Neurath/Stock Boston **896,** David Young-Wolff/Tony Stone Images **931,** Robert Severi/Gamma Liaison **946,** © Scott Teven **963,** Robert Severi/Gamma Liaison **966,** AP/Wide World Photos **978,** Frances M. Roberts **994,** Owen Franken/Stock Boston

LIBRARY OF CONGRESS CATALOGING-IN-PUBLICATION DATA

Bittinger, Marvin L.
Developmental mathematics/Marvin L. Bittinger, Judith A. Beecher.—5th ed.
p. cm.
ISBN 0-201-34027-5
1. Arithmetic. 2. Algebra. I. Beecher, Judith A. II. Title.
QA107.B54 1999
513'.14—dc21 99-40860
CIP

Developmental Mathematics, 5th Edition, TASP Version ISBN 0-201-34028-3

1 2 3 4 5 6 7 8 9 10—VH—03020100

Contents

Appendixes

Preface

Intended for use by students needing a review in arithmetic skills before covering introductory algebra topics, this text begins with a review of arithmetic concepts, and then develops statistics, geometry, and introductory algebra. It is part of a series of texts that includes the following:

Bittinger/Ellenbogen: *Prealgebra,* Third Edition

Bittinger: *Basic Mathematics,* Eighth Edition

Bittinger: *Fundamental Mathematics,* Second Edition

Bittinger: *Introductory Algebra,* Eighth Edition

Bittinger: *Intermediate Algebra,* Eighth Edition

Bittinger/Beecher: *Developmental Mathematics,* Fifth Edition

Bittinger/Beecher: *Introductory and Intermediate Algebra: A Combined Approach*

All of the topics on the Texas Academic Skills Program test (TASP), and the majority of the topics on the state-level mathematics tests, including the CSU Entry Level Mathematics Test (ELM) and the CUNY Mathematics Skills Assessment Test, are incorporated in this edition. Guidelines from many states and educational institutions were considered while planning the revision, including those for the Florida CLAST test, the Alabama Junior College Curriculum, the Tennessee State Board of Regents, and the mathematics requirements of the New York Commissioner's Regulations. Many of the skills required by these guidelines are covered in *Developmental Mathematics,* Fifth Edition.

Developmental Mathematics, Fifth Edition, is a significant revision of the Fourth Edition, particularly with respect to design, art program, pedagogy, features, and supplements package. Its unique approach, which has been developed and refined over five editions, continues to blend the following elements in order to bring students success:

- ***Real data.*** Real-data applications aid in motivating students by connecting the mathematics to their everyday lives. Extensive research was conducted to find new applications that relate mathematics to the real world.

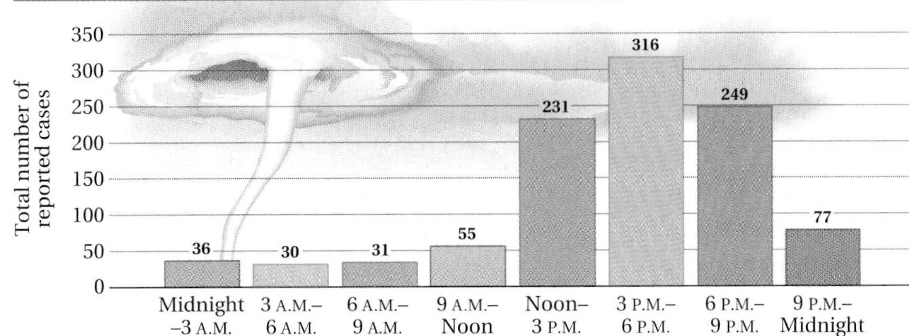

Source: National Weather Service

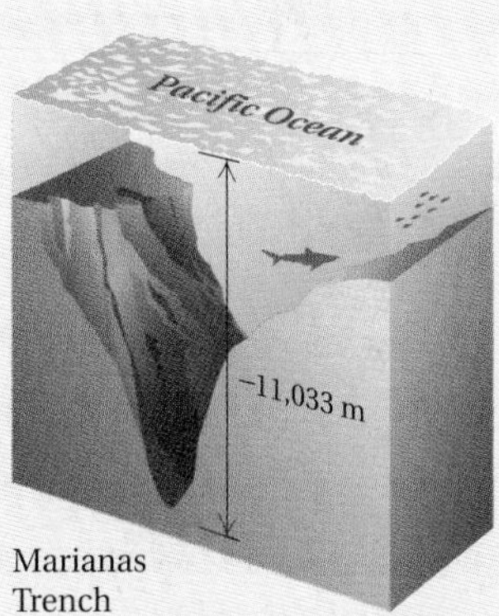

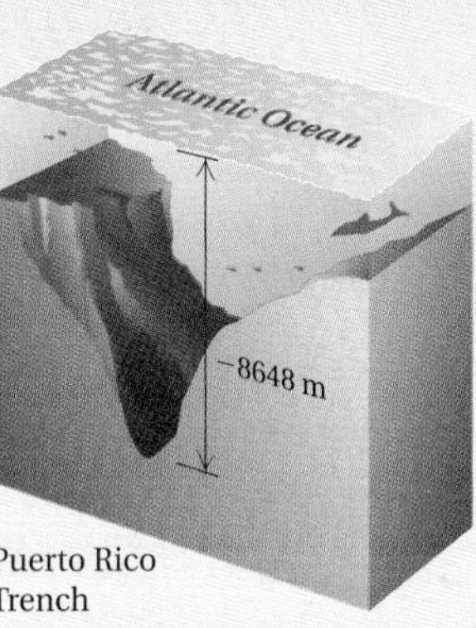

- ***Problem-solving approach.*** The basis for solving problems and real-data applications is a five-step process (*Familiarize, Translate, Solve, Check,* and *State*) introduced early in the text and used consistently throughout. This problem-solving approach provides students with a consistent framework for solving applications. (See pages 43, 489, and 778.)
- ***Writing style.*** The authors write in a clear easy-to-read style that helps students progress from concepts through examples and margin exercises to section exercises.
- ***Art program.*** The art program has been expanded to improve the visualization of mathematical concepts and to enhance the real-data applications.
- ***Reviewer feedback.*** The authors solicit feedback from reviewers and students to help fulfill student and instructor needs.
- ***Accuracy.*** The manuscript is subjected to an extensive accuracy-checking process to eliminate errors.
- ***Supplements package.*** All ancillary materials are closely tied with the text and created by members of the author team to provide a complete and consistent package for both students and instructors.

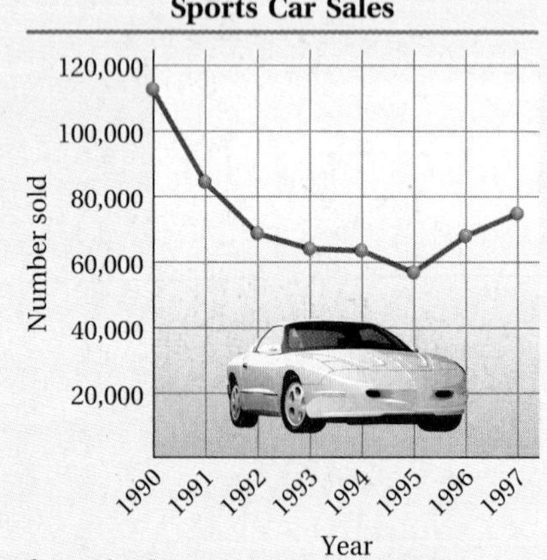

Source: Autodata

What's New in the Fifth Edition?

The style, format, and approach of the Fourth Edition have been strengthened in this new edition in a number of ways.

Updated Applications Extensive research has been done to make the applications in the Fifth Edition even more up to date and realistic. A large number of the applications are new to this edition, and many are drawn from the fields of business and economics, life and physical sciences, social sciences, and areas of general interest such as sports and daily life. To encourage students to understand the relevance of mathematics, many applications are enhanced by graphs and drawings similar to those found in today's newspapers and magazines. Many applications are also titled for quick and easy reference, and most real-data applications are credited with a source line. (See pages 191, 249, 278, 495, 534, and 978.)

Elephant Population

Cameroon
Zimbabwe
Sudan
Zaire
Tanzania
Botswana
= 10,000 elephants

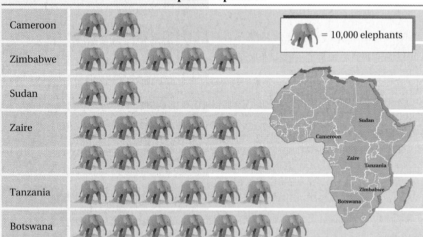

Source: National Geographic

Improving Your Math Study Skills Referenced in the table of contents, these mini-lessons provide students with concrete techniques to improve studying and test-taking. These features can be covered in their entirety at the beginning of the course, encouraging good study habits early on, or they can be used as they occur in the text, allowing students to learn them gradually. These features can also be used in conjunction with Marvin L. Bittinger's "Math Study Skills" Videotape, which is free to adopters. Please contact your Addison Wesley Longman sales consultant for details on how to obtain this videotape. (See pages 16, 160, and 540.)

Calculator Spotlights ▦, ▣ Designed specifically for the beginning developmental-mathematics student, these optional features include scientific- and graphing-calculator instruction and practice exercises (see pages 57, 221, 845, and 952). Answers to all Calculator Spotlight exercises appear at the back of the text.

New Art and Design To enhance the greater emphasis on real data and applications, we have extensively increased the number of pieces of technical and situational art (see pages 3, 186, 404, 491, and 872). The use of color has been carried out in a methodical and precise manner so that its use carries a consistent meaning, which enhances the readability of the text. For example, when perimeter is considered, figures have a red border to emphasize the perimeter. When area is considered, figures are outlined in black and screened with amber to emphasize the area. Similarly, when volume is considered, figures are three-dimensional and air-brushed blue.

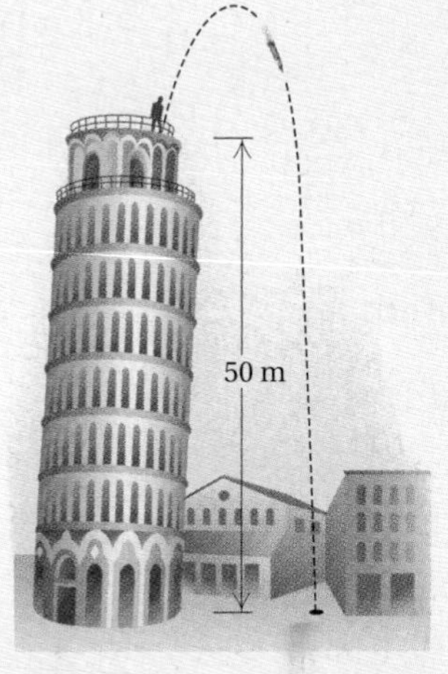

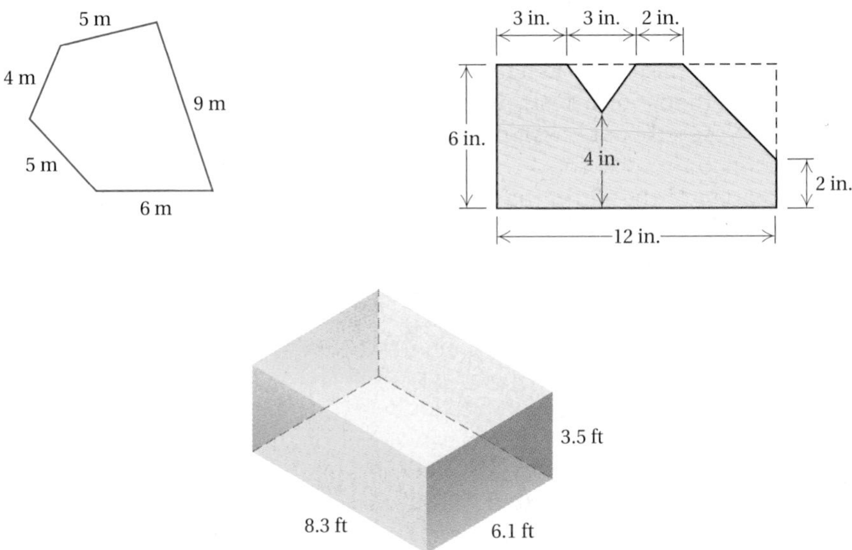

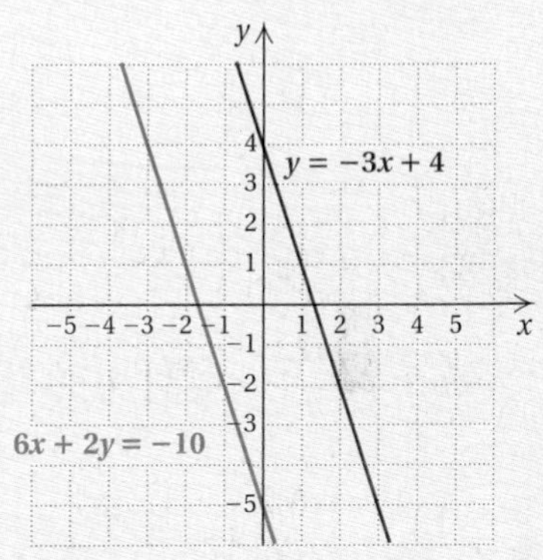

The use of both red and blue in mathematical art increases understanding of the concepts. When two lines are graphed using the same set of axes, one is usually red and the other blue. Note that equation labels are the same color as the corresponding line to aid in understanding.

In addition, the margin color has been improved so that student annotations can be read more clearly. Answer lines have also been deleted from all section exercise sets to allow room for more exercises and additional art to better illustrate the exercises.

World Wide Web Integration World Wide Web The World Wide Web is a powerful resource available to more and more people every day. In an effort to get students more involved in using this source of information, we have added a World Wide Web address (www.mathmax.com) to every chapter opener (see pages 83, 203, 533, and 931). Students can go to this page on the World Wide Web to further explore the subject matter of the chapter-opening application. Selected exercise sets, marked on the first page of the exercise set with an icon (see pages 107, 259, 431, and 555), have additional practice-problem worksheets that can be downloaded from this site. Additional, more extensive, Summary and Review pages for each chapter, as well as other supplementary material, can also be downloaded for instructor and student use.

Algebraic–Graphical Connections AG To give students a better visual understanding of algebra, we have included algebraic–graphical connections in the Fifth Edition (see pages 552, 765, 840, and 933). This feature gives the algebra more meaning by connecting the algebra to a graphical interpretation.

Collaborative Learning Features An icon located at the end of an exercise set signals the existence of a Collaborative Learning Activity correlating to that section in Irene Doo's *Collaborative Learning Activities Manual* (see pages 54, 108, 788, and 822). Please contact your Addison Wesley Longman sales consultant for details on ordering this supplement.

Exercises The deletion of answer lines in the exercise sets has allowed us to include more exercises in the Fifth Edition. Exercises are paired, meaning that each even-numbered exercise is very much like the odd-numbered one that precedes it. This gives the instructor several options: If an instructor wants the student to have answers available, the odd-numbered exercises are assigned; if an instructor wants the student to practice (perhaps for a test), with no answers available, then the even-numbered exercises are assigned. In this way, each exercise set actually serves as two exercise sets. Answers to all odd-numbered exercises, with the exception of the Thinking and Writing exercises, and *all* Skill Maintenance exercises are provided at the back of the text. If an instructor wants the student to have access to all the answers, a complete answer book is available.

Skill Maintenance Exercises The Skill Maintenance exercises have been enhanced by the inclusion of 65% more exercises in this edition. These exercises focus on the four Objectives for Retesting listed at the beginning of each chapter, but they also review concepts from other sections of the text in order to prepare students for a final examination. Section and objective codes appear next to each Skill Maintenance exercise for easy reference. Answers to all Skill Maintenance exercises appear at the back of the book (see pages 126, 172, 629, and 746).

Synthesis Exercises These exercises now appear in every exercise set, Summary and Review, and Chapter Test. Synthesis exercises help build critical thinking skills by requiring students to synthesize or combine learning objectives from the section being studied as well as preceding sections in the book (see pages 214, 708, 830, and 904).

Thinking and Writing Exercises ◆ Two Thinking and Writing exercises (denoted by the maze icon) have been added to the Synthesis section of every exercise set and Summary and Review. Designed to develop comprehension of critical concepts, these exercises encourage students to both think and write about key mathematical ideas in the chapter (see pages 162, 214, 808, and 848).

Content We have made the following improvements to the content of *Developmental Mathematics.*

- The topic of estimation is spiraled throughout the text. It is used to approximate the results of operations, to check possible solutions to applied problems, and as a basic skill. Estimation has been expanded in this edition to include fractions and mixed numerals. (See pages 29, 129, and 179.)

- In Chapter 5 (*Data Analysis, Graphs, and Statistics*), a new section on data analysis and predictions has been added. Students will learn to compare two sets of data using their means and to make predictions from a set of data using interpolation and extrapolation. This chapter is also filled with new real-data applications.

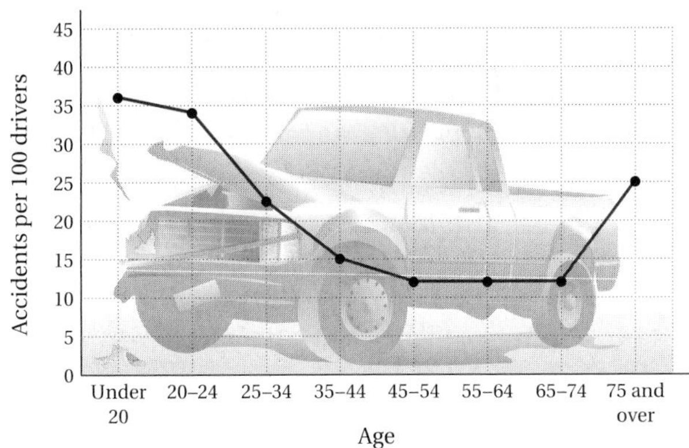

- Graphing, formerly located in Chapter 12, has been divided into two chapters. Basic graphing of equations is now covered in Chapter 9 (*Graphs of Equations; Data Analysis*) and the concept of slope appears in Chapter 13 (*Graphs, Slope, and Applications*). This earlier introduction to graphing allows expanded coverage and integration of graphing in the rest of the text.
- Section 9.4 ("Applications and Data Analysis with Graphs") on data analysis and predictions is optional new content to the Fifth Edition. Students will learn to compare two sets of data using their means and to make predictions from a set of data using interpolation and extrapolation.
- Section 13.2 ("Equations of Lines") has been improved by emphasizing the use of $y = mx + b$ in determining equations of lines and by placing all of these topics in one section.
- To provide expanded preparation for later courses, the topic of functions is introduced in Chapter 16.
- Two new appendixes (*Multistep Applications* and *Sample TASP Exam*) have been added to the Fifth Edition.

Learning Aids

Interactive Worktext Approach The pedagogy of this text is designed to provide an interactive learning experience between the student and the exposition, annotated examples, art, margin exercises, and exercise sets. This approach provides students with a clear set of learning objectives, involves them with the development of the material, and provides immediate and continual reinforcement and assessment.

Section objectives are keyed by letter not only to section subheadings, but also to exercises in the exercise sets and Summary and Review, as well as answers to the Pretest, and Chapter Test questions. This enables students to easily find appropriate review material if they are unable to work a particular exercise.

Throughout the text, students are directed to numerous *margin exercises,* which provide immediate reinforcement of the concepts covered in each section.

Review Material The Fifth Edition of *Developmental Mathematics* continues to provide many opportunities for students to prepare for final assessment.

Now in a two-column format, the *Summary and Review* appears at the end of each chapter and provides an extensive set of review exercises. Reference codes beside each exercise or direction line preceding it allow the student to easily return to the objective being reviewed (see pages 133, 263, and 831).

Objectives for Retesting are covered in each Summary and Review and Chapter Test, and are also included in the Skill Maintenance exercises and in the Printed Test Bank (see pages 204, 666, and 800).

Objectives

a Given the coordinates of two points on a line, find the slope of the line.

b Find the slope of a line from an equation.

c Find the slope or rate of change in an applied problem involving slope.

For Extra Help

TAPE 25

MAC
WIN

CD-ROM

For Extra Help Many valuable study aids accompany this text. Below the list of objectives found at the beginning of each section are references to appropriate videotape, tutorial software, and CD-ROM programs to make it easy for the student to find the correct support materials.

Testing The following assessment opportunities exist in the text.

Chapter Pretests can then be used to place students in a specific section of the chapter, allowing them to concentrate on topics with which they have particular difficulty (see pages 466, 534, and 882).

Chapter Tests allow students to review and test comprehension of chapter skills, as well as the four Objectives for Retesting from earlier chapters (see pages 315, 581, and 985).

Answers to all Chapter Pretest and Chapter Test questions are found at the back of the book, along with appropriate section and objective references.

Supplements for the Instructor

Instructor's Solutions Manual

0-201-63678-6

The *Instructor's Solutions Manual* by Judith A. Penna contains brief worked-out solutions to all even-numbered exercises in the exercise sets and answers to all Thinking and Writing exercises.

Printed Test Bank/Instructor's Resource Guide

by Donna DeSpain

0-201-63684-0

The test-bank section of this supplement contains the following:

- Two alternate test forms for each chapter, modeled after the Chapter Tests in the text
- Two alternate test forms for each chapter, designed for a 50-minute class period
- Two multiple-choice versions of each Chapter Test

- Six final examinations: two with questions organized by chapter, two with questions scrambled, and two with multiple-choice questions
- Answers for the Chapter Tests and the final examinations
- TASP Practice Test

The resource-guide section contains the following:

- Extra practice exercises (with answers) for the most difficult topics in the text
- Black-line masters of grids and number lines for transparency masters or test preparation
- Index to the videotapes that accompany the text
- Three-column chapter Summary and Review listing objectives, brief procedures, worked-out examples, multiple-choice problems similar to the example, and the answers to those problems

Collaborative Learning Activities Manual

0-201-70003-4

The *Collaborative Learning Activities Manual,* written by Irene Doo of Austin Community College, features group activities that are tied to sections of the text via an icon Collaborative Learning Manual. Instructions for classroom setup are also included in the manual.

TestGen-EQ/QuizMaster-EQ CD-ROM

0-201-64615-3

This test generation software is available in Windows and Macintosh versions. TestGen-EQ's friendly graphical interface enables instructors to easily view, edit, and add questions, transfer questions to tests, and print tests in a variety of fonts and forms. Search and sort features help the instructor quickly locate questions and arrange them in a preferred order. Six question formats are available, including short-answer, true–false, multiple-choice, essay, matching, and bimodal formats. A built-in question editor gives the instructor the ability to create graphs, import graphics, insert mathematical symbols and templates, and insert variable numbers or text. Computerized testbanks include algorithmically defined problems organized according to each textbook. An "Export to HTML" feature lets instructors create practice tests for the World Wide Web. TestGen-EQ is free to qualifying adopters.

QuizMaster-EQ enables instructors to create and save tests and quizzes using TestGen-EQ so students can take them on a computer network. Instructors can set preferences for how and when tests are administered. QuizMaster-EQ automatically grades the exams and allows the instructor to view or print a variety of reports for individual students, classes, or courses. This software is available for both Windows and Macintosh and is fully networkable. QuizMaster-EQ is free to qualifying adopters.

Supplements for the Student

Student's Solutions Manual

0-201-63681-6

The *Student's Solutions Manual* by Judith A. Penna contains fully worked-out solutions with step-by-step annotations for all the odd-numbered exercises in the exercise sets in the text, with the exception of the Thinking and Writing exercises. It may be purchased by your students from Addison Wesley Longman.

"Steps to Success" Videotapes

0-201-64809-1

Steps to Success is a complete revision of the existing series of videotapes, based on extensive input from both students and instructors. These videotapes feature an engaging team of mathematics teachers who present comprehensive coverage of each section of the text in a student-interactive format. The lecturers' presentations include examples and problems from the text and support an approach that emphasizes visualization and problem solving. A video icon at the beginning of each section references the appropriate videotape number. The videotapes are free to qualifying adopters.

"Math Study Skills for Students" Videotape

0-201-88039-3

Designed to help students make better use of their math study time, this videotape help students improve retention of concepts and procedures taught in classes from basic mathematics through intermediate algebra. Through carefully-crafted graphics and comprehensive on-camera explanations, Marvin L. Bittinger helps viewers focus on study skills that are commonly overlooked.

InterAct Math Tutorial Software CD-ROM

0-201-64596-3

InterAct Math Tutorial Software has been developed and designed by professional software engineers working closely with a team of experienced developmental-math teachers. This software includes exercises that are linked one-to-one with the odd-numbered exercises in the text and require the same computational and problem-solving skills as their companion exercises in the text. Each exercise has an example and an interactive guided solution that are designed to involve students in the solution process and to help them identify precisely where they are having trouble. In addition, the software recognizes common student errors and provides students with appropriate customized feedback. With its sophisticated answer recognition capabilities, *InterAct Math Tutorial Software* recognizes equivalent forms of the same answer for any kind of input. It also tracks for each section student activity and scores that can then be printed out. A CD-ROM icon at the beginning of each section identifies section coverage. Available for Windows and Macintosh computers, this software is free to qualifying adopters or can be bundled with books for sale to students.

World Wide Web Supplement (www.mathmax.com)

This on-line supplement provides additional practice and learning resources for the student of introductory algebra. For each book chapter, students can find additional practice exercises, Web links for further exploration, and expanded Summary and Review pages that review and reinforce the concepts and skills learned throughout the chapter. In addition, students can download a plug-in for Addison Wesley Longman's *InterAct Math Tutorial Software* that allows them to access additional tutorial problems directly through their Web browser. Students and instructors can also learn about the other supplements available for the MathMax series.

MathMax Multimedia CD-ROM for Developmental Mathematics

0-201-66217-5

The Developmental Mathematics CD provides an interactive environment using graphics, animations, and audio narration to build on some of the unique and proven features of the MathMax series. Highlighting key concepts from the book, the content of the CD is tightly and consistently integrated with the *Developmental Mathematics* text and retains references to the *Developmental Mathematics* numbering scheme so that students can move smoothly between the CD and other *Developmental Mathematics* supplements. The CD includes Addison Wesley Longman's *InterAct Math Tutorial Software* so that students can practice additional tutorial problems. An interactive Chapter Review section allows students to review and practice what they have learned in each chapter; and multimedia presentations reiterate important study skills described throughout the book. A CD-ROM icon at the beginning of each section indicates section coverage. The Developmental Mathematics CD is available for both Windows and Macintosh computers. Contact your Addison Wesley Longman sales consultant for a demonstration.

Your authors and their team have committed themselves to publishing an accessible, clear, accomplishable, error-free book and supplements package that will provide the student with a successful learning experience and will foster appreciation and enjoyment of mathematics. As part of our continual effort to accomplish this goal, we welcome your comments and suggestions at the following email addresses:

Marv Bittinger
exponent@aol.com

Judy Beecher
jabeecher@worldnet.att.net

Acknowledgments

Many of you have helped to shape the Fifth Edition by reviewing, participating in telephone surveys and focus groups, filling out questionnaires, and spending time with us on your campuses. Our deepest appreciation to all of you and in particular to the following:

Courtney Adams, *Louisiana State University—Eunice*
Andrea Adlman, *Ventura College*
Dolores Anenson, *Merced College*
Michelle Bach, *Kansas City Kansas Community College*
Martin Baker, *Parks College*
Linda Balfour, *Schoolcraft College*
Sharon Balk, *Northeast Iowa Community College*
William Bordeaux, Jr., *Huntingdon College*
Ted Bright, *College of Marin*
Kim Brown, *Tarrant County Junior College, Northeast*
Richard Burns, *Springfield Technical Community College*
Mark Campbell, *Slippery Rock University*
Debra Caplinger Cross, *Alabama Southern Community College*
Tyrone Clinton, *St. Petersburg Junior College*
Barbara Conway, *Berkshire Community College*
Dan Cortney, *Glendale Community College*
Sherry Crabtree, *Northwest Shoals Community College*
Nancy Desilet, *Carroll Community College*
Irene Doo, *Austin Community College*
Irene Duranczyk, *Eastern Michigan University*
M. R. Eisfelder, *McHenry County College*
Peter Embalabala, *Lincolnland Community College*
Rebecca Farrow, *John Tyler Community College*
Kathy Fenimore, *Frederick Community College*
Jacqueline Fesq, *Raritan Valley Community College*
Margaret Finster, *Erie Community College—South Campus*
David Frankenreiter, *Jefferson College*
Jimy Fulford, *Gulf Coast Community College*
Dewey Furnass, *Ricks College*
Linda Galloway, *Macon College*
Bob Grant, *Mesa Community College*
Catherine Green, *Lawson State Community College*
Phil Green, *Skagit Valley College*
Don Griffin, *Greenville Technical College*
Janet Guynn, *Blue Ridge Community College*
Linda Hurst, *Central Texas College*
Charlotte Hutt, *Rogue Community College*
Mary Indelicado, *Normandale Community College*

Kathy Janoviak, *Mid-Michigan Community College*
Yvonne Jessee, *Mountain Empire Community College*
Juan Carlos Jiminez, *Springfield Technical Community College*
Kathy Johnson, *Volunteer State Community College*
Joe Jordan, *John Tyler Community College*
Robert Kaiden, *Lorain County Community College*
Joanne Kendall, *College of the Mainland*
Roxann King, *Prince Georges Community College*
Steve Kinholt, *Green River Community College*
Paulette Kirkpatrick, *Wharton County Junior College*
Karen Knight, *Jefferson College*
Evelyn Kral, *Morgan Community College*
Jeanne Kubier, *Oklahoma State University—Oklahoma City*
Lee M. Lacey, *Glendale Community College—Upper Keys*
Ira Lansing, *College of Marin*
Christine Ledwith, *Florida Keys Community College—Upper Keys Center*
Linda Long, *Ricks College*
Lynn Maracek, *Rancho Santiago College*
Lawrence Marler, *Olympic College*
C. Vernon Marlin, *Southeastern Community College*
Bob Martin, *Tarrant County Junior College, Northeast*
Ray Maruca, *Delaware County Community College*
Hubert McClure, *Tri-County Technical College*
Sharon McKindrick, *New Mexico State University—Grants Beach*
John Menzie, *Barstow College*
Rebecca Metzger
Sandra Miller, *Harrisburg Area Community College*
Molly Misko, *Gadsden State Community College*
Mike Montano, *Riverside Community College—City Campus*
Charlie Montgomery, *Alabama Southern Community College*
Frank Mulvaney, *Delaware County Community College*
Ethel Muter, *Raritan Valley Community College*
Ellen O'Connell, *Triton College*
Linda Padilla, *Joliet Junior College*
Julie Pendleton, *Brookhaven College*
Charles Perry, *Southern Union State Junior College*
Marilyn Platt, *Gatson College*
George Podorski, *Jefferson College—Hillsboro*
Elizabeth Polen, *County College of Morris*
Mary Pusch, *Rogue Community College*
Carol Rardin, *Central Wyoming College*
Eugena Rohrberg, *Los Angeles Valley College*
Patricio Rojas, *New Mexico State University—Grants*
James Ronner, *Southwestern Michigan College*
Suzanne Rosenberger, *Harrisburg Area Community College*
Pat Roux, *Delgado Community College*
Martin Sade, *Pima Community College*
Ned Schillow, *Lehigh Carbon Community College*
Radha Shrinivas, *St. Louis Community College—Forest Park*
Minnie Shuler, *Gulf Coast Community College*
Lynn Siedenstrang, *Gray's Harbor College*
Maxine Smith, *Greenville Technical College*
Larry Smyrski, *Henry Ford Community College*
Lee Ann Spahr, *Durham Technical Community College*
Helen Stewart, *Brunswick Community College*
Tom Swiersz, *St. Petersburg Junior College*

Sharon Testone, *Onondaga Community College*
Jane Thieling, *Dyersburg State Community College*
Victor Thomas, *Holyoke Community College*
Tami Wellick, *McHenry County College*
Joyce Wellington, *Southeastern Community College*
Elaine Werner, *University of the Pacific*
Steve Wittel, *Augusta College*
Lily Yang, *Brevard Community College*
Ben Zandy, *Fullerton College*

We also wish to recognize the following people who wrote scripts, presented lessons on camera, and checked the accuracy of the videotapes:

Beth Burkenstock
Donna DeSpain, *Benedictine University*
Margaret Donlan, *University of Delaware*
David J. Ellenbogen, *Community College of Vermont*
Barbara Johnson, *Indiana University—Purdue University at Indianapolis*
Judith A. Penna, *Indiana University—Purdue University at Indianapolis*
Clen Vance, *Houston Community College*

We wish to thank Jason Jordan, our publisher and friend at Addison Wesley Longman, for his encouragement, for his marketing insight, and for providing us with the environment of creative freedom. The unwavering support of the Developmental Math group and the endless hours of hard work by Martha Morong and Janet Theurer have led to products of which we are immensely proud.

We also want to thank Judy Penna for writing the Student's and the Instructor's Solutions Manuals and for her strong leadership in the preparation of the printed supplements, videotapes, and interactive CD-ROM. Other strong support has come from Donna DeSpain for the Printed Test Bank; Irene Doo for the Collaborative Learning Activities Manual; and Irene Doo, Pat Ewert, Barbara Johnson, Linda Long, Judy Penna, Vera Preston, Peggy Reijto, and Patty Slipher for their accuracy checking.

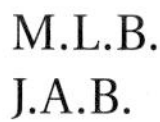

M.L.B.
J.A.B.

1

Operations on the Whole Numbers

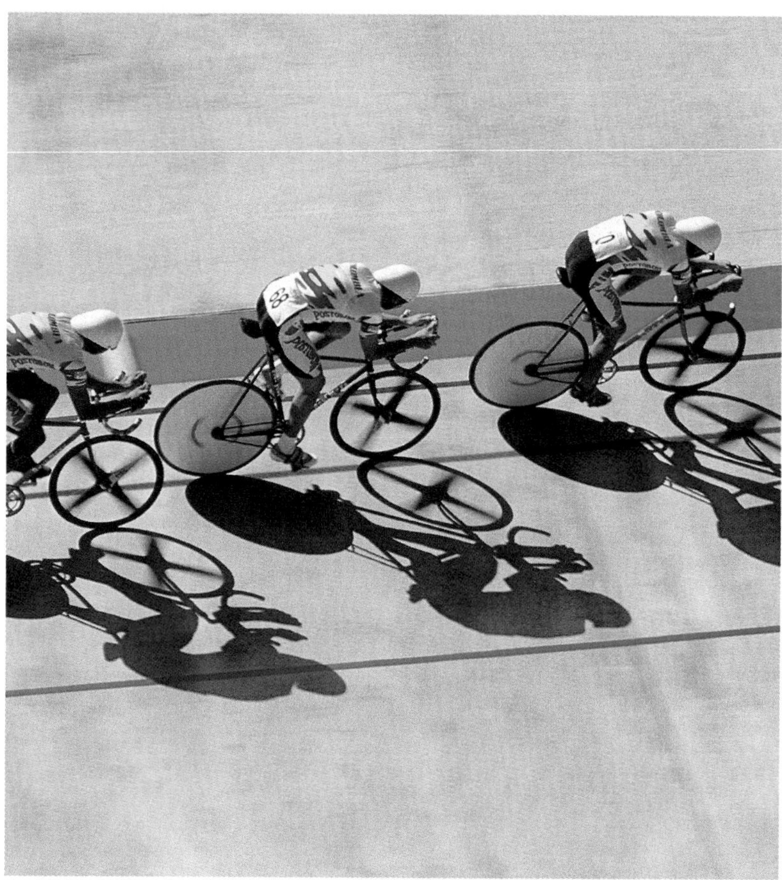

An Application

Total sales, in millions of dollars, of bicycles and related sporting supplies were $2973 in 1992, $3534 in 1993, $3470 in 1994, and $3435 in 1995 (***Source*:** National Sporting Goods Association). Find the total sales for the entire four-year period.

This problem appears as Exercise 4 in Exercise Set 1.5.

The Mathematics

We let T = the total sales. Since we are combining sales, addition can be used. We translate the problem to the equation

$$\underbrace{2973 + 3534 + 3470 + 3435 = T.}$$

This is how addition can occur in applications and problem solving.

World Wide Web For more information, visit us at www.mathmax.com

Introduction

In this chapter, we consider addition, subtraction, multiplication, and division of whole numbers. Then we study the solving of simple equations and apply all these skills to the solving of problems.

Factorizations, divisibility, and least common multiples are also studied.

Pretest: Chapter 1

1. Write a word name: 3,078,059.

2. Write expanded notation: 6987.

3. Write standard notation: Two billion, forty-seven million, three hundred ninety-eight thousand, five hundred eighty-nine.

4. What does the digit 6 mean in 2,967,342?

Use either < or > for ▒ to write a true sentence.

5. 346 ▒ 364

6. 54 ▒ 45

7. Add.

$$\begin{array}{r} 7\,3\,1\,2 \\ +\,2\,9\,0\,4 \\ \hline \end{array}$$

8. Subtract.

$$\begin{array}{r} 7\,0\,1\,2 \\ -\,2\,9\,0\,4 \\ \hline \end{array}$$

9. Multiply: $359 \cdot 64$.

10. Divide: $23{,}149 \div 46$.

Solve.

11. $326 \cdot 17 = m$

12. $y = 924 \div 42$

13. $19 + x = 53$

14. $34 \cdot n = 850$

Solve.

15. Anna weighs 121 lb and Kari weighs 109 lb. How much more does Anna weigh?

16. How many 12-jar cases can be filled with 1512 jars of spaghetti sauce?

17. *Population.* The population of Illinois is 11,830,000. The population of Ohio is 11,151,000. (***Source:*** U.S. Bureau of the Census) What is the total population of Illinois and Ohio?

18. A lot measures 48 ft by 54 ft. A pool that is 15 ft by 20 ft is constructed on the lot. How much area is left over?

19. Evaluate: 4^3.

20. Find the LCM of 15 and 24.

Simplify.

21. $8^2 \div 8 \cdot 2 - (2 + 2 \cdot 7)$

22. $108 \div 9 - \{4 \cdot [18 - (5 \cdot 3)]\}$

23. Determine whether 59 is prime, composite, or neither.

24. Find the prime factorization of 140.

25. Determine whether 1503 is divisible by 9.

26. Determine whether 788 is divisible by 8.

1.1 Standard Notation; Order

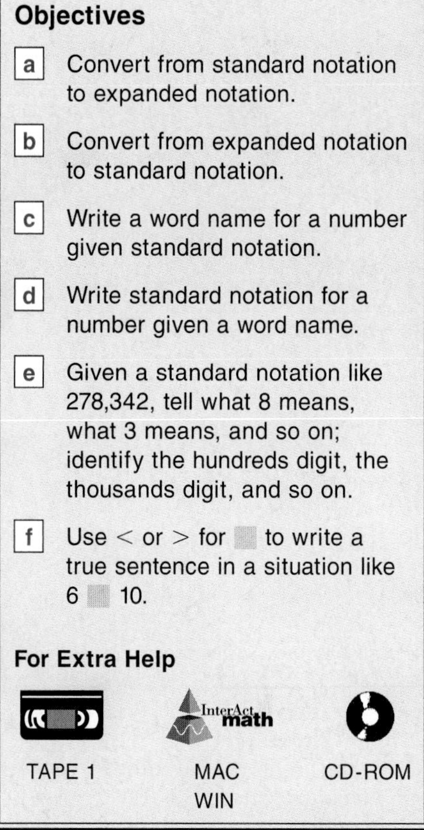

Objectives

a Convert from standard notation to expanded notation.

b Convert from expanded notation to standard notation.

c Write a word name for a number given standard notation.

d Write standard notation for a number given a word name.

e Given a standard notation like 278,342, tell what 8 means, what 3 means, and so on; identify the hundreds digit, the thousands digit, and so on.

f Use < or > for ▒ to write a true sentence in a situation like 6 ▒ 10.

For Extra Help

TAPE 1 | MAC WIN | CD-ROM

We study mathematics in order to be able to solve problems. In this chapter, we learn how to use operations on the whole numbers. We begin by studying how numbers are named.

a From Standard Notation to Expanded Notation

To answer questions such as "How many?", "How much?", and "How tall?", we use whole numbers. The set, or collection, of **whole numbers** is

0, 1, 2, 3, 4, 5, 6, 7, 8, 9, 10, 11, 12, ...

The set goes on indefinitely. There is no largest whole number, and the smallest whole number is 0. Each whole number can be named using various notations. The set 1, 2, 3, 4, 5, ..., without 0, is called the set of **natural numbers**.

As examples, we use data from the bar graph shown here.

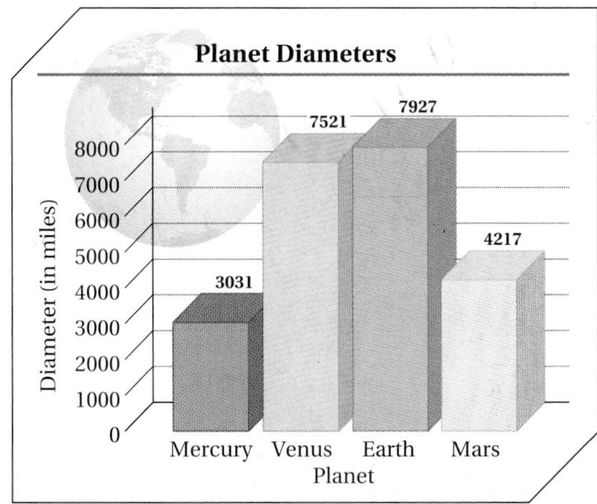

Note that the diameter of Mars is 4217 miles (mi). **Standard notation** for this number is 4217. We find **expanded notation** for 4217 as follows:

4217 = 4 thousands + 2 hundreds + 1 ten + 7 ones.

Example 1 Write expanded notation for 3031 mi, the diameter of Mercury.

3031 = 3 thousands + 0 hundreds + 3 tens + 1 one

Example 2 Write expanded notation for 54,567.

54,567 = 5 ten thousands + 4 thousands
+ 5 hundreds + 6 tens + 7 ones

Do Exercises 1 and 2 (in the margin at the right).

Write expanded notation.

1. 1805

2. 36,223

Answers on page A-1

Write expanded notation.

3. 3210

4. 2009

5. 5700

Write standard notation.

6. 5 thousands + 6 hundreds + 8 tens + 9 ones

7. 8 ten thousands + 7 thousands + 1 hundred + 2 tens + 8 ones

8. 9 thousands + 3 ones

Write a word name.

9. 57

10. 29

11. 88

Answers on page A-1

Example 3 Write expanded notation for 3400.

3400 = 3 thousands + 4 hundreds + 0 tens + 0 ones, or
3 thousands + 4 hundreds

Do Exercises 3–5.

b From Expanded Notation to Standard Notation

Example 4 Write standard notation for 2 thousands + 5 hundreds + 7 tens + 5 ones.

Standard notation is 2575.

Example 5 Write standard notation for 9 ten thousands + 6 thousands + 7 hundreds + 1 ten + 8 ones.

Standard notation is 96,718.

Example 6 Write standard notation for 2 thousands + 3 tens.

Standard notation is 2030.

Do Exercises 6–8.

c Word Names

"Three," "two hundred one," and "forty-two" are **word names** for numbers. When we write word names for two-digit numbers like 42, 76, and 91, we use hyphens. For example, U.S. Olympic team pitcher Michelle Granger can pitch a softball at a speed of 72 mph. A word name for 72 is "seventy-two."

Examples Write a word name.

7. 43 Forty-three

8. 91 Ninety-one

Do Exercises 9–11.

For large numbers, digits are separated into groups of three, called **periods**. Each period has a name: *ones, thousands, millions, billions,* and so on. When we write or read a large number, we start at the left with the largest period. The number named in the period is followed by the name of the period; then a comma is written and the next period is named. Recently, the U.S. national debt was $5,103,040,000,000. We can use a **place-value** chart to illustrate how to use periods to read the number 5,103,040,000,000.

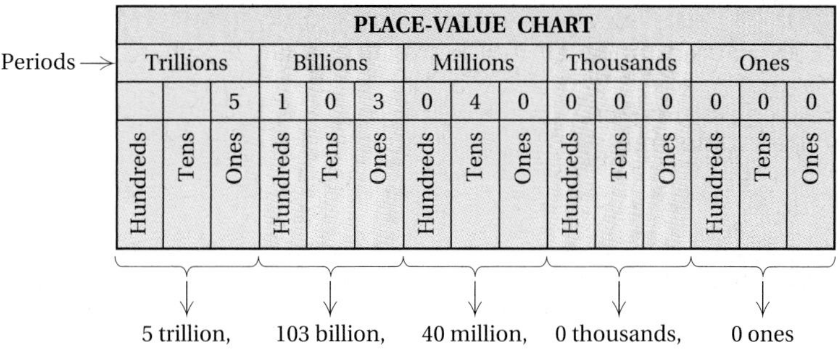

PLACE-VALUE CHART
Periods
Trillions
Billions
Millions
Thousands
Ones
5 1 0 3 0 4 0 0 0 0 0 0 0
Hundreds Tens Ones Hundreds Tens Ones Hundreds Tens Ones Hundreds Tens Ones Hundreds Tens Ones
5 trillion,
103 billion,
40 million,
0 thousands,
0 ones

Example 9 Write a word name for 46,605,314,732.

Forty-six billion,
six hundred five million,
three hundred fourteen thousand,
seven hundred thirty-two

The word "and" *should not* appear in word names for whole numbers. Although we commonly hear such expressions as "two hundred *and* one," the use of "and" is not, strictly speaking, correct in word names for whole numbers. For decimal notation, it is appropriate to use "and" for the decimal point. For example, 317.4 is read as "three hundred seventeen *and* four tenths."

Do Exercises 12–15.

d From Word Names to Standard Notation

Example 10 Write standard notation.

Five hundred six million,
three hundred forty-five thousand,
two hundred twelve

Standard notation is 506,345,212.

Do Exercise 16.

e Digits

A **digit** is a number 0, 1, 2, 3, 4, 5, 6, 7, 8, or 9 that names a place-value location.

Examples What does the digit 8 mean in each case?

11. 278,342 — 8 thousands
12. 872,342 — 8 hundred thousands
13. 28,343,399,223 — 8 billions

Do Exercises 17–20.

Write a word name.

12. 204

13. 79,204

14. 1,879,204

15. 22,301,879,204

16. Write standard notation.

Two hundred thirteen million, one hundred five thousand, three hundred twenty-nine

What does the digit 2 mean in each case?

17. 526,555

18. 265,789

19. 42,789,654

20. 24,789,654

Answers on page A-1

Golf Balls. On an average day, Americans buy 486,575 golf balls. In 486,575, what digit tells the number of:

21. Thousands?

22. Ten thousands?

23. Ones?

24. Hundreds?

Use $<$ or $>$ for $\square$ to write a true sentence. Draw a number line if necessary.

25. 8 $\square$ 12

26. 12 $\square$ 8

27. 76 $\square$ 64

28. 64 $\square$ 76

29. 217 $\square$ 345

30. 345 $\square$ 217

Answers on page A-1

Example 14 *Dunkin Donuts.* On an average day about 2,739,726 Dunkin Donuts are served in the United States. In 2,739,726, what digits tells the number of:

a) Hundred thousands 7

b) Thousands 9

Do Exercises 21–24.

f Order

We know that 2 is not the same as 5. We express this by the sentence $2 \neq 5$. We also know that 2 is less than 5. We can see this order on a number line: 2 is to the left of 5. The number 0 is the smallest whole number.

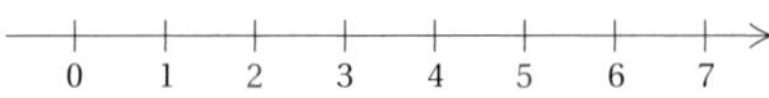

For any whole numbers a and b:

1. $a < b$ (read "a is less than b") is true when a is to the left of b on a number line.
2. $a > b$ (read "a is greater than b") is true when a is to the right of b on a number line.

We call $<$ and $>$ **inequality symbols.**

Example 15 Use $<$ or $>$ for $\square$ to write a true sentence: 7 $\square$ 11.

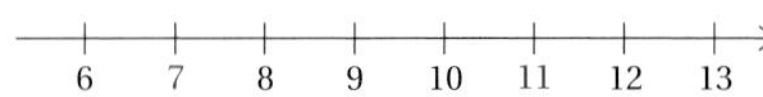

Since 7 is to the left of 11, $7 < 11$.

Example 16 Use $<$ or $>$ for $\square$ to write a true sentence: 92 $\square$ 87.

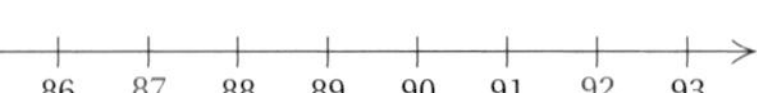

Since 92 is to the right of 87, $92 > 87$.

A sentence like $8 + 5 = 13$ is called an **equation**. A sentence like $7 < 11$ is called an **inequality**. The sentence $7 < 11$ is a true inequality. The sentence $23 > 69$ is a false inequality.

Do Exercises 25–30.

Exercise Set 1.1

Always review the objectives before doing an exercise set. See page 3. Note how the objectives are keyed to the exercises.

a Write expanded notation.

1. 5742

2. 3897

3. 27,342

4. 93,986

5. 5609

6. 9990

7. 2300

8. 7020

b Write standard notation.

9. 2 thousands + 4 hundreds + 7 tens + 5 ones

10. 7 thousands + 9 hundreds + 8 tens + 3 ones

11. 6 ten thousands + 8 thousands + 9 hundreds + 3 tens + 9 ones

12. 1 ten thousand + 8 thousands + 4 hundreds + 6 tens + 1 one

13. 7 thousands + 3 hundreds + 0 tens + 4 ones

14. 8 thousands + 0 hundreds + 2 tens + 0 ones

15. 1 thousand + 9 ones

16. 2 thousands + 4 hundreds + 5 tens

c Write a word name.

17. 85

18. 48

19. 88,000

20. 45,987

21. 123,765

22. 111,013

23. 7,754,211,577

24. 43,550,651,808

Write a word name for the number in the sentence.

25. *NBA Salaries.* In a recent year, the average salary of a player in the NBA was $1,867,000.

26. The area of the Pacific Ocean is about 64,186,000 square miles.

27. *Population.* The population of South Asia is about 1,583,141,000.

28. *Monopoly.* In a recent Monopoly game sponsored by McDonald's restaurants, the odds of winning the grand prize was estimated to be 467,322,388 to 1.

d Write standard notation.

29. Two million, two hundred thirty-three thousand, eight hundred twelve

30. Three hundred fifty-four thousand, seven hundred two

31. Eight billion

32. Seven hundred million

Write standard notation for the number in the sentence.

33. *Light Distance.* Light travels nine trillion, four hundred sixty billion kilometers in one year.

34. *Pluto.* The distance from the sun to Pluto is three billion, six hundred sixty-four million miles.

35. *Area of Greenland.* The area of Greenland is two million, nine hundred seventy-four thousand, six hundred square kilometers.

36. *Memory Space.* On computer hard drives, one gigabyte is actually one billion, seventy-three million, seven hundred forty-one thousand, eight hundred twenty-four bytes of memory.

e What does the digit 5 mean in each case?

37. 235,888

38. 253,888

39. 488,526

40. 500,346

In 89,302, what digit tells the number of:

41. Hundreds?

42. Thousands?

43. Tens?

44. Ones?

f Use < or > for ▒ to write a true sentence. Draw a number line if necessary.

45. 0 ▒ 17

46. 32 ▒ 0

47. 34 ▒ 12

48. 28 ▒ 18

49. 1000 ▒ 1001

50. 77 ▒ 117

51. 133 ▒ 132

52. 999 ▒ 997

53. 460 ▒ 17

54. 345 ▒ 456

55. 37 ▒ 11

56. 12 ▒ 32

Synthesis

Exercises designated as *Synthesis exercises* differ from those found in the main body of the exercise set. The icon ◆ denotes synthesis exercises that are writing exercises. Writing exercises are meant to be answered in one or more complete sentences. Because answers to writing exercises often vary, they are not listed at the back of the book.

Exercises marked with a ▦ are meant to be solved using a calculator. These and the other synthesis exercises will often challenge you to put together two or more objectives at once.

57. ◆ Write an English sentence in which the number 260,000,000 is used.

58. ◆ Explain why we use commas when writing large numbers.

59. ▦ What is the largest number that you can name on your calculator? How many digits does that number have? How many periods?

60. How many whole numbers between 100 and 400 contain the digit 2 in their standard notation?

1.2 Addition and Subtraction

a Addition and the Real World

Addition of whole numbers corresponds to combining or putting things together. Let's look at various situations in which addition applies.

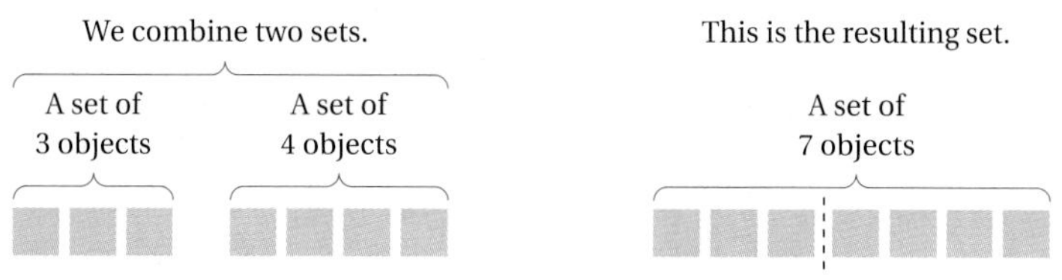

The addition that corresponds to the figure above is

$3 + 4 = 7.$

The number of objects in a set can be found by counting. We count and find that the two sets have 3 members and 4 members, respectively. After combining, we count and find that there are 7 objects. We say that the **sum** of 3 and 4 is 7. The numbers added are called **addends**.

$$\underbrace{3}_{\text{Addend}} + \underbrace{4}_{\text{Addend}} = \underbrace{7}_{\text{Sum}}$$

Example 1 Write an addition sentence that corresponds to this situation.

A student has \$3 and earns \$10 more. How much money does the student have?

An addition that corresponds is \$3 + \$10 = \$13.

Do Exercise 1.

Addition also corresponds to combining distances or lengths.

Example 2 Write an addition sentence that corresponds to this situation.

A car is driven 44 mi from San Francisco to San Jose. It is then driven 42 mi from San Jose to Oakland. How far is it from San Francisco to Oakland along the same route?

44 mi + 42 mi = 86 mi

Do Exercises 2 and 3.

When we find the sum of the distances around an object, we are finding its **perimeter.** Addition also corresponds to combining areas and volumes.

Objectives

a Write an addition sentence that corresponds to a situation.

b Add whole numbers.

c Write a subtraction sentence that corresponds to a situation involving "take away."

d Given a subtraction sentence, write a related addition sentence; and given an addition sentence, write two related subtraction sentences.

e Write a subtraction sentence that corresponds to a situation involving "how much more."

f Subtract whole numbers.

For Extra Help

TAPE 1 | MAC WIN | CD-ROM

Write an addition sentence that corresponds to the situation.

1. John has 8 music CD-ROMs in his backpack. Then he buys 2 educational CD-ROMs at the bookstore. How many CD-ROMs does John have in all?

Write an addition sentence that corresponds to the situation.

2. A car is driven 100 mi from Austin to Waco. It is then driven 93 mi from Waco to Dallas. How far is it from Austin to Dallas along the same route?

3. A coaxial cable 5 ft (feet) long is connected to a cable 7 ft long. How long is the resulting cable?

Answers on page A-1

Write an addition sentence that corresponds to the situation.

4. Find the perimeter of (distance around) the figure.

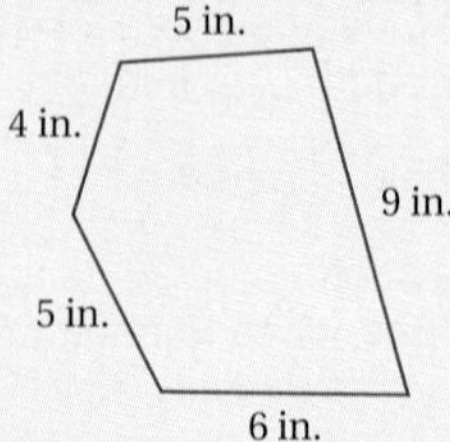

5. Find the perimeter of (distance around) the figure.

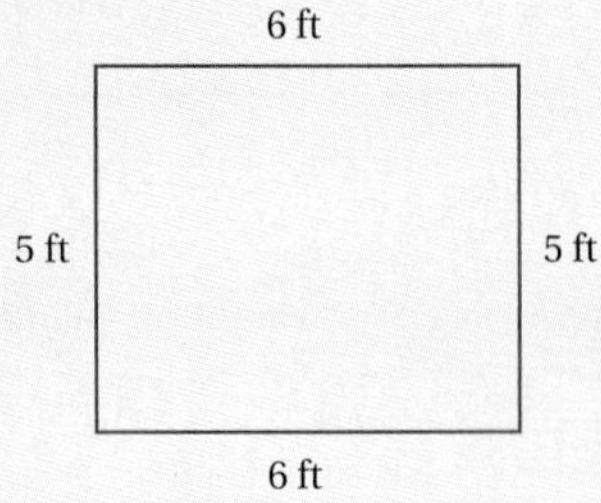

Write an addition sentence that corresponds to the situation.

6. The front parking lot of Sparks Electronics contains 30,000 square feet (sq ft) of parking space. The back lot contains 40,000 sq ft. What is the total area of the two parking lots?

7. You own a small rug that contains 8 square yards (sq yd) of fabric. You buy another rug that contains 9 sq yd. What is the area of the floor covered by both rugs?

Answers on page A-1

Example 3 Write an addition sentence that corresponds to this situation.

A computer sales rep travels the following route to visit various electronics stores. How long is the route?

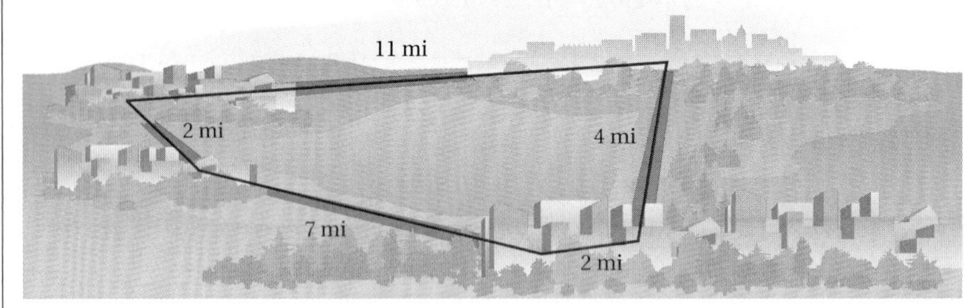

2 mi + 7 mi + 2 mi + 4 mi + 11 mi = 26 mi

Do Exercises 4 and 5.

Addition also corresponds to combining areas.

Example 4 Write an addition sentence that corresponds to this situation.

The area of a standard large index card is 40 square inches (sq in.). The area of a standard small index card is 15 sq in. Altogether, what is the total area of a large and a small card?

The area of the large index card is 40 sq in.		The area of the small index card is 15 sq in.		The total area of the two cards is 55 sq in.
40 sq in.	+	15 sq in.	=	55 sq in.

Do Exercises 6 and 7.

Addition corresponds to combining volumes as well.

Example 5 Write an addition sentence that corresponds to this situation.

Two trucks haul dirt to a construction site. One hauls 5 cubic yards (cu yd) and the other hauls 7 cu yd. Altogether, how many cubic yards of dirt have they hauled to the site?

5 cu yd + 7 cu yd = 12 cu yd

Do Exercises 8 and 9 on the following page.

b Addition of Whole Numbers

To add numbers, we add the ones digits first, then the tens, then the hundreds, and so on.

Example 6 Add: 7312 + 2504.

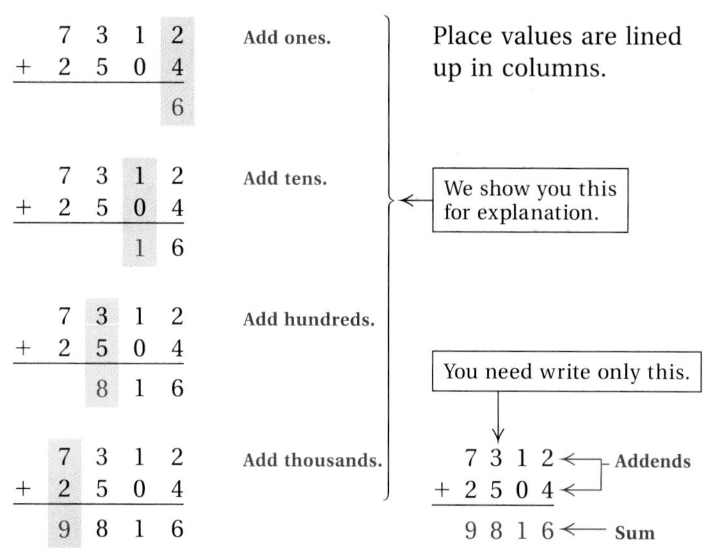

Do Exercise 10.

Example 7 Add: 2391 + 3276 + 8789 + 1498.

```
      2
  2 3 9 1
  3 2 7 6
  8 7 8 9
+ 1 4 9 8
---------
        4
```

Add ones: We get 24, so we have 2 tens + 4 ones. Write 4 in the ones column and 2 above the tens. This is called *carrying*, or *regrouping*.

```
    3 2
  2 3 9 1
  3 2 7 6
  8 7 8 9
+ 1 4 9 8
---------
      5 4
```

Add tens: We get 35 tens, so we have 30 tens + 5 tens. This is also 3 hundreds + 5 tens. Write 5 in the tens column and 3 above the hundreds.

```
  1 3 2
  2 3 9 1
  3 2 7 6
  8 7 8 9
+ 1 4 9 8
---------
    9 5 4
```

Add hundreds: We get 19 hundreds, or 1 thousand + 9 hundreds. Write 9 in the hundreds column and 1 above the thousands.

```
  1 3 2
  2 3 9 1
  3 2 7 6
  8 7 8 9
+ 1 4 9 8
---------
1 5 9 5 4
```

Add thousands: We get 15 thousands.

Do Exercises 11–13.

Write an addition sentence that corrresponds to the situation.

8. Two trucks haul sand to a construction site to use in a driveway. One hauls 6 cu yd and the other hauls 8 cu yd. Altogether, how many cubic yards of sand have they hauled to the site?

9. A football fan drives to all college football games using a motor home. On one trip the fan buys 80 gallons (gal) of gasoline and on another, 56 gal. How many gallons were bought in all?

10. Add.

```
  6 2 0 3
+ 3 5 4 2
```

Add.

11.
```
  7 9 6 8
+ 5 4 9 7
```

12.
```
  9 8 0 4
+ 6 3 7 8
```

13. Add.

```
  1 9 3 2
  6 7 2 3
  9 8 7 8
+ 8 9 4 1
```

Answers on page A-1

c Subtraction and the Real World: Take Away

Subtraction of whole numbers corresponds to two kinds of situations. The first one is called "take away."

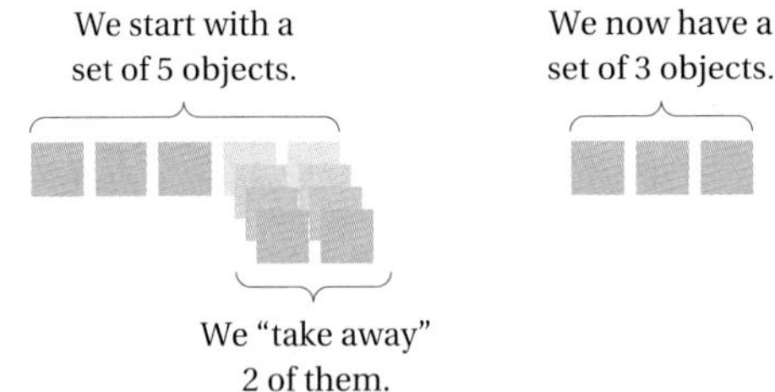

We start with a set of 5 objects.
We now have a set of 3 objects.
We "take away" 2 of them.

The subtraction that corresponds to the figure above is as follows.

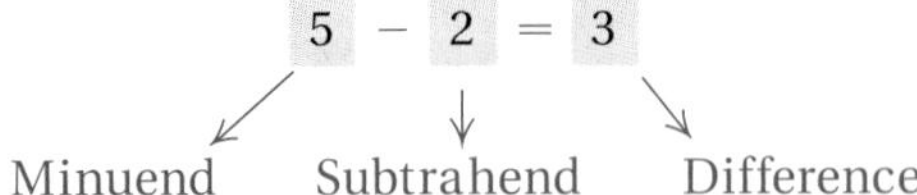

A **subtrahend** is the number being subtracted. A **difference** is the result of subtracting one number from another. That is, it is the result of subtracting the subtrahend from the **minuend**.

Examples Write a subtraction sentence that corresponds to the situation.

8. Juan goes to a music store and chooses 10 CDs to take to the listening station. He rejects 7 of them, but buys the rest. How many CDs did Juan buy?

There are 10 CDs to begin with.
He rejects 7 of them.
He buys the remaining 3.
10 − 7 = 3

9. A student has $300 and spends $85 for office supplies. How much money is left?

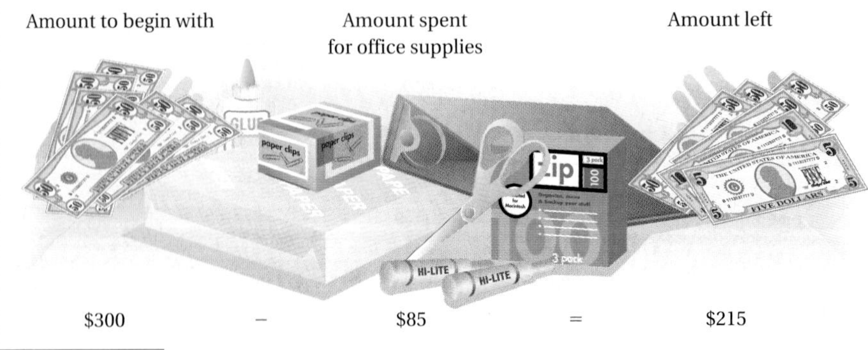

$300 − $85 = $215

Do Exercises 14 and 15.

Write a subtraction sentence that corresponds to the situation.

14. A contractor removes 5 cu yd of sand from a pile containing 67 cu yd. How many cubic yards of sand are left in the pile?

15. Sparks Electronics owns a field next door that has an area of 20,000 sq ft. Deciding they need more room for parking, the owners have 12,000 sq ft paved. How many square feet of field are left unpaved?

Answers on page A-1

d Related Sentences

Subtraction is defined in terms of addition. For example, $5 - 2$ is that number which when added to 2 gives 5. Thus for the subtraction sentence

$5 - 2 = 3,$ **Taking away 2 from 5 gives 3.**

there is a *related addition* sentence

$5 = 3 + 2.$ **Putting back the 2 gives 5 again.**

In fact, we know answers to subtractions are correct only because of the related addition, which provides a handy way to check a subtraction.

Example 10 Write a related addition sentence: $8 - 5 = 3$.

$8 - 5 = 3$

This number gets added (after 3).

$8 = 3 + 5$

By the commutative law of addition, there is also another addition sentence:

$8 = 5 + 3.$

The related addition sentence is $8 = 3 + 5$.

Do Exercises 16 and 17.

Write a related addition sentence.

16. $7 - 5 = 2$

17. $17 - 8 = 9$

Example 11 Write two related subtraction sentences: $4 + 3 = 7$.

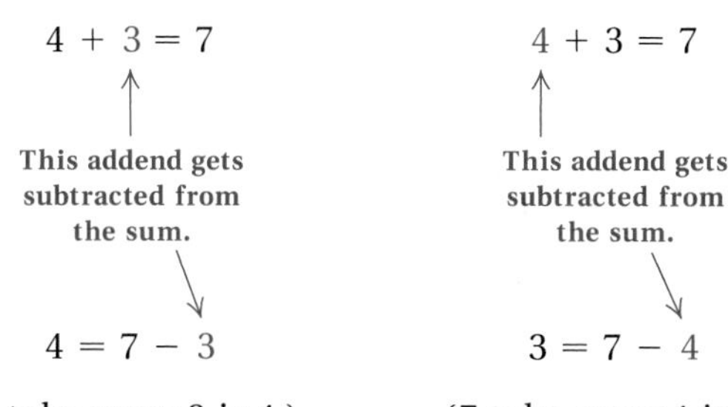
4 + 3 = 7
This addend gets subtracted from the sum.
4 = 7 − 3
4 + 3 = 7
This addend gets subtracted from the sum.
3 = 7 − 4

(7 take away 3 is 4.) (7 take away 4 is 3.)

The related subtraction sentences are $4 = 7 - 3$ and $3 = 7 - 4$.

Do Exercises 18 and 19.

Write two related subtraction sentences.

18. $5 + 8 = 13$

19. $11 + 3 = 14$

e How Much More?

The second kind of situation to which subtraction corresponds is called "how much more"? We need the concept of a missing addend for "how-much-more" problems. From the related sentences, we see that finding a *missing addend* is the same as finding a *difference.*

Missing addend → $12 = 3 + \square$ Difference → $12 - 3 = \square$

Examples Write a subtraction sentence that corresponds to the situation.

12. A student has \$47 and wants to buy a graphing calculator that costs \$89. How much more is needed to buy the calculator?

Answers on page A-1

To find the subtraction sentence, we first consider addition.

Amount that the student has	plus	Amount needed	is	Cost of the calculator
$47	+	□	=	$89

Now we write a related subtraction sentence:

$47 + \square = 89$

$\square = 89 - 47.$ **The addend 47 gets subtracted.**

13. Cathy is reading *True Success: A New Philosophy of Excellence,* by Tom Morris, as part of her philosophy class. It contains 288 pages. She has read 126 pages. How many more pages must she read?

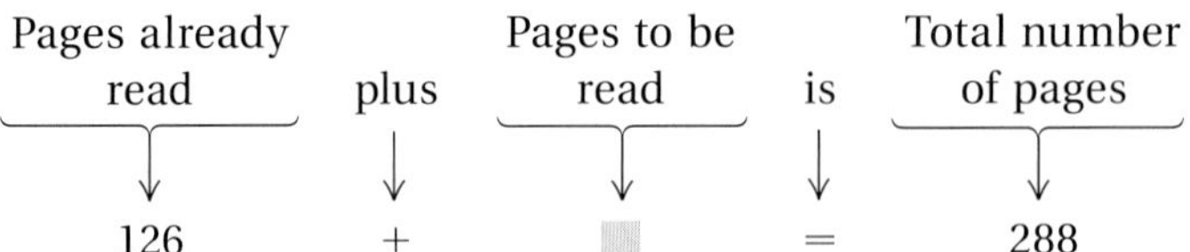

Pages already read
plus
Pages to be read
is
Total number of pages
126
+
=
288

Now we write a related subtraction sentence:

$126 + \square = 288$

$\square = 288 - 126.$ **126 gets subtracted.**

Do Exercises 20 and 21.

f Subtraction of Whole Numbers

To subtract numbers, we subtract the ones digits first, then the tens, then the hundreds, and so on.

Example 14 Subtract: 9768 − 4320.

$$\begin{array}{r} 9\,7\,6\,8 \\ -\,4\,3\,2\,0 \\ \hline 8 \end{array}$$ **Subtract ones.**

$$\begin{array}{r} 9\,7\,6\,8 \\ -\,4\,3\,2\,0 \\ \hline 4\,8 \end{array}$$ **Subtract tens.**

$$\begin{array}{r} 9\,7\,6\,8 \\ -\,4\,3\,2\,0 \\ \hline 4\,4\,8 \end{array}$$ **Subtract hundreds.**

$$\begin{array}{r} 9\,7\,6\,8 \\ -\,4\,3\,2\,0 \\ \hline 5\,4\,4\,8 \end{array}$$ **Subtract thousands.**

← This is for explanation.

$$\begin{array}{r} 9\,7\,6\,8 \\ -\,4\,3\,2\,0 \\ \hline 5\,4\,4\,8 \end{array}$$

← You should write only this.

Do Exercise 22.

Write an addition sentence and a related subtraction sentence corresponding to the situation. You need not carry out the subtraction.

20. It is 348 mi from Miami to Jacksonville. Alice has driven 67 mi from Miami to West Palm Beach on the way to Jacksonville. How much farther does she have to drive to get to Jacksonville?

21. A bricklayer estimates that it will take 1200 bricks to complete the side of a building but he has only 800 bricks on the job site. How many more bricks will be needed?

22. Subtract.

$$\begin{array}{r} 7\,8\,9\,3 \\ -\,4\,0\,9\,2 \\ \hline \end{array}$$

Answers on page A-1

Sometimes we need to borrow.

Example 15 Subtract: 6246 − 1879.

$$\begin{array}{r} \scriptstyle 3\ 16 \\ 6\,2\,\not{4}\,\not{6} \\ -\ 1\,8\,7\,9 \\ \hline 7 \end{array}$$

We cannot subtract 9 ones from 6 ones, but we can subtract 9 ones from 16 ones. We borrow 1 ten to get 16 ones.

$$\begin{array}{r} \scriptstyle 13 \\ \scriptstyle 1\ \not{3}\ 16 \\ 6\,\not{2}\,\not{4}\,\not{6} \\ -\ 1\,8\,7\,9 \\ \hline 6\,7 \end{array}$$

We cannot subtract 7 tens from 3 tens, but we can subtract 7 tens from 13 tens. We borrow 1 hundred to get 13 tens.

$$\begin{array}{r} \scriptstyle 11\ 13 \\ \scriptstyle 5\ \not{1}\ \not{3}\ 16 \\ \not{6}\,\not{2}\,\not{4}\,\not{6} \\ -\ 1\,8\,7\,9 \\ \hline 4\,3\,6\,7 \end{array}$$

We cannot subtract 8 hundreds from 1 hundred, but we can subtract 8 hundreds from 11 hundreds. We borrow 1 thousand to get 11 hundreds.

We can always check the answer by adding it to the number being subtracted.

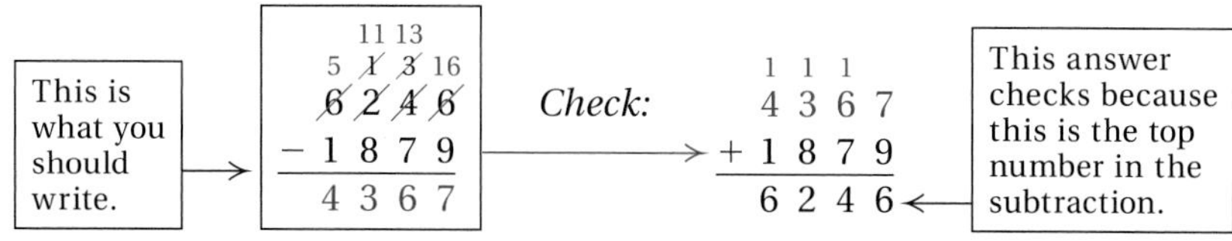

This is what you should write.
11 13
5 1 3 16
6 2 4 6
− 1 8 7 9
4 3 6 7
Check:
1 1 1
4 3 6 7
+ 1 8 7 9
6 2 4 6
This answer checks because this is the top number in the subtraction.

Do Exercises 23 and 24.

Example 16 Subtract: 902 − 477.

$$\begin{array}{r} \scriptstyle 8\ 9\ 12 \\ \not{9}\,\not{0}\,\not{2} \\ -\ 4\,7\,7 \\ \hline 4\,2\,5 \end{array}$$

We cannot subtract 7 ones from 2 ones. We have 9 hundreds, or 90 tens. We borrow 1 ten to get 12 ones. We then have 89 tens.

Do Exercises 25 and 26.

Example 17 Subtract: 8003 − 3667.

$$\begin{array}{r} \scriptstyle 7\ 9\ 9\ 13 \\ \not{8}\,\not{0}\,\not{0}\,\not{3} \\ -\ 3\,6\,6\,7 \\ \hline 4\,3\,3\,6 \end{array}$$

We have 8 thousands, or 800 tens. We borrow 1 ten to get 13 ones. We then have 799 tens.

Examples

18. Subtract: 6000 − 3762.

$$\begin{array}{r} \scriptstyle 5\ 9\ 9\ 10 \\ \not{6}\,\not{0}\,\not{0}\,\not{0} \\ -\ 3\,7\,6\,2 \\ \hline 2\,2\,3\,8 \end{array}$$

19. Subtract: 6024 − 2968.

$$\begin{array}{r} \scriptstyle 11 \\ \scriptstyle 5\ 9\ \not{1}\ 14 \\ \not{6}\,\not{0}\,\not{2}\,\not{4} \\ -\ 2\,9\,6\,8 \\ \hline 3\,0\,5\,6 \end{array}$$

Do Exercises 27–29.

Subtract. Check by adding.

23. $\begin{array}{r} 8\,6\,8\,6 \\ -\ 2\,3\,5\,8 \\ \hline \end{array}$ **24.** $\begin{array}{r} 7\,1\,4\,5 \\ -\ 2\,3\,9\,8 \\ \hline \end{array}$

Subtract.

25. $\begin{array}{r} 7\,0 \\ -\ 1\,4 \\ \hline \end{array}$ **26.** $\begin{array}{r} 5\,0\,3 \\ -\ 2\,9\,8 \\ \hline \end{array}$

Subtract.

27. $\begin{array}{r} 7\,0\,0\,7 \\ -\ 6\,3\,4\,9 \\ \hline \end{array}$ **28.** $\begin{array}{r} 6\,0\,0\,0 \\ -\ 3\,1\,4\,9 \\ \hline \end{array}$

29. $\begin{array}{r} 9\,0\,3\,5 \\ -\ 7\,4\,8\,9 \\ \hline \end{array}$

Answers on page A-1

Improving Your Math Study Skills

Tips for Using This Textbook

Throughout this textbook, you will find a feature called "Improving Your Math Study Skills." At least one such topic is included in each chapter. Each topic title is listed in the table of contents beginning on p. iii.

One of the most important ways to improve your math study skills is to learn the proper use of the textbook. Here we highlight a few points that we consider most helpful.

- **Be sure to note the special symbols a, b, c, and so on, that correspond to the objectives you are to be able to perform.** They appear in many places throughout the text. The first time you see them is in the margin at the beginning of each section. The second time is in the subheadings of each section, and the third time is in the exercise set. You will also find them next to the skill maintenance exercises in each exercise set and in the review exercises at the end of the chapter, as well as in the answers to the chapter tests and the cumulative reviews. These objective symbols allow you to refer back whenever you need to review a topic.
- **Note the symbols in the margin under the list of objectives at the beginning of each section.** These refer to the many distinctive study aids that accompany the book.
- **Read and study each step of each example.** The examples include important side comments that explain each step. These carefully chosen examples and notes prepare you for success in the exercise set.
- **Stop and do the margin exercises as you study a section.** When our students come to us troubledabout how they are doing in the course, the first question we ask is "Are you doing the margin exercises when directed to do so?" This is one of the most effective ways to enhance your ability to learn mathematics from this text. Don't deprive yourself of its benefits!
- **When you study the book, don't mark the points that you think are important, but mark the points you do not understand!** This book includes many design features that highlight important points. Use your efforts to mark where you are having trouble. Then when you go to class, a math lab, or a tutoring session, you will be prepared to ask questions that home in on your difficulties rather than spending time going over what you already understand.
- **If you are having trouble, consider using the *Student's Solutions Manual*, which contains worked-out solutions to the odd-numbered exercises in the exercise sets.**
- **Try to keep one section ahead of your syllabus.** If you study ahead of your lectures, you can concentrate on what is being explained in them, rather than trying to write everything down. You can then take notes only of special points or of questions related to what is happening in class.

Exercise Set 1.2

a Write an addition sentence that corresponds to the situation.

1. Isabel receives 7 e-mail messages on Tuesday and 8 on Wednesday. How many e-mail messages did she receive altogether on the two days?

2. At a construction site, there are two gasoline containers to be used by earth-moving vehicles. One contains 400 gal and the other 200 gal. How many gallons do both contain altogether?

Find the perimeter of (distance around) the figure.

3.

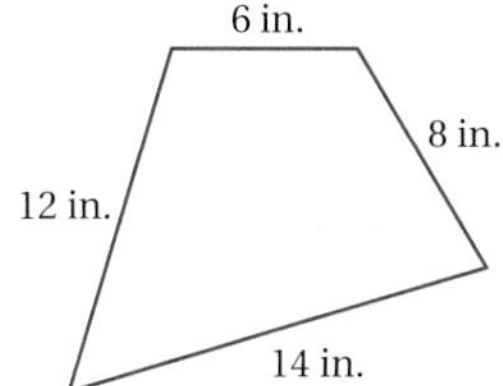

4.

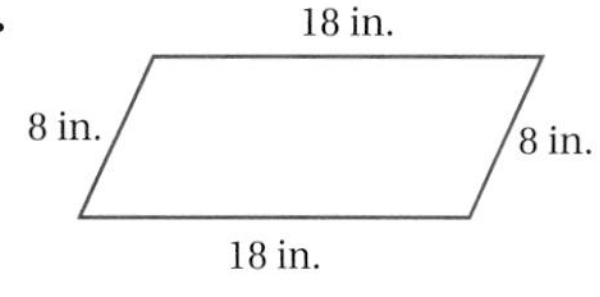

5.

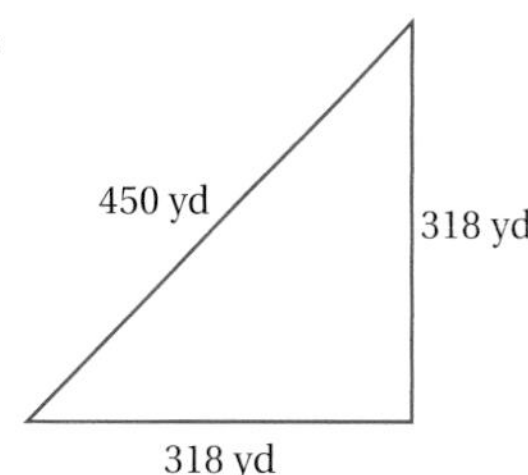

6.

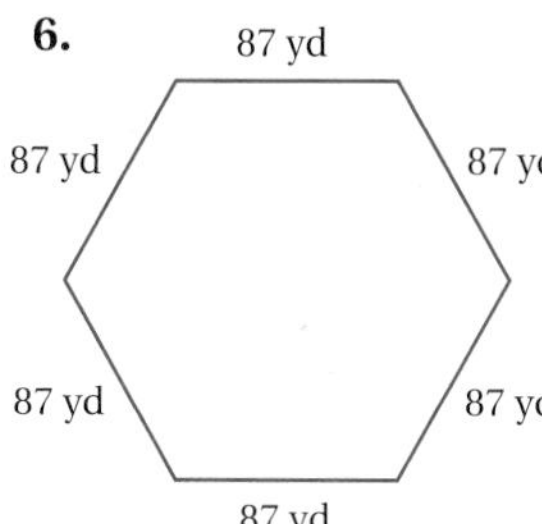

b Add.

7. $\begin{array}{r} 364 \\ +\ \ 23 \\ \hline \end{array}$

8. $\begin{array}{r} 1521 \\ +\ \ 348 \\ \hline \end{array}$

9. $\begin{array}{r} 1716 \\ +3482 \\ \hline \end{array}$

10. $\begin{array}{r} 7503 \\ +2683 \\ \hline \end{array}$

11. 909 + 101

12. 707 + 909

13. 356 + 4910

14. 280 + 34,702

15. $\begin{array}{r} 5093 \\ +3217 \\ \hline \end{array}$

16. $\begin{array}{r} 3654 \\ +2700 \\ \hline \end{array}$

17. $\begin{array}{r} 23,443 \\ +10,989 \\ \hline \end{array}$

18. $\begin{array}{r} 67,654 \\ +98,786 \\ \hline \end{array}$

19. $\begin{array}{r} 327 \\ 428 \\ 569 \\ 787 \\ +209 \\ \hline \end{array}$

20. $\begin{array}{r} 989 \\ 566 \\ 834 \\ 920 \\ +703 \\ \hline \end{array}$

21. $\begin{array}{r} 3420 \\ 8719 \\ 4312 \\ +6203 \\ \hline \end{array}$

22. $\begin{array}{r} 2003 \\ 149 \\ 58 \\ +3426 \\ \hline \end{array}$

c Write a subtraction sentence that corresponds to the situation. You need not carry out the subtraction.

23. *Chocolate Cake.* One slice of chocolate cake with fudge frosting contains 564 calories. One cup of hot cocoa made with skim milk contains 188 calories. How many more calories are in the cake than in the cocoa?

24. *Frozen Yogurt.* A dispenser at a frozen yogurt store contains 126 ounces (oz) of strawberry yogurt. A 13-oz cup is sold to a customer. How much is left in the dispenser?

d Write a related addition sentence.

25. $7 - 4 = 3$

26. $12 - 5 = 7$

27. $13 - 8 = 5$

28. $9 - 9 = 0$

Write two related subtraction sentences.

29. $6 + 9 = 15$

30. $7 + 9 = 16$

31. $8 + 7 = 15$

32. $8 + 0 = 8$

e Write an addition sentence and a related subtraction sentence corresponding to the situation. You need not carry out the subtraction.

33. *Kangaroos.* There are 32 million kangaroos in Australia and 17 million people. How many more kangaroos are there than people?

34. Marv needs to bowl a score of 223 in order to beat his opponent. His score with one frame to go is 195. How many pins does Marv need in the last frame to beat his opponent?

f Subtract.

35. $86 - 47$

36. $73 - 28$

37. $625 - 327$

38. $726 - 509$

39. $866 - 333$

40. $526 - 323$

41. $3982 - 2489$

42. $7650 - 1765$

43. $12{,}647 - 4{,}899$

44. $95{,}654 - 48{,}985$

45. $140 - 56$

46. $2300 - 109$

47. $7000 - 2794$

48. $8001 - 6543$

49. $48{,}000 - 37{,}695$

50. $17{,}043 - 11{,}598$

Skill Maintenance

The exercises that follow begin an important feature called *skill maintenance exercises.* These exercises provide an ongoing review of any preceding objective in the book. You will see them in virtually every exercise set. It has been found that this kind of extensive review can significantly improve your performance on a final examination.

51. Write standard notation for 7 thousands + 9 hundreds + 9 tens + 2 ones. [1.1b]

52. Write a word name for the number in the following sentence: [1.1c]

In a recent year, the gross revenue of the NBA was $924,600,000 (***Source:*** *Wall Street Journal*).

Synthesis

53. ◆ Describe a situation that corresponds to the addition 80 sq ft + 140 sq ft. (See Examples 2–5.)

54. ◆ Describe a situation that corresponds to the subtraction $20 − $17. (See Examples 9 and 12.)

1.3 Multiplication and Division; Rounding and Estimating

a Multiplication and the Real World

Multiplication of whole numbers corresponds to two kinds of situations.

Repeated Addition

The multiplication 3×5 corresponds to this repeated addition:

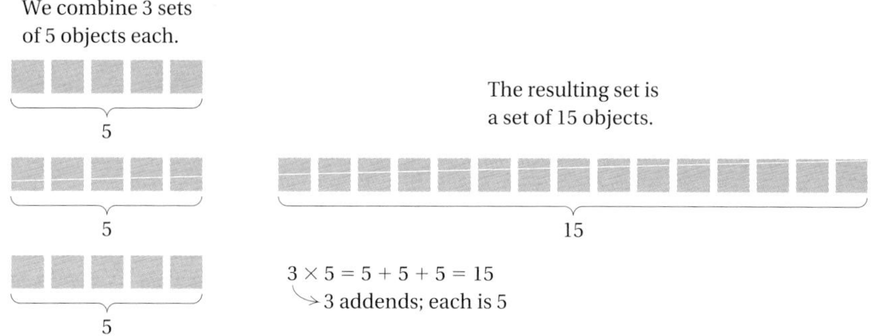

We say that the *product* of 3 and 5 is 15. The numbers 3 and 5 are called *factors.* The numbers that we multiply can be called **factors**. The result of the multiplication is a number called a **product**.

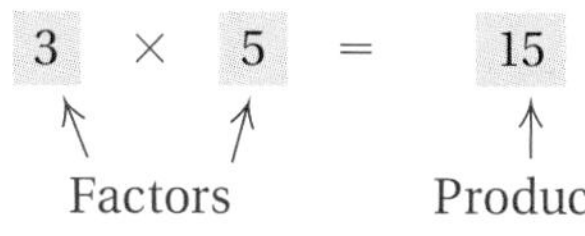

Rectangular Arrays

The multiplication 3×5 corresponds to this rectangular array. When you write a multiplication sentence corresponding to a real-world situation, you should think of either a rectangular array or repeated addition. In some cases, it may help to think both ways.

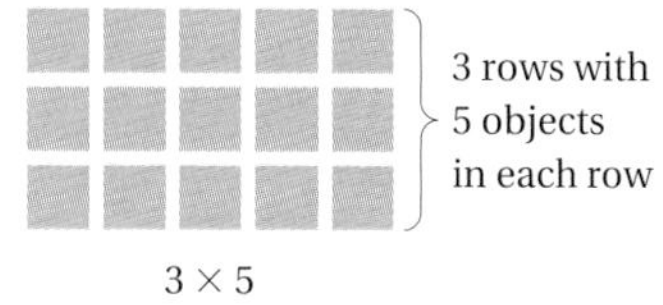

We have used an "×" to denote multiplication. A dot "·" is also commonly used. (Use of the dot is attributed to the German mathematician Gottfried Wilhelm von Leibniz in 1698.) Parentheses are also used to denote multiplication—for example, $(3)(5) = 15$, or $3(5) = 15$.

Example 1 Write a multiplication sentence that corresponds to this situation.

It is known that Americans drink 24 million gal of soft drinks per day. What quantity of soft drinks is consumed every 5 days?

We make a drawing. Repeated addition fits best in this case.

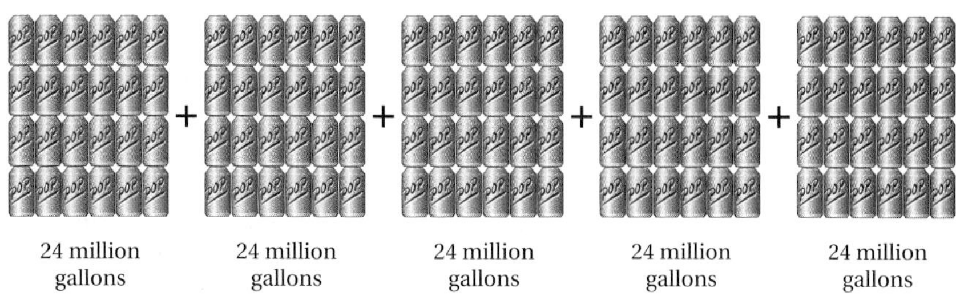

$5 \cdot 24$ million gallons $= 120$ million gallons

 = 1 million gallons

Objectives

a Write a multiplication sentence that corresponds to a situation.

b Multiply whole numbers.

c Write a division sentence that corresponds to a situation.

d Given a division sentence, write a related multiplication sentence; and given a multiplication sentence, write two related division sentences.

e Divide whole numbers.

f Round to the nearest ten, hundred, or thousand; and estimate sums, differences, and products by rounding.

For Extra Help

TAPE 1

MAC
WIN

CD-ROM

Write a multiplication sentence that corresponds to the situation.

1. Marv practices for the U.S. Open bowling tournament. He bowls 8 games a day for 7 days. How many games does he play altogether for practice?

2. A lab technician pours 75 milliliters (mL) of acid into each of 10 beakers. How much acid is poured in all?

3. *Checkerboard.* A checkerboard consists of 8 rows with 8 squares in each row. How many squares in all are there on a checkerboard?

4. What is the area of this pool table?

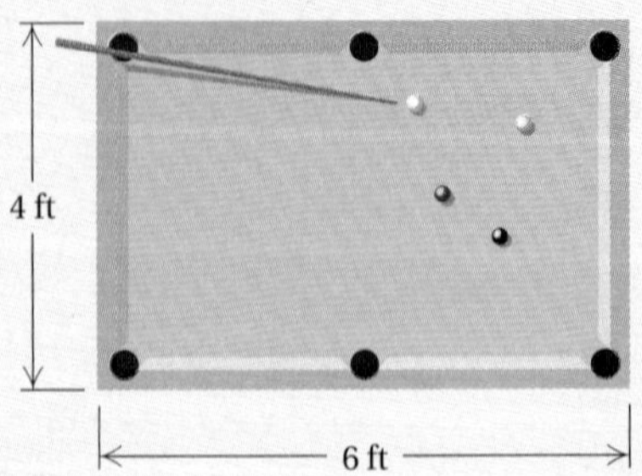

Answers on page A-2

Example 2 Write a multiplication sentence that corresponds to this situation.

One side of a building has 6 floors with 7 windows on each floor. How many windows are there on that side of the building?

We have a rectangular array and can easily make a drawing.

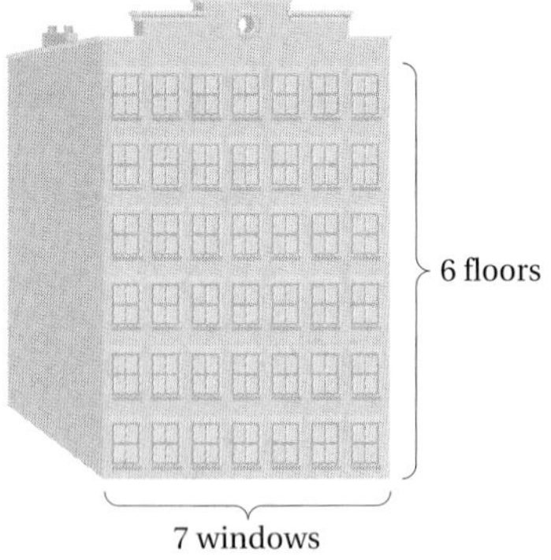

$6 \cdot 7 = 42$

Do Exercises 1–3.

Area

The area of a rectangular region is often considered to be the number of square units needed to fill it. Here is a rectangle 4 cm (centimeters) long and 3 cm wide. It takes 12 square centimeters (sq cm) to fill it.

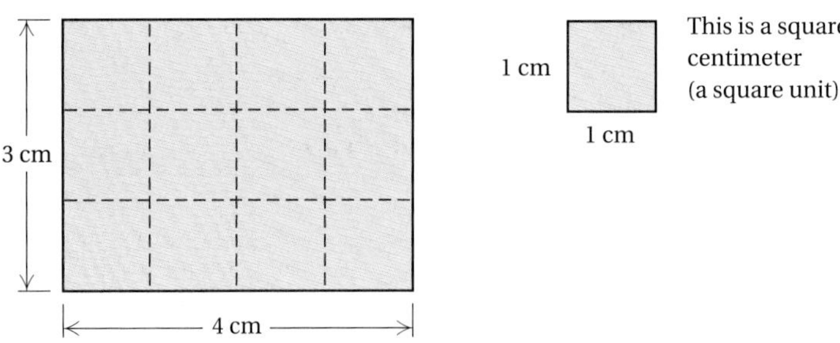

In this case, we have a rectangular array. The number of square units is $3 \cdot 4$, or 12.

Example 3 Write a multiplication sentence that corresponds to this situation.

A rectangular floor is 10 ft long and 8 ft wide. Find its area.

We make a drawing.

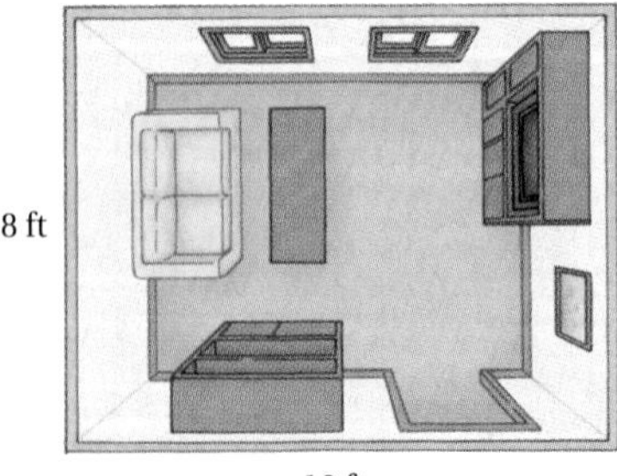

If we think of filling the rectangle with square feet, we have a rectangular array. The length $l = 10$ ft, and the width $w = 8$ ft. The area A is given by the formula

$$A = l \cdot w = 10 \times 8 = 80 \text{ sq ft.}$$

Do Exercise 4.

b Multiplication of Whole Numbers

Let's find the product

$$\begin{array}{r} 54 \\ \times\ 32 \\ \hline \end{array}$$

To do this, we multiply 54 by 2, then 54 by 30, and then add.

$$\begin{array}{r} 54 \\ \times\ \ 2 \\ \hline 108 \end{array} \qquad \begin{array}{r} \scriptstyle 1 \\ 54 \\ \times\ \ 30 \\ \hline 1620 \end{array}$$

Since we are going to add the results, let's write the work this way.

$$\begin{array}{rl} 54 & \\ \times\ 32 & \\ \hline 108 & \text{Multiplying by 2} \\ 1620 & \text{Multiplying by 30} \\ \hline 1728 & \text{Adding to obtain the product} \end{array}$$

Example 4 Multiply: 43×57.

$$\begin{array}{rl} \scriptstyle 2\ \ & \\ 57 & \\ \times\ 43 & \\ \hline 171 & \text{Multiplying by 3} \end{array}$$

$$\begin{array}{rl} \scriptstyle 2\ \ & \\ \scriptstyle 2\ \ & \\ 57 & \\ \times\ 43 & \\ \hline 171 & \text{Multiplying by 40. (We write a 0} \\ 2280 & \text{and then multiply 57 by 4.)} \end{array}$$

You may have learned that such a 0 does not have to be written. You may omit it if you wish. If you do omit it, remember, when multiplying by tens, to put the answer in the tens place.

$$\begin{array}{rl} \scriptstyle 2\ \ & \\ \scriptstyle 2\ \ & \\ 57 & \\ \times\ 43 & \\ \hline 171 & \\ 2280 & \\ \hline 2451 & \text{Adding to obtain the product} \end{array}$$

Do Exercises 5 and 6.

Multiply.

5. $\begin{array}{r} 45 \\ \times\ 23 \\ \hline \end{array}$

6. 48×63

Answers on page A-2

Multiply.

7.
```
  7 4 6
×   6 2
```

8. 245 × 837

Multiply.

9.
```
  4 7 2
× 3 0 6
```

10. 408 × 704

11.
```
  2 3 4 4
× 6 0 0 5
```

Multiply.

12.
```
  4 7 2
× 8 3 0
```

13.
```
  2 3 4 4
× 7 4 0 0
```

14. 100 × 562

15. 1000 × 562

Answers on page A-2

Example 5 Multiply: 457 × 683.

```
     5  2
     6  8  3
  ×  4  5  7
  ----------
  4  7  8  1      Multiplying 683 by 7

     4  1
     5  2
     6  8  3
  ×  4  5  7
  ----------
   4  7  8  1
3  4  1  5  0     Multiplying 683 by 50

      3  1
      4  1
      5  2
      6  8  3
   ×  4  5  7
  -----------
      4  7  8  1
   3  4  1  5  0
2  7  3  2  0  0  Multiplying 683 by 400
----------------
3  1  2, 1  3  1  Adding
```

Do Exercises 7 and 8.

Zeros in Multiplication

Example 6 Multiply: 306 × 274.

Note that 306 = 3 hundreds + 6 ones.

```
      2 7 4
    × 3 0 6
    -------
    1 6 4 4     Multiplying by 6
  8 2 2 0 0     Multiplying by 3 hundreds. (We write 00
  ---------     and then multiply 274 by 3.)
  8 3,8 4 4     Adding
```

Do Exercises 9–11.

Example 7 Multiply: 360 × 274.

Note that 360 = 3 hundreds + 6 tens.

```
      2 7 4
  ×   3 6 0
  ---------
  1 6 4 4 0     Multiplying by 6 tens. (We write 0
                and then multiply 274 by 6.)
  8 2 2 0 0     Multiplying by 3 hundreds. (We write 00
  ---------     and then multiply 274 by 3.)
  9 8,6 4 0     Adding
```

Do Exercises 12–15.

C Division and the Real World

Division of whole numbers corresponds to two kinds of situations. In the first, consider the division $20 \div 5$, read "20 divided by 5." We can think of 20 objects arranged in a rectangular array. We ask "How many rows, each with 5 objects, are there?"

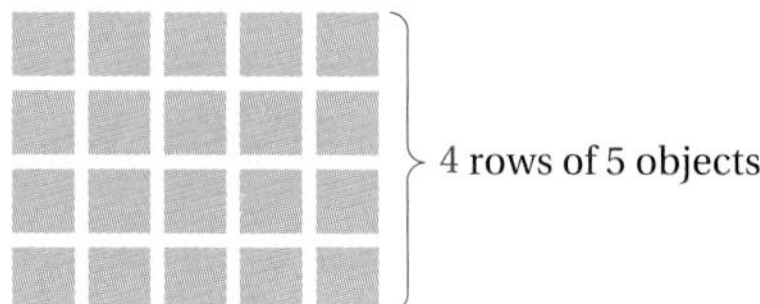

Since there are 4 rows of 5 objects each, we have

$$20 \div 5 = 4.$$

In the second situation, we can ask, "If we make 5 rows, how many objects will there be in each row?"

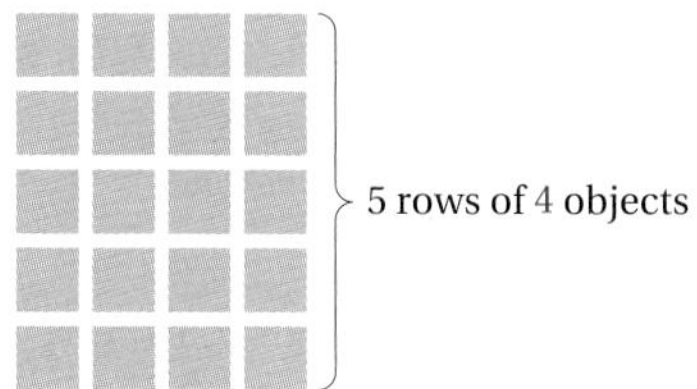

Since there are 4 objects in each of the 5 rows, we have

$$20 \div 5 = 4.$$

We say that the **dividend** is 20, the **divisor** is 5, and the **quotient** is 4.

$$\underset{\text{Dividend}}{\underset{\downarrow}{20}} \quad \div \quad \underset{\text{Divisor}}{\underset{\downarrow}{5}} \quad = \quad \underset{\text{Quotient}}{\underset{\downarrow}{4}}$$

The *dividend* is what we are dividing into. The result of the division is the *quotient*.

We also write a division such as $20 \div 5$ as

$$20/5 \quad \text{or} \quad \frac{20}{5} \quad \text{or} \quad 5\overline{)20}.$$

Example 8 Write a division sentence that corresponds to this situation.

A parent gives $24 to 3 children, with each child getting the same amount. How much does each child get?

We think of an array with 3 rows. Each row will go to a child. How many dollars will be in each row?

3 rows with 8 in each row

$$24 \div 3 = 8$$

Write a division sentence that corresponds to the situation. You need not carry out the division.

16. There are 112 students in a college band, and they are marching with 14 in each row. How many rows are there?

17. A college band is in a rectangular array. There are 112 students in the band, and they are marching in 8 rows. How many students are there in each row?

Answers on page A-2

Example 9 Write a division sentence that corresponds to this situation. You need not carry out the division.

How many mailboxes that cost $45 each can be purchased for $495?

We think of an array with 45 one-dollar bills in each row. The money in each row will buy a mailbox. How many rows will there be?

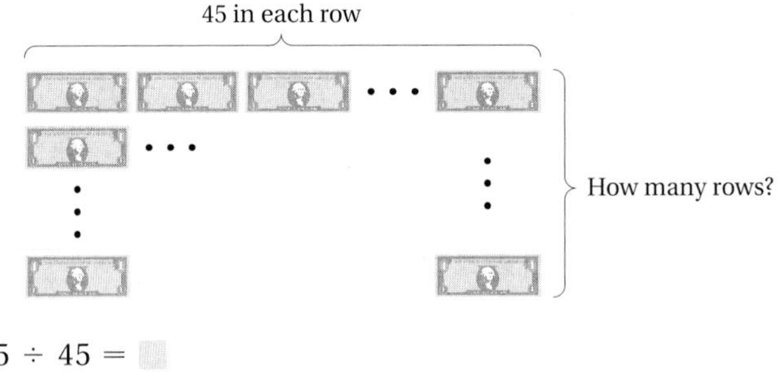

$495 \div 45 = \square$

> Whenever we have a rectangular array, we know the following:
>
> (The total number) ÷ (The number of rows) = (The number in each row).
>
> Also:
>
> (The total number) ÷ (The number in each row) = (The number of rows).

Do Exercises 16 and 17.

d Related Sentences

By looking at rectangular arrays, we can see how multiplication and division are related. The following array shows that $4 \cdot 5 = 20$.

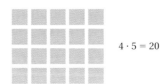

The array also shows the following:

$20 \div 5 = 4$ and $20 \div 4 = 5$.

Division is actually defined in terms of multiplication. For example, $20 \div 5$ is defined to be the number that when multiplied by 5 gives 20. Thus, for every division sentence, there is a related multiplication sentence.

$20 \div 5 = 4$ **Division sentence**

$20 = 4 \cdot 5$ **Related multiplication sentence**

> To get the related multiplication sentence, we use
> Dividend = Quotient · Divisor.

Example 10 Write a related multiplication sentence: $12 \div 6 = 2$.

We have

$12 \div 6 = 2$ **Division sentence**

$12 = 2 \cdot 6.$ **Related multiplication sentence**

The related multiplication sentence is $12 = 2 \cdot 6$.

By the commutative law of multiplication, there is also another multiplication sentence: $12 = 6 \cdot 2$.

Do Exercises 18 and 19.

For every multiplication sentence, we can write related divisions, as we can see from the preceding array.

Example 11 Write two related division sentences: $7 \cdot 8 = 56$.

We have

$7 \cdot 8 = 56$ — This factor becomes a divisor. — $7 = 56 \div 8.$

$7 \cdot 8 = 56$ — This factor becomes a divisor. — $8 = 56 \div 7.$

The related division sentences are $7 = 56 \div 8$ and $8 = 56 \div 7$.

Do Exercises 20 and 21.

e Division of Whole Numbers

Multiplication can be thought of as repeated addition. Division can be thought of as repeated subtraction. Compare.

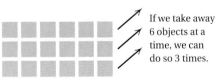

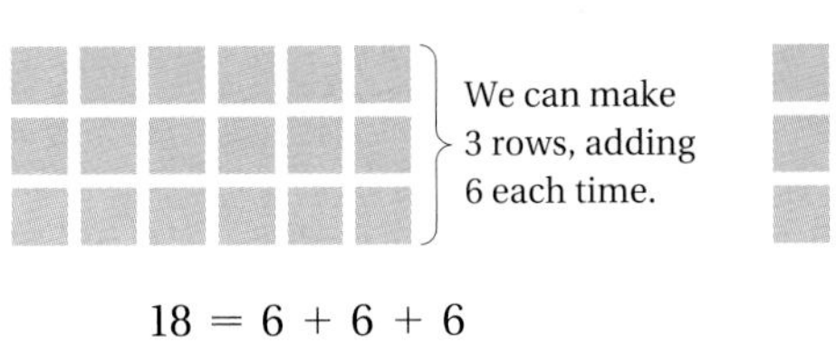
We can make
3 rows, adding
6 each time.

$$18 = 6 + 6 + 6$$
$$= 3 \cdot 6$$

If we take away
6 objects at a
time, we can
do so 3 times.

$$18 - 6 - 6 - 6 = 0$$

3 times

$$18 \div 6 = 3$$

Write a related multiplication sentence.

18. $15 \div 3 = 5$

19. $72 \div 8 = 9$

Write two related division sentences.

20. $6 \cdot 2 = 12$

21. $7 \cdot 6 = 42$

Answers on page A-2

Divide by repeated subtraction. Then check.

22. $54 \div 9$

23. $61 \div 9$

24. $53 \div 12$

25. $157 \div 24$

Answers on page A-2

To divide by repeated subtraction, we keep track of the number of times we subtract.

Example 12 Divide by repeated subtraction: $20 \div 4$.

$$\begin{array}{r} 20 \\ -\ \ 4 \\ \hline 16 \\ -\ \ 4 \\ \hline 12 \\ -\ \ 4 \\ \hline 8 \\ -\ \ 4 \\ \hline 4 \\ -\ \ 4 \\ \hline 0 \end{array}$$

We subtracted 5 times, so $20 \div 4 = 5$.

Example 13 Divide by repeated subtraction: $23 \div 5$.

$$\begin{array}{r} 23 \\ -\ \ 5 \\ \hline 18 \\ -\ \ 5 \\ \hline 13 \\ -\ \ 5 \\ \hline 8 \\ -\ \ 5 \\ \hline 3 \end{array}$$

We subtracted 4 times.

We have 3 left. This number is called the *remainder*.

We write

$$23 \div 5 = 4 \text{ R } 3$$

Dividend (23), Divisor (5), Quotient (4), Remainder (3)

CHECKING DIVISIONS. To check a division, we multiply. Suppose we divide 98 by 2 and get 49:

$$98 \div 2 = 49.$$

To check, we think of the related multiplication sentence $49 \cdot 2 = \square$. We multiply 49 by 2 and see if we get 98.

If there is a remainder, we add it after multiplying.

Example 14 Check the division in Example 13.

We found that $23 \div 5 = 4 \text{ R } 3$. To check, we multiply 5 by 4. This gives us 20. Then we add 3 to get 23. The dividend is 23, so the answer checks.

Do Exercises 22–25.

When we use the process of long division, we are doing repeated subtraction, even though we are going about it in a different way.

To divide, we start from the digit of highest place value in the dividend and work down to the lowest through the remainders. At each step we ask if there are multiples of the divisor in the quotient.

Example 15 Divide and check: 3642 ÷ 5.

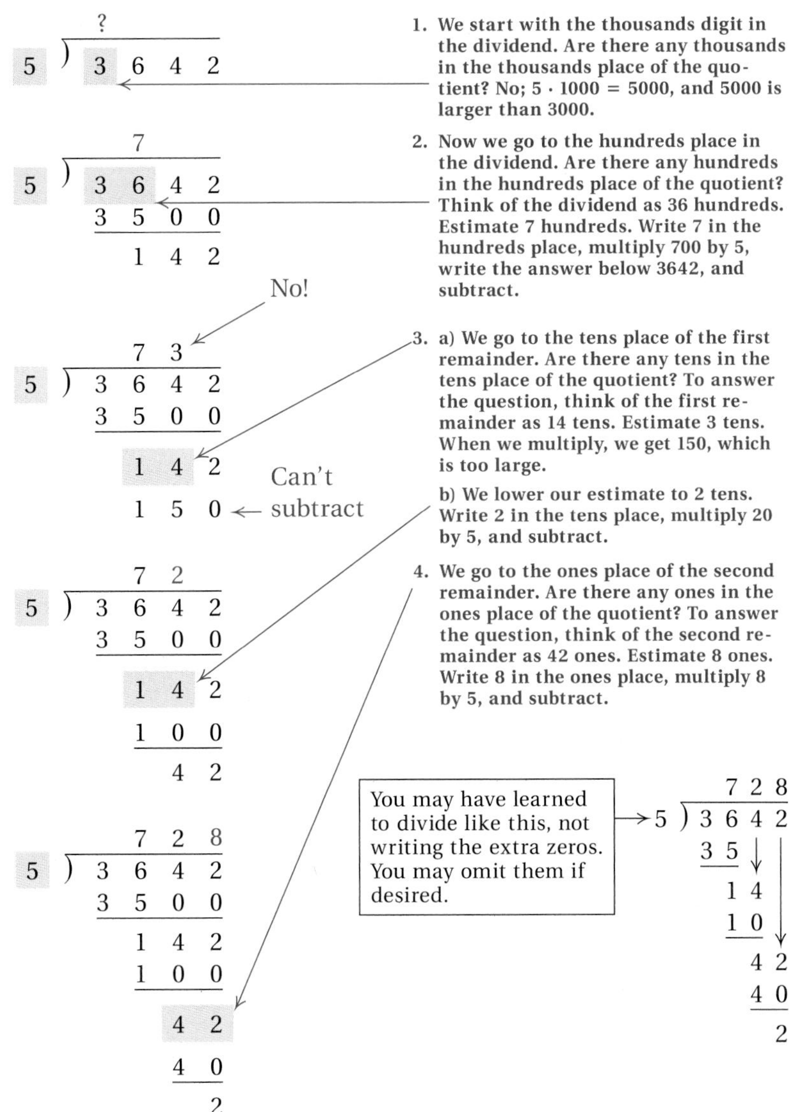

1. We start with the thousands digit in the dividend. Are there any thousands in the thousands place of the quotient? No; 5 · 1000 = 5000, and 5000 is larger than 3000.
2. Now we go to the hundreds place in the dividend. Are there any hundreds in the hundreds place of the quotient? Think of the dividend as 36 hundreds. Estimate 7 hundreds. Write 7 in the hundreds place, multiply 700 by 5, write the answer below 3642, and subtract.
3. a) We go to the tens place of the first remainder. Are there any tens in the tens place of the quotient? To answer the question, think of the first remainder as 14 tens. Estimate 3 tens. When we multiply, we get 150, which is too large.

   b) We lower our estimate to 2 tens. Write 2 in the tens place, multiply 20 by 5, and subtract.
4. We go to the ones place of the second remainder. Are there any ones in the ones place of the quotient? To answer the question, think of the second remainder as 42 ones. Estimate 8 ones. Write 8 in the ones place, multiply 8 by 5, and subtract.

You may have learned to divide like this, not writing the extra zeros. You may omit them if desired.

The answer is 728 R 2. To check, we multiply the quotient 728 by the divisor 5. This gives us 3640. Then we add 2 to get 3642. The dividend is 3642, so the answer checks.

Do Exercises 26–28.

Divide and check.

26. 4) 2 3 9

27. 6) 8 8 5 5

28. 5) 5 0 7 5

Answers on page A-2

Divide.

29. $6\overline{)4846}$

30. $7\overline{)7616}$

Divide.

31. $27\overline{)9724}$

32. $56\overline{)44{,}847}$

Answers on page A-2

Zeros in Quotients

Example 16 Divide: 6341 ÷ 7.

```
        9
   ________
7 ) 6 3 4 1  ←  Think: 63 hundreds ÷ 7.
    6 3 0 0     Estimate 9 hundreds.
    -------
        4 1
```

```
        9 0
   ________
7 ) 6 3 4 1
    6 3 0 0
    -------
        4 1  ←  Think: 4 tens ÷ 7. There are no tens
                in the quotient (other than the tens in 900).
                We write a 0 to show this.
```

```
        9 0 5
   ________
7 ) 6 3 4 1
    6 3 0 0
    -------
        4 1  ←  Think: 41 ones ÷ 7.
        3 5     Estimate 5 ones.
        ---
          6
```

The answer is 905 R 6.

Do Exercises 29 and 30.

Example 17 Divide: 8889 ÷ 37.

We round 37 to 40.

```
          2
     ________
3 7 ) 8 8 8 9  ←  Think: 37 ≈ 40; 88 hundreds ÷ 40.
      7 4 0 0     Estimate 2 hundreds, but write
      -------     2 × 37 = 74.
      1 4 8 9
```

```
          2 4
     ________
3 7 ) 8 8 8 9
      7 4 0 0
      -------
      1 4 8 9  ←  Think: 148 tens ÷ 40.
      1 4 8 0     Estimate 4 tens, but write
      -------     4 × 37 = 148.
            9
```

```
          2 4 0
     ________
3 7 ) 8 8 8 9
      7 4 0 0
      -------
      1 4 8 9
      1 4 8 0
      -------
            9  ←  Think: 9 ones ÷ 40.
                  There are no ones in the quotient.
```

The answer is 240 R 9.

Do Exercises 31 and 32.

f Rounding and Estimating

Rounding

We round numbers in various situations if we do not need an exact answer. For example, we might round to check if an answer to a problem is reasonable or to check a calculation done by hand or on a calculator. We might also round to see if we are being charged the correct amount in a store.

To understand how to round, we first look at some examples using number lines, even though this is not the way we normally do rounding.

Example 18 Round 47 to the nearest ten.

Here is a part of a number line; 47 is between 40 and 50.

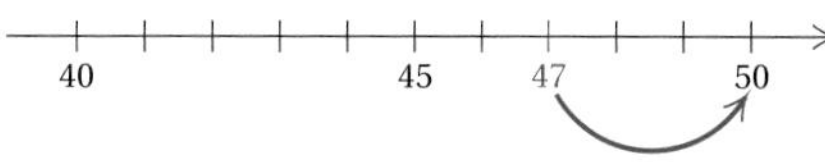

Since 47 is closer to 50, we round up to 50.

Example 19 Round 42 to the nearest ten.

42 is between 40 and 50.

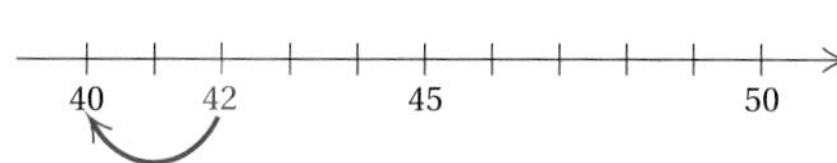

Since 42 is closer to 40, we round down to 40.

Do Exercises 33–36.

Example 20 Round 45 to the nearest ten.

45 is halfway between 40 and 50.

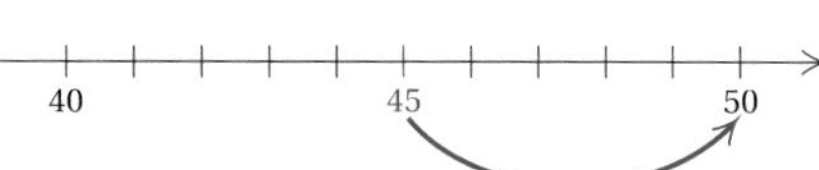

We could round 45 down to 40 or up to 50. We agree to round up to 50.

> When a number is halfway between rounding numbers, round up.

Do Exercises 37–39.

Here is a rule for rounding.

> To round to a certain place:
>
> 1. Locate the digit in that place.
> 2. Consider the next digit to the right.
> 3. If the digit to the right is 5 or higher, round up; if the digit to the right is 4 or lower, round down.
> 4. Change all digits to the right of the rounding location to zeros.

Round to the nearest ten.

33. 37

34. 52

35. 73

36. 98

Round to the nearest ten.

37. 35

38. 75

39. 85

Answers on page A-2

Round to the nearest ten.

40. 137

41. 473

42. 235

43. 285

Round to the nearest hundred.

44. 641

45. 759

46. 750

47. 9325

Round to the nearest thousand.

48. 7896

49. 8459

50. 19,343

51. 68,500

Answers on page A-2

Example 21 Round 6485 to the nearest (a) ten; (b) hundred; (c) thousand.

a) Locate the digit in the tens place. It is 8.

6 4 8 5
    ↑

The next digit to the right is 5, so we round up. The answer is 6490.

b) Locate the digit in the hundreds place. It is 4.

6 4 8 5
  ↑

The next digit to the right is 8, so we round up. The answer is 6500.

c) Locate the digit in the thousands place. It is 6.

6 4 8 5
↑

The next digit to the right is 4, so we round down. The answer is 6000.

CAUTION! 7000 is not a correct answer to Example 21(c). It is incorrect to round from the ones digit over, as follows:

6485, 6490, 6500, 7000.

Do Exercises 40–51.

There are many methods of rounding. For example, in computer applications, the rounding of 8563 to the nearest hundred might be done using a different rule called **truncating**, meaning that we simply change all digits to the right of the rounding location to zeros. Thus, 8563 would round to 8500, which is not the same answer that we would get using the rule discussed in this section.

Estimating

Estimating is used to simplify a problem so that it can then be solved easily or mentally. Rounding is used when estimating. There are many ways to estimate.

Example 22 Michelle earned $21,791 as a consultant and $17,239 as an instructor in a recent year. Estimate Michelle's yearly earnings.

One way to estimate is to round each number to the nearest thousand and then add.

$$\begin{array}{r} 21,791 \\ +\ 17,239 \\ \hline \end{array} \qquad \begin{array}{r} 22,000 \\ +\ 17,000 \\ \hline 39,000 \end{array} \leftarrow \text{Estimated answer}$$

We often use the symbol $\approx$ instead of $=$ when approximating or estimating. The symbol $\approx$ means **"is approximately equal to."** In a situation such as the one in Example 22, we might write

$$\$21,791 + \$17,239 \approx \$39,000.$$

Example 23 Estimate this sum by first rounding to the nearest hundred:

850 + 674 + 986 + 839.

We have

```
  8 5 0        9 0 0
  6 7 4        7 0 0
  9 8 6      1 0 0 0
+ 8 3 9    +   8 0 0
           ---------
             3 4 0 0
```

Example 24 Estimate the difference by first rounding to the nearest thousand: 9324 − 2849.

We have

```
  9 3 2 4      9 0 0 0
− 2 8 4 9    − 3 0 0 0
             ---------
               6 0 0 0
```

Do Exercises 52–54.

Example 25 Estimate the following product by first rounding to the nearest ten and to the nearest hundred: 683 × 457.

```
Nearest ten     Nearest hundred     Exact
      6 8 0            7 0 0           6 8 3
×     4 6 0     ×      5 0 0     ×     4 5 7
  4 0 8 0 0     3 5 0 0 0 0          4 7 8 1
2 7 2 0 0 0                        3 4 1 5 0
3 1 2 8 0 0                      2 7 3 2 0 0
                                 3 1 2 1 3 1
```

Note in Example 25 that the estimate, having been rounded to the nearest ten, is

312,800.

The estimate, having been rounded to the nearest hundred, is

350,000.

Note how the estimates compare to the exact answer,

312,131.

Why does rounding give a larger answer than the exact one?

Do Exercise 55.

52. Estimate the sum by first rounding to the nearest hundred. Show your work.

```
  6 5 0
  6 8 5
  2 3 8
+ 1 6 8
```

53. Estimate the difference by first rounding to the nearest hundred. Show your work.

```
  9 2 8 5
− 6 7 3 9
```

54. Estimate the difference by first rounding to the nearest thousand. Show your work.

```
  2 3,2 7 8
− 1 1,6 9 8
```

55. Estimate the product by first rounding to the nearest ten and the nearest hundred. Show your work.

```
  8 3 7
× 2 4 5
```

Answers on page A-2

Improving Your Math Study Skills

Getting Started in a Math Class: The First-Day Handout or Syllabus

There are many ways in which to improve your math study skills. We have already considered some tips on using this book (see Section 1.2). We now consider some more general tips.

- **Textbook.** On the first day of class, most instructors distribute a handout that lists the textbook and other materials needed in the course. If possible, call the instructor or the department office before the term begins to find out which textbook you will be using and visit the bookstore to pick it up. This way, you can purchase the book before class starts and be ready to begin studying. Delay in obtaining a copy of the textbook may cause you to fall behind in your homework.
- **Attendance.** The handout may also describe the attendance policy for your class. Some instructors take attendance at every class, while others use different methods to track students' attendance. Regardless of the policy, you should plan to attend class every time. Missing even one class can cause you to fall behind. If attendance counts toward your course grade, find out if there is a way to make up for missed days. In general, missing a class is not as catastrophic if you put in the effort to catch up by studying the material on your own.

  If you do miss a class, call the instructor as soon as possible to find out what material was covered and what was assigned for the next class. If you have a study partner, call this person; ask if you can make a copy of his or her notes and find out what the homework assignment was. It is a good idea to meet with your instructor in person to clarify any concepts that you do not understand. This way, when you do return to class, you will be able to follow along with the rest of the group.
- **Homework.** The first-day handout may also detail how homework is handled. Find out when, and how often, homework will be assigned, whether homework is collected or graded, and whether there will be quizzes over the homework material.

  If the homework will be graded, find out what part of the final grade it will determine. Ask what the policy is for late homework: Some instructors are willing to accept homework after the deadline, while others are more strict. If you do miss a homework deadline, be sure to do the assigned homework anyway, as this is the best way to learn the material.
- **Grading.** The handout may also provide information on how your grade will be calculated at the end of the term. Typically, there will be tests during the term and a final exam at the end of the term. Frequently, homework is counted as part of the grade calculation, as are the quizzes. Find out how many tests will be given, if there is an option for make-up tests, or if any test grades will be dropped at the end of the term.

  Some instructors keep the class grades on a computer. If this is the case, find out if you can receive current grade reports throughout the term. This will help you focus on what is needed to obtain the desired grade in the course. Although a good grade should not be your only goal in this class, most students find it motivational to know what their grade is at any time during the term.
- **Get to know your classmates.** It can be a big help in a math class to get to know your fellow students. You might consider forming a study group. If you do so, find out their phone numbers and schedules so that you can coordinate study time for homework or tests.
- **Get to know your instructor.** It can, of course, help immensely to get to know your instructor. Trivial though it may seem, get basic information like his or her name, how he or she can be contacted outside of class, and where the office is.

  Learn about your instructor's teaching style and try to adapt your learning to it. Does he or she use an overhead projector or the board? Will there be frequent in-class questions?

Exercise Set 1.3

a Write a multiplication sentence that corresponds to the situation.

1. The *Los Angeles Sunday Times* crossword puzzle is arranged rectangularly with squares in 21 rows and 21 columns. How many squares does the puzzle have altogether?

2. *Pixels.* A computer screen consists of small rectangular dots called *pixels.* How many pixels are there on a screen that has 600 rows with 800 pixels in each row?

What is the area of the region?

3.

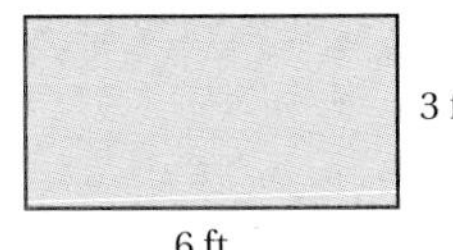

4.

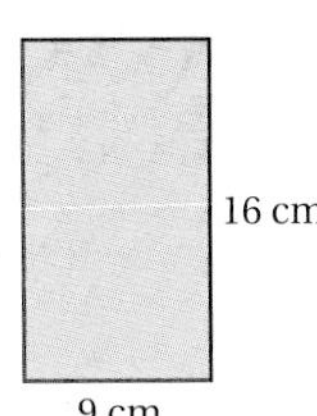

5.

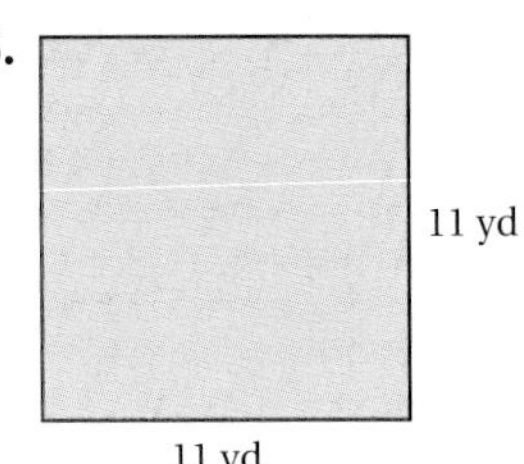

6.

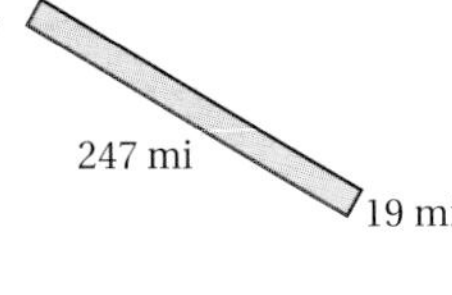

b Multiply.

7. $\begin{array}{r} 100 \\ \times\ 96 \\ \hline \end{array}$

8. $\begin{array}{r} 2340 \\ \times\ 1000 \\ \hline \end{array}$

9. $\begin{array}{r} 94 \\ \times\ 6 \\ \hline \end{array}$

10. $\begin{array}{r} 76 \\ \times\ 9 \\ \hline \end{array}$

11. $3 \cdot 509$

12. $7 \cdot 806$

13. $7(9229)$

14. $4(7867)$

15. $90(53)$

16. $60(78)$

17. $(47)(85)$

18. $(34)(87)$

19. $\begin{array}{r} 640 \\ \times\ 72 \\ \hline \end{array}$

20. $\begin{array}{r} 666 \\ \times\ 66 \\ \hline \end{array}$

21. $\begin{array}{r} 444 \\ \times\ 33 \\ \hline \end{array}$

22. $\begin{array}{r} 509 \\ \times\ 88 \\ \hline \end{array}$

23. $\begin{array}{r} 509 \\ \times\ 408 \\ \hline \end{array}$

24. $\begin{array}{r} 432 \\ \times\ 375 \\ \hline \end{array}$

25. $\begin{array}{r} 853 \\ \times\ 936 \\ \hline \end{array}$

26. $\begin{array}{r} 346 \\ \times\ 650 \\ \hline \end{array}$

27. $\begin{array}{r} 6428 \\ \times\ 3224 \\ \hline \end{array}$

28. $\begin{array}{r} 8928 \\ \times\ 3172 \\ \hline \end{array}$

29. $\begin{array}{r} 3482 \\ \times\ 104 \\ \hline \end{array}$

30. $\begin{array}{r} 6408 \\ \times\ 6064 \\ \hline \end{array}$

31. $\begin{array}{r} 5006 \\ \times 4008 \\ \hline \end{array}$

32. $\begin{array}{r} 6789 \\ \times 2330 \\ \hline \end{array}$

33. $\begin{array}{r} 5608 \\ \times 4500 \\ \hline \end{array}$

34. $\begin{array}{r} 4560 \\ \times 7890 \\ \hline \end{array}$

c Write a division sentence that corresponds to the situation. You need not carry out the division.

35. *Canyonlands.* The trail boss for a trip into Canyonlands National Park divides 760 pounds (lb) of equipment among 4 mules. How many pounds does each mule carry?

36. *Surf Expo.* In a swimwear showing at Surf Expo, a trade show for retailers of beach supplies, each swimsuit test takes 8 minutes (min). If the show runs for 240 min, how many tests can be scheduled?

d Write a related multiplication sentence.

37. $18 \div 3 = 6$

38. $72 \div 9 = 8$

39. $22 \div 22 = 1$

40. $32 \div 1 = 32$

Write two related division sentences.

41. $9 \times 5 = 45$

42. $2 \cdot 7 = 14$

43. $37 \cdot 1 = 37$

44. $4 \cdot 12 = 48$

e Divide.

45. $277 \div 5$

46. $699 \div 3$

47. $864 \div 8$

48. $869 \div 8$

49. $4\overline{)1228}$

50. $3\overline{)2124}$

51. $738 \div 8$

52. $881 \div 6$

53. $5\overline{)8515}$

54. $3\overline{)6027}$

55. $30\overline{)875}$

56. $40\overline{)987}$

57. $85\overline{)7672}$

58. $54\overline{)2729}$

59. $111\overline{)3219}$

60. $102\overline{)5612}$

61. $24\overline{)8880}$

62. $36\overline{)7563}$

63. $28\overline{)17{,}067}$

64. $36\overline{)28{,}929}$

f Round to the nearest ten.

65. 48

66. 17

67. 67

68. 99

69. 731

70. 532

71. 895

72. 798

Round to the nearest hundred.

73. 146

74. 874

75. 957

76. 650

77. 9079

78. 4645

79. 32,850

80. 198,402

Round to the nearest thousand.

81. 5876

82. 4500

83. 7500

84. 2001

85. 45,340

86. 735,562

87. 373,405

88. 6,713,855

Estimate the sum or difference by first rounding to the nearest ten. Show your work.

89. $\begin{array}{r} 78 \\ +\ 97 \\ \hline \end{array}$

90. $\begin{array}{r} 62 \\ 97 \\ 46 \\ +\ 88 \\ \hline \end{array}$

91. $\begin{array}{r} 8074 \\ -\ 2347 \\ \hline \end{array}$

92. $\begin{array}{r} 673 \\ -\ \ 28 \\ \hline \end{array}$

Estimate the sum or difference by first rounding to the nearest hundred. Show your work.

93. $\begin{array}{r} 7348 \\ +\ 9247 \\ \hline \end{array}$

94. $\begin{array}{r} 568 \\ 472 \\ 938 \\ +\ 402 \\ \hline \end{array}$

95. $\begin{array}{r} 6852 \\ -\ 1748 \\ \hline \end{array}$

96. $\begin{array}{r} 9438 \\ -\ 2787 \\ \hline \end{array}$

Estimate the sum or difference by first rounding to the nearest thousand. Show your work.

97. $\begin{array}{r} 9643 \\ 4821 \\ 8943 \\ +\ 7004 \\ \hline \end{array}$

98. $\begin{array}{r} 7648 \\ 9348 \\ 7842 \\ +\ 2222 \\ \hline \end{array}$

99. $\begin{array}{r} 92{,}149 \\ -\ 22{,}555 \\ \hline \end{array}$

100. $\begin{array}{r} 84{,}890 \\ -\ 11{,}110 \\ \hline \end{array}$

Estimate the product by first rounding to the nearest ten. Show your work.

101. $\begin{array}{r} 45 \\ \times 67 \\ \hline \end{array}$

102. $\begin{array}{r} 51 \\ \times 78 \\ \hline \end{array}$

103. $\begin{array}{r} 34 \\ \times 29 \\ \hline \end{array}$

104. $\begin{array}{r} 63 \\ \times 54 \\ \hline \end{array}$

Estimate the product by first rounding to the nearest hundred. Show your work.

105. $\begin{array}{r} 876 \\ \times 345 \\ \hline \end{array}$

106. $\begin{array}{r} 355 \\ \times 299 \\ \hline \end{array}$

107. $\begin{array}{r} 432 \\ \times 199 \\ \hline \end{array}$

108. $\begin{array}{r} 789 \\ \times 434 \\ \hline \end{array}$

Estimate the product by first rounding to the nearest thousand. Show your work.

109. $\begin{array}{r} 5608 \\ \times 4576 \\ \hline \end{array}$

110. $\begin{array}{r} 2344 \\ \times 6123 \\ \hline \end{array}$

111. $\begin{array}{r} 7888 \\ \times 6224 \\ \hline \end{array}$

112. $\begin{array}{r} 6501 \\ \times 3449 \\ \hline \end{array}$

Skill Maintenance

113. Write expanded notation for 7882. [1.1a]

114. Use < or > for ▒ to write a true sentence: [1.1f]

888 ▒ 788.

Write a related addition sentence. [1.2d]

115. $21 - 16 = 5$

116. $56 - 14 = 42$

Write two related subtraction sentences. [1.2d]

117. $47 + 9 = 56$

118. $350 + 64 = 414$

Synthesis

119. ◆ Describe a situation that corresponds to the division $1180 \div 295$. (See Examples 8 and 9.)

120. ◆ Is division associative? Why or why not?

121. A group of 1231 college students is going to take buses for a field trip. Each bus can hold only 42 students. How many buses are needed?

122. 🖩 Fill in the missing digits to make the equation true:

34,584,132 ÷ 76▒ = 4▒,386.

123. 🖩 An 18-story office building is box-shaped. Each floor measures 172 ft by 84 ft with a 20-ft by 35-ft rectangular area lost to an elevator and a stairwell. How much area is available as office space?

Collaborative Learning Manual

Multiply numbers using the Russian peasant multiplication method.

1.4 Solving Equations

a Solutions by Trial

Let's find a number that we can put in the blank to make this sentence true:

$9 = 3 + \square$.

We are asking "9 is 3 plus what number?" The answer is 6.

$9 = 3 + \boxed{6}$

Do Exercises 1 and 2.

A sentence with = is called an **equation.** A **solution** of an equation is a number that makes the sentence true. Thus, 6 is a solution of

$9 = 3 + \square$ because $9 = 3 + \boxed{6}$ is true.

However, 7 is not a solution of

$9 = 3 + \square$ because $9 = 3 + \boxed{7}$ is false.

Do Exercises 3 and 4.

We can use a letter instead of a blank. For example,

$9 = 3 + x$.

We call x a **variable** because it can represent any number.

> A **solution** is a replacement for the variable that makes the equation true. When we find all the solutions, we say that we have **solved** the equation.

Example 1 Solve $x + 12 = 27$ by trial.

We replace x with several numbers.

If we replace x with 13, we get a false equation: $13 + 12 = 27$.
If we replace x with 14, we get a false equation: $14 + 12 = 27$.
If we replace x with 15, we get a true equation: $15 + 12 = 27$.

No other replacement makes the equation true, so the solution is 15.

Examples Solve.

2. $7 + n = 22$
(7 plus what number is 22?)
The solution is 15.

3. $8 \cdot 23 = y$
(8 times 23 is what?)
The solution is 184.

Note, as in Example 3, that when the variable is alone on one side of the equation, the other side shows us what calculations to do in order to find the solution.

Do Exercises 5–8.

Objectives

a Solve simple equations by trial.

b Solve equations like $t + 28 = 54$, $28 \cdot x = 168$, and $98 \div 2 = y$.

For Extra Help

TAPE 2

MAC
WIN

CD-ROM

Find a number that makes the sentence true.

1. $8 = 1 + \square$

2. $\square + 2 = 7$

3. Determine whether 7 is a solution of $\square + 5 = 9$.

4. Determine whether 4 is a solution of $\square + 5 = 9$.

Solve by trial.

5. $n + 3 = 8$

6. $x - 2 = 8$

7. $45 \div 9 = y$

8. $10 + t = 32$

Answers on page A-2

Solve.

9. $346 \times 65 = y$

10. $x = 2347 + 6675$

11. $4560 \div 8 = t$

12. $x = 6007 - 2346$

Solve.

13. $x + 9 = 17$

14. $77 = m + 32$

Answers on page A-2

b Solving Equations

We now begin to develop more efficient ways to solve certain equations. When an equation has a variable alone on one side, it is easy to see the solution or to compute it. For example, the solution of

$$x = 12$$

is 12. When a calculation is on one side and the variable is alone on the other, we can find the solution by carrying out the calculation.

Example 4 Solve: $x = 245 \times 34$.

To solve the equation, we carry out the calculation.

$$\begin{array}{r} 2\,4\,5 \\ \times \quad 3\,4 \\ \hline 9\,8\,0 \\ 7\,3\,5\,0 \\ \hline 8\,3\,3\,0 \end{array}$$

The solution is 8330.

Do Exercises 9–12.

Look at the equation

$$x + 12 = 27.$$

We can get x alone on one side of the equation by writing a related subtraction sentence:

$x = 27 - 12$ **12 gets subtracted to find the related subtraction sentence.**

$x = 15.$ **Doing the subtraction**

It is useful in our later study of algebra to think of this as "subtracting 12 *on both sides.*" Thus,

$x + 12 - 12 = 27 - 12$ **Subtracting 12 on both sides**

$x + 0 = 15$ **Carrying out the subtraction**

$x = 15.$

> To solve $x + a = b$, subtract a on both sides.

If we can get an equation in a form with the variable alone on one side, we can "see" the solution.

Example 5 Solve: $t + 28 = 54$.

We have

$t + 28 = 54$

$t + 28 - 28 = 54 - 28$ **Subtracting 28 on both sides**

$t + 0 = 26$

$t = 26.$

The solution is 26.

Do Exercises 13 and 14.

Example 6 Solve: $182 = 65 + n$.

We have

$$
\begin{aligned}
182 &= 65 + n \\
182 - 65 &= 65 + n - 65 && \textbf{Subtracting 65 on both sides} \\
117 &= 0 + n && \textbf{65 plus } n \textbf{ minus 65 is } 0 + n. \\
117 &= n.
\end{aligned}
$$

The solution is 117.

15. Solve: $155 = t + 78$.

Do Exercise 15.

Example 7 Solve: $7381 + x = 8067$.

We have

$$
\begin{aligned}
7381 + x &= 8067 \\
7381 + x - 7381 &= 8067 - 7381 && \textbf{Subtracting 7381 on both sides} \\
x &= 686.
\end{aligned}
$$

The solution is 686.

Solve.

16. $4566 + x = 7877$

17. $8172 = h + 2058$

Do Exercises 16 and 17.

We now learn to solve equations like $8 \cdot n = 96$. Look at

$8 \cdot n = 96$.

We can get n alone by writing a related division sentence:

$$
\begin{aligned}
n = 96 \div 8 &= \frac{96}{8} && \textbf{96 is divided by 8.} \\
&= 12. && \textbf{Doing the division}
\end{aligned}
$$

Note that $n = 12$ is easier to solve than $8 \cdot n = 96$. This is because we see easily that if we replace n on the left side with 12, we get a true sentence: $12 = 12$. The solution of $n = 12$ is 12, which is also the solution of $8 \cdot n = 96$.

It is useful in our later study of algebra to think of the preceding as "dividing by 8 *on both sides.*" Thus,

$$
\begin{aligned}
\frac{8 \cdot n}{8} &= \frac{96}{8} && \textbf{Dividing by 8 on both sides} \\
n &= 12. && \textbf{8 times } n \textbf{ divided by 8 is } n.
\end{aligned}
$$

> To solve $a \cdot x = b$, divide by a on both sides.

Answers on page A-2

Solve.

18. $8 \cdot x = 64$

19. $144 = 9 \cdot n$

20. Solve: $5152 = 8 \cdot t$.

21. Solve: $18 \cdot y = 1728$.

22. Solve: $n \cdot 48 = 4512$.

Answers on page A-2

Example 8 Solve: $10 \cdot x = 240$.

We have

$$10 \cdot x = 240$$

$$\frac{10 \cdot x}{10} = \frac{240}{10} \quad \text{Dividing by 10 on both sides}$$

$$x = 24.$$

The solution is 24.

Do Exercises 18 and 19.

Example 9 Solve: $5202 = 9 \cdot t$.

We have

$$5202 = 9 \cdot t$$

$$\frac{5202}{9} = \frac{9 \cdot t}{9} \quad \text{Dividing by 9 on both sides}$$

$$578 = t.$$

The solution is 578.

Do Exercise 20.

Example 10 Solve: $14 \cdot y = 1092$.

We have

$$14 \cdot y = 1092$$

$$\frac{14 \cdot y}{14} = \frac{1092}{14} \quad \text{Dividing by 14 on both sides}$$

$$y = 78.$$

The solution is 78.

Do Exercise 21.

Example 11 Solve: $n \cdot 56 = 4648$.

We have

$$n \cdot 56 = 4648$$

$$\frac{n \cdot 56}{56} = \frac{4648}{56} \quad \text{Dividing by 56 on both sides}$$

$$n = 83.$$

The solution is 83.

Do Exercise 22.

Exercise Set 1.4

a Solve by trial.

1. $x + 0 = 14$

2. $x - 7 = 18$

3. $y \cdot 17 = 0$

4. $56 \div m = 7$

b Solve.

5. $13 + x = 42$

6. $15 + t = 22$

7. $12 = 12 + m$

8. $16 = t + 16$

9. $3 \cdot x = 24$

10. $6 \cdot x = 42$

11. $112 = n \cdot 8$

12. $162 = 9 \cdot m$

13. $45 \times 23 = x$

14. $23 \times 78 = y$

15. $t = 125 \div 5$

16. $w = 256 \div 16$

17. $p = 908 - 458$

18. $9007 - 5667 = m$

19. $x = 12{,}345 + 78{,}555$

20. $5678 + 9034 = t$

21. $3 \cdot m = 96$

22. $4 \cdot y = 96$

23. $715 = 5 \cdot z$

24. $741 = 3 \cdot t$

25. $10 + x = 89$

26. $20 + x = 57$

27. $61 = 16 + y$

28. $53 = 17 + w$

29. $6 \cdot p = 1944$

30. $4 \cdot w = 3404$

31. $5 \cdot x = 3715$

32. $9 \cdot x = 1269$

33. $47 + n = 84$

34. $56 + p = 92$

35. $x + 78 = 144$

36. $z + 67 = 133$

37. $165 = 11 \cdot n$

38. $660 = 12 \cdot n$

39. $624 = t \cdot 13$

40. $784 = y \cdot 16$

41. $x + 214 = 389$

42. $x + 221 = 333$

43. $567 + x = 902$

44. $438 + x = 807$

45. $18 \cdot x = 1872$

46. $19 \cdot x = 6080$

47. $40 \cdot x = 1800$

48. $20 \cdot x = 1500$

49. $2344 + y = 6400$

50. $9281 = 8322 + t$

51. $8322 + 9281 = x$

52. $9281 - 8322 = y$

53. $234 \times 78 = y$

54. $10{,}534 \div 458 = q$

55. $58 \cdot m = 11{,}890$

56. $233 \cdot x = 22{,}135$

Skill Maintenance

57. Write two related subtraction sentences: $7 + 8 = 15$. [1.2d]

58. Write two related division sentences: $6 \cdot 8 = 48$. [1.3d]

Use $>$ or $<$ for $\square$ to write a true sentence. [1.1f]

59. $123 \ \square \ 789$

60. $342 \ \square \ 339$

Divide. [1.3e]

61. $1283 \div 9$

62. $17 \overline{)5689}$

Synthesis

63. ◆ Describe a procedure that can be used to convert any equation of the form $a + b = c$ to a related subtraction equation.

64. ◆ Describe a procedure that can be used to convert any equation of the form $a \cdot b = c$ to a related division equation.

Solve.

65. 🖩 $23{,}465 \cdot x = 8{,}142{,}355$

66. 🖩 $48{,}916 \cdot x = 14{,}332{,}388$

1.5 Applications and Problem Solving

Objective

a Solve applied problems involving addition, subtraction, multiplication, or division of whole numbers.

For Extra Help

TAPE 2 — InterAct math MAC WIN — CD-ROM

a Applications and problem solving are the main uses of mathematics. To solve a problem using the operations on the whole numbers, we first look at the situation. We try to translate the problem to an equation. Then we solve the equation. We check to see if the solution of the equation is a solution of the original problem. Thus we are using the following five-step strategy.

FIVE STEPS FOR PROBLEM SOLVING

1. *Familiarize* yourself with the situation. If it is described in words, as in a textbook, *read carefully.* In any case, think about the situation. Draw a picture whenever it makes sense to do so. Choose a letter, or *variable,* to represent the unknown quantity to be solved for.
2. *Translate* the problem to an equation.
3. *Solve* the equation.
4. *Check* the answer in the original wording of the problem.
5. *State* the answer to the problem clearly with appropriate units.

Example 1 *Minivan Sales.* Recently, sales of minivans have soared. The bar graph at right shows the number of Chrysler Town & Country LXi minivans sold in recent years. Find the total number of minivans sold during those years.

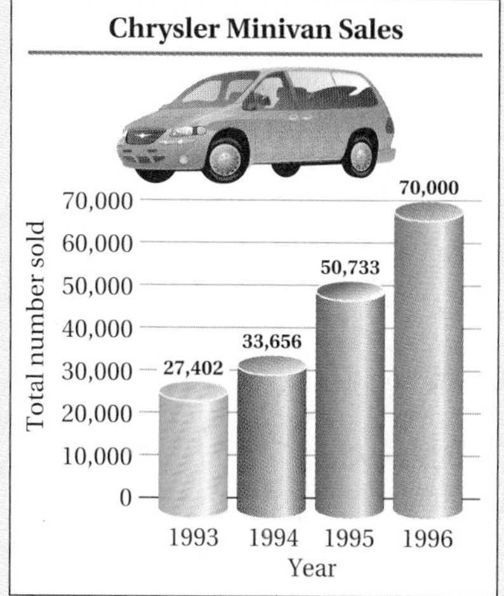

Source: Chrysler Corporation

1. **Familiarize.** We can make a drawing or at least visualize the situation.

27,402	+	33,656	+	50,733	+	70,000	=	n
in 1993		in 1994		in 1995		in 1996		Total sold

Since we are combining objects, addition can be used. First we define the unknown. We let $n =$ the total number of minivans sold.

2. **Translate.** We translate to an equation:

$$27{,}402 + 33{,}656 + 50{,}733 + 70{,}000 = n.$$

3. **Solve.** We solve the equation by carrying out the addition.

$$\begin{array}{r} \scriptstyle 1\ 1\ \ \ 1\ \ \ \\ 27{,}402 \\ 33{,}656 \\ 50{,}733 \\ +\ 70{,}000 \\ \hline 181{,}791 \end{array}$$

Thus, $181{,}791 = n$, or $n = 181{,}791$.

4. **Check.** We check 181,791 in the original problem. There are many ways in which this can be done. For example, we can repeat the calculation. (We leave this to the student.) Another way is to check the reasonableness of the answer. In this case, we would expect the answer to be larger than the sales in any of the individual years, which it is. We can also estimate by rounding. Here, we round to the nearest thousand:

$$\begin{aligned} &27{,}402 + 33{,}656 + 50{,}733 + 70{,}000 \\ &\approx 27{,}000 + 34{,}000 + 51{,}000 + 70{,}000 \\ &= 182{,}000. \end{aligned}$$

1. *Teacher needs in 2005.* The data in the table show the estimated number of new jobs for teachers in the year 2005. The reason is an expected boom in the number of youngsters under the age of 18. Find the total number of jobs available for teachers in 2005.

Type of Teacher	Number of New Jobs
Secondary	386,000
Aides	364,000
Childcare workers	248,000
Elementary	220,000
Special education	206,000

Source: Bureau of Labor Statistics

2. *Checking Account.* You have \$756 in your checking account. You write a check for \$387 to pay for a VCR for your campus apartment. How much is left in your checking account?

Answers on page A-2

Since 181,791 $\approx$ 182,000, we have a partial check. If we had an estimate like 236,000 or 580,000, we might be suspicious that our calculated answer is incorrect. Since our estimated answer is close to our calculation, we are further convinced that our answer checks.

5. **State.** The total number of minivans sold during these years is 181,791.

Do Exercise 1.

Example 2 *Hard-Drive Space.* The hard drive on your computer has 572 megabytes (MB) of storage space available. You install a software package called Microsoft® Office, which uses 84 MB of space. How much storage space do you have left after the installation?

1. **Familiarize.** We first make a drawing or at least visualize the situation. We let M = the amount of space left.

572 MB 84 MB

2. **Translate.** We see that this is a "take-away" situation. We translate to an equation.

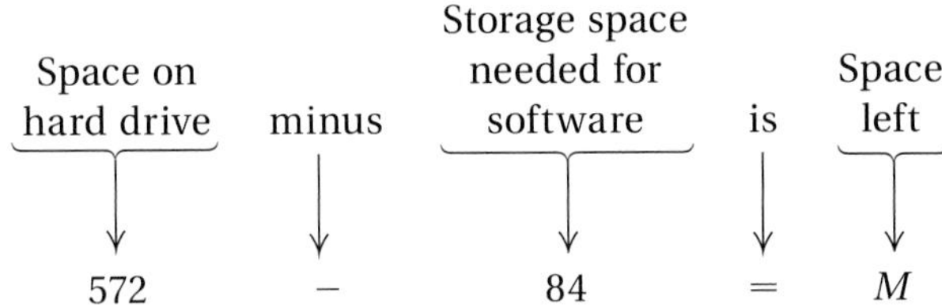

3. **Solve.** This sentence tells us what to do. We subtract.

$$\begin{array}{r} \scriptstyle 16 \\ \scriptstyle 4\ \not{6}\ 12 \\ \not{5}\,\not{7}\,\not{2} \\ -\quad 8\ 4 \\ \hline 4\ 8\ 8 \end{array}$$

Thus, $488 = M$, or $M = 488$.

4. **Check.** We check our answer of 488 MB by repeating the calculation. We note that the answer should be less than the original amount of memory, 572 MB, which it is. We can also add the difference, 488, to the subtrahend, 84: $84 + 488 = 572$. We can also estimate:

$$572 - 84 \approx 600 - 100 = 500 \approx 488.$$

5. **State.** There is 488 MB of memory left.

Do Exercise 2.

In the real world, problems may not be stated in written words. You must still become familiar with the situation before you can solve the problem.

Example 3 *Travel Distance.* Vicki is driving from Indianapolis to Salt Lake City to work during summer vacation. The distance from Indianapolis to Salt Lake City is 1634 mi. She travels 1154 mi to Denver. How much farther must she travel?

1. **Familiarize.** We first make a drawing or at least visualize the situation. We let x = the remaining distance to Salt Lake City.

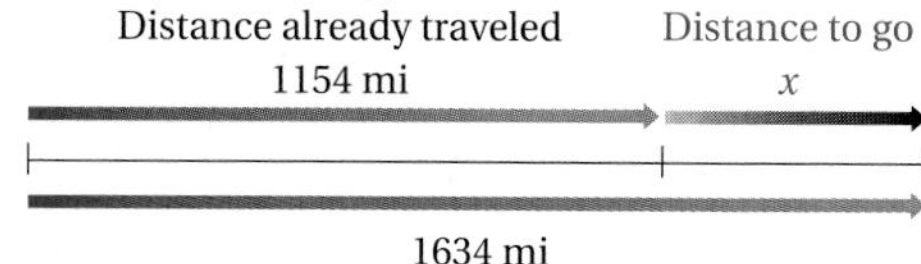

2. **Translate.** We see that this is a "how-much-more" situation. We translate to an equation.

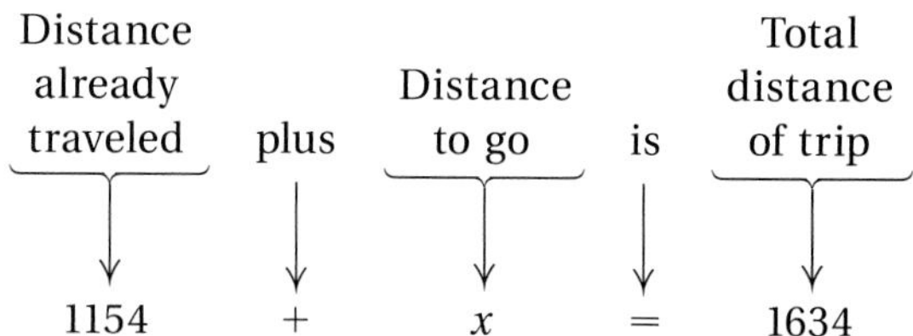

3. **Solve.** We solve the equation.

$$1154 + x = 1634$$
$$1154 + x - 1154 = 1634 - 1154 \quad \text{Subtracting 1154 on both sides}$$
$$x = 480$$

$$\begin{array}{r} \overset{5\ 13}{1\,\not{6}\,\not{3}\,4} \\ -\ 1\,1\,5\,4 \\ \hline 4\,8\,0 \end{array}$$

4. **Check.** We check our answer of 480 mi in the original problem. This number should be less than the total distance, 1634 mi, which it is. We can add the difference, 480, to the subtrahend, 1154: $1154 + 480 = 1634$. We can also estimate:

$$1634 - 1154 \approx 1600 - 1200$$
$$= 400 \approx 480.$$

The answer, 480 mi, checks.

5. **State.** Vicki must travel 480 mi farther to Salt Lake City.

Do Exercise 3.

3. *Calculator Purchase.* Bernardo has $76. He wants to purchase a graphing calculator for $94. How much more does he need?

Answer on page A-2

4. *Total Cost of Laptop Computers.* What is the total cost of 12 laptop computers with CD-ROM drives and matrix color if each one costs $3249?

Answer on page A-2

Example 4 *Total Cost of VCRs.* What is the total cost of 5 four-head VCRs if each one costs $289?

1. **Familiarize.** We first make a drawing or at least visualize the situation. We let n = the cost of 5 VCRs. Repeated addition works well here.

2. **Translate.** We translate to an equation.

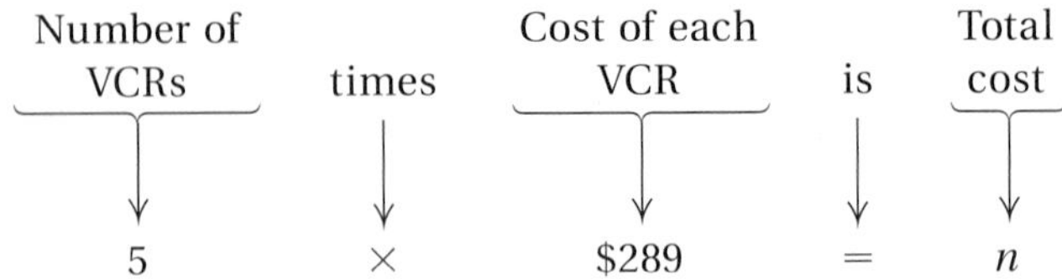

3. **Solve.** This sentence tells us what to do. We multiply.

$$\begin{array}{r} {}^{4\,4} \\ 2\,8\,9 \\ \times \quad 5 \\ \hline 1\,4\,4\,5 \end{array}$$

Thus, $n = 1445$.

4. **Check.** We have an answer that is much larger than the cost of any individual VCR, which is reasonable. We can repeat our calculation. We can also check by estimating:

$$5 \times 289 \approx 5 \times 300 = 1500 \approx 1445.$$

The answer checks.

5. **State.** The total cost of 5 VCRs is $1445.

Do Exercise 4.

Example 5 *Bed Sheets.* The dimensions of a sheet for a king-size bed are 108 in. by 102 in. What is the area of the sheet? (The dimension labels on sheets list width × length.)

1. **Familiarize.** We first make a drawing. We let A = the area.

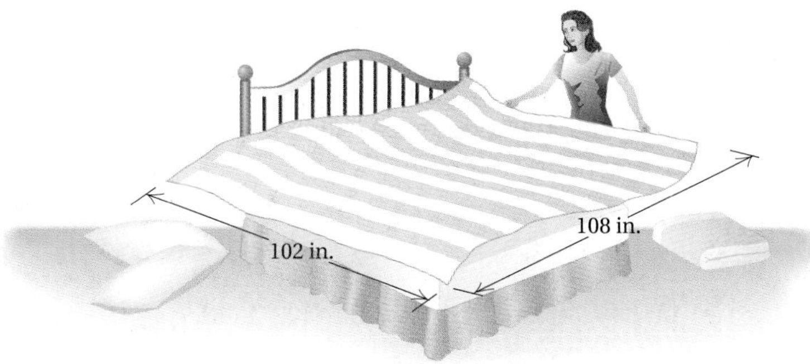

2. **Translate.** Using a formula for area, we have

$$A = \text{length} \cdot \text{width} = l \cdot w = 102 \cdot 108.$$

3. **Solve.** We carry out the multiplication.

$$\begin{array}{r} 1\,0\,8 \\ \times \quad 1\,0\,2 \\ \hline 2\,1\,6 \\ 1\,0\,8\,0\,0 \\ \hline 1\,1\,0\,1\,6 \end{array}$$

Thus, $A = 11{,}016$.

4. **Check.** We repeat our calculation. We also note that the answer is larger than either the length or the width, which it should be. (This might not be the case if we were using decimals.) The answer checks.
5. **State.** The area of a king-size bed sheet is 11,016 sq in.

Do Exercise 5.

Example 6 *Diet Cola Packaging.* Diet Cola has become very popular in the quest to control our weight. A bottling company produces 2203 cans of cola. How many 8-can packages can be filled? How many cans will be left over?

1. **Familiarize.** We first draw a picture. We let n = the number of 8-can packages to be filled.

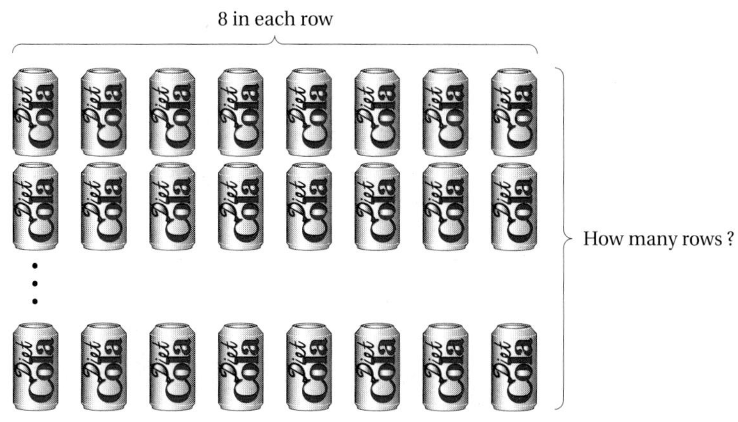

5. *Bed Sheets.* The dimensions of a sheet for a queen-size bed are 90 in. by 102 in. What is the area of the sheet?

Answer on page A-2

6. *Diet Cola Packaging.* The bottling company also uses 6-can packages. How many 6-can packages can be filled with 2269 cans of cola? How many cans will be left over?

Answer on page A-2

2. **Translate.** We can translate to an equation as follows.

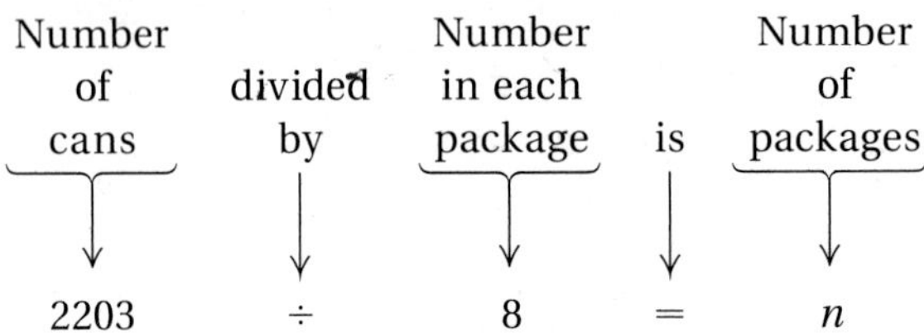

Number of cans
divided by
Number in each package
is
Number of packages
2203
÷
8
=
n

3. **Solve.** We solve the equation by carrying out the division.

```
      2 7 5
  8 ) 2 2 0 3
      1 6 0 0
        6 0 3
        5 6 0
          4 3
          4 0
            3
```

4. **Check.** We can check by multiplying the number of packages by 8 and adding the remainder, 3:

$8 \cdot 275 = 2200, \qquad 2200 + 3 = 2203.$

5. **State.** Thus, 275 8-can packages can be filled. There will be 3 cans left over.

Do Exercise 6.

Example 7 *Automobile Mileage.* The Chrysler Town & Country LXi minivan featured in Example 1 gets 18 miles to the gallon (mpg) in city driving. How many gallons will it use in 4932 mi of city driving?

1. **Familiarize.** We first make a drawing. It is often helpful to be descriptive about how you define a variable. In this example, we let g = the number of gallons (g comes from "gallons").

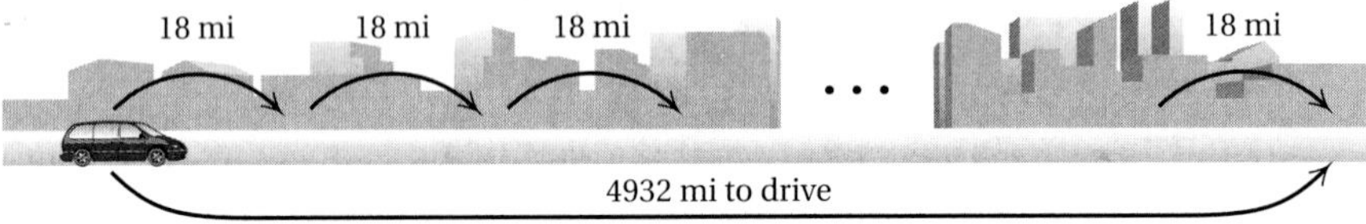

2. **Translate.** Repeated addition applies here. Thus the following multiplication corresponds to the situation.

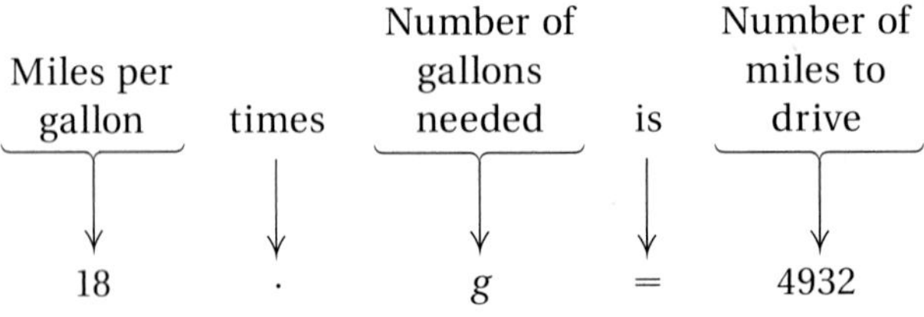

Miles per gallon
times
Number of gallons needed
is
Number of miles to drive
18
·
g
=
4932

3. Solve. To solve the equation, we divide by 18 on both sides.

$$18 \cdot g = 4932$$

$$\frac{18 \cdot g}{18} = \frac{4932}{18}$$

$$g = 274$$

$$\begin{array}{r} 274 \\ 18 \overline{)4932} \\ \underline{3600} \\ 1332 \\ \underline{1260} \\ 72 \\ \underline{72} \\ 0 \end{array}$$

4. Check. To check, we multiply 274 by 18: $18 \cdot 274 = 4932$.

5. State. The minivan will use 274 gal.

Do Exercise 7.

7. *Automobile Mileage.* The Chrysler Town & Country LXi minivan gets 24 miles to the gallon (mpg) in country driving. How many gallons will it use in 888 mi of country driving?

Answer on page A-2

Multistep Problems

Sometimes we must use more than one operation to solve a problem, as in the following example.

Example 8 *Weight Loss.* Many Americans exercise for weight control. It is known that one must burn off about 3500 calories in order to lose one pound. The chart shown here details how many calories are burned by certain activities. How long would an individual have to run at a brisk pace in order to lose one pound?

To burn off 100 calories, you must:
• Run for 8 min at a brisk pace, or
• Swim for 2 min at a brisk pace, or
• Bicycle for 15 min at 9 mph, or
• Do aerobic exercises for 15 min.

1. Familiarize. We first make a chart.

ONE POUND			
3500 calories			
100 cal 8 min	100 cal 8 min	... ...	100 cal 8 min

2. Translate. Repeated addition applies here. Thus the following multiplication corresponds to the situation. We must find out how many 100's there are in 3500. We let x = the number of 100's in 3500.

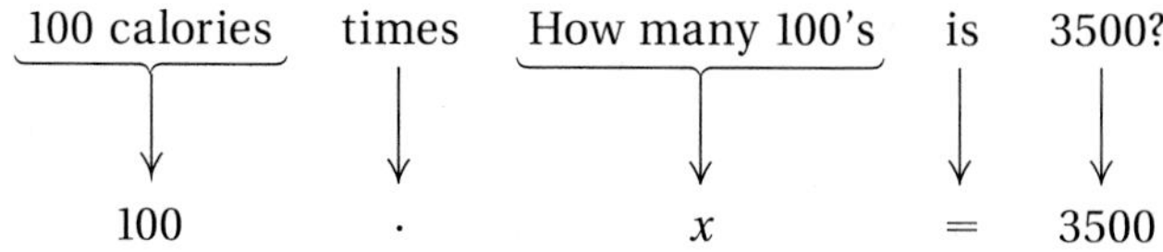

8. *Weight Loss.* Using the chart for Example 8, determine how long an individual must swim in order to lose one pound.

9. *Bones in the Hands and Feet.* There are 27 bones in each human hand and 26 bones in each human foot. How many bones are there in all in the hands and feet?

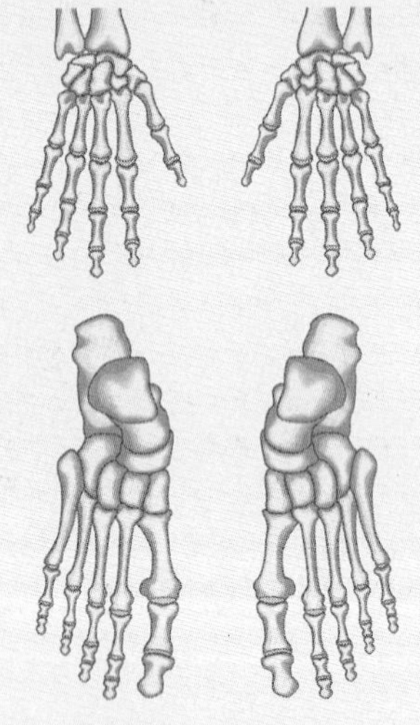

Answers on page A-2

3. *Solve.* To solve the equation, we divide by 100 on both sides.

$$100 \cdot x = 3500$$

$$\frac{100 \cdot x}{100} = \frac{3500}{100}$$

$$x = 35$$

$$\begin{array}{r} 35 \\ 100\overline{)3500} \\ \underline{3000} \\ 500 \\ \underline{500} \\ 0 \end{array}$$

We know that running for 8 min will burn off 100 calories. To do this 35 times will burn off one pound, so you must run for 35 times 8 minutes in order to burn off one pound. We let $t =$ the time it takes to run off one pound.

$$35 \times 8 = t$$

$$280 = t$$

$$\begin{array}{r} 35 \\ \times\ \ 8 \\ \hline 280 \end{array}$$

4. Check. Suppose you run for 280 min. If we divide 280 by 8, we get 35, and 35 times 100 is 3500, the number of calories it takes to lose one pound.

5. State. It will take 280 min, or 4 hr, 40 min, of running to lose one pound.

Do Exercises 8 and 9.

As you consider the following exercises, here are some words and phrases that may be helpful to look for when you are translating problems to equations.

Addition:	sum, total, increase, altogether, plus
Subtraction:	difference, minus, how much more?, how many more?, decrease, deducted, how many left?
Multiplication:	given rows and columns, how many in all?, product, total from a repeated addition, area, of
Division:	how many in each row?, how many rows?, how many pieces?, how many parts in a whole?, quotient, divisible

Exercise Set 1.5

a Solve.

1. During the first four months of a recent year, Campus Depot Business Machine Company reported the following sales:

January	\$3572
February	2718
March	2809
April	3177

What were the total sales over this time period?

2. A family travels the following miles during a five-day trip:

Monday	568
Tuesday	376
Wednesday	424
Thursday	150
Friday	224

How many miles did they travel altogether?

Bicycle Sales. The bar graph below shows the total sales, in millions of dollars, for bicycles and related supplies in recent years. Use this graph for Exercises 3–6.

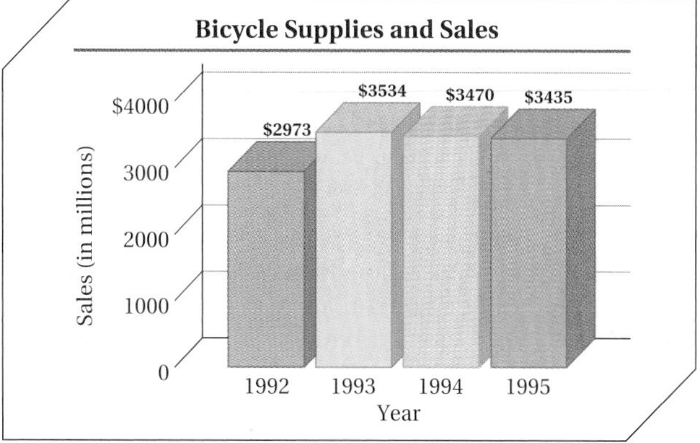

Source: National Sporting Goods Association

3. What were the total sales for 1993 and 1994?

4. What were the total sales for 1992 through 1995?

5. How much more were the sales in 1993 than in 1994?

6. How much more were the sales in 1994 than in 1992?

7. *Longest Rivers.* The longest river in the world is the Nile, which has a length of 4145 mi. It is 138 mi longer than the next longest river, which is the Amazon in South America. How long is the Amazon?

8. *Largest Lakes.* The largest lake in the world is the Caspian Sea, which has an area of 317,000 square kilometers (sq km). The Caspian is 288,900 sq km larger than the second largest lake, which is Lake Superior. What is the area of Lake Superior?

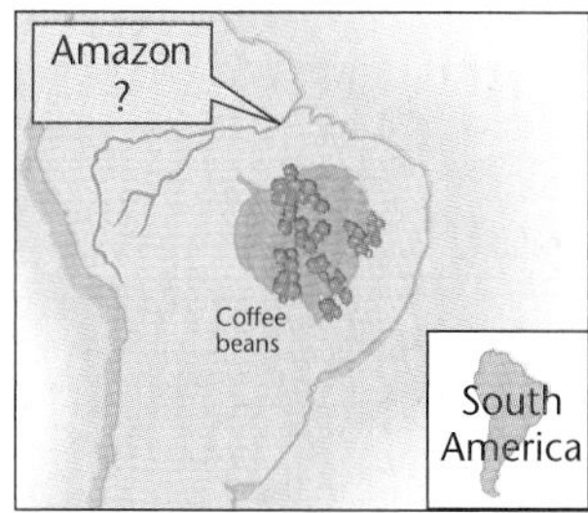

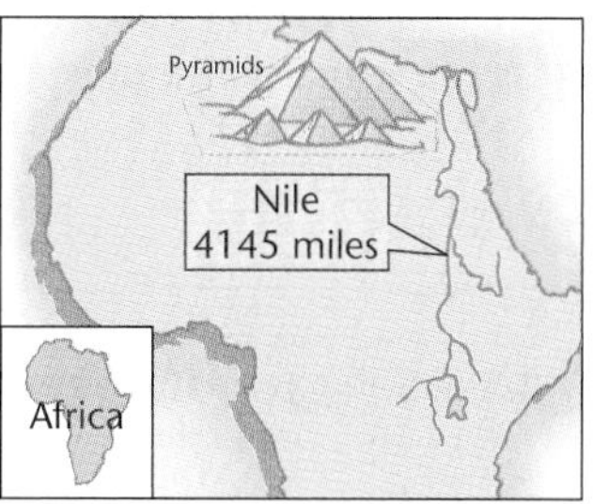

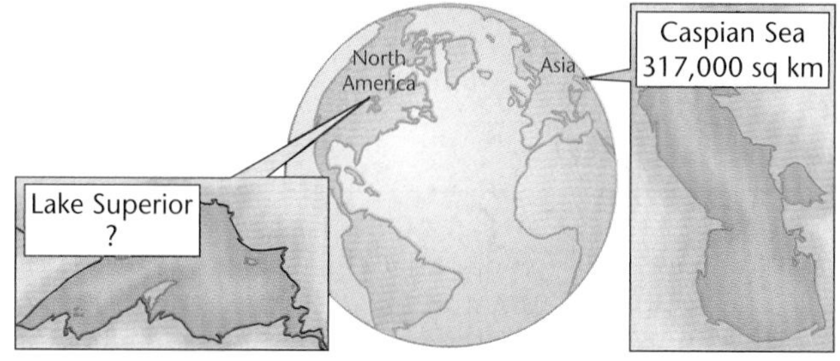

9. *Sheet Perimeter.* The dimensions of a sheet for a queen-size bed are 90 in. by 102 in. What is the perimeter of the sheet?

10. *Sheet Perimeter.* The dimensions of a sheet for a king-size bed are 108 in. by 102 in. What is the perimeter of the sheet?

11. *Paper Quantity.* A ream of paper contains 500 sheets. How many sheets are in 9 reams?

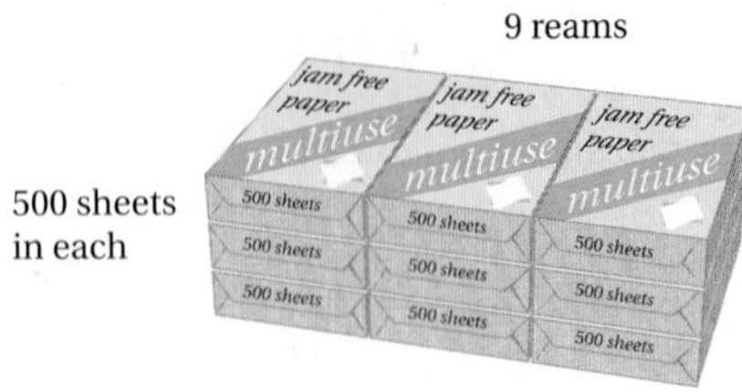

12. *Reading Rate.* Cindy's reading rate is 205 words per minute. How many words can she read in 30 min?

13. *Elvis Impersonators.* When Elvis Presley died in 1977, there were already 48 professional Elvis impersonators (***Source*:** *Chance Magazine* 9, no. 1, Winter 1996). In 1995, there were 7328. How many more were there in 1995?

14. *LAV Vehicle.* A combat-loaded U.S. Light Armed Vehicle 25 (LAV-25) weighs 3930 lb more than its empty curb weight. The loaded LAV-25 weighs 28,400 lb. (***Source*:** *Car & Driver* 42, no. 1, July 1996: 153–155) What is its curb weight?

15. Dana borrows $5928 for a used car. The loan is to be paid off in 24 equal monthly payments. How much is each payment (excluding interest)?

16. A family borrows $4824 to build a sunroom on the back of their house. The loan is to be paid off in equal monthly payments of $134 (excluding interest). How many months will it take to pay off the loan?

17. *Cheers Episodes.* *Cheers* is the longest-running comedy in the history of television, with 271 episodes created. A local station picks up the syndicated reruns. If the station runs 5 episodes per week, how many full weeks will pass before it must start over with past episodes? How many episodes will be left for the last week?

18. A lab technician separates a vial containing 70 cubic centimeters (cc) of blood into test tubes, each of which contain 3 cc of blood. How many test tubes can be filled? How much blood is left over?

19. There are 24 hours (hr) in a day and 7 days in a week. How many hours are there in a week?

20. There are 60 min in an hour and 24 hr in a day. How many minutes are there in a day?

21. You have $568 in your checking account. You write checks for $46, $87, and $129. Then you deposit $94 back in the account upon the return of some books. How much is left in your account?

22. The balance in your checking account is $749. You write checks for $34 and $65. Then you make a deposit of $123 from your paycheck. What is your new balance?

23. *NBA Court.* The standard basketball court used by college and NBA players has dimensions of 50 ft by 94 ft (***Source*:** National Basketball Association).

a) What is its area?
b) What is its perimeter?

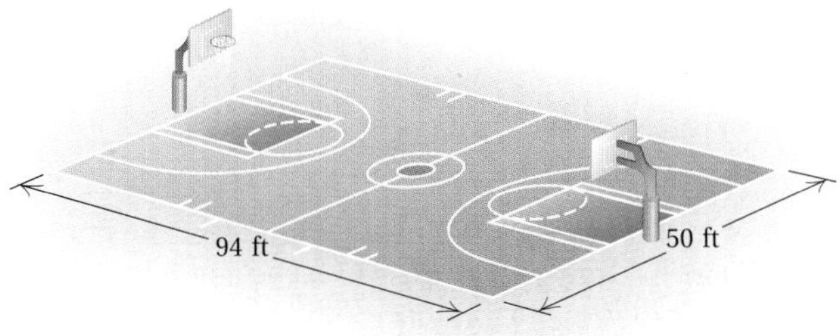

24. *High School Court.* The standard basketball court used by high school players has dimensions of 50 ft by 84 ft.

a) What is its area? What is its perimeter?
b) How much larger is the area of an NBA court than a high school court? (See Exercise 23.)

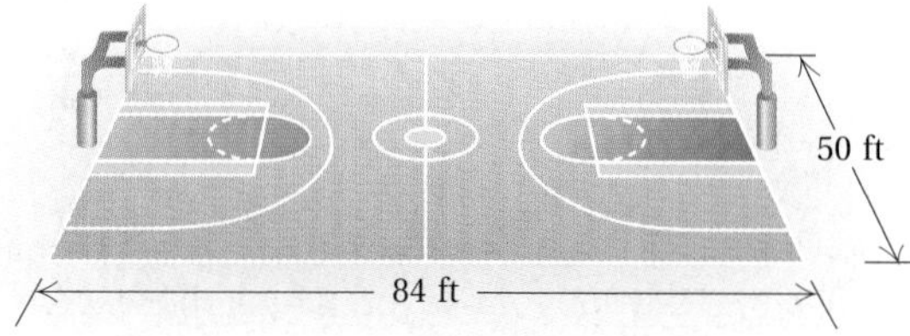

25. Copies of this book are generally shipped from the warehouse in cartons containing 24 books each. How many cartons are needed to ship 840 books?

26. Sixteen-ounce bottles of catsup are generally shipped in cartons containing 12 bottles each. How many cartons are needed to ship 528 bottles of catsup?

27. Copies of this book are generally shipped from the warehouse in cartons containing 24 books each. How many cartons are needed to ship 1355 books? How many books are left over?

28. Sixteen-ounce bottles of catsup are generally shipped in cartons containing 12 bottles each. How many cartons are needed to ship 1033 bottles of catsup? How many bottles are left over?

29. *Map Drawing.* A map has a scale of 64 mi to the inch. How far apart *in reality* are two cities that are 25 in. apart on the map? How far apart *on the map* are two cities that, in reality, are 1728 mi apart?

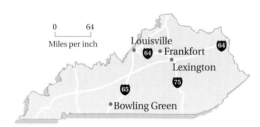

30. *Map Drawing.* A map has a scale of 25 mi to the inch. How far apart *on the map* are two cities that, in reality, are 2200 mi apart? How far apart *in reality* are two cities that are 13 in. apart on the map?

31. A carpenter drills 216 holes in a rectangular array in a pegboard. There are 12 holes in each row. How many rows are there?

32. Lou works as a CPA. He arranges 504 entries on a spreadsheet in a rectangular array that has 36 rows. How many entries are in each row?

33. Elaine buys 5 video games at \$44 each and pays for them with \$10 bills. How many \$10 bills did it take?

34. Lowell buys 5 video games at \$44 each and pays for them with \$20 bills. How many \$20 bills did it take?

35. Before going back to college, David buys 4 shirts at $59 each and 6 pairs of pants at $78 each. What was the total cost of this clothing?

36. Ann buys office supplies at Office Depot. One day she buys 8 reams of paper at $24 each and 16 pens at $3 each. How much did she spend?

37. *Weight Loss.* Use the information from the chart on page 67. How long must you do aerobic exercises in order to lose one pound?

38. *Weight Loss.* Use the information from the chart on page 67. How long must you bicycle at 9 mph in order to lose one pound?

39. *Index Cards.* Index cards of dimension 3 in. by 5 in. are normally shipped in packages containing 100 cards each. How much writing area is available if one uses the front and back sides of a package of these cards?

40. An office for adjunct instructors at a community college has 6 bookshelves, each of which is 3 ft long. The office is moved to a new location that has dimensions of 16 ft by 21 ft. Is it possible for the bookshelves to be put side by side on the 16-ft wall?

Skill Maintenance

Round 234,562 to the nearest: [1.3f]

41. Hundred.

42. Thousand.

Estimate the computation by rounding to the nearest thousand. [1.3f]

43. 2783 + 4602 + 5797 + 8111

44. 28,430 − 11,977

Estimate the product by rounding to the nearest hundred. [1.3f]

45. 787 · 363

46. 887 · 799

Synthesis

47. ◈ Of the five problem-solving steps listed at the beginning of this section, which is the most difficult for you? Why?

48. ◈ Write a problem for a classmate to solve. Design the problem so that the solution is "The driver still has 329 mi to travel."

49. ▦ *Speed of Light.* Light travels about 186,000 miles per second (mi/sec) in a vacuum as in outer space. In ice it travels about 142,000 mi/sec, and in glass it travels about 109,000 mi/sec. In 18 sec, how many more miles will light travel in a vacuum than in ice? in glass?

50. Carney Community College has 1200 students. Each professor teaches 4 classes and each student takes 5 classes. There are 30 students and 1 teacher in each classroom. How many professors are there at Carney Community College?

Make a budget for a road trip to your favorite destination.

1.6 Exponential Notation and Order of Operations

a Exponential Notation

Consider the product $3 \cdot 3 \cdot 3 \cdot 3$. Such products occur often enough that mathematicians have found it convenient to create a shorter notation, called **exponential notation,** explained as follows.

$\underbrace{3 \cdot 3 \cdot 3 \cdot 3}_{\text{4 factors}}$ is shortened to 3^4 ← exponent; 3 is the base

We read 3^4 as "three to the fourth power," 5^3 as "five to the third power," or "five cubed," and 5^2 as "five squared." The latter comes from the fact that a square of side s has area A given by $A = s^2$.

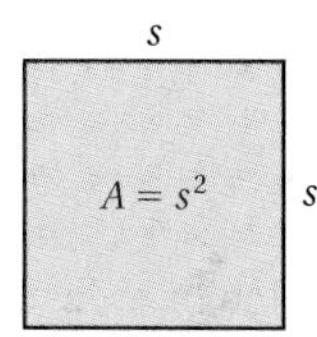

Example 1 Write exponential notation for $10 \cdot 10 \cdot 10 \cdot 10 \cdot 10$.

Exponential notation is 10^5. **5 is the *exponent*. 10 is the *base*.**

Example 2 Write exponential notation for $2 \cdot 2 \cdot 2$.

Exponential notation is 2^3.

Do Exercises 1–4.

b Evaluating Exponential Notation

We evaluate exponential notation by rewriting it as a product and computing the product.

Example 3 Evaluate: 10^3.

$10^3 = 10 \cdot 10 \cdot 10 = 1000$

Example 4 Evaluate: 5^4.

$5^4 = 5 \cdot 5 \cdot 5 \cdot 5 = 625$

Caution! 5^4 does not mean $5 \cdot 4$.

Do Exercises 5–8.

Objectives

a Write exponential notation for products such as $4 \cdot 4 \cdot 4$.

b Evaluate exponential notation.

c Simplify expressions using the rules for order of operations.

d Remove parentheses within parentheses.

For Extra Help

TAPE 2

MAC WIN

CD-ROM

Write exponential notation.

1. $5 \cdot 5 \cdot 5 \cdot 5$

2. $5 \cdot 5 \cdot 5 \cdot 5 \cdot 5$

3. $10 \cdot 10$

4. $10 \cdot 10 \cdot 10 \cdot 10$

Evaluate.

5. 10^4

6. 10^2

7. 8^3

8. 2^5

Answers on page A-2

Simplify.

9. $93 - 14 \cdot 3$

10. $104 \div 4 + 4$

11. $25 \cdot 26 - (56 + 10)$

12. $75 \div 5 + (83 - 14)$

Simplify and compare.

13. $64 \div (32 \div 2)$ and $(64 \div 32) \div 2$

14. $(28 + 13) + 11$ and $28 + (13 + 11)$

Answers on page A-2

c Simplifying Expressions

Suppose we have a calculation like the following:

$$3 + 4 \cdot 8.$$

How do we find the answer? Do we add 3 to 4 and then multiply by 8, or do we multiply 4 by 8 and then add 3? In the first case, the answer is 56. In the second, the answer is 35. We agree to compute as in the second case.

Consider the calculation

$$7 \cdot 14 - (12 + 18).$$

What do the parentheses mean? To deal with these questions, we must make some agreement regarding the order in which we perform operations. The rules are as follows.

RULES FOR ORDER OF OPERATIONS

1. Do all calculations within parentheses ( ), brackets [ ], or braces { } before operations outside.
2. Evaluate all exponential expressions.
3. Do all multiplications and divisions in order from left to right.
4. Do all additions and subtractions in order from left to right.

It is worth noting that these are the rules that a computer uses to do computations. In order to program a computer, you must know these rules.

Example 5 Simplify: $16 \div 8 \times 2$.

There are no parentheses or exponents, so we start with the third step.

$16 \div 8 \times 2 = 2 \times 2$ — **Doing all multiplications and divisions in order from left to right**

$= 4$

Example 6 Simplify: $7 \cdot 14 - (12 + 18)$.

$7 \cdot 14 - (12 + 18) = 7 \cdot 14 - 30$ — **Carrying out operations inside parentheses**

$= 98 - 30$ — **Doing all multiplications and divisions**

$= 68$ — **Doing all additions and subtractions**

Do Exercises 9–12.

Example 7 Simplify and compare: $23 - (10 - 9)$ and $(23 - 10) - 9$.

We have

$$23 - (10 - 9) = 23 - 1 = 22;$$
$$(23 - 10) - 9 = 13 - 9 = 4.$$

We can see that $23 - (10 - 9)$ and $(23 - 10) - 9$ represent different numbers. Thus subtraction is not associative.

Do Exercises 13 and 14.

Example 8 Simplify: $7 \cdot 2 - (12 + 0) \div 3 - (5 - 2)$.

$7 \cdot 2 - (12 + 0) \div 3 - (5 - 2) = 7 \cdot 2 - 12 \div 3 - 3$ — **Carrying out operations inside parentheses**

$= 14 - 4 - 3$ — **Doing all multiplications and divisions in order from left to right**

$= 7$ — **Doing all additions and subtractions in order from left to right**

Do Exercise 15.

Example 9 Simplify: $15 \div 3 \cdot 2 \div (10 - 8)$.

$15 \div 3 \cdot 2 \div (10 - 8) = 15 \div 3 \cdot 2 \div 2$ — **Carrying out operations inside parentheses**

$= 5 \cdot 2 \div 2$ — **Doing all multiplications and divisions in order from left to right**

$= 10 \div 2$

$= 5$

Do Exercises 16–18.

Example 10 Simplify: $4^2 \div (10 - 9 + 1)^3 \cdot 3 - 5$.

$4^2 \div (10 - 9 + 1)^3 \cdot 3 - 5$

$= 4^2 \div (1 + 1)^3 \cdot 3 - 5$ — **Carrying out operations inside parentheses**

$= 4^2 \div 2^3 \cdot 3 - 5$ — **Adding inside parentheses**

$= 16 \div 8 \cdot 3 - 5$ — **Evaluating exponential expressions**

$= 2 \cdot 3 - 5$

$= 6 - 5$ — **Doing all multiplications and divisions in order from left to right**

$= 1$

Do Exercises 19–21.

Calculator Spotlight

Calculators often have an [x^y], [a^x], or [^] key for raising a base to a power. To find 3^5 with such a key, we press [3] [x^y] [5] [=]. The result is 243.

1. Find 4^5. **2.** Find 7^9. **3.** Find 2^{20}.

To determine whether a calculator is programmed to follow the rules for order of operations, press [3] [+] [4] [×] [2] [=]. If the result is 11, that particular calculator follows the rules. If the result is 14, the calculator performs operations as they are entered. To compensate for the latter case, we would press [4] [×] [2] [=] [+] [3] [=].

4. Find $84 - 5 \cdot 7$. **5.** Find $80 + 50 \div 10$.

When a calculator has parentheses, [(] and [)], expressions like $5(4 + 3)$ can be found without first entering the addition. We simply press [5] [×] [(] [4] [+] [3] [)] [=]. The result is 35.

6. Find $9(7 + 8)$. **7.** Find $8[4 + 3(7 - 1)]$.

15. Simplify:

$9 \times 4 - (20 + 4$

22.

Simplify.

16. $5 \cdot 5 \cdot 5 + 26 \cdot 71 - (16 + 25 \cdot 3)$

17. $30 \div 5 \cdot 2 + 10 \cdot 20 + 8 \cdot 8 - 23$

18. $95 - 2 \cdot 2 \cdot 2 \cdot 5 \div (24 - 4)$

Simplify.

19. $5^3 + 26 \cdot 71 - (16 + 25 \cdot 3)$

20. $(1 + 3)^3 + 10 \cdot 20 + 8^2 - 23$

21. $95 - 2^3 \cdot 5 \div (24 - 4)$

Answers on page A-2

BA Tall Men. The heights, in inches, of several of the tallest players in the NBA are given in the bar graph below. Find the average height of these players.

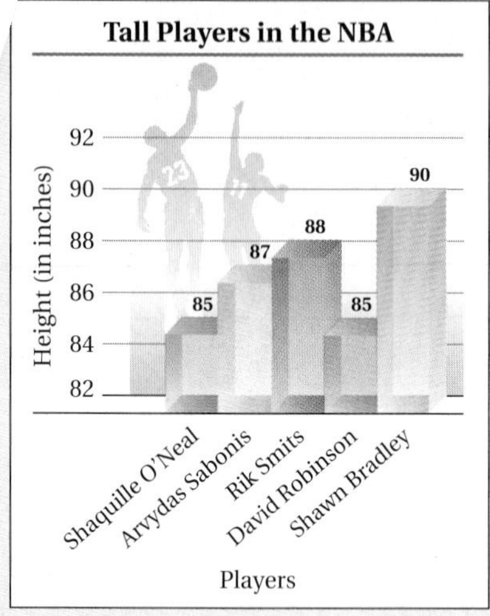

Source: NBA

Example 11 *Average Height of Waterfalls.* The heights of the four highest waterfalls in the world are given in the bar graph at right. Find the average height of all four.
To find the **average** of a set of numbers, we first add the numbers and then divide by the number of addends.

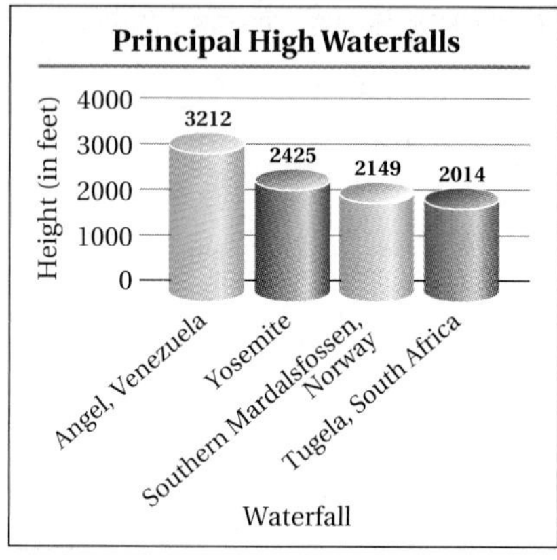

Source: World Almanac

The average is given by

$$(3212 + 2425 + 2149 + 2014) \div 4.$$

To find the average, we carry out the computation using the rules for order of operations:

$$\begin{aligned}(3212 + 2425 + 2149 + 2014) \div 4 &= 9800 \div 4 \\ &= 2450.\end{aligned}$$

Thus the average height of the four highest waterfalls is 2450 ft.

Do Exercise 22.

d Parentheses Within Parentheses

When parentheses occur within parentheses, we can make them different shapes, such as [ ] (also called "brackets") and { } (also called "braces"). All of these have the same meaning. When parentheses occur within parentheses, computations in the innermost ones are to be done first.

Example 12 Simplify: $16 \div 2 + \{40 - [13 - (4 + 2)]\}$.

$16 \div 2 + \{40 - [13 - (4 + 2)]\}$

$= 16 \div 2 + \{40 - [13 - 6]\}$ **Doing the calculations in the innermost parentheses first**

$= 16 \div 2 + \{40 - 7\}$ **Again, doing the calculations in the innermost parentheses**

$= 16 \div 2 + 33$

$= 8 + 33$ **Doing all multiplications and divisions in order from left to right**

$= 41$ **Doing all additions and subtractions in order from left to right**

Example 13 Simplify: $[25 - (4 + 3) \times 3] \div (11 - 7)$.

$$\begin{aligned}[25 - (4 + 3) \times 3] \div (11 - 7) &= [25 - 7 \times 3] \div (11 - 7) \\ &= [25 - 21] \div (11 - 7) \\ &= 4 \div 4 \\ &= 1\end{aligned}$$

Do Exercises 23 and 24.

Simplify.

23. $9 \times 5 + \{6 \div [14 - (5 + 3)]\}$

24. $[18 - (2 + 7) \div 3] - (31 - 10 \times 2)$

Answers on page A-2

Exercise Set 1.6

a Write exponential notation.

1. $3 \cdot 3 \cdot 3 \cdot 3$

2. $2 \cdot 2 \cdot 2 \cdot 2 \cdot 2$

3. $5 \cdot 5$

4. $13 \cdot 13 \cdot 13$

5. $7 \cdot 7 \cdot 7 \cdot 7 \cdot 7$

6. $10 \cdot 10$

7. $10 \cdot 10 \cdot 10$

8. $1 \cdot 1 \cdot 1 \cdot 1$

b Evaluate.

9. 7^2

10. 5^3

11. 9^3

12. 10^2

13. 12^4

14. 10^5

15. 11^2

16. 6^3

c Simplify.

17. $12 + (6 + 4)$

18. $(12 + 6) + 18$

19. $52 - (40 - 8)$

20. $(52 - 40) - 8$

21. $1000 \div (100 \div 10)$

22. $(1000 \div 100) \div 10$

23. $(256 \div 64) \div 4$

24. $256 \div (64 \div 4)$

25. $(2 + 5)^2$

26. $2^2 + 5^2$

27. $(11 - 8)^2 - (18 - 16)^2$

28. $(32 - 27)^3 + (19 + 1)^3$

29. $16 \cdot 24 + 50$

30. $23 + 18 \cdot 20$

31. $83 - 7 \cdot 6$

32. $10 \cdot 7 - 4$

33. $10 \cdot 10 - 3 \cdot 4$

34. $90 - 5 \cdot 5 \cdot 2$

35. $4^3 \div 8 - 4$

36. $8^2 - 8 \cdot 2$

37. $17 \cdot 20 - (17 + 20)$

38. $1000 \div 25 - (15 + 5)$

39. $6 \cdot 10 - 4 \cdot 10$

40. $3 \cdot 8 + 5 \cdot 8$

41. $300 \div 5 + 10$

42. $144 \div 4 - 2$

43. $3 \cdot (2 + 8)^2 - 5 \cdot (4 - 3)^2$

44. $7 \cdot (10 - 3)^2 - 2 \cdot (3 + 1)^2$

45. $4^2 + 8^2 \div 2^2$

46. $6^2 - 3^4 \div 3^3$

47. $10^3 - 10 \cdot 6 - (4 + 5 \cdot 6)$

48. $7^2 + 20 \cdot 4 - (28 + 9 \cdot 2)$

49. $6 \cdot 11 - (7 + 3) \div 5 - (6 - 4)$

50. $8 \times 9 - (12 - 8) \div 4 - (10 - 7)$

51. $120 - 3^3 \cdot 4 \div (5 \cdot 6 - 6 \cdot 4)$

52. $80 - 2^4 \cdot 15 \div (7 \cdot 5 - 45 \div 3)$

53. Find the average of \$64, \$97, and \$121.

54. Find the average of four test grades of 86, 92, 80, and 78.

d Simplify.

55. $8 \times 13 + \{42 \div [18 - (6 + 5)]\}$

56. $72 \div 6 - \{2 \times [9 - (4 \times 2)]\}$

57. $[14 - (3 + 5) \div 2] - [18 \div (8 - 2)]$

58. $[92 \times (6 - 4) \div 8] + [7 \times (8 - 3)]$

59. $(82 - 14) \times [(10 + 45 \div 5) - (6 \cdot 6 - 5 \cdot 5)]$

60. $(18 \div 2) \cdot \{[(9 \cdot 9 - 1) \div 2] - [5 \cdot 20 - (7 \cdot 9 - 2)]\}$

61. $4 \times \{(200 - 50 \div 5) - [(35 \div 7) \cdot (35 \div 7) - 4 \times 3]\}$

62. $\{[18 - 2 \cdot 6] - [40 \div (17 - 9)]\} + \{48 - 13 \times 3 + [(50 - 7 \cdot 5) + 2]\}$

Skill Maintenance

Solve. [1.4b]

63. $x + 341 = 793$

64. $7 \cdot x = 91$

Solve. [1.5a]

65. *Colorado.* The state of Colorado is roughly the shape of a rectangle that is 270 mi by 380 mi. What is its area?

66. On a long four-day trip, a family bought the following amounts of gasoline for their motor home:

23 gallons, 24 gallons,
26 gallons, 25 gallons.

How much gasoline did they buy in all?

Synthesis

67. ◆ The expression $9 - (4 \times 2)$ contains parentheses. Are they necessary? Why or why not?

68. ◆ The expression $(3 \cdot 4)^2$ contains parentheses. Are they necessary? Why or why not?

Simplify.

69. 🖩 $15(23 - 4 \cdot 2)^3 \div (3 \cdot 25)$

70. 🖩 $(19 - 2^4)^5 - (141 \div 47)^2$

Each of the expressions in Exercises 71–73 is incorrect. First find the correct answer. Then place as many parentheses as needed in the expression in order to make the incorrect answer correct.

71. $1 + 5 \cdot 4 + 3 = 36$

72. $12 \div 4 + 2 \cdot 3 - 2 = 2$

73. $12 \div 4 + 2 \cdot 3 - 2 = 4$

74. Write out the symbols $+, -, \times, \div$, and ( ) and one occurrence each of 1, 2, 3, 4, 5, 6, 7, 8, and 9 to represent 100.

Collaborative Learning Manual

Use the order of operations to simplify expressions.

1.7 Factorizations

In this chapter, we begin our work with fractions. Certain skills make such work easier. For example, in order to simplify

$$\frac{12}{32},$$

it is important that we be able to *factor* the 12 and the 32, as follows:

$$\frac{12}{32} = \frac{4 \cdot 3}{4 \cdot 8}.$$

Then we "remove" a factor of 1:

$$\frac{4 \cdot 3}{4 \cdot 8} = \frac{4}{4} \cdot \frac{3}{8} = 1 \cdot \frac{3}{8} = \frac{3}{8}.$$

Thus factoring is an important skill in working with fractions.

Objectives

a Find the factors of a number.

b Find some multiples of a number, and determine whether a number is divisible by another.

c Given a number from 1 to 100, tell whether it is prime, composite, or neither.

d Find the prime factorization of a composite number.

For Extra Help

TAPE 3 | InterAct math MAC WIN | CD-ROM

a Factors and Factorization

In Sections 1.7 and 1.8, we consider only the **natural numbers** 1, 2, 3, and so on.

Let's look at the product $3 \cdot 4 = 12$. We say that 3 and 4 are **factors** of 12. Since $12 = 12 \cdot 1$, we also know that 12 and 1 are factors of 12.

> A **factor** of a given number is a number multiplied in a product.
>
> A **factorization** of a number is an expression for the number that shows it as a product of natural numbers.

For example, each of the following gives a factorization of 12.

$12 = 4 \cdot 3$	This factorization shows that 4 and 3 are factors of 12.
$12 = 12 \cdot 1$	This factorization shows that 12 and 1 are factors of 12.
$12 = 6 \cdot 2$	This factorization shows that 6 and 2 are factors of 12.
$12 = 2 \cdot 3 \cdot 2$	This factorization shows that 2 and 3 are factors of 12.

Since $n = n \cdot 1$, every number has a factorization, and every number has factors even if its only factors are itself and 1.

Example 1 Find all the factors of 24.

We first find some factorizations.

$24 = 1 \cdot 24$ $\quad 24 = 3 \cdot 8$

$24 = 2 \cdot 12$ $\quad 24 = 4 \cdot 6$

Note that all but one of the factors of a natural number are *less* than the number.

Factors: 1, 2, 3, 4, 6, 8, 12, 24.

Do Exercises 1–4.

Find all the factors of the number. (*Hint*: Find some factorizations of the number.)

1. 6

2. 8

3. 10

4. 32

Answers on page A-2

5. Show that each of the numbers 5, 45, and 100 is a multiple of 5.

6. Show that each of the numbers 10, 60, and 110 is a multiple of 10.

7. Multiply by 1, 2, 3, and so on, to find ten multiples of 5.

Calculator Spotlight

Rather than list remainders, most calculators display quotients using decimal notation. Although decimal notation is not studied until Chapter 3, it is still possible for us to now check for divisibility using a calculator.

To determine whether a number, like 551, is divisible by another number, like 19, we simply press [5] [5] [1] [÷] [1] [9] [=]. If the resulting quotient contains no digits to the right of the decimal point, the first number is divisible by the second. Thus, since 551 ÷ 19 = 29, we know that 551 is divisible by 19. On the other hand, since 551 ÷ 20 = 27.55, we know that 551 is not divisible by 20.

Exercises

For each pair of numbers, determine whether the first number is divisible by the second number.

1. 731, 17
2. 1502, 79
3. 1053, 36
4. 4183, 47

Answers on page A-2

b Multiples and Divisibility

A **multiple** of a natural number is a product of it and some natural number. For example, some multiples of 2 are:

2 (because $2 = 1 \cdot 2$);
4 (because $4 = 2 \cdot 2$);
6 (because $6 = 3 \cdot 2$);
8 (because $8 = 4 \cdot 2$);
10 (because $10 = 5 \cdot 2$).

Note that all but one of the multiples of a number are *larger* than the number.

We find multiples of 2 by counting by twos: 2, 4, 6, 8, and so on. We can find multiples of 3 by counting by threes: 3, 6, 9, 12, and so on.

Example 2 Show that each of the numbers 3, 6, 9, and 15 is a multiple of 3.

$3 = 1 \cdot 3 \qquad 6 = 2 \cdot 3 \qquad 9 = 3 \cdot 3 \qquad 15 = 5 \cdot 3$

Do Exercises 5 and 6.

Example 3 Multiply by 1, 2, 3, and so on, to find ten multiples of 7.

$1 \cdot 7 = 7$	$6 \cdot 7 = 42$
$2 \cdot 7 = 14$	$7 \cdot 7 = 49$
$3 \cdot 7 = 21$	$8 \cdot 7 = 56$
$4 \cdot 7 = 28$	$9 \cdot 7 = 63$
$5 \cdot 7 = 35$	$10 \cdot 7 = 70$

Do Exercise 7.

A number b is said to be **divisible** by another number a if b is a multiple of a (a is a factor of b).

Thus,

4 is divisible by 2 because 4 is a multiple of 2 ($4 = 2 \cdot 2$);
27 is divisible by 3 because 27 is a multiple of 3 ($27 = 9 \cdot 3$);
100 is divisible by 25 because 100 is a multiple of 25 ($100 = 4 \cdot 25$).

A number b is **divisible** by another number a if division of b by a results in a remainder of zero. We sometimes say that a divides b "evenly."

Example 4 Determine whether 24 is divisible by 3.

We divide 24 by 3:

$$\begin{array}{r} 8 \\ 3\overline{)24} \\ \underline{24} \\ 0 \end{array}$$

Because the remainder is 0, 24 is divisible by 3.

Do Exercises 8–10 on the following page.

c Prime and Composite Numbers

> A natural number that has exactly two *different* factors, only itself and 1, is called a **prime number**.

Example 5 Tell whether the numbers 2, 3, 5, 7, and 11 are prime.

The number 2 is prime. It has only the factors 1 and 2.

The number 5 is prime. It has only the factors 1 and 5.

The numbers 3, 7, and 11 are also prime.

Some natural numbers are not prime.

Example 6 Tell whether the numbers 4, 6, 8, 10, 63, and 1 are prime.

The number 4 is not prime. It has the factors 1, 2, and 4.

The numbers 6, 8, 10, and 63 are not prime. Each has more than two different factors.

The number 1 is not prime. It does not have two *different* factors.

> A natural number, other than 1, that is not prime is called **composite**.

In other words, if a number can be factored into a product of natural numbers, some of which are not the number itself or 1, it is composite. Thus, 2, 3, 5, 7, and 11 are prime; 4, 6, 8, 10, and 63 are composite; 1 is neither prime nor composite. The number 0 is also neither prime nor composite, but 0 is *not* a natural number and thus is not considered here. We are considering only natural numbers.

Do Exercise 11.

d Prime Factorizations

To factor a composite number into a product of primes is to find a **prime factorization** of the number. To do this, we consider the primes 2, 3, 5, 7, 11, 13, 17, 19, 23, and so on, and determine whether a given number is divisible by the primes.

Example 7 Find the prime factorization of 39.

a) We divide by the first prime, 2.

$$\begin{array}{r} 19 \quad R = 1 \\ 2\overline{)39} \\ \underline{38} \\ 1 \end{array}$$

Because the remainder is not 0, 2 is not a factor of 39.

b) We divide by the next prime, 3.

$$\begin{array}{r} 13 \quad R = 0 \\ 3\overline{)39} \end{array}$$

Because 13 is a prime, we are finished. The prime factorization is

$39 = 3 \cdot 13.$

8. Determine whether 16 is divisible by 2.

9. Determine whether 125 is divisible by 5.

10. Determine whether 125 is divisible by 6.

11. Tell whether each number is prime, composite, or neither.

1, 4, 6, 8, 13, 19, 41

The following is a table of the prime numbers from 2 to 157. There are more extensive tables, but these prime numbers will be the most helpful to you in this text.

A TABLE OF PRIMES

2, 3, 5, 7, 11, 13, 17, 19, 23, 29, 31, 37, 41, 43, 47, 53, 59, 61, 67, 71, 73, 79, 83, 89, 97, 101, 103, 107, 109, 113, 127, 131, 137, 139, 149, 151, 157

Answers on page A-2

Find the prime factorization of the number.

12. 6

13. 12

14. 45

15. 98

16. 126

17. 144

Answers on page A-2

Example 8 Find the prime factorization of 76.

a) We divide by the first prime, 2.

$$2\overline{)76}\quad 38 \quad R = 0$$

b) Because 38 is composite, we start with 2 again:

$$2\overline{)38}\quad 19 \quad R = 0$$

Because 19 is a prime, we are finished. The prime factorization is

$$76 = 2 \cdot 2 \cdot 19.$$

We abbreviate our procedure as follows.

$$\begin{array}{r} 19 \\ 2\overline{)38} \\ 2\overline{)76} \end{array}$$

$$76 = 2 \cdot 2 \cdot 19$$

Multiplication is commutative so a factorization such as $2 \cdot 2 \cdot 19$ could also be expressed as $2 \cdot 19 \cdot 2$ or $19 \cdot 2 \cdot 2$ (or in exponential notation, as $2^2 \cdot 19$ or $19 \cdot 2^2$), but the prime factors are still the same. For this reason, we agree that any of these is "the" prime factorization of 76.

Example 9 Find the prime factorization of 72.

We can do divisions "up" as follows.

$$\begin{array}{r} 3 \\ 3\overline{)9} \\ 2\overline{)18} \\ 2\overline{)36} \\ 2\overline{)72} \end{array} \longleftarrow \text{Begin here.}$$

$$72 = 2 \cdot 2 \cdot 2 \cdot 3 \cdot 3$$

Example 10 Find the prime factorization of 189.

We can use a string of successive divisions.

$$\begin{array}{r} 7 \\ 3\overline{)21} \\ 3\overline{)63} \\ 3\overline{)189} \end{array}$$

189 is not divisible by 2. We move to 3.
63 is not divisible by 2. We move to 3.
21 is not divisible by 2. We move to 3.

$$189 = 3 \cdot 3 \cdot 3 \cdot 7$$

Example 11 Find the prime factorization of 65.

We can use a string of successive divisions.

$$\begin{array}{r} 13 \\ 5\overline{)65} \end{array}$$

$$65 = 5 \cdot 13$$

Do Exercises 12–17.

Exercise Set 1.7

a Find all the factors of the number.

1. 18
2. 16
3. 54
4. 48
5. 4
6. 9
7. 7
8. 11
9. 1
10. 3
11. 98
12. 100

b Multiply by 1, 2, 3, and so on, to find ten multiples of the number.

13. 4
14. 11
15. 20
16. 50
17. 3
18. 5
19. 12
20. 13
21. 10
22. 6
23. 9
24. 14
25. Determine whether 26 is divisible by 6.
26. Determine whether 29 is divisible by 9.
27. Determine whether 1880 is divisible by 8.
28. Determine whether 4227 is divisible by 3.
29. Determine whether 256 is divisible by 16.
30. Determine whether 102 is divisible by 4.
31. Determine whether 4227 is divisible by 9.
32. Determine whether 200 is divisible by 25.
33. Determine whether 8650 is divisible by 16.
34. Determine whether 4143 is divisible by 7.

c Determine whether the number is prime, composite, or neither.

35. 1
36. 2
37. 9
38. 19
39. 11
40. 27
41. 29
42. 49

d Find the prime factorization of the number.

43. 8 **44.** 16 **45.** 14 **46.** 15

47. 42 **48.** 32 **49.** 25 **50.** 40

51. 50 **52.** 62 **53.** 169 **54.** 140

55. 100 **56.** 110 **57.** 35 **58.** 70

59. 72 **60.** 86 **61.** 77 **62.** 99

63. 2884 **64.** 484 **65.** 51 **66.** 91

Skill Maintenance

Multiply. [1.3b]

67. $2 \cdot 13$ **68.** $8 \cdot 32$ **69.** $17 \cdot 25$ **70.** $25 \cdot 168$

Divide. [1.3e]

71. $0 \div 22$ **72.** $22 \div 1$ **73.** $22 \div 22$ **74.** $66 \div 22$

Solve. [1.5a]

75. Find the total cost of 7 shirts at \$48 each and 4 pairs of pants at \$69 each.

76. Sandy can type 62 words per minute. How long will it take her to type 12,462 words?

Synthesis

77. ◆ Explain a method for finding a composite number that contains exactly two factors other than itself and 1.

78. ◆ Is every natural number a multiple of 1? Why or why not?

79. *Factors and Sums.* To *factor* a number is to express it as a product. Since $15 = 5 \cdot 3$, we say that 15 is *factored* and that 5 and 3 are *factors* of 15. In the table below, the top number in each column has been factored in such a way that the sum of the factors is the bottom number in the column. For example, in the first column, 56 has been factored as $7 \cdot 8$, and $7 + 8 = 15$, the bottom number. Such thinking will be important in understanding the meaning of a factor and in algebra.

Product	56	63	36	72	140	96		168	110			
Factor	7									9	24	3
Factor	8						8	8		10	18	
Sum	15	16	20	38	24	20	14		21			24

Find the missing numbers in the table.

Find all the prime numbers less than 200, using the Sieve of Eratosthenes.

1.8 Divisibility

Suppose you are asked to find the simplest fractional notation for

$$\frac{117}{225}.$$

Since the numbers are quite large, you might feel that the task is difficult. However, both the numerator and the denominator have 9 as a factor. If you knew this, you could factor and simplify quickly as follows:

$$\frac{117}{225} = \frac{9 \cdot 13}{9 \cdot 25} = \frac{9}{9} \cdot \frac{13}{25} = 1 \cdot \frac{13}{25} = \frac{13}{25}.$$

How did we know that both numbers have 9 as a factor? There are fast tests for such determinations. If the sum of the digits of a number is divisible by 9, then the number is divisible by 9; that is, it has 9 as a factor. Since $1 + 1 + 7 = 9$ and $2 + 2 + 5 = 9$, both numbers have 9 as a factor.

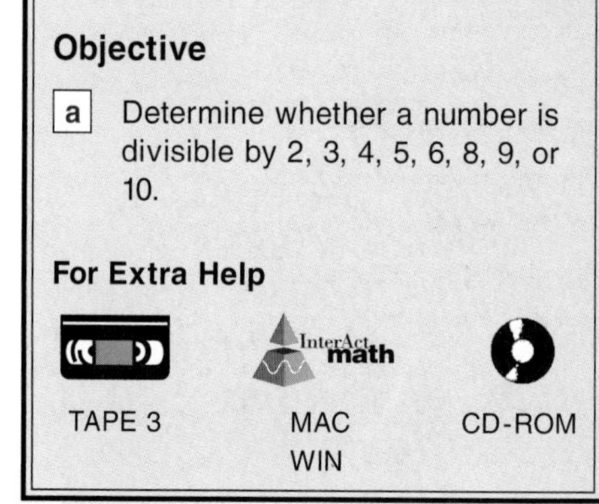

a Rules for Divisibility

In this section, we learn fast ways of determining whether numbers are divisible by 2, 3, 4, 5, 6, 8, 9, and 10. This will make simplifying fractional notation much easier.

Divisibility by 2

You may already know the test for divisibility by 2.

> A number is divisible by 2 (is *even*) if it has a ones digit of 0, 2, 4, 6, or 8 (that is, it has an even ones digit).

Let's see why. Consider 354, which is

3 hundreds + 5 tens + 4.

Hundreds and tens are both multiples of 2. If the last digit is a multiple of 2, then the entire number is a multiple of 2.

Examples Determine whether the number is divisible by 2.

1. 355 is not a multiple of 2; 5 is *not* even.
2. 4786 is a multiple of 2; 6 is even.
3. 8990 is a multiple of 2; 0 is even.
4. 4261 is not a multiple of 2; 1 is *not* even.

Do Exercises 1–4.

Determine whether the number is divisible by 2.

1. 84

2. 59

3. 998

4. 2225

Answers on page A-3

Divisibility by 3

> A number is divisible by 3 if the sum of its digits is divisible by 3.

Examples Determine whether the number is divisible by 3.

5. 18 $1 + 8 = 9$
6. 93 $9 + 3 = 12$
7. 201 $2 + 0 + 1 = 3$

All divisible by 3 because the sums of their digits are divisible by 3.

8. 256 $2 + 5 + 6 = 13$

The sum, 13, is not divisible by 3, so 256 is not divisible by 3.

Do Exercises 5–8.

Divisibility by 6

A number divisible by 6 is a multiple of 6. But $6 = 2 \cdot 3$, so the number is also a multiple of 2 and 3. Thus we have the following.

> A number is divisible by 6 if its ones digit is 0, 2, 4, 6, or 8 (is even) and the sum of its digits is divisible by 3.

Examples Determine whether the number is divisible by 6.

9. 720

Because 720 is even, it is divisible by 2. Also, $7 + 2 + 0 = 9$, so 720 is divisible by 3. Thus, 720 is divisible by 6.

720 ↑ Even $7 + 2 + 0 = 9$ ↑ Divisible by 3

10. 73

73 is *not* divisible by 6 because it is *not* even.

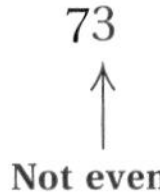

73 ↑ Not even

11. 256

256 is *not* divisible by 6 because the sum of its digits is *not* divisible by 3.

$2 + 5 + 6 = 13$ ↑ Not divisible by 3

Do Exercises 9–12.

Determine whether the number is divisible by 3.

5. 111

6. 1111

7. 309

8. 17,216

Determine whether the number is divisible by 6.

9. 420

10. 106

11. 321

12. 444

Answers on page A-3

Divisibility by 9

The test for divisibility by 9 is similar to the test for divisibility by 3.

> A number is divisible by 9 if the sum of its digits is divisible by 9.

Example 12 The number 6984 is divisible by 9 because

$$6 + 9 + 8 + 4 = 27$$

and 27 is divisible by 9.

Example 13 The number 322 is *not* divisible by 9 because

$$3 + 2 + 2 = 7$$

and 7 is not divisible by 9.

Do Exercises 13–16.

Divisibility by 10

> A number is divisible by 10 if its ones digit is 0.

We know that this test works because the product of 10 and *any* number has a ones digit of 0.

Examples Determine whether the number is divisible by 10.

14. 3440 is divisible by 10 because the ones digit is 0.

15. 3447 is *not* divisible by 10 because the ones digit is not 0.

Do Exercises 17–20.

Divisibility by 5

> A number is divisible by 5 if its ones digit is 0 or 5.

Examples Determine whether the number is divisible by 5.

16. 220 is divisible by 5 because the ones digit is 0.

17. 475 is divisible by 5 because the ones digit is 5.

18. 6514 is *not* divisible by 5 because the ones digit is neither a 0 nor a 5.

Do Exercises 21–24.

Let's see why the test for 5 works. Consider 7830:

$$7830 = 10 \cdot 783 = 5 \cdot 2 \cdot 783.$$

Since 7830 is divisible by 10 and 5 is a factor of 10, 7830 is divisible by 5.

Determine whether the number is divisible by 9.

13. 16

14. 117

15. 930

16. 29,223

Determine whether the number is divisible by 10.

17. 305

18. 300

19. 847

20. 8760

Determine whether the number is divisible by 5.

21. 5780

22. 3427

23. 34,678

24. 7775

Answers on page A-3

Determine whether the number is divisible by 4.

25. 216

26. 217

27. 5865

28. 23,524

Determine whether the number is divisible by 8.

29. 7564

30. 7864

31. 17,560

32. 25,716

Answers on page A-3

Consider 6734:

$$6734 = 673 \text{ tens} + 4.$$

Tens are multiples of 5, so the only number that must be checked is the ones digit. If the last digit is a multiple of 5, the entire number is. In this case, 4 is not a multiple of 5, so 6734 is not divisible by 5.

Divisibility by 4

The test for divisibility by 4 is similar to the test for divisibility by 2.

> A number is divisible by 4 if the number named by its last *two* digits is divisible by 4.

Examples Determine whether the number is divisible by 4.

19. 8212 is divisible by 4 because 12 is divisible by 4.

20. 5216 is divisible by 4 because 16 is divisible by 4.

21. 8211 is *not* divisible by 4 because 11 is *not* divisible by 4.

22. 7515 is *not* divisible by 4 because 15 is *not* divisible by 4.

Do Exercises 25–28.

To see why the test for divisibility by 4 works, consider 516:

$$516 = 5 \text{ hundreds} + 16.$$

Hundreds are multiples of 4. If the number named by the last two digits is a multiple of 4, then the entire number is a multiple of 4.

Divisibility by 8

The test for divisibility by 8 is an extension of the tests for divisibility by 2 and 4.

> A number is divisible by 8 if the number named by its last *three* digits is divisible by 8.

Examples Determine whether the number is divisible by 8.

23. 5648 is divisible by 8 because 648 is divisible by 8.

24. 96,088 is divisible by 8 because 88 is divisible by 8.

25. 7324 is *not* divisible by 8 because 324 is *not* divisible by 8.

26. 13,420 is *not* divisible by 8 because 420 is *not* divisible by 8.

Do Exercises 29–32.

A Note About Divisibility by 7

There are several tests for divisibility by 7, but all of them are more complicated than simply dividing by 7. So if you want to test for divisibility by 7, simply divide by 7.

Exercise Set 1.8

a To answer Exercises 1–8, consider the following numbers.

46	300	85	256
224	36	711	8064
19	45,270	13,251	1867
555	4444	254,765	21,568

1. Which of the above are divisible by 2?

2. Which of the above are divisible by 3?

3. Which of the above are divisible by 4?

4. Which of the above are divisible by 5?

5. Which of the above are divisible by 6?

6. Which of the above are divisible by 8?

7. Which of the above are divisible by 9?

8. Which of the above are divisible by 10?

To answer Exercises 9–16, consider the following numbers.

56	200	75	35
324	42	812	402
784	501	2345	111,111
55,555	3009	2001	1005

9. Which of the above are divisible by 3?

10. Which of the above are divisible by 2?

11. Which of the above are divisible by 5?

12. Which of the above are divisible by 4?

13. Which of the above are divisible by 9?

14. Which of the above are divisible by 6?

15. Which of the above are divisible by 10?

16. Which of the above are divisible by 8?

Skill Maintenance

Solve. [1.4b]

17. $56 + x = 194$

18. $y + 124 = 263$

19. $18 \cdot t = 1008$

20. $24 \cdot m = 624$

Divide. [1.3e]

21. $2106 \div 9$

22. $45\overline{)180{,}135}$

Solve. [1.5a]

23. An automobile with a 5-speed transmission gets 33 mpg in city driving. How many gallons of gas will it use to travel 1485 mi?

24. There are 60 min in 1 hr. How many minutes are there in 72 hr?

Synthesis

25. ◆ Are the divisibility tests useful for finding prime factorizations? Why or why not?

26. ◆ Which of the years from 1990 to 2010, if any, also happen to be prime numbers? Explain at least two ways in which you might go about solving this problem.

Find the prime factorization of the number.

27. 7800

28. 2520

29. 2772

30. 1998

31. Using each of the digits 1, 2, 3, . . . , 7, find the largest seven-digit number for which the sum of any two consecutive digits is a prime number.

32. A passenger in a taxicab asks for the driver's company number. The driver says abruptly, "Sure—you can have my number. Work it out: If you divide it by 2, 3, 4, 5, or 6, you will get a remainder of 1. If you divide it by 11, the remainder will be 0 and no driver has a company number that is smaller than this one." Determine the number.

Use the divisibility rules and properties of numbers to discover an unknown number.

1.9 Least Common Multiples

In this chapter, we study addition and subtraction using fractional notation. Suppose we want to add $\frac{2}{3}$ and $\frac{1}{2}$. To do so, we find the least common multiple of the denominators: $\frac{2}{3} + \frac{1}{2} = \frac{4}{6} + \frac{3}{6}$. Then we add the numerators and keep the common denominator, 6. Before we do this, though, we study finding the **least common denominator (LCD),** or **least common multiple (LCM),** of the denominators.

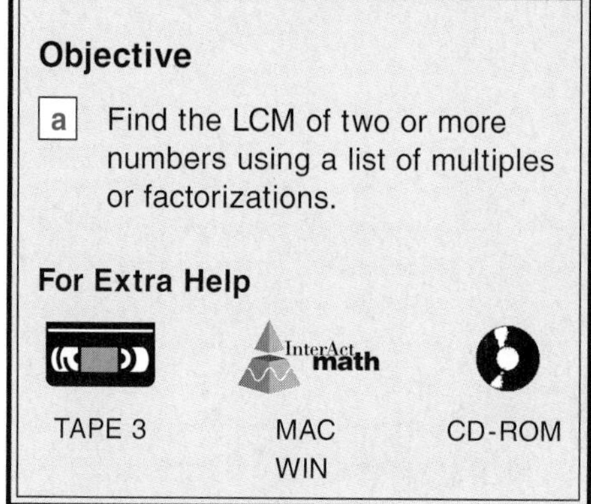

Objective
a Find the LCM of two or more numbers using a list of multiples or factorizations.
For Extra Help
InterAct math
TAPE 3
MAC
WIN
CD-ROM

a Finding Least Common Multiples

> The **least common multiple,** or LCM, of two natural numbers is the smallest number that is a multiple of both.

Example 1 Find the LCM of 20 and 30.

a) First list some multiples of 20 by multiplying 20 by 1, 2, 3, and so on:

20, 40, 60, 80, 100, 120, 140, 160, 180, 200, 220, 240, ...

b) Then list some multiples of 30 by multiplying 30 by 1, 2, 3, and so on:

30, 60, 90, 120, 150, 180, 210, 240, ...

c) Now list the numbers *common* to both lists, the common multiples:

60, 120, 180, 240, ...

d) These are the common multiples of 20 and 30. Which is the smallest? The LCM of 20 and 30 is 60.

Do Exercise 1.

Next we develop two methods that are more efficient for finding LCMs. You may choose to learn either method (consult with your instructor), or both, but if you are going on to a study of algebra, you should definitely learn method 2.

Method 1: Finding LCMs Using One List of Multiples

Method 1. To find the LCM of a set of numbers (9, 12):

a) Determine whether the largest number is a multiple of the others. If it is, it is the LCM. That is, if the largest number has the others as factors, the LCM is that number.

(12 is not a multiple of 9)

b) If not, check multiples of the largest number until you get one that is a multiple of the others.

($2 \cdot 12 = 24$, not a multiple of 9)

($3 \cdot 12 = 36$, a multiple of 9)

c) That number is the LCM.

LCM = 36

1. By examining lists of multiples, find the LCM of 9 and 15.

Answer on page A-3

2. By examining lists of multiples, find the LCM of 8 and 10.

Find the LCM.

3. 10, 15

4. 6, 8

Find the LCM.

5. 5, 10

6. 20, 40, 80

Answers on page A-3

Example 2 Find the LCM of 12 and 15.

a) 15 is not a multiple of 12.

b) Check multiples:

$2 \cdot 15 = 30,$ Not a multiple of 12

$3 \cdot 15 = 45,$ Not a multiple of 12

$4 \cdot 15 = 60.$ A multiple of 12

c) The LCM = 60.

Do Exercise 2.

Example 3 Find the LCM of 4 and 14.

a) 14 is not a multiple of 4.

b) Check multiples:

$2 \cdot 14 = 28.$ A multiple of 4

c) The LCM = 28.

Do Exercises 3 and 4.

Example 4 Find the LCM of 8 and 32.

a) 32 is a multiple of 8, so it is the LCM.

c) The LCM = 32.

Example 5 Find the LCM of 10, 100, and 1000.

a) 1000 is a multiple of 10 and 100, so it is the LCM.

c) The LCM = 1000.

Do Exercises 5 and 6.

Method 2: Finding LCMs Using Factorizations

A second method for finding LCMs uses prime factorizations. Consider again 20 and 30. Their prime factorizations are

$$20 = 2 \cdot 2 \cdot 5 \quad \text{and} \quad 30 = 2 \cdot 3 \cdot 5.$$

Let's look at these prime factorizations in order to find the LCM. Any multiple of 20 will have to have *two* 2's as factors and *one* 5 as a factor. Any multiple of 30 will have to have *one* 2, *one* 3, and *one* 5 as factors. The smallest number satisfying these conditions is

Two 2's, one 5; makes 20 a factor

$$2 \cdot 2 \cdot 3 \cdot 5.$$

One 2, one 3, one 5; makes 30 a factor

The LCM must have all the factors of 20 and all the factors of 30, but the factors need not be repeated when they are common to both numbers.

The greatest number of times a 2 occurs as a factor of either 20 or 30 is two, and the LCM has 2 as a factor twice. The greatest number of times a 3 occurs as a factor of either 20 or 30 is one, and the LCM has 3 as a factor once. The greatest number of times that 5 occurs as a factor of either 20 or 30 is one, and the LCM has 5 as a factor once. The LCM is the product $2 \cdot 2 \cdot 3 \cdot 5$, or 60.

Method 2. To find the LCM of a set of numbers using prime factorizations:

a) Find the prime factorization of each number.

b) Create a product of factors, using each factor the greatest number of times that it occurs in any one factorization.

Example 6 Find the LCM of 6 and 8.

a) Find the prime factorization of each number.

$6 = 2 \cdot 3, \qquad 8 = 2 \cdot 2 \cdot 2$

b) Create a product by writing factors, using each the greatest number of times that it occurs in any one factorization.

Consider the factor 2. The greatest number of times that 2 occurs in any one factorization is three. We write 2 as a factor three times.

$2 \cdot 2 \cdot 2 \cdot ?$

Consider the factor 3. The greatest number of times that 3 occurs in any one factorization is one. We write 3 as a factor one time.

$2 \cdot 2 \cdot 2 \cdot 3 \cdot ?$

Since there are no other prime factors in either factorization, the

LCM is $2 \cdot 2 \cdot 2 \cdot 3$, or 24.

Example 7 Find the LCM of 24 and 36.

a) Find the prime factorization of each number.

$24 = 2 \cdot 2 \cdot 2 \cdot 3, \qquad 36 = 2 \cdot 2 \cdot 3 \cdot 3$

b) Create a product by writing factors, using each the greatest number of times that it occurs in any one factorization.

Consider the factor 2. The greatest number of times that 2 occurs in any one factorization is three. We write 2 as a factor three times:

$2 \cdot 2 \cdot 2 \cdot ?$

Consider the factor 3. The greatest number of times that 3 occurs in any one factorization is two. We write 3 as a factor two times:

$2 \cdot 2 \cdot 2 \cdot 3 \cdot 3 \cdot ?$

Since there are no other prime factors in either factorization, the

LCM is $2 \cdot 2 \cdot 2 \cdot 3 \cdot 3$, or 72.

Do Exercises 7–9.

Use prime factorizations to find the LCM.

7. 8, 10

8. 18, 40

9. 32, 54

Answers on page A-3

10. Find the LCM of 24, 35, and 45.

Find the LCM.

11. 3, 18

12. 12, 24

Find the LCM.

13. 4, 9

14. 5, 6, 7

Answer on page A-3

Example 8 Find the LCM of 27, 90, and 84.

a) Find the prime factorization of each number.

$27 = 3 \cdot 3 \cdot 3, \qquad 90 = 2 \cdot 3 \cdot 3 \cdot 5, \qquad 84 = 2 \cdot 2 \cdot 3 \cdot 7$

b) Create a product by writing factors, using each the greatest number of times that it occurs in any one factorization.

Consider the factor 2. The greatest number of times that 2 occurs in any one factorization is two. We write 2 as a factor two times:

$2 \cdot 2 \cdot ?$

Consider the factor 3. The greatest number of times that 3 occurs in any one factorization is three. We write 3 as a factor three times:

$2 \cdot 2 \cdot 3 \cdot 3 \cdot 3 \cdot ?$

Consider the factor 5. The greatest number of times that 5 occurs in any one factorization is one. We write 5 as a factor one time:

$2 \cdot 2 \cdot 3 \cdot 3 \cdot 3 \cdot 5 \cdot ?$

Consider the factor 7. The greatest number of times that 7 occurs in any one factorization is one. We write 7 as a factor one time:

$2 \cdot 2 \cdot 3 \cdot 3 \cdot 3 \cdot 5 \cdot 7 \cdot ?$

Since no other prime factors are possible in any of the factorizations, the

LCM is $2 \cdot 2 \cdot 3 \cdot 3 \cdot 3 \cdot 5 \cdot 7$, or 3780.

Do Exercise 10.

Example 9 Find the LCM of 7 and 21.

Note that 7 is a factor of 21. We stated earlier that if one number is a factor of another, the LCM is the larger of the numbers. Thus the LCM is 21. When you notice this at the outset, you can find the LCM quickly without using factorizations.

Do Exercises 11 and 12.

Example 10 Find the LCM of 8 and 9.

The two numbers 8 and 9 have no common prime factor. When this happens, the LCM is just the product of the two numbers. Thus the LCM is $8 \cdot 9$, or 72.

Do Exercises 13 and 14.

Let's compare the two methods considered for finding LCMs: the multiples method and the factorization method.

Method 1, the **multiples method,** can be longer than the factorization method when the LCM is large or when there are more than two numbers. But this method is faster and easier to use mentally for two numbers.

Method 2, the **factorization method,** works well for several numbers. It is just like a method used in algebra. If you are going to study algebra, you should definitely learn the factorization method.

Exercise Set 1.9

[a] Find the LCM of the set of numbers.

1. 2, 4

2. 3, 15

3. 10, 25

4. 10, 15

5. 20, 40

6. 8, 12

7. 18, 27

8. 9, 11

9. 30, 50

10. 24, 36

11. 30, 40

12. 21, 27

13. 18, 24

14. 12, 18

15. 60, 70

16. 35, 45

17. 16, 36

18. 18, 20

19. 32, 36

20. 36, 48

21. 2, 3, 5

22. 5, 18, 3

23. 3, 5, 7

24. 6, 12, 18

25. 24, 36, 12

26. 8, 16, 22

27. 5, 12, 15

28. 12, 18, 40

29. 9, 12, 6

30. 8, 16, 12

31. 180, 100, 450

32. 18, 30, 50, 48

33. 8, 48

34. 16, 32

35. 5, 50

36. 12, 72

37. 11, 13

38. 13, 14

39. 12, 35

40. 23, 25

41. 54, 63

42. 56, 72

43. 81, 90

44. 75, 100

Applications of LCMs: Planet Orbits. The earth, Jupiter, Saturn, and Uranus all revolve around the sun. The earth takes 1 yr, Jupiter 12 yr, Saturn 30 yr, and Uranus 84 yr to make a complete revolution. On a certain night, you look at those three distant planets and wonder how many years it will take before they have the same position again. (*Hint*: To find out, you find the LCM of 12, 30, and 84. It will be that number of years.)

45. How often will Jupiter and Saturn appear in the same direction in the night sky as seen from the earth?

46. How often will Jupiter, Saturn, and Uranus appear in the same direction in the night sky as seen from the earth?

Skill Maintenance

47. Use $<$ or $>$ for ▢ to write a true sentence: [1.1f]
9001 ▢ 10,001.

48. A performing arts center was sold out for a musical. Its seats sell for $13 each. Total receipts were $3250. How many seats does this auditorium contain? [1.5a]

49. Multiply: $23 \cdot 345$. [1.3b]

50. Subtract: $10{,}007 - 3068$. [1.2f]

51. Write expanded notation for 24,605. [1.1a]

52. Write a word name for 102,960. [1.1c]

Synthesis

53. ◆ Is the LCM of two prime numbers always their product? Why or why not?

54. ◆ Is the LCM of two numbers always at least twice as large as the larger of the two numbers? Why or why not?

▦ Use a calculator and the multiples method to find the LCM of each pair of numbers.

55. 288, 324

56. 2700, 7800

57. Find the LCM of 27, 90, 84, 210, 108, and 50.

58. Find the LCM of 18, 21, 24, 36, 63, 56, and 20.

59. A pencil company uses two sizes of boxes, 5 in. by 6 in. and 5 in. by 8 in. These boxes are packed in bigger cartons for shipping. Find the width and the length of the smallest carton that will accommodate boxes of either size without any room left over. (Each carton can contain only one type of box and all boxes must point in the same direction.)

60. Consider 8 and 12. Determine whether each of the following is the LCM of 8 and 12. Tell why or why not.

a) $2 \cdot 2 \cdot 3 \cdot 3$
b) $2 \cdot 2 \cdot 3$
c) $2 \cdot 3 \cdot 3$
d) $2 \cdot 2 \cdot 2 \cdot 3$

Find the least common multiple of two or more numbers using shaped markers.

Summary and Review Exercises: Chapter 1

The review exercises that follow are for practice. Answers are given at the back of the book. If you miss an exercise, restudy the objective indicated in blue next to the exercise or direction line that precedes it.

Write expanded notation. [1.1a]

1. 2793

2. 56,078

3. Write standard notation. [1.1b]

9 ten thousands + 8 hundreds + 4 tens + 4 ones

Write a word name. [1.1c]

4. 67,819

5. 2,781,427

6. Write standard notation. [1.1d]

San Francisco International. The total number of passengers passing through San Francisco International Airport in a recent year was thirty-six million, two hundred sixty thousand, sixty-four.

7. What does the digit 8 mean in 4,678,952? [1.1e]

Use < or > for ▒ to write a true sentence. [1.1f]

8. 67 ▒ 56

9. 1 ▒ 23

10. Write an addition sentence that corresponds to the situation. [1.2a]

Tony has $406 in her checking account. She is paid $78 for a part-time job and deposits that in her checking account. How much is then in the account?

11. Find the perimeter. [1.2a]

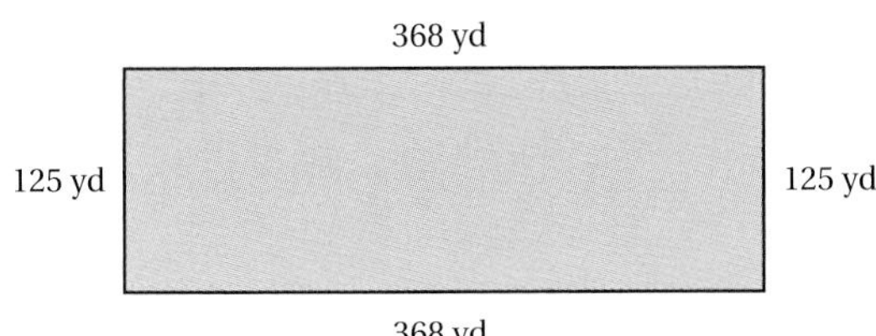

Add. [1.2b]

12. 7304 + 6968

13. 27,609 + 38,415

14. 2743 + 4125 + 6274 + 8956

15. $\begin{array}{r} 91,426 \\ +\ 7,495 \\ \hline \end{array}$

Write a subtraction sentence that corresponds to the situation. [1.2c, e]

16. By exercising daily, you lose 12 lb in one month. If you weighed 151 lb at the beginning of the month, what is your weight now?

17. Natosha has $196 and wants to buy a fax machine for $340. How much more does she need?

18. Write a related addition sentence: [1.2d]

$10 - 6 = 4.$

19. Write two related subtraction sentences: [1.2d]

$8 + 3 = 11.$

Subtract. [1.2f]

20. 8045 − 2897

21. 8465 − 7312

22. 6003 − 3729

23. $\begin{array}{r} 37,405 \\ -\ 19,648 \\ \hline \end{array}$

Round 345,759 to the nearest: [1.3f]

24. Hundred.

25. Ten.

26. Thousand.

Estimate the sum, difference, or product by first rounding to the nearest hundred. Show your work. [1.3f]

27. 41,348 + 19,749

28. 38,652 − 24,549

29. $396 \cdot 748$

30. Write a multiplication sentence that corresponds to the situation. [1.3a]

A farmer plants apple trees in a rectangular array. He plants 15 rows with 32 trees in each row. How many apple trees does he have altogether?

31. Find the area of the rectangle in Exercise 11. [1.3a]

Multiply. [1.3b]

32. $700 \cdot 600$

33. $7846 \cdot 800$

34. $726 \cdot 698$

35. $\begin{array}{r} 8305 \\ \times\ 642 \\ \hline \end{array}$

36. Write a division sentence that corresponds to the situation. [1.3c]

A cheese factory made 176 lb of Monterey Jack cheese. The cheese was placed in 4-lb boxes. How many boxes were filled?

37. Write a related multiplication sentence: [1.3d]

$56 \div 8 = 7.$

38. Write two related division sentences: [1.3d]

$13 \cdot 4 = 52.$

Divide. [1.3e]

39. $63 \div 5$

40. $80 \div 16$

41. $60\overline{)286}$

42. $4266 \div 79$

43. $38\overline{)17{,}176}$

44. $52{,}668 \div 12$

Solve. [1.4b]

45. $46 \cdot n = 368$

46. $47 + x = 92$

47. $x = 782 - 236$

Solve. [1.5a]

48. An apartment builder bought 3 electric ranges at $299 each and 4 dishwashers at $379 each. What was the total cost?

49. *Lincoln-Head Pennies.* In 1909, the first Lincoln-head pennies were minted. Seventy-three years later, these pennies were first minted with a decreased copper content. In what year was the copper content reduced?

50. A family budgets $4950 for food and clothing and $3585 for entertainment. The yearly income of the family was $28,283. How much of this income remained after these two allotments?

51. A chemist has 2753 mL of alcohol. How many 20-mL beakers can be filled? How much will be left over?

52. Write exponential notation: $4 \cdot 4 \cdot 4$. [1.6a]

Evaluate. [1.6b]

53. 10^4

54. 6^2

Simplify. [1.6c, d]

55. $8 \cdot 6 + 17$

56. $7 + (4 + 3)^2$

57. $10 \cdot 24 - (18 + 2) \div 4 - (9 - 7)$

58. $(80 \div 16) \times [(20 - 56 \div 8) + (8 \cdot 8 - 5 \cdot 5)]$

59. Find the average of 157, 170, and 168. [1.6c]

Find the prime factorization of the number. [1.7d]

60. 70

61. 30

62. 45

63. 150

Determine whether: [1.8a]

64. 2432 is divisible by 6.

65. 182 is divisible by 4.

66. 4344 is divisible by 9.

67. 4344 is divisible by 8.

68. Determine whether 37 is prime, composite, or neither. [1.7c]

Find the LCM. [1.9a]

69. 12 and 18

70. 26, 36, and 54

Synthesis

71. ◆ Write a problem for a classmate to solve. Design the problem so that the solution is "Each of the 144 bottles will contain 8 oz of hot sauce." [1.5a]

72. 🖩 Determine the missing digits a and b. [1.3e]

$$\begin{array}{r} 9\,a\,1 \\ 2\,b\,1\overline{)\,236{,}421} \end{array}$$

73. A prime number that becomes a prime number when its digits are reversed is called a **palindrome prime**. For example, 17 is a palindrome prime because both 17 and 71 are primes. Which of the following numbers are palindrome primes? [1.7c]

13, 91, 16, 11, 15, 24, 29, 101, 201

Test: Chapter 1

1. Write expanded notation: 8843.

2. Write a word name: 38,403,277.

3. In the number 546,789, which digit tells the number of hundred thousands?

Add.

4. $\begin{array}{r} 6811 \\ +3178 \\ \hline \end{array}$

5. $\begin{array}{r} 45{,}889 \\ +17{,}902 \\ \hline \end{array}$

6. $\begin{array}{r} 12 \\ 8 \\ 3 \\ 7 \\ +\ \ 4 \\ \hline \end{array}$

7. $\begin{array}{r} 6203 \\ +4312 \\ \hline \end{array}$

Subtract.

8. $\begin{array}{r} 7983 \\ -4353 \\ \hline \end{array}$

9. $\begin{array}{r} 2974 \\ -1935 \\ \hline \end{array}$

10. $\begin{array}{r} 8907 \\ -2059 \\ \hline \end{array}$

11. $\begin{array}{r} 23{,}067 \\ -17{,}892 \\ \hline \end{array}$

Multiply.

12. $\begin{array}{r} 4568 \\ \times\ \ \ \ 9 \\ \hline \end{array}$

13. $\begin{array}{r} 8876 \\ \times\ 600 \\ \hline \end{array}$

14. $\begin{array}{r} 65 \\ \times 37 \\ \hline \end{array}$

15. $\begin{array}{r} 678 \\ \times 788 \\ \hline \end{array}$

Divide.

16. $15 \div 4$

17. $420 \div 6$

18. $89\overline{)8633}$

19. $44\overline{)35{,}428}$

Solve.

20. *James Dean.* James Dean was 24 yr old when he died. He was born in 1931. In what year did he die?

21. A beverage company produces 739 cans of soda. How many 8-can packages can be filled? How many cans will be left over?

22. *Area of New England.* Listed below are the areas, in square miles, of the New England states (***Source:*** U.S. Bureau of the Census). What is the total area of New England?

Maine	30,865
Massachusetts	7,838
New Hampshire	8,969
Vermont	9,249
Connecticut	4,845
Rhode Island	1,045

23. A rectangular lot measures 200 m by 600 m. What is the area of the lot? What is the perimeter of the lot?

600 m

200 m

24. A sack of oranges weighs 27 lb. A sack of apples weighs 32 lb. Find the total weight of 16 bags of oranges and 43 bags of apples.

25. A box contains 5000 staples. How many staplers can be filled from the box if each stapler holds 250 staples?

Answers

1. ______
2. ______
3. ______
4. ______
5. ______
6. ______
7. ______
8. ______
9. ______
10. ______
11. ______
12. ______
13. ______
14. ______
15. ______
16. ______
17. ______
18. ______
19. ______
20. ______
21. ______
22. ______
23. ______
24. ______
25. ______

Answers

26. ____
27. ____
28. ____
29. ____
30. ____
31. ____
32. ____
33. ____
34. ____
35. ____
36. ____
37. ____
38. ____
39. ____
40. ____
41. ____
42. ____
43. ____
44. ____
45. ____
46. ____
47. ____
48. ____
49. ____
50. ____
51. ____
52. ____
53. ____
54. ____

Solve.

26. $28 + x = 74$

27. $169 \div 13 = n$

28. $38 \cdot y = 532$

Round 34,578 to the nearest:

29. Thousand.

30. Ten.

31. Hundred.

Estimate the sum, difference, or product by first rounding to the nearest hundred. Show your work.

32. $\begin{array}{r} 23{,}649 \\ +\,54{,}746 \\ \hline \end{array}$

33. $\begin{array}{r} 54{,}751 \\ -\,23{,}649 \\ \hline \end{array}$

34. $\begin{array}{r} 824 \\ \times\,489 \\ \hline \end{array}$

Use < or > for ▒ to write a true sentence.

35. 34 ▒ 17

36. 117 ▒ 157

37. Write exponential notation: $12 \cdot 12 \cdot 12 \cdot 12$.

Evaluate.

38. 7^3

39. 2^3

Simplify.

40. $(10 - 2)^2$

41. $10^2 - 2^2$

42. $(25 - 15) \div 5$

43. $8 \times \{(20 - 11) \cdot [(12 + 48) \div 6 - (9 - 2)]\}$

44. $2^4 + 24 \div 12$

45. Find the average of 97, 98, 87, and 86.

Find the prime factorization of the number.

46. 18

47. 60

48. Determine whether 1784 is divisible by 8.

49. Determine whether 784 is divisible by 9.

50. Find the LCM of 12 and 16.

Synthesis

51. An open cardboard shoe box is 8 in. wide, 12 in. long, and 6 in. high. How many square inches of cardboard are used?

52. Cara spends $229 a month to repay her student loan. If she has already paid $9160 on the 10-yr loan, how many payments remain?

53. Jennie scores three 90's, four 80's, and a 74 on her eight quizzes. Find her average.

54. Use trials to find the single digit number a for which

$$359 - 46 + a \div 3 \times 25 - 7^2 = 339.$$

2

Fractional Notation

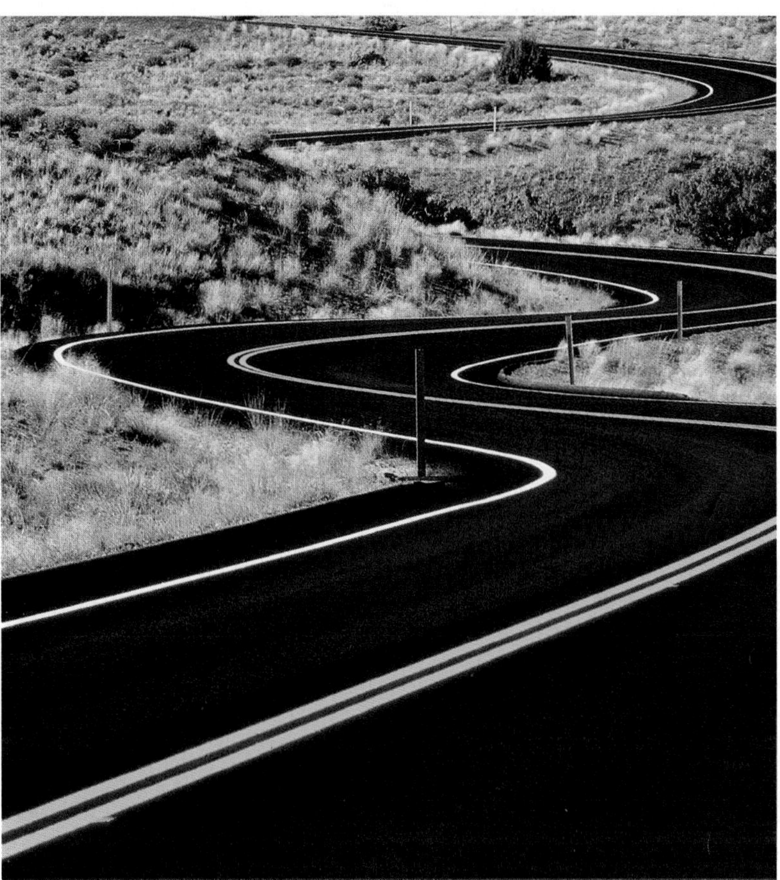

Introduction

In this chapter, we consider addition, subtraction, multiplication, and division using fractional notation. Also discussed are addition, subtraction, multiplication, and division using mixed numerals. We then apply all these operations to rules for order of operations, estimating, and applied problems.

An Application

Liz and her road crew paint the lines in the middle and on the sides of a highway. They average about $\frac{5}{16}$ mi each hour. How long will it take them to paint the lines on 70 mi of highway?

This problem appears as Exercise 46 in the Summary and Review Exercises.

The Mathematics

We let t = the time it takes to paint 70 mi of highway. The problem then translates to this equation:

$$\underbrace{t = 70 \div \frac{5}{16}}.$$

Here is how division using fractional notation occurs in applied problems.

World Wide Web For more information, visit us at www.mathmax.com

retest: Chapter 2

Simplify.

1. $\frac{57}{57}$

2. $\frac{68}{1}$

3. $\frac{0}{50}$

4. $\frac{8}{32}$

5. Use < or > for ■ to write a true sentence:

$\frac{7}{9}$ ■ $\frac{4}{5}$.

6. Find the reciprocal: $\frac{7}{8}$.

7. Convert to fractional notation: $7\frac{5}{8}$.

8. Convert to a mixed numeral: $\frac{11}{2}$.

9. Add. Write a mixed numeral for the answer.

$$\begin{array}{r} 8\frac{11}{12} \\ +\,2\frac{3}{5} \\ \hline \end{array}$$

10. Divide. Write a mixed numeral for the answer.

$5\frac{5}{12} \div 3\frac{1}{4}$

11. Multiply and simplify: $\frac{1}{3} \cdot \frac{18}{5}$.

12. Subtract and simplify: $\frac{5}{12} - \frac{1}{4}$.

Solve.

13. $\frac{2}{3} + x = \frac{8}{9}$

14. $\frac{7}{10} \cdot x = 21$

15. A piece of tubing $\frac{5}{8}$ m long is to be cut into 15 pieces of the same length. What is the length of each piece?

16. At a summer camp, the cook bought 100 lb of potatoes and used $78\frac{3}{4}$ lb. How many pounds were left?

17. A courier drove $214\frac{3}{10}$ mi one day and $136\frac{9}{10}$ mi the next. How far did she travel in all?

18. A cake recipe calls for $3\frac{3}{4}$ cups of flour. How much flour would be used to make 6 cakes?

Estimate each of the following as 0, $\frac{1}{2}$, or 1.

19. $\frac{2}{41}$

20. $\frac{15}{29}$

Estimate each of the following as a whole number or as a mixed numeral where the fractional part is $\frac{1}{2}$.

21. $10\frac{2}{17}$

22. $\frac{1}{10} + \frac{7}{8} + \frac{41}{39}$

Objectives for Retesting

The objectives to be tested in addition to the material in this chapter are as follows.

[1.2f] Subtract whole numbers.

[1.3e] Divide whole numbers.

[1.4b] Solve equations like $t + 28 = 54$, $28 \cdot x = 268$, and $98 \div 2 = y$.

[1.5a] Solve applied problems involving addition, subtraction, multiplication, or division of whole numbers.

2.1 Fractional Notation and Simplifying

The study of arithmetic begins with the set of whole numbers

0, 1, 2, 3, 4, 5, 6, 7, 8, 9, 10, 11, and so on.

The need soon arises for fractional parts of numbers such as halves, thirds, fourths, and so on. Here are some examples:

$\frac{1}{25}$ of the parking spaces in a commercial area in the state of Indiana are to be marked for the handicapped.

$\frac{1}{11}$ of all women develop breast cancer.

$\frac{1}{4}$ of the minimum daily requirement of calcium is provided by a cup of frozen yogurt.

$\frac{43}{100}$ of all corporate travel money is spent on airfares.

The following are some additional examples of fractions:

$$\frac{1}{2}, \quad \frac{3}{4}, \quad \frac{8}{5}, \quad \frac{11}{23}.$$

This way of writing number names is called **fractional notation.** The top number is called the **numerator** and the bottom number is called the **denominator**.

Objectives

a. Write fractional notation for part of an object or part of a set of objects.

b. Simplify fractional notation like n/n to 1, $0/n$ to 0, and $n/1$ to n.

c. Multiply using fractional notation.

d. Find another name for a number, but having a new denominator. Use multiplying by 1.

e. Simplify fractional notation.

For Extra Help

TAPE 4

MAC
WIN

CD-ROM

a Fractions and the Real World

Example 1 What part is shaded?

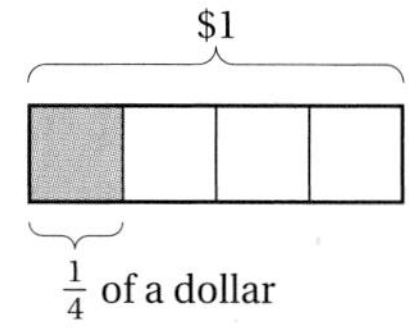

When an object is divided into 4 parts of the same size, each of these parts is $\frac{1}{4}$ of the object. Thus, $\frac{1}{4}$ (*one-fourth*) is shaded.

Do Exercises 1–4.

Example 2 What part is shaded?

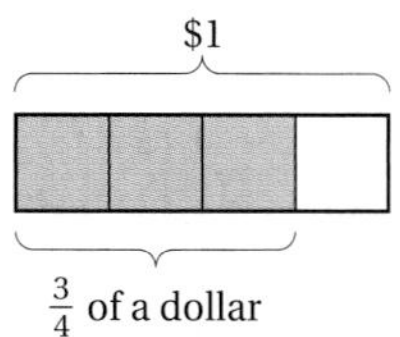

The object is divided into 4 parts of the same size, and 3 of them are shaded. This is $3 \cdot \frac{1}{4}$, or $\frac{3}{4}$. Thus, $\frac{3}{4}$ (*three-fourths*) of the object is shaded.

Do Exercises 5–8 on the following page.

What part is shaded?

1. $1

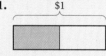

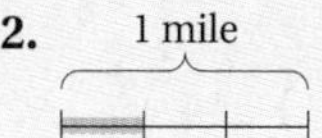

2. 1 mile

3.

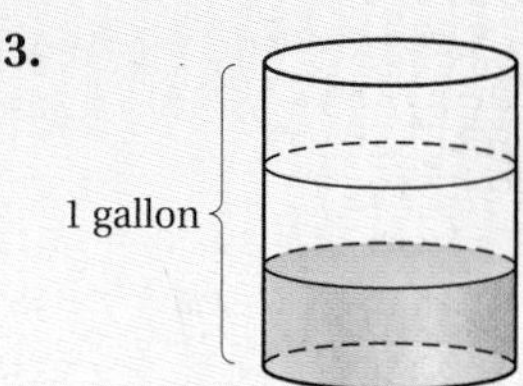

4.

Answers on page A-4

What part is shaded?

5.

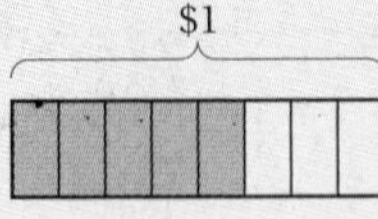

6.

1 mile

7.

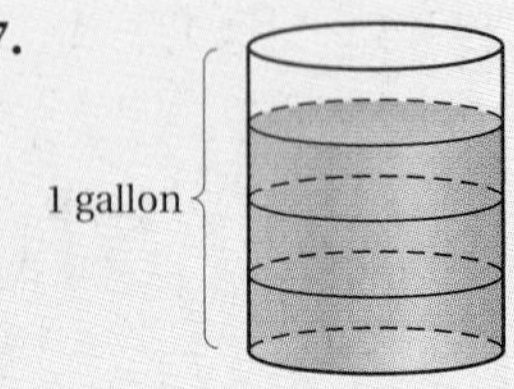

8.

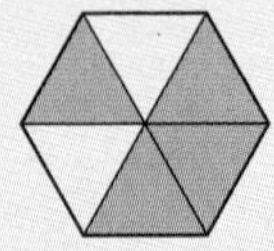

What part is shaded?

9.

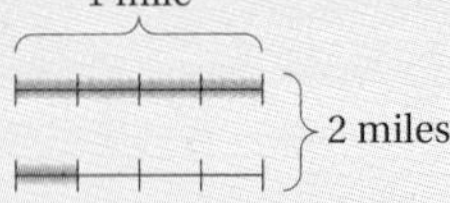

10.

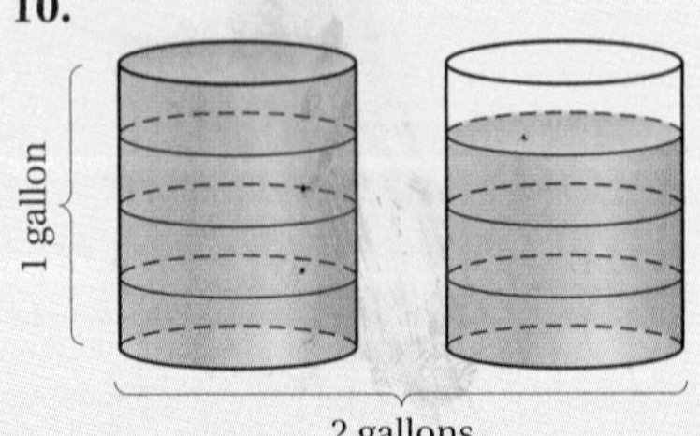

Answers on page A-4

Fractions greater than 1 correspond to situations like the following.

Example 3 What part is shaded?

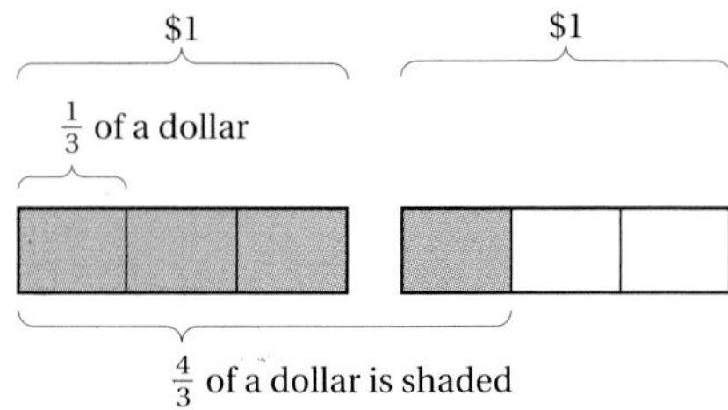

We divide the two objects into 3 parts each and take 4 of those parts. We have more than one whole object. In this case, it is $4 \cdot \frac{1}{3}$, or $\frac{4}{3}$.

Do Exercises 9 and 10.

Circle graphs, or pie charts, are often used to illlustrate the relationships of fractional parts of a whole. The following graph shows color preferences of bicycles.

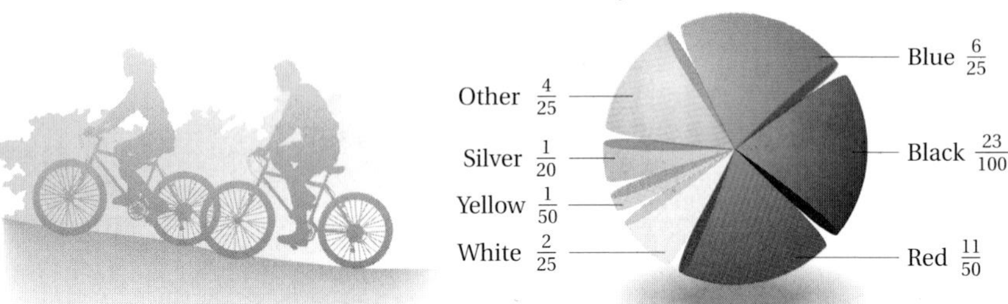

b Some Fractional Notation for Whole Numbers

Fractional Notation for 1

The number 1 corresponds to situations like those shown here.

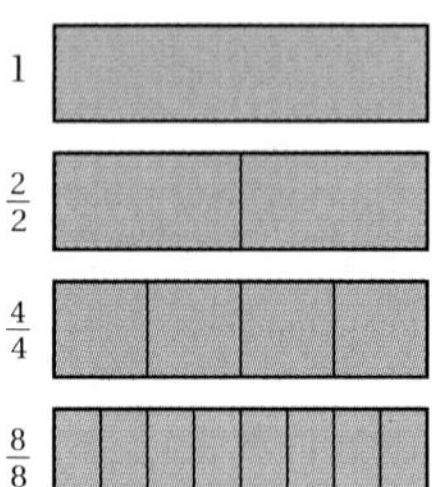

If we divide an object into n parts and take n of them, we get all of the object (1 whole object).

$$\frac{n}{n} = 1, \quad \text{for any whole number } n \text{ that is not 0.}$$

Examples Simplify.

4. $\frac{5}{5} = 1$ **5.** $\frac{9}{9} = 1$ **6.** $\frac{23}{23} = 1$

Do Exercises 11–16 on the following page.

Fractional Notation for 0

Consider $\frac{0}{4}$. This corresponds to dividing an object into 4 parts and taking none of them. We get 0.

> $\frac{0}{n} = 0$, for any whole number n that is not 0.

Examples Simplify.

7. $\frac{0}{1} = 0$ **8.** $\frac{0}{9} = 0$ **9.** $\frac{0}{23} = 0$

Fractional notation with a denominator of 0, such as $n/0$, is meaningless because we cannot speak of an object divided into *zero* parts. (If it is not divided at all, then we say that it is undivided and remains in one part.)

> $\frac{n}{0}$ is not defined for any whole number n.

Do Exercises 17–22.

Other Whole Numbers

Consider $\frac{4}{1}$. This corresponds to taking 4 objects and dividing each into 1 part. (We do not divide them.) We have 4 objects.

> Any whole number divided by 1 is the whole number. That is,
>
> $\frac{n}{1} = n$, for any whole number n.

Examples Simplify.

10. $\frac{2}{1} = 2$ **11.** $\frac{9}{1} = 9$ **12.** $\frac{34}{1} = 34$

Do Exercises 23–26.

c Multiplication Using Fractional Notation

We find a product such as $\frac{9}{7} \cdot \frac{3}{4}$ as follows.

> To multiply a fraction by a fraction,
>
> a) multiply the numerators to get the new numerator, and
>
> b) multiply the denominators to get the new denominator.
>
> $$\frac{9}{7} \cdot \frac{3}{4} = \frac{9 \cdot 3}{7 \cdot 4} = \frac{27}{28}$$

Simplify.

11. $\frac{1}{1}$ **12.** $\frac{4}{4}$

13. $\frac{34}{34}$ **14.** $\frac{100}{100}$

15. $\frac{2347}{2347}$ **16.** $\frac{103}{103}$

Simplify, if possible.

17. $\frac{0}{1}$ **18.** $\frac{0}{8}$

19. $\frac{0}{107}$ **20.** $\frac{4-4}{567}$

21. $\frac{15}{0}$ **22.** $\frac{0}{3-3}$

Simplify.

23. $\frac{8}{1}$ **24.** $\frac{10}{1}$

25. $\frac{346}{1}$ **26.** $\frac{24-1}{23-22}$

Answers on page A-4

Multiply.

27. $\frac{3}{8} \cdot \frac{5}{7}$

28. $\frac{4}{3} \times \frac{8}{5}$

29. $\frac{3}{10} \cdot \frac{1}{10}$

30. $7 \cdot \frac{2}{3}$

31. Draw diagrams like those in the text to show how the multiplication $\frac{1}{3} \cdot \frac{4}{5}$ corresponds to a real-world situation.

Answers on page A-4

Examples Multiply.

13. $\frac{5}{6} \times \frac{7}{4} = \frac{5 \times 7}{6 \times 4} = \frac{35}{24}$

Skip this step whenever you can.

14. $\frac{3}{5} \cdot \frac{7}{8} = \frac{3 \cdot 7}{5 \cdot 8} = \frac{21}{40}$

15. $\frac{3}{5} \cdot \frac{3}{4} = \frac{9}{20}$

16. $\frac{1}{4} \cdot \frac{1}{3} = \frac{1}{12}$

17. $6 \cdot \frac{4}{5} = \frac{6}{1} \cdot \frac{4}{5} = \frac{24}{5}$

Do Exercises 27–30.

Unless one of the factors is a whole number, multiplication using fractional notation does not correspond to repeated addition. Let's see how multiplication of fractions corresponds to situations in the real world. We consider the multiplication

$$\frac{3}{5} \cdot \frac{3}{4}.$$

We first consider some object and take $\frac{3}{4}$ of it. We divide it into 4 vertical parts, or columns of the same area, and take 3 of them. That is shown in the shading below.

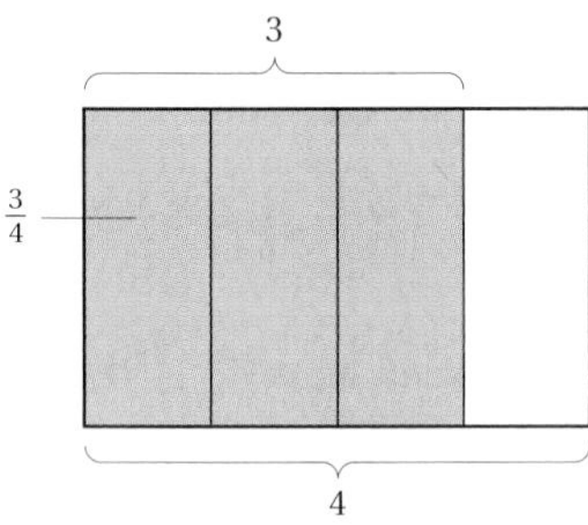

Next, we take $\frac{3}{5}$ of the result. We divide the shaded part into 5 horizontal parts, or rows of the same area, and take 3 of them. That is shown below.

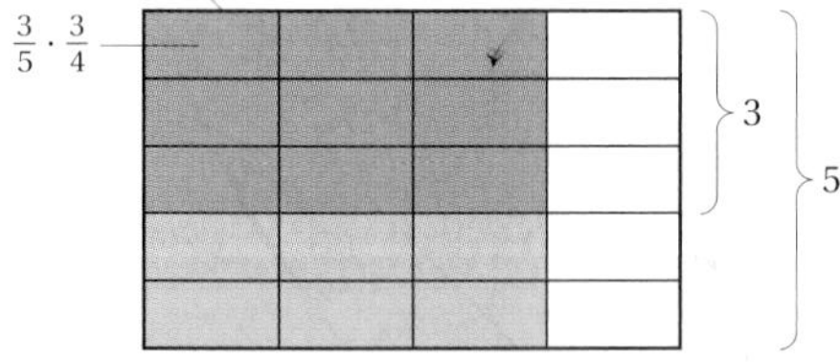

The entire object has been divided into 20 parts, and we have shaded 9 of them for a second time:

$$\frac{3}{5} \cdot \frac{3}{4} = \frac{3 \cdot 3}{5 \cdot 4} = \frac{9}{20}.$$

The figure above shows a rectangular array inside a rectangular array. The number of pieces in the entire array is $5 \cdot 4$ (the product of the denominators). The number of pieces shaded a second time is $3 \cdot 3$ (the product of the numerators). For the answer, we take 9 pieces out of a set of 20 to get $\frac{9}{20}$.

Do Exercise 31.

d Multiplying by 1

Recall the following:

$$1 = \frac{1}{1} = \frac{2}{2} = \frac{3}{3} = \frac{4}{4} = \frac{10}{10} = \frac{45}{45} = \frac{100}{100} = \frac{n}{n}.$$

Any nonzero number divided by itself is 1.

> When we multiply a number by 1, we get the same number.
>
> $$\frac{3}{5} = \frac{3}{5} \cdot 1 = \frac{3}{5} \cdot \frac{4}{4} = \frac{12}{20}$$

Since $\frac{3}{5} \cdot 1 = \frac{12}{20}$, we know that $\frac{3}{5}$ and $\frac{12}{20}$ are two names for the same number. We also say that $\frac{3}{5}$ and $\frac{12}{20}$ are **equivalent**.

Do Exercises 32–35.

Suppose we want to find a name for $\frac{2}{3}$, but one that has a denominator of 9. We can multiply by 1 to find equivalent fractions:

$$\frac{2}{3} = \frac{2}{3} \cdot \frac{3}{3} = \frac{2 \cdot 3}{3 \cdot 3} = \frac{6}{9}.$$

We chose $\frac{3}{3}$ for 1 in order to get a denominator of 9.

Example 18 Find a name for $\frac{1}{4}$ with a denominator of 24.

Since $4 \cdot 6 = 24$, we multiply by $\frac{6}{6}$:

$$\frac{1}{4} = \frac{1}{4} \cdot \frac{6}{6} = \frac{1 \cdot 6}{4 \cdot 6} = \frac{6}{24}.$$

Example 19 Find a name for $\frac{2}{5}$ with a denominator of 35.

Since $5 \cdot 7 = 35$, we multiply by $\frac{7}{7}$:

$$\frac{2}{5} = \frac{2}{5} \cdot \frac{7}{7} = \frac{2 \cdot 7}{5 \cdot 7} = \frac{14}{35}.$$

Do Exercises 36–40.

e Simplifying

All of the following are names for three-fourths:

$$\frac{3}{4}, \frac{6}{8}, \frac{9}{12}, \frac{12}{16}, \frac{15}{20}.$$

We say that $\frac{3}{4}$ is **simplest** because it has the smallest numerator and the smallest denominator. That is, the numerator and the denominator have no common factor other than 1.

Multiply.

32. $\frac{1}{2} \cdot \frac{8}{8}$

33. $\frac{3}{5} \cdot \frac{10}{10}$

34. $\frac{13}{25} \cdot \frac{4}{4}$

35. $\frac{8}{3} \cdot \frac{25}{25}$

Find another name for the number, but with the denominator indicated. Use multiplying by 1.

36. $\frac{4}{3} = \frac{?}{9}$

37. $\frac{3}{4} = \frac{?}{24}$

38. $\frac{9}{10} = \frac{?}{100}$

39. $\frac{3}{15} = \frac{?}{45}$

40. $\frac{8}{7} = \frac{?}{49}$

Answers on page A-4

Simplify.

41. $\frac{2}{8}$

42. $\frac{10}{12}$

43. $\frac{40}{8}$

44. $\frac{24}{18}$

Answers on page A-4

To simplify, we reverse the process of multiplying by 1.

$$\frac{12}{18} = \frac{2 \cdot 6}{3 \cdot 6} \quad \begin{array}{l} \leftarrow \text{Factoring the numerator} \\ \leftarrow \text{Factoring the denominator} \end{array}$$

$$= \frac{2}{3} \cdot \frac{6}{6} \qquad \text{Factoring the fraction}$$

$$= \frac{2}{3} \cdot 1 \qquad \frac{6}{6} = 1$$

$$= \frac{2}{3} \qquad \text{Removing a factor of 1: } \frac{2}{3} \cdot 1 = \frac{2}{3}$$

Examples Simplify.

20. $\frac{8}{20} = \frac{2 \cdot 4}{5 \cdot 4} = \frac{2}{5} \cdot \frac{4}{4} = \frac{2}{5}$

21. $\frac{2}{6} = \frac{1 \cdot 2}{3 \cdot 2} = \frac{1}{3} \cdot \frac{2}{2} = \frac{1}{3}$

The number 1 allows for pairing of factors in the numerator and the denominator.

22. $\frac{30}{6} = \frac{5 \cdot 6}{1 \cdot 6} = \frac{5}{1} \cdot \frac{6}{6} = \frac{5}{1} = 5$

We could also simplify $\frac{30}{6}$ by doing the division $30 \div 6$. That is, $\frac{30}{6} = 30 \div 6 = 5$.

Do Exercises 41–44.

The use of prime factorizations can be helpful for simplifying when numerators and/or denominators are larger numbers.

Example 23 Simplify: $\frac{90}{84}$.

$$\frac{90}{84} = \frac{2 \cdot 3 \cdot 3 \cdot 5}{2 \cdot 2 \cdot 3 \cdot 7} \qquad \text{Factoring the numerator and the denominator into primes}$$

$$= \frac{2 \cdot 3 \cdot 3 \cdot 5}{2 \cdot 3 \cdot 2 \cdot 7} \qquad \text{Changing the order so that like primes are above and below each other}$$

$$= \frac{2}{2} \cdot \frac{3}{3} \cdot \frac{3 \cdot 5}{2 \cdot 7} \qquad \text{Factoring the fraction}$$

$$= 1 \cdot 1 \cdot \frac{3 \cdot 5}{2 \cdot 7}$$

$$= \frac{3 \cdot 5}{2 \cdot 7} \qquad \text{Removing factors of 1}$$

$$= \frac{15}{14}$$

We could have shortened the preceding example had we recalled our tests for divisibility (Section 1.8) and noted that 6 is a factor of both the numerator and the denominator. Then

$$\frac{90}{84} = \frac{6 \cdot 15}{6 \cdot 14} = \frac{6}{6} \cdot \frac{15}{14} = \frac{15}{14}.$$

The tests for divisibility are very helpful in simplifying.

Example 24 Simplify: $\frac{603}{207}$.

At first glance this looks difficult. But note, using the test for divisibility by 9 (sum of digits divisible by 9), that both the numerator and the denominator are divisible by 9. Thus we can factor 9 from both numbers:

$$\frac{603}{207} = \frac{9 \cdot 67}{9 \cdot 23} = \frac{9}{9} \cdot \frac{67}{23} = \frac{67}{23}.$$

Do Exercises 45–49.

Simplify.

45. $\frac{35}{40}$

46. $\frac{801}{702}$

47. $\frac{24}{21}$

48. $\frac{75}{300}$

49. Simplify each fraction in this circle graph.

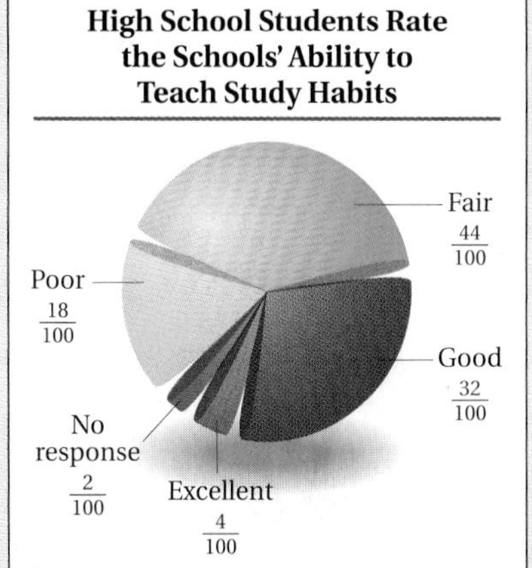

CANCELING Canceling is a shortcut that you may have used for removing a factor of 1 when working with fractional notation. With *great* concern, we mention it as a possibility for speeding up your work. Canceling may be done only when removing common factors in numerators and denominators. Each common factor allows us to remove a factor of 1 in a product.

Our concern is that canceling be done with care and understanding. In effect, slashes are used to indicate factors of 1 that have been removed. For instance, Example 6 might have been done faster as follows:

$$\frac{90}{84} = \frac{2 \cdot 3 \cdot 3 \cdot 5}{2 \cdot 2 \cdot 3 \cdot 7}$$ **Factoring the numerator and the denominator**

$$= \frac{\not{2} \cdot \not{3} \cdot 3 \cdot 5}{2 \cdot \not{2} \cdot \not{3} \cdot 7}$$ **When a factor of 1 is noted, it is "canceled" as shown: $\frac{2 \cdot 3}{2 \cdot 3} = 1$.**

$$= \frac{3 \cdot 5}{2 \cdot 7} = \frac{15}{14}.$$

CAUTION! The difficulty with canceling is that it is often applied incorrectly in situations like the following:

$$\frac{\not{2} + 3}{\not{2}} = 3; \quad \frac{\not{4} + 1}{\not{4} + 2} = \frac{1}{2}; \quad \frac{1\not{5}}{\not{5}4} = \frac{1}{4}.$$

Wrong! Wrong! Wrong!

The correct answers are

$$\frac{2 + 3}{2} = \frac{5}{2}; \quad \frac{4 + 1}{4 + 2} = \frac{5}{6}; \quad \frac{15}{54} = \frac{5}{18}.$$

In each situation, the number canceled was not a factor of 1. Factors are parts of products. For example, in $2 \cdot 3$, 2 and 3 are factors, but in $2 + 3$, 2 and 3 are *not* factors. Canceling may not be done when sums or differences are in numerators or denominators, as shown here.

> If you cannot factor, do not cancel! If in doubt, do not cancel!

Calculator Spotlight

Fraction calculators are equipped with a key, often labeled [$a\,b/c$], that allows for simplification with fractional notation. To simplify

$$\frac{208}{256}$$

with such a fraction calculator, the following keystrokes can be used.

[2] [0] [8] [$a\,b/c$] [2] [5] [6] [=].

The display that appears

13 ⌟ 16.

represents simplified fractional notation $\frac{13}{16}$.

Exercises

Use a fraction calculator to simplify each of the following.

1. $\frac{84}{90}$ **2.** $\frac{35}{40}$ **3.** $\frac{690}{835}$ **4.** $\frac{42}{150}$

Answers on page A-4

Improving Your Math Study Skills

Classwork: Before and During Class

Before Class

Textbook

- Check your syllabus (or ask your instructor) to find out which sections will be covered during the next class. Then be sure to read these sections *before* class. Although you may not understand all the concepts, you will at least be familiar with the material, which will help you follow the discussion during class.
- This book makes use of color, shading, and design elements to highlight important concepts, so you do not need to highlight these. Instead, it is more productive for you to note trouble spots with either a highlighter or Post-It notes. Then use these marked points as possible questions for clarification by your instructor at the appropriate time.

Homework

- Review the previous day's homework just before class. This will refresh your memory on the concepts covered in the last class, and again provide you with possible questions to ask your instructor.

During Class

Class Seating

- If possible, choose a seat at the front of the class. In most classes, the more serious students tend to sit there so you will probably be able to concentrate better if you do the same. You should also avoid sitting next to noisy or distracting students.
- If your instructor uses an overhead projector, consider choosing a seat that will give you an unobstructed view of the screen.

Taking Notes

- This textbook has been written and laid out so that it represents a quality set of notes at the same time that it teaches. Thus you might not need to take notes in class. Just watch, listen, and ask yourself questions as the class moves along, rather than racing to keep up your note-taking.

  However, if you still feel more comfortable taking your own notes, consider using the following two-column method. Divide your page in half vertically so that you have two columns side by side. Write down what is on the board in the left column; then, in the right column, write clarifying comments or questions.
- If you have any difficulty keeping up with the instructor, use abbreviations to speed up your note-taking. Consider standard abbreviations like "Ex" for "Example," "$\approx$" for "approximately equal to," or "$\therefore$" for "therefore." Create your own abbreviations as well.
- Another shortcut for note-taking is to write only the beginning of a word, leaving space for the rest. Be sure you write enough of the word to know what it means later on!

Exercise Set 2.1

a What part of the object or set of objects is shaded?

1.

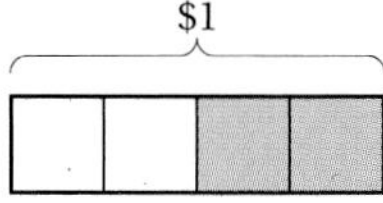

2.

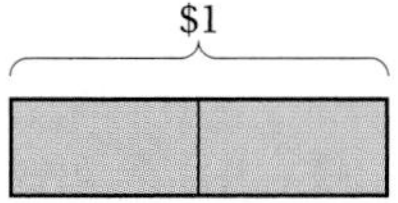

3.

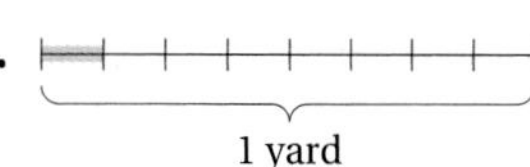

4.

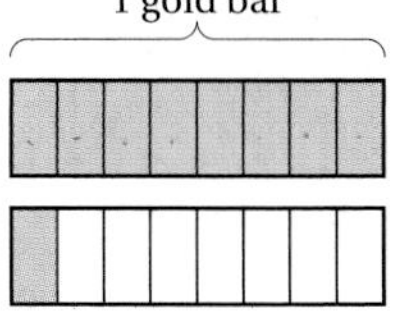

5.

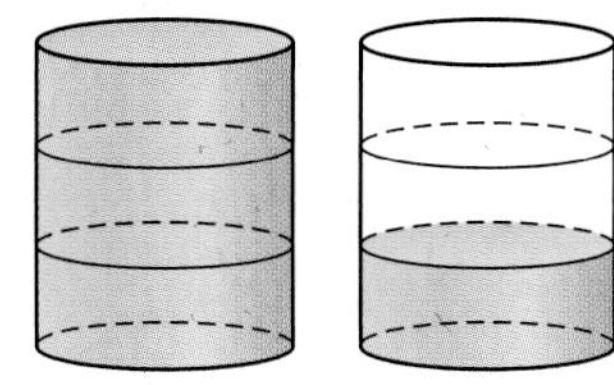

6.

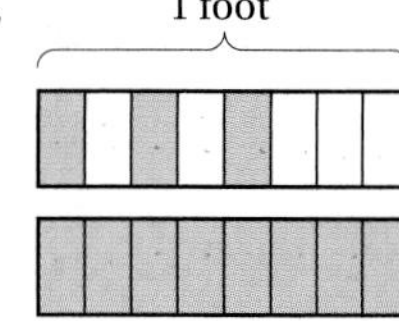

7.

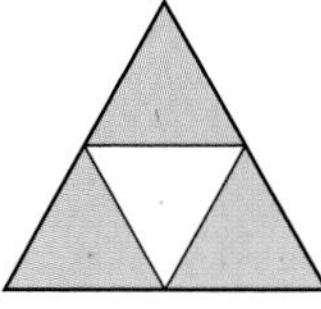

8.

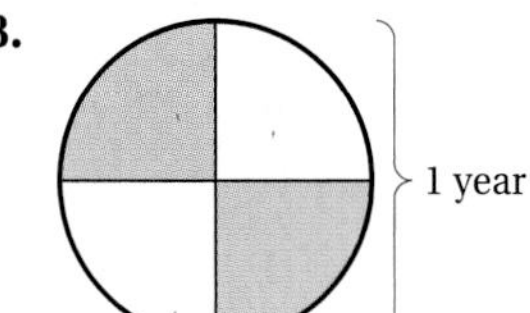

9.

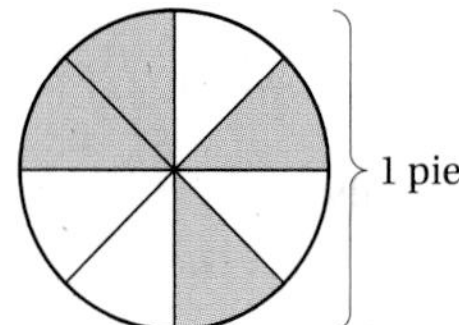

10.

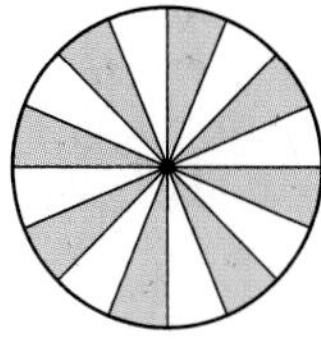

11.

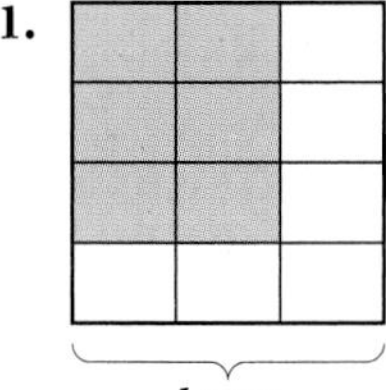

12.

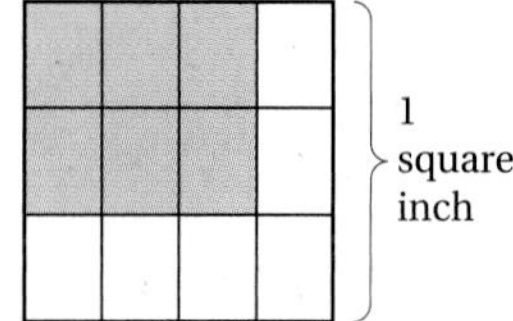

b Simplify.

13. $\frac{18}{1}$

14. $\frac{0}{16}$

15. $\frac{0}{8}$

16. $\frac{16}{1}$

17. $\frac{20}{20}$

18. $\frac{3}{3}$

19. $\frac{5}{6-6}$

20. $\frac{238}{1}$

21. $\frac{729}{0}$

22. $\frac{8-8}{1247}$

23. $\frac{87}{87}$

24. $\frac{1317}{0}$

c Multiply.

25. $\frac{1}{2} \cdot \frac{1}{3}$

26. $\frac{1}{6} \cdot \frac{1}{4}$

27. $5 \times \frac{1}{8}$

28. $4 \times \frac{1}{5}$

29. $\frac{2}{3} \times \frac{1}{5}$

30. $\frac{3}{5} \times \frac{1}{5}$

31. $\frac{2}{5} \cdot \frac{2}{3}$

32. $\frac{3}{4} \cdot \frac{3}{5}$

33. $\frac{3}{4} \cdot \frac{3}{4}$

34. $\frac{3}{7} \cdot \frac{4}{5}$

35. $\frac{2}{3} \cdot \frac{7}{13}$

36. $\frac{3}{11} \cdot \frac{4}{5}$

37. $7 \cdot \frac{3}{4}$

38. $7 \cdot \frac{2}{5}$

39. $\frac{7}{8} \cdot \frac{7}{8}$

40. $\frac{3}{10} \cdot \frac{7}{100}$

d Find another name for the given number, but with the denominator indicated. Use multiplying by 1.

41. $\frac{1}{2} = \frac{?}{10}$

42. $\frac{1}{6} = \frac{?}{18}$

43. $\frac{5}{8} = \frac{?}{32}$

44. $\frac{2}{9} = \frac{?}{18}$

45. $\frac{5}{3} = \frac{?}{45}$

46. $\frac{11}{5} = \frac{?}{30}$

47. $\frac{7}{22} = \frac{?}{132}$

48. $\frac{10}{21} = \frac{?}{126}$

e Simplify.

49. $\frac{6}{8}$

50. $\frac{8}{12}$

51. $\frac{3}{15}$

52. $\frac{8}{10}$

53. $\frac{24}{8}$

54. $\frac{36}{9}$

55. $\frac{18}{24}$

56. $\frac{42}{48}$

57. $\frac{14}{16}$

58. $\frac{15}{25}$

59. $\frac{17}{51}$

60. $\frac{425}{525}$

61. $\frac{150}{25}$

62. $\frac{19}{76}$

Synthesis

63. ◆ Explain in your own words when it *is* possible to "cancel" and when it *is not* possible to "cancel."

64. ◆ Can fractional notation be simplified if its numerator and its denominator are two different prime numbers? Why or why not?

Use fraction bars to represent equivalent fractions.

2.2 Multiplication and Division

Objectives

a Multiply and simplify using fractional notation.

b Find the reciprocal of a number.

c Divide and simplify using fractional notation.

d Solve equations of the type $a \cdot x = b$ and $x \cdot a = b$, where a and b may be fractions.

For Extra Help

TAPE 4

MAC WIN

CD-ROM

a Simplifying After Multiplying

We usually simplify after we multiply. To make such simplifying easier, it is generally best not to carry out the products in the numerator and the denominator, but to factor and simplify before multiplying. Consider the product

$$\frac{3}{8} \cdot \frac{4}{9}.$$

We proceed as follows:

$$\frac{3}{8} \cdot \frac{4}{9} = \frac{3 \cdot 4}{8 \cdot 9}$$ We write the products in the numerator and the denominator, but we do not carry them out.

$$= \frac{3 \cdot 2 \cdot 2}{2 \cdot 2 \cdot 2 \cdot 3 \cdot 3}$$ Factoring the numerator and the denominator

$$= \frac{3 \cdot 2 \cdot 2}{3 \cdot 2 \cdot 2} \cdot \frac{1}{2 \cdot 3}$$ Factoring the fraction

$$= 1 \cdot \frac{1}{2 \cdot 3}$$

$$= \frac{1}{2 \cdot 3}$$ Removing a factor of 1

$$= \frac{1}{6}.$$

The procedure could have been shortened had we noticed that 4 is a factor of the 8 in the denominator:

$$\frac{3}{8} \cdot \frac{4}{9} = \frac{3 \cdot 4}{8 \cdot 9} = \frac{3 \cdot 4}{4 \cdot 2 \cdot 3 \cdot 3} = \frac{3 \cdot 4}{3 \cdot 4} \cdot \frac{1}{2 \cdot 3} = 1 \cdot \frac{1}{2 \cdot 3} = \frac{1}{2 \cdot 3} = \frac{1}{6}.$$

To multiply and simplify:

a) Write the products in the numerator and the denominator, but do not carry out the products.

b) Factor the numerator and the denominator.

c) Factor the fraction to remove factors of 1.

d) Carry out the remaining products.

Examples Multiply and simplify.

1. $\frac{2}{3} \cdot \frac{9}{4} = \frac{2 \cdot 9}{3 \cdot 4} = \frac{2 \cdot 3 \cdot 3}{3 \cdot 2 \cdot 2} = \frac{2 \cdot 3}{2 \cdot 3} \cdot \frac{3}{2} = 1 \cdot \frac{3}{2} = \frac{3}{2}$

2. $\frac{6}{7} \cdot \frac{5}{3} = \frac{6 \cdot 5}{7 \cdot 3} = \frac{3 \cdot 2 \cdot 5}{7 \cdot 3} = \frac{3}{3} \cdot \frac{2 \cdot 5}{7} = 1 \cdot \frac{2 \cdot 5}{7} = \frac{2 \cdot 5}{7} = \frac{10}{7}$

3. $40 \cdot \frac{7}{8} = \frac{40 \cdot 7}{8} = \frac{8 \cdot 5 \cdot 7}{8 \cdot 1} = \frac{8}{8} \cdot \frac{5 \cdot 7}{1} = 1 \cdot \frac{5 \cdot 7}{1} = \frac{5 \cdot 7}{1} = 35$

Multiply and simplify.

1. $\frac{2}{3} \cdot \frac{7}{8}$

2. $\frac{4}{5} \cdot \frac{5}{12}$

3. $16 \cdot \frac{3}{8}$

4. $\frac{5}{8} \cdot 4$

Find the reciprocal.

5. $\frac{2}{5}$

6. $\frac{10}{7}$

7. 9

8. $\frac{1}{5}$

Answers on page A-4

CAUTION! Canceling can be used as follows for these examples.

1. $\frac{2}{3} \cdot \frac{9}{4} = \frac{2 \cdot 9}{3 \cdot 4} = \frac{\not{2} \cdot \not{3} \cdot 3}{\not{3} \cdot \not{2} \cdot 2} = \frac{3}{2}$ Removing a factor of 1: $\frac{2 \cdot 3}{2 \cdot 3} = 1$

2. $\frac{6}{7} \cdot \frac{5}{3} = \frac{6 \cdot 5}{7 \cdot 3} = \frac{\not{3} \cdot 2 \cdot 5}{7 \cdot \not{3}} = \frac{2 \cdot 5}{7} = \frac{10}{7}$ Removing a factor of 1: $\frac{3}{3} = 1$

3. $40 \cdot \frac{7}{8} = \frac{40 \cdot 7}{8} = \frac{\not{8} \cdot 5 \cdot 7}{\not{8} \cdot 1} = \frac{5 \cdot 7}{1} = 35$ Removing a factor of 1: $\frac{8}{8} = 1$

Remember, if you can't factor, you can't cancel!

Do Exercises 1–4.

b Reciprocals

Look at these products:

$$8 \cdot \frac{1}{8} = \frac{8 \cdot 1}{8} = \frac{8}{8} = 1; \qquad \frac{2}{3} \cdot \frac{3}{2} = \frac{2 \cdot 3}{3 \cdot 2} = \frac{6}{6} = 1.$$

If the product of two numbers is 1, we say that they are **reciprocals** of each other. To find a reciprocal, interchange the numerator and the denominator.

Number $\longrightarrow \frac{3}{4} \quad \frac{4}{3} \longleftarrow$ Reciprocal

Examples Find the reciprocal.

4. The reciprocal of $\frac{4}{5}$ is $\frac{5}{4}$. $\quad \frac{4}{5} \cdot \frac{5}{4} = \frac{20}{20} = 1$

5. The reciprocal of $\frac{8}{7}$ is $\frac{7}{8}$. $\quad \frac{8}{7} \cdot \frac{7}{8} = \frac{56}{56} = 1$

6. The reciprocal of 8 is $\frac{1}{8}$. $\quad$ Think of 8 as $\frac{8}{1}$: $\frac{8}{1} \cdot \frac{1}{8} = \frac{8}{8} = 1$.

7. The reciprocal of $\frac{1}{3}$ is 3. $\quad \frac{1}{3} \cdot 3 = \frac{3}{3} = 1$

Do Exercises 5–8.

Does 0 have a reciprocal? If it did, it would have to be a number x such that

$$0 \cdot x = 1.$$

But 0 times any number is 0. Thus,

The number 0, or $\frac{0}{n}$, has no reciprocal. $\left(\text{Recall that } \frac{n}{0} \text{ is not defined.}\right)$

c Division

Recall that $a \div b$ is the number that when multiplied by b gives a. Consider the division $\frac{3}{4} \div \frac{1}{8}$. We are asking how many $\frac{1}{8}$'s are in $\frac{3}{4}$. We can answer this by looking at the figure below.

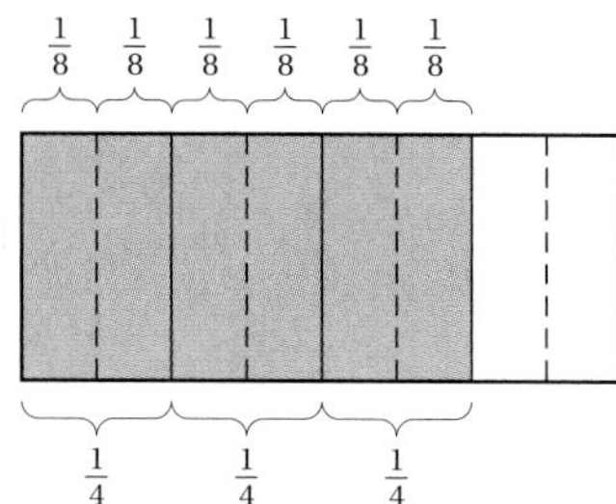

We see that there are six $\frac{1}{8}$'s in $\frac{3}{4}$. Thus,

$$\frac{3}{4} \div \frac{1}{8} = 6.$$

We can check this by multiplying:

$$6 \cdot \frac{1}{8} = \frac{6}{8} = \frac{3}{4}.$$

Here is a faster way to divide.

> To divide fractions, multiply the dividend by the reciprocal of the divisor:
>
> $$\frac{2}{5} \div \frac{3}{4} = \frac{2}{5} \cdot \frac{4}{3} = \frac{2 \cdot 4}{5 \cdot 3} = \frac{8}{15}.$$
>
> Multiply by the reciprocal of the divisor.

Examples Divide and simplify.

8. $\frac{5}{6} \div \frac{2}{3} = \frac{5}{6} \cdot \frac{3}{2} = \frac{5 \cdot 3}{6 \cdot 2} = \frac{5 \cdot 3}{3 \cdot 2 \cdot 2} = \frac{3}{3} \cdot \frac{5}{2 \cdot 2} = \frac{5}{2 \cdot 2} = \frac{5}{4}$

9. $\frac{3}{4} \div \frac{1}{8} = \frac{3}{4} \cdot 8 = \frac{3 \cdot 8}{4} = \frac{3 \cdot 4 \cdot 2}{4 \cdot 1} = \frac{4}{4} \cdot \frac{3 \cdot 2}{1} = \frac{3 \cdot 2}{1} = 6$

10. $\frac{2}{5} \div 6 = \frac{2}{5} \cdot \frac{1}{6} = \frac{2 \cdot 1}{5 \cdot 6} = \frac{2 \cdot 1}{5 \cdot 2 \cdot 3} = \frac{2}{2} \cdot \frac{1}{5 \cdot 3} = \frac{1}{5 \cdot 3} = \frac{1}{15}$

CAUTION! Canceling can be used as follows for Examples 8–10.

8. $\frac{5}{6} \div \frac{2}{3} = \frac{5}{6} \cdot \frac{3}{2} = \frac{5 \cdot 3}{6 \cdot 2} = \frac{5 \cdot \cancel{3}}{\cancel{3} \cdot 2 \cdot 2} = \frac{5}{2 \cdot 2} = \frac{5}{4}$ **Removing a factor of 1:** $\frac{3}{3} = 1$

9. $\frac{3}{4} \div \frac{1}{8} = \frac{3}{4} \cdot 8 = \frac{3 \cdot 8}{4} = \frac{3 \cdot \cancel{4} \cdot 2}{\cancel{4} \cdot 1} = \frac{3 \cdot 2}{1} = 6$ **Removing a factor of 1:** $\frac{4}{4} = 1$

10. $\frac{2}{5} \div 6 = \frac{2}{5} \cdot \frac{1}{6} = \frac{2 \cdot 1}{5 \cdot 6} = \frac{\cancel{2} \cdot 1}{5 \cdot \cancel{2} \cdot 3} = \frac{1}{5 \cdot 3} = \frac{1}{15}$ **Removing a factor of 1:** $\frac{2}{2} = 1$

Remember, if you can't factor, you can't cancel!

Do Exercises 9–13.

Divide and simplify.

9. $\frac{6}{7} \div \frac{3}{4}$

10. $\frac{2}{3} \div \frac{1}{4}$

11. $\frac{4}{5} \div 8$

12. $60 \div \frac{3}{5}$

13. $\frac{3}{5} \div \frac{3}{5}$

Answers on page A-4

14. Divide by multiplying by 1:

$$\frac{\frac{4}{5}}{\frac{6}{7}}.$$

Why do we multiply by a reciprocal when dividing? To see this, let's consider $\frac{2}{3} \div \frac{7}{5}$. We will multiply by 1. The name for 1 that we will use is $(5/7)/(5/7)$; it comes from the reciprocal of $\frac{7}{5}$.

$$\frac{2}{3} \div \frac{7}{5} = \frac{\frac{2}{3}}{\frac{7}{5}} \qquad \text{Writing fractional notation for the division}$$

$$= \frac{\frac{2}{3}}{\frac{7}{5}} \cdot 1 \qquad \text{Multiplying by 1}$$

$$= \frac{\frac{2}{3}}{\frac{7}{5}} \cdot \frac{\frac{5}{7}}{\frac{5}{7}} \qquad \text{Multiplying by 1; } \tfrac{5}{7} \text{ is the reciprocal of } \tfrac{7}{5} \text{ and } \frac{\frac{5}{7}}{\frac{5}{7}} = 1$$

$$= \frac{\frac{2}{3} \cdot \frac{5}{7}}{\frac{7}{5} \cdot \frac{5}{7}} \qquad \text{Multiplying the numerators and the denominators}$$

$$= \frac{\frac{2}{3} \cdot \frac{5}{7}}{1} = \frac{2}{3} \cdot \frac{5}{7} = \frac{10}{21}.$$

After we multiplied, we got 1 for the denominator. The numerator (in color) shows the multiplication by the reciprocal.

Do Exercise 14.

Solve.

15. $\frac{5}{6} \cdot y = \frac{2}{3}$

16. $\frac{3}{4} \cdot n = 24$

d Solving Equations

Now let's solve equations $a \cdot x = b$ and $x \cdot a = b$, where a and b may be fractions. Proceeding as we have before, we divide by a on both sides.

Example 11 Solve: $\frac{4}{3} \cdot x = \frac{6}{7}$.

$$\frac{4}{3} \cdot x = \frac{6}{7}$$

$$x = \frac{6}{7} \div \frac{4}{3} \qquad \text{Dividing by } \tfrac{4}{3} \text{ on both sides}$$

$$= \frac{6}{7} \cdot \frac{3}{4} \qquad \text{Multiplying by the reciprocal}$$

$$= \frac{2 \cdot 3 \cdot 3}{7 \cdot 2 \cdot 2} = \frac{2}{2} \cdot \frac{3 \cdot 3}{7 \cdot 2} = \frac{3 \cdot 3}{7 \cdot 2} = \frac{9}{14}$$

The solution is $\frac{9}{14}$.

Example 12 Solve: $t \cdot \frac{4}{5} = 80$.

Dividing by $\frac{4}{5}$ on both sides, we get

$$t = 80 \div \frac{4}{5} = 80 \cdot \frac{5}{4} = \frac{80 \cdot 5}{4} = \frac{4 \cdot 20 \cdot 5}{4 \cdot 1} = \frac{4}{4} \cdot \frac{20 \cdot 5}{1} = \frac{20 \cdot 5}{1} = 100.$$

The solution is 100.

Do Exercises 15 and 16.

Answers on page A-4

Exercise Set 2.2

a Multiply and simplify.

Don't forget to simplify!

1. $\frac{2}{3} \cdot \frac{1}{2}$

2. $\frac{3}{8} \cdot \frac{1}{3}$

3. $\frac{1}{4} \cdot \frac{2}{3}$

4. $\frac{4}{6} \cdot \frac{1}{6}$

5. $\frac{12}{5} \cdot \frac{9}{8}$

6. $\frac{16}{15} \cdot \frac{5}{4}$

7. $\frac{10}{9} \cdot \frac{7}{5}$

8. $\frac{25}{12} \cdot \frac{4}{3}$

9. $9 \cdot \frac{1}{9}$

10. $4 \cdot \frac{1}{4}$

11. $\frac{7}{5} \cdot \frac{5}{7}$

12. $\frac{2}{11} \cdot \frac{11}{2}$

13. $24 \cdot \frac{1}{6}$

14. $16 \cdot \frac{1}{2}$

15. $12 \cdot \frac{3}{4}$

16. $18 \cdot \frac{5}{6}$

17. $\frac{7}{10} \cdot 28$

18. $\frac{5}{8} \cdot 34$

19. $240 \cdot \frac{1}{8}$

20. $150 \cdot \frac{1}{5}$

21. $\frac{4}{10} \cdot \frac{5}{10}$

22. $\frac{7}{10} \cdot \frac{34}{150}$

23. $\frac{8}{10} \cdot \frac{45}{100}$

24. $\frac{3}{10} \cdot \frac{8}{10}$

25. $\frac{11}{24} \cdot \frac{3}{5}$

26. $\frac{15}{22} \cdot \frac{4}{7}$

27. $\frac{10}{21} \cdot \frac{3}{4}$

28. $\frac{17}{18} \cdot \frac{3}{5}$

b Find the reciprocal.

29. $\frac{5}{6}$

30. $\frac{7}{8}$

31. 6

32. 4

33. $\frac{1}{6}$

34. $\frac{1}{4}$

35. $\frac{10}{3}$

36. $\frac{17}{4}$

c Divide and simplify.

Don't forget to simplify!

37. $\frac{3}{5} \div \frac{3}{4}$

38. $\frac{2}{3} \div \frac{3}{4}$

39. $\frac{3}{5} \div \frac{9}{4}$

40. $\frac{6}{7} \div \frac{3}{5}$

41. $\frac{4}{3} \div \frac{1}{3}$

42. $\frac{10}{9} \div \frac{1}{3}$

43. $\frac{1}{3} \div \frac{1}{6}$

44. $\frac{1}{4} \div \frac{1}{5}$

45. $\frac{3}{8} \div 3$

46. $\frac{5}{6} \div 5$

47. $\frac{12}{7} \div 4$

48. $\frac{18}{5} \div 2$

49. $12 \div \frac{3}{2}$

50. $24 \div \frac{3}{8}$

51. $28 \div \frac{4}{5}$

52. $40 \div \frac{2}{3}$

53. $\frac{5}{8} \div \frac{5}{8}$

54. $\frac{2}{5} \div \frac{2}{5}$

55. $\frac{8}{15} \div \frac{4}{5}$

56. $\frac{6}{13} \div \frac{3}{26}$

57. $\frac{9}{5} \div \frac{4}{5}$

58. $\frac{5}{12} \div \frac{25}{36}$

59. $120 \div \frac{5}{6}$

60. $360 \div \frac{8}{7}$

d Solve.

61. $\frac{4}{5} \cdot x = 60$

62. $\frac{3}{2} \cdot t = 90$

63. $\frac{5}{3} \cdot y = \frac{10}{3}$

64. $\frac{4}{9} \cdot m = \frac{8}{3}$

65. $x \cdot \frac{25}{36} = \frac{5}{12}$

66. $p \cdot \frac{4}{5} = \frac{8}{15}$

67. $n \cdot \frac{8}{7} = 360$

68. $y \cdot \frac{5}{6} = 120$

Skill Maintenance

Divide. [1.3e]

69. $268 \div 4$

70. $268 \div 8$

71. $6842 \div 24$

72. $8765 \div 85$

Solve. [1.4b]

73. $4 \cdot x = 268$

74. $4 + x = 268$

75. $y + 502 = 9001$

76. $56 \cdot 78 = T$

Synthesis

77. ◆ Without performing the division, explain why $5 \div \frac{1}{7}$ is a greater number than $5 \div \frac{2}{3}$.

78. ◆ A student incorrectly insists that $\frac{2}{5} \div \frac{3}{4}$ is $\frac{15}{8}$. What mistake is the student probably making?

79. If $\frac{1}{3}$ of a number is $\frac{1}{4}$, what is $\frac{1}{2}$ of the number?

2.3 Addition and Subtraction; Order

a Like Denominators

Addition using fractional notation corresponds to combining or putting like things together, just as addition with whole numbers does. For example,

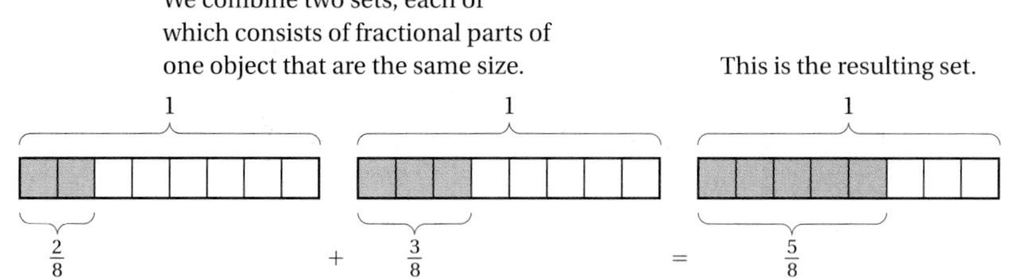

2 eighths + 3 eighths = 5 eighths,

or $\quad 2 \cdot \frac{1}{8} + 3 \cdot \frac{1}{8} = 5 \cdot \frac{1}{8}, \quad$ or $\quad \frac{2}{8} + \frac{3}{8} = \frac{5}{8}.$

We see that to add when denominators are the same, we add the numerators, keep the denominator, and simplify, if possible.

Do Exercise 1.

> To add when denominators are the same,
>
> a) add the numerators,
>
> b) keep the denominator, and
>
> c) simplify, if possible.
>
> $$\frac{2}{6} + \frac{5}{6} = \frac{2+5}{6} = \frac{7}{6}$$

Examples Add and simplify.

1. $\frac{2}{4} + \frac{1}{4} = \frac{2+1}{4} = \frac{3}{4}$ **No simplifying is possible.**

2. $\frac{11}{6} + \frac{3}{6} = \frac{11+3}{6} = \frac{14}{6} = \frac{2 \cdot 7}{2 \cdot 3} = \frac{2}{2} \cdot \frac{7}{3} = 1 \cdot \frac{7}{3} = \frac{7}{3}$ **Here we simplified.**

3. $\frac{3}{12} + \frac{5}{12} = \frac{3+5}{12} = \frac{8}{12} = \frac{4 \cdot 2}{4 \cdot 3} = \frac{4}{4} \cdot \frac{2}{3} = 1 \cdot \frac{2}{3} = \frac{2}{3}$

Do Exercises 2–4.

b Addition Using the LCD: Different Denominators

What do we do when denominators are different? We try to find a common denominator. We can do this by multiplying by 1. Consider adding $\frac{1}{6}$ and $\frac{3}{4}$. There are several common denominators that can be obtained. Let's look at two possibilities.

Objectives

a Add using fractional notation when denominators are the same.

b Add using fractional notation when denominators are different, by multiplying by 1 to find the least common denominator.

c Subtract using fractional notation.

d Use $<$ or $>$ with fractional notation to write a true sentence.

e Solve equations of the type $x + a = b$ and $a + x = b$, where a and b may be fractions.

For Extra Help

TAPE 4 | MAC WIN | CD-ROM

1. Find $\frac{1}{5} + \frac{3}{5}$.

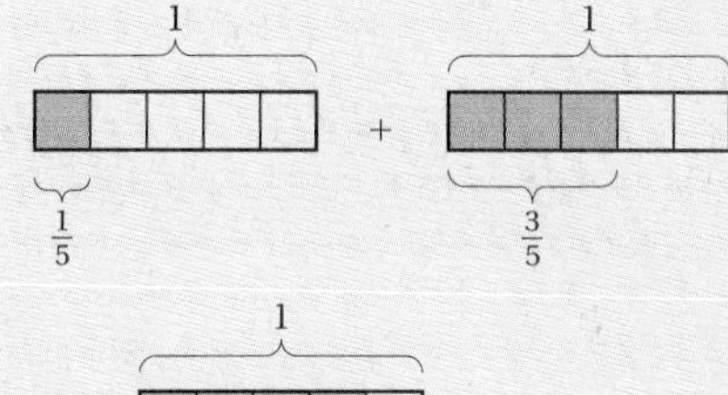

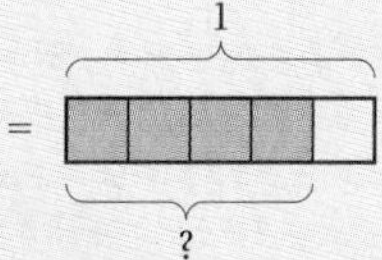

Add and simplify.

2. $\frac{1}{3} + \frac{2}{3}$

3. $\frac{5}{12} + \frac{1}{12}$

4. $\frac{9}{16} + \frac{3}{16}$

Answers on page A-4

5. Add. (Find the least common denominator.)

$$\frac{2}{3} + \frac{1}{6}$$

6. Add: $\frac{3}{8} + \frac{5}{6}$.

Answers on page A-4

A. $\frac{1}{6} + \frac{3}{4} = \frac{1}{6} \cdot 1 + \frac{3}{4} \cdot 1$
$= \frac{1}{6} \cdot \frac{4}{4} + \frac{3}{4} \cdot \frac{6}{6}$
$= \frac{4}{24} + \frac{18}{24}$
$= \frac{22}{24}$
$= \frac{11}{12}$

B. $\frac{1}{6} + \frac{3}{4} = \frac{1}{6} \cdot 1 + \frac{3}{4} \cdot 1$
$= \frac{1}{6} \cdot \frac{2}{2} + \frac{3}{4} \cdot \frac{3}{3}$
$= \frac{2}{12} + \frac{9}{12}$
$= \frac{11}{12}$

We had to simplify in (A). We didn't have to simplify in (B). In (B), we used the least common multiple of the denominators, 12. That number is called the **least common denominator,** or **LCD**.

To add when denominators are different:

a) Find the least common multiple of the denominators. That number is the least common denominator, LCD.

b) Multiply by 1, using an appropriate notation, n/n, to express each number in terms of the LCD.

c) Add the numerators, keeping the same denominator.

d) Simplify, if possible.

Example 4 Add: $\frac{3}{4} + \frac{1}{8}$.

The LCD is 8. **4 is a factor of 8 so the LCM of 4 and 8 is 8.**

$\frac{3}{4} + \frac{1}{8} = \frac{3}{4} \cdot 1 + \frac{1}{8}$ ← **This fraction already has the LCD as its denominator.**

$= \frac{3}{4} \cdot \frac{2}{2} + \frac{1}{8}$ **Think: 4 × ■ = 8. The answer is 2, so we multiply by 1, using $\frac{2}{2}$.**

$= \frac{6}{8} + \frac{1}{8}$

$= \frac{7}{8}$ **No simplification is necessary.**

Do Exercise 5.

Example 5 Add: $\frac{1}{9} + \frac{5}{6}$.

The LCD is 18. **9 = 3 · 3 and 6 = 2 · 3, so the LCM of 9 and 6 is 2 · 3 · 3, or 18.**

$\frac{1}{9} + \frac{5}{6} = \frac{1}{9} \cdot 1 + \frac{5}{6} \cdot 1 = \frac{1}{9} \cdot \frac{2}{2} + \frac{5}{6} \cdot \frac{3}{3}$

Think: 6 × ■ = 18. The answer is 3, so we multiply by 1 using $\frac{3}{3}$.

Think: 9 × ■ = 18. The answer is 2, so we multiply by 1, using $\frac{2}{2}$.

$= \frac{2}{18} + \frac{15}{18} = \frac{17}{18}$

Do Exercise 6.

Example 6 Add: $\frac{5}{9} + \frac{11}{18}$.

The LCD is 18.

$$\frac{5}{9} + \frac{11}{18} = \frac{5}{9} \cdot \frac{2}{2} + \frac{11}{18}$$
$$= \frac{10}{18} + \frac{11}{18}$$
$$= \frac{21}{18}$$
$$= \frac{7}{6}$$

We may still have to simplify, but it is usually easier if we have used the LCD.

Do Exercise 7.

Example 7 Add: $\frac{1}{10} + \frac{3}{100} + \frac{7}{1000}$.

Since 10 and 100 are factors of 1000, the LCD is 1000. Then

$$\frac{1}{10} + \frac{3}{100} + \frac{7}{1000} = \frac{1}{10} \cdot \frac{100}{100} + \frac{3}{100} \cdot \frac{10}{10} + \frac{7}{1000}$$
$$= \frac{100}{1000} + \frac{30}{1000} + \frac{7}{1000} = \frac{137}{1000}.$$

Look back over this example. Try to think it out so that you can do it mentally.

Example 8 Add: $\frac{13}{70} + \frac{11}{21} + \frac{6}{15}$.

We have

$$\frac{13}{70} + \frac{11}{21} + \frac{6}{15} = \frac{13}{2 \cdot 5 \cdot 7} + \frac{11}{3 \cdot 7} + \frac{6}{3 \cdot 5}. \quad \text{Factoring denominators}$$

The LCD is $2 \cdot 3 \cdot 5 \cdot 7$, or 210. Then

$$\frac{13}{70} + \frac{11}{21} + \frac{6}{15} = \frac{13}{2 \cdot 5 \cdot 7} \cdot \frac{3}{3} + \frac{11}{3 \cdot 7} \cdot \frac{2 \cdot 5}{2 \cdot 5} + \frac{6}{3 \cdot 5} \cdot \frac{7 \cdot 2}{7 \cdot 2}$$
$$= \frac{13 \cdot 3}{2 \cdot 5 \cdot 7 \cdot 3} + \frac{11 \cdot 2 \cdot 5}{3 \cdot 7 \cdot 2 \cdot 5} + \frac{6 \cdot 7 \cdot 2}{3 \cdot 5 \cdot 7 \cdot 2}$$
$$= \frac{39}{3 \cdot 5 \cdot 7 \cdot 2} + \frac{110}{3 \cdot 5 \cdot 7 \cdot 2} + \frac{84}{3 \cdot 5 \cdot 7 \cdot 2}$$
$$= \frac{233}{3 \cdot 5 \cdot 7 \cdot 2}$$
$$= \frac{233}{210}. \quad \text{We left 210 factored until we knew we could not simplify.}$$

The LCD of 70, 21, and 15 is $2 \cdot 3 \cdot 5 \cdot 7$. In each case, we multiply by 1 to obtain the LCD.

Do Exercises 8–10.

7. Add: $\frac{1}{6} + \frac{7}{18}$.

Add.

8. $\frac{4}{10} + \frac{1}{100} + \frac{3}{1000}$

9. $\frac{7}{10} + \frac{5}{100} + \frac{9}{1000}$
(Try to do this one mentally.)

10. $\frac{7}{10} + \frac{2}{21} + \frac{1}{7}$

Answers on page A-4

Subtract and simplify.

11. $\frac{7}{8} - \frac{3}{8}$

12. $\frac{10}{16} - \frac{4}{16}$

13. $\frac{8}{10} - \frac{3}{10}$

Answers on page A-4

c Subtraction

Like Denominators

We can consider the difference $\frac{4}{8} - \frac{3}{8}$ as we did before, as either "take away" or "how much more." Let's consider "take away."

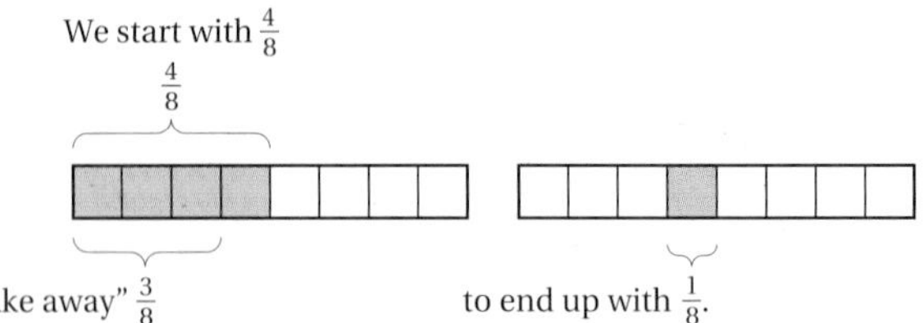

We start with 4 eighths and take away 3 eighths:

4 eighths − 3 eighths = 1 eighth,

or $4 \cdot \frac{1}{8} - 3 \cdot \frac{1}{8} = \frac{1}{8}$, or $\frac{4}{8} - \frac{3}{8} = \frac{1}{8}$.

> To subtract when denominators are the same,
> a) subtract the numerators,
> b) keep the denominator, and
> c) simplify, if possible.
>
> $$\frac{7}{10} - \frac{4}{10} = \frac{7-4}{10} = \frac{3}{10}$$

Examples Subtract and simplify.

9. $\frac{7}{10} - \frac{3}{10} = \frac{7-3}{10} = \frac{4}{10} = \frac{2 \cdot 2}{5 \cdot 2} = \frac{2}{5} \cdot \frac{2}{2} = \frac{2}{5} \cdot 1 = \frac{2}{5}$

10. $\frac{8}{9} - \frac{2}{9} = \frac{8-2}{9} = \frac{6}{9} = \frac{2 \cdot 3}{3 \cdot 3} = \frac{2}{3} \cdot \frac{3}{3} = \frac{2}{3} \cdot 1 = \frac{2}{3}$

11. $\frac{32}{12} - \frac{25}{12} = \frac{32-25}{12} = \frac{7}{12}$

Do Exercises 11–13.

Different Denominators

> To subtract when denominators are different:
> a) Find the least common multiple of the denominators. That number is the least common denominator, LCD.
> b) Multiply by 1, using an appropriate notation, n/n, to express each number in terms of the LCD.
> c) Subtract the numerators, keeping the same denominator.
> d) Simplify, if possible.

Example 12 Subtract: $\frac{5}{6} - \frac{7}{12}$.

Since 6 is a factor of 12, the LCM of 6 and 12 is 12. The LCD is 12.

$$\frac{5}{6} - \frac{7}{12} = \frac{5}{6} \cdot \frac{2}{2} - \frac{7}{12}$$
$$= \frac{10}{12} - \frac{7}{12}$$
$$= \frac{10 - 7}{12} = \frac{3}{12}$$
$$= \frac{3 \cdot 1}{3 \cdot 4} = \frac{3}{3} \cdot \frac{1}{4}$$
$$= \frac{1}{4}$$

Do Exercises 14 and 15.

Example 13 Subtract: $\frac{17}{24} - \frac{4}{15}$.

We have

$$\frac{17}{24} - \frac{4}{15} = \frac{17}{3 \cdot 2 \cdot 2 \cdot 2} - \frac{4}{5 \cdot 3}.$$

The LCD is $3 \cdot 2 \cdot 2 \cdot 2 \cdot 5$, or 120. Then

$$\frac{17}{24} - \frac{4}{15} = \frac{17}{3 \cdot 2 \cdot 2 \cdot 2} \cdot \frac{5}{5} - \frac{4}{5 \cdot 3} \cdot \frac{2 \cdot 2 \cdot 2}{2 \cdot 2 \cdot 2}$$
$$= \frac{17 \cdot 5}{3 \cdot 2 \cdot 2 \cdot 2 \cdot 5} - \frac{4 \cdot 2 \cdot 2 \cdot 2}{5 \cdot 3 \cdot 2 \cdot 2 \cdot 2}$$
$$= \frac{85}{120} - \frac{32}{120} = \frac{53}{120}.$$

The LCD of 24 and 15 is $2 \cdot 2 \cdot 2 \cdot 3 \cdot 5$. In each case, we multiply by 1 to obtain the LCD.

Do Exercise 16.

d Order

We see from this figure that $\frac{4}{5} > \frac{3}{5}$. That is, $\frac{4}{5}$ is greater than $\frac{3}{5}$.

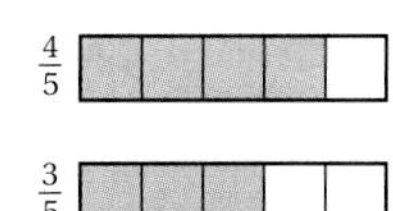

To determine which of two numbers is greater when there is a common denominator, compare the numerators:

$$\frac{4}{5}, \frac{3}{5}, \quad 4 > 3 \quad \frac{4}{5} > \frac{3}{5}.$$

Do Exercises 17 and 18.

Subtract.

14. $\frac{5}{6} - \frac{1}{9}$

15. $\frac{4}{5} - \frac{3}{10}$

16. Subtract: $\frac{11}{28} - \frac{5}{16}$.

17. Use < or > for ▒ to write a true sentence:

$$\frac{3}{8} \;▒\; \frac{5}{8}.$$

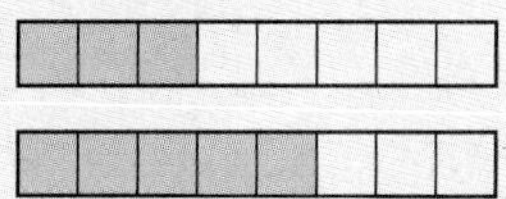

18. Use < or > for ▒ to write a true sentence:

$$\frac{7}{10} \;▒\; \frac{6}{10}.$$

Answers on page A-4

Use < or > for ▒ to write a true sentence.

19. $\frac{2}{3} \;\blacksquare\; \frac{5}{8}$

20. $\frac{3}{4} \;\blacksquare\; \frac{8}{12}$

21. $\frac{5}{6} \;\blacksquare\; \frac{7}{8}$

Solve.

22. $x + \frac{2}{3} = \frac{5}{6}$

23. $\frac{3}{5} + t = \frac{7}{8}$

Answers on page A-4

When denominators are different, we cannot compare numerators. We multiply by 1 to make the denominators the same.

Example 14 Use < or > for ▒ to write a true sentence:

$$\frac{2}{5} \;\blacksquare\; \frac{3}{4}.$$

We have

$$\frac{2}{5} \cdot \frac{4}{4} = \frac{8}{20};$$ We multiply by 1 using $\frac{4}{4}$ to get the LCD.

$$\frac{3}{4} \cdot \frac{5}{5} = \frac{15}{20}.$$ We multiply by 1 using $\frac{5}{5}$ to get the LCD.

Now that the denominators are the same, 20, we can compare the numerators. Since $8 < 15$, it follows that $\frac{8}{20} < \frac{15}{20}$, so

$$\frac{2}{5} < \frac{3}{4}.$$

Example 15 Use < or > for ▒ to write a true sentence:

$$\frac{9}{10} \;\blacksquare\; \frac{89}{100}.$$

The LCD is 100.

$$\frac{9}{10} \cdot \frac{10}{10} = \frac{90}{100}$$ We multiply by $\frac{10}{10}$ to get the LCD.

Since $90 > 89$, it follows that $\frac{90}{100} > \frac{89}{100}$, so

$$\frac{9}{10} > \frac{89}{100}.$$

Do Exercises 19–21.

e Solving Equations

Now let's solve equations of the form $x + a = b$ or $a + x = b$, where a and b may be fractions. Proceeding as we have before, we subtract a on both sides of the equation.

Example 16 Solve: $x + \frac{1}{4} = \frac{3}{5}$.

$$x + \frac{1}{4} - \frac{1}{4} = \frac{3}{5} - \frac{1}{4}$$ Subtracting $\frac{1}{4}$ on both sides

$$x + 0 = \frac{3}{5} \cdot \frac{4}{4} - \frac{1}{4} \cdot \frac{5}{5}$$ The LCD is 20. We multiply by 1 to get the LCD.

$$x = \frac{12}{20} - \frac{5}{20} = \frac{7}{20}$$

Do Exercises 22 and 23.

Exercise Set 2.3

a, b Add and simplify.

1. $\frac{7}{8}+\frac{1}{8}$
2. $\frac{2}{5}+\frac{3}{5}$
3. $\frac{1}{8}+\frac{5}{8}$
4. $\frac{3}{10}+\frac{3}{10}$
5. $\frac{2}{3}+\frac{5}{6}$
6. $\frac{5}{6}+\frac{1}{9}$
7. $\frac{1}{8}+\frac{1}{6}$
8. $\frac{1}{6}+\frac{3}{4}$
9. $\frac{4}{5}+\frac{7}{10}$
10. $\frac{3}{4}+\frac{1}{12}$
11. $\frac{5}{12}+\frac{3}{8}$
12. $\frac{7}{8}+\frac{1}{16}$
13. $\frac{3}{20}+\frac{3}{4}$
14. $\frac{2}{15}+\frac{2}{5}$
15. $\frac{5}{6}+\frac{7}{9}$
16. $\frac{3}{16}+\frac{1}{12}$
17. $\frac{7}{8}+\frac{0}{1}$
18. $\frac{0}{1}+\frac{5}{6}$
19. $\frac{3}{8}+\frac{1}{6}$
20. $\frac{5}{8}+\frac{1}{6}$
21. $\frac{5}{8}+\frac{5}{6}$
22. $\frac{3}{10}+\frac{1}{100}$
23. $\frac{9}{10}+\frac{3}{100}$
24. $\frac{5}{12}+\frac{4}{15}$
25. $\frac{5}{12}+\frac{7}{24}$
26. $\frac{1}{18}+\frac{7}{12}$
27. $\frac{8}{10}+\frac{7}{100}+\frac{4}{1000}$
28. $\frac{1}{10}+\frac{2}{100}+\frac{3}{1000}$
29. $\frac{3}{8}+\frac{5}{12}+\frac{8}{15}$
30. $\frac{1}{2}+\frac{3}{8}+\frac{1}{4}$
31. $\frac{15}{24}+\frac{7}{36}+\frac{91}{48}$
32. $\frac{5}{7}+\frac{25}{52}+\frac{7}{4}$

c Subtract and simplify.

33. $\frac{5}{6}-\frac{1}{6}$
34. $\frac{5}{8}-\frac{3}{8}$
35. $\frac{11}{12}-\frac{2}{12}$
36. $\frac{17}{18}-\frac{11}{18}$
37. $\frac{3}{4}-\frac{1}{8}$
38. $\frac{2}{3}-\frac{1}{9}$

39. $\frac{1}{8} - \frac{1}{12}$

40. $\frac{1}{6} - \frac{1}{8}$

41. $\frac{4}{3} - \frac{5}{6}$

42. $\frac{3}{4} - \frac{4}{16}$

43. $\frac{5}{12} - \frac{2}{15}$

44. $\frac{9}{10} - \frac{11}{16}$

45. $\frac{6}{10} - \frac{7}{100}$

46. $\frac{9}{10} - \frac{3}{100}$

47. $\frac{7}{15} - \frac{3}{25}$

48. $\frac{5}{6} - \frac{2}{3}$

49. $\frac{5}{12} - \frac{3}{8}$

50. $\frac{7}{12} - \frac{2}{9}$

51. $\frac{7}{8} - \frac{1}{16}$

52. $\frac{5}{12} - \frac{5}{16}$

53. $\frac{17}{25} - \frac{4}{15}$

54. $\frac{11}{18} - \frac{7}{24}$

55. $\frac{23}{25} - \frac{112}{150}$

56. $\frac{89}{90} - \frac{53}{120}$

d Use < or > for ▒ to write a true sentence.

57. $\frac{5}{8}$ ▒ $\frac{6}{8}$

58. $\frac{7}{9}$ ▒ $\frac{5}{9}$

59. $\frac{1}{3}$ ▒ $\frac{1}{4}$

60. $\frac{1}{8}$ ▒ $\frac{1}{6}$

61. $\frac{2}{3}$ ▒ $\frac{5}{7}$

62. $\frac{3}{5}$ ▒ $\frac{4}{7}$

63. $\frac{4}{5}$ ▒ $\frac{5}{6}$

64. $\frac{3}{2}$ ▒ $\frac{7}{5}$

65. $\frac{19}{20}$ ▒ $\frac{4}{5}$

66. $\frac{5}{6}$ ▒ $\frac{13}{16}$

67. $\frac{19}{20}$ ▒ $\frac{9}{10}$

68. $\frac{3}{4}$ ▒ $\frac{11}{15}$

69. $\frac{31}{21}$ ▒ $\frac{41}{13}$

70. $\frac{12}{7}$ ▒ $\frac{132}{49}$

e Solve.

71. $x + \frac{1}{30} = \frac{1}{10}$

72. $y + \frac{9}{12} = \frac{11}{12}$

73. $\frac{2}{3} + t = \frac{4}{5}$

74. $\frac{2}{3} + p = \frac{7}{8}$

75. $m + \frac{5}{6} = \frac{9}{10}$

76. $x + \frac{1}{3} = \frac{5}{6}$

Skill Maintenance

Subtract. [1.2f]

77. $\begin{array}{r} 9060 \\ -\,4387 \\ \hline \end{array}$

78. $\begin{array}{r} 7800 \\ -\,2462 \\ \hline \end{array}$

Divide. [1.3e]

79. $35\overline{)7140}$

80. $13{,}602 \div 7$

Synthesis

81. ◆ Explain the role of multiplication when adding using fractional notation with different denominators.

82. ◆ Explain how one could use pictures to convince someone that $\frac{7}{29}$ is larger than $\frac{13}{57}$.

83. A mountain climber, beginning at sea level, climbs $\frac{3}{5}$ km, descends $\frac{1}{4}$ km, climbs $\frac{1}{3}$ km, and then descends $\frac{1}{7}$ km. At what elevation does the climber finish?

Arrange sockets and drill bits in fractional sizes from smallest to largest.

2.4 Mixed Numerals

a What Is a Mixed Numeral?

A symbol like $2\frac{3}{4}$ is called a **mixed numeral.**

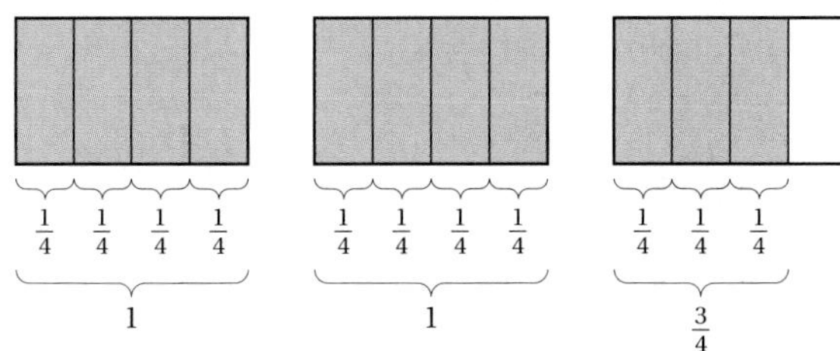

$$2\frac{3}{4} \quad \text{means} \quad 2 + \frac{3}{4}$$

This is a whole number. This is a fraction less than 1.

Examples Convert to a mixed numeral.

1. $7 + \frac{2}{5} = 7\frac{2}{5}$

2. $4 + \frac{3}{10} = 4\frac{3}{10}$

Do Exercises 1–3.

The notation $2\frac{3}{4}$ has a plus sign left out. To aid in understanding, we sometimes write the missing plus sign.

Examples Convert to fractional notation.

3. $2\frac{3}{4} = 2 + \frac{3}{4}$ Inserting the missing plus sign

$= \frac{2}{1} + \frac{3}{4}$ $2 = \frac{2}{1}$

$= \frac{2}{1} \cdot \frac{4}{4} + \frac{3}{4}$ Finding a common denominator

$= \frac{8}{4} + \frac{3}{4} = \frac{11}{4}$

4. $4\frac{3}{10} = 4 + \frac{3}{10} = \frac{4}{1} + \frac{3}{10} = \frac{4}{1} \cdot \frac{10}{10} + \frac{3}{10} = \frac{40}{10} + \frac{3}{10} = \frac{43}{10}$

Do Exercises 4 and 5.

Let's now consider a faster method for converting a mixed numeral to fractional notation.

To convert from a mixed numeral to fractional notation:

(a) Multiply: $4 \cdot 10 = 40$.
(b) Add: $40 + 3 = 43$.
(c) Keep the denominator.

(b) (a) $4\frac{3}{10} = \frac{43}{10}$

Objectives

a Convert from mixed numerals to fractional notation and from fractional notation to mixed numerals.
b Add using mixed numerals.
c Subtract using mixed numerals.
d Multiply using mixed numerals.
e Divide using mixed numerals.

For Extra Help

TAPE 5 MAC WIN CD-ROM

1. $1 + \frac{2}{3} = \square$ —— Convert to a mixed numeral.

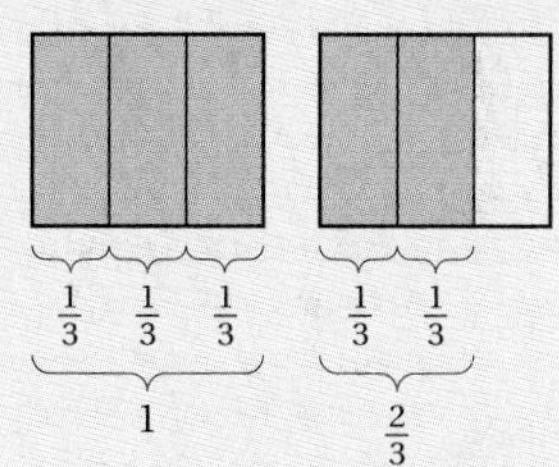

Convert to a mixed numeral.

2. $8 + \frac{3}{4}$

3. $12 + \frac{2}{3}$

Convert to fractional notation.

4. $4\frac{2}{5}$

5. $6\frac{1}{10}$

Answers on page A-4

Convert to fractional notation. Use the faster method.

6. $4\frac{5}{6}$

7. $9\frac{1}{4}$

8. $20\frac{2}{3}$

Convert to a mixed numeral.

9. $\frac{7}{3}$

10. $\frac{11}{10}$

11. $\frac{110}{6}$

Answers on page A-4

Examples Convert to fractional notation.

5. $6\frac{2}{3} = \frac{20}{3}$ **6.** $8\frac{2}{9} = \frac{74}{9}$ **7.** $10\frac{7}{8} = \frac{87}{8}$

Do Exercises 6–8.

We can find a mixed numeral for $\frac{5}{3}$ as follows:

$$\frac{5}{3} = \frac{3}{3} + \frac{2}{3} = 1 + \frac{2}{3} = 1\frac{2}{3}.$$

Fractional symbols like $\frac{5}{3}$ also indicate division. Let's divide the numerator by the denominator.

$$\begin{array}{r} 1 \\ 3\overline{)5} \\ \underline{3} \\ 2 \end{array} \longleftarrow 3\overline{)2}^{\frac{2}{3}} \quad \text{or} \quad 2 \div 3 = \frac{2}{3}$$

Thus, $\frac{5}{3} = 1\frac{2}{3}$.

In terms of objects, we can think of $\frac{5}{3}$ as $\frac{3}{3}$, or 1, plus $\frac{2}{3}$, as shown below.

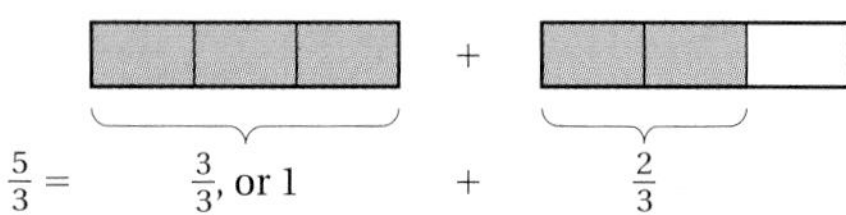

$\frac{5}{3} = \frac{3}{3}$, or 1 $+ \frac{2}{3}$

To convert from fractional notation to a mixed numeral, divide.

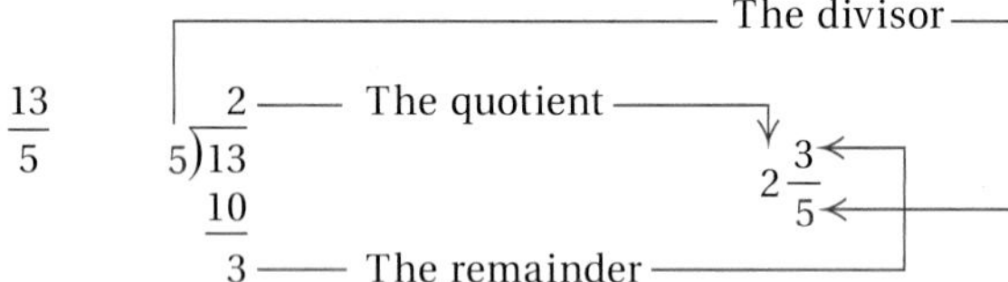

Examples Convert to a mixed numeral.

8. $\frac{8}{5}$ $\quad \begin{array}{r} 1 \\ 5\overline{)8} \\ \underline{5} \\ 3 \end{array} \quad \frac{8}{5} = 1\frac{3}{5}$

A fraction larger than 1, such as $\frac{8}{5}$, is sometimes referred to as an "improper" fraction. We have intentionally avoided such terminology. The use of such notation as $\frac{8}{5}$, $\frac{69}{10}$, and so on, is quite proper and very common in algebra.

9. $\frac{122}{8}$ $\quad \begin{array}{r} 15 \\ 8\overline{)122} \\ \underline{80} \\ 42 \\ \underline{40} \\ 2 \end{array} \quad \frac{122}{8} = 15\frac{2}{8} = 15\frac{1}{4}$

Do Exercises 9–11.

b Addition

To find the sum $1\frac{5}{8} + 3\frac{1}{8}$, we first add the fractions. Then we add the whole numbers.

$$\begin{array}{rl} 1\frac{5}{8} & = \\ +\,3\frac{1}{8} & = \\ \hline \frac{6}{8} & \end{array} \qquad \begin{array}{r} 1\frac{5}{8} \\ +\,3\frac{1}{8} \\ \hline 4\frac{6}{8} = 4\frac{3}{4} \end{array}$$

Simplifying

Add the fractions. Add the whole numbers.

Do Exercise 12.

Example 10 Add: $5\frac{2}{3} + 3\frac{5}{6}$. Write a mixed numeral for the answer.

The LCD is 6.

$$\begin{array}{rl} 5\frac{2}{3}\cdot\frac{2}{2} & = \quad 5\frac{4}{6} \\ +\,3\frac{5}{6} & = +\,3\frac{5}{6} \\ \hline & \quad 8\frac{9}{6} = 8 + \frac{9}{6} \\ & \quad = 8 + 1\frac{1}{2} \\ & \quad = 9\frac{1}{2} \end{array}$$

To find a mixed numeral for $\frac{9}{6}$, we divide:

$$6\overline{)9} = 1 \text{ R } 3 \qquad \frac{9}{6} = 1\frac{3}{6} = 1\frac{1}{2}$$

$\frac{19}{2}$ is also a correct answer, but it is not a mixed numeral, which is what we are working with in Section 2.4.

Do Exercise 13.

Example 11 Add: $10\frac{5}{6} + 7\frac{3}{8}$.

The LCD is 24.

$$\begin{array}{rl} 10\frac{5}{6}\cdot\frac{4}{4} & = \quad 10\frac{20}{24} \\ +\,7\frac{3}{8}\cdot\frac{3}{3} & = +\,7\frac{9}{24} \\ \hline & \quad 17\frac{29}{24} = 18\frac{5}{24} \end{array}$$

Do Exercise 14.

12. Add.

$$\begin{array}{r} 2\frac{3}{10} \\ +\,5\frac{1}{10} \\ \hline \end{array}$$

13. Add.

$$\begin{array}{r} 8\frac{2}{5} \\ +\,3\frac{7}{10} \\ \hline \end{array}$$

14. Add.

$$\begin{array}{r} 9\frac{3}{4} \\ +\,3\frac{5}{6} \\ \hline \end{array}$$

Answers on page A-4

Subtract.

15. $\begin{array}{r} 10\frac{7}{8} \\ -\ 9\frac{3}{8} \\ \hline \end{array}$

16. $\begin{array}{r} 8\frac{2}{3} \\ -\ 5\frac{1}{2} \\ \hline \end{array}$

17. Subtract.

$\begin{array}{r} 5\frac{1}{12} \\ -\ 1\frac{3}{4} \\ \hline \end{array}$

18. Subtract.

$\begin{array}{r} 5 \\ -\ 1\frac{1}{3} \\ \hline \end{array}$

Answers on page A-4

c Subtraction

Example 12 Subtract: $7\frac{3}{4} - 2\frac{1}{4}$.

$$\begin{array}{rcr} 7\frac{3}{4} & = & 7\frac{3}{4} \\ -\ 2\frac{1}{4} & = & -\ 2\frac{1}{4} \\ \hline \frac{2}{4} & & 5\frac{2}{4} = 5\frac{1}{2} \end{array}$$

Subtract the fractions. Subtract the whole numbers. Simplifying

Example 13 Subtract: $9\frac{4}{5} - 3\frac{1}{2}$.

The LCD is 10.

$$\begin{array}{rcr} 9\frac{4}{5}\cdot\frac{2}{2} & = & 9\frac{8}{10} \\ -\ 3\frac{1}{2}\cdot\frac{5}{5} & = & -\ 3\frac{5}{10} \\ \hline & & 6\frac{3}{10} \end{array}$$

Do Exercises 15 and 16.

Example 14 Subtract: $7\frac{1}{6} - 2\frac{1}{4}$.

The LCD is 12.

$$\begin{array}{rcr} 7\frac{1}{6}\cdot\frac{2}{2} & = & 7\frac{2}{12} \\ -\ 2\frac{1}{4}\cdot\frac{3}{3} & = & -\ 2\frac{3}{12} \\ \hline \end{array}$$

We cannot subtract $\frac{3}{12}$ from $\frac{2}{12}$.
We borrow 1, or $\frac{12}{12}$, from 7:
$7\frac{2}{12} = 6 + 1 + \frac{2}{12} = 6 + \frac{12}{12} + \frac{2}{12} = 6\frac{14}{12}$.

We can write this as

$$\begin{array}{rcr} 7\frac{2}{12} & = & 6\frac{14}{12} \\ -\ 2\frac{3}{12} & = & -\ 2\frac{3}{12} \\ \hline & & 4\frac{11}{12} \end{array}$$

Do Exercise 17.

Example 15 Subtract: $12 - 9\frac{3}{8}$.

$$\begin{array}{rcr} 12 & = & 11\frac{8}{8} \\ -\ 9\frac{3}{8} & = & -\ 9\frac{3}{8} \\ \hline & & 2\frac{5}{8} \end{array}$$

$12 = 11 + 1 = 11 + \frac{8}{8} = 11\frac{8}{8}$

Do Exercise 18.

d Multiplication

Carrying out addition and subtraction with mixed numerals is easier if the numbers are left as mixed numerals. With multiplication and division, however, it is easier to convert the numbers first to fractional notation.

> To multiply using mixed numerals, first convert to fractional notation. Then multiply with fractional notation and convert the answer back to a mixed numeral, if appropriate.

Example 16 Multiply: $6 \cdot 2\frac{1}{2}$.

$$6 \cdot 2\frac{1}{2} = \frac{6}{1} \cdot \frac{5}{2} = \frac{6 \cdot 5}{1 \cdot 2} = \frac{2 \cdot 3 \cdot 5}{2 \cdot 1} = \frac{2}{2} \cdot \frac{3 \cdot 5}{1} = 15$$

Here we write fractional notation.

Do Exercise 19.

Example 17 Multiply: $3\frac{1}{2} \cdot \frac{3}{4}$.

$$3\frac{1}{2} \cdot \frac{3}{4} = \frac{7}{2} \cdot \frac{3}{4} = \frac{21}{8} = 2\frac{5}{8}$$

Note that fractional notation is needed to carry out the multiplication.

Do Exercise 20.

Example 18 Multiply: $8 \cdot 4\frac{2}{3}$.

$$8 \cdot 4\frac{2}{3} = \frac{8}{1} \cdot \frac{14}{3} = \frac{112}{3} = 37\frac{1}{3}$$

Do Exercise 21.

Example 19 Multiply: $2\frac{1}{4} \cdot 3\frac{2}{5}$.

$$2\frac{1}{4} \cdot 3\frac{2}{5} = \frac{9}{4} \cdot \frac{17}{5} = \frac{153}{20} = 7\frac{13}{20}$$

CAUTION! $2\frac{1}{4} \cdot 3\frac{2}{5} \neq 6\frac{2}{20}$. A common error is to multiply the whole numbers and then the fractions. This does not give the correct answer, $7\frac{13}{20}$, which is found by converting first to fractional notation.

Do Exercise 22.

19. Multiply: $6 \cdot 3\frac{1}{3}$.

20. Multiply: $2\frac{1}{2} \cdot \frac{3}{4}$.

21. Multiply: $2 \cdot 6\frac{2}{5}$.

22. Multiply: $3\frac{1}{3} \cdot 2\frac{1}{2}$.

Answers on page A-4

23. Divide: $84 \div 5\frac{1}{4}$.

24. Divide: $26 \div 3\frac{1}{2}$.

Divide.

25. $2\frac{1}{4} \div 1\frac{1}{5}$

26. $1\frac{3}{4} \div 2\frac{1}{2}$

Answers on page A-4

e Division

The division $1\frac{1}{2} \div \frac{1}{6}$ is shown here.

$$1\frac{1}{2} \div \frac{1}{6} = \frac{3}{2} \div \frac{1}{6}$$

$$= \frac{3}{2} \cdot 6 = \frac{3 \cdot 6}{2} = \frac{3 \cdot 3 \cdot 2}{2 \cdot 1} = \frac{3 \cdot 3}{1} \cdot \frac{2}{2} = \frac{3 \cdot 3}{1} \cdot 1 = 9$$

To divide using mixed numerals, first write fractional notation. Then divide with fractional notation and convert the answer back to a mixed numeral, if appropriate.

Example 20 Divide: $32 \div 3\frac{1}{5}$.

$$32 \div 3\frac{1}{5} = \frac{32}{1} \div \frac{16}{5}$$

$$= \frac{32}{1} \cdot \frac{5}{16} = \frac{32 \cdot 5}{1 \cdot 16} = \frac{2 \cdot 16 \cdot 5}{1 \cdot 16} = \frac{16}{16} \cdot \frac{2 \cdot 5}{1} = 10$$

Remember to multiply by the reciprocal.

Do Exercise 23.

Example 21 Divide: $35 \div 4\frac{1}{3}$.

$$35 \div 4\frac{1}{3} = \frac{35}{1} \div \frac{13}{3} = \frac{35}{1} \cdot \frac{3}{13} = \frac{105}{13} = 8\frac{1}{13}$$

Do Exercise 24.

Example 22 Divide: $2\frac{1}{3} \div 1\frac{3}{4}$.

$$2\frac{1}{3} \div 1\frac{3}{4} = \frac{7}{3} \div \frac{7}{4} = \frac{7}{3} \cdot \frac{4}{7} = \frac{7 \cdot 4}{7 \cdot 3} = \frac{7}{7} \cdot \frac{4}{3} = 1 \cdot \frac{4}{3} = \frac{4}{3} = 1\frac{1}{3}$$

CAUTION! The reciprocal of $1\frac{3}{4}$ is *not* $1\frac{4}{3}$!

Example 23 Divide: $1\frac{3}{5} \div 3\frac{1}{3}$.

$$1\frac{3}{5} \div 3\frac{1}{3} = \frac{8}{5} \div \frac{10}{3} = \frac{8}{5} \cdot \frac{3}{10} = \frac{2 \cdot 4 \cdot 3}{5 \cdot 2 \cdot 5} = \frac{2}{2} \cdot \frac{4 \cdot 3}{5 \cdot 5} = 1 \cdot \frac{4 \cdot 3}{5 \cdot 5} = \frac{12}{25} = \frac{12}{25}$$

Do Exercises 25 and 26.

Exercise Set 2.4

a Convert to fractional notation.

1. $5\frac{2}{3}$ **2.** $20\frac{1}{5}$ **3.** $9\frac{5}{6}$ **4.** $1\frac{3}{5}$ **5.** $12\frac{3}{4}$ **6.** $33\frac{1}{3}$

Convert to a mixed numeral.

7. $\frac{18}{5}$ **8.** $\frac{17}{4}$ **9.** $\frac{57}{10}$ **10.** $\frac{50}{8}$ **11.** $\frac{345}{8}$ **12.** $\frac{467}{100}$

b Add. Write a mixed numeral for the answer.

13. $2\frac{7}{8} + 3\frac{5}{8}$

14. $4\frac{5}{6} + 3\frac{5}{6}$

15. $1\frac{1}{4} + 1\frac{2}{3}$

16. $4\frac{1}{3} + 5\frac{2}{9}$

17. $8\frac{3}{4} + 5\frac{5}{6}$

18. $4\frac{3}{8} + 6\frac{5}{12}$

19. $12\frac{4}{5} + 8\frac{7}{10}$

20. $15\frac{5}{8} + 11\frac{3}{4}$

21. $14\frac{5}{8} + 13\frac{1}{4}$

22. $16\frac{1}{4} + 15\frac{7}{8}$

23. $7\frac{1}{8} + 9\frac{2}{3} + 10\frac{3}{4}$

24. $45\frac{2}{3} + 31\frac{3}{5} + 12\frac{1}{4}$

c Subtract. Write a mixed numeral for the answer.

25. $4\frac{1}{5} - 2\frac{3}{5}$

26. $5\frac{1}{8} - 2\frac{3}{8}$

27. $6\frac{3}{5} - 2\frac{1}{2}$

28. $7\frac{2}{3} - 6\frac{1}{2}$

29. $34 - 18\frac{5}{8}$

30. $23 - 19\frac{3}{4}$

31. $21\frac{1}{6} - 13\frac{3}{4}$

32. $42\frac{1}{10} - 23\frac{7}{12}$

33. $14\frac{1}{8} - \frac{3}{4}$

34. $28\frac{1}{6} - 5$

35. $25\frac{1}{9} - 13\frac{5}{6}$

36. $23\frac{5}{16} - 14\frac{7}{12}$

d Multiply. Write a mixed numeral for the answer.

37. $8 \cdot 2\frac{5}{6}$

38. $5 \cdot 3\frac{3}{4}$

39. $3\frac{5}{8} \cdot \frac{2}{3}$

40. $6\frac{2}{3} \cdot \frac{1}{4}$

41. $3\frac{1}{2} \cdot 2\frac{1}{3}$

42. $4\frac{1}{5} \cdot 5\frac{1}{4}$

43. $3\frac{2}{5} \cdot 2\frac{7}{8}$

44. $2\frac{3}{10} \cdot 4\frac{2}{5}$

45. $4\frac{7}{10} \cdot 5\frac{3}{10}$

46. $6\frac{3}{10} \cdot 5\frac{7}{10}$

47. $20\frac{1}{2} \cdot 10\frac{1}{5} \cdot 4\frac{2}{3}$

48. $21\frac{1}{3} \cdot 11\frac{1}{3} \cdot 3\frac{5}{8}$

e Divide. Write a mixed numeral for the answer.

49. $20 \div 3\frac{1}{5}$

50. $18 \div 2\frac{1}{4}$

51. $8\frac{2}{5} \div 7$

52. $3\frac{3}{8} \div 3$

53. $4\frac{3}{4} \div 1\frac{1}{3}$

54. $5\frac{4}{5} \div 2\frac{1}{2}$

55. $1\frac{7}{8} \div 1\frac{2}{3}$

56. $4\frac{3}{8} \div 2\frac{5}{6}$

57. $5\frac{1}{10} \div 4\frac{3}{10}$

58. $4\frac{1}{10} \div 2\frac{1}{10}$

59. $20\frac{1}{4} \div 90$

60. $12\frac{1}{2} \div 50$

Skill Maintenance

Solve. [1.5a]

61. A dairy produced 4578 oz of milk one week. How many 16-oz cartons were filled? How much milk was left over?

62. A pet-care service cut 29,824 lb of hair in 8 yr of operation. On the average, how many pounds did it cut each year?

Synthesis

63. ◆ Write a problem for a classmate to solve. Design the problem so that its solution is found by performing the multiplication $4\frac{1}{2} \cdot 33\frac{1}{3}$.

64. ◆ Under what circumstances is a pair of mixed numerals more easily added than multiplied?

Collaborative Learning Manual — Add and subtract mixed numerals using fraction bars.

2.5 Applications and Problem Solving

Objective

a Solve applied problems involving addition, subtraction, multiplication, and division using fractional notation and mixed numerals.

For Extra Help

TAPE 5 MAC WIN CD-ROM

a We solve applied problems using fractional notation and mixed numerals in the same way that we do when using whole numbers. The five steps for problem solving on p. 43 should be reviewed.

Most applied problems that can be solved by multiplying fractions can be thought of in terms of rectangular arrays.

Example 1 A rancher owns a square mile of land. He gives $\frac{4}{5}$ of it to his daughter and she gives $\frac{2}{3}$ of her share to her son. How much land goes to the son?

1. **Familiarize.** We first make a drawing to help solve the problem. The land may not be square. It could be in a shape like A or B below, or it could even be in more than one piece. But to think out the problem, we can think of it as a square, as shown by shape C.

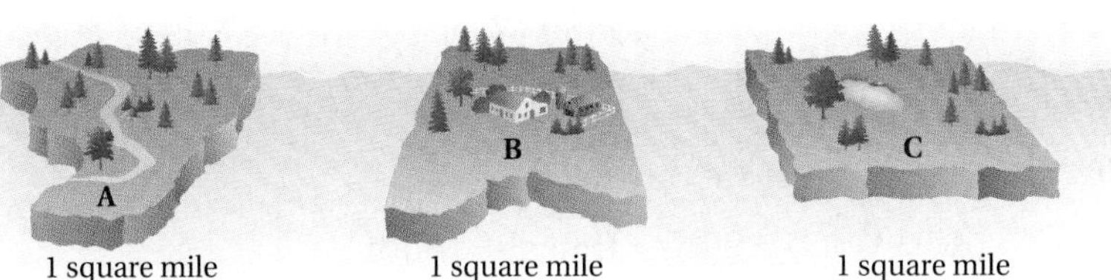

The daughter gets $\frac{4}{5}$ of the land. We shade $\frac{4}{5}$.

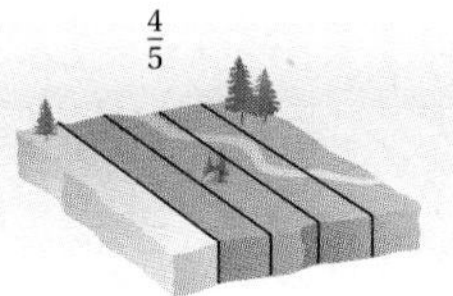

Her son gets $\frac{2}{3}$ of her part. We shade that.

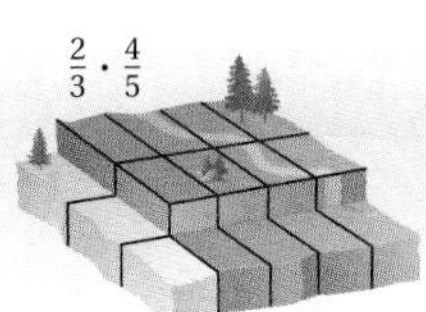

2. **Translate.** We let n = the part of the land that goes to the son. We are taking "two-thirds of four-fifths." The word "of" corresponds to multiplication. Thus the following multiplication sentence corresponds to the situation:

$$\frac{2}{3} \cdot \frac{4}{5} = n.$$

3. **Solve.** The number sentence tells us what to do. We multiply:

$$\frac{2}{3} \cdot \frac{4}{5} = \frac{8}{15}.$$

1. A resort hotel uses $\frac{3}{4}$ of its extra land for recreational purposes. Of that, $\frac{1}{2}$ is used for swimming pools. What part of the land is used for swimming pools?

2. *Area of Fax Key.* The length of a button on a fax machine is $\frac{9}{10}$ cm. The width is $\frac{7}{10}$ cm. What is the area?

Answers on page A-4

4. **Check.** We can check partially by noting that the answer is smaller than the original area, 1, which we expect since the rancher is giving parts of the land away. Thus, $\frac{8}{15}$ is a reasonable answer. We can also check this in the figure above, where we see that 8 of 15 parts have been shaded a second time.
5. **State.** The son gets $\frac{8}{15}$ of a square mile of land.

Do Exercise 1.

Example 1 and the preceding discussion indicate that the area of a rectangular region can be found by multiplying length by width. That is true whether length and width are whole numbers or not. Remember, the area of a rectangular region is given by the formula

$$A = l \cdot w.$$

Example 2 *Area of Calculator Key.* The length of a rectangular key on a calculator is $\frac{7}{10}$ cm. The width is $\frac{3}{10}$ cm. What is the area?

1. **Familiarize.** Recall that area is length times width. We draw a picture, letting A = the area of the calculator key.

2. **Translate.** Then we translate.

Area	is	Length	times	Width
↓	↓	↓	↓	↓
A	$=$	$\frac{7}{10}$	$\times$	$\frac{3}{10}$

3. **Solve.** The sentence tells us what to do. We multiply:

$$\frac{7}{10} \cdot \frac{3}{10} = \frac{7 \cdot 3}{10 \cdot 10} = \frac{21}{100}.$$

4. **Check.** We check by repeating the calculation. This is left to the student.
5. **State.** The area is $\frac{21}{100}$ cm^2.

Do Exercise 2.

Example 3 *Test Tubes.* How many test tubes, each containing $\frac{3}{5}$ mL, can a nursing student fill from a container of 60 mL?

1. **Familiarize.** Repeated addition will apply here. We let n = the number of test tubes in all. We make a drawing.

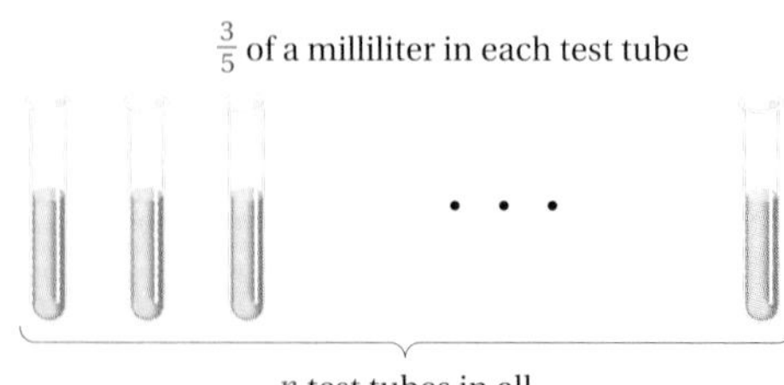

2. **Translate.** The equation that corresponds to the situation is

$$n \cdot \frac{3}{5} = 60.$$

3. **Solve.** We solve the equation by dividing by $\frac{3}{5}$ on both sides and carrying out the division:

$$n = 60 \div \frac{3}{5} = 60 \cdot \frac{5}{3} = \frac{60 \cdot 5}{3} = \frac{3 \cdot 20 \cdot 5}{3 \cdot 1} = \frac{3}{3} \cdot \frac{20 \cdot 5}{1} = 100.$$

4. **Check.** We check by repeating the calculation.
5. **State.** Thus, 100 test tubes can be filled.

Do Exercise 3.

3. Each loop in a spring uses $\frac{3}{8}$ in. of wire. How many loops can be made from 120 in. of wire?

Example 4 After driving 210 mi, $\frac{5}{6}$ of a sales trip was completed. How long was the total trip?

1. **Familiarize.** We make a drawing or at least visualize the situation. We let n = the length of the trip.

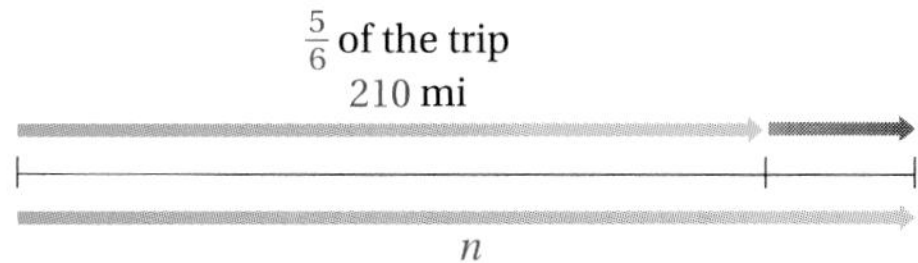

2. **Translate.** We translate to an equation.

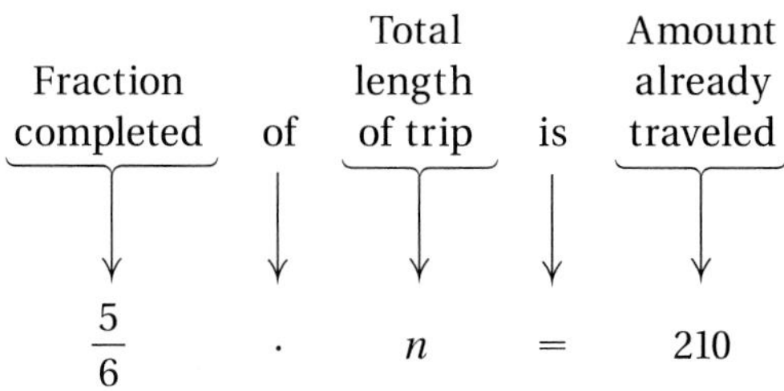

$$\frac{5}{6} \cdot n = 210$$

3. **Solve.** The equation that corresponds to the situation is $\frac{5}{6} \cdot n = 210$. We divide by $\frac{5}{6}$ on both sides and carry out the division:

$$n = 210 \div \frac{5}{6} = 210 \cdot \frac{6}{5} = \frac{210 \cdot 6}{5} = \frac{5 \cdot 42 \cdot 6}{5 \cdot 1} = \frac{5}{5} \cdot \frac{42 \cdot 6}{1} = 252.$$

4. **Check.** We check by repeating the calculation.
5. **State.** The total trip was 252 mi.

Do Exercise 4.

4. A service station tank had 175 gal of oil when it was $\frac{7}{8}$ full. How much could the tank hold altogether?

Example 5 A jogger has run $\frac{2}{3}$ mi and will stop running when she has run $\frac{7}{8}$ mi. How much farther does the jogger have to go?

1. **Familiarize.** We first make a drawing or at least visualize the situation. We let d = the distance to go.

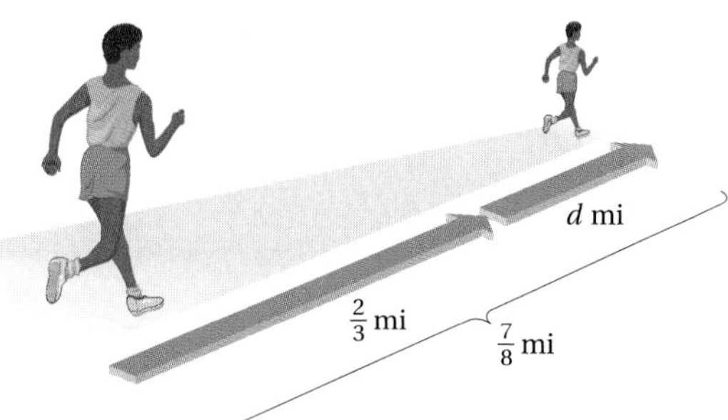

Answers on page A-4

5. There is $\frac{1}{4}$ cup of olive oil in a measuring cup. How much oil must be added to make a total of $\frac{4}{5}$ cup of oil in the measuring cup?

2. **Translate.** We see that this is a "how much more" situation. Now we translate to an equation.

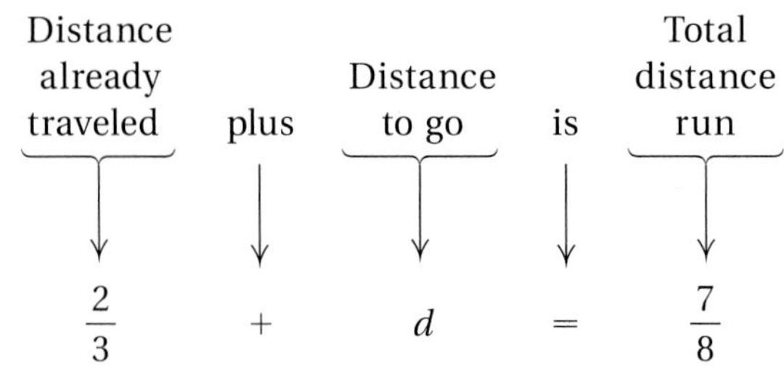

3. **Solve.** To solve the equation, we subtract $\frac{2}{3}$ on both sides:

$$\frac{2}{3} + d - \frac{2}{3} = \frac{7}{8} - \frac{2}{3}$$ **Subtracting $\frac{2}{3}$ on both sides**

$$d + 0 = \frac{7}{8} \cdot \frac{3}{3} - \frac{2}{3} \cdot \frac{8}{8}$$ **The LCD is 24. We multiply by 1 to obtain the LCD.**

$$d = \frac{21}{24} - \frac{16}{24} = \frac{5}{24}.$$

4. **Check.** To check, we return to the original problem and add:

$$\frac{2}{3} + \frac{5}{24} = \frac{2}{3} \cdot \frac{8}{8} + \frac{5}{24} = \frac{16}{24} + \frac{5}{24} = \frac{21}{24} = \frac{7}{8} \cdot \frac{3}{3} = \frac{7}{8}.$$

5. **State.** The jogger has $\frac{5}{24}$ mi to go.

Do Exercise 5.

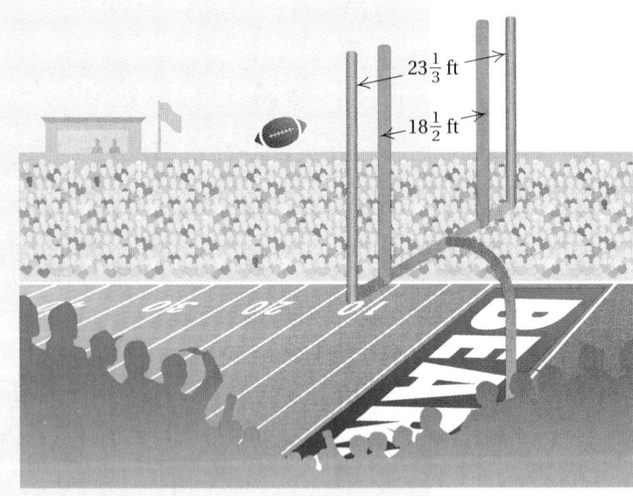

Source: NCAA

Example 6 *NCAA Football Goalposts.* Recently, in college football, the distance between goalposts was reduced from $23\frac{1}{3}$ ft to $18\frac{1}{2}$ ft. How much was it reduced?

1. **Familiarize.** We let d = the amount of reduction and make a drawing to illustrate the situation.

2. **Translate.** We translate as follows.

Former distance	−	New distance	=	Amount of reduction
↓	↓	↓	↓	↓
$23\frac{1}{3}$	−	$18\frac{1}{2}$	=	d

3. **Solve.** To solve the equation, we carry out the subtraction. The LCD is 6.

$$\begin{array}{rcrcrcr} 23\frac{1}{3} & = & 23\frac{1}{3} \cdot \frac{2}{2} & = & 23\frac{2}{6} & = & 22\frac{8}{6} \\ -\ 18\frac{1}{2} & = & -\ 18\frac{1}{2} \cdot \frac{3}{3} & = & -18\frac{3}{6} & = & -\ 18\frac{3}{6} \\ \hline & & & & & & 4\frac{5}{6} \end{array}$$

Thus, $d = 4\frac{5}{6}$ ft.

Answers on page A-4

4. Check. To check, we add the reduction to the new distance:

$$18\frac{1}{2} + 4\frac{5}{6} = 18\frac{3}{6} + 4\frac{5}{6} = 22\frac{8}{6} = 23\frac{2}{6} = 23\frac{1}{3}.$$

This checks.

5. State. The reduction in the goalpost distance was $4\frac{5}{6}$ ft.

Do Exercise 6.

6. *Damascus Blade.* The Damascus blade of a pearl-handled folding knife is $3\frac{3}{4}$ in. long. The same blade in an ATS-34 is $4\frac{1}{8}$ in. long (***Source:*** *Blade Magazine* 23, no. 10, October 1996: 26–27). How many inches longer is the ATS-34 blade?

Multistep Problems

Example 7 *Intel Stock Price.* One morning, the stock of Intel Corporation opened at a price of $\$100\frac{3}{8}$ per share. By noon, the price had risen $\$4\frac{7}{8}$. At the end of the day, it had fallen $\$10\frac{3}{4}$ from the price at noon. What was the closing price?

1. Familiarize. We first make a drawing or at least visualize the situation. We let p = the price at noon, after the rise, and c = the price at the close, after the drop.

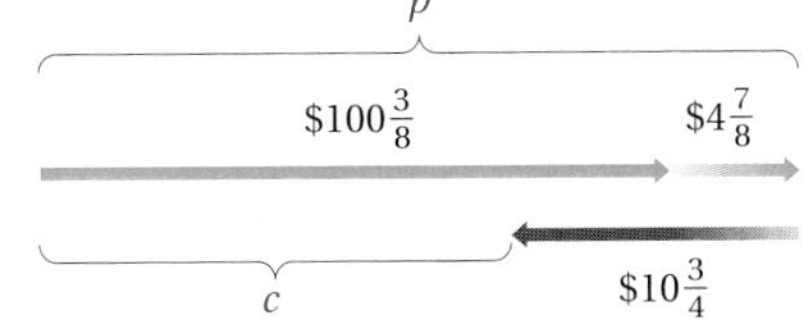

2. Translate. From the figure, we see that the price at the close is the price at noon minus the amount of the drop. Thus,

$$c = p - \$10\frac{3}{4} = \left(\$100\frac{3}{8} + \$4\frac{7}{8}\right) - \$10\frac{3}{4}.$$

3. Solve. This is a two-step problem.

a) We first add $\$4\frac{7}{8}$ to $\$100\frac{3}{8}$ to find the price p of the stock at noon.

$$\begin{array}{r} 100\frac{3}{8} \\ +\ \ 4\frac{7}{8} \\ \hline 104\frac{10}{8} = 105\frac{1}{4} = p \end{array}$$

b) Next we subtract $\$10\frac{3}{4}$ from $\$105\frac{1}{4}$ to find the price c of the stock at closing.

$$\begin{array}{rcr} 105\frac{1}{4} & = & 104\frac{5}{4} \\ -\ 10\frac{3}{4} & = & -\ 10\frac{3}{4} \\ \hline & & 94\frac{2}{4} = 94\frac{1}{2} = c \end{array}$$

4. Check. We check by repeating the calculation.

5. State. The price of the stock at closing is $\$94\frac{1}{2}$.

Do Exercise 7.

7. There are $20\frac{1}{3}$ gal of water in a barrel; $5\frac{3}{4}$ gal are poured out and $8\frac{2}{3}$ gal are poured back in. How many gallons of water are then in the barrel?

Answers on page A-4

8. A room is $22\frac{1}{2}$ ft by $15\frac{1}{2}$ ft. A 9-ft by 12-ft Oriental rug is placed in the center of the room. How much area is not covered by the rug?

Answer on page A-4

Example 8 An L-shaped room consists of a rectangle that is $8\frac{1}{2}$ by 11 ft and one that is $6\frac{1}{2}$ by $7\frac{1}{2}$ ft. What is the total area of a carpet that covers the floor?

1. **Familiarize.** We make a drawing of the situation. We let a = the total floor area.

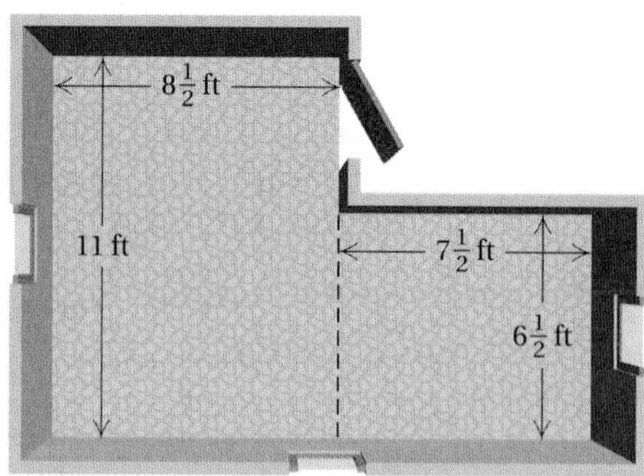

2. **Translate.** The total area is the sum of the areas of the two rectangles. This gives us the following equation:

$$a = 8\frac{1}{2} \cdot 11 + 7\frac{1}{2} \cdot 6\frac{1}{2}.$$

3. **Solve.** This is a multistep problem. We perform each multiplication and then add. This follows the rules for order of operations:

$$\begin{aligned} a &= 8\frac{1}{2} \cdot 11 + 7\frac{1}{2} \cdot 6\frac{1}{2} \\ &= \frac{17}{2} \cdot 11 + \frac{15}{2} \cdot \frac{13}{2} \\ &= \frac{17 \cdot 11}{2} + \frac{15 \cdot 13}{2 \cdot 2} \\ &= \frac{187}{2} + \frac{195}{4} \\ &= 93\frac{1}{2} + 48\frac{3}{4} \\ &= 93\frac{2}{4} + 48\frac{3}{4} \\ &= 141\frac{5}{4} \\ &= 142\frac{1}{4}. \end{aligned}$$

4. **Check.** We perform a partial check by estimating the total area as $11 \cdot 9 + 7 \cdot 7 = 99 + 49 = 148$ ft^2. Our answer, $142\frac{1}{4}$ ft^2, seems reasonable.

5. **State.** The total area of the carpet is $142\frac{1}{4}$ ft^2.

Do Exercise 8.

Exercise Set 2.5

a Solve.

1. A rectangular table top measures $\frac{4}{5}$ m long by $\frac{3}{5}$ m wide. What is its area?

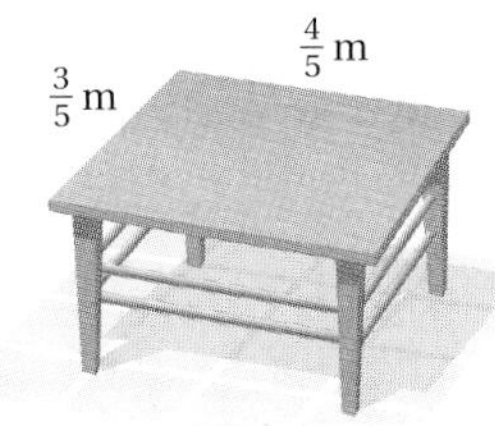

2. If each piece of pie is $\frac{1}{6}$ of a pie, how much of the pie is $\frac{1}{2}$ of a piece?

3. Anna receives \$36 for working a full day doing inventory at a hardware store. How much will she receive for working $\frac{3}{4}$ of the day?

4. After Jack completes 60 hr of teacher training in college, he can earn \$45 for working a full day as a substitute teacher. How much will he receive for working $\frac{1}{5}$ of a day?

5. *Map Scaling.* On a map, 1 in. represents 240 mi. How much does $\frac{2}{3}$ in. represent?

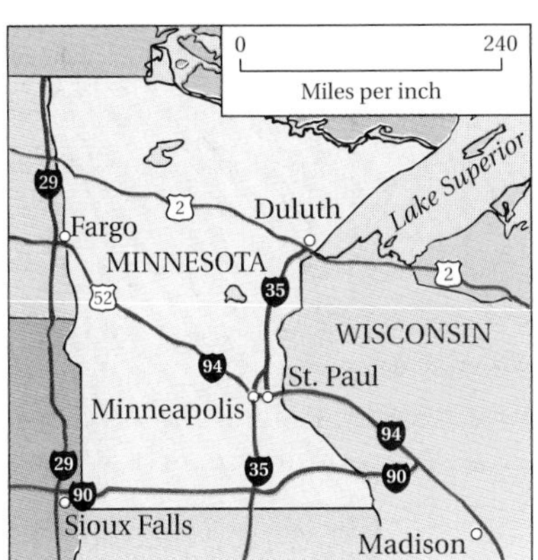

6. *Map Scaling.* On a map, 1 in. represents 120 mi. How much does $\frac{3}{4}$ in. represent?

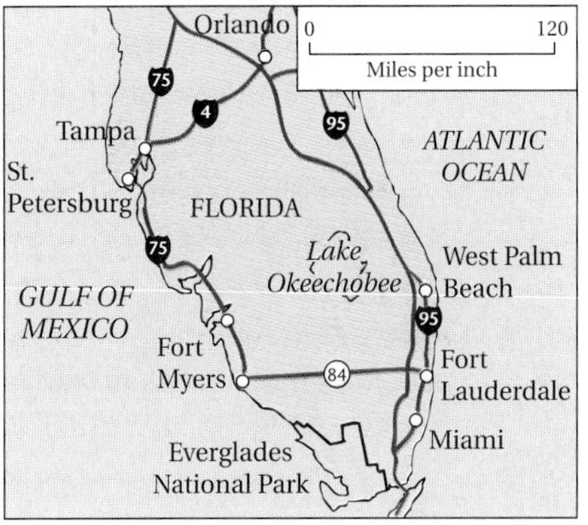

7. A pair of basketball shorts requires $\frac{3}{4}$ yd of nylon. How many pairs of shorts can be made from 24 yd of nylon?

8. A child's baseball shirt requires $\frac{5}{6}$ yd of a certain fabric. How many shirts can be made from 25 yd of the fabric?

9. After driving 180 kilometers (km), $\frac{5}{8}$ of a trip is completed. How long is the total trip? How many kilometers are left to drive?

10. A road crew repaves $\frac{1}{12}$ mi of road each day. How long will it take the crew to repave a $\frac{3}{4}$-mi stretch of road?

11. Russ walked $\frac{7}{6}$ mi to a friend's dormitory, and then $\frac{3}{4}$ mi to class. How far did he walk?

12. Elaine walked $\frac{7}{8}$ mi to the student union, and then $\frac{2}{5}$ mi to class. How far did she walk?

13. *Concrete Mix.* A cubic meter of concrete mix contains 420 kilograms (kg) of cement, 150 kg of stone, and 120 kg of sand. What is the total weight of the cubic meter of concrete mix? What part is cement? stone? sand? Add these amounts. What is the result?

14. *Punch Recipe.* A recipe for strawberry punch calls for $\frac{1}{5}$ quart (qt) of ginger ale and $\frac{3}{5}$ qt of strawberry soda. How much liquid is needed? If the recipe is doubled, how much liquid is needed? If the recipe is halved, how much liquid is needed?

15. Monica spent $\frac{3}{4}$ hr listening to tapes of Beethoven and Brahms. She spent $\frac{1}{3}$ hr listening to Beethoven. How many hours were spent listening to Brahms?

16. From a $\frac{4}{5}$-lb wheel of cheese, a $\frac{1}{4}$-lb piece was served. How much cheese remained on the wheel?

17. *Tire Tread.* A new long-life tire has a tread depth of $\frac{3}{8}$ in. instead of the more typical $\frac{11}{32}$ in. (***Source:*** *Popular Science*). How much deeper is the new tread depth?

18. As part of a fitness program, Deb swims $\frac{1}{2}$ mi every day. She has already swum $\frac{1}{5}$ mi. How much farther should Deb swim?

19. Tricia is 66 in. tall and her son is $59\frac{7}{12}$ in. tall. How much taller is Tricia?

20. Nicholas is $73\frac{2}{3}$ in. tall and his daughter is $71\frac{5}{16}$ in. tall. How much taller is Nicholas?

21. *Toys "R" Us Stock.* During a recent year, the price of one share of stock in Toys "R" Us varied between a low of $\$20\frac{1}{2}$ and a high of $\$37\frac{5}{8}$ (***Source:*** Toys "R" Us annual report). What was the difference between the high and the low?

22. *Nike, Inc., Stock.* During a recent year, the lowest price of one share of Nike, Inc., stock was $\$31\frac{3}{4}$. Its highest price was $\$22\frac{1}{4}$ more than its lowest price (***Source:*** Nike, Inc., annual report). What was the highest price?

23. A butcher sold packages of hamburger weighing $1\frac{2}{3}$ lb and $5\frac{3}{4}$ lb. What was the total weight of the meat?

24. A plumber uses pipes of length $10\frac{5}{16}$ ft and $8\frac{3}{4}$ ft in the installation of a sink. How much pipe was used?

Find the perimeter of (distance around) the figure.

25.

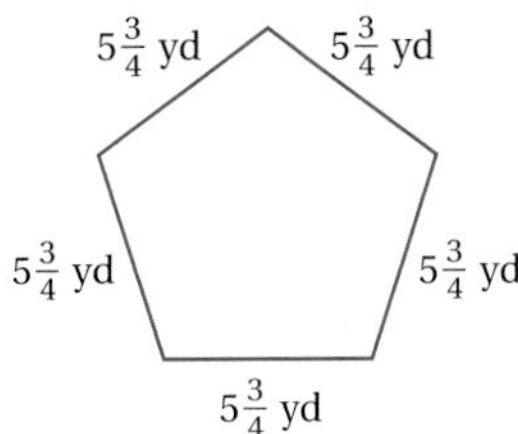

26.

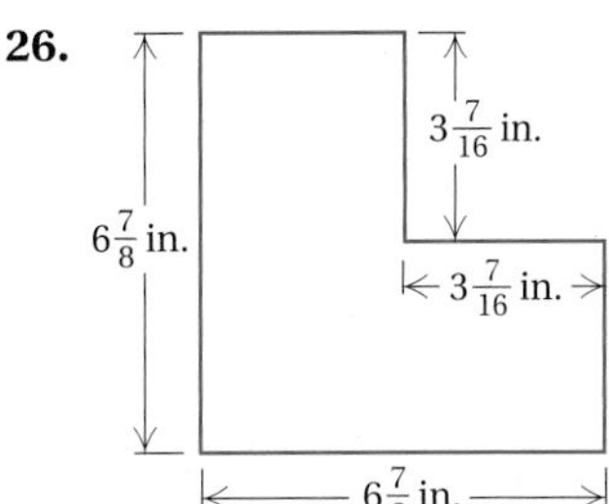

27. Sue, an interior designer, worked $10\frac{1}{2}$ hr over a three-day period. If Sue worked $2\frac{1}{2}$ hr on the first day and $4\frac{1}{5}$ hr on the second, how many hours did Sue work on the third day?

28. A painter had $3\frac{1}{2}$ gal of paint. It took $2\frac{3}{4}$ gal for a family room. It was estimated that it would take $2\frac{1}{4}$ gal to paint the living room. How much more paint was needed?

29. Find the length d in the figure.

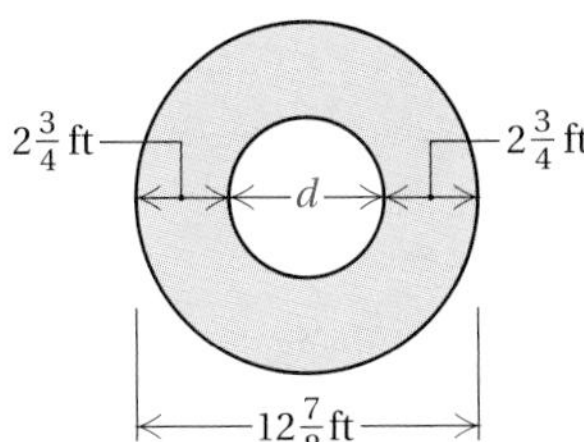

30. Find the smallest length of a bolt that will pass through a piece of tubing with an outside diameter of $\frac{1}{2}$ in., a washer $\frac{1}{16}$ in. thick, a piece of tubing with a $\frac{3}{4}$-in. outside diameter, another washer, and a nut $\frac{3}{16}$ in. thick.

31. The tape in a VCR operating in the short-play mode travels at a rate of $1\frac{3}{8}$ in. per second. How many inches of tape are used to record for 60 sec in the short-play mode?

32. *Temperature.* Fahrenheit temperature can be obtained from Celsius (centigrade) temperature by multiplying by $1\frac{4}{5}$ and adding 32°. What Fahrenheit temperature corresponds to the Celsius temperature of boiling water, which is 100°?

33. *Weight of Water.* The weight of water is $62\frac{1}{2}$ lb per cubic foot. How many cubic feet would be occupied by 250 lb of water?

34. *Turkey Servings.* Turkey contains $1\frac{1}{3}$ servings per pound. How many pounds are needed for 32 servings?

35. *Shuttle Orbits.* Most space shuttles orbit the earth once every $1\frac{1}{2}$ hr. How many orbits are made every 24 hr?

36. A car traveled 385 mi on $15\frac{4}{10}$ gal of gas. How many miles per gallon did it get?

37. A car traveled 213 mi on $14\frac{2}{10}$ gal of gas. How many miles per gallon did it get?

38. A bicycle wheel makes $66\frac{2}{3}$ revolutions per minute. If it rotates for 21 min, how many revolutions does it make?

39. Irene wants to build a bookcase to hold her collection of favorite videocassette movies. Each shelf in the bookcase will be 27 in. long and each videocassette is $1\frac{1}{8}$ in. thick. How many cassettes can she place on each shelf?

40. *Sodium Consumption.* The average American woman consumes $1\frac{1}{3}$ tsp of sodium each day (***Source:*** *Nutrition Action Health Letter,* March 1994, p. 6. 1875 Connecticut Ave., N.W., Washington, DC 20009-5728). How much sodium do 10 average American women consume in one day?

Find the area of the shaded region.

41.

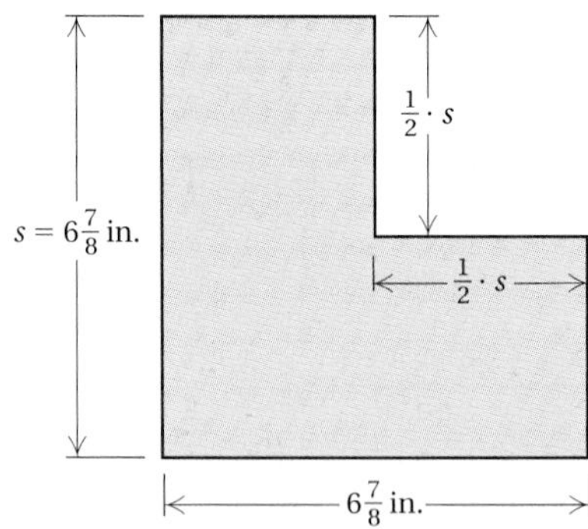

42.

$10\frac{1}{2}$ ft

$8\frac{1}{2}$ ft

4 ft

$10\frac{1}{2}$ ft

43. A rectangular lot has dimensions of $302\frac{1}{2}$ ft by $205\frac{1}{4}$ ft. A building with dimensions of 100 ft by $25\frac{1}{2}$ ft is built on the lot. How much area is left over?

44. *Word Processing.* Kelly wants to create a table using Microsoft® Word software for word processing. She needs to have two columns, each $1\frac{1}{2}$ in. wide, and five columns, each $\frac{3}{4}$ in. wide. Will this table fit on a piece of standard paper that is $8\frac{1}{2}$ in. wide? If so, how wide will each margin be if her margins on each side are to be of equal width?

Skill Maintenance

Solve. [1.5a]

45. A playing field is 78 ft long and 64 ft wide. What is its area?

46. A landscaper buys 13 small maple trees and 17 small oak trees for a project. A maple costs \$23 and an oak costs \$37. How much is spent altogether for the trees?

Subtract. [1.2f]

47. $34 - 23$

48. $50 - 18$

49. $803 - 617$

50. $8344 - 5607$

Solve. [1.4b]

51. $30 \cdot x = 150$

52. $5280 = 1760 + t$

Synthesis

53. A VCR can record up to 6 hr on one tape. It can also fill that same tape in either 4 hr or 2 hr when running at faster speeds. A tape is placed in the machine, which records for $\frac{1}{2}$ hr at the 4-hr speed and $\frac{3}{4}$ hr at the 2-hr speed. How much time is left on the tape to record at the 6-hr speed?

54. As part of a rehabilitation program, an athlete must swim and then walk a total of $\frac{9}{10}$ km each day. If one lap in the swimming pool is $\frac{3}{80}$ km, how far must the athlete walk after swimming 10 laps?

Analyze stock market prices.

2.6 Order of Operations; Estimation

a Order of Operations: Fractional Notation and Mixed Numerals

The rules for order of operations that we use with whole numbers (see Section 1.6) apply when we are simplifying expressions involving fractional notation and mixed numerals. For review, these rules are listed below.

RULES FOR ORDER OF OPERATIONS

1. Do all calculations within parentheses before operations outside.
2. Evaluate all exponential expressions.
3. Do all multiplications and divisions in order from left to right.
4. Do all additions and subtractions in order from left to right.

Example 1 Simplify: $\frac{2}{3} \div \frac{1}{2} \cdot \frac{5}{8} + \frac{1}{6}$.

$\frac{2}{3} \div \frac{1}{2} \cdot \frac{5}{8} + \frac{1}{6} = \frac{2}{3} \cdot \frac{2}{1} \cdot \frac{5}{8} + \frac{1}{6}$ Doing the division first by multiplying by the reciprocal of $\frac{1}{2}$

$= \frac{4}{3} \cdot \frac{5}{8} + \frac{1}{6}$

$= \frac{4 \cdot 5}{3 \cdot 8} + \frac{1}{6}$ Doing the multiplication

$= \frac{\cancel{4} \cdot 5}{3 \cdot \cancel{4} \cdot 2} + \frac{1}{6}$ Factoring in order to simplify

$= \frac{5}{3 \cdot 2} + \frac{1}{6}$ Removing a factor of 1: $\frac{4}{4} = 1$

$= \frac{5}{6} + \frac{1}{6}$

$= \frac{6}{6}$, or 1 Doing the addition

Do Exercises 1 and 2.

Example 2 Simplify: $\frac{2}{3} \cdot 24 - 11\frac{1}{2}$.

$\frac{2}{3} \cdot 24 - 11\frac{1}{2} = \frac{2 \cdot 24}{3} - 11\frac{1}{2}$ Doing the multiplication first

$= \frac{2 \cdot \cancel{3} \cdot 8}{\cancel{3}} - 11\frac{1}{2}$ Factoring the numerator

$= 2 \cdot 8 - 11\frac{1}{2}$ Removing a factor of 1: $\frac{3}{3} = 1$

$= 16 - 11\frac{1}{2}$ Completing the multiplication

$= 4\frac{1}{2}$, or $\frac{9}{2}$ Doing the subtraction

Do Exercise 3.

Objectives

a Simplify expressions using the rules for order of operations.

b Estimate with fractional and mixed-numeral notation.

For Extra Help

TAPE 5

MAC
WIN

CD-ROM

Simplify.

1. $\frac{2}{5} \cdot \frac{5}{8} + \frac{1}{4}$

2. $\frac{1}{3} \cdot \frac{3}{4} \div \frac{5}{8} - \frac{1}{10}$

3. Simplify: $\frac{3}{4} \cdot 16 + 8\frac{2}{3}$.

Answers on page A-5

4. Find the average of

$$\frac{1}{2}, \frac{1}{3}, \text{ and } \frac{5}{6}.$$

5. Find the average of $\frac{3}{4}$ and $\frac{4}{5}$.

6. Simplify:

$$\left(\frac{2}{3} + \frac{3}{4}\right) \div 2\frac{1}{3} - \left(\frac{1}{2}\right)^3.$$

Answers on page A-5

Example 3 To find the **average** of a set of numbers, we add the numbers and then divide by the number of addends. Find the average of $\frac{1}{2}$, $\frac{3}{4}$, and $\frac{7}{8}$.

The average is given by

$$\left(\frac{1}{2} + \frac{3}{4} + \frac{7}{8}\right) \div 3.$$

To find the average, we carry out the computation using the rules for order of operations:

$$\left(\frac{1}{2} + \frac{3}{4} + \frac{7}{8}\right) \div 3 = \left(\frac{4}{8} + \frac{6}{8} + \frac{7}{8}\right) \div 3$$ **Doing the operations inside parentheses first: adding by finding a common denominator**

$$= \frac{17}{8} \div 3$$ **Adding**

$$= \frac{17}{8} \cdot \frac{1}{3}$$ **Dividing by multiplying by the reciprocal**

$$= \frac{17}{24}$$ **Multiplying**

The average is $\frac{17}{24}$.

Do Exercises 4 and 5.

Example 4 Simplify: $\left(\frac{7}{8} - \frac{1}{3}\right) \times 48 + \left(13 + \frac{4}{5}\right)^2$.

$$\left(\frac{7}{8} - \frac{1}{3}\right) \times 48 + \left(13 + \frac{4}{5}\right)^2$$

$$= \left(\frac{7}{8} \cdot \frac{3}{3} - \frac{1}{3} \cdot \frac{8}{8}\right) \times 48 + \left(13 \cdot \frac{5}{5} + \frac{4}{5}\right)^2$$ **Carrying out operations inside parentheses first. To do so, we first multiply by 1 to obtain the LCD.**

$$= \left(\frac{21}{24} - \frac{8}{24}\right) \times 48 + \left(\frac{65}{5} + \frac{4}{5}\right)^2$$

$$= \frac{13}{24} \times 48 + \left(\frac{69}{5}\right)^2$$ **Completing the operations within parentheses**

$$= \frac{13}{24} \times 48 + \frac{4761}{25}$$ **Evaluating exponential expressions next**

$$= 26 + \frac{4761}{25}$$ **Doing the multiplication**

$$= 26 + 190\frac{11}{25}$$ **Converting to a mixed numeral**

$$= 216\frac{11}{25}, \text{ or } \frac{5411}{25}$$ **Adding**

Answers can be given using either fractional notation or mixed numerals as desired. Consult with your instructor.

Do Exercise 6.

b Estimation with Fractional Notation and Mixed Numerals

We now estimate with fractional notation and mixed numerals.

Examples Estimate each of the following as 0, $\frac{1}{2}$, or 1.

5. $\dfrac{2}{17}$

A fraction is very close to 0 when the numerator is very small in comparison to the denominator. Thus, 0 is an estimate for $\frac{2}{17}$ because 2 is very small in comparison to 17. Thus, $\frac{2}{17} \approx 0$.

6. $\dfrac{11}{23}$

A fraction is very close to $\frac{1}{2}$ when the denominator is about twice the numerator. Thus, $\frac{1}{2}$ is an estimate for $\frac{11}{23}$ because $2 \cdot 11 = 22$ and 22 is close to 23. Thus, $\frac{11}{23} \approx \frac{1}{2}$.

7. $\dfrac{37}{38}$

A fraction is very close to 1 when the numerator is nearly equal to the denominator. Thus, 1 is an estimate for $\frac{37}{38}$ because 37 is nearly equal to 38. Thus, $\frac{37}{38} \approx 1$.

8. $\dfrac{43}{41}$

As in the preceding example, the numerator 43 is very close to the denominator 41. Thus, $\frac{43}{41} \approx 1$.

Do Exercises 7–10.

Example 9 Find a number for the blank so that $\dfrac{9}{\blacksquare}$ is close to but less than 1. Answers may vary.

If the number in the blank were 9, we would have 1, so we increase 9 to 10. The answer is 10; $\frac{9}{10}$ is close to 1. The number 11 would also be a correct answer; $\frac{9}{11}$ is close to 1.

Do Exercises 11 and 12.

Example 10 Estimate $16\frac{8}{9} + 11\frac{2}{13} - 4\frac{22}{43}$ as a whole number or as a mixed number where the fractional part is $\frac{1}{2}$.

We estimate each fraction as 0, $\frac{1}{2}$, or 1. Then we calculate:

$$16\frac{8}{9} + 11\frac{2}{13} - 4\frac{22}{43} \approx 17 + 11 - 4\frac{1}{2} = 23\frac{1}{2}.$$

Do Exercises 13–15.

Estimate each of the following as 0, $\frac{1}{2}$, or 1.

7. $\dfrac{3}{59}$

8. $\dfrac{61}{59}$

9. $\dfrac{29}{59}$

10. $\dfrac{57}{59}$

Find a number for the blank so that the fraction is close to but less than 1.

11. $\dfrac{11}{\blacksquare}$

12. $\dfrac{\blacksquare}{33}$

Estimate each of the following as a whole number or as a mixed numeral where the fractional part is $\frac{1}{2}$.

13. $5\dfrac{9}{10} + 26\dfrac{1}{2} - 10\dfrac{3}{29}$

14. $10\dfrac{7}{8} \cdot \left(25\dfrac{11}{13} - 14\dfrac{1}{9}\right)$

15. $\left(10\dfrac{4}{5} + 7\dfrac{5}{9}\right) \div \dfrac{17}{30}$

Answers on page A-5

Calculator Spotlight

Mixed Numerals on a Calculator. Fraction calculators are equipped with a key, often labeled [a b/c], that allows for computations with fractional notation and mixed numerals. To calculate

$$\frac{2}{3}+\frac{4}{5}$$

with such a fraction calculator, the following keystrokes can be used:

[2] [a b/c] [3] [+] [4] [a b/c] [5] [=].

The display that appears,

1 ⌟ 7 ⌟ 15 ,

represents the mixed numeral $1\frac{7}{15}$.

To express the answer in fractional notation, we use the following keystrokes:

[Shift] [d/c].

The display that appears,

22 ⌟ 15 ,

represents the fraction $\frac{22}{15}$.

To enter a mixed numeral like $3\frac{2}{5}$ on a fraction calculator equipped with an [a b/c] key, we press

[3] [a b/c] [2] [a b/c] [5].

The calculator's display is in the form

3 ⌟ 2 ⌟ 5 .

Some calculators are capable of displaying mixed numerals in the way in which we write them, as shown below.

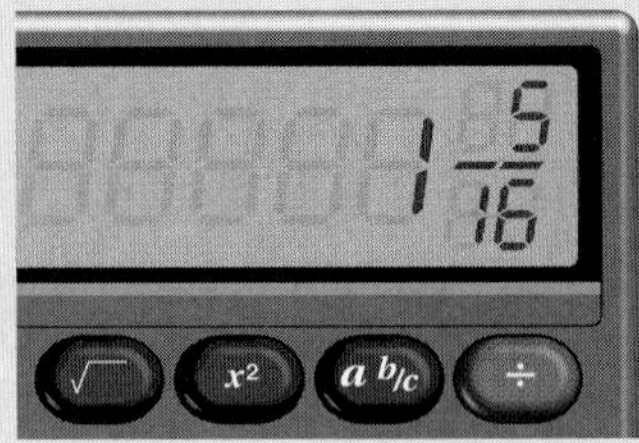

Exercises

Calculate using a fraction calculator. Give the answer in fractional notation.

1. $\frac{3}{8}+\frac{1}{4}$
2. $\frac{5}{12}+\frac{7}{10}-\frac{5}{12}$
3. $\frac{15}{7}\cdot\frac{1}{3}$
4. $\frac{19}{20}\div\frac{17}{35}$
5. $\frac{29}{30}-\frac{18}{25}\cdot\frac{2}{3}$
6. $\frac{1}{2}+\frac{13}{29}\cdot\frac{3}{4}$

Calculate using a fraction calculator. Give the answer in mixed numerals.

7. $4\frac{1}{2}\cdot 5\frac{3}{7}$
8. $7\frac{2}{3}\div 9\frac{4}{5}$
9. $8\frac{3}{7}+5\frac{2}{9}$
10. $13\frac{4}{9}-7\frac{5}{8}$
11. $13\frac{1}{4}-2\frac{1}{5}\cdot 4\frac{3}{8}$
12. $2\frac{5}{6}+5\frac{1}{6}\cdot 3\frac{1}{4}$

Exercise Set 2.6

a Simplify.

1. $\frac{1}{2} \cdot \frac{1}{3} \cdot \frac{1}{4}$

2. $\frac{5}{6} \div \frac{3}{4} \div \frac{2}{5}$

3. $\frac{5}{8} \div \frac{1}{4} - \frac{2}{3} \cdot \frac{4}{5}$

4. $\frac{3}{4} \div \frac{1}{2} \cdot \left(\frac{8}{9} - \frac{2}{3}\right)$

5. $28\frac{1}{8} - 5\frac{1}{4} + 3\frac{1}{2}$

6. $10\frac{3}{5} - 4\frac{1}{10} - 1\frac{1}{2}$

7. $\frac{7}{8} \div \frac{1}{2} \cdot \frac{1}{4}$

8. $\frac{7}{10} \cdot \frac{4}{5} \div \frac{2}{3}$

9. $\left(\frac{2}{3}\right)^2 - \frac{1}{3} \cdot 1\frac{1}{4}$

10. $\left(\frac{3}{4}\right)^2 + 3\frac{1}{2} \div 1\frac{1}{4}$

11. $\frac{1}{2} - \left(\frac{1}{2}\right)^2 + \left(\frac{1}{2}\right)^3$

12. $1 + \frac{1}{4} + \left(\frac{1}{4}\right)^2 - \left(\frac{1}{4}\right)^3$

13. Find the average of $\frac{1}{6}$, $\frac{1}{8}$, and $\frac{3}{4}$.

14. Find the average of $\frac{4}{5}$, $\frac{1}{2}$, and $\frac{1}{10}$.

15. Find the average of $3\frac{1}{2}$ and $9\frac{3}{8}$.

16. Find the average of $10\frac{2}{3}$ and $24\frac{5}{6}$.

Simplify.

17. $\left(\frac{1}{2} + \frac{1}{3}\right)^2 \cdot 144 - \frac{5}{8} \div 10\frac{1}{2}$

18. $\left(3\frac{1}{2} - 2\frac{1}{3}\right)^2 + 6 \cdot 2\frac{1}{2} \div 32$

b Estimate each of the following as 0, $\frac{1}{2}$, or 1.

19. $\frac{2}{47}$

20. $\frac{4}{5}$

21. $\frac{7}{100}$

22. $\frac{5}{9}$

23. $\frac{6}{11}$

24. $\frac{10}{13}$

Find a number for the blank so that the fraction is close to but greater than $\frac{1}{2}$. Answers may vary.

25. $\frac{\square}{11}$

26. $\frac{\square}{8}$

27. $\frac{\square}{23}$

28. $\frac{7}{\square}$

29. $\frac{8}{\square}$

30. $\frac{51}{\square}$

Find a number for the blank so that the fraction is close to but greater than 1. Answers may vary.

31. $\frac{7}{\square}$ **32.** $\frac{11}{\square}$ **33.** $\frac{13}{\square}$ **34.** $\frac{\square}{9}$ **35.** $\frac{\square}{15}$ **36.** $\frac{\square}{100}$

Estimate each part of the following as a whole number or as a mixed numeral where the fractional part is $\frac{1}{2}$.

37. $2\frac{7}{8}$

38. $26\frac{6}{13}$

39. $\frac{2}{3} + \frac{7}{13} + \frac{5}{9}$

40. $\frac{8}{9} + \frac{4}{5} + \frac{11}{12}$

41. $24 \div 7\frac{8}{9}$

42. $43\frac{16}{17} \div 11\frac{2}{13}$

43. $76\frac{3}{14} + 23\frac{19}{20}$

44. $76\frac{13}{14} \cdot 23\frac{17}{20}$

45. $\frac{43}{100} + \frac{1}{10} - \frac{11}{1000}$

46. $\frac{23}{24} + \frac{37}{39} + \frac{51}{50}$

47. $7\frac{29}{60} + 10\frac{12}{13} \cdot 24\frac{2}{17}$

48. $5\frac{13}{14} - 1\frac{5}{8} + 1\frac{23}{28} \cdot 6\frac{35}{74}$

49. $16\frac{1}{5} \div 2\frac{1}{11} + 25\frac{9}{10} - 4\frac{11}{23}$

50. $96\frac{2}{13} \div 5\frac{19}{20} + 3\frac{1}{7} \cdot 5\frac{18}{21}$

Skill Maintenance

Solve. [1.5a]

51. A shopper has $3458 in a checking account and writes checks for $329 and $52. How much is left in the account?

52. What is the total cost of 5 sweaters at $89 each and 6 shirts at $49 each?

Subtract. [1.2f]

53. 2037 − 1189

54. 9001 − 6798

55. 67,113 − 29,874

Synthesis

56. ◆ A student insists that $3\frac{2}{5} \cdot 1\frac{3}{7} = 3\frac{6}{35}$. What mistake is the student making and how should he have proceeded?

57. ◆ A student insists that $5 \cdot 3\frac{2}{7} = (5 \cdot 3) \cdot \left(5 \cdot \frac{2}{7}\right)$. What mistake is the student making and how should she have proceeded?

58. Find r if

$$\frac{1}{r} = \frac{1}{100} + \frac{1}{150} + \frac{1}{200}.$$

59. Use a standard calculator. Arrange the following in order from smallest to largest.

$$\frac{3}{4}, \frac{17}{21}, \frac{13}{15}, \frac{7}{9}, \frac{15}{17}, \frac{13}{12}, \frac{19}{22}$$

Summary and Review Exercises: Chapter 2

Beginning with this chapter, certain objectives, from four particular sections of preceding chapters, will be retested on the chapter test. The objectives to be tested in addition to the material in this chapter are [1.2f], [1.3e], [1.4b], and [1.5a].

1. What fractional part is shaded? [2.1a]

2. Simplify, if possible, the fractions on this circle graph. [2.1e]

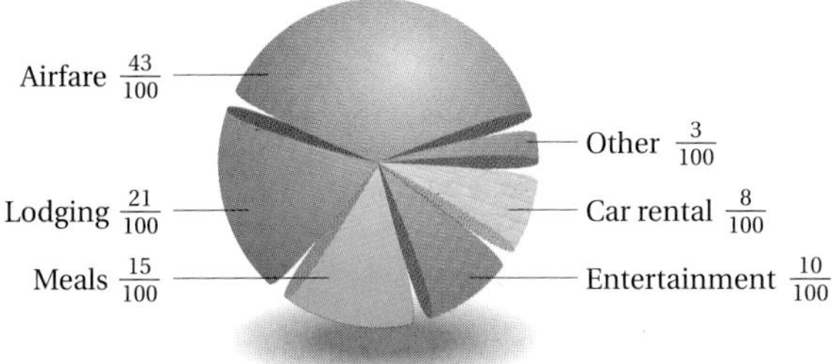

Simplify. [2.1b], [2.1e]

3. $\frac{0}{4}$ **4.** $\frac{23}{23}$ **5.** $\frac{48}{1}$

6. $\frac{48}{8}$ **7.** $\frac{12}{30}$ **8.** $\frac{18}{0}$

9. $\frac{9}{27}$ **10.** $\frac{7}{28}$ **11.** $\frac{10}{15}$

Multiply and simplify. [2.2a]

12. $4 \cdot \frac{3}{8}$ **13.** $\frac{6}{5} \cdot 20$

14. $\frac{5}{7} \cdot \frac{1}{10}$ **15.** $\frac{3}{7} \cdot \frac{14}{9}$

Find the reciprocal. [2.2b]

16. $\frac{4}{5}$ **17.** 3

Divide and simplify. [2.2c]

18. $\frac{3}{14} \div \frac{6}{7}$ **19.** $\frac{1}{4} \div \frac{1}{9}$

20. $180 \div \frac{3}{5}$ **21.** $\frac{2}{3} \div \frac{3}{2}$

Add and simplify. [2.3b]

22. $\frac{6}{5} + \frac{3}{8}$ **23.** $\frac{5}{16} + \frac{1}{12}$

24. $\frac{6}{5} + \frac{11}{15}$ **25.** $\frac{5}{16} + \frac{1}{8}$

Subtract and simplify. [2.3c]

26. $\frac{5}{9} - \frac{2}{9}$ **27.** $\frac{7}{8} - \frac{3}{4}$

28. $\frac{11}{27} - \frac{2}{9}$ **29.** $\frac{5}{6} - \frac{2}{9}$

Use $<$ or $>$ for ▢ to write a true sentence. [2.3d]

30. $\frac{4}{7}$ ▢ $\frac{5}{9}$ **31.** $\frac{8}{9}$ ▢ $\frac{11}{13}$

Convert to fractional notation. [2.4a]

32. $7\frac{1}{2}$ **33.** $8\frac{3}{8}$

Convert to a mixed numeral. [2.4a]

34. $\frac{7}{3}$ **35.** $\frac{27}{4}$

Add, subtract, multiply, or divide. Write a mixed numeral for the answer. [2.4b, c, d, e]

36. $5\frac{3}{5} + 4\frac{4}{5}$

37. $8\frac{1}{3} + 3\frac{2}{5}$

38. $10\frac{1}{4} - 6\frac{1}{10}$

39. $24 - 10\frac{5}{8}$

40. $6 \cdot 2\frac{2}{3}$

41. $5\frac{1}{4} \cdot \frac{2}{3}$

42. $2\frac{2}{5} \div 1\frac{7}{10}$

43. $3\frac{1}{4} \div 26$

Solve. [2.2d], [2.3e]

44. $\frac{5}{4} \cdot t = \frac{3}{8}$

45. $\frac{1}{2} + y = \frac{9}{10}$

Solve. [2.5a]

46. Liz and her road crew paint the lines in the middle and on the sides of a highway. They average about $\frac{5}{16}$ of a mile each day. How long will it take to paint the lines on 70 mi of highway?

47. A recipe calls for $\frac{2}{3}$ cup of diced bell peppers. In making $\frac{1}{2}$ of this recipe, how much diced pepper should be used?

48. Bernardo usually earns $42 for working a full day. How much does he receive for working $\frac{1}{7}$ of a day?

49. A curtain requires $2\frac{3}{5}$ yd of material. How many curtains can be made from 39 yd of material?

50. *Alcoa Stock Price.* On the first day of trading on the stock market, stock in Alcoa opened at $\$67\frac{3}{4}$ and rose by $\$2\frac{5}{8}$ at the close of trading. What was the stock's closing price?

51. A board $\frac{9}{10}$ in. thick is glued to a board $\frac{8}{10}$ in. thick. The glue is $\frac{3}{100}$ in. thick. How thick is the result?

52. Simplify this expression using the rules for order of operations: [2.6a]

$$\frac{1}{8} \div \frac{1}{4} + \frac{1}{2}.$$

53. Find the average of $\frac{1}{2}$, $\frac{1}{4}$, $\frac{1}{3}$, and $\frac{1}{5}$. [2.6a]

Estimate each of the following as 0, $\frac{1}{2}$, or 1. [2.6b]

54. $\frac{29}{59}$

55. $\frac{2}{59}$

56. $\frac{61}{59}$

Estimate each of the following as a whole number or as a mixed numeral where the fractional part is $\frac{1}{2}$. [2.6b]

57. $6\frac{7}{8}$

58. $\frac{3}{10} + \frac{5}{6} + \frac{31}{29}$

Skill Maintenance

Solve. [1.4b]

59. $17 \cdot x = 408$

60. $765 + t = 1234$

Solve. [1.5a]

61. The balance in your checking account is $789. After writing checks for $78, $97, and $102 and making a deposit of $400, what is your new balance?

62. An economy car gets 43 mpg on the highway. How far can the car be driven on a full tank of 18 gal of gasoline?

63. Divide: [1.3e]

$36 \overline{)14{,}697}$

64. Subtract: [1.2f]

$5604 - 1997$

Synthesis

65. ◆ Write, in your own words, a series of steps that can be used when simplifying fractional notation. [2.1e]

66. ◆ A student claims that "taking $\frac{1}{2}$ of a number is the same as dividing by $\frac{1}{2}$." Explain the error in this reasoning. [2.1c], [2.2c]

67. ▦ In the division below, find a and b. [2.2c]

$$\frac{19}{24} \div \frac{a}{b} = \frac{187{,}853}{268{,}224}$$

Test: Chapter 2

1. What part is shaded?

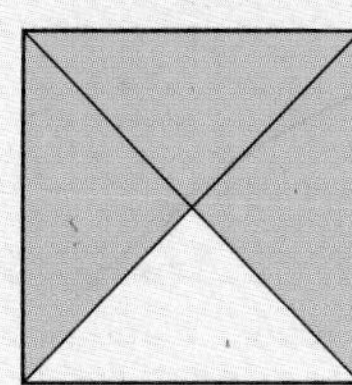

2. Use $<$ or $>$ for ■ to write a true sentence:

$$\frac{6}{7} \;\blacksquare\; \frac{21}{25}.$$

Simplify.

3. $\dfrac{12}{12}$

4. $\dfrac{0}{16}$

5. $\dfrac{2}{28}$

6. $\dfrac{9}{0}$

Add and simplify.

7. $\dfrac{1}{2} + \dfrac{5}{2}$

8. $\dfrac{7}{8} + \dfrac{2}{3}$

9. $\dfrac{7}{10} + \dfrac{9}{100}$

Subtract and simplify.

10. $\dfrac{5}{6} - \dfrac{3}{6}$

11. $\dfrac{5}{6} - \dfrac{3}{4}$

12. $\dfrac{17}{24} - \dfrac{5}{8}$

Multiply and simplify.

13. $\dfrac{4}{3} \cdot 24$

14. $\dfrac{2}{3} \cdot \dfrac{15}{4}$

15. $\dfrac{3}{5} \cdot \dfrac{1}{6}$

Find the reciprocal.

16. $\dfrac{5}{8}$

17. 18

Divide and simplify.

18. $\dfrac{3}{8} \div \dfrac{5}{4}$

19. $\dfrac{1}{5} \div \dfrac{1}{8}$

20. $12 \div \dfrac{2}{3}$

Solve.

21. $\dfrac{7}{8} \cdot x = 56$

22. $x + \dfrac{2}{3} = \dfrac{11}{12}$

23. Convert to a mixed numeral: $\dfrac{74}{9}$.

24. Convert to fractional notation: $3\dfrac{1}{2}$.

Answers

1. ______
2. ______
3. ______
4. ______
5. ______
6. ______
7. ______
8. ______
9. ______
10. ______
11. ______
12. ______
13. ______
14. ______
15. ______
16. ______
17. ______
18. ______
19. ______
20. ______
21. ______
22. ______
23. ______
24. ______

Answers

25. ____
26. ____
27. ____
28. ____
29. ____
30. ____
31. ____
32. ____
33. ____
34. ____
35. ____
36. ____
37. ____
38. ____
39. ____
40. ____
41. ____
42. ____
43. ____
44. ____
45. ____
46. ____
47. ____

Add, subtract, multiply or divide. Write a mixed numeral for the answer in Exercises 25–28.

25. $\begin{array}{r} 6\frac{2}{5} \\ +\ 7\frac{4}{5} \\ \hline \end{array}$

26. $\begin{array}{r} 10\frac{1}{6} \\ -\ 5\frac{7}{8} \\ \hline \end{array}$

27. $6\frac{3}{4} \cdot \frac{2}{3}$

28. $2\frac{1}{3} \div 1\frac{1}{6}$

Solve.

29. It takes $\frac{7}{8}$ lb of salt to use in the ice of one batch of homemade ice cream. How much salt is required for 32 batches?

30. A strip of taffy $\frac{9}{10}$ m long is cut into 12 equal pieces. What is the length of each piece?

31. *Turkey Loaf.* A low-cholesterol turkey loaf recipe calls for $3\frac{1}{2}$ cups of turkey breast. How much turkey is needed for 5 recipes?

32. An order of books for a math course weighs 220 lb. Each book weighs $2\frac{3}{4}$ lb. How many books are in the order?

33. The weights of two students are $183\frac{2}{3}$ lb and $176\frac{3}{4}$ lb. What is their total weight?

34. A standard piece of paper is $8\frac{1}{2}$ in. by 11 in. By how much does the length exceed the width?

35. Simplify: $\frac{2}{3} + 1\frac{1}{3} \cdot 2\frac{1}{8}$.

36. Find the average of $\frac{2}{5}$, $\frac{3}{4}$, and $\frac{1}{2}$.

Estimate each of the following as 0, $\frac{1}{2}$, or 1.

37. $\frac{3}{82}$

38. $\frac{93}{91}$

Estimate each of the following as a whole number or as a mixed numeral where the fractional part is $\frac{1}{2}$.

39. $3\frac{8}{9}$

40. $256 \div 15\frac{19}{21}$

Skill Maintenance

Solve.

41. $x + 198 = 2003$

42. $47 \cdot t = 4747$

43. It is 2060 mi from San Francisco to Winnipeg, Canada. It is 1575 mi from Winnipeg to Atlanta. What is the total length of a route from San Francisco to Winnipeg to Atlanta?

44. Divide: $24\overline{)9127}$

45. Subtract: $\begin{array}{r} 8001 \\ -\ 3567 \\ \hline \end{array}$

Synthesis

46. A recipe for a batch of buttermilk pancakes calls for $\frac{3}{4}$ teaspoon (tsp) of salt. Jacqueline plans to cut the amount of salt in half for each of 5 batches of pancakes. How much salt will she need?

47. Dolores runs 17 laps at her health club. Terence runs 17 laps at his health club. If the track at Dolores's health club is $\frac{1}{7}$ mi long, and the track at Terence's is $\frac{1}{8}$ mi long, who runs farther? How much farther?

3

Decimal Notation

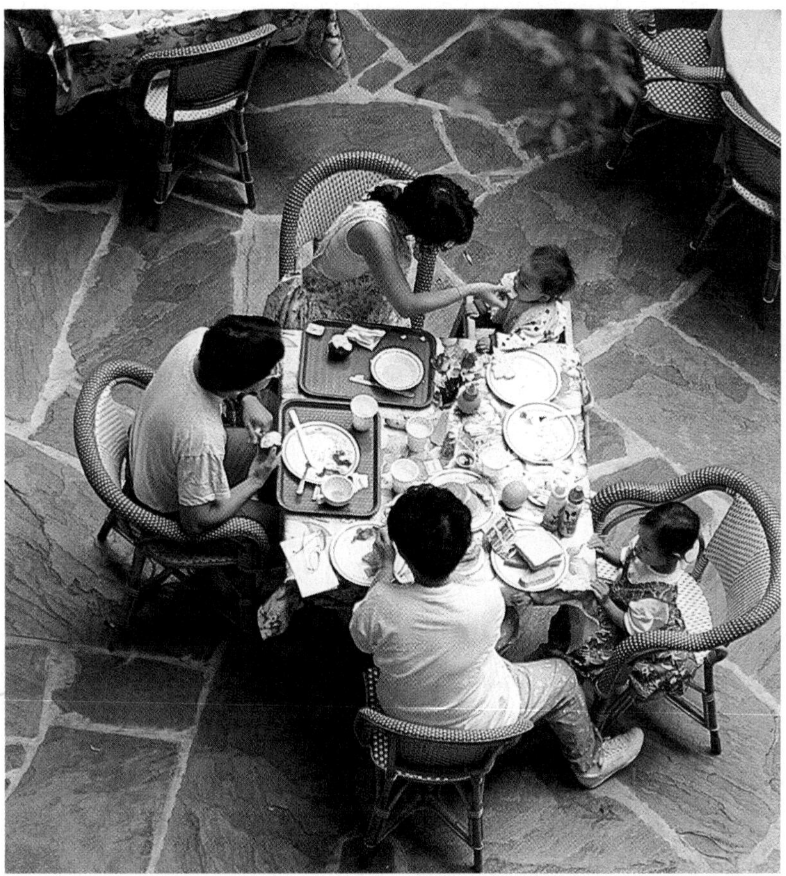

An Application

As life becomes busier, Americans are eating many more meals outside the home. In 1995, the average check for a casual meal eaten out was $28.90 and in 1996, it was $39.51 (***Source***: Sandelman and Associates, Brea CA). How much more is the average check for 1996 than that for 1995?

This problem appears as Example 1 in Section 3.7.

The Mathematics

We let c = the additional amount spent in 1996. The problem then translates to the equation

$$\underbrace{28.90 + c = 39.51.}$$

This equation involves decimal notation, which arises often in applied problems.

World Wide Web For more information, visit us at www.mathmax.com

Introduction

In this chapter, we consider the operations of addition, subtraction, multiplication, and division with decimal notation. This will allow us to solve applied problems like the one below. We will also study estimating sums, differences, products, and quotients. Conversion between fractional and decimal notation in which the decimal notation may be repeating will also be discussed.

Pretest: Chapter 3

1. Write a word name: 2.347.

2. Write a word name, as on a check, for $3264.78.

Write fractional notation.

3. 0.21

4. 5.408

Write decimal notation.

5. $\frac{379}{1000}$

6. $28\frac{439}{1000}$

Which number is larger?

7. 3.2, 0.321

8. 0.099, 0.091

Round 21.0448 to the nearest:

9. Tenth.

10. Thousandth.

11. Add:
$$\begin{array}{r} 601.3 \\ 5.81 \\ +\ 0.109 \\ \hline \end{array}$$

12. Subtract:
$$\begin{array}{r} 40.0 \\ -\ 0.9099 \\ \hline \end{array}$$

Multiply.

13.
$$\begin{array}{r} 0.835 \\ \times\ 0.74 \\ \hline \end{array}$$

14. 0.001×324.56

Divide.

15. $6.6\,\overline{)\,200.64}$

16. $\frac{576.98}{1000}$

Solve.

17. $9.6 \cdot y = 808.896$

18. $54.96 + q = 6400.117$

Solve.

19. *Travel Distance.* On a three-day trip, a traveler drove these distances: 432.6 mi, 179.2 mi, and 469.8 mi. What is the total number of miles driven?

20. A checking account contained $434.19. After a $148.24 check was drawn, how much was left in the account?

21. What is the cost of 6 videotapes at $14.95 each?

22. *Land Purchase.* A developer paid $47,567.89 for 14 acres of land. How much was paid for 1 acre? Round to the nearest cent.

23. Estimate the product 6.92×32.458 by rounding to the nearest one.

Find decimal notation. Use multiplying by 1.

24. $\frac{7}{5}$

25. $\frac{37}{40}$

Find decimal notation. Use division.

26. $\frac{11}{4}$

27. $\frac{29}{7}$

Round the answer to Exercise 27 to the nearest:

28. Tenth.

29. Hundredth.

30. Thousandth.

31. Convert from cents to dollars: 949 cents.

32. Convert to standard notation: 490 trillion.

Calculate.

33. $(1 - 0.06)^2 + 8[5(12.1 - 7.8) + 20(17.3 - 8.7)]$

34. $\frac{2}{3} \times 89.95 - \frac{5}{9} \times 3.234$

Objectives for Retesting

The objectives to be tested in addition to the material in this chapter are as follows.

[1.7d] Find the prime factorization of a composite number.
[1.9a] Find the LCM of two or more numbers using a list of multiples or factorizations.
[2.1e] Simplify fractional notation.
[2.4b, c] Add and subtract using mixed numerals.

3.1 Decimal Notation, Order, and Rounding

Objectives

a Given decimal notation, write a word name, and write a word name for an amount of money.

b Convert from decimal notation to fractional notation.

c Convert from fractional notation and mixed numerals to decimal notation.

d Given a pair of numbers in decimal notation, tell which is larger.

e Round to the nearest thousandth, hundredth, tenth, one, ten, hundred, or thousand.

For Extra Help

TAPE 6

MAC WIN

CD-ROM

The set of **arithmetic numbers,** or **nonnegative rational numbers,** consists of the whole numbers 0, 1, 2, 3, 4, 5, 6, 7, 8, 9, 10, and so on, and fractions like $\frac{1}{2}$, $\frac{2}{3}$, $\frac{7}{8}$, $\frac{17}{10}$, and so on. We studied the use of fractional notation for arithmetic numbers in Chapters 2 and 3. In Chapter 3, we will study the use of *decimal notation.* Although we are using different notation, we are still considering the same set of numbers. For example, instead of using fractional notation for $\frac{7}{8}$, we use decimal notation, 0.875.

a Decimal Notation and Word Names

Decimal notation for the women's shotput record is 74.249 ft. To understand what 74.249 means, we use a **place-value chart.** The value of each place is $\frac{1}{10}$ as large as the one to its left.

PLACE-VALUE CHART							
Hundreds	Tens	Ones	Ten*ths*	Hundred*ths*	Thousand*ths*	Ten-Thousand*ths*	Hundred-Thousand*ths*
100	10	1	$\frac{1}{10}$	$\frac{1}{100}$	$\frac{1}{1000}$	$\frac{1}{10,000}$	$\frac{1}{100,000}$
	7	4 .	2	4	9		

The decimal notation 74.249 means

$$7 \text{ tens} + 4 \text{ ones} + 2 \text{ tenths} + 4 \text{ hundredths} + 9 \text{ thousandths},$$

or $$7 \cdot 10 + 4 \cdot 1 + 2 \cdot \frac{1}{10} + 4 \cdot \frac{1}{100} + 9 \cdot \frac{1}{1000},$$

or $$70 + 4 + \frac{2}{10} + \frac{4}{100} + \frac{9}{1000}.$$

A mixed numeral for 74.249 is $74\frac{249}{1000}$. We read 74.249 as "seventy-four and two hundred forty-nine thousandths." When we come to the decimal point, we read "and." We can also read 74.249 as "seven four *point* two four nine."

To write a word name from decimal notation,

a) write a word name for the whole number (the number named to the left of the decimal point), — 397.685 → Three hundred ninety-seven

b) write the word "and" for the decimal point, and — 397.685 → Three hundred ninety-seven and

c) write a word name for the number named to the right of the decimal point, followed by the place value of the last digit. — 397.685 → Three hundred ninety-seven and six hundred eighty-five *thousandths*

Example 1 Write a word name for the number in this sentence: Each person consumes an average of 41.2 gallons of water per year.

Forty-one and two tenths

Example 2 Write a word name for 410.87.

Four hundred ten and eighty-seven hundredths

Example 3 Write a word name for the number in this sentence: The world record in the men's marathon is 2.1833 hours.

Two and one thousand eight hundred thirty-three ten-thousandths

Example 4 Write a word name for 1788.405.

One thousand, seven hundred eighty-eight and four hundred five thousandths

Do Exercises 1–4.

Write a word name for the number.

1. Each person in this country consumes an average of 21.1 gallons of coffee per year (***Source:*** Department of Agriculture).
2. The racehorse *Swale* won the Belmont Stakes in a time of 2.4533 minutes.
3. 245.89
4. 31,079.764

Decimal notation is also used with money. It is common on a check to write "and ninety-five cents" as "and $\frac{95}{100}$ dollars."

Example 5 Write a word name for the amount on the check, \$5876.95.

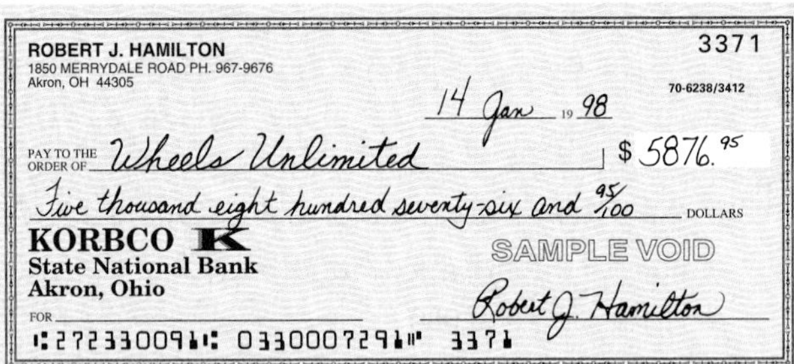

Five thousand, eight hundred seventy-six and $\frac{95}{100}$ dollars

Do Exercises 5 and 6.

Write a word name as on a check.

5. \$4217.56
6. \$13.98

b Converting from Decimal Notation to Fractional Notation

We can find fractional notation as follows:

$$9.875 = 9 + \frac{8}{10} + \frac{7}{100} + \frac{5}{1000}$$

$$= 9 \cdot \frac{1000}{1000} + \frac{8}{10} \cdot \frac{100}{100} + \frac{7}{100} \cdot \frac{10}{10} + \frac{5}{1000}$$

$$= \frac{9000}{1000} + \frac{800}{1000} + \frac{70}{1000} + \frac{5}{1000} = \frac{9875}{1000}.$$

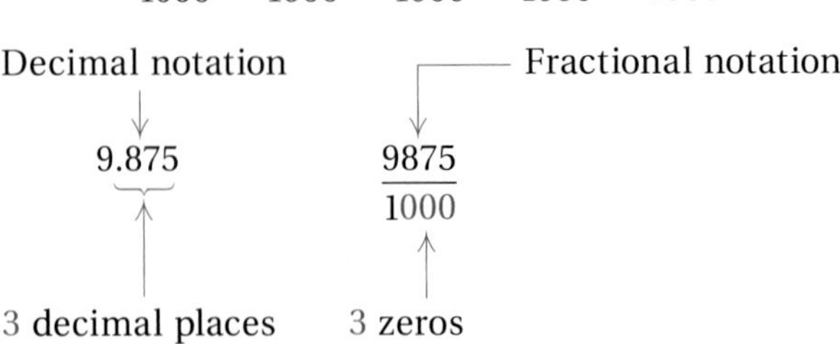

Answers on page A-5

To convert from decimal to fractional notation,

a) count the number of decimal places, 4.98 — 2 places

b) move the decimal point that many places to the right, and 4.98. Move 2 places.

c) write the answer over a denominator with a 1 followed by that number of zeros. $\frac{498}{100}$ 2 zeros

Example 6 Write fractional notation for 0.876. Do not simplify.

0.876 0.876. $0.876 = \frac{876}{1000}$

3 places 3 zeros

For a number like 0.876, we generally write a 0 before the decimal point to avoid forgetting or omitting it.

Example 7 Write fractional notation for 56.23. Do not simplify.

56.23 56.23. $56.23 = \frac{5623}{100}$

2 places 2 zeros

Example 8 Write fractional notation for 1.5018. Do not simplify.

1.5018 1.5018. $1.5018 = \frac{15{,}018}{10{,}000}$

4 places 4 zeros

Do Exercises 7–10.

c Converting from Fractional Notation and Mixed Numerals to Decimal Notation

If fractional notation has a denominator that is a power of ten, such as 10, 100, 1000, and so on, we reverse the procedure we used before.

To convert from fractional notation to decimal notation when the denominator is 10, 100, 1000, and so on,

a) count the number of zeros, and $\frac{8679}{1000}$ — 3 zeros

b) move the decimal point that number of places to the left. Leave off the denominator. 8.679. Move 3 places.

$\frac{8679}{1000} = 8.679$

Write fractional notation.

7. 0.896

8. 23.78

9. 5.6789

10. 1.9

Answers on page A-5

Write decimal notation.

11. $\frac{743}{100}$

12. $\frac{406}{1000}$

13. $\frac{67{,}089}{10{,}000}$

14. $\frac{9}{10}$

15. $\frac{57}{1000}$

16. $\frac{830}{10{,}000}$

Write decimal notation.

17. $4\frac{3}{10}$

18. $283\frac{71}{100}$

19. $456\frac{13}{1000}$

Answers on page A-5

Example 9 Write decimal notation for $\frac{47}{10}$.

$\frac{47}{10}$ ← 1 zero 4.7. (1 place) $\frac{47}{10} = 4.7$

Example 10 Write decimal notation for $\frac{123{,}067}{10{,}000}$.

$\frac{123{,}067}{10{,}000}$ ← 4 zeros 12.3067. (4 places) $\frac{123{,}067}{10{,}000} = 12.3067$

Example 11 Write decimal notation for $\frac{13}{1000}$.

$\frac{13}{1000}$ ← 3 zeros 0.013. (3 places) $\frac{13}{1000} = 0.013$

Example 12 Write decimal notation for $\frac{570}{100{,}000}$.

$\frac{570}{100{,}000}$ ← 5 zeros 0.00570. (5 places) $\frac{570}{100{,}000} = 0.0057$

Do Exercises 11–16.

When denominators are numbers other than 10, 100, and so on, we will use another method for conversion. It will be considered in Section 3.5.

If a mixed numeral has a fractional part with a denominator that is a power of ten, such as 10, 100, or 1000, and so on, we first write the mixed numeral as a sum of a whole number and a fraction. Then we convert to decimal notation.

Example 13 Write decimal notation for $23\frac{59}{100}$.

$$23\frac{59}{100} = 23 + \frac{59}{100} = 23 \text{ and } \frac{59}{100} = 23.59$$

Example 14 Write decimal notation for $772\frac{129}{10{,}000}$.

$$772\frac{129}{10{,}000} = 772 + \frac{129}{10{,}000} = 772 \text{ and } \frac{129}{10{,}000} = 772.0129$$

Do Exercises 17–19.

d Order

To understand how to compare numbers in decimal notation, consider 0.85 and 0.9. First note that $0.9 = 0.90$ because $\frac{9}{10} = \frac{90}{100}$. Then $0.85 = \frac{85}{100}$ and $0.90 = \frac{90}{100}$. Since $\frac{85}{100} < \frac{90}{100}$, it follows that $0.85 < 0.90$. This leads us to a quick way to compare two numbers in decimal notation.

> To compare two numbers in decimal notation, start at the left and compare corresponding digits moving from left to right. If two digits differ, the number with the larger digit is the larger of the two numbers. To ease the comparison, extra zeros can be written to the right of the last decimal place.

Example 15 Which of 2.109 and 2.1 is larger?

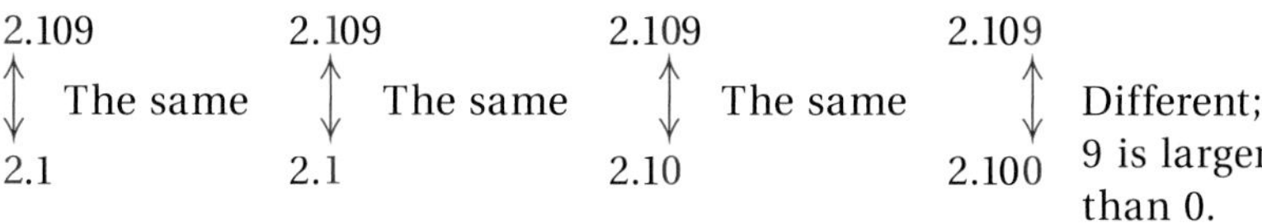

Thus, 2.109 is larger.

Example 16 Which of 0.09 and 0.108 is larger?

Thus, 0.108 is larger.

Do Exercises 20–25.

e Rounding

Rounding is done as for whole numbers. To understand, we first consider an example using a number line. It might help to review Section 1.4.

Example 17 Round 0.37 to the nearest tenth.

Here is part of a number line.

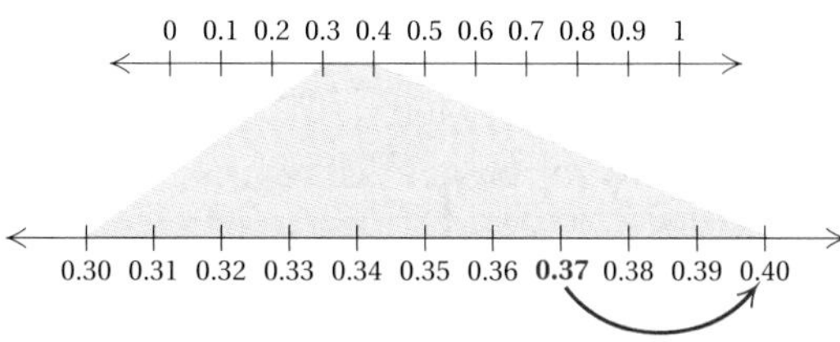

We see that 0.37 is closer to 0.40 than to 0.30. Thus, 0.37 rounded to the nearest tenth is 0.4.

Which number is larger?

20. 2.04, 2.039

21. 0.06, 0.008

22. 0.5, 0.58

23. 1, 0.9999

24. 0.8989, 0.09898

25. 21.006, 21.05

Answers on page A-5

Round to the nearest tenth.

26. 2.76 **27.** 13.85

28. 234.448 **29.** 7.009

Round to the nearest hundredth.

30. 0.636 **31.** 7.834

32. 34.675 **33.** 0.025

Round to the nearest thousandth.

34. 0.9434 **35.** 8.0038

36. 43.1119 **37.** 37.4005

Round 7459.3548 to the nearest:

38. Thousandth.

39. Hundredth.

40. Tenth.

41. One.

42. Ten. (*Caution*: "Tens" are not "tenths.")

43. Hundred.

44. Thousand.

Answers on page A-5

To round to a certain place:

a) Locate the digit in that place.

b) Consider the next digit to the right.

c) If the digit to the right is 5 or higher, round up; if the digit to the right is 4 or lower, round down.

Example 18 Round 3872.2459 to the nearest tenth.

a) Locate the digit in the tenths place.

3 8 7 2.2 4 5 9
↑

b) Consider the next digit to the right.

3 8 7 2.2 4 5 9
↑

c) Since that digit, 4, is less than 5, round down.

3 8 7 2.2 ⟵ **This is the answer.**

CAUTION! 3872.3 is not a correct answer to Example 18. It is incorrect to round from the ten-thousandths digit over to the tenths digit, as follows:

3872.246, 3872.25, 3872.3.

Example 19 Round 3872.2459 to the nearest thousandth, hundredth, tenth, one, ten, hundred, and thousand.

Thousandth:	3872.246	Ten:	3870
Hundredth:	3872.25	Hundred:	3900
Tenth:	3872.2	Thousand:	4000
One:	3872		

Example 20 Round 14.8973 to the nearest hundredth.

a) Locate the digit in the hundredths place. 1 4.8 9 7 3
↑

b) Consider the next digit to the right. 1 4.8 9 7 3
↑

c) Since that digit, 7, is 5 or higher, round up. When we make the hundredths digit a 10, we carry 1 to the tenths place.

The answer is 14.90. Note that the 0 in 14.90 indicates that the answer is correct to the nearest hundredth.

Example 21 Round 0.008 to the nearest tenth.

a) Locate the digit in the tenths place. 0.0 0 8
↑

b) Consider the next digit to the right. 0.0 0 8
↑

c) Since that digit, 0, is less than 5, round down.

The answer is 0.0.

Do Exercises 26–44.

Exercise Set 3.1

a Write a word name for the number in the sentence.

1. The largest pumpkin ever grown weighed 449.06 kilograms (***Source***: *Guinness Book of Records,* 1997).

2. The average loss of daylight in October in Anchorage, Alaska, is 5.63 min per day.

3. Recently, one British pound was worth about \$1.5599 in U.S. currency.

4. The cost of a fast modem for a computer is about \$289.95.

Write a word name.

5. 34.891

6. 27.1245

Write a word name as on a check.

7. \$326.48

8. \$125.99

9. \$36.72

10. \$0.67

b Write fractional notation. Do not simplify.

11. 8.3

12. 0.17

13. 3.56

14. 203.6

15. 46.03

16. 1.509

17. 0.00013

18. 0.0109

19. 1.0008

20. 2.0114

21. 20.003

22. 4567.2

c Write decimal notation.

23. $\frac{8}{10}$

24. $\frac{51}{10}$

25. $\frac{889}{100}$

26. $\frac{92}{100}$

27. $\frac{3798}{1000}$

28. $\frac{780}{1000}$

29. $\frac{78}{10,000}$

30. $\frac{56,788}{100,000}$

31. $\frac{19}{100,000}$

32. $\frac{2173}{100}$

33. $\frac{376,193}{1,000,000}$

34. $\frac{8,953,074}{1,000,000}$

35. $99\frac{44}{100}$

36. $4\frac{909}{1000}$

37. $3\frac{798}{1000}$

38. $67\frac{83}{100}$

39. $2\frac{1739}{10,000}$

40. $9243\frac{1}{10}$

41. $8\frac{953,073}{1,000,000}$

42. $2256\frac{3059}{10,000}$

d Which number is larger?

43. 0.06, 0.58

44. 0.008, 0.8

45. 0.905, 0.91

46. 42.06, 42.1

47. 0.0009, 0.001

48. 7.067, 7.054

49. 234.07, 235.07

50. 0.99999, 1

51. 0.004, $\frac{4}{100}$

52. $\frac{73}{10}$, 0.73

53. 0.432, 0.4325

54. 0.8437, 0.84384

e Round to the nearest tenth.

55. 0.11

56. 0.85

57. 0.49

58. 0.5794

59. 2.7449 **60.** 4.78 **61.** 123.65 **62.** 36.049

Round to the nearest hundredth.

63. 0.893 **64.** 0.675 **65.** 0.6666 **66.** 6.529

67. 0.995 **68.** 207.9976 **69.** 0.094 **70.** 11.4246

Round to the nearest thousandth.

71. 0.3246 **72.** 0.6666 **73.** 17.0015 **74.** 123.4562

75. 10.1011 **76.** 0.1161 **77.** 9.9989 **78.** 67.100602

Round 809.4732 to the nearest:

79. Hundred. **80.** Tenth. **81.** Thousandth.

82. Hundredth. **83.** One. **84.** Ten.

Round 34.54389 to the nearest:

85. Ten-thousandth. **86.** Thousandth. **87.** Hundredth.

88. Tenth. **89.** One. **90.** Ten.

Skill Maintenance

Round 6172 to the nearest: [1.3f]

91. Ten. **92.** Hundred. **93.** Thousand.

94. Find the LCM of 18, 27, and 54. [1.9a]

95. Subtract and simplify: $24 - 17\frac{2}{5}$. [2.4c]

Synthesis

96. ◆ Describe in your own words a procedure for converting from decimal notation to fractional notation.

97. ◆ A fellow student rounds 236.448 to the nearest one and gets 237. Explain the possible error.

There are other methods of rounding decimal notation. A computer often uses a method called **truncating**. To round using truncating, drop off all decimal places past the rounding place, which is the same as changing all digits to the right to zeros. For example, rounding 6.78093456285102 to the ninth decimal place, using truncating, gives us 6.780934562. Use truncating to round each of the following to the fifth decimal place, that is, the nearest hundred thousandth.

98. 6.78346123 **99.** 6.783461902 **100.** 99.999999999 **101.** 0.030303030303

3.2 Addition and Subtraction with Decimal Notation

Objectives

a Add using decimal notation.

b Subtract using decimal notation.

c Solve equations of the type $x + a = b$ and $a + x = b$, where a and b may be in decimal notation.

For Extra Help

TAPE 6

MAC WIN

CD-ROM

a Addition

Adding with decimal notation is similar to adding whole numbers. First we line up the decimal points so that we can add corresponding place-value digits. Then we add digits from the right. For example, we add the thousandths, then the hundredths, and so on, carrying if necessary. If desired, we can write extra zeros to the right of the decimal point so that the number of places is the same.

Example 1 Add: 56.314 + 17.78.

$$\begin{array}{r} 56.314 \\ +\ 17.780 \\ \hline \end{array}$$

Lining up the decimal points in order to add

Writing an extra zero to the right of the decimal point

$$\begin{array}{r} 56.314 \\ +\ 17.780 \\ \hline 4 \end{array}$$

Adding thousandths

$$\begin{array}{r} 56.314 \\ +\ 17.780 \\ \hline 94 \end{array}$$

Adding hundredths

$$\begin{array}{r} \overset{1}{5}6.314 \\ +\ 17.780 \\ \hline .094 \end{array}$$

Adding tenths

Write a decimal point in the answer.

We get 10 tenths = 1 one + 0 tenths, so we carry the 1 to the ones column.

$$\begin{array}{r} \overset{1}{5}\overset{1}{6}.314 \\ +\ 17.780 \\ \hline 4.094 \end{array}$$

Adding ones

We get 14 ones = 1 ten + 4 ones, so we carry the 1 to the tens column.

$$\begin{array}{r} \overset{1}{5}\overset{1}{6}.314 \\ +\ 17.780 \\ \hline 74.094 \end{array}$$

Adding tens

Do Exercises 1 and 2.

Add.

1. $\begin{array}{r} 0.847 \\ +\ 10.07 \\ \hline \end{array}$

2. $\begin{array}{r} 2.1 \\ 0.739 \\ +\ 31.3689 \\ \hline \end{array}$

Remember, we can write extra zeros to the right of the decimal point to get the same number of decimal places.

Example 2 Add: 3.42 + 0.237 + 14.1.

$$\begin{array}{r} 3.420 \\ 0.237 \\ +\ 14.100 \\ \hline 17.757 \end{array}$$

Lining up the decimal points and writing extra zeros

Adding

Do Exercises 3–5.

Add.

3. 0.02 + 4.3 + 0.649

4. 0.12 + 3.006 + 0.4357

5. 0.4591 + 0.2374 + 8.70894

Answers on page A-6

Add.

6. 789 + 123.67

7. 45.78 + 2467 + 1.993

Subtract.

8. 37.428 − 26.674

9.
```
  0.3 4 7
- 0.0 0 8
```

Answers on page A-6

Consider the addition 3456 + 19.347. Keep in mind that a whole number, such as 3456, has an "unwritten" decimal point at the right, with 0 fractional parts. When adding, we can always write in that decimal point and extra zeros if desired.

Example 3 Add: 3456 + 19.347.

```
     1
  3 4 5 6.0 0 0    Writing in the decimal point and extra zeros
+       1 9.3 4 7  Lining up the decimal points
  3 4 7 5.3 4 7    Adding
```

Do Exercises 6 and 7.

b Subtraction

Subtracting with decimal notation is similar to subtracting whole numbers. First we line up the decimal points so that we can subtract corresponding place-value digits. Then we subtract digits from the right. For example, we subtract the thousandths, then the hundredths, the tenths, and so on, borrowing if necessary.

Example 4 Subtract: 56.314 − 17.78.

```
  5 6.3 1 4   Lining up the decimal points in order to subtract
- 1 7.7 8 0   Writing an extra 0

  5 6.3 1 4   Subtracting thousandths
- 1 7.7 8 0
          4

       2 11
  5 6.3 1 4   Borrowing tenths to subtract hundredths
- 1 7.7 8 0
        3 4

      12
     5 2 11
  5 6.3 1 4   Borrowing ones to subtract tenths
- 1 7.7 8 0
      .5 3 4  Writing a decimal point

    15 12
   4 5 2 11
  5 6.3 1 4   Borrowing tens to subtract ones
- 1 7.7 8 0
    8.5 3 4

    15 12
   4 5 2 11
  5 6.3 1 4   Subtracting tens
- 1 7.7 8 0
  3 8.5 3 4
```

CHECK:
```
    1 1 1
  3 8.5 3 4
+ 1 7.7 8 0
  5 6.3 1 4
```

Do Exercises 8 and 9.

Example 5 Subtract: 13.07 − 9.205.

```
   12
   2 10 6 10
  1 3.0 7 0     Writing an extra zero
−   9.2 0 5
    3.8 6 5     Subtracting
```

Example 6 Subtract: 23.08 − 5.0053.

```
 1 13    7 9 10
 2 3.0 8 0 0     Writing two extra zeros
−  5.0 0 5 3
 1 8.0 7 4 7     Subtracting
```

Do Exercises 10–12.

When subtraction involves a whole number, again keep in mind that there is an "unwritten" decimal point that can be written in if desired. Extra zeros can also be written in to the right of the decimal point.

Example 7 Subtract: 456 − 2.467.

```
      5 9 9 10
 4 5 6.0 0 0     Writing in the decimal point and extra zeros
−      2.4 6 7
 4 5 3.5 3 3     Subtracting
```

Do Exercises 13 and 14.

c Solving Equations

Now let's solve equations $x + a = b$ and $a + x = b$, where a and b may be in decimal notation. Proceeding as we have before, we subtract a on both sides.

Example 8 Solve: $x + 28.89 = 74.567$.

We have

$$x + 28.89 - 28.89 = 74.567 - 28.89$$ Subtracting 28.89 on both sides

$$x = 45.677.$$

```
 6 13 14 16
 7 4.5 6 7
− 2 8.8 9 0
  4 5.6 7 7
```

The solution is 45.677.

Subtract.

10. 1.2345 − 0.7

11. 0.9564 − 0.4392

12. 7.37 − 0.00008

Subtract.

13. 1277 − 82.78

14. 5 − 0.0089

Answers on page A-6

Solve.

15. $x + 17.78 = 56.314$

16. $8.906 + t = 23.07$

17. Solve: $241 + y = 2374.5$.

Answers on page A-6

Example 9 Solve: $0.8879 + y = 9.0026$.

We have

$$0.8879 + y - 0.8879 = 9.0026 - 0.8879 \quad \text{Subtracting 0.8879 on both sides}$$

$$y = 8.1147.$$

$$\begin{array}{r} \overset{8}{\cancel{9}}.\overset{9}{\cancel{0}}\,\overset{9}{\cancel{0}}\,\overset{11}{\cancel{2}}\,\overset{16}{\cancel{6}} \\ -\ 0.8\ 8\ 7\ 9 \\ \hline 8.1\ 1\ 4\ 7 \end{array}$$

The solution is 8.1147.

Do Exercises 15 and 16.

Example 10 Solve: $120 + x = 4380.6$.

We have

$$120 + x - 120 = 4380.6 - 120 \quad \text{Subtracting 120 on both sides}$$

$$x = 4260.6$$

$$\begin{array}{r} 4\ 3\ 8\ 0.6 \\ -\ \ 1\ 2\ 0.0 \\ \hline 4\ 2\ 6\ 0.6 \end{array}$$

The solution is 4260.6.

Do Exercise 17.

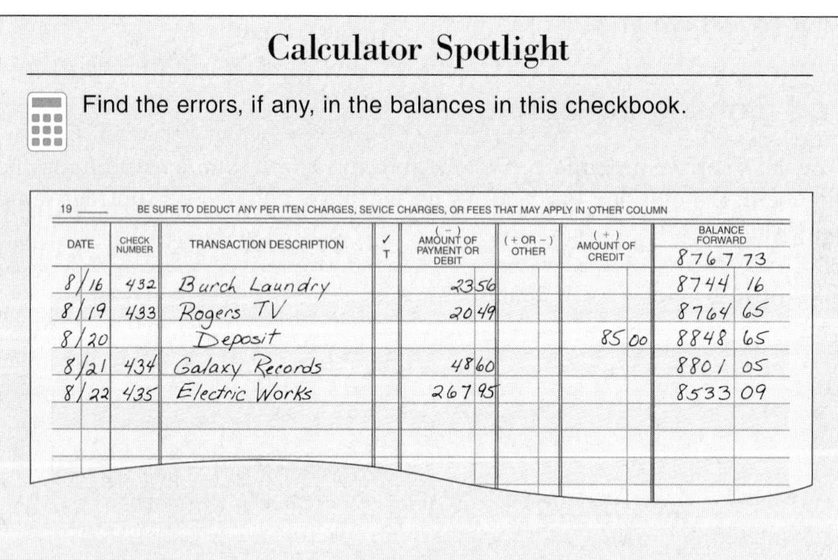

Calculator Spotlight

Find the errors, if any, in the balances in this checkbook.

19 ____ BE SURE TO DEDUCT ANY PER ITEN CHARGES, SEVICE CHARGES, OR FEES THAT MAY APPLY IN 'OTHER' COLUMN

DATE	CHECK NUMBER	TRANSACTION DESCRIPTION	✓ T	(−) AMOUNT OF PAYMENT OR DEBIT	(+ OR −) OTHER	(+) AMOUNT OF CREDIT	BALANCE FORWARD
							8767 73
8/16	432	Burch Laundry		23 56			8744 16
8/19	433	Rogers TV		20 49			8764 65
8/20		Deposit				85 00	8848 65
8/21	434	Galaxy Records		48 60			8801 05
8/22	435	Electric Works		267 95			8533 09

Exercise Set 3.2

a Add.

1. $$\begin{array}{r} 316.25 \\ +\ 18.12 \\ \hline \end{array}$$

2. $$\begin{array}{r} 641.803 \\ +\ 14.935 \\ \hline \end{array}$$

3. $$\begin{array}{r} 659.403 \\ +916.812 \\ \hline \end{array}$$

4. $$\begin{array}{r} 4203.28 \\ +\ 3.39 \\ \hline \end{array}$$

5. $$\begin{array}{r} 9.104 \\ +123.456 \\ \hline \end{array}$$

6. $$\begin{array}{r} 6.1528 \\ +5.2777 \\ \hline \end{array}$$

7. $$\begin{array}{r} 81.008 \\ +\ 3.409 \\ \hline \end{array}$$

8. $0.8096 + 0.7856$

9. $20.0124 + 30.0124$

10. $0.687 + 0.9$

11. $39 + 1.007$

12. $0.845 + 10.02$

13. $0.34 + 3.5 + 0.127 + 768$

14. $2.3 + 0.729 + 23$

15. $17 + 3.24 + 0.256 + 0.3689$

16. $$\begin{array}{r} 47.8 \\ 219.852 \\ 43.59 \\ +666.713 \\ \hline \end{array}$$

17. $$\begin{array}{r} 2.703 \\ 78.33 \\ 28.0009 \\ +118.4341 \\ \hline \end{array}$$

18. $$\begin{array}{r} 13.72 \\ 9.112 \\ 6542.7908 \\ +\ 23.901 \\ \hline \end{array}$$

19. $99.6001 + 7285.18 + 500.042 + 870$

20. $65.987 + 9.4703 + 6744.02 + 1.0003 + 200.895$

b Subtract.

21. $\begin{array}{r} 5.2 \\ -\ 3.9 \\ \hline \end{array}$

22. $\begin{array}{r} 44.345 \\ -\ 3.105 \\ \hline \end{array}$

23. $\begin{array}{r} 51.31 \\ -\ 2.29 \\ \hline \end{array}$

24. $\begin{array}{r} 87.46 \\ -\ 6.32 \\ \hline \end{array}$

25. $\begin{array}{r} 48.76 \\ -\ 3.15 \\ \hline \end{array}$

26. $\begin{array}{r} 97.01 \\ -\ 3.15 \\ \hline \end{array}$

27. $\begin{array}{r} 92.341 \\ -\ 6.42 \\ \hline \end{array}$

28. $\begin{array}{r} 0.8468 \\ -\ 0.034 \\ \hline \end{array}$

29. $\begin{array}{l} 2.5 \\ -\ 0.0025 \\ \hline \end{array}$

30. $\begin{array}{r} 39.0 \\ -\ 0.28 \\ \hline \end{array}$

31. $\begin{array}{l} 3.4 \\ -\ 0.003 \\ \hline \end{array}$

32. $\begin{array}{l} 2.8 \\ -\ 2.08 \\ \hline \end{array}$

33. 28.2 − 19.35

34. 100.16 − 0.118

35. 34.07 − 30.7

36. 36.2 − 16.28

37. 8.45 − 7.405

38. 3.801 − 2.81

39. 6.003 − 2.3

40. 9.087 − 8.807

41. 1 − 0.0098

42. 2 − 1.0908

43. 100 − 0.34

44. 624 − 18.79

45. 7.48 − 2.6

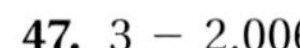

46. 18.4 − 5.92

47. 3 − 2.006

48. 263.7 − 102.08

49. 19 − 1.198

50. 2548.98 − 2.007

51. 65 − 13.87

52. 45 − 0.999

53. 3.907 − 1.416

54. 70.0009 − 23.0567

55. $\begin{array}{r} 32.7978 \\ -\ \ 0.0592 \\ \hline \end{array}$

56. $\begin{array}{r} 0.49634 \\ -0.12678 \\ \hline \end{array}$

57. $\begin{array}{r} 3.0074 \\ -1.3408 \\ \hline \end{array}$

58. $\begin{array}{l} \ \ 6.07 \\ -2.0078 \\ \hline \end{array}$

59. $\begin{array}{r} 2345.90786 \\ -\ \ \ \ \ \ \ \ 0.999\ \ \ \\ \hline \end{array}$

60. $\begin{array}{l} \ \ 1.0 \\ -0.9999 \\ \hline \end{array}$

c Solve.

61. $x + 17.5 = 29.15$

62. $t + 50.7 = 54.07$

63. $3.205 + m = 22.456$

64. $4.26 + q = 58.32$

65. $17.95 + p = 402.63$

66. $w + 1.3004 = 47.8$

67. $13{,}083.3 = x + 12{,}500.33$

68. $100.23 = 67.8 + z$

69. $x + 2349 = 17{,}684.3$

70. $1830.4 + t = 23{,}067$

Skill Maintenance

71. Find the LCM of 32 and 85. [1.9a]

72. Find the prime factorization of 228. [1.7d]

Subtract.

73. $\frac{13}{24} - \frac{3}{8}$ [2.3c]

74. $\frac{8}{9} - \frac{2}{15}$ [2.3c]

75. $8805 - 2639$ [1.2f]

76. $8005 - 2639$ [1.2f]

Solve.

77. A serving of filleted fish is generally considered to be about $\frac{1}{3}$ lb. How many servings can be prepared from $5\frac{1}{2}$ lb of flounder fillet? [2.5a]

78. A photocopier technician drove $125\frac{7}{10}$ mi away from Scottsdale for a repair call. The next day he drove $65\frac{1}{2}$ mi back toward Scottsdale for another service call. How far was the technician from Scottsdale? [2.5a]

Synthesis

79. ◆ Explain the error in the following:

Add.

$$\begin{array}{r} 1\,3.0\,7 \\ +\quad 9.2\,0\,5 \\ \hline 1\,0.5\,1\,2 \end{array}$$

80. ◆ Explain the error in the following:

Subtract.

$$\begin{array}{r} 7\,3.0\,8\,9 \\ -\,5.0\,0\,6\,1 \\ \hline 2.3\,0\,2\,8 \end{array}$$

81. A student presses the wrong button when using a calculator and adds 235.7 instead of subtracting it. The incorrect answer is 817.2. What is the correct answer?

3.3 Multiplication with Decimal Notation

Objectives

a Multiply using decimal notation.

b Convert from notation like 45.7 million to standard notation, and from dollars to cents and cents to dollars.

For Extra Help

TAPE 6 MAC WIN CD-ROM

a Multiplication

Let's find the product

$$2.3 \times 1.12.$$

To understand how we find such a product, we first convert each factor to fractional notation. Next, we multiply the whole numbers 23 and 112, and then divide by 1000.

$$2.3 \times 1.12 = \frac{23}{10} \times \frac{112}{100} = \frac{23 \times 112}{10 \times 100} = \frac{2576}{1000} = 2.576$$

Note the number of decimal places.

$$\begin{array}{rl} 1.1\,2 & \text{(2 decimal places)} \\ \times \quad 2.3 & \text{(1 decimal place)} \\ \hline 2.5\,7\,6 & \text{(3 decimal places)} \end{array}$$

Now consider

$$0.011 \times 15.0002 = \frac{11}{1000} \times \frac{150{,}002}{10{,}000} = \frac{1{,}650{,}022}{10{,}000{,}000} = 0.1650022.$$

Note the number of decimal places.

$$\begin{array}{rl} 1\,5.0\,0\,0\,2 & \text{(4 decimal places)} \\ \times \quad 0.0\,1\,1 & \text{(3 decimal places)} \\ \hline 0.1\,6\,5\,0\,0\,2\,2 & \text{(7 decimal places)} \end{array}$$

To multiply using decimals:

a) Ignore the decimal points and multiply as though both factors were whole numbers.

b) Then place the decimal point in the result. The number of decimal places in the product is the sum of the numbers of places in the factors (count places from the right).

0.8×0.43

$$\begin{array}{rl} \overset{2}{0.4\,3} & \\ \times \quad 0.8 & \text{Ignore the decimal} \\ \hline 3\,4\,4 & \text{points for now.} \end{array}$$

$$\begin{array}{rl} 0.4\,3 & \text{(2 decimal places)} \\ \times \quad 0.8 & \text{(1 decimal place)} \\ \hline 0.3\,4\,4 & \text{(3 decimal places)} \end{array}$$

1. Multiply.

$$\begin{array}{r} 85.4 \\ \times \quad 6.2 \\ \hline \end{array}$$

Multiply.

2. $\begin{array}{r} 1234 \\ \times 0.0041 \\ \hline \end{array}$

3. $\begin{array}{r} 42.65 \\ \times 0.804 \\ \hline \end{array}$

Answers on page A-6

Example 1 Multiply: 8.3×74.6.

a) Ignore the decimal points and multiply as though factors were whole numbers:

$$\begin{array}{r} \overset{3}{\underset{1}{7}}\,\overset{4}{\underset{1}{4}}.6 \\ \times \quad 8.3 \\ \hline 2238 \\ 59680 \\ \hline 61918 \end{array}$$

b) Place the decimal point in the result. The number of decimal places in the product is the sum, $1 + 1$, of the number of places in the factors.

$$\begin{array}{rl} 74.6 & \text{(1 decimal place)} \\ \times \quad 8.3 & \text{(1 decimal place)} \\ \hline 2238 & \downarrow \\ 59680 & \\ \hline 619.18 & \text{(2 decimal places)} \end{array}$$

Do Exercise 1.

Example 2 Multiply: 0.0032×2148.

As we catch on to the skill, we can combine the two steps.

$$\begin{array}{rl} 2148 & \text{(0 decimal places)} \\ \times 0.0032 & \text{(4 decimal places)} \\ \hline 4296 & \downarrow \\ 64440 & \\ \hline 6.8736 & \text{(4 decimal places)} \end{array}$$

Example 3 Multiply: 0.14×0.867.

$$\begin{array}{rl} 0.867 & \text{(3 decimal places)} \\ \times \quad 0.14 & \text{(2 decimal places)} \\ \hline 3468 & \downarrow \\ 8670 & \\ \hline 0.12138 & \text{(5 decimal places)} \end{array}$$

Do Exercises 2 and 3.

Now let's consider some special kinds of products. The first involves multiplying by a tenth, hundredth, thousandth, or ten-thousandth. Let's look at those products.

$$0.1 \times 38 = \frac{1}{10} \times 38 = \frac{38}{10} = 3.8$$

$$0.01 \times 38 = \frac{1}{100} \times 38 = \frac{38}{100} = 0.38$$

$$0.001 \times 38 = \frac{1}{1000} \times 38 = \frac{38}{1000} = 0.038$$

$$0.0001 \times 38 = \frac{1}{10{,}000} \times 38 = \frac{38}{10{,}000} = 0.0038$$

Note in each case that the product is *smaller* than 38.

To multiply any number by a tenth, hundredth, or thousandth,

a) count the number of decimal places in the tenth, hundredth, or thousandth, and

0.001 × 34.45678
→ 3 places

b) move the decimal point that many places to the left.

0.001 × 34.45678 = 0.034.45678

Move 3 places to the left.

0.001 × 34.45678 = 0.03445678

Examples Multiply.

4. 0.1 × 14.605 = 1.4605 (1.4.605)

5. 0.01 × 14.605 = 0.14605

6. 0.001 × 14.605 = 0.014605
We write an extra zero.

7. 0.0001 × 14.605 = 0.0014605
We write two extra zeros.

Do Exercises 4–7.

Now let's consider multiplying by a power of ten, such as 10, 100, 1000, and so on. Let's look at those products.

10 × 97.34 = 973.4

100 × 97.34 = 9734.

1000 × 97.34 = 97,340

10,000 × 97.34 = 973,400

Note in each case that the product is *larger* than 97.34.

Multiply.

4. 0.1 × 3.48

5. 0.01 × 3.48

6. 0.001 × 3.48

7. 0.0001 × 3.48

Answers on page A-6

Multiply.

8. 10×3.48

9. 100×3.48

10. 1000×3.48

11. $10{,}000 \times 3.48$

Answers on page A-6

To multiply any number by a power of ten, such as 10, 100, 1000, and so on,

a) count the number of zeros, and

b) move the decimal point that many places to the right.

1000×34.45678 → 3 zeros

$1000 \times 34.45678 = 34.456.78$

Move 3 places to the right.

$1000 \times 34.45678 = 34{,}456.78$

Examples Multiply.

8. $10 \times 14.605 = 146.05$ 14.6.05

9. $100 \times 14.605 = 1460.5$

10. $1000 \times 14.605 = 14{,}605$

11. $10{,}000 \times 14.605 = 146{,}050$ 14.6050.

Do Exercises 8–11.

b Applications Using Multiplication with Decimal Notation

Naming Large Numbers

We often see notation like the following in newspapers and magazines and on television.

O'Hare International Airport handles 67.3 million passengers per year.

Americans eat $1.1 billion worth of lettuce each year.

The population of the world is 6.6 billion.

To understand such notation, it helps to consider the following table.

1 hundred = 100 = 10^2 → 2 zeros

1 thousand = 1000 = 10^3 → 3 zeros

1 million = 1,000,000 = 10^6 → 6 zeros

1 billion = 1,000,000,000 = 10^9 → 9 zeros

1 trillion = 1,000,000,000,000 = 10^{12} → 12 zeros

To convert to standard notation, we proceed as follows.

Example 12 Convert the number in this sentence to standard notation: O'Hare's International Airport handles 67.3 million passengers per year.

$$67.3 \text{ million} = 67.3 \times 1 \text{ million}$$
$$= 67.3 \times 1{,}000{,}000$$
$$= 67{,}300{,}000$$

Do Exercises 12 and 13.

Convert the number in the sentence to standard notation.

12. In a recent year, the total payroll of major league baseball was $938 million.

13. In a recent year, the U.S. trade deficit with Japan was $44.1 billion.

Money Conversion

Converting from dollars to cents is like multiplying by 100. To see why, consider $19.43.

$19.43 = 19.43 × $1	We think of $19.43 as 19.43 × 1 dollar, or 19.43 × $1.
= 19.43 × 100¢	Substituting 100¢ for $1: $1 = 100¢
= 1943¢	Multiplying

> To convert from dollars to cents, move the decimal point two places to the right and change from the $ sign in front to the ¢ sign at the end.

Examples Convert from dollars to cents.

13. $189.64 = 18,964¢

14. $0.75 = 75¢

Do Exercises 14 and 15.

Convert from dollars to cents.

14. $15.69

15. $0.17

Converting from cents to dollars is like multiplying by 0.01. To see why, consider 65¢.

65¢ = 65 × 1¢	We think of 65¢ as 65 × 1 cent, or 65 × 1¢.
= 65 × $0.01	Substituting $0.01 for 1¢: 1¢ = $0.01
= $0.65	Multiplying

> To convert from cents to dollars, move the decimal point two places to the left and change from the ¢ sign at the end to the $ sign in front.

Examples Convert from cents to dollars.

15. 395¢ = $3.95

16. 8503¢ = $85.03

Do Exercises 16 and 17.

Convert from cents to dollars.

16. 35¢

17. 577¢

Answers on page A-6

Improving Your Math Study Skills

Learning Resources and Time Management

Two other topics to consider in enhancing your math study skills are learning resources and time management.

Learning Resources

- **Textbook supplements.** Are you aware of all the supplements that exist for this textbook? Many details are given in the preface. Now that you are more familiar with the book, let's discuss them.

  1. The *Student's Solutions Manual* contains worked-out solutions to the odd-numbered exercises in the exercise sets. Consider obtaining a copy if you are having trouble. It should be your first choice if you can make an additional purchase.
  2. An extensive set of *videotapes* supplement this text. These may be available to you on your campus at a learning center or math lab. Check with your instructor.
  3. *Tutorial software* also accompanies the text. If not available in the campus learning center, you might order it by calling the number 1-800-322-1377.

- **The Internet.** Our on-line World Wide Web supplement provides additional practice resources. If you have internet access, you can reach this site through the address:

  http://www.mathmax.com

  It contains many helpful ideas as well as many links to other resources for learning mathematics.

- **Your college or university.** Your own college or university probably has resources to enhance your math learning.

  1. For example, is there a learning lab or tutoring center for drop-in tutoring?
  2. Are there special lab classes or formal tutoring sessions tailored for the specific course you are taking?
  3. Perhaps there is a bulletin board or network where you can locate the names of experienced private tutors.

- **Your instructor.** Although it may seem obvious, students neglect to consider the most underused resource available to them: their instructor. Find out your instructor's office hours and make it a point to visit when you need additional help.

Time Management

- **Juggling time.** Have reasonable expectations about the time you need to study math. Unreasonable expectations may lead to lower grades and frustrations. Working 40 hours per week and taking 12 hours of credit is equivalent to working two full-time jobs. Can you handle such a load? As a rule of thumb, your ratio of work hours to credit load should be about 40/3, 30/6, 20/9, 10/12, and 5/14. Budget about 2–3 hours of homework and studying per hour of class.
- **Daily schedule.** Make an hour-by-hour schedule of your typical week. Include work, college, home, personal, sleep, study, and leisure times. Be realistic about the amount of time needed for sleep and home duties. If possible, try to schedule time for study when you are most alert.

Exercise Set 3.3

a Multiply.

1. 8.6×7

2. 5.7×0.8

3. 0.84×8

4. 9.4×0.6

5. 6.3×0.04

6. 9.8×0.08

7. 87×0.006

8. 18.4×0.07

9. 10×23.76

10. 100×3.8798

11. 1000×583.686852

12. 0.34×1000

13. 7.8×100

14. 0.00238×10

15. 0.1×89.23

16. 0.01×789.235

17. 0.001×97.68

18. 8976.23×0.001

19. 78.2×0.01

20. 0.0235×0.1

21. 32.6×16

22. 9.28×8.6

23. 0.984×3.3

24. 8.489×7.4

25. 374×2.4

26. 865×1.08

27. 749×0.43

28. 978×20.5

29. 0.87×64

30. 7.25×60

31. 46.50×75

32. 8.24×703

33. 81.7×0.612

34. 31.82×7.15

35. 10.105×11.324

36. 151.2×4.555

37. 12.3×1.08

38. 7.82×0.024

39. 32.4×2.8

40. 8.09×0.0075

41. $\begin{array}{r} 0.00342 \\ \times \quad 0.84 \\ \hline \end{array}$

42. $\begin{array}{r} 2.0056 \\ \times \quad 3.8 \\ \hline \end{array}$

43. $\begin{array}{r} 0.347 \\ \times \quad 2.09 \\ \hline \end{array}$

44. $\begin{array}{r} 2.532 \\ \times 1.067 \\ \hline \end{array}$

45. $\begin{array}{r} 3.005 \\ \times 0.623 \\ \hline \end{array}$

46. $\begin{array}{r} 16.34 \\ \times 0.000512 \\ \hline \end{array}$

47. 1000×45.678

48. 0.001×45.678

b Convert from dollars to cents.

49. \$28.88

50. \$67.43

51. \$0.66

52. \$1.78

Convert from cents to dollars.

53. 34¢

54. 95¢

55. 3445¢

56. 933¢

Convert the number in the sentence to standard notation.

57. In a recent year, the net sales of Morton International, Inc., were \$3.6 billion.

58. Annual production of sugarcane is 1.075 billion tons.

59. The total surface area of the earth is 196.8 million square miles.

60. Annual sales of *Sports Illustrated* magazine is 3.2 million copies per year.

Skill Maintenance

61. Multiply: $2\frac{1}{3} \cdot 4\frac{4}{5}$. [2.4d]

62. Divide: $2\frac{1}{3} \div 4\frac{4}{5}$. [2.4e]

Divide. [1.3e]

63. $24\overline{)8208}$

64. $4\overline{)348}$

65. $7\overline{)31{,}962}$

66. $18\overline{)22{,}626}$

Synthesis

67. ◆ If two rectangles have the same perimeter, will they also have the same area? Why?

68. ◆ A student insists that 346.708×0.1 is 3467.08. How could you convince the student that a mistake has been made?

Express as a power of 10.

69. (1 trillion) · (1 billion)

70. (1 million) · (1 billion)

3.4 Division with Decimal Notation

a Division

Whole-Number Divisors

Compare these divisions by a whole number.

$$\frac{588}{7} = 84$$

$$\frac{58.8}{7} = 8.4$$

$$\frac{5.88}{7} = 0.84$$

$$\frac{0.588}{7} = 0.084$$

When we are dividing by a whole number, the number of decimal places in the *quotient* is the same as the number of decimal places in the *dividend.*

These examples lead us to the following method for dividing by a whole number.

> To divide by a whole number,
>
> a) place the decimal point directly above the decimal point in the dividend, and
>
> b) divide as though dividing whole numbers.

```
      0.8 4  <— Quotient
  7 ) 5.8 8
      5 6 0
        2 8
        2 8
          0  <— Remainder
```

Example 1 Divide: 379.2 ÷ 8.

```
         4 7.4     <— Place the decimal point.
    8 ) 3 7 9.2
        3 2 0 0
          5 9 2
          5 6 0    Divide as though dividing whole numbers.
            3 2
            3 2
              0
```

Example 2 Divide: 82.08 ÷ 24.

```
            3.4 2     <— Place the decimal point.
    2 4 ) 8 2.0 8
          7 2 0 0
          1 0 0 8
            9 6 0     Divide as though dividing whole numbers.
              4 8
              4 8
                0
```

Do Exercises 1–3.

Objectives

a Divide using decimal notation.

b Solve equations of the type $a \cdot x = b$, where a and b may be in decimal notation.

c Simplify expressions using the rules for order of operations.

For Extra Help

TAPE 6

MAC
WIN

CD-ROM

Divide.

1. $9\overline{)5.4}$

2. $15\overline{)22.5}$

3. $82\overline{)38.54}$

Answers on page A-6

Divide.

4. $25\overline{)8}$

5. $4\overline{)15}$

6. $86\overline{)21.5}$

Answers on page A-6

Sometimes it helps to write some extra zeros to the right of the decimal point. They don't change the number.

Example 3 Divide: 30 ÷ 8.

$$\begin{array}{r} 3. \\ 8\,\overline{)\,30.} \\ \underline{24} \\ 6 \end{array}$$

Place the decimal point and divide to find how many ones.

$$\begin{array}{r} 3. \\ 8\,\overline{)\,30.0} \\ \underline{24} \\ 6\,0 \end{array}$$

Write an extra zero.

$$\begin{array}{r} 3.7 \\ 8\,\overline{)\,30.0} \\ \underline{24} \\ 6\,0 \\ \underline{5\,6} \\ 4 \end{array}$$

Divide to find how many tenths.

$$\begin{array}{r} 3.7 \\ 8\,\overline{)\,30.00} \\ \underline{24} \\ 6\,0 \\ \underline{5\,6} \\ 4\,0 \end{array}$$

Write an extra zero.

$$\begin{array}{r} 3.75 \\ 8\,\overline{)\,30.00} \\ \underline{24} \\ 6\,0 \\ \underline{5\,6} \\ 4\,0 \\ \underline{4\,0} \\ 0 \end{array}$$

Divide to find how many hundredths.

Example 4 Divide: 4 ÷ 25.

$$\begin{array}{r} 0.16 \\ 25\,\overline{)\,4.00} \\ \underline{2\,5} \\ 1\,5\,0 \\ \underline{1\,5\,0} \\ 0 \end{array}$$

Do Exercises 4–6.

Divisors That Are Not Whole Numbers

Consider the division

$0.24\overline{)8.208}$

We write the division as $\frac{8.208}{0.24}$. Then we multiply by 1 to change to a whole-number divisor:

$$\frac{8.208}{0.24} = \frac{8.208}{0.24} \times \frac{100}{100} = \frac{820.8}{24}.$$

The divisor is now a whole number. The division

$0.24\overline{)8.208}$

is the same as

$24\overline{)820.8}$

To divide when the divisor is not a whole number,

a) move the decimal point (multiply by 10, 100, and so on) to make the divisor a whole number;

$0.24\overline{)8.208}$

Move 2 places to the right.

b) move the decimal point (multiply the same way) in the dividend the same number of places; and

$0.24\overline{)8.208}$

Move 2 places to the right.

c) place the decimal point directly above the new decimal point in the dividend and divide as though dividing whole numbers.

$$\begin{array}{r} 34.2 \\ 0.24\overline{)8.20_{\wedge}8} \\ 7200 \\ \hline 1008 \\ 960 \\ \hline 48 \\ 48 \\ \hline 0 \end{array}$$

(The new decimal point in the dividend is indicated by a caret.)

Example 5 Divide: 5.848 ÷ 8.6.

$8.6\overline{)5.848}$

Multiply the divisor by 10 (move the decimal point 1 place). Multiply the same way in the dividend (move 1 place).

$$\begin{array}{r} 0.68 \\ 8.6\overline{)5.8_{\wedge}48} \\ 5160 \\ \hline 688 \\ 688 \\ \hline 0 \end{array}$$

Then divide.

Note: $\frac{5.848}{8.6} = \frac{5.848}{8.6} \cdot \frac{10}{10} = \frac{58.48}{86}.$

Do Exercises 7–9.

7. a) Complete.

$$\frac{3.75}{0.25} = \frac{3.75}{0.25} \times \frac{100}{100} = \frac{(\quad)}{25}$$

b) Divide.

$0.25\overline{)3.75}$

Divide.

8. $0.83\overline{)4.067}$

9. $3.5\overline{)44.8}$

Answers on page A-6

10. Divide.

$1.6\overline{)25}$

Answer on page A-6

Example 6 Divide: 12 ÷ 0.64.

$0.64\overline{)12.}$ Put a decimal point at the end of the whole number.

$0.64\overline{)12.00}$ Multiply the divisor by 100 (move the decimal point 2 places). Multiply the same way in the dividend (move 2 places).

$$\begin{array}{r} 18.75 \\ 0.64_\wedge\overline{)12.00_\wedge00} \\ \underline{6\,4\,0} \\ 5\,6\,0 \\ \underline{5\,1\,2} \\ 4\,8\,0 \\ \underline{4\,4\,8} \\ 3\,2\,0 \\ \underline{3\,2\,0} \\ 0 \end{array}$$

Then divide.

Do Exercise 10.

It is often helpful to be able to divide quickly by a ten, hundred, or thousand, or by a tenth, hundredth, or thousandth. The procedure we use is based on multiplying by 1. Consider the following examples:

$$\frac{23.789}{1000} = \frac{23.789}{1000} \cdot \frac{1000}{1000} = \frac{23{,}789}{1{,}000{,}000} = 0.023789.$$

We are dividing by a number greater than 1: The result is *smaller* than 23.789.

$$\frac{23.789}{0.01} = \frac{23.789}{0.01} \cdot \frac{100}{100} = \frac{2378.9}{1} = 2378.9.$$

We are dividing by a number less than 1: The result is *larger* than 23.789.

We use the following procedure.

> To divide by a power of ten, such as 10, 100, or 1000, and so on,
>
> a) count the number of zeros in the divisor, and $\frac{713.49}{100}$ → 2 zeros
>
> b) move the decimal point that number of places to the left.
>
> $\frac{713.49}{100}$, 7.13.49 (2 places to the left) $\frac{713.49}{100} = 7.1349$
>
> To divide by a tenth, hundredth, or thousandth,
>
> a) count the number of decimal places in the divisor, and $\frac{713.49}{0.001}$ → 3 places
>
> b) move the decimal point that number of places to the right.
>
> $\frac{713.49}{0.001}$, 713.490. (3 places to the right) $\frac{713.49}{0.001} = 713{,}490$

Example 7 Divide: $\dfrac{0.0104}{10}$.

$\dfrac{0.0104}{\underbrace{10}}$, 0.0.0104, $\dfrac{0.0104}{10} = 0.00104$

1 zero — 1 place to the left to change 10 to 1

Example 8 Divide: $\dfrac{23.738}{0.001}$.

$\dfrac{23.738}{\underbrace{0.001}}$, 23.738. $\dfrac{23.738}{0.001} = 23{,}738$

3 places — 3 places to the right to change 0.001 to 1

Do Exercises 11–14.

b Solving Equations

Now let's solve equations of the type $a \cdot x = b$, where a and b may be in decimal notation. Proceeding as before, we divide by a on both sides.

Example 9 Solve: $8 \cdot x = 27.2$.

We have

$$\frac{8 \cdot x}{8} = \frac{27.2}{8} \quad \text{Dividing by 8 on both sides}$$

$$x = 3.4.$$

```
      3.4
8 ) 2 7.2
    2 4 0
      3 2
      3 2
        0
```

The solution is 3.4.

Example 10 Solve: $2.9 \cdot t = 0.14616$.

We have

$$\frac{2.9 \cdot t}{2.9} = \frac{0.14616}{2.9} \quad \text{Dividing by 2.9 on both sides}$$

$$t = 0.0504.$$

```
          0.0 5 0 4
2.9 ) 0.1^4 6 1 6
        1 4 5 0 0
            1 1 6
            1 1 6
                0
```

The solution is 0.0504.

Do Exercises 15 and 16.

Divide.

11. $\dfrac{0.1278}{0.01}$

12. $\dfrac{0.1278}{100}$

13. $\dfrac{98.47}{1000}$

14. $\dfrac{6.7832}{0.1}$

Solve.

15. $100 \cdot x = 78.314$

16. $0.25 \cdot y = 276.4$

Answers on page A-6

c Order of Operations: Decimal Notation

The same rules for order of operations used with whole numbers and fractional notation apply when simplifying expressions with decimal notation.

RULES FOR ORDER OF OPERATIONS

1. Do all calculations within parentheses before operations outside.
2. Evaluate all exponential expressions.
3. Do all multiplications and divisions in order from left to right.
4. Do all additions and subtractions in order from left to right.

Example 11 Simplify: $(5 - 0.06) \div 2 + 3.42 \times 0.1$.

$(5 - 0.06) \div 2 + 3.42 \times 0.1 = 4.94 \div 2 + 3.42 \times 0.1$ Carrying out operations inside parentheses

$= 2.47 + 0.342$ Doing all multiplications and divisions in order from left to right

$= 2.812$

Example 12 Simplify: $10^2 \times \{[(3 - 0.24) \div 2.4] - (0.21 - 0.092)\}$.

$10^2 \times \{[(3 - 0.24) \div 2.4] - (0.21 - 0.092)\}$

$= 10^2 \times \{[2.76 \div 2.4] - 0.118\}$ Doing the calculations in the innermost parentheses first

$= 10^2 \times \{1.15 - 0.118\}$ Again, doing the calculations in the innermost parentheses

$= 10^2 \times 1.032$ Subtracting inside the parentheses

$= 100 \times 1.032$ Evaluating the exponential expression

$= 103.2$

Do Exercises 17 and 18.

Example 13 *Average Movie Revenue.* The bar graph shows movie box-office revenue (money taken in), in billions, in each of the four years from 1993 to 1996. Find the average revenue.

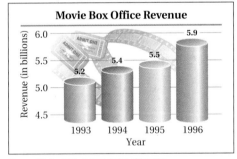

Movie Box Office Revenue

Revenue (in billions): 6.0, 5.5, 5.0, 4.5

1993: 5.2; 1994: 5.4; 1995: 5.5; 1996: 5.9

Year

Source: Motion Picture Association of America

To find the average of a set of numbers, we add them. Then we divide by the number of addends. In this case, we are finding the average of 5.2, 5.4, 5.5, and 5.9. The average is given by

$$(5.2 + 5.4 + 5.5 + 5.9) \div 4.$$

Thus,

$$(5.2 + 5.4 + 5.5 + 5.9) \div 4 = 22 \div 4 = 5.5.$$

The average box-office revenue was $5.5 billion.

Do Exercise 19.

Simplify.

17. $0.25 \cdot (1 + 0.08) - 0.0274$

18. $20^2 - 3.4^2 + \{2.5[20(9.2 - 5.6)] + 5(10 - 5)\}$

19. *Tickets Sold at the Movies.* The number of tickets sold at the movies in each of the four years from 1993 to 1996 is shown in the bar graph below. Find the average number of tickets sold.

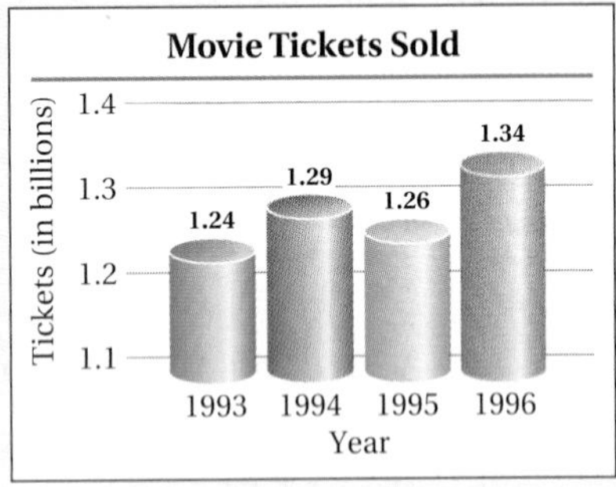

Source: Motion Picture Association of America

Answers on page A-6

Exercise Set 3.4

a Divide.

1. $2\overline{)5.98}$

2. $5\overline{)18}$

3. $4\overline{)95.12}$

4. $8\overline{)25.92}$

5. $12\overline{)89.76}$

6. $23\overline{)25.07}$

7. $33\overline{)237.6}$

8. $12.4 \div 4$

9. $9.144 \div 8$

10. $4.5 \div 9$

11. $12.123 \div 3$

12. $7\overline{)5.6}$

13. $5\overline{)0.35}$

14. $0.04\overline{)1.68}$

15. $0.12\overline{)8.4}$

16. $0.36\overline{)2.88}$

17. $3.4\overline{)68}$

18. $0.25\overline{)5}$

19. $15\overline{)6}$

20. $12\overline{)1.8}$

21. $36\overline{)14.76}$

22. $52\overline{)119.6}$

23. $3.2\overline{)27.2}$

24. $8.5\overline{)27.2}$

25. $4.2\overline{)39.06}$

26. $4.8\overline{)0.1104}$

27. $8\overline{)5}$

28. $8\overline{)3}$

29. $0.47 \overline{)0.1222}$

30. $1.08 \overline{)0.54}$

31. $4.8 \overline{)75}$

32. $0.28 \overline{)63}$

33. $0.032 \overline{)0.07488}$

34. $0.017 \overline{)1.581}$

35. $82 \overline{)38.54}$

36. $34 \overline{)0.1462}$

37. $\dfrac{213.4567}{1000}$

38. $\dfrac{213.4567}{100}$

39. $\dfrac{213.4567}{10}$

40. $\dfrac{100.7604}{0.1}$

41. $\dfrac{1.0237}{0.001}$

42. $\dfrac{1.0237}{0.01}$

b Solve.

43. $4.2 \cdot x = 39.06$

44. $36 \cdot y = 14.76$

45. $1000 \cdot y = 9.0678$

46. $789.23 = 0.25 \cdot q$

47. $1048.8 = 23 \cdot t$

48. $28.2 \cdot x = 423$

c Simplify.

49. $14 \times (82.6 + 67.9)$

50. $(26.2 - 14.8) \times 12$

51. $0.003 + 3.03 \div 0.01$

52. $9.94 + 4.26 \div (6.02 - 4.6) - 0.9$

53. $42 \times (10.6 + 0.024)$

54. $(18.6 - 4.9) \times 13$

55. $4.2 \times 5.7 + 0.7 \div 3.5$

56. $123.3 - 4.24 \times 1.01$

57. $9.0072 + 0.04 \div 0.1^2$

58. $12 \div 0.03 - 12 \times 0.03^2$

59. $(8 - 0.04)^2 \div 4 + 8.7 \times 0.4$

60. $(5 - 2.5)^2 \div 100 + 0.1 \times 6.5$

61. $86.7 + 4.22 \times (9.6 - 0.03)^2$

62. $2.48 \div (1 - 0.504) + 24.3 - 11 \times 2$

63. $4 \div 0.4 + 0.1 \times 5 - 0.1^2$

64. $6 \times 0.9 + 0.1 \div 4 - 0.2^3$

65. $5.5^2 \times [(6 - 4.2) \div 0.06 + 0.12]$

66. $12^2 \div (12 + 2.4) - [(2 - 1.6) \div 0.8]$

67. $200 \times \{[(4 - 0.25) \div 2.5] - (4.5 - 4.025)\}$

68. $0.03 \times \{1 \times 50.2 - [(8 - 7.5) \div 0.05]\}$

69. Find the average of \$1276.59, \$1350.49, \$1123.78, and \$1402.56.

70. Find the average weight of two wrestlers who weigh 308 lb and 296.4 lb.

Global Warming. The following table lists the global average temperature for the years 1984 through 1994. Use the table for Exercises 71 and 72.

YEAR	1986	1987	1988	1989	1990	1991	1992	1993	1994	1995	1996
Global temperature (in degrees Fahrenheit)	59.29°	59.58°	59.63°	59.45°	59.85°	59.74°	59.23°	59.36°	59.56°	59.72°	59.58°

Source: Lester R. Brown et al., *Vital Signs 1997*.

71. Find the average temperature for the years 1992 through 1996.

72. Find the average temperature for the years 1987 through 1991.

Skill Maintenance

73. Add: $10\frac{1}{2} + 4\frac{5}{8}$. [2.4b]

74. Subtract: $10\frac{1}{2} - 4\frac{5}{8}$. [2.4c]

75. Simplify: $\frac{36}{42}$. [2.1e]

76. Find the prime factorization of 162. [1.7d]

77. Find the prime factorization of 684. [1.7d]

78. Simplify: $\frac{56}{64}$. [2.1e]

Synthesis

79. ◆ A student insists that $0.247 \div 0.1$ is 0.0247. How could you convince this student that a mistake has been made?

80. ◆ A student insists that $0.247 \div 10$ is 2.47. How could you convince this student that a mistake has been made?

Simplify.

81. ▦ $9.0534 - 2.041^2 \times 0.731 \div 1.043^2$

82. ▦ $23.042(7 - 4.037 \times 1.46 - 0.932^2)$

In Exercises 83–86, find the missing value.

83. $439.57 \times 0.01 \div 1000 \times \square = 4.3957$

84. $5.2738 \div 0.01 \times 1000 \div \square = 52.738$

85. $0.0329 \div 0.001 \times 10^4 \div \square = 3290$

86. $0.0047 \times 0.01 \div 10^4 \times \square = 4.7$

3.5 Converting from Fractional Notation to Decimal Notation

a Fractional Notation to Decimal Notation

When a denominator has no prime factors other than 2's and 5's, we can find decimal notation by multiplying by 1. We multiply to get a denominator that is a power of ten, like 10, 100, or 1000.

Example 1 Find decimal notation for $\frac{3}{5}$.

$$\frac{3}{5} = \frac{3}{5} \cdot \frac{2}{2} = \frac{6}{10} = 0.6$$ We use $\frac{2}{2}$ for 1 to get a denominator of 10.

Example 2 Find decimal notation for $\frac{7}{20}$.

$$\frac{7}{20} = \frac{7}{20} \cdot \frac{5}{5} = \frac{35}{100} = 0.35$$ We use $\frac{5}{5}$ for 1 to get a denominator of 100.

Example 3 Find decimal notation for $\frac{9}{40}$.

$$\frac{9}{40} = \frac{9}{40} \cdot \frac{25}{25} = \frac{225}{1000} = 0.225$$ We use $\frac{25}{25}$ for 1 to get a denominator of 1000.

Example 4 Find decimal notation for $\frac{87}{25}$.

$$\frac{87}{25} = \frac{87}{25} \cdot \frac{4}{4} = \frac{348}{100} = 3.48$$ We use $\frac{4}{4}$ for 1 to get a denominator of 100.

Do Exercises 1–4.

We can also divide to find decimal notation.

Example 5 Find decimal notation for $\frac{3}{5}$.

$$\frac{3}{5} = 3 \div 5 \qquad \begin{array}{r} 0.6 \\ 5\overline{)\,3.0} \\ \underline{3\,0} \\ 0 \end{array} \qquad \frac{3}{5} = 0.6$$

Example 6 Find decimal notation for $\frac{7}{8}$.

$$\frac{7}{8} = 7 \div 8 \qquad \begin{array}{r} 0.875 \\ 8\overline{)\,7.000} \\ \underline{6\,4} \\ 6\,0 \\ \underline{5\,6} \\ 4\,0 \\ \underline{4\,0} \\ 0 \end{array} \qquad \frac{7}{8} = 0.875$$

Do Exercises 5 and 6.

Objectives

a Convert from fractional notation to decimal notation.

b Round numbers named by repeating decimals.

c Calculate using fractional and decimal notation together.

For Extra Help

TAPE 7 | MAC WIN | CD-ROM

Find decimal notation. Use multiplying by 1.

1. $\frac{4}{5}$

2. $\frac{9}{20}$

3. $\frac{11}{40}$

4. $\frac{33}{25}$

Find decimal notation.

5. $\frac{2}{5}$

6. $\frac{3}{8}$

Answers on page A-6

Find decimal notation.

7. $\frac{1}{6}$

8. $\frac{2}{3}$

Answers on page A-6

In Examples 5 and 6, the division *terminated,* meaning that eventually we got a remainder of 0. A **terminating decimal** occurs when the denominator has only 2's or 5's, or both, as factors, as in $\frac{17}{25}$, $\frac{5}{8}$, or $\frac{83}{100}$. This assumes that the fractional notation has been simplified.

Consider a different situation:

$$\frac{5}{6}, \quad \text{or} \quad \frac{5}{2 \cdot 3}.$$

Since 6 has a 3 as a factor, the division will not terminate. Although we can still use division to get decimal notation, the answer will be a **repeating decimal,** as follows.

Example 7 Find decimal notation for $\frac{5}{6}$.

We have

$$\frac{5}{6} = 5 \div 6 \qquad \begin{array}{r} 0.8\,3\,3 \\ 6\,\overline{)\,5.0\,0\,0} \\ \underline{4\,8} \\ 2\,0 \\ \underline{1\,8} \\ 2\,0 \\ \underline{1\,8} \\ 2 \end{array}$$

Since 2 keeps reappearing as a remainder, the digits repeat and will continue to do so; therefore,

$$\frac{5}{6} = 0.83333\ldots.$$

The dots indicate an endless sequence of digits in the quotient. When there is a repeating pattern, the dots are often replaced by a bar to indicate the repeating part—in this case, only the 3:

$$\frac{5}{6} = 0.8\overline{3}.$$

Do Exercises 7 and 8.

Example 8 Find decimal notation for $\frac{4}{11}$.

$$\frac{4}{11} = 4 \div 11 \qquad \begin{array}{r} 0.3\,6\,3\,6 \\ 11\,\overline{)\,4.0\,0\,0\,0} \\ \underline{3\,3} \\ 7\,0 \\ \underline{6\,6} \\ 4\,0 \\ \underline{3\,3} \\ 7\,0 \\ \underline{6\,6} \\ 4 \end{array}$$

Since 7 and 4 keep reappearing as remainders, the sequence of digits "36" repeats in the quotient, and

$$\frac{4}{11} = 0.363636\ldots, \quad \text{or} \quad 0.\overline{36}.$$

Do Exercises 9 and 10.

Example 9 Find decimal notation for $\frac{5}{7}$.

We have

$$\begin{array}{r} 0.714285 \\ 7\overline{)5.000000} \\ \underline{4\,9} \\ 10 \\ \underline{7} \\ 30 \\ \underline{28} \\ 20 \\ \underline{14} \\ 60 \\ \underline{56} \\ 40 \\ \underline{35} \\ 5 \end{array}$$

Since 5 appears as a remainder, the sequence of digits "714285" repeats in the quotient, and

$$\frac{5}{7} = 0.714285714285\ldots, \quad \text{or} \quad 0.\overline{714285}.$$

The length of a repeating part can be very long—too long to find on a calculator. An example is $\frac{5}{97}$, which has a repeating part of 96 digits.

Do Exercise 11.

b Rounding in Problem Solving

In applied problems, repeating decimals are rounded to get approximate answers.

Examples Round each to the nearest tenth, hundredth, and thousandth.

	Nearest tenth	*Nearest hundredth*	*Nearest thousandth*
10. $0.8\overline{3} = 0.83333\ldots$	0.8	0.83	0.833
11. $0.\overline{09} = 0.090909\ldots$	0.1	0.09	0.091
12. $0.\overline{714285} = 0.714285714285\ldots$	0.7	0.71	0.714

Do Exercises 12–14.

Find decimal notation.

9. $\frac{5}{11}$

10. $\frac{12}{11}$

11. Find decimal notation for $\frac{3}{7}$.

Round each to the nearest tenth, hundredth, and thousandth.

12. $0.\overline{6}$

13. $0.\overline{808}$

14. $6.\overline{245}$

Answers on page A-6

15. Calculate: $\frac{5}{6} \times 0.864$.

Calculate.

16. $\frac{1}{3} \times 0.384 + \frac{5}{8} \times 0.6784$

17. $\frac{5}{6} \times 0.864 + 14.3 \div \frac{8}{5}$

Answers on page A-6

c Calculations with Fractional and Decimal Notation Together

In certain kinds of calculations, fractional and decimal notation might occur together. In such cases, there are at least three ways in which we might proceed.

Example 13 Calculate: $\frac{2}{3} \times 0.576$.

METHOD 1. One way to do this calculation is to convert the decimal notation to fractional notation so that both numbers are in fractional notation. The answer can be left in fractional notation and simplified, or we can convert back to decimal notation and round, if appropriate.

$$\begin{aligned}\frac{2}{3} \times 0.576 &= \frac{2}{3} \cdot \frac{576}{1000} = \frac{2 \cdot 576}{3 \cdot 1000} \\ &= \frac{2 \cdot 2 \cdot 2 \cdot 2 \cdot 2 \cdot 2 \cdot 2 \cdot 3 \cdot 3}{2 \cdot 2 \cdot 2 \cdot 3 \cdot 5 \cdot 5 \cdot 5} \\ &= \frac{2 \cdot 2 \cdot 2 \cdot 3}{2 \cdot 2 \cdot 2 \cdot 3} \cdot \frac{2 \cdot 2 \cdot 2 \cdot 2 \cdot 3}{5 \cdot 5 \cdot 5} \\ &= 1 \cdot \frac{2 \cdot 2 \cdot 2 \cdot 2 \cdot 3}{5 \cdot 5 \cdot 5} \\ &= \frac{2 \cdot 2 \cdot 2 \cdot 2 \cdot 3}{5 \cdot 5 \cdot 5} = \frac{48}{125}, \text{ or } 0.384\end{aligned}$$

METHOD 2. A second way to do this calculation is to convert the fractional notation to decimal notation so that both numbers are in decimal notation. Since $\frac{2}{3}$ converts to repeating decimal notation, it is first rounded to some chosen decimal place. We choose three decimal places. Then, using decimal notation, we multiply. Note that the answer is not as accurate as that found by method 1, due to the rounding.

$$\frac{2}{3} \times 0.576 = 0.\overline{6} \times 0.576 \approx 0.667 \times 0.576 = 0.384192$$

METHOD 3. A third way to do this calculation is to treat 0.576 as $\frac{0.576}{1}$. Then we multiply 0.576 by 2, and divide the result by 3.

$$\frac{2}{3} \times 0.576 = \frac{2}{3} \times \frac{0.576}{1} = \frac{2 \times 0.576}{3} = \frac{1.152}{3} = 0.384$$

Do Exercise 15.

Example 14 Calculate: $\frac{2}{3} \times 0.576 + 3.287 \div \frac{4}{5}$.

We use the rules for order of operations, doing first the multiplication and then the division. Then we add.

$$\begin{aligned}\frac{2}{3} \times 0.576 + 3.287 \div \frac{4}{5} &= 0.384 + 3.287 \cdot \frac{5}{4} \\ &= 0.384 + 4.10875 \\ &= 4.49275\end{aligned}$$

Do Exercises 16 and 17.

Exercise Set 3.5

a Find decimal notation.

1. $\frac{3}{5}$
2. $\frac{19}{20}$
3. $\frac{13}{40}$
4. $\frac{3}{16}$
5. $\frac{1}{5}$
6. $\frac{3}{20}$
7. $\frac{17}{20}$
8. $\frac{9}{40}$
9. $\frac{19}{40}$
10. $\frac{81}{40}$
11. $\frac{39}{40}$
12. $\frac{31}{40}$
13. $\frac{13}{25}$
14. $\frac{61}{125}$
15. $\frac{2502}{125}$
16. $\frac{181}{200}$
17. $\frac{1}{4}$
18. $\frac{1}{2}$
19. $\frac{23}{40}$
20. $\frac{11}{20}$
21. $\frac{18}{25}$
22. $\frac{37}{25}$
23. $\frac{19}{16}$
24. $\frac{5}{8}$
25. $\frac{4}{15}$
26. $\frac{7}{9}$
27. $\frac{1}{3}$
28. $\frac{1}{9}$
29. $\frac{4}{3}$
30. $\frac{8}{9}$
31. $\frac{7}{6}$
32. $\frac{7}{11}$
33. $\frac{4}{7}$
34. $\frac{14}{11}$
35. $\frac{11}{12}$
36. $\frac{5}{12}$

b

37.–47. Round each answer of the odd-numbered Exercises 25–35 to the nearest tenth, hundredth, and thousandth.

38.–48. Round each answer of the even-numbered Exercises 26–36 to the nearest tenth, hundredth, and thousandth.

Round each to the nearest tenth, hundredth, and thousandth.

49. $0.\overline{18}$

50. $0.\overline{83}$

51. $0.2\overline{7}$

52. $3.5\overline{4}$

c Calculate.

53. $\frac{7}{8} \times 12.64$

54. $\frac{4}{5} \times 384.8$

55. $2\frac{3}{4} + 5.65$

56. $4\frac{4}{5} + 3.25$

57. $\frac{47}{9} \times 79.95$

58. $\frac{7}{11} \times 2.7873$

59. $\frac{1}{2} - 0.5$

60. $3\frac{1}{8} - 2.75$

61. $4.875 - 2\frac{1}{16}$

62. $55\frac{3}{5} - 12.22$

63. $\frac{5}{6} \times 0.0765 + \frac{5}{4} \times 0.1124$

64. $\frac{3}{5} \times 6384.1 - \frac{3}{8} \times 156.56$

65. $\frac{4}{5} \times 384.8 + 24.8 \div \frac{8}{3}$

66. $102.4 \div \frac{2}{5} - 12 \times \frac{5}{6}$

67. $\frac{7}{8} \times 0.86 - 0.76 \times \frac{3}{4}$

68. $17.95 \div \frac{5}{8} + \frac{3}{4} \times 16.2$

69. $3.375 \times 5\frac{1}{3}$

70. $2.5 \times 3\frac{5}{8}$

71. $6.84 \div 2\frac{1}{2}$

72. $8\frac{1}{2} \div 2.125$

Skill Maintenance

73. Multiply: $9 \cdot 2\frac{1}{3}$. [2.4d]

74. Divide: $84 \div 8\frac{2}{5}$. [2.4e]

75. Subtract: $20 - 16\frac{3}{5}$. [2.4c]

76. Add: $14\frac{3}{5} + 16\frac{1}{10}$. [2.4b]

Solve. [2.5a]

77. A recipe for bread calls for $\frac{2}{3}$ cup of water, $\frac{1}{4}$ cup of milk, and $\frac{1}{8}$ cup of oil. How many cups of liquid ingredients does the recipe call for?

78. A board $\frac{9}{10}$ in. thick is glued to a board $\frac{8}{10}$ in. thick. The glue is $\frac{3}{100}$ in. thick. How thick is the result?

Synthesis

79. ◈ When is long division *not* the fastest way of converting a fraction to decimal notation?

80. ◈ Examine Example 13 of this section. How could the problem be changed so that method 2 would give a result that is completely accurate?

Find decimal notation.

81. $\frac{1}{7}$

82. $\frac{2}{7}$

83. $\frac{3}{7}$

84. $\frac{4}{7}$

85. $\frac{5}{7}$

86. From the pattern of Exercises 81–85, guess the decimal notation for $\frac{6}{7}$. Check on your calculator.

Find decimal notation.

87. $\frac{1}{9}$

88. $\frac{1}{99}$

89. $\frac{1}{999}$

90. From the pattern of Exercises 87–89, guess the decimal notation for $\frac{1}{9999}$. Check on your calculator.

3.6 Estimating

a Estimating Sums, Differences, Products and Quotients

Estimating has many uses. It can be done before a problem is even attempted in order to get an idea of the answer. It can be done afterward as a check, even when we are using a calculator. In many situations, an estimate is all we need. We usually estimate by rounding the numbers so that there are one or two nonzero digits. Consider the following advertisements for Examples 1–4.

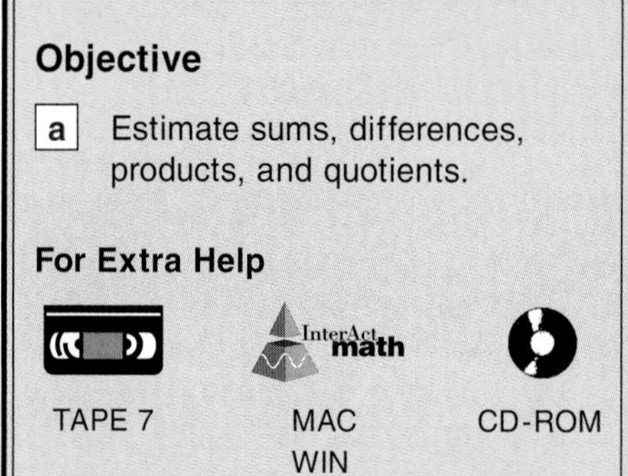
Objective

a Estimate sums, differences, products, and quotients.

For Extra Help

TAPE 7 MAC WIN CD-ROM

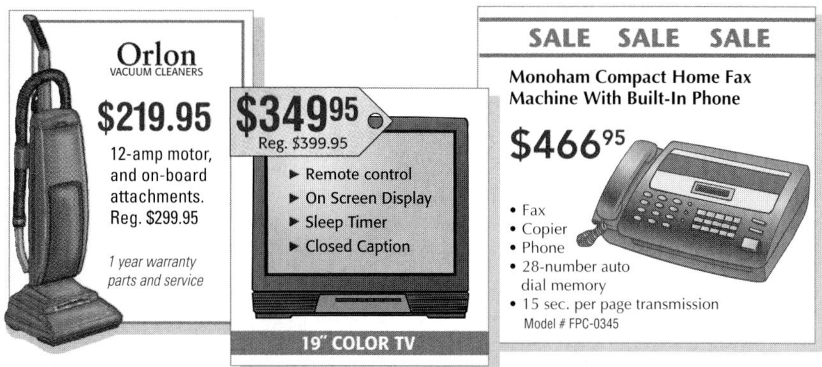

Example 1 Estimate to the nearest ten the total cost of one fax machine and one TV.

We are estimating the sum

$466.95 + $349.95 = Total cost.

The estimate to the nearest ten is

$470 + $350 = $820. (Estimated total cost)

We rounded $466.95 to the nearest ten and $349.95 to the nearest ten. The estimated sum is $820.

Do Exercise 1.

Example 2 About how much more does the fax machine cost than the TV? Estimate to the nearest ten.

We are estimating the difference

$466.95 − $349.95 = Price difference.

The estimate to the nearest ten is

$470 − $350 = $120. (Estimated price difference)

Do Exercise 2.

1. Estimate to the nearest ten the total cost of one TV and one vacuum cleaner. Which of the following is an appropriate estimate?

a) $5700 b) $570
c) $790 d) $57

2. About how much more does the TV cost than the vacuum cleaner? Estimate to the nearest ten. Which of the following is an appropriate estimate?

a) $130 b) $1300
c) $580 d) $13

Answers on page A-7

Example 3 Estimate the total cost of 4 vacuum cleaners.

We are estimating the product

$$4 \times \$219.95 = \text{Total cost.}$$

The estimate is found by rounding $219.95 to the nearest ten:

$$4 \times \$220 = \$880.$$

Do Exercise 3.

Example 4 About how many fax machines can be bought for $1580?

We estimate the quotient

$$\$1580 \div \$466.95.$$

Since we want a whole-number estimate, we choose our rounding appropriately. Rounding $466.95 to the nearest hundred, we get $500. Since $1580 is close to $1500, which is a multiple of 500, we estimate

$$\$1500 \div \$500,$$

so the answer is 3.

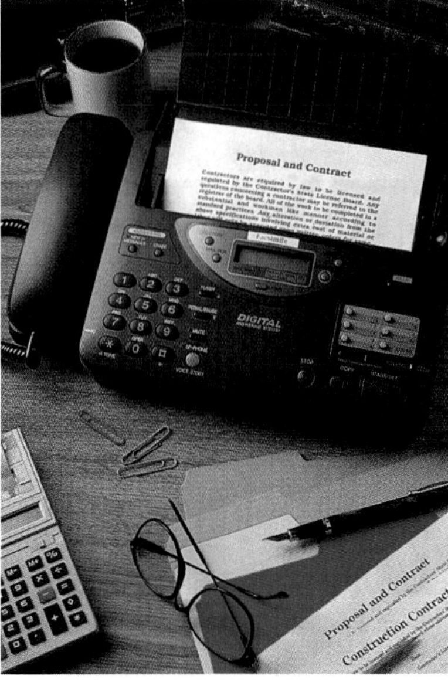

Do Exercise 4.

Example 5 Estimate: 4.8×52. Do not find the actual product. Which of the following is an appropriate estimate?

a) 25 b) 250 c) 2500 d) 360

We have

$$5 \times 50 = 250. \quad \text{(Estimated product)}$$

We rounded 4.8 to the nearest one and 52 to the nearest ten. Thus an appropriate estimate is (b).

Compare these estimates for the product 4.94×38:

$$5 \times 40 = 200, \quad 5 \times 38 = 190, \quad 4.9 \times 40 = 196.$$

The first estimate was the easiest. You could probably do it mentally. The others had more nonzero digits.

Do Exercises 5–10.

3. Estimate the total cost of 6 fax machines. Which of the following is an appropriate estimate?

a) $4400 b) $300
c) $30,000 d) $3000

4. About how many vacuum cleaners can be bought for $1100? Which of the following is an appropriate estimate?

a) 8 b) 5
c) 11 d) 124

Estimate the product. Do not find the actual product. Which of the following is an appropriate estimate?

5. 2.4×8

a) 16 b) 34
c) 125 d) 5

6. 24×0.6

a) 200 b) 5
c) 110 d) 20

7. 0.86×0.432

a) 0.04 b) 0.4
c) 1.1 d) 4

8. 0.82×0.1

a) 800 b) 8
c) 0.08 d) 80

9. 0.12×18.248

a) 180 b) 1.8
c) 0.018 d) 18

10. 24.234×5.2

a) 200 b) 125
c) 12.5 d) 234

Answers on page A-7

Example 6 Estimate: $82.08 \div 24$. Which of the following is an appropriate estimate?

a) 400 b) 16 c) 40 d) 4

This is about $80 \div 20$, so the answer is about 4. Thus an appropriate estimate is (d).

Example 7 Estimate: $94.18 \div 3.2$. Which of the following is an appropriate estimate?

a) 30 b) 300 c) 3 d) 60

This is about $90 \div 3$, so the answer is about 30. Thus an appropriate estimate is (a).

Example 8 Estimate: $0.0156 \div 1.3$. Which of the following is an appropriate estimate?

a) 0.2 b) 0.002 c) 0.02 d) 20

This is about $0.02 \div 1$, so the answer is about 0.02. Thus an appropriate estimate is (c).

Do Exercises 11–13.

In some cases, it is easier to estimate a quotient directly rather than by rounding the divisor and the dividend.

Example 9 Estimate: $0.0074 \div 0.23$. Which of the following is an appropriate estimate?

a) 0.3 b) 0.03 c) 300 d) 3

We estimate 3 for a quotient. We check by multiplying.

$$0.23 \times 3 = 0.69$$

We make the estimate smaller. We estimate 0.3 and check by multiplying.

$$0.23 \times 0.3 = 0.069$$

We make the estimate smaller. We estimate 0.03 and check by multiplying.

$$0.23 \times 0.03 = 0.0069$$

This is about 0.0074, so the quotient is about 0.03. Thus an appropriate estimate is (b).

Do Exercise 14.

Estimate the quotient. Which of the following is an appropriate estimate?

11. $59.78 \div 29.1$

a) 200 b) 20
c) 2 d) 0.2

12. $82.08 \div 2.4$

a) 40 b) 4.0
c) 400 d) 0.4

13. $0.1768 \div 0.08$

a) 8 b) 10
c) 2 d) 20

14. Estimate: $0.0069 \div 0.15$. Which of the following is an appropriate estimate?

a) 0.5 b) 50
c) 0.05 d) 23.4

Answers on page A-7

Calculator Spotlight

Calculators can perform calculations so quickly that repeated experimental trials are not particularly time-consuming.

1. Use one of +, −, ×, and ÷ in each blank to make a true sentence.

a) (0.37 ▢ 18.78) ▢ 2^{13} = 156,876.8

b) 2.56 ▢ 6.4 ▢ 51.2 ▢ 17.4 = 312.84

2. In the subtraction below, *a* and *b* are digits. Find *a* and *b*.

$$\begin{array}{r} b876.a4321 \\ -1234.a678b \\ \hline 8641.b7a32 \end{array}$$

3. Look for a pattern in the following list, and find the missing numbers.

$22.22, $33.34, $44.46, $55.58, ____, ____, ____, ____, ____, ____, ____.

4. Look for a pattern in the following list, and find the missing numbers.

$2344.78, $2266, $2187.22, $2108.44, ____, ____, ____, ____.

Each of the following is called a *magic square.* The sum along each row, column, or diagonal is the same. Find the missing numbers.

5.

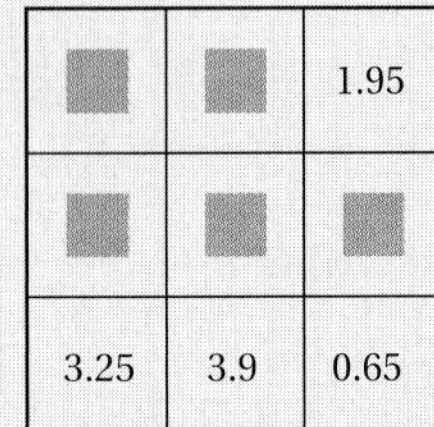

▢	▢	1.95
▢	▢	▢
3.25	3.9	0.65

Magic sum = ▢

6.

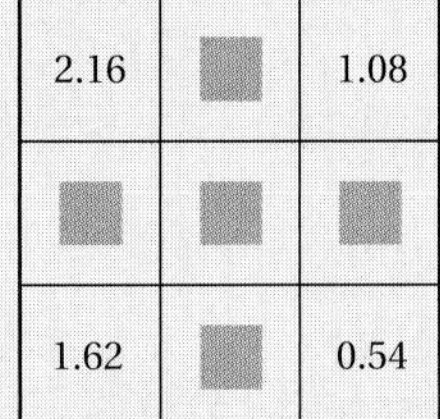

2.16	▢	1.08
▢	▢	▢
1.62	▢	0.54

Magic sum = 4.05

7.

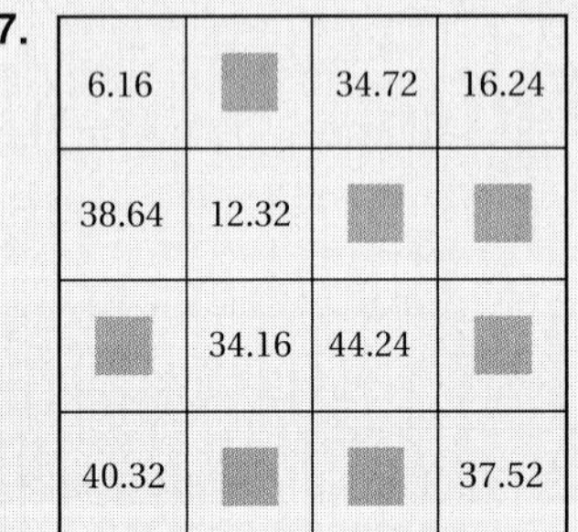

6.16	▢	34.72	16.24
38.64	12.32	▢	▢
▢	34.16	44.24	▢
40.32	▢	▢	37.52

Magic sum = 100.24

Exercise Set 3.6

a Consider the following advertisements for Exercises 1–8. Estimate the sums, differences, products, or quotients involved in these problems. Indicate which of the choices is an appropriate estimate.

1. Estimate the total cost of one entertainment center and one sound system.
a) $36 b) $72 c) $3.60 d) $360

2. Estimate the total cost of one entertainment center and one TV.
a) $410 b) $820 c) $41 d) $4.10

3. About how much more does the TV cost than the sound system?
a) $500 b) $80 c) $50 d) $5

4. About how much more does the TV cost than the entertainment center?
a) $100 b) $190 c) $250 d) $150

5. Estimate the total cost of 9 TVs.
a) $2700 b) $27 c) $270 d) $540

6. Estimate the total cost of 16 sound systems.
a) $5010 b) $4000 c) $40 d) $410

7. About how many TVs can be bought for $1700?
a) 600 b) 72 c) 6 d) 60

8. About how many sound systems can be bought for $1300?
a) 10 b) 5 c) 50 d) 500

Estimate by rounding as directed.

9. 0.02 + 1.31 + 0.34; nearest tenth

10. 0.88 + 2.07 + 1.54; nearest one

11. 6.03 + 0.007 + 0.214; nearest one

12. 1.11 + 8.888 + 99.94; nearest one

13. 52.367 + 1.307 + 7.324 nearest one

14. 12.9882 + 1.0115; nearest tenth

15. 2.678 − 0.445; nearest tenth

16. 12.9882 − 1.0115; nearest one

17. 198.67432 − 24.5007; nearest ten

Estimate. Choose a rounding digit that gives one or two nonzero digits. Indicate which of the choices is an appropriate estimate.

18. 234.12321 − 200.3223

a) 600 b) 60
c) 300 d) 30

19. 49×7.89

a) 400 b) 40
c) 4 d) 0.4

20. 7.4×8.9

a) 95 b) 63
c) 124 d) 6

21. 98.4×0.083

a) 80 b) 12
c) 8 d) 0.8

22. 78×5.3

a) 400 b) 800
c) 40 d) 8

23. $3.6 \div 4$

a) 10 b) 1
c) 0.1 d) 0.01

24. $0.0713 \div 1.94$

a) 4 b) 0.4
c) 0.04 d) 40

25. $74.68 \div 24.7$

a) 9 b) 3
c) 12 d) 120

26. $914 \div 0.921$

a) 9 b) 90
c) 900 d) 0.9

27. *Movie Revenue.* Total summer box-office revenue (money taken in) for the movie *Eraser* was $53.6 million (***Source***: *Hollywood Reporter Magazine*). Each theater showing the movie averaged $6716 in revenue. Estimate how many screens were showing this movie.

28. *Nintendo and the Sears Tower.* The Nintendo Game Boy portable video game is 4.5 in. (0.375 ft) tall (***Source***: Nintendo of America). Estimate how many game units it would take to reach the top of the Sears Tower, which is 1454 ft tall. Round to the nearest one.

Skill Maintenance

Find the prime factorization. [1.7d]

29. 108 **30.** 400 **31.** 325 **32.** 666

Simplify. [2.1e]

33. $\frac{125}{400}$ **34.** $\frac{3225}{6275}$ **35.** $\frac{72}{81}$ **36.** $\frac{325}{625}$

Synthesis

37. ◆ A roll of fiberglass insulation costs $21.95. Describe two situations involving estimating and the cost of fiberglass insulation. Devise one situation so that $21.95 is rounded to $22. Devise the other situation so that $21.95 is rounded to $20.

38. ◆ Describe a situation in which an estimation is made by rounding to the nearest 10,000 and then multiplying.

The following were done on a calculator. Estimate to see if the decimal point was placed correctly.

39. $178.9462 \times 61.78 = 11{,}055.29624$

40. $14{,}973.35 \div 298.75 = 501.2$

41. $19.7236 - 1.4738 \times 4.1097 = 1.366672414$

42. $28.46901 \div 4.9187 - 2.5081 = 3.279813473$

Estimate the cost of food for a catered party.

3.7 Applications and Problem Solving

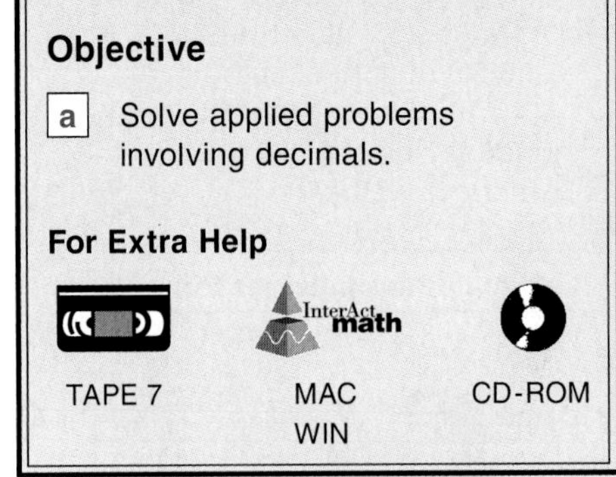

a Solving applied problems with decimals is like solving applied problems with whole numbers. We translate first to an equation that corresponds to the situation. Then we solve the equation.

Example 1 *Eating Out.* More and more Americans are eating meals outside the home. The following graph compares the average check for meals of various types for the years 1995 and 1996. How much more is the average check for casual dining in 1996 than in 1995?

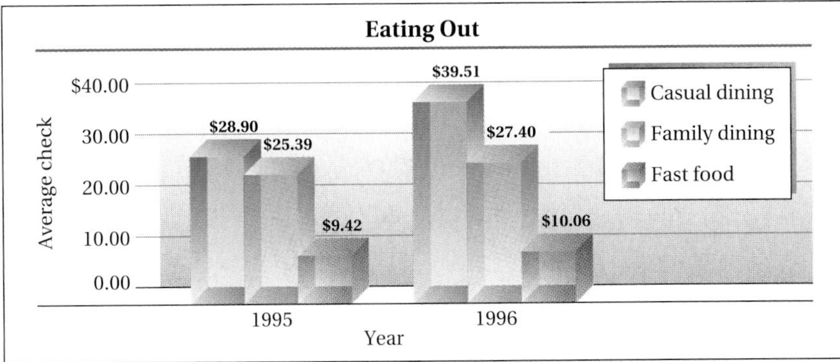

Source: Sandelman and Associates, Brea, California

1. **Familiarize.** We use the bar graph to visualize the situation and to obtain the appropriate data. We let $c =$ the additional amount spent in 1996.
2. **Translate.** This is a "how-much-more" situation. We translate as follows, using the data from the bar graph.

Average check in 1995	plus	Additional amount	is	Average check in 1996
↓	↓	↓	↓	↓
\$28.90	$+$	c	$=$	\$39.51

3. **Solve.** We solve the equation, first subtracting 28.90 from both sides:

$$28.90 + c - 28.90 = 39.51 - 28.90$$
$$c = 10.61.$$

$$\begin{array}{r} \overset{8\ 15}{3\ \cancel{9}.\cancel{5}\ 1} \\ -\ 2\ 8.9\ 0 \\ \hline 1\ 0.6\ 1 \end{array}$$

4. **Check.** We can check by adding 10.61 to 28.90 to get 39.51.
5. **State.** The average check for casual dining in 1996 was \$10.61 more than in 1995.

Do Exercise 1.

1. *Body Temperature.* Normal body temperature is 98.6°F. When fevered, most people will die if their bodies reach 107°F. This is a rise of how many degrees?

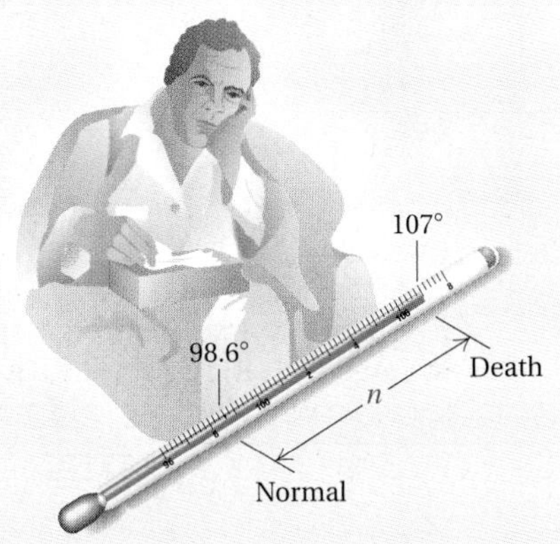

Answer on page A-7

2. Each year, the average American drinks about 49.0 gal of soft drinks, 41.2 gal of water, 25.3 gal of milk, 24.8 gal of coffee, and 7.8 gal of fruit juice. What is the total amount that the average American drinks?

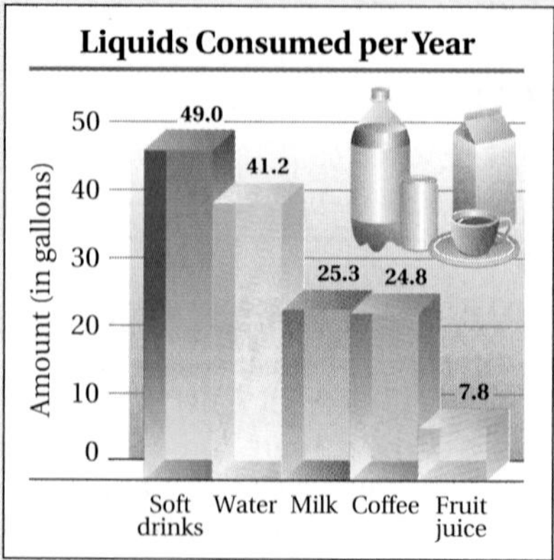

Source: U.S. Department of Agriculture

Example 2 *Injections of Medication.* A patient was given injections of 2.8 mL, 1.35 mL, 2.0 mL, and 1.88 mL over a 24-hr period. What was the total amount of the injections?

1. **Familiarize.** We make a drawing or at least visualize the situation. We let $t =$ the amount of the injections.

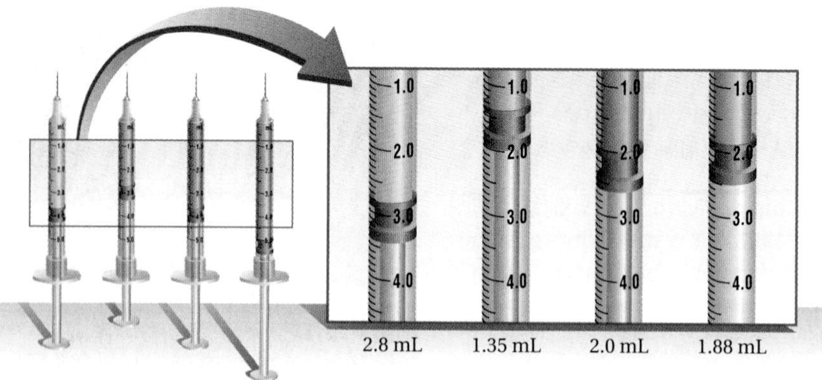

2. **Translate.** Amounts are being combined. We translate to an equation:

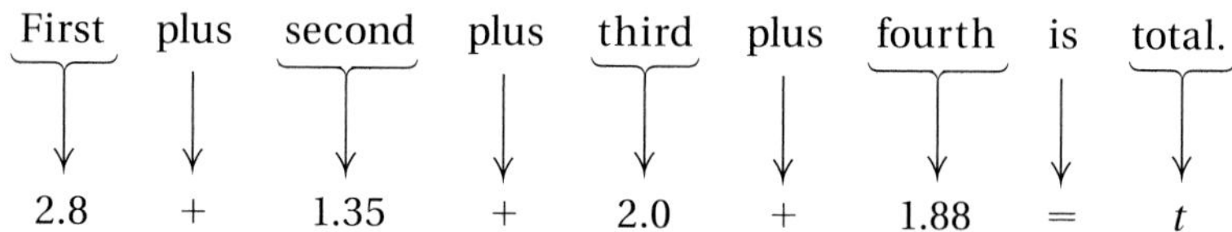

$$2.8 + 1.35 + 2.0 + 1.88 = t$$

3. **Solve.** To solve, we carry out the addition.

$$\begin{array}{r} {}^{2\ 1} \\ 2.8\ 0 \\ 1.3\ 5 \\ 2.0\ 0 \\ +\ 1.8\ 8 \\ \hline 8.0\ 3 \end{array}$$

Thus, $t = 8.03$.

4. **Check.** We can check by repeating our addition. We can also see whether our answer is reasonable by first noting that it is indeed larger than any of the numbers being added. We can also check by rounding:

$$\begin{aligned} 2.8 + 1.35 + 2.0 + 1.88 &\approx 3 + 1 + 2 + 2 \\ &= 8 \approx 8.03. \end{aligned}$$

If we had gotten an answer like 80.3 or 0.803, then our estimate, 8, would have told us that we did something wrong, like not lining up the decimal points.

5. **State.** The total amount of the injections was 8.03 mL.

Do Exercise 2.

Answer on page A-7

Example 3 *IRS Driving Allowance.* In a recent year, the Internal Revenue Service allowed a tax deduction of 31¢ per mile for mileage driven for business purposes. What deduction, in dollars, would be allowed for driving 127 mi?

1. **Familiarize.** We first make a drawing or at least visualize the situation. Repeated addition fits this situation. We let d = the deduction, in dollars, allowed for driving 127 mi.

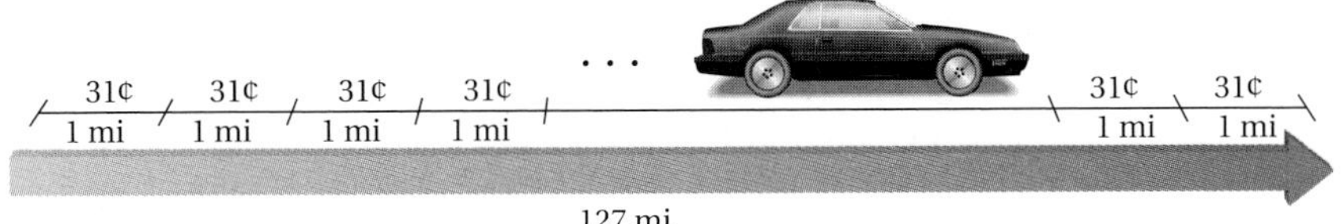

2. **Translate.** We translate as follows.

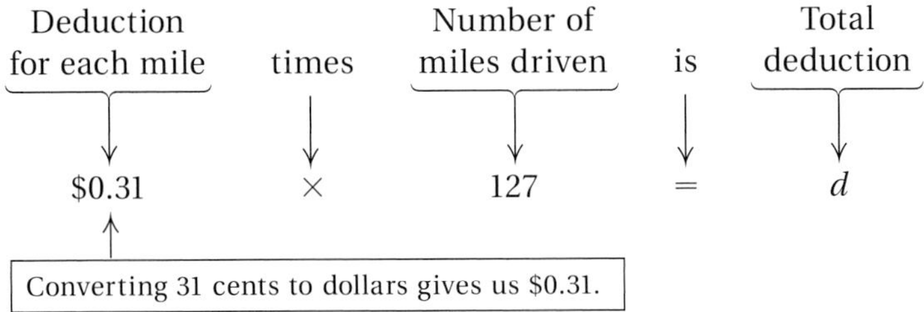

3. **Solve.** To solve the equation, we carry out the multiplication.

$$\begin{array}{r} 127 \\ \times\ 0.31 \\ \hline 127 \\ 3810 \\ \hline 39.37 \end{array}$$

Thus, $d = 39.37$.

4. **Check.** We can obtain a partial check by rounding and estimating:

$$127 \times 0.31 \approx 130 \times 0.3$$
$$= 39 \approx 39.37.$$

5. **State.** The total allowable deduction would be $39.37.

Do Exercise 3.

Example 4 *Loan Payments.* A car loan of $7382.52 is to be paid off in 36 monthly payments. How much is each payment?

1. **Familiarize.** We first make a drawing. We let n = the amount of each payment.

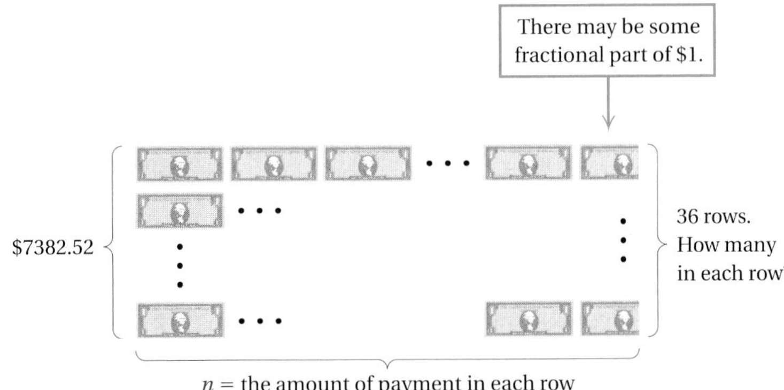

3. *Printing Costs.* At a printing company, the cost of copying is 12 cents per page. How much, in dollars, would it cost to make 466 copies?

Answer on page A-7

4. *Loan Payments.* A loan of \$4425 is to be paid off in 12 monthly payments. How much is each payment?

2. **Translate.** The problem can be translated to the following equation, thinking that

(Total loan) ÷ (Number of payments) = Amount of each payment

$$\$7382.52 \div 36 = n.$$

3. **Solve.** To solve the equation, we carry out the division.

$$\begin{array}{r} 205.07 \\ 36\,\overline{)\,7382.52} \\ 720000 \\ \hline 18252 \\ 18000 \\ \hline 252 \\ 252 \\ \hline 0 \end{array}$$

Thus, $n = 205.07$.

4. **Check.** A partial check can be obtained by estimating the quotient: \$7382.56 ÷ 36 ≈ 8000 ÷ 40 = 200 ≈ 205.07. The estimate checks.

5. **State.** Each payment is \$205.07.

Do Exercise 4.

The area of a rectangular region is given by the formula *Area* = *Length* · *Width,* or $A = l \cdot w$. We can use this formula with decimal notation.

Example 5 *Poster Area.* A rectangular poster measures 73.2 cm by 61.8 cm. Find the area.

1. **Familiarize.** We first make a drawing, letting A = the area.

2. **Translate.** Then we use the formula $A = l \cdot w$ and translate:

$A = 73.2 \times 61.8.$

3. **Solve.** We solve by carrying out the multiplication.

$$\begin{array}{r} 73.2 \\ \times \quad 61.8 \\ \hline 5856 \\ 7320 \\ 439200 \\ \hline 4523.76 \end{array}$$

Thus, $A = 4523.76$.

Answer on page A-7

4. **Check.** We obtain a partial check by estimating the product:

$$73.2 \times 61.8 \approx 70 \times 60 = 4200 \approx 4523.76.$$

Since this estimate is not too close, we might repeat our calculation or change our estimate to be more certain. We leave this to the student. We see that 4523.76 checks.

5. **State.** The area is 4523.76 cm^2.

Do Exercise 5.

Example 6 *Cost of Crabmeat.* One pound of crabmeat makes 3 servings at the Key West Seafood Restaurant. It costs $14.98 per pound. What is the cost per serving? Round to the nearest cent.

1. **Familiarize.** We let $c =$ the cost per serving.
2. **Translate.** We translate as follows.

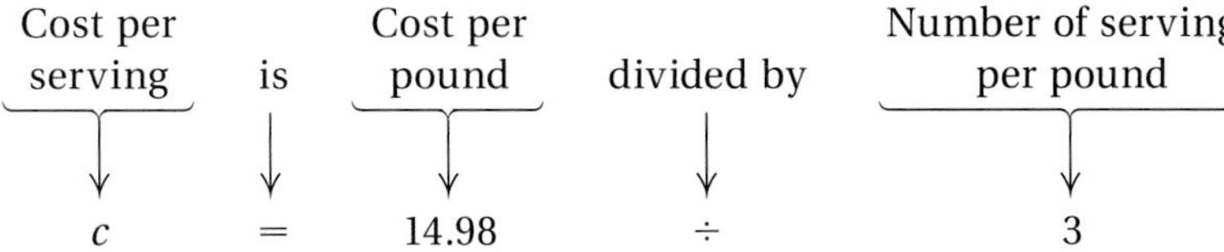

Cost per serving
is
Cost per pound
divided by
Number of servings per pound
c
=
14.98
÷
3

3. **Solve.** To solve, we carry out the division.

$$\begin{array}{r} 4.9933 \\ 3\,\overline{)\,14.9800} \\ \underline{12} \\ 2\,9 \\ \underline{2\,7} \\ 28 \\ \underline{27} \\ 10 \\ \underline{9} \\ 10 \end{array} \qquad c = 4.99\overline{3}$$

4. **Check.** We check by estimating the quotient:

$$14.98 \div 3 \approx 15 \div 3 = 5 \approx 4.99\overline{3}.$$

In this case, our check provides a good estimate.

5. **State.** We round $4.99\overline{3}$ and find the cost per serving to be about $4.99.

Do Exercise 6.

5. A standard-size index card measures 12.7 cm by 7.6 cm. Find its area.

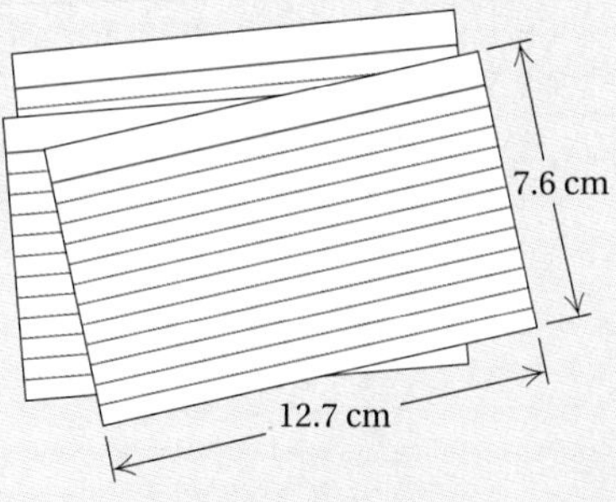

6. One pound of lean boneless ham contains 4.5 servings. It costs $3.99 per pound. What is the cost per serving? Round to the nearest cent.

Answers on page A-7

7. *Gas Mileage.* A driver filled the gasoline tank and noted that the odometer read 38,320.8. After the next filling, the odometer read 38,735.5. It took 14.5 gal to fill the tank. How many miles per gallon did the driver get?

Answer on page A-7

Multistep Problems

Example 7 *Gas Mileage.* A driver filled the gasoline tank and noted that the odometer read 67,507.8. After the next filling, the odometer read 68,006.1. It took 16.5 gal to fill the tank. How many miles per gallon did the driver get?

1. **Familiarize.** We first make a drawing.

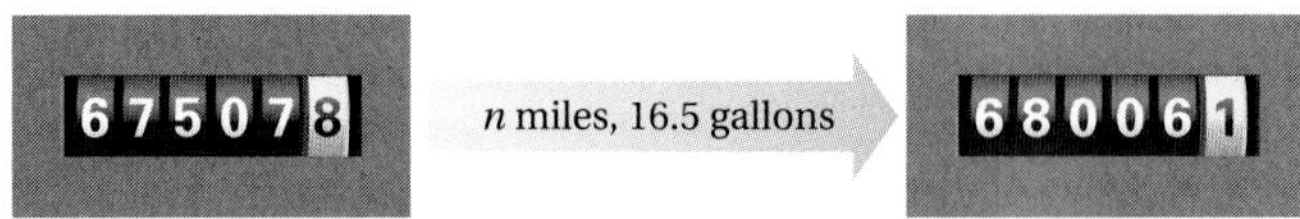

This is a two-step problem. First, we find the number of miles that have been driven between fillups. We let n = the number of miles driven.

2., 3. **Translate** and **Solve.** This is a "how-much-more" situation. We translate and solve as follows.

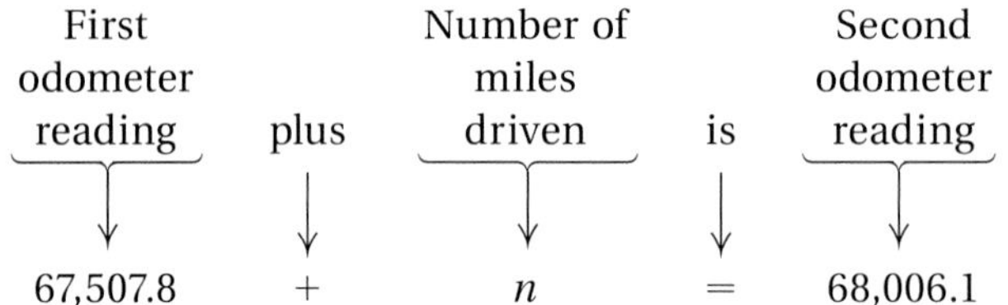

First odometer reading
plus
Number of miles driven
is
Second odometer reading
67,507.8
+
n
=
68,006.1

To solve the equation, we subtract 67,507.8 on both sides:

$$n = 68{,}006.1 - 67{,}507.8 = 498.3.$$

$$\begin{array}{r} 68{,}006.1 \\ -\ 67{,}507.8 \\ \hline 498.3 \end{array}$$

Second, we divide the total number of miles driven by the number of gallons. This gives us m = the number of miles per gallon—that is, the mileage. The division that corresponds to the situation is

$$498.3 \div 16.5 = m.$$

To find the number m, we divide.

$$\begin{array}{r} 30.2 \\ 16.5\overline{)\,498.30} \\ \underline{495\ 0} \\ 3\ 30 \\ \underline{3\ 30} \\ 0 \end{array}$$

Thus, $m = 30.2$.

4. **Check.** To check, we first multiply the number of miles per gallon times the number of gallons:

$$16.5 \times 30.2 = 498.3.$$

Then we add 498.3 to 67,507.8:

$$67{,}507.8 + 498.3 = 68{,}006.1.$$

The mileage 30.2 checks.

5. **State.** The driver got 30.2 miles per gallon.

Do Exercise 7.

Example 8 *Home-Cost Comparison.* Suppose you own a home like the one shown here and it is valued at $250,000 in Indianapolis, Indiana. What would it cost to buy a similar (replacement) home in Beverly Hills, California? To find out, we can use an index table prepared by Coldwell Banker Real Estate Corporation (***Source*:** Coldwell Banker Real Estate Corporation. For a complete index table, contact your local representative.). We use the following formula:

$$\begin{pmatrix}\text{Cost of your}\\ \text{home in new city}\end{pmatrix} = \begin{pmatrix}\text{Value of}\\ \text{your home}\end{pmatrix} \div \begin{pmatrix}\text{Index of}\\ \text{your city}\end{pmatrix} \times \begin{pmatrix}\text{Index of}\\ \text{new city}\end{pmatrix}$$

Find the cost of your Indianapolis home in Beverly Hills.

State	City	Index
California	San Francisco	286
	Beverly Hills	376
	Fresno	82
Indiana	Indianapolis	79
	Fort Wayne	69
Arizona	Phoenix	90
	Tucson	79
Illinois	Chicago	214
	Naperville	101
Texas	Austin	89
	Dallas	70
	Houston	61
Florida	Miami	85
	Orlando	76
	Tampa	72
Massachusetts	Wellesley	231
	Cape Cod	84
Georgia	Atlanta	81
New York	Queens	179
	Albany	89

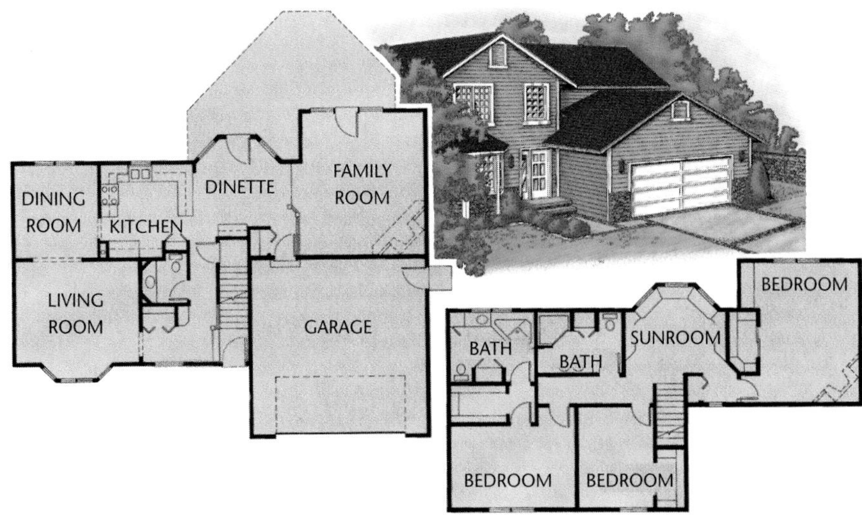

8. *Home-cost Comparison.* Find the replacement cost of a \$250,000 home in Indianapolis if you were to try to replace it when moving to Dallas.

9. Find the replacement cost of a \$250,000 home in Phoenix if you were to try to replace it when moving to Chicago.

Answers on page A-7

1. **Familiarize.** We let C = the cost of the home in Beverly Hills. We use the table and look up the indexes of the city in which you now live and the city to which you are moving.
2. **Translate.** Using the formula, we translate to the following equation:

$$C = \$250{,}000 \div 79 \times 376.$$

3. **Solve.** To solve, we carry out the computations using the rules for order of operations (see Section 3.4):

$$\begin{aligned} C &= \$250{,}000 \div 79 \times 376 \\ &= \$3164.557 \times 376 && \text{Carrying out the division first} \\ &\approx \$1{,}189{,}873. && \text{Carrying out the multiplication and rounding to the nearest one} \end{aligned}$$

On a calculator, the computation could be done in one step.

4. **Check.** We can repeat our computations or round and estimate as we learned in Section 3.6:

$$\begin{aligned} C &= \$250{,}000 \div 79 \times 376 \\ &\approx \$250{,}000 \div 100 \times 400 \\ &= \$1{,}000{,}000. \end{aligned}$$

Since $\$1{,}000{,}000 \approx \$1{,}189{,}873$, we have a partial check.

5. **State.** A home selling for \$250,000 in Indianapolis would cost about \$1,189,873 in Beverly Hills.

Do Exercises 8 and 9.

Exercise Set 3.7

a Solve.

1. What is the cost of 8 pairs of socks at \$4.95 per pair?

2. What is the cost of 7 shirts at \$32.98 each?

3. *Gasoline Cost.* What is the cost, in dollars, of 17.7 gal of gasoline at 119.9 cents per gallon? (119.9 cents = \$1.199) Round the answer to the nearest cent.

4. *Gasoline Cost.* What is the cost, in dollars, of 20.4 gal of gasoline at 149.9 cents per gallon? Round the answer to the nearest cent.

5. Roberto bought a CD for \$16.99 and paid with a \$20 bill. How much change did he receive?

6. Madeleine buys a book for \$44.68 and pays with a \$50 bill. How much change does she receive?

7. *Body Temperature.* Normal body temperature is 98.6°F. During an illness, a patient's temperature rose 4.2°. What was the new temperature?

8. *Blood Test.* A medical assistant draws 9.85 mL of blood and uses 4.68 mL in a blood test. How much is left?

9. *Lottery Winnings.* In Texas, one of the state lotteries is called "Cash 5." In a recent weekly game, the lottery prize of \$127,315 was shared equally by 6 winners. How much was each winner's share? Round to the nearest cent.

10. A group of 4 students pays \$40.76 for lunch. What is each person's share?

11. A rectangular parking lot measures 800.4 ft by 312.6 ft. What is its area?

12. A rectangular fenced yard measures 40.3 yd by 65.7 yd. What is its area?

13. *Odometer Reading.* A family checked the odometer before starting a trip. It read 22,456.8 and they know that they will be driving 234.7 mi. What will the odometer read at the end of the trip?

14. *Miles Driven.* Petra bought gasoline when the odometer read 14,296.3. At the next gasoline purchase, the odometer read 14,515.8. How many miles had been driven?

15. *Eating Habits.* Each year, Americans eat 24.8 billion hamburgers and 15.9 billion hot dogs. How many more hamburgers than hot dogs do Americans eat?

16. *Gas Mileage.* A driver wants to estimate gas mileage per gallon. At 36,057.1 mi, the tank is filled with 10.7 gal. At 36,217.6 mi, the tank is filled with 11.1 gal. Find the mileage per gallon. Round to the nearest tenth.

17. *Jet-Powered Car.* A jet-powered car was measured on a computer to go from a speed of mach 0.85 to mach 1.15 (mach 1.0 is the speed of sound). What was the difference in these speeds?

18. *Fat Content.* There is 0.8 g of fat in one serving $\left(3\frac{1}{2}\text{ oz}\right)$ of raw scallops. In one serving of oysters, there is 2.5 g of fat. How much more fat is in one serving of oysters than in one serving of scallops?

19. *Gas Mileage.* Peggy filled her van's gas tank and noted that the odometer read 26,342.8. After the next filling, the odometer read 26,736.7. It took 19.5 gal to fill the tank. How many miles per gallon did the van get?

20. *Gas Mileage.* Peter filled his Honda's gas tank and noted that the odometer read 18,943.2. After the next filling, the odometer read 19,306.2. It took 13.2 gal to fill the tank. How many miles per gallon did the car get?

21. The water in a filled tank weighs 748.45 lb. One cubic foot of water weighs 62.5 lb. How many cubic feet of water does the tank hold?

22. *Highway Routes.* You can drive from home to work using either of two routes:

    Route A: Via interstate highway, 7.6 mi, with a speed limit of 65 mph.
    Route B: Via a country road, 5.6 mi, with a speed limit of 50 mph.

    Assuming you drive at the posted speed limit, which route takes less time? (Use the formula *Distance* = *Speed* × *Time*.)

23. *Cost of Video Game.* The average video game costs 25 cents and runs for 1.5 min. Assuming a player does not win any free games and plays continuously, how much money, in dollars, does it cost to play a video game for 1 hr?

24. *Property Taxes.* The Colavitos own a house with an assessed value of $124,500. For every $1000 of assessed value, they pay $7.68 in taxes. How much do they pay in taxes?

Find the distance around (perimeter of) the figure.

25.

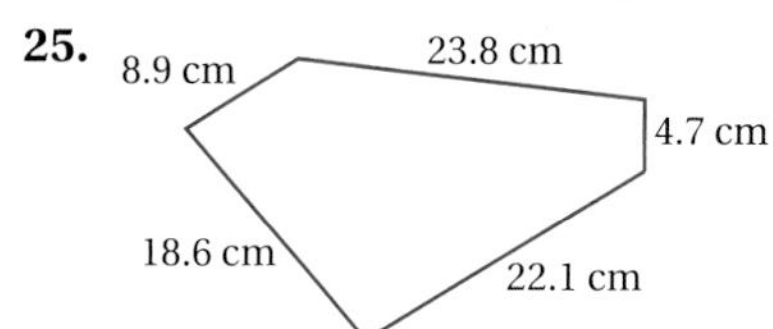

26.

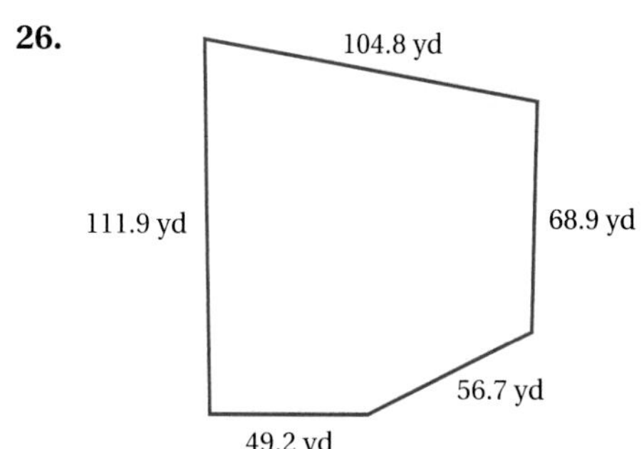

Find the length d in the figure.

27.

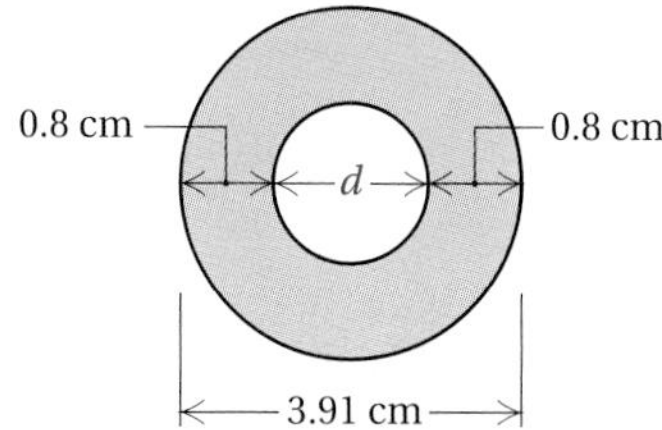

28.

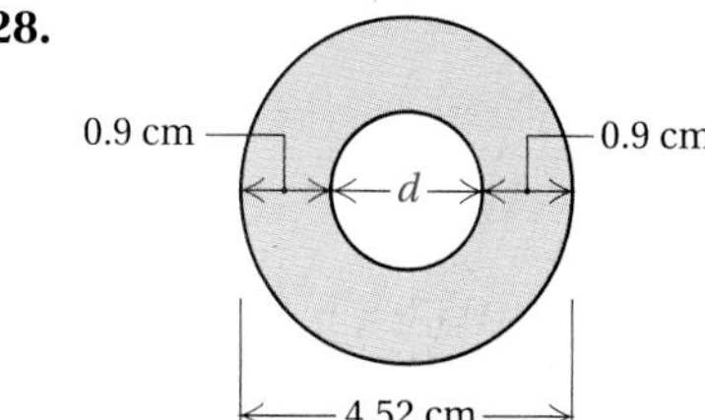

29. *Calories Burned Mowing.* A person weighing 150 lb burns 7.3 calories per minute while mowing a lawn with a power lawnmower (***Source:*** *Hanely Science Answer Book*). How many calories would be burned in 2 hr of mowing?

30. Lot A measures 250.1 ft by 302.7 ft. Lot B measures 389.4 ft by 566.2 ft. What is the total area of the two lots?

31. Holly had $1123.56 in her checking account. She wrote checks of $23.82, $507.88, and $98.32 to pay some bills. She then deposited a bonus check of $678.20. How much is in her account after these changes?

32. Natalie Clad had $185.00 to spend for fall clothes: $44.95 was spent for shoes, $71.95 for a jacket, and $55.35 for pants. How much was left?

33. A rectangular yard is 20 ft by 15 ft. The yard is covered with grass except for an 8.5-ft square flower garden. How much grass is in the yard?

34. Rita earns a gross paycheck (before deductions) of $495.72. Her deductions are $59.60 for federal income tax, $29.00 for FICA, and $29.00 for medical insurance. What is her take-home paycheck?

35. *Batting Average.* In a recent year, Bernie Williams of the New York Yankees got 168 hits in 551 times at bat. What part of his at-bats were hits? Give decimal notation to the nearest thousandth. (This is a player's *batting average.*)

36. *Batting Average.* In a recent year, Chipper Jones of the Atlanta Braves got 185 hits in 598 times at bat. What was his batting average? Give decimal notation to the nearest thousandth. (See Exercise 35.)

37. It costs $24.95 a day plus 27 cents per mile to rent a compact car at Shuttles Rent-a-Car. How much, in dollars, would it cost to drive the car 120 mi in 1 day?

38. Zachary worked 53 hr during a week one summer. He earned $6.50 per hour for the first 40 hr and $9.75 per hour for overtime (hours exceeding 40). How much did Zachary earn during the week?

39. A family of five can save $6.72 per week by eating cooked cereal instead of ready-to-eat cereal. How much will they save in 1 year? Use 52 weeks for 1 year.

40. A medical assistant prepares 200 injections, each with 2.5 mL of penicillin. How much penicillin is used in all?

41. A restaurant owner bought 20 dozen eggs for $13.80. Find the cost of each egg to the nearest tenth of a cent (thousandth of a dollar).

42. *Weight Loss.* A person weighing 170 lb burns 8.6 calories per minute while mowing a lawn. One must burn about 3500 calories in order to lose 1 lb. How many pounds would be lost by mowing for 2 hr? Round to the nearest tenth.

43. *Soccer Field.* The dimensions of a World Cup soccer field are 114.9 yd by 74.4 yd. The dimensions of a standard football field are 120 yd by 53.3 yd. How much greater is the area of a soccer field?

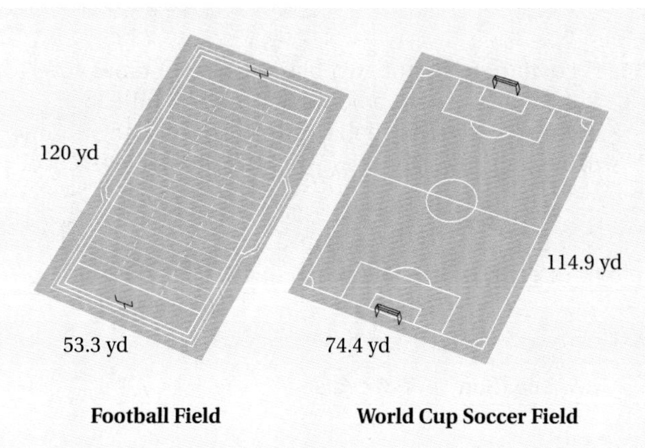

44. *Construction Pay.* A construction worker is paid \$13.50 per hour for the first 40 hr of work, and time and a half, or \$20.25 per hour, for any overtime exceeding 40 hr per week. One week she works 46 hr. How much is her pay?

45. *Loan Payment.* In order to make money on loans, financial institutions are paid back more money than they loan. You borrow \$120,000 to buy a house and agree to make monthly payments of \$880.52 for 30 yr. How much do you pay back altogether? How much more do you pay back than the amount of the loan?

46. *Car-Rental Cost.* Enterprise Rent-A-Car charges \$59.99 per day plus \$0.25 per mile for a luxury sedan (***Source*:** Enterprise Rent-A-Car). How much is the rental charge for a 4-day trip of 876 mi?

Airport Passengers. The following graph shows the number of passengers in a recent year who traveled through the country's busiest airports. (Use the graph for Exercises 47–50.)

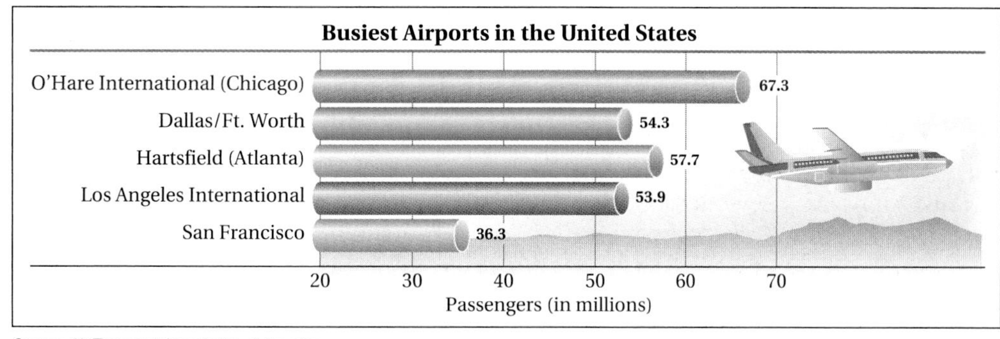

Source: Air Transport Association of America

47. How many more passengers does O'Hare handle than San Francisco?

48. How many more passengers does Hartsfield handle than Los Angeles?

49. How many passengers do Dallas/Ft. Worth and San Francisco handle altogether?

50. How many passengers do all these airports handle altogether?

51. *Body Temperature.* Normal body temperature is 98.6°F. A baby's bath water should be 100°F. How many degrees above normal body temperature is this?

52. *Body Temperature.* Normal body temperature is 98.6°F. The lowest temperature at which a patient has survived is 69°F. How many degrees below normal is this?

53. A student used a $20 bill to buy a poster for $10.75. The change was a five-dollar bill, three one-dollar bills, a dime, and two nickels. Was the change correct?

54. A customer bought two blank cassette tapes for $13.88. They were paid for with a $20 bill. The change was a five-dollar bill, a one-dollar bill, one dime, and two pennies. Was the change correct?

Home-Cost Comparison. Use the table and formula from Example 8. In each of the following cases, find the value of the house in the new location.

	Value	Present Location	New Location	New Value
55.	$125,000	Fresno	San Francisco	
56.	$180,000	Chicago	Beverly Hills	
57.	$96,000	Indianapolis	Tampa	
58.	$300,000	Miami	Queens	
59.	$240,000	San Francisco	Atlanta	
60.	$160,000	Cape Cod	Phoenix	

Skill Maintenance

Add.

61. 4569 + 1766 [1.2b]

62. $\frac{2}{3} + \frac{5}{8}$ [2.3b]

63. $4\frac{1}{3} + 2\frac{1}{2}$ [2.4b]

64. $\frac{5}{6} + \frac{7}{10}$ [2.3b]

65. 8099 + 5667 [1.2b]

Subtract.

66. 4569 − 1766 [1.2f]

67. $\frac{2}{3} - \frac{5}{8}$ [2.3c]

68. $4\frac{1}{3} - 2\frac{1}{2}$ [2.4c]

69. $\frac{5}{6} - \frac{7}{10}$ [2.3c]

70. 8099 − 5667 [1.2f]

Solve. [2.5a]

71. If a water wheel made 469 revolutions at a rate of $16\frac{3}{4}$ revolutions per minute, how long did it rotate?

72. If a bicycle wheel made 480 revolutions at a rate of $66\frac{2}{3}$ revolutions per minute, how long did it rotate?

Synthesis

73. ◆ Write a problem for a classmate to solve. Design the problem so that the solution is "Mona's Buick got 23.5 mpg."

74. ◆ Write a problem for a classmate to solve. Design the problem so that the solution is "The larger field is 200 m^2 bigger."

75. You buy a half-dozen packs of basketball cards with a dozen cards in each pack. The cost is a dozen cents for each half-dozen cards. How much do you pay for the cards?

Summary and Review Exercises: Chapter 3

The objectives to be tested in addition to the material in this chapter are [1.7d], [1.9a], [2.1e], and [2.4b, c].

Write a word name. [3.1a]

1. 3.47

2. 0.031

Write a word name as on a check. [3.1a]

3. $597.25

4. $0.96

Write fractional notation. [3.1b]

5. 0.09

6. 4.561

7. 0.089

8. 3.0227

Write decimal notation. [3.1c]

9. $\frac{34}{1000}$

10. $\frac{42{,}603}{10{,}000}$

11. $27\frac{91}{100}$

12. $867\frac{6}{1000}$

Which number is larger? [3.1d]

13. 0.034, 0.0185

14. 0.91, 0.19

15. 0.741, 0.6943

16. 1.038, 1.041

Round 17.4287 to the nearest: [3.1e]

17. Tenth.

18. Hundredth.

19. Thousandth.

20. One.

Add. [3.2a]

21.
$$\begin{array}{r} 2.048 \\ 65.371 \\ +\ 507.1 \\ \hline \end{array}$$

22.
$$\begin{array}{r} 0.6 \\ 0.004 \\ 0.07 \\ +\ 0.0098 \\ \hline \end{array}$$

23. 219.3 + 2.8 + 7

24. 0.41 + 4.1 + 41 + 0.041

Subtract. [3.2b]

25.
$$\begin{array}{r} 30.0 \\ -\ 0.7908 \\ \hline \end{array}$$

26.
$$\begin{array}{r} 845.08 \\ -\ 54.79 \\ \hline \end{array}$$

27. 37.645 − 8.497

28. 70.8 − 0.0109

Multiply. [3.3a]

29.
$$\begin{array}{r} 48 \\ \times\ 0.27 \\ \hline \end{array}$$

30.
$$\begin{array}{r} 0.174 \\ \times\ 0.83 \\ \hline \end{array}$$

31. 100×0.043

32. 0.001×24.68

Divide. [3.4a]

33. $8\,\overline{)\,60}$

34. $52\,\overline{)\,23.4}$

35. $2.6\,\overline{)\,117.52}$

36. $2.14\,\overline{)\,2.18708}$

37. $\frac{276.3}{1000}$

38. $\frac{13.892}{0.01}$

Solve. [3.2c], [3.4b]

39. $x + 51.748 = 548.0275$

40. $3 \cdot x = 20.85$

41. $10 \cdot y = 425.4$

42. $0.0089 + y = 5$

Solve. [3.7a]

43. Katrina earned $310.37 during a 40-hr week. What was her hourly wage? Round to the nearest cent.

44. Derek had $6274.53 in his checking account. He wrote a check for $385.79 to buy a fax-modem for his computer. How much was left in his account?

45. *Tea Consumption.* The average person drinks about 3.48 cups of tea per day. How many cups of tea does the average person drink in a week? in a 30-day month?

46. *Software Purchase.* Four software disks containing video instruction are purchased by a learning lab at a cost of $59.95 each. How much is spent altogether?

47. *Telephone Poles.* In the United States, there are 51.81 telephone poles for every 100 people. In Canada, there are 40.65. How many more telephone poles for every 100 people are there in the United States?

48. Antarctica Agricultural College has four subterranean cornfields. One year the harvest in each field was 1419.3 bushels, 1761.8 bushels, 1095.2 bushels, and 2088.8 bushels. What was the total harvest?

Estimate each of the following. [3.6a]

49. The product 7.82×34.487 by rounding to the nearest one

50. The difference $219.875 - 4.478$ by rounding to the nearest one

51. The quotient $82.304 \div 17.287$ by rounding to the nearest ten

52. The sum $45.78 + $78.99 by rounding to the nearest one

Find decimal notation. Use multiplying by 1. [3.5a]

53. $\frac{13}{5}$ 54. $\frac{32}{25}$ 55. $\frac{11}{4}$

Find decimal notation. Use division. [3.5a]

56. $\frac{13}{4}$ 57. $\frac{7}{6}$ 58. $\frac{17}{11}$

Round the answer to Exercise 58 to the nearest: [3.5b]

59. Tenth. 60. Hundredth. 61. Thousandth.

Convert from cents to dollars. [3.3b]

62. 8273 cents 63. 487 cents

Convert from dollars to cents. [3.3b]

64. $24.93 65. $9.86

Convert the number in the sentence to standard notation. [3.3b]

66. The statutory debt limit is $5.5 trillion.

67. Your blood travels 1.2 million miles in a week.

Calculate. [3.5c]

68. $(8 - 1.23) \div 4 + 5.6 \times 0.02$

69. $(1 + 0.07)^2 + 10^3 \div 10^2$
$+ [4(10.1 - 5.6) + 8(11.3 - 7.8)]$

70. $\frac{3}{4} \times 20.85$

71. $\frac{1}{3} \times 123.7 + \frac{4}{9} \times 0.684$

Skill Maintenance

72. Add: $12\frac{1}{2} + 7\frac{3}{10}$. [2.4b]

73. Subtract: $24 - 17\frac{2}{5}$. [2.4c]

74. Simplify: $\frac{28}{56}$. [2.1e]

75. Find the prime factorization of 192. [1.7d]

76. Find the LCM of 20, 33, and 75. [1.9a]

Synthesis

77. ◈ Consider finding decimal notation for $\frac{44}{125}$. Discuss as many ways as you can for finding such notation and give the answer. [3.5a]

78. ◈ Explain how we can use fractional notation to understand why we count decimal places when multiplying with decimal notation. [3.3a]

79. ▦ In each of the following, use one of +, −, ×, and ÷ in each blank to make a true sentence. [3.4c]

a) 2.56 ▒ 6.4 ▒ 51.2 ▒ 17.4 ▒ 89.7 = 72.62
b) (0.37 ▒ 18.78) ▒ 2^{13} = 156,876.8

80. Find repeating decimal notation for 1 and explain. Use the following hints. [3.5a]

$$\frac{1}{3} = 0.33333333\ldots,$$

$$\frac{2}{3} = 0.66666666\ldots$$

81. Find repeating decimal notation for 2. [3.5a]

Test: Chapter 3

1. Write a word name: 2.34.

2. Write a word name, as on a check, for $1234.78.

Write fractional notation.

3. 0.91

4. 2.769

Write decimal notation.

5. $\frac{74}{1000}$

6. $\frac{37,047}{10,000}$

7. $756\frac{9}{100}$

8. $91\frac{703}{1000}$

Which number is larger?

9. 0.07, 0.162

10. 0.078, 0.06

11. 0.09, 0.9

Round 5.6783 to the nearest:

12. One.

13. Hundredth.

14. Thousandth.

15. Tenth.

Add.

16. $\begin{array}{r} 402.3 \\ 2.81 \\ +\quad 0.109 \\ \hline \end{array}$

17. $\begin{array}{r} 0.7 \\ 0.08 \\ 0.009 \\ +\ 0.0012 \\ \hline \end{array}$

18. $102.4 + 6.1 + 78$

19. $0.93 + 9.3 + 93 + 930$

Subtract.

20. $\begin{array}{r} 52.678 \\ -\quad 4.321 \\ \hline \end{array}$

21. $\begin{array}{r} 20.0 \\ -\quad 0.9099 \\ \hline \end{array}$

22. $2 - 0.0054$

23. $234.6788 - 81.7854$

Multiply.

24. $\begin{array}{r} 0.125 \\ \times\quad 0.24 \\ \hline \end{array}$

25. $\begin{array}{r} 32 \\ \times\ 0.25 \\ \hline \end{array}$

26. 0.001×213.45

27. 1000×73.962

Divide.

28. $4\,\overline{)\,19}$

29. $42\,\overline{)\,10.08}$

30. $3.3\,\overline{)\,100.32}$

31. $82\,\overline{)\,15.58}$

32. $\frac{346.89}{1000}$

33. $\frac{346.89}{0.01}$

Solve.

34. $4.8 \cdot y = 404.448$

35. $x + 0.018 = 9$

Answers

1. ______
2. ______
3. ______
4. ______
5. ______
6. ______
7. ______
8. ______
9. ______
10. ______
11. ______
12. ______
13. ______
14. ______
15. ______
16. ______
17. ______
18. ______
19. ______
20. ______
21. ______
22. ______
23. ______
24. ______
25. ______
26. ______
27. ______
28. ______
29. ______
30. ______
31. ______
32. ______
33. ______
34. ______
35. ______

Answers

36. ____
37. ____
38. ____
39. ____
40. ____
41. ____
42. ____
43. ____
44. ____
45. ____
46. ____
47. ____
48. ____
49. ____
50. ____
51. ____
52. ____
53. ____
54. ____
55. ____
56. ____
57. ____
58. ____
59. ____
60. ____
61. ____
62. ____
63. ____

Solve.

36. A marathon runner ran 24.85 km in 5 hr. How far did she run in 1 hr?

37. Carla has a balance of $10,200 in her checking account before writing checks of $123.89, $56.68, and $3446.98. What was the balance after she had written the checks?

38. Ben Westlund paid $23,457 for 14 acres of land adjoining his ranch. How much did he pay for 1 acre? Round to the nearest cent.

39. A government agency bought 6 new flags at $79.95 each. How much was spent altogether?

Estimate each of the following.

40. The product 8.91×22.457 by rounding to the nearest one

41. The quotient $78.2209 \div 16.09$ by rounding to the nearest ten

Find decimal notation. Use multiplying by 1.

42. $\frac{8}{5}$

43. $\frac{22}{25}$

44. $\frac{21}{4}$

Find decimal notation. Use division.

45. $\frac{3}{4}$

46. $\frac{11}{9}$

47. $\frac{15}{7}$

Round the answer to Exercise 47 to the nearest:

48. Tenth.

49. Hundredth.

50. Thousandth.

51. Convert from cents to dollars: 949 cents.

52. Convert to standard notation: Procter & Gamble spent $2.8 billion on advertising in a recent year.

Calculate.

53. $256 \div 3.2 \div 2 - 1.56 + 78.325 \times 0.02$

54. $(1 - 0.08)^2 + 6[5(12.1 - 8.7) + 10(14.3 - 9.6)]$

55. $\frac{7}{8} \times 345.6$

56. $\frac{2}{3} \times 79.95 - \frac{7}{9} \times 1.235$

Skill Maintenance

57. Add: $2\frac{3}{16} + \frac{1}{2}$.

58. Subtract: $28\frac{2}{3} - 2\frac{1}{6}$.

59. Simplify: $\frac{33}{54}$.

60. Find the LCM of 15, 36, and 40.

61. Find the prime factorization of 360.

Synthesis

62. The Fit Fiddle health club generally charges a $79 membership fee and $42.50 a month. Allise has a coupon that will allow her to join the club for $299 for six months. How much will Allise save if she uses the coupon?

63. 🖩 Arrange from smallest to largest.

$\frac{2}{3}, \frac{15}{19}, \frac{11}{13}, \frac{5}{7}, \frac{13}{15}, \frac{17}{20}$

4

Percent Notation

Introduction

This chapter introduces a new kind of notation for numbers: percent notation. We will see that $\frac{3}{8}$, 0.375, and 37.5% are all names for the same number. Then we will use percent notation and equations to solve applied problems.

Also introduced are the topics of ratio and proportion.

An Application

In a treadmill test, a doctor's goal is to get the patient to reach his or her *maximum heart rate,* in beats per minute, which is found by subtracting the patient's age from 220 and taking 85% of the result. The author of this text took such a test at age 55. What was his maximum heart rate?

This problem appears as Exercise 11 in Section 4.6.

The Mathematics

We let x = the maximum heart rate. This problem translates to the equation

$$x = \underbrace{85\%}_{\uparrow} \cdot (220 - 55).$$

This is percent notation.

World Wide Web For more information, visit us at www.mathmax.com

Pretest: Chapter 4

Write fractional notation for the ratio.

1. 35 to 43

2. 0.079 to 1.043

Solve.

3. $\frac{5}{6} = \frac{x}{27}$

4. What is the rate in miles per gallon?
   408 miles, 16 gallons

5. Find decimal notation for 87%.

6. Find percent notation for 0.537.

7. Find percent notation for $\frac{3}{4}$.

8. Find fractional notation for 37%.

9. Translate to an equation. Then solve.
   What is 60% of 75?

10. Translate to a proportion. Then solve.
    What percent of 50 is 35?

Solve.

11. *Weight of Muscles.* The weight of muscles in a human body is 40% of total body weight. A person weighs 225 lb. What do the muscles weigh?

12. The population of a town increased from 3000 to 3600. Find the percent of increase in population.

13. *Tax Rate in Massachusetts.* The sales tax rate in Massachusetts is 5%. How much tax is charged on a purchase of $286? What is the total price?

14. A salesperson's commission rate is 28%. What is the commission from the sale of $18,400 worth of merchandise?

15. The marked price of a stereo is $450. The stereo is on sale at Lowland Appliances for 25% off. What are the discount and the sale price?

16. What is the simple interest on $1200 principal at the interest rate of 8.3% for 1 year?

17. What is the simple interest on $500 at 8% for $\frac{1}{2}$ year?

18. Interest is compounded annually. Find the amount in an account if $6000 is invested at 9% for 2 years.

Objectives for Retesting

The objectives to be tested in addition to the material in this chapter are as follows.

[2.4a] Convert from mixed numerals to fractional notation, and vice versa.
[3.1b] Convert from decimal notation to fractional notation.
[3.4b] Solve equations of the type $a \cdot x = b$, where a and b may be in decimal notation.
[3.5a] Convert from fractional notation to decimal notation.

4.1 Ratio and Proportion

a Ratios

> A **ratio** is the quotient of two quantities.

For example, each day in this country about 5200 people die. Of these, 1070 die of cancer. The *ratio* of those who die of cancer to those who die is shown by the fractional notation

$$\frac{1070}{5200} \quad \text{or by the notation} \quad 1070{:}5200.$$

We read such notation as "the ratio of 1070 to 5200," listing the numerator first and the denominator second.

> The **ratio** of a to b is given by $\frac{a}{b}$, where a is the numerator and b is the denominator, or by $a{:}b$.

Example 1 Find the ratio of 31.4 to 100.

The ratio is $\frac{31.4}{100}$, or $31.4{:}100$.

In most of our work, we will use fractional notation for ratios.

Example 2 Hank Aaron hit 755 home runs in 12,364 at-bats. Find the ratio of at-bats to home runs.

The ratio is $\frac{12{,}364}{755}$.

Do Exercises 1–4.

Example 3 In the triangle at right:

a) What is the ratio of the length of the longest side to the length of the shortest side?

$$\frac{5}{3}$$

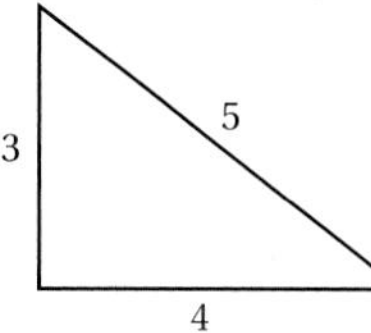

b) What is the ratio of the length of the shortest side to the length of the longest side?

$$\frac{3}{5}$$

Do Exercise 5.

Objectives

a Find fractional notation for ratios.

b Give the ratio of two different kinds of measure as a rate.

c Determine whether two pairs of numbers are proportional.

d Solve proportions.

e Solve applied problems involving proportions.

For Extra Help

TAPE 8 | MAC WIN | CD-ROM

1. Find the ratio of 5 to 11.

2. Find the ratio of 57.3 to 86.1.

3. Find the ratio of $6\frac{3}{4}$ to $7\frac{2}{5}$.

4. *Fat Content.* In one serving $\left(3\frac{1}{2}\text{ oz}\right)$ of raw scallops, there is 0.8 g of fat. In one serving of oysters, there is 2.5 g of fat. Find the ratio of fat in one serving of oysters to that in one serving of scallops.

5. In the triangle below, what is the ratio of the length of the shortest side to the length of the longest side?

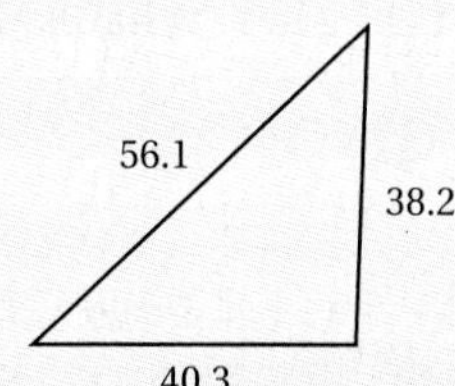

Answers on page A-8

6. Find the ratio of 18 to 27. Then simplify and find two other numbers in the same ratio.

7. Find the ratio of 3.6 to 12. Then simplify and find two other numbers in the same ratio.

8. A standard television screen with a length of 16 in. has a width, or height, of 12 in. What is the ratio of length to width?

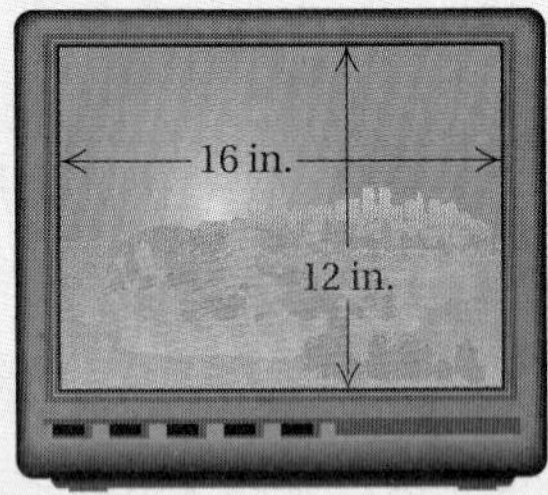

What is the rate, or speed, in miles per hour?

9. 45 mi, 9 hr

10. 120 mi, 10 hr

11. 3 mi, 10 hr

What is the rate, or speed, in feet per second?

12. 2200 ft, 2 sec

13. 52 ft, 13 sec

14. 232 ft, 16 sec

15. A well-hit golf ball can travel 500 ft in 2 sec. What is the rate, or speed, of the golf ball in feet per second?

16. A leaky faucet can lose 14 gal of water in a week. What is the rate in gallons per day?

Answers on page A-8

Sometimes a ratio can be simplified. This provides a means of finding other numbers with the same ratio.

Example 4 Find the ratio of 2.4 to 10. Then simplify and find two other numbers in the same ratio.

We first write the ratio. Next, we multiply by 1 to clear the decimal from the numerator. Then we simplify:

$$\frac{2.4}{10} = \frac{2.4}{10} \cdot \frac{10}{10} = \frac{24}{100} = \frac{4 \cdot 6}{4 \cdot 25} = \frac{4}{4} \cdot \frac{6}{25} = \frac{6}{25}.$$

Thus, 2.4 is to 10 as 6 is to 25.

Do Exercises 6–8.

b Rates

When a ratio is used to compare two different kinds of measure, we call it a **rate.** Suppose that a car is driven 200 km in 4 hr. The ratio

$$\frac{200 \text{ km}}{4 \text{ hr}}, \quad \text{or } 50\frac{\text{km}}{\text{hr}}, \quad \text{or 50 kilometers per hour,} \quad \text{or 50 km/h}$$

Recall that "per" means "division," or "for each."

is the rate traveled in kilometers per hour, which is the division of the number of kilometers by the number of hours. A ratio of distance traveled to time is also called **speed.**

Example 5 A European driver travels 145 km on 2.5 L of gas. What is the rate in kilometers per liter?

$$\frac{145 \text{ km}}{2.5 \text{ L}}, \quad \text{or} \quad 58\frac{\text{km}}{\text{L}}$$

Do Exercises 9–14.

Example 6 It takes 60 oz of grass seed to seed 3000 sq ft of lawn. What is the rate in ounces per square foot?

$$\frac{60 \text{ oz}}{3000 \text{ sq ft}} = \frac{1}{50}\frac{\text{oz}}{\text{sq ft}}, \quad \text{or} \quad 0.02\frac{\text{oz}}{\text{sq ft}}$$

Do Exercises 15 and 16.

c Proportion

Suppose we want to compare $\frac{3}{6}$ and $\frac{2}{4}$. We find a common denominator and compare numerators. To do this, we multiply by 1:

$$\left.\begin{aligned} \frac{3}{6} &= \frac{3}{6} \cdot \frac{4}{4} = \frac{3 \cdot 4}{6 \cdot 4} = \frac{12}{24} \\ \frac{2}{4} &= \frac{2}{4} \cdot \frac{6}{6} = \frac{2 \cdot 6}{4 \cdot 6} = \frac{12}{24} \end{aligned}\right\} \quad \text{We see that } \frac{3}{6} = \frac{2}{4}.$$

We need only check the products $3 \cdot 4$ and $6 \cdot 2$ to test for equality.

We multiply these two numbers: $3 \cdot 4$. $\quad \dfrac{3}{6} \quad \dfrac{2}{4} \quad$ We multiply these two numbers: $6 \cdot 2$.

Since $3 \cdot 4 = 6 \cdot 2$, we know that $\frac{3}{6} = \frac{2}{4}$. We call $3 \cdot 4$ and $6 \cdot 2$ *cross products.*

When two pairs of numbers (such as 3, 2 and 6, 4) have the same ratio, we say that they are **proportional**. The equation

$$\frac{3}{2} = \frac{6}{4}$$

states that the pairs 3, 2 and 6, 4 are proportional. Such an equation is called a **proportion**. We sometimes read $\frac{3}{2} = \frac{6}{4}$ as "3 is to 2 as 6 is to 4."

Example 7 Determine whether 1, 2 and 3, 6 are proportional.

We can use cross products:

$$1 \cdot 6 = 6 \qquad \frac{1}{2} \quad \frac{3}{6} \qquad 2 \cdot 3 = 6.$$

Since the cross products are the same, $6 = 6$, we know that $\frac{1}{2} = \frac{3}{6}$, so the numbers are proportional.

Example 8 Determine whether 2, 5 and 4, 7 are proportional.

We can use cross products:

$$2 \cdot 7 = 14 \qquad \frac{2}{5} \quad \frac{4}{7} \qquad 5 \cdot 4 = 20.$$

Since the cross products are not the same, $14 \neq 20$, we know that $\frac{2}{5} \neq \frac{4}{7}$, so the numbers are not proportional.

Do Exercises 17–19.

d Solving Proportions

Let's now look at solving proportions. Consider the proportion

$$\frac{x}{3} = \frac{4}{6}.$$

One way to solve a proportion is to use cross products. Then we can divide on both sides to get the variable alone:

$x \cdot 6 = 3 \cdot 4$ — **Equating cross products (finding cross products and setting them equal)**

$x = \dfrac{3 \cdot 4}{6}$ — **Dividing by 6 on both sides**

$= \dfrac{12}{6}$ — **Multiplying**

$= 2.$ — **Dividing**

Determine whether the two pairs of numbers are proportional.

17. 3, 4 and 6, 8

18. 1, 4 and 10, 39

19. 1, 2 and 20, 39

Answers on page A-8

20. Solve: $\frac{x}{63} = \frac{2}{9}$.

21. Solve: $\frac{x}{9} = \frac{5}{4}$.

22. Solve: $\frac{21}{5} = \frac{n}{2.5}$.

Answers on page A-8

We can check that 2 is the solution by replacing x with 2 and using cross products:

$$2 \cdot 6 = 12 \qquad \frac{2}{3} \quad \frac{4}{6} \qquad 3 \cdot 4 = 12.$$

Since the cross products are the same, it follows that $\frac{2}{3} = \frac{4}{6}$ so the numbers 2, 3 and 4, 6 are proportional, and 2 is the solution of the equation.

> To solve $\frac{x}{a} = \frac{c}{d}$, equate cross products and divide on both sides to get x alone.

Do Exercise 20.

Example 9 Solve: $\frac{x}{7} = \frac{5}{3}$. Write a mixed numeral for the answer.

We have

$$\frac{x}{7} = \frac{5}{3}$$

$$3 \cdot x = 7 \cdot 5 \qquad \text{Equating cross products}$$

$$x = \frac{7 \cdot 5}{3} \qquad \text{Dividing by 3}$$

$$= \frac{35}{3}, \text{ or } 11\frac{2}{3}.$$

The solution is $11\frac{2}{3}$.

Do Exercise 21.

Example 10 Solve: $\frac{7.7}{15.4} = \frac{y}{2.2}$.

We have

$$\frac{7.7}{15.4} = \frac{y}{2.2}$$

$$7.7 \times 2.2 = 15.4 \times y \qquad \text{Equating cross products}$$

$$\frac{7.7 \times 2.2}{15.4} = y \qquad \text{Dividing by 15.4}$$

$$\frac{16.94}{15.4} = y \qquad \text{Multiplying}$$

$$1.1 = y. \qquad \text{Dividing:}$$

```
          1.1
 15.4 ) 16.94
        15 4 0
          1 5 4
          1 5 4
              0
```

The solution is 1.1.

Do Exercise 22.

Example 11 Solve: $\frac{8}{x} = \frac{5}{3}$. Write decimal notation for the answer.

We have

$$\frac{8}{x} = \frac{5}{3}$$

$8 \cdot 3 = x \cdot 5$ Equating cross products

$\frac{8 \cdot 3}{5} = x$ Dividing by 5

$\frac{24}{5} = x$ Multiplying

$4.8 = x.$ Simplifying

The solution is 4.8.

Do Exercise 23.

Example 12 Solve: $\frac{3.4}{4.93} = \frac{10}{n}$.

We have

$$\frac{3.4}{4.93} = \frac{10}{n}$$

$3.4 \times n = 4.93 \times 10$ Equating cross products

$n = \frac{4.93 \times 10}{3.4}$ Dividing by 3.4

$= \frac{49.3}{3.4}$ Multiplying

$= 14.5.$ Dividing

The solution is 14.5.

Do Exercise 24.

e Applications of Proportions

Example 13 *Predicting Medication.* To control a fever, a doctor suggests that a child who weighs 28 kg be given 420 mg of Tylenol. If the dosage is proportional to the child's weight, how much Tylenol is recommended for a child who weighs 35 kg?

1. **Familiarize.** We let t = the number of milligrams of Tylenol.
2. **Translate.** We translate to a proportion, keeping the amount of Tylenol in the numerators.

Tylenol suggested → 420 / Child's weight → 28 = t ← Tylenol suggested / 35 ← Child's weight

$$\frac{420}{28} = \frac{t}{35}$$

3. **Solve.** Next, we solve the proportion:

$420 \cdot 35 = 28 \cdot t$ Equating cross products

$\frac{420 \cdot 35}{28} = t$ Dividing by 28 on both sides

$525 = t.$ Multiplying and dividing

23. Solve: $\frac{6}{x} = \frac{25}{11}$.

24. Solve: $\frac{0.4}{0.9} = \frac{4.8}{t}$.

Calculator Spotlight

You may have noticed in Examples 9–12 that after equating cross products, we divided on both sides. Since this is always the case when solving proportions, calculators can be useful. For instance, to solve Example 12 with a calculator, we could press

[4] [.] [9] [3] [×] [1] [0] [÷] [3] [.] [4] [=].

Exercises

1. Use a calculator to solve Examples 9–12.
2. Use a calculator to check your answers to Margin Exercises 20–24.

Answers on page A-8

25. *Predicting Paint Needs.* Lowell and Chris run a summer painting company to support their college expenses. They can paint 1700 ft^2 of clapboard with 4 gal of paint. How much paint would be needed for a building with 6000 ft^2 of clapboard?

26. *Construction Plans.* In Example 14, the length of the actual deck is 28.5 ft. What is the length of the deck on the blueprints?

Answers on page A-8

4. **Check.** We substitute into the proportion and check cross products:

$$\frac{420}{28} = \frac{525}{35};$$

$$420 \cdot 35 = 14{,}700; \quad 28 \cdot 525 = 14{,}700.$$

The cross products are the same.

5. **State.** The dosage for a child who weighs 35 kg is 525 mg.

Do Exercise 25.

Example 14 *Construction Plans.* Architects make blueprints of projects being constructed. These are scale drawings in which lengths are in proportion to actual sizes. The Hennesseys are constructing a rectangular deck just outside their house. The architectural blueprints are rendered such that $\frac{3}{4}$ in. on the drawing is actually 2.25 ft on the deck. The width of the deck on the drawing is 4.3 in. How wide is the deck in reality?

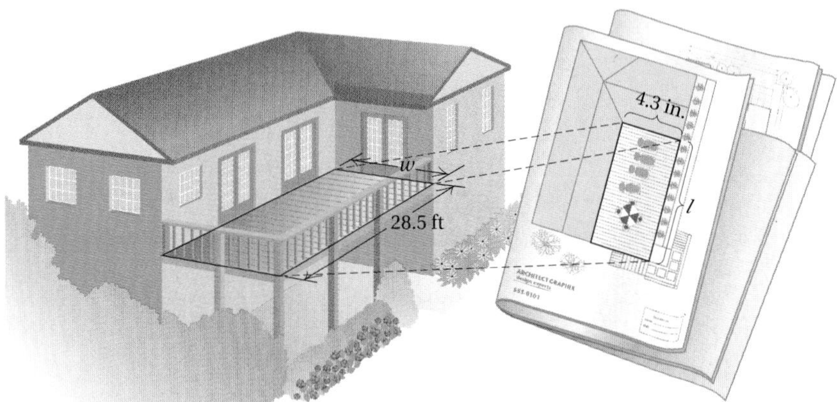

1. **Familiarize.** We let w = the width of the deck.
2. **Translate.** Then we translate to a proportion, using 0.75 for $\frac{3}{4}$ in.

$$\frac{\text{Measure on drawing} \rightarrow 0.75}{\text{Measure on deck} \rightarrow 2.25} = \frac{4.3 \leftarrow \text{Width on drawing}}{w \leftarrow \text{Width on deck}}$$

3. **Solve.** Next, we solve the proportion:

$$0.75 \cdot w = 2.25 \cdot 4.3 \quad \text{Equating cross products}$$

$$w = \frac{2.25 \cdot 4.3}{0.75} \quad \text{Dividing by 0.75 on both sides}$$

$$= 12.9.$$

4. **Check.** We substitute into the proportion and check cross products:

$$\frac{0.75}{2.25} = \frac{4.3}{12.9};$$

$$0.75 \times 12.9 = 9.675; \quad 2.25 \times 4.3 = 9.675.$$

The cross products are the same.

5. **State.** The width of the deck is 12.9 ft.

Do Exercise 26.

Exercise Set 4.1

a Find fractional notation for the ratio. You need not simplify.

1. 4 to 5

2. 329 to 967

3. 56.78 to 98.35

4. $10\frac{1}{2}$ to $43\frac{1}{4}$

5. *Corvette Accidents.* Of every 5 fatal accidents involving a Corvette, 4 do not involve another vehicle (***Source:*** *Harper's Magazine*). Find the ratio of fatal accidents involving just a Corvette to those involving a Corvette and at least one other vehicle.

6. *New York Commuters.* Of every 5 people who commute to work in New York City, 2 spend more than 90 min a day commuting (***Source:*** *The Amicus Journal*). Find the ratio of people whose daily commute to New York exceeds 90 min a day to those whose commute is 90 min or less.

Simplify the ratio.

7. 18 to 24

8. 5.6 to 10

9. 0.48 to 0.64

10. 0.32 to 0.96

11. In this right triangle, find the ratio of shortest length to longest length and simplify.

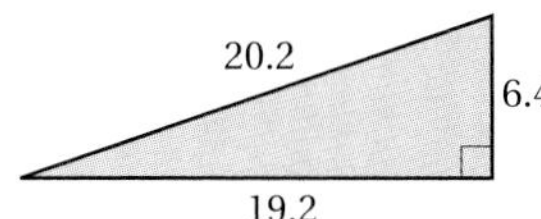

12. In this rectangle, find the ratio of width to length and simplify.

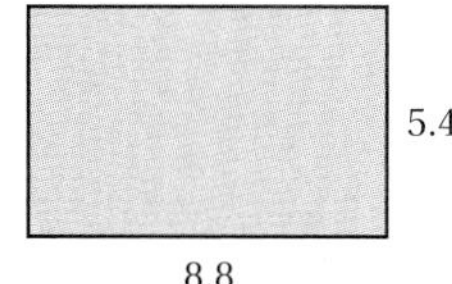

b In Exercises 13–18, find the rate as a ratio of distance to time.

13. 120 km, 3 hr

14. 18 mi, 9 hr

15. 440 m, 40 sec

16. 200 mi, 25 sec

17. 342 yd, 2.25 days

18. 492 m, 60 sec

19. A long-distance telephone call between two cities costs $5.75 for 10 min. What is the rate in cents per minute?

20. An 8-lb boneless ham contains 36 servings of meat. What is the ratio in servings per pound?

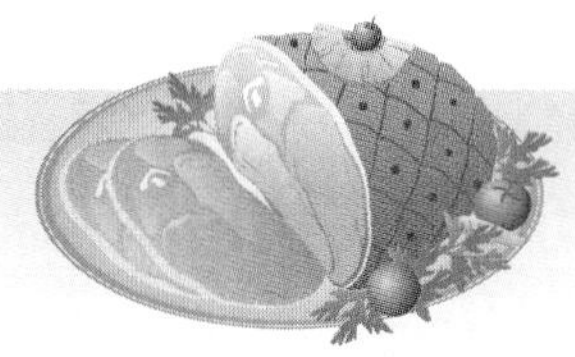

21. To water a lawn adequately requires 623 gal of water for every 1000 ft^2. What is the rate in gallons per square foot?

22. A car is driven 200 km on 40 L of gasoline. What is the rate in kilometers per liter?

23. Impulses in nerve fibers travel 310 km in 2.5 hr. What is the rate, or speed, in kilometers per hour?

24. A black racer snake can travel 4.6 km in 2 hr. What is its rate, or speed, in kilometers per hour?

25. A jet flew 2660 mi in 4.75 hr. What was its speed?

26. A turtle traveled 0.42 mi in 2.5 hr. What was its speed?

c Determine whether the two pairs of numbers are proportional.

27. 5, 6 and 7, 9

28. 7, 5 and 6, 4

29. 1, 2 and 10, 20

30. 7, 3 and 21, 9

31. 2.4, 3.6 and 1.8, 2.7

32. 4.5, 3.8 and 6.7, 5.2

33. $5\frac{1}{3}, 8\frac{1}{4}$ and $2\frac{1}{5}, 9\frac{1}{2}$

34. $2\frac{1}{3}, 3\frac{1}{2}$ and 14, 21

d Solve.

35. $\frac{18}{4} = \frac{x}{10}$

36. $\frac{x}{45} = \frac{20}{25}$

37. $\frac{t}{12} = \frac{5}{6}$

38. $\frac{12}{4} = \frac{x}{3}$

39. $\frac{2}{5} = \frac{8}{n}$

40. $\frac{10}{6} = \frac{5}{x}$

41. $\frac{16}{12} = \frac{24}{x}$

42. $\frac{7}{11} = \frac{2}{x}$

43. $\frac{t}{0.16} = \frac{0.15}{0.40}$

44. $\frac{x}{11} = \frac{7.1}{2}$

45. $\frac{100}{25} = \frac{20}{n}$

46. $\frac{35}{125} = \frac{7}{m}$

47. $\frac{\frac{1}{4}}{\frac{1}{2}} = \frac{\frac{1}{2}}{x}$

48. $\frac{5\frac{1}{5}}{6\frac{1}{6}} = \frac{y}{3\frac{1}{2}}$

49. $\frac{1.28}{3.76} = \frac{4.28}{y}$

50. $\frac{10.4}{12.4} = \frac{6.76}{t}$

e Solve.

51. *Travel Distance.* Monica bicycled 234 mi in 14 days. At this rate, how far would Monica travel in 42 days?

52. *Gasoline Mileage.* Chuck's van traveled 84 mi on 6.5 gal of gasoline. At this rate, how many gallons would be needed to travel 126 mi?

53. *Coffee.* Coffee beans from 14 trees are required to produce the 17 lb of coffee that the average person in the United States drinks each year. How many trees are required to produce 375 lb of coffee?

54. *Turkey Servings.* An 8-lb turkey breast contains 36 servings of meat. How many pounds of turkey breast would be needed for 54 servings?

55. *Quality Control.* A quality-control inspector examined 200 lightbulbs and found 18 of them to be defective. At this rate, how many defective bulbs will there be in a lot of 22,000?

56. If 2 bars of soap cost $0.89, how many bars of soap can be purchased with $6.50?

57. *Deck Sealant.* Bonnie can waterproof 450 ft^2 of decking with 2 gal of sealant. How many gallons should Bonnie buy for a 1200-ft^2 deck?

58. *Grass-Seed Coverage.* It takes 60 oz of grass seed to seed 3000 ft^2 of lawn. At this rate, how much would be needed for 5000 ft^2 of lawn?

59. In the rectangular paintings below, the ratio of length to height is the same. Find the height of the larger painting.

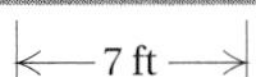

60. *Estimating a Wildlife Population.* To determine the number of fish in a lake, a conservationist catches 225 fish, tags them, and throws them back into the lake. Later, 108 fish are caught, and it is found that 15 of them are tagged. Estimate how many fish are in the lake.

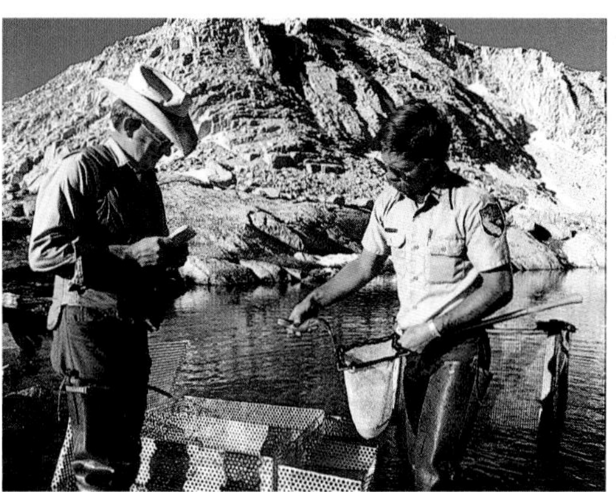

61. *Snow to Water.* Under typical conditions, $1\frac{1}{2}$ ft of snow will melt to 2 in. of water. To how many inches of water will $5\frac{1}{2}$ ft of snow melt?

62. *Tire Wear.* Tires are often priced according to the number of miles that they are expected to be driven. Suppose a tire priced at $59.76 is expected to be driven 35,000 mi. How much would you pay for a tire that is expected to be driven 40,000 mi?

Earned Run Average. In baseball, the average number of earned runs given up by a pitcher in 9 innings is the pitcher's *earned run average,* or *ERA*. For example, John Smoltz of the Atlanta Braves gave up 83 earned runs in $253\frac{2}{3}$ innings during the year in which he earned the Cy Young award as the finest pitcher in the National League. His earned average is found by solving the following proportion:

$$\frac{\text{ERA}}{9} = \frac{83}{253\frac{2}{3}}$$

$$\text{ERA} = \frac{9 \cdot 83}{253\frac{2}{3}} \approx 2.94.$$

Complete the following table to find the ERA of each National League pitcher in the same year.

	Player	Team	Earned Runs	Innings Pitched	ERA
	John Smoltz	Atlanta Braves	83	$253\frac{2}{3}$	2.94
63.	Greg Maddux	Atlanta Braves	74	245	
64.	Jaime Navarro	Chicago Cubs	103	$236\frac{2}{3}$	
65.	Kevin Ritz	Colorado Rockies	125	213	
66.	Hideo Nomo	Los Angeles Dodgers	81	$228\frac{1}{3}$	

Skill Maintenance

Solve. [3.7a]

67. Dallas, Texas, receives an average of 31.1 in. (78.994 cm) of rain and 2.6 in. (6.604 cm) of snow each year (***Source*:** National Oceanic and Atmospheric Administration).

a) What is the total amount of precipitation in inches?
b) What is the total amount of precipitation in centimeters?

68. The distance, by air, from New York to St. Louis is 876 mi (1401.6 km) and from St. Louis to Los Angeles is 1562 mi (2499.2 km).

a) How far, in miles, is it from New York to Los Angeles?
b) How far, in kilometers, is it from New York to Los Angeles?

Synthesis

69. ◆ Polly solved Example 13 by forming the proportion $\frac{t}{420} = \frac{35}{28}$, whereas Rudy wrote $\frac{420}{35} = \frac{28}{t}$. Are both approaches valid? Why or why not?

70. ◆ An instructor predicts that a student's test grade will be proportional to the amount of time the student spends studying. What is meant by this? Write an example of a proportion that involves the grades of two students and their study times.

71. ▦ Carney College is expanding from 850 to 1050 students. To avoid any rise in the student-to-faculty ratio, the faculty of 69 professors must also increase. How many new faculty positions should be created?

72. ▦ In recognition of her outstanding work, Sheri's salary has been increased from \$26,000 to \$29,380. Tim is earning \$23,000 and is requesting a proportional raise. How much more should he ask for?

73. Sue can paint 950 ft^2 with 2 gal of paint. How many gallons should Sue buy in order to paint a 30-ft by 100-ft wall?

74. Cy Young, one of the greatest baseball pitchers of all time, had an earned run average of 2.63. He pitched more innings, 7356, than anyone in the history of baseball. How many earned runs did he give up?

Analyze the ratios of different-colored M&M candies.
Use proportions to estimate your college's student population.

4.2 Percent Notation

a Understanding Percent Notation

Of all wood harvested, 35% of it is used for paper production. What does this mean? It means that, on average, of every 100 tons of wood harvested, 35 tons is used to produce paper. Thus, 35% is a ratio of 35 to 100, or $\frac{35}{100}$.

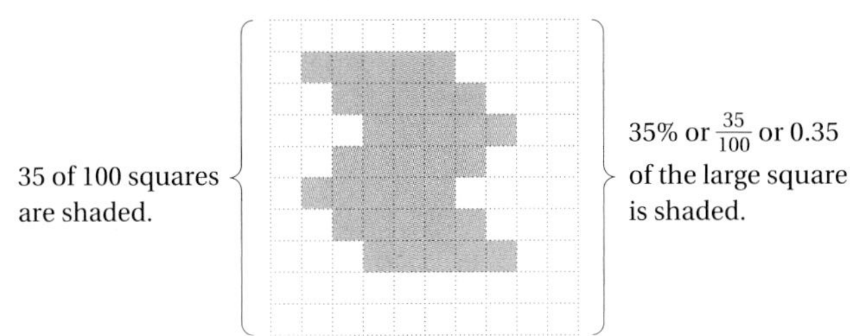

Percent notation is used extensively in our lives. Here are some examples:

Astronauts lose 1% of their bone mass for each month of weightlessness.

95% of hair spray is alcohol.

55% of all baseball merchandise sold is purchased by women.

62.4% of all aluminum cans were recycled in a recent year.

56% of all fruit juice purchased is orange juice.

45.8% of us sleep between 7 and 8 hours per night.

74% of the times a major-league baseball player strikes out swinging, the pitch was out of the strike zone.

Percent notation is often represented by pie charts to show how the parts of a quantity are related. For example, the chart below relates the amounts of different kinds of juices that are sold.

Juices Sold

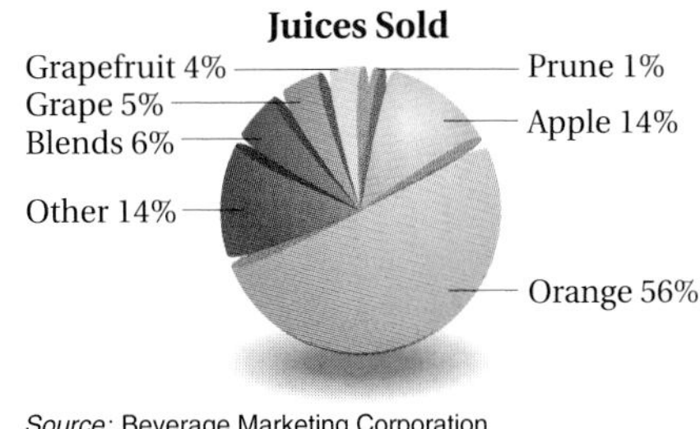

Source: Beverage Marketing Corporation

> The notation ***n*%** means "n per hundred."

Objectives

- a Write three kinds of notation for a percent.
- b Convert from percent notation to decimal notation.
- c Convert from decimal notation to percent notation.

For Extra Help

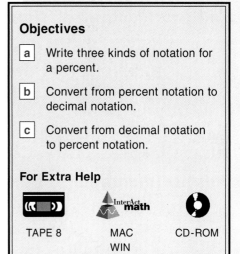

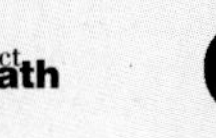

TAPE 8 | MAC WIN | CD-ROM

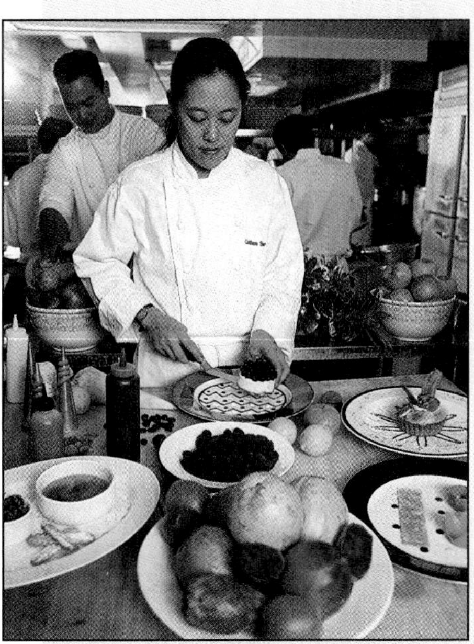

The number of jobs for professional chefs will increase by 43% from the year 1992 to the year 2005.

Write three kinds of notation as in Examples 1 and 2.

1. 70%

2. 23.4%

3. 100%

It is thought that the Roman Emperor Augustus began percent notation by taxing goods sold at a rate of $\frac{1}{100}$. In time, the symbol "%" evolved by interchanging the parts of the symbol "100" to "0/0" and then to "%".

Answers on page A-8

This definition leads us to the following equivalent ways of defining percent notation.

> **Percent notation, *n*%, is defined using:**
>
> ratio → $n\% = $ the ratio of n to 100 $= \frac{n}{100}$;
>
> fractional notation → $n\% = n \times \frac{1}{100}$;
>
> decimal notation → $n\% = n \times 0.01$.

Example 1 Write three kinds of notation for 35%.

Using ratio: $35\% = \frac{35}{100}$ A ratio of 35 to 100

Using fractional notation: $35\% = 35 \times \frac{1}{100}$ Replacing % with $\times \frac{1}{100}$

Using decimal notation: $35\% = 35 \times 0.01$ Replacing % with $\times$ 0.01

Example 2 Write three kinds of notation for 67.8%.

Using ratio: $67.8\% = \frac{67.8}{100}$ A ratio of 67.8 to 100

Using fractional notation: $67.8\% = 67.8 \times \frac{1}{100}$ Replacing % with $\times \frac{1}{100}$

Using decimal notation: $67.8\% = 67.8 \times 0.01$ Replacing % with $\times$ 0.01

Do Exercises 1–3.

b Converting from Percent Notation to Decimal Notation

Consider 78%. To convert to decimal notation, we can think of percent notation as a ratio and write

$78\% = \frac{78}{100}$ Using the definition of percent as a ratio

$= 0.78.$ Converting to decimal notation

Similarly,

$4.9\% = \frac{4.9}{100}$ Using the definition of percent as a ratio

$= 0.049.$ Converting to decimal notation

We could also convert 78% to decimal notation by replacing "%" with "$\times$ 0.01" and write

$78\% = 78 \times 0.01$ Replacing % with $\times$ 0.01

$= 0.78.$ Multiplying

Similarly,

$4.9\% = 4.9 \times 0.01$ Replacing % with $\times$ 0.01

$= 0.049.$ Multiplying

Dividing by 100 amounts to moving the decimal point two places to the left, which is the same as multiplying by 0.01. This leads us to a quick way to convert from percent notation to decimal notation: We drop the percent symbol and move the decimal point two places to the left.

To convert from percent notation to decimal notation,	36.5%	
a) replace the percent symbol % with × 0.01, and	36.5 × 0.01	
b) multiply by 0.01, which means move the decimal point two places to the left.	0.36.5 36.5% = 0.365	Move 2 places to the left.

Find decimal notation.

4. 34%

5. 78.9%

Example 3 Find decimal notation for 99.44%.

a) Replace the percent symbol with × 0.01. 99.44 × 0.01

b) Move the decimal point two places to the left. 0.99.44

Thus, 99.44% = 0.9944.

Example 4 The interest rate on a $2\frac{1}{2}$-year certificate of deposit is 5.1%. Find decimal notation for 5.1%.

a) Replace the percent symbol with × 0.01. 5.1 × 0.01

b) Move the decimal point two places to the left. 0.05.1

Thus, 5.1% = 0.051.

Find decimal notation for the percent notation in the sentence.

6. People forget 83% of all names they learn.

7. Soft drink sales in the United States have grown 4.2% annually over the past decade.

Do Exercises 4–7.

c Converting from Decimal Notation to Percent Notation

To convert 0.38 to percent notation, we can first write fractional notation, as follows:

$$0.38 = \frac{38}{100}$$ Converting to fractional notation

$$= 38\%.$$ Using the definition of percent as a ratio

Note that 100% = 100 × 0.01 = 1. Thus to convert 0.38 to percent notation, we can multiply by 1, using 100% as a symbol for 1. Then

$$0.38 = 0.38 \times 1$$
$$= 0.38 \times 100\%$$
$$= 0.38 \times 100 \times 0.01$$
$$= 38 \times 0.01$$
$$= 38\%.$$ Replacing "× 0.01" with the % symbol

Even more quickly, since 0.38 = 0.38 × 100%, we can simply multiply 0.38 by 100 and write the % symbol.

Answers on page A-8

Find percent notation.

8. 0.24

9. 3.47

10. 1

Find percent notation for the decimal notation in the sentence.

11. Blood is 0.9 water.

12. Of those accidents requiring medical attention, 0.108 of them occur on roads.

Answers on page A-8

To convert from decimal notation to percent notation, multiply b41 100%—that is, move the decimal point two places to the right and write a percent symbol.

To convert from decimal notation to percent notation, multiply by 100%. That is,	$0.675 = 0.675 \times 100\%$	
a) move the decimal point two places to the right, and	0.67.5	Move 2 places to the right.
b) write a % symbol.	67.5% $0.675 = 67.5\%$	

Example 5 Find percent notation for 1.27.

a) Move the decimal point two places to the right. 1.27.

b) Write a % symbol. 127%

Thus, 1.27 = 127%.

Example 6 Television sets are on 0.25 of the time. Find percent notation for 0.25.

a) Move the decimal point two places to the right. 0.25.

b) Write a % symbol. 25%

Thus, 0.25 = 25%.

Example 7 Find percent notation for 5.6.

a) Move the decimal point two places to the right, adding an extra zero. 5.60.

b) Write a % symbol. 560%

Thus, 5.6 = 560%.

Do Exercises 8–12.

Exercise Set 4.2

a Write three kinds of notation as in Examples 1 and 2 on p. 216.

1. 90%

2. 58.7%

3. 12.5%

4. 130%

b Find decimal notation.

5. 67%

6. 17%

7. 45.6%

8. 76.3%

9. 59.01%

10. 30.02%

11. 10%

12. 40%

13. 1%

14. 100%

15. 200%

16. 300%

17. 0.1%

18. 0.4%

19. 0.09%

20. 0.12%

21. 0.18%

22. 5.5%

23. 23.19%

24. 87.99%

Find decimal notation for the percent notation in the sentence.

25. On average, about 40% of the body weight of an adult male is muscle.

26. On average, about 23% of the body weight of an adult female is muscle.

27. A person's brain is 2.5% of his or her body weight.

28. It is known that 16% of all dessert orders in restaurants is for pie.

29. It is known that 62.2% of us think Monday is the worst day of the week.

30. Of all 18-year-olds, 68.4% have a driver's license.

c Find percent notation.

31. 0.47

32. 0.87

33. 0.03

34. 0.01

35. 8.7

36. 4

37. 0.334

38. 0.889

39. 0.75

40. 0.99

41. 0.4

42. 0.5

43. 0.006

44. 0.008

45. 0.017

46. 0.024

47. 0.2718

48. 0.8911

49. 0.0239

50. 0.00073

Find percent notation for the decimal notation in the sentence.

51. Around the fourth of July, about 0.000104 of all children aged 15 to 19 suffer injuries from fireworks.

52. About 0.144 of all children are cared for by relatives.

53. It is known that 0.24 of us go to the movies once a month.

54. It is known that 0.458 of us sleep between 7 and 8 hours.

55. Of all CDs purchased, 0.581 of them are pop/rock.

56. About 0.026 of all college football players go on to play professional football.

Skill Maintenance

Convert to a mixed numeral. [2.4a]

57. $\frac{100}{3}$

58. $\frac{75}{2}$

59. $\frac{75}{8}$

60. $\frac{297}{16}$

Convert to decimal notation. [3.5a]

61. $\frac{2}{3}$

62. $\frac{1}{3}$

63. $\frac{5}{6}$

64. $\frac{17}{12}$

Synthesis

65. ◆ ▦ What would you do to an entry on a calculator in order to get percent notation? Explain.

66. ◆ ▦ What would you do to percent notation on a calculator in order to get decimal notation? Explain.

4.3 Percent Notation and Fractional Notation

a Converting from Fractional Notation to Percent Notation

Consider the fractional notation $\frac{7}{8}$. To convert to percent notation, we use two skills we already have. We first find decimal notation by dividing:

$$\frac{7}{8} = 0.875 \qquad \begin{array}{r} 0.875 \\ 8\,\overline{)\,7.000} \\ \underline{6\,4} \\ 60 \\ \underline{56} \\ 40 \\ \underline{40} \\ 0 \end{array}$$

Then we convert the decimal notation to percent notation. We move the decimal point two places to the right

0.87.5

and write a % symbol:

$$\frac{7}{8} = 87.5\%, \text{ or } 87\frac{1}{2}\%.$$

To convert from fractional notation to percent notation,	$\frac{3}{5}$ Fractional notation
a) find decimal notation by division, and	$\begin{array}{r} 0.6 \\ 5\,\overline{)\,3.0} \\ \underline{3\,0} \\ 0 \end{array}$
b) convert the decimal notation to percent notation.	$0.6 = 0.60 = 60\%$ Percent notation $\frac{3}{5} = 60\%$

Example 1 Find percent notation for $\frac{3}{8}$.

a) Find decimal notation by division.

$$\begin{array}{r} 0.375 \\ 8\,\overline{)\,3.000} \\ \underline{2\,4} \\ 60 \\ \underline{56} \\ 40 \\ \underline{40} \\ 0 \end{array} \qquad \frac{3}{8} = 0.375$$

Objectives

a Convert from fractional notation to percent notation.

b Convert from percent notation to fractional notation.

For Extra Help

TAPE 8 MAC WIN CD-ROM

Calculator Spotlight

Conversion. Calculators are often used when we are converting fractional notation to percent notation. We simply perform the division on the calculator and then convert the decimal notation to percent notation. For example, percent notation for $\frac{17}{40}$ can be found by pressing

[1] [7] [÷] [4] [0] [=]

and then converting the result, 0.425, to percent notation, 42.5%.

Exercises

Find percent notation. Round to the nearest hundredth of a percent.

1. $\frac{13}{25}$ **2.** $\frac{5}{13}$

3. $\frac{42}{39}$ **4.** $\frac{12}{7}$

5. $\frac{217}{364}$ **6.** $\frac{2378}{8401}$

Find percent notation.

1. $\frac{1}{4}$

2. $\frac{5}{8}$

3. The human body is $\frac{2}{3}$ water. Find percent notation for $\frac{2}{3}$.

4. Find percent notation: $\frac{5}{6}$.

Find percent notation.

5. $\frac{57}{100}$

6. $\frac{19}{25}$

Answers on page A-8

b) Convert the decimal notation to percent notation. Move the decimal point two places to the right, and write a % symbol.

$$0.37.5$$

$$\frac{3}{8} = 37.5\%, \text{ or } 37\frac{1}{2}\%$$

Don't forget the % symbol.

Do Exercises 1 and 2.

Example 2 Of all meals, $\frac{1}{3}$ are eaten outside the home. Find percent notation for $\frac{1}{3}$.

a) Find decimal notation by division.

$$\begin{array}{r} 0.3\,3\,3 \\ 3\,\overline{)\,1.0\,0\,0} \\ \underline{9} \\ 1\,0 \\ \underline{9} \\ 1\,0 \\ \underline{9} \\ 1 \end{array}$$

We get a repeating decimal: $0.33\overline{3}$.

b) Convert the answer to percent notation.

$$0.33.\overline{3}$$

$$\frac{1}{3} = 33.\overline{3}\%, \text{ or } 33\frac{1}{3}\%$$

Do Exercises 3 and 4.

In some cases, division is not the fastest way to convert. The following are some optional ways in which conversion might be done.

Example 3 Find percent notation for $\frac{69}{100}$.

We use the definition of percent as a ratio.

$$\frac{69}{100} = 69\%$$

Example 4 Find percent notation for $\frac{17}{20}$.

We multiply by 1 to get 100 in the denominator. We think of what we have to multiply 20 by in order to get 100. That number is 5, so we multiply by 1 using $\frac{5}{5}$.

$$\frac{17}{20} \cdot \frac{5}{5} = \frac{85}{100} = 85\%$$

Note that this shortcut works only when the denominator is a factor of 100.

Do Exercises 5 and 6.

b Converting from Percent Notation to Fractional Notation

To convert from percent notation to fractional notation,	30%	Percent notation
a) use the definition of percent as a ratio, and	$\frac{30}{100}$	
b) simplify, if possible.	$\frac{3}{10}$	Fractional notation

Example 5 Find fractional notation for 75%.

$$75\% = \frac{75}{100} \qquad \text{Using the definition of percent}$$

$$= \frac{3 \cdot 25}{4 \cdot 25} = \frac{3}{4} \cdot \frac{25}{25} = \frac{3}{4} \qquad \text{Simplifying}$$

Example 6 Find fractional notation for 62.5%.

$$62.5\% = \frac{62.5}{100} \qquad \text{Using the definition of percent}$$

$$= \frac{62.5}{100} \times \frac{10}{10} \qquad \text{Multiplying by 1 to eliminate the decimal point in the numerator}$$

$$= \frac{625}{1000}$$

$$= \frac{5 \cdot 125}{8 \cdot 125} = \frac{5}{8} \cdot \frac{125}{125} = \frac{5}{8} \qquad \text{Simplifying}$$

Example 7 Find fractional notation for $16\frac{2}{3}\%$.

$$16\frac{2}{3}\% = \frac{50}{3}\% \qquad \text{Converting from the mixed numeral to fractional notation}$$

$$= \frac{50}{3} \times \frac{1}{100} \qquad \text{Using the definition of percent}$$

$$= \frac{50 \cdot 1}{3 \cdot 50 \cdot 2} = \frac{1}{6} \cdot \frac{50}{50} = \frac{1}{6} \qquad \text{Simplifying}$$

Do Exercises 7–10.

The table on the inside front cover lists decimal, fractional, and percent equivalents used so often that it would speed up your work if you learned them. For example, $\frac{1}{3} = 0.\overline{3}$, so we say that the **decimal equivalent** of $\frac{1}{3}$ is $0.\overline{3}$, or that $0.\overline{3}$ has the **fractional equivalent** $\frac{1}{3}$.

Find fractional notation.

7. 60%

8. 3.25%

9. $66\frac{2}{3}\%$

10. Complete this table.

Fractional Notation	$\frac{1}{5}$		
Decimal Notation		$0.8\overline{33}$	
Percent Notation			$37\frac{1}{2}\%$

Answers on page A-8

Calculator Spotlight

Applications of Ratio and Percent: The Price–Earnings Ratio and Stock Yields

The Price–Earnings Ratio

If the total earnings of a company one year were \$5,000,000 and 100,000 shares of stock were issued, the earnings per share was \$50. At one time, the price per share of Coca-Cola was $\$48\frac{5}{8}$ and the earnings per share was \$1.35. The **price–earnings ratio, *P/E*,** is the price of the stock divided by the earnings per share. For the Coca-Cola stock, the price–earnings ratio, *P/E*, is given by

$$\frac{P}{E} = \frac{48\frac{5}{8}}{1.35}$$

$$= \frac{48.625}{1.35} \quad \text{Converting to decimal notation}$$

$$\approx 36.02. \quad \text{Dividing, using a calculator, and rounding to the nearest tenth}$$

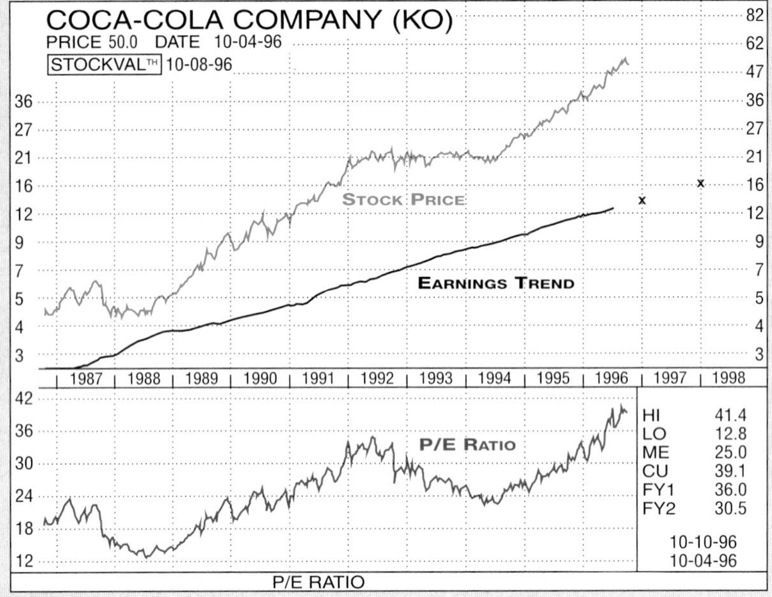

Source: C. H. Dean

Stock Yields

At one time, the price per share of Coca-Cola stock was $\$48\frac{5}{8}$ and the company was paying a yearly dividend of \$0.50 per share. It is helpful to those interested in stocks to know what percent the dividend is of the price of the stock. The percent is called the **yield**. For the Coca-Cola stock, the yield is given by

$$\text{Yield} = \frac{\text{Dividend}}{\text{Price per share}}$$

$$= \frac{0.50}{48\frac{5}{8}}$$

$$= \frac{0.50}{48.625} \quad \text{Converting to decimal notation}$$

$$\approx 0.0103 \quad \text{Dividing and rounding to the nearest thousandth}$$

$$\approx 1.03\%. \quad \text{Converting to percent notation}$$

Exercises

Compute the price–earnings ratio and the yield for each stock.

Stock	*Price per Share*	*Earnings*	*Dividend*
1. Monsanto	$\$35\frac{3}{4}$	\$1.41	\$0.60
2. K-Mart	$10\frac{7}{8}$	0.91	0.00
3. Rubbermaid	24	0.42	0.60
4. AT&T	$38\frac{5}{8}$	0.98	1.32

Exercise Set 4.3

a Find percent notation.

1. $\frac{41}{100}$
2. $\frac{36}{100}$
3. $\frac{5}{100}$
4. $\frac{1}{100}$
5. $\frac{2}{10}$
6. $\frac{7}{10}$
7. $\frac{3}{10}$
8. $\frac{9}{10}$
9. $\frac{1}{2}$
10. $\frac{3}{4}$
11. $\frac{5}{8}$
12. $\frac{1}{8}$
13. $\frac{4}{5}$
14. $\frac{2}{5}$
15. $\frac{2}{3}$
16. $\frac{1}{3}$
17. $\frac{1}{6}$
18. $\frac{5}{6}$
19. $\frac{4}{25}$
20. $\frac{17}{25}$
21. $\frac{1}{20}$
22. $\frac{31}{50}$
23. $\frac{17}{50}$
24. $\frac{3}{20}$

Find percent notation for the fractional notation in the sentence.

25. Bread is $\frac{9}{25}$ water.

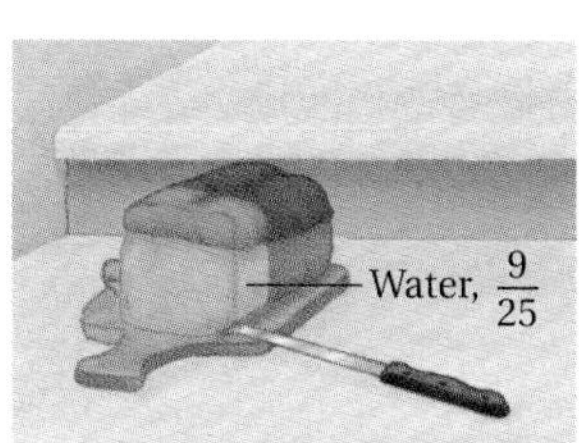

26. Milk is $\frac{7}{8}$ water.

Write percent notation for the fractions in this pie chart.

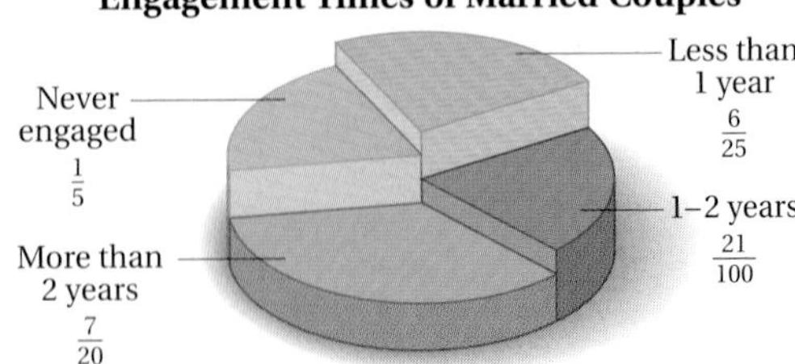

27. $\frac{21}{100}$

28. $\frac{1}{5}$

29. $\frac{6}{25}$

30. $\frac{7}{20}$

b Find fractional notation. Simplify.

31. 85%

32. 55%

33. 62.5%

34. 12.5%

35. $33\frac{1}{3}\%$

36. $83\frac{1}{3}\%$

37. $16.\overline{6}\%$

38. $66.\overline{6}\%$

39. 7.25%

40. 4.85%

41. 0.8%

42. 0.2%

43. $25\frac{3}{8}\%$

44. $48\frac{7}{8}\%$

45. $78\frac{2}{9}\%$

46. $16\frac{5}{9}\%$

47. $64\frac{7}{11}\%$

48. $73\frac{3}{11}\%$

49. 150%

50. 110%

51. 0.0325%

52. 0.419%

53. $33.\overline{3}\%$

54. $83.\overline{3}\%$

Find fractional notation for the percents in this bar graph.

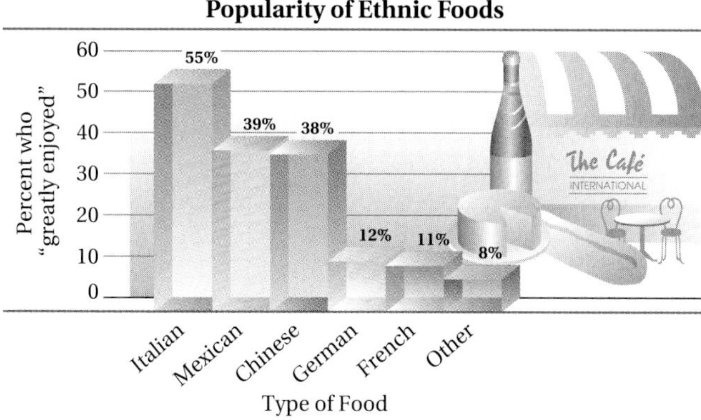

Note that there can be multiple responses.

55. 55%

56. 39%

57. 38%

58. 12%

59. 11%

60. 8%

Find fractional notation for the percent notation in the sentence.

61. A 30-g cup of Wheaties supplies 25% of the minimum daily requirement of vitamin A.

62. A 30-g cup of Wheaties supplies 45% of the minimum daily requirement of iron.

63. The interest rate on a 5-yr certificate of deposit was 5.69%.

64. One year the sales of *USA Today* increased 7.2%.

Complete the table.

65.

Fractional Notation	Decimal Notation	Percent Notation
$\frac{1}{8}$		$12\frac{1}{2}\%$, or 12.5%
$\frac{1}{6}$		
		20%
	0.25	
		$33\frac{1}{3}\%$, or $33.\overline{3}\%$
		$37\frac{1}{2}\%$, or 37.5%
		40%
$\frac{1}{2}$	0.5	50%

66.

Fractional Notation	Decimal Notation	Percent Notation
$\frac{3}{5}$		
	0.625	
$\frac{2}{3}$		
	0.75	75%
$\frac{4}{5}$		
$\frac{5}{6}$		$83\frac{1}{3}\%$, or $83.\overline{3}\%$
$\frac{7}{8}$		$87\frac{1}{2}\%$, or 87.5%
		100%

67.

Fractional Notation	Decimal Notation	Percent Notation
	0.5	
$\frac{1}{3}$		
		25%
		$16\frac{2}{3}\%$, or $16.\overline{6}\%$
	0.125	
$\frac{3}{4}$		
	$0.8\overline{3}$	
$\frac{3}{8}$		

68.

Fractional Notation	Decimal Notation	Percent Notation
		40%
		$62\frac{1}{2}\%$, or 62.5%
	0.875	
$\frac{1}{1}$		
	0.6	
	$0.\overline{6}$	
$\frac{1}{5}$		

Skill Maintenance

Solve.

69. $13 \cdot x = 910$ [1.4b]

70. $15 \cdot y = 75$ [1.4b]

71. $0.05 \times b = 20$ [3.4b]

72. $3 = 0.16 \times b$ [3.4b]

73. $\frac{1}{2} \cdot x = 2$ [2.2d]

74. $4 \cdot x = \frac{3}{11}$ [2.2d]

Convert to a mixed numeral. [2.4a]

75. $\frac{40}{31}$

76. $\frac{63}{12}$

77. $\frac{250}{3}$

78. $\frac{123}{6}$

79. $\frac{345}{8}$

80. $\frac{373}{6}$

81. $\frac{75}{4}$

82. $\frac{67}{9}$

Synthesis

83. ◆ Is it always best to convert from fractional notation to percent notation by first finding decimal notation? Why or why not?

84. ◆ Athletes sometimes speak of "giving 110%" effort. Does this make sense? Why or why not?

Find percent notation.

85. ▦ $\frac{41}{369}$

86. ▦ $\frac{54}{999}$

Find decimal notation.

87. $\frac{14}{9}\%$

88. $\frac{19}{12}\%$

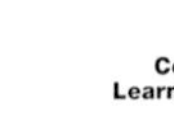

Use percent squares to develop a number sense for percents.

4.4 Solving Percent Problems Using Equations

a Translating to Equations

To solve a problem involving percents, it is helpful to translate first to an equation.

Example 1 Translate:

23%	of	5	is	what?
↓	↓	↓	↓	↓
23%	$\cdot$	5	=	a

> "**Of**" translates to "·", or "×". "**Is**" translates to "=".
>
> "**What**" translates to any letter. **%** translates to "$\times \frac{1}{100}$" or "× 0.01".

Example 2 Translate:

What	is	11%	of	49?
↓	↓	↓	↓	↓
a	=	11%	$\cdot$	49

Any letter can be used.

Do Exercises 1 and 2.

Example 3 Translate:

3	is	10%	of	what?
↓	↓	↓	↓	↓
3	=	10%	$\cdot$	b

Example 4 Translate:

45%	of	what	is	23?
↓	↓	↓	↓	↓
45%	×	b	=	23

Do Exercises 3 and 4.

Example 5 Translate:

10	is	what percent	of	20?
↓	↓	↓	↓	↓
10	=	n	×	20

Example 6 Translate:

What percent	of	50	is	7?
↓	↓	↓	↓	↓
n	$\cdot$	50	=	7

Do Exercises 5 and 6.

Objectives

a Translate percent problems to equations.

b Solve basic percent problems.

For Extra Help

TAPE 8 MAC WIN CD-ROM

Translate to an equation. Do not solve.

1. 12% of 50 is what?

2. What is 40% of 60?

Translate to an equation. Do not solve.

3. 45 is 20% of what?

4. 120% of what is 60?

Translate to an equation. Do not solve.

5. 16 is what percent of 40?

6. What percent of 84 is 10.5?

Answers on page A-9

7. Solve:

What is 12% of 50?

Answer on page A-9

b Solving Percent Problems

In solving percent problems, we use the *Translate* and *Solve* steps in the problem-solving strategy used throughout this text.

Percent problems are actually of three different types. Although the method we present does *not* require that you be able to identify which type we are studying, it is helpful to know them.

We know that

15 is 25% of 60, or

$15 = 25\% \times 60.$

We can think of this as:

> Amount = Percent number × Base.

Each of the three types of percent problems depends on which of the three pieces of information is missing.

1. **Finding the *amount* (the result of taking the percent)**

   Example: What is 25% of 60?

   Translation: $y = 25\% \cdot 60$

2. **Finding the *base* (the number you are taking the percent of)**

   Example: 15 is 25% of what number?

   Translation: $15 = 25\% \cdot y$

3. **Finding the *percent number* (the percent itself)**

   Example: 15 is what percent of 60?

   Translation: $15 = y \cdot 60$

Finding the Amount

Example 7 What is 11% of 49?

Translate: $a = 11\% \times 49.$

Solve: The letter is by itself. To solve the equation, we just convert 11% to decimal notation and multiply.

$$\begin{array}{r} 49 \\ \times\ 0.11 \\ \hline 49 \\ 490 \\ \hline a = 5.39 \end{array} \qquad 11\% = 0.11$$

> A way of checking answers is by estimating as follows:
>
> $11\% \times 49 \approx 10\% \times 50$
> $= 0.10 \times 50 = 5.$
>
> Since 5 is close to 5.39, our answer is reasonable.

Thus, 5.39 is 11% of 49. The answer is 5.39.

Do Exercise 7.

Example 8 120% of \$42 is what?

Translate: $120\% \times 42 = a$.

Solve: The letter is by itself. To solve the equation, we carry out the calculation.

$$\begin{array}{r} 4\,2 \\ \times\ 1.2 \\ \hline 8\,4 \\ 4\,2\,0 \\ \hline a = 5\,0.4 \end{array} \qquad 120\% = 1.20 = 1.2$$

Thus, 120% of \$42 is \$50.40. The answer is \$50.40.

Do Exercise 8.

8. Solve:

64% of \$55 is what?

Finding the Base

Example 9 5% of what is 20?

Translate: $5\% \times b = 20$.

Solve: This time the letter is *not* by itself. To solve the equation, we divide by 5% on both sides:

$$\frac{5\% \times b}{5\%} = \frac{20}{5\%} \qquad \text{Dividing by 5\% on both sides}$$

$$b = \frac{20}{0.05} \qquad 5\% = 0.05$$

$$= 400.$$

$$\begin{array}{r} 4\,0\,0. \\ 0.0\,5\,\overline{)\,2\,0.0\,0_\wedge} \\ 2\,0\,0\,0 \\ \hline 0 \end{array}$$

Thus, 5% of 400 is 20. The answer is 400.

Example 10 \$3 is 16% of what?

Translate: \$3 is 16% of what?

$$\begin{array}{ccccc} \$3 & \text{is} & 16\% & \text{of} & \text{what?} \\ \downarrow & \downarrow & \downarrow & \downarrow & \downarrow \\ 3 & = & 16\% & \times & b. \end{array}$$

Solve: Again, the letter is not by itself. To solve the equation, we divide by 16% on both sides:

$$\frac{3}{16\%} = \frac{16\% \times b}{16\%} \qquad \text{Dividing by 16\% on both sides}$$

$$\frac{3}{0.16} = b \qquad 16\% = 0.16$$

$$18.75 = b.$$

$$\begin{array}{r} 1\,8.7\,5 \\ 0.1\,6\,\overline{)\,3.0\,0_\wedge0\,0} \\ 1\,6 \\ \hline 1\,4\,0 \\ 1\,2\,8 \\ \hline 1\,2\,0 \\ 1\,1\,2 \\ \hline 8\,0 \\ 8\,0 \\ \hline 0 \end{array}$$

Thus, \$3 is 16% of \$18.75. The answer is \$18.75.

Do Exercises 9 and 10.

Solve.

9. 20% of what is 45?

10. \$60 is 120% of what?

Answers on page A-9

11. Solve:

16 is what percent of 40?

Finding the Percent Number

In solving these problems, you *must* remember to convert to percent notation after you have solved the equation.

Example 11 10 is what percent of 20?

Translate: 10 is what percent of 20?

$$10 = n \times 20.$$

Solve: To solve the equation, we divide by 20 on both sides and convert the result to percent notation:

$$n \cdot 20 = 10$$

$$\frac{n \cdot 20}{20} = \frac{10}{20} \qquad \text{Dividing by 20 on both sides}$$

$$n = 0.50 = 50\%. \qquad \text{Converting to percent notation}$$

Thus, 10 is 50% of 20. The answer is 50%.

Do Exercise 11.

Example 12 What percent of \$50 is \$16?

Translate: What percent of \$50 is \$16?

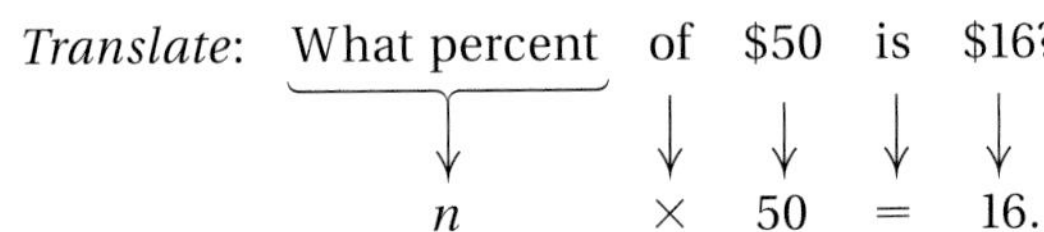

$$n \times 50 = 16.$$

12. Solve:

What percent of \$84 is \$10.50?

Solve: To solve the equation, we divide by 50 on both sides and convert the answer to percent notation:

$$\frac{n \times 50}{50} = \frac{16}{50} \qquad \text{Dividing by 50 on both sides}$$

$$n = \frac{16}{50}$$

$$= \frac{16}{50} \cdot \frac{2}{2}$$

$$= \frac{32}{100}$$

$$= 32\%. \qquad \text{Converting to percent notation}$$

Thus, 32% of \$50 is \$16. The answer is 32%.

Do Exercise 12.

CAUTION! When a question asks "what percent?", be sure to give the answer in percent notation.

Answers on page A-9

Exercise Set 4.4

a Translate to an equation. Do not solve.

1. What is 32% of 78?

2. 98% of 57 is what?

3. 89 is what percent of 99?

4. What percent of 25 is 8?

5. 13 is 25% of what?

6. 21.4% of what is 20?

b Solve.

7. What is 85% of 276?

8. What is 74% of 53?

9. 150% of 30 is what?

10. 100% of 13 is what?

11. What is 6% of $300?

12. What is 4% of $45?

13. 3.8% of 50 is what?

14. $33\frac{1}{3}$% of 480 is what? (*Hint:* $33\frac{1}{3}\% = \frac{1}{3}$.)

15. $39 is what percent of $50?

16. $16 is what percent of $90?

17. 20 is what percent of 10?

18. 60 is what percent of 20?

19. What percent of $300 is $150?

20. What percent of $50 is $40?

21. What percent of 80 is 100?

22. What percent of 60 is 15?

23. 20 is 50% of what?

24. 57 is 20% of what?

25. 40% of what is \$16?

26. 100% of what is \$74?

27. 56.32 is 64% of what?

28. 71.04 is 96% of what?

29. 70% of what is 14?

30. 70% of what is 35?

31. What is $62\frac{1}{2}\%$ of 10?

32. What is $35\frac{1}{4}\%$ of 1200?

33. What is 8.3% of \$10,200?

34. What is 9.2% of \$5600?

Skill Maintenance

Write fractional notation. [3.1b]

35. 0.09

36. 1.79

37. 0.875

38. 0.9375

Write decimal notation. [3.5a]

39. $\frac{89}{100}$

40. $\frac{7}{100}$

41. $\frac{3}{10}$

42. $\frac{17}{1000}$

Synthesis

43. ◆ Write a question that could be translated to the equation

$25 = 4\% \times b.$

44. ◆ To calculate a 15% tip on a \$24 bill, a customer adds \$2.40 and half of \$2.40, or \$1.20, to get \$3.60. Is this procedure valid? Why or why not?

Solve.

45. 🖩 What is 7.75% of \$10,880?
Estimate ______________
Calculate ______________

46. 🖩 50,951.775 is what percent of 78,995?
Estimate ______________
Calculate ______________

47. *Recyclables.* It is estimated that 40% to 50% of all trash is recyclable. If a community produces 270 tons of trash, how much of their trash is recyclable?

48. 40% of $18\frac{3}{4}\%$ of \$25,000 is what?

4.5 Solving Percent Problems Using Proportions*

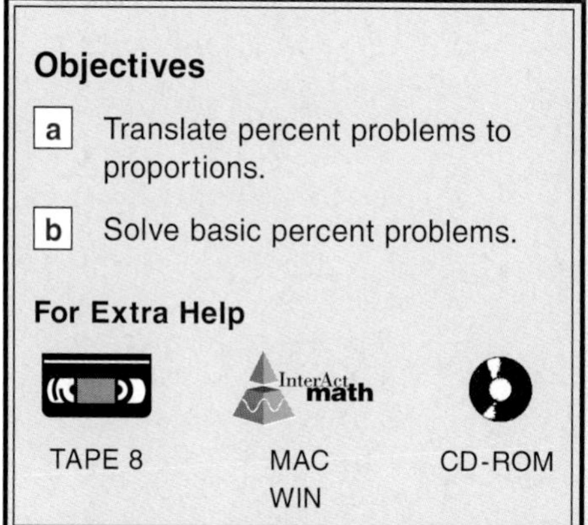

Objectives
a Translate percent problems to proportions.
b Solve basic percent problems.
For Extra Help
InterAct math
TAPE 8
MAC
WIN
CD-ROM

a Translating to Proportions

A percent is a ratio of some number to 100. For example, 75% is the ratio $\frac{75}{100}$. The numbers 3 and 4 have the same ratio as 75 and 100. Thus,

$$75\% = \frac{75}{100} = \frac{3}{4}.$$

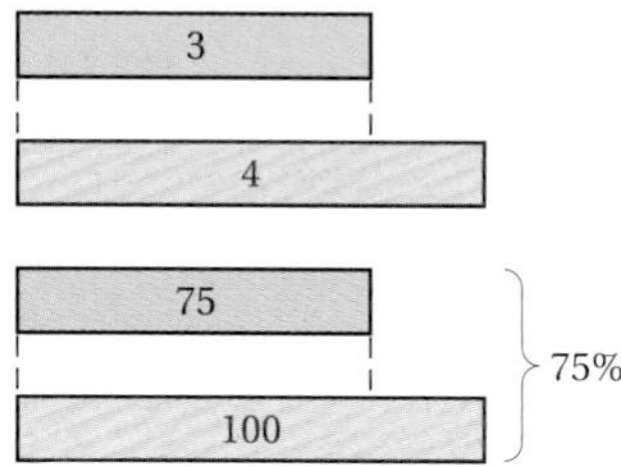

To solve a percent problem using a proportion, we translate as follows:

$$\begin{array}{l} \text{Number} \to \\ 100 \longrightarrow \end{array} \frac{N}{100} = \frac{a}{b} \begin{array}{l} \leftarrow \text{Amount} \\ \leftarrow \text{Base} \end{array}$$

You might find it helpful to read this as "part is to whole as part is to whole."

For example,

60% of 25 is 15

translates to

$$\frac{60}{100} = \frac{15}{25}. \begin{array}{l} \leftarrow \text{Amount} \\ \leftarrow \text{Base} \end{array}$$

A clue in translating is that the base, b, corresponds to 100 and usually follows the wording "percent of." Also, $N\%$ always translates to $N/100$. Another aid in translating is to make a comparison drawing. To do this, we start with the percent side and list 0% at the top and 100% near the bottom. Then we estimate where the specified percent—in this case, 60%—is located. The corresponding quantities are then filled in. The base—in this case, 25—always corresponds to 100% and the amount—in this case, 15—corresponds to the specified percent.

Percents	Quantities
0%	0
100%	

Percents	Quantities
0%	0
60%	
100%	

Percents	Quantities
0%	0
60%	15
100%	25

The proportion can then be read easily from the drawing.

**Note*: *This section presents an alternative method for solving basic percent problems. You can use either equations or proportions to solve percent problems, but you might prefer one method over the other, or your instructor may direct you to use one method over the other.*

Translate to a proportion. Do not solve.

1. 12% of 50 is what?

2. What is 40% of 60?

3. 130% of 72 is what?

Translate to a proportion. Do not solve.

4. 45 is 20% of what?

5. 120% of what is 60?

Answers on page A-9

Example 1 Translate to a proportion.

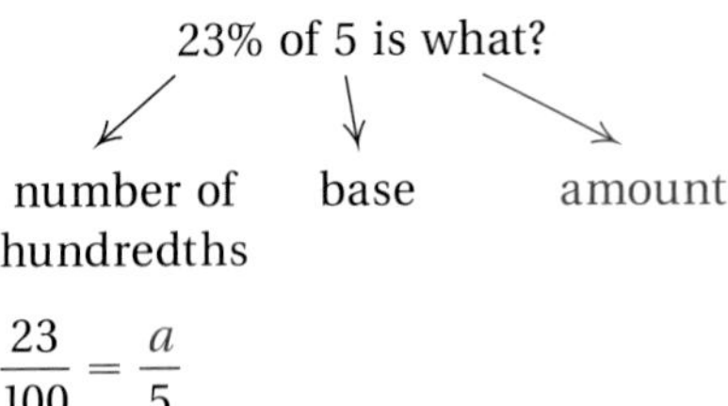

$$\frac{23}{100} = \frac{a}{5}$$

Percents	Quantities
0%	0
23%	a
100%	5

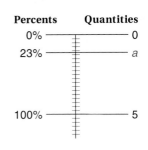

Example 2 Translate to a proportion.

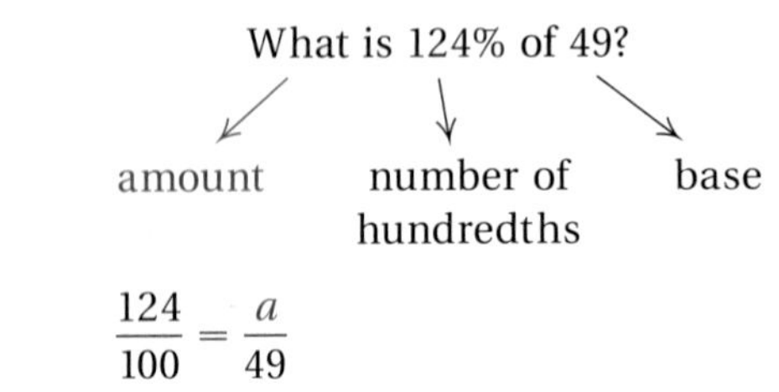

$$\frac{124}{100} = \frac{a}{49}$$

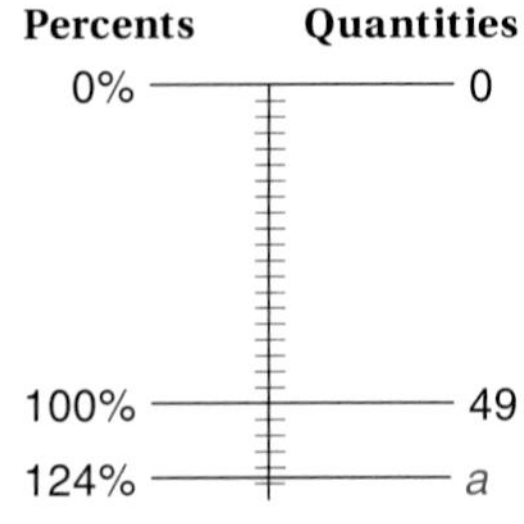

Do Exercises 1–3.

Example 3 Translate to a proportion.

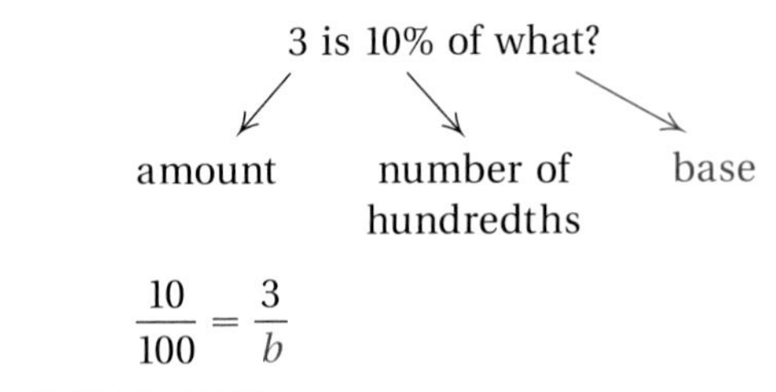

$$\frac{10}{100} = \frac{3}{b}$$

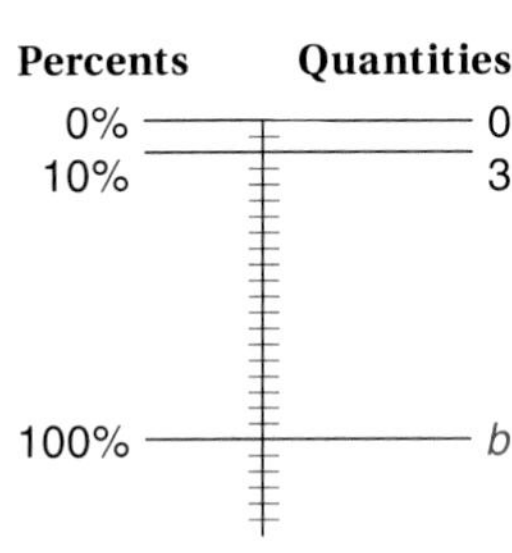

Example 4 Translate to a proportion.

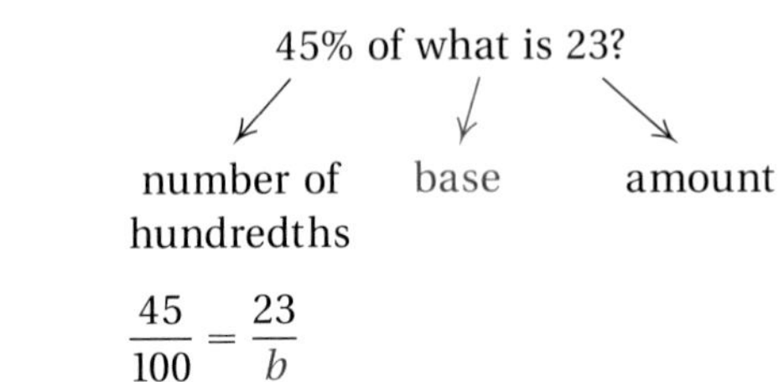

$$\frac{45}{100} = \frac{23}{b}$$

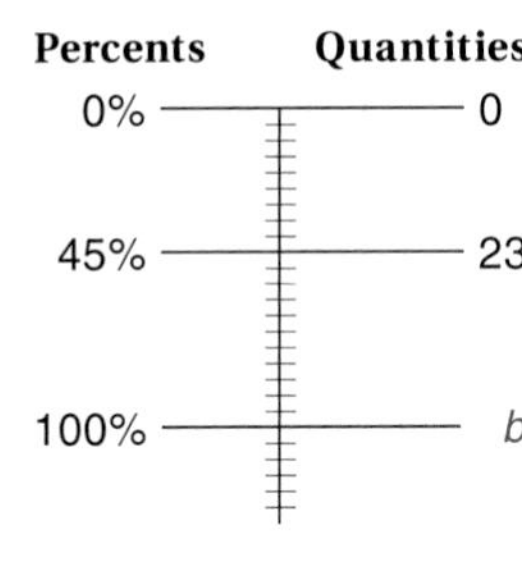

Do Exercises 4 and 5.

Example 5 Translate to a proportion.

10 is what percent of 20?

amount — number of hundredths — base

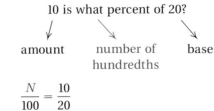

$$\frac{N}{100} = \frac{10}{20}$$

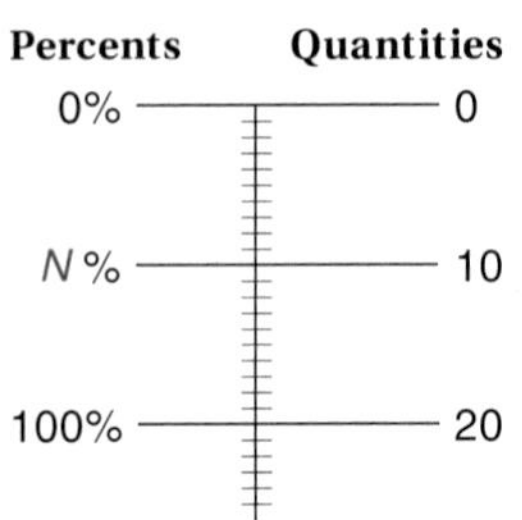

Example 6 Translate to a proportion.

What percent of 50 is 7?

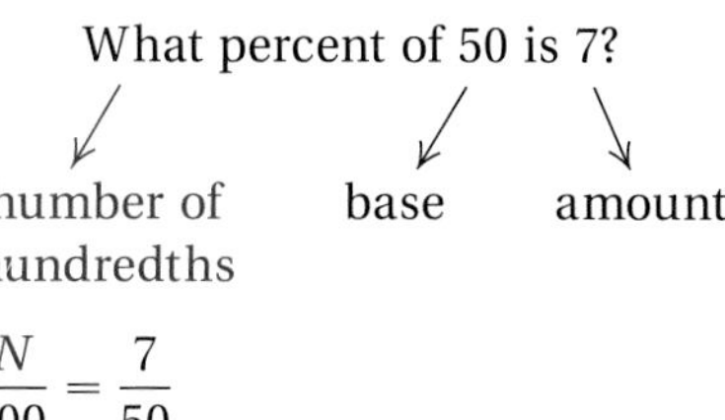

$$\frac{N}{100} = \frac{7}{50}$$

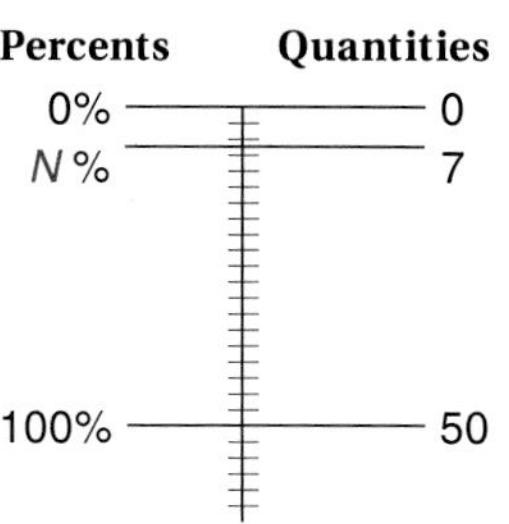

Do Exercises 6 and 7.

Translate to a proportion. Do not solve.

6. 16 is what percent of 40?

7. What percent of 84 is 10.5?

b Solving Percent Problems

After a percent problem has been translated to a proportion, we solve as in Section 4.1.

Example 7 5% of what is $20?

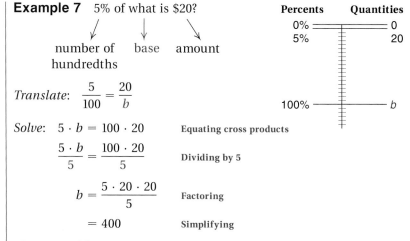

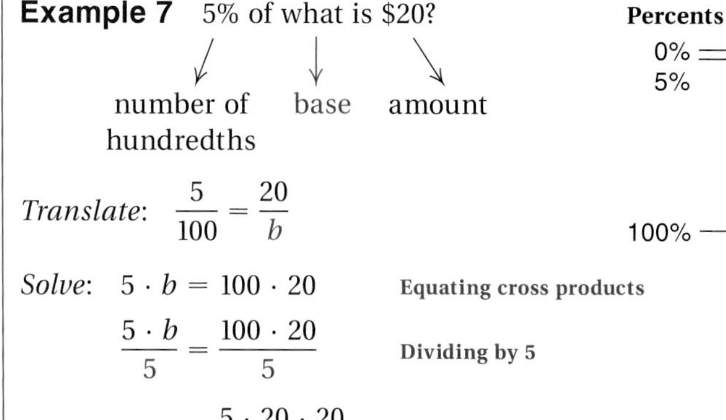

Translate: $\frac{5}{100} = \frac{20}{b}$

Solve: $5 \cdot b = 100 \cdot 20$ **Equating cross products**

$\frac{5 \cdot b}{5} = \frac{100 \cdot 20}{5}$ **Dividing by 5**

$b = \frac{5 \cdot 20 \cdot 20}{5}$ **Factoring**

$= 400$ **Simplifying**

Thus, 5% of $400 is $20. The answer is $400.

Do Exercise 8.

8. Solve:

20% of what is $45?

Example 8 120% of 42 is what?

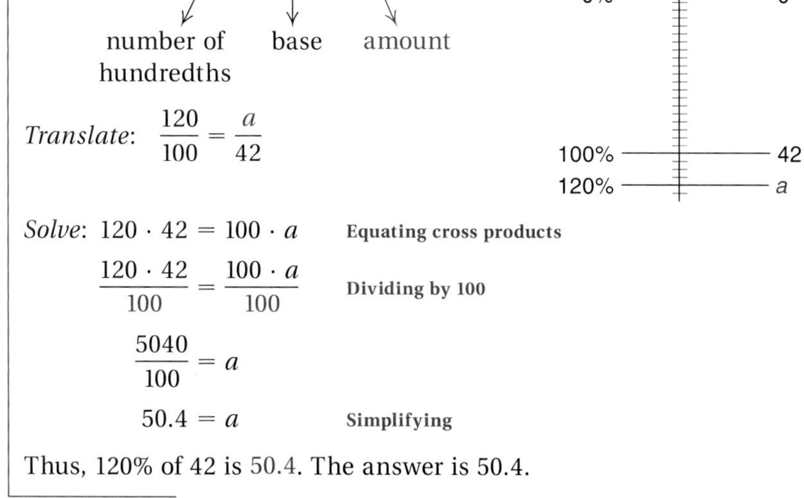

Translate: $\frac{120}{100} = \frac{a}{42}$

Solve: $120 \cdot 42 = 100 \cdot a$ **Equating cross products**

$\frac{120 \cdot 42}{100} = \frac{100 \cdot a}{100}$ **Dividing by 100**

$\frac{5040}{100} = a$

$50.4 = a$ **Simplifying**

Thus, 120% of 42 is 50.4. The answer is 50.4.

Do Exercises 9 and 10.

Solve.

9. 64% of 55 is what?

10. What is 12% of 50?

Answers on page A-9

11. Solve:

60 is 120% of what?

12. Solve:

\$12 is what percent of \$40?

13. Solve:

What percent of 84 is 10.5?

Answers on page A-9

Example 9 3 is 16% of what?

amount — 3; number of hundredths — 16; base — what

Percents	Quantities
0%	0
16%	3
100%	b

Translate: $\frac{3}{b} = \frac{16}{100}$

Solve:

$3 \cdot 100 = b \cdot 16$ **Equating cross products**

$\frac{3 \cdot 100}{16} = \frac{b \cdot 16}{16}$ **Dividing by 16**

$\frac{3 \cdot 100}{16} = b$

$18.75 = b$ **Multiplying and dividing**

Thus, 3 is 16% of 18.75. The answer is 18.75.

Do Exercise 11.

Example 10 \$10 is what percent of \$20?

amount — \$10; number of hundredths — what percent; base — \$20

Percents	Quantities
0%	0
N%	\$10
100%	\$20

Translate: $\frac{10}{20} = \frac{N}{100}$

Solve:

$10 \cdot 100 = 20 \cdot N$ **Equating cross products**

$\frac{10 \cdot 100}{20} = \frac{20 \cdot N}{20}$ **Dividing by 20**

$\frac{10 \cdot 100}{20} = N$

$50 = N$ **Multiplying and dividing**

Thus, \$10 is 50% of \$20. The answer is 50%.

Do Exercise 12.

Example 11 What percent of 50 is 16?

number of hundredths — what percent; base — 50; amount — 16

Percents	Quantities
0%	0
N%	16
100%	50

Translate: $\frac{N}{100} = \frac{16}{50}$

Solve:

$50 \cdot N = 100 \cdot 16$ **Equating cross products**

$\frac{50 \cdot N}{50} = \frac{100 \cdot 16}{50}$ **Dividing by 50**

$N = \frac{100 \cdot 16}{50}$

$= 32$ **Multiplying and dividing**

Thus, 32% of 50 is 16. The answer is 32%.

Do Exercise 13.

Exercise Set 4.5

a Translate to a proportion. Do not solve.

1. What is 37% of 74?

2. 66% of 74 is what?

3. 4.3 is what percent of 5.9?

4. What percent of 6.8 is 5.3?

5. 14 is 25% of what?

6. 133% of what is 40?

b Solve.

7. What is 76% of 90?

8. What is 32% of 70?

9. 70% of 660 is what?

10. 80% of 920 is what?

11. What is 4% of 1000?

12. What is 6% of 2000?

13. 4.8% of 60 is what?

14. 63.1% of 80 is what?

15. \$24 is what percent of \$96?

16. \$14 is what percent of \$70?

17. 102 is what percent of 100?

18. 103 is what percent of 100?

19. What percent of \$480 is \$120?

20. What percent of \$80 is \$60?

21. What percent of 160 is 150?

22. What percent of 33 is 11?

23. \$18 is 25% of what?

24. \$75 is 20% of what?

25. 60% of what is 54?

26. 80% of what is 96?

27. 65.12 is 74% of what?

28. 63.7 is 65% of what?

29. 80% of what is 16?

30. 80% of what is 10?

31. What is $62\frac{1}{2}\%$ of 40?

32. What is $43\frac{1}{4}\%$ of 2600?

33. What is 9.4% of \$8300?

34. What is 8.7% of \$76,000?

Skill Maintenance

Solve. [4.1d]

35. $\frac{x}{188} = \frac{2}{47}$

36. $\frac{15}{x} = \frac{3}{800}$

37. $\frac{4}{7} = \frac{x}{14}$

38. $\frac{612}{t} = \frac{72}{244}$

39. $\frac{5000}{t} = \frac{3000}{60}$

40. $\frac{75}{100} = \frac{n}{20}$

41. $\frac{x}{1.2} = \frac{36.2}{5.4}$

42. $\frac{y}{1\frac{1}{2}} = \frac{2\frac{3}{4}}{22}$

Solve. [2.5a]

43. A recipe for muffins calls for $\frac{1}{2}$ qt of buttermilk, $\frac{1}{3}$ qt of skim milk, and $\frac{1}{16}$ qt of oil. How many quarts of liquid ingredients does the recipe call for?

44. The Ferristown School District purchased $\frac{3}{4}$ ton (T) of clay. If the clay is to be shared equally among the district's 6 art departments, how much will each art department receive?

Synthesis

45. ◆ In your own words, list steps that a classmate could use to solve any percent problem in this section.

46. ◆ In solving Example 10, a student simplifies $\frac{10}{20}$ before solving. Is this a good idea? Why or why not?

Solve.

47. 🖩 What is 8.85% of \$12,640?
Estimate ______________
Calculate ______________

48. 🖩 78.8% of what is 9809.024?
Estimate ______________
Calculate ______________

4.6 Applications of Percent

a Applied Problems Involving Percent

Applied problems involving percent are not always stated in a manner easily translated to an equation. In such cases, it is helpful to rephrase the problem before translating. Sometimes it also helps to make a drawing.

Example 1 *Paper Recycling.* In a recent year, the United States generated 73.3 million tons of paper waste, of which 20.5 million tons were recycled (***Source*:** Environmental Protection Agency). What percent of paper waste was recycled?

1. **Familiarize.** The question asks for a percent. We know that 10% of 73.3 is 7.33. Since $20.5 \approx 3 \times 7.33$, we expect the answer to be close to 30%. We let $n =$ the percent of paper waste that was recycled.
2. **Translate.** We can rephrase the question and translate as follows:

20.5 million	is	what percent	of	73.3 million?
↓	↓	↓	↓	↓
20,500,000	$=$	n	$\times$	73,300,000

3. **Solve.** We solve as we did in Section 4.4:*

$$20{,}500{,}000 = n \times 73{,}300{,}000$$

$$\frac{20{,}500{,}000}{73{,}300{,}000} = \frac{n \times 73{,}300{,}000}{73{,}300{,}000} \quad \text{Dividing by 73,300,000 on both sides}$$

$$0.28 \approx n \quad \text{Rounding to the nearest hundredth}$$

$$28\% = n. \quad \text{Remember to find percent notation.}$$

4. **Check.** To check, we note that the answer, 28%, is close to 30%, as predicted in the *Familiarize* step.
5. **State.** About 28% of the paper waste was recycled.

Do Exercise 1.

*We can also use the proportion method of Section 4.5 and solve:

$$\frac{N}{100} = \frac{20{,}500{,}000}{73{,}300{,}000}.$$

Objectives

a Solve applied problems involving percent.

b Solve applied problems involving percent of increase or decrease.

For Extra Help

TAPE 8 | MAC WIN | CD-ROM

1. *Desserts.* If a restaurant sells 250 desserts in an evening, it is typical that 40 of them will be pie. What percent of the desserts sold will be pie?

Answer on page A-9

2. *Desserts.* Of all desserts sold in restaurants, 20% of them are chocolate cake. One evening a restaurant sells 250 desserts. How many were chocolate cake?

Answer on page A-9

Example 2 *Junk Mail.* The U.S. Postal Service estimates that we read 78% of the junk mail we receive. Suppose that a business sends out 9500 advertising brochures. How many brochures can the business expect to be opened and read?

1. **Familiarize.** We can draw a pie chart to help familiarize ourselves with the problem. We let $a =$ the number of brochures that are opened and read.

Mail Advertising Opened

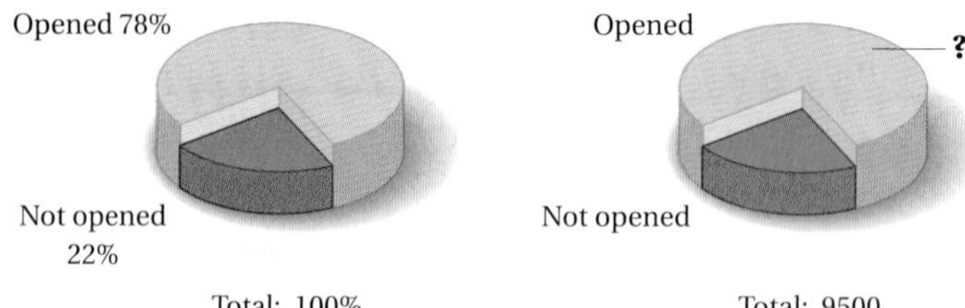

2. **Translate.** The question can be rephrased and translated as follows.

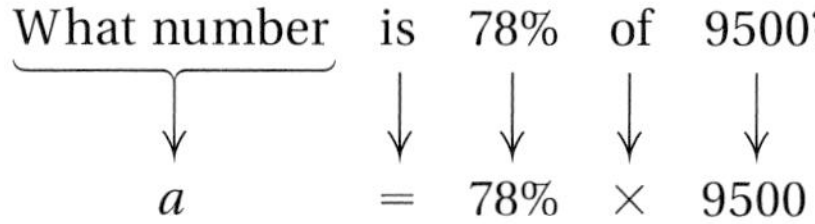

What number is 78% of 9500?

$$a = 78\% \times 9500$$

3. **Solve.** We convert 78% to decimal notation and multiply:*

$$a = 78\% \times 9500 = 0.78 \times 9500 = 7410.$$

4. **Check.** To check, we can repeat the calculation. We can also think about our answer. Since we are taking 78% of 9500, we would expect 7410 to be smaller than 9500 and about three-fourths of 9500, which it is.

5. **State.** The business can expect 7410 of its brochures to be opened and read.

Do Exercise 2.

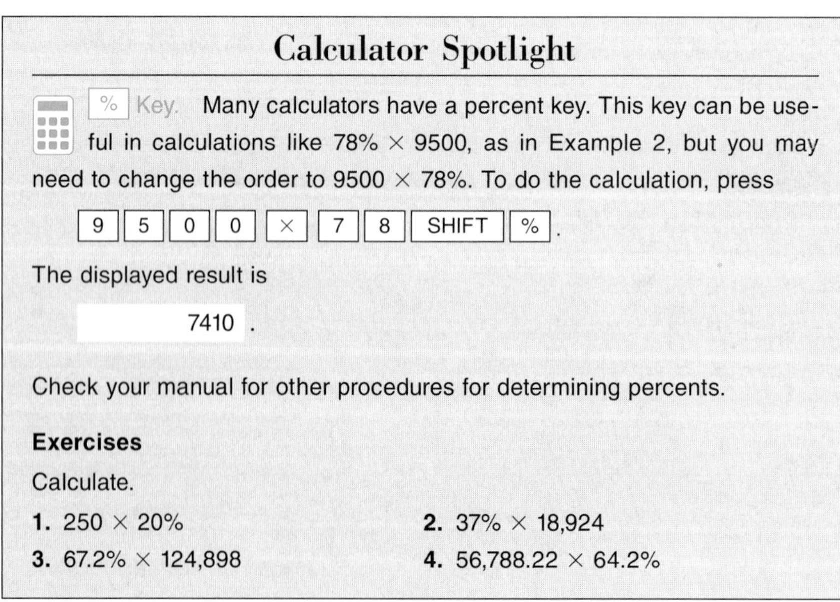

Calculator Spotlight

% Key. Many calculators have a percent key. This key can be useful in calculations like 78% × 9500, as in Example 2, but you may need to change the order to 9500 × 78%. To do the calculation, press

9 5 0 0 × 7 8 SHIFT %.

The displayed result is

7410.

Check your manual for other procedures for determining percents.

Exercises

Calculate.

1. 250 × 20%

2. 37% × 18,924

3. 67.2% × 124,898

4. 56,788.22 × 64.2%

*We can also use the proportion method of Section 4.5 and solve:

$$\frac{N}{9500} = \frac{78}{100}.$$

b Percent of Increase or Decrease

Percent is often used to state increases or decreases. For example, the average salary of an NBA basketball player increased from \$1.558 million in 1994 to \$1.867 million in 1995. To find the *percent of increase* in salary, we first subtract to find out how much more the salary was in 1995:

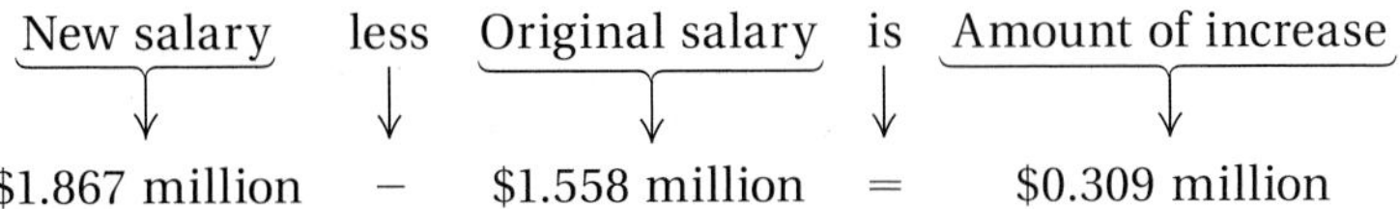

\$1.867 million − \$1.558 million = \$0.309 million

Let's first look at this with a drawing.

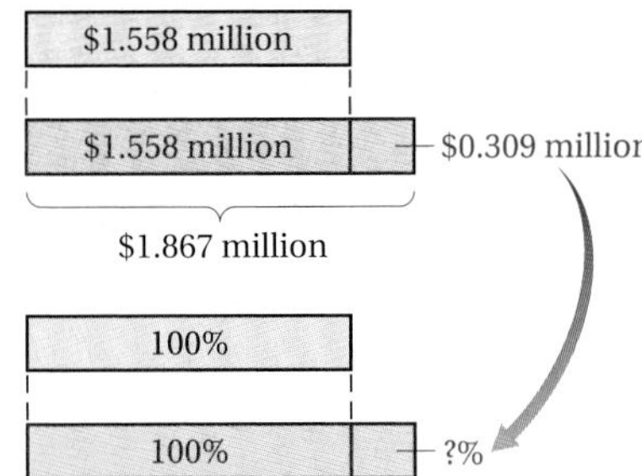

Then we determine what percent of the original amount the increase was. Since \$0.309 million = \$309,000, we are asking

\$309,000 is what percent of \$1,558,000?

This translates to the following:

$309{,}000 = x \cdot 1{,}558{,}000?$

This is an equation of the type studied in Sections 4.4 and 4.5. Solving the equation, we can confirm that \$0.309 million, or \$309,000, is about 19.8% of \$1.558 million. Thus the percent of increase in salary was 19.8%.

> To find a percent of increase or decrease:
>
> a) Find the amount of increase or decrease.
>
> b) Then determine what percent this is of the original amount.

3. *Automobile Price.* The price of an automobile increased from \$15,800 to \$17,222. What was the percent of increase?

Answer on page A-9

Example 3 *Digital-Camera Screen Size.* The diagonal of the display screen of a digital camera was recently increased from 1.8 in. to 2.5 in. What was the percent of increase in the diagonal?

1. **Familiarize.** We note that the increase in the diagonal was $2.5 - 1.8$, or 0.7 in. A drawing can help us to visualize the situation. We let $n =$ the percent of increase.

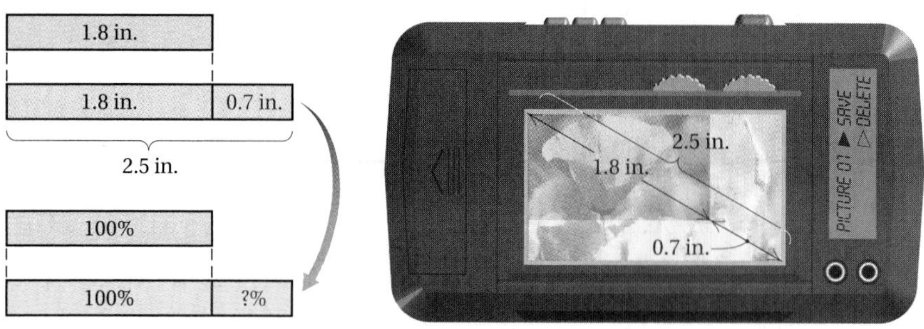

2. **Translate.** We rephrase the question and translate.

$$\underbrace{0.7\text{ in.}}_{0.7}\ \underbrace{\text{is}}_{=}\ \underbrace{\text{what percent}}_{n}\ \underbrace{\text{of}}_{\times}\ \underbrace{1.8\text{ in.?}}_{1.8}$$

3. **Solve.** To solve the equation, we divide by 1.8 on both sides:*

$$\frac{0.7}{1.8} = \frac{n \times 1.8}{1.8}$$

$$0.389 \approx n \qquad \textbf{Rounded to the nearest thousandth}$$

$$38.9\% \approx n. \qquad \textbf{Remember to find percent notation.}$$

4. **Check.** To check, we take 38.9% of 1.8:

$$38.9\% \times 1.8 = 0.389 \times 1.8 = 0.7002.$$

Since we rounded the percent, this approximation is close enough to 0.7 to be a good check.

5. **State.** The percent of increase of the screen diagonal is 38.9%.

Do Exercise 3.

What do we mean when we say that the price of Swiss cheese has decreased 8%? If the price was \$5.00 per pound and it went down to \$4.60 per pound, then the decrease is \$0.40, which is 8% of the original price. We can see this in the following figure.

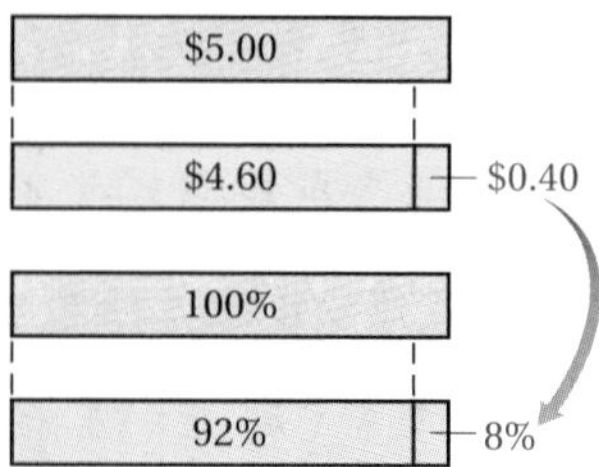

*We can also use the proportion method of Section 4.5 and solve:

$$\frac{0.7}{1.8} = \frac{N}{100}.$$

Example 4 *Fuel Bill.* With proper furnace maintenance, a family that pays a monthly fuel bill of \$78.00 can reduce their bill to \$70.20. What is the percent of decrease?

1. **Familiarize.** We find the amount of decrease and then make a drawing.

7 8.0 0	Original bill
− 7 0.2 0	New bill
7.8 0	Decrease

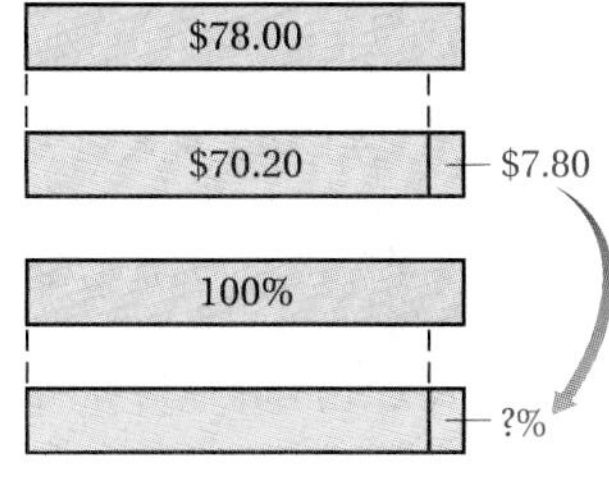

We let n = the percent of decrease.

2. **Translate.** We rephrase and translate as follows.*

7.80	is	what percent	of	78.00?
↓	↓	↓	↓	↓
7.80	=	n	×	78.00

3. **Solve.** To solve the equation, we divide by 78 on both sides:

$$\frac{7.80}{78.00} = \frac{n \times 78.00}{78.00}$$ Dividing by 78 on both sides

$$0.1 = n$$ You may have noticed earlier that 7.8 is 10% of 78.

$$10\% = n.$$ Changing from decimal to percent notation

4. **Check.** To check, we note that, with a 10% decrease, the reduced bill should be 90% of the original bill. Since 90% of 78 = 0.9 × 78 = 70.20, our answer checks.

5. **State.** The percent of decrease of the fuel bill is 10%.

Do Exercise 4.

Example 5 A part-time teacher's aide earns \$9700 one year and receives a 6% raise the next. What is the new salary?

1. **Familiarize.** We make a drawing.

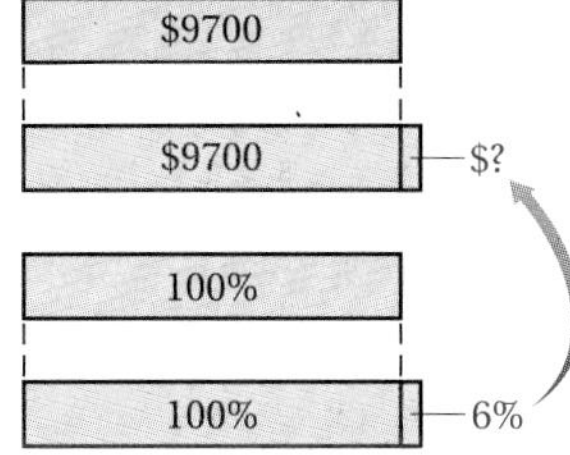

This is a two-step problem. First, we find the increase. We let a = the salary raise.

---

*We can also use the proportion method of Section 4.5 and solve:

$$\frac{7.8}{78} = \frac{N}{100}.$$

4. *Fuel Bill.* By using only cold water in the washing machine, a household with a monthly fuel bill of \$78.00 can reduce their bill to \$74.88. What is the percent of decrease?

Answer on page A-9

5. A part-time salesperson earns $9800 one year and gets a 9% raise the next. What is the new salary?

Answer on page A-9

2. **Translate.** We rephrase the question and translate as follows.

What	is	6%	of	9700?
↓	↓	↓	↓	↓
a	=	6%	×	9700

3. **Solve.** We convert 6% to a decimal notation and multiply:

$$a = 0.06 \times 9700 = 582.$$

Next, we add $582 to the old salary:

$$9700 + 582 = 10{,}282.$$

4. **Check.** To check, we can repeat the calculation. We can also check by estimating. The old salary, $9700, is approximately $10,000, and 6% of $10,000 is 0.06 × 10,000, or $600. The new salary would be about 9700 + 600, or $10,300. Since $10,282 is close to $10,300, we have a partial check.

5. **State.** The new salary is $10,282.

Do Exercise 5.

Calculator Spotlight

The [%] Key and Percent of Increase or Decrease. On a calculator with a percent key, there may be a fast way to find the result of adding or subtracting a percent from a number. In Example 5, the result of taking 6% of $9700 and adding it to $9700 might be found by pressing

[9] [7] [0] [0] [×] [6] [SHIFT] [%] [+].

The displayed result would be

10,282 .

If the salary had been reduced by 6%, the computation would be

[9] [7] [0] [0] [×] [6] [SHIFT] [%] [−].

The displayed result would be

9118 .

Check your manual for other procedures for determining percents.

Exercises

Use a calculator with a [%] key.

1. Find the result of Margin Exercise 5.
2. Find the result of Margin Exercise 5 if the salary were decreased by 9%.

Exercise Set 4.6

a Solve.

1. *Left-handed Professional Bowlers.* It has been determined by sociologists that 17% of the population is left-handed. Each tournament conducted by the Professional Bowlers Association has 120 entrants. How many would you expect to be left-handed? not left-handed? Round to the nearest one.

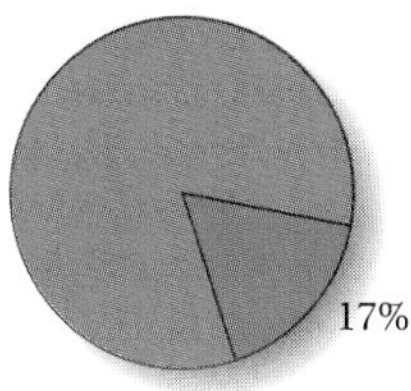

2. *Advertising Budget.* A common guideline for businesses is to use 5% of their operating budget for advertising. Ariel Electronics has an operating budget of $8000 per week. How much should it spend each week for advertising? for other expenses?

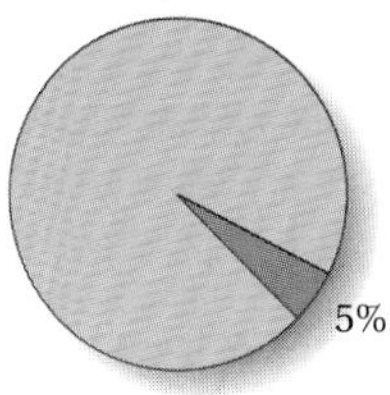

3. Of all moviegoers, 67% are in the 12–29 age group. A theater held 800 people for a showing of *Star Trek-18*. How many were in the 12–29 age group? not in this age group?

4. Deming, New Mexico, claims to have the purest drinking water in the world. It is 99.9% pure. If you had 240 L of water from Deming, how much of it, in liters, would be pure? impure?

5. A baseball player gets 13 hits in 40 at-bats. What percent are hits? not hits?

6. On a test of 80 items, Erika had 76 correct. What percent were correct? incorrect?

7. A lab technician has 680 mL of a solution of water and acid; 3% is acid. How many milliliters are acid? water?

8. A lab technician has 540 mL of a solution of alcohol and water; 8% is alcohol. How many milliliters are alcohol? water?

9. *TV Usage.* Of the 8760 hr in a year, most television sets are on for 2190 hr. What percent is this?

10. *Colds from Kissing.* In a medical study, it was determined that if 800 people kiss someone who has a cold, only 56 will actually catch a cold. What percent is this?

11. *Maximum Heart Rate.* Treadmill tests are often administered to diagnose heart ailments. A guideline in such a test is to try to get you to reach your *maximum heart rate,* in beats per minute. The maximum heart rate is found by subtracting your age from 220 and then multiplying by 85%. What is the maximum heart rate of someone whose age is 25? 36? 48? 55? 76? Round to the nearest one.

12. It costs an oil company $40,000 a day to operate two refineries. Refinery A accounts for 37.5% of the cost, and refinery B for the rest of the cost.

a) What percent of the cost does it take to run refinery B?

b) What is the cost of operating refinery A? refinery B?

b Solve.

13. The amount in a savings account increased from $200 to $216. What was the percent of increase?

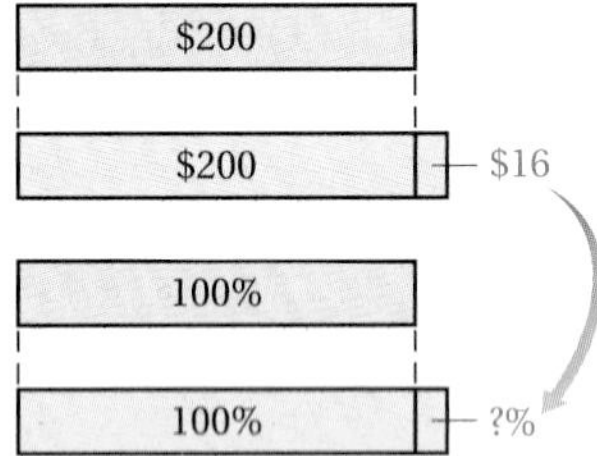

14. The population of a small mountain town increased from 840 to 882. What was the percent of increase?

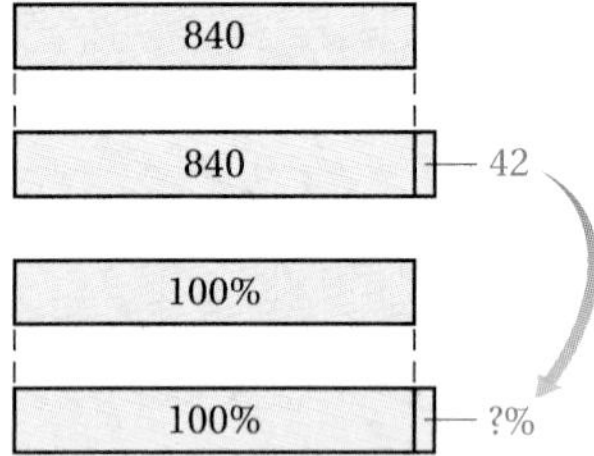

15. During a sale, a dress decreased in price from $90 to $72. What was the percent of decrease?

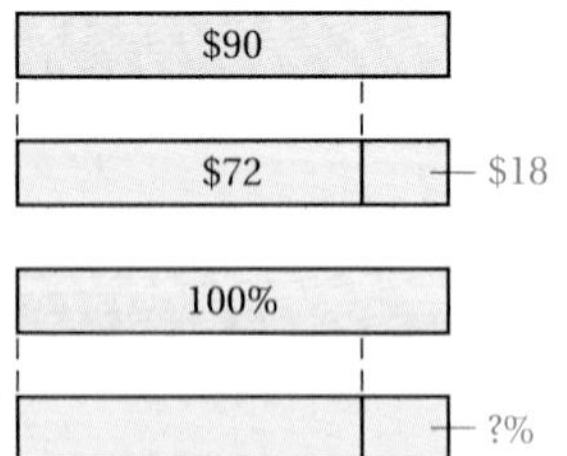

16. A person on a diet goes from a weight of 125 lb to a weight of 110 lb. What is the percent of decrease?

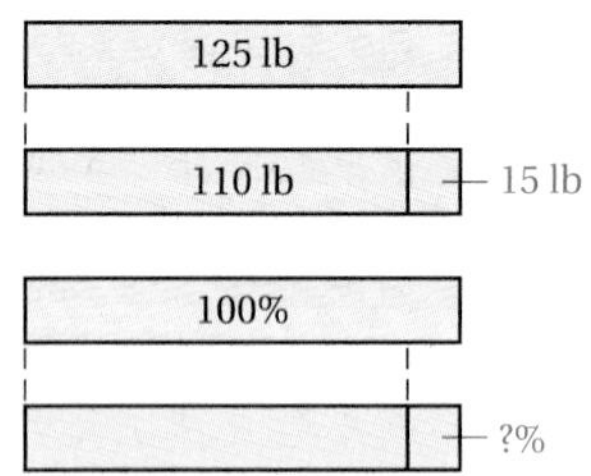

17. A person earns $28,600 one year and receives a 5% raise in salary. What is the new salary?

18. A person earns $20,400 one year and receives an 8% raise in salary. What is the new salary?

19. The value of a car typically decreases by 30% in the first year. A car is bought for $18,000. What is its value one year later?

20. One year the pilots of an airline shocked the business world by taking an 11% pay cut. The former salary was $55,000. What was the reduced salary?

21. *World Population.* World population is increasing by 1.6% each year. In 1999, it was 6.0 billion. How much will it be in 2000? 2001? 2002?

22. *Cooling Costs.* By increasing the thermostat from 72° to 78°, a family can reduce its cooling bill by 50%. If the cooling bill was $106.00, what would the new bill be? By what percent has the temperature been increased?

23. *Car Depreciation.* A car generally depreciates 30% of its original value in the first year. A car is worth $25,480 after the first year. What was its original cost?

24. *Car Depreciation.* Given normal use, an American-made car will depreciate 30% of its original cost the first year and 14% of its remaining value in the second year. What is the value of a car at the end of the second year if its original cost was $36,400? $28,400? $26,800?

25. *Tipping.* Diners frequently add a 15% tip when charging a meal to a credit card. What is the total amount charged if the cost of the meal, without tip, is $15? $34? $49?

26. *Two-by-Four.* A cross-section of a standard or nominal "two-by-four" board actually measures $1\frac{1}{2}$ in. by $3\frac{1}{2}$ in. The rough board is 2 in. by 4 in. but is planed and dried to the finished size. What percent of the wood is removed in planing and drying?

27. *MADD.* Despite efforts by groups such as MADD (Mothers Against Drunk Driving), the number of alcohol-related deaths is rising after many years of decline. The data in the table shows the number of deaths from 1986 to 1995.

a) What is the percent of increase in the number of alcohol-related deaths from 1994 to 1995?
b) What is the percent of decrease in the number of alcohol-related deaths from 1986 to 1994?

Alcohol-Related Traffic Deaths Back on the Increase!!

Year	Deaths
1986	24,045
1987	23,641
1988	23,626
1989	22,436
1990	22,084
1991	19,887
1992	17,859
1993	17,473
1994	16,589
1995	17,274

Source: National Highway Traffic Safety Administration

28. *Fetal Acoustic Stimulation.* Each year there are about 4 million births in the United States. Of these, about 120,000 births occur in breech position (delivery of a fetus with the buttocks or feet appearing first). A new technique, called *fetal acoustic stimulation (FAS),* uses sound directed through a mother's abdomen in order to stimulate movement of the fetus to a safer position. In a recent study of this low-risk and low-cost procedure, FAS enabled doctors to turn the baby in 34 of 38 cases (***Source*:** Johnson and Elliott, "Fetal Acoustic Stimulation, an Adjunct to External Cephalic Versions: A Blinded, Randomized Crossover Study," *American Journal of Obstetrics & Gynecology* **173**, no. 5 (1995): 1369–1372).

a) What percent of U.S. births are breech?
b) What percent (rounded to the nearest tenth) of cases showed success with FAS?
c) About how many breech babies yearly might be turned if FAS could be implemented in all births in the United States?
d) Breech position is one reason for performing Caesarean section (or C-section) birth surgery. Researchers expect that FAS alone can eliminate the need for about 2000 C-sections yearly in the United States. Given this information, how many yearly C-sections are due to breech position alone?

29. *Strike Zone.* In baseball, the *strike zone* is normally a 17-in. by 40-in. rectangle. Some batters give the pitcher an advantage by swinging at pitches thrown out of the strike zone. By what percent is the area of the strike zone increased if a 2-in. border is added to the outside?

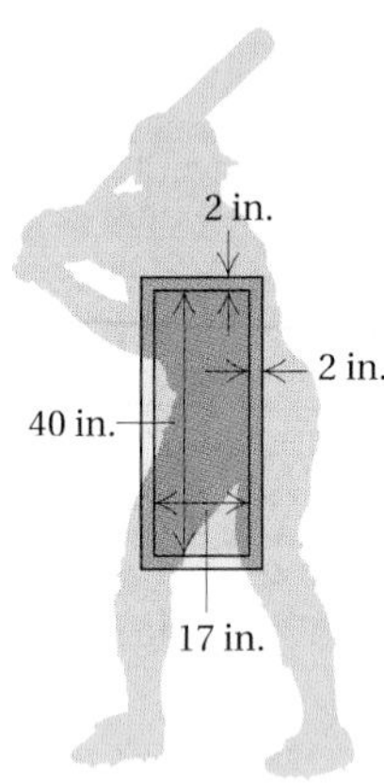

30. Tony is planting grass on a 24-ft by 36-ft area in his back yard. He installs a 6-ft by 8-ft garden. By what percent has he reduced the area he has to mow?

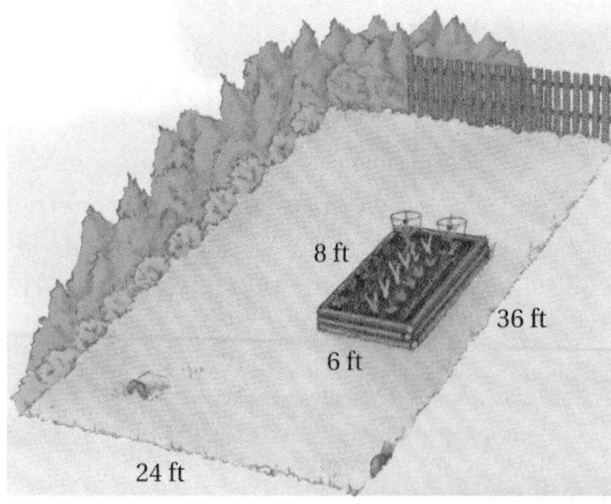

Skill Maintenance

Convert to decimal notation. [3.1c], [3.5a]

31. $\frac{25}{11}$

32. $\frac{11}{25}$

33. $\frac{27}{8}$

34. $\frac{43}{9}$

35. $\frac{23}{25}$

36. $\frac{20}{24}$

37. $\frac{14}{32}$

38. $\frac{2317}{1000}$

39. $\frac{34,809}{10,000}$

40. $\frac{27}{40}$

Synthesis

41. ◆ Which is better for a wage earner, and why: a 10% raise followed by a 5% raise a year later, or a 5% raise followed by a 10% raise a year later?

42. ◆ Write a problem for a classmate to solve. Design the problem so that the solution is "Jackie's raise was $7\frac{1}{2}$%."

43. ▦ A worker receives raises of 3%, 6%, and then 9%. By what percent has the original salary increased?

44. ◆ ▦ A workers' union is offered either a 5% "across-the-board" raise in which all salaries would increase 5%, or a flat $1650 raise for each worker. If the total payroll for the 123 workers is $4,213,365, which offer should the union select? Why?

45. *Adult Height.* It has been determined that at the age of 10, a girl has reached 84.4% of her final adult growth. Cynthia is 4 ft, 8 in. at the age of 10. What will be her final adult height?

46. *Adult Height.* It has been determined that at the age of 15, a boy has reached 96.1% of his final adult height. Claude is 6 ft, 4 in. at the age of 15. What will be his final adult height?

47. If p is 120% of q, then q is what percent of p?

48. A coupon allows a couple to have dinner and then have $10 subtracted from the bill. Before subtracting $10, however, the restaurant adds a tip of 15%. If the couple is presented with a bill for $44.05, how much would the dinner (without tip) have cost without the coupon?

4.7 Consumer Applications

a Sales Tax

Sales tax computations represent a special type of percent of increase problem. The sales tax rate in Arkansas is 3%. This means that the tax is 3% of the purchase price. Suppose the purchase price on a coat is $124.95. The sales tax is then

3% of $124.95, or 0.03×124.95,

or

3.7485, or about $3.75.

The total that you pay is the price plus the sales tax:

$124.95 + $3.75, or $128.70.

> **Sales tax** = Sales tax rate × Purchase price
>
> **Total price** = Purchase price + Sales tax

Example 1 *Florida.* The sales tax rate in Florida is 6%. How much tax is charged on the purchase of 3 CDs at $13.95 each? What is the total price?

a) We first find the cost of the CDs. It is

$3 \times \$13.95 = \$41.85.$

b) The sales tax on items costing $41.85 is

$$\underbrace{\text{Sales tax rate}}_{\downarrow\ 6\%} \times \underbrace{\text{Purchase price}}_{\downarrow\ \$41.85},$$

or 0.06×41.85, or 2.511. Thus the tax is $2.51.

c) The total price is given by the purchase price plus the sales tax:

$41.85 + $2.51, or $44.36.

Do Exercises 1 and 2.

Example 2 The sales tax is $32 on the purchase of an $800 sofa. What is the sales tax rate?

Rephrase: Sales tax is what percent of purchase price?

Translate: $32 = r \times 800$

Solve:

$$\frac{32}{800} = \frac{r \times 800}{800} \qquad \text{Dividing by 800 on both sides}$$

$$0.04 = r$$

$$4\% = r.$$

The sales tax rate is 4%.

Do Exercise 3.

Objectives

- a Solve applied problems involving sales tax and percent.
- b Solve applied problems involving commission and percent.
- c Solve applied problems involving discount and percent.
- d Solve applied problems involving simple interest and percent.
- e Solve applied problems involving compound interest.

For Extra Help

TAPE 9 | MAC WIN | CD-ROM

1. *Connecticut.* The sales tax rate in Connecticut is 8%. How much tax is charged on the purchase of a refrigerator that sells for $668.95? What is the total price?

2. *New Jersey.* Morris buys 5 blank audiocassettes in New Jersey, where the sales tax rate is 7%. If each tape costs $2.95, how much tax will be charged? What is the total price?

3. The sales tax is $33 on the purchase of a $550 washing machine. What is the sales tax rate?

Answers on page A-10

4. Raul's commission rate is 30%. What is the commission from the sale of $18,760 worth of air conditioners?

5. Ben's commission rate is 16%. He receives a commission of $268 from sales of clothing. How many dollars worth of clothing were sold?

Answers on page A-10

b Commission

When you work for a **salary**, you receive the same amount of money each week or month. When you work for a **commission**, you are paid a percentage of the total sales for which you are responsible.

> **Commission** = Commission rate × Sales

Example 3 A salesperson's commission rate is 20%. What is the commission from the sale of $25,560 worth of stereophonic equipment?

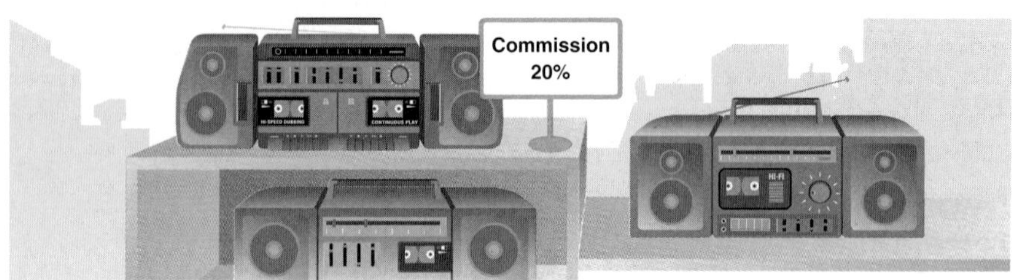

Commission	=	*Commission rate*	×	*Sales*
C	=	20%	×	25,560

This tells us what to do. We multiply.

$$C = 20\% \times 25{,}560 = 0.2 \times 25{,}560 = 5112.$$

The commission is $5112.

Do Exercise 4.

Example 4 Joyce's commission rate is 25%. She receives a commission of $425 on the sale of a motorbike. How much did the motorbike cost?

Commission	=	*Commission rate*	×	*Sales*
425	=	25%	×	S

To solve this equation, we divide by 0.25 on both sides:

$$\frac{425}{0.25} = \frac{0.25 \times S}{0.25}$$

$$1700 = S.$$

The motorbike cost $1700.

Do Exercise 5.

c Discount

Suppose that the regular price of a rug is \$60, and the rug is on sale at 25% off. Since 25% of \$60 is \$15, the sale price is \$60 − \$15, or \$45. We call \$60 the **original**, or **marked price,** 25% the **rate of discount,** \$15 the **discount**, and \$45 the **sale price.** Note that discount problems are a type of percent of decrease problem.

> **Discount** = Rate of discount × Original price
>
> **Sale price** = Original price − Discount

Example 5 A rug marked \$240 is on sale at 25% off. What is the discount? the sale price?

a) *Discount* = *Rate of discount* × *Original price*

$D = 25\% \times 240$

This tells us what to do. We convert 25% to decimal notation and multiply:

$D = 0.25 \times 240 = 60.$

The discount is \$60.

b) *Sale price* = *Marked price* − *Discount*

$S = 240 - 60$

This tells us what to do. We subtract.

$S = 240 - 60\text{-}180.$

The sale price is \$180.

Do Exercise 6.

6. A suit marked \$140 is on sale at 24% off. What is the discount? the sale price?

Answer on page A-10

7. What is the interest on \$4300 invested at an interest rate of 14% for 1 year?

8. What is the interest on a principal of \$4300 invested at an interest rate of 14% for $\frac{3}{4}$ year?

Answers on page A-10

d Simple Interest

Suppose you put \$100 into an investment for 1 year. The \$100 is called the **principal**. If the **interest rate** is 8%, in addition to the principal, you get back 8% of the principal, which is

8% of \$100, or 0.08×100, or \$8.00.

The \$8.00 is called the **simple interest**. It is, in effect, the price that a financial institution pays for the use of the money over time.

> The **simple interest** I on principal P, invested for t years at interest rate r, is given by
>
> $$I = P \cdot r \cdot t.$$

Example 6 What is the interest on \$2500 invested at an interest rate of 6% for 1 year?

We use the formula $I = P \cdot r \cdot t$:

$$\begin{aligned} I = P \cdot r \cdot t &= \$2500 \times 6\% \times 1 \\ &= \$2500 \times 0.06 \\ &= \$150. \end{aligned}$$

```
   2 5 0 0
×    0.0 6
----------
 1 5 0.0 0
```

The interest for 1 year is \$150.

Do Exercise 7.

Example 7 What is the interest on a principal of \$2500 invested at an interest rate of 6% for $\frac{1}{4}$ year?

We use the formula $I = P \cdot r \cdot t$:

$$\begin{aligned} I = P \cdot r \cdot t &= \$2500 \times 6\% \times \frac{1}{4} \\ &= \frac{\$2500 \times 0.06}{4} \\ &= \$37.50. \end{aligned}$$

```
       3 7.5
4 ) 1 5 0.0
    1 2 0
    -----
      3 0
      2 8
      ---
        2 0
        2 0
        ---
          0
```

> We could have instead found $\frac{1}{4}$ of 6% and then multiplied by 2500.

The interest for $\frac{1}{4}$ year is \$37.50.

Do Exercise 8.

When time is given in days, we usually divide it by 365 to express the time as a fractional part of a year.

Example 8 To pay for a shipment of tee shirts, New Wave Designs borrows $8000 at 9% for 60 days. Find (a) the amount of simple interest that is due and (b) the total amount that must be paid after 60 days.

a) We express 60 days as a fractional part of a year:

$$I = P \cdot r \cdot t = \$8000 \times 9\% \times \frac{60}{365}$$

$$= \$8000 \times 0.09 \times \frac{60}{365}$$

$$\approx \$118.36. \qquad \text{Usng a calculator}$$

The interest due for 60 days is $118.36.

b) The total amount to be paid after 60 days is the principal plus the interest:

$$\$8000 + \$118.36 = \$8118.36.$$

The total amount due is $8118.36.

Do Exercise 9.

9. The Glass Nook borrows $4800 at 7% for 30 days. Find (a) the amount of simple interest due and (b) the total amount that must be paid after 30 days.

e Compound Interest

When interest is paid *on interest,* we call it **compound interest.** This is the type of interest usually paid on investments. Suppose you have $5000 in a savings account at 6%. In 1 year, the account will contain the original $5000 plus 6% of $5000. Thus the total in the account after 1 year will be

106% of $5000, or $1.06 \times \$5000$, or $5300.

Now suppose that the total of $5300 remains in the account for another year. At the end of this second year, the account will contain the $5300 plus 6% of $5300. The total in the account would thus be

106% of $5300, or $1.06 \times \$5300$, or $5618.

Note that in the second year, interest is earned on the first year's interest. When this happens, we say that interest is **compounded annually.**

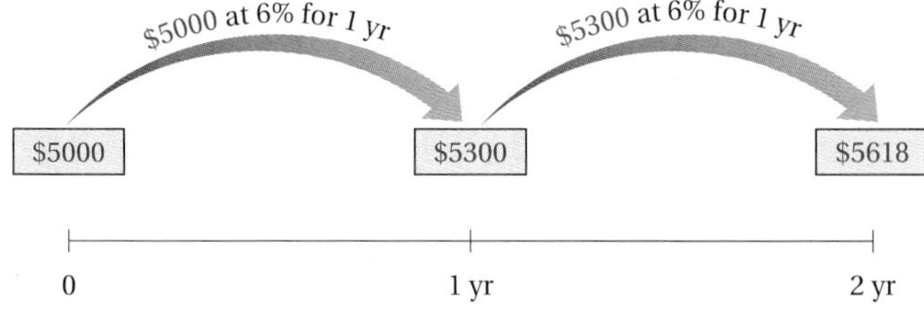

Answer on page A-10

10. Find the amount in an account if \$2000 is invested at 11%, compounded annually, for 2 years.

Example 9 Find the amount in an account if \$2000 is invested at 8%, compounded annually, for 2 years.

a) After 1 year, the account will contain 108% of \$2000:

$1.08 \times \$2000 = \$2160.$

$$\begin{array}{r} 2\,0\,0\,0 \\ \times \quad 1.0\,8 \\ \hline 1\,6\,0\,0\,0 \\ 0\,0\,0\,0 \\ 2\,0\,0\,0 \\ \hline 2\,1\,6\,0.0\,0 \end{array}$$

b) At the end of the second year, the account will contain 108% of \$2160:

$1.08 \times \$2160 = \$2332.80.$

$$\begin{array}{r} 2\,1\,6\,0 \\ \times \quad 1.0\,8 \\ \hline 1\,7\,2\,8\,0 \\ 0\,0\,0\,0 \\ 2\,1\,6\,0 \\ \hline 2\,3\,3\,2.8\,0 \end{array}$$

The amount in the account after 2 years is \$2332.80.

Do Exercise 10.

Suppose that the interest in Example 9 were **compounded semiannually**—that is, every half year. Interest would then be calculated twice a year at a rate of $8\% \div 2$, or 4%, each time. The approach used in Example 4 can then be adapted, as follows.

After the first $\frac{1}{2}$ year, the account will contain 104% of \$2000:

$1.04 \times \$2000 = \$2080.$ These calculations can be confirmed with a calculator.

After a second $\frac{1}{2}$ year (1 full year), the account will contain 104% of \$2080:

$1.04 \times \$2080 = \$2163.20.$

After a third $\frac{1}{2}$ year $\left(1\frac{1}{2}\text{ full years}\right)$, the account will contain 104% of \$2163.20:

$1.04 \times \$2163.20 = \$2249.728$

$\approx \$2249.73.$ Rounding to the nearest cent

Finally, after a fourth $\frac{1}{2}$ year (2 full years), the account will contain 104% of \$2249.73:

$1.04 \times \$2249.73 = \$2339.7192$

$\approx \$2339.72.$ Rounding to the nearest cent

Note that each multiplication was by 1.04 and that

$\$2000 \times 1.04^4 = \$2339.72.$ Using a calculator and rounding to the nearest cent

Answer on page A-10

We have illustrated the following result.

> If a principal P has been invested at interest rate r, compounded n times a year, in t years it will grow to an amount A given by
> $$A = P \cdot \left(1 + \frac{r}{n}\right)^{n \cdot t}.$$

Example 10 The Ibsens invest \$4000 in an account paying 8%, compounded quarterly. Find the amount in the account after $2\frac{1}{2}$ years.

We substitute \$4000 for P, 8% for r, 4 for n, and $2\frac{1}{2}$ fot t and solve for A:

$$\begin{aligned} A &= P \cdot \left(1 + \frac{r}{n}\right)^{n \cdot t} \\ &= 4000 \cdot \left(1 + \frac{0.08}{4}\right)^{4 \cdot (5/2)} && 8\% = 0.08;\ 2\tfrac{1}{2} = \tfrac{5}{2} \\ &= 4000 \cdot (1 + 0.02)^{10} \\ &= 4000 \cdot (1.02)^{10} \\ &\approx 4875.98. && \text{Using a calculator} \end{aligned}$$

The amount in the account after $2\frac{1}{2}$ years is \$4875.98.

Do Exercise 11.

11. A couple invests \$7000 in an account paying 10%, compounded semiannually. Find the amount in the account after $1\frac{1}{2}$ years.

Answer on page A-10

Calculator Spotlight

When using a calculator for interest computations, it is important to remember the order in which operations are performed and to minimize "round-off error." For example, to find the amount due on a \$20,000 loan made for 25 days at 11%, compounded daily, we press the following sequence of keys:

[2] [0] [0] [0] [0] [×] [(] [1] [+] [0] [.] [1] [1] [÷] [3] [6] [5] [)] [x^y] [2] [5] [=].

This key may appear differently. See p. 57.

Without parentheses keys, we would press

[1] [+] [0] [.] [1] [1] [÷] [3] [6] [5] [=] [x^y] [2] [5] [=] [×] [2] [0] [0] [0] [0] [=].

Note that in both sequences of keystrokes we raise $1 + \frac{0.11}{365}$, not just $\frac{0.11}{365}$, to the power.

Exercises

1. Find the amount due on a \$16,000 loan made for 62 days at 13%, compounded daily.
2. An investment of \$12,500 is made for 90 days at 8.5%, compounded daily. How much is the investment worth after 90 days?

Improving Your Math Study Skills

Homework

Before Doing Your Homework

- **Setting.** Consider doing your homework as soon as possible after class, before you forget what you learned in the lecture. Research has shown that after 24 hours, most people forget about half of what is in their short-term memory. To avoid this "automatic" forgetting, you need to transfer the knowledge into long-term memory. The best way to do this with math concepts is to perform practice exercises repeatedly. This is the "drill-and-practice" part of learning math that comes when you do your homework. It cannot be overlooked if you want to succeed in your study of math.

  Try to set a specific time for your homework. Then choose a location that is quiet and uninterrupted. Some students find it helpful to listen to music when doing homework. Research has shown that classical music creates the best atmosphere for studying: Give it a try!
- **Reading.** Before you begin doing the homework exercises, you should reread the assigned material in the textbook. You may also want to look over your class notes again and rework some of the examples given in class.

  You should not read a math textbook as you would a novel or history textbook. Math texts are not meant to be read passively. Be sure to stop and do the margin exercises when directed. Also be sure to reread any paragraphs as you see the need.

While Doing Your Homework

- **Study groups.** For some students, forming a study group can be helpful. Many times, two heads are better than one. Also, it is true that "to teach is to learn again." Thus, when you explain a concept to your classmate, you often gain a deeper understanding of the concept yourself. If you do study in a group, resist the temptation to waste time by socializing.

  If you work regularly with someone, be careful not to become dependent on that person. Work on your own some of the time so that you do not rely heavily on others and are able to learn even when they are not available.
- **Notebook.** When doing your homework, consider using notebook paper in a spiral or three-ring binder. You want to be able to go over your homework when studying for a test. Therefore, you need to be able to easily access any problem in your homework notebook. Write legibly in your notebook so you can check over your work. Label each section and each exercise clearly, and show all steps. Your clear writing will also be appreciated by your instructor should your homework be collected. Also, tutors and instructors can be more helpful if they can see and understand all the steps in your work.

  When you are finished with your homework, check the answers to the odd-numbered exercises at the back of the book or in the *Student's Solutions Manual* and make corrections. If you do not understand why an answer is wrong, put a star by it so you can ask questions in class or during the instructor's office hours.

After Doing Your Homework

- **Review.** If you complete your homework several days before the next class, review your work every day. This will keep the material fresh in your mind. You should also review the work immediately before the next class so that you can ask questions as needed.

Exercise Set 4.7

a Solve.

1. *Illinois.* The sales tax rate in Illinois is 6.25%. How much tax is charged on a purchase of 5 telephones at \$53 apiece? What is the total price?

2. *New York City.* The sales tax rate in New York City is 8.25%. How much tax is charged on photo equipment costing \$248? What is the total price?

3. The sales tax is \$48 on the purchase of a dining room set that sells for \$960. What is the sales tax rate?

4. The sales tax is \$15 on the purchase of a diamond ring that sells for \$500. What is the sales tax rate?

5. The sales tax is \$35.80 on the purchase of a refrigerator–freezer that sells for \$895. What is the sales tax rate?

6. The sales tax is \$9.12 on the purchase of a patio set that sells for \$456. What is the sales tax rate?

7. The sales tax on a used car is \$100 and the sales tax rate is 5%. Find the purchase price (the price before taxes are added).

8. The sales tax on a stereo is \$66 and the sales tax rate is 5.5%. Find the purchase price.

9. The sales tax rate in Dallas is 1% for the city and 6% for the state. Find the total amount paid for 2 shower units at \$332.50 apiece.

10. The sales tax rate in Omaha is 1.5% for the city and 5% for the state. Find the total amount paid for 3 air conditioners at \$260 apiece.

11. The sales tax is \$1030.40 on an automobile purchase of \$18,400. What is the sales tax rate?

12. The sales tax is \$979.60 on an automobile purchase of \$15,800. What is the sales tax rate?

b Solve.

13. Sondra's commission rate is 6%. What is the commission from the sale of $45,000 worth of furnaces?

14. Jose's commission rate is 32%. What is the commission from the sale of $12,500 worth of sailboards?

15. Vince earns $120 selling $2400 worth of television sets. What is the commission rate?

16. Donna earns $408 selling $3400 worth of shoes. What is the commission rate?

17. An art gallery's commission rate is 40%. They receive a commission of $392. How many dollars worth of artwork were sold?

18. A real estate agent's commission rate is 7%. She receives a commission of $5600 on the sale of a home. How much did the home sell for?

19. A real estate commission is 6%. What is the commission on the sale of a $98,000 home?

20. A real estate commission is 8%. What is the commission on the sale of a piece of land for $68,000?

21. Bonnie earns $280.80 selling $2340 worth of tee shirts. What is the commission rate?

22. Chuck earns $1147.50 selling $7650 worth of ski passes. What is the commission rate?

23. Miguel's commission is increased according to how much he sells. He receives a commission of 5% for the first $2000 and 8% on the amount over $2000. What is the total commission on sales of $6000?

24. Lucinda earns a salary of $500 a month, plus a 2% commission on sales. One month, she sold $990 worth of encyclopedias. What were her wages that month?

c Solve.

25. Find the marked price and the rate of discount for the camcorder in this ad.

REDUCED
$83
Power Zoom
X12
Palmaster
VHS-C Camcorder
• Large Video Head Cylinder for Jitter-free, Crisp Pictures
• 12:1 Variable Speed Power Zoom
• Lens Cover Opens Automatically when Camera is Turned On
$377

26. Find the discount and the rate of discount for the calculator in this ad.

Calc-U-Sure C96
Graphing Calculator
• 8 line × 16 character display
• Pull-down menus
• Uses 3 "AAA" batteries
• Sliding plastic cover
• Model C96
• Mfr. List $115.00
69 98

Find what is missing.

27.

Marked Price	Rate of Discount	Discount	Sale Price
$300	10%		

28.

Marked Price	Rate of Discount	Discount	Sale Price
$2000	40%		

29.

Marked Price	Rate of Discount	Discount	Sale Price
$17.00	15%		

30.

Marked Price	Rate of Discount	Discount	Sale Price
$20.00	25%		

31.

Marked Price	Rate of Discount	Discount	Sale Price
	10%	$12.50	

32.

Marked Price	Rate of Discount	Discount	Sale Price
	15%	$65.70	

33.

Marked Price	Rate of Discount	Discount	Sale Price
$600		$240	

34.

Marked Price	Rate of Discount	Discount	Sale Price
$12,800		$1920	

d Find the *simple* interest.

	Principal	*Rate of interest*	*Time*
35.	$200	13%	1 year
36.	$450	18%	1 year
37.	$2000	12.4%	$\frac{1}{2}$ year
38.	$200	7.7%	$\frac{1}{2}$ year
39.	$4300	14%	$\frac{1}{4}$ year
40.	$2000	15%	$\frac{1}{4}$ year

Solve. Assume that simple interest is being calculated in each case.

41. Animal Instinct, a pet supply shop, borrows $6500 at 8% for 90 days. Find (a) the amount of interest due and (b) the total amount that must be paid after 90 days.

42. Andante's Cafe borrows $4500 at 9% for 60 days. Find (a) the amount of interest due and (b) the total amount that must be paid after 60 days.

e Interest is compounded annually. Find the amount in the account after the given length of time. Round to the nearest cent.

	Principal	*Rate of interest*	*Time*
43.	$400	10%	2 years
44.	$400	7.7%	2 years
45.	$200	8.8%	2 years
46.	$1000	15%	2 years

Interest is compounded semiannually. Find the amount in the account after the given length of time. Round to the nearest cent.

	Principal	*Rate of interest*	*Time*
47.	$4000	7%	1 year
48.	$1000	5%	1 year
49.	🖩 $2000	9%	3 years
50.	🖩 $5000	8%	30 months

Solve.

51. 🖩 A family invests $4000 in an account paying 6%, compounded monthly. How much is in the account after 5 months?

52. 🖩 The O'Hares invest $6000 in an account paying 8%, compounded quarterly. How much is in the account after 18 months?

Skill Maintenance

53. Write fractional notation: 0.93. [3.1b]

54. Solve: $2.3 \times y = 85.1$. [3.4b]

55. Convert to decimal notation: $\frac{13}{11}$. [3.5a]

56. Convert to a mixed numeral: $\frac{29}{11}$. [2.4a]

Synthesis

57. ◆ Which is a better investment and why: $1000 invested at $14\frac{3}{4}\%$ simple interest for 1 year, or $1000 invested at 14% compounded monthly for 1 year?

58. ◆ A firm must choose between borrowing $5000 at 10% for 30 days and borrowing $10,000 at 8% for 60 days. Give arguments in favor of and against each option.

Calculate the costs associated with the purchase of a car or truck.
Prepare an amortization table for a car loan.

Summary and Review Exercises: Chapter 4

Important Properties and Formulas

Commission = Commission rate × Sales

Discount = Rate of discount × Original price

Sale price = Original price − Discount

Simple Interest: $I = P \cdot r \cdot t$

Compounded Interest: $A = P \cdot \left(1 + \frac{r}{n}\right)^{n \cdot t}$

The objectives to be tested in addition to the material in this chapter are [2.4a], [3.1b], [3.4b], and [3.5a].

Write fractional notation for the ratio. Do not simplify. [4.1a]

1. 47 to 84

2. 46 to 1.27

3. What is the rate in miles per hour? [4.1b]

117.7 miles, 5 hours

4. A lawn requires 319 gal of water for every 500 ft^2. What is the rate in gallons per square foot? [4.1b]

5. *Turkey Servings.* A 25-lb turkey serves 18 people. What is the rate in servings per pound? [4.1b]

Determine whether the two pairs of numbers are proportional. [4.1c]

6. 9, 15 and 36, 59

7. 24, 37 and 40, 46.25

Solve. [4.1d]

8. $\frac{8}{9} = \frac{x}{36}$

9. $\frac{4.5}{120} = \frac{0.9}{x}$

Solve. [4.1e]

10. If 3 dozen eggs cost $2.67, how much will 5 dozen eggs cost?

11. *Quality Control.* A factory manufacturing computer circuits found 39 defective circuits in a lot of 65 circuits. At this rate, how many defective circuits can be expected in a lot of 585 circuits?

12. A train travels 448 mi in 7 hr. At this rate, how far will it travel in 13 hr?

13. *Garbage Production.* It is known that 5 people produce 13 kg of garbage in one day. San Diego, California, has 1,150,000 people. How many kilograms of garbage are produced in San Diego in one day?

Find percent notation. [4.2c]

14. 0.483

15. 0.36

Find percent notation. [4.3a]

16. $\frac{3}{8}$

17. $\frac{1}{3}$

Find decimal notation. [4.2b]

18. 73.5%

19. $6\frac{1}{2}\%$

Find fractional notation. [4.3b]

20. 24%

21. 6.3%

Translate to an equation. Then solve. [4.4a, b]

22. 30.6 is what percent of 90?

23. 63 is 84 percent of what?

24. What is $38\frac{1}{2}\%$ of 168?

Translate to a proportion. Then solve. [4.5a, b]

25. 24 percent of what is 16.8?

26. 42 is what percent of 30?

27. What is 10.5% of 84?

Solve. [4.6a, b]

28. Food expenses account for 26% of the average family's budget. A family makes $2300 one month. How much do they spend for food?

29. The price of a television set was reduced from $350 to $308. Find the percent of decrease in price.

30. The price of a box of cookies increased from $1.70 to $2.04. What was the percent of increase in the price?

31. Carney College has a student body of 960 students. Of these, 17.5% are seniors. How many students are seniors?

Solve. [4.7a, b, c]

32. A state charges a meals tax of $4\frac{1}{2}$%. What is the meals tax charged on a dinner party costing $320?

33. In a certain state, a sales tax of $378 is collected on the purchase of a used car for $7560. What is the sales tax rate?

34. Kim earns $753.50 selling $6850 worth of televisions. What is the commission rate?

35. An air conditioner has a marked price of $350. It is placed on sale at 12% off. What are the discount and the sale price?

36. An insurance salesperson receives a 7% commission. If $42,000 worth of life insurance is sold, what is the commission?

Solve. [4.7d, e]

37. The Dress Shack borrows $24,000 at 10% simple interest for 60 days. Find (a) the amount of interest due and (b) the total amount that must be paid after 60 days.

38. What is the simple interest on $2200 principal at the interest rate of 5.5% for 1 year?

39. The Kleins invest $7500 in an investment account paying 12%, compounded monthly. How much is in the account after 3 months?

40. Find the amount in an investment account if $8000 is invested at 9%, compounded annually, for 2 years.

Skill Maintenance

Write fractional notation. [3.1b]

41. 3.107

42. 0.29

Solve. [3.4b]

43. $10.4 \times y = 665.6$

44. $100 \cdot x = 761.23$

Convert to decimal notation. [3.5a]

45. $\frac{11}{3}$

46. $\frac{11}{7}$

Convert to a mixed numeral. [2.4a]

47. $\frac{11}{3}$

48. $\frac{121}{7}$

Synthesis

49. ◈ Ollie buys a microwave oven during a 10%-off sale. The sale price that Ollie paid was $162. To find the original price, Ollie calculates 10% of $162 and adds that to $162. Is this correct? Why or why not? [4.7c]

50. ◈ Which is a better deal for a consumer and why: a discount of 40% or a discount of 20% followed by another of 22%? [4.7c]

51. It takes Yancy Martinez 10 min to type two-thirds of a page of his term paper. At this rate, how long will it take him to type a 7-page term paper? [4.1e]

52. A $200 coat is marked up 20%. After 30 days, it is marked down 30% and sold. What was the final selling price of the coat? [4.7c]

Test: Chapter 4

Write fractional notation for the ratio. Do not simplify.

1. 85 to 97

2. 0.34 to 124

3. What is the rate in feet per second?

   10 feet, 16 seconds

4. *Ham Servings.* A 12-lb shankless ham contains 16 servings. What is the rate in servings per pound?

Determine whether the two pairs of numbers are proportional.

5. 7, 8 and 63, 72

6. 1.3, 3.4 and 5.6, 15.2

Solve.

7. $\frac{9}{4} = \frac{27}{x}$

8. $\frac{150}{2.5} = \frac{x}{6}$

Solve.

9. *Time Loss.* A watch loses 2 min in 10 hr. At this rate, how much will it lose in 24 hr?

10. *Map Scaling.* On a map, 3 in. represents 225 mi. If two cities are 7 in. apart on the map, how far apart are they in reality?

11. Find decimal notation for 89%.

12. Find percent notation for 0.674.

13. Find percent notation for $\frac{11}{8}$.

14. Find fractional notation for 65%.

15. Translate to an equation. Then solve.

    What is 40% of 55?

16. Translate to a proportion. Then solve.

    What percent of 80 is 65?

Solve.

17. *Weight of Muscles.* The weight of muscles in a human body is 40% of total body weight. A person weighs 125 lb. What do the muscles weigh?

18. The population of Rippington increased from 1500 to 3600. Find the percent of increase in population.

Answers

1. ______
2. ______
3. ______
4. ______
5. ______
6. ______
7. ______
8. ______
9. ______
10. ______
11. ______
12. ______
13. ______
14. ______
15. ______
16. ______
17. ______
18. ______

Answers

19. ______

20. ______

21. ______

22. ______

23. ______

24. ______

25. ______

26. ______

27. ______

28. ______

29. ______

30. ______

31. ______

32. ______

19. *Arizona Tax Rate.* The sales tax rate in Arizona is 5%. How much tax is charged on a purchase of \$324? What is the total price?

20. Gwen's commission rate is 15%. What is the commission from the sale of \$4200 worth of merchandise?

21. The marked price of a CD player is \$200 and the item is on sale at 20% off. What are the discount and the sale price?

22. What is the simple interest on a principal of \$120 at the interest rate of 7.1% for 1 year?

23. The Burnham Parents–Teachers Association invests \$5200 at 6% simple interest. How much is in the account after $\frac{1}{2}$ year?

24. Find the amount in an account if \$1000 is invested at 5%, compounded annually, for 2 years.

25. The Suarez family invests \$10,000 at 9%, compounded monthly. How much is in the account after 3 months?

26. Find the discount and the discount rate of the bed in this ad.

WHITE IRON DAYBED
WITH BRASS ACCENTS
100 TO SELL
FANTASTIC VALUE!
MARKET VALUE
\$249.95
Choice of finish!
\$118
Springs Included!

Skill Maintenance

27. Solve: $8.4 \times y = 1864.8$.

28. Write fractional notation for 44.7.

29. Convert to decimal notation: $\frac{17}{12}$.

30. Convert to a mixed numeral: $\frac{153}{44}$.

Synthesis

31. By selling a home without using a realtor, Juan and Marie can avoid paying a 7.5% commission. They receive an offer of \$109,000 from a potential buyer. In order to give a comparable offer, for what price would a realtor need to sell the house? Round to the nearest hundred.

32. Nancy Morano-Smith wants to win a season football ticket from the local bookstore. Her goal is to guess the number of marbles in an 8-gal jar. She knows that there are 128 oz in a gallon. She goes home and fills an 8-oz jar with 46 marbles. How many marbles should she guess are in the jar?

5

Data Analysis, Graphs, and Statistics

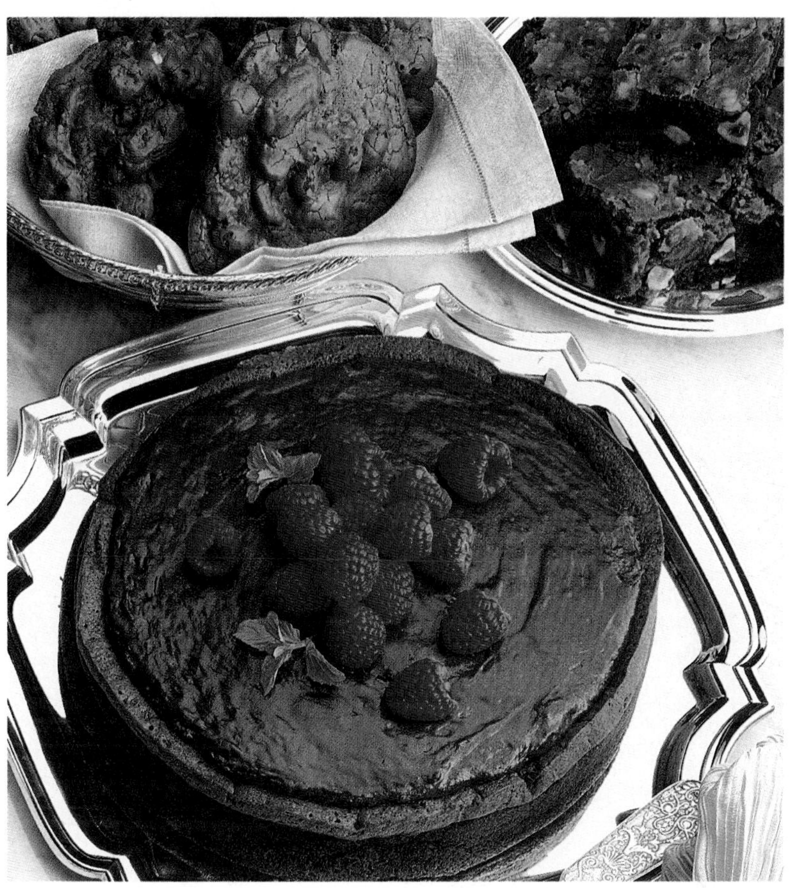

An Application

Tricia adds one slice of chocolate cake with fudge frosting (560 calories) to her diet each day for one year (365 days) and makes no other changes in her eating or exercise habits. The consumption of 3500 extra calories will add about 1 lb to her body weight. How many pounds will she gain?

This problem appears as Exercise 12 in Exercise Set 5.3.

The Mathematics

We see that Tricia will consume

365×560, or $204{,}400$ calories

in 1 year. Thus her weight gain is

$$\frac{204{,}400}{3500} \approx 58 \text{ lb.}$$

World Wide Web For more information, visit us at www.mathmax.com

Introduction

There are many ways in which we can analyze or describe data. One is to organize the data into a table or a graph. Another is to look at certain *statistics* related to the data. After such organization, many problems can be solved, including those of making predictions from the data. We will consider many kinds of graphs: pictographs, bar graphs, line graphs, and circle graphs. Statistics we will consider are *means, medians,* and *modes.*

Pretest: Chapter 5

In Questions 1–3, find (a) the average, (b) the median, and (c) the mode.

1. 46, 50, 53, 55

2. 5, 4, 3, 2, 1

3. 4, 17, 4, 18, 4, 17, 18, 20

4. A car was driven 660 mi in 12 hr. What was the average number of miles per hour?

5. To get a C in chemistry, Delia must average 70 on four tests. Scores on the first three tests were 68, 71, and 65. What is the lowest score that she can make on the last test and still get a C?

6. *Reasons for Exercising.* The following data show the percentage of women selecting a particular reason for exercising. Make a circle graph to show the data.

Health: 51%
Lose weight: 38%
Relieve stress: 11%

7. *Cost of Life Insurance.* The following table shows the comparison of the cost of a $100,000 life insurance policy for female smokers and nonsmokers at certain ages.

a) How much does it cost a female nonsmoker, age 32, for insurance?
b) How much more does it cost a female smoker, age 35, than a nonsmoker at the same age?

8. Using the data in Question 7, draw a vertical bar graph showing the cost of insurance for a female smoker at various ages. Use age on the horizontal scale and cost on the vertical scale.

9. Using the data in Question 7, draw a line graph showing the cost of insurance for a female smoker at various ages. Use age on the horizontal scale and cost on the vertical scale.

LIFE INSURANCE: FEMALE

Age	Cost (Smoker)	Cost (Nonsmoker)
31	$254	$201
32	273	208
33	294	221
34	319	236
35	341	249

Source: State Farm Insurance

Risk of Heart Disease. The line graph below shows the relationship between blood cholesterol level and risk of coronary heart disease.

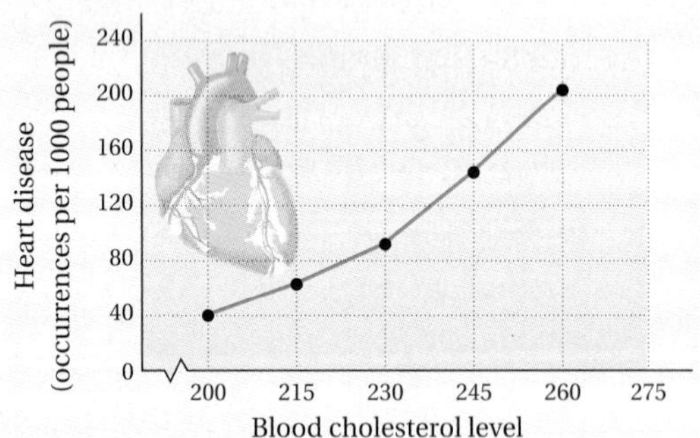

10. At what cholesterol level is the risk highest?

11. About how much higher is the risk at 260 than at 200?

12. *Study Time vs. Grades.* An English instructor asked his students to keep track of how much time each spent studying for a chapter test. He collected the information together with the test scores. The data are given in the table below.

Study Time (in hours)	Test Grade (in percent)
9	75
11	83
13	80
15	85
17	80
18	86
21	87
23	92
24	?

a) Draw a line graph of the data.
b) Estimate the missing data value.

Objectives for Retesting

The objectives to be tested in addition to the material in this chapter are as follows.

[2.2c] Divide and simplify using fractional notation.
[4.1e] Solve applied problems involving proportions.
[4.4b], [4.5b] Solve basic percent problems.
[4.6a] Solve applied problems involving percent.

5.1 Averages, Medians, and Modes

Data are often available regarding some kind of application involving mathematics. We can use tables and graphs of various kinds to show information about the data and to extract information from the data that can lead us to make analyses and predictions. Graphs allow us to communicate a message from the data.

For example, the following show data regarding credit-card spending between Thanksgiving and Christmas in recent years. Examine each method of presentation. Which method, if any, do you like the best and why? Which do you like the least and why?

Objectives

a. Find the average of a set of numbers and solve applied problems involving averages.

b. Find the median of a set of numbers and solve applied problems involving medians.

c. Find the mode of a set of numbers and solve applied problems involving modes.

For Extra Help

TAPE 10

MAC
WIN

CD-ROM

PARAGRAPH FORM

The National Credit Counseling Services has recently released data regarding credit-card spending between Thanksgiving and Christmas for various years. In 1991, spending was $59.8 billion; in 1992, it was $66.8 billion; in 1993, it was $79.1 billion; in 1994, it was $96.9 billion; in 1995, it was $116.3 billion; and finally, in 1996, it was $131.4 billion.

TABLE

Year	Credit-Card Spending from Thanksgiving to Christmas (in billions)
1991	$ 59.8
1992	66.8
1993	79.1
1994	96.9
1995	116.3
1996	131.4

Source: RAM Research Group, National Credit Counseling Services

PICTOGRAPH

Credit-Card Spending from Thanksgiving to Christmas

[card icon] = $10 billion dollars

1991

1992

1993

1994

1995

1996

BAR GRAPH

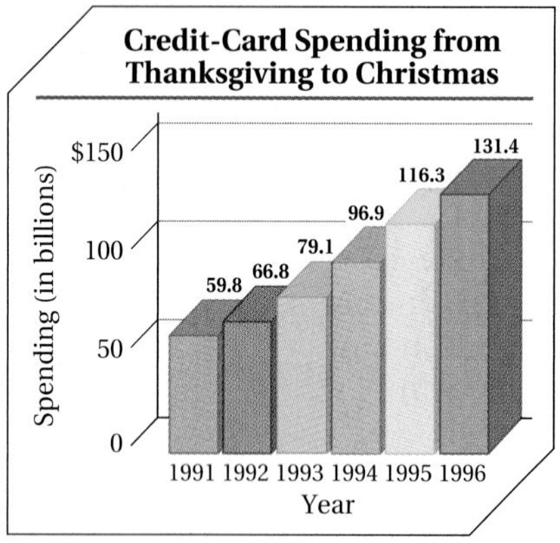

LINE GRAPH

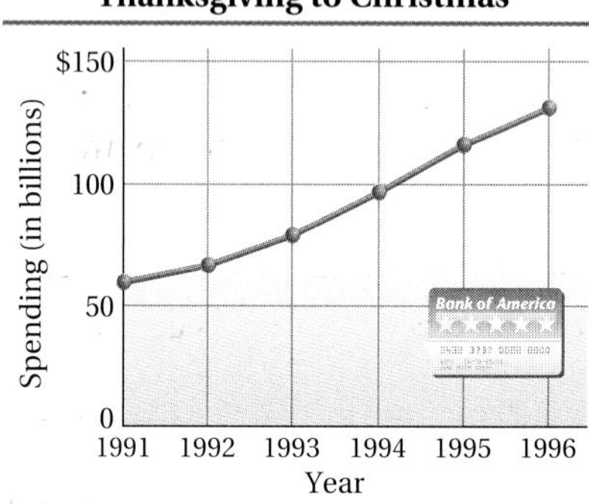

CIRCLE, OR PIE, GRAPH

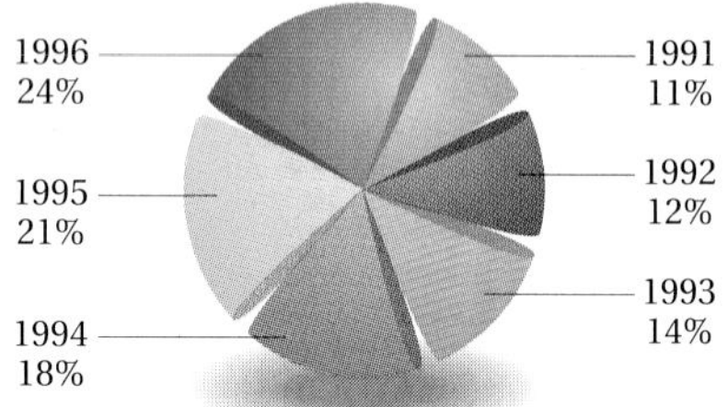

Most people would not find the paragraph method for displaying the data most useful. It takes time to read, and it is hard to look for a trend and make predictions. The circle, or pie, graph might be used to compare what part of the entire amount of spending over the six years each individual year represents, but that comparison is not the same comparison as those presented by the bar and line graphs. The bar and line graphs might be more worthwhile if we want to see the trend of increased spending and to make predictions about the years 1997 and beyond.

In this chapter, we will learn not only how to extract information from various kinds of tables and graphs, but also how to create various kinds of graphs.

a Averages

A **statistic** is a number describing a set of data. One statistic is a *center point* that characterizes the data. The most common kind of center point is the *mean,* or *average,* of a set of numbers. We first considered averages in Section 1.9.

Let's consider the data on credit-card spending (given in billions):

$59.8, $66.8, $79.1, $96.9, $116.3, $131.4.

What is the *average* of the numbers? First, we add the numbers:

$$59.8 + 66.8 + 79.1 + 96.9 + 116.3 + 131.4 = 550.3.$$

Next, we divide by the number of data items, 6:

$$\frac{550.3}{6} \approx \$91.7. \quad \textbf{Rounding to the nearest tenth}$$

Note that

$$91.7 + 91.7 + 91.7 + 91.7 + 91.7 + 91.7 = 550.2 \approx 550.3.$$

The number 91.7 is called the **average** of the set of numbers. It is also called the **arithmetic** (pronounced ăr′ ĭth-mĕt′-ĭk) **mean** or simply the **mean**.

> To find the **average** of a set of numbers, add the numbers and then divide by the number of items of data.

Example 1 On a 4-day trip, a car was driven the following number of miles each day: 240, 302, 280, 320. What was the average number of miles per day?

$$\frac{240 + 302 + 280 + 320}{4} = \frac{1142}{4}, \quad \text{or} \quad 285.5$$

The car was driven an average of 285.5 mi per day. Had the car been driven exactly 285.5 mi each day, the same total distance (1142 mi) would have been traveled.

Do Exercises 1–4.

Example 2 *Food Waste.* Courtney is a typical American consumer. In the course of 1 yr, she discards 100 lb of food waste. What is the average number of pounds of food waste discarded each week? Round to the nearest tenth.

We already know the total amount of food waste for the year. Since there are 52 weeks in a year, we divide by 52 and round:

$$\frac{100}{52} \approx 1.9.$$

On average, Courtney discards 1.9 lb of food waste per week.

Do Exercise 5.

Find the average.

1. 14, 175, 36

2. 75, 36.8, 95.7, 12.1

3. A student scored the following on five tests: 68, 85, 82, 74, 96. What was the average score?

4. In the first five games, a basketball player scored points as follows: 26, 21, 13, 14, 23. Find the average number of points scored per game.

5. *Food Waste.* Courtney also composts (converts to dirt) 5 lb of food waste each year. How much, on average, does Courtney compost per month? Round to the nearest tenth.

Answers on page A-11

6. *Gas Mileage.* According to recent EPA estimates, a Toyota Camry LE can be expected to travel 209 mi (city) on 11 gal of gasoline (***Source:*** *Popular Science Magazine*). What is the average number of miles expected per gallon?

7. *GPA.* Jennifer earned the following grades one semester.

Grade	Number of Credit Hours in Course
B	3
C	4
C	4
A	2

What was Jennifer's grade point average? Assume that the grade point values are 4.0 for an A, 3.0 for a B, and so on. Round to the nearest tenth.

Answers on page A-11

Example 3 *Gas Mileage.* According to recent EPA estimates, an Oldsmobile Aurora can be expected to travel 204 mi (city) on 12 gal of gasoline. What is the average number of miles expected per gallon?

We divide the total number of miles, 204, by the number of gallons, 12:

$$\frac{204}{12} = 17 \text{ mpg.}$$

The Aurora's expected average is 17 miles per gallon.

Do Exercise 6.

Example 4 *GPA.* In most colleges, students are assigned grade point values for grades obtained. The **grade point average,** or **GPA,** is the average of the grade point values for each credit hour taken. At many colleges, grade point values are assigned as follows:

A: 4.0
B: 3.0
C: 2.0
D: 1.0
F: 0.0

Tom earned the following grades for one semester. What was his grade point average?

Course	Grade	Number of Credit Hours in Course
History	B	4
Basic mathematics	A	5
English	A	5
French	C	3
Physical education	F	1

To find the GPA, we first multiply the grade point value (in color below) by the number of credit hours in the course and then add, as follows:

History	$3.0 \cdot 4 =$	12
Basic mathematics	$4.0 \cdot 5 =$	20
English	$4.0 \cdot 5 =$	20
French	$2.0 \cdot 3 =$	6
Physical education	$0.0 \cdot 1 =$	0
		58 (Total)

The total number of credit hours taken is $4 + 5 + 5 + 3 + 1$, or 18. We divide 58 by 18 and round to the nearest tenth:

$$\text{GPA} = \frac{58}{18} \approx 3.2.$$

Tom's grade point average was 3.2.

Do Exercise 7.

Example 5 To get a B in math, Geraldo must score an average of 80 on the tests. On the first four tests, his scores were 79, 88, 64, and 78. What is the lowest score that Geraldo can get on the last test and still get a B?

We can find the total of the five scores needed as follows:

$$80 + 80 + 80 + 80 + 80 = 5 \cdot 80, \quad \text{or} \quad 400.$$

The total of the scores on the first four tests is

$$79 + 88 + 64 + 78 = 309.$$

Thus Geraldo needs to get at least

$$400 - 309, \quad \text{or} \quad 91$$

in order to get a B. We can check this as follows:

$$\frac{79 + 88 + 64 + 78 + 91}{5} = \frac{400}{5}, \quad \text{or} \quad 80.$$

Do Exercise 8.

8. To get an A in math, Rosa must score an average of 90 on the tests. On the first three tests, her scores were 80, 100, and 86. What is the lowest score that Rosa can get on the last test and still get an A?

b Medians

Another type of center-point statistic is the *median.* Medians are useful when we wish to de-emphasize unusually extreme scores. For example, suppose a small class scored as follows on an exam.

Phil:	78	Pat:	56
Jill:	81	Olga:	84
Matt:	82		

Let's first list the scores in order from smallest to largest:

56, 78, 81, 82, 84.

↑ Middle score (81)

The middle score—in this case, 81—is called the **median.** Note that because of the extremely low score of 56, the average of the scores is 76.2. In this example, the median may be a more appropriate center-point statistic.

Example 6 What is the median of this set of numbers?

99, 870, 91, 98, 106, 90, 98

We first rearrange the numbers in order from smallest to largest. Then we locate the middle number, 98.

90, 91, 98, 98, 99, 106, 870

↑ Middle number (98)

The median is 98.

Do Exercises 9–11.

Find the median.

9. 17, 13, 18, 14, 19

10. 20, 14, 13, 19, 16, 18, 17

11. 78, 81, 83, 91, 103, 102, 122, 119, 88

> Once a set of data is listed in order, from smallest to largest, the **median** is the middle number if there is an odd number of data items. If there is an even number of items, the median is the number that is the average of the two middle numbers.

Calculator Spotlight

Averages can be easily computed on a calculator if we remember the order in which operations are performed. For example, to calculate

$$\frac{85 + 92 + 79}{3}$$

on most calculators, we press

[8] [5] [+] [9] [2] [+] [7] [9] [=] [÷] [3] [=],

or

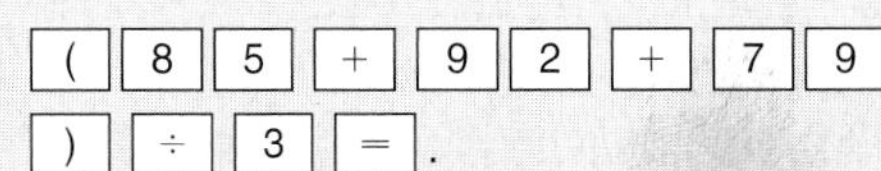

Exercises

1. What would the result have been if we had not used parentheses in the latter sequence of keystrokes?
2. Use a calculator to solve Examples 1–5.

Answers on page A-11

Find the median.

12. \$1300, \$2000, \$1900, \$1600, \$1800, \$1400

13. 68, 34, 67, 69, 34, 70

Find the modes of these data.

14. 23, 45, 45, 45, 78

15. 34, 34, 67, 67, 68, 70

16. 13, 24, 27, 28, 67, 89

17. In a lab, Gina determined the mass, in grams, of each of five eggs:

15 g, 19 g, 19 g, 14 g, 18 g.

a) What is the mean?

b) What is the median?

c) What is the mode?

Answers on page A-11

Example 7 What is the median of this set of numbers?

69, 80, 61, 63, 62, 65

We first rearrange the numbers in order from smallest to largest. There is an even number of numbers. We look for the middle two, which are 63 and 65. The median is halfway between 63 and 65, the number 64.

61, 62, 63, 65, 69, 80

The average of the middle numbers is $\frac{63 + 65}{2}$, or 64.

The median is 64.

Example 8 What is the median of this set of yearly salaries?

\$35,000, \$500,000, \$28,000, \$34,000

We rearrange the numbers in order from smallest to largest. The two middle numbers are \$34,000 and \$35,000. Thus the median is halfway between \$34,000 and \$35,000 (the average of \$34,000 and \$35,000):

\$28,000, \$34,000, \$35,000, \$500,000

$$\text{Median} = \frac{\$34{,}000 + \$35{,}000}{2} = \frac{\$69{,}000}{2} = \$34{,}500.$$

Do Exercises 12 and 13.

c Modes

The final type of center-point statistic is the **mode**.

> The **mode** of a set of data is the number or numbers that occur most often. If each number occurs the same number of times, there is *no* mode.

Example 9 Find the mode of these data.

13, 14, 17, 17, 18, 19

The number that occurs most often is 17. Thus the mode is 17.

A set of data has just one average (mean) and just one median, but it can have more than one mode. It is also possible for a set of data to have no mode—when all numbers are equally represented. For example, the set of data 5, 7, 11, 13, 19 has no mode.

Example 10 Find the modes of these data.

33, 34, 34, 34, 35, 36, 37, 37, 37, 38, 39, 40

There are two numbers that occur most often, 34 and 37. Thus the modes are 34 and 37.

Do Exercises 14–17.

Exercise Set 5.1

a, b, c For each set of numbers, find the average, the median, and any modes that exist.

1. 16, 18, 29, 14, 29, 19, 15

2. 72, 83, 85, 88, 92

3. 5, 30, 20, 20, 35, 5, 25

4. 13, 32, 25, 27, 13

5. 1.2, 4.3, 5.7, 7.4, 7.4

6. 13.4, 13.4, 12.6, 42.9

7. 234, 228, 234, 229, 234, 278

8. $29.95, $28.79, $30.95, $29.95

9. The following temperatures were recorded for seven days in Hartford:

43°, 40°, 23°, 38°, 54°, 35°, 47°.

What was the average temperature? the median? the mode?

10. Lauri Merten, a professional golfer, scored 71, 71, 70, and 68 to win the U.S. Women's Open in a recent year. What was the average score? the median? the mode?

11. *Gas Mileage.* According to recent EPA estimates, an Achieva can be expected to travel 297 mi (highway) on 9 gal of gasoline (***Source:*** *Motor Trend Magazine*). What is the average number of miles expected per gallon?

12. *Gas Mileage.* According to recent EPA estimates an Aurora can be expected to travel 192 mi (highway) on 8 gal of gasoline (***Source:*** *Motor Trend Magazine*). What is the average number of miles expected per gallon?

GPA. In Exercises 13 and 14 are the grades of a student for one semester. In each case, find the grade point average. Assume that the grade point values are 4.0 for an A, 3.0 for a B, and so on. Round to the nearest tenth.

13.

Grades	Number of Credit Hours in Course
B	4
B	5
B	3
C	4

14.

Grades	Number of Credit Hours in Course
A	5
B	4
B	3
C	5

15. The following prices per pound of Atlantic salmon were found at five fish markets:

$7.99, $9.49, $9.99, $7.99, $10.49.

What was the average price per pound? the median price? the mode?

16. The following prices per pound of Vermont cheddar cheese were found at five supermarkets:

$4.99, $5.79, $4.99, $5.99, $5.79.

What was the average price per pound? the median price? the mode?

17. To get a B in math, Rich must score an average of 80 on five tests. Scores on the first four tests were 80, 74, 81, and 75. What is the lowest score that Rich can get on the last test and still receive a B?

18. To get an A in math, Cybil must score an average of 90 on five tests. Scores on the first four tests were 90, 91, 81, and 92. What is the lowest score that Cybil can get on the last test and still receive an A?

19. Marta was pregnant 270 days, 259 days, and 272 days for her first three pregnancies. In order for Marta's average pregnancy to equal the worldwide average of 266 days, how long must her fourth pregnancy last? (***Source*:** David Crystal (ed.), *The Cambridge Factfinder.* Cambridge CB2 1RP: Cambridge University Press, 1993, p. 84.)

20. Jason's brothers are 174 cm, 180 cm, 179 cm, and 172 cm tall. The average male is 176.5 cm tall. How tall is Jason if he and his brothers have an average height of 176.5 cm?

Skill Maintenance

Multiply.

21. $14 \cdot 14$ [1.3b]

22. $\frac{2}{3} \cdot \frac{2}{3}$ [2.2a]

23. 1.4×1.4 [3.3a]

24. 1.414×1.414 [3.3a]

Solve. [4.1e]

25. Four software CDs cost $239.80. How much would 19 comparable CDs cost?

26. A car is driven 700 mi in 5 days. At this rate, how far will it have been driven in 24 days?

Synthesis

27. ◆ You are applying for an entry-level job at a large firm. You can be informed of the mean, median, or mode salary. Which of the three figures would you request? Why?

28. ◆ Is it possible for a driver to average 20 mph on a 30-mi trip and still receive a ticket for driving 75 mph? Why or why not?

Bowling Averages. Bowling averages are always computed by rounding down to the nearest integer. For example, suppose a bowler gets a total of 599 for 3 games. To find the average, we divide 599 by 3 and drop the amount to the right of the decimal point:

$$\frac{599}{3} \approx 199.67. \quad \text{The bowler's average is 199.}$$

In each case, find the bowling average.

29. 547 in 3 games

30. 4621 in 27 games

31. *Hank Aaron.* Hank Aaron averaged $34\frac{7}{22}$ home runs per year over a 22-yr career. After 21 yr, Aaron had averaged $35\frac{10}{21}$ home runs per year. How many home runs did Aaron hit in his final year?

32. The ordered set of data 18, 21, 24, a, 36, 37, b has a median of 30 and an average of 32. Find a and b.

Perform a statistical analysis of pulse rates.

5.2 Tables and Pictographs

a Reading and Interpreting Tables

A **table** is often used to present data in rows and columns.

Example 1 *Cereal Data.* Let's assume that you generally have a 2-cup bowl of cereal each morning. The following table lists nutritional information for five name-brand cereals. (It does not consider the use of milk, sugar, or sweetener.) The data have been determined by doubling the information given for a 1-cup serving that is found in the Nutrition Facts panel on a box of cereal.

Cereal	Calories	Fat	Total Carbohydrate	Sodium
Ralston Rice Chex	240	0 g	54 g	460 mg
Kellogg's Complete Bran Flakes	240	1.3 g	64 g	613.3 mg
Kellogg's Special K	220	0 g	44 g	500 mg
Honey Nut Cheerios	240	3 g	48 g	540 mg
Wheaties	220	2 g	48 g	440 mg

a) Which cereal has the least amount of sodium per serving?

b) Which cereal has the greatest amount of fat?

c) Which cereal has the least amount of fat?

d) Find the average total carbohydrate in the cereals.

Careful examination of the table will give the answers.

a) To determine which cereal has the least amount of sodium, look down the column headed "Sodium" until you find the smallest number. That number is 440 mg. Then look across that row to find the brand of cereal, Wheaties.

b) To determine which cereal has the greatest amount of fat, look down the column headed "Fat" until you find the largest number. That number is 3 g. Then look across that row to find the cereal, Honey Nut Cheerios.

c) To determine which cereal has the least amount of fat, look down the column headed "Fat" until you find the smallest number. There are two listings of 0 g. Then look across those rows to find the cereals, Ralston Rice Chex and Kellogg's Special K.

d) Find the average of all the numbers in the column headed "Total Carbohydrate":

$$\frac{54 + 64 + 44 + 48 + 48}{5} = 51.6.$$

The average total carbohydrate content is 51.6 g.

Do Exercises 1–7. (Exercises 5–7 are on the following page.)

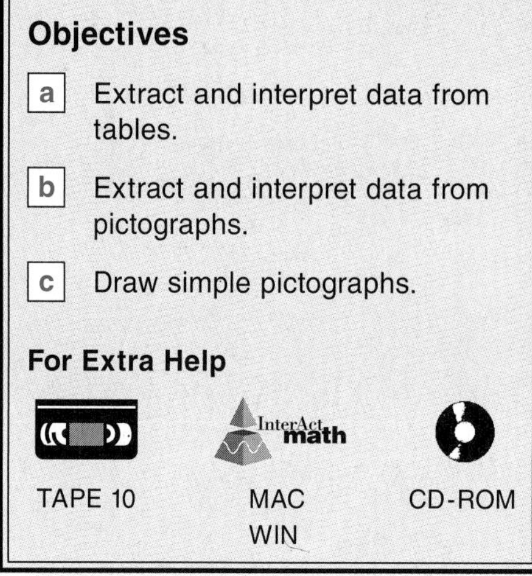

Objectives
a Extract and interpret data from tables.
b Extract and interpret data from pictographs.
c Draw simple pictographs.
For Extra Help
TAPE 10
InterAct math
MAC
WIN
CD-ROM

Use the table in Example 1 to answer each of the following.

1. Which cereal has the most total carbohydrate?

2. Which cereal has the least total carbohydrate?

3. Which cereal has the least number of calories?

4. Which cereal has the greatest number of calories?

Answers on page A-11

Example 2 *Wheaties Nutrition Facts.* Most foods are required by law to provide factual information regarding nutrition, as shown in the following table of Nutrition Facts from a box of Wheaties cereal. Although this can be very helpful to the consumer, one must be careful in interpreting the data. The % Daily Value figures shown here are based on a 2000-calorie diet. Your daily values may be higher or lower, depending on your calorie needs or intake.

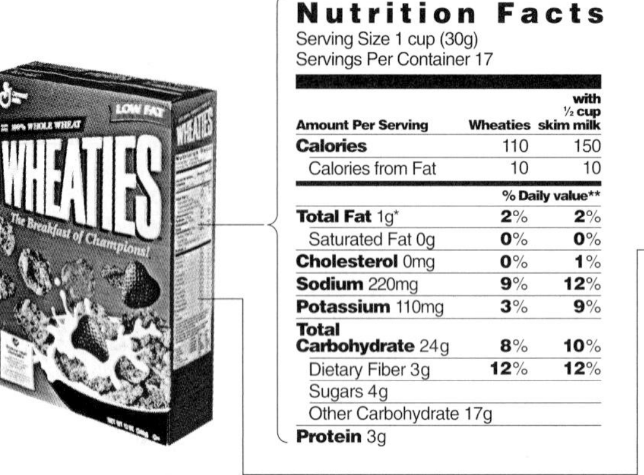

Nutrition Facts

Serving Size 1 cup (30g)
Servings Per Container 17

Amount Per Serving	Wheaties	with ½ cup skim milk
Calories	110	150
Calories from Fat	10	10
	% Daily value**	
Total Fat 1g*	**2%**	**2%**
Saturated Fat 0g	**0%**	**0%**
Cholesterol 0mg	**0%**	**1%**
Sodium 220mg	**9%**	**12%**
Potassium 110mg	**3%**	**9%**
Total Carbohydrate 24g	**8%**	**10%**
Dietary Fiber 3g	**12%**	**12%**
Sugars 4g		
Other Carbohydrate 17g		
Protein 3g		

Vitamin A	25%	30%
Vitamin C	25%	25%
Calcium	0%	15%
Iron	45%	45%
Vitamin D	10%	25%
Thiamin	25%	30%
Riboflavin	25%	35%
Niacin	25%	25%
Vitamin B_6	25%	25%
Folic Acid	25%	25%
Phosphorus	10%	20%
Magnesium	8%	10%
Zinc	4%	8%
Copper	4%	4%

*Amount in Cereal. A serving of cereal plus skim milk provides 1g fat, less than 5mg cholesterol, 280mg sodium, 310mg potassium, 30g carbohydrate (10g sugars), and 7g protein.

**Percent Daily Values are based on a 2,000 calorie diet. Your daily values may be higher or lower depending on your calorie needs.

Suppose your morning bowl of cereal consists of 2 cups of Wheaties together with 1 cup of skim milk, with artificial sweetener containing 0 calories.

a) How many calories have you consumed?

b) What percent of the daily value of total fat have you consumed?

c) A nutritionist recommends that you look for foods that provide 10% or more of the daily value for vitamin C. Do you get that with your bowl of Wheaties?

d) Suppose you are trying to limit your daily caloric intake to 2500 calories. How many bowls of cereal would it take to exceed the 2500 calories, even though you probably would not eat just cereal?

Careful examination of the table of nutrition facts will give the answers.

a) Look at the column marked "with ½ cup skim milk" and note that 1 cup of cereal with ½ cup skim milk contains 150 calories. Since you are having twice that amount, you are consuming

$$2 \times 150, \quad \text{or} \quad 300 \text{ calories.}$$

b) Read across from "Total Fat" and note that in 1 cup of cereal with ½ cup skim milk, you get 2% of the daily value of fat. Since you are doubling that, you get 4% of the daily value of fat.

c) Find the row labeled "Vitamin C" on the left and look under the column labeled "with ½ cup skim milk." Note that you get 25% of the daily value for "1 cup with ½ cup of skim milk," and since you are doubling that, you are more than satisfying the 10% requirement.

d) From part (a), we know that you are consuming 300 calories per bowl. Dividing 2500 by 300 gives $\frac{2500}{300} \approx 8.33$. Thus if you eat 9 bowls of cereal in this manner, you will exceed the 2500 calories.

Do Exercises 8–12.

5. Find the average amount of sodium in the cereals.

6. Find the median of the amount of sodium in the cereals.

7. Find the mean, the median, and the mode of the number of calories in the cereals.

Use the Nutrition Facts data from the Wheaties box and the bowl of cereal described in Example 2 to answer each of the following.

8. How many calories from fat are in your bowl of cereal?

9. A nutritionist recommends that you look for foods that provide 10% or more of the daily value for iron. Do you get that with your bowl of Wheaties?

10. How much sodium have you consumed?

11. What daily value of sodium have you consumed?

12. How much protein have you consumed?

Answers on page A-11

b Reading and Interpreting Pictographs

Pictographs (or *picture graphs*) are another way to show information. Instead of actually listing the amounts to be considered, a **pictograph** uses symbols to represent the amounts. In addition, a *key* is given telling what each symbol represents.

Example 3 *Elephant Population.* The following pictograph shows the elephant population of various countries in Africa. Located on the graph is a key that tells you that each symbol represents 10,000 elephants.

Elephant Population

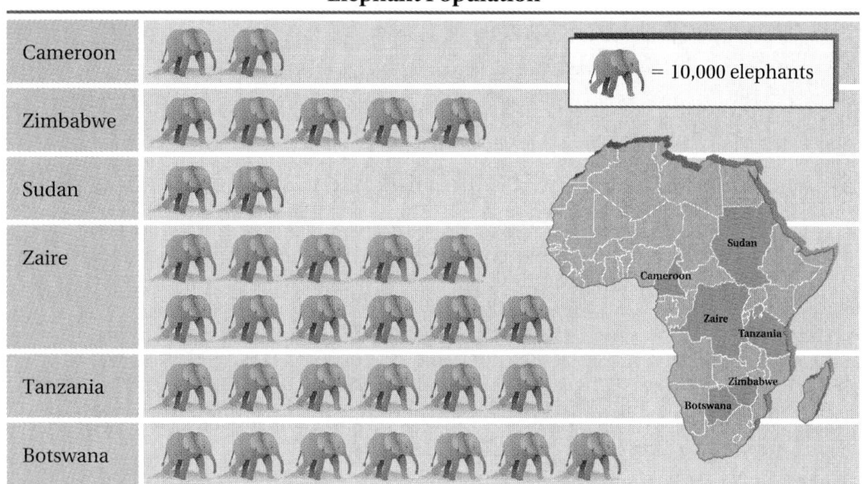

Source: National Geographic

a) Which country has the greatest number of elephants?

b) Which country has the least number of elephants?

c) How many more elephants are there in Zaire than in Botswana?

We can compute the answers by first reading the pictograph.

a) The country with the most symbols has the greatest number of elephants: Zaire, with $11 \times 10{,}000$, or 110,000 elephants.

b) The countries with the fewest symbols have the least number of elephants: Cameroon and Sudan, each with $2 \times 10{,}000$, or 20,000 elephants.

c) From part (a), we know that there are 110,000 elephants in Zaire. In Botswana there are $7 \times 10{,}000$, or 70,000 elephants. Thus there are $110{,}000 - 70{,}000$, or 40,000 more elephants in Zaire than in Botswana.

Do Exercises 13–15.

You have probably noticed that, although they seem to be very easy to read, pictographs are difficult to draw accurately because whole symbols reflect loose approximations due to significant rounding. In pictographs, you also need to use some mathematics to find the actual amounts.

Use the pictograph in Example 3 to answer each of the following.

13. How many elephants are there in Tanzania?

14. How does the elephant population of Zimbabwe compare to that of Cameroon?

15. What is the average number of elephants in these six countries?

Answers on page A-11

Use the pictograph in Example 4 to answer each of the following.

16. Determine the approximate coffee consumption per capita of France.

17. Determine the approximate coffee consumption per capita of Italy.

18. The approximate coffee consumption of Finland is about the same as the combined coffee consumptions of Switzerland and the United States. What is the approximate coffee consumption of Finland?

Answers on page A-11

Example 4 *Coffee Consumption.* For selected countries, the following pictograph shows approximately how many cups of coffee each person (per capita) drinks annually.

Coffee Consumption

Germany

United States

Switzerland

France

Italy

= 100 cups

Source: Beverage Marketing Corporation

a) Determine the approximate annual coffee consumption per capita of Germany.

b) Which two countries have the greatest difference in coffee consumption? Estimate that difference.

We use the data from the pictograph as follows.

a) Germany's consumption is represented by 11 whole symbols (1100 cups) and, though it is visually debatable, about $\frac{1}{8}$ of another symbol (about 13 cups), for a total of 1113 cups.

b) Visually, we see that Switzerland has the most consumption and that the United States has the least consumption. Switzerland's annual coffee consumption per capita is represented by 12 whole symbols (1200 cups) and about $\frac{1}{5}$ of another symbol (20 cups), for a total of 1220 cups. U.S. consumption is represented by 6 whole symbols (600 cups) and about $\frac{1}{10}$ of another symbol (10 cups), for a total of 610 cups. The difference between these amounts is $1220 - 610$, or 610 cups.

One advantage of pictographs is that the appropriate choice of a symbol will tell you, at a glance, the kind of measurement being made. Another advantage is that the comparison of amounts represented in the graph can be expressed more easily by just counting symbols. For instance, in Example 3, the ratio of elephants in Zaire to those in Cameroon is 11:2.

One disadvantage of pictographs is that, to make a pictograph easy to read, the amounts must be rounded significantly to the unit that a symbol represents. This makes it difficult to accurately represent an amount. Another problem is that it is difficult to determine very accurately how much a partial symbol represents. A third disadvantage is that you must use some mathematics to finally compute the amount represented, since there is usually no explicit statement of the amount.

Do Exercises 16–18.

c Drawing Pictographs

Example 5 *Concert Revenue.* The following list shows the top five concert acts in a recent year and their total gross revenue (money taken in) (***Source:*** *Pollstar Magazine*). Draw a pictograph to represent the data. Let the symbol [dollar bill symbol] represent \$10,000,000.

Kiss	\$43,600,000
Garth Brooks	\$34,500,000
Neil Diamond	\$32,200,000
Rod Stewart	\$29,100,000
Bob Seger	\$26,300,000

Some computation is necessary before we can draw the pictograph.

Kiss: Note that $43{,}600{,}000 = 4.36 \times 10{,}000{,}000$. Thus we need 4 whole symbols and 0.36 of another symbol. Now 0.36 is hard to draw, but we estimate it to be about 33%, or $\frac{1}{3}$, of a symbol.

Garth Brooks: Note that $34{,}500{,}000 = 3.45 \times 10{,}000{,}000$. Thus we need 3 whole symbols and 0.45, or about half, of another symbol.

Neil Diamond: Note that $32{,}200{,}000 = 3.22 \times 10{,}000{,}000$. Thus we need 3 whole symbols and 0.22, or about 20% or $\frac{1}{5}$, of another symbol.

Rod Stewart: Note that $29{,}100{,}000 = 2.91 \times 10{,}000{,}000$. Thus we need 2 whole symbols and 0.91 of another, or about 3 whole symbols.

Bob Seger: Note that $26{,}300{,}000 = 2.63 \times 10{,}000{,}000$. Thus we need 2 whole symbols and 0.63, or about 60% or $\frac{3}{5}$, of another symbol.

The pictograph can now be drawn as follows. We list the concert act or performer in one column, draw the monetary amounts with their symbols, and title the overall graph "Total Gross Revenue."

Total Gross Revenue

Kiss

Garth Brooks

Neil Diamond

Rod Stewart

Bob Seger

[symbol] = \$10,000,000

Do Exercise 19.

19. *Concert Revenue.* The following is a list of the next five concert acts for the same year and their total gross revenues. Draw a pictograph to represent the data.

Jimmy Buffett	\$26,200,000
Reba McEntire	\$26,100,000
Alanis Morissette	\$23,200,000
Hootie & the Blowfish	\$21,400,000
Ozzy Osbourne	\$21,300,000

Answer on page A-11

Improving Your Math Study Skills

Forming Math Study Groups, by James R. Norton

Dr. James Norton has taught at the University of Phoenix and Scottsdale Community College. He has extensive experience with the use of study groups to learn mathematics.

The use of math study groups for learning has become increasingly more common in recent years. Some instructors regard them as a primary source of learning, while others let students form groups on their own.

A study group generally consists of study partners who help each other learn the material and do the homework. You will probably meet outside of class at least once or twice a week. Here are some do's and don'ts to make your study group more valuable.

- DO make the group up of no more than four or five people. Research has shown clearly that this size works best.
- DO trade phone numbers so that you can get in touch with each other for help between team meetings.
- DO make sure that everyone in the group has a chance to contribute.
- DON'T let a group member copy from others without contributing. If this should happen, one member should speak with that student privately; if the situation continues, that student should be asked to leave the group.
- DON'T let the "A" students drop the ball. The group needs them! The benefits to even the best students are twofold: (1) Other students will benefit from their expertise and (2) the bright students will learn the material better by teaching it to someone else.
- DON'T let the slower students drop the ball either. *Everyone* can contribute something, and being in a group will actually improve their self-esteem as well as their performance.

How do you form study groups if the instructor has not already done so? A good place to begin is to get together with three or four friends and arrange a study time. If you don't know anyone, start getting acquainted with other people in the class during the first week of the semester.

What should you look for in a study partner?

- Do you live near each other to make it easy to get together?
- What are your class schedules like? Are you both on campus? Do you have free time?
- What about work schedules, athletic practice, and other out-of-school commitments that you might have to work around?

Making use of a study group is not a form of "cheating." You are merely helping each other learn. So long as everyone in the group is both contributing and doing the work, this method will bring you great success!

Exercise Set 5.2

a *Planets.* Use the following table, which lists information about the planets, for Exercises 1–10.

Planet	Average Distance from Sun (in miles)	Diameter (in miles)	Length of Planet's Day in Earth Time (in days)	Time of Revolution in Earth Time (in years)
Mercury	35,983,000	3,031	58.82	0.24
Venus	67,237,700	7,520	224.59	0.62
Earth	92,955,900	7,926	1.00	1.00
Mars	141,634,800	4,221	1.03	1.88
Jupiter	483,612,200	88,846	0.41	11.86
Saturn	888,184,000	74,898	0.43	29.46
Uranus	1,782,000,000	31,763	0.45	84.01
Neptune	2,794,000,000	31,329	0.66	164.78
Pluto	3,666,000,000	1,423	6.41	248.53

Source: *Handy Science Answer Book*, Gale Research, Inc.

1. Find the average distance from the sun to Jupiter.

2. How long is a day on Venus?

3. Which planet has a time of revolution of 164.78 yr?

4. Which planet has a diameter of 4221 mi?

5. Which planets have an average distance from the sun that is greater than 1,000,000 mi?

6. Which planets have a diameter that is less than 100,000 mi?

7. About how many earth diameters would it take to equal one Jupiter diameter?

8. How much longer is the longest time of revolution than the shortest?

9. What are the average, the median, and the mode of the diameters of the planets?

10. What are the average, the median, and the mode of the average distances from the sun of the planets?

Heat Index. In warm weather, a person can feel hotter due to reduced heat loss from the skin caused by higher humidity. The **temperature–humidity index,** or **apparent temperature,** is what the temperature would have to be with no humidity in order to give the same heat effect. The following table lists the apparent temperatures for various actual temperatures and relative humidities. Use this table for Exercises 11–22.

Actual Temperature (°F)	Relative Humidity									
	10%	20%	30%	40%	50%	60%	70%	80%	90%	100%
	Apparent Temperature (°F)									
75°	75	77	79	80	82	84	86	88	90	92
80°	80	82	85	87	90	92	94	97	99	102
85°	85	88	91	94	97	100	103	106	108	111
90°	90	93	97	100	104	107	111	114	118	121
95°	95	99	103	107	111	115	119	123	127	131
100°	100	105	109	114	118	123	127	132	137	141
105°	105	110	115	120	125	131	136	141	146	151

In Exercises 11–14, find the apparent temperature for the given actual temperature and humidity combinations.

11. 80°, 60%

12. 90°, 70%

13. 85°, 90%

14. 95°, 80%

15. How many temperature–humidity combinations give an apparent temperature of 100°?

16. How many temperature–humidity combinations give an apparent temperature of 111°?

17. At a relative humidity of 50%, what actual temperatures give an apparent temperature above 100°?

18. At a relative humidity of 90%, what actual temperatures give an apparent temperature above 100°?

19. At an actual temperature of 95°, what relative humidities give an apparent temperature above 100°?

20. At an actual temperature of 85°, what relative humidities give an apparent temperature above 100°?

21. At an actual temperature of 85°, by how much would the humidity have to increase in order to raise the apparent temperature from 97° to 111°?

22. At an actual temperature of 80°, by how much would the humidity have to increase in order to raise the apparent temperature from 87° to 102°?

Global Warming. Ecologists are increasingly concerned about global warming, that is, the trend of average global temperatures to rise over recent years. One possible effect is the melting of the polar icecaps. Use the following table for Exercises 23–26.

Year	Average Global Temperature (°F)
1986	59.29°
1987	59.58°
1988	59.63°
1989	59.45°
1990	59.85°
1991	59.74°
1992	59.23°
1993	59.36°
1994	59.56°
1995	59.72°
1996	59.58°

Source: Vital Signs, 1997

23. Find the average global temperatures in 1986 and 1987. What was the percent of increase in the temperature from 1986 to 1987?

24. Find the average global temperatures in 1992 and 1993. What was the percent of increase in the temperature from 1992 to 1993?

25. Find the average of the average global temperatures for the years 1986 to 1988. Find the average of the average global temperatures for the years 1994 to 1996. By how many degrees does the latter average exceed the former?

26. Find the average of the average global temperatures for the years 1994 to 1996. Find the ten-year average of the average global temperatures for the years 1987 to 1996. By how many degrees does the former average exceed the latter?

b *World Population Growth.* The following pictograph shows world population in various years. Use the pictograph for Exercises 27–34.

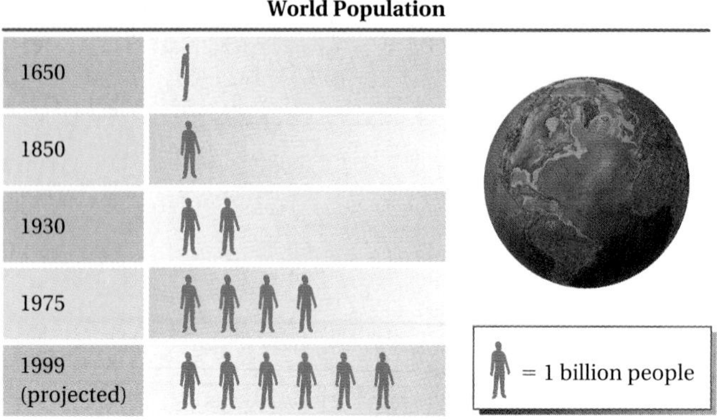

27. What was the world population in 1850?

28. What was the world population in 1975?

29. In which year will the population be the greatest?

30. In which year was the population the least?

31. Between which two years was the amount of growth the least?

32. Between which two years was the amount of growth the greatest?

33. How much greater will the world population in 1999 be than in 1975? What is the percent of increase?

34. How much greater will the world population be in 1999 than in 1930? What is the percent of increase?

Mountain Bikes. The following pictograph shows sales of mountain bikes for a bicycle company for six consecutive years. Use the pictograph for Exercises 35–42.

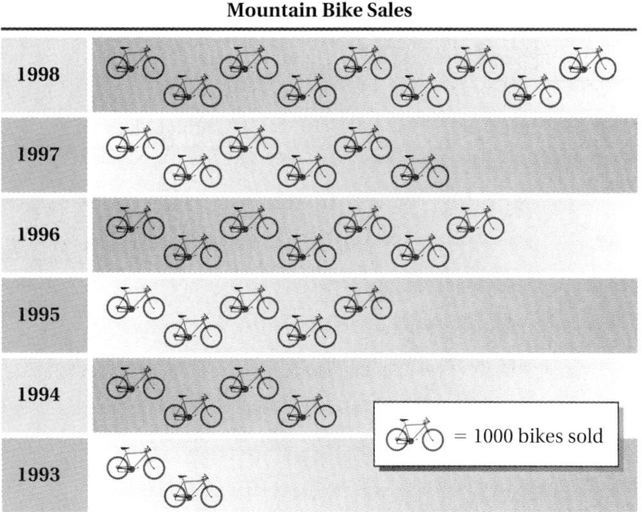

35. In which year was the greatest number of bikes sold?

36. Between which two consecutive years was there the greatest growth?

37. Between which two years did the least amount of positive growth occur?

38. How many sales does one bike symbol represent?

39. Approximately how many bikes were sold in 1996?

40. Approximately how many more bikes were sold in 1998 than in 1993?

41. In which year was there actually a decline in the number of bikes sold?

42. The sales for 1998 were how many times the sales for 1993?

c

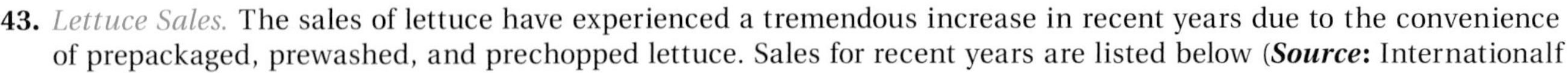

43. *Lettuce Sales.* The sales of lettuce have experienced a tremendous increase in recent years due to the convenience of prepackaged, prewashed, and prechopped lettuce. Sales for recent years are listed below (***Source*:** Internationalf Fresh-Cut Produce Association). Draw a pictograph to represent lettuce sales for these years. Use the symbol  to represent $100,000,000.

1992	$168,000,000
1993	$312,000,000
1994	$577,000,000
1995	$889,000,000
1996	$1,100,000,000

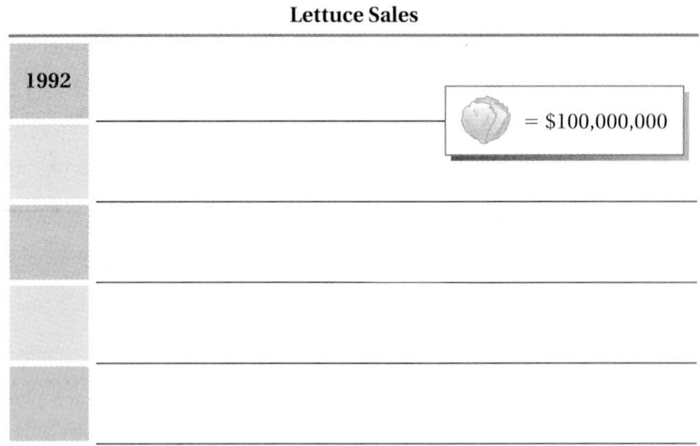

Skill Maintenance

Solve.

44. A football team has won 3 of its first 4 games. At this rate, how many games will it win in a 16-game season? [4.1e]

45. The state of Maine is 90% forest. The area of Maine is 30,955 mi^2. How many square miles of Maine are forest? [4.6a]

Find fractional notation for the percent notation in the sentence. [4.3b]

46. The United States uses 24% of the world's energy.

47. The United States has 4.8% of the world's population.

Synthesis

48. ◆ Loreena is drawing a pictograph in which dollar bills are used as symbols to represent the tuition at various private colleges. Should each dollar bill represent $8000, $4000, or $400? Why?

49. ◆ What advantage(s) does a table have over a pictograph?

50. Redraw the pictograph appearing in Example 4 as one in which each symbol represents 150 cups of coffee.

5.3 Bar Graphs and Line Graphs

A **bar graph** is convenient for showing comparisons because you can tell at a glance which amount represents the largest or smallest quantity. Of course, since a bar graph is a more abstract form of pictograph, this is true of pictographs as well. However, with bar graphs, a *second scale* is usually included so that a more accurate determination of the amount can be made.

Objectives

a Extract and interpret data from bar graphs.

b Draw bar graphs.

c Extract and interpret data from line graphs.

d Draw simple line graphs.

For Extra Help

TAPE 10 | InterAct math MAC WIN | CD-ROM

a Reading and Interpreting Bar Graphs

Example 1 *Fat Content in Fast Foods.* Wendy's Hamburgers is a national food franchise. The following bar graph shows the fat content of various sandwiches sold by Wendy's.

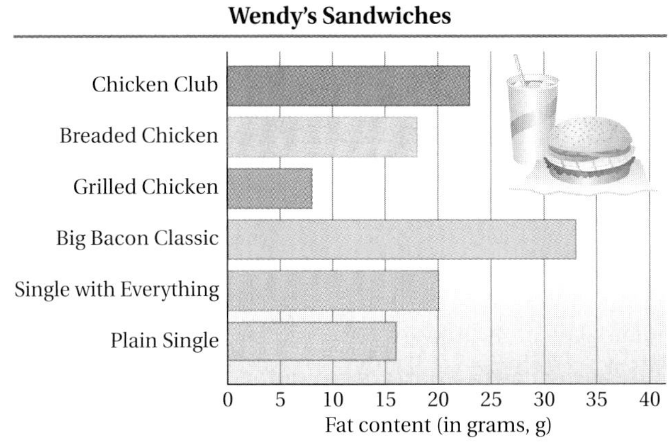

Source: Wendy's International

a) About how much fat is in a chicken club sandwich?

b) Which sandwich contains the least amount of fat?

c) Which sandwich contains about 20 g of fat?

We look at the graph to answer the questions.

a) We move to the right along the bar representing chicken club sandwiches. We can read, fairly accurately, that there is approximately 23 g of fat in the chicken club sandwich.

b) The shortest bar is for the grilled chicken sandwich. Thus that sandwich contains the least amount of fat.

c) We locate the line representing 20 g and then go up until we reach a bar that ends at approximately 20 g. We then go across to the left and read the name of the sandwich, which is the "Single with Everything."

Do Exercises 1–3.

Use the bar graph in Example 1 to answer each of the following.

1. About how much fat is in the plain single sandwich?

2. Which sandwich contains the greatest amount of fat?

3. Which sandwiches contain 20 g or more of fat?

Answers on page A-12

Bar graphs are often drawn vertically and sometimes a double bar graph is used to make comparisons.

Example 2 *Breast Cancer.* The following graph indicates the incidence and mortality rates of breast cancer for women of various age groups.

When Breast Cancer Strikes

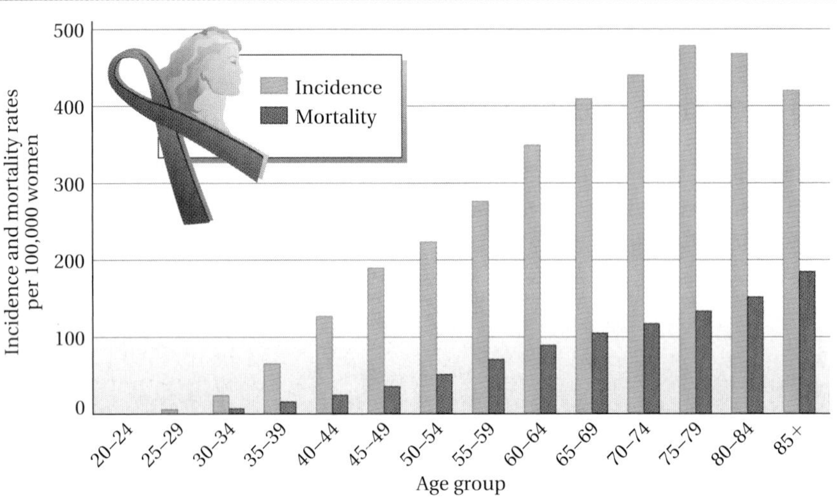

Source: National Cancer Institute

a) Approximately how many women, per 100,000, develop breast cancer between the ages of 40 and 44?

b) In what age range is the mortality rate for breast cancer approximately 100 for every 100,000 women?

c) In what age range is the incidence of breast cancer the highest?

d) Does the incidence of breast cancer seem to increase from the youngest to the oldest age group?

We look at the graph to answer the questions.

a) We go to the right, across the bottom, to the green bar above the age group 40–44. Next, we go up to the top of that bar and, from there, back to the left to read approximately 130 on the vertical scale. About 130 out of every 100,000 women develop breast cancer between the ages of 40 and 44.

b) We read up the vertical scale to the number 100. From there we move to the right until we come to the top of a red bar. Moving down that bar, we find that in the 65–69 age group, about 100 out of every 100,000 women die of breast cancer.

c) We look for the tallest green bar and read the age range below it. The incidence of breast cancer is highest for women in the 75–79 age group.

d) Looking at the heights of the bars, we see that the incidence of breast cancer increases to a high point in the 75–79 age group and then decreases.

Do Exercises 4–7.

Use the bar graph in Example 2 to answer each of the following.

4. Approximately how many women, per 100,000, develop breast cancer between the ages of 35 and 39?

5. In what age group is the mortality rate the highest?

6. In what age group do about 350 out of every 100,000 women develop breast cancer?

7. Does the breast-cancer mortality rate seem to increase from the youngest to the oldest age group?

Answers on page A-12

b Drawing Bar Graphs

Example 3 *Heights of NBA Centers.* Listed below are the heights of some of the tallest centers in the NBA (***Source*:** National Basketball Association). Make a vertical bar graph of the data.

Shaquille O'Neal:	7′1″ (85 in.)
Shawn Bradley:	7′6″ (90 in.)
Rik Smits:	7′4″ (88 in.)
Gheorghe Muresan:	7′7″ (91 in.)
Arvydas Sabonis:	7′3″ (87 in.)
David Robinson:	7′1″ (85 in.)

First, we indicate on the base or horizontal scale in six equally spaced intervals the different names of the players and give the horizontal scale the title "Players." (See the figure on the left below.) Then we label the vertical scale with "Height (in inches)." We note that the largest number (in inches) is 91 and the smallest is 85. We could start the vertical scaling at 0, but then the bars would be very high. We decide to start at 83, using the jagged line to indicate the missing numbers. We label the marks by 1's from 83 to 91. Finally, we draw vertical bars to show the various heights (in inches), as shown in the figure on the right below. We give the graph the overall title "NBA Centers."

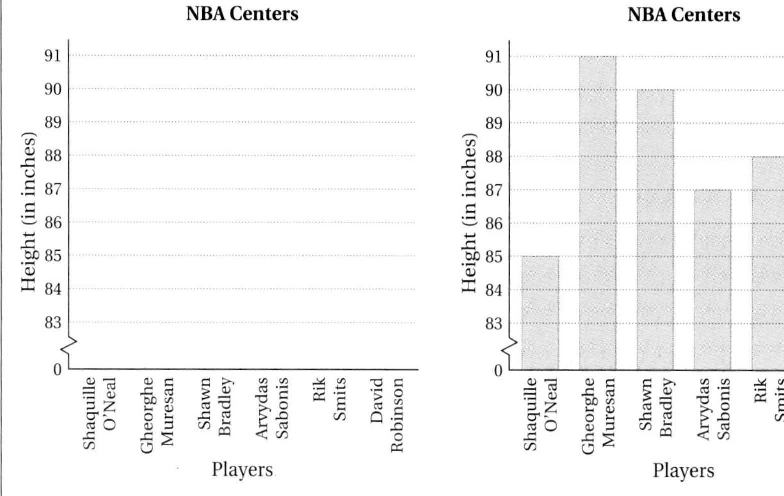

Do Exercise 8.

8. *Planetary Moons.* Make a horizontal bar graph to show the number of moons orbiting the various planets.

Planet	Number of Moons
Earth	1
Mars	2
Jupiter	16
Saturn	18
Uranus	15
Neptune	8
Pluto	1

Answer on page A-12

Use the line graph in Example 4 to answer each of the following.

9. For which month were new home sales lowest?

10. Between which months did new home sales decrease?

11. For which months were new home sales below 700 thousand?

Answers on page A-12

c Reading and Interpreting Line Graphs

Line graphs are often used to show a change over time as well as to indicate patterns or trends.

Example 4 *New Home Sales.* The following line graph shows the number of new home sales, in thousands, over a twelve-month period. The jagged line at the base of the vertical scale indicates an unnecessary portion of the scale. Note that the vertical scale differs from the horizontal scale so that the data can be shown reasonably.

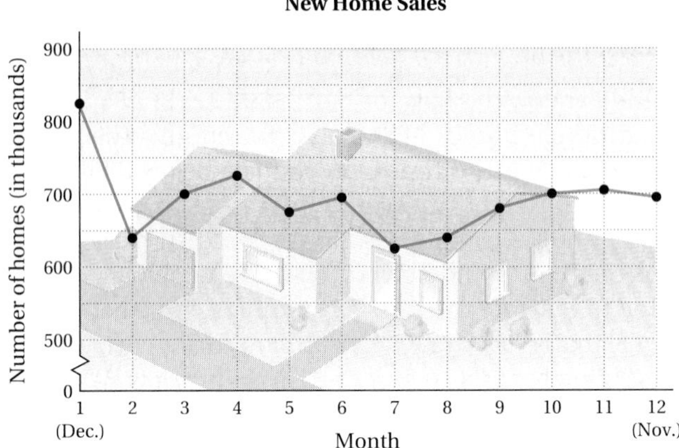

Source: U.S. Department of Commerce

a) For which month were new home sales the greatest?

b) Between which months did new home sales increase?

c) For which months were new home sales about 700 thousand?

We look at the graph to answer the questions.

a) The greatest number of new home sales was about 825 thousand in month 1.

b) Reading the graph from left to right, we see that new home sales increased from month 2 to month 3, from month 3 to month 4, from month 5 to month 6, from month 7 to month 8, from month 8 to month 9, from month 9 to month 10, and from month 10 to month 11.

c) We look from left to right along the line at 700.

We see that points are closest to 700 thousand at months 3, 6, 10, 11, and 12.

Do Exercises 9–11.

Example 5 *Monthly Loan Payment.* Suppose that you borrow \$110,000 at an interest rate of 9% to buy a home. The following graph shows the monthly payment required to pay off the loan, depending on the length of the loan. (*Caution*: A low monthly payment means that you will pay more interest over the duration of the loan.)

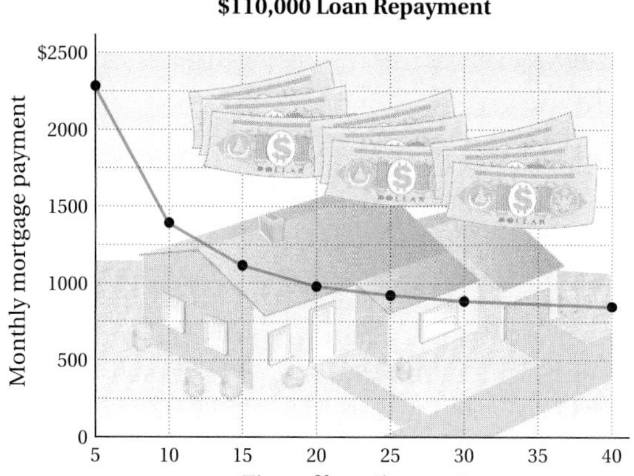

Note that there is no value for 35 yr. We will consider the reason for this in Section 5.5.

a) Estimate the monthly payment for a loan of 15 yr.

b) What time period corresponds to a monthly payment of about \$1400?

c) By how much does the monthly payment decrease when the loan period is increased from 10 yr to 20 yr?

We look at the graph to answer the questions.

a) We find the time period labeled "15" on the bottom scale and move up from that point to the line. We then go straight across to the left and find that the monthly payment is about \$1100.

b) We locate \$1400 on the vertical axis. Then we move to the right until we hit the line. The point \$1400 crosses the line at the 10-yr time period.

c) The graph shows that the monthly payment for 10 yr is about \$1400; for 20 yr, it is about \$990. Thus the monthly payment is decreased by \$1400 − \$990, or \$410. (It should be noted that you will pay back \$990 · 20 · 12 − \$1400 · 10 · 12, or \$69,600, more in interest for a 20-yr loan.)

Do Exercises 12–14.

Use the line graph in Example 5 to answer each of the following.

12. Estimate the monthly payment for a loan of 25 yr.

13. What time period corresponds to a monthly payment of about \$850?

14. By how much does the monthly payment decrease when the loan period is increased from 5 yr to 20 yr?

Answers on page A-12

15. *SAT Scores.* Draw a line graph to show how the average combined verbal–math SAT score has changed over a period of 6 yr. Use the following data (***Source*:** The College Board).

1991: 999
1992: 1001
1993: 1003
1994: 1003
1995: 1010
1996: 1013

Answer on page A-12

d Drawing Line Graphs

Example 6 *Movie Releases.* Draw a line graph to show how the number of movies released each year has changed over a period of 6 yr. Use the following data (***Source*:** Motion Picture Association of America).

1991: 164 movies
1992: 150 movies
1993: 161 movies
1994: 184 movies
1995: 234 movies
1996: 260 movies

First, we indicate on the horizontal scale the different years and title it "Year." (See the graph below.) Then we mark the vertical scale appropriately by 50's to show the number of movies released and title it "Number per Year." We also give the overall title "Movies Released" to the graph.

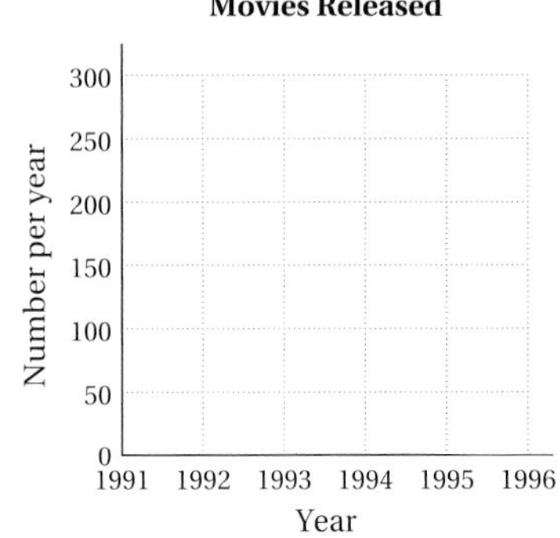

Next, we mark at the appropriate level above each year the points that indicate the number of movies released. Then we draw line segments connecting the points. The change over time can now be observed easily from the graph.

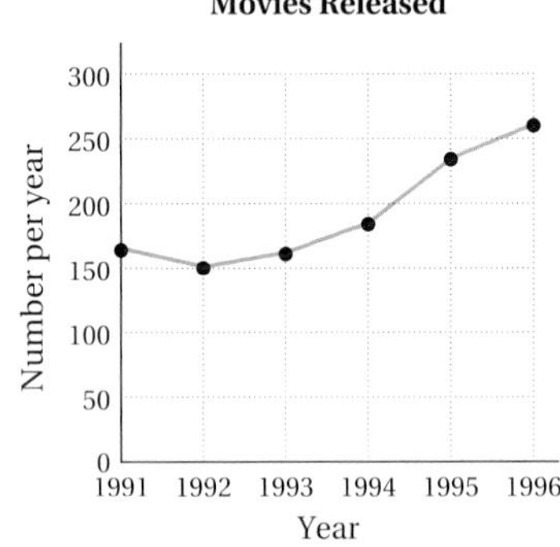

Do Exercise 15.

Exercise Set 5.3

a *Chocolate Desserts.* The following horizontal bar graph shows the average caloric content of various kinds of chocolate desserts. Use the bar graph for Exercises 1–12.

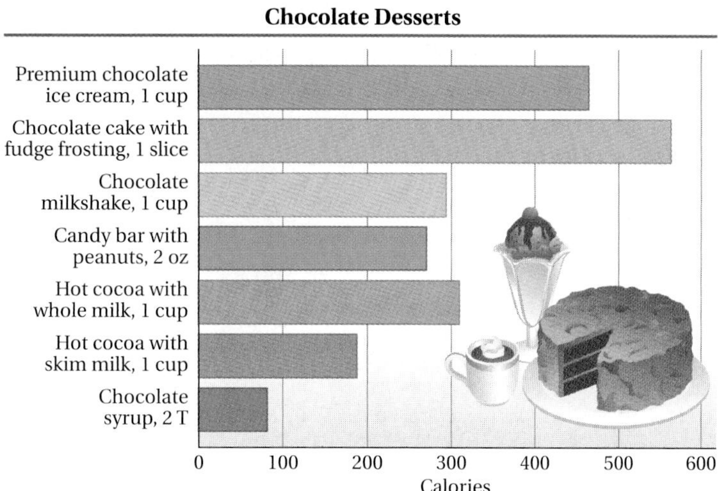

Source: *Better Homes and Gardens*, December 1996

1. Estimate how many calories there are in 1 cup of hot cocoa with skim milk.

2. Estimate how many calories there are in 1 cup of premium chocolate ice cream.

3. Which dessert has the highest caloric content?

4. Which dessert has the lowest caloric content?

5. Which dessert contains about 460 calories?

6. Which desserts contain about 300 calories?

7. How many more calories are there in 1 cup of hot cocoa made with whole milk than in 1 cup of hot cocoa made with skim milk?

8. Fred generally drinks a 4-cup chocolate milkshake. How many calories does he consume?

9. Kristin likes to eat 2 cups of premium chocolate ice cream at bedtime. How many calories does she consume?

10. Barney likes to eat a 6-oz chocolate bar with peanuts for lunch. How many calories does he consume?

11. Paul adds a 2-oz chocolate bar with peanuts to his diet each day for 1 yr (365 days) and makes no other changes in his eating or exercise habits. Consumption of 3500 extra calories will add about 1 lb to his body weight. How many pounds will he gain?

12. Tricia adds one slice of chocolate cake with fudge frosting to her diet each day for 1 yr (365 days) and makes no other changes in her eating or exercise habits. Consumption of 3500 extra calories will add about 1 lb to her body weight. How many pounds will she gain?

Deforestation. The world is gradually losing its tropical forests. The following vertical triple bar graph shows the amount of forested land of three tropical regions in the years 1980 and 1990. Use the bar graph for Exercises 13–20.

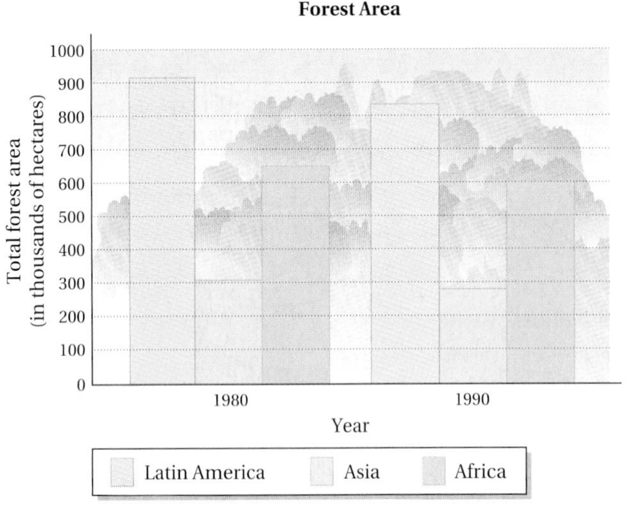

Source: World Resources Institute

13. What was the forest area of Latin America in 1980?

14. What was the forest area of Africa in 1990?

15. Which region experienced the greatest loss of forest area from 1980 to 1990?

16. Which region experienced the smallest loss of forest area from 1980 to 1990?

17. Which region had a forest area of about 600 thousand hectares in 1990?

18. Which region had a forest area of about 300 thousand hectares in 1980?

19. What was the average forest area in Latin America for the two years?

20. What was the average forest area in Asia for the two years?

b

21. *Commuting Time.* The following table lists the average commuting time in metropolitan areas with more than 1 million people. Make a vertical bar graph to illustrate the data.

City	Commuting Time (in minutes)
New York	30.6
Los Angeles	26.4
Phoenix	23.0
Dallas	24.1
Indianapolis	21.9
Orlando	22.9

Source: Census Bureau

Use the data and the bar graph in Exercise 21 to do Exercises 22–25.

22. Which city has the greatest commuting time?

23. Which city has the least commuting time?

24. What was the average commuting time for all six cities?

25. What was the median commuting time for the six cities?

26. *Deaths from Driving Incidents.* The following table lists for various years the number of driving incidents that resulted in death. Make a horizontal bar graph illustrating the data.

Year	Number of Incidents Causing Death
1990	1129
1991	1297
1992	1478
1993	1555
1994	1669
1995	1708

Source: AAA Foundation

Use the data and the bar graph in Exercise 26 to do Exercises 27–30.

27. Between which two years was the greatest increase in the number of incidents causing death?

28. Between which two years was the smallest increase in the number of incidents causing death?

29. What was the average number of incidents causing death over the 6-yr period?

30. What was the median number of incidents causing death over the 6-yr period?

c *Average Salary of Major-League Baseball Players.* The following graph shows the average salary of major-league baseball players over a recent 7-yr period. Use the graph for Exercises 31–36.

Average Salary of Major League Baseball Players

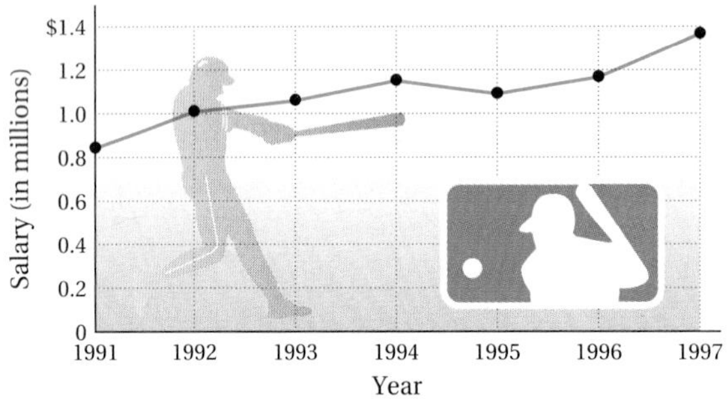

31. In which year was the average salary the highest?

32. In which year was the average salary the lowest?

33. What was the difference in salary between the highest and lowest salaries?

34. Between which two years was the increase in salary the greatest?

35. Between which two years did the salary decrease?

36. What was the percent of increase in salary between 1991 and 1996?

d Make a line graph of the data in the following tables (Exercises 37 and 42), using time on the horizontal scale.

37. *Ozone Layer.*

Year	Ozone Level (in parts per billion)
1991	2981
1992	3133
1993	3148
1994	3138
1995	3124

Source: National Oceanic and Atmospheric Administration

Use the data and the line graph in Exercise 37 to do Exercises 38–41.

38. Between which two years was the increase in the ozone level the greatest?

39. Between which two years was the decrease in the ozone level the greatest?

40. What was the average ozone level over the 5-yr period?

41. What was the median ozone level over the 5-yr period?

42. *Motion Picture Expense.*

Year	Average Expense per Picture (in millions)
1991	$38.2
1992	42.4
1993	44.0
1994	50.4
1995	54.1
1996	61.0

Source: Motion Picture Association of America

Use the data and the line graph in Exercise 42 to do Exercises 43–46.

43. Between which two years was the increase in motion-picture expense the greatest?

44. Between which two years was the increase in motion-picture expense the least?

45. What was the average motion-picture expense over the 6-yr period?

46. What was the median motion-picture expense over the 6-yr period?

47. What was the average motion-picture expense from 1991 through 1993?

48. What was the average motion-picture expense from 1994 through 1996?

Skill Maintenance

Solve.

49. A clock loses 3 min every 12 hr. At this rate, how much time will the clock lose in 72 hr? [4.1e]

50. It is known to operators of pizza restaurants that if 50 pizzas are ordered in an evening, people will request extra cheese on 9 of them. What percent of the pizzas sold are ordered with extra cheese? [4.6a]

51. 110% of 75 is what? [4.4b], [4.5b]

52. 34 is what percent of 51? [4.4b], [4.5b]

Synthesis

53. ◆ Compare bar graphs and line graphs. Discuss why you might use one over the other to graph a particular set of data.

54. ◆ Can bar graphs always, sometimes, or never be converted to line graphs? Why?

55. Referring to Exercise 42, what do you think was the average expense per picture in 1997? Justify your answer. How could you tell for sure?

Analyze class grades using line graphs and bar graphs.

5.4 Circle Graphs

We often use **circle graphs,** also called *pie charts,* to show the percent of a quantity used in different categories. Circle graphs can also be used very effectively to show visually the *ratio* of one category to another. In either case, it is quite often necessary to use mathematics to find the actual amounts represented for each specific category.

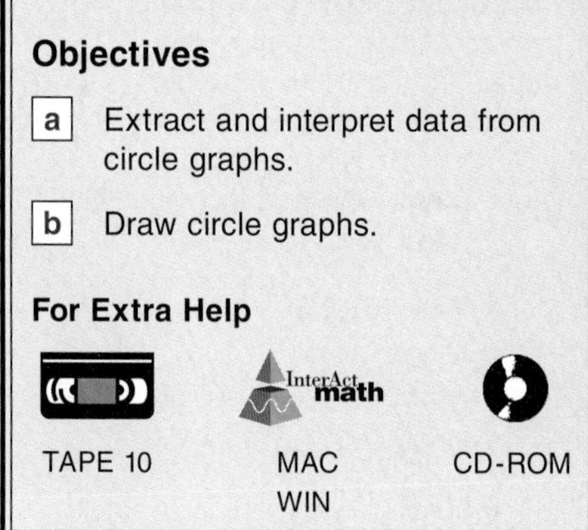

a Reading and Interpreting Circle Graphs

Example 1 *Costs of Owning a Dog.* The following circle graph shows the relative costs of raising a dog from birth to death.

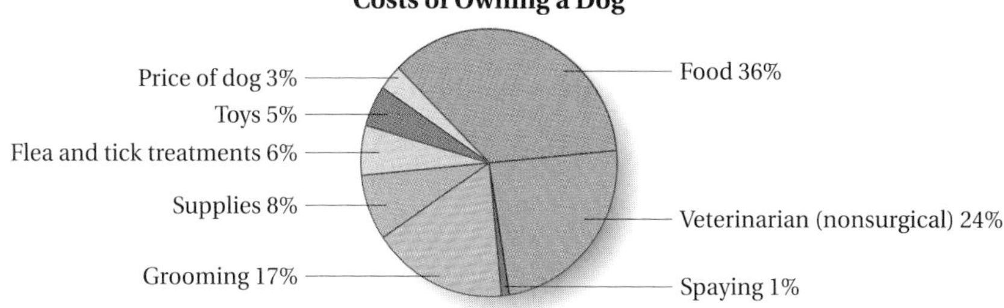

Source: The American Pet Products Manufacturers Association

a) Which item costs the most?

b) What percent of the total cost is spent on grooming?

c) Which item involves 24% of the cost?

d) The American Pet Products Manufacturers Association estimates that the total cost of owning a dog for its lifetime is $6600. How much of that amount is spent for food?

e) What percent of the expense is for grooming and flea and tick treatments?

We look at the sections of the graph to find the answers.

a) The largest section (or sector) of the graph, 36%, is for food.

b) We see that grooming is 17% of the cost.

c) Nonsurgical veterinarian bills account for 24% of the cost.

d) The section of the graph representing food costs is 36%; 36% of $6600 is $2376.

e) In a circle graph, we can add percents for questions like this. Therefore,

$$17\% \text{ (grooming)} + 6\% \text{ (flea and tick treatments)} = 23\%.$$

Do Exercises 1–4.

Use the circle graph in Example 1 to answer each of the following.

1. Which item costs the least?

2. What percent of the total cost is spent on toys?

3. How much of the $6600 lifetime cost of owning a dog is for grooming?

4. What part of the expense is for supplies and for buying the dog?

Answers on page A-12

5. *Lengths of Engagement of Married Couples.* The data below relate the percent of married couples who were engaged for a certain time period before marriage (***Source:*** Bruskin Goldring Research). Use this information to draw a circle graph.

Less than 1 yr:	24%
1–2 yr:	21%
More than 2 yr:	35%
Never engaged:	20%

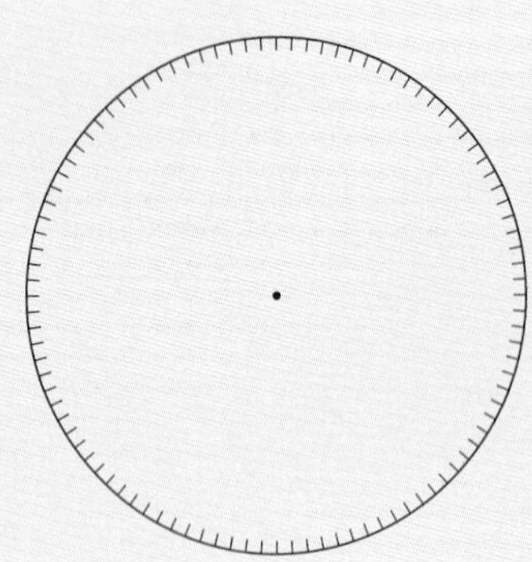

Answers on page A-12

b Drawing Circle Graphs

To draw a circle graph, or pie chart, like the one in Example 1, think of a pie cut into 100 equally sized pieces. We would then shade in a wedge equal in size to 36 of these pieces to represent 36% for food. We shade a wedge equal in size to 5 of these pieces to represent 5% for toys, and so on.

Example 2 *Fruit Juice Sales.* The percents of various kinds of fruit juice sold are given in the list at right (***Source:*** Beverage Marketing Corporation). Use this information to draw a circle graph.

Apple:	14%
Orange:	56%
Blends:	6%
Grape:	5%
Grapefruit:	4%
Prune:	1%
Other:	14%

Using a circle marked with 100 equally spaced ticks, we start with the 14% given for apple juice. We draw a line from the center to any one tick. Then we count off 14 ticks and draw another line. We shade the wedge with a color—in this case, red—and label the wedge as shown in the figure on the left below.

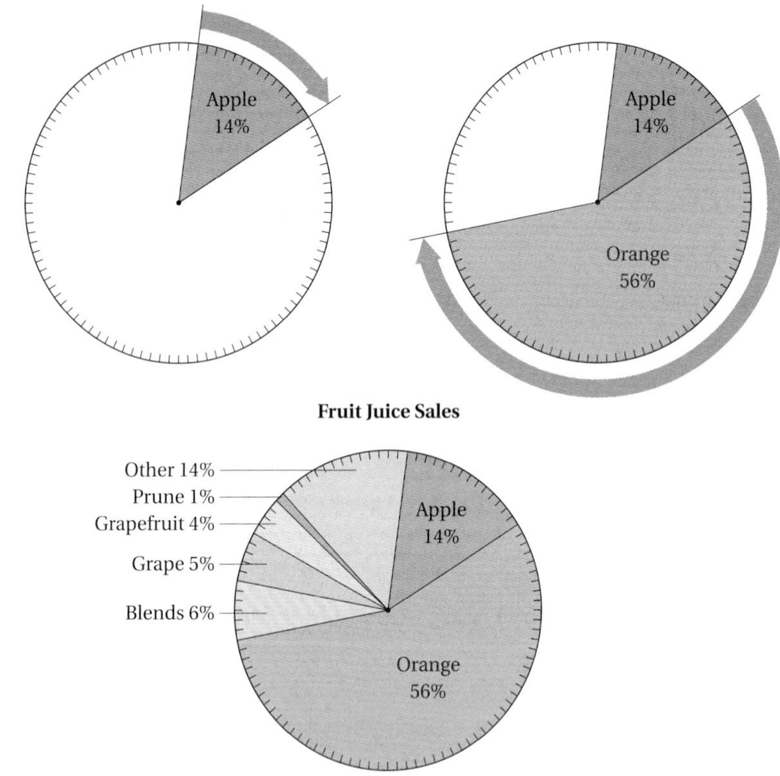

To shade a wedge for orange juice, at 56%, we count off 56 ticks and draw another line. We shade the wedge with a different color—in this case, orange—and label the wedge as shown in the figure on the right above. Continuing in this manner and choosing different colors, we obtain the graph shown above. Finally, we give the graph the overall title "Fruit Juice Sales."

Do Exercise 5.

Exercise Set 5.4

a *Musical Recordings.* This circle graph, in the shape of a CD, shows music preferences of customers on the basis of music store sales. Use the graph for Exercises 1–6.

Musical Recordings

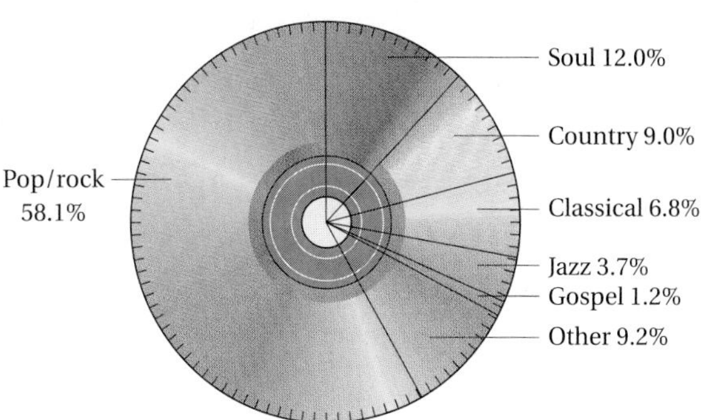

Source: National Association of Recording Merchandisers

1. What percent of all recordings sold are jazz?

2. Together, what percent of all recordings sold are either soul or pop/rock?

3. Lou's Music Store sells 3000 recordings a month. How many are country?

4. Al's Music Store sells 2500 recordings a month. How many are gospel?

5. What percent of all recordings sold are classical?

6. Together, what percent of all recordings sold are either classical or jazz?

Family Expenses. This circle graph shows expenses as a percent of income in a family of four. (*Note*: Due to rounding, the sum of the percents is 101% instead of 100%.) Use the graph for Exercises 7–10.

Family Expenses

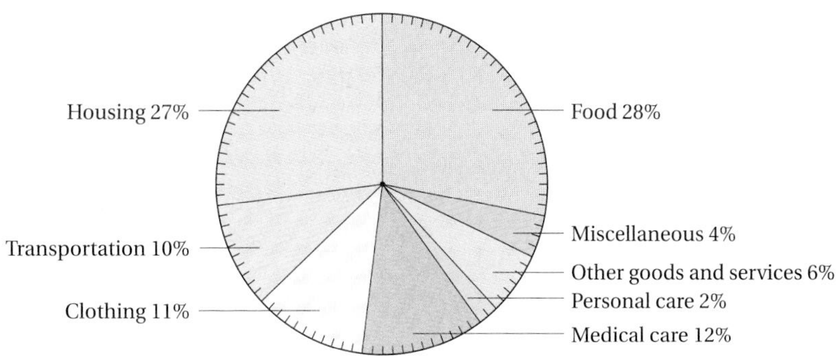

Source: Bureau of Labor Statistics

7. Which item accounts for the greatest expense?

8. In a family with a $2000 monthly income, how much is spent for transportation?

9. Some surveys combine medical care with personal care. What percent would be spent on those two items combined?

10. In a family with a $2000 monthly income, what is the ratio of the amount spent on medical care to the amount spent on personal care?

b Use the given information to complete a circle graph. Note that each circle is divided into 100 sections. (Circle graphs for Exercises 15 and 16 should be drawn on a separate sheet of paper.)

11. *Where Homebuyers Prefer to Live.*

City:	4%
Rural:	30%
Outlying suburbs:	34%
Nearby suburbs:	30%
Other:	2%

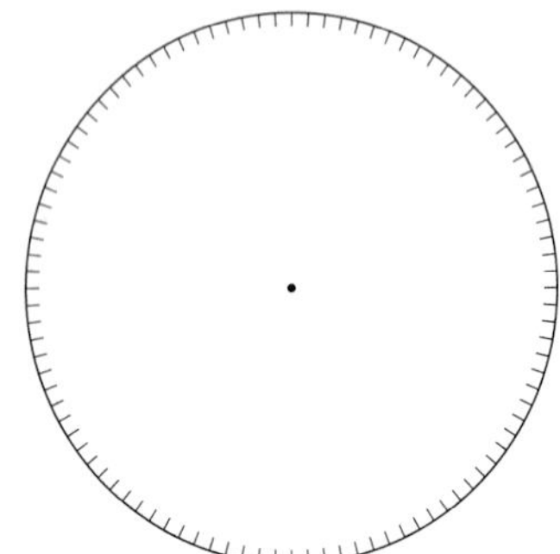

12. *How Vacation Money is Spent.*

Transportation:	15%
Meals:	20%
Lodging:	32%
Recreation:	18%
Other:	15%

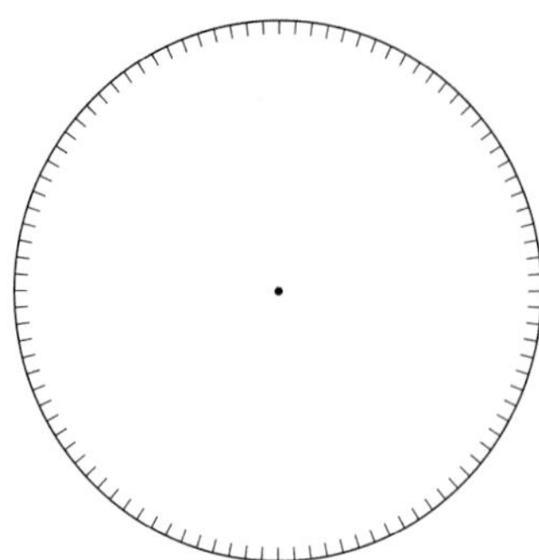

13. *Pilots' Ages.*

20–29:	6%
30–39:	32%
40–49:	36%
50–59:	26%

Source: Federal Aviation Administration

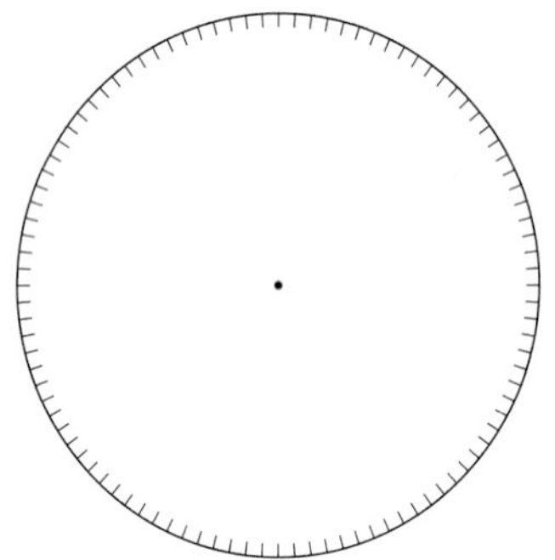

14. *Sources of Water.*

Drinking water:	29%
Tea and coffee:	24%
Vegetables:	9%
Milk/dairy products:	9%
Soft drinks:	8%
Other:	21%

Source: U.S. Department of Agriculture

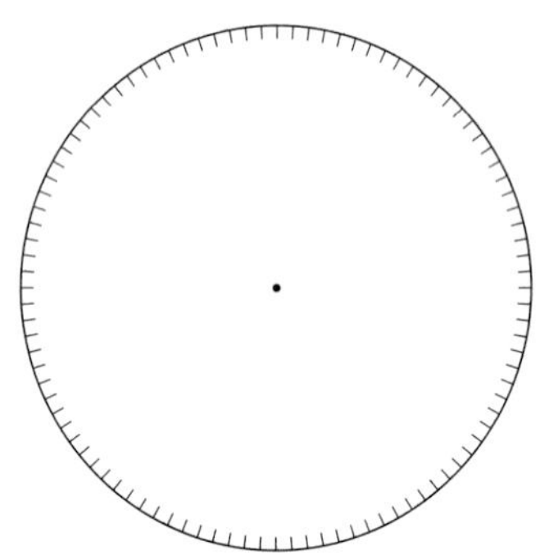

15. *Holiday Gift Giving by Men.*

More than 30 gifts:	14%
21–30 gifts:	13%
11–20 gifts:	32%
6–10 gifts:	24%
1–5 gifts:	13%
0 gifts:	4%

Source: Maritz AmeriPoll

16. *Reasons for Drinking Coffee.*

To get going in the morning:	32%
Like the taste:	33%
Not sure:	2%
To relax:	4%
As a pick-me-up:	10%
A habit:	19%

Source: LMK Associates survey for Au Bon Pain Co., Inc.

Skill Maintenance

Solve. [4.4b], [4.5b]

17. What is 45% of 668?

18. 16 is what percent of 64?

19. 23 is 20 percent of what?

Use a circle graph to show household expenses.

5.5 Data Analysis and Predictions

a Comparing Two Sets of Data

We have seen how to organize, display, and interpret data using graphs and how to calculate averages, medians, and modes from data. Now we learn data analysis for the purpose of solving applied problems.

One goal of analyzing two sets of data is to make a determination about which of two groups is "better." One way to do so is by comparing the means.

Example 1 *Battery Testing.* An experiment is performed to compare battery quality. Two kinds of battery were tested to see how long, in hours, they kept a portable CD player running. On the basis of this test, which battery is better?

Battery A: EternReady Times (in hours)	Battery B: SturdyCell Times (in hours)
27.9 28.3 27.4 27.6 27.9 28.0 26.8 27.7 28.1 28.2 26.9 27.4	28.3 27.6 27.8 27.4 27.6 27.9 26.9 27.8 28.1 27.9 28.7 27.6

Note that it is difficult to analyze the data at a glance because the numbers are close together. We need a way to compare the two groups. Let's compute the average of each set of data.

Battery A: Average

$$= \frac{27.9 + 28.3 + 27.4 + 27.6 + 27.9 + 28.0 + 26.8 + 27.7 + 28.1 + 28.2 + 26.9 + 27.4}{12}$$

$$= \frac{332.2}{12} \approx 27.68$$

Battery B: Average

$$= \frac{28.3 + 27.6 + 27.8 + 27.4 + 27.6 + 27.9 + 26.9 + 27.8 + 28.1 + 27.9 + 28.7 + 27.6}{12}$$

$$= \frac{333.6}{12} = 27.8$$

We see that the average of battery B is higher than that of battery A and thus conclude that battery B is "better." (It should be noted that statisticians might question whether these differences are what they call "significant." The answer to that question belongs to a later math course.)

Do Exercise 1.

Objectives

a Compare two sets of data using their means.

b Make predictions from a set of data using interpolation or extrapolation.

For Extra Help

TAPE 10 | MAC WIN | CD-ROM

1. *Growth of Wheat.* Rudy experiments to see which of two kinds of wheat is better. (In this situation, the shorter wheat is considered "better.") He grows both kinds under similar conditions and measures stalk heights, in inches, as follows. Which kind is better?

Wheat A Stalk Heights (in inches)			
16.2	42.3	19.5	25.7
25.6	18.0	15.6	41.7
22.6	26.4	18.4	12.6
41.5	13.7	42.0	21.6
Wheat B Stalk Heights (in inches)			
19.7	18.4	19.7	17.2
19.7	14.6	32.0	25.7
14.0	21.6	42.5	32.6
22.6	10.9	26.7	22.8

Answer on page A-13

2. *World Bicycle Production.* Use interpolation to estimate world bicycle production in 1994 from the information in the following table.

Year	World Bicycle Production (in millions)
1989	95
1990	90
1991	96
1992	103
1993	108
1994	?
1995	114

Source: United Nations Interbike Directory

Answer on page A-13

b Making Predictions

Sometimes we use data to make predictions or estimates of missing data points. One process for doing so is called **interpolation**. Let's return to some data first considered in Section 5.3.

Example 2 *Monthly Loan Payment.* The following table lists monthly repayments on a loan of $110,000 at 9% interest. Note that we have no data point for a 35-yr loan. Use interpolation to estimate its value.

Year	Monthly Payment
5	$2283.42
10	1393.43
15	1115.69
20	989.70
25	923.12
30	885.08
35	?
40	848.50

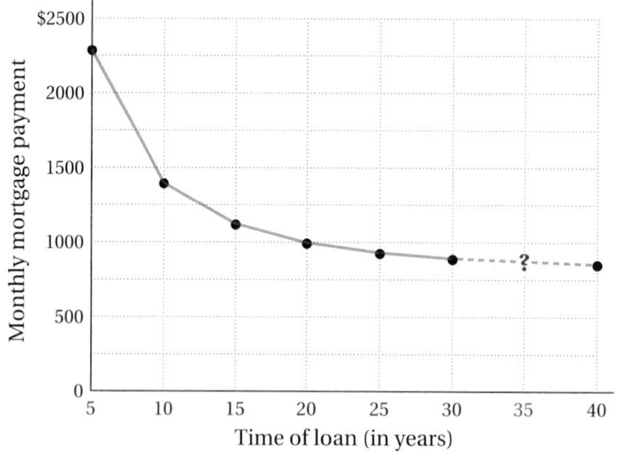

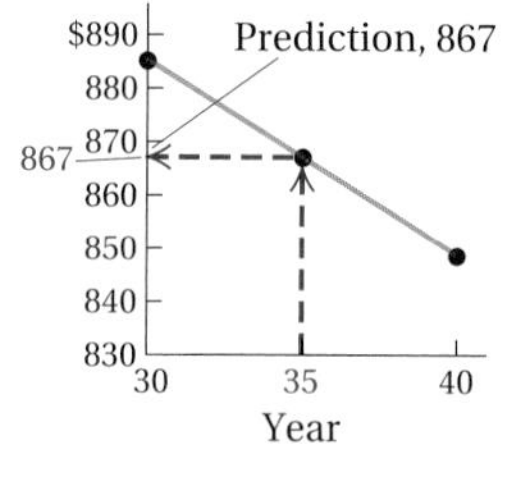

Refer to the figures shown above. First, we analyze the data and look for trends. Note that the monthly mortgage payments are decreasing and the data points resemble a straight line. It seems reasonable that we can draw a line between the points for 30 and 40. We draw a "zoomed-in" graph. Then we consider a vertical line up from the point for 35 and see where the vertical line crosses the line between the data points. We move to the left and read off a value—about $867. We can also estimate this value by taking the average of the data values $885.08 and $848.50:

$$\frac{\$885.08 + \$848.50}{2} = \$866.79.$$

When we estimate in this way to find an "in-between value," we are using a process called *interpolation*. Real-world information about the data might tell us that an estimate found in this way is unreliable. For example, data from the stock market might be very erratic.

Do Exercise 2.

We often analyze data with the view of going "beyond" the data. One process for doing so is called **extrapolation**.

Example 3 *Movies Released.* The data in the following table and graphs show the number of movie releases over a period of years. Use extrapolation to estimate the number of movies released in 1997.

Year	Movies Released
1991	164
1992	150
1993	161
1994	184
1995	234
1996	260
1997	?

Source: Motion Picture Association of America

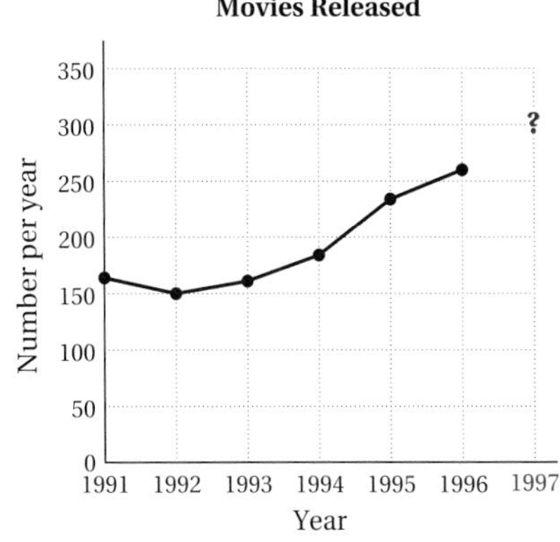

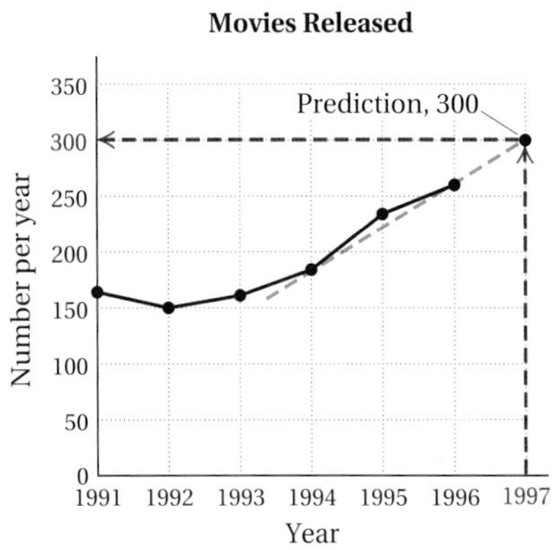

First, we analyze the data and note that they tend to follow a straight line past 1994. Keeping this trend in mind, we draw a "representative" line through the data and beyond. To estimate a value for 1997, we draw a vertical line up from 1997 until it hits the representative line. We go to the left and read off a value—about 300. When we estimate in this way to find a "go-beyond value," we are using a process called *extrapolation*. Answers found with this method can vary greatly depending on the points chosen to determine the "representative" line.

Do Exercise 3.

3. *Study Time and Test Scores.* A professor gathered the following data comparing study time and test scores. Use extrapolation to estimate the test score received when studying for 23 hr.

Study Time (in hours)	Test Grade (in percent)
19	83
20	85
21	88
22	91
23	?

Answer on page A-13

Improving Your Math Study Skills

How Many Women Have Won the Ultimate Math Contest?

Although this Study Skill feature does not contain specific tips on studying mathematics, we hope that you will find this article both challenging and encouraging.

Every year on college campuses across the United States and Canada, the most brilliant math students face the ultimate challenge. For six hours, they struggle with problems from the merely intractable to the seemingly impossible.

Every spring, five are chosen winners of the William Lowell Putnam Mathematical Competition, the Olympics of college mathematics. Every year for 56 years, all have been men.

Until this year.

This spring, Ioana Dumitriu (pronounced yo-AHN-na doo-mee-TREE-oo), 20, a New York University sophomore from Romania, became the first woman to win the award.

Ms. Dumitriu, the daughter of two electrical engineering professors in Romania, who as a girl solved math puzzles for fun, was identified as a math talent early in her schooling in Bucharest. At 11, Ms. Dumitriu was steered into years of math training camps as preparation for the Romanian entry in the International Mathematics Olympiad.

It was this training, and a handsome young coach, that led her to New York City. He was several years older. They fell in love. He chose N.Y.U. for its graduate school in mathematics, and at 19 she joined him in New York.

The test Ms. Dumitriu won is dauntingly difficult, even for math majors. About half of the 2,407 test-takers scored 2 or less of a possible 120, and a third scored 0. Some students simply walk out after staring at the questions for a while.

Ms. Dumitriu said that in the six hours allotted, she had time to do 8 of the 12 problems, each worth a maximum of 10 points. The last one she did in 10 minutes. This year, Ms. Dumitriu and her five co-winners (there was a tie for fifth place) scored between 76 and 98. She does not know her exact score or rank because the organizers do not announce them.

"I didn't ever tell myself that I was unlikely to win, that no woman before had ever won and therefore I couldn't," she said. "It is not that I forget that I'm a woman. It's just that I don't see it as an obstacle or a —."

Her English is near-perfect, but she paused because she could not find the right word. "The mathematics community is made up of persons, and that is what I am primarily."

Prof. Joel Spencer, who was a Putnam winner himself, said her work for his class in problem solving last year was remarkable. "What really got me was her fearlessness," he said. "To be good at math, you have to go right at it and start playing around with it, and she had that from the start."

In the graduate lounge in the Courant Institute of Mathematical Sciences at N.Y.U., Ms. Dumitriu, a tall, striking redhead, stands out. Instead of jeans and T-shirts, she wears gray pin-striped slacks and a rust-colored turtleneck and vest.

"There is a social perception of women and math, a stereotype," Ms. Dumitriu said during an interview. "What's happening right now is that the stereotype is defied. It starts breaking."

Still, even as women began to flock to sciences, math has remained largely a male bastion.

"Math remains the bottom line of sex differences for many," said Sheila Tobias, author of "Overcoming Math Anxiety" (W.W. Norton & Company, 1994). "It's one thing for women to write books, negotiate bills through Congress, litigate, fire missiles; quite another for them to do math."

Besides collecting the $1,000 awarded to each Putnam fellow, Ms. Dumitriu also won the $500 Elizabeth Lowell Putman prize for the top woman finisher for the second year in a row, a prize created five years ago to encourage women to take the test. This year 414 did.

In her view, there are never too many problems, never too much practice.

Besides, each new problem holds its own allure: "When you have all the pieces and you put them together and you see the puzzle, that moment always amazes me."

Exercise Set 5.5

a

1. *Light-Bulb Testing.* An experiment is performed to compare the lives of two types of light bulb. Several bulbs of each type were tested and the results are listed in the following table. On the basis of this test, which bulb is better?

Bulb A: HotLight Times (in hours)			Bulb B: BrightBulb Times (in hours)		
983	964	1214	979	1083	1344
1417	1211	1521	984	1445	975
1084	1075	892	1492	1325	1283
1423	949	1322	1325	1352	1432

2. *Cola Testing.* An experiment is conducted to determine which of two colas tastes better. Students drank each cola and gave it a rating from 1 to 10. The results are given in the following table. On the basis of this test, which cola tastes better?

Cola A: Vervcola				Cola B: Cola-cola			
6	8	10	7	10	9	9	6
7	9	9	8	8	8	10	7
5	10	9	10	8	7	4	3
9	4	7	6	7	8	10	9

b Use interpolation or extrapolation to find the missing data values.

3. *Study Time vs. Grades.* A math instructor asked her students to keep track of how much time each spent studying the chapter on percent notation in her basic mathematics course. They collected the information together with test scores from that chapter's test. The data are given in the following table. Estimate the missing data value.

Study Time (in hours)	Test Grade (in percent)
9	75
11	93
13	80
15	85
16	85
17	80
18	?
21	86
23	91

4. *Maximum Heart Rate.* A person's maximum heart rate depends on his or her gender, age, and resting heart rate. The following table relates resting heart rate and maximum heart rate for a 20-yr-old man. Estimate the missing data value.

Resting Heart Rate (in beats per minute)	Maximum Heart Rate (in beats per minute)
50	166
60	168
65	?
70	170
80	172

Source: American Heart Association

Estimate the missing data value in each of the following tables.

5. *Ozone Layer.*

Year	Ozone Level (in parts per billion)
1991	2981
1992	3133
1993	3148
1994	3138
1995	3124
1996	?

Source: National Oceanic and Atmospheric Administration

6. *Motion Picture Expense.*

Year	Average Expense per Picture (in millions)
1991	$38.2
1992	42.4
1993	44.0
1994	50.4
1995	54.1
1996	61.0
1997	?

Source: Motion Picture Association of America

7. *Credit-Card Spending.*

Year	Credit-Card Spending from Thanksgiving to Christmas (in billions)
1991	$ 59.8
1992	66.8
1993	79.1
1994	96.9
1995	116.3
1996	131.4
1997	?

Source: RAM Research Group, National Credit Counseling Services

8. *U.S. Book-Buying Growth.*

Year	Book Sales (in billions)
1992	$21
1993	23
1994	24
1995	25
1996	26
1997	?

Source: Book Industry Trends 1995

Skill Maintenance

Solve. [4.1e]

9. The building costs on a 2200-ft^2 house are $118,000. Using this rate, find the building costs on a 2400-ft^2 house.

10. A glaucoma medication is mixed in the ratio of 25 parts of medicine to 400 parts of saline solution. How many cubic centimeters of medicine should be added to 10 mL of saline solution? (1 cubic centimeter = 1 milliliter)

11. Four software CDs cost $239.80. How much would 23 comparable CDs cost?

12. A car is driven 700 mi in 5 days. At this rate, how far will it have been driven in 36 days?

Divide. [2.2c]

13. $\frac{5}{6} \div \frac{7}{18}$

14. $256 \div \frac{6}{11}$

15. $\frac{17}{25} \div 1000$

16. $\frac{1}{12} \div \frac{1}{11}$

Synthesis

17. ◆ Discuss how you might test the estimates that you found in Exercises 3–8.

18. ◆ Compare and contrast the processes of interpolation and extrapolation.

Summary and Review Exercises: Chapter 5

The objectives to be tested in addition to the material in this chapter are [2.2c], [4.1e], [4.4b], [4.5b], and [4.6a].

FedEx Mailing Costs. Federal Express has three types of delivery service for packages of various weights, as shown in the following table. Use this table for Exercises 1–6. [5.2a]

Delivery by 10:00 a.m. next business day → FedEx Priority Overnight®
Delivery by 3:00 p.m. next business day → FedEx Standard Overnight®
Delivery by 4:30 p.m. second business day → FedEx 2Day®

		FedEx Priority Overnight®	FedEx Standard Overnight®	FedEx 2Day®
FedEx Letter up to 8 oz.		$ 13.25	$ 11.50	$ n/a
All other packaging/Weight in lbs.	1 lb.	$ 18.30	$ 16.00	$ 9.25
	2 lbs.	19.20	17.00	9.95
	3	21.00	18.00	11.00
	4	22.80	19.00	12.00
	5	24.90	20.00	13.00
	6	27.30	21.75	14.25
	7	29.70	23.50	15.25
	8	31.80	25.25	16.25
	9	34.20	27.00	17.25
	10	36.80	28.75	18.25
	11	37.80	30.75	19.25

Source: Federal Express Corporation

1. Find the cost of a 3-lb FedEx Priority Overnight delivery.

2. Find the cost of a 10-lb FedEx Standard Overnight delivery.

3. How much would you save by sending the package listed in Exercise 1 by FedEx 2Day delivery?

4. How much would you save by sending the package in Exercise 2 by FedEx 2Day delivery?

5. Is there any difference in price between sending a 5-oz package FedEx Priority Overnight and sending an 8-oz package in the same way?

6. An author has a 4-lb manuscript to send by FedEx Standard Overnight delivery to her publisher. She calls and the package is picked up. Later that day she completes work on another part of her manuscript that weighs 5 lb. She calls and sends it by FedEx Standard Overnight delivery to the same address. How much could she have saved if she had waited and sent both packages as one?

TV Sales. This pictograph shows the projected number of television sets to be sold by a company. Use it for Exercises 7–10. [5.2b]

Television Sets Sold

7. About how many TV sets will be sold in 1999?

8. In which year does the company sell the least number of TV sets?

9. In which year does the company sell the greatest number of TV sets?

10. Estimate the average number of TV sets sold per year over the 4-yr period.

Find the average. [5.1a]

11. 26, 34, 43, 51

12. 7, 11, 14, 17, 18

13. 0.2, 1.7, 1.9, 2.4

14. 700, 900, 1900, 2700, 3000

15. \$2, \$14, \$17, \$17, \$21, \$29

16. 20, 190, 280, 470, 470, 500

Find the median. [5.1b]

17. 26, 34, 43, 51

18. 7, 11, 14, 17, 18

19. 0.2, 1.7, 1.9, 2.4

20. 700, 900, 1900, 2700, 3000

21. \$2, \$14, \$17, \$17, \$21, \$29

22. 20, 190, 280, 470, 470, 500

Find the mode. [5.1c]

23. 26, 34, 43, 26, 51

24. 7, 11, 11, 14, 17, 17, 18

25. 0.2, 0.2, 0.2, 1.7, 1.9, 2.4

26. 700, 700, 800, 2700, 800

27. \$2, \$14, \$17, \$17, \$21, \$29

28. 20, 20, 20, 20, 20, 500

29. One summer, a student earned the following amounts over a four-week period: \$102, \$112, \$130, and \$98. What was the average amount earned per week? the median? [5.1a, b]

30. The following temperatures were recorded in St. Louis every four hours on a certain day in June: 63°, 58°, 66°, 72°, 71°, 67°. What was the average temperature for that day? [5.1a]

31. To get an A in math, a student must score an average of 90 on four tests. Scores on the first three tests were 94, 78, and 92. What is the lowest score that the student can make on the last test and still get an A? [5.1a]

Calorie Content in Fast Foods. Wendy's Hamburgers is a national food franchise. The following bar graph shows the caloric content of various sandwiches sold by Wendy's. Use the bar graph for Exercises 32–39. [5.3a]

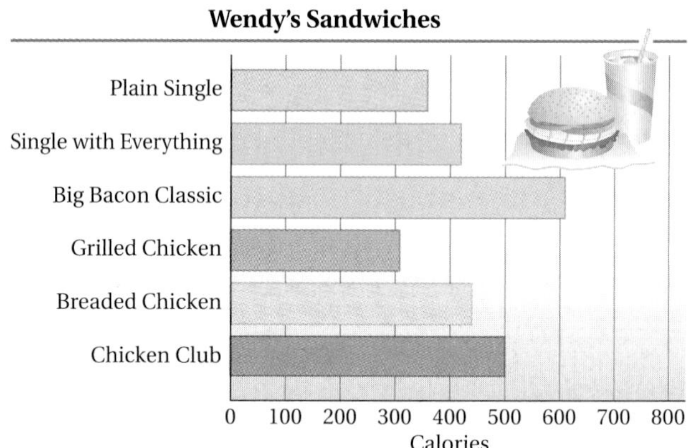

Source: Wendy's International

32. How many calories are in a Single with Everything?

33. How many calories are in a breaded chicken sandwich?

34. Which sandwich has the highest caloric content?

35. Which sandwich has the lowest caloric content?

36. Which sandwich contains about 360 calories?

37. Which sandwich contains about 500 calories?

38. How many more calories are in a chicken club than in a Single with Everything?

39. How many more calories are in a Big Bacon Classic than in a plain single?

Accidents of Drivers by Age. The following line graph shows the number of accidents per 100 drivers, by age. Use the graph for Exercises 40–45. [5.3c]

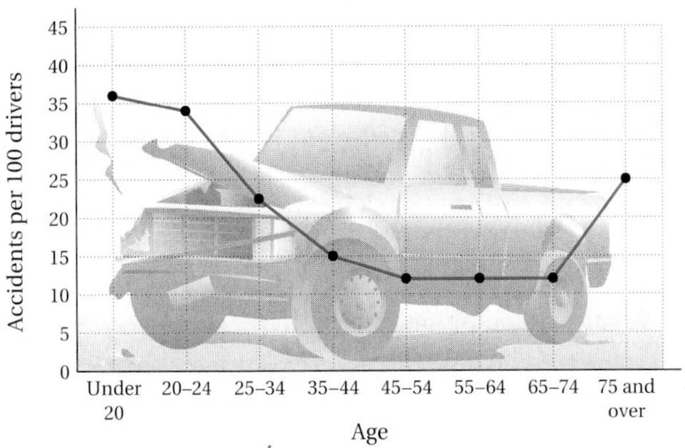

40. Which age group has the most accidents per 100 drivers?

41. What is the fewest number of accidents per 100 in any age group?

42. How many more accidents do people over 75 yr of age have than those in the age range of 65–74?

43. Between what ages does the number of accidents stay basically the same?

44. How many fewer accidents do people 25–34 yr of age have than those 20–24 yr of age?

45. Which age group has accidents more than three times as often as people 55–64 yr of age?

Hotel Preferences. This circle graph shows hotel preferences for travelers. Use the graph for Exercises 46–49. [5.4a]

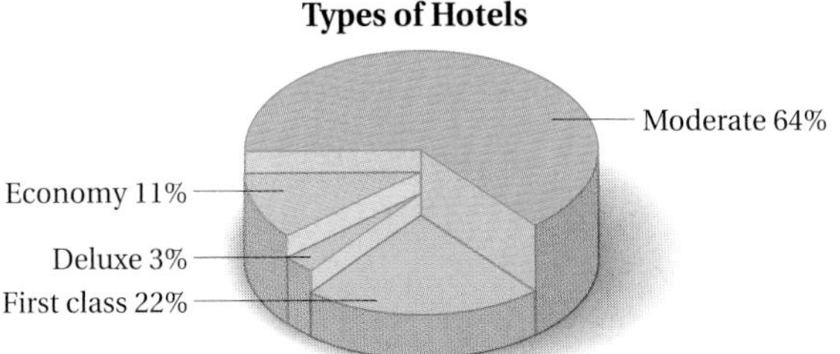

46. What percent of travelers prefer a first-class hotel?

47. What percent of travelers prefer an economy hotel?

48. Suppose 2500 travelers arrive in a city one day. How many of them might seek a moderate room?

49. What percent of travelers prefer either a first-class or a deluxe hotel?

Coca-Cola. The following table shows the total yearly revenue (money taken in), in millions, of the Coca-Cola Company from 1991 to 1996. Use the table for Exercises 50–53.

Year	Total Revenue (in millions)
1991	$465
1992	656
1993	687
1994	724
1995	762
1996	?

Source: Coca-Cola Bottling Consolidated Annual Report

50. Make a vertical bar graph of the data. [5.3b]

51. Make a line graph of the data. [5.3d]

52. Use extrapolation to find the missing data value for 1996. [5.5b]

53. Suppose you know that the revenue in 1997 was $850 million. Use this information, the other data in the table, and interpolation to estimate the missing data value for 1996. [5.5b]

54. *Battery Testing.* An experiment is performed to compare battery quality. Two kinds of battery were tested to see how long, in hours, they kept a hand radio running. On the basis of this test, which battery is better? [5.5a]

Battery A: Times (in hours)		
38.9	39.3	40.4
53.1	41.7	38.0
36.8	47.7	48.1
38.2	46.9	47.4

Battery B: Times (in hours)		
39.3	38.6	38.8
37.4	47.6	37.9
46.9	37.8	38.1
47.9	50.1	38.2

Skill Maintenance

Solve.

55. A company car was driven 4200 mi in the first 4 months of a year. At this rate, how far will it be driven in 12 months? [4.1e]

56. 92% of the world population does not have a telephone. The population is about 5.9 billion. How many do not have a telephone? [4.6a]

57. 789 is what percent of 355.05? [4.4b], [4.5b]

58. What percent of 98 is 49? [4.4b], [4.5b]

Divide and simplify. [2.2c]

59. $\frac{3}{4} \div \frac{5}{6}$

60. $\frac{5}{8} \div \frac{3}{2}$

Synthesis

61. ◆ Compare and contrast averages, medians, and modes. Discuss why you might use one over the others to analyze a set of data. [5.1a, b, c]

62. ◆ Find a real-world situation that fits this equation: [5.1a]

$$\frac{(20{,}500 + 22{,}800 + 23{,}400 + 26{,}000)}{4}.$$

63. The ordered set of data 298, 301, 305, a, 323, b, 390 has a median of 316 and an average of 326. Find a and b. [5.1a, b]

Test: Chapter 5

Answers

Retirement Savings. The following table lists estimates of the type of retirement savings a person should have, based on his or her household yearly income, age, gender, and marital status. Use the table for Exercises 1–6.

Household Yearly Income	Age 35	45	55	65
		Couple		
$ 50,000	$ 2,756	$ 34,443	$117,739	$187,593
$100,000	$28,850	$101,462	$261,139	$474,590
$150,000	$60,538	$200,825	$468,837	$820,215
		Single Male		
$ 50,000	$ 2,558	$ 38,939	$125,420	$180,953
$100,000	$26,345	$115,816	$275,744	$472,326
$150,000	$53,519	$209,960	$468,259	$779,456
		Single Female		
$ 50,000	$ 35,158	$ 69,391	$121,242	$181,577
$100,000	$ 90,601	$193,985	$341,413	$504,500
$150,000	$152,725	$326,846	$565,817	$831,025

Source: Merrill Lynch

1. What is the recommended retirement savings for a 55-year-old single female with an annual income of $100,000?

2. What is the recommended retirement savings for a 35-year-old single male with an annual income of $50,000?

3. What type of person(s) needs a retirement savings of $474,590?

4. What type of person(s) needs a retirement savings of $326,846?

5. How much more retirement savings does a 45-year-old single female with an income of $100,000 need than a comparable single male?

6. How much more retirement savings does a 65-year-old couple with an income of $100,000 need than a comparable single male?

Shampoo Sales. The following pictograph shows projected sales of shampoo for a soap company for six consecutive years. Use the pictograph for Exercises 7–10.

Shampoo Sales

2003

2002

2001

2000

1999

1998

= 1000 bottles sold

7. In which year will the greatest number of bottles be sold?

8. Between which two consecutive years will there be the greatest growth?

9. How many more bottles will be sold in 2003 than in 1998?

10. In which year will there actually be a decline in the number of bottles sold?

1. ______

2. ______

3. ______

4. ______

5. ______

6. ______

7. ______

8. ______

9. ______

10. ______

Answers

11. ______

12. ______

13. ______

14. ______

15. ______

16. ______

17. ______

18. ______

19. ______

20. ______

21. ______

22. ______

23. ______

24. ______

Find the average.

11. 45, 49, 52, 54

12. 1, 2, 3, 4, 5

13. 3, 17, 17, 18, 18, 20

Find the median and the mode.

14. 45, 49, 52, 54

15. 1, 2, 3, 4, 5

16. 3, 17, 17, 18, 18, 20

17. A car is driven 754 km in 13 hr. What is the average number of kilometers per hour?

18. To get a C in chemistry, a student must score an average of 70 on four tests. Scores on the first three tests were 68, 71, and 65. What is the lowest score that the student can make on the last test and still get a C?

Nike, Inc. The following line graph shows the revenues of Nike, Inc. Use the graph for Exercises 19–24.

Revenue of Nike, Inc.

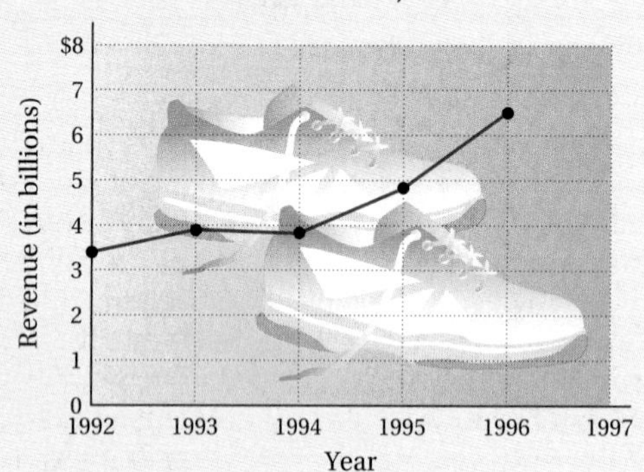

Source: Nike, Inc., annual report

19. How much revenue was earned in 1996?

20. What was the average of the revenues for the five years?

21. How much more revenue was earned in 1996 than in 1992?

22. In which year was the increase in revenue the greatest?

23. What was the median of the revenues for the five years?

24. Use extrapolation to estimate the revenue in 1997.

25. *Animal Speeds.* The following table lists maximum running speeds for various animals, in miles per hour, compared to the speed of the fastest human. Make a vertical bar graph of the data.

Animal	Speed (in miles per hour)
Antelope	61
Bear	30
Cheetah	70
Fastest Human	28
Greyhound	39
Lion	50
Zebra	40

Refer to the table and graph in Question 25 for Questions 26–29.

26. By how much does the fastest speed exceed the slowest speed?

27. Does a human have a chance of outrunning a lion? Explain.

28. Find the average of all the speeds.

29. Find the median of all the speeds.

30. *Popularity of Sport Utility Vehicles.* Use the following information to make a circle graph showing the reasons people buy sport utility vehicles.

Sporty look:	11%
Status symbol:	27%
Drives well in bad weather:	32%
Hauling or carrying capacity:	21%
Off-road capability:	6%
Other:	3%

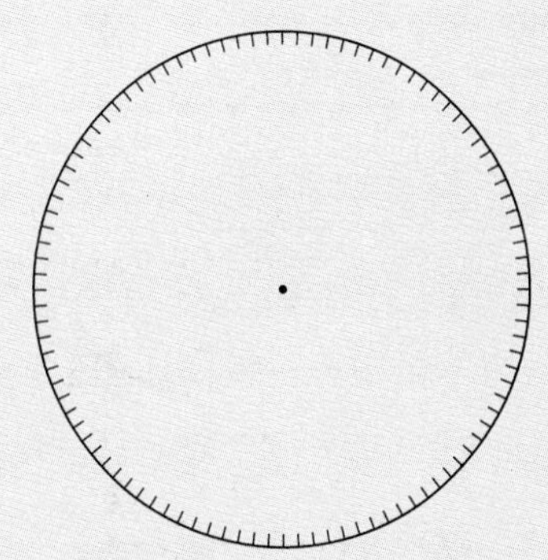

Answers

25. ____________

26. ____________

27. ____________

28. ____________

29. ____________

30. ____________

Answers

31. ______

32. ______

33. ______

34. ______

35. ______

36. ______

37. ______

38. ______

Pepsico. Pepsico makes soft drinks and snack foods and operates restaurants such as Taco Bell, Pizza Hut, and KFC. The following table shows the total revenue, in billions, of Pepsico for various years. Use the table for Questions 31 and 32.

Year	Total Revenue (in billions)
1992	$22.0
1993	25.0
1994	28.5
1995	30.4
1996	31.6
1997	?

Source: Pepsico Annual Report

31. Make a line graph of the data.

32. Suppose you know that the revenue in 1998 was $37.2 billion. Use this information, the other data in the table, and interpolation to estimate the missing data value for 1997.

33. *Chocolate Bars.* An experiment is performed to compare the quality of new Swiss chocolate bars being introduced in the United States. People were asked to taste the candies and rate them on a scale of 1 to 10. On the basis of this test, which chocolate bar is better?

Bar A: Swiss Pecan	Bar B: Swiss Hazelnut
9 10 8	10 6 8
10 9 7	9 10 10
6 9 10	8 7 6
7 8 8	9 10 8

Skill Maintenance

34. Divide and simplify: $\frac{3}{5} \div \frac{12}{125}$.

35. 17 is 25% of what number?

36. On a particular Sunday afternoon, 78% of the television sets that were on were tuned to one of the major networks. Suppose 20,000 TV sets in a town are being watched. How many are tuned to a major network?

37. A baseball player gets 7 hits in the first 20 times at bat. At this rate, how many times at bat will it take to get 119 hits?

Synthesis

38. The ordered set of data 69, 71, 73, a, 78, 98, b has a median of 74 and a mean of 82. Find a and b.

6

Geometry

An Application

A standard-sized slow-pitch softball diamond is a square with sides of length 65 ft. What is the perimeter of this softball diamond? (This is the distance you would have to run if you hit a home run.)

This problem appears as Exercise 16 in Exercise Set 6.2.

The Mathematics

We find the perimeter by finding the distance around the square:

$$\begin{aligned} P &= 65 \text{ ft} + 65 \text{ ft} + 65 \text{ ft} + 65 \text{ ft} \\ &= 4 \cdot (65 \text{ ft}) = \underbrace{260 \text{ ft}}. \end{aligned}$$

This is the perimeter.

For more information, visit us at www.mathmax.com

Introduction

In this chapter, we introduce basic geometric figures, such as segments, rays, lines, and angles. Measures considered in this chapter are perimeter, area, and volume. Relationships between angle measures, congruent and similar triangles, and properties of parallelograms are also studied.

Pretest: Chapter 6

1. Find the perimeter.

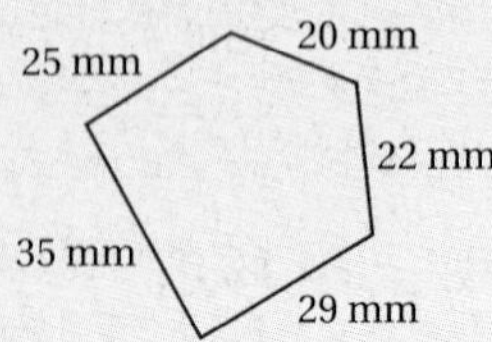

2. Find the area of the shaded region.

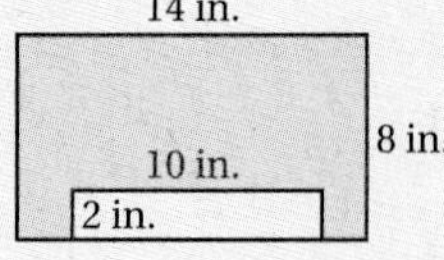

Find the area.

3.

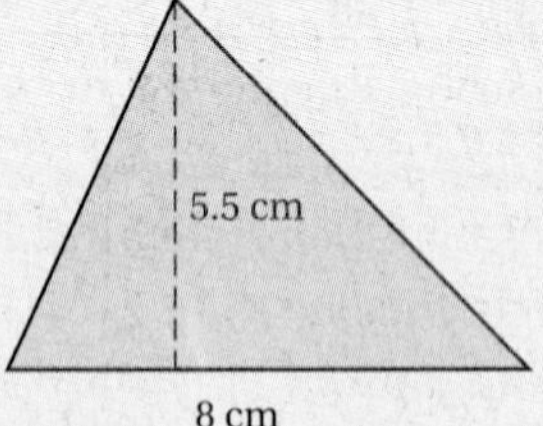

4.

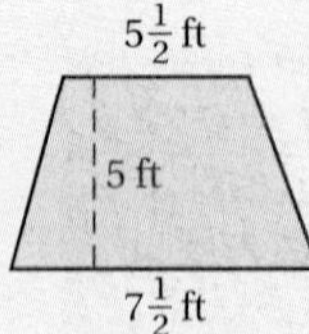

5.

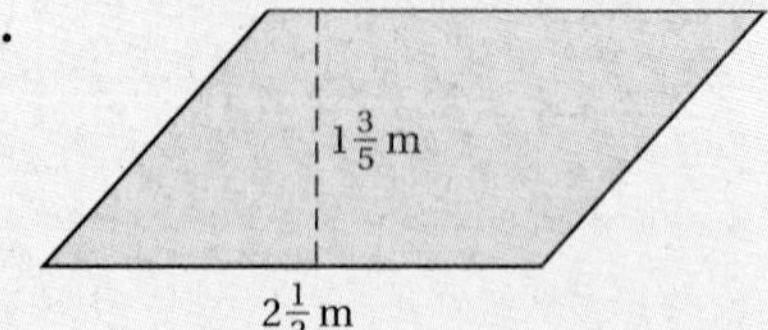

6. Find the length of a diameter of a circle with a radius of 4.8 m.

7. Find the circumference and the area of the circle in Question 6. Use 3.14 for π.

8. Find the volume and the surface area.

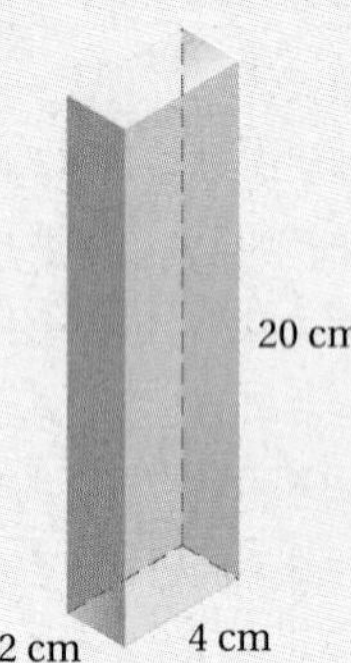

9. Find the volume. Use 3.14 for π.

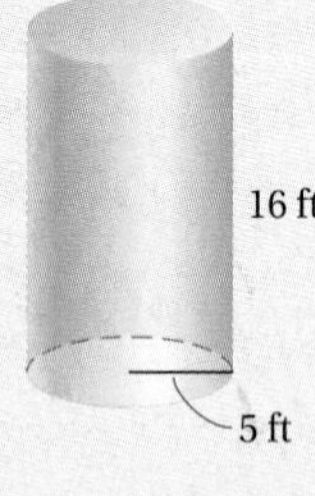

10. If $m \parallel n$ and $m\angle 8 = 29°$, what are the measures of the other angles?

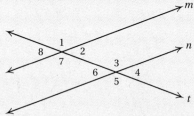

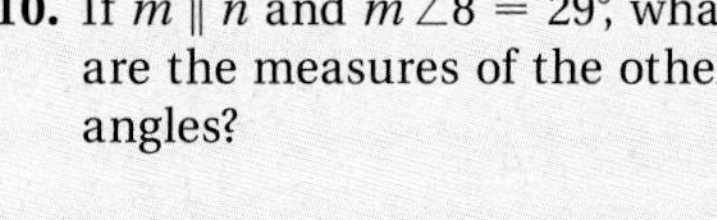

11. Given that $\triangle PQR \cong \triangle STV$, list the congruent corresponding parts.

12. Given that $\triangle MAC \sim \triangle GET$, find MA and GT.

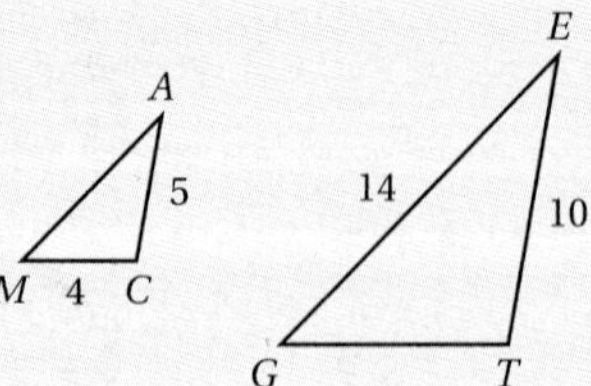

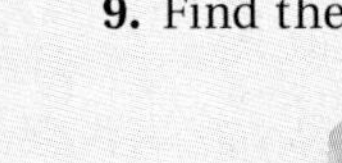

Objectives for Retesting

The objectives to be tested in addition to the material in this chapter are as follows.

[1.6b] Evaluate exponential notation.
[2.4d] Multiply using mixed numerals.
[3.3a] Multiply using decimal notation.
[4.2b, c] Convert from percent notation to decimal notation and from decimal notation to percent notation.
[4.3a, b] Convert from fractional notation to percent notation and from percent notation to fractional notation.

6.1 Basic Geometric Figures

In geometry we study sets of points. A **geometric figure** (or *figure*) is simply a set of points. Thus a figure can be a set with one point, a set with two points, or sets that look like those below.

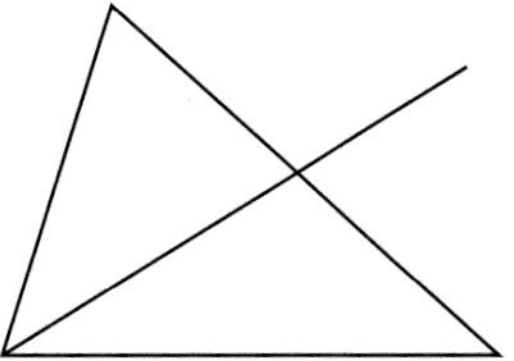

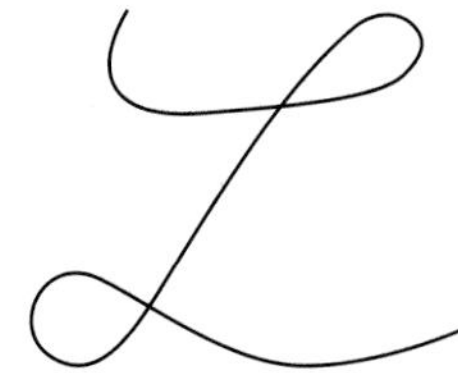

Objectives

a Draw and name segments, rays, and lines. Also, identify endpoints, if they exist.

b Name a given angle in four different ways and given an angle, measure it with a protractor.

c Classify a given angle as right, straight, acute, or obtuse.

d Identify perpendicular lines.

e Classify a triangle as equilateral, isosceles, or scalene and as right, obtuse, or acute. Given a polygon of twelve, ten, or fewer sides, classify it as a dodecagon, decagon, and so on.

f Given a polygon of n sides, find the sum of its angle measures using the formula $(n - 2) \cdot 180°$.

For Extra Help

TAPE 11

MAC WIN

CD-ROM

a Segments, Rays, and Lines

A **segment** is a geometric figure consisting of two points, called *endpoints,* and all points between them. The segment whose endpoints are A and B is shown below. It can be named $\overline{AB}$ or $\overline{BA}$.

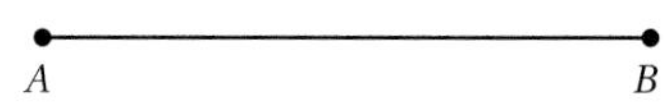

Do Exercise 1.

We get an idea of a geometric figure called a ray by thinking of a ray of light. A **ray** consists of a segment, say $\overline{AB}$, and all points X such that B is between A and X: that is, $\overline{AB}$ and all points "beyond" B.

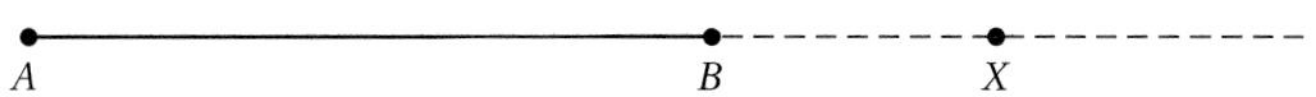

A ray is usually drawn as shown below. It has just one endpoint. The arrow indicates that it extends forever in one direction.

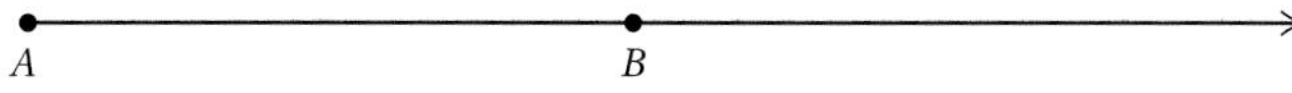

A ray is named $\overrightarrow{AB}$, where B is some point on the ray other than A. The endpoint is always listed first. Thus rays $\overrightarrow{AB}$ and $\overrightarrow{BA}$ are different.

Do Exercises 2–5.

Two rays such as $\overrightarrow{PQ}$ and $\overrightarrow{QP}$ make up what is known as a **line**. A line can be named with a smaller letter m, as shown below, or it can be named by two points P and Q on the line as $\overleftrightarrow{PQ}$.

Do Exercises 6–11 on the following page.

1. a) Draw a segment.

b) Label its endpoints E and F.

c) Name this segment in two ways.

2. Draw two points P and Q.

3. Draw $\overline{PQ}$.

4. Draw $\overrightarrow{PQ}$. What is its endpoint?

5. Use a colored pencil to draw $\overrightarrow{QP}$. What is its endpoint?

Answers on page A-14

6. Draw two points R and S.

7. Draw $\overline{RS}$. What are its endpoints?

8. Draw $\overrightarrow{RS}$. What is its endpoint?

9. Draw $\overrightarrow{SR}$. What is its endpoint?

10. Draw $\overleftrightarrow{RS}$. What are its endpoints?

11. Name this line in seven different ways.

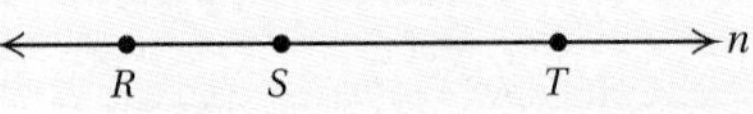

Name the angle in four different ways.

12.

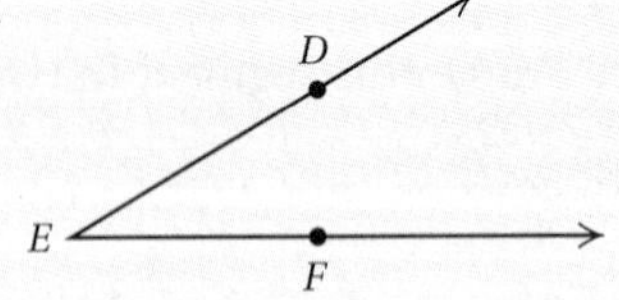

13.

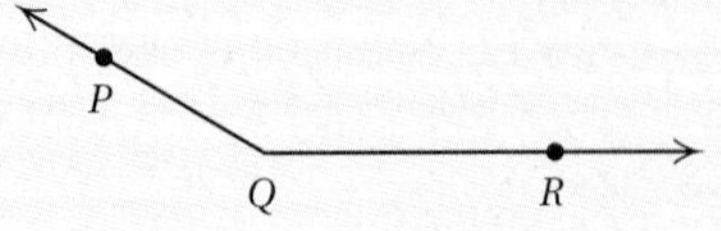

Answers on page A-14

Lines in the same plane are called **coplanar**. Coplanar lines that do not intersect are called **parallel**. For example, lines l and m below are *parallel* ($l \parallel m$).

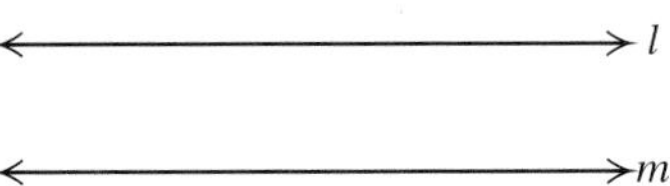

The figure below shows two lines that cross. Their *intersection* is D. They are also called **intersecting lines.**

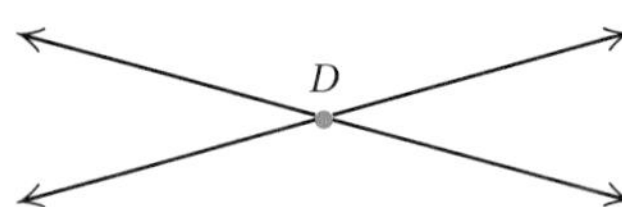

b Angles

An **angle** is a set of points consisting of two rays with a common endpoint. The endpoint, B, is called the **vertex**.

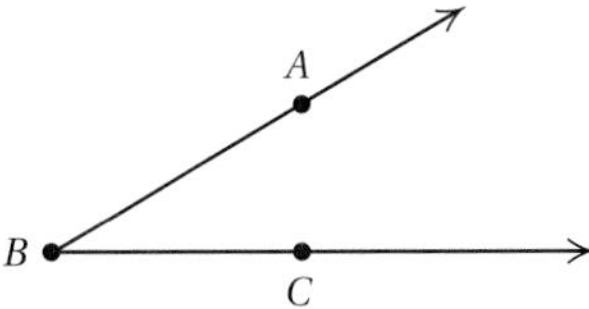

The rays are called the *sides*. The angle above can be named

angle ABC, $\angle ABC$, $\angle CBA$, or $\angle B$.

Note that the name of the vertex either is in the middle or is listed by itself.

Do Exercises 12 and 13.

Measuring angles is similar to measuring segments. To measure angles, we start with some arbitrary angle and assign to it a measure of 1. We call it a *unit angle.* Suppose that $\angle ABC$, below, is a unit angle. Let's then measure $\angle DEF$. If we made 3 copies of $\angle ABC$, they would "fill up" $\angle DEF$. Thus the measure of $\angle DEF$ is 3.

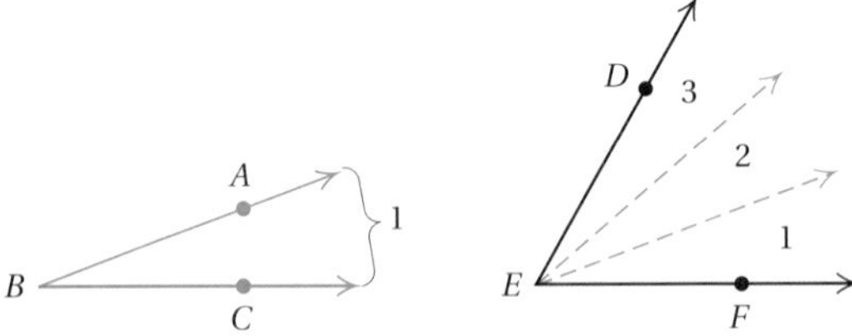

The unit most commonly used for angle measure is the degree. Below is such a unit. Its measure is 1 degree, or 1°.

Here are some other angles with their degree measures.

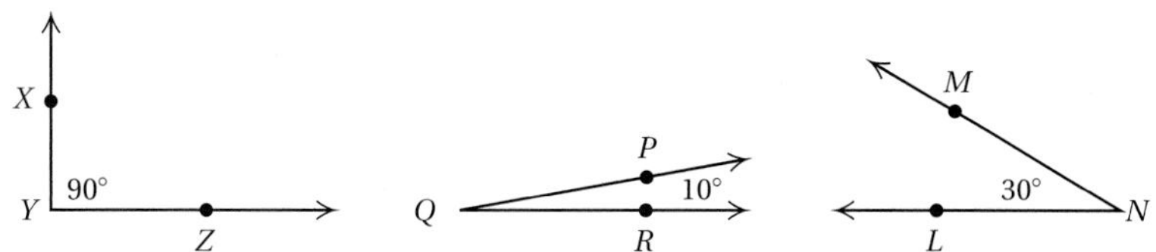

To indicate the measure of $\angle XYZ$, we write $m\angle XYZ = 90°$.

We can use a device called a *protractor* to measure angles. Note the two scales in the figure below. Let's find the measure of $\angle PQR$. We place the △ at the vertex and line up one of the sides at 0°. Then we check where the other side crosses the scale. Since 0° is on the inside scale in this case, we check where the side crosses the inside scale. Thus, $m\angle PQR = 145°$.

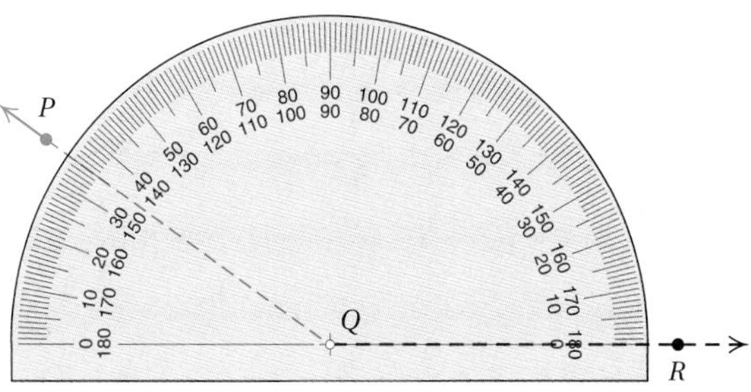

Do Exercise 14.

Let's find the measure of $\angle ABC$. This time we will use the 0° on the outside scale. We see that $m\angle ABC = 42°$.

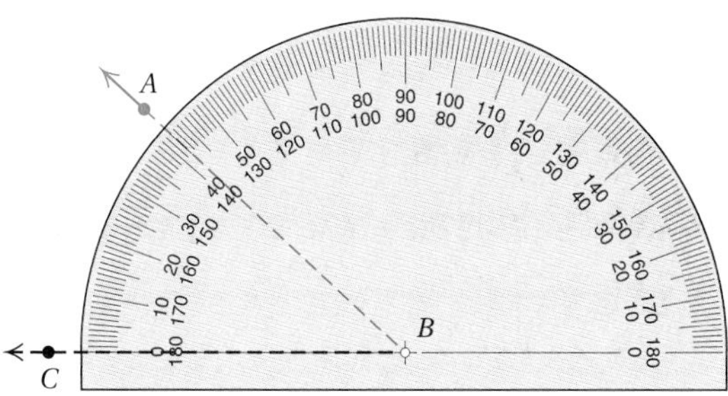

Do Exercise 15.

14. Use a protractor to measure this angle.

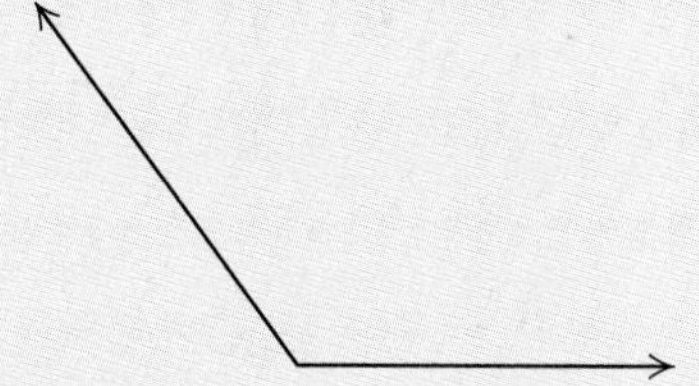

15. Use a protractor to measure this angle.

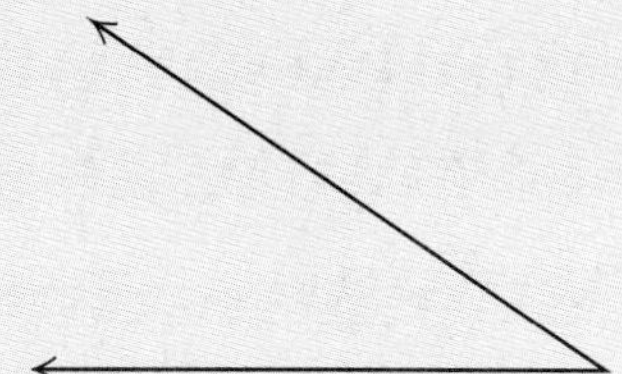

Answers on page A-14

Classify the angle as right, straight, acute, or obtuse. Use a protractor if necessary.

16. **17.**

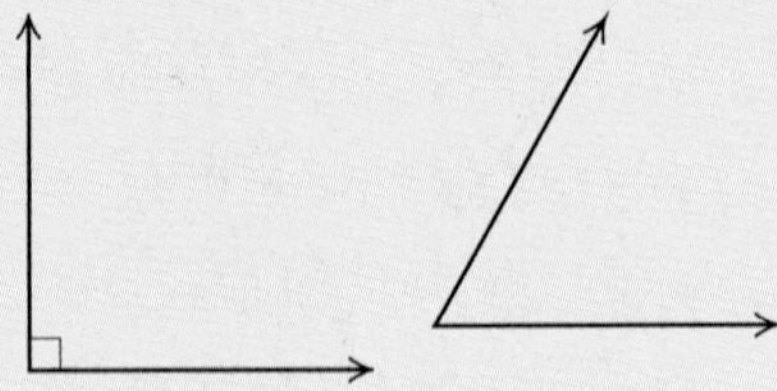

18.

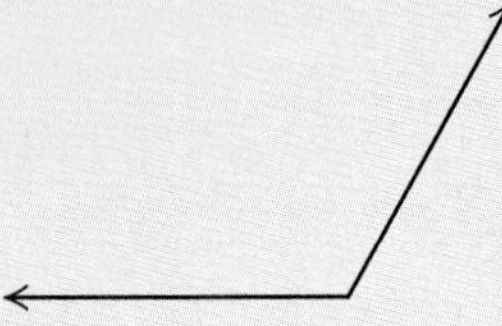

19. C D E

Determine whether the pair of lines is perpendicular. Use a protractor.

20.

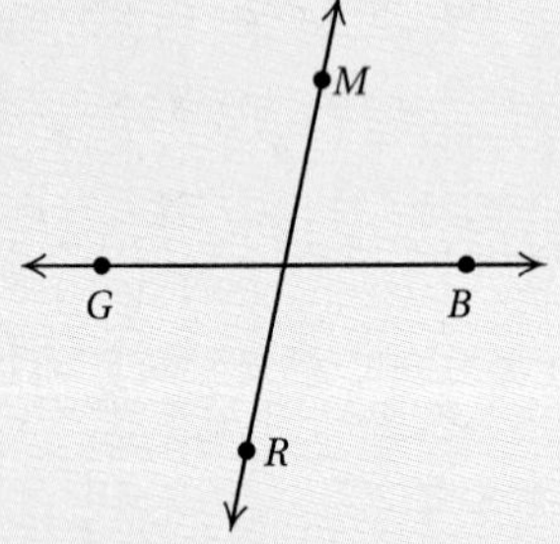

21.

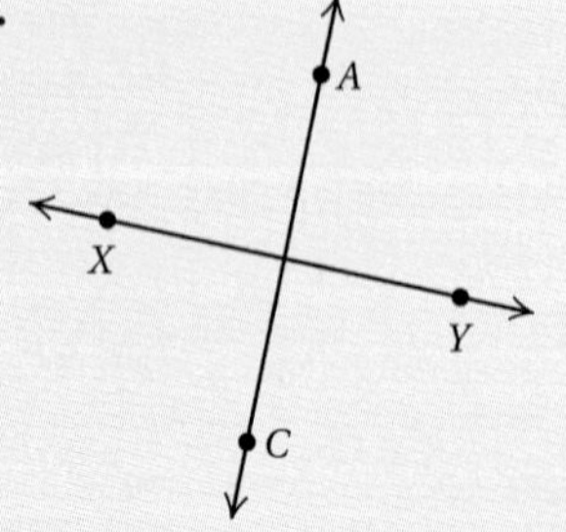

Answers on page A-14

c Classifying Angles

The following are ways in which we classify angles.

> *Right angles*: Angles whose measure is 90°.
> *Straight angles*: Angles whose measure is 180°.
> *Acute angles*: Angles whose measure is greater than 0° and less than 90°.
> *Obtuse angles*: Angles whose measure is greater than 90° and less than 180°.

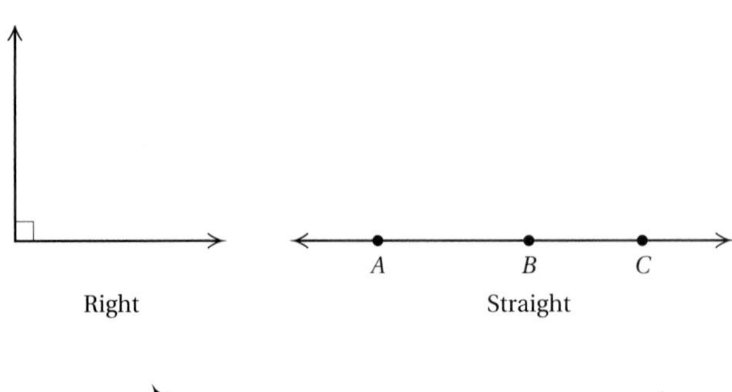

Right Straight

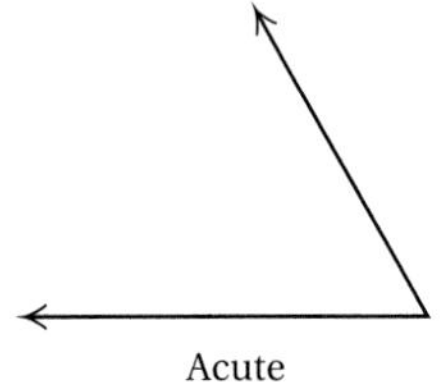

Acute

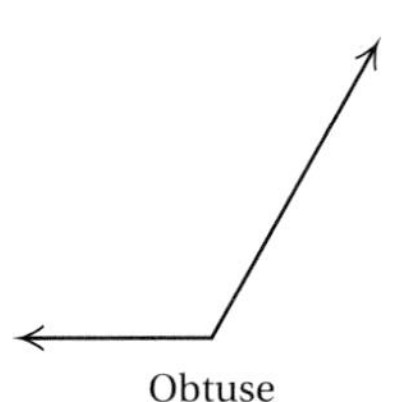

Obtuse

Do Exercises 16–19.

d Perpendicular Lines

Two lines are **perpendicular** if they intersect to form a right angle.

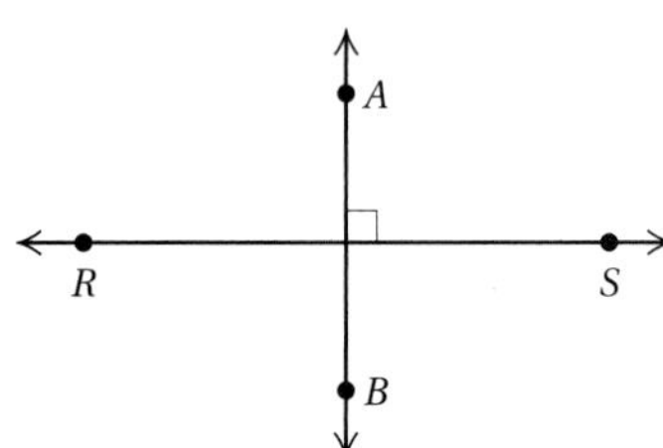

To say that $\overleftrightarrow{AB}$ is perpendicular to $\overleftrightarrow{RS}$, we write $\overleftrightarrow{AB} \perp \overleftrightarrow{RS}$. If two lines intersect to form one right angle, they form four right angles.

Do Exercises 20 and 21.

e Polygons

The figures below are examples of **polygons**.

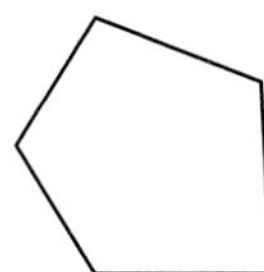
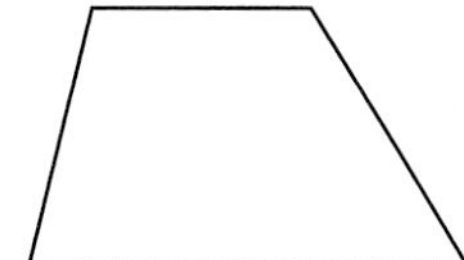

A **triangle** is a polygon made up of three segments, or sides. Consider these triangles. The triangle with vertices *A*, *B*, and *C* can be named $\triangle ABC$.

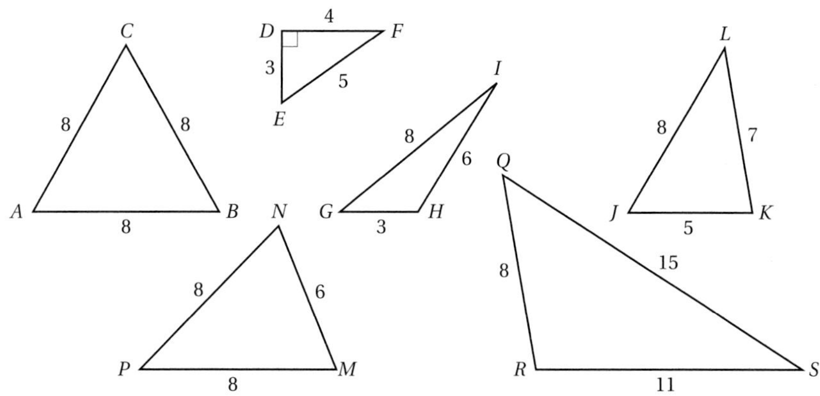

We can classify triangles according to sides and according to angles.

> *Equilateral triangle*: All sides are the same length.
> *Isosceles triangle*: Two or more sides are the same length.
> *Scalene triangle*: All sides are of different lengths.
> *Right triangle*: One angle is a right angle.
> *Obtuse triangle*: One angle is an obtuse angle.
> *Acute triangle*: All three angles are acute.

Do Exercises 22–25.

We can further classify polygons as follows.

Number of Sides	Polygon	Number of Sides	Polygon
4	Quadrilateral	8	Octagon
5	Pentagon	9	Nonagon
6	Hexagon	10	Decagon
7	Heptagon	12	Dodecagon

Do Exercises 26–31. (Exercise 31 is on the following page.)

22. Which triangles on this page are:

a) equilateral?

b) isosceles?

c) scalene?

23. Are all equilateral triangles isosceles?

24. Are all isosceles triangles equilateral?

25. Which triangles on this page are:

a) right triangles?

b) obtuse triangles?

c) acute triangles?

Classify the polygon by name.

26.

27.

28.

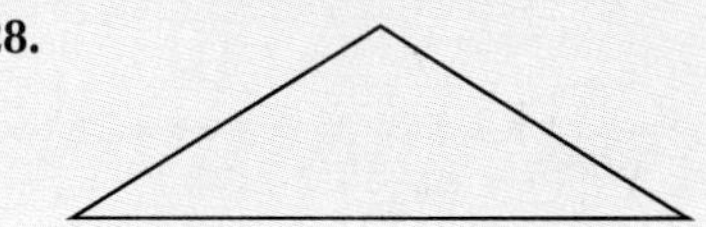

29.

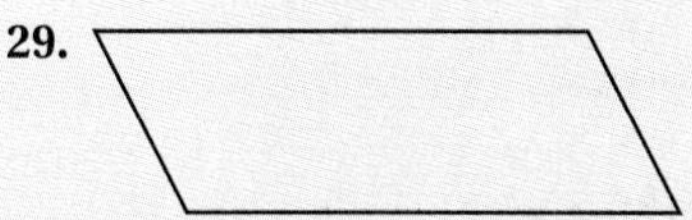

30.

Answers on page A-14

31.

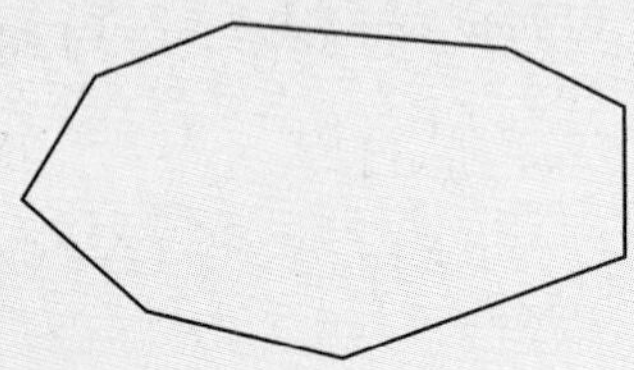

32. Find the missing angle measure.

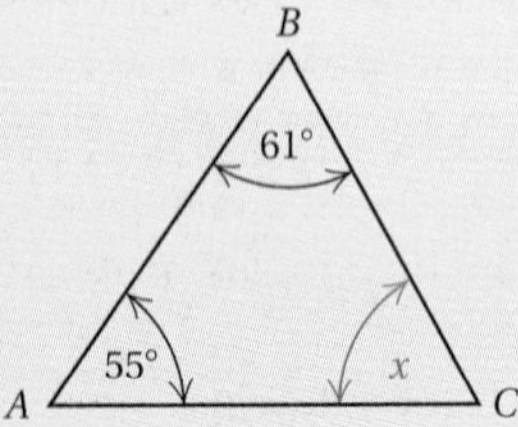

33. Consider a five-sided figure:

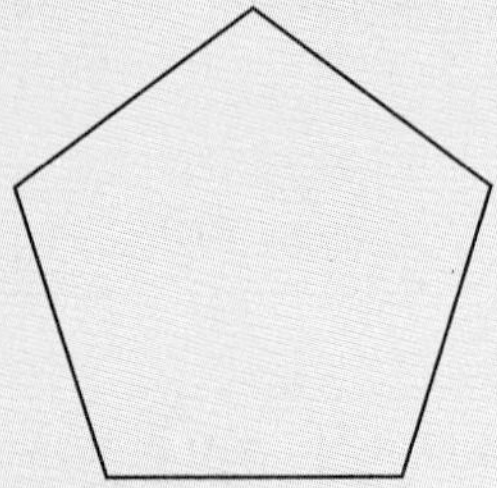

Complete.

a) The figure can be divided into ____ triangles.

b) The sum of the angle measures of each triangle is ____.

c) The sum of the angle measures of the polygon is ____ · 180°, or ____.

34. What is the sum of the angle measures of an octagon?

35. What is the sum of the angle measures of a 25-sided figure?

Answers on page A-14

f Sum of the Angle Measures of a Polygon

The measures of the angles of a triangle add up to 180°. Thus if we know the measures of two angles of a triangle, we can calculate the third.

> In any triangle, the sum of the measures of the angles is 180°:
> $m(\angle A) + m(\angle B) + m(\angle C) = 180°$.

Example 1 Find the missing angle measure.

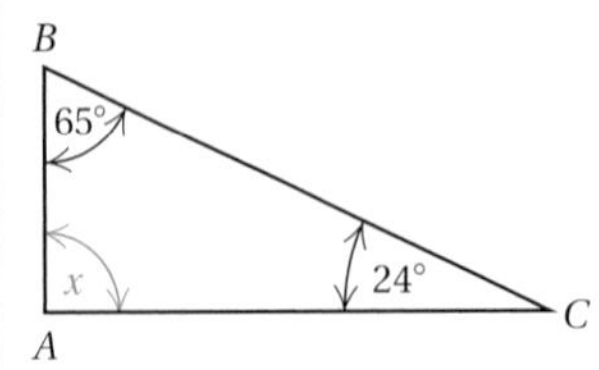

$$\begin{aligned} m(\angle A) + m(\angle B) + m(\angle C) &= 180° \\ x + 65° + 24° &= 180° \\ x + 89° &= 180° \\ x &= 180° - 89° \\ x &= 91° \end{aligned}$$

Do Exercise 32.

Now let's use this idea to find the sum of the measures of the angles of a polygon of n sides. First let's consider a four-sided figure:

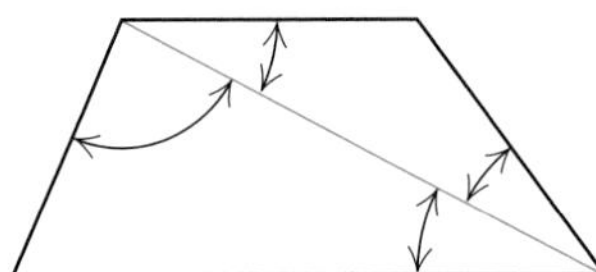

We can divide the figure into two triangles. The sum of the angle measures of each triangle is 180°. We have two triangles, so the sum of the angle measures of the figure is $2 \cdot 180°$, or 360°.

Do Exercise 33.

If a polygon has n sides, it can be divided into $n - 2$ triangles, each having 180° as the sum of its angle measures. Thus the sum of the angle measures of the polygon is $(n - 2) \cdot 180°$.

> If a polygon has n sides, then the sum of its angle measures is $(n - 2) \cdot 180°$.

Example 2 What is the sum of the angle measures of a hexagon?

A hexagon has 6 sides. We use the formula $(n - 2) \cdot 180°$:

$$\begin{aligned} (n - 2) \cdot 180° &= (6 - 2) \cdot 180° \\ &= 4 \cdot 180° \\ &= 720°. \end{aligned}$$

Do Exercises 34 and 35.

Exercise Set 6.1

1. Draw the segment whose endpoints are G and H. Name the segment in two ways.

2. Draw the segment whose endpoints are C and D. Name the segment in two ways.

C D

3. Draw the ray with endpoint Q. Name the ray.

Q D

4. Draw the ray with endpoint D. Name the ray.

Q D

Name the line in seven different ways.

5.

6. m J K T

b Name the angle in four different ways.

7.

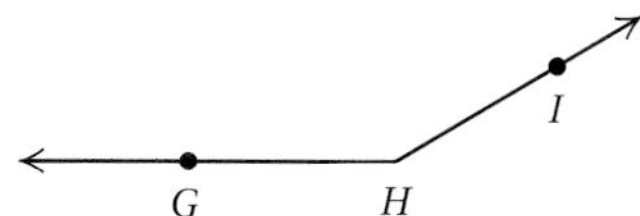

8.

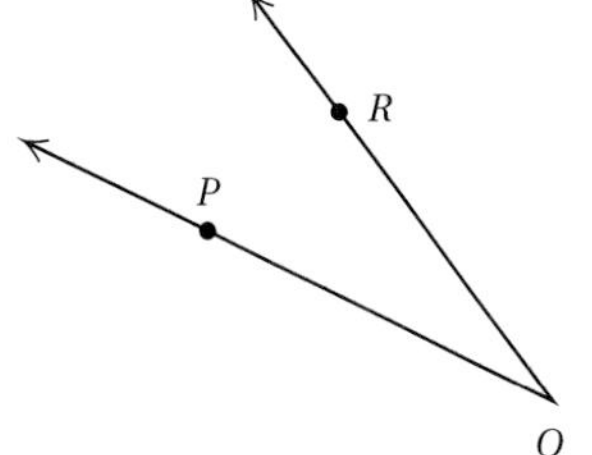

Use a protractor and measure the angle.

9.

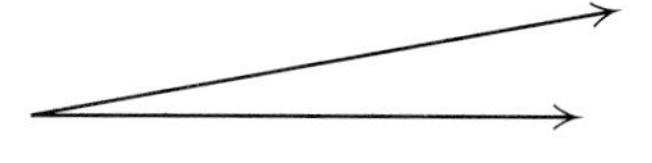

10.

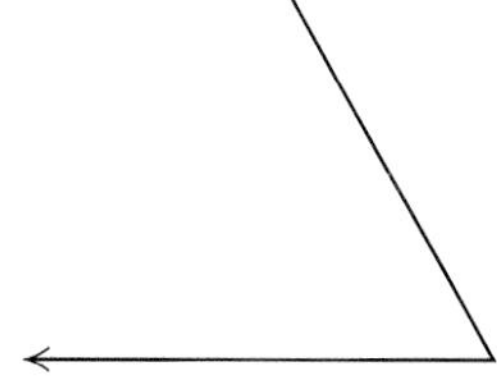

11.

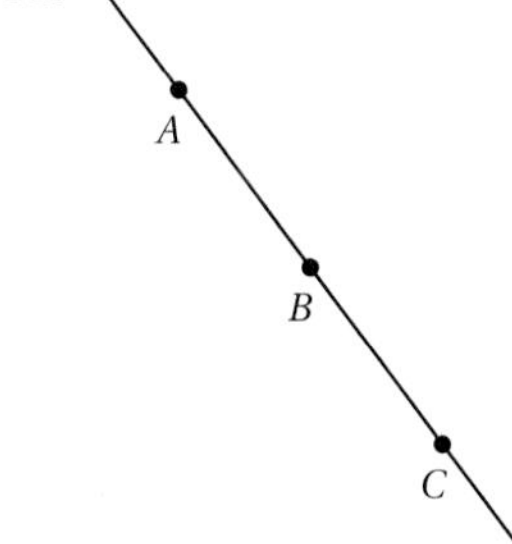

12.

13.

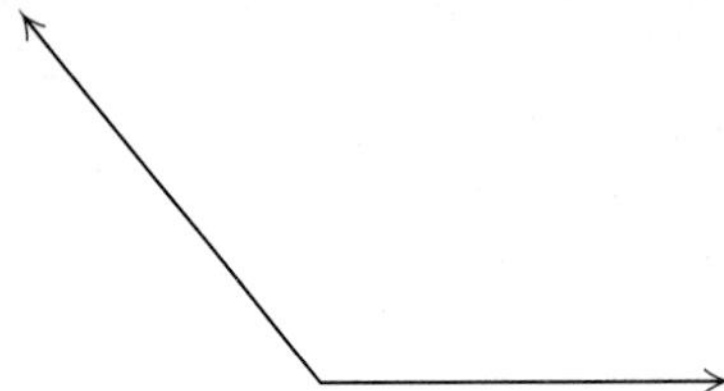

14.

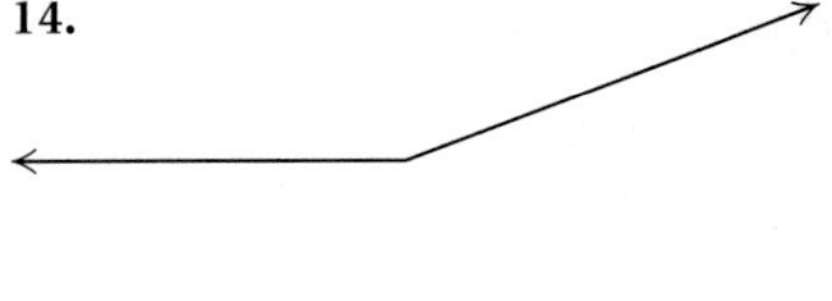

c

15.–20. Classify each of the angles in Exercises 9–14 as right, straight, acute, or obtuse.

d Determine whether the pair of lines is perpendicular. Use a protractor.

21.

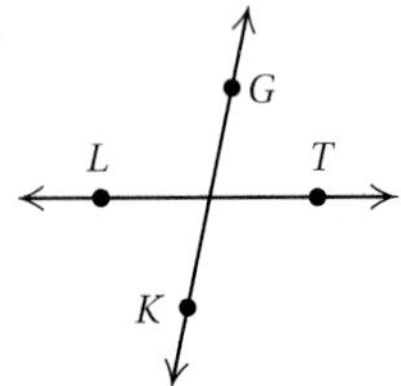

22.

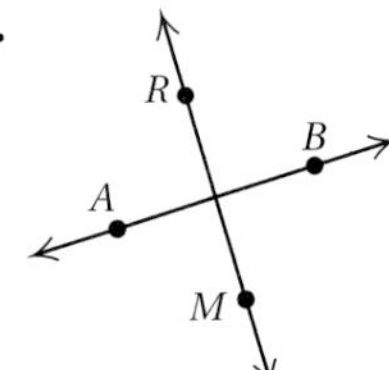

23.

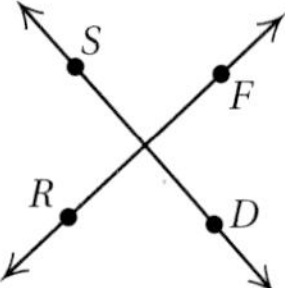

24.

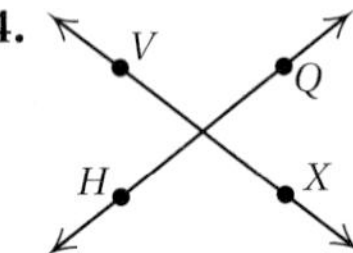

e Classify the triangle as equilateral, isosceles, or scalene. Then classify it as right, obtuse, or acute.

25.

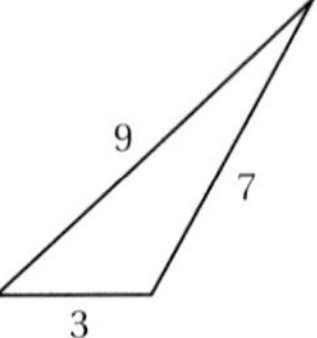

26.

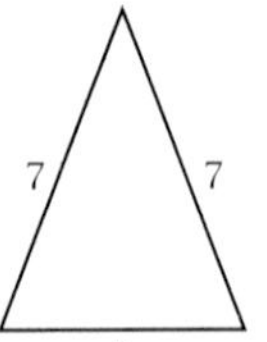

27.

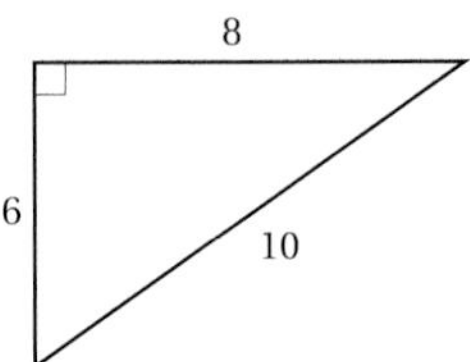

28. 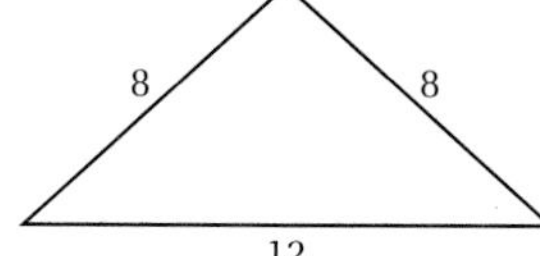

29. 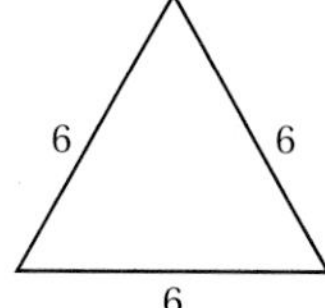

30. 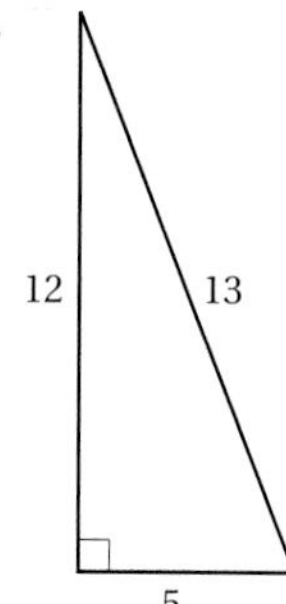

31. 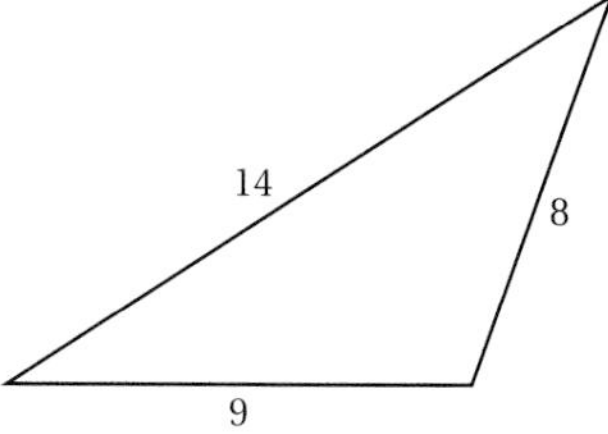

32. 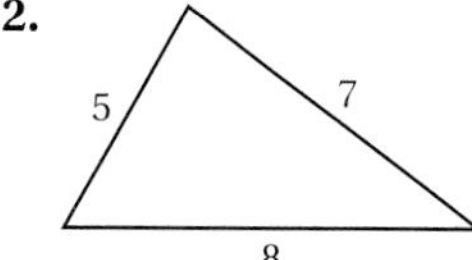

Classify the polygon by name.

33. 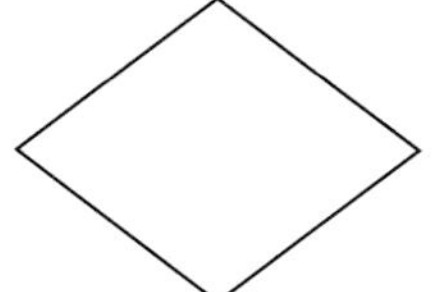

34. 

35. 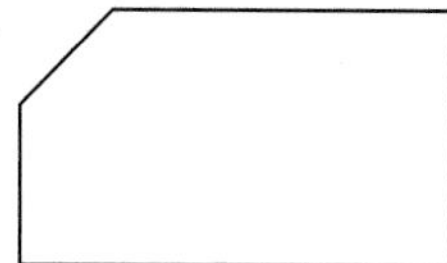

36. 

37. 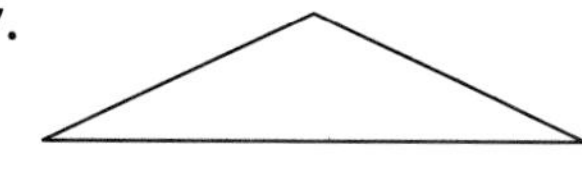

38. 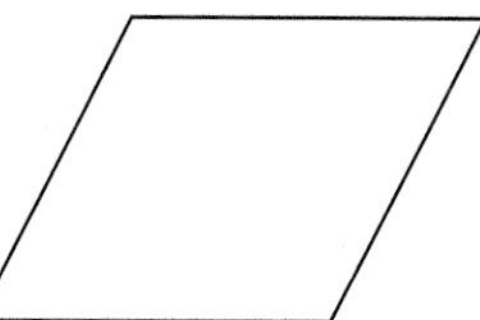

39. 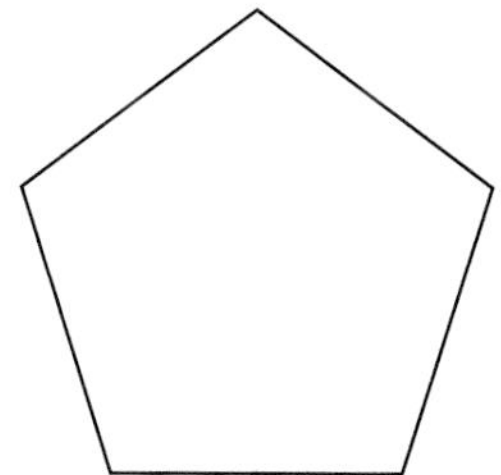

40. 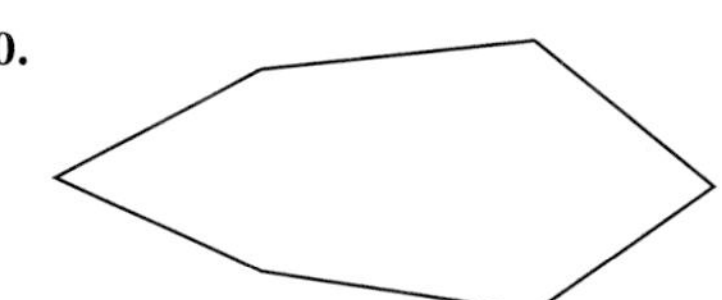

41.

42.

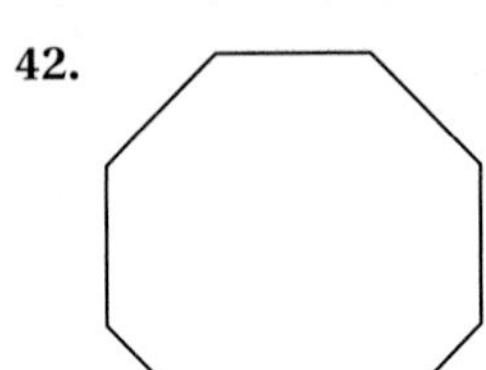

f Find the sum of the angle measures of each of the following.

43. A decagon

44. A quadrilateral

45. A heptagon

46. A nonagon

47. A 14-sided polygon

48. A 17-sided polygon

49. A 20-sided polygon

50. A 32-sided polygon

Find the missing angle measure.

51.

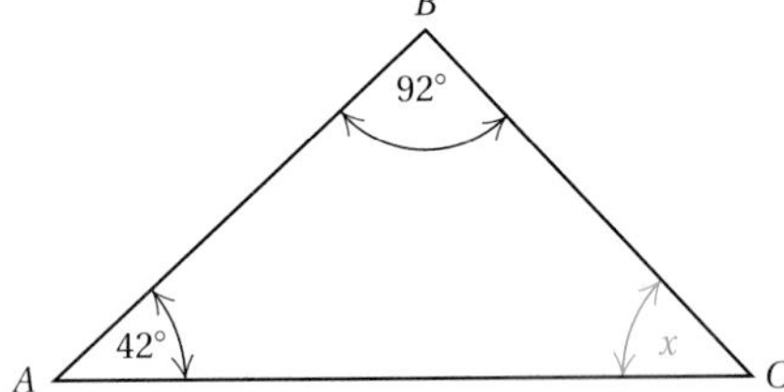

52.

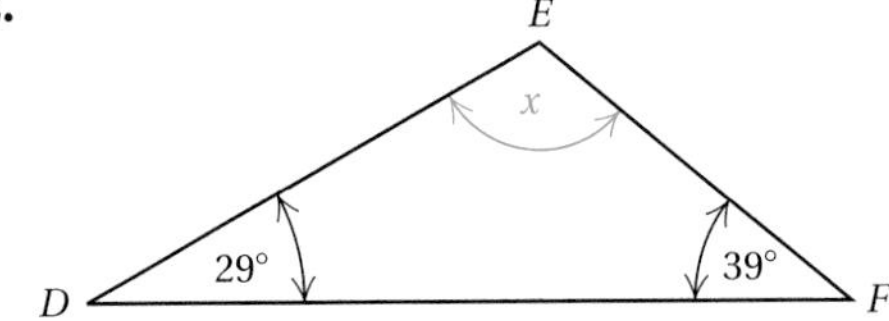

53. In $\triangle RST$, $m(\angle S) = 58°$ and $m(\angle T) = 79°$. Find $m(\angle R)$.

54. In $\triangle KNP$, $m(\angle K) = 137°$ and $m(\angle P) = 12°$. Find $m(\angle N)$.

Skill Maintenance

Divide. Find decimal notation for the answer. [3.4a]

55. $21 \div 12$

56. $23.4 \div 10$

57. $23.4 \div 100$

58. $23.4 \div 1000$

59. Multiply 3.14×4.41. Round to the nearest hundredth. [3.3a], [3.1e]

60. Multiply: $4 \times 20\frac{1}{8}$. [2.4d]

61. Multiply: $48 \times \frac{1}{12}$. [2.2a]

Synthesis

62. ◆ Determine whether the following statement is true or false and explain your answer.

All equilateral triangles are isosceles, but not all isosceles triangles are equilateral.

63. ◆ Sara is incorrect when she says that $\triangle ABC$ is a right triangle with $m(\angle A) = 141°$. Explain why she is wrong.

6.2 Perimeter

a Finding Perimeters

A **polygon** is a geometric figure with three or more sides. The **perimeter of a polygon** is the distance around it, or the sum of the lengths of its sides.

Example 1 Find the perimeter of this polygon.

We add the lengths of the sides. Since all the units are the same, we add the numbers, keeping meters (m) as the unit.

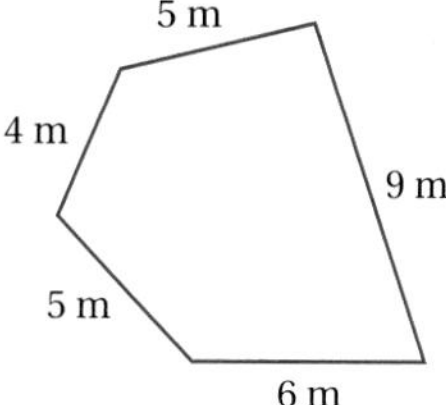

Perimeter = 6 m + 5 m + 4 m + 5 m + 9 m
= (6 + 5 + 4 + 5 + 9) m
= 29 m

Do Exercises 1 and 2.

A **rectangle** is a figure with four sides and four 90°-angles, like the one shown in Example 2.

Example 2 Find the perimeter of a rectangle that is 3 cm by 4 cm.

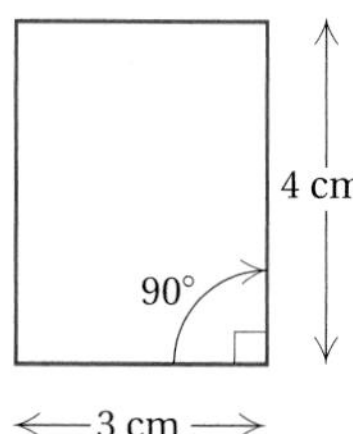

Perimeter = 3 cm + 3 cm + 4 cm + 4 cm
= (3 + 3 + 4 + 4) cm
= 14 cm

Do Exercise 3.

The **perimeter of a rectangle** is twice the sum of the length and the width, or 2 times the length plus 2 times the width:

$P = 2 \cdot (l + w)$, or $P = 2 \cdot l + 2 \cdot w$.

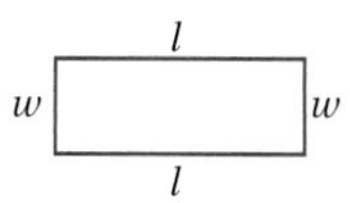

Example 3 Find the perimeter of a rectangle that is 4.3 ft by 7.8 ft.

$P = 2 \cdot (l + w) = 2 \cdot (4.3\text{ ft} + 7.8\text{ ft}) = 2 \cdot (12.1\text{ ft}) = 24.2\text{ ft}$

Do Exercises 4 and 5.

A **square** is a rectangle with all sides the same length.

Example 4 Find the perimeter of a square whose sides are 9 mm long.

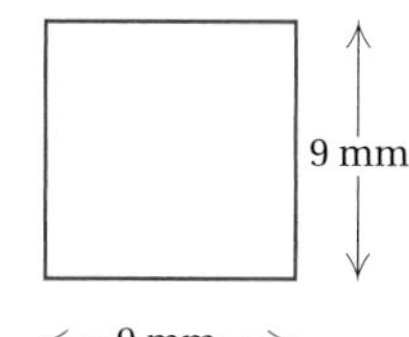

P = 9 mm + 9 mm + 9 mm + 9 mm
= (9 + 9 + 9 + 9) mm = 36 mm

Objectives

a Find the perimeter of a polygon.

b Solve applied problems involving perimeter.

For Extra Help

TAPE 11 | MAC WIN | CD-ROM

Find the perimeter of the polygon.

1.

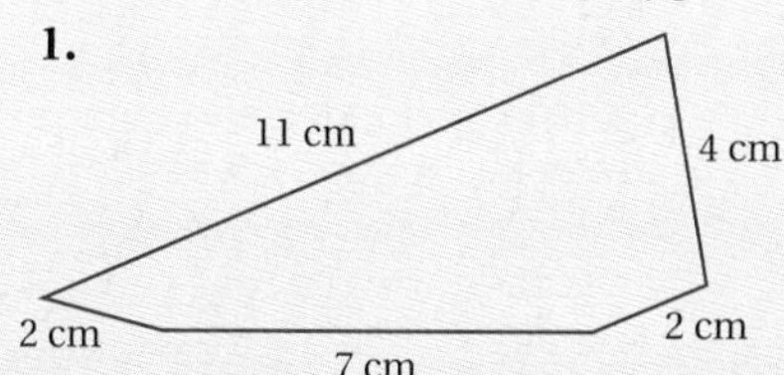

2.

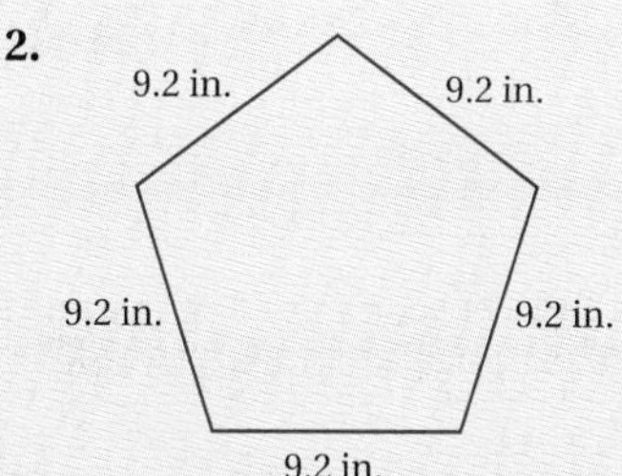

3. Find the perimeter of a rectangle that is 2 cm by 4 cm.

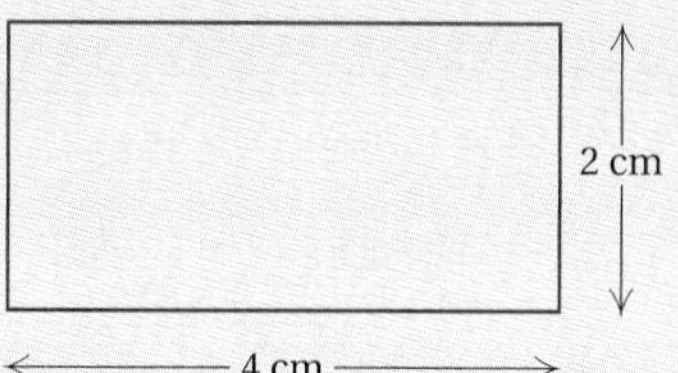

4. Find the perimeter of a rectangle that is 5.25 yd by 3.5 yd.

5. Find the perimeter of a rectangle that is 8 km by 8 km.

Answers on page A-15

6. Find the perimeter of a square with sides of length 10 km.

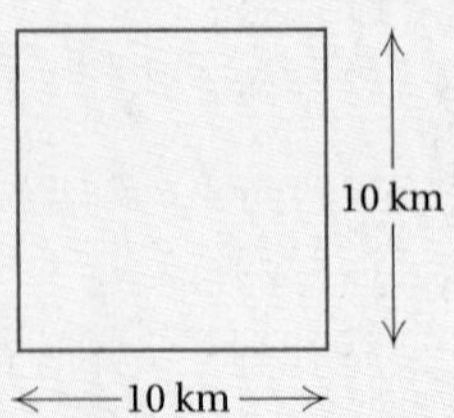

7. Find the perimeter of a square with sides of length $5\frac{1}{4}$ yd.

8. Find the perimeter of a square with sides of length 7.8 km.

9. A play area is 25 ft by 10 ft. A fence is to be built around the play area. How many feet of fencing will be needed? If fencing costs $4.95 per foot, what will the fencing cost?

Answers on page A-15

Do Exercise 6.

The **perimeter of a square** is four times the length of a side:

$P = 4 \cdot s.$

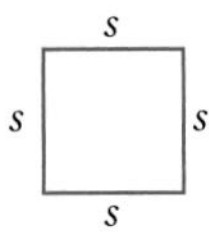

Example 5 Find the perimeter of a square whose sides are $20\frac{1}{8}$ in. long.

$$P = 4 \cdot s = 4 \cdot 20\frac{1}{8} \text{ in.}$$
$$= 4 \cdot \frac{161}{8} \text{ in.} = \frac{4 \cdot 161}{4 \cdot 2} \text{ in.}$$
$$= \frac{161}{2} \cdot \frac{4}{4} \text{ in.} = 80\frac{1}{2} \text{ in.}$$

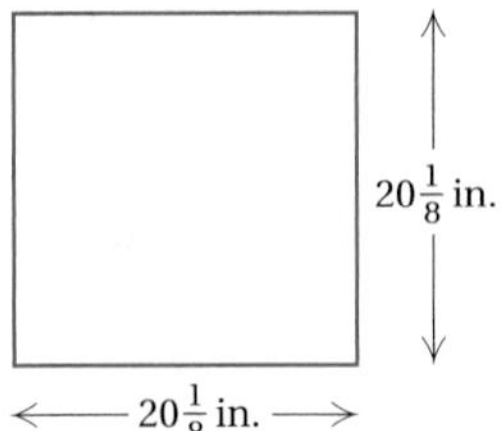

Do Exercises 7 and 8.

b Solving Applied Problems

Example 6 A vegetable garden is 20 ft by 15 ft. A fence is to be built around the garden. How many feet of fence will be needed? If fencing sells for $2.95 per foot, what will the fencing cost?

1. Familiarize. We make a drawing and let P = the perimeter.

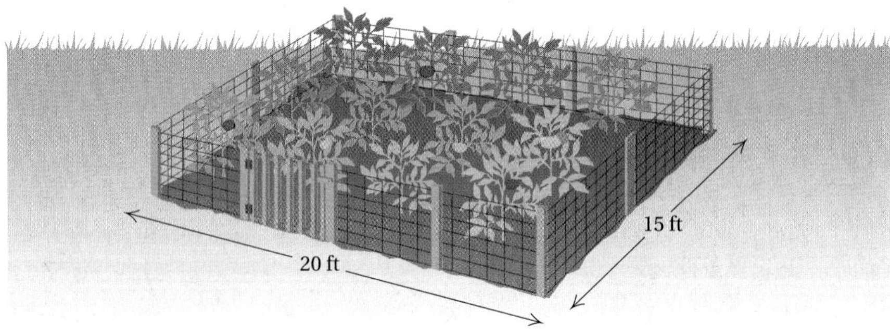

2. Translate. The perimeter of the garden is given by

$$P = 2 \cdot (l + w) = 2 \cdot (20 \text{ ft} + 15 \text{ ft}).$$

3. Solve. We calculate the perimeter as follows:

$$P = 2 \cdot (20 \text{ ft} + 15 \text{ ft}) = 2 \cdot (35 \text{ ft}) = 70 \text{ ft}.$$

Then we multiply by $2.95 to find the cost of the fencing:

$$\text{Cost} = \$2.95 \times \text{Perimeter} = \$2.95 \times 70 \text{ ft} = \$206.50.$$

4. Check. The check is left to the student.

5. State. The 70 ft of fencing that is needed will cost $206.50.

Do Exercise 9.

Exercise Set 6.2

a Find the perimeter of the polygon.

1.

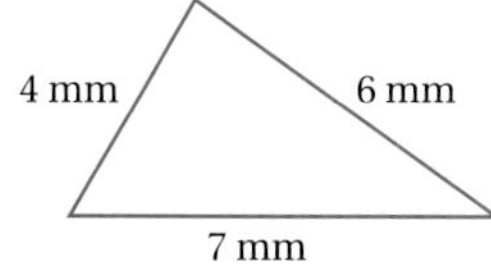

2.

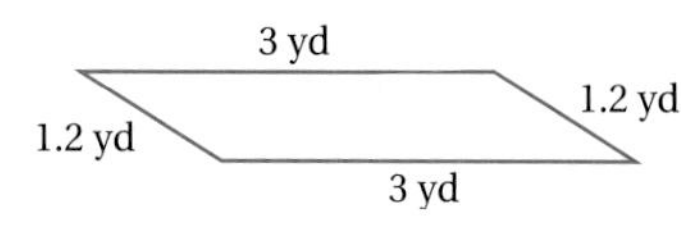

3.

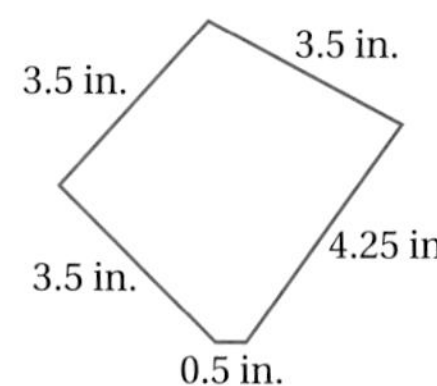

4.

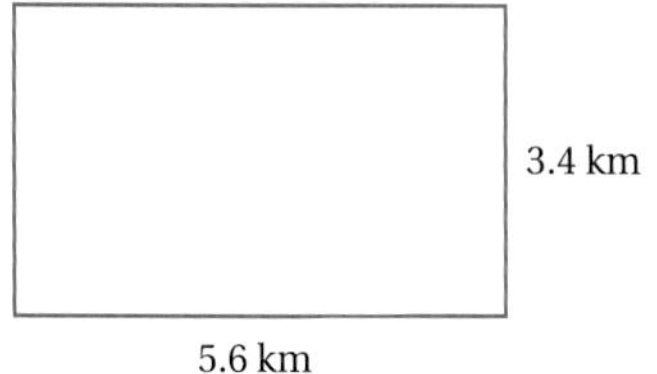

5.

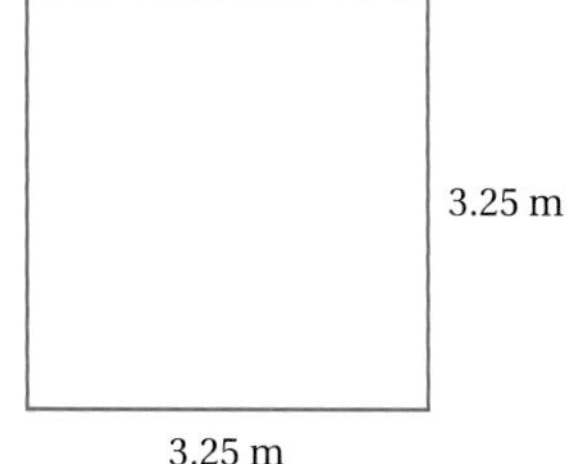

6.

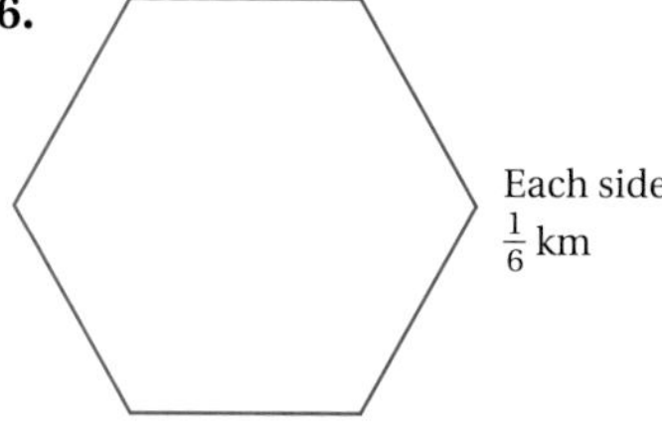

Find the perimeter of the rectangle.

7. 5 ft by 10 ft

8. 2.5 m by 100 m

9. 34.67 cm by 4.9 cm

10. $3\frac{1}{2}$ yd by $4\frac{1}{2}$ yd

Find the perimeter of the square.

11. 22 ft on a side

12. 56.9 km on a side

13. 45.5 mm on a side

14. $3\frac{1}{8}$ yd on a side

b Solve.

15. A security fence is to be built around a 173-m by 240-m field. What is the perimeter of the field? If fence wire costs $1.45 per meter, what will the fencing cost?

16. *Softball Diamond.* A standard-sized slow-pitch softball diamond is a square with sides of length 65 ft. What is the perimeter of this softball diamond? (This is the distance you would have to run if you hit a home run.)

17. A piece of flooring tile is a square with sides of length 30.5 cm. What is the perimeter of a piece of tile?

18. A posterboard is 61.8 cm by 87.9 cm. What is the perimeter of the board?

19. A rain gutter is to be installed around the house shown in the figure.

a) Find the perimeter of the house.
b) If the gutter costs $4.59 per foot, what is the total cost of the gutter?

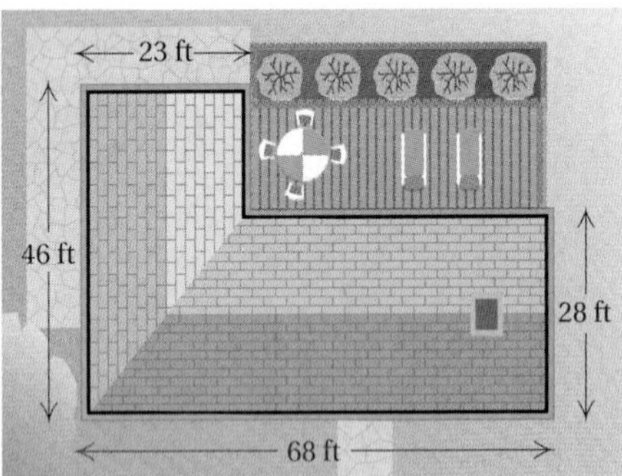

20. A carpenter is to build a fence around a 9-m by 12-m garden.

a) The posts are 3 m apart. How many posts will be needed?
b) The posts cost $2.40 each. How much will the posts cost?
c) The fence will surround all but 3 m of the garden, which will be a gate. How long will the fence be?
d) The fence costs $2.85 per meter. What will the cost of the fence be?
e) The gate costs $9.95. What is the total cost of the materials?

Skill Maintenance

21. Convert to decimal notation: 56.1%. [4.2b]

22. Convert to percent notation: 0.6734. [4.2c]

23. Convert to percent notation: $\frac{9}{8}$. [4.3a]

Evaluate. [1.6b]

24. 5^2

25. 10^2

26. 31^2

Convert the number in the sentence to standard notation. [3.3b]

27. It is estimated that 4.7 million fax machines were sold in a recent year.

28. In a recent year, 4.3 billion CDs were sold.

Synthesis

29. ◆ Create for a fellow student a development of the formula

$$P = 2 \cdot (l + w) = 2 \cdot l + 2 \cdot w$$

for the perimeter of a rectangle.

30. ◆ Create for a fellow student a development of the formula

$$P = 4 \cdot s$$

for the perimeter of a square.

Find the perimeter of the figure in feet.

31.

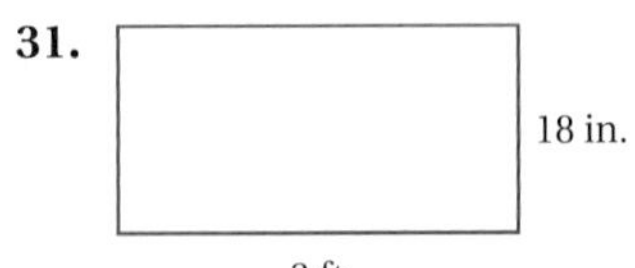

32.

78 in.

5.5 yd

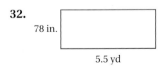

6.3 Area: Rectangles and Squares

a Rectangles and Squares

A polygon and its interior form a plane region. We can find the area of a *rectangular region* by filling it with square units. Two such units, a *square inch* and a *square centimeter,* are shown below.

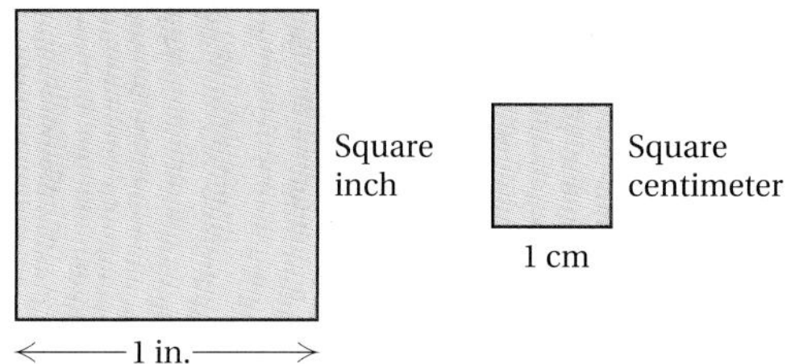

Objectives

a Find the area of a rectangle or a square.

b Solve applied problems involving areas of rectangles or squares.

For Extra Help

TAPE 11 | MAC WIN | CD-ROM

Example 1 What is the area of this region?

We have a rectangular array. Since the region is filled with 12 square centimeters, its area is 12 square centimeters (sq cm), or 12 cm^2. The number of units is 3×4, or 12.

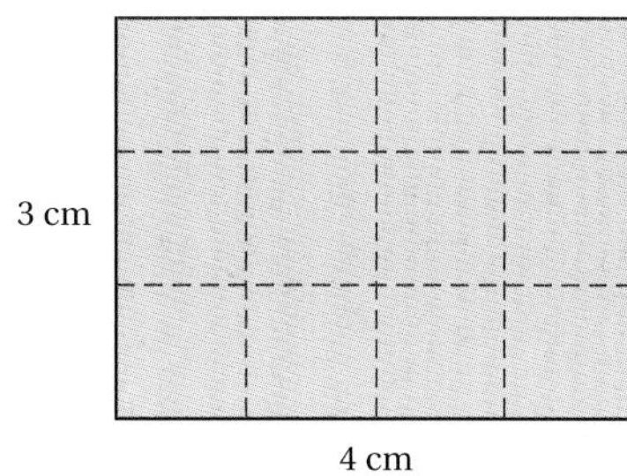

Do Exercise 1.

1. What is the area of this region? Count the number of square centimeters.

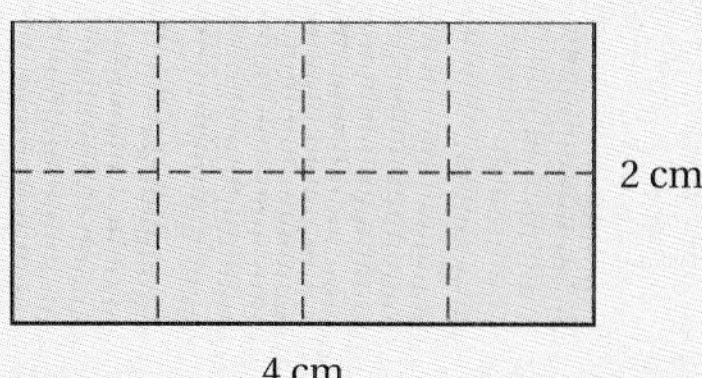

The **area of a rectangular region** is the product of the length l and the width w:

$$A = l \cdot w.$$

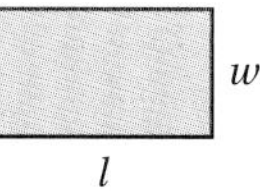

Example 2 Find the area of a rectangle that is 7 yd by 4 yd.

$$A = l \cdot w = 7 \text{ yd} \cdot 4 \text{ yd} = 7 \cdot 4 \cdot \text{yd} \cdot \text{yd} = 28 \text{ yd}^2$$

We think of yd · yd as $(\text{yd})^2$ and denote it yd^2. Thus we read "28 yd^2" as "28 square yards."

Do Exercises 2 and 3.

2. Find the area of a rectangle that is 7 km by 8 km.

3. Find the area of a rectangle that is $5\frac{1}{4}$ yd by $3\frac{1}{2}$ yd.

Example 3 Find the area of a square with sides of length 9 mm.

$$A = (9 \text{ mm}) \cdot (9 \text{ mm})$$
$$= 9 \cdot 9 \cdot \text{mm} \cdot \text{mm}$$
$$= 81 \text{ mm}^2$$

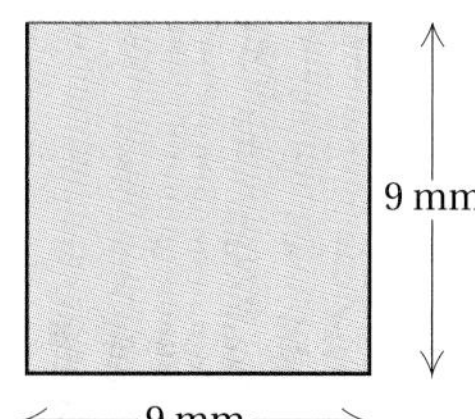

Do Exercise 4.

4. Find the area of a square with sides of length 12 km.

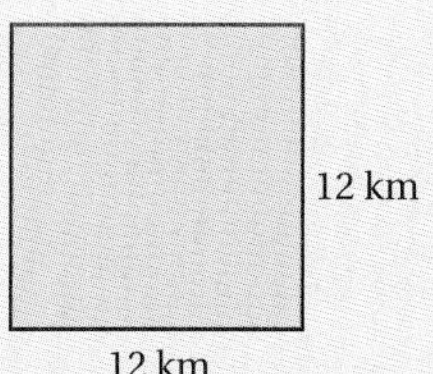

Answers on page A-15

5. Find the area of a square with sides of length 10.9 m.

6. Find the area of a square with sides of length $3\frac{1}{2}$ yd.

7. A square flower bed 3.5 m on a side is dug on a 30-m by 22.4-m lawn. How much area is left over? Draw a picture first.

Answers on page A-15

The **area of a square region** is the square of the length of a side:

$A = s \cdot s$, or $A = s^2$.

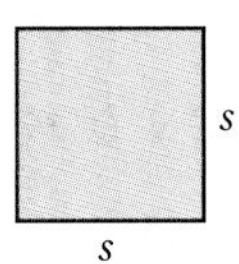

Example 4 Find the area of a square with sides of length 20.3 m.

$$A = s \cdot s = 20.3 \text{ m} \times 20.3 \text{ m}$$
$$= 20.3 \times 20.3 \times \text{m} \times \text{m} = 412.09 \text{ m}^2$$

Do Exercises 5 and 6.

b Solving Applied Problems

Example 5 *Mowing Cost.* A square sandbox 1.5 m on a side is placed on a 20-m by 31.2-m lawn. It costs \$0.04 per square meter to have the lawn mowed. What is the total cost of mowing?

1. **Familiarize.**
   We make a drawing.

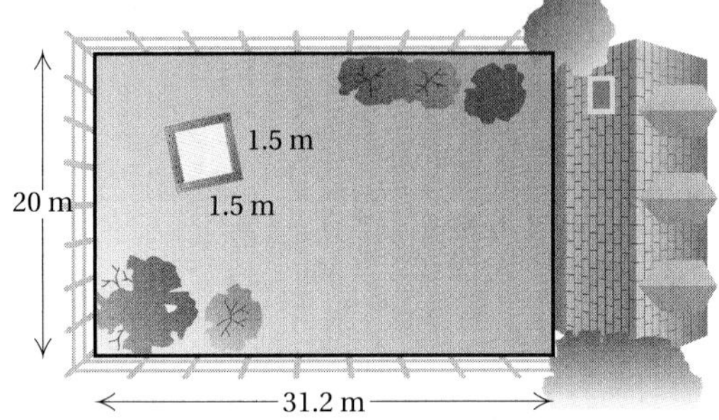

2. **Translate.** This is a two-step problem. We first find the area left over after the area of the sandbox has been subtracted. Then we multiply by the cost per square meter. We let A = the area left over.

Area left over	is	Area of lawn	minus	Area of sandbox
↓	↓	↓	↓	↓
A	$=$	$(20 \text{ m}) \times (31.2 \text{ m})$	$-$	$(1.5 \text{ m}) \times (1.5 \text{ m})$

3. **Solve.** The area of the lawn is

$$(20 \text{ m}) \times (31.2 \text{ m}) = 20 \times 31.2 \times \text{m} \times \text{m} = 624 \text{ m}^2.$$

The area of the sandbox is

$$(1.5 \text{ m}) \times (1.5 \text{ m}) = 1.5 \times 1.5 \times \text{m} \times \text{m} = 2.25 \text{ m}^2.$$

The area left over is

$$A = 624 \text{ m}^2 - 2.25 \text{ m}^2 = 621.75 \text{ m}^2.$$

Then we multiply by \$0.04:

$$\$0.04 \times 621.75 = \$24.87.$$

4. **Check.** The check is left to the student.
5. **State.** The total cost of mowing the lawn is \$24.87.

Do Exercise 7.

Exercise Set 6.3

a Find the area.

1. 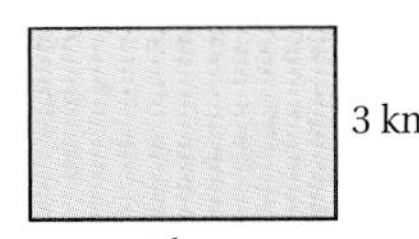

2. 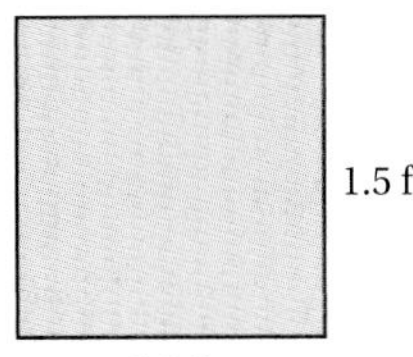

3. 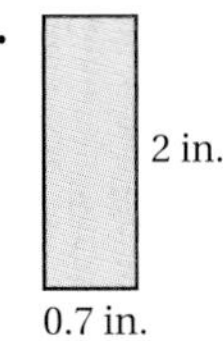

4. 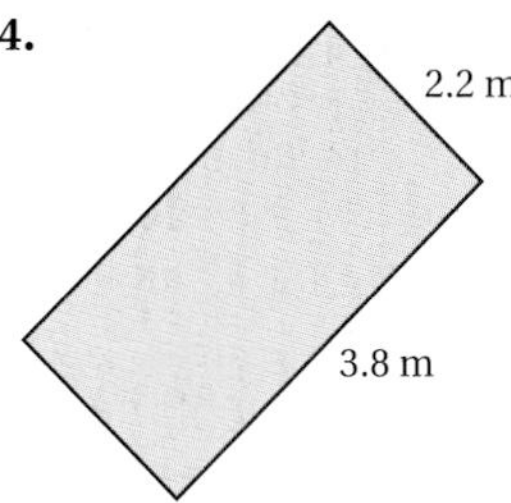

5. 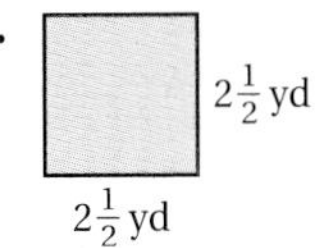

6. 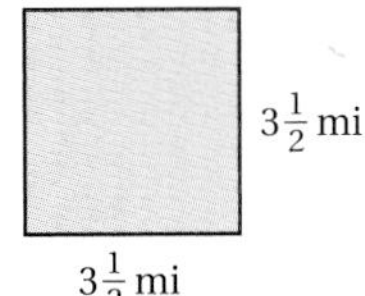

7. 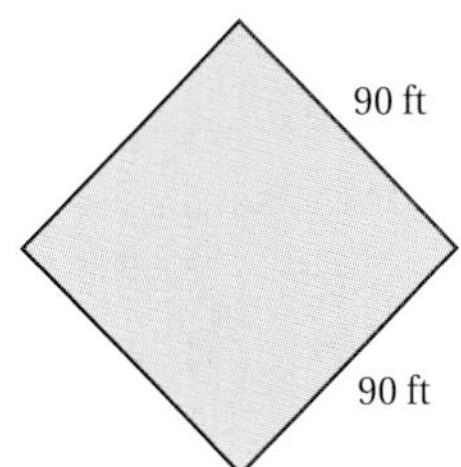

8. 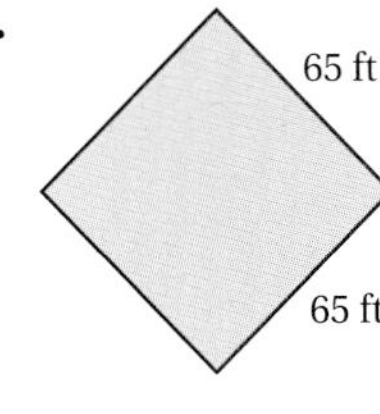

Find the area of the rectangle.

9. 5 ft by 10 ft

10. 14 yd by 8 yd

11. 34.67 cm by 4.9 cm

12. 2.45 km by 100 km

13. $4\frac{2}{3}$ in. by $8\frac{5}{6}$ in.

14. $10\frac{1}{3}$ mi by $20\frac{2}{3}$ mi

Find the area of the square.

15. 22 ft on a side

16. 18 yd on a side

17. 56.9 km on a side

18. 45.5 m on a side

19. $5\frac{3}{8}$ yd on a side

20. $7\frac{2}{3}$ ft on a side

b Solve.

21. A lot is 40 m by 36 m. A house 27 m by 9 m is built on the lot. How much area is left over for a lawn?

22. A field is 240.8 m by 450.2 m. Part of the field, 160.4 m by 90.6 m, is paved for a parking lot. How much area is unpaved?

23. Franklin Construction Company builds a sidewalk around two sides of the Municipal Trust Bank building, as shown in the figure. What is the area of the sidewalk?

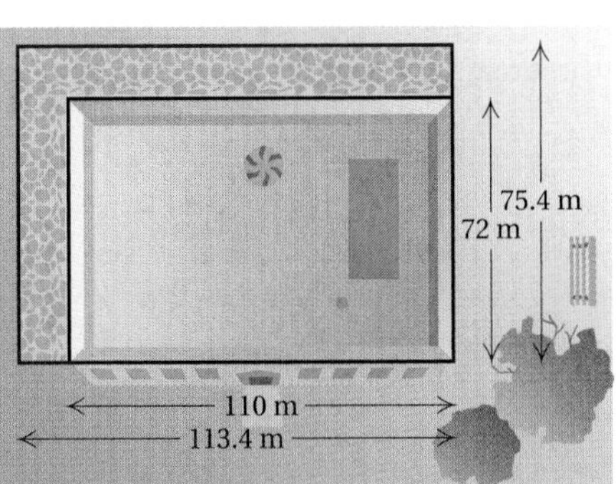

24. A standard sheet of typewriter paper is $8\frac{1}{2}$ in. by 11 in. We generally type on a $7\frac{1}{2}$-in. by 9-in. area of the paper. What is the area of the margin?

25. A room is 15 ft by 20 ft. The ceiling is 8 ft above the floor. There are two windows in the room, each 3 ft by 4 ft. The door is $2\frac{1}{2}$ ft by $6\frac{1}{2}$ ft.

a) What is the total area of the walls and the ceiling?

b) A gallon of paint will cover 86.625 ft^2. How many gallons of paint are needed for the room, including the ceiling?

c) Paint costs $17.95 a gallon. How much will it cost to paint the room?

26. A restaurant owner wants to carpet a 15-yd by 20-yd room.

a) How many square yards of carpeting are needed?

b) The carpeting they want is $18.50 per square yard. How much will it cost to carpet the room?

Find the area of the region.

27.

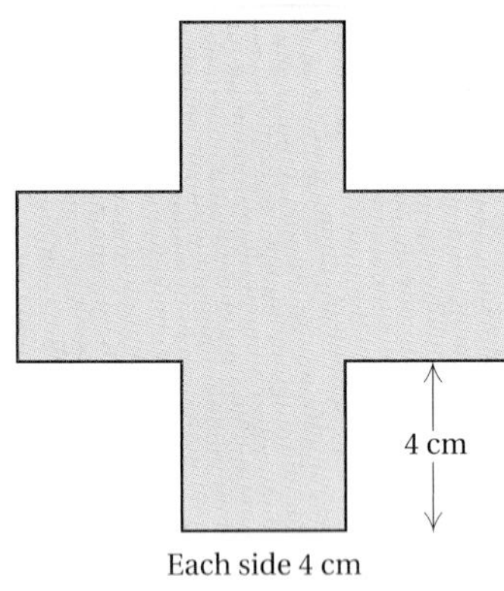

28.

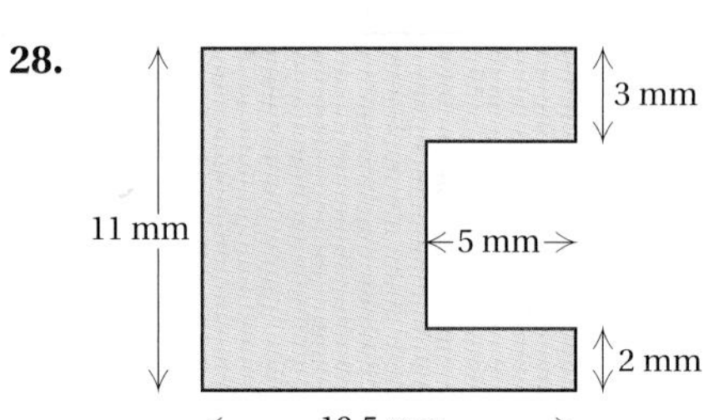

Skill Maintenance

Convert to percent notation. [4.2c], [4.3a]

29. 0.452

30. $\frac{1}{3}$

31. $\frac{11}{20}$

32. $\frac{22}{25}$

33. *Tourist Spending.* Foreign tourists spend $13.1 billion in this country annually. The most money, $2.7 billion, is spent in Florida. What is the ratio of amount spent in Florida to total amount spent? What is the ratio of total amount spent to amount spent in Florida? [4.1a]

34. One person in four plays a musical instrument. In a given group of people, what is the ratio of those who play an instrument to total number of people? What is the ratio of those who do not play an instrument to total number of people? [4.1a]

Synthesis

35. ◆ The length and the width of one rectangle are each three times the length and the width of another rectangle. Is the area of the first rectangle three times the area of the other rectangle? Why or why not?

36. ◆ Create for a fellow student a development of the formula

$$A = l \cdot w$$

for the area of a rectangle.

37. Find the area, in square inches, of the shaded region.

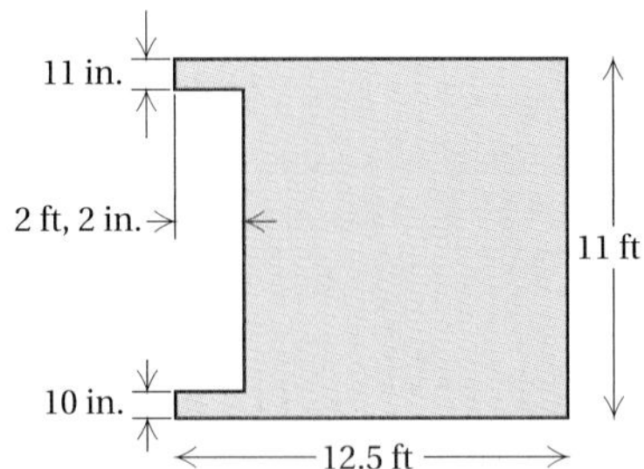

38. Find the area, in square feet, of the shaded region.

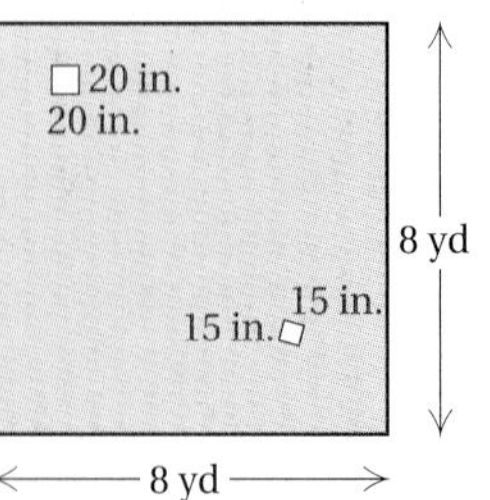

Collaborative Learning Manual

Prepare a budget for redecorating the classroom.

6.4 Area: Parallelograms, Triangles, and Trapezoids

Objectives

a Find the area of a parallelogram, a triangle, and a trapezoid.

b Solve applied problems involving areas of parallelograms, triangles, and trapezoids.

For Extra Help

TAPE 12

MAC
WIN

CD-ROM

a Finding Other Areas

Parallelograms

A **parallelogram** is a four-sided figure with two pairs of parallel sides, as shown below.

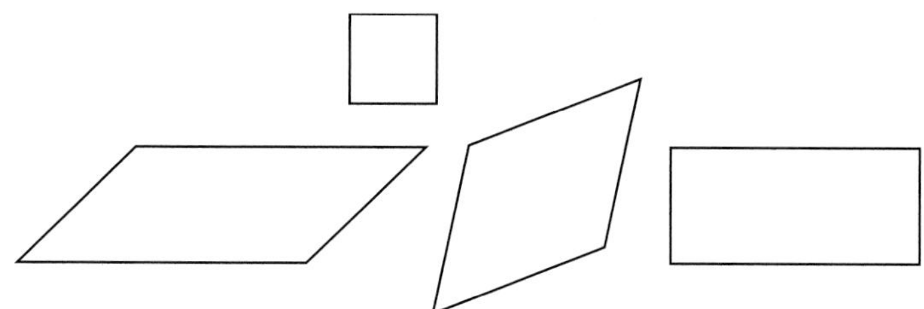

To find the area of a parallelogram, consider the one below.

If we cut off a piece and move it to the other end, we get a rectangle.

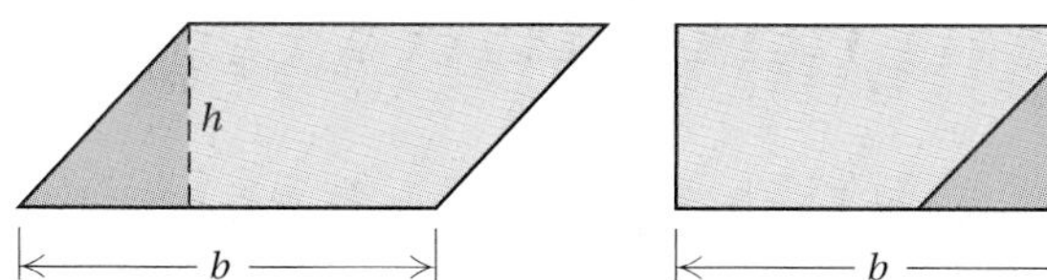

We can find the area by multiplying the length b, called a **base**, by h, called the **height**.

> The **area of a parallelogram** is the product of the length of a base b and the height h:
>
> $A = b \cdot h.$
>
> (figure: parallelogram with height h and base b)

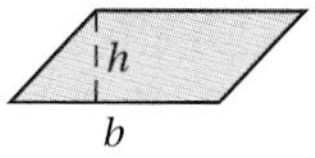

Example 1 Find the area of this parallelogram.

$A = b \cdot h$

$= 7 \text{ km} \cdot 5 \text{ km}$

$= 35 \text{ km}^2$

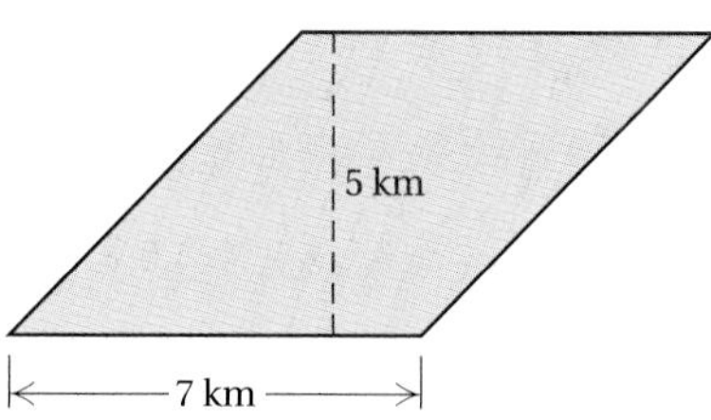

Find the area.

1.

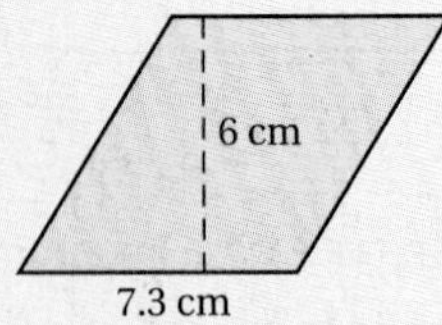

2.

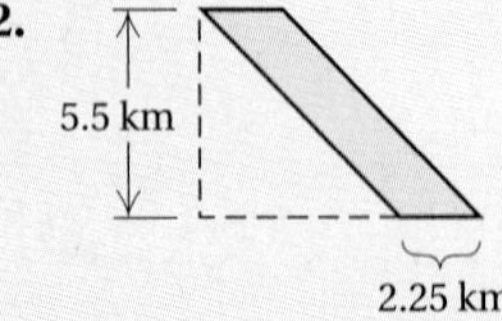

Example 2 Find the area of this parallelogram.

$$A = b \cdot h$$
$$= (1.2 \text{ m}) \times (6 \text{ m})$$
$$= 7.2 \text{ m}^2$$

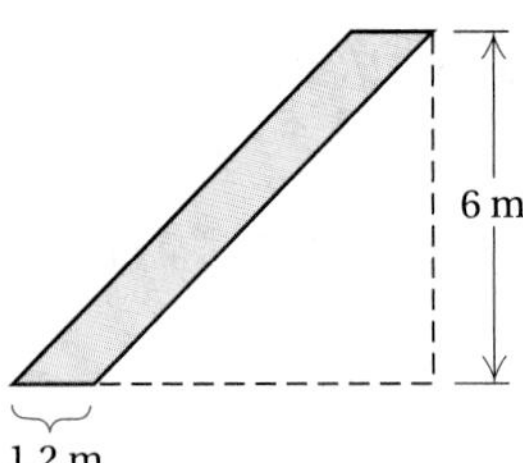

Do Exercises 1 and 2.

Triangles

To find the area of a triangle, think of cutting out another just like it.

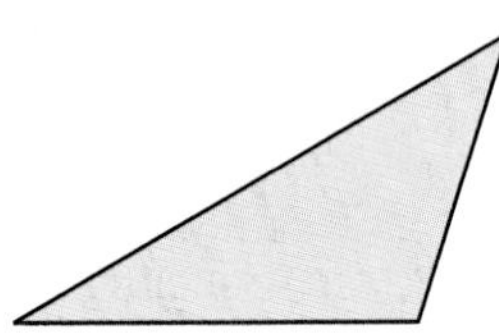

Then place the second one like this.

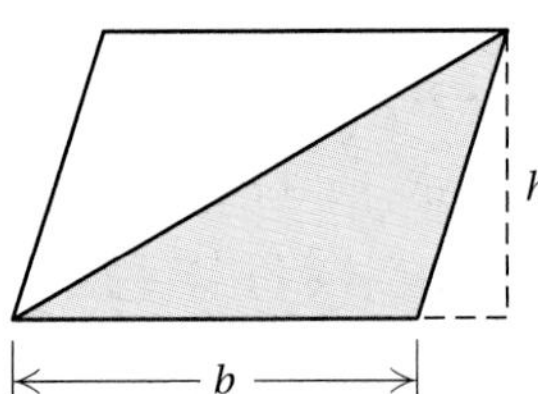

The resulting figure is a parallelogram whose area is

$$b \cdot h.$$

The triangle we started with has half the area of the parallelogram, or

$$\frac{1}{2} \cdot b \cdot h.$$

The **area of a triangle** is half the length of the base times the height:

$$A = \frac{1}{2} \cdot b \cdot h.$$

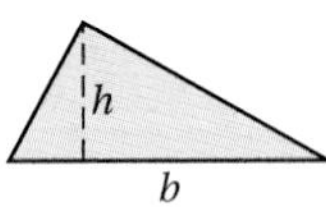

Example 3 Find the area of this triangle.

$$A = \frac{1}{2} \cdot b \cdot h$$
$$= \frac{1}{2} \cdot 9 \text{ m} \cdot 6 \text{ m}$$
$$= \frac{9 \cdot 6}{2} \text{ m}^2$$
$$= 27 \text{ m}^2$$

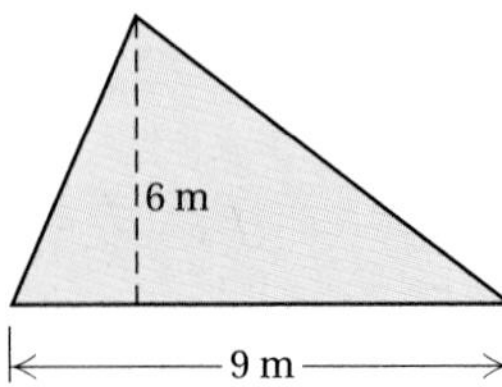

Answers on page A-15

Example 4 Find the area of this triangle.

$$A = \frac{1}{2} \cdot b \cdot h$$
$$= \frac{1}{2} \times 6.25 \text{ cm} \times 5.5 \text{ cm}$$
$$= 0.5 \times 6.25 \times 5.5 \text{ cm}^2$$
$$= 17.1875 \text{ cm}^2$$

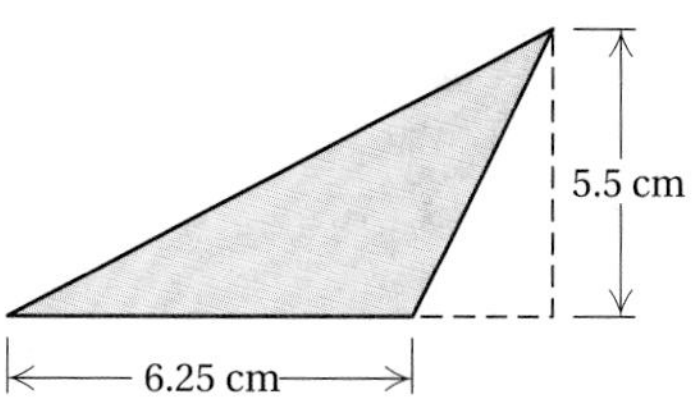

Do Exercises 3 and 4.

Find the area.

3.

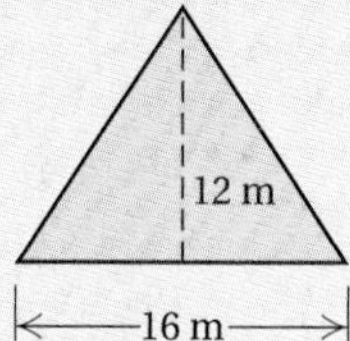

4.

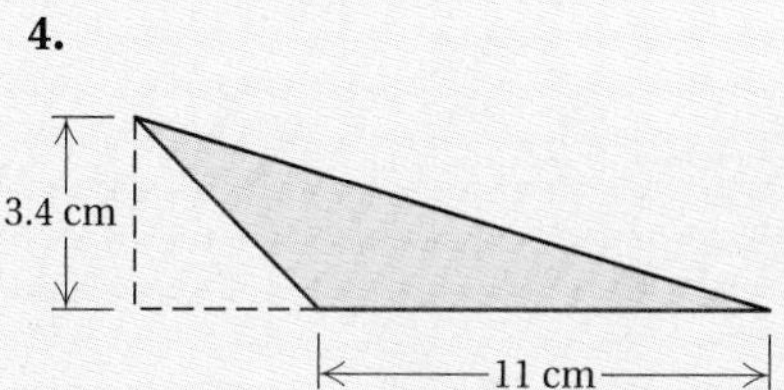

Answers on page A-15

Trapezoids

A **trapezoid** is a polygon with four sides, two of which, the **bases**, are parallel to each other.

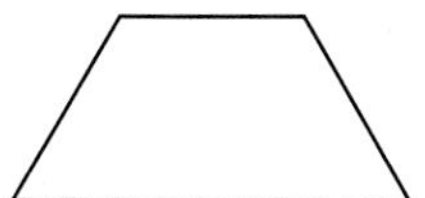

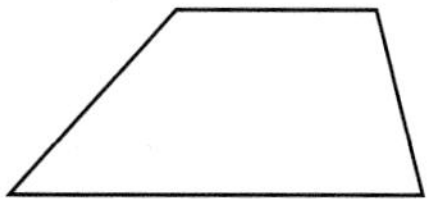

To find the area of a trapezoid, think of cutting out another just like it.

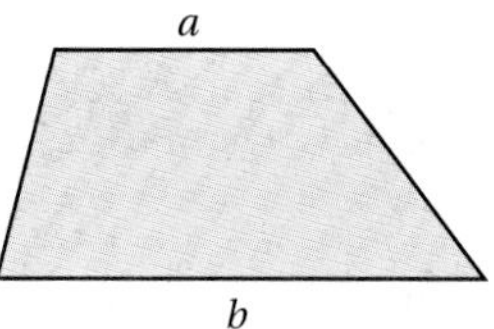

Then place the second one like this.

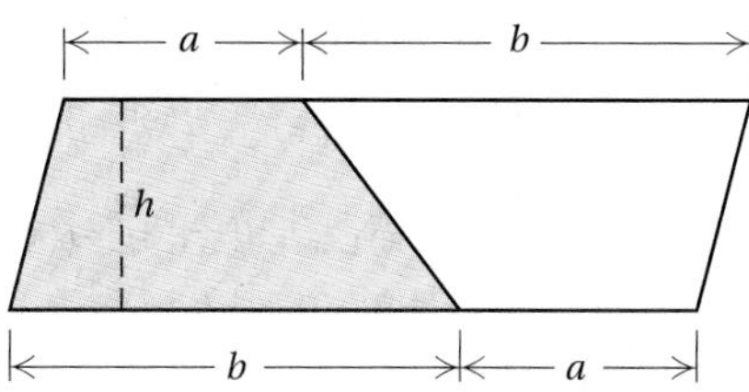

The resulting figure is a parallelogram whose area is

$h \cdot (a + b)$. **The base is $a + b$.**

The trapezoid we started with has half the area of the parallelogram, or

$$\frac{1}{2} \cdot h \cdot (a + b).$$

> The **area of a trapezoid** is half the product of the height and the sum of the lengths of the parallel sides, or the product of the height and the average length of the bases:
>
> $$A = \frac{1}{2} \cdot h \cdot (a + b) = h \cdot \frac{a + b}{2}.$$
>
> 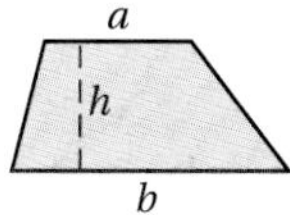
>

Find the area.

5.

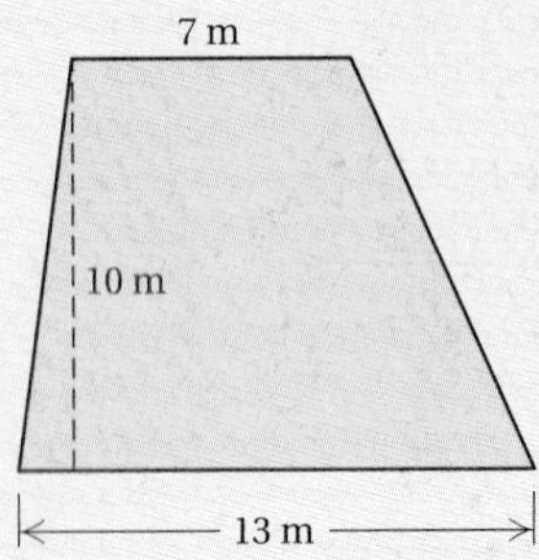

6.

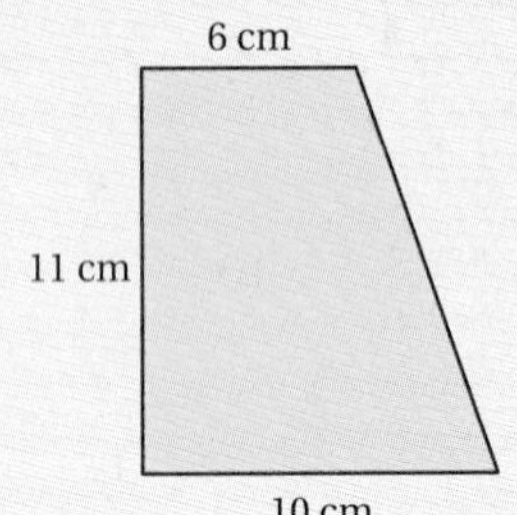

7. Find the area.

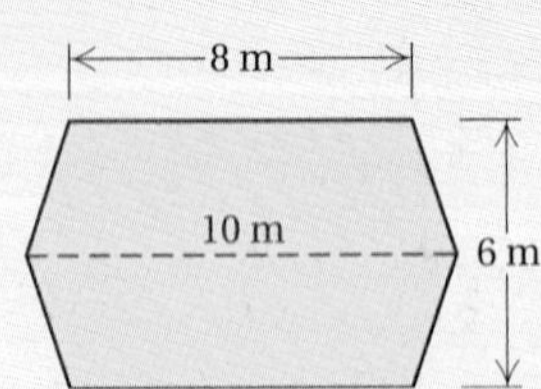

Answers on page A-15

Example 5 Find the area of this trapezoid.

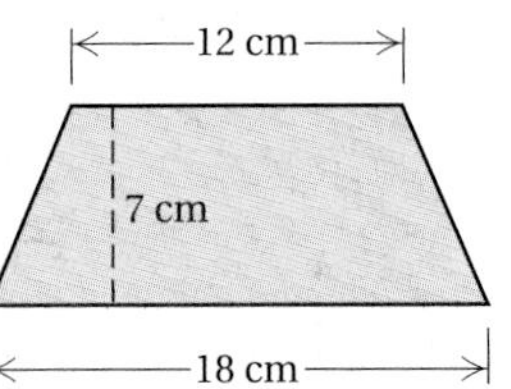

$$A = \frac{1}{2} \cdot h \cdot (a + b)$$

$$= \frac{1}{2} \cdot 7 \text{ cm} \cdot (12 + 18) \text{ cm}$$

$$= \frac{7 \cdot 30}{2} \cdot \text{cm}^2 = \frac{7 \cdot 15 \cdot 2}{1 \cdot 2} \text{ cm}^2$$

$$= \frac{7 \cdot 15}{1} \cdot \frac{2}{2} \text{ cm}^2$$

$$= 105 \text{ cm}^2$$

Do Exercises 5 and 6.

b Solving Applied Problems

Example 6 Find the area of this kite.

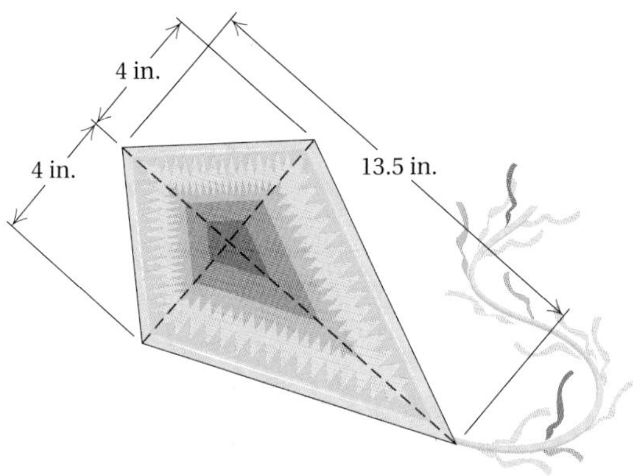

1. **Familiarize.** We look for the kinds of figures whose areas we can calculate using area formulas that we already know.
2. **Translate.** The kite consists of two triangles, each with a base of 13.5 in. and a height of 4 in. We can apply the formula $A = \frac{1}{2} \cdot b \cdot h$ for the area of a triangle and then multiply by 2.
3. **Solve.** We have

   $$A = \frac{1}{2} \cdot (13.5 \text{ in.}) \cdot (4 \text{ in.}) = 27 \text{ in}^2.$$

   Then we multiply by 2:

   $$2 \cdot 27 \text{ in}^2 = 54 \text{ in}^2.$$

4. **Check.** We can check by repeating the calculations.
5. **State.** The area of the kite is 54 in^2.

Do Exercise 7.

Exercise Set 6.4

a Find the area.

1.

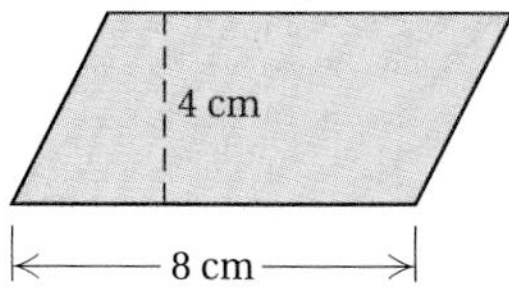

2.

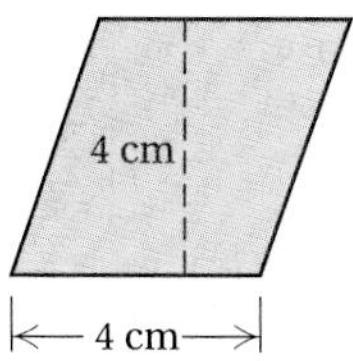

3.

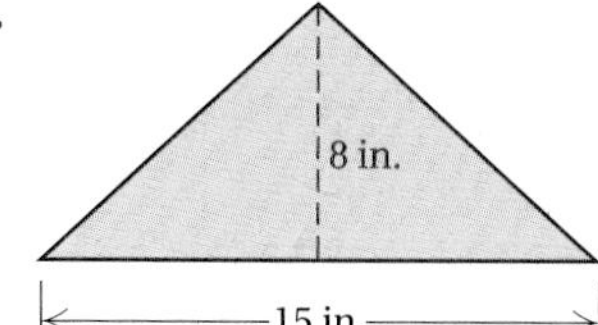

4.

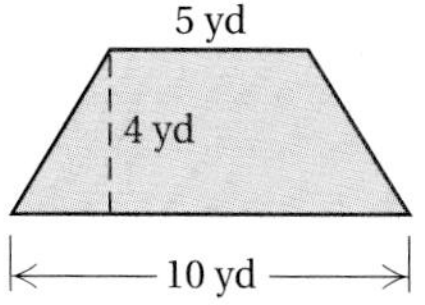

5.

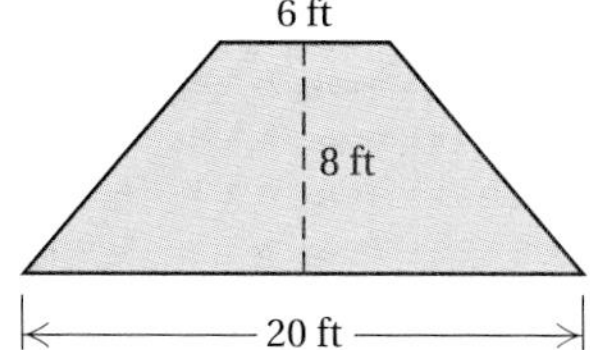

6.

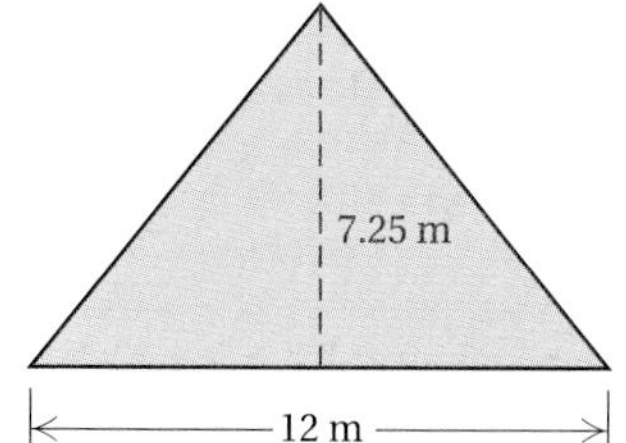

7.

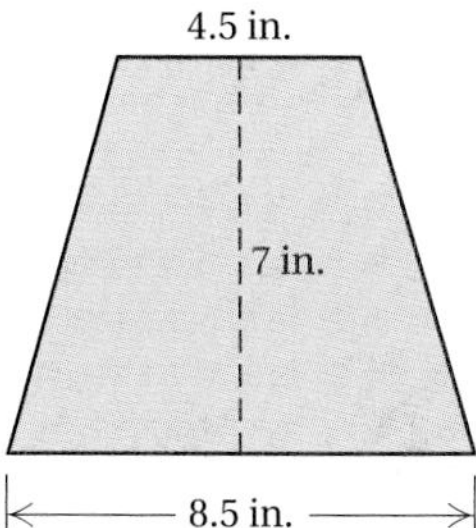

8.

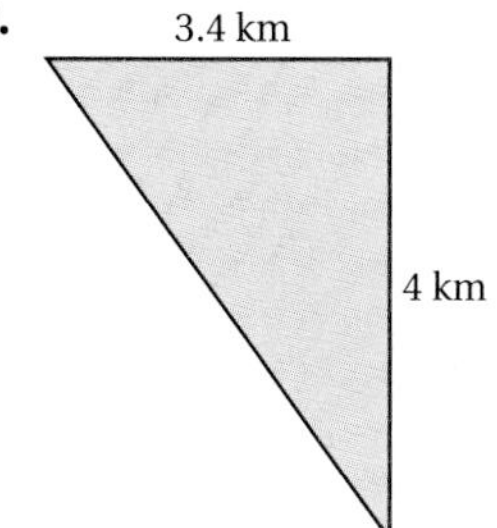

9.

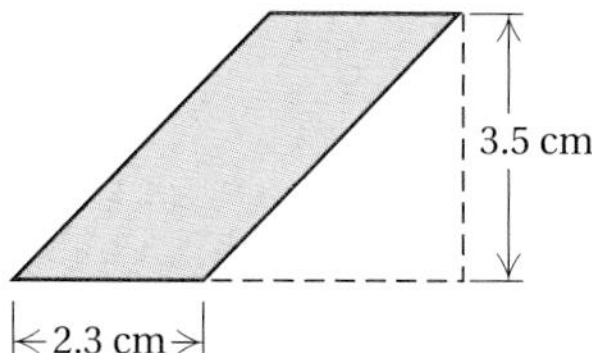

10.

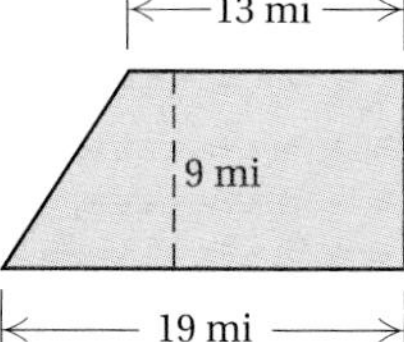

11.

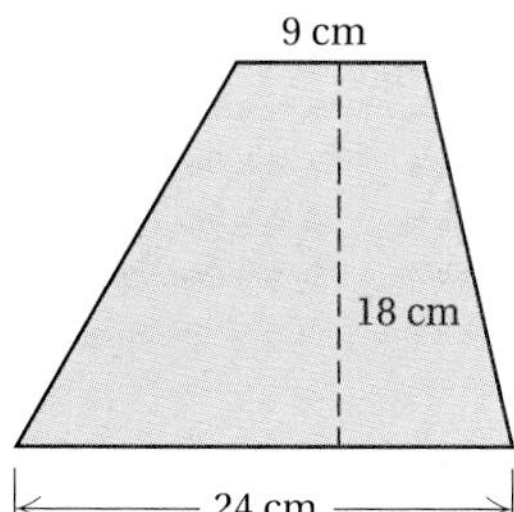

12.

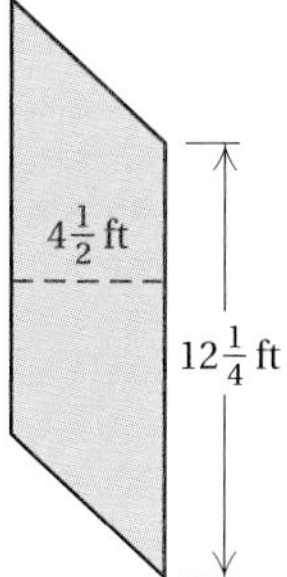

13.

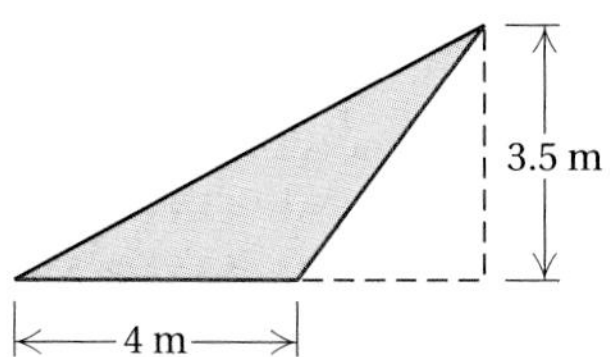

14.

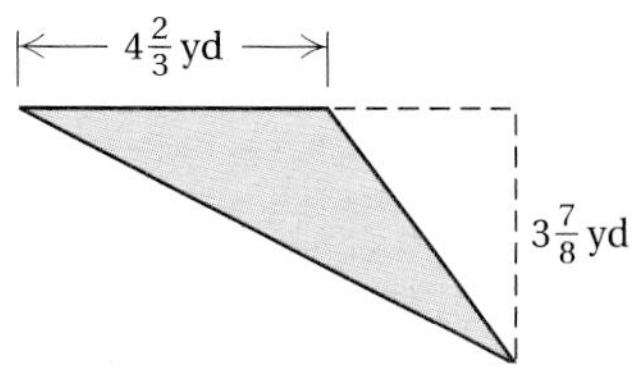

b Find the area of the shaded region.

15.

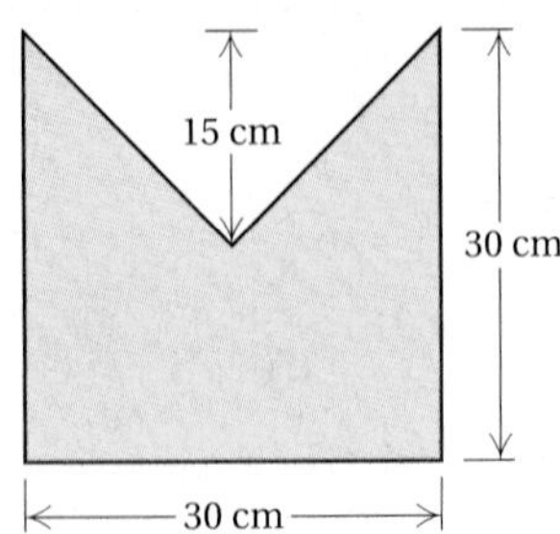

16.

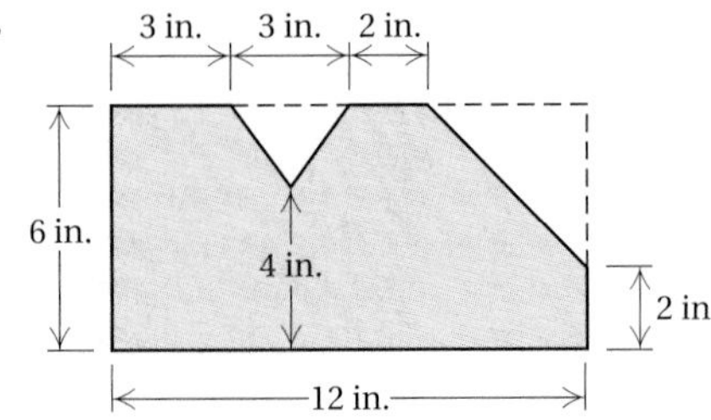

17.

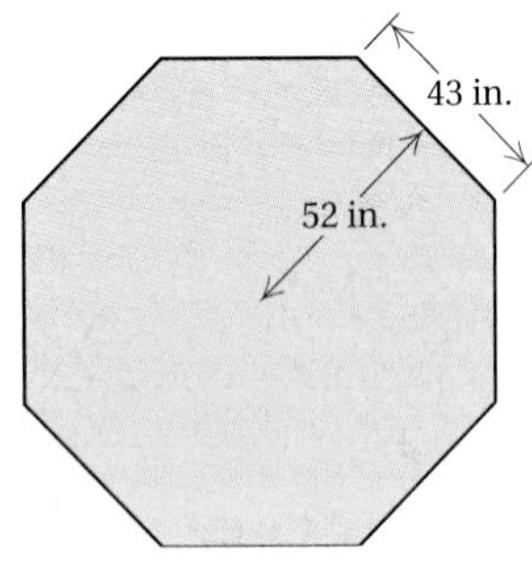

18.

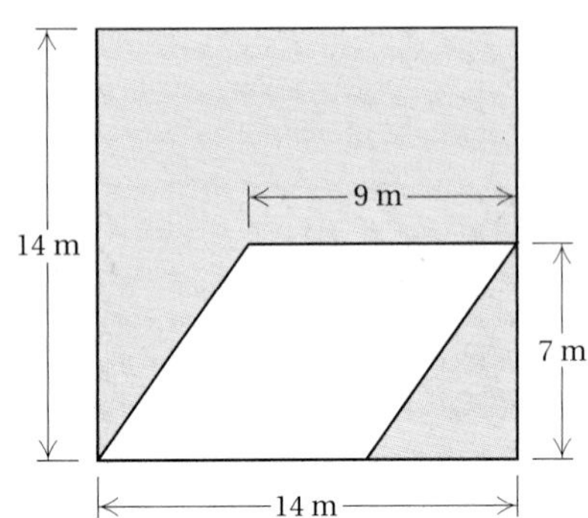

19. A rectangular piece of sailcloth is 36 ft by 24 ft. A triangular area with a height of 4.6 ft and a base of 5.2 ft is cut from the sailcloth. How much area is left over?

20. Find the total area of the sides and ends of the building.

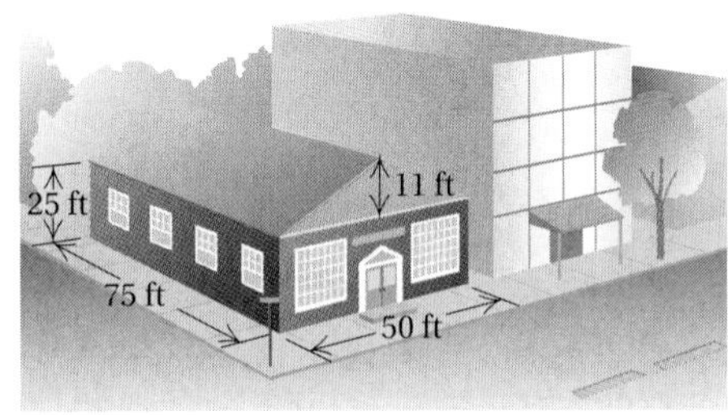

Skill Maintenance

Convert to fractional notation. [4.3b]

21. 35%

22. 85.5%

23. $37\frac{1}{2}\%$

24. $66.\overline{6}\%$

25. $83.\overline{3}\%$

26. $16\frac{2}{3}\%$

Solve. [1.5a]

27. A ream of paper contains 500 sheets. How many sheets are there in 15 reams?

28. A lab technician separates a vial containing 140 cc of blood into test tubes, each of which contains 3 cc of blood. How many test tubes can be filled? How much blood is left over?

Synthesis

29. ◆ Explain how the area of a parallelogram can be found by considering the area of a rectangle.

30. ◆ Explain how the area of a triangle can be found by considering the area of a parallelogram.

Verify the formulas for the area of a parallelogram, a triangle, and a trapezoid.

6.5 Circles

a Radius and Diameter

At the right is a circle with center O. Segment $\overline{AC}$ is a *diameter.* A **diameter** is a segment that passes through the center of the circle and has endpoints on the circle. Segment $\overline{OB}$ is called a *radius.* A **radius** is a segment with one endpoint on the center and the other endpoint on the circle.

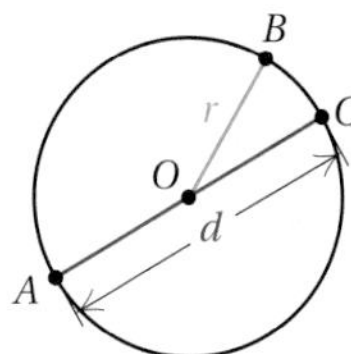

> Suppose that d is the diameter of a circle and r is the radius. Then
>
> $$d = 2 \cdot r \quad \text{and} \quad r = \frac{d}{2}.$$

Example 1 Find the length of a radius of this circle.

$$r = \frac{d}{2} = \frac{12\text{ m}}{2} = 6\text{ m}$$

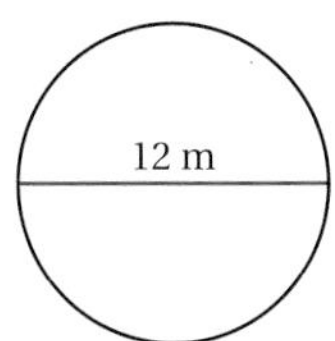

The radius is 6 m.

Example 2 Find the length of a diameter of this circle.

$$d = 2 \cdot r = 2 \cdot \frac{1}{4}\text{ ft} = \frac{1}{2}\text{ ft}$$

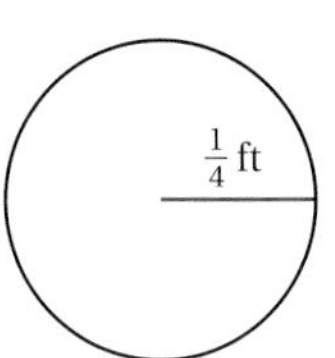

The diameter is $\frac{1}{2}$ ft.

Do Exercises 1 and 2.

b Circumference

The **circumference** of a circle is the distance around it. Calculating circumference is similar to finding the perimeter of a polygon.

To find a formula for the circumference of any circle given its diameter, we first need to consider the ratio C/d. Take a 12-oz soda can and measure the circumference C with a tape measure. Also measure the diameter d. The results are shown in the figure. Then

$$\frac{C}{d} = \frac{7.8\text{ in.}}{2.5\text{ in.}} \approx 3.1.$$

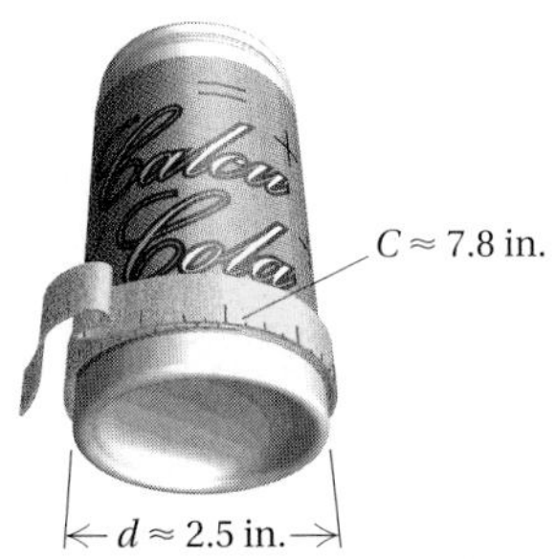

Objectives

a Find the length of a radius of a circle given the length of a diameter, and find the length of a diameter given the length of a radius.

b Find the circumference of a circle given the length of a diameter or a radius.

c Find the area of a circle given the length of a radius.

d Solve applied problems involving circles.

For Extra Help

TAPE 12

MAC
WIN

CD-ROM

1. Find the length of a radius.

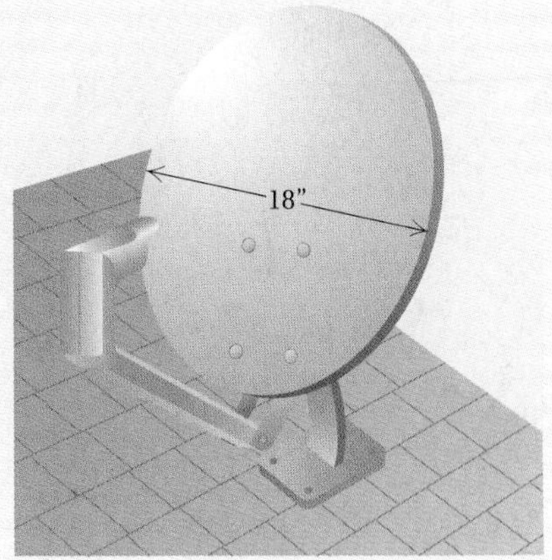

2. Find the length of a diameter.

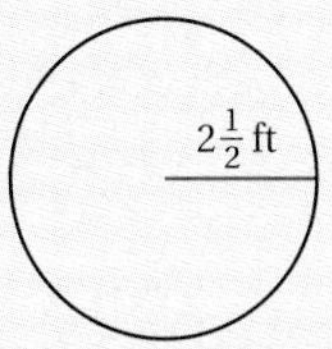

Answers on page A-15

Suppose we did this with cans and circles of several sizes. We would get a number close to 3.1. For any circle, if we divide the circumference C by the diameter d, we get the same number. We call this number π (pi).

> $\frac{C}{d} = \pi$ or $C = \pi \cdot d$. The number π is about 3.14, or about $\frac{22}{7}$.

Example 3 Find the circumference of this circle. Use 3.14 for π.

$$C = \pi \cdot d$$
$$\approx 3.14 \times 6 \text{ cm}$$
$$\approx 18.84 \text{ cm}$$

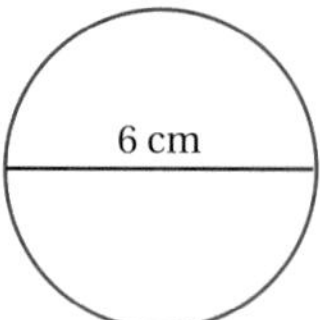

The circumference is about 18.84 cm.

Do Exercise 3.

Since $d = 2 \cdot r$, where r is the length of a radius, it follows that

$$C = \pi \cdot d = \pi \cdot (2 \cdot r).$$

> $C = 2 \cdot \pi \cdot r$

Example 4 Find the circumference of this circle. Use $\frac{22}{7}$ for π.

$$C = 2 \cdot \pi \cdot r$$
$$\approx 2 \cdot \frac{22}{7} \cdot 70 \text{ in.}$$
$$\approx 2 \cdot 22 \cdot \frac{70}{7} \text{ in.}$$
$$\approx 44 \cdot 10 \text{ in.}$$
$$\approx 440 \text{ in.}$$

The circumference is about 440 in.

Example 5 Find the perimeter of this figure. Use 3.14 for π.

We let $P =$ the perimeter. We see that we have half a circle attached to a square. Thus we add half the circumference to the lengths of the three line segments.

$$P = 3 \times 9.4 \text{ km} + \frac{1}{2} \times 2 \times \pi \times 4.7 \text{ km}$$
$$\approx 28.2 \text{ km} + 3.14 \times 4.7 \text{ km}$$
$$\approx 28.2 \text{ km} + 14.758 \text{ km}$$
$$\approx 42.958 \text{ km}$$

The perimeter is about 42.958 km.

Do Exercises 4 and 5.

3. Find the circumference of this circle. Use 3.14 for π.

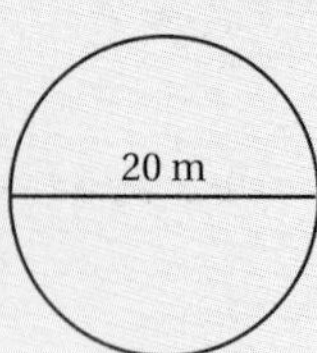

4. Find the circumference of this circle. Use $\frac{22}{7}$ for π.

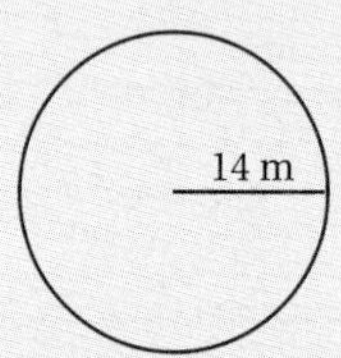

5. Find the perimeter of this figure. Use 3.14 for π.

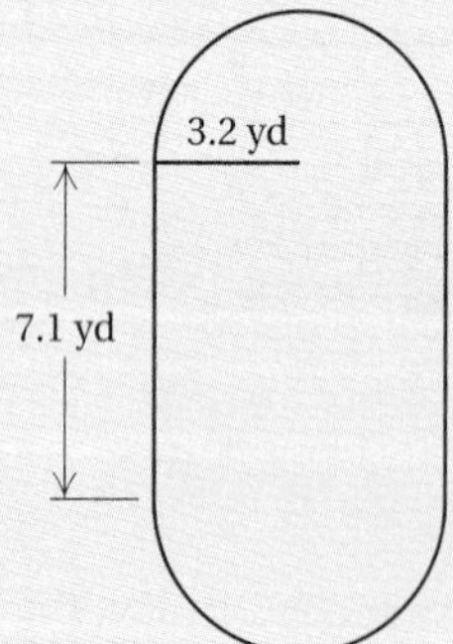

Answers on page A-15

c Area

Below is a circle of radius r.

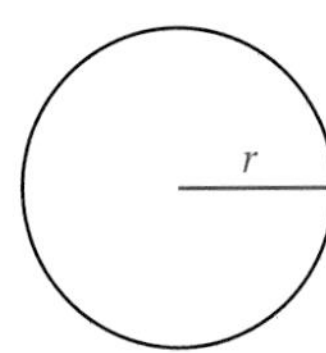

Think of cutting half the circular region into small pieces and arranging them as shown below.

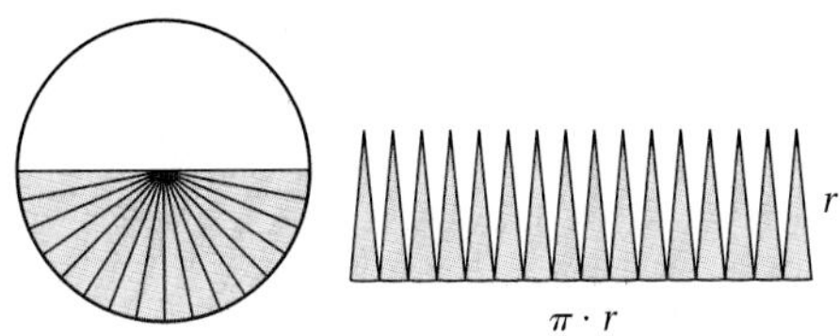

Then imagine cutting the other half of the circular region and arranging the pieces in with the others as shown below.

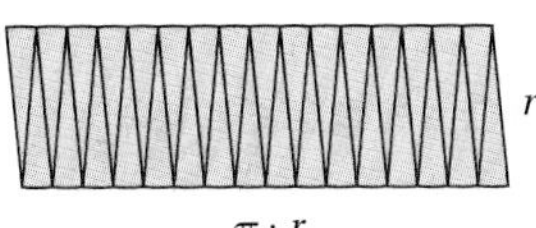

This is almost a parallelogram. The base has length $\frac{1}{2} \cdot 2 \cdot \pi \cdot r$, or $\pi \cdot r$ (half the circumference) and the height is r. Thus the area is

$$(\pi \cdot r) \cdot r.$$

This is the area of a circle.

The **area of a circle** with radius of length r is given by

$$A = \pi \cdot r \cdot r, \quad \text{or} \quad A = \pi \cdot r^2.$$

Example 6 Find the area of this circle. Use $\frac{22}{7}$ for π.

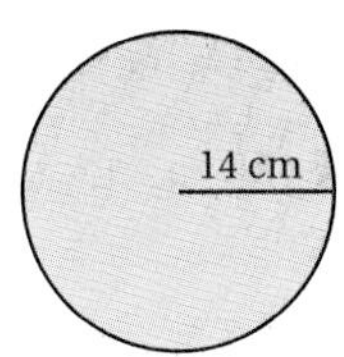

$$A = \pi \cdot r \cdot r$$

$$\approx \frac{22}{7} \cdot 14\text{ cm} \cdot 14\text{ cm}$$

$$\approx \frac{22}{7} \cdot 196\text{ cm}^2$$

$$\approx 616\text{ cm}^2$$

The area is about 616 cm^2.

Do Exercise 6.

6. Find the area of this circle. Use $\frac{22}{7}$ for π.

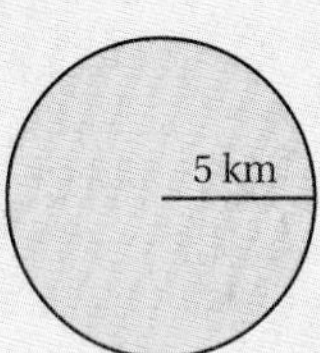

Calculator Spotlight

On certain calculators, there is a pi key, $\boxed{\pi}$. You can use a $\boxed{\pi}$ key for most computations instead of stopping to round the value of π. Rounding, if necessary, is done at the end.

Exercises

1. If you have a $\boxed{\pi}$ key on your calculator, to how many places does this key give the value of π?
2. Find the circumference and the area of a circle with a radius of 225.68 in.
3. Find the area of a circle with a diameter of $46\frac{12}{13}$ in.
4. Find the area of a large irrigated farming circle with a diameter of 400 ft.

Answer on page A-15

7. Find the area of this circle. Use 3.14 for π.

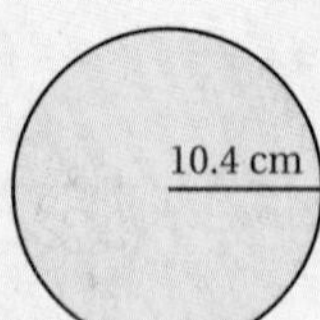

CAUTION!

Circumference $= \pi \cdot d = \pi \cdot (r + r) = \pi \cdot (2 \cdot r)$,

Area $= \pi \cdot r^2 = \pi \cdot (r \cdot r)$,

and

$r^2 \neq 2 \cdot r$.

Example 7 Find the area of this circle. Use 3.14 for π. Round to the nearest hundredth.

$$\begin{aligned} A &= \pi \cdot r \cdot r \\ &\approx 3.14 \times 2.1 \text{ m} \times 2.1 \text{ m} \\ &\approx 3.14 \times 4.41 \text{ m}^2 \\ &\approx 13.8474 \text{ m}^2 \\ &\approx 13.85 \text{ m}^2 \end{aligned}$$

2.1 m

The area is about 13.85 m^2.

Do Exercise 7.

d Solving Applied Problems

Example 8 *Area of Pizza Pan.* Which makes a larger pizza and by how much: a 16-in. square pizza pan or a 16-in. diameter circular pizza pan?

First, we make a drawing of each.

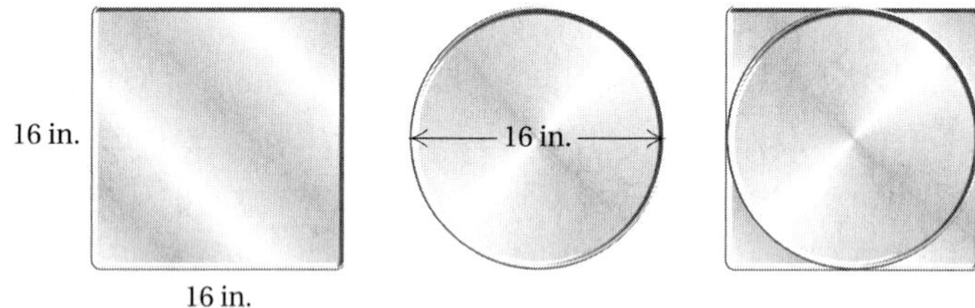

8. Which is larger and by how much: a 10-ft square flower bed or a 12-ft diameter flower bed?

Then we compute areas.

The area of the square is

$$\begin{aligned} A &= s \cdot s \\ &= 16 \text{ in.} \times 16 \text{ in.} = 256 \text{ in}^2. \end{aligned}$$

The diameter of the circle is 16 in., so the radius is 16 in./2, or 8 in. The area of the circle is

$$\begin{aligned} A &= \pi \cdot r \cdot r \\ &\approx 3.14 \times 8 \text{ in.} \times 8 \text{ in.} \approx 200.96 \text{ in}^2. \end{aligned}$$

We see that the square pizza pan is larger by about

$$256 \text{ in}^2 - 200.96 \text{ in}^2, \quad \text{or} \quad 55.04 \text{ in}^2.$$

Thus the square pan makes the larger pizza, by about 55.04 in^2.

Do Exercise 8.

Answers on page A-15

Exercise Set 6.5

a, b, c For each circle, find the length of a diameter, the circumference, and the area. Use $\frac{22}{7}$ for π.

1.

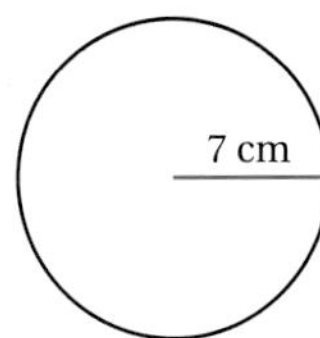

2.

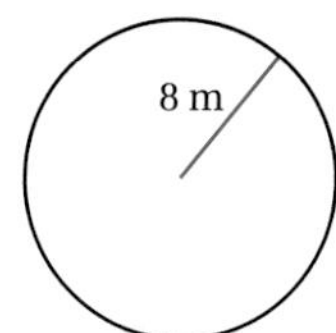

3.

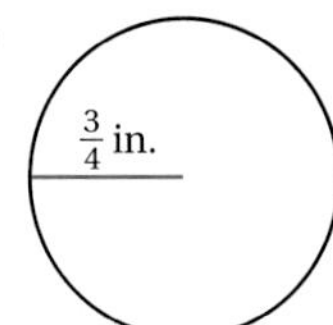

4.

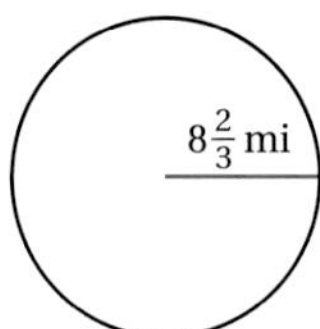

For each circle, find the length of a radius, the circumference, and the area. Use 3.14 for π.

5.

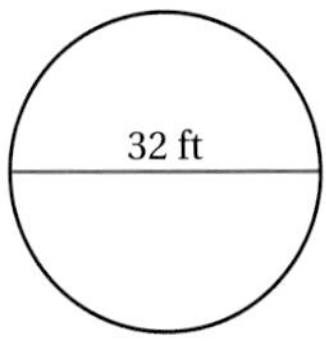

6.

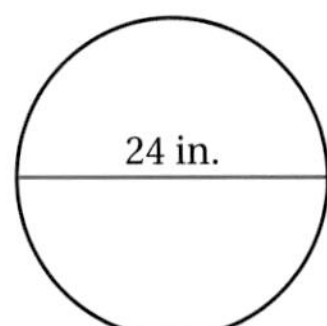

7.

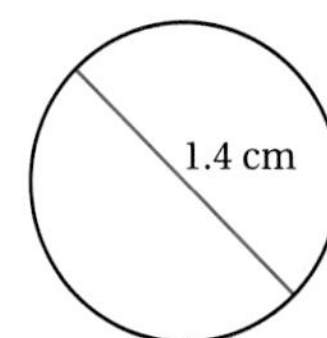

8.

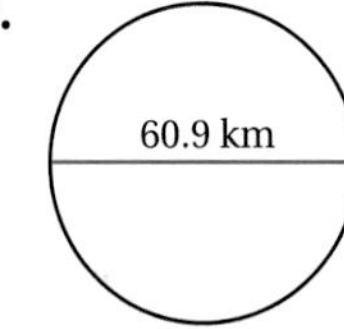

d Solve. Use 3.14 for π.

9. The top of a soda can has a 6-cm diameter. What is its radius? its circumference? its area?

10. A penny has a 1-cm radius. What is its diameter? its circumference? its area?

11. A radio station is allowed by the FCC to broadcast over an area with a radius of 220 mi. How much area is this?

12. *Pizza Areas.* Which is larger and by how much: a 12-in. circular pizza or a 12-in. square pizza?

13. *Dimensions of a Quarter.* The circumference of a quarter is 7.85 cm. What is the diameter? the radius? the area?

14. *Dimensions of a Dime.* The circumference of a dime is 2.23 in. What is the diameter? the radius? the area?

15. *Gypsy-Moth Tape.* To protect an elm tree in your backyard, you need to attach gypsy moth caterpillar tape around the trunk. The tree has a 1.1-ft diameter. What length of tape is needed?

16. *Silo.* A silo has a 10-m diameter. What is its circumference?

17. *Swimming-Pool Walk.* You want to install a 1-yd–wide walk around a circular swimming pool. The diameter of the pool is 20 yd. What is the area of the walk?

18. *Roller-Rink Floor.* A roller rink floor is shown below. What is its area? If hardwood flooring costs \$10.50 per square meter, how much will the flooring cost?

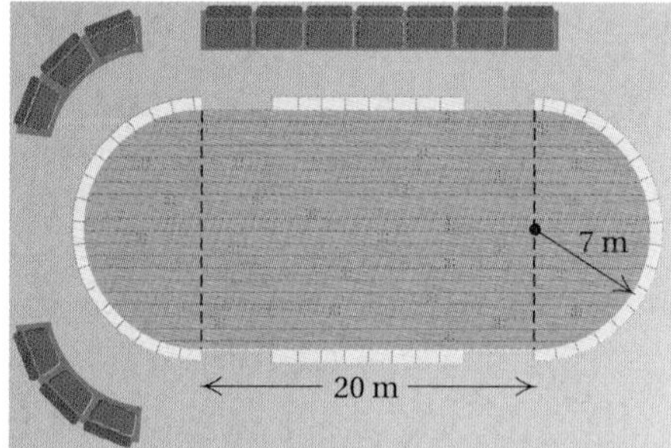

Find the perimeter. Use 3.14 for π.

19.

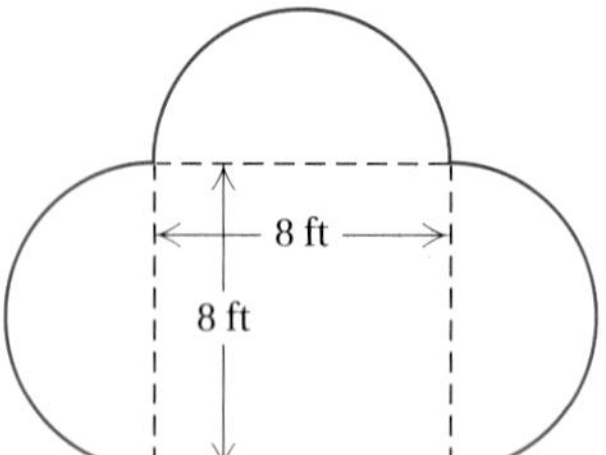

20.

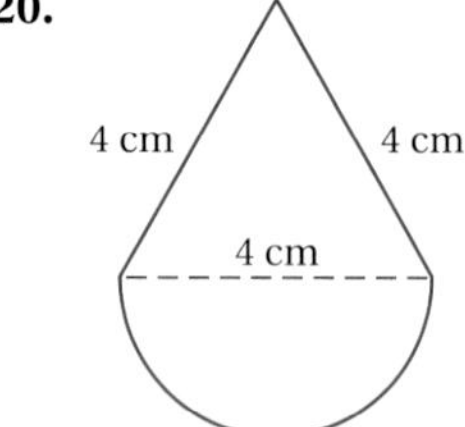

21.

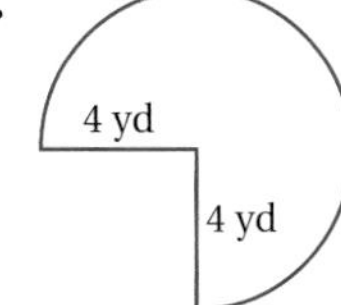

22.

8 in. 8 in. 8 in. 8 in.

23.

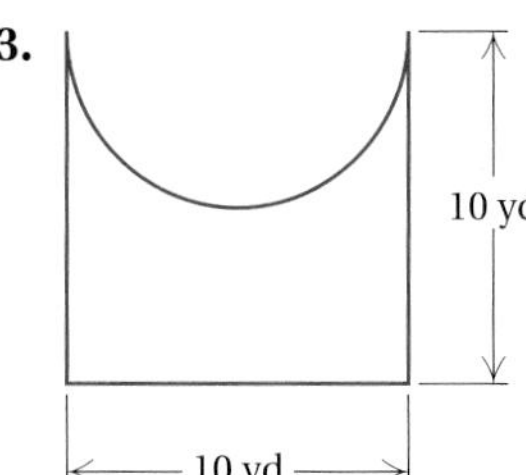

24.

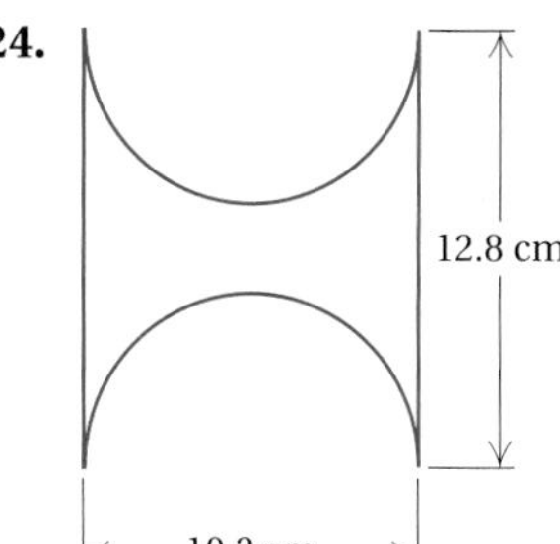

Find the area of the shaded region. Use 3.14 for π.

25.

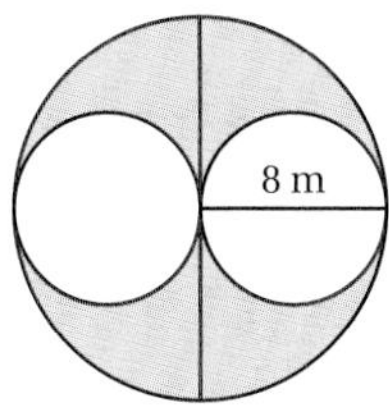

26.

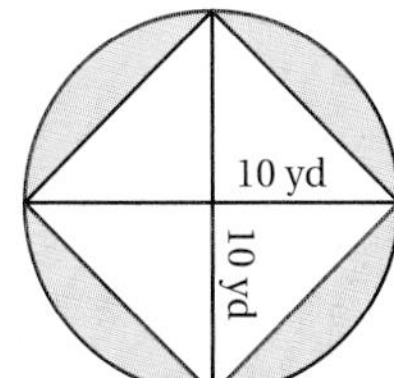

27.

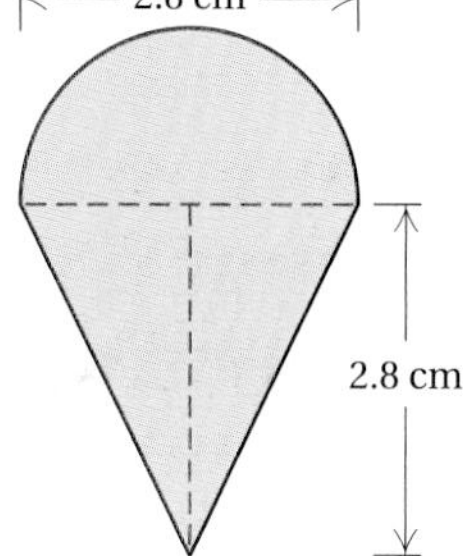

28.

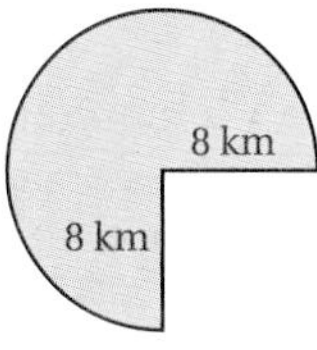

29.

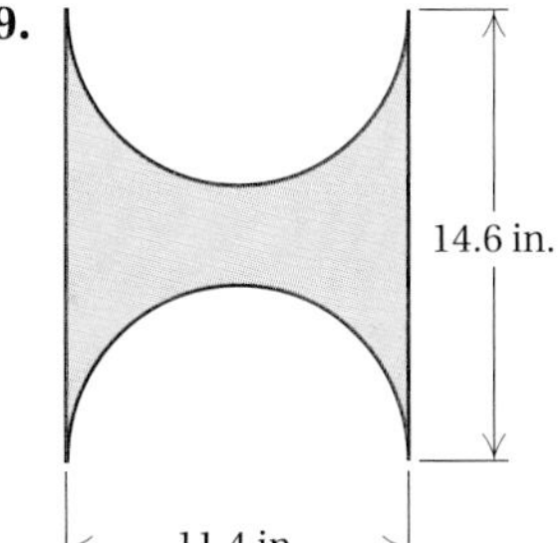

30.

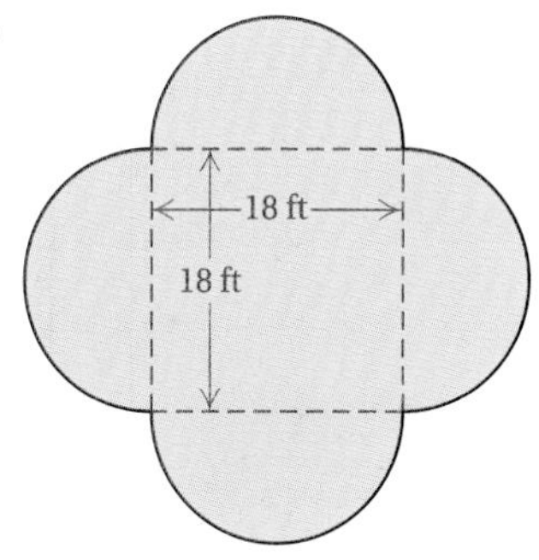

Skill Maintenance

Convert to percent notation. [4.2c]

31. 0.875

32. 0.58

33. $0.\overline{6}$

34. 0.4361

Convert to percent notation. [4.3a]

35. $\frac{3}{8}$

36. $\frac{5}{8}$

37. $\frac{2}{3}$

38. $\frac{1}{5}$

Estimate each of the following as a whole number or as a mixed number where the fractional part is $\frac{1}{2}$. [2.6b]

39. $3\frac{7}{8}$

40. $8\frac{1}{3}$

41. $13\frac{1}{6}$

42. $39\frac{7}{13}$

43. $\frac{4}{5} + 3\frac{7}{8}$

44. $\frac{1}{11} \cdot \frac{7}{15}$

45. $\frac{2}{3} + \frac{7}{15} + \frac{8}{9}$

46. $\frac{8}{9} + \frac{4}{5} + \frac{13}{14}$

47. $\frac{57}{100} - \frac{1}{10} + \frac{9}{1000}$

48. $\frac{23}{24} + \frac{38}{39} + \frac{61}{60}$

49. $11\frac{29}{80} + 10\frac{14}{15} \cdot 24\frac{2}{17}$

50. $\frac{13}{14} + 9\frac{5}{8} - 1\frac{23}{28} \cdot 1\frac{36}{73}$

Synthesis

51. ◆ Explain why a 16-in.–diameter pizza that costs \$16.25 is a better buy than a 10-in.–diameter pizza that costs \$7.85.

52. ◆ The radius of one circle is twice the size of another circle's radius. Is the area of the first circle twice the area of the other circle? Why or why not?

53. ▦ $\pi \approx \frac{3927}{1250}$ is another approximation for π. Find decimal notation using a calculator. Round to the nearest thousandth.

54. ▦ The distance from Kansas City to Indianapolis is 500 mi. A car was driven this distance using tires with a radius of 14 in. How many revolutions of each tire occurred on the trip? Use $\frac{22}{7}$ for π.

55. *Tennis Balls.* Tennis balls are usually packed vertically three in a can, one on top of another. Suppose the diameter of a tennis ball is d. Find the height of the stack of balls. Find the circumference of one ball. Which is greater? Explain.

Collaborative Learning Manual

Estimate the value of π.

6.6 Volume and Surface Area

Objectives

[a] Find the volume and the surface area of a rectangular solid.

[b] Given the radius and the height, find the volume of a circular cylinder.

[c] Given the radius, find the volume of a sphere.

[d] Given the radius and the height, find the volume of a circular cone.

For Extra Help

TAPE 12 | InterAct math MAC WIN | CD-ROM

[a] Rectangular Solids

The **volume** of a **rectangular solid** is the number of unit cubes needed to fill it.

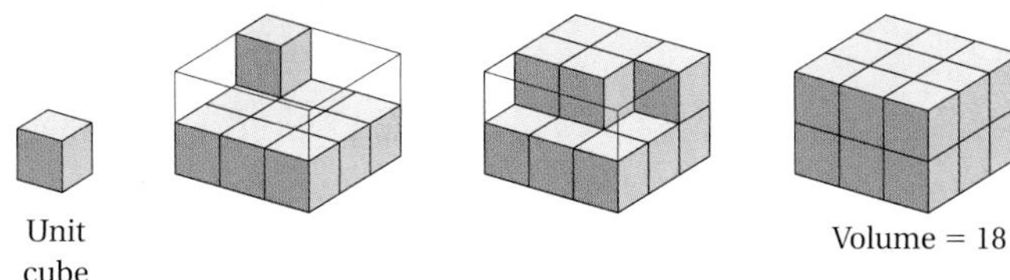

Two other units are shown below.

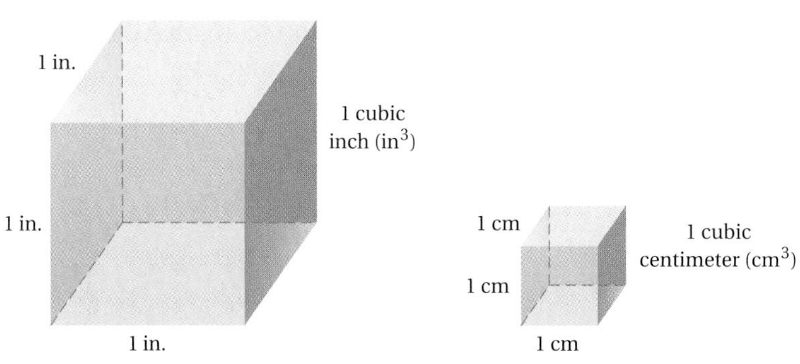

Example 1 Find the volume.

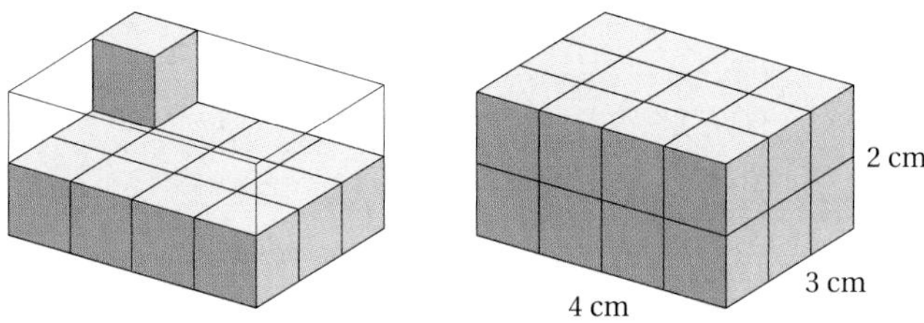

The figure is made up of 2 layers of 12 cubes each, so its volume is 24 cubic centimeters (cm^3).

Do Exercise 1.

The **volume of a rectangular solid** is found by multiplying length by width by height:

$$V = l \cdot w \cdot h.$$

h

w *l*

1. Find the volume.

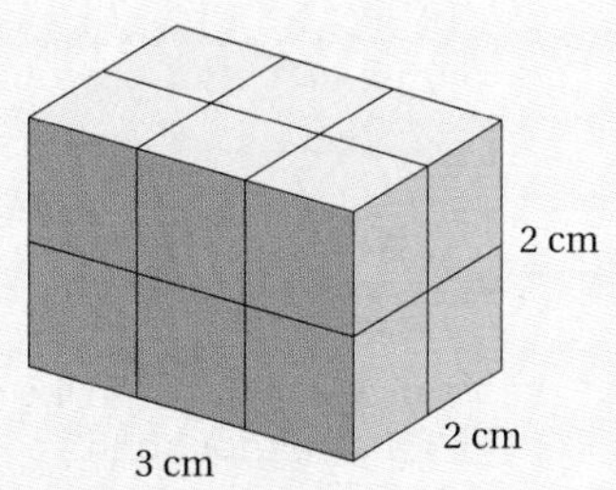

Answer on page A-15

2. In a recent year, people in the United States bought enough unpopped popcorn to provide every person in the country with a bag of popped corn measuring 2 ft by 2 ft by 5 ft. Find the volume of such a bag.

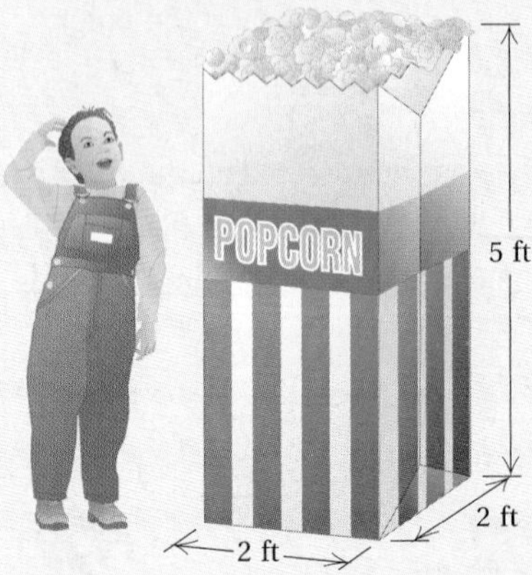

3. *Cord of Wood.* A cord of wood is 4 ft by 4 ft by 8 ft. What is the volume of a cord of wood?

Find the volume and the surface area of the rectangular solid.

4.

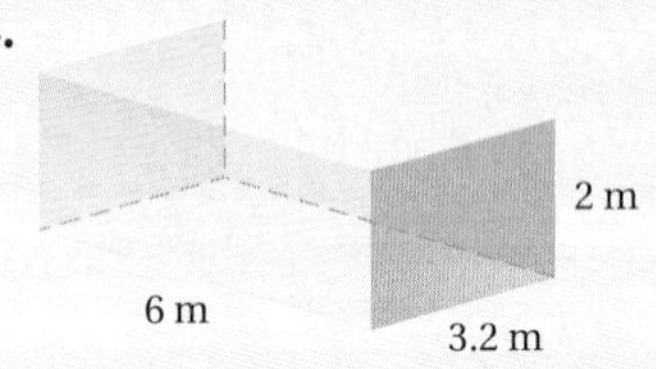

5.

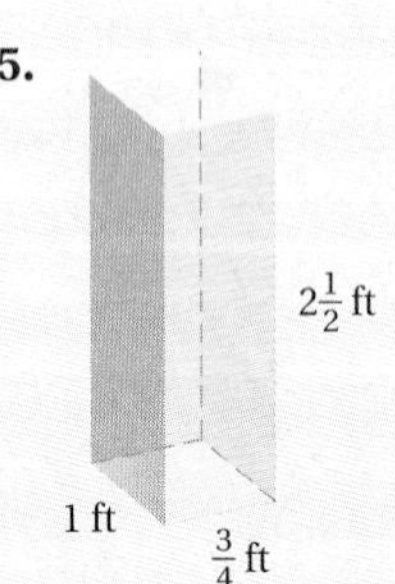

Answers on page A-15

Example 2 The largest piece of luggage that you can carry on an airplane measures 23 in. by 10 in. by 13 in. Find the volume of this solid.

$$\begin{aligned} V &= l \cdot w \cdot h \\ &= 23 \text{ in.} \cdot 10 \text{ in.} \cdot 13 \text{ in.} \\ &= 230 \cdot 13 \text{ in}^3 \\ &= 2990 \text{ in}^3 \end{aligned}$$

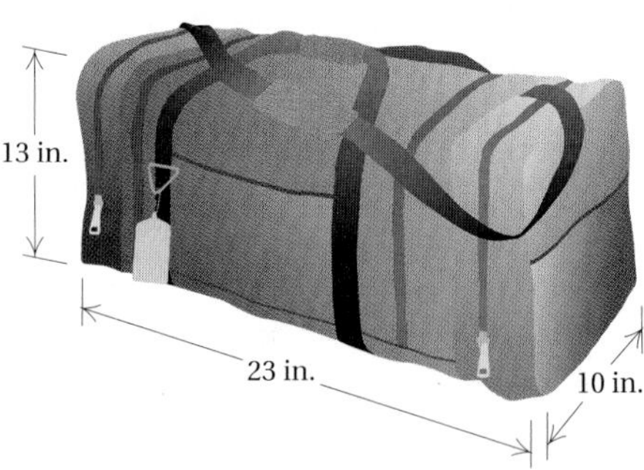

Do Exercises 2 and 3.

The **surface area** of a rectangular solid is the total area of the six rectangles that form the surface of the solid. For the rectangular solid below, we can show the six rectangles with a diagram.

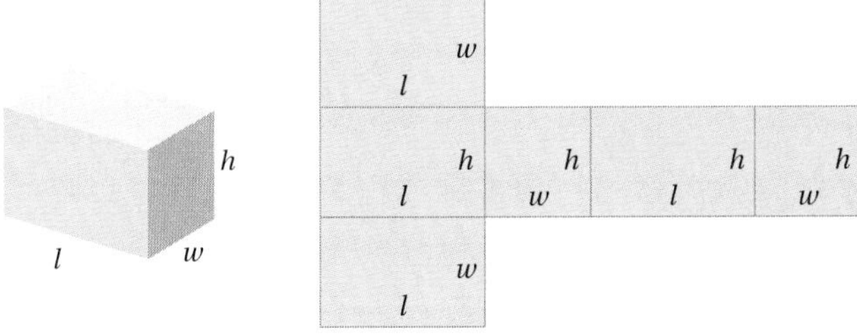

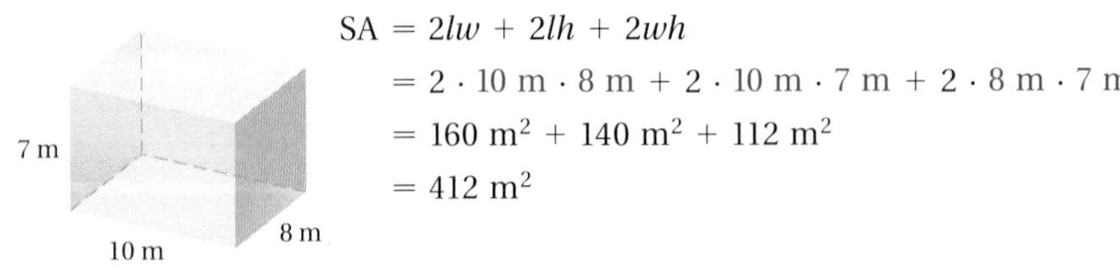

$$\begin{aligned} \text{SA} &= lw + lw + lh + wh + lh + wh \\ &= 2lw + 2lh + 2wh, \quad \text{or} \quad 2(lw + lh + wh) \end{aligned}$$

The surface area of a rectangular solid with length l, width w, and height h is given by the formula

$$\text{SA} = 2lw + 2lh + 2wh, \quad \text{or} \quad 2(lw + lh + wh).$$

Example 3 Find the surface area of this rectangular solid.

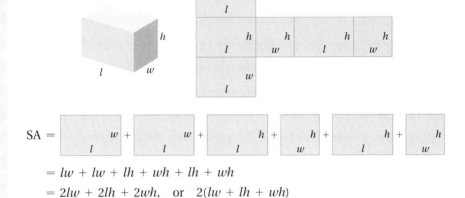

$$\begin{aligned} \text{SA} &= 2lw + 2lh + 2wh \\ &= 2 \cdot 10 \text{ m} \cdot 8 \text{ m} + 2 \cdot 10 \text{ m} \cdot 7 \text{ m} + 2 \cdot 8 \text{ m} \cdot 7 \text{ m} \\ &= 160 \text{ m}^2 + 140 \text{ m}^2 + 112 \text{ m}^2 \\ &= 412 \text{ m}^2 \end{aligned}$$

7 m

10 m

8 m

The units used for area are square units.
The units used for volume are cubic units.

Do Exercises 4 and 5.

b Cylinders

A rectangular solid is shown below. Note that we can think of the volume as the product of the area of the base times the height:

$$\begin{aligned} V &= l \cdot w \cdot h \\ &= (l \cdot w) \cdot h \\ &= (\text{Area of the base}) \cdot h \\ &= B \cdot h, \end{aligned}$$

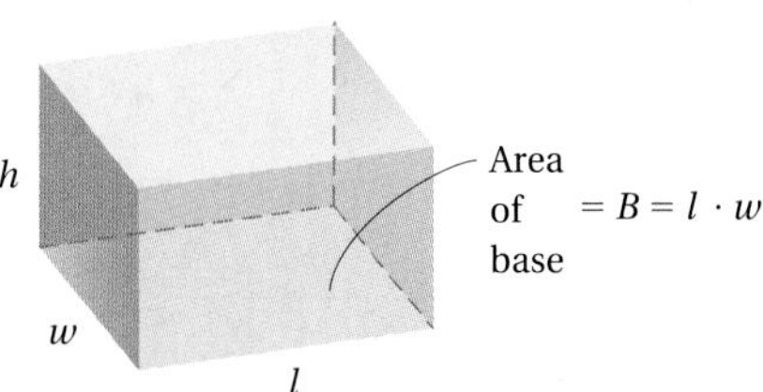

where B represents the area of the base.

Like rectangular solids, **circular cylinders** have bases of equal area that lie in parallel planes. The bases of circular cylinders are circular regions.

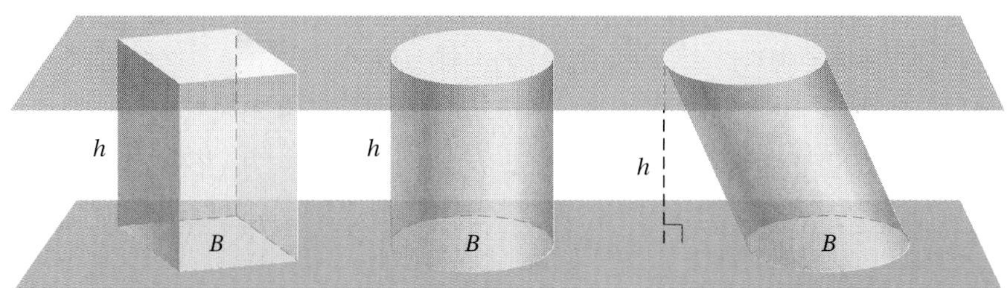

The volume of a circular cylinder is found in a manner similar to finding the volume of a rectangular solid. The volume is the product of the area of the base times the height. The height is always measured perpendicular to the base.

> The **volume of a circular cylinder** is the product of the area of the base B and the height h:
>
> $V = B \cdot h$, or $V = \pi \cdot r^2 \cdot h$.

Example 4 Find the volume of this circular cylinder. Use 3.14 for π.

$$\begin{aligned} V &= Bh = \pi \cdot r^2 \cdot h \\ &\approx 3.14 \times 4 \text{ cm} \times 4 \text{ cm} \times 12 \text{ cm} \\ &\approx 602.88 \text{ cm}^3 \end{aligned}$$

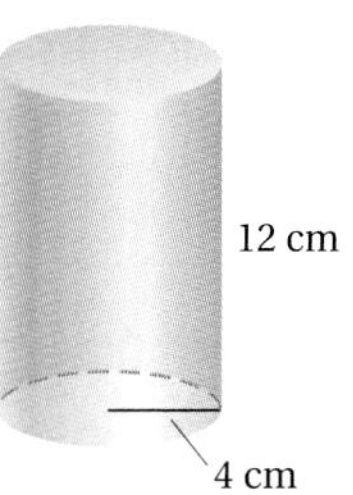

Do Exercises 6 and 7.

c Spheres

A **sphere** is the three-dimensional counterpart of a circle. It is the set of all points in space that are a given distance (the radius) from a given point (the center).

6. Find the volume of the cylinder. Use 3.14 for π.

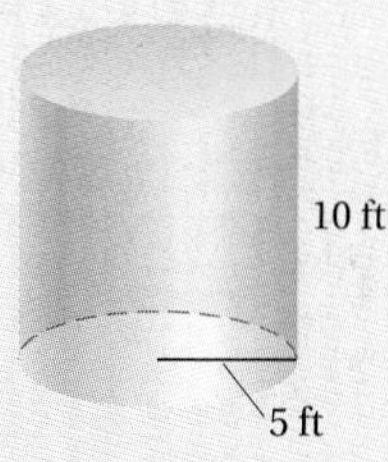

7. Find the volume of the cylinder. Use $\frac{22}{7}$ for π.

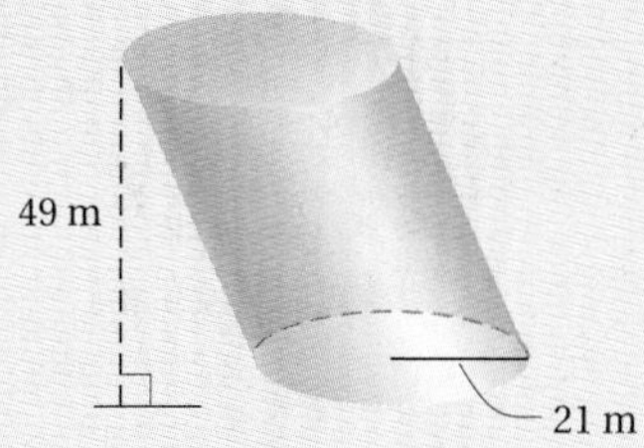

Answers on page A-15

8. Find the volume of the sphere. Use $\frac{22}{7}$ for π.

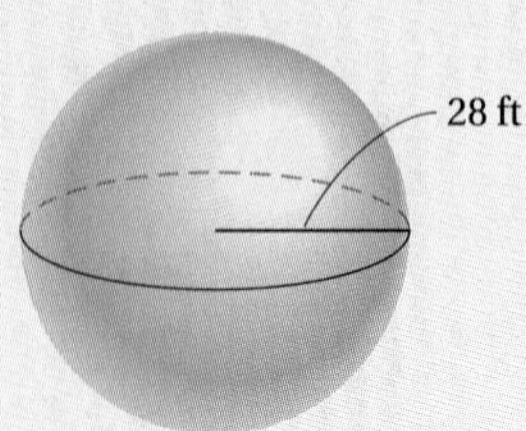

9. The radius of a standard-sized golf ball is 2.1 cm. Find its volume. Use 3.14 for π.

10. Find the volume of this cone. Use 3.14 for π.

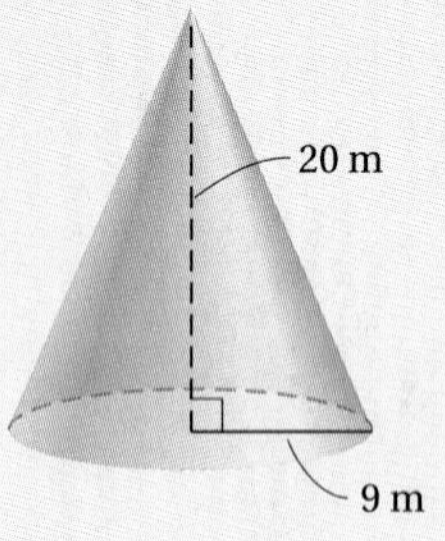

11. Find the volume of this cone. Use $\frac{22}{7}$ for π.

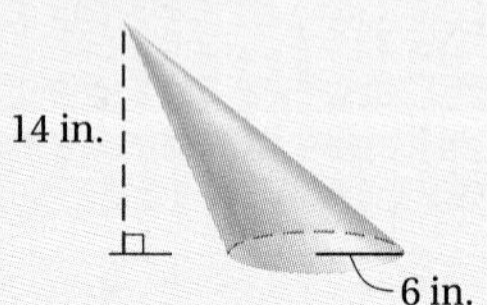

Answers on page A-15

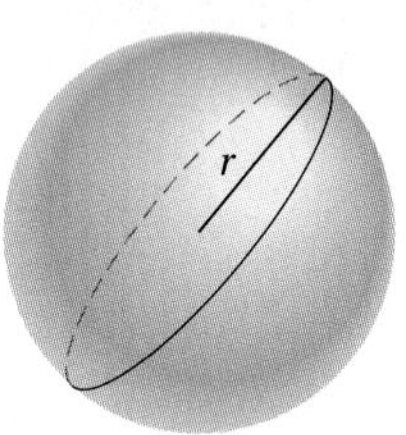

We find the volume of a sphere as follows.

> The **volume of a sphere** of radius r is given by
>
> $$V = \frac{4}{3} \cdot \pi \cdot r^3.$$

Example 5 *Bowling Ball.* The radius of a standard-sized bowling ball is 4.2915 in. Find the volume of a standard-sized bowling ball. Round to the nearest hundredth of a cubic inch. Use 3.14 for π.

$$V = \tfrac{4}{3} \cdot \pi \cdot r^3 \approx \tfrac{4}{3} \times 3.14 \times (4.2915 \text{ in.})^3$$
$$\approx \frac{4 \times 3.14 \times 79.0364 \text{ in}^3}{3} \approx 330.90 \text{ in}^3$$

Do Exercises 8 and 9.

d Cones

Consider a circle in a plane and choose any point P not in the plane. The circular region, together with the set of all segments connecting P to a point on the circle, is called a **circular cone.**

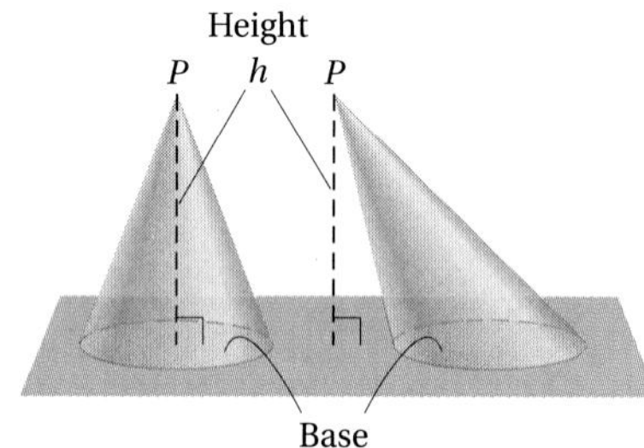

We find the volume of a cone as follows.

> The **volume of a circular cone** with base radius r is one-third the product of the base area and the height:
>
> $$V = \frac{1}{3} \cdot B \cdot h = \frac{1}{3} \pi \cdot r^2 \cdot h.$$

Example 6 Find the volume of this circular cone. Use 3.14 for π.

$$V = \tfrac{1}{3}\pi \cdot r^2 \cdot h$$
$$\approx \tfrac{1}{3} \times 3.14 \times 3 \text{ cm} \times 3 \text{ cm} \times 7 \text{ cm}$$
$$\approx 65.94 \text{ cm}^3$$

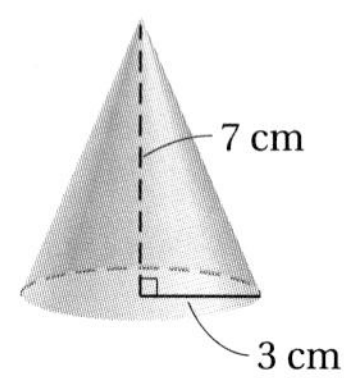

Do Exercises 10 and 11.

Exercise Set 6.6

a Find the volume and the surface area of the rectangular solid.

1.

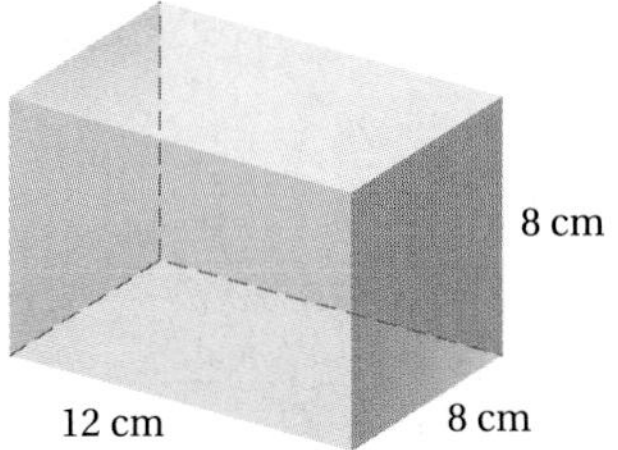

2.

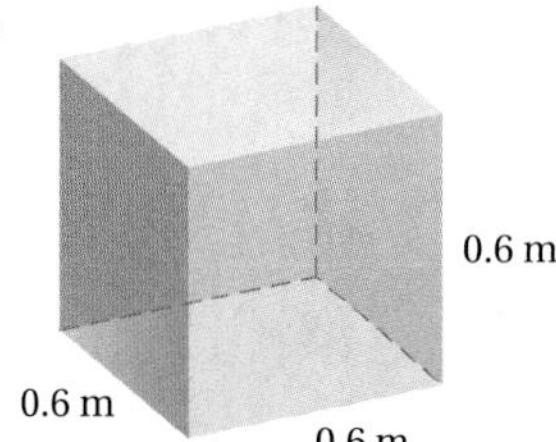

3.

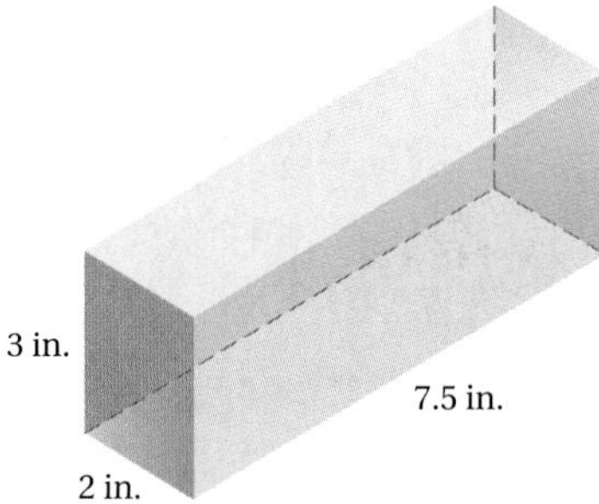

4.

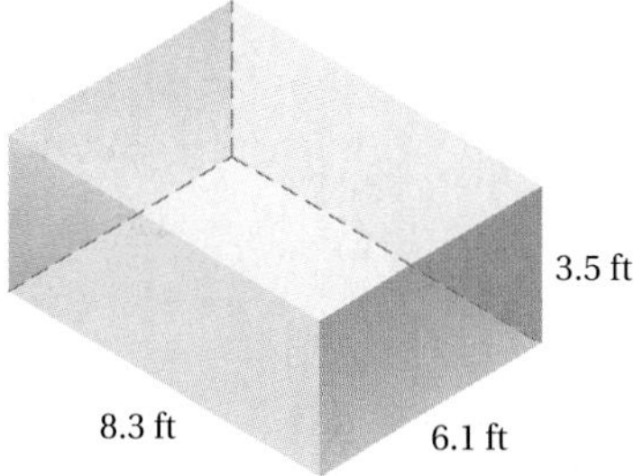

5.

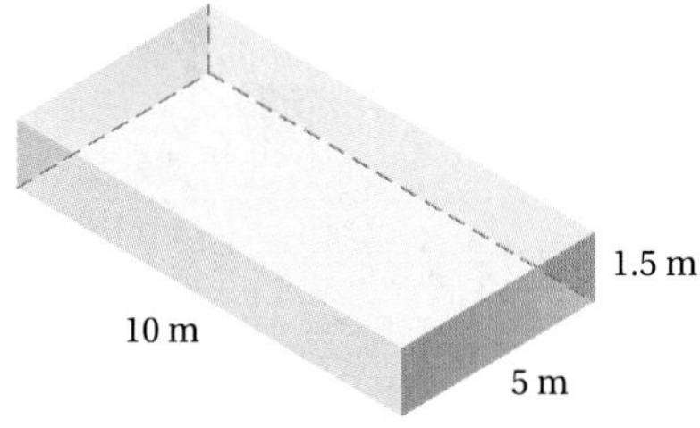

6.

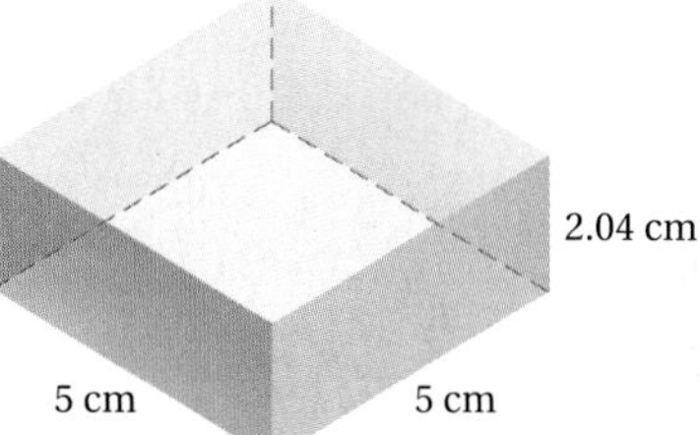

7.

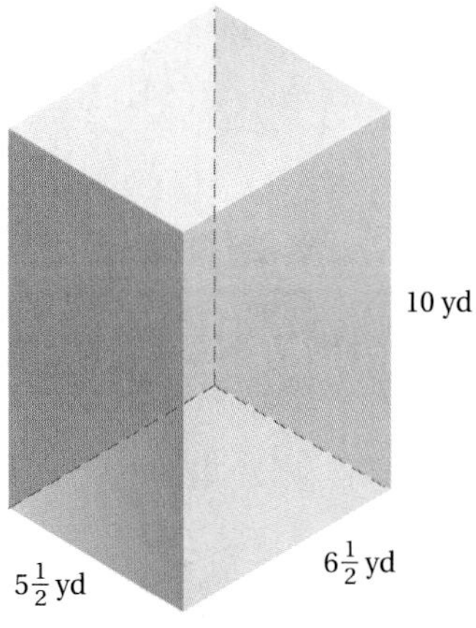

8.

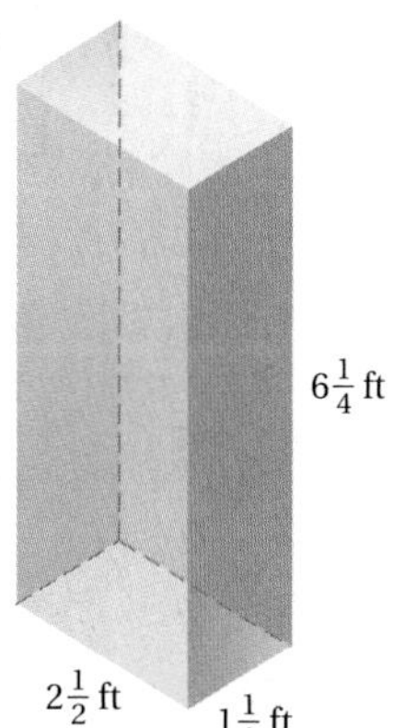

b Find the volume of the circular cylinder. Use 3.14 for π in Exercises 9–12. Use $\frac{22}{7}$ for π in Exercises 13 and 14.

9.

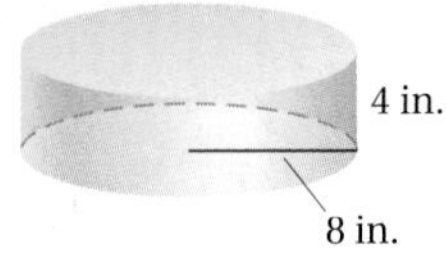

10.

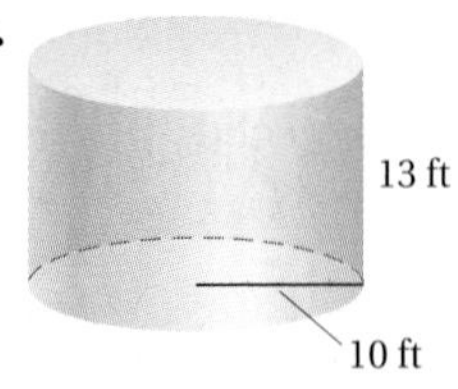

11.

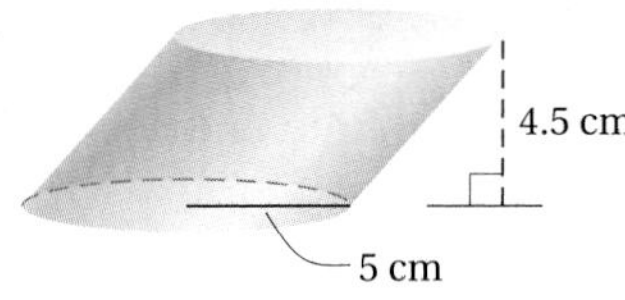

12.

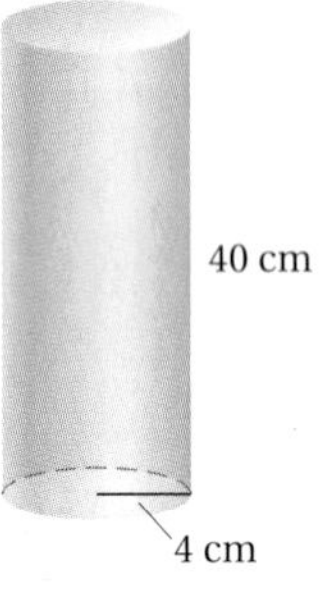

13.

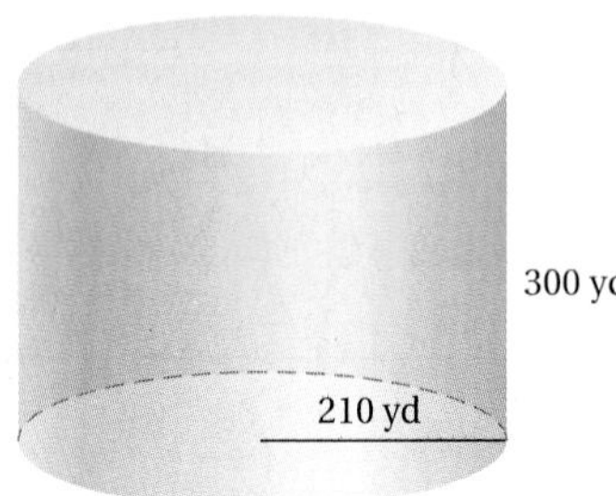

14.

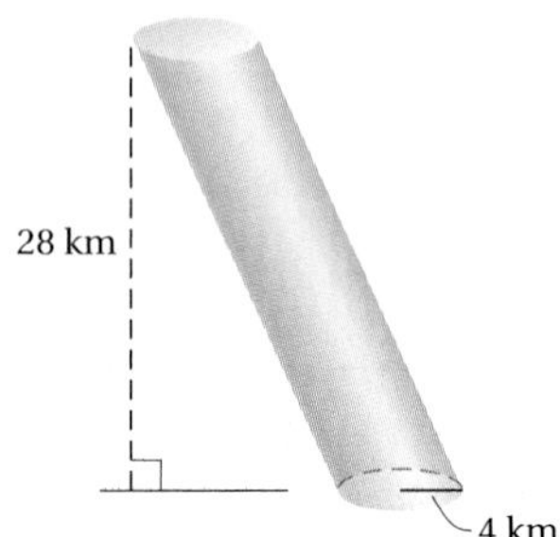

c Find the volume of the sphere. Use 3.14 for π in Exercises 15–18. Use $\frac{22}{7}$ for π in Exercises 19 and 20.

15.

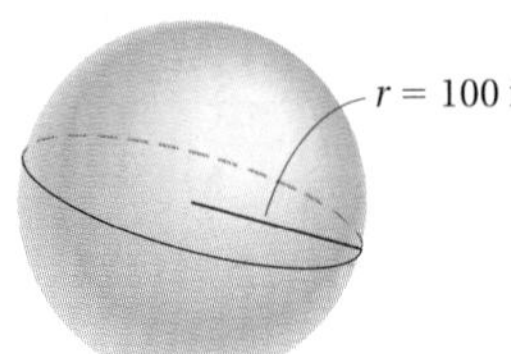

16.

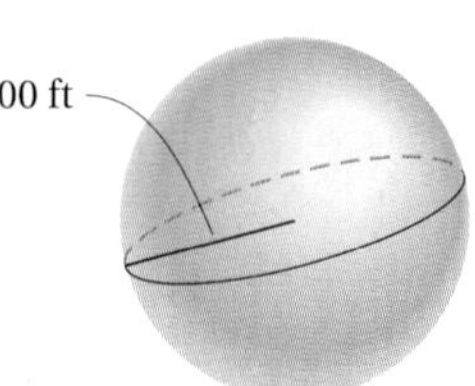

17.

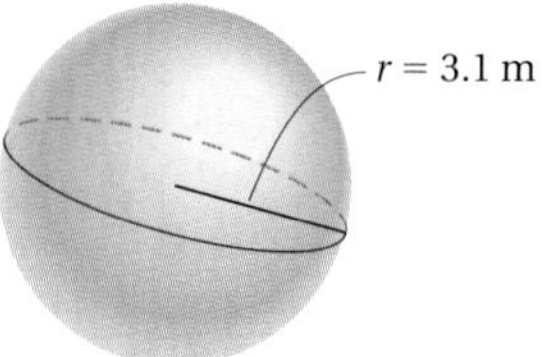

18.

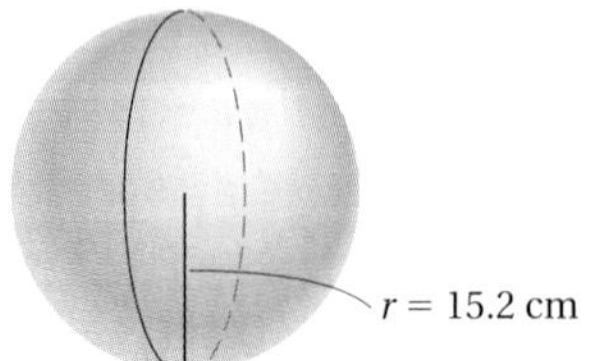

19.

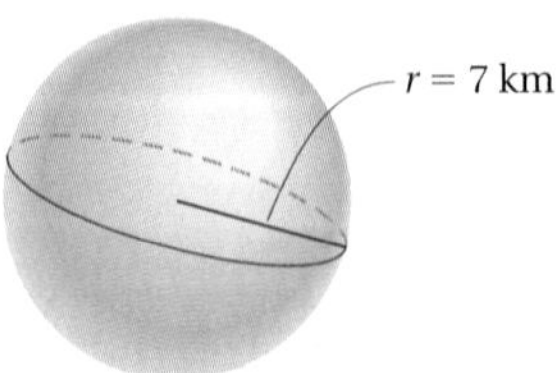

20.

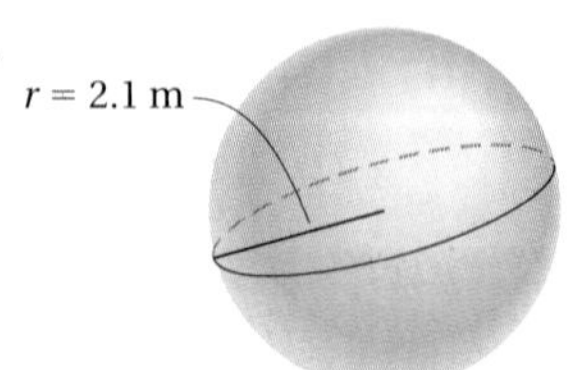

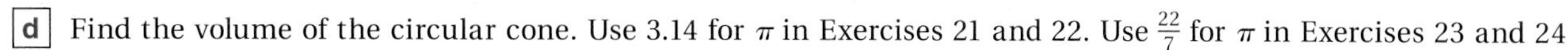

 Find the volume of the circular cone. Use 3.14 for π in Exercises 21 and 22. Use $\frac{22}{7}$ for π in Exercises 23 and 24.

21.

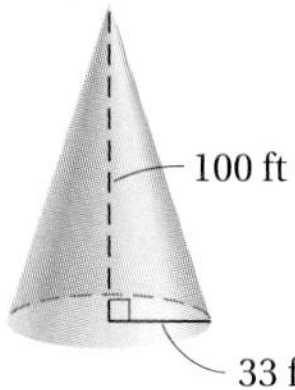

22.

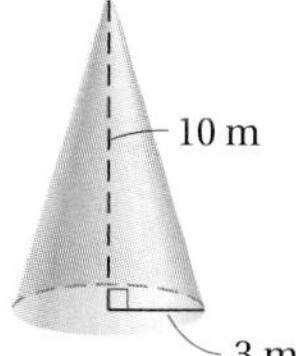

23.

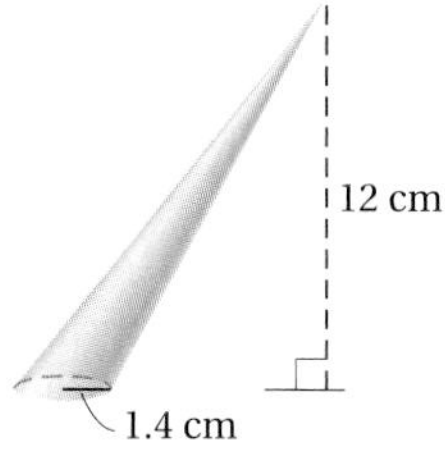

24.

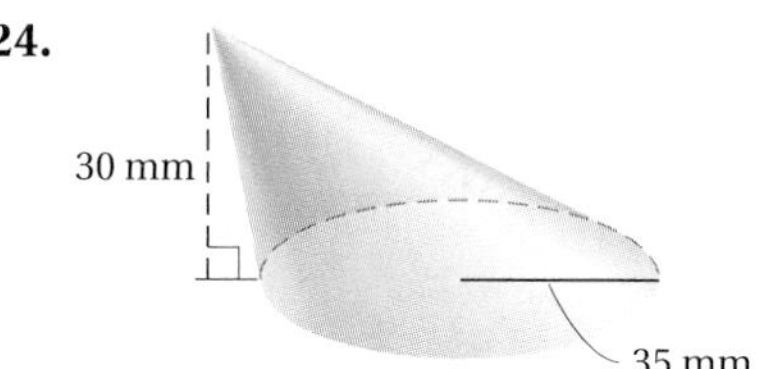

b, c Solve.

25. The diameter of the base of a circular cylinder is 14 yd. The height is 220 yd. Find the volume. Use $\frac{22}{7}$ for π.

26. A rung of a ladder is 2 in. in diameter and 16 in. long. Find the volume. Use 3.14 for π.

27. *Barn Silo.* A barn silo, excluding the top, is a circular cylinder. The silo is 6 m in diameter and the height is 13 m. Find the volume. Use 3.14 for π.

28. A log of wood has a diameter of 12 cm and a height of 42 cm. Find the volume. Use 3.14 for π.

29. *Tennis Ball.* The diameter of a tennis ball is 6.5 cm. Find the volume. Use 3.14 for π.

30. *Spherical Gas Tank.* The diameter of a spherical gas tank is 6 m. Find the volume. Use 3.14 for π.

31. *Volume of Earth.* The diameter of the earth is about 3980 mi. Find the volume of the earth. Use 3.14 for π. Round to the nearest ten thousand cubic miles.

32. The volume of a ball is 36π cm^3. Find the dimensions of a rectangular box that is just large enough to hold the ball.

Skill Maintenance

33. Find the simple interest on \$600 at 6.4% for $\frac{1}{2}$ yr. [4.7d]

34. Find the simple interest on \$600 at 8% for 2 yr. [4.7d]

Evaluate. [1.6b]

35. 10^3

36. 15^2

37. 7^2

38. 4^3

Solve.

39. *Sales Tax.* In a certain state, a sales tax of \$878 is collected on the purchase of a car for \$17,560. What is the sales tax rate? [4.7a]

40. *Commission Rate.* Rich earns \$1854.60 selling \$16,860 worth of cellular phones. What is the commission rate? [4.7b]

Synthesis

41. ◆ The design of a modern home includes a cylindrical tower that will be capped with either a 10-ft–high dome or a 10-ft–high cone. Which type of cap will be more energy-efficient and why?

42. ◆ A 2-cm–wide stream of water passes through a 30-m garden hose. At the instant that the water is turned off, how many liters of water are in the hose?

43. ▦ The width of a dollar bill is 2.3125 in., the length is 6.0625 in., and the thickness is 0.0041 in. Find the volume occupied by one million one-dollar bills.

© 1998 AL SATTERWHITE

44. ▦ Audio cassette cases are typically 7 cm by 10.75 cm by 1.5 cm and contain 74 min of music. Compact-disc cases are typically 12.4 cm by 14.1 cm by 1 cm and contain 50 min of music. Which container holds the most music per cubic centimeter?

45. ▦ A hot water tank is a right circular cylinder with a base of diameter 16 in. and height 5 ft. Find the volume of the tank in cubic feet. One cubic foot of water is about 7.5 gal. About how many gallons will the tank hold?

46. ▦ Find the volume.

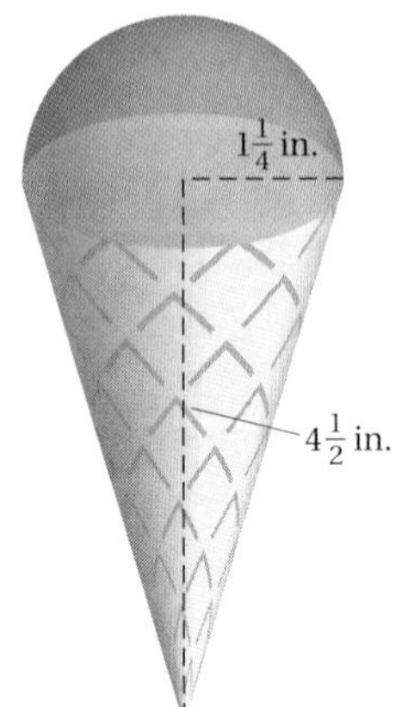

6.7 Relationships Between Angle Measures

a Complementary and Supplementary Angles

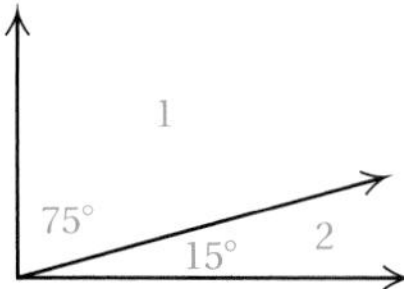

$\angle 1$ and $\angle 2$ above are **complementary** angles.

$$m\angle 1 + m\angle 2 = 90°$$
$$75° + 15° = 90°$$

> Two angles are **complementary** if and only if the sum of their measures is 90°. Each angle is called a complement of the other.

If two angles are complementary, each is an acute angle.

Example 1 Identify each pair of complementary angles.

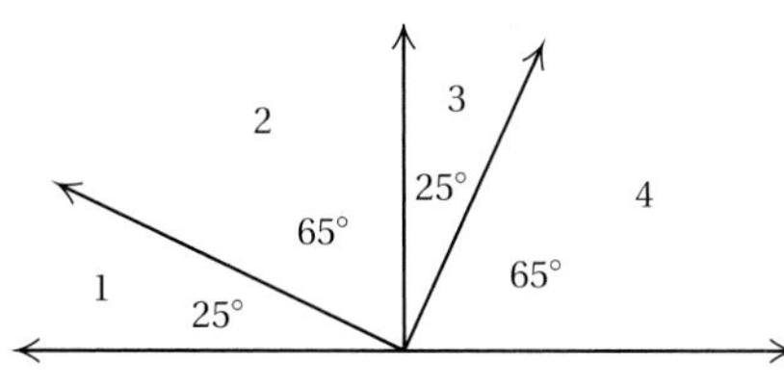

$\angle 1$ and $\angle 2$ **25° + 65° = 90°**

$\angle 1$ and $\angle 4$

$\angle 2$ and $\angle 3$

$\angle 3$ and $\angle 4$

Do Exercise 1.

Example 2 Find the measure of a complement of an angle of 39°.

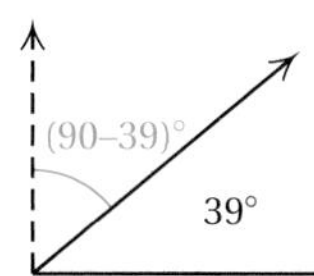

$90° - 39° = 51°$

The measure of a complement is 51°.

Do Exercises 2–4.

Objectives

a Identify complementary and supplementary angles and find the measure of a complement or a supplement of a given angle.

b Determine whether segments are congruent and whether angles are congruent.

c Use the Vertical Angle Property to find measures of angles.

d Identify pairs of corresponding angles, interior angles, and alternate interior angles and apply properties of transversals and parallel lines to find measures of angles.

For Extra Help

TAPE 13

MAC WIN

CD-ROM

1. Identify each pair of complementary angles.

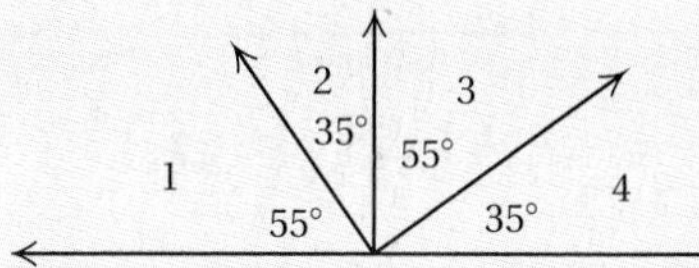

Find the measure of a complement of the angle.

2.

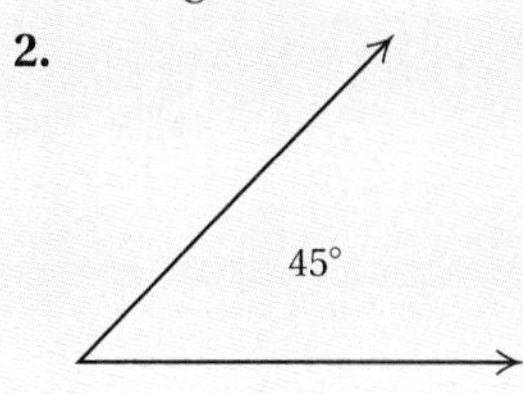

3.

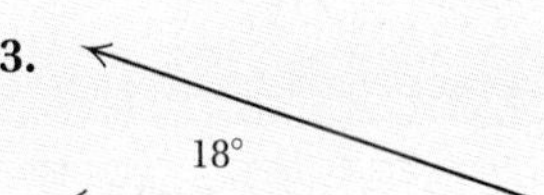

4.

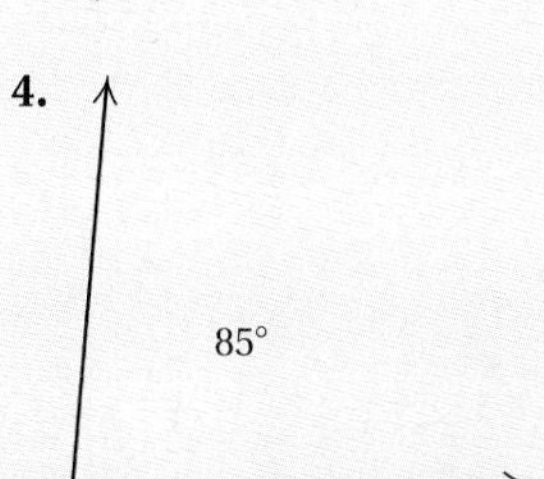

Answers on page A-15

$\angle 1$ and $\angle 2$ above are **supplementary** angles.

$$m\angle 1 + m\angle 2 = 180^\circ$$
$$30^\circ + 150^\circ = 180^\circ$$

> Two angles are **supplementary** if and only if the sum of their measures is 180°. Each angle is called a supplement of the other.

Example 3 Identify each pair of supplementary angles.

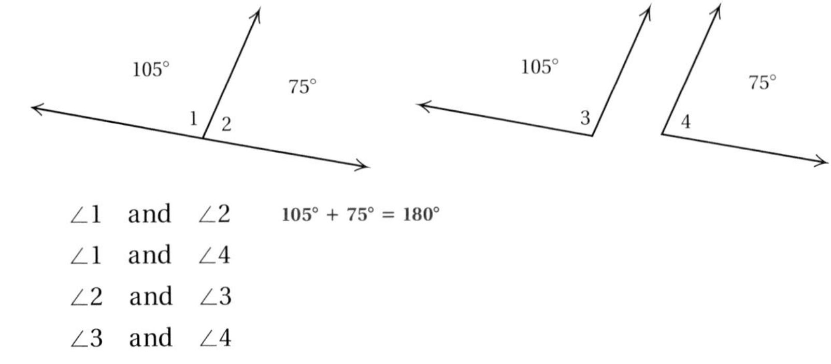

$\angle 1$ and $\angle 2$ $105^\circ + 75^\circ = 180^\circ$

$\angle 1$ and $\angle 4$

$\angle 2$ and $\angle 3$

$\angle 3$ and $\angle 4$

Do Exercise 5.

Example 4 Find the measure of a supplement of an angle of 112°.

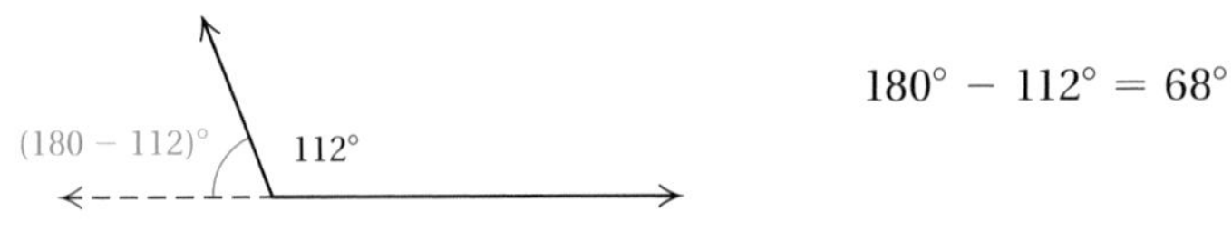

$$180^\circ - 112^\circ = 68^\circ$$

The measure of a supplement is 68°.

Do Exercises 6–8.

5. Identify each pair of supplementary angles.

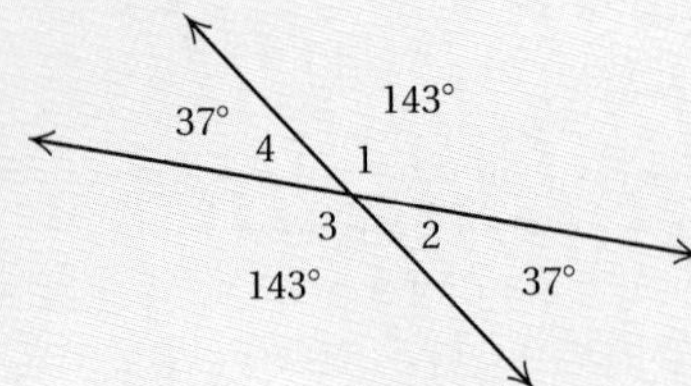

Find the measure of a supplement of an angle with the given measure.

6. 38°

7. 157°

8. 90°

Answers on page A-15

b Congruent Segments and Angles

Congruent figures have the same size and shape. They fit together exactly.

> Two segments are **congruent** if and only if they have the same length.

Example 5 Use a ruler to show that $\overline{PQ}$ and $\overline{RS}$ are congruent.

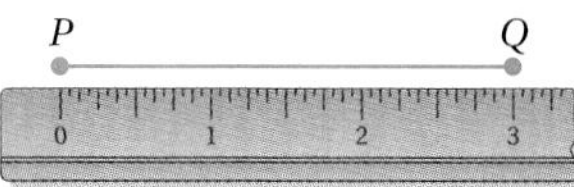

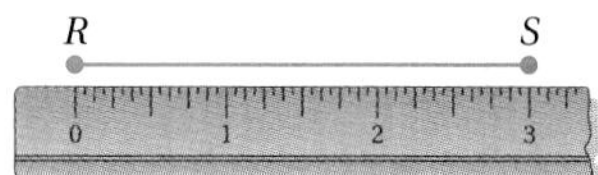

Since both segments have the same length, $\overline{PQ}$ and $\overline{RS}$ are congruent. To say that $\overline{PQ}$ and $\overline{RS}$ are congruent, we write

$$\overline{PQ} \cong \overline{RS}.$$

Example 6 Which pairs of segments are congruent? Use a ruler.

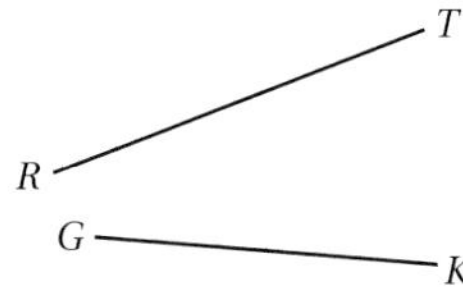

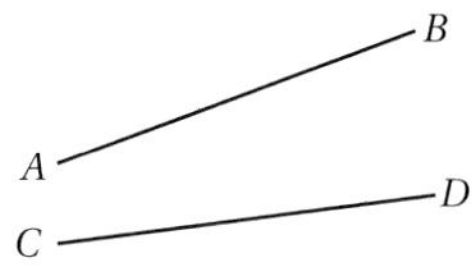

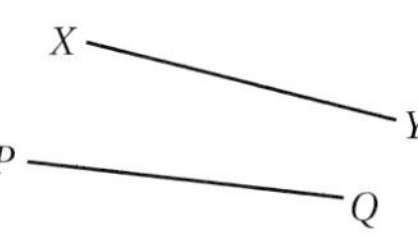

$\overline{AB} \cong \overline{CD}$ and $\overline{PQ} \cong \overline{XY}$.

Do Exercises 9 and 10.

> Two angles are **congruent** if and only if they have the same measure.

Example 7 Use a protractor to show that $\angle P$ and $\angle Q$ are congruent.

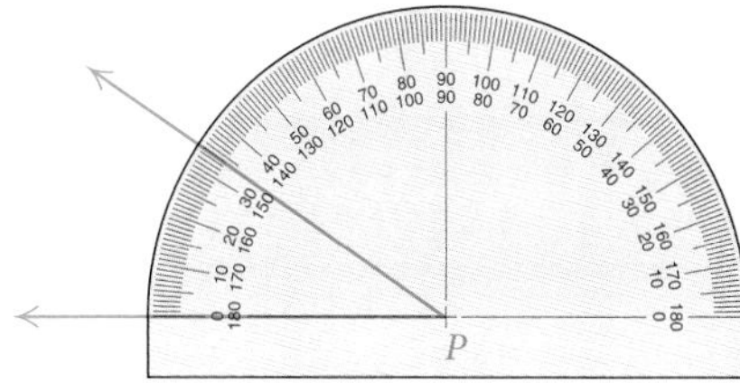

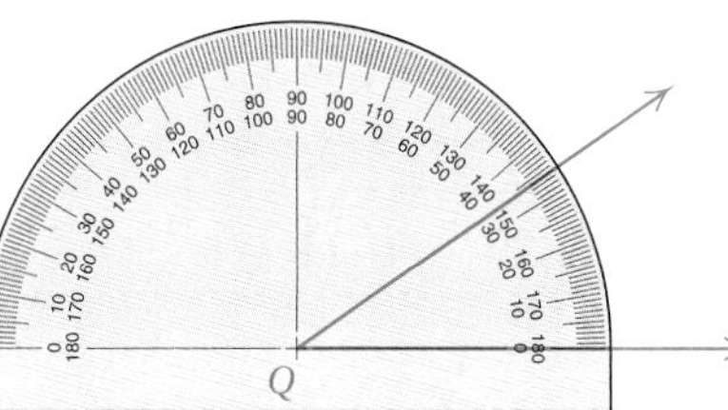

Since $m\angle P = m\angle Q = 34°$, $\angle P$ and $\angle Q$ are congruent. To say that $\angle P$ and $\angle Q$ are congruent, we write

$$\angle P \cong \angle Q.$$

Which pairs of segments are congruent? Use a ruler.

9.

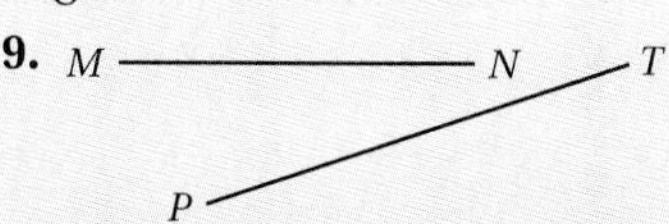

10.

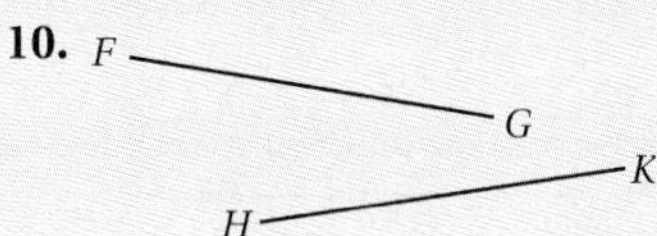

Answers on page A-15

Which pairs of angles are congruent? Use a protractor.

11.

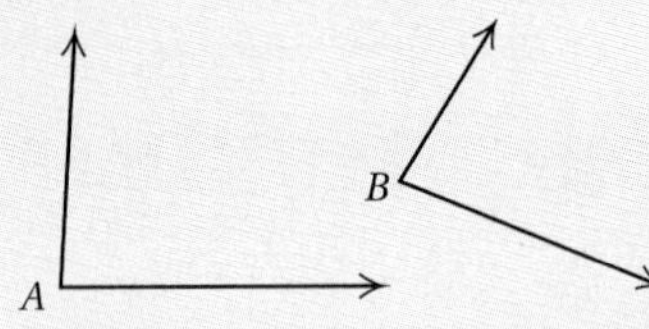

12.

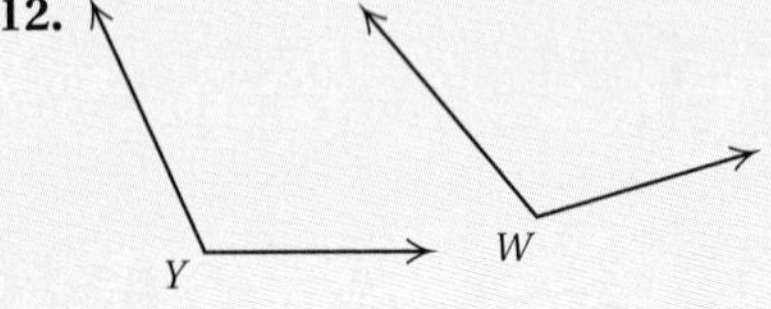

13. In the figure below, $m\angle 2 = 41°$ and $m\angle 4 = 10°$. Find $m\angle 1$, $m\angle 3$, $m\angle 5$, and $m\angle 6$.

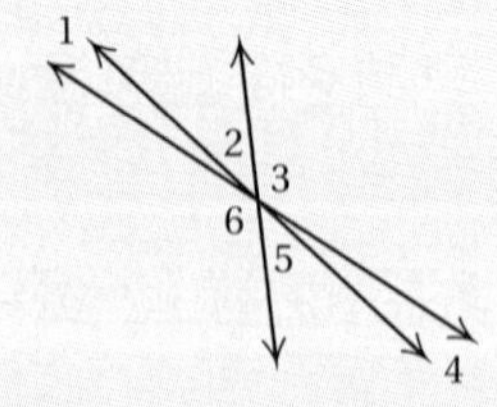

Answers on page A-15

Example 8 Which pairs of angles are congruent? Use a protractor.

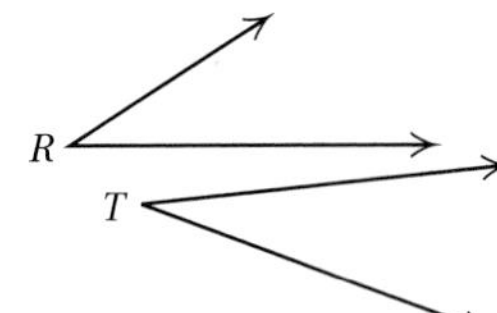

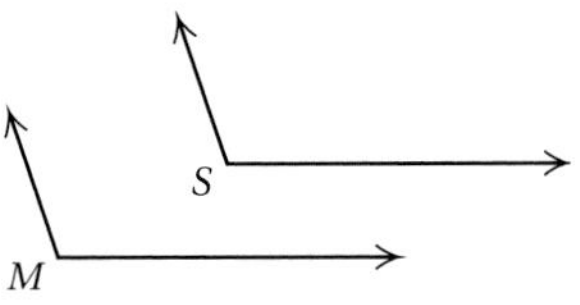

$$\angle M \cong \angle S \quad \text{since} \quad m\angle M = m\angle S = 108°.$$

Do Exercises 11 and 12.

If two angles are congruent, then their supplements are congruent and their complements are congruent.

c Vertical Angles

When $\overleftrightarrow{RT}$ intersects $\overleftrightarrow{SQ}$ at P, four angles are formed:

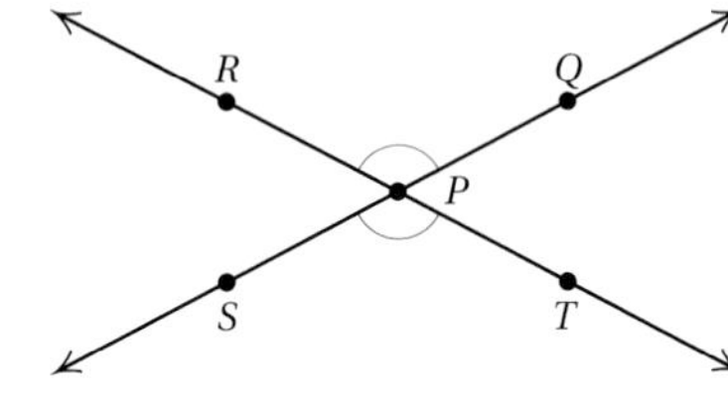

$\angle SPT$

$\angle RPQ$

$\angle SPR$

$\angle QPT$

Pairs of angles such as $\angle RPQ$ and $\angle SPT$ are called **vertical angles.**

> Two nonstraight angles are **vertical angles** if and only if their sides form two pairs of opposite rays.

Vertical angles are supplements of the same angle. Thus they are congruent.

> **THE VERTICAL ANGLE PROPERTY**
>
> Vertical angles are congruent.

Example 9 In the figure below, $m\angle 1 = 23°$ and $m\angle 3 = 34°$. Find $m\angle 2$, $m\angle 4$, $m\angle 5$, and $m\angle 6$.

Since $\angle 1$ and $\angle 4$ are vertical angles, $m\angle 4 = 23°$. Likewise, $\angle 3$ and $\angle 6$ are vertical angles, so $m\angle 6 = 34°$.

$$m\angle 1 + m\angle 2 + m\angle 3 = 180$$
$$23 + m\angle 2 + 34 = 180 \quad \text{Substituting}$$
$$m\angle 2 = 180 - 57$$
$$m\angle 2 = 123°$$

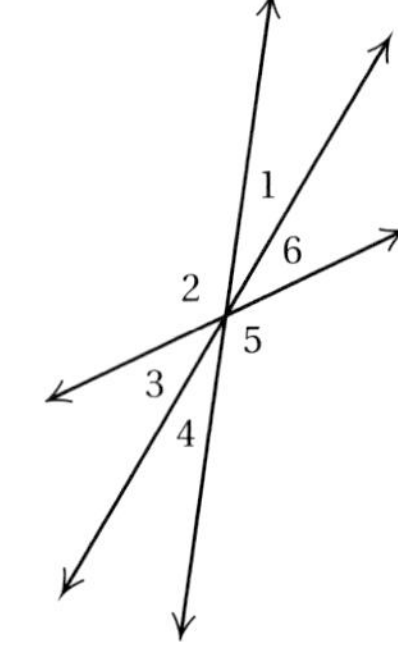

Since $\angle 2$ and $\angle 5$ are vertical angles, $m\angle 5 = 123°$.

Do Exercise 13.

d Transversals and Angles

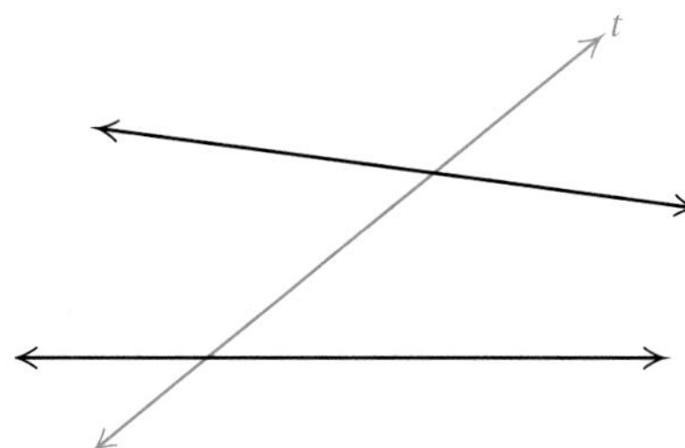

> A **transversal** is a line that intersects two or more coplanar lines in different points.

When a transversal intersects a pair of lines, eight angles are formed. Certain pairs of these angles have special names.

Corresponding Angles

∠2 and ∠6
∠3 and ∠7
∠1 and ∠5
∠4 and ∠8

Interior Angles

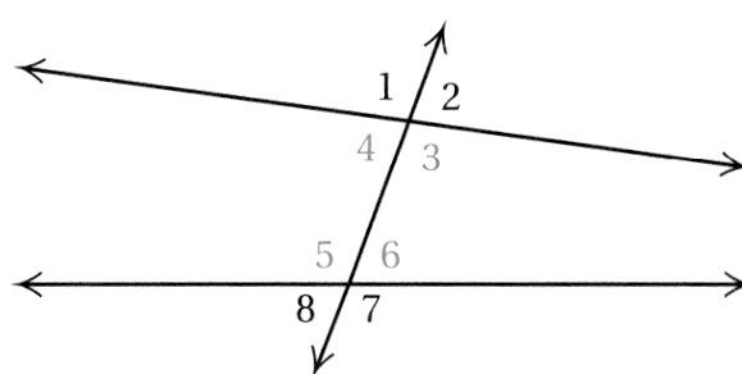

∠3, ∠4, ∠5, and ∠6

Alternate Interior Angles

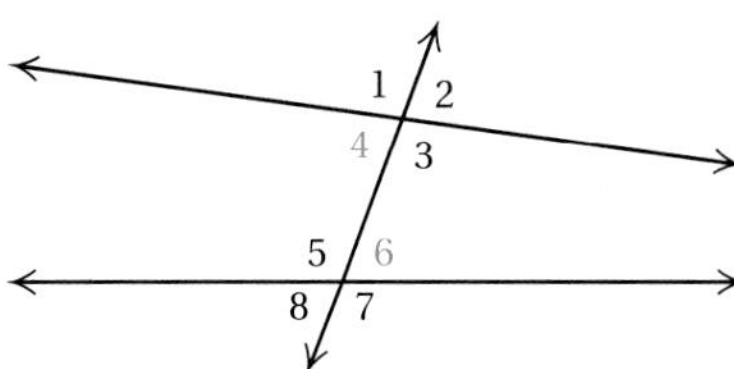

∠4 and ∠6
∠3 and ∠5

Use the following figure to answer Margin Exercises 14–16.

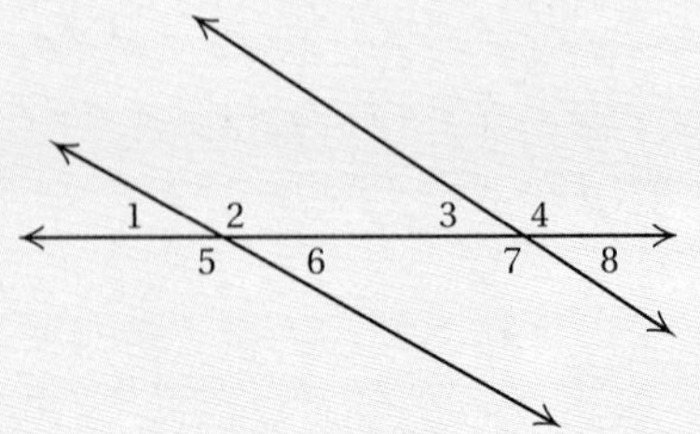

14. Identify all pairs of corresponding angles.

15. Identify all interior angles.

16. Identify all pairs of alternate interior angles.

Answers on page A-15

Example 10 Identify all pairs of corresponding angles, all interior angles, and all pairs of alternate interior angles.

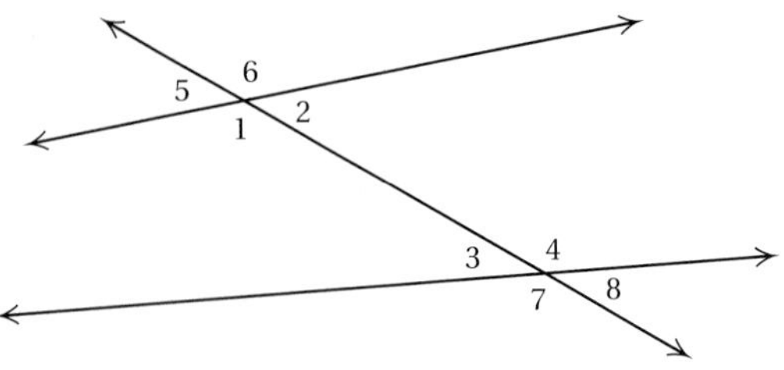

Corresponding angles:	$\angle 6$ and $\angle 4$, $\angle 2$ and $\angle 8$, $\angle 5$ and $\angle 3$, $\angle 1$ and $\angle 7$
Interior angles:	$\angle 1$, $\angle 2$, $\angle 3$, $\angle 4$
Alternate interior angles:	$\angle 1$ and $\angle 4$, $\angle 2$ and $\angle 3$

Do Exercises 14–16.

Given a line l and a point P not on l, there is at most one line that contains P and is parallel to l.

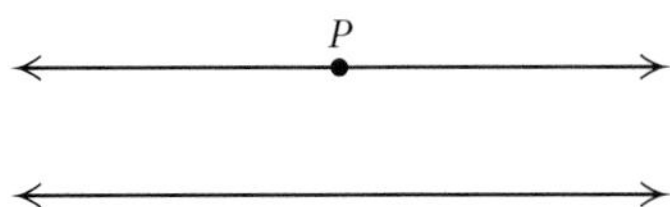

If two lines are parallel, the following relations hold.

Properties of Parallel Lines

1. If a transversal intersects two parallel lines, then the corresponding angles are congruent.

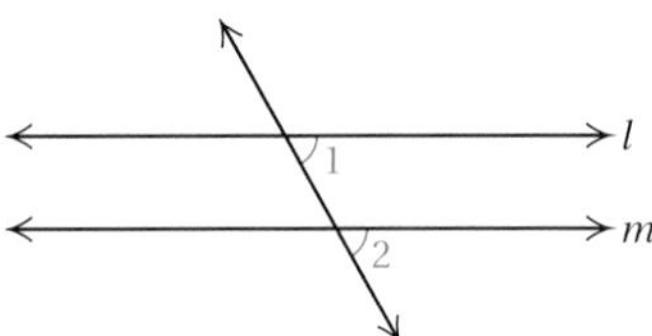

If $l \parallel m$, then $\angle 1 \cong \angle 2$.

2. If a transversal intersects two parallel lines, then the alternate interior angles are congruent.

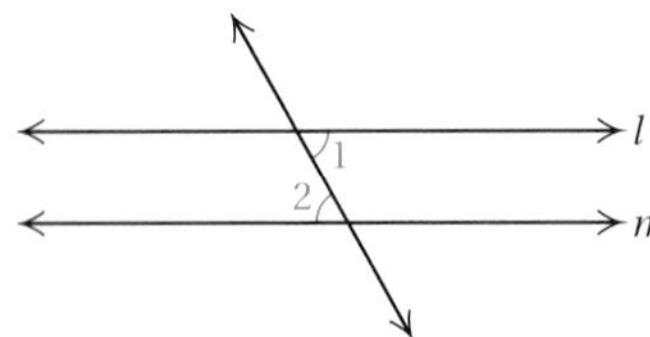

If $l \parallel m$, then $\angle 1 \cong \angle 2$.

3. In a plane, if two lines are parallel to a third line, then the two lines are parallel to each other.

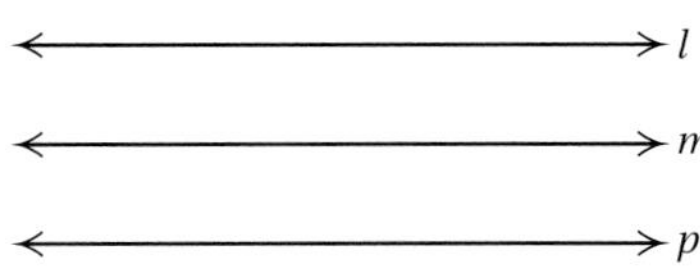

If $l \parallel p$ and $m \parallel p$, then $l \parallel m$.

4. If a transversal intersects two parallel lines, then the interior angles on the same side of the transversal are supplementary.

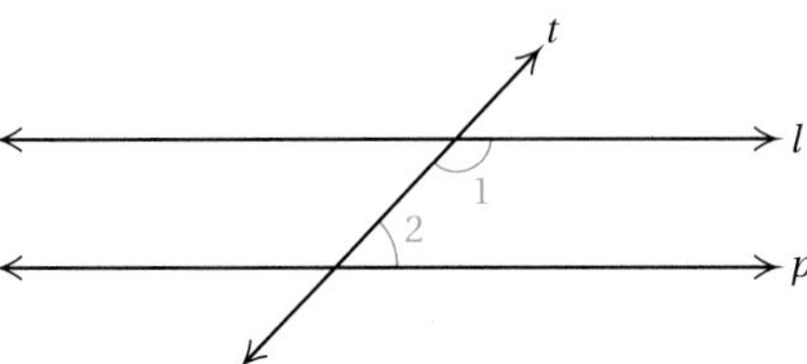

If $l \parallel p$, then $m\angle 1 + m\angle 2 = 180°$.

5. If a transversal is perpendicular to one of two parallel lines, then it is perpendicular to the other.

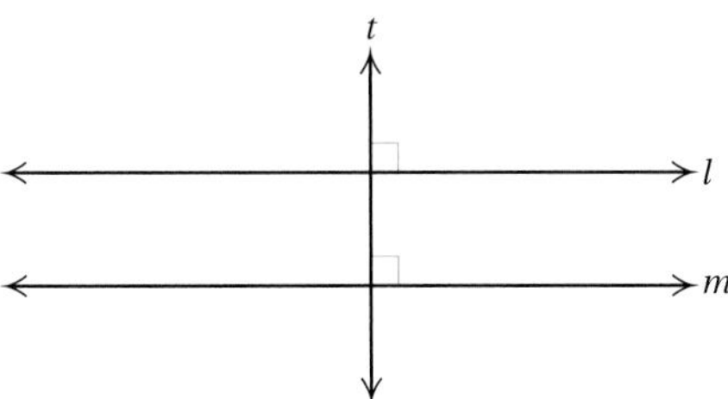

Example 11 If $l \parallel m$ and $m\angle 1 = 40°$, what are the measures of the other angles?

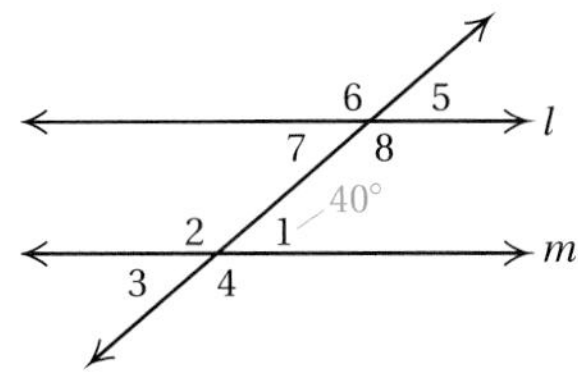

$m\angle 7 = 40°$	**Using Property 2**
$m\angle 5 = 40°$	**Using Property 1**
$m\angle 8 = 140°$	**Using Property 4**
$m\angle 3 = 40°$	**$\angle 1$ and $\angle 3$ are vertical angles**
$m\angle 4 = 140°$	**Using Property 1 and $m\angle 8 = 140°$**
$m\angle 2 = 140°$	**$\angle 2$ and $\angle 4$ are vertical angles and $m\angle 4 = 140°$**
$m\angle 6 = 140°$	**$\angle 6$ and $\angle 8$ are vertical angles and $m\angle 8 = 140°$**

Do Exercise 17.

17. If $l \parallel m$ and $m\angle 3 = 51°$, what are the measures of the other angles?

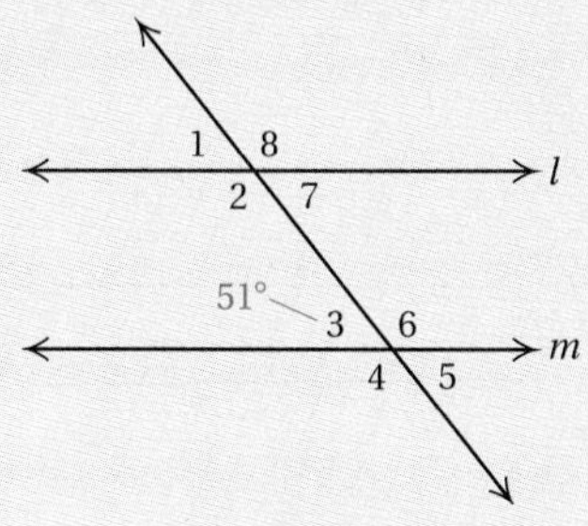

18. If $\overline{AB} \parallel \overline{CD}$, which pairs of angles are congruent?

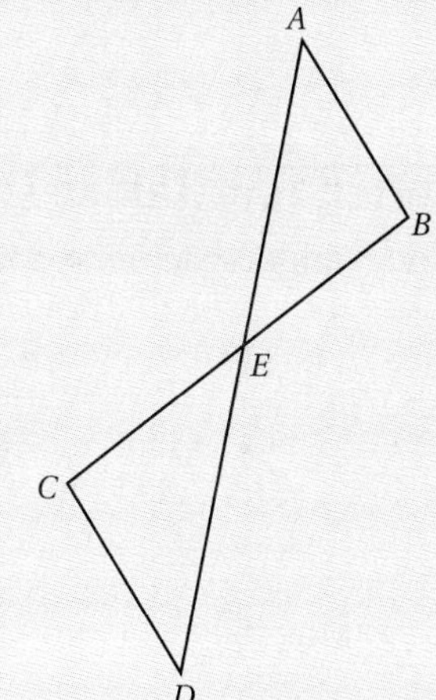

Answer on page A-15

19. If $\overline{PQ} \parallel \overline{RS}$, which pairs of angles are congruent?

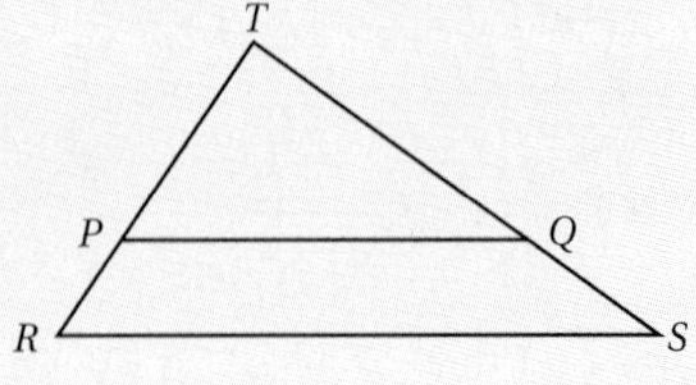

Answer on page A-15

Example 12 If $\overline{PT} \parallel \overline{SR}$, which pairs of angles are congruent?

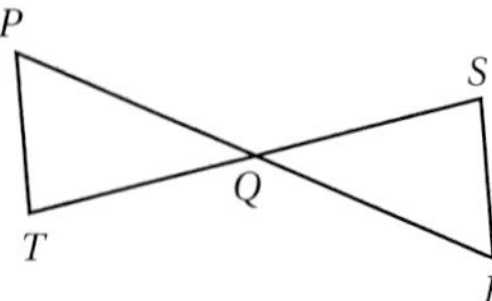

$\angle TPQ \cong \angle SRQ$ and $\angle PTQ \cong \angle RSQ$ Using Property 2
$\angle PQT \cong \angle RQS$ and $\angle PQS \cong \angle RQT$ Vertical angles

Do Exercise 18 on the preceding page.

Example 13 If $\overline{DE} \parallel \overline{BC}$, which pairs of angles are congruent?

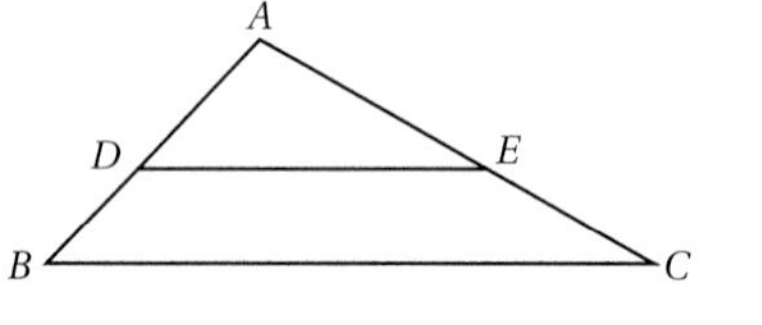

$\angle ADE \cong \angle ABC$ and $\angle AED \cong \angle ACB$ Using Property 1

Do Exercise 19.

Improving Your Math Study Skills

Better Test Taking

How often do you make the following statement after taking a test: "I was able to do the homework, but I froze during the test"? Instructors have heard this comment for years, and in most cases, it is merely a coverup for a lack of proper study habits. Here are two related tips, however, to help you with this difficulty. Both are intended to make test taking less stressful by getting you to practice good test-taking habits on a daily basis.

- **Treat *every* homework exercise as if it were a test question.** If you had to work a problem at your job with no backup answer provided, what would you do? You would probably work it very deliberately, checking and rechecking every step. You might work it more than one time, or you might try to work it another way to check the result. Try to use this approach when doing your homework. Treat every exercise as though it were a test question and no answer were provided at the back of the book.
- **Be sure that you do questions without answers as part of every homework assignment whether or not the instructor has assigned them!** One reason a test may seem such a different task is that questions on a test lack answers. That is the reason for taking a test: to see if you can do the questions without assistance. As part of your test preparation, be sure you do some exercises for which you do not have the answers. Thus when you take a test, you are doing a more familiar task.

The purpose of doing your homework using these approaches is to give you more test-taking practice beforehand. Let's make a sports analogy here. At a basketball game, the players take lots of practice shots before the game. They play the first half, go to the locker room, and come out for the second half. What do they do before the second half, even though they have just played 20 minutes of basketball? They shoot baskets again! We suggest the same approach here. Create more and more situations in which you practice taking test questions by treating each homework exercise like a test question and by doing exercises for which you have no answers. Good luck! Please send me an e-mail (exponent@aol.com) and let me know how it works for you.

Exercise Set 6.7

a Find the measure of a complement of an angle with the given measure.

1. 11°

2. 83°

3. 67°

4. 5°

Find the measure of a supplement of an angle with the given measure.

5. 3°

6. 54°

7. 139°

8. 13°

b Determine if the pair of segments is congruent. Use a ruler.

9.

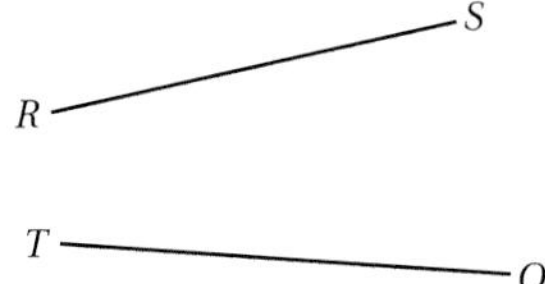

10.

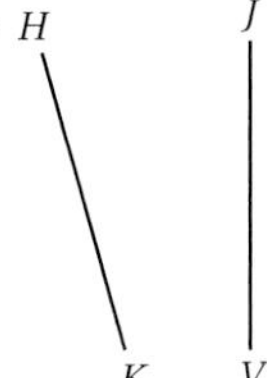

Determine if the pair of angles is congruent. Use a protractor.

11.

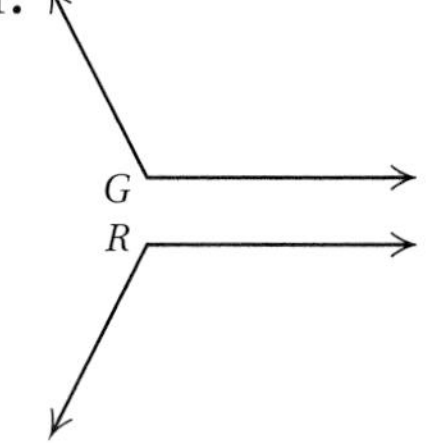

12.

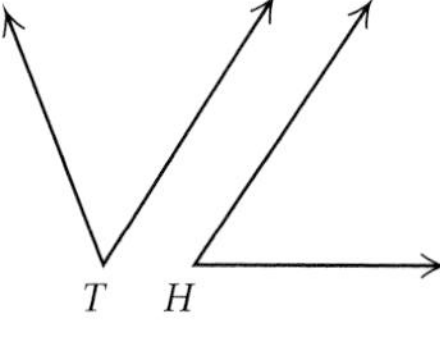

c

13. In the figure, $m\angle 1 = 80°$ and $m\angle 5 = 67°$. Find $m\angle 2$, $m\angle 3$, $m\angle 4$, and $m\angle 6$.

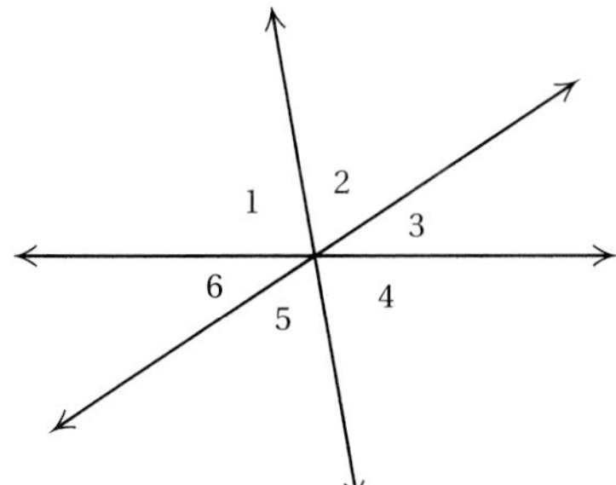

14. In the figure, $m\angle 2 = 42°$ and $m\angle 4 = 56°$. Find $m\angle 1$, $m\angle 3$, $m\angle 5$, and $m\angle 6$.

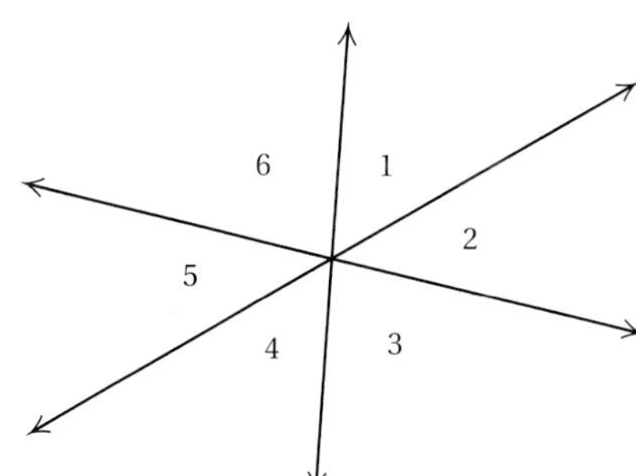

d In Exercises 15 and 16, (a) identify all pairs of corresponding angles, (b) identify all interior angles, and (c) identify all pairs of alternate interior angles.

15.

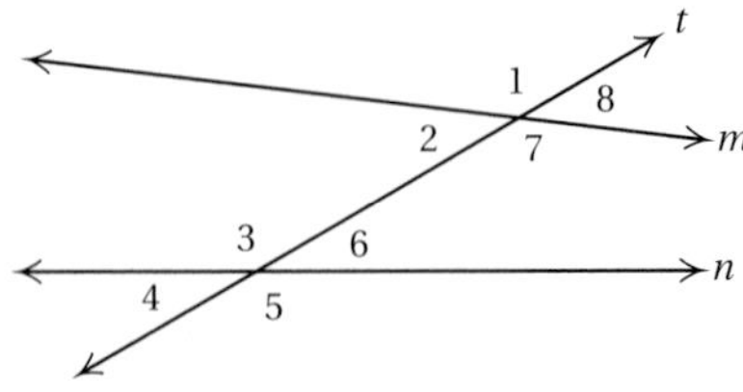

Lines m and n
Transversal t

16.

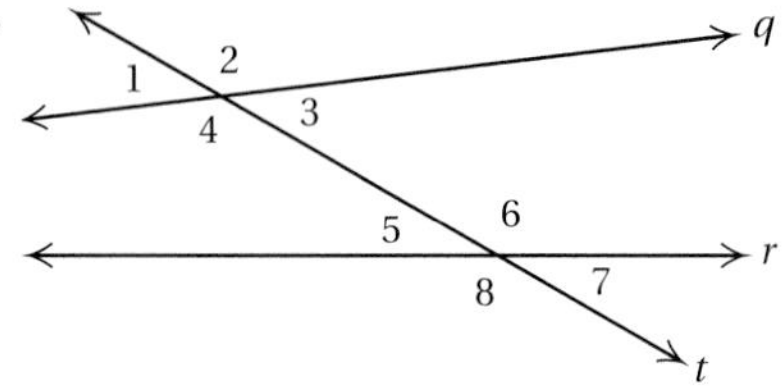

Lines q and r
Transversal t

17. If $m \parallel n$ and $m\angle 4 = 125°$, what are the measures of the other angles?

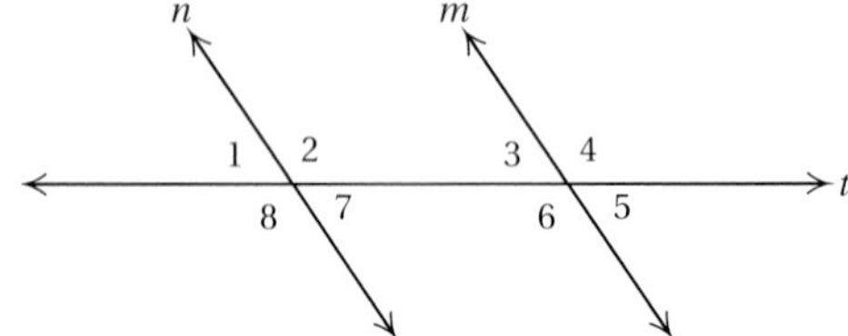

18. If $m \parallel n$ and $m\angle 8 = 34°$, what are the measures of the other angles?

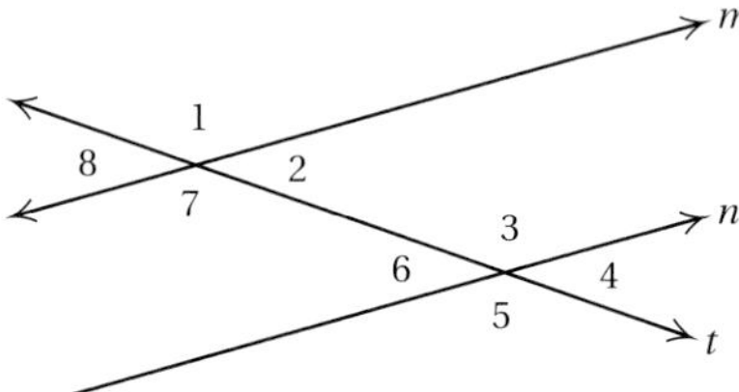

In each figure, $\overline{AB} \parallel \overline{CD}$. Identify pairs of congruent angles. When possible, give the measures of the angles.

19.

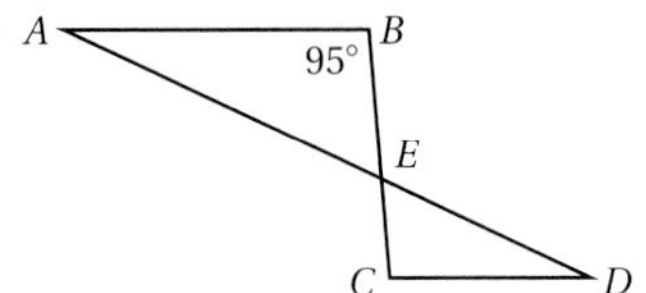

20.

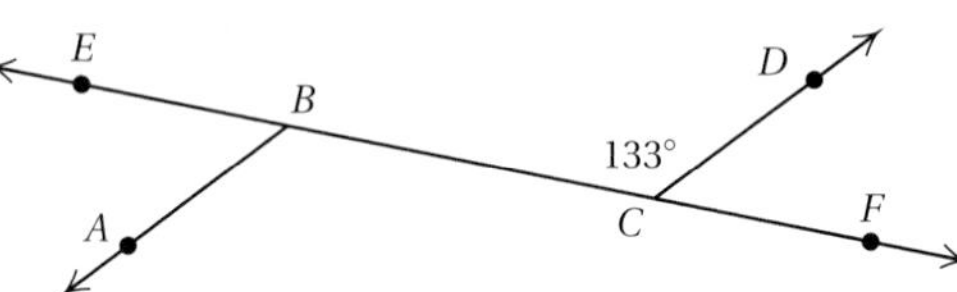

21.

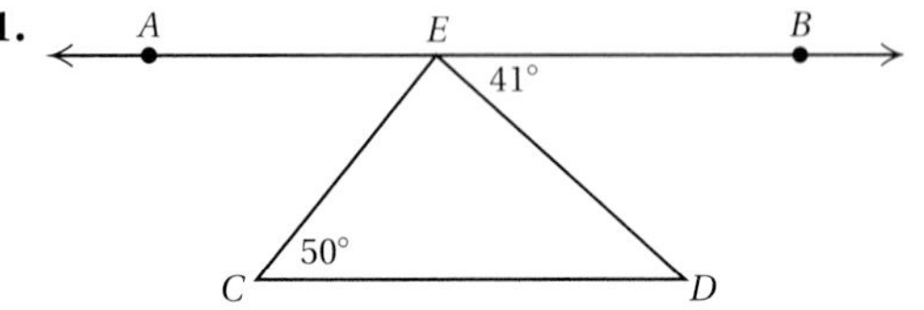

22.

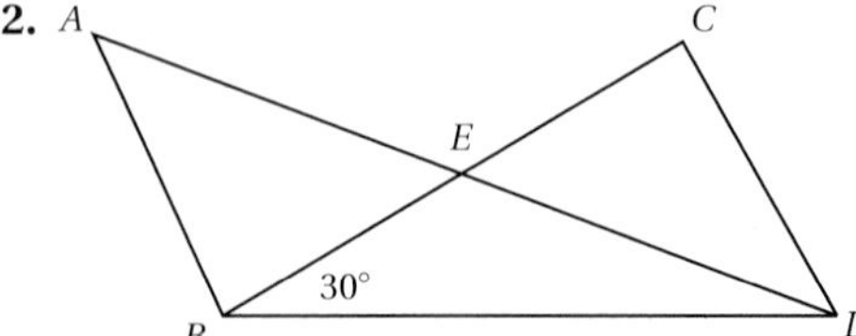

6.8 Congruent Triangles and Properties of Parallelograms

Objectives

a. Identify the corresponding parts of congruent triangles and show why triangles are congruent using SAS, SSS, and ASA.

b. Use properties of parallelograms to find lengths of sides and measures of angles of parallelograms.

For Extra Help

TAPE 13 | MAC WIN | CD-ROM

a Congruent Triangles

Triangles can be classified by their angles.

Acute: All angles acute
Right: One right angle
Obtuse: One obtuse angle
Equiangular: All angles congruent

Triangles can also be classified by their sides.

Equilateral: All sides congruent
Isosceles: At least two sides congruent
Scalene: No sides congruent

We know that congruent figures fit together exactly.

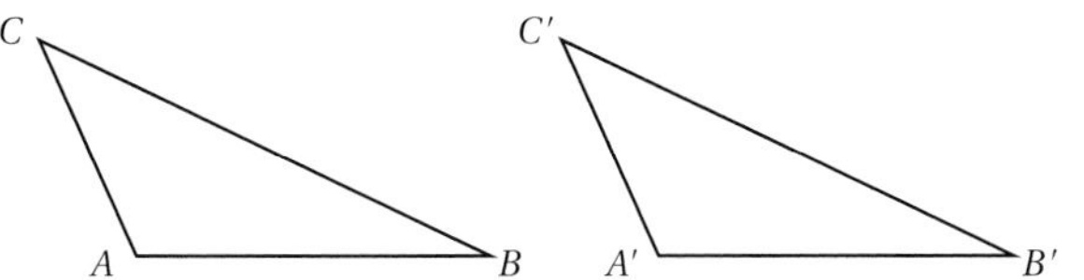

B' is read "B prime."

These triangles will fit together exactly if we match A with A', B with B', and C with C'. On the other hand, if we match A with B', B with C', and C with A', the triangles will not fit together exactly. The matching of vertices determines corresponding sides and angles.

Examples Consider $\triangle ABC$ and $\triangle A'B'C'$ above.

1. If we match A with A', B with B', and C with C', what are the corresponding sides?

$\overline{AB} \leftrightarrow \overline{A'B'}$
$\overline{BC} \leftrightarrow \overline{B'C'}$
$\overline{AC} \leftrightarrow \overline{A'C'}$

$\leftrightarrow$ means "corresponds to."

2. If we match A with B', B with C', and C with A', what are the corresponding angles?

$\angle A \leftrightarrow \angle B' \qquad \angle B \leftrightarrow \angle C' \qquad \angle C \leftrightarrow \angle A'$

If $A \leftrightarrow A'$, $B \leftrightarrow B'$, and $C \leftrightarrow C'$, then we write $ABC \leftrightarrow A'B'C'$.

> Two triangles are **congruent** if and only if their vertices can be matched so that the corresponding angles and sides are congruent.

The corresponding sides and angles of two congruent triangles are called *corresponding parts* of congruent triangles. Corresponding parts of congruent triangles are always congruent.

1. Suppose that $\triangle ABC \cong \triangle DEF$. What are the congruent corresponding parts?

2. Name the corresponding parts of these congruent triangles.

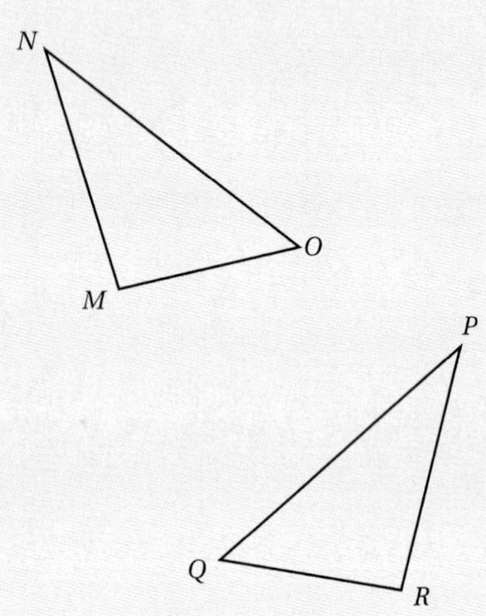

Answers on page A-16

We write $\triangle ABC \cong \triangle A'B'C'$ to say that $\triangle ABC$ and $\triangle A'B'C'$ are congruent. We agree that this symbol also tells us the way in which the vertices are matched.

$$\triangle ABC \cong \triangle A'B'C'$$

$\triangle ABC \cong \triangle A'B'C'$ means that

$$\angle A \cong \angle A' \quad \text{and} \quad \overline{AB} \cong \overline{A'B'}$$
$$\angle B \cong \angle B' \qquad \overline{AC} \cong \overline{A'C'}$$
$$\angle C \cong \angle C' \qquad \overline{BC} \cong \overline{B'C'}.$$

Example 3 Suppose that $\triangle PQR \cong \triangle STV$. What are the congruent corresponding parts?

Angles	*Sides*
$\angle P \cong \angle S$	$\overline{PQ} \cong \overline{ST}$
$\angle Q \cong \angle T$	$\overline{PR} \cong \overline{SV}$
$\angle R \cong \angle V$	$\overline{QR} \cong \overline{TV}$

Do Exercise 1.

Example 4 Name the corresponding parts of these congruent triangles.

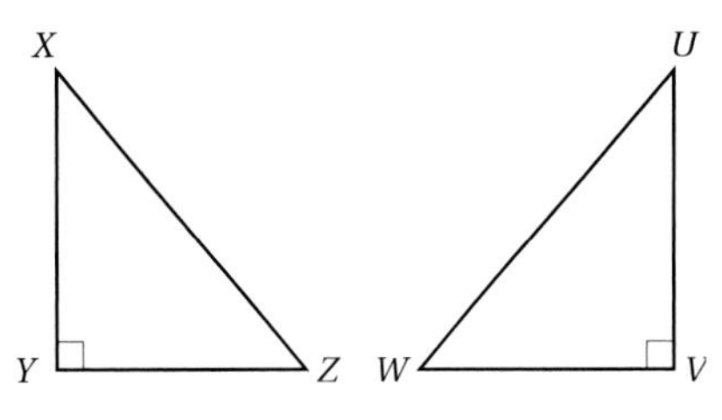

Angles	*Sides*
$\angle X \cong \angle U$	$\overline{XY} \cong \overline{UV}$
$\angle Y \cong \angle V$	$\overline{YZ} \cong \overline{VW}$
$\angle Z \cong \angle W$	$\overline{ZX} \cong \overline{WU}$

Do Exercise 2.

Sometimes we can show that triangles are congruent without already knowing that all six corresponding parts are congruent.

On a full sheet of paper, draw $\triangle ABC$. On another sheet of paper, make a copy of $\angle A$. Label the copy $\angle D$. On the sides of $\angle D$, copy $\overline{AB}$ and $\overline{AC}$. Label the copy $\overline{DE}$ and $\overline{DF}$. Draw $\overline{EF}$. Cut out $\triangle DEF$ and $\triangle ABC$ and place them together. What do you conclude?

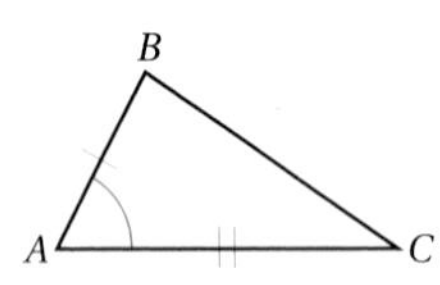

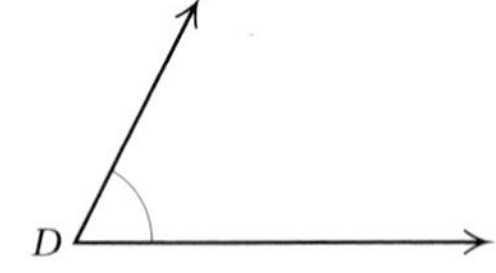

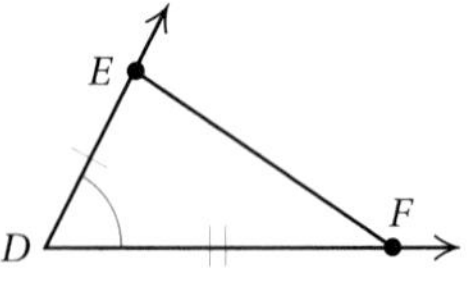

THE SIDE–ANGLE–SIDE (SAS) PROPERTY

Two triangles are congruent if two sides and the included angle of one triangle are congruent to two sides and the included angle of the other triangle.

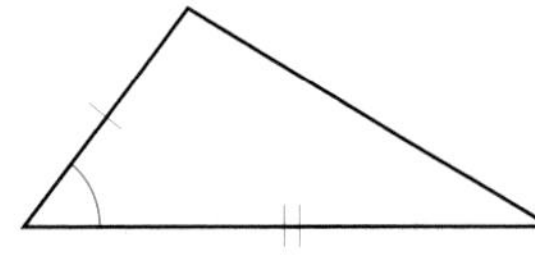

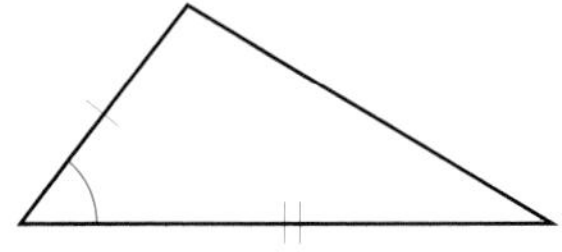

Example 5 Which pairs of triangles are congruent by the SAS property?

a)

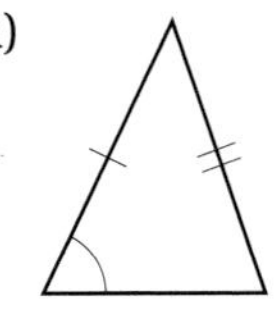

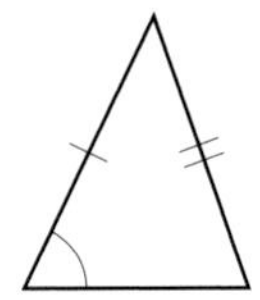

b)

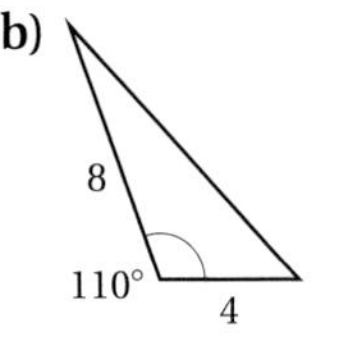

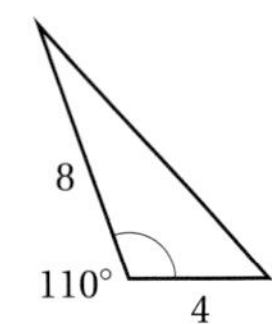

c)

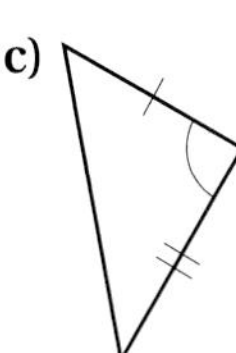

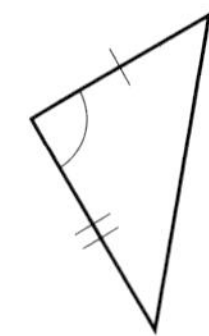

d)

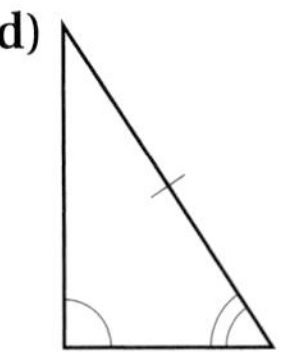

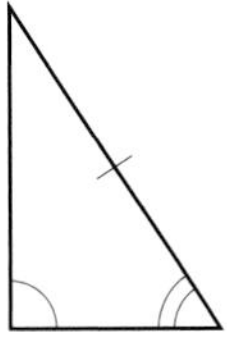

Pairs (b) and (c) are congruent by the SAS property.

Do Exercise 3.

On a sheet of paper, draw a triangle. Then copy this triangle by copying each of its sides. Cut both triangles out and place them together. This suggests the following property.

> **THE SIDE–SIDE–SIDE (SSS) PROPERTY**
>
> If three sides of one triangle are congruent to three sides of another triangle, then the triangles are congruent.

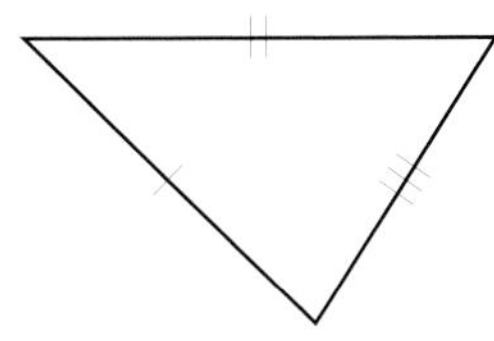

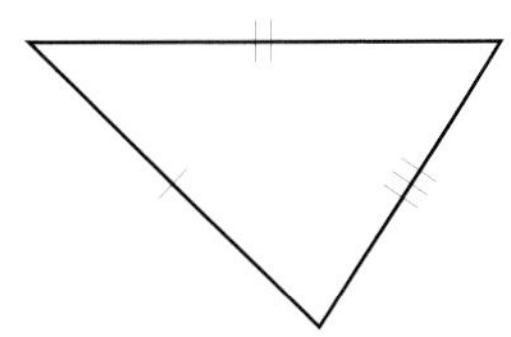

Example 6 Which pairs of triangles are congruent by the SSS property?

a)

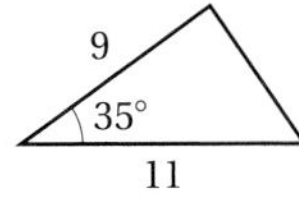

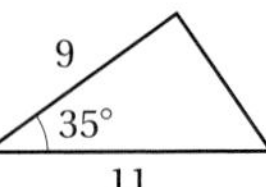

b)

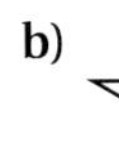

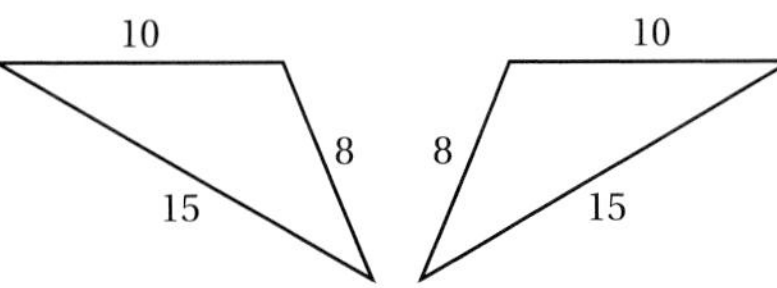

c)

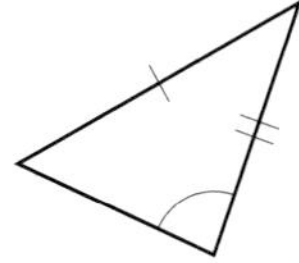

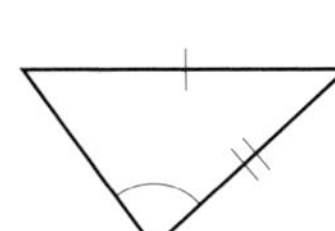

d)

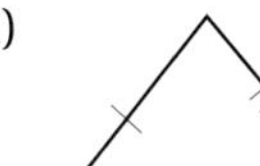

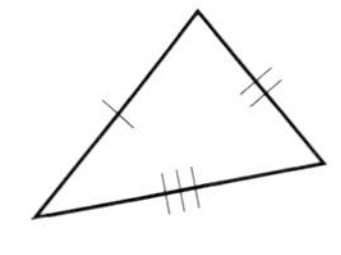

Pairs (b) and (d) are congruent by the SSS property.

Do Exercise 4.

We have shown triangles to be congruent using SAS and SSS. A third way to show congruence is shown on the following page.

3. Which pairs of triangles are congruent by the SAS property?

a)

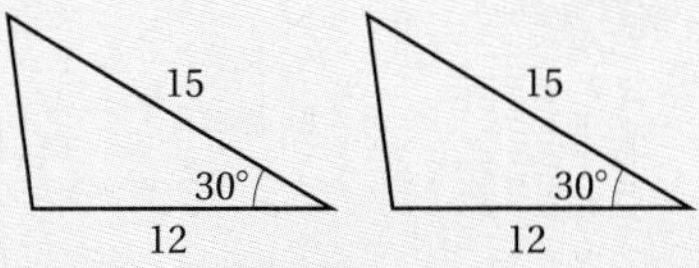

b)

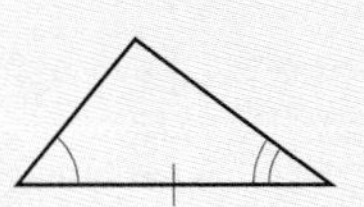

c)

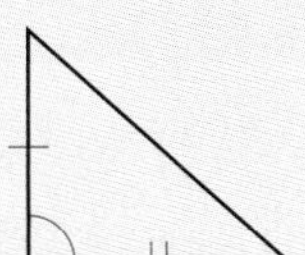

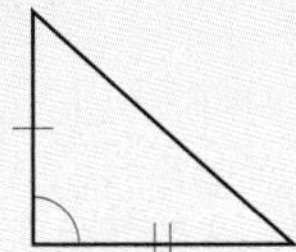

d)

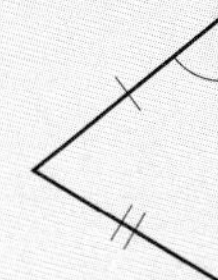

4. Which pairs of triangles are congruent by the SSS property?

a)

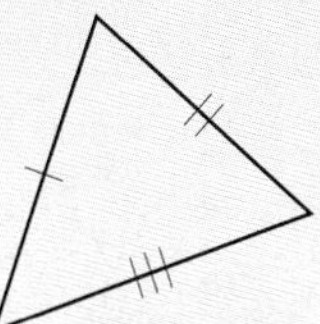

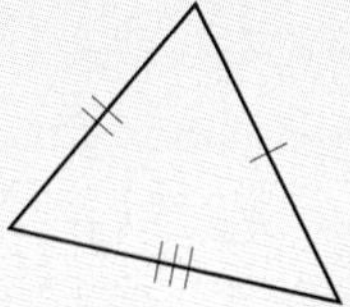

b)

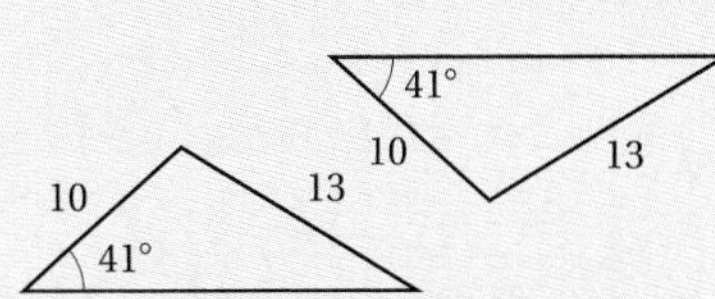

Answers on page A-16

5. Which pairs of triangles are congruent by the ASA property?

a)

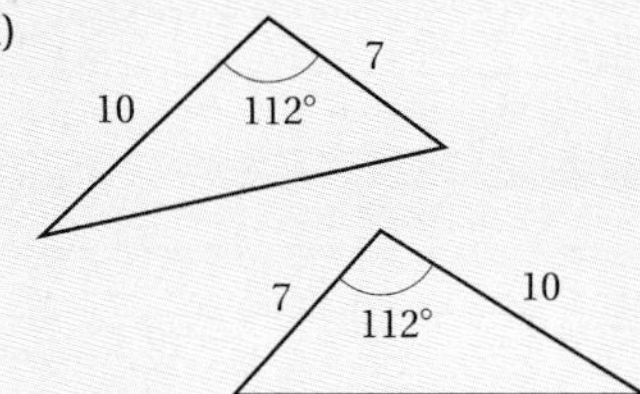

b)

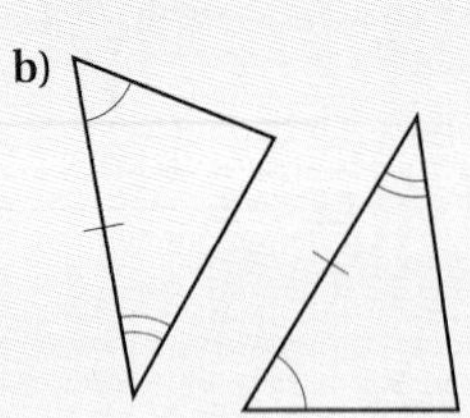

c)

Which property (if any) should be used to show that these pairs of triangles are congruent?

6.

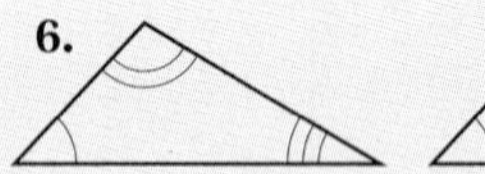

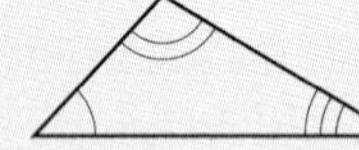

7.

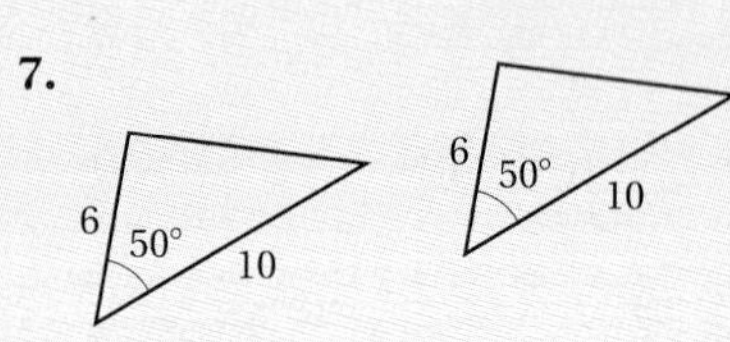

8.

9.

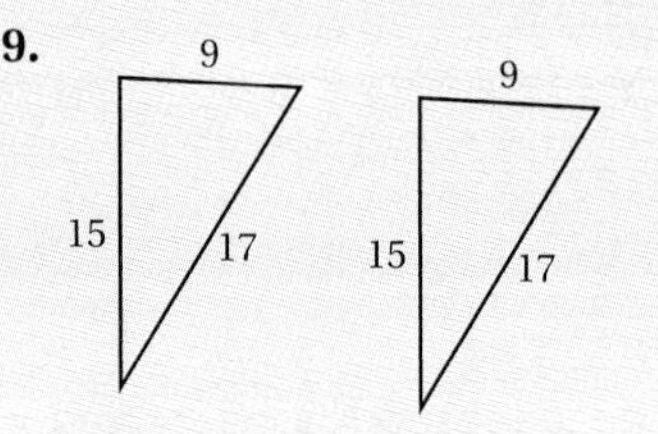

Answers on page A-16

On a full sheet of paper, draw a triangle, $\triangle ABC$. On another sheet of paper, draw a segment $\overline{DE}$ so that $DE = AB$*. At D, make a copy of $\angle A$. At E, make a copy of $\angle B$. Label the third vertex of the copy F. Cut out $\triangle ABC$ and $\triangle DEF$ and place them together. What do you conclude?

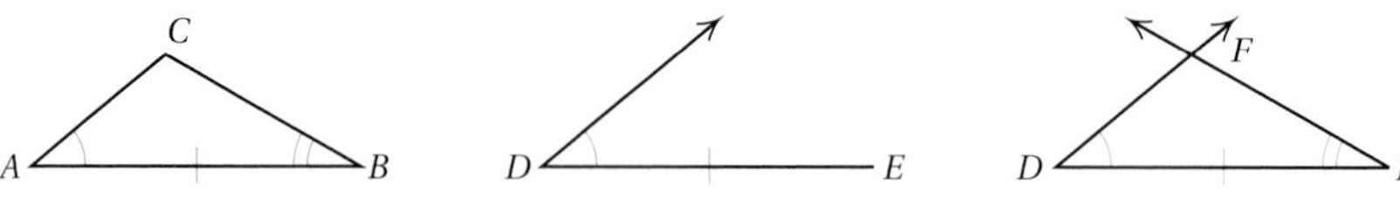

THE ANGLE–SIDE–ANGLE (ASA) PROPERTY

If two angles and the included side of a triangle are congruent to two angles and the included side of another triangle, then the triangles are congruent.

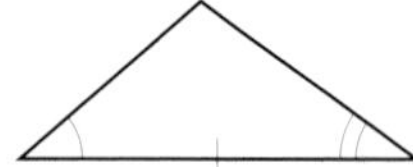

Example 7 Which pairs of triangles are congruent by the ASA property?

a) b) c)

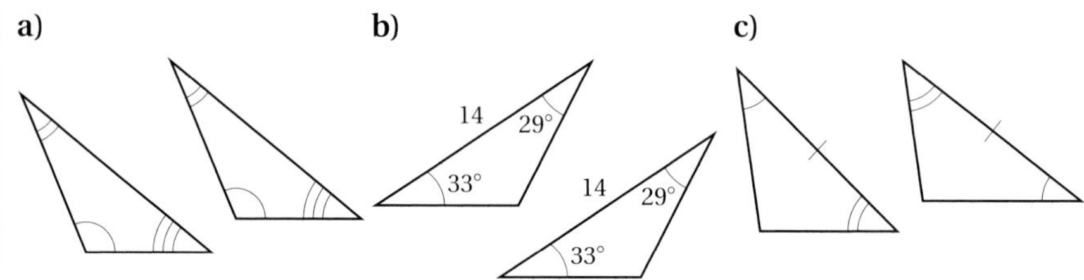

Pairs (b) and (c) are congruent by the ASA property.

Do Exercise 5.

Examples Which property (if any) should be used to show that these pairs of triangles are congruent?

8.

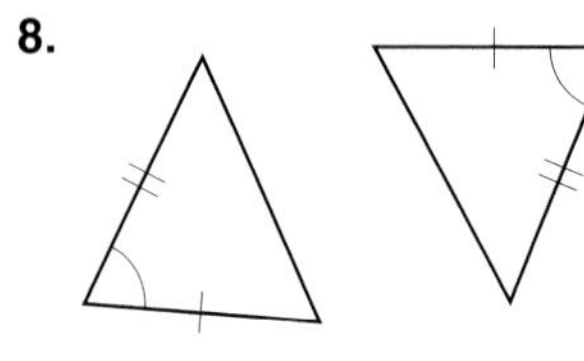

Use SAS.

9.

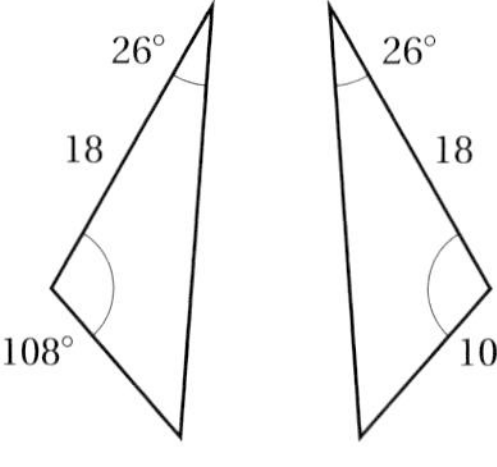

Use ASA.

10.

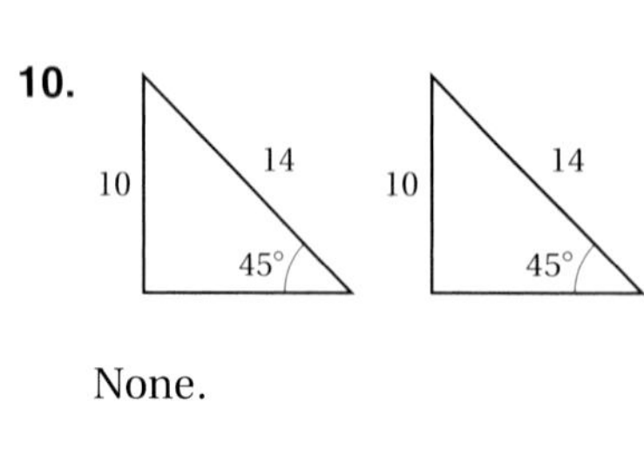

None.

11.

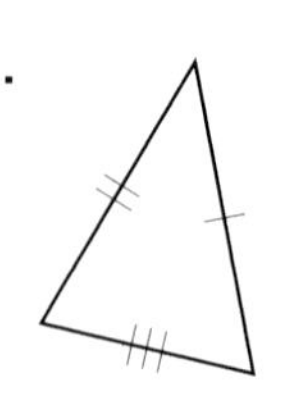

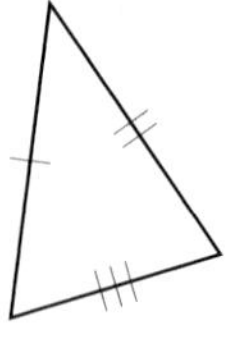

Use SSS.

Do Exercises 6–9.

---

*$\overline{DE}$ denotes the segment with endpoints D and E. DE denotes the length of $\overline{DE}$.

It is important to be able to explain why triangles are congruent.

Example 12 In $\triangle ABC$ and $\triangle DEF$, $\overline{AB} \cong \overline{DE}$, $\overline{AC} \cong \overline{DF}$, and $\angle A \cong \angle D$. Explain why the triangles are congruent.

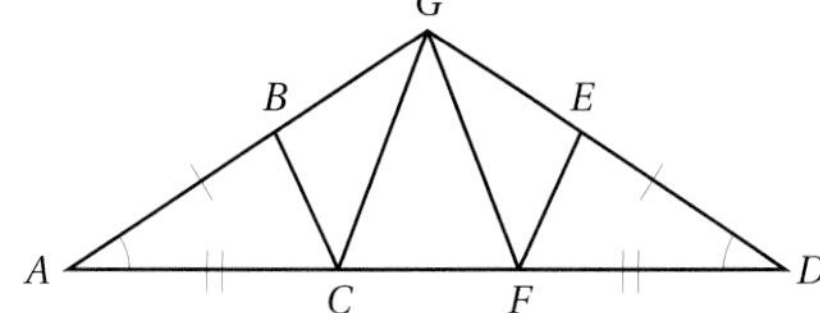

We have two sides and an included angle of $\triangle ABC$ congruent to the corresponding parts of $\triangle DEF$. Thus, $\triangle ABC \cong \triangle DEF$ by SAS.

Example 13 In $\triangle CPD$ and $\triangle EQD$, $\overline{CP} \perp \overline{QP}$ and $\overline{EQ} \perp \overline{QP}$. Also, $\angle QDE \cong \angle PDC$ and D is the midpoint of $\overline{QP}$. Explain why $\triangle CPD \cong \triangle EQD$.

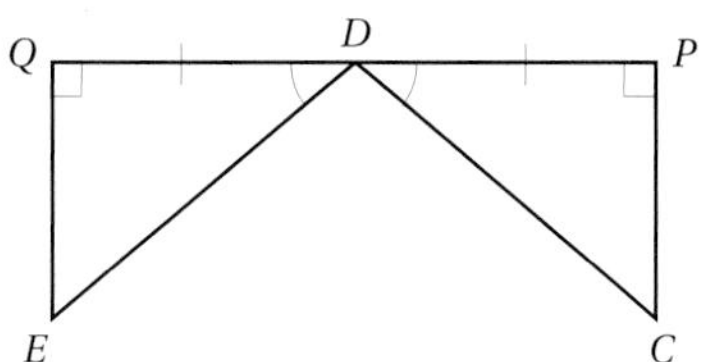

The perpendicular sides form right angles, which are congruent. Since D is the midpoint of $\overline{QP}$, we know that $\overline{QD} \cong \overline{PD}$. With $\angle QDE \cong \angle PDC$, we have $\triangle CPD \cong \triangle EQD$ by ASA.

Do Exercise 10.

Sometimes we can conclude that angles and segments are congruent by first showing that triangles are congruent.

Example 14 $\overline{AB} \cong \overline{BC}$ and $\overline{EB} \cong \overline{DB}$. What can you conclude?

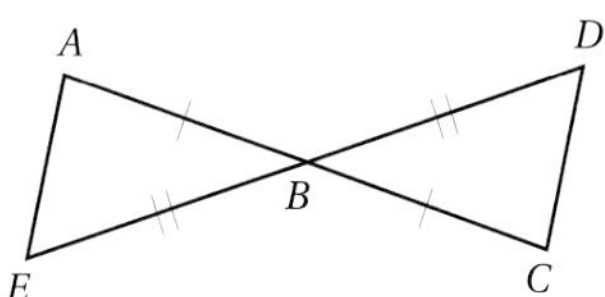

Since $\angle ABE$ and $\angle CBD$ are vertical angles, $\angle ABE \cong \angle CBD$. Thus, $\triangle ABE \cong \triangle CBD$ by SAS. As corresponding parts, $\overline{AE} \cong \overline{CD}$, $\angle A \cong \angle C$, and $\angle E \cong \angle D$.

Example 15 Explain how you can use congruent triangles to find the distance across a marsh.

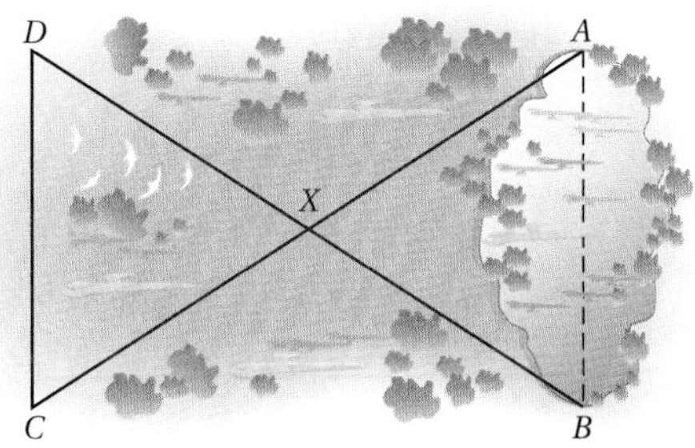

Mark off distances AX and BX. Extend $\overline{AX}$ and $\overline{BX}$ so that point X becomes the midpoint of $\overline{AC}$ and $\overline{BD}$. Then $\triangle ABX \cong \triangle CDX$ by SAS. Thus, $\overline{DC} \cong \overline{AB}$ as corresponding parts. Then we can measure $\overline{DC}$ knowing that $DC = AB$.

Do Exercises 11 and 12.

10. In this figure, $\overline{AB} \perp \overline{ED}$ and B is the midpoint of $\overline{ED}$. Explain why $\triangle ABD \cong \triangle ABE$.

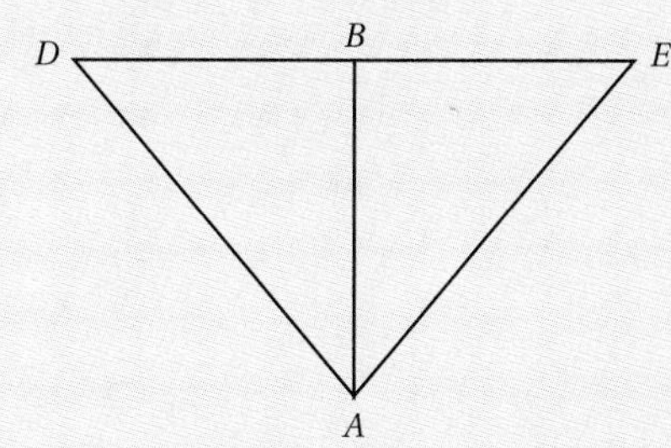

11. $\angle R \cong \angle T$, $\angle W \cong \angle V$, and $\overline{RW} \cong \overline{TV}$. What can you conclude about this figure?

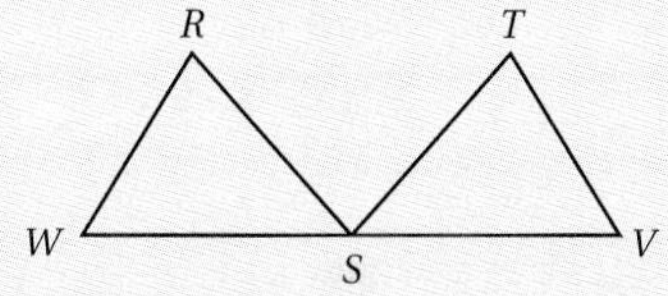

12. On a pair of pinking shears, the indicated angles and sides are congruent. How do you know that P is the midpoint of $\overline{GR}$?

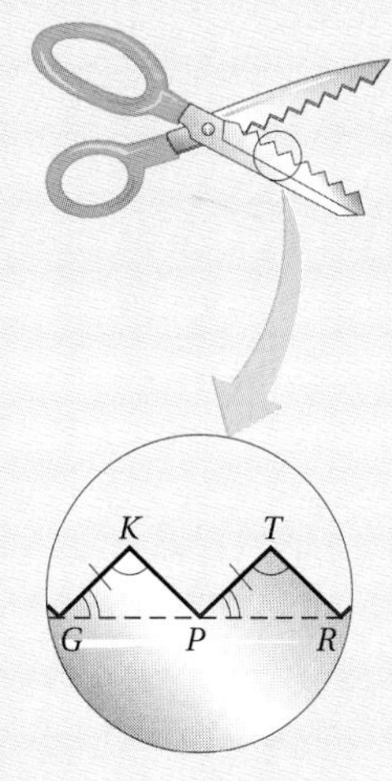

Answers on page A-16

b Properties of Parallelograms

A quadrilateral is a polygon with four sides. A **diagonal** of a quadrilateral is a segment that joins two opposite vertices.

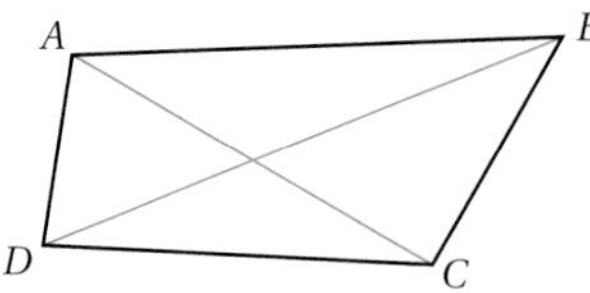

$\overline{AC}$ and $\overline{BD}$ are diagonals.

The sum of the measures of the angles of a quadrilateral is 360°.

A parallelogram is a quadrilateral with two pairs of parallel sides.

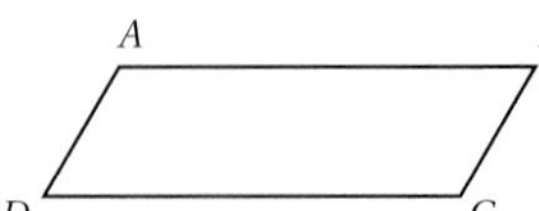

$\overline{AB} \parallel \overline{DC}$

$\overline{AD} \parallel \overline{BC}$

Draw two pairs of parallel lines to form parallelogram *ABCD*. Compare the lengths of opposite sides. Compare the measures of opposite angles. Compare the measures of consecutive angles. Draw diagonal $\overline{AC}$. How are $\triangle ADC$ and $\triangle CBA$ related? Draw diagonal $\overline{BD}$, intersecting $\overline{AC}$ at point *E*. What is special about point *E*?

Using the comparisons and the fact that corresponding parts of congruent triangles are congruent, we can list the following properties of parallelograms.

PROPERTIES OF PARALLELOGRAMS

1. A diagonal of a parallelogram determines two congruent triangles.
2. The opposite angles of a parallelogram are congruent.
3. The opposite sides of a parallelogram are congruent.
4. Consecutive angles of a parallelogram are supplementary.
5. The diagonals of a parallelogram bisect each other.

Example 16 If $m\angle A = 120°$, find the measures of the other angles of parallelogram *ABCD*.

$m\angle C = 120°$ Using Property 2

$m\angle B = 60°$ Using Property 4

$m\angle D = 60°$ Using Property 2

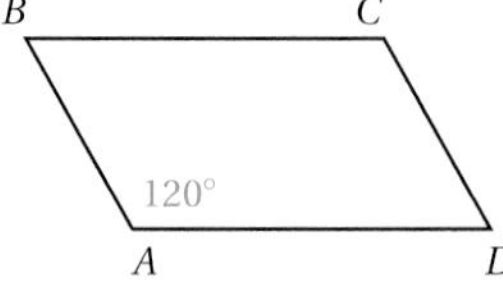

Example 17 Find *AB* and *BC*.

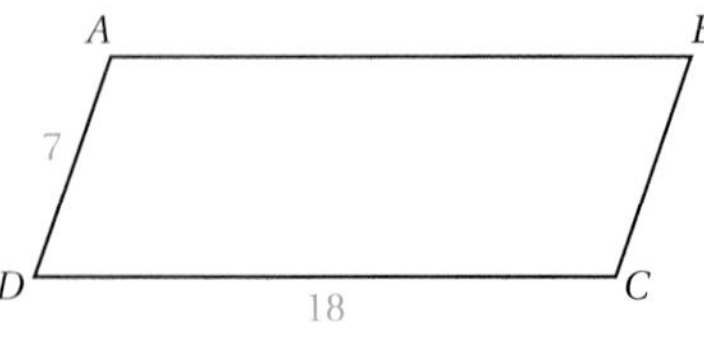

$AB = 18$ and $BC = 7$ Using Property 3

Do Exercises 13–16.

Find the measure of each angle.

13.

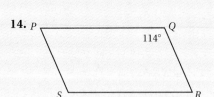

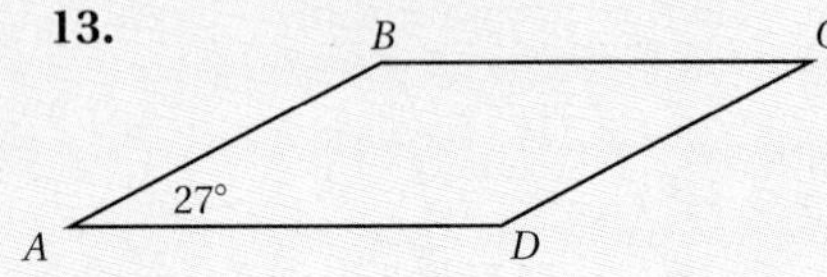

14. P Q 114° S R

Find the length of each side.

15.

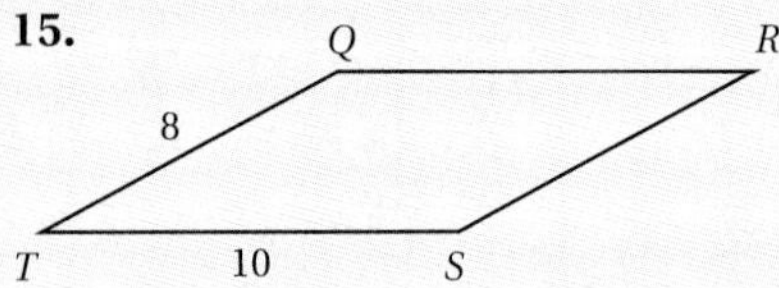

16. The perimeter of ▱*DEFG* is 68.

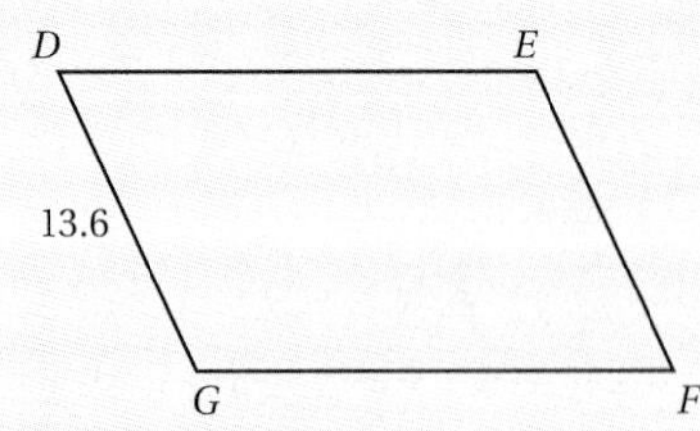

Answers on page A-16

Exercise Set 6.8

a Name the corresponding parts of the congruent triangles.

1. $\triangle ABC \cong \triangle RST$

2. $\triangle MNQ \cong \triangle HJK$

3. $\triangle DEF \cong \triangle GHK$

4. $\triangle ABC \cong \triangle ABC$

5. $\triangle XYZ \cong \triangle UVW$

6. $\triangle ABC \cong \triangle ACB$

Name the corresponding parts of the congruent triangles.

7. 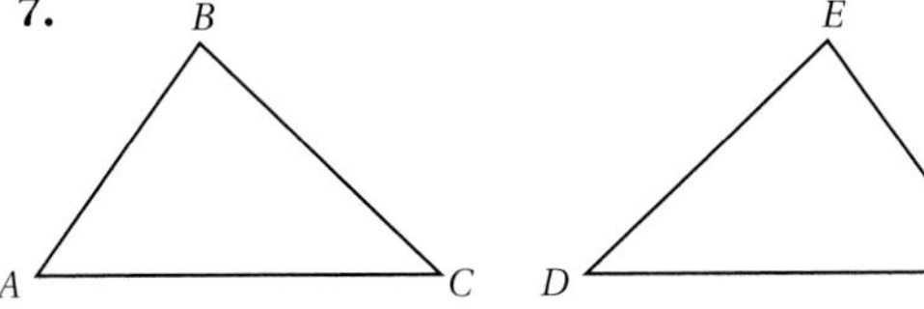

8. 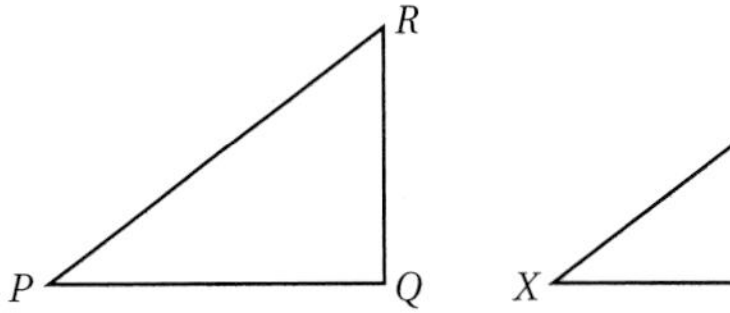

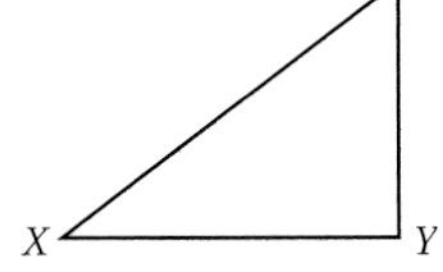

9. 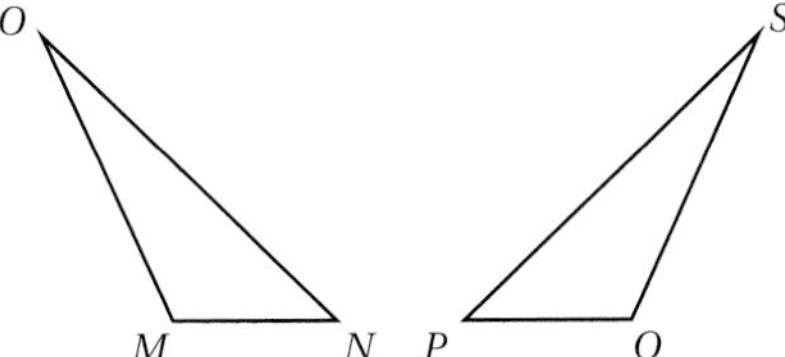

10. 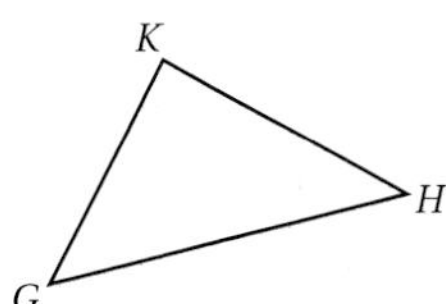

Determine whether the pair of triangles is congruent by the SAS property.

11. 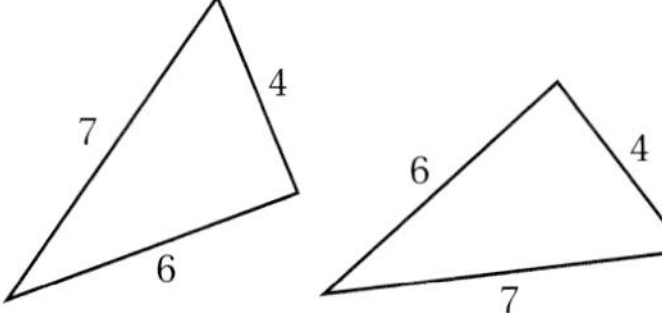

12. 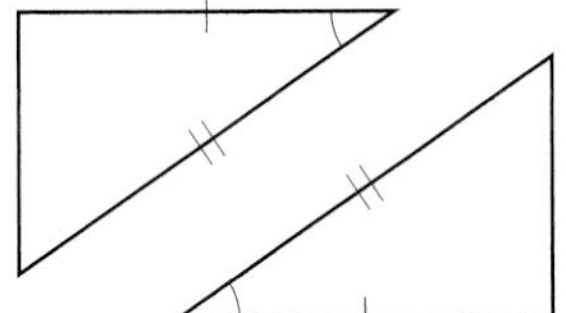

13. 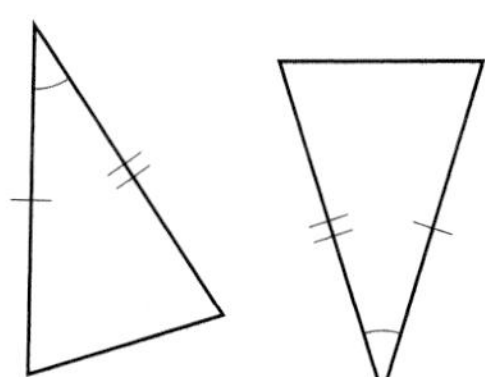

14. 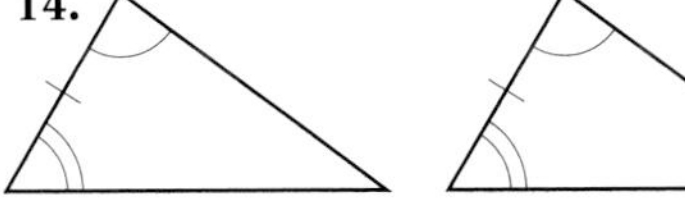

15. 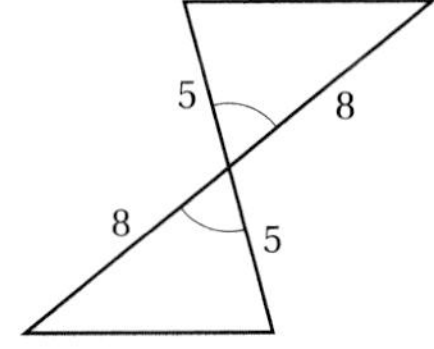

16. 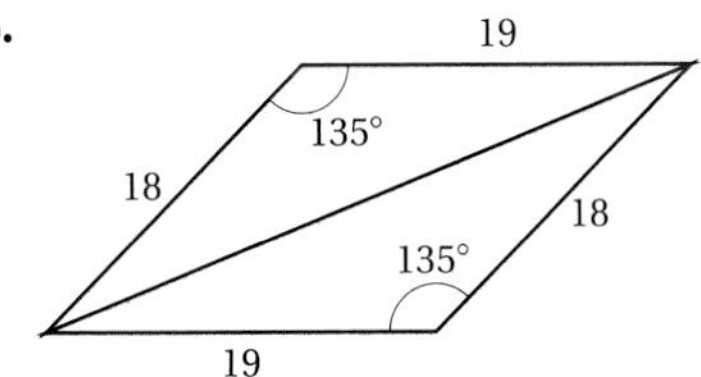

Determine whether the pair of triangles is congruent by the SSS property.

17.

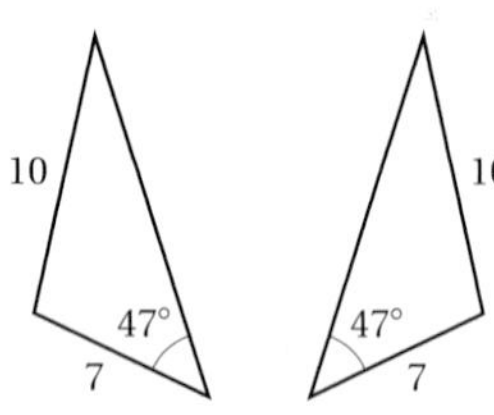

18.

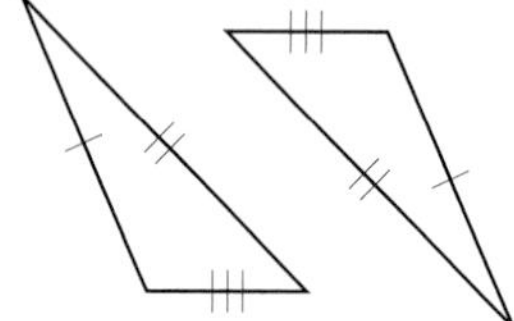

19.

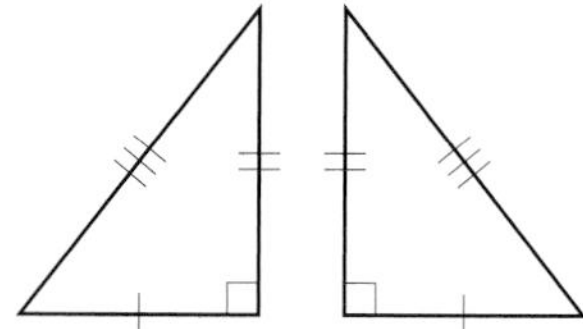

20.

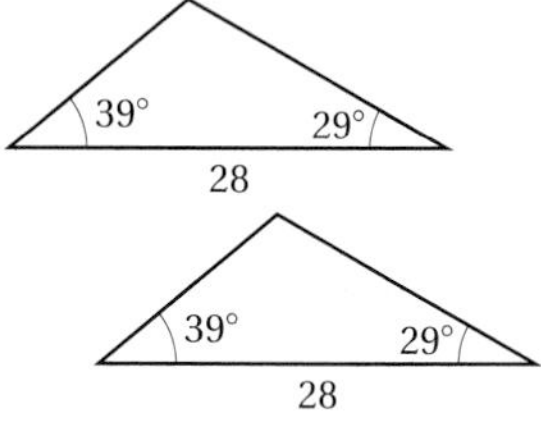

21.

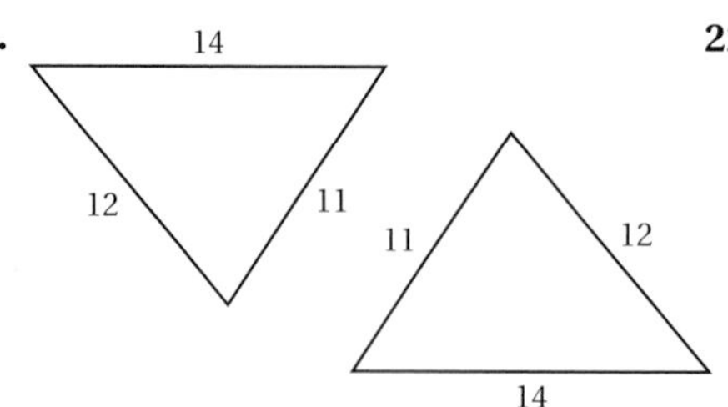

22.

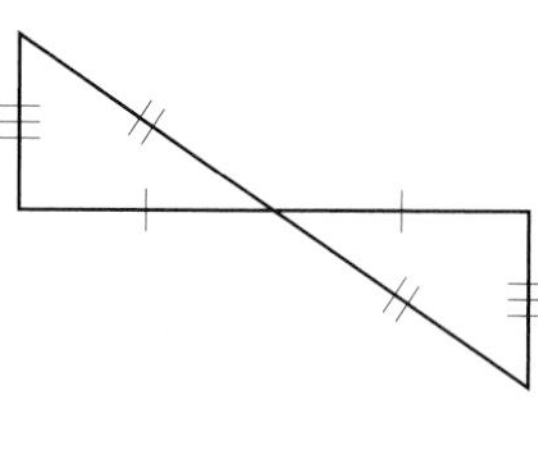

Determine whether the pair of triangles is congruent by the ASA property.

23.

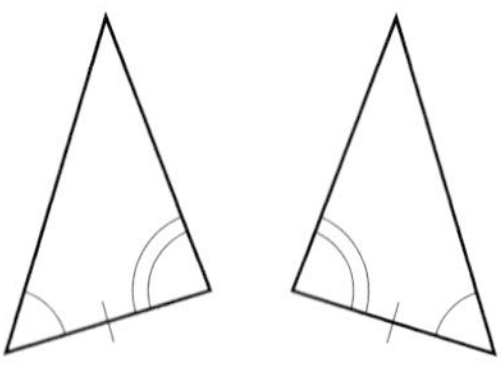

24.

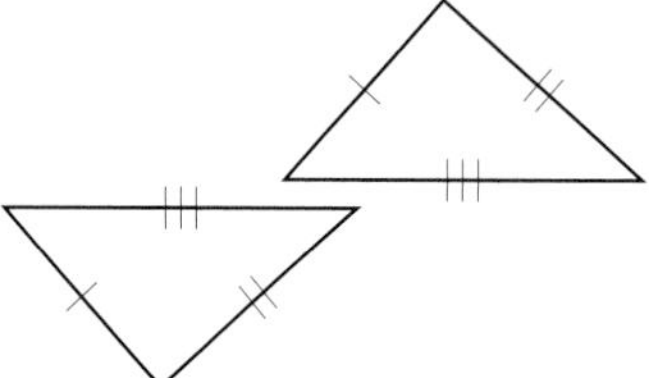

25.

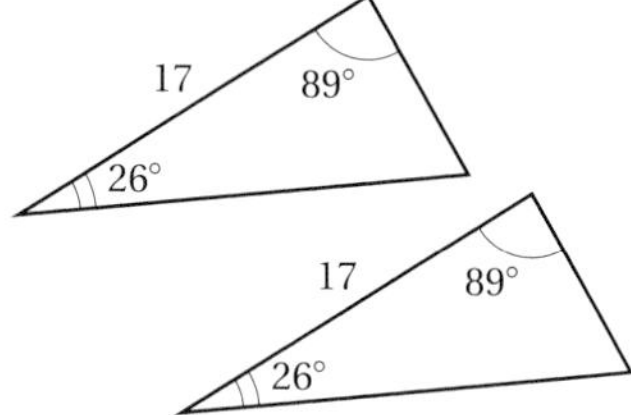

26.

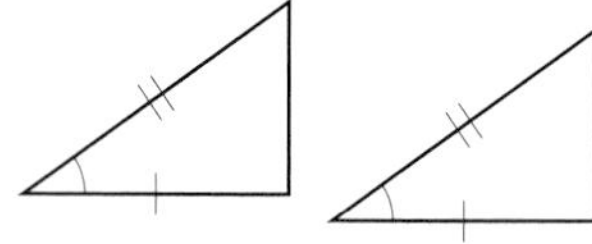

27.

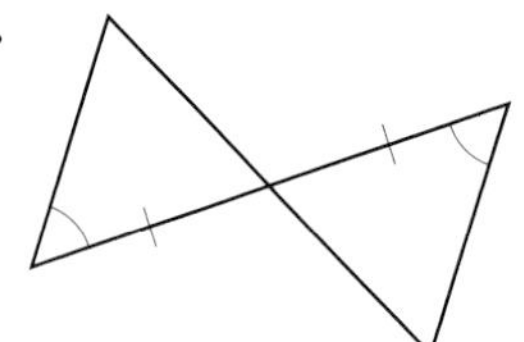

28.

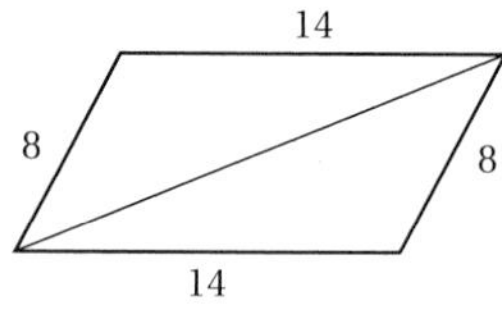

Which property (if any) should be used to show that the pair of triangles is congruent?

29.

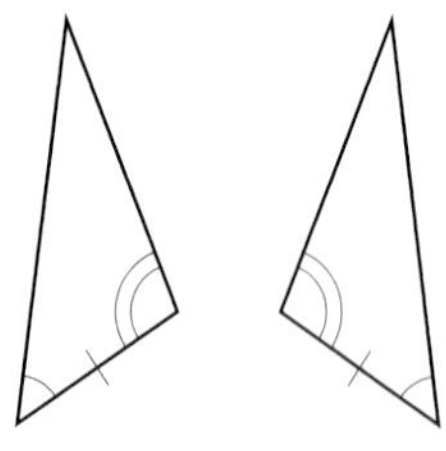

30.

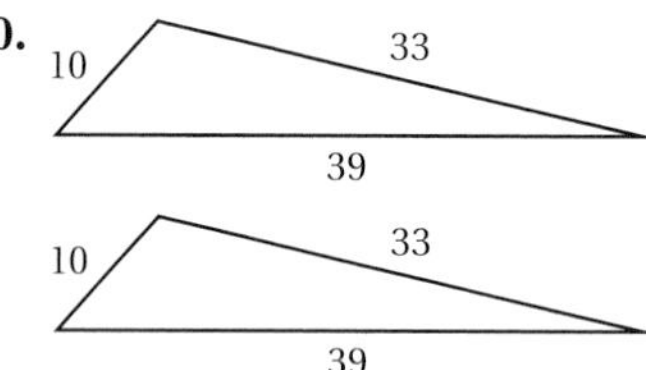

31.

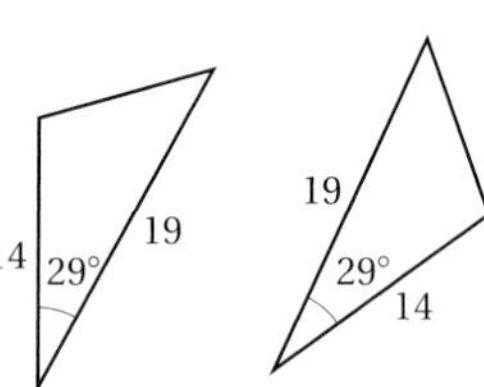

32.

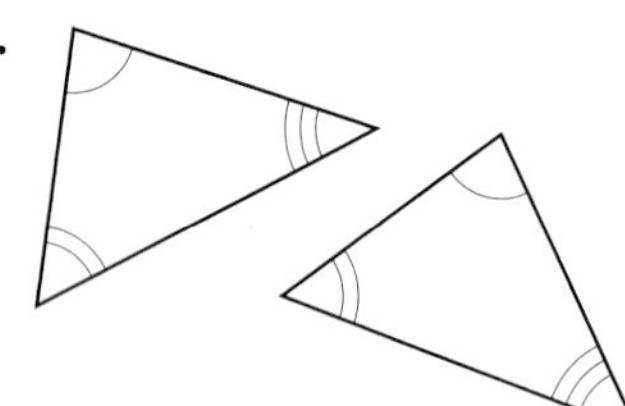

33.

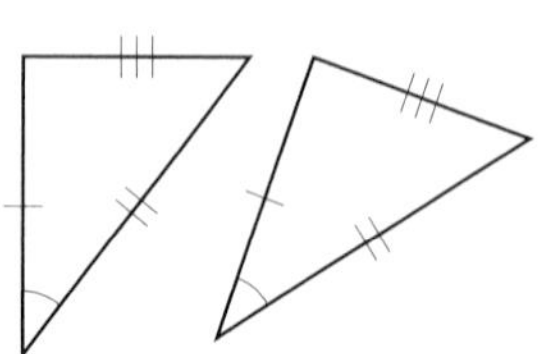

34.

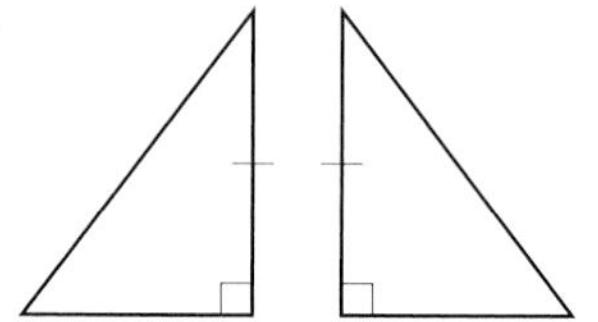

Explain why the triangles indicated in parentheses are congruent.

35. R is the midpoint of both $\overline{PT}$ and $\overline{QS}$. ($\triangle PRQ \cong \triangle TRS$)

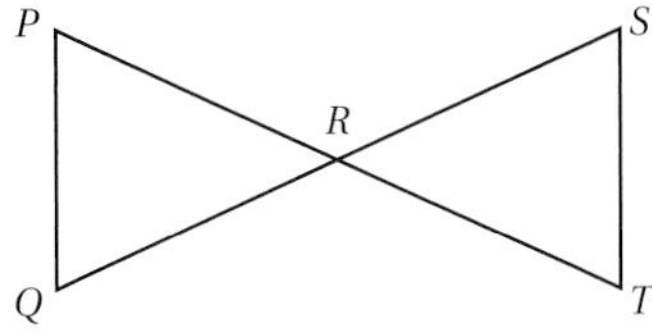

36. $\angle 1$ and $\angle 2$ are right angles, X is the midpoint of $\overline{AY}$, and $\overline{XB} \cong \overline{YZ}$. ($\triangle ABX \cong \triangle XZY$)

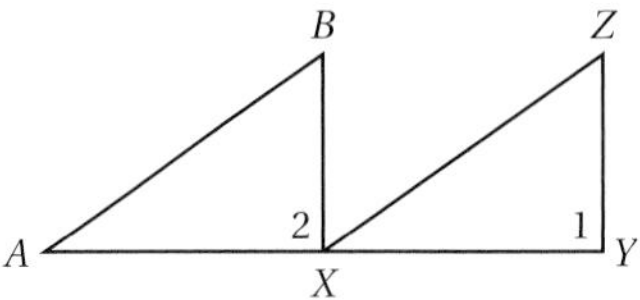

37. L is the midpoint of $\overline{KM}$ and $\overline{GL} \perp \overline{KM}$. ($\triangle KLG \cong \triangle MLG$)

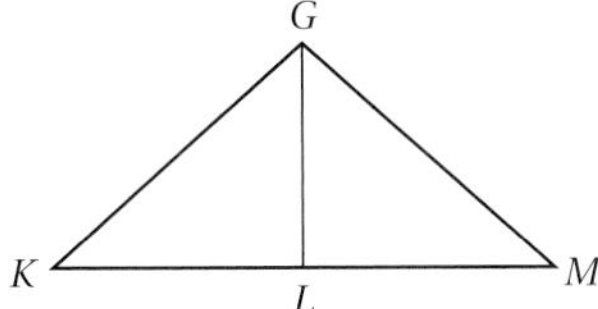

38. X is the midpoint of $\overline{QS}$ and $\overline{RP}$ with $RQ = SP$. ($\triangle RQX \cong \triangle PSX$)

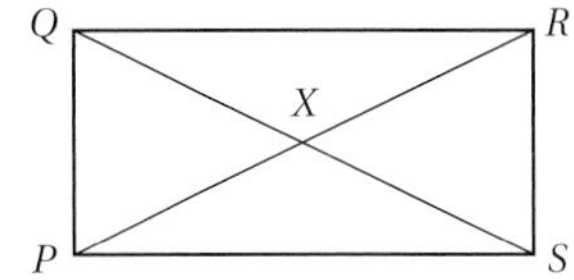

39. $\triangle AEB$ and $\triangle CDB$ are isosceles with $\overline{AE} \cong \overline{AB} \cong \overline{CB} \cong \overline{CD}$. Also, B is the midpoint of $\overline{ED}$. ($\triangle AEB \cong \triangle CDB$)

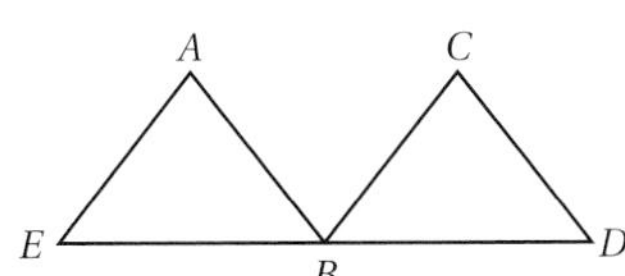

40. $\overline{AB} \perp \overline{BE}$ and $\overline{DE} \perp \overline{BE}$. $\overline{AB} \cong \overline{DE}$ and $\angle BAC \cong \angle EDC$. ($\triangle ABC \cong \triangle DEC$)

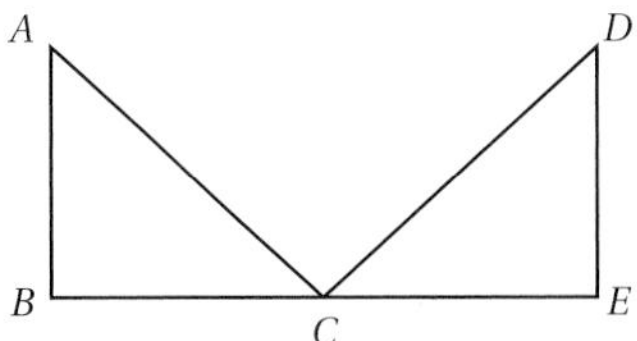

What can you conclude about each figure using the given information?

41. $\overline{GK} \perp \overline{LJ}$, $\overline{HK} \cong \overline{KJ}$, and $\overline{GK} \cong \overline{LK}$

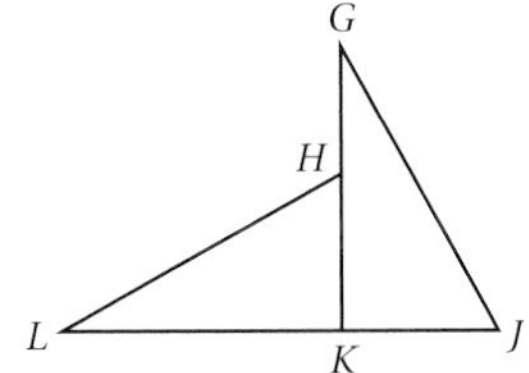

42. $\overline{AB} \cong \overline{DC}$ and $\angle BAC \cong \angle DCA$

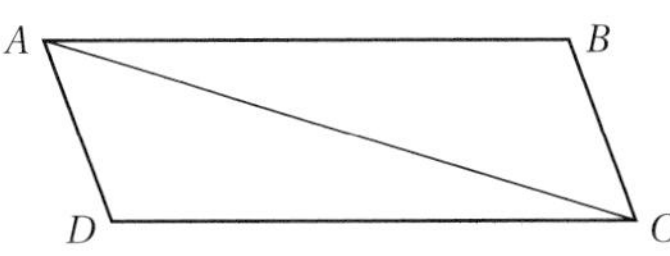

Use corresponding parts to solve Exercises 43 and 44.

43. On this national flag, the indicated segments and angles are congruent. Explain why P is the midpoint of $\overline{EF}$.

44. The indicated sides of a kite are congruent. Explain how you know that $\angle 1 \cong \angle 2$.

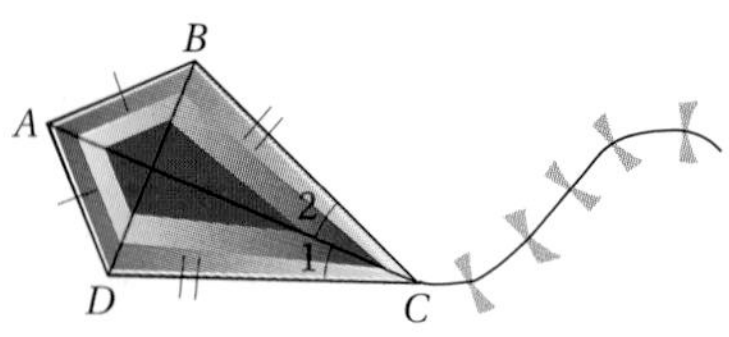

b Find the measures of the angles of the parallelogram.

45.

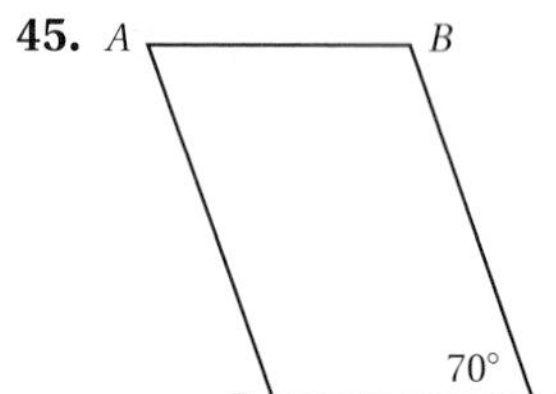

46. E F 110° H G

47.

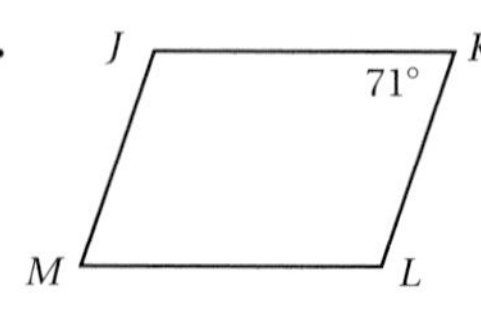

48.

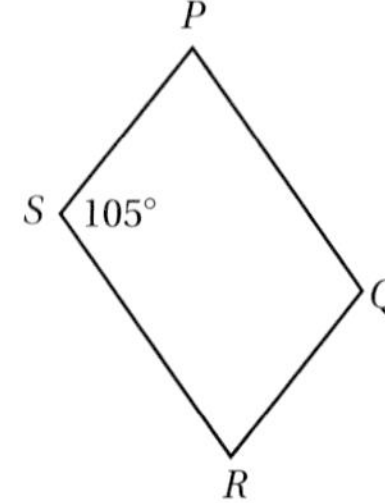

Find the lengths of the sides of the parallelogram.

49.

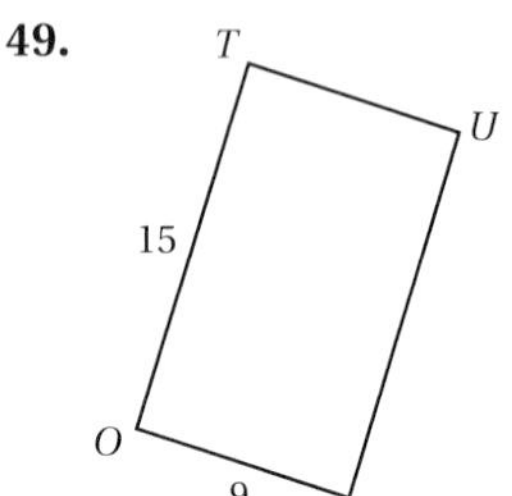

50.

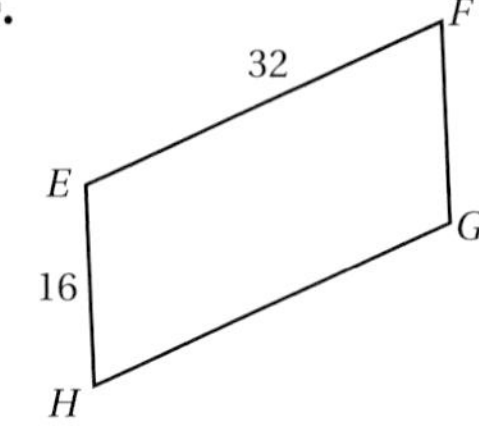

51. The perimeter of ▱ $JKLM$ is 22.

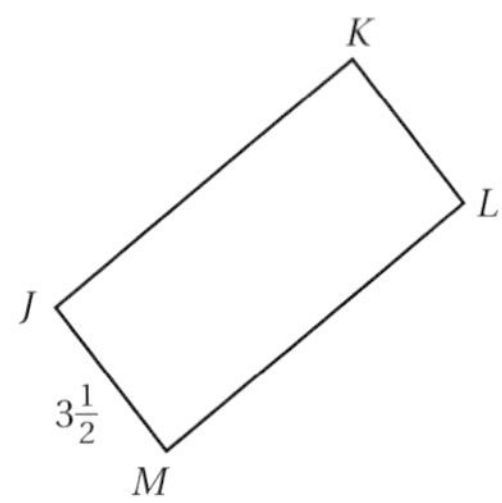

52. The perimeter of ▱ $WXYZ$ is 248.

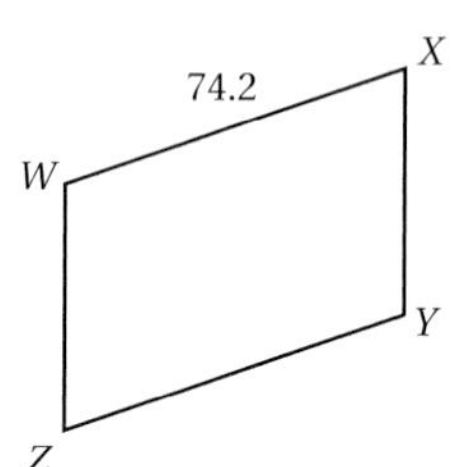

53. $AB = 14$ and $BD = 19$. Find the length of each diagonal.

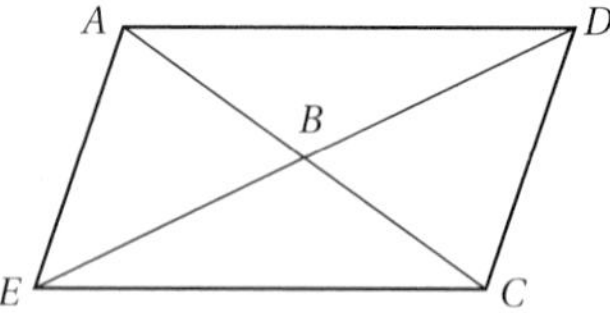

54. $EJ = 23$ and $GJ = 13$. Find the length of each diagonal.

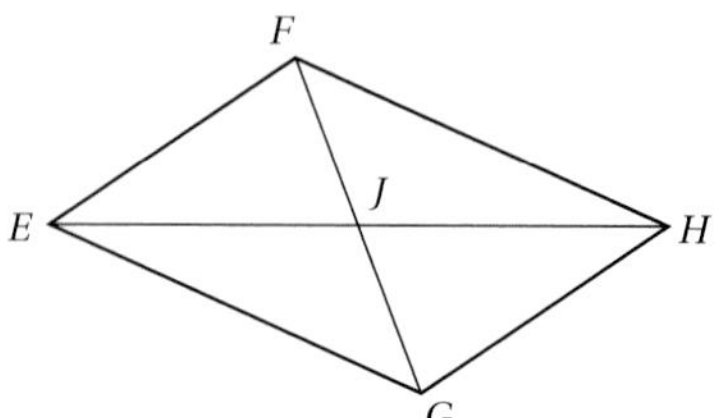

6.9 Similar Triangles

a Proportions and Similar Triangles

We know that congruent figures have the same shape and size. *Similar figures* have the same shape, but are not necessarily the same size.

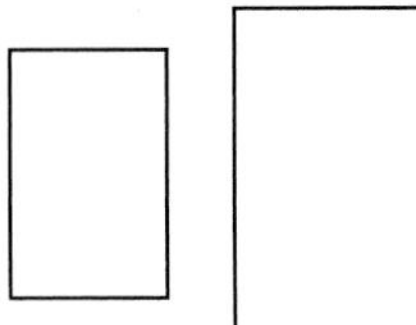
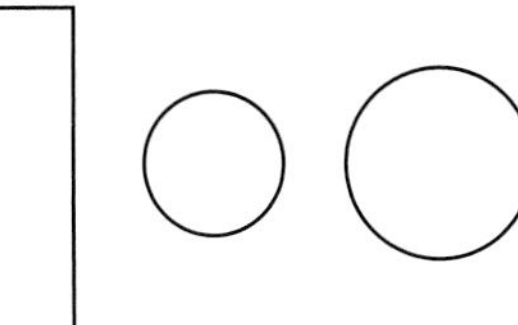
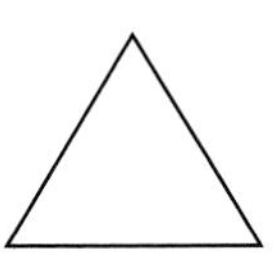
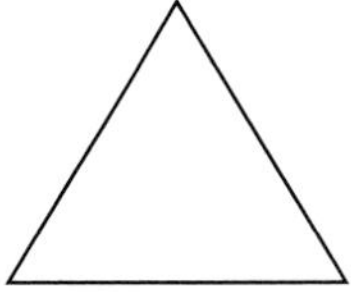

Similar figures

Example 1 Which pairs of triangles appear to be similar?

a)

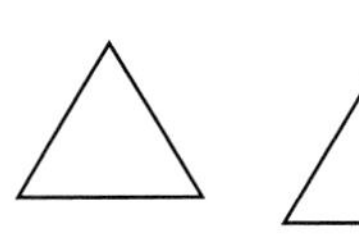

b)

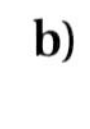
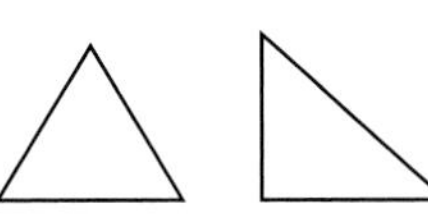

c)

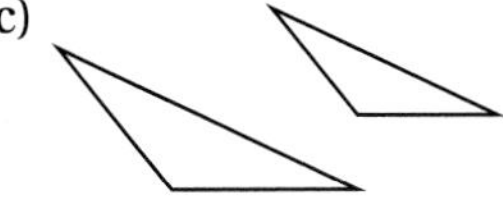

d)

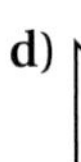
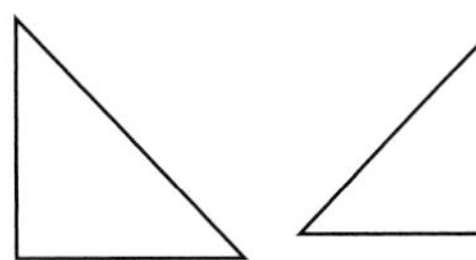

Pairs (a), (c), and (d) appear to be similar.

Do Exercise 1.

Similar triangles have corresponding sides and angles.

Example 2 $\triangle ABC$ and $\triangle DEF$ are similar. Name their corresponding sides and angles.

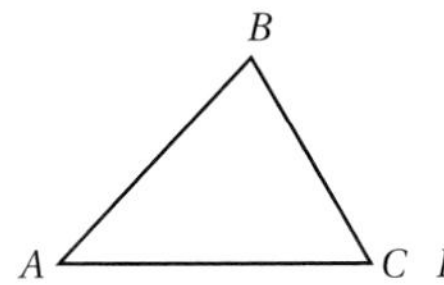

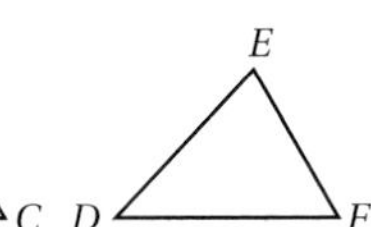

$\overline{AB} \leftrightarrow \overline{DE}$ $\angle A \leftrightarrow \angle D$

$\overline{AC} \leftrightarrow \overline{DF}$ $\angle B \leftrightarrow \angle E$

$\overline{BC} \leftrightarrow \overline{EF}$ $\angle C \leftrightarrow \angle F$

Do Exercise 2.

Objectives

a Identify the corresponding parts of similar triangles and determine which sides of a given pair of triangles have lengths that are proportional.

b Find lengths of sides of similar triangles using proportions.

For Extra Help

TAPE 13

MAC
WIN

CD-ROM

1. Which pairs of triangles appear to be similar?

a)

b)

c)

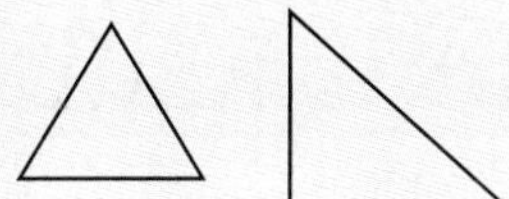

d)

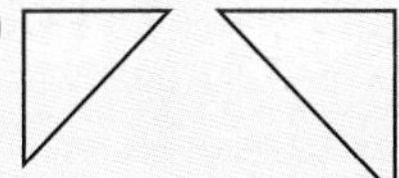

2. $\triangle PQR$ and $\triangle GHK$ are similar. Name their corresponding sides and angles.

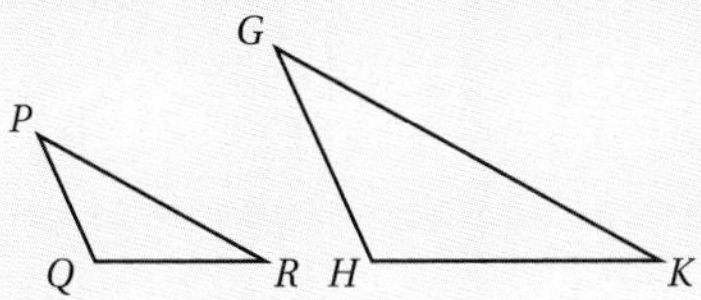

Answers on page A-16

3. Suppose that $\triangle JKL \sim \triangle ABC$. Which angles are congruent? Which sides are proportional?

4. These triangles are similar. Which sides are proportional?

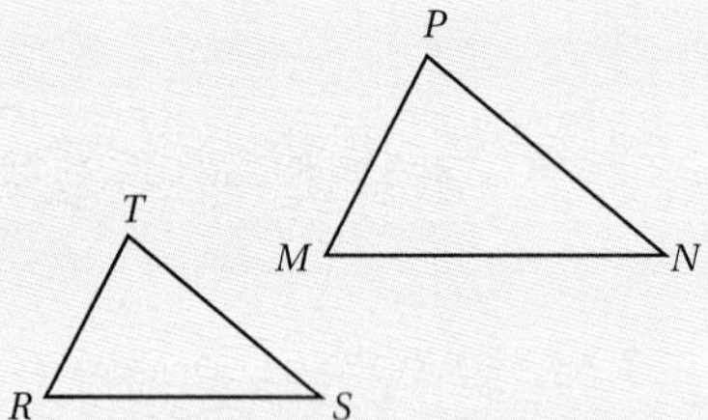

Answers on page A-16

Two triangles are **similar** if and only if their vertices can be matched so that the corresponding angles are congruent and the lengths of corresponding sides are proportional.

To say that $\triangle ABC$ and $\triangle DEF$ are similar, we write "$\triangle ABC \sim \triangle DEF$." We will agree that this symbol also tells us the way in which the vertices are matched.

$$\triangle ABC \sim \triangle DEF$$

Thus, $\triangle ABC \sim \triangle DEF$ means that

$$\angle A \cong \angle D \qquad \angle B \cong \angle E \qquad \angle C \cong \angle F \quad \text{and} \quad \frac{AB}{DE} = \frac{AC}{DF} = \frac{BC}{EF}.$$

Example 3 Suppose that $\triangle PQR \sim \triangle STV$. Which angles are congruent? Which sides are proportional?

$$\angle P \cong \angle S \qquad \angle Q \cong \angle T \qquad \angle R \cong \angle V \quad \text{and} \quad \frac{PQ}{ST} = \frac{PR}{SV} = \frac{QR}{TV}$$

Do Exercise 3.

Example 4 These triangles are similar. Which sides are proportional?

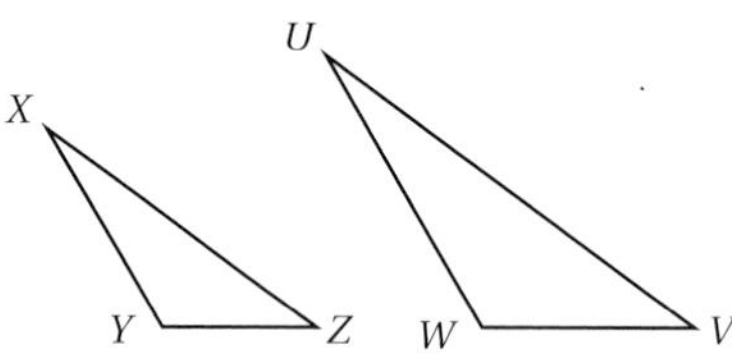

It appears that if we match X with U, Y with W, and Z with V, the corresponding angles will be congruent. Thus,

$$\frac{XY}{UW} = \frac{XZ}{UV} = \frac{YZ}{WV}.$$

Do Exercise 4.

b Proportions and Similar Triangles

We can find lengths of sides in similar triangles.

Example 5 If $\triangle RAE \sim \triangle GQL$, find QL and GL.

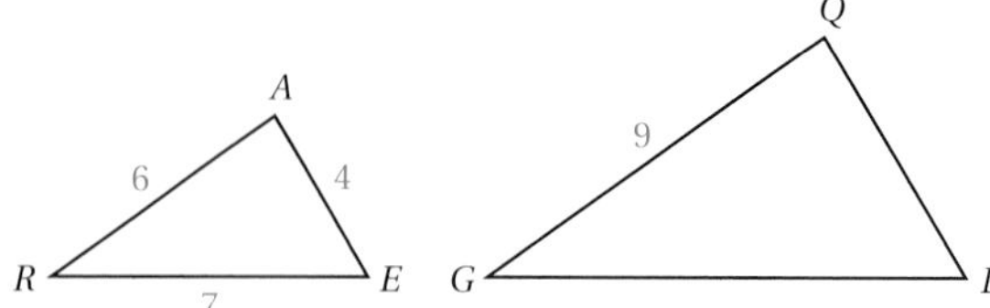

Since $\triangle RAE \sim \triangle GQL$, the corresponding sides are proportional. Thus,

$$\frac{6}{9} = \frac{4}{QL}$$

$$6(QL) = 9 \cdot 4 \qquad \textbf{Equating cross products}$$

$$6(QL) = 36$$

$$QL = 6 \qquad \textbf{Dividing by 6 on both sides}$$

and

$$\frac{6}{9} = \frac{7}{GL}$$

$$6(GL) = 9 \cdot 7$$

$$6(GL) = 63$$

$$GL = 10\tfrac{1}{2}.$$

Do Exercise 5.

Example 6 If $\overline{AB} \parallel \overline{CD}$, find CD.

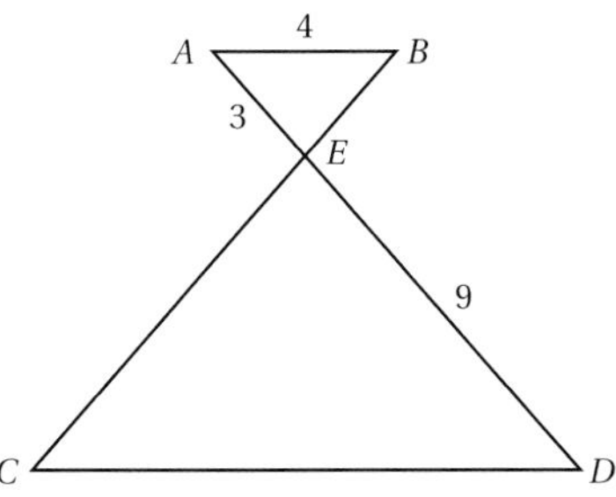

Recall that if a transversal intersects two parallel lines, then the alternate interior angles are congruent (Section 6.7). Thus,

$$\angle A \cong \angle D \quad \text{and} \quad \angle C \cong \angle B,$$

because they are pairs of alternate interior angles. Since $\angle AEB$ and $\angle DEC$ are vertical angles, they are congruent. Thus by definition

$$\triangle AEB \sim \triangle DEC$$

and the lengths of the corresponding sides are proportional. Thus,

$$\frac{AE}{DE} = \frac{AB}{CD}.$$

Solve: $$\frac{3}{9} = \frac{4}{CD} \qquad \textbf{Substituting}$$

$$3(CD) = 9 \cdot 4 \qquad \textbf{Equating cross products}$$

$$3(CD) = 36$$

$$CD = 12. \qquad \textbf{Dividing by 3 on both sides}$$

Do Exercise 6.

Similar triangles and proportions can often be used to find lengths that would ordinarily be difficult to measure. For example, we could find the height of a flagpole without climbing it or the distance across a river without crossing it.

5. If $\triangle WNE \sim \triangle CBT$, find BT and CT.

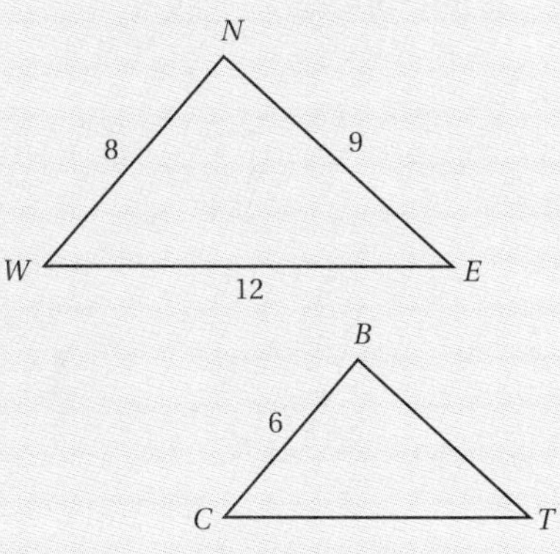

6. If $\overline{QR} \parallel \overline{ST}$, find QR.

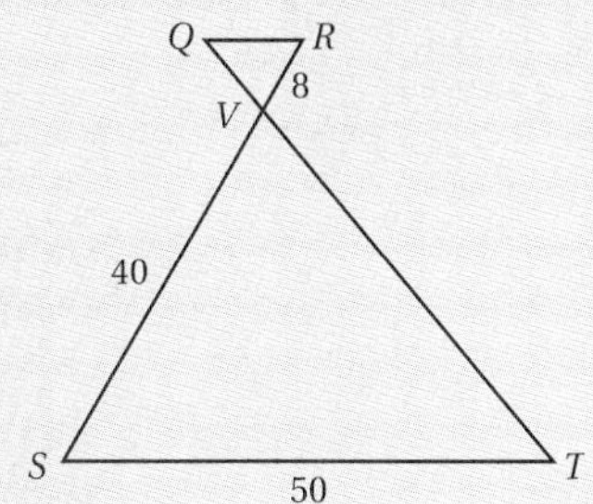

Answers on page A-16

7. How high is a flagpole that casts a 45-ft shadow at the same time that a 5.5-ft woman casts a 10-ft shadow?

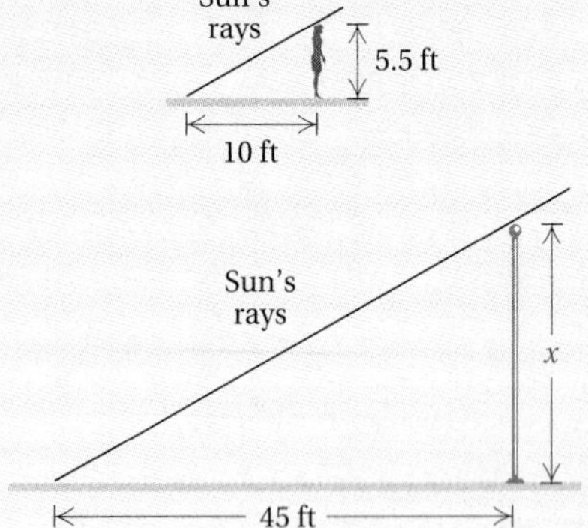

Example 7 How high is a flagpole that casts a 56-ft shadow at the same time that a 6-ft man casts a 5-ft shadow?

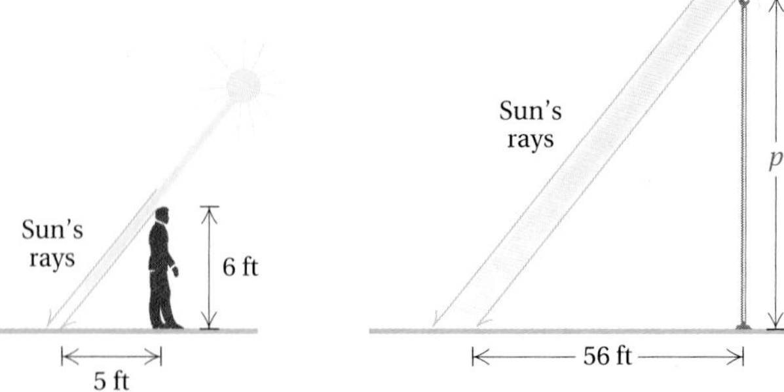

If we use the sun's rays to represent the third side of the triangle in our drawing of the situation, we see that we have similar triangles. Let p = the height of the flagpole. The ratio of 6 to p is the same as the ratio of 5 to 56. Thus we have the proportion

$$\text{Height of man} \rightarrow \frac{6}{p} = \frac{5}{56}. \leftarrow \text{Length of shadow of man}$$
$$\text{Height of pole} \rightarrow \qquad \leftarrow \text{Length of shadow of pole}$$

Solve: $6 \cdot 56 = 5 \cdot p$ Equating cross products

$\frac{6 \cdot 56}{5} = p$ Dividing by 5 on both sides

$67.2 = p$ Simplifying

The height of the flagpole is 67.2 ft.

Do Exercise 7.

8. *F-106 Blueprint.* Referring to Example 3, find the length x of the wing.

Example 8 *F-106 Blueprint.* A blueprint for an F-106 Delta Dart fighter plane is a scale drawing. Each wing of the plane has a triangular shape. The blueprint shows similar triangles. Find the length of side a of the wing.

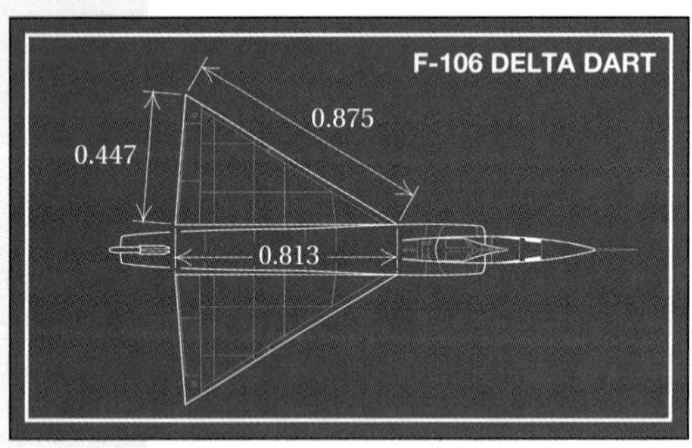

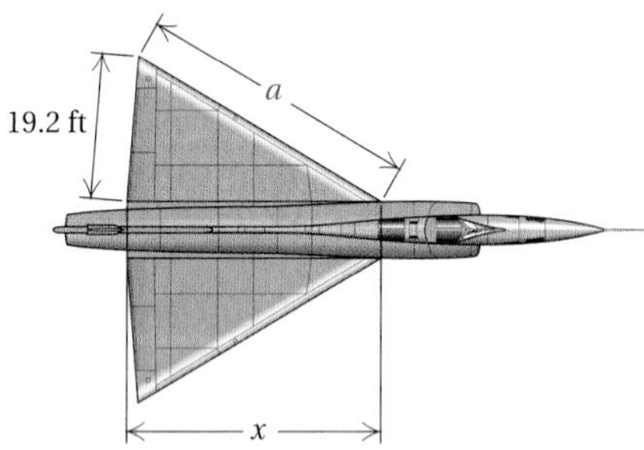

We let a = the length of the wing. Thus we have the proportion

$$\text{Length on the blueprint} \rightarrow \frac{0.447}{19.2} = \frac{0.875}{a}. \leftarrow \text{Length on the blueprint}$$
$$\text{Length of the wing} \rightarrow \qquad \leftarrow \text{Length of the wing}$$

Solve: $0.447 \cdot a = 19.2 \cdot 0.875$ Equating cross products

$a = \frac{19.2 \cdot 0.875}{0.447}$ Dividing by 0.447 on both sides

≈ 37.6 ft

The length of side a of the wing is about 37.6 ft.

Do Exercise 8.

Answers on page A-16

Exercise Set 6.9

a For each pair of similar triangles, name the corresponding sides and angles.

1.

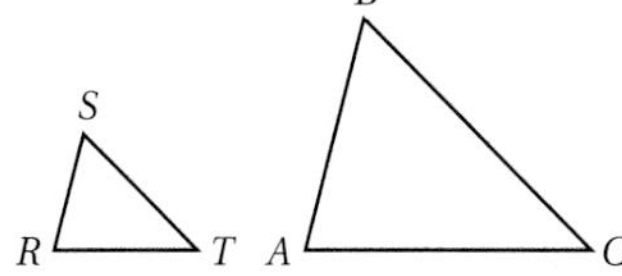

2.

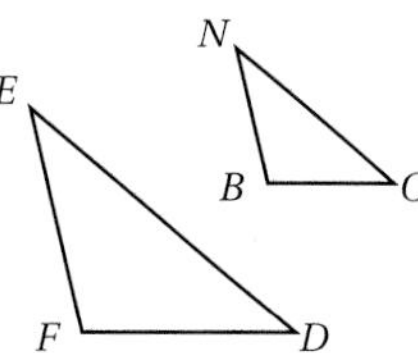

3.

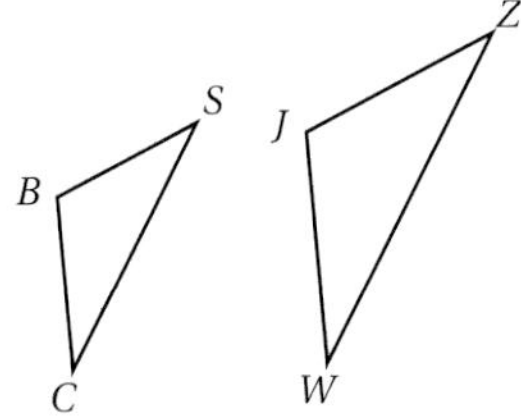

4.

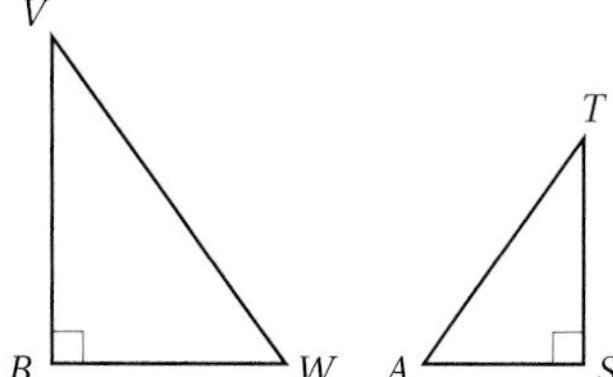

For each pair of similar triangles, name the congruent angles and proportional sides.

5. $\triangle ABC \sim \triangle RST$

6. $\triangle PQR \sim \triangle STV$

7. $\triangle MES \sim \triangle CLF$

8. $\triangle SMH \sim \triangle WLK$

Name the proportional sides in these similar triangles.

9.

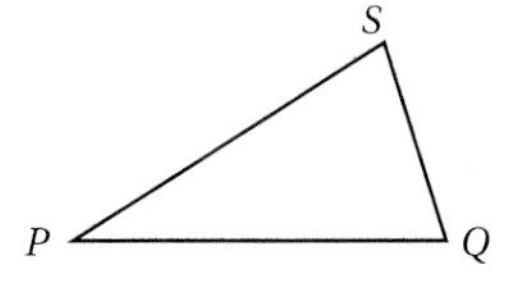

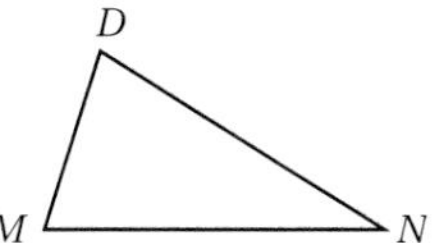

10.

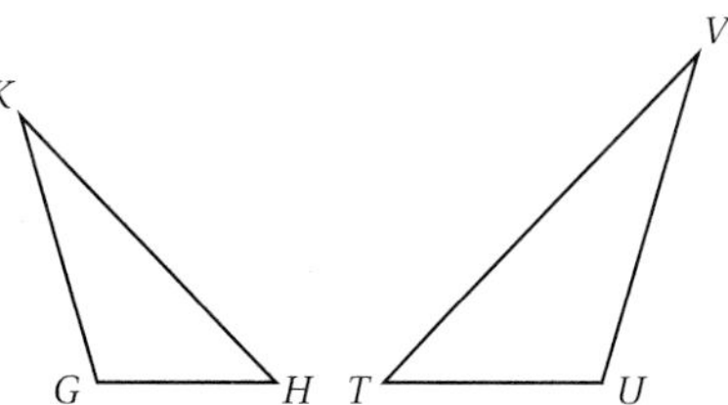

11.

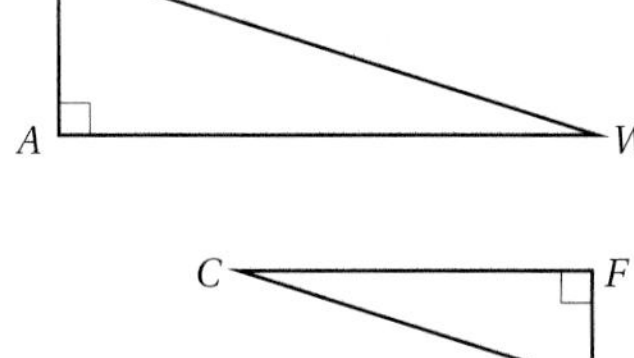

12.

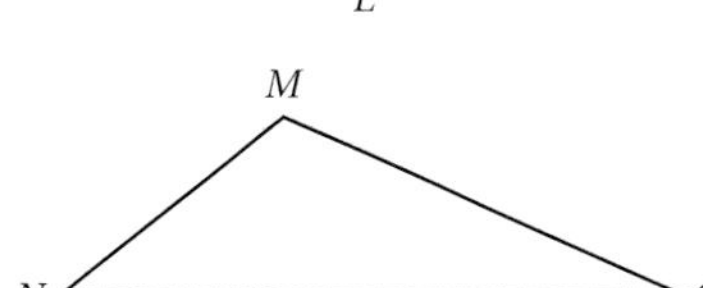

b Find the missing lengths.

13. If $\triangle ABC \sim \triangle PQR$, find QR and PR.

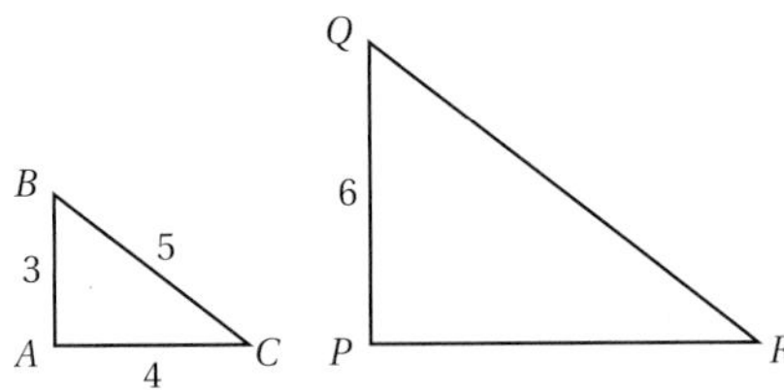

14. If $\triangle MAC \sim \triangle GET$, find AM and GT.

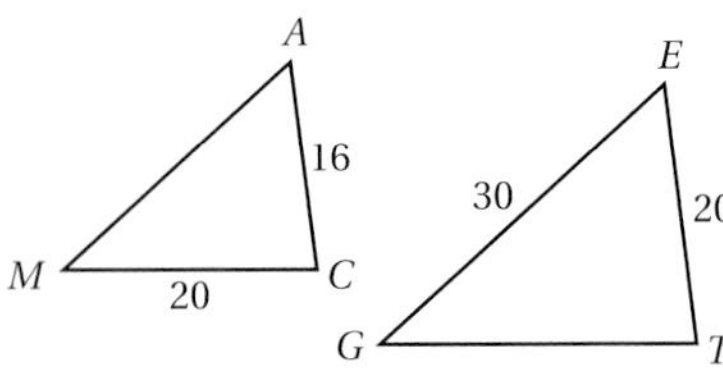

15. If $\overline{AD} \parallel \overline{CB}$, find EC.

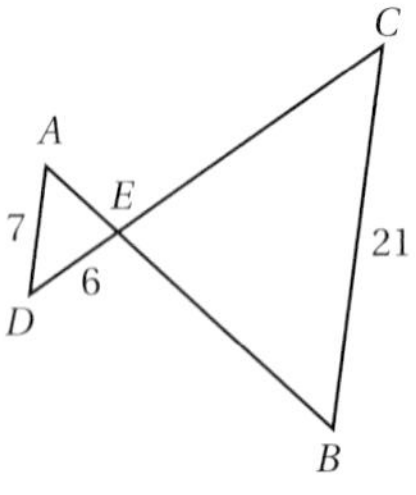

16. If $\overline{LN} \parallel \overline{PM}$, find QM.

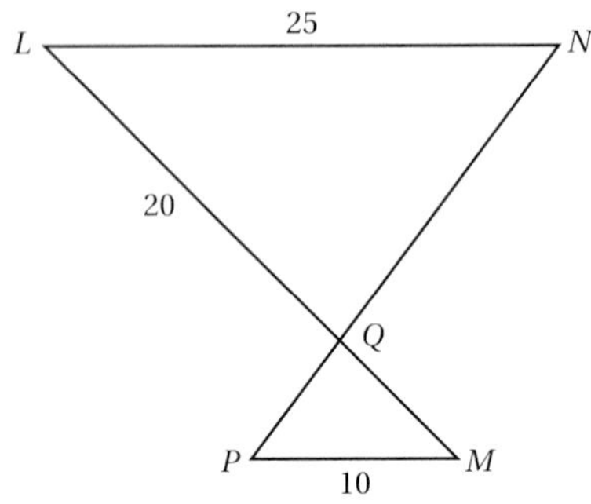

17. How high is a tree that casts a 27-ft shadow at the same time that a 4-ft fence post casts a 3-ft shadow?

18. How high is a flagpole that casts a 42-ft shadow at the same time that a $5\frac{1}{2}$-ft woman casts a 7-ft shadow?

19. Find the distance across the river. Assume that the ratio of d to 25 ft is the same as the ratio of 40 ft to 10 ft.

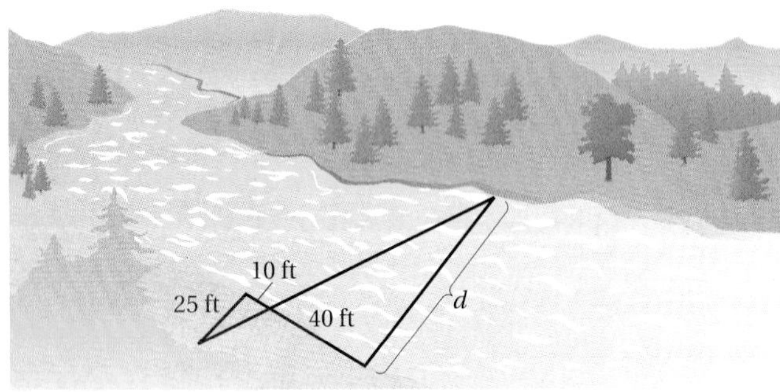

20. To measure the height of a hill, a string is drawn tight from level ground to the top of the hill. A 3-ft yardstick is placed under the string, touching it at point P, a distance of 5 ft from point G, where the string touches the ground. The string is then detached and found to be 120 ft long. How high is the hill?

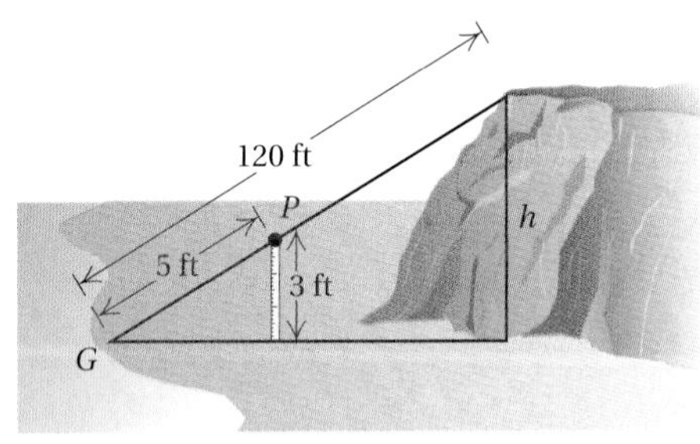

Skill Maintenance

Multiply. [2.4d], [3.3a]

21. $2\frac{4}{5} \times 10\frac{1}{2}$

22. 3.05×0.08

23. $8 \times 9\frac{3}{4}$

24. 10.01×6.11

Synthesis

25. ◆ Is it possible for two triangles to have two pairs of sides that are proportional without the triangles being similar? Why or why not?

26. ◆ Design for a classmate a problem involving similar triangles for which

$$\frac{18}{128.95} = \frac{x}{789.89}.$$

Summary and Review Exercises: Chapter 6

Important Properties and Formulas

Perimeter of a Rectangle:	$P = 2 \cdot (l + w)$, or $P = 2 \cdot l + 2 \cdot w$
Perimeter of a Square:	$P = 4 \cdot s$
Area of a Rectangle:	$A = l \cdot w$
Area of a Square:	$A = s \cdot s$, or $A = s^2$
Area of a Parallelogram:	$A = b \cdot h$
Area of a Triangle:	$A = \frac{1}{2} \cdot b \cdot h$
Area of a Trapezoid:	$A = \frac{1}{2} \cdot h \cdot (a + b)$
Radius and Diameter of a Circle:	$d = 2 \cdot r$, or $r = \frac{d}{2}$
Circumference of a Circle:	$C = \pi \cdot d$, or $C = 2 \cdot \pi \cdot r$
Area of a Circle:	$A = \pi \cdot r \cdot r$, or $A = \pi \cdot r^2$
Volume of a Rectangular Solid:	$V = l \cdot w \cdot h$
Volume of a Circular Cylinder:	$V = \pi \cdot r^2 \cdot h$
Volume of a Sphere:	$V = \frac{4}{3} \cdot \pi \cdot r^3$
Volume of a Cone:	$V = \frac{1}{3} \cdot \pi \cdot r^2 \cdot h$

The objectives to be tested in addition to the material in this chapter are [1.6b], [2.4d], [3.3a], [4.2b, c], and [4.3a, b].

Find the perimeter. [6.2a]

1.

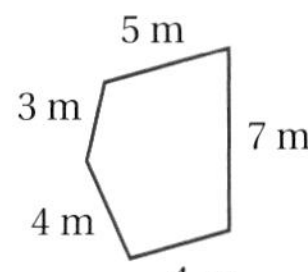

2. The dimensions of a standard-sized tennis court are 78 ft by 36 ft. Find the perimeter and the area of the tennis court. [6.2b], [6.3b]

Find the area. [6.3a], [6.4a]

3.

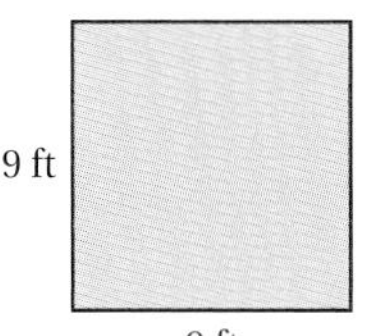

4.

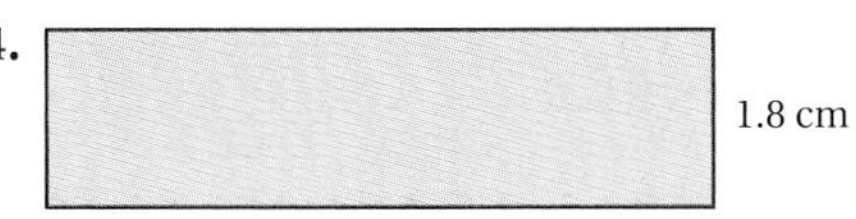

5.

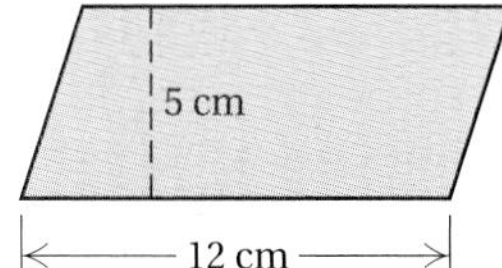

6.

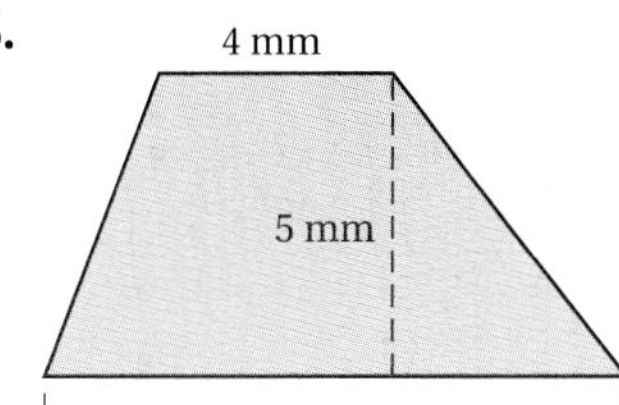

7.

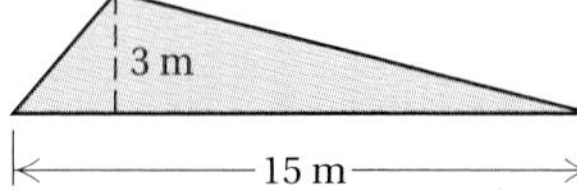

8.

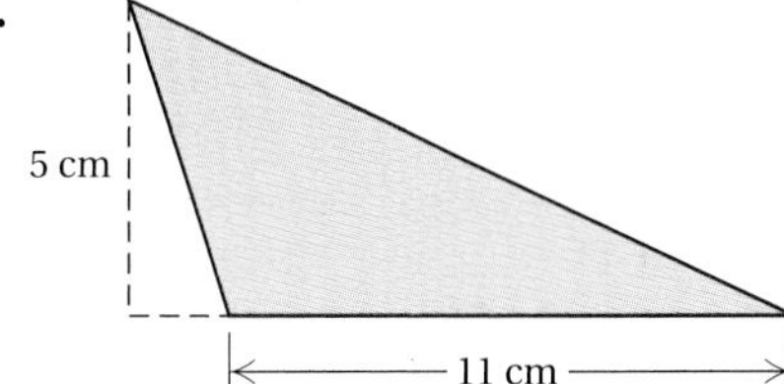

9. A grassy area is to be seeded around three sides of a building and has equal width on the three sides, as shown below. What is the seeded area? [6.3b]

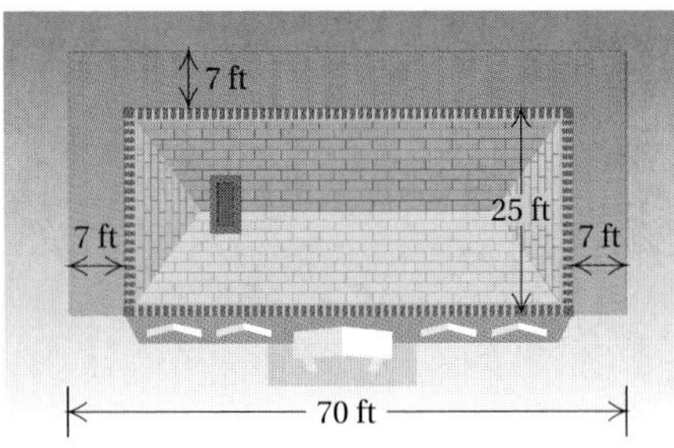

Find the length of a radius of the circle. [6.5a]

10.

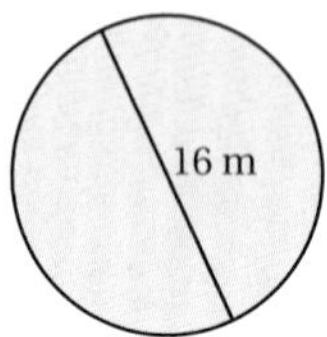

11.

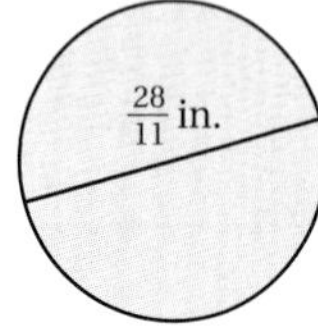

Find the length of a diameter of the circle. [6.5a]

12.

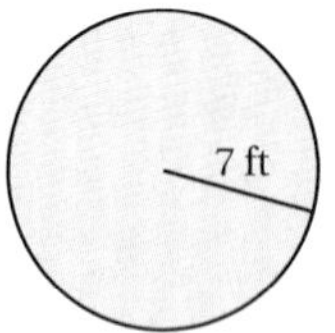

13.

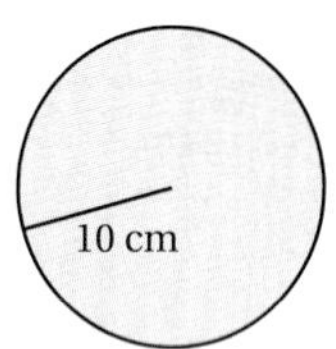

14. Find the circumference of the circle in Exercise 10. Use 3.14 for π. [6.5b]

15. Find the circumference of the circle in Exercise 11. Use $\frac{22}{7}$ for π. [6.5b]

16. Find the area of the circle in Exercise 10. Use 3.14 for π. [6.5c]

17. Find the area of the circle in Exercise 11. Use $\frac{22}{7}$ for π. [6.5c]

Find the volume and the surface area. [6.6a]

18.

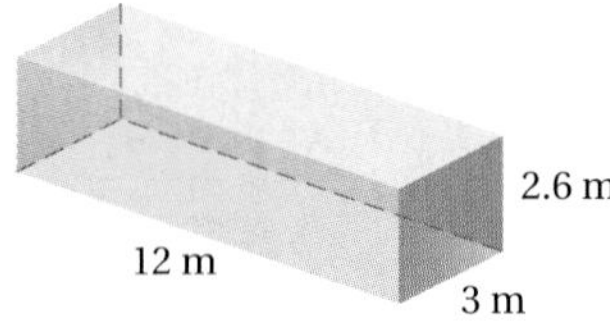

19.

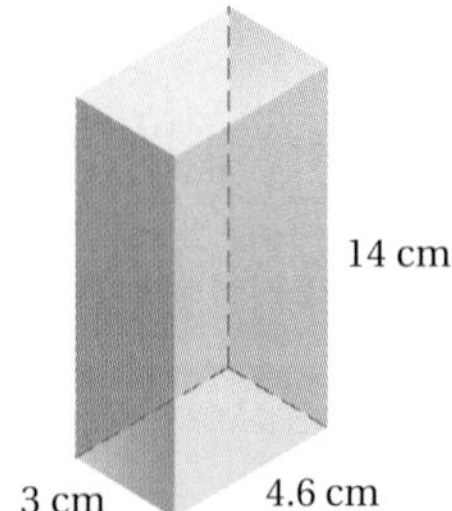

Find the volume. Use 3.14 for π.

20. [6.6b]

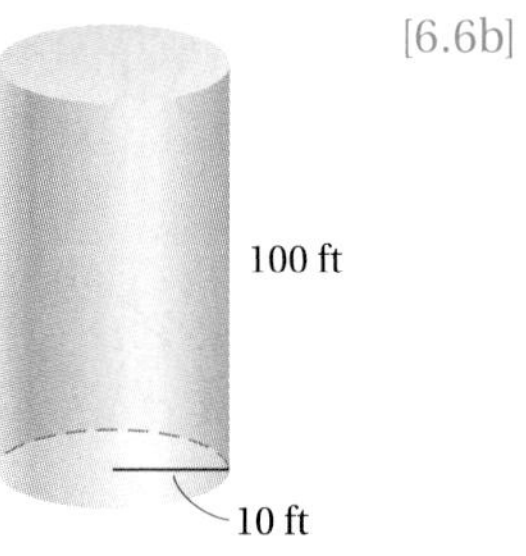

21. [6.6c]

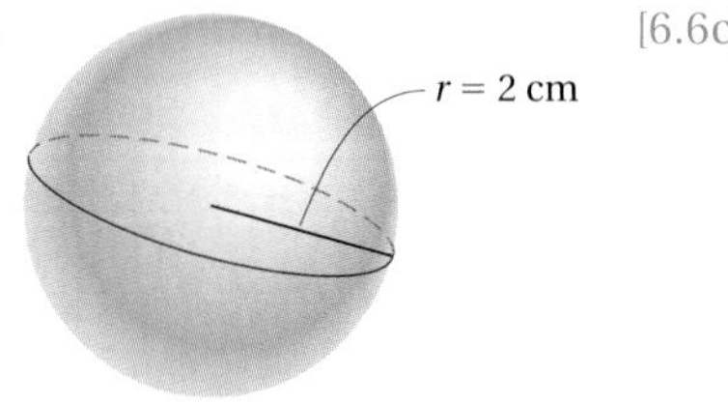

22. [6.6d]

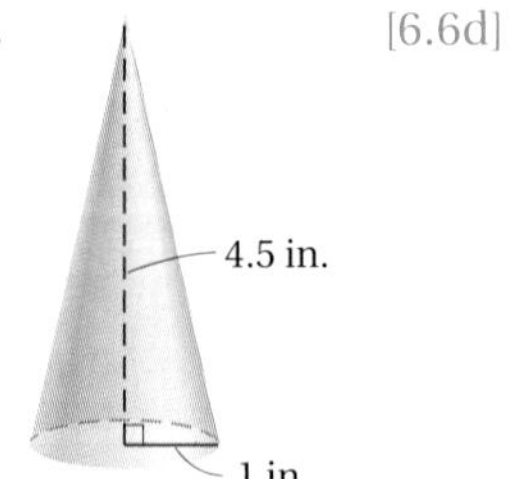

Use this figure for Questions 23–25.

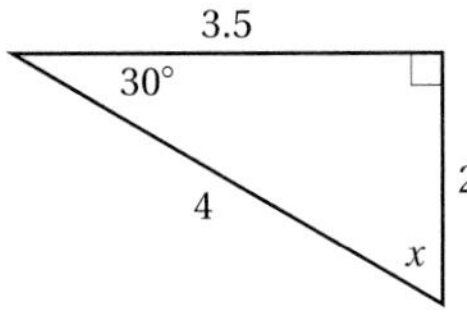

23. Find the missing angle measure. [6.1f]

24. Classify the triangle as equilateral, isosceles, or scalene. [6.1e]

25. Classify the triangle as right, obtuse, or acute. [6.1e]

26. Find the sum of the angle measures of a hexagon. [6.1f]

Find the measure of a complement of an angle with the given measure. [6.7a]

27. 82°

28. 5°

Find the measure of a supplement of an angle with the given measure. [6.7a]

29. 33°

30. 133°

31. In this figure, $m\angle 1 = 38°$ and $m\angle 5 = 105°$. Find $m\angle 2$, $m\angle 3$, $m\angle 4$, and $m\angle 6$. [6.7c]

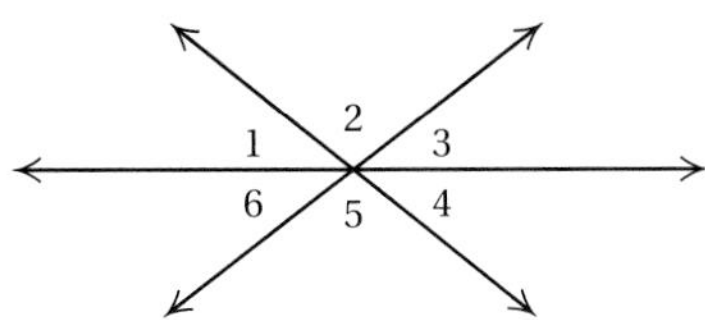

32. In this figure, identify (a) all pairs of corresponding angles, (b) all interior angles, and (c) all pairs of alternate interior angles. [6.7d]

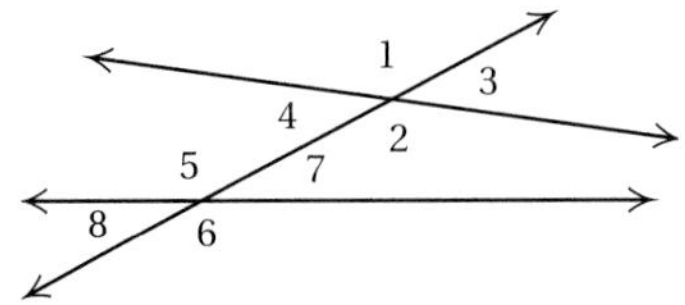

33. If $m \parallel n$ and $m\angle 4 = 135°$, what are the measures of the other angles? [6.7d]

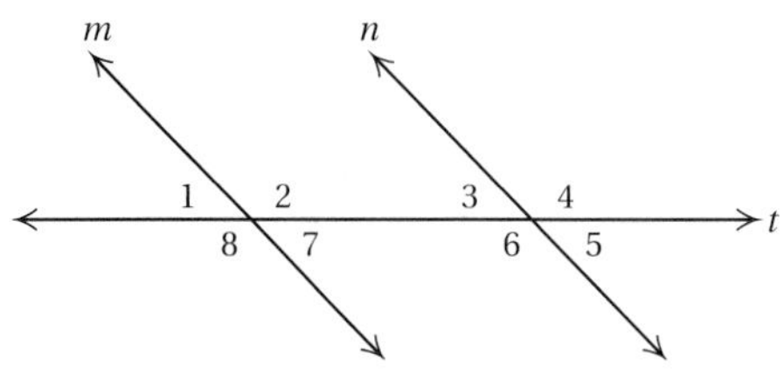

Name the corresponding parts of these congruent triangles. [6.8a]

34. $\triangle DHJ \cong \triangle RZK$

35.

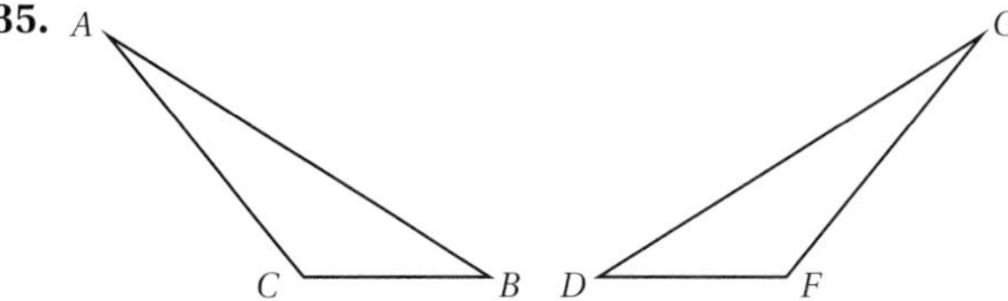

Which property (if any) should be used to show that the following pairs of triangles are congruent? [6.8a]

36.

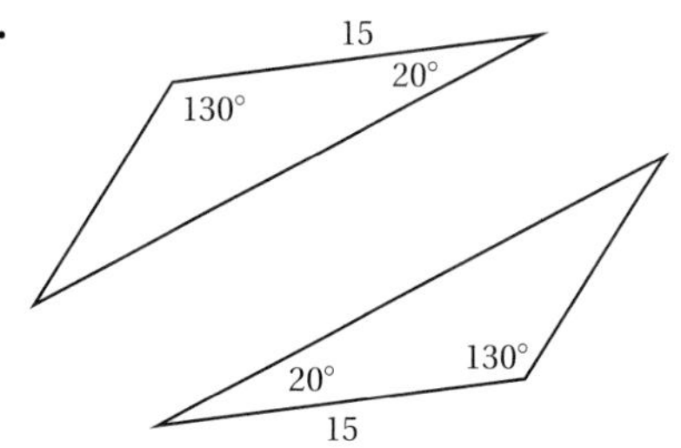

37.

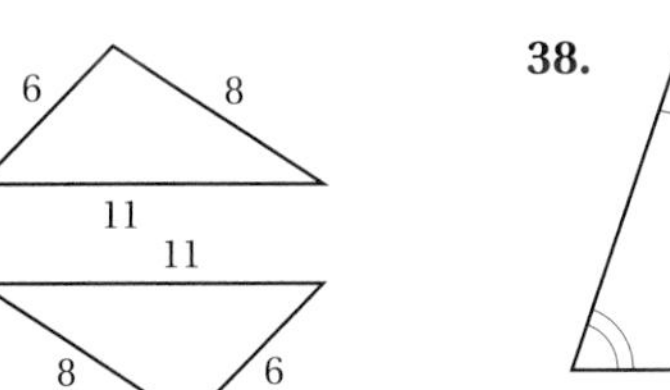

38.

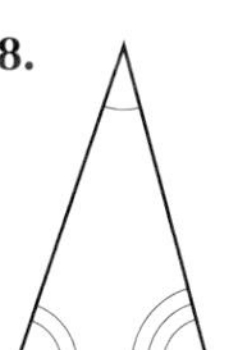

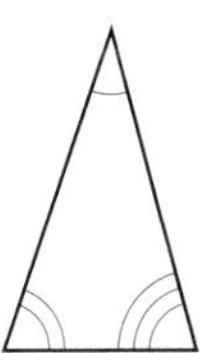

39. J is the midpoint of $\overline{IK}$ and $\overline{HI} \parallel \overline{KL}$. Explain why $\triangle JIH \cong \triangle JKL$. [6.8a]

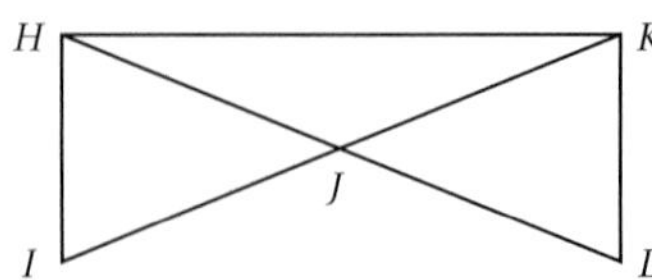

40. Find the measures of the angles and the lengths of the sides of this parallelogram. [6.8b]

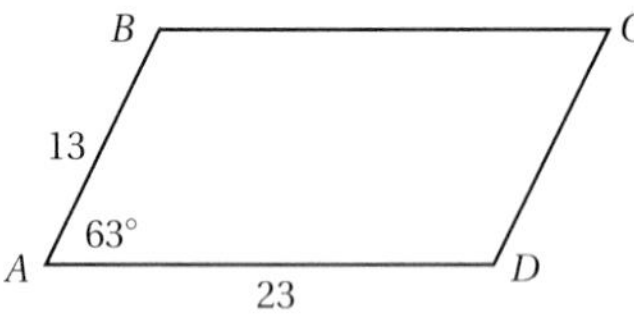

41. If $\triangle CQW \sim \triangle FAS$, name the congruent angles and the proportional sides. [6.9a]

42. If $\triangle NMO \sim \triangle STR$, find MO. [6.9b]

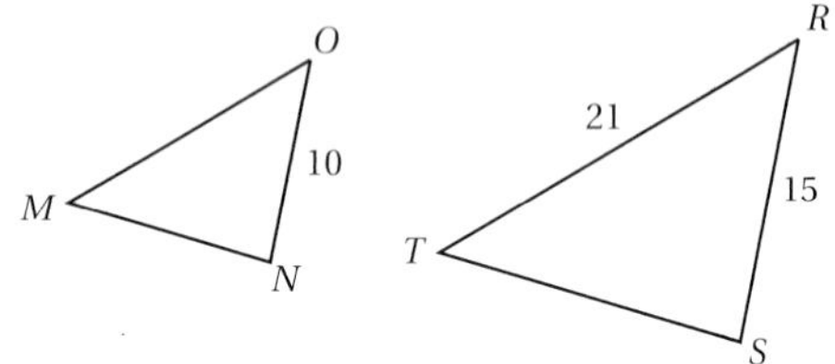

Skill Maintenance

43. Multiply: $5\frac{3}{4} \times 9\frac{1}{2}$. [2.4d]

Evaluate. [1.6b]

44. 4.7^3

45. $\left(\frac{1}{2}\right)^4$

46. Convert to fractional notation: 73%. [4.3b]

47. Convert to percent notation: 0.47. [4.2c]

48. Convert to percent notation: $\frac{23}{25}$. [4.3a]

Synthesis

49. ◈ Describe the difference among linear, area, and volume units of measure. [6.2a], [6.3a], [6.6a]

50. ◈ List and describe all the volume formulas that you have learned in this chapter. [6.6a, b, c, d]

51. Find the area of the shaded region. Use 3.14 for π. [6.5d]

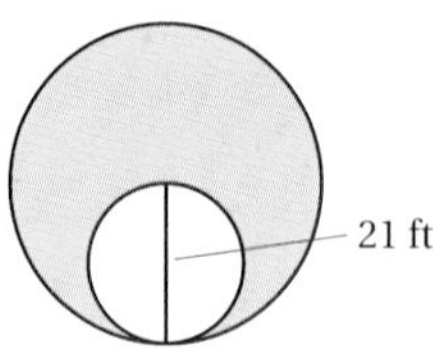

52. A square is cut in half so that the perimeter of each of the resulting rectangles is 30 ft. Find the area of the original square. [6.2a], [6.3a]

53. Find the area, in square meters, of the shaded region. [6.3b]

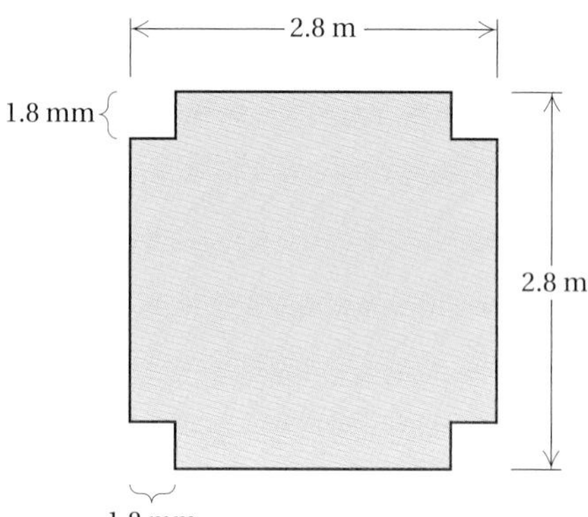

54. Find the area, in square centimeters, of the shaded region. [6.4b]

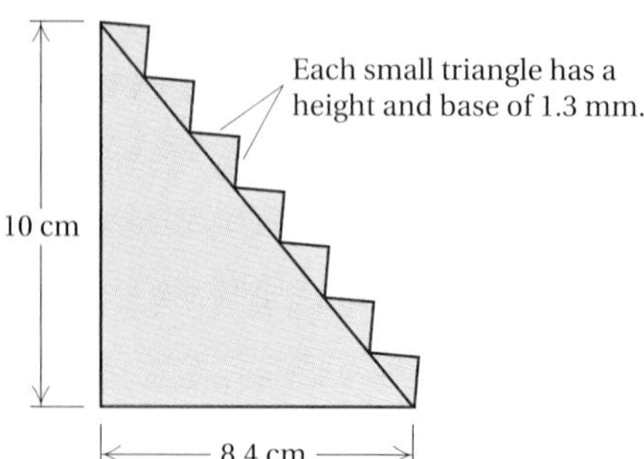

Test: Chapter 6

Answers

1. Find the perimeter.

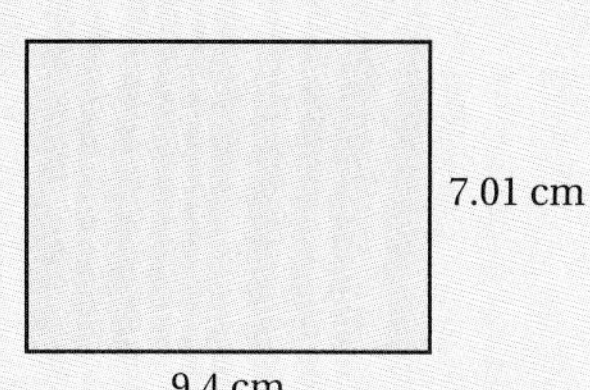

2. The dimensions of a doormat are $2\frac{1}{2}$ ft by $4\frac{1}{2}$ ft. Find the perimeter and the area of the mat.

Find the area.

3.

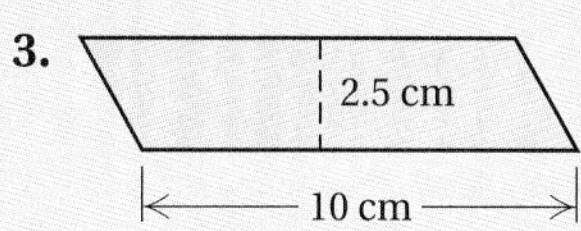

4.

3 m

8 m

5.

4 ft

3 ft

8 ft

6.

25 m

25 m

7. Find the area of the shaded region.

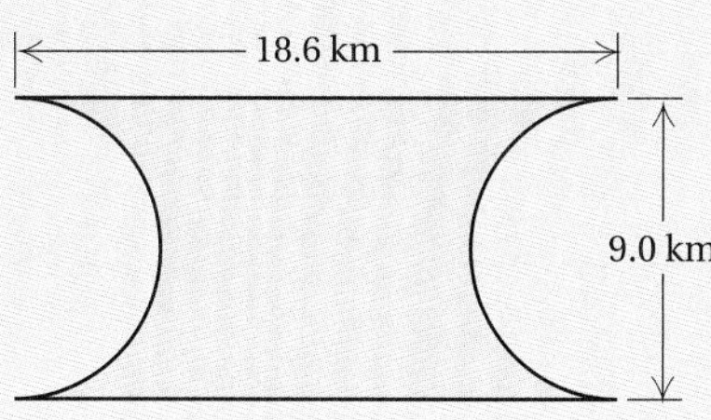

8. Find the length of a diameter of this circle.

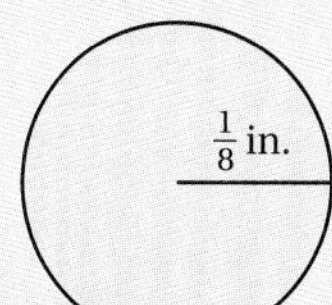

9. Find the length of a radius of this circle.

18 cm

1. ____________

2. ____________

3. ____________

4. ____________

5. ____________

6. ____________

7. ____________

8. ____________

9. ____________

Answers

10. ____________________

11. ____________________

12. ____________________

13. ____________________

14. ____________________

15. ____________________

16. ____________________

17. ____________________

18. ____________________

10. Find the circumference of the circle in Question 8. Use $\frac{22}{7}$ for π.

11. Find the area of the circle in Question 9. Use 3.14 for π.

12. Find the volume and the surface area.

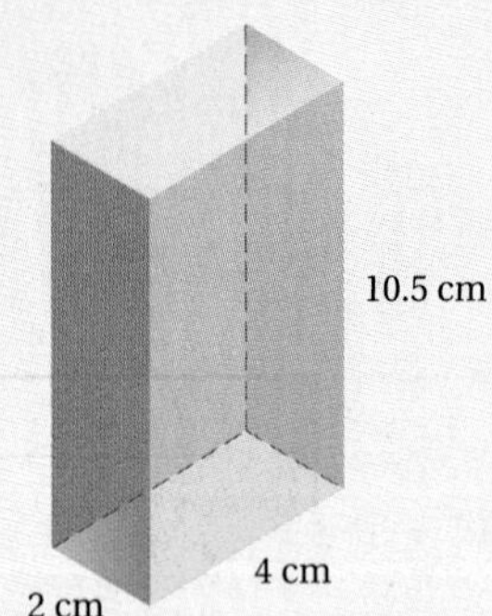

Find the volume. Use 3.14 for π.

13.

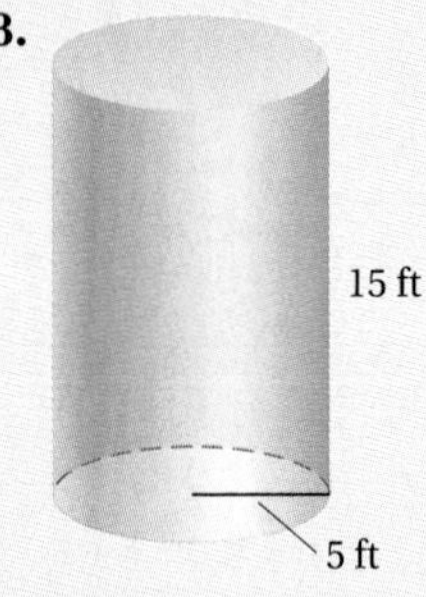

14.

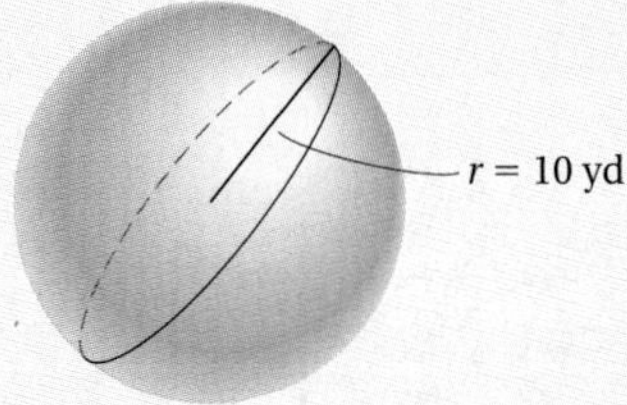

15.

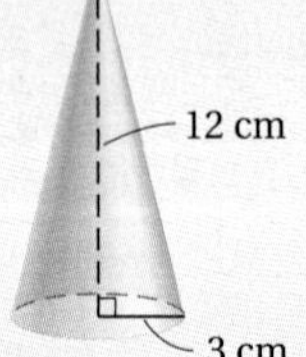

Use the following triangle for Questions 16–18.

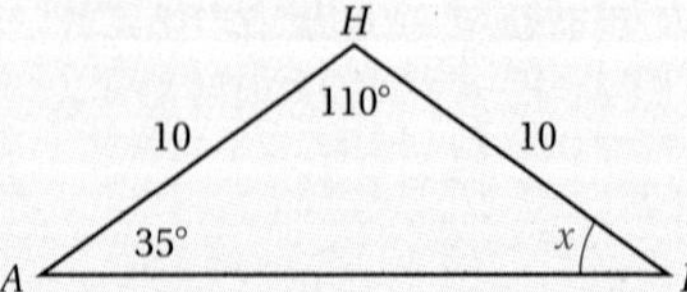

16. Find the missing angle measure.

17. Classify the triangle as equilateral, isosceles, or scalene.

18. Classify the triangle as right, obtuse, or acute.

19. Find the sum of the angle measures of a pentagon.

20. Find the measure of a supplement of an angle of 31°.

21. Find the measure of a complement of an angle of 79°.

22. In the figure, $m\angle 1 = 62°$ and $m\angle 5 = 110°$. Find $m\angle 2$, $m\angle 3$, $m\angle 4$, and $m\angle 6$.

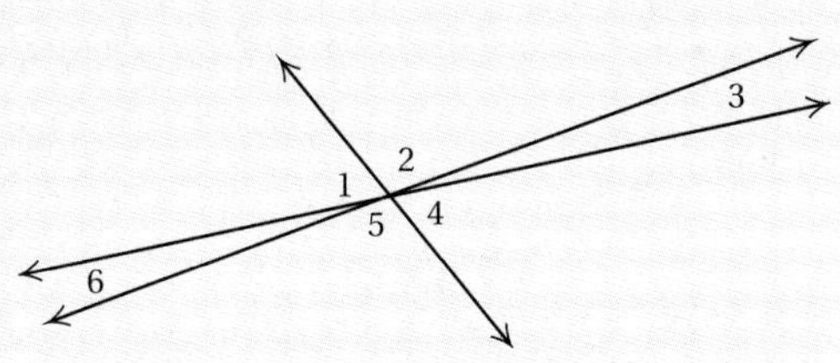

23. If $m \parallel n$ and $m\angle 4 = 120°$, what are the measures of the other angles?

24. Name the corresponding parts of these congruent triangles: $\triangle CWS \cong \triangle ATZ$.

Which property (if any) would you use to show that $\triangle RST \cong \triangle DEF$ with the given information?

25. $\overline{RS} \cong \overline{DE}$, $\overline{RT} \cong \overline{DF}$, and $\angle R \cong \angle D$

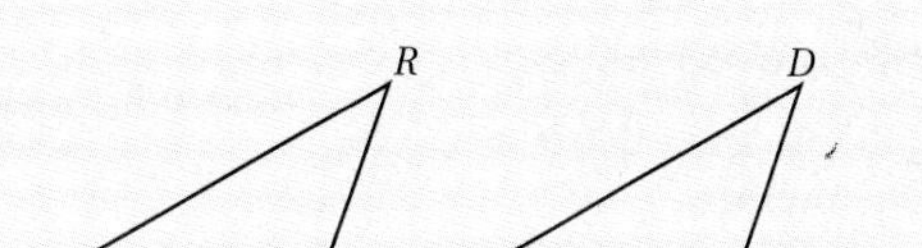

26. $\angle R \cong \angle D$, $\angle S \cong \angle E$, and $\angle T \cong \angle F$

27. $\overline{RS} \cong \overline{DE}$, $\angle R \cong \angle D$, and $\angle S \cong \angle E$

28. $\angle R \cong \angle D$, $\overline{RT} \cong \overline{DF}$, and $\overline{ST} \cong \overline{EF}$

Answers

19. ______________

20. ______________

21. ______________

22. ______________

23. ______________

24. ______________

25. ______________

26. ______________

27. ______________

28. ______________

Answers

29. ______

30. ______

31. ______

32. ______

33. ______

34. ______

35. ______

36. ______

37. ______

38. ______

39. ______

40. ______

41. ______

29. The perimeter of $\square DEFG$ is 62. Find the measures of the angles and the lengths of the sides.

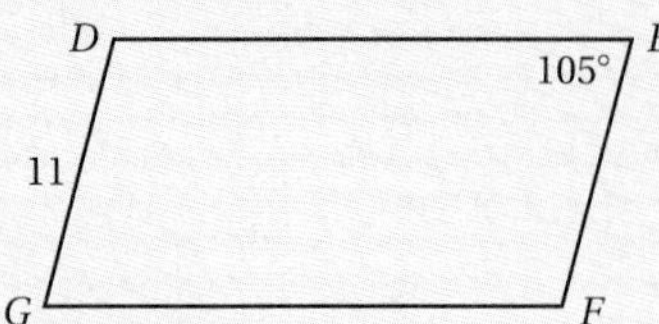

30. In $\square JKLM$, $JN = 3.2$ and $KN = 3$. Find the lengths of the diagonals, $\overline{LJ}$ and $\overline{KM}$.

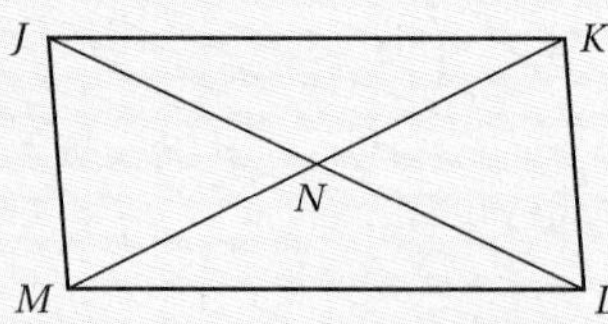

31. If $\triangle ERS \sim \triangle TGF$, name the congruent angles and the proportional sides.

32. If $\triangle GTR \sim \triangle ZEK$, find EK and ZK.

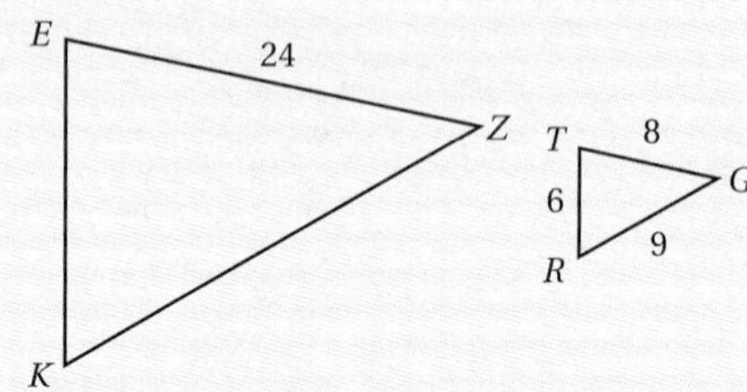

Skill Maintenance

Multiply.

33. 4.6×2.31

34. $8\frac{1}{4} \times 2\frac{2}{3}$

Evaluate.

35. 10^3

36. $\left(\frac{1}{4}\right)^2$

37. Convert to percent notation: $\frac{13}{16}$.

38. Convert to decimal notation: 93.2%.

39. Convert to fractional notation: $33\frac{1}{3}\%$.

Synthesis

40. Find the area of the shaded region. Give the answer in square feet.

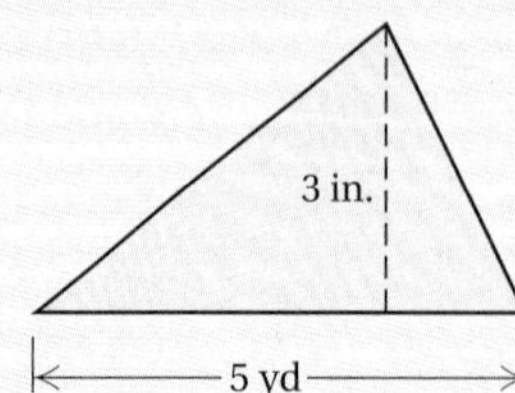

41. Find the volume of the solid. Give the answer in cubic feet. (Note that the solid is not drawn in perfect proportion.)

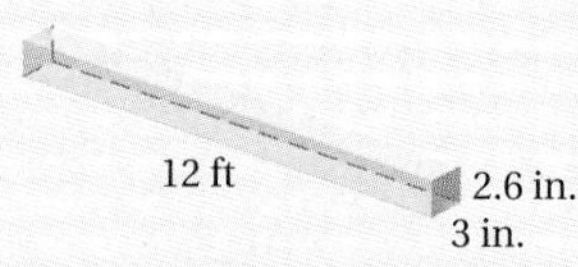

7

Introduction to Real Numbers and Algebraic Expressions

Introduction

In this chapter, we consider the number system used most often in algebra. It is called the real-number system. We will learn to add, subtract, multiply, and divide real numbers and to manipulate certain expressions. Such manipulation will be important when we solve equations and applied problems in Chapter 8.

An Application

The casino game of blackjack makes use of many card-counting systems to give players a winning edge if the count becomes negative. One such system is called *High–Low*, first developed by Harvey Dubner in 1963. Each card counts as -1, 0, or 1 as follows:

2, 3, 4, 5, 6	count as $+1$;
7, 8, 9	count as 0;
10, J, Q, K, A	count as -1.

Find the final count on the sequence of cards

K, A, 2, 4, 5, 10, J, 8, Q, K, 5.

Source: Patterson, Jerry L., *Casino Gambling.* New York: Perigee, 1982

The Mathematics

We add the following numbers:

$(-1) + (-1) + 1 + 1 + 1 + (-1) + (-1) + 0 + (-1) + (-1) + 1 = -2.$

The numbers in red are negative numbers.

This problem appears as Exercise 120 in Exercise Set 7.4.

World Wide Web For more information, visit us at www.mathmax.com

Pretest: Chapter 7

1. Evaluate $x/2y$ for $x = 5$ and $y = 8$.

2. Write an algebraic expression: Seventy-eight percent of some number.

3. Find the area of a rectangle when the length is 22.5 ft and the width is 16 ft.

4. Find $-x$ when $x = -12$.

Use either < or > for ▒ to write a true sentence.

5. 0 ▒ -5

6. 10 ▒ -5

7. -35 ▒ -45

8. $-\frac{2}{3}$ ▒ $\frac{4}{5}$

Find the absolute value.

9. $|-12|$

10. $|2.3|$

11. $|0|$

Find the opposite, or additive inverse.

12. 5.4

13. $-\frac{2}{3}$

Find the reciprocal.

14. 10

15. $-\frac{2}{3}$

Compute and simplify.

16. $-9 + (-8)$

17. $20.2 - (-18.4)$

18. $-\frac{5}{6} - \frac{3}{10}$

19. $-11.5 + 6.5$

20. $-9(-7)$

21. $\frac{5}{8}\left(-\frac{2}{3}\right)$

22. $-19.6 \div 0.2$

23. $-56 \div (-7)$

24. $12 - (-6) + 14 - 8$

25. $20 - 10 \div 5 + 2^3$

Multiply.

26. $9(z - 2)$

27. $-2(2a + b - 5c)$

Factor.

28. $4x - 12$

29. $6y - 9z - 18$

Simplify.

30. $3y - 7 - 2(2y + 3)$

31. $\{2[3(y + 1) - 4] - [5(y - 3) - 5]\}$

32. Write an inequality with the same meaning as $x > 12$.

Objectives for Retesting

The objectives to be tested in addition to the material in this chapter are as follows.

[1.6b] Evaluate exponential notation.
[1.9a] Find the LCM of two or more numbers.
[2.2c] Divide and simplify using fractional notation.
[4.3a] Convert from fractional notation to percent notation.

7.1 Introduction to Algebra

Many types of problems require the use of equations in order to be solved effectively. The study of algebra involves the use of equations to solve problems. Equations are constructed from algebraic expressions. The purpose of this section is to introduce you to the types of expressions encountered in algebra.

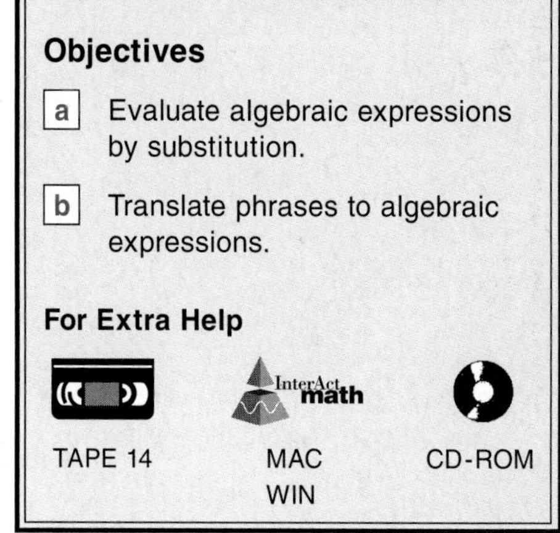

Objectives
a Evaluate algebraic expressions by substitution.
b Translate phrases to algebraic expressions.
For Extra Help
InterAct math
TAPE 14
MAC
WIN
CD-ROM

a Evaluating Algebraic Expressions

In arithmetic, you have worked with expressions such as

$$49 + 75, \quad 8 \times 6.07, \quad 29 - 14, \quad \text{and} \quad \frac{5}{6}.$$

In algebra, we use certain letters for numbers and work with *algebraic expressions* such as

$$x + 75, \quad 8 \times y, \quad 29 - t, \quad \text{and} \quad \frac{a}{b}.$$

Sometimes a letter can represent various numbers. In that case, we call the letter a **variable**. Let $a =$ your age. Then a is a variable since a changes from year to year. Sometimes a letter can stand for just one number. In that case, we call the letter a **constant**. Let $b =$ your date of birth. Then b is a constant.

Where do algebraic expressions occur? Most often we encounter them when we are solving applied problems. For example, consider the bar graph shown at right, one that we might find in a book or magazine. Suppose we want to know how much longer the diameter of Earth is than the diameter of Mars.

In algebra, we translate the problem into an equation. It might be done as follows.

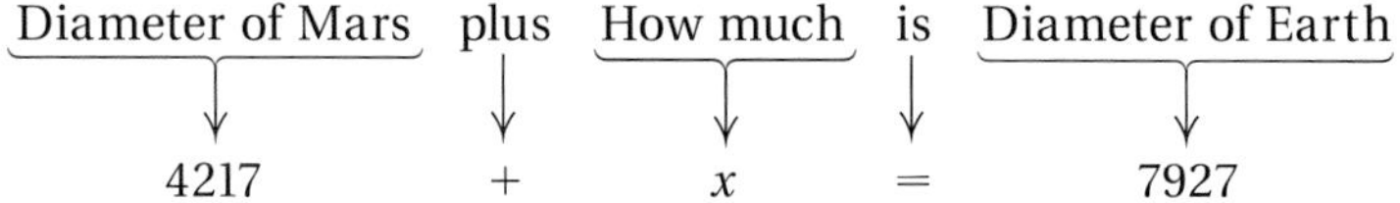

Note that we have an algebraic expression on the left of the equals sign. To find the number x, we can subtract 4217 on both sides of the equation:

$$4217 + x = 7927$$
$$4217 + x - 4217 = 7927 - 4217$$
$$x = 3710.$$

The value of x gives us the answer, 3710 miles.

In arithmetic, you probably would do this subtraction right away without considering an equation. In algebra, more complex problems are difficult to solve without first solving an equation.

Do Exercise 1.

1. Translate this problem to an equation. Use the graph below.

How much longer is the diameter of Venus than the diameter of Pluto?

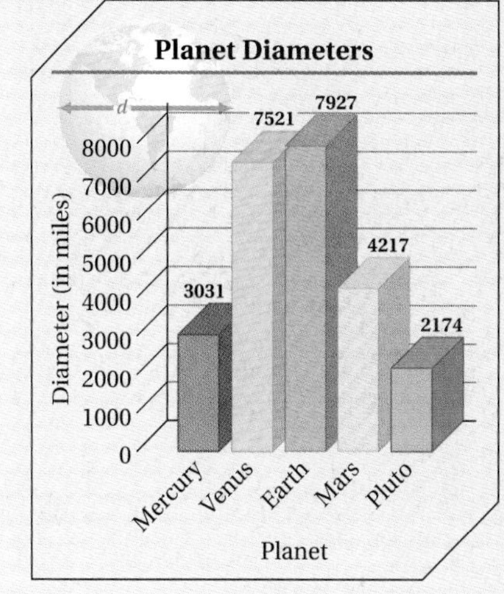

Answer on page A-17

2. Evaluate $a + b$ for $a = 38$ and $b = 26$.

3. Evaluate $x - y$ for $x = 57$ and $y = 29$.

4. Evaluate $4t$ for $t = 15$.

5. Find the area of a rectangle when l is 24 ft and w is 8 ft.

Calculator Spotlight

Evaluating Algebraic Expressions. To evaluate the expression $12m/n$ for $m = 8$ and $n = 16$, we first write (or think)

$$\frac{12 \cdot 8}{16}.$$

Then we press

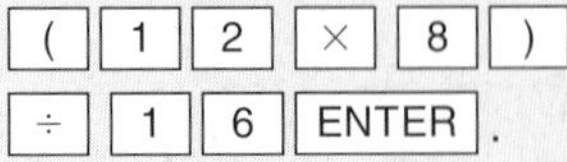

The answer is 6.

Exercises

Evaluate.

1. $\frac{12m}{n}$, for $m = 82.3$ and $n = 16.56$
2. $a + b$, for $a = 8.2$ and $b = 3.7$
3. $b + a$, for $a = 8.2$ and $b = 3.7$
4. $27xy$, for $x = 12.7$ and $y = 100.4$
5. $3a + 2b$, for $a = 8.2$ and $b = 3.7$
6. $2a + 3b$, for $a = 8.2$ and $b = 3.7$

Answers on page A-17

An **algebraic expression** consists of variables, constants, numerals, and operation signs. When we replace a variable with a number, we say that we are **substituting** for the variable. This process is called **evaluating the expression.**

Example 1 Evaluate $x + y$ for $x = 37$ and $y = 29$.

We substitute 37 for x and 29 for y and carry out the addition:

$$x + y = 37 + 29 = 66.$$

The number 66 is called the **value** of the expression.

Algebraic expressions involving multiplication can be written in several ways. For example, "8 times a" can be written as $8 \times a$, $8 \cdot a$, $8(a)$, or simply $8a$. Two letters written together without an operation symbol, such as ab, also indicates a multiplication.

Example 2 Evaluate $3y$ for $y = 14$.

$$3y = 3(14) = 42$$

Do Exercises 2–4.

Example 3 The area A of a rectangle of length l and width w is given by the formula $A = lw$. Find the area when l is 24.5 in. and w is 16 in.

We substitute 24.5 in. for l and 16 in. for w and carry out the multiplication:

$$\begin{aligned} A = lw &= (24.5 \text{ in.})(16 \text{ in.}) \\ &= (24.5)(16)(\text{in.})(\text{in.}) \\ &= 392 \text{ in}^2, \text{ or } 392 \text{ square inches.} \end{aligned}$$

w

l

Do Exercise 5.

Algebraic expressions involving division can also be written in several ways. For example, "8 divided by t" can be written as $8 \div t$, $\frac{8}{t}$, or $8/t$, where the fraction bar is a division symbol.

Example 4 Evaluate $\frac{a}{b}$ for $a = 63$ and $b = 9$.

We substitute 63 for a and 9 for b and carry out the division:

$$\frac{a}{b} = \frac{63}{9} = 7.$$

Example 5 Evaluate $\frac{12m}{n}$ for $m = 8$ and $n = 16$.

$$\frac{12m}{n} = \frac{12 \cdot 8}{16} = \frac{96}{16} = 6$$

Do Exercises 6 and 7 on the following page.

Example 6 *Motorcycle Travel.* Ed takes a trip on his motorcycle. He wants to travel 660 mi on a particular day. The time t, in hours, that it takes to travel 660 mi is given by

$$t = \frac{660}{r},$$

where r is the speed of Ed's motorcycle. Find the time of travel if the speed r is 60 mph.

We substitute 60 for r and carry out the division:

$$t = \frac{660}{r} = \frac{660}{60} = 11 \text{ hr}.$$

Do Exercise 8.

b Translating to Algebraic Expressions

In algebra, we translate problems to equations. The different parts of an equation are translations of word phrases to algebraic expressions. It is easier to translate if we know that certain words often translate to certain operation symbols.

KEY WORDS			
Addition (+)	**Subtraction (−)**	**Multiplication (·)**	**Division (÷)**
add sum plus more than increased by	subtract difference minus less than decreased by take from	multiply product times twice of	divide quotient divided by

Example 7 Translate to an algebraic expression:

Twice (or two times) some number.

Think of some number, say, 8. What number is twice 8? It is 16. How did you get 16? You multiplied by 2. Do the same thing using a variable. We can use any variable we wish, such as x, y, m, or n. Let's use y to stand for some number. If we multiply by 2, we get an expression

$$y \times 2, \quad 2 \times y, \quad 2 \cdot y, \quad \text{or} \quad 2y.$$

Example 8 Translate to an algebraic expression:

Thirty-eight percent of some number.

The word "of" translates to a multiplication symbol, so we get the following expressions as a translation:

$$38\% \cdot n, \quad 0.38 \times n, \quad \text{or} \quad 0.38n.$$

6. Evaluate a/b for $a = 200$ and $b = 8$.

7. Evaluate $10p/q$ for $p = 40$ and $q = 25$.

8. *Motorcycle Travel.* Find the time it takes to travel 660 mi if the speed is 55 mph.

Answers on page A-17

Translate to an algebraic expression.

9. Eight less than some number

10. Eight more than some number

11. Four less than some number

12. Half of a number

13. Six more than eight times some number

14. The difference of two numbers

15. Fifty-nine percent of some number

16. Two hundred less than the product of two numbers

17. The sum of two numbers

Answers on page A-17

Example 9 Translate to an algebraic expression:

Seven less than some number.

We let

x represent the number.

Now if the number were 23, then the translation would be $23 - 7$. If we knew the number to be 345, then the translation would be $345 - 7$. If the number is x, then the translation is

$$x - 7.$$

CAUTION! Note that $7 - x$ is *not* a correct translation of the expression in Example 9. The expression $7 - x$ is a translation of "seven minus some number" or "some number less than seven."

Example 10 Translate to an algebraic expression:

Eighteen more than a number.

We let

$t =$ the number.

Now if the number were 26, then the translation would be $26 + 18$. If we knew the number to be 174, then the translation would be $174 + 18$. If the number is t, then the translation is

$$t + 18, \quad \text{or} \quad 18 + t.$$

Example 11 Translate to an algebraic expression:

A number divided by 5.

We let

$m =$ the number.

Now if the number were 76, then the translation would be $76 \div 5$, or $76/5$, or $\frac{76}{5}$. If the number were 213, then the translation would be $213 \div 5$, or $213/5$, or $\frac{213}{5}$. If the number is m, then the translation is

$$m \div 5, \quad m/5, \quad \text{or} \quad \frac{m}{5}.$$

Example 12 Translate each of the following phrases to an algebraic expression.

Phrase	Algebraic Expression
Five more than some number	$n + 5$, or $5 + n$
Half of a number	$\frac{1}{2}t$, $\frac{t}{2}$, or $t/2$
Five more than three times some number	$3p + 5$, or $5 + 3p$
The difference of two numbers	$x - y$
Six less than the product of two numbers	$mn - 6$
Seventy-six percent of some number	$76\%z$, or $0.76z$

Do Exercises 9–17.

Exercise Set 7.1

a Substitute to find values of the expressions in each of the following applied problems.

1. *Enrollment Costs.* At Emmett Community College, it costs \$600 to enroll in the 8 A.M. section of Elementary Algebra. Suppose that the variable n stands for the number of students who enroll. Then $600n$ stands for the total amount of money collected for this course. How much is collected if 34 students enroll? 78 students? 250 students?

2. *Commuting Time.* It takes Erin 24 min less time to commute to work than it does George. Suppose that the variable x stands for the time it takes George to get to work. Then $x - 24$ stands for the time it takes Erin to get to work. How long does it take Erin to get to work if it takes George 56 min? 93 min? 105 min?

3. The area A of a triangle with base b and height h is given by $A = \frac{1}{2}bh$. Find the area when $b = 45$ m (meters) and $h = 86$ m.

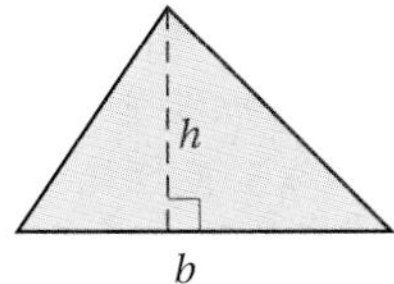

4. The area A of a parallelogram with base b and height h is given by $A = bh$. Find the area of the parallelogram when the height is 15.4 cm (centimeters) and the base is 6.5 cm.

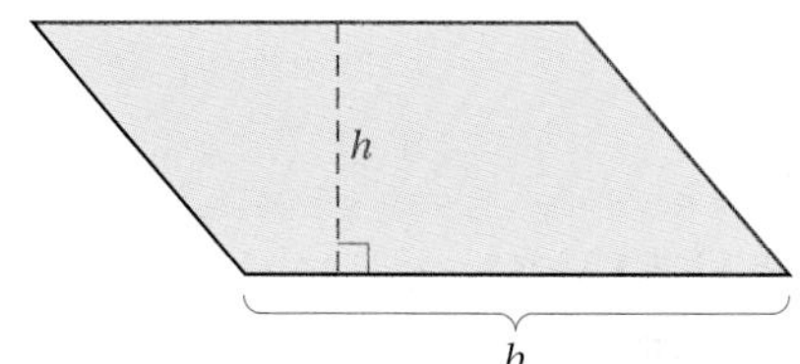

5. *Distance Traveled.* A driver who drives at a speed of r mph for t hr will travel a distance d mi given by $d = rt$ mi. How far will a driver travel at a speed of 65 mph for 4 hr?

6. *Simple Interest.* The simple interest I on a principal of P dollars at interest rate r for time t, in years, is given by $I = Prt$. Find the simple interest on a principal of \$4800 at 9% for 2 yr. (*Hint*: 9% = 0.09.)

Evaluate.

7. $8x$, for $x = 7$

8. $6y$, for $y = 7$

9. $\frac{a}{b}$, for $a = 24$ and $b = 3$

10. $\frac{p}{q}$, for $p = 16$ and $q = 2$

11. $\frac{3p}{q}$, for $p = 2$ and $q = 6$

12. $\frac{5y}{z}$, for $y = 15$ and $z = 25$

13. $\frac{x + y}{5}$, for $x = 10$ and $y = 20$

14. $\frac{p + q}{2}$, for $p = 2$ and $q = 16$

15. $\frac{x - y}{8}$, for $x = 20$ and $y = 4$

16. $\frac{m - n}{5}$, for $m = 16$ and $n = 6$

b Translate to an algebraic expression.

17. 7 more than b

18. 9 more than t

19. 12 less than c

20. 14 less than d

21. 4 increased by q

22. 13 increased by z

23. b more than a

24. c more than d

25. x less than y

26. c less than h

27. x added to w

28. s added to t

29. m subtracted from n

30. p subtracted from q

31. The sum of r and s

32. The sum of a and b

33. Twice z

34. Three times q

35. 3 multiplied by m

36. The product of 8 and t

37. The product of 89% and some number

38. 67% of some number

39. A driver drove at a speed of 55 mph for t hours. How far did the driver travel?

40. An executive assistant has d dollars before going to an office supply store. He bought some fax paper for \$18.95. How much did he have after the purchase?

Skill Maintenance

Find the prime factorization. [1.7d]

41. 54

42. 32

43. 108

44. 192

Find the LCM. [1.9a]

45. 6, 18

46. 6, 24, 32

47. 10, 20, 30

48. 16, 24

Synthesis

49. ◆ If the length of a rectangle is doubled, does the area double? Why or why not?

50. ◆ If the height and the base of a triangle are doubled, what happens to the area? Explain.

Translate to an algebraic expression.

51. Some number x plus three times y

52. Some number a plus 2 plus b

53. A number that is 3 less than twice x

54. Your age in 5 years, if you are a years old now

7.2 The Real Numbers

A **set** is a collection of objects. For our purposes, we will most often be considering sets of numbers. One way to name a set uses what is called **roster notation.** For example, roster notation for the set containing the numbers 0, 2, and 5 is {0, 2, 5}.

Sets that are parts of other sets are called **subsets**. In this section, we become acquainted with the set of *real numbers* and its various subsets.

Two important subsets of the real numbers are listed below using roster notation.

> **Natural numbers** = {1, 2, 3, ...}. These are the numbers used for counting.

> **Whole numbers** = {0, 1, 2, 3, ...}. This is the set of natural numbers with 0 included.

We can represent these sets on a number line. The natural numbers are those to the right of zero. The whole numbers are the natural numbers and zero.

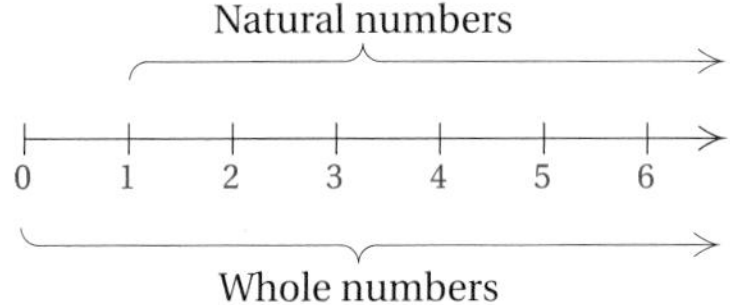

We create a new set, called the *integers,* by starting with the whole numbers, 0, 1, 2, 3, and so on. For each natural number 1, 2, 3, and so on, we obtain a new number to the left of zero on the number line:

For the number 1, there will be an *opposite* number -1 (negative 1).

For the number 2, there will be an *opposite* number -2 (negative 2).

For the number 3, there will be an *opposite* number -3 (negative 3), and so on.

The **integers** consist of the whole numbers and these new numbers. We picture them on a number line as follows.

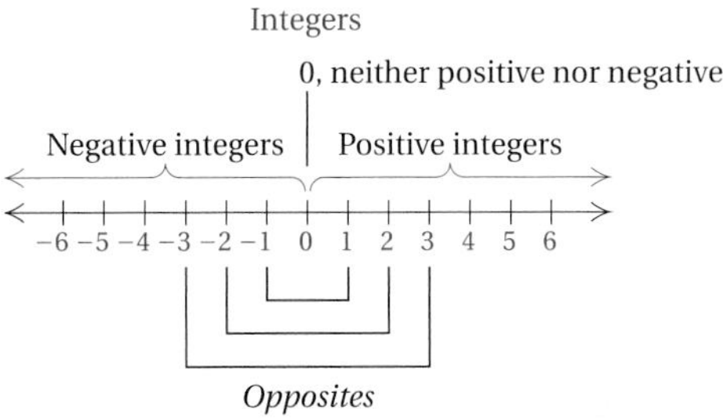

We call these new numbers to the left of 0 **negative integers.** The natural numbers are also called **positive integers.** Zero is neither positive nor negative. We call -1 and 1 **opposites** of each other. Similarly, -2 and 2 are opposites, -3 and 3 are opposites, -100 and 100 are opposites, and 0 is its own opposite. Pairs of opposite numbers like -3 and 3 are equidistant from 0. The integers extend infinitely on the number line to the left and right of zero.

Objectives

a Name the integer that corresponds to a real-world situation.

b Graph rational numbers on a number line.

c Convert from fractional notation to decimal notation for a rational number.

d Determine which of two real numbers is greater and indicate which, using $<$ or $>$; given an inequality like $a < b$, write another inequality with the same meaning. Determine whether an inequality like $-3 \le 5$ is true or false.

e Find the absolute value of a real number.

For Extra Help

TAPE 14 MAC WIN CD-ROM

The set of **integers** = {..., −5, −4, −3, −2, −1, 0, 1, 2, 3, 4, 5, ...}.

a Integers and the Real World

Integers correspond to many real-world problems and situations. The following examples will help you get ready to translate problem situations that involve integers to mathematical language.

Example 1 Tell which integer corresponds to this situation: The temperature is 3 degrees below zero.

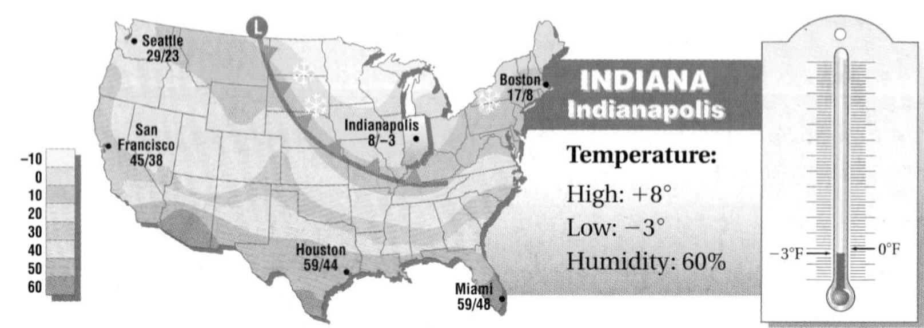

The integer −3 corresponds to the situation. The temperature is −3°.

Example 2 *Jeopardy.* Tell which integer corresponds to this situation: A contestant missed a $600 question on the television game show "Jeopardy."

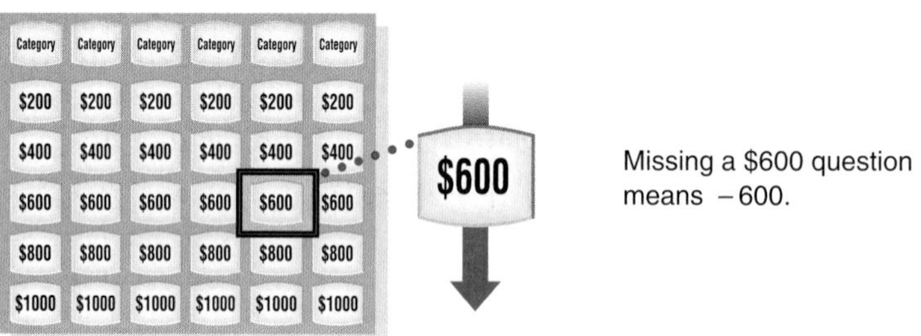

Missing a $600 question causes a $600 loss on the score—that is, the contestant earns −600 dollars.

Example 3 *Elevation.* Tell which integer corresponds to this situation: The lowest point in New Orleans is 8 ft below sea level.

The integer −8 corresponds to the situation. The elevation is −8 ft.

Example 4 *Dow Jones Industrial Average.* Tell which integers correspond to this situation: The largest daily decrease in the Dow Jones Industrial Average was 508 points; the largest increase was 187 (***Source:*** *The Guinness Book of Records*).

The integer -508 corresponds to a decrease in the average. The integer 187 corresponds to an increase in the average.

Do Exercises 1–4.

Tell which integers correspond to the given situation.

1. The halfback gained 8 yd on the first down. The quarterback was sacked for a 5-yd loss on the second down.

2. The highest temperature ever recorded in the United States was 134° in Death Valley on July 10, 1913. The coldest temperature ever recorded in the United States was 80° below zero in Prospect Creek, Alaska, in January 1971.

3. At 10 sec before liftoff, ignition occurs. At 156 sec after liftoff, the first stage is detached from the rocket.

4. A submarine dove 120 ft, rose 50 ft, and then dove 80 ft.

b The Rational Numbers

We created the set of integers by obtaining a negative number for each natural number. To create a larger number system, called the set of **rational numbers,** we consider quotients of integers with nonzero divisors. The following are rational numbers:

$$\frac{2}{3},\quad -\frac{2}{3},\quad \frac{7}{1},\quad 4,\quad -3,\quad 0,\quad \frac{23}{-8},\quad 2.4,\quad -0.17,\quad 10\frac{1}{2}.$$

The number $-\frac{2}{3}$ (read "negative two-thirds") can also be named $\frac{2}{-3}$ or $\frac{-2}{3}$. The number 2.4 can be named $\frac{24}{10}$ or $\frac{12}{5}$, and -0.17 can be named $-\frac{17}{100}$.

Note that this new set of numbers, the rational numbers, contains the whole numbers, the integers, and the arithmetic numbers (also called the nonnegative rational numbers). We can describe the set of rational numbers using **set-builder notation,** as follows.

> The set of **rational numbers** $= \left\{\frac{a}{b} \,\middle|\, a \text{ and } b \text{ are integers and } b \neq 0\right\}$.
>
> (This is read "the set of numbers $\frac{a}{b}$, where a and b are integers and $b \neq 0$.")

We picture the rational numbers on a number line as follows. There is a point on the line for every rational number.

Negative numbers | 0 | Positive numbers

-2.7, $-\frac{3}{2}$, 1.5, 4

-4 -3 -2 -1 0 1 2 3 4

To **graph** a number means to find and mark its point on the number line. Some rational numbers are graphed in the preceding figure.

Example 5 Graph: $\frac{5}{2}$.

The number $\frac{5}{2}$ can be named $2\frac{1}{2}$, or 2.5. Its graph is halfway between 2 and 3.

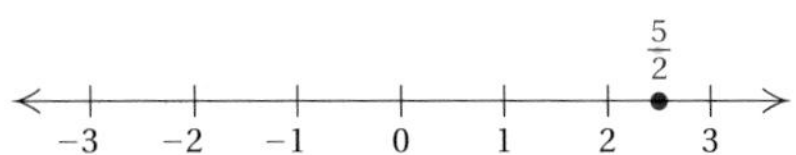

Answers on page A-17

Graph on a number line.

5. $-\frac{7}{2}$

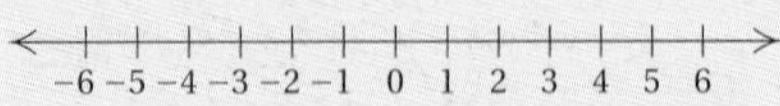

6. -1.4

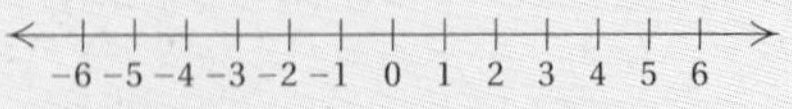

7. $\frac{11}{4}$

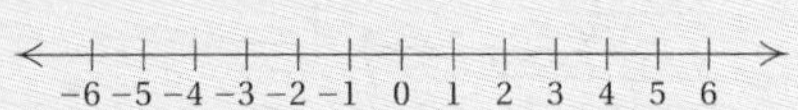

Calculator Spotlight

Negative Numbers on a Calculator. To enter a negative number on a graphing calculator, we use an opposite key (−). Note that this is different from the subtraction key. To enter −5, we press

The display then reads

−5 .

To enter $-\frac{7}{8}$, we press

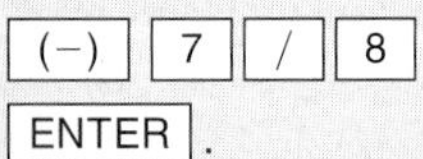

The number is displayed in decimal notation:

−.875 .

Exercises

Press the appropriate keys so that your calculator displays the given number.

1. −3
2. −508
3. −0.17
4. $-\frac{5}{8}$

Answers on page A-17

Example 6 Graph: −3.2.

The graph of −3.2 is $\frac{2}{10}$ of the way from −3 to −4.

Example 7 Graph: $\frac{13}{8}$.

The number $\frac{13}{8}$ can be named $1\frac{5}{8}$, or 1.625. The graph is about $\frac{6}{10}$ of the way from 1 to 2.

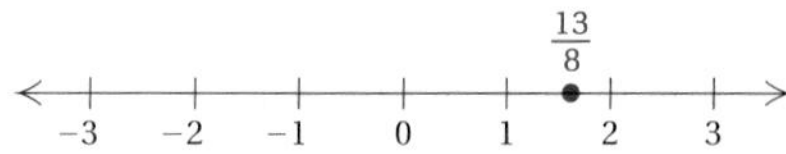

Do Exercises 5–7.

c Notation for Rational Numbers

Each rational number can be named using fractional or decimal notation.

Example 8 Convert to decimal notation: $-\frac{5}{8}$.

We first find decimal notation for $\frac{5}{8}$. Since $\frac{5}{8}$ means 5 ÷ 8, we divide.

```
     0.6 2 5
 8 ) 5.0 0 0
     4 8
       2 0
       1 6
         4 0
         4 0
           0
```

Thus, $\frac{5}{8} = 0.625$, so $-\frac{5}{8} = -0.625$.

Decimal notation for $-\frac{5}{8}$ is −0.625. We consider −0.625 to be a **terminating decimal.** Decimal notation for some numbers repeats.

Example 9 Convert to decimal notation: $\frac{7}{11}$.

We divide.

```
        0.6 3 6 3...
 1 1 ) 7.0 0 0 0
       6 6
         4 0
         3 3
           7 0
           6 6
             4 0
             3 3
               7
```

$$\frac{7}{11} = 0.\overline{63}$$

We can abbreviate repeating decimal notation by writing a bar over the repeating part—in this case, $0.\overline{63}$.

The following are other examples to show how each rational number can be named using fractional or decimal notation:

$$0 = \frac{0}{8}, \quad \frac{27}{100} = 0.27, \quad -8\frac{3}{4} = -8.75, \quad \frac{-13}{6} = -2.1\overline{6}.$$

Do Exercises 8–10.

Convert to decimal notation.

8. $-\frac{3}{8}$

9. $-\frac{6}{11}$

10. $\frac{4}{3}$

d The Real Numbers and Order

Every rational number has a point on the number line. However, there are some points on the line for which there is no rational number. These points correspond to what are called **irrational numbers.**

What kinds of numbers are irrational? One example is the number π, which is used in finding the area and the circumference of a circle: $A = \pi r^2$ and $C = 2\pi r$.

Another example of an irrational number is the square root of 2, named $\sqrt{2}$.

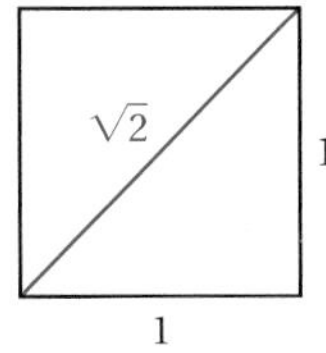

It is the length of the diagonal of a square with sides of length 1. It is also the number that when multiplied by itself gives 2. There is no rational number that can be multiplied by itself to get 2. But the following are rational *approximations*:

1.4 is an approximation of $\sqrt{2}$ because $(1.4)^2 = 1.96$;

1.41 is a better approximation because $(1.41)^2 = 1.9881$;

1.4142 is an even better approximation because $(1.4142)^2 = 1.99996164$.

We can find rational approximations for square roots using a calculator.

Decimal notation for rational numbers *either* terminates *or* repeats. Decimal notation for irrational numbers *neither* terminates *nor* repeats. Some other examples of irrational numbers are $\sqrt{3}$, $-\sqrt{8}$, $\sqrt{11}$, and 0.121221222122221.... Whenever we take the square root of a number that is not a perfect square, we will get an irrational number.

The rational numbers and the irrational numbers together correspond to all the points on a number line and make up what is called the **real-number system.**

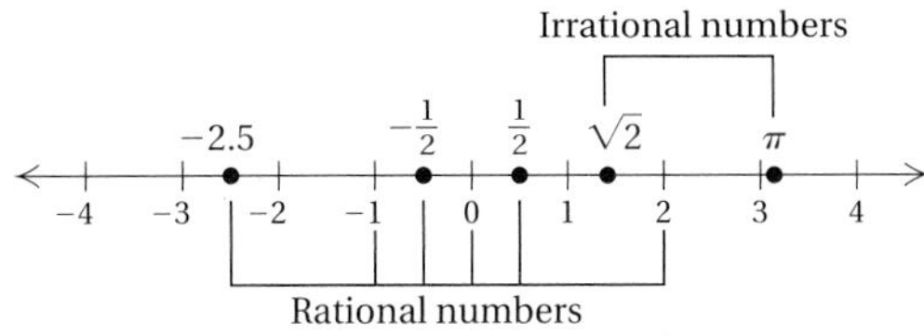

> The set of **real numbers** = The set of all numbers corresponding to points on the number line.

Calculator Spotlight

Approximating Square Roots and π. Square roots are found by pressing [2nd] [√‾]. (√‾ is the second operation associated with the [x^2] key.)

To find an approximation for $\sqrt{48}$, we press

[2nd] [√‾] [4] [8] [ENTER].

The approximation 6.92820323 is displayed.

To find $8 \cdot \sqrt{13}$, we press

[8] [2nd] [√‾] [1] [3] [ENTER].

The approximation 28.8444102 is displayed.

The number π is used widely enough to have its own key. (π is the second operation associated with the [^] key.)

To approximate π, we press

[2nd] [π] [ENTER].

The approximation 3.141592654 is displayed.

Exercises

Approximate.

1. $\sqrt{76}$
2. $\sqrt{317}$
3. $15 \cdot \sqrt{20}$
4. $29 + \sqrt{42}$
5. π
6. $29 \cdot \pi$
7. $\pi \cdot 13^2$
8. $5 \cdot \pi + 8 \cdot \sqrt{237}$

Answers on page A-17

The real numbers consist of the rational numbers and the irrational numbers. The following figure shows the relationships among various kinds of numbers.

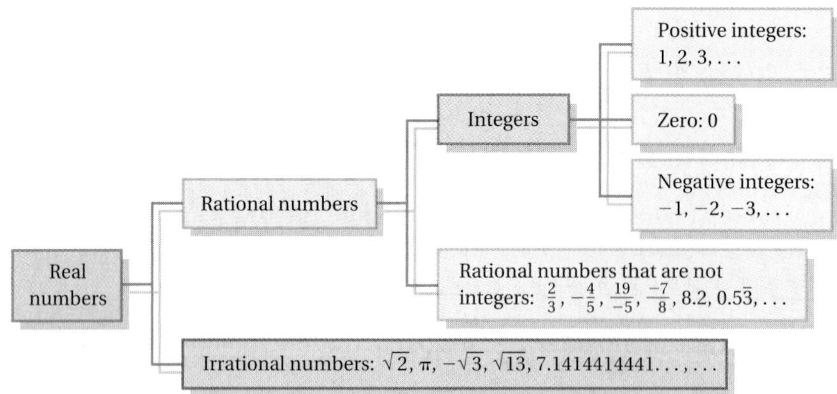

Order

Real numbers are named in order on the number line, with larger numbers named farther to the right. For any two numbers on the line, the one to the left is less than the one to the right.

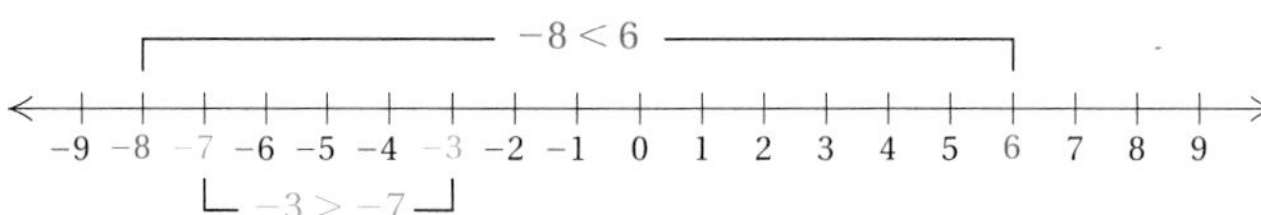

We use the symbol **<** to mean "**is less than.**" The sentence $-8 < 6$ means "-8 is less than 6." The symbol **>** means "**is greater than.**" The sentence $-3 > -7$ means "-3 is greater than -7." The sentences $-8 < 6$ and $-3 > -7$ are **inequalities**.

Examples Use either < or > for □ to write a true sentence.

10. 2 □ 9 — Since 2 is to the left of 9, 2 is less than 9, so $2 < 9$.

11. −7 □ 3 — Since −7 is to the left of 3, we have $-7 < 3$.

12. 6 □ −12 — Since 6 is to the right of −12, then $6 > -12$.

13. −18 □ −5 — Since −18 is to the left of −5, we have $-18 < -5$.

14. -2.7 □ $-\frac{3}{2}$ — The answer is $-2.7 < -\frac{3}{2}$.

15. 1.5 □ −2.7 — The answer is $1.5 > -2.7$.

16. 1.38 □ 1.83 — The answer is $1.38 < 1.83$.

17. −3.45 □ 1.32 — The answer is $-3.45 < 1.32$.

18. −4 □ 0 — The answer is $-4 < 0$.

19. 5.8 □ 0 — The answer is $5.8 > 0$.

20. $\frac{5}{8}$ □ $\frac{7}{11}$ — We convert to decimal notation: $\frac{5}{8} = 0.625$ and $\frac{7}{11} = 0.6363\ldots$. Thus, $\frac{5}{8} < \frac{7}{11}$.

Do Exercises 11–18.

Use either < or > for □ to write a true sentence.

11. −3 □ 7

12. −8 □ −5

13. 7 □ −10

14. 3.1 □ −9.5

15. $-\frac{2}{3}$ □ -1

16. $-\frac{11}{8}$ □ $\frac{23}{15}$

17. $-\frac{2}{3}$ □ $-\frac{5}{9}$

18. −4.78 □ −5.01

Answers on page A-17

Note that both $-8 < 6$ and $6 > -8$ are true. Every true inequality yields another true inequality when we interchange the numbers or variables and reverse the direction of the inequality sign.

> $a < b$ also has the meaning $b > a$.

Examples Write another inequality with the same meaning.

21. $a < -5$ The inequality $-5 > a$ has the same meaning.

22. $-3 > -8$ The inequality $-8 < -3$ has the same meaning.

A helpful mental device is to think of an inequality sign as an "arrow" with the arrow pointing to the smaller number.

Do Exercises 19 and 20.

Note that all positive real numbers are greater than zero and all negative real numbers are less than zero.

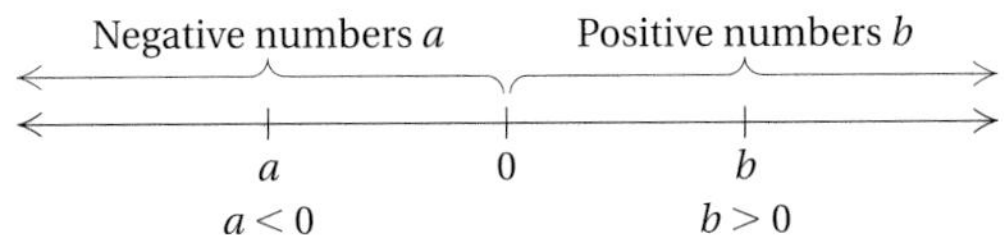

> If b is a positive real number, then $b > 0$.
> If a is a negative real number, then $a < 0$.

Expressions like $a \le b$ and $b \ge a$ are also inequalities. We read $a \le b$ as "***a* is less than or equal to *b*.**" We read $a \ge b$ as "***a* is greater than or equal to *b*.**"

Examples Write true or false for each statement.

23. $-3 \le 5.4$ True since $-3 < 5.4$ is true

24. $-3 \le -3$ True since $-3 = -3$ is true

25. $-5 \ge 1\frac{2}{3}$ False since neither $-5 > 1\frac{2}{3}$ nor $-5 = 1\frac{2}{3}$ is true

Do Exercises 21–23.

e Absolute Value

From the number line, we see that numbers like 4 and -4 are the same distance from zero. Distance is always a nonnegative number. We call the distance from zero on a number line the **absolute value** of the number.

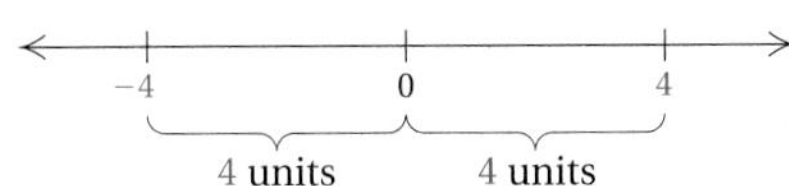

> The **absolute value** of a number is its distance from zero on a number line. We use the symbol $|x|$ to represent the absolute value of a number x.

Write another inequality with the same meaning.

19. $-5 < 7$

20. $x > 4$

Write true or false.

21. $-4 \le -6$

22. $7.8 \ge 7.8$

23. $-2 \le \dfrac{3}{8}$

Answers on page A-17

To find absolute value:

a) If a number is negative, make it positive.

b) If a number is positive or zero, leave it alone.

Examples Find the absolute value.

26. $|-7|$ The distance of -7 from 0 is 7, so $|-7| = 7$.

27. $|12|$ The distance of 12 from 0 is 12, so $|12| = 12$.

28. $|0|$ The distance of 0 from 0 is 0, so $|0| = 0$.

29. $\left|\frac{3}{2}\right| = \frac{3}{2}$

30. $|-2.73| = 2.73$

Do Exercises 24–28.

Find the absolute value.

24. $|8|$

25. $|0|$

26. $|-9|$

27. $\left|-\frac{2}{3}\right|$

28. $|5.6|$

Answers on page A-17

Exercise Set 7.2

a Tell which integers correspond to the situation.

1. *Elevation.* The Dead Sea, between Jordan and Israel, is 1286 ft below sea level; Mt. Rainier in Washington State is 13,804 ft above sea level.

2. Amy's golf score was 3 under par; Juan's was 7 over par.

3. On Wednesday, the temperature was 24° above zero. On Thursday, it was 2° below zero.

4. A student deposited her tax refund of \$750 in a savings account. Two weeks later, she withdrew \$125 to pay sorority fees.

5. *U.S. Public Debt.* Recently, the total public debt of the United States was about \$5.2 trillion (***Source:*** U.S. Department of the Treasury).

6. *Birth and Death Rates.* Recently, the world birth rate was 27 per thousand. The death rate was 9.7 per thousand. (***Source:*** United Nations Population Fund)

b Graph the number on the number line.

7. $\frac{10}{3}$

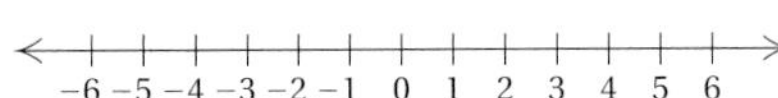

8. $-\frac{17}{4}$

−6 −5 −4 −3 −2 −1 0 1 2 3 4 5 6

9. -5.2

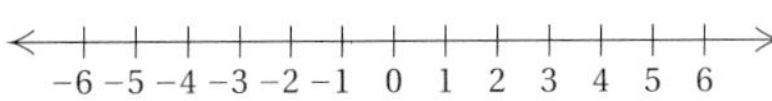

10. 4.78

−6 −5 −4 −3 −2 −1 0 1 2 3 4 5 6

c Convert to decimal notation.

11. $-\frac{7}{8}$ **12.** $-\frac{1}{8}$ **13.** $\frac{5}{6}$ **14.** $\frac{5}{3}$ **15.** $\frac{7}{6}$ **16.** $\frac{5}{12}$

17. $\frac{2}{3}$ **18.** $\frac{1}{4}$ **19.** $-\frac{1}{2}$ **20.** $\frac{5}{8}$ **21.** $\frac{1}{10}$ **22.** $-\frac{7}{20}$

d Use either $<$ or $>$ for ☐ to write a true sentence.

23. 8 ☐ 0 **24.** 3 ☐ 0 **25.** −8 ☐ 3 **26.** 6 ☐ −6

27. −8 ☐ 8 **28.** 0 ☐ −9 **29.** −8 ☐ −5 **30.** −4 ☐ −3

31. −5 ☐ −11 **32.** −3 ☐ −4 **33.** −6 ☐ −5 **34.** −10 ☐ −14

35. 2.14 ☐ 1.24 **36.** −3.3 ☐ −2.2 **37.** −14.5 ☐ 0.011 **38.** 17.2 ☐ −1.67

39. $-12.88 \;\square\; -6.45$

40. $-14.34 \;\square\; -17.88$

41. $\frac{5}{12} \;\square\; \frac{11}{25}$

42. $-\frac{13}{16} \;\square\; -\frac{5}{9}$

Write true or false.

43. $-3 \geq -11$

44. $5 \leq -5$

45. $0 \geq 8$

46. $-5 \leq 7$

Write an inequality with the same meaning.

47. $-6 > x$

48. $x < 8$

49. $-10 \leq y$

50. $12 \geq t$

e Find the absolute value.

51. $|-3|$

52. $|-7|$

53. $|10|$

54. $|11|$

55. $|0|$

56. $|-4|$

57. $|-24|$

58. $|325|$

59. $\left|-\frac{2}{3}\right|$

60. $\left|-\frac{10}{7}\right|$

61. $\left|\frac{0}{4}\right|$

62. $|14.8|$

Skill Maintenance

Convert to decimal notation. [4.2b]

63. 63%

64. 8.3%

65. 110%

66. 22.76%

Convert to percent notation. [4.3a]

67. $\frac{3}{4}$

68. $\frac{5}{8}$

69. $\frac{5}{6}$

70. $\frac{19}{32}$

Synthesis

71. ◆ ▦ When Jennifer's calculator gives a decimal approximation for $\sqrt{2}$ and that approximation is promptly squared, the result is 2. Yet, when that same approximation is entered by hand and then squared, the result is not exactly 2. Why do you suppose this happens?

72. ◆ How many rational numbers are there between 0 and 1? Why?

List in order from the least to the greatest.

73. $-\frac{2}{3}, \frac{1}{2}, -\frac{3}{4}, -\frac{5}{6}, \frac{3}{8}, \frac{1}{6}$

74. $-8\frac{7}{8}, 7^1, -5, |-6|, 4, |3|, -8\frac{5}{8}, -100, 0, 1^7, \frac{14}{4}, \frac{-67}{8}$

Given that $0.3\overline{3} = \frac{1}{3}$ and $0.6\overline{6} = \frac{2}{3}$, express each of the following as a quotient or ratio of two integers.

75. $0.1\overline{1}$

76. $0.9\overline{9}$

77. $5.5\overline{5}$

7.3 Addition of Real Numbers

In this section, we consider addition of real numbers. First, to gain an understanding, we add using a number line. Then we consider rules for addition.

Addition of numbers can be illustrated on a number line. To do the addition $a + b$, we start at a, and then move according to b.

a) If b is positive, we move to the right.

b) If b is negative, we move to the left.

c) If b is 0, we stay at a.

Example 1 Add: $3 + (-5)$.

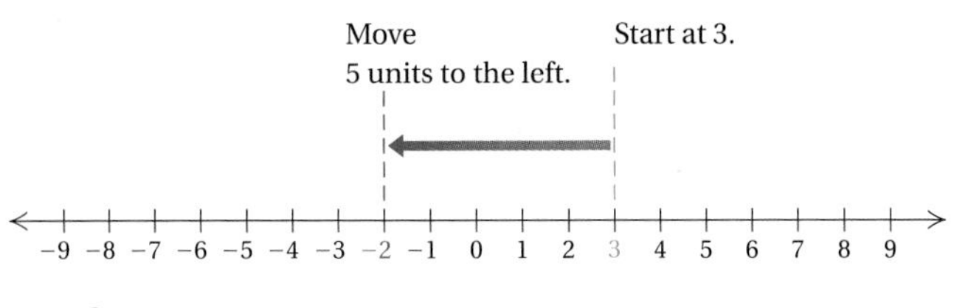

$3 + (-5) = -2$

Example 2 Add: $-4 + (-3)$.

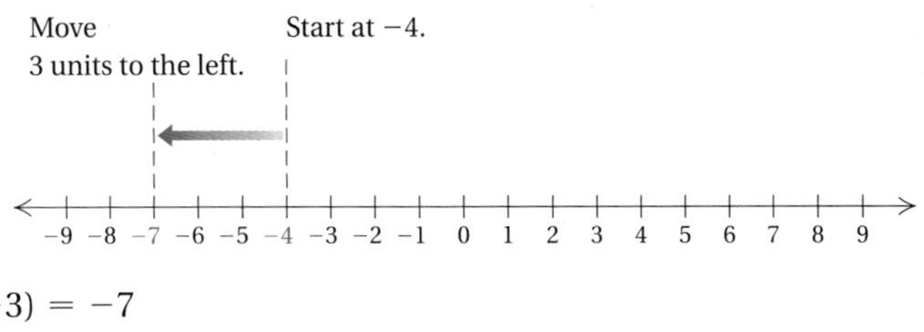

$-4 + (-3) = -7$

Example 3 Add: $-4 + 9$.

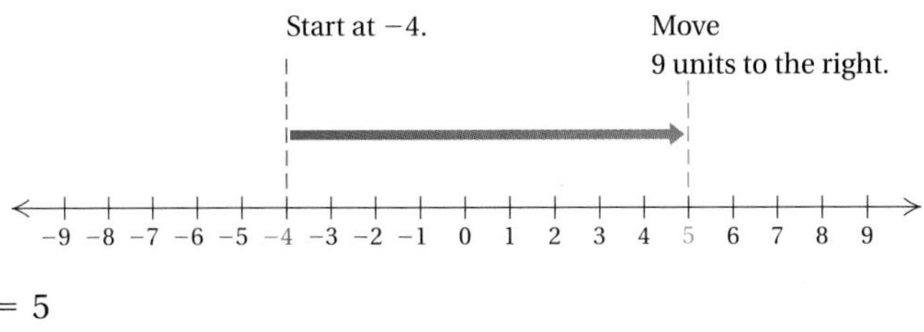

$-4 + 9 = 5$

Example 4 Add: $-5.2 + 0$.

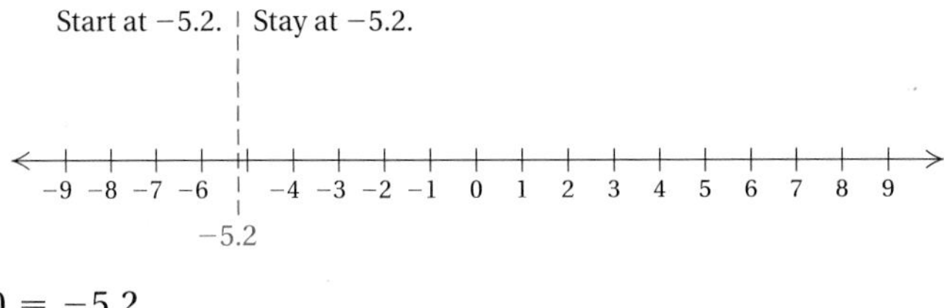

$-5.2 + 0 = -5.2$

Do Exercises 1–6.

Objectives

a Add real numbers without using a number line.

b Find the opposite, or additive inverse, of a real number.

For Extra Help

TAPE 14 | MAC WIN | CD-ROM

Add using a number line.

1. $0 + (-6)$

2. $1 + (-4)$

3. $-3 + (-5)$

4. $-3 + 7$

5. $-5.4 + 5.4$

6. $-\frac{5}{2} + \frac{1}{2}$

Answers on page A-17

Add without using a number line.

7. $-5 + (-6)$

8. $-9 + (-3)$

9. $-4 + 6$

10. $-7 + 3$

11. $5 + (-7)$

12. $-20 + 20$

13. $-11 + (-11)$

14. $10 + (-7)$

15. $-0.17 + 0.7$

16. $-6.4 + 8.7$

17. $-4.5 + (-3.2)$

18. $-8.6 + 2.4$

19. $\frac{5}{9} + \left(-\frac{7}{9}\right)$

20. $-\frac{1}{5} + \left(-\frac{3}{4}\right)$

Answers on page A-17

a Adding Without a Number Line

You may have noticed some patterns in the preceding examples. These lead us to rules for adding without using a number line that are more efficient for adding larger numbers.

RULES FOR ADDITION OF REAL NUMBERS

1. *Positive numbers*: Add the same as arithmetic numbers. The answer is positive.
2. *Negative numbers*: Add absolute values. The answer is negative.
3. *A positive and a negative number*: Subtract the smaller absolute value from the larger. Then:
   a) If the positive number has the greater absolute value, the answer is positive.
   b) If the negative number has the greater absolute value, the answer is negative.
   c) If the numbers have the same absolute value, the answer is 0.
4. *One number is zero*: The sum is the other number.

Rule 4 is known as the **identity property of 0.** It says that for any real number a, $a + 0 = a$.

Examples Add without using a number line.

5. $-12 + (-7) = -19$ — Two negatives. *Think*: Add the absolute values, 12 and 7, getting 19. Make the answer *negative*, -19.

6. $-1.4 + 8.5 = 7.1$ — The absolute values are 1.4 and 8.5. The difference is 7.1. The positive number has the larger absolute value, so the answer is *positive*, 7.1.

7. $-36 + 21 = -15$ — The absolute values are 36 and 21. The difference is 15. The negative number has the larger absolute value, so the answer is *negative*, -15.

8. $1.5 + (-1.5) = 0$ — The numbers have the same absolute value. The sum is 0.

9. $-\frac{7}{8} + 0 = -\frac{7}{8}$ — One number is zero. The sum is $-\frac{7}{8}$.

10. $-9.2 + 3.1 = -6.1$

11. $-\frac{3}{2} + \frac{9}{2} = \frac{6}{2} = 3$

12. $-\frac{2}{3} + \frac{5}{8} = -\frac{16}{24} + \frac{15}{24} = -\frac{1}{24}$

Do Exercises 7–20.

Suppose we want to add several numbers, some positive and some negative, as follows. How can we proceed?

$$15 + (-2) + 7 + 14 + (-5) + (-12)$$

We can change grouping and order as we please when adding. For instance, we can group the positive numbers together and the negative numbers together and add them separately. Then we add the two results.

Example 13 Add: $15 + (-2) + 7 + 14 + (-5) + (-12)$.

a) $15 + 7 + 14 = 36$ **Adding the positive numbers**

b) $-2 + (-5) + (-12) = -19$ **Adding the negative numbers**

c) $36 + (-19) = 17$ **Adding the results**

We can also add the numbers in any other order we wish, say, from left to right as follows:

$$\begin{aligned} 15 + (-2) + 7 + 14 + (-5) + (-12) &= 13 + 7 + 14 + (-5) + (-12) \\ &= 20 + 14 + (-5) + (-12) \\ &= 34 + (-5) + (-12) \\ &= 29 + (-12) \\ &= 17 \end{aligned}$$

Do Exercises 21–24.

b Opposites, or Additive Inverses

Suppose we add two numbers that are **opposites**, such as 6 and -6. The result is 0. When opposites are added, the result is always 0. Such numbers are also called **additive inverses.** Every real number has an opposite, or additive inverse.

> Two numbers whose sum is 0 are called **opposites**, or **additive inverses,** of each other.

Examples Find the opposite of each number.

14. 34   The opposite of 34 is -34 because $34 + (-34) = 0$.

15. -8 The opposite of -8 is 8 because $-8 + 8 = 0$.

16. 0   The opposite of 0 is 0 because $0 + 0 = 0$.

17. $-\frac{7}{8}$ The opposite of $-\frac{7}{8}$ is $\frac{7}{8}$ because $-\frac{7}{8} + \frac{7}{8} = 0$.

Do Exercises 25–30.

To name the opposite, we use the symbol $-$, as follows.

> The opposite, or additive inverse, of a number a can be named $-a$ (read "the opposite of a," or "the additive inverse of a").

Note that if we take a number, say, 8, and find its opposite, -8, and then find the opposite of the result, we will have the original number, 8, again.

Add.

21. $(-15) + (-37) + 25 + 42 + (-59) + (-14)$

22. $42 + (-81) + (-28) + 24 + 18 + (-31)$

23. $-2.5 + (-10) + 6 + (-7.5)$

24. $-35 + 17 + 14 + (-27) + 31 + (-12)$

Find the opposite.

25. -4

26. 8.7

27. -7.74

28. $-\frac{8}{9}$

29. 0

30. 12

Answers on page A-17

Find $-x$ and $-(-x)$ when x is each of the following.

31. 14

32. 1

33. -19

34. -1.6

35. $\frac{2}{3}$

36. $-\frac{9}{8}$

Find the opposite. (Change the sign.)

37. -4

38. -13.4

39. 0

40. $\frac{1}{4}$

Answers on page A-17

> The opposite of the opposite of a number is the number itself. (The additive inverse of the additive inverse of a number is the number itself.) That is, for any number a,
>
> $$-(-a) = a.$$

Example 18 Find $-x$ and $-(-x)$ when $x = 16$.

If $x = 16$, then $-x = -16$. The opposite of 16 is -16.

If $x = 16$, then $-(-x) = -(-16) = 16$. The opposite of the opposite of 16 is 16.

Example 19 Find $-x$ and $-(-x)$ when $x = -3$.

If $x = -3$, then $-x = -(-3) = 3$.

If $x = -3$, then $-(-x) = -(-(-3)) = -3$.

Note that in Example 19 we used a second set of parentheses to show that we are substituting the negative number -3 for x. Symbolism like $--x$ is not considered meaningful.

Do Exercises 31–36.

A symbol such as -8 is usually read "negative 8." It could be read "the additive inverse of 8," because the additive inverse of 8 is negative 8. It could also be read "the opposite of 8," because the opposite of 8 is -8. Thus a symbol like -8 can be read in more than one way. A symbol like $-x$, which has a variable, should be read "the opposite of x" or "the additive inverse of x" and *not* "negative x," because we do not know whether x represents a positive number, a negative number, or 0. You can check this in Examples 18 and 19.

We can use the symbolism $-a$ to restate the definition of opposite, or additive inverse.

> For any real number a, the **opposite**, or **additive inverse**, of a, which is $-a$, is such that
>
> $$a + (-a) = (-a) + a = 0.$$

Signs of Numbers

A negative number is sometimes said to have a "negative sign." A positive number is said to have a "positive sign." When we replace a number with its opposite, we can say that we have "changed its sign."

Examples Find the opposite. (Change the sign.)

20. -3 $\quad -(-3) = 3$ The opposite of -3 is 3.

21. -10 $\quad -(-10) = 10$

22. 0 $\quad -(0) = 0$

23. 14 $\quad -(14) = -14$

Do Exercises 37–40.

Exercise Set 7.3

a Add. Do not use a number line except as a check.

1. $2 + (-9)$

2. $-5 + 2$

3. $-11 + 5$

4. $4 + (-3)$

5. $-6 + 6$

6. $8 + (-8)$

7. $-3 + (-5)$

8. $-4 + (-6)$

9. $-7 + 0$

10. $-13 + 0$

11. $0 + (-27)$

12. $0 + (-35)$

13. $17 + (-17)$

14. $-15 + 15$

15. $-17 + (-25)$

16. $-24 + (-17)$

17. $18 + (-18)$

18. $-13 + 13$

19. $-28 + 28$

20. $11 + (-11)$

21. $8 + (-5)$

22. $-7 + 8$

23. $-4 + (-5)$

24. $10 + (-12)$

25. $13 + (-6)$

26. $-3 + 14$

27. $-25 + 25$

28. $50 + (-50)$

29. $53 + (-18)$

30. $75 + (-45)$

31. $-8.5 + 4.7$

32. $-4.6 + 1.9$

33. $-2.8 + (-5.3)$

34. $-7.9 + (-6.5)$

35. $-\frac{3}{5} + \frac{2}{5}$

36. $-\frac{4}{3} + \frac{2}{3}$

37. $-\frac{2}{9} + \left(-\frac{5}{9}\right)$

38. $-\frac{4}{7} + \left(-\frac{6}{7}\right)$

39. $-\frac{5}{8} + \frac{1}{4}$

40. $-\frac{5}{6} + \frac{2}{3}$

41. $-\frac{5}{8} + \left(-\frac{1}{6}\right)$

42. $-\frac{5}{6} + \left(-\frac{2}{9}\right)$

43. $-\frac{3}{8} + \frac{5}{12}$

44. $-\frac{7}{16} + \frac{7}{8}$

45. $76 + (-15) + (-18) + (-6)$

46. $29 + (-45) + 18 + 32 + (-96)$

47. $-44 + \left(-\frac{3}{8}\right) + 95 + \left(-\frac{5}{8}\right)$

48. $24 + 3.1 + (-44) + (-8.2) + 63$

49. $98 + (-54) + 113 + (-998) + 44 + (-612)$

50. $-458 + (-124) + 1025 + (-917) + 218$

b Find the opposite, or additive inverse.

51. 24

52. -64

53. -26.9

54. 48.2

Find $-x$ when x is each of the following.

55. 8

56. -27

57. $-\frac{13}{8}$

58. $\frac{1}{236}$

Find $-(-x)$ when x is each of the following.

59. -43

60. 39

61. $\frac{4}{3}$

62. -7.1

Find the opposite. (Change the sign.)

63. -24

64. -12.3

65. $-\frac{3}{8}$

66. 10

Skill Maintenance

Convert to decimal notation. [4.2b]

67. 57%

68. 49%

69. 52.9%

70. 71.3%

Convert to percent notation. [4.3a]

71. $\frac{5}{4}$

72. $\frac{1}{8}$

73. $\frac{13}{25}$

74. $\frac{13}{32}$

Synthesis

75. ◆ Without actually performing the addition, explain why the sum of all integers from -50 to 50 is 0.

76. ◆ Explain in your own words why the sum of two negative numbers is always negative.

77. For what numbers x is $-x$ negative?

78. For what numbers x is $-x$ positive?

Add.

79. ▦ $-3496 + (-2987)$

80. ▦ $2708 + (-3749)$

Tell whether the sum is positive, negative, or zero.

81. If a is positive and b is negative, $-a + b$ is ________.

82. If $a = b$ and a and b are negative, $-a + (-b)$ is ________.

Add integers using a variety of methods.

7.4 Subtraction of Real Numbers

a Subtraction

We now consider subtraction of real numbers. Subtraction is defined as follows.

> The difference $a - b$ is the number that when added to b gives a.

For example, $45 - 17 = 28$ because $28 + 17 = 45$. Let's consider an example whose answer is a negative number.

Example 1 Subtract: $5 - 8$.

Think: $5 - 8$ is the number that when added to 8 gives 5. What number can we add to 8 to get 5? The number must be negative. The number is -3:

$$5 - 8 = -3.$$

That is, $5 - 8 = -3$ because $5 = -3 + 8$.

Do Exercises 1–3.

The definition above does *not* provide the most efficient way to do subtraction. From that definition, however, we can develop a faster way to subtract. Look for a pattern in the following examples.

Subtractions	Adding an Opposite
$5 - 8 = -3$	$5 + (-8) = -3$
$-6 - 4 = -10$	$-6 + (-4) = -10$
$-7 - (-10) = 3$	$-7 + 10 = 3$
$-7 - (-2) = -5$	$-7 + 2 = -5$

Do Exercises 4–7.

Perhaps you have noticed that we can subtract by adding the opposite of the number being subtracted. This can always be done.

> For any real numbers a and b,
>
> $a - b = a + (-b)$.
>
> (To subtract, add the opposite, or additive inverse, of the number being subtracted.)

This is the method generally used for quick subtraction of real numbers.

Objectives

a Subtract real numbers and simplify combinations of additions and subtractions.

b Solve applied problems involving addition and subtraction of real numbers.

For Extra Help

TAPE 14

MAC WIN

CD-ROM

Subtract.

1. $-6 - 4$

   Think: What number can be added to 4 to get -6?

2. $-7 - (-10)$

   Think: What number can be added to -10 to get -7?

3. $-7 - (-2)$

   Think: What number can be added to -2 to get -7?

Complete the addition and compare with the subtraction.

4. $4 - 6 = -2$;
   $4 + (-6) =$ ______

5. $-3 - 8 = -11$;
   $-3 + (-8) =$ ______

6. $-5 - (-9) = 4$;
   $-5 + 9 =$ ______

7. $-5 - (-3) = -2$;
   $-5 + 3 =$ ______

Answers on page A-18

Examples Subtract.

2. $2 - 6 = 2 + (-6) = -4$

The opposite of 6 is −6. We change the subtraction to addition and add the opposite.

3. $4 - (-9) = 4 + 9 = 13$

The opposite of −9 is 9. We change the subtraction to addition and add the opposite.

4. $-4.2 - (-3.6) = -4.2 + 3.6 = -0.6$

Adding the opposite. *Check*: $-0.6 + (-3.6) = -4.2$.

5. $-\frac{1}{2} - \left(-\frac{3}{4}\right) = -\frac{1}{2} + \frac{3}{4} = \frac{1}{4}$

Adding the opposite. *Check*: $\frac{1}{4} + \left(-\frac{3}{4}\right) = -\frac{1}{2}$.

Do Exercises 8–13.

Examples Read each of the following. Then subtract by adding the opposite of the number being subtracted.

6. $3 - 5$;
$3 - 5 = 3 + (-5) = -2$

Read "three minus five is three plus the opposite of five"

7. $\frac{1}{8} - \frac{7}{8}$;
$\frac{1}{8} - \frac{7}{8} = \frac{1}{8} + \left(-\frac{7}{8}\right) = -\frac{6}{8}$, or $-\frac{3}{4}$

Read "one-eighth minus seven-eighths is one-eighth plus the opposite of seven-eighths"

8. $-4.6 - (-9.8)$;
$-4.6 - (-9.8) = -4.6 + 9.8 = 5.2$

Read "negative four point six minus negative nine point eight is negative four point six plus the opposite of negative nine point eight"

9. $-\frac{3}{4} - \frac{7}{5}$;
$-\frac{3}{4} - \frac{7}{5} = -\frac{3}{4} + \left(-\frac{7}{5}\right) = -\frac{15}{20} + \left(-\frac{28}{20}\right) = -\frac{43}{20}$

Read "negative three-fourths minus seven-fifths is negative three-fourths plus the opposite of seven-fifths"

Do Exercises 14–18.

When several additions and subtractions occur together, we can make them all additions.

Examples Simplify.

10. $8 - (-4) - 2 - (-4) + 2 = 8 + 4 + (-2) + 4 + 2$
$= 16$

Adding the opposites where subtraction is indicated

11. $8.2 - (-6.1) + 2.3 - (-4) = 8.2 + 6.1 + 2.3 + 4$
$= 20.6$

Do Exercises 19–21.

Subtract.

8. $2 - 8$

9. $-6 - 10$

10. $12.4 - 5.3$

11. $-8 - (-11)$

12. $-8 - (-8)$

13. $\frac{2}{3} - \left(-\frac{5}{6}\right)$

Read each of the following. Then subtract by adding the opposite of the number being subtracted.

14. $3 - 11$

15. $12 - 5$

16. $-12 - (-9)$

17. $-12.4 - 10.9$

18. $-\frac{4}{5} - \left(-\frac{4}{5}\right)$

Simplify.

19. $-6 - (-2) - (-4) - 12 + 3$

20. $9 - (-6) + 7 - 11 - 14 - (-20)$

21. $-9.6 + 7.4 - (-3.9) - (-11)$

Answers on page A-18

b Applications and Problem Solving

Let's now see how we can use addition and subtraction of real numbers to solve applied problems.

Example 12 *Home-Run Differential.* In baseball the difference between the number of home runs hit by a team's players and the number given up by its pitchers is called the *home-run differential,* that is,

$$\text{Home run differential} = \frac{\text{Number of}}{\text{home runs}} - \frac{\text{Number of home}}{\text{runs allowed}}.$$

Teams strive for a positive home-run differential.

a) In a recent year, Atlanta hit 197 home runs and gave up 120. Find its home-run differential.

b) In a recent year, San Francisco hit 153 home runs and gave up 194. Find its home-run differential.

We solve as follows.

a) We subtract 120 from 197 to find the home-run differential for Atlanta:

Home-run differential = 197 − 120 = 77.

b) We subtract 194 from 153 to find the home-run differential for San Francisco:

Home-run differential = 153 − 194 = −41.

Do Exercises 22 and 23.

22. *Home-Run Differential.* Complete the following table to find the home-run differentials for all the major-league baseball teams.

National League

	HRs	HRs allowed	Diff.
Atlanta	197	120	+77
Florida	150	113	
Los Angeles	150	125	
Cincinnati	191	167	
Colorado	221	198	
San Diego	147	138	
Montreal	148	152	
Cubs	175	184	
Mets	147	159	
Houston	129	154	
Philadelphia	132	160	
St. Louis	142	173	
San Francisco	153	194	−41
Pittsburgh	138	183	

American League

	HRs	HRs allowed	Diff.
Texas	221	168	
Baltimore	257	209	
Cleveland	218	173	
Oakland	243	205	
Seattle	245	216	
Boston	209	185	
White Sox	195	174	
Yankees	162	143	
Toronto	177	187	
California	192	219	
Milwaukee	178	213	
Detroit	204	241	
Kansas City	123	176	
Minnesota	118	233	

23. *Temperature Extremes.* In Churchill, Manitoba, Canada, the average daily low temperature in January is −31°C. The average daily low temperature in Key West, Florida, is 19°C. How much higher is the average daily low temperature in Key West, Florida?

Answers on page A-18

Improving Your Math Study Skills

Classwork: During and After Class

During Class

Asking Questions

Many students are afraid to ask questions in class. You will find that most instructors are not only willing to answer questions during class, but often encourage students to ask questions. In fact, some instructors would like more questions than are offered. Probably your question is one that other students in the class might have been afraid to ask!

Consider waiting for an appropriate time to ask questions. Some instructors will pause to ask the class if they have questions. Use this opportunity to get clarification on any concept you do not understand.

After Class

Restudy Examples and Class Notes

As soon as possible after class, find some time to go over your notes. Read the appropriate sections from the textbook and try to correlate the text with your class notes. You may also want to restudy the examples in the textbook for added comprehension.

Often students make the mistake of doing the homework exercises without reading their notes or textbook. This is not a good idea, since you may lose the opportunity for a complete understanding of the concepts. Simply being able to work the exercises does not ensure that you know the material well enough to work problems on a test.

Videotapes

If you can find the time, visit the library, math lab, or media center to view the videotapes on the textbook. Look on the first page of each section in the textbook for the appropriate tape reference.

The videotapes provide detailed explanations of each objective and they may give you a different presentation than the one offered by your instructor. Being able to pause the tape while you take notes or work the examples or replay the tape as many times as you need are additional advantages to using the videos.

Also, consider studying the special tapes *Math Problem Solving in the Real World* and *Math Study Skills* prepared by the author. If these are not available in the media center, contact your instructor.

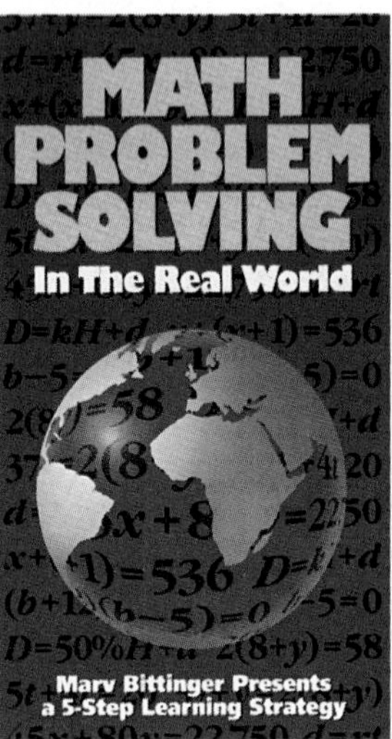

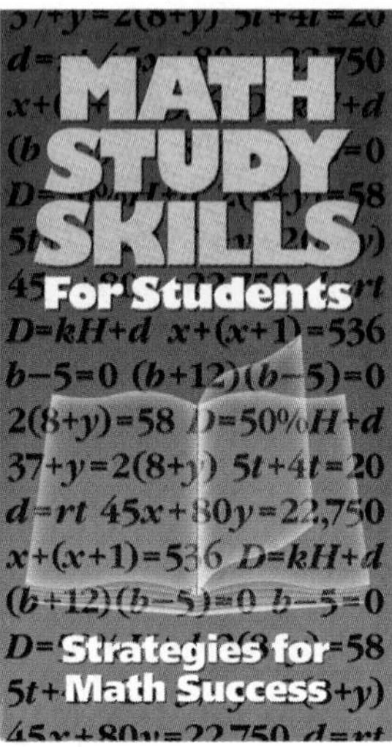

Software

If you would like additional practice on any section of the textbook, you can use the accompanying Interact Math Tutorial Software. This software can generate many different versions of the basic odd-numbered exercises for added practice. You can also ask the software to work out each problem step by step.

Ask your instructor about the availability of this software.

Exercise Set 7.4

a Subtract.

1. $2 - 9$
2. $3 - 8$
3. $0 - 4$
4. $0 - 9$
5. $-8 - (-2)$
6. $-6 - (-8)$
7. $-11 - (-11)$
8. $-6 - (-6)$
9. $12 - 16$
10. $14 - 19$
11. $20 - 27$
12. $30 - 4$
13. $-9 - (-3)$
14. $-7 - (-9)$
15. $-40 - (-40)$
16. $-9 - (-9)$
17. $7 - 7$
18. $9 - 9$
19. $7 - (-7)$
20. $4 - (-4)$
21. $8 - (-3)$
22. $-7 - 4$
23. $-6 - 8$
24. $6 - (-10)$
25. $-4 - (-9)$
26. $-14 - 2$
27. $1 - 8$
28. $2 - 8$
29. $-6 - (-5)$
30. $-4 - (-3)$
31. $8 - (-10)$
32. $5 - (-6)$
33. $0 - 10$
34. $0 - 18$
35. $-5 - (-2)$
36. $-3 - (-1)$
37. $-7 - 14$
38. $-9 - 16$
39. $0 - (-5)$
40. $0 - (-1)$
41. $-8 - 0$
42. $-9 - 0$
43. $7 - (-5)$
44. $7 - (-4)$

45. $2 - 25$

46. $18 - 63$

47. $-42 - 26$

48. $-18 - 63$

49. $-71 - 2$

50. $-49 - 3$

51. $24 - (-92)$

52. $48 - (-73)$

53. $-50 - (-50)$

54. $-70 - (-70)$

55. $-\frac{3}{8} - \frac{5}{8}$

56. $\frac{3}{9} - \frac{9}{9}$

57. $\frac{3}{4} - \frac{2}{3}$

58. $\frac{5}{8} - \frac{3}{4}$

59. $-\frac{3}{4} - \frac{2}{3}$

60. $-\frac{5}{8} - \frac{3}{4}$

61. $-\frac{5}{8} - \left(-\frac{3}{4}\right)$

62. $-\frac{3}{4} - \left(-\frac{2}{3}\right)$

63. $6.1 - (-13.8)$

64. $1.5 - (-3.5)$

65. $-2.7 - 5.9$

66. $-3.2 - 5.8$

67. $0.99 - 1$

68. $0.87 - 1$

69. $-79 - 114$

70. $-197 - 216$

71. $0 - (-500)$

72. $500 - (-1000)$

73. $-2.8 - 0$

74. $6.04 - 1.1$

75. $7 - 10.53$

76. $8 - (-9.3)$

77. $\frac{1}{6} - \frac{2}{3}$

78. $-\frac{3}{8} - \left(-\frac{1}{2}\right)$

79. $-\frac{4}{7} - \left(-\frac{10}{7}\right)$

80. $\frac{12}{5} - \frac{12}{5}$

81. $-\frac{7}{10} - \frac{10}{15}$

82. $-\frac{4}{18} - \left(-\frac{2}{9}\right)$

83. $\frac{1}{5} - \frac{1}{3}$

84. $-\frac{1}{7} - \left(-\frac{1}{6}\right)$

Simplify.

85. $18 - (-15) - 3 - (-5) + 2$

86. $22 - (-18) + 7 + (-42) - 27$

87. $-31 + (-28) - (-14) - 17$

88. $-43 - (-19) - (-21) + 25$

89. $-34 - 28 + (-33) - 44$

90. $39 + (-88) - 29 - (-83)$

91. $-93 - (-84) - 41 - (-56)$

92. $84 + (-99) + 44 - (-18) - 43$

93. $-5 - (-30) + 30 + 40 - (-12)$

94. $14 - (-50) + 20 - (-32)$

95. $132 - (-21) + 45 - (-21)$

96. $81 - (-20) - 14 - (-50) + 53$

b Solve.

97. *Ocean Depth.* The deepest point in the Pacific Ocean is the Marianas Trench, with a depth of 11,033 m. The deepest point in the Atlantic Ocean is the Puerto Rico Trench, with a depth of 8648 m. What is the difference in the elevation of the two trenches?

98. *Depth of Offshore Oil Wells.* In 1993, the elevation of the world's deepest offshore oil well was −2860 ft. By 1998, the deepest well is expected to be 360 ft deeper. (***Source:*** *New York Times,* 12/7/94, p. D1.) What will be the elevation of the deepest well in 1998?

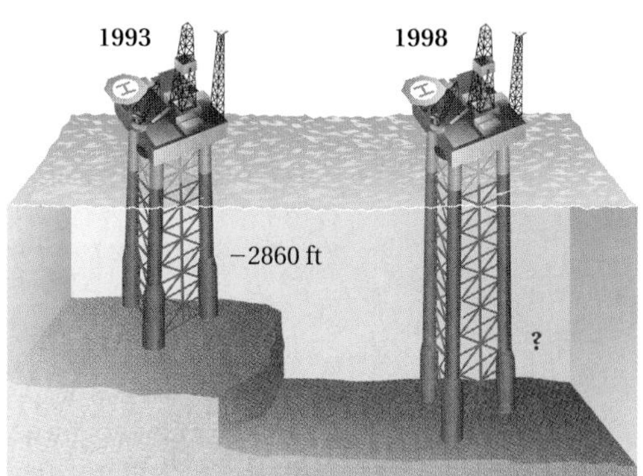

99. Laura has a charge of \$476.89 on her credit card, but she then returns a sweater that cost \$128.95. How much does she now owe on her credit card?

100. Chris has \$720 in a checking account. He writes a check for \$970 to pay for a sound system. What is the balance in his checking account?

101. *Temperature Records.* The greatest recorded temperature change in one day occurred in Browning, Montana, where the temperature fell from 44°F to −56°F (***Source:*** *The Guinness Book of Records*). How much did the temperature drop?

102. *Low Points on Continents.* The lowest point in Africa is Lake Assal, which is 515 ft below sea level. The lowest point in South America is the Valdes Peninsula, which is 132 ft below sea level. How much lower is Lake Assal than the Valdes Peninsula?

Skill Maintenance

103. Evaluate: 5^3. [1.6b]

104. Find the prime factorization of 864. [1.7d]

105. Simplify: $256 \div 64 \div 2^3 + 100$. [1.6c]

106. Simplify: $5 \cdot 6 + (7 \cdot 2)^2$. [1.6c]

107. Convert to decimal notation: 58.3%. [4.2b]

108. Simplify: $\frac{164}{256}$. [2.1e]

SYNTHESIS

109. ◈ If a negative number is subtracted from a positive number, will the result always be positive? Why or why not?

110. ◈ Write a problem for a classmate to solve. Design the problem so that the solution is "The temperature dropped to −9°."

Subtract.

111. 🖩 123,907 − 433,789

112. 🖩 23,011 − (−60,432)

Tell whether the statement is true or false for all integers a and b. If false, show why.

113. $a - 0 = 0 - a$

114. $0 - a = a$

115. If $a \neq b$, then $a - b \neq 0$.

116. If $a = -b$, then $a + b = 0$.

117. If $a + b = 0$, then a and b are opposites.

118. If $a - b = 0$, then $a = -b$.

119. Maureen is a stockbroker. She kept track of the changes in the stock market over a period of 5 weeks. By how many points had the market risen or fallen over this time?

Week 1	Week 2	Week 3	Week 4	Week 5
Down 13 pts	Down 16 pts	Up 36 pts	Down 11 pts	Up 19 pts

120. *Blackjack Counting System.* The casino game of blackjack makes use of many card-counting systems to give players a winning edge if the count becomes negative. One such system is called *High–Low,* first developed by Harvey Dubner in 1963. Each card counts as −1, 0, or 1 as follows:

2, 3, 4, 5, 6 count as +1;
7, 8, 9 count as 0;
10, J, Q, K, A count as −1.

(***Source:*** Patterson, Jerry L., *Casino Gambling.* New York: Perigee, 1982)

a) Find the final count on the sequence of cards

K, A, 2, 4, 5, 10, J, 8, Q, K, 5.

b) Does the player have a winning edge?

Subtract integers using tiles.

7.5 Multiplication of Real Numbers

a Multiplication

Multiplication of real numbers is very much like multiplication of arithmetic numbers. The only difference is that we must determine whether the answer is positive or negative.

Multiplication of a Positive Number and a Negative Number

To see how to multiply a positive number and a negative number, consider the pattern of the following.

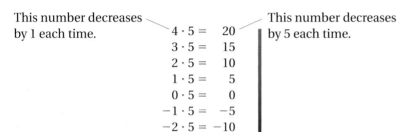

This number decreases by 1 each time. This number decreases by 5 each time.

$$\begin{aligned}
4 \cdot 5 &= 20\\
3 \cdot 5 &= 15\\
2 \cdot 5 &= 10\\
1 \cdot 5 &= 5\\
0 \cdot 5 &= 0\\
-1 \cdot 5 &= -5\\
-2 \cdot 5 &= -10\\
-3 \cdot 5 &= -15
\end{aligned}$$

Do Exercise 1.

According to this pattern, it looks as though the product of a negative number and a positive number is negative. That is the case, and we have the first part of the rule for multiplying numbers.

> To multiply a positive number and a negative number, multiply their absolute values. The answer is negative.

Examples Multiply.

1. $8(-5) = -40$ **2.** $-\frac{1}{3} \cdot \frac{5}{7} = -\frac{5}{21}$ **3.** $(-7.2)5 = -36$

Do Exercises 2–7.

Multiplication of Two Negative Numbers

How do we multiply two negative numbers? Again, we look for a pattern.

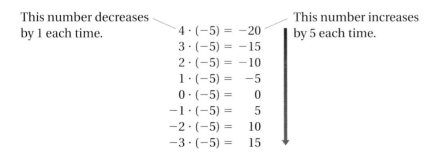

This number decreases by 1 each time. This number increases by 5 each time.

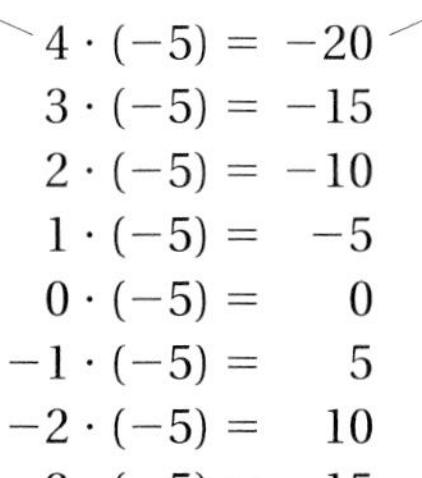

$$\begin{aligned}
4 \cdot (-5) &= -20\\
3 \cdot (-5) &= -15\\
2 \cdot (-5) &= -10\\
1 \cdot (-5) &= -5\\
0 \cdot (-5) &= 0\\
-1 \cdot (-5) &= 5\\
-2 \cdot (-5) &= 10\\
-3 \cdot (-5) &= 15
\end{aligned}$$

Do Exercise 8.

Objective

a Multiply real numbers.

For Extra Help

TAPE 14 MAC WIN CD-ROM

InterAct math

1. Complete, as in the example.

$$\begin{aligned}
4 \cdot 10 &= 40\\
3 \cdot 10 &= 30\\
2 \cdot 10 &=\\
1 \cdot 10 &=\\
0 \cdot 10 &=\\
-1 \cdot 10 &=\\
-2 \cdot 10 &=\\
-3 \cdot 10 &=
\end{aligned}$$

Multiply.

2. $-3 \cdot 6$

3. $20 \cdot (-5)$

4. $4 \cdot (-20)$

5. $-\frac{2}{3} \cdot \frac{5}{6}$

6. $-4.23(7.1)$

7. $\frac{7}{8}\left(-\frac{4}{5}\right)$

8. Complete, as in the example.

$$\begin{aligned}
3 \cdot (-10) &= -30\\
2 \cdot (-10) &= -20\\
1 \cdot (-10) &=\\
0 \cdot (-10) &=\\
-1 \cdot (-10) &=\\
-2 \cdot (-10) &=\\
-3 \cdot (-10) &=
\end{aligned}$$

Answers on page A-18

Multiply.

9. $-9 \cdot (-3)$

10. $-16 \cdot (-2)$

11. $-7 \cdot (-5)$

12. $-\frac{4}{7}\left(-\frac{5}{9}\right)$

13. $-\frac{3}{2}\left(-\frac{4}{9}\right)$

14. $-3.25(-4.14)$

Multiply.

15. $5(-6)$

16. $(-5)(-6)$

17. $(-3.2) \cdot 0$

18. $\left(-\frac{4}{5}\right)\left(\frac{10}{3}\right)$

Answers on page A-18

According to the pattern, it appears that the product of two negative numbers is positive. That is actually so, and we have the second part of the rule for multiplying real numbers.

> To multiply two negative numbers, multiply their absolute values. The answer is positive.

Do Exercises 9–14.

The following is another way to consider the rules we have for multiplication.

> To multiply two real numbers:
>
> a) Multiply the absolute values.
>
> b) If the signs are the same, the answer is positive.
>
> c) If the signs are different, the answer is negative.

Multiplication by Zero

The only case that we have not considered is multiplying by zero. As with other numbers, the product of any real number and 0 is 0.

> **THE MULTIPLICATION PROPERTY OF ZERO**
>
> For any real number a,
>
> $$a \cdot 0 = 0.$$
>
> (The product of 0 and any real number is 0.)

Examples Multiply.

4. $(-3)(-4) = 12$
5. $-1.6(2) = -3.2$
6. $-19 \cdot 0 = 0$
7. $\left(-\frac{5}{6}\right)\left(-\frac{1}{9}\right) = \frac{5}{54}$

Do Exercises 15–18.

Multiplying More Than Two Numbers

When multiplying more than two real numbers, we can choose order and grouping as we please.

Examples Multiply.

8. $-8 \cdot 2(-3) = -16(-3)$ **Multiplying the first two numbers**
$= 48$

9. $-8 \cdot 2(-3) = 24 \cdot 2$ **Multiplying the negatives. Every pair of negative numbers gives a positive product.**
$= 48$

10. $-3(-2)(-5)(4) = 6(-5)(4)$ **Multiplying the first two numbers**
$= (-30)4$
$= -120$

11. $\left(-\frac{1}{2}\right)(8)\left(-\frac{2}{3}\right)(-6) = (-4)4$ **Multiplying the first two numbers and the last two numbers**
$= -16$

12. $-5 \cdot (-2) \cdot (-3) \cdot (-6) = 10 \cdot 18$
$= 180$

13. $(-3)(-5)(-2)(-3)(-6) = (-30)(18)$
$= -540$

We can see the following pattern in the results of Examples 12 and 13.

> The product of an even number of negative numbers is positive.
> The product of an odd number of negative numbers is negative.

Do Exercises 19–24.

Let's compare the expressions $(-x)^2$ and $-x^2$.

Example 14 Evaluate $(-x)^2$ and $-x^2$ for $x = 5$.

$(-x)^2 = (-5)^2 = (-5)(-5) = 25;$ **Substitute 5 for x. Then evaluate the power.**

$-x^2 = -(5)^2 = -25$ **Substitute 5 for x. Evaluate the power. Then find the opposite.**

The expressions $(-x)^2$ and $-x^2$ are *not* equivalent. That is, they do not have the same value for every allowable replacement of the variable by a real number. To find $(-x)^2$, we take the opposite and then square. To find $-x^2$, we find the square and then take the opposite.

Example 15 Evaluate $2x^2$ for $x = 3$ and $x = -3$.

$2x^2 = 2(3)^2 = 2(9) = 18;$

$2x^2 = 2(-3)^2 = 2(9) = 18$

Do Exercises 25–27.

Multiply.

19. $5 \cdot (-3) \cdot 2$

20. $-3 \times (-4.1) \times (-2.5)$

21. $-\frac{1}{2} \cdot \left(-\frac{4}{3}\right) \cdot \left(-\frac{5}{2}\right)$

22. $-2 \cdot (-5) \cdot (-4) \cdot (-3)$

23. $(-4)(-5)(-2)(-3)(-1)$

24. $(-1)(-1)(-2)(-3)(-1)(-1)$

25. Evaluate $(-x)^2$ and $-x^2$ for $x = 2$.

26. Evaluate $(-x)^2$ and $-x^2$ for $x = 3$.

27. Evaluate $3x^2$ for $x = 4$ and for $x = -4$.

Answers on page A-18

Improving Your Math Study Skills

Studying for Tests and Making the Most of Tutoring Sessions

This math study skill feature focuses on the very important task of test preparation.

Test-Taking Tips

- **Make up your own test questions as you study.** You have probably become accustomed by now to the section and objective codes that appear throughout the book. After you have done your homework over a particular objective, write one or two questions on your own that you think might be on a test. You will be amazed at the insight this will provide. You are actually carrying out a task similar to what a teacher does in preparing an exam.
- **Do an overall review of the chapter focusing on the objectives and the examples.** This should be accompanied by a study of any class notes you may have taken.
- **Do the review exercises at the end of the chapter.** Check your answers at the back of the book. If you have trouble with an exercise, use the objective symbol as a guide to go back and do further study of that objective. These review exercises are very much like a sample test.
- **Do the chapter test at the end of the chapter.** This is like taking a second sample test. Check the answers and objective symbols at the back of the book.
- **Ask former students for old exams.** Working such exams can be very helpful and allows you to see what various professors think is important.
- **When taking a test, read each question carefully and try to do all the questions the first time through, but pace yourself.** Answer all the questions, and mark those to recheck if you have time at the end. Very often, your first hunch will be correct.
- **Try to write your test in a neat and orderly manner.** Very often, your instructor tries to give you partial credit when grading an exam. If your test paper is sloppy and disorderly, it is difficult to verify the partial credit. Doing your work neatly can ease such a task for the instructor. Try using an erasable pen to make your writing darker and therefore more readable.
- **What about the student who says, "I could do the work at home, but on the test I made silly mistakes"?** Yes, all of us, including instructors, make silly computational mistakes in class, on homework, and on tests. But your instructor, if he or she has taught for some time, is probably aware that 90% of students who make such comments in truth do not have the required depth of knowledge of the subject matter, and such silly mistakes often are a sign that the student has not mastered the material. There is no way we can make that analysis for you. It will have to be unraveled by some careful soul searching on your part or by a conference with your instructor.

Making the Most of Tutoring and Help Sessions

Often you will determine that a tutoring session would be helpful. The following comments may help you to make the most of such sessions.

- **Work on the topics before you go to the help or tutoring session. Do not go to such sessions viewing yourself as an empty cup and the tutor as a magician who will pour in the learning.** The primary source of your ability to learn is within you. We have seen so many students over the years go to help or tutoring sessions with no advanced preparation. You are often wasting your time and perhaps your money if you are paying for such sessions. Go to class, study the textbook, and mark trouble spots. Then use the help and tutoring sessions to deal with these difficulties most efficiently.
- **Do not be afraid to ask questions in these sessions!** The more you talk to your tutor, the more the tutor can help you with your difficulties.
- **Try being a "tutor" yourself.** Explaining a topic to someone else—a classmate, your instructor—is often the best way to learn it.

Exercise Set 7.5

a Multiply.

1. $-4 \cdot 2$

2. $-3 \cdot 5$

3. $-8 \cdot 6$

4. $-5 \cdot 2$

5. $8 \cdot (-3)$

6. $9 \cdot (-5)$

7. $-9 \cdot 8$

8. $-10 \cdot 3$

9. $-8 \cdot (-2)$

10. $-2 \cdot (-5)$

11. $-7 \cdot (-6)$

12. $-9 \cdot (-2)$

13. $15 \cdot (-8)$

14. $-12 \cdot (-10)$

15. $-14 \cdot 17$

16. $-13 \cdot (-15)$

17. $-25 \cdot (-48)$

18. $39 \cdot (-43)$

19. $-3.5 \cdot (-28)$

20. $97 \cdot (-2.1)$

21. $9 \cdot (-8)$

22. $7 \cdot (-9)$

23. $4 \cdot (-3.1)$

24. $3 \cdot (-2.2)$

25. $-5 \cdot (-6)$

26. $-6 \cdot (-4)$

27. $-7 \cdot (-3.1)$

28. $-4 \cdot (-3.2)$

29. $\frac{2}{3} \cdot \left(-\frac{3}{5}\right)$

30. $\frac{5}{7} \cdot \left(-\frac{2}{3}\right)$

31. $-\frac{3}{8} \cdot \left(-\frac{2}{9}\right)$

32. $-\frac{5}{8} \cdot \left(-\frac{2}{5}\right)$

33. -6.3×2.7

34. -4.1×9.5

35. $-\frac{5}{9} \cdot \frac{3}{4}$

36. $-\frac{8}{3} \cdot \frac{9}{4}$

37. $7 \cdot (-4) \cdot (-3) \cdot 5$

38. $9 \cdot (-2) \cdot (-6) \cdot 7$

39. $-\frac{2}{3} \cdot \frac{1}{2} \cdot \left(-\frac{6}{7}\right)$

40. $-\frac{1}{8} \cdot \left(-\frac{1}{4}\right) \cdot \left(-\frac{3}{5}\right)$

41. $-3 \cdot (-4) \cdot (-5)$

42. $-2 \cdot (-5) \cdot (-7)$

43. $-2 \cdot (-5) \cdot (-3) \cdot (-5)$

44. $-3 \cdot (-5) \cdot (-2) \cdot (-1)$

45. $\frac{1}{5}\left(-\frac{2}{9}\right)$

46. $-\frac{3}{5}\left(-\frac{2}{7}\right)$

47. $-7 \cdot (-21) \cdot 13$

48. $-14 \cdot (34) \cdot 12$

49. $-4 \cdot (-1.8) \cdot 7$

50. $-8 \cdot (-1.3) \cdot (-5)$

51. $-\frac{1}{9}\left(-\frac{2}{3}\right)\left(\frac{5}{7}\right)$

52. $-\frac{7}{2}\left(-\frac{5}{7}\right)\left(-\frac{2}{5}\right)$

53. $4 \cdot (-4) \cdot (-5) \cdot (-12)$

54. $-2 \cdot (-3) \cdot (-4) \cdot (-5)$

55. $0.07 \cdot (-7) \cdot 6 \cdot (-6)$

56. $80 \cdot (-0.8) \cdot (-90) \cdot (-0.09)$

57. $\left(-\frac{5}{6}\right)\left(\frac{1}{8}\right)\left(-\frac{3}{7}\right)\left(-\frac{1}{7}\right)$

58. $\left(\frac{4}{5}\right)\left(-\frac{2}{3}\right)\left(-\frac{15}{7}\right)\left(\frac{1}{2}\right)$

59. $(-14) \cdot (-27) \cdot 0$

60. $7 \cdot (-6) \cdot 5 \cdot (-4) \cdot 3 \cdot (-2) \cdot 1 \cdot 0$

61. $(-8)(-9)(-10)$

62. $(-7)(-8)(-9)(-10)$

63. $(-6)(-7)(-8)(-9)(-10)$

64. $(-5)(-6)(-7)(-8)(-9)(-10)$

65. Evaluate $(-3x)^2$ and $-3x^2$ for $x = 7$.

66. Evaluate $(-2x)^2$ and $-2x^2$ for $x = 3$.

67. Evaluate $5x^2$ for $x = 2$ and for $x = -2$.

68. Evaluate $2x^2$ for $x = 5$ and for $x = -5$.

Skill Maintenance

69. Find the LCM of 36 and 60. [1.9a]

70. Find the prime factorization of 4608. [1.7d]

Simplify. [2.1e]

71. $\frac{26}{39}$

72. $\frac{48}{54}$

73. $\frac{264}{484}$

74. $\frac{1025}{6625}$

Synthesis

75. ◆ Multiplication can be thought of as repeated addition. Using this concept and a number line, explain why $3 \cdot (-5) = -15$.

76. ◆ What rule have we developed that would tell you the sign of $(-7)^8$ and $(-7)^{11}$ without doing the computations? Explain.

77. After diving 95 m below the surface, a diver rises at a rate of 7 meters per minute for 9 min. What is the diver's new elevation?

78. Jo wrote seven checks for $13 each. If she began with a balance of $68 in her account, what was her balance after having written the checks?

79. What must be true of a and b if $-ab$ is to be (a) positive? (b) zero? (c) negative?

80. Evaluate $-6(3x - 5y) + z$ for $x = -2$, $y = -4$, and $z = 5$.

7.6 Division of Real Numbers

We now consider division of real numbers. The definition of division results in rules for division that are the same as those for multiplication.

Objectives

a Divide integers.

b Find the reciprocal of a real number.

c Divide real numbers.

For Extra Help

TAPE 15

MAC WIN

CD-ROM

a Division of Integers

> The quotient $\frac{a}{b}$ (or $a \div b$) is the number, if there is one, that when multiplied by b gives a.

Let's use the definition to divide integers.

Examples Divide, if possible. Check your answer.

1. $14 \div (-7) = -2$ — *Think*: What number multiplied by -7 gives 14? That number is -2. *Check*: $(-2)(-7) = 14$.

2. $\frac{-32}{-4} = 8$ — *Think*: What number multiplied by -4 gives -32? That number is 8. *Check*: $8(-4) = -32$.

3. $\frac{-10}{7} = -\frac{10}{7}$ — *Think*: What number multiplied by 7 gives -10? That number is $-\frac{10}{7}$. *Check*: $-\frac{10}{7} \cdot 7 = -10$.

4. $\frac{-17}{0}$ is **undefined**. — *Think*: What number multiplied by 0 gives -17? There is no such number because the product of 0 and *any* number is 0.

The rules for division are the same as those for multiplication.

> To multiply or divide two real numbers:
> a) Multiply or divide the absolute values.
> b) If the signs are the same, the answer is positive.
> c) If the signs are different, the answer is negative.

Do Exercises 1–6.

Divide.

1. $6 \div (-3)$

Think: What number multiplied by -3 gives 6?

2. $\frac{-15}{-3}$

Think: What number multiplied by -3 gives -15?

3. $-24 \div 8$

Think: What number multiplied by 8 gives -24?

4. $\frac{-48}{-6}$

5. $\frac{30}{-5}$

6. $\frac{30}{-7}$

Division by Zero

Example 4 shows why we cannot divide -17 by 0. We can use the same argument to show why we cannot divide any nonzero number b by 0. Consider $b \div 0$. We look for a number that when multiplied by 0 gives b. There is no such number because the product of 0 and any number is 0. Thus we cannot divide a nonzero number b by 0.

On the other hand, if we divide 0 by 0, we look for a number r such that $0 \cdot r = 0$. But $0 \cdot r = 0$ for any number r. Thus it appears that $0 \div 0$ could be any number we choose. Getting any answer we want when we divide 0 by 0 would be very confusing. Thus we agree that division by zero is undefined.

> Division by 0 is undefined.
>
> $a \div 0$ is undefined for all real numbers a.
>
> 0 divided by a nonzero number a is 0.
>
> $0 \div a = 0, \quad a \neq 0.$

Answers on page A-18

For example, $\frac{0}{4} = 0$, $\frac{4}{0}$ is undefined, and $\frac{0}{0}$ is undefined.

Do Exercises 7 and 8.

b Reciprocals

When two numbers like $\frac{1}{2}$ and 2 are multiplied, the result is 1. Such numbers are called **reciprocals** of each other. Every nonzero real number has a reciprocal, also called a **multiplicative inverse.**

> Two numbers whose product is 1 are called **reciprocals**, or **multiplicative inverses,** of each other.

Examples Find the reciprocal.

5. $\frac{7}{8}$ The reciprocal of $\frac{7}{8}$ is $\frac{8}{7}$ because $\frac{7}{8} \cdot \frac{8}{7} = 1$.

6. -5 The reciprocal of -5 is $-\frac{1}{5}$ because $-5\left(-\frac{1}{5}\right) = 1$.

7. 3.9 The reciprocal of 3.9 is $\frac{1}{3.9}$ because $3.9\left(\frac{1}{3.9}\right) = 1$.

8. $-\frac{1}{2}$ The reciprocal of $-\frac{1}{2}$ is -2 because $\left(-\frac{1}{2}\right)(-2) = 1$.

9. $-\frac{2}{3}$ The reciprocal of $-\frac{2}{3}$ is $-\frac{3}{2}$ because $\left(-\frac{2}{3}\right)\left(-\frac{3}{2}\right) = 1$.

10. $\frac{1}{3/4}$ The reciprocal of $\frac{1}{3/4}$ is $\frac{3}{4}$ because $\left(\frac{1}{3/4}\right)\left(\frac{3}{4}\right) = 1$.

> For $a \neq 0$, the reciprocal of a can be named $\frac{1}{a}$ and the reciprocal of $\frac{1}{a}$ is a.
>
> The reciprocal of a nonzero number $\frac{a}{b}$ can be named $\frac{b}{a}$.
>
> The number 0 has no reciprocal.

Do Exercises 9–14.

The reciprocal of a positive number is also a positive number, because their product must be the positive number 1. The reciprocal of a negative number is also a negative number, because their product must be the positive number 1.

> The reciprocal of a number has the same sign as the number itself.

CAUTION! It is important *not* to confuse *opposite* with *reciprocal.* Keep in mind that the opposite, or additive inverse, of a number is what we add to the number to get 0. The reciprocal, or multiplicative inverse, is what we multiply the number by to get 1.

Divide, if possible.

7. $\frac{-5}{0}$

8. $\frac{0}{-3}$

Find the reciprocal.

9. $\frac{2}{3}$

10. $-\frac{5}{4}$

11. -3

12. $-\frac{1}{5}$

13. 1.6

14. $\frac{1}{2/3}$

Answers on page A-18

Compare the following.

Number	Opposite (Change the Sign.)	Reciprocal (Invert But Do Not Change the Sign.)
$-\frac{3}{8}$	$\frac{3}{8}$	$-\frac{8}{3}$
19	−19	$\frac{1}{19}$
$\frac{18}{7}$	$-\frac{18}{7}$	$\frac{7}{18}$
−7.9	7.9	$-\frac{1}{7.9}$, or $-\frac{10}{79}$
0	0	Undefined

$\left(-\frac{3}{8}\right)\left(-\frac{8}{3}\right) = 1$

$-\frac{3}{8} + \frac{3}{8} = 0$

Do Exercise 15.

15. Complete the following table.

Number	Opposite	Reciprocal
$\frac{2}{3}$		
$-\frac{5}{4}$		
0		
1		
−8		
−4.5		

c Division of Real Numbers

We know that we can subtract by adding an opposite. Similarly, we can divide by multiplying by a reciprocal.

For any real numbers a and b, $b \neq 0$,

$$a \div b = \frac{a}{b} = a \cdot \frac{1}{b}.$$

(To divide, we can multiply by the reciprocal of the divisor.)

Examples Rewrite the division as a multiplication.

11. $-4 \div 3$ $\quad$ $-4 \div 3$ is the same as $-4 \cdot \frac{1}{3}$

12. $\frac{6}{-7}$ $\quad$ $\frac{6}{-7} = 6\left(-\frac{1}{7}\right)$

13. $\frac{x+2}{5}$ $\quad$ $\frac{x+2}{5} = (x+2)\frac{1}{5}$ $\quad$ **Parentheses are necessary here.**

14. $\frac{-17}{1/b}$ $\quad$ $\frac{-17}{1/b} = -17 \cdot b$

15. $\frac{3}{5} \div \left(-\frac{9}{7}\right)$ $\quad$ $\frac{3}{5} \div \left(-\frac{9}{7}\right) = \frac{3}{5}\left(-\frac{7}{9}\right)$

Do Exercises 16–20.

Rewrite the division as a multiplication.

16. $\frac{4}{7} \div \left(-\frac{3}{5}\right)$

17. $\frac{5}{-8}$

18. $\frac{a-b}{7}$

19. $\frac{-23}{1/a}$

20. $-5 \div 7$

When actually doing division calculations, we sometimes multiply by a reciprocal and we sometimes divide directly. With fractional notation, it is usually better to multiply by a reciprocal. With decimal notation, it is usually better to divide directly.

Answers on page A-18

Divide by multiplying by the reciprocal of the divisor.

21. $\frac{4}{7} \div \left(-\frac{3}{5}\right)$

22. $-\frac{8}{5} \div \frac{2}{3}$

23. $-\frac{12}{7} \div \left(-\frac{3}{4}\right)$

24. Divide: $21.7 \div (-3.1)$.

Find two equal expressions for the number with negative signs in different places.

25. $\frac{-5}{6}$

26. $-\frac{8}{7}$

27. $\frac{10}{-3}$

Answers on page A-18

Examples Divide by multiplying by the reciprocal of the divisor.

16. $\frac{2}{3} \div \left(-\frac{5}{4}\right) = \frac{2}{3} \cdot \left(-\frac{4}{5}\right) = -\frac{8}{15}$

17. $-\frac{5}{6} \div \left(-\frac{3}{4}\right) = -\frac{5}{6} \cdot \left(-\frac{4}{3}\right) = \frac{20}{18} = \frac{10 \cdot 2}{9 \cdot 2} = \frac{10}{9} \cdot \frac{2}{2} = \frac{10}{9}$

Caution! Be careful not to change the sign when taking a reciprocal!

18. $-\frac{3}{4} \div \frac{3}{10} = -\frac{3}{4} \cdot \left(\frac{10}{3}\right) = -\frac{30}{12} = -\frac{5}{2} \cdot \frac{6}{6} = -\frac{5}{2}$

With decimal notation, it is easier to carry out long division than to multiply by the reciprocal.

Examples Divide.

19. $-27.9 \div (-3) = \frac{-27.9}{-3} = 9.3$ Do the long division $3\overline{)27.9}$ (quotient 9.3). The answer is positive.

20. $-6.3 \div 2.1 = -3$ Do the long division $2.1\overline{)6.3}$ (quotient 3.). The answer is negative.

Do Exercises 21–24.

Consider the following:

1. $\frac{2}{3} = \frac{2}{3} \cdot 1 = \frac{2}{3} \cdot \frac{-1}{-1} = \frac{2(-1)}{3(-1)} = \frac{-2}{-3}$. Thus, $\frac{2}{3} = \frac{-2}{-3}$.

2. $-\frac{2}{3} = -1 \cdot \frac{2}{3} = \frac{-1}{1} \cdot \frac{2}{3} = \frac{-1 \cdot 2}{1 \cdot 3} = \frac{-2}{3}$. Thus, $-\frac{2}{3} = \frac{-2}{3}$.

$\frac{-2}{3} = \frac{-2}{3} \cdot 1 = \frac{-2}{3} \cdot \frac{-1}{-1} = \frac{-2(-1)}{3(-1)} = \frac{2}{-3}$. Thus, $\frac{-2}{3} = \frac{2}{-3}$.

We can use the following properties to make sign changes in fractional notation.

For any numbers a and b, $b \neq 0$:

1. $\frac{-a}{-b} = \frac{a}{b}$

(The opposite of a number a divided by the opposite of another number b is the same as the quotient of the two numbers a and b.)

2. $\frac{-a}{b} = \frac{a}{-b} = -\frac{a}{b}$

(The opposite of a number a divided by another number b is the same as the number a divided by the opposite of the number b, and both are the same as the opposite of *a divided by b*.)

Do Exercises 25–27.

Exercise Set 7.6

a Divide, if possible. Check each answer.

1. $48 \div (-6)$

2. $\frac{42}{-7}$

3. $\frac{28}{-2}$

4. $24 \div (-12)$

5. $\frac{-24}{8}$

6. $-18 \div (-2)$

7. $\frac{-36}{-12}$

8. $-72 \div (-9)$

9. $\frac{-72}{9}$

10. $\frac{-50}{25}$

11. $-100 \div (-50)$

12. $\frac{-200}{8}$

13. $-108 \div 9$

14. $\frac{-63}{-7}$

15. $\frac{200}{-25}$

16. $-300 \div (-16)$

17. $\frac{75}{0}$

18. $\frac{0}{-5}$

19. $\frac{-23}{-2}$

20. $\frac{-23}{0}$

b Find the reciprocal.

21. $\frac{15}{7}$

22. $\frac{3}{8}$

23. $-\frac{47}{13}$

24. $-\frac{31}{12}$

25. 13

26. -10

27. 4.3

28. -8.5

29. $\frac{1}{-7.1}$

30. $\frac{1}{-4.9}$

31. $\frac{p}{q}$

32. $\frac{s}{t}$

33. $\frac{1}{4y}$

34. $\frac{-1}{8a}$

35. $\frac{2a}{3b}$

36. $\frac{-4y}{3x}$

c Rewrite the division as a multiplication.

37. $4 \div 17$

38. $5 \div (-8)$

39. $\frac{8}{-13}$

40. $-\frac{13}{47}$

41. $\frac{13.9}{-1.5}$

42. $-\frac{47.3}{21.4}$

43. $\frac{x}{\frac{1}{y}}$

44. $\frac{13}{x}$

45. $\dfrac{3x + 4}{5}$

46. $\dfrac{4y - 8}{-7}$

47. $\dfrac{5a - b}{5a + b}$

48. $\dfrac{2x + x^2}{x - 5}$

Divide.

49. $\dfrac{3}{4} \div \left(-\dfrac{2}{3}\right)$

50. $\dfrac{7}{8} \div \left(-\dfrac{1}{2}\right)$

51. $-\dfrac{5}{4} \div \left(-\dfrac{3}{4}\right)$

52. $-\dfrac{5}{9} \div \left(-\dfrac{5}{6}\right)$

53. $-\dfrac{2}{7} \div \left(-\dfrac{4}{9}\right)$

54. $-\dfrac{3}{5} \div \left(-\dfrac{5}{8}\right)$

55. $-\dfrac{3}{8} \div \left(-\dfrac{8}{3}\right)$

56. $-\dfrac{5}{8} \div \left(-\dfrac{6}{5}\right)$

57. $-6.6 \div 3.3$

58. $-44.1 \div (-6.3)$

59. $\dfrac{-11}{-13}$

60. $\dfrac{-1.9}{20}$

61. $\dfrac{48.6}{-3}$

62. $\dfrac{-17.8}{3.2}$

63. $\dfrac{-9}{17 - 17}$

64. $\dfrac{-8}{-5 + 5}$

Skill Maintenance

65. Simplify: $\dfrac{264}{468}$. [2.1e]

66. Convert to decimal notation: 47.7%. [4.2b]

67. Simplify: $2^3 - 5 \cdot 3 + 8 \cdot 10 \div 2$. [1.6c]

68. Add and simplify: $\dfrac{2}{3} + \dfrac{5}{6}$. [2.3b]

69. Convert to percent notation: $\dfrac{7}{8}$. [4.3a]

70. Simplify: $\dfrac{40}{60}$. [2.1e]

71. Divide and simplify: $\dfrac{12}{25} \div \dfrac{32}{75}$. [2.2c]

72. Multiply and simplify: $\dfrac{12}{25} \cdot \dfrac{32}{75}$. [2.2a]

Synthesis

73. ◆ Explain how multiplication can be used to justify why a negative number divided by a positive number is negative.

74. ◆ Explain how multiplication can be used to justify why a negative number divided by a negative number is positive.

75. ◆ ▦ Find the reciprocal of -10.5. What happens if you take the reciprocal of the result?

76. Determine those real numbers a for which the opposite of a is the same as the reciprocal of a.

Tell whether the expression represents a positive number or a negative number when a and b are negative.

77. $\dfrac{-a}{b}$

78. $\dfrac{-a}{-b}$

79. $-\left(\dfrac{a}{-b}\right)$

80. $-\left(\dfrac{-a}{b}\right)$

81. $-\left(\dfrac{-a}{-b}\right)$

7.7 Properties of Real Numbers

a Equivalent Expressions

In solving equations and doing other kinds of work in algebra, we manipulate expressions in various ways. For example, instead of

$$x + x,$$

we might write

$$2x,$$

knowing that the two expressions represent the same number for any allowable replacement of x. In that sense, the expressions $x + x$ and $2x$ are **equivalent**, as are $3/x$ and $3x/x^2$, even though 0 is not an allowable replacement because division by 0 is undefined.

> Two expressions that have the same value for all allowable replacements are called **equivalent**.

The expressions $x + 3x$ and $5x$ are *not* equivalent.

Do Exercises 1 and 2.

In this section, we will consider several laws of real numbers that will allow us to find equivalent expressions. The first two laws are the *identity properties of 0 and 1.*

> **THE IDENTITY PROPERTY OF 0**
>
> For any real number a,
>
> $$a + 0 = 0 + a = a.$$
>
> (The number 0 is the *additive identity.*)

> **THE IDENTITY PROPERTY OF 1**
>
> For any real number a,
>
> $$a \cdot 1 = 1 \cdot a = a.$$
>
> (The number 1 is the *multiplicative identity.*)

We often refer to the use of the identity property of 1 as "multiplying by 1." We can use this method to find equivalent fractional expressions. Recall from arithmetic that to multiply with fractional notation, we multiply numerators and denominators. (See also Section 2.2.)

Objectives

a Find equivalent fractional expressions and simplify fractional expressions.

b Use the commutative and associative laws to find equivalent expressions.

c Use the distributive laws to multiply expressions like 8 and $x - y$.

d Use the distributive laws to factor expressions like $4x - 12 + 24y$.

e Collect like terms.

For Extra Help

TAPE 15 MAC WIN CD-ROM

Complete the table by evaluating each expression for the given values.

1.

	$x + x$	$2x$
$x = 3$		
$x = -6$		
$x = 4.8$		

2.

	$x + 3x$	$5x$
$x = 2$		
$x = -6$		
$x = 4.8$		

Answers on page A-18

3. Write a fractional expression equivalent to $\frac{3}{4}$ with a denominator of 8.

4. Write a fractional expression equivalent to $\frac{3}{4}$ with a denominator of $4t$.

Simplify.

5. $\frac{3y}{4y}$

6. $-\frac{16m}{12m}$

7. Evaluate $x + y$ and $y + x$ for $x = -2$ and $y = 3$.

8. Evaluate xy and yx for $x = -2$ and $y = 5$.

Answers on page A-18

Example 1 Write a fractional expression equivalent to $\frac{2}{3}$ with a denominator of $3x$.

Note that $3x = 3 \cdot x$. We want fractional notation for $\frac{2}{3}$ that has a denominator of $3x$, but the denominator 3 is missing a factor of x. Thus we multiply by 1, using x/x as an equivalent expression for 1:

$$\frac{2}{3} = \frac{2}{3} \cdot 1 = \frac{2}{3} \cdot \frac{x}{x} = \frac{2x}{3x}.$$

The expressions 2/3 and $2x/3x$ are equivalent. They have the same value for any allowable replacement. Note that $2x/3x$ is undefined for a replacement of 0, but for all nonzero real numbers, the expressions 2/3 and $2x/3x$ have the same value.

Do Exercises 3 and 4.

In algebra, we consider an expression like 2/3 to be "simplified" from $2x/3x$. To find such simplified expressions, we use the identity property of 1 to remove a factor of 1. (See also Section 2.1.)

Example 2 Simplify: $-\frac{20x}{12x}$.

$$-\frac{20x}{12x} = -\frac{5 \cdot 4x}{3 \cdot 4x}$$ We look for the largest factor common to both the numerator and the denominator and factor each.

$$= -\frac{5}{3} \cdot \frac{4x}{4x}$$ Factoring the fractional expression

$$= -\frac{5}{3} \cdot 1$$ $\frac{4x}{4x} = 1$

$$= -\frac{5}{3}$$ Removing a factor of 1 using the identity property of 1

Do Exercises 5 and 6.

b The Commutative and Associative Laws

The Commutative Laws

Let's examine the expressions $x + y$ and $y + x$, as well as xy and yx.

Example 3 Evaluate $x + y$ and $y + x$ for $x = 4$ and $y = 3$.

We substitute 4 for x and 3 for y in both expressions:

$$x + y = 4 + 3 = 7; \qquad y + x = 3 + 4 = 7.$$

Example 4 Evaluate xy and yx for $x = 23$ and $y = 12$.

We substitute 23 for x and 12 for y in both expressions:

$$xy = 23 \cdot 12 = 276; \qquad yx = 12 \cdot 23 = 276.$$

Do Exercises 7 and 8.

Note that the expressions

$$x + y \quad \text{and} \quad y + x$$

have the same values no matter what the variables stand for. Thus they are equivalent. Therefore, when we add two numbers, the order in which we add does not matter. Similarly, the expressions xy and yx are equivalent. They also have the same values, no matter what the variables stand for. Therefore, when we multiply two numbers, the order in which we multiply does not matter.

The following are examples of general patterns or laws.

THE COMMUTATIVE LAWS

Addition. For any numbers a and b,

$$a + b = b + a.$$

(We can change the order when adding without affecting the answer.)

Multiplication. For any numbers a and b,

$$ab = ba.$$

(We can change the order when multiplying without affecting the answer.)

Using a commutative law, we know that $x + 2$ and $2 + x$ are equivalent. Similarly, $3x$ and $x(3)$ are equivalent. Thus, in an algebraic expression, we can replace one with the other and the result will be equivalent to the original expression.

Example 5 Use the commutative laws to write an expression equivalent to $y + 5$, ab, and $7 + xy$.

An expression equivalent to $y + 5$ is $5 + y$ by the commutative law of addition.

An expression equivalent to ab is ba by the commutative law of multiplication.

An expression equivalent to $7 + xy$ is $xy + 7$ by the commutative law of addition. Another expression equivalent to $7 + xy$ is $7 + yx$ by the commutative law of multiplication.

Do Exercises 9–11.

Use a commutative law to write an equivalent expression.

9. $x + 9$

10. pq

11. $xy + t$

The Associative Laws

Now let's examine the expressions $a + (b + c)$ and $(a + b) + c$. Note that these expressions involve the use of parentheses as *grouping* symbols, and they also involve three numbers. Calculations within parentheses are to be done first.

Example 6 Calculate and compare: $3 + (8 + 5)$ and $(3 + 8) + 5$.

$$3 + (8 + 5) = 3 + 13 \qquad \text{Calculating within parentheses first; adding the 8 and 5}$$
$$= 16;$$

$$(3 + 8) + 5 = 11 + 5 \qquad \text{Calculating within parentheses first; adding the 3 and 8}$$
$$= 16$$

Answers on page A-18

12. Calculate and compare:

$8 + (9 + 2)$ and $(8 + 9) + 2$.

13. Calculate and compare:

$10 \cdot (5 \cdot 3)$ and $(10 \cdot 5) \cdot 3$.

Use an associative law to write an equivalent expression.

14. $r + (s + 7)$

15. $9(ab)$

Answers on page A-18

The two expressions in Example 6 name the same number. Moving the parentheses to group the additions differently does not affect the value of the expression.

Example 7 Calculate and compare: $3 \cdot (4 \cdot 2)$ and $(3 \cdot 4) \cdot 2$.

$3 \cdot (4 \cdot 2) = 3 \cdot 8 = 24;\qquad (3 \cdot 4) \cdot 2 = 12 \cdot 2 = 24$

Do Exercises 12 and 13.

You may have noted that when only addition is involved, parentheses can be placed any way we please without affecting the answer. When only multiplication is involved, parentheses also can be placed any way we please without affecting the answer.

THE ASSOCIATIVE LAWS

Addition. For any numbers a, b, and c,

$$a + (b + c) = (a + b) + c.$$

(Numbers can be grouped in any manner for addition.)

Multiplication. For any numbers a, b, and c,

$$a \cdot (b \cdot c) = (a \cdot b) \cdot c.$$

(Numbers can be grouped in any manner for multiplication.)

Example 8 Use an associative law to write an expression equivalent to $(y + z) + 3$ and $8(xy)$.

An equivalent expression is $y + (z + 3)$ by the associative law of addition.

An equivalent expression is $(8x)y$ by the associative law of multiplication.

Do Exercises 14 and 15.

The associative laws say parentheses can be placed any way we please when only additions or only multiplications are involved. Thus we often omit them. For example,

$$x + (y + 2) \text{ means } x + y + 2, \quad \text{and} \quad (lw)h \text{ means } lwh.$$

Using the Commutative and Associative Laws Together

Example 9 Use the commutative and associative laws to write at least three expressions equivalent to $(x + 5) + y$.

a) $(x + 5) + y = x + (5 + y)$
$= x + (y + 5)$ — **Using the associative law first and then using the commutative law**

b) $(x + 5) + y = y + (x + 5)$
$= y + (5 + x)$ — **Using the commutative law first and then the commutative law again**

c) $(x + 5) + y = (5 + x) + y$
$= 5 + (x + y)$ — **Using the commutative law first and then the associative law**

Example 10 Use the commutative and associative laws to write at least three expressions equivalent to $(3x)y$.

a) $(3x)y = 3(xy)$ **Using the associative law first and then using the commutative law**

$= 3(yx)$

b) $(3x)y = y(3x)$ **Using the commutative law twice**

$= y(x3)$

c) $(3x)y = (x3)y$ **Using the commutative law, and then the associative law, and then the commutative law again**

$= x(3y)$

$= x(y3)$

Do Exercises 16 and 17.

Use the commutative and associative laws to write at least three equivalent expressions.

16. $4(tu)$

17. $r + (2 + s)$

c The Distributive Laws

The *distributive laws* are the basis of many procedures in both arithmetic and algebra. They are probably the most important laws that we use to manipulate algebraic expressions. The distributive law of multiplication over addition involves two operations: addition and multiplication.

Let's begin by considering a multiplication problem from arithmetic:

```
    4 5
  ×   7
  -----
    3 5 ←This is 7 · 5.
  2 8 0 ←This is 7 · 40.
  -----
  3 1 5 ←This is the sum 7 · 40 + 7 · 5.
```

To carry out the multiplication, we actually added two products. That is,

$$7 \cdot 45 = 7(40 + 5) = 7 \cdot 40 + 7 \cdot 5.$$

Let's examine this further. If we wish to multiply a sum of several numbers by a factor, we can either add and then multiply, or multiply and then add.

Example 11 Compute in two ways: $5 \cdot (4 + 8)$.

a) $5 \cdot (4 + 8)$ **Adding within parentheses first, and then multiplying**

$= 5 \cdot 12$

$= 60$

b) $(5 \cdot 4) + (5 \cdot 8)$ **Distributing the multiplication to terms within parentheses first and then adding**

$= 20 + 40$

$= 60$

Do Exercises 18–20.

Compute.

18. a) $7 \cdot (3 + 6)$

b) $(7 \cdot 3) + (7 \cdot 6)$

19. a) $2 \cdot (10 + 30)$

b) $(2 \cdot 10) + (2 \cdot 30)$

20. a) $(2 + 5) \cdot 4$

b) $(2 \cdot 4) + (5 \cdot 4)$

Answers on page A-18

THE DISTRIBUTIVE LAW OF MULTIPLICATION OVER ADDITION

For any numbers a, b, and c,

$$a(b + c) = ab + ac.$$

Calculate.

21. a) $4(5 - 3)$

b) $4 \cdot 5 - 4 \cdot 3$

22. a) $-2 \cdot (5 - 3)$

b) $-2 \cdot 5 - (-2) \cdot 3$

23. a) $5 \cdot (2 - 7)$

b) $5 \cdot 2 - 5 \cdot 7$

What are the terms of the expression?

24. $5x - 8y + 3$

25. $-4y - 2x + 3z$

Multiply.

26. $3(x - 5)$

27. $5(x + 1)$

28. $\frac{3}{5}(p + q - t)$

Answers on page A-18

In the statement of the distributive law, we know that in an expression such as $ab + ac$, the multiplications are to be done first according to the rules for order of operations. (See Section 1.6.) So, instead of writing $(4 \cdot 5) + (4 \cdot 7)$, we can write $4 \cdot 5 + 4 \cdot 7$. However, in $a(b + c)$, we cannot omit the parentheses. If we did, we would have $ab + c$, which means $(ab) + c$. For example, $3(4 + 2) = 18$, but $3 \cdot 4 + 2 = 14$.

There is another distributive law that relates multiplication and subtraction. This law says that to multiply by a difference, we can either subtract and then multiply, or multiply and then subtract.

THE DISTRIBUTIVE LAW OF MULTIPLICATION OVER SUBTRACTION

For any numbers a, b, and c,

$$a(b - c) = ab - ac.$$

We often refer to "*the* distributive law" when we mean *either* or *both* of these laws.

Do Exercises 21–23.

What do we mean by the *terms* of an expression? **Terms** are separated by addition signs. If there are subtraction signs, we can find an equivalent expression that uses addition signs.

Example 12 What are the terms of $3x - 4y + 2z$?

We have

$$3x - 4y + 2z = 3x + (-4y) + 2z. \quad \text{Separating parts with + signs}$$

The terms are $3x$, $-4y$, and $2z$.

Do Exercises 24 and 25.

The distributive laws are a basis for a procedure in algebra called **multiplying**. In an expression like $8(a + 2b - 7)$, we multiply each term inside the parentheses by 8:

$$8(a + 2b - 7) = 8 \cdot a + 8 \cdot 2b - 8 \cdot 7 = 8a + 16b - 56.$$

Examples Multiply.

13. $9(x - 5) = 9x - 9(5)$ Using the distributive law of multiplication over subtraction

$= 9x - 45$

14. $\frac{2}{3}(w + 1) = \frac{2}{3} \cdot w + \frac{2}{3} \cdot 1$ Using the distributive law of multiplication over addition

$= \frac{2}{3}w + \frac{2}{3}$

15. $\frac{4}{3}(s - t + w) = \frac{4}{3}s - \frac{4}{3}t + \frac{4}{3}w$ Using both distributive laws

Do Exercises 26–28.

Example 16 Multiply: $-4(x - 2y + 3z)$.

$$-4(x - 2y + 3z) = -4 \cdot x - (-4)(2y) + (-4)(3z) \quad \text{Using both distributive laws}$$
$$= -4x - (-8y) + (-12z) \quad \text{Multiplying}$$
$$= -4x + 8y - 12z$$

We can also do this problem by first finding an equivalent expression with all plus signs and then multiplying:

$$-4(x - 2y + 3z) = -4[x + (-2y) + 3z]$$
$$= -4 \cdot x + (-4)(-2y) + (-4)(3z)$$
$$= -4x + 8y - 12z.$$

Do Exercises 29–31.

Multiply.

29. $-2(x - 3)$

30. $5(x - 2y + 4z)$

31. $-5(x - 2y + 4z)$

d Factoring

Factoring is the reverse of multiplying. To factor, we can use the distributive laws in reverse:

$$ab + ac = a(b + c) \quad \text{and} \quad ab - ac = a(b - c).$$

> To **factor** an expression is to find an equivalent expression that is a product.

Look at Example 13. To *factor* $9x - 45$, we find an equivalent expression that is a product, $9(x - 5)$. When all the terms of an expression have a factor in common, we can "factor it out" using the distributive laws. Note the following.

$9x$ has the factors $9, -9, 3, -3, 1, -1, x, -x, 3x, -3x, 9x, -9x$;

-45 has the factors $1, -1, 3, -3, 5, -5, 9, -9, 15, -15, 45, -45$

We generally remove the largest common factor. In this case, that factor is 9. Thus,

$$9x - 45 = 9 \cdot x - 9 \cdot 5$$
$$= 9(x - 5).$$

Remember that an expression has been factored when we have found an equivalent expression that is a product.

Examples Factor.

17. $5x - 10 = 5 \cdot x - 5 \cdot 2$ Try to do this step mentally.

$= 5(x - 2)$ You can check by multiplying.

18. $ax - ay + az = a(x - y + z)$

19. $9x + 27y - 9 = 9 \cdot x + 9 \cdot 3y - 9 \cdot 1 = 9(x + 3y - 1)$

Answers on page A-18

Factor.

32. $6x - 12$

33. $3x - 6y + 9$

34. $bx + by - bz$

35. $16a - 36b + 42$

36. $\frac{3}{8}x - \frac{5}{8}y + \frac{7}{8}$

37. $-12x + 32y - 16z$

Collect like terms.

38. $6x - 3x$

39. $7x - x$

40. $x - 9x$

41. $x - 0.41x$

42. $5x + 4y - 2x - y$

43. $3x - 7x - 11 + 8y + 4 - 13y$

44. $-\frac{2}{3} - \frac{3}{5}x + y + \frac{7}{10}x - \frac{2}{9}y$

Answers on page A-19

Examples Factor. Try to write just the answer, if you can.

20. $5x - 5y = 5(x - y)$

21. $-3x + 6y - 9z = -3(x - 2y + 3z)$

We usually factor out a negative when the first term is negative. The way we factor can depend on the situation in which we are working. We might also factor the expression in Example 21 as follows:

$$-3x + 6y - 9z = 3(-x + 2y - 3z).$$

22. $18z - 12x - 24 = 6(3z - 2x - 4)$

23. $\frac{1}{2}x + \frac{3}{2}y - \frac{1}{2} = \frac{1}{2}(x + 3y - 1)$

Remember that you can always check factoring by multiplying. Keep in mind that an expression is factored when it is written as a product.

Do Exercises 32–37.

e Collecting Like Terms

Terms such as $5x$ and $-4x$, whose variable factors are exactly the same, are called **like terms.** Similarly, numbers, such as -7 and 13, are like terms. Also, $3y^2$ and $9y^2$ are like terms because the variables are raised to the same power. Terms such as $4y$ and $5y^2$ are not like terms, and $7x$ and $2y$ are not like terms.

The process of **collecting like terms** is also based on the distributive laws. We can apply the distributive law when a factor is on the right because of the commutative law of multiplication.

Examples Collect like terms. Try to write just the answer, if you can.

24. $4x + 2x = (4 + 2)x = 6x$ Factoring out the x using a distributive law

25. $2x + 3y - 5x - 2y = 2x - 5x + 3y - 2y$
$= (2 - 5)x + (3 - 2)y = -3x + y$

26. $3x - x = (3 - 1)x = 2x$

27. $x - 0.24x = 1 \cdot x - 0.24x = (1 - 0.24)x = 0.76x$

28. $x - 6x = 1 \cdot x - 6 \cdot x = (1 - 6)x = -5x$

29. $4x - 7y + 9x - 5 + 3y - 8 = 13x - 4y - 13$

30. $\frac{2}{3}a - b + \frac{4}{5}a + \frac{1}{4}b - 10 = \frac{2}{3}a - 1 \cdot b + \frac{4}{5}a + \frac{1}{4}b - 10$
$= \left(\frac{2}{3} + \frac{4}{5}\right)a + \left(-1 + \frac{1}{4}\right)b - 10$
$= \left(\frac{10}{15} + \frac{12}{15}\right)a + \left(-\frac{4}{4} + \frac{1}{4}\right)b - 10$
$= \frac{22}{15}a - \frac{3}{4}b - 10$

Do Exercises 38–44.

Exercise Set 7.7

a Find an equivalent expression with the given denominator.

1. $\frac{3}{5}$; $5y$

2. $\frac{5}{8}$; $8t$

3. $\frac{2}{3}$; $15x$

4. $\frac{6}{7}$; $14y$

Simplify.

5. $-\frac{24a}{16a}$

6. $-\frac{42t}{18t}$

7. $-\frac{42ab}{36ab}$

8. $-\frac{64pq}{48pq}$

b Write an equivalent expression. Use a commutative law.

9. $y + 8$

10. $x + 3$

11. mn

12. ab

13. $9 + xy$

14. $11 + ab$

15. $ab + c$

16. $rs + t$

Write an equivalent expression. Use an associative law.

17. $a + (b + 2)$

18. $3(vw)$

19. $(8x)y$

20. $(y + z) + 7$

21. $(a + b) + 3$

22. $(5 + x) + y$

23. $3(ab)$

24. $(6x)y$

Use the commutative and associative laws to write three equivalent expressions.

25. $(a + b) + 2$

26. $(3 + x) + y$

27. $5 + (v + w)$

28. $6 + (x + y)$

29. $(xy)3$

30. $(ab)5$

31. $7(ab)$

32. $5(xy)$

c Multiply.

33. $2(b + 5)$

34. $4(x + 3)$

35. $7(1 + t)$

36. $4(1 + y)$

37. $6(5x + 2)$

38. $9(6m + 7)$

39. $7(x + 4 + 6y)$

40. $4(5x + 8 + 3p)$

41. $7(x-3)$

42. $15(y-6)$

43. $-3(x-7)$

44. $1.2(x-2.1)$

45. $\frac{2}{3}(b-6)$

46. $\frac{5}{8}(y+16)$

47. $7.3(x-2)$

48. $5.6(x-8)$

49. $-\frac{3}{5}(x-y+10)$

50. $-\frac{2}{3}(a+b-12)$

51. $-9(-5x-6y+8)$

52. $-7(-2x-5y+9)$

53. $-4(x-3y-2z)$

54. $8(2x-5y-8z)$

55. $3.1(-1.2x+3.2y-1.1)$

56. $-2.1(-4.2x-4.3y-2.2)$

List the terms of the expression.

57. $4x+3z$

58. $8x-1.4y$

59. $7x+8y-9z$

60. $8a+10b-18c$

d Factor. Check by multiplying.

61. $2x+4$

62. $5y+20$

63. $30+5y$

64. $7x+28$

65. $14x+21y$

66. $18a+24b$

67. $5x+10+15y$

68. $9a+27b+81$

69. $8x-24$

70. $10x-50$

71. $32-4y$

72. $24-6m$

73. $8x + 10y - 22$

74. $9a + 6b - 15$

75. $ax - a$

76. $by - 9b$

77. $ax - ay - az$

78. $cx + cy - cz$

79. $18x - 12y + 6$

80. $-14x + 21y + 7$

81. $\frac{2}{3}x - \frac{5}{3}y + \frac{1}{3}$

82. $\frac{3}{5}a + \frac{4}{5}b - \frac{1}{5}$

e Collect like terms.

83. $9a + 10a$

84. $12x + 2x$

85. $10a - a$

86. $-16x + x$

87. $2x + 9z + 6x$

88. $3a - 5b + 7a$

89. $7x + 6y^2 + 9y^2$

90. $12m^2 + 6q + 9m^2$

91. $41a + 90 - 60a - 2$

92. $42x - 6 - 4x + 2$

93. $23 + 5t + 7y - t - y - 27$

94. $45 - 90d - 87 - 9d + 3 + 7d$

95. $\frac{1}{2}b + \frac{1}{2}b$

96. $\frac{2}{3}x + \frac{1}{3}x$

97. $2y + \frac{1}{4}y + y$

98. $\frac{1}{2}a + a + 5a$

99. $11x - 3x$

100. $9t - 17t$

101. $6n - n$

102. $10t - t$

103. $y - 17y$

104. $3m - 9m + 4$

105. $-8 + 11a - 5b + 6a - 7b + 7$

106. $8x - 5x + 6 + 3y - 2y - 4$

107. $9x + 2y - 5x$

108. $8y - 3z + 4y$

109. $11x + 2y - 4x - y$

110. $13a + 9b - 2a - 4b$

111. $2.7x + 2.3y - 1.9x - 1.8y$

112. $6.7a + 4.3b - 4.1a - 2.9b$

113. $\frac{13}{2}a + \frac{9}{5}b - \frac{2}{3}a - \frac{3}{10}b - 42$

114. $\frac{11}{4}x + \frac{2}{3}y - \frac{4}{5}x - \frac{1}{6}y + 12$

Skill Maintenance

115. Add and simplify: $\frac{11}{12} + \frac{15}{16}$. [2.3b]

116. Subtract and simplify: $\frac{7}{8} - \frac{2}{3}$. [2.3c]

117. Find the LCM of 16, 18, and 24. [1.9a]

118. Convert to percent notation: $\frac{3}{10}$. [4.3a]

119. Subtract and simplify: $\frac{1}{8} - \frac{1}{3}$. [2.3c]

120. Find the LCM of 12, 15, and 20. [1.9a]

Synthesis

121. ◆ The distributive law was introduced before the discussion on collecting like terms. Why do you think this was done?

122. ◆ Find two different expressions for the total area of the two rectangles shown below. Explain the equivalence of the expressions in terms of the distributive law.

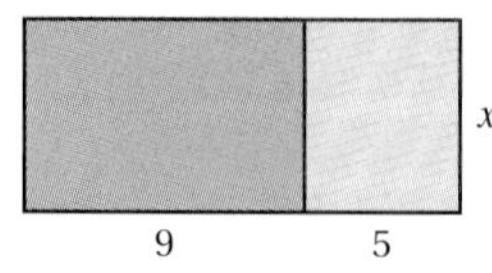

Tell whether the expressions are equivalent. Explain.

123. $3t + 5$ and $3 \cdot 5 + t$

124. $4x$ and $x + 4$

125. $5m + 6$ and $6 + 5m$

126. $(x + y) + z$ and $z + (x + y)$

Collect like terms, if possible, and factor the result.

127. $q + qr + qrs + qrst$

128. $21x + 44xy + 15y - 16x - 8y - 38xy + 2y + xy$

Collaborative Learning Manual

Use the commutative and associative laws to add a series of numbers.

7.8 Simplifying Expressions; Order of Operations

We now expand our ability to manipulate expressions by first considering opposites of sums and differences. Then we simplify expressions involving parentheses.

Objectives

a Find an equivalent expression for an opposite without parentheses, where an expression has several terms.

b Simplify expressions by removing parentheses and collecting like terms.

c Simplify expressions with parentheses inside parentheses.

d Simplify expressions using rules for order of operations.

For Extra Help

TAPE 15 | MAC WIN | CD-ROM

a Opposites of Sums

What happens when we multiply a real number by -1? Consider the following products:

$$-1(7) = -7, \qquad -1(-5) = 5, \qquad -1(0) = 0.$$

From these examples, it appears that when we multiply a number by -1, we get the opposite, or additive inverse, of that number.

> **THE PROPERTY OF -1**
>
> For any real number a,
>
> $$-1 \cdot a = -a.$$
>
> (Negative one times a is the opposite, or additive inverse, of a.)

The property of -1 enables us to find certain expressions equivalent to opposites of sums.

Examples Find an equivalent expression without parentheses.

1. $-(3 + x) = -1(3 + x)$ — **Using the property of -1**

$= -1 \cdot 3 + (-1)x$ — **Using a distributive law, multiplying each term by -1**

$= -3 + (-x)$ — **Using the property of -1**

$= -3 - x$

2. $-(3x + 2y + 4) = -1(3x + 2y + 4)$ — **Using the property of -1**

$= -1(3x) + (-1)(2y) + (-1)4$ — **Using a distributive law**

$= -3x - 2y - 4$ — **Using the property of -1**

Do Exercises 1 and 2.

Find an equivalent expression without parentheses.

1. $-(x + 2)$

2. $-(5x + 2y + 8)$

Suppose we want to remove parentheses in an expression like

$$-(x - 2y + 5).$$

We can first rewrite any subtractions inside the parentheses as additions. Then we take the opposite of each term:

$$-(x - 2y + 5) = -[x + (-2y) + 5]$$
$$= -x + 2y - 5.$$

The most efficient method for removing parentheses is to replace each term in the parentheses with its opposite ("change the sign of every term"). Doing so for $-(x - 2y + 5)$, we obtain $-x + 2y - 5$ as an equivalent expression.

Answers on page A-19

Find an equivalent expression without parentheses. Try to do this in one step.

3. $-(6 - t)$

4. $-(x - y)$

5. $-(-4a + 3t - 10)$

6. $-(18 - m - 2n + 4z)$

Remove parentheses and simplify.

7. $5x - (3x + 9)$

8. $5y - 2 - (2y - 4)$

Remove parentheses and simplify.

9. $6x - (4x + 7)$

10. $8y - 3 - (5y - 6)$

11. $(2a + 3b - c) - (4a - 5b + 2c)$

Answers on page A-19

Examples Find an equivalent expression without parentheses.

3. $-(5 - y) = -5 + y$ Changing the sign of each term

4. $-(2a - 7b - 6) = -2a + 7b + 6$

5. $-(-3x + 4y + z - 7w - 23) = 3x - 4y - z + 7w + 23$

Do Exercises 3–6.

b Removing Parentheses and Simplifying

When a sum is added, as in $5x + (2x + 3)$, we can simply remove, or drop, the parentheses and collect like terms because of the associative law of addition:

$$5x + (2x + 3) = 5x + 2x + 3 = 7x + 3.$$

On the other hand, when a sum is subtracted, as in $3x - (4x + 2)$, no "associative" law applies. However, we can subtract by adding an opposite. We then remove parentheses by changing the sign of each term inside the parentheses and collecting like terms.

Example 6 Remove parentheses and simplify.

$3x - (4x + 2) = 3x + [-(4x + 2)]$ Adding the opposite of $(4x + 2)$

$= 3x + (-4x - 2)$ Changing the sign of each term inside the parentheses

$= 3x - 4x - 2$

$= -x - 2$ Collecting like terms

Do Exercises 7 and 8.

In practice, the first three steps of Example 6 are usually combined by changing the sign of each term in parentheses and then collecting like terms.

Examples Remove parentheses and simplify.

7. $5y - (3y + 4) = 5y - 3y - 4$ Removing parentheses by changing the sign of every term inside the parentheses

$= 2y - 4$ Collecting like terms

8. $3y - 2 - (2y - 4) = 3y - 2 - 2y + 4 = y + 2$

9. $(3a + 4b - 5) - (2a - 7b + 4c - 8)$

$= 3a + 4b - 5 - 2a + 7b - 4c + 8$

$= a + 11b - 4c + 3$

Do Exercises 9–11.

Next, consider subtracting an expression consisting of several terms multiplied by a number other than 1 or -1.

Example 10 Remove parentheses and simplify.

$x - 3(x + y) = x + [-3(x + y)]$ Adding the opposite of $3(x + y)$

$= x + [-3x - 3y]$ Multiplying $x + y$ by -3

$= x - 3x - 3y$

$= -2x - 3y$ Collecting like terms

Examples Remove parentheses and simplify.

11. $3y - 2(4y - 5) = 3y - 8y + 10$ Multiplying each term in parentheses by -2
$= -5y + 10$

12. $(2a + 3b - 7) - 4(-5a - 6b + 12)$
$= 2a + 3b - 7 + 20a + 24b - 48$
$= 22a + 27b - 55$

13. $2y - \frac{1}{3}(9y - 12) = 2y - 3y + 4$
$= -y + 4$

Do Exercises 12–15.

c Parentheses Within Parentheses

In addition to parentheses, some expressions contain other grouping symbols such as brackets [ ] and braces { }.

> When more than one kind of grouping symbol occurs, do the computations in the innermost ones first. Then work from the inside out.

Examples Simplify.

14. $[3 - (7 + 3)] = [3 - 10]$ Computing $7 + 3$
$= -7$

15. $\{8 - [9 - (12 + 5)]\} = \{8 - [9 - 17]\}$ Computing $12 + 5$
$= \{8 - [-8]\}$ Computing $9 - 17$
$= 8 + 8$
$= 16$

16. $\left[(-4) \div \left(-\frac{1}{4}\right)\right] \div \frac{1}{4} = [(-4) \cdot (-4)] \div \frac{1}{4}$ Working within the brackets computing $(-4) \div \left(-\frac{1}{4}\right)$
$= 16 \div \frac{1}{4}$
$= 16 \cdot 4$
$= 64$

17. $4(2 + 3) - \{7 - [4 - (8 + 5)]\}$
$= 4 \cdot 5 - \{7 - [4 - 13]\}$ Working with the innermost parentheses first
$= 20 - \{7 - [-9]\}$ Computing $4 \cdot 5$ and $4 - 13$
$= 20 - 16$ Computing $7 - [-9]$
$= 4$

Do Exercises 16–19.

Example 18 Simplify.

$[5(x + 2) - 3x] - [3(y + 2) - 7(y - 3)]$
$= [5x + 10 - 3x] - [3y + 6 - 7y + 21]$ Working with the innermost parentheses first
$= [2x + 10] - [-4y + 27]$ Collecting like terms within brackets
$= 2x + 10 + 4y - 27$ Removing brackets
$= 2x + 4y - 17$ Collecting like terms

Do Exercise 20.

Remove parentheses and simplify.

12. $y - 9(x + y)$

13. $5a - 3(7a - 6)$

14. $4a - b - 6(5a - 7b + 8c)$

15. $5x - \frac{1}{4}(8x + 28)$

Simplify.

16. $12 - (8 + 2)$

17. $\{9 - [10 - (13 + 6)]\}$

18. $[24 \div (-2)] \div (-2)$

19. $5(3 + 4) - \{8 - [5 - (9 + 6)]\}$

20. Simplify:

$[3(x + 2) + 2x] -$
$[4(y + 2) - 3(y - 2)]$.

Answers on page A-19

d Order of Operations

When several operations are to be done in a calculation or a problem, we apply the same rules that we did in Section 1.6. We repeat them here for review. (If you did not study that section earlier, you should do so now.)

RULES FOR ORDER OF OPERATIONS

1. Do all calculations within parentheses before operations outside.
2. Evaluate all exponential expressions.
3. Do all multiplications and divisions in order from left to right.
4. Do all additions and subtractions in order from left to right.

These rules are consistent with the way in which most computers and scientific calculators perform calculations.

Example 19 Simplify: $-34 \cdot 56 - 17$.

There are no parentheses or powers, so we start with the third step.

$$-34 \cdot 56 - 17 = -1904 - 17$$ Carrying out all multiplications and divisions in order from left to right

$$= -1921$$ Carrying out all additions and subtractions in order from left to right

Example 20 Simplify: $2^4 + 51 \cdot 4 - (37 + 23 \cdot 2)$.

$$2^4 + 51 \cdot 4 - (37 + 23 \cdot 2)$$

$$= 2^4 + 51 \cdot 4 - (37 + 46)$$ Following the rules for order of operations within the parentheses first

$$= 2^4 + 51 \cdot 4 - 83$$ Completing the addition inside parentheses

$$= 16 + 51 \cdot 4 - 83$$ Evaluating exponential expressions

$$= 16 + 204 - 83$$ Doing all multiplications

$$= 220 - 83$$ Doing all additions and subtractions in order from left to right

$$= 137$$

A fraction bar can play the role of a grouping symbol, although such a symbol is not as evident as the others.

Example 21 Simplify: $\dfrac{-64 \div (-16) \div (-2)}{2^3 - 3^2}$.

An equivalent expression with brackets as grouping symbols is

$$[-64 \div (-16) \div (-2)] \div [2^3 - 3^2].$$

This shows, in effect, that we can do the calculations in the numerator and then in the denominator, and divide the results:

$$\frac{-64 \div (-16) \div (-2)}{2^3 - 3^2} = \frac{4 \div (-2)}{8 - 9} = \frac{-2}{-1} = 2.$$

Do Exercises 21–24.

Simplify.

21. $23 - 42 \cdot 30$

22. $32 \div 8 \cdot 2$

23. $52 \cdot 5 + 5^3 - (4^2 - 48 \div 4)$

24. $\dfrac{5 - 10 - 5 \cdot 23}{2^3 + 3^2 - 7}$

Answers on page A-19

Calculator Spotlight

Calculations Using Order of Operations. To do a calculation like $-8 - (-2.3)$, we press the following keys:

[(−)] [8] [−] [(−)] [2] [.] [3] [ENTER].

The answer is -5.7.

Note that we did not need to key in grouping symbols. Sometimes we do need the parenthesis keys [(] and [)]. For example, to do a calculation like $-7(2 - 9) - 20$, we press the following keys:

[(−)] [7] [(] [2] [−] [9] [)] [−]
[2] [0] [ENTER].

The answer is 29.

To enter a power like $(-39)^4$, we press

[(] [(−)] [3] [9] [)] [^] [4] [ENTER].

The answer is 2,313,441.

To find -39^4, think of the expression as -1×39^4. Then we press

[(−)] [3] [9] [^] [4] [ENTER].

The answer is $-2{,}313{,}441$.

To simplify an expression like

$$\frac{38 + 142}{2 - 47},$$

we first think of it using grouping symbols as

$$(38 + 142) \div (2 - 47).$$

We then press

[(] [3] [8] [+] [1] [4] [2] [)] [÷]
[(] [2] [−] [4] [7] [)] [ENTER].

The answer is -4.

Exercises

Evaluate.

1. $-8 + 4(7 - 9) + 5$
2. $-3[2 + (-5)]$
3. $7[4 - (-3)] + 5[3^2 - (-4)]$
4. $(-7)^6$
5. $(-17)^5$
6. $(-104)^3$
7. -7^6
8. -17^5
9. -104^3

Calculate.

10. $\dfrac{38 - 178}{5 + 30}$

11. $\dfrac{311 - 17^2}{2 - 13}$

12. $785 - \dfrac{285 - 5^4}{17 + 3 \cdot 51}$

In Exercises 13 and 14, place one of $+$, $-$, $\times$, and $\div$ in each blank to make a true sentence.

13. $-32 \ \square \ (88 \ \square \ 29) = -1888$
14. $3^5 \ \square \ 10^2 \ \square \ 5^2 = -2257$
15. Consider the numbers 2, 4, 6, and 8. Assume that each can be placed in a blank in the following.

    $\square + \square \cdot \square - \square = ?$

    What placement of the numbers in the blanks yields the largest number? Explain why there are two answers.
16. Consider the numbers 3, 5, 7, and 9. Assume that each can be placed in a blank in the following.

    $\square + \square^2 \cdot \square - \square = ?$

    What placement of the numbers in the blanks yields the largest number? Explain why, unlike Exercise 15, there is just one answer.

Improving Your Math Study Skills

Study Tips for Trouble Spots

By now you have probably encountered certain topics that gave you more difficulty than others. It is important to know that this happens to every person who studies mathematics. Unfortunately, frustration is often part of the learning process and it is important not to give up when difficulty arises.

One source of frustration for many students is not being able to set aside sufficient time for studying. Family commitments, work schedules, and athletics are just a few of the time demands that many students face. Couple these demands with a math lesson that seems to require a greater than usual amount of study time, and it is no wonder that many students often feel frustrated. Below are some study tips that might be useful if and when troubles arise.

- **Realize that everyone—even your instructor—has been stymied at times when studying math.** You are not the first person, nor will you be the last, to encounter a "roadblock."
- **Whether working alone or with a classmate, try to allow enough study time so that you won't need to constantly glance at a clock.** Difficult material is best mastered when your mind is completely focused on the subject matter. Thus, if you are tired, it is usually best to study early the next morning or to take a ten-minute "power-nap" in order to make the most productive use of your time.
- **Talk about your trouble spot with a classmate.** It is possible that she or he is also having difficulty with the same material. If that is the case, perhaps the majority of your class is confused and your instructor's coverage of the topic is not yet finished. If your classmate *does* understand the topic that is troubling you, patiently allow him or her to explain it to you. By verbalizing the math in question, your classmate may help clarify the material for both of you. Perhaps you will be able to return the favor for your classmate when he or she is struggling with a topic that you understand.
- **Try to study in a "controlled" environment.** What we mean by this is that you can often put yourself in a setting that will enable you to maximize your powers of concentration. For example, whereas some students may succeed in studying at home or in a dorm room, for many these settings are filled with distractions. Consider a trip to a library, classroom building, or perhaps the attic or basement if such a setting is more conducive to studying. If you plan on working with a classmate, try to find a location in which conversation will not be bothersome to others.
- **When working on difficult material, it is often helpful to first "back up" and review the most recent material that *did* make sense.** This can build your confidence and create a momentum that can often carry you through the roadblock. Sometimes a small piece of information that appeared in a previous section is all that is needed for your problem spot to disappear. When the difficult material is finally mastered, try to make use of what is fresh in your mind by taking a "sneak preview" of what your next topic for study will be.

Exercise Set 7.8

a Find an equivalent expression without parentheses.

1. $-(2x + 7)$

2. $-(8x + 4)$

3. $-(5x - 8)$

4. $-(4x - 3)$

5. $-(4a - 3b + 7c)$

6. $-(x - 4y - 3z)$

7. $-(6x - 8y + 5)$

8. $-(4x + 9y + 7)$

9. $-(3x - 5y - 6)$

10. $-(6a - 4b - 7)$

11. $-(-8x - 6y - 43)$

12. $-(-2a + 9b - 5c)$

b Remove parentheses and simplify.

13. $9x - (4x + 3)$

14. $4y - (2y + 5)$

15. $2a - (5a - 9)$

16. $12m - (4m - 6)$

17. $2x + 7x - (4x + 6)$

18. $3a + 2a - (4a + 7)$

19. $2x - 4y - 3(7x - 2y)$

20. $3a - 9b - 1(4a - 8b)$

21. $15x - y - 5(3x - 2y + 5z)$

22. $4a - b - 4(5a - 7b + 8c)$

23. $(3x + 2y) - 2(5x - 4y)$

24. $(-6a - b) - 5(2b + a)$

25. $(12a - 3b + 5c) - 5(-5a + 4b - 6c)$

26. $(-8x + 5y - 12) - 6(2x - 4y - 10)$

c Simplify.

27. $[9 - 2(5 - 4)]$

28. $[6 - 5(8 - 4)]$

29. $8[7 - 6(4 - 2)]$

30. $10[7 - 4(7 - 5)]$

31. $[4(9 - 6) + 11] - [14 - (6 + 4)]$

32. $[7(8 - 4) + 16] - [15 - (7 + 8)]$

33. $[10(x + 3) - 4] + [2(x - 1) + 6]$

34. $[9(x + 5) - 7] + [4(x - 12) + 9]$

35. $[7(x + 5) - 19] - [4(x - 6) + 10]$

36. $[6(x + 4) - 12] - [5(x - 8) + 14]$

37. $3\{[7(x - 2) + 4] - [2(2x - 5) + 6]\}$

38. $4\{[8(x - 3) + 9] - [4(3x - 2) + 6]\}$

39. $4\{[5(x - 3) + 2] - 3[2(x + 5) - 9]\}$

40. $3\{[6(x - 4) + 5] - 2[5(x + 8) - 3]\}$

d Simplify.

41. $8 - 2 \cdot 3 - 9$

42. $8 - (2 \cdot 3 - 9)$

43. $(8 - 2 \cdot 3) - 9$

44. $(8 - 2)(3 - 9)$

45. $[(-24) \div (-3)] \div \left(-\frac{1}{2}\right)$

46. $[32 \div (-2)] \div (-2)$

47. $16 \cdot (-24) + 50$

48. $10 \cdot 20 - 15 \cdot 24$

49. $2^4 + 2^3 - 10$

50. $40 - 3^2 - 2^3$

51. $5^3 + 26 \cdot 71 - (16 + 25 \cdot 3)$

52. $4^3 + 10 \cdot 20 + 8^2 - 23$

53. $4 \cdot 5 - 2 \cdot 6 + 4$

54. $4 \cdot (6 + 8)/(4 + 3)$

55. $4^3/8$

56. $5^3 - 7^2$

57. $8(-7) + 6(-5)$

58. $10(-5) + 1(-1)$

59. $19 - 5(-3) + 3$

60. $14 - 2(-6) + 7$

61. $9 \div (-3) + 16 \div 8$

62. $-32 - 8 \div 4 - (-2)$

63. $6 - 4^2$

64. $(2 - 5)^2$

65. $(3 - 8)^2$

66. $3 - 3^2$

67. $12 - 20^3$

68. $20 + 4^3 \div (-8)$

69. $2 \cdot 10^3 - 5000$

70. $-7(3^4) + 18$

71. $6[9 - (3 - 4)]$

72. $8[(6 - 13) - 11]$

73. $-1000 \div (-100) \div 10$

74. $256 \div (-32) \div (-4)$

75. $8 - (7 - 9)$

76. $(8 - 7) - 9$

77. $\dfrac{10 - 6^2}{9^2 + 3^2}$

78. $\dfrac{5^2 - 4^3 - 3}{9^2 - 2^2 - 1^5}$

79. $\dfrac{3(6-7)-5\cdot 4}{6\cdot 7-8(4-1)}$

80. $\dfrac{20(8-3)-4(10-3)}{10(2-6)-2(5+2)}$

81. $\dfrac{2^3-3^2+12\cdot 5}{-32\div(-16)\div(-4)}$

82. $\dfrac{|3-5|^2-|7-13|}{|12-9|+|11-14|}$

Skill Maintenance

83. Find the prime factorization of 236. [1.7d]

84. Find the LCM of 28 and 36. [1.9a]

85. Divide and simplify: $\dfrac{2}{3}\div\dfrac{5}{12}$. [2.2c]

86. Multiply and simplify: $\dfrac{2}{3}\cdot\dfrac{5}{12}$. [2.2a]

Evaluate. [1.6b]

87. 3^4

88. 10^3

89. 10^2

90. 15^2

Synthesis

91. ◆ Some students use the memory device PEMDAS ("Please Excuse My Dear Aunt Sally") to remember the rules for the order of operations. Explain how this can be done.

92. ◆ Determine whether $|-x|$ and $|x|$ are equivalent. Explain.

Find an equivalent expression by enclosing the last three terms in parentheses preceded by a minus sign.

93. $6y+2x-3a+c$

94. $x-y-a-b$

95. $6m+3n-5m+4b$

Simplify.

96. $z-\{2z-[3z-(4z-5z)-6z]-7z\}-8z$

97. $\{x-[f-(f-x)]+[x-f]\}-3x$

98. $x-\{x-1-[x-2-(x-3-\{x-4-[x-5-(x-6)]\})]\}$

99. ▦ Use your calculator to do the following.

a) Evaluate x^2+3 for $x=7$, for $x=-7$, and for $x=-5.013$.

b) Evaluate $1-x^2$ for $x=5$, for $x=-5$, and for $x=-10.455$.

100. Express $3^3+3^3+3^3$ as a power of 3.

Use the order of operations as a group to simplify expressions.

Summary and Review Exercises: Chapter 7

Important Properties and Formulas

Properties of the Real-Number System

The Commutative Laws:	$a + b = b + a, \quad ab = ba$
The Associative Laws:	$a + (b + c) = (a + b) + c, \quad a(bc) = (ab)c$
The Identity Properties:	For every real number a, $a + 0 = a$ and $a \cdot 1 = a$.
The Inverse Properties:	For each real number a, there is an opposite $-a$, such that $a + (-a) = 0$.
	For each nonzero real number a, there is a reciprocal $\frac{1}{a}$, such that $a\left(\frac{1}{a}\right) = 1$.
The Distributive Laws:	$a(b + c) = ab + ac, \quad a(b - c) = ab - ac$

The review exercises that follow are for practice. Answers are at the back of the book. If you miss an exercise, restudy the objective indicated in blue after the exercise or the direction line that precedes it. Beginning with this chapter, certain objectives, from four particular sections of preceding chapters, will be retested on the chapter test. The objectives to be tested in addition to the material in this chapter are [1.6b], [1.9a], [2.2c], and [4.3a].

1. Evaluate $\frac{x - y}{3}$ for $x = 17$ and $y = 5$. [7.1a]

2. Translate to an algebraic expression: [7.1b]
Nineteen percent of some number.

3. Tell which integers correspond to this situation: [7.2a]
David has a debt of \$45 and Joe has \$72 in his savings account.

4. Find: $|-38|$. [7.2e]

Graph the number on a number line. [7.2b]

5. -2.5

6. $\frac{8}{9}$

Use either $<$ or $>$ for ▒ to write a true sentence. [7.2d]

7. -3 ▒ 10

8. -1 ▒ -6

9. 0.126 ▒ -12.6

10. $-\frac{2}{3}$ ▒ $-\frac{1}{10}$

Find the opposite. [7.3b]

11. 3.8

12. $-\frac{3}{4}$

Find the reciprocal. [7.6b]

13. $\frac{3}{8}$

14. -7

15. Find $-x$ when $x = -34$. [7.3b]

16. Find $-(-x)$ when $x = 5$. [7.3b]

Compute and simplify.

17. $4 + (-7)$ [7.3a]

18. $6 + (-9) + (-8) + 7$ [7.3a]

19. $-3.8 + 5.1 + (-12) + (-4.3) + 10$ [7.3a]

20. $-3 - (-7)$ [7.4a]

21. $-\frac{9}{10} - \frac{1}{2}$ [7.4a]

22. $-3.8 - 4.1$ [7.4a]

23. $-9 \cdot (-6)$ [7.5a]

24. $-2.7(3.4)$ [7.5a]

25. $\frac{2}{3} \cdot \left(-\frac{3}{7}\right)$ [7.5a]

26. $3 \cdot (-7) \cdot (-2) \cdot (-5)$ [7.5a]

27. $35 \div (-5)$ [7.6a]

28. $-5.1 \div 1.7$ [7.6c]

29. $-\frac{3}{11} \div \left(-\frac{4}{11}\right)$ [7.6c]

30. $(-3.4 - 12.2) - 8(-7)$ [7.8d]

31. $\frac{-12(-3) - 2^3 - (-9)(-10)}{3 \cdot 10 + 1}$ [7.8d]

Solve. [7.4b]

32. On the first, second, and third downs, a football team had these gains and losses: 5-yd gain, 12-yd loss, and 15-yd gain, respectively. Find the total gain (or loss).

33. Kaleb's total assets are \$170. He borrows \$300. What are his total assets now?

Multiply. [7.7c]

34. $5(3x - 7)$

35. $-2(4x - 5)$

36. $10(0.4x + 1.5)$

37. $-8(3 - 6x)$

Factor. [7.7d]

38. $2x - 14$

39. $6x - 6$

40. $5x + 10$

41. $12 - 3x$

Collect like terms. [7.7e]

42. $11a + 2b - 4a - 5b$

43. $7x - 3y - 9x + 8y$

44. $6x + 3y - x - 4y$

45. $-3a + 9b + 2a - b$

Remove parentheses and simplify.

46. $2a - (5a - 9)$ [7.8b]

47. $3(b + 7) - 5b$ [7.8b]

48. $3[11 - 3(4 - 1)]$ [7.8c]

49. $2[6(y - 4) + 7]$ [7.8c]

50. $[8(x + 4) - 10] - [3(x - 2) + 4]$ [7.8c]

51. $5\{[6(x - 1) + 7] - [3(3x - 4) + 8]\}$ [7.8c]

Write true or false. [7.2d]

52. $-9 \leq 11$

53. $-11 \geq -3$

54. Write another inequality with the same meaning as $-3 < x$. [7.2d]

Skill Maintenance

55. Divide and simplify: $\frac{11}{12} \div \frac{7}{10}$. [2.2c]

56. Compute and simplify: $\frac{5^3 - 2^4}{5 \cdot 2 + 2^3}$. [1.6c]

57. Evaluate: $(2.5)^2$. [1.6b]

58. Convert to percent notation: $\frac{5}{8}$. [4.3a]

59. Convert to decimal notation: 5.67%. [4.2b]

60. Find the LCM of 15, 27, and 30. [1.9a]

Synthesis

Simplify. [7.2e], [7.4a], [7.6a], [7.8d]

61. $-\left|\frac{7}{8} - \left(-\frac{1}{2}\right) - \frac{3}{4}\right|$

62. $(|2.7 - 3| + 3^2 - |-3|) \div (-3)$

63. $2000 - 1990 + 1980 - 1970 + \cdots + 20 - 10$

64. Find a formula for the perimeter of the following figure. [7.7e]

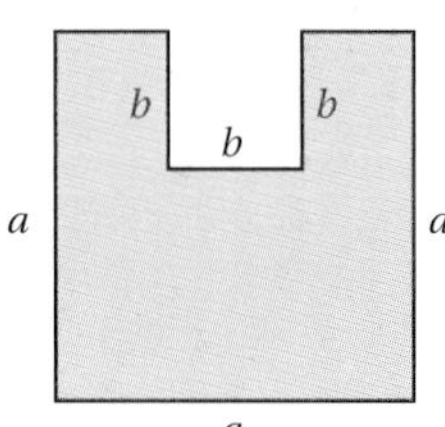

Test: Chapter 7

1. Evaluate $\frac{3x}{y}$ for $x = 10$ and $y = 5$.

2. Write an algebraic expression: Nine less than some number.

3. Find the area of a triangle when the height h is 30 ft and the base b is 16 ft.

Use either $<$ or $>$ for ▒ to write a true sentence.

4. -4 ▒ 0

5. -3 ▒ -8

6. -0.78 ▒ -0.87

7. $-\frac{1}{8}$ ▒ $\frac{1}{2}$

Find the absolute value.

8. $|-7|$

9. $\left|\frac{9}{4}\right|$

10. $|-2.7|$

Find the opposite.

11. $\frac{2}{3}$

12. -1.4

13. Find $-x$ when $x = -8$.

Find the reciprocal.

14. -2

15. $\frac{4}{7}$

Compute and simplify.

16. $3.1 - (-4.7)$

17. $-8 + 4 + (-7) + 3$

18. $-\frac{1}{5} + \frac{3}{8}$

19. $2 - (-8)$

20. $3.2 - 5.7$

21. $\frac{1}{8} - \left(-\frac{3}{4}\right)$

Answers

1. ________

2. ________

3. ________

4. ________

5. ________

6. ________

7. ________

8. ________

9. ________

10. ________

11. ________

12. ________

13. ________

14. ________

15. ________

16. ________

17. ________

18. ________

19. ________

20. ________

21. ________

Answers

22. ______

23. ______

24. ______

25. ______

26. ______

27. ______

28. ______

29. ______

30. ______

31. ______

32. ______

33. ______

34. ______

35. ______

36. ______

37. ______

38. ______

39. ______

40. ______

41. ______

42. ______

43. ______

44. ______

45. ______

46. ______

22. $4 \cdot (-12)$

23. $-\frac{1}{2} \cdot \left(-\frac{3}{8}\right)$

24. $-45 \div 5$

25. $-\frac{3}{5} \div \left(-\frac{4}{5}\right)$

26. $4.864 \div (-0.5)$

27. $-2(16) - |2(-8) - 5^3|$

28. *Antarctica Highs and Lows.* The continent of Antarctica, which lies in the southern hemisphere, experiences winter in July. The average high temperature is $-67°$F and the average low temperature is $-81°$F. How much higher is the average high than the average low?

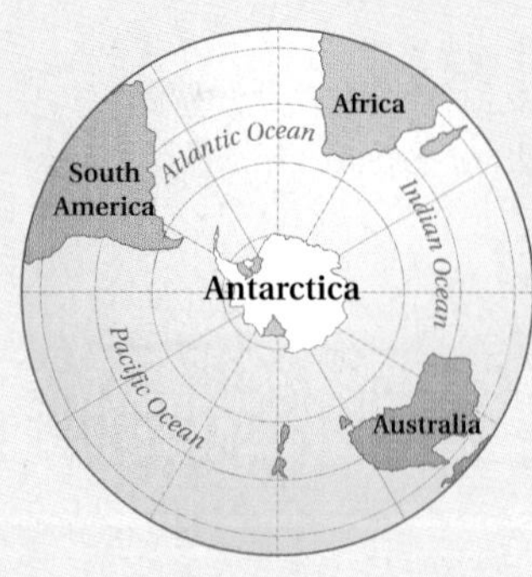

Multiply.

29. $3(6 - x)$

30. $-5(y - 1)$

Factor.

31. $12 - 22x$

32. $7x + 21 + 14y$

Simplify.

33. $6 + 7 - 4 - (-3)$

34. $5x - (3x - 7)$

35. $4(2a - 3b) + a - 7$

36. $4\{3[5(y - 3) + 9] + 2(y + 8)\}$

37. $256 \div (-16) \div 4$

38. $2^3 - 10[4 - (-2 + 18)3]$

39. Write an inequality with the same meaning as $x \leq -2$.

Skill Maintenance

40. Evaluate: $(1.2)^3$.

41. Convert to percent notation: $\frac{1}{8}$.

42. Divide and simplify: $\frac{3}{8} \div \frac{3}{16}$.

43. Find the LCM of 16, 20, and 30.

Synthesis

Simplify.

44. $|-27 - 3(4)| - |-36| + |-12|$

45. $a - \{3a - [4a - (2a - 4a)]\}$

46. Find a formula for the perimeter of the following figure.

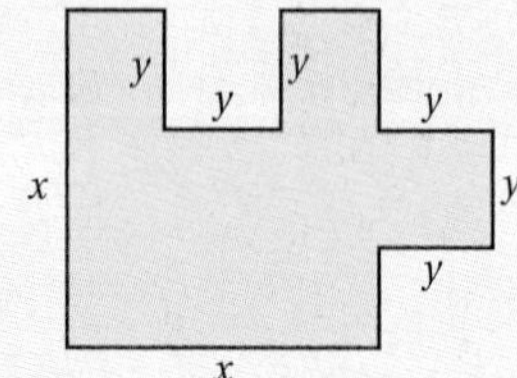

8

Solving Equations and Inequalities

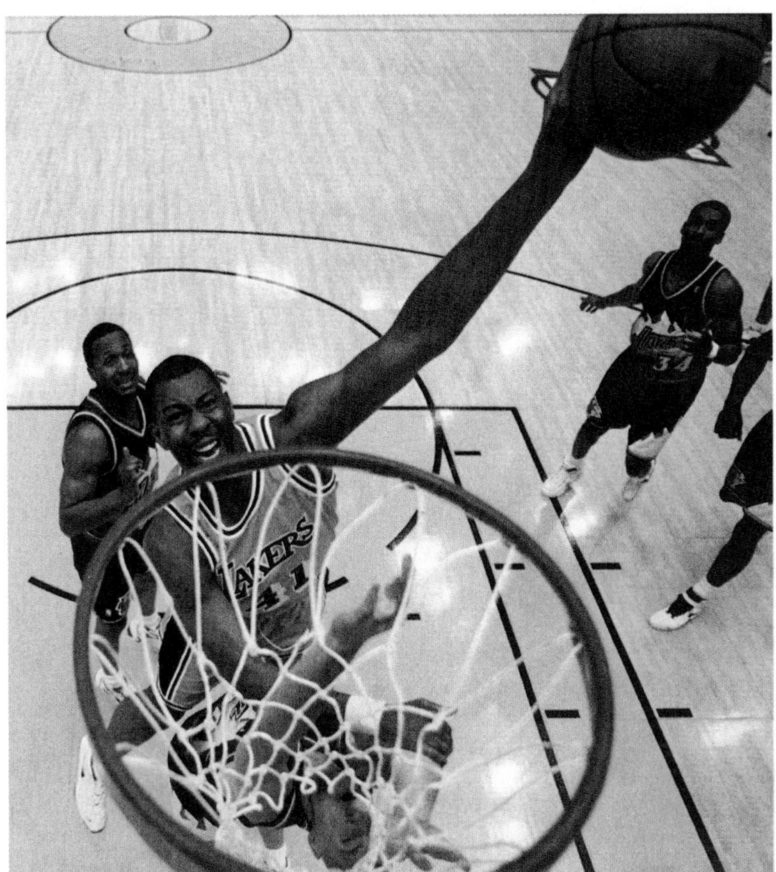

Introduction

In this chapter, we use the manipulations considered in Chapter 7 to solve equations and inequalities. We then use equations and inequalities to solve applied problems.

An Application

The perimeter of an NBA basketball court is 288 ft. The length is 44 ft longer than the width. (***Source*:** National Basketball Association) Find the dimensions of the court.

This problem appears as Example 5 in Section 8.4.

The Mathematics

The perimeter P of a rectangle is given by the formula $2l + 2w = P$, where $l =$ the length and $w =$ the width. To translate the problem, we substitute $w + 44$ for l and 288 for P:

$$2l + 2w = P$$
$$\underbrace{2(w + 44) + 2w = 288.}$$

To find the dimensions, we first solve this equation.

World Wide Web For more information, visit us at www.mathmax.com

Pretest: Chapter 8

Solve.

1. $-7x = 49$

2. $4y + 9 = 2y + 7$

3. $6a - 2 = 10$

4. $4 + x = 12$

5. $7 - 3(2x - 1) = 40$

6. $\frac{4}{9}x - 1 = \frac{7}{8}$

7. $1 + 2(a + 3) = 3(2a - 1) + 6$

8. $-3x \leq 18$

9. $y + 5 > 1$

10. $5 - 2a < 7$

11. $3x + 4 \geq 2x + 7$

12. $8y < -18$

13. Solve for G: $P = 3KG$.

14. Solve for a: $A = \frac{3a - b}{b}$.

Solve.

15. The perimeter of the ornate frame of an oil painting is 146 in. The width is 5 in. less than the length. Find the dimensions.

16. Money is invested in a savings account at 4.25% simple interest. After 1 year, there is $479.55 in the account. How much was originally invested?

17. The sum of three consecutive integers is 246. Find the integers.

18. When 18 is added to six times a number, the result is less than 120. For what numbers is this possible?

Graph on a number line.

19. $x > -3$

20. $x \leq 4$

Objectives for Retesting

The objectives to be tested in addition to the material in this chapter are as follows.

[7.1a] Evaluate algebraic expressions by substitution.
[7.1b] Translate phrases to algebraic expressions.
[7.3a] Add real numbers.
[7.8b] Simplify expressions by removing parentheses and collecting like terms.

8.1 Solving Equations: The Addition Principle

Objectives

a. Determine whether a given number is a solution of a given equation.

b. Solve equations using the addition principle.

For Extra Help

TAPE 16 | MAC WIN | CD-ROM

a Equations and Solutions

In order to solve problems, we must learn to solve equations.

> An **equation** is a number sentence that says that the expressions on either side of the equals sign, =, represent the same number.

Here are some examples:

$$3 + 2 = 5, \quad 14 - 10 = 1 + 3, \quad x + 6 = 13, \quad 3x - 2 = 7 - x.$$

Equations have expressions on each side of the equals sign. The sentence "$14 - 10 = 1 + 3$" asserts that the expressions $14 - 10$ and $1 + 3$ name the same number.

Some equations are true. Some are false. Some are neither true nor false.

Examples Determine whether the equation is true, false, or neither.

1. $3 + 2 = 5$ The equation is *true.*

2. $7 - 2 = 4$ The equation is *false.*

3. $x + 6 = 13$ The equation is *neither* true nor false, because we do not know what number x represents.

Do Exercises 1–3.

Determine whether the equation is true, false, or neither.

1. $5 - 8 = -4$

2. $12 + 6 = 18$

3. $x + 6 = 7 - x$

> Any replacement for the variable that makes an equation true is called a **solution** of the equation. To solve an equation means to find *all* of its solutions.

One way to determine whether a number is a solution of an equation is to evaluate the expression on each side of the equals sign by substitution. If the values are the same, then the number is a solution.

Example 4 Determine whether 7 is a solution of $x + 6 = 13$.

We have

$x + 6 = 13$ **Writing the equation**

$7 + 6 \ ? \ 13$ **Substituting 7 for x**

$13 \mid$ TRUE

Since the left-hand and the right-hand sides are the same, we have a solution. No other number makes the equation true, so the only solution is the number 7.

Answers on page A-20

Determine whether the given number is a solution of the given equation.

4. 8; $x + 4 = 12$

5. 0; $x + 4 = 12$

6. -3; $7 + x = -4$

7. Solve using the addition principle:

$x + 2 = 11.$

Answers on page A-20

Example 5 Determine whether 19 is a solution of $7x = 141$.

We have

$7x = 141$	Writing the equation
$7(19) \ ? \ 141$	Substituting 19 for x
$133 \mid$	FALSE

Since the left-hand and the right-hand sides are not the same, we do not have a solution.

Do Exercises 4–6.

b Using the Addition Principle

Consider the equation

$$x = 7.$$

We can easily see that the solution of this equation is 7. If we replace x with 7, we get

$$7 = 7, \quad \text{which is true.}$$

Now consider the equation of Example 4:

$$x + 6 = 13.$$

In Example 4, we discovered that the solution of this equation is also 7, but the fact that 7 is the solution is not as obvious. We now begin to consider principles that allow us to start with an equation and end up with an *equivalent equation*, like $x = 7$, in which the variable is alone on one side and for which the solution is easy to find.

> Equations with the same solutions are called **equivalent equations.**

One of the principles that we use in solving equations involves adding. An equation $a = b$ says that a and b stand for the same number. Suppose this is true, and we add a number c to the number a. We get the same answer if we add c to b, because a and b are the same number.

> **THE ADDITION PRINCIPLE**
>
> For any real numbers a, b, and c,
>
> $a = b$ is equivalent to $a + c = b + c$.

Let's again solve the equation $x + 6 = 13$ using the addition principle. We want to get x alone on one side. To do so, we use the addition principle, choosing to add -6 because $6 + (-6) = 0$:

$x + 6 = 13$	
$x + 6 + (-6) = 13 + (-6)$	Using the addition principle: adding -6 on both sides
$x + 0 = 7$	Simplifying
$x = 7.$	Identity property of 0: $x + 0 = x$

Do Exercise 7.

When we use the addition principle, we sometimes say that we "add the same number on both sides of the equation." This is also true for subtraction, since we can express every subtraction as an addition. That is, since

$$a - c = b - c \quad \text{is equivalent to} \quad a + (-c) = b + (-c),$$

the addition principle tells us that we can "subtract the same number on both sides of the equation."

Example 6 Solve: $x + 5 = -7$.

We have

$$x + 5 = -7$$

$$x + 5 - 5 = -7 - 5 \qquad \text{Using the addition principle: adding } -5 \text{ on both sides or subtracting 5 on both sides}$$

$$x + 0 = -12 \qquad \text{Simplifying}$$

$$x = -12. \qquad \text{Identity property of 0}$$

We can see that the solution of $x = -12$ is the number -12. To check the answer, we substitute -12 in the original equation.

CHECK:

$$\begin{array}{c|c} x + 5 & -7 \\ \hline -12 + 5 \ ? & -7 \\ -7 & \end{array} \qquad \text{TRUE}$$

The solution of the original equation is -12.

In Example 6, to get x alone, we used the addition principle and subtracted 5 on both sides. This eliminated the 5 on the left. We started with $x + 5 = -7$, and, using the addition principle, we found a simpler equation $x = -12$ for which it was easy to "*see*" the solution. The equations $x + 5 = -7$ and $x = -12$ are *equivalent.*

Do Exercise 8.

Now we solve an equation with a subtraction using the addition principle.

Example 7 Solve: $a - 4 = 10$.

We have

$$a - 4 = 10$$

$$a - 4 + 4 = 10 + 4 \qquad \text{Using the addition principle: adding 4 on both sides}$$

$$a + 0 = 14 \qquad \text{Simplifying}$$

$$a = 14. \qquad \text{Identity property of 0}$$

CHECK:

$$\begin{array}{c|c} a - 4 & 10 \\ \hline 14 - 4 \ ? & 10 \\ 10 & \end{array} \qquad \text{TRUE}$$

The solution is 14.

Do Exercise 9.

8. Solve using the addition principle, subtracting 7 on both sides:

$x + 7 = 2.$

9. Solve: $t - 3 = 19$.

Answers on page A-20

Solve.

10. $8.7 = n - 4.5$

11. $y + 17.4 = 10.9$

Solve.

12. $x + \frac{1}{2} = -\frac{3}{2}$

13. $t - \frac{13}{4} = \frac{5}{8}$

Answers on page A-20

Example 8 Solve: $-6.5 = y - 8.4$.

We have

$$-6.5 = y - 8.4$$
$$-6.5 + 8.4 = y - 8.4 + 8.4 \qquad \text{Using the addition principle: adding 8.4 to eliminate } -8.4 \text{ on the right}$$
$$1.9 = y.$$

CHECK:

$$\begin{array}{c|c} -6.5 & y - 8.4 \\ \hline -6.5 \ ? & 1.9 - 8.4 \\ & -6.5 \end{array} \qquad \text{TRUE}$$

The solution is 1.9.

Note that equations are reversible. That is, if $a = b$ is true, then $b = a$ is true. Thus when we solve $-6.5 = y - 8.4$, we can reverse it and solve $y - 8.4 = -6.5$ if we wish.

Do Exercises 10 and 11.

Example 9 Solve: $-\frac{2}{3} + x = \frac{5}{2}$.

We have

$$-\frac{2}{3} + x = \frac{5}{2}$$
$$\frac{2}{3} - \frac{2}{3} + x = \frac{2}{3} + \frac{5}{2} \qquad \text{Adding } \frac{2}{3}$$
$$x = \frac{2}{3} \cdot \frac{2}{2} + \frac{5}{2} \cdot \frac{3}{3} \qquad \text{Multiplying by 1 to obtain equivalent fractional expressions with the least common denominator 6}$$
$$= \frac{4}{6} + \frac{15}{6}$$
$$= \frac{19}{6}.$$

CHECK:

$$\begin{array}{c|c} -\frac{2}{3} + x & \frac{5}{2} \\ \hline -\frac{2}{3} + \frac{19}{6} \ ? & \frac{5}{2} \\ -\frac{4}{6} + \frac{19}{6} & \\ \frac{15}{6} & \\ \frac{5}{2} & \end{array} \qquad \text{TRUE}$$

The solution is $\frac{19}{6}$.

Do Exercises 12 and 13.

Exercise Set 8.1

a Determine whether the given number is a solution of the given equation.

1. 15; $x + 17 = 32$

2. 35; $t + 17 = 53$

3. 21; $x - 7 = 12$

4. 36; $a - 19 = 17$

5. -7; $6x = 54$

6. -9; $8y = -72$

7. 30; $\frac{x}{6} = 5$

8. 49; $\frac{y}{8} = 6$

9. 19; $5x + 7 = 107$

10. 9; $9x + 5 = 86$

11. -11; $7(y - 1) = 63$

12. -18; $x + 3 = 3 + x$

b Solve using the addition principle. Don't forget to check!

13. $x + 2 = 6$

CHECK: $x + 2 = 6$?

14. $y + 4 = 11$

CHECK: $y + 4 = 11$?

15. $x + 15 = -5$

CHECK: $x + 15 = -5$?

16. $t + 10 = 44$

CHECK: $t + 10 = 44$?

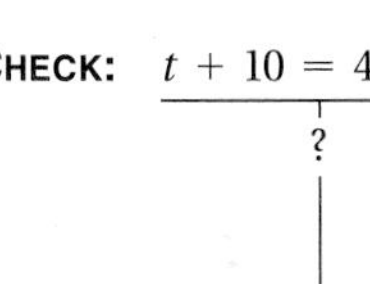

17. $x + 6 = -8$

CHECK: $x + 6 = -8$?

18. $z + 9 = -14$

19. $x + 16 = -2$

20. $m + 18 = -13$

21. $x - 9 = 6$

22. $x - 11 = 12$

23. $x - 7 = -21$

24. $x - 3 = -14$

25. $5 + t = 7$

26. $8 + y = 12$

27. $-7 + y = 13$

28. $-8 + y = 17$

29. $-3 + t = -9$

30. $-8 + t = -24$

31. $x + \frac{1}{2} = 7$

32. $24 = -\frac{7}{10} + r$

33. $12 = a - 7.9$

34. $2.8 + y = 11$

35. $r + \frac{1}{3} = \frac{8}{3}$

36. $t + \frac{3}{8} = \frac{5}{8}$

37. $m + \frac{5}{6} = -\frac{11}{12}$

38. $x + \frac{2}{3} = -\frac{5}{6}$

39. $x - \frac{5}{6} = \frac{7}{8}$

40. $y - \frac{3}{4} = \frac{5}{6}$

41. $-\frac{1}{5} + z = -\frac{1}{4}$

42. $-\frac{1}{8} + y = -\frac{3}{4}$

43. $7.4 = x + 2.3$

44. $8.4 = 5.7 + y$

45. $7.6 = x - 4.8$

46. $8.6 = x - 7.4$

47. $-9.7 = -4.7 + y$

48. $-7.8 = 2.8 + x$

49. $5\frac{1}{6} + x = 7$

50. $5\frac{1}{4} = 4\frac{2}{3} + x$

51. $q + \frac{1}{3} = -\frac{1}{7}$

52. $52\frac{3}{8} = -84 + x$

Skill Maintenance

53. Add: $-3 + (-8)$. [7.3a]

54. Subtract: $-3 - (-8)$. [7.4a]

55. Multiply: $-\frac{2}{3} \cdot \frac{5}{8}$. [7.5a]

56. Divide: $-\frac{3}{7} \div \left(-\frac{9}{7}\right)$. [7.6c]

57. Divide: $\frac{2}{3} \div \left(-\frac{4}{9}\right)$. [7.6c]

58. Add: $-8.6 + 3.4$. [7.3a]

Translate to an algebraic expression. [7.1b]

59. Liza had \$50 before paying x dollars for a pizza. How much does she have left?

60. Donnie drove his S-10 pickup truck 65 mph for t hours. How far did he drive?

Synthesis

61. ◆ Explain the difference between equivalent expressions and equivalent equations.

62. ◆ When solving an equation using the addition principle, how do you determine which number to add or subtract on both sides of the equation?

Solve.

63. 🖩 $-356.788 = -699.034 + t$

64. $-\frac{4}{5} + \frac{7}{10} = x - \frac{3}{4}$

65. $x + \frac{4}{5} = -\frac{2}{3} - \frac{4}{15}$

66. $8 - 25 = 8 + x - 21$

67. $16 + x - 22 = -16$

68. $x + x = x$

69. $x + 3 = 3 + x$

70. $x + 4 = 5 + x$

71. $-\frac{3}{2} + x = -\frac{5}{17} - \frac{3}{2}$

72. $|x| = 5$

73. $|x| + 6 = 19$

8.2 Solving Equations: The Multiplication Principle

Objective

a Solve equations using the multiplication principle.

For Extra Help

TAPE 16 | MAC WIN | CD-ROM

a Using the Multiplication Principle

Suppose that $a = b$ is true, and we multiply a by some number c. We get the same answer if we multiply b by c, because a and b are the same number.

THE MULTIPLICATION PRINCIPLE

For any real numbers a, b, and c, $c \neq 0$,

$$a = b \quad \text{is equivalent to} \quad a \cdot c = b \cdot c.$$

When using the multiplication principle, we sometimes say that we "multiply on both sides of the equation by the same number."

Example 1 Solve: $5x = 70$.

To get x alone, we multiply by the *multiplicative inverse,* or *reciprocal,* of 5. Then we get the *multiplicative identity* 1 times x, or $1 \cdot x$, which simplifies to x. This allows us to eliminate 5 on the left.

$$5x = 70 \qquad \text{The reciprocal of 5 is } \tfrac{1}{5}.$$

$$\frac{1}{5} \cdot 5x = \frac{1}{5} \cdot 70 \qquad \text{Multiplying by } \tfrac{1}{5} \text{ to get } 1 \cdot x \text{ and eliminate 5 on the left}$$

$$1 \cdot x = 14 \qquad \text{Simplifying}$$

$$x = 14 \qquad \text{Identity property of 1: } 1 \cdot x = x$$

CHECK:

$$\begin{array}{r|l} 5x & = 70 \\ \hline 5 \cdot 14 & ?\ 70 \\ 70 & \end{array} \quad \text{TRUE}$$

The solution is 14.

The multiplication principle also tells us that we can "divide on both sides of the equation by a nonzero number." This is because division is the same as multiplying by a reciprocal. That is,

$$\frac{a}{c} = \frac{b}{c} \quad \text{is equivalent to} \quad a \cdot \frac{1}{c} = b \cdot \frac{1}{c}, \quad \text{when } c \neq 0.$$

In an expression like $5x$ in Example 1, the number 5 is called the **coefficient**. Example 1 could be done as follows, dividing by 5, the coefficient of x, on both sides.

1. Solve. Multiply on both sides.

$6x = 90$

2. Solve. Divide on both sides.

$4x = -7$

3. Solve: $-6x = 108$.

4. Solve: $-x = -10$.

Answers on page A-20

Example 2 Solve: $5x = 70$.

We have

$$5x = 70$$

$$\frac{5x}{5} = \frac{70}{5} \quad \text{Dividing by 5 on both sides}$$

$$1 \cdot x = 14 \quad \text{Simplifying}$$

$$x = 14. \quad \text{Identity property of 1}$$

Do Exercises 1 and 2.

Example 3 Solve: $-4x = 92$.

We have

$$-4x = 92$$

$$\frac{-4x}{-4} = \frac{92}{-4} \quad \text{Using the multiplication principle. Dividing by } -4 \text{ on both sides is the same as multiplying by } -\tfrac{1}{4}.$$

$$1 \cdot x = -23 \quad \text{Simplifying}$$

$$x = -23. \quad \text{Identity property of 1}$$

CHECK:

$$-4x = 92$$

$$-4(-23) \ ? \ 92$$

$$92 \quad \text{TRUE}$$

The solution is -23.

Do Exercise 3.

Example 4 Solve: $-x = 9$.

We have

$$-x = 9$$

$$-1 \cdot x = 9 \quad \text{Using the property of } -1\text{: } -x = -1 \cdot x$$

$$\frac{-1 \cdot x}{-1} = \frac{9}{-1} \quad \text{Dividing by } -1$$

$$1 \cdot x = -9$$

$$x = -9.$$

CHECK:

$$-x = 9$$

$$-(-9) \ ? \ 9$$

$$9 \quad \text{TRUE}$$

The solution is -9.

Do Exercise 4.

In practice, it is generally more convenient to "divide" on both sides of the equation if the coefficient of the variable is in decimal notation or is an integer. If the coefficient is in fractional notation, it is more convenient to "multiply" by a reciprocal.

Example 5 Solve: $\frac{3}{8} = -\frac{5}{4}x$.

$$\frac{3}{8} = -\frac{5}{4}x$$

The reciprocal of $-\frac{5}{4}$ is $-\frac{4}{5}$. There is no sign change.

$$-\frac{4}{5} \cdot \frac{3}{8} = -\frac{4}{5} \cdot \left(-\frac{5}{4}x\right)$$ Multiplying by $-\frac{4}{5}$ to get $1 \cdot x$ and eliminate $-\frac{5}{4}$ on the right

$$-\frac{12}{40} = 1 \cdot x$$

$$-\frac{3}{10} = 1 \cdot x$$ Simplifying

$$-\frac{3}{10} = x$$ Identity property of 1

CHECK:

$$\frac{3}{8} = -\frac{5}{4}x$$

$$\frac{3}{8} \; ? \; -\frac{5}{4}\left(-\frac{3}{10}\right)$$

$$\frac{3}{8}$$ TRUE

The solution is $-\frac{3}{10}$.

Note that equations are reversible. That is, if $a = b$ is true, then $b = a$ is true. Thus when we solve $\frac{3}{8} = -\frac{5}{4}x$, we can reverse it and solve $-\frac{5}{4}x = \frac{3}{8}$ if we wish.

Do Exercise 5.

Example 6 Solve: $1.16y = 9744$.

$$1.16y = 9744$$

$$\frac{1.16y}{1.16} = \frac{9744}{1.16}$$ Dividing by 1.16

$$y = \frac{9744}{1.16}$$

$$= 8400$$

CHECK:

$$1.16y = 9744$$

$$1.16(8400) \; ? \; 9744$$

$$9744$$ TRUE

The solution is 8400.

Do Exercises 6 and 7.

5. Solve: $\frac{2}{3} = -\frac{5}{6}y$.

Solve.

6. $1.12x = 8736$

7. $6.3 = -2.1y$

Answers on page A-20

8. Solve: $-14 = \frac{-y}{2}$.

Answer on page A-20

Now we solve an equation with a division using the multiplication principle. Consider an equation like $-y/9 = 14$. In Chapter 7, we learned that a division can be expressed as multiplication by the reciprocal of the divisor. Thus,

$$\frac{-y}{9} \text{ is equivalent to } \frac{1}{9}(-y).$$

The reciprocal of $\frac{1}{9}$ is 9. Then, using the multiplication principle, we multiply by 9 on both sides. This is shown in the following example.

Example 7 Solve: $\frac{-y}{9} = 14$.

$$\frac{-y}{9} = 14$$

$$\frac{1}{9}(-y) = 14$$

$$9 \cdot \frac{1}{9}(-y) = 9 \cdot 14 \qquad \text{Multiplying by 9 on both sides}$$

$$-y = 126$$

$$-1 \cdot (-y) = -1 \cdot 126 \qquad \text{Multiplying by } -1\text{, or dividing by } -1\text{, on both sides}$$

$$y = -126$$

CHECK:

$\frac{-y}{9} = 14$	
$\frac{-(-126)}{9}$	? 14
$\frac{126}{9}$	
14	

TRUE

The solution is -126.

Do Exercise 8.

Exercise Set 8.2

a Solve using the multiplication principle. Don't forget to check!

1. $6x = 36$

CHECK: $6x = 36$ | ?

2. $3x = 51$

CHECK: $3x = 51$ | ?

3. $5x = 45$

CHECK: $5x = 45$ | ?

4. $8x = 72$

CHECK: $8x = 72$ | ?

5. $84 = 7x$

6. $63 = 9x$

7. $-x = 40$

8. $53 = -x$

9. $-x = -1$

10. $-47 = -t$

11. $7x = -49$

12. $8x = -56$

13. $-12x = 72$

14. $-15x = 105$

15. $-21x = -126$

16. $-13x = -104$

17. $\frac{t}{7} = -9$

18. $\frac{y}{-8} = 11$

19. $\frac{3}{4}x = 27$

20. $\frac{4}{5}x = 16$

21. $\frac{-t}{3} = 7$

22. $\frac{-x}{6} = 9$

23. $-\frac{m}{3} = \frac{1}{5}$

24. $\frac{1}{8} = -\frac{y}{5}$

25. $-\frac{3}{5}r = \frac{9}{10}$

26. $\frac{2}{5}y = -\frac{4}{15}$

27. $-\frac{3}{2}r = -\frac{27}{4}$

28. $-\frac{3}{8}x = -\frac{15}{16}$

29. $6.3x = 44.1$

30. $2.7y = 54$

31. $-3.1y = 21.7$

32. $-3.3y = 6.6$

33. $38.7m = 309.6$

34. $29.4m = 235.2$

35. $-\frac{2}{3}y = -10.6$

36. $-\frac{9}{7}y = 12.06$

Skill Maintenance

Collect like terms. [7.7e]

37. $3x + 4x$

38. $6x + 5 - 7x$

39. $-4x + 11 - 6x + 18x$

40. $8y - 16y - 24y$

Remove parentheses and simplify. [7.8b]

41. $3x - (4 + 2x)$

42. $2 - 5(x + 5)$

43. $8y - 6(3y + 7)$

44. $-2a - 4(5a - 1)$

Translate to an algebraic expression. [7.1b]

45. Patty drives her van for 8 hr at a speed of r mph. How far does she drive?

46. A triangle has a height of 10 meters and a base of b meters. What is the area of the triangle?

Synthesis

47. ◆ When solving an equation using the multiplication principle, how do you determine by what number to multiply or divide on both sides of the equation?

48. ◆ Are the equations $x = 5$ and $x^2 = 25$ equivalent? Why or why not?

Solve.

49. $-0.2344m = 2028.732$

50. $0 \cdot x = 0$

51. $0 \cdot x = 9$

52. $4|x| = 48$

53. $2|x| = -12$

Solve for x.

54. $ax = 5a$

55. $3x = \frac{b}{a}$

56. $cx = a^2 + 1$

57. $\frac{a}{b}x = 4$

58. A student makes a calculation and gets an answer of 22.5. On the last step, the student multiplies by 0.3 when a division by 0.3 should have been done. What is the correct answer?

8.3 Using the Principles Together

a Applying Both Principles

Consider the equation $3x + 4 = 13$. It is more complicated than those we discussed in the preceding two sections. In order to solve such an equation, we first isolate the x-term, $3x$, using the addition principle. Then we apply the multiplication principle to get x by itself.

Example 1 Solve: $3x + 4 = 13$.

$$3x + 4 = 13$$

$$3x + 4 - 4 = 13 - 4 \qquad \text{Using the addition principle: subtracting 4 on both sides}$$

First, isolate the x-term. → $3x = 9$ Simplifying

$$\frac{3x}{3} = \frac{9}{3} \qquad \text{Using the multiplication principle: dividing by 3 on both sides}$$

Then isolate x. → $x = 3$ Simplifying

CHECK:

$3x + 4 = 13$	
$3 \cdot 3 + 4$?	13
$9 + 4$	
13	TRUE

We use the rules for order of operations to carry out the check. We find the product $3 \cdot 3$. Then we add 4.

The solution is 3.

Do Exercise 1.

Example 2 Solve: $-5x - 6 = 16$.

$$-5x - 6 = 16$$

$$-5x - 6 + 6 = 16 + 6 \qquad \text{Adding 6 on both sides}$$

$$-5x = 22$$

$$\frac{-5x}{-5} = \frac{22}{-5} \qquad \text{Dividing by } -5 \text{ on both sides}$$

$$x = -\frac{22}{5}, \text{ or } -4\frac{2}{5} \qquad \text{Simplifying}$$

CHECK:

$-5x - 6 = 16$	
$-5\left(-\frac{22}{5}\right) - 6$?	16
$22 - 6$	
16	TRUE

The solution is $-\frac{22}{5}$.

Do Exercises 2 and 3.

Objectives

a Solve equations using both the addition and the multiplication principles.

b Solve equations in which like terms may need to be collected.

c Solve equations by first removing parentheses and collecting like terms.

For Extra Help

TAPE 16

MAC WIN

CD-ROM

1. Solve: $9x + 6 = 51$.

Solve.

2. $8x - 4 = 28$

3. $-\frac{1}{2}x + 3 = 1$

Answers on page A-20

4. Solve: $-18 - m = -57$.

Solve.

5. $-4 - 8x = 8$

6. $41.68 = 4.7 - 8.6y$

Solve.

7. $4x + 3x = -21$

8. $x - 0.09x = 728$

Answers on page A-20

Example 3 Solve: $45 - t = 13$.

$$45 - t = 13$$
$$-45 + 45 - t = -45 + 13 \quad \text{Adding } -45 \text{ on both sides}$$
$$-t = -32$$
$$-1 \cdot t = -32 \quad \text{Using the property of } -1\text{: } -t = -1 \cdot t$$
$$\frac{-1 \cdot t}{-1} = \frac{-32}{-1}$$
Dividing by -1 on both sides (You could have multiplied by -1 on both sides instead. That would also change the sign on both sides.)
$$t = 32$$

The number 32 checks and is the solution.

Do Exercise 4.

Example 4 Solve: $16.3 - 7.2y = -8.18$.

$$16.3 - 7.2y = -8.18$$
$$-16.3 + 16.3 - 7.2y = -16.3 + (-8.18) \quad \text{Adding } -16.3 \text{ on both sides}$$
$$-7.2y = -24.48$$
$$\frac{-7.2y}{-7.2} = \frac{-24.48}{-7.2} \quad \text{Dividing by } -7.2 \text{ on both sides}$$
$$y = 3.4$$

CHECK:

$16.3 - 7.2y$	$= -8.18$	
$16.3 - 7.2(3.4)$	? -8.18	
$16.3 - 24.48$		
-8.18		TRUE

The solution is 3.4.

Do Exercises 5 and 6.

b Collecting Like Terms

If there are like terms on one side of the equation, we collect them before using the addition or the multiplication principle.

Example 5 Solve: $3x + 4x = -14$.

$$3x + 4x = -14$$
$$7x = -14 \quad \text{Collecting like terms}$$
$$\frac{7x}{7} = \frac{-14}{7} \quad \text{Dividing by 7 on both sides}$$
$$x = -2$$

The number -2 checks, so the solution is -2.

Do Exercises 7 and 8.

If there are like terms on opposite sides of the equation, we get them on the same side by using the addition principle. Then we collect them. In other words, we get all terms with a variable on one side and all numbers on the other.

Example 6 Solve: $2x - 2 = -3x + 3$.

$$2x - 2 = -3x + 3$$

$$2x - 2 + 2 = -3x + 3 + 2 \qquad \text{Adding 2}$$

$$2x = -3x + 5 \qquad \text{Collecting like terms}$$

$$2x + 3x = -3x + 3x + 5 \qquad \text{Adding } 3x$$

$$5x = 5 \qquad \text{Simplifying}$$

$$\frac{5x}{5} = \frac{5}{5} \qquad \text{Dividing by 5}$$

$$x = 1 \qquad \text{Simplifying}$$

CHECK:

$2x - 2 = -3x + 3$		
$2 \cdot 1 - 2$?	$-3 \cdot 1 + 3$	
$2 - 2$	$-3 + 3$	
0	0	TRUE

The solution is 1.

Do Exercise 9.

In Example 6, we used the addition principle to get all terms with a variable on one side and all numbers on the other side. Then we collected like terms and proceeded as before. If there are like terms on one side at the outset, they should be collected before proceeding.

Example 7 Solve: $6x + 5 - 7x = 10 - 4x + 3$.

$$6x + 5 - 7x = 10 - 4x + 3$$

$$-x + 5 = 13 - 4x \qquad \text{Collecting like terms}$$

$$4x - x + 5 = 13 - 4x + 4x \qquad \text{Adding } 4x \text{ to get all terms with a variable on one side}$$

$$3x + 5 = 13 \qquad \text{Simplifying; that is, collecting like terms}$$

$$3x + 5 - 5 = 13 - 5 \qquad \text{Subtracting 5}$$

$$3x = 8 \qquad \text{Simplifying}$$

$$\frac{3x}{3} = \frac{8}{3} \qquad \text{Dividing by 3}$$

$$x = \frac{8}{3} \qquad \text{Simplifying}$$

The number $\frac{8}{3}$ checks, so it is the solution.

Do Exercises 10–12.

9. Solve: $7y + 5 = 2y + 10$.

Solve.

10. $5 - 2y = 3y - 5$

11. $7x - 17 + 2x = 2 - 8x + 15$

12. $3x - 15 = 5x + 2 - 4x$

Answers on page A-20

13. Solve: $\frac{7}{8}x - \frac{1}{4} + \frac{1}{2}x = \frac{3}{4} + x.$

Answer on page A-20

Clearing Fractions and Decimals

In general, equations are easier to solve if they do not contain fractions or decimals. Consider, for example,

$$\frac{1}{2}x + 5 = \frac{3}{4} \quad \text{and} \quad 2.3x + 7 = 5.4.$$

If we multiply by 4 on both sides of the first equation and by 10 on both sides of the second equation, we have

$$4\left(\frac{1}{2}x + 5\right) = 4 \cdot \frac{3}{4} \quad \text{and} \quad 10(2.3x + 7) = 10 \cdot 5.4$$

or

$$4 \cdot \frac{1}{2}x + 4 \cdot 5 = 4 \cdot \frac{3}{4} \quad \text{and} \quad 10 \cdot 2.3x + 10 \cdot 7 = 10 \cdot 5.4$$

or

$$2x + 20 = 3 \quad \text{and} \quad 23x + 70 = 54.$$

The first equation has been "cleared of fractions" and the second equation has been "cleared of decimals." Both resulting equations are equivalent to the original equations and are easier to solve. *It is your choice* whether to clear fractions or decimals, but doing so often eases computations.

The easiest way to clear an equation of fractions is to multiply *every term on both sides* by the **least common multiple of all the denominators.**

Example 8 Solve: $\frac{2}{3}x - \frac{1}{6} + \frac{1}{2}x = \frac{7}{6} + 2x.$

The number 6 is the least common multiple of all the denominators. We multiply by 6 on both sides.

$$6\left(\frac{2}{3}x - \frac{1}{6} + \frac{1}{2}x\right) = 6\left(\frac{7}{6} + 2x\right) \qquad \text{Multiplying by 6 on both sides}$$

$$6 \cdot \frac{2}{3}x - 6 \cdot \frac{1}{6} + 6 \cdot \frac{1}{2}x = 6 \cdot \frac{7}{6} + 6 \cdot 2x$$

Using the distributive law (*Caution*! Be sure to multiply *all* the terms by 6.)

$$4x - 1 + 3x = 7 + 12x \qquad \text{Simplifying. Note that the fractions are cleared.}$$

$$7x - 1 = 7 + 12x \qquad \text{Collecting like terms}$$

$$7x - 1 - 12x = 7 + 12x - 12x \qquad \text{Subtracting } 12x$$

$$-5x - 1 = 7 \qquad \text{Collecting like terms}$$

$$-5x - 1 + 1 = 7 + 1 \qquad \text{Adding 1}$$

$$-5x = 8 \qquad \text{Collecting like terms}$$

$$\frac{-5x}{-5} = \frac{8}{-5} \qquad \text{Dividing by } -5$$

$$x = -\frac{8}{5}$$

The number $-\frac{8}{5}$ checks, so it is the solution.

Do Exercise 13.

To illustrate clearing decimals, we repeat Example 4, but this time we clear the equation of decimals first. Compare both methods.

To clear an equation of decimals, we count the greatest number of decimal places in any one number. If the greatest number of decimal places is 1, we multiply by 10; if it is 2, we multiply by 100; and so on.

Example 9 Solve: $16.3 - 7.2y = -8.18$.

The greatest number of decimal places in any one number is *two*. Multiplying by 100, which has *two* 0's, will clear all decimals.

$$100(16.3 - 7.2y) = 100(-8.18) \quad \text{Multiplying by 100 on both sides}$$

$$100(16.3) - 100(7.2y) = 100(-8.18) \quad \text{Using the distributive law}$$

$$1630 - 720y = -818 \quad \text{Simplifying}$$

$$1630 - 720y - 1630 = -818 - 1630 \quad \text{Subtracting 1630 on both sides}$$

$$-720y = -2448 \quad \text{Collecting like terms}$$

$$\frac{-720y}{-720} = \frac{-2448}{-720} \quad \text{Dividing by } -720 \text{ on both sides}$$

$$y = 3.4$$

The number 3.4 checks, so it is the solution.

Do Exercise 14.

14. Solve: $41.68 = 4.7 - 8.6y$.

c Equations Containing Parentheses

To solve certain kinds of equations that contain parentheses, we first use the distributive laws to remove the parentheses. Then we proceed as before.

Example 10 Solve: $4x = 2(12 - 2x)$.

$$4x = 2(12 - 2x)$$

$$4x = 24 - 4x \quad \text{Using the distributive law to multiply and remove parentheses}$$

$$4x + 4x = 24 - 4x + 4x \quad \text{Adding } 4x \text{ to get all the } x\text{-terms on one side}$$

$$8x = 24 \quad \text{Collecting like terms}$$

$$\frac{8x}{8} = \frac{24}{8} \quad \text{Dividing by 8}$$

$$x = 3$$

CHECK:

$4x$	$= 2(12 - 2x)$
$4 \cdot 3$?	$2(12 - 2 \cdot 3)$
12	$2(12 - 6)$
	$2 \cdot 6$
	12

We use the rules for order of operations to carry out the calculations on each side of the equation.

TRUE

The solution is 3.

Do Exercises 15 and 16.

Solve.

15. $2(2y + 3) = 14$

16. $5(3x - 2) = 35$

Answers on page A-20

Solve.

17. $3(7 + 2x) = 30 + 7(x - 1)$

18. $4(3 + 5x) - 4 = 3 + 2(x - 2)$

Calculator Spotlight

Checking Possible Solutions. To check possible solutions on a calculator, we substitute and carry out the calculations on each side. Let's check the solution of Example 11. We first substitute -2 on the left:

$$2 - 5(-2 + 5).$$

We carry out this computation as shown in the Calculator Spotlight in Section 7.8. We get -13. Then we substitute -2 on the right:

$$3(-2 - 2) - 1.$$

Carrying out the calculations, we again get -13.

Exercises

1. Check the solution of Margin Exercise 17.
2. Check the solution of Margin Exercise 18.
3. Check the solution of Example 9.

Answers on page A-20

Here is a procedure for solving the types of equation discussed in this section.

AN EQUATION-SOLVING PROCEDURE

1. Multiply on both sides to clear the equation of fractions or decimals. (This is optional, but it can ease computations.)
2. If parentheses occur, multiply to remove them using the *distributive laws.*
3. Collect like terms on each side, if necessary.
4. Get all terms with variables on one side and all numbers (constant terms) on the other side, using the *addition principle.*
5. Collect like terms again, if necessary.
6. Multiply or divide to solve for the variable, using the *multiplication principle.*
7. Check all possible solutions in the original equation.

Example 11 Solve: $2 - 5(x + 5) = 3(x - 2) - 1$.

$$2 - 5(x + 5) = 3(x - 2) - 1$$

$$2 - 5x - 25 = 3x - 6 - 1 \quad \text{Using the distributive laws to multiply and remove parentheses}$$

$$-5x - 23 = 3x - 7 \quad \text{Collecting like terms}$$

$$-5x - 23 + 5x = 3x - 7 + 5x \quad \text{Adding } 5x$$

$$-23 = 8x - 7 \quad \text{Collecting like terms}$$

$$-23 + 7 = 8x - 7 + 7 \quad \text{Adding 7}$$

$$-16 = 8x \quad \text{Collecting like terms}$$

$$\frac{-16}{8} = \frac{8x}{8} \quad \text{Dividing by 8}$$

$$-2 = x$$

CHECK:

$2 - 5(x + 5)$	$= 3(x - 2) - 1$	
$2 - 5(-2 + 5)$	? $3(-2 - 2) - 1$	
$2 - 5(3)$	$3(-4) - 1$	
$2 - 15$	$-12 - 1$	
-13	-13	TRUE

The solution is -2.

Do Exercises 17 and 18.

Exercise Set 8.3

a Solve. Don't forget to check!

1. $5x + 6 = 31$

CHECK: $5x + 6 = 31$?

2. $7x + 6 = 13$

CHECK: $7x + 6 = 13$?

3. $8x + 4 = 68$

CHECK: $8x + 4 = 68$?

4. $4y + 10 = 46$

CHECK: $4y + 10 = 46$?

5. $4x - 6 = 34$

6. $5y - 2 = 53$

7. $3x - 9 = 33$

8. $4x - 19 = 5$

9. $7x + 2 = -54$

10. $5x + 4 = -41$

11. $-45 = 3 + 6y$

12. $-91 = 9t + 8$

13. $-4x + 7 = 35$

14. $-5x - 7 = 108$

15. $-7x - 24 = -129$

16. $-6z - 18 = -132$

b Solve.

17. $5x + 7x = 72$

CHECK: $5x + 7x = 72$?

18. $8x + 3x = 55$

CHECK: $8x + 3x = 55$?

19. $8x + 7x = 60$

CHECK: $8x + 7x = 60$?

20. $8x + 5x = 104$

CHECK: $8x + 5x = 104$?

21. $4x + 3x = 42$

22. $7x + 18x = 125$

23. $-6y - 3y = 27$

24. $-5y - 7y = 144$

f **25.** $-7y - 8y = -15$

26. $-10y - 3y = -39$

27. $x + \frac{1}{3}x = 8$

28. $x + \frac{1}{4}x = 10$

29. $10.2y - 7.3y = -58$

30. $6.8y - 2.4y = -88$

31. $8y - 35 = 3y$

32. $4x - 6 = 6x$

33. $8x - 1 = 23 - 4x$

34. $5y - 2 = 28 - y$

35. $2x - 1 = 4 + x$

36. $4 - 3x = 6 - 7x$

37. $6x + 3 = 2x + 11$

38. $14 - 6a = -2a + 3$

39. $5 - 2x = 3x - 7x + 25$

40. $-7z + 2z - 3z - 7 = 17$

41. $4 + 3x - 6 = 3x + 2 - x$

42. $5 + 4x - 7 = 4x - 2 - x$

43. $4y - 4 + y + 24 = 6y + 20 - 4y$

44. $5y - 7 + y = 7y + 21 - 5y$

Solve. Clear fractions or decimals first.

45. $\frac{7}{2}x + \frac{1}{2}x = 3x + \frac{3}{2} + \frac{5}{2}x$

46. $\frac{7}{8}x - \frac{1}{4} + \frac{3}{4}x = \frac{1}{16} + x$

47. $\frac{2}{3} + \frac{1}{4}t = \frac{1}{3}$

48. $-\frac{3}{2} + x = -\frac{5}{6} - \frac{4}{3}$

49. $\frac{2}{3} + 3y = 5y - \frac{2}{15}$

50. $\frac{1}{2} + 4m = 3m - \frac{5}{2}$

51. $\frac{5}{3} + \frac{2}{3}x = \frac{25}{12} + \frac{5}{4}x + \frac{3}{4}$

52. $1 - \frac{2}{3}y = \frac{9}{5} - \frac{y}{5} + \frac{3}{5}$

53. $2.1x + 45.2 = 3.2 - 8.4x$

54. $0.96y - 0.79 = 0.21y + 0.46$

55. $1.03 - 0.62x = 0.71 - 0.22x$

56. $1.7t + 8 - 1.62t = 0.4t - 0.32 + 8$

57. $\frac{2}{7}x - \frac{1}{2}x = \frac{3}{4}x + 1$

58. $\frac{5}{16}y + \frac{3}{8}y = 2 + \frac{1}{4}y$

c Solve.

59. $3(2y - 3) = 27$

60. $8(3x + 2) = 30$

61. $40 = 5(3x + 2)$

62. $9 = 3(5x - 2)$

63. $2(3 + 4m) - 9 = 45$

64. $5x + 5(4x - 1) = 20$

65. $5r - (2r + 8) = 16$

66. $6b - (3b + 8) = 16$

67. $6 - 2(3x - 1) = 2$

68. $10 - 3(2x - 1) = 1$

69. $5(d + 4) = 7(d - 2)$

70. $3(t - 2) = 9(t + 2)$

71. $8(2t + 1) = 4(7t + 7)$

72. $7(5x - 2) = 6(6x - 1)$

73. $3(r - 6) + 2 = 4(r + 2) - 21$

74. $5(t + 3) + 9 = 3(t - 2) + 6$

75. $19 - (2x + 3) = 2(x + 3) + x$

76. $13 - (2c + 2) = 2(c + 2) + 3c$

77. $2[4 - 2(3 - x)] - 1 = 4[2(4x - 3) + 7] - 25$

78. $5[3(7 - t) - 4(8 + 2t)] - 20 = -6[2(6 + 3t) - 4]$

79. $0.7(3x + 6) = 1.1 - (x + 2)$

80. $0.9(2x + 8) = 20 - (x + 5)$

81. $a + (a - 3) = (a + 2) - (a + 1)$

82. $0.8 - 4(b - 1) = 0.2 + 3(4 - b)$

Skill Maintenance

83. Divide: $-22.1 \div 3.4$. [7.6c]

84. Factor: $7x - 21 - 14y$. [7.7d]

85. Use < or > for ▒ to write a true sentence: [7.2d]
-15 ▒ -13.

86. Find $-(-x)$ when $x = -14$. [7.3b]

87. Add: $-22.1 + 3.4$. [7.3a]

88. Subtract: $-22.1 - 3.4$. [7.4a]

Translate to an algebraic expression. [7.1b]

89. A number c is divided by 8.

90. A parallelogram with height h has a base length of 13.4. What is the area of the parallelogram?

Synthesis

91. ◆ What procedure would you follow to solve an equation like $0.23x + \frac{17}{3} = -0.8 + \frac{3}{4}x$? Could your procedure be streamlined? If so, how?

92. ◆ Consider any equation of the form $ax + b = c$. Describe a procedure that can be used to solve for x.

Solve.

93. ▦ $0.008 + 9.62x - 42.8 = 0.944x + 0.0083 - x$

94. $\frac{y - 2}{3} = \frac{2 - y}{5}$

95. $0 = y - (-14) - (-3y)$

96. $3x = 4x$

97. $\frac{5 + 2y}{3} = \frac{25}{12} + \frac{5y + 3}{4}$

98. ▦ $0.05y - 1.82 = 0.708y - 0.504$

99. $-2y + 5y = 6y$

100. $\frac{1}{4}(8y + 4) - 17 = -\frac{1}{2}(4y - 8)$

101. $\frac{1}{3}(6x + 24) - 20 = -\frac{1}{4}(12x - 72)$

102. $\frac{2}{3}\left(\frac{7}{8} - 4x\right) - \frac{5}{8} = \frac{3}{8}$

103. $\frac{3}{4}\left(3x - \frac{1}{2}\right) - \frac{2}{3} = \frac{1}{3}$

104. $\frac{4 - 3x}{7} = \frac{2 + 5x}{49} - \frac{x}{14}$

105. Solve the equation $4x - 8 = 32$ by first using the addition principle. Then solve it by first using the multiplication principle.

Solve linear equations as a group.

8.4 Applications and Problem Solving

Objective

a Solve applied problems by translating to equations.

For Extra Help

TAPE 16 | MAC WIN | CD-ROM

a Five Steps for Solving Problems

We have studied many new equation-solving tools in this chapter. We now use them for applications and problem solving. The following five-step strategy can be very helpful in solving problems.

FIVE STEPS FOR PROBLEM SOLVING IN ALGEBRA

1. *Familiarize* yourself with the problem situation.
2. *Translate* the problem to an equation.
3. *Solve* the equation.
4. *Check* the answer in the original problem.
5. *State* the answer to the problem clearly.

Of the five steps, the most important is probably the first one: becoming familiar with the problem situation. The table in the margin lists some hints for familiarization.

To familiarize yourself with a problem situation:

- If a problem is given in words, read it carefully. Reread the problem, perhaps aloud. Try to verbalize the problem as if you were explaining it to someone else.
- Make a drawing and label it with known information, using specific units if given. Also, indicate unknown information.
- Choose a variable (or variables) to represent the unknown and clearly state what the variable represents. Be descriptive! For example, let L = length, d = distance, and so on.
- Find further information. Look up formulas or definitions with which you are not familiar. (Geometric formulas appear on the inside front cover of this text.) Consult a reference librarian or an expert in the field.
- Create a table that lists all the information you have available. Look for patterns that may help in the translation to an equation.
- Guess or estimate the answer.

Example 1 *Subway Sandwich.* Subway is a national restaurant firm that serves sandwiches prepared in buns of length 18 in. (***Source:*** Subway Restaurants). Suppose Jenny, Demi, and Sarah buy one of these sandwiches and take it back to their room. Since they have different appetites, Jenny cuts the sandwich in such a way that Demi gets half of what Jenny gets and Sarah gets three-fourths of what Jenny gets. Find the length of each person's sandwich.

1. **Familiarize.** We first make a drawing. Because the sandwich lengths are expressed in terms of Jenny's sandwich, we let

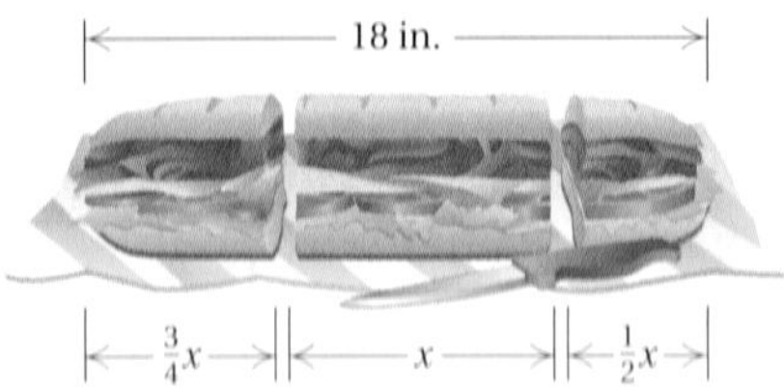

x = the length of Jenny's sandwich.

Then $\frac{1}{2}x$ = the length of Demi's sandwich

and $\frac{3}{4}x$ = the length of Sarah's sandwich.

2. **Translate.** From the statement of the problem and the drawing, we see that the lengths add up to 18 in. That gives us our translation:

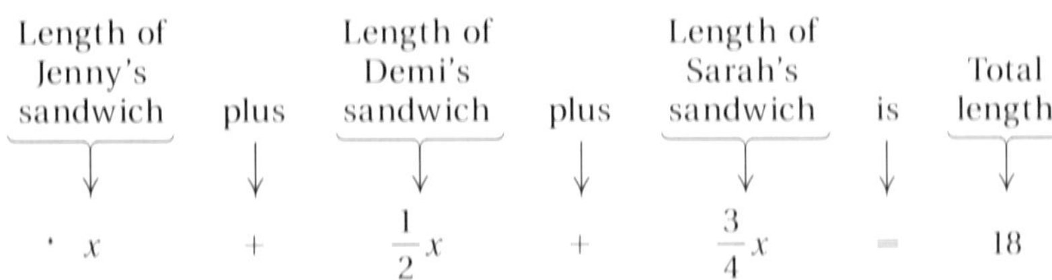

Length of Jenny's sandwich	plus	Length of Demi's sandwich	plus	Length of Sarah's sandwich	is	Total length
x	$+$	$\frac{1}{2}x$	$+$	$\frac{3}{4}x$	$=$	18

1. *Rocket Sections.* A rocket is divided into three sections: the payload and navigation section in the top, the fuel section in the middle, and the rocket engine section in the bottom. The top section is one-sixth the length of the bottom section. The middle section is one-half the length of the bottom section. The total length is 240 ft. Find the length of each section.

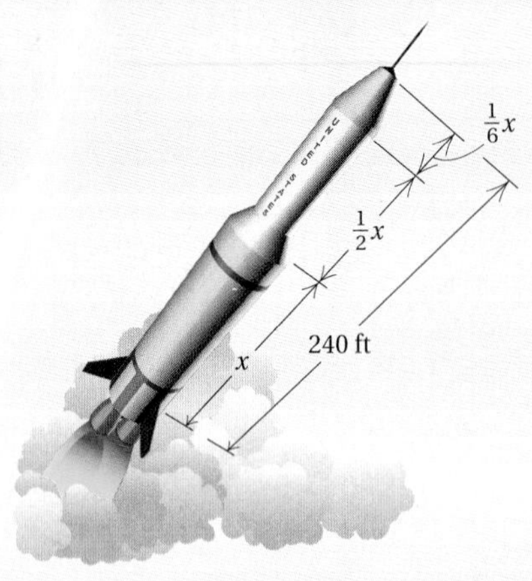

3. **Solve.** We solve the equation by clearing fractions as follows:

$$x + \frac{1}{2}x + \frac{3}{4}x = 18 \qquad \text{The LCM of all the denominators is 4.}$$

$$4\left(x + \frac{1}{2}x + \frac{3}{4}x\right) = 4 \cdot 18 \qquad \text{Multiplying by the LCM, 4}$$

$$4 \cdot x + 4 \cdot \frac{1}{2}x + 4 \cdot \frac{3}{4}x = 4 \cdot 18 \qquad \text{Using the distributive law}$$

$$4x + 2x + 3x = 72 \qquad \text{Simplifying}$$

$$9x = 72 \qquad \text{Collecting like terms}$$

$$\frac{9x}{9} = \frac{72}{9} \qquad \text{Dividing by 9}$$

$$x = 8.$$

4. **Check.** Do we have an answer to the *problem*? If the length of Jenny's sandwich is 8 in., then the length of Demi's sandwich is $\frac{1}{2} \cdot 8$ in., or 4 in., and the length of Sarah's sandwich is $\frac{3}{4} \cdot 8$ in., or 6 in. These lengths add up to 18 in. Our answer checks.
5. **State.** The length of Jenny's sandwich is 8 in., the length of Demi's sandwich is 4 in., and the length of Sarah's sandwich is 6 in.

Do Exercise 1.

Example 2 Five plus three more than a number is nineteen. What is the number?

1. **Familiarize.** Let $x =$ the number. Then "three more than a number" translates to $x + 3$, and "five plus three more than a number" translates to $5 + (x + 3)$.
2. **Translate.** The familiarization leads us to the following translation:

Five	plus	Three more than a number	is	Nineteen.
↓	↓	↓	↓	↓
5	+	$(x + 3)$	=	19.

3. **Solve.** We solve the equation:

$$5 + (x + 3) = 19$$

$$x + 8 = 19 \qquad \text{Collecting like terms}$$

$$x + 8 - 8 = 19 - 8 \qquad \text{Subtracting 8}$$

$$x = 11.$$

4. **Check.** Be sure to check your answer in the original wording of the problem, not in the equation that you solved. This will enable you to check for errors in the translation as well. Three more than 11 is 14. Adding 5 to 14, we get 19. This checks.
5. **State.** The number is 11.

Do Exercise 2.

2. If 5 is subtracted from three times a certain number, the result is 10. What is the number?

Recall that the

Set of integers $= \{\ldots, -5, -4, -3, -2, -1, 0, 1, 2, 3, 4, 5, \ldots\}$.

Before we solve the next problem, we need to learn some additional terminology regarding integers.

Answers on page A-20

The following are examples of **consecutive integers:** 16, 17, 18, 19, 20; and −31, −30, −29, −28. Note that consecutive integers can be represented in the form x, $x + 1$, $x + 2$, and so on.

The following are examples of **consecutive even integers:** 16, 18, 20, 22, 24; and −52, −50, −48, −46. Note that consecutive even integers can be represented in the form x, $x + 2$, $x + 4$, and so on.

The following are examples of **consecutive odd integers:** 21, 23, 25, 27, 29; and −71, −69, −67, −65. Note that consecutive odd integers can be represented in the form x, $x + 2$, $x + 4$, and so on.

Example 3 *Interstate Mile Markers.* If you are traveling on a U.S. interstate highway, you will notice numbered markers every mile to tell your location in case of an accident or other emergency. In many states, the numbers on the markers increase from west to east. (***Source*:** Federal Highway Administration, Ed Rotalewski) The sum of two consecutive mile markers on I-70 in Kansas is 559. Find the numbers on the markers.

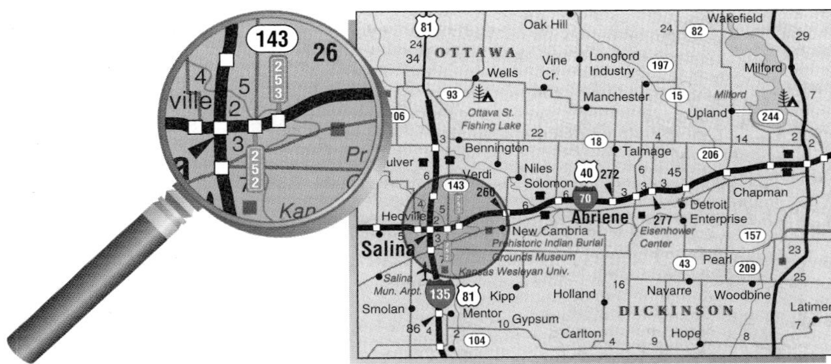

1. **Familiarize.** The numbers on the mile markers are consecutive positive integers. Thus if we let $x =$ the smaller number, then $x + 1 =$ the larger number.

   To become familiar with the problem, we can make a table. First, we guess a value for x; then we find $x + 1$. Finally, we add the two numbers and check the sum. From the table, we see that the first marker should be between 252 and 302. You might actually solve the problem this way, but let's work on developing our algebra skills.

x	$x + 1$	Sum of x and $x + 1$
114	115	229
252	253	505
302	303	605

2. **Translate.** We reword the problem and translate as follows.

First integer	plus	Second integer	is	559	**Rewording**
↓	↓	↓	↓	↓	
x	$+$	$(x + 1)$	$=$	559	**Translating**

3. **Solve.** We solve the equation:

$$x + (x + 1) = 559$$
$$2x + 1 = 559 \quad \text{Collecting like terms}$$
$$2x + 1 - 1 = 559 - 1 \quad \text{Subtracting 1}$$
$$2x = 558$$
$$\frac{2x}{2} = \frac{558}{2} \quad \text{Dividing by 2}$$
$$x = 279.$$

If x is 279, then $x + 1$ is 280.

3. *Interstate Mile Markers.* The sum of two consecutive mile markers on I-90 in upstate New York is 627 (***Source*:** New York State Department of Transportation). (On I-90 in New York, the marker numbers *increase* from east to west.) Find the numbers on the markers.

Answer on page A-20

4. *IKON Copiers.* The law firm in Example 4 decides to raise its budget to \$2400 for the 3-month period. How many copies can they make for \$2400?

Answer on page A-20

4. **Check.** Our possible answers are 279 and 280. These are consecutive positive integers and $279 + 280 = 559$, so the answers check.
5. **State.** The mile markers are 279 and 280.

Do Exercise 3 on the preceding page.

Example 4 *IKON Copiers.* IKON Office Solutions rents a Canon GP30F copier for \$240 per month plus 1.8¢ per copy. A law firm needs to lease a copy machine for use during a special case that they anticipate will take 3 months. If they allot a budget of \$1500, how many copies can they make?

Source: IKON Office Solutions, Keith Palmer

1. **Familiarize.** Suppose that the law firm makes 20,000 copies. Then the cost is

Monthly charges plus Copy charges

or

$$\underbrace{3(\$240)}_{\$720} \quad + \quad \underbrace{\text{Cost per copy}}_{\$0.018} \quad \cdot \quad \underbrace{\text{Number of copies}}_{20{,}000},$$

which is \$1080. This process familiarizes us with the way in which a calculation is made. Note that we convert 1.8¢ to \$0.018 so that all information is in the same unit, dollars. Otherwise, we will not get the correct answer.

We let $c =$ the number of copies that can be made for \$1500.

2. **Translate.** We reword the problem and translate as follows.

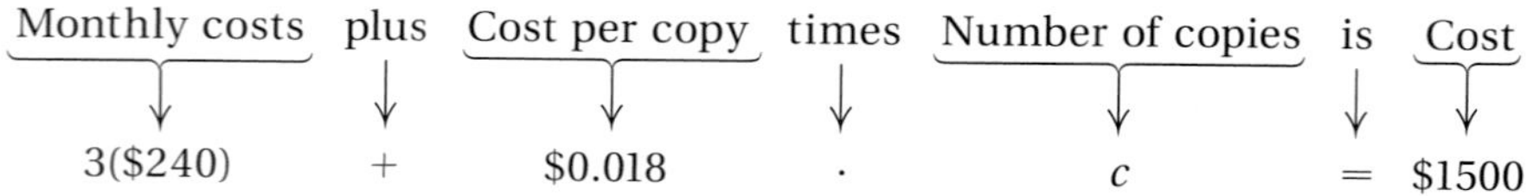

$$\underbrace{\text{Monthly costs}}_{3(\$240)} \; + \; \underbrace{\text{Cost per copy}}_{\$0.018} \; \cdot \; \underbrace{\text{Number of copies}}_{c} \; = \; \underbrace{\text{Cost}}_{\$1500}$$

3. **Solve.** We solve the equation:

$$\begin{aligned}
3(240) + 0.018c &= 1500 \\
720 + 0.018c &= 1500 \\
1000(720 + 0.018c) &= 1000 \cdot 1500 && \text{Multiplying by 1000 on both sides to clear decimals} \\
1000(720) + 1000(0.018c) &= 1{,}500{,}000 && \text{Using the distributive law} \\
720{,}000 + 18c &= 1{,}500{,}000 && \text{Simplifying} \\
720{,}000 + 18c - 720{,}000 &= 1{,}500{,}000 - 720{,}000 && \text{Subtracting 720,000} \\
18c &= 780{,}000 \\
\frac{18c}{18} &= \frac{780{,}000}{18} && \text{Dividing by 18} \\
c &\approx 43{,}333. && \text{Rounding to the nearest one. “}\approx\text{” means “is approximately equal to.”}
\end{aligned}$$

4. **Check.** We check in the original problem. The cost for 43,333 pages is 43,333(\$0.018) = \$779.994. The rental for 3 months is 3(\$240) = \$720. The total cost is then \$779.994 + \$720 ≈ \$1499.99, which is just about the \$1500 allotted.

5. **State.** The law firm can make 43,333 copies on the copy rental allotment of \$1500.

Do Exercise 4 on the preceding page.

Example 5 *Perimeter of NBA Court.* The perimeter of an NBA basketball court is 288 ft. The length is 44 ft longer than the width. (***Source***: National Basketball Association) Find the dimensions of the court.

1. **Familiarize.** We first make a drawing.

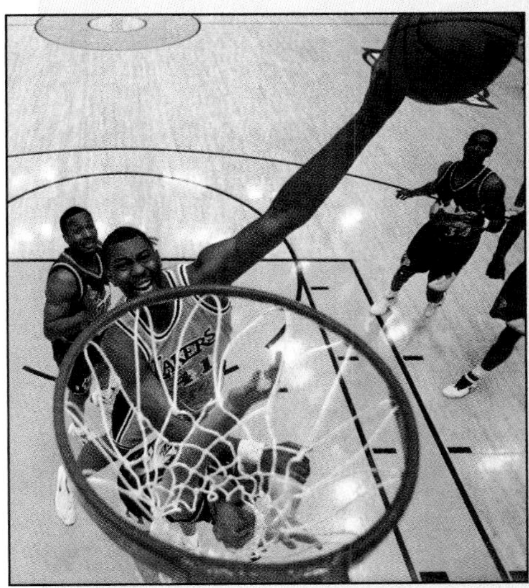

We let w = the width of the rectangle. Then $w + 44$ = the length. The perimeter P of a rectangle is the distance around the rectangle and is given by the formula $2l + 2w = P$, where

l = the length and w = the width.

2. **Translate.** To translate the problem, we substitute $w + 44$ for l and 288 for P:

$$2l + 2w = P$$
$$2(w + 44) + 2w = 288.$$

3. **Solve.** We solve the equation:

$$\begin{aligned} 2(w + 44) + 2w &= 288 \\ 2 \cdot w + 2 \cdot 44 + 2w &= 288 &&\text{Using the distributive law} \\ 4w + 88 &= 288 &&\text{Collecting like terms} \\ 4w + 88 - 88 &= 288 - 88 &&\text{Subtracting 88} \\ 4w &= 200 \\ \frac{4w}{4} &= \frac{200}{4} &&\text{Dividing by 4} \\ w &= 50. \end{aligned}$$

Thus possible dimensions are

$w = 50$ ft and $l = w + 44 = 50 + 44$, or 94 ft.

4. **Check.** If the width is 50 ft and the length is 94 ft, then the perimeter is 2(50 ft) + 2(94 ft), or 288 ft. This checks.

5. **State.** The width is 50 ft and the length is 94 ft.

Do Exercise 5.

5. *Perimeter of High School Basketball Court.* The perimeter of a standard high school basketball court is 268 ft. The length is 34 ft longer than the width. Find the dimensions of the court.

Answer on page A-20

6. The second angle of a triangle is three times as large as the first. The third angle measures 30° more than the first angle. Find the measures of the angles.

Answer on page A-20

Example 6 *Cross Section of a Roof.* In a triangular cross section of a roof, the second angle is twice as large as the first angle. The measure of the third angle is 20° greater than that of the first angle. How large are the angles?

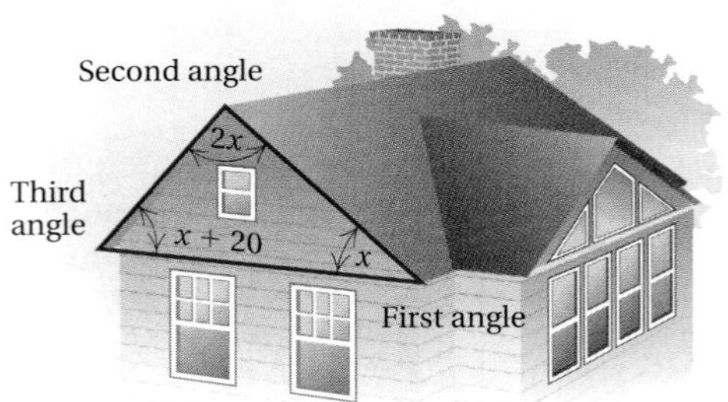

1. **Familiarize.** We first make a drawing as shown above. We let

$$\text{measure of first angle} = x.$$

Then $\quad \text{measure of second angle} = 2x$

and $\quad \text{measure of third angle} = x + 20.$

2. **Translate.** To translate, we need to recall a geometric fact. (You might, as part of step 1, look it up in a geometry book or in the list of formulas on the inside front cover.) Remember, the measures of the angles of a triangle total 180°.

Measure of first angle	plus	Measure of second angle	plus	Measure of third angle	is	180°
↓	↓	↓	↓	↓	↓	↓
x	$+$	$2x$	$+$	$(x + 20)$	$=$	180°

3. **Solve.** We solve the equation:

$$\begin{aligned} x + 2x + (x + 20) &= 180 \\ 4x + 20 &= 180 \\ 4x + 20 - 20 &= 180 - 20 \\ 4x &= 160 \\ \frac{4x}{4} &= \frac{160}{4} \\ x &= 40. \end{aligned}$$

Possible measures for the angles are as follows:

First angle: $x = 40°$;
Second angle: $2x = 2(40) = 80°$;
Third angle: $x + 20 = 40 + 20 = 60°$.

4. **Check.** Consider our answers: 40°, 80°, and 60°. The second is twice the first and the third is 20° greater than the first. The sum is 180°. The angles check.

5. **State.** The measures of the angles are 40°, 80°, and 60°.

Caution! Units are important in answers. Remember to include them, where appropriate.

Do Exercise 6.

Example 7 *Nike, Inc.* The equation

$$y = 0.69606x + 1.68722$$

can be used to approximate the total revenue y, in billions of dollars, of Nike, Inc., in year x, where

$x = 0$ corresponds to 1990,
$x = 1$ corresponds to 1991,
$x = 2$ corresponds to 1992,
$x = 10$ corresponds to 2000,
and so on.

(***Source***: Nike, Inc.) (This equation was developed from a procedure called *regression*. Its discussion belongs to a later course.)

a) Find the total revenue in 1999 and 2008.

b) In what year will the total revenue be about \$12.82418 billion?

Since a formula has been given, we will not use the five-step problem-solving strategy.

a) To find the total revenue for 1999, note first that $1999 - 1990 = 9$. We substitute 9 for x:

$$\begin{aligned} y &= 0.69606x + 1.68722 \\ &= 0.69606(9) + 1.68722 \\ &= \$7.95176. \end{aligned}$$

To find the total revenue for 2008, note that $2008 - 1990 = 18$. We substitute 18 for x:

$$\begin{aligned} y &= 0.69606x + 1.68722 \\ &= 0.69606(18) + 1.68722 \\ &= \$14.2163. \end{aligned}$$

Thus the total revenue will be \$7.95176 billion in 1999 and \$14.2163 billion in 2008.

b) To determine the year in which the total revenue will be about \$12.82418 billion, we first substitute 12.82418 for y. Then we solve for x. Note that we have not cleared decimals because the numbers have the same number of decimal places. (You may choose to do so.)

$$\begin{aligned} y &= 0.69606x + 1.68722 \\ 12.82418 &= 0.69606x + 1.68722 \\ 12.82418 - 1.68722 &= 0.69606x + 1.68722 - 1.68722 \\ 11.13696 &= 0.69606x \\ \frac{11.13696}{0.69606} &= \frac{0.69606x}{0.69606} \\ 16 &= x \end{aligned}$$

The number 16 is the number of years *after* 1990. To find that year, we add 16 to 1990: $1990 + 16 = 2006$. Thus, assuming the equation continues to be valid, the total revenue of Nike, Inc., will be \$12.82418 billion in 2006.

Do Exercise 7.

7. *Nike, Inc.* Referring to Example 7:

a) Find the total revenue in 2002 and 2010.

b) In what year will the total revenue be about \$9.34388 billion?

Answer on page A-20

Improving Your Math Study Skills

Extra Tips on Problem Solving

The following tips, which are focused on problem solving, summarize some points already considered and propose some new ones.

- Get in the habit of using all five steps for problem solving.

1. ***Familiarize*** **yourself with the problem situation.** Some suggestions for this are given on p. 489.
2. ***Translate*** **the problem to an equation.** As you study more mathematics, you will find that the translation may be to some other kind of mathematical language, such as an inequality.
3. ***Solve*** **the equation.** If the translation is to some other kind of mathematical language, you would carry out some kind of mathematical manipulation—in the case of an inequality, you would solve it.
4. ***Check*** **the answer in the original problem.** This does not mean to check in the translated equation. It means to go back to the original worded problem.
5. ***State*** **the answer to the problem clearly.**

For Step 4, some further comment on checking is appropriate. *You may be able to translate to an equation and to solve the equation, but you may find that none of the solutions of the equation is the solution of the original problem.* To see how this can happen, consider this example.

Example

The sum of two consecutive even integers is 537. Find the integers.

1. ***Familiarize.*** Suppose we let x = the first number. Then $x + 2$ = the second number.
2. ***Translate.*** The problem can be translated to the following equation: $x + (x + 2) = 537$.
3. ***Solve.*** We solve the equation as follows:

$$2x + 2 = 537$$
$$2x = 535$$
$$x = \frac{535}{2}, \text{ or } 267.5.$$

4. ***Check.*** Then $x + 2 = 269.5$. However, the numbers are not only not even, but they are not integers.
5. ***State.*** The problem has no solution.

The following are some other tips.

- **To be good at problem solving, do lots of problems.** Learning to solve problems is similar to learning other skills such as golf. At first you may not be successful, but the more you practice and work at improving your skills, the more successful you will become. For problem solving, do more than just two or three odd-numbered assigned problems. Do them all, and if you have time, do the even-numbered problems as well. Then find another book on the same subject and do problems in that book.
- **Look for patterns when solving problems.** You will eventually see patterns in similar kinds of problems. For example, there is a pattern in the way that you solve problems involving consecutive integers.
- **When translating to an equation, or some other mathematical language, consider the dimensions of the variables and the constants in the equation.** The variables that represent length should all be in the same unit, those that represent money should all be in dollars or in cents, and so on.

Exercise Set 8.4

a Solve.

1. Two times a number added to 85 is 117. Find the number.

2. Eight times a number plus 7 is 2559. Find the number.

3. Three less than twice a number is -4. Find the number.

4. Seven less than four times a number is -27. Find the number.

5. When 17 is subtracted from four times a certain number, the result is 211. What is the number?

6. When 36 is subtracted from five times a certain number, the result is 374. What is the number?

7. A 240-in. pipe is cut into two pieces. One piece is three times the length of the other. Find the lengths of the pieces.

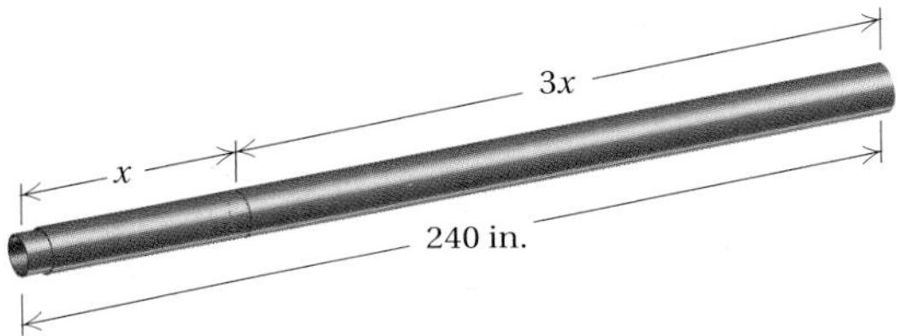

8. A 72-in. board is cut into two pieces. One piece is 2 in. longer than the other. Find the lengths of the pieces.

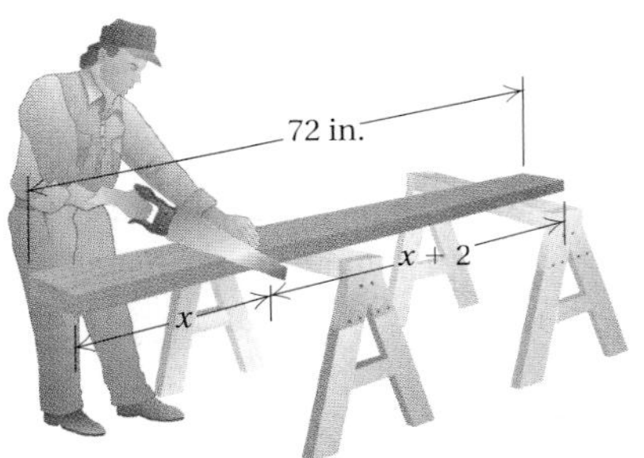

9. *Statue of Liberty.* The height of the Eiffel Tower is 974 ft, which is about 669 ft higher than the Statue of Liberty. What is the height of the Statue of Liberty?

10. *Area of Lake Ontario.* The area of Lake Superior is about four times the area of Lake Ontario. The area of Lake Superior is 30,172 mi^2. What is the area of Lake Ontario?

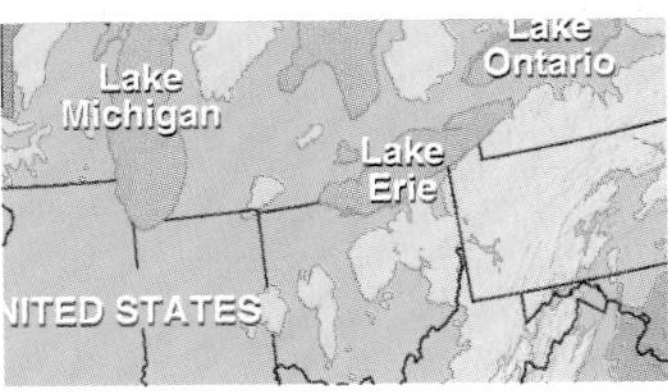

11. *Wheaties.* Recently, the cost of four 18-oz boxes of Wheaties cereal was \$11.56. What was the cost of one box?

12. *Women's Dresses.* In a recent year, the total amount spent on women's blouses was \$6.5 billion. This was \$0.2 billion more than what was spent on women's dresses. How much was spent on women's dresses?

13. If you double a certain number and then add 16, you get $\frac{2}{3}$ of the original number. What is the original number?

14. If you double a certain number and then add 85, you get $\frac{3}{4}$ of the original number. What is the original number?

15. *Iditarod Race.* The Iditarod sled dog race extends for 1049 mi from Anchorage to Nome (***Source*:** Iditarod Trail Commission). If a musher is twice as far from Anchorage as from Nome, how much of the race has the musher completed?

16. *Home Remodeling.* In a recent year, Americans spent a total of \$35 billion to remodel bathrooms and kitchens. Twice as much was spent on kitchens as bathrooms. How much was spent on each?

17. *Consecutive Page Numbers.* The sum of the page numbers on the facing pages of a book is 573. What are the page numbers?

18. *Consecutive Post Office Box Numbers.* The sum of the numbers on two consecutive post office boxes is 547. What are the numbers?

19. The numbers on Sam's three raffle tickets are consecutive integers. The sum of the numbers is 126. What are the numbers?

20. The ages of Whitney, Wesley, and Wanda are consecutive integers. The sum of their ages is 108. What are their ages?

21. The sum of three consecutive odd integers is 189. What are the integers?

22. Three consecutive integers are such that the first plus one-half the second plus seven less than twice the third is 2101. What are the integers?

23. *Standard Billboard Sign.* A standard rectangular highway billboard sign has a perimeter of 124 ft. The length is 6 ft more than three times the width. Find the dimensions.

24. *Two-by-Four.* The perimeter of a cross section of a "two-by-four" piece of lumber is $10\frac{1}{2}$ in. The length is twice the width. Find the actual dimensions of the cross section of a two-by-four.

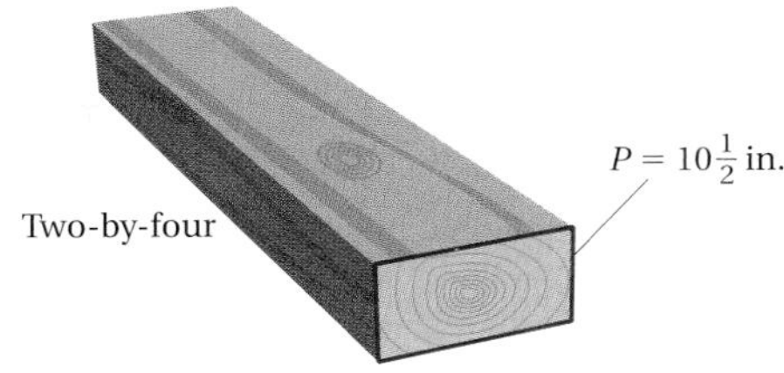

25. *Parking Costs.* A hospital parking lot charges \$1.50 for the first hour or part thereof, and \$1.00 for each additional hour or part thereof. A weekly pass costs \$27.00 and allows unlimited parking for 7 days. Suppose that each visit Ed makes to the hospital lasts $1\frac{1}{2}$ hr. What is the minimum number of times that Ed would have to visit per week to make it worthwhile for him to buy the pass?

26. *Van Rental.* Value Rent-A-Car rents vans at a daily rate of \$84.95 plus 60 cents per mile. Molly rents a van to deliver electrical parts to her customers. She is allotted a daily budget of \$320. How many miles can she drive for \$320?

27. The second angle of a triangular field is three times as large as the first angle. The third angle is 40° greater than the first angle. How large are the angles?

28. *Triangular Parking Lot.* The second angle of a triangular parking lot is four times as large as the first angle. The third angle is 45° less than the sum of the other two angles. How large are the angles?

29. *Triangular Backyard.* A home has a triangular backyard. The second angle of the triangle is 5° more than the first angle. The third angle is 10° more than three times the first angle. Find the angles of the triangular yard.

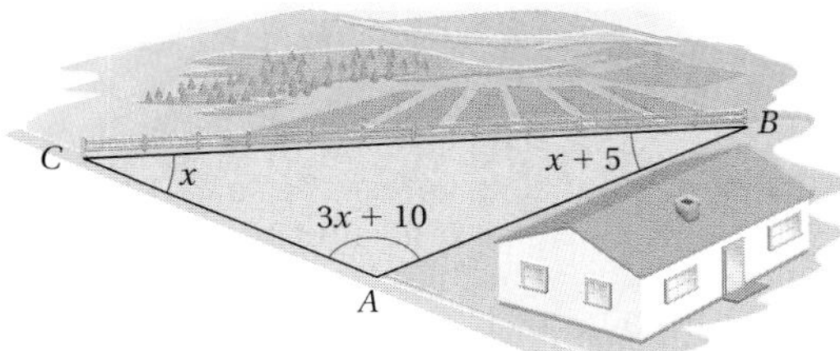

30. *Boarding Stable.* A rancher needs to form a triangular horse pen using ropes next to a stable. The second angle is three times the first angle. The third angle is 15° less than the first angle. Find the angles of the triangular pen.

31. *Coca-Cola Co.* The equation

$$y = 66.2x + 460.2$$

can be used to approximate the total revenue y, in millions of dollars, of the Coca-Cola Co., in year x, where

$x = 0$ corresponds to 1990,

$x = 2$ corresponds to 1992,

$x = 10$ corresponds to 2000,

and so on.

(***Source:*** Coca-Cola Bottling Consolidated)

a) Find the total revenue in 1999, 2000, and 2010.
b) In what year will the total revenue be about $1254.6 million?

32. *Running Records in the 200-m Dash.* The equation

$$R = -0.028t + 20.8$$

can be used to predict the world record in the 200-m dash, where R = the record in seconds, and t = the number of years since 1920 (***Source:*** International Amateur Athletic Federation).

a) Predict the record in 2000 and 2010.
b) In what year will the record be 18.0 sec?

Skill Maintenance

Calculate.

33. $-\frac{4}{5} - \frac{3}{8}$ [7.4a]

34. $-\frac{4}{5} + \frac{3}{8}$ [7.3a]

35. $-\frac{4}{5} \cdot \frac{3}{8}$ [7.5a]

36. $-\frac{4}{5} \div \frac{3}{8}$ [7.6c]

37. $-25.6 \div (-16)$ [7.6c]

38. $-25.6(-16)$ [7.5a]

39. $-25.6 - (-16)$ [7.4a]

40. $-25.6 + (-16)$ [7.3a]

Synthesis

41. ◆ A fellow student claims to be able to solve most of the problems in this section by guessing. Is there anything wrong with this approach? Why or why not?

42. ◆ Write a problem for a classmate to solve so that it can be translated to the equation

$\frac{2}{3}x + (x + 5) + x = 375.$

43. Apples are collected in a basket for six people. One-third, one-fourth, one-eighth, and one-fifth are given to four people, respectively. The fifth person gets ten apples with one apple remaining for the sixth person. Find the original number of apples in the basket.

44. A student scored 78 on a test that had 4 seven-point fill-ins and 24 three-point multiple-choice questions. The student had one fill-in wrong. How many multiple-choice questions did the student answer correctly?

45. The area of this triangle is 2.9047 in^2. Find x.

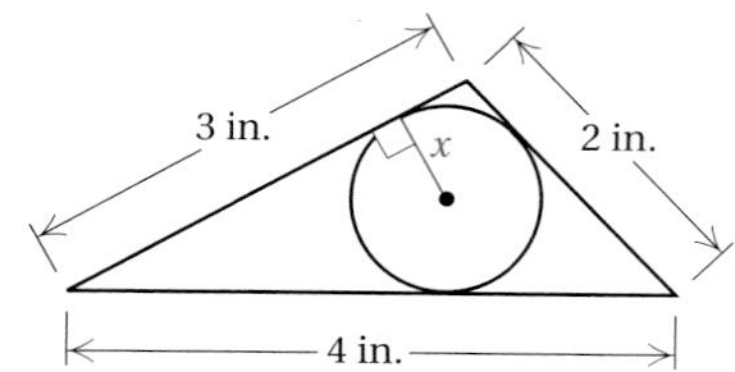

46. A storekeeper goes to the bank to get $10 worth of change. She requests twice as many quarters as half dollars, twice as many dimes as quarters, three times as many nickels as dimes, and no pennies or dollars. How many of each coin did the storekeeper get?

8.5 Applications with Percents

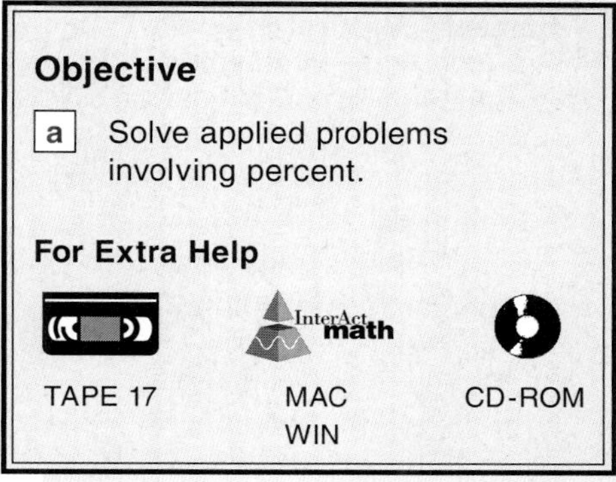

a Many applied problems involve percents. We can use our knowledge of equations and the problem-solving process to solve such problems. For background on percent notation, see Sections 4.2–4.4.

Example 1 What percent of 45 is 15?

1. **Familiarize.** This type of problem is stated so explicitly that we can proceed directly to the translation. We first let x = the percent.
2. **Translate.** We translate as follows:

$$\underbrace{\text{What percent}}_{x} \quad \text{of} \quad 45 \quad \text{is} \quad 15?$$
$$x \cdot 45 = 15.$$

3. **Solve.** We solve the equation:

$$x \cdot 45 = 15$$
$$\frac{x \cdot 45}{45} = \frac{15}{45} \quad \text{Dividing by 45}$$
$$x = \frac{1}{3} \quad \text{Simplifying}$$
$$= 33\frac{1}{3}\%. \quad \text{Changing fractional notation to percent notation}$$

4. **Check.** We check by finding $33\frac{1}{3}\%$ of 45:

$$33\frac{1}{3}\% \cdot 45 = \frac{1}{3} \cdot 45 = 15.$$

5. **State.** The answer is $33\frac{1}{3}\%$.

Do Exercises 1 and 2.

Solve.

1. What percent of 50 is 16?

2. 15 is what percent of 60?

Example 2 3 is 16% of what number?

1. **Familiarize.** This problem is stated so explicitly that we can proceed directly to the translation. We let y = the number that we are taking 16% of.
2. **Translate.** The translation is as follows:

$$\text{3 is 16\% of what?}$$
$$3 = 16\% \cdot y$$

3. **Solve.** We solve the equation:

$$3 = 16\% \cdot y$$
$$3 = 0.16y \quad \text{Converting to decimal notation}$$
$$0.16y = 3 \quad \text{Reversing the equation}$$
$$\frac{0.16y}{0.16} = \frac{3}{0.16} \quad \text{Dividing by 0.16}$$
$$y = 18.75.$$

Answers on page A-20

Solve.

3. 45 is 20 percent of what number?

4. 120 percent of what number is 60?

Solve.

5. What is 23% of 48?

6. Referring to Example 3, determine how many deaths by lightning occurred near telephone poles.

Answers on page A-20

4. **Check.** We check by finding 16% of 18.75:

$$16\% \times 18.75 = 0.16 \times 18.75 = 3.$$

5. **State.** The answer is 18.75.

Do Exercises 3 and 4.

Perhaps you have noticed that to handle percents in problems such as those in Examples 1 and 2, you can convert to decimal notation before continuing.

Example 3 *Locations of Deaths by Lightning.* The circle graph below shows the various locations of people who are struck and killed by lightning. It is known that in the United States, 3327 people were killed by lightning from 1959 to 1996. How many were killed in fields or ballparks?

Number of Deaths Due to Lightning

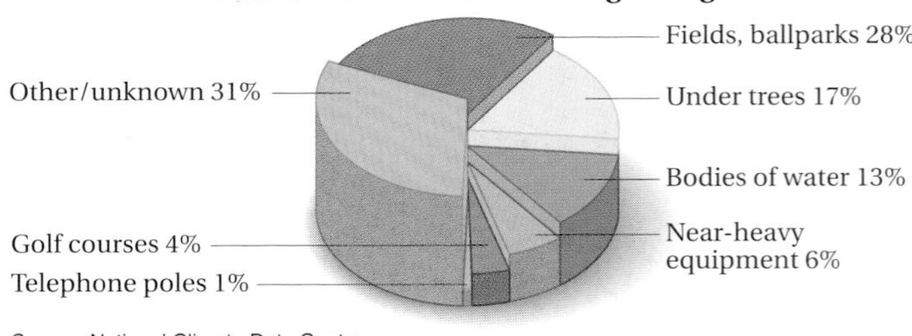

Source: National Climate Data Center

1. **Familiarize.** We first write down the information.

Total number of lightning strikes: 3327

Percent killed in fields or ballparks: 28%

We let x = the number of people killed in fields or ballparks. It seems reasonable that we would take 28% of 3327. This leads us to rewording and translating the problem.

2. **Translate.** The translation is as follows.

28%	of	3327	is	what?	**Rewording**
↓	↓	↓	↓	↓	
28%	$\cdot$	3327	=	x	**Translating**

3. **Solve.** We solve the equation:

$$28\% \cdot 3327 = x$$

$$0.28 \times 3327 = x \quad \text{Converting 28\% to decimal notation}$$

$$932 \approx x. \quad \text{Multiplying and rounding to the nearest one}$$

4. **Check.** The check is actually the computation we use to solve the equation:

$$28\% \cdot 3327 = 0.28 \times 3327 = 931.56 \approx 932.$$

5. **State.** About 932 lightning deaths occurred in fields or ballparks.

Do Exercises 5–7. (Exercise 7 is on the following page.)

Example 4 *Simple Interest.* An investment is made at 6% simple interest for 1 year. It grows to \$768.50. How much was originally invested (the principal)?

1. **Familiarize.** Suppose that \$100 was invested. Recalling the formula for simple interest, $I = Prt$, we know that the interest for 1 year on \$100 at 6% simple interest is given by $I = \$100 \cdot 6\% \cdot 1 = \$6$. Then, at the end of the year, the amount in the account is found by adding the principal and the interest:

$$\begin{array}{ccccc} \text{Principal} & + & \text{Interest} & = & \text{Amount} \\ \downarrow & & \downarrow & & \downarrow \\ \$100 & + & \$6 & = & \$106. \end{array}$$

In this problem, we are working backward. We are trying to find the principal, which is the original investment. We let $x =$ the principal.

2. **Translate.** We reword the problem and then translate.

$$\begin{array}{ccccc} \text{Principal} & + & \text{Interest} & = & \text{Amount} \\ \downarrow & & \downarrow & & \downarrow \\ x & + & 6\%x & = & 768.50 \end{array}$$

Interest is 6% of the principal.

3. **Solve.** We solve the equation:

$x + 6\%x = 768.50$

$x + 0.06x = 768.50$ **Converting to decimal notation**

$1x + 0.06x = 768.50$ **Identity property of 1**

$1.06x = 768.50$ **Collecting like terms**

$\dfrac{1.06x}{1.06} = \dfrac{768.50}{1.06}$ **Dividing by 1.06**

$x = 725.$

4. **Check.** We check by taking 6% of \$725 and adding it to \$725:

$6\% \times \$725 = 0.06 \times 725 = \$43.50.$

Then $\$725 + \$43.50 = \$768.50$, so \$725 checks.

5. **State.** The original investment was \$725.

Do Exercise 8.

7. The area of Arizona is 19% of the area of Alaska. The area of Alaska is 586,400 mi^2. What is the area of Arizona?

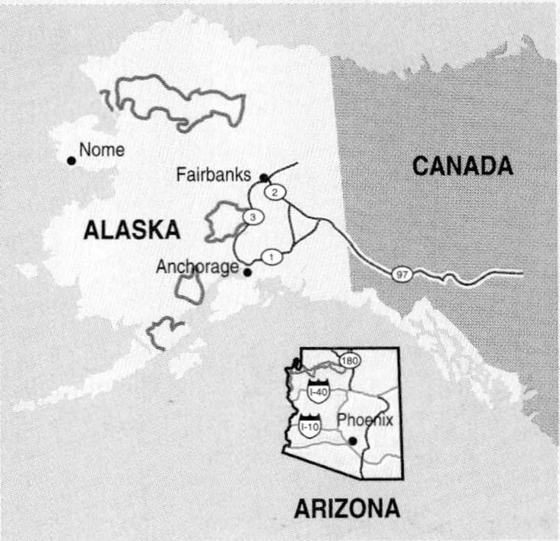

8. An investment is made at 7% simple interest for 1 year. It grows to \$8988. How much was originally invested (the principal)?

Answers on page A-20

FOR SALE
Sales Contract
Seller will pay Realtor 7.5% Commission

Example 5 *Selling a Home.* The Fowlers are selling their home. They want to clear \$115,625 after paying a $7\frac{1}{2}\%$ commission to a realtor. For how much must they sell the house?

1. **Familiarize.** Suppose the Fowlers sold the house for \$120,000. We can determine the $7\frac{1}{2}\%$ commission by taking $7\frac{1}{2}\%$ of \$120,000:

$$7\tfrac{1}{2}\% \text{ of } \$120{,}000 = 0.075(\$120{,}000) = \$9000.$$

Subtracting this commission from \$120,000 would leave the Fowlers with

$$\$120{,}000 - \$9000 = \$111{,}000.$$

This shows us that in order for the Fowlers to clear \$115,625, the house must be sold for more than \$120,000. To determine exactly what the sale price would need to be, we could check more guesses. Instead, let's take advantage of our algebra skills. We let $x =$ the selling price of the house. Because the commission is $7\frac{1}{2}\%$, the realtor receives $7\frac{1}{2}\%x$.

2. **Translate.** We reword and translate.

Selling price	minus	Commission	is	Amount cleared	**Rewording**
↓	↓	↓	↓	↓	
x	$-$	$7\frac{1}{2}\%x$	$=$	115,625	**Translating**

3. **Solve.** We solve the equation:

$$\begin{aligned} x - 7\tfrac{1}{2}\%x &= 115{,}625 \\ 1x - 0.075x &= 115{,}625 && \text{Converting to decimal notation} \\ (1 - 0.075)x &= 115{,}625 && \text{Collecting like terms} \\ 0.925x &= 115{,}625 \\ \frac{0.925x}{0.925} &= \frac{115{,}625}{0.925} && \text{Dividing by 0.925} \\ x &= 125{,}000. \end{aligned}$$

4. **Check.** To check, we first find $7\frac{1}{2}\%$ of \$125,000, calculating as we did in the *Familiarize* step:

$$7\tfrac{1}{2}\% \text{ of } \$125{,}000 = 0.075(\$125{,}000) = \$9375. \quad \text{This is the commission.}$$

Then we subtract the commission to find the amount cleared:

$$\$125{,}000 - \$9375 = \$115{,}625.$$

Since, after the commission, the Fowlers are left with \$115,625, our answer checks. Note that the sale price of \$125,000 is greater than \$120,000, as predicted in the *Familiarize* step.

5. **State.** The Fowlers need to sell their house for \$125,000.

CAUTION! The problem in Example 5 is easy to solve with algebra. Without algebra, it is not. A common error in such a problem is to take $7\frac{1}{2}\%$ of the sale price and then subtract or add. Note that $7\frac{1}{2}\%$ of the selling price $\left(7\frac{1}{2}\% \cdot \$125{,}000 = \$9375\right)$ is not equal to $7\frac{1}{2}\%$ of the price that the Fowlers wanted to clear $\left(7\frac{1}{2}\% \cdot \$115{,}625 \approx \$8671.88\right)$.

Do Exercise 9.

9. The price of a suit was decreased to a sale price of \$526.40. This was a 20% reduction. What was the former price?

Answer on page A-20

Exercise Set 8.5

a Solve.

1. What percent of 180 is 36?

2. What percent of 76 is 19?

3. 45 is 30% of what number?

4. 20.4 is 24% of what number?

5. What number is 65% of 840?

6. What number is 1% of 1,000,000?

7. 30 is what percent of 125?

8. 57 is what percent of 300?

9. 12% of what number is 0.3?

10. 7 is 175% of what number?

11. 2 is what percent of 40?

12. 40 is 2% of what number?

National Hamburger Sales. The circle graph below shows hamburger sales by various restaurants in 1996. The total sales were $39 billion.

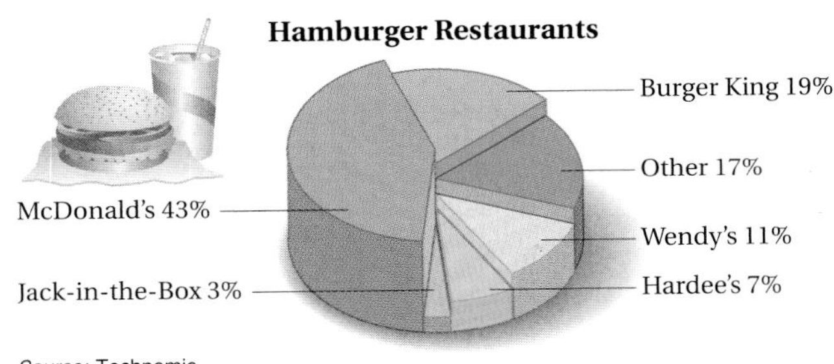

13. What was the total amount of hamburger sales, in dollars, by McDonald's?

14. What was the total amount of hamburger sales, in dollars, by Wendy's?

15. *Junk Mail.* The U.S. Postal Service reports that we open and read 78% of the junk mail that we receive (***Source***: U.S. Postal Service). A sports instructional videotape company sends out 10,500 advertising brochures.
    a) How many of the brochures can it expect to be opened and read?
    b) The company sells videos to 189 of the people who receive the brochure. What percent of the 10,500 people who receive the brochure buy the video?

16. *FBI Applications.* The FBI annually receives 16,000 applications for agents. It accepts 600 of these applicants. (***Source***: Federal Bureau of Investigation) What percent does it accept?

17. Leon left a $4 tip for a meal that cost $25.

a) What percent of the cost of the meal was the tip?

b) What was the total cost of the meal including the tip?

18. Selena left a $12.76 tip for a meal that cost $58.

a) What percent of the cost of the meal was the tip?

b) What was the total cost of the meal including the tip?

19. Leon left a 15% tip for a meal that cost $25.

a) How much was the tip?

b) What was the total cost of the meal including the tip?

20. Selena left a 15% tip for a meal that cost $58.

a) How much was the tip?

b) What was the total cost of the meal including the tip?

21. Leon left a 15% tip of $4.32 for a meal.

a) What was the cost of the meal before the tip?

b) What was the total cost of the meal including the tip?

22. Selena left a 15% tip of $8.40 for a meal.

a) What was the cost of the meal before the tip?

b) What was the total cost of the meal including the tip?

23. Leon left a 15% tip for a meal. The total cost of the meal, including the tip, was $41.40. What was the cost of the meal before the tip was added?

24. Selena left an 18% tip for a meal. The total cost of the meal, including the tip, was $40.71. What was the cost of the meal before the tip was added?

25. In a medical study of a group of pregnant women with "poor" diets, 16 of the women, or 8%, had babies who were in good or excellent health. How many women were in the original study?

26. In a medical study of a group of pregnant women with "good-to-excellent" diets, 285 of the women, or 95%, had babies who were in good or excellent health. How many women were in the original study?

27. *Life Insurance for Smokers vs. Nonsmokers.* The premium for a \$100,000 life insurance policy for a female nonsmoker, age 22, is about \$166 per year. The premium for a smoker is 170% of the premium for a nonsmoker. What is the premium for a smoker?

28. *Catching Colds from Kissing.* In a medical study, it was determined that if 800 people kiss someone else who has a cold, only 56 will actually catch the cold. What percent is this?

29. *Body Fat.* The author of this text exercises regularly at a local YMCA that recently offered a body-fat percentage test to its members. The device used measures the passage of a very low voltage of electricity through the body. The author's body-fat percentage was found to be 19.8% and he weighs 214 lb. What part, in pounds, of his body weight is fat?

30. *Calories Burned.* The author of this text exercises regularly at a local YMCA. The readout on a stairmaster machine tells him that if he exercises for 24 min, he will burn 356 calories. He decides to increase his time on the machine in order to lose more weight.

a) By what percent has he increased his time on the stairmaster if he exercises for 30 min?

b) How many calories does he burn if he exercises for 30 min? 40 min?

31. *Lightning Deaths Under Trees.* Referring to Example 3, determine how many deaths by lightning occurred under trees from 1959 to 1996.

32. *Lightning Deaths on Golf Courses.* Referring to Example 3, determine how many deaths by lightning occurred on golf courses from 1959 to 1996.

33. *Major League Baseball.* In 1997, the Boston Red Sox made the highest increase in the average price of a ticket to a baseball game. The price was \$17.69 and represented an increase of 15% from the preceding season. (***Source*:** Major League Baseball) What was the average price of a ticket the preceding season?

34. *Major League Baseball.* The Atlanta Braves opened a new ballpark, Turner Field, in 1997. Attending a game represented the costliest outing in major league baseball that year. To take a family of four to a game, buy four small soft drinks, two small beers, and four hot dogs, and add in the cost of parking, two game programs, and two twill caps, the average cost would be \$129.16, an increase of 6% over the year before. (***Source*:** Major League Baseball) What did this outing cost the year before?

35. An investment was made at 6% simple interest for 1 year. It grows to \$8268. How much was originally invested?

36. Money is borrowed at 6.2% simple interest. After 1 year, \$6945.48 pays off the loan. How much was originally borrowed?

37. After a 40% reduction, a shirt is on sale for \$34.80. What was the original price (that is, the price before reduction)?

38. After a 34% reduction, a blouse is on sale for \$42.24. What was the original price?

Skill Maintenance

Compute.

39. $9.076 \div 0.05$ [3.4a]

40. 9.076×0.05 [3.3a]

41. $1.089 + 10.89 + 0.1089$ [3.2a]

42. $1000.23 - 156.0893$ [3.2b]

Evaluate. [7.1a]

43. $x - y$, for $x = 58$ and $y = 42$

44. $8t$, for $t = 23.7$

45. $\frac{6a}{b}$, for $a = 25$ and $b = 15$

46. $\frac{a + b}{8}$, for $a = 45.6$ and $b = 102.3$

Synthesis

47. ◈ Comment on the following quote by Yogi Berra, a famous Major League Hall of Fame baseball player: "Ninety percent of hitting is mental. The other half is physical."

48. ◈ Erin returns a tent that she bought during a storewide 35% off sale that has ended. She is offered store credit for 125% of what she paid (not to be used on sale items). Is this fair to Erin? Why or why not?

49. It has been determined that at the age of 15, a boy has reached 96.1% of his final adult height. Jaraan is 6 ft, 4 in. at the age of 15. What will his final adult height be?

50. It has been determined that at the age of 10, a girl has reached 84.4% of her final adult height. Dana is 4 ft, 8 in. at the age of 10. What will her final adult height be?

51. In one city, a sales tax of 9% was added to the price of gasoline as registered on the pump. Suppose a driver asked for \$10 worth of gas. The attendant filled the tank until the pump read \$9.10 and charged the driver \$10. Something was wrong. Use algebra to correct the error.

Collaborative Learning Manual

Calculate the sale price and the original price of discounted items.

8.6 Formulas

a Evaluating and Solving Formulas

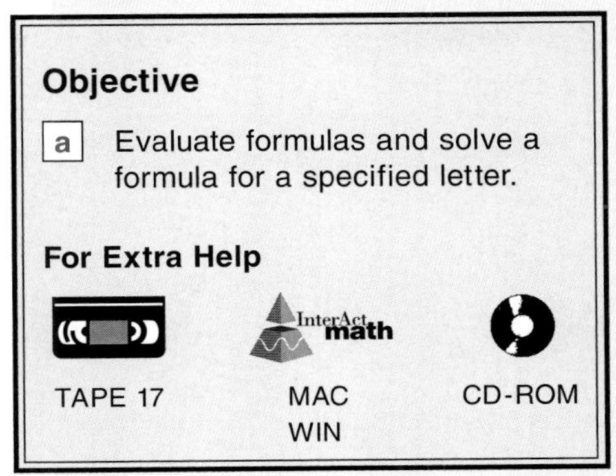

Objective
a Evaluate formulas and solve a formula for a specified letter.
For Extra Help
InterAct math
TAPE 17
MAC WIN
CD-ROM

A **formula** is a "recipe" for doing a certain type of calculation. Formulas are often given as equations. Here is an example of a formula that has to do with weather: $M = \frac{1}{5}n$. You see a flash of lightning. After a few seconds you hear the thunder associated with that flash. How far away was the lightning?

Your distance from the storm is M miles. You can find that distance by counting the number of seconds n that it takes the sound of the thunder to reach you and then multiplying by $\frac{1}{5}$.

Example 1 *Storm Distance.* Consider the formula $M = \frac{1}{5}n$. It takes 10 sec for the sound of thunder to reach you after you have seen a flash of lightning. How far away is the storm?

We substitute 10 for n and calculate M: $M = \frac{1}{5}n = \frac{1}{5}(10) = 2$. The storm is 2 mi away.

Do Exercise 1.

1. Suppose that it takes the sound of thunder 14 sec to reach you. How far away is the storm?

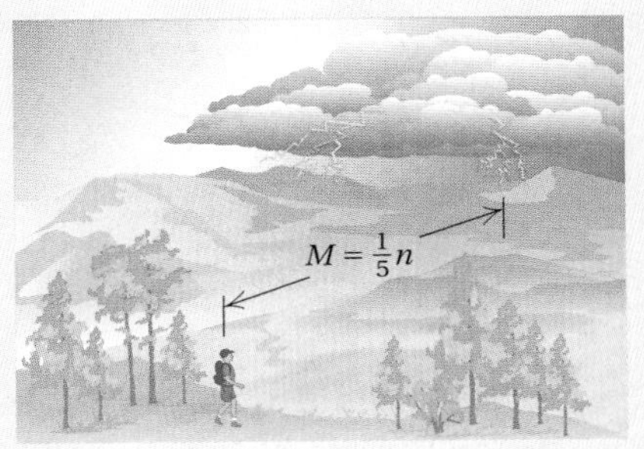

Suppose that we think we know how far we are from the storm and want to check by calculating the number of seconds it should take the sound of the thunder to reach us. We could substitute a number for M—say, 2—and solve for n:

$$2 = \frac{1}{5}n$$

$$10 = n. \quad \text{Multiplying by 5}$$

However, if we wanted to do this repeatedly, it might be easier to solve for n by getting it alone on one side. We "solve" the formula for n.

Example 2 Solve for n: $M = \frac{1}{5}n$.

We have

$$M = \frac{1}{5}n \quad \text{We want this letter alone.}$$

$$5 \cdot M = 5 \cdot \frac{1}{5}n \quad \text{Multiplying by 5 on both sides}$$

$$5M = n.$$

In the above situation for $M = 2$, $n = 5(2)$, or 10.

Do Exercise 2.

2. Solve for I: $E = IR$.
(This is a formula from electricity relating voltage E, current I, and resistance R.)

To see how the addition and multiplication principles apply to formulas, compare the following.

A. Solve.

$$5x + 2 = 12$$
$$5x = 12 - 2$$
$$5x = 10$$
$$x = \frac{10}{5} = 2$$

B. Solve.

$$5x + 2 = 12$$
$$5x = 12 - 2$$
$$x = \frac{12 - 2}{5}$$

C. Solve for x.

$$ax + b = c$$
$$ax = c - b$$
$$x = \frac{c - b}{a}$$

In (A), we solved as we did before. In (B), we did not carry out the calculations. In (C), we could not carry out the calculations because we had unknown numbers.

Answers on page A-21

3. Solve for D: $C = \pi D$.
(This is a formula for the circumference C of a circle of diameter D.)

4. *Averages.* Solve for c:

$$A = \frac{a + b + c + d}{4}.$$

5. Use the formula of Example 5.
a) Estimate the weight of a yellow tuna that is 7 ft long and has a girth of about 54 in.
b) Solve the formula for L.

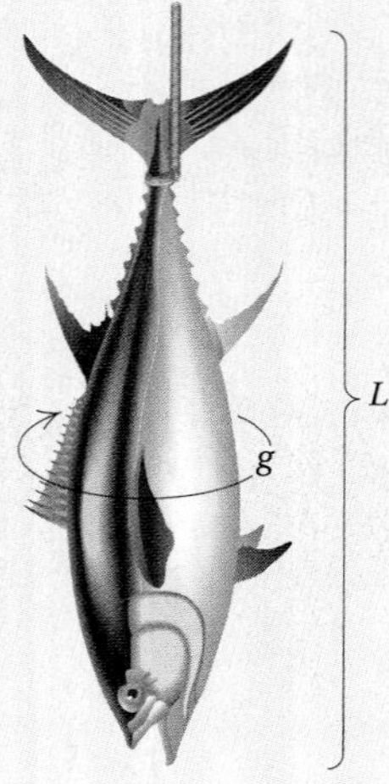

Answers on page A-21

Example 3 *Circumference.* Solve for r: $C = 2\pi r$. This is a formula for the circumference C of a circle of radius r.

$$C = 2\pi r \quad \text{We want this letter alone.}$$

$$\frac{C}{2\pi} = \frac{2\pi r}{2\pi} \quad \text{Dividing by } 2\pi$$

$$\frac{C}{2\pi} = r$$

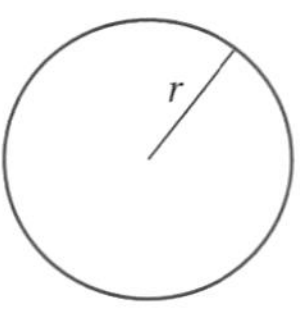

To solve a formula for a given letter, identify the letter and:

1. Multiply on both sides to clear fractions or decimals, if that is needed.
2. Collect like terms on each side, if necessary.
3. Get all terms with the letter to be solved for on one side of the equation and all other terms on the other side.
4. Collect like terms again, if necessary.
5. Solve for the letter in question.

Example 4 Solve for a: $A = \dfrac{a + b + c}{3}$. This is a formula for the average A of three numbers a, b, and c.

$$A = \frac{a + b + c}{3} \quad \text{We want the letter } a \text{ alone.}$$

$$3A = a + b + c \quad \text{Multiplying by 3 to clear the fraction}$$

$$3A - b - c = a \quad \text{Subtracting } b \text{ and } c$$

Do Exercises 3 and 4.

Example 5 *Estimating the Weight of a Fish.* An ancient fisherman's formula for estimating the weight of a fish is

$$W = \frac{Lg^2}{800},$$

where W is the weight in pounds, L is the length in inches, and g is the girth (distance around the midsection) in inches.

a) Estimate the weight of a great bluefin tuna that is 8 ft long and has a girth of about 76 in.
b) Solve the formula for g^2.

We solve as follows:

a) We substitute 96 for L (8 ft = 96 in.) and 76 for g. Then we calculate W.

$$W = \frac{Lg^2}{800} = \frac{96 \cdot 76^2}{800} \approx 693 \text{ lb}$$

The tuna weighs about 693 lb.

b) $$W = \frac{Lg^2}{800} \quad \text{We want to get } g^2 \text{ alone.}$$

$$800W = Lg^2 \quad \text{Multiplying by 800}$$

$$\frac{800W}{L} = g^2 \quad \text{Dividing by } L$$

Do Exercise 5.

Exercise Set 8.6

a Solve for the given letter.

1. *Area of a Parallelogram*:

$A = bh$, for h

(Area A, base b, height h)

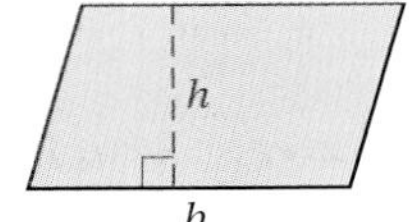

2. *Distance Formula*:

$d = rt$, for r

(Distance d, speed r, time t)

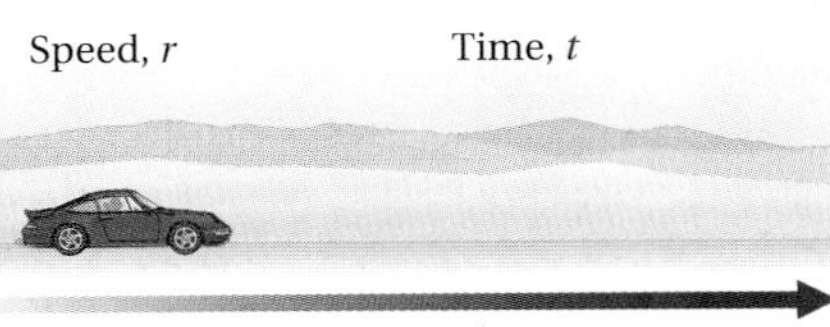

3. *Perimeter of a Rectangle*:

$P = 2l + 2w$, for w

(Perimeter P, length l, width w)

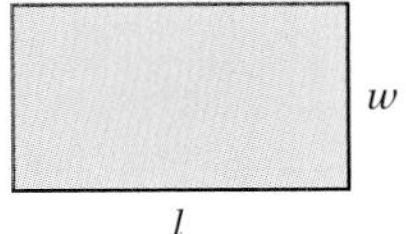

4. *Area of a Circle*:

$A = \pi r^2$, for r^2

(Area A, radius r)

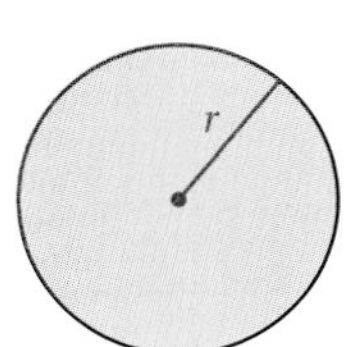

5. *Average of Two Numbers*:

$A = \dfrac{a + b}{2}$, for a

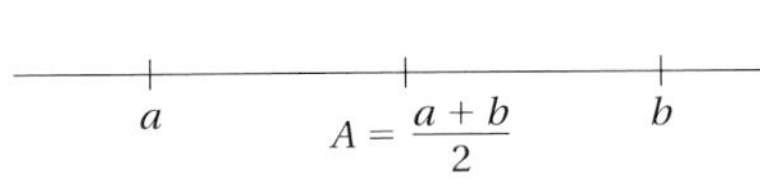

6. *Area of a Triangle*:

$A = \dfrac{1}{2}bh$, for b

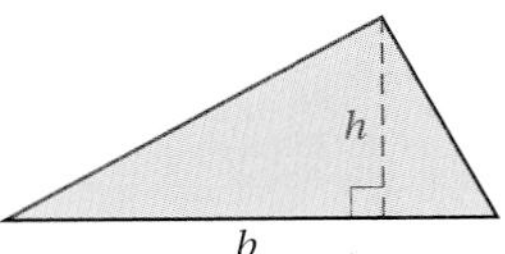

7. *Force*:

$F = ma$, for a

(Force F, mass m, acceleration, a)

8. *Simple Interest*:

$I = Prt$, for P

(Interest I, principal P, interest rate r, time t)

9. *Relativity*:

$E = mc^2$, for c^2

(Energy E, mass m, speed of light c)

10. $Q = \dfrac{p - q}{2}$, for p

11. $Ax + By = c$, for x

12. $Ax + By = c$, for y

13. $v = \dfrac{3k}{t}$, for t

14. $P = \dfrac{ab}{c}$, for c

15. *Furnace Output.* The formula

$b = 30a$

is used in New England to estimate the minimum furnace output, b, in Btu's, for a modern house with a square feet of flooring.

a) Determine the minimum furnace output for a 1900-ft^2 modern house.

b) Solve the formula for a.

16. *Surface Area of a Cube.* The surface area of a cube with side s is given by

$A = 6s^2$.

a) Determine the surface area of a cube with 3-in. sides.

b) Solve the formula for s^2.

17. *Full-Time Equivalent Students.* Colleges accommodate students who need to take different total-credit-hour loads. They determine the number of "full-time-equivalent" students, F, using the formula

$$F = \frac{n}{15},$$

where n is the total number of credits students enroll in for a given semester.

a) Determine the number of full-time equivalent students on a campus in which students register for 21,345 credits.
b) Solve the formula for n.

18. *Young's Rule in Medicine.* Young's rule for determining the amount of a medicine dosage for a child is given by the formula

$$c = \frac{ad}{a + 12},$$

where a is the child's age and d is the usual adult dosage. (*Warning*! Do not apply this formula without checking with a physician!)

a) The usual adult dosage of medication for an adult is 250 mg. Find the dosage for a child of age 2.
b) Solve the formula for d. (***Source*:** Olsen, June L., et al., *Medical Dosage Calculations*, 6th ed. Reading, MA: Addison Wesley Longman, p. A-31.)

19. *Female Caloric Needs.* The number of calories K needed each day by a moderately active woman who weighs w pounds, is h inches tall, and is a years old can be estimated by the formula

$$K = 917 + 6(w + h - a).$$

(***Source*:** Parker, M., *She Does Math.* Mathematical Association of America, p. 96.)

a) Elaine is moderately active, weighs 120 lb, is 67 in. tall, and is 23 yr old. What are her caloric needs?
b) Solve the formula for a, for h, and for w.

20. *Male Caloric Needs.* The number of calories K needed each day by a moderately active man who weighs w kilograms, is h centimeters tall, and is a years old can be estimated by the formula

$$K = 19.18w + 7h - 9.52a + 92.4.$$

(***Source*:** Parker, M., *She Does Math.* Mathematical Association of America, p. 96.)

a) Marv is moderately active, weighs 97 kg, is 185 cm tall, and is 55 yr old. What are his caloric needs?
b) Solve the formula for a, for h, and for w.

Skill Maintenance

21. Convert to decimal notation: $\frac{23}{25}$. [3.5a]

22. Add: $-23 + (-67)$. [7.3a]

23. Subtract: $-45.8 - (-32.6)$. [7.4a]

24. Remove parentheses and simplify: [7.8b]
$4a - 8b - 5(5a - 4b)$.

25. Add: $-\frac{2}{3} + \frac{5}{6}$. [7.3a]

26. Subtract: $-\frac{2}{3} - \frac{5}{6}$. [7.4a]

Synthesis

27. ◆ Devise an application in which it would be useful to solve the equation $d = rt$ for r. (See Exercise 2.)

28. ◆ The equations

$$P = 2l + 2w \quad \text{and} \quad w = \frac{P}{2} - l$$

are equivalent formulas involving the perimeter P, the length l, and the width w of a rectangle. Devise a problem for which the second of the two formulas would be more useful.

Solve.

29. $A = \frac{1}{2}ah + \frac{1}{2}bh$, for b; for h

30. $P = 4m + 7mn$, for m

31. In $A = lw$, l and w both double. What is the effect on A?

32. In $P = 2a + 2b$, P doubles. Do a and b necessarily both double?

33. In $A = \frac{1}{2}bh$, b increases by 4 units and h does not change. What happens to A?

34. Solve for F:

$$D = \frac{1}{E + F}.$$

8.7 Solving Inequalities

We now extend our equation-solving principles to the solving of inequalities.

Objectives

- a Determine whether a given number is a solution of an inequality.
- b Graph an inequality on a number line.
- c Solve inequalities using the addition principle.
- d Solve inequalities using the multiplication principle.
- e Solve inequalities using the addition and multiplication principles together.

For Extra Help

TAPE 17   MAC WIN   CD-ROM

a Solutions of Inequalities

In Section 7.2, we defined the symbols $>$ (greater than), $<$ (less than), $\geq$ (greater than or equal to), and $\leq$ (less than or equal to). For example, $3 \leq 4$ and $3 \leq 3$ are both true, but $-3 \leq -4$ and $0 \geq 2$ are both false.

An **inequality** is a number sentence with $>$, $<$, $\geq$, or $\leq$ as its verb—for example,

$$-4 > t, \quad x < 3, \quad 2x + 5 \geq 0, \quad \text{and} \quad -3y + 7 \leq -8.$$

Some replacements for a variable in an inequality make it true and some make it false.

> A replacement that makes an inequality true is called a **solution**. The set of all solutions is called the **solution set.** When we have found the set of all solutions of an inequality, we say that we have **solved** the inequality.

Examples Determine whether the number is a solution of $x < 2$.

1. -2.7 Since $-2.7 < 2$ is true, -2.7 is a solution.

2. 2 Since $2 < 2$ is false, 2 is not a solution.

Examples Determine whether the number is a solution of $y \geq 6$.

3. 6 Since $6 \geq 6$ is true, 6 is a solution.

4. $-\frac{4}{3}$ Since $-\frac{4}{3} \geq 6$ is false, $-\frac{4}{3}$ is not a solution.

Do Exercises 1 and 2.

Determine whether each number is a solution of the inequality.

1. $x > 3$

a) 2 b) 0

c) -5 d) 15.4

e) 3 f) $-\frac{2}{5}$

2. $x \leq 6$

a) 6 b) 0

c) -4.3 d) 25

e) -6 f) $\frac{5}{8}$

Answers on page A-21

b Graphs of Inequalities

Some solutions of $x < 2$ are 0.45, -8.9, $-\pi$, $\frac{5}{8}$, and so on. In fact, there are infinitely many real numbers that are solutions. Because we cannot list them all individually, it is helpful to make a drawing that represents all the solutions.

A **graph** of an inequality is a drawing that represents its solutions. An inequality in one variable can be graphed on a number line. An inequality in two variables can be graphed on a coordinate plane; we will study such graphs in Chapter 13.

We first graph inequalities in one variable on a number line.

Example 5 Graph: $x < 2$.

The solutions of $x < 2$ are all those numbers less than 2. They are shown on the graph by shading all points to the left of 2. The open circle at 2 indicates that 2 is not part of the graph.

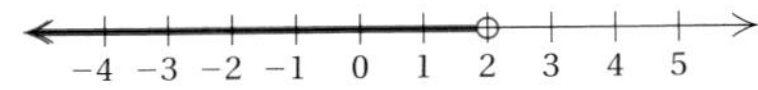

Graph.

3. $x \leq 4$

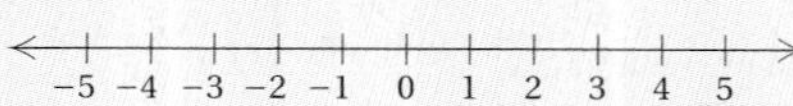

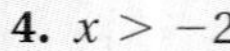

4. $x > -2$

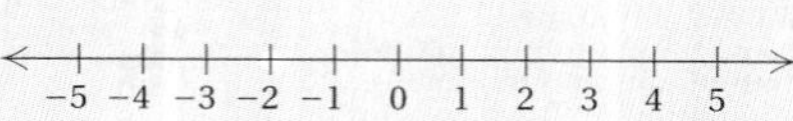

5. $-2 < x \leq 4$

−5 −4 −3 −2 −1 0 1 2 3 4 5

Answers on page A-21

Example 6 Graph: $x \geq -3$.

The solutions of $x \geq -3$ are shown on the number line by shading the point for -3 and all points to the right of -3. The closed circle at -3 indicates that -3 *is* part of the graph.

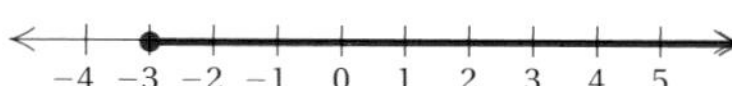

Example 7 Graph: $-3 \leq x < 2$.

The inequality $-3 \leq x < 2$ is read "-3 is less than or equal to x *and* x is less than 2," or "x is greater than or equal to -3 *and* x is less than 2." In order to be a solution of this inequality, a number must be a solution of both $-3 \leq x$ and $x < 2$. The number 1 is a solution, as are -1.7, 0, 1.5, and $\frac{3}{8}$. We can see from the graphs below that the solution set consists of the numbers that overlap in the two solution sets in Examples 5 and 6:

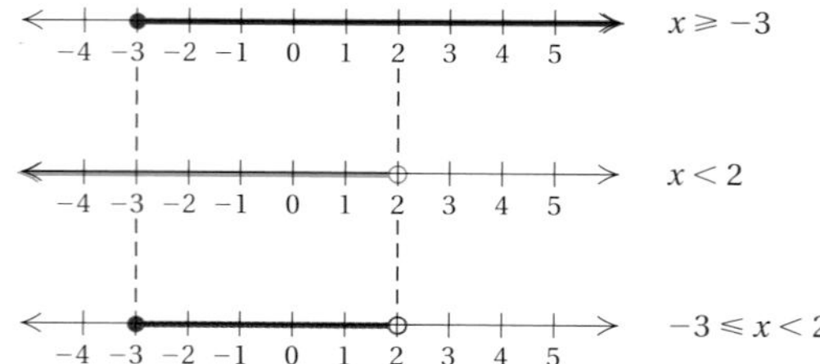

The open circle at 2 means that 2 is *not* part of the graph. The closed circle at -3 means that -3 *is* part of the graph. The other solutions are shaded.

Do Exercises 3–5.

c Solving Inequalities Using the Addition Principle

Consider the true inequality $3 < 7$. If we add 2 on both sides, we get another true inequality:

$$3 + 2 < 7 + 2, \quad \text{or} \quad 5 < 9.$$

Similarly, if we add -4 on both sides of $x + 4 < 10$, we get an *equivalent* inequality:

$$x + 4 + (-4) < 10 + (-4),$$

or

$$x < 6.$$

To say that $x + 4 < 10$ and $x < 6$ are **equivalent** is to say that they have the same solution set. For example, the number 3 is a solution of $x + 4 < 10$. It is also a solution of $x < 6$. The number -2 is a solution of $x < 6$. It is also a solution of $x + 4 < 10$. Any solution of one is a solution of the other—they are equivalent.

> **THE ADDITION PRINCIPLE FOR INEQUALITIES**
>
> For any real numbers a, b, and c:
>
> $a < b$ is equivalent to $a + c < b + c$;
>
> $a > b$ is equivalent to $a + c > b + c$;
>
> $a \leq b$ is equivalent to $a + c \leq b + c$;
>
> $a \geq b$ is equivalent to $a + c \geq b + c$.
>
> In other words, when we add or subtract the same number on both sides of an inequality, the direction of the inequality symbol is not changed.

As with equation solving, when solving inequalities, our goal is to isolate the variable on one side. Then it is easier to determine the solution set.

Example 8 Solve: $x + 2 > 8$. Then graph.

We use the addition principle, subtracting 2 on both sides:

$$x + 2 - 2 > 8 - 2$$
$$x > 6.$$

From the inequality $x > 6$, we can determine the solutions directly. Any number greater than 6 makes the last sentence true and is a solution of that sentence. Any such number is also a solution of the original sentence. Thus the inequality is solved. The graph is as follows:

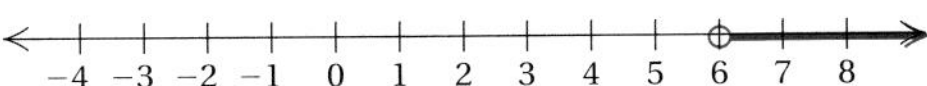

We cannot check all the solutions of an inequality by substitution, as we can check solutions of equations, because there are too many of them. A partial check can be done by substituting a number greater than 6—say, 7—into the original inequality:

$x + 2 > 8$		
$7 + 2$	8	
9		TRUE

Since $9 > 8$ is true, 7 is a solution. Any number greater than 6 is a solution.

Example 9 Solve: $3x + 1 \leq 2x - 3$. Then graph.

We have

$3x + 1 \leq 2x - 3$	
$3x + 1 - 1 \leq 2x - 3 - 1$	**Subtracting 1**
$3x \leq 2x - 4$	**Simplifying**
$3x - 2x \leq 2x - 4 - 2x$	**Subtracting $2x$**
$x \leq -4.$	**Simplifying**

The graph is as follows:

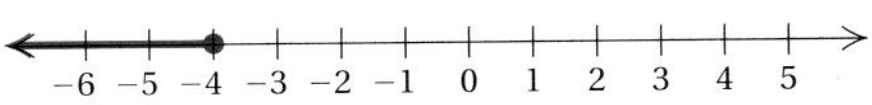

Solve. Then graph.

6. $x + 3 > 5$

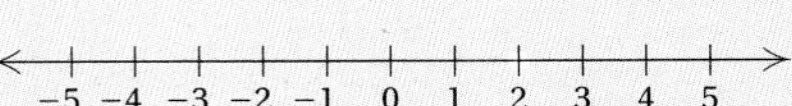

7. $x - 1 \leq 2$

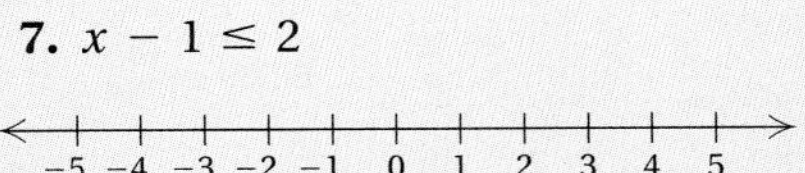

8. $5x + 1 < 4x - 2$

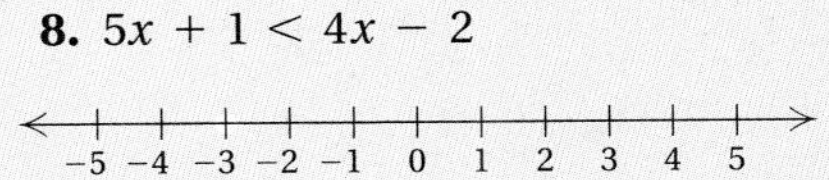

Answers on page A-21

Solve.

9. $x + \frac{2}{3} \geq \frac{4}{5}$

10. $5y + 2 \leq -1 + 4y$

Answers on page A-21

In Example 9, any number less than or equal to -4 is a solution. The following are some solutions:

$$-4, \quad -5, \quad -6, \quad -\frac{13}{3}, \quad -204.5, \quad \text{and} \quad -18\pi.$$

Besides drawing a graph, we can also describe all the solutions of an inequality using **set notation.** We could just begin to list them in a set using roster notation (see p. 403), as follows:

$$\{-4, -5, -6, -4.1, -204.5, -18\pi, \ldots\}.$$

We can never list them all this way, however. Seeing this set without knowing the inequality makes it difficult for us to know what real numbers we are considering. There is, however, another kind of notation that we can use. It is

$$\{x \mid x \leq -4\},$$

which is read

"The set of all x such that x is less than or equal to -4."

This shorter notation for sets is called **set-builder notation** (see Section 7.2). From now on, we will use this notation when solving inequalities.

Do Exercises 6–8 on the preceding page.

Example 10 Solve: $x + \frac{1}{3} > \frac{5}{4}$.

We have

$$x + \frac{1}{3} > \frac{5}{4}$$

$$x + \frac{1}{3} - \frac{1}{3} > \frac{5}{4} - \frac{1}{3} \qquad \text{Subtracting } \tfrac{1}{3}$$

$$x > \frac{5}{4} \cdot \frac{3}{3} - \frac{1}{3} \cdot \frac{4}{4} \qquad \text{Multiplying by 1 to obtain a common denominator}$$

$$x > \frac{15}{12} - \frac{4}{12}$$

$$x > \frac{11}{12}.$$

Any number greater than $\frac{11}{12}$ is a solution. The solution set is

$$\left\{x \mid x > \frac{11}{12}\right\},$$

which is read

"The set of all x such that x is greater than $\frac{11}{12}$."

When solving inequalities, you may obtain an answer like $7 < x$. Recall from Chapter 7 that this has the same meaning as $x > 7$. Thus the solution set can be described as $\{x \mid 7 < x\}$ or as $\{x \mid x > 7\}$. The latter is used most often.

Do Exercises 9 and 10.

d Solving Inequalities Using the Multiplication Principle

There is a multiplication principle for inequalities that is similar to that for equations, but it must be modified. When we are multiplying on both sides by a negative number, the direction of the inequality symbol must be changed. Let's see what happens. Consider the true inequality $3 < 7$. If we multiply on both sides by a *positive* number, like 2, we get another true inequality:

$3 \cdot 2 < 7 \cdot 2$, or $6 < 14$. **True**

If we multiply on both sides by a *negative* number, like -2, and we do not change the direction of the inequality symbol, we get a *false* inequality:

$3 \cdot (-2) < 7 \cdot (-2)$, or $-6 < -14$. **False**

The fact that $6 < 14$ is true but $-6 < -14$ is false stems from the fact that the negative numbers, in a sense, mirror the positive numbers. That is, whereas 14 is to the *right* of 6 on a number line, the number -14 is to the *left* of -6. Thus, if we reverse (change the direction of) the inequality symbol, we get a *true* inequality: $-6 > -14$.

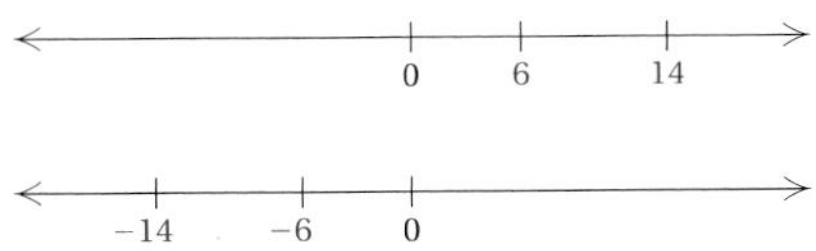

THE MULTIPLICATION PRINCIPLE FOR INEQUALITIES

For any real numbers a and b, and any *positive* number c:

$a < b$ is equivalent to $ac < bc$;

$a > b$ is equivalent to $ac > bc$.

For any real numbers a and b, and any *negative* number c:

$a < b$ is equivalent to $ac > bc$;

$a > b$ is equivalent to $ac < bc$.

Similar statements hold for $\leq$ and $\geq$.

In other words, when we multiply or divide by a positive number on both sides of an inequality, the direction of the inequality symbol stays the same. When we multiply or divide by a negative number on both sides of an inequality, the direction of the inequality symbol is reversed.

Example 11 Solve: $4x < 28$. Then graph.

We have

$4x < 28$

$\frac{4x}{4} < \frac{28}{4}$ **Dividing by 4**

The symbol stays the same.

$x < 7$. **Simplifying**

The solution set is $\{x \mid x < 7\}$. The graph is as follows:

Do Exercises 11 and 12.

Solve. Then graph.

11. $8x < 64$

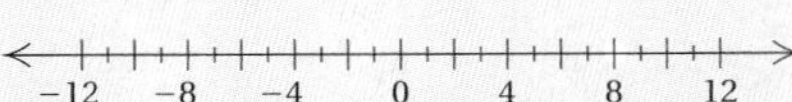

12. $5y \geq 160$

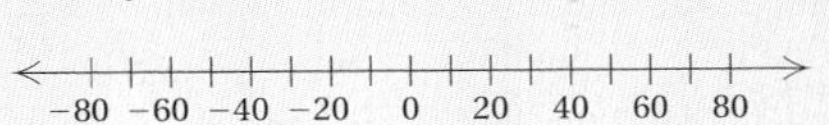

Answers on page A-21

Solve.

13. $-4x \leq 24$

14. $-5y > 13$

15. Solve: $7 - 4x < 8$.

Answers on page A-21

Example 12 Solve: $-2y < 18$. Then graph.

We have

$$-2y < 18$$

$$\frac{-2y}{-2} > \frac{18}{-2} \quad \text{Dividing by } -2$$

The symbol must be reversed!

$$y > -9. \quad \text{Simplifying}$$

The solution set is $\{y \mid y > -9\}$. The graph is as follows:

−10 −9 −8 −7 −6 −5 −4 −3 −2 −1 0 1 2 3 4 5 6 7 8

Do Exercises 13 and 14.

e Using the Principles Together

All of the equation-solving techniques used in Sections 8.1–8.3 can be used with inequalities provided we remember to reverse the inequality symbol when multiplying or dividing on both sides by a negative number.

Example 13 Solve: $6 - 5y > 7$.

We have

$$6 - 5y > 7$$

$$-6 + 6 - 5y > -6 + 7 \quad \text{Adding } -6\text{. The symbol stays the same.}$$

$$-5y > 1 \quad \text{Simplifying}$$

$$\frac{-5y}{-5} < \frac{1}{-5} \quad \text{Dividing by } -5$$

The symbol must be reversed.

$$y < -\frac{1}{5}. \quad \text{Simplifying}$$

The solution set is $\{y \mid y < -\frac{1}{5}\}$.

Do Exercise 15.

Example 14 Solve: $8y - 5 > 17 - 5y$.

$$-17 + 8y - 5 > -17 + 17 - 5y \quad \text{Adding } -17\text{. The symbol stays the same.}$$

$$8y - 22 > -5y \quad \text{Simplifying}$$

$$-8y + 8y - 22 > -8y - 5y \quad \text{Adding } -8y$$

$$-22 > -13y \quad \text{Simplifying}$$

$$\frac{-22}{-13} < \frac{-13y}{-13} \quad \text{Dividing by } -13$$

The symbol must be reversed.

$$\frac{22}{13} < y.$$

The solution set is $\{y \mid \frac{22}{13} < y\}$, or $\{y \mid y > \frac{22}{13}\}$.

We can often solve inequalities in such a way as to avoid having to reverse the inequality symbol. We add so that after like terms have been collected, the coefficient of the variable term is positive. We show this by solving the inequality in Example 14 a different way.

Example 15 Solve: $8y - 5 > 17 - 5y$.

Note that if we add $5y$ on both sides, the coefficient of the y-term will be positive after like terms have been collected.

$$8y - 5 + 5y > 17 - 5y + 5y \quad \text{Adding } 5y$$
$$13y - 5 > 17 \quad \text{Simplifying}$$
$$13y - 5 + 5 > 17 + 5 \quad \text{Adding 5}$$
$$13y > 22 \quad \text{Simplifying}$$
$$\frac{13y}{13} > \frac{22}{13} \quad \text{Dividing by 13}$$
$$y > \frac{22}{13}$$

The solution set is $\{y \mid y > \frac{22}{13}\}$.

Do Exercises 16 and 17.

Example 16 Solve: $3(x - 2) - 1 < 2 - 5(x + 6)$.

$$3(x - 2) - 1 < 2 - 5(x + 6)$$
$$3x - 6 - 1 < 2 - 5x - 30 \quad \text{Using the distributive law to multiply and remove parentheses}$$
$$3x - 7 < -5x - 28 \quad \text{Simplifying}$$
$$3x + 5x < -28 + 7 \quad \text{Adding } 5x \text{ and 7 to get all } x\text{-terms on one side and all other terms on the other side}$$
$$8x < -21 \quad \text{Simplifying}$$
$$x < \frac{-21}{8}, \text{ or } -\frac{21}{8} \quad \text{Dividing by 8}$$

The solution set is $\{x \mid x < -\frac{21}{8}\}$.

Do Exercise 18.

Example 17 Solve: $16.3 - 7.2p \leq -8.18$.

The greatest number of decimal places in any one number is *two*. Multiplying by 100, which has two 0's, will clear decimals. Then we proceed as before.

$$16.3 - 7.2p \leq -8.18$$
$$100(16.3 - 7.2p) \leq 100(-8.18) \quad \text{Multiplying by 100}$$
$$100(16.3) - 100(7.2p) \leq 100(-8.18) \quad \text{Using the distributive law}$$
$$1630 - 720p \leq -818 \quad \text{Simplifying}$$
$$1630 - 720p - 1630 \leq -818 - 1630 \quad \text{Subtracting 1630}$$
$$-720p \leq -2448 \quad \text{Simplifying}$$
$$\frac{-720p}{-720} \geq \frac{-2448}{-720} \quad \text{Dividing by } -720$$

The symbol must be reversed.

$$p \geq 3.4$$

The solution set is $\{p \mid p \geq 3.4\}$.

16. Solve: $24 - 7y \leq 11y - 14$.

17. Solve. Use a method like the one used in Example 15.

$24 - 7y \leq 11y - 14$

18. Solve:

$3(7 + 2x) \leq 30 + 7(x - 1)$.

Answers on page A-21

19. Solve:

$2.1x + 43.2 \geq 1.2 - 8.4x.$

20. Solve:

$\frac{3}{4} + x < \frac{7}{8}x - \frac{1}{4} + \frac{1}{2}x.$

Answers on page A-21

Do Exercise 19.

Example 18 Solve: $\frac{2}{3}x - \frac{1}{6} + \frac{1}{2}x > \frac{7}{6} + 2x.$

The number 6 is the least common multiple of all the denominators. Thus we multiply by 6 on both sides.

$$\frac{2}{3}x - \frac{1}{6} + \frac{1}{2}x > \frac{7}{6} + 2x$$

$$6\left(\frac{2}{3}x - \frac{1}{6} + \frac{1}{2}x\right) > 6\left(\frac{7}{6} + 2x\right) \quad \text{Multiplying by 6 on both sides}$$

$$6 \cdot \frac{2}{3}x - 6 \cdot \frac{1}{6} + 6 \cdot \frac{1}{2}x > 6 \cdot \frac{7}{6} + 6 \cdot 2x \quad \text{Using the distributve law}$$

$$4x - 1 + 3x > 7 + 12x \quad \text{Simplifying}$$

$$7x - 1 > 7 + 12x \quad \text{Collecting like terms}$$

$$7x - 1 - 12x > 7 + 12x - 12x \quad \text{Subtracting } 12x$$

$$-5x - 1 > 7 \quad \text{Collecting like terms}$$

$$-5x - 1 + 1 > 7 + 1 \quad \text{Adding 1}$$

$$-5x > 8 \quad \text{Simplifying}$$

$$\frac{-5x}{-5} < \frac{8}{-5} \quad \text{Dividing by } -5$$

↑ The symbol must be reversed.

$$x < -\frac{8}{5}$$

The solution set is $\{x \mid x < -\frac{8}{5}\}$.

Do Exercise 20.

Exercise Set 8.7

a Determine whether each number is a solution of the given inequality.

1. $x > -4$

a) 4
b) 0
c) -4
d) 6
e) 5.6

2. $x \leq 5$

a) 0
b) 5
c) -1
d) -5
e) $7\frac{1}{4}$

3. $x \geq 6.8$

a) -6
b) 0
c) 6
d) 8
e) $-3\frac{1}{2}$

4. $x < 8$

a) 8
b) -10
c) 0
d) 11
e) -4.7

b Graph on a number line.

5. $x > 4$

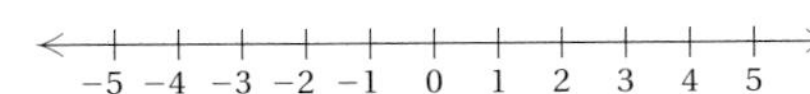

6. $x < 0$

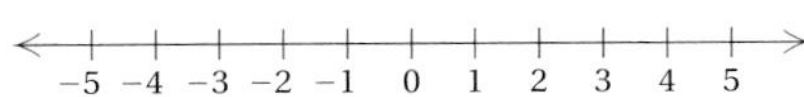

7. $t < -3$

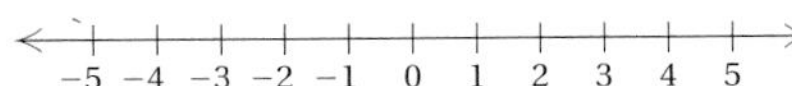

8. $y > 5$

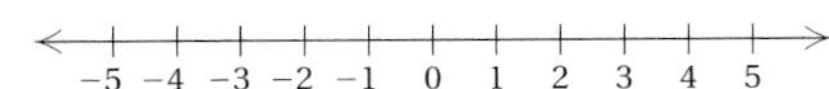

9. $m \geq -1$

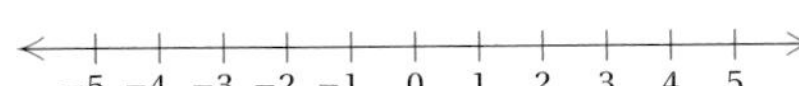

10. $x \leq -2$

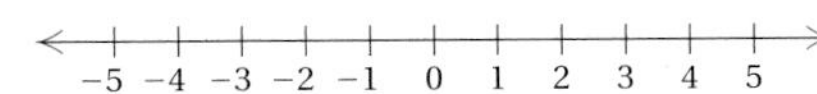

11. $-3 < x \leq 4$

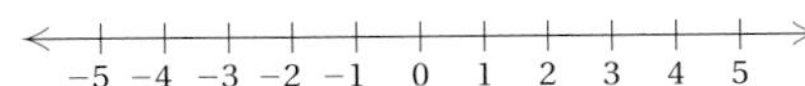

12. $-5 \leq x < 2$

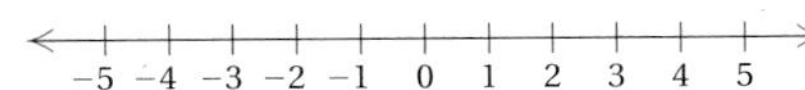

13. $0 < x < 3$

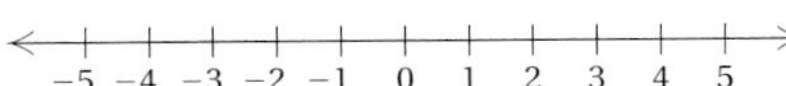

14. $-5 \leq x \leq 0$

c Solve using the addition principle. Then graph.

15. $x + 7 > 2$

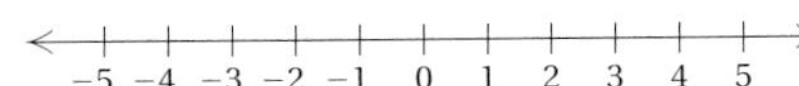

16. $x + 5 > 2$

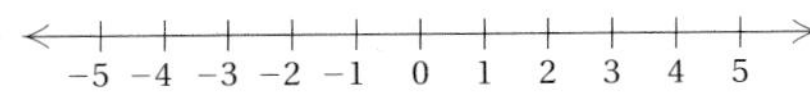

17. $x + 8 \leq -10$

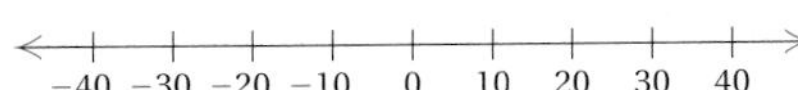

18. $x + 8 \leq -11$

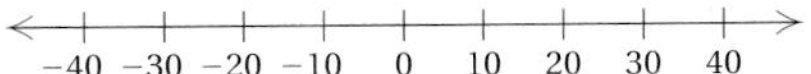

Solve using the addition principle.

19. $y - 7 > -12$

20. $y - 9 > -15$

21. $2x + 3 > x + 5$

22. $2x + 4 > x + 7$

23. $3x + 9 \leq 2x + 6$

24. $3x + 18 \leq 2x + 16$

25. $5x - 6 < 4x - 2$

26. $9x - 8 < 8x - 9$

27. $-9 + t > 5$

28. $-8 + p > 10$

29. $y + \frac{1}{4} \leq \frac{1}{2}$

30. $x - \frac{1}{3} \leq \frac{5}{6}$

31. $x - \frac{1}{3} > \frac{1}{4}$

32. $x + \frac{1}{8} > \frac{1}{2}$

d Solve using the multiplication principle. Then graph.

33. $5x < 35$

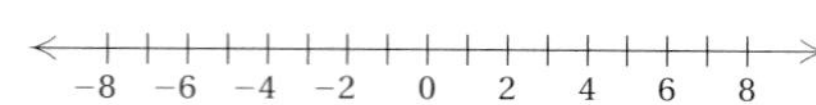

34. $8x \geq 32$

−5 −4 −3 −2 −1 0 1 2 3 4 5

35. $-12x > -36$

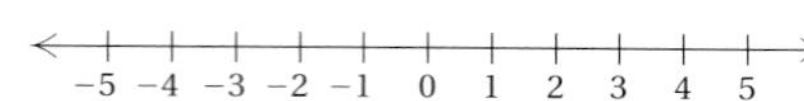

36. $-16x > -64$

−5 −4 −3 −2 −1 0 1 2 3 4 5

Solve using the multiplication principle.

37. $5y \geq -2$

38. $3x < -4$

39. $-2x \leq 12$

40. $-3x \leq 15$

41. $-4y \geq -16$

42. $-7x < -21$

43. $-3x < -17$

44. $-5y > -23$

45. $-2y > \frac{1}{7}$

46. $-4x \leq \frac{1}{9}$

47. $-\frac{6}{5} \leq -4x$

48. $-\frac{7}{9} > 63x$

e Solve using the addition and multiplication principles.

49. $4 + 3x < 28$

50. $3 + 4y < 35$

51. $3x - 5 \leq 13$

52. $5y - 9 \leq 21$

53. $13x - 7 < -46$

54. $8y - 6 < -54$

55. $30 > 3 - 9x$

56. $48 > 13 - 7y$

57. $4x + 2 - 3x \leq 9$

58. $15x + 5 - 14x \leq 9$

59. $-3 < 8x + 7 - 7x$

60. $-8 < 9x + 8 - 8x - 3$

61. $6 - 4y > 4 - 3y$

62. $9 - 8y > 5 - 7y + 2$

63. $5 - 9y \leq 2 - 8y$

64. $6 - 18x \leq 4 - 12x - 5x$

65. $19 - 7y - 3y < 39$

66. $18 - 6y - 4y < 63 + 5y$

67. $2.1x + 45.2 > 3.2 - 8.4x$

68. $0.96y - 0.79 \leq 0.21y + 0.46$

69. $\frac{x}{3} - 2 \leq 1$

70. $\frac{2}{3} + \frac{x}{5} < \frac{4}{15}$

71. $\frac{y}{5} + 1 \leq \frac{2}{5}$

72. $\frac{3x}{4} - \frac{7}{8} \geq -15$

73. $3(2y - 3) < 27$

74. $4(2y - 3) > 28$

75. $2(3 + 4m) - 9 \geq 45$

76. $3(5 + 3m) - 8 \leq 88$

77. $8(2t + 1) > 4(7t + 7)$

78. $7(5y - 2) > 6(6y - 1)$

79. $3(r - 6) + 2 < 4(r + 2) - 21$

80. $5(x + 3) + 9 \leq 3(x - 2) + 6$

81. $0.8(3x + 6) \geq 1.1 - (x + 2)$

82. $0.4(2x + 8) \geq 20 - (x + 5)$

83. $\frac{5}{3} + \frac{2}{3}x < \frac{25}{12} + \frac{5}{4}x + \frac{3}{4}$

84. $1 - \frac{2}{3}y \geq \frac{9}{5} - \frac{y}{5} + \frac{3}{5}$

Skill Maintenance

Add or subtract. [7.3a], [7.4a]

85. $-56 + (-18)$

86. $-2.3 + 7.1$

87. $-\frac{3}{4} + \frac{1}{8}$

88. $8.12 - 9.23$

89. $-56 - (-18)$

90. $-\frac{3}{4} - \frac{1}{8}$

91. $-2.3 - 7.1$

92. $-8.12 + 9.23$

Simplify. [7.8b, d]

93. $5 - 3^2 + (8 - 2)^2 \cdot 4$

94. $10 \div 2 \cdot 5 - 3^2 + (-5)^2$

95. $5(2x - 4) - 3(4x + 1)$

96. $9(3 + 5x) - 4(7 + 2x)$

Synthesis

97. ◈ Are the inequalities $3x - 4 < 10 - 4x$ and $2(x - 5) > 3(2x - 6)$ equivalent? Why or why not?

98. ◈ Explain in your own words why it is necessary to reverse the inequality symbol when multiplying on both sides of an inequality by a negative number.

99. Determine whether each number is a solution of the inequality $|x| < 3$.

a) 0
b) -2
c) -3
d) 4
e) 3
f) 1.7
g) -2.8

100. Graph $|x| < 3$ on a number line.

−5 −4 −3 −2 −1 0 1 2 3 4 5

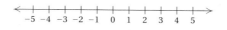

Solve.

101. $x + 3 \leq 3 + x$

102. $x + 4 < 3 + x$

Solve linear inequalities as a group.

8.8 Applications and Problem Solving with Inequalities

We can use inequalities to solve certain types of problems.

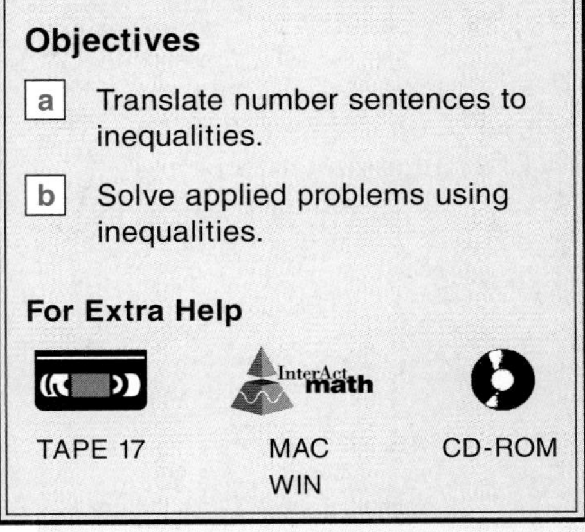

Objectives
a Translate number sentences to inequalities.
b Solve applied problems using inequalities.
For Extra Help
TAPE 17
InterAct math
MAC
WIN
CD-ROM

a Translating to Inequalities

First let's practice translating sentences to inequalities.

Examples Translate to an inequality.

1. A number is less than 5.

$x < 5$

2. A number is greater than or equal to $3\frac{1}{2}$.

$y \geq 3\frac{1}{2}$

3. He can earn, at most, \$34,000.

$E \leq \$34{,}000$

4. The number of compact disc players sold in this city in a year is at least 2700.

$C \geq 2700$

5. 12 more than twice a number is less than 37.

$2x + 12 < 37$

Do Exercises 1–5.

Translate.

1. A number is less than or equal to 8.

2. A number is greater than -2.

3. That car can be driven at most 180 mph.

4. The price of that car is at least \$5800.

5. Twice a number minus 32 is greater than 5.

Answers on page A-21

b Solving Problems

Example 6 *Test Scores.* A pre-med student is taking a chemistry course in which four tests are to be given. To get an A, she must average at least 90 on the four tests. The student got scores of 91, 86, and 89 on the first three tests. Determine (in terms of an inequality) what scores on the last test will allow her to get an A.

1. Familiarize. Let's try some guessing. Suppose the student gets a 92 on the last test. The average of the four scores is their sum divided by the number of tests, 4, and is given by

$$\frac{91 + 86 + 89 + 92}{4} = 89.5.$$

In order for this average to be *at least* 90, it must be greater than or equal to 90. Since $89.5 \geq 90$ is false, a score of 92 will not give the student an A. But there are scores that will give an A. To find them, we translate to an inequality and solve. Let $x =$ the student's score on the last test.

2. Translate. The average of the four scores must be *at least* 90. This means that it must be greater than or equal to 90. Thus we can translate the problem to the inequality

$$\frac{91 + 86 + 89 + x}{4} \geq 90.$$

6. *Test Scores.* A student is taking a literature course in which four tests are to be given. To get a B, he must average at least 80 on the four tests. The student got scores of 82, 76, and 78 on the first three tests. Determine (in terms of an inequality) what scores on the last test will allow him to get at least a B.

7. *Gold Temperatures.* Gold stays solid at Fahrenheit temperatures below 1945.4°. Determine (in terms of an inequality) those Celsius temperatures for which gold stays solid. Use the formula given in Example 7.

Answers on page A-21

3. **Solve.** We solve the inequality. We first multiply by 4 to clear the fraction.

$$4\left(\frac{91 + 86 + 89 + x}{4}\right) \geq 4 \cdot 90 \qquad \text{Multiplying by 4}$$
$$91 + 86 + 89 + x \geq 360$$
$$266 + x \geq 360 \qquad \text{Collecting like terms}$$
$$x \geq 94 \qquad \text{Subtracting 266}$$

The solution set is $\{x \mid x \geq 94\}$.

4. **Check.** We can obtain a partial check by substituting a number greater than or equal to 94. We leave it to the student to try 95 in a manner similar to what was done in the *Familiarize* step.

5. **State.** Any score that is at least 94 will give the student an A.

Do Exercise 6.

Example 7 *Butter Temperatures.* Butter stays solid at Fahrenheit temperatures below 88°. The formula

$$F = \tfrac{9}{5}C + 32$$

can be used to convert Celsius temperatures C to Fahrenheit temperatures F. Determine (in terms of an inequality) those Celsius temperatures for which butter stays solid.

1. **Familiarize.** Let's make a guess. We try a Celsius temperature of 40°. We substitute and find F:

$$F = \tfrac{9}{5}C + 32 = \tfrac{9}{5}(40) + 32 = 72 + 32 = 104°.$$

This is higher than 88°, so 40° is *not* a solution. To find the solutions, we need to solve an inequality.

2. **Translate.** The Fahrenheit temperature F is to be less than 88. We have the inequality

$$F < 88.$$

To find the Celsius temperatures C that satisfy this condition, we substitute $\tfrac{9}{5}C + 32$ for F, which gives us the following inequality:

$$\tfrac{9}{5}C + 32 < 88.$$

3. **Solve.** We solve the inequality:

$$\tfrac{9}{5}C + 32 < 88$$
$$5\left(\tfrac{9}{5}C + 32\right) < 5(88) \qquad \text{Multiplying by 5 to clear the fraction}$$
$$5\left(\tfrac{9}{5}C\right) + 5(32) < 440 \qquad \text{Using a distributive law}$$
$$9C + 160 < 440 \qquad \text{Simplifying}$$
$$9C < 280 \qquad \text{Subtracting 160}$$
$$C < \frac{280}{9} \qquad \text{Dividing by 9}$$
$$C < 31.1. \qquad \text{Dividing and rounding to the nearest tenth}$$

The solution set of the inequality is $\{C \mid C < 31.1°\}$.

4. **Check.** The check is left to the student.

5. **State.** Butter stays solid at Celsius temperatures below 31.1°.

Do Exercise 7.

Exercise Set 8.8

a Translate to an inequality.

1. A number is greater than 8.

2. A number is less than 5.

3. A number is less than or equal to -4.

4. A number is greater than or equal to 18.

5. The number of people is at least 1300.

6. The cost is at most \$4857.95.

7. The amount of acid is not to exceed 500 liters.

8. The cost of gasoline is no less than 94 cents per gallon.

9. Two more than three times a number is less than 13.

10. Five less than one-half a number is greater than 17.

b Solve.

11. *Test Scores.* Your quiz grades are 73, 75, 89, and 91. Determine (in terms of an inequality) what scores on the last quiz will allow you to get an average quiz grade of at least 85.

12. *Body Temperatures.* The human body is considered to be fevered when its temperature is higher than 98.6°F. Using the formula given in Example 7, determine (in terms of an inequality) those Celsius temperatures for which the body is fevered.

13. *World Records in the 1500-m Run.* The formula

$$R = -0.075t + 3.85$$

can be used to predict the world record in the 1500-m run t years after 1930. Determine (in terms of an inequality) those years for which the world record will be less than 3.5 min.

14. *World Records in the 200-m Dash.* The formula

$$R = -0.028t + 20.8$$

can be used to predict the world record in the 200-m dash t years after 1920. Determine (in terms of an inequality) those years for which the world record will be less than 19.0 sec.

15. *Sizes of Envelopes.* Rhetoric Advertising is a direct-mail company. It determines that for a particular campaign, it can use any envelope with a fixed width of $3\frac{1}{2}$ in. and an area of at least $17\frac{1}{2}$ in^2. Determine (in terms of an inequality) those lengths that will satisfy the company constraints.

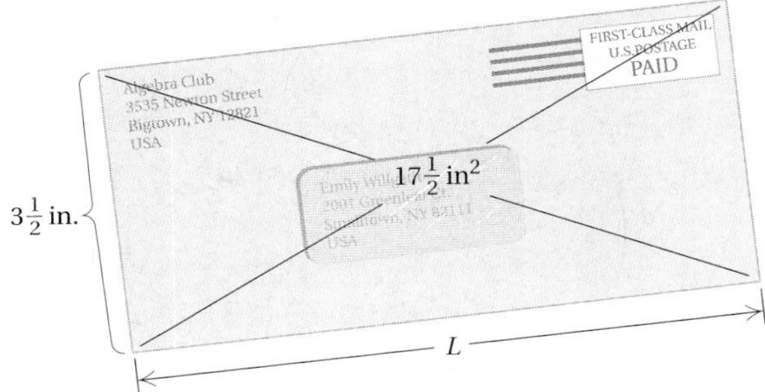

16. *Sizes of Packages.* An overnight delivery service accepts packages of up to 165 in. in length and girth combined. (Girth is the distance around the package.) A package has a fixed girth of 53 in. Determine (in terms of an inequality) those lengths for which a package is acceptable.

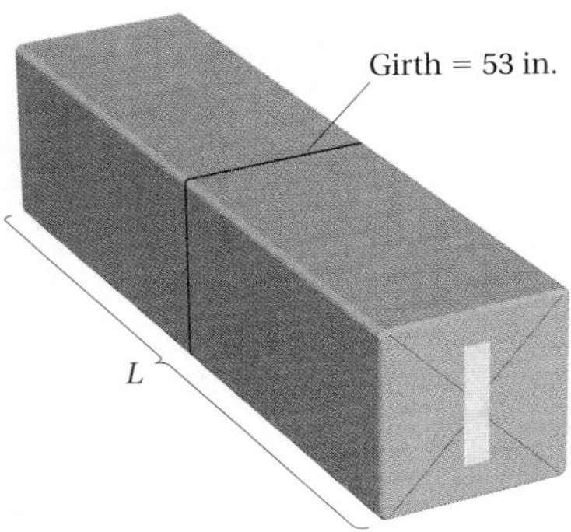

17. Find all numbers such that the sum of the number and 15 is less than four times the number.

18. Find all numbers such that three times the number minus ten times the number is greater than or equal to eight times the number.

19. *Black Angus Calves.* Black Angus calves weigh about 75 lb at birth and gain about 2 lb per day for the first few weeks. Determine (in terms of an inequality) those days for which the calf's weight is more than 125 lb.

20. *IKON Copiers.* IKON Office Solutions rents a Canon GP30F copier for $240 per month plus 1.8¢ per copy (***Source:*** Ikon Office Solutions, Keith Palmer). A catalog publisher needs to lease a copy machine for use during a special project that they anticipate will take 3 months. They decide to rent the copier, but must stay within a budget of $5400 for copies. Determine (in terms of an inequality) the number of copies they can make per month and still remain within budget.

21. One side of a triangle is 2 cm shorter than the base. The other side is 3 cm longer than the base. What lengths of the base will allow the perimeter to be greater than 19 cm?

22. The perimeter of a rectangular swimming pool is not to exceed 70 ft. The length is to be twice the width. What widths will meet these conditions?

23. Dirk's Electric made 17 customer calls last week and 22 calls this week. How many calls must be made next week in order to maintain an average of at least 20 calls for the three-week period?

24. Ginny and Jill do volunteer work at a hospital. Jill worked 3 hr more than Ginny, and together they worked more than 27 hr. What possible number of hours did each work?

25. A family's air conditioner needs freon. The charge for a service call is a flat fee of $70 plus $60 an hour. The freon costs $35. The family has at most $150 to pay for the service call. Determine (in terms of an inequality) those lengths of time of the call that will allow the family to stay within its $150 budget.

26. A student is shopping for a new pair of jeans and two sweaters of the same kind. He is determined to spend no more than $120.00 for the outfit. He buys jeans for $21.95. What is the most that the student can spend for each sweater?

27. *Skippy Reduced-Fat Peanut Butter.* In order for a food to be advertised as "reduced fat," it must have at least 25% less fat than the regular food of that type. Reduced-fat Skippy Peanut Butter contains 12 g of fat per serving. What can you conclude about how much fat is in regular Skippy peanut butter?

28. A landscaping company is laying out a triangular flower bed. The height of the triangle is 16 ft. What lengths of the base will make the area at least 200 ft^2?

Skill Maintenance

Simplify.

29. $-3 + 2(-5)^2(-3) - 7$ [7.8d]

30. $3x + 2[4 - 5(2x - 1)]$ [7.8c]

31. $23(2x - 4) - 15(10 - 3x)$ [7.8b]

32. $256 \div 64 \div 4^2$ [7.8d]

Synthesis

33. ◆ Chassman and Bem booksellers offers a preferred customer card for $25. The card entitles a customer to a 10% discount on all purchases for a period of 1 year. Under what circumstances would an individual save money by buying a card?

34. ◆ After 9 quizzes, Brenda's average is 84. Is it possible for her to improve her average by two points with the next quiz? Why or why not?

Summary and Review Exercises: Chapter 8

Important Properties and Formulas

The Addition Principle for Equations: For any real numbers a, b, and c: $a = b$ is equivalent to $a + c = b + c$.

The Multiplication Principle for Equations: For any real numbers a, b, and c, $c \neq 0$: $a = b$ is equivalent to $a \cdot c = b \cdot c$.

The Addition Principle for Inequalities: For any real numbers a, b, and c:
$a < b$ is equivalent to $a + c < b + c$;
$a > b$ is equivalent to $a + c > b + c$;
$a \leq b$ is equivalent to $a + c \leq b + c$;
$a \geq b$ is equivalent to $a + c \geq b + c$.

The Multiplication Principle for Inequalities: For any real numbers a and b, and any *positive* number c:
$a < b$ is equivalent to $ac < bc$; $a > b$ is equivalent to $ac > bc$.
For any real numbers a and b, and any *negative* number c:
$a < b$ is equivalent to $ac > bc$; $a > b$ is equivalent to $ac < bc$.

The objectives to be tested in addition to the material in this chapter are [7.1a], [7.1b], [7.3a], and [7.8b].

Solve. [8.1b]

1. $x + 5 = -17$

2. $n - 7 = -6$

3. $x - 11 = 14$

4. $y - 0.9 = 9.09$

Solve. [8.2a]

5. $-\frac{2}{3}x = -\frac{1}{6}$

6. $-8x = -56$

7. $-\frac{x}{4} = 48$

8. $15x = -35$

9. $\frac{4}{5}y = -\frac{3}{16}$

Solve. [8.3a]

10. $5 - x = 13$

11. $\frac{1}{4}x - \frac{5}{8} = \frac{3}{8}$

Solve. [8.3b]

12. $5t + 9 = 3t - 1$

13. $7x - 6 = 25x$

14. $14y = 23y - 17 - 10$

15. $0.22y - 0.6 = 0.12y + 3 - 0.8y$

16. $\frac{1}{4}x - \frac{1}{8}x = 3 - \frac{1}{16}x$

Solve. [8.3c]

17. $4(x + 3) = 36$

18. $3(5x - 7) = -66$

19. $8(x - 2) = 5(x + 4)$

20. $-5x + 3(x + 8) = 16$

Determine whether the given number is a solution of the inequality $x \leq 4$. [8.7a]

21. -3

22. 7

23. 4

Solve. Write set notation for the answers. [8.7c, d, e]

24. $y + \frac{2}{3} \geq \frac{1}{6}$

25. $9x \geq 63$

26. $2 + 6y > 14$

27. $7 - 3y \geq 27 + 2y$

28. $3x + 5 < 2x - 6$

29. $-4y < 28$

30. $3 - 4x < 27$

31. $4 - 8x < 13 + 3x$

32. $-3y \geq -21$

33. $-4x \leq \frac{1}{3}$

Graph on a number line. [8.7b, e]

34. $4x - 6 < x + 3$

35. $-2 < x \leq 5$

36. $y > 0$

Solve. [8.6a]

37. $C = \pi d$, for d

38. $V = \frac{1}{3}Bh$, for B

39. $A = \frac{a + b}{2}$, for a

Solve. [8.4a]

40. *Dimensions of Wyoming.* The state of Wyoming is roughly in the shape of a rectangle whose perimeter is 1280 mi. The length is 90 mi more than the width. Find the dimensions.

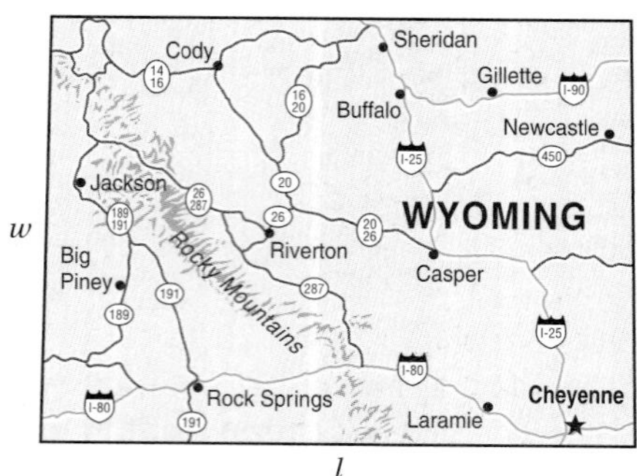

41. If 14 is added to a certain number, the result is 41. Find the number.

42. *Interstate Mile Markers.* The sum of two mile markers on I-5 in California is 691. Find the numbers on the markers.

43. An entertainment center sold for \$2449 in June. This was \$332 more than the cost in February. Find the cost in February.

44. Ty is paid a commission of \$4 for each appliance he sells. One week, he received \$108 in commissions. How many appliances did he sell?

45. The measure of the second angle of a triangle is 50° more than that of the first angle. The measure of the third angle is 10° less than twice the first angle. Find the measures of the angles.

Solve. [8.5a]

46. After a 30% reduction, a bread maker is on sale for \$154. What was the marked price (the price before the reduction)?

47. A hotel manager's salary is \$30,000, which is a 15% increase over the previous year's salary. What was the previous salary (to the nearest dollar)?

48. A tax-exempt charity received a bill of \$145.90 for a sump pump. The bill incorrectly included sales tax of 5%. How much does the charity actually owe?

Solve.

49. *Test Scores.* Your test grades are 71, 75, 82, and 86. What is the lowest grade that you can get on the next test and still have an average test score of at least 80? [8.8b]

50. The length of a rectangle is 43 cm. What widths will make the perimeter greater than 120 cm? [8.8b]

51. *Estimating the Weight of a Fish.* An ancient fisherman's formula for estimating the weight of a fish is

$$W = \frac{Lg^2}{800},$$

where W is the weight in pounds, L is the length in inches, and g is the girth (distance around the midsection) in inches. [8.6a]

a) Estimate the weight of a salmon that is 3 ft long and has a girth of about 13.5 in.
b) Solve the formula for L.

Skill Maintenance

52. Evaluate $\dfrac{a+b}{4}$ for $a = 16$ and $b = 25$. [7.1a]

53. Translate to an algebraic expression: [7.1b]
Tricia drives her car at 58 mph for t hours. How far has she driven?

54. Add: $-12 + 10 + (-19) + (-24)$. [7.3a]

55. Remove parentheses and simplify: $5x - 8(6x - y)$. [7.8b]

Synthesis

56. ◆ Would it be better to receive a 5% raise and then an 8% raise or the other way around? Why?

57. ◆ Are the inequalities $x > -5$ and $-x < 5$ equivalent? Why or why not?

Solve.

58. $2|x| + 4 = 50$ [7.2e], [8.3a]

59. $|3x| = 60$ [7.2e], [8.2a]

60. $y = 2a - ab + 3$, for a [8.6a]

Test: Chapter 8

Solve.

1. $x + 7 = 15$

2. $t - 9 = 17$

3. $3x = -18$

4. $-\frac{4}{7}x = -28$

5. $3t + 7 = 2t - 5$

6. $\frac{1}{2}x - \frac{3}{5} = \frac{2}{5}$

7. $8 - y = 16$

8. $-\frac{2}{5} + x = -\frac{3}{4}$

9. $3(x + 2) = 27$

10. $-3x - 6(x - 4) = 9$

11. $0.4p + 0.2 = 4.2p - 7.8 - 0.6p$

Solve. Write set notation for the answers.

12. $x + 6 \leq 2$

13. $14x + 9 > 13x - 4$

14. $12x \leq 60$

15. $-2y \geq 26$

16. $-4y \leq -32$

17. $-5x \geq \frac{1}{4}$

18. $4 - 6x > 40$

19. $5 - 9x \geq 19 + 5x$

Graph on a number line.

20. $y \leq 9$

−12 −8 −4 0 4 8 12

21. $6x - 3 < x + 2$

−5 −4 −3 −2 −1 0 1 2 3 4 5

22. $-2 \leq x \leq 2$

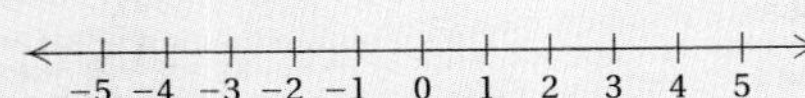

Answers

1. ______
2. ______
3. ______
4. ______
5. ______
6. ______
7. ______
8. ______
9. ______
10. ______
11. ______
12. ______
13. ______
14. ______
15. ______
16. ______
17. ______
18. ______
19. ______
20. ______
21. ______
22. ______

Answers

23. ____

24. ____

25. ____

26. ____

27. ____

28. ____

29. a) ____

b) ____

30. ____

31. ____

32. ____

33. ____

34. ____

35. ____

36. ____

37. ____

38. ____

Solve.

23. The perimeter of a rectangular photograph is 36 cm. The length is 4 cm greater than the width. Find the width and the length.

24. If you triple a number and then subtract 14, you get two-thirds of the original number. What is the original number?

25. The numbers on three raffle tickets are consecutive integers whose sum is 7530. Find the integers.

26. Money is invested in a savings account at 5% simple interest. After 1 year, there is \$924 in the account. How much was originally invested?

27. An 8-m board is cut into two pieces. One piece is 2 m longer than the other. How long are the pieces?

28. Solve $A = 2\pi rh$ for r.

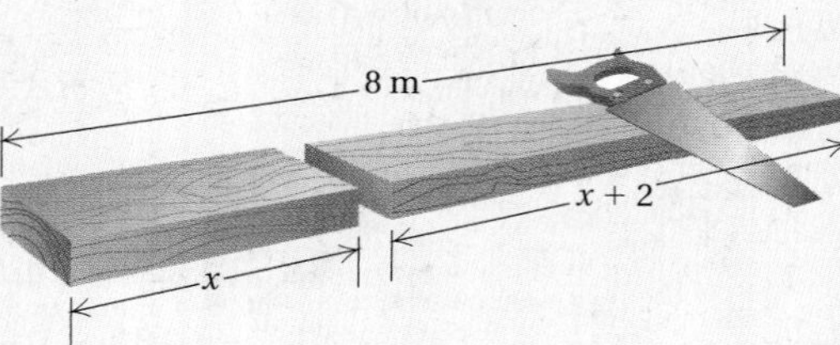

29. *Male Caloric Needs.* The number of calories K needed each day by a moderately active man who weighs w kilograms, is h centimeters tall, and is a years old can be estimated by the formula

$$K = 19.18w + 7h - 9.52a + 92.4.$$

a) David is moderately active, weighs 89 kg, is 180 cm tall, and is 43 yr old. What are his caloric needs?

b) Solve the formula for w.

Source: Parker, M., *She Does Math.* Mathematical Association of America, p. 96.

30. Find all numbers such that six times the number is greater than the number plus 30.

31. The width of a rectangle is 96 yd. Find all possible lengths such that the perimeter of the rectangle will be at least 540 yd.

Skill Maintenance

32. Add: $\frac{2}{3} + \left(-\frac{8}{9}\right)$.

33. Evaluate $\frac{4x}{y}$ for $x = 2$ and $y = 3$.

34. Translate to an algebraic expression: Seventy-three percent of p.

35. Simplify: $2x - 3y - 5(4x - 8y)$.

Synthesis

36. Solve $c = \frac{1}{a - d}$ for d.

37. Solve: $3|w| - 8 = 37$.

38. A movie theater had a certain number of tickets to give away. Five people got the tickets. The first got one-third of the tickets, the second got one-fourth of the tickets, and the third got one-fifth of the tickets. The fourth person got eight tickets, and there were five tickets left for the fifth person. Find the total number of tickets given away.

9

Graphs of Equations; Data Analysis

An Application

The cost y, in dollars, of mailing a FedEx Priority Overnight package weighing 1 lb or more is given by the equation

$$y = 2.085x + 15.08,$$

where x = the number of pounds (***Source***: Federal Express Corporation). Graph the equation and then use the graph to estimate the cost of mailing a $6\frac{1}{2}$-lb package.

This problem appears as Example 8 in Section 9.2.

The Mathematics

The graph is shown below. It appears that the cost of mailing a $6\frac{1}{2}$-lb package is about \$29.

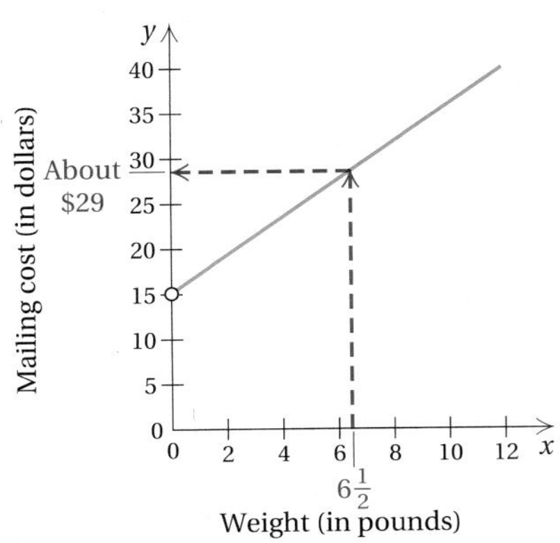

For more information, visit us at www.mathmax.com

Introduction

We now begin a study of graphs. First, we examine graphs as they commonly appear in newspapers or magazines and develop some terminology. Following that, we study graphs of certain equations. Finally, we use graphs with applications and data analysis.

Pretest: Chapter 9

Graph on a plane.

1. $y = -x$

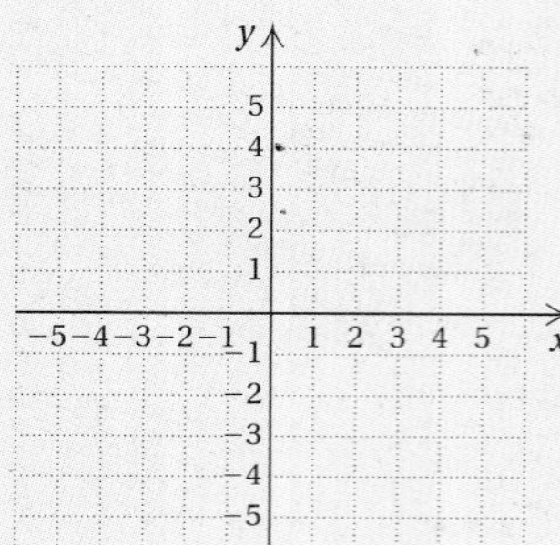

2. $x = -4$

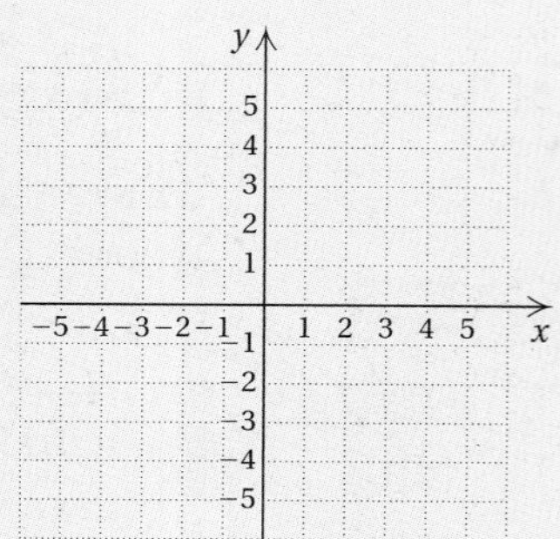

3. $4x - 5y = 20$

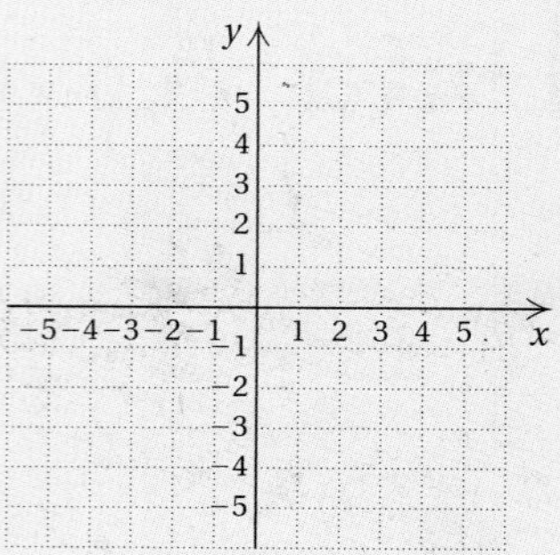

4. $y = \frac{2}{3}x - 1$

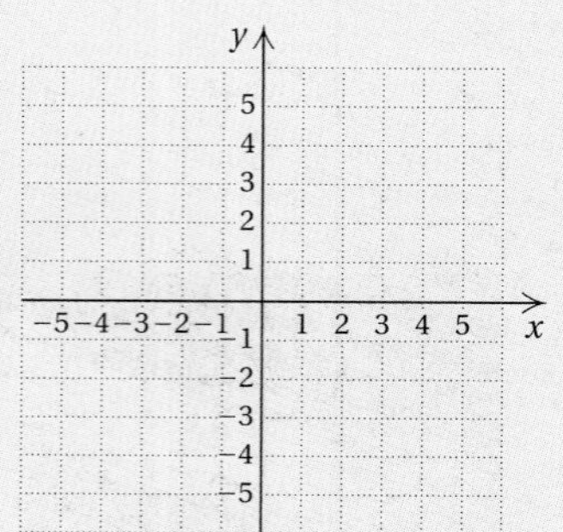

5. In which quadrant is the point $(-4, -1)$ located?

6. Determine whether the ordered pair $(-4, -1)$ is a solution of $4x - 5y = 20$.

7. Find the intercepts of the graph of $4x - 5y = 20$.

8. Find the y-intercept of $y = 3x - 8$.

9. *Price of Printing.* The price P, in cents, of a photocopied and bound lab manual is given by

$$P = \frac{7}{2}n + 20,$$

where n = the number of pages in the manual. Graph the equation and then use the graph to estimate the price of an 85-page manual.

10. *Blood Alcohol Levels of Drivers in Fatal Accidents.* Find the mean, the median, and the mode of this set of data:

0.18, 0.17, 0.21, 0.16, 0.18.

11. *Height of Girls.* Use extrapolation to estimate the missing data.

Age	Height (in centimeters)
2	95.7
4	103.2
6	115.9
8	128.0
10	138.6
12	151.9
14	159.6
16	162.2
18	?

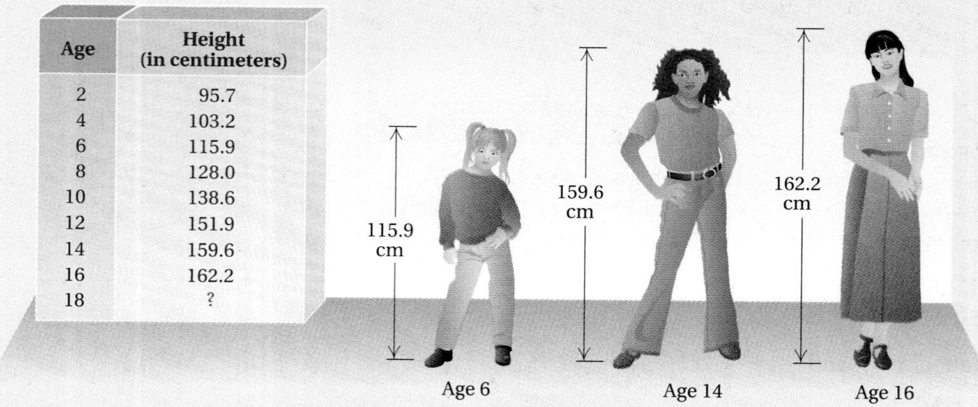

Source: Kempe, C. Henry, et al (eds.), *Current Pediatric Diagnosis & Treatment 1987.* Norwalk, CT: Appleton & Lange, 1987

Objectives for Retesting

The objectives to be retested in addition to the material in this chapter are as follows.

[3.1e] Round numbers to a specified decimal place.
[7.2c] Convert from fractional notation to decimal notation for a rational number.
[7.2e] Find the absolute value of a real number.
[8.5a] Solve applied problems involving percent.

9.1 Graphs and Applications

Often data are available regarding an application in mathematics that we are reviewing. We can use graphs to show the data and extract information about the data that can lead to making analyses and predictions.

Today's print and electronic media make extensive use of graphs. This is due in part to the ease with which some graphs can be prepared by computer and in part to the large quantity of information that a graph can display. We first consider applications with circle, bar, and line graphs.

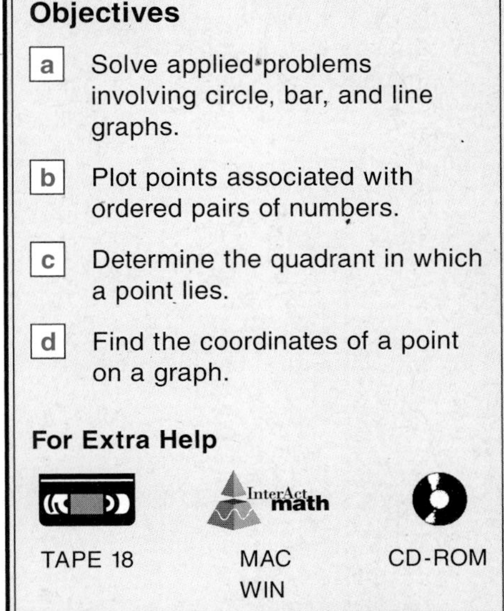

Objectives

a Solve applied problems involving circle, bar, and line graphs.

b Plot points associated with ordered pairs of numbers.

c Determine the quadrant in which a point lies.

d Find the coordinates of a point on a graph.

For Extra Help

TAPE 18 | MAC WIN | CD-ROM

a Applications with Graphs

Circle Graphs

Circle graphs and *pie graphs,* or *charts,* are often used to show what percent of a whole each particular item in a group represents.

Example 1 *U.S. Soft-Drink Retail Sales.* The following circle graph shows the percentages of sales of various soft drinks in the United States in a recent year.

U.S. Soft Drink Retail Sales

Other 11%
Coca-Cola 43%
Dr. Pepper/7Up 15%
Pepsi 31%

Source: Pepsico, Inc.

Total soft-drink sales in the United States that year reached \$54 billion. What were the sales of Pepsi?

1. **Familiarize.** The graph shows that 31% of the soft-drink sales were of Pepsi. We let

   $y =$ the amount spent on Pepsi.

2. **Translate.** We reword and translate the problem as follows.

   What is 31% of \$54? **Rewording**

   $y = 31\% \cdot 54$ **Translating**

3. **Solve.** We solve the equation by carrying out the computation on the right:

   $y = 31\% \cdot 54 = 0.31 \cdot 54 = \$16.74.$

4. **Check.** We leave the check to the student.
5. **State.** Pepsi accounted for \$16.74 billion of the soft-drink sales that particular year.

Do Exercise 1.

1. Referring to Example 1, determine the soft-drink sales of Coca-Cola.

Answer on page A-22

2. *Tornado Touchdowns.* Referring to Example 2, determine the following.

a) During which interval did the smallest number of touchdowns occur?

b) During which intervals was the number of touchdowns less than 60?

Answers on page A-22

Bar Graphs

Bar graphs are convenient for showing comparisons. In every bar graph, certain categories are paired with certain numbers. Example 2 pairs intervals of time with the total number of reported cases of tornado touchdowns.

Example 2 *Tornado Touchdowns.* The following bar graph shows the total number of tornado touchdowns by time of day in Indiana from 1950–1994.

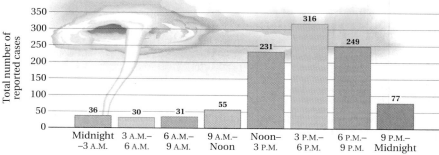

Tornado Touchdowns in Indiana by Time of Day (1950–1994)

Time of day	Total number of reported cases
Midnight–3 A.M.	36
3 A.M.–6 A.M.	30
6 A.M.–9 A.M.	31
9 A.M.–Noon	55
Noon–3 P.M.	231
3 P.M.–6 P.M.	316
6 P.M.–9 P.M.	249
9 P.M.–Midnight	77

Source: National Weather Service

a) During which interval of time did the greatest number of tornado touchdowns occur?

b) During which intervals was the number of tornado touchdowns greater than 200?

We solve as follows.

a) In this bar graph, the values are written at the top of the bars. We see that 316 is the greatest number. We look at the bottom of that bar on the horizontal scale and see that the time interval of greatest occurrence is 3 P.M.–6 P.M.

b) We locate 200 on the vertical scale and move across the graph or draw a horizontal line. We note that the value on three bars exceeds 200. Then we look down at the horizontal scale and see that the corresponding time intervals are noon–3 P.M., 3 P.M.–6 P.M., and 6 P.M.–9 P.M.

Do Exercise 2.

Line Graphs

Line graphs are often used to show change over time. Certain points are plotted to represent given information. When segments are drawn to connect the points, a line graph is formed.

Sometimes it is impractical to begin the listing of horizontal or vertical values with zero. When this happens, as in Example 3, the symbol ⌇ is used to indicate a break in the list of values.

Example 3 *Exercise and Pulse Rate.* The following line graph shows the relationship between a person's resting pulse rate and months of regular exercise.

Exercise to Improve Your Heart Rate

Beats per minute

100
90
80
70
60
50
0

1 2 3 4 5 6 7 8 9 10

Months of regular exercise

Source: Hughes, Martin, *Body Clock*. New York: Facts on File, Inc., p. 60

a) How many months of regular exercise are required to lower the pulse rate to its lowest point?

b) How many months of regular exercise are needed to achieve a pulse rate of 65 beats per minute?

We solve as follows.

a) The lowest point on the graph occurs above the number 6. Thus after 6 months of regular exercise, the pulse rate has been lowered as much as possible.

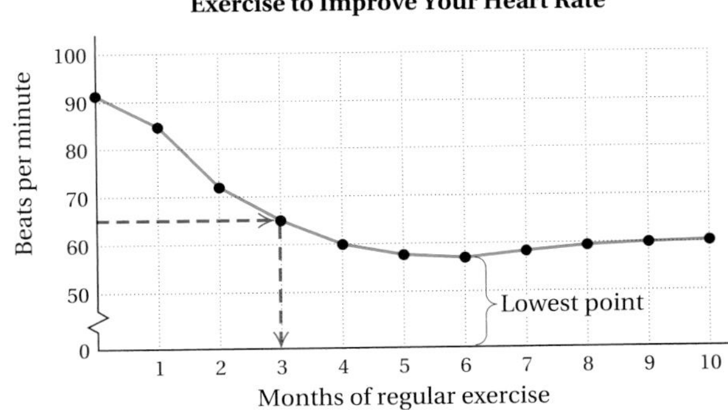

b) We locate 65 on the vertical scale and then move right until we reach the line. At that point, we move down to the horizontal scale and read the information we are seeking. The pulse rate is 65 beats per minute after 3 months of regular exercise.

Do Exercise 3.

3. *Exercise and Pulse Rate.* Referring to Example 3, determine the following.

a) About how many months of regular exercise are needed to achieve a pulse rate of about 72 beats per minute?

b) What pulse rate has been achieved after 10 months of exercise?

Answers on page A-22

Plot these points on the graph below.

4. (4, 5) **5.** (5, 4)

6. (−2, 5) **7.** (−3, −4)

8. (5, −3) **9.** (−2, −1)

10. (0, −3) **11.** (2, 0)

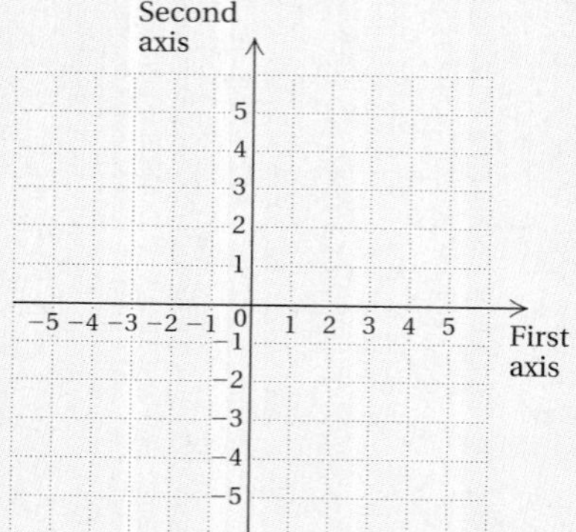

Answers on page A-22

b Plotting Ordered Pairs

The line graph in Example 3 is formed from a collection of points. Each point pairs a number of months of exercise with a pulse rate.

In Chapter 8, we graphed numbers and inequalities in one variable on a line. To enable us to graph an equation that contains two variables, we now learn to graph number pairs on a plane.

On a number line, each point is the graph of a number. On a plane, each point is the graph of a number pair. We use two perpendicular number lines called **axes**. They cross at a point called the **origin**. The arrows show the positive directions.

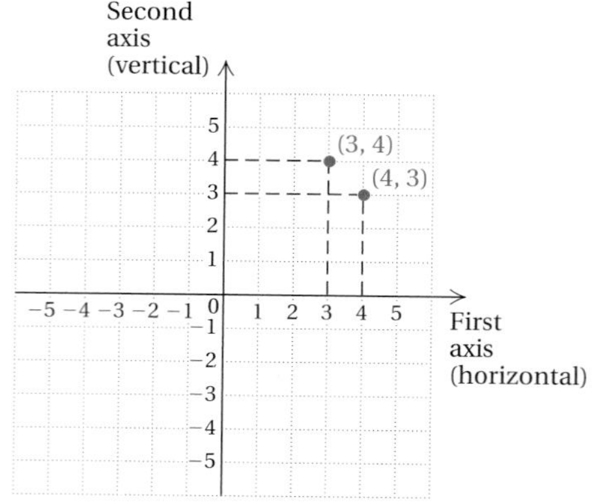

Consider the ordered pair (3, 4). The numbers in an ordered pair are called **coordinates**. In (3, 4), the **first coordinate** is 3 and the **second coordinate** is 4. To plot (3, 4), we start at the origin and move horizontally to the 3. Then we move up vertically 4 units and make a "dot."

The point (4, 3) is also plotted. Note that (3, 4) and (4, 3) give different points. The order of the numbers in the pair is indeed important. They are called **ordered pairs** because it makes a difference which number comes first. The coordinates of the origin are (0, 0) even though it is usually labeled either with the number 0 or not at all.

Example 4 Plot the point (−5, 2).

The first number, −5, is negative. Starting at the origin, we move −5 units in the horizontal direction (5 units to the left). The second number, 2, is positive. We move 2 units in the vertical direction (up).

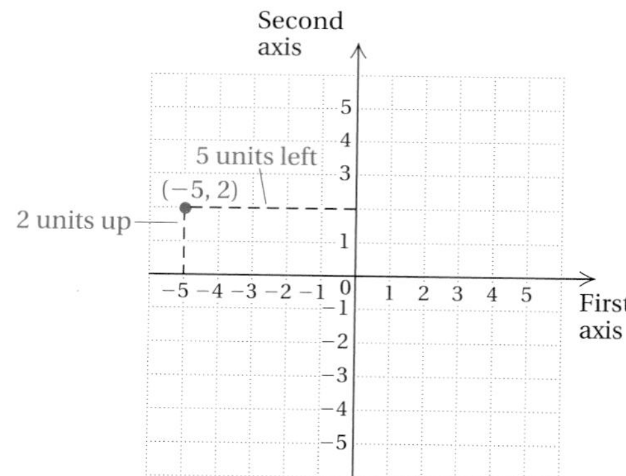

Do Exercises 4–11.

c Quadrants

This figure shows some points and their coordinates. In region I (the *first quadrant*), both coordinates of any point are positive. In region II (the *second quadrant*), the first coordinate is negative and the second positive. In region III (the *third quadrant*), both coordinates are negative. In region IV (the *fourth quadrant*), the first coordinate is positive and the second is negative.

Example 5 In which quadrant, if any, are the points (−4, 5), (5, −5), (2, 4), (−2, −5), and (−5, 0) located?

The point (−4, 5) is in the second quadrant. The point (5, −5) is in the fourth quadrant. The point (2, 4) is in the first quadrant. The point (−2, −5) is in the third quadrant. The point (−5, 0) is on an axis and is *not* in any quadrant.

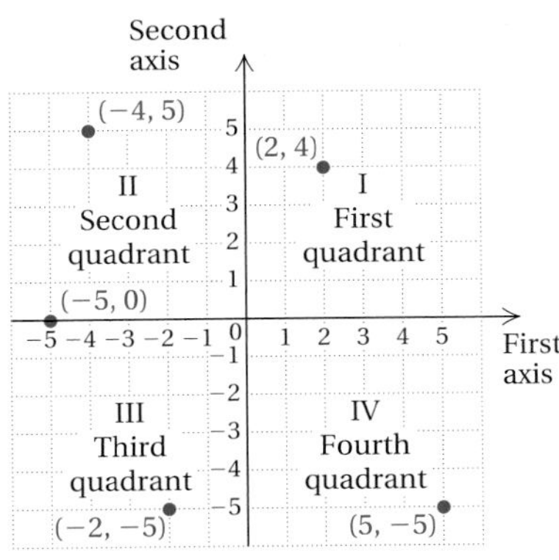

Do Exercises 12–18.

12. What can you say about the coordinates of a point in the third quadrant?

13. What can you say about the coordinates of a point in the fourth quadrant?

In which quadrant, if any, is the point located?

14. (5, 3)

15. (−6, −4)

16. (10, −14)

17. (−13, 9)

18. (0, −3)

d Finding Coordinates

To find the coordinates of a point, we see how far to the right or left of zero it is located and how far up or down.

Example 6 Find the coordinates of points *A*, *B*, *C*, *D*, *E*, *F*, and *G*.

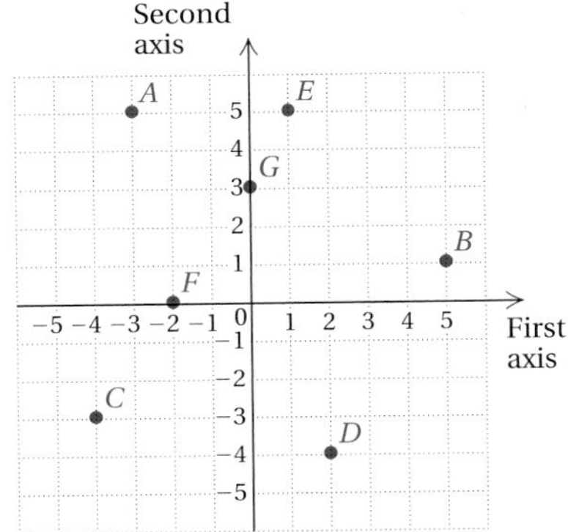

Point *A* is 3 units to the left (horizontal direction) and 5 units up (vertical direction). Its coordinates are (−3, 5). The coordinates of the other points are as follows:

B: (5, 1); *C*: (−4, −3); *D*: (2, −4);
E: (1, 5); *F*: (−2, 0); *G*: (0, 3).

Do Exercise 19.

19. Find the coordinates of points *A*, *B*, *C*, *D*, *E*, *F*, and *G* on the graph below.

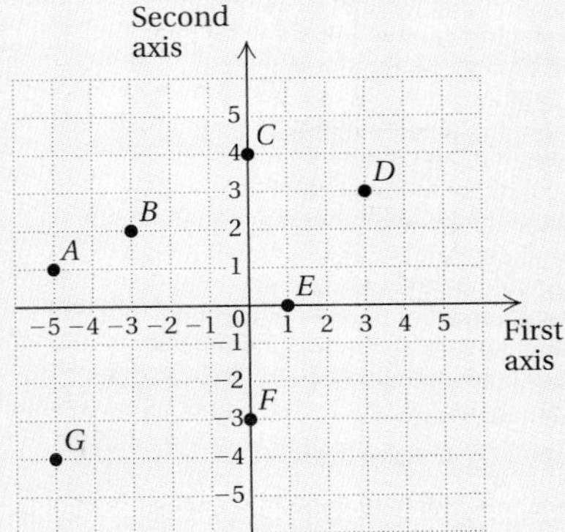

Answers on page A-23

Improving Your Math Study Skills

A Checklist of Your Study Skills

You are now about halfway through this textbook as well as the course. How are you doing? If you are struggling, are you making full use of the study skills we have suggested to you in these inserts? To decide, go through the following checklist, marking the questions "Yes" or "No." This list is a review of many of the study skill suggestions considered so far in the text.

Study Skill Questions	Yes	No
1. Are you stopping to work the margin exercises when directed to do so?		
2. Are you doing your homework as soon as possible after class?		
3. Are you doing your homework at a specified time and in a quiet setting?		
4. Have you found a study group in which to work?		
5. Are you consistently trying to apply the five-step problem-solving strategy?		
6. If you are going to help or tutoring sessions, have you studied enough to generate questions to ask the tutor, rather than having the tutor teach you from scratch?		
7. Are you doing lots of even-numbered exercises for which answers are not available?		
8. Are you keeping one section ahead on your syllabus?		
9. Are you using the book supplements, such as the *Student's Solutions Manual* and the *InterAct Math Tutorial Software*?		
10. When you study the book, are you marking the points you do not understand as a source for in-class questions?		
11. Are you reading and studying each step of each example?		
12. Are you noting and using the objective code symbols [a], [b], [c], and so on, that appear at the beginning of each section, throughout the section, in the exercise sets, the summary–reviews, and the tests?		

If you have 7 or more "No" answers to these questions, and are struggling in the course, you now have many suggestions for improvement as you progress to the end of the course. A consultation with your instructor is strongly advised. Good luck!

Exercise Set 9.1

a Solve.

Driving While Intoxicated (DWI). State laws have determined that a blood alcohol level of at least 0.10% or higher indicates that an individual has consumed too much alcohol to drive safely. The following bar graph shows the number of drinks that a person of a certain weight would need to consume in order to reach a blood alcohol level of 0.10%. A 12-oz beer, a 5-oz glass of wine, or a cocktail containing $1\frac{1}{2}$ oz of distilled liquor all count as one drink. Use the bar graph for Exercises 1–6.

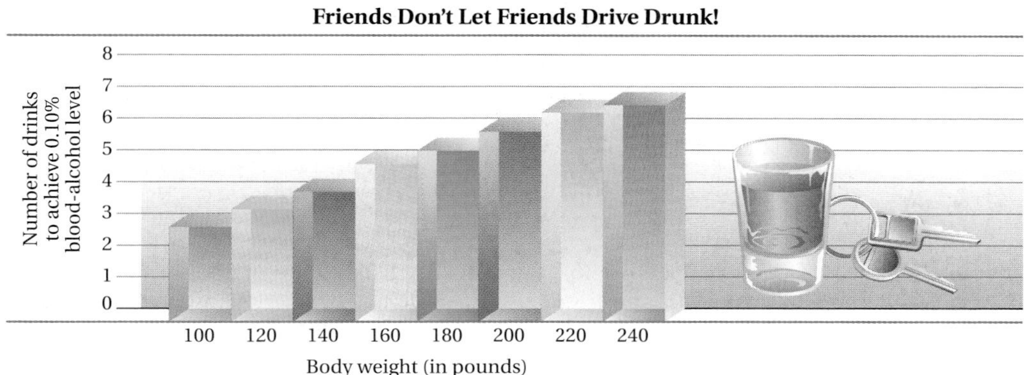

Source: *Neighborhood Digest*, 7, no. 12

1. Approximately how many drinks would a 200-lb person have consumed if he or she had a blood alcohol level of 0.10%?

2. What can be concluded about the weight of someone who can consume 4 drinks without reaching a blood alcohol level of 0.10%?

3. What can be concluded about the weight of someone who can consume 6 drinks without reaching a blood alcohol level of 0.10%?

4. Approximately how many drinks would a 160-lb person have consumed if he or she had a blood alcohol level of 0.10%?

5. What can be concluded about the weight of someone who has consumed $3\frac{1}{2}$ drinks without reaching a blood alcohol level of 0.10%?

6. What can be concluded about the weight of someone who has consumed $4\frac{1}{2}$ drinks without reaching a blood alcohol level of 0.10%?

Cost of Raising a Child. A family is in a $32,800–$55,500 income bracket. The following pie chart shows the various costs involved for a family in a $32,800–$55,500 income bracket in raising a child to the age of 18. Use the pie chart for Exercises 7–10.

Cost of Raising a Child to the Age of 18

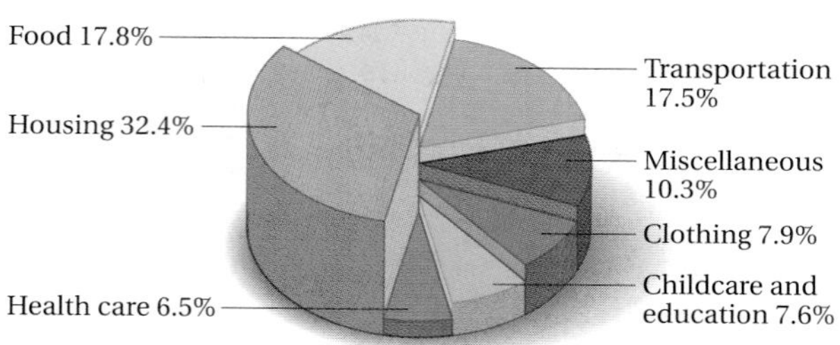

Source: U.S. Department of Agriculture, Food, Nutrition, and Consumer Service

7. What percent of the total expense is for housing?

8. What percent of the total expense is for health care?

9. It costs a total of about $136,320 to raise a child to the age of 18. How much of this cost is for child care and education?

10. It costs a total of about $136,320 to raise a child to the age of 18. How much of this cost is for transportation?

MADD (Mothers Against Drunk Driving). Despite efforts by groups such as MADD, the number of alcohol-related deaths is rising after many years of decline. The data in the following graph show the number of deaths from 1989 to 1995. Use this graph for Exercises 11–16.

Number of Alcohol-Related Traffic Deaths

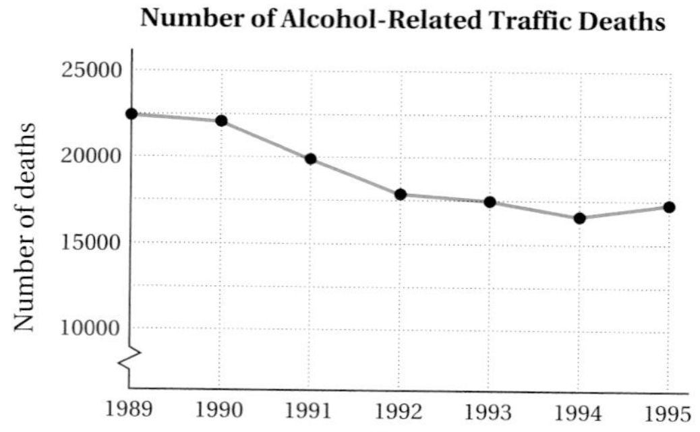

Source: National Highway Traffic Safety Administration

11. About how many alcohol-related deaths occurred in 1991?

12. About how many alcohol-related deaths occurred in 1995?

13. In what year did the lowest number of deaths occur?

14. In what years did fewer than 18,000 deaths occur?

15. By how much did the number of alcohol-related deaths increase from 1994 to 1995?

16. By how much did the number of alcohol-related deaths decrease from 1989 to 1994?

b

17. Plot these points.

(2, 5) (−1, 3) (3, −2) (−2, −4)
(0, 4) (0, −5) (5, 0) (−5, 0)

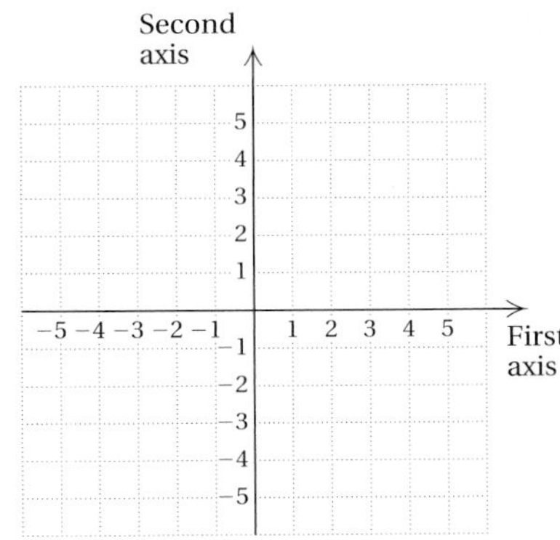

18. Plot these points.

(4, 4) (−2, 4) (5, −3) (−5, −5)
(0, 4) (0, −4) (3, 0) (−4, 0)

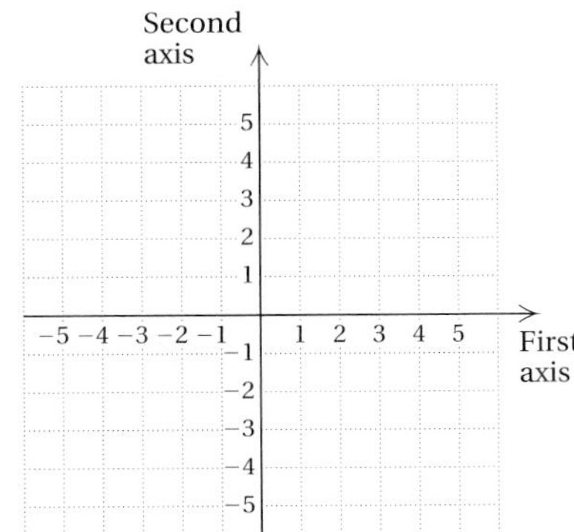

c In which quadrant is the point located?

19. (−5, 3)

20. (1, −12)

21. (100, −1)

22. (−2.5, 35.6)

23. (−6, −29)

24. (3.6, 105.9)

25. (3.8, 9.2)

26. (−895, −492)

27. $\left(-\frac{1}{3}, \frac{15}{7}\right)$

28. $\left(-\frac{2}{3}, -\frac{9}{8}\right)$

29. $\left(12\frac{7}{8}, -1\frac{1}{2}\right)$

30. $\left(23\frac{5}{8}, 81.74\right)$

31. In quadrant III, first coordinates are always __________ and second coordinates are always __________.

32. In quadrant II, __________ coordinates are always positive and __________ coordinates are always negative.

33. In quadrant IV, __________ coordinates are always negative and __________ coordinates are always positive.

34. In quadrant I, first coordinates are always __________ and second coordinates are always __________.

In Exercises 35–38, tell in which quadrant(s) the point can be located.

35. The first coordinate is positive.

36. The second coordinate is negative.

37. The first and second coordinates are equal.

38. The first coordinate is the additive inverse of the second coordinate.

d

39. Find the coordinates of points A, B, C, D, and E.

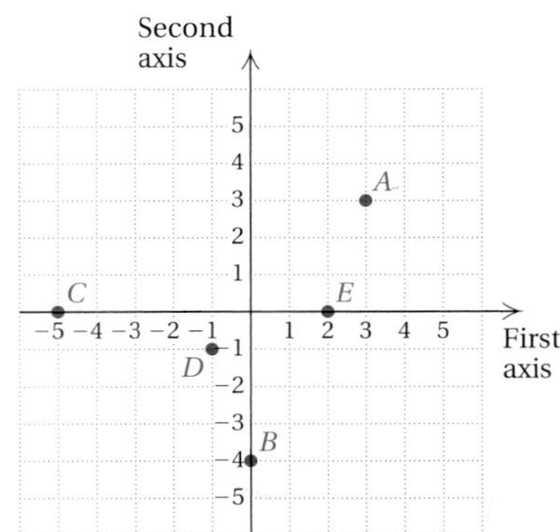

40. Find the coordinates of points A, B, C, D, and E.

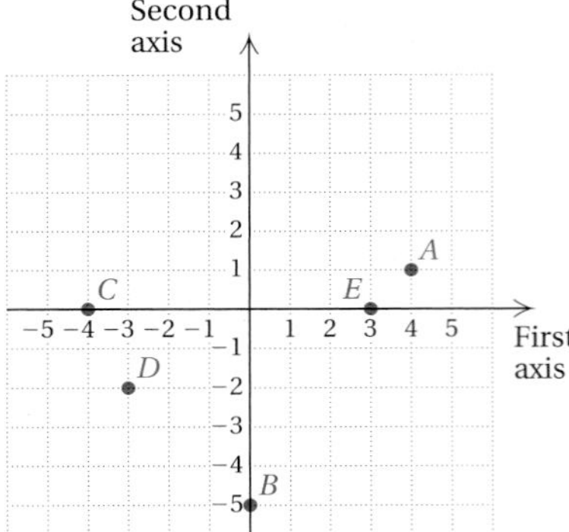

Skill Maintenance

Find the absolute value. [7.2e]

41. $|-12|$

42. $|4.89|$

43. $|0|$

44. $\left|-\frac{4}{5}\right|$

Solve. [8.5a]

45. *Baseball Salaries.* In 1997, the total amount spent on the salaries of major-league baseball players soared to $1.06 billion. This was a 17.7% increase over the total amount spent in 1996. (***Source:*** Major League Baseball) How much was spent in 1996?

46. Erin left a 15% tip for a meal. The total cost of the meal, including the tip, was $21.16. What was the cost of the meal before the tip was added?

Synthesis

47. ◈ The sales of snow skis are highest in the winter months and lowest in the summer months. Sketch a line graph that might show the sales of a ski store and explain how an owner might use such a graph in decision making.

48. ◈ The graph in Example 3 tends to flatten out. Explain why the graph does not continue to decrease downward.

49. The points $(-1, 1)$, $(4, 1)$, and $(4, -5)$ are three vertices of a rectangle. Find the coordinates of the fourth vertex.

50. Three parallelograms share the vertices $(-2, -3)$, $(-1, 2)$, and $(4, -3)$. Find the fourth vertex of each parallelogram.

51. Graph eight points such that the sum of the coordinates in each pair is 6.

52. Graph eight points such that the first coordinate minus the second coordinate is 1.

53. Find the perimeter of a rectangle whose vertices have coordinates $(5, 3)$, $(5, -2)$, $(-3, -2)$, and $(-3, 3)$.

54. Find the area of a triangle whose vertices have coordinates $(0, 9)$, $(0, -4)$, and $(5, -4)$.

Practice finding and plotting ordered pairs by playing a variation of the game Battleship.

9.2 Graphing Linear Equations

We have seen how circle, bar, and line graphs can be used to represent the data in an application. Now we begin to learn how graphs can be used to represent solutions of equations.

a Solutions of Equations

When an equation contains two variables, the solutions of the equation are *ordered pairs* in which each number in the pair corresponds to a letter in the equation. Unless stated otherwise, to determine whether a pair is a solution, we use the first number in each pair to replace the variable that occurs first alphabetically.

Example 1 Determine whether each of the following pairs is a solution of $4q - 3p = 22$: (2, 7) and (−1, 6).

For (2, 7), we substitute 2 for p and 7 for q (using alphabetical order of variables):

$$\begin{array}{r|l} 4q - 3p & 22 \\ \hline 4 \cdot 7 - 3 \cdot 2 \ ? & 22 \\ 28 - 6 & \\ 22 & \quad \text{TRUE} \end{array}$$

Thus, (2, 7) is a solution of the equation.

For (−1, 6), we substitute −1 for p and 6 for q:

$$\begin{array}{r|l} 4q - 3p & 22 \\ \hline 4 \cdot 6 - 3 \cdot -1 \ ? & 22 \\ 24 + 3 & \\ 27 & \quad \text{FALSE} \end{array}$$

Thus, (−1, 6) is *not* a solution of the equation.

Do Exercises 1 and 2.

Example 2 Show that the pairs (3, 7), (0, 1), and (−3, −5) are solutions of $y = 2x + 1$. Then graph the three points and use the graph to determine another pair that is a solution.

To show that a pair is a solution, we substitute, replacing x with the first coordinate and y with the second coordinate of each pair:

$$\begin{array}{r|l} y & 2x + 1 \\ \hline 7 \ ? & 2 \cdot 3 + 1 \\ & 6 + 1 \\ & 7 \quad \text{TRUE} \end{array} \qquad \begin{array}{r|l} y & 2x + 1 \\ \hline 1 \ ? & 2 \cdot 0 + 1 \\ & 0 + 1 \\ & 1 \quad \text{TRUE} \end{array}$$

$$\begin{array}{r|l} y & 2x + 1 \\ \hline -5 \ ? & 2(-3) + 1 \\ & -6 + 1 \\ & -5 \quad \text{TRUE} \end{array}$$

In each of the three cases, the substitution results in a true equation. Thus the pairs are all solutions.

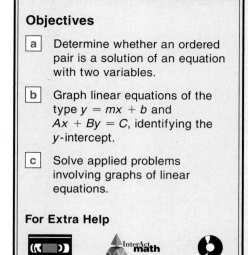

Objectives

a Determine whether an ordered pair is a solution of an equation with two variables.

b Graph linear equations of the type $y = mx + b$ and $Ax + By = C$, identifying the y-intercept.

c Solve applied problems involving graphs of linear equations.

For Extra Help

TAPE 18 MAC WIN CD-ROM

1. Determine whether (2, −4) is a solution of $4q - 3p = 22$.

2. Determine whether (2, −4) is a solution of $7a + 5b = -6$.

Answers on page A-23

3. Use the graph in Example 2 to find at least two more points that are solutions of $y = 2x + 1$.

Answer on page A-23

We plot the points as shown at right. The order of the points follows the alphabetical order of the variables. That is, x comes before y, so x-values are first coordinates and y-values are second coordinates. Similarly, we also label the horizontal axis as the x-axis and the vertical axis as the y-axis.

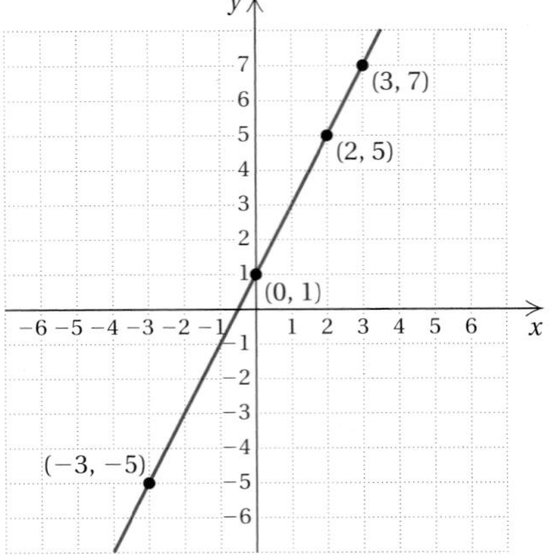

Note that the three points appear to "line up." That is, they appear to be on a straight line. Will other points that line up with these points also represent solutions of $y = 2x + 1$? To find out, we use a straightedge and lightly sketch a line passing through (3, 7), (0, 1), and (−3, −5).

The line appears to also pass through (2, 5). Let's see if this pair is a solution of $y = 2x + 1$:

$$\begin{array}{c|l} y & = 2x + 1 \\ \hline 5 & ?\ 2 \cdot 2 + 1 \\ & 4 + 1 \\ & 5 \qquad \text{TRUE} \end{array}$$

Thus, (2, 5) is a solution.

Do Exercise 3.

Example 2 leads us to suspect that any point on the line that passes through (3, 7), (0, 1), and (−3, −5) represents a solution of $y = 2x + 1$. In fact, every solution of $y = 2x + 1$ is represented by a point on that line and every point on that line represents a solution. The line is said to be the *graph* of the equation.

> The **graph** of an equation is a drawing that represents all its solutions.

b Graphs of Linear Equations

Equations like $y = 2x + 1$ and $4p - 3p = 22$ are said to be **linear** because the graph of each equation is a straight line. In general, any equation equivalent to one of the form $y = mx + b$ or $Ax + By = C$, where m, b, A, B, and C are constants (not variables) and A and B are not both 0, is linear.

> To graph a linear equation:
>
> 1. Select a value for one variable and calculate the corresponding value of the other variable. Form an ordered pair using alphabetical order as indicated by the variables.
> 2. Repeat step (1) to obtain at least two other ordered pairs. Two points are essential to determine a straight line. A third point serves as a check.
> 3. Plot the ordered pairs and draw a straight line passing through the points.

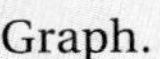

In general, calculating three (or more) ordered pairs is not difficult for equations of the form $y = mx + b$. We simply substitute values for x and calculate the corresponding values for y.

Example 3 Graph: $y = 2x$.

First, we find some ordered pairs that are solutions. We choose *any* number for x and then determine y by substitution. Since $y = 2x$, we find y by doubling x. Suppose that we choose 3 for x. Then

$$y = 2x = 2 \cdot 3 = 6.$$

We get a solution: the ordered pair (3, 6).

Suppose that we choose 0 for x. Then

$$y = 2x = 2 \cdot 0 = 0.$$

We get another solution: the ordered pair (0, 0).

For a third point, we make a negative choice for x. We now have enough points to plot the line, but if we wish, we can compute more. If a number takes us off the graph paper, we either do not use it or we use larger paper or rescale the axes. Continuing in this manner, we create a table like the one shown below.

Now we plot these points. We draw the line, or graph, with a straightedge and label it $y = 2x$.

x	y $y = 2x$	(x, y)
3	6	(3, 6)
1	2	(1, 2)
0	0	(0, 0)
−2	−4	(−2, −4)
−3	−6	(−3, −6)

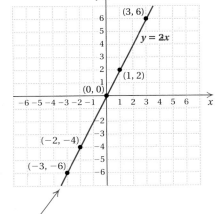

(1) Choose x.
(2) Compute y.
(3) Form the pair (x, y).
(4) Plot the points.

Do Exercises 4 and 5.

Graph.

4. $y = -2x$

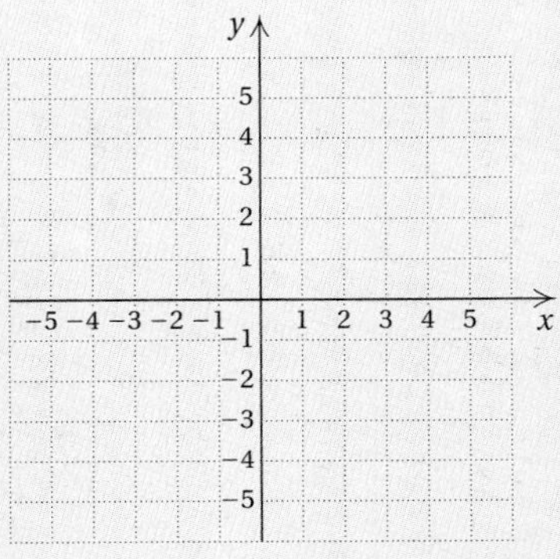

5. $y = \frac{1}{2}x$

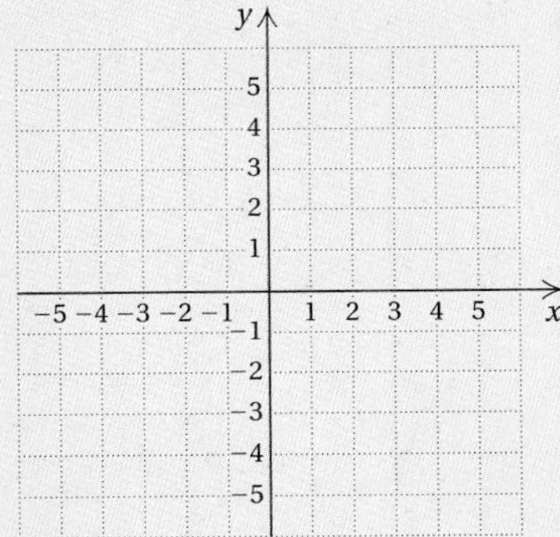

Answers on page A-23

Graph.

6. $y = 2x + 3$

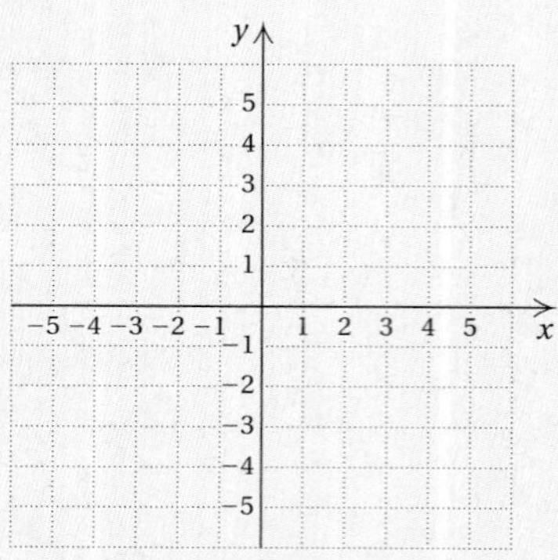

Example 4 Graph: $y = -3x + 1$.

We select a value for x, compute y, and form an ordered pair. Then we repeat the process for other choices of x.

If $x = 2$, then $y = -3 \cdot 2 + 1 = -5$, and $(2, -5)$ is a solution.
If $x = 0$, then $y = -3 \cdot 0 + 1 = 1$, and $(0, 1)$ is a solution.
If $x = -1$, then $y = -3 \cdot (-1) + 1 = 4$, and $(-1, 4)$ is a solution.

Results are often listed in a table, as shown below. The points corresponding to each pair are then plotted.

x	y $y = -3x + 1$	(x, y)
2	-5	$(2, -5)$
0	1	$(0, 1)$
-1	4	$(-1, 4)$

(1) Choose x.
(2) Compute y.
(3) Form the pair (x, y).
(4) Plot the points.

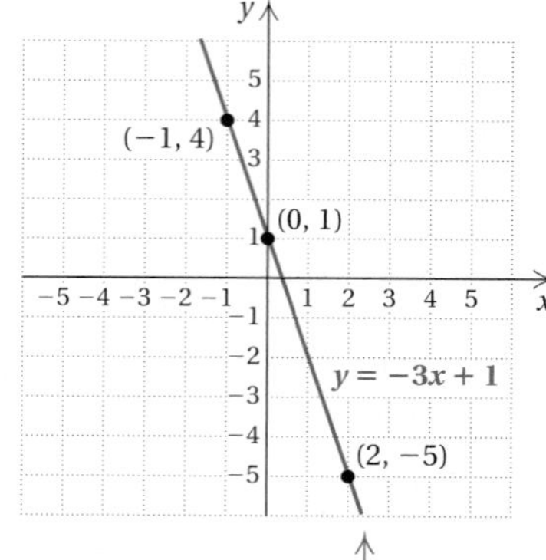

Note that all three points line up. If they did not, we would know that we had made a mistake. When only two points are plotted, a mistake is harder to detect. We use a ruler or other straightedge to draw a line through the points. Every point on the line represents a solution of $y = -3x + 1$.

7. $y = -\frac{1}{2}x - 3$

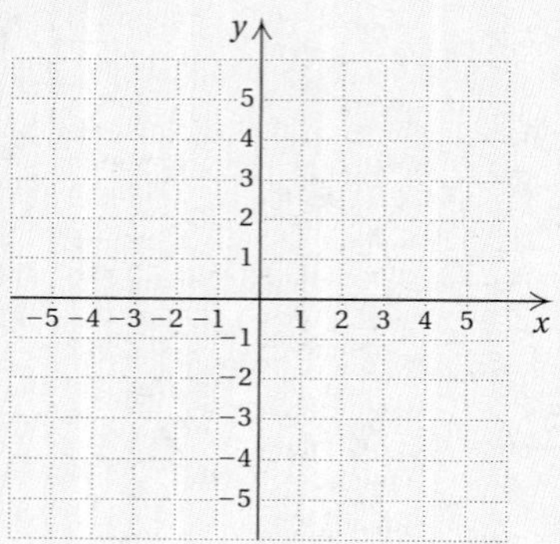

Do Exercises 6 and 7.

In Example 3, we saw that $(0, 0)$ is a solution of $y = 2x$. It is also the point at which the graph crosses the y-axis. Similarly, in Example 4, we saw that $(0, 1)$ is a solution of $y = -3x + 1$. It is also the point at which the graph crosses the y-axis. A generalization can be made: If x is replaced with 0 in the equation $y = mx + b$, then the corresponding y-value is $m \cdot 0 + b$, or b. Thus any equation of the form $y = mx + b$ has a graph that passes through the point $(0, b)$. Since $(0, b)$ is the point at which the graph crosses the y-axis, it is called the **y-intercept**. Sometimes, for convenience, we simply refer to b as the y-intercept.

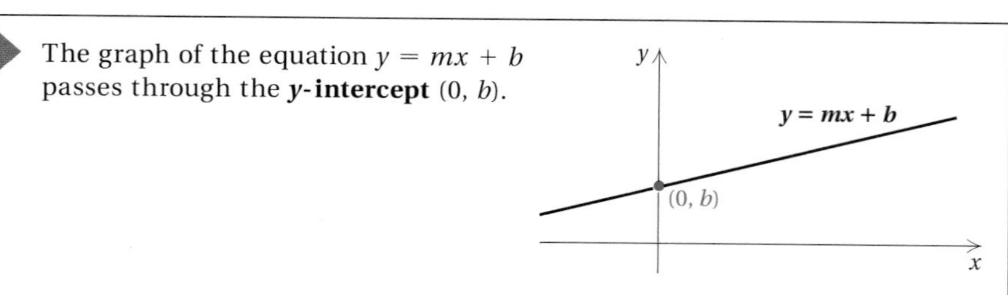

The graph of the equation $y = mx + b$ passes through the **y-intercept** $(0, b)$.

Answers on page A-23

Example 5 Graph $y = \frac{2}{5}x + 4$ and identify the y-intercept.

We select a value for x, compute y, and form an ordered pair. Then we repeat the process for other choices of x. In this case, using multiples of 5 avoids fractions.

If $x = 0$, then $y = \frac{2}{5} \cdot 0 + 4 = 4$, and $(0, 4)$ is a solution.

If $x = 5$, then $y = \frac{2}{5} \cdot 5 + 4 = 6$, and $(5, 6)$ is a solution.

If $x = -5$, then $y = \frac{2}{5} \cdot (-5) + 4 = 2$, and $(-5, 2)$ is a solution.

The following table lists these solutions. Next, we plot the points and see that they form a line. Finally, we draw and label the line.

x	y $y = \frac{2}{5}x + 4$	(x, y)
0	4	(0, 4)
5	6	(5, 6)
−5	2	(−5, 2)

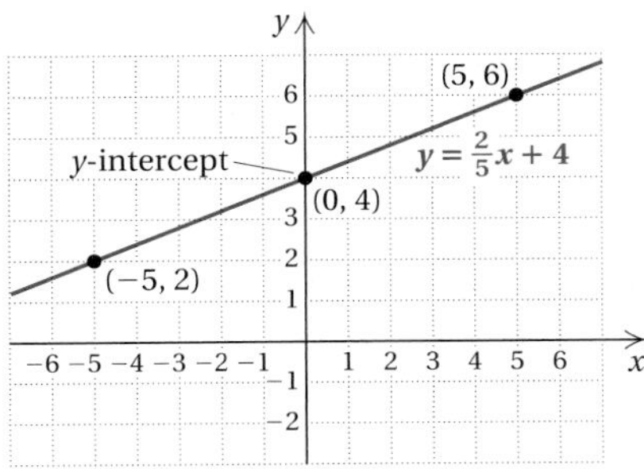

We see that $(0, 4)$ is a solution of $y = \frac{2}{5}x + 4$. It is the y-intercept. Because the equation is in the form $y = mx + b$, we can read the y-intercept directly from the equation as follows:

$y = \frac{2}{5}x + 4$ $(0, 4)$ is the y-intercept.

Do Exercises 8 and 9.

Calculating ordered pairs is generally easiest when y is isolated on one side of the equation, as in $y = mx + b$. To graph an equation in which y is not isolated, we can use the addition and multiplication principles to solve for y (see Sections 8.3 and 8.6).

Example 6 Graph $3y + 5x = 0$ and identify the y-intercept.

To find an equivalent equation in the form $y = mx + b$, we solve for y:

$$3y + 5x = 0$$

$$3y + 5x - 5x = 0 - 5x \quad \text{Subtracting } 5x$$

$$3y = -5x \quad \text{Collecting like terms}$$

$$\frac{3y}{3} = \frac{-5x}{3} \quad \text{Dividing by 3}$$

$$y = -\frac{5}{3}x.$$

Graph the equation and identify the y-intercept.

8. $y = \frac{3}{5}x + 2$

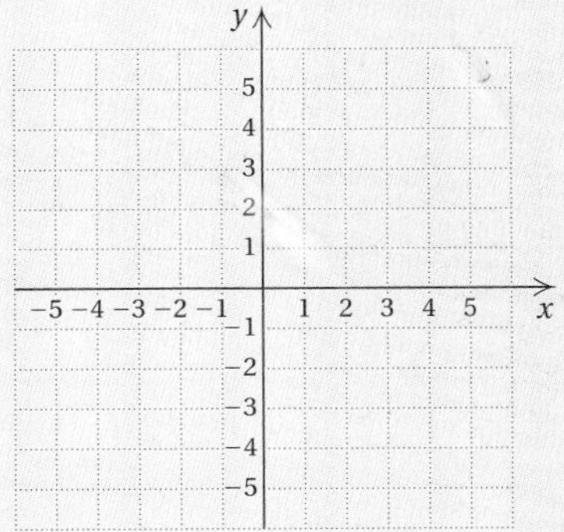

9. $y = -\frac{3}{5}x - 1$

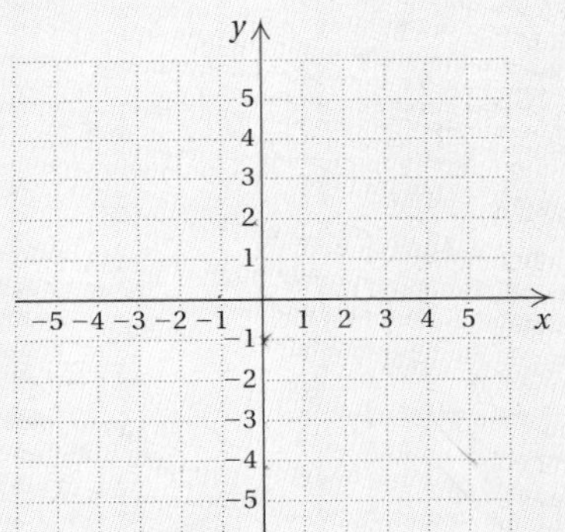

Answers on page A-23

Graph the equation and identify the y-intercept.

10. $5y + 4x = 0$

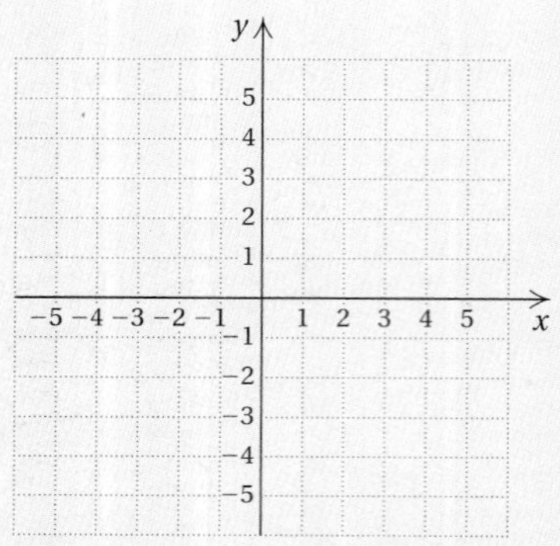

11. $4y = 3x$

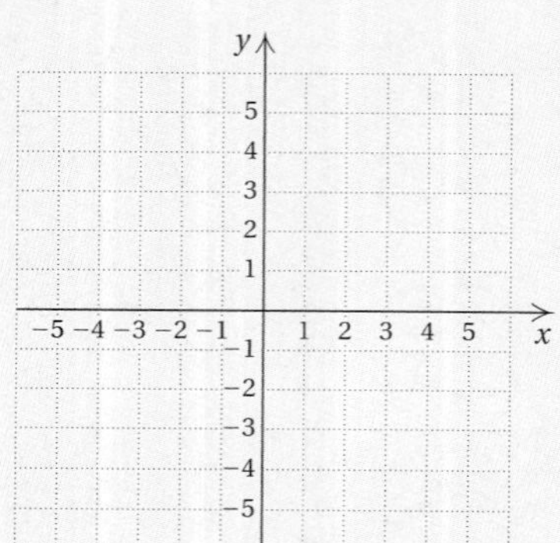

Answers on page A-23

Because all the equations above are equivalent, we can use $y = -\frac{5}{3}x$ to draw the graph of $3y + 5x = 0$. To graph $y = -\frac{5}{3}x$, we select x-values and compute y-values. In this case, if we select multiples of 3, we can avoid fractions.

$$\text{If } x = 0, \quad \text{then } y = -\frac{5}{3} \cdot 0 = 0.$$

$$\text{If } x = 3, \quad \text{then } y = -\frac{5}{3} \cdot 3 = -5.$$

$$\text{If } x = -3, \quad \text{then } y = -\frac{5}{3} \cdot (-3) = 5.$$

We list these solutions in a table. Next, we plot the points and see that they form a line. Finally, we draw and label the line. The y-intercept is (0, 0).

x	y
0	0
3	−5
−3	5

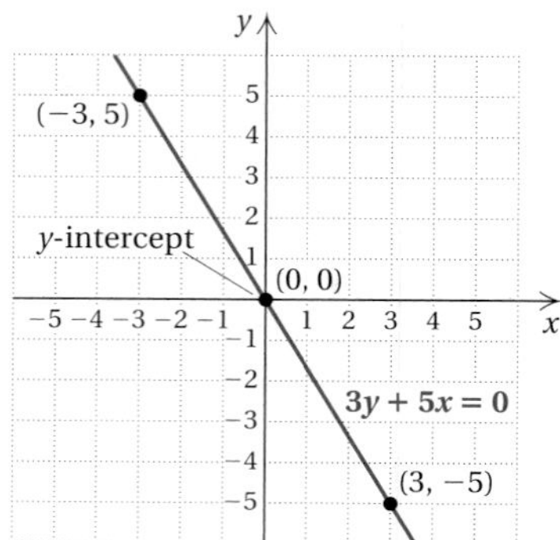

Do Exercises 10 and 11.

Example 7 Graph $4y + 3x = -8$ and identify the y-intercept.

To find an equivalent equation in the form $y = mx + b$, we solve for y:

$$\begin{aligned}
4y + 3x &= -8 \\
4y + 3x - 3x &= -8 - 3x && \text{Subtracting } 3x \\
4y &= -3x - 8 && \text{Simplifying} \\
\frac{1}{4} \cdot 4y &= \frac{1}{4} \cdot (-3x - 8) && \text{Multiplying by } \tfrac{1}{4} \text{ or dividing by 4} \\
y &= \frac{1}{4} \cdot (-3x) - \frac{1}{4} \cdot 8 && \text{Using the distributive law} \\
&= -\frac{3}{4}x - 2. && \text{Simplifying}
\end{aligned}$$

Thus, $4y + 3x = -8$ is equivalent to $y = -\frac{3}{4}x - 2$. The y-intercept is $(0, -2)$. We find two other pairs using multiples of 4 for x to avoid fractions. We then complete and label the graph as shown.

x	y
0	−2
4	−5
−4	1

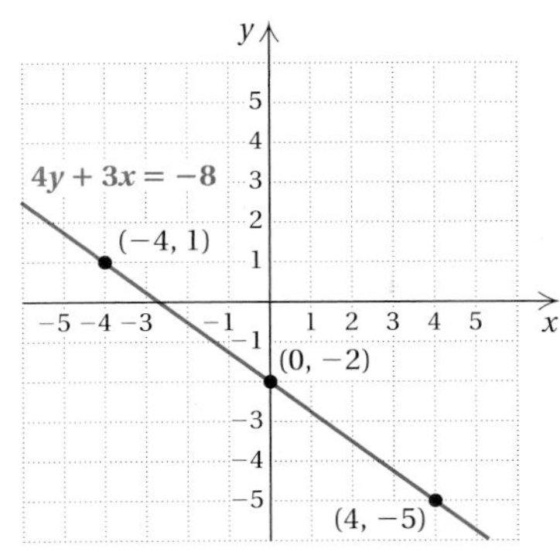

Do Exercises 12 and 13.

Graph the equation and identify the y-intercept.

12. $5y - 3x = -10$

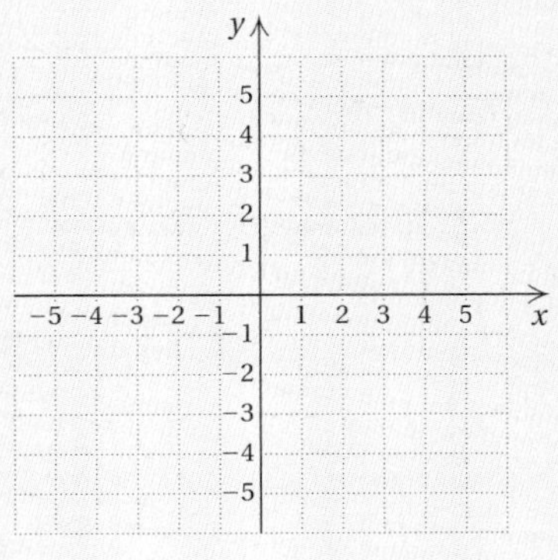

13. $5y + 3x = 20$

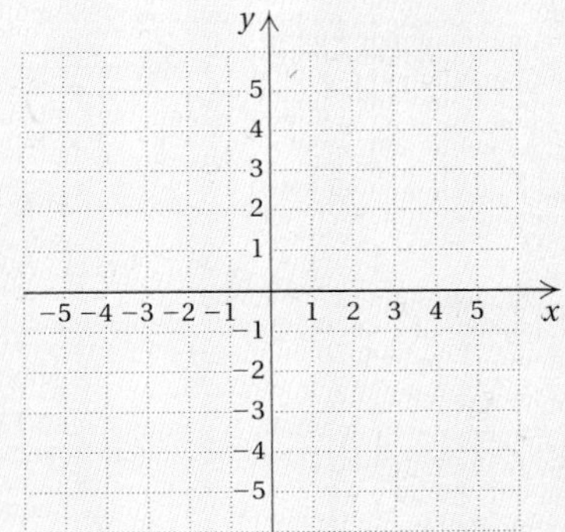

Answers on page A-23

c Applications of Linear Equations

Mathematical concepts become more understandable through visualization. Throughout this text, you will occasionally see the heading AG Algebraic–Graphical Connection, as in Example 8, which follows. In this feature, the algebraic approach is enhanced and expanded with a graphical connection. Relating a solution of an equation to a graph can often give added meaning to the algebraic solution.

Example 8 *FedEx Mailing Costs.* The cost y, in dollars, of mailing a FedEx Priority Overnight package weighing 1 lb or more is given by the equation

$$y = 2.085x + 15.08,$$

where x = the number of pounds (***Source***: Federal Express Corporation).

a) Find the cost of mailing packages weighing 2 lb, 5 lb, and 7 lb.

b) Graph the equation and then use the graph to estimate the cost of mailing a $6\frac{1}{2}$-lb package.

c) If a package costs \$177.71 to mail, how much does it weigh?

14. *Value of a Color Copier.* The value of Dupliographic's color copier is given by

$$v = -0.68t + 3.4,$$

where v = the value, in thousands of dollars, t years from the date of purchase.

a) Find the value after 1 yr, 2 yr, 4 yr, and 5 yr.

b) Graph the equation and use the graph to estimate the value of the copier after $2\frac{1}{2}$ yr.

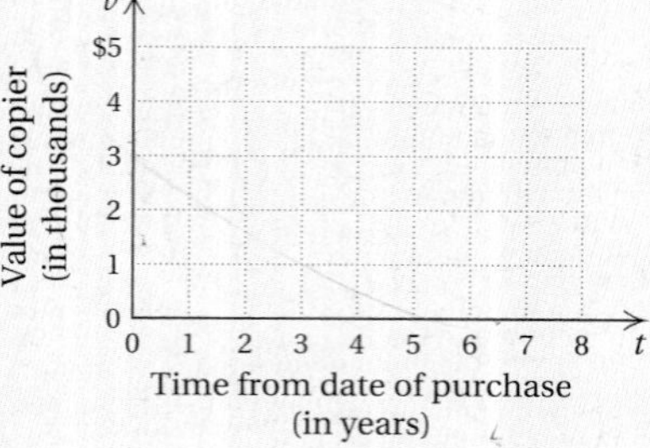

c) After what amount of time is the value of the copier $1500?

Answers on page A-23

We solve as follows.

a) We substitute 2, 5, and 7 for x and then calculate y:

If $x = 2$, then $y = 2.085(2) + 15.08 = \$19.25$.
If $x = 5$, then $y = 2.085(5) + 15.08 \approx \$25.51$.
If $x = 7$, then $y = 2.085(7) + 15.08 \approx \$29.68$.

AG Algebraic–Graphical Connection

b) We have three ordered pairs from (a). We plot these points and see that they line up. Thus our calculations are probably correct. Since zero and negative x-values have no meaning in this problem, we use an open circle at (0, 15.08) when drawing the graph.

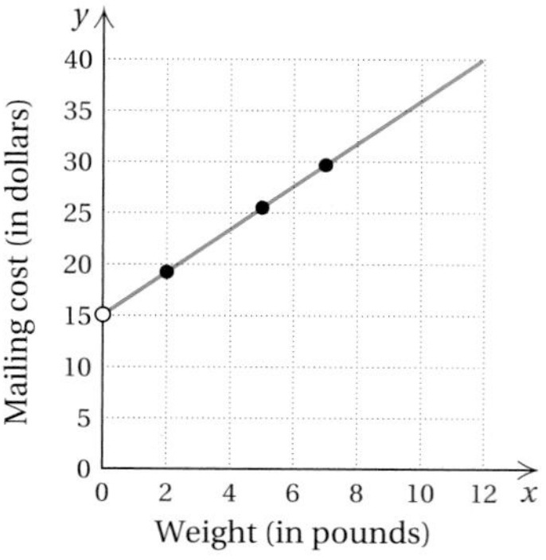

To estimate the cost of mailing a $6\frac{1}{2}$-lb package, we need to determine what y-value is paired with $x = 6\frac{1}{2}$. We locate the point on the line that is above $6\frac{1}{2}$ and then find the value on the y-axis that corresponds to that point. It appears that the cost of mailing a $6\frac{1}{2}$-lb package is about $29.

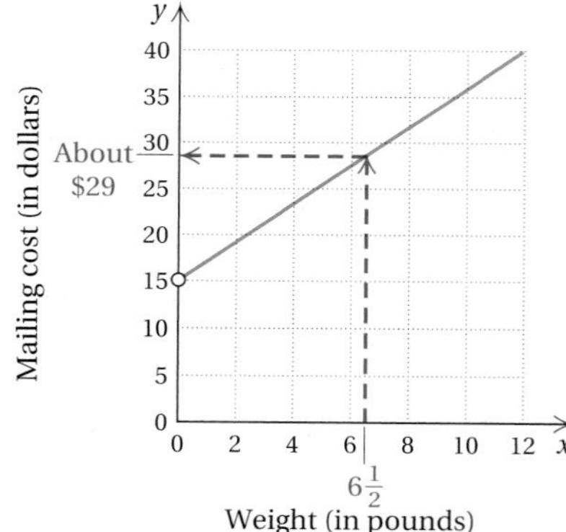

To obtain a more accurate cost, we can simply substitute into the equation:

$$y = 2.085(6.5) + 15.08 \approx \$28.63.$$

AG

c) We substitute \$177.71 for y and then solve for x:

$$\begin{aligned} y &= 2.085x + 15.08 \\ 177.71 &= 2.085x + 15.08 && \textbf{Substituting} \\ 162.63 &= 2.085x && \textbf{Subtracting 15.08} \\ 78 &= x. && \textbf{Dividing by 2.085} \end{aligned}$$

Do Exercise 14 on the preceding page.

Many equations in two variables have graphs that are not straight lines. Three such graphs are shown below. As before, each graph represents the solutions of the given equation. We are not going to develop methods of doing such graphing at this time, although such *nonlinear graphs* can be created very easily using a graphing calculator. We will cover such graphs in the optional Calculator Spotlights.

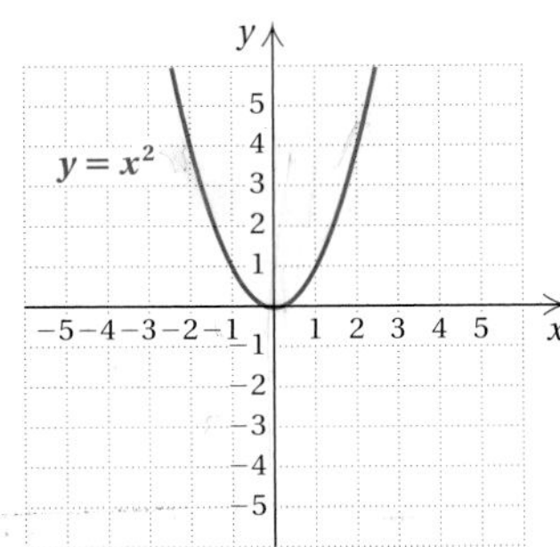

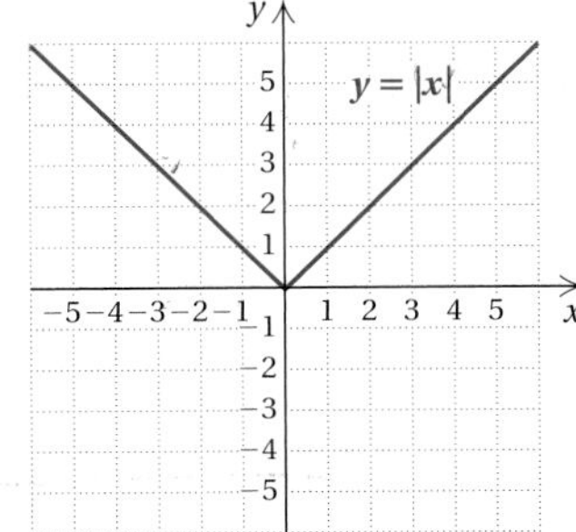

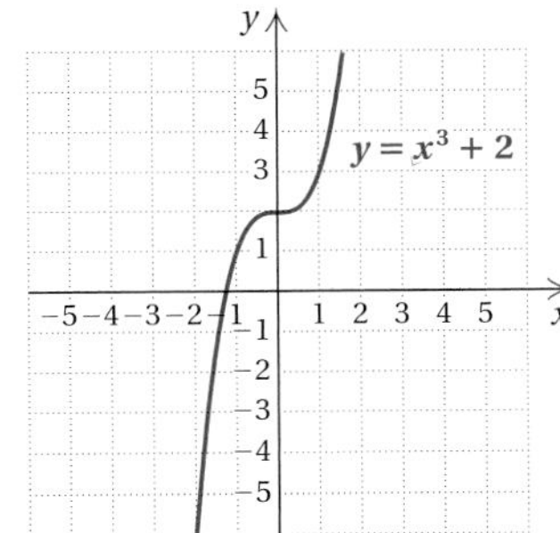

Calculator Spotlight

Introduction to the Use of a Graphing Calculator: Windows and Graphs

Viewing Windows. One feature common to all graphers is the **viewing window.** Windows are described by four numbers, [**L**, **R**, **B**, **T**], which represent the **L**eft and **R**ight endpoints of the x-axis and the **B**ottom and **T**op endpoints of the y-axis. A WINDOW feature is used to set these dimensions. Below is a window setting of $[-20, 20, -5, 5]$ with axis scaling denoted as Xscl = 5 and Yscl = 1. The notation Xres = 1 indicates the number of pixels (black rectangular dots). We will usually leave it at 1 and not refer to it unless needed. On some graphers, a setting of $[-10, 10, -10, 10]$, Xscl = 1, Yscl = 1 is considered **standard**.

```
WINDOW
  Xmin = -20
  Xmax = 20
  Xscl = 5
  Ymin = -5
  Ymax = 5
  Yscl = 1
  Xres = 1
```

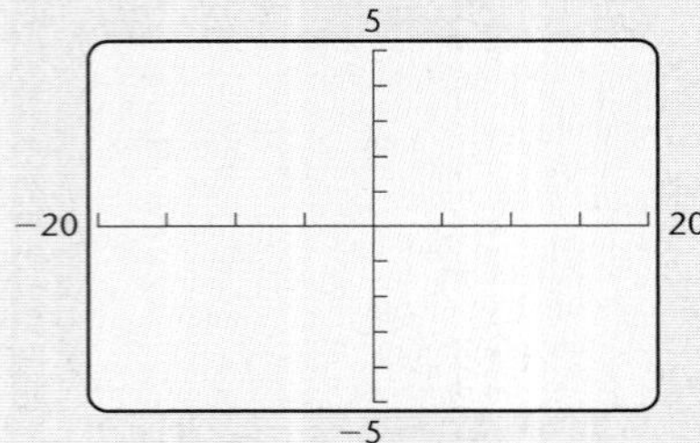

The primary use for a grapher is to graph equations. For example, let's graph the equation $y = x^2 - 3x - 5$. The equation can be entered using the $\boxed{y=}$ key.

$y = x^2 - 3x - 5$

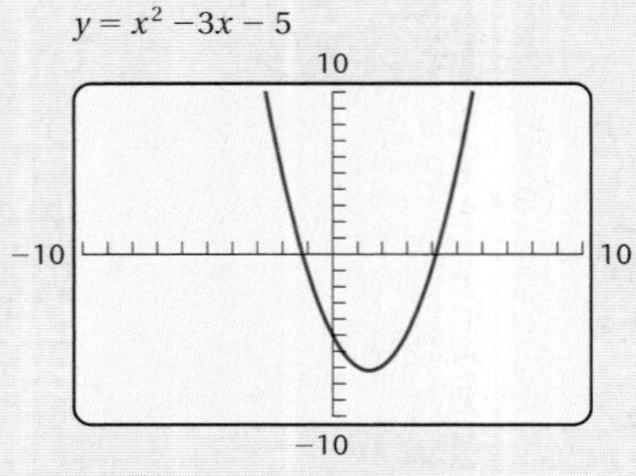

To graph an equation like $4y + 3x = -8$, most graphers require that the equation be solved for y, that is, "$y = \ldots$." Thus we must rewrite the equation $4y + 3x = -8$ as

$$y = \frac{-3x - 8}{4}, \quad \text{or} \quad y = -\frac{3}{4}x - 2,$$

as we did in Example 7. We then enter this equation as $y = -(3/4)x - 2$.

$y = -\frac{3}{4}x - 2$

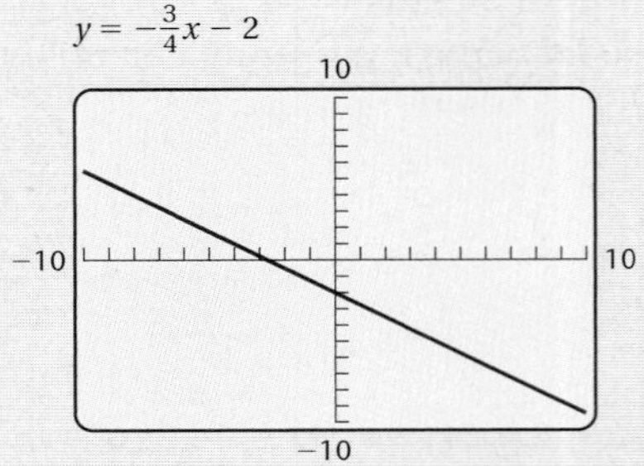

Exercises

Use a grapher to graph each of the following equations. Select the standard window $[-10, 10, -10, 10]$ and axis scaling Xscl = 1, Yscl = 1.

1. $y = 2x + 1$

2. $y = -3x + 1$

3. $y = \frac{2}{5}x + 4$

4. $y = -\frac{3}{5}x - 1$

5. $y = 2.085x + 15.08$

6. $y = -\frac{4}{5}x + \frac{13}{7}$

7. $2x + 3y = 18$

8. $5y + 3x = 4$

9. $y = x^2$

10. $y = 0.5x^2$

11. $y = 8 - x^2$

12. $y = 4 - 3x - x^2$

13. $y = 5x^2 - 3x - 10$

14. $y = x^3 + 2$

15. $y = |x|$ (On most graphers, this is entered as $y = \text{abs}(x)$.)

16. $y = |x - 5|$

17. $y = |x| - 5$

18. $y = 8 - |x|$

Exercise Set 9.2

a Determine whether the given point is a solution of the equation.

1. $(2, 9)$; $y = 3x - 1$

2. $(1, 7)$; $y = 2x + 5$

3. $(4, 2)$; $2x + 3y = 12$

4. $(0, 5)$; $5x - 3y = 15$

5. $(3, -1)$; $3a - 4b = 13$

6. $(-5, 1)$; $2p - 3q = -13$

In Exercises 7–12, an equation and two ordered pairs are given. Show that each pair is a solution. Then use the graph of the two points to determine another solution. Answers may vary.

7. $y = x - 5$; $(4, -1)$ and $(1, -4)$

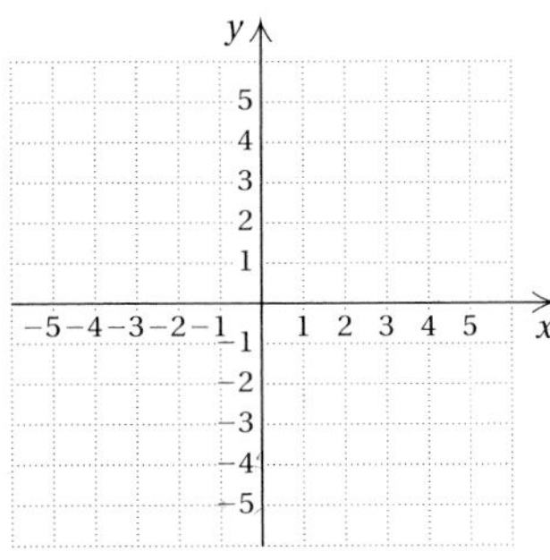

8. $y = x + 3$; $(-1, 2)$ and $(3, 6)$

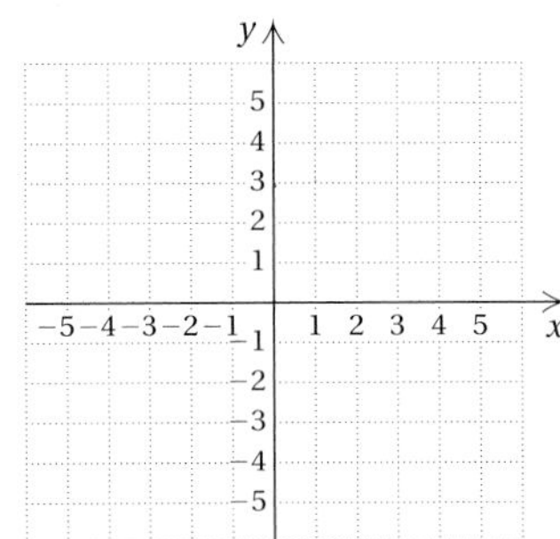

9. $y = \frac{1}{2}x + 3$; $(4, 5)$ and $(-2, 2)$

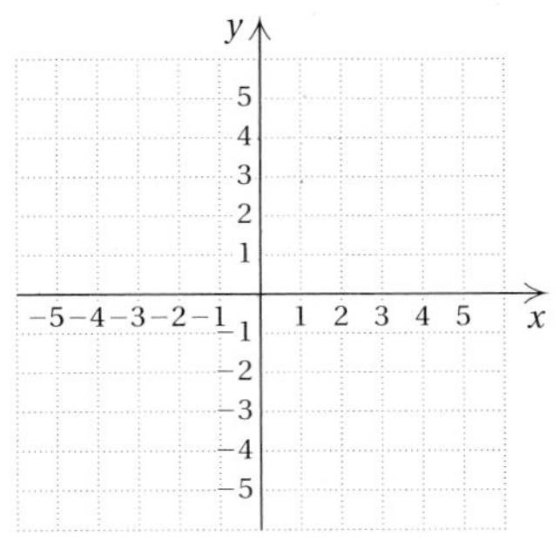

10. $3x + y = 7$; $(2, 1)$ and $(4, -5)$

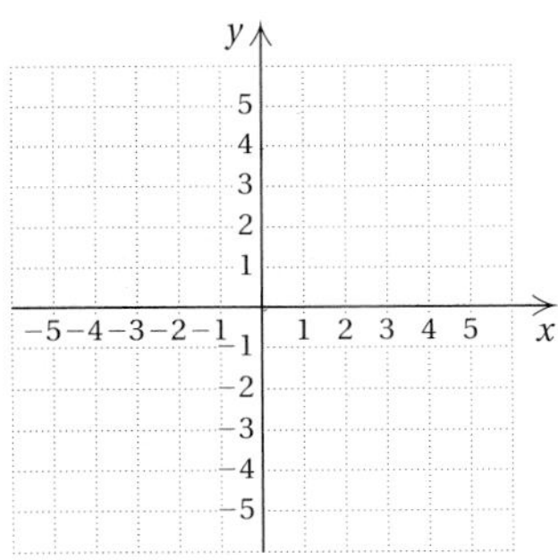

11. $4x - 2y = 10$; $(0, -5)$ and $(4, 3)$

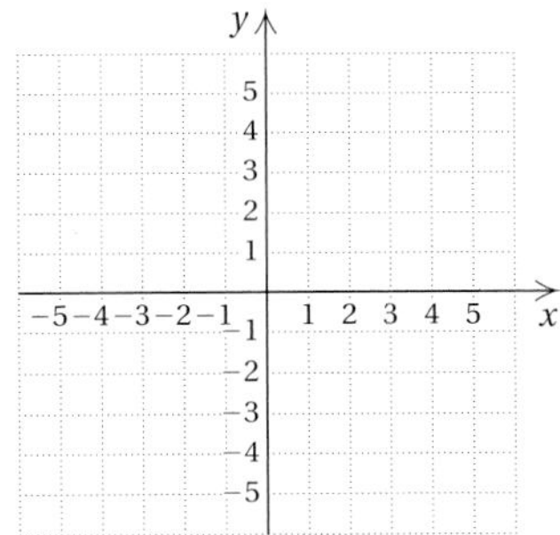

12. $6x - 3y = 3$; $(1, 1)$ and $(-1, -3)$

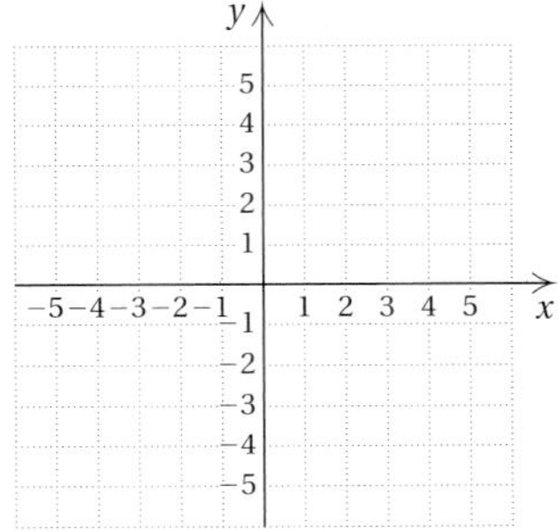

b Graph the equation and identify the y-intercept.

13. $y = x + 1$

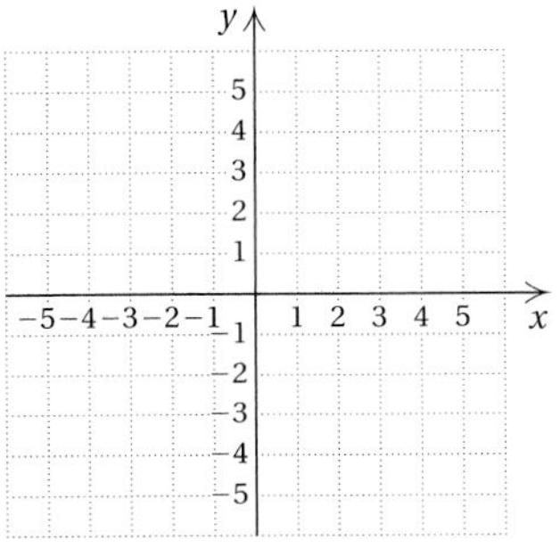

14. $y = x - 1$

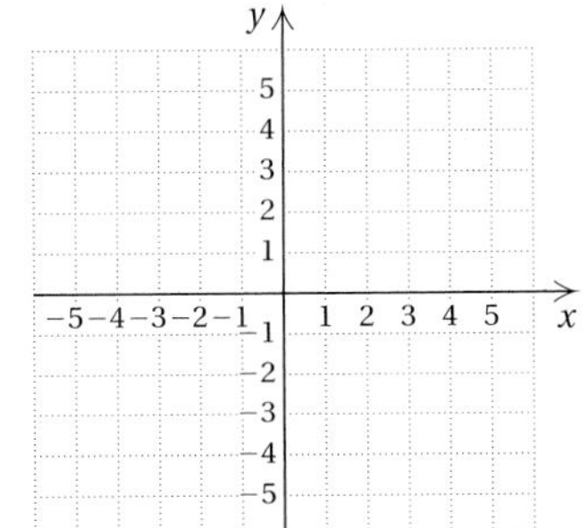

15. $y = x$

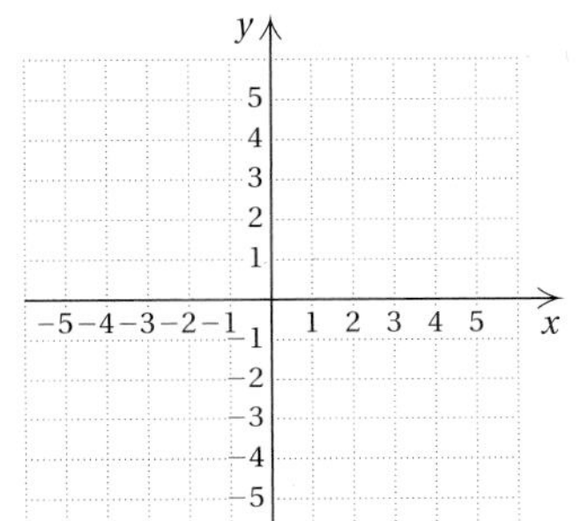

16. $y = -x$

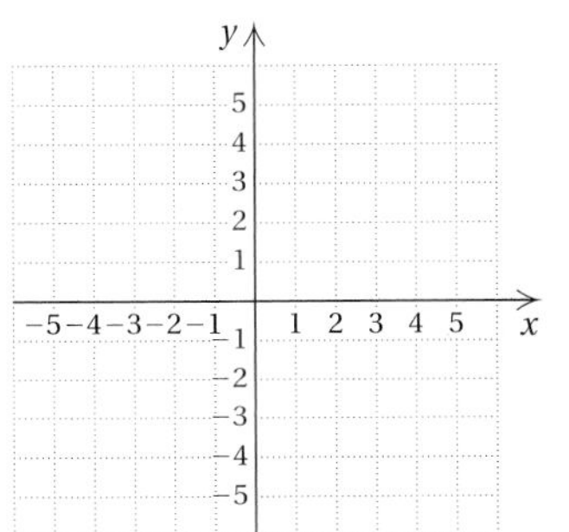

17. $y = \frac{1}{2}x$

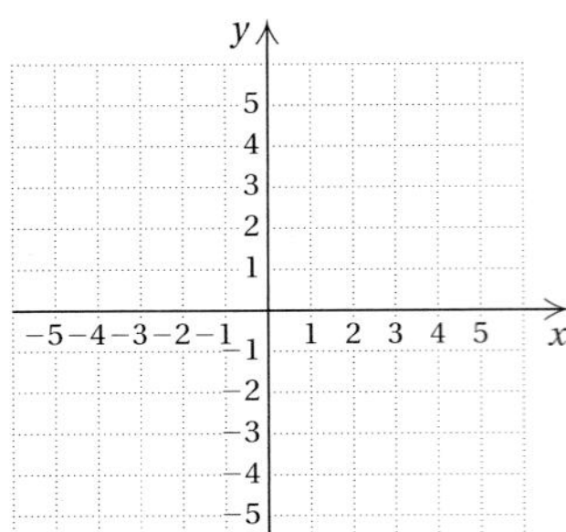

18. $y = \frac{1}{3}x$

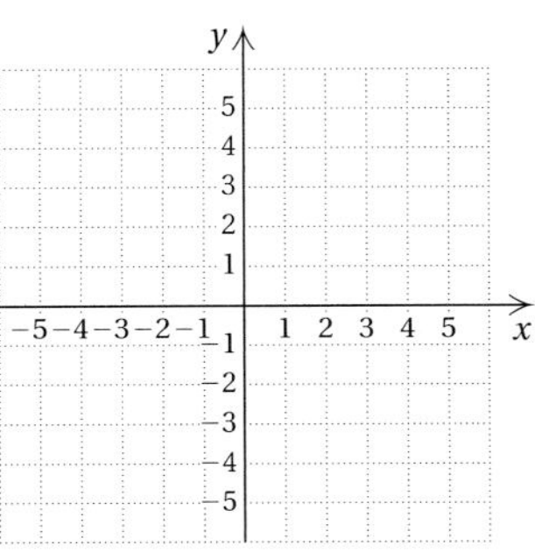

19. $y = x - 3$

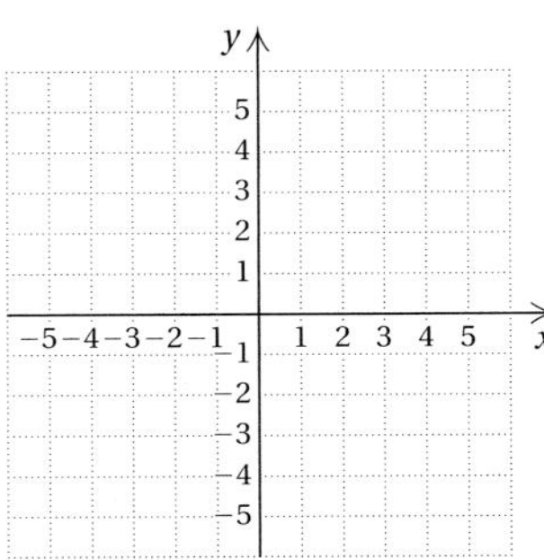

20. $y = x + 3$

21. $y = 3x - 2$

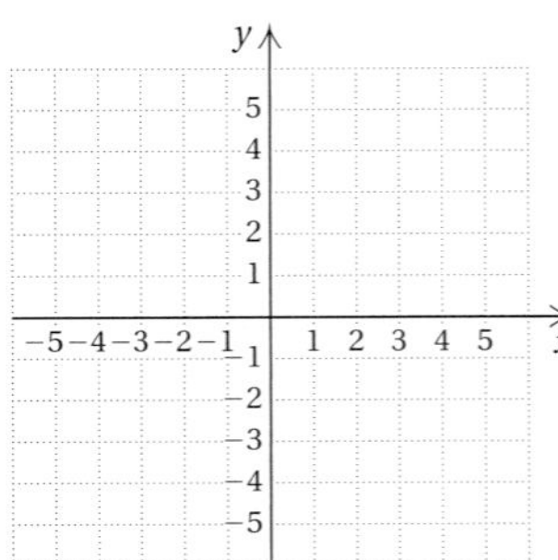

22. $y = 2x + 2$

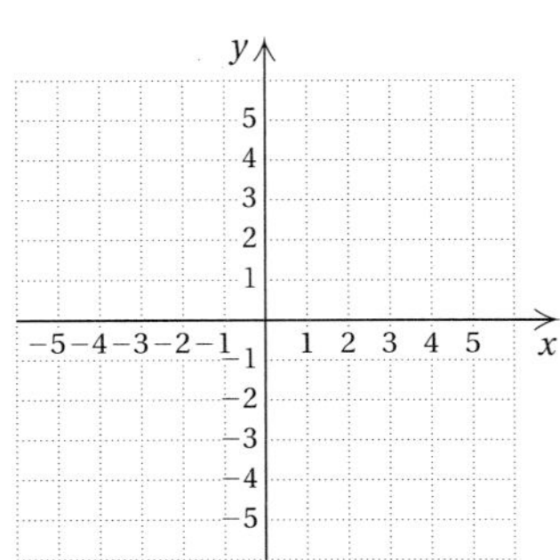

23. $y = \frac{1}{2}x + 1$

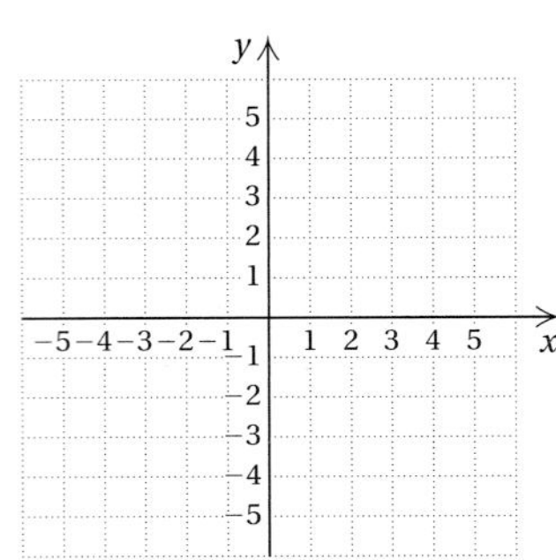

24. $y = \frac{1}{3}x - 4$

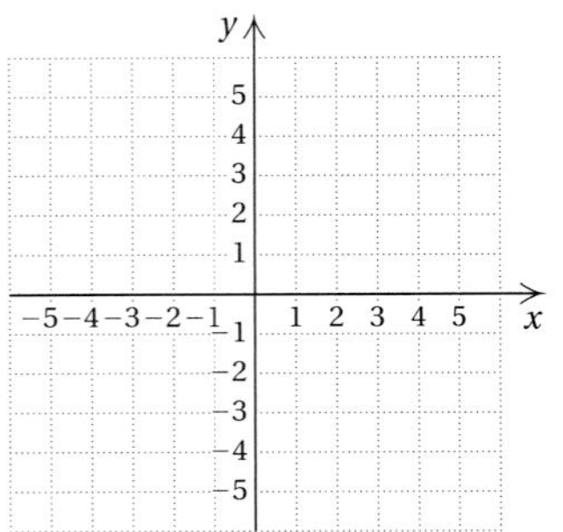

25. $x + y = -5$

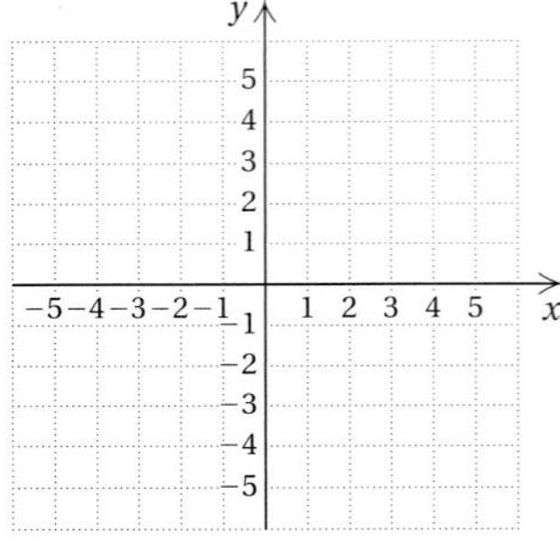

26. $x + y = 4$

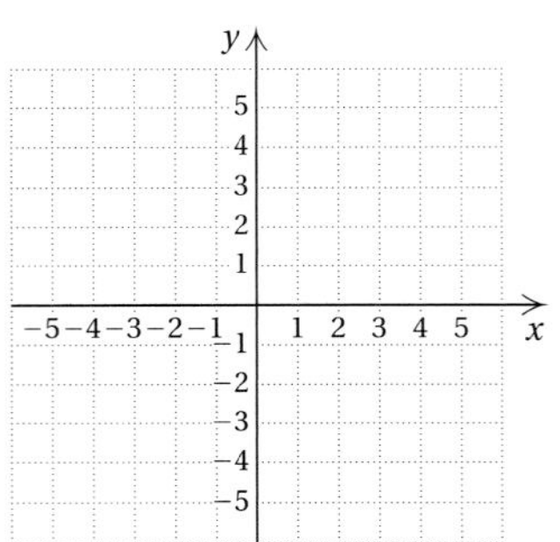

27. $y = \frac{5}{3}x - 2$

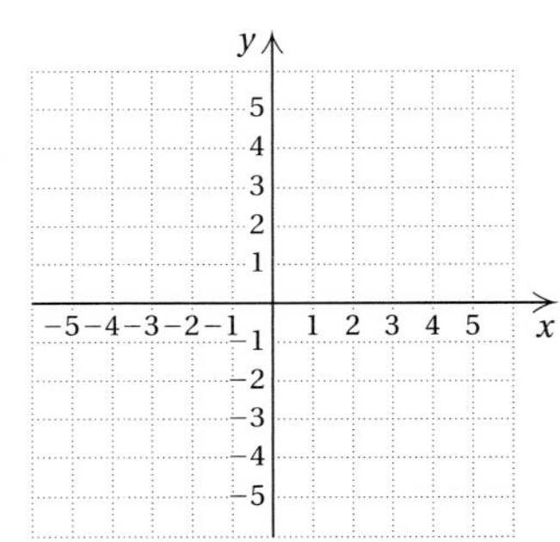

28. $y = \frac{5}{2}x + 3$

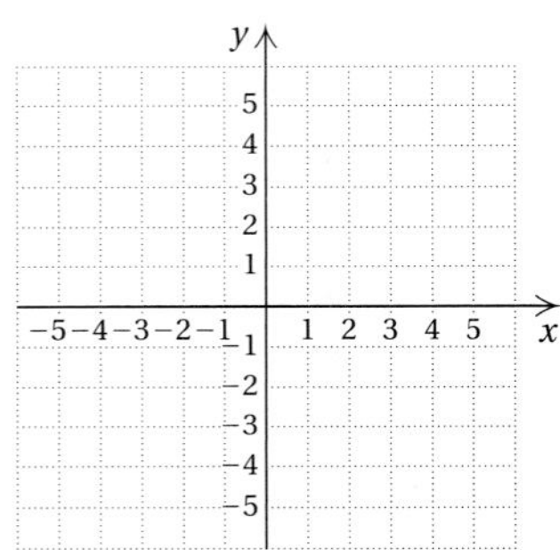

29. $x + 2y = 8$

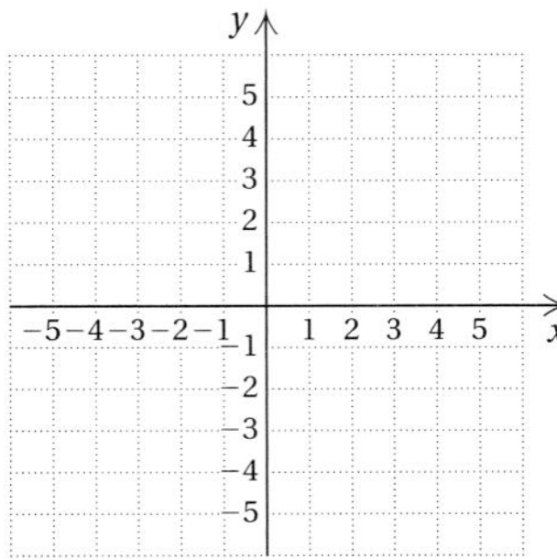

30. $x + 2y = -6$

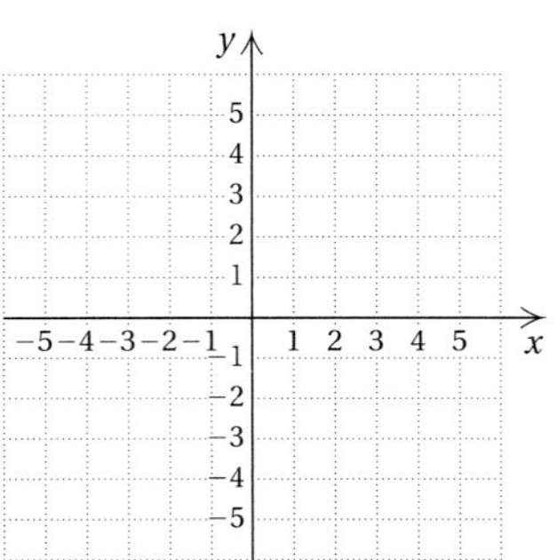

31. $y = \frac{3}{2}x + 1$

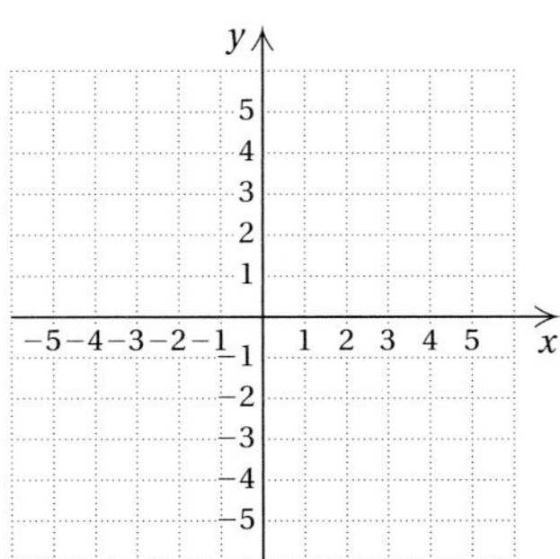

32. $y = -\frac{1}{2}x - 3$

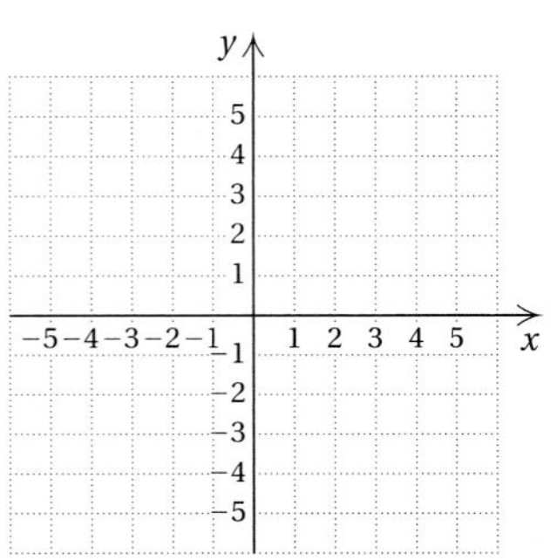

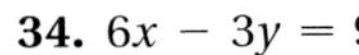

33. $8x - 2y = -10$

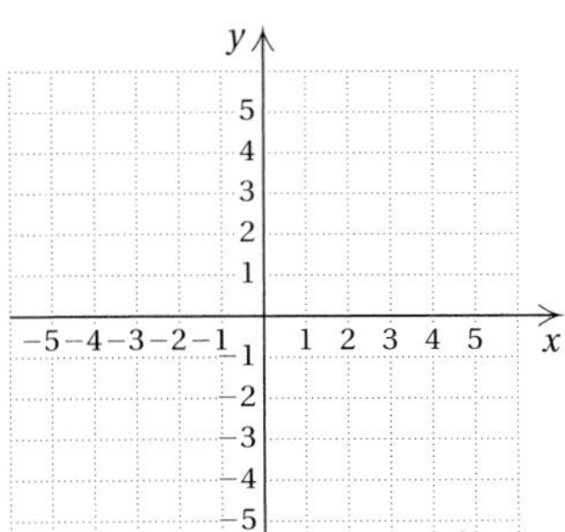

34. $6x - 3y = 9$

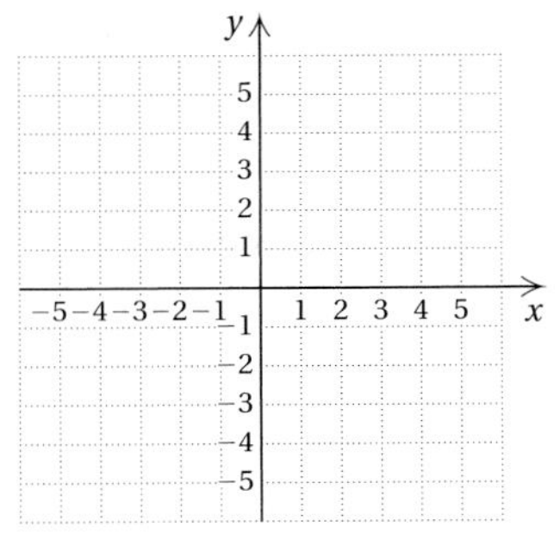

35. $8y + 2x = -4$

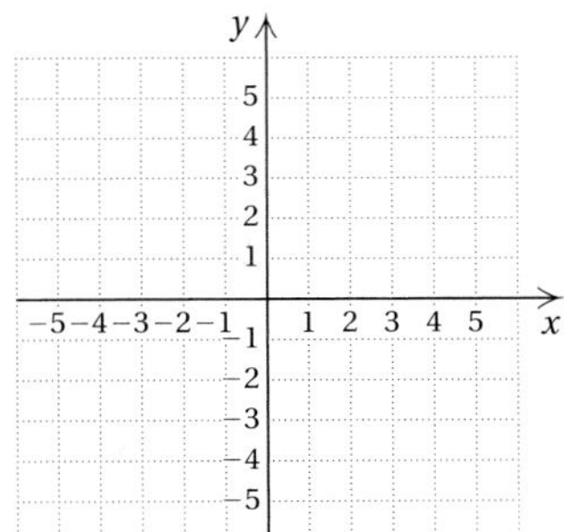

36. $6y + 2x = 8$

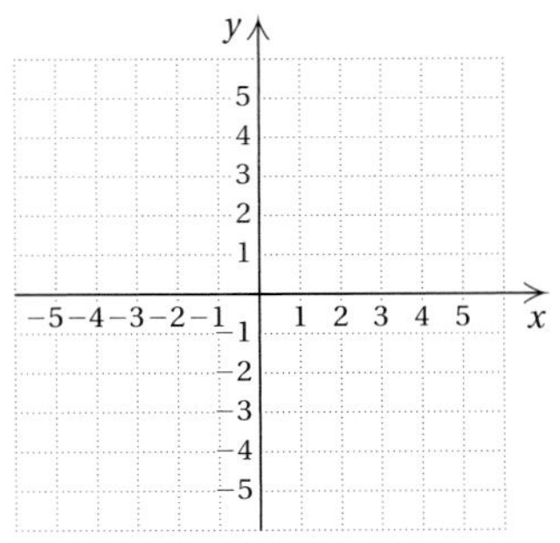

c Solve.

37. *Value of Computer Software.* The value V, in dollars, of a shopkeeper's inventory software program is given by

$$V = -50t + 300,$$

where $t =$ the number of years since the shopkeeper first bought the program.

a) Find the value of the software after 0 yr, 4 yr, and 6 yr.

b) Graph the equation and then use the graph to estimate the value of the software after 5 yr.

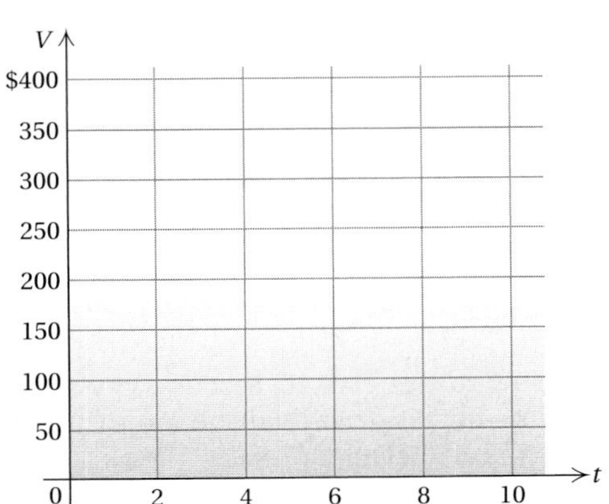

c) After how many years is the value of the software \$150?

38. *College Costs.* The cost T, in dollars, of tuition and fees at many community colleges can be approximated by

$$T = 120c + 100,$$

where $c =$ the number of credits for which a student registers (***Source:*** Community College of Vermont).

a) Find the cost of tuition for a student who takes 8 hr, 12 hr, and 15 hr.

b) Graph the equation and then use the graph to estimate the cost of tuition for a student who takes 9 hr.

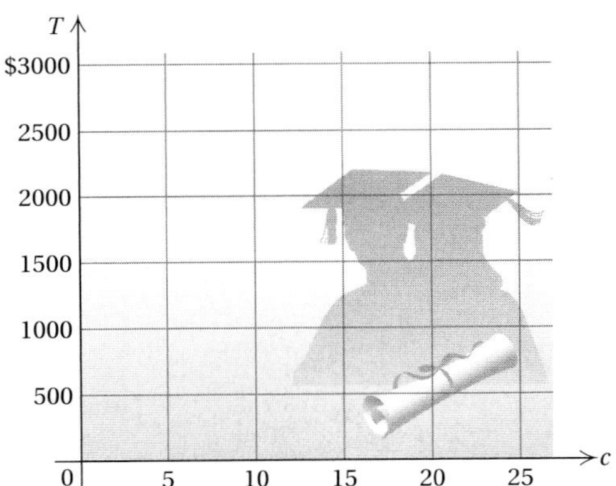

c) Estimate how many hours a student can take for \$1420.

39. *Tea Consumption.* The number of gallons N of tea consumed each year by the average U.S. consumer can be approximated by

$$N = 0.1d + 7,$$

where d = the number of years since 1991 (***Source:*** Statistical Abstract of the United States).

a) Find the number of gallons of tea consumed in 1992 ($n = 1$), 1995 ($n = 4$), 1999 ($n = 8$), and 2001 ($n = 10$).

b) Graph the equation and use the graph to estimate what the tea consumption was in 1997.

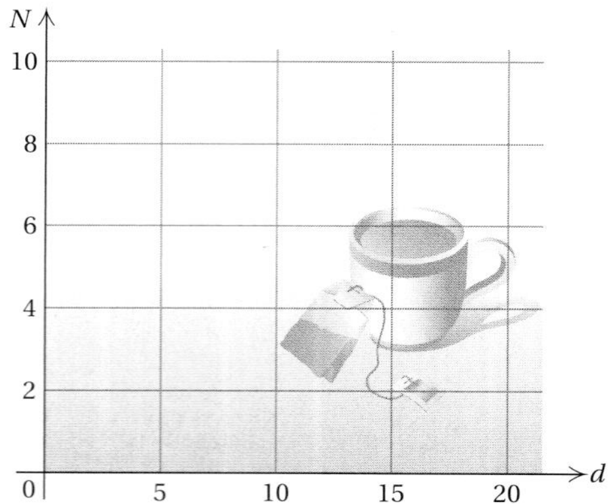

c) In what year will tea consumption be about 9 gal?

40. *Record Temperature Drop.* On 22 January 1943, the temperature T, in degrees Fahrenheit, in Spearfish, South Dakota, could be approximated by

$$T = -2.15m + 54,$$

where m = the number of minutes since 9:00 that morning (***Source:*** *Information Please Almanac*, 1996).

a) Find the temperature at 9:01 A.M., 9:08 A.M., and 9:20 A.M.

b) Graph the equation and use the graph to estimate the temperature at 9:15 A.M.

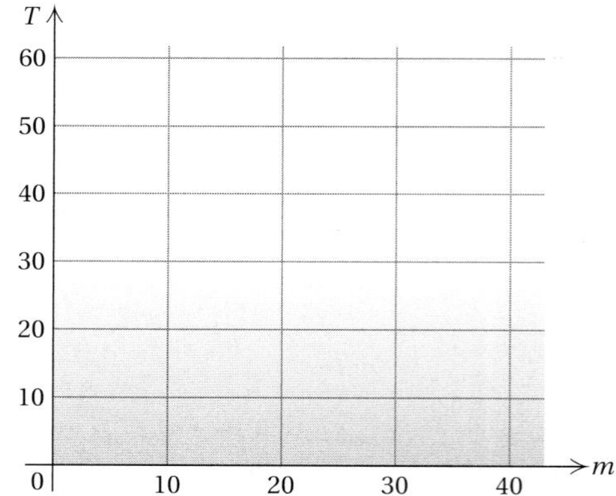

c) The temperature stopped dropping when it reached $-4°$. At what time did this occur? (*Note*: The linear equation could not be used after that time.)

Skill Maintenance

Round to the nearest thousand. [1.3f], [3.1e]

41. 2567.03

42. 124,748

43. 293.4572

44. 6,078,124

Convert to decimal notation. [7.2c]

45. $-\frac{7}{8}$

46. $\frac{23}{32}$

47. $\frac{117}{64}$

48. $-\frac{27}{12}$

Synthesis

49. ◈ The equations $3x + 4y = 8$ and $y = -\frac{3}{4}x + 2$ are equivalent. Which equation is easier to graph and why?

50. ◈ Referring to Exercise 40, discuss why the linear equation no longer described the temperature after the temperature reached $-4°$.

In Exercises 51–54, find an equation for the graph shown.

51.

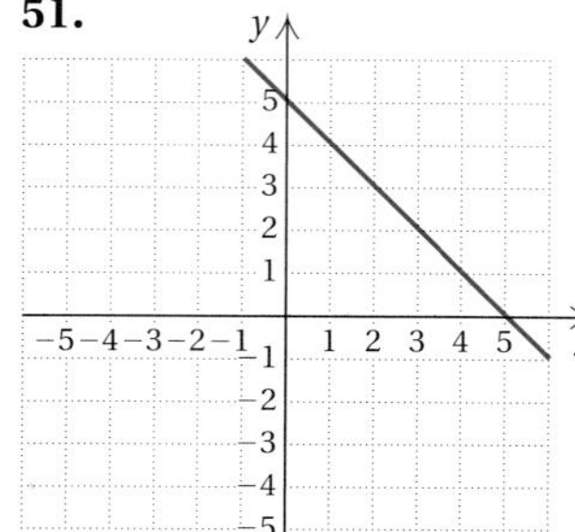

52.

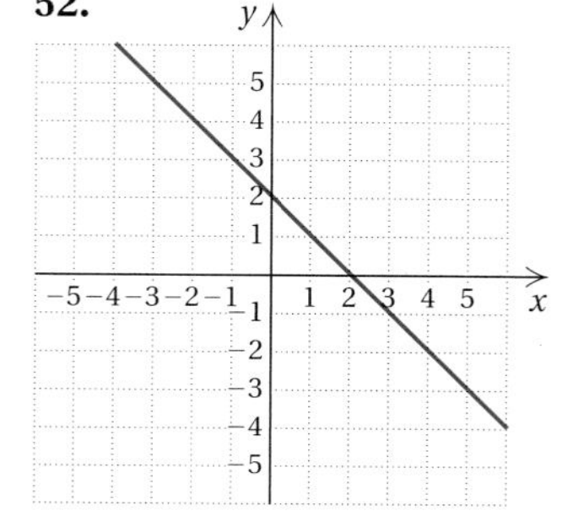

53.

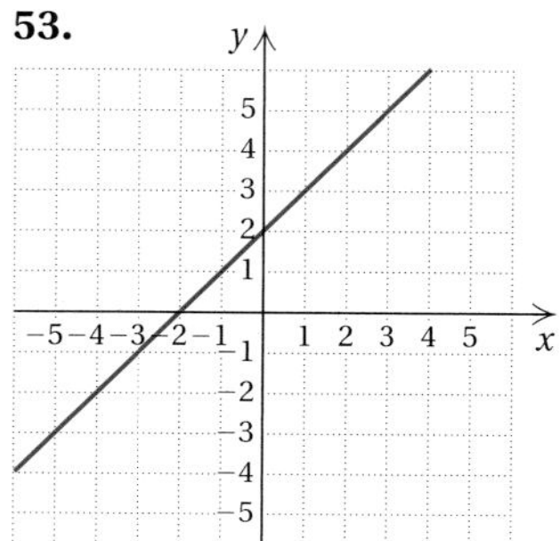

54.

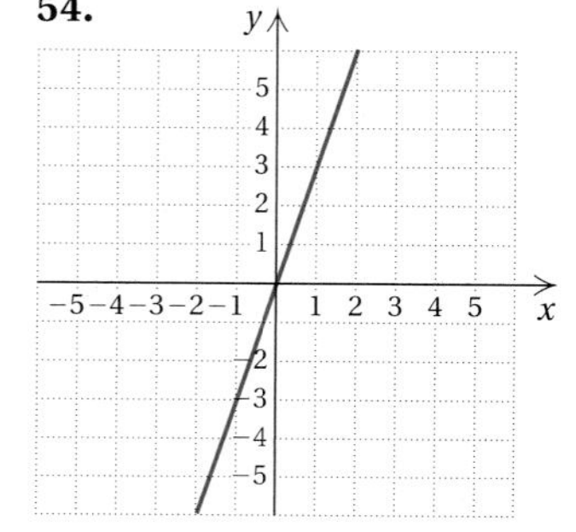

9.3 More with Graphing and Intercepts

a Graphing Using Intercepts

In Section 9.2, we graphed linear equations of the form $Ax + By = C$ by first solving for y to find an equivalent equation in the form $y = mx + b$. We did so because it is then easier to calculate the y-value that corresponds to a given x-value. Another convenient way to graph $Ax + By = C$ is to use **intercepts**. Look at the graph of $-2x + y = 4$ shown below.

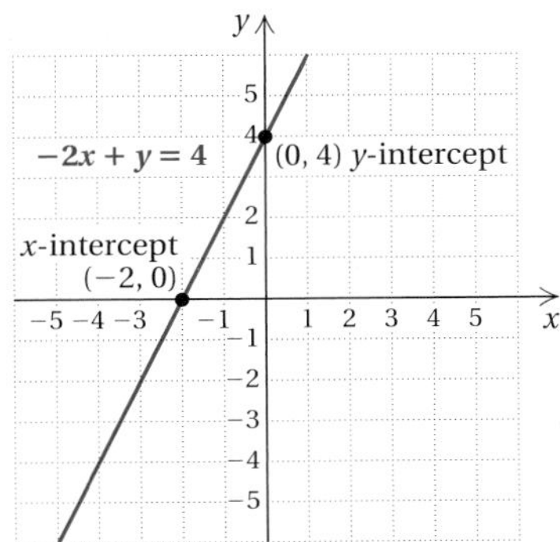

The y-intercept is (0, 4). It occurs where the line crosses the y-axis and thus will always have 0 as the first coordinate. The x-intercept is (−2, 0). It occurs where the line crosses the x-axis and thus will always have 0 as the second coordinate.

Do Exercise 1.

We find intercepts as follows.

> The **y-intercept** is $(0, b)$. To find b, let $x = 0$ and solve the original equation for y.
>
> The **x-intercept** is $(a, 0)$. To find a, let $y = 0$ and solve the original equation for x.

Now let's draw a graph using intercepts.

Example 1 Consider $4x + 3y = 12$. Find the intercepts. Then graph the equation using the intercepts.

To find the y-intercept, we let $x = 0$. Then we solve for y:

$$4 \cdot 0 + 3y = 12$$
$$3y = 12$$
$$y = 4.$$

Thus, (0, 4) is the y-intercept. Note that finding this intercept amounts to covering up the x-term and solving the rest of the equation.

To find the x-intercept, we let $y = 0$. Then we solve for x:

$$4x + 3 \cdot 0 = 12$$
$$4x = 12$$
$$x = 3.$$

Objectives

a Find the intercepts of a linear equation, and graph using intercepts.

b Graph equations equivalent to those of the type $x = a$ and $y = b$.

For Extra Help

TAPE 18 | MAC WIN | CD-ROM

1. Look at the graph shown below.

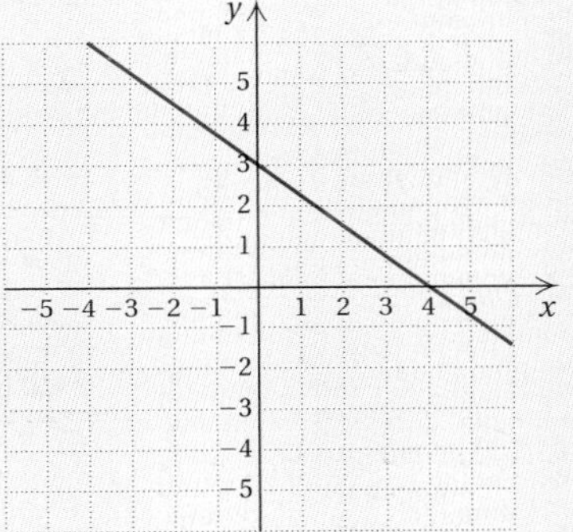

a) Find the coordinates of the y-intercept.

b) Find the coordinates of the x-intercept.

Answers on page A-25

For each equation, find the intercepts. Then graph the equation using the intercepts.

2. $2x + 3y = 6$

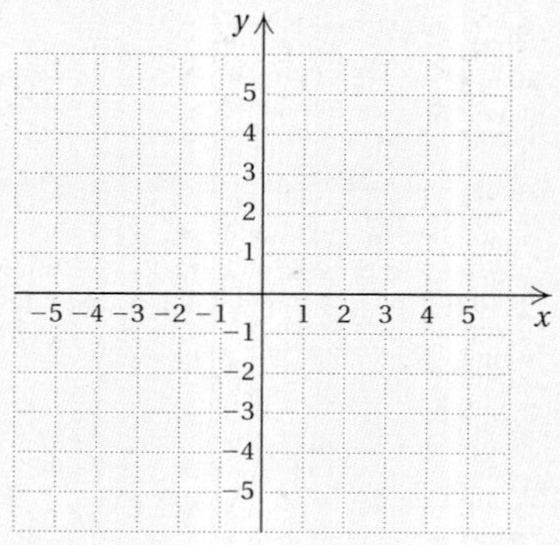

3. $3y - 4x = 12$

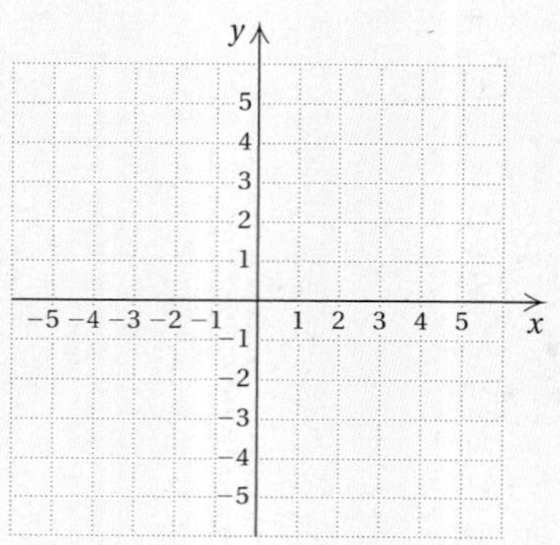

Graph.

4. $x = 5$

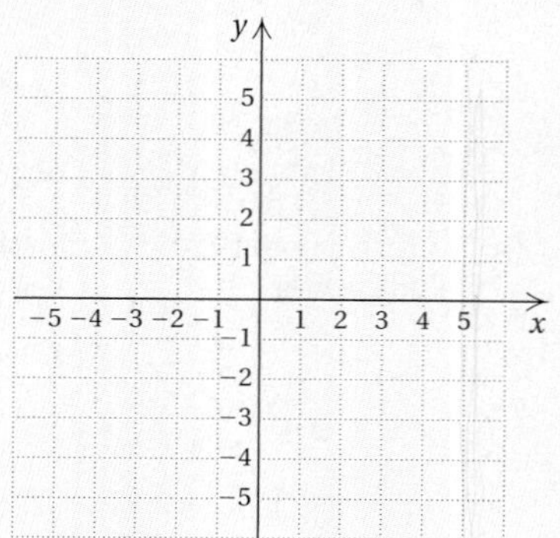

Answers on page A-25

Thus, (3, 0) is the x-intercept. Note that finding this intercept amounts to covering up the y-term and solving the rest of the equation.

We plot these points and draw the line, or graph.

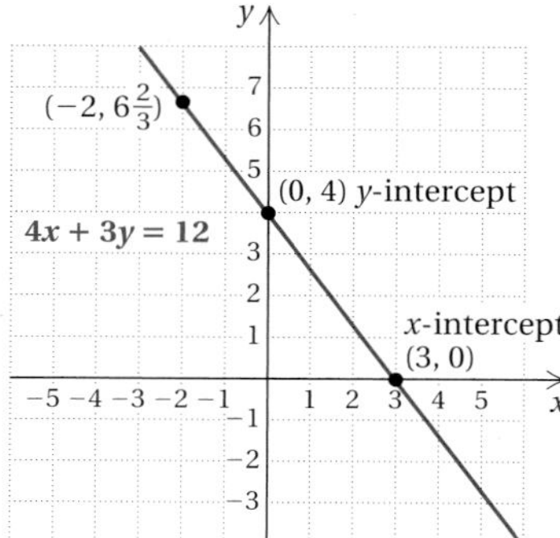

A third point should be used as a check. We substitute any convenient value for x and solve for y. In this case, we choose $x = -2$. Then

$$4(-2) + 3y = 12 \quad \text{Substituting } -2 \text{ for } x$$
$$-8 + 3y = 12$$
$$3y = 12 + 8 = 20$$
$$y = \frac{20}{3}, \text{ or } 6\frac{2}{3}. \quad \text{Solving for } y$$

It appears that the point $\left(-2, 6\frac{2}{3}\right)$ is on the graph, though graphing fractional values can be inexact. The graph is probably correct.

Graphs of equations of the type $y = mx$ pass through the origin. Thus the x-intercept and the y-intercept are the same, (0, 0). In such cases, we must calculate another point in order to complete the graph. Another point would also have to be calculated if a check is desired.

Do Exercises 2 and 3.

b Equations Whose Graphs Are Horizontal or Vertical Lines

Example 2 Graph: $y = 3$.

Consider $y = 3$. We can also think of this equation as $0 \cdot x + y = 3$. No matter what number we choose for x, we find that y is 3. We make up a table with all 3's in the y-column.

x	y
	3
	3
	3

y must be 3.

Choose any number for x. →

x	y
−2	3
0	3
4	3

← y-intercept (row 0, 3)

When we plot the ordered pairs $(-2, 3)$, $(0, 3)$, and $(4, 3)$ and connect the points, we will obtain a horizontal line. Any ordered pair $(x, 3)$ is a solution. So the line is parallel to the x-axis with y-intercept $(0, 3)$.

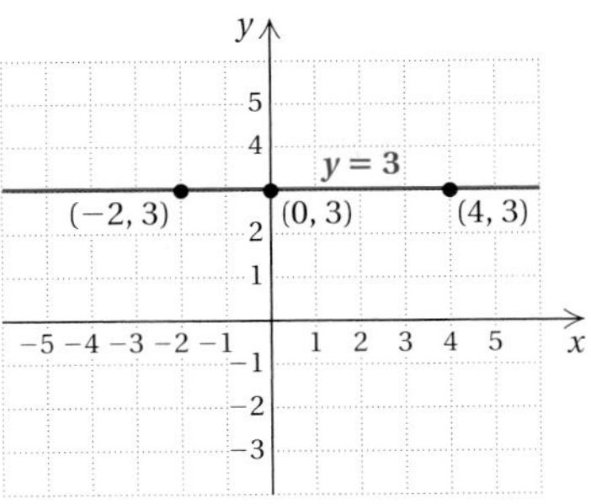

Example 3 Graph: $x = -4$.

Consider $x = -4$. We can also think of this equation as $x + 0 \cdot y = -4$. We make up a table with all -4's in the x-column.

x	y
−4	
−4	
−4	
−4	

Choose any number for y. →

x must be -4.

x-intercept →

x	y
−4	−5
−4	1
−4	3
−4	0

When we plot the ordered pairs $(-4, -5)$, $(-4, 1)$, $(-4, 3)$, and $(-4, 0)$ and connect the points, we will obtain a vertical line. Any ordered pair $(-4, y)$ is a solution. So the line is parallel to the y-axis with x-intercept $(-4, 0)$.

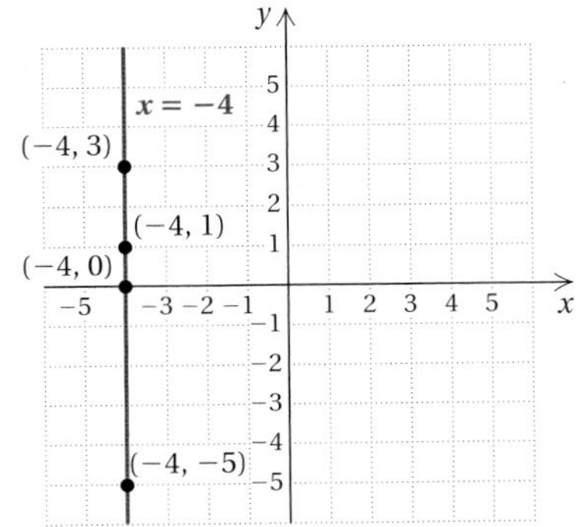

The graph of $y = b$ is a **horizontal line.** The y-intercept is $(0, b)$.

The graph of $x = a$ is a **vertical line.** The x-intercept is $(a, 0)$.

Do Exercises 4–7. (Exercise 4 is on the preceding page.)

Graph.

5. $y = -2$

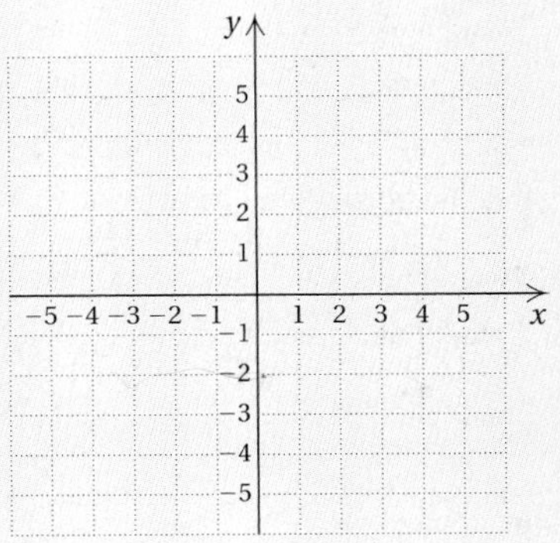

6. $x = 0$

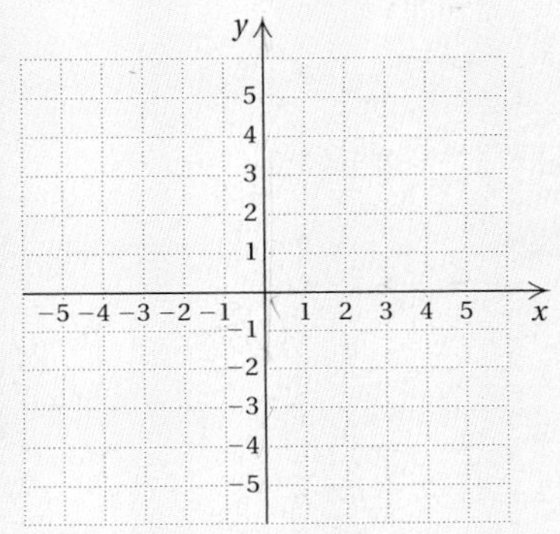

7. $x = -3$

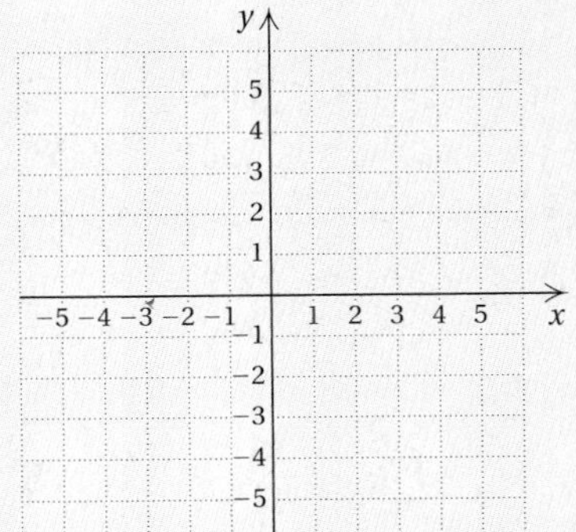

Answers on page A-25

Calculator Spotlight

Viewing the Intercepts. Graph the equation $y = -x + 15$ using the standard viewing window.

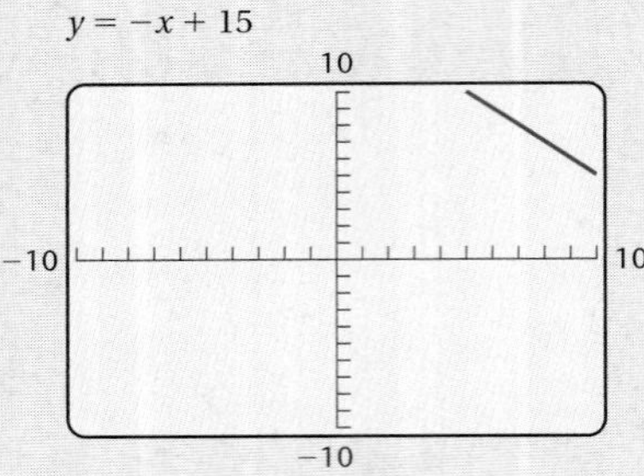

Note that the graph is barely visible in the upper right-hand corner, and neither intercept can be seen. To better view the intercepts, we can try different window settings.

$y = -x + 15$

50

−50 50

−50

Exercises

Find the intercepts of the equation using algebra (algebraically). Then graph the equation and adjust the window and tick mark settings so that the intercepts can be clearly seen on both axes.

1. $y = -7.2x - 15$
2. $y - 2.13x = 27$
3. $5x + 6y = 84$
4. $2x - 7y = 150$
5. $3x + 2y = 50$
6. $y = 0.2x - 9$
7. $y = 1.3x - 15$
8. $25x - 20y = 1$

The following is a general procedure for graphing linear equations.

Graphing Linear Equations

1. If the equation is of the type $x = a$ or $y = b$, the graph will be a line parallel to an axis; $x = a$ is vertical and $y = b$ is horizontal.

   Examples.

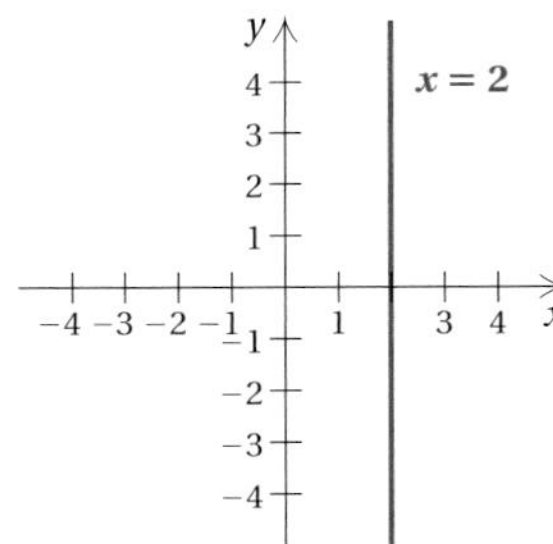

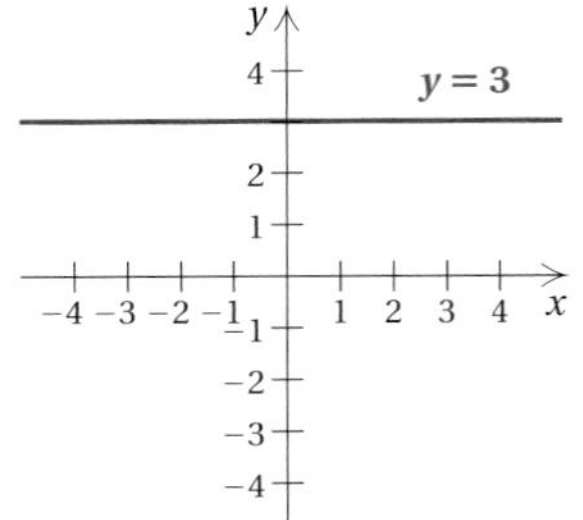

2. If the equation is of the type $y = mx$, both intercepts are the origin, (0, 0). Plot (0, 0) and two other points.

   Example.

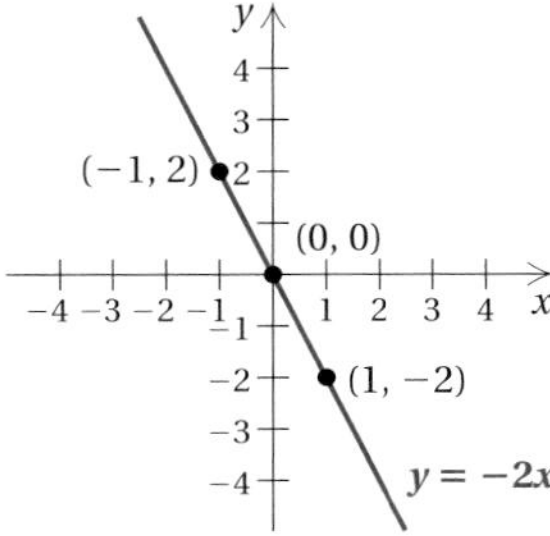

3. If the equation is of the type $y = mx + b$, plot the y-intercept $(0, b)$ and two other points.

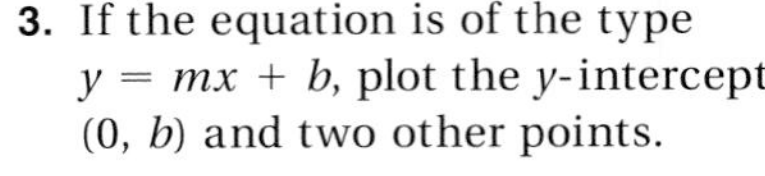

   Example.

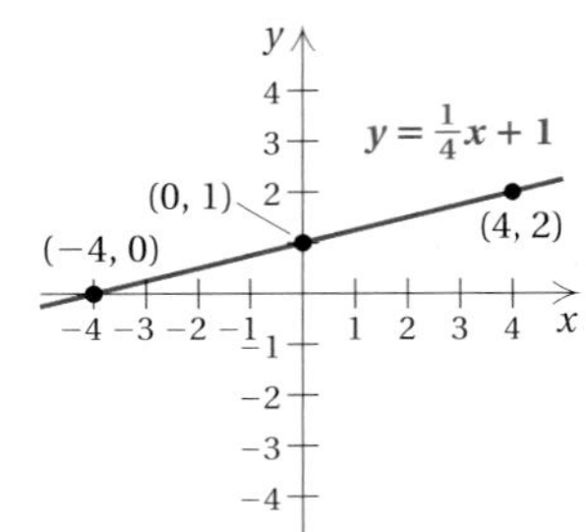

4. If the equation is of the type $Ax + By = C$, but not of the type $x = a$, $y = b$, $y = mx$, or $y = mx + b$, then either solve for y and proceed as with the equation $y = mx + b$, or graph using intercepts. If the intercepts are too close together, choose another point or points farther from the origin.

   Examples.

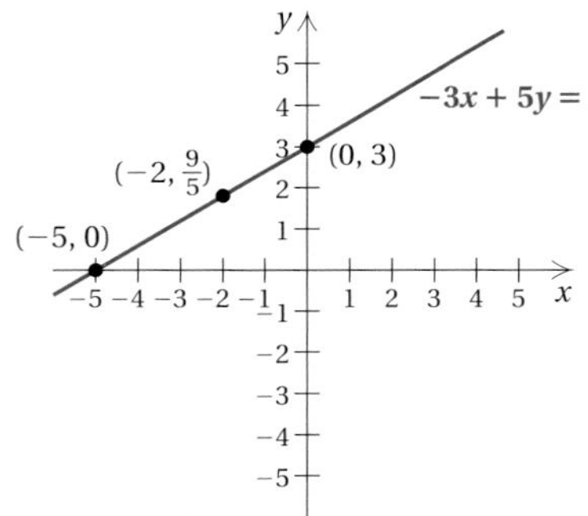

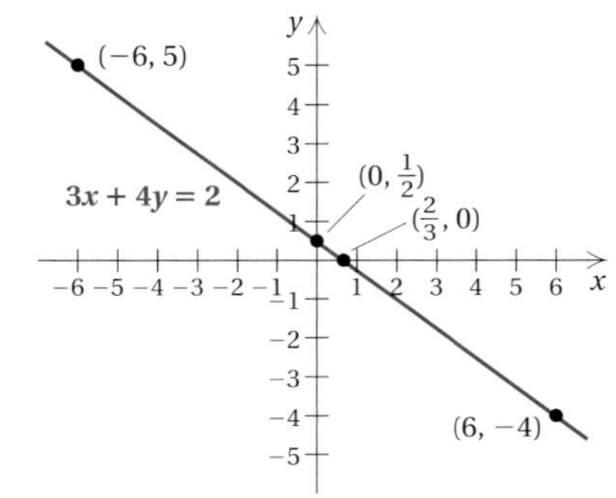

Exercise Set 9.3

a For Exercises 1–4, find (a) the coordinates of the y-intercept and (b) the coordinates of the x-intercept.

1.

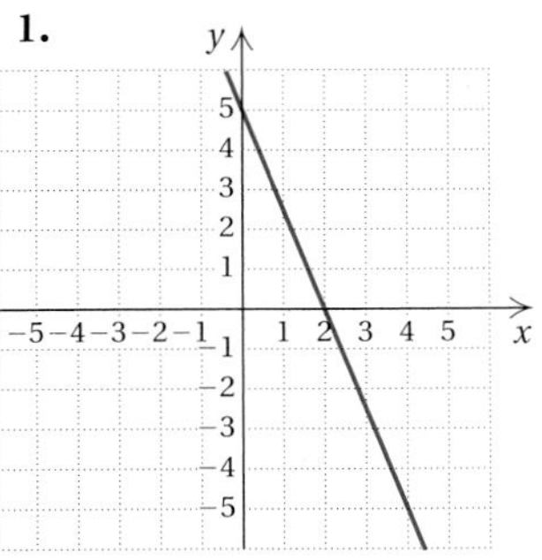

2.

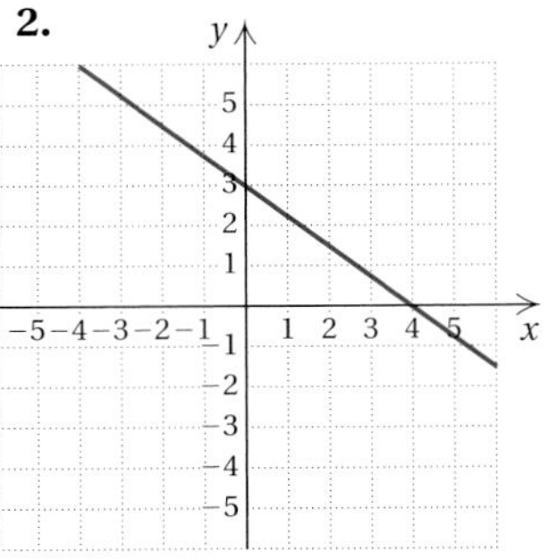

3.

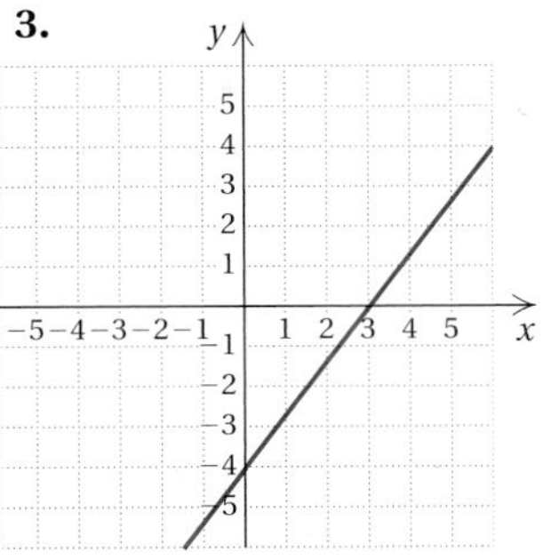

4.

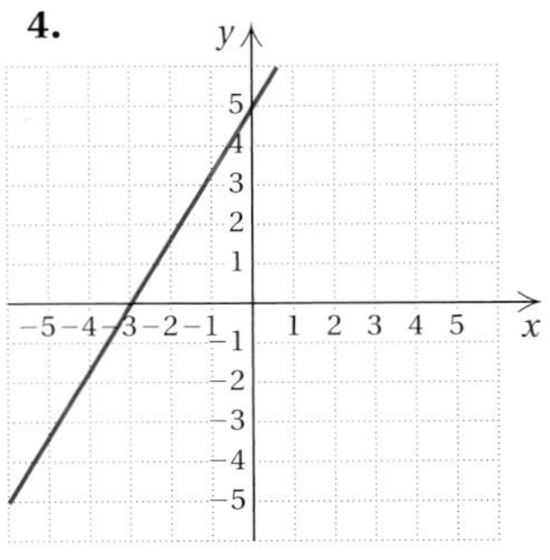

For Exercises 5–12, find (a) the coordinates of the y-intercept and (b) the coordinates of the x-intercept. Do not graph.

5. $3x + 5y = 15$

6. $5x + 2y = 20$

7. $7x - 2y = 28$

8. $3x - 4y = 24$

9. $-4x + 3y = 10$

10. $-2x + 3y = 7$

11. $6x - 3 = 9y$

12. $4y - 2 = 6x$

For each equation, find the intercepts. Then use the intercepts to graph the equation.

13. $x + 3y = 6$

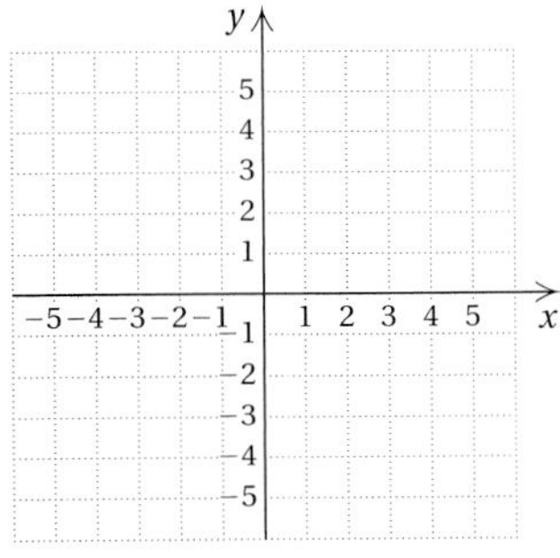

14. $x + 2y = 2$

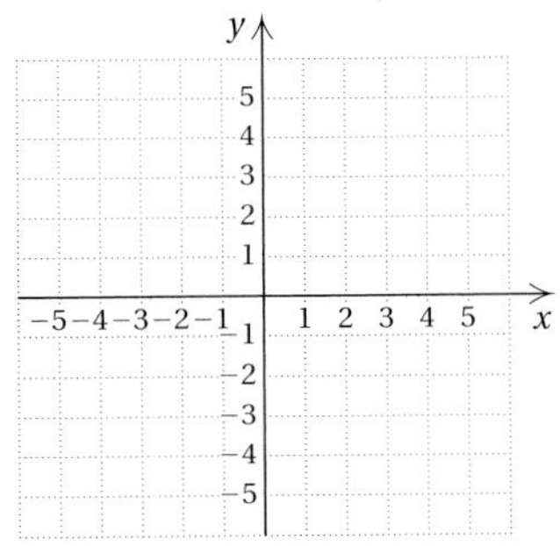

15. $-x + 2y = 4$

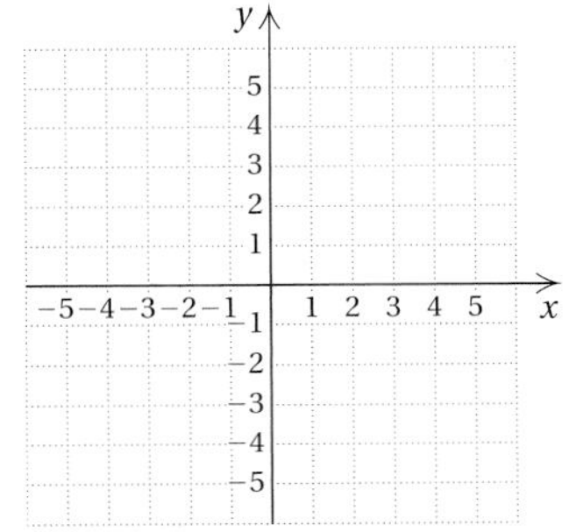

16. $-x + y = 5$

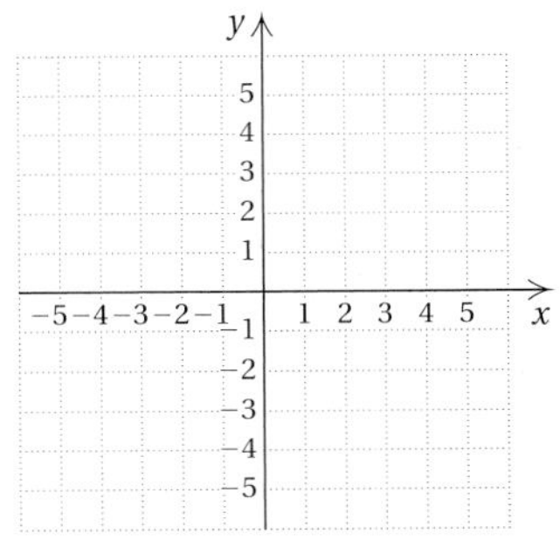

17. $3x + y = 6$

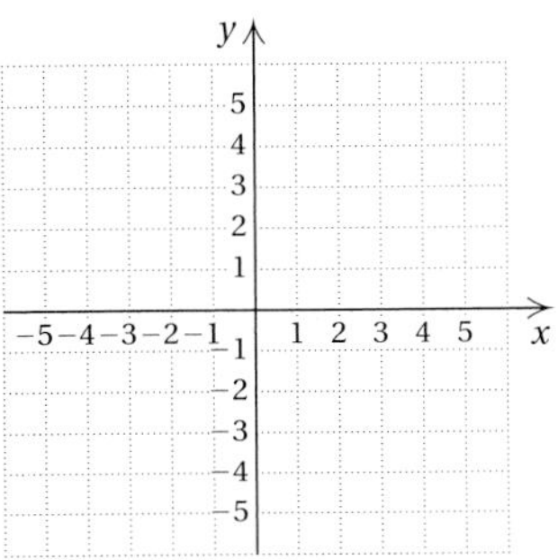

18. $2x + y = 6$

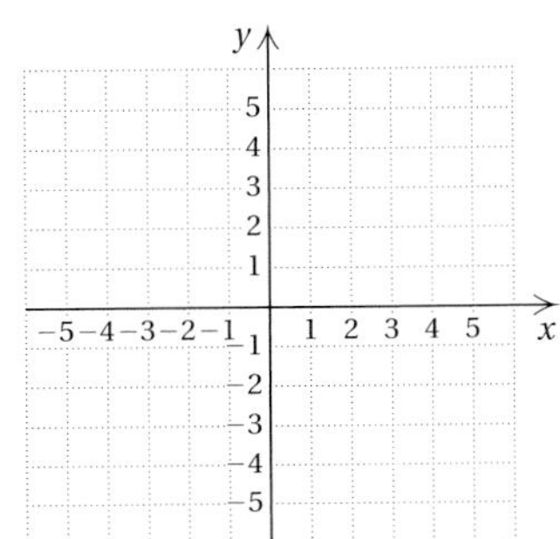

19. $2y - 2 = 6x$

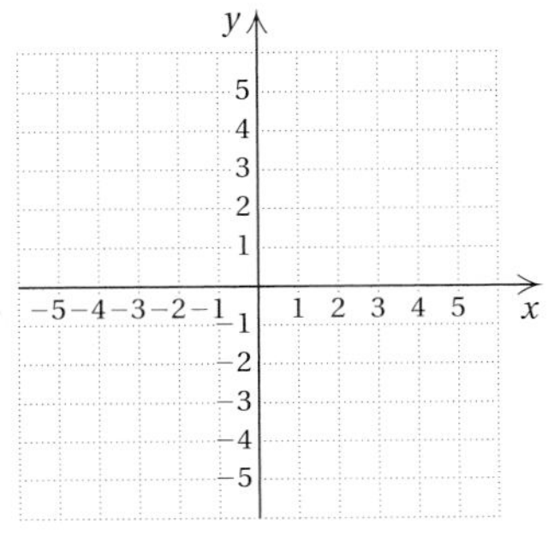

20. $3y - 6 = 9x$

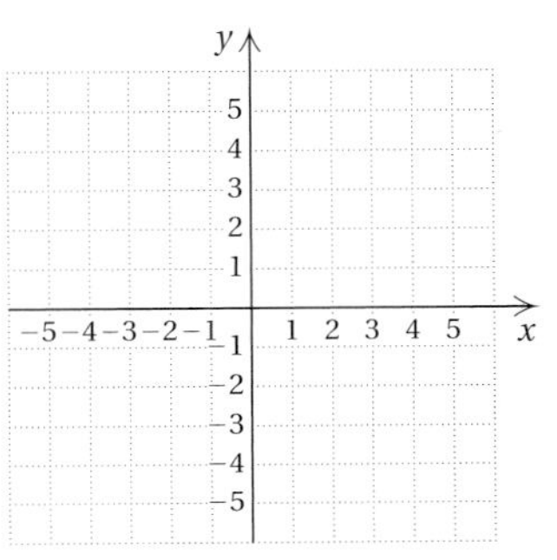

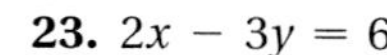

21. $3x - 9 = 3y$

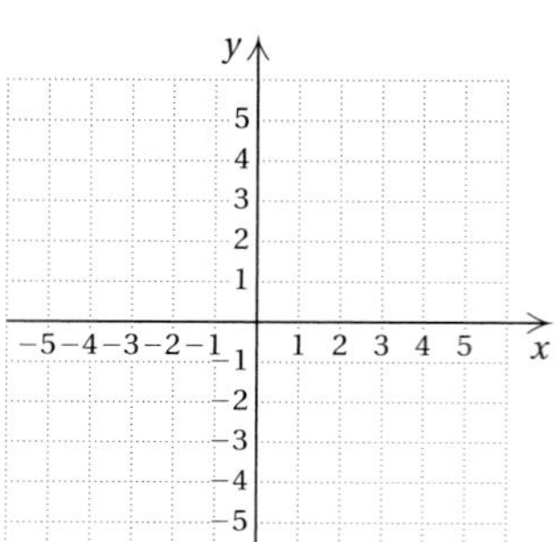

22. $5x - 10 = 5y$

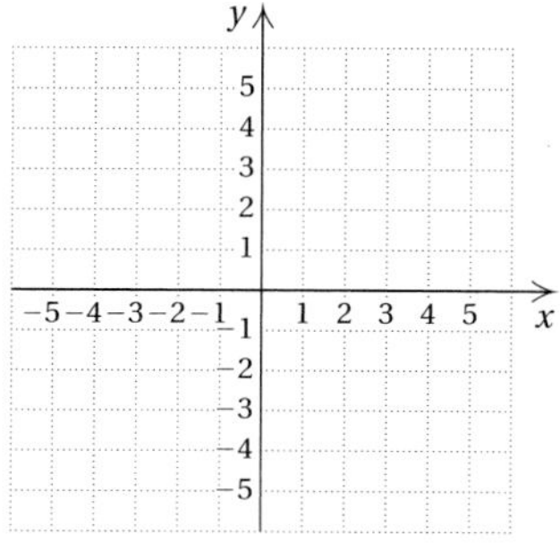

23. $2x - 3y = 6$

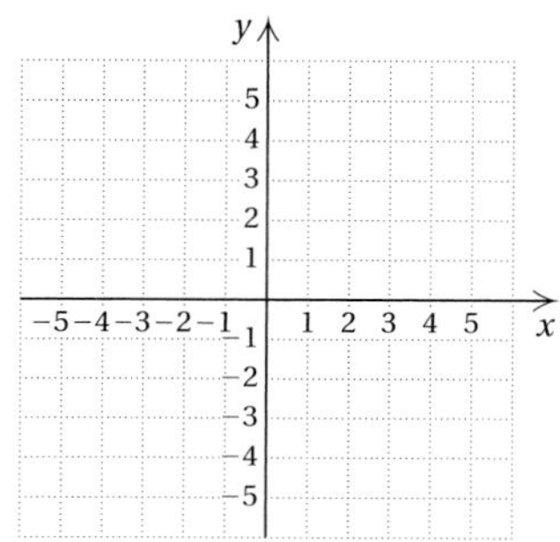

24. $2x - 5y = 10$

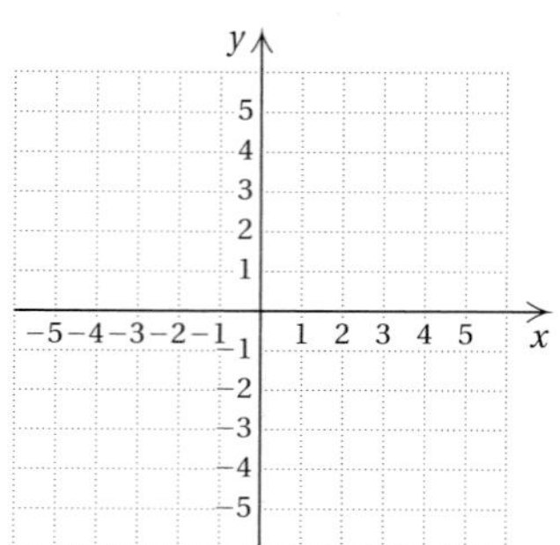

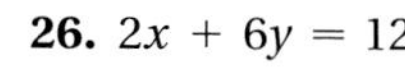

25. $4x + 5y = 20$

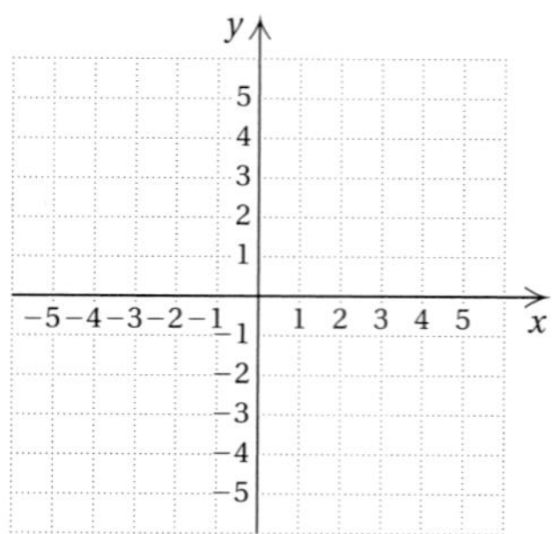

26. $2x + 6y = 12$

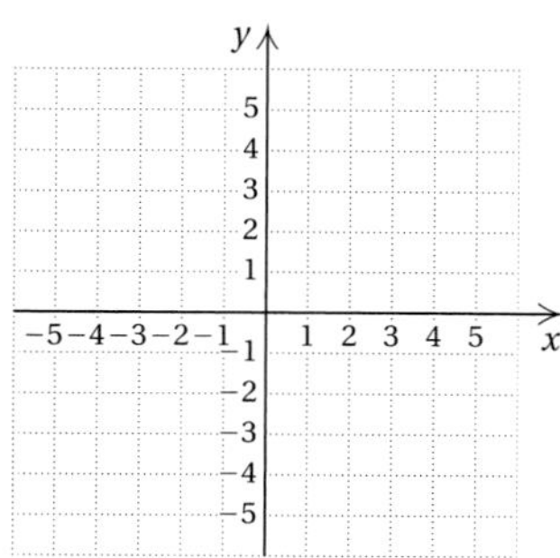

27. $2x + 3y = 8$

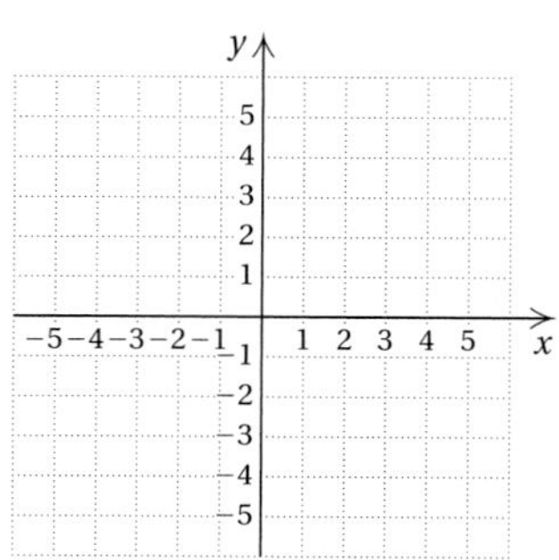

28. $x - 1 = y$

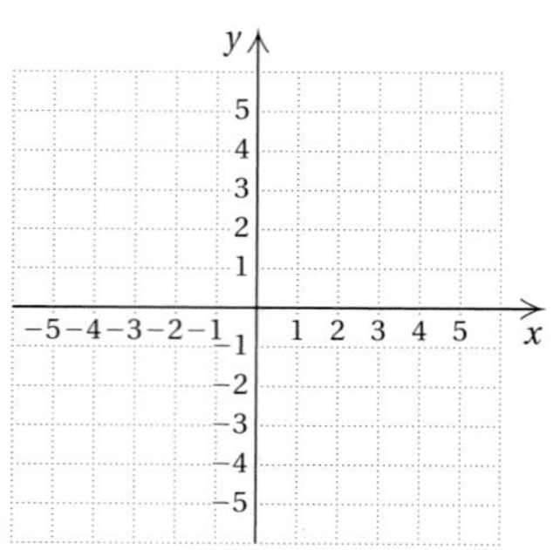

29. $x - 3 = y$

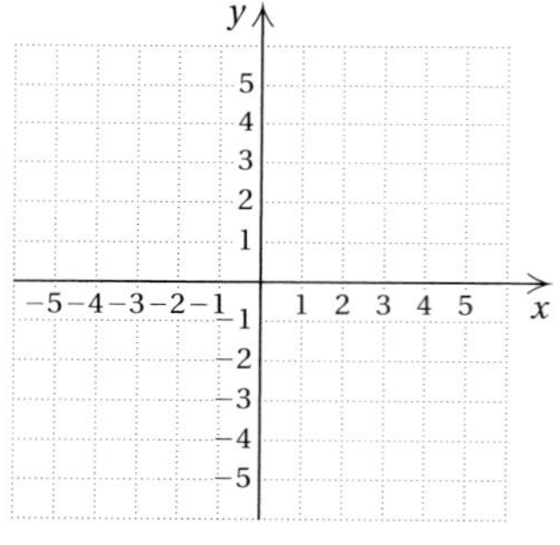

30. $2x - 1 = y$

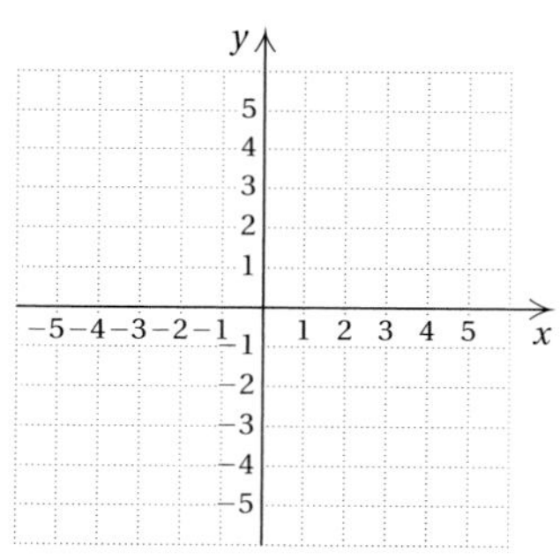

31. $3x - 2 = y$

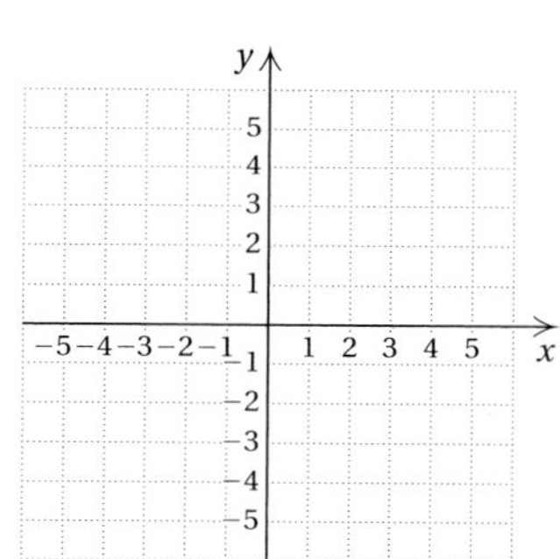

32. $4x - 3y = 12$

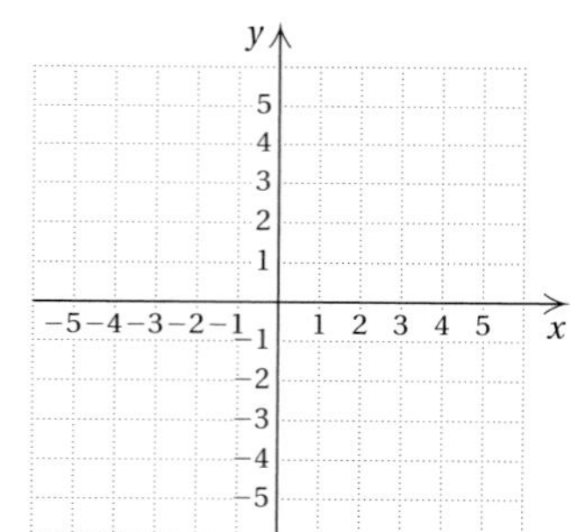

33. $6x - 2y = 12$

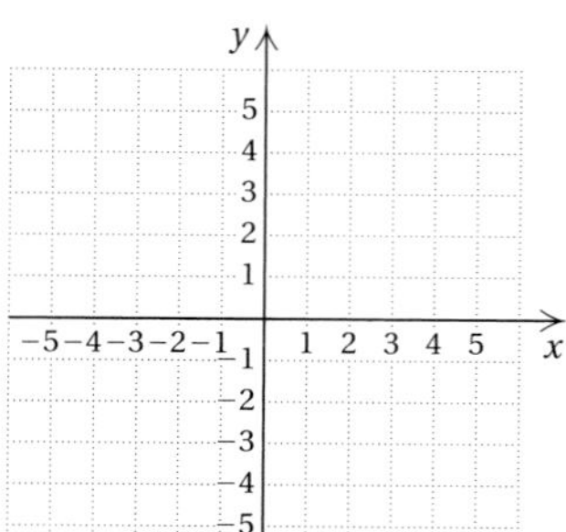

34. $7x + 2y = 6$

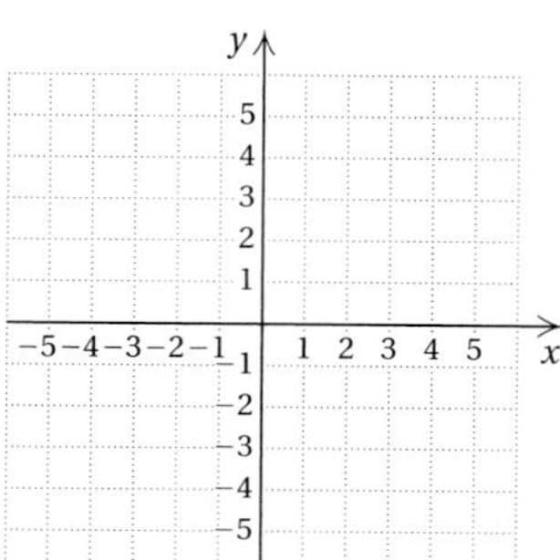

35. $3x + 4y = 5$

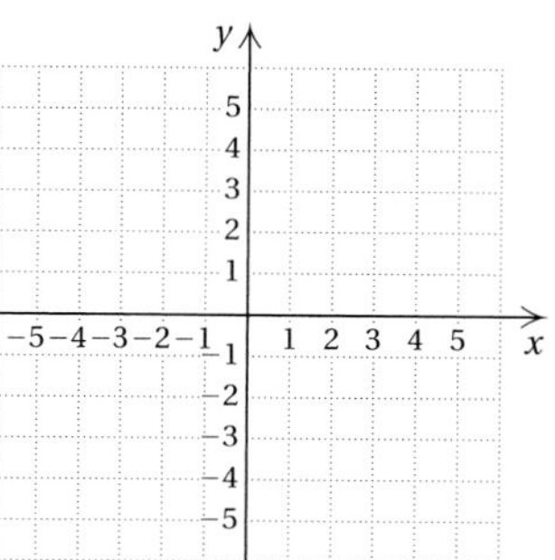

36. $y = -4 - 4x$

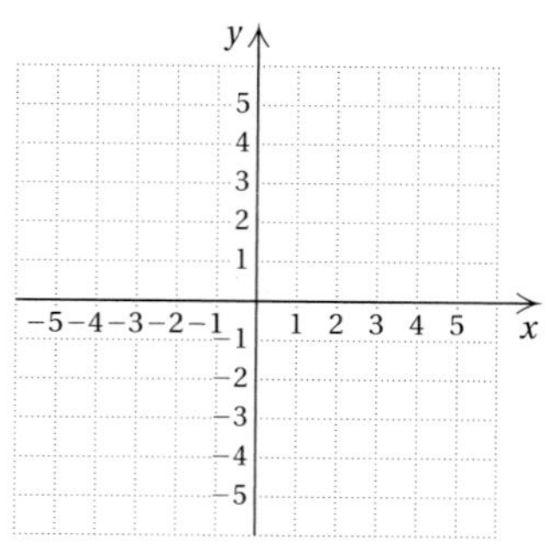

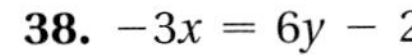

37. $y = -3 - 3x$

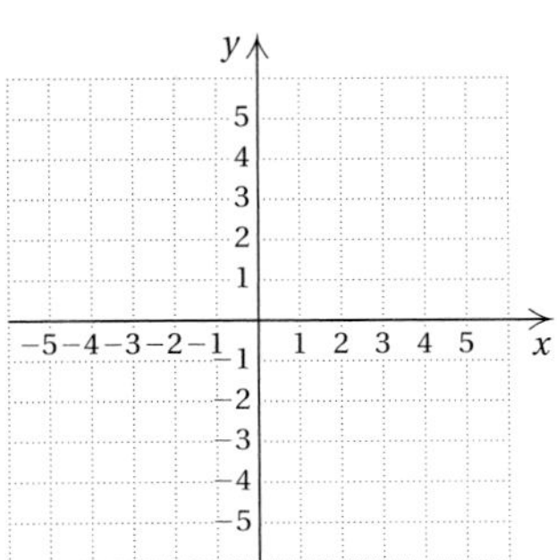

38. $-3x = 6y - 2$

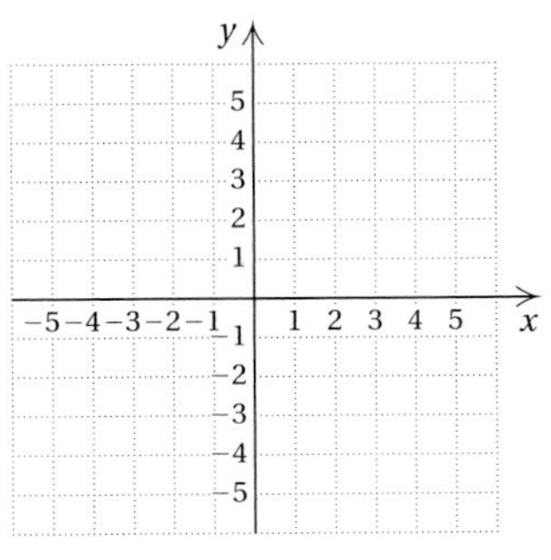

39. $y - 3x = 0$

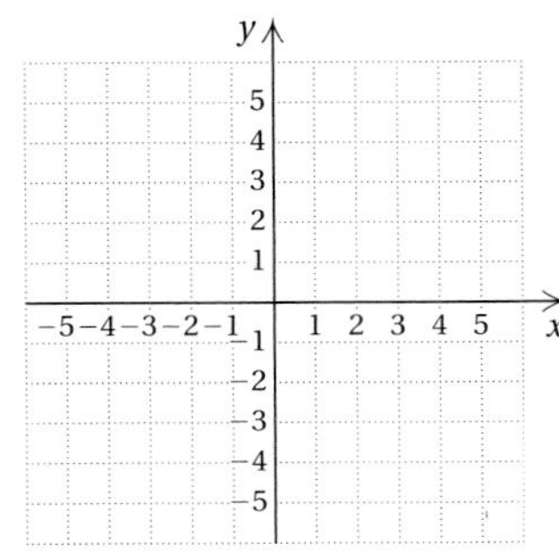

40. $x + 2y = 0$

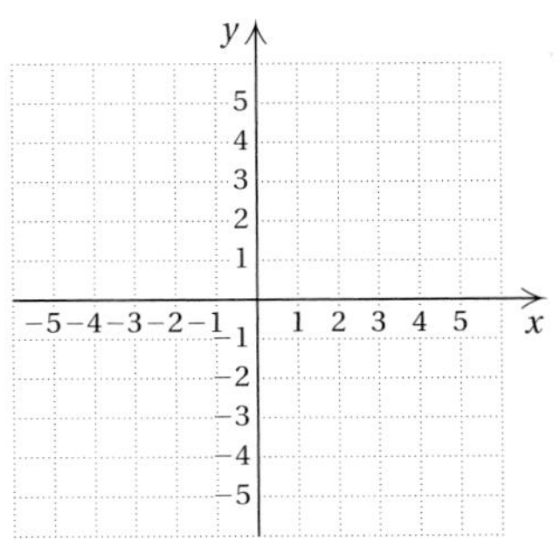

b Graph.

41. $x = -2$

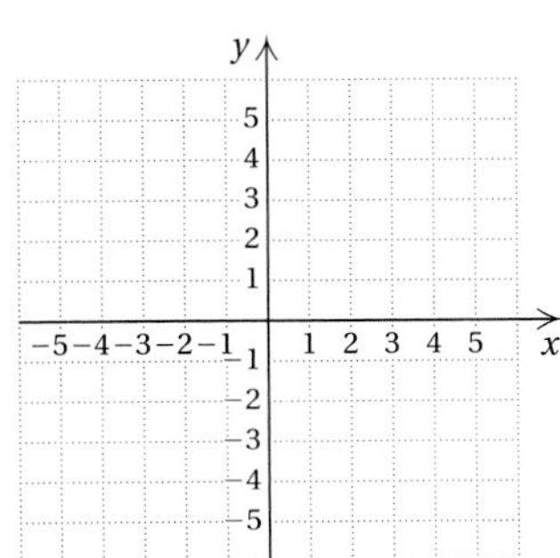

42. $x = 1$

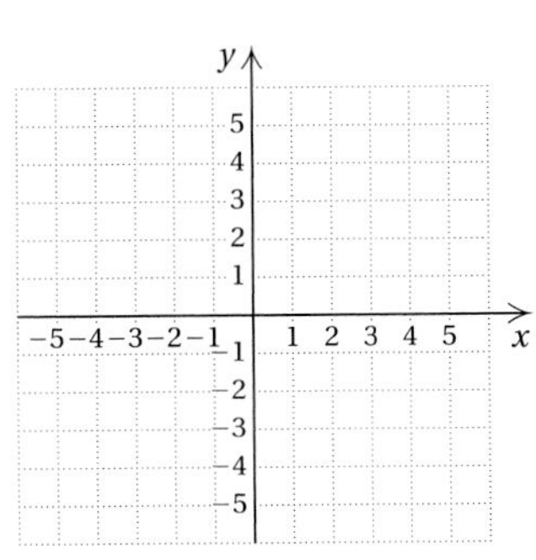

43. $y = 2$

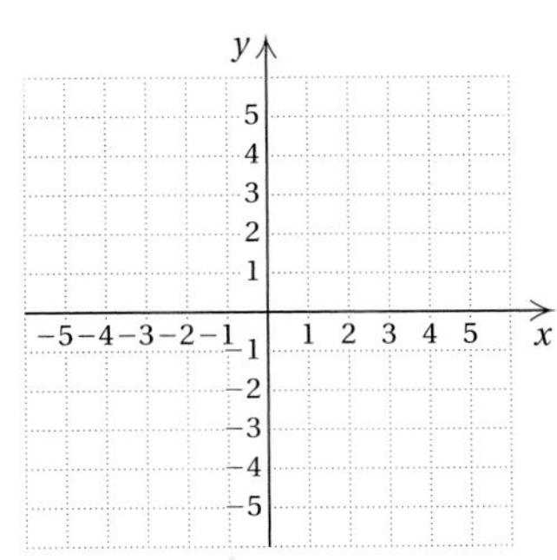

44. $y = -4$

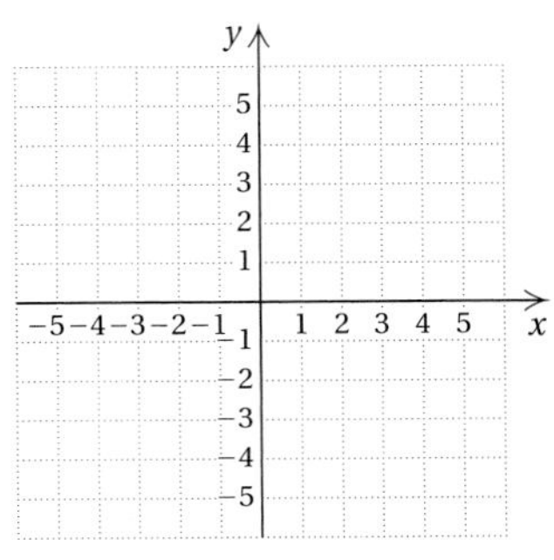

45. $x = 2$

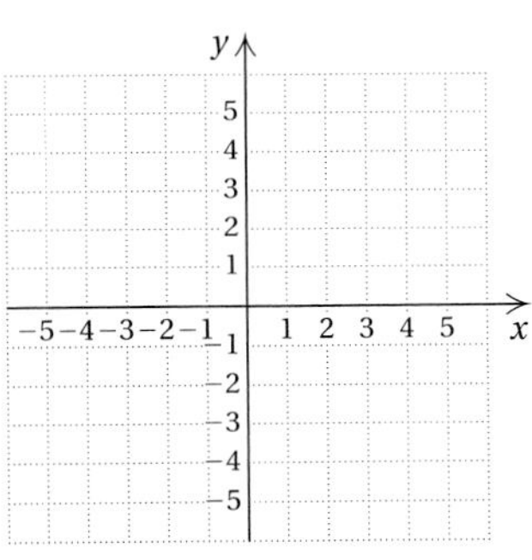

46. $x = 3$

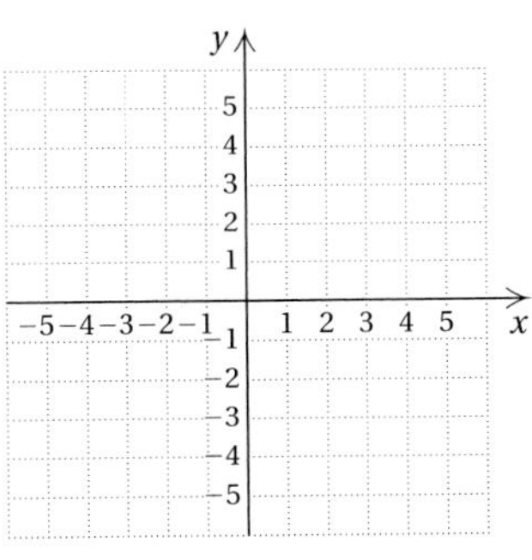

47. $y = 0$

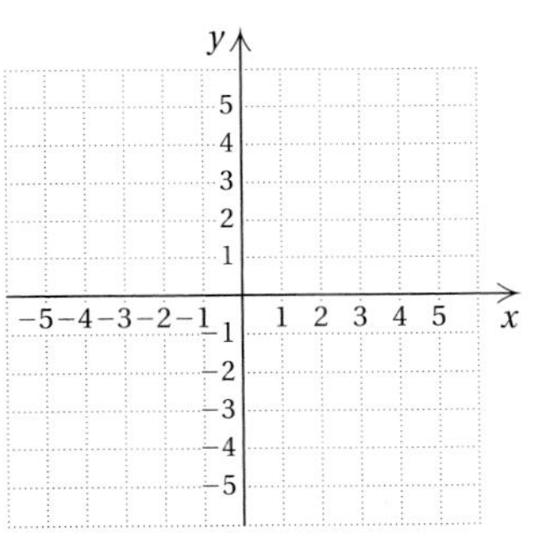

48. $y = -1$

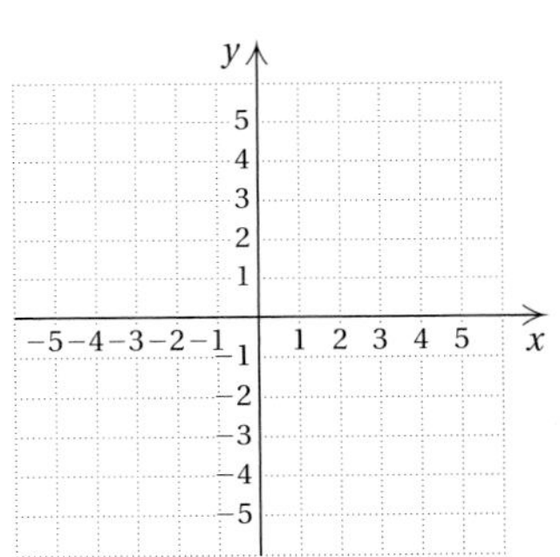

49. $x = \frac{3}{2}$

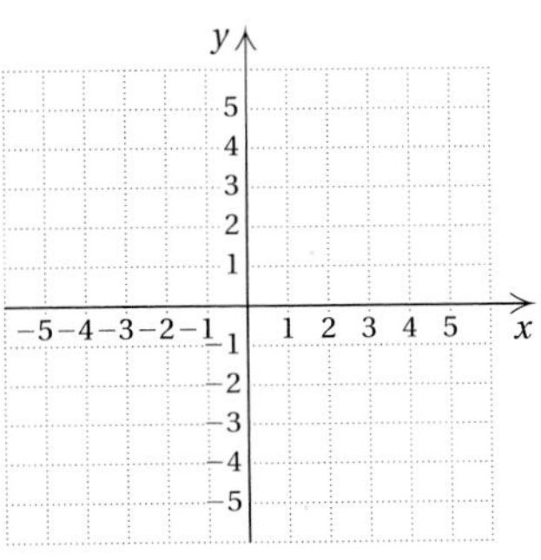

50. $x = -\frac{5}{2}$

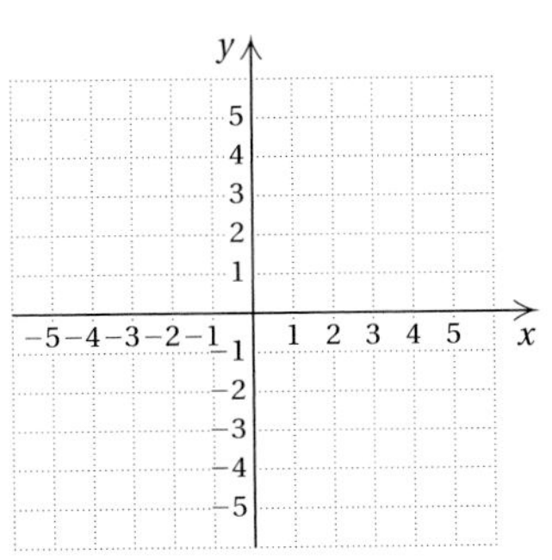

51. $3y = -5$

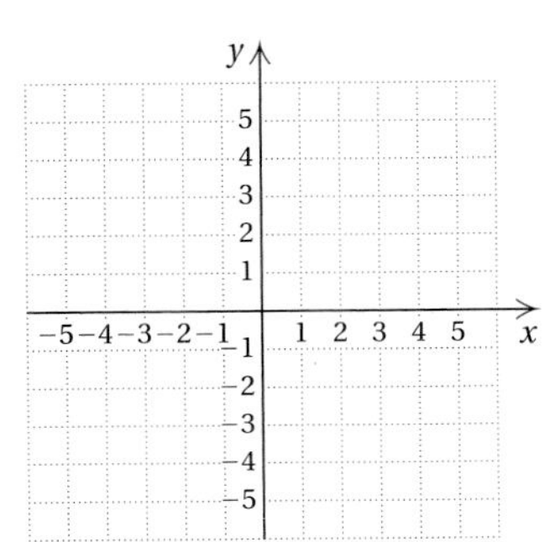

52. $12y = 45$

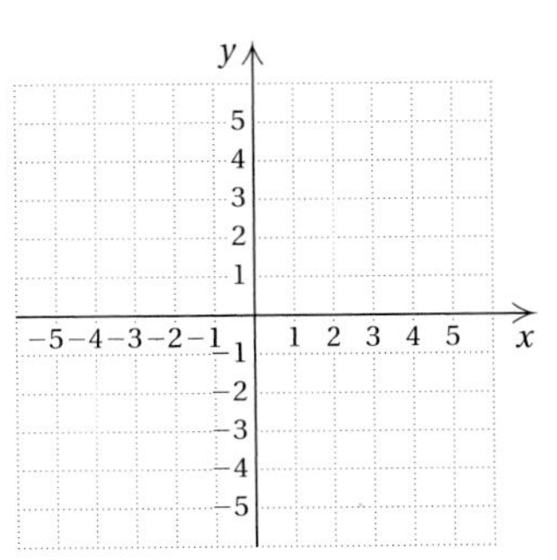

53. $4x + 3 = 0$

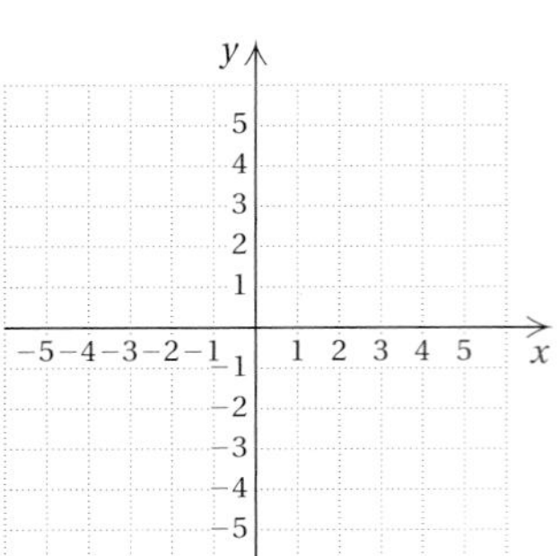

54. $-3x + 12 = 0$

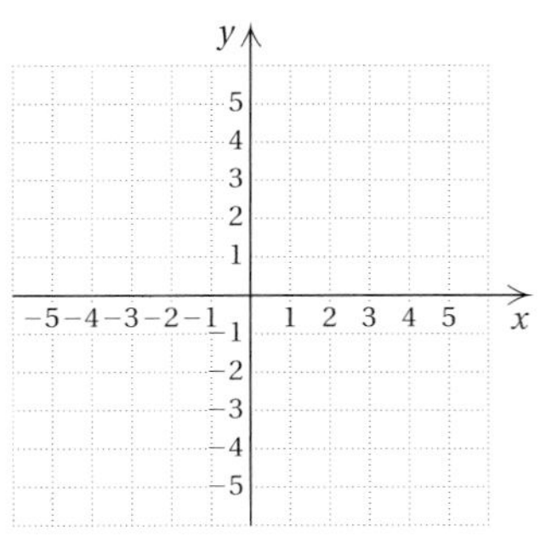

55. $48 - 3y = 0$

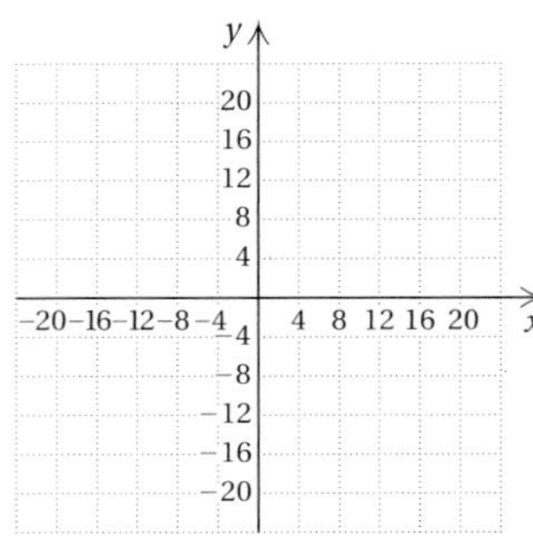

56. $63 + 7y = 0$

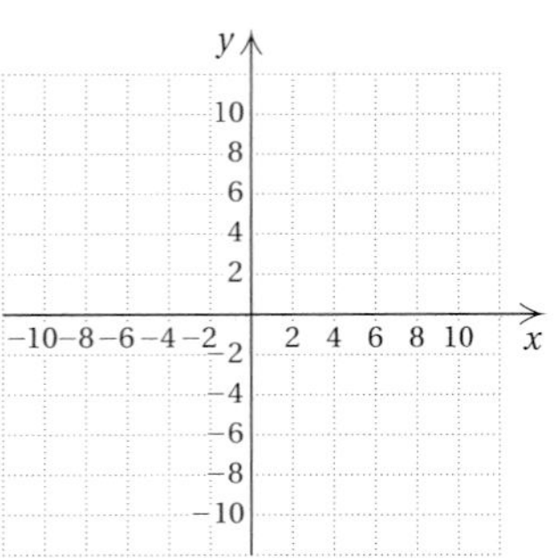

Write an equation for the graph shown.

57.

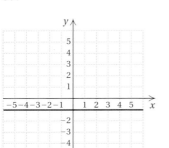

58.

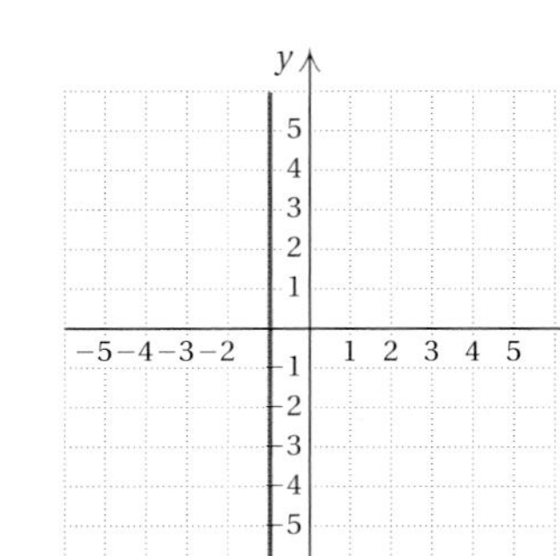

59.

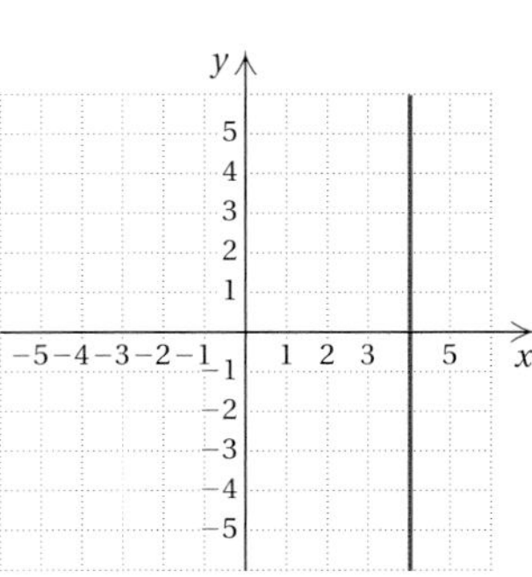

60.

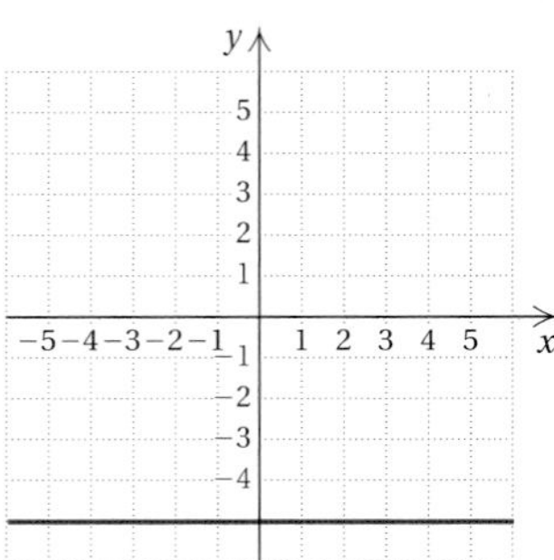

Skill Maintenance

Solve. [8.5a]

61. *Desserts.* If a restaurant sells 250 desserts in an evening, it is typical that 40 of them will be pie. What percent of the desserts sold will be pie?

62. *Desserts.* Of all desserts sold in restaurants, 20% of them are chocolate cake. One evening a restaurant sells 350 desserts. How many were chocolate cake?

63. Harry left a 20% tip of $6.50 for a meal. What was the cost of the meal before the tip?

64. Rambeau paid $27.60 for a taxi ride. This included a 20% tip. How much was the fare before the tip?

Solve. [8.7d, e]

65. $-1.6x < 64$

66. $-12x - 71 \geq 13$

67. $x + (x - 1) < (x + 2) - (x + 1)$

68. $6 - 18x \leq 4 - 12x - 5x$

Synthesis

69. ◆ If the graph of the equation $Ax + By = C$ is a horizontal line, what can you conclude about A? Why?

70. ◆ Explain in your own words why the graph of $x = 7$ is a vertical line.

71. Write an equation for the y-axis.

72. Write an equation for the x-axis.

73. Write an equation of a line parallel to the x-axis and passing through $(-3, -4)$.

74. Find the value of m such that the graph of $y = mx + 6$ has an x-intercept of $(2, 0)$.

75. Find the value of k such that the graph of $3x + k = 5y$ has an x-intercept of $(-4, 0)$.

76. Find the value of k such that the graph of $4x = k - 3y$ has a y-intercept of $(0, -8)$.

9.4 Applications and Data Analysis with Graphs

Objectives

a Find the mean (average), the median, and the mode of a set of data and solve related applied problems.

b Compare two sets of data using their means.

c Make predictions from a set of data using interpolation or extrapolation.

For Extra Help

TAPE 18 | MAC WIN | CD-ROM

a Mean, Median, and Mode

One way to analyze data is to look for a single representative number, called a *center point* or *measure of central tendency.* Those most often used are the *mean* (or *average*), the *median,* and the *mode.*

Mean

Let's first consider the *mean,* or average.

> The **mean**, or **average**, of a set of numbers is the sum of the numbers divided by the number of addends.

Example 1 Consider the following data on revenue, in billions of dollars, at McDonald's restaurants in five recent years:

\$12.5, \$13.2, \$14.2, \$14.9, \$15.9.

(***Source*:** McDonalds Corporation). What is the mean of the numbers?

First, we add the numbers:

$$12.5 + 13.2 + 14.2 + 14.9 + 15.9 = 70.7.$$

Then we divide by the number of addends, 5:

$$\frac{(12.5 + 13.2 + 14.2 + 14.9 + 15.9)}{5} = \frac{70.7}{5} = 14.14.$$

The mean, or average, revenue of McDonald's for those five years is \$14.14 billion.

Note that

$$14.14 + 14.14 + 14.14 + 14.14 + 14.14 = 70.7.$$

If we use this center point, 14.14, repeatedly as the addend, we get the same sum that we do when adding the individual data numbers.

Do Exercises 1–3.

Find the mean. Round to the nearest tenth.

1. 28, 103, 39

2. 85, 46, 105.7, 22.1

3. A student scored the following on five tests:

78, 95, 84, 100, 82.

What was the average score?

Answers on page A-27

Median

The *median* is useful when we wish to de-emphasize extreme scores. For example, suppose five workers in a technology company manufactured the following numbers of computers during one day's work:

Sarah:	88	Jen:	94
Matt:	92	Mark:	91
Pat:	66		

Let's first list the scores in order from smallest to largest:

66 88 91 92 94.

↑ Middle number (91)

The middle number—in this case, 91—is the **median.**

Find the median.

4. 17, 13, 18, 14, 19

5. 17, 18, 16, 19, 13, 14

6. 122, 102, 103, 91, 83, 81, 78, 119, 88

Find any modes that exist.

7. 33, 55, 55, 88, 55

8. 90, 54, 88, 87, 87, 54

9. 23.7, 27.5, 54.9, 17.2, 20.1

10. In conducting laboratory tests, Carole discovers bacteria in different lab dishes grew to the following areas, in square millimeters:

25, 19, 29, 24, 28.

a) What is the mean?

b) What is the median?

c) What is the mode?

Answers on page A-27

> Once a set of data has been arranged from smallest to largest, the **median** of the set of data is the middle number if there is an odd number of data numbers. If there is an even number of data numbers, then there are two middle numbers and the median is the *average* of the two middle numbers.

Example 2 What is the median of the following set of yearly salaries?

\$76,000, \$58,000, \$87,000, \$32,500, \$64,800, \$62,500

We first rearrange the numbers in order from smallest to largest.

\$32,500 \$58,000 \$62,500 \$64,800 \$76,000 \$87,000

↑ Median

There is an even number of numbers. We look for the middle two, which are \$62,500 and \$64,800. In this case, the median is the average of \$62,500 and \$64,800:

$$\frac{\$62{,}500 + \$64{,}800}{2} = \$63{,}650.$$

Do Exercises 4–6.

Mode

The last center point we consider is called the **mode**. A number that occurs most often in a set of data can be considered a representative number or center point.

> The **mode** of a set of data is the number or numbers that occur most often. If each number occurs the same number of times, there is *no* mode.

Example 3 Find the mode of the following data:

23, 24, 27, 18, 19, 27

The number that occurs most often is 27. Thus the mode is 27.

Example 4 Find the mode of the following data:

83, 84, 84, 84, 85, 86, 87, 87, 87, 88, 89, 90.

There are two numbers that occur most often, 84 and 87. Thus the modes are 84 and 87.

Example 5 Find the mode of the following data:

115, 117, 211, 213, 219.

Each number occurs the same number of times. The set of data has *no* mode.

Do Exercises 7–10.

b Comparing Two Sets of Data

We have seen how data are displayed and interpreted using graphs, and we have calculated the mean, the median, and the mode from data. Now we look into using data analysis to solve applied problems.

One goal of analyzing two sets of data is to make a determination about which of two groups is "better." One way to do so is by comparing the means.

Example 6 *Light-Bulb Testing.* An experiment is performed to compare the lives of two types of light bulb. Several bulbs of each type were tested and the results are listed in the following table. On the basis of this test, which bulb is better?

Bulb A: HotLight Life Times (in hours)		
983	964	1214
1417	1211	1521
1084	1075	892
1423	949	

Bulb B: BrightBulb Life Times (in hours)		
979	1083	1344
984	1445	975
1492	1325	1283
1325	1352	1432

Note that it is difficult to analyze the data at a glance because the numbers are close together and there is a different number of data points in each set. We need a way to compare the two groups. Let's compute the average of each set of data.

Bulb A: Average

$$= \frac{(983 + 964 + 1214 + 1417 + 1211 + 1521 + 1084 + 1075 + 892 + 1423 + 949)}{11}$$

$$= \frac{12{,}733}{11} \approx 1157.55;$$

Bulb B: Average

$$= \frac{(979 + 1083 + 1344 + 984 + 1445 + 975 + 1492 + 1325 + 1283 + 1325 + 1352 + 1432)}{12}$$

$$= \frac{15{,}019}{12} \approx 1251.58.$$

We see that the average life of bulb B is higher than that of bulb A and thus conclude that bulb B is "better." (It should be noted that statisticians might question whether these differences are what they call "significant." The answer to that question belongs to a later math course.)

Do Exercise 11.

11. *Quality of Baseballs.* Lauri experiments to see which of two kinds of baseball is better by determining which bounces highest. She drops balls of each kind from a height of 6 ft onto concrete and measures how high they bounce, in inches. Which kind of ball is better?

Ball A Bouncing Heights (in inches)			
16.2	22.3	19.5	15.7
19.6	18.0	15.6	21.7
19.8	16.4	18.4	16.6
21.5	18.7	22.0	18.3

Ball B Bouncing Heights (in inches)			
19.7	18.4	19.7	17.2
19.7	14.6	22.0	23.7
16.5	21.6	22.5	19.8
22.6	17.9	18.7	

Answer on page A-27

c Making Predictions

Sometimes we use data to make predictions or estimates of missing data points. One process for doing so is called *interpolation.* It uses graphs and/or averages to guess a missing data point between two data points.

Example 7 *World Bicycle Production.* The following table shows how world bicycle production has grown in recent years. Note that there is no data for 1994. Use interpolation to estimate the missing value.

Year	World Bicycle Production (in millions)
1989	95
1990	90
1991	96
1992	103
1993	108
1994	?
1995	114

Source: United Nations Interbike Directory

First, we analyze the data and look for trends. Note that production decreases for the early years, but in more recent years it increases more like a straight line. It seems reasonable that we can draw a line between the points for 1993 and 1995. We zoom in on that portion of the graph, as shown below.

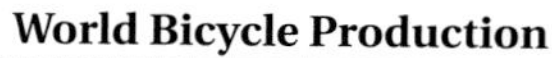

World Bicycle Production

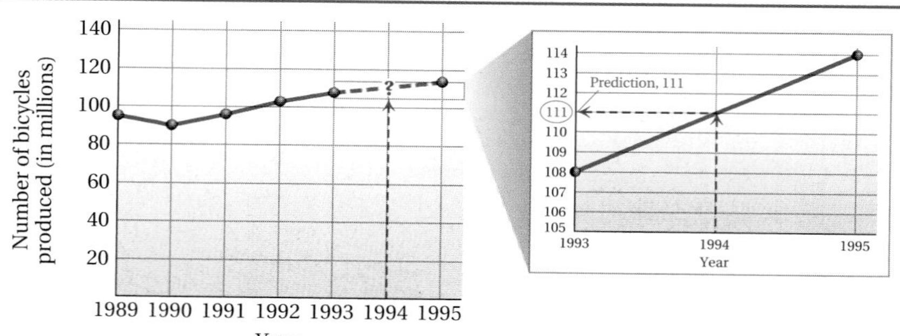

Then we visualize a vertical line up from the point for 1994 and see where the vertical line crosses the line between the data points (1993, 108) and (1995, 114). We move to the left and read off a value—about 111. We can also estimate this value by taking the average of the data values 108 and 114:

$$\frac{(108 + 114)}{2} = 111.$$

When we estimate in this way to find an "in-between value," we are using a process called **interpolation.** Real-world information about the data might tell us that an estimate found in this way is unreliable. For example, data from the stock market might be very erratic.

Do Exercise 12.

12. *Monthly Loan Payment.* The following table lists monthly payments on a loan of $110,000 at 9% interest. Note that there is no data point for a 35-yr loan. Use interpolation to estimate the missing value.

Number of Years	Monthly Payment (in dollars)
5	$2283.42
10	1393.43
15	1115.69
20	989.70
25	923.12
30	885.08
35	?
40	848.50

Answer on page A-27

We often analyze data with the view of going "beyond" the data. One process for doing so is called *extrapolation.*

Example 8 *World Bicycle Production Extended.* Let's now consider that we know the world production of bicycles in 1994 to be 111 million and add it to the table of data. Use extrapolation to estimate world bicycle production in 1996.

Year	World Bicycle Production (in millions)
1989	95
1990	90
1991	96
1992	103
1993	108
1994	111
1995	114
1996	?

Source: United Nations Interbike Directory

World Bicycle Production

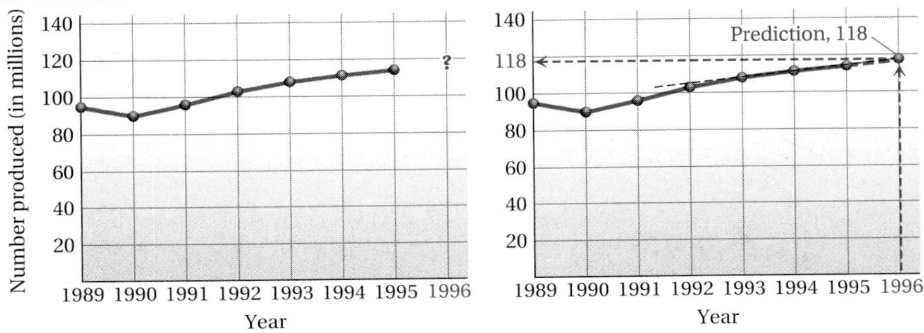

First, we analyze the data and note that they tend to follow a straight line past 1993. Keeping this trend in mind, we draw a "representative" line through the data and beyond. To estimate a value for 1996, we draw a vertical line up from 1996 until it hits the representative line. We go to the left and read off a value—about 118. When we estimate in this way to find a "go-beyond value," we are using a process called **extrapolation.** Answers found with this method can vary greatly depending on the points chosen to determine the "representative" line.

Do Exercise 13.

13. *Study Time and Test Scores.* A professor gathered the following data comparing study time and test scores. Use extrapolation to estimate the test score received when studying for 22 hr.

Study Time (in hours)	Test Grade (in percent)
18	84
19	86
20	89
21	92
22	?

Answer on page A-27

Calculator Spotlight

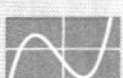

Trace and Table Features

Trace Feature. There are two ways in which we can determine the coordinates of points on a graph drawn by a grapher. One approach is to use a TRACE key.

Let's consider the equation for the FedEx mailing costs considered in Example 8 of Section 9.2.

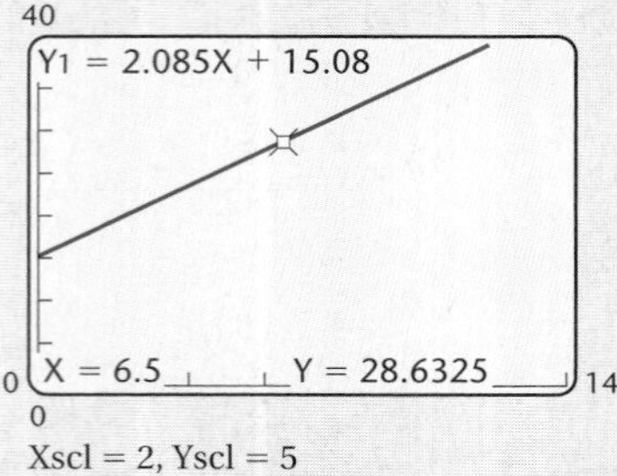

The cursor on the line means that the TRACE feature has been activated. The coordinates at the bottom indicate that the cursor is at the point with coordinates (6.5, 28.6325). By using the arrow keys, we can obtain coordinates of other points. For example, if we press the left arrow key ◁ seven times, we move the cursor to the location shown below, obtaining a point on the graph with coordinates (5.5319149, 26.614043).

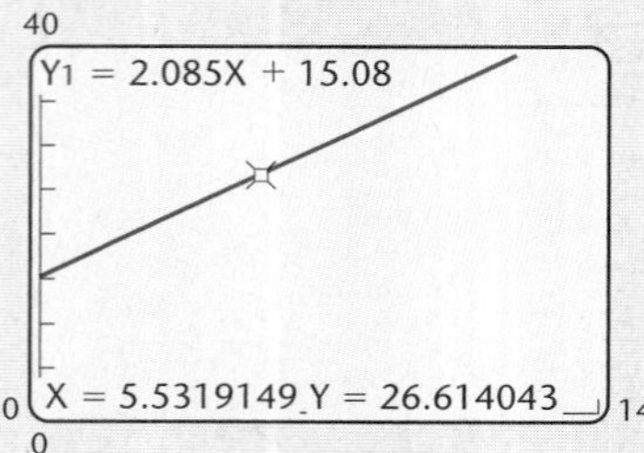

Table Feature. Another way to find the coordinates of solutions of equations makes use of the TABLE feature. We first press 2nd TBLSET and set TblStart = 0 and $\triangle$Tbl = 10.

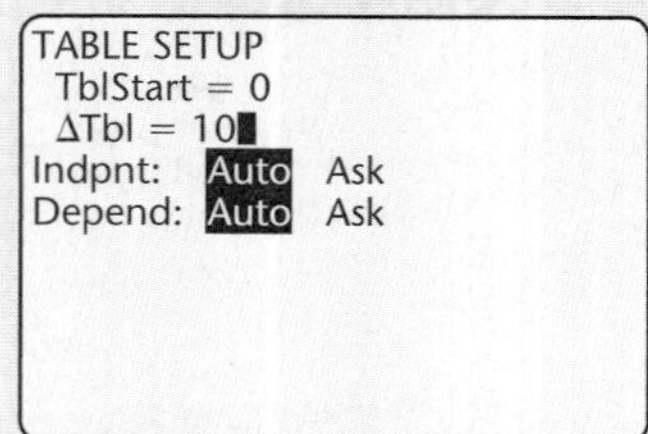

This means that the table's x-values will start at 0 and increase by 10. By setting Indpnt and Depend to Auto, we obtain the following when we press 2nd TABLE: The arrow keys allow us to scroll up and down the table and extend it to other values not initially shown.

X	Y1
0	15.08
10	35.93
20	56.78
30	77.63
40	98.48
50	119.33
60	140.18

X = 0

X	Y1
20	56.78
30	77.63
40	98.48
50	119.33
60	140.18
70	161.03
80	181.88

X = 80

Exercises

1. Use the TRACE feature to find five different ordered-pair solutions of the equation $y = 2.085x + 15.08$.
2. Use the TABLE feature to construct a table of solutions of the equation $y = 2.085x + 15.08$. Set TblStart = 100 with $\triangle$Tbl = 50. Find the value of y when x is 100. Then find the value of y when x is 150.
3. Again, use the equation $y = 2.085x + 15.08$ and adjust the table settings to Indpnt: Ask. How does the table change? Enter a number of your choice and see what happens. Use this setting to find the value of y when x is 187.
4. *Value of an Office Machine.* (Refer to Margin Exercise 14 in Section 9.2.) The value of Dupliographic's color copier is given by $v = -0.68t + 3.4$, where v = the value, in thousands of dollars, t years from the date of purchase.
   a) Graph the equation, choosing an appropriate viewing window that shows the intercepts.
   b) Does the equation have meaning for negative x-values?
   c) Find the x-intercept.
   d) For what x-values does the equation have meaning?
   e) Use the TRACE feature to find five ordered-pair solutions for values of x between 0 and 5.
   f) After what amount of time is the value of the copier about $1500?
   g) Using the TABLE feature, find the value of the copier after 1.7 yr, 2.3 yr, 4.1 yr, and 5 yr.

Exercise Set 9.4

a For each set of numbers, find the mean, the median, and any modes that exist. Round to the nearest tenth.

1. 15, 40, 30, 30, 45, 15, 25

2. 26, 28, 39, 24, 39, 29, 25

3. 81, 93, 96, 98, 102, 94

4. 23.4, 23.4, 22.6, 52.9

5. 23, 42, 35, 37, 23

6. 101.2, 104.3, 107.4, 105.7, 107.4

7. *Coffee Consumption.* The following lists the annual coffee consumption, in cups per person, for various countries.

Germany	1113
United States	610
Switzerland	1215
France	798
Italy	750

(***Source***: Beverage Marketing Corporation)

8. *Calories in Cereal.* The following lists the caloric content of a 2-cup bowl of certain cereals.

Ralston Rice Chex	240
Kellogg's Complete Bran Flakes	240
Kellogg's Special K	220
Honey Nut Cheerios	240
Wheaties	220

9. *NBA Tall Men.* The following is a list of the heights, in inches, of the tallest men in the NBA in a recent year.

Shaquille O'Neal	85
Gheorghe Muresan	91
Shawn Bradley	90
Priest Lauderdale	88
Rik Smits	88
David Robinson	85
Arvydas Sabonis	87

(***Source***: National Basketball Association)

10. *Movie Tickets Sold.* The following lists the number of movie tickets sold, in billions, in six recent years.

1990	1.19
1991	1.14
1992	1.17
1993	1.24
1994	1.29
1995	1.26

(***Source***: Motion Picture Association of America)

b Compare the set of data using their means.

11. *Battery Testing.* An experiment is performed to compare battery quality. Two kinds of battery were tested to see how long, in hours, they kept a portable CD player running. On the basis of this test, which battery is better?

Battery A: EternReady Times (in hours)			Battery B: SturdyCell Times (in hours)		
27.9	28.3	27.4	28.3	27.6	27.8
27.6	27.9	28.0	27.4		27.9
26.8	27.7	28.1	26.9	27.8	28.1
28.2	26.9	27.4	27.9	28.7	27.6

12. *Growth of Wheat.* A farmer experiments to see which of two kinds of wheat is better. (In this situation, the shorter wheat is considered "better.") He grows both kinds under similar conditions and measures stalk heights, in inches, as follows. Which kind is better?

Wheat A Stalk Heights (in inches)				Wheat B Stalk Heights (in inches)			
16.2	42.3	19.5	25.7	19.7	18.4	32.0	25.7
25.6	18.0	15.6	41.7	19.7	21.6	42.5	32.6
22.6	26.4	18.4	12.6	14.0	10.9	26.7	22.8
41.5	13.7	42.0	21.6	22.6	19.7	17.2	

c Use interpolation or extrapolation to estimate the missing data values. Answers found using expolation can vary greatly.

13. *Height of Girls.*

Age	Height (in centimeters)
2	95.7
4	103.2
6	115.9
8	128.0
10	138.6
12	151.9
14	159.6
16	162.2
17	?
18	162.5

Source: Kempe, C. Henry, M.D., et al., eds., *Current Pediatric Diagnosis & Treatment 1987.* Norwalk, CT: Appleton & Lange, 1987.

14. *Height of Boys.*

Age	Height (in centimeters)
2	96.2
4	103.4
6	117.5
8	130.0
10	140.3
12	149.6
14	162.7
15	?
16	171.6
18	174.5

Source: Kempe, C. Henry, M.D., et al., eds., *Current Pediatric Diagnosis & Treatment 1987.* Norwalk, CT: Appleton & Lange, 1987.

15. *World Population.*

Population (in billions)	Year in Which Population Is Reached
1	1804
2	1927
3	1960
4	1974
5	1987
6	1998
7	?

Source: U.S. Bureau of the Census

16. *SAT Scores.*

Year	Average SAT Score, Math and Verbal
1991	999
1992	1001
1993	1003
1994	1003
1995	1010
1996	1013
1997	?

Source: The College Board

17. *Movie Tickets Sold.*

Year	Tickets Sold (in billions)
1991	1.19
1992	1.14
1993	1.17
1994	1.24
1995	1.29
1996	1.26
1997	?
2000	?

Source: Motion Picture Association of America

18. *McDonald's Restaurant Revenue in the United States.*

Year	Revenue (in billions)
1990	$12.3
1991	12.5
1992	13.2
1993	14.2
1994	14.9
1995	15.9
1996	?
2000	?

Source: McDonald's Corporation

19. *Average Price of a 30-Second Super Bowl Commercial.*

Year	Cost
1991	$ 800,000
1992	850,000
1993	850,000
1994	900,000
1995	1,000,000
1996	1,300,000
1997	?
2000	?

Source: National Football League

20. *Retail Revenue from Lettuce.*

Year	Revenue (in millions)
1991	$ 106
1992	168
1993	312
1994	577
1995	889
1996	1100
1997	?
2000	?

Source: International Fresh-Cut Produce Association, Information Resources

21. *Study Time vs. Grades.* A mathematics instructor asked her students to keep track of how much time each spent studying the chapter on percent notation in her basic mathematics course. She collected the information together with test scores from that chapter's test. The data are given in the following table. Estimate the missing data value.

Study Time (in hours)	Test Grade (in percent)
9	76
11	94
13	81
15	86
16	87
17	81
18	?
19	87
20	92

22. *Maximum Heart Rate.* A person's maximum heart rate depends on his or her gender, age, and resting heart rate. The following table relates resting heart rate and maximum heart rate for a 20-yr-old man. Estimate the missing data value.

Resting Heart Rate (in beats per minute)	Maximum Heart Rate (in beats per minute)
50	166
60	168
70	170
75	?
80	172

Source: American Heart Association

Skill Maintenance

Convert to fractional notation. [4.3b]

23. 16%

24. $33\frac{1}{3}\%$

25. 37.5%

26. 75%

Solve. [8.5a]

27. Jennifer left an $8.50 tip for a meal that cost $42.50. What percent of the cost of the meal was the tip?

28. Kristen left an 18% tip of $3.24 for a meal. What was the cost of the meal before the tip?

29. Juan left a 15% tip for a meal. The total cost of the meal, including the tip, was $51.92. What was the cost of the meal before the tip was added?

30. After a 25% reduction, a sweater is on sale for $41.25. What was the original price?

Synthesis

31. ◆ In a recent year, the average salary of all players in baseball's American League was $1.3 million and the median was $400,000 (***Source:*** Major League Baseball). Discuss the merits of each value as a measure of central tendency.

32. ◆ Discuss how you might test the estimates that you found in Exercises 13–20.

Graph the equation using the standard viewing window. Then construct a table of y-values for x-values starting at $x = -10$ with $\triangle\text{Tbl} = 0.1$.

33. $y = 0.35x - 7$

34. $y = 5.6 - x^2$

35. $y = x^3 - 5$

36. $y = 4 + 3x - x^2$

Collaborative Learning Manual

Perform a statistical analysis of pulse rates. Make predictions from a set of data.

Summary and Review Exercises: Chapter 9

The objectives to be tested in addition to the material in this chapter are [3.1e], [7.2c], [7.2e], and [8.5a].

1. *Federal Spending.* The following pie chart shows how our federal income tax dollars are used. As a freelance graphic artist, Jennifer pays $3525 in taxes. How much of Jennifer's tax payment goes toward defense? toward social programs? [9.1a]

Where Your Tax Dollars Are Spent

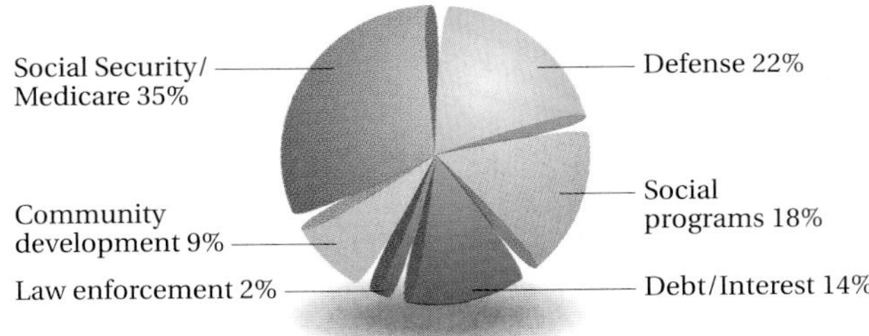

Source: U.S. Department of the Treasury

Chicken Consumption. The following line graph shows average chicken consumption from 1980 to 2000. (The value for the year 2000 is projected.) [9.1a]

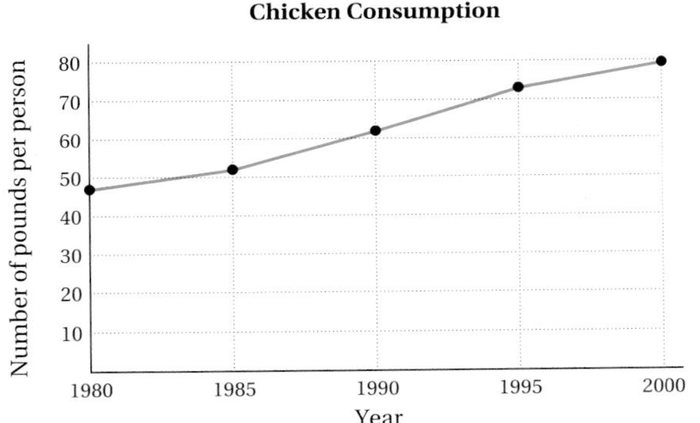

2. About how many pounds of chicken were consumed per person in 1980?

3. About how many pounds of chicken will be consumed per person in 2000?

4. By what amount did chicken consumption increase from 1980 to 2000?

5. In what year did the consumption of chicken exceed 70 lb per person?

6. In what 5-yr period was the difference in consumption the greatest?

Water Usage. The following bar graph shows water usage, in gallons, for various tasks. [9.1a]

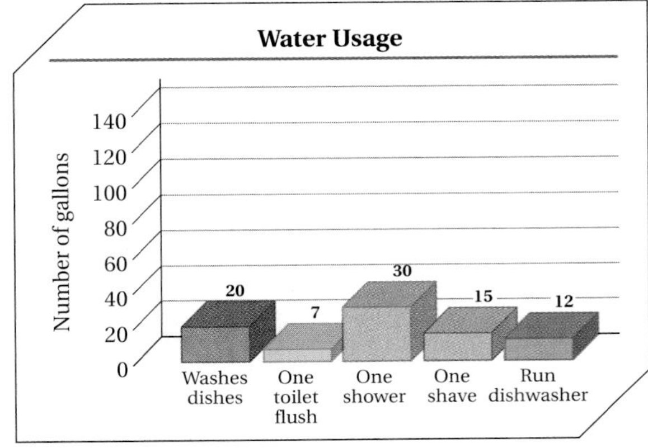

Source: American Water Works Association

7. Which task requires the most water?

8. Which task requires the least water?

9. Which tasks require 15 or more gallons?

10. Which task requires 7 gallons?

Find the coordinates of the point. [9.1d]

11. A **12.** B **13.** C

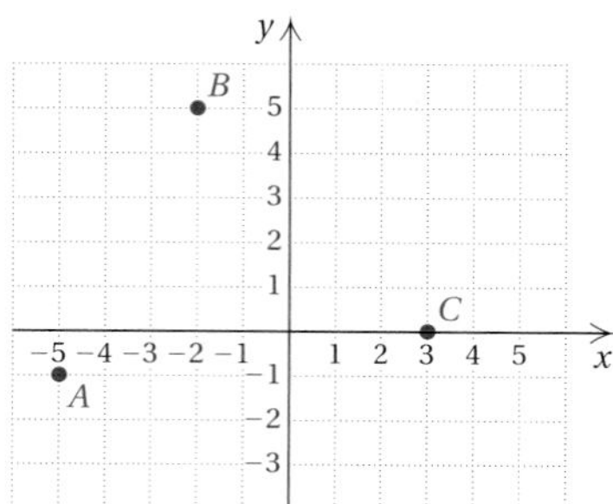

Plot the point. [9.1b]

14. (2, 5) **15.** (0, −3) **16.** (−4, −2)

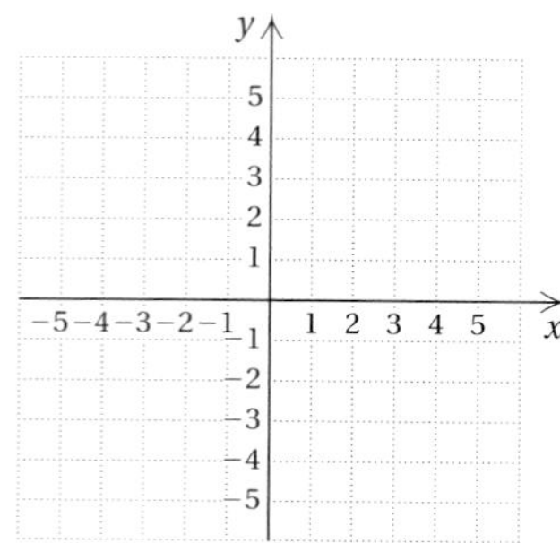

In which quadrant is the point located? [9.1c]

17. (3, −8) **18.** (−20, −14) **19.** (4.9, 1.3)

Determine whether the point is a solution of $2y - x = 10$. [9.2a]

20. (2, −6) **21.** (0, 5)

22. Show that the ordered pairs (0, −3) and (2, 1) are solutions of the equation $2x - y = 3$. Then use the graph of the two points to determine another solution. Answers may vary. [9.2a]

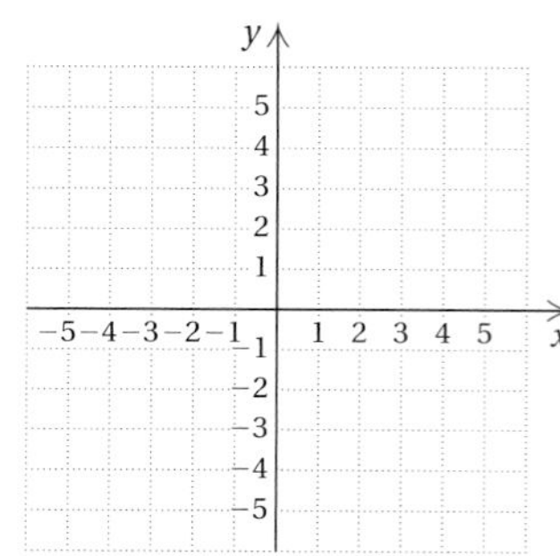

Graph the equation, identifying the y-intercept. [9.2b]

23. $y = 2x - 5$

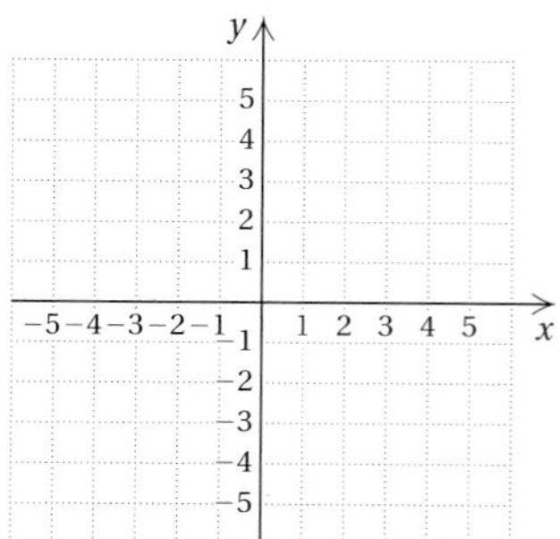

24. $y = -\frac{3}{4}x$

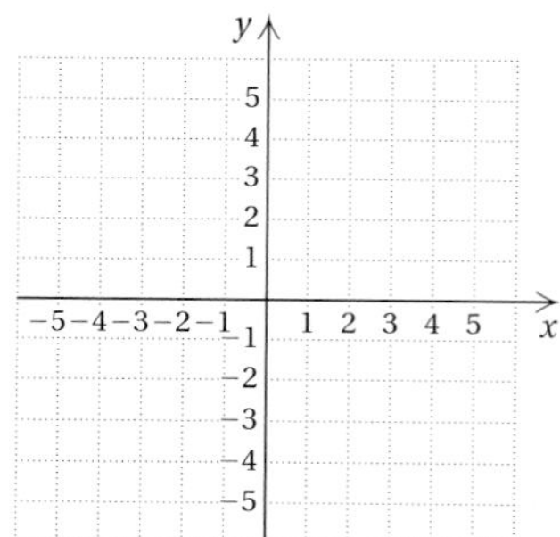

25. $y = -x + 4$

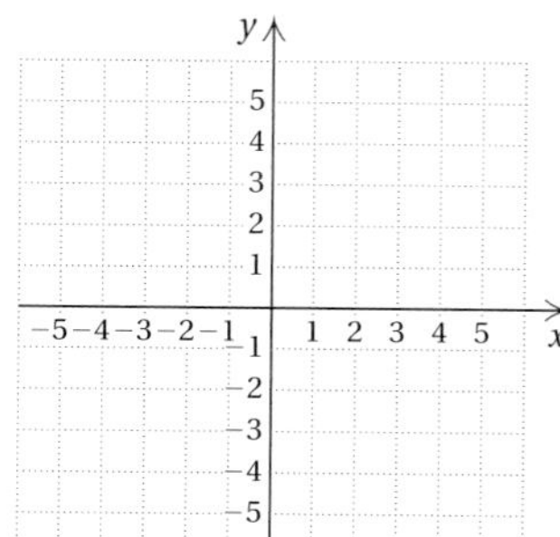

26. $y = 3 - 4x$

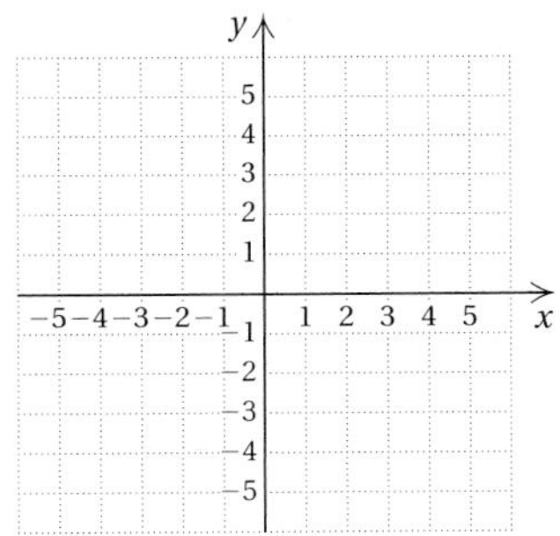

Graph the equation. [9.3b]

27. $y = 3$

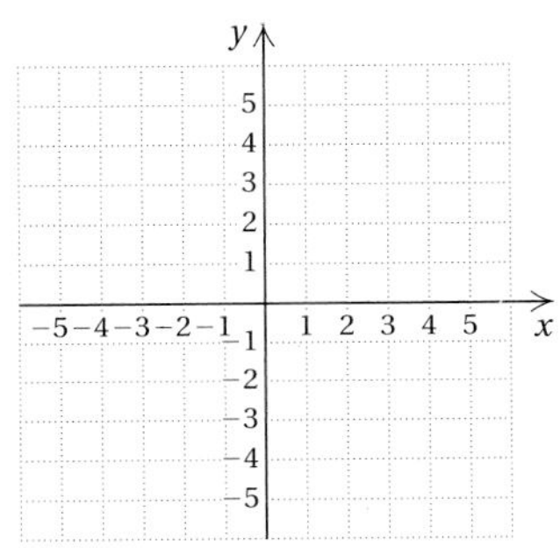

28. $5x - 4 = 0$

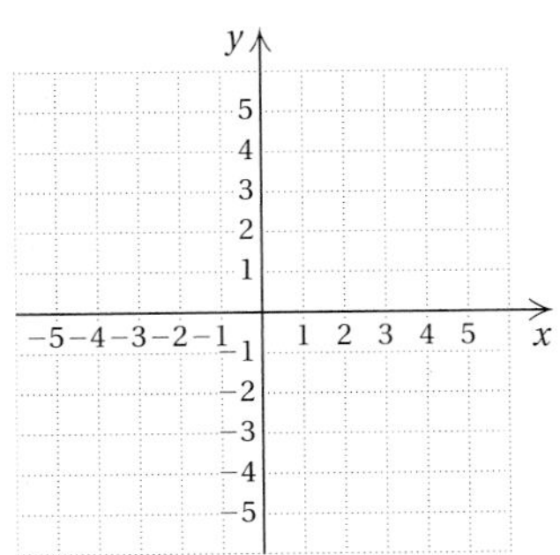

Find the intercepts of the equation. Then graph the equation. [9.3a]

29. $x - 2y = 6$

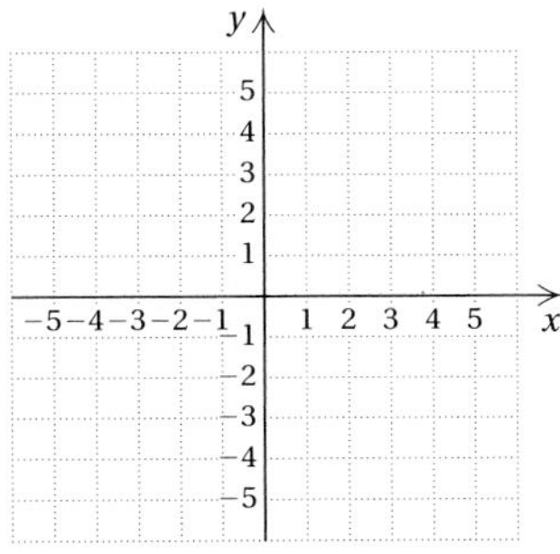

30. $5x - 2y = 10$

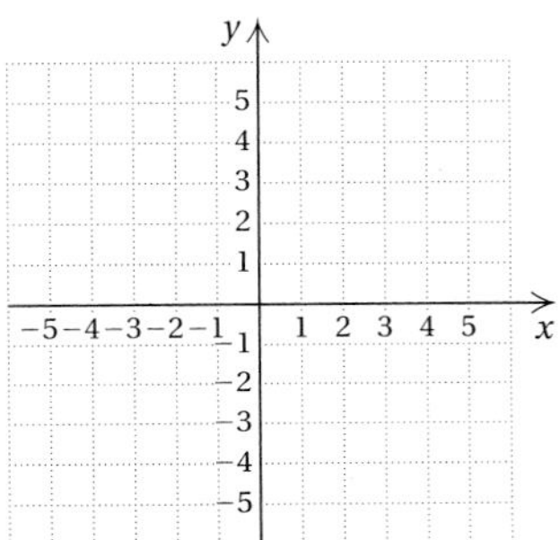

Solve. [9.2c]

31. *Kitchen Design.* Kitchen designers recommend that a refrigerator be selected on the basis of the number of people n in the household. The appropriate size S, in cubic feet, is given by

$$S = \frac{3}{2}n + 13.$$

a) Determine the recommended size of a refrigerator if the number of people is 1, 2, 5, and 10.

b) Graph the equation and use the graph to estimate the recommended size of a refrigerator for 3 people sharing an apartment.

c) A refrigerator is 22 ft^3. For how many residents is it the recommended size?

Find the mean, the median, and the mode. [9.4a]

32. 27, 35, 44, 52

33. 8, 12, 15, 17, 19, 12, 17, 19

34. *Blood Alcohol Levels of Drivers in Fatal Accidents.* 0.18, 0.17, 0.21, 0.16

35. *Bowling Scores of the Author in a Recent Tournament.* 215, 259, 215, 223, 237

36. *Movie Production Costs (in millions of dollars) per Film in Five Recent Years.* \$42.3, \$44.0, \$50.4, \$54.1, \$59.7

37. *Popcorn Testing.* An experiment is performed to compare the quality of two types of popcorn. The tester puts 200 kernels in a pan, pops them, and counts the number of unpopped kernels. The results are shown in the following table. Determine which type of popcorn is better by comparing their means. [9.4b]

Popcorn A Unpopped Kernels		
20	23	35
10	12	18
18	24	11
19	21	

Popcorn B Unpopped Kernels		
19	25	32
8	22	14
19	13	24
15	22	10

Estimate the missing data values using interpolation or extrapolation. [9.4c]

38. *Height of Girls.*

Age	Height (in centimeters)
2	95.7
4	103.2
5	?
6	115.9
8	128.0
10	138.6
12	151.9
14	159.6
16	162.2
18	162.5

Source: Kempe, C. Henry, M.D., et al., eds., *Current Pediatric Diagnosis & Treatment 1987.* Norwalk, CT: Appleton & Lange, 1987.

39. *Movie Production Costs.*

Year	Cost per Film (in millions)
1992	\$42.3
1993	44.0
1994	50.4
1995	54.1
1996	59.7
1997	?
2000	?

Source: Motion Picture Association of America

Skill Maintenance

Convert to decimal notation. [7.2c]

40. $-\dfrac{11}{32}$

41. $\dfrac{8}{9}$

Find the absolute value. [7.2e]

42. $|-3.2|$

43. $\left|\dfrac{17}{19}\right|$

Round to the nearest hundredth. [3.1e]

44. 42.705

45. 112.5278

Solve. [8.5a]

46. An investment was made at 6% simple interest for 1 year. It grows to \$10,340.40. How much was originally invested?

47. After a 20% reduction, a pair of slacks is on sale for \$63.96. What was the original price (that is, the price before reduction)?

Synthesis

48. ◆ Describe two ways in which a small business might make use of graphs. [9.1a], [9.2c]

49. ◆ Explain why the first coordinate of the y-intercept is always 0. [9.2b]

50. Find the value of m in $y = mx + 3$ such that $(-2, 5)$ is on the graph. [9.2a]

51. Find the area and the perimeter of a rectangle for which $(-2, 2)$, $(7, 2)$, and $(7, -3)$ are three of the vertices. [9.1b]

Test: Chapter 9

Answers

Toothpaste Sales. The following pie chart shows the percentages of sales of various toothpaste brands in the United States. In a recent year, total sales of toothpaste were $1,500,000,000.

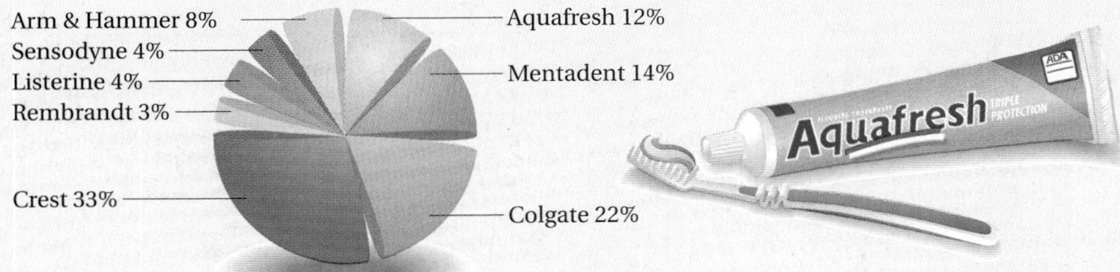

1. What were the total sales of Crest?

2. Which two brands together accounted for over half the sales?

3. Which brand had the greatest sales?

4. Which brand had sales of $120,000,000?

Tornado Touchdowns. The following bar graph shows the total number of tornado touchdowns by month in Indiana from 1950–1994.

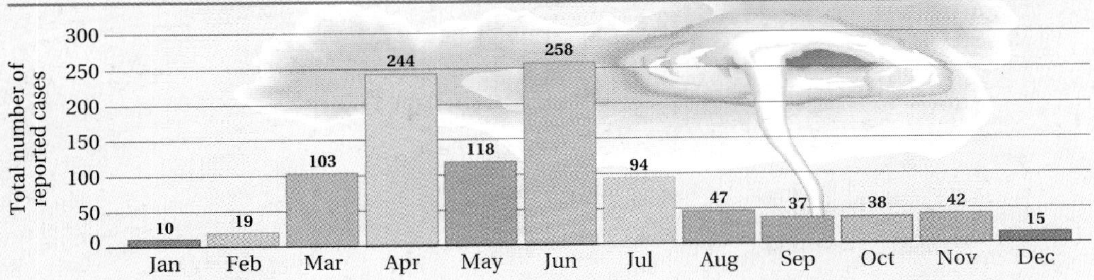

Source: National Weather Service

5. In which month did the greatest number of touchdowns occur?

6. In which month did the least number of touchdowns occur?

7. In which months was the number of touchdowns greater than 90?

8. In which month were there 47 touchdowns?

1. ______
2. ______
3. ______
4. ______
5. ______
6. ______
7. ______
8. ______

Answers

9. ______

10. ______

11. ______

12. ______

13. ______

14. ______

15. ______

16. ______

17. ______

18. ______

19. ______

20. ______

21. ______

Average Salary of Major-League Baseball Players. The line graph at right shows the average salary of major-league baseball players over a recent seven-year period. Use the graph for Exercises 9–14.

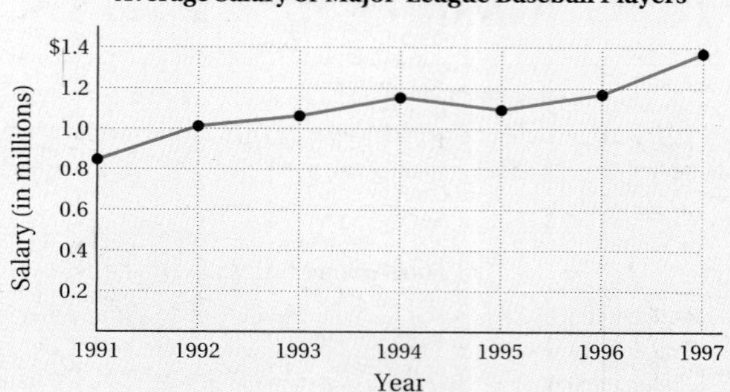

9. In which year was the average salary the highest?

10. In which year was the average salary the lowest?

11. What was the difference in salary between the highest and lowest salaries?

12. Between which two years was the increase in salary the greatest?

13. Between which two years did the salary decrease?

14. By how much did salaries increase between 1991 and 1997?

In which quadrant is the point located?

15. $\left(-\frac{1}{2}, 7\right)$

16. $(-5, -6)$

Find the coordinates of the point.

17. A

18. B

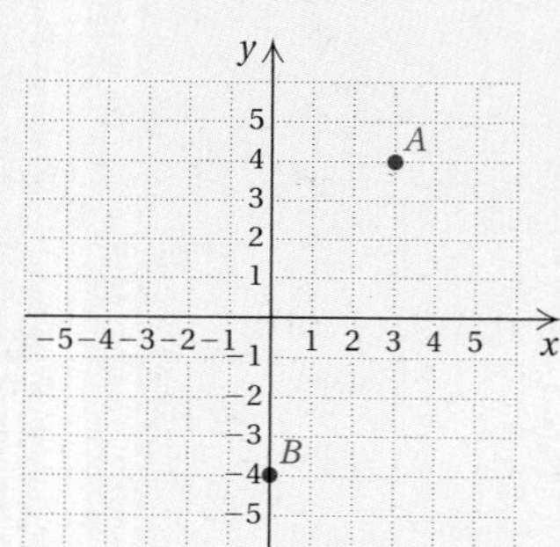

19. Show that the ordered pairs $(-4, -3)$ and $(-1, 3)$ are solutions of the equation $y - 2x = 5$. Then use the graph of the two points to determine another solution. Answers may vary.

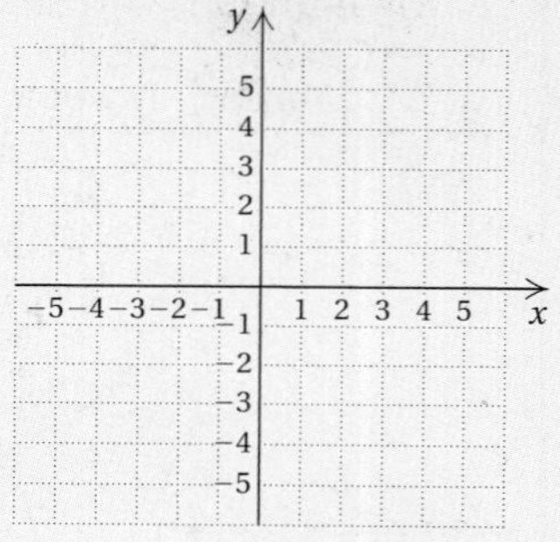

Graph the equation. Identify the y-intercept.

20. $y = 2x - 1$

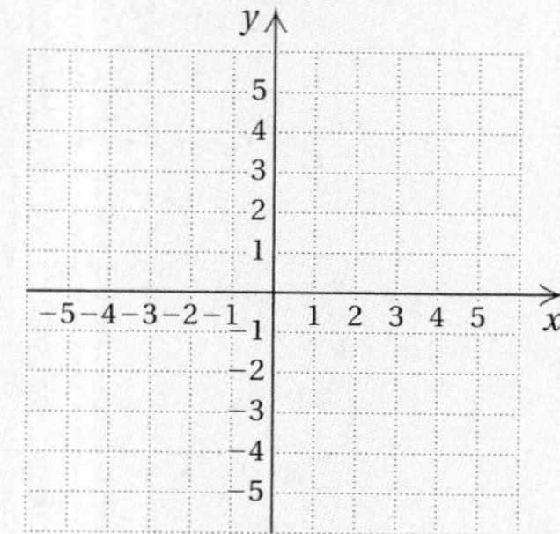

21. $y = -\frac{3}{2}x$

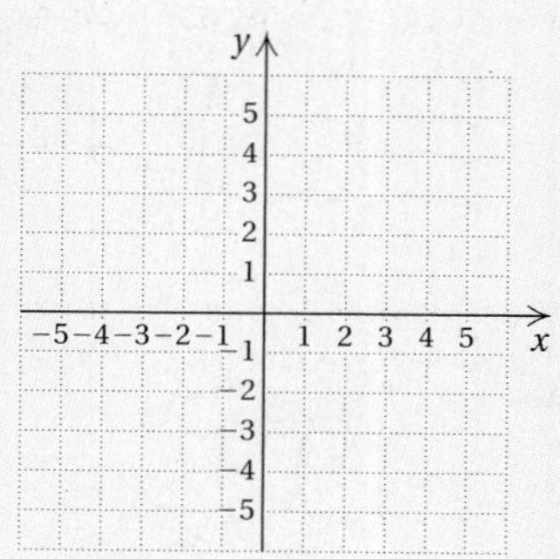

Graph the equation.

22. $2x + 8 = 0$

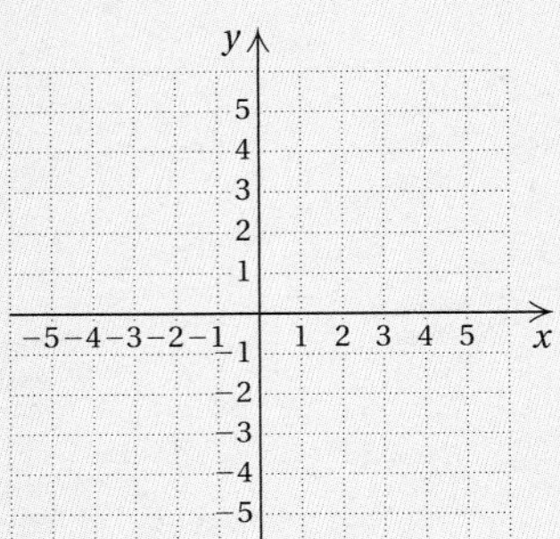

23. $y = 5$

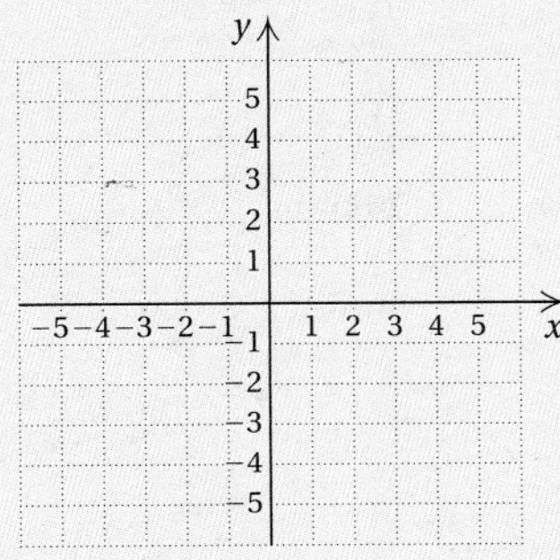

Find the intercepts of the equation. Then graph the equation.

24. $2x - 4y = -8$

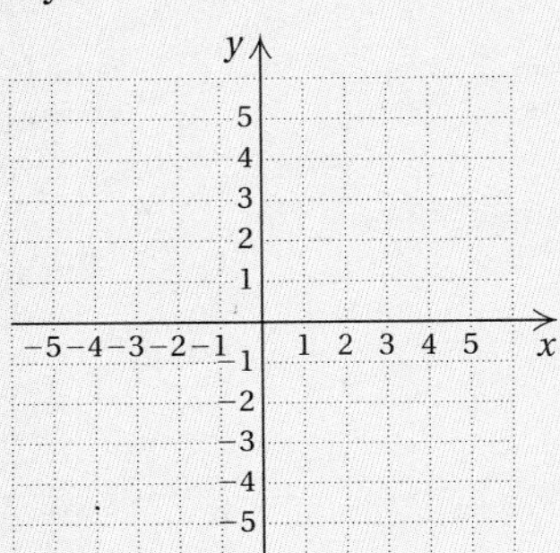

25. $2x - y = 3$

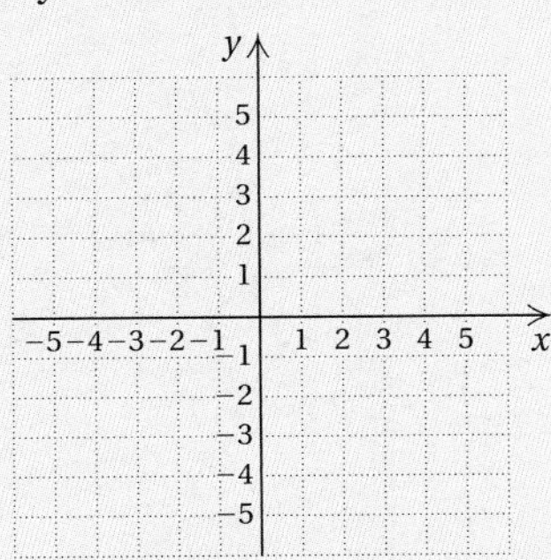

26. *Private-College Costs.* The cost T, in thousands of dollars, of tuition and fees at a private college (all expenses) can be approximated by

$$T = \frac{4}{5}n + 17,$$

where $n =$ the number of years since 1992 (***Source:*** Statistical Abstract of the United States). That is, $n = 0$ corresponds to 1992, $n = 7$ corresponds to 1999, and so on.

a) Find the cost of tuition in 1992, 1995, 1999, and 2001.

b) Graph the equation and then use the graph to estimate the cost of tuition in 2005.

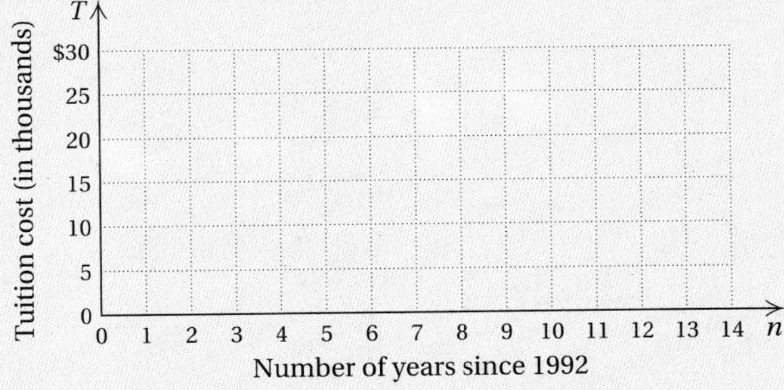

c) Estimate the year in which the cost of tuition will be $25,000.

Find the mean, the median, and the mode.

27. 46, 50, 53, 55

28. 2, 3, 4, 5, 6, 5

29. 4, 19, 20, 18, 19, 18

30. *Animal Speeds.*

Animal	Speed (in miles per hour)
Antelope	61
Bear	30
Cheetah	70
Fastest human	28
Greyhound	39
Lion	50
Zebra	40

Answers

22. ____

23. ____

24. ____

25. ____

26. a) ____

b) ____

c) ____

27. ____

28. ____

29. ____

30. ____

Answers

31. ______

32. ______

33. ______

34. ______

35. ______

36. ______

37. ______

38. ______

39. ______

40. ______

41. ______

42. ______

43. ______

31. *Quality of Golf Balls.* A golf pro experiments to see which of two kinds of golf ball is better. He drops each type of ball from a height of 8 ft onto concrete and measures how high they bounce, in inches. Determine which type of ball is better by comparing their means.

Ball A Bouncing Heights (in inches)				Ball B Bouncing Heights (in inches)			
59.7	58.4	59.7	57.2	56.2	62.3	59.5	65.7
59.7	64.6	62.0	63.7	59.6	58.0	61.6	62.7
66.5	61.6	62.5	59.8	59.8	56.4	58.4	66.6
61.6	57.9	58.7		61.5	58.7	62.0	68.3

Estimate the missing data values using interpolation or extrapolation.

32. *Height of Boys.*

Age	Height (in centimeters)
2	96.2
4	103.4
6	117.5
8	130.0
9	?
10	140.3
12	149.6
14	162.7
16	171.6
18	174.5

Source: Kempe, C. Henry, M.D., et al., eds., *Current Pediatric Diagnosis & Treatment 1987.* Norwalk, CT: Appleton & Lange, 1987.

33. *Deaths from Driving Incidents.* The following table lists for several years the number of driving incidents that resulted in death.

Year	Incidents
1990	1129
1991	1297
1992	1478
1993	1555
1994	1669
1995	1708
1996	?
2000	?

Source: AAA Foundation

Skill Maintenance

Convert to decimal notation.

34. $\frac{39}{40}$

35. $-\frac{13}{12}$

Find the absolute value.

36. $|71.2|$

37. $\left|-\frac{13}{47}\right|$

Round to the nearest thousandth.

38. 42.7047

39. 112.52702

Solve.

40. After a 24% reduction, a software game is on sale for $64.22. What was the original price (that is, the price before reduction)?

41. An investment was made at 7% simple interest for 1 year. It grows to $38,948. How much was originally invested?

Synthesis

42. A diagonal of a square connects the points $(-3, -1)$ and $(2, 4)$. Find the area and the perimeter of the square.

43. Write an equation of a line parallel to the x-axis and 3 units above it.

10

Polynomials: Operations

An Application

The Olympic flame at the 1992 Summer Olympics was lit by a flaming arrow. As the arrow moved d feet horizontally from the archer, its height h, in feet, could be approximated by the polynomial

$$-0.002d^2 + 0.8d + 6.6.$$

Use the polynomial to approximate the height of the arrow after it has traveled 100 ft horizontally.

This problem appears as Exercise 19 in Exercise Set 10.3.

The Mathematics

We substitute 100 for d and evaluate:

$$\underbrace{-0.002d^2 + 0.8d + 6.6}_{\uparrow} = -0.002(100)^2 + 0.8(100) + 6.6 = 66.6 \text{ ft.}$$

This is a polynomial.

World Wide Web For more information, visit us at www.mathmax.com

Introduction

Algebraic expressions like

$$-0.002d^2 + 0.8d + 6.6$$

and

$$x^3 - 5x^2 + 4x - 11$$

are called *polynomials*. One of the most important parts of introductory algebra is the study of polynomials. In this chapter, we learn to add, subtract, multiply, and divide polynomials.

Of particular importance here is the study of the quick ways to multiply certain polynomials. These *special products* will be helpful not only in this text but also in more advanced mathematics.

Pretest: Chapter 10

1. Multiply: $x^{-3} \cdot x^5$.

2. Divide: $\dfrac{x^{-2}}{x^5}$.

3. Simplify: $(-4x^2y^{-3})^2$.

4. Express using a positive exponent: p^{-3}.

5. Convert to scientific notation: 0.000347.

6. Convert to decimal notation: 3.4×10^6.

7. Identify the degree of each term and the degree of the polynomial:

$2x^3 - 4x^2 + 3x - 5$.

8. Collect like terms:

$2a^3b - a^2b^2 + ab^3 + 9 - 5a^3b - a^2b^2 + 12b^3$.

9. Add:

$(5x^2 - 7x + 8) + (6x^2 + 11x - 19)$.

10. Subtract:

$(5x^2 - 7x + 8) - (6x^2 + 11x - 19)$.

Multiply.

11. $5x^2(3x^2 - 4x + 1)$

12. $(x + 5)^2$

13. $(x - 5)(x + 5)$

14. $(x^3 + 6)(4x^3 - 5)$

15. $(2x - 3y)(2x - 3y)$

16. Divide: $(x^3 - x^2 + x + 2) \div (x - 2)$.

Objectives for Retesting

The objectives to be retested in addition to the material in this chapter are as follows.

[7.4a] Subtract real numbers and simplify combinations of additions and subtractions.

[7.7d] Use the distributive laws to factor expressions like $4x - 12 + 24y$.

[8.3b, c] Solve equations in which like terms may need to be collected, and solve equations by first removing parentheses and collecting like terms.

[8.4a] Solve applied problems by translating to equations.

10.1 Integers as Exponents

We introduced integer exponents of 2 or higher in Section 1.6. Here we consider 0 and 1, as well as negative integers, as exponents.

Objectives

a Tell the meaning of exponential notation.

b Evaluate exponential expressions with exponents of 0 and 1.

c Evaluate algebraic expressions containing exponents.

d Use the product rule to multiply exponential expressions with like bases.

e Use the quotient rule to divide exponential expressions with like bases.

f Express an exponential expression involving negative exponents with positive exponents.

For Extra Help

TAPE 19 | MAC WIN | CD-ROM

a Exponential Notation

An exponent of 2 or greater tells how many times the base is used as a factor. For example,

$$a \cdot a \cdot a \cdot a = a^4.$$

In this case, the **exponent** is 4 and the **base** is a. An expression for a power is called **exponential notation.**

$a^n \leftarrow$ This is the exponent.

$\uparrow$

This is the base.

Example 1 What is the meaning of 3^5? of n^4? of $(2n)^3$? of $50x^2$?

3^5 means $3 \cdot 3 \cdot 3 \cdot 3 \cdot 3$; $\quad n^4$ means $n \cdot n \cdot n \cdot n$;

$(2n)^3$ means $2n \cdot 2n \cdot 2n$; $\quad 50x^2$ means $50 \cdot x \cdot x$

Do Exercises 1–4.

We read exponential notation as follows:

a^n is read the ***n*th power of *a*,** or simply ***a* to the *n*th,** or ***a* to the *n*.**

We often read x^2 as "***x*-squared.**" The reason for this is that the area of a square of side x is $x \cdot x$, or x^2. We often read x^3 as "***x*-cubed.**" The reason for this is that the volume of a cube with length, width, and height x is $x \cdot x \cdot x$, or x^3.

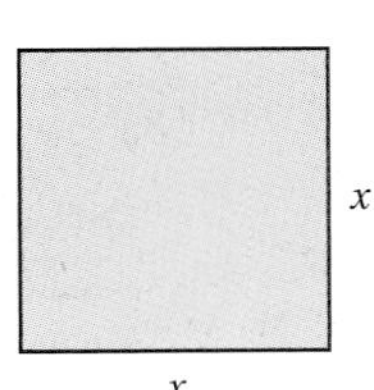

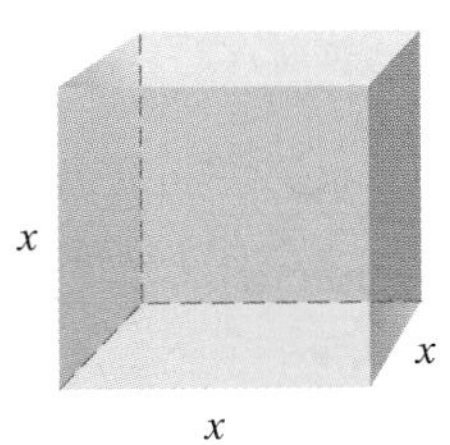

What is the meaning of each of the following?

1. 5^4

2. x^5

3. $(3t)^2$

4. $3t^2$

Answers on page A-29

b One and Zero as Exponents

Look for a pattern in the following:

On each side, we divide by 8 at each step.

$$8 \cdot 8 \cdot 8 \cdot 8 = 8^4$$
$$8 \cdot 8 \cdot 8 = 8^3$$
$$8 \cdot 8 = 8^2$$
$$8 = 8^?$$
$$1 = 8^?.$$

On this side, the exponents decrease by 1.

To continue the pattern, we would say that

$$8 = 8^1$$

and $\quad 1 = 8^0.$

Evaluate.

5. 6^1

6. 7^0

7. $(8.4)^1$

8. 8654^0

Answers on page A-29

We make the following definition.

> $a^1 = a$, for any number a;
> $a^0 = 1$, for any nonzero number a.

We consider 0^0 to be undefined. We will explain why later in this section.

Example 2 Evaluate 5^1, 8^1, 3^0, $(-7.3)^0$, and $(186{,}892{,}046)^0$.

$5^1 = 5$; $8^1 = 8$; $3^0 = 1$;

$(-7.3)^0 = 1$; $(186{,}892{,}046)^0 = 1$

Do Exercises 5–8.

c Evaluating Algebraic Expressions

Algebraic expressions can involve exponential notation. For example, the following are algebraic expressions:

$$x^4, \quad (3x)^3 - 2, \quad a^2 + 2ab + b^2.$$

We evaluate algebraic expressions by replacing variables with numbers and following the rules for order of operations.

Example 3 Evaluate x^4 for $x = 2$.

$x^4 = 2^4$ **Substituting**

$= 2 \cdot 2 \cdot 2 \cdot 2 = 16$

Example 4 *Area of a Compact Disc.* The standard compact disc used for software and music has a radius of 6 cm. Find the area of such a CD (ignoring the hole in the middle).

$A = \pi r^2$

$= \pi \cdot (6 \text{ cm})^2$

$\approx 3.14 \times 36 \text{ cm}^2$

$\approx 113.04 \text{ cm}^2$

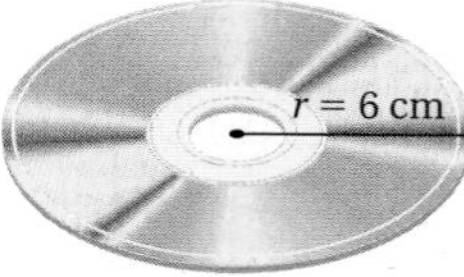

In Example 4, "cm^2" means "square centimeters" and "$\approx$" means "is approximately equal to."

Example 5 Evaluate $(5x)^3$ for $x = -2$.

When we evaluate with a negative number, we often use extra parentheses to show the substitution.

$(5x)^3 = [5 \cdot (-2)]^3$ **Substituting**

$= [-10]^3$ **Multiplying within brackets first**

$= -1000$ **Evaluating the power**

Example 6 Evaluate $5x^3$ for $x = -2$.

$$5x^3 = 5 \cdot (-2)^3 \quad \text{Substituting}$$
$$= 5(-8) \quad \text{Evaluating the power first}$$
$$= -40$$

Recall that two expressions are equivalent if they have the same value for all meaningful replacements. Note that Examples 5 and 6 show that $(5x)^3$ and $5x^3$ are *not* equivalent—that is, $(5x)^3 \neq 5x^3$.

Do Exercises 9–13.

d Multiplying Powers with Like Bases

There are several rules for manipulating exponential notation to obtain equivalent expressions. We first consider multiplying powers with like bases:

$$a^3 \cdot a^2 = \underbrace{(a \cdot a \cdot a)}_{\text{3 factors}}\underbrace{(a \cdot a)}_{\text{2 factors}} = \underbrace{a \cdot a \cdot a \cdot a \cdot a}_{\text{5 factors}} = a^5.$$

Since an integer exponent greater than 1 tells how many times we use a base as a factor, then $(a \cdot a \cdot a)(a \cdot a) = a \cdot a \cdot a \cdot a \cdot a = a^5$ by the associative law. Note that the exponent in a^5 is the sum of those in $a^3 \cdot a^2$. That is, $3 + 2 = 5$. Likewise,

$$b^4 \cdot b^3 = (b \cdot b \cdot b \cdot b)(b \cdot b \cdot b) = b^7, \quad \text{where} \quad 4 + 3 = 7.$$

Adding the exponents gives the correct result.

THE PRODUCT RULE

For any number a and any positive integers m and n,

$$a^m \cdot a^n = a^{m+n}.$$

(When multiplying with exponential notation, if the bases are the same, keep the base and add the exponents.)

Examples Multiply and simplify. By simplify, we mean write the expression as one number to a nonnegative power.

7. $8^4 \cdot 8^3 = 8^{4+3}$ Adding exponents: $a^m \cdot a^n = a^{m+n}$
$= 8^7$

8. $x^2 \cdot x^9 = x^{2+9}$
$= x^{11}$

9. $m^5 m^{10} m^3 = m^{5+10+3}$
$= m^{18}$

10. $x \cdot x^8 = x^1 \cdot x^8 = x^{1+8}$
$= x^9$

11. $(a^3b^2)(a^3b^5) = (a^3a^3)(b^2b^5)$
$= a^6b^7$

Do Exercises 14–18.

9. Evaluate t^3 for $t = 5$.

10. Find the area of a circle when $r = 32$ cm. Use 3.14 for π.

11. Evaluate $200 - a^4$ for $a = 3$.

12. Evaluate $t^1 - 4$ and $t^0 - 4$ for $t = 7$.

13. a) Evaluate $(4t)^2$ for $t = -3$.

b) Evaluate $4t^2$ for $t = -3$.

c) Determine whether $(4t)^2$ and $4t^2$ are equivalent.

Multiply and simplify.

14. $3^5 \cdot 3^5$

15. $x^4 \cdot x^6$

16. $p^4p^{12}p^8$

17. $x \cdot x^4$

18. $(a^2b^3)(a^7b^5)$

Answers on page A-29

Divide and simplify.

19. $\dfrac{4^5}{4^2}$

20. $\dfrac{y^6}{y^2}$

21. $\dfrac{p^{10}}{p}$

22. $\dfrac{a^7b^6}{a^3b^4}$

Answers on page A-29

e Dividing Powers with Like Bases

The following suggests a rule for dividing powers with like bases, such as a^5/a^2:

$$\frac{a^5}{a^2} = \frac{a \cdot a \cdot a \cdot a \cdot a}{a \cdot a} = \frac{a \cdot a \cdot a \cdot a \cdot a}{1 \cdot a \cdot a} = \frac{a \cdot a \cdot a}{1} \cdot \frac{a \cdot a}{a \cdot a} = \frac{a \cdot a \cdot a}{1} \cdot 1$$
$$= a \cdot a \cdot a = a^3.$$

Note that the exponent in a^3 is the difference of those in $a^5 \div a^2$. If we subtract exponents, we get $5 - 2$, which is 3.

> **THE QUOTIENT RULE**
>
> For any nonzero number a and any positive integers m and n,
>
> $$\frac{a^m}{a^n} = a^{m-n}.$$
>
> (When dividing with exponential notation, if the bases are the same, keep the base and subtract the exponent of the denominator from the exponent of the numerator.)

Examples Divide and simplify. By simplify, we mean write the expression as one number to a nonnegative power.

12. $\dfrac{6^5}{6^3} = 6^{5-3}$ Subtracting exponents
$= 6^2$

13. $\dfrac{x^8}{x^2} = x^{8-2}$
$= x^6$

14. $\dfrac{t^{12}}{t} = \dfrac{t^{12}}{t^1} = t^{12-1}$
$= t^{11}$

15. $\dfrac{p^5q^7}{p^2q^5} = \dfrac{p^5}{p^2} \cdot \dfrac{q^7}{q^5} = p^{5-2}q^{7-5}$
$= p^3q^2$

The quotient rule can also be used to explain the definition of 0 as an exponent. Consider the expression a^4/a^4, where a is nonzero:

$$\frac{a^4}{a^4} = \frac{a \cdot a \cdot a \cdot a}{a \cdot a \cdot a \cdot a} = 1.$$

This is true because the numerator and the denominator are the same. Now suppose we apply the rule for dividing powers with the same base:

$$\frac{a^4}{a^4} = a^{4-4} = a^0 = 1.$$

Since both expressions a^4/a^4 and a^{4-4} are equivalent to 1, it follows that $a^0 = 1$, when $a \neq 0$.

We can explain why we do not define 0^0 using the quotient rule. We know that 0^0 is 0^{1-1}. But 0^{1-1} is also equal to $0/0$. We have already seen that division by 0 is undefined, so 0^0 is also undefined.

Do Exercises 19–22.

f Negative Integers as Exponents

We can use the rule for dividing powers with like bases to lead us to a definition of exponential notation when the exponent is a negative integer. Consider $5^3/5^7$ and first simplify it using procedures we have learned for working with fractions:

$$\frac{5^3}{5^7} = \frac{5 \cdot 5 \cdot 5}{5 \cdot 5 \cdot 5 \cdot 5 \cdot 5 \cdot 5 \cdot 5} = \frac{5 \cdot 5 \cdot 5 \cdot 1}{5 \cdot 5 \cdot 5 \cdot 5 \cdot 5 \cdot 5 \cdot 5}$$

$$= \frac{5 \cdot 5 \cdot 5}{5 \cdot 5 \cdot 5} \cdot \frac{1}{5 \cdot 5 \cdot 5 \cdot 5} = \frac{1}{5^4}.$$

Now we apply the rule for dividing powers with the same bases. Then

$$\frac{5^3}{5^7} = 5^{3-7} = 5^{-4}.$$

From these two expressions for $5^3/5^7$, it follows that

$$5^{-4} = \frac{1}{5^4}.$$

This leads to our definition of negative exponents:

> For any real number a that is nonzero and any integer n,
>
> $$a^{-n} = \frac{1}{a^n}.$$

In fact, the numbers a^n and a^{-n} are reciprocals of each other because

$$a^n \cdot a^{-n} = a^n \cdot \frac{1}{a^n} = \frac{a^n}{a^n} = 1.$$

Examples Express using positive exponents. Then simplify.

16. $4^{-2} = \frac{1}{4^2} = \frac{1}{16}$

17. $(-3)^{-2} = \frac{1}{(-3)^2} = \frac{1}{(-3)(-3)} = \frac{1}{9}$

18. $m^{-3} = \frac{1}{m^3}$

19. $ab^{-1} = a\left(\frac{1}{b^1}\right) = a\left(\frac{1}{b}\right) = \frac{a}{b}$

20. $\frac{1}{x^{-3}} = x^{-(-3)} = x^3$

21. $3c^{-5} = 3\left(\frac{1}{c^5}\right) = \frac{3}{c^5}$

CAUTION! Note in Example 16 that

$$4^{-2} \neq -16 \quad \text{and} \quad 4^{-2} \neq -\frac{1}{16}.$$

Do Exercises 23–28.

The rules for multiplying and dividing powers with like bases still hold when exponents are 0 or negative. We will state them in a summary at the end of this section.

Express with positive exponents. Then simplify.

23. 4^{-3}

24. 5^{-2}

25. 2^{-4}

26. $(-2)^{-3}$

27. $4p^{-3}$

28. $\frac{1}{x^{-2}}$

Answers on page A-29

Simplify.

29. $5^{-2} \cdot 5^4$

30. $x^{-3} \cdot x^{-4}$

31. $\frac{7^{-2}}{7^3}$

32. $\frac{b^{-2}}{b^{-3}}$

33. $\frac{t}{t^{-5}}$

Answers on page A-29

Examples Simplify. By simplify, we generally mean write the expression as one number to a nonnegative power.

22. $7^{-3} \cdot 7^6 = 7^{-3+6} = 7^3$ **Adding exponents**

23. $x^4 \cdot x^{-3} = x^{4+(-3)} = x^1 = x$

24. $\frac{5^4}{5^{-2}} = 5^{4-(-2)} = 5^{4+2} = 5^6$ **Subtracting exponents**

25. $\frac{x}{x^7} = x^{1-7} = x^{-6} = \frac{1}{x^6}$

26. $\frac{b^{-4}}{b^{-5}} = b^{-4-(-5)} = b^{-4+5} = b^1 = b$

27. $y^{-4} \cdot y^{-8} = y^{-4+(-8)} = y^{-12} = \frac{1}{y^{12}}$

In Examples 24–26 (division with exponents), it may help to think as follows: After writing the base, write the top exponent. Then write a subtraction sign. Next write the bottom exponent. Then do the subtraction by adding the opposite. For example,

$$\frac{x^{-3}}{x^{-5}} = x^{-3-(-5)} = x^{-3+5} = x^2$$

(1) Write the base.
(2) Write the top exponent.
(3) Write a subtraction sign.
(4) Write the bottom exponent.

Do Exercises 29–33.

The following is another way to arrive at the definition of negative exponents.

On each side, we divide by 5 at each step.

$$5 \cdot 5 \cdot 5 \cdot 5 = 5^4$$
$$5 \cdot 5 \cdot 5 = 5^3$$
$$5 \cdot 5 = 5^2$$
$$5 = 5^1$$
$$1 = 5^0$$
$$\frac{1}{5} = 5^?$$
$$\frac{1}{25} = 5^?$$

On this side, the exponents decrease by 1.

To continue the pattern, it should follow that

$$\frac{1}{5} = \frac{1}{5^1} = 5^{-1} \quad \text{and} \quad \frac{1}{25} = \frac{1}{5^2} = 5^{-2}.$$

The following is a summary of the definitions and rules for exponents that we have considered in this section.

DEFINITIONS AND RULES FOR EXPONENTS

1 as an exponent:	$a^1 = a$;
0 as an exponent:	$a^0 = 1,\ a \neq 0$;
Negative integers as exponents:	$a^{-n} = \frac{1}{a^n},\ \frac{1}{a^{-n}} = a^n;\ a \neq 0$
Product Rule:	$a^m \cdot a^n = a^{m+n}$;
Quotient Rule:	$\frac{a^m}{a^n} = a^{m-n},\ a \neq 0$

Calculator Spotlight

Checking Equivalent Expressions. Let's look at the expressions $x^2 \cdot x^3$ and x^5. We know from the product rule, $x^m \cdot x^n = x^{m+n}$, that these expressions are equivalent. In this case, $x^2 \cdot x^3 = x^{2+3} = x^5$ is true for any real-number substitution. This use of the product rule is an algebraic check of the correctness of the statement $x^2 \cdot x^3 = x^5$. How can we check the result using a grapher? We can do it *graphically* by looking at graphs and *numerically* by looking at a table of values.

Graphical Check. Let's first do a graphical check of $x^2 \cdot x^3 = x^5$. We consider each expression separately and form two equations to be graphed: $y_1 = x^2 \cdot x^3$ and $y_2 = x^5$. The "y_1" is read "y sub 1" and refers simply to a "first" equation. Similarly, "y_2" is read "y sub 2" and refers to a "second" equation. We enter these equations into the grapher using the [y=] key and then graph them, as shown on the left below. Note that the graphs appear to coincide. This is a partial check that the expressions are equivalent. We say "partial check" because most graphs cannot be drawn completely so there is always an element of uncertainty.

$y_1 = x^2 \cdot x^3,\ y_2 = x^5$

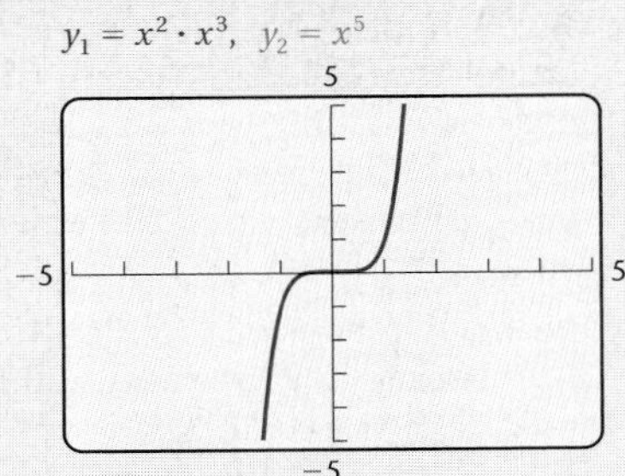

X	Y1	Y2
6	7776	7776
7	16807	16807
8	32768	32768
9	59049	59049
10	100000	100000
11	161051	161051
12	248832	248832

Y1 = 32768

Numerical Check. Now let's use the TABLE feature to check $x^2 \cdot x^3 = x^5$. We already have the equations $y_1 = x^2 \cdot x^3$ and $y_2 = x^5$ entered. The TABLE feature allows us to compare y-values for various x-values. Note in the table on the right above that the y_1- and y_2-values agree. Thus we have a partial check that we have an identity. We say "partial check" because it is impossible to compute all possible y-values and there may be some that disagree.

Let's now consider the equation $x^2 \cdot x^3 = x^6$. Is this a correct result? It seems to violate the product rule, $x^m \cdot x^n = x^{m+n}$. Let's check the equation both graphically and numerically.

Graphical Check. We graph $y_1 = x^2 \cdot x^3$ and $y_2 = x^6$, as shown on the left below. On the TI-83, there is a way to choose a graphing style so that the graphs look different when graphed in the same window. See the window in the middle below. It is obvious that the graphs are different. Thus the equation is not correct.

$y_1 = x^2 \cdot x^3,\ y_2 = x^6$

$y_1 = x^2 \cdot x^3,\ y_2 = x^6$

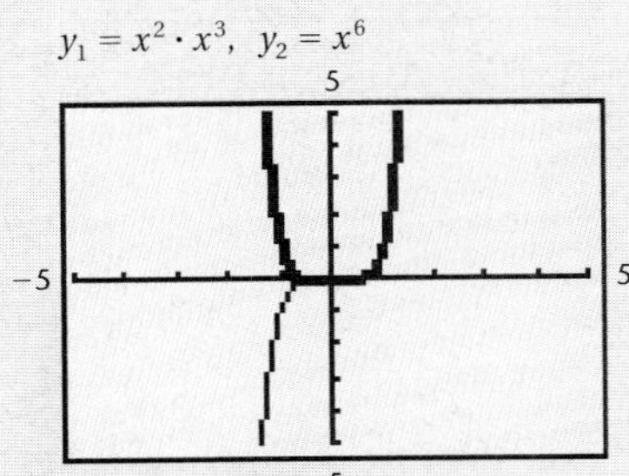

X	Y1	Y2
5	3125	15625
6	7776	46656
7	16807	117649
8	32768	262144
9	59049	531441
10	100000	1E6
11	161051	1.77E6

X = 5

Numerical Check. Let's check a table of y-values. See the table on the right above. Here we note that the y_1- and y_2-values are not the same. Thus, $x^2 \cdot x^3 = x^6$ is not correct.

Exercises Determine whether each of the following equations is correct.

1. $x \cdot x^2 = x^3$
2. $x \cdot x^2 = x^2$
3. $\dfrac{x^3}{x^2} = x^5$
4. $\dfrac{x^5}{x^2} = x^3$
5. $\left(\dfrac{x}{3}\right)^2 = \dfrac{x^2}{9}$
6. $(5x)^2 = 25x^2$
7. $(x + 2)^2 = x^2 + 4$
8. $(x + 2)^2 = x^2 + 4x + 4$
9. $x + 3 = 3 + x$
10. $3(x - 1) = 3x - 3$
11. $5x - 5 = 5(x - 5)$
12. $10x + 20 = 5(2x + 4)$
13. $2 + (3 + x) = (2 + 3) + x$
14. $5(2x) = 5x$

Improving Your Math Study Skills

Tips from a Former Student

A former student of Professor Bittinger, Mike Rosenborg earned a master's degree in mathematics and now teaches mathematics. Here are some of his study tips.

- Because working problems is the best way to learn math, instructors generally assign lots of problems. Never let yourself get behind in your math homework.
- If you are struggling with a math concept, do not give up. Ask for help from your friends and your instructor. Since each concept is built on previous concepts, any gaps in your understanding will follow you through the entire course, so make sure you understand each concept as you go along.
- Math contains many rules that cannot be "bent." Don't try inventing your own rules and still expect to get correct answers. Although there is usually more than one way to solve a problem, each method must follow the established rules.
- Read your textbook! It will often contain the tips and help you need to solve any problem with which you're struggling. It may also bring out points that you missed in class or that your instructor may not have covered.
- Learn to use scratch paper to jot down your thoughts and to draw pictures. Don't try to figure everything out "in your head." You will think more clearly and accurately this way.
- When preparing for a test, it is often helpful to work at least two problems per section as practice: one easy and one difficult. Write out all the new rules and procedures your test will cover, and then read through them twice. Doing so will enable you to both learn and retain them better.
- Some people like to work in study groups, while others prefer solitary study. Although it's important to be flexible, it's more important that you be comfortable with your study method, so consider trying both. You may find one or the other or a combination of both effective.
- Most schools have classrooms set up where you can get free help from math tutors. Take advantage of this, but be sure you do the work first. Don't let your tutor do all the work for you—otherwise you'll never learn the material.
- In math, as in many other areas of life, patience and persistence are virtues—cultivate them. "Cramming" for an exam will not help you learn and retain the material.
- Do your work neatly and in pencil. Then if you make a mistake, it will be relatively easy to find and correct. Write out each step in the problem's solution; don't skip steps or take shortcuts. Each step should follow clearly from the preceding step, and the entire solution should be easy to follow. If you understand the concepts and get a wrong answer, the first thing you should look for is a "small" mistake, like writing a "+" instead of a "−."

Exercise Set 10.1

a What is the meaning of each of the following?

1. 3^4

2. 4^3

3. $(1.1)^5$

4. $(87.2)^6$

5. $\left(\frac{2}{3}\right)^4$

6. $\left(-\frac{5}{8}\right)^3$

7. $(7p)^2$

8. $(11c)^3$

9. $8k^3$

10. $17x^2$

b Evaluate.

11. a^0, $a \neq 0$

12. t^0, $t \neq 0$

13. b^1

14. c^1

15. $\left(\frac{2}{3}\right)^0$

16. $\left(-\frac{5}{8}\right)^0$

17. 8.38^0

18. 8.38^1

19. $(ab)^1$

20. $(ab)^0$, $a, b \neq 0$

21. ab^1

22. ab^0

c Evaluate.

23. m^3, for $m = 3$

24. x^6, for $x = 2$

25. p^1, for $p = 19$

26. x^{19}, for $x = 0$

27. x^4, for $x = 4$

28. y^{15}, for $y = 1$

29. $y^2 - 7$, for $y = -10$

30. $z^5 + 5$, for $z = -2$

31. $x^1 + 3$ and $x^0 + 3$, for $x = 7$

32. $y^0 - 8$ and $y^1 - 8$, for $y = -3$

33. Find the area of a circle when $r = 34$ ft. Use 3.14 for π.

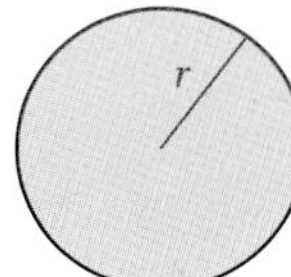

34. The area A of a square with sides of length s is given by $A = s^2$. Find the area of a square with sides of length 24 m.

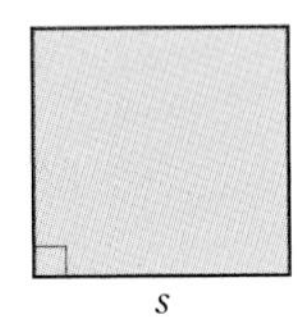

f Express using positive exponents. Then simplify.

35. 3^{-2}

36. 2^{-3}

37. 10^{-3}

38. 5^{-4}

39. 7^{-3}

40. 5^{-2}

41. a^{-3}

42. x^{-2}

43. $\frac{1}{8^{-2}}$

44. $\frac{1}{2^{-5}}$

45. $\frac{1}{y^{-4}}$

46. $\frac{1}{t^{-7}}$

47. $\frac{1}{z^{-n}}$

48. $\frac{1}{h^{-n}}$

Express using negative exponents.

49. $\frac{1}{4^3}$

50. $\frac{1}{5^2}$

51. $\frac{1}{x^3}$

52. $\frac{1}{y^2}$

53. $\frac{1}{a^5}$

54. $\frac{1}{b^7}$

d, **f** Multiply and simplify.

55. $2^4 \cdot 2^3$

56. $3^5 \cdot 3^2$

57. $8^5 \cdot 8^9$

58. $n^3 \cdot n^{20}$

59. $x^4 \cdot x^3$

60. $y^7 \cdot y^9$

61. $9^{17} \cdot 9^{21}$

62. $t^0 \cdot t^{16}$

63. $(3y)^4(3y)^8$

64. $(2t)^8(2t)^{17}$

65. $(7y)^1(7y)^{16}$

66. $(8x)^0(8x)^1$

67. $3^{-5} \cdot 3^8$

68. $5^{-8} \cdot 5^9$

69. $x^{-2} \cdot x$

70. $x \cdot x^{-1}$

71. $x^{14} \cdot x^3$

72. $x^9 \cdot x^4$

73. $x^{-7} \cdot x^{-6}$

74. $y^{-5} \cdot y^{-8}$

75. $a^{11} \cdot a^{-3} \cdot a^{-18}$

76. $a^{-11} \cdot a^{-3} \cdot a^{-7}$

77. $t^8 \cdot t^{-8}$

78. $m^{10} \cdot m^{-10}$

e, **f** Divide and simplify.

79. $\frac{7^5}{7^2}$

80. $\frac{5^8}{5^6}$

81. $\frac{8^{12}}{8^6}$

82. $\frac{8^{13}}{8^2}$

83. $\frac{y^9}{y^5}$

84. $\frac{x^{11}}{x^9}$

85. $\frac{16^2}{16^8}$

86. $\frac{7^2}{7^9}$

87. $\frac{m^6}{m^{12}}$

88. $\frac{a^3}{a^4}$

89. $\frac{(8x)^6}{(8x)^{10}}$

90. $\frac{(8t)^4}{(8t)^{11}}$

91. $\frac{(2y)^9}{(2y)^9}$

92. $\frac{(6y)^7}{(6y)^7}$

93. $\frac{x}{x^{-1}}$

94. $\frac{y^8}{y}$

95. $\frac{x^7}{x^{-2}}$

96. $\frac{t^8}{t^{-3}}$

97. $\frac{z^{-6}}{z^{-2}}$

98. $\frac{x^{-9}}{x^{-3}}$

99. $\frac{x^{-5}}{x^{-8}}$

100. $\frac{y^{-2}}{y^{-9}}$

101. $\frac{m^{-9}}{m^{-9}}$

102. $\frac{x^{-7}}{x^{-7}}$

Simplify.

103. 5^2, 5^{-2}, $\left(\frac{1}{5}\right)^2$, $\left(\frac{1}{5}\right)^{-2}$, -5^2, and $(-5)^2$

104. 8^2, 8^{-2}, $\left(\frac{1}{8}\right)^2$, $\left(\frac{1}{8}\right)^{-2}$, -8^2, and $(-8)^2$

Skill Maintenance

105. Translate to an algebraic expression: Sixty-four percent of t. [7.1b]

106. Evaluate $3x/y$ for $x = 4$ and $y = 12$. [7.1a]

107. Divide: $1555.2 \div 24.3$. [7.6c]

108. Add: $1555.2 + 24.3$. [7.3a]

109. Solve: $3x - 4 + 5x - 10x = x - 8$. [8.3b]

110. Factor: $8x - 56$. [7.7d]

Solve. [8.4a]

111. A 12-in. submarine sandwich is cut into two pieces. One piece is twice as long as the other. How long are the pieces?

112. A book is opened. The sum of the page numbers on the facing pages is 457. Find the page numbers.

Synthesis

113. ◆ Under what conditions does a^n represent a negative number? Why?

114. ◆ Explain the errors in each of the following.

a) $2^{-3} = \frac{1}{-8}$

b) $m^{-2}m^5 = m^{10}$

Determine whether each of the following is correct.

115. $(x + 1)^2 = x^2 + 1$

116. $(x - 1)^2 = x^2 - 2x + 1$

117. $(5x)^0 = 5x^0$

118. $\frac{x^3}{x^5} = x^2$

Simplify.

119. $(y^{2x})(y^{3x})$

120. $a^{5k} \div a^{3k}$

121. $\frac{a^{6t}(a^{7t})}{a^{9t}}$

122. $\frac{\left(\frac{1}{2}\right)^4}{\left(\frac{1}{2}\right)^5}$

123. $\frac{(0.8)^5}{(0.8)^3(0.8)^2}$

124. Determine whether $(a + b)^2$ and $a^2 + b^2$ are equivalent. (*Hint*: Choose values for a and b and evaluate.)

Use $>$, $<$, or $=$ for ▭ to write a true sentence.

125. 3^5 ▭ 3^4

126. 4^2 ▭ 4^3

127. 4^3 ▭ 5^3

128. 4^3 ▭ 3^4

Find a value of the variable that shows that the two expressions are *not* equivalent.

129. $3x^2$; $(3x)^2$

130. $\frac{x + 2}{2}$; x

10.2 Exponents and Scientific Notation

We now enhance our ability to manipulate exponential expressions by considering three more rules. The rules are also applied to a new way to name numbers called *scientific notation.*

Objectives

a. Use the power rule to raise powers to powers.

b. Raise a product to a power and a quotient to a power.

c. Convert between scientific notation and decimal notation.

d. Multiply and divide using scientific notation.

e. Solve applied problems using scientific notation.

For Extra Help

TAPE 19

MAC
WIN

CD-ROM

a Raising Powers to Powers

Consider an expression like $(3^2)^4$. We are raising 3^2 to the fourth power:

$$\begin{aligned}(3^2)^4 &= (3^2)(3^2)(3^2)(3^2)\\ &= (3\cdot 3)(3\cdot 3)(3\cdot 3)(3\cdot 3)\\ &= 3\cdot 3\cdot 3\cdot 3\cdot 3\cdot 3\cdot 3\cdot 3\\ &= 3^8.\end{aligned}$$

Note that in this case we could have multiplied the exponents:

$$(3^2)^4 = 3^{2\cdot 4} = 3^8.$$

Likewise, $(y^8)^3 = (y^8)(y^8)(y^8) = y^{24}$. Once again, we get the same result if we multiply the exponents:

$$(y^8)^3 = y^{8\cdot 3} = y^{24}.$$

THE POWER RULE

For any real number a and any integers m and n,

$$(a^m)^n = a^{mn}.$$

(To raise a power to a power, multiply the exponents.)

Calculator Spotlight

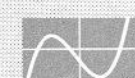

Exercises

Determine both graphically and numerically whether each of the following is correct. That is, use the GRAPH and TABLE features on your grapher.

1. $(x^2)^3 = x^6$
2. $(x^2)^3 = x^5$

Examples Simplify. Express the answers using positive exponents.

1. $(3^5)^4 = 3^{5\cdot 4}$ **Multiplying exponents**
$\quad = 3^{20}$

2. $(2^2)^5 = 2^{2\cdot 5} = 2^{10}$

3. $(y^{-5})^7 = y^{-5\cdot 7} = y^{-35} = \dfrac{1}{y^{35}}$

4. $(x^4)^{-2} = x^{4(-2)} = x^{-8} = \dfrac{1}{x^8}$

5. $(a^{-4})^{-6} = a^{(-4)(-6)} = a^{24}$

Do Exercises 1–4.

Simplify. Express the answers using positive exponents.

1. $(3^4)^5$

2. $(x^{-3})^4$

3. $(y^{-5})^{-3}$

4. $(x^4)^{-8}$

Answers on page A-29

b Raising a Product or a Quotient to a Power

When an expression inside parentheses is raised to a power, the inside expression is the base. Let's compare $2a^3$ and $(2a)^3$:

$2a^3 = 2\cdot a\cdot a\cdot a;$ **The base is a.**

$$\begin{aligned}(2a)^3 &= (2a)(2a)(2a) && \textbf{The base is } 2a.\\ &= (2\cdot 2\cdot 2)(a\cdot a\cdot a) && \textbf{Using the associative and commutative laws of multiplication to regroup the factors}\\ &= 2^3a^3\\ &= 8a^3.\end{aligned}$$

We see that $2a^3$ and $(2a)^3$ are *not* equivalent. We also see that we can evaluate the power $(2a)^3$ by raising each factor to the power 3. This leads us to the following rule for raising a product to a power.

Simplify.

5. $(2x^5y^{-3})^4$

6. $(5x^5y^{-6}z^{-3})^2$

7. $[(-x)^{37}]^2$

8. $(3y^{-2}x^{-5}z^8)^3$

Simplify.

9. $\left(\frac{x^6}{5}\right)^2$

10. $\left(\frac{2t^5}{w^4}\right)^3$

11. $\left(\frac{x^4}{3}\right)^{-2}$

Calculator Spotlight

Exercises

Determine both graphically and numerically whether each of the following is correct. That is, use the GRAPH and TABLE features on your grapher.

1. $(3x)^2 = 9x^2$
2. $\left(\frac{x}{4}\right)^2 = \frac{x^2}{16}$
3. $(2x)^3 = 2x^3$
4. $\left(\frac{x}{5}\right)^2 = \frac{x^2}{5}$

Answers on page A-29

RAISING A PRODUCT TO A POWER

For any real numbers a and b and any integer n,

$$(ab)^n = a^nb^n.$$

(To raise a product to the nth power, raise each factor to the nth power.)

Examples

6. $(4x^2)^3 = 4^3 \cdot (x^2)^3$ **Raising each factor to the third power**
$= 64x^6$

7. $(5x^3y^5z^2)^4 = 5^4(x^3)^4(y^5)^4(z^2)^4$ **Raising each factor to the fourth power**
$= 625x^{12}y^{20}z^8$

8. $(-5x^4y^3)^3 = (-5)^3(x^4)^3(y^3)^3$
$= -125x^{12}y^9$

9. $[(-x)^{25}]^2 = (-x)^{50}$ **Using the power rule**
$= (-1 \cdot x)^{50}$ **Using the property of −1 (Section 7.8)**
$= (-1)^{50}x^{50}$
$= 1 \cdot x^{50}$ **The product of an even number of negative factors is positive.**
$= x^{50}$

10. $(5x^2y^{-2})^3 = 5^3(x^2)^3(y^{-2})^3 = 125x^6y^{-6}$ **Be sure to raise *each* factor to the third power.**
$= \frac{125x^6}{y^6}$

11. $(3x^3y^{-5}z^2)^4 = 3^4(x^3)^4(y^{-5})^4(z^2)^4$
$= 81x^{12}y^{-20}z^8 = \frac{81x^{12}z^8}{y^{20}}$

Do Exercises 5–8.

There is a similar rule for raising a quotient to a power.

RAISING A QUOTIENT TO A POWER

For any real numbers a and b, $b \neq 0$, and any integer n,

$$\left(\frac{a}{b}\right)^n = \frac{a^n}{b^n}.$$

(To raise a quotient to the nth power, raise both the numerator and the denominator to the nth power.)

Examples Simplify.

12. $\left(\frac{x^2}{4}\right)^3 = \frac{(x^2)^3}{4^3} = \frac{x^6}{64}$

13. $\left(\frac{3a^4}{b^3}\right)^2 = \frac{(3a^4)^2}{(b^3)^2} = \frac{3^2(a^4)^2}{b^{3\cdot 2}} = \frac{9a^8}{b^6}$

14. $\left(\frac{y^3}{5}\right)^{-2} = \frac{(y^3)^{-2}}{5^{-2}} = \frac{y^{-6}}{5^{-2}} = \frac{\frac{1}{y^6}}{\frac{1}{5^2}} = \frac{1}{y^6} \div \frac{1}{5^2} = \frac{1}{y^6} \cdot \frac{5^2}{1} = \frac{25}{y^6}$

Do Exercises 9–11.

c Scientific Notation

There are many kinds of symbols, or notation, for numbers. You are already familiar with fractional notation, decimal notation, and percent notation. Now we study another, **scientific notation,** which is especially useful when calculations involve very large or very small numbers. The following are examples of scientific notation:

Niagara Falls: On the Canadian side, during the summer the amount of water that spills over the falls in 1 min is about

1.3088×10^8 L = 130,880,000 L.

The mass of a hydrogen atom:

1.7×10^{-24} g = 0.0000000000000000000000017 g.

> **Scientific notation** for a number is an expression of the type
>
> $M \times 10^n$,
>
> where n is an integer, M is greater than or equal to 1 and less than 10 ($1 \le M < 10$), and M is expressed in decimal notation. 10^n is also considered to be scientific notation when $M = 1$.

You should try to make conversions to scientific notation mentally as much as possible. Here is a handy mental device.

> A positive exponent in scientific notation indicates a large number (greater than 1) and a negative exponent indicates a small number (less than 1).

Examples Convert to scientific notation.

15. $78{,}000 = 7.8 \times 10^4$ 7.8,000. (4 places)

Large number, so the exponent is positive.

16. $0.0000057 = 5.7 \times 10^{-6}$ 0.000005.7 (6 places)

Small number, so the exponent is negative.

Each of the following is *not* scientific notation.

12.46×10^7 — This number is greater than 10.

0.347×10^{-5} — This number is less than 1.

Do Exercises 12 and 13.

Examples Convert mentally to decimal notation.

17. $7.893 \times 10^5 = 789{,}300$ 7.89300. (5 places)

Positive exponent, so the answer is a large number.

18. $4.7 \times 10^{-8} = 0.000000047$ 0.00000004.7 (8 places)

Negative exponent, so the answer is a small number.

Convert to scientific notation.

12. 0.000517

13. 523,000,000

Convert to decimal notation.

14. 6.893×10^{11}

15. 5.67×10^{-5}

Answers on page A-29

Calculator Spotlight

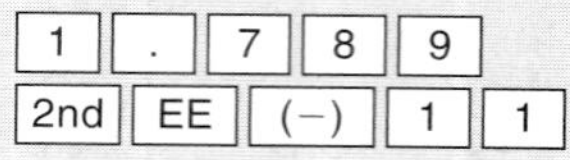 To enter a number in scientific notation into a graphing calculator, we first enter the decimal portion of the number and then press [2nd] [EE] followed by the exponent.

For example, to enter 1.789×10^{-11}, we press

[1] [.] [7] [8] [9]
[2nd] [EE] [(−)] [1] [1]

The display reads,

1.789 E ⁻11 .

Exercises

Enter in scientific notation.

1. 260,000,000
2. 0.00000000006709

Multiply and write scientific notation for the result.

16. $(1.12 \times 10^{-8})(5 \times 10^{-7})$

17. $(9.1 \times 10^{-17})(8.2 \times 10^{3})$

Answers on page A-29

Do Exercises 14 and 15 on the preceding page.

d Multiplying and Dividing Using Scientific Notation

Multiplying

Consider the product

$$400 \cdot 2000 = 800{,}000.$$

In scientific notation, this is

$$(4 \times 10^2) \cdot (2 \times 10^3) = (4 \cdot 2)(10^2 \cdot 10^3) = 8 \times 10^5.$$

By applying the commutative and associative laws, we can find this product by multiplying $4 \cdot 2$, to get 8, and $10^2 \cdot 10^3$, to get 10^5 (we do this by adding the exponents).

Example 19 Multiply: $(1.8 \times 10^6) \cdot (2.3 \times 10^{-4})$.

We apply the commutative and associative laws to get

$$\begin{aligned}(1.8 \times 10^6) \cdot (2.3 \times 10^{-4}) &= (1.8 \cdot 2.3) \times (10^6 \cdot 10^{-4})\\ &= 4.14 \times 10^{6+(-4)} \quad \text{Adding exponents}\\ &= 4.14 \times 10^2.\end{aligned}$$

Example 20 Multiply: $(3.1 \times 10^5) \cdot (4.5 \times 10^{-3})$.

We have

$$\begin{aligned}(3.1 \times 10^5) \cdot (4.5 \times 10^{-3}) &= (3.1 \times 4.5)(10^5 \cdot 10^{-3})\\ &= 13.95 \times 10^2.\end{aligned}$$

The answer at this stage is 13.95×10^2, but this is *not* scientific notation, because 13.95 is not a number between 1 and 10. To find scientific notation for the product, we convert 13.95 to scientific notation and simplify:

$$\begin{aligned}13.95 \times 10^2 &= (1.395 \times 10^1) \times 10^2 \quad \text{Substituting } 1.395 \times 10^1 \text{ for } 13.95\\ &= 1.395 \times (10^1 \times 10^2) \quad \text{Associative law}\\ &= 1.395 \times 10^3. \quad \text{Adding exponents}\end{aligned}$$

The answer is

$$1.395 \times 10^3.$$

Do Exercises 16 and 17.

Dividing

Consider the quotient

$$800{,}000 \div 400 = 2000.$$

In scientific notation, this is

$$(8 \times 10^5) \div (4 \times 10^2) = \frac{8 \times 10^5}{4 \times 10^2} = \frac{8}{4} \times \frac{10^5}{10^2} = 2 \times 10^3.$$

We can find this product by dividing 8 by 4, to get 2, and 10^5 by 10^2, to get 10^3 (we do this by subtracting the exponents).

Example 21 Divide: $(3.41 \times 10^5) \div (1.1 \times 10^{-3})$.

$$\begin{aligned}(3.41 \times 10^5) \div (1.1 \times 10^{-3}) &= \frac{3.41 \times 10^5}{1.1 \times 10^{-3}} \\ &= \frac{3.41}{1.1} \times \frac{10^5}{10^{-3}} \\ &= 3.1 \times 10^{5-(-3)} \\ &= 3.1 \times 10^8\end{aligned}$$

Example 22 Divide: $(6.4 \times 10^{-7}) \div (8.0 \times 10^6)$.

We have

$$\begin{aligned}(6.4 \times 10^{-7}) \div (8.0 \times 10^6) &= \frac{6.4 \times 10^{-7}}{8.0 \times 10^6} \\ &= \frac{6.4}{8.0} \times \frac{10^{-7}}{10^6} \\ &= 0.8 \times 10^{-7-6} \\ &= 0.8 \times 10^{-13}.\end{aligned}$$

The answer at this stage is

$$0.8 \times 10^{-13},$$

but this is *not* scientific notation, because 0.8 is not a number between 1 and 10. To find scientific notation for the quotient, we convert 0.8 to scientific notation and simplify:

$$\begin{aligned}0.8 \times 10^{-13} &= (8.0 \times 10^{-1}) \times 10^{-13} && \text{Substituting } 8.0 \times 10^{-1} \text{ for } 0.8 \\ &= 8.0 \times (10^{-1} \times 10^{-13}) && \text{Associative law} \\ &= 8.0 \times 10^{-14}. && \text{Adding exponents}\end{aligned}$$

The answer is

$$8.0 \times 10^{-14}.$$

Do Exercises 18 and 19.

e Applications with Scientific Notation

Example 23 *Distance from the Sun to Earth.* Light from the sun traveling at a rate of 300,000 kilometers per second (km/s) reaches Earth in 499 sec. Find the distance, expressed in scientific notation, from the sun to Earth.

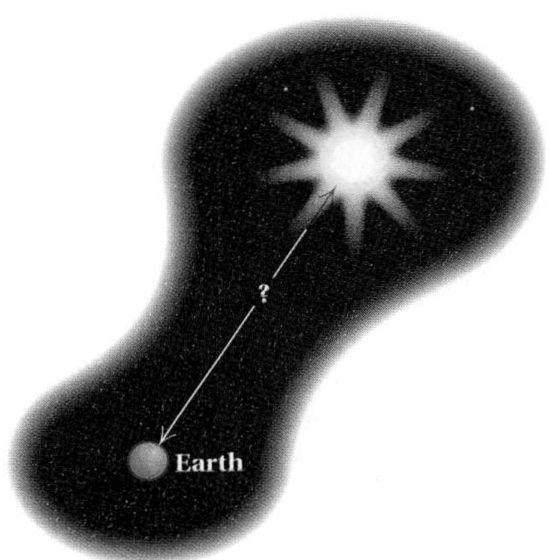

Divide and write scientific notation for the result.

18. $\dfrac{4.2 \times 10^5}{2.1 \times 10^2}$

19. $\dfrac{1.1 \times 10^{-4}}{2.0 \times 10^{-7}}$

Answers on page A-29

20. *Niagara Falls Water Flow.* On the Canadian side, during the summer the amount of water that spills over the falls in 1 min is about

1.3088×10^8 L.

How much water spills over the falls in one day? Express the answer in scientific notation.

21. *Earth vs. Saturn.* The mass of Earth is about 6×10^{21} metric tons. The mass of Saturn is about 5.7×10^{23} metric tons. About how many times the mass of Earth is the mass of Saturn? Express the answer in scientific notation.

Answers on page A-29

The time t that it takes for light to reach Earth from the sun is 4.99×10^2 sec (s). The speed is 3.0×10^5 km/s. Recall that distance can be expressed in terms of speed and time as

$$\text{Distance} = \text{Speed} \cdot \text{Time}$$
$$d = rt.$$

We substitute 3.0×10^5 for r and 4.99×10^2 for t:

$$\begin{aligned} d &= rt \\ &= (3.0 \times 10^5)(4.99 \times 10^2) && \text{Substituting} \\ &= 14.97 \times 10^7 \\ &= 1.497 \times 10^8 \text{ km.} && \text{Converting to scientific notation} \end{aligned}$$

Thus the distance from the sun to Earth is 1.497×10^8 km.

Do Exercise 20.

Example 24 *Earth vs. Jupiter.* The mass of Earth is about 6×10^{21} metric tons. The mass of Jupiter is about 1.908×10^{24} metric tons. About how many times the mass of Earth is the mass of Jupiter? Express the answer in scientific notation.

To determine how many times the mass of Jupiter is of the mass of Earth, we divide the mass of Jupiter by the mass of Earth:

$$\begin{aligned} \frac{1.908 \times 10^{24}}{6 \times 10^{21}} &= \frac{1.908}{6} \times \frac{10^{24}}{10^{21}} \\ &= 0.318 \times 10^3 \\ &= (3.18 \times 10^{-1}) \times 10^3 \\ &= 3.18 \times 10^2. \end{aligned}$$

Thus the mass of Jupiter is 3.18×10^2, or 318, times the mass of Earth.

Do Exercise 21.

The following is a summary of the definitions and rules for exponents that we have considered in this section and the preceding one.

DEFINITIONS AND RULES FOR EXPONENTS

Exponent of 1:	$a^1 = a$
Exponent of 0:	$a^0 = 1,\ a \neq 0$
Negative exponents:	$a^{-n} = \dfrac{1}{a^n},\ a \neq 0$
Product Rule:	$a^m \cdot a^n = a^{m+n}$
Quotient Rule:	$\dfrac{a^m}{a^n} = a^{m-n},\ a \neq 0$
Power Rule:	$(a^m)^n = a^{mn}$
Raising a product to a power:	$(ab)^n = a^n b^n$
Raising a quotient to a power:	$\left(\dfrac{a}{b}\right)^n = \dfrac{a^n}{b^n},\ b \neq 0$
Scientific notation:	$M \times 10^n$, or 10^n, where $1 \leq M < 10$

Exercise Set 10.2

a, **b** Simplify.

1. $(2^3)^2$

2. $(5^2)^4$

3. $(5^2)^{-3}$

4. $(7^{-3})^5$

5. $(x^{-3})^{-4}$

6. $(a^{-5})^{-6}$

7. $(4x^3)^2$

8. $4(x^3)^2$

9. $(x^4y^5)^{-3}$

10. $(t^5x^3)^{-4}$

11. $(x^{-6}y^{-2})^{-4}$

12. $(x^{-2}y^{-7})^{-5}$

13. $(3x^3y^{-8}z^{-3})^2$

14. $(2a^2y^{-4}z^{-5})^3$

15. $\left(\frac{a^2}{b^3}\right)^4$

16. $\left(\frac{x^3}{y^4}\right)^5$

17. $\left(\frac{y^3}{2}\right)^2$

18. $\left(\frac{a^5}{3}\right)^3$

19. $\left(\frac{y^2}{2}\right)^{-3}$

20. $\left(\frac{a^4}{3}\right)^{-2}$

21. $\left(\frac{x^2y}{z}\right)^3$

22. $\left(\frac{m}{n^4p}\right)^3$

23. $\left(\frac{a^2b}{cd^3}\right)^{-2}$

24. $\left(\frac{2a^2}{3b^4}\right)^{-3}$

c Convert to scientific notation.

25. 28,000,000,000

26. 4,900,000,000,000

27. 907,000,000,000,000,000

28. 168,000,000,000,000

29. 0.00000304

30. 0.000000000865

31. 0.000000018

32. 0.00000000002

33. 100,000,000,000

34. 0.0000001

Convert the number in the sentence to scientific notation.

35. *Niagara Falls Water Flow.* On the American side, during the summer the amount of water that spills over the falls in 1 min is about 11.35 million L (1 million $= 10^6$).

36. *Proctor & Gamble.* In a recent year, Proctor & Gamble led the nation's advertisers by spending \$2.777 billion on advertising (***Source:*** *Advertising Age*) (1 billion $= 10^9$).

Convert to decimal notation.

37. 8.74×10^7

38. 1.85×10^8

39. 5.704×10^{-8}

40. 8.043×10^{-4}

41. 10^7

42. 10^6

43. 10^{-5}

44. 10^{-8}

d Multiply or divide and write scientific notation for the result.

45. $(3 \times 10^4)(2 \times 10^5)$

46. $(3.9 \times 10^8)(8.4 \times 10^{-3})$

47. $(5.2 \times 10^5)(6.5 \times 10^{-2})$

48. $(7.1 \times 10^{-7})(8.6 \times 10^{-5})$

49. $(9.9 \times 10^{-6})(8.23 \times 10^{-8})$

50. $(1.123 \times 10^4) \times 10^{-9}$

51. $\frac{8.5 \times 10^{8}}{3.4 \times 10^{-5}}$

52. $\frac{5.6 \times 10^{-2}}{2.5 \times 10^{5}}$

53. $(3.0 \times 10^{6}) \div (6.0 \times 10^{9})$

54. $(1.5 \times 10^{-3}) \div (1.6 \times 10^{-6})$

55. $\frac{7.5 \times 10^{-9}}{2.5 \times 10^{12}}$

56. $\frac{4.0 \times 10^{-3}}{8.0 \times 10^{20}}$

e Solve.

57. *Total Income of Two-Person Households.* In 1993, there were about 31.2 million two-person households in the United States. The average income of these households was about $42,400. (***Source*:** Statistical Abstract of the United States) Find the total income generated by two-person households in 1993. Express the answer in scientific notation.

58. *Niagara Falls Water Flow.* On the American side, during the summer the amount of water that spills over the falls in 1 min is about 11.35 million L (1 million $= 10^{6}$). How much water spills over the falls in 1 yr? (Use 365 days for 1 yr.) Express the answer in scientific notation.

59. *Stars.* It is estimated that there are 10 billion trillion stars in the known universe. Express the number of stars in scientific notation.

60. *Closest Star.* Excluding the sun, the closest star to Earth is Proxima Centauri, which is 4.3 light-years away (one light-year $= 5.88 \times 10^{12}$ mi). How far, in miles, is Proxima Centauri from Earth? Express the answer in scientific notation.

61. *Earth vs. Sun.* The mass of Earth is about 6×10^{21} metric tons. The mass of the sun is about 1.998×10^{27} metric tons. About how times the mass of Earth is the mass of the sun? Express the answer in scientific notation.

62. *Red Light.* The wavelength of light is given by the velocity divided by the frequency. The velocity of red light is 300,000,000 m/sec, and its frequency is 400,000,000,000,000 cycles per second. What is the wavelength of red light? Express the answer in scientific notation.

Space Travel. Use the following information for Exercises 63 and 64.

Earth $d = 240{,}000$ mi Moon

Approximate Distance from Earth to:	
Moon	240,000 mi
Mars	35,000,000 mi
Pluto	2,670,000,000 mi

63. *Time to Reach Mars.* Suppose that it takes about 3 days for a space vehicle to travel from Earth to the moon. About how long would it take the same space vehicle traveling at the same speed to reach Mars? Express the answer in scientific notation.

64. *Time to Reach Pluto.* Suppose that it takes about 3 days for a space vehicle to travel from Earth to the moon. About how long would it take the same space vehicle traveling at the same speed to reach Pluto? Express the answer in scientific notation.

Skill Maintenance

Factor. [7.7d]

65. $9x - 36$

66. $4x - 2y + 16$

67. $3s + 3t + 24$

68. $-7x - 14$

Solve. [8.3b]

69. $2x - 4 - 5x + 8 = x - 3$

70. $8x + 7 - 9x = 12 - 6x + 5$

Solve. [8.3c]

71. $8(2x + 3) - 2(x - 5) = 10$

72. $4(x - 3) + 5 = 6(x + 2) - 8$

Graph. [9.2b], [9.3a]

73. $y = x - 5$

74. $2x + y = 8$

Synthesis

75. ◆ Using the quotient rule, explain why 9^0 is defined to be 1.

76. ◆ Explain in your own words when exponents should be added and when they should be multiplied.

77. 🖩 Carry out the indicated operations. Express the result in scientific notation.

$$\frac{(5.2 \times 10^6)(6.1 \times 10^{-11})}{1.28 \times 10^{-3}}$$

78. Find the reciprocal and express it in scientific notation.

$$6.25 \times 10^{-3}$$

Simplify.

79. $\dfrac{(5^{12})^2}{5^{25}}$

80. $\dfrac{a^{22}}{(a^2)^{11}}$

81. $\dfrac{(3^5)^4}{3^5 \cdot 3^4}$

82. $\dfrac{49^{18}}{7^{35}}$

(*Hint*: Study Exercise 80.)

83. $\left(\dfrac{1}{a}\right)^{-n}$

84. $\dfrac{(0.4)^5}{[(0.4)^3]^2}$

Determine whether each of the following is true for any pairs of integers m and n and any positive numbers x and y.

85. $x^m \cdot y^n = (xy)^{mn}$

86. $x^m \cdot y^m = (xy)^{2m}$

87. $(x - y)^m = x^m - y^m$

Collaborative Learning Manual

Use exponential and scientific notation to represent the salary for a job.

10.3 Introduction to Polynomials

We have already learned to evaluate and to manipulate certain kinds of algebraic expressions. We will now consider algebraic expressions called *polynomials.*

The following are examples of *monomials in one variable*:

$$3x^2, \quad 2x, \quad -5, \quad 37p^4, \quad 0.$$

Each expression is a constant or a constant times some variable to a nonnegative integer power.

> A **monomial** is an expression of the type ax^n, where a is a real-number constant and n is a nonnegative integer.

Algebraic expressions like the following are **polynomials:**

$$\tfrac{3}{4}y^5, \quad -2, \quad 5y + 3, \quad 3x^2 + 2x - 5, \quad -7a^3 + \tfrac{1}{2}a, \quad 6x, \quad 37p^4, \quad x, \quad 0.$$

> A **polynomial** is a monomial or a combination of sums and/or differences of monomials.

The following algebraic expressions are *not* polynomials:

$$\textbf{(1)}\ \frac{x+3}{x-4}, \qquad \textbf{(2)}\ 5x^3 - 2x^2 + \frac{1}{x}, \qquad \textbf{(3)}\ \frac{1}{x^3 - 2}.$$

Expressions (1) and (3) are not polynomials because they represent quotients, not sums. Expression (2) is not a polynomial because

$$\frac{1}{x} = x^{-1},$$

and this is not a monomial because the exponent is negative.

Do Exercise 1.

Objectives

- a Evaluate a polynomial for a given value of the variable.
- b Identify the terms of a polynomial.
- c Identify the like terms of a polynomial.
- d Identify the coefficients of a polynomial.
- e Collect the like terms of a polynomial.
- f Arrange a polynomial in descending order, or collect the like terms and then arrange in descending order.
- g Identify the degree of each term of a polynomial and the degree of the polynomial.
- h Identify the missing terms of a polynomial.
- i Classify a polynomial as a monomial, binomial, trinomial, or none of these.

For Extra Help

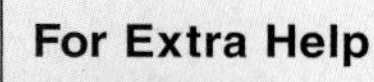

TAPE 19 | MAC WIN | CD-ROM

1. Write three polynomials.

a Evaluating Polynomials and Applications

When we replace the variable in a polynomial with a number, the polynomial then represents a number called a **value** of the polynomial. Finding that number, or value, is called **evaluating the polynomial.** We evaluate a polynomial using the rules for order of operations (Section 7.8).

Example 1 Evaluate the polynomial for $x = 2$.

a) $3x + 5 = 3 \cdot 2 + 5$
$\quad = 6 + 5$
$\quad = 11$

b) $2x^2 - 7x + 3 = 2 \cdot 2^2 - 7 \cdot 2 + 3$
$\quad = 2 \cdot 4 - 7 \cdot 2 + 3$
$\quad = 8 - 14 + 3$
$\quad = -3$

Answer on page A-29

Evaluate the polynomial for $x = 3$.

2. $-4x - 7$

3. $-5x^3 + 7x + 10$

Evaluate the polynomial for $x = -4$.

4. $5x + 7$

5. $2x^2 + 5x - 4$

6. Referring to Example 3, what is the total number of games to be played in a league of 12 teams?

7. *Perimeter of Baseball Diamond.* The perimeter of a square of side x is given by the polynomial $4x$.

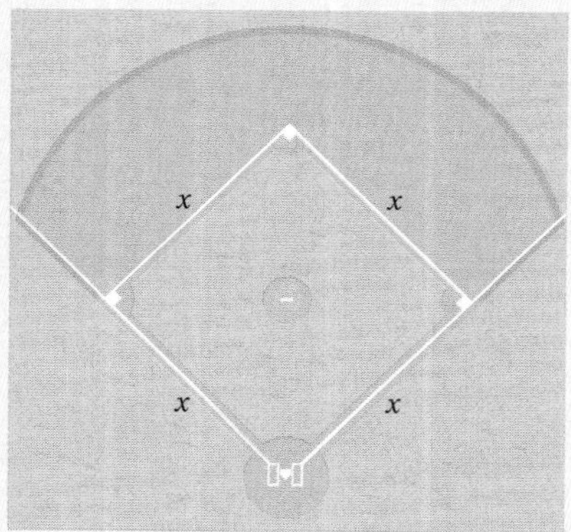

A baseball diamond is a square 90 ft on a side. Find the perimeter of a baseball diamond.

8. Use *only* the graph shown in Example 4 to evaluate the polynomial $2x - 2$ for $x = 4$ and for $x = -1$.

Answers on page A-29

Example 2 Evaluate the polynomial for $x = -4$.

a) $2 - x^3 = 2 - (-4)^3 = 2 - (-64)$
$= 2 + 64$
$= 66$

b) $-x^2 - 3x + 1 = -(-4)^2 - 3(-4) + 1$
$= -16 + 12 + 1$
$= -3$

Do Exercises 2–5.

Polynomials occur in many real-world situations.

Example 3 *Games in a Sports League.* In a sports league of n teams in which each team plays every other team twice, the total number of games to be played is given by the polynomial

$$n^2 - n.$$

A women's slow-pitch softball league has 10 teams. What is the total number of games to be played?

We evaluate the polynomial for $n = 10$:

$$n^2 - n = 10^2 - 10 = 100 - 10 = 90.$$

The league plays 90 games.

Do Exercises 6 and 7.

AG Algebraic–Graphical Connection

An equation like $y = 2x - 2$, which has a polynomial on one side and y on the other, is called a **polynomial equation.** We will here and in many places throughout the book connect graphs to related concepts.

Recall from Chapter 9 that in order to plot points before graphing an equation, we choose values for x and compute the corresponding y-values. If the equation has y on one side and a polynomial involving x on the other, then determining y is the same as evaluating the polynomial. Once the graph of such an equation has been drawn, we can evaluate the polynomial for a given x-value by finding the y-value that is paired with it on the graph.

Example 4 Use *only* the given graph of $y = 2x - 2$ to evaluate the polynomial $2x - 2$ for $x = 3$.

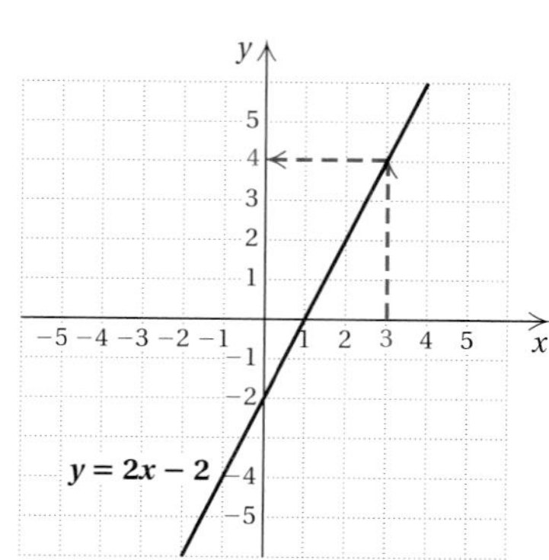

First, we locate 3 on the x-axis. From there we move vertically to the graph of the equation and then horizontally to the y-axis. There we locate the y-value that is paired with 3. Although our drawing may not be precise, it appears that the y-value 4 is paired with 3. Thus the value of $2x - 2$ is 4 when $x = 3$.

Do Exercise 8.

AG

Example 5 *Medical Dosage.* The concentration C, in parts per million, of a certain antibiotic in the bloodstream after t hours is given by the polynomial equation

$$C = -0.05t^2 + 2t + 2.$$

Find the concentration after 2 hr.

To find the concentration after 2 hr, we evaluate the polynomial for $t = 2$:

$$\begin{aligned} -0.05t^2 + 2t + 2 &= -0.05(2)^2 + 2(2) + 2 \\ &= -0.05(4) + 2(2) + 2 \\ &= -0.2 + 4 + 2 \\ &= -0.2 + 6 \\ &= 5.8. \end{aligned}$$

Carrying out the calculation using the rules for order of operations

The concentration after 2 hr is 5.8 parts per million.

Do Exercise 9.

AG Algebraic–Graphical Connection

The polynomial equation in Example 5 can be graphed if we evaluate the polynomial for several values of t. We list the values in a table and show the graph below. Note that the concentration peaks at the 20-hr mark and after a bit more than 40 hr, the concentration is 0. Since neither time nor concentration can be negative, our graph uses only the first quadrant.

t	$-0.05t^2 + 2t + 2$
0	2
2	5.8
10	17
20	22
30	17

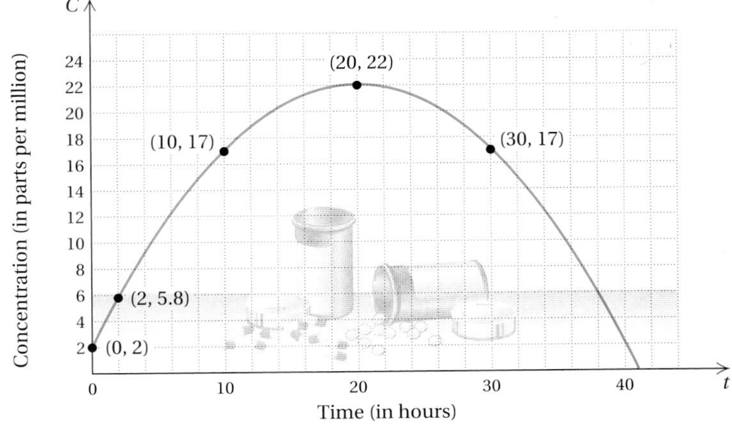

Do Exercise 10.

9. Referring to Example 5, find the concentration after 3 hr.

10. Use *only* the graph showing medical dosage to estimate the value of the polynomial for $t = 26$.

Answers on page A-29

Calculator Spotlight

Evaluating Polynomials. One way to evaluate a polynomial like $-x^2 - 3x + 1$ for $x = -4$ (see Example 2b) is to graph $y_1 = -x^2 - 3x + 1$.

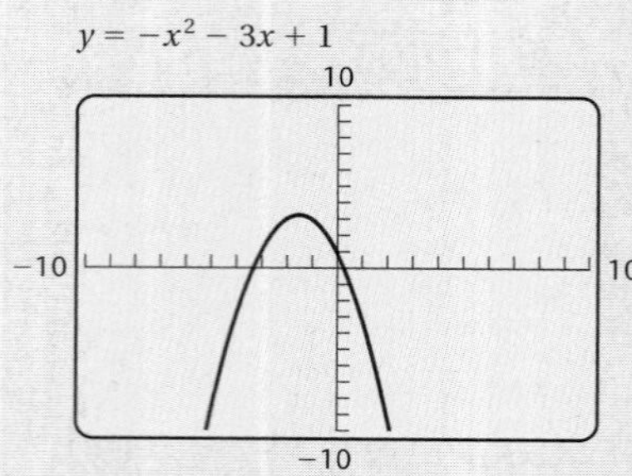

We can also adjust the window to obtain a better view of the graph.

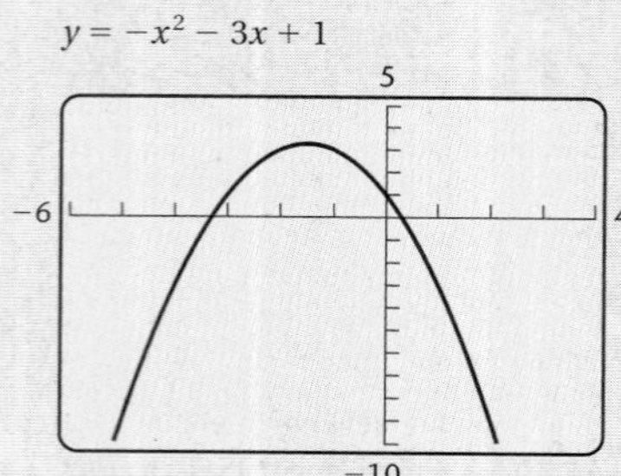

To evaluate the polynomial, we then use the CALC feature and choose VALUE.

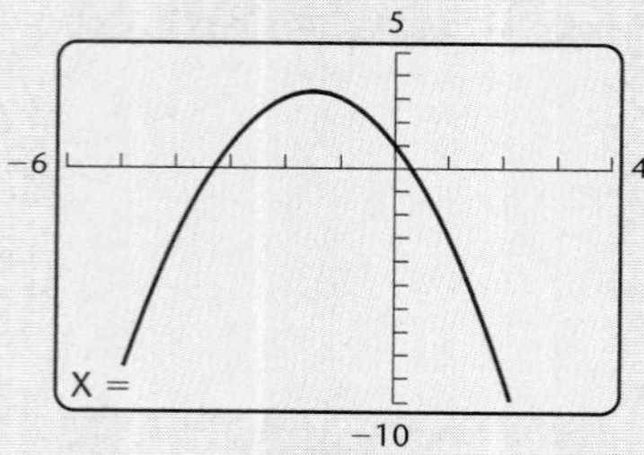

We enter $x = -4$. The y-value, -3, is shown together with a TRACE indicator showing the point $(-4, -3)$.

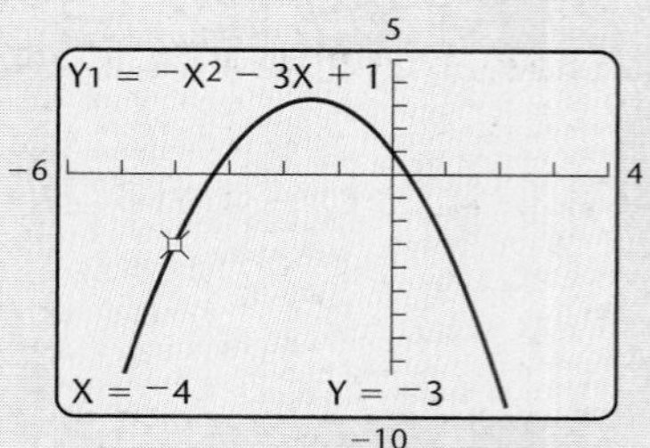

Polynomials can also be evaluated using the TABLE feature as described in Section 9.4.

Exercises

1. Evaluate the polynomial $-x^2 - 3x + 1$ for $x = -1$, for $x = -0.3$, and for $x = 1.7$.
2. Evaluate the polynomial $-0.05x^2 + 2x + 2$ for $x = 0$, for $x = 10$, for $x = 23$, and for $x = 36.4$. Use the viewing window [0, 41, 0, 25].
3. Evaluate the polynomial $2x^2 - x - 8$ for $x = -3$, for $x = -2$, for $x = 0$, for $x = 1.8$, and for $x = 3$.

b Identifying Terms

As we saw in Section 7.4, subtractions can be rewritten as additions. For any polynomial that has some subtractions, we can find an equivalent polynomial using only additions.

Examples Find an equivalent polynomial using only additions.

6. $-5x^2 - x = -5x^2 + (-x)$

7. $4x^5 - 2x^6 - 4x + 7 = 4x^5 + (-2x^6) + (-4x) + 7$

Do Exercises 11 and 12.

When a polynomial has only additions, the monomials being added are called **terms**. In Example 6, the terms are $-5x^2$ and $-x$. In Example 7, the terms are $4x^5$, $-2x^6$, $-4x$, and 7.

Example 8 Identify the terms of the polynomial

$$4x^7 + 3x + 12 + 8x^3 + 5x.$$

Terms: $4x^7$, $3x$, 12, $8x^3$, and $5x$.

If there are subtractions, you can *think* of them as additions without rewriting.

Example 9 Identify the terms of the polynomial

$$3t^4 - 5t^6 - 4t + 2.$$

Terms: $3t^4$, $-5t^6$, $-4t$, and 2.

Do Exercises 13 and 14.

c Like Terms

When terms have the same variable and the variable is raised to the same power, we say that they are **like terms,** or **similar terms.**

Examples Identify the like terms in the polynomials.

10. $4x^3 + 5x - 4x^2 + 2x^3 + x^2$

Like terms: $4x^3$ and $2x^3$ — Same variable and exponent

Like terms: $-4x^2$ and x^2 — Same variable and exponent

11. $6 - 3a^2 + 8 - a - 5a$

Like terms: 6 and 8 — Constant terms are like terms because $6 = 6x^0$ and $8 = 8x^0$.

Like terms: $-a$ and $-5a$

Do Exercises 15–17.

d Coefficients

The coefficient of the term $5x^3$ is 5. In the following polynomial, the color numbers are the **coefficients**:

$$3x^5 - 2x^3 + 5x + 4.$$

Find an equivalent polynomial using only additions.

11. $-9x^3 - 4x^5$

12. $-2y^3 + 3y^7 - 7y$

Identify the terms of the polynomial.

13. $3x^2 + 6x + \frac{1}{2}$

14. $-4y^5 + 7y^2 - 3y - 2$

Identify the like terms in the polynomial.

15. $4x^3 - x^3 + 2$

16. $4t^4 - 9t^3 - 7t^4 + 10t^3$

17. $5x^2 + 3x - 10 + 7x^2 - 8x + 11$

Answers on page A-29

18. Identify the coefficient of each term in the polynomial

$2x^4 - 7x^3 - 8.5x^2 + 10x - 4.$

Collect like terms.

19. $3x^2 + 5x^2$

20. $4x^3 - 2x^3 + 2 + 5$

21. $\frac{1}{2}x^5 - \frac{3}{4}x^5 + 4x^2 - 2x^2$

22. $24 - 4x^3 - 24$

23. $5x^3 - 8x^5 + 8x^5$

24. $-2x^4 + 16 + 2x^4 + 9 - 3x^5$

Collect like terms.

25. $7x - x$

26. $5x^3 - x^3 + 4$

27. $\frac{3}{4}x^3 + 4x^2 - x^3 + 7$

28. $8x^2 - x^2 + x^3 - 1 - 4x^2 + 10$

Answers on page A-29

Example 12 Identify the coefficient of each term in the polynomial

$$3x^4 - 4x^3 + 7x^2 + x - 8.$$

The coefficient of the first term is 3.

The coefficient of the second term is -4.

The coefficient of the third term is 7.

The coefficient of the fourth term is 1.

The coefficient of the fifth term is -8.

Do Exercise 18.

e Collecting Like Terms

We can often simplify polynomials by **collecting like terms,** or **combining similar terms.** To do this, we use the distributive laws. We factor out the variable expression and add or subtract the coefficients. We try to do this mentally as much as possible.

Examples Collect like terms.

13. $2x^3 - 6x^3 = (2 - 6)x^3 = -4x^3$ **Using a distributive law**

14. $5x^2 + 7 + 4x^4 + 2x^2 - 11 - 2x^4 = (5 + 2)x^2 + (4 - 2)x^4 + (7 - 11)$
$= 7x^2 + 2x^4 - 4$

Note that using the distributive laws in this manner allows us to collect like terms by adding or subtracting the coefficients. Often the middle step is omitted and we add or subtract mentally, writing just the answer. In collecting like terms, we may get 0.

Examples Collect like terms.

15. $5x^3 - 5x^3 = (5 - 5)x^3 = 0x^3 = 0$

16. $3x^4 + 2x^2 - 3x^4 + 8 = (3 - 3)x^4 + 2x^2 + 8$
$= 0x^4 + 2x^2 + 8 = 2x^2 + 8$

Do Exercises 19–24.

Multiplying a term of a polynomial by 1 does not change the term, but it may make it easier to factor or collect like terms.

Examples Collect like terms.

17. $5x^2 + x^2 = 5x^2 + 1x^2$ **Replacing x^2 with $1x^2$**
$= (5 + 1)x^2$ **Using a distributive law**
$= 6x^2$

18. $5x^4 - 6x^3 - x^4 = 5x^4 - 6x^3 - 1x^4$ $x^4 = 1x^4$
$= (5 - 1)x^4 - 6x^3$
$= 4x^4 - 6x^3$

19. $\frac{2}{3}x^4 - x^3 - \frac{1}{6}x^4 + \frac{2}{5}x^3 - \frac{3}{10}x^3 = \left(\frac{2}{3} - \frac{1}{6}\right)x^4 + \left(-1 + \frac{2}{5} - \frac{3}{10}\right)x^3$
$= \left(\frac{4}{6} - \frac{1}{6}\right)x^4 + \left(-\frac{10}{10} + \frac{4}{10} - \frac{3}{10}\right)x^3$
$= \frac{3}{6}x^4 - \frac{9}{10}x^3 = \frac{1}{2}x^4 - \frac{9}{10}x^3$

Do Exercises 25–28.

f Descending and Ascending Order

Note in the following polynomial that the exponents decrease from left to right. We say that the polynomial is arranged in **descending order:**

$$2x^4 - 8x^3 + 5x^2 - x + 3.$$

The term with the largest exponent is first. The term with the next largest exponent is second, and so on. The associative and commutative laws allow us to arrange the terms of a polynomial in descending order.

Examples Arrange the polynomial in descending order.

20. $6x^5 + 4x^7 + x^2 + 2x^3 = 4x^7 + 6x^5 + 2x^3 + x^2$

21. $\frac{2}{3} + 4x^5 - 8x^2 + 5x - 3x^3 = 4x^5 - 3x^3 - 8x^2 + 5x + \frac{2}{3}$

We usually arrange polynomials in descending order, but not always. The opposite order is called **ascending order.** Generally, if an exercise is written in a certain order, we give the answer in that same order.

Do Exercises 29–31.

Example 22 Collect like terms and then arrange in descending order:

$$2x^2 - 4x^3 + 3 - x^2 - 2x^3.$$

We have

$$2x^2 - 4x^3 + 3 - x^2 - 2x^3 = x^2 - 6x^3 + 3 \quad \text{Collecting like terms}$$
$$= -6x^3 + x^2 + 3 \quad \text{Arranging in descending order}$$

Do Exercises 32 and 33.

g Degrees

The **degree** of a term is the exponent of the variable. The degree of the term $5x^3$ is 3.

Example 23 Identify the degree of each term of $8x^4 + 3x + 7$.

The degree of $8x^4$ is 4.

The degree of $3x$ is 1. Recall that $x = x^1$.

The degree of 7 is 0. Think of 7 as $7x^0$. Recall that $x^0 = 1$.

The **degree of a polynomial** is the largest of the degrees of the terms, unless it is the polynomial 0. The polynomial 0 is a special case. We agree that it has *no* degree either as a term or as a polynomial. This is because we can express 0 as $0 = 0x^5 = 0x^7$, and so on, using any exponent we wish.

Example 24 Identify the degree of the polynomial $5x^3 - 6x^4 + 7$.

We have

$$5x^3 - 6x^4 + 7. \quad \text{The largest exponent is 4.}$$

The degree of the polynomial is 4.

Do Exercise 34.

Arrange the polynomial in descending order.

29. $x + 3x^5 + 4x^3 + 5x^2 + 6x^7 - 2x^4$

30. $4x^2 - 3 + 7x^5 + 2x^3 - 5x^4$

31. $-14 + 7t^2 - 10t^5 + 14t^7$

Collect like terms and then arrange in descending order.

32. $3x^2 - 2x + 3 - 5x^2 - 1 - x$

33. $-x + \frac{1}{2} + 14x^4 - 7x - 1 - 4x^4$

34. Identify the degree of each term and the degree of the polynomial

$$-6x^4 + 8x^2 - 2x + 9.$$

Answers on page A-30

Identify the missing terms in the polynomial.

35. $2x^3 + 4x^2 - 2$

36. $-3x^4$

37. $x^3 + 1$

38. $x^4 - x^2 + 3x + 0.25$

Classify the polynomial as a monomial, binomial, trinomial, or none of these.

39. $5x^4$

40. $4x^3 - 3x^2 + 4x + 2$

41. $3x^2 + x$

42. $3x^2 + 2x - 4$

Answers on page A-30

Let's summarize the terminology that we have learned, using the polynomial

$$3x^4 - 8x^3 + 5x^2 + 7x - 6.$$

Term	Coefficient	Degree of the Term	Degree of the Polynomial
$3x^4$	3	4	
$-8x^3$	-8	3	
$5x^2$	5	2	4
$7x$	7	1	
-6	-6	0	

h Missing Terms

If a coefficient is 0, we generally do not write the term. We say that we have a **missing term.**

Example 25 Identify the missing terms in the polynomial

$$8x^5 - 2x^3 + 5x^2 + 7x + 8.$$

There is no term with x^4. We say that the x^4-term (or the *fourth-degree term*) is missing.

For certain skills or manipulations, we can write missing terms with zero coefficients or leave space. For example, we can write the polynomial $3x^2 + 9$ as

$$3x^2 + 0x + 9 \quad \text{or} \quad 3x^2 + \qquad 9.$$

Do Exercises 35–38.

i Classifying Polynomials

Polynomials with just one term are called **monomials**. Polynomials with just two terms are called **binomials**. Those with just three terms are called **trinomials**. Those with more than three terms are generally not specified with a name.

Example 26

Monomials	Binomials	Trinomials	None of These
$4x^2$	$2x + 4$	$3x^3 + 4x + 7$	$4x^3 - 5x^2 + x - 8$
9	$3x^5 + 6x$	$6x^7 - 7x^2 + 4$	
$-23x^{19}$	$-9x^7 - 6$	$4x^2 - 6x - \frac{1}{2}$	

Do Exercises 39–42.

Exercise Set 10.3

a Evaluate the polynomial for $x = 4$ and for $x = -1$.

1. $-5x + 2$

2. $-8x + 1$

3. $2x^2 - 5x + 7$

4. $3x^2 + x - 7$

5. $x^3 - 5x^2 + x$

6. $7 - x + 3x^2$

Evaluate the polynomial for $x = -2$ and for $x = 0$.

7. $3x + 5$

8. $8 - 4x$

9. $x^2 - 2x + 1$

10. $5x + 6 - x^2$

11. $-3x^3 + 7x^2 - 3x - 2$

12. $-2x^3 + 5x^2 - 4x + 3$

13. *Skydiving.* During the first 13 sec of a jump, the number of feet that a skydiver falls in t seconds can be approximated by the polynomial

$11.12t^2$.

Approximately how far has a skydiver fallen 10 sec after having jumped from a plane?

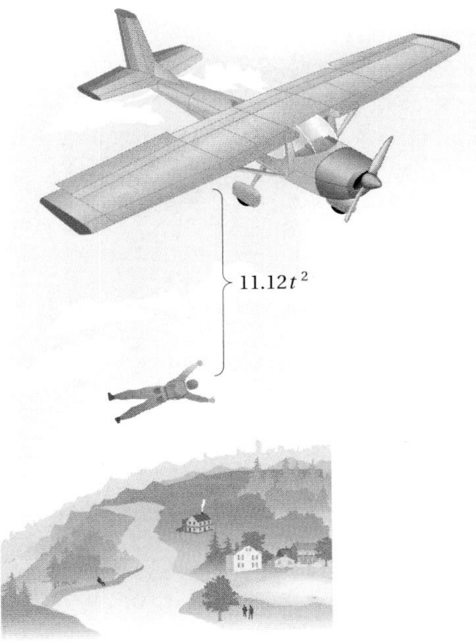

14. *Skydiving.* For jumps that exceed 13 sec, the polynomial

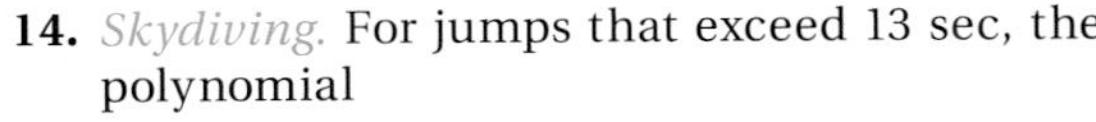

$173t - 369$

can be used to approximate the distance, in feet, that a skydiver has fallen in t seconds. Approximately how far has a skydiver fallen 20 sec after having jumped from a plane?

15. *Total Revenue.* Hadley Electronics is marketing a new kind of high-density TV. The firm determines that when it sells x TVs, its total revenue (the total amount of money taken in) will be

$280x - 0.4x^2$ dollars.

What is the total revenue from the sale of 75 TVs? 100 TVs?

16. *Total Cost.* Hadley Electronics determines that the total cost of producing x high-density TVs is given by

$5000 + 0.6x^2$ dollars.

What is the total cost of producing 500 TVs? 650 TVs?

17. The graph of the polynomial equation $y = 5 - x^2$ is shown below. Use *only* the graph to estimate the value of the polynomial for $x = -3$, for $x = -1$, for $x = 0$, for $x = 1.5$, and for $x = 2$.

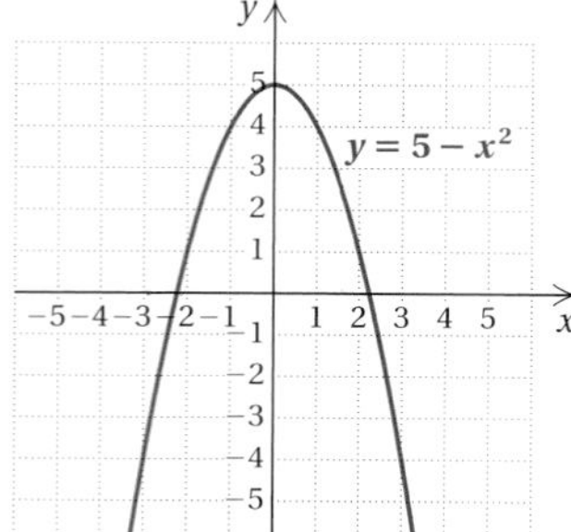

18. The graph of the polynomial equation $y = 6x^3 - 6x$ is shown below. Use *only* the graph to estimate the value of the polynomial for $x = -1$, for $x = -0.5$, for $x = 0.5$, for $x = 1$, and for $x = 1.1$.

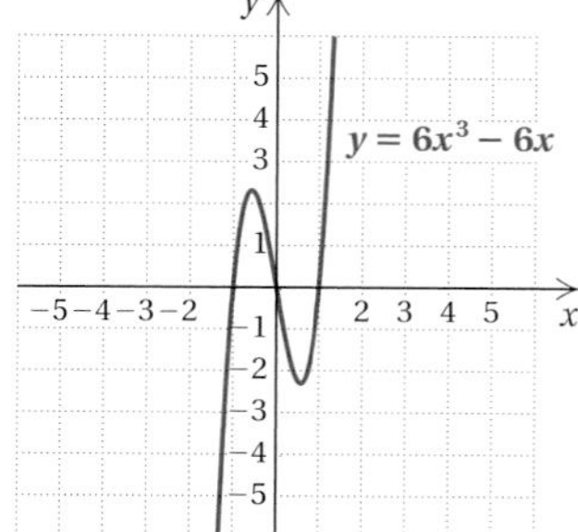

19. *Path of the Olympic Arrow.* The Olympic flame at the 1992 Summer Olympics was lit by a flaming arrow. As the arrow moved d feet horizontally from the archer, its height h, in feet, could be approximated by the polynomial equation

$h = -0.002d^2 + 0.8d + 6.6.$

The graph of this equation is shown at right. Use either the graph or the polynomial to approximate the height of the arrow after it has traveled horizontally for 100 ft, 200 ft, 300 ft, and 350 ft.

$h = -0.002d^2 + 0.8d + 6.6$

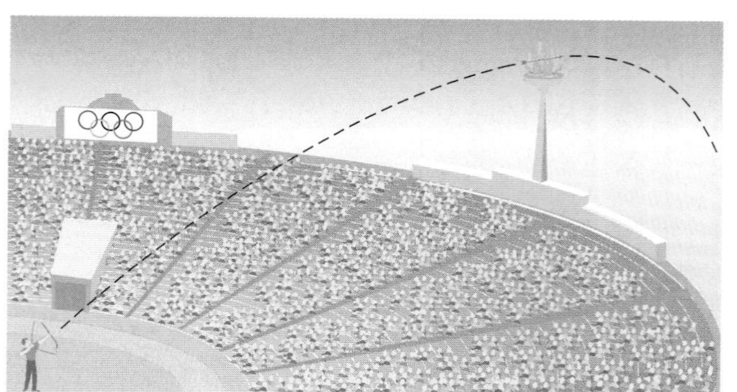

20. *Hearing-Impaired Americans.* The number N, in millions, of hearing-impaired Americans of age x can be approximated by the polynomial equation

$$N = -0.00006x^3 + 0.006x^2 - 0.1x + 1.9$$

(***Source***: American Speech-Language Hearing Association). The graph of this equation is shown at right. Use either the graph or the polynomial to approximate the number of hearing-impaired Americans of ages 20, 40, 50, and 60.

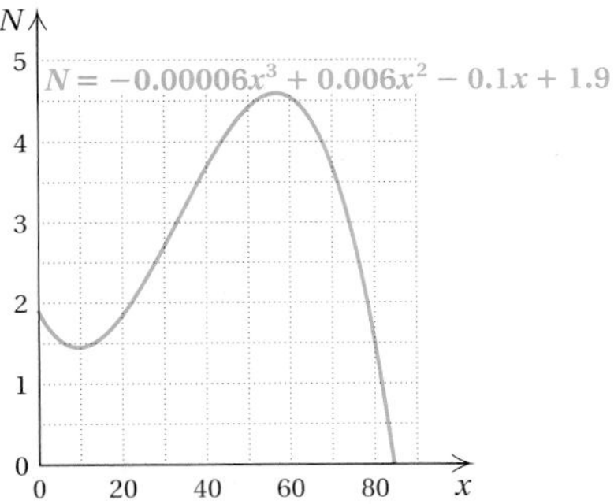

b Identify the terms of the polynomial.

21. $2 - 3x + x^2$

22. $2x^2 + 3x - 4$

c Identify the like terms in the polynomial.

23. $5x^3 + 6x^2 - 3x^2$

24. $3x^2 + 4x^3 - 2x^2$

25. $2x^4 + 5x - 7x - 3x^4$

26. $-3t + t^3 - 2t - 5t^3$

27. $3x^5 - 7x + 8 + 14x^5 - 2x - 9$

28. $8x^3 + 7x^2 - 11 - 4x^3 - 8x^2 - 29$

d Identify the coefficient of each term of the polynomial.

29. $-3x + 6$

30. $2x - 4$

31. $5x^2 + 3x + 3$

32. $3x^2 - 5x + 2$

33. $-5x^4 + 6x^3 - 3x^2 + 8x - 2$

34. $7x^3 - 4x^2 - 4x + 5$

e Collect like terms.

35. $2x - 5x$

36. $2x^2 + 8x^2$

37. $x - 9x$

38. $x - 5x$

39. $5x^3 + 6x^3 + 4$

40. $6x^4 - 2x^4 + 5$

41. $5x^3 + 6x - 4x^3 - 7x$

42. $3a^4 - 2a + 2a + a^4$

43. $6b^5 + 3b^2 - 2b^5 - 3b^2$

44. $2x^2 - 6x + 3x + 4x^2$

45. $\frac{1}{4}x^5 - 5 + \frac{1}{2}x^5 - 2x - 37$

46. $\frac{1}{3}x^3 + 2x - \frac{1}{6}x^3 + 4 - 16$

47. $6x^2 + 2x^4 - 2x^2 - x^4 - 4x^2$

48. $8x^2 + 2x^3 - 3x^3 - 4x^2 - 4x^2$

49. $\frac{1}{4}x^3 - x^2 - \frac{1}{6}x^2 + \frac{3}{8}x^3 + \frac{5}{16}x^3$

50. $\frac{1}{5}x^4 + \frac{1}{5} - 2x^2 + \frac{1}{10} - \frac{3}{15}x^4 + 2x^2 - \frac{3}{10}$

f Arrange the polynomial in descending order.

51. $x^5 + x + 6x^3 + 1 + 2x^2$

52. $3 + 2x^2 - 5x^6 - 2x^3 + 3x$

53. $5y^3 + 15y^9 + y - y^2 + 7y^8$

54. $9p - 5 + 6p^3 - 5p^4 + p^5$

Collect like terms and then arrange in descending order.

55. $3x^4 - 5x^6 - 2x^4 + 6x^6$

56. $-1 + 5x^3 - 3 - 7x^3 + x^4 + 5$

57. $-2x + 4x^3 - 7x + 9x^3 + 8$

58. $-6x^2 + x - 5x + 7x^2 + 1$

59. $3x + 3x + 3x - x^2 - 4x^2$

60. $-2x - 2x - 2x + x^3 - 5x^3$

61. $-x + \frac{3}{4} + 15x^4 - x - \frac{1}{2} - 3x^4$

62. $2x - \frac{5}{6} + 4x^3 + x + \frac{1}{3} - 2x$

g Identify the degree of each term of the polynomial and the degree of the polynomial.

63. $2x - 4$

64. $6 - 3x$

65. $3x^2 - 5x + 2$

66. $5x^3 - 2x^2 + 3$

67. $-7x^3 + 6x^2 + 3x + 7$

68. $5x^4 + x^2 - x + 2$

69. $x^2 - 3x + x^6 - 9x^4$

70. $8x - 3x^2 + 9 - 8x^3$

71. Complete the following table for the polynomial $-7x^4 + 6x^3 - 3x^2 + 8x - 2$.

Term	Coefficient	Degree of the Term	Degree of the Polynomial
$6x^3$	6		
		2	
$8x$		1	
	-2		

72. Complete the following table for the polynomial $3x^2 + 8x^5 - 46x^3 + 6x - 2.4 - \frac{1}{2}x^4$.

Term	Coefficient	Degree of the Term	Degree of the Polynomial
		5	
$-\frac{1}{2}x^4$		4	
	-46		
$3x^2$		2	
	6		
-2.4			

h Identify the missing terms in the polynomial.

73. $x^3 - 27$

74. $x^5 + x$

75. $x^4 - x$

76. $5x^4 - 7x + 2$

77. $2x^3 - 5x^2 + x - 3$

78. $-6x^3$

i Classify the polynomial as a monomial, binomial, trinomial, or none of these.

79. $x^2 - 10x + 25$

80. $-6x^4$

81. $x^3 - 7x^2 + 2x - 4$

82. $x^2 - 9$

83. $4x^2 - 25$

84. $2x^4 - 7x^3 + x^2 + x - 6$

85. $40x$

86. $4x^2 + 12x + 9$

Skill Maintenance

87. Three tired campers stopped for the night. All they had to eat was a bag of apples. During the night, one awoke and ate one-third of the apples. Later, a second camper awoke and ate one-third of the apples that remained. Much later, the third camper awoke and ate one-third of those apples yet remaining after the other two had eaten. When they got up the next morning, 8 apples were left. How many apples did they begin with? [8.4a]

Subtract. [7.4a]

88. $1 - 20$

89. $\frac{1}{8} - \frac{5}{6}$

90. $\frac{3}{8} - \left(-\frac{1}{4}\right)$

91. $5.6 - 8.2$

92. Solve: $3(x + 2) = 5x - 9$. [8.3c]

93. Solve $cx = ab - r$ for b. [8.6a]

94. A nut dealer has 1800 lb of peanuts, 1500 lb of cashews, and 700 lb of almonds. What percent of the total is peanuts? cashews? almonds? [8.5a]

95. Factor: $3x - 15y + 63$. [7.7d]

Synthesis

96. ◆ Is it better to evaluate a polynomial before or after like terms have been collected? Why?

97. ◆ Explain why an understanding of the rules of order of operations is essential when evaluating polynomials.

Collect like terms.

98. $\frac{9}{2}x^8 + \frac{1}{9}x^2 + \frac{1}{2}x^9 + \frac{9}{2}x^1 + \frac{9}{2}x^9 + \frac{8}{9}x^2 + \frac{1}{2}x - \frac{1}{2}x^8$

99. $(3x^2)^3 + 4x^2 \cdot 4x^4 - x^4(2x)^2 + ((2x)^2)^3 - 100x^2(x^2)^2$

100. Construct a polynomial in x (meaning that x is the variable) of degree 5 with four terms and coefficients that are integers.

101. What is the degree of $(5m^5)^2$?

102. A polynomial in x has degree 3. The coefficient of x^2 is 3 less than the coefficient of x^3. The coefficient of x is three times the coefficient of x^2. The remaining coefficient is 2 more than the coefficient of x^3. The sum of the coefficients is -4. Find the polynomial.

Use the CALC feature and choose VALUE on your grapher to find the values in each of the following.

103. Exercise 17

104. Exercise 18

105. Exercise 19

106. Exercise 20

10.4 Addition and Subtraction of Polynomials

a Addition of Polynomials

To add two polynomials, we can write a plus sign between them and then collect like terms. Depending on the situation, you may see polynomials written in descending order, ascending order, or neither. Generally, if an exercise is written in a particular order, we write the answer in that same order.

Example 1 Add: $(-3x^3 + 2x - 4) + (4x^3 + 3x^2 + 2)$.

$(-3x^3 + 2x - 4) + (4x^3 + 3x^2 + 2)$

$= (-3 + 4)x^3 + 3x^2 + 2x + (-4 + 2)$ Collecting like terms (*No* signs are changed.)

$= x^3 + 3x^2 + 2x - 2$

Example 2 Add:

$\left(\frac{2}{3}x^4 + 3x^2 - 2x + \frac{1}{2}\right) + \left(-\frac{1}{3}x^4 + 5x^3 - 3x^2 + 3x - \frac{1}{2}\right)$.

We have

$\left(\frac{2}{3}x^4 + 3x^2 - 2x + \frac{1}{2}\right) + \left(-\frac{1}{3}x^4 + 5x^3 - 3x^2 + 3x - \frac{1}{2}\right)$

$= \left(\frac{2}{3} - \frac{1}{3}\right)x^4 + 5x^3 + (3 - 3)x^2 + (-2 + 3)x + \left(\frac{1}{2} - \frac{1}{2}\right)$ Collecting like terms

$= \frac{1}{3}x^4 + 5x^3 + x.$

We can add polynomials as we do because they represent numbers. After some practice, you will be able to add mentally.

Do Exercises 1–4.

Example 3 Add: $(3x^2 - 2x + 2) + (5x^3 - 2x^2 + 3x - 4)$.

$(3x^2 - 2x + 2) + (5x^3 - 2x^2 + 3x - 4)$

$= 5x^3 + (3 - 2)x^2 + (-2 + 3)x + (2 - 4)$ You might do this step mentally.

$= 5x^3 + x^2 + x - 2$ Then you would write only this.

Do Exercises 5 and 6.

We can also add polynomials by writing like terms in columns.

Example 4 Add: $9x^5 - 2x^3 + 6x^2 + 3$ and $5x^4 - 7x^2 + 6$ and $3x^6 - 5x^5 + x^2 + 5$.

We arrange the polynomials with the like terms in columns.

$$\begin{array}{rrrrrr} & 9x^5 & & -\ 2x^3 & +\ 6x^2 & +\ 3 \\ & & 5x^4 & & -\ 7x^2 & +\ 6 \\ 3x^6 & -\ 5x^5 & & & +\ x^2 & +\ 5 \\ \hline 3x^6 & +\ 4x^5 & +\ 5x^4 & -\ 2x^3 & & +\ 14 \end{array}$$

We leave spaces for missing terms.

Adding

We write the answer as $3x^6 + 4x^5 + 5x^4 - 2x^3 + 14$ without the space.

Objectives

a. Add polynomials.

b. Find the opposite of a polynomial.

c. Subtract polynomials.

d. Use polynomials to represent perimeter and area.

For Extra Help

TAPE 19   MAC WIN   CD-ROM

Add.

1. $(3x^2 + 2x - 2) + (-2x^2 + 5x + 5)$

2. $(-4x^5 + x^3 + 4) + (7x^4 + 2x^2)$

3. $(31x^4 + x^2 + 2x - 1) + (-7x^4 + 5x^3 - 2x + 2)$

4. $(17x^3 - x^2 + 3x + 4) + \left(-15x^3 + x^2 - 3x - \frac{2}{3}\right)$

Add mentally. Try to write just the answer.

5. $(4x^2 - 5x + 3) + (-2x^2 + 2x - 4)$

6. $(3x^3 - 4x^2 - 5x + 3) + \left(5x^3 + 2x^2 - 3x - \frac{1}{2}\right)$

Answers on page A-30

Add.

7. $\begin{array}{r} -2x^3+5x^2-2x+4 \\ x^4 \quad\quad\quad +6x^2+7x-10 \\ -9x^4+6x^3+x^2 \quad\quad -2 \\ \hline \end{array}$

8. $-3x^3 + 5x + 2$ and
$x^3 + x^2 + 5$ and
$x^3 - 2x - 4$

Find two equivalent expressions for the opposite of the polynomial.

9. $12x^4 - 3x^2 + 4x$

10. $-4x^4 + 3x^2 - 4x$

11. $-13x^6 + 2x^4 - 3x^2 + x - \frac{5}{13}$

12. $-7y^3 + 2y^2 - y + 3$

Simplify.

13. $-(4x^3 - 6x + 3)$

14. $-(5x^4 + 3x^2 + 7x - 5)$

15. $-\left(14x^{10} - \frac{1}{2}x^5 + 5x^3 - x^2 + 3x\right)$

Answers on page A-30

Do Exercises 7 and 8.

b Opposites of Polynomials

We now look at subtraction of polynomials. To do so, we first consider the opposite, or additive inverse, of a polynomial.

We know that two numbers are opposites of each other if their sum is zero. For example, 5 and -5 are opposites, since $5 + (-5) = 0$. The same definition holds for polynomials. Two polynomials are **opposites**, or **additive inverses,** of each other if their sum is zero.

To find a way to determine an opposite, look for a pattern in the following examples:

a) $2x + (-2x) = 0$;

b) $-6x^2 + 6x^2 = 0$;

c) $(5t^3 - 2) + (-5t^3 + 2) = 0$;

d) $(7x^3 - 6x^2 - x + 4) + (-7x^3 + 6x^2 + x - 4) = 0$.

Since $(5t^3 - 2) + (-5t^3 + 2) = 0$, we know that the opposite of $(5t^3 - 2)$ is $(-5t^3 + 2)$. To say the same thing with purely algebraic symbolism, consider

$$\underbrace{\text{The opposite of}}_{\downarrow} \quad \underbrace{(5t^3-2)}_{\downarrow} \quad \underset{\downarrow}{\text{is}} \quad \underbrace{-5t^3+2.}_{\downarrow}$$
$$- \quad (5t^3 - 2) \quad = \quad -5t^3 + 2.$$

> We can find an equivalent polynomial for the opposite, or additive inverse, of a polynomial by replacing each term with its opposite—that is, *changing the sign of every term.*

Example 5 Find two equivalent expressions for the opposite of

$$4x^5 - 7x^3 - 8x + \tfrac{5}{6}.$$

The opposite of $4x^5 - 7x^3 - 8x + \frac{5}{6}$ is

$-\left(4x^5 - 7x^3 - 8x + \frac{5}{6}\right)$, or

$-4x^5 + 7x^3 + 8x - \frac{5}{6}$. **Changing the sign of every term**

Thus, $-\left(4x^5 - 7x^3 - 8x + \frac{5}{6}\right)$ is equivalent to $-4x^5 + 7x^3 + 8x - \frac{5}{6}$, and each is the opposite of the original polynomial $4x^5 - 7x^3 - 8x + \frac{5}{6}$.

Do Exercises 9–12.

Example 6 Simplify: $-\left(-7x^4 - \frac{5}{9}x^3 + 8x^2 - x + 67\right)$.

$$-\left(-7x^4 - \tfrac{5}{9}x^3 + 8x^2 - x + 67\right) = 7x^4 + \tfrac{5}{9}x^3 - 8x^2 + x - 67$$

Do Exercises 13–15.

c Subtraction of Polynomials

Recall that we can subtract a real number by adding its opposite, or additive inverse: $a - b = a + (-b)$. This allows us to find an equivalent expression for the difference of two polynomials.

Example 7 Subtract:

$$(9x^5 + x^3 - 2x^2 + 4) - (2x^5 + x^4 - 4x^3 - 3x^2).$$

We have

$(9x^5 + x^3 - 2x^2 + 4) - (2x^5 + x^4 - 4x^3 - 3x^2)$

$= 9x^5 + x^3 - 2x^2 + 4 + [-(2x^5 + x^4 - 4x^3 - 3x^2)]$ — Adding the opposite

$= 9x^5 + x^3 - 2x^2 + 4 - 2x^5 - x^4 + 4x^3 + 3x^2$ — Finding the opposite by changing the sign of *each* term

$= 7x^5 - x^4 + 5x^3 + x^2 + 4.$ — Collecting like terms

Do Exercises 16 and 17.

As with similar work in Section 7.8, we combine steps by changing the sign of each term of the polynomial being subtracted and collecting like terms. Try to do this mentally as much as possible.

Example 8 Subtract: $(9x^5 + x^3 - 2x) - (-2x^5 + 5x^3 + 6)$.

$(9x^5 + x^3 - 2x) - (-2x^5 + 5x^3 + 6)$

$= 9x^5 + x^3 - 2x + 2x^5 - 5x^3 - 6$ — Finding the opposite by changing the sign of each term

$= 11x^5 - 4x^3 - 2x - 6$ — Collecting like terms

Do Exercises 18 and 19.

We can use columns to subtract. We replace coefficients with their opposites, as shown in Example 7.

Example 9 Write in columns and subtract:

$$(5x^2 - 3x + 6) - (9x^2 - 5x - 3).$$

a) $\begin{array}{r} 5x^2 - 3x + 6 \\ -(9x^2 - 5x - 3) \\ \hline \end{array}$ — Writing similar terms in columns

b) $\begin{array}{r} 5x^2 - 3x + 6 \\ -9x^2 + 5x + 3 \\ \hline \end{array}$ — Changing signs

c) $\begin{array}{r} 5x^2 - 3x + 6 \\ -9x^2 + 5x + 3 \\ \hline -4x^2 + 2x + 9 \end{array}$ — Adding

If you can do so without error, you can arrange the polynomials in columns and write just the answer.

Subtract.

16. $(7x^3 + 2x + 4) - (5x^3 - 4)$

17. $(-3x^2 + 5x - 4) - (-4x^2 + 11x - 2)$

Subtract.

18. $(-6x^4 + 3x^2 + 6) - (2x^4 + 5x^3 - 5x^2 + 7)$

19. $\left(\frac{3}{2}x^3 - \frac{1}{2}x^2 + 0.3\right) - \left(\frac{1}{2}x^3 + \frac{1}{2}x^2 + \frac{4}{3}x + 1.2\right)$

Answers on page A-30

Write in columns and subtract.

20. $(4x^3 + 2x^2 - 2x - 3) - (2x^3 - 3x^2 + 2)$

21. $(2x^3 + x^2 - 6x + 2) - (x^5 + 4x^3 - 2x^2 - 4x)$

22. Find a polynomial for the sum of the perimeters and the areas of the rectangles.

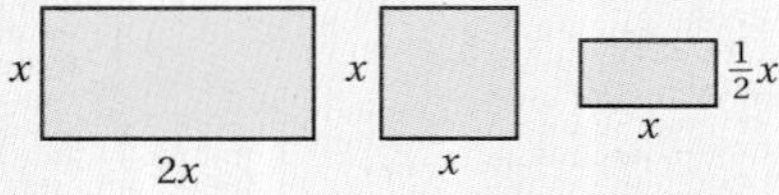

23. Find a polynomial for the shaded area.

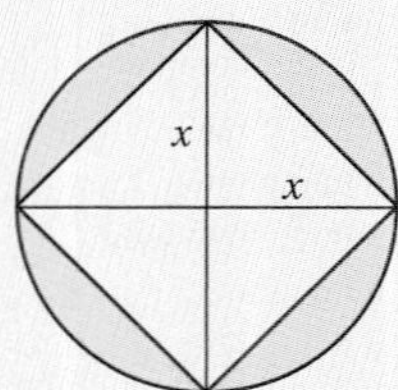

Answers on page A-30

Example 10 Write in columns and subtract:

$$(x^3 + x^2 + 2x - 12) - (-2x^3 + x^2 - 3x).$$

We have

$$\begin{array}{rl} x^3 + x^2 + 2x - 12 & \\ -2x^3 + x^2 - 3x & \quad \textbf{Changing signs} \\ \hline 3x^3 + 5x - 12. & \quad \textbf{Adding} \end{array}$$

Do Exercises 20 and 21.

d Polynomials and Geometry

Example 11 Find a polynomial for the sum of the areas of these rectangles.

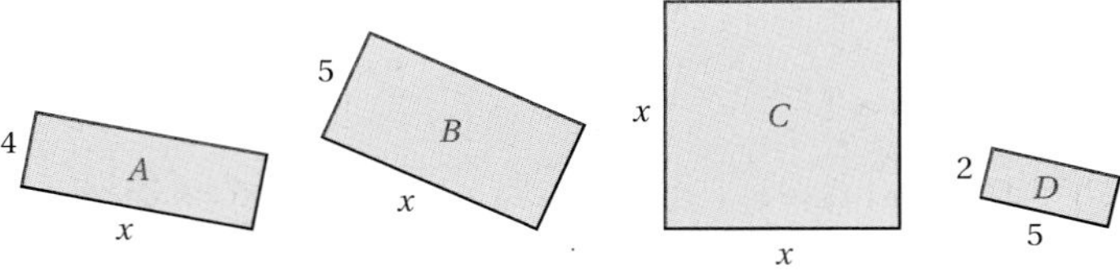

Recall that the area of a rectangle is the product of the length and the width. The sum of the areas is a sum of products. We find these products and then collect like terms.

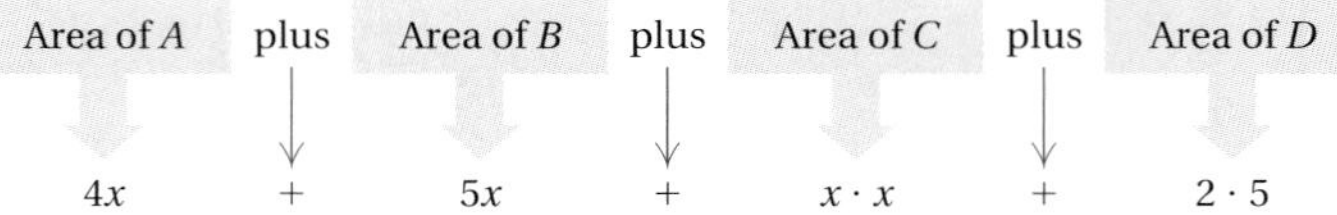

Area of A
plus
Area of B
plus
Area of C
plus
Area of D
4x
+
5x
+
x · x
+
2 · 5

We collect like terms:

$$4x + 5x + x^2 + 10 = x^2 + 9x + 10.$$

Do Exercise 22.

Example 12 A water fountain with a 4-ft by 4-ft square base is placed on a square grassy park area that is x ft on a side. To determine the amount of grass seed needed for the lawn, find a polynomial for the grassy area.

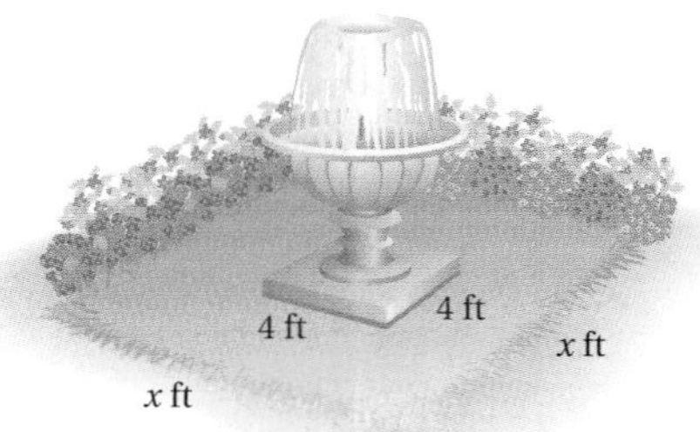

We draw a picture of the situation as shown here. We then reword the problem and write the polynomial as follows.

Area of park	$-$	Area of base of fountain	$=$	Area left over
$x \cdot x$	$-$	$4 \cdot 4$	$=$	Area left over

Then $x^2 - 16 =$ Area left over.

Do Exercise 23.

Exercise Set 10.4

a Add.

1. $(3x + 2) + (-4x + 3)$

2. $(6x + 1) + (-7x + 2)$

3. $(-6x + 2) + (x^2 + x - 3)$

4. $(x^2 - 5x + 4) + (8x - 9)$

5. $(x^2 - 9) + (x^2 + 9)$

6. $(x^3 + x^2) + (2x^3 - 5x^2)$

7. $(3x^2 - 5x + 10) + (2x^2 + 8x - 40)$

8. $(6x^4 + 3x^3 - 1) + (4x^2 - 3x + 3)$

9. $(1.2x^3 + 4.5x^2 - 3.8x) + (-3.4x^3 - 4.7x^2 + 23)$

10. $(0.5x^4 - 0.6x^2 + 0.7) + (2.3x^4 + 1.8x - 3.9)$

11. $(1 + 4x + 6x^2 + 7x^3) + (5 - 4x + 6x^2 - 7x^3)$

12. $(3x^4 - 6x - 5x^2 + 5) + (6x^2 - 4x^3 - 1 + 7x)$

13. $\left(\frac{1}{4}x^4 + \frac{2}{3}x^3 + \frac{5}{8}x^2 + 7\right) + \left(-\frac{3}{4}x^4 + \frac{3}{8}x^2 - 7\right)$

14. $\left(\frac{1}{3}x^9 + \frac{1}{5}x^5 - \frac{1}{2}x^2 + 7\right) +$
$\left(-\frac{1}{5}x^9 + \frac{1}{4}x^4 - \frac{3}{5}x^5 + \frac{3}{4}x^2 + \frac{1}{2}\right)$

15. $(0.02x^5 - 0.2x^3 + x + 0.08) +$
$(-0.01x^5 + x^4 - 0.8x - 0.02)$

16. $(0.03x^6 + 0.05x^3 + 0.22x + 0.05) +$
$\left(\frac{7}{100}x^6 - \frac{3}{100}x^3 + 0.5\right)$

17. $(9x^8 - 7x^4 + 2x^2 + 5) + (8x^7 + 4x^4 - 2x) +$
$(-3x^4 + 6x^2 + 2x - 1)$

18. $(4x^5 - 6x^3 - 9x + 1) + (6x^3 + 9x^2 + 9x) +$
$(-4x^3 + 8x^2 + 3x - 2)$

19.
$$\begin{array}{rrrrr} 0.15x^4 & + 0.10x^3 & - \ 0.9x^2 & & \\ & - 0.01x^3 & + 0.01x^2 & + x & \\ 1.25x^4 & & + 0.11x^2 & & + 0.01 \\ & 0.27x^3 & & & + 0.99 \\ -0.35x^4 & & + \ 15x^2 & & - 0.03 \\ \hline \end{array}$$

20.
$$\begin{array}{rrrrr} 0.05x^4 & + 0.12x^3 & - \ 0.5x^2 & & \\ & - 0.02x^3 & + 0.02x^2 & + 2x & \\ 1.5x^4 & & + 0.01x^2 & & + 0.15 \\ & 0.25x^3 & & & + 0.85 \\ -0.25x^4 & & + \ 10x^2 & & - 0.04 \\ \hline \end{array}$$

b Find two equivalent expressions for the opposite of the polynomial.

21. $-5x$

22. $x^2 - 3x$

23. $-x^2 + 10x - 2$

24. $-4x^3 - x^2 - x$

25. $12x^4 - 3x^3 + 3$

26. $4x^3 - 6x^2 - 8x + 1$

Simplify.

27. $-(3x - 7)$

28. $-(-2x + 4)$

29. $-(4x^2 - 3x + 2)$

30. $-(-6a^3 + 2a^2 - 9a + 1)$

31. $-(-4x^4 + 6x^2 + \frac{3}{4}x - 8)$

32. $-(-5x^4 + 4x^3 - x^2 + 0.9)$

c Subtract.

33. $(3x + 2) - (-4x + 3)$

34. $(6x + 1) - (-7x + 2)$

35. $(-6x + 2) - (x^2 + x - 3)$

36. $(x^2 - 5x + 4) - (8x - 9)$

37. $(x^2 - 9) - (x^2 + 9)$

38. $(x^3 + x^2) - (2x^3 - 5x^2)$

39. $(6x^4 + 3x^3 - 1) - (4x^2 - 3x + 3)$

40. $(-4x^2 + 2x) - (3x^3 - 5x^2 + 3)$

41. $(1.2x^3 + 4.5x^2 - 3.8x) - (-3.4x^3 - 4.7x^2 + 23)$

42. $(0.5x^4 - 0.6x^2 + 0.7) - (2.3x^4 + 1.8x - 3.9)$

43. $\left(\frac{5}{8}x^3 - \frac{1}{4}x - \frac{1}{3}\right) - \left(-\frac{1}{8}x^3 + \frac{1}{4}x - \frac{1}{3}\right)$

44. $\left(\frac{1}{5}x^3 + 2x^2 - 0.1\right) - \left(-\frac{2}{5}x^3 + 2x^2 + 0.01\right)$

45. $(0.08x^3 - 0.02x^2 + 0.01x) - (0.02x^3 + 0.03x^2 - 1)$

46. $(0.8x^4 + 0.2x - 1) - \left(\frac{7}{10}x^4 + \frac{1}{5}x - 0.1\right)$

Subtract.

47. $\begin{array}{r} x^2 + 5x + 6 \\ \underline{x^2 + 2x \quad\;\;} \end{array}$

48. $\begin{array}{l} x^3 \qquad\quad + 1 \\ \underline{x^3 + x^2 \quad\;\;} \end{array}$

49. $\begin{array}{r} 5x^4 + 6x^3 - 9x^2 \qquad\qquad \\ \underline{-6x^4 - 6x^3 \qquad\quad + 8x + 9} \end{array}$

50. $\begin{array}{r} 5x^4 \qquad + 6x^2 - 3x + 6 \\ \underline{6x^3 + 7x^2 - 8x - 9} \end{array}$

51. $\begin{array}{r} x^5 \qquad\qquad\qquad\qquad\quad - 1 \\ \underline{x^5 - x^4 + x^3 - x^2 + x - 1} \end{array}$

52. $\begin{array}{r} x^5 + x^4 - x^3 + x^2 - x + 2 \\ \underline{x^5 - x^4 + x^3 - x^2 - x + 2} \end{array}$

d Solve.

53. Find a polynomial for the sum of the areas of these rectangles.

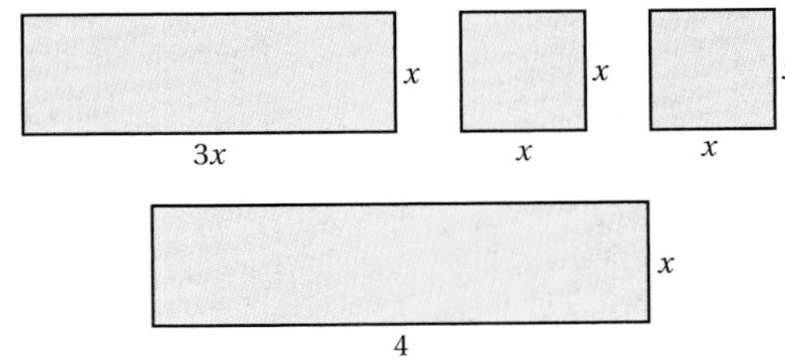

54. Find a polynomial for the sum of the areas of these circles.

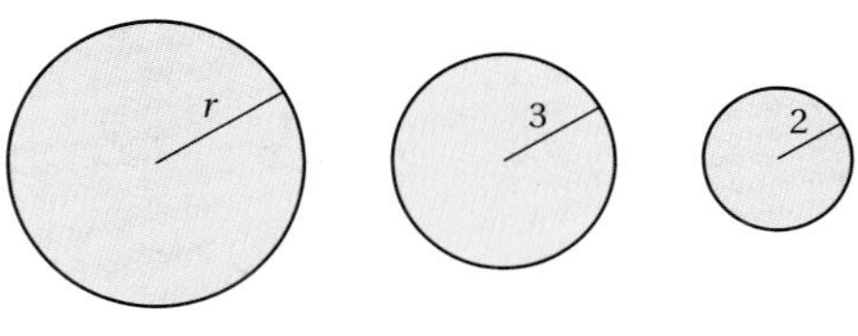

Find a polynomial for the perimeter of the figure.

55.

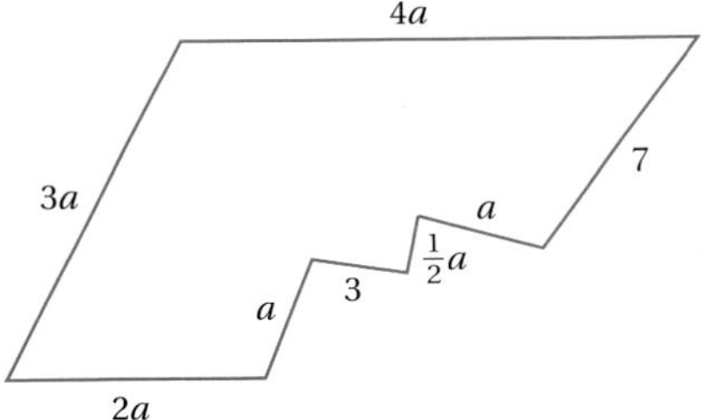

56.

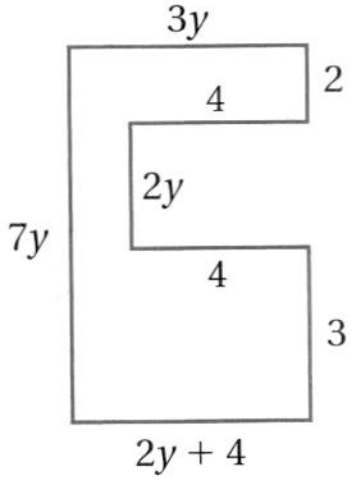

Skill Maintenance

Solve. [8.3b]

57. $8x + 3x = 66$

58. $5x - 7x = 38$

59. $\frac{3}{8}x + \frac{1}{4} - \frac{3}{4}x = \frac{11}{16} + x$

60. $5x - 4 = 26 - x$

61. $1.5x - 2.7x = 22 - 5.6x$

62. $3x - 3 = -4x + 4$

Solve. [8.3c]

63. $6(y - 3) - 8 = 4(y + 2) + 5$

64. $8(5x + 2) = 7(6x - 3)$

Solve. [8.7e]

65. $3x - 7 \leq 5x + 13$

66. $2(x - 4) > 5(x - 3) + 7$

Synthesis

67. ◆ Is the sum of two binomials ever a trinomial? Why or why not?

68. ◆ Which, if any, of the commutative, associative, and distributive laws are needed for adding polynomials? Why?

Find two algebraic expressions for the area of the figure.

69.
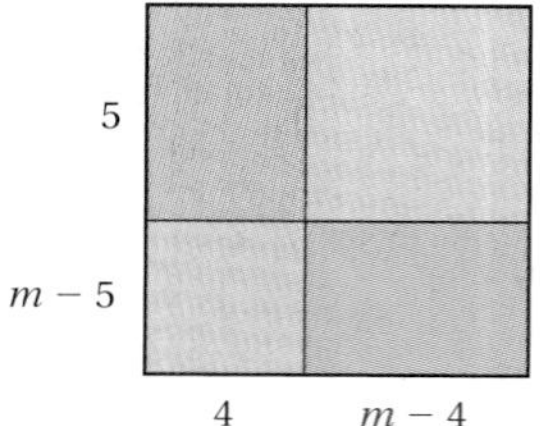

70.
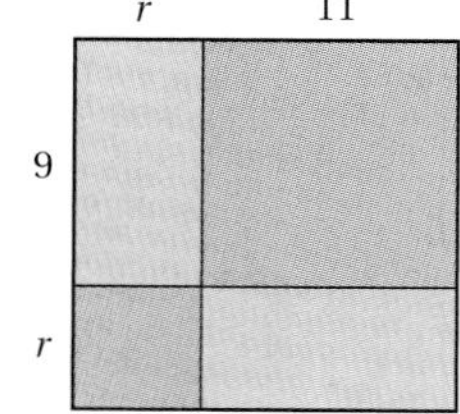

Find a polynomial for the shaded area of the figure.

71.
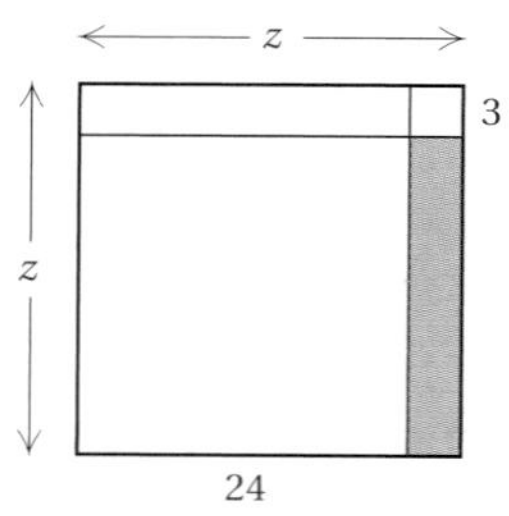

72.
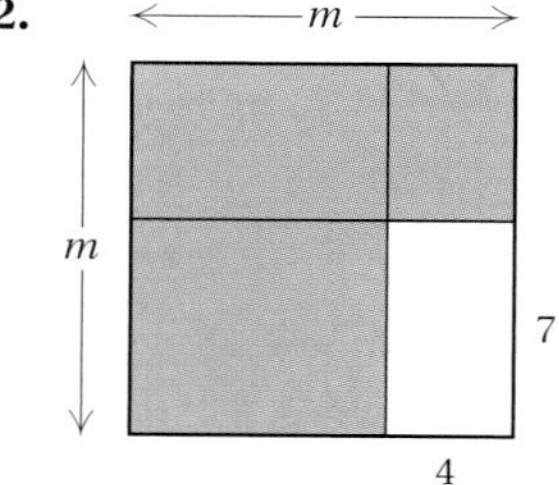

73. Find $(y - 2)^2$ using the four parts of this square.

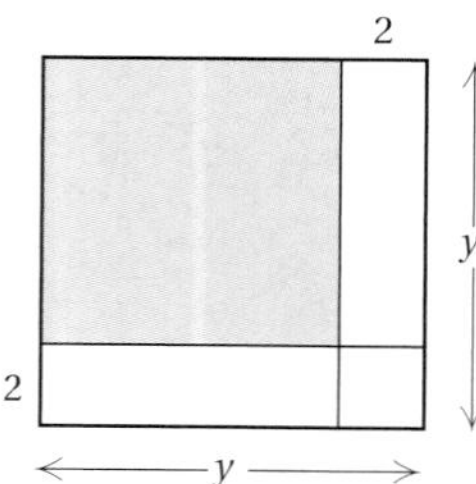

74. Find a polynomial for the surface area of this right rectangular solid.

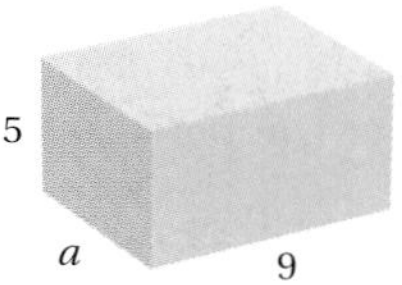

Simplify.

75. $(3x^2 - 4x + 6) - (-2x^2 + 4) + (-5x - 3)$

76. $(7y^2 - 5y + 6) - (3y^2 + 8y - 12) + (8y^2 - 10y + 3)$

77. $(-4 + x^2 + 2x^3) - (-6 - x + 3x^3) - (-x^2 - 5x^3)$

78. $(-y^4 - 7y^3 + y^2) + (-2y^4 + 5y - 2) - (-6y^3 + y^2)$

The TABLE feature can be used as a partial check that polynomials have been added or subtracted correctly. To check Example 3, we enter

$y_1 = (3x^2 - 2x + 2) + (5x^3 - 2x^2 + 3x - 4)$
and $y_2 = 5x^3 + x^2 + x - 2$.

If our addition was correct, the y_1- and y_2-values should be the same, regardless of the table settings used.

X	Y1	Y2
4	338	338
5	653	653
6	1120	1120
7	1769	1769
8	2630	2630
9	3733	3733
10	5108	5108

X = 4

79. Use the TABLE feature to check Exercise 7.

80. Use the TABLE feature to check Exercise 9.

10.5 Multiplication of Polynomials

We now multiply polynomials using techniques based, for the most part, on the distributive laws, but also on the associative and commutative laws. As we proceed in this chapter, we will develop special ways to find certain products.

Objectives

a. Multiply monomials.

b. Multiply a monomial and any polynomial.

c. Multiply two binomials.

d. Multiply any two polynomials.

For Extra Help

TAPE 20

MAC
WIN

CD-ROM

a Multiplying Monomials

Consider $(3x)(4x)$. We multiply as follows:

$$
\begin{aligned}
(3x)(4x) &= 3 \cdot x \cdot 4 \cdot x && \text{By the associative law of multiplication} \\
&= 3 \cdot 4 \cdot x \cdot x && \text{By the commutative law of multiplication} \\
&= (3 \cdot 4)(x \cdot x) && \text{By the associative law} \\
&= 12x^2. && \text{Using the product rule for exponents}
\end{aligned}
$$

> To find an equivalent expression for the product of two monomials, multiply the coefficients and then multiply the variables using the product rule for exponents.

Examples Multiply.

1. $5x \cdot 6x = (5 \cdot 6)(x \cdot x)$ By the associative and commutative laws
$= 30x^2$ Multiplying the coefficients and multiplying the variables

2. $(3x)(-x) = (3x)(-1x)$
$= (3)(-1)(x \cdot x)$
$= -3x^2$

3. $(-7x^5)(4x^3) = (-7 \cdot 4)(x^5 \cdot x^3)$
$= -28x^{5+3}$ Adding the exponents
$= -28x^8$ Simplifying

After some practice, you can do this mentally. Multiply the coefficients and then the variables by keeping the base and adding the exponents. Write only the answer.

Do Exercises 1–8.

Multiply.

1. $(3x)(-5)$

2. $(-x) \cdot x$

3. $(-x)(-x)$

4. $(-x^2)(x^3)$

5. $3x^5 \cdot 4x^2$

6. $(4y^5)(-2y^6)$

7. $(-7y^4)(-y)$

8. $7x^5 \cdot 0$

Answers on page A-30

b Multiplying a Monomial and Any Polynomial

To find an equivalent expression for the product of a monomial, such as $2x$, and a binomial, such as $5x + 3$, we use a distributive law and multiply each term of $5x + 3$ by $2x$.

Example 4 Multiply: $2x(5x + 3)$.

$$
\begin{aligned}
2x(5x + 3) &= (2x)(5x) + (2x)(3) && \text{Using a distributive law} \\
&= 10x^2 + 6x && \text{Multiplying the monomials}
\end{aligned}
$$

Multiply.

9. $4x(2x + 4)$

10. $3t^2(-5t + 2)$

11. $5x^3(x^3 + 5x^2 - 6x + 8)$

Multiply.

12. $(x + 8)(x + 5)$

13. $(x + 5)(x - 4)$

Answers on page A-30

Example 5 Multiply: $5x(2x^2 - 3x + 4)$.

$$\begin{aligned} 5x(2x^2 - 3x + 4) &= (5x)(2x^2) - (5x)(3x) + (5x)(4) \\ &= 10x^3 - 15x^2 + 20x \end{aligned}$$

> To multiply a monomial and a polynomial, multiply each term of the polynomial by the monomial.

Example 6 Multiply: $2x^2(x^3 - 7x^2 + 10x - 4)$.

$$2x^2(x^3 - 7x^2 + 10x - 4) = 2x^5 - 14x^4 + 20x^3 - 8x^2$$

Do Exercises 9–11.

c Multiplying Two Binomials

To find an equivalent expression for the product of two binomials, we use the distributive laws more than once. In Example 7, we use a distributive law three times.

Example 7 Multiply: $(x + 5)(x + 4)$.

$$\begin{aligned} (x + 5)(x + 4) &= x(x + 4) + 5(x + 4) && \text{Using a distributive law} \\ &= x \cdot x + x \cdot 4 + 5 \cdot x + 5 \cdot 4 && \text{Using a distributive law on each part} \\ &= x^2 + 4x + 5x + 20 && \text{Multiplying the monomials} \\ &= x^2 + 9x + 20 && \text{Collecting like terms} \end{aligned}$$

To visualize the product in Example 7, consider a rectangle of length $x + 5$ and width $x + 4$.

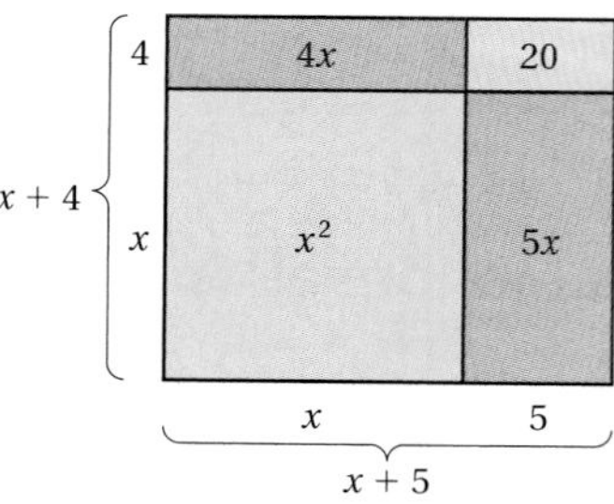

The total area can be expressed as $(x + 5)(x + 4)$ or, by adding the four smaller areas, $x^2 + 5x + 4x + 20$.

Do Exercises 12 and 13.

Example 8 Multiply: $(4x + 3)(x - 2)$.

$$(4x + 3)(x - 2) = 4x(x - 2) + 3(x - 2) \quad \text{Using a distributive law}$$
$$= 4x \cdot x - 4x \cdot 2 + 3 \cdot x - 3 \cdot 2 \quad \text{Using a distributive law on each part}$$
$$= 4x^2 - 8x + 3x - 6 \quad \text{Multiplying the monomials}$$
$$= 4x^2 - 5x - 6 \quad \text{Collecting like terms}$$

Do Exercises 14 and 15.

Multiply.

14. $(5x + 3)(x - 4)$

15. $(2x - 3)(3x - 5)$

d Multiplying Any Two Polynomials

Let's consider the product of a binomial and a trinomial. We use a distributive law four times. You may see ways to skip some steps and do the work mentally.

Example 9 Multiply: $(x^2 + 2x - 3)(x^2 + 4)$.

$$(x^2 + 2x - 3)(x^2 + 4) = x^2(x^2 + 4) + 2x(x^2 + 4) - 3(x^2 + 4)$$
$$= x^2 \cdot x^2 + x^2 \cdot 4 + 2x \cdot x^2 + 2x \cdot 4 - 3 \cdot x^2 - 3 \cdot 4$$
$$= x^4 + 4x^2 + 2x^3 + 8x - 3x^2 - 12$$
$$= x^4 + 2x^3 + x^2 + 8x - 12$$

Do Exercises 16 and 17.

Multiply.

16. $(x^2 + 3x - 4)(x^2 + 5)$

17. $(3y^2 - 7)(2y^3 - 2y + 5)$

Perhaps you have discovered the following in the preceding examples.

> To multiply two polynomials P and Q, select one of the polynomials—say, P. Then multiply each term of P by every term of Q and collect like terms.

We can use columns for long multiplications. We multiply each term at the top by every term at the bottom. We write like terms in columns, and then we add the results. Such multiplication is like multiplying with whole numbers:

4 5 7	4 5 7	$= 400 + 50 + 7$
× 6 3	× 6 3	$= 60 + 3$
1 3 7 1	1200 + 150 + 21	$= 3(457) = 3(400 + 50 + 7)$
2 7 4 2 0	24000 + 3000 + 420	$= 60(457) = 60(400 + 50 + 7)$
2 8 7 9 1	24000 + 4200 + 570 + 21	$= 28{,}791$

Example 10 Multiply: $(4x^2 - 2x + 3)(x + 2)$.

$$\begin{array}{r} 4x^2 - 2x + 3 \\ x + 2 \\ \hline 8x^2 - 4x + 6 \\ 4x^3 - 2x^2 + 3x \quad \\ \hline 4x^3 + 6x^2 - x + 6 \end{array}$$

Multiplying the top row by 2

Multiplying the top row by x

Collecting like terms

Line up like terms in columns.

Answers on page A-30

Multiply.

18.
$$\begin{array}{r} 3x^2 - 2x + 4 \\ x + 5 \\ \hline \end{array}$$

19.
$$\begin{array}{r} -5x^2 + 4x + 2 \\ -4x^2 - 8 \\ \hline \end{array}$$

20. Multiply.

$$\begin{array}{r} 3x^2 - 2x - 5 \\ 2x^2 + x - 2 \\ \hline \end{array}$$

Example 11 Multiply: $(5x^3 - 3x + 4)(-2x^2 - 3)$.

When missing terms occur, it helps to leave spaces for them and align like terms as we multiply.

$$\begin{array}{rrrrrl} & 5x^3 & & -3x & +4 & \\ & & -2x^2 & & -3 & \\ \hline & -15x^3 & & +9x & -12 & \text{Multiplying by } -3 \\ -10x^5 & +6x^3 & -8x^2 & & & \text{Multiplying by } -2x^2 \\ \hline -10x^5 & -9x^3 & -8x^2 & +9x & -12 & \text{Collecting like terms} \end{array}$$

Do Exercises 18 and 19.

Example 12 Multiply: $(2x^2 + 3x - 4)(2x^2 - x + 3)$.

$$\begin{array}{rrrrrl} & & 2x^2 & +3x & -4 & \\ & & 2x^2 & -x & +3 & \\ \hline & & 6x^2 & +9x & -12 & \text{Multiplying by 3} \\ & -2x^3 & -3x^2 & +4x & & \text{Multiplying by } -x \\ 4x^4 & +6x^3 & -8x^2 & & & \text{Multiplying by } 2x^2 \\ \hline 4x^4 & +4x^3 & -5x^2 & +13x & -12 & \text{Collecting like terms} \end{array}$$

Do Exercise 20.

Calculator Spotlight

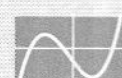 Checking Multiplications with a Table or Graph

Table. The TABLE feature can be used as a partial check that polynomials have been multiplied correctly. To check whether

$$(x + 5)(x - 4) = x^2 + 9x - 20$$

is correct, we enter

$$y_1 = (x + 5)(x - 4) \quad \text{and} \quad y_2 = x^2 + 9x - 20.$$

X	Y1	Y2
−23	486	302
−13	136	32
−3	−14	−38
7	36	92
17	286	422
27	736	952
37	1386	1682
X = −23		

If our multiplication is correct, the y_1- and y_2-values should be the same, regardless of the table settings used. We see that y_1 and y_2 are not the same, so the multiplication is not correct.

Graph. Multiplication of polynomials can also be checked with the GRAPH feature. In this case, we see that the graphs differ, so the multiplication is not correct.

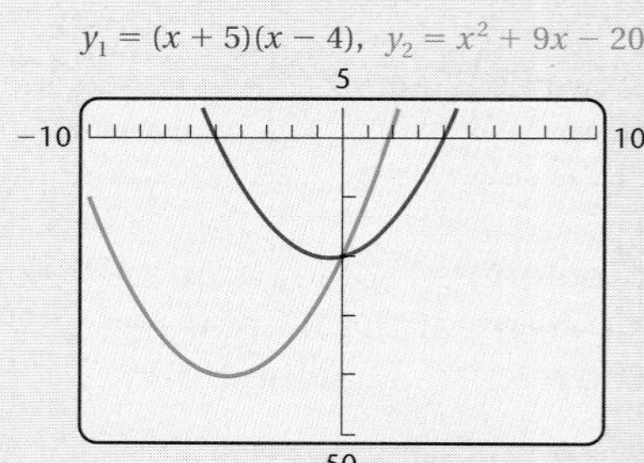

Exercises

Use the TABLE or GRAPH feature to check whether each of the following is correct.

1. $(x + 5)(x + 4) = x^2 + 9x + 20$ (Example 7)
2. $(4x + 3)(x - 2) = 4x^2 - 5x - 6$ (Example 8)
3. $(5x + 3)(x - 4) = 5x^2 + 17x - 12$
4. $(2x - 3)(3x - 5) = 6x^2 - 19x - 15$
5. $(x - 3)(x - 3) = x^2 - 9$
6. $(x - 3)(x + 3) = x^2 - 9$

Answers on page A-30

Exercise Set 10.5

a Multiply.

1. $(8x^2)(5)$

2. $(4x^2)(-2)$

3. $(-x^2)(-x)$

4. $(-x^3)(x^2)$

5. $(8x^5)(4x^3)$

6. $(10a^2)(2a^2)$

7. $(0.1x^6)(0.3x^5)$

8. $(0.3x^4)(-0.8x^6)$

9. $\left(-\frac{1}{5}x^3\right)\left(-\frac{1}{3}x\right)$

10. $\left(-\frac{1}{4}x^4\right)\left(\frac{1}{5}x^8\right)$

11. $(-4x^2)(0)$

12. $(-4m^5)(-1)$

13. $(3x^2)(-4x^3)(2x^6)$

14. $(-2y^5)(10y^4)(-3y^3)$

b Multiply.

15. $2x(-x + 5)$

16. $3x(4x - 6)$

17. $-5x(x - 1)$

18. $-3x(-x - 1)$

19. $x^2(x^3 + 1)$

20. $-2x^3(x^2 - 1)$

21. $3x(2x^2 - 6x + 1)$

22. $-4x(2x^3 - 6x^2 - 5x + 1)$

23. $(-6x^2)(x^2 + x)$

24. $(-4x^2)(x^2 - x)$

25. $(3y^2)(6y^4 + 8y^3)$

26. $(4y^4)(y^3 - 6y^2)$

c Multiply.

27. $(x + 6)(x + 3)$

28. $(x + 5)(x + 2)$

29. $(x + 5)(x - 2)$

30. $(x + 6)(x - 2)$

31. $(x - 4)(x - 3)$

32. $(x - 7)(x - 3)$

33. $(x + 3)(x - 3)$

34. $(x + 6)(x - 6)$

35. $(5 - x)(5 - 2x)$

36. $(3 + x)(6 + 2x)$

37. $(2x + 5)(2x + 5)$

38. $(3x - 4)(3x - 4)$

39. $\left(x - \frac{5}{2}\right)\left(x + \frac{2}{5}\right)$

40. $\left(x + \frac{4}{3}\right)\left(x + \frac{3}{2}\right)$

41. $(x - 2.3)(x + 4.7)$

42. $(2x + 0.13)(2x - 0.13)$

d Multiply.

43. $(x^2 + x + 1)(x - 1)$

44. $(x^2 + x - 2)(x + 2)$

45. $(2x + 1)(2x^2 + 6x + 1)$

46. $(3x - 1)(4x^2 - 2x - 1)$

47. $(y^2 - 3)(3y^2 - 6y + 2)$

48. $(3y^2 - 3)(y^2 + 6y + 1)$

49. $(x^3 + x^2)(x^3 + x^2 - x)$

50. $(x^3 - x^2)(x^3 - x^2 + x)$

51. $(-5x^3 - 7x^2 + 1)(2x^2 - x)$

52. $(-4x^3 + 5x^2 - 2)(5x^2 + 1)$

53. $(1 + x + x^2)(-1 - x + x^2)$

54. $(1 - x + x^2)(1 - x + x^2)$

55. $(2t^2 - t - 4)(3t^2 + 2t - 1)$

56. $(3a^2 - 5a + 2)(2a^2 - 3a + 4)$

57. $(x - x^3 + x^5)(x^2 - 1 + x^4)$

58. $(x - x^3 + x^5)(3x^2 + 3x^6 + 3x^4)$

59. $(x^3 + x^2 + x + 1)(x - 1)$

60. $(x + 2)(x^3 - x^2 + x - 2)$

Skill Maintenance

Simplify.

61. $-\frac{1}{4} - \frac{1}{2}$ [7.4a]

62. $-3.8 - (-10.2)$ [7.4a]

63. $(10 - 2)(10 + 2)$ [7.8d]

64. $10 - 2 + (-6)^2 \div 3 \cdot 2$ [7.8d]

Factor. [7.7d]

65. $15x - 18y + 12$

66. $16x - 24y + 36$

67. $-9x - 45y + 15$

68. $100x - 100y + 1000a$

69. Graph: $y = \frac{1}{2}x - 3$. [9.2b]

70. Solve: $4(x - 3) = 5(2 - 3x) + 1$. [8.3c]

Synthesis

71. ◈ Under what conditions will the product of two binomials be a trinomial?

72. ◈ Is it possible to understand polynomial multiplication without first understanding the distributive law? Why or why not?

73. Find a polynomial for the shaded area.

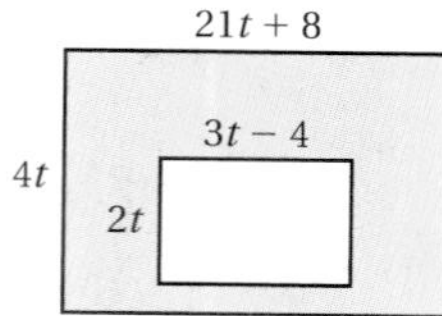

74. A box with a square bottom is to be made from a 12-in.-square piece of cardboard. Squares with side x are cut out of the corners and the sides are folded up. Find polynomials for the volume and the outside surface area of the box.

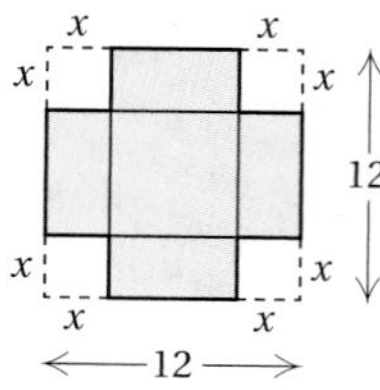

75. The height of a triangle is 4 ft longer than its base. Find a polynomial for the area.

Compute and simplify.

76. $(x + 3)(x + 6) + (x + 3)(x + 6)$

77. $(x - 2)(x - 7) - (x - 2)(x - 7)$

Visualize polynomial multiplication using rectangles.

10.6 Special Products

We encounter certain products so often that it is helpful to have faster methods of computing. We now consider special ways of multiplying any two binomials. Such techniques are called *special products.*

Objectives

- a Multiply two binomials mentally using the FOIL method.
- b Multiply the sum and the difference of two terms mentally.
- c Square a binomial mentally.
- d Find special products when polynomial products are mixed together.

For Extra Help

TAPE 20 MAC WIN CD-ROM

a Products of Two Binomials Using FOIL

To multiply two binomials, we can select one binomial and multiply each term of that binomial by every term of the other. Then we collect like terms. Consider the product $(x + 5)(x + 4)$:

$$\begin{aligned}(x + 5)(x + 4) &= x \cdot x + 5 \cdot x + x \cdot 4 + 5 \cdot 4\\ &= x^2 + 5x + 4x + 20\\ &= x^2 + 9x + 20.\end{aligned}$$

We can rewrite the first line of this product to show a special technique for finding the product of two binomials:

$$(x + 5)(x + 4) = \underbrace{x \cdot x}_{\text{First terms}} + \underbrace{4 \cdot x}_{\text{Outside terms}} + \underbrace{5 \cdot x}_{\text{Inside terms}} + \underbrace{5 \cdot 4}_{\text{Last terms}}.$$

To remember this method of multiplying, we use the initials **FOIL**.

THE FOIL METHOD

To multiply two binomials, $A + B$ and $C + D$, multiply the First terms AC, the Outside terms AD, the Inside terms BC, and then the Last terms BD. Then collect like terms, if possible.

$$(A + B)(C + D) = AC + AD + BC + BD$$

1. Multiply **F**irst terms: AC.
2. Multiply **O**utside terms: AD.
3. Multiply **I**nside terms: BC.
4. Multiply **L**ast terms: BD.

FOIL

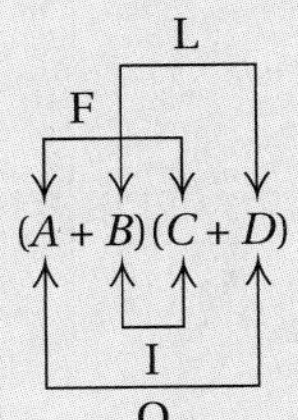

Example 1 Multiply: $(x + 8)(x^2 - 5)$.

We have

$$\begin{aligned}&\quad\ \ \ \text{F}\quad\ \text{O}\quad\ \ \text{I}\quad\ \ \text{L}\\ (x + 8)(x^2 - 5) &= x^3 - 5x + 8x^2 - 40\\ &= x^3 + 8x^2 - 5x - 40.\end{aligned}$$

Since each of the original binomials is in descending order, we write the product in descending order, as is customary, but this is not a "must."

Multiply mentally, if possible. If you need extra steps, be sure to use them.

1. $(x + 3)(x + 4)$

2. $(x + 3)(x - 5)$

3. $(2x - 1)(x - 4)$

4. $(2x^2 - 3)(x - 2)$

5. $(6x^2 + 5)(2x^3 + 1)$

6. $(y^3 + 7)(y^3 - 7)$

7. $(t + 5)(t + 3)$

8. $(2x^4 + x^2)(-x^3 + x)$

Multiply.

9. $\left(x + \frac{4}{5}\right)\left(x - \frac{4}{5}\right)$

10. $(x^3 - 0.5)(x^2 + 0.5)$

11. $(2 + 3x^2)(4 - 5x^2)$

12. $(6x^3 - 3x^2)(5x^2 - 2x)$

Answers on page A-31

Often we can collect like terms after we have multiplied.

Examples Multiply.

2. $(x + 6)(x - 6) = x^2 - 6x + 6x - 36$ **Using FOIL**
$= x^2 - 36$ **Collecting like terms**

3. $(x + 7)(x + 4) = x^2 + 4x + 7x + 28$
$= x^2 + 11x + 28$

4. $(y - 3)(y - 2) = y^2 - 2y - 3y + 6$
$= y^2 - 5y + 6$

5. $(x^3 - 5)(x^3 + 5) = x^6 + 5x^3 - 5x^3 - 25$
$= x^6 - 25$

Do Exercises 1–8.

Examples Multiply.

6. $(4t^3 + 5)(3t^2 - 2) = 12t^5 - 8t^3 + 15t^2 - 10$

7. $\left(x - \frac{2}{3}\right)\left(x + \frac{2}{3}\right) = x^2 + \frac{2}{3}x - \frac{2}{3}x - \frac{4}{9}$
$= x^2 - \frac{4}{9}$

8. $(x^2 - 0.3)(x^2 - 0.3) = x^4 - 0.3x^2 - 0.3x^2 + 0.09$
$= x^4 - 0.6x^2 + 0.09$

9. $(3 - 4x)(7 - 5x^3) = 21 - 15x^3 - 28x + 20x^4$
$= 21 - 28x - 15x^3 + 20x^4$

(*Note*: If the original polynomials are in ascending order, it is natural to write the product in ascending order, but this is not a "must.")

10. $(5x^4 + 2x^3)(3x^2 - 7x) = 15x^6 - 35x^5 + 6x^5 - 14x^4$
$= 15x^6 - 29x^5 - 14x^4$

Do Exercises 9–12.

We can show the FOIL method geometrically as follows.

The area of the large rectangle is $(A + B)(C + D)$.

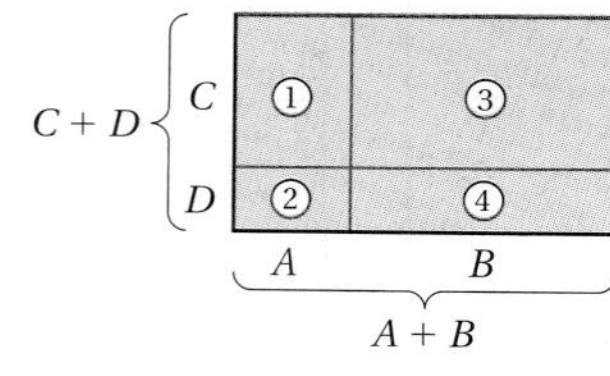

The area of rectangle ① is AC.
The area of rectangle ② is AD.
The area of rectangle ③ is BC.
The area of rectangle ④ is BD.

The area of the large rectangle is the sum of the areas of the smaller rectangles. Thus,

$$(A + B)(C + D) = AC + AD + BC + BD.$$

b Multiplying Sums and Differences of Two Terms

Consider the product of the sum and the difference of the same two terms, such as

$$(x + 2)(x - 2).$$

Since this is the product of two binomials, we can use FOIL. This type of product occurs so often, however, that it would be valuable if we could use an even faster method. To find a faster way to compute such a product, look for a pattern in the following:

a) $(x + 2)(x - 2) = x^2 - 2x + 2x - 4$
$= x^2 - 4;$

b) $(3x - 5)(3x + 5) = 9x^2 + 15x - 15x - 25$
$= 9x^2 - 25.$

Do Exercises 13 and 14.

Perhaps you discovered in each case that when you multiply the two binomials, two terms are opposites, or additive inverses, which add to 0 and "drop out."

> The product of the sum and the difference of the same two terms is the square of the first term minus the square of the second term:
>
> $(A + B)(A - B) = A^2 - B^2.$

It is helpful to memorize this rule in both words and symbols. (If you do forget it, you can, of course, use FOIL.)

Examples Multiply. (Carry out the rule and say the words as you go.)

$\mathbf{(A + B)(A - B) = A^2 - B^2}$

11. $(x + 4)(x - 4) = x^2 - 4^2$ "The square of the first term, x^2, minus the square of the second, 4^2"
$= x^2 - 16$ Simplifying

12. $(5 + 2w)(5 - 2w) = 5^2 - (2w)^2$
$= 25 - 4w^2$

13. $(3x^2 - 7)(3x^2 + 7) = (3x^2)^2 - 7^2$
$= 9x^4 - 49$

14. $(-4x - 10)(-4x + 10) = (-4x)^2 - 10^2$
$= 16x^2 - 100$

15. $\left(x + \frac{3}{8}\right)\left(x - \frac{3}{8}\right) = x^2 - \left(\frac{3}{8}\right)^2 = x^2 - \frac{9}{64}$

Do Exercises 15–19.

c Squaring Binomials

Consider the square of a binomial, such as $(x + 3)^2$. This can be expressed as $(x + 3)(x + 3)$. Since this is the product of two binomials, we can again use FOIL. But again, this type of product occurs so often that we would like to use an even faster method. Look for a pattern in the following:

a) $(x + 3)^2 = (x + 3)(x + 3)$
$= x^2 + 3x + 3x + 9$
$= x^2 + 6x + 9;$

b) $(5 + 3p)^2 = (5 + 3p)(5 + 3p)$
$= 25 + 15p + 15p + 9p^2$
$= 25 + 30p + 9p^2;$

Multiply.

13. $(x + 5)(x - 5)$

14. $(2x - 3)(2x + 3)$

Multiply.

15. $(x + 2)(x - 2)$

16. $(x - 7)(x + 7)$

17. $(6 - 4y)(6 + 4y)$

18. $(2x^3 - 1)(2x^3 + 1)$

19. $\left(x - \frac{2}{5}\right)\left(x + \frac{2}{5}\right)$

Answers on page A-31

Multiply.

20. $(x + 8)(x + 8)$

21. $(x - 5)(x - 5)$

c) $(x - 3)^2 = (x - 3)(x - 3)$
$= x^2 - 3x - 3x + 9$
$= x^2 - 6x + 9;$

d) $(3x - 5)^2 = (3x - 5)(3x - 5)$
$= 9x^2 - 15x - 15x + 25$
$= 9x^2 - 30x + 25.$

Do Exercises 20 and 21.

When squaring a binomial, we multiply a binomial by itself. Perhaps you noticed that two terms are the same and when added give twice their product. The other two terms are squares.

> The square of a sum or a difference of two terms is the square of the first term, plus or minus twice the product of the two terms, plus the square of the last term:
>
> $(A + B)^2 = A^2 + 2AB + B^2;$
>
> $(A - B)^2 = A^2 - 2AB + B^2.$

It is helpful to memorize this rule in both words and symbols.

Examples Multiply. (Carry out the rule and say the words as you go.)

$$(A + B)^2 = A^2 + 2 \cdot A \cdot B + B^2$$

16. $(x + 3)^2 = x^2 + 2 \cdot x \cdot 3 + 3^2$ "x^2 plus 2 times x times 3 plus 3^2"
$= x^2 + 6x + 9$

$$(A - B)^2 = A^2 - 2 \cdot A \cdot B + B^2$$

17. $(t - 5)^2 = t^2 - 2 \cdot t \cdot 5 + 5^2$ "t^2 minus 2 times t times 5 plus 5^2"
$= t^2 - 10t + 25$

18. $(2x + 7)^2 = (2x)^2 + 2 \cdot 2x \cdot 7 + 7^2$
$= 4x^2 + 28x + 49$

19. $(5x - 3x^2)^2 = (5x)^2 - 2 \cdot 5x \cdot 3x^2 + (3x^2)^2$
$= 25x^2 - 30x^3 + 9x^4$

20. $(2.3 - 5.4m)^2 = 2.3^2 - 2(2.3)(5.4m) + (5.4m)^2$
$= 5.29 - 24.84m + 29.16m^2$

Do Exercises 22–27.

Multiply.

22. $(x + 2)^2$

23. $(a - 4)^2$

24. $(2x + 5)^2$

25. $(4x^2 - 3x)^2$

26. $(7.8 + 1.2y)(7.8 + 1.2y)$

27. $(3x^2 - 5)(3x^2 - 5)$

CAUTION! Note carefully in these examples that the square of a sum is *not* the sum of the squares:

The middle term $2AB$ is missing.

$$(A + B)^2 \neq A^2 + B^2.$$

To see this, note that

$$(20 + 5)^2 = 25^2 = 625,$$

but

$$20^2 + 5^2 = 400 + 25 = 425 \quad \text{and} \quad 425 \neq 625.$$

However, $20^2 + 2(20)(5) + 5^2 = 625$, which illustrates that

$$(A + B)^2 = A^2 + 2AB + B^2.$$

Answers on page A-31

We can look at the rule for finding $(A + B)^2$ geometrically as follows. The area of the large square is

$$(A + B)(A + B) = (A + B)^2.$$

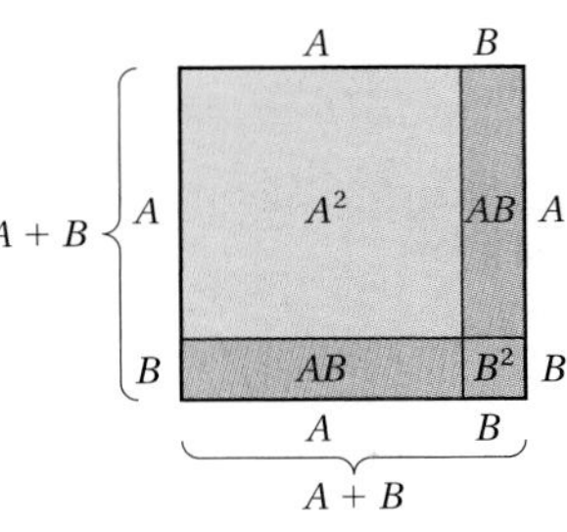

This is equal to the sum of the areas of the smaller rectangles:

$$A^2 + AB + AB + B^2 = A^2 + 2AB + B^2.$$

Thus,

$$(A + B)^2 = A^2 + 2AB + B^2.$$

d Multiplication of Various Types

We have considered how to quickly multiply certain kinds of polynomials. Let's now try several types of multiplications mixed together so that we can learn to sort them out. When you multiply, first see what kind of multiplication you have. Then use the best method. The formulas you should know and the questions you should ask yourself are as follows.

MULTIPLYING TWO POLYNOMIALS

1. Is the product the square of a binomial? If so, use the following:

   $$(A + B)(A + B) = (A + B)^2 = A^2 + 2AB + B^2,$$

   or $$(A - B)(A - B) = (A - B)^2 = A^2 - 2AB + B^2.$$

   The square of a binomial is the square of the first term, plus or minus *twice* the product of the two terms, plus the square of the last term.

   [The answer has 3 terms.]

   Example: $(x + 7)(x + 7) = (x + 7)^2$
   $= x^2 + 2 \cdot x \cdot 7 + 7^2 = x^2 + 14x + 49$

2. Is it the product of the sum and the difference of the *same* two terms? If so, use the following:

   $$(A + B)(A - B) = A^2 - B^2.$$

   The product of the sum and the difference of the same two terms is the difference of the squares.

   [The answer has 2 terms.]

   Example: $(x + 7)(x - 7) = x^2 - 7^2 = x^2 - 49$

3. Is it the product of two binomials other than those above? If so, use FOIL.

   [The answer will have 3 or 4 terms.]

   Example: $(x + 7)(x - 4) = x^2 - 4x + 7x - 28 = x^2 + 3x - 28$

4. Is it the product of a monomial and a polynomial? If so, multiply each term of the polynomial by the monomial.

   Example: $5x(x + 7) = 5x \cdot x + 5x \cdot 7 = 5x^2 + 35x$

5. Is it the product of two polynomials other than those above? If so, multiply each term of one by every term of the other. Use columns if you wish.

   [The answer will have 2 or more terms, usually more than 2 terms.]

   Example: $(x^2 - 3x + 2)(x + 7) = x^2(x + 7) - 3x(x + 7) + 2(x + 7)$
   $= x^2 \cdot x + x^2 \cdot 7 - 3x \cdot x - 3x \cdot 7$
   $+ 2 \cdot x + 2 \cdot 7$
   $= x^3 + 7x^2 - 3x^2 - 21x + 2x + 14$
   $= x^3 + 4x^2 - 19x + 14$

Remember that FOIL will *always* work for two binomials. You can use it instead of either of the first two rules, but those rules will make your work go faster.

Multiply.

28. $(x+5)(x+6)$

29. $(t-4)(t+4)$

30. $4x^2(-2x^3+5x^2+10)$

31. $(9x^2+1)^2$

32. $(2a-5)(2a+8)$

33. $\left(5x+\frac{1}{2}\right)^2$

34. $\left(2x-\frac{1}{2}\right)^2$

35. $(x^2-x+4)(x-2)$

Answers on page A-31

Example 21 Multiply: $(x+3)(x-3)$.

$(x+3)(x-3) = x^2 - 9$ Using method 2 (the product of the sum and the difference of two terms)

Example 22 Multiply: $(t+7)(t-5)$.

$(t+7)(t-5) = t^2 + 2t - 35$ Using method 3 (the product of two binomials, but neither the square of a binomial nor the product of the sum and the difference of two terms)

Example 23 Multiply: $(x+6)(x+6)$.

$$(x+6)(x+6) = x^2 + 2(6)x + 36$$ Using method 1 (the square of a binomial sum)

$$= x^2 + 12x + 36$$

Example 24 Multiply: $2x^3(9x^2+x-7)$.

$2x^3(9x^2+x-7) = 18x^5 + 2x^4 - 14x^3$ Using method 4 (the product of a monomial and a trinomial; multiplying each term of the trinomial by the monomial)

Example 25 Multiply: $(5x^3-7x)^2$.

$$(5x^3-7x)^2 = 25x^6 - 2(5x^3)(7x) + 49x^2$$ Using method 1 (the square of a binomial difference)

$$= 25x^6 - 70x^4 + 49x^2$$

Example 26 Multiply: $\left(3x+\frac{1}{4}\right)^2$.

$$\left(3x+\tfrac{1}{4}\right)^2 = 9x^2 + 2(3x)\left(\tfrac{1}{4}\right) + \tfrac{1}{16}$$ Using method 1 (the square of a binomial sum. To get the middle term, we multiply $3x$ by $\frac{1}{4}$ and double.)

$$= 9x^2 + \tfrac{3}{2}x + \tfrac{1}{16}$$

Example 27 Multiply: $\left(4x-\frac{3}{4}\right)^2$.

$$\left(4x-\tfrac{3}{4}\right)^2 = 16x^2 - 2(4x)\left(\tfrac{3}{4}\right) + \tfrac{9}{16}$$ Using method 1 (the square of a binomial difference)

$$= 16x^2 - 6x + \tfrac{9}{16}$$

Example 28 Multiply: $(p+3)(p^2+2p-1)$.

$$\begin{array}{r} p^2 + 2p - 1 \\ p + 3 \\ \hline 3p^2 + 6p - 3 \\ p^3 + 2p^2 - p \\ \hline p^3 + 5p^2 + 5p - 3 \end{array}$$

Using method 5 (the product of two polynomials)

Multiplying by 3

Multiplying by p

Do Exercises 28–35.

Exercise Set 10.6

a Multiply. Try to write only the answer. If you need more steps, be sure to use them.

1. $(x + 1)(x^2 + 3)$

2. $(x^2 - 3)(x - 1)$

3. $(x^3 + 2)(x + 1)$

4. $(x^4 + 2)(x + 10)$

5. $(y + 2)(y - 3)$

6. $(a + 2)(a + 3)$

7. $(3x + 2)(3x + 2)$

8. $(4x + 1)(4x + 1)$

9. $(5x - 6)(x + 2)$

10. $(x - 8)(x + 8)$

11. $(3t - 1)(3t + 1)$

12. $(2m + 3)(2m + 3)$

13. $(4x - 2)(x - 1)$

14. $(2x - 1)(3x + 1)$

15. $\left(p - \frac{1}{4}\right)\left(p + \frac{1}{4}\right)$

16. $\left(q + \frac{3}{4}\right)\left(q + \frac{3}{4}\right)$

17. $(x - 0.1)(x + 0.1)$

18. $(x + 0.3)(x - 0.4)$

19. $(2x^2 + 6)(x + 1)$

20. $(2x^2 + 3)(2x - 1)$

21. $(-2x + 1)(x + 6)$

22. $(3x + 4)(2x - 4)$

23. $(a + 7)(a + 7)$

24. $(2y + 5)(2y + 5)$

25. $(1 + 2x)(1 - 3x)$

26. $(-3x - 2)(x + 1)$

27. $(x^2 + 3)(x^3 - 1)$

28. $(x^4 - 3)(2x + 1)$

29. $(3x^2 - 2)(x^4 - 2)$

30. $(x^{10} + 3)(x^{10} - 3)$

31. $(2.8x - 1.5)(4.7x + 9.3)$

32. $\left(x - \frac{3}{8}\right)\left(x + \frac{4}{7}\right)$

33. $(3x^5 + 2)(2x^2 + 6)$

34. $(1 - 2x)(1 + 3x^2)$

35. $(8x^3 + 1)(x^3 + 8)$

36. $(4 - 2x)(5 - 2x^2)$

37. $(4x^2 + 3)(x - 3)$

38. $(7x - 2)(2x - 7)$

39. $(4y^4 + y^2)(y^2 + y)$

40. $(5y^6 + 3y^3)(2y^6 + 2y^3)$

b Multiply mentally, if possible. If you need extra steps, be sure to use them.

41. $(x + 4)(x - 4)$

42. $(x + 1)(x - 1)$

43. $(2x + 1)(2x - 1)$

44. $(x^2 + 1)(x^2 - 1)$

45. $(5m - 2)(5m + 2)$

46. $(3x^4 + 2)(3x^4 - 2)$

47. $(2x^2 + 3)(2x^2 - 3)$

48. $(6x^5 - 5)(6x^5 + 5)$

49. $(3x^4 - 4)(3x^4 + 4)$

50. $(t^2 - 0.2)(t^2 + 0.2)$

51. $(x^6 - x^2)(x^6 + x^2)$

52. $(2x^3 - 0.3)(2x^3 + 0.3)$

53. $(x^4 + 3x)(x^4 - 3x)$

54. $\left(\frac{3}{4} + 2x^3\right)\left(\frac{3}{4} - 2x^3\right)$

55. $(x^{12} - 3)(x^{12} + 3)$

56. $(12 - 3x^2)(12 + 3x^2)$

57. $(2y^8 + 3)(2y^8 - 3)$

58. $\left(m - \frac{2}{3}\right)\left(m + \frac{2}{3}\right)$

59. $\left(\frac{5}{8}x - 4.3\right)\left(\frac{5}{8}x + 4.3\right)$

60. $(10.7 - x^3)(10.7 + x^3)$

c Multiply mentally, if possible. If you need extra steps, be sure to use them.

61. $(x + 2)^2$

62. $(2x - 1)^2$

63. $(3x^2 + 1)^2$

64. $\left(3x + \frac{3}{4}\right)^2$

65. $\left(a - \frac{1}{2}\right)^2$

66. $\left(2a - \frac{1}{5}\right)^2$

67. $(3 + x)^2$

68. $(x^3 - 1)^2$

69. $(x^2 + 1)^2$

70. $(8x - x^2)^2$

71. $(2 - 3x^4)^2$

72. $(6x^3 - 2)^2$

73. $(5 + 6t^2)^2$

74. $(3p^2 - p)^2$

75. $\left(x - \frac{5}{8}\right)^2$

76. $(0.3y + 2.4)^2$

d Multiply mentally, if possible.

77. $(3 - 2x^3)^2$

78. $(x - 4x^3)^2$

79. $4x(x^2 + 6x - 3)$

80. $8x(-x^5 + 6x^2 + 9)$

81. $\left(2x^2 - \frac{1}{2}\right)\left(2x^2 - \frac{1}{2}\right)$

82. $(-x^2 + 1)^2$

83. $(-1 + 3p)(1 + 3p)$

84. $(-3q + 2)(3q + 2)$

85. $3t^2(5t^3 - t^2 + t)$

86. $-6x^2(x^3 + 8x - 9)$

87. $(6x^4 + 4)^2$

88. $(8a + 5)^2$

89. $(3x + 2)(4x^2 + 5)$

90. $(2x^2 - 7)(3x^2 + 9)$

91. $(8 - 6x^4)^2$

92. $\left(\frac{1}{5}x^2 + 9\right)\left(\frac{3}{5}x^2 - 7\right)$

93. $(t - 1)(t^2 + t + 1)$

94. $(y + 5)(y^2 - 5y + 25)$

Compute each of the following and compare.

95. $3^2 + 4^2$; $(3 + 4)^2$

96. $6^2 + 7^2$; $(6 + 7)^2$

97. $9^2 - 5^2$; $(9 - 5)^2$

98. $11^2 - 4^2$; $(11 - 4)^2$

Skill Maintenance

99. In apartment 3B, lamps, an air conditioner, and a television set are all operating at the same time. The lamps use 10 times as many watts of electricity as the television set, and the air conditioner uses 40 times as many watts as the television set. The total wattage used in the apartment is 2550. How many watts are used by each appliance? [8.4a]

Solve. [8.3c]

100. $3x - 8x = 4(7 - 8x)$

101. $3(x - 2) = 5(2x + 7)$

102. $5(2x - 3) - 2(3x - 4) = 20$

Solve. [8.6a]

103. $3x - 2y = 12$, for y

104. $ab - cd = 4$, for a

Synthesis

105. ◈ Under what conditions is the product of two binomials a binomial?

106. ◈ Todd feels that since the FOIL method can be used to find the product of any two binomials, he needn't study the other special products. What advice would you give him?

Multiply.

107. $5x(3x - 1)(2x + 3)$

108. $[(2x - 3)(2x + 3)](4x^2 + 9)$

109. $[(a - 5)(a + 5)]^2$

110. $(a - 3)^2(a + 3)^2$
(*Hint*: Examine Exercise 109.)

111. $(3t^4 - 2)^2(3t^4 + 2)^2$
(*Hint*: Examine Exercise 109.)

112. $[3a - (2a - 3)][3a + (2a - 3)]$

Solve.

113. $(x + 2)(x - 5) = (x + 1)(x - 3)$

114. $(2x + 5)(x - 4) = (x + 5)(2x - 4)$

115. *Factors and Sums.* To *factor* a number is to express it as a product. Since $12 = 4 \cdot 3$, we say that 12 is *factored* and that 4 and 3 are *factors* of 12. In the following table, the top number has been factored in such a way that the sum of the factors is the bottom number. For example, in the first column, 40 has been factored as $5 \cdot 8$, and $5 + 8 = 13$, the bottom number. Such thinking is important in algebra when we factor trinomials of the type $x^2 + bx + c$. Find the missing numbers in the table.

Product	40	63	36	72	−140	−96	48	168	110			
Factor	5									−9	−24	−3
Factor	8									−10	18	
Sum	13	16	−20	−38	−4	4	−14	−29	−21			18

Find the total shaded area.

116.

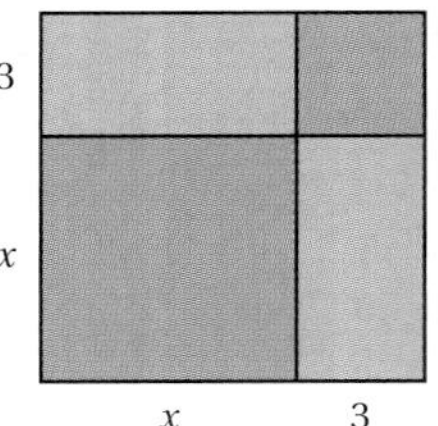

117.

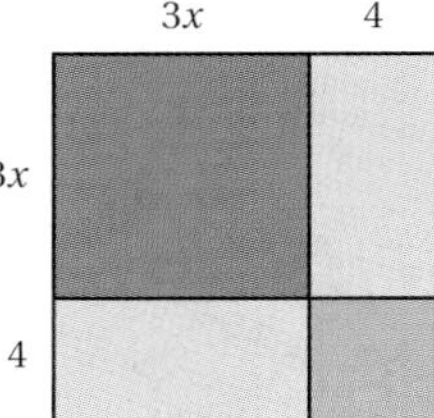

118. A factored polynomial for the shaded area in this rectangle is $(A + B)(A - B)$.

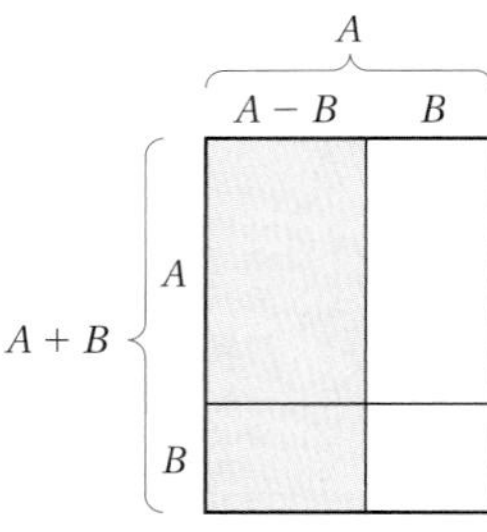

a) Find a polynomial for the area of the entire rectangle.
b) Find a polynomial for the sum of the areas of the two small unshaded rectangles.
c) Find a polynomial for the area in part (a) minus the area in part (b).
d) Find a polynomial for the area of the shaded region and compare this with the polynomial found in part (c).

Use the TABLE or GRAPH feature to check whether each of the following is correct.

119. $(x - 1)^2 = x^2 - 2x + 1$

120. $(x - 2)^2 = x^2 - 4x - 4$

121. $(x - 3)(x + 3) = x^2 - 6$

122. $(x - 3)(x + 2) = x^2 - x - 6$

Collaborative Learning Manual

Derive the special-product formulas.

10.7 Operations with Polynomials in Several Variables

The polynomials that we have been studying have only one variable. A **polynomial in several variables** is an expression like those you have already seen, but with more than one variable. Here are two examples:

$$3x + xy^2 + 5y + 4, \qquad 8xy^2z - 2x^3z - 13x^4y^2 + 15.$$

a Evaluating Polynomials

Example 1 Evaluate the polynomial $4 + 3x + xy^2 + 8x^3y^3$ for $x = -2$ and $y = 5$.

We replace x with -2 and y with 5:

$$\begin{aligned} 4 + 3x + xy^2 + 8x^3y^3 &= 4 + 3(-2) + (-2) \cdot 5^2 + 8(-2)^3 \cdot 5^3 \\ &= 4 - 6 - 50 - 8000 \\ &= -8052. \end{aligned}$$

Example 2 *Male Caloric Needs.* The number of calories needed each day by a moderately active man who weighs w kilograms, is h centimeters tall, and is a years old can be estimated by the polynomial

$$19.18w + 7h - 9.52a + 92.4$$

(***Source:*** Parker, M., *She Does Math.* Mathematical Association of America, p. 96). The author of this text is moderately active, weighs 97 kg, is 185 cm tall, and is 55 yr old. What are his daily caloric needs?

We evaluate the polynomial for $w = 97$, $h = 185$, and $a = 55$:

$$\begin{aligned} &19.18w + 7h - 9.52a + 92.4 \\ &= 19.18(97) + 7(185) - 9.52(55) + 92.4 \qquad \text{Substituting} \\ &= 2724.26. \end{aligned}$$

His daily caloric need is about 2724 calories.

Do Exercises 1–3.

Objectives

- a Evaluate a polynomial in several variables for given values of the variables.
- b Identify the coefficients and the degrees of the terms of a polynomial and the degree of a polynomial.
- c Collect like terms of a polynomial.
- d Add polynomials.
- e Subtract polynomials.
- f Multiply polynomials.

For Extra Help

TAPE 20

MAC WIN

CD-ROM

1. Evaluate the polynomial

$$4 + 3x + xy^2 + 8x^3y^3$$

for $x = 2$ and $y = -5$.

2. Evaluate the polynomial

$$8xy^2 - 2x^3z - 13x^4y^2 + 5$$

for $x = -1$, $y = 3$, and $z = 4$.

3. *Female Caloric Needs.* The number of calories needed each day by a moderately active woman who weighs w pounds, is h inches tall, and is a years old can be estimated by the polynomial

$$917 + 6w + 6h - 6a$$

(***Source:*** Parker, M., *She Does Math.* Mathematical Association of America, p. 96). Christine Poolos, a Project Manager for Addison Wesley Longman Publishing Co., is moderately active, weighs 125 lb, is 64 in. tall, and is 27 yr old. What are her daily caloric needs?

Answers on page A-31

4. Identify the coefficient of each term:

$-3xy^2 + 3x^2y - 2y^3 + xy + 2.$

5. Identify the degree of each term and the degree of the polynomial

$4xy^2 + 7x^2y^3z^2 - 5x + 2y + 4.$

Collect like terms.

6. $4x^2y + 3xy - 2x^2y$

7. $-3pq - 5pqr^3 - 12 + 8pq + 5pqr^3 + 4$

Answers on page A-31

b Coefficients and Degrees

The **degree** of a term is the sum of the exponents of the variables. The **degree of a polynomial** is the degree of the term of highest degree.

Example 3 Identify the coefficient and the degree of each term and the degree of the polynomial

$$9x^2y^3 - 14xy^2z^3 + xy + 4y + 5x^2 + 7.$$

Term	Coefficient	Degree	Degree of the Polynomial
$9x^2y^3$	9	5	
$-14xy^2z^3$	-14	6	6
xy	1	2	
$4y$	4	1	Think: $4y = 4y^1$.
$5x^2$	5	2	
7	7	0	Think: $7 = 7x^0$, or $7x^0y^0z^0$.

Do Exercises 4 and 5.

c Collecting Like Terms

Like terms (or **similar terms**) have exactly the same variables with exactly the same exponents. For example,

$3x^2y^3$ and $-7x^2y^3$ are like terms;

$9x^4z^7$ and $12x^4z^7$ are like terms.

But

$13xy^5$ and $-2x^2y^5$ are *not* like terms, because the x-factors have different exponents;

and

$3xyz^2$ and $4xy$ are *not* like terms, because there is no factor of z^2 in the second expression.

Collecting like terms is based on the distributive laws.

Examples Collect like terms.

4. $5x^2y + 3xy^2 - 5x^2y - xy^2 = (5 - 5)x^2y + (3 - 1)xy^2 = 2xy^2$
5. $8a^2 - 2ab + 7b^2 + 4a^2 - 9ab - 17b^2 = 12a^2 - 11ab - 10b^2$
6. $7xy - 5xy^2 + 3xy^2 - 7 + 6x^3 + 9xy - 11x^3 + y - 1$
$= -2xy^2 + 16xy - 5x^3 + y - 8$

Do Exercises 6 and 7.

d Addition

We can find the sum of two polynomials in several variables by writing a plus sign between them and then collecting like terms.

Example 7 Add: $(-5x^3 + 3y - 5y^2) + (8x^3 + 4x^2 + 7y^2)$.

$$(-5x^3 + 3y - 5y^2) + (8x^3 + 4x^2 + 7y^2)$$
$$= (-5 + 8)x^3 + 4x^2 + 3y + (-5 + 7)y^2$$
$$= 3x^3 + 4x^2 + 3y + 2y^2$$

Example 8 Add:

$$(5xy^2 - 4x^2y + 5x^3 + 2) + (3xy^2 - 2x^2y + 3x^3y - 5).$$

We first look for like terms. They are $5xy^2$ and $3xy^2$, $-4x^2y$ and $-2x^2y$, and 2 and -5. We collect these. Since there are no more like terms, the answer is

$$8xy^2 - 6x^2y + 5x^3 + 3x^3y - 3.$$

Do Exercises 8–10.

e Subtraction

We subtract a polynomial by adding its opposite, or additive inverse. The opposite of the polynomial

$$4x^2y - 6x^3y^2 + x^2y^2 - 5y$$

can be represented by

$$-(4x^2y - 6x^3y^2 + x^2y^2 - 5y).$$

We find an equivalent expression for the opposite of a polynomial by replacing each coefficient with its opposite, or by changing the sign of each term. Thus,

$$-(4x^2y - 6x^3y^2 + x^2y^2 - 5y) = -4x^2y + 6x^3y^2 - x^2y^2 + 5y.$$

Example 9 Subtract:

$$(4x^2y + x^3y^2 + 3x^2y^3 + 6y + 10) - (4x^2y - 6x^3y^2 + x^2y^2 - 5y - 8).$$

We have

$$(4x^2y + x^3y^2 + 3x^2y^3 + 6y + 10) - (4x^2y - 6x^3y^2 + x^2y^2 - 5y - 8)$$
$$= 4x^2y + x^3y^2 + 3x^2y^3 + 6y + 10 - 4x^2y + 6x^3y^2 - x^2y^2 + 5y + 8$$

Finding the opposite by changing the sign of each term

$$= 7x^3y^2 + 3x^2y^3 - x^2y^2 + 11y + 18.$$

Collecting like terms. (Try to write just the answer!)

Do Exercises 11 and 12.

Add.

8. $(4x^3 + 4x^2 - 8y - 3) + (-8x^3 - 2x^2 + 4y + 5)$

9. $(13x^3y + 3x^2y - 5y) + (x^3y + 4x^2y - 3xy + 3y)$

10. $(-5p^2q^4 + 2p^2q^2 + 3q) + (6pq^2 + 3p^2q + 5)$

Subtract.

11. $(-4s^4t + s^3t^2 + 2s^2t^3) - (4s^4t - 5s^3t^2 + s^2t^2)$

12. $(-5p^4q + 5p^3q^2 - 3p^2q^3 - 7q^4 - 2) - (4p^4q - 4p^3q^2 + p^2q^3 + 2q^4 - 7)$

Answers on page A-31

Multiply.

13. $(x^2y^3 + 2x)(x^3y^2 + 3x)$

14. $(p^4q - 2p^3q^2 + 3q^3)(p + 2q)$

Multiply.

15. $(3xy + 2x)(x^2 + 2xy^2)$

16. $(x - 3y)(2x - 5y)$

17. $(4x + 5y)^2$

18. $(3x^2 - 2xy^2)^2$

19. $(2xy^2 + 3x)(2xy^2 - 3x)$

20. $(3xy^2 + 4y)(-3xy^2 + 4y)$

21. $(3y + 4 - 3x)(3y + 4 + 3x)$

22. $(2a + 5b + c)(2a - 5b - c)$

Answers on page A-31

f Multiplication

To multiply polynomials in several variables, we can multiply each term of one by every term of the other. We can use columns for long multiplications as with polynomials in one variable. We multiply each term at the top by every term at the bottom. We write like terms in columns, and then we add the results.

Example 10 Multiply: $(3x^2y - 2xy + 3y)(xy + 2y)$.

$$\begin{array}{rrrrl}
 & 3x^2y & -\ 2xy & +\ 3y & \\
 & & xy & +\ 2y & \\
\hline
 & 6x^2y^2 & -\ 4xy^2 & +\ 6y^2 & \textbf{Multiplying by } 2y \\
3x^3y^2 & -\ 2x^2y^2 & +\ 3xy^2 & & \textbf{Multiplying by } xy \\
\hline
3x^3y^2 & +\ 4x^2y^2 & -\ xy^2 & +\ 6y^2 & \textbf{Adding}
\end{array}$$

Do Exercises 13 and 14.

Where appropriate, we use the special products that we have learned.

Examples Multiply.

$$\text{F} \quad \text{O} \quad \text{I} \quad \text{L}$$

11. $(x^2y + 2x)(xy^2 + y^2) = x^3y^3 + x^2y^3 + 2x^2y^2 + 2xy^2$

12. $(p + 5q)(2p - 3q) = 2p^2 - 3pq + 10pq - 15q^2$
$= 2p^2 + 7pq - 15q^2$

$(A + B)^2 = A^2 + 2 \cdot A \cdot B + B^2$

13. $(3x + 2y)^2 = (3x)^2 + 2(3x)(2y) + (2y)^2$
$= 9x^2 + 12xy + 4y^2$

$(A - B)^2 = A^2 - 2 \cdot A \cdot B + B^2$

14. $(2y^2 - 5x^2y)^2 = (2y^2)^2 - 2(2y^2)(5x^2y) + (5x^2y)^2$
$= 4y^4 - 20x^2y^3 + 25x^4y^2$

$(A + B)(A - B) = A^2 - B^2$

15. $(3x^2y + 2y)(3x^2y - 2y) = (3x^2y)^2 - (2y)^2$
$= 9x^4y^2 - 4y^2$

16. $(-2x^3y^2 + 5t)(2x^3y^2 + 5t) = (5t - 2x^3y^2)(5t + 2x^3y^2)$
$= (5t)^2 - (2x^3y^2)^2$
$= 25t^2 - 4x^6y^4$

$(A - B)(A + B) = A^2 - B^2$

17. $(\,2x + 3\, - 2y)(\,2x + 3\, + 2y) = (\,2x + 3\,)^2 - (2y)^2$
$= 4x^2 + 12x + 9 - 4y^2$

Do Exercises 15–22.

Exercise Set 10.7

a Evaluate the polynomial for $x = 3$, $y = -2$, and $z = -5$.

1. $x^2 - y^2 + xy$

2. $x^2 + y^2 - xy$

3. $x^2 - 3y^2 + 2xy$

4. $x^2 - 4xy + 5y^2$

5. $8xyz$

6. $-3xyz^2$

7. $xyz^2 - z$

8. $xy - xz + yz$

Lung Capacity. The polynomial

$$0.041h - 0.018A - 2.69$$

can be used to estimate the lung capacity, in liters, of a female of height h, in centimeters, and age A, in years.

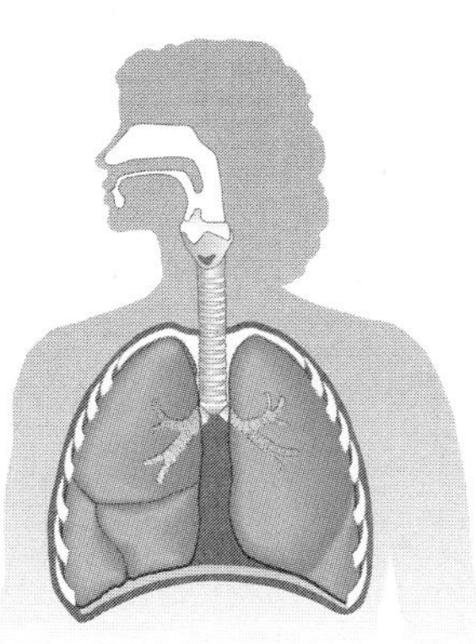

9. Find the lung capacity of a 20-yr-old woman who is 165 cm tall.

10. Find the lung capacity of a 50-yr-old woman who is 160 cm tall.

Altitude of a Launched Object. The altitude, in meters, of a launched object is given by the polynomial

$$h + vt - 4.9t^2,$$

where h is the height, in meters, from which the launch occurs, v is the initial upward speed (or velocity), in meters per second (m/s), and t is the number of seconds for which the object is airborne.

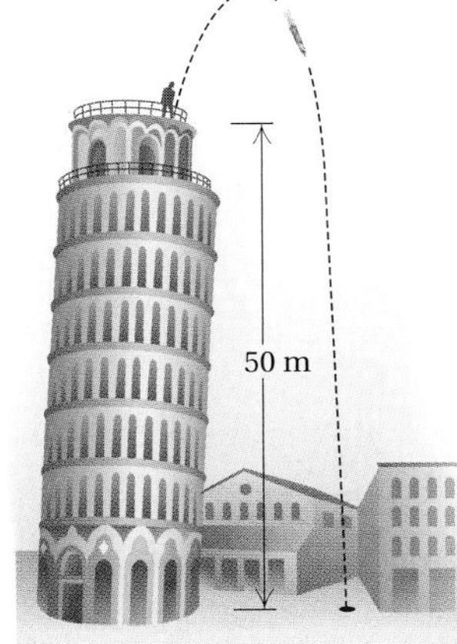

11. A model rocket is launched from the top of the Leaning Tower of Pisa, 50 m above the ground. The upward speed is 40 m/s. How high will the rocket be 2 sec after the blastoff?

12. A golf ball is thrown upward with an initial speed of 30 m/s by a golfer atop the Washington Monument, which is 160 m above the ground. How high above the ground will the ball be after 3 sec?

Surface Area of a Right Circular Cylinder. The area of a right circular cylinder is given by the polynomial

$$2\pi rh + 2\pi r^2,$$

where h is the height and r is the radius of the base.

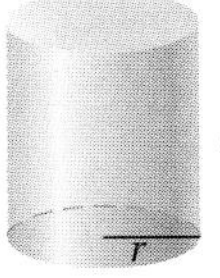

13. A 16-oz beverage can has a height of 6.3 in. and a radius of 1.2 in. Evaluate the polynomial for $h = 6.3$ and $r = 1.2$ to find the area of the can. Use 3.14 for π.

14. A 26-oz coffee can has a height of 6.5 in. and a radius of 2.5 in. Evaluate the polynomial for $h = 6.5$ and $r = 2.5$ to find the area of the can. Use 3.14 for π.

b Identify the coefficient and the degree of each term of the polynomial. Then find the degree of the polynomial.

15. $x^3y - 2xy + 3x^2 - 5$

16. $5y^3 - y^2 + 15y + 1$

17. $17x^2y^3 - 3x^3yz - 7$

18. $6 - xy + 8x^2y^2 - y^5$

c Collect like terms.

19. $a + b - 2a - 3b$

20. $y^2 - 1 + y - 6 - y^2$

21. $3x^2y - 2xy^2 + x^2$

22. $m^3 + 2m^2n - 3m^2 + 3mn^2$

23. $6au + 3av + 14au + 7av$

24. $3x^2y - 2z^2y + 3xy^2 + 5z^2y$

25. $2u^2v - 3uv^2 + 6u^2v - 2uv^2$

26. $3x^2 + 6xy + 3y^2 - 5x^2 - 10xy - 5y^2$

d Add.

27. $(2x^2 - xy + y^2) + (-x^2 - 3xy + 2y^2)$

28. $(2z - z^2 + 5) + (z^2 - 3z + 1)$

29. $(r - 2s + 3) + (2r + s) + (s + 4)$

30. $(ab - 2a + 3b) + (5a - 4b) + (3a + 7ab - 8b)$

31. $(b^3a^2 - 2b^2a^3 + 3ba + 4) + (b^2a^3 - 4b^3a^2 + 2ba - 1)$

32. $(2x^2 - 3xy + y^2) + (-4x^2 - 6xy - y^2) + (x^2 + xy - y^2)$

e Subtract.

33. $(a^3 + b^3) - (a^2b - ab^2 + b^3 + a^3)$

34. $(x^3 - y^3) - (-2x^3 + x^2y - xy^2 + 2y^3)$

35. $(xy - ab - 8) - (xy - 3ab - 6)$

36. $(3y^4x^2 + 2y^3x - 3y - 7) - (2y^4x^2 + 2y^3x - 4y - 2x + 5)$

37. $(-2a + 7b - c) - (-3b + 4c - 8d)$

38. Find the sum of $2a + b$ and $3a - b$. Then subtract $5a + 2b$.

f Multiply.

39. $(3z - u)(2z + 3u)$

40. $(a - b)(a^2 + b^2 + 2ab)$

41. $(a^2b - 2)(a^2b - 5)$

42. $(xy + 7)(xy - 4)$

43. $(a^3 + bc)(a^3 - bc)$

44. $(m^2 + n^2 - mn)(m^2 + mn + n^2)$

45. $(y^4x + y^2 + 1)(y^2 + 1)$

46. $(a - b)(a^2 + ab + b^2)$

47. $(3xy - 1)(4xy + 2)$

48. $(m^3n + 8)(m^3n - 6)$

49. $(3 - c^2d^2)(4 + c^2d^2)$

50. $(6x - 2y)(5x - 3y)$

51. $(m^2 - n^2)(m + n)$

52. $(pq + 0.2)(0.4pq - 0.1)$

53. $(xy + x^5y^5)(x^4y^4 - xy)$

54. $(x - y^3)(2y^3 + x)$

55. $(x + h)^2$

56. $(3a + 2b)^2$

57. $(r^3t^2 - 4)^2$

58. $(3a^2b - b^2)^2$

59. $(p^4 + m^2n^2)^2$

60. $(2ab - cd)^2$

61. $\left(2a^3 - \frac{1}{2}b^3\right)^2$

62. $-3x(x + 8y)^2$

63. $3a(a - 2b)^2$

64. $(a^2 + b + 2)^2$

65. $(2a - b)(2a + b)$

66. $(x - y)(x + y)$

67. $(c^2 - d)(c^2 + d)$

68. $(p^3 - 5q)(p^3 + 5q)$

69. $(ab + cd^2)(ab - cd^2)$

70. $(xy + pq)(xy - pq)$

71. $(x + y - 3)(x + y + 3)$

72. $(p + q + 4)(p + q - 4)$

73. $[x + y + z][x - (y + z)]$

74. $[a + b + c][a - (b + c)]$

75. $(a + b + c)(a - b - c)$

76. $(3x + 2 - 5y)(3x + 2 + 5y)$

Skill Maintenance

In which quadrant is the point located? [9.1c]

77. $(2, -5)$

78. $(-8, -9)$

79. $(16, 23)$

80. $(-3, 2)$

Graph. [9.3b]

81. $2x = -10$

82. $y = -4$

83. $8y - 16 = 0$

84. $x = 4$

Find the mean, the median, and the mode, if it exists. [9.4a]

85. 23, 31, 24, 31, 25, 28, 31

86. 5.2, 5.6, 5.8, 6.1, 5.6, 5.2, 6.3

Synthesis

87. ◆ Is it possible for a polynomial in four variables to have a degree less than 4? Why or why not?

88. ◆ Can the sum of two trinomials in several variables be a trinomial in one variable? Why or why not?

Find a polynomial for the shaded area. (Leave results in terms of π where appropriate.)

89.

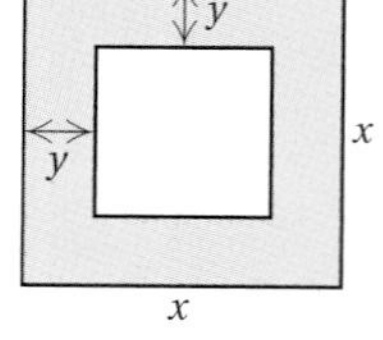

90.

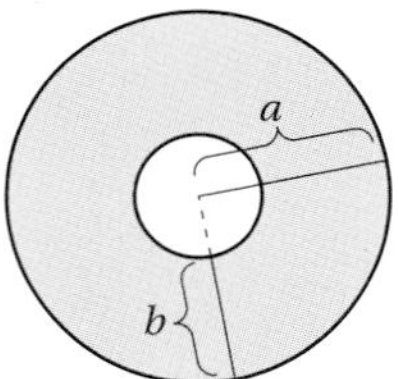

91.

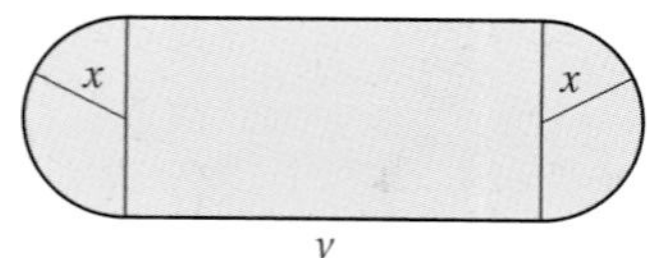

92.

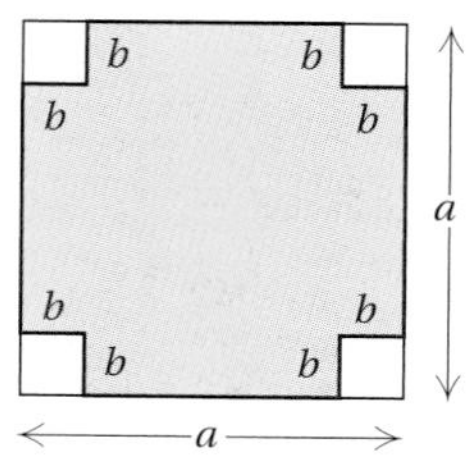

Find a formula for the surface area of the solid object. Leave results in terms of π.

93.

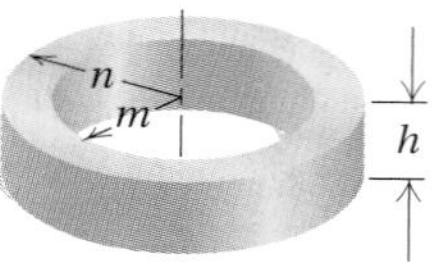

94.

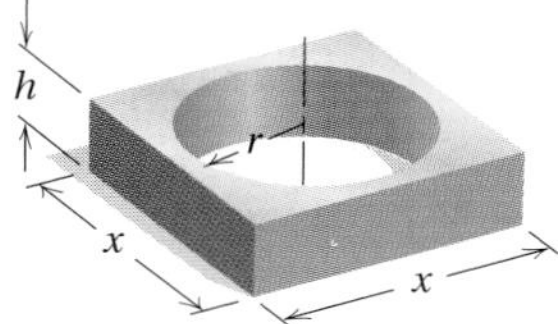

10.8 Division of Polynomials

In this section, we consider division of polynomials. You will see that such division is similar to what is done in arithmetic.

Objectives

a Divide a polynomial by a monomial.

b Divide a polynomial by a divisor that is not a monomial.

For Extra Help

TAPE 20 | MAC WIN | CD-ROM

a Divisor a Monomial

We first consider division by a monomial. When dividing a monomial by a monomial, we use the quotient rule of Section 10.1 to subtract exponents when the bases are the same. We also divide the coefficients.

Examples Divide.

1. $\dfrac{10x^2}{2x} = \dfrac{10}{2} \cdot \dfrac{x^2}{x} = 5x^{2-1} = 5x$

2. $\dfrac{x^9}{3x^2} = \dfrac{1x^9}{3x^2} = \dfrac{1}{3} \cdot \dfrac{x^9}{x^2} = \dfrac{1}{3}x^{9-2} = \dfrac{1}{3}x^7$

3. $\dfrac{-18x^{10}}{3x^3} = \dfrac{-18}{3} \cdot \dfrac{x^{10}}{x^3} = -6x^{10-3} = -6x^7$

4. $\dfrac{42a^2b^5}{-3ab^2} = \dfrac{42}{-3} \cdot \dfrac{a^2}{a} \cdot \dfrac{b^5}{b^2} = -14a^{2-1}b^{5-2} = -14ab^3$

Do Exercises 1–4.

Divide.

1. $\dfrac{20x^3}{5x}$

2. $\dfrac{-28x^{14}}{4x^3}$

3. $\dfrac{-56p^5q^7}{2p^2q^6}$

4. $\dfrac{x^5}{4x}$

Answers on page A-32

To divide a polynomial by a monomial, we note that since

$$\frac{A}{C} + \frac{B}{C} = \frac{A + B}{C},$$

it follows that

$$\frac{A + B}{C} = \frac{A}{C} + \frac{B}{C}.$$

This is actually the procedure we use when performing divisions like $86 \div 2$. Although we might write

$$\frac{86}{2} = 43,$$

we could also calculate as follows:

$$\frac{86}{2} = \frac{80 + 6}{2} = \frac{80}{2} + \frac{6}{2} = 40 + 3 = 43.$$

Similarly, to divide a polynomial by a monomial, we divide each term by the monomial.

Example 5 Divide: $(9x^8 + 12x^6) \div 3x^2$.

We have

$$(9x^8 + 12x^6) \div 3x^2 = \frac{9x^8 + 12x^6}{3x^2}$$

$$= \frac{9x^8}{3x^2} + \frac{12x^6}{3x^2}.$$ **To see this, add and get the original expression.**

5. Divide: $(28x^7 + 32x^5) \div 4x^3$. Check the result.

6. Divide: $(2x^3 + 6x^2 + 4x) \div 2x$. Check the result.

7. Divide: $(6x^2 + 3x - 2) \div 3$. Check the result.

Divide and check.

8. $(8x^2 - 3x + 1) \div 2$

9. $\dfrac{2x^4y^6 - 3x^3y^4 + 5x^2y^3}{x^2y^2}$

Answers on page A-32

We now perform the separate divisions:

$$\frac{9x^8}{3x^2} + \frac{12x^6}{3x^2} = \frac{9}{3} \cdot \frac{x^8}{x^2} + \frac{12}{3} \cdot \frac{x^6}{x^2}$$
$$= 3x^{8-2} + 4x^{6-2}$$
$$= 3x^6 + 4x^4.$$

Caution! The coefficients are *divided*, but the exponents are *subtracted.*

To check, we multiply the quotient $3x^6 + 4x^4$ by the divisor $3x^2$:

$$(3x^6 + 4x^4)3x^2 = (3x^6)(3x^2) + (4x^4)(3x^2) = 9x^8 + 12x^6.$$

Do Exercises 5–7.

Example 6 Divide and check: $(10a^5b^4 - 2a^3b^2 + 6a^2b) \div (2a^2b)$.

$$\frac{10a^5b^4 - 2a^3b^2 + 6a^2b}{2a^2b} = \frac{10a^5b^4}{2a^2b} - \frac{2a^3b^2}{2a^2b} + \frac{6a^2b}{2a^2b}$$
$$= \frac{10}{2}a^{5-2}b^{4-1} - \frac{2}{2}a^{3-2}b^{2-1} + \frac{6}{2}$$
$$= 5a^3b^3 - ab + 3$$

CHECK:

$$\begin{array}{r} 5a^3b^3 - ab + 3 \\ 2a^2b \\ \hline 10a^5b^4 - 2a^3b^2 + 6a^2b \end{array}$$

We multiply.

The answer checks.

To divide a polynomial by a monomial, divide each term by the monomial.

Do Exercises 8 and 9.

b Divisor Not a Monomial

Let's first consider long division as it is performed in arithmetic. When we divide, we repeat the following procedure.

LONG DIVISION

1. Divide,
2. Multiply,
3. Subtract, and
4. Bring down the next term.

We review this by considering the division $3711 \div 8$.

$$\begin{array}{r} 4 \\ 8\overline{)3711} \\ 32 \\ \hline 51 \end{array}$$

(1) Divide: $37 \div 8 \approx 4$.

(2) Multiply: $4 \times 8 = 32$.

(3) Subtract: $37 - 32 = 5$.

(4) Bring down the 1.

$$\begin{array}{r} 463 \\ 8\overline{)3711} \\ 32 \\ \hline 51 \\ 48 \\ \hline 31 \\ 24 \\ \hline 7 \end{array}$$

Next, we repeat the process two more times. We obtain the complete division as shown on the right above. The quotient is 463. The remainder is 7, expressed as R = 7. We write the answer as

$$463 \text{ R } 7 \quad \text{or} \quad 463 + \frac{7}{8} = 463\frac{7}{8}.$$

We check by multiplying the quotient, 463, by the divisor, 8, and adding the remainder, 7:

$$8 \cdot 463 + 7 = 3704 + 7 = 3711.$$

Now let's look at long division with polynomials. We use this procedure when the divisor is not a monomial. We write polynomials in descending order and then write in missing terms.

Example 7 Divide $x^2 + 5x + 6$ by $x + 2$.

We have

$$\begin{array}{r|l} x & \leftarrow \text{Divide the first term by the first term: } x^2/x = x. \text{ Ignore the term 2.} \\ x + 2\overline{)x^2 + 5x + 6} & \\ x^2 + 2x & \leftarrow \text{Multiply } x \text{ above by the divisor, } x + 2. \\ \hline 3x & \leftarrow \text{Subtract: } (x^2 + 5x) - (x^2 + 2x) = x^2 + 5x - x^2 - 2x = 3x. \end{array}$$

We now "bring down" the next term of the dividend—in this case, 6.

$$\begin{array}{r|l} x + 3 & \leftarrow \text{Divide the first term by the first term: } 3x/x = 3. \\ x + 2\overline{)x^2 + 5x + 6} & \\ x^2 + 2x & \\ \hline 3x + 6 & \leftarrow \text{The 6 has been "brought down."} \\ 3x + 6 & \leftarrow \text{Multiply 3 by the divisor, } x + 2. \\ \hline 0 & \leftarrow \text{Subtract: } (3x + 6) - (3x + 6) = 3x + 6 - 3x - 6 = 0. \end{array}$$

The quotient is $x + 3$. The remainder is 0, expressed as R = 0. A remainder of 0 is generally not listed in an answer.

To check, we multiply the quotient by the divisor and add the remainder, if any, to see if we get the dividend:

$$\underbrace{(x + 2)}_{\text{Divisor}} \cdot \underbrace{(x + 3)}_{\text{Quotient}} + \underbrace{0}_{\text{Remainder}} = \underbrace{x^2 + 5x + 6}_{\text{Dividend}}. \quad \text{The division checks.}$$

Do Exercise 10.

10. Divide and check:

$(x^2 + x - 6) \div (x + 3)$.

Example 8 Divide and check: $(x^2 + 2x - 12) \div (x - 3)$.

We have

$$\begin{array}{r|l} x & \leftarrow \text{Divide the first term by the first term: } x^2/x = x. \\ x - 3\overline{)x^2 + 2x - 12} & \\ x^2 - 3x & \leftarrow \text{Multiply } x \text{ above by the divisor, } x - 3. \\ \hline 5x & \leftarrow \text{Subtract: } (x^2 + 2x) - (x^2 - 3x) = x^2 + 2x - x^2 + 3x = 5x. \end{array}$$

We now "bring down" the next term of the dividend—in this case, -12.

$$\begin{array}{r|l} x + 5 & \leftarrow \text{Divide the first term by the first term: } 5x/x = 5. \\ x - 3\overline{)x^2 + 2x - 12} & \\ x^2 - 3x & \\ \hline 5x - 12 & \leftarrow \text{Bring down the } -12. \\ 5x - 15 & \leftarrow \text{Multiply 5 above by the divisor, } x - 3. \\ \hline 3 & \leftarrow \text{Subtract: } (5x - 12) - (5x - 15) = 5x - 12 - 5x + 15 = 3. \end{array}$$

Answer on page A-32

11. Divide and check:

$x - 2\overline{)x^2 + 2x - 8}$.

The answer is $x + 5$ with $R = 3$, or

$$x + 5 + \frac{3}{x - 3}$$

Quotient: $x + 5$; Remainder: 3; Divisor: $x - 3$

(This is the way answers will be given at the back of the book.)

CHECK: We can check by multiplying the divisor by the quotient and adding the remainder, as follows:

$$(x - 3)(x + 5) + 3 = x^2 + 2x - 15 + 3$$
$$= x^2 + 2x - 12.$$

When dividing, an answer may "come out even" (that is, have a remainder of 0, as in Example 7), or it may not (as in Example 8). If a remainder is not 0, we continue dividing until the degree of the remainder is less than the degree of the divisor. Check this in each of Examples 7 and 8.

Do Exercises 11 and 12.

Divide and check.

12. $x + 3\overline{)x^2 + 7x + 10}$

Example 9 Divide and check: $(x^3 + 1) \div (x + 1)$.

$$\begin{array}{rl}
 & x^2 - x + 1 \\
x + 1 & \overline{)x^3 + 0x^2 + 0x + 1} \leftarrow \text{Fill in the missing terms (see Section 4.3).} \\
 & \underline{x^3 + x^2} \\
 & -x^2 + 0x \quad \text{This subtraction is } x^3 - (x^3 + x^2). \\
 & \underline{-x^2 - x} \\
 & x + 1 \quad \text{This subtraction is } -x^2 - (-x^2 - x). \\
 & \underline{x + 1} \\
 & 0
\end{array}$$

The answer is $x^2 - x + 1$. The check is left to the student.

13. $(x^3 - 1) \div (x - 1)$

Example 10 Divide and check: $(x^4 - 3x^2 + 1) \div (x - 4)$.

$$\begin{array}{rl}
 & x^3 + 4x^2 + 13x + 52 \\
x - 4 & \overline{)x^4 + 0x^3 - 3x^2 + 0x + 1} \leftarrow \text{Fill in the missing terms.} \\
 & \underline{x^4 - 4x^3} \\
 & 4x^3 - 3x^2 \quad x^4 - (x^4 - 4x^3) \\
 & \underline{4x^3 - 16x^2} \\
 & 13x^2 + 0x \leftarrow (4x^3 - 3x^2) - (4x^3 - 16x^2) \\
 & \underline{13x^2 - 52x} \\
 & 52x + 1 \\
 & \underline{52x - 208} \\
 & 209
\end{array}$$

The answer is $x^3 + 4x^2 + 13x + 52$, with $R = 209$, or

$$x^3 + 4x^2 + 13x + 52 + \frac{209}{x - 4}.$$

CHECK: $(x - 4)(x^3 + 4x^2 + 13x + 52) + 209$

$$= -4x^3 - 16x^2 - 52x - 208 + x^4 + 4x^3 + 13x^2 + 52x + 209$$
$$= x^4 - 3x^2 + 1$$

Do Exercise 13.

Answers on page A-32

Exercise Set 10.8

a Divide and check.

1. $\frac{24x^4}{8}$

2. $\frac{-2u^2}{u}$

3. $\frac{25x^3}{5x^2}$

4. $\frac{16x^7}{-2x^2}$

5. $\frac{-54x^{11}}{-3x^8}$

6. $\frac{-75a^{10}}{3a^2}$

7. $\frac{64a^5b^4}{16a^2b^3}$

8. $\frac{-34p^{10}q^{11}}{-17pq^9}$

9. $\frac{24x^4 - 4x^3 + x^2 - 16}{8}$

10. $\frac{12a^4 - 3a^2 + a - 6}{6}$

11. $\frac{u - 2u^2 - u^5}{u}$

12. $\frac{50x^5 - 7x^4 + x^2}{x}$

13. $(15t^3 + 24t^2 - 6t) \div (3t)$

14. $(25t^3 + 15t^2 - 30t) \div (5t)$

15. $(20x^6 - 20x^4 - 5x^2) \div (-5x^2)$

16. $(24x^6 + 32x^5 - 8x^2) \div (-8x^2)$

17. $(24x^5 - 40x^4 + 6x^3) \div (4x^3)$

18. $(18x^6 - 27x^5 - 3x^3) \div (9x^3)$

19. $\frac{18x^2 - 5x + 2}{2}$

20. $\frac{15x^2 - 30x + 6}{3}$

21. $\frac{12x^3 + 26x^2 + 8x}{2x}$

22. $\frac{2x^4 - 3x^3 + 5x^2}{x^2}$

23. $\frac{9r^2s^2 + 3r^2s - 6rs^2}{3rs}$

24. $\frac{4x^4y - 8x^6y^2 + 12x^8y^6}{4x^4y}$

b Divide.

25. $(x^2 + 4x + 4) \div (x + 2)$

26. $(x^2 - 6x + 9) \div (x - 3)$

27. $(x^2 - 10x - 25) \div (x - 5)$

28. $(x^2 + 8x - 16) \div (x + 4)$

29. $(x^2 + 4x - 14) \div (x + 6)$

30. $(x^2 + 5x - 9) \div (x - 2)$

31. $\dfrac{x^2 - 9}{x + 3}$

32. $\dfrac{x^2 - 25}{x - 5}$

33. $\dfrac{x^5 + 1}{x + 1}$

34. $\dfrac{x^5 - 1}{x - 1}$

35. $\dfrac{8x^3 - 22x^2 - 5x + 12}{4x + 3}$

36. $\dfrac{2x^3 - 9x^2 + 11x - 3}{2x - 3}$

37. $(x^6 - 13x^3 + 42) \div (x^3 - 7)$

38. $(x^6 + 5x^3 - 24) \div (x^3 - 3)$

39. $(x^4 - 16) \div (x - 2)$

40. $(x^4 - 81) \div (x - 3)$

41. $(t^3 - t^2 + t - 1) \div (t - 1)$

42. $(t^3 - t^2 + t - 1) \div (t + 1)$

Skill Maintenance

Subtract. [7.4a]

43. $17 - 45$

44. $-14 - 45$

45. $-2.3 - (-9.1)$

46. $-\dfrac{5}{8} - \dfrac{3}{4}$

Solve. [8.4a]

47. The perimeter of a rectangle is 640 ft. The length is 15 ft more than the width. Find the area of the rectangle.

48. The first angle of a triangle is 24° more than the second. The third angle is twice the first. Find the measures of the angles of the triangle.

Solve. [8.3c]

49. $-6(2 - x) + 10(5x - 7) = 10$

50. $-10(x - 4) = 5(2x + 5) - 7$

Factor. [7.7d]

51. $4x - 12 + 24y$

52. $256 - 2a - 4b$

Synthesis

53. ◆ Explain how the equation

$$(2x + 3)(3x - 1) = 6x^2 + 7x - 3$$

can be used to write two equations involving division.

54. ◆ Can the quotient of two binomials be a trinomial? Why or why not?

Divide.

55. $(x^4 + 9x^2 + 20) \div (x^2 + 4)$

56. $(y^4 + a^2) \div (y + a)$

57. $(5a^3 + 8a^2 - 23a - 1) \div (5a^2 - 7a - 2)$

58. $(15y^3 - 30y + 7 - 19y^2) \div (3y^2 - 2 - 5y)$

59. $(6x^5 - 13x^3 + 5x + 3 - 4x^2 + 3x^4) \div (3x^3 - 2x - 1)$

60. $(5x^7 - 3x^4 + 2x^2 - 10x + 2) \div (x^2 - x + 1)$

61. $(a^6 - b^6) \div (a - b)$

62. $(x^5 + y^5) \div (x + y)$

If the remainder is 0 when one polynomial is divided by another, the divisor is a *factor* of the dividend. Find the value(s) of c for which $x - 1$ is a factor of the polynomial.

63. $x^2 + 4x + c$

64. $2x^2 + 3cx - 8$

65. $c^2x^2 - 2cx + 1$

Summary and Review Exercises: Chapter 10

Important Properties and Formulas

FOIL: $(A + B)(C + D) = AC + AD + BC + BD$

Square of a Sum: $(A + B)(A + B) = (A + B)^2 = A^2 + 2AB + B^2$

Square of a Difference: $(A - B)(A - B) = (A - B)^2 = A^2 - 2AB + B^2$

Product of a Sum and a Difference: $(A + B)(A - B) = A^2 - B^2$

Definitions and Rules for Exponents

See p. 592.

The objectives to be tested in addition to the material in this chapter are [7.4a], [7.7d], [8.3b, c], and [8.4a].

Multiply and simplify. [10.1d, f]

1. $7^2 \cdot 7^{-4}$

2. $y^7 \cdot y^3 \cdot y$

3. $(3x)^5 \cdot (3x)^9$

4. $t^8 \cdot t^0$

Divide and simplify. [10.1e, f]

5. $\dfrac{4^5}{4^2}$

6. $\dfrac{a^5}{a^8}$

7. $\dfrac{(7x)^4}{(7x)^4}$

Simplify.

8. $(3t^4)^2$ [10.2a, b]

9. $(2x^3)^2(-3x)^2$ [10.1d], [10.2a, b]

10. $\left(\dfrac{2x}{y}\right)^{-3}$ [10.2b]

11. Express using a negative exponent: $\dfrac{1}{t^5}$. [10.1f]

12. Express using a positive exponent: y^{-4}. [10.1f]

13. Convert to scientific notation: 0.0000328. [10.2c]

14. Convert to decimal notation: 8.3×10^6. [10.2c]

Multiply or divide and write scientific notation for the result. [10.2d]

15. $(3.8 \times 10^4)(5.5 \times 10^{-1})$

16. $\dfrac{1.28 \times 10^{-8}}{2.5 \times 10^{-4}}$

17. *Diet-Drink Consumption.* It has been estimated that there will be 275 million people in the United States by the year 2000 and that on average, each of them will drink 15.3 gal of diet drinks that year (***Source***: U.S. Department of Agriculture). How many gallons of diet drinks will be consumed by the entire population in 2000? Express the answer in scientific notation. [10.2e]

18. Evaluate the polynomial $x^2 - 3x + 6$ for $x = -1$. [10.3a]

19. Identify the terms of the polynomial $-4y^5 + 7y^2 - 3y - 2$. [10.3b]

20. Identify the missing terms in $x^3 + x$. [10.3h]

21. Identify the degree of each term and the degree of the polynomial $4x^3 + 6x^2 - 5x + \frac{5}{3}$. [10.3g]

Classify the polynomial as a monomial, binomial, trinomial, or none of these. [10.3i]

22. $4x^3 - 1$

23. $4 - 9t^3 - 7t^4 + 10t^2$

24. $7y^2$

Collect like terms and then arrange in descending order. [10.3f]

25. $3x^2 - 2x + 3 - 5x^2 - 1 - x$

26. $-x + \frac{1}{2} + 14x^4 - 7x^2 - 1 - 4x^4$

Add. [10.4a]

27. $(3x^4 - x^3 + x - 4) + (x^5 + 7x^3 - 3x^2 - 5) + (-5x^4 + 6x^2 - x)$

28. $(3x^5 - 4x^4 + x^3 - 3) + (3x^4 - 5x^3 + 3x^2) + (-5x^5 - 5x^2) + (-5x^4 + 2x^3 + 5)$

Subtract. [10.4c]

29. $(5x^2 - 4x + 1) - (3x^2 + 1)$

30. $(3x^5 - 4x^4 + 3x^2 + 3) - (2x^5 - 4x^4 + 3x^3 + 4x^2 - 5)$

31. Find a polynomial for the perimeter and for the area. [10.4d], [10.5b]

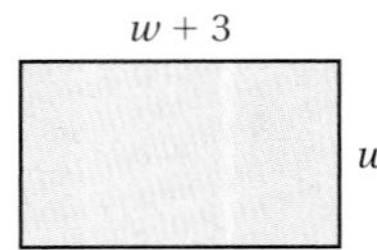

Multiply.

32. $\left(x + \frac{2}{3}\right)\left(x + \frac{1}{2}\right)$ [10.6a]

33. $(7x + 1)^2$ [10.6c]

34. $(4x^2 - 5x + 1)(3x - 2)$ [10.5d]

35. $(3x^2 + 4)(3x^2 - 4)$ [10.6b]

36. $5x^4(3x^3 - 8x^2 + 10x + 2)$ [10.5b]

37. $(x + 4)(x - 7)$ [10.6a]

38. $(3y^2 - 2y)^2$ [10.6c]

39. $(2t^2 + 3)(t^2 - 7)$ [10.6a]

40. Evaluate the polynomial

$$2 - 5xy + y^2 - 4xy^3 + x^6$$

for $x = -1$ and $y = 2$. [10.7a]

41. Identify the coefficient and the degree of each term of the polynomial

$$x^5y - 7xy + 9x^2 - 8.$$

Then find the degree of the polynomial. [10.7b]

Collect like terms. [10.7c]

42. $y + w - 2y + 8w - 5$

43. $m^6 - 2m^2n + m^2n^2 + n^2m - 6m^3 + m^2n^2 + 7n^2m$

44. Add: [10.7d]

$(5x^2 - 7xy + y^2) + (-6x^2 - 3xy - y^2) +$
$(x^2 + xy - 2y^2).$

45. Subtract: [10.7e]

$(6x^3y^2 - 4x^2y - 6x) - (-5x^3y^2 + 4x^2y + 6x^2 - 6).$

Multiply. [10.7f]

46. $(p - q)(p^2 + pq + q^2)$ **47.** $\left(3a^4 - \frac{1}{3}b^3\right)^2$

Divide.

48. $(10x^3 - x^2 + 6x) \div (2x)$ [10.8a]

49. $(6x^3 - 5x^2 - 13x + 13) \div (2x + 3)$ [10.8b]

50. The graph of the polynomial equation $y = 10x^3 - 10x$ is shown below. Use *only* the graph to estimate the value of the polynomial for $x = -1$, for $x = -0.5$, for $x = 0.5$, for $x = 1$, and for $x = 1.1$. [10.3a]

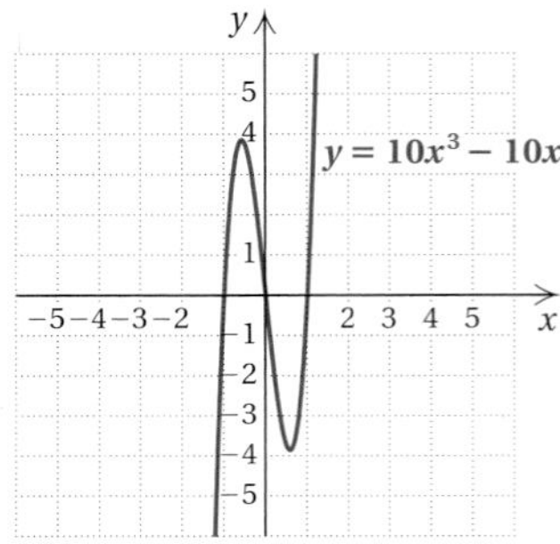

Skill Maintenance

51. Factor: $25t - 50 + 100m$. [7.7d]

52. Solve: $7x + 6 - 8x = 11 - 5x + 4$. [8.3b]

53. Solve: $3(x - 2) + 6 = 5(x + 3) + 9$. [8.3c]

54. Subtract: $-3.4 - 7.8$. [7.4a]

55. The perimeter of a rectangle is 540 m. The width is 19 m less than the length. Find the width and the length. [8.4a]

Synthesis

56. ◈ Explain why the expression 578.6×10^{-7} is not in scientific notation. [10.2c]

57. ◈ Write a short explanation of the difference between a monomial, a binomial, a trinomial, and a general polynomial. [10.3i]

Find a polynomial for the shaded area. [10.4d], [10.6b]

58.

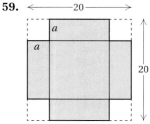

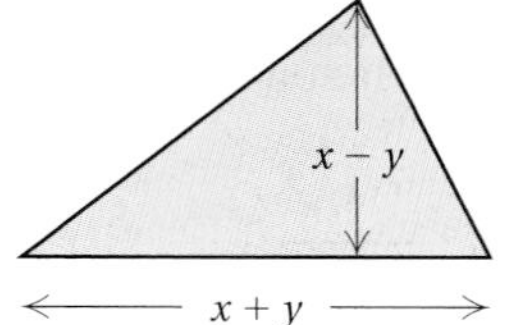

59.

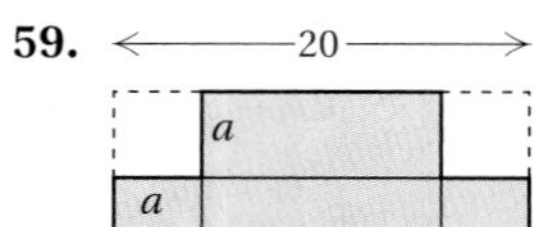

60. Collect like terms: [10.1d], [10.2a], [10.3e]

$-3x^5 \cdot 3x^3 - x^6(2x)^2 + (3x^4)^2 + (2x^2)^4 - 40x^2(x^3)^2.$

61. Solve: [10.6a]

$$(x - 7)(x + 10) = (x - 4)(x - 6).$$

62. The product of two polynomials is $x^5 - 1$. One of the polynomials is $x - 1$. Find the other. [10.8b]

Test: Chapter 10

Multiply and simplify.

1. $6^{-2} \cdot 6^{-3}$

2. $x^6 \cdot x^2 \cdot x$

3. $(4a)^3 \cdot (4a)^8$

Divide and simplify.

4. $\frac{3^5}{3^2}$

5. $\frac{x^3}{x^8}$

6. $\frac{(2x)^5}{(2x)^5}$

Simplify.

7. $(x^3)^2$

8. $(-3y^2)^3$

9. $(2a^3b)^4$

10. $\left(\frac{ab}{c}\right)^3$

11. $(3x^2)^3(-2x^5)^3$

12. $3(x^2)^3(-2x^5)^3$

13. $2x^2(-3x^2)^4$

14. $(2x)^2(-3x^2)^4$

15. Express using a positive exponent: 5^{-3}.

16. Express using a negative exponent: $\frac{1}{y^8}$.

17. Convert to scientific notation: 3,900,000,000.

18. Convert to decimal notation: 5×10^{-8}.

Multiply or divide and write scientific notation for the answer.

19. $\frac{5.6 \times 10^6}{3.2 \times 10^{-11}}$

20. $(2.4 \times 10^5)(5.4 \times 10^{16})$

21. A CD-ROM can contain about 600 million pieces of information (bytes). How many sound files, each containing 40,000 bytes, can a CD-ROM hold? Express the answer in scientific notation.

22. Evaluate the polynomial $x^5 + 5x - 1$ for $x = -2$.

23. Identify the coefficient of each term of the polynomial $\frac{1}{3}x^5 - x + 7$.

24. Identify the degree of each term and the degree of the polynomial $2x^3 - 4 + 5x + 3x^6$.

25. Classify the polynomial $7 - x$ as a monomial, binomial, trinomial, or none of these.

Collect like terms.

26. $4a^2 - 6 + a^2$

27. $y^2 - 3y - y + \frac{3}{4}y^2$

Answers

1. ______
2. ______
3. ______
4. ______
5. ______
6. ______
7. ______
8. ______
9. ______
10. ______
11. ______
12. ______
13. ______
14. ______
15. ______
16. ______
17. ______
18. ______
19. ______
20. ______
21. ______
22. ______
23. ______
24. ______
25. ______
26. ______
27. ______

Answers

28. ______
29. ______
30. ______
31. ______
32. ______
33. ______
34. ______
35. ______
36. ______
37. ______
38. ______
39. ______
40. ______
41. ______
42. ______
43. ______
44. ______
45. ______
46. ______
47. ______
48. ______
49. ______
50. ______
51. ______
52. ______
53. ______

28. Collect like terms and then arrange in descending order:
$3 - x^2 + 2x^3 + 5x^2 - 6x - 2x + x^5$.

Add.

29. $(3x^5 + 5x^3 - 5x^2 - 3) +$
$(x^5 + x^4 - 3x^3 - 3x^2 + 2x - 4)$

30. $\left(x^4 + \frac{2}{3}x + 5\right) + \left(4x^4 + 5x^2 + \frac{1}{3}x\right)$

Subtract.

31. $(2x^4 + x^3 - 8x^2 - 6x - 3) -$
$(6x^4 - 8x^2 + 2x)$

32. $(x^3 - 0.4x^2 - 12) -$
$(x^5 + 0.3x^3 + 0.4x^2 + 9)$

Multiply.

33. $-3x^2(4x^2 - 3x - 5)$

34. $\left(x - \frac{1}{3}\right)^2$

35. $(3x + 10)(3x - 10)$

36. $(3b + 5)(b - 3)$

37. $(x^6 - 4)(x^8 + 4)$

38. $(8 - y)(6 + 5y)$

39. $(2x + 1)(3x^2 - 5x - 3)$

40. $(5t + 2)^2$

41. Collect like terms: $x^3y - y^3 + xy^3 + 8 - 6x^3y - x^2y^2 + 11$.

42. Subtract: $(8a^2b^2 - ab + b^3) - (-6ab^2 - 7ab - ab^3 + 5b^3)$.

43. Multiply: $(3x^5 - 4y^5)(3x^5 + 4y^5)$.

Divide.

44. $(12x^4 + 9x^3 - 15x^2) \div (3x^2)$

45. $(6x^3 - 8x^2 - 14x + 13) \div (3x + 2)$

46. The graph of the polynomial equation $y = x^3 - 5x - 1$ is shown at right. Use *only* the graph to estimate the value of the polynomial for $x = -1$, for $x = -0.5$, for $x = 0.5$, for $x = 1$, and for $x = 1.1$.

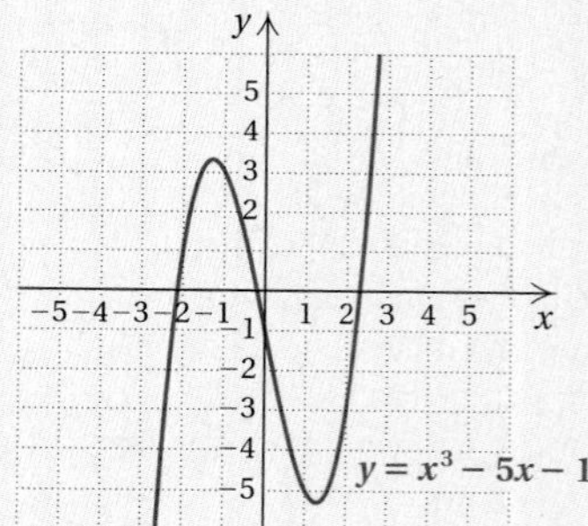

Skill Maintenance

47. Solve: $7x - 4x - 2 = 37$.

48. Solve: $4(x + 2) - 21 = 3(x - 6) + 2$.

49. Factor: $64t - 32m + 16$.

50. Subtract: $\frac{2}{5} - \left(-\frac{3}{4}\right)$.

51. The first angle of a triangle is four times as large as the second. The measure of the third angle is 30° greater than that of the second. How large are the angles?

Synthesis

52. The height of a box is 1 less than its length, and the length is 2 more than its width. Find the volume in terms of the length.

53. Solve: $(x - 5)(x + 5) = (x + 6)^2$.

11

Polynomials: Factoring

An Application

A ladder of length 13 ft is placed against a building in such a way that the distance from the top of the ladder to the ground is 7 ft more than the distance from the bottom of the ladder to the building. Find both distances.

This problem appears as Example 6 in Section 11.8.

The Mathematics

If we visualize this as a triangle, we can let x = the length of the side (leg) across the bottom. Then $x + 7$ = the length of the other side (leg). The hypotenuse has length 13 ft. Using the Pythagorean theorem, we translate the problem to

$$x^2 + (x + 7)^2 = 13^2$$
$$x^2 + x^2 + 14x + 49 = 169$$
$$\underbrace{2x^2 + 14x - 120 = 0.}$$

This is a second-degree, or quadratic, equation.

World Wide Web For more information, visit us at www.mathmax.com

Introduction

Factoring is the reverse of multiplying. To *factor* a polynomial, or other algebraic expression, is to find an equivalent expression that is a product. In this chapter, we study factoring polynomials. To learn to factor quickly, we use the quick methods for multiplication that we learned in Chapter 10.

At the end of this chapter, we find the payoff for learning to factor. We can solve certain new equations containing second-degree polynomials. This in turn allows us to solve applied problems, like the one below, that we could not have solved before.

Pretest: Chapter 11

1. Find three factorizations of $-20x^6$.

Factor.

2. $2x^2 + 4x + 2$

3. $x^2 + 6x + 8$

4. $8a^5 + 4a^3 - 20a$

5. $-6 + 5x^2 - 13x$

6. $81 - z^4$

7. $y^6 - 4y^3 + 4$

8. $3x^3 + 2x^2 + 12x + 8$

9. $p^2 - p - 30$

10. $x^4y^2 - 64$

11. $2p^2 + 7pq - 4q^2$

Solve.

12. $x^2 - 5x = 0$

13. $(x - 4)(5x - 3) = 0$

14. $3x^2 + 10x - 8 = 0$

Solve.

15. Six less than the square of a number is five times the number. Find all such numbers.

16. The height of a triangle is 3 cm longer than the base. The area of the triangle is 44 cm^2. Find the base and the height.

Objectives for Retesting

The objectives to be tested in addition to the material in this chapter are as follows.

[7.6c] Divide real numbers.
[8.7e] Solve inequalities using the addition and multiplication principles together.
[9.3a] Find the intercepts of a linear equation, and graph using intercepts.
[10.6d] Find special products when polynomial products are mixed together.

11.1 Introduction to Factoring

To solve certain types of algebraic equations involving polynomials of second degree, we must learn to factor polynomials.

Consider the product $15 = 3 \cdot 5$. We say that 3 and 5 are **factors** of 15 and that $3 \cdot 5$ is a **factorization** of 15. Since $15 = 15 \cdot 1$, we also know that 15 and 1 are factors of 15 and that $15 \cdot 1$ is a factorization of 15.

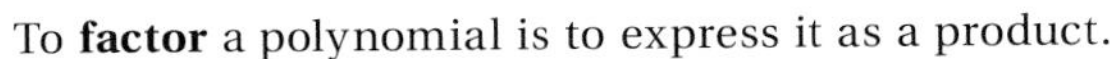

> To **factor** a polynomial is to express it as a product.
>
> A **factor** of a polynomial P is a polynomial that can be used to express P as a product.
>
> A **factorization** of a polynomial is an expression that names that polynomial as a product.

Objectives

- a Factor monomials.
- b Factor polynomials when the terms have a common factor, factoring out the largest common factor.
- c Factor certain expressions with four terms using factoring by grouping.

For Extra Help

TAPE 21 | InterAct math MAC WIN | CD-ROM

a Factoring Monomials

To factor a monomial, we find two monomials whose product is equivalent to the original monomial. Compare.

	Multiplying	*Factoring*
a)	$(4x)(5x) = 20x^2$	$20x^2 = (4x)(5x)$
b)	$(2x)(10x) = 20x^2$	$20x^2 = (2x)(10x)$
c)	$(-4x)(-5x) = 20x^2$	$20x^2 = (-4x)(-5x)$
d)	$(x)(20x) = 20x^2$	$20x^2 = (x)(20x)$

You can see that the monomial $20x^2$ has many factorizations. There are still other ways to factor $20x^2$.

Do Exercises 1 and 2.

Example 1 Find three factorizations of $15x^3$.

a) $15x^3 = (3 \cdot 5)(x \cdot x^2)$
$= (3x)(5x^2)$

b) $15x^3 = (3 \cdot 5)(x^2 \cdot x)$
$= (3x^2)(5x)$

c) $15x^3 = (-15)(-1)x^3$
$= (-15)(-x^3)$

Do Exercises 3–5.

b Factoring When Terms Have a Common Factor

To factor polynomials quickly, we consider the special-product rules studied in Chapter 10, but we first factor out the largest common factor.

To multiply a monomial and a polynomial with more than one term, we multiply each term of the polynomial by the monomial using the distributive laws,

$$a(b + c) = ab + ac \quad \text{and} \quad a(b - c) = ab - ac.$$

1. a) Multiply: $(3x)(4x)$.

b) Factor: $12x^2$.

2. a) Multiply: $(2x)(8x^2)$.

b) Factor: $16x^3$.

Find three factorizations of the monomial.

3. $8x^4$

4. $21x^2$

5. $6x^5$

Answers on page A-33

6. a) Multiply: $3(x + 2)$.

b) Factor: $3x + 6$.

7. a) Multiply: $2x(x^2 + 5x + 4)$.

b) Factor: $2x^3 + 10x^2 + 8x$.

Answers on page A-33

To factor, we do the reverse. We express a polynomial as a product using the distributive laws in reverse:

$$ab + ac = a(b + c) \quad \textbf{and} \quad ab - ac = a(b - c).$$

Compare.

Multiply	*Factor*
$3x(x^2 + 2x - 4)$	$3x^3 + 6x^2 - 12x$
$= 3x \cdot x^2 + 3x \cdot 2x - 3x \cdot 4$	$= 3x \cdot x^2 + 3x \cdot 2x - 3x \cdot 4$
$= 3x^3 + 6x^2 - 12x$	$= 3x(x^2 + 2x - 4)$

Do Exercises 6 and 7.

CAUTION! Consider the following:

$$3x^3 + 6x^2 - 12x = 3 \cdot x \cdot x \cdot x + 2 \cdot 3 \cdot x \cdot x - 2 \cdot 2 \cdot 3 \cdot x.$$

The terms of the polynomial, $3x^3$, $6x^2$, and $-12x$, have been factored but the polynomial itself has not been factored. This is not what we mean by a factorization of the polynomial. The *factorization* is

$$3x(x^2 + 2x - 4).$$

The expressions $3x$ and $x^2 + 2x - 4$ are *factors* of $3x^3 + 6x^2 - 12x$.

To factor, we first try to find a factor common to all terms. There may not always be one other than 1. When there is, we generally use the factor with the largest possible coefficient and the largest possible exponent.

Example 2 Factor: $7x^2 + 14$.

We have

$$7x^2 + 14 = 7 \cdot x^2 + 7 \cdot 2 \quad \text{Factoring each term}$$
$$= 7(x^2 + 2). \quad \text{Factoring out the common factor 7}$$

CHECK: We multiply to check:

$$7(x^2 + 2) = 7 \cdot x^2 + 7 \cdot 2 = 7x^2 + 14.$$

Example 3 Factor: $16x^3 + 20x^2$.

$$16x^3 + 20x^2 = (4x^2)(4x) + (4x^2)(5) \quad \text{Factoring each term}$$
$$= 4x^2(4x + 5) \quad \text{Factoring out the common factor } 4x^2$$

Suppose in Example 3 that you had not recognized the largest common factor and removed only part of it, as follows:

$$16x^3 + 20x^2 = (2x^2)(8x) + (2x^2)(10)$$
$$= 2x^2(8x + 10).$$

Note that $8x + 10$ still has a common factor of 2. You need not begin again. Just continue factoring out common factors, as follows, until finished:

$$= 2x^2[2(4x + 5)]$$
$$= 4x^2(4x + 5).$$

Example 4 Factor: $15x^5 - 12x^4 + 27x^3 - 3x^2$.

$$15x^5 - 12x^4 + 27x^3 - 3x^2 = (3x^2)(5x^3) - (3x^2)(4x^2) + (3x^2)(9x) - (3x^2)(1)$$
$$= 3x^2(5x^3 - 4x^2 + 9x - 1) \quad \text{Factoring out } 3x^2$$

CAUTION! Don't forget the term -1.

CHECK: We multiply to check:

$$3x^2(5x^3 - 4x^2 + 9x - 1)$$
$$= (3x^2)(5x^3) - (3x^2)(4x^2) + (3x^2)(9x) - (3x^2)(1)$$
$$= 15x^5 - 12x^4 + 27x^3 - 3x^2.$$

As you become more familiar with factoring, you will be able to spot the largest common factor without factoring each term. Then you can write just the answer.

Examples Factor.

5. $8m^3 - 16m = 8m(m^2 - 2)$

6. $14p^2y^3 - 8py^2 + 2py = 2py(7py^2 - 4y + 1)$

7. $\frac{4}{5}x^2 + \frac{1}{5}x + \frac{2}{5} = \frac{1}{5}(4x^2 + x + 2)$

8. $2.4x^2 + 1.2x - 3.6 = 1.2(2x^2 + x - 3)$

Do Exercises 8–13.

There are two important points to keep in mind as we study this chapter.

- Before doing any other kind of factoring, first try to factor out the largest common factor.
- Always check the result of factoring by multiplying.

c Factoring by Grouping: Four Terms

Certain polynomials with four terms can be factored using a method called *factoring by grouping*.

Example 9 Factor: $x^2(x + 1) + 2(x + 1)$.

The binomial $x + 1$ is common to both terms:

$$x^2(x + 1) + 2(x + 1) = (x^2 + 2)(x + 1).$$

The factorization is $(x^2 + 2)(x + 1)$.

Do Exercises 14 and 15.

Factor. Check by multiplying.

8. $x^2 + 3x$

9. $3y^6 - 5y^3 + 2y^2$

10. $9x^4 - 15x^3 + 3x^2$

11. $\frac{3}{4}t^3 + \frac{5}{4}t^2 + \frac{7}{4}t + \frac{1}{4}$

12. $35x^7 - 49x^6 + 14x^5 - 63x^3$

13. $8.4x^2 - 5.6x + 2.8$

Factor.

14. $x^2(x + 7) + 3(x + 7)$

15. $x^2(a + b) + 2(a + b)$

Answers on page A-33

Factor by grouping.

16. $x^3 + 7x^2 + 3x + 21$

17. $8t^3 + 2t^2 + 12t + 3$

18. $3m^5 - 15m^3 + 2m^2 - 10$

19. $3x^3 - 6x^2 - x + 2$

20. $4x^3 - 6x^2 - 6x + 9$

21. $y^4 - 2y^3 - 2y - 10$

Answers on page A-33

Consider the four-term polynomial

$$x^3 + x^2 + 2x + 2.$$

There is no factor other than 1 that is common to all the terms. We can, however, factor $x^3 + x^2$ and $2x + 2$ separately:

$x^3 + x^2 = x^2(x + 1);$ **Factoring $x^3 + x^2$**

$2x + 2 = 2(x + 1).$ **Factoring $2x + 2$**

We have grouped certain terms and factored each polynomial separately:

$$\begin{aligned} x^3 + x^2 + 2x + 2 &= (x^3 + x^2) + (2x + 2) \\ &= x^2(x + 1) + 2(x + 1) \\ &= (x^2 + 2)(x + 1), \end{aligned}$$

as in Example 9. This method is called **factoring by grouping.** We began with a polynomial with four terms. After grouping and removing common factors, we obtained a polynomial with two parts, each having a common factor $x + 1$. Not all polynomials with four terms can be factored by this procedure, but it does give us a method to try.

Examples Factor by grouping.

10. $6x^3 - 9x^2 + 4x - 6$

$= (6x^3 - 9x^2) + (4x - 6)$

$= 3x^2(2x - 3) + 2(2x - 3)$ **Factoring each binomial**

$= (3x^2 + 2)(2x - 3)$ **Factoring out the common factor $2x - 3$**

We think through this process as follows:

$$6x^3 - 9x^2 + 4x - 6 = 3x^2(2x - 3) \;\square\; (2x - 3)$$

(1) Factor the first two terms.

(2) This factor, $2x - 3$, gives us a hint to the factorization on the right.

(3) Now we ask ourselves, "What needs to be here to enable us to get $4x - 6$ when we multiply?"

11. $x^3 + x^2 + x + 1 = (x^3 + x^2) + (x + 1)$

$= x^2(x + 1) + 1(x + 1)$ **Factoring each binomial**

$= (x^2 + 1)(x + 1)$ **Factoring out the common factor $x + 1$**

12. $2x^3 - 6x^2 - x + 3$

$= (2x^3 - 6x^2) + (-x + 3)$

$= 2x^2(x - 3) - 1(x - 3)$ ***Check:*** **$-1(x - 3) = -x + 3$.**

$= (2x^2 - 1)(x - 3)$ **Factoring out the common factor $x - 3$**

13. $12x^5 + 20x^2 - 21x^3 - 35 = 4x^2(3x^3 + 5) - 7(3x^3 + 5)$

$= (4x^2 - 7)(3x^3 + 5)$

14. $x^3 + x^2 + 2x - 2 = x^2(x + 1) + 2(x - 1)$

This polynomial is not factorable using factoring by grouping. It may be factorable, but not by methods that we will consider in this text.

Do Exercises 16–21.

Exercise Set 11.1

a Find three factorizations for the monomial.

1. $8x^3$ **2.** $6x^4$ **3.** $-10a^6$ **4.** $-8y^5$ **5.** $24x^4$ **6.** $15x^5$

b Factor. Check by multiplying.

7. $x^2 - 6x$ **8.** $x^2 + 5x$ **9.** $2x^2 + 6x$

10. $8y^2 - 8y$ **11.** $x^3 + 6x^2$ **12.** $3x^4 - x^2$

13. $8x^4 - 24x^2$ **14.** $5x^5 + 10x^3$ **15.** $2x^2 + 2x - 8$

16. $8x^2 - 4x - 20$ **17.** $17x^5y^3 + 34x^3y^2 + 51xy$ **18.** $16p^6q^4 + 32p^5q^3 - 48pq^2$

19. $6x^4 - 10x^3 + 3x^2$ **20.** $5x^5 + 10x^2 - 8x$ **21.** $x^5y^5 + x^4y^3 + x^3y^3 - x^2y^2$

22. $x^9y^6 - x^7y^5 + x^4y^4 + x^3y^3$ **23.** $2x^7 - 2x^6 - 64x^5 + 4x^3$ **24.** $8y^3 - 20y^2 + 12y - 16$

25. $1.6x^4 - 2.4x^3 + 3.2x^2 + 6.4x$ **26.** $2.5x^6 - 0.5x^4 + 5x^3 + 10x^2$

27. $\frac{5}{3}x^6 + \frac{4}{3}x^5 + \frac{1}{3}x^4 + \frac{1}{3}x^3$ **28.** $\frac{5}{9}x^7 + \frac{2}{9}x^5 - \frac{4}{9}x^3 - \frac{1}{9}x$

c Factor.

29. $x^2(x + 3) + 2(x + 3)$ **30.** $3z^2(2z + 1) + (2z + 1)$

31. $5a^3(2a - 7) - (2a - 7)$ **32.** $m^4(8 - 3m) - 7(8 - 3m)$

Factor by grouping.

33. $x^3 + 3x^2 + 2x + 6$

34. $6z^3 + 3z^2 + 2z + 1$

35. $2x^3 + 6x^2 + x + 3$

36. $3x^3 + 2x^2 + 3x + 2$

37. $8x^3 - 12x^2 + 6x - 9$

38. $10x^3 - 25x^2 + 4x - 10$

39. $12x^3 - 16x^2 + 3x - 4$

40. $18x^3 - 21x^2 + 30x - 35$

41. $5x^3 - 5x^2 - x + 1$

42. $7x^3 - 14x^2 - x + 2$

43. $x^3 + 8x^2 - 3x - 24$

44. $2x^3 + 12x^2 - 5x - 30$

45. $2x^3 - 8x^2 - 9x + 36$

46. $20g^3 - 4g^2 - 25g + 5$

Skill Maintenance

Solve.

47. $-2x < 48$ [8.7d]

48. $4x - 8x + 16 \geq 6(x - 2)$ [8.7e]

49. Divide: $\frac{-108}{-4}$. [7.6a]

50. Solve $A = \frac{p + q}{2}$ for p. [8.6a]

Multiply. [10.6d]

51. $(y + 5)(y + 7)$

52. $(y + 7)^2$

53. $(y + 7)(y - 7)$

54. $(y - 7)^2$

Find the intercepts of the equation. Then graph the equation. [9.3a]

55. $x + y = 4$

56. $x - y = 3$

57. $5x - 3y = 15$

58. $y - 3x = 6$

Synthesis

59. ◆ Josh says that there is no need to print answers for Exercises 1–46 at the back of the book. Is he correct in saying this? Why or why not?

60. ◆ Explain how one could construct a polynomial with four terms that can be factored by grouping.

Factor.

61. $4x^5 + 6x^3 + 6x^2 + 9$

62. $x^6 + x^4 + x^2 + 1$

63. $x^{12} + x^7 + x^5 + 1$

64. $x^3 - x^2 - 2x + 5$

65. $p^3 + p^2 - 3p + 10$

11.2 Factoring Trinomials of the Type $x^2 + bx + c$

a We now begin a study of the factoring of trinomials. We first factor trinomials like

$x^2 + 5x + 6$ and $x^2 + 3x - 10$

by a refined *trial-and-error* process. In this section, we restrict our attention to trinomials of the type $ax^2 + bx + c$, where $a = 1$. The coefficient a is often called the **leading coefficient.**

Constant Term Positive

Recall the FOIL method of multiplying two binomials:

$$\begin{aligned}(x+2)(x+5) &= \overset{F}{x^2} + \overset{O}{5x} + \overset{I}{2x} + \overset{L}{10}\\ &= x^2 + 7x + 10.\end{aligned}$$

The product above is a trinomial. The term of highest degree, x^2, called the leading term, has a coefficient of 1. The constant term, 10, is positive. To factor $x^2 + 7x + 10$, we think of FOIL in reverse. We multiplied x times x to get the first term of the trinomial, so we know that the first term of each binomial factor is x. Next, we look for numbers p and q such that

$x^2 + 7x + 10 = (x + p)(x + q)$.

To get the middle term and the last term of the trinomial, we look for two numbers p and q whose product is 10 and whose sum is 7. Those numbers are 2 and 5. Thus the factorization is

$(x + 2)(x + 5)$.

Example 1 Factor: $x^2 + 5x + 6$.

Think of FOIL in reverse. The first term of each factor is x: $(x + \square)(x + \square)$. Next, we look for two numbers whose product is 6 and whose sum is 5. All the pairs of factors of 6 are shown in the table on the left below. Since both the product, 6, and the sum, 5, of the pair of numbers must be positive, we need consider only the positive factors, listed in the table on the right.

Pairs of Factors	Sums of Factors
1, 6	7
−1, −6	−7
2, 3	5
−2, −3	−5

Pairs of Factors	Sums of Factors
1, 6	7
2, 3	5

↑ The numbers we need are 2 and 3.

The factorization is $(x + 2)(x + 3)$. We can check by multiplying to see whether we get the original trinomial.

CHECK: $(x + 2)(x + 3) = x^2 + 3x + 2x + 6 = x^2 + 5x + 6$.

Do Exercises 1 and 2.

Objective

a Factor trinomials of the type $x^2 + bx + c$ by examining the constant term c.

For Extra Help

TAPE 21 MAC WIN CD-ROM

1. Consider the trinomial $x^2 + 7x + 12$.

a) Complete the following table.

Pairs of Factors	Sums of Factors
1, 12	13
−1, −12	
2, 6	
−2, −6	
3, 4	
−3, −4	

b) Explain why you need to consider only positive factors, as in the following table.

Pairs of Factors	Sums of Factors
1, 12	
2, 6	
3, 4	

c) Factor: $x^2 + 7x + 12$.

2. Factor: $x^2 + 13x + 36$.

Answers on page A-33

3. Explain why you would not consider the pairs of factors listed below in factoring $y^2 - 8y + 12$.

Pairs of Factors	Sums of Factors
1, 12	
2, 6	
3, 4	

Factor.

4. $x^2 - 8x + 15$

$(x - 5)(x - 3)$

5. $t^2 - 9t + 20$

$(t - 5)(t - 4)$

Answers on page A-33

Consider this multiplication:

$$(x - 2)(x - 5) = \overset{F}{x^2} \overset{O}{- 5x} \overset{I}{- 2x} \overset{L}{+ 10}$$
$$= x^2 - 7x + 10.$$

When the constant term of a trinomial is positive, we look for two numbers with the same sign (both negative or both positive). The sign is that of the middle term:

$$x^2 - 7x + 10 = (x - 2)(x - 5), \quad \text{or} \quad x^2 + 7x + 10 = (x + 2)(x + 5).$$

Example 2 Factor: $y^2 - 8y + 12$.

Since the constant term, 12, is positive and the coefficient of the middle term, -8, is negative, we look for a factorization of 12 in which both factors are negative. Their sum must be -8.

Pairs of Factors	Sums of Factors
−1, −12	−13
−2, −6	−8 ← The numbers we need are −2 and −6.
−3, −4	−7

The factorization is $(y - 2)(y - 6)$.

Do Exercises 3–5.

Constant Term Negative

Sometimes when we use FOIL, the product has a negative constant term. Consider these multiplications:

a) $(x - 5)(x + 2) = \overset{F}{x^2} \overset{O}{+ 2x} \overset{I}{- 5x} \overset{L}{- 10}$
$$= x^2 - 3x - 10;$$

b) $(x + 5)(x - 2) = \overset{F}{x^2} \overset{O}{- 2x} \overset{I}{+ 5x} \overset{L}{- 10}$
$$= x^2 + 3x - 10.$$

Reversing the signs of the factors changes the sign of the middle term.

When the constant term of a trinomial is negative, we look for two factors whose product is negative. One of them must be positive and the other negative. Their sum must be the coefficient of the middle term:

$$x^2 - 3x - 10 = (x - 5)(x + 2), \quad \text{or} \quad x^2 + 3x - 10 = (x + 5)(x - 2).$$

Example 3 Factor: $x^3 - 8x^2 - 20x$.

Always look first for a common factor. This time there is one, x. We first factor it out: $x^3 - 8x^2 - 20x = x(x^2 - 8x - 20)$. Now consider the expression $x^2 - 8x - 20$. Since the constant term, -20, is negative, we look for a factorization of -20 in which one factor is positive and one factor is negative. The sum must be -8, so the negative factor must have the larger absolute value. Thus we consider only pairs of factors in which the negative factor has the larger absolute value.

Pairs of Factors	Sums of Factors
1, −20	−19
2, −10	−8 ← The numbers we need are 2 and −10.
4, −5	−1

The factorization of $x^2 - 8x - 20$ is $(x + 2)(x - 10)$. But we must also remember to include the common factor. The factorization of the original polynomial is

$$x(x + 2)(x - 10).$$

Do Exercise 6.

6. Explain why you would not consider the pairs of factors listed below in factoring $x^2 - 8x - 20$.

Pairs of Factors	Sums of Factors
−1, 20	
−2, 10	
−4, 5	

Example 4 Factor: $t^2 - 24 + 5t$.

It helps to first write the trinomial in descending order: $t^2 + 5t - 24$. Since the constant term, -24, is negative, we look for a factorization of -24 in which one factor is positive and one factor is negative. Their sum must be 5, so the positive factor must have the larger absolute value. Thus we consider only pairs of factors in which the positive term has the larger absolute value.

Pairs of Factors	Sums of Factors
−1, 24	23
−2, 12	10
−3, 8	5 ← The numbers we need are −3 and 8.
−4, 6	2

The factorization is $(t - 3)(t + 8)$.

Do Exercise 7.

7. Explain why you would not consider the pairs of factors listed below in factoring $t^2 + 5t - 24$.

Pairs of Factors	Sums of Factors
1, −24	
2, −12	
3, −8	
4, −6	

Example 5 Factor: $x^4 - x^2 - 110$.

Consider this trinomial as $(x^2)^2 - x^2 - 110$. We look for numbers p and q such that

$$x^4 - x^2 - 110 = (x^2 + p)(x^2 + q).$$

Since the constant term, -110, is negative, we look for a factorization of -110 in which one factor is positive and one factor is negative. Their sum must be -1. The middle-term coefficient, -1, is small compared to -110. This tells us that the desired factors are close to each other in absolute value. The numbers we want are 10 and -11. The factorization is

$$(x^2 + 10)(x^2 - 11).$$

Answers on page A-33

Factor.

8. $x^3 + 4x^2 - 12x$

9. $y^2 - 12 - 4y$

10. $t^4 + 5t^2 - 14$

11. $p^2 - pq - 3pq^2$

12. $x^2 + 2x + 7$

13. Factor: $x^2 + 8x + 16$.

Answers on page A-33

Example 6 Factor: $a^2 + 4ab - 21b^2$.

We consider the trinomial in the equivalent form

$$a^2 + 4ba - 21b^2.$$

We think of $4b$ as a "coefficient" of a. Then we look for factors of $-21b^2$ whose sum is $4b$. Those factors are $-3b$ and $7b$. The factorization is

$$(a - 3b)(a + 7b).$$

There are polynomials that are not factorable.

Example 7 Factor: $x^2 - x + 5$.

Since 5 has very few factors, we can easily check all possibilities.

Pairs of Factors	Sums of Factors
5, 1	6
−5, −1	−6

There are no factors whose sum is −1. Thus the polynomial is *not* factorable into binomials.

Do Exercises 8–12.

We can factor a trinomial that is a perfect square using this method.

Example 8 Factor: $x^2 - 10x + 25$.

Since the constant term, 25, is positive and the coefficient of the middle term, −10, is negative, we look for a factorization of 25 in which both factors are negative. Their sum must be −10.

Pairs of Factors	Sums of Factors
−25, −1	−26
−5, −5	−10 ← The numbers we need are −5 and −5.

The factorization is $(x - 5)(x - 5)$, or $(x - 5)^2$.

Do Exercise 13.

The following is a summary of our procedure for factoring $x^2 + bx + c$.

To factor $x^2 + bx + c$:

1. First arrange in descending order.
2. Use a trial-and-error process that looks for factors of c whose sum is b.
3. If c is positive, the signs of the factors are the same as the sign of b.
4. If c is negative, one factor is positive and the other is negative. If the sum of two factors is the opposite of b, changing the sign of each factor will give the desired factors whose sum is b.
5. Check by multiplying.

Exercise Set 11.2

a Factor. Remember that you can check by multiplying.

1. $x^2 + 8x + 15$

2. $x^2 + 5x + 6$

3. $x^2 + 7x + 12$

4. $x^2 + 9x + 8$

5. $x^2 - 6x + 9$

6. $y^2 - 11y + 28$

7. $x^2 + 9x + 14$

8. $a^2 + 11a + 30$

9. $b^2 + 5b + 4$

10. $z^2 - 8z + 7$

11. $x^2 + \frac{2}{3}x + \frac{1}{9}$

12. $x^2 - \frac{2}{5}x + \frac{1}{25}$

13. $d^2 - 7d + 10$

14. $t^2 - 12t + 35$

15. $y^2 - 11y + 10$

16. $x^2 - 4x - 21$

17. $x^2 + x - 42$

18. $x^2 + 2x - 15$

19. $x^2 - 7x - 18$

20. $y^2 - 3y - 28$

21. $x^3 - 6x^2 - 16x$

22. $x^3 - x^2 - 42x$

23. $y^3 - 4y^2 - 45y$

24. $x^3 - 7x^2 - 60x$

25. $-2x - 99 + x^2$

26. $x^2 - 72 + 6x$

27. $c^4 + c^2 - 56$

28. $b^4 + 5b^2 - 24$

29. $a^4 + 2a^2 - 35$

30. $x^4 - x^2 - 6$

31. $x^2 + x + 1$

32. $x^2 + 5x + 3$

33. $7 - 2p + p^2$

34. $11 - 3w + w^2$

35. $x^2 + 20x + 100$

36. $a^2 + 19a + 88$

37. $x^4 - 21x^3 - 100x^2$

38. $x^4 - 20x^3 + 96x^2$

39. $x^2 - 21x - 72$

40. $4x^2 + 40x + 100$

41. $x^2 - 25x + 144$

42. $y^2 - 21y + 108$

43. $a^2 + a - 132$

44. $a^2 + 9a - 90$

45. $120 - 23x + x^2$

46. $96 + 22d + d^2$

47. $108 - 3x - x^2$

48. $112 + 9y - y^2$

49. $y^2 - 0.2y - 0.08$

50. $t^2 - 0.3t - 0.10$

51. $p^2 + 3pq - 10q^2$

52. $a^2 + 2ab - 3b^2$

53. $m^2 + 5mn + 4n^2$

54. $x^2 + 11xy + 24y^2$

55. $s^2 - 2st - 15t^2$

56. $p^2 + 5pq - 24q^2$

Skill Maintenance

Multiply. [10.6d]

57. $8x(2x^2 - 6x + 1)$

58. $(7w + 6)(4w - 11)$

59. $(7w + 6)^2$

60. $(4w - 11)^2$

61. $(4w - 11)(4w + 11)$

62. Simplify: $(3x^4)^3$. [10.2b]

Solve. [8.3a]

63. $3x - 8 = 0$

64. $2x + 7 = 0$

Solve.

65. *Arrests for Counterfeiting.* In a recent year, 29,200 people were arrested for counterfeiting. This number was down 1.2% from the preceding year. How many people were arrested the preceding year? [8.5a]

66. The first angle of a triangle is four times as large as the second. The measure of the third angle is 30° greater than that of the second. Find the angle measures. [8.4a]

Synthesis

67. ◆ Without doing the multiplication $(x - 17)(x - 18)$, explain why it cannot possibly be a factorization of $x^2 + 35x + 306$.

68. ◆ When searching for a factorization of $x^2 + bx + c$, why do we list pairs of numbers with the specified product c instead of pairs of numbers with the specified sum b?

69. Find all integers m for which $y^2 + my + 50$ can be factored.

70. Find all integers b for which $a^2 + ba - 50$ can be factored.

Factor completely.

71. $x^2 - \frac{1}{2}x - \frac{3}{16}$

72. $x^2 - \frac{1}{4}x - \frac{1}{8}$

73. $x^2 + \frac{30}{7}x - \frac{25}{7}$

74. $\frac{1}{3}x^3 + \frac{1}{3}x^2 - 2x$

75. $b^{2n} + 7b^n + 10$

76. $a^{2m} - 11a^m + 28$

Find a polynomial in factored form for the shaded area. (Leave answers in terms of π.)

77.

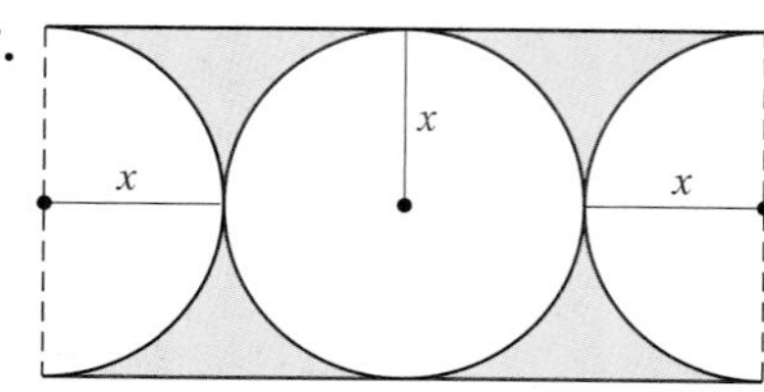

78.

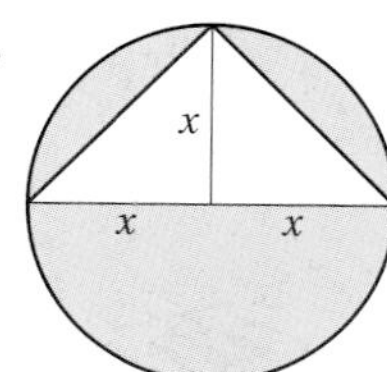

11.3 Factoring $ax^2 + bx + c$, $a \neq 1$, Using FOIL

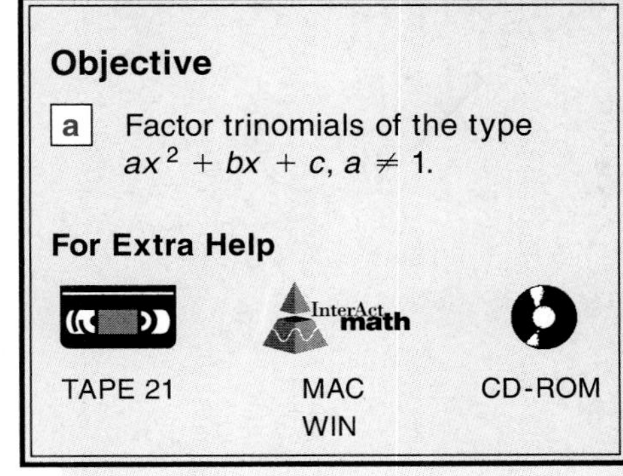

Objective
a Factor trinomials of the type $ax^2 + bx + c$, $a \neq 1$.
For Extra Help
InterAct math
TAPE 21
MAC
WIN
CD-ROM

In Section 11.2, we learned a trial-and-error method to factor trinomials of the type $x^2 + bx + c$. In this section, we factor trinomials in which the coefficient of the leading term x^2 is not 1. The procedure we learn is a refined trial-and-error method. (In Section 11.4, we will consider an alternative method for the same kind of factoring. It involves *factoring by grouping*.)

a We want to factor trinomials of the type $ax^2 + bx + c$. Consider the following multiplication:

$$
\begin{array}{rcccccccc}
 & & F & & O & & I & & L \\
(2x+5)(3x+4) & = & 6x^2 & + & 8x & + & 15x & + & 20 \\
 & = & 6x^2 & + & & 23x & & + & 20
\end{array}
$$

F	O + I		L
$2 \cdot 3$	$2 \cdot 4$	$5 \cdot 3$	$5 \cdot 4$

To factor $6x^2 + 23x + 20$, we reverse the above multiplication, using what we might call an "unFOIL" process. We look for two binomials $rx + p$ and $sx + q$ whose product is this trinomial. The product of the First terms must be $6x^2$. The product of the Outside terms plus the product of the Inside terms must be $23x$. The product of the Last terms must be 20. We know from the preceding discussion that the answer is $(2x + 5)(3x + 4)$. Generally, however, finding such an answer is a refined trial-and-error process. It turns out that $(-2x - 5)(-3x - 4)$ is also a correct answer, but we usually choose an answer in which the first coefficients are positive.

We will use the following trial-and-error method.

To factor $ax^2 + bx + c$, $a \neq 1$, using FOIL:

1. Factor out the largest common factor, if any.
2. Find the First terms whose product is ax^2.

   $(\blacksquare x + \quad)(\blacksquare x + \quad) = ax^2 + bx + c.$

   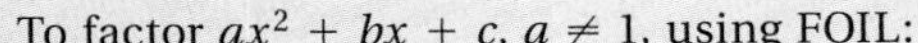
   FOIL

3. Find two Last terms whose product is c:

   $(\quad x + \blacksquare)(\quad x + \blacksquare) = ax^2 + bx + c.$

   FOIL

4. Repeat steps (2) and (3) until a combination is found for which the sum of the Outer and Inner products is bx:

   $(\blacksquare x + \blacksquare)(\blacksquare x + \blacksquare) = ax^2 + bx + c.$

   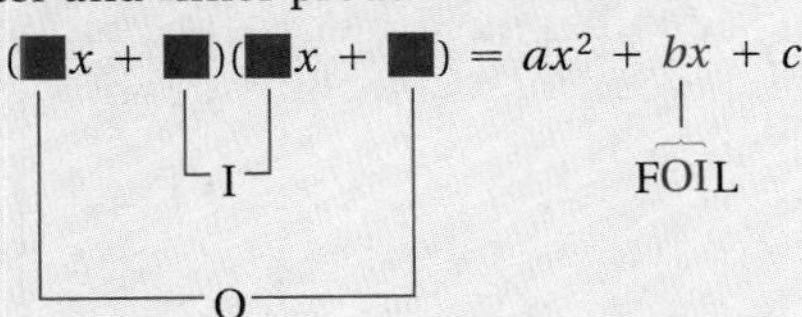

Factor.

1. $2x^2 - x - 15$

2. $12x^2 - 17x - 5$

Answers on page A-34

Example 1 Factor: $3x^2 - 10x - 8$.

1) First, we check for a common factor. Here there is none (other than 1 or -1).

2) Find two **F**irst terms whose product is $3x^2$.

The only possibilities for the **F**irst terms are $3x$ and x, so any factorization must be of the form

$$(3x + \square)(x + \square).$$

3) Find two **L**ast terms whose product is -8.

Possible factorizations of -8 are

$(-8) \cdot 1,\quad 8 \cdot (-1),\quad (-2) \cdot 4,\quad$ and $\quad 2 \cdot (-4)$.

Since the First terms are not identical, we must also consider

$1 \cdot (-8),\quad (-1) \cdot 8,\quad 4 \cdot (-2),\quad$ and $\quad (-4) \cdot 2$.

4) Inspect the **O**uter and **I**nner products resulting from steps (2) and (3). Look for a combination in which the sum of the products is the middle term, $-10x$:

Trial	*Product*	
$(3x - 8)(x + 1)$	$3x^2 + 3x - 8x - 8$ $= 3x^2 - 5x - 8$	← **Wrong middle term**
$(3x + 8)(x - 1)$	$3x^2 - 3x + 8x - 8$ $= 3x^2 + 5x - 8$	← **Wrong middle term**
$(3x - 2)(x + 4)$	$3x^2 + 12x - 2x - 8$ $= 3x^2 + 10x - 8$	← **Wrong middle term**
$(3x + 2)(x - 4)$	$3x^2 - 12x + 2x - 8$ $= 3x^2 - 10x - 8$	← **Correct middle term!**
$(3x + 1)(x - 8)$	$3x^2 - 24x + x - 8$ $= 3x^2 - 23x - 8$	← **Wrong middle term**
$(3x - 1)(x + 8)$	$3x^2 + 24x - x - 8$ $= 3x^2 + 23x - 8$	← **Wrong middle term**
$(3x + 4)(x - 2)$	$3x^2 - 6x + 4x - 8$ $= 3x^2 - 2x - 8$	← **Wrong middle term**
$(3x - 4)(x + 2)$	$3x^2 + 6x - 4x - 8$ $= 3x^2 + 2x - 8$	← **Wrong middle term**

The correct factorization is $(3x + 2)(x - 4)$.

CHECK: $(3x + 2)(x - 4) = 3x^2 - 10x - 8$.

Two observations can be made from Example 1. First, we listed all possible trials even though we could have stopped after having found the correct factorization. We did this to show that each trial differs only in the middle term of the product. Second, note that as in Section 11.2, only the sign of the middle term changes when the signs in the binomials are reversed.

Do Exercises 1 and 2.

Example 2 Factor: $24x^2 - 76x + 40$.

1) First, we factor out the largest common factor, 4:

$$4(6x^2 - 19x + 10).$$

Now we factor the trinomial $6x^2 - 19x + 10$.

2) Because $6x^2$ can be factored as $3x \cdot 2x$ or $6x \cdot x$, we have these possibilities for factorizations:

$$(3x + \square)(2x + \square) \quad \text{or} \quad (6x + \square)(x + \square).$$

3) There are four pairs of factors of 10 and they each can be listed in two ways:

$$10, 1 \qquad -10, -1 \qquad 5, 2 \qquad -5, -2$$

and

$$1, 10 \qquad -1, -10 \qquad 2, 5 \qquad -2, -5.$$

4) The two possibilities from step (2) and the eight possibilities from step (3) give $2 \cdot 8$, or 16 possibilities for factorizations. We look for **O**uter and **I**nner products resulting from steps (2) and (3) for which the sum is the middle term, $-19x$. Since the sign of the middle term is negative, but the sign of the last term, 10, is positive, the two factors of 10 must both be negative. This means only four pairings from step (3) need be considered. We first try these factors with $(3x + \square)(2x + \square)$. If none gives the correct factorization, we will consider $(6x + \square)(x + \square)$.

Trial	*Product*	
$(3x - 10)(2x - 1)$	$6x^2 - 3x - 20x + 10$ $= 6x^2 - 23x + 10$	← **Wrong middle term**
$(3x - 1)(2x - 10)$	$6x^2 - 30x - 2x + 10$ $= 6x^2 - 32x + 10$	← **Wrong middle term**
$(3x - 5)(2x - 2)$	$6x^2 - 6x - 10x + 10$ $= 6x^2 - 16x + 10$	← **Wrong middle term**
$(3x - 2)(2x - 5)$	$6x^2 - 15x - 4x + 10$ $= 6x^2 - 19x + 10$	← **Correct middle term!**

Since we have a correct factorization, we need not consider

$$(6x + \square)(x + \square).$$

The factorization of $6x^2 - 19x + 10$ is $(3x - 2)(2x - 5)$, but *do not forget the common factor*! We must include it in order to factor the original trinomial:

$$24x^2 - 76x + 40 = 4(6x^2 - 19x + 10)$$
$$= 4(3x - 2)(2x - 5).$$

CAUTION! When factoring any polynomial, always look for a common factor. Failure to do so is such a common error that this caution bears repeating.

Factor.

3. $3x^2 - 19x + 20$

4. $20x^2 - 46x + 24$

5. Factor: $6x^2 + 7x + 2$.

Answers on page A-34

In Example 2, look again at the possibility $(3x - 5)(2x - 2)$. Without multiplying, we can reject such a possibility. To see why, consider the following:

$$(3x - 5)(2x - 2) = 2(3x - 5)(x - 1).$$

The expression $2x - 2$ has a common factor, 2. But we removed the *largest* common factor in the first step. If $2x - 2$ were one of the factors, then 2 would have to be a common factor in addition to the original 4. Thus, $(2x - 2)$ cannot be part of the factorization of the original trinomial.

Given that the largest common factor is factored out at the outset, we need not consider factorizations that have a common factor.

Do Exercises 3 and 4.

Example 3 Factor: $10x^2 + 37x + 7$.

1) There is no common factor (other than 1 or -1).

2) Because $10x^2$ factors as $10x \cdot x$ or $5x \cdot 2x$, we have these possibilities for factorizations:

$$(10x + \quad)(x + \quad) \quad \text{or} \quad (5x + \quad)(2x + \quad).$$

3) There are two pairs of factors of 7 and they each can be listed in two ways:

$$1, 7 \qquad -1, -7 \qquad \text{and} \qquad 7, 1 \qquad -7, -1.$$

4) From steps (2) and (3), we see that there are 8 possibilities for factorizations. Look for **O**uter and **I**nner products for which the sum is the middle term. Because all coefficients in $10x^2 + 37x + 7$ are positive, we need consider only positive factors of 7. The possibilities are

$$(10x + 1)(x + 7) = 10x^2 + 71x + 7,$$
$$(10x + 7)(x + 1) = 10x^2 + 17x + 7,$$
$$(5x + 7)(2x + 1) = 10x^2 + 19x + 7,$$
$$(5x + 1)(2x + 7) = 10x^2 + 37x + 7.$$

The factorization is $(5x + 1)(2x + 7)$.

Keep in mind that this method of factoring trinomials of the type $ax^2 + bx + c$ involves *trial and error.* As you practice, you will find that you can make better and better guesses.

Do Exercise 5.

TIPS FOR FACTORING $ax^2 + bx + c$, $a \neq 1$

1. Always factor out the largest common factor, if one exists. Once the common factor has been factored out of the original trinomial, no binomial factor can contain a common factor (other than 1 or -1).
2. If c is positive, then the signs in both binomial factors must match the sign of b. (This assumes that $a > 0$.)
3. Reversing the signs in the binomials reverses the sign of the middle term of their product.
4. Be systematic about your trials. Keep track of those pairs you have tried and those you have not.
5. Always check by multiplying.

Example 4 Factor: $6p^2 - 13pq - 28q^2$.

1) Factor out a common factor, if any. There is none (other than 1 or -1).
2) Factor the first term, $6p^2$. Possibilities are $2p$, $3p$ and $6p$, p. We have these as possibilities for factorizations:

$$(2p + \quad)(3p + \quad) \quad \text{or} \quad (6p + \quad)(p + \quad).$$

3) Factor the last term, $-28q^2$, which has a negative coefficient. The possibilities are $-14q$, $2q$ and $14q$, $-2q$; $-28q$, q and $28q$, $-q$; and $-7q$, $4q$ and $7q$, $-4q$.
4) The coefficient of the middle term is negative, so we look for combinations of factors from steps (2) and (3) such that the sum of their products has a negative coefficient. We try some possibilities:

$$(2p + q)(3p - 28q) = 6p^2 - 53pq - 28q^2,$$
$$(2p - 7q)(3p + 4q) = 6p^2 - 13pq - 28q^2.$$

The factorization of $6p^2 - 13pq - 28q^2$ is $(2p - 7q)(3p + 4q)$.

Do Exercises 6 and 7.

Factor.

6. $6a^2 - 5ab + b^2$

7. $6x^2 + 15xy + 9y^2$

Calculator Spotlight

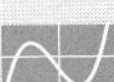 Checking Factorizations with a Table or a Graph

Table. The TABLE feature can be used as a partial check that polynomials have been factored correctly. To check whether the factoring of Example 1,

$$3x^2 - 10x - 8 = (3x + 2)(x - 4),$$

is correct, we enter

$$y_1 = 3x^2 - 10x - 8 \quad \text{and} \quad y_2 = (3x + 2)(x - 4).$$

If our factoring is correct, the y_1- and y_2-values should be the same, regardless of the table settings used.

X	Y1	Y2
0	−8	−8
1	−15	−15
2	−16	−16
3	−11	−11
4	0	0
5	17	17
6	40	40
X = 0		

We see that y_1 and y_2 are the same, so the factoring seems to be correct. Remember, though, that this is only a partial check.

Graph. This factorization of Example 1 can also be checked with the GRAPH feature. We see that the graphs of y_1 and y_2 are the same, so the factoring seems to be correct.

$y_1 = 3x^2 - 10x - 8$, $y_2 = (3x + 2)(x - 4)$

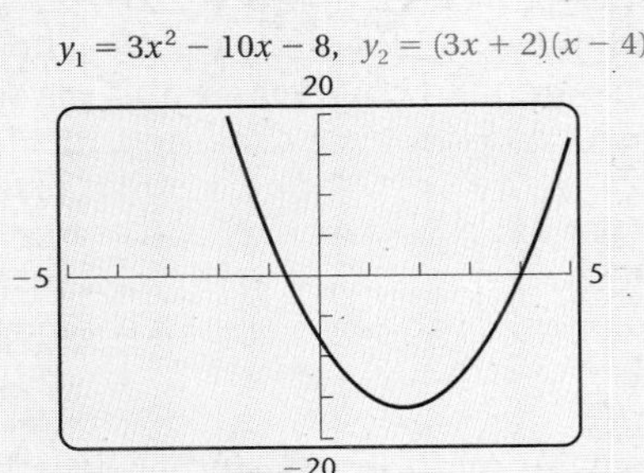

Exercises

Use the TABLE or the GRAPH feature to check whether the factorization is correct.

1. $24x^2 - 76x + 40 = 4(3x - 2)(2x - 5)$ (Example 2)
2. $4x^2 - 5x - 6 = (4x + 3)(x - 2)$
3. $5x^2 + 17x - 12 = (5x + 3)(x - 4)$
4. $10x^2 + 37x + 7 = (5x - 1)(2x + 7)$
5. $12x^2 - 17x - 5 = (6x + 1)(2x - 5)$
6. $12x^2 - 17x - 5 = (4x + 1)(3x - 5)$
7. $x^2 - 4 = (x - 2)(x - 2)$
8. $x^2 - 4 = (x + 2)(x - 2)$

Answers on page A-34

Improving Your Math Study Skills

On Reading and Writing Mathematics

Mike Rosenborg is a former student of Marv Bittinger. He went on to receive a Master's Degree in mathematics and is now a math teacher. Here are some of his study tips regarding the reading and writing of mathematics.

Why read your math text? This is a legitimate question when you consider that the instructor usually covers most of the material in the text. I have a reason: I don't want to be spoon-fed the material; I want to learn on my own. It's a lot more fun, and it builds my self-confidence to know that I can learn the material without the need for a teacher or a classroom.

It's a good idea to read your math text regularly for several reasons. When you read mathematics, you rely exclusively on the written word, and this is where mathematics derives much of its power. Mathematics is very precise, and it depends on writing to maintain and communicate this precision. Definitions and theorems in mathematics are stated in precise terms, and mathematical manipulations (such as solving equations) are performed by writing in a precise way.

In general, math texts develop the concepts in mathematics in a clear, tightly reasoned format, showing many examples along the way. Remember: The authors are mathematicians, and the way they write reflects their extensive mathematical training and thought processes. If you carefully read through the text, you will experience what it is like to think in a mathematical, rigorous, and precise way. This will not happen if you rely exclusively on oral lectures, because oral presentations are intrinsically "loose."

Reading your math text has other benefits. You will often find how to solve a difficult problem in the exercise set by looking at the text; in fact, there may be an example developed for you in the text that is much like your problem. Often your instructor will not have the time to cover everything in the text, or may want to cover something a little different. In these cases, reading your text will fill in the gaps.

But how do you read a math text? There is, of course, a difference between reading mathematics and a novel. Mathematics is like a chain with each link being developed in sequence and in order, and each link demands careful thought and attention before one can proceed to the next link—each link depends on the link before it. This is why you will experience troubles throughout an entire math course if you miss a single concept. Here, then, is a list of math reading tips and techniques:

- **Always read with a pencil and a piece of paper nearby.** When you find a section in your text that is difficult to understand, stop and work it out on your scratch paper.
- **Make notes in the margins of your text.** For instance, if you come across a word you don't understand, look it up and write its definition in the margin. Also, if the book refers to something covered previously that you have forgotten, find where it was originally covered, write the page number down in the margin, and go back and review the word.
- **Proceed slowly and carefully, making sure you understand what you read before continuing.** If there are worked-out examples in the book, read one and then try the next ones on your own. If you make a mistake, the details on the worked examples in the text will enable you to find your mistake quickly.
- **Do the Thinking and Writing Exercises.** The benefits of writing mathematics are that writing forces you to think through what you are writing about in a step-by-step manner, the act of writing itself helps to reinforce the concepts in your mind, and you have a ready, easy-to-read reference for further study or review. For instance, when studying for a test, if you have written up all your homework problems in a complete way, it will be easy for you to study for the test directly from your homework.

Exercise Set 11.3

a Factor.

1. $2x^2 - 7x - 4$
2. $3x^2 - x - 4$
3. $5x^2 - x - 18$
4. $4x^2 - 17x + 15$
5. $6x^2 + 23x + 7$
6. $6x^2 - 23x + 7$
7. $3x^2 + 4x + 1$
8. $7x^2 + 15x + 2$
9. $4x^2 + 4x - 15$
10. $9x^2 + 6x - 8$
11. $2x^2 - x - 1$
12. $15x^2 - 19x - 10$
13. $9x^2 + 18x - 16$
14. $2x^2 + 5x + 2$
15. $3x^2 - 5x - 2$
16. $18x^2 - 3x - 10$
17. $12x^2 + 31x + 20$
18. $15x^2 + 19x - 10$
19. $14x^2 + 19x - 3$
20. $35x^2 + 34x + 8$
21. $9x^2 + 18x + 8$
22. $6 - 13x + 6x^2$
23. $49 - 42x + 9x^2$
24. $16 + 36x^2 + 48x$
25. $24x^2 + 47x - 2$
26. $16p^2 - 78p + 27$
27. $35x^2 - 57x - 44$
28. $9a^2 + 12a - 5$
29. $20 + 6x - 2x^2$
30. $15 + x - 2x^2$
31. $12x^2 + 28x - 24$
32. $6x^2 + 33x + 15$
33. $30x^2 - 24x - 54$
34. $18t^2 - 24t + 6$
35. $4y + 6y^2 - 10$
36. $-9 + 18x^2 - 21x$
37. $3x^2 - 4x + 1$
38. $6t^2 + 13t + 6$
39. $12x^2 - 28x - 24$
40. $6x^2 - 33x + 15$
41. $-1 + 2x^2 - x$
42. $-19x + 15x^2 + 6$
43. $9x^2 - 18x - 16$
44. $14y^2 + 35y + 14$
45. $15x^2 - 25x - 10$
46. $18x^2 + 3x - 10$
47. $12p^3 + 31p^2 + 20p$
48. $15x^3 + 19x^2 - 10x$

49. $14x^4 + 19x^3 - 3x^2$

50. $70x^4 + 68x^3 + 16x^2$

51. $168x^3 - 45x^2 + 3x$

52. $144x^5 + 168x^4 + 48x^3$

53. $15x^4 - 19x^2 + 6$

54. $9x^4 + 18x^2 + 8$

55. $25t^2 + 80t + 64$

56. $9x^2 - 42x + 49$

57. $6x^3 + 4x^2 - 10x$

58. $18x^3 - 21x^2 - 9x$

59. $25x^2 + 79x + 64$

60. $9y^2 + 42y + 47$

61. $6x^2 - 19x - 5$

62. $2x^2 + 11x - 9$

63. $12m^2 - mn - 20n^2$

64. $12a^2 - 17ab + 6b^2$

65. $6a^2 - ab - 15b^2$

66. $3p^2 - 16pq - 12q^2$

67. $9a^2 + 18ab + 8b^2$

68. $10s^2 + 4st - 6t^2$

69. $35p^2 + 34pq + 8q^2$

70. $30a^2 + 87ab + 30b^2$

71. $18x^2 - 6xy - 24y^2$

72. $15a^2 - 5ab - 20b^2$

Skill Maintenance

Solve. [8.6a]

73. $A = pq - 7$, for q

74. $y = mx + b$, for x

75. $3x + 2y = 6$, for y

76. $p - q + r = 2$, for q

Solve. [8.7e]

77. $5 - 4x < -11$

78. $2x - 4(x + 3x) \geq 6x - 8 - 9x$

79. Graph: $y = \frac{2}{5}x - 1$. [9.2b]

80. Divide: $\frac{y^{12}}{y^4}$. [10.1e]

Multiply. [10.6d]

81. $(3x - 5)(3x + 5)$

82. $(4a - 3)^2$

Synthesis

83. ◈ Explain how the factoring in Exercise 21 can be used to aid the factoring in Exercise 67.

84. ◈ A student presents the following work:

$$4x^2 + 28x + 48 = (2x + 6)(2x + 8)$$
$$= 2(x + 3)(x + 4).$$

Is it correct? Explain.

Factor.

85. $20x^{2n} + 16x^n + 3$

86. $-15x^{2m} + 26x^m - 8$

87. $3x^{6a} - 2x^{3a} - 1$

88. $x^{2n+1} - 2x^{n+1} + x$

11.4 Factoring $ax^2 + bx + c, a \neq 1$, Using Grouping

Objective

a Factor trinomials of the type $ax^2 + bx + c, a \neq 1$, by splitting the middle term and using grouping.

For Extra Help

TAPE 21 | MAC WIN | CD-ROM

a Another method of factoring trinomials of the type $ax^2 + bx + c$, $a \neq 1$, is known as the **grouping method.** It involves factoring by grouping. We know how to factor the trinomial $x^2 + 5x + 6$. We look for factors of the constant term, 6, whose sum is the coefficient of the middle term, 5:

$$x^2 + 5x + 6.$$

(1) Factor: $6 = 2 \cdot 3$

(2) Sum: $2 + 3 = 5$

What happens when the leading coefficient is not 1? To factor a trinomial like $3x^2 - 10x - 8$, we can use a method similar to what we used for the preceding trinomial, but we need two more steps. That method is outlined as follows.

To factor $ax^2 + bx + c, a \neq 1$, using the grouping method:

1. Factor out a common factor, if any.
2. Multiply the leading coefficient a and the constant c.
3. Try to factor the product ac so that the sum of the factors is b. That is, find integers p and q such that $pq = ac$ and $p + q = b$.
4. Split the middle term. That is, write it as a sum using the factors found in step (3).
5. Then factor by grouping.

Example 1 Factor: $3x^2 - 10x - 8$.

1) First, we factor out a common factor, if any. There is none (other than 1 or -1).

2) We multiply the leading coefficient, 3, and the constant, -8:

$$3(-8) = -24.$$

3) Then we look for a factorization of -24 in which the sum of the factors is the coefficient of the middle term, -10.

Pairs of Factors	Sums of Factors
−1, 24	23
1, −24	−23
−2, 12	10
2, −12	−10 ← $2 + (-12) = -10$
−3, 8	5
3, −8	−5
−4, 6	2
4, −6	−2

4) Next, we split the middle term as a sum or a difference using the factors found in step (3):

$$-10x = 2x - 12x.$$

Factor.

1. $6x^2 + 7x + 2$

2. $12x^2 - 17x - 5$

5) Finally, we factor by grouping, as follows:

$$3x^2 - 10x - 8 = 3x^2 + 2x - 12x - 8 \quad \text{Substituting } 2x - 12x \text{ for } -10x$$
$$= x(3x + 2) - 4(3x + 2) \quad \text{Factoring by grouping; see Section 11.1}$$
$$= (x - 4)(3x + 2).$$

We can also split the middle term as $-12x + 2x$. We still get the same factorization, although the factors may be in a different order. Note the following:

$$3x^2 - 10x - 8 = 3x^2 - 12x + 2x - 8 \quad \text{Substituting } -12x + 2x \text{ for } -10x$$
$$= 3x(x - 4) + 2(x - 4) \quad \text{Factoring by grouping; see Section 11.1}$$
$$= (3x + 2)(x - 4).$$

Check by multiplying: $(3x + 2)(x - 4) = 3x^2 - 10x - 8$.

Do Exercises 1 and 2.

Factor.

3. $6x^2 + 15x + 9$

4. $20x^2 - 46x + 24$

Example 2 Factor: $8x^2 + 8x - 6$.

1) First, we factor out a common factor, if any. The number 2 is common to all three terms, so we factor it out:

$2(4x^2 + 4x - 3)$.

2) Next, we factor the trinomial $4x^2 + 4x - 3$. We multiply the leading coefficient and the constant, 4 and -3:

$4(-3) = -12$.

3) We try to factor -12 so that the sum of the factors is 4.

Pairs of Factors	Sums of Factors
−1, 12	11
1, −12	−11
−2, 6	4 ← $-2 + 6 = 4$
2, −6	−4
−3, 4	1
3, −4	−1

4) Then we split the middle term, $4x$, as follows:

$4x = -2x + 6x$.

5) Finally, we factor by grouping:

$$4x^2 + 4x - 3 = 4x^2 - 2x + 6x - 3 \quad \text{Substituting } -2x + 6x \text{ for } 4x$$
$$= 2x(2x - 1) + 3(2x - 1) \quad \text{Factoring by grouping}$$
$$= (2x + 3)(2x - 1).$$

The factorization of $4x^2 + 4x - 3$ is $(2x + 3)(2x - 1)$. But don't forget the common factor! We must include it to get a factorization of the original trinomial:

$$8x^2 + 8x - 6 = 2(2x + 3)(2x - 1).$$

Do Exercises 3 and 4.

Answers on page A-34

Exercise Set 11.4

a Factor. Note that the middle term has already been split.

1. $x^2 + 2x + 7x + 14$

2. $x^2 + 3x + x + 3$

3. $x^2 - 4x - x + 4$

4. $a^2 + 5a - 2a - 10$

5. $6x^2 + 4x + 9x + 6$

6. $3x^2 - 2x + 3x - 2$

7. $3x^2 - 4x - 12x + 16$

8. $24 - 18y - 20y + 15y^2$

9. $35x^2 - 40x + 21x - 24$

10. $8x^2 - 6x - 28x + 21$

11. $4x^2 + 6x - 6x - 9$

12. $2x^4 - 6x^2 - 5x^2 + 15$

13. $2x^4 + 6x^2 + 5x^2 + 15$

14. $9x^4 - 6x^2 - 6x^2 + 4$

Factor by grouping.

15. $2x^2 - 7x - 4$

16. $5x^2 - x - 18$

17. $3x^2 + 4x - 15$

18. $3x^2 + x - 4$

19. $6x^2 + 23x + 7$

20. $6x^2 + 13x + 6$

21. $3x^2 + 4x + 1$

22. $7x^2 + 15x + 2$

23. $4x^2 + 4x - 15$

24. $9x^2 + 6x - 8$

25. $2x^2 + x - 1$

26. $15x^2 + 19x - 10$

27. $9x^2 - 18x - 16$

28. $2x^2 - 5x + 2$

29. $3x^2 + 5x - 2$

30. $18x^2 + 3x - 10$

31. $12x^2 - 31x + 20$

32. $15x^2 - 19x - 10$

33. $14x^2 + 19x - 3$

34. $35x^2 + 34x + 8$

35. $9x^2 + 18x + 8$

36. $6 - 13x + 6x^2$

37. $49 - 42x + 9x^2$

38. $25x^2 + 40x + 16$

39. $24x^2 + 47x - 2$

40. $16a^2 + 78a + 27$

41. $35x^5 - 57x^4 - 44x^3$

42. $18a^3 + 24a^2 - 10a$

43. $60x + 18x^2 - 6x^3$

44. $60x + 4x^2 - 8x^3$

45. $15x^3 + 33x^4 + 6x^5$

46. $8x^2 + 2x + 6x^3$

Skill Maintenance

Solve. [8.7d, e]

47. $-10x > 1000$

48. $-3.8x \leq -824.6$

49. $6 - 3x \geq -18$

50. $3 - 2x - 4x > -9$

51. $\frac{1}{2}x - 6x + 10 \leq x - 5x$

52. $-2(x + 7) > -4(x - 5)$

53. $3x - 6x + 2(x - 4) > 2(9 - 4x)$

54. $-6(x - 4) + 8(4 - x) \leq 3(x - 7)$

Synthesis

55. ◆ If you have studied both the FOIL and the grouping methods of factoring $ax^2 + bx + c$, $a \neq 1$, decide which method you think is better and explain why.

56. ◆ Explain factoring $ax^2 + bx + c$, $a \neq 1$, by grouping as though you were teaching a fellow student.

Factor.

57. $9x^{10} - 12x^5 + 4$

58. $24x^{2n} + 22x^n + 3$

59. $16x^{10} + 8x^5 + 1$

60. $(a + 4)^2 - 2(a + 4) + 1$

61.–70. Use the TABLE feature to check the factoring in Exercises 15–24.

11.5 Factoring Trinomial Squares and Differences of Squares

In this section, we first learn to factor trinomials that are squares of binomials. Then we factor binomials that are differences of squares.

Objectives

a. Recognize trinomial squares.

b. Factor trinomial squares.

c. Recognize differences of squares.

d. Factor differences of squares, being careful to factor completely.

For Extra Help

TAPE 22 | MAC WIN | CD-ROM

a Recognizing Trinomial Squares

Some trinomials are squares of binomials. For example, the trinomial $x^2 + 10x + 25$ is the square of the binomial $x + 5$. To see this, we can calculate $(x + 5)^2$. It is $x^2 + 2 \cdot x \cdot 5 + 5^2$, or $x^2 + 10x + 25$. A trinomial that is the square of a binomial is called a **trinomial square.**

In Chapter 10, we considered squaring binomials as special-product rules:

$$(A + B)^2 = A^2 + 2AB + B^2;$$
$$(A - B)^2 = A^2 - 2AB + B^2.$$

We can use these equations in reverse to factor trinomial squares.

$$A^2 + 2AB + B^2 = (A + B)^2;$$
$$A^2 - 2AB + B^2 = (A - B)^2$$

How can we recognize when an expression to be factored is a trinomial square? Look at $A^2 + 2AB + B^2$ and $A^2 - 2AB + B^2$. In order for an expression to be a trinomial square:

a) Two terms, A^2 and B^2, must be squares, such as

$$4, \quad x^2, \quad 25x^4, \quad 16t^2.$$

When the coefficient is a perfect square and the power(s) of the variable(s) is (are) even, then the expression is a perfect square.

b) There must be no minus sign before A^2 or B^2.

c) If we multiply A and B (expressions whose squares are A^2 and B^2) and double the result, we get either the remaining term $2 \cdot A \cdot B$, or its opposite, $-2 \cdot A \cdot B$.

Example 1 Determine whether $x^2 + 6x + 9$ is a trinomial square.

a) We know that x^2 and 9 are squares.

b) There is no minus sign before x^2 or 9.

c) If we multiply the square roots, x and 3, and double the product, we get the remaining term: $2 \cdot x \cdot 3 = 6x$.

Thus, $x^2 + 6x + 9$ is the square of a binomial. In fact, $x^2 + 6x + 9 = (x + 3)^2$.

Example 2 Determine whether $x^2 + 6x + 11$ is a trinomial square.

The answer is no, because only one term is a square.

Determine whether each is a trinomial square. Write "yes" or "no."

1. $x^2 + 8x + 16$

2. $25 - x^2 + 10x$

3. $t^2 - 12t + 4$

4. $25 + 20y + 4y^2$

5. $5x^2 + 16 - 14x$

6. $16x^2 + 40x + 25$

7. $p^2 + 6p - 9$

8. $25a^2 + 9 - 30a$

Factor.

9. $x^2 + 2x + 1$

10. $1 - 2x + x^2$

11. $4 + t^2 + 4t$

12. $25x^2 - 70x + 49$

13. $49 - 56y + 16y^2$

Answers on page A-34

Example 3 Determine whether $16x^2 + 49 - 56x$ is a trinomial square.

It helps to first write the trinomial in descending order:

$$16x^2 - 56x + 49.$$

a) We know that $16x^2$ and 49 are squares.

b) There is no minus sign before $16x^2$ or 49.

c) If we multiply the square roots, $4x$ and 7, and double the product, we get the opposite of the remaining term: $2 \cdot 4x \cdot 7 = 56x$; $56x$ is the opposite of $-56x$.

Thus, $16x^2 + 49 - 56x$ is a trinomial square. In fact, $16x^2 - 56x + 49 = (4x - 7)^2$.

Do Exercises 1–8.

b Factoring Trinomial Squares

We can use the trial-and-error or grouping methods from Sections 11.2–11.4 to factor such trinomial squares, but there is a faster method using the following equations:

$$A^2 + 2AB + B^2 = (A + B)^2;$$
$$A^2 - 2AB + B^2 = (A - B)^2.$$

We consider 3 to be a square root of 9 because $3^2 = 9$. Similarly, A is a square root of A^2. We use square roots of the squared terms and the sign of the remaining term to factor a trinomial square.

Example 4 Factor: $x^2 + 6x + 9$.

$$x^2 + 6x + 9 = x^2 + 2 \cdot x \cdot 3 + 3^2 = (x + 3)^2$$
$$A^2 + 2\ A\ B + B^2 = (A + B)^2$$

The sign of the middle term is positive.

Example 5 Factor: $x^2 + 49 - 14x$.

$$x^2 + 49 - 14x = x^2 - 14x + 49$$ **Changing order**

$$= x^2 - 2 \cdot x \cdot 7 + 7^2$$ **The sign of the middle term is negative.**

$$= (x - 7)^2$$

Example 6 Factor: $16x^2 - 40x + 25$.

$$16x^2 - 40x + 25 = (4x)^2 - 2 \cdot 4x \cdot 5 + 5^2 = (4x - 5)^2$$
$$A^2 - 2\ A\ B + B^2 = (A - B)^2$$

Do Exercises 9–13.

Example 7 Factor: $t^4 + 20t^2 + 100$.

$$t^4 + 20t^2 + 100 = (t^2)^2 + 2(t^2)(10) + 10^2$$
$$= (t^2 + 10)^2$$

Example 8 Factor: $75m^3 + 210m^2 + 147m$.

Always look first for a common factor. This time there is one, $3m$:

$$75m^3 + 210m^2 + 147m = 3m[25m^2 + 70m + 49]$$
$$= 3m[(5m)^2 + 2(5m)(7) + 7^2]$$
$$= 3m(5m + 7)^2.$$

Example 9 Factor: $4p^2 - 12pq + 9q^2$.

$$4p^2 - 12pq + 9q^2 = (2p)^2 - 2(2p)(3q) + (3q)^2$$
$$= (2p - 3q)^2$$

Do Exercises 14–17.

Factor.

14. $48m^2 + 75 + 120m$

15. $p^4 + 18p^2 + 81$

16. $4z^5 - 20z^4 + 25z^3$

17. $9a^2 + 30ab + 25b^2$

Answers on page A-34

c Recognizing Differences of Squares

The following polynomials are *differences of squares*:

$$x^2 - 9, \quad 4t^2 - 49, \quad a^2 - 25b^2.$$

To factor a difference of squares such as $x^2 - 9$, think about the formula we used in Chapter 10:

$$(A + B)(A - B) = A^2 - B^2.$$

Equations are reversible, so we also know that

$$A^2 - B^2 = (A + B)(A - B).$$

Thus,

$$x^2 - 9 = (x + 3)(x - 3).$$

To use this formula, we must be able to recognize when it applies. A **difference of squares** is an expression like the following:

$$A^2 - B^2.$$

How can we recognize such expressions? Look at $A^2 - B^2$. In order for a binomial to be a difference of squares:

a) There must be two expressions, both squares, such as

$$4x^2, \quad 9, \quad 25t^4, \quad 1, \quad x^6, \quad 49y^8.$$

b) The terms must have different signs.

Determine whether each is a difference of squares. Write "yes" or "no."

18. $x^2 - 25$

19. $t^2 - 24$

20. $y^2 + 36$

21. $4x^2 - 15$

22. $16x^4 - 49$

23. $9w^6 - 1$

24. $-49 + 25t^2$

Answers on page A-34

Example 10 Is $9x^2 - 64$ a difference of squares?

a) The first expression is a square: $9x^2 = (3x)^2$.
The second expression is a square: $64 = 8^2$.

b) The terms have different signs.

Thus we have a difference of squares, $(3x)^2 - 8^2$.

Example 11 Is $25 - t^3$ a difference of squares?

a) The expression t^3 is not a square.

The expression is not a difference of squares.

Example 12 Is $-4x^2 + 16$ a difference of squares?

a) The expressions $4x^2$ and 16 are squares: $4x^2 = (2x)^2$ and $16 = 4^2$.

b) The terms have different signs.

Thus we have a difference of squares. We can also see this by rewriting in the equivalent form: $16 - 4x^2$.

Do Exercises 18–24.

d Factoring Differences of Squares

To factor a difference of squares, we use the following equation:

$$A^2 - B^2 = (A + B)(A - B).$$

To factor a difference of squares $A^2 - B^2$, we find A and B, which are square roots of the expressions A^2 and B^2. We then use A and B to form two factors. One is the sum $A + B$, and the other is the difference $A - B$.

Example 13 Factor: $x^2 - 4$.

$$\begin{array}{ccccc} x^2 - 4 = & x^2 - 2^2 & = & (x + 2)(x - 2) \\ & \uparrow \quad \uparrow & & \uparrow \quad \uparrow \uparrow \quad \uparrow \\ & \mathbf{A^2 - B^2} & = & \mathbf{(A + B)(A - B)} \end{array}$$

Example 14 Factor: $9 - 16t^4$.

$$\begin{array}{ccccc} 9 - 16t^4 = & 3^2 - (4t^2)^2 & = & (3 + 4t^2)(3 - 4t^2) \\ & \uparrow \quad\quad \uparrow & & \uparrow \quad \uparrow \quad \uparrow \quad \uparrow \\ & \mathbf{A^2 - \quad B^2} & = & \mathbf{(A + B)\ (A - B)} \end{array}$$

Example 15 Factor: $m^2 - 4p^2$.

$$m^2 - 4p^2 = m^2 - (2p)^2 = (m + 2p)(m - 2p)$$

Example 16 Factor: $x^2 - \frac{1}{9}$.

$$x^2 - \frac{1}{9} = x^2 - \left(\frac{1}{3}\right)^2 = \left(x + \frac{1}{3}\right)\left(x - \frac{1}{3}\right)$$

Example 17 Factor: $18x^2 - 50x^6$.

Always look first for a factor common to all terms. This time there is one, $2x^2$.

$$\begin{aligned} 18x^2 - 50x^6 &= 2x^2(9 - 25x^4) \\ &= 2x^2[3^2 - (5x^2)^2] \\ &= 2x^2(3 + 5x^2)(3 - 5x^2) \end{aligned}$$

Example 18 Factor: $49x^4 - 9x^6$.

$$49x^4 - 9x^6 = x^4(49 - 9x^2) = x^4(7 + 3x)(7 - 3x)$$

Do Exercises 25–29.

CAUTION! Note carefully in these examples that a difference of squares is *not* the square of the difference; that is,

$$A^2 - B^2 \neq (A - B)^2 = A^2 - 2AB + B^2.$$

For example,

$$(45 - 5)^2 = 40^2 = 1600,$$

but

$$45^2 - 5^2 = 2025 - 25 = 2000.$$

Factoring Completely

If a factor with more than one term can still be factored, you should do so. When no factor can be factored further, you have **factored completely.** Always factor completely whenever told to factor.

Example 19 Factor: $p^4 - 16$.

$$\begin{aligned} p^4 - 16 &= (p^2)^2 - 4^2 \\ &= (p^2 + 4)(p^2 - 4) && \text{Factoring a difference of squares} \\ &= (p^2 + 4)(p + 2)(p - 2) && \text{Factoring further. The factor } p^2 - 4 \text{ is a difference of squares.} \end{aligned}$$

The polynomial $p^2 + 4$ cannot be factored further into polynomials with real coefficients.

CAUTION! If the greatest common factor has been removed, then you cannot factor a sum of squares further. In particular,

$$(A + B)^2 \neq A^2 + B^2.$$

Consider $25x^2 + 100$. This is a case in which we have a sum of squares, but there is a common factor, 25. Factoring, we get $25(x^2 + 4)$. Now $x^2 + 4$ cannot be factored further.

Example 20 Factor: $y^4 - 16x^{12}$.

$$\begin{aligned} y^4 - 16x^{12} &= (y^2 + 4x^6)(y^2 - 4x^6) && \text{Factoring a difference of squares} \\ &= (y^2 + 4x^6)(y + 2x^3)(y - 2x^3) && \text{Factoring further. The factor } y^2 - 4x^6 \text{ is a difference of squares.} \end{aligned}$$

Factor.

25. $x^2 - 9$

26. $64 - 4t^2$

27. $a^2 - 25b^2$

28. $64x^4 - 25x^6$

29. $5 - 20t^6$
[*Hint*: $1 = 1^2$, $t^6 = (t^3)^2$.]

Answers on page A-34

Factor completely.

30. $81x^4 - 1$

31. $49p^4 - 25q^6$

Answers on page A-34

FACTORING HINTS

1. Always look first for a common factor. If there is one, factor out the largest common factor.
2. Always factor completely.
3. Check by multiplying.

Do Exercises 30 and 31.

Exercise Set 11.5

a Determine whether each of the following is a trinomial square.

1. $x^2 - 14x + 49$

2. $x^2 - 16x + 64$

3. $x^2 + 16x - 64$

4. $x^2 - 14x - 49$

5. $x^2 - 2x + 4$

6. $x^2 + 3x + 9$

7. $9x^2 - 36x + 24$

8. $36x^2 - 24x + 16$

b Factor completely. Remember to look first for a common factor and to check by multiplying.

9. $x^2 - 14x + 49$

10. $x^2 - 20x + 100$

11. $x^2 + 16x + 64$

12. $x^2 + 20x + 100$

13. $x^2 - 2x + 1$

14. $x^2 + 2x + 1$

15. $4 + 4x + x^2$

16. $4 + x^2 - 4x$

17. $q^4 - 6q^2 + 9$

18. $64 + 16a^2 + a^4$

19. $49 + 56y + 16y^2$

20. $75 + 48a^2 - 120a$

21. $2x^2 - 4x + 2$

22. $2x^2 - 40x + 200$

23. $x^3 - 18x^2 + 81x$

24. $x^3 + 24x^2 + 144x$

25. $12q^2 - 36q + 27$

26. $20p^2 + 100p + 125$

27. $49 - 42x + 9x^2$

28. $64 - 112x + 49x^2$

29. $5y^4 + 10y^2 + 5$

30. $a^4 + 14a^2 + 49$

31. $1 + 4x^4 + 4x^2$

32. $1 - 2a^5 + a^{10}$

33. $4p^2 + 12pq + 9q^2$

34. $25m^2 + 20mn + 4n^2$

35. $a^2 - 6ab + 9b^2$

36. $x^2 - 14xy + 49y^2$

37. $81a^2 - 18ab + b^2$

38. $64p^2 + 16pq + q^2$

39. $36a^2 + 96ab + 64b^2$

40. $16m^2 - 40mn + 25n^2$

c Determine whether each of the following is a difference of squares.

41. $x^2 - 4$

42. $x^2 - 36$

43. $x^2 + 25$

44. $x^2 + 9$

45. $x^2 - 45$

46. $x^2 - 80y^2$

47. $16x^2 - 25y^2$

48. $-1 + 36x^2$

d Factor completely. Remember to look first for a common factor.

49. $y^2 - 4$

50. $q^2 - 1$

51. $p^2 - 9$

52. $x^2 - 36$

53. $-49 + t^2$

54. $-64 + m^2$

55. $a^2 - b^2$

56. $p^2 - q^2$

57. $25t^2 - m^2$

58. $w^2 - 49z^2$

59. $100 - k^2$

60. $81 - w^2$

61. $16a^2 - 9$

62. $25x^2 - 4$

63. $4x^2 - 25y^2$

64. $9a^2 - 16b^2$

65. $8x^2 - 98$

66. $24x^2 - 54$

67. $36x - 49x^3$

68. $16x - 81x^3$

69. $49a^4 - 81$

70. $25a^4 - 9$

71. $a^4 - 16$

72. $y^4 - 1$

73. $5x^4 - 405$

74. $4x^4 - 64$

75. $1 - y^8$

76. $x^8 - 1$

77. $x^{12} - 16$

78. $x^8 - 81$

79. $y^2 - \frac{1}{16}$

80. $x^2 - \frac{1}{25}$

81. $25 - \frac{1}{49}x^2$

82. $\frac{1}{4} - 9q^2$

83. $16m^4 - t^4$

84. $p^4q^4 - 1$

Skill Maintenance

Divide. [7.6c]

85. $(-110) \div 10$

86. $-1000 \div (-2.5)$

87. $\left(-\frac{2}{3}\right) \div \frac{4}{5}$

88. $8.1 \div (-9)$

89. $-64 \div (-32)$

90. $-256 \div 1.6$

Find a polynomial for the shaded area. (Leave results in terms of π where appropriate.) [10.4d]

91.

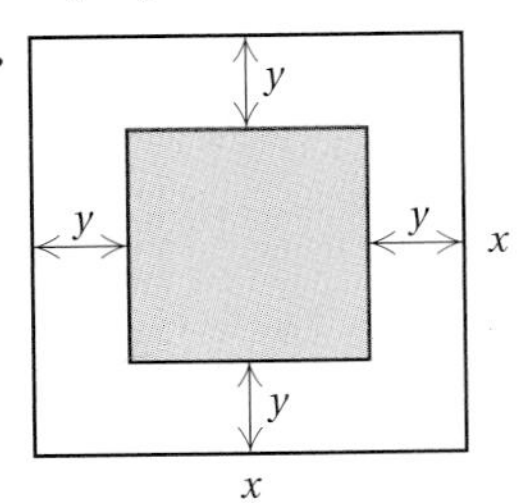

92.

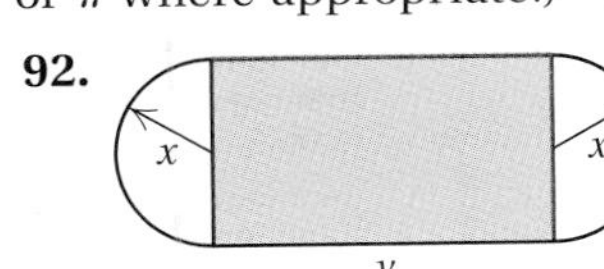

Simplify.

93. $y^5 \cdot y^7$ [10.1d]

94. $(5a^2b^3)^2$ [10.2b]

Find the intercepts. Then graph the equation. [9.3a]

95. $y - 6x = 6$

96. $3x - 5y = 15$

Synthesis

97. ◆ Explain in your own words how to determine whether a polynomial is a trinomial square.

98. ◆ A student concludes that since $x^2 - 9 = (x - 3)(x + 3)$, it must follow that $x^2 + 9 = (x + 3)(x + 3)$. What mistake is the student making? How would you go about correcting the misunderstanding?

Factor completely, if possible.

99. $49x^2 - 216$

100. $27x^3 - 13x$

101. $x^2 + 22x + 121$

102. $x^2 - 5x + 25$

103. $18x^3 + 12x^2 + 2x$

104. $162x^2 - 82$

105. $x^8 - 2^8$

106. $4x^4 - 4x^2$

107. $3x^5 - 12x^3$

108. $3x^2 - \frac{1}{3}$

109. $18x^3 - \frac{8}{25}x$

110. $x^2 - 2.25$

111. $0.49p - p^3$

112. $3.24x^2 - 0.81$

113. $0.64x^2 - 1.21$

114. $1.28x^2 - 2$

115. $(x + 3)^2 - 9$

116. $(y - 5)^2 - 36q^2$

117. $x^2 - \left(\frac{1}{x}\right)^2$

118. $a^{2n} - 49b^{2n}$

119. $81 - b^{4k}$

120. $9x^{18} + 48x^9 + 64$

121. $9b^{2n} + 12b^n + 4$

122. $(x + 7)^2 - 4x - 24$

123. $(y + 3)^2 + 2(y + 3) + 1$

124. $49(x + 1)^2 - 42(x + 1) + 9$

Find c such that the polynomial is the square of a binomial.

125. $cy^2 + 6y + 1$

126. $cy^2 - 24y + 9$

Use the TABLE feature to determine whether the factorization is correct.

127. $x^2 + 9 = (x + 3)(x + 3)$

128. $x^2 - 49 = (x - 7)(x + 7)$

129. $x^2 + 9 = (x + 3)^2$

130. $x^2 - 49 = (x - 7)^2$

11.6 Factoring: A General Strategy

a We now combine all of our factoring techniques and consider a general strategy for factoring polynomials. Here we will encounter polynomials of all the types we have considered, in random order, so you will have the opportunity to determine which method to use.

Objective

a Factor polynomials completely using any of the methods considered in this chapter.

For Extra Help

TAPE 22 MAC WIN CD-ROM

To factor a polynomial:

a) Always look first for a common factor. If there is one, factor out the largest common factor.

b) Then look at the number of terms.

Two terms: Determine whether you have a difference of squares. Do not try to factor a sum of squares: $A^2 + B^2$.

Three terms: Determine whether the trinomial is a square. If it is, you know how to factor. If not, try trial and error, using FOIL or grouping.

Four terms: Try factoring by grouping.

c) *Always factor completely.* If a factor with more than one term can still be factored, you should factor it. When no factor can be factored further, you have finished.

Example 1 Factor: $5t^4 - 80$.

a) We look for a common factor:

$$5t^4 - 80 = 5(t^4 - 16).$$

b) The factor $t^4 - 16$ has only two terms. It is a difference of squares: $(t^2)^2 - 4^2$. We factor it, being careful to include the common factor:

$$5(t^2 + 4)(t^2 - 4).$$

c) We see that one of the factors is again a difference of squares. We factor it:

$$5(t^2 + 4)(t + 2)(t - 2).$$

This is a sum of squares. It cannot be factored!

We have factored completely because no factor with more than one term can be factored further.

Example 2 Factor: $2x^3 + 10x^2 + x + 5$.

a) We look for a common factor. There isn't one.

b) There are four terms. We try factoring by grouping:

$2x^3 + 10x^2 + x + 5$

$= (2x^3 + 10x^2) + (x + 5)$ — **Separating into two binomials**

$= 2x^2(x + 5) + 1(x + 5)$ — **Factoring each binomial**

$= (2x^2 + 1)(x + 5).$ — **Factoring out the common factor $x + 5$**

c) None of these factors can be factored further, so we have factored completely.

Factor.

1. $3m^4 - 3$

2. $x^6 + 8x^3 + 16$

3. $2x^4 + 8x^3 + 6x^2$

4. $3x^3 + 12x^2 - 2x - 8$

5. $8x^3 - 200x$

Answers on page A-35

Example 3 Factor: $x^5 - 2x^4 - 35x^3$.

a) We look first for a common factor. This time there is one, x^3:

$$x^5 - 2x^4 - 35x^3 = x^3(x^2 - 2x - 35).$$

b) The factor $x^2 - 2x - 35$ has three terms, but it is not a trinomial square. We factor it using trial and error (FOIL or grouping):

$$x^5 - 2x^4 - 35x^3 = x^3(x^2 - 2x - 35) = x^3(x - 7)(x + 5).$$

Don't forget to include the common factor in the final answer!

c) No factor with more than one term can be factored further, so we have factored completely.

Example 4 Factor: $x^4 - 10x^2 + 25$.

a) We look first for a common factor. There isn't one.

b) There are three terms. We see that this polynomial is a trinomial square. We factor it:

$$x^4 - 10x^2 + 25 = (x^2)^2 - 2 \cdot x^2 \cdot 5 + 5^2 = (x^2 - 5)^2.$$

c) Since $x^2 - 5$ cannot be factored further, we have factored completely.

Do Exercises 1–5.

Example 5 Factor: $6x^2y^4 - 21x^3y^5 + 3x^2y^6$.

a) We look first for a common factor:

$$6x^2y^4 - 21x^3y^5 + 3x^2y^6 = 3x^2y^4(2 - 7xy + y^2).$$

b) There are three terms in $2 - 7xy + y^2$. We determine whether the trinomial is a square. Since only y^2 is a square, we do not have a trinomial square. Can the trinomial be factored by trial and error? A key to the answer is that x is only in the term $-7xy$. The polynomial might be in a form like $(1 - y)(2 + y)$, but there would be no x in the middle term. Thus, $2 - 7xy + y^2$ cannot be factored.

c) Have we factored completely? Yes, because no factor with more than one term can be factored further.

Example 6 Factor: $(p + q)(x + 2) + (p + q)(x + y)$.

a) We look for a common factor:

$$(p + q)(x + 2) + (p + q)(x + y) = (p + q)[(x + 2) + (x + y)]$$
$$= (p + q)(2x + y + 2).$$

b) There are three terms in $2x + y + 2$, but this trinomial cannot be factored further.

c) Neither factor can be factored further, so we have factored completely.

Example 7 Factor: $px + py + qx + qy$.

a) We look first for a common factor. There isn't one.

b) There are four terms. We try factoring by grouping:

$$\begin{aligned} px + py + qx + qy &= p(x + y) + q(x + y) \\ &= (p + q)(x + y). \end{aligned}$$

c) Have we factored completely? Since neither factor can be factored further, we have factored completely.

Example 8 Factor: $25x^2 + 20xy + 4y^2$.

a) We look first for a common factor. There isn't one.

b) There are three terms. We determine whether the trinomial is a square. The first term and the last term are squares:

$$25x^2 = (5x)^2 \quad \text{and} \quad 4y^2 = (2y)^2.$$

Since twice the product of $5x$ and $2y$ is the other term,

$$2 \cdot 5x \cdot 2y = 20xy,$$

the trinomial is a perfect square.

We factor by writing the square roots of the square terms and the sign of the middle term:

$$25x^2 + 20xy + 4y^2 = (5x + 2y)^2.$$

We can check by squaring $5x + 2y$.

c) Since $5x + 2y$ cannot be factored further, we have factored completely.

Example 9 Factor: $p^2q^2 + 7pq + 12$.

a) We look first for a common factor. There isn't one.

b) There are three terms. We determine whether the trinomial is a square. The first term is a square, but neither of the other terms is a square, so we do not have a trinomial square. We use the trial-and-error or grouping method, thinking of the product pq as a single variable. We consider this possibility for factorization:

$$(pq + \square)(pq + \square).$$

We factor the last term, 12. All the signs are positive, so we consider only positive factors. Possibilities are 1, 12 and 2, 6 and 3, 4. The pair 3, 4 gives a sum of 7 for the coefficient of the middle term. Thus,

$$p^2q^2 + 7pq + 12 = (pq + 3)(pq + 4).$$

c) No factor with more than one term can be factored further, so we have factored completely.

Factor.

6. $x^4y^2 + 2x^3y + 3x^2y$

7. $10p^6q^2 + 4p^5q^3 + 2p^4q^4$

8. $(a - b)(x + 5) + (a - b)(x + y^2)$

9. $ax^2 + ay + bx^2 + by$

10. $x^4 + 2x^2y^2 + y^4$

11. $x^2y^2 + 5xy + 4$

12. $p^4 - 81q^4$

Answers on page A-35

Example 10 Factor: $8x^4 - 20x^2y - 12y^2$.

a) We look first for a common factor:

$$8x^4 - 20x^2y - 12y^2 = 4(2x^4 - 5x^2y - 3y^2).$$

b) There are three terms in $2x^4 - 5x^2y - 3y^2$. We determine whether the trinomial is a square. Since none of the terms is a square, we do not have a trinomial square. We factor $2x^4$. Possibilities are $2x^2$, x^2 and $2x$, x^3 and others. We also factor the last term, $-3y^2$. Possibilities are $3y$, $-y$ and $-3y$, y and others. We look for factors such that the sum of their products is the middle term. We try some possibilities:

$$(2x - y)(x^3 + 3y) = 2x^4 + 6xy - x^3y - 3y^2,$$
$$(2x^2 - y)(x^2 + 3y) = 2x^4 + 5x^2y - 3y^2,$$
$$(2x^2 + y)(x^2 - 3y) = 2x^4 - 5x^2y - 3y^2.$$

c) No factor with more than one term can be factored further, so we have factored completely. The factorization, including the common factor, is

$$4(2x^2 + y)(x^2 - 3y).$$

Example 11 Factor: $a^4 - 16b^4$.

a) We look first for a common factor. There isn't one.

b) There are two terms. Since $a^4 = (a^2)^2$ and $16b^4 = (4b^2)^2$, we see that we do have a difference of squares. Thus,

$$a^4 - 16b^4 = (a^2 + 4b^2)(a^2 - 4b^2).$$

c) The last factor can be factored further. It is also a difference of squares. Thus,

$$a^4 - 16b^4 = (a^2 + 4b^2)(a + 2b)(a - 2b).$$

Do Exercises 6–12.

Exercise Set 11.6

a Factor completely.

1. $3x^2 - 192$

2. $2t^2 - 18$

3. $a^2 + 25 - 10a$

4. $y^2 + 49 + 14y$

5. $2x^2 - 11x + 12$

6. $8y^2 - 18y - 5$

7. $x^3 + 24x^2 + 144x$

8. $x^3 - 18x^2 + 81x$

9. $x^3 + 3x^2 - 4x - 12$

10. $x^3 - 5x^2 - 25x + 125$

11. $48x^2 - 3$

12. $50x^2 - 32$

13. $9x^3 + 12x^2 - 45x$

14. $20x^3 - 4x^2 - 72x$

15. $x^2 + 4$

16. $t^2 + 25$

17. $x^4 + 7x^2 - 3x^3 - 21x$

18. $m^4 + 8m^3 + 8m^2 + 64m$

19. $x^5 - 14x^4 + 49x^3$

20. $2x^6 + 8x^5 + 8x^4$

21. $20 - 6x - 2x^2$

22. $45 - 3x - 6x^2$

23. $x^2 - 6x + 1$

24. $x^2 + 8x + 5$

25. $4x^4 - 64$

26. $5x^5 - 80x$

27. $1 - y^8$

28. $t^8 - 1$

29. $x^5 - 4x^4 + 3x^3$

30. $x^6 - 2x^5 + 7x^4$

31. $\frac{1}{81}x^6 - \frac{8}{27}x^3 + \frac{16}{9}$

32. $36a^2 - 15a + \frac{25}{16}$

33. $mx^2 + my^2$

34. $12p^2 + 24q^3$

35. $9x^2y^2 - 36xy$

36. $x^2y - xy^2$

37. $2\pi rh + 2\pi r^2$

38. $10p^4q^4 + 35p^3q^3 + 10p^2q^2$

39. $(a + b)(x - 3) + (a + b)(x + 4)$

40. $5c(a^3 + b) - (a^3 + b)$

41. $(x - 1)(x + 1) - y(x + 1)$

42. $3(p - q) - q^2(p - q)$

43. $n^2 + 2n + np + 2p$

44. $a^2 - 3a + ay - 3y$

45. $6q^2 - 3q + 2pq - p$

46. $2x^2 - 4x + xy - 2y$

47. $4b^2 + a^2 - 4ab$

48. $x^2 + y^2 - 2xy$

49. $16x^2 + 24xy + 9y^2$

50. $9c^2 + 6cd + d^2$

51. $49m^4 - 112m^2n + 64n^2$

52. $4x^2y^2 + 12xyz + 9z^2$

53. $y^4 + 10y^2z^2 + 25z^4$

54. $0.01x^4 - 0.1x^2y^2 + 0.25y^4$

55. $\frac{1}{4}a^2 + \frac{1}{3}ab + \frac{1}{9}b^2$

56. $4p^2q + pq^2 + 4p^3$

57. $a^2 - ab - 2b^2$

58. $3b^2 - 17ab - 6a^2$

59. $2mn - 360n^2 + m^2$

60. $15 + x^2y^2 + 8xy$

61. $m^2n^2 - 4mn - 32$

62. $p^2q^2 + 7pq + 6$

63. $a^2b^6 + 4ab^5 - 32b^4$

64. $p^5q^2 + 3p^4q - 10p^3$

65. $a^5 + 4a^4b - 5a^3b^2$

66. $2s^6t^2 + 10s^3t^3 + 12t^4$

67. $a^2 - \frac{1}{25}b^2$

68. $p^2 - \frac{1}{49}b^2$

69. $x^2 - y^2$

70. $p^2q^2 - r^2$

71. $16 - p^4q^4$

72. $15a^4 - 15b^4$

73. $1 - 16x^{12}y^{12}$

74. $81a^4 - b^4$

75. $q^3 + 8q^2 - q - 8$

76. $m^3 - 7m^2 - 4m + 28$

77. $112xy + 49x^2 + 64y^2$

78. $4ab^5 - 32b^4 + a^2b^6$

Skill Maintenance

Sports-Car Sales. The sales of sports cars rise and fall over the years due often to new or redesigned models, such as the 1997 Corvette. Sales for recent years are shown in the following line graph. Use the graph for Exercises 79–84. [9.1a]

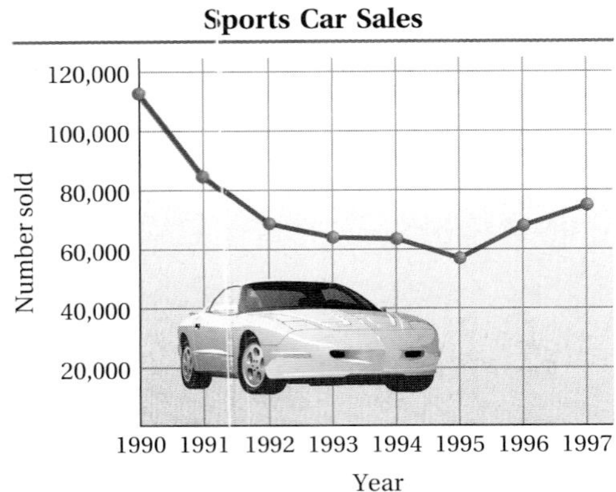

Source: Autodata

79. In which year were sports-car sales the highest?

80. In which year were sports-car sales the lowest?

81. In which year were sports-car sales about 68,000?

82. What were sports-car sales in 1997?

83. By how much did sales increase from 1995 to 1997?

84. By how much did sales decrease from 1990 to 1995?

85. Divide: $\frac{7}{5} \div \left(-\frac{11}{10}\right)$. [7.6c]

86. Multiply: $(5x - t)^2$. [10.6d], [10.7f]

87. Solve $A = aX + bX - 7$ for X. [8.6a]

88. Solve: $4(x - 9) - 2(x + 7) < 14$. [8.7e]

Synthesis

89. ◆ Kelly factored $16 - 8x + x^2$ as $(x - 4)^2$, while Tony factored it as $(4 - x)^2$. Evaluate each expression for several values of x. Then explain why both answers are correct.

90. ◆ Describe in your own words a strategy that can be used to factor polynomials.

Factor completely.

91. $a^4 - 2a^2 + 1$

92. $x^4 + 9$

93. $12.25x^2 - 7x + 1$

94. $\frac{1}{5}x^2 - x + \frac{4}{5}$

95. $5x^2 + 13x + 7.2$

96. $x^3 - (x - 3x^2) - 3$

97. $18 + y^3 - 9y - 2y^2$

98. $-(x^4 - 7x^2 - 18)$

99. $a^3 + 4a^2 + a + 4$

100. $x^3 + x^2 - (4x + 4)$

101. $x^4 - 7x^2 - 18$

102. $3x^4 - 15x^2 + 12$

103. $x^3 - x^2 - 4x + 4$

104. $y^2(y + 1) - 4y(y + 1) - 21(y + 1)$

105. $y^2(y - 1) - 2y(y - 1) + (y - 1)$

106. $6(x - 1)^2 + 7y(x - 1) - 3y^2$

107. $(y + 4)^2 + 2x(y + 4) + x^2$

108. $a^4 - 81$

Collaborative Learning Manual — Create polynomials for factoring.

11.7 Solving Quadratic Equations by Factoring

In this section, we introduce a new equation-solving method and use it along with factoring to solve certain equations like $x^2 + x - 156 = 0$.

> A **quadratic equation** is an equation equivalent to an equation of the type
>
> $$ax^2 + bx + c = 0, \quad \text{where } a > 0.$$
>
> The trinomial on the left is of second degree.

Objectives

a. Solve equations (already factored) using the principle of zero products.

b. Solve quadratic equations by factoring and then using the principle of zero products.

For Extra Help

TAPE 22 | MAC WIN | CD-ROM

a The Principle of Zero Products

The product of two numbers is 0 if one or both of the numbers is 0. Furthermore, *if any product is* 0, *then a factor must be* 0. For example:

If $7x = 0$, then we know that $x = 0$.

If $x(2x - 9) = 0$, then we know that $x = 0$ or $2x - 9 = 0$.

If $(x + 3)(x - 2) = 0$, then we know that $x + 3 = 0$ or $x - 2 = 0$.

In a product such as $ab = 24$, we cannot conclude with certainty that a is 24 or that b is 24, but if $ab = 0$, we can conclude that $a = 0$ or $b = 0$.

Example 1 Solve: $(x + 3)(x - 2) = 0$.

We have a product of 0. This equation will be true when either factor is 0. Thus it is true when

$$x + 3 = 0 \quad \text{or} \quad x - 2 = 0.$$

Here we have two simple equations that we know how to solve:

$$x = -3 \quad \text{or} \quad x = 2.$$

Each of the numbers -3 and 2 is a solution of the original equation, as we can see in the following checks.

CHECK: For -3:

$$\begin{array}{c|c} (x + 3)(x - 2) & 0 \\ \hline (-3 + 3)(-3 - 2) & ?\ 0 \\ 0(-5) & \\ 0 & \end{array} \quad \text{TRUE}$$

For 2:

$$\begin{array}{c|c} (x + 3)(x - 2) & 0 \\ \hline (2 + 3)(2 - 2) & ?\ 0 \\ 5(0) & \\ 0 & \end{array} \quad \text{TRUE}$$

We now have a principle to help in solving quadratic equations.

> **THE PRINCIPLE OF ZERO PRODUCTS**
>
> An equation $ab = 0$ is true if and only if $a = 0$ is true or $b = 0$ is true, or both are true. (A product is 0 if and only if one or both of the factors is 0.)

Solve using the principle of zero products.

1. $(x - 3)(x + 4) = 0$

2. $(x - 7)(x - 3) = 0$

3. $(4t + 1)(3t - 2) = 0$

4. Solve: $y(3y - 17) = 0$.

Answers on page A-35

Calculator Spotlight

Checking Solutions on a Grapher. We can check solutions using the TABLE or the CALC feature. Consider the equation $(x + 3)(x - 2) = 0$ of Example 1. We enter $y_1 = (x + 3)(x - 2)$ into the grapher.

Table. First, we set up the table with TblStart $= -4$ and ΔTbl $= 1$. By scrolling through the table, we see that when $x = -3$, $y = 0$ and when $x = 2$, $y = 0$.

X	Y1	
−4	6	
−3	0	
−2	−4	
−1	−6	
0	−6	
1	−4	
2	0	
X = −4		

Calc. We graph the equation $y = (x + 3)(x - 2)$ and use the CALC and VALUE features to check the solutions to the equation $(x + 3)(x - 2) = 0$. By entering -3 for x, we see that when $x = -3$, $y = 0$ (that is, when $x = -3$, the graph crosses the x-axis). It can also be shown that when $x = 2$, $y = 0$.

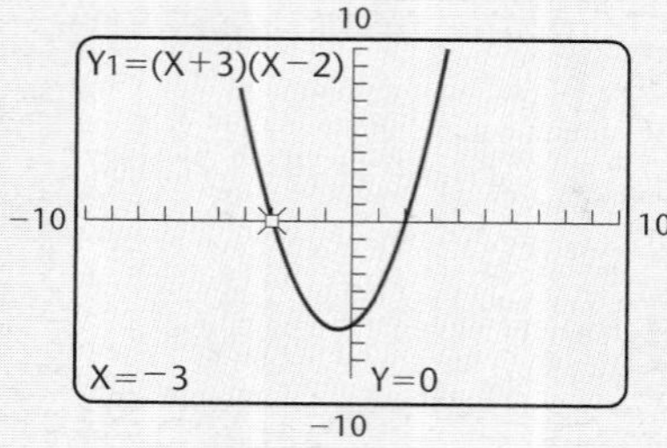

Exercises

1. Check the solutions of the equations in Examples 1–3.

Example 2 Solve: $(5x + 1)(x - 7) = 0$.

$$\begin{aligned} (5x + 1)(x - 7) &= 0 \\ 5x + 1 = 0 \quad &or \quad x - 7 = 0 \qquad \text{Using the principle of zero products} \\ 5x = -1 \quad &or \quad x = 7 \qquad \text{Solving the two equations separately} \\ x = -\tfrac{1}{5} \quad &or \quad x = 7 \end{aligned}$$

CHECK: For $-\frac{1}{5}$:

$$\begin{array}{c|c} (5x + 1)(x - 7) & 0 \\ \hline \left(5\left(-\frac{1}{5}\right) + 1\right)\left(-\frac{1}{5} - 7\right) & ?\ 0 \\ (-1 + 1)\left(-7\frac{1}{5}\right) & \\ 0\left(-7\frac{1}{5}\right) & \\ 0 & \end{array} \quad \text{TRUE}$$

For 7:

$$\begin{array}{c|c} (5x + 1)(x - 7) & 0 \\ \hline (5(7) + 1)(7 - 7) & ?\ 0 \\ (35 + 1) \cdot 0 & \\ 36 \cdot 0 & \\ 0 & \end{array} \quad \text{TRUE}$$

The solutions are $-\frac{1}{5}$ and 7.

When you solve an equation using the principle of zero products, you may wish to check by substitution, as in Examples 1 and 2. Such a check will detect errors in solving.

Do Exercises 1–3 on the preceding page.

When some factors have only one term, you can still use the principle of zero products.

Example 3 Solve: $x(2x - 9) = 0$.

$$\begin{aligned} x(2x - 9) &= 0 \\ x = 0 \quad &or \quad 2x - 9 = 0 \qquad \text{Using the principle of zero products} \\ x = 0 \quad &or \quad 2x = 9 \\ x = 0 \quad &or \quad x = \frac{9}{2} \end{aligned}$$

The solutions are 0 and $\frac{9}{2}$. The check is left to the student.

Do Exercise 4 on the preceding page.

b Using Factoring to Solve Equations

Using factoring and the principle of zero products, we can solve some new kinds of equations. Thus we have extended our equation-solving abilities.

Example 4 Solve: $x^2 + 5x + 6 = 0$.

Compare this equation to those that we know how to solve from Chapter 8. There are no like terms to collect, and we have a squared term. We first factor the polynomial. Then we use the principle of zero products.

$$\begin{aligned} x^2 + 5x + 6 &= 0 \\ (x + 2)(x + 3) &= 0 \qquad \text{Factoring} \\ x + 2 = 0 \quad &or \quad x + 3 = 0 \qquad \text{Using the principle of zero products} \\ x = -2 \quad &or \quad x = -3 \end{aligned}$$

CHECK: For -2:

$$\begin{array}{r|l} x^2 + 5x + 6 & 0 \\ \hline (-2)^2 + 5(-2) + 6 & ?\ 0 \\ 4 - 10 + 6 & \\ -6 + 6 & \\ 0 & \quad \text{TRUE} \end{array}$$

For -3:

$$\begin{array}{r|l} x^2 + 5x + 6 & 0 \\ \hline (-3)^2 + 5(-3) + 6 & ?\ 0 \\ 9 - 15 + 6 & \\ -6 + 6 & \\ 0 & \quad \text{TRUE} \end{array}$$

The solutions are -2 and -3.

CAUTION! Keep in mind that you *must* have 0 on one side of the equation before you can use the principle of zero products. Get all nonzero terms on one side and 0 on the other.

Do Exercise 5.

Example 5 Solve: $x^2 - 8x = -16$.

We first add 16 to get a 0 on one side:

$$x^2 - 8x = -16$$

$$x^2 - 8x + 16 = 0 \quad \text{Adding 16}$$

$$(x - 4)(x - 4) = 0 \quad \text{Factoring}$$

$$x - 4 = 0 \quad or \quad x - 4 = 0 \quad \text{Using the principle of zero products}$$

$$x = 4 \quad or \quad x = 4$$

There is only one solution, 4. The check is left to the student.

Do Exercises 6 and 7.

Example 6 Solve: $x^2 + 5x = 0$.

$$x^2 + 5x = 0$$

$$x(x + 5) = 0 \quad \text{Factoring out a common factor}$$

$$x = 0 \quad or \quad x + 5 = 0 \quad \text{Using the principle of zero products}$$

$$x = 0 \quad or \quad x = -5$$

The solutions are 0 and -5. The check is left to the student.

Example 7 Solve: $4x^2 = 25$.

$$4x^2 = 25$$

$$4x^2 - 25 = 0 \quad \text{Subtracting 25 on both sides to get 0 on one side}$$

$$(2x - 5)(2x + 5) = 0 \quad \text{Factoring a difference of squares}$$

$$2x - 5 = 0 \quad or \quad 2x + 5 = 0$$

$$2x = 5 \quad or \quad 2x = -5$$

$$x = \frac{5}{2} \quad or \quad x = -\frac{5}{2}$$

The solutions are $\frac{5}{2}$ and $-\frac{5}{2}$.

Do Exercises 8 and 9.

5. Solve: $x^2 - x - 6 = 0$.

Solve.

6. $x^2 - 3x = 28$

7. $x^2 = 6x - 9$

Solve.

8. $x^2 - 4x = 0$

9. $9x^2 = 16$

Answers on page A-35

10. Solve: $(x + 1)(x - 1) = 8$.

11. Find the x-intercepts of the graph shown below.

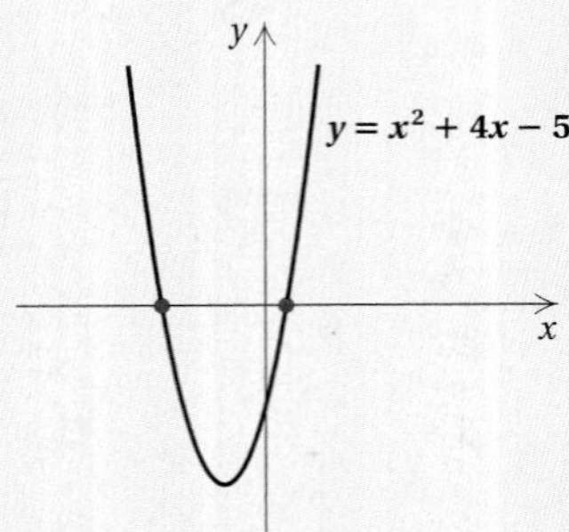

12. Use *only* the graph shown below to solve $3x - x^2 = 0$.

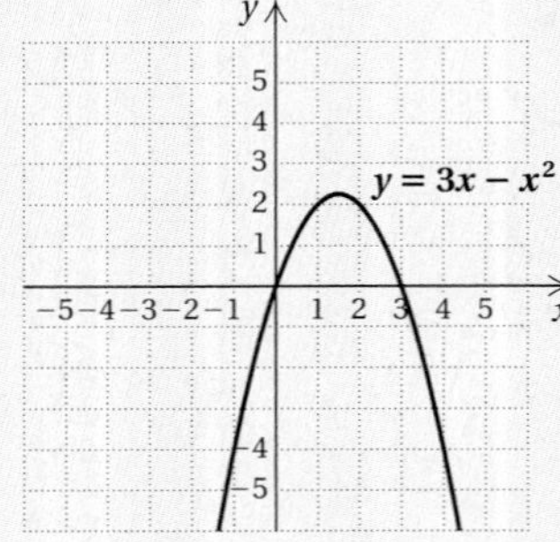

Answer on page A-35

Example 8 Solve: $(x + 2)(x - 2) = 5$.

Be careful with an equation like this one! It might be tempting to set each factor equal to 5. Remember: We must have a 0 on one side. We first carry out the product on the left. Then we subtract 5 on both sides to get 0 on one side. Then we proceed with the principle of zero products.

$$(x + 2)(x - 2) = 5$$

$$x^2 - 4 = 5 \quad \text{Multiplying on the left}$$

$$x^2 - 4 - 5 = 5 - 5 \quad \text{Subtracting 5}$$

$$x^2 - 9 = 0 \quad \text{Simplifying}$$

$$(x + 3)(x - 3) = 0 \quad \text{Factoring}$$

$$x + 3 = 0 \quad or \quad x - 3 = 0 \quad \text{Using the principle of zero products}$$

$$x = -3 \quad or \quad x = 3$$

The solutions are -3 and 3. The check is left to the student.

Do Exercise 10.

AG Algebraic–Graphical Connection

In Chapter 9, we graphed linear equations of the type $y = mx + b$ and $Ax + By = C$. Recall that to find the x-intercept, we replaced y with 0 and solved for x. This procedure can also be used to find the x-intercepts when an equation of the form $y = ax^2 + bx + c$, $a \neq 0$, is to be graphed. Although the details of creating such graphs will be left to Chapter 16, we consider them briefly here from the standpoint of finding the x-intercepts. The graphs are shaped like the following curves. Note that each x-intercept represents a solution of $ax^2 + bx + c = 0$.

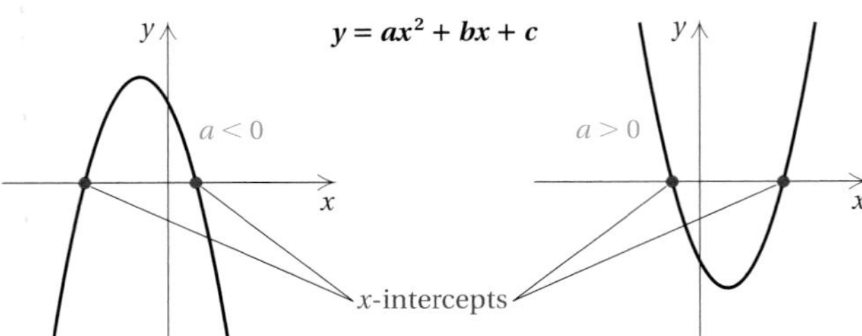

Example 9 Find the x-intercepts of the graph of $y = x^2 - 4x - 5$ shown at right.

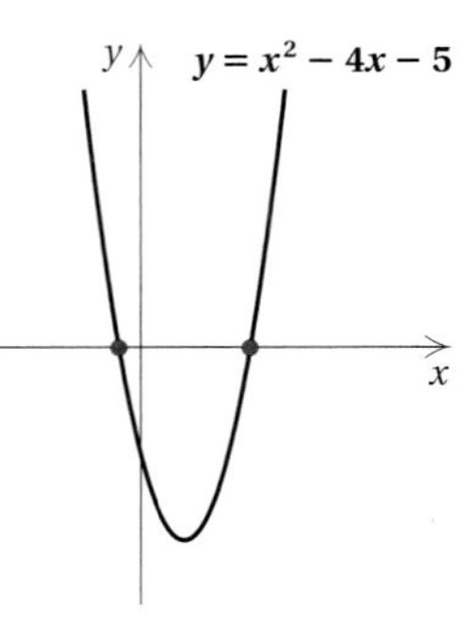

To find the x-intercepts, we let $y = 0$ and solve for x:

$$0 = x^2 - 4x - 5 \quad \text{Substituting 0 for } y$$

$$0 = (x - 5)(x + 1) \quad \text{Factoring}$$

$$x - 5 = 0 \quad or \quad x + 1 = 0 \quad \text{Using the principle of zero products}$$

$$x = 5 \quad or \quad x = -1.$$

The x-intercepts are $(5, 0)$ and $(-1, 0)$.

Do Exercises 11 and 12.

Exercise Set 11.7

a Solve using the principle of zero products.

1. $(x + 4)(x + 9) = 0$

2. $(x + 2)(x - 7) = 0$

3. $(x + 3)(x - 8) = 0$

4. $(x + 6)(x - 8) = 0$

5. $(x + 12)(x - 11) = 0$

6. $(x - 13)(x + 53) = 0$

7. $x(x + 3) = 0$

8. $y(y + 5) = 0$

9. $0 = y(y + 18)$

10. $0 = x(x - 19)$

11. $(2x + 5)(x + 4) = 0$

12. $(2x + 9)(x + 8) = 0$

13. $(5x + 1)(4x - 12) = 0$

14. $(4x + 9)(14x - 7) = 0$

15. $(7x - 28)(28x - 7) = 0$

16. $(13x + 14)(6x - 5) = 0$

17. $2x(3x - 2) = 0$

18. $55x(8x - 9) = 0$

19. $\left(\frac{1}{5} + 2x\right)\left(\frac{1}{9} - 3x\right) = 0$

20. $\left(\frac{7}{4}x - \frac{1}{16}\right)\left(\frac{2}{3}x - \frac{16}{15}\right) = 0$

21. $(0.3x - 0.1)(0.05x + 1) = 0$

22. $(0.1x + 0.3)(0.4x - 20) = 0$

23. $9x(3x - 2)(2x - 1) = 0$

24. $(x + 5)(x - 75)(5x - 1) = 0$

b Solve by factoring and using the principle of zero products. Remember to check.

25. $x^2 + 6x + 5 = 0$

26. $x^2 + 7x + 6 = 0$

27. $x^2 + 7x - 18 = 0$

28. $x^2 + 4x - 21 = 0$

29. $x^2 - 8x + 15 = 0$

30. $x^2 - 9x + 14 = 0$

31. $x^2 - 8x = 0$

32. $x^2 - 3x = 0$

33. $x^2 + 18x = 0$

34. $x^2 + 16x = 0$

35. $x^2 = 16$

36. $100 = x^2$

37. $9x^2 - 4 = 0$

38. $4x^2 - 9 = 0$

39. $0 = 6x + x^2 + 9$

40. $0 = 25 + x^2 + 10x$

41. $x^2 + 16 = 8x$

42. $1 + x^2 = 2x$

43. $5x^2 = 6x$

44. $7x^2 = 8x$

45. $6x^2 - 4x = 10$

46. $3x^2 - 7x = 20$

47. $12y^2 - 5y = 2$

48. $2y^2 + 12y = -10$

49. $t(3t + 1) = 2$

50. $x(x - 5) = 14$

51. $100y^2 = 49$

52. $64a^2 = 81$

53. $x^2 - 5x = 18 + 2x$

54. $3x^2 + 8x = 9 + 2x$

55. $10x^2 - 23x + 12 = 0$

56. $12x^2 + 17x - 5 = 0$

Find the x-intercepts for the graph of the equation.

57.

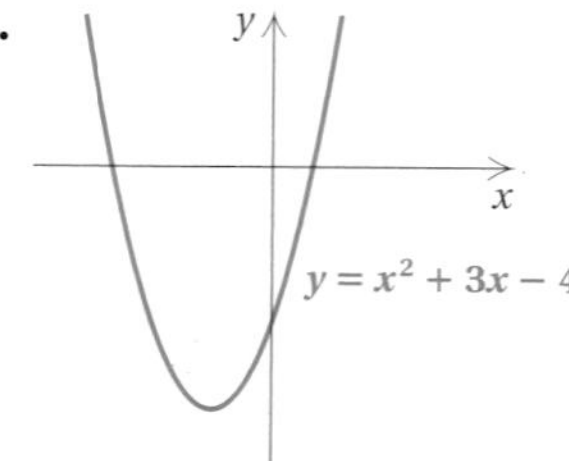

58.

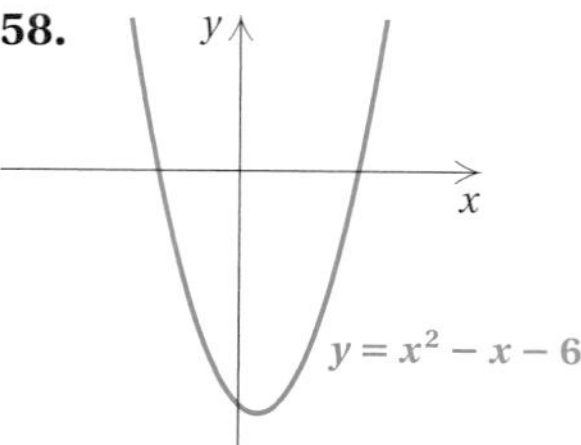

59.

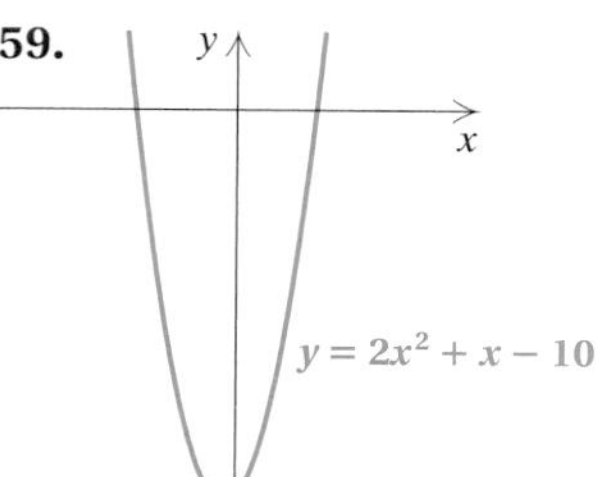

60.

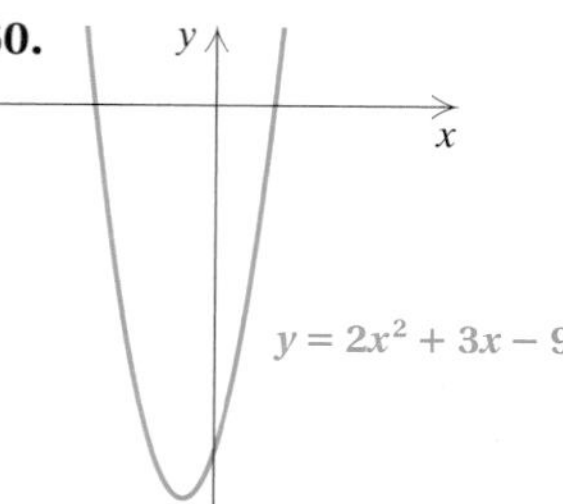

Skill Maintenance

Translate to an algebraic expression. [7.1b]

61. The square of the sum of a and b

62. The sum of the squares of a and b

Divide. [7.6c]

63. $144 \div (-9)$

64. $-24.3 \div 5.4$

65. $-\frac{5}{8} \div \frac{3}{16}$

66. $-\frac{3}{16} \div \left(-\frac{5}{8}\right)$

Synthesis

67. ◆ What is wrong with the following? Explain the correct method of solution.

$$(x - 3)(x + 4) = 8$$
$$x - 3 = 8 \quad or \quad x + 4 = 8$$
$$x = 11 \quad or \quad x = 4$$

68. ◆ What is incorrect about solving $x^2 = 3x$ by dividing by x on both sides?

Solve.

69. $b(b + 9) = 4(5 + 2b)$

70. $y(y + 8) = 16(y - 1)$

71. $(t - 3)^2 = 36$

72. $(t - 5)^2 = 2(5 - t)$

73. $x^2 - \frac{1}{64} = 0$

74. $x^2 - \frac{25}{36} = 0$

75. $\frac{5}{16}x^2 = 5$

76. $\frac{27}{25}x^2 = \frac{1}{3}$

77. Find an equation that has the given numbers as solutions. For example, 3 and −2 are solutions to $x^2 - x - 6 = 0$.

a) −3, 4 b) −3, −4 c) $\frac{1}{2}, \frac{1}{2}$ d) 5, −5 e) 0, 0.1, $\frac{1}{4}$

Create and solve quadratic equations.

11.8 Applications and Problem Solving

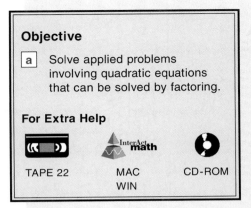

a We can now use our new method for solving quadratic equations and the five steps for solving problems.

Example 1 One more than a number times one less than the number is 8. Find all such numbers.

1. **Familiarize.** Let's make a guess. Try 5. One more than 5 is 6. One less than the number is 4. The product of one more than the number and one less than the number is 6(4), or 24, which is too large. We could continue to guess, but let's use our algebraic skills to find the numbers. Let x = the number (there could be more than one).
2. **Translate.** From the familiarization, we can translate as follows:

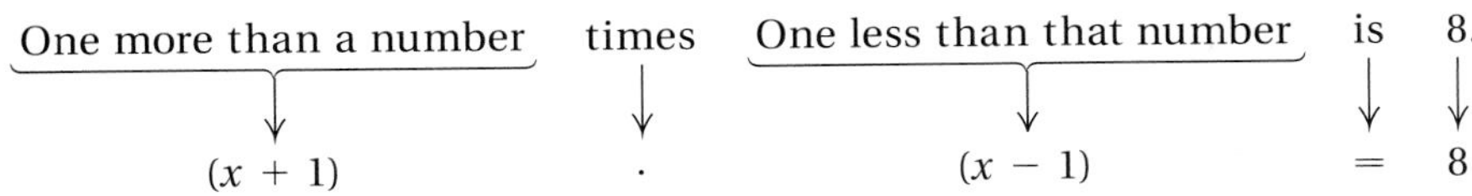

One more than a number
times
One less than that number
is
8.
$(x + 1)$
$\cdot$
$(x - 1)$
$=$
8

3. **Solve.** We solve the equation as follows:

$$(x + 1)(x - 1) = 8$$
$$x^2 - 1 = 8 \quad \text{Multiplying}$$
$$x^2 - 1 - 8 = 8 - 8 \quad \text{Subtracting 8 on both sides to get 0 on one side}$$
$$x^2 - 9 = 0 \quad \text{Simplifying}$$
$$(x - 3)(x + 3) = 0 \quad \text{Factoring}$$
$$x - 3 = 0 \quad or \quad x + 3 = 0 \quad \text{Using the principle of zero products}$$
$$x = 3 \quad or \quad x = -3.$$

4. **Check.** One more than 3 (this is 4) times one less than 3 (this is 2) is 8. Thus, 3 checks. One more than -3 (this is -2) times one less than -3 (this is -4) is 8. Thus, -3 also checks.
5. **State.** There are two such numbers, 3 and -3.

Do Exercises 1 and 2.

Example 2 The square of a number minus twice the number is 48. Find all such numbers.

1. **Familiarize.** Let's make a guess to help understand the problem and ease the translation. Try 6. The square of 6 is 36, and twice the number is 12. Then $36 - 12 = 24$, so 6 is not a number we want. We find the numbers using our algebraic skills. Let x = the number, or numbers.
2. **Translate.** We translate as follows:

The square of a number	minus	Twice the number	is	48.
↓	↓	↓	↓	↓
x^2	$-$	$2x$	$=$	48

3. **Solve.** We solve the equation as follows:

$$x^2 - 2x = 48$$
$$x^2 - 2x - 48 = 48 - 48 \quad \text{Subtracting 48 to get 0 on one side}$$
$$x^2 - 2x - 48 = 0 \quad \text{Simplifying}$$
$$(x - 8)(x + 6) = 0 \quad \text{Factoring}$$
$$x - 8 = 0 \quad or \quad x + 6 = 0 \quad \text{Using the principle of zero products}$$
$$x = 8 \quad or \quad x = -6.$$

Objective

a Solve applied problems involving quadratic equations that can be solved by factoring.

For Extra Help

TAPE 22 | InterAct math MAC WIN | CD-ROM

1. One more than a number times one less than the number is 24. Find all such numbers.

2. Seven less than a number times eight less than the number is 0. Find all such numbers.

Answers on page A-35

3. The square of a number minus the number is 20. Find all such numbers.

4. The width of a rectangle is 2 cm less than the length. The area is 15 cm^2. Find the length and the width.

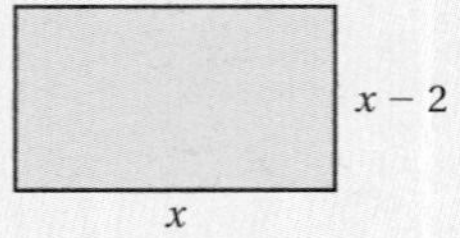

Answers on page A-35

4. **Check.** The square of 8 is 64, and twice the number 8 is 16. Then $64 - 16$ is 48, so 8 checks. The square of -6 is $(-6)^2$, or 36, and twice -6 is -12. Then $36 - (-12)$ is 48, so -6 checks.
5. **State.** There are two such numbers, 8 and -6.

Do Exercise 3.

Example 3 *Sailing.* The height of a triangular foresail on a racing yacht is 7 ft more than the base. The area of the triangle is 30 ft^2. Find the height and the base.

1. **Familiarize.** We first make a drawing. If you don't remember the formula for the area of a triangle, look it up on the inside front cover of this book or in a geometry book. The area is $\frac{1}{2}$(base)(height).

   We let b = the base of the triangle. Then $b + 7$ = the height.

2. **Translate.** It helps to reword this problem before translating:

$\frac{1}{2}$	times	Base	times	Height	is	30.	Rewording
↓	↓	↓	↓	↓	↓	↓	
$\frac{1}{2}$	$\cdot$	b	$\cdot$	$(b + 7)$	$=$	30	Translating

3. **Solve.** We solve the equation as follows:

$$\frac{1}{2} \cdot b \cdot (b + 7) = 30$$
$$\frac{1}{2}(b^2 + 7b) = 30 \quad \text{Multiplying}$$
$$2 \cdot \frac{1}{2}(b^2 + 7b) = 2 \cdot 30 \quad \text{Multiplying by 2}$$
$$b^2 + 7b = 60 \quad \text{Simplifying}$$
$$b^2 + 7b - 60 = 60 - 60 \quad \text{Subtracting 60 to get 0 on one side}$$
$$b^2 + 7b - 60 = 0$$
$$(b + 12)(b - 5) = 0 \quad \text{Factoring}$$
$$b + 12 = 0 \quad or \quad b - 5 = 0 \quad \text{Using the principle of zero products}$$
$$b = -12 \quad or \quad b = 5.$$

4. **Check.** The base of a triangle cannot have a negative length, so -12 cannot be a solution. Suppose the base is 5 ft. Then the height is 7 ft more than the base, so the height is 12 ft and the area is $\frac{1}{2}(5)(12)$, or 30 ft^2. These numbers check in the original problem.
5. **State.** The height is 12 ft and the base is 5 ft.

Do Exercise 4.

Example 4 *Games in a Sports League.* In a sports league of n teams in which each team plays every other team twice, the total number N of games to be played is given by

$$n^2 - n = N.$$

If a basketball league plays a total of 240 games, how many teams are in the league?

1., 2. Familiarize and **Translate.** We are given that n = the number of teams in a league and N = the number of games. To familiarize yourself with this problem, reread Example 3 in Section 10.3 where we first considered it. To find the number of teams n in a league in which 240 games are played, we substitute 240 for N in the equation:

$n^2 - n = 240.$ Substituting 240 for N

3. Solve. We solve the equation as follows:

$$n^2 - n = 240$$
$$n^2 - n - 240 = 240 - 240 \quad \text{Subtracting 240 to get 0 on one side}$$
$$n^2 - n - 240 = 0$$
$$(n - 16)(n + 15) = 0 \quad \text{Factoring}$$
$$n - 16 = 0 \quad or \quad n + 15 = 0 \quad \text{Using the principle of zero products}$$
$$n = 16 \quad or \quad n = -15.$$

4. Check. The solutions of the equation are 16 and -15. Since the number of teams cannot be negative, -15 cannot be a solution. But 16 checks, since $16^2 - 16 = 256 - 16 = 240$.

5. State. There are 16 teams in the league.

Do Exercise 5.

Example 5 The product of the numbers of two consecutive entrants in a marathon race is 156. Find the numbers.

1. Familiarize. The numbers are consecutive integers. Recall that consecutive integers are next to each other, such as 49 and 50, or -6 and -5. Let x = the smaller integer; then $x + 1$ = the larger integer.

2. Translate. It helps to reword the problem before translating:

First integer	times	Second integer	is	156.	Rewording
↓	↓	↓	↓	↓	
x	$\cdot$	$(x + 1)$	$=$	156	Translating

3. Solve. We solve the equation as follows:

$$x(x + 1) = 156$$
$$x^2 + x = 156 \quad \text{Multiplying}$$
$$x^2 + x - 156 = 156 - 156 \quad \text{Subtracting 156 to get 0 on one side}$$
$$x^2 + x - 156 = 0 \quad \text{Simplifying}$$
$$(x - 12)(x + 13) = 0 \quad \text{Factoring}$$
$$x - 12 = 0 \quad or \quad x + 13 = 0 \quad \text{Using the principle of zero products}$$
$$x = 12 \quad or \quad x = -13.$$

4. Check. The solutions of the equation are 12 and -13. When x is 12, then $x + 1$ is 13, and $12 \cdot 13 = 156$. The numbers 12 and 13 are consecutive integers that are solutions to the problem. When x is -13, then $x + 1$ is -12, and $(-13)(-12) = 156$. The numbers -13 and -12 are also consecutive integers, but they are not solutions of the problem because negative numbers are not used as entry numbers.

5. State. The entry numbers are 12 and 13.

5. Use $N = n^2 - n$ for the following.

a) *Volleyball League.* A women's volleyball league has 19 teams. What is the total number of games to be played?

b) *Softball League.* A slow-pitch softball league plays a total of 72 games. How many teams are in the league?

Answers on page A-35

6. The product of the page numbers on two facing pages of a book is 506. Find the page numbers.

7. The length of one leg of a right triangle is 1 m longer than the other. The length of the hypotenuse is 5 m. Find the lengths of the legs.

Answers on page A-35

Do Exercise 6.

The following example involves the Pythagorean theorem, which relates the lengths of the sides of a right triangle. A **right triangle** has a 90° angle. The side opposite the 90° angle is called the **hypotenuse**. The other sides are called **legs**.

> **THE PYTHAGOREAN THEOREM**
>
> The sum of the squares of the legs of a right triangle is equal to the square of the hypotenuse:
>
> $$a^2 + b^2 = c^2.$$

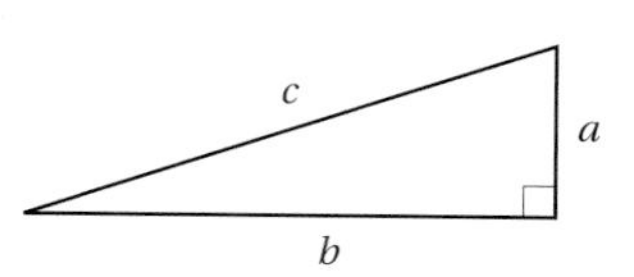

Example 6 *Ladder Settings.* A ladder of length 13 ft is placed against a building in such a way that the distance from the top of the ladder to the ground is 7 ft more than the distance from the bottom of the ladder to the building. Find both distances.

1. **Familiarize.** We first make a drawing. The ladder and the missing distances form the hypotenuse and legs of a right triangle. We let x = the length of the side (leg) across the bottom. Then $x + 7$ = the length of the other side (leg). The hypotenuse has length 13 ft.

2. **Translate.** Since a right triangle is formed, we can use the Pythagorean theorem:

$$a^2 + b^2 = c^2$$
$$x^2 + (x + 7)^2 = 13^2. \quad \text{Substituting}$$

3. **Solve.** We solve the equation as follows:

$$x^2 + (x^2 + 14x + 49) = 169 \quad \text{Squaring the binomial and 13}$$
$$2x^2 + 14x + 49 = 169 \quad \text{Collecting like terms}$$
$$2x^2 + 14x + 49 - 169 = 169 - 169 \quad \text{Subtracting 169 to get 0 on one side}$$
$$2x^2 + 14x - 120 = 0 \quad \text{Simplifying}$$
$$2(x^2 + 7x - 60) = 0 \quad \text{Factoring out a common factor}$$
$$x^2 + 7x - 60 = 0 \quad \text{Dividing by 2}$$
$$(x + 12)(x - 5) = 0 \quad \text{Factoring}$$
$$x + 12 = 0 \quad or \quad x - 5 = 0$$
$$x = -12 \quad or \quad x = 5.$$

4. **Check.** The negative integer -12 cannot be the length of a side. When $x = 5$, $x + 7 = 12$, and $5^2 + 12^2 = 13^2$. So 5 and 12 check.
5. **State.** The distance from the top of the ladder to the ground is 12 ft. The distance from the bottom of the ladder to the building is 5 ft.

Do Exercise 7.

Exercise Set 11.8

a Solve.

1. If 7 is added to the square of a number, the result is 32. Find all such numbers.

2. If you subtract a number from four times its square, the result is 3. Find all such numbers.

3. Fifteen more than the square of a number is eight times the number. Find all such numbers.

4. Eight more than the square of a number is six times the number. Find all such numbers.

5. *Calculator Dimensions.* The length of a rectangular calculator is 5 cm greater than the width. The area of the calculator is 84 cm^2. Find the length and the width.

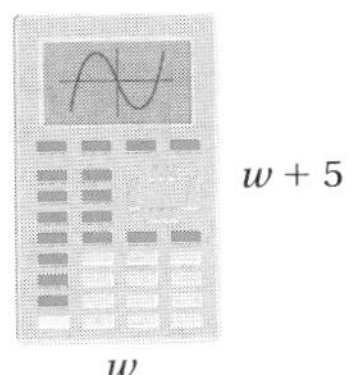

6. *Garden Dimensions.* The length of a rectangular garden is 4 m greater than the width. The area of the garden is 96 m^2. Find the length and the width.

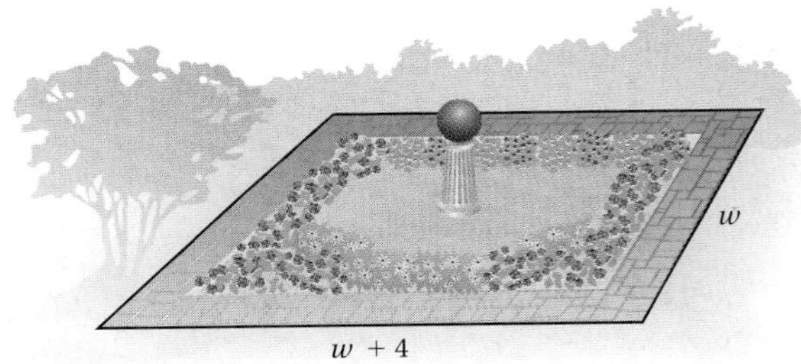

7. *Consecutive Page Numbers.* The product of the page numbers on two facing pages of a book is 210. Find the page numbers.

8. *Consecutive Page Numbers.* The product of the page numbers on two facing pages of a book is 420. Find the page numbers.

9. The product of two consecutive even integers is 168. Find the integers.

10. The product of two consecutive even integers is 224. Find the integers.

11. The product of two consecutive odd integers is 255. Find the integers.

12. The product of two consecutive odd integers is 143. Find the integers.

13. The area of a square bookcase is 5 more than the perimeter. Find the length of a side.

14. The perimeter of a square porch is 3 more than the area. Find the length of a side.

15. *Sharks' Teeth.* Sharks' teeth are shaped like triangles. The height of a tooth of a great white shark is 1 cm longer than the base. The area is 15 cm^2. Find the height and the base.

16. The base of a triangle is 6 cm greater than twice the height. The area is 28 cm^2. Find the height and the base.

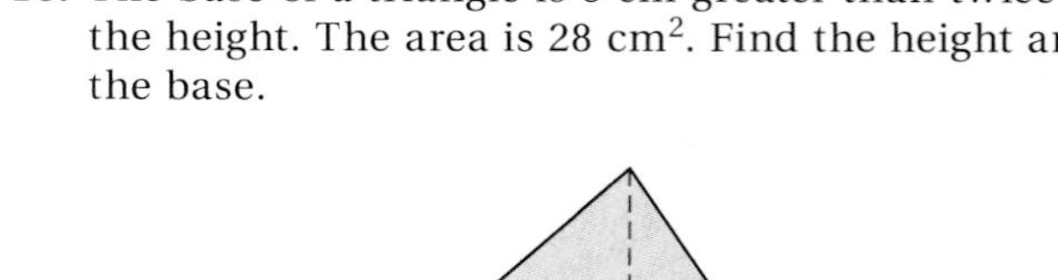

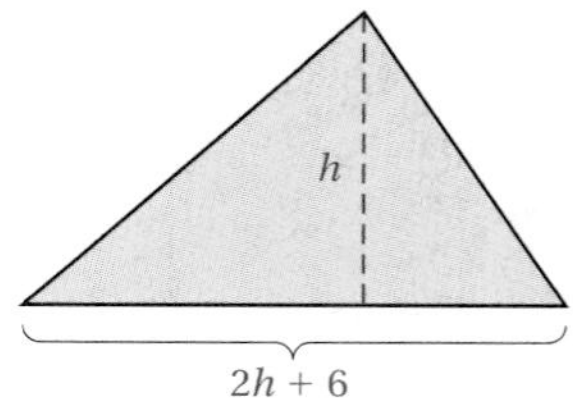

17. If the sides of a square are lengthened by 3 km, the area becomes 81 km^2. Find the length of a side of the original square.

18. The base and the height of a triangle are the same length. If the length of the base is increased by 4 in., the area becomes 96 in^2. Find the length of the base of the original triangle.

Rocket Launch. A model water rocket is launched with an initial velocity of 180 ft/sec. Its height h, in feet, after t seconds is given by the formula

$$h = 180t - 16t^2.$$

Use this formula for Exercises 19 and 20.

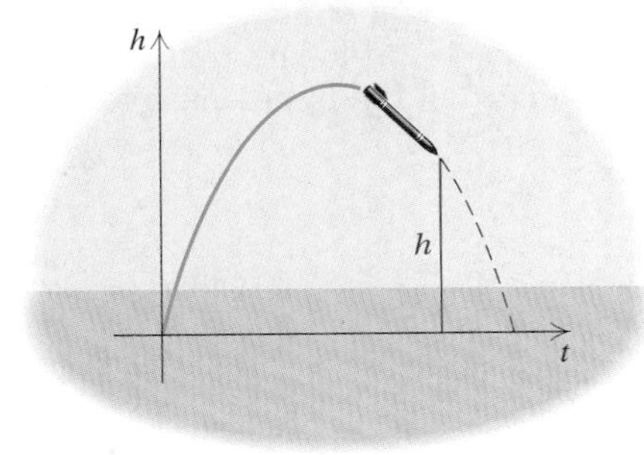

19. After how many seconds will the rocket first reach a height of 464 ft?

20. After how many seconds will the rocket again be at that same height of 464 ft? (See Exercise 19.)

21. The sum of the squares of two consecutive odd positive integers is 74. Find the integers.

22. The sum of the squares of two consecutive odd positive integers is 130. Find the integers.

Games in a League. Use $n^2 - n = N$ for Exercises 23–26.

23. A chess league has 14 teams. What is the total number of games to be played?

24. A women's volleyball league has 23 teams. What is the total number of games to be played?

25. A slow-pitch softball league plays a total of 132 games. How many teams are in the league?

26. A basketball league plays a total of 90 games. How many teams are in the league?

Handshakes. A researcher wants to investigate the potential spread of germs by contact. She knows that the number of possible handshakes within a group of n people is given by

$$N = \frac{1}{2}(n^2 - n).$$

27. There are 100 people at a party. How many handshakes are possible?

28. There are 40 people at a meeting. How many handshakes are possible?

29. Everyone at a meeting shook hands. There were 300 handshakes in all. How many people were at the meeting?

30. Everyone at a party shook hands. There were 190 handshakes in all. How many people were at the party?

31. The length of one leg of a right triangle is 8 ft. The length of the hypotenuse is 2 ft longer than the other leg. Find the length of the hypotenuse and the other leg.

32. The length of one leg of a right triangle is 24 ft. The length of the other leg is 16 ft shorter than the hypotenuse. Find the length of the hypotenuse and the other leg.

Skill Maintenance

Multiply. [10.6d], [10.7f]

33. $(3x - 5y)(3x + 5y)$

34. $(3x - 5y)^2$

35. $(3x + 5y)^2$

36. $(3x - 5y)(2x + 7y)$

Find the intercepts of the equation. [9.3a]

37. $4x - 16y = 64$

38. $4x + 16y = 64$

39. $x - 1.3y = 6.5$

40. $\frac{2}{3}x + \frac{5}{8}y = \frac{5}{12}$

Synthesis

41. ◆ Write a problem in which a quadratic equation must be solved.

42. ◆ Write a problem for a classmate to solve such that only one of the two solutions of a quadratic equation can be used as an answer.

43. A cement walk of constant width is built around a 20-ft by 40-ft rectangular pool. The total area of the pool and the walk is 1500 ft^2. Find the width of the walk.

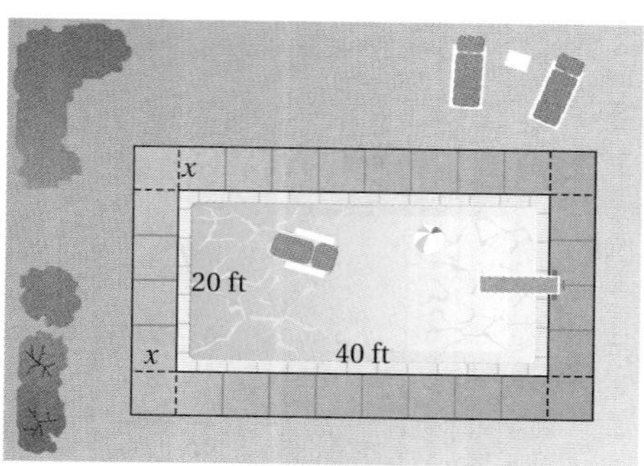

44. An open rectangular gutter is made by turning up the sides of a piece of metal 20 in. wide. The area of the cross-section of the gutter is 50 in^2. Find the depth of the gutter.

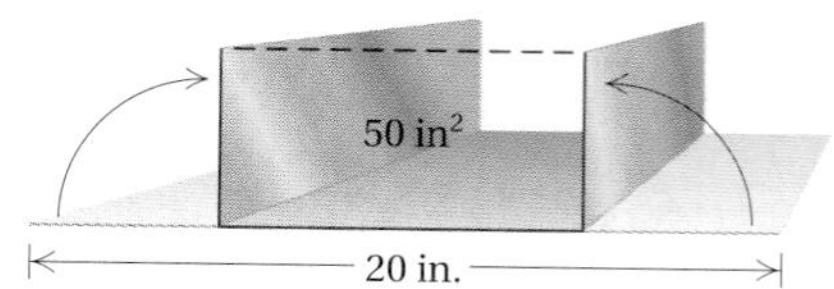

45. The ones digit of a number less than 100 is 4 greater than the tens digit. The sum of the number and the product of the digits is 58. Find the number.

46. The total surface area of a closed box is 350 m^2. The box is 9 m high and has a square base and lid. Find the length of the side of the base.

47. A rectangular piece of cardboard is twice as long as it is wide. A 4-cm square is cut out of each corner, and the sides are turned up to make a box with an open top. The volume of the box is 616 cm^3. Find the original dimensions of the cardboard.

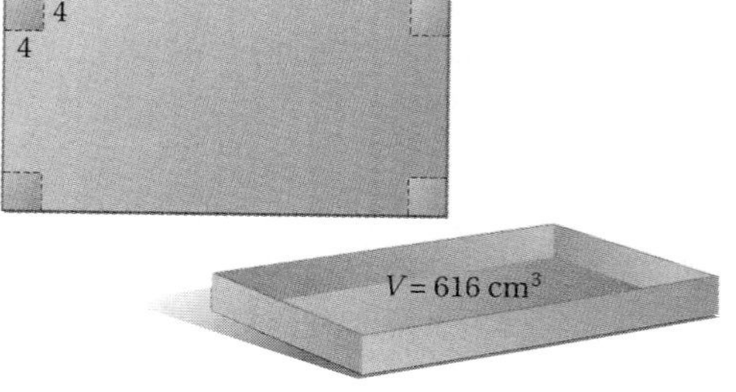

Summary and Review Exercises: Chapter 11

Important Properties and Formulas

Factoring Formulas: $A^2 - B^2 = (A + B)(A - B)$, $A^2 + 2AB + B^2 = (A + B)^2$, $A^2 - 2AB + B^2 = (A - B)^2$

The Principle of Zero Products: **An equation $ab = 0$ is true if and only if $a = 0$ is true or $b = 0$ is true, or both are true.**

The objectives to be tested in addition to the material in this chapter are [7.6c], [8.7e], [9.3a], and [10.6d].

Find three factorizations of the monomial. [11.1a]

1. $-10x^2$

2. $36x^5$

Factor completely. [11.6a]

3. $5 - 20x^6$

4. $x^2 - 3x$

5. $9x^2 - 4$

6. $x^2 + 4x - 12$

7. $x^2 + 14x + 49$

8. $6x^3 + 12x^2 + 3x$

9. $x^3 + x^2 + 3x + 3$

10. $6x^2 - 5x + 1$

11. $x^4 - 81$

12. $9x^3 + 12x^2 - 45x$

13. $2x^2 - 50$

14. $x^4 + 4x^3 - 2x - 8$

15. $16x^4 - 1$

16. $8x^6 - 32x^5 + 4x^4$

17. $75 + 12x^2 + 60x$

18. $x^2 + 9$

19. $x^3 - x^2 - 30x$

20. $4x^2 - 25$

21. $9x^2 + 25 - 30x$

22. $6x^2 - 28x - 48$

23. $x^2 - 6x + 9$

24. $2x^2 - 7x - 4$

25. $18x^2 - 12x + 2$

26. $3x^2 - 27$

27. $15 - 8x + x^2$

28. $25x^2 - 20x + 4$

29. $49b^{10} + 4a^8 - 28a^4b^5$

30. $x^2y^2 + xy - 12$

31. $12a^2 + 84ab + 147b^2$

32. $m^2 + 5m + mt + 5t$

33. $32x^4 - 128y^4z^4$

Solve. [11.7a], [11.7b]

34. $(x - 1)(x + 3) = 0$

35. $x^2 + 2x - 35 = 0$

36. $x^2 + x - 12 = 0$

37. $3x^2 + 2 = 5x$

38. $2x^2 + 5x = 12$

39. $16 = x(x - 6)$

Solve. [11.8a]

40. The square of a number is 6 more than the number. Find all such numbers.

41. The product of two consecutive even integers is 288. Find the integers.

42. Twice the square of a number is 10 more than the number. Find all such numbers.

43. The product of two consecutive odd integers is 323. Find the integers.

44. *House Plan.* An architect has allocated a rectangular space of 264 ft^2 for a square dining room and a 10-ft wide kitchen. Find the dimensions of each room.

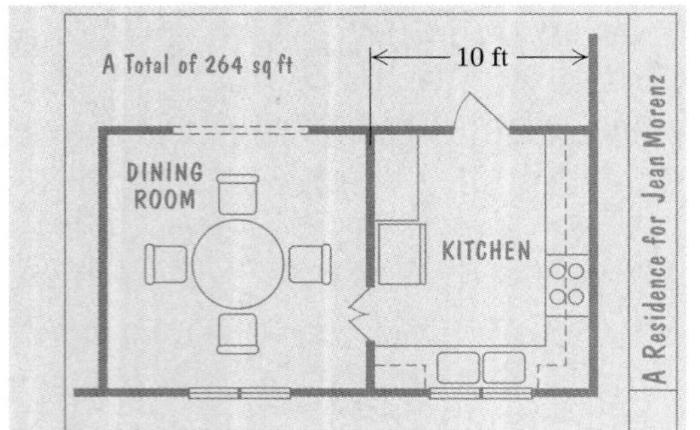

45. *Antenna Guy Wire.* The guy wires for a television antenna are 1 m longer than the height of the antenna. The guy wires are anchored 3 m from the foot of the antenna. How tall is the antenna?

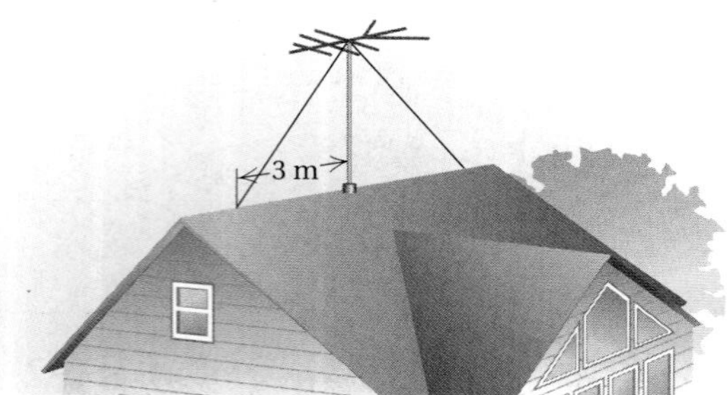

Find the x-intercepts for the graph of the equation. [11.7b]

46. $y = x^2 + 9x + 20$

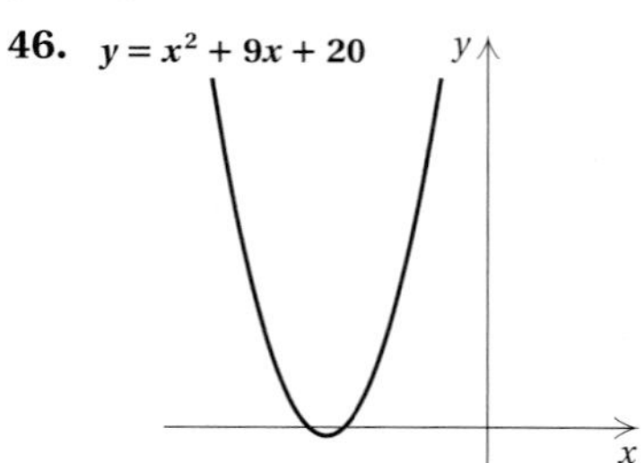

47. $y = 2x^2 - 7x - 15$

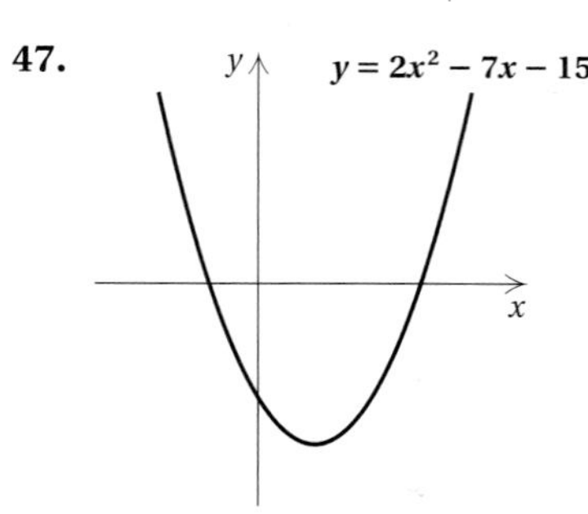

Skill Maintenance

48. Divide: $-\frac{12}{25} \div \left(-\frac{21}{10}\right)$. [7.6c]

49. Solve: $20 - (3x + 2) \geq 2(x + 5) + x$. [8.7e]

50. Multiply: $(2a - 3)(2a + 3)$. [10.6d]

51. Find the intercepts. Then graph the equation. [9.3a]

$$3y - 4x = -12$$

Synthesis

52. ◆ Compare the types of equations that we are able to solve after having studied this chapter with those we have studied earlier. [8.3a, b, c], [11.7a, b]

53. ◆ Describe as many procedures as you can for checking the result of factoring a polynomial. [11.3a], [11.7a]

Solve. [11.8a]

54. The pages of a book measure 15 cm by 20 cm. Margins of equal width surround the printing on each page and constitute one-half of the area of the page. Find the width of the margins.

55. The cube of a number is the same as twice the square of the number. Find all such numbers.

56. The length of a rectangle is two times its width. When the length is increased by 20 and the width decreased by 1, the area is 160. Find the original length and width.

Solve. [11.7b]

57. $x^2 + 25 = 0$

58. $(x - 2)(x + 3)(2x - 5) = 0$

59. For each equation in group A, find an equivalent equation in group B. [11.6a]

A. a) $3x^2 - 4x + 8 = 0$
b) $(x - 6)(x + 3) = 0$
c) $x^2 + 2x + 9 = 0$
d) $(2x - 5)(x + 4) = 0$
e) $5x^2 - 5 = 0$
f) $x^2 + 10x - 2 = 0$

B. g) $4x^2 + 8x + 36 = 0$
h) $(2x + 8)(2x - 5) = 0$
i) $9x^2 - 12x + 24 = 0$
j) $(x + 1)(5x - 5) = 0$
k) $x^2 - 3x - 18 = 0$
l) $2x^2 + 20x - 4 = 0$

60. Which is greater, $2^{90} + 2^{90}$, or 2^{100}? Why? [11.1b]

Test: Chapter 11

1. Find three factorizations of $4x^3$.

Factor completely.

2. $x^2 - 7x + 10$

3. $x^2 + 25 - 10x$

4. $6y^2 - 8y^3 + 4y^4$

5. $x^3 + x^2 + 2x + 2$

6. $x^2 - 5x$

7. $x^3 + 2x^2 - 3x$

8. $28x - 48 + 10x^2$

9. $4x^2 - 9$

10. $x^2 - x - 12$

11. $6m^3 + 9m^2 + 3m$

12. $3w^2 - 75$

13. $60x + 45x^2 + 20$

14. $3x^4 - 48$

15. $49x^2 - 84x + 36$

16. $5x^2 - 26x + 5$

17. $x^4 + 2x^3 - 3x - 6$

18. $80 - 5x^4$

19. $4x^2 - 4x - 15$

20. $6t^3 + 9t^2 - 15t$

21. $3m^2 - 9mn - 30n^2$

Answers

1. ______
2. ______
3. ______
4. ______
5. ______
6. ______
7. ______
8. ______
9. ______
10. ______
11. ______
12. ______
13. ______
14. ______
15. ______
16. ______
17. ______
18. ______
19. ______
20. ______
21. ______

Answers

22. ______

23. ______

24. ______

25. ______

26. ______

27. ______

28. ______

29. ______

30. ______

31. ______

32. ______

33. ______

34. ______

35. ______

36. ______

Solve.

22. $x^2 - x - 20 = 0$

23. $2x^2 + 7x = 15$

24. $x(x - 3) = 28$

Solve.

25. The square of a number is 24 more than five times the number. Find all such numbers.

26. *Dimensions of a Sail.* The height of the jib sail on a Lightning sailboat is 5 ft greater than the length of its "foot." If the area of the sail is 42 ft^2, find the length of the foot and the height of the sail.

Find the x-intercepts for the graph of the equation.

27.

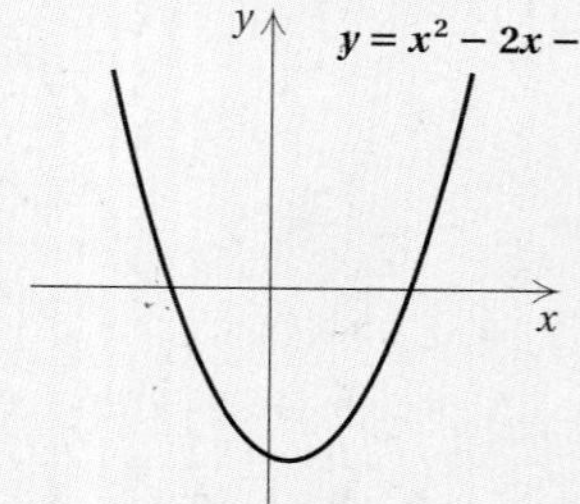

28.

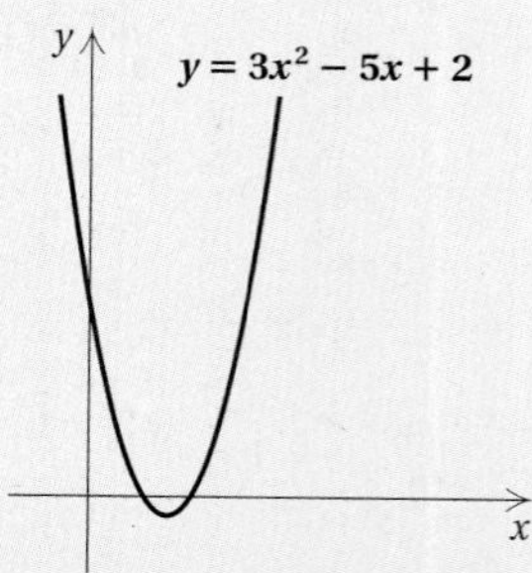

Skill Maintenance

29. Divide: $\frac{5}{8} \div \left(-\frac{11}{16}\right)$.

30. Solve: $10(x - 3) < 4(x + 2)$.

31. Find the intercepts. Then graph the equation.

$2y - 5x = 10$

32. Multiply: $(5x^2 - 7)^2$.

Synthesis

33. The length of a rectangle is five times its width. When the length is decreased by 3 and the width is increased by 2, the area of the new rectangle is 60. Find the original length and width.

34. Factor: $(a + 3)^2 - 2(a + 3) - 35$.

35. If $x^2 - 4 = (14)(18)$, then one possibility for x is which of the following?

a) 12 b) 14
c) 16 d) 18

36. If $x + y = 4$ and $x - y = 6$, then $x^2 - y^2 = ?$

a) 2 b) 10
c) 34 d) 24

12

Rational Expressions and Equations

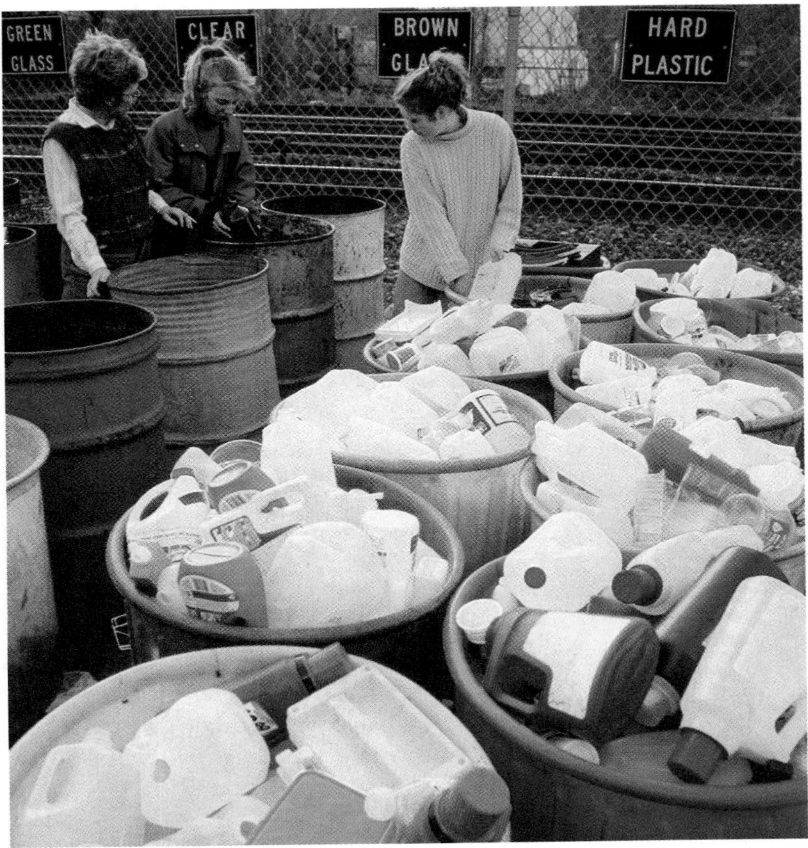

An Application

Erin and Nick work as volunteers at a community recycling depot. Erin can sort a morning's accumulation of recyclables in 4 hr, while Nick requires 6 hr to do the same job. How long would it take them, working together, to sort the recyclables?

This problem appears as Example 3 in Section 12.7.

The Mathematics

We let t = the time it takes them, working together, to sort the recyclables. The problem then translates to the equation

These are *rational expressions.*

$$\frac{t}{4} + \frac{t}{6} = 1.$$

This is a *rational equation.*

World Wide Web For more information, visit us at www.mathmax.com

Introduction

In this chapter, we learn to manipulate rational expressions. We learn to simplify rational expressions as well as to add, subtract, multiply, and divide them. Then we use these skills to solve equations, formulas, and applied problems.

Pretest: Chapter 12

1. Find the LCM of $x^2 + 5x + 6$ and $x^2 + 6x + 9$.

Perform the indicated operations and simplify.

2. $\dfrac{b-1}{2-b} + \dfrac{b^2-3}{b^2-4}$

3. $\dfrac{4y-4}{y^2-y-2} - \dfrac{3y-5}{y^2-y-2}$

4. $\dfrac{4}{a+2} + \dfrac{3}{a}$

5. $\dfrac{x}{x+1} - \dfrac{x}{x-1} + \dfrac{2x^2}{x^2-1}$

6. $\dfrac{4x+8}{x+1} \cdot \dfrac{x^2-2x-3}{2x^2-8}$

7. $\dfrac{x+3}{x^2-9} \div \dfrac{x+3}{x^2-6x+9}$

8. Simplify: $\dfrac{\dfrac{1}{x}+\dfrac{1}{y}}{\dfrac{1}{x}-\dfrac{1}{y}}$.

Solve.

9. $\dfrac{1}{x+4} = \dfrac{5}{x}$

10. $\dfrac{3}{x-2} + \dfrac{x}{2} = \dfrac{6}{2x-4}$

11. Solve $R = \dfrac{1}{3}M(a-b)$ for M.

12. It takes 6 hr for a paper carrier to deliver 200 papers. At this rate, how long would it take to deliver 350 papers?

13. One data-entry clerk can key in a report in 6 hr. Another can key in the same report in 5 hr. How long would it take them, working together, to key in the same report?

14. One car travels 20 mph faster than another. While one car travels 300 mi, the other travels 400 mi. Find their speeds.

Objectives for Retesting

The objectives to be tested in addition to the material in this chapter are as follows.

[10.2b] Raise a product to a power and a quotient to a power.
[10.4c] Subtract polynomials.
[11.6a] Factor polynomials completely.
[11.8a] Solve applied problems involving quadratic equations that can be solved by factoring.

12.1 Multiplying and Simplifying Rational Expressions

a Rational Expressions and Replacements

Rational numbers are quotients of integers. Some examples are

$$\frac{2}{3}, \quad \frac{4}{-5}, \quad \frac{-8}{17}, \quad \frac{563}{1}.$$

The following are called **rational expressions** or **fractional expressions.** They are quotients, or ratios, of polynomials:

$$\frac{3}{4}, \quad \frac{z}{6}, \quad \frac{5}{x+2}, \quad \frac{t^2+3t-10}{7t^2-4}.$$

A rational expression is also a division. For example,

$$\frac{3}{4} \text{ means } 3 \div 4 \quad \text{and} \quad \frac{x-8}{x+2} \text{ means } (x-8) \div (x+2).$$

Because rational expressions indicate division, we must be careful to avoid denominators of zero. When a variable is replaced with a number that produces a denominator equal to zero, the rational expression is undefined. For example, in the expression

$$\frac{x-8}{x+2},$$

when x is replaced with -2, the denominator is 0, and the expression is undefined:

$$\frac{x-8}{x+2} = \frac{-2-8}{-2+2} = \frac{-10}{0} \longleftarrow \text{Undefined}$$

When x is replaced with a number other than -2, such as 3, the expression *is* defined because the denominator is nonzero:

$$\frac{x-8}{x+2} = \frac{3-8}{3+2} = \frac{-5}{5} = -1.$$

Example 1 Find all numbers for which the rational expression

$$\frac{x+4}{x^2-3x-10}$$

is undefined.

To determine which numbers make the rational expression undefined, we set the denominator equal to 0 and solve:

$$x^2 - 3x - 10 = 0$$

$$(x-5)(x+2) = 0 \qquad \text{Factoring}$$

$$x - 5 = 0 \quad or \quad x + 2 = 0 \qquad \text{Using the principle of zero products}$$

$$x = 5 \quad or \quad x = -2.$$

The expression is undefined for the replacement numbers 5 and -2.

Do Exercises 1–3.

Objectives

- a Find all numbers for which a rational expression is undefined.
- b Multiply a rational expression by 1, using an expression such as A/A.
- c Simplify rational expressions by factoring the numerator and the denominator and removing factors of 1.
- d Multiply rational expressions and simplify.

For Extra Help

TAPE 23

MAC WIN

CD-ROM

Find all numbers for which the rational expression is undefined.

1. $\dfrac{16}{x-3}$

2. $\dfrac{2x-7}{x^2+5x-24}$

3. $\dfrac{x+5}{8}$

Answers on page A-36

Multiply.

4. $\dfrac{2x + 1}{3x - 2} \cdot \dfrac{x}{x}$

5. $\dfrac{x + 1}{x - 2} \cdot \dfrac{x + 2}{x + 2}$

6. $\dfrac{x - 8}{x - y} \cdot \dfrac{-1}{-1}$

Answers on page A-36

b Multiplying by 1

We multiply rational expressions in the same way that we multiply fractional notation in arithmetic. For a review, see Section 2.2. We saw there that

$$\frac{3}{7} \cdot \frac{2}{5} = \frac{3 \cdot 2}{7 \cdot 5} = \frac{6}{35}.$$

> To multiply rational expressions, multiply the numerators and multiply the denominators.

For example,

$$\frac{x - 2}{3} \cdot \frac{x + 2}{x + 7} = \frac{(x - 2)(x + 2)}{3(x + 7)}.$$ **Multiplying the numerators and the denominators**

Note that we leave the numerator, $(x - 2)(x + 2)$, and the denominator, $3(x + 7)$, in factored form because it is easier to simplify if we do not multiply. In order to learn to simplify, we first need to consider multiplying the rational expression by 1.

Any rational expression with the same numerator and denominator is a symbol for 1:

$$\frac{19}{19} = 1, \qquad \frac{x + 8}{x + 8} = 1, \qquad \frac{3x^2 - 4}{3x^2 - 4} = 1, \qquad \frac{-1}{-1} = 1.$$

> Expressions that have the same value for all allowable (or meaningful) replacements are called **equivalent expressions.**

We can multiply by 1 to obtain an equivalent expression. At this point, we select expressions for 1 arbitrarily. Later, we will have a system for our choices when we add and subtract.

Examples Multiply.

2. $\dfrac{3x + 2}{x + 1} \cdot 1 = \dfrac{3x + 2}{x + 1} \cdot \dfrac{2x}{2x} = \dfrac{(3x + 2)2x}{(x + 1)2x}$ **Identity property of 1**

3. $\dfrac{x + 2}{x - 7} \cdot \dfrac{x + 3}{x + 3} = \dfrac{(x + 2)(x + 3)}{(x - 7)(x + 3)}$

4. $\dfrac{2 + x}{2 - x} \cdot \dfrac{-1}{-1} = \dfrac{(2 + x)(-1)}{(2 - x)(-1)}$

Do Exercises 4–6.

c Simplifying Rational Expressions

Simplifying rational expressions is similar to simplifying fractional expressions in arithmetic. For a review, see Section 2.1. We saw there, for example, that an expression like $\frac{15}{40}$ can be simplified as follows:

$$\frac{15}{40} = \frac{3 \cdot 5}{8 \cdot 5}$$ Factoring the numerator and the denominator. Note the common factor of 5.

$$= \frac{3}{8} \cdot \frac{5}{5}$$ Factoring the fractional expression

$$= \frac{3}{8} \cdot 1$$ $\frac{5}{5} = 1$

$$= \frac{3}{8}.$$ Using the identity property of 1, or "removing a factor of 1"

In algebra, instead of simplifying

$$\frac{15}{40},$$

we may need to simplify an expression like

$$\frac{x^2 - 16}{x + 4}.$$

Just as factoring is important in simplifying in arithmetic, so too is it important in simplifying rational expressions. The factoring we use most is the factoring of polynomials, which we studied in Chapter 11.

To simplify, we can do the reverse of multiplying. We factor the numerator and the denominator and "remove" a factor of 1.

Example 5 Simplify by removing a factor of 1: $\frac{8x^2}{24x}$.

$$\frac{8x^2}{24x} = \frac{8 \cdot x \cdot x}{3 \cdot 8 \cdot x}$$ Factoring the numerator and the denominator

$$= \frac{8x}{8x} \cdot \frac{x}{3}$$ Factoring the rational expression

$$= 1 \cdot \frac{x}{3}$$ $\frac{8x}{8x} = 1$

$$= \frac{x}{3}$$ We removed a factor of 1.

Do Exercises 7 and 8.

Examples Simplify by removing a factor of 1.

6. $$\frac{5a + 15}{10} = \frac{5(a + 3)}{5 \cdot 2}$$ Factoring the numerator and the denominator

$$= \frac{5}{5} \cdot \frac{a + 3}{2}$$ Factoring the rational expression

$$= 1 \cdot \frac{a + 3}{2}$$ $\frac{5}{5} = 1$

$$= \frac{a + 3}{2}$$ Removing a factor of 1

Simplify by removing a factor of 1.

7. $\frac{5y}{y}$

8. $\frac{9x^2}{36x}$

Answers on page A-36

Simplify by removing a factor of 1.

9. $\dfrac{2x^2 + x}{3x^2 + 2x}$

10. $\dfrac{x^2 - 1}{2x^2 - x - 1}$

11. $\dfrac{7x + 14}{7}$

12. $\dfrac{12y + 24}{48}$

Answers on page A-36

7. $\dfrac{6a + 12}{7a + 14} = \dfrac{6(a + 2)}{7(a + 2)}$ Factoring the numerator and the denominator

$= \dfrac{6}{7} \cdot \dfrac{a + 2}{a + 2}$ Factoring the rational expression

$= \dfrac{6}{7} \cdot 1$ $\quad \dfrac{a+2}{a+2} = 1$

$= \dfrac{6}{7}$ Removing a factor of 1

8. $\dfrac{6x^2 + 4x}{2x^2 + 2x} = \dfrac{2x(3x + 2)}{2x(x + 1)}$ Factoring the numerator and the denominator

$= \dfrac{2x}{2x} \cdot \dfrac{3x + 2}{x + 1}$ Factoring the rational expression

$= 1 \cdot \dfrac{3x + 2}{x + 1}$ $\quad \dfrac{2x}{2x} = 1$

$= \dfrac{3x + 2}{x + 1}$ Removing a factor of 1. Note in this step that you *cannot* remove the x's because x is not a factor of the entire numerator and the entire denominator.

9. $\dfrac{x^2 + 3x + 2}{x^2 - 1} = \dfrac{(x + 2)(x + 1)}{(x + 1)(x - 1)}$

$= \dfrac{x + 1}{x + 1} \cdot \dfrac{x + 2}{x - 1}$

$= 1 \cdot \dfrac{x + 2}{x - 1}$

$= \dfrac{x + 2}{x - 1}$

Canceling

You may have encountered canceling when working with rational expressions. With great concern, we mention it as a possible way to speed up your work. Our concern is that canceling be done with care and understanding. Example 9 might have been done faster as follows:

$\dfrac{x^2 + 3x + 2}{x^2 - 1} = \dfrac{(x + 2)(x + 1)}{(x + 1)(x - 1)}$ Factoring the numerator and the denominator

$= \dfrac{(x + 2)\cancel{(x + 1)}}{\cancel{(x + 1)}(x - 1)}$ When a factor of 1 is noted, it is canceled, as shown: $\dfrac{x+1}{x+1} = 1$.

$= \dfrac{x + 2}{x - 1}$. Simplifying

CAUTION! The difficulty with canceling is that it is often applied incorrectly, as in the following situations:

$\dfrac{\cancel{x} + 3}{\cancel{x}} = 3$; $\quad \dfrac{\cancel{4} + 1}{\cancel{4} + 2} = \dfrac{1}{2}$; $\quad \dfrac{1\cancel{5}}{\cancel{5}4} = \dfrac{1}{4}$.

Wrong! Wrong! Wrong!

In each of these situations, the expressions canceled were *not* factors of 1. Factors are parts of products. For example, in $2 \cdot 3$, 2 and 3 are factors, but in $2 + 3$, 2 and 3 are *not* factors. **If you can't factor, you can't cancel. If in doubt, don't cancel!**

Do Exercises 9–12.

Factors That Are Opposites

Consider

$$\frac{x-4}{4-x}.$$

At first glance, the numerator and the denominator do not appear to have any common factors other than 1. But $x - 4$ and $4 - x$ are opposites, or additive inverses, of each other. Thus we can rewrite one as the opposite of the other by factoring out a -1.

Example 10 Simplify: $\dfrac{x-4}{4-x}$.

$$\begin{aligned}\frac{x-4}{4-x} = \frac{x-4}{-(-4+x)} &= \frac{x-4}{-1(x-4)}\\ &= -1 \cdot \frac{x-4}{x-4}\\ &= -1 \cdot 1\\ &= -1\end{aligned}$$

Do Exercises 13–15.

Simplify.

13. $\dfrac{x-8}{8-x}$

14. $\dfrac{c-d}{d-c}$

15. $\dfrac{-x-7}{x+7}$

d Multiplying and Simplifying

We try to simplify after we multiply. That is why we leave the numerator and the denominator in factored form.

Example 11 Multiply and simplify: $\dfrac{5a^3}{4} \cdot \dfrac{2}{5a}$.

$$\frac{5a^3}{4} \cdot \frac{2}{5a} = \frac{5a^3(2)}{4(5a)}$$ Multiplying the numerators and the denominators

$$= \frac{2 \cdot 5 \cdot a \cdot a \cdot a}{2 \cdot 2 \cdot 5 \cdot a}$$ Factoring the numerator and the denominator

$$= \frac{\cancel{2} \cdot \cancel{5} \cdot \cancel{a} \cdot a \cdot a}{\cancel{2} \cdot 2 \cdot \cancel{5} \cdot \cancel{a}}$$ Removing a factor of 1: $\dfrac{2 \cdot 5 \cdot a}{2 \cdot 5 \cdot a} = 1$

$$= \frac{a^2}{2}$$ Simplifying

Example 12 Multiply and simplify: $\dfrac{x^2+6x+9}{x^2-4} \cdot \dfrac{x-2}{x+3}$.

$$\frac{x^2+6x+9}{x^2-4} \cdot \frac{x-2}{x+3} = \frac{(x^2+6x+9)(x-2)}{(x^2-4)(x+3)}$$ Multiplying the numerators and the denominators

$$= \frac{(x+3)(x+3)(x-2)}{(x+2)(x-2)(x+3)}$$ Factoring the numerator and the denominator

$$= \frac{\cancel{(x+3)}(x+3)\cancel{(x-2)}}{(x+2)\cancel{(x-2)}\cancel{(x+3)}}$$ Removing a factor of 1: $\dfrac{(x+3)(x-2)}{(x+3)(x-2)} = 1$

$$= \frac{x+3}{x+2}$$ Simplifying

Do Exercise 16.

16. Multiply and simplify:

$$\frac{a^2-4a+4}{a^2-9} \cdot \frac{a+3}{a-2}.$$

Answers on page A-36

17. Multiply and simplify:

$$\frac{x^2-25}{6}\cdot\frac{3}{x+5}.$$

Example 13 Multiply and simplify: $\dfrac{x^2+x-2}{15}\cdot\dfrac{5}{2x^2-3x+1}.$

$$\frac{x^2+x-2}{15}\cdot\frac{5}{2x^2-3x+1}=\frac{(x^2+x-2)5}{15(2x^2-3x+1)}$$ **Multiplying the numerators and the denominators**

$$=\frac{(x+2)(x-1)5}{5(3)(x-1)(2x-1)}$$ **Factoring the numerator and the denominator**

$$=\frac{(x+2)\cancel{(x-1)5}}{\cancel{5}(3)\cancel{(x-1)}(2x-1)}$$ **Removing a factor of 1:** $\frac{(x-1)5}{(x-1)5}=1$

$$=\frac{x+2}{3(2x-1)}$$ **Simplifying**

You need not carry out this multiplication.

Do Exercise 17.

Calculator Spotlight

Checking Simplifications Using a Table. We can use the TABLE feature as a partial check that rational expressions have been multiplied and/or simplified correctly. To check whether the simplifying of Example 9,

$$\frac{x^2+3x+2}{x^2-1}=\frac{x+2}{x-1},$$

is correct, we enter

$$y_1=\frac{x^2+3x+2}{x^2-1}\quad\text{as}\quad y_1=(x^2+3x+2)/(x^2-1)$$

and

$$y_2=\frac{x+2}{x-1}\quad\text{as}\quad y_2=(x+2)/(x-1).$$

If our simplifying is correct, the y_1- and y_2-values should be the same for all allowable replacements, regardless of the table settings used. Let TblStart = −4 and ΔTbl = 1.

X	Y1	Y2
−4	.4	.4
−3	.25	.25
−2	0	0
−1	ERROR	−.5
0	−2	−2
1	ERROR	ERROR
2	4	4

X = −4

Note that −1 and 1 are not allowable replacements in the first expression, and 1 is not an allowable replacement in the second. These facts are indicated by the ERROR messages. For all other numbers, we see that y_1 and y_2 are the same, so the simplifying seems to be correct. Remember, this is only a partial check.

Exercises

Use the TABLE feature to check whether each of the following is correct.

1. $\dfrac{8x^2}{24x}=\dfrac{x}{3}$
2. $\dfrac{5a+15}{10}=\dfrac{a+3}{2}$
3. $\dfrac{x+3}{x}=3$
4. $\dfrac{x^2-3x+2}{x^2-1}=\dfrac{x+2}{x-1}$
5. $\dfrac{x^2-9}{x^2-3}=3$
6. $\dfrac{x^2+6x+9}{x^2-4}\cdot\dfrac{x-2}{x+3}=\dfrac{x+3}{x+2}$
7. $\dfrac{x^2-25}{6}\cdot\dfrac{3}{x+5}=\dfrac{x-5}{3}$

Answer on page A-36

Exercise Set 12.1

a Find all numbers for which the rational expression is undefined.

1. $\frac{-3}{2x}$

2. $\frac{24}{-8y}$

3. $\frac{5}{x - 8}$

4. $\frac{y - 4}{y + 6}$

5. $\frac{3}{2y + 5}$

6. $\frac{x^2 - 9}{4x - 12}$

7. $\frac{x^2 + 11}{x^2 - 3x - 28}$

8. $\frac{p^2 - 9}{p^2 - 7p + 10}$

9. $\frac{m^3 - 2m}{m^2 - 25}$

10. $\frac{7 - 3x + x^2}{49 - x^2}$

11. $\frac{x - 4}{3}$

12. $\frac{x^2 - 25}{14}$

b Multiply. Do not simplify. Note that in each case you are multiplying by 1.

13. $\frac{4x}{4x} \cdot \frac{3x^2}{5y}$

14. $\frac{5x^2}{5x^2} \cdot \frac{6y^3}{3z^4}$

15. $\frac{2x}{2x} \cdot \frac{x - 1}{x + 4}$

16. $\frac{2a - 3}{5a + 2} \cdot \frac{a}{a}$

17. $\frac{3 - x}{4 - x} \cdot \frac{-1}{-1}$

18. $\frac{x - 5}{5 - x} \cdot \frac{-1}{-1}$

19. $\frac{y + 6}{y + 6} \cdot \frac{y - 7}{y + 2}$

20. $\frac{x^2 + 1}{x^3 - 2} \cdot \frac{x - 4}{x - 4}$

c Simplify.

21. $\frac{8x^3}{32x}$

22. $\frac{4x^2}{20x}$

23. $\frac{48p^7q^5}{18p^5q^4}$

24. $\frac{-76x^8y^3}{-24x^4y^3}$

25. $\frac{4x - 12}{4x}$

26. $\frac{5a - 40}{5}$

27. $\dfrac{3m^2 + 3m}{6m^2 + 9m}$

28. $\dfrac{4y^2 - 2y}{5y^2 - 5y}$

29. $\dfrac{a^2 - 9}{a^2 + 5a + 6}$

30. $\dfrac{t^2 - 25}{t^2 + t - 20}$

31. $\dfrac{a^2 - 10a + 21}{a^2 - 11a + 28}$

32. $\dfrac{x^2 - 2x - 8}{x^2 - x - 6}$

33. $\dfrac{x^2 - 25}{x^2 - 10x + 25}$

34. $\dfrac{x^2 + 8x + 16}{x^2 - 16}$

35. $\dfrac{a^2 - 1}{a - 1}$

36. $\dfrac{t^2 - 1}{t + 1}$

37. $\dfrac{x^2 + 1}{x + 1}$

38. $\dfrac{m^2 + 9}{m + 3}$

39. $\dfrac{6x^2 - 54}{4x^2 - 36}$

40. $\dfrac{8x^2 - 32}{4x^2 - 16}$

41. $\dfrac{6t + 12}{t^2 - t - 6}$

42. $\dfrac{4x + 32}{x^2 + 9x + 8}$

43. $\dfrac{2t^2 + 6t + 4}{4t^2 - 12t - 16}$

44. $\dfrac{3a^2 - 9a - 12}{6a^2 + 30a + 24}$

45. $\dfrac{t^2 - 4}{(t + 2)^2}$

46. $\dfrac{m^2 - 10m + 25}{m^2 - 25}$

47. $\dfrac{6-x}{x-6}$

48. $\dfrac{t-3}{3-t}$

49. $\dfrac{a-b}{b-a}$

50. $\dfrac{y-x}{-x+y}$

51. $\dfrac{6t-12}{2-t}$

52. $\dfrac{5a-15}{3-a}$

53. $\dfrac{x^2-1}{1-x}$

54. $\dfrac{a^2-b^2}{b^2-a^2}$

d Multiply and simplify.

55. $\dfrac{4x^3}{3x} \cdot \dfrac{14}{x}$

56. $\dfrac{18}{x^3} \cdot \dfrac{5x^2}{6}$

57. $\dfrac{3c}{d^2} \cdot \dfrac{4d}{6c^3}$

58. $\dfrac{3x^2y}{2} \cdot \dfrac{4}{xy^3}$

59. $\dfrac{x^2-3x-10}{x^2-4x+4} \cdot \dfrac{x-2}{x-5}$

60. $\dfrac{t^2}{t^2-4} \cdot \dfrac{t^2-5t+6}{t^2-3t}$

61. $\dfrac{a^2-9}{a^2} \cdot \dfrac{a^2-3a}{a^2+a-12}$

62. $\dfrac{x^2+10x-11}{x^2-1} \cdot \dfrac{x+1}{x+11}$

63. $\dfrac{4a^2}{3a^2-12a+12} \cdot \dfrac{3a-6}{2a}$

64. $\dfrac{5v+5}{v-2} \cdot \dfrac{v^2-4v+4}{v^2-1}$

65. $\dfrac{t^4-16}{t^4-1} \cdot \dfrac{t^2+1}{t^2+4}$

66. $\dfrac{x^4-1}{x^4-81} \cdot \dfrac{x^2+9}{x^2+1}$

67. $\dfrac{(x+4)^3}{(x+2)^3} \cdot \dfrac{x^2+4x+4}{x^2+8x+16}$

68. $\dfrac{(t-2)^3}{(t-1)^3} \cdot \dfrac{t^2-2t+1}{t^2-4t+4}$

69. $\dfrac{5a^2-180}{10a^2-10} \cdot \dfrac{20a+20}{2a-12}$

70. $\dfrac{2t^2-98}{4t^2-4} \cdot \dfrac{8t+8}{16t-112}$

Skill Maintenance

Solve.

71. The product of two consecutive even integers is 360. Find the integers. [11.8a]

72. About 5 L of oxygen can be dissolved in 100 L of water at 0°C. This is 1.6 times the amount that can be dissolved in the same volume of water at 20°C. How much oxygen can be dissolved in 100 L at 20°C? [8.4a]

Factor. [11.6a]

73. $x^2 - x - 56$

74. $a^2 - 16a + 64$

75. $x^5 - 2x^4 - 35x^3$

76. $2y^3 - 10y^2 + y - 5$

77. $16 - t^4$

78. $10x^2 + 80x + 70$

79. $x^2 - 9x + 14$

80. $x^2 + x + 7$

81. $16x^2 - 40xy + 25y^2$

82. $a^2 - 9ab + 14b^2$

Synthesis

83. ◆ How is the process of canceling related to the identity property of 1?

84. ◆ Explain how a rational expression can be formed for which −3 and 4 are not allowable replacements.

Simplify.

85. $\dfrac{x^4-16y^4}{(x^2+4y^2)(x-2y)}$

86. $\dfrac{(a-b)^2}{b^2-a^2}$

87. $\dfrac{t^4-1}{t^4-81} \cdot \dfrac{t^2-9}{t^2+1} \cdot \dfrac{(t-9)^2}{(t+1)^2}$

88. $\dfrac{(t+2)^3}{(t+1)^3} \cdot \dfrac{t^2+2t+1}{t^2+4t+4} \cdot \dfrac{t+1}{t+2}$

89. $\dfrac{x^2-y^2}{(x-y)^2} \cdot \dfrac{x^2-2xy+y^2}{x^2-4xy-5y^2}$

90. $\dfrac{x-1}{x^2+1} \cdot \dfrac{x^4-1}{(x-1)^2} \cdot \dfrac{x^2-1}{x^4-2x^2+1}$

12.2 Division and Reciprocals

There is a similarity between what we do with rational expressions and what we do with rational numbers. In fact, after variables have been replaced with rational numbers, a rational expression represents a rational number.

Objectives

a. Find the reciprocal of a rational expression.

b. Divide rational expressions and simplify.

For Extra Help

TAPE 23

MAC WIN

CD-ROM

a Finding Reciprocals

Two expressions are reciprocals of each other if their product is 1. The reciprocal of a rational expression is found by interchanging the numerator and the denominator.

Examples

1. The reciprocal of $\frac{2}{5}$ is $\frac{5}{2}$. $\left(\text{This is because } \frac{2}{5} \cdot \frac{5}{2} = \frac{10}{10} = 1.\right)$

2. The reciprocal of $\frac{2x^2 - 3}{x + 4}$ is $\frac{x + 4}{2x^2 - 3}$.

3. The reciprocal of $x + 2$ is $\frac{1}{x + 2}$. $\left(\text{Think of } x + 2 \text{ as } \frac{x + 2}{1}.\right)$

Do Exercises 1–4.

Find the reciprocal.

1. $\frac{7}{2}$

2. $\frac{x^2 + 5}{2x^3 - 1}$

3. $x - 5$

4. $\frac{1}{x^2 - 3}$

b Division

We divide rational expressions in the same way that we divide fractional notation in arithmetic. For a review, see Section 2.2.

> To divide rational expressions, multiply by the reciprocal of the divisor. Then factor and simplify the result.

Examples Divide.

4. $\frac{3}{4} \div \frac{2}{5} = \frac{3}{4} \cdot \frac{5}{2}$ **Multiplying by the reciprocal of the divisor**

$= \frac{3 \cdot 5}{4 \cdot 2}$

$= \frac{15}{8}$

5. $\frac{2}{x} \div \frac{x}{3} = \frac{2}{x} \cdot \frac{3}{x}$ **Multiplying by the reciprocal of the divisor**

$= \frac{2 \cdot 3}{x \cdot x}$

$= \frac{6}{x^2}$

Do Exercises 5 and 6.

Divide.

5. $\frac{3}{5} \div \frac{7}{2}$

6. $\frac{x}{8} \div \frac{5}{x}$

Answers on page A-37

Example 6 Divide: $\frac{x+1}{x+2} \div \frac{x-1}{x+3}$.

$$\frac{x+1}{x+2} \div \frac{x-1}{x+3} = \frac{x+1}{x+2} \cdot \frac{x+3}{x-1}$$ Multiplying by the reciprocal of the divisor

$$= \frac{(x+1)(x+3)}{(x+2)(x-1)}$$

We usually do not carry out the multiplication in the numerator or the denominator. It is not wrong to do so, but the factored form is often more useful.

Do Exercise 7.

Example 7 Divide and simplify: $\frac{x+1}{x^2-1} \div \frac{x+1}{x^2-2x+1}$.

$$\frac{x+1}{x^2-1} \div \frac{x+1}{x^2-2x+1}$$

$$= \frac{x+1}{x^2-1} \cdot \frac{x^2-2x+1}{x+1}$$ Multiplying by the reciprocal

$$= \frac{(x+1)(x^2-2x+1)}{(x^2-1)(x+1)}$$

$$= \frac{(x+1)(x-1)(x-1)}{(x-1)(x+1)(x+1)}$$ Factoring the numerator and the denominator

$$= \frac{\cancel{(x+1)}\cancel{(x-1)}(x-1)}{\cancel{(x-1)}\cancel{(x+1)}(x+1)}$$ Removing a factor of 1: $\frac{(x+1)(x-1)}{(x+1)(x-1)} = 1$

$$= \frac{x-1}{x+1}$$

Example 8 Divide and simplify: $\frac{x^2-2x-3}{x^2-4} \div \frac{x+1}{x+5}$.

$$\frac{x^2-2x-3}{x^2-4} \div \frac{x+1}{x+5}$$

$$= \frac{x^2-2x-3}{x^2-4} \cdot \frac{x+5}{x+1}$$ Multiplying by the reciprocal

$$= \frac{(x^2-2x-3)(x+5)}{(x^2-4)(x+1)}$$

$$= \frac{(x-3)(x+1)(x+5)}{(x-2)(x+2)(x+1)}$$ Factoring the numerator and the denominator

$$= \frac{(x-3)\cancel{(x+1)}(x+5)}{(x-2)(x+2)\cancel{(x+1)}}$$ Removing a factor of 1: $\frac{x+1}{x+1} = 1$

$$= \frac{(x-3)(x+5)}{(x-2)(x+2)}$$

You need not carry out the multiplications in the numerator and the denominator.

Do Exercises 8–10.

7. Divide:

$$\frac{x-3}{x+5} \div \frac{x+5}{x-2}.$$

Divide and simplify.

8. $\frac{x-3}{x+5} \div \frac{x+2}{x+5}$

9. $\frac{x^2-5x+6}{x+5} \div \frac{x+2}{x+5}$

10. $\frac{y^2-1}{y+1} \div \frac{y^2-2y+1}{y+1}$

Calculator Spotlight

Use the TABLE feature to check the divisions in Examples 5–8.

Answers on page A-37

Exercise Set 12.2

a Find the reciprocal.

1. $\frac{4}{x}$
2. $\frac{a+3}{a-1}$
3. $x^2 - y^2$
4. $x^2 - 5x + 7$
5. $\frac{1}{a+b}$
6. $\frac{x^2}{x^2-3}$
7. $\frac{x^2+2x-5}{x^2-4x+7}$
8. $\frac{(a-b)(a+b)}{(a+4)(a-5)}$

b Divide and simplify.

9. $\frac{2}{5} \div \frac{4}{3}$
10. $\frac{3}{10} \div \frac{3}{2}$
11. $\frac{2}{x} \div \frac{8}{x}$
12. $\frac{t}{3} \div \frac{t}{15}$
13. $\frac{a}{b^2} \div \frac{a^2}{b^3}$
14. $\frac{x^2}{y} \div \frac{x^3}{y^3}$
15. $\frac{a+2}{a-3} \div \frac{a-1}{a+3}$
16. $\frac{x-8}{x+9} \div \frac{x+2}{x-1}$
17. $\frac{x^2-1}{x} \div \frac{x+1}{x-1}$
18. $\frac{4y-8}{y+2} \div \frac{y-2}{y^2-4}$
19. $\frac{x+1}{6} \div \frac{x+1}{3}$
20. $\frac{a}{a-b} \div \frac{b}{a-b}$
21. $\frac{5x-5}{16} \div \frac{x-1}{6}$
22. $\frac{4y-12}{12} \div \frac{y-3}{3}$
23. $\frac{-6+3x}{5} \div \frac{4x-8}{25}$
24. $\frac{-12+4x}{4} \div \frac{-6+2x}{6}$
25. $\frac{a+2}{a-1} \div \frac{3a+6}{a-5}$
26. $\frac{t-3}{t+2} \div \frac{4t-12}{t+1}$
27. $\frac{x^2-4}{x} \div \frac{x-2}{x+2}$
28. $\frac{x+y}{x-y} \div \frac{x^2+y}{x^2-y^2}$
29. $\frac{x^2-9}{4x+12} \div \frac{x-3}{6}$
30. $\frac{a-b}{2a} \div \frac{a^2-b^2}{8a^3}$
31. $\frac{c^2+3c}{c^2+2c-3} \div \frac{c}{c+1}$
32. $\frac{y+5}{2y} \div \frac{y^2-25}{4y^2}$

33. $\dfrac{2y^2 - 7y + 3}{2y^2 + 3y - 2} \div \dfrac{6y^2 - 5y + 1}{3y^2 + 5y - 2}$

34. $\dfrac{x^2 + x - 20}{x^2 - 7x + 12} \div \dfrac{x^2 + 10x + 25}{x^2 - 6x + 9}$

35. $\dfrac{x^2 - 1}{4x + 4} \div \dfrac{2x^2 - 4x + 2}{8x + 8}$

36. $\dfrac{5t^2 + 5t - 30}{10t + 30} \div \dfrac{2t^2 - 8}{6t^2 + 36t + 54}$

Skill Maintenance

Solve.

37. Bonnie is taking an astronomy course. In order to receive an A, she must average at least 90 after four exams. Bonnie scored 96, 98, and 89 on the first three tests. Determine (in terms of an inequality) what scores on the last test will earn her an A. [8.8b]

38. Sixteen more than the square of a number is eight times the number. Find the number. [11.8a]

Subtract. [10.4c]

39. $(8x^3 - 3x^2 + 7) - (8x^2 + 3x - 5)$

40. $(3p^2 - 6pq + 7q^2) - (5p^2 - 10pq + 11q^2)$

Simplify. [10.2b]

41. $(2x^{-3}y^4)^2$

42. $(5x^6y^{-4})^3$

43. $\left(\dfrac{2x^3}{y^5}\right)^2$

44. $\left(\dfrac{a^{-3}}{b^4}\right)^5$

Synthesis

45. ◆ Explain why 5, −1, and 7 are *not* allowable replacements in the division

$$\frac{x + 3}{x - 5} \div \frac{x - 7}{x + 1}.$$

46. ◆ Is the reciprocal of a product the product of the reciprocals? Why or why not?

Simplify.

47. $\dfrac{3a^2 - 5ab - 12b^2}{3ab + 4b^2} \div (3b^2 - ab)$

48. $\dfrac{3x + 3y + 3}{9x} \div \left(\dfrac{x^2 + 2xy + y^2 - 1}{x^4 + x^2}\right)$

49. The volume of this rectangular solid is $x - 3$. What is its height?

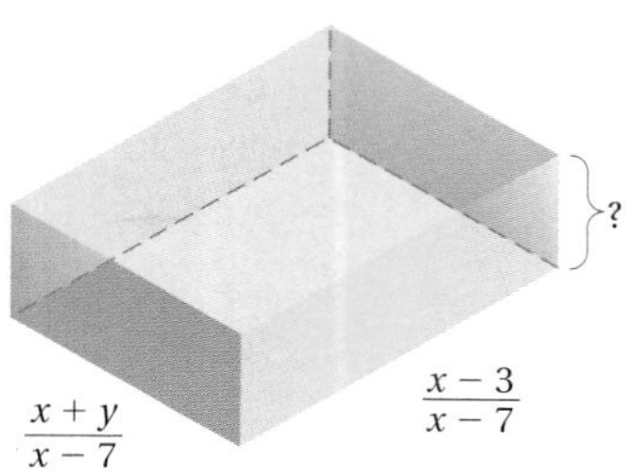

12.3 Least Common Multiples and Denominators

a Least Common Multiples

To add when denominators are different, we first find a common denominator. For a review, see Section 2.3. We saw there, for example, that to add $\frac{5}{12}$ and $\frac{7}{30}$, we first look for the **least common multiple, LCM,** of both 12 and 30. That number becomes the **least common denominator, LCD.** To find the LCM of 12 and 30, we factor:

$$12 = 2 \cdot 2 \cdot 3;$$
$$30 = 2 \cdot 3 \cdot 5.$$

The LCM is the number that has 2 as a factor twice, 3 as a factor once, and 5 as a factor once:

$$\text{LCM} = 2 \cdot 2 \cdot 3 \cdot 5, \text{ or } 60.$$

> To find the LCM, use each factor the greatest number of times that it appears in any one factorization.

Example 1 Find the LCM of 24 and 36.

$$\left.\begin{aligned} 24 &= 2 \cdot 2 \cdot 2 \cdot 3 \\ 36 &= 2 \cdot 2 \cdot 3 \cdot 3 \end{aligned}\right\} \quad \text{LCM} = 2 \cdot 2 \cdot 2 \cdot 3 \cdot 3, \text{ or } 72$$

Do Exercises 1–4.

b Adding Using the LCD

Let's finish adding $\frac{5}{12}$ and $\frac{7}{30}$:

$$\frac{5}{12} + \frac{7}{30} = \frac{5}{2 \cdot 2 \cdot 3} + \frac{7}{2 \cdot 3 \cdot 5}.$$

The least common denominator, LCD, is $2 \cdot 2 \cdot 3 \cdot 5$. To get the LCD in the first denominator, we need a 5. To get the LCD in the second denominator, we need another 2. We get these numbers by multiplying by 1:

$$\frac{5}{12} + \frac{7}{30} = \frac{5}{2 \cdot 2 \cdot 3} \cdot \frac{5}{5} + \frac{7}{2 \cdot 3 \cdot 5} \cdot \frac{2}{2} \qquad \text{Multiplying by 1}$$

$$= \frac{25}{2 \cdot 2 \cdot 3 \cdot 5} + \frac{14}{2 \cdot 3 \cdot 5 \cdot 2} \qquad \text{The denominators are now the LCD.}$$

$$= \frac{39}{2 \cdot 2 \cdot 3 \cdot 5} \qquad \text{Adding the numerators and keeping the LCD}$$

$$= \frac{\cancel{3} \cdot 13}{2 \cdot 2 \cdot \cancel{3} \cdot 5} \qquad \text{Factoring the numerator and removing a factor of 1: } \frac{3}{3} = 1$$

$$= \frac{13}{20}. \qquad \text{Simplifying}$$

Objectives

a Find the LCM of several numbers by factoring.

b Add fractions, first finding the LCD.

c Find the LCM of algebraic expressions by factoring.

For Extra Help

TAPE 23 MAC WIN CD-ROM

Find the LCM by factoring.

1. 16, 18

2. 6, 12

3. 2, 5

4. 24, 30, 20

Answers on page A-37

Add, first finding the LCD. Simplify, if possible.

5. $\frac{3}{16} + \frac{1}{18}$

6. $\frac{1}{6} + \frac{1}{12}$

7. $\frac{1}{2} + \frac{3}{5}$

8. $\frac{1}{24} + \frac{1}{30} + \frac{3}{20}$

Find the LCM.

9. $12xy^2, \quad 15x^3y$

10. $y^2 + 5y + 4, \quad y^2 + 2y + 1$

11. $t^2 + 16, \quad t - 2, \quad 7$

12. $x^2 + 2x + 1, \quad 3x^2 - 3x, \quad x^2 - 1$

Answers on page A-37

Example 2 Add: $\frac{5}{12} + \frac{11}{18}$.

$$\left.\begin{aligned} 12 &= 2 \cdot 2 \cdot 3 \\ 18 &= 2 \cdot 3 \cdot 3 \end{aligned}\right\} \quad \text{LCD} = 2 \cdot 2 \cdot 3 \cdot 3, \text{ or } 36$$

$$\frac{5}{12} + \frac{11}{18} = \frac{5}{2 \cdot 2 \cdot 3} \cdot \frac{3}{3} + \frac{11}{2 \cdot 3 \cdot 3} \cdot \frac{2}{2} = \frac{15 + 22}{2 \cdot 2 \cdot 3 \cdot 3} = \frac{37}{36}$$

Do Exercises 5–8.

c LCMs of Algebraic Expressions

To find the LCM of two or more algebraic expressions, we factor them. Then we use each factor the greatest number of times that it occurs in any one expression.

Example 3 Find the LCM of $12x$, $16y$, and $8xyz$.

$$\left.\begin{aligned} 12x &= 2 \cdot 2 \cdot 3 \cdot x \\ 16y &= 2 \cdot 2 \cdot 2 \cdot 2 \cdot y \\ 8xyz &= 2 \cdot 2 \cdot 2 \cdot x \cdot y \cdot z \end{aligned}\right\} \quad \begin{aligned} \text{LCM} &= 2 \cdot 2 \cdot 2 \cdot 2 \cdot 3 \cdot x \cdot y \cdot z \\ &= 48xyz \end{aligned}$$

Example 4 Find the LCM of $x^2 + 5x - 6$ and $x^2 - 1$.

$$\left.\begin{aligned} x^2 + 5x - 6 &= (x + 6)(x - 1) \\ x^2 - 1 &= (x + 1)(x - 1) \end{aligned}\right\} \quad \text{LCM} = (x + 6)(x - 1)(x + 1)$$

Example 5 Find the LCM of $x^2 + 4$, $x + 1$, and 5.

These expressions do not share a common factor other than 1, so the LCM is their product:

$5(x^2 + 4)(x + 1)$.

Example 6 Find the LCM of $x^2 - 25$ and $2x - 10$.

$$\left.\begin{aligned} x^2 - 25 &= (x + 5)(x - 5) \\ 2x - 10 &= 2(x - 5) \end{aligned}\right\} \quad \text{LCM} = 2(x + 5)(x - 5)$$

Example 7 Find the LCM of $x^2 - 4y^2$, $x^2 - 4xy + 4y^2$, and $x - 2y$.

$$\left.\begin{aligned} x^2 - 4y^2 &= (x - 2y)(x + 2y) \\ x^2 - 4xy + 4y^2 &= (x - 2y)(x - 2y) \\ x - 2y &= x - 2y \end{aligned}\right\} \quad \begin{aligned} \text{LCM} &= (x + 2y)(x - 2y)(x - 2y) \\ &= (x + 2y)(x - 2y)^2 \end{aligned}$$

Do Exercises 9–12.

Exercise Set 12.3

a Find the LCM.

1. 12, 27

2. 10, 15

3. 8, 9

4. 12, 18

5. 6, 9, 21

6. 8, 36, 40

7. 24, 36, 40

8. 4, 5, 20

9. 10, 100, 500

10. 28, 42, 60

b Add, first finding the LCD. Simplify, if possible.

11. $\frac{7}{24} + \frac{11}{18}$

12. $\frac{7}{60} + \frac{2}{25}$

13. $\frac{1}{6} + \frac{3}{40}$

14. $\frac{5}{24} + \frac{3}{20}$

15. $\frac{1}{20} + \frac{1}{30} + \frac{2}{45}$

16. $\frac{2}{15} + \frac{5}{9} + \frac{3}{20}$

c Find the LCM.

17. $6x^2, \quad 12x^3$

18. $2a^2b, \quad 8ab^3$

19. $2x^2, \quad 6xy, \quad 18y^2$

20. $p^3q, \quad p^2q, \quad pq^2$

21. $2(y - 3), \quad 6(y - 3)$

22. $5(m + 2), \quad 15(m + 2)$

23. $t, \quad t + 2, \quad t - 2$

24. $y, \quad y - 5, \quad y + 5$

25. $x^2 - 4, \quad x^2 + 5x + 6$

26. $x^2 - 4, \quad x^2 - x - 2$

27. $t^3 + 4t^2 + 4t, \quad t^2 - 4t$

28. $m^4 - m^2, \quad m^3 - m^2$

29. $a + 1, \quad (a - 1)^2, \quad a^2 - 1$

30. $a^2 - 2ab + b^2, \quad a^2 - b^2, \quad 3a + 3b$

31. $m^2 - 5m + 6, \quad m^2 - 4m + 4$

32. $2x^2 + 5x + 2, \quad 2x^2 - x - 1$

33. $2 + 3x, \quad 4 - 9x^2, \quad 2 - 3x$

34. $9 - 4x^2, \quad 3 + 2x, \quad 3 - 2x$

35. $10v^2 + 30v, \quad 5v^2 + 35v + 60$

36. $12a^2 + 24a, \quad 4a^2 + 20a + 24$

37. $9x^3 - 9x^2 - 18x, \quad 6x^5 - 24x^4 + 24x^3$

38. $x^5 - 4x^3, \quad x^3 + 4x^2 + 4x$

39. $x^5 + 4x^4 + 4x^3, \quad 3x^2 - 12, \quad 2x + 4$

40. $x^5 + 2x^4 + x^3, \quad 2x^3 - 2x, \quad 5x - 5$

Skill Maintenance

Factor. [11.6a]

41. $x^2 - 6x + 9$

42. $6x^2 + 4x$

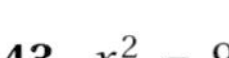
43. $x^2 - 9$

44. $x^2 + 4x - 21$

45. $x^2 + 6x + 9$

46. $x^2 - 4x - 21$

Divorce Rate. The graph at right is that of the equation

$$D = 0.00509x^2 - 19.17x + 18{,}065.305$$

for values of x ranging from 1900 to 2010. It shows the percentage of couples who are married in a given year, x, whose marriages, it is predicted, will end in divorce. Use *only* the graph to answer the questions in Exercises 47–52. [9.1a]

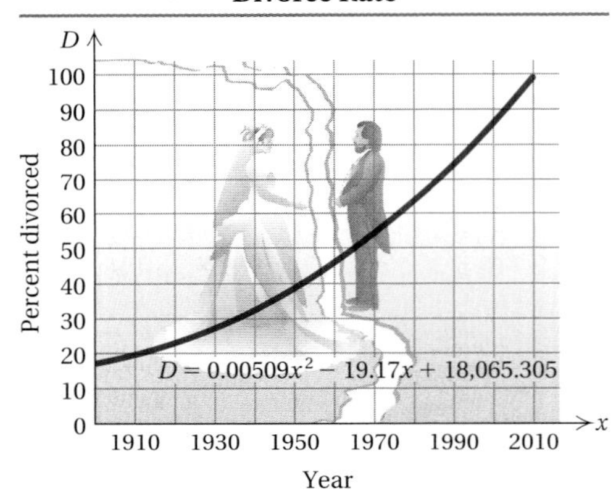

Source: Gottman, John, *What Predicts Divorce: The Relationship Between Marital Processes and Marital Outcomes*. New Jersey: Lawrence Erlbaum Associates, 1993.

47. Estimate the divorce percentage of those married in 1970.

48. Estimate the divorce percentage of those married in 1980.

49. Estimate the divorce percentage of those married in 1990.

50. Estimate the divorce percentage of those married' in 2010.

51. In what year was the divorce percentage about 50%?

52. In what year was the divorce percentage about 84%?

Synthesis

53. ◆ If the LCM of a binomial and a trinomial is the trinomial, what relationship exists between the two expressions?

54. ◆ Explain how you might find the LCD of these two expressions:

$$\frac{x + 1}{x^2 - 4}, \quad \frac{x - 2}{x^2 + 5x + 6}.$$

12.4 Adding Rational Expressions

a We add rational expressions as we do rational numbers.

> To add when the denominators are the same, add the numerators and keep the same denominator.

Examples Add.

1. $\dfrac{x}{x+1} + \dfrac{2}{x+1} = \dfrac{x+2}{x+1}$

2. $\dfrac{2x^2+3x-7}{2x+1} + \dfrac{x^2+x-8}{2x+1} = \dfrac{(2x^2+3x-7)+(x^2+x-8)}{2x+1}$
$= \dfrac{3x^2+4x-15}{2x+1}$

3. $\dfrac{x-5}{x^2-9} + \dfrac{2}{x^2-9} = \dfrac{(x-5)+2}{x^2-9} = \dfrac{x-3}{x^2-9}$
$= \dfrac{x-3}{(x-3)(x+3)}$ Factoring
$= \dfrac{\cancel{x-3}}{\cancel{(x-3)}(x+3)}$ Removing a factor of 1: $\dfrac{x-3}{x-3} = 1$
$= \dfrac{1}{x+3}$ Simplifying

As in Example 3, simplifying should be done if possible after adding.

Do Exercises 1–3.

When denominators are not the same, we multiply by 1 to obtain equivalent expressions with the same denominator. When one denominator is the opposite of the other, we can first multiply either expression by 1 using $-1/-1$.

Examples

4. $\dfrac{x}{2} + \dfrac{3}{-2} = \dfrac{x}{2} + \dfrac{3}{-2} \cdot \dfrac{-1}{-1}$ Multiplying by 1 using $\dfrac{-1}{-1}$
$= \dfrac{x}{2} + \dfrac{-3}{2}$ The denominators are now the same.
$= \dfrac{x+(-3)}{2} = \dfrac{x-3}{2}$

5. $\dfrac{3x+4}{x-2} + \dfrac{x-7}{2-x} = \dfrac{3x+4}{x-2} + \dfrac{x-7}{2-x} \cdot \dfrac{-1}{-1}$

We could have chosen to multiply this expression by $-1/-1$. We multiply only one expression, *not* both.

$= \dfrac{3x+4}{x-2} + \dfrac{-x+7}{x-2}$ *Note:* $(2-x)(-1) = -2+x = x-2.$
$= \dfrac{(3x+4)+(-x+7)}{x-2} = \dfrac{2x+11}{x-2}$

Do Exercises 4 and 5.

Objective

a Add rational expressions.

For Extra Help

TAPE 23

MAC WIN

CD-ROM

Add.

1. $\dfrac{5}{9} + \dfrac{2}{9}$

2. $\dfrac{3}{x-2} + \dfrac{x}{x-2}$

3. $\dfrac{4x+5}{x-1} + \dfrac{2x-1}{x-1}$

Add.

4. $\dfrac{x}{4} + \dfrac{5}{-4}$

5. $\dfrac{2x+1}{x-3} + \dfrac{x+2}{3-x}$

Answers on page A-37

Add.

6. $\frac{3x}{16} + \frac{5x^2}{24}$

7. $\frac{3}{16x} + \frac{5}{24x^2}$

Answers on page A-37

When denominators are different, we find the least common denominator, LCD. The procedure we will use is as follows.

To add rational expressions with different denominators:

1. Find the LCM of the denominators. This is the least common denominator (LCD).
2. For each rational expression, find an equivalent expression with the LCD. To do so, multiply by 1 using an expression for 1 made up of factors of the LCD that are missing from the original denominator.
3. Add the numerators. Write the sum over the LCD.
4. Simplify, if possible.

Example 6 Add: $\frac{5x^2}{8} + \frac{7x}{12}$.

First, we find the LCD:

$$\left.\begin{aligned} 8 &= 2 \cdot 2 \cdot 2 \\ 12 &= 2 \cdot 2 \cdot 3 \end{aligned}\right\} \quad \text{LCD} = 2 \cdot 2 \cdot 2 \cdot 3, \text{ or } 24.$$

Compare the factorization $8 = 2 \cdot 2 \cdot 2$ with the factorization of the LCD, $24 = 2 \cdot 2 \cdot 2 \cdot 3$. The factor of the LCD missing from 8 is 3. Compare $12 = 2 \cdot 2 \cdot 3$ and $24 = 2 \cdot 2 \cdot 2 \cdot 3$. The factor of the LCD missing from 12 is 2. We multiply by 1 to get the LCD in each expression, and then add and simplify, if possible:

$$\begin{aligned} \frac{5x^2}{8} + \frac{7x}{12} &= \frac{5x^2}{2 \cdot 2 \cdot 2} + \frac{7x}{2 \cdot 2 \cdot 3} \\ &= \frac{5x^2}{2 \cdot 2 \cdot 2} \cdot \frac{3}{3} + \frac{7x}{2 \cdot 2 \cdot 3} \cdot \frac{2}{2} \qquad \text{Multiplying by 1 to get the same denominators} \\ &= \frac{15x^2}{24} + \frac{14x}{24} \\ &= \frac{15x^2 + 14x}{24}. \end{aligned}$$

Example 7 Add: $\frac{3}{8x} + \frac{5}{12x^2}$.

First, we find the LCD:

$$\left.\begin{aligned} 8x &= 2 \cdot 2 \cdot 2 \cdot x \\ 12x^2 &= 2 \cdot 2 \cdot 3 \cdot x \cdot x \end{aligned}\right\} \quad \text{LCD} = 2 \cdot 2 \cdot 2 \cdot 3 \cdot x \cdot x, \text{ or } 24x^2.$$

The factors of the LCD missing from $8x$ are 3 and x. The factor of the LCD missing from $12x^2$ is 2. We multiply by 1 to get the LCD in each expression, and then add and simplify, if possible:

$$\begin{aligned} \frac{3}{8x} + \frac{5}{12x^2} &= \frac{3}{8x} \cdot \frac{3 \cdot x}{3 \cdot x} + \frac{5}{12x^2} \cdot \frac{2}{2} \\ &= \frac{9x}{24x^2} + \frac{10}{24x^2} \\ &= \frac{9x + 10}{24x^2}. \end{aligned}$$

Do Exercises 6 and 7.

Example 8 Add: $\frac{2a}{a^2 - 1} + \frac{1}{a^2 + a}$.

First, we find the LCD:

$$\left.\begin{aligned} a^2 - 1 &= (a - 1)(a + 1) \\ a^2 + a &= a(a + 1) \end{aligned}\right\} \quad \text{LCD} = a(a - 1)(a + 1).$$

We multiply by 1 to get the LCD in each expression, and then add and simplify:

$$\frac{2a}{(a - 1)(a + 1)} \cdot \frac{a}{a} + \frac{1}{a(a + 1)} \cdot \frac{a - 1}{a - 1}$$

$$= \frac{2a^2}{a(a - 1)(a + 1)} + \frac{a - 1}{a(a - 1)(a + 1)}$$

$$= \frac{2a^2 + a - 1}{a(a - 1)(a + 1)}$$

$$= \frac{(a + 1)(2a - 1)}{a(a - 1)(a + 1)} \quad \text{Factoring the numerator in order to simplify}$$

$$= \frac{\cancel{(a + 1)}(2a - 1)}{a(a - 1)\cancel{(a + 1)}} \quad \text{Removing a factor of 1: } \frac{a + 1}{a + 1} = 1$$

$$= \frac{2a - 1}{a(a - 1)}.$$

Do Exercise 8.

Example 9 Add: $\frac{x + 4}{x - 2} + \frac{x - 7}{x + 5}$.

First, we find the LCD. It is just the product of the denominators:

$$\text{LCD} = (x - 2)(x + 5).$$

We multiply by 1 to get the LCD in each expression, and then add and simplify:

$$\frac{x + 4}{x - 2} \cdot \frac{x + 5}{x + 5} + \frac{x - 7}{x + 5} \cdot \frac{x - 2}{x - 2} = \frac{(x + 4)(x + 5)}{(x - 2)(x + 5)} + \frac{(x - 7)(x - 2)}{(x - 2)(x + 5)}$$

$$= \frac{x^2 + 9x + 20}{(x - 2)(x + 5)} + \frac{x^2 - 9x + 14}{(x - 2)(x + 5)}$$

$$= \frac{x^2 + 9x + 20 + x^2 - 9x + 14}{(x - 2)(x + 5)}$$

$$= \frac{2x^2 + 34}{(x - 2)(x + 5)}.$$

Do Exercise 9.

8. Add:

$$\frac{3}{x^3 - x} + \frac{4}{x^2 + 2x + 1}.$$

9. Add:

$$\frac{x - 2}{x + 3} + \frac{x + 7}{x + 8}.$$

Calculator Spotlight

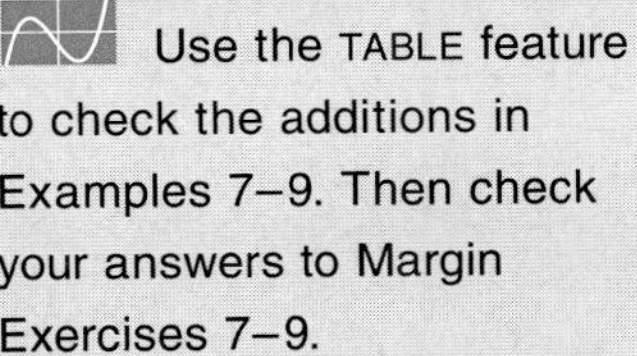

Use the TABLE feature to check the additions in Examples 7–9. Then check your answers to Margin Exercises 7–9.

Answers on page A-37

10. Add:

$$\frac{5}{x^2 + 17x + 16} + \frac{3}{x^2 + 9x + 8}.$$

11. Add:

$$\frac{x + 3}{x^2 - 16} + \frac{5}{12 - 3x}.$$

Answers on page A-37

Example 10 Add: $\dfrac{x}{x^2 + 11x + 30} + \dfrac{-5}{x^2 + 9x + 20}$.

$$\frac{x}{x^2 + 11x + 30} + \frac{-5}{x^2 + 9x + 20}$$

$$= \frac{x}{(x + 5)(x + 6)} + \frac{-5}{(x + 5)(x + 4)}$$

Factoring the denominators in order to find the LCD. The LCD is $(x + 4)(x + 5)(x + 6)$.

$$= \frac{x}{(x + 5)(x + 6)} \cdot \frac{x + 4}{x + 4} + \frac{-5}{(x + 5)(x + 4)} \cdot \frac{x + 6}{x + 6}$$

Multiplying by 1

$$= \frac{x(x + 4) + (-5)(x + 6)}{(x + 4)(x + 5)(x + 6)} = \frac{x^2 + 4x - 5x - 30}{(x + 4)(x + 5)(x + 6)}$$

$$= \frac{x^2 - x - 30}{(x + 4)(x + 5)(x + 6)}$$

$$= \frac{(x - 6)\cancel{(x + 5)}}{(x + 4)\cancel{(x + 5)}(x + 6)}$$

$$= \frac{(x - 6)}{(x + 4)(x + 6)}$$

→ **Always simplify at the end if possible: $\dfrac{x + 5}{x + 5} = 1$.**

Do Exercise 10.

Suppose that after we factor to find the LCD, we find factors that are opposites. There are several ways to handle this, but the easiest is to first go back and multiply by $-1/-1$ appropriately to change factors so that they are not opposites.

Example 11 Add: $\dfrac{x}{x^2 - 25} + \dfrac{3}{10 - 2x}$.

First, we factor as though we are going to find the LCD:

$x^2 - 25 = (x - 5)(x + 5)$;

$10 - 2x = 2(5 - x)$.

We note that there is an $x - 5$ as one factor and a $5 - x$ as another factor. If the denominator of the second expression were $2x - 10$, this situation would not occur. To rewrite the second expression with a denominator of $2x - 10$, we multiply by 1 using $-1/-1$, and then continue as before:

$$\frac{x}{x^2 - 25} + \frac{3}{10 - 2x} = \frac{x}{(x - 5)(x + 5)} + \frac{3}{10 - 2x} \cdot \frac{-1}{-1}$$

$$= \frac{x}{(x - 5)(x + 5)} + \frac{-3}{2x - 10}$$

$$= \frac{x}{(x - 5)(x + 5)} + \frac{-3}{2(x - 5)}$$

LCD = $2(x - 5)(x + 5)$

$$= \frac{x}{(x - 5)(x + 5)} \cdot \frac{2}{2} + \frac{-3}{2(x - 5)} \cdot \frac{x + 5}{x + 5}$$

$$= \frac{2x - 3(x + 5)}{2(x - 5)(x + 5)} = \frac{2x - 3x - 15}{2(x - 5)(x + 5)}$$

$$= \frac{-x - 15}{2(x - 5)(x + 5)}.$$

Collecting like terms

Do Exercise 11.

Exercise Set 12.4

a Add. Simplify, if possible.

1. $\frac{5}{8} + \frac{3}{8}$

2. $\frac{3}{16} + \frac{5}{16}$

3. $\frac{1}{3 + x} + \frac{5}{3 + x}$

4. $\frac{4x + 6}{2x - 1} + \frac{5 - 8x}{-1 + 2x}$

5. $\frac{x^2 + 7x}{x^2 - 5x} + \frac{x^2 - 4x}{x^2 - 5x}$

6. $\frac{4}{x + y} + \frac{9}{y + x}$

7. $\frac{7}{8} + \frac{5}{-8}$

8. $\frac{5}{-3} + \frac{11}{3}$

9. $\frac{3}{t} + \frac{4}{-t}$

10. $\frac{5}{-a} + \frac{8}{a}$

11. $\frac{2x + 7}{x - 6} + \frac{3x}{6 - x}$

12. $\frac{2x - 7}{5x - 8} + \frac{6 + 10x}{8 - 5x}$

13. $\frac{y^2}{y - 3} + \frac{9}{3 - y}$

14. $\frac{t^2}{t - 2} + \frac{4}{2 - t}$

15. $\frac{b - 7}{b^2 - 16} + \frac{7 - b}{16 - b^2}$

16. $\frac{a - 3}{a^2 - 25} + \frac{a - 3}{25 - a^2}$

17. $\frac{a^2}{a - b} + \frac{b^2}{b - a}$

18. $\frac{x^2}{x - 7} + \frac{49}{7 - x}$

19. $\frac{x + 3}{x - 5} + \frac{2x - 1}{5 - x} + \frac{2(3x - 1)}{x - 5}$

20. $\frac{3(x - 2)}{2x - 3} + \frac{5(2x + 1)}{2x - 3} + \frac{3(x + 1)}{3 - 2x}$

21. $\frac{2(4x + 1)}{5x - 7} + \frac{3(x - 2)}{7 - 5x} + \frac{-10x - 1}{5x - 7}$

22. $\frac{5(x - 2)}{3x - 4} + \frac{2(x - 3)}{4 - 3x} + \frac{3(5x + 1)}{4 - 3x}$

23. $\dfrac{x+1}{(x+3)(x-3)} + \dfrac{4(x-3)}{(x-3)(x+3)} + \dfrac{(x-1)(x-3)}{(3-x)(x+3)}$

24. $\dfrac{2(x+5)}{(2x-3)(x-1)} + \dfrac{3x+4}{(2x-3)(1-x)} + \dfrac{x-5}{(3-2x)(x-1)}$

25. $\dfrac{2}{x} + \dfrac{5}{x^2}$

26. $\dfrac{3}{y^2} + \dfrac{6}{y}$

27. $\dfrac{5}{6r} + \dfrac{7}{8r}$

28. $\dfrac{13}{18x} + \dfrac{7}{24x}$

29. $\dfrac{4}{xy^2} + \dfrac{6}{x^2y}$

30. $\dfrac{8}{ab^3} + \dfrac{3}{a^2b}$

31. $\dfrac{2}{9t^3} + \dfrac{1}{6t^2}$

32. $\dfrac{5}{c^2d^3} + \dfrac{-4}{7cd^2}$

33. $\dfrac{x+y}{xy^2} + \dfrac{3x+y}{x^2y}$

34. $\dfrac{2c-d}{c^2d} + \dfrac{c+d}{cd^2}$

35. $\dfrac{3}{x-2} + \dfrac{3}{x+2}$

36. $\dfrac{2}{y+1} + \dfrac{2}{y-1}$

37. $\dfrac{3}{x+1} + \dfrac{2}{3x}$

38. $\dfrac{4}{5y} + \dfrac{7}{y-2}$

39. $\dfrac{2x}{x^2-16} + \dfrac{x}{x-4}$

40. $\dfrac{4x}{x^2 - 25} + \dfrac{x}{x + 5}$

41. $\dfrac{5}{z + 4} + \dfrac{3}{3z + 12}$

42. $\dfrac{t}{t - 3} + \dfrac{5}{4t - 12}$

43. $\dfrac{3}{x - 1} + \dfrac{2}{(x - 1)^2}$

44. $\dfrac{8}{(y + 3)^2} + \dfrac{5}{y + 3}$

45. $\dfrac{4a}{5a - 10} + \dfrac{3a}{10a - 20}$

46. $\dfrac{9x}{6x - 30} + \dfrac{3x}{4x - 20}$

47. $\dfrac{x + 4}{x} + \dfrac{x}{x + 4}$

48. $\dfrac{a}{a - 3} + \dfrac{a - 3}{a}$

49. $\dfrac{4}{a^2 - a - 2} + \dfrac{3}{a^2 + 4a + 3}$

50. $\dfrac{a}{a^2 - 2a + 1} + \dfrac{1}{a^2 - 5a + 4}$

51. $\dfrac{x + 3}{x - 5} + \dfrac{x - 5}{x + 3}$

52. $\dfrac{3x}{2y - 3} + \dfrac{2x}{3y - 2}$

53. $\dfrac{a}{a^2 - 1} + \dfrac{2a}{a^2 - a}$

54. $\dfrac{3x + 2}{3x + 6} + \dfrac{x - 2}{x^2 - 4}$

55. $\dfrac{6}{x - y} + \dfrac{4x}{y^2 - x^2}$

56. $\dfrac{a - 2}{3 - a} + \dfrac{4 - a^2}{a^2 - 9}$

57. $\dfrac{4 - a}{25 - a^2} + \dfrac{a + 1}{a - 5}$

58. $\dfrac{x + 2}{x - 7} + \dfrac{3 - x}{49 - x^2}$

59. $\dfrac{2}{t^2 + t - 6} + \dfrac{3}{t^2 - 9}$

60. $\dfrac{10}{a^2 - a - 6} + \dfrac{3a}{a^2 + 4a + 4}$

Skill Maintenance

Subtract. [10.4c]

61. $(x^2 + x) - (x + 1)$

62. $(4y^3 - 5y^2 + 7y - 24) - (-9y^3 + 9y^2 - 5y + 49)$

Simplify. [10.2b]

63. $(2x^4y^3)^{-3}$

64. $\left(\dfrac{x^3}{5y}\right)^2$

65. $\left(\dfrac{x^{-4}}{y^7}\right)^3$

66. $(5x^{-2}y^{-3})^2$

Graph.

67. $y = \dfrac{1}{2}x - 5$ [9.2b]

68. $2y + x + 10 = 0$ [9.2b], [9.3a]

69. $y = 3$ [9.3b]

70. $x = -5$ [9.3b]

Solve.

71. $3x - 7 = 5x + 9$ [8.3b]

72. $2a + 8 = 13 - 4a$ [8.3b]

73. $x^2 - 8x + 15 = 0$ [11.7b]

74. $x^2 - 7x = 18$ [11.7b]

Synthesis

75. ◆ Explain why the expressions

$$\frac{1}{3 - x} \text{ and } \frac{1}{x - 3}$$

are opposites.

76. ◆ Why is it better to use the *least* common denominator, rather than *any* common denominator, when adding rational expressions?

Find the perimeter and the area of the figure.

77.

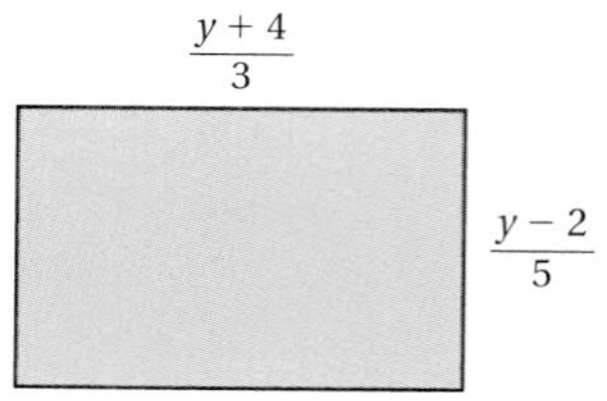

78.

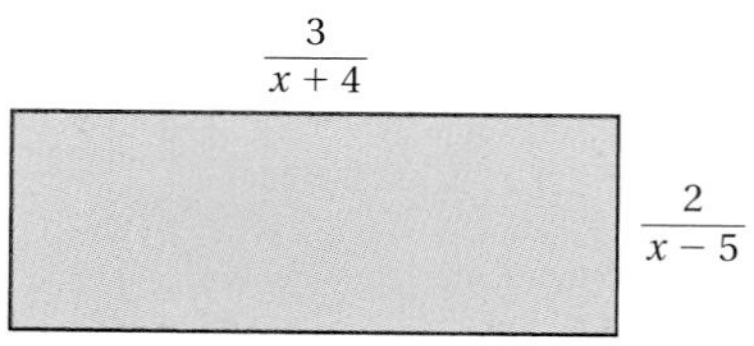

Add. Simplify, if possible.

79. $\dfrac{5}{z + 2} + \dfrac{4z}{z^2 - 4} + 2$

80. $\dfrac{-2}{y^2 - 9} + \dfrac{4y}{(y - 3)^2} + \dfrac{6}{3 - y}$

81. $\dfrac{3z^2}{z^4 - 4} + \dfrac{5z^2 - 3}{2z^4 + z^2 - 6}$

82. Find an expression equivalent to

$$\frac{a - 3b}{a - b}$$

that is a sum of two fractional expressions. Answers may vary.

83.–88. Use the TABLE feature to check the additions in Exercises 47–52.

12.5 Subtracting Rational Expressions

a We subtract rational expressions as we do rational numbers.

> To subtract when the denominators are the same, subtract the numerators and keep the same denominator.

Example 1 Subtract: $\frac{8}{x} - \frac{3}{x}$.

$$\frac{8}{x} - \frac{3}{x} = \frac{8-3}{x} = \frac{5}{x}$$

Example 2 Subtract: $\frac{3x}{x+2} - \frac{x-2}{x+2}$.

$$\frac{3x}{x+2} - \frac{x-2}{x+2} = \frac{3x-(x-2)}{x+2}$$

The parentheses are important to make sure that you subtract the entire numerator.

$$= \frac{3x-x+2}{x+2} = \frac{2x+2}{x+2}$$

Do Exercises 1–3.

When one denominator is the opposite of the other, we can first multiply one expression by $-1/-1$ to obtain a common denominator.

Example 3 Subtract: $\frac{x}{5} - \frac{3x-4}{-5}$.

$$\frac{x}{5} - \frac{3x-4}{-5} = \frac{x}{5} - \frac{3x-4}{-5}\cdot\frac{-1}{-1}$$

Multiplying by 1 using $\frac{-1}{-1}$

This is equal to 1 (not -1).

$$= \frac{x}{5} - \frac{(3x-4)(-1)}{(-5)(-1)}$$

$$= \frac{x}{5} - \frac{4-3x}{5}$$

Remember the parentheses!

$$= \frac{x-(4-3x)}{5}$$

$$= \frac{x-4+3x}{5} = \frac{4x-4}{5}$$

Example 4 Subtract: $\frac{5y}{y-5} - \frac{2y-3}{5-y}$.

$$\frac{5y}{y-5} - \frac{2y-3}{5-y} = \frac{5y}{y-5} - \frac{2y-3}{5-y}\cdot\frac{-1}{-1}$$

$$= \frac{5y}{y-5} - \frac{(2y-3)(-1)}{(5-y)(-1)} = \frac{5y}{y-5} - \frac{3-2y}{y-5}$$

Remember the parentheses!

$$= \frac{5y-(3-2y)}{y-5}$$

Objectives

a Subtract rational expressions.

b Simplify combined additions and subtractions of rational expressions.

For Extra Help

TAPE 23

MAC WIN

CD-ROM

Subtract.

1. $\frac{7}{11} - \frac{3}{11}$

2. $\frac{7}{y} - \frac{2}{y}$

3. $\frac{2x^2+3x-7}{2x+1} - \frac{x^2+x-8}{2x+1}$

Answers on page A-38

Subtract.

4. $\dfrac{x}{3} - \dfrac{2x - 1}{-3}$

5. $\dfrac{3x}{x - 2} - \dfrac{x - 3}{2 - x}$

6. Subtract:

$$\frac{x - 2}{3x} - \frac{2x - 1}{5x}.$$

Answers on page A-38

Then $= \dfrac{5y - 3 + 2y}{y - 5}$

$= \dfrac{7y - 3}{y - 5}.$

Do Exercises 4 and 5.

To subtract rational expressions with different denominators, we use a procedure similar to what we used for addition, except that we subtract numerators and write the difference over the LCD.

> To subtract rational expressions with different denominators:
>
> 1. Find the LCM of the denominators. This is the least common denominator (LCD).
> 2. For each rational expression, find an equivalent expression with the LCD. To do so, multiply by 1 using a symbol for 1 made up of factors of the LCD that are missing from the original denominator.
> 3. Subtract the numerators. Write the difference over the LCD.
> 4. Simplify, if possible.

Example 5 Subtract: $\dfrac{x + 2}{x - 4} - \dfrac{x + 1}{x + 4}.$

The LCD $= (x - 4)(x + 4)$.

$$\frac{x + 2}{x - 4} \cdot \frac{x + 4}{x + 4} - \frac{x + 1}{x + 4} \cdot \frac{x - 4}{x - 4} \quad \text{Multiplying by 1}$$

$$= \frac{(x + 2)(x + 4)}{(x - 4)(x + 4)} - \frac{(x + 1)(x - 4)}{(x - 4)(x + 4)}$$

$$= \frac{x^2 + 6x + 8}{(x - 4)(x + 4)} - \frac{x^2 - 3x - 4}{(x - 4)(x + 4)}$$

$$= \frac{x^2 + 6x + 8 - (x^2 - 3x - 4)}{(x - 4)(x + 4)} \quad \text{Subtracting this numerator. Don't forget the parentheses.}$$

$$= \frac{x^2 + 6x + 8 - x^2 + 3x + 4}{(x - 4)(x + 4)}$$

$$= \frac{9x + 12}{(x - 4)(x + 4)}$$

Do Exercise 6.

Example 6 Subtract: $\dfrac{x}{x^2 + 5x + 6} - \dfrac{2}{x^2 + 3x + 2}.$

$$\frac{x}{x^2 + 5x + 6} - \frac{2}{x^2 + 3x + 2}$$

$$= \frac{x}{(x + 2)(x + 3)} - \frac{2}{(x + 2)(x + 1)} \quad \text{LCD} = (x + 1)(x + 2)(x + 3)$$

$$= \frac{x}{(x + 2)(x + 3)} \cdot \frac{x + 1}{x + 1} - \frac{2}{(x + 2)(x + 1)} \cdot \frac{x + 3}{x + 3}$$

$$= \frac{x^2 + x}{(x + 1)(x + 2)(x + 3)} - \frac{2x + 6}{(x + 1)(x + 2)(x + 3)}$$

Then

$$= \frac{x^2 + x - (2x + 6)}{(x + 1)(x + 2)(x + 3)}$$ Subtracting this numerator. Don't forget the parentheses.

$$= \frac{x^2 + x - 2x - 6}{(x + 1)(x + 2)(x + 3)}$$

$$= \frac{x^2 - x - 6}{(x + 1)(x + 2)(x + 3)}$$

$$= \frac{(x + 2)(x - 3)}{(x + 1)(x + 2)(x + 3)}$$

$$= \frac{\cancel{(x + 2)}(x - 3)}{(x + 1)\cancel{(x + 2)}(x + 3)}$$ Simplifying by removing a factor of 1: $\frac{x + 2}{x + 2} = 1$

$$= \frac{x - 3}{(x + 1)(x + 3)}.$$

Do Exercise 7.

Suppose that after we factor to find the LCD, we find factors that are opposites. Then we multiply by $-1/-1$ appropriately to change factors so that they are not opposites.

Example 7 Subtract: $\frac{p}{64 - p^2} - \frac{5}{p - 8}$.

Factoring $64 - p^2$, we get $(8 - p)(8 + p)$. Note that the factors $8 - p$ in the first denominator and $p - 8$ in the second denominator are opposites. We multiply the first expression by $-1/-1$ to avoid this situation. Then we proceed as before.

$$\frac{p}{64 - p^2} - \frac{5}{p - 8} = \frac{p}{64 - p^2} \cdot \frac{-1}{-1} - \frac{5}{p - 8}$$

$$= \frac{-p}{p^2 - 64} - \frac{5}{p - 8}$$

$$= \frac{-p}{(p - 8)(p + 8)} - \frac{5}{p - 8}$$ LCD = $(p - 8)(p + 8)$

$$= \frac{-p}{(p - 8)(p + 8)} - \frac{5}{p - 8} \cdot \frac{p + 8}{p + 8}$$

$$= \frac{-p}{(p - 8)(p + 8)} - \frac{5p + 40}{(p - 8)(p + 8)}$$

$$= \frac{-p - (5p + 40)}{(p - 8)(p + 8)}$$ Subtracting this numerator. Don't forget the parentheses.

$$= \frac{-p - 5p - 40}{(p - 8)(p + 8)}$$

$$= \frac{-6p - 40}{(p - 8)(p + 8)}$$

Do Exercise 8.

7. Subtract:

$$\frac{x}{x^2 + 15x + 56} - \frac{6}{x^2 + 13x + 42}.$$

8. Subtract:

$$\frac{y}{16 - y^2} - \frac{7}{y - 4}.$$

Calculator Spotlight

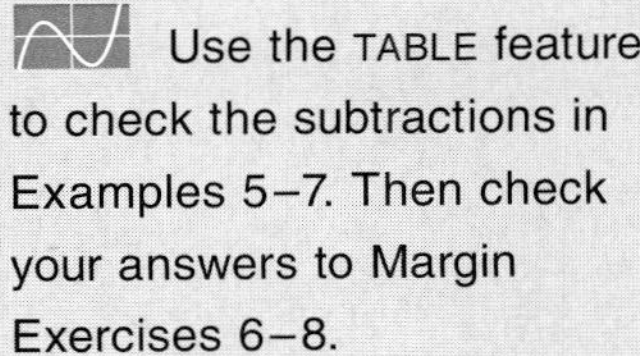

Use the TABLE feature to check the subtractions in Examples 5–7. Then check your answers to Margin Exercises 6–8.

Answers on page A-38

9. Perform the indicated operations and simplify:

$$\frac{x+2}{x^2-9} - \frac{x-7}{9-x^2} + \frac{-8-x}{x^2-9}.$$

10. Perform the indicated operations and simplify:

$$\frac{1}{x} - \frac{5}{3x} + \frac{2x}{x+1}.$$

Answers on page A-38

b Combined Additions and Subtractions

Now let's look at some combined additions and subtractions.

Example 8 Perform the indicated operations and simplify:

$$\frac{x+9}{x^2-4} + \frac{5-x}{4-x^2} - \frac{2+x}{x^2-4}.$$

We have

$$\begin{aligned}\frac{x+9}{x^2-4} + \frac{5-x}{4-x^2} - \frac{2+x}{x^2-4} &= \frac{x+9}{x^2-4} + \frac{5-x}{4-x^2}\cdot\frac{-1}{-1} - \frac{2+x}{x^2-4}\\ &= \frac{x+9}{x^2-4} + \frac{x-5}{x^2-4} - \frac{2+x}{x^2-4}\\ &= \frac{(x+9)+(x-5)-(2+x)}{x^2-4}\\ &= \frac{x+9+x-5-2-x}{x^2-4}\\ &= \frac{x+2}{x^2-4}\\ &= \frac{\cancel{(x+2)}\cdot 1}{\cancel{(x+2)}(x-2)} \qquad \frac{x+2}{x+2}=1\\ &= \frac{1}{x-2}.\end{aligned}$$

Do Exercise 9.

Example 9 Perform the indicated operations and simplify:

$$\frac{1}{x} - \frac{1}{x^2} + \frac{2}{x+1}.$$

The LCD $= x\cdot x(x+1)$, or $x^2(x+1)$.

$$\begin{aligned}&\frac{1}{x}\cdot\frac{x(x+1)}{x(x+1)} - \frac{1}{x^2}\cdot\frac{(x+1)}{(x+1)} + \frac{2}{x+1}\cdot\frac{x^2}{x^2}\\ &= \frac{x(x+1)}{x^2(x+1)} - \frac{x+1}{x^2(x+1)} + \frac{2x^2}{x^2(x+1)}\\ &= \frac{x(x+1)-(x+1)+2x^2}{x^2(x+1)} \qquad \text{Subtracting this numerator. Don't forget the parentheses.}\\ &= \frac{x^2+x-x-1+2x^2}{x^2(x+1)}\\ &= \frac{3x^2-1}{x^2(x+1)}\end{aligned}$$

Do Exercise 10.

Exercise Set 12.5

a Subtract. Simplify, if possible.

1. $\frac{7}{x} - \frac{3}{x}$

2. $\frac{5}{a} - \frac{8}{a}$

3. $\frac{y}{y - 4} - \frac{4}{y - 4}$

4. $\frac{t^2}{t + 5} - \frac{25}{t + 5}$

5. $\frac{2x - 3}{x^2 + 3x - 4} - \frac{x - 7}{x^2 + 3x - 4}$

6. $\frac{x + 1}{x^2 - 2x + 1} - \frac{5 - 3x}{x^2 - 2x + 1}$

7. $\frac{11}{6} - \frac{5}{-6}$

8. $\frac{5}{9} - \frac{7}{-9}$

9. $\frac{5}{a} - \frac{8}{-a}$

10. $\frac{8}{x} - \frac{3}{-x}$

11. $\frac{4}{y - 1} - \frac{4}{1 - y}$

12. $\frac{5}{a - 2} - \frac{3}{2 - a}$

13. $\frac{3 - x}{x - 7} - \frac{2x - 5}{7 - x}$

14. $\frac{t^2}{t - 2} - \frac{4}{2 - t}$

15. $\frac{a - 2}{a^2 - 25} - \frac{6 - a}{25 - a^2}$

16. $\frac{x - 8}{x^2 - 16} - \frac{x - 8}{16 - x^2}$

17. $\frac{4 - x}{x - 9} - \frac{3x - 8}{9 - x}$

18. $\frac{4x - 6}{x - 5} - \frac{7 - 2x}{5 - x}$

19. $\dfrac{2(x-1)}{2x-3} - \dfrac{3(x+2)}{2x-3} - \dfrac{x-1}{3-2x}$

20. $\dfrac{5(2y+1)}{2y-3} - \dfrac{3(y-1)}{3-2y} - \dfrac{3(y-2)}{2y-3}$

21. $\dfrac{a-2}{10} - \dfrac{a+1}{5}$

22. $\dfrac{y+3}{2} - \dfrac{y-4}{4}$

23. $\dfrac{4z-9}{3z} - \dfrac{3z-8}{4z}$

24. $\dfrac{a-1}{4a} - \dfrac{2a+3}{a}$

25. $\dfrac{4x+2t}{3xt^2} - \dfrac{5x-3t}{x^2t}$

26. $\dfrac{5x+3y}{2x^2y} - \dfrac{3x+4y}{xy^2}$

27. $\dfrac{5}{x+5} - \dfrac{3}{x-5}$

28. $\dfrac{3t}{t-1} - \dfrac{8t}{t+1}$

29. $\dfrac{3}{2t^2-2t} - \dfrac{5}{2t-2}$

30. $\dfrac{11}{x^2-4} - \dfrac{8}{x+2}$

31. $\dfrac{2s}{t^2-s^2} - \dfrac{s}{t-s}$

32. $\dfrac{3}{12+x-x^2} - \dfrac{2}{x^2-9}$

33. $\dfrac{y-5}{y} - \dfrac{3y-1}{4y}$

34. $\dfrac{3x-2}{4x} - \dfrac{3x+1}{6x}$

35. $\dfrac{a}{x+a} - \dfrac{a}{x-a}$

36. $\dfrac{a}{a-b} - \dfrac{a}{a+b}$

37. $\frac{5x}{x^2 - 9} - \frac{4}{3 - x}$

38. $\frac{8x}{16 - x^2} - \frac{5}{x - 4}$

39. $\frac{t^2}{2t^2 - 2t} - \frac{1}{2t - 2}$

40. $\frac{4}{5a^2 - 5a} - \frac{2}{5a - 5}$

41. $\frac{x}{x^2 + 5x + 6} - \frac{2}{x^2 + 3x + 2}$

42. $\frac{a}{a^2 + 11a + 30} - \frac{5}{a^2 + 9a + 20}$

b Perform the indicated operations and simplify.

43. $\frac{3(2x + 5)}{x - 1} - \frac{3(2x - 3)}{1 - x} + \frac{6x - 1}{x - 1}$

44. $\frac{a - 2b}{b - a} - \frac{3a - 3b}{a - b} + \frac{2a - b}{a - b}$

45. $\frac{x - y}{x^2 - y^2} + \frac{x + y}{x^2 - y^2} - \frac{2x}{x^2 - y^2}$

46. $\frac{x - 3y}{2(y - x)} + \frac{x + y}{2(x - y)} - \frac{2x - 2y}{2(x - y)}$

47. $\frac{10}{2y - 1} - \frac{6}{1 - 2y} + \frac{y}{2y - 1} + \frac{y - 4}{1 - 2y}$

48. $\frac{(x + 1)(2x - 1)}{(2x - 3)(x - 3)} - \frac{(x - 3)(x + 1)}{(3 - x)(3 - 2x)} + \frac{(2x + 1)(x + 3)}{(3 - 2x)(x - 3)}$

49. $\frac{a + 6}{4 - a^2} - \frac{a + 3}{a + 2} + \frac{a - 3}{2 - a}$

50. $\frac{4t}{t^2 - 1} - \frac{2}{t} - \frac{2}{t + 1}$

51. $\frac{2z}{1-2z} + \frac{3z}{2z+1} - \frac{3}{4z^2-1}$

52. $\frac{1}{x-y} - \frac{2x}{x^2-y^2} + \frac{1}{x+y}$

53. $\frac{1}{x+y} - \frac{1}{x-y} + \frac{2x}{x^2-y^2}$

54. $\frac{2b}{a^2-b^2} - \frac{1}{a+b} + \frac{1}{a-b}$

Skill Maintenance

Simplify.

55. $\frac{x^8}{x^3}$ [10.1e]

56. $3x^4 \cdot 10x^8$ [10.1d]

57. $(a^2b^{-5})^{-4}$ [10.2b]

58. $\frac{54x^{10}}{3x^7}$ [10.1e]

59. $\frac{66x^2}{11x^5}$ [10.1e]

60. $5x^{-7} \cdot 2x^4$ [10.1d]

Find a polynomial for the shaded area of the figure. [10.4d]

61.

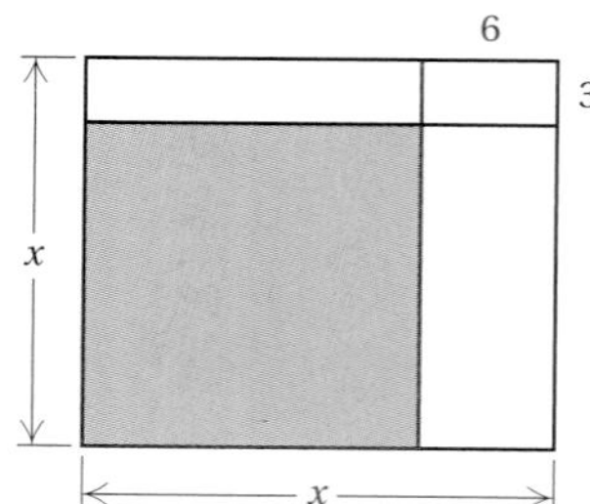

62.

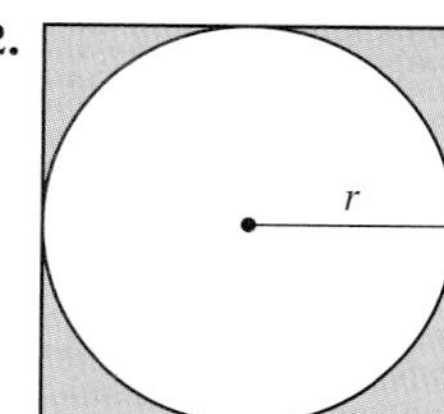

Synthesis

63. ◈ Are parentheses as important when adding rational expressions as they are when subtracting? Why or why not?

64. ◈ Is it possible to add or subtract rational expressions without knowing how to factor? Why or why not?

Perform the indicated operations and simplify.

65. $\frac{2x+11}{x-3} \cdot \frac{3}{x+4} + \frac{2x+1}{4+x} \cdot \frac{3}{3-x}$

66. $\frac{x^2}{3x^2-5x-2} - \frac{2x}{3x+1} \cdot \frac{1}{x-2}$

67. $\frac{x}{x^4-y^4} - \left(\frac{1}{x+y}\right)^2$

68. $\left(\frac{a}{a-b} + \frac{b}{a+b}\right)\left(\frac{1}{3a+b} + \frac{2a+6b}{9a^2-b^2}\right)$

69. The perimeter of the following right triangle is $2a + 5$. Find the length of the missing side and the area.

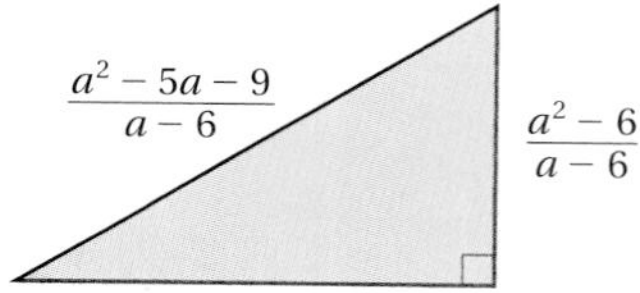

70.–75. Use the TABLE feature to check the subtractions in Exercises 29–34.

12.6 Solving Rational Equations

a Rational Equations

In Sections 12.1–12.5, we studied operations with *rational expressions.* These expressions have no equals signs. We can perform the operations and simplify, but we cannot solve if there are no equals signs—as, for example, in

$$\frac{x^2 + 6x + 9}{x^2 - 4} \cdot \frac{x - 2}{x + 3}, \quad \frac{x + y}{x - y} \div \frac{x^2 + y}{x^2 - y^2}, \quad \text{and} \quad \frac{a + 3}{a^2 - 16} + \frac{5}{12 - 3a}.$$

Operation signs occur. There are no equals signs!

Most often, the result of our calculation is another rational expression that has not been cleared of fractions.

Equations *do have* equals signs, and we can clear them of fractions as we did in Section 8.3. A **rational**, or **fractional, equation** is an equation containing one or more rational expressions. Here are some examples:

$$\frac{2}{3} + \frac{5}{6} = \frac{x}{9}, \quad x + \frac{6}{x} = -5, \quad \text{and} \quad \frac{x^2}{x - 1} = \frac{1}{x - 1}.$$

There are equals signs as well as operation signs.

> To solve a rational equation, the first step is to clear the equation of fractions. To do this, multiply both sides of the equation by the LCM of all the denominators. Then carry out the equation-solving process as we learned it in Chapter 8.

When clearing an equation of fractions, we use the terminology LCM instead of LCD because we are *not* adding or subtracting rational expressions.

Example 1 Solve: $\frac{2}{3} + \frac{5}{6} = \frac{x}{9}$.

The LCM of all denominators is $2 \cdot 3 \cdot 3$, or 18. We multiply by 18 on both sides:

$$18\left(\frac{2}{3} + \frac{5}{6}\right) = 18 \cdot \frac{x}{9} \quad \text{Multiplying by the LCM on both sides}$$

$$18 \cdot \frac{2}{3} + 18 \cdot \frac{5}{6} = 18 \cdot \frac{x}{9} \quad \text{Multiplying to remove parentheses}$$

When clearing an equation of fractions, be sure to multiply *each* term by the LCM.

$$12 + 15 = 2x \quad \text{Simplifying. Note that we have now cleared fractions.}$$

$$27 = 2x$$

$$\frac{27}{2} = x.$$

The solution is $\frac{27}{2}$.

Do Exercise 1.

Objective

a Solve rational equations.

For Extra Help

TAPE 24 | MAC WIN | CD-ROM

1. Solve: $\frac{3}{4} + \frac{5}{8} = \frac{x}{12}$.

Answer on page A-38

2. Solve: $\frac{1}{x} = \frac{1}{6 - x}$.

Example 2 Solve: $\frac{1}{x} = \frac{1}{4 - x}$.

The LCM is $x(4 - x)$. We multiply by $x(4 - x)$ on both sides:

$$\frac{1}{x} = \frac{1}{4 - x}$$

$$x(4 - x) \cdot \frac{1}{x} = x(4 - x) \cdot \frac{1}{4 - x} \quad \text{Multiplying by the LCM on both sides}$$

$$4 - x = x \quad \text{Simplifying}$$

$$4 = 2x$$

$$x = 2.$$

CHECK:

$\frac{1}{x}$	$= \frac{1}{4 - x}$	
$\frac{1}{2}$	$\frac{1}{4 - 2}$	
	$\frac{1}{2}$	TRUE

This checks, so the solution is 2.

Do Exercise 2.

3. Solve: $\frac{x}{4} - \frac{x}{6} = \frac{1}{8}$.

Example 3 Solve: $\frac{x}{6} - \frac{x}{8} = \frac{1}{12}$.

The LCM is 24. We multiply by 24 on both sides:

$$\frac{x}{6} - \frac{x}{8} = \frac{1}{12}$$

$$24\left(\frac{x}{6} - \frac{x}{8}\right) = 24 \cdot \frac{1}{12} \quad \text{Multiplying by the LCM on both sides}$$

$$24 \cdot \frac{x}{6} - 24 \cdot \frac{x}{8} = 24 \cdot \frac{1}{12} \quad \text{Multiplying to remove parentheses}$$

Be sure to multiply *each* term by the LCM.

$$4x - 3x = 2 \quad \text{Simplifying}$$

$$x = 2.$$

CHECK:

$\frac{x}{6} - \frac{x}{8}$	$= \frac{1}{12}$	
$\frac{2}{6} - \frac{2}{8}$	$\frac{1}{12}$	
$\frac{1}{3} - \frac{1}{4}$		
$\frac{4}{12} - \frac{3}{12}$		
$\frac{1}{12}$		TRUE

This checks, so the solution is 2.

Do Exercise 3.

Answers on page A-38

AG Algebraic–Graphical Connection

We can obtain a visual check of the solutions of a rational equation by graphing. For example, consider the equation

$$\frac{x}{4} + \frac{x}{2} = 6.$$

We can examine the solution by graphing the equations

$$y = \frac{x}{4} + \frac{x}{2} \quad \text{and} \quad y = 6$$

using the same set of axes.

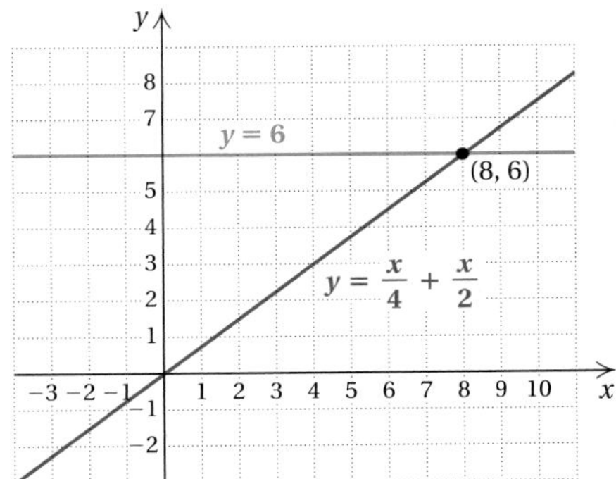

The y-values for each equation will be the same where the graphs intersect. The x-value of that point will yield that value, so it will be the solution of the equation. It appears from the graph that when $x = 8$, the value of $x/4 + x/2$ is 6. We can check by substitution:

$$\frac{x}{4} + \frac{x}{2} = \frac{8}{4} + \frac{8}{2} = 2 + 4 = 6.$$

Thus the solution is 8.

Example 4 Solve: $\frac{2}{3x} + \frac{1}{x} = 10.$

The LCM is $3x$. We multiply by $3x$ on both sides:

$$\frac{2}{3x} + \frac{1}{x} = 10$$

$$3x\left(\frac{2}{3x} + \frac{1}{x}\right) = 3x \cdot 10 \qquad \text{Multiplying by the LCM on both sides}$$

$$3x \cdot \frac{2}{3x} + 3x \cdot \frac{1}{x} = 3x \cdot 10 \qquad \text{Multiplying to remove parentheses}$$

$$2 + 3 = 30x \qquad \text{Simplifying}$$

$$5 = 30x$$

$$\frac{5}{30} = x$$

$$\frac{1}{6} = x.$$

We leave the check to the student. The solution is $\frac{1}{6}$.

Do Exercise 4.

4. Solve: $\frac{1}{2x} + \frac{1}{x} = -12.$

Answer on page A-38

5. Solve: $x + \frac{1}{x} = 2.$

Answer on page A-38

Example 5 Solve: $x + \frac{6}{x} = -5.$

The LCM is x. We multiply by x on both sides:

$$x + \frac{6}{x} = -5$$

$$x\left(x + \frac{6}{x}\right) = -5x \qquad \text{Multiplying by } x \text{ on both sides}$$

$$x \cdot x + x \cdot \frac{6}{x} = -5x \qquad \text{Note that each rational expression on the left is now multiplied by } x.$$

$$x^2 + 6 = -5x \qquad \text{Simplifying}$$

$$x^2 + 5x + 6 = 0 \qquad \text{Adding } 5x \text{ to get a 0 on one side}$$

$$(x + 3)(x + 2) = 0 \qquad \text{Factoring}$$

$$x + 3 = 0 \quad or \quad x + 2 = 0 \qquad \text{Using the principle of zero products}$$

$$x = -3 \quad or \quad x = -2.$$

CHECK: For -3:

$x + \frac{6}{x} = -5$	
$-3 + \frac{6}{-3}$	-5
$-3 - 2$	
-5	

TRUE

For -2:

$x + \frac{6}{x} = -5$	
$-2 + \frac{6}{-2}$	-5
$-2 - 3$	
-5	

TRUE

Both of these check, so there are two solutions, -3 and -2.

Do Exercise 5.

Calculator Spotlight

Checking Solutions Graphically. A grapher can be used to check the solutions of the equation in Example 5:

$$x + \frac{6}{x} = -5.$$

To do so, we graph

$$y_1 = x + \frac{6}{x} \quad \text{and} \quad y_2 = -5.$$

$y_1 = x + \frac{6}{x}$, $y_2 = -5$

10, −10, 10, −10

We see that the graphs appear to cross each other in the third quadrant. To get a better look, we change window settings and obtain the following graph.

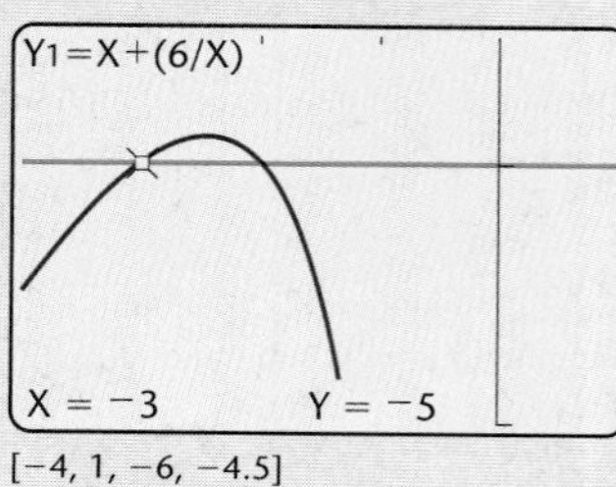

[−4, 1, −6, −4.5]

Next, we can use the CALC and VALUE features to confirm that the points of intersection occur at $x = -3$ and $x = -2$.

Exercises

Use a grapher to check the solutions in each of the following.

1. Example 1
2. Margin Exercise 1
3. Example 3
4. Margin Exercise 5

When we multiply on both sides of an equation by the LCM, the resulting equation might have solutions that are *not* solutions of the original equation. Thus we must *always* check possible solutions in the original equation.

1. If you have carried out all algebraic procedures correctly, you need only check to see if a number makes a denominator 0 in the original equation. If it does make a denominator 0, it is *not* a solution.
2. To be sure that no computational errors have been made and that you indeed have a solution, a complete check is necessary, as we did in Chapter 8.

The next example illustrates the importance of checking all possible solutions.

Example 6 Solve: $\dfrac{x^2}{x-1} = \dfrac{1}{x-1}$.

The LCM is $x - 1$. We multiply by $x - 1$ on both sides:

$$\frac{x^2}{x-1} = \frac{1}{x-1}$$

$$(x-1)\cdot\frac{x^2}{x-1} = (x-1)\cdot\frac{1}{x-1} \qquad \textbf{Multiplying by } x-1 \textbf{ on both sides}$$

$$x^2 = 1 \qquad \textbf{Simplifying}$$

$$x^2 - 1 = 0 \qquad \textbf{Subtracting 1 to get a 0 on one side}$$

$$(x-1)(x+1) = 0 \qquad \textbf{Factoring}$$

$$x - 1 = 0 \quad \text{or} \quad x + 1 = 0 \qquad \textbf{Using the principle of zero products}$$

$$x = 1 \quad \text{or} \quad x = -1.$$

The numbers 1 and -1 are possible solutions. We look at the original equation and see that 1 makes a denominator 0 and is therefore not a solution. The number -1 checks and is a solution.

Do Exercise 6.

Example 7 Solve: $\dfrac{3}{x-5} + \dfrac{1}{x+5} = \dfrac{2}{x^2-25}$.

The LCM is $(x-5)(x+5)$. We multiply by $(x-5)(x+5)$ on both sides:

$$(x-5)(x+5)\left(\frac{3}{x-5} + \frac{1}{x+5}\right) = (x-5)(x+5)\left(\frac{2}{x^2-25}\right)$$

Multiplying on both sides by the LCM

$$(x-5)(x+5)\cdot\frac{3}{x-5} + (x-5)(x+5)\cdot\frac{1}{x+5} = (x-5)(x+5)\cdot\frac{2}{x^2-25}$$

$$3(x+5) + (x-5) = 2 \qquad \textbf{Simplifying}$$

$$3x + 15 + x - 5 = 2 \qquad \textbf{Removing parentheses}$$

$$4x + 10 = 2$$

$$4x = -8$$

$$x = -2.$$

The check is left to the student. The number -2 checks and is the solution.

Do Exercise 7.

6. Solve: $\dfrac{x^2}{x+2} = \dfrac{4}{x+2}$.

7. Solve: $\dfrac{4}{x-2} + \dfrac{1}{x+2} = \dfrac{26}{x^2-4}$.

Calculator Spotlight

Use a grapher to check the solution to Example 6.

CAUTION! We have introduced a new use of the LCM in this section. We previously used the LCM in adding or subtracting rational expressions. *Now* we have equations with equals signs. We clear fractions by multiplying on both sides of the equation by the LCM. This eliminates the denominators. Do *not* make the mistake of trying to clear fractions when you do not have an equation.

Answers on page A-38

Improving Your Math Study Skills

Are You Calculating or Solving?

At the beginning of this section, we noted that one of the common difficulties with this chapter is knowing for sure the task at hand. Are you combining expressions using operations to get another *rational expression,* or are you solving equations for which the results are numbers that are *solutions* of an equation? To learn to make these decisions, complete the following list by writing in the blank the type of answer you should get: "Rational expression" or "Solutions." You do not need to complete the mathematical operations.

Task	Answer (Just write "Rational expression" or "Solutions.")
1. Add: $\frac{4}{x-2} + \frac{1}{x+2}$.	
2. Solve: $\frac{4}{x-2} = \frac{1}{x+2}$.	
3. Subtract: $\frac{4}{x-2} - \frac{1}{x+2}$.	
4. Multiply: $\frac{4}{x-2} \cdot \frac{1}{x+2}$.	
5. Divide: $\frac{4}{x-2} \div \frac{1}{x+2}$.	
6. Solve: $\frac{4}{x-2} + \frac{1}{x+2} = \frac{26}{x^2-4}$.	
7. Perform the indicated operations and simplify: $\frac{4}{x-2} + \frac{1}{x+2} - \frac{26}{x^2-4}$.	
8. Solve: $\frac{x^2}{x-1} = \frac{1}{x-1}$.	
9. Solve: $\frac{2}{y^2-25} = \frac{3}{y-5} + \frac{1}{y-5}$.	
10. Solve: $\frac{x}{x+4} - \frac{4}{x-4} = \frac{x^2+16}{x^2-16}$.	
11. Perform the indicated operations and simplify: $\frac{x}{x+4} - \frac{4}{x-4} - \frac{x^2+16}{x^2-16}$.	
12. Solve: $\frac{5}{y-3} - \frac{30}{y^2-9} = 1$.	
13. Add: $\frac{5}{y-3} + \frac{30}{y^2-9} + 1$.	

Exercise Set 12.6

a Solve. Don't forget to check!

1. $\frac{4}{5} - \frac{2}{3} = \frac{x}{9}$

2. $\frac{x}{20} = \frac{3}{8} - \frac{4}{5}$

3. $\frac{3}{5} + \frac{1}{8} = \frac{1}{x}$

4. $\frac{2}{3} + \frac{5}{6} = \frac{1}{x}$

5. $\frac{3}{8} + \frac{4}{5} = \frac{x}{20}$

6. $\frac{3}{5} + \frac{2}{3} = \frac{x}{9}$

7. $\frac{1}{x} = \frac{2}{3} - \frac{5}{6}$

8. $\frac{1}{x} = \frac{1}{8} - \frac{3}{5}$

9. $\frac{1}{6} + \frac{1}{8} = \frac{1}{t}$

10. $\frac{1}{8} + \frac{1}{12} = \frac{1}{t}$

11. $x + \frac{4}{x} = -5$

12. $\frac{10}{x} - x = 3$

13. $\frac{x}{4} - \frac{4}{x} = 0$

14. $\frac{x}{5} - \frac{5}{x} = 0$

15. $\frac{5}{x} = \frac{6}{x} - \frac{1}{3}$

16. $\frac{4}{x} = \frac{5}{x} - \frac{1}{2}$

17. $\frac{5}{3x} + \frac{3}{x} = 1$

18. $\frac{5}{2y} + \frac{8}{y} = 1$

19. $\frac{t-2}{t+3} = \frac{3}{8}$

20. $\frac{x-7}{x+2} = \frac{1}{4}$

21. $\frac{2}{x+1} = \frac{1}{x-2}$

22. $\frac{8}{y-3} = \frac{6}{y+4}$

23. $\frac{x}{6} - \frac{x}{10} = \frac{1}{6}$

24. $\frac{x}{8} - \frac{x}{12} = \frac{1}{8}$

25. $\frac{t+2}{5} - \frac{t-2}{4} = 1$

26. $\frac{x+1}{3} - \frac{x-1}{2} = 1$

27. $\frac{5}{x-1} = \frac{3}{x+2}$

28. $\frac{x-7}{x-9} = \frac{2}{x-9}$

29. $\frac{a-3}{3a+2} = \frac{1}{5}$

30. $\frac{x+7}{8x-5} = \frac{2}{3}$

31. $\dfrac{x-1}{x-5} = \dfrac{4}{x-5}$

32. $\dfrac{y+11}{y+8} = \dfrac{3}{y+8}$

33. $\dfrac{2}{x+3} = \dfrac{5}{x}$

34. $\dfrac{6}{y} = \dfrac{5}{y-8}$

35. $\dfrac{x-2}{x-3} = \dfrac{x-1}{x+1}$

36. $\dfrac{t+5}{t-2} = \dfrac{t-2}{t+4}$

37. $\dfrac{1}{x+3} + \dfrac{1}{x-3} = \dfrac{1}{x^2-9}$

38. $\dfrac{4}{x-3} + \dfrac{2x}{x^2-9} = \dfrac{1}{x+3}$

39. $\dfrac{x}{x+4} - \dfrac{4}{x-4} = \dfrac{x^2+16}{x^2-16}$

40. $\dfrac{5}{y-3} - \dfrac{30}{y^2-9} = 1$

41. $\frac{4-a}{8-a} = \frac{4}{a-8}$

42. $\frac{3}{x-7} = \frac{x+10}{x-7}$

43. $2 - \frac{a-2}{a+3} = \frac{a^2-4}{a+3}$

44. $\frac{5}{x-1} + x + 1 = \frac{5x+4}{x-1}$

Skill Maintenance

Simplify.

45. $(a^2b^5)^{-3}$ [10.2b]

46. $(x^{-2}y^{-3})^{-4}$ [10.2b]

47. $\left(\frac{2x}{t^2}\right)^4$ [10.2b]

48. $\left(\frac{y^3}{w^2}\right)^{-2}$ [10.2b]

49. $4x^{-5} \cdot 8x^{11}$ [10.1d]

50. $(8x^5y^{-4})^2$ [10.2b]

Find the intercepts. Then graph the equation. [9.3a]

51. $5x + 10y = 20$

52. $2x - 4y = 8$

53. $10y - 4x = -20$

54. $y - 5x = 5$

Synthesis

55. ◆ Why is it especially important to check the possible solutions to a rational equation?

56. ◆ How can a graph be used to determine how many solutions an equation has?

Solve.

57. $\frac{4}{y-2} - \frac{2y-3}{y^2-4} = \frac{5}{y+2}$

58. $\frac{x}{x^2+3x-4} + \frac{x+1}{x^2+6x+8} = \frac{2x}{x^2+x-2}$

59. $\frac{x+1}{x+2} = \frac{x+3}{x+4}$

60. $\frac{x^2}{x^2-4} = \frac{x}{x+2} - \frac{2x}{2-x}$

61. $4a - 3 = \frac{a+13}{a+1}$

62. $\frac{3x-9}{x-3} = \frac{5x-4}{2}$

63. $\frac{y^2-4}{y+3} = 2 - \frac{y-2}{y+3}$

64. $\frac{3a-5}{a^2+4a+3} + \frac{2a+2}{a+3} = \frac{a-3}{a+1}$

65. Use a grapher to check the solutions to Exercises 1–4.

66. Use a grapher to check the solutions to Exercises 13, 15, and 25.

12.7 Applications, Proportions, and Problem Solving

a Solving Applied Problems

Example 1 If 2 is subtracted from a number and then the reciprocal is found, the result is twice the reciprocal of the number itself. What is the number?

1. **Familiarize.** Let's try to guess such a number. Try 10: $10 - 2$ is 8, and the reciprocal of 8 is $\frac{1}{8}$. Two times the reciprocal of 10 is $2\left(\frac{1}{10}\right)$, or $\frac{1}{5}$. Since $\frac{1}{8} \neq \frac{1}{5}$, the number 10 does not check, but the process helps us understand the translation. Let $x =$ the number.
2. **Translate.** From the *Familiarize* step, we get the following translation. Subtracting 2 from the number gives us $x - 2$. Twice the reciprocal of the original number is $2(1/x)$.

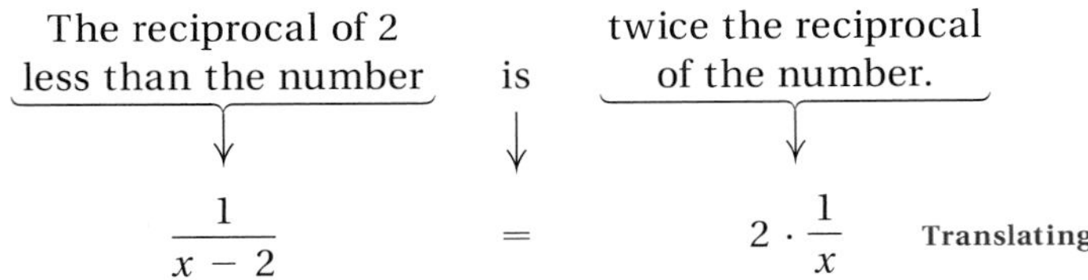

$$\frac{1}{x-2} = 2 \cdot \frac{1}{x} \qquad \text{Translating}$$

3. **Solve.** We solve the equation. The LCM is $x(x - 2)$.

$$x(x-2) \cdot \frac{1}{x-2} = x(x-2) \cdot \frac{2}{x} \qquad \text{Multiplying by the LCM}$$
$$x = 2(x-2) \qquad \text{Simplifying}$$
$$x = 2x - 4$$
$$-x = -4$$
$$x = 4$$

4. **Check.** We go back to the original problem. The number to be checked is 4. Two from 4 is 2. The reciprocal of 2 is $\frac{1}{2}$. The reciprocal of the number itself is $\frac{1}{4}$. Since $\frac{1}{2}$ is twice $\frac{1}{4}$, the conditions are satisfied.
5. **State.** The number is 4.

Do Exercise 1.

Example 2 *Animal Speeds.* A cheetah can run 20 mph faster than a lion. A cheetah can run 7 mi in the same time that a lion can run 5 mi. Find the speed of each animal.

1. **Familiarize.** We first make a drawing. Let $r =$ the speed of the lion. Then $r + 20 =$ the speed of the cheetah.

Objectives

a Solve applied problems using rational equations.

b Solve proportion problems.

For Extra Help

TAPE 24 — InterAct math MAC WIN — CD-ROM

1. The reciprocal of 2 more than a number is three times the reciprocal of the number. Find the number.

Answer on page A-39

2. *Car Speeds.* One car travels 20 km/h faster than another. While one car travels 240 km, the other travels 160 km. Find the speed of each car.

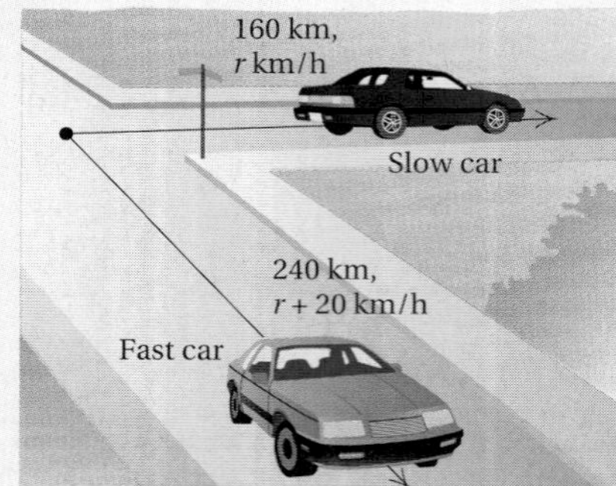

Answer on page A-39

Recall that sometimes we need to find a formula in order to solve an application. A formula that relates the notions of distance, speed, and time is $d = rt$, or

Distance = Speed · Time.

(Indeed, you may need to look up such a formula.)

Since each animal travels the same length of time, we can use just t for time. We organize the information in a chart, as follows.

	d = Distance	r · Speed	t Time	
Lion	5	r	t	$\rightarrow 5 = rt$
Cheetah	7	$r + 20$	t	$\rightarrow 7 = (r + 20)t$

2. Translate. We can apply the formula $d = rt$ along the rows of the table to obtain two equations:

$$5 = rt, \quad (1)$$
$$7 = (r + 20)t. \quad (2)$$

We know that the animals travel for the same length of time. Thus if we solve each equation for t and set the results equal to each other, we get an equation in terms of r.

Solving $5 = rt$ for t: $\quad t = \frac{5}{r}$

Solving $7 = (r + 20)t$ for t: $\quad t = \frac{7}{r + 20}$

Since the times are the same, we have the following equation:

$$\frac{5}{r} = \frac{7}{r + 20}.$$

3. Solve. To solve the equation, we first multiply on both sides by the LCM, which is $r(r + 20)$:

$$r(r + 20) \cdot \frac{5}{r} = r(r + 20) \cdot \frac{7}{r + 20} \quad \text{Multiplying on both sides by the LCM, which is } r(r + 20)$$
$$5(r + 20) = 7r \quad \text{Simplifying}$$
$$5r + 100 = 7r \quad \text{Removing parentheses}$$
$$100 = 2r$$
$$50 = r.$$

We now have a possible solution. The speed of the lion is 50 mph, and the speed of the cheetah is $r = 50 + 20$, or 70 mph.

4. Check. We first reread the problem to see what we were to find. We check the speeds of 50 for the lion and 70 for the cheetah. The cheetah does travel 20 mph faster than the lion and will travel farther than the lion, which runs at a slower speed. If the cheetah runs 7 mi at 70 mph, the time it has traveled is $\frac{7}{70}$, or $\frac{1}{10}$ hr. If the lion runs 5 mi at 50 mph, the time it has traveled is $\frac{5}{50}$, or $\frac{1}{10}$ hr. Since the times are the same, the speeds check.

5. State. The speed of the lion is 50 mph and the speed of the cheetah is 70 mph.

Do Exercise 2.

Example 3 *Recyclable Work.* Erin and Nick work as volunteers at a community recycling depot. Erin can sort a morning's accumulation of recyclables in 4 hr, while Nick requires 6 hr to do the same job. How long would it take them, working together, to sort the recyclables?

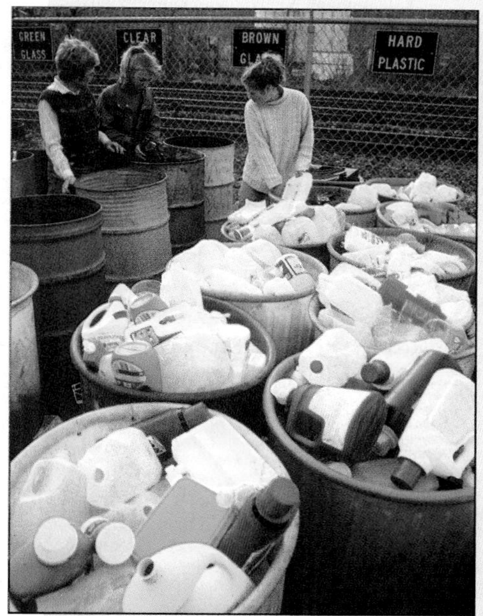

1. **Familiarize.** We familiarize ourselves with the problem by considering two *incorrect* ways of translating the problem to mathematical language.

   a) A common *incorrect* way to translate the problem is to add the two times: 4 hr + 6 hr = 10 hr. Let's think about this. Erin can do the job alone in 4 hr. If Erin and Nick work together, whatever time it takes them should be *less* than 4 hr. Thus we reject 10 hr as a solution, but we do have a partial check on any answer we get. The answer should be less than 4 hr.

   b) Another *incorrect* way to translate the problem is as follows. Suppose the two people split up the sorting job in such a way that Erin does half the sorting and Nick does the other half. Then

   Erin sorts $\frac{1}{2}$ the recyclables in $\frac{1}{2}$(4 hr), or 2 hr,

   and Nick sorts $\frac{1}{2}$ the recyclables in $\frac{1}{2}$(6 hr), or 3 hr.

   But time is wasted since Erin would finish 1 hr earlier than Nick. In effect, they have not worked together to get the job done as fast as possible. If Erin helps Nick after completing her half, the entire job could be done in a time somewhere between 2 hr and 3 hr.

We proceed to a translation by considering how much of the job is finished in 1 hr, 2 hr, 3 hr, and so on. It takes Erin 4 hr to do the sorting job alone. Then, in 1 hr, she can do $\frac{1}{4}$ of the job. It takes Nick 6 hr to do the job alone. Then, in 1 hr, he can do $\frac{1}{6}$ of the job. Working together, they can do

$$\frac{1}{4} + \frac{1}{6}, \text{ or } \frac{5}{12} \text{ of the job in 1 hr.}$$

In 2 hr, Erin can do $2\left(\frac{1}{4}\right)$ of the job and Nick can do $2\left(\frac{1}{6}\right)$ of the job. Working together, they can do

$$2\left(\frac{1}{4}\right) + 2\left(\frac{1}{6}\right), \text{ or } \frac{5}{6} \text{ of the job in 2 hr.}$$

Continuing this reasoning, we can create a table like the following one.

	Fraction of the Job Completed		
Time	**Erin**	**Nick**	**Together**
1 hr	$\frac{1}{4}$	$\frac{1}{6}$	$\frac{1}{4} + \frac{1}{6}$, or $\frac{5}{12}$
2 hr	$2\left(\frac{1}{4}\right)$	$2\left(\frac{1}{6}\right)$	$2\left(\frac{1}{4}\right) + 2\left(\frac{1}{6}\right)$, or $\frac{5}{6}$
3 hr	$3\left(\frac{1}{4}\right)$	$3\left(\frac{1}{6}\right)$	$3\left(\frac{1}{4}\right) + 3\left(\frac{1}{6}\right)$, or $1\frac{1}{4}$
t hr	$t\left(\frac{1}{4}\right)$	$t\left(\frac{1}{6}\right)$	$t\left(\frac{1}{4}\right) + t\left(\frac{1}{6}\right)$

3. By checking work records, a contractor finds that it takes Eduardo 6 hr to construct a wall of a certain size. It takes Yolanda 8 hr to construct the same wall. How long would it take if they worked together?

Answer on page A-39

Calculator Spotlight

 Work Problems on a Calculator.

Using a reciprocal key $\boxed{x^{-1}}$ on a calculator can make it easier to solve problems when using the work formula.

Example. It takes Bianca 6.75 hr, working alone, to keyboard a financial statement. It takes Arturo 7 hr, working alone, and Alani 8 hr, working alone, to produce the same statement. How long does it take if they all work together?

Let $t =$ the time it takes them to do the job working together. Then

$$\frac{t}{6.75} + \frac{t}{7} + \frac{t}{8} = 1, \quad \text{or}$$

$$\frac{1}{6.75} + \frac{1}{7} + \frac{1}{8} = \frac{1}{t}.$$

We calculate $1/t$ using the reciprocal key:

$$\frac{1}{t} = 6.75^{-1} + 7^{-1} + 8^{-1}$$

$$\frac{1}{t} = 0.416005291.$$

This is $1/t$, so we press the reciprocal key $\boxed{x^{-1}}$ again to obtain

$$t = 2.40381558.$$

The financial statement will take about 2.4 hr to complete if they all work together.

Exercises

Solve.

1. Irwin can edge a lawn in 45 min if he works alone. It takes Hannah 38 min to edge the same lawn if she works alone. How long would it take if they work together?
2. Five sorority sisters work on invitations to a dance. Working alone, it would take them, respectively, 2.8, 3.1, 2.2, 3.4, and 2.4 hr to do the job. How long would it take if they work together?

From the table, we see that if they work 3 hr, the fraction of the job completed is $1\frac{1}{4}$, which is more of the job than needs to be done. We see again that the answer is somewhere between 2 hr and 3 hr. What we want is a number t such that the fraction of the job that gets completed is 1; that is, the job is just completed.

2. **Translate.** From the table, we see that the time we want is some number t for which

$$t\left(\frac{1}{4}\right) + t\left(\frac{1}{6}\right) = 1, \quad \text{or} \quad \frac{t}{4} + \frac{t}{6} = 1,$$

where 1 represents the idea that the entire job is completed in time t.

3. **Solve.** We solve the equation:

$$12\left(\frac{t}{4} + \frac{t}{6}\right) = 12 \cdot 1 \qquad \text{Multiplying by the LCM, which is } 2 \cdot 2 \cdot 3, \text{ or } 12$$

$$12 \cdot \frac{t}{4} + 12 \cdot \frac{t}{6} = 12$$

$$3t + 2t = 12$$

$$5t = 12$$

$$t = \frac{12}{5}, \text{ or } 2\frac{2}{5} \text{ hr.}$$

4. **Check.** The check can be done by recalculating:

$$\frac{12}{5}\left(\frac{1}{4}\right) + \frac{12}{5}\left(\frac{1}{6}\right) = \frac{3}{5} + \frac{2}{5} = \frac{5}{5} = 1.$$

We also have another check in what we learned from the *Familiarize* step. The answer, $2\frac{2}{5}$ hr, is between 2 hr and 3 hr (see the table), and it is less than 4 hr, the time it takes Erin working alone.

5. **State.** It takes $2\frac{2}{5}$ hr for them to do the sorting, working together.

THE WORK PRINCIPLE

Suppose $a =$ the time it takes A to do a job, $b =$ the time it takes B to do the same job, and $t =$ the time it takes them to do the same job working together. Then

$$\frac{t}{a} + \frac{t}{b} = 1, \quad \text{or} \quad \frac{1}{a} + \frac{1}{b} = \frac{1}{t}.$$

Do Exercise 3 on the preceding page.

b Applications Involving Proportions

We now consider applications with proportions. A **proportion** involves ratios. A **ratio** of two quantities is their quotient. For example, 73% is the ratio of 73 to 100, $\frac{73}{100}$. The ratio of two different kinds of measure is called a **rate**. Suppose an animal travels 720 ft in 2.5 hr. Its **rate**, or **speed**, is then

$$\frac{720 \text{ ft}}{2.5 \text{ hr}} = 288 \frac{\text{ft}}{\text{hr}}.$$

Do Exercises 4–7 on the following page.

> An equality of ratios, $A/B = C/D$, is called a **proportion**. The numbers named in a proportion are said to be **proportional**.

Proportions can be used to solve applications by expressing a single ratio in two ways.

Example 4 *Gas Mileage.* A Ford Taurus can travel 135 mi of city driving on 6 gal of gas (***Source:*** Ford Motor Company). How much gas would be required for 360 mi of city driving?

1. **Familiarize.** We know that the Taurus can travel 135 mi on 6 gal of gas. Thus we can set up ratios, letting x = the amount of gas required to drive 360 mi.

2. **Translate.** We assume that the car uses gas at the same rate throughout the 360 miles. Thus the ratios are the same and we can write a proportion. Note that the units of *mileage* are in the numerators and the units of *gasoline* are in the denominators.

$$\begin{matrix}\text{Miles} \longrightarrow \\ \text{Gas} \longrightarrow\end{matrix} \frac{135}{6} = \frac{360}{x} \begin{matrix}\longleftarrow \text{Miles} \\ \longleftarrow \text{Gas}\end{matrix}$$

3. **Solve.** To solve for x, we multiply on both sides by the LCM, which is $6x$:

$$6x \cdot \frac{135}{6} = 6x \cdot \frac{360}{x}$$

$$135x = 2160 \quad \text{Simplifying}$$

$$\frac{135x}{135} = \frac{2160}{135} \quad \text{Dividing by 135}$$

$$x = 16. \quad \text{Simplifying}$$

We can also use **cross products** to solve the proportion:

$$\frac{135}{6} = \frac{360}{x} \quad 135x \text{ and } 6 \cdot 360 \text{ are called cross products.}$$

$$135x = 6 \cdot 360 \quad \text{Equating the cross products}$$

$$\frac{135x}{135} = \frac{6 \cdot 360}{135} \quad \text{Dividing by 135}$$

$$x = 16.$$

4. **Check.** We leave the check to the student.

5. **State.** The Taurus will require 16 gal of gas for 360 mi of city driving.

Do Exercise 8.

4. Find the ratio of 145 km to 2.5 liters (L).

5. *Batting Average.* Recently, a baseball player got 7 hits in 25 times at bat. What was the rate, or batting average, in hits per times at bat?

6. Impulses in nerve fibers travel 310 km in 2.5 hr. What is the rate, or speed, in kilometers per hour?

7. A lake of area 550 yd^2 contains 1320 fish. What is the population density of the lake in fish per square yard?

8. *Gas Mileage.* An Oldsmobile Achieva can travel 576 mi of interstate driving on 18 gal of gas (***Source:*** General Motors Corporation). How much gas would be required for 2592 mi of interstate driving?

Answers on page A-39

9. In 1997, Mark McGwire of the Oakland Athletics (and later with the St. Louis Cardinals) had 27 home runs after 77 games.

a) At this rate, how many home runs could McGwire hit in 162 games?

b) Could it be predicted that he would break Maris's record? (McGwire actually completed the season hitting a major-league high of 58 home runs.) (***Source***: Major League Baseball)

10. A sample of 184 light bulbs contained 6 defective bulbs. How many would you expect to find in a sample of 1288 bulbs?

Answers on page A-39

Proportions can be used in many types of applications. In the following example, we predict whether an important home-run record can be broken.

Example 5 *Home-Run Record.* Baseball fans enjoy speculating about records being broken. Roger Maris hit 61 home runs in 1961 to claim the major-league season home-run record. In 1997, Ken Griffey, Jr., had 20 home runs after 44 games. The season consists of 162 games. At this rate, could it be predicted that Griffey would break Maris's record? (***Source***: Major League Baseball)

1. **Familiarize.** Let's assume that Griffey's rate of hitting 20 home runs in 44 games will continue for the 162-game season. We let $H =$ the number of home runs that Griffey can hit in 162 games.

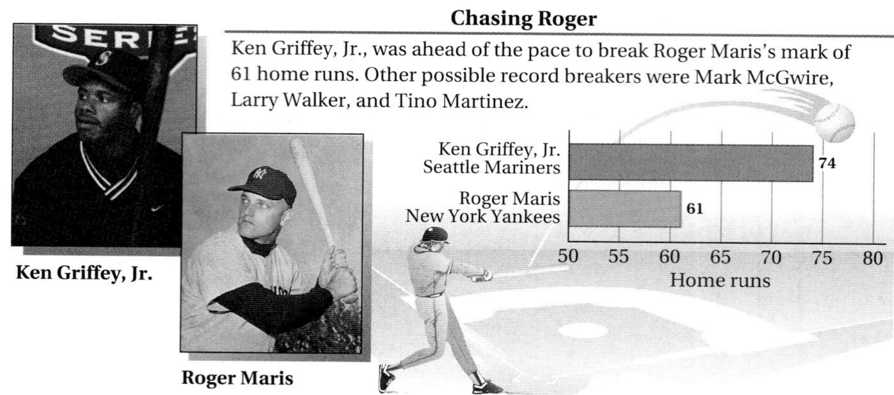
Chasing Roger
Ken Griffey, Jr., was ahead of the pace to break Roger Maris's mark of 61 home runs. Other possible record breakers were Mark McGwire, Larry Walker, and Tino Martinez.
Ken Griffey, Jr.
Seattle Mariners
74
Roger Maris
New York Yankees
61
50 55 60 65 70 75 80
Home runs
Ken Griffey, Jr.
Roger Maris

2. **Translate.** Assuming the rate of hitting home runs continues, the ratios are the same, and we have the proportion

$$\begin{array}{r} \text{Number of home runs} \rightarrow \\ \text{Number of games} \rightarrow \end{array} \frac{H}{162} = \frac{20}{44}. \begin{array}{l} \leftarrow \text{Number of home runs} \\ \leftarrow \text{Number of games} \end{array}$$

3. **Solve.** We solve the equation:

$$\frac{H}{162} = \frac{20}{44}$$

$$44H = 162 \cdot 20 \qquad \text{Equating cross products}$$

$$\frac{44H}{44} = \frac{162 \cdot 20}{44} \qquad \text{Dividing by 44}$$

$$H \approx 73.64.$$

4. **Check.** We leave the check to the student.

5. **State.** We can indeed predict that Griffey, Jr., will hit about 74 home runs and break Maris's record. (Griffey actually completed the season with 56 home runs, having hit only 8 home runs in June and July.)

Do Exercises 9 and 10.

Example 6 *Estimating Wildlife Populations.* To determine the number of fish in a lake, a park ranger catches 225 fish, tags them, and throws them back into the lake. Later, 108 fish are caught, and 15 of them are found to be tagged. Estimate how many fish are in the lake.

1. **Familiarize.** The ratio of fish tagged to the total number of fish in the lake, F, is $\frac{225}{F}$. Of the 108 fish caught later, 15 fish were tagged. The ratio of fish tagged to fish caught is $\frac{15}{108}$.
2. **Translate.** Assuming that the two ratios are the same, we can translate to a proportion.

$$\begin{array}{r} \text{Fish tagged originally} \longrightarrow \\ \text{Fish in lake} \longrightarrow \end{array} \frac{225}{F} = \frac{15}{108} \begin{array}{l} \longleftarrow \text{Tagged fish caught later} \\ \longleftarrow \text{Fish caught later} \end{array}$$

3. **Solve.** We solve the proportion. We multiply by the LCM, which is $108F$:

$$108F \cdot \frac{225}{F} = 108F \cdot \frac{15}{108} \qquad \text{Multiplying by } 108F$$

$$108 \cdot 225 = F \cdot 15$$

$$\frac{108 \cdot 225}{15} = F \qquad \text{Dividing by 15}$$

$$1620 = F.$$

4. **Check.** We leave the check to the student.
5. **State.** We estimate that there are about 1620 fish in the lake.

Do Exercise 11.

Similar Triangles

Proportions also occur geometrically with *similar triangles.* Although similar triangles have the same shape, their sizes may be different.

$$\frac{a}{r} = \frac{b}{s} = \frac{c}{t}$$

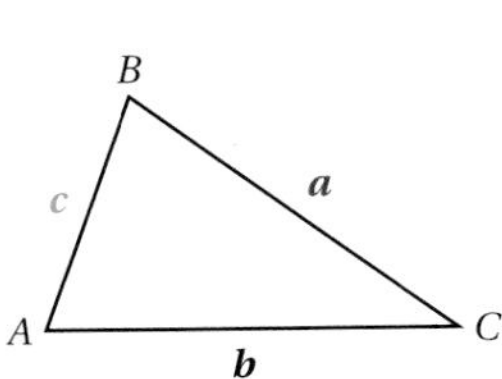

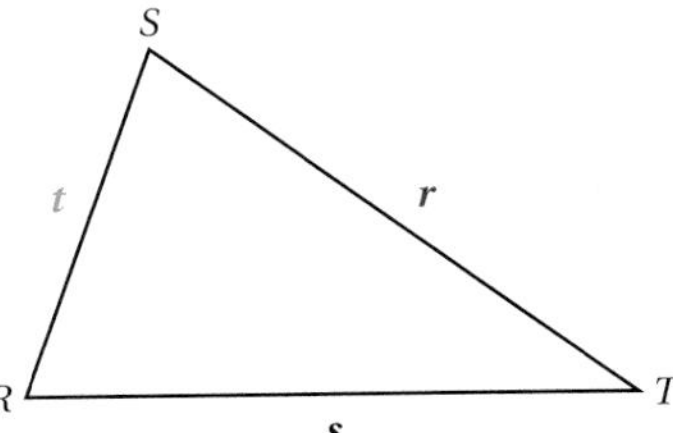

In **similar triangles,** corresponding angles have the same measure and the lengths of corresponding sides are proportional.

11. To determine the number of deer in a forest, a conservationist catches 612 deer, tags them, and lets them loose. Later, 244 deer are caught, 72 of which are tagged. Estimate how many deer are in the forest.

Answer on page A-39

12. *Height of a Flagpole.* How high is a flagpole that casts a 45-ft shadow at the same time that a 5.5-ft woman casts a 10-ft shadow?

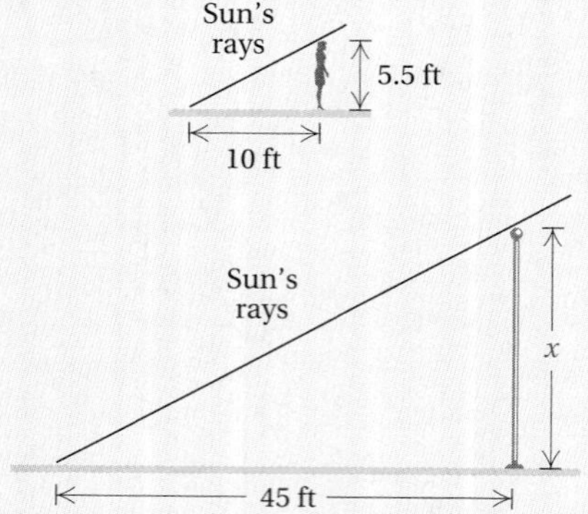

13. *F-106 Blueprint.* Referring to Example 8, find the length x on the plane.

Answers on page A-39

Example 7 Triangles ABC and XYZ below are similar triangles. Solve for z if $x = 10$, $a = 8$, and $c = 5$.

We make a sketch, write a proportion, and then solve. Note that side a is always opposite angle A, side x is always opposite angle X, and so on.

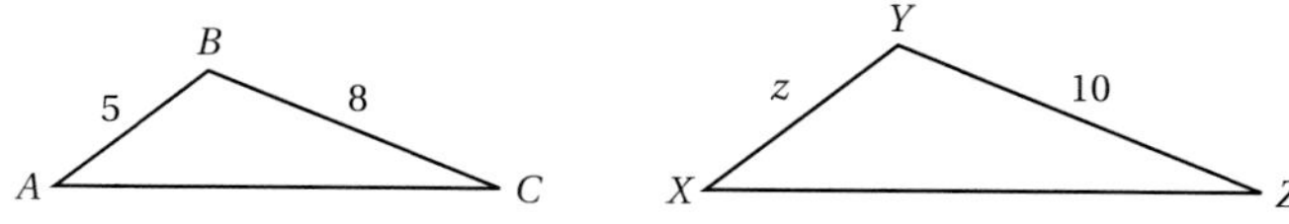

We have

$$\frac{z}{5} = \frac{10}{8}$$ The proportion $\frac{5}{z} = \frac{8}{10}$ could also be used.

$$40 \cdot \frac{z}{5} = 40 \cdot \frac{10}{8}$$ Multiplying by 40

$$8z = 50$$

$$z = \frac{50}{8} \text{ or } 6.25.$$ Dividing by 8

Example 8 *F-106 Blueprint.* A blueprint for an F-106 Delta Dart fighter plane is a scale drawing, as shown below. Each wing has a triangular shape. The blueprint shows similar triangles. Find the length of side a of the wing.

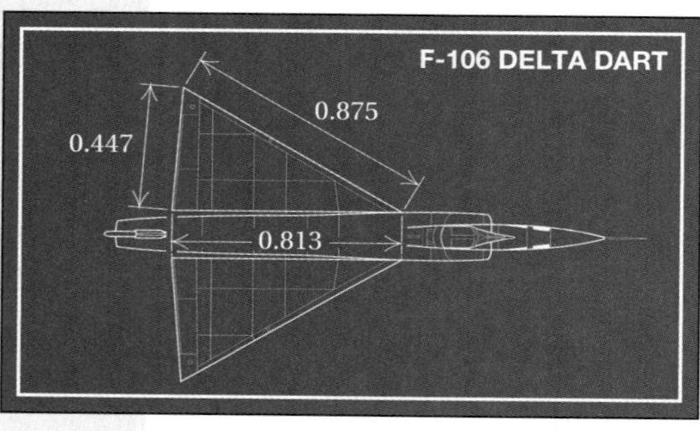

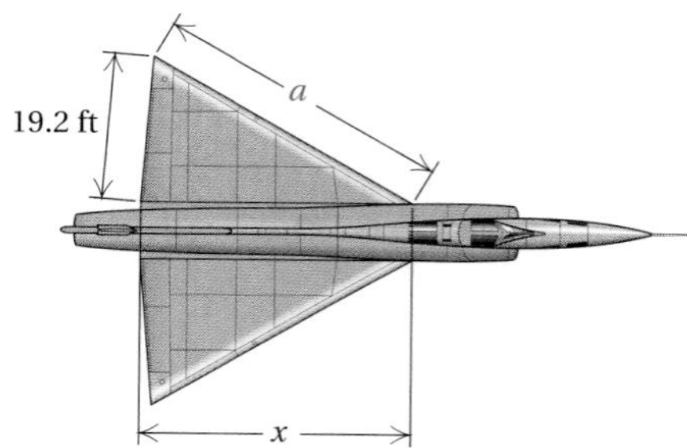

We let $a =$ the length of the wing. Thus we have the proportion

$$\begin{array}{r} \text{Length on the blueprint} \rightarrow \\ \text{Length of the wing} \rightarrow \end{array} \frac{0.447}{19.2} = \frac{0.875}{a} \begin{array}{l} \leftarrow \text{Length on the blueprint} \\ \leftarrow \text{Length of the wing} \end{array}$$

Solve: $0.447 \cdot a = 19.2 \cdot 0.875$ Equating cross products

$$a = \frac{19.2 \cdot 0.875}{0.447}$$ Dividing by 0.447

$$a \approx 37.6 \text{ ft}.$$

The length of side a of the wing is about 37.6 ft.

Do Exercises 12 and 13.

Exercise Set 12.7

a Solve.

1. The reciprocal of 6 plus the reciprocal of 8 is the reciprocal of what number?

2. The reciprocal of 5 plus the reciprocal of 4 is the reciprocal of what number?

3. One number is 5 more than another. The quotient of the larger divided by the smaller is $\frac{4}{3}$. Find the numbers.

4. One number is 4 more than another. The quotient of the larger divided by the smaller is $\frac{5}{2}$. Find the numbers.

5. *Car Speeds.* Rick drives his four-wheel-drive truck 40 km/h faster than Sarah drives her Saturn. While Sarah travels 150 km, Rick travels 350 km. Find their speeds.

   Complete this table and the equations as part of the *Familiarize* step.

	d Distance	$=$ r Speed	$\cdot$ t Time	
Car	150	r		$\rightarrow 150 = r(\quad)$
Truck	350		t	$\rightarrow 350 = (\quad)t$

6. *Car Speeds.* A passenger car travels 30 km/h faster than a delivery truck. While the car goes 400 km, the truck goes 250 km. Find their speeds.

7. *Train Speeds.* The speed of a freight train is 14 mph slower than the speed of a passenger train. The freight train travels 330 mi in the same time that it takes the passenger train to travel 400 mi. Find the speed of each train.

   Complete this table and the equations as part of the *Familiarize* step.

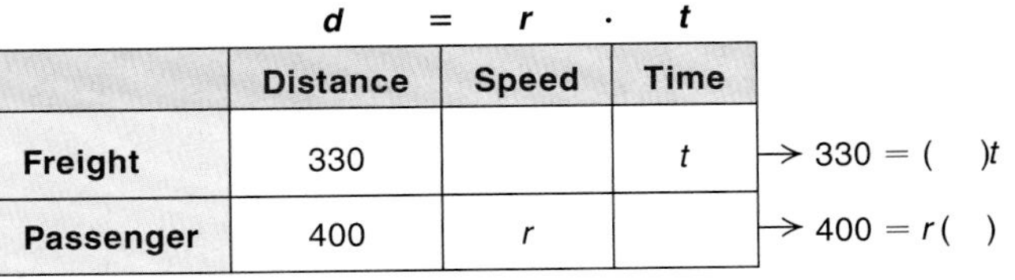

	d Distance	$=$ r Speed	$\cdot$ t Time	
Freight	330		t	$\rightarrow 330 = (\quad)t$
Passenger	400	r		$\rightarrow 400 = r(\quad)$

8. *Train Speeds.* The speed of a freight train is 15 mph slower than the speed of a passenger train. The freight train travels 390 mi in the same time that it takes the passenger train to travel 480 mi. Find the speed of each train.

9. A long-distance trucker traveled 120 mi in one direction during a snowstorm. The return trip in rainy weather was accomplished at double the speed and took 3 hr less time. Find the speed going.

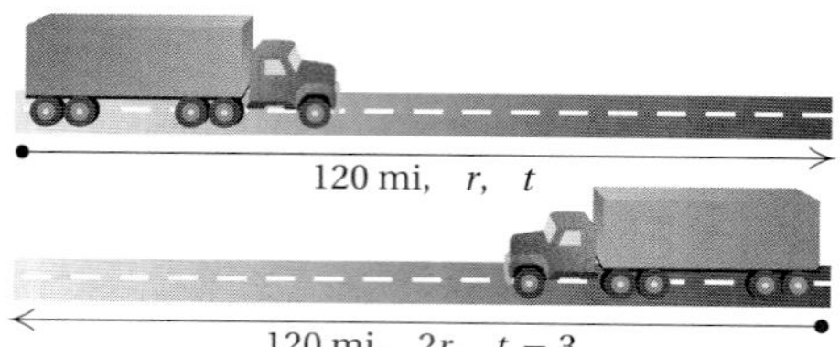

10. After making a trip of 126 mi, a person found that the trip would have taken 1 hr less time by increasing the speed by 8 mph. What was the actual speed?

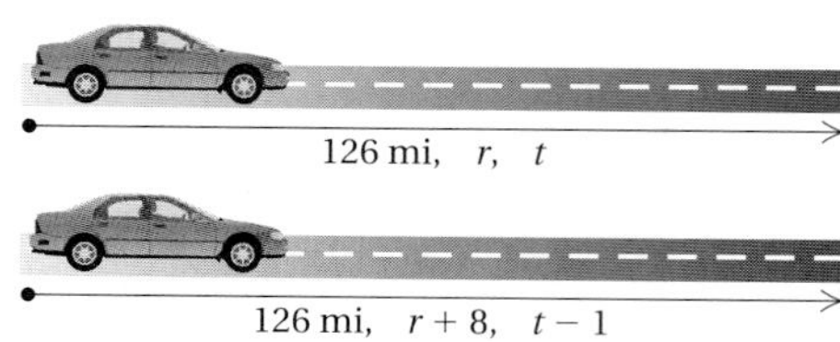

11. The Brother MFC4500 can fax a year-end report in 10 min while the Xerox 850 can fax the same report in 8 min. How long would it take the two machines, working together, to fax the report? (Assume that the recipient has two machines for incoming faxes.)

12. Zack mows the backyard in 40 min, while Angela can mow the same yard in 50 min. How long would it take them, working together with two mowers, to mow the yard?

13. By checking work records, a plumber finds that Rory can fit a kitchen in 12 hr. Mira can do the same job in 9 hr. How long would it take if they worked together?

14. Morgan can proofread 25 pages in 40 min. Shelby can proofread the same 25 pages in 30 min. How long would it take them, working together, to proofread 25 pages?

b Find the ratio of the following. Simplify, if possible.

15. 54 days, 6 days

16. 800 mi, 50 gal

17. A black racer snake travels 4.6 km in 2 hr. What is the speed in kilometers per hour?

18. *Speed of Light.* Light travels 558,000 mi in 3 sec. What is the speed in miles per second?

Solve.

19. A 120-lb person should eat a minimum of 44 g of protein each day. How much protein should a 180-lb person eat each day?

20. *Coffee Beans.* The coffee beans from 14 trees are required to produce 7.7 kg of coffee (this is the average amount that each person in the United States drinks each year). How many trees are required to produce 320 kg of coffee?

21. A student traveled 234 km in 14 days. At this same rate, how far would the student travel in 42 days?

22. In a potato bread recipe, the ratio of milk to flour is $\frac{3}{13}$. If 5 cups of milk are used, how many cups of flour are used?

23. A sample of 144 firecrackers contained 9 "duds." How many duds would you expect in a sample of 3200 firecrackers?

24. *Grass Seed.* It takes 60 oz of grass seed to seed 3000 ft^2 of lawn. At this rate, how much would be needed to seed 5000 ft^2 of lawn?

25. *Home Runs.* In 1997, Tino Martinez of the New York Yankees had 17 home runs after 44 games (***Source:*** Major League Baseball).

a) At this rate, how many home runs could Martinez hit in 162 games?

b) Could it be predicted that Martinez would break Maris's record of 61 home runs in a season?

26. *Home Runs.* In 1997, Larry Walker of the Colorado Rockies had 14 home runs after 40 games (***Source:*** Major League Baseball).

a) At this rate, how many home runs could Walker hit in 162 games?

b) Could it be predicted that Walker would break Maris's record of 61 home runs in a season?

27. *Estimating Whale Population.* To determine the number of blue whales in the world's oceans, marine biologists tag 500 blue whales in various parts of the world. Later, 400 blue whales are checked, and it is found that 20 of them are tagged. Estimate the blue whale population.

28. *Estimating Trout Population.* To determine the number of trout in a lake, a conservationist catches 112 trout, tags them, and throws them back into the lake. Later, 82 trout are caught; 32 of them are tagged. Estimate the number of trout in the lake.

29. *Weight on Mars.* The ratio of the weight of an object on Mars to the weight of an object on Earth is 0.4 to 1.

a) How much would a 12-ton rocket weigh on Mars?

b) How much would a 120-lb astronaut weigh on Mars?

30. *Weight on Moon.* The ratio of the weight of an object on the moon to the weight of an object on Earth is 0.16 to 1.

a) How much would a 12-ton rocket weigh on the moon?

b) How much would a 180-lb astronaut weigh on the moon?

31. A basketball team has 12 more games to play. They have won 25 of the 36 games they have played. How many more games must they win in order to finish with a 0.750 record?

32. Simplest fractional notation for a rational number is $\frac{9}{17}$. Find an equal ratio in which the sum of the numerator and the denominator is 104.

For each pair of similar triangles, find the length of the indicated letter.

33. b:

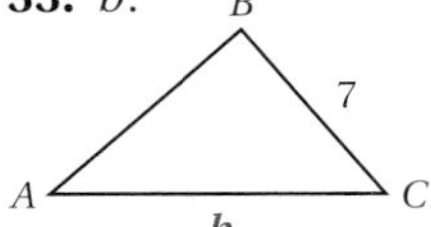

Y 4 X Z 6

34. a:

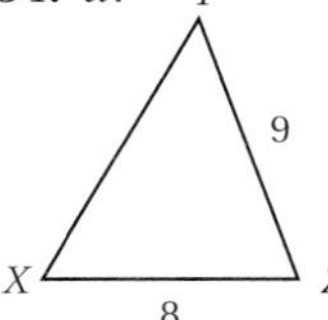

B a A C 6

35. f:

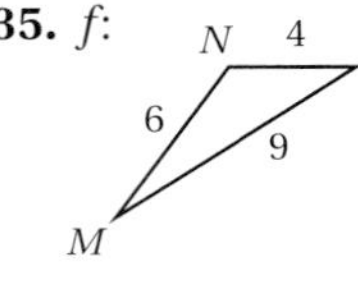

G f H 4 6 F

36. r:

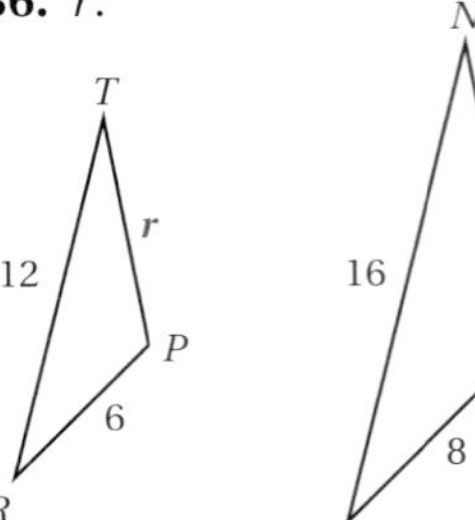

37. h:

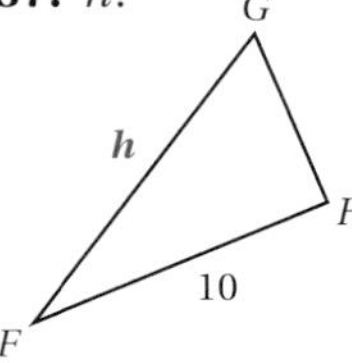

S 7 T 6 R

38. n:

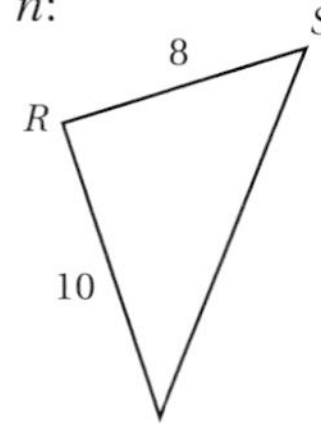

M 5 N n P

Skill Maintenance

Simplify. [10.1d]

39. $x^5 \cdot x^6$

40. $x^{-5} \cdot x^6$

41. $x^{-5} \cdot x^{-6}$

42. $x^5 \cdot x^{-6}$

Synthesis

43. ◆ Explain why it is incorrect to assume that two workers can complete a task twice as quickly as one person working alone.

44. ◆ Write a problem similar to Example 3 or Margin Exercise 3 for a classmate to solve. Design the problem so that the translation step is

$$\frac{t}{7} + \frac{t}{5} = 1.$$

45. Larry, Moe, and Curly are accountants who can complete a financial report together in 3 days. Larry can do the job in 8 days and Moe can do it in 10 days. How many days will it take Curly to complete the job?

46. Ann and Betty work together and complete a sales report in 4 hr. It would take Betty 6 hr longer, working alone, to do the job than it would Ann. How long would it take each of them to do the job working alone?

47. The denominator of a fraction is 1 more than the numerator. If 2 is subtracted from both the numerator and the denominator, the resulting fraction is $\frac{1}{2}$. Find the original fraction.

48. Express 100 as the sum of two numbers for which the ratio of one number, increased by 5, to the other number, decreased by 5, is 4.

49. How soon after 5 o'clock will the hands on a clock first be together?

50. Rachel allows herself 1 hr to reach a sales appointment 50 mi away. After she has driven 30 mi, she realizes that she must increase her speed by 15 mph in order to get there on time. What was her speed for the first 30 mi?

12.8 Formulas and Applications

a The use of formulas is important in many applications of mathematics. We use the following procedure to solve a rational formula for a letter.

> To solve a rational formula for a given letter, identify the letter, and:
>
> 1. Multiply on both sides to clear fractions or decimals, if that is needed.
> 2. Multiply to remove parentheses, if necessary.
> 3. Get all terms with the letter to be solved for on one side of the equation and all other terms on the other side, using the addition principle.
> 4. Factor out the unknown.
> 5. Solve for the letter in question, using the multiplication principle.

Example 1 *Gravitational Force.* The gravitational force f between planets of mass M and m, at a distance d from each other, is given by

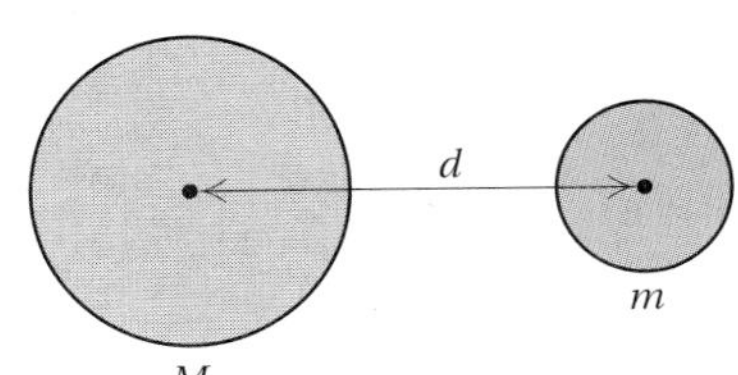

$$f = \frac{kMm}{d^2},$$

where k represents a fixed number constant. Solve for m.

We have

$$f \cdot d^2 = \frac{kMm}{d^2} \cdot d^2 \qquad \text{Multiplying by the LCM, } d^2$$

$$fd^2 = kMm \qquad \text{Simplifying}$$

$$\frac{fd^2}{kM} = m. \qquad \text{Dividing by } kM$$

Do Exercise 1.

Example 2 *Area of a Trapezoid.* The area A of a trapezoid is half the product of the height h and the sum of the lengths b_1 and b_2 of the parallel sides. Solve for b_2.

$$A = \frac{1}{2}h(b_1 + b_2)$$

We consider b_1 and b_2 to be different variables (or constants). The letter b_1 represents the length of the first parallel side and b_2 represents the length of the second parallel side. The small numbers 1 and 2 are called **subscripts**. Subscripts are used to identify different variables with related meanings.

$$2 \cdot A = 2 \cdot \frac{1}{2}h(b_1 + b_2) \qquad \text{Multiplying by 2 to clear fractions}$$

$$2A = h(b_1 + b_2) \qquad \text{Simplifying}$$

Objective

a Solve a formula for a letter.

For Extra Help

TAPE 24 | InterAct Math MAC WIN | CD-ROM

1. Solve for M: $f = \dfrac{kMm}{d^2}$.

Answer on page A-39

2. Solve for b_1: $A = \frac{1}{2}h(b_1 + b_2)$.

3. Solve for f: $\frac{1}{p} + \frac{1}{q} = \frac{1}{f}$.
(This is an optics formula.)

4. Solve for b: $Q = \frac{a - b}{2b}$.

Answers on page A-39

Then $\quad 2A = hb_1 + hb_2 \quad$ Using a distributive law to remove parentheses

$$2A - hb_1 = hb_2 \quad \text{Subtracting } hb_1$$

$$\frac{2A - hb_1}{h} = b_2. \quad \text{Dividing by } h$$

Do Exercise 2.

Example 3 *A Work Formula.* The following work formula was considered in Section 12.7. Solve it for t.

$$\frac{t}{a} + \frac{t}{b} = 1$$

We multiply by the LCM, which is ab:

$$ab \cdot \left(\frac{t}{a} + \frac{t}{b}\right) = ab \cdot 1 \quad \text{Multiplying by } ab$$

$$ab \cdot \frac{t}{a} + ab \cdot \frac{t}{b} = ab \quad \text{Using a distributive law to remove parentheses}$$

$$bt + at = ab \quad \text{Simplifying}$$

$$(b + a)t = ab \quad \text{Factoring out } t$$

$$t = \frac{ab}{b + a}. \quad \text{Dividing by } b + a$$

Do Exercise 3.

In Examples 1 and 2, the letter for which we solved was on the right side of the equation. In Example 3, the letter was on the left. Since all equations are reversible, the location of the letter is a matter of choice.

> **TIP FOR FORMULA SOLVING**
>
> The variable to be solved for should be alone on one side of the equation, with *no* occurrence of that variable on the other side.

Example 4 Solve for b: $S = \frac{a + b}{3b}$.

We multiply by the LCM, which is $3b$:

$$3b \cdot S = 3b \cdot \frac{a + b}{3b} \quad \text{Multiplying by } 3b$$

$$3bS = a + b \quad \text{Simplifying}$$

If we divide by $3S$, we will have b alone on the left, but we will still have a term with b on the right.

$$3bS - b = a \quad \text{Subtracting } b \text{ to get all terms involving } b \text{ on one side}$$

$$b(3S - 1) = a$$

$$b = \frac{a}{3S - 1}. \quad \text{Dividing by } 3S - 1$$

Do Exercise 4.

Exercise Set 12.8

a Solve.

1. $S = 2\pi rh$, for r

2. $A = P(1 + rt)$, for t
(An interest formula)

3. $A = \frac{1}{2}bh$, for b
(The area of a triangle)

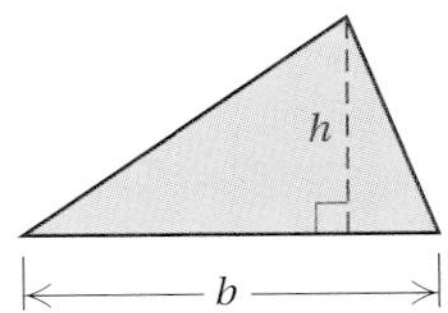

4. $s = \frac{1}{2}gt^2$, for g

5. $S = 180(n - 2)$, for n

6. $S = \frac{n}{2}(a + l)$, for a

7. $V = \frac{1}{3}k(B + b + 4M)$, for b

8. $A = P + Prt$, for P
(*Hint*: Factor the right-hand side.)

9. $S(r - 1) = rl - a$, for r

10. $T = mg - mf$, for m
(*Hint*: Factor the right-hand side.)

11. $A = \frac{1}{2}h(b_1 + b_2)$, for h

12. $S = 2\pi r(r + h)$, for h
(The surface area of a right circular cylinder)

13. $\frac{A - B}{AB} = Q$, for B

14. $L = \frac{Mt + g}{t}$, for t

15. $\frac{1}{p} + \frac{1}{q} = \frac{1}{f}$, for p

16. $\frac{1}{a} + \frac{1}{b} = \frac{1}{t}$, for b

17. $\frac{A}{P} = 1 + r$, for A

18. $\frac{2A}{h} = a + b$, for h

19. $\frac{1}{R} = \frac{1}{r_1} + \frac{1}{r_2}$, for R
(An electricity formula)

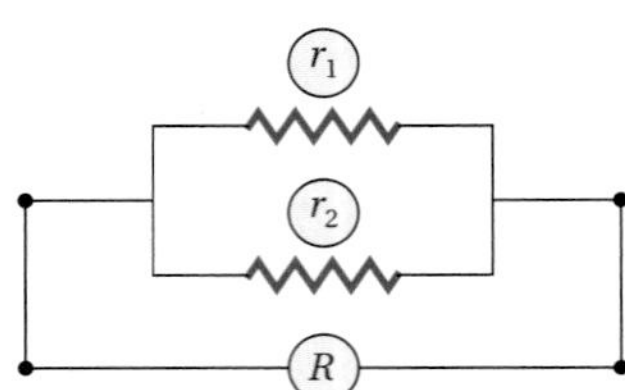

20. $\frac{1}{R} = \frac{1}{r_1} + \frac{1}{r_2}$, for r_1

21. $\frac{A}{B} = \frac{C}{D}$, for D

22. $q = \frac{VQ}{I}$, for I
(An engineering formula)

23. $h_1 = q\left(1 + \frac{h_2}{p}\right)$, for h_2

24. $S = \frac{a - ar^n}{1 - r}$, for a

25. $C = \frac{Ka - b}{a}$, for a

26. $Q = \frac{Pt - h}{t}$, for t

Skill Maintenance

Subtract. [10.4c]

27. $(5x^3 - 7x^2 + 9) - (8x^3 - 2x^2 + 4)$

28. $(5x^4 - 6x^3 + 23x^2 - 79x + 24) - (-18x^4 - 56x^3 + 84x - 17)$

Factor. [11.6a]

29. $x^2 - 4$

30. $30y^4 + 9y^2 - 12$

31. $49m^2 - 112mn + 64n^2$

32. $y^2 + 2y - 35$

33. $y^4 - 1$

34. $a^2 - 100b^2$

Divide and check. [10.8b]

35. $(x^3 + 4x - 4) \div (x - 2)$

36. $(x^4 - 6x^2 + 9) \div (x^2 - 3)$

Synthesis

37. ◆ Describe a situation in which the result of Example 3,

$$t = \frac{ab}{a + b},$$

would be especially useful.

38. ◆ Which of the following is easier to solve for x?

$$\frac{1}{23} + \frac{1}{25} = \frac{1}{x} \quad \text{or} \quad \frac{1}{a} + \frac{1}{b} = \frac{1}{x}$$

Explain the reasons for your choice.

Solve.

39. $u = -F\left(E - \frac{P}{T}\right)$, for T

40. $l = a + (n - 1)d$, for d

41. The formula

$$N = \frac{(b + d)f_1 - v}{(b - v)f_2}$$

is used when monitoring the water in fisheries. Solve for v.

42. In

$$N = \frac{a}{c},$$

what is the effect on N when c increases? when c decreases? Assume that a, c, and N are positive.

Collaborative Learning Manual

Develop a formula for calculating the time required to complete a task when two or more people are working together.

12.9 Complex Rational Expressions

a A **complex rational expression,** or **complex fractional expression,** is a rational expression that has one or more rational expressions within its numerator or denominator. Here are some examples:

$$\frac{1+\frac{2}{x}}{3}, \quad \frac{\frac{x+y}{2}}{\frac{2x}{x+1}}, \quad \frac{\frac{1}{3}+\frac{1}{5}}{\frac{2}{x}-\frac{x}{y}}.$$

These are rational expressions within the complex rational expression.

There are two methods to simplify complex rational expressions. We will consider them both. Use the one that works best for you or the one that your instructor directs you to use.

Objective

a Simplify complex rational expressions.

For Extra Help

TAPE 24 | MAC WIN | CD-ROM

Multiplying by the LCM of All the Denominators: Method 1

METHOD 1

To simplify a complex rational expression:

1. First, find the LCM of all the denominators of all the rational expressions occurring *within* both the numerator and the denominator of the complex rational expression.
2. Then multiply by 1 using LCM/LCM.
3. If possible, simplify by removing a factor of 1.

Example 1 Simplify: $\dfrac{\frac{1}{2}+\frac{3}{4}}{\frac{5}{6}-\frac{3}{8}}$.

We have

$$\frac{\frac{1}{2}+\frac{3}{4}}{\frac{5}{6}-\frac{3}{8}}$$

The denominators *within* the complex rational expression are 2, 4, 6, and 8. The LCM of these denominators is 24. We multiply by 1 using $\frac{24}{24}$.

$$= \frac{\frac{1}{2}+\frac{3}{4}}{\frac{5}{6}-\frac{3}{8}} \cdot \frac{24}{24}$$ Multiplying by 1

$$= \frac{\left(\frac{1}{2}+\frac{3}{4}\right)24}{\left(\frac{5}{6}-\frac{3}{8}\right)24}$$

← Multiplying the numerator by 24

← Multiplying the denominator by 24

1. Simplify. Use method 1.

$$\frac{\frac{1}{3}+\frac{4}{5}}{\frac{7}{8}-\frac{5}{6}}$$

2. Simplify. Use method 1.

$$\frac{\frac{x}{2}+\frac{2x}{3}}{\frac{1}{x}-\frac{x}{2}}$$

3. Simplify. Use method 1.

$$\frac{1+\frac{1}{x}}{1-\frac{1}{x^2}}$$

Answers on page A-39

Using the distributive laws, we carry out the multiplications:

$$= \frac{\frac{1}{2}(24)+\frac{3}{4}(24)}{\frac{5}{6}(24)-\frac{3}{8}(24)}$$

$$= \frac{12+18}{20-9} \quad \textbf{Simplifying}$$

$$= \frac{30}{11}.$$

Multiplying in this manner has the effect of clearing fractions in both the top and the bottom of the complex rational expression.

Do Exercise 1.

Example 2 Simplify: $\dfrac{\frac{3}{x}+\frac{1}{2x}}{\frac{1}{3x}-\frac{3}{4x}}$.

The denominators within the complex expression are x, $2x$, $3x$, and $4x$. The LCM of these denominators is $12x$. We multiply by 1 using $12x/12x$.

$$\frac{\frac{3}{x}+\frac{1}{2x}}{\frac{1}{3x}-\frac{3}{4x}}\cdot\frac{12x}{12x} = \frac{\left(\frac{3}{x}+\frac{1}{2x}\right)12x}{\left(\frac{1}{3x}-\frac{3}{4x}\right)12x} = \frac{\frac{3}{x}(12x)+\frac{1}{2x}(12x)}{\frac{1}{3x}(12x)-\frac{3}{4x}(12x)}$$

$$= \frac{36+6}{4-9} = -\frac{42}{5}$$

Do Exercise 2.

Example 3 Simplify: $\dfrac{1-\frac{1}{x}}{1-\frac{1}{x^2}}$.

The denominators within the complex expression are x and x^2. The LCM of these denominators is x^2. We multiply by 1 using x^2/x^2. Then, after obtaining a single rational expression, we simplify:

$$\frac{1-\frac{1}{x}}{1-\frac{1}{x^2}}\cdot\frac{x^2}{x^2} = \frac{\left(1-\frac{1}{x}\right)x^2}{\left(1-\frac{1}{x^2}\right)x^2} = \frac{1(x^2)-\frac{1}{x}(x^2)}{1(x^2)-\frac{1}{x^2}(x^2)} = \frac{x^2-x}{x^2-1}$$

$$= \frac{x\cancel{(x-1)}}{(x+1)\cancel{(x-1)}} = \frac{x}{x+1}.$$

Do Exercise 3.

Adding in the Numerator and the Denominator: Method 2

METHOD 2

To simplify a complex rational expression:

1. Add or subtract, as necessary, to get a single rational expression in the numerator.
2. Add or subtract, as necessary, to get a single rational expression in the denominator.
3. Divide the numerator by the denominator.
4. If possible, simplify by removing a factor of 1.

We will redo Examples 1–3 using this method.

Example 4 Simplify: $\dfrac{\frac{1}{2}+\frac{3}{4}}{\frac{5}{6}-\frac{3}{8}}$.

We have

$$\frac{\frac{1}{2}+\frac{3}{4}}{\frac{5}{6}-\frac{3}{8}} = \frac{\frac{1}{2}\cdot\frac{2}{2}+\frac{3}{4}}{\frac{5}{6}\cdot\frac{4}{4}-\frac{3}{8}\cdot\frac{3}{3}}$$

Multiplying the $\frac{1}{2}$ by 1 to get a common denominator

Multiplying the $\frac{5}{6}$ and the $\frac{3}{8}$ by 1 to get a common denominator

$$= \frac{\frac{2}{4}+\frac{3}{4}}{\frac{20}{24}-\frac{9}{24}}$$

$$= \frac{\frac{5}{4}}{\frac{11}{24}}$$ Adding in the numerator; subtracting in the denominator

$$= \frac{5}{4}\cdot\frac{24}{11}$$ Multiplying by the reciprocal of the divisor

$$= \frac{5\cdot 3\cdot 2\cdot 2\cdot 2}{2\cdot 2\cdot 11}$$ Factoring

$$= \frac{5\cdot 3\cdot 2\cdot \cancel{2}\cdot \cancel{2}}{\cancel{2}\cdot \cancel{2}\cdot 11}$$ Removing a factor of 1: $\frac{2\cdot 2}{2\cdot 2}=1$

$$= \frac{30}{11}.$$

Do Exercise 4.

4. Simplify. Use method 2.

$$\frac{\frac{1}{3}+\frac{4}{5}}{\frac{7}{8}-\frac{5}{6}}$$

Answer on page A-39

5. Simplify. Use method 2.

$$\frac{\frac{x}{2}+\frac{2x}{3}}{\frac{1}{x}-\frac{x}{2}}$$

6. Simplify. Use method 2.

$$\frac{1+\frac{1}{x}}{1-\frac{1}{x^2}}$$

Answers on page A-39

Example 5 Simplify: $\dfrac{\frac{3}{x}+\frac{1}{2x}}{\frac{1}{3x}-\frac{3}{4x}}$.

We have

$$\frac{\frac{3}{x}+\frac{1}{2x}}{\frac{1}{3x}-\frac{3}{4x}}=\frac{\frac{3}{x}\cdot\frac{2}{2}+\frac{1}{2x}}{\frac{1}{3x}\cdot\frac{4}{4}-\frac{3}{4x}\cdot\frac{3}{3}}$$

Finding the LCD, $2x$, and multiplying by 1 in the numerator

Finding the LCD, $12x$, and multiplying by 1 in the denominator

$$=\frac{\frac{6}{2x}+\frac{1}{2x}}{\frac{4}{12x}-\frac{9}{12x}}=\frac{\frac{7}{2x}}{\frac{-5}{12x}}$$

Adding in the numerator and subtracting in the denominator

$$=\frac{7}{2x}\cdot\frac{12x}{-5}$$

Multiplying by the reciprocal of the divisor

$$=\frac{7}{2x}\cdot\frac{6(2x)}{-5}$$

Factoring

$$=\frac{7}{\cancel{2x}}\cdot\frac{6(\cancel{2x})}{-5}$$

Removing a factor of 1: $\frac{2x}{2x}=1$

$$=\frac{42}{-5}=-\frac{42}{5}.$$

Do Exercise 5.

Example 6 Simplify: $\dfrac{1-\frac{1}{x}}{1-\frac{1}{x^2}}$.

We have

$$\frac{1-\frac{1}{x}}{1-\frac{1}{x^2}}=\frac{\frac{x}{x}-\frac{1}{x}}{\frac{x^2}{x^2}-\frac{1}{x^2}}$$

Finding the LCD, x, and multiplying by 1 in the numerator

Finding the LCD, x^2, and multiplying by 1 in the denominator

$$=\frac{\frac{x-1}{x}}{\frac{x^2-1}{x^2}}$$

Subtracting in the numerator and subtracting in the denominator

$$=\frac{x-1}{x}\cdot\frac{x^2}{x^2-1}$$

Multiplying by the reciprocal of the divisor

$$=\frac{(x-1)x\cdot x}{x(x-1)(x+1)}$$

Factoring

$$=\frac{\cancel{(x-1)}\cancel{x}\cdot x}{\cancel{x}\cancel{(x-1)}(x+1)}$$

Removing a factor of 1: $\frac{x(x-1)}{x(x-1)}=1$

$$=\frac{x}{x+1}.$$

Do Exercise 6.

Exercise Set 12.9

a Simplify.

1. $\dfrac{1+\dfrac{9}{16}}{1-\dfrac{3}{4}}$

2. $\dfrac{6-\dfrac{3}{8}}{4+\dfrac{5}{6}}$

3. $\dfrac{1-\dfrac{3}{5}}{1+\dfrac{1}{5}}$

4. $\dfrac{2+\dfrac{2}{3}}{2-\dfrac{2}{3}}$

5. $\dfrac{\dfrac{1}{2}+\dfrac{3}{4}}{\dfrac{5}{8}-\dfrac{5}{6}}$

6. $\dfrac{\dfrac{3}{4}+\dfrac{7}{8}}{\dfrac{2}{3}-\dfrac{5}{6}}$

7. $\dfrac{\dfrac{1}{x}+3}{\dfrac{1}{x}-5}$

8. $\dfrac{2-\dfrac{1}{a}}{4+\dfrac{1}{a}}$

9. $\dfrac{4-\dfrac{1}{x^2}}{2-\dfrac{1}{x}}$

10. $\dfrac{\dfrac{2}{y}+\dfrac{1}{2y}}{y+\dfrac{y}{2}}$

11. $\dfrac{8+\dfrac{8}{d}}{1+\dfrac{1}{d}}$

12. $\dfrac{3+\dfrac{2}{t}}{3-\dfrac{2}{t}}$

13. $\dfrac{\dfrac{x}{8}-\dfrac{8}{x}}{\dfrac{1}{8}+\dfrac{1}{x}}$

14. $\dfrac{\dfrac{2}{m}+\dfrac{m}{2}}{\dfrac{m}{3}-\dfrac{3}{m}}$

15. $\dfrac{1+\dfrac{1}{y}}{1-\dfrac{1}{y^2}}$

16. $\dfrac{\dfrac{1}{q^2}-1}{\dfrac{1}{q}+1}$

17. $\dfrac{\dfrac{1}{5}-\dfrac{1}{a}}{\dfrac{5-a}{5}}$

18. $\dfrac{\dfrac{4}{t}}{4+\dfrac{1}{t}}$

19. $\dfrac{\frac{1}{a}+\frac{1}{b}}{\frac{1}{a^2}-\frac{1}{b^2}}$

20. $\dfrac{\frac{1}{x^2}-\frac{1}{y^2}}{\frac{2}{x}-\frac{2}{y}}$

21. $\dfrac{\frac{p}{q}+\frac{q}{p}}{\frac{1}{p}+\frac{1}{q}}$

22. $\dfrac{x-3+\frac{2}{x}}{x-4+\frac{3}{x}}$

Skill Maintenance

Add. [10.4a]

23. $(2x^3 - 4x^2 + x - 7) + (4x^4 + x^3 + 4x^2 + x)$

24. $(2x^3 - 4x^2 + x - 7) + (-2x^3 + 4x^2 - x + 7)$

Factor. [11.6a]

25. $p^2 - 10p + 25$

26. $p^2 + 10p + 25$

27. $50p^2 - 100$

28. $5p^2 - 40p - 100$

Solve. [11.8a]

29. The length of a rectangle is 3 yd greater than the width. The area of the rectangle is 10 yd^2. Find the perimeter.

30. A ladder of length 13 ft is placed against a building in such a way that the distance from the top of the ladder to the ground is 7 ft more than the distance from the bottom of the ladder to the building. Find these distances.

Synthesis

31. ◆ Why is factoring an important skill when simplifying complex rational expressions?

32. ◆ Why is the distributive law especially important when using method 1 of this section?

33. Find the reciprocal of $\dfrac{2}{x-1} - \dfrac{1}{3x-2}$.

Simplify.

34. $\dfrac{\frac{a}{b}+\frac{c}{d}}{\frac{b}{a}+\frac{d}{c}}$

35. $\dfrac{\frac{a}{b}-\frac{c}{d}}{\frac{b}{a}-\frac{d}{c}}$

36. $\left[\dfrac{\frac{x+1}{x-1}+1}{\frac{x+1}{x-1}-1}\right]^5$

37. $1+\dfrac{1}{1+\dfrac{1}{1+\dfrac{1}{1+\frac{1}{x}}}}$

38. $\dfrac{\dfrac{z}{1-\frac{z}{2+2z}}-2z}{\dfrac{2z}{5z-2}-3}$

Summary and Review Exercises: Chapter 12

The objectives to be tested in addition to the material in this chapter are [10.2b], [10.4c], [11.6a], and [11.8a].

Find all numbers for which the rational expression is undefined. [12.1a]

1. $\dfrac{3}{x}$

2. $\dfrac{4}{x-6}$

3. $\dfrac{x+5}{x^2-36}$

4. $\dfrac{x^2-3x+2}{x^2+x-30}$

5. $\dfrac{-4}{(x+2)^2}$

6. $\dfrac{x-5}{x^3-8x^2+15x}$

Simplify. [12.1c]

7. $\dfrac{4x^2-8x}{4x^2+4x}$

8. $\dfrac{14x^2-x-3}{2x^2-7x+3}$

9. $\dfrac{(y-5)^2}{y^2-25}$

Multiply and simplify. [12.1d]

10. $\dfrac{a^2-36}{10a} \cdot \dfrac{2a}{a+6}$

11. $\dfrac{6t-6}{2t^2+t-1} \cdot \dfrac{t^2-1}{t^2-2t+1}$

Divide and simplify. [12.2b]

12. $\dfrac{10-5t}{3} \div \dfrac{t-2}{12t}$

13. $\dfrac{4x^4}{x^2-1} \div \dfrac{2x^3}{x^2-2x+1}$

Find the LCM. [12.3c]

14. $3x^2,\ 10xy,\ 15y^2$

15. $a-2,\ 4a-8$

16. $y^2-y-2,\ y^2-4$

Add and simplify. [12.4a]

17. $\dfrac{x+8}{x+7} + \dfrac{10-4x}{x+7}$

18. $\dfrac{3}{3x-9} + \dfrac{x-2}{3-x}$

19. $\dfrac{2a}{a+1} + \dfrac{4a}{a^2-1}$

20. $\dfrac{d^2}{d-c} + \dfrac{c^2}{c-d}$

Subtract and simplify. [12.5a]

21. $\dfrac{6x-3}{x^2-x-12} - \dfrac{2x-15}{x^2-x-12}$

22. $\dfrac{3x-1}{2x} - \dfrac{x-3}{x}$

23. $\dfrac{x+3}{x-2} - \dfrac{x}{2-x}$

24. $\dfrac{1}{x^2-25} - \dfrac{x-5}{x^2-4x-5}$

25. Perform the indicated operations and simplify: [12.5b]

$$\frac{3x}{x+2} - \frac{x}{x-2} + \frac{8}{x^2-4}.$$

Simplify. [12.9a]

26. $\dfrac{\dfrac{1}{z}+1}{\dfrac{1}{z^2}-1}$

27. $\dfrac{\dfrac{c}{d}-\dfrac{d}{c}}{\dfrac{1}{c}+\dfrac{1}{d}}$

Solve. [12.6a]

28. $\dfrac{3}{y} - \dfrac{1}{4} = \dfrac{1}{y}$

29. $\dfrac{15}{x} - \dfrac{15}{x+2} = 2$

Solve. [12.7a]

30. In checking records, a contractor finds that crew A can pave a certain length of highway in 9 hr, while crew B can do the same job in 12 hr. How long would it take if they worked together?

31. *Train Speeds.* A manufacturer is testing two high-speed trains. One train travels 40 km/h faster than the other. While one train travels 70 km, the other travels 60 km. Find the speed of each train.

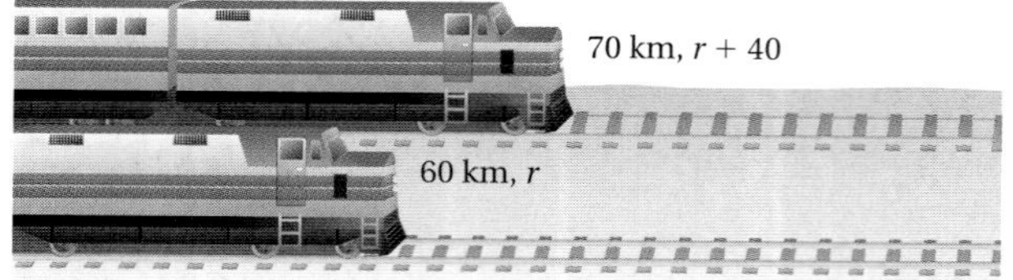

32. The reciprocal of 1 more than a number is twice the reciprocal of the number itself. What is the number?

33. *Airplane Speeds.* One plane travels 80 mph faster than another. While one travels 1750 mi, the other travels 950 mi. Find the speed of each plane.

Solve. [12.7b]

34. A sample of 250 calculators contained 8 defective calculators. How many defective calculators would you expect to find in a sample of 5000?

35. It is known that 10 cm^3 of a normal specimen of human blood contains 1.2 g of hemoglobin. How many grams of hemoglobin would 16 cm^3 of the same blood contain?

36. Triangles ABC and XYZ below are similar. Find the value of x.

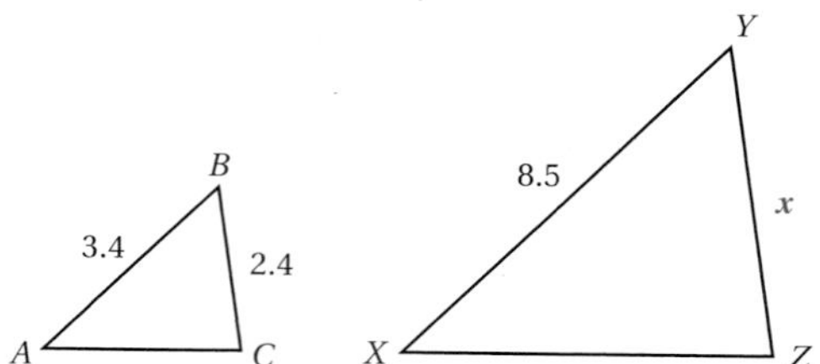

Solve for the letter indicated. [12.8a]

37. $\frac{1}{r} + \frac{1}{s} = \frac{1}{t}$, for s

38. $F = \frac{9C + 160}{5}$, for C

39. $V = \frac{4}{3}\pi r^3$, for r^3
(The volume of a sphere)

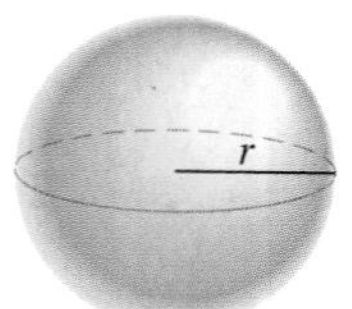

Skill Maintenance

40. Factor: $5x^3 + 20x^2 - 3x - 12$. [11.6a]

41. Simplify: $(5x^3y^2)^{-3}$. [10.2b]

42. Subtract: [10.4c]
$(5x^3 - 4x^2 + 3x - 4) - (7x^3 - 7x^2 - 9x + 14)$.

43. The width of a rectangle is 2 cm less than the length. The area is 15 cm^2. Find the dimensions and the perimeter of the rectangle. [11.8a]

Synthesis

Carry out the direction for each of the following. Explain the use of the LCM in each case.

44. Add: $\frac{4}{x-2} + \frac{1}{x+2}$. [12.4a]

45. Subtract: $\frac{4}{x-2} - \frac{1}{x+2}$. [12.5a]

46. Solve: $\frac{4}{x-2} + \frac{1}{x+2} = \frac{26}{x^2-4}$. [12.6a]

47. Simplify: $\frac{1 - \frac{2}{x}}{1 + \frac{x}{4}}$. [12.9a]

Simplify.

48. $\frac{2a^2 + 5a - 3}{a^2} \cdot \frac{5a^3 + 30a^2}{2a^2 + 7a - 4} \div \frac{a^2 + 6a}{a^2 + 7a + 12}$
[12.1d], [12.2b]

49. $\frac{12a}{(a-b)(b-c)} - \frac{2a}{(b-a)(c-b)}$ [12.5a]

50. Compare
$$\frac{A+B}{B} = \frac{C+D}{D}$$
with the proportion
$$\frac{A}{B} = \frac{C}{D}.$$
[12.7b]

Test: Chapter 12

Find all numbers for which the rational expression is undefined.

1. $\frac{8}{2x}$

2. $\frac{5}{x + 8}$

3. $\frac{x - 7}{x^2 - 49}$

4. $\frac{x^2 + x - 30}{x^2 - 3x + 2}$

5. $\frac{11}{(x - 1)^2}$

6. $\frac{x + 2}{x^3 + 8x^2 + 15x}$

7. Simplify:

$$\frac{6x^2 + 17x + 7}{2x^2 + 7x + 3}.$$

8. Multiply and simplify:

$$\frac{a^2 - 25}{6a} \cdot \frac{3a}{a - 5}.$$

9. Divide and simplify:

$$\frac{25x^2 - 1}{9x^2 - 6x} \div \frac{5x^2 + 9x - 2}{3x^2 + x - 2}.$$

10. Find the LCM:

$y^2 - 9,\ y^2 + 10y + 21,\ y^2 + 4y - 21.$

Add or subtract. Simplify, if possible.

11. $\frac{16 + x}{x^3} + \frac{7 - 4x}{x^3}$

12. $\frac{5 - t}{t^2 + 1} - \frac{t - 3}{t^2 + 1}$

13. $\frac{x - 4}{x - 3} + \frac{x - 1}{3 - x}$

14. $\frac{x - 4}{x - 3} - \frac{x - 1}{3 - x}$

15. $\frac{5}{t - 1} + \frac{3}{t}$

16. $\frac{1}{x^2 - 16} - \frac{x + 4}{x^2 - 3x - 4}$

17. $\frac{1}{x - 1} + \frac{4}{x^2 - 1} - \frac{2}{x^2 - 2x + 1}$

18. Simplify: $\dfrac{9 - \frac{1}{y^2}}{3 - \frac{1}{y}}.$

Answers

1. ____________

2. ____________

3. ____________

4. ____________

5. ____________

6. ____________

7. ____________

8. ____________

9. ____________

10. ____________

11. ____________

12. ____________

13. ____________

14. ____________

15. ____________

16. ____________

17. ____________

18. ____________

Answers

19. ____

20. ____

21. ____

22. ____

23. ____

24. ____

25. ____

26. ____

27. ____

28. ____

29. ____

30. ____

31. ____

Solve.

19. $\frac{7}{y} - \frac{1}{3} = \frac{1}{4}$

20. $\frac{15}{x} - \frac{15}{x-2} = -2$

Solve.

21. The reciprocal of 3 less than a number is four times the reciprocal of the number itself. What is the number?

22. A sample of 125 spark plugs contained 4 defective spark plugs. How many defective spark plugs would you expect to find in a sample of 500?

23. One car travels 20 mph faster than another on a freeway. While one goes 225 mi, the other goes 325 mi. Find the speed of each car.

24. Solve $L = \frac{Mt - g}{t}$ for t.

25. This pair of triangles is similar. Find the missing length x.

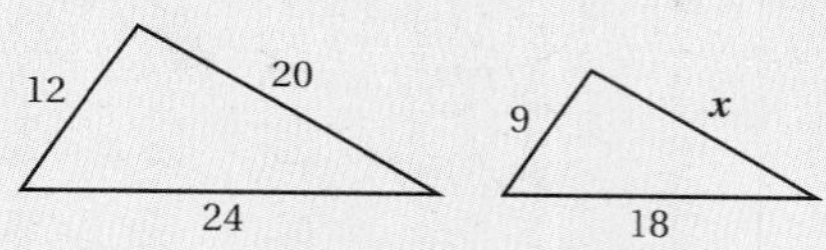

Skill Maintenance

26. Factor: $16a^2 - 49$.

27. Simplify: $\left(\frac{3x^2}{y^3}\right)^{-4}$.

28. Subtract:
$(5x^2 - 19x + 34) - (-8x^2 + 10x - 42)$.

29. The product of two consecutive integers is 462. Find the integers.

Synthesis

30. Team A and team B work together and complete a job in $2\frac{6}{7}$ hr. It would take team B 6 hr longer, working alone, to do the job than it would team A. How long would it take each of them to do the job working alone?

31. Simplify: $1 + \cfrac{1}{1 + \cfrac{1}{1 + \cfrac{1}{a}}}$.

13

Graphs, Slope, and Applications

An Application

In order to meet federal standards, a wheelchair ramp must not rise more than 1 ft over a horizontal distance of 12 ft. Express this slope as a grade.

This problem appears as Exercise 40 in Exercise Set 13.1.

The Mathematics

The slope, or grade, is the vertical change in distance divided by the horizontal change in distance, or

$$m = \frac{1}{12} \approx 0.083 \approx 8.3\%.$$

World Wide Web For more information, visit us at www.mathmax.com

Introduction

We began our study of graphs in Chapter 9, where we focused on linear equations and intercepts. Here we expand our study of linear equations to consider the concept of *slope*. Slope is a number that describes the way in which a line slants. We will also consider applications such as variation and the graphing of inequalities in two variables.

Pretest: Chapter 13

Find the slope, if it exists, of the line.

1. $-4x + y = 6$

2. $y = 3$

3. Find the slope and the y-intercept of the line $x - 3y = 7$.

4. Find the slope, if it exists, of the line containing the points (3, 0) and (3, 6).

5. Find an equation of the line containing the points $(3, -1)$ and $(1, -3)$.

6. Find an equation of the line containing the point $(-1, 3)$ and having slope 4.

7. Find an equation of variation in which y varies directly as x and $y = 10$ when $x = 4$.

8. Find an equation of variation in which y varies inversely as x and $y = 10$ when $x = 4$.

Graph on a plane.

9. $y < x + 2$

10. $2y - 3x \geq 6$

Determine whether the graphs of the equations are parallel, perpendicular, or neither.

11. $y - 3x = 9,$
$y - 3x = 7$

12. $-x + 2y = 7,$
$2x + y = 4$

13. $y = \frac{2}{3}x - 5,$
$y = -\frac{3}{2}x + 4$

14. Determine whether the ordered pair $(-3, 4)$ is a solution of $2x + 5y < 17$.

15. *Consumer Spending on Software.* The line graph at right describes the amount of consumer spending S, in billions, on software in recent years.

a) Find an equation of the line.

b) What is the rate of change in software spending?

c) Use the equation to predict consumer spending on software in 2002.

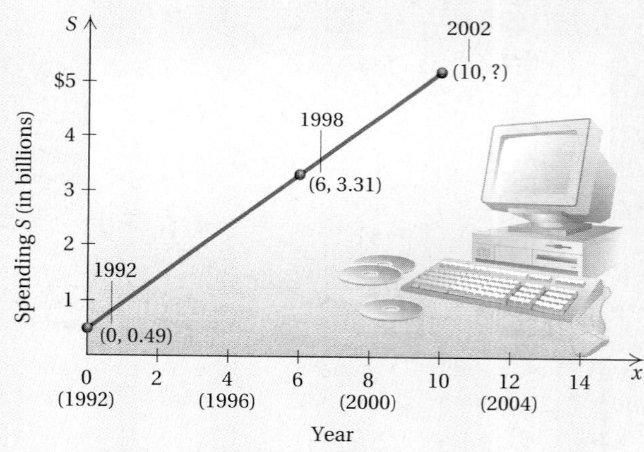

Source: Veronis, Suhler & Associates, PC Data

Objectives for Retesting

The objectives to be tested in addition to the material in this chapter are as follows.

[7.8d] Simplify expressions using rules for order of operations.
[11.7b] Solve quadratic equations by factoring and then using the principle of zero products.
[12.6a] Solve rational equations.
[12.7a] Solve applied problems using rational equations.

13.1 Slope and Applications

Objectives

a Given the coordinates of two points on a line, find the slope of the line.

b Find the slope of a line from an equation.

c Find the slope or rate of change in an applied problem involving slope.

For Extra Help

TAPE 25 | MAC WIN | CD-ROM

a Slope

In Chapter 9, we considered two forms of a linear equation,

$$Ax + By = C \quad \text{and} \quad y = mx + b.$$

We found that from the form of the equation $y = mx + b$, we know certain information, namely, that the y-intercept of the line is $(0, b)$.

$y = mx + b$

? ← The y-intercept is $(0, b)$.

What about the constant m? Does it give us certain information about the line? Look at the following graphs and see if you can make any connection between the constant m and the "slant" of the line.

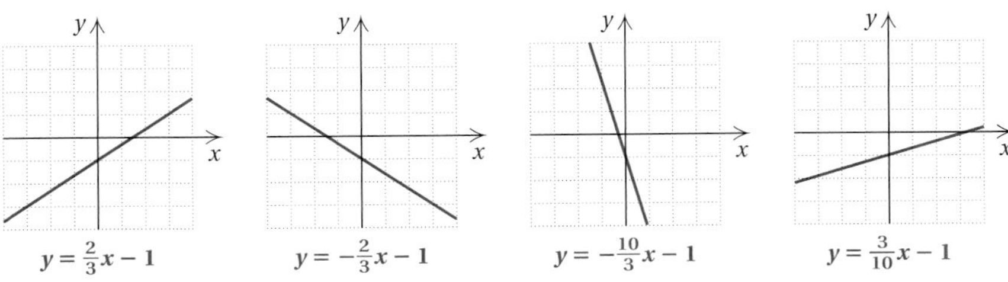

The graphs of some linear equations slant upward from left to right. Others slant downward. Some are vertical and some are horizontal. Some slant more steeply than others. We now look for a way to describe such possibilities with numbers.

Consider a line with two points marked P and Q. As we move from P to Q, the y-coordinate changes from 1 to 3 and the x-coordinate changes from 2 to 6. The change in y is $3 - 1$, or 2. The change in x is $6 - 2$, or 4.

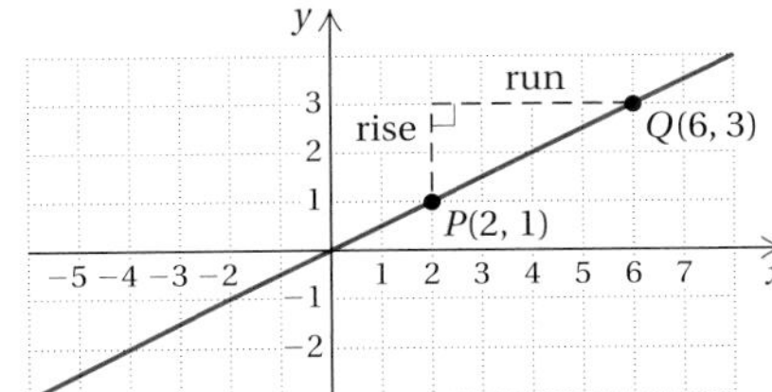

We call the change in y the **rise** and the change in x the **run**. The ratio rise/run is the same for any two points on a line. We call this ratio the **slope**. Slope describes the slant of a line. The slope of the line in the graph above is given by

$$\frac{\text{rise}}{\text{run}} = \frac{\text{the change in } y}{\text{the change in } x}, \text{ or } \frac{2}{4}, \text{ or } \frac{1}{2}.$$

The **slope** of a line containing points (x_1, y_1) and (x_2, y_2) is given by

$$m = \frac{\text{rise}}{\text{run}} = \frac{\text{the change in } y}{\text{the change in } x} = \frac{y_2 - y_1}{x_2 - x_1}.$$

In the definition above, (x_1, y_1) and (x_2, y_2)—read "x sub-one, y sub-one and x sub-two, y sub-two"—represent two different points on a line. It does not matter which point is considered (x_1, y_1) and which is considered (x_2, y_2) so long as coordinates are subtracted in the same order in both the numerator and the denominator.

Graph the line containing the points and find the slope in two different ways.

1. $(-2, 3)$ and $(3, 5)$

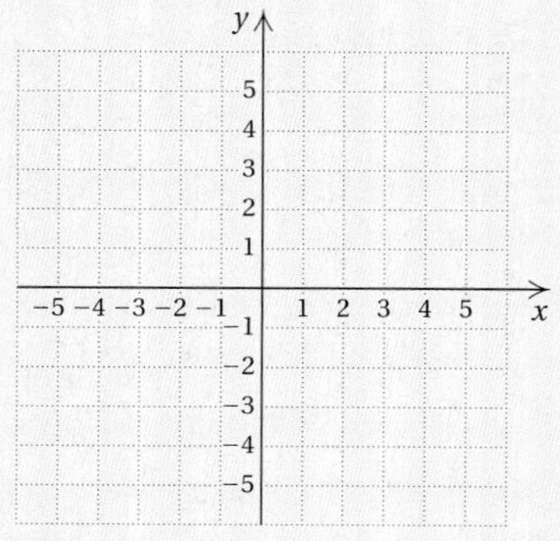

2. $(0, -3)$ and $(-3, 2)$

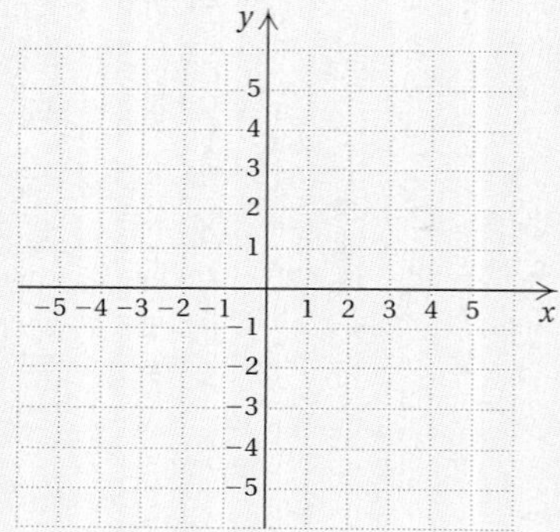

Answers on page A-40

Example 1 Graph the line containing the points $(-4, 3)$ and $(2, -6)$ and find the slope.

The graph is shown below. From $(-4, 3)$ and $(2, -6)$, we see that the change in y, or the rise, is $-6 - 3$, or -9. The change in x, or the run, is $2 - (-4)$, or 6. We consider (x_1, y_1) to be $(-4, 3)$ and (x_2, y_2) to be $(2, -6)$:

$$\text{Slope} = \frac{\text{rise}}{\text{run}} = \frac{\text{change in } y}{\text{change in } x}$$

$$= \frac{y_2 - y_1}{x_2 - x_1}$$

$$= \frac{-6 - 3}{2 - (-4)}$$

$$= \frac{-9}{6} = -\frac{9}{6}, \text{ or } -\frac{3}{2}.$$

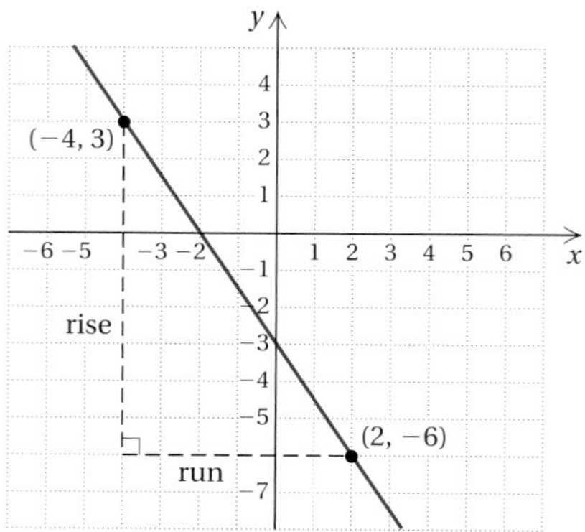

When we use the formula

$$m = \frac{y_2 - y_1}{x_2 - x_1},$$

we can subtract in two ways. We must remember, however, to subtract the y-coordinates in the same order that we subtract the x-coordinates. Let's redo Example 1, where we consider (x_1, y_1) to be $(2, -6)$ and (x_2, y_2) to be $(-4, 3)$:

$$\text{Slope} = \frac{\text{change in } y}{\text{change in } x} = \frac{3 - (-6)}{-4 - 2} = \frac{9}{-6} = -\frac{3}{2}.$$

The slope of a line tells how it slants. A line with positive slope slants up from left to right. The larger the slope, the steeper the slant. A line with negative slope slants downward from left to right.

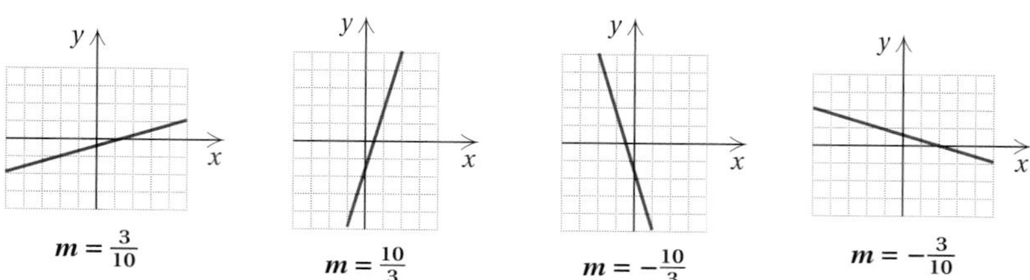

Do Exercises 1 and 2.

b Finding the Slope from an Equation

It is possible to find the slope of a line from its equation. Let's consider the equation $y = 2x + 3$, which is in the form $y = mx + b$. We can find two points by choosing convenient values for x—say, 0 and 1—and substituting to find the corresponding y-values. We find the two points on the line to be (0, 3) and (1, 5). The slope of the line is found using the definition of slope:

$$m = \frac{\text{change in } y}{\text{change in } x} = \frac{5 - 3}{1 - 0} = \frac{2}{1} = 2.$$

The slope is 2. Note that this is also the coefficient of the x-term in the equation $y = 2x + 3$.

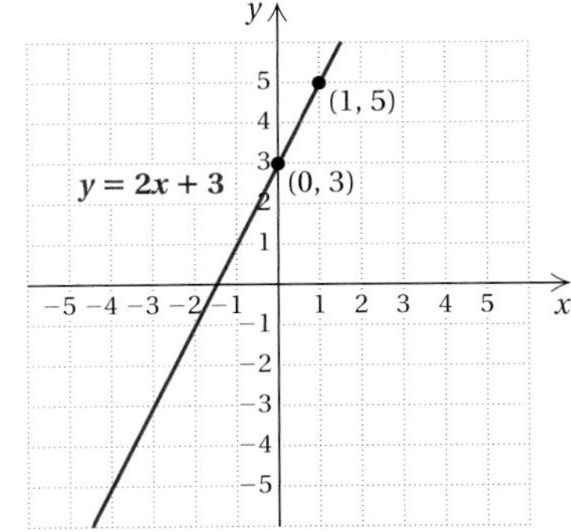

> The slope of the line $y = mx + b$ is m. To find the slope of a nonvertical line, solve the linear equation in x and y for y and get the resulting equation in the form $y = mx + b$. The coefficient of the x-term, m, is the slope of the line.

Examples Find the slope of the line.

2. $y = -3x + \frac{2}{9}$
→ $m = -3 = \text{Slope}$

3. $y = \frac{4}{5}x$
→ $m = \frac{4}{5} = \text{Slope}$

4. $y = x + 6$
→ $m = 1 = \text{Slope}$

5. $y = -0.6x - 3.5$
→ $m = -0.6 = \text{Slope}$

Do Exercises 3–6.

To find slope from an equation, we may have to first find an equivalent form of the equation.

Example 6 Find the slope of the line $2x + 3y = 7$.

We solve for y to get the equation in the form $y = mx + b$:

$$2x + 3y = 7$$
$$3y = -2x + 7$$
$$y = \frac{-2x + 7}{3}$$
$$y = -\frac{2}{3}x + \frac{7}{3}. \quad \text{This is } y = mx + b.$$

The slope is $-\frac{2}{3}$.

Do Exercises 7 and 8.

Calculator Spotlight

Visualizing Slope

Exercises

Graph each of the following sets of equations using the window settings $[-6, 6, -4, 4]$, Xscl = 1, Yscl = 1.

1. $y = x$, $y = 2x$,
$y = 5x$, $y = 10x$

What do you think the graph of $y = 123x$ will look like?

2. $y = x$, $y = \frac{3}{4}x$,
$y = 0.38x$, $y = \frac{5}{32}x$

What do you think the graph of $y = 0.000043x$ will look like?

Find the slope of the line.

3. $y = 4x + 11$

4. $y = -17x + 8$

5. $y = -x + \frac{1}{2}$

6. $y = \frac{2}{3}x - 1$

Find the slope of the line.

7. $4x + 4y = 7$

8. $5x - 4y = 8$

Answers on page A-40

Calculator Spotlight

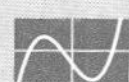 Visualizing Slope

Exercises

Graph each of the following sets of equations using the window settings [−6, 6, −4, 4], Xscl = 1, Yscl = 1.

1. $y = -x$, $y = -2x$, $y = -5x$, $y = -10x$

What do you think the graph of $y = -123x$ will look like?

2. $y = -x$, $y = -\frac{3}{4}x$, $y = -0.38x$, $y = -\frac{5}{32}x$

What do you think the graph of $y = -0.000043x$ will look like?

Find the slope, if it exists, of the line.

9. $x = 7$

10. $y = -5$

Answers on page A-40

What about the slope of a horizontal or a vertical line?

Example 7 Find the slope of the line $y = 5$.

We can think of $y = 5$ as $y = 0x + 5$. Then from this equation, we see that $m = 0$. Consider the points $(-3, 5)$ and $(4, 5)$, which are on the line. The change in $y = 5 - 5$, or 0. The change in $x = -3 - 4$, or -7. We have

$$m = \frac{5 - 5}{-3 - 4} = \frac{0}{-7} = 0.$$

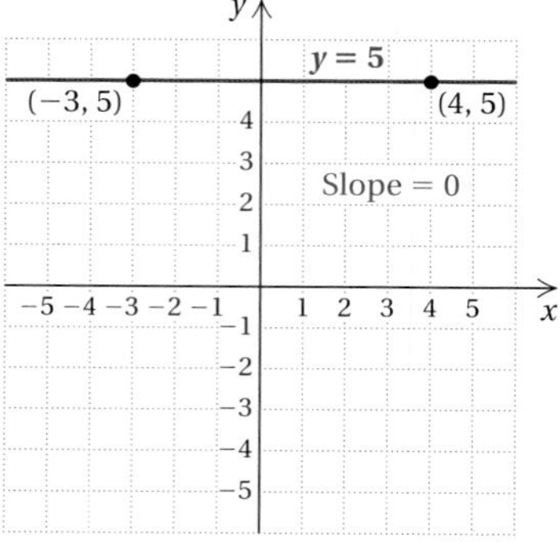

Any two points on a horizontal line have the same y-coordinate. Thus the change in y is 0.

Example 8 Find the slope of the line $x = -4$.

Consider the points $(-4, 3)$ and $(-4, -2)$, which are on the line. The change in $y = 3 - (-2)$, or 5. The change in $x = -4 - (-4)$, or 0. We have

$$m = \frac{3 - (-2)}{-4 - (-4)} = \frac{5}{0}. \quad \textbf{Undefined}$$

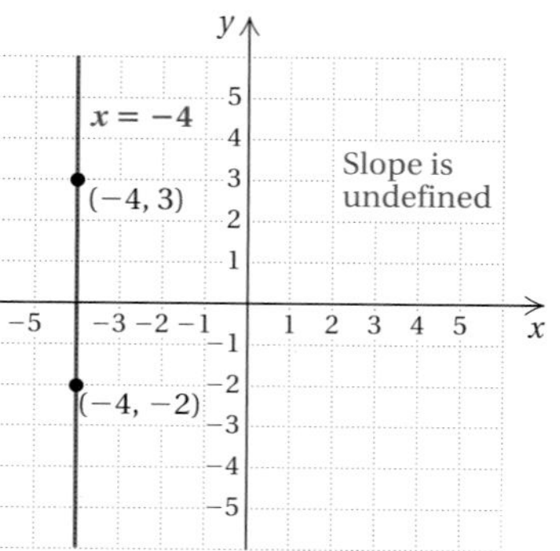

Since division by 0 is undefined, the slope of this line is undefined. The answer in this example is "The slope of this line is undefined."

> A horizontal line has slope 0. The slope of a vertical line is undefined.

Do Exercises 9 and 10.

c Applications of Slope

Slope has many real-world applications. For example, numbers like 2%, 3%, and 6% are often used to represent the *grade* of a road, a measure of how steep a road on a hill or mountain is. For example, a 3% grade $\left(3\% = \frac{3}{100}\right)$ means that for every horizontal distance of 100 ft, the road rises 3 ft, and a −3% grade means that for every horizontal distance of 100 ft, the road drops 3 ft. The concept of grade also occurs in skiing or snowboarding, where a 4% grade is considered very tame, but a 40% grade is considered extremely steep. And in cardiology, a physician may change the grade of a treadmill to measure its effect on heartbeat.

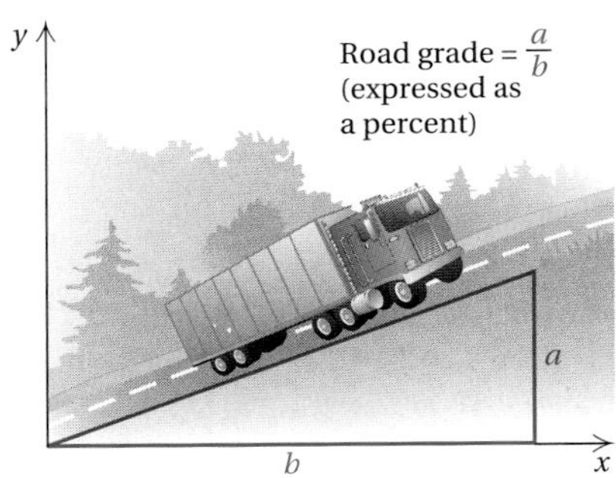

Architects and carpenters use slope when designing and building stairs, ramps, or roof pitches. Another application occurs in hydrology. When a river flows, the strength or force of the river depends on how far the river falls vertically compared to how far it flows horizontally. Slope can also be considered as a **rate of change.**

11. *Headwall Ski Slope.* One of the steepest ski slopes in North America, the Headwall on Mount Washington in New Hampshire, drops 720 ft over a horizontal distance of 900 ft. Find the grade of the Headwall.

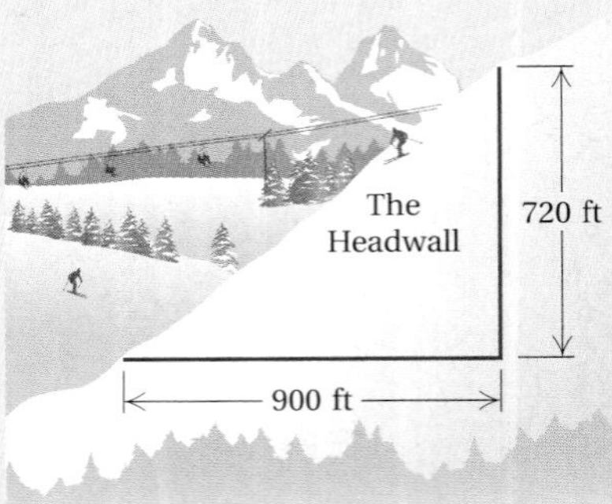

Answer on page A-40

AG Algebraic–Graphical Connection

Example 9 *Cost of a Formal Wedding.* The cost of a formal wedding has increased over the years, as shown in the following graph. Find the rate of change of the cost.

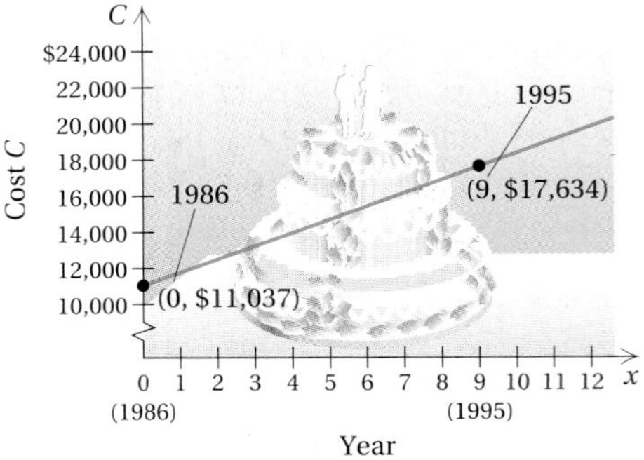

Source: *Modern Bride Magazine*

We determine the coordinates of two points on the graph. In this case, they are given as (0, $11,037) and (9, $17,634). Then we compute the slope, or rate of change:

$$\text{Slope} = \text{Rate of change} = \frac{\text{change in } y}{\text{change in } x}$$

$$= \frac{\$17{,}634 - \$11{,}037}{9 - 0} = \frac{\$6597}{9} = 733\,\frac{\$}{\text{yr}}.$$

What this means is that each year the cost of a formal wedding is $733 more than it was the preceding year.

Do Exercise 11.

Exercise Set 13.1

a Find the slope, if it exists, of the line.

1.

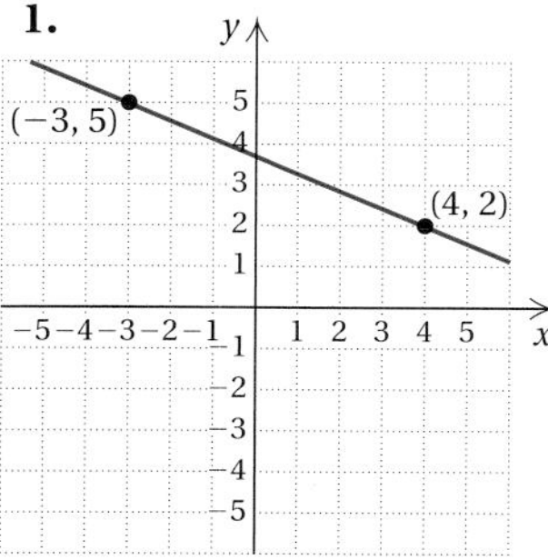

2.

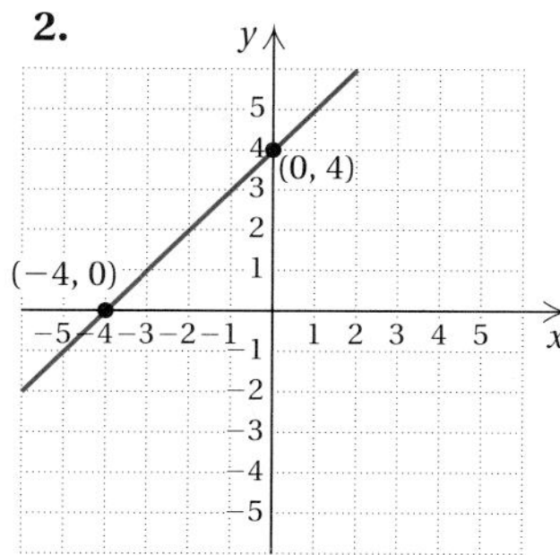

3.

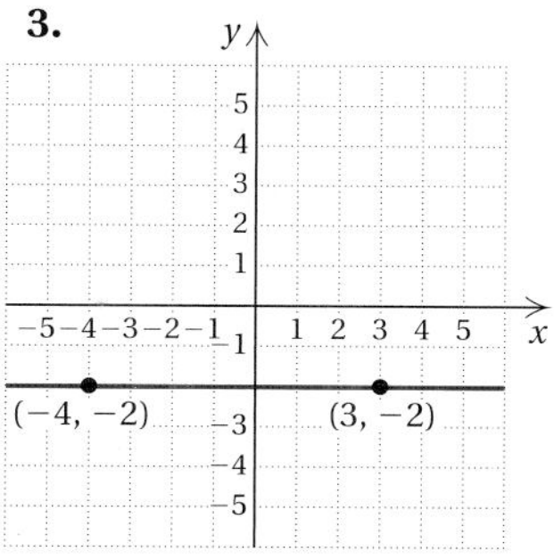

4.

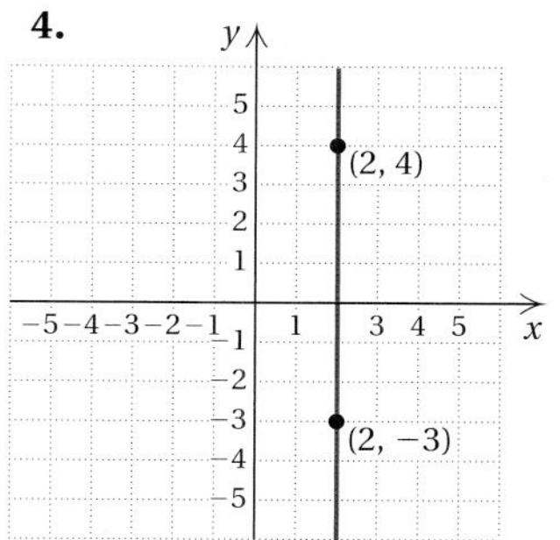

Graph the line containing the given pair of points and find the slope.

5. $(-2, 4)$, $(3, 0)$

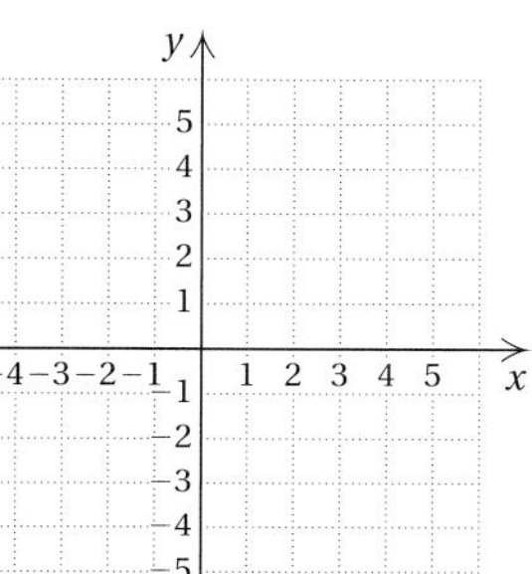

6. $(2, -4)$, $(-3, 2)$

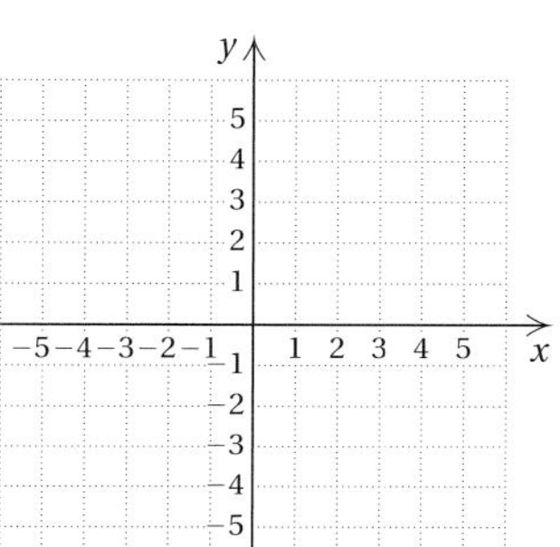

7. $(-4, 0)$, $(-5, -3)$

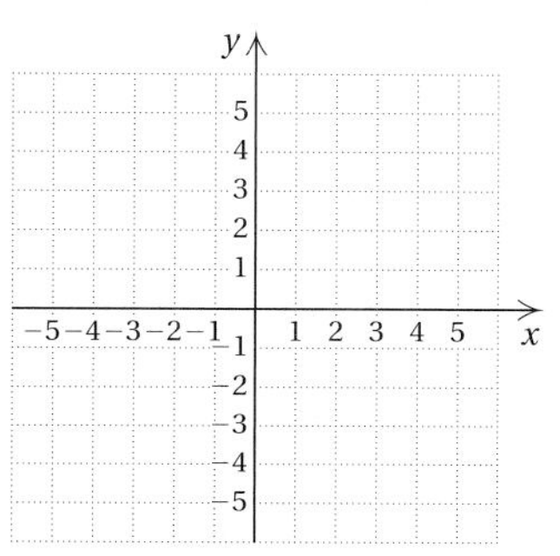

8. $(-3, 0)$, $(-5, -2)$

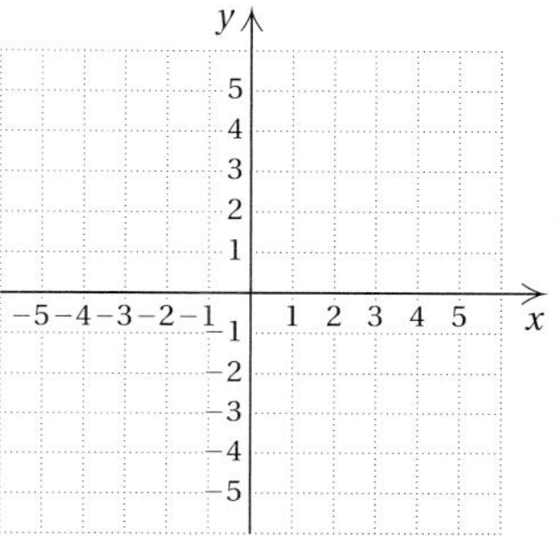

Find the slope, if it exists, of the line containing the given pair of points.

9. $\left(2, -\frac{1}{2}\right)$, $\left(5, \frac{3}{2}\right)$

10. $\left(\frac{2}{3}, -1\right)$, $\left(\frac{5}{3}, 2\right)$

11. $(4, -2)$, $(4, 3)$

12. $(4, -3)$, $(-2, -3)$

b Find the slope, if it exists.

13. $y = -10x + 7$

14. $y = \frac{10}{3}x - \frac{5}{7}$

15. $y = 3.78x - 4$

16. $y = -\frac{3}{5}x + 28$

17. $3x - y = 4$

18. $-2x + y = 8$

19. $x + 5y = 10$

20. $x - 4y = 8$

21. $3x + 2y = 6$

22. $2x - 4y = 8$

23. $5x - 7y = 14$

24. $3x - 6y = 10$

25. $y = -2.74x$

26. $y = \frac{219}{298}x - 6.7$

27. $9x = 3y + 5$

28. $4y = 9x - 7$

29. $5x - 4y + 12 = 0$

30. $16 + 2x - 8y = 0$

31. $y = 4$

32. $x = -3$

c Find the slope (or rate of change) in each exercise.

33. Find the slope (or pitch) of the roof.

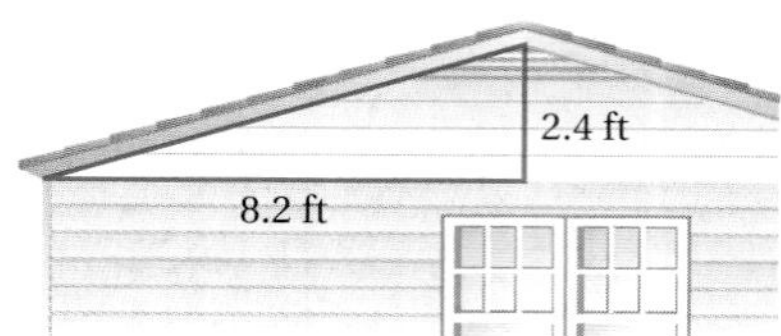

34. Find the grade of the road.

35. Find the slope of the river.

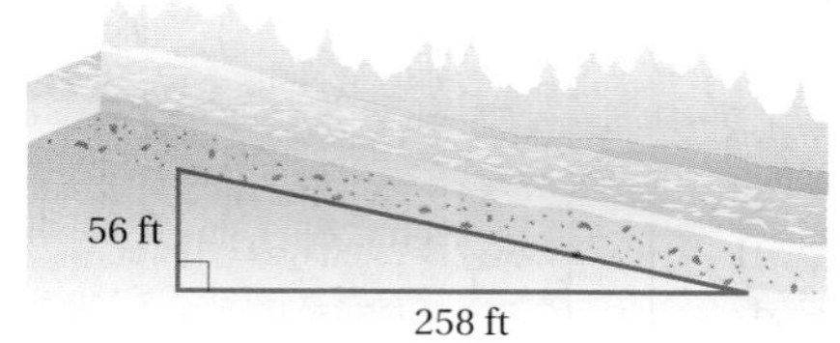

36. Find the slope (or grade) of the treadmill.

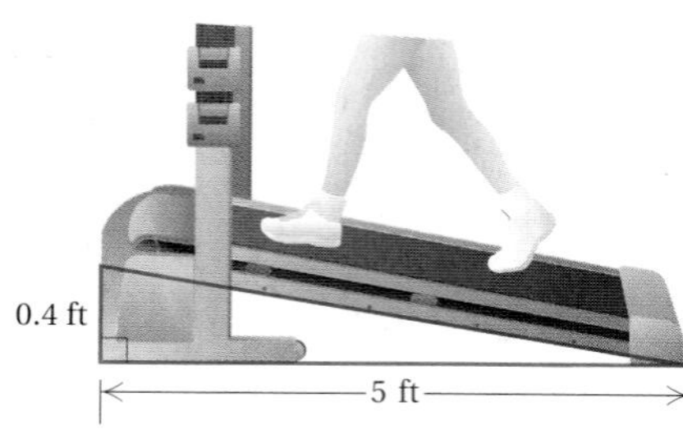

37. Find the rate at which a runner burns calories.

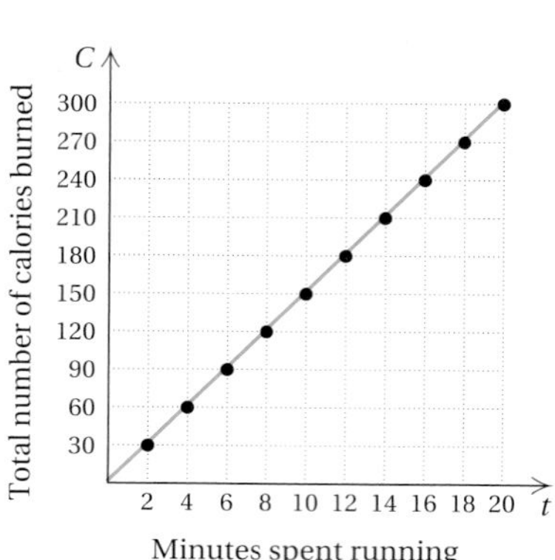

38. Find the rate of change in the number of U.S. farms.

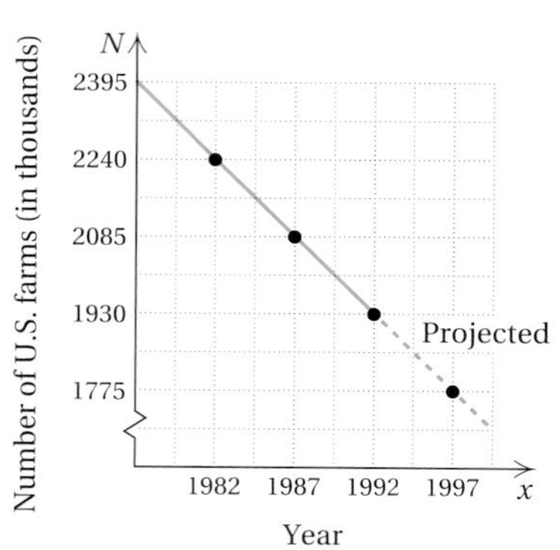

Source: Statistical Abstract of the United States

39. *Slope of Long's Peak.* From a base elevation of 9600 ft, Long's Peak in Colorado rises to a summit elevation of 14,255 ft over a horizontal distance of 15,840 ft. Find the grade of Long's Peak.

40. *Ramps for the Disabled.* In order to meet federal standards, a wheelchair ramp must not rise more than 1 ft over a horizontal distance of 12 ft. Express this slope as a grade.

Skill Maintenance

Simplify. [7.8d]

41. $11 \cdot 6 \div 3 \cdot 2 \div 7$

42. $2^4 - 2^4 \div 2^2 - 2$

43. $10 - 3(7 - 2)$

44. $5^3 - 4^2 + 6(5 \cdot 7 + 4 \cdot 3)$

45. $\dfrac{4^3 + 2^2}{5^3 - 4^2}$

46. $(4^3 + 2^2) \cdot (5^3 - 4^2)$

47. $1000 \div 100 \div 10 \div 2$

48. $3^{10} \div 3^2 \div 3^4 \div 9$

Synthesis

49. ◈ If one line has a slope of -3 and another has a slope of 2, which line is steeper? Why?

50. ◈ Graph several equations that have the same slope. How are they related?

Collaborative Learning Manual

Verify that m and b represent the slope and the y-intercept of the line $y = mx + b$.

13.2 Equations of Lines

a Finding an Equation of a Line When the Slope and the y-Intercept Are Given

We know that in the equation $y = mx + b$ the slope is m and the y-intercept is $(0, b)$. Thus we call the equation $y = mx + b$ the **slope–intercept equation.**

> **THE SLOPE–INTERCEPT EQUATION: $y = mx + b$**
>
> The equation $y = mx + b$ is called the **slope–intercept equation.** The slope is m and the y-intercept is $(0, b)$.

Example 1 Find the slope and the y-intercept of $2x - 3y = 8$.

We first solve for y:

$$2x - 3y = 8$$

$$-3y = -2x + 8 \quad \text{Subtracting } 2x$$

$$\frac{-3y}{-3} = \frac{-2x + 8}{-3} \quad \text{Dividing by } -3$$

$$y = \frac{-2x}{-3} + \frac{8}{-3}$$

$$y = \frac{2}{3}x - \frac{8}{3}$$

The slope is $\frac{2}{3}$. The y-intercept is $\left(0, -\frac{8}{3}\right)$.

Do Exercises 1–5.

Example 2 A line has slope -2.4 and y-intercept $(0, 11)$. Find an equation of the line.

We use the slope–intercept equation and substitute -2.4 for m and 11 for b:

$$y = mx + b$$

$$y = -2.4x + 11. \quad \text{Substituting}$$

Do Exercise 6.

b Finding an Equation of a Line When the Slope and a Point Are Given

Suppose we know the slope of a line and a certain point on that line. We can use the slope–intercept equation $y = mx + b$ to find an equation of the line. To write an equation in this form, we need to know the slope (m) and the y-intercept (b).

Objectives

a Given an equation in the form $y = mx + b$, find the slope and the y-intercept; and find an equation of a line when the slope and the y-intercept are given.

b Find an equation of a line when the slope and a point on the line are given.

c Find an equation of a line when two points on the line are given.

For Extra Help

TAPE 25 | MAC WIN | CD-ROM

Find the slope and the y-intercept.

1. $y = 5x$

2. $y = -\frac{3}{2}x - 6$

3. $3x + 4y = 15$

4. $2y = 4x - 17$

5. $-7x - 5y = 22$

6. A line has slope 3.5 and y-intercept $(0, -23)$. Find an equation of the line.

Answers on page A-40

Find an equation of the line that contains the given point and has the given slope.

7. $(4, 2)$, $m = 5$

8. $(-2, 1)$, $m = -3$

9. $(3, 5)$, $m = 6$

10. $(1, 4)$, $m = -\frac{2}{3}$

Find an equation of the line containing the given points.

11. $(2, 4)$ and $(3, 5)$

12. $(-1, 2)$ and $(-3, -2)$

Answers on page A-41

Example 3 Find the equation of the line with slope 3 that contains the point (4, 1).

We know that the slope is 3, so the equation is $y = 3x + b$. Using the point (4, 1), we substitute 4 for x and 1 for y in $y = 3x + b$. Then we solve for b:

$$y = 3x + b$$
$$1 = 3(4) + b \quad \text{Substituting}$$
$$-11 = b. \quad \text{Solving for } b\text{, the } y\text{-intercept}$$

We use the equation $y = mx + b$ and substitute 3 for m and -11 for b:

$$y = 3x - 11.$$

Example 4 Find an equation of the line with slope -5 that contains the point $(-2, 3)$.

We know that the slope is -5, so the equation is $y = -5x + b$. Using the point $(-2, 3)$, we substitute -2 for x and 3 for y in $y = -5x + b$. Then we solve for b:

$$y = -5x + b$$
$$3 = -5(-2) + b \quad \text{Substituting}$$
$$3 = 10 + b$$
$$-7 = b. \quad \text{Solving for } b$$

We use the equation $y = mx + b$ and substitute -5 for m and -7 for b:

$$y = -5x - 7.$$

Do Exercises 7–10.

c Finding an Equation of a Line When Two Points Are Given

We can also use the slope–intercept equation to find an equation of a line when two points are given.

Example 5 Find an equation of the line containing the points (2, 3) and $(-6, 1)$.

First, we find the slope:

$$m = \frac{3 - 1}{2 - (-6)} = \frac{2}{8}, \text{ or } \frac{1}{4}.$$

Thus, $y = \frac{1}{4}x + b$. We then proceed as we did in Example 4, using either point to find b. We choose (2, 3) and substitute 2 for x and 3 for y:

$$y = \tfrac{1}{4}x + b$$
$$3 = \tfrac{1}{4} \cdot 2 + b \quad \text{Substituting}$$
$$3 = \tfrac{1}{2} + b$$
$$\tfrac{5}{2} = b. \quad \text{Solving for } b$$

We use the equation $y = mx + b$ and substitute $\frac{1}{4}$ for m and $\frac{5}{2}$ for b:

$$y = \tfrac{1}{4}x + \tfrac{5}{2}.$$

Do Exercises 11 and 12.

Exercise Set 13.2

a Find the slope and the y-intercept.

1. $y = -4x - 9$

2. $y = -2x + 3$

3. $y = 1.8x$

4. $y = -27.4x$

5. $-8x - 7y = 21$

6. $-2x - 8y = 16$

7. $4x = 9y + 7$

8. $5x + 4y = 12$

9. $-6x = 4y + 2$

10. $4.8x - 1.2y = 36$

11. $y = -17$

12. $y = 28$

Find an equation of the line with the given slope and y-intercept.

13. Slope $= -7$, y-intercept $= (0, -13)$

14. Slope $= 73$, y-intercept $= (0, 54)$

15. Slope $= 1.01$, y-intercept $= (0, -2.6)$

16. Slope $= -\frac{3}{8}$, y-intercept $= \left(0, \frac{7}{11}\right)$

b Find an equation of the line containing the given point and having the given slope.

17. $(-3, 0)$, $m = -2$

18. $(2, 5)$, $m = 5$

19. $(2, 4)$, $m = \dfrac{3}{4}$

20. $\left(\dfrac{1}{2}, 2\right)$, $m = -1$

21. $(2, -6)$, $m = 1$

22. $(4, -2)$, $m = 6$

23. $(0, 3)$, $m = -3$

24. $(-2, -4)$, $m = 0$

c Find an equation of the line that contains the given pair of points.

25. $(12, 16)$ and $(1, 5)$

26. $(-6, 1)$ and $(2, 3)$

27. $(0, 4)$ and $(4, 2)$

28. $(0, 0)$ and $(4, 2)$

29. (3, 2) and (1, 5)

30. (−4, 1) and (−1, 4)

31. (−4, 5) and (−2, −3)

32. (−2, −4) and (2, −1)

33. *Aerobic Exercise.* The line graph below describes the *target heart rate, T,* in beats per minute, of a person of age *a*, who is exercising. The goal is to get the number of heart beats per minute to this target level.

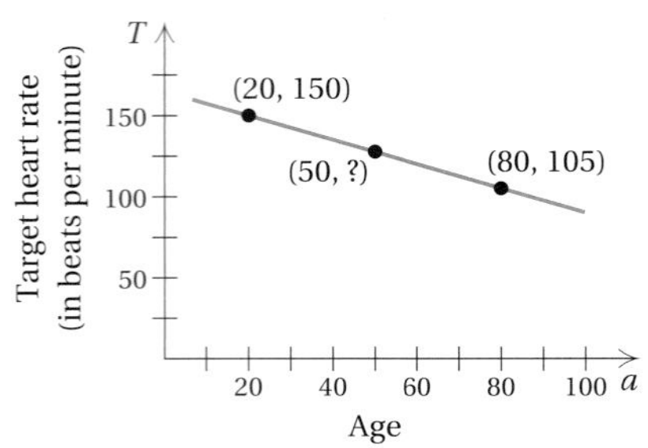

a) Find an equation of the line.
b) What is the rate of change in the target heart rate?
c) Use the equation to calculate the target heart rate of a person of age 50.

34. *Diabetes Cases.* The line graph below describes the number *N*, in millions, of cases of diabetes in this country in years *x* since 1983.

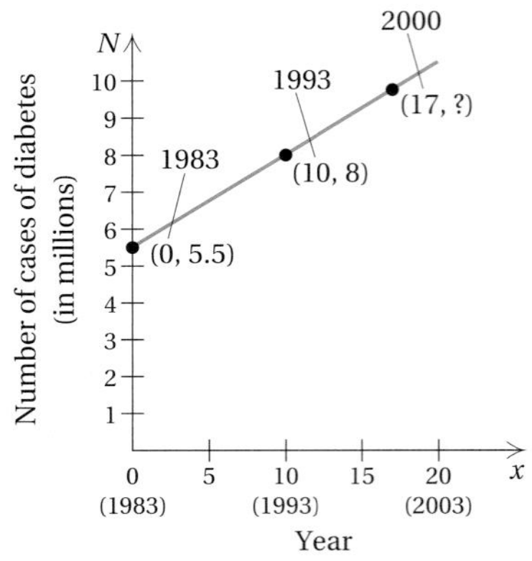

Source: U.S. National Center for Health Statistics

a) Find an equation of the line.
b) What is the rate of change of the number of cases of diabetes?
c) Use the equation to predict the number of cases of diabetes in 2000.

Skill Maintenance

Solve. [11.7b]

35. $2x^2 + 6x = 0$

36. $x^2 - 49 = 0$

37. $x^2 - x - 6 = 0$

38. $x^2 + 4x - 5 = 0$

39. $2x^2 + 11x = 21$

40. $5x^2 = 14x + 24$

41. $x^2 + 5x - 14 = 0$

42. $12x^2 + 16x - 16 = 0$

Solve. [8.3c]

43. $3x - 4(9 - x) = 17$

44. $2(5 + 2y) + 4y = 13$

45. $40(2x - 7) = 50(4 - 6x)$

46. $\frac{2}{3}(x - 5) = \frac{3}{8}(x + 5)$

Synthesis

47. ◆ Do all graphs of linear equations have y-intercepts? Why or why not?

48. ◆ Do all graphs of linear equations have x-intercepts? Why or why not?

49. Find an equation of the line that contains the point (2, −3) and has the same slope as the line $3x - y + 4 = 0$.

50. Find an equation of the line that has the same y-intercept as the line $x - 3y = 6$ and contains the point (5, −1).

51. Find an equation of the line with the same slope as the line $3x - 2y = 8$ and the same y-intercept as the line $2y + 3x = -4$.

13.3 Parallel and Perpendicular Lines

When we graph a pair of linear equations, there are three possibilities:

1. The graphs are the same.
2. The graphs intersect at exactly one point.
3. The graphs are parallel (they do not intersect).

Objectives

a Determine whether the graphs of two linear equations are parallel.

b Determine whether the graphs of two linear equations are perpendicular.

For Extra Help

TAPE 25 MAC WIN CD-ROM

a Parallel Lines

The graphs shown at right are of the linear equations

$$y = 2x + 5$$

and $\quad y = 2x - 3.$

The slope of each line is 2. The y-intercepts are $(0, 5)$ and $(0, -3)$ and are different. The lines do not intersect and are parallel.

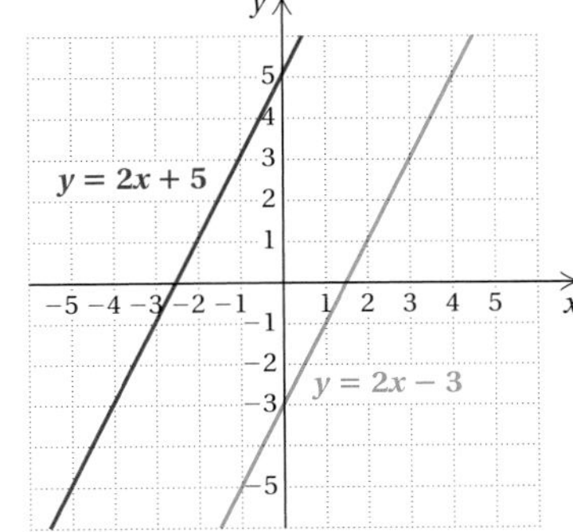

PARALLEL LINES

- Parallel nonvertical lines have the *same* slope, $m_1 = m_2$, and *different* y-intercepts, $b_1 \neq b_2$.
- Parallel horizontal lines have equations $y = p$ and $y = q$, where $p \neq q$.
- Parallel vertical lines have equations $x = p$ and $x = q$, where $p \neq q$.

Determine whether the graphs of the pair of equations are parallel.

1. $y - 3x = 1,$
$-2y = 3x + 2$

2. $3x - y = -5,$
$y - 3x = -2$

By simply graphing, we may find it difficult to determine whether lines are parallel. Sometimes they may intersect only very far from the origin. We can use the preceding statements about slopes, y-intercepts, and parallel lines to determine for certain whether lines are parallel.

Example 1 Determine whether the graphs of the lines $y = -3x + 4$ and $6x + 2y = -10$ are parallel.

The graphs of these equations are shown below, but they are not necessary in order to determine whether the lines are parallel.

We first solve each equation for y. In this case, the first equation is already solved for y.

a) $y = -3x + 4$

b) $6x + 2y = -10$

$$2y = -6x - 10$$

$$y = \frac{1}{2}(-6x - 10)$$

$$y = -3x - 5$$

The slope of each line is -3. The y-intercepts are $(0, 4)$ and $(0, -5)$ and are different. The lines are parallel.

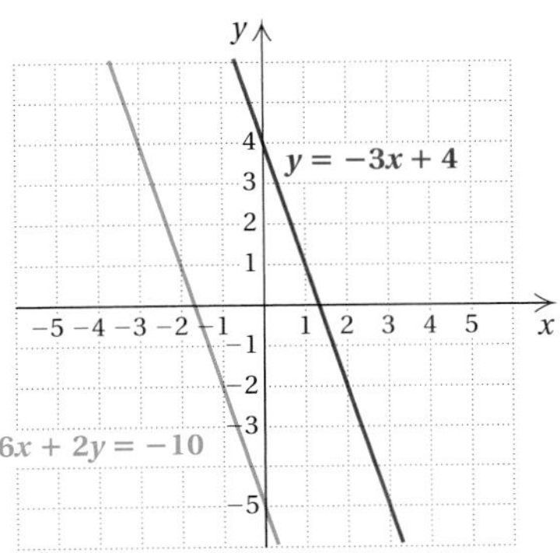

Do Exercises 1 and 2.

Calculator Spotlight

Visualizing Parallel Lines

Exercises

Graph each pair of equations in Margin Exercises 1 and 2 using the window settings $[-6, 6, -4, 4]$, Xscl = 1, Yscl = 1. Check visually whether they appear to be parallel.

Answers on page A-41

Determine whether the graphs of the pair of equations are perpendicular.

3. $y = -\frac{3}{4}x + 7,$
$y = \frac{4}{3}x - 9$

4. $4x - 5y = 8,$
$6x + 9y = -12$

b Perpendicular Lines

Perpendicular lines in a plane are lines that intersect at a right angle. The measure of a right angle is 90°. The lines whose graphs are shown below are perpendicular. You can check this approximately by using a protractor or placing a rectangular piece of paper at the intersection.

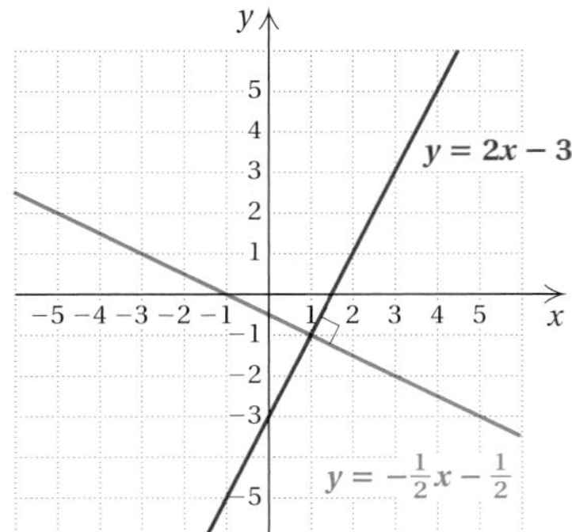

The slopes of the lines are 2 and $-\frac{1}{2}$. Note that $2\left(-\frac{1}{2}\right) = -1$. That is, the product of the slopes is -1.

PERPENDICULAR LINES

- Two nonvertical lines are perpendicular if the product of their slopes is -1, $m_1 \cdot m_2 = -1$. (If one line has slope m, the slope of the line perpendicular to it is $-1/m$.)
- If one equation in a pair of perpendicular lines is vertical, then the other is horizontal. These equations are of the form $x = a$ and $y = b$.

Example 2 Determine whether the graphs of the lines $3y = 9x + 3$ and $6y + 2x = 6$ are perpendicular.

The graphs are shown below, but they are not necessary in order to determine whether the lines are perpendicular.

We first solve each equation for y in order to determine the slopes:

a) $3y = 9x + 3$
$y = \frac{1}{3}(9x + 3)$
$y = 3x + 1;$

b) $6y + 2x = 6$
$6y = -2x + 6$
$y = \frac{1}{6}(-2x + 6)$
$y = -\frac{1}{3}x + 1.$

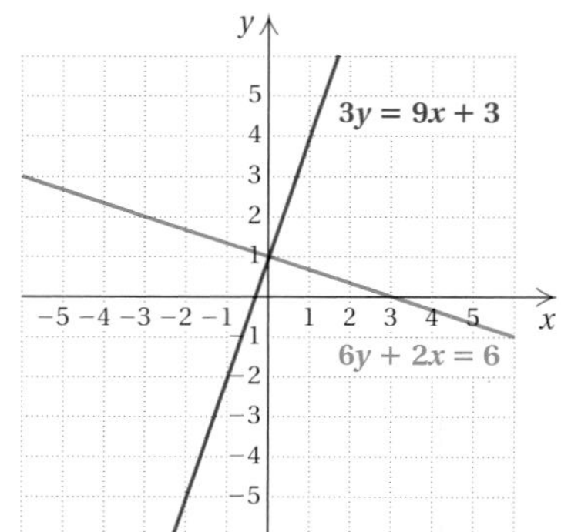

The slopes are 3 and $-\frac{1}{3}$. The product of the slopes is $3\left(-\frac{1}{3}\right) = -1$. The lines are perpendicular.

Do Exercises 3 and 4.

Calculator Spotlight

 Visualizing Perpendicular Lines

Exercises

Graph each pair of equations in Margin Exercises 3 and 4 using the window settings $[-9, 9, -6, 6]$, Xscl = 1, Yscl = 1. Check visually whether they appear to be perpendicular.

Answers on page A-41

Exercise Set 13.3

a Determine whether the graphs of the equations are parallel lines.

1. $x + 4 = y,$
$y - x = -3$

2. $3x - 4 = y,$
$y - 3x = 8$

3. $y + 3 = 6x,$
$-6x - y = 2$

4. $y = -4x + 2,$
$-5 = -2y + 8x$

5. $10y + 32x = 16.4,$
$y + 3.5 = 0.3125x$

6. $y = 6.4x + 8.9,$
$5y - 32x = 5$

7. $y = 2x + 7,$
$5y + 10x = 20$

8. $y + 5x = -6,$
$3y + 5x = -15$

9. $3x - y = -9,$
$2y - 6x = -2$

10. $y - 6 = -6x,$
$-2x + y = 5$

11. $x = 3,$
$x = 4$

12. $y = 1,$
$y = -2$

b Determine whether the graphs of the equations are perpendicular lines.

13. $y = -4x + 3,$
$4y + x = -1$

14. $y = -\frac{2}{3}x + 4,$
$3x + 2y = 1$

15. $x + y = 6,$
$4y - 4x = 12$

16. $2x - 5y = -3,$
$5x + 2y = 6$

17. $y = -0.3125x + 11,$
$y - 3.2x = -14$

18. $y = -6.4x - 7,$
$64y - 5x = 32$

19. $y = -x + 8,$
$x - y = -1$

20. $2x + 6y = -3,$
$12y = 4x + 20$

21. $\frac{3}{8}x - \frac{y}{2} = 1,$
$\frac{4}{3}x - y + 1 = 0$

22. $\frac{1}{2}x + \frac{3}{4}y = 6,$
$-\frac{3}{2}x + y = 4$

23. $x = 0,$
$y = -2$

24. $x = -3,$
$y = 5$

Skill Maintenance

Solve. [12.7a]

25. A train leaves a station and travels west at 70 km/h. Two hours later, a second train leaves on a parallel track and travels west at 90 km/h. When will it overtake the first train?

26. One car travels 10 km/h faster than another. While one car travels 130 km, the other travels 140 km. What is the speed of each car?

Solve. [12.6a]

27. $\frac{x^2}{x+4} = \frac{16}{x+4}$

28. $\frac{2}{3} - \frac{5}{6} = \frac{1}{x}$

29. $\frac{t}{3} + \frac{t}{10} = 1$

30. $\frac{5}{x-4} = \frac{3}{x+2}$

31. $\frac{4}{x-2} + \frac{7}{x-3} = \frac{10}{x^2 - 5x + 6}$

32. $\frac{3}{x-5} + \frac{4}{x+5} = \frac{2}{x^2 - 25}$

Synthesis

33. ◆ Consider two equations of the type $Ax + By = C$. Explain how you would go about showing that their graphs are perpendicular.

34. ◆ Consider two equations of the type $Ax + By = C$. Explain how you would go about showing that their graphs are parallel.

35.–40. Check the results of Exercises 1–6 by graphing each pair of equations using the window settings $[-6, 6, -4, 4]$, Xscl = 1, Yscl = 1.

41.–46. Check the results of Exercises 13–18 by graphing each pair of equations using the window settings $[-24, 24, -16, 16]$, Xscl = 1, Yscl = 1.

47. Find an equation of a line that contains the point $(0, 6)$ and is parallel to $y - 3x = 4$.

48. Find an equation of the line that contains the point $(-2, 4)$ and is parallel to $y = 2x - 3$.

49. Find an equation of the line that contains the point $(0, 2)$ and is perpendicular to $3y - x = 0$.

50. Find an equation of the line that contains the point $(1, 0)$ and is perpendicular to $2x + y = -4$.

51. Find an equation of the line that has x-intercept $(-2, 0)$ and is parallel to $4x - 8y = 12$.

52. Find the value of k such that $4y = kx - 6$ and $5x + 20y = 12$ are parallel.

53. Find the value of k such that $4y = kx - 6$ and $5x + 20y = 12$ are perpendicular.

The lines in the graphs in Exercises 54 and 55 are perpendicular and the lines in the graph in Exercise 56 are parallel. Find an equation of each line.

54.

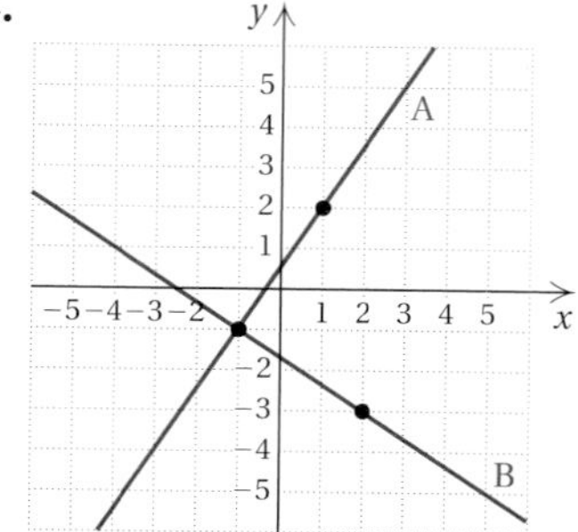

55.

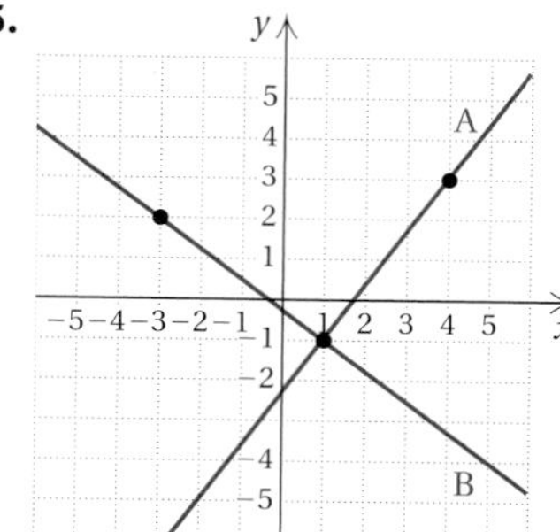

56.

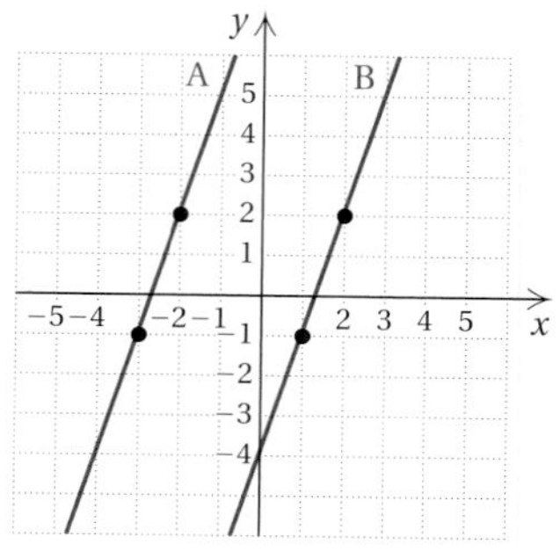

13.4 Graphing Inequalities in Two Variables

A graph of an inequality is a drawing that represents its solutions. An inequality in one variable can be graphed on a number line. An inequality in two variables can be graphed on a coordinate plane.

Objectives

a Determine whether an ordered pair of numbers is a solution of an inequality in two variables.

b Graph linear inequalities.

For Extra Help

TAPE 25

MAC WIN

CD-ROM

a Solutions of Inequalities in Two Variables

The solutions of inequalities in two variables are ordered pairs.

Example 1 Determine whether $(-3, 2)$ is a solution of $5x + 4y < 13$.

We use alphabetical order to replace x with -3 and y with 2.

$$\begin{array}{r|l} 5x + 4y & < 13 \\ \hline 5(-3) + 4 \cdot 2 & ?\ 13 \\ -15 + 8 & \\ -7 & \end{array} \quad \text{TRUE}$$

Since $-7 < 13$ is true, $(-3, 2)$ is a solution.

Example 2 Determine whether $(6, 8)$ is a solution of $5x + 4y < 13$.

We use alphabetical order to replace x with 6 and y with 8.

$$\begin{array}{r|l} 5x + 4y & < 13 \\ \hline 5(6) + 4(8) & ?\ 13 \\ 30 + 32 & \\ 62 & \end{array} \quad \text{FALSE}$$

Since $62 < 13$ is false, $(6, 8)$ is not a solution.

Do Exercises 1 and 2.

1. Determine whether $(4, 3)$ is a solution of $3x - 2y < 1$.

2. Determine whether $(2, -5)$ is a solution of $4x + 7y \geq 12$.

b Graphing Inequalities in Two Variables

Example 3 Graph: $y > x$.

We first graph the line $y = x$. Every solution of $y = x$ is an ordered pair like $(3, 3)$. The first and second coordinates are the same. We draw the line $y = x$ dashed because its points (as shown on the left below) are *not* solutions of $y > x$.

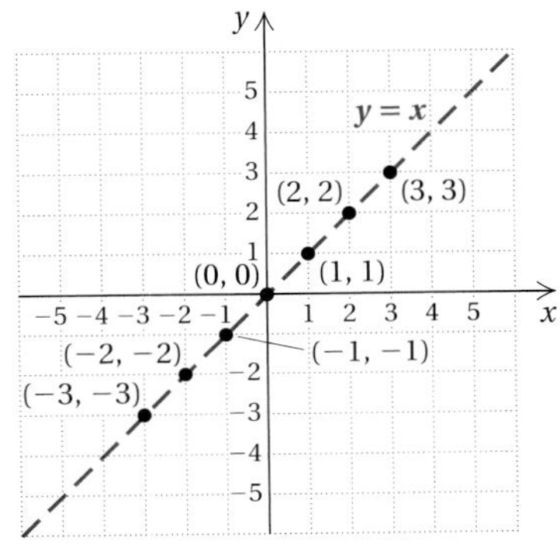

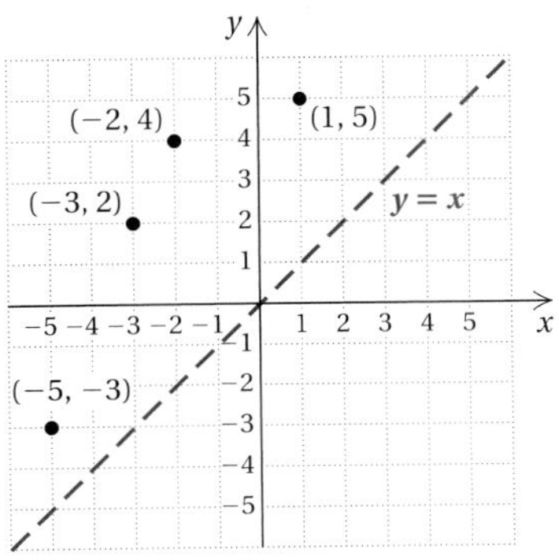

Answers on page A-41

3. Graph: $y < x$.

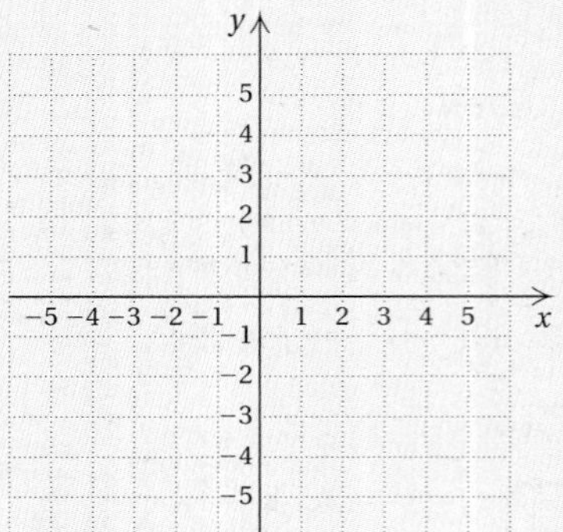

Now look at the graph on the right on the preceding page. Several ordered pairs are plotted in the half-plane above the line $y = x$. Each is a solution of $y > x$.

We can check a pair such as $(-2, 4)$ as follows:

$$\begin{array}{c|c} y & x \\ \hline 4 & -2 \end{array} \quad y > x: \; 4 \; ? \; -2 \quad \text{TRUE}$$

It turns out that any point on the same side of $y = x$ as $(-2, 4)$ is also a solution. *If we know that one point in a half-plane is a solution, then all points in that half-plane are solutions.* The graph of $y > x$ is shown below. (Solutions are indicated by color shading throughout.) We shade the half-plane above $y = x$.

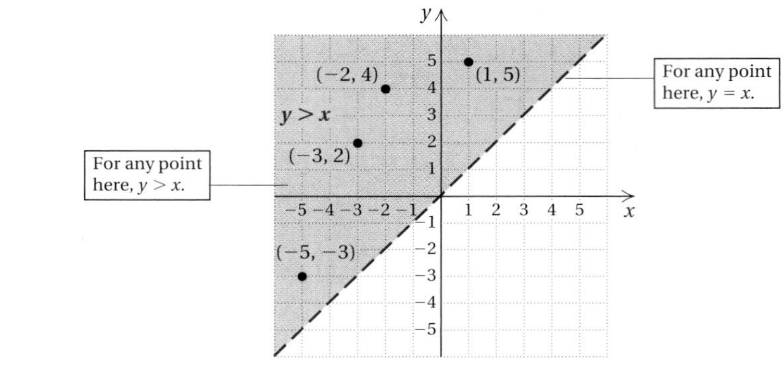

Do Exercise 3.

A **linear inequality** is one that we can get from a linear equation by changing the equals symbol to an inequality symbol. Every linear equation has a graph that is a straight line. The graph of a linear inequality is a half-plane, sometimes including the line along the edge.

> To graph an inequality in two variables:
>
> **1.** Replace the inequality symbol with an equals sign and graph this related equation.
>
> **2.** If the inequality symbol is $<$ or $>$, draw the line dashed. If the inequality symbol is $\le$ or $\ge$, draw the line solid.
>
> **3.** The graph consists of a half-plane, either above or below or left or right of the line, and, if the line is solid, the line as well. To determine which half-plane to shade, choose a point not on the line as a test point. Substitute to find whether that point is a solution of the inequality. If it is, shade the half-plane containing that point. If it is not, shade the half-plane on the opposite side of the line.

Example 4 Graph: $5x - 2y < 10$.

1. We first graph the line $5x - 2y = 10$. The intercepts are $(0, -5)$ and $(2, 0)$. This line forms the boundary of the solutions of the inequality.

2. Since the inequality contains the $<$ symbol, points on the line are not solutions of the inequality, so we draw a dashed line.

Answer on page A-41

3. To determine which half-plane to shade, we consider a test point *not* on the line. We try $(3, -2)$ and substitute:

$$\begin{array}{c|c} 5x - 2y < 10 & \\ \hline 5(3) - 2(-2) \ ? \ 10 & \\ 15 + 4 & \\ 19 & \text{FALSE} \end{array}$$

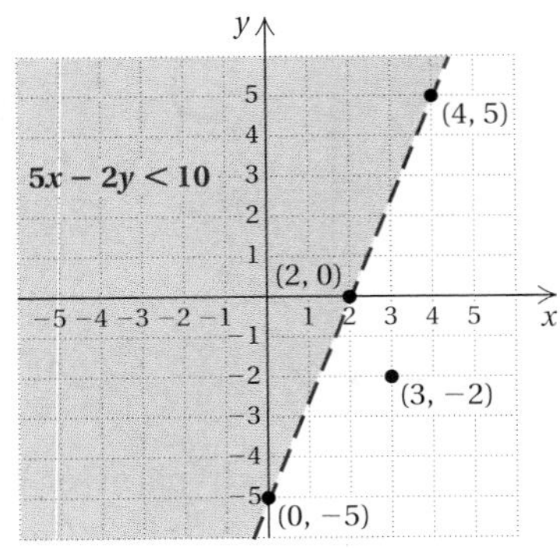

Since this inequality is false, the point $(3, -2)$ is *not* a solution; no point in the half-plane containing $(3, -2)$ is a solution. Thus the points in the opposite half-plane are solutions. The graph is shown above.

Do Exercise 4.

Example 5 Graph: $2x + 3y \leq 6$.

First, we graph the line $2x + 3y = 6$. The intercepts are $(0, 2)$ and $(3, 0)$. Since the inequality contains the $\leq$ symbol, we draw the line solid to indicate that any pair on the line is a solution. Next, we choose a test point that does not belong to the line. We substitute to determine whether this point is a solution. The origin $(0, 0)$ is generally an easy one to use:

$$\begin{array}{c|c} 2x + 3y \leq 6 & \\ \hline 2 \cdot 0 + 3 \cdot 0 \ ? \ 6 & \\ 0 & \text{TRUE} \end{array}$$

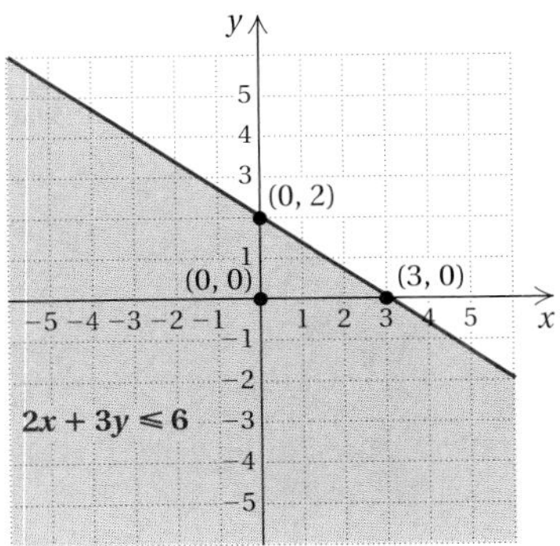

We see that $(0, 0)$ is a solution, so we shade the lower half-plane. Had the substitution given us a false inequality, we would have shaded the other half-plane.

Do Exercises 5 and 6.

Example 6 Graph $x < 3$ on a plane.

There is a missing variable in this inequality. Thus we rewrite this inequality as $x + 0y < 3$. We use the same technique that we have used with the other examples. First, we graph the related equation $x = 3$ on the plane and draw the graph with a dashed line since the inequality symbol is $<$.

4. Graph: $2x + 4y < 8$.

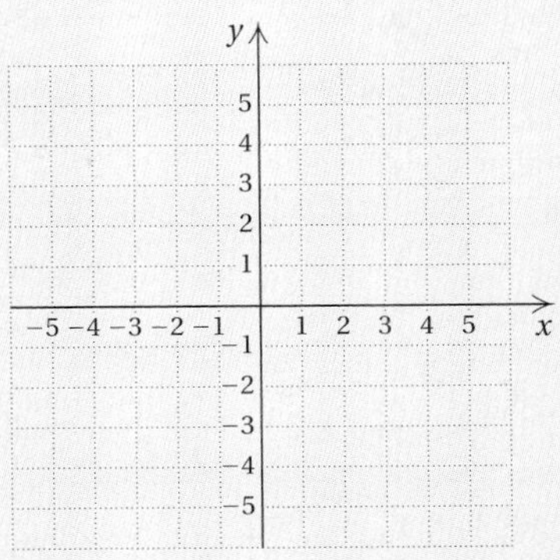

Graph.

5. $3x - 5y < 15$

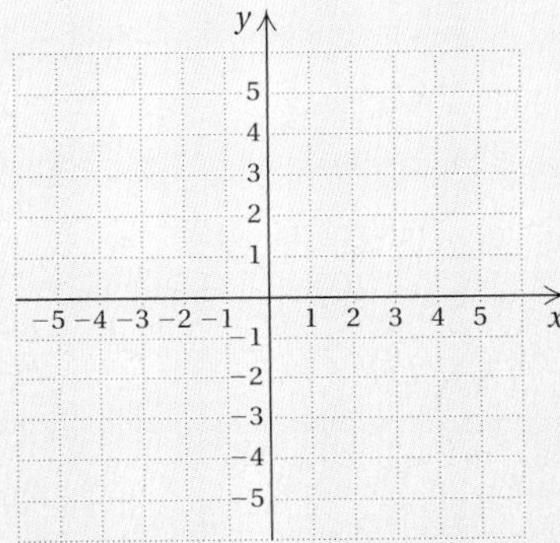

6. $2x + 3y \geq 12$

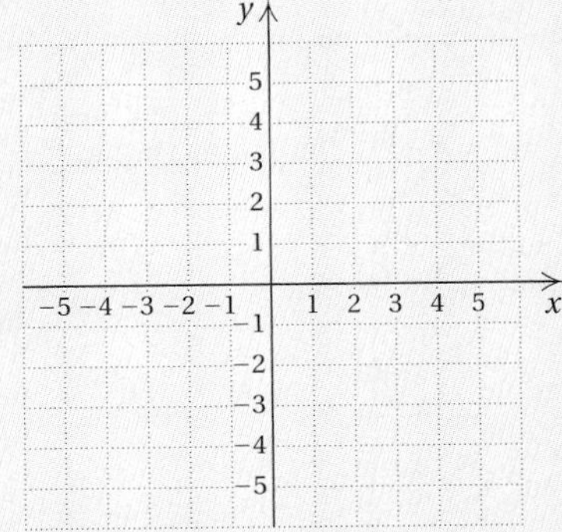

Answers on page A-41

Graph.

7. $x > -3$

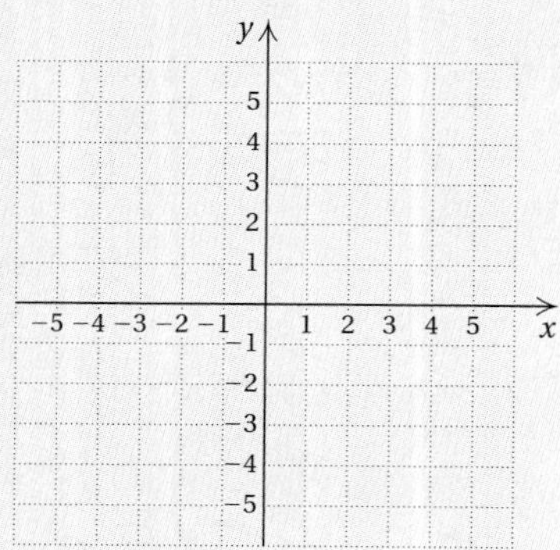

8. $y \le 4$

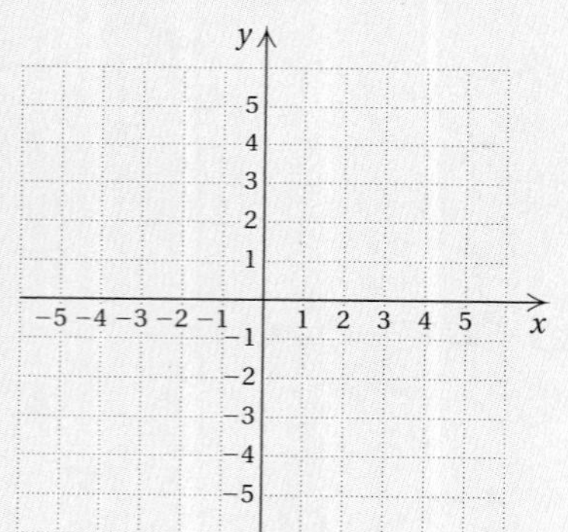

Answers on page A-41

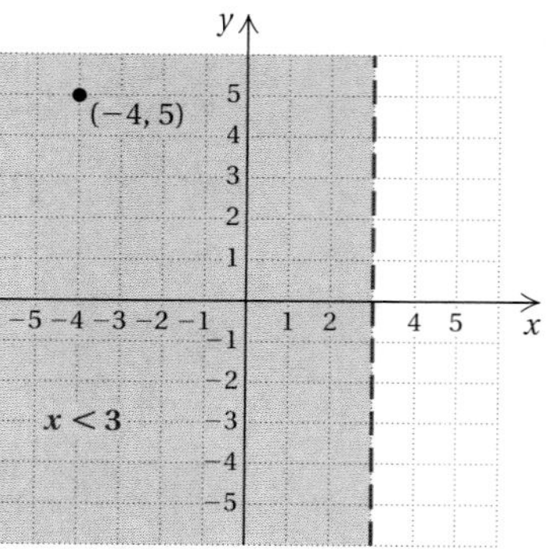

The graph is a half-plane either to the left or to the right of the line $x = 3$. To determine which, we consider a test point, $(-4, 5)$:

$$\begin{array}{c|c} x + 0y & < 3 \\ \hline -4 + 0(5) & ?\ 3 \\ -4 & \end{array} \quad \text{TRUE}$$

We see that $(-4, 5)$ is a solution, so all the pairs in the half-plane containing $(-4, 5)$ are solutions. We shade that half-plane.

We see from the graph that the solutions of $x < 3$ are all those ordered pairs whose first coordinates are less than 3.

If we graph the inequality in Example 6 on a line rather than on a plane, its graph is as follows:

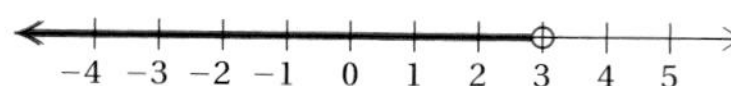

Example 7 Graph $y \ge -4$ on a plane.

We first graph $y = -4$ using a solid line to indicate that all points on the line are solutions. We then use (2, 3) as a test point and substitute:

$$\begin{array}{c|c} 0x + y & \ge -4 \\ \hline 0(2) + 3 & ?\ -4 \\ 3 & \end{array} \quad \text{TRUE}$$

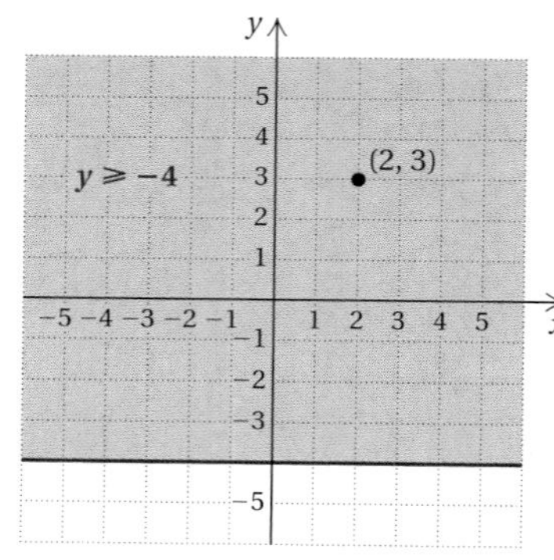

Since (2, 3) is a solution, all points in the half-plane containing (2, 3) are solutions. Note that this half-plane consists of all ordered pairs whose second coordinate is greater than or equal to -4.

Do Exercises 7 and 8.

Exercise Set 13.4

a

1. Determine whether $(-3, -5)$ is a solution of $-x - 3y < 18$.

2. Determine whether $(2, -3)$ is a solution of $5x - 4y \geq 1$.

3. Determine whether $\left(\frac{1}{2}, -\frac{1}{4}\right)$ is a solution of $7y - 9x \leq -3$.

4. Determine whether $(-8, 5)$ is a solution of $x + 0 \cdot y > 4$.

b Graph on a plane.

5. $x > 2y$

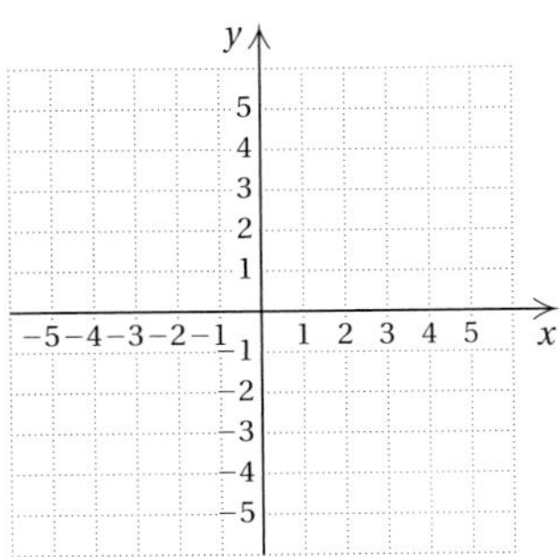

6. $x > 3y$

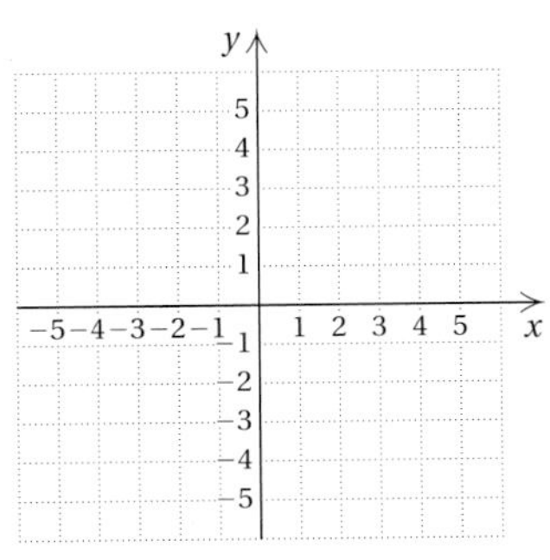

7. $y \leq x - 3$

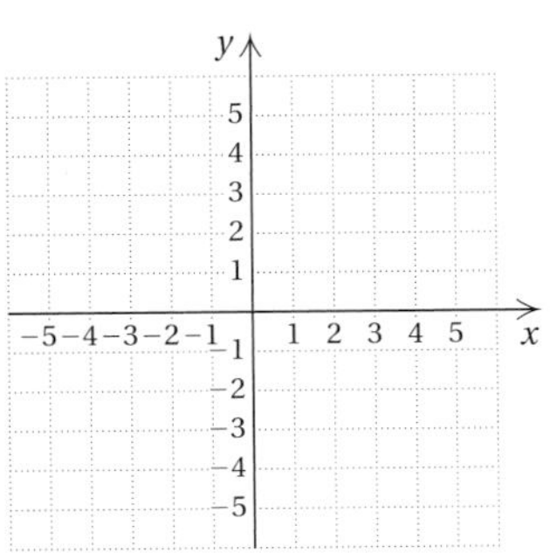

8. $y \leq x - 5$

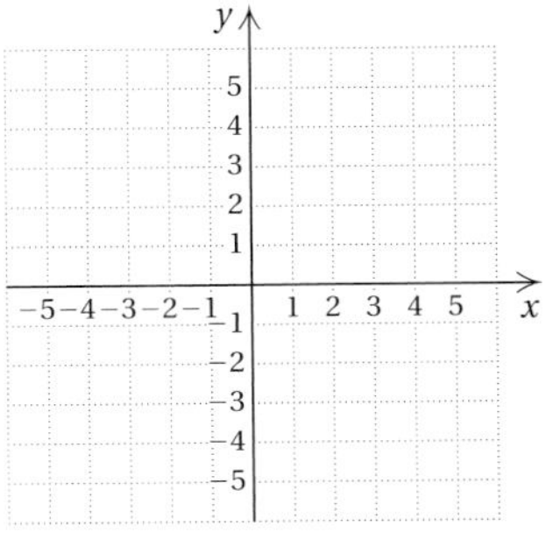

9. $y < x + 1$

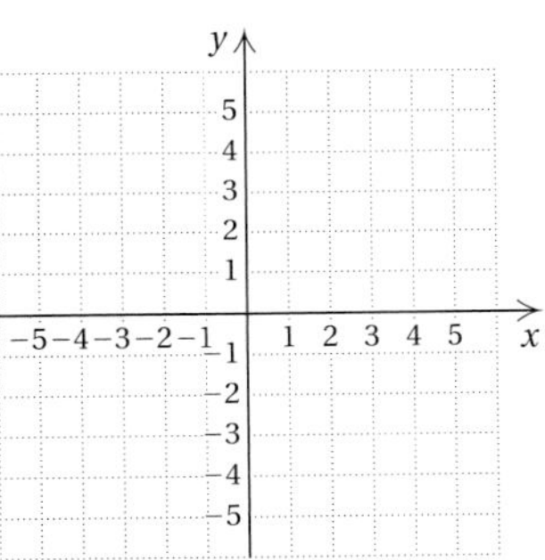

10. $y < x + 4$

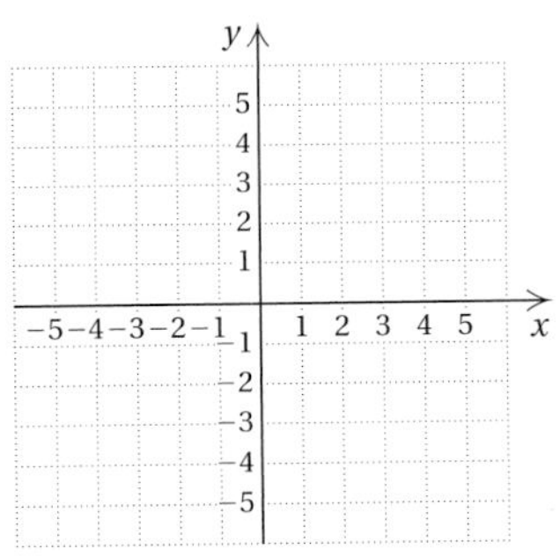

11. $y \geq x - 2$

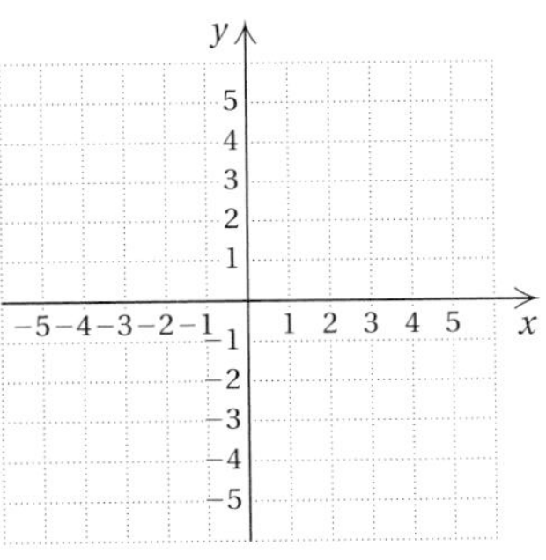

12. $y \geq x - 1$

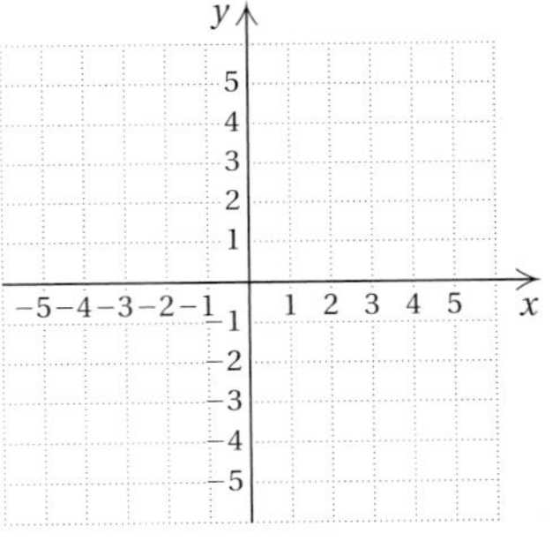

13. $y \leq 2x - 1$

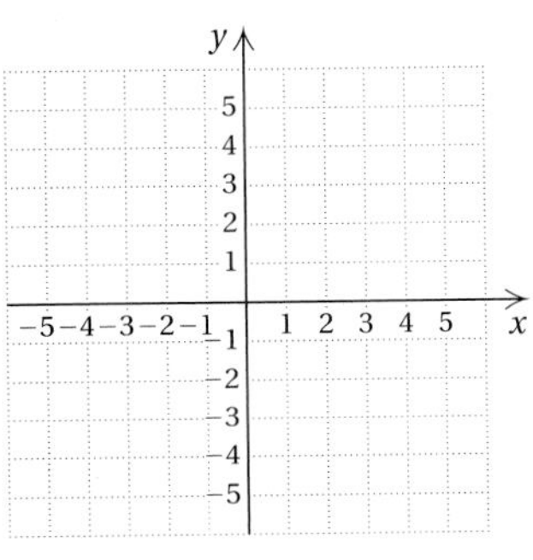

14. $y \leq 3x + 2$

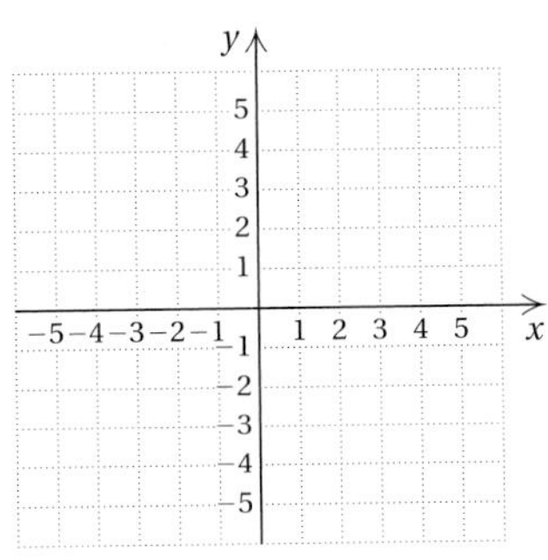

15. $x + y \leq 3$

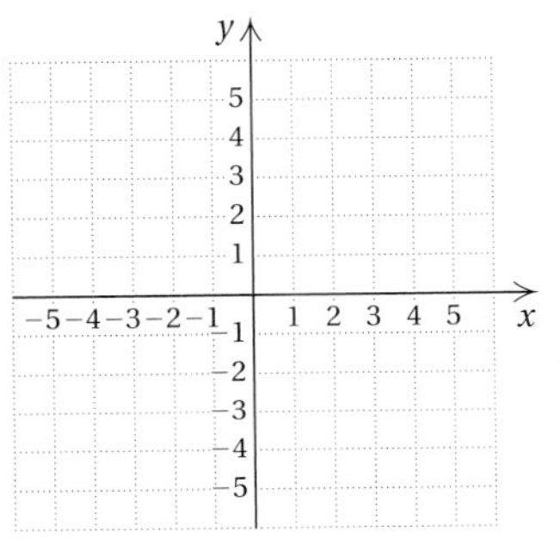

16. $x + y \leq 4$

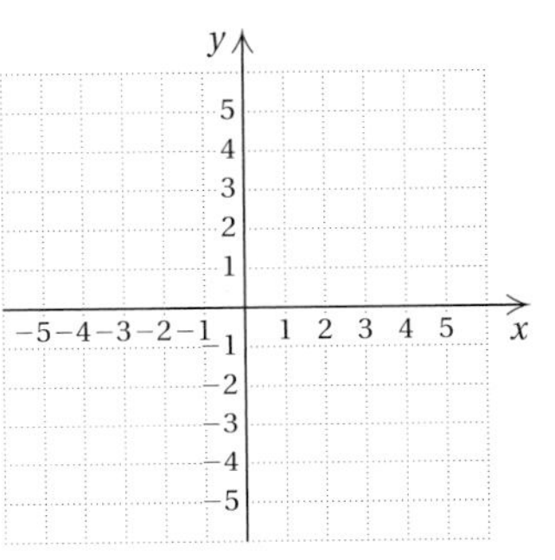

17. $x - y > 7$

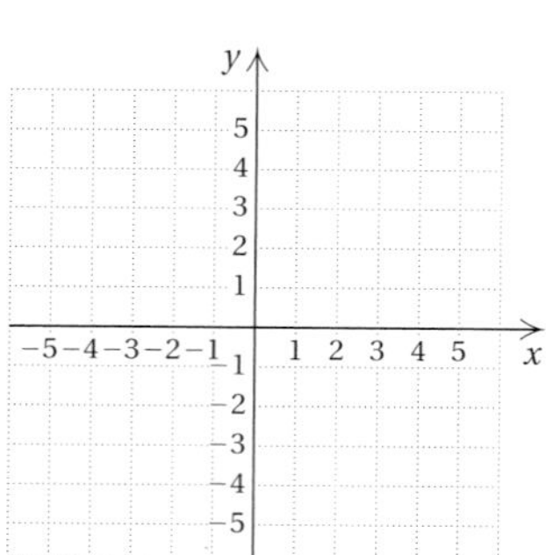

18. $x - y > -2$

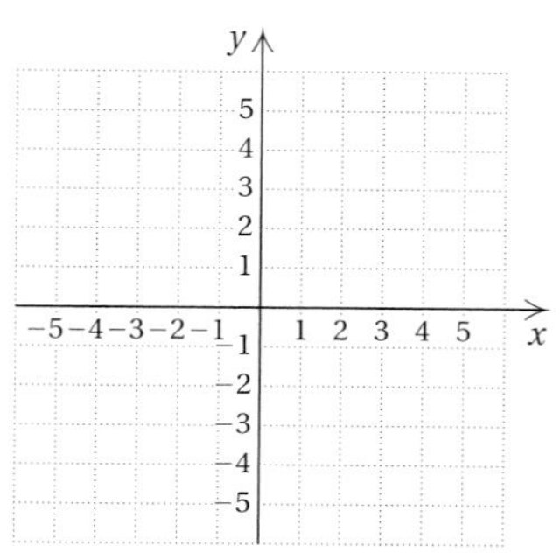

19. $2x + 3y \leq 12$

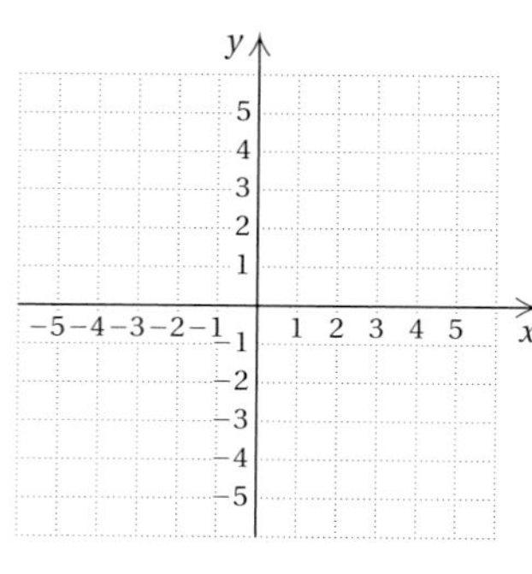

20. $5x + 4y \geq 20$

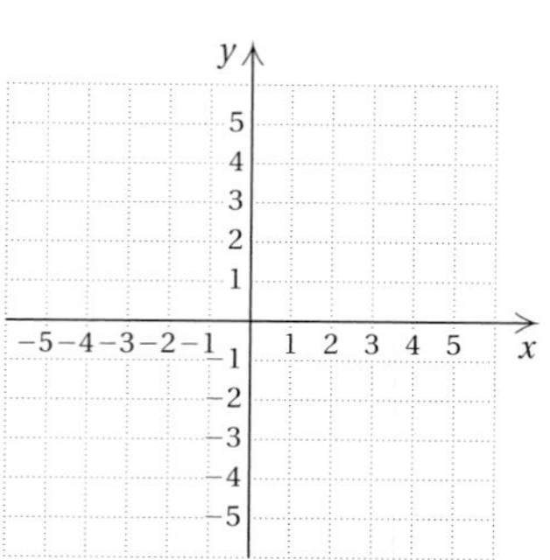

21. $y \geq 1 - 2x$

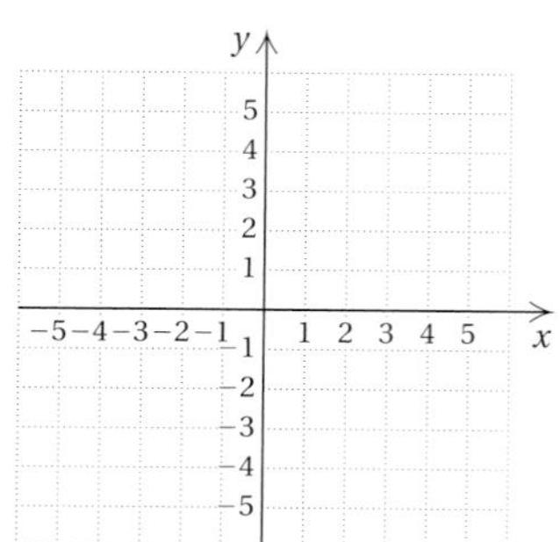

22. $y - 2x \leq -1$

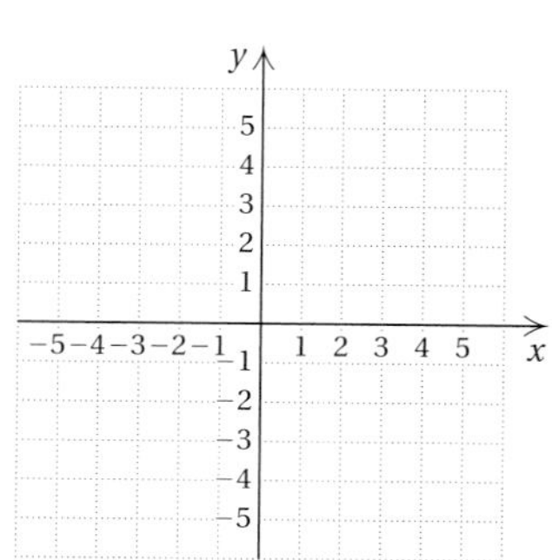

23. $2x - 3y > 6$

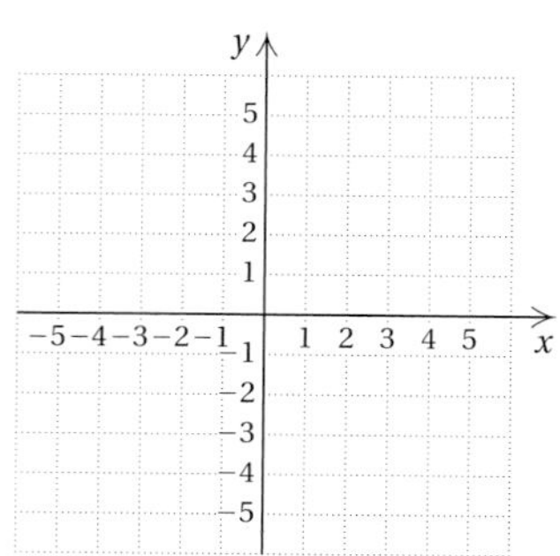

24. $5y - 2x \leq 10$

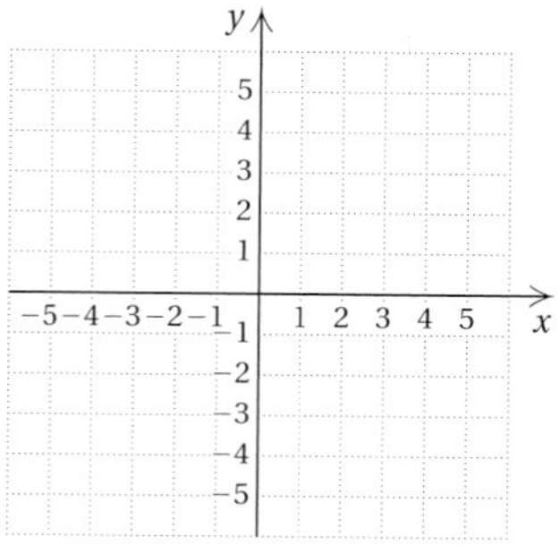

25. $y \leq 3$

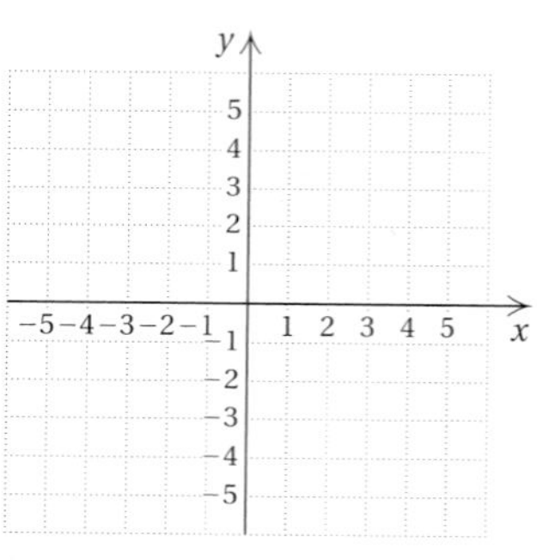

26. $y > -1$

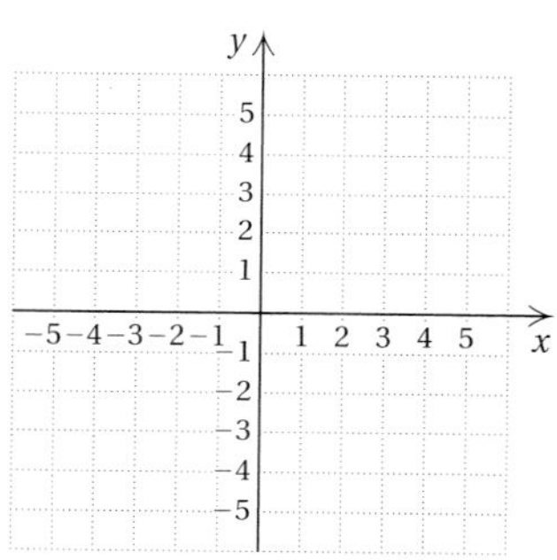

27. $x \geq -1$

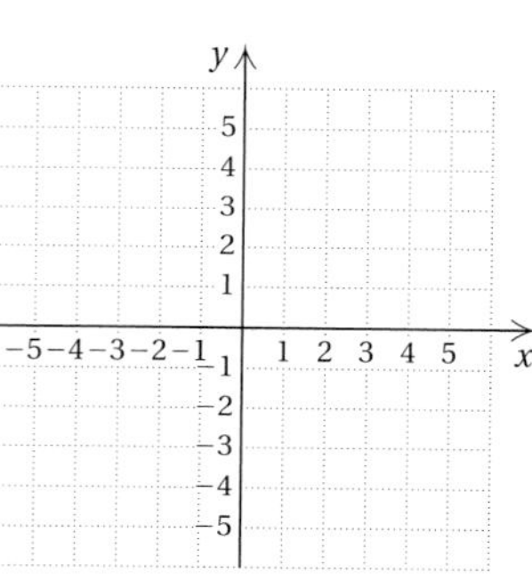

28. $x < 0$

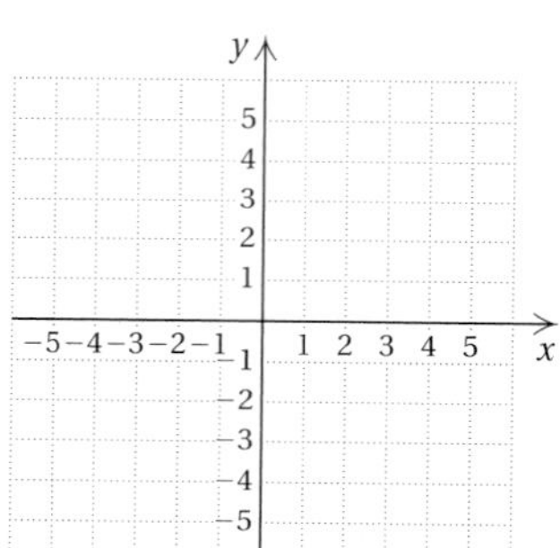

Skill Maintenance

Solve. [12.6a]

29. $\dfrac{12}{x} = \dfrac{48}{x + 9}$

30. $x + 5 = -\dfrac{6}{x}$

Solve. [11.7b]

31. $x^2 + 16 = 8x$

32. $12x^2 + 17x = 5$

Synthesis

33. ◆ Why is (0, 0) such a "convenient" test point?

34. ◆ Is the graph of any inequality in the form $y > mx + b$ shaded *above* the line $y = mx + b$? Why or why not?

35. *Elevators.* Many elevators have a capacity of 1 metric ton (1000 kg). Suppose c children, each weighing 35 kg, and a adults, each weighing 75 kg, are on an elevator. Find and graph an inequality that asserts that the elevator is overloaded.

36. *Hockey Wins and Losses.* A hockey team determines that it needs at least 60 points for the season in order to make the playoffs. A win w is worth 2 points and a tie t is worth 1 point. Find and graph an inequality that describes the situation.

Collaborative Learning Manual

Practice graphing linear inequalities by playing a variation of the game Battleship.

13.5 Direct and Inverse Variation

a Equations of Direct Variation

A bicycle is traveling at a speed of 15 km/h. In 1 hr, it goes 15 km; in 2 hr, it goes 30 km; in 3 hr, it goes 45 km; and so on. We can form a set of ordered pairs using the number of hours as the first coordinate and the number of kilometers traveled as the second coordinate. These determine a set of ordered pairs:

$$(1, 15),\ (2, 30),\ (3, 45),\ (4, 60),\ \text{and so on.}$$

Note that the ratio of the second coordinate to the first is the same number:

$$\frac{15}{1} = 15,\quad \frac{30}{2} = 15,\quad \frac{45}{3} = 15,\quad \frac{60}{4} = 15,\quad \text{and so on.}$$

Whenever a situation produces pairs of numbers in which the *ratio is constant,* we say that there is **direct variation.** Here the distance varies directly as the time:

$$\frac{d}{t} = 15 \text{ (a constant)},\quad \text{or}\quad d = 15t.$$

The equation is an **equation of direct variation.** The coefficient—in this case, 15—is called the **variation constant.** The graph of $d = 15t$ is shown at right.

> **DIRECT VARIATION**
>
> If a situation translates to an equation described by $y = kx$, where k is a positive constant, $y = kx$ is called an **equation of direct variation,** and k is called the **variation constant.** We say that y varies directly as x.

The terminologies "y varies as x," "y is directly proportional to x," and "y is proportional to x" also imply direct variation and are used in many situations. The constant k is often referred to as a **constant of proportionality.**

When there is direct variation $y = kx$, the variation constant can be found if one pair of values of x and y is known. Then other values can be found.

Example 1 Find an equation of variation in which y varies directly as x and $y = 7$ when $x = 25$.

We first substitute to find k:

$$y = kx$$
$$7 = k \cdot 25$$
$$\frac{7}{25} = k,\quad \text{or } k = 0.28.$$

Then the equation of variation is

$$y = 0.28x.$$

Note that the answer is an *equation.*

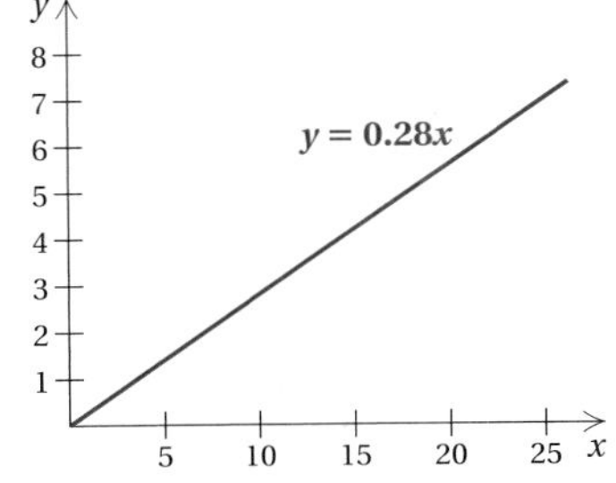

Do Exercises 1 and 2 on the following page.

Objectives

a Find an equation of direct variation given a pair of values of the variables.

b Solve applied problems involving direct variation.

c Find an equation of inverse variation given a pair of values of the variables.

d Solve applied problems involving inverse variation.

For Extra Help

TAPE 25 MAC WIN CD-ROM

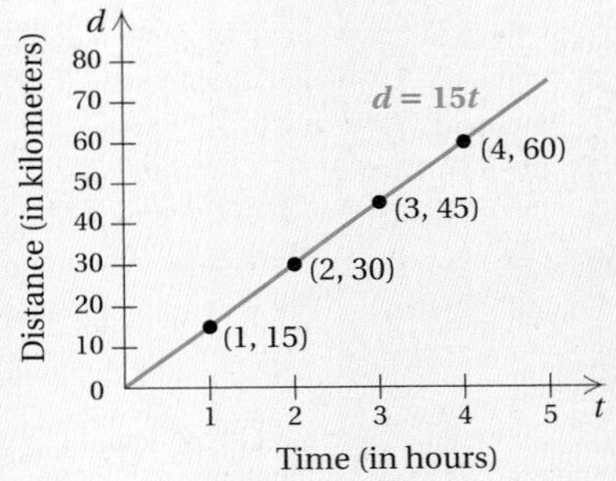

b Applications of Direct Variation

Example 2 *Karat Ratings of Gold Objects.* It is known that the karat rating K of a gold object varies directly as the actual percentage P of gold in the object. A 14-karat gold ring is 58.25% gold. What is the percentage of gold in a 10-karat chain?

1., 2. Familiarize and **Translate.** The problem states that we have direct variation between the variables K and P. Thus an equation $K = kP$, $k > 0$, applies. As the percentage of gold increases, the karat rating increases. The letters K and k represent different quantities.

3. Solve. The mathematical manipulation has two parts. First, we determine the equation of variation by substituting known values for K and P to find the variation constant k. Second, we compute the percentage of gold in a 10-karat chain.

a) First, we find an equation of variation:

$$K = kP$$

$$14 = k(0.5825) \quad \text{Substituting 14 for } K \text{ and 58.25\%, or 0.5825, for } P$$

$$\frac{14}{0.5825} = k$$

$$24.03 \approx k. \quad \text{Dividing and rounding to the nearest hundredth}$$

The equation of variation is $K = 24.03P$.

b) We then use the equation to find the percentage of gold in a 10-karat chain:

$$K = 24.03P$$

$$10 = 24.03P \quad \text{Substituting 10 for } K$$

$$\frac{10}{24.03} = P$$

$$0.416 \approx P$$

$$41.6\% \approx P.$$

4. Check. The check might be done by repeating the computations. You might also do some reasoning about the answer. The karat rating decreased from 14 to 10. Similarly, the percentage decreased from 58.25% to 41.6%.

5. State. A 10-karat chain is 41.6% gold.

Do Exercises 3 and 4.

Let's consider direct variation from the standpoint of a graph. The graph of $y = kx$, $k > 0$, always goes through the origin and rises from left to right. Note that as x increases, y increases; and as x decreases, y decreases. This is why the terminology "direct" is used. What one variable does, the other does as well.

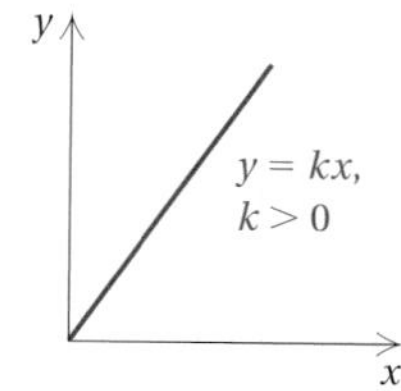

Find an equation of variation in which y varies directly as x and the following is true.

1. $y = 84$ when $x = 12$

2. $y = 50$ when $x = 80$

3. *Electricity Costs.* The cost C of operating a television varies directly as the number n of hours that it is in operation. It costs \$14.00 to operate a standard-size color TV continuously for 30 days. At this rate, how much would it cost to operate the TV for 1 day? for 1 hour?

4. *Weight on Venus.* The weight V of an object on Venus varies directly as its weight E on Earth. A person weighing 165 lb on Earth would weigh 145.2 lb on Venus. How much would a person weighing 198 lb on Earth weigh on Venus?

Answers on page A-42

c Equations of Inverse Variation

A car is traveling a distance of 20 mi. At a speed of 5 mph, it will take 4 hr; at 20 mph, it will take 1 hr; at 40 mph, it will take $\frac{1}{2}$ hr; and so on. We use speed as the first coordinate and the time as the second coordinate. These determine a set of ordered pairs:

$$(5, 4),\ (20, 1),\ \left(40, \tfrac{1}{2}\right),\ \left(60, \tfrac{1}{3}\right),\ \text{and so on.}$$

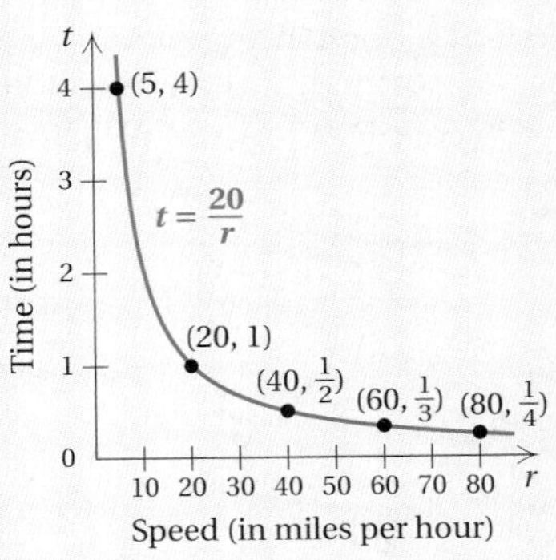

Note that the products of the coordinates in each ordered pair are all the same number:

$$5 \cdot 4 = 20,\quad 20 \cdot 1 = 20,\quad 40 \cdot \tfrac{1}{2} = 20,\quad 60 \cdot \tfrac{1}{3} = 20,\quad \text{and so on.}$$

Whenever a situation produces pairs of numbers in which the *product is constant,* we say that there is **inverse variation.** Here the time varies inversely as the speed:

$$rt = 20 \text{ (a constant)}, \quad \text{or} \quad t = \frac{20}{r}.$$

The equation is an **equation of inverse variation.** The coefficient—in this case, 20—is called the **variation constant.** Note that as the first number gets larger, the second number gets smaller.

> **INVERSE VARIATION**
>
> If a situation translates to an equation described by $y = k/x$, where k is a positive constant, $y = k/x$ is called an **equation of inverse variation.** We say that y varies inversely as x.

The terminology "y is inversely proportional to x" also implies inverse variation and is used in some situations.

Example 3 Find an equation of variation in which y varies inversely as x and $y = 145$ when $x = 0.8$.

We first substitute to find k:

$$y = \frac{k}{x}$$

$$145 = \frac{k}{0.8}$$

$$(0.8)145 = k$$

$$116 = k.$$

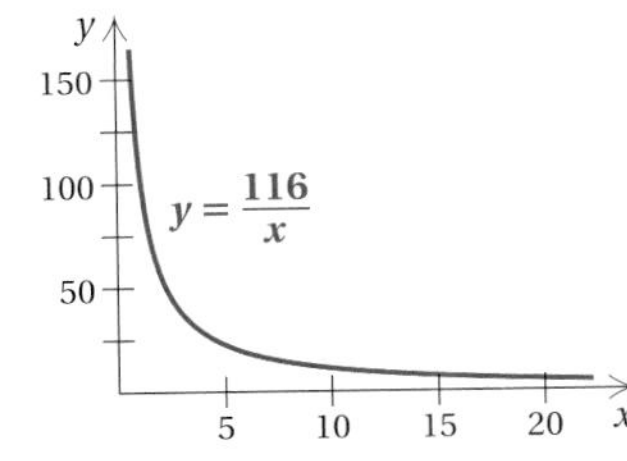

The equation of variation is $y = \dfrac{116}{x}$.

Do Exercises 5 and 6.

Find an equation of variation in which y varies inversely as x and the following is true.

5. $y = 105$ when $x = 0.6$

6. $y = 45$ when $x = 20$

The graph of $y = k/x$, $k > 0$, is shaped like the figure at right for positive values of x. (You need not know how to graph such equations at this time.) Note that as x increases, y decreases; and as x decreases, y increases. This is why the terminology "inverse" is used. One variable does the opposite of what the other does.

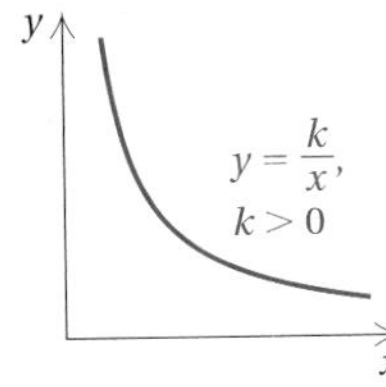

Answers on page A-42

7. Referring to Example 4, determine how long it would take 10 people to do the job.

8. The time required to drive a fixed distance varies inversely as the speed r. It takes 5 hr at 60 km/h to drive a fixed distance. How long would it take at 40 km/h?

Answers on page A-42

d Applications of Inverse Variation

Often in an applied situation we must decide which kind of variation, if any, might apply to the problem.

Example 4 *Work Time.* Molly is a maintenance supervisor. She notes that it takes 4 hr for 20 people to wash and wax the floors in a building. How long would it then take 25 people to do the job?

1. **Familiarize.** Think about the problem situation. What kind of variation would be used? It seems reasonable that the more people there are working on the job, the less time it will take to finish. (One might argue that too many people in a crowded area would be counterproductive, but we will disregard that possibility.) Thus inverse variation might apply. We let T = the time to do the job, in hours, and N = the number of people. Assuming inverse variation, we know that an equation $T = k/N$, $k > 0$, applies. As the number of people increases, the time it takes to do the job decreases.
2. **Translate.** We write an equation of variation:

$$T = \frac{k}{N}.$$

Time varies inversely as the number of people involved.

3. **Solve.** The mathematical manipulation has two parts. First, we find the equation of variation by substituting known values for T and N to find k. Second, we compute the amount of time it would take 25 people to do the job.

a) First, we find an equation of variation:

$$T = \frac{k}{N}$$

$$4 = \frac{k}{20} \quad \text{Substituting 4 for } T \text{ and 20 for } N$$

$$20 \cdot 4 = k$$

$$80 = k.$$

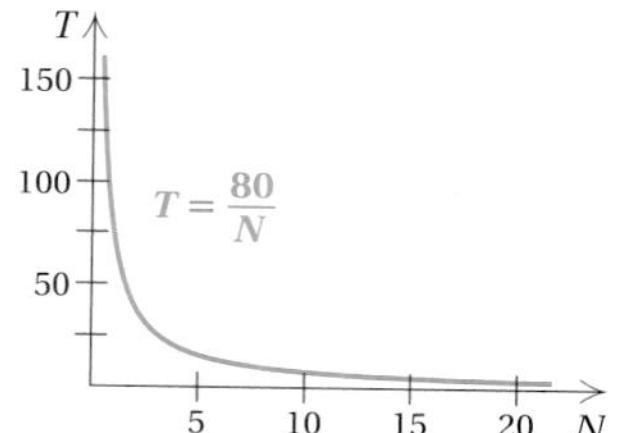

The equation of variation is $T = \dfrac{80}{N}$.

b) We then use the equation to find the amount of time that it takes 25 people to do the job:

$$T = \frac{80}{N}$$

$$= \frac{80}{25} \quad \text{Substituting 25 for } N$$

$$= 3.2.$$

4. **Check.** The check might be done by repeating the computations. We might also analyze the results. The number of people increased from 20 to 25. Did the time decrease? It did, and this confirms what we expect with inverse variation.
5. **State.** It should take 3.2 hr for 25 people to complete the job.

Do Exercises 7 and 8.

Exercise Set 13.5

a Find an equation of variation in which y varies directly as x and the following are true.

1. $y = 36$ when $x = 9$

2. $y = 60$ when $x = 16$

3. $y = 0.8$ when $x = 0.5$

4. $y = 0.7$ when $x = 0.4$

5. $y = 630$ when $x = 175$

6. $y = 400$ when $x = 125$

7. $y = 500$ when $x = 60$

8. $y = 200$ when $x = 300$

b Solve.

9. A person's paycheck P varies directly as the number H of hours worked. For working 15 hr, the pay is \$84. Find the pay for 35 hr of work.

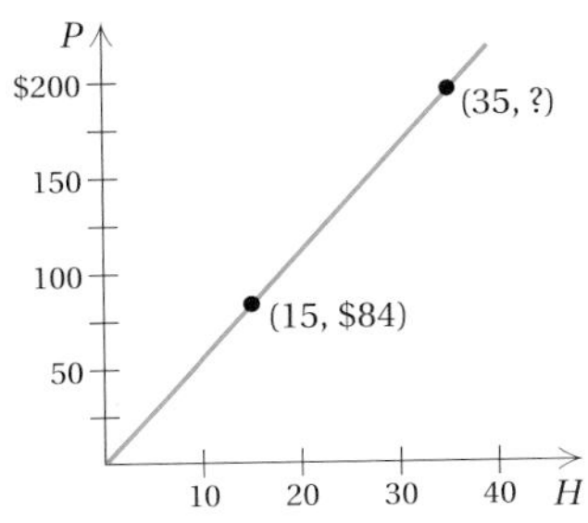

10. The interest I earned in 1 yr on a fixed principal varies directly as the interest rate r. An investment earns \$53.55 at an interest rate of 4.25%. How much will the investment earn at a rate of 5.75%?

11. The cost C to fill a sandbox varies directly as the depth S of the sand. Lucinda checks at her local hardware store and finds that it would cost \$75 to fill the box with 6 in. of sand. She decides to fill the sandbox to a depth of 8 in. How much will the sand cost Lucinda?

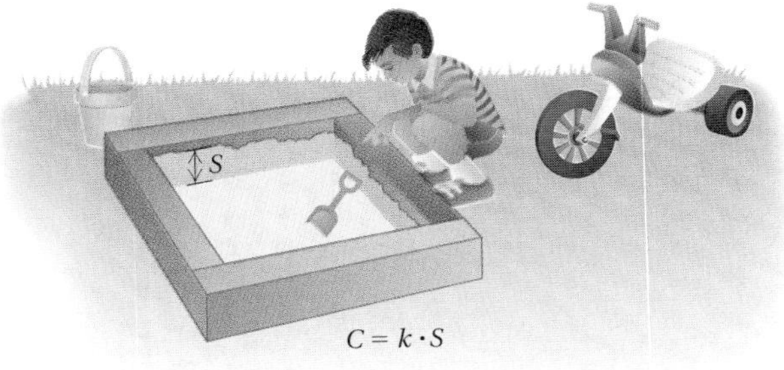

12. The cost C of cement needed to pave a driveway varies directly as the depth D of the driveway. John checks at his local building materials store and finds that it costs \$500 to install his driveway with a depth of 8 in. He decides to build a stronger driveway at a depth of 12 in. How much will it cost?

13. A chef is planning meals in a refreshment tent at a golf tournament. The number of servings S of meat that can be obtained from a turkey varies directly as its weight W. From a turkey weighing 14 kg, one can get 40 servings of meat. How many servings can be obtained from an 8-kg turkey?

14. The number of servings S of meat that can be obtained from round steak varies directly as the weight W. From 9 kg of round steak, one can get 70 servings of meat. How many servings can one get from 12 kg of round steak?

15. *Weight on Moon.* The weight M of an object on the moon varies directly as its weight E on Earth. A person who weighs 171.6 lb on Earth weighs 28.6 lb on the moon. How much would a 220-lb person weigh on the moon?

16. *Weight on Mars.* The weight M of an object on Mars varies directly as its weight E on Earth. A person who weighs 209 lb on Earth weighs 79.42 lb on Mars. How much would a 176-lb person weigh on Mars?

17. *Computer Megahertz.* The number of computer instructions N per second varies directly as the speed S of its internal processor. A processor with a speed of 25 megahertz can perform 2,000,000 instructions per second. How many instructions will the same processor perform if it is running at a speed of 200 megahertz?

18. *Water in Human Body.* The number of kilograms W of water in a human body varies directly as the total body weight B. A person who weighs 75 kg contains 54 kg of water. How many kilograms of water are in a person who weighs 95 kg?

c Find an equation of variation in which y varies inversely as x and the following are true.

19. $y = 3$ when $x = 25$

20. $y = 2$ when $x = 45$

21. $y = 10$ when $x = 8$

22. $y = 10$ when $x = 7$

23. $y = 6.25$ when $x = 0.16$

24. $y = 0.125$ when $x = 8$

25. $y = 50$ when $x = 42$

26. $y = 25$ when $x = 42$

27. $y = 0.2$ when $x = 0.3$

28. $y = 0.4$ when $x = 0.6$

b, **d** Solve.

29. A production line produces 15 compact disc players every 8 hr. How many players can it produce in 37 hr?

a) What kind of variation might apply to this situation?
b) Solve the problem.

30. A person works for 15 hr and makes $93.75. How much will the person make by working 35 hr?

a) What kind of variation might apply to this situation?
b) Solve the problem.

31. It takes 4 hr for 9 cooks to prepare the food for a wedding rehearsal dinner. How long will it take 8 cooks to prepare the dinner?

a) What kind of variation might apply to this situation?
b) Solve the problem.

32. It takes 16 hr for 2 people to resurface a tennis court. How long will it take 6 people to do the job?

a) What kind of variation might apply to this situation?
b) Solve the problem.

33. *Miles per Gallon.* To travel a fixed distance, the number of gallons N of gasoline needed is inversely proportional to the miles-per-gallon rating P of the car. A car that gets 20 miles per gallon (mpg) needs 14 gal to travel the distance. How much gas will be needed for a car that gets 28 mpg?

34. *Miles per Gallon.* To travel a fixed distance, the number of gallons N of gasoline needed is inversely proportional to the miles-per-gallon rating P of the car. A car that gets 25 miles per gallon (mpg) needs 12 gal to travel the distance. How much gas will be needed for a car that gets 20 mpg?

35. *Electrical Current.* The current I in an electrical conductor varies inversely as the resistance R of the conductor. The current is 96 amperes when the resistance is 20 ohms. What is the current when the resistance is 60 ohms?

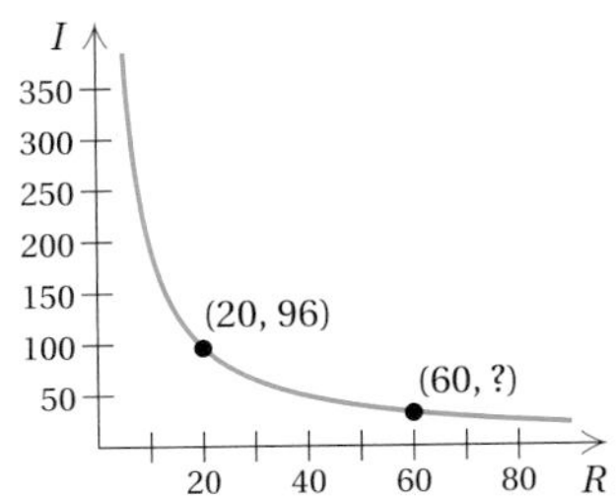

36. *Gas Volume.* The volume V of a gas varies inversely as the pressure P on it. The volume of a gas is 200 cm^3 under a pressure of 32 kg/cm^2. What will be its volume under a pressure of 20 kg/cm^2?

37. The number of files N of the same size that a computer's hard drive will hold varies inversely as the size of the files. Loretta's hard drive will hold 1600 files if each is 50,000 bytes long. How many files will the drive hold if each is 125,000 bytes long?

38. The time t required to empty a tank varies inversely as the rate r of pumping. A pump can empty a tank in 90 min at a rate of 1200 L/min. How long will it take the pump to empty the tank at a rate of 2000 L/min?

39. The apparent size A of an object varies inversely as the distance d of the object from the eye. A flagpole 30 ft from an observer appears to be 27.5 ft tall. How tall will the same flagpole appear to be if it is 100 ft from the eye?

40. The time t required to drive a fixed distance varies inversely as the speed r. It takes 5 hr at 55 mph to drive a fixed distance. How long would it take at 40 mph?

Skill Maintenance

Solve. [12.6a]

41. $\frac{x+2}{x+5} = \frac{x-4}{x-6}$

42. $\frac{x-3}{x-5} = \frac{x+5}{x+1}$

Solve. [11.7b]

43. $x^2 - 25x + 144 = 0$

44. $t^2 + 21t + 108 = 0$

45. $35x^2 + 8 = 34x$

46. $14x^2 - 19x - 3 = 0$

Calculate. [7.8d]

47. $3^7 \div 3^4 \div 3^3 \div 3$

48. $\frac{37 - 5(4-6)}{2 \cdot 6 + 8}$

Synthesis

In Exercises 49–52, determine whether the situation represents direct variation, inverse variation, or neither. Give a reason for your answer.

49. ◆ The cost of mailing a first-class letter in the United States and the distance that it travels

50. ◆ The number of hours that a student watches TV per week and the student's grade point average

51. ◆ The weight of a turkey and the cooking time

52. ◆ The number of plays that it takes to go 80 yd for a touchdown and the average gain per play

53. Graph the equation that corresponds to Exercise 12. Then use the TABLE feature to create a table with TblStart = 1 and ΔTbl = 1. What happens to the y-values as the x-values become larger?

54. Graph the equation that corresponds to Exercise 13. Then use the TABLE feature to create a table with TblStart = 1 and ΔTbl = 1. What happens to the y-values as the x-values become larger?

Write an equation of variation for the situation.

55. The square of the pitch P of a vibrating string varies directly as the tension t on the string.

56. In a stream, the amount S of salt carried varies directly as the sixth power of the speed V of the stream.

57. The power P in a windmill varies directly as the cube of the wind speed V.

58. The volume V of a sphere varies directly as the cube of the radius r.

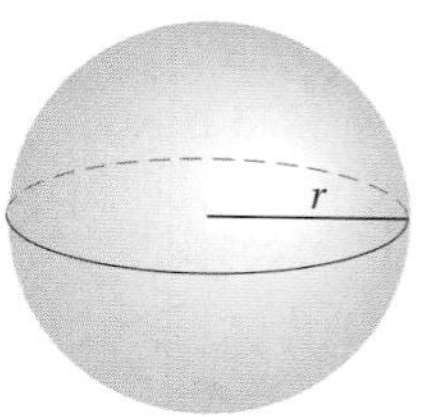

Summary and Review Exercises: Chapter 13

Important Properties and Formulas

Slope $= m = \dfrac{y_2 - y_1}{x_2 - x_1}$

Slope–Intercept Equation:	$y = mx + b$
Parallel Lines:	Slopes equal, y-intercepts different
Perpendicular Lines:	Product of slopes $= -1$
Equation of Direct Variation:	$y = kx$
Equation of Inverse Variation:	$y = \dfrac{k}{x}$

The objectives to be tested in addition to the material in this chapter are [7.8d], [11.7b], [12.6a], and [12.7a].

Find the slope, if it exists, of the line. [13.1a]

1.

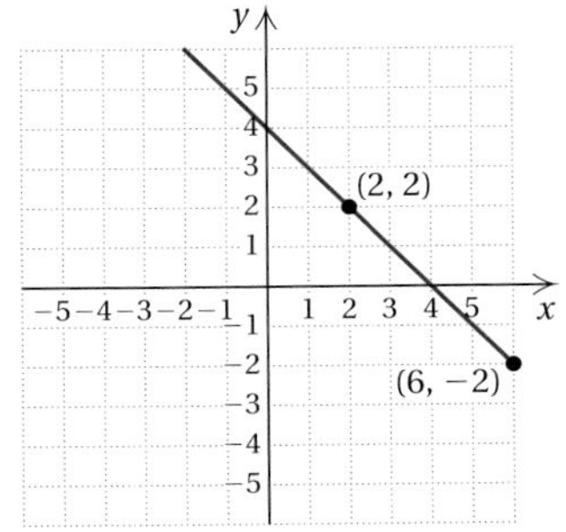

2.

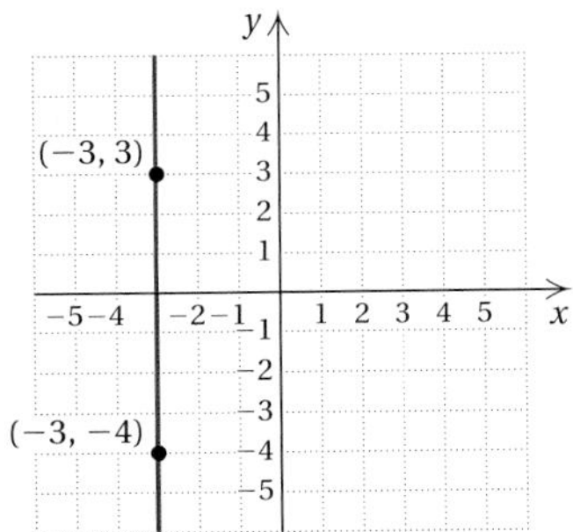

Graph the line containing the given pair of points and find the slope. [13.1a]

3. $(5, -2)$ and $(-3, 4)$

4. $(0, -3)$ and $(5, -3)$

Find the slope, if it exists, of the line containing the given pair of points. [13.1a]

5. $(6, 8)$ and $(-2, -4)$

6. $(-8.3, 4.6)$ and $(-9.9, 1.4)$

Find the slope of the line, if it exists. [13.1b]

7. $y = -6$

8. $x = 90$

9. $4x + 3y = -20$

Find the slope and the y-intercept. [13.2a]

10. $y = -9x + 46$

11. $x + y = 9$

12. $3x - 5y = 4$

Find an equation of the line with the given slope and y-intercept. [13.2a]

13. Slope $= -2.8$; y-intercept: $(0, 19)$

14. Slope $= \frac{5}{8}$; y-intercept: $\left(0, -\frac{7}{8}\right)$

Find an equation of the line containing the given point and with the given slope. [13.2b]

15. $(1, 2)$, $m = 3$

16. $(-2, -5)$, $m = \frac{2}{3}$

17. $(0, -4)$, $m = -2$

Find an equation of the line containing the given pair of points. [13.2c]

18. $(5, 7)$ and $(-1, 1)$

19. $(2, 0)$ and $(-4, -3)$

Solve. [13.2c]

20. *Median Age of Cars.* People are driving cars for longer periods of time. The line graph below describes the *median age of cars A*, in years, for years since 1990.

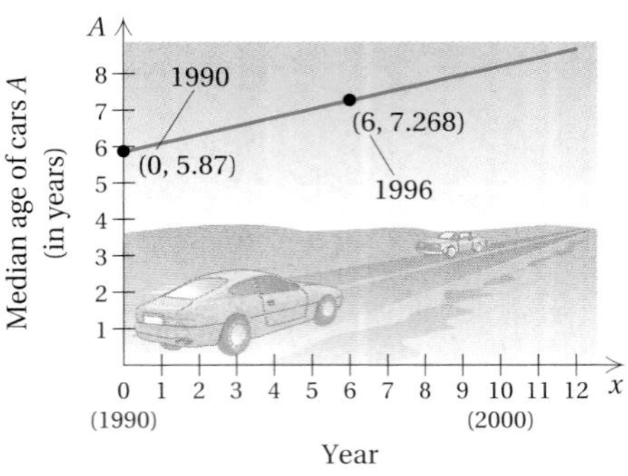

Source: The Polk Co.

a) Find an equation of the line.
b) What is the rate of change in the median age of cars?
c) Use the equation to predict the median age of cars in 2000.

Determine whether the graphs of the equations are parallel, perpendicular, or neither. [13.3a, b]

21. $4x + y = 6,$
$4x + y = 8$

22. $2x + y = 10,$
$y = \frac{1}{2}x - 4$

23. $x + 4y = 8,$
$x = -4y - 10$

24. $3x - y = 6,$
$3x + y = 8$

Determine whether the given point is a solution of the inequality $x - 2y > 1$. [13.4a]

25. $(0, 0)$

26. $(1, 3)$

27. $(4, -1)$

Graph on a plane. [13.4b]

28. $x < y$

29. $x + 2y \geq 4$

30. $x > -2$

Find an equation of variation in which y varies directly as x and the following are true. [13.5a]

31. $y = 12$ when $x = 4$

32. $y = 4$ when $x = 8$

33. $y = 0.4$ when $x = 0.5$

Find an equation of variation in which y varies inversely as x and the following are true. [13.5c]

34. $y = 5$ when $x = 6$

35. $y = 0.5$ when $x = 2$

36. $y = 1.3$ when $x = 0.5$

Solve.

37. A person's paycheck P varies directly as the number H of hours worked. The pay is $165.00 for working 20 hr. Find the pay for 35 hr of work. [13.5b]

38. It takes 5 hr for 2 washing machines to wash a fixed amount of laundry. How long would it take 10 washing machines to do the same job? (The number of hours varies inversely as the number of washing machines.) [13.5d]

Skill Maintenance

39. Judd can paint a shed alone in 5 hr. Bud can paint the same shed in 10 hr. How long would it take both of them, working together, to paint the fence? [12.7a]

40. Compute: $13 \cdot 6 \div 3 \cdot 26 \div 13$. [7.8d]

Solve.

41. $\frac{x^2}{x - 4} = \frac{16}{x - 4}$ [12.6a]

42. $a^2 + 6a - 55 = 0$ [11.7b]

Synthesis

43. ◈ Briefly describe the concept of slope. [13.1a]

44. ◈ Graph $x < 1$ on both a number line and a plane, and explain the difference between the graphs. [13.4b]

45. In chess, the knight can move to any of the eight squares shown by a, b, c, d, e, f, g, and h below. If lines are drawn from the beginning to the end of the move, what slopes are possible for these lines? [13.1a]

Test: Chapter 13

Find the slope, if it exists, of the line.

1.

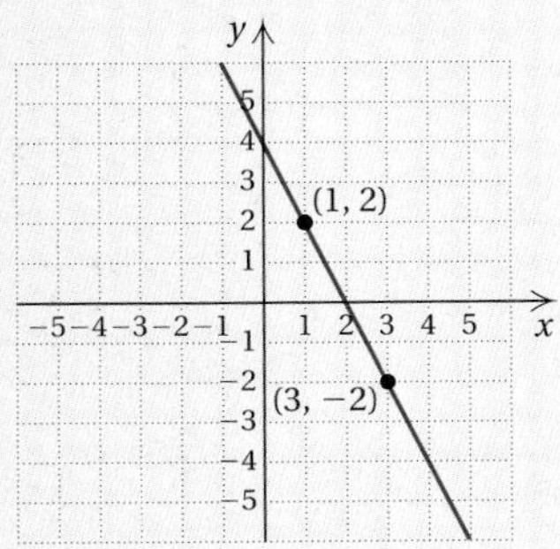

2.

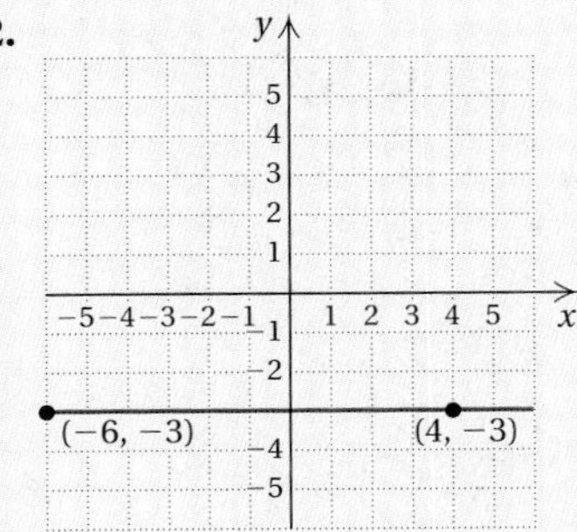

3. Graph the line containing the given pair of points and find the slope.

(5, −4) and (−2, −2)

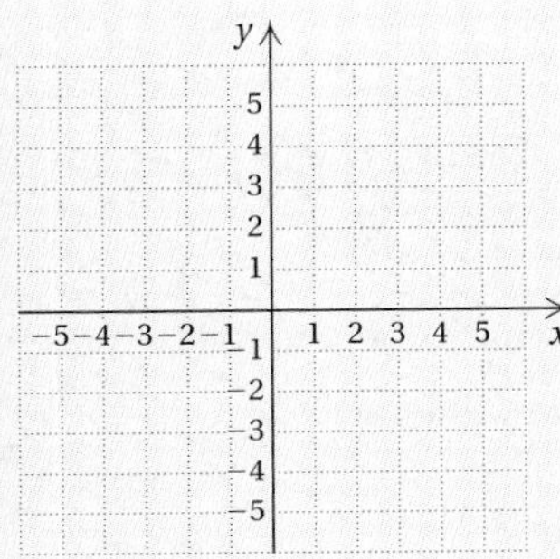

Find the slope, if it exists, of the line containing the given pair of points.

4. (4, 7) and (4, −1)

5. (9, 2) and (−3, −5)

Find the slope of the line, if it exists.

6. $y = -7$

7. $x = 6$

Find the slope and the y-intercept.

8. $y = 2x - \frac{1}{4}$

9. $-4x + 3y = -6$

Find an equation of the line with the given slope and y-intercept.

10. Slope = 1.8; y-intercept: (0, −7)

11. Slope $= -\frac{3}{8}$; y-intercept: $\left(0, -\frac{1}{8}\right)$

Find an equation of the line containing the given point and with the given slope.

12. (3, 5), $m = 1$

13. (−2, 0), $m = -3$

Find an equation of the line containing the given pair of points.

14. (1, 1) and (2, −2)

15. (4, −1) and (−4, −3)

16. *Cancer Research.* Increasing amounts of money are being spent each year on cancer research. The line graph at right describes the amount spent on cancer research M, in millions of dollars, for years since 1992.

a) Find an equation of the line.

b) What is the rate of change in the amount spent on cancer research?

c) Use the equation to predict the amount spent on cancer research in 2000.

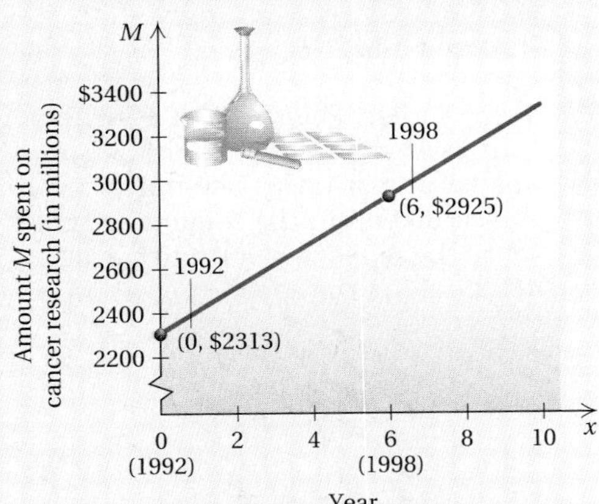

Source: The New England Journal of Medicine

Determine whether the graphs of the equations are parallel, perpendicular, or neither.

17. $2x + y = 8,$
$2x + y = 4$

18. $2x + 5y = 2,$
$y = 2x + 4$

19. $x + 2y = 8,$
$-2x + y = 8$

Answers

1. ______
2. ______
3. ______
4. ______
5. ______
6. ______
7. ______
8. ______
9. ______
10. ______
11. ______
12. ______
13. ______
14. ______
15. ______
16. a) ______
b) ______
c) ______
17. ______
18. ______
19. ______

Answers

20. __________

21. __________

22. __________

23. __________

24. __________

25. __________

26. __________

27. __________

28. __________

29. __________

30. __________

31. __________

32. __________

33. __________

34. __________

35. __________

Determine whether the given point is a solution of the inequality $3y - 2x < -2$.

20. (0, 0)

21. (−4, −10)

Graph on a plane.

22. $y > x - 1$

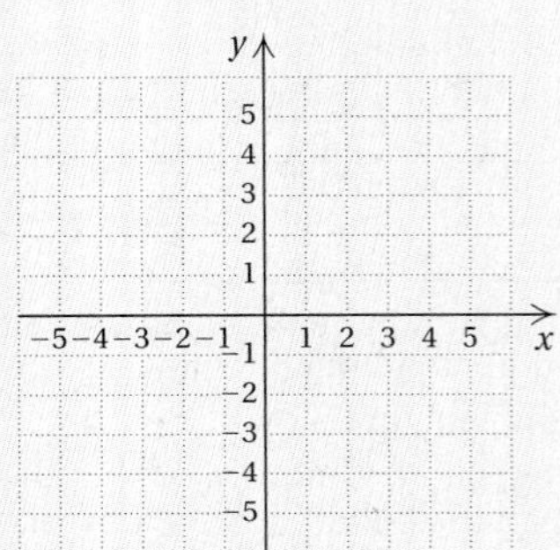

23. $2x - y \leq 4$

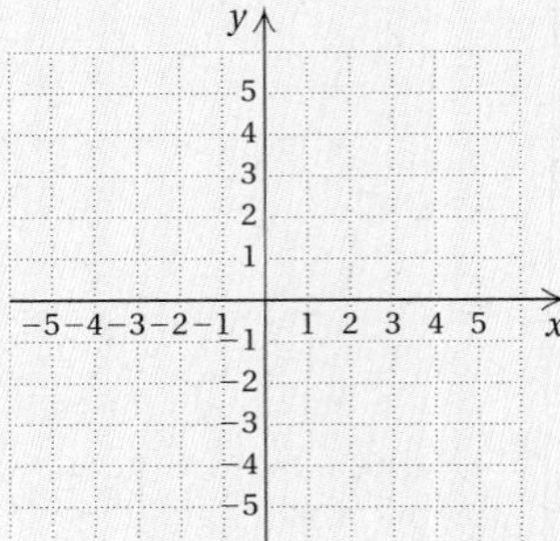

Find an equation of variation in which y varies directly as x and the following are true.

24. $y = 6$ when $x = 3$

25. $y = 1.5$ when $x = 3$

Find an equation of variation in which y varies inversely as x and the following are true.

26. $y = 6$ when $x = 3$

27. $y = 11$ when $x = 2$

Solve.

28. The distance d traveled by a train varies directly as the time t that it travels. The train travels 60 km in $\frac{1}{2}$ hr. How far will it travel in 2 hr?

29. It takes 3 hr for 2 concrete mixers to mix a fixed amount of concrete. The number of hours varies inversely as the number of concrete mixers used. How long would it take 5 concrete mixers to do the same job?

Skill Maintenance

30. *Train Speeds.* The speed of a freight train is 15 mph slower than the speed of a passenger train. The freight train travels 360 mi in the same time that it takes the passenger train to travel 420 mi. Find the speed of each train.

31. Compute: $\dfrac{3^2 - 2^3}{2^2 + 3 - 12 \div 2}$.

Solve.

32. $\dfrac{x^2}{x + 10} = \dfrac{100}{x + 10}$

33. $a^2 + 3a - 28 = 0$

Synthesis

34. Find the value of k such that $3x + 7y = 14$ and $ky - 7x = -3$ are perpendicular.

35. Find the slope–intercept equation of the line that contains the point $(-4, 1)$ and has the same slope as the line $2x - 3y = -6$.

14

Systems of Equations

An Application

In most areas of the United States, gas stations offer three grades of gasoline, indicated by octane ratings on the pumps, such as 87, 89, and 93. When a tanker delivers gas, it brings only two grades of gasoline, the highest and lowest, filling two large underground tanks. If you purchase the middle grade, the pump's computer mixes the other two grades appropriately. How much 87-octane gas and 93-octane gas should be blended in order to make 18 gal of 89-octane gas?

This problem appears as Exercise 37 in Exercise Set 14.4.

The Mathematics

We let x = the number of gallons of the 87-octane gasoline and y = the number of gallons of the 93-octane gasoline. Then we can translate the problem to this pair of equations:

$$\left.\begin{aligned} x + y &= 18, \\ 87x + 93y &= 89 \cdot 114. \end{aligned}\right\}$$

This is a *system of equations*.

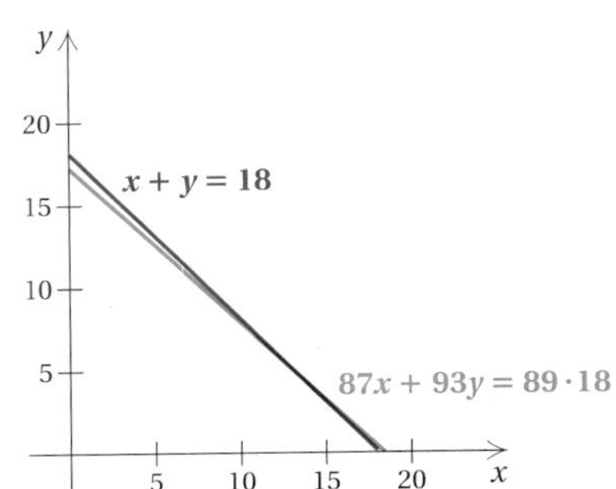

World Wide Web For more information, visit us at www.mathmax.com

Introduction

We now consider how the graphs of two linear equations might intersect. Such a point of intersection is a solution of what is called a *system of equations*. Many applications and problems involve two facts about two quantities and are easier to solve by translating to a system of two equations in two variables. Systems of equations have extensive applications in many fields such as psychology, sociology, business, education, engineering, and science.

Pretest: Chapter 14

1. Determine whether the ordered pair $(-1, 1)$ is a solution of the system of equations

$2x + y = -1,$

$3x - 2y = -5.$

2. Solve this system by graphing.

$2x = y + 1,$

$2x - y = 5$

Solve by the substitution method.

3. $x + y = 7,$
$x = 2y + 1$

4. $2x - 3y = 7,$
$x + y = 1$

Solve by the elimination method.

5. $2x - y = 1,$
$2x + y = 2$

6. $2x - 3y = -4,$
$3x - 4y = -7$

7. $\frac{3}{5}x - \frac{1}{4}y = 4,$
$\frac{1}{5}x + \frac{3}{4}y = 8$

8. Find two numbers whose sum is 74 and whose difference is 26.

9. Two angles are complementary. (Complementary angles are angles whose sum is 90°.) One angle is 15° more than twice the other. Find the angles.

10. A train leaves a station and travels north at 96 mph. Two hours later, a second train leaves on a parallel track and travels north at 120 mph. When will it overtake the first train?

Objectives for Retesting

The objectives to be tested in addition to the material in this chapter are as follows.

[9.3a]	Find the intercepts of a linear equation, and graph using intercepts.
[10.1d, e, f]	Use the product and quotient rules to multiply and divide exponential expressions with like bases, and express exponential expressions involving negative exponents with positive exponents.
[12.1c]	Simplify rational expressions by factoring the numerator and the denominator and removing factors of 1.
[12.5a]	Subtract rational expressions.

14.1 Systems of Equations in Two Variables

a Systems of Equations and Solutions

Many problems can be solved more easily by translating to two equations in two variables. The following is such a **system of equations:**

$$x + y = 8,$$
$$2x - y = 1.$$

> A **solution** of a system of two equations is an ordered pair that makes both equations true.

Objectives

a. Determine whether an ordered pair is a solution of a system of equations.

b. Solve systems of two linear equations in two variables by graphing.

For Extra Help

TAPE 26 | MAC WIN | CD-ROM

Look at the graphs shown at right. Recall that a graph of an equation is a drawing that represents its solution set. Each point on the graph corresponds to a solution of that equation. Which points (ordered pairs) are solutions of *both* equations?

The graph shows that there is only one. It is the point P where the graphs cross. This point looks as if its coordinates are (3, 5). We check to see if (3, 5) is a solution of *both* equations, substituting 3 for x and 5 for y.

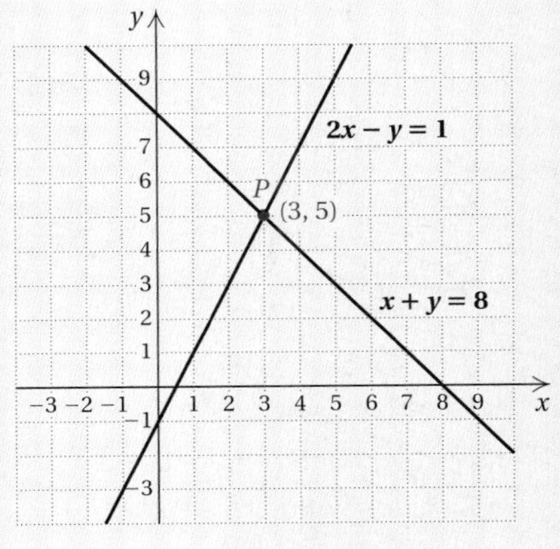

CHECK:

$$\begin{array}{c|c} x + y & 8 \\ \hline 3 + 5 & 8 \\ 8 & \end{array} \quad \text{TRUE}$$

$$\begin{array}{c|c} 2x - y & 1 \\ \hline 2 \cdot 3 - 5 & 1 \\ 6 - 5 & \\ 1 & \end{array} \quad \text{TRUE}$$

There is just one solution of the system of equations. It is (3, 5). In other words, $x = 3$ and $y = 5$.

Example 1 Determine whether (1, 2) is a solution of the system

$$y = x + 1,$$
$$2x + y = 4.$$

We check by substituting alphabetically 1 for x and 2 for y.

CHECK:

$$\begin{array}{c|c} y & x + 1 \\ \hline 2 & 1 + 1 \\ & 2 \end{array} \quad \text{TRUE}$$

$$\begin{array}{c|c} 2x + y & 4 \\ \hline 2 \cdot 1 + 2 & 4 \\ 2 + 2 & \\ 4 & \end{array} \quad \text{TRUE}$$

This checks, so (1, 2) is a solution of the system.

Example 2 Determine whether (−3, 2) is a solution of the system

$$p + q = -1,$$
$$q + 3p = 4.$$

We check by substituting alphabetically −3 for p and 2 for q.

CHECK:

$$\begin{array}{c|c} p + q & -1 \\ \hline -3 + 2 & -1 \\ -1 & \end{array} \quad \text{TRUE}$$

$$\begin{array}{c|c} q + 3p & 4 \\ \hline 2 + 3(-3) & 4 \\ 2 - 9 & \\ -7 & \end{array} \quad \text{FALSE}$$

The point (−3, 2) is not a solution of $q + 3p = 4$. Thus it is not a solution of the system.

Determine whether the given ordered pair is a solution of the system of equations.

1. $(2, -3)$; $x = 2y + 8,$
$2x + y = 1$

2. $(20, 40)$; $a = \frac{1}{2}b,$
$b - a = 60$

3. Solve this system by graphing:

$2x + y = 1,$
$x = 2y + 8.$

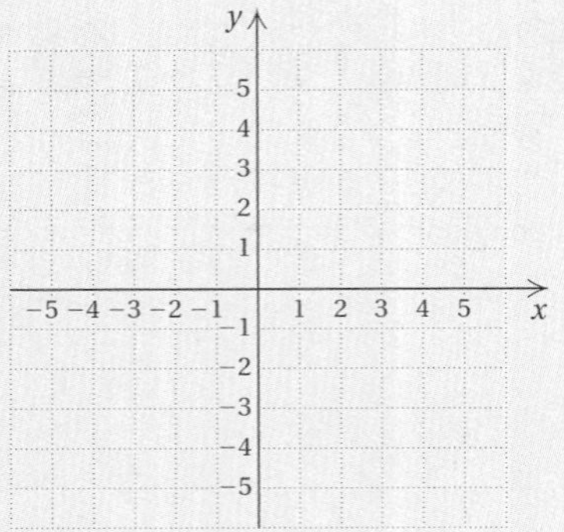

Calculator Spotlight

Checking with the TABLE feature. We can use the TABLE feature to check a possible solution of a system. Consider the system in Example 3. We enter each equation in the "$y =$" form:

$y_1 = 6 - x,$
$y_2 = x - 2.$

We set TblStart = 1 and ΔTbl = 1 and obtain the following table.

X	Y1	Y2
1	5	−1
2	4	0
3	3	1
4	2	2
5	1	3
6	0	4
7	−1	5
X = 4		

Note that when $x = 4$, $y_1 = 2$ and $y_2 = 2$.

Answers on page A-43

Example 2 illustrates that an ordered pair may be a solution of one equation but *not both*. If that is the case, it is *not* a solution of the system.

Do Exercises 1 and 2.

b Graphing Systems of Equations

Recall that the **graph** of an equation is a drawing that represents its solution set. If the graph of an equation is a line, then every point on the line corresponds to an ordered pair that is a solution of the equation. If we graph a **system** of two linear equations, we graph both equations and find the coordinates of the points of intersection, if any exist.

Example 3 Solve this system of equations by graphing:

$x + y = 6,$
$x = y + 2.$

We graph the equations using any of the methods studied in Chapter 9. Point P with coordinates (4, 2) looks as if it is the solution.

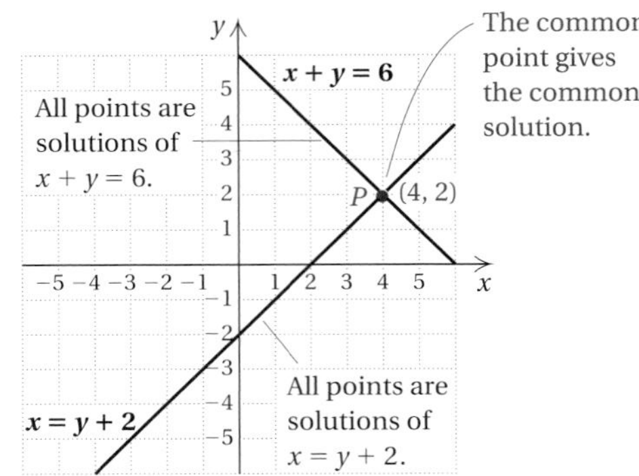

We check the pair as follows.

CHECK:

$$\begin{array}{c|c} x + y & 6 \\ \hline 4 + 2 & ? \ 6 \\ 6 & \end{array} \quad \text{TRUE} \qquad \begin{array}{c|c} x & y + 2 \\ \hline 4 & ? \ 2 + 2 \\ & 4 \end{array} \quad \text{TRUE}$$

The solution is (4, 2).

Do Exercise 3.

Example 4 Solve this system of equations by graphing:

$x = 2,$
$y = -3.$

The graph of $x = 2$ is a vertical line, and the graph of $y = -3$ is a horizontal line. They intersect at the point $(2, -3)$.

The solution is $(2, -3)$.

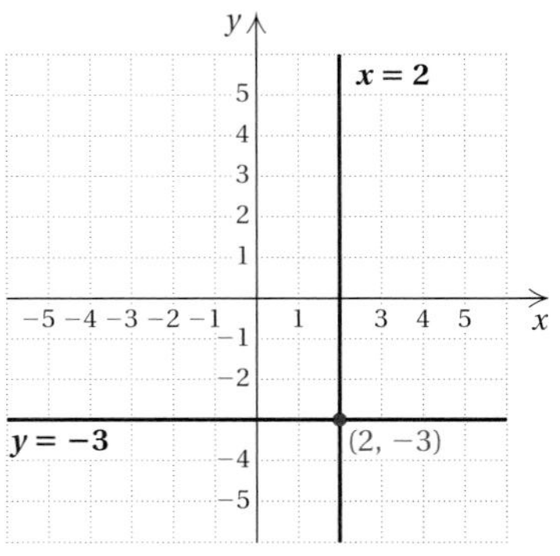

Do Exercise 4 on the following page.

Sometimes the equations in a system have graphs that are parallel lines.

Example 5 Solve this system of equations by graphing:

$$y = 3x + 4,$$
$$y = 3x - 3.$$

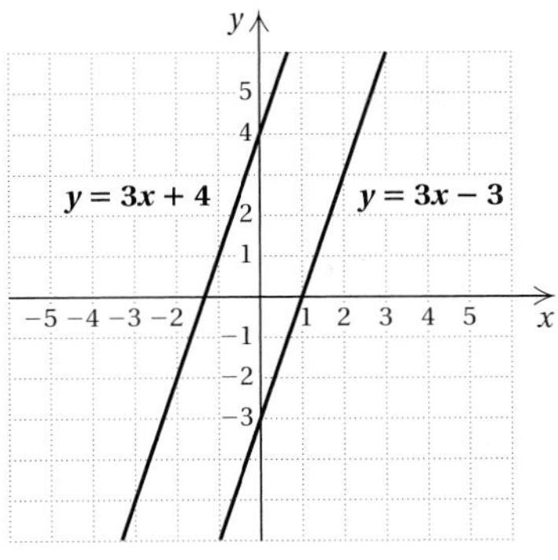

We graph the equations, again using any of the methods studied in Chapter 9. The lines have the same slope, 3, and different y-intercepts, (0, 4) and (0, −3), so they are parallel.

There is no point at which the lines cross, so the system has no solution. The solution set is the empty set, denoted $\varnothing$, or { }.

Do Exercise 5.

Sometimes the equations in a system have the same graph.

Example 6 Solve this system of equations by graphing:

$$2x + 3y = 6,$$
$$-8x - 12y = -24.$$

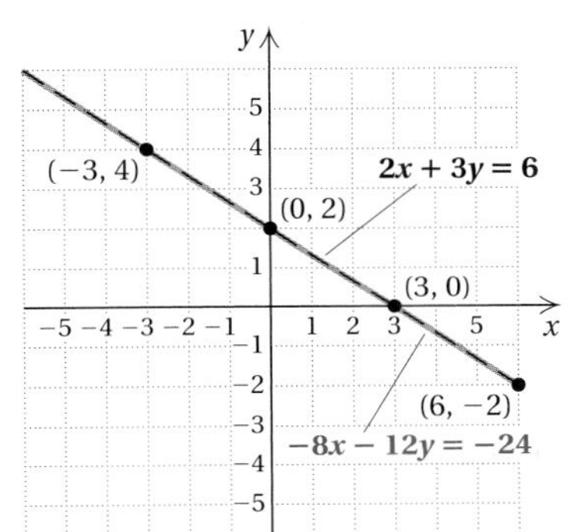

We graph the equations and see that the graphs are the same. Thus any solution of one of the equations is a solution of the other. Each equation has an infinite number of solutions, some of which are indicated on the graph.

We check one such solution, (0, 2): the y-intercept of each equation.

CHECK:

$2x + 3y = 6$			$-8x - 12y = -24$		
$2(0) + 3(2)$	?	6	$-8(0) - 12(2)$	?	-24
$0 + 6$			$0 - 24$		
6		TRUE	-24		TRUE

We leave it to the student to check that (−3, 4) is also a solution of the system. If (0, 2) and (−3, 4) are solutions, then all points on the line containing them are solutions. The system has an infinite number of solutions.

Do Exercise 6.

When we graph a system of two equations in two variables, we obtain one of the following three results.

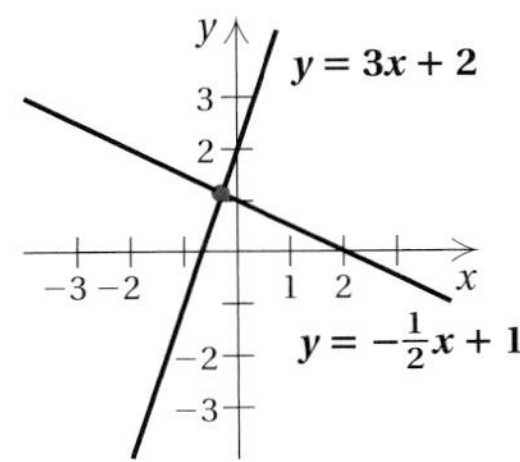

One solution.
Graphs intersect.

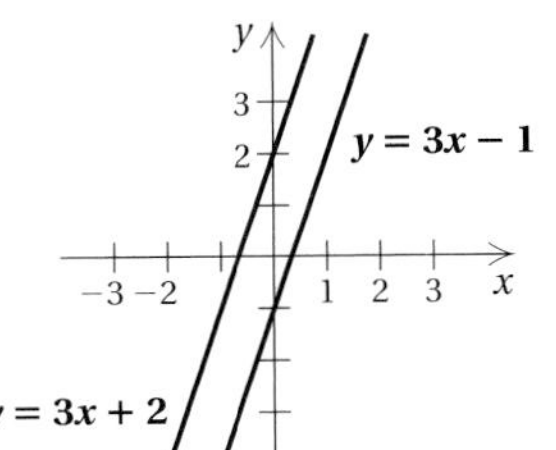

No solution.
Graphs are parallel.

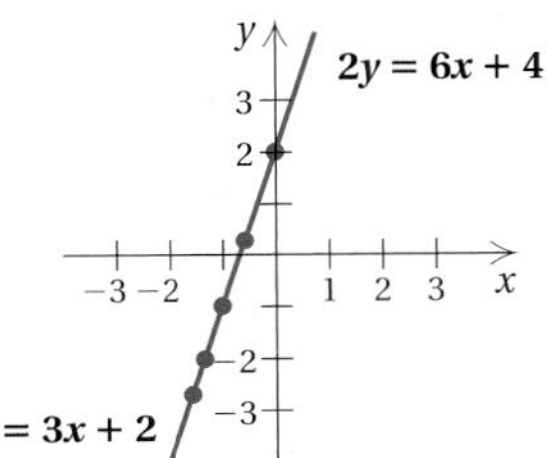

Infinitely many solutions.
Equations have the same graph.

4. Solve this system by graphing:

$$x = -4,$$
$$y = 3.$$

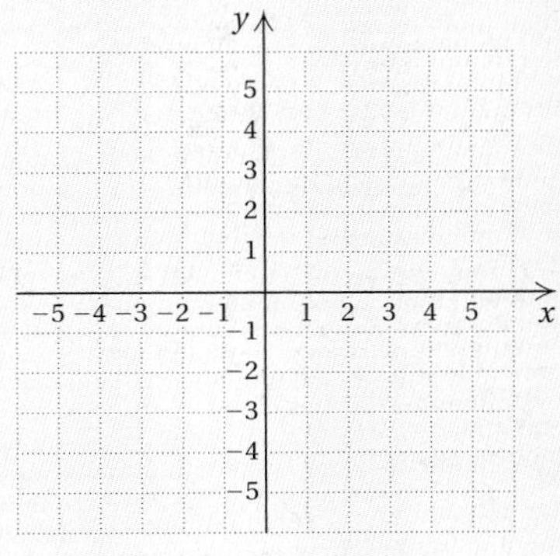

5. Solve this system by graphing:

$$y + 4 = x,$$
$$x - y = -2.$$

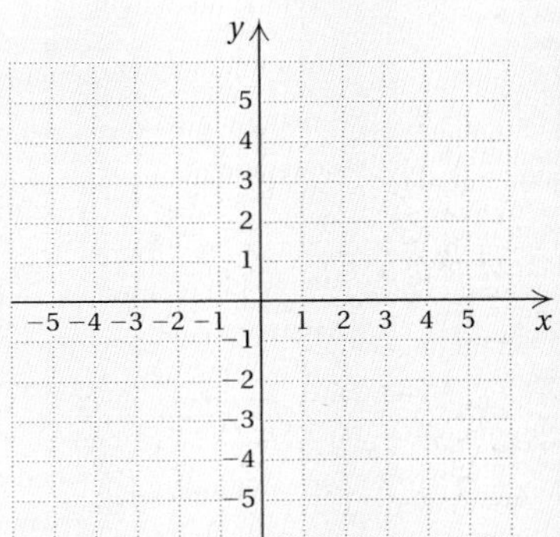

6. Solve this system by graphing:

$$2x + y = 4,$$
$$-6x - 3y = -12.$$

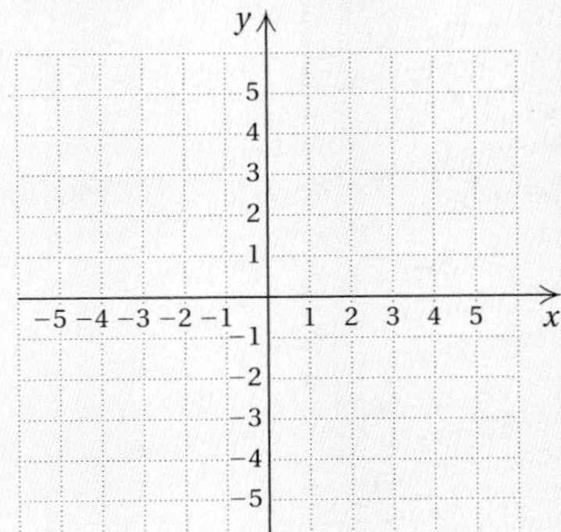

Answers on page A-43

7. a) Solve $2x - 1 = 8 - x$ algebraically.

b) Solve $2x - 1 = 8 - x$ graphically using method 1.

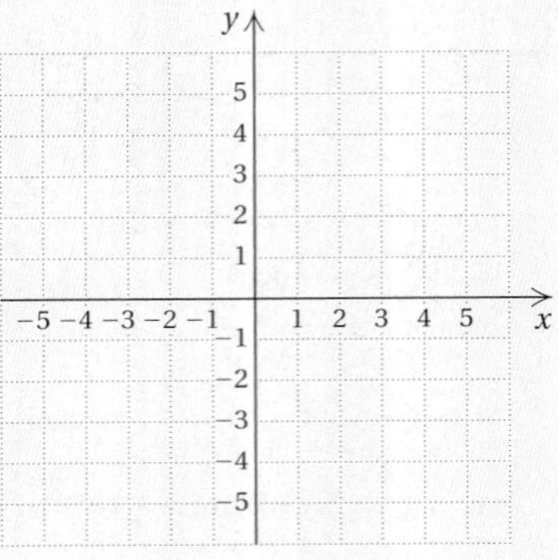

c) Compare your answers to parts (a) and (b).

8. a) Solve $2x - 1 = 8 - x$ graphically using method 2.

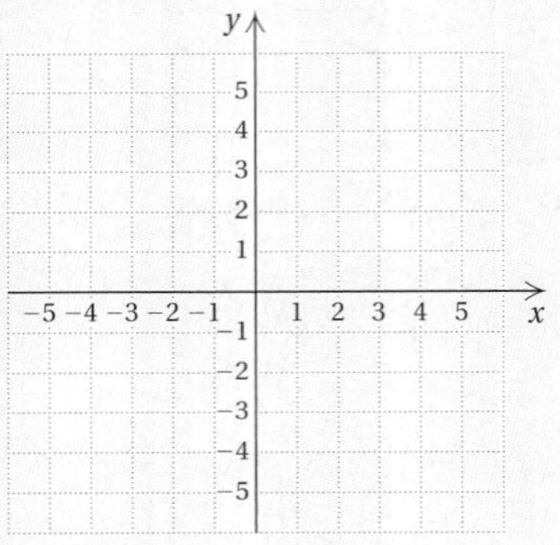

b) Compare your answers to 7(a), 7(b), and 8(a).

Answers on page A-43

AG Algebraic–Graphical Connection

To bring together the concepts of Chapters 7–14, let's take an algebraic–graphical look at equation solving. Such interpretation is useful when using a graphing calculator or computer graphing software.

Consider the equation $6 - x = x - 2$. Let's solve it algebraically as we did in Chapter 8:

$$6 - x = x - 2$$

$$6 = 2x - 2 \quad \text{Adding } x$$

$$8 = 2x \quad \text{Adding 2}$$

$$4 = x. \quad \text{Dividing by 2}$$

Can we also solve the equation graphically? We can, as we see in the following two methods.

METHOD 1 Solve $6 - x = x - 2$ graphically.

We let $y = 6 - x$ and $y = x - 2$. Graphing the system of equations as we did in Example 3 gives us the following. The point of intersection is $(4, 2)$. Note that the x-coordinate of the intersection is 4. This value for x is *also* the solution of the equation $6 - x = x - 2$.

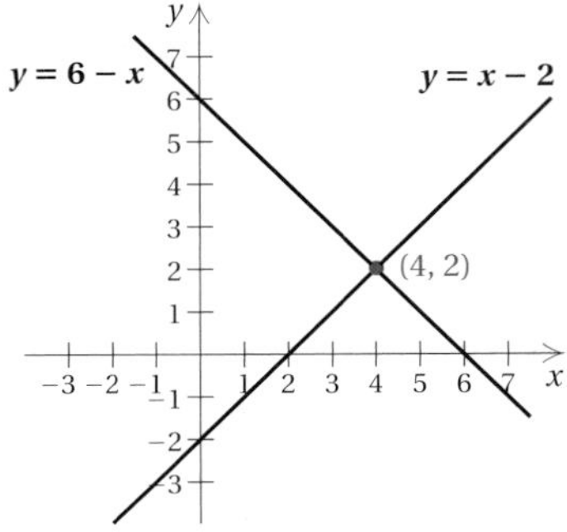

Do Exercise 7.

METHOD 2 Solve $6 - x = x - 2$ graphically.

Adding x and -6 on both sides, we obtain the form $0 = 2x - 8$. In this case, we let $y = 0$ and $y = 2x - 8$. Since $y = 0$ is the x-axis, we need only graph $y = 2x - 8$ and see where it crosses the x-axis. Note that the x-intercept of $y = 2x - 8$ is $(4, 0)$, or just 4. This x-value is *also* the solution of the equation $6 - x = x - 2$.

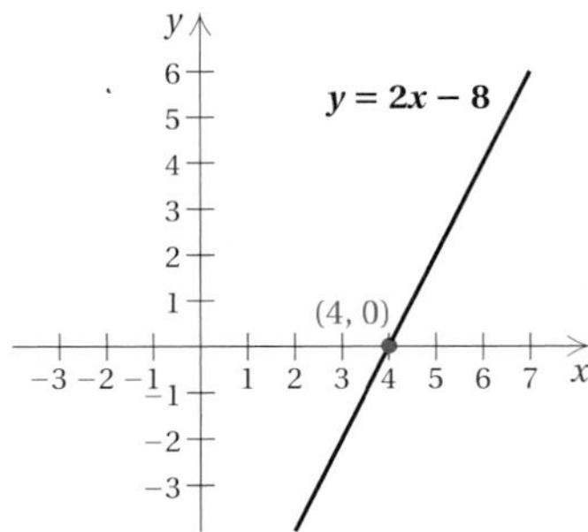

Do Exercise 8.

Let's compare the two methods. Using method 1, we graph two equations. The solution of the original equation is the x-coordinate of the point of intersection. Using method 2, we find that the solution of the original equation is the x-intercept of the graph.

Exercise Set 14.1

a Determine whether the given ordered pair is a solution of the system of equations. Use alphabetical order of the variables.

1. $(1, 5)$; $5x - 2y = -5,$
$3x - 7y = -32$

2. $(3, 2)$; $2x + 3y = 12,$
$x - 4y = -5$

3. $(4, 2)$; $3b - 2a = -2,$
$b + 2a = 8$

4. $(6, -6)$; $t + 2s = 6,$
$t - s = -12$

5. $(15, 20)$; $3x - 2y = 5,$
$6x - 5y = -10$

6. $(-1, -5)$; $4r + s = -9,$
$3r = 2 + s$

7. $(-1, 1)$; $x = -1,$
$x - y = -2$

8. $(-3, 4)$; $2x = -y - 2,$
$y = -4$

9. $(18, 3)$; $y = \frac{1}{6}x,$
$2x - y = 33$

10. $(-3, 1)$; $y = -\frac{1}{3}x,$
$3y = -5x - 12$

b Solve the system of equations by graphing.

11. $x - y = 2,$
$x + y = 6$

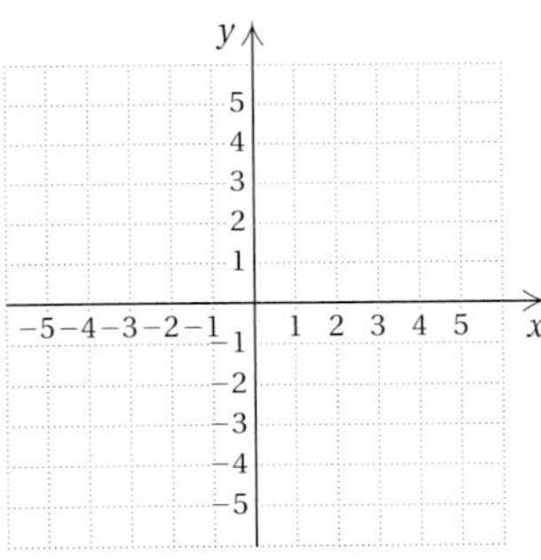

12. $x + y = 3,$
$x - y = 1$

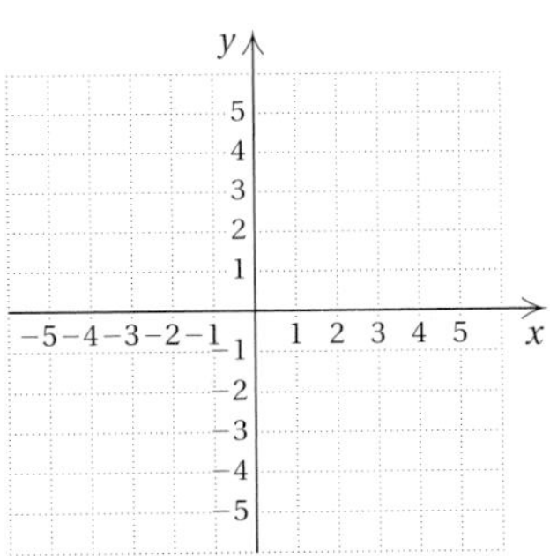

13. $8x - y = 29,$
$2x + y = 11$

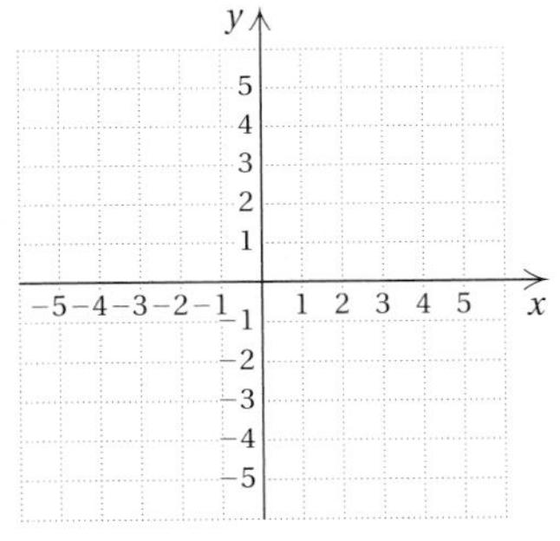

14. $4x - y = 10,$
$3x + 5y = 19$

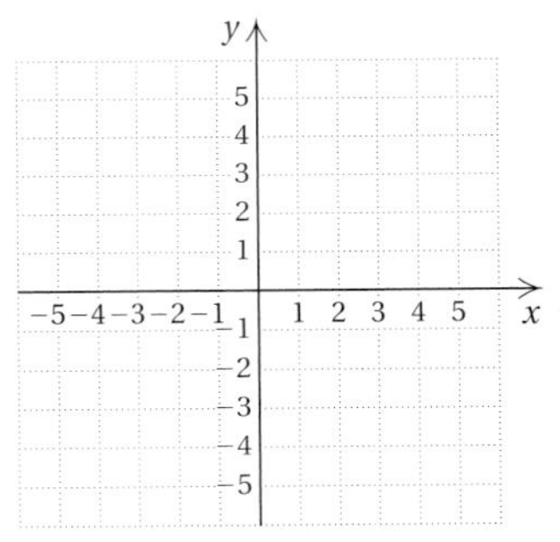

15. $u = v,$
$4u = 2v - 6$

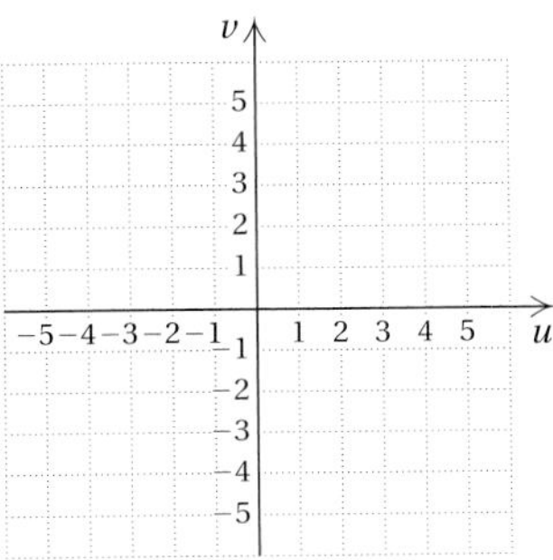

16. 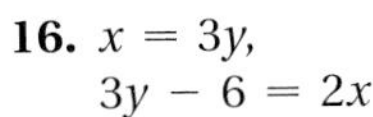
$x = 3y,$
$3y - 6 = 2x$

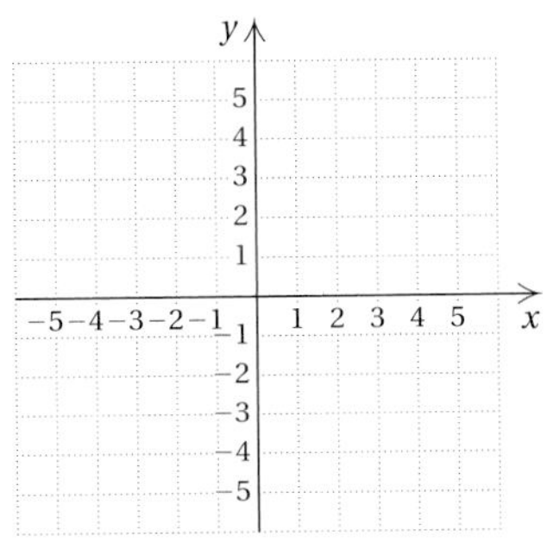

17. $x = -y,$
$x + y = 4$

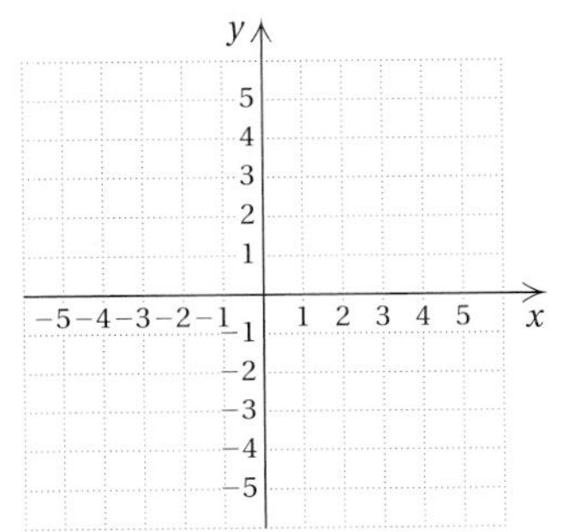

18. 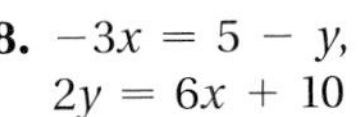
$-3x = 5 - y,$
$2y = 6x + 10$

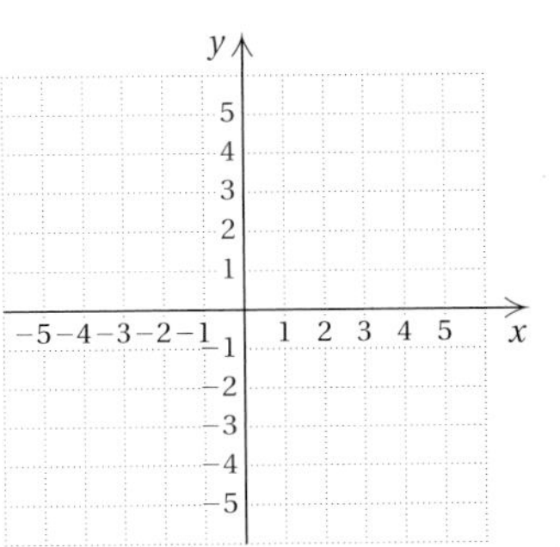

19. $a = \frac{1}{2}b + 1,$
$a - 2b = -2$

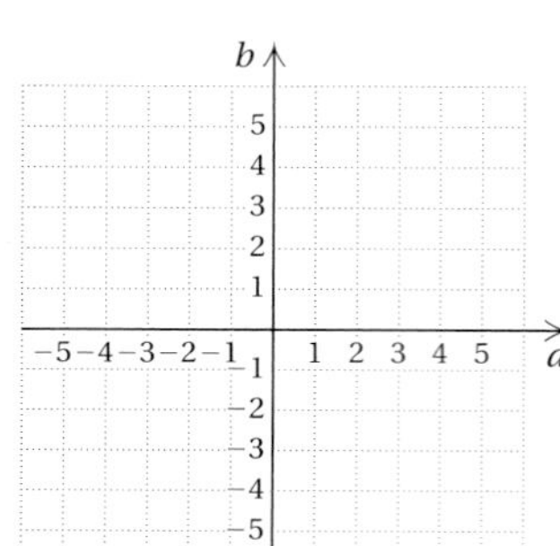

20. $x = \frac{1}{3}y + 2,$
$-2x - y = 1$

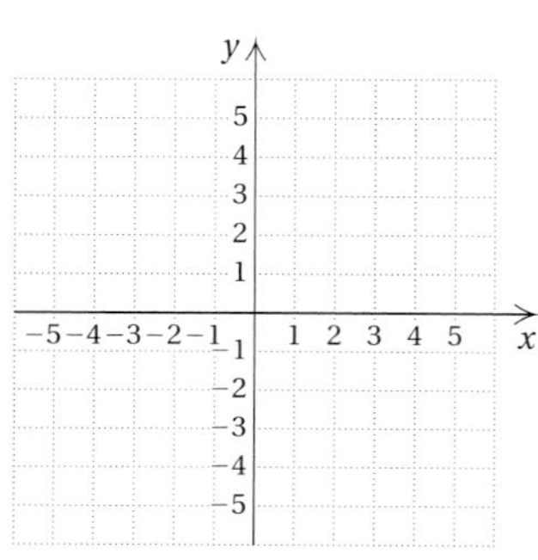

21. $y - 2x = 0,$
$y = 6x - 2$

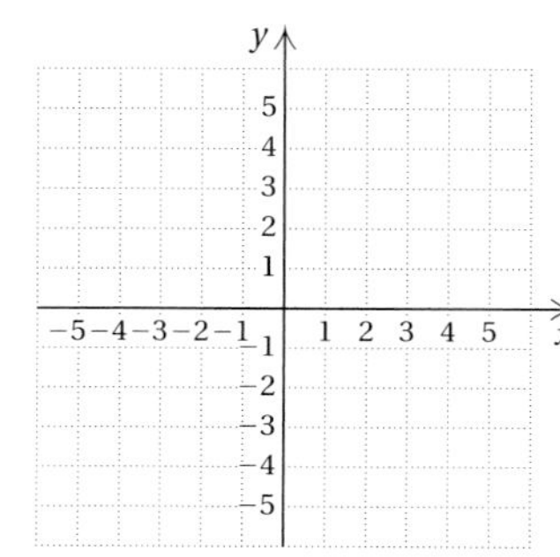

22. $y = 3x,$
$y = -3x + 2$

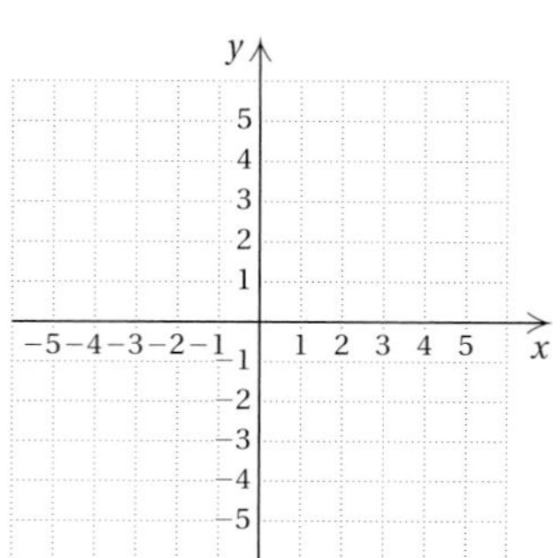

23. $x + y = 9,$
$3x + 3y = 27$

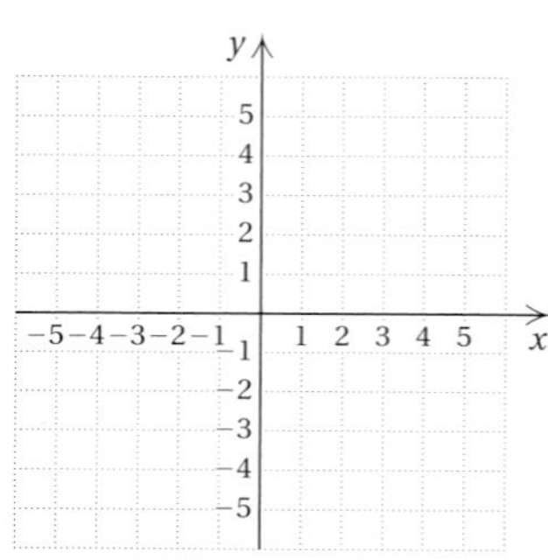

24. $x + y = 4,$
$x + y = -4$

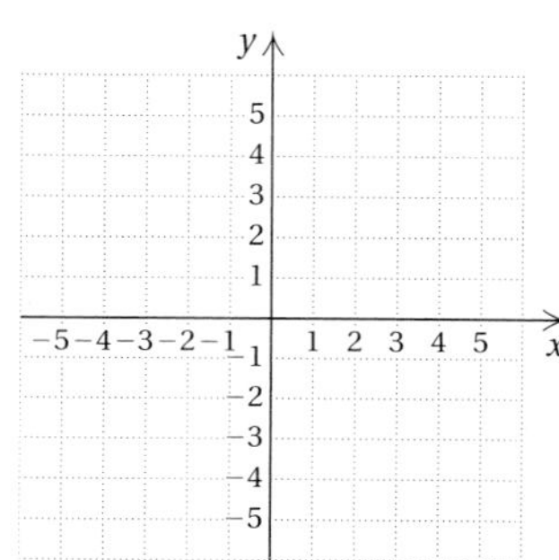

25. $x = 5,$
$y = -3$

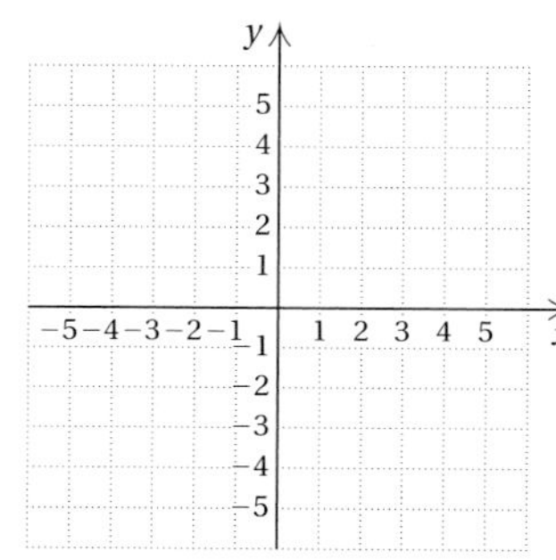

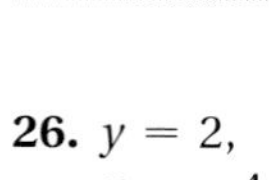

26. $y = 2,$
$y = -4$

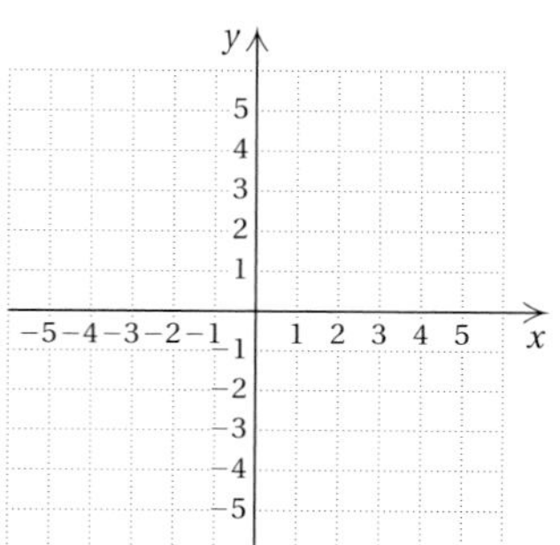

Skill Maintenance

27. Multiply: $(9x^{-5})(12x^{-8})$. [10.1d, f]

28. Divide: $\frac{9x^{-5}}{3x^{-8}}$. [10.1e, f]

Simplify.

29. $\frac{1}{x} - \frac{1}{x^2} + \frac{1}{x + 1}$ [12.5b]

30. $\frac{3 - x}{x - 2} - \frac{x - 7}{2 - x}$ [12.5a]

31. $\frac{x + 2}{x - 4} - \frac{x + 1}{x + 4}$ [12.5a]

32. $\frac{2x^2 - x - 15}{x^2 - 9}$ [12.1c]

Classify the polynomial as a monomial, binomial, trinomial, or none of these. [10.3i]

33. $5x^2 - 3x + 7$

34. $4x^3 - 2x^2$

35. $1.8x^5$

36. $x^3 + 2x^2 - 3x + 1$

Synthesis

37. ◆ Suppose you have shown that the solution of the equation $3x - 1 = 9 - 2x$ is 2. How can this result be used to determine where the graphs of $y = 3x - 1$ and $y = 9 - 2x$ intersect?

38. ◆ Graph this system of equations. What happens when you try to determine a solution from the graph?

$x - 2y = 6,$
$3x + 2y = 4$

39. The solution of the following system is $(2, -3)$. Find A and B.

$Ax - 3y = 13,$
$x - By = 8$

40. Find an equation to go with $5x + 2y = 11$ such that the solution of the system is $(3, -2)$. Answers may vary.

41. Find a system of equations with $(6, -2)$ as a solution. Answers may vary.

42.–49. Use the TABLE feature to check your answers to Exercises 11–18.

14.2 The Substitution Method

Consider the following system of equations:

$$3x + 7y = 5,$$
$$6x - 7y = 1.$$

Suppose we try to solve this system graphically. We obtain this graph.

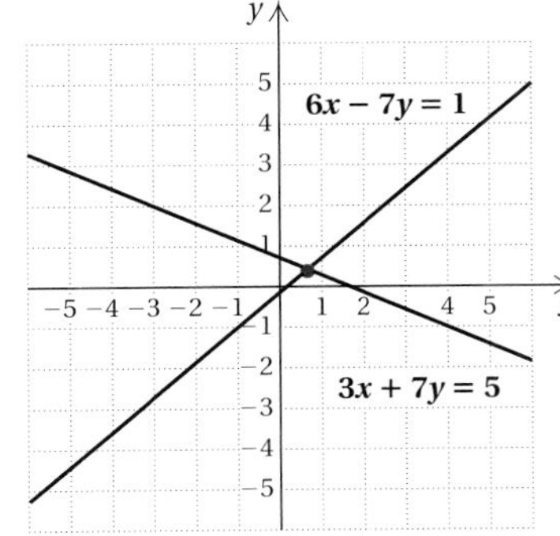

What is the solution? It is rather difficult to tell exactly. It would appear that fractions are involved. It turns out that the solution is $\left(\frac{2}{3}, \frac{3}{7}\right)$. We need techniques involving algebra to determine the solution exactly. Graphing helps us picture the solution of a system of equations, but solving by graphing, though practical in many applications, is not always fast or accurate in cases where solutions are not integers. We now learn other methods using algebra. Because they use algebra, they are called **algebraic**.

Objectives

a Solve a system of two equations in two variables by the substitution method when one of the equations has a variable alone on one side.

b Solve a system of two equations in two variables by the substitution method when neither equation has a variable alone on one side.

c Solve applied problems by translating to a system of two equations and then solving using the substitution method.

For Extra Help

TAPE 26 MAC WIN CD-ROM

a Solving by the Substitution Method

One nongraphical method for solving systems is known as the **substitution method.** In Example 1, we use the substitution method to solve a system we graphed in Example 3 of Section 14.1.

Example 1 Solve the system

$$x + y = 6, \quad (1)$$
$$x = y + 2. \quad (2)$$

Equation (2) says that x and $y + 2$ name the same thing. Thus in equation (1), we can substitute $y + 2$ for x:

$x + y = 6$ Equation (1)

$(y + 2) + y = 6.$ Substituting $y + 2$ for x

This last equation has only one variable. We solve it:

$y + 2 + y = 6$ Removing parentheses

$2y + 2 = 6$ Collecting like terms

$2y + 2 - 2 = 6 - 2$ Subtracting 2 on both sides

$2y = 4$ Simplifying

$\frac{2y}{2} = \frac{4}{2}$ Dividing by 2

$y = 2.$ Simplifying

We have found the y-value of the solution. To find the x-value, we return to the original pair of equations. Substituting into either equation will give us the x-value.

1. Solve by the substitution method. Do not graph.

$$x + y = 5,$$
$$x = y + 1$$

We choose equation (2) because it has x alone on one side:

$$\begin{aligned} x &= y + 2 && \text{Equation (2)} \\ &= 2 + 2 && \text{Substituting 2 for } y \\ &= 4. \end{aligned}$$

The ordered pair (4, 2) may be a solution. We check.

CHECK:

$x + y = 6$		
$4 + 2$? 6		
6		TRUE

$x = y + 2$		
4 ? $2 + 2$		
	4	TRUE

Since (4, 2) checks, we have the solution. We could also express the answer as $x = 4$, $y = 2$.

Note in Example 1 that substituting 2 for y in equation (1) will also give us the x-value of the solution:

$$\begin{aligned} x + y &= 6 \\ x + 2 &= 6 \\ x &= 4. \end{aligned}$$

Note also that we are using alphabetical order in listing the coordinates in an ordered pair. That is, since x precedes y, we list 4 before 2 in the pair (4, 2).

Do Exercise 1.

Calculator Spotlight

Use the TABLE feature to check the possible solutions of the systems in Example 1 and Margin Exercise 1.

Example 2 Solve the system

$$t = 1 - 3s, \quad (1)$$
$$s - t = 11. \quad (2)$$

We substitute $1 - 3s$ for t in equation (2):

$$\begin{aligned} s - t &= 11 && \text{Equation (2)} \\ s - (1 - 3s) &= 11. && \text{Substituting } 1 - 3s \text{ for } t \end{aligned}$$

Remember to use parentheses when you substitute.

Now we solve for s:

$$\begin{aligned} s - 1 + 3s &= 11 && \text{Removing parentheses} \\ 4s - 1 &= 11 && \text{Collecting like terms} \\ 4s &= 12 && \text{Adding 1} \\ s &= 3. && \text{Dividing by 4} \end{aligned}$$

Next, we substitute 3 for s in equation (1) of the original system:

$$\begin{aligned} t &= 1 - 3s && \text{Equation (1)} \\ &= 1 - 3 \cdot 3 && \text{Substituting 3 for } s \\ &= -8. \end{aligned}$$

The pair $(3, -8)$ checks and is the solution. Remember: We list the answer in alphabetical order, (s, t). That is, since s comes before t in the alphabet, 3 is listed first and -8 second.

Do Exercise 2.

2. Solve by the substitution method:

$$a - b = 4,$$
$$b = 2 - a.$$

Answers on page A-43

b Solving for the Variable First

Sometimes neither equation of a pair has a variable alone on one side. Then we solve one equation for one of the variables and proceed as before, substituting into the *other* equation. If possible, we solve in either equation for a variable that has a coefficient of 1.

Example 3 Solve the system

$$x - 2y = 6, \quad (1)$$
$$3x + 2y = 4. \quad (2)$$

We solve one equation for one variable. Since the coefficient of x is 1 in equation (1), it is easier to solve that equation for x:

$x - 2y = 6$ Equation (1)

$x = 6 + 2y.$ Adding $2y$ (3)

We substitute $6 + 2y$ for x in equation (2) of the original pair and solve for y:

$3x + 2y = 4$	Equation (2)
$3(6 + 2y) + 2y = 4$	Substituting $6 + 2y$ for x
$18 + 6y + 2y = 4$	Removing parentheses
$18 + 8y = 4$	Collecting like terms
$8y = -14$	Subtracting 18
$y = \frac{-14}{8}$, or $-\frac{7}{4}$.	Dividing by 8

To find x, we go back to either of the original equations (1) or (2) or to equation (3), which we solved for x. It is generally easier to use an equation like equation (3) where we have solved for a specific variable. We substitute $-\frac{7}{4}$ for y in equation (3) and compute x:

$x = 6 + 2y$	Equation (3)
$= 6 + 2\left(-\frac{7}{4}\right)$	Substituting $-\frac{7}{4}$ for y
$= 6 - \frac{7}{2} = \frac{5}{2}.$	

We check the ordered pair $\left(\frac{5}{2}, -\frac{7}{4}\right)$.

CHECK:

$x - 2y = 6$		$3x + 2y = 4$	
$\frac{5}{2} - 2\left(-\frac{7}{4}\right) \ ? \ 6$		$3 \cdot \frac{5}{2} + 2\left(-\frac{7}{4}\right) \ ? \ 4$	
$\frac{5}{2} + \frac{7}{2}$		$\frac{15}{2} - \frac{7}{2}$	
$\frac{12}{2}$		$\frac{8}{2}$	
6	TRUE	4	TRUE

Since $\left(\frac{5}{2}, -\frac{7}{4}\right)$ checks, it is the solution.

This solution would have been difficult to find graphically because it involves fractions.

Do Exercise 3.

3. Solve:

$$x - 2y = 8,$$
$$2x + y = 8.$$

Calculator Spotlight

Solving Systems. Graphers can be used to solve systems of equations, provided each equation has been solved first for y. Consider the system in Example 3. We put each equation in the "$y =$" form by solving for y:

$y_1 = (1/2)x - 3,$ **(1)**

$y_2 = (-3/2)x + 2.$ **(2)**

We then enter the equations.

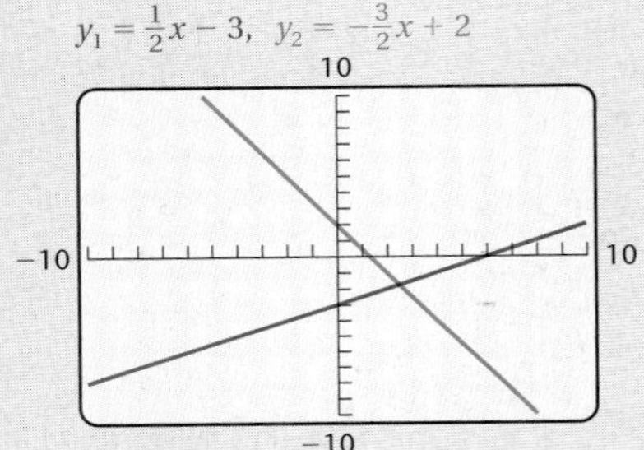

To find the solution, we use the CALC-INTERSECT feature.

$y_1 = \frac{1}{2}x - 3,\ y_2 = -\frac{3}{2}x + 2$

Intersection
X=2.5   Y=−1.75

The solution is listed at the bottom of the screen.

Exercises

Use a grapher to solve each of the following systems.

1. Margin Exercise 3

2. $6x - 7y = 1,$
   $3x + 7y = 5$

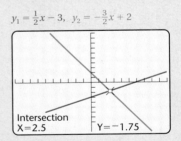

Answers on page A-43

c Solving Applied Problems

Now let's solve an applied problem using systems of equations and the substitution method.

4. *Perimeter of High School Court.* The perimeter of a standard high school basketball court is 268 ft. The length is 34 ft longer than the width. Find the dimensions of the court.

Answer on page A-43

Example 4 *Perimeter of NBA Court.* The perimeter of an NBA-sized basketball court is 288 ft. The length is 44 ft longer than the width. (***Source:*** National Basketball Association) Find the dimensions of the court.

1. **Familiarize.** We make a drawing and label it, calling the length l and the width w.

2. **Translate.** The perimeter of the rectangle is $2l + 2w$. We translate the first statement.

$$\underbrace{\text{The perimeter}}_{2l + 2w} \quad \underbrace{\text{is}}_{=} \quad \underbrace{288 \text{ ft.}}_{288}$$

We translate the second statement:

$$\underbrace{\text{The length}}_{l} \quad \underbrace{\text{is}}_{=} \quad \underbrace{\text{44 ft longer than the width.}}_{44 + w}$$

We now have a system of equations:

$$2l + 2w = 288, \qquad (1)$$
$$l = 44 + w. \qquad (2)$$

3. **Solve.** We solve the system. To begin, we substitute $44 + w$ for l in the first equation and solve:

$$2(44 + w) + 2w = 288 \qquad \text{Substituting } 44 + w \text{ for } l \text{ in equation (1)}$$
$$2 \cdot 44 + 2 \cdot w + 2w = 288 \qquad \text{Using a distributive law}$$
$$4w + 88 = 288 \qquad \text{Collecting like terms}$$
$$4w + 88 - 88 = 288 - 88 \qquad \text{Subtracting 88}$$
$$4w = 200$$
$$w = 50. \qquad \text{Dividing by 4}$$

We go back to the original equations and substitute 50 for w:

$$l = 44 + w = 44 + 50 = 94. \qquad \text{Substituting in equation (2)}$$

4. **Check.** If the width is 50 ft and the length is 94 ft, then the length is 44 ft more than the width and the perimeter is 2(50 ft) + 2(94 ft), or 288 ft. This checks.

5. **State.** The width is 50 ft and the length is 94 ft.

The problem in Example 4 illustrates that many problems that can be solved by translating to *one* equation in *one* variable are actually easier to solve by translating to *two* equations in *two* variables.

Do Exercise 4.

Exercise Set 14.2

a Solve using the substitution method.

1. $x + y = 10,$
$y = x + 8$

2. $x + y = 4,$
$y = 2x + 1$

3. $y = x - 6,$
$x + y = -2$

4. $y = x + 1,$
$2x + y = 4$

5. $y = 2x - 5,$
$3y - x = 5$

6. $y = 2x + 1,$
$x + y = -2$

7. $x = -2y,$
$x + 4y = 2$

8. $r = -3s,$
$r + 4s = 10$

b Solve using the substitution method. First, solve one equation for one variable.

9. $x - y = 6,$
$x + y = -2$

10. $s + t = -4,$
$s - t = 2$

11. $y - 2x = -6,$
$2y - x = 5$

12. $x - y = 5,$
$x + 2y = 7$

13. $2x + 3y = -2,$
$2x - y = 9$

14. $x + 2y = 10,$
$3x + 4y = 8$

15. $x - y = -3,$
$2x + 3y = -6$

16. $3b + 2a = 2,$
$-2b + a = 8$

17. $r - 2s = 0,$
$4r - 3s = 15$

18. $y - 2x = 0,$
$3x + 7y = 17$

c Solve.

19. The sum of two numbers is 37. One number is 5 more than the other. Find the numbers.

20. The sum of two numbers is 26. One number is 12 more than the other. Find the numbers.

21. Find two numbers whose sum is 52 and whose difference is 28.

22. Find two numbers whose sum is 63 and whose difference is 5.

23. The difference between two numbers is 12. Two times the larger is five times the smaller. What are the numbers?

24. The difference between two numbers is 18. Twice the smaller number plus three times the larger is 74. What are the numbers?

25. *Dimensions of Colorado.* The state of Colorado is roughly in the shape of a rectangle whose perimeter is 1300 mi. The width is 110 mi less than the length. Find the length and the width.

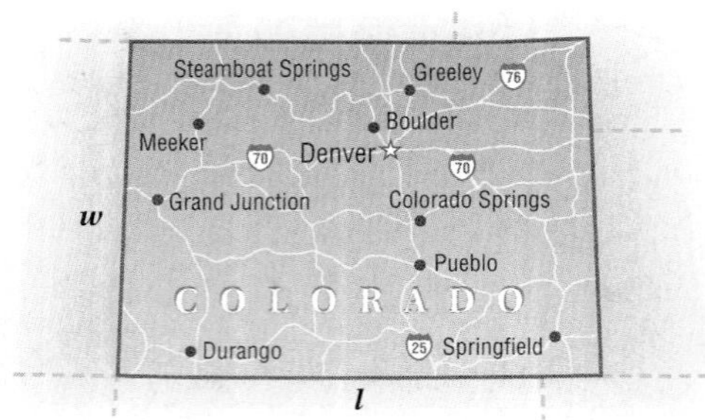

26. *Standard Billboard.* A standard rectangular highway billboard has a perimeter of 124 ft. The length is 6 ft more than three times the width. Find the length and the width.

27. *Two-by-Four.* The perimeter of a cross section of a "two-by-four" piece of lumber is $10\frac{1}{2}$ in. The length is twice the width. Find the actual dimensions of the cross section of a two-by-four.

28. The perimeter of a rectangular rose garden is 400 m. The length is 3 m more than twice the width. Find the length and the width.

Skill Maintenance

Graph. [9.3a, b]

29. $2x - 3y = 6$

30. $2x + 3y = 6$

31. $y = 2x - 5$

32. $y = 4$

Factor completely. [11.6a]

33. $6x^2 - 13x + 6$

34. $4p^2 - p - 3$

35. $4x^2 + 3x + 2$

36. $9a^2 - 25$

Synthesis

37. ◈ Joel solves every system of two equations (in x and y) by first solving for y in the first equation and then substituting into the second equation. Is he using the best approach, given that he always uses substitution? Why or why not?

38. ◈ Compare the solution of Example 4 in this section with the solution of the same problem in Example 5 of Section 8.4. Discuss the merits of using a system of equations to solve the problem.

 Solve using a grapher and its CALC-INTERSECT feature.

39. $x - y = 5,$
$x + 2y = 7$

40. $y - 2x = -6,$
$2y - x = 5$

41. $y - 2.35x = -5.97,$
$2.14y - x = 4.88$

42. $y = 1.2x - 32.7,$
$y = -0.7x + 46.15$

14.3 The Elimination Method

a Solving by the Elimination Method

The **elimination method** for solving systems of equations makes use of the *addition principle.* Some systems are much easier to solve using this method. Trying to solve the system in Example 1 by substitution would necessitate the use of fractions and extra steps because no variable has a coefficient of 1. Instead we use the elimination method.

Example 1 Solve the system

$$2x + 3y = 13, \quad (1)$$
$$4x - 3y = 17. \quad (2)$$

The key to the advantage of the elimination method for solving this system involves the $3y$ in one equation and the $-3y$ in the other. The terms are opposites. If we add the terms on the sides of the equations, the y-terms will add to 0, and in effect, the variable y will be eliminated.

We will use the addition principle for equations. According to equation (2), $4x - 3y$ and 17 are the same number. Thus we can use a vertical form and add $4x - 3y$ to the left side of equation (1) and 17 to the right side—in effect, adding the same number on both sides of equation (1):

$$\begin{aligned} 2x + 3y &= 13 && (1) \\ 4x - 3y &= 17 && (2) \\ \hline 6x + 0y &= 30. && \text{Adding} \end{aligned}$$

We have "eliminated" one variable. This is why we call this the **elimination method.** We now have an equation with just one variable that can be solved for x:

$$\begin{aligned} 6x &= 30 \\ x &= 5. \end{aligned}$$

Next, we substitute 5 for x in either of the original equations:

$$\begin{aligned} 2x + 3y &= 13 && \text{Equation (1)} \\ 2(5) + 3y &= 13 && \text{Substituting 5 for } x \\ 10 + 3y &= 13 \\ 3y &= 3 \\ y &= 1. && \text{Solving for } y \end{aligned}$$

We check the ordered pair (5, 1).

CHECK:

$2x + 3y = 13$		
$2(5) + 3(1)$	?	13
$10 + 3$		
13		TRUE

$4x - 3y = 17$		
$4(5) - 3(1)$	?	17
$20 - 3$		
17		TRUE

Since (5, 1) checks, it is the solution.

Do Exercises 1 and 2.

Objectives

a Solve a system of two equations in two variables using the elimination method when no multiplication is necessary.

b Solve a system of two equations in two variables using the elimination method when multiplication is necessary.

c Solve applied problems by translating to a system of two equations and then solving using the elimination method.

For Extra Help

TAPE 26

MAC WIN

CD-ROM

Solve using the elimination method.

1. $x + y = 5,$
$2x - y = 4$

2. $-2x + y = -4,$
$2x - 5y = 12$

Calculator Spotlight

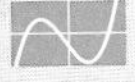
Use the TABLE feature to check the possible solutions of the systems in Example 1 and Margin Exercises 1 and 2.

Answers on page A-44

3. Solve. Multiply one equation by -1 first.

$5x + 3y = 17,$
$5x - 2y = -3$

4. Solve the system

$3x - 2y = -30,$
$5x - 2y = -46.$

Calculator Spotlight

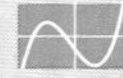 Use the CALC-INTERSECT feature to solve the systems in Examples 2 and 3 and Margin Exercises 3 and 4.

Answers on page A-44

b Using the Multiplication Principle First

The elimination method allows us to eliminate a variable. We may need to multiply by certain numbers first, however, so that terms become opposites.

Example 2 Solve the system

$$2x + 3y = 8, \quad (1)$$
$$x + 3y = 7. \quad (2)$$

If we add, we will not eliminate a variable. However, if the $3y$ were $-3y$ in one equation, we could eliminate y. Thus we multiply by -1 on both sides of equation (2) and then add, using a vertical form:

$$\begin{array}{rl} 2x + 3y = 8 & \text{Equation (1)} \\ -x - 3y = -7 & \text{Multiplying equation (2) by } -1 \\ \hline x \quad\quad = 1. & \text{Adding} \end{array}$$

Next, we substitute 1 for x in one of the original equations:

$$\begin{array}{rl} x + 3y = 7 & \text{Equation (2)} \\ 1 + 3y = 7 & \text{Substituting 1 for } x \\ 3y = 6 & \\ y = 2. & \text{Solving for } y \end{array}$$

We check the ordered pair (1, 2).

CHECK:

$$\begin{array}{r|l} 2x + 3y & 8 \\ \hline 2 \cdot 1 + 3 \cdot 2 \ ? & 8 \\ 2 + 6 & \\ 8 & \end{array} \quad \text{TRUE} \qquad \begin{array}{r|l} x + 3y & 7 \\ \hline 1 + 3 \cdot 2 \ ? & 7 \\ 1 + 6 & \\ 7 & \end{array} \quad \text{TRUE}$$

Since (1, 2) checks, it is the solution.

Do Exercises 3 and 4.

In Example 2, we used the multiplication principle, multiplying by -1. However, we often need to multiply by something other than -1.

Example 3 Solve the system

$$3x + 6y = -6, \quad (1)$$
$$5x - 2y = 14. \quad (2)$$

Looking at the terms with variables, we see that if $-2y$ were $-6y$, we would have terms that are opposites. We can achieve this by multiplying by 3 on both sides of equation (2). Then we add and solve for x:

$$\begin{array}{rl} 3x + 6y = -6 & \text{Equation (1)} \\ 15x - 6y = 42 & \text{Multiplying equation (2) by 3} \\ \hline 18x \quad\quad = 36 & \text{Adding} \\ x = 2. & \text{Solving for } x \end{array}$$

Next, we go back to equation (1) and substitute 2 for x:

$$\begin{aligned} 3 \cdot 2 + 6y &= -6 && \text{Substituting} \\ 6 + 6y &= -6 \\ 6y &= -12 \\ y &= -2. && \text{Solving for } y \end{aligned}$$

We check the ordered pair $(2, -2)$.

CHECK:

$$\begin{array}{c|l} 3x + 6y & = -6 \\ \hline 3 \cdot 2 + 6 \cdot (-2) & ?\ -6 \\ 6 + (-12) & \\ -6 & \quad \text{TRUE} \end{array} \qquad \begin{array}{c|l} 5x - 2y & = 14 \\ \hline 5 \cdot 2 - 2 \cdot (-2) & ?\ 14 \\ 10 - (-4) & \\ 14 & \quad \text{TRUE} \end{array}$$

Since $(2, -2)$ checks, it is the solution.

Do Exercises 5 and 6.

Part of the strategy in using the elimination method is making a decision about which variable to eliminate. So long as the algebra has been carried out correctly, the solution can be found by eliminating *either* variable. We multiply so that terms involving the variable to be eliminated are opposites. It is helpful to first get each equation in a form equivalent to $Ax + By = C$.

Example 4 Solve the system

$$\begin{aligned} 3y + 1 + 2x &= 0, && (1) \\ 5x &= 7 - 4y. && (2) \end{aligned}$$

We first rewrite each equation in a form equivalent to $Ax + By = C$:

$$\begin{aligned} 2x + 3y &= -1, && (1) && \text{Subtracting 1 on both sides and rearranging terms} \\ 5x + 4y &= 7. && (2) && \text{Adding } 4y \text{ on both sides} \end{aligned}$$

We decide to eliminate the x-term. We do this by multiplying by 5 on both sides of equation (1) and by -2 on both sides of equation (2). Then we add and solve for y:

$$\begin{aligned} 10x + 15y &= -5 && \text{Multiplying by 5 on both sides of equation (1)} \\ -10x - 8y &= -14 && \text{Multiplying by } -2 \text{ on both sides of equation (2)} \\ \hline 7y &= -19 && \text{Adding} \\ y &= \frac{-19}{7}, \text{ or } -\frac{19}{7}. && \text{Dividing by 7} \end{aligned}$$

Next, we substitute $-\frac{19}{7}$ for y in one of the original equations:

$$\begin{aligned} 2x + 3y &= -1 && \text{Equation (1)} \\ 2x + 3\left(-\tfrac{19}{7}\right) &= -1 && \text{Substituting } -\tfrac{19}{7} \text{ for } y \\ 2x - \tfrac{57}{7} &= -1 \\ 2x &= -1 + \tfrac{57}{7} \\ 2x &= -\tfrac{7}{7} + \tfrac{57}{7} \\ 2x &= \tfrac{50}{7} \\ x &= \tfrac{50}{7} \cdot \tfrac{1}{2}, \text{ or } \tfrac{25}{7}. && \text{Solving for } x \end{aligned}$$

We check the ordered pair $\left(\frac{25}{7}, -\frac{19}{7}\right)$.

Solve the system.

5. $4a + 7b = 11,$
$2a + 3b = 5$

6. $3x - 8y = 2,$
$5x + 2y = -12$

CAUTION! Solving a *system* of equations in two variables requires finding an ordered *pair* of numbers. Once you have solved for one variable, don't forget the other, and remember to list the ordered-pair solution using alphabetical order.

Answers on page A-44

7. Solve the system

$$3x = 5 + 2y,$$
$$2x + 3y - 1 = 0.$$

8. Solve the system

$$2x + y = 15,$$
$$4x + 2y = 23.$$

9. Solve the system

$$5x - 2y = 3,$$
$$-15x + 6y = -9.$$

Calculator Spotlight

Exercises

Use a viewing window of $[-6, 6, -4, 4]$ for the following.

1. Check the results of Example 5 on a grapher. What happens when you solve each equation for y?
2. Check the results of Example 6 on a grapher. What happens when you solve each equation for y?

Answers on page A-44

CHECK:

$3y + 1 + 2x = 0$		
$3\left(-\frac{19}{7}\right) + 1 + 2\left(\frac{25}{7}\right)$	?	0
$-\frac{57}{7} + \frac{7}{7} + \frac{50}{7}$		
0		TRUE

$5x = 7 - 4y$		
$5\left(\frac{25}{7}\right)$	?	$7 - 4\left(-\frac{19}{7}\right)$
$\frac{125}{7}$		$\frac{49}{7} + \frac{76}{7}$
		$\frac{125}{7}$ TRUE

The solution is $\left(\frac{25}{7}, -\frac{19}{7}\right)$.

Do Exercise 7.

Let's consider a system with no solution and see what happens when we apply the elimination method.

Example 5 Solve the system

$$y - 3x = 2, \quad (1)$$
$$y - 3x = 1. \quad (2)$$

We multiply by -1 on both sides of equation (2) and then add:

$$y - 3x = 2$$
$$-y + 3x = -1 \quad \text{Multiplying by } -1$$
$$0 = 1. \quad \text{Adding}$$

We obtain a false equation, $0 = 1$, so there is *no solution*. The slope–intercept forms of these equations are

$$y = 3x + 2,$$
$$y = 3x + 1.$$

The slopes are the same and the y-intercepts are different. Thus the lines are parallel. They do not intersect.

Do Exercise 8.

Sometimes there is an infinite number of solutions. Let's look at a system that we graphed in Example 6 of Section 14.1.

Example 6 Solve the system

$$2x + 3y = 6, \quad (1)$$
$$-8x - 12y = -24. \quad (2)$$

We multiply by 4 on both sides of equation (1) and then add the two equations:

$$8x + 12y = 24 \quad \text{Multiplying by 4}$$
$$-8x - 12y = -24$$
$$0 = 0. \quad \text{Adding}$$

We have eliminated both variables, and what remains, $0 = 0$, is an equation easily seen to be true. If this happens when we use the elimination method, we have an infinite number of solutions.

Do Exercise 9.

When decimals or fractions appear, we first multiply to clear them. Then we proceed as before.

Example 7 Solve the system

$$\frac{1}{3}x + \frac{1}{2}y = -\frac{1}{6}, \qquad (1)$$

$$\frac{1}{2}x + \frac{2}{5}y = \frac{7}{10}. \qquad (2)$$

The number 6 is a multiple of all the denominators of equation (1). The number 10 is a multiple of all the denominators of equation (2). We multiply by 6 on both sides of equation (1) and by 10 on both sides of equation (2):

$$6\left(\frac{1}{3}x + \frac{1}{2}y\right) = 6\left(-\frac{1}{6}\right) \qquad 10\left(\frac{1}{2}x + \frac{2}{5}y\right) = 10\left(\frac{7}{10}\right)$$

$$6 \cdot \frac{1}{3}x + 6 \cdot \frac{1}{2}y = -1 \qquad 10 \cdot \frac{1}{2}x + 10 \cdot \frac{2}{5}y = 7$$

$$2x + 3y = -1; \qquad 5x + 4y = 7.$$

The resulting system is

$$2x + 3y = -1,$$

$$5x + 4y = 7.$$

As we saw in Example 4, the solution of this system is $\left(\frac{25}{7}, -\frac{19}{7}\right)$.

Do Exercises 10 and 11.

The following is a summary that compares the graphical, substitution, and elimination methods for solving systems of equations.

Method	Strengths	Weaknesses
Graphical	Can "see" solution.	Inexact when solution involves numbers that are not integers or are very large and off the graph.
Substitution	Works well when solutions are not integers. Easy to use when a variable is alone on one side.	Introduces extensive computations with fractions for more complicated systems where coefficients are not 1 or −1. Cannot "see" solution.
Elimination	Works well when solutions are not integers, when coefficients are not 1 or −1, and when coefficients involve decimals or fractions.	Cannot "see" solution.

When deciding which method to use, consider the preceding chart and directions from your instructor. The situation is like having a piece of wood to cut and three saws with which to cut it. The saw you use depends on the type of wood, the type of cut you are making, and how you want the wood to turn out.

Solve the system.

10. $\frac{1}{2}x + \frac{3}{10}y = \frac{1}{5},$

$\frac{3}{5}x + y = -\frac{2}{5}$

11. $3.3x + 6.6y = -6.6,$

$0.1x - 0.04y = 0.28$

Answers on page A-44

12. *Car Rental.* Budget Rent-A-Car rents a car at a daily rate of \$41.95 plus 43 cents per mile. Speedo Rentzit rents a car for \$44.95 plus 39 cents per mile. For what mileage are the costs the same?

Answer on page A-44

c Solving Applied Problems

We now use the elimination method to solve an applied problem.

Example 8 *Truck Rental.* At one time, Value Rent-A-Car rented pickup trucks at a daily rate of \$43.95 plus 40 cents per mile. Thrifty Rent-A-Car rented the same type of pickup trucks at a daily rate of \$42.95 plus 42 cents per mile. For what mileage are the costs the same?

1. **Familiarize.** To become familiar with the problem, we make a guess. Suppose a person rents a pickup truck from each rental agency and drives it 100 mi. The total cost at Value is \$43.95 + \$0.40(100) = \$43.95 + \$40.00, or \$83.95. The total cost at Thrifty is \$42.95 + \$0.42(100) = \$42.95 + \$42.00, or \$84.95. Note that we converted all of the money units to dollars. The resulting costs are very nearly the same, so our guess is close. We can, of course, refine our guess. Instead, we will use algebra to solve the problem. We let M = the number of miles driven and C = the total cost of the truck rental.

2. **Translate.** We translate the first statement, using \$0.40 for 40 cents. It helps to reword the problem before translating.

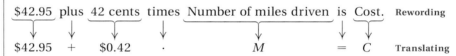

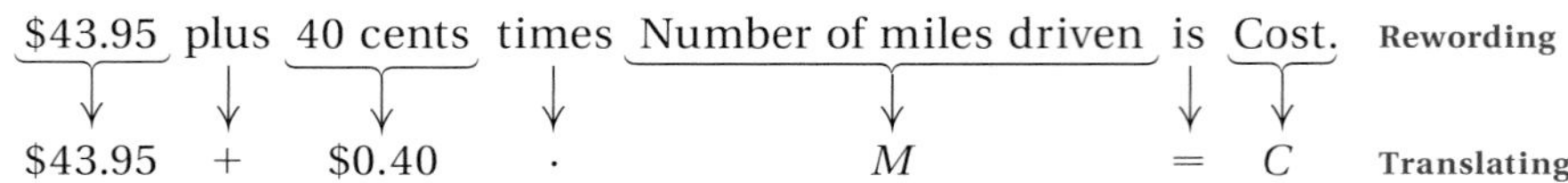

\$43.95 plus 40 cents times Number of miles driven is Cost. Rewording
\$43.95 + \$0.40 · M = C Translating

We translate the second statement, but again it helps to reword it first.

\$42.95 plus 42 cents times Number of miles driven is Cost. **Rewording**

$\$42.95 + \$0.42 \cdot M = C$ **Translating**

We have now translated to a system of equations:

$$43.95 + 0.40M = C,$$
$$42.95 + 0.42M = C.$$

3. **Solve.** We solve the system of equations. We clear the system of decimals by multiplying by 100 on both sides. Then we multiply the second equation by -1 and add:

$$\begin{aligned} 4395 + 40M &= 100C \\ -4295 - 42M &= -100C \\ \hline 100 - 2M &= 0 \\ 100 &= 2M \\ 50 &= M. \end{aligned}$$

4. **Check.** For 50 mi, the cost of the Value truck is \$43.95 + \$0.40(50), or \$43.95 + \$20, or \$63.95, and the cost of the Thrifty truck is \$42.95 + \$0.42(50), or \$42.95 + \$21, or \$63.95. Thus the costs are the same when the mileage is 50.

5. **State.** When the trucks are driven 50 mi, the costs will be the same.

Do Exercise 12.

Exercise Set 14.3

a Solve using the elimination method.

1. $x - y = 7,$
   $x + y = 5$

2. $x + y = 11,$
   $x - y = 7$

3. $x + y = 8,$
   $-x + 2y = 7$

4. $x + y = 6,$
   $-x + 3y = -2$

5. $5x - y = 5,$
   $3x + y = 11$

6. $2x - y = 8,$
   $3x + y = 12$

7. $4a + 3b = 7,$
   $-4a + b = 5$

8. $7c + 5d = 18,$
   $c - 5d = -2$

9. $8x - 5y = -9,$
   $3x + 5y = -2$

10. $3a - 3b = -15,$
    $-3a - 3b = -3$

11. $4x - 5y = 7,$
    $-4x + 5y = 7$

12. $2x + 3y = 4,$
    $-2x - 3y = -4$

b Solve using the multiplication principle first. Then add.

13. $x + y = -7,$
    $3x + y = -9$

14. $-x - y = 8,$
    $2x - y = -1$

15. $3x - y = 8,$
    $x + 2y = 5$

16. $x + 3y = 19,$
    $x - y = -1$

17. $x - y = 5,$
$4x - 5y = 17$

18. $x + y = 4,$
$5x - 3y = 12$

19. $2w - 3z = -1,$
$3w + 4z = 24$

20. $7p + 5q = 2,$
$8p - 9q = 17$

21. $2a + 3b = -1,$
$3a + 5b = -2$

22. $3x - 4y = 16,$
$5x + 6y = 14$

23. $x = 3y,$
$5x + 14 = y$

24. $5a = 2b,$
$2a + 11 = 3b$

25. $2x + 5y = 16,$
$3x - 2y = 5$

26. $3p - 2q = 8,$
$5p + 3q = 7$

27. $p = 32 + q,$
$3p = 8q + 6$

28. $3x = 8y + 11,$
$x + 6y - 8 = 0$

29. $3x - 2y = 10,$
$-6x + 4y = -20$

30. $2x + y = 13,$
$4x + 2y = 23$

31. $0.06x + 0.05y = 0.07,$
$0.4x - 0.3y = 1.1$

32. $1.8x - 2y = 0.9,$
$0.04x + 0.18y = 0.15$

33. $\frac{1}{3}x + \frac{3}{2}y = \frac{5}{4},$
$\frac{3}{4}x - \frac{5}{6}y = \frac{3}{8}$

34. $x - \frac{3}{2}y = 13,$
$\frac{3}{2}x - y = 17$

35. $-4.5x + 7.5y = 6,$
$-x + 1.5y = 5$

36. $0.75x + 0.6y = -0.3,$
$3.9x + 5.2y = 96.2$

c Solve.

37. *Van Rental.* A family plans to rent a van to move a daughter to college. Quick-Haul rents a 10-ft moving van at a daily rate of \$19.95 plus 39 cents per mile. Another company rents the same size van for \$39.95 plus 29 cents per mile. For what mileage are the costs the same?

38. *Car Rental.* Elite Rent-A-Car rents a basic car at a daily rate of \$45.95 plus 40 cents per mile. Another company rents a basic car for \$46.95 plus 20 cents per mile. For what mileage are the costs the same?

39. Two angles are supplementary. (**Supplementary angles** are angles whose sum is 180°.) One is 30° more than two times the other. Find the angles.

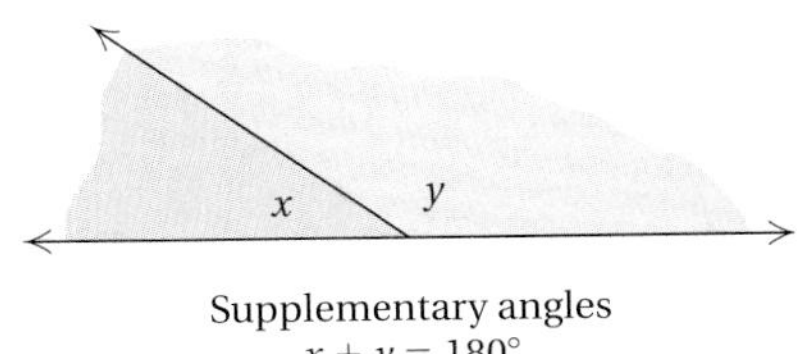

Supplementary angles
$x + y = 180°$

40. Two angles are supplementary. One is 8° less than three times the other. Find the angles.

41. Two angles are complementary. (**Complementary angles** are angles whose sum is 90°.) Their difference is 34°. Find the angles.

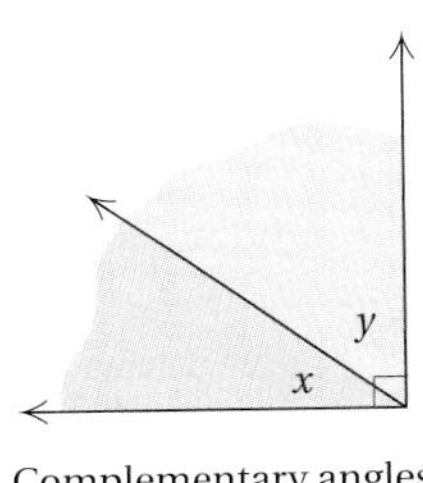

Complementary angles
$x + y = 90°$

42. Two angles are complementary. One angle is 42° more than one-half the other. Find the angles.

43. The Rolling Velvet Horse Farm allots 650 hectares to plant hay and oats. The owners know that their needs are best met if they plant 180 hectares more of hay than of oats. How many hectares of each should they plant?

44. In a vineyard, a vintner uses 820 hectares to plant Chardonnay and Riesling grapes. The vintner knows that the profits will be greatest by planting 140 hectares more of Chardonnay than of Riesling. How many hectares of each grape should be planted?

Skill Maintenance

Simplify. [10.1d, e, f]

45. $x^{-2} \cdot x^{-5}$

46. $x^{-2} \cdot x^{5}$

47. $x^{2} \cdot x^{-5}$

48. $x^{2} \cdot x^{5}$

49. $\dfrac{x^{-2}}{x^{-5}}$

50. $\dfrac{x^{2}}{x^{-5}}$

51. $(a^2b^{-3})(a^5b^{-6})$

52. $\dfrac{a^2b^{-3}}{a^5b^{-6}}$

Simplify. [12.1c]

53. $\dfrac{x^2 - 5x + 6}{x^2 - 4}$

54. $\dfrac{x^2 - 25}{x^2 - 10x + 25}$

Subtract. [12.5a]

55. $\dfrac{x-2}{x+3} - \dfrac{2x-5}{x-4}$

56. $\dfrac{x+7}{x^2-1} - \dfrac{3}{x+1}$

SYNTHESIS

57. ◆ The following lists the steps a student uses to solve a system of equations, but an error occurs. Find and describe the error and correct the answer.

$$\begin{aligned} 3x - y &= 4 \\ 2x + y &= 16 \\ \hline 5x \quad &= 20 \\ x &= 4 \end{aligned}$$

$$\begin{aligned} 3x - y &= 4 \\ 3(4) - y &= 4 \\ y &= 4 - 12 \\ y &= -8 \end{aligned}$$

The solution is $(4, -8)$.

58. ◆ Explain how the addition and multiplication principles are used in this section. Then count the number of times that these principles are used in Example 4.

59.–68. Use the TABLE feature to check the possible solutions to Exercises 1–10.

69.–78. Use a grapher and the CALC-INTERSECT feature to solve the systems in Exercises 21–30.

79. Will's age is 20% of his father's age. Twenty years from now, Will's age will be 52% of his father's age. How old are Will and his father now?

80. If 5 is added to a woman's age and the total is divided by 5, the result will be her daughter's age. Five years ago, the woman's age was eight times her daughter's age. Find their present ages.

Solve using either the substitution or the elimination method.

81. $3(x - y) = 9,$
$x + y = 7$

82. $2(x - y) = 3 + x,$
$x = 3y + 4$

83. $2(5a - 5b) = 10,$
$-5(6a + 2b) = 10$

84. $\dfrac{x}{3} + \dfrac{y}{2} = 1\dfrac{1}{3},$
$x + 0.05y = 4$

85. Several ancient Chinese books included problems that can be solved by translating to systems of equations. *Arithmetical Rules in Nine Sections* is a book of 246 problems compiled by a Chinese mathematician, Chang Tsang, who died in 152 B.C. One of the problems is: Suppose there are a number of rabbits and pheasants confined in a cage. In all, there are 35 heads and 94 feet. How many rabbits and how many pheasants are there? Solve the problem.

Collaborative Learning Manual

Compare the three methods for solving systems of equations in two variables.

14.4 Applications and Problem Solving

a We now use systems of equations to solve applied problems that involve two equations in two variables.

Example 1 *Pizza and Soda Prices.* A campus vendor charges \$3.50 for one slice of pizza and one medium soda and \$9.15 for three slices of pizza and two medium sodas. Determine the price of one medium soda and the price of one slice of pizza.

1. **Familiarize.** We let p = the price of one slice of pizza and s = the price of one medium soda.
2. **Translate.** The price of one slice of pizza and one medium soda is \$3.50. This gives us one equation:

   $$p + s = 3.50.$$

   The price of three slices of pizza and two medium sodas is \$9.15. This gives us another equation:

   $$3p + 2s = 9.15.$$

3. **Solve.** We solve the system of equations

   $$p + s = 3.50, \quad (1)$$
   $$3p + 2s = 9.15. \quad (2)$$

   Which method should we use? As we discussed in Section 14.3, any method can be used. Each has its advantages and disadvantages. We decide to proceed with the elimination method, because we see that if we multiply each side of equation (1) by -2 and add, the s-terms can be eliminated:

   $$-2p - 2s = -7.00 \quad \text{Multiplying equation (1) by } -2$$
   $$3p + 2s = 9.15 \quad \text{Equation (2)}$$
   $$p = 2.15. \quad \text{Adding}$$

   Next, we substitute 2.15 for p in equation (1) and solve for s:

   $$p + s = 3.50$$
   $$2.15 + s = 3.50$$
   $$s = 1.35.$$

4. **Check.** The sum of the prices for one slice of pizza and one medium soda is

   $$\$2.15 + \$1.35, \quad \text{or} \quad \$3.50.$$

   Three times the price of one slice of pizza plus twice the price of a medium soda is

   $$3(\$2.15) + 2(\$1.35), \quad \text{or} \quad \$9.15.$$

   The prices check.

5. **State.** The price of one slice of pizza is \$2.15, and the price of one medium soda is \$1.35.

Do Exercise 1.

Objective

a Solve applied problems by translating to a system of two equations in two variables.

For Extra Help

TAPE 26

MAC WIN

CD-ROM

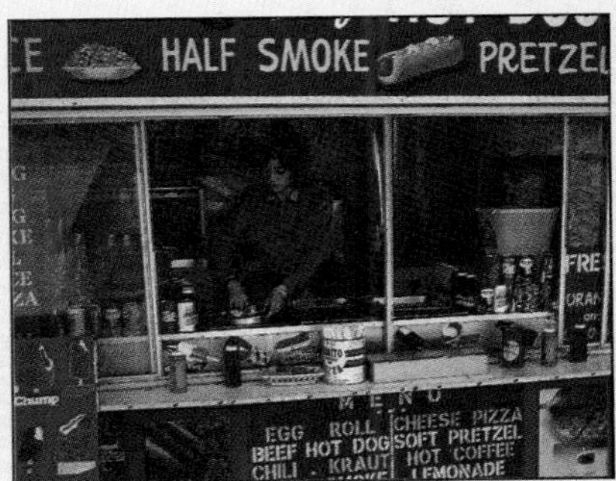

1. *Chicken and Hamburger Prices.* Fast Rick's Burger restaurant decides to include chicken on its menu. It offers a special two-and-one promotion. The price of one hamburger and two pieces of chicken is \$5.39, and the price of two hamburgers and one piece of chicken is \$5.68. Find the price of one hamburger and the price of one piece of chicken.

Answer on page A-44

2. *Ages.* Sarah is 26 yr older than Malcolm. In 5 yr, Sarah will be twice as old as Malcolm. How old are they now?

Complete the following table to aid with the familiarization.

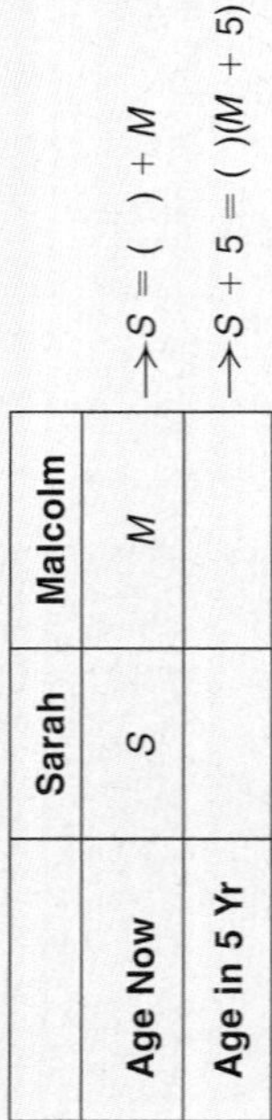

	Sarah	Malcolm	
Age Now	S	M	$\rightarrow S = (\quad) + M$
Age in 5 Yr			$\rightarrow S + 5 = (\quad)(M + 5)$

Example 2 *Ages.* Caleb is 21 yr older than Tanya. In 6 yr, Caleb will be twice as old as Tanya. How old are they now?

1. **Familiarize.** Let's consider some conditions of the problem. We let C = Caleb's age now and T = Tanya's age now. How will the ages relate in 6 yr? In 6 yr, Tanya will be $T + 6$ and Caleb will be $C + 6$. We make a table to organize our information.

	Caleb	Tanya	
Age Now	C	T	$\rightarrow C = 21 + T$
Age in 6 Yr	$C + 6$	$T + 6$	$\rightarrow C + 6 = 2(T + 6)$

2. **Translate.** From the present ages, we get the following rewording and translation.

Caleb's age	is	21	more than	Tanya's age.	**Rewording**
↓	↓	↓	↓	↓	
C	$=$	21	$+$	T	**Translating**

From their ages in 6 yr, we get the following rewording and translation.

Caleb's age in 6 yr	will be	twice	Tanya's age in 6 yr	**Rewording**
↓	↓	↓	↓	
$C + 6$	$=$	$2 \cdot$	$(T + 6)$	**Translating**

The problem has been translated to the following system of equations:

$$C = 21 + T, \qquad \textbf{(1)}$$
$$C + 6 = 2(T + 6). \qquad \textbf{(2)}$$

3. **Solve.** We solve the system. This time we use the substitution method since there is a variable alone on one side. We substitute $21 + T$ for C in equation (2) and solve for T:

$$\begin{aligned} C + 6 &= 2(T + 6) \\ (21 + T) + 6 &= 2(T + 6) \\ T + 27 &= 2T + 12 \\ 15 &= T. \end{aligned}$$

We find C by substituting 15 for T in the first equation:

$$\begin{aligned} C &= 21 + T \\ &= 21 + 15 \\ &= 36. \end{aligned}$$

4. **Check.** Caleb's age is 36, which is 21 more than 15, Tanya's age. In 6 yr, when Caleb will be 42 and Tanya 21, Caleb's age will be twice Tanya's age.

5. **State.** Caleb is now 36 and Tanya is 15.

Do Exercise 2.

Answer on page A-44

Example 3 *Imax Movie Prices.* There were 270 people at a recent showing of the IMAX 3D movie *Antarctica.* Admission was $8.00 each for adults and $4.75 each for children, and receipts totaled $2088.50. How many adults and how many children attended?

1. **Familiarize.** There are many ways in which to familiarize ourselves with a problem situation. This time, let's make a guess and do some calculations. The total number of people at the movie was 270, so we choose numbers that total 270. Let's try

220 adults and
50 children.

How much money was taken in? The problem says that adults paid $8.00 each, so the total amount of money collected from the adults was

220($8), or $1760.

Children paid $4.75 each, so the total amount of money collected from the children was

50($4.75), or $237.50.

This makes the total receipts $1760 + $237.50, or $1997.50.

Our guess is not the answer to the problem because the total taken in, according to the problem, was $2088.50. If we were to continue guessing, we would need to add more adults and fewer children, since our first guess gave us an amount of total receipts that was lower than $2088.50. The steps we have used to see if our guesses are correct help us to understand the actual steps involved in solving the problem.

Let's list the information in a table. That usually helps in the familiarization process. We let a = the number of adults and c = the number of children.

	Adults	Children	Total	
Admission	$8.00	$4.75		
Number Attending	a	c	270	→ $a + c = 270$
Money Taken in	$8.00a$	$4.75c$	$2088.50	→ $8.00a + 4.75c = 2088.50$

2. **Translate.** The total number of people attending was 270, so

$$a + c = 270.$$

The amount taken in from the adults was $8.00a$, and the amount taken in from the children was $4.75c$. These amounts are in dollars. The total was $2088.50, so we have

$$8.00a + 4.75c = 2088.50.$$

We can multiply by 100 on both sides to clear decimals. Thus we have a translation to a system of equations:

$$a + c = 270, \quad (1)$$

$$800a + 475c = 208{,}850. \quad (2)$$

3. *Game Admissions.* There were 166 paid admissions to a game. The price was \$3.10 each for adults and \$1.75 each for children. The amount taken in was \$459.25. How many adults and how many children attended?

Complete the following table to aid with the familiarization.

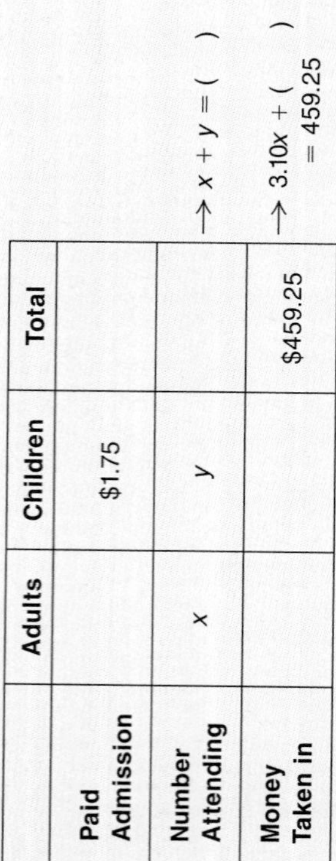

	Adults	Children	Total	
Paid Admission		\$1.75		
Number Attending	x	y		$\rightarrow x + y = (\quad)$
Money Taken in			\$459.25	$\rightarrow 3.10x + (\quad) = 459.25$

Answer on page A-44

3. **Solve.** We solve the system. We use the elimination method since the equations are both in the form $Ax + By = C$. (A case can certainly be made for using the substitution method since we can solve for one of the variables quite easily in the first equation. Very often a decision is just a matter of choice.) We multiply by -475 on both sides of equation (1) and then add and solve for a:

$$\begin{aligned} -475a - 475c &= -128{,}250 && \text{Multiplying by } -475 \\ 800a + 475c &= 208{,}850 \\ \hline 325a \qquad &= 80{,}600 && \text{Adding} \\ a &= \frac{80{,}600}{325} && \text{Dividing by 325} \\ a &= 248. \end{aligned}$$

Next, we go back to equation (1), substituting 248 for a, and solve for c:

$$\begin{aligned} a + c &= 270 \\ 248 + c &= 270 \\ c &= 22. \end{aligned}$$

4. **Check.** We leave the check to the student. It is similar to what we did in the *Familiarize* step.
5. **State.** Attending the showing were 248 adults and 22 children.

Do Exercise 3.

Example 4 *Mixture of Solutions.* A chemist has one solution that is 80% acid (that is, 8 parts are acid and 2 parts are water) and another solution that is 30% acid. What is needed is 200 L of a solution that is 62% acid. The chemist will prepare it by mixing the two solutions. How much of each should be used?

1. **Familiarize.** We can make a drawing of the situation. The chemist uses x liters of the first solution and y liters of the second solution.

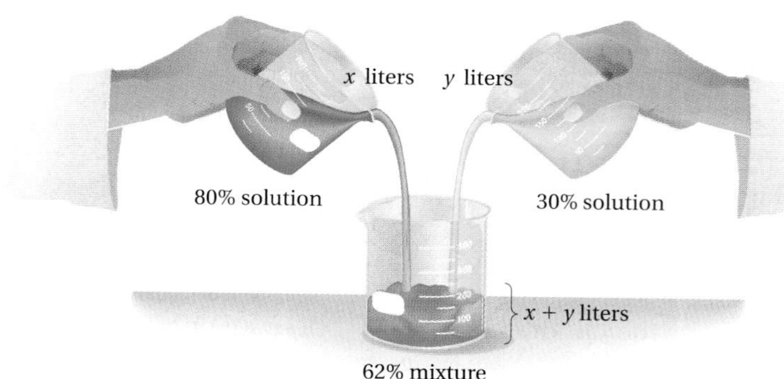

We can also arrange the information in a table.

Type of Solution	First	Second	Mixture	
Amount of Solution	x	y	200 L	$\rightarrow x + y = 200$
Percent of Acid	80%	30%	62%	
Amount of Acid in Solution	$80\%x$	$30\%y$	$62\% \times 200$, or 124 L	$\rightarrow 80\%x + 30\%y = 124$

2. **Translate.** The chemist uses x liters of the first solution and y liters of the second. Since the total is to be 200 L, we have

Total amount of solution: $x + y = 200$.

The amount of acid in the new mixture is to be 62% of 200 L, or 124 L. The amounts of acid from the two solutions are $80\%x$ and $30\%y$. Thus,

Total amount of acid: $80\%x + 30\%y = 124$

or $0.8x + 0.3y = 124$.

We clear decimals by multiplying by 10 on both sides:

$$10(0.8x + 0.3y) = 10 \cdot 124$$
$$8x + 3y = 1240.$$

Thus we have a translation to a system of equations:

$$x + y = 200, \quad (1)$$
$$8x + 3y = 1240. \quad (2)$$

3. **Solve.** We solve the system. We use the elimination method, again because equations are in the form $Ax + By = C$ and a multiplication in one equation will allow us to eliminate a variable, but substitution would also work. We multiply by -3 on both sides of equation (1) and then add and solve for x:

$$-3x - 3y = -600 \quad \text{Multiplying by } -3$$
$$8x + 3y = 1240$$
$$5x = 640 \quad \text{Adding}$$
$$x = \frac{640}{5} \quad \text{Dividing by 5}$$
$$x = 128.$$

Next, we go back to equation (1) and substitute 128 for x:

$$x + y = 200$$
$$128 + y = 200$$
$$y = 72.$$

The solution is $x = 128$ and $y = 72$.

4. **Check.** The sum of 128 and 72 is 200. Also, 80% of 128 is 102.4 and 30% of 72 is 21.6. These add up to 124.

5. **State.** The chemist should use 128 L of the 80%-acid solution and 72 L of the 30%-acid solution.

Do Exercise 4.

Example 5 *Candy Mixtures.* A bulk wholesaler wishes to mix some candy worth 45 cents per pound and some worth 80 cents per pound to make 350 lb of a mixture worth 65 cents per pound. How much of each type of candy should be used?

1. **Familiarize.** Arranging the information in a table will help. We let x = the amount of 45-cents candy and y = the amount of 80-cents candy.

4. *Mixture of Solutions.* One solution is 50% alcohol and a second is 70% alcohol. How much of each should be mixed in order to make 30 L of a solution that is 55% alcohol?

Complete the following table to aid in the familiarization.

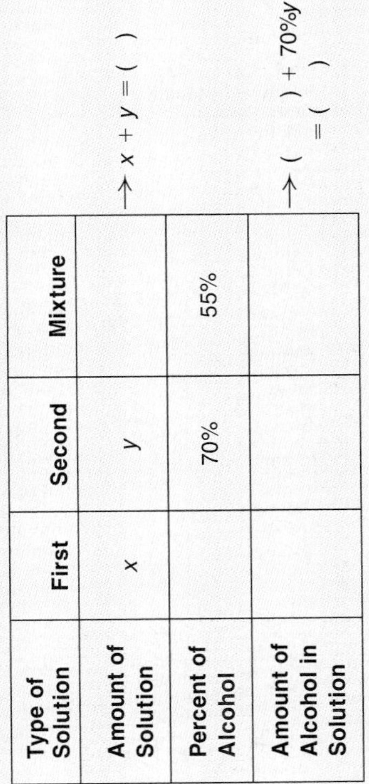

Type of Solution	First	Second	Mixture	
Amount of Solution	x	y		→ x + y = ( )
Percent of Alcohol		70%	55%	
Amount of Alcohol in Solution				→ ( ) + 70%y = ( )

Answer on page A-44

5. *Mixture of Grass Seeds.* Grass seed A is worth \$1.40 per pound and seed B is worth \$1.75 per pound. How much of each should be mixed in order to make 50 lb of a mixture worth \$1.54 per pound?

Complete the following table to aid in the familiarization.

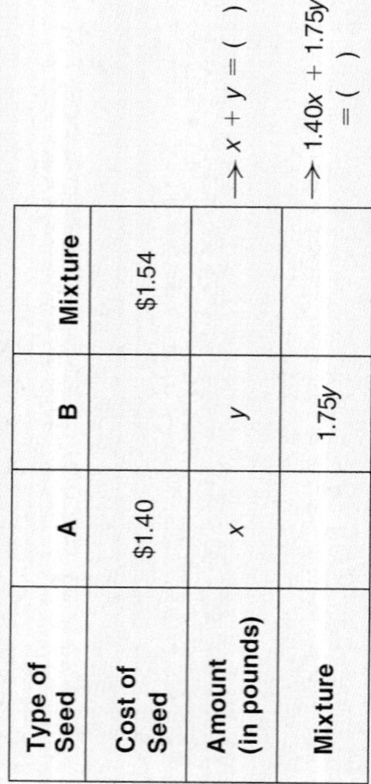

Type of Seed	A	B	Mixture	
Cost of Seed	\$1.40		\$1.54	
Amount (in pounds)	x	y		$\rightarrow x + y = (\ \ )$
Mixture		$1.75y$		$\rightarrow 1.40x + 1.75y = (\ \ )$

Answer on page A-44

Type of Candy	Inexpensive Candy	Expensive Candy	Mixture	
Cost of Candy	45 cents	80 cents	65 cents	
Amount (in pounds)	x	y	350	$\rightarrow x + y = 350$
Total Cost	$45x$	$80y$	65 cents · (350), or 22,750 cents	$\rightarrow 45x + 80y = 22{,}750$

Note the similarity of this problem to Example 3. Here we consider types of candy instead of groups of people.

2. Translate. We translate as follows. From the second row of the table, we find that

Total amount of candy: $x + y = 350.$

Our second equation will come from the costs. The value of the inexpensive candy, in cents, is $45x$ (x pounds at 45 cents per pound). The value of the expensive candy is $80y$, and the value of the mixture is 65×350, or 22,750 cents. Thus we have

Total cost of mixture: $45x + 80y = 22{,}750.$

Remember the problem-solving tip about dimension symbols. In this last equation, all expressions are given in cents. We could have expressed them all in dollars, but we do not want some in cents and some in dollars. Thus we have a translation to a system of equations:

$$x + y = 350, \quad (1)$$
$$45x + 80y = 22{,}750. \quad (2)$$

3. Solve. We solve the system using the elimination method again. We multiply by -45 on both sides of equation (1) and then add and solve for y:

$$\begin{aligned} -45x - 45y &= -15{,}750 && \text{Multiplying by } -45 \\ 45x + 80y &= 22{,}750 \\ \hline 35y &= 7{,}000 && \text{Adding} \\ y &= \frac{7{,}000}{35} \\ y &= 200. \end{aligned}$$

Next, we go back to equation (1), substituting 200 for y, and solve for x:

$$\begin{aligned} x + y &= 350 \\ x + 200 &= 350 \\ x &= 150. \end{aligned}$$

4. Check. We consider $x = 150$ lb and $y = 200$ lb. The sum is 350 lb. The value of the candy is $45(150) + 80(200)$, or 22,750 cents and each pound of the mixture is worth $22{,}750 \div 350$, or 65 cents. These values check.

5. State. The grocer should mix 150 lb of the 45-cents candy with 200 lb of the 80-cents candy.

Do Exercise 5.

Example 6 *Coin Value.* A student assistant at the university copy center has some nickels and dimes to use for change when students make copies. The value of the coins is \$7.40. There are 26 more dimes than nickels. How many of each kind of coin are there?

1. **Familiarize.** We let d = the number of dimes and n = the number of nickels.
2. **Translate.** We have one equation at once:

   $$d = n + 26.$$

   The value of the nickels, in cents, is $5n$, since each coin is worth 5 cents. The value of the dimes, in cents, is $10d$, since each coin is worth 10 cents. The total value is given as \$7.40. Since we have the values of the nickels and dimes *in cents,* we must use cents for the total value. This is 740. This gives us another equation:

   $$10d + 5n = 740.$$

   We now have a system of equations:

   $$d = n + 26, \qquad (1)$$
   $$10d + 5n = 740. \qquad (2)$$

3. **Solve.** Since we have d alone on one side of one equation, we use the substitution method. We substitute $n + 26$ for d in equation (2):

   $$\begin{aligned} 10d + 5n &= 740 \\ 10(n + 26) + 5n &= 740 && \text{Substituting } n + 26 \text{ for } d \\ 10n + 260 + 5n &= 740 && \text{Removing parentheses} \\ 15n + 260 &= 740 && \text{Collecting like terms} \\ 15n &= 480 && \text{Subtracting 260} \\ n &= \frac{480}{15}, \text{ or } 32. && \text{Dividing by 15} \end{aligned}$$

   Next, we substitute 32 for n in either of the original equations to find d. We use equation (1):

   $$\begin{aligned} d &= n + 26 \\ &= 32 + 26 \\ &= 58. \end{aligned}$$

4. **Check.** We have 58 dimes and 32 nickels. There are 26 more dimes than nickels. The value of the coins is 58(\$0.10) + 32(\$0.05), which is \$7.40. This checks.
5. **State.** The student assistant has 58 dimes and 32 nickels.

Do Exercise 6.

6. *Coin Value.* On a table are 20 coins, quarters and dimes. Their value is \$3.05. How many of each kind of coin are there?

Answer on page A-44

You should look back over Examples 3–6. The problems are quite similar in their structure. Compare them and try to see the similarities. The problems in Examples 3–6 are often called *mixture problems.* These problems provide a pattern, or model, for many related problems.

PROBLEM-SOLVING TIP

When solving problems, see if they are patterned or modeled after other problems that you have studied.

Exercise Set 14.4

a Solve.

1. *Basketball Scoring.* Shaquille O'Neill once scored 36 points on 22 shots in an NBA game, shooting only two-pointers and foul shots (one point) (***Source*:** National Basketball Association). How many of each type of shot did he make?

2. *Household Trash.* The Perezes generate twice as much trash as their neighbors, the Willises. Together, the two households produce 14 bags of trash each month. How much trash does each household produce?

3. The Kuyatts' house is twice as old as the Marconis' house. Eight years ago, the Kuyatts' house was three times as old as the Marconis' house. How old is each house?

4. David is twice as old as his daughter. In 4 yr, David's age will be three times what his daughter's age was 6 yr ago. How old are they now?

5. Randy is four times as old as Mandy. In 12 yr, Mandy's age will be half of Randy's. How old are they now?

6. Jennifer is twice as old as Ramon. The sum of their ages 7 yr ago was 13. How old are they now?

7. *Coffee Blends.* Cafebucks coffee shop mixes Brazilian coffee worth \$19 per pound with Turkish coffee worth \$22 per pound. The mixture is to sell for \$20 per pound. How much of each type of coffee should be used in order to make a 300-lb mixture? Complete the following table to aid in the familiarization.

Type of Coffee	Brazilian	Turkish	Mixture	
Cost of Coffee	\$19		\$20	
Amount (in pounds)	x	y	300	$\longrightarrow x + y = (\quad)$
Mixture		$22y$	20(300), or \$6000	$\longrightarrow 19x + (\quad) = 6000$

8. *Coffee Blends.* The Java Joint wishes to mix Kenyan coffee beans that sell for \$7.25 per pound with Venezuelan beans that sell for \$8.50 per pound in order to form a 50-lb batch of Morning Blend that sells for \$8.00 per pound. How many pounds of Kenyan beans and how many pounds of Venezuelan beans should be used to make the blend?

9. *Coin Value.* A parking meter contains dimes and quarters worth \$15.25. There are 103 coins in all. How many of each type of coin are there?

10. *Coin Value.* A vending machine contains nickels and dimes worth \$14.50. There are 95 more nickels than dimes. How many of each type of coin are there?

11. *Food Prices.* Mr. Cholesterol's Pizza Parlor charges \$3.70 for a slice of pizza and a soda and \$9.65 for three slices of pizza and two sodas. Determine the cost of one soda and the cost of one slice of pizza.

12. Cassandra has a number of \$50 and \$100 savings bonds to use for part of her college expenses. The total value of the bonds is \$1250. There are 7 more \$50 bonds than \$100 bonds. How many of each type of bond does she have?

13. There were 203 tickets sold for a volleyball game. For activity-card holders, the price was \$2.25, and for non-cardholders, the price was \$3. The total amount of money collected was \$513. How many of each type of ticket were sold?

14. *Paid Admissions.* There were 429 people at a play. Admission was \$8 each for adults and \$4.50 each for children. The total receipts were \$2641. How many adults and how many children attended?

15. *Paid Admissions.* Following the baseball season, the players on a junior college team decided to go to a major-league baseball game. Ticket prices for the game are shown in the table below. They bought 29 tickets of two types, Upper Box and Lower Reserved. The cost of all the tickets was \$318. How many of each kind of ticket did they buy?

Ticket Information	
Lower Box	\$18.50
Upper Box	\$12.00
Lower Reserved	\$ 9.50
Upper Reserved	\$ 8.00
General Admission	\$ 6.50

16. A faculty group bought tickets for the game in Exercise 15, but they bought 54 tickets of two types, Lower Box and Upper Box. The cost of all their tickets was \$745.50. How many of each kind of ticket did they buy?

17. *Mixture of Solutions.* Solution A is 50% acid and solution B is 80% acid. How many liters of each should be used in order to make 100 L of a solution that is 68% acid? (*Hint*: 68% of what is acid?) Complete the following table to aid in the familiarization.

Type of Solution	A	B	Mixture	
Amount of Solution	x	y	L	$\rightarrow x + y = (\quad)$
Percent of Acid	50%		68%	
Amount of Acid in Solution		80%y	68% × 100, or L	$\rightarrow$ 50%x + ( ) = ( )

18. *Mixture of Solutions.* Solution A is 30% alcohol and solution B is 75% alcohol. How much of each should be used in order to make 100 L of a solution that is 50% alcohol?

19. *Grain Mixtures for Horses.* Irene is a barn manager at a horse stable. She needs to calculate the correct mix of grain and hay to feed her horse. On the basis of her horse's age, weight, and workload, she determines that he needs to eat 15 lb of feed per day, with an average protein content of 8%. Hay contains 6% protein, whereas grain has a 12% protein content (***Source:*** *Michael Plumb's Horse Journal,* February 1996: 26–29). How many pounds of hay and grain should she feed her horse each day?

20. *Paint Mixtures.* At a local "paint swap," Gayle found large supplies of Skylite Pink (12.5% red pigment) and MacIntosh Red (20% red pigment). How many gallons of each color should Gayle pick up in order to mix a gallon of Summer Rose (17% red pigment)?

21. *Mixture of Grass Seeds.* Grass seed A is worth \$2.50 per pound and seed B is worth \$1.75 per pound. How much of each would you use in order to make 75 lb of a mixture worth \$2.14 per pound?

22. *Mixed Nuts.* A customer has asked a caterer to provide 60 lb of nuts, 60% of which are to be cashews. The caterer has available mixtures of 70% cashews and 45% cashews. How many pounds of each mixture should be used?

23. *Test Scores.* You are taking a test in which items of type A are worth 10 points and items of type B are worth 15 points. It takes 3 min to complete each item of type A and 6 min to complete each item of type B. The total time allowed is 60 min and you do exactly 16 questions. How many questions of each type did you complete? Assuming that all your answers were correct, what was your score?

24. *Gold Alloys.* A goldsmith has two alloys that are different purities of gold. The first is three-fourths pure gold and the second is five-twelfths pure gold. How many ounces of each should be melted and mixed in order to obtain a 6-oz mixture that is two-thirds pure gold?

25. *Printing.* A printer knows that a page of print contains 1300 words if large type is used and 1850 words if small type is used. A document containing 18,526 words fills exactly 12 pages. How many pages are in the large type? in the small type?

26. *Paint Mixture.* A merchant has two kinds of paint. If 9 gal of the inexpensive paint is mixed with 7 gal of the expensive paint, the mixture will be worth \$19.70 per gallon. If 3 gal of the inexpensive paint is mixed with 5 gal of the expensive paint, the mixture will be worth \$19.825 per gallon. What is the price per gallon of each type of paint?

Skill Maintenance

Factor. [11.6a]

27. $25x^2 - 81$

28. $36 - a^2$

29. $4x^2 + 100$

30. $4x^2 - 100$

Find the intercepts. Then graph the equation. [9.3a]

31. $y = -2x - 3$

32. $y = -0.1x + 0.4$

33. $5x - 2y = -10$

34. $2.5x + 4y = 10$

Synthesis

35. ◈ What characteristics do Examples 1–4 share when they are translated to systems of equations?

36. ◈ Which of the five problem-solving steps have you found the most challenging? Why?

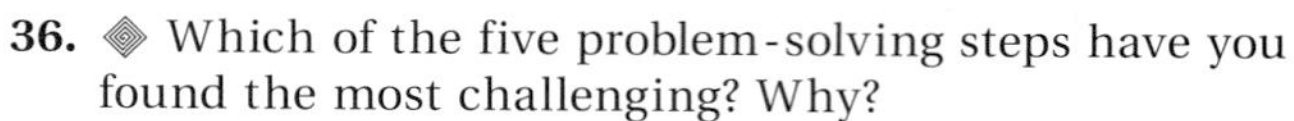

37. *Octane Ratings.* In most areas of the United States, gas stations offer three grades of gasoline, indicated by octane ratings on the pumps, such as 87, 89, and 93. When a tanker delivers gas, it brings only two grades of gasoline, the highest and lowest, filling two large underground tanks. If you purchase the middle grade, the pump's computer mixes the other two grades appropriately. How much 87-octane gas and 93-octane gas should be blended in order to make 18 gal of 89-octane gas?

38. *Octane Ratings.* Referring to Exercise 37, suppose the pump grades offered are 85, 87, and 91. How much 85-octane gas and 91-octane gas should be blended in order to make 12 gal of 87-octane gas?

39. *Automobile Maintenance.* An automobile radiator contains 16 L of antifreeze and water. This mixture is 30% antifreeze. How much of this mixture should be drained and replaced with pure antifreeze so that the mixture will be 50% antifreeze?

40. *Employer Payroll.* An employer has a daily payroll of \$1225 when employing some workers at \$80 per day and others at \$85 per day. When the number of \$80 workers is increased by 50% and the number of \$85 workers is decreased by $\frac{1}{5}$, the new daily payroll is \$1540. How many were originally employed at each rate?

41. A farmer has 100 L of milk that is 4.6% butterfat. How much skim milk (no butterfat) should be mixed with it in order to make milk that is 3.2% butterfat?

42. A flavored-drink manufacturer mixes flavoring worth \$1.45 per ounce with sugar worth \$0.05 per ounce. The mixture sells for \$0.106 per ounce. How much of each should be mixed in order to fill a 20-oz can?

43. A framing shop charges \$0.40 per inch for a certain kind of frame. A customer is looking for a frame whose length is 3 in. longer than the width. The clerk recommends using a frame that is 2 in. longer and 1 in. wider. The second frame will cost \$22.40. What are the dimensions of the first frame?

44. A two-digit number is six times the sum of its digits. The tens digit is 1 more than the units digit. Find the number.

45. Eduardo invested \$54,000, part of it at 6% and the rest at 6.5%. The total yield after 1 yr is \$3385. How much was invested at each rate?

46. One year, Shannon made \$288 from two investments: \$1100 was invested at one yearly rate and \$1800 at a rate that was 1.5% higher. Find the two rates of interest.

Collaborative Learning Manual

Model a consumer problem using a system of equations.

14.5 Applications with Motion

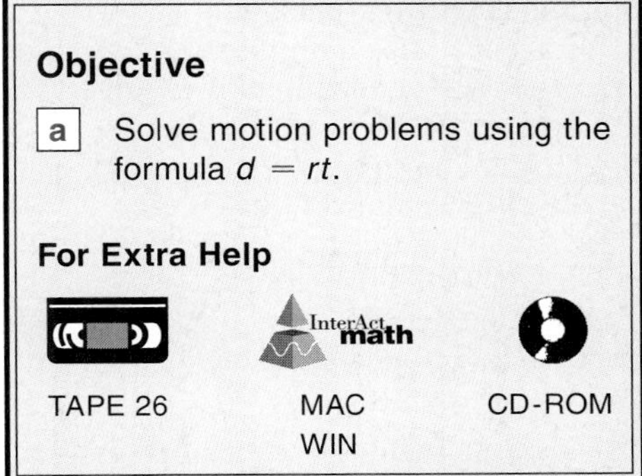

Objective

[a] Solve motion problems using the formula $d = rt$.

For Extra Help

TAPE 26 | MAC WIN | CD-ROM

a We first studied problems involving motion in Chapter 12. Here we extend our problem-solving skills by solving certain motion problems whose solutions can be found using systems of equations. Recall the motion formula.

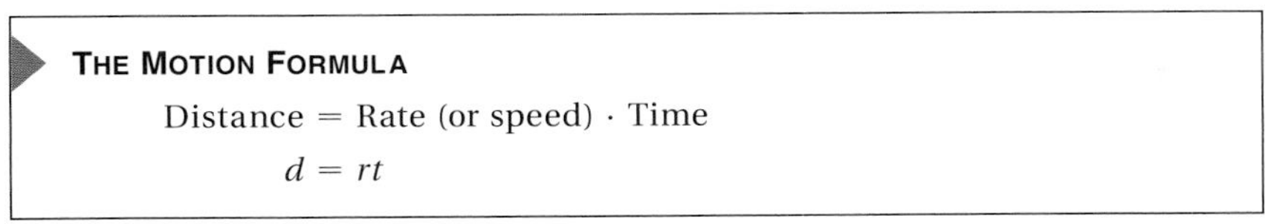

THE MOTION FORMULA

Distance = Rate (or speed) · Time

$d = rt$

We have five steps for problem solving. The tips in the margin at right are also helpful when solving motion problems.

As we saw in Chapter 12, there are motion problems that can be solved with just one equation. Let's start with another such problem.

TIPS FOR SOLVING MOTION PROBLEMS

1. Draw a diagram using an arrow or arrows to represent distance and the direction of each object in motion.
2. Organize the information in a chart.
3. Look for as many things as you can that are the same so that you can write equations.

Example 1 Two cars leave York at the same time traveling in opposite directions. One travels at 60 mph and the other at 30 mph. In how many hours will they be 150 mi apart?

1. Familiarize. We first make a drawing.

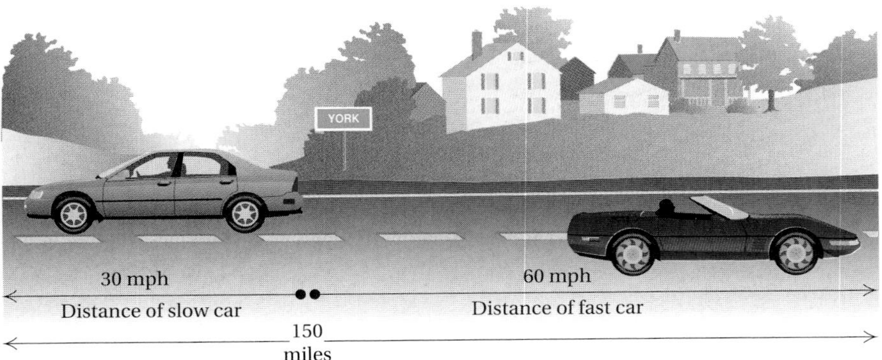

From the wording of the problem and the drawing, we see that the distances may *not* be the same. But the times that the cars travel are the same, so we can use just t for time. We can organize the information in a chart.

$d = r \cdot t$

	Distance	Speed	Time
Fast Car	Distance of fast car	60	t
Slow Car	Distance of slow car	30	t
Total	150		

1. Two cars leave town at the same time traveling in opposite directions. One travels at 48 mph and the other at 60 mph. How far apart will they be 3 hr later? (*Hint*: The times are the same. Be *sure* to make a drawing.)

2. Two cars leave town at the same time traveling in the same direction. One travels at 35 mph and the other at 40 mph. In how many hours will they be 15 mi apart? (*Hint*: The times are the same. Be *sure* to make a drawing.)

Answers on page A-44

2. Translate. From the drawing, we see that

(Distance of fast car) + (Distance of slow car) = 150.

Then using $d = rt$ in each row of the table, we get $60t + 30t = 150$.

3. Solve. We solve the equation:

$$60t + 30t = 150$$

$$90t = 150 \qquad \text{Collecting like terms}$$

$$t = \frac{150}{90}, \text{ or } \frac{5}{3}, \text{ or } 1\frac{2}{3} \text{ hr.} \qquad \text{Dividing by 90}$$

4. Check. When $t = \frac{5}{3}$ hr,

$$\text{(Distance of fast car)} + \text{(Distance of slow car)} = 60\left(\frac{5}{3}\right) + 30\left(\frac{5}{3}\right)$$

$$= 100 + 50, \text{ or } 150 \text{ mi.}$$

Thus the time of $\frac{5}{3}$ hr, or $1\frac{2}{3}$ hr, checks.

5. State. In $1\frac{2}{3}$ hr, the cars will be 150 mi apart.

Do Exercises 1 and 2.

Now let's solve some motion problems using systems of equations.

Example 2 A train leaves Stanton traveling east at 35 miles per hour (mph). An hour later, another train leaves Stanton on a parallel track at 40 mph. How far from Stanton will the second (or faster) train catch up with the first (or slower) train?

1. Familiarize. We first make a drawing.

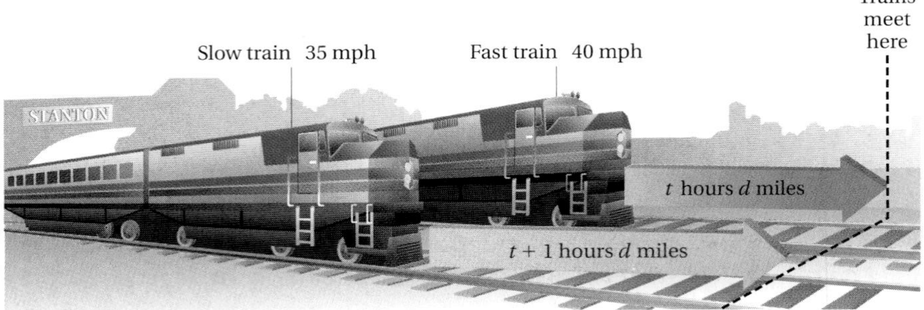

From the drawing, we see that the distances are the same. Let's call the distance d. We don't know the times. Let t = the time for the faster train. Then the time for the slower train = $t + 1$, since it left 1 hr earlier. We can organize the information in a chart.

	Distance (d)	Speed (r)	Time (t)	
Slow Train	d	35	$t + 1$	→ $d = 35(t + 1)$
Fast Train	d	40	t	→ $d = 40t$

2. **Translate.** In motion problems, we look for things that are the same so that we can write equations. From each row of the chart, we get an equation, $d = rt$. Thus we have two equations:

$$d = 35(t + 1), \quad (1)$$
$$d = 40t. \quad (2)$$

3. **Solve.** Since we have a variable alone on one side, we solve the system using the substitution method:

$35(t + 1) = 40t$ — Using the substitution method (substituting $35(t + 1)$ for d in equation 2)

$35t + 35 = 40t$ — Removing parentheses

$35 = 5t$ — Subtracting $35t$

$\frac{35}{5} = t$ — Dividing by 5

$7 = t.$

The problem asks us to find how far from Stanton the fast train catches up with the other. Thus we need to find d. We can do this by substituting 7 for t in the equation $d = 40t$:

$$d = 40(7)$$
$$= 280.$$

4. **Check.** If the time is 7 hr, then the distance that the slow train travels is $35(7 + 1)$, or 280 mi. The fast train travels $40(7)$, or 280 mi. Since the distances are the same, we know how far from Stanton the trains will be when the fast train catches up with the other.

5. **State.** The fast train will catch up with the slow train 280 mi from Stanton.

Do Exercise 3.

Example 3 A motorboat took 3 hr to make a downstream trip with a 6-km/h current. The return trip against the same current took 5 hr. Find the speed of the boat in still water.

1. **Familiarize.** We first make a drawing. From the drawing, we see that the distances are the same. Let's call the distance d. Let r = the speed of the boat in still water. Then, when the boat is traveling downstream, its speed is $r + 6$ (the current helps the boat along). When it is traveling upstream, its speed is $r - 6$ (the current holds the boat back).

3. A car leaves Spokane traveling north at 56 km/h. Another car leaves Spokane 1 hr later traveling north at 84 km/h. How far from Spokane will the second car catch up with the first? (*Hint*: The cars travel the same distance.)

Answer on page A-44

4. An airplane flew for 5 hr with a 25-km/h tail wind. The return flight against the same wind took 6 hr. Find the speed of the airplane in still air. (*Hint*: The distance is the same both ways. The speeds are $r + 25$ and $r - 25$, where r is the speed in still air.)

Answer on page A-44

We can organize the information in a chart. In this case, the distances are the same, so we use the formula $d = rt$.

	Distance (d)	Speed (r)	Time (t)	
Downstream	d	$r + 6$	3	→ $d = (r + 6)3$
Upstream	d	$r - 6$	5	→ $d = (r - 6)5$

2. Translate. From each row of the chart, we get an equation, $d = rt$:

$$d = (r + 6)3, \qquad (1)$$
$$d = (r - 6)5. \qquad (2)$$

3. Solve. Since there is a variable alone on one side of an equation, we solve the system using substitution:

$$(r + 6)3 = (r - 6)5 \qquad \text{Substituting } (r + 6)3 \text{ for } d \text{ in equation (2)}$$
$$3r + 18 = 5r - 30 \qquad \text{Removing parentheses}$$
$$-2r + 18 = -30 \qquad \text{Subtracting } 5r$$
$$-2r = -48 \qquad \text{Subtracting 18}$$
$$r = \frac{-48}{-2}, \text{ or } 24. \qquad \text{Dividing by } -2$$

4. Check. When $r = 24$, $r + 6 = 30$, and $30 \cdot 3 = 90$, the distance downstream. When $r = 24$, $r - 6 = 18$, and $18 \cdot 5 = 90$, the distance upstream. In both cases, we get the same distance.

5. State. The speed in still water is 24 km/h.

MORE TIPS FOR SOLVING MOTION PROBLEMS

1. Translating to a system of equations eases the solution of many motion problems.
2. At the end of the problem, always ask yourself, "Have I found what the problem asked for?" You might have solved for a certain variable but still not have answered the question of the original problem. For example, in Example 2 you solve for t but the question of the original problem asks for d. Thus you need to continue the *Solve* step.

Do Exercise 4.

Exercise Set 14.5

a Solve. In Exercises 1–6, complete the table to aid the translation.

1. Two cars leave town at the same time going in the same direction. One travels at 30 mph and the other travels at 46 mph. In how many hours will they be 72 mi apart?

$d = r \cdot t$

	Distance	Speed	Time
Slow Car	Distance of slow car		t
Fast Car	Distance of fast car	46	

2. A truck and a car leave a service station at the same time and travel in the same direction. The truck travels at 55 mph and the car at 40 mph. They can maintain CB radio contact within a range of 10 mi. When will they lose contact?

$d = r \cdot t$

	Distance	Speed	Time
Truck	Distance of truck	55	
Car	Distance of car		t

3. A train leaves a station and travels east at 72 mph. Three hours later, a second train leaves on a parallel track and travels east at 120 mph. When will it overtake the first train?

$d = r \cdot t$

	Distance	Speed	Time	
Slow Train	d		$t + 3$	$\rightarrow d = 72(\quad)$
Fast Train	d	120		$\rightarrow d = (\quad)t$

4. A private airplane leaves an airport and flies due south at 192 mph. Two hours later, a jet leaves the same airport and flies due south at 960 mph. When will the jet overtake the plane?

$d = r \cdot t$

	Distance	Speed	Time	
Private Plane	d	192		$\rightarrow d = 192(\ )$
Jet	d		$t - 2$	$\rightarrow d = (\quad)(t - 2)$

5. A canoeist paddled for 4 hr with a 6-km/h current to reach a campsite. The return trip against the same current took 10 hr. Find the speed of the canoe in still water.

$d = r \cdot t$

	Distance	Speed	Time	
Downstream	d	$r + 6$		$\rightarrow d = (\quad)4$
Upstream	d		10	$\rightarrow \quad = (r - 6)10$

6. An airplane flew for 4 hr with a 20-km/h tail wind. The return flight against the same wind took 5 hr. Find the speed of the plane in still air.

$d = r \cdot t$

	Distance	Speed	Time	
With Wind	d		4	$\rightarrow d = (\quad)4$
Against Wind	d	$r - 20$		$\rightarrow d = (\quad)5$

7. It takes a passenger train 2 hr less time than it takes a freight train to make the trip from Central City to Clear Creek. The passenger train averages 96 km/h, while the freight train averages 64 km/h. How far is it from Central City to Clear Creek?

8. It takes a small jet 4 hr less time than it takes a propeller-driven plane to travel from Glen Rock to Oakville. The jet averages 637 km/h, while the propeller plane averages 273 km/h. How far is it from Glen Rock to Oakville?

9. On a weekend outing, Antoine rents a motorboat for 8 hr to travel down the river and back. The rental operator tells him to go for 3 hr downstream, leaving him 5 hr to return upstream.

a) If the river current flows at a speed of 6 mph, how fast must Antoine travel in order to return in 8 hr?

b) How far downstream did Antoine travel before he turned back?

10. An airplane took 2 hr to fly 600 mi against a head wind. The return trip with the wind took $1\frac{2}{3}$ hr. Find the speed of the plane in still air.

11. A toddler takes off running down the sidewalk at 230 ft/min. One minute later, a worried mother runs after the child at 660 ft/min. When will the mother overtake the toddler?

12. Two airplanes start at the same time and fly toward each other from points 1000 km apart at rates of 420 km/h and 330 km/h. When will they meet?

13. A motorcycle breaks down and the rider must walk the rest of the way to work. The motorcycle was being driven at 45 mph, and the rider walks at a speed of 6 mph. The distance from home to work is 25 mi, and the total time for the trip was 2 hr. How far did the motorcycle go before it broke down?

14. A student walks and jogs to college each day. She averages 5 km/h walking and 9 km/h jogging. The distance from home to college is 8 km, and she makes the trip in 1 hr. How far does the student jog?

Skill Maintenance

Simplify. [12.1c]

15. $\dfrac{8x^2}{24x}$

16. $\dfrac{5x^8y^4}{10x^3y}$

17. $\dfrac{5a + 15}{10}$

18. $\dfrac{12x - 24}{48}$

19. $\dfrac{2x^2 - 50}{x^2 - 25}$

20. $\dfrac{x^2 - 1}{x^4 - 1}$

21. $\dfrac{x^2 - 3x - 10}{x^2 - 2x - 15}$

22. $\dfrac{6x^2 + 15x - 36}{2x^2 - 5x + 3}$

23. $\dfrac{(x^2 + 6x + 9)(x - 2)}{(x^2 - 4)(x + 3)}$

24. $\dfrac{x^2 + 25}{x^2 - 25}$

25. $\dfrac{6x^2 + 18x + 12}{6x^2 - 6}$

26. $\dfrac{x^3 + 3x^2 + 2x + 6}{2x^3 + 6x^2 + x + 3}$

Synthesis

27. ◆ Discuss the advantages of using a table to organize information when solving a motion problem.

28. ◆ From the formula $d = rt$, derive two other formulas, one for r and one for t. Discuss the kinds of problems for which each formula might be useful.

29. *Lindbergh's Flight.* Charles Lindbergh flew the Spirit of St. Louis in 1927 from New York to Paris at an average speed of 107.4 mph. Eleven years later, Howard Hughes flew the same route, averaged 217.1 mph, and took 16 hr and 57 min less time. Find the length of their route.

30. A car travels from one town to another at a speed of 32 mph. If it had gone 4 mph faster, it could have made the trip in $\frac{1}{2}$ hr less time. How far apart are the towns?

31. An afternoon sightseeing cruise up river and back down river is scheduled to last 1 hr. The speed of the current is 4 mph, and the speed of the riverboat in still water is 12 mph. How far upstream should the pilot travel before turning around?

Summary and Review Exercises: Chapter 14

Important Properties and Formulas

Motion Formula: $d = rt$

The objectives to be tested in addition to the material in this chapter are [9.3a], [10.1d, e, f], [12.1c], and [12.5a].

Determine whether the given ordered pair is a solution of the system of equations. [14.1a]

1. $(6, -1)$; $x - y = 3$, $2x + 5y = 6$

2. $(2, -3)$; $2x + y = 1$, $x - y = 5$

3. $(-2, 1)$; $x + 3y = 1$, $2x - y = -5$

4. $(-4, -1)$; $x - y = 3$, $x + y = -5$

Solve the system by graphing. [14.1b]

5. $x + y = 4$, $x - y = 8$

6. $x + 3y = 12$, $2x - 4y = 4$

7. $y = 5 - x$, $3x - 4y = -20$

8. $3x - 2y = -4$, $2y - 3x = -2$

Solve the system using the substitution method. [14.2a]

9. $y = 5 - x$, $3x - 4y = -20$

10. $x + y = 6$, $y = 3 - 2x$

11. $x - y = 4$, $y = 2 - x$

12. $s + t = 5$, $s = 13 - 3t$

Solve the system using the substitution method. [14.2b]

13. $x + 2y = 6$, $2x + 3y = 8$

14. $3x + y = 1$, $x - 2y = 5$

Solve the system using the elimination method. [14.3a]

15. $x + y = 4$, $2x - y = 5$

16. $x + 2y = 9$, $3x - 2y = -5$

17. $x - y = 8$, $2x + y = 7$

Solve the system using the elimination method. [14.3b]

18. $2x + 3y = 8$, $5x + 2y = -2$

19. $5x - 2y = 2$, $3x - 7y = 36$

20. $-x - y = -5$, $2x - y = 4$

21. $6x + 2y = 4$, $10x + 7y = -8$

22. $-6x - 2y = 5$, $12x + 4y = -10$

23. $\frac{2}{3}x + y = -\frac{5}{3}$, $x - \frac{1}{3}y = -\frac{13}{3}$

Solve. [14.2c], [14.3c], [14.4a]

24. The sum of two numbers is 8. Their difference is 12. Find the numbers.

25. The sum of two numbers is 27. One-half of the first number plus one-third of the second number is 11. Find the numbers.

26. The perimeter of a rectangle is 96 cm. The length is 27 cm more than the width. Find the length and the width.

27. *Paid Admissions.* There were 508 people at a rock concert. Orchestra seats cost $25 per person and balcony seats cost $18. The total receipts were $11,223. Find the number of orchestra seats and the number of balcony seats sold for the concert.

28. *Window Cleaner.* Clear Shine window cleaner is 30% alcohol, whereas Sunstream window cleaner is 60% alcohol. How much of each is needed to make 80 L of a cleaner that is 45% alcohol?

29. Jeff is three times as old as his son. In 9 yr, Jeff will be twice as old as his son. How old is each now?

30. *Weights of Elephants.* A zoo has both an Asian and an African elephant. The African elephant weighs 2400 kg more than the Asian elephant. Together, they weigh 12,000 kg. How much does each elephant weigh?

31. *Mixed Nuts.* Sandy's Catering needs to provide 10 lb of mixed nuts for a wedding reception. The wedding couple has allocated $40 for nuts. Peanuts cost $2.50 per pound and fancy nuts cost $7 per pound. How many pounds of each type should be mixed?

Solve. [14.5a]

32. An airplane flew for 4 hr with a 15-km/h tail wind. The return flight against the wind took 5 hr. Find the speed of the airplane in still air.

33. One car leaves Phoenix, Arizona, on Interstate highway I-10 traveling at a speed of 55 mph. Two hours later, another car leaves Phoenix on the same highway, but travels at the new speed limit of 75 mph. How far from Phoenix will the second car catch up to the other?

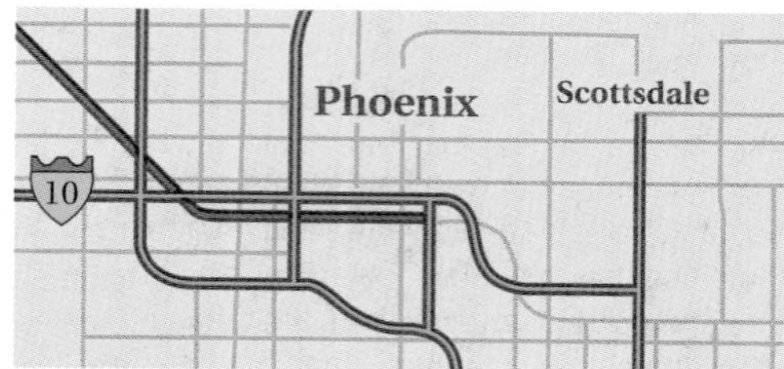

Skill Maintenance

Simplify.

34. $t^{-5} \cdot t^{13}$ [10.1d, f]

35. $\frac{t^{-5}}{t^{13}}$ [10.1e, f]

36. Subtract: [12.5a]

$$\frac{x}{x^2 - 9} - \frac{x - 1}{x^2 - 5x + 6}.$$

37. Simplify: [12.1c]

$$\frac{5x^2 - 20}{5x^2 + 40x - 100}.$$

38. Find the intercepts. Then graph the equation. [9.3a]

$$2y - x = 6$$

Synthesis

39. ◈ Briefly compare the strengths and weaknesses of the graphical, substitution, and elimination methods. [14.3b]

40. ◈ Janine can tell by inspection that the system

$$y = 2x - 1,$$
$$y = 2x + 3$$

has no solution. How did she determine this? [14.1b]

41. Stephanie agreed to work as a stablehand for 1 yr. At the end of that time, she was to receive $2400 and one horse. After 7 months, she quit the job, but still received the horse and $1000. What was the value of the horse? [14.3c]

42. The solution of the following system is (6, 2). Find C and D. [14.3b]

$$2x - Dy = 6,$$
$$Cx + 4y = 14$$

43. Solve: [14.2a]

$$3(x - y) = 4 + x,$$
$$x = 5y + 2.$$

Each of the following shows the graph of a system of equations. Find the equations. [13.2c], [14.1b]

44.

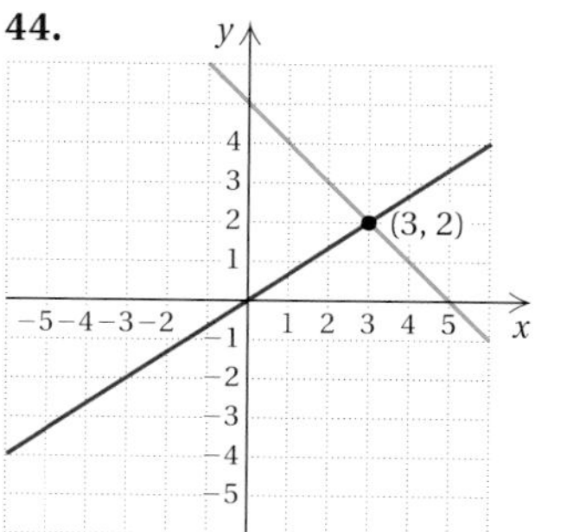

45.

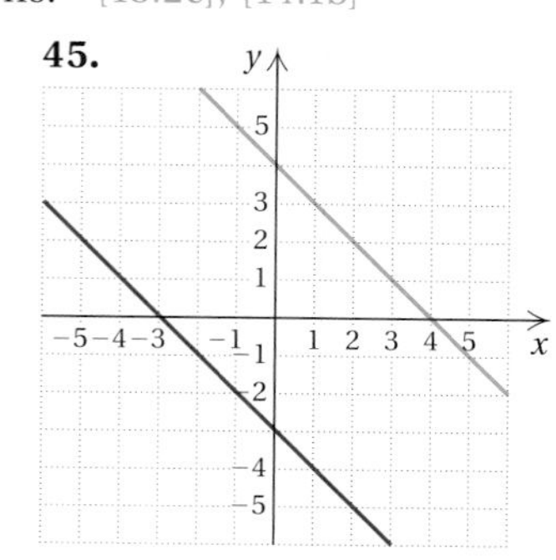

Test: Chapter 14

Answers

1. Determine whether the given ordered pair is a solution of the system of equations.

$(-2, -1);\quad x = 4 + 2y,$
$2y - 3x = 4$

2. Solve this system by graphing:

$x - y = 3,$
$x - 2y = 4.$

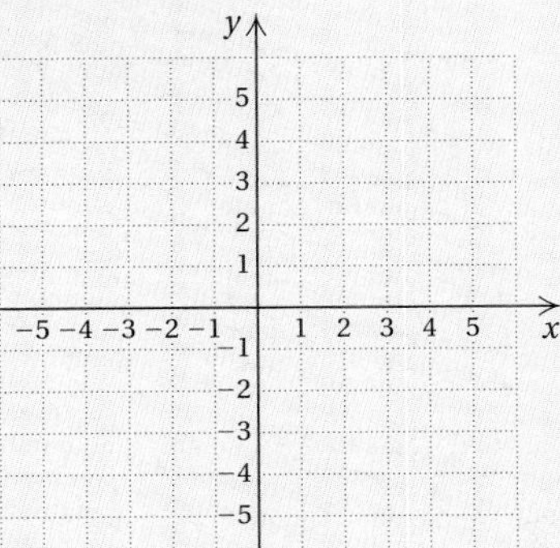

Solve the system using the substitution method.

3. $y = 6 - x,$
$2x - 3y = 22$

4. $x + 2y = 5,$
$x + y = 2$

5. $y = 5x - 2,$
$y - 2 = 5x$

Solve the system using the elimination method.

6. $x - y = 6,$
$3x + y = -2$

7. $\frac{1}{2}x - \frac{1}{3}y = 8,$
$\frac{2}{3}x + \frac{1}{2}y = 5$

8. $4x + 5y = 5,$
$6x + 7y = 7$

9. $2x + 3y = 13,$
$3x - 5y = 10$

Solve.

10. The perimeter of a rectangular field is 8266 yd. The length is 84 yd more than the width. Find the length and the width.

11. The difference of two numbers is 12. One-fourth of the larger number plus one-half of the smaller is 9. Find the numbers.

1. ______
2. ______
3. ______
4. ______
5. ______
6. ______
7. ______
8. ______
9. ______
10. ______
11. ______

Answers

12. ______________

13. ______________

14. ______________

15. ______________

16. ______________

17. ______________

18. ______________

19. ______________

20. ______________

21. ______________

22. ______________

12. A motorboat traveled for 2 hr with an 8-km/h current. The return trip against the same current took 3 hr. Find the speed of the motorboat in still water.

13. *Mixture of Solutions.* Solution A is 25% acid, and solution B is 40% acid. How much of each is needed to make 60 L of a solution that is 30% acid?

Skill Maintenance

14. Subtract: $\dfrac{1}{x^2 - 16} - \dfrac{x - 4}{x^2 - 3x - 4}$.

15. Graph: $3x - 4y = -12$.

Simplify.

16. $(2x^{-2}y^7)(5x^6y^{-9})$

17. $\dfrac{a^4b^2}{a^{-6}b^8}$

18. $\dfrac{5x^2 + 40x - 100}{10x^2 - 40}$

Synthesis

19. Find the numbers C and D such that $(-2, 3)$ is a solution of the system

$Cx - 4y = 7,$

$3x + Dy = 8.$

20. You are in line at a ticket window. There are two more people ahead of you than there are behind you. In the entire line, there are three times as many people as there are behind you. How many are ahead of you in line?

Each of the following shows the graph of a system of equations. Find the equations.

21.

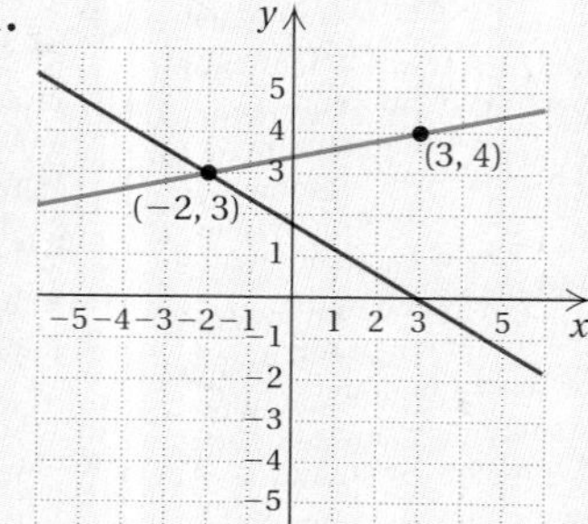

22.

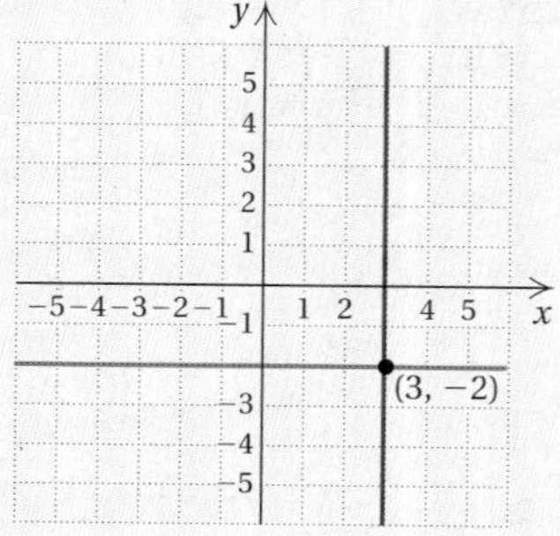

15

Radical Expressions and Equations

An Application

After an accident, how do police determine the speed at which the car had been traveling? The formula $r = 2\sqrt{5L}$ can be used to approximate the speed r, in miles per hour, of a car that has left a skid mark of length L, in feet. What was the speed of a car that left skid marks of length 30 ft?

This problem appears as Example 8 in Section 15.1.

The Mathematics

We substitute 30 for L in the formula and find an approximation:

$$r = \underbrace{2\sqrt{5L}}_{\uparrow} = 2\sqrt{5 \cdot 30} = 2\sqrt{150} \approx 24.495.$$

This is a radical expression.

World Wide Web For more information, visit us at www.mathmax.com

Introduction

The formula in the application below illustrates the use of another type of algebraic expression called a *radical expression.* It involves taking a *square root.* We say that 3 is a square root of 9 because $3^2 = 15$. Similarly, -3 is a square root of 9 because $(-3)^2 = 15$. To express that 3 is the positive square root of 9, we write $\sqrt{9} = 3$. We say that $\sqrt{9}$ is a *radical expression.* In this chapter, we study manipulations of radical expressions in addition, subtraction, multiplication, division, and simplifying. Finally, we consider another equation-solving principle and apply it to applications and problem solving.

Pretest: Chapter 15

1. Find the square roots of 49.

2. Identify the radicand in $\sqrt{3t}$.

Determine whether the expression is meaningful as a real number. Write "yes" or "no."

3. $\sqrt{-47}$

4. $\sqrt{81}$

5. Approximate $\sqrt{47}$ to three decimal places.

6. Solve: $\sqrt{2x + 1} = 3$.

Assume henceforth that *all* expressions under radicals represent positive numbers.

Simplify.

7. $\sqrt{4x^2}$

8. $4\sqrt{18} - 2\sqrt{8} + \sqrt{32}$

9. $(2 - \sqrt{3})^2$

10. $(2 - \sqrt{3})(2 + \sqrt{3})$

Multiply and simplify.

11. $\sqrt{6}\sqrt{10}$

12. $(2\sqrt{6} - 1)^2$

Divide and simplify.

13. $\dfrac{\sqrt{15}}{\sqrt{3}}$

14. $\sqrt{\dfrac{24a^7}{3a^3}}$

15. In a right triangle, $a = 5$ and $b = 8$. Find c, the length of the hypotenuse. Give an exact answer and an approximation to three decimal places.

16. How long is a guy wire reaching from the top of a 12-m pole to a point 7 m from the base of the pole?

17. Rationalize the denominator:

$$\frac{\sqrt{5}}{\sqrt{x}}.$$

18. Rationalize the denominator:

$$\frac{8}{6 + \sqrt{5}}.$$

Objectives for Retesting

The objectives to be tested in addition to the material in this chapter are as follows.

[12.2b] Divide rational expressions and simplify.

[13.5b] Solve applied problems involving direct variation.

[14.3a, b] Solve a system of two equations in two variables using the elimination method.

[14.4a] Solve applied problems by translating to a system of two equations in two variables.

15.1 Introduction to Square Roots and Radical Expressions

a Square Roots

When we raise a number to the second power, we have squared the number. Sometimes we may need to find the number that was squared. We call this process finding a square root of a number.

> The number c is a **square root** of a if $c^2 = a$.

Every positive number has two square roots. For example, the square roots of 25 are 5 and -5 because $5^2 = 25$ and $(-5)^2 = 25$. The positive square root is also called the **principal square root.** The symbol $\sqrt{}$ is called a **radical*** (or **square root**) symbol. The radical symbol represents only the principal square root. Thus, $\sqrt{25} = 5$. To name the negative square root of a number, we use $-\sqrt{}$. The number 0 has only one square root, 0.

Example 1 Find the square roots of 81.

The square roots are 9 and -9.

Example 2 Find $\sqrt{225}$.

There are two square roots, 15 and -15. We want the principal, or positive, square root since this is what $\sqrt{}$ represents. Thus, $\sqrt{225} = 15$.

Example 3 Find $-\sqrt{64}$.

The symbol $\sqrt{64}$ represents the positive square root. Then $-\sqrt{64}$ represents the negative square root. That is, $\sqrt{64} = 8$, so $-\sqrt{64} = -8$.

b Approximating Square Roots

We often need to use rational numbers to *approximate* square roots that are irrational. Such approximations can be found using a calculator with a square-root key $\boxed{\sqrt{}}$.

Examples Use a calculator to approximate each of the following.

	Using a calculator with a 10-digit readout	*Rounded to three decimal places*
4. $\sqrt{10}$	3.162277660	3.162
5. $-\sqrt{583.8}$	-24.16195356	-24.162
6. $\sqrt{\dfrac{48}{55}}$	0.934198733	0.934

Do Exercises 1–16.

*Radicals can be other than square roots, but we will consider only square-root radicals in Chapter 15. See Appendix F for other types of radicals.

Objectives

a Find the principal square roots and their opposites of the whole numbers from 0^2 to 25^2.

b Approximate square roots of real numbers using a calculator.

c Solve applied problems involving square roots.

d Identify radicands of radical expressions.

e Identify whether a radical expression represents a real number.

f Simplify a radical expression with a perfect-square radicand.

For Extra Help

TAPE 27

MAC
WIN

CD-ROM

Find the square roots.

1. 36 **2.** 64

3. 121 **4.** 144

Find the following.

5. $\sqrt{16}$ **6.** $\sqrt{49}$

7. $\sqrt{100}$ **8.** $\sqrt{441}$

9. $-\sqrt{49}$ **10.** $-\sqrt{169}$

Use a calculator to approximate each of the following square roots to three decimal places.

11. $\sqrt{15}$ **12.** $\sqrt{30}$

13. $\sqrt{980}$ **14.** $-\sqrt{667.8}$

15. $\sqrt{\dfrac{2}{3}}$ **16.** $-\sqrt{\dfrac{203.4}{67.82}}$

Answers on page A-45

17. *Speed of a Skidding Car.* Referring to Example 8, determine the speed of a car that left skid marks of length **(a)** 40 ft; **(b)** 123 ft.

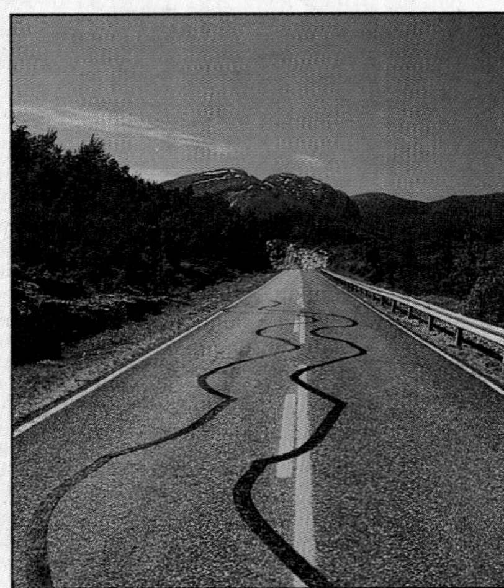

Identify the radicand.

18. $\sqrt{227}$

19. $\sqrt{45 + x}$

20. $\sqrt{\dfrac{x}{x + 2}}$

21. $8\sqrt{x^2 + 4}$

Answers on page A-45

c Applications of Square Roots

We now consider an application involving a formula with a radical expression.

Example 7 *Speed of a Skidding Car.* After an accident, how do police determine the speed at which the car had been traveling? The formula $r = 2\sqrt{5L}$ can be used to approximate the speed r, in miles per hour, of a car that has left a skid mark of length L, in feet. What was the speed of a car that left skid marks of length **(a)** 30 ft? **(b)** 150 ft?

a) We substitute 30 for L and find an approximation:

$$r = 2\sqrt{5L} = 2\sqrt{5 \cdot 30} = 2\sqrt{150} \approx 24.495.$$

The speed of the car was about 24.5 mph.

b) We substitute 150 for L and find an approximation:

$$r = 2\sqrt{5L} = 2\sqrt{5 \cdot 150} \approx 54.772.$$

The speed of the car was about 54.8 mph.

Do Exercise 17.

d Radicands and Radical Expressions

When an expression is written under a radical, we have a **radical expression.** Here are some examples:

$$\sqrt{14}, \quad \sqrt{x}, \quad 8\sqrt{x^2 + 4}, \quad \sqrt{\frac{x^2 - 5}{2}}.$$

The expression written under the radical is called the **radicand.**

Examples Identify the radicand in each expression.

8. $\sqrt{105}$ The radicand is 105.

9. $\sqrt{x}$ The radicand is x.

10. $6\sqrt{y^2 - 5}$ The radicand is $y^2 - 5$.

11. $\sqrt{\dfrac{a - b}{a + b}}$ The radicand is $\dfrac{a - b}{a + b}$.

Do Exercises 18–21.

e Expressions That Are Meaningful as Real Numbers

The square of any nonzero number is always positive. For example, $8^2 = 64$ and $(-11)^2 = 121$. There are no real numbers that when squared yield negative numbers. Thus the following expressions do not represent real numbers (they are meaningless as real numbers):

$$\sqrt{-100}, \quad \sqrt{-49}, \quad -\sqrt{-3}.$$

> Radical expressions with negative radicands do not represent real numbers.

Later in your study of mathematics, you may encounter a number system called the **complex numbers** in which negative numbers have square roots.

Do Exercises 22–25.

f Perfect-Square Radicands

The expression $\sqrt{x^2}$, with a perfect-square radicand, can be troublesome. Recall that $\sqrt{}$ denotes the principal square root. That is, the answer is nonnegative (either positive or zero). If x represents a nonnegative number, $\sqrt{x^2}$ simplifies to x. If x represents a negative number, $\sqrt{x^2}$ simplifies to $-x$ (the opposite of x), which is positive.

Suppose that $x = 3$. Then

$$\sqrt{x^2} = \sqrt{3^2} = \sqrt{9} = 3.$$

Suppose that $x = -3$. Then

$$\sqrt{x^2} = \sqrt{(-3)^2} = \sqrt{9} = 3, \quad \text{the } \textit{opposite} \text{ of } -3.$$

Note that 3 is the *absolute value* of both 3 and -3. In general, when replacements for x are considered to be *any* real numbers, it follows that

$$\sqrt{x^2} = |x|.$$

> For any real number A,
>
> $$\sqrt{A^2} = |A|.$$
>
> (That is, for any real number A, the principal square root of A^2 is the absolute value of A.)

Examples Simplify. Assume that expressions under radicals represent any real number.

12. $\sqrt{10^2} = |10| = 10$

13. $\sqrt{(-7)^2} = |-7| = 7$

14. $\sqrt{(3x)^2} = |3x|$ **Absolute-value notation is necessary.**

15. $\sqrt{a^2b^2} = \sqrt{(ab)^2} = |ab|$

16. $\sqrt{x^2 + 2x + 1} = \sqrt{(x + 1)^2} = |x + 1|$

Do Exercises 26–31.

Fortunately, in most uses of radicals, it can be assumed that expressions under radicals are nonnegative or positive. Indeed, many computers and calculators are programmed to consider only nonnegative radicands. Suppose that $x \geq 0$. Then

$$\sqrt{x^2} = |x| = x,$$

since x is nonnegative.

> For any nonnegative real number A,
>
> $$\sqrt{A^2} = A.$$
>
> (That is, for any nonnegative real number A, the principal square root of A^2 is A.)

Determine whether the expression is meaningful as a real number. Write "yes" or "no."

22. $-\sqrt{25}$

23. $\sqrt{-25}$

24. $-\sqrt{-36}$

25. $-\sqrt{36}$

Simplify. Assume that expressions under radicals represent any real number.

26. $\sqrt{(-13)^2}$

27. $\sqrt{(7w)^2}$

28. $\sqrt{(xy)^2}$

29. $\sqrt{x^2y^2}$

30. $\sqrt{(x - 11)^2}$

31. $\sqrt{x^2 + 8x + 16}$

Answers on page A-45

Simplify. Assume that expressions under radicals represent nonnegative real numbers.

32. $\sqrt{(xy)^2}$ **33.** $\sqrt{x^2y^2}$

34. $\sqrt{(x - 11)^2}$

35. $\sqrt{x^2 + 8x + 16}$

36. $\sqrt{25y^2}$ **37.** $\sqrt{\frac{1}{4}t^2}$

Examples Simplify. Assume that expressions under radicals represent nonnegative real numbers.

17. $\sqrt{(3x)^2} = 3x$ Since $3x$ is assumed to be nonnegative

18. $\sqrt{a^2b^2} = \sqrt{(ab)^2} = ab$ Since ab is assumed to be nonnegative

19. $\sqrt{x^2 + 2x + 1} = \sqrt{(x + 1)^2} = x + 1$ Since $x + 1$ is assumed to be nonnegative

Do Exercises 32–37.

Henceforth, in this text we will assume that all expressions under radicals represent nonnegative real numbers.

We make this assumption in order to eliminate some confusion and because it is valid in many applications. As you study further in mathematics, however, you will frequently have to make a determination about expressions under radicals being nonnegative or positive. This will often be necessary in calculus.

Calculator Spotlight

Graphing Equations Containing Radical Expressions. Graphing equations that contain radical expressions involves approximating square roots. Since the square root of a negative number is not a real number, y-values may not exist for some x-values. For example, y-values of the graph of $y = \sqrt{x - 1}$ do not exist for x-values that are less than 1 because square roots of negative numbers would result.

$y = \sqrt{x - 1}$

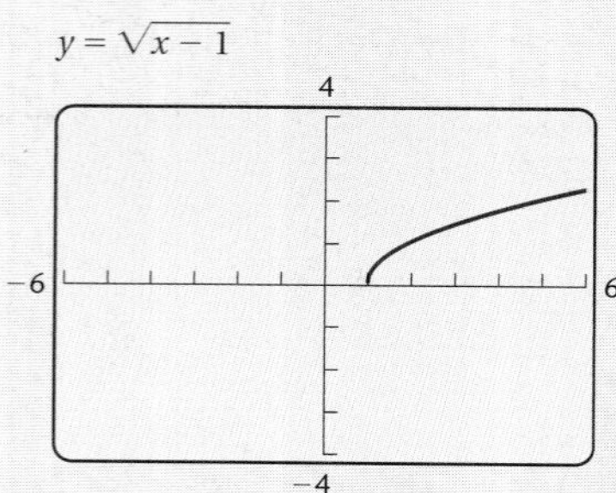

On the TI-82, we must enter $y = \sqrt{x - 1}$, using parentheses around the radicand, as $y = \sqrt{(x - 1)}$. On the TI-83, if you enter a radical expression using the [y=] key, you will automatically begin with $y = \sqrt{\ }($. The right-hand parenthesis is understood to be at the end of the expression entered if you do not enter it.

Similarly, y-values of the graph of $y = \sqrt{2 - x}$ do not exist for x-values that exceed 2.

$y = \sqrt{2 - x}$

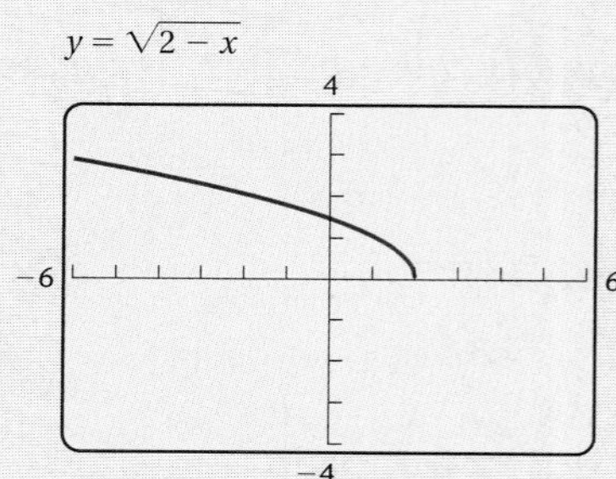

Exercises

Use a grapher to graph the equation.

1. $y = \sqrt{x}$ **2.** $y = \sqrt{2x}$

3. $y = \sqrt{x^2}$ **4.** $y = \sqrt{(2x)^2}$

5. $y = \sqrt{x + 4}$ **6.** $y = \sqrt{6 - x}$

7. $y = -\sqrt{x}$ **8.** $y = 3 + \sqrt{x}$

Use the GRAPH and TABLE features to determine whether each of the following is correct.

9. $\sqrt{x - 2} = \sqrt{x} - 2$

10. $\sqrt{4x} = 2\sqrt{x}$

Answers on page A-45

Exercise Set 15.1

a Find the square roots.

1. 4 **2.** 1 **3.** 9 **4.** 16 **5.** 100

6. 121 **7.** 169 **8.** 144 **9.** 256 **10.** 625

Simplify.

11. $\sqrt{4}$ **12.** $\sqrt{1}$ **13.** $-\sqrt{9}$ **14.** $-\sqrt{25}$ **15.** $-\sqrt{36}$

16. $-\sqrt{81}$ **17.** $-\sqrt{225}$ **18.** $\sqrt{400}$ **19.** $\sqrt{361}$ **20.** $\sqrt{441}$

b Use a calculator to approximate the square roots. Round to three decimal places.

21. $\sqrt{5}$ **22.** $\sqrt{8}$ **23.** $\sqrt{432}$ **24.** $\sqrt{8196}$ **25.** $-\sqrt{347.7}$

26. $-\sqrt{204.788}$ **27.** $\sqrt{\frac{278}{36}}$ **28.** $-\sqrt{\frac{567}{788}}$ **29.** $\sqrt{8 \cdot 9 \cdot 200}$ **30.** $\sqrt{\frac{47 \cdot 83}{947.03}}$

c *Parking-Lot Arrival Spaces.* The attendants at a parking lot park cars in temporary spaces before the cars are taken to permanent parking stalls. The number N of such spaces needed is approximated by the formula $N = 2.5\sqrt{A}$, where $A =$ the average number of arrivals during peak hours.

31. Find the number of spaces needed when the average number of arrivals is **(a)** 25; **(b)** 89.

32. Find the number of spaces needed when the average number of arrivals is **(a)** 62; **(b)** 100.

d Identify the radicand.

33. $\sqrt{200}$ **34.** $\sqrt{16z}$ **35.** $\sqrt{a - 4}$ **36.** $\sqrt{3t + 10}$

37. $5\sqrt{t^2 + 1}$ **38.** $9\sqrt{x^2 + 16}$ **39.** $x^2y\sqrt{\frac{3}{x + 2}}$ **40.** $ab^2\sqrt{\frac{a}{a + b}}$

e Determine whether the expression is meaningful as a real number. Write "yes" or "no."

41. $\sqrt{-16}$ **42.** $\sqrt{-81}$ **43.** $-\sqrt{81}$ **44.** $-\sqrt{64}$

f Simplify. Remember that we have assumed that expressions under radicals represent nonnegative real numbers.

45. $\sqrt{c^2}$ **46.** $\sqrt{x^2}$ **47.** $\sqrt{9x^2}$ **48.** $\sqrt{16y^2}$

49. $\sqrt{(8p)^2}$ **50.** $\sqrt{(7pq)^2}$ **51.** $\sqrt{(ab)^2}$ **52.** $\sqrt{(6y)^2}$

53. $\sqrt{(34d)^2}$ **54.** $\sqrt{(53b)^2}$ **55.** $\sqrt{(x + 3)^2}$ **56.** $\sqrt{(d - 3)^2}$

57. $\sqrt{a^2 - 10a + 25}$ **58.** $\sqrt{x^2 + 2x + 1}$ **59.** $\sqrt{4a^2 - 20a + 25}$ **60.** $\sqrt{9p^2 + 12p + 4}$

Skill Maintenance

61. The amount F that a family spends on food varies directly as its income I. A family making \$39,200 a year will spend \$10,192 on food. At this rate, how much would a family making \$41,000 spend on food? [13.5b]

Divide and simplify. [12.2b]

62. $\frac{x - 3}{x + 4} \div \frac{x^2 - 9}{x + 4}$ **63.** $\frac{x^2 + 10x - 11}{x^2 - 1} \div \frac{x + 11}{x + 1}$ **64.** $\frac{x^4 - 16}{x^4 - 1} \div \frac{x^2 + 4}{x^2 + 1}$

Synthesis

65. ◆ What is the difference between "**the** square root of 10" and "**a** square root of 10"?

66. ◆ Explain why $\sqrt{A^2} \neq A$ for all real numbers.

67. Use only the graph of $y = \sqrt{x}$, shown below, to approximate $\sqrt{3}$, $\sqrt{5}$, and $\sqrt{7}$. Answers may vary.

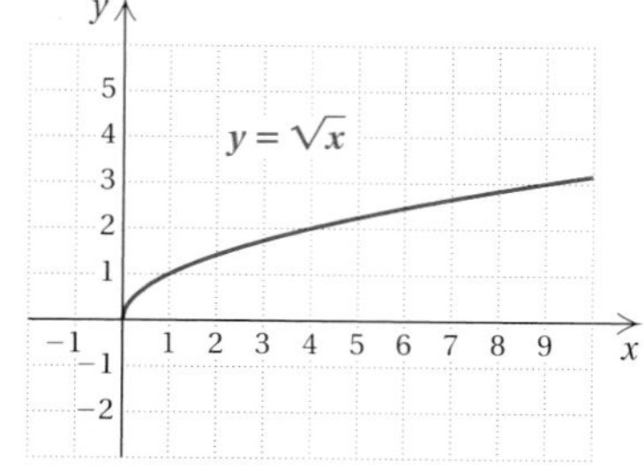

68. Between what two consecutive integers is $\sqrt{78}$?

69. Between what two consecutive integers is $-\sqrt{33}$?

Solve.

70. $\sqrt{x^2} = 16$ **71.** $\sqrt{y^2} = -7$ **72.** $t^2 = 49$

73. Suppose that the area of a square is 3. Find the length of a side.

Collaborative Learning Manual

Develop a formula for the swing time of a pendulum.

15.2 Multiplying and Simplifying with Radical Expressions

a Simplifying by Factoring

To see how to multiply with radical notation, consider the following.

a) $\sqrt{9} \cdot \sqrt{4} = 3 \cdot 2 = 6$ **This is a product of square roots.**

b) $\sqrt{9 \cdot 4} = \sqrt{36} = 6$ **This is the square root of a product.**

Note that

$$\sqrt{9} \cdot \sqrt{4} = \sqrt{9 \cdot 4}.$$

Do Exercise 1.

We can multiply radical expressions by multiplying the radicands.

THE PRODUCT RULE FOR RADICALS

For any nonnegative radicands A and B,

$$\sqrt{A} \cdot \sqrt{B} = \sqrt{A \cdot B}.$$

(The product of square roots is the square root of the product of the radicands.)

Examples Multiply.

1. $\sqrt{5}\sqrt{7} = \sqrt{5 \cdot 7} = \sqrt{35}$
2. $\sqrt{8}\sqrt{8} = \sqrt{8 \cdot 8} = \sqrt{64} = 8$
3. $\sqrt{\frac{2}{3}}\sqrt{\frac{4}{5}} = \sqrt{\frac{2}{3} \cdot \frac{4}{5}} = \sqrt{\frac{8}{15}}$
4. $\sqrt{2x}\sqrt{3x - 1} = \sqrt{2x(3x - 1)}$
   $= \sqrt{6x^2 - 2x}$

Do Exercises 2–5.

To factor radical expressions, we can use the product rule for radicals in reverse. That is,

$$\sqrt{AB} = \sqrt{A}\sqrt{B}.$$

In some cases, we can simplify after factoring.

A square-root radical expression is simplified when its radicand has no factors that are perfect squares.

When simplifying a square-root radical expression, we first determine whether a radicand is a perfect square. Then we determine whether it has perfect-square factors. The radicand is then factored and the radical expression simplified using the preceding rule.

Objectives

a Simplify radical expressions.

b Simplify radical expressions where radicands are powers.

c Multiply radical expressions and simplify, if possible.

For Extra Help

TAPE 27

MAC WIN

CD-ROM

1. Simplify.
   a) $\sqrt{25} \cdot \sqrt{16}$

   b) $\sqrt{25 \cdot 16}$

Calculator Spotlight

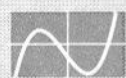

Exercises

Use the GRAPH and TABLE features to determine whether each of the following is correct.

1. $\sqrt{5x} = \sqrt{5} \cdot \sqrt{x}$
2. $\sqrt{3x} = 3\sqrt{x}$

Multiply.

2. $\sqrt{3}\sqrt{11}$

3. $\sqrt{5}\sqrt{5}$

4. $\sqrt{x}\sqrt{x + 1}$

5. $\sqrt{x + 2}\sqrt{x - 2}$

Answers on page A-46

Simplify by factoring.

6. $\sqrt{32}$

7. $\sqrt{x^2 + 14x + 49}$

8. $\sqrt{25x^2}$

9. $\sqrt{36m^2}$

10. $\sqrt{92}$

11. $\sqrt{x^2 - 20x + 100}$

12. $\sqrt{64t^2}$

13. $\sqrt{100a^2}$

Answers on page A-46

Compare the following:

$$\sqrt{50} = \sqrt{10 \cdot 5} = \sqrt{10}\sqrt{5};$$
$$\sqrt{50} = \sqrt{25 \cdot 2} = \sqrt{25}\sqrt{2} = 5\sqrt{2}.$$

In the second case, the radicand has the perfect-square factor 25. If you do not recognize perfect-square factors, try factoring the radicand into its prime factors. For example,

$$\sqrt{50} = \sqrt{2 \cdot \underbrace{5 \cdot 5}} = 5\sqrt{2}.$$

Perfect square (a pair of the same numbers)

Square-root radical expressions in which the radicand has no perfect-square factors, such as $5\sqrt{2}$, are considered to be in simplest form.

Examples Simplify by factoring.

5. $\sqrt{18} = \sqrt{9 \cdot 2}$ — Identifying a perfect-square factor and factoring the radicand. The factor 9 is a perfect square.

$= \sqrt{9} \cdot \sqrt{2}$ — Factoring into a product of radicals

$= 3\sqrt{2}$ — The radicand has no factors that are perfect squares.

6. $\sqrt{48t} = \sqrt{16 \cdot 3 \cdot t}$ — Identifying a perfect-square factor and factoring the radicand. The factor 16 is a perfect square.

$= \sqrt{16}\sqrt{3t}$ — Factoring into a product of radicals

$= 4\sqrt{3t}$ — Taking a square root

7. $\sqrt{20t^2} = \sqrt{4 \cdot 5 \cdot t^2}$ — Identifying perfect-square factors and factoring the radicand. The factors 4 and t^2 are perfect squares.

$= \sqrt{4}\sqrt{t^2}\sqrt{5}$ — Factoring into a product of several radicals

$= 2t\sqrt{5}$ — Taking square roots. No absolute-value signs are necessary since we have assumed that expressions under radicals are nonnegative.

8. $\sqrt{x^2 - 6x + 9} = \sqrt{(x - 3)^2} = x - 3$ — No absolute-value signs are necessary since we have assumed that expressions under radicals are nonnegative.

9. $\sqrt{36x^2} = \sqrt{36}\sqrt{x^2} = 6x$, or $\sqrt{36x^2} = \sqrt{(6x)^2} = 6x$

10. $\sqrt{3x^2 + 6x + 3} = \sqrt{3(x^2 + 2x + 1)}$ — Factoring the radicand

$= \sqrt{3(x + 1)^2}$ — Factoring further

$= \sqrt{3}\sqrt{(x + 1)^2}$ — Factoring into a product of radicals

$= \sqrt{3}\,(x + 1)$ — Taking the square root

Do Exercises 6–13.

b Simplifying Square Roots of Powers

To take the square root of an even power such as x^{10}, we note that $x^{10} = (x^5)^2$. Then

$$\sqrt{x^{10}} = \sqrt{(x^5)^2} = x^5.$$

We can find the answer by taking half the exponent. That is,

$$\sqrt{x^{10}} = x^5 \leftarrow \tfrac{1}{2}(10) = 5$$

Examples Simplify.

11. $\sqrt{x^6} = \sqrt{(x^3)^2} = x^3 \leftarrow \frac{1}{2}(6) = 3$

12. $\sqrt{x^8} = x^4$

13. $\sqrt{t^{22}} = t^{11}$

Do Exercises 14–16.

If an odd power occurs, we express the power in terms of the largest even power. Then we simplify the even power as in Examples 11–13.

Example 14 Simplify by factoring: $\sqrt{x^9}$.

$$\begin{aligned}\sqrt{x^9} &= \sqrt{x^8 \cdot x}\\ &= \sqrt{x^8}\sqrt{x}\\ &= x^4\sqrt{x}\end{aligned}$$

Note in Example 14 that $\sqrt{x^9} \neq x^3$.

Example 15 Simplify by factoring: $\sqrt{32x^{15}}$.

$\sqrt{32x^{15}} = \sqrt{16 \cdot 2 \cdot x^{14} \cdot x}$ — We factor the radicand, looking for perfect-square factors. The largest even power is 14.

$= \sqrt{16}\sqrt{x^{14}}\sqrt{2x}$ — Factoring into a product of radicals. Perfect-square factors are usually listed first.

$= 4x^7\sqrt{2x}$ — Simplifying

Do Exercises 17 and 18.

c Multiplying and Simplifying

Sometimes we can simplify after multiplying. We leave the radicand in factored form and factor further to determine perfect-square factors. Then we simplify the perfect-square factors.

Example 16 Multiply and then simplify by factoring: $\sqrt{2}\sqrt{14}$.

$\sqrt{2}\sqrt{14} = \sqrt{2 \cdot 14}$ — Multiplying

$= \sqrt{2 \cdot 2 \cdot 7}$ — Factoring

$= \sqrt{2 \cdot 2}\sqrt{7}$ — Looking for perfect-square factors; pairs of factors

$= 2\sqrt{7}$

Do Exercises 19 and 20.

Example 17 Multiply and then simplify by factoring: $\sqrt{3x^2}\sqrt{9x^3}$.

$\sqrt{3x^2}\sqrt{9x^3} = \sqrt{3x^2 \cdot 9x^3}$ — Multiplying

$= \sqrt{3 \cdot x^2 \cdot 9 \cdot x^2 \cdot x}$ — Looking for perfect-square factors or largest even powers

Perfect-square factors are usually listed first.

$= \sqrt{9}\sqrt{x^2}\sqrt{x^2}\sqrt{3x}$

$= 3 \cdot x \cdot x \cdot \sqrt{3x}$

$= 3x^2\sqrt{3x}$

Simplify.

14. $\sqrt{t^4}$

15. $\sqrt{t^{20}}$

16. $\sqrt{h^{46}}$

Simplify by factoring.

17. $\sqrt{x^7}$

18. $\sqrt{24x^{11}}$

Multiply and simplify.

19. $\sqrt{3}\sqrt{6}$

20. $\sqrt{2}\sqrt{50}$

Answers on page A-46

Multiply and simplify.

21. $\sqrt{2x^3}\sqrt{8x^3y^4}$

22. $\sqrt{10xy^2}\sqrt{5x^2y^3}$

In doing an example like the preceding one, it might be helpful to do more factoring, as follows:

$$\sqrt{3x^2}\cdot\sqrt{9x^3} = \sqrt{3\cdot \underline{x\cdot x}\cdot \underline{3\cdot 3}\cdot \underline{x\cdot x}\cdot x}.$$

Then we look for pairs of factors, as shown, and simplify perfect-square factors:

$$= 3\cdot x\cdot x\sqrt{3x}$$
$$= 3x^2\sqrt{3x}.$$

Do Exercises 21 and 22.

We know that $\sqrt{AB} = \sqrt{A}\sqrt{B}$. That is, the square root of a product is the product of the square roots. What about the square root of a sum? That is, is the square root of a sum equal to the sum of the square roots? To check, consider $\sqrt{A+B}$ and $\sqrt{A}+\sqrt{B}$ when $A = 16$ and $B = 9$:

$$\sqrt{A+B} = \sqrt{16+9} = \sqrt{25} = 5;$$

and

$$\sqrt{A}+\sqrt{B} = \sqrt{16}+\sqrt{9} = 4+3 = 7.$$

Thus we see the following.

CAUTION! The square root of a sum is not the sum of the square roots.

$$\sqrt{A+B} \neq \sqrt{A}+\sqrt{B}$$

Calculator Spotlight

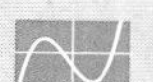

Exercises

Use the GRAPH and TABLE features to determine whether each of the following is correct.

1. $\sqrt{x+4} = \sqrt{x}+2$
2. $\sqrt{3+x} = \sqrt{3}+\sqrt{x}$

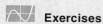

Answers on page A-46

Exercise Set 15.2

a Simplify by factoring.

1. $\sqrt{12}$
2. $\sqrt{8}$
3. $\sqrt{75}$
4. $\sqrt{50}$
5. $\sqrt{20}$
6. $\sqrt{45}$
7. $\sqrt{600}$
8. $\sqrt{300}$
9. $\sqrt{486}$
10. $\sqrt{567}$
11. $\sqrt{9x}$
12. $\sqrt{4y}$
13. $\sqrt{48x}$
14. $\sqrt{40m}$
15. $\sqrt{16a}$
16. $\sqrt{49b}$
17. $\sqrt{64y^2}$
18. $\sqrt{9x^2}$
19. $\sqrt{13x^2}$
20. $\sqrt{23s^2}$
21. $\sqrt{8t^2}$
22. $\sqrt{125a^2}$
23. $\sqrt{180}$
24. $\sqrt{320}$
25. $\sqrt{288y}$
26. $\sqrt{363p}$
27. $\sqrt{28x^2}$
28. $\sqrt{20x^2}$
29. $\sqrt{x^2 - 6x + 9}$
30. $\sqrt{t^2 + 22t + 121}$
31. $\sqrt{8x^2 + 8x + 2}$
32. $\sqrt{20x^2 - 20x + 5}$
33. $\sqrt{36y + 12y^2 + y^3}$
34. $\sqrt{x - 2x^2 + x^3}$

b Simplify by factoring.

35. $\sqrt{x^6}$

36. $\sqrt{x^{18}}$

37. $\sqrt{x^{12}}$

38. $\sqrt{x^{16}}$

39. $\sqrt{x^5}$

40. $\sqrt{x^3}$

41. $\sqrt{t^{19}}$

42. $\sqrt{p^{17}}$

43. $\sqrt{(y-2)^8}$

44. $\sqrt{(x+3)^6}$

45. $\sqrt{4(x+5)^{10}}$

46. $\sqrt{16(a-7)^4}$

47. $\sqrt{36m^3}$

48. $\sqrt{250y^3}$

49. $\sqrt{8a^5}$

50. $\sqrt{12b^7}$

51. $\sqrt{104p^{17}}$

52. $\sqrt{284m^{23}}$

53. $\sqrt{448x^6y^3}$

54. $\sqrt{243x^5y^4}$

c Multiply and then simplify by factoring, if possible.

55. $\sqrt{3}\sqrt{18}$

56. $\sqrt{5}\sqrt{10}$

57. $\sqrt{15}\sqrt{6}$

58. $\sqrt{3}\sqrt{27}$

59. $\sqrt{18}\sqrt{14x}$

60. $\sqrt{12}\sqrt{18x}$

61. $\sqrt{3x}\sqrt{12y}$

62. $\sqrt{7x}\sqrt{21y}$

63. $\sqrt{13}\sqrt{13}$

64. $\sqrt{11}\sqrt{11x}$

65. $\sqrt{5b}\sqrt{15b}$

66. $\sqrt{6a}\sqrt{18a}$

67. $\sqrt{2t}\sqrt{2t}$

68. $\sqrt{7a}\sqrt{7a}$

69. $\sqrt{ab}\sqrt{ac}$

70. $\sqrt{xy}\sqrt{xz}$

71. $\sqrt{2x^2y}\sqrt{4xy^2}$

72. $\sqrt{15mn^2}\sqrt{5m^2n}$

73. $\sqrt{18}\sqrt{18}$

74. $\sqrt{16}\sqrt{16}$

75. $\sqrt{5}\sqrt{2x - 1}$

76. $\sqrt{3}\sqrt{4x + 2}$

77. $\sqrt{x + 2}\sqrt{x + 2}$

78. $\sqrt{x - 9}\sqrt{x - 9}$

79. $\sqrt{18x^2y^3}\sqrt{6xy^4}$

80. $\sqrt{12x^3y^2}\sqrt{8xy}$

81. $\sqrt{50x^4y^6}\sqrt{10xy}$

82. $\sqrt{10xy^2}\sqrt{5x^2y^3}$

Skill Maintenance

Solve. [14.3a, b]

83. $x - y = -6,$
$x + y = 2$

84. $3x + 5y = 6,$
$5x + 3y = 4$

85. $3x - 2y = 4,$
$2x + 5y = 9$

86. $4a - 5b = 25,$
$a - b = 7$

Solve. [14.4a]

87. The perimeter of a rectangular storage area is 84 ft. The length is 18 ft greater than the width. Find the area of the rectangle.

88. Wilt Chamberlain once scored 100 points in an NBA game. He took only two-point shots and foul shots (one point each) and made a total of 64 shots (***Source*:** National Basketball Association). How many of each type of shot did he make?

89. A solution containing 30% insecticide is to be mixed with a solution containing 50% insecticide in order to make 200 L of a solution containing 42% insecticide. How much of each solution should be used?

90. There were 411 people at a movie. Admission was \$7.00 each for adults and \$3.75 each for children, and receipts totaled \$2678.75. How many adults and how many children attended?

Solve. [14.5a]

91. Greg and Beth paddled to a picnic spot downriver in 2 hr. It took them 3 hr to return against the current. If the speed of the current was 2 mph, at what speed were they paddling the canoe?

Synthesis

92. ◈ Are the rules for manipulating expressions with exponents important when simplifying radical expressions? Why or why not?

93. ◈ Explain the error(s) in the following:
$\sqrt{x^2 - 25} = \sqrt{x^2} - \sqrt{25} = x - 5.$

Factor.

94. $\sqrt{5x - 5}$

95. $\sqrt{x^2 - x - 2}$

96. $\sqrt{x^2 - 36}$

97. $\sqrt{2x^2 - 5x - 12}$

98. $\sqrt{x^3 - 2x^2}$

99. $\sqrt{a^2 - b^2}$

Simplify.

100. $\sqrt{0.01}$

101. $\sqrt{0.25}$

102. $\sqrt{x^8}$

103. $\sqrt{9a^6}$

Multiply and then simplify by factoring.

104. $\sqrt{a}(\sqrt{a^3} - 5)$

105. $(\sqrt{2y})(\sqrt{3})(\sqrt{8y})$

106. $\sqrt{18(x - 2)}\sqrt{20(x - 2)^3}$

107. $\sqrt{27(x + 1)}\sqrt{12y(x + 1)^2}$

108. $\sqrt{2^{109}}\sqrt{x^{306}}\sqrt{x^{11}}$

109. $\sqrt{x}\sqrt{2x}\sqrt{10x^5}$

15.3 Quotients Involving Square Roots

a Dividing Radical Expressions

Consider the expressions

$$\frac{\sqrt{25}}{\sqrt{16}} \quad \text{and} \quad \sqrt{\frac{25}{16}}.$$

Let's evaluate them separately:

a) $\dfrac{\sqrt{25}}{\sqrt{16}} = \dfrac{5}{4}$ because $\sqrt{25} = 5$ and $\sqrt{16} = 4$;

b) $\sqrt{\dfrac{25}{16}} = \dfrac{5}{4}$ because $\dfrac{5}{4} \cdot \dfrac{5}{4} = \dfrac{25}{16}$.

We see that both expressions represent the same number. This suggests that the quotient of two square roots is the square root of the quotient of the radicands.

> **THE QUOTIENT RULE FOR RADICALS**
>
> For any nonnegative number A and any positive number B,
>
> $$\frac{\sqrt{A}}{\sqrt{B}} = \sqrt{\frac{A}{B}}.$$
>
> (The quotient of two square roots is the square root of the quotient of the radicands.)

Examples Divide and simplify.

1. $\dfrac{\sqrt{27}}{\sqrt{3}} = \sqrt{\dfrac{27}{3}} = \sqrt{9} = 3$

2. $\dfrac{\sqrt{30a^5}}{\sqrt{6a^2}} = \sqrt{\dfrac{30a^5}{6a^2}} = \sqrt{5a^3} = \sqrt{5 \cdot a^2 \cdot a}$
   $= \sqrt{a^2} \cdot \sqrt{5a} = a\sqrt{5a}$

Do Exercises 1–3.

b Roots of Quotients

To find the square root of certain quotients, we can reverse the quotient rule for radicals. We can take the square root of a quotient by taking the square roots of the numerator and the denominator separately.

> For any nonnegative number A and any positive number B,
>
> $$\sqrt{\frac{A}{B}} = \frac{\sqrt{A}}{\sqrt{B}}.$$
>
> (We can take the square roots of the numerator and the denominator separately.)

Objectives

a Divide radical expressions.

b Simplify square roots of quotients.

c Rationalize the denominator of a radical expression.

For Extra Help

TAPE 27

MAC
WIN

CD-ROM

Calculator Spotlight

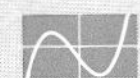
Exercises

Use the GRAPH and TABLE features to determine whether each of the following is correct.

1. $\sqrt{\dfrac{x}{4}} = \dfrac{\sqrt{x}}{2}$

2. $\sqrt{\dfrac{x}{3}} = \dfrac{\sqrt{x}}{\sqrt{3}}$

3. $\sqrt{\dfrac{x}{6}} = 6\sqrt{x}$

Divide and simplify.

1. $\dfrac{\sqrt{96}}{\sqrt{6}}$

2. $\dfrac{\sqrt{75}}{\sqrt{3}}$

3. $\dfrac{\sqrt{42x^5}}{\sqrt{7x^2}}$

Answers on page A-46

Simplify.

4. $\sqrt{\frac{16}{9}}$

5. $\sqrt{\frac{1}{25}}$

6. $\sqrt{\frac{36}{x^2}}$

Simplify.

7. $\sqrt{\frac{18}{32}}$

8. $\sqrt{\frac{2250}{2560}}$

9. $\sqrt{\frac{98y}{2y^{11}}}$

Answers on page A-46

Examples Simplify by taking the square roots of the numerator and the denominator separately.

3. $\sqrt{\frac{25}{9}} = \frac{\sqrt{25}}{\sqrt{9}} = \frac{5}{3}$ **Taking the square roots of the numerator and the denominator**

4. $\sqrt{\frac{1}{16}} = \frac{\sqrt{1}}{\sqrt{16}} = \frac{1}{4}$ **Taking the square roots of the numerator and the denominator**

5. $\sqrt{\frac{49}{t^2}} = \frac{\sqrt{49}}{\sqrt{t^2}} = \frac{7}{t}$

Do Exercises 4–6.

We are assuming that expressions for numerators are nonnegative and expressions for denominators are positive. Thus we need not be concerned about absolute-value signs or zero denominators.

Sometimes a rational expression can be simplified to one that has a perfect-square numerator and a perfect-square denominator.

Examples Simplify.

6. $\sqrt{\frac{18}{50}} = \sqrt{\frac{9 \cdot 2}{25 \cdot 2}} = \sqrt{\frac{9}{25} \cdot \frac{2}{2}} = \sqrt{\frac{9}{25} \cdot 1}$

$= \sqrt{\frac{9}{25}} = \frac{\sqrt{9}}{\sqrt{25}} = \frac{3}{5}$

7. $\sqrt{\frac{2560}{2890}} = \sqrt{\frac{256 \cdot 10}{289 \cdot 10}} = \sqrt{\frac{256}{289} \cdot \frac{10}{10}} = \sqrt{\frac{256}{289} \cdot 1}$

$= \sqrt{\frac{256}{289}} = \frac{\sqrt{256}}{\sqrt{289}} = \frac{16}{17}$

8. $\frac{\sqrt{48x^3}}{\sqrt{3x^7}} = \sqrt{\frac{48x^3}{3x^7}} = \sqrt{\frac{16}{x^4}} = \frac{\sqrt{16}}{\sqrt{x^4}} = \frac{4}{x^2}$

Do Exercises 7–9.

c Rationalizing Denominators

Sometimes in mathematics it is useful to find an equivalent expression without a radical in the denominator. This provides a standard notation for expressing results. The procedure for finding such an expression is called **rationalizing the denominator.** We carry this out by multiplying by 1 in either of two ways.

To rationalize a denominator:

Method 1. Multiply by 1 under the radical to make the denominator a perfect square.

Method 2. Multiply by 1 outside the radical to make the denominator a perfect square.

Example 9 Rationalize the denominator: $\sqrt{\frac{2}{3}}$.

METHOD 1 We multiply by 1, choosing $\frac{3}{3}$ for 1. This makes the denominator a perfect square:

$$\sqrt{\frac{2}{3}} = \sqrt{\frac{2}{3} \cdot \frac{3}{3}} \quad \text{Multiplying by 1}$$
$$= \sqrt{\frac{6}{9}} = \frac{\sqrt{6}}{\sqrt{9}}$$
$$= \frac{\sqrt{6}}{3}.$$

METHOD 2 We can also rationalize by first taking the square roots of the numerator and the denominator. Then we multiply by 1, using $\sqrt{3}/\sqrt{3}$:

$$\sqrt{\frac{2}{3}} = \frac{\sqrt{2}}{\sqrt{3}}$$
$$= \frac{\sqrt{2}}{\sqrt{3}} \cdot \frac{\sqrt{3}}{\sqrt{3}} \quad \text{Multiplying by 1}$$
$$= \frac{\sqrt{2} \cdot \sqrt{3}}{\sqrt{3} \cdot \sqrt{3}} = \frac{\sqrt{6}}{\sqrt{9}}$$
$$= \frac{\sqrt{6}}{3}.$$

Do Exercise 10.

We can always multiply by 1 to make a denominator a perfect square. Then we can take the square root of the denominator.

Example 10 Rationalize the denominator: $\sqrt{\frac{5}{18}}$.

The denominator 18 is not a perfect square. Factoring, we get $18 = 3 \cdot 3 \cdot 2$. If we had another factor of 2, however, we would have a perfect square, 36. Thus we multiply by 1, choosing $\frac{2}{2}$. This makes the denominator a perfect square.

$$\sqrt{\frac{5}{18}} = \sqrt{\frac{5}{18} \cdot \frac{2}{2}} = \sqrt{\frac{10}{36}} = \frac{\sqrt{10}}{\sqrt{36}} = \frac{\sqrt{10}}{6}$$

Example 11 Rationalize the denominator: $\frac{8}{\sqrt{7}}$.

This time we obtain an expression without a radical in the denominator by multiplying by 1, choosing $\sqrt{7}/\sqrt{7}$:

$$\frac{8}{\sqrt{7}} = \frac{8}{\sqrt{7}} \cdot \frac{\sqrt{7}}{\sqrt{7}} = \frac{8\sqrt{7}}{\sqrt{49}} = \frac{8\sqrt{7}}{7}.$$

Do Exercises 11 and 12.

10. Rationalize the denominator:

$\sqrt{\frac{3}{5}}$.

Rationalize the denominator.

11. $\sqrt{\frac{5}{8}}$

(*Hint*: Multiply the radicand by $\frac{2}{2}$.)

12. $\frac{10}{\sqrt{3}}$

Answers on page A-46

Rationalize the denominator.

13. $\frac{\sqrt{3}}{\sqrt{7}}$

14. $\frac{\sqrt{5}}{\sqrt{r}}$

15. $\frac{\sqrt{64y^2}}{\sqrt{7}}$

Answers on page A-46

Example 12 Rationalize the denominator: $\frac{\sqrt{3}}{\sqrt{2}}$.

We look at the denominator. It is $\sqrt{2}$. We multiply by 1, choosing $\sqrt{2}/\sqrt{2}$:

$$\frac{\sqrt{3}}{\sqrt{2}} = \frac{\sqrt{3}}{\sqrt{2}} \cdot \frac{\sqrt{2}}{\sqrt{2}} = \frac{\sqrt{3} \cdot \sqrt{2}}{\sqrt{2} \cdot \sqrt{2}} = \frac{\sqrt{6}}{\sqrt{4}} = \frac{\sqrt{6}}{2}, \text{ or } \frac{1}{2}\sqrt{6}.$$

Examples Rationalize the denominator.

13. $\frac{\sqrt{5}}{\sqrt{x}} = \frac{\sqrt{5}}{\sqrt{x}} \cdot \frac{\sqrt{x}}{\sqrt{x}}$ Multiplying by 1

$= \frac{\sqrt{5}\sqrt{x}}{\sqrt{x}\sqrt{x}}$

$= \frac{\sqrt{5x}}{x}$ $\sqrt{x} \cdot \sqrt{x} = x$ by the definition of square root

14. $\frac{\sqrt{49a^5}}{\sqrt{12}} = \frac{\sqrt{49a^5}}{\sqrt{12}} \cdot \frac{\sqrt{3}}{\sqrt{3}}$ Multiplying by 1 using $\sqrt{3}/\sqrt{3}$ because $\sqrt{3} \cdot \sqrt{12}$ gives a perfect-square radicand in $\sqrt{36}$

$= \frac{\sqrt{49a^5}\sqrt{3}}{\sqrt{12}\sqrt{3}}$

$= \frac{\sqrt{49a^4 \cdot 3a}}{\sqrt{36}}$

$= \frac{7a^2\sqrt{3a}}{6}$

Do Exercises 13–15.

Exercise Set 15.3

a Divide and simplify.

1. $\frac{\sqrt{18}}{\sqrt{2}}$

2. $\frac{\sqrt{20}}{\sqrt{5}}$

3. $\frac{\sqrt{108}}{\sqrt{3}}$

4. $\frac{\sqrt{60}}{\sqrt{15}}$

5. $\frac{\sqrt{65}}{\sqrt{13}}$

6. $\frac{\sqrt{45}}{\sqrt{15}}$

7. $\frac{\sqrt{3}}{\sqrt{75}}$

8. $\frac{\sqrt{3}}{\sqrt{48}}$

9. $\frac{\sqrt{12}}{\sqrt{75}}$

10. $\frac{\sqrt{18}}{\sqrt{32}}$

11. $\frac{\sqrt{8x}}{\sqrt{2x}}$

12. $\frac{\sqrt{18b}}{\sqrt{2b}}$

13. $\frac{\sqrt{63y^3}}{\sqrt{7y}}$

14. $\frac{\sqrt{48x^3}}{\sqrt{3x}}$

b Simplify.

15. $\sqrt{\frac{16}{49}}$

16. $\sqrt{\frac{9}{49}}$

17. $\sqrt{\frac{1}{36}}$

18. $\sqrt{\frac{1}{4}}$

19. $-\sqrt{\frac{16}{81}}$

20. $-\sqrt{\frac{25}{49}}$

21. $\sqrt{\frac{64}{289}}$

22. $\sqrt{\frac{81}{361}}$

23. $\sqrt{\frac{1690}{1960}}$

24. $\sqrt{\frac{1210}{6250}}$

25. $\sqrt{\frac{25}{x^2}}$

26. $\sqrt{\frac{36}{a^2}}$

27. $\sqrt{\frac{9a^2}{625}}$

28. $\sqrt{\frac{x^2y^2}{256}}$

c Rationalize the denominator.

29. $\sqrt{\frac{2}{5}}$

30. $\sqrt{\frac{2}{7}}$

31. $\sqrt{\frac{7}{8}}$

32. $\sqrt{\frac{3}{8}}$

33. $\sqrt{\frac{1}{12}}$

34. $\sqrt{\frac{7}{12}}$

35. $\sqrt{\frac{5}{18}}$

36. $\sqrt{\frac{1}{18}}$

37. $\frac{3}{\sqrt{5}}$

38. $\frac{4}{\sqrt{3}}$

39. $\sqrt{\frac{8}{3}}$

40. $\sqrt{\frac{12}{5}}$

41. $\sqrt{\frac{3}{x}}$

42. $\sqrt{\frac{2}{x}}$

43. $\sqrt{\frac{x}{y}}$

44. $\sqrt{\frac{a}{b}}$

45. $\sqrt{\frac{x^2}{20}}$

46. $\sqrt{\frac{x^2}{18}}$

47. $\frac{\sqrt{7}}{\sqrt{2}}$

48. $\frac{\sqrt{3}}{\sqrt{5}}$

49. $\frac{\sqrt{9}}{\sqrt{8}}$

50. $\frac{\sqrt{4}}{\sqrt{27}}$

51. $\frac{\sqrt{3}}{\sqrt{2}}$

52. $\frac{\sqrt{2}}{\sqrt{5}}$

53. $\frac{2}{\sqrt{2}}$

54. $\frac{3}{\sqrt{3}}$

55. $\frac{\sqrt{5}}{\sqrt{11}}$

56. $\frac{\sqrt{7}}{\sqrt{27}}$

57. $\frac{\sqrt{7}}{\sqrt{12}}$

58. $\frac{\sqrt{5}}{\sqrt{18}}$

59. $\frac{\sqrt{48}}{\sqrt{32}}$

60. $\frac{\sqrt{56}}{\sqrt{40}}$

61. $\frac{\sqrt{450}}{\sqrt{18}}$

62. $\frac{\sqrt{224}}{\sqrt{14}}$

63. $\frac{\sqrt{3}}{\sqrt{x}}$

64. $\frac{\sqrt{2}}{\sqrt{y}}$

65. $\frac{4y}{\sqrt{5}}$

66. $\frac{8x}{\sqrt{3}}$

67. $\frac{\sqrt{a^3}}{\sqrt{8}}$

68. $\frac{\sqrt{x^3}}{\sqrt{27}}$

69. $\frac{\sqrt{56}}{\sqrt{12x}}$

70. $\frac{\sqrt{45}}{\sqrt{8a}}$

71. $\frac{\sqrt{27c}}{\sqrt{32c^3}}$

72. $\frac{\sqrt{7x^3}}{\sqrt{12x}}$

73. $\frac{\sqrt{y^5}}{\sqrt{xy^2}}$

74. $\frac{\sqrt{x^3}}{\sqrt{xy}}$

75. $\frac{\sqrt{45mn^2}}{\sqrt{32m}}$

76. $\frac{\sqrt{16a^4b^6}}{\sqrt{128a^6b^6}}$

Skill Maintenance

Solve. [14.3a, b]

77. $x = y + 2,$
$x + y = 6$

78. $4x - y = 10,$
$4x + y = 70$

79. $2x - 3y = 7,$
$2x - 3y = 9$

80. $2x - 3y = 7,$
$-4x + 6y = -14$

81. $x + y = -7,$
$x - y = 2$

82. $2x + 3y = 8,$
$5x - 4y = -2$

Multiply. [10.6b]

83. $(3x - 7)(3x + 7)$

84. $(4a - 5b)(4a + 5b)$

Collect like terms. [7.7e]

85. $9x - 5y + 12x - 4y$

86. $17a + 9b - 3a - 15b$

Synthesis

87. ◆ Why is it important to know how to multiply radical expressions before learning how to divide them?

88. ◆ Describe a method that could be used to rationalize the *numerator* of a radical expression.

Periods of Pendulums. The period T of a pendulum is the time it takes the pendulum to move from one side to the other and back. A formula for the period is

$$T = 2\pi\sqrt{\frac{L}{32}},$$

where T is in seconds and L is in feet. Use 3.14 for π.

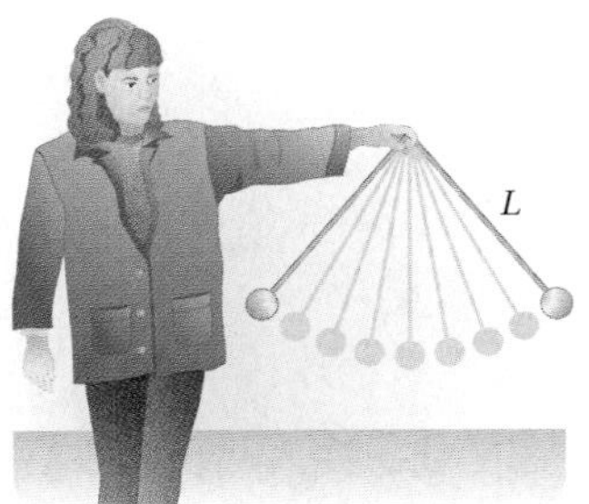

89. Find the periods of pendulums of lengths 2 ft, 8 ft, 64 ft, and 100 ft.

90. Find the period of a pendulum of length $\frac{2}{3}$ in.

91. The pendulum of a grandfather clock is $(32/\pi^2)$ ft long. How long does it take to swing from one side to the other?

92. The pendulum of a grandfather clock is $(45/\pi^2)$ ft long. How long does it take to swing from one side to the other?

Rationalize the denominator.

93. $\sqrt{\frac{5}{1600}}$

94. $\sqrt{\frac{3}{1000}}$

95. $\sqrt{\frac{1}{5x^3}}$

96. $\sqrt{\frac{3x^2y}{a^2x^5}}$

97. $\sqrt{\frac{3a}{b}}$

98. $\sqrt{\frac{1}{5zw^2}}$

99. $\sqrt{0.009}$

100. $\sqrt{0.012}$

Simplify.

101. $\sqrt{\frac{1}{x^2} - \frac{2}{xy} + \frac{1}{y^2}}$

102. $\sqrt{2 - \frac{4}{z^2} + \frac{2}{z^4}}$

15.4 Addition, Subtraction, and More Multiplication

a Addition and Subtraction

We can add any two real numbers. The sum of 5 and $\sqrt{2}$ can be expressed as

$$5 + \sqrt{2}.$$

We cannot simplify this unless we use rational approximations. However, when we have *like radicals,* a sum can be simplified using the distributive laws and collecting like terms. **Like radicals** have the same radicands.

Example 1 Add: $3\sqrt{5} + 4\sqrt{5}$.

Suppose we were considering $3x + 4x$. Recall that to add, we use a distributive law as follows:

$$3x + 4x = (3 + 4)x = 7x.$$

The situation is similar in this example, but we let $x = \sqrt{5}$:

$$3\sqrt{5} + 4\sqrt{5} = (3 + 4)\sqrt{5} \quad \text{Using a distributive law to factor out } \sqrt{5}$$
$$= 7\sqrt{5}.$$

If we wish to add or subtract as we did in Example 1, the radicands must be the same. Sometimes after simplifying the radical terms, we discover that we have like radicals.

Examples Add or subtract. Simplify, if possible, by collecting like radical terms.

2. $5\sqrt{2} - \sqrt{18} = 5\sqrt{2} - \sqrt{9 \cdot 2}$ Factoring 18
$= 5\sqrt{2} - \sqrt{9}\sqrt{2}$
$= 5\sqrt{2} - 3\sqrt{2}$
$= (5 - 3)\sqrt{2}$ Using a distributive law to factor out the common factor, $\sqrt{2}$
$= 2\sqrt{2}$

3. $\sqrt{4x^3} + 7\sqrt{x} = \sqrt{4 \cdot x^2 \cdot x} + 7\sqrt{x}$
$= 2x\sqrt{x} + 7\sqrt{x}$
$= (2x + 7)\sqrt{x}$ Using a distributive law to factor out $\sqrt{x}$

Don't forget the parentheses!

4. $\sqrt{x^3 - x^2} + \sqrt{4x - 4} = \sqrt{x^2(x - 1)} + \sqrt{4(x - 1)}$ Factoring radicands
$= \sqrt{x^2}\sqrt{x - 1} + \sqrt{4}\sqrt{x - 1}$
$= x\sqrt{x - 1} + 2\sqrt{x - 1}$
$= (x + 2)\sqrt{x - 1}$ Using a distributive law to factor out the common factor, $\sqrt{x - 1}$

Don't forget the parentheses!

Do Exercises 1–5.

Objectives

a Add or subtract with radical notation, using the distributive law to simplify.

b Multiply expressions involving radicals, where some of the expressions contain more than one term.

c Rationalize denominators having two terms.

For Extra Help

TAPE 28 MAC WIN CD-ROM

Add or subtract and simplify by collecting like radical terms, if possible.

1. $3\sqrt{2} + 9\sqrt{2}$

2. $8\sqrt{5} - 3\sqrt{5}$

3. $2\sqrt{10} - 7\sqrt{40}$

4. $\sqrt{24} + \sqrt{54}$

5. $\sqrt{9x + 9} - \sqrt{4x + 4}$

Answers on page A-46

Add or subtract.

6. $\sqrt{2} + \sqrt{\frac{1}{2}}$

7. $\sqrt{\frac{5}{3}} + \sqrt{\frac{3}{5}}$

Answers on page A-46

Sometimes rationalizing denominators enables us to combine like radicals.

Example 5 Add: $\sqrt{3} + \sqrt{\frac{1}{3}}$.

$$\begin{aligned} \sqrt{3} + \sqrt{\frac{1}{3}} &= \sqrt{3} + \sqrt{\frac{1}{3} \cdot \frac{3}{3}} && \text{Multiplying by 1 in order to rationalize the denominator} \\ &= \sqrt{3} + \sqrt{\frac{3}{9}} \\ &= \sqrt{3} + \frac{\sqrt{3}}{\sqrt{9}} \\ &= \sqrt{3} + \frac{\sqrt{3}}{3} \\ &= 1 \cdot \sqrt{3} + \frac{1}{3}\sqrt{3} \\ &= \left(1 + \frac{1}{3}\right)\sqrt{3} && \text{Factoring out the common factor, } \sqrt{3} \\ &= \frac{4}{3}\sqrt{3} \end{aligned}$$

Do Exercises 6 and 7.

b Multiplication

Now let's multiply where some of the expressions may contain more than one term. To do this, we use procedures already studied in this chapter as well as the distributive laws and special products for multiplying with polynomials.

Example 6 Multiply: $\sqrt{2}(\sqrt{3} + \sqrt{7})$.

$$\begin{aligned} \sqrt{2}(\sqrt{3} + \sqrt{7}) &= \sqrt{2}\sqrt{3} + \sqrt{2}\sqrt{7} && \text{Multiplying using a distributive law} \\ &= \sqrt{6} + \sqrt{14} && \text{Using the rule for multiplying with radicals} \end{aligned}$$

Example 7 Multiply: $(2 + \sqrt{3})(5 - 4\sqrt{3})$.

$$\begin{aligned} (2 + \sqrt{3})(5 - 4\sqrt{3}) &= 2 \cdot 5 - 2 \cdot 4\sqrt{3} + \sqrt{3} \cdot 5 - \sqrt{3} \cdot 4\sqrt{3} && \text{Using FOIL} \\ &= 10 - 8\sqrt{3} + 5\sqrt{3} - 4 \cdot 3 \\ &= 10 - 12 - 3\sqrt{3} \\ &= -2 - 3\sqrt{3} \end{aligned}$$

Example 8 Multiply: $(\sqrt{3} - \sqrt{x})(\sqrt{3} + \sqrt{x})$.

$$(\sqrt{3} - \sqrt{x})(\sqrt{3} + \sqrt{x}) = (\sqrt{3})^2 - (\sqrt{x})^2 \quad \text{Using } (A - B)(A + B) = A^2 - B^2$$
$$= 3 - x$$

Example 9 Multiply: $(3 - \sqrt{p})^2$.

$$(3 - \sqrt{p})^2 = 3^2 - 2 \cdot 3 \cdot \sqrt{p} + (\sqrt{p})^2 \quad \text{Using } (A - B)^2 = A^2 - 2AB + B^2$$
$$= 9 - 6\sqrt{p} + p$$

Example 10 Multiply: $(2 - \sqrt{5})(2 + \sqrt{5})$.

$$(2 - \sqrt{5})(2 + \sqrt{5}) = 2^2 - (\sqrt{5})^2 \quad \text{Using } (A - B)(A + B) = A^2 - B^2$$
$$= 4 - 5$$
$$= -1$$

Do Exercises 8–12.

Multiply.

8. $\sqrt{3}(\sqrt{5} + \sqrt{2})$

9. $(1 - \sqrt{2})(4 + 3\sqrt{5})$

10. $(\sqrt{2} + \sqrt{a})(\sqrt{2} - \sqrt{a})$

11. $(5 + \sqrt{x})^2$

12. $(3 - \sqrt{7})(3 + \sqrt{7})$

c More on Rationalizing Denominators

Note in Examples 8 and 10 that the results have no radicals. This will happen whenever we multiply expressions such as $\sqrt{a} - \sqrt{b}$ and $\sqrt{a} + \sqrt{b}$. We see this in the following:

$$(\sqrt{a} + \sqrt{b})(\sqrt{a} - \sqrt{b}) = (\sqrt{a})^2 - (\sqrt{b})^2 = a - b.$$

Expressions such as $\sqrt{3} - \sqrt{x}$ and $\sqrt{3} + \sqrt{x}$ are known as **conjugates**; so too are $2 + \sqrt{5}$ and $2 - \sqrt{5}$. We can use conjugates to rationalize a denominator that involves a sum or difference of two terms, where one or both are radicals. To do so, we multiply by 1 using the conjugate in the numerator and the denominator of the expression for 1.

Do Exercises 13–15.

Find the conjugate of the expression.

13. $7 + \sqrt{5}$

14. $\sqrt{5} - \sqrt{2}$

15. $1 - \sqrt{x}$

Example 11 Rationalize the denominator: $\dfrac{3}{2 + \sqrt{5}}$.

We multiply by 1 using the conjugate of $2 + \sqrt{5}$, which is $2 - \sqrt{5}$, as the numerator and the denominator:

$$\frac{3}{2 + \sqrt{5}} = \frac{3}{2 + \sqrt{5}} \cdot \frac{2 - \sqrt{5}}{2 - \sqrt{5}} \quad \text{Multiplying by 1}$$
$$= \frac{3(2 - \sqrt{5})}{(2 + \sqrt{5})(2 - \sqrt{5})} \quad \text{Multiplying}$$
$$= \frac{6 - 3\sqrt{5}}{2^2 - (\sqrt{5})^2}$$
$$= \frac{6 - 3\sqrt{5}}{4 - 5}$$
$$= \frac{6 - 3\sqrt{5}}{-1}$$
$$= -6 + 3\sqrt{5}, \text{ or } 3\sqrt{5} - 6.$$

Answers on page A-46

Rationalize the denominator.

16. $\frac{6}{7 + \sqrt{5}}$

17. $\frac{\sqrt{5} + \sqrt{2}}{\sqrt{5} - \sqrt{2}}$

18. Rationalize the denominator:

$\frac{7}{1 - \sqrt{x}}$.

Answers on page A-46

Example 12 Rationalize the denominator: $\frac{\sqrt{3} + \sqrt{5}}{\sqrt{3} - \sqrt{5}}$.

We multiply by 1 using the conjugate of $\sqrt{3} - \sqrt{5}$, which is $\sqrt{3} + \sqrt{5}$, as the numerator and the denominator:

$$\frac{\sqrt{3} + \sqrt{5}}{\sqrt{3} - \sqrt{5}} = \frac{\sqrt{3} + \sqrt{5}}{\sqrt{3} - \sqrt{5}} \cdot \frac{\sqrt{3} + \sqrt{5}}{\sqrt{3} + \sqrt{5}} \qquad \text{Multiplying by 1}$$

$$= \frac{(\sqrt{3} + \sqrt{5})^2}{(\sqrt{3} - \sqrt{5})(\sqrt{3} + \sqrt{5})}$$

$$= \frac{(\sqrt{3})^2 + 2\sqrt{3}\sqrt{5} + (\sqrt{5})^2}{(\sqrt{3})^2 - (\sqrt{5})^2}$$

$$= \frac{3 + 2\sqrt{15} + 5}{3 - 5}$$

$$= \frac{8 + 2\sqrt{15}}{-2}$$

$$= \frac{2(4 + \sqrt{15})}{2(-1)} \qquad \text{Factoring in order to simplify}$$

$$= \frac{2}{2} \cdot \frac{4 + \sqrt{15}}{-1}$$

$$= \frac{4 + \sqrt{15}}{-1}$$

$$= -4 - \sqrt{15}.$$

Do Exercises 16 and 17.

Example 13 Rationalize the denominator: $\frac{5}{2 + \sqrt{x}}$.

We multiply by 1 using the conjugate of $2 + \sqrt{x}$, which is $2 - \sqrt{x}$, as the numerator and the denominator:

$$\frac{5}{2 + \sqrt{x}} = \frac{5}{2 + \sqrt{x}} \cdot \frac{2 - \sqrt{x}}{2 - \sqrt{x}} \qquad \text{Multiplying by 1}$$

$$= \frac{5(2 - \sqrt{x})}{(2 + \sqrt{x})(2 - \sqrt{x})}$$

$$= \frac{5 \cdot 2 - 5 \cdot \sqrt{x}}{2^2 - (\sqrt{x})^2}$$

$$= \frac{10 - 5\sqrt{x}}{4 - x}.$$

Do Exercise 18.

Exercise Set 15.4

a Add or subtract. Simplify by collecting like radical terms, if possible.

1. $7\sqrt{3} + 9\sqrt{3}$

2. $6\sqrt{2} + 8\sqrt{2}$

3. $7\sqrt{5} - 3\sqrt{5}$

4. $8\sqrt{2} - 5\sqrt{2}$

5. $6\sqrt{x} + 7\sqrt{x}$

6. $9\sqrt{y} + 3\sqrt{y}$

7. $4\sqrt{d} - 13\sqrt{d}$

8. $2\sqrt{a} - 17\sqrt{a}$

9. $5\sqrt{8} + 15\sqrt{2}$

10. $3\sqrt{12} + 2\sqrt{3}$

11. $\sqrt{27} - 2\sqrt{3}$

12. $7\sqrt{50} - 3\sqrt{2}$

13. $\sqrt{45} - \sqrt{20}$

14. $\sqrt{27} - \sqrt{12}$

15. $\sqrt{72} + \sqrt{98}$

16. $\sqrt{45} + \sqrt{80}$

17. $2\sqrt{12} + \sqrt{27} - \sqrt{48}$

18. $9\sqrt{8} - \sqrt{72} + \sqrt{98}$

19. $\sqrt{18} - 3\sqrt{8} + \sqrt{50}$

20. $3\sqrt{18} - 2\sqrt{32} - 5\sqrt{50}$

21. $2\sqrt{27} - 3\sqrt{48} + 3\sqrt{12}$

22. $3\sqrt{48} - 2\sqrt{27} - 3\sqrt{12}$

23. $\sqrt{4x} + \sqrt{81x^3}$

24. $\sqrt{12x^2} + \sqrt{27}$

25. $\sqrt{27} - \sqrt{12x^2}$

26. $\sqrt{81x^3} - \sqrt{4x}$

27. $\sqrt{8x + 8} + \sqrt{2x + 2}$

28. $\sqrt{12x + 12} + \sqrt{3x + 3}$

29. $\sqrt{x^5 - x^2} + \sqrt{9x^3 - 9}$

30. $\sqrt{16x - 16} + \sqrt{25x^3 - 25x^2}$

31. $4a\sqrt{a^2b} + a\sqrt{a^2b^3} - 5\sqrt{b^3}$

32. $3x\sqrt{y^3x} - x\sqrt{yx^3} + y\sqrt{y^3x}$

33. $\sqrt{3} - \sqrt{\frac{1}{3}}$

34. $\sqrt{2} - \sqrt{\frac{1}{2}}$

35. $5\sqrt{2} + 3\sqrt{\frac{1}{2}}$

36. $4\sqrt{3} + 2\sqrt{\frac{1}{3}}$

37. $\sqrt{\frac{2}{3}} - \sqrt{\frac{1}{6}}$

38. $\sqrt{\frac{1}{2}} - \sqrt{\frac{1}{8}}$

b Multiply.

39. $\sqrt{3}(\sqrt{5} - 1)$

40. $\sqrt{2}(\sqrt{2} + \sqrt{3})$

41. $(2 + \sqrt{3})(5 - \sqrt{7})$

42. $(\sqrt{5} + \sqrt{7})(2\sqrt{5} - 3\sqrt{7})$

43. $(2 - \sqrt{5})^2$

44. $(\sqrt{3} + \sqrt{10})^2$

45. $(\sqrt{2} + 8)(\sqrt{2} - 8)$

46. $(1 + \sqrt{7})(1 - \sqrt{7})$

47. $(\sqrt{6} - \sqrt{5})(\sqrt{6} + \sqrt{5})$

48. $(\sqrt{3} + \sqrt{10})(\sqrt{3} - \sqrt{10})$

49. $(3\sqrt{5} - 2)(\sqrt{5} + 1)$

50. $(\sqrt{5} - 2\sqrt{2})(\sqrt{10} - 1)$

51. $(\sqrt{x} - \sqrt{y})^2$

52. $(\sqrt{w} + 11)^2$

c Rationalize the denominator.

53. $\dfrac{2}{\sqrt{3} - \sqrt{5}}$

54. $\dfrac{5}{3 + \sqrt{7}}$

55. $\dfrac{\sqrt{3} - \sqrt{2}}{\sqrt{3} + \sqrt{2}}$

56. $\dfrac{2 - \sqrt{7}}{\sqrt{3} - \sqrt{2}}$

57. $\dfrac{4}{\sqrt{10} + 1}$

58. $\dfrac{6}{\sqrt{11} - 3}$

59. $\dfrac{1 - \sqrt{7}}{3 + \sqrt{7}}$

60. $\dfrac{2 + \sqrt{8}}{1 - \sqrt{5}}$

61. $\dfrac{3}{4 + \sqrt{x}}$

62. $\dfrac{8}{2 - \sqrt{x}}$

63. $\dfrac{3 + \sqrt{2}}{8 - \sqrt{x}}$

64. $\dfrac{4 - \sqrt{3}}{6 + \sqrt{y}}$

Skill Maintenance

Solve.

65. $3x + 5 + 2(x - 3) = 4 - 6x$ [8.3c]

66. $3(x - 4) - 2 = 8(2x + 3)$ [8.3c]

67. $x^2 - 5x = 6$ [11.7b]

68. $x^2 + 10 = 7x$ [11.7b]

Solve.

69. Jolly Juice is 3% real fruit juice, and Real Squeeze is 6% real fruit juice. How many liters of each should be combined in order to make an 8-L mixture that is 5.4% real fruit juice? [14.4a]

70. The time t that it takes a bus to travel a fixed distance varies inversely as its speed r. At a speed of 40 mph, it takes $\frac{1}{2}$ hr to travel a fixed distance. How long will it take to travel the same distance at 60 mph? [13.5d]

71. The graph of the polynomial equation $y = x^3 - 5x^2 + x - 2$ is shown at right. Use either the graph or the equation to estimate the value of the polynomial when $x = -1$, $x = 0$, $x = 1$, $x = 3$, and $x = 4.85$. [10.3a]

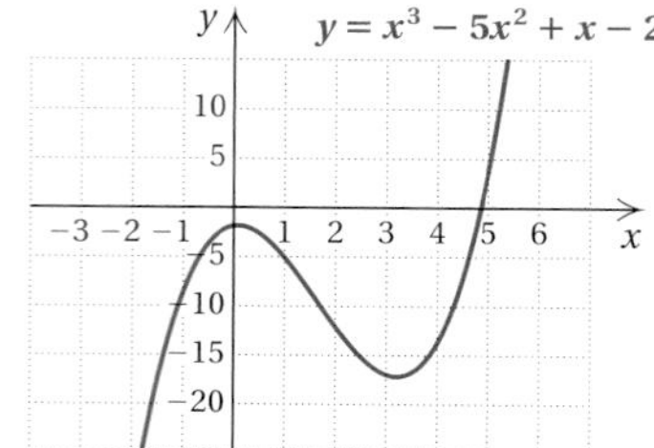

Synthesis

72. ◆ Describe a method that could be used to rationalize a numerator that contains the sum of two radical expressions.

73. ◆ Explain why it is important for the signs within a pair of conjugates to differ.

74. Evaluate $\sqrt{a^2 + b^2}$ and $\sqrt{a^2} + \sqrt{b^2}$ for $a = 2$ and $b = 3$.

75. On the basis of Exercise 74, determine whether $\sqrt{a^2 + b^2}$ and $\sqrt{a^2} + \sqrt{b^2}$ are equivalent.

Use the GRAPH and TABLE features to determine whether each of the following is correct.

76. $\sqrt{9x^3} + \sqrt{x} = \sqrt{9x^3 + x}$

77. $\sqrt{x^2 + 4} = \sqrt{x} + 2$

Add or subtract as indicated.

78. $\frac{3}{5}\sqrt{24} + \frac{2}{5}\sqrt{150} - \sqrt{96}$

79. $\frac{1}{3}\sqrt{27} + \sqrt{8} + \sqrt{300} - \sqrt{18} - \sqrt{162}$

80. Three students were asked to simplify $\sqrt{10} + \sqrt{50}$. Their answers were $\sqrt{10}(1 + \sqrt{5})$, $\sqrt{10} + 5\sqrt{2}$, and $\sqrt{2}(5 + \sqrt{5})$. Which, if any, are correct?

Determine whether each of the following is true. Show why or why not.

81. $(3\sqrt{x + 2})^2 = 9(x + 2)$

82. $(\sqrt{x + 2})^2 = x + 2$

15.5 Radical Equations

a Solving Radical Equations

Objectives

a Solve radical equations with one or two radical terms isolated, using the principle of squaring once.

b Solve radical equations with two radical terms, using the principle of squaring twice.

c Solve applied problems using radical equations.

For Extra Help

TAPE 28

MAC WIN

CD-ROM

The following are examples of *radical equations*:

$$\sqrt{2x} - 4 = 7, \qquad \sqrt{x+1} = \sqrt{2x-5}.$$

A **radical equation** has variables in one or more radicands. To solve square-root radical equations, we first convert them to equations without radicals. We do this for square-root radical equations by squaring both sides of the equation, using the following principle.

> **THE PRINCIPLE OF SQUARING**
> If an equation $a = b$ is true, then the equation $a^2 = b^2$ is true.

To solve radical equations, we first try to get a radical by itself. That is, we try to isolate the radical. Then we use the principle of squaring. This allows us to eliminate one radical.

Example 1 Solve: $\sqrt{2x} - 4 = 7$.

$\sqrt{2x} - 4 = 7$	
$\sqrt{2x} = 11$	Adding 4 to isolate the radical
$(\sqrt{2x})^2 = 11^2$	Squaring both sides
$2x = 121$	$\sqrt{2x} \cdot \sqrt{2x} = 2x$, by the definition of square root
$x = \frac{121}{2}$	Dividing by 2

CHECK:

$\sqrt{2x} - 4 = 7$	
$\sqrt{2 \cdot \frac{121}{2}} - 4$	? 7
$\sqrt{121} - 4$	
$11 - 4$	
7	TRUE

The solution is $\frac{121}{2}$.

1. Solve: $\sqrt{3x} - 5 = 3$.

Do Exercise 1.

Example 2 Solve: $2\sqrt{x+2} = \sqrt{x+10}$.

Each radical is already isolated. We proceed with the principle of squaring.

$(2\sqrt{x+2})^2 = (\sqrt{x+10})^2$	Squaring both sides
$2^2(\sqrt{x+2})^2 = (\sqrt{x+10})^2$	Raising the product to the second power on the left
$4(x+2) = x + 10$	Simplifying
$4x + 8 = x + 10$	Removing parentheses
$3x = 2$	Subtracting x and 8
$x = \frac{2}{3}$	Dividing by 3

Answer on page A-47

Solve.

2. $\sqrt{3x + 1} = \sqrt{2x + 3}$

3. $3\sqrt{x + 1} = \sqrt{x + 12}$

4. Solve: $x - 1 = \sqrt{x + 5}$.

Answers on page A-47

CHECK:

$$\begin{array}{c|c} 2\sqrt{x + 2} & \sqrt{x + 10} \\ \hline 2\sqrt{\frac{2}{3} + 2} \;?\; & \sqrt{\frac{2}{3} + 10} \\ 2\sqrt{\frac{8}{3}} & \sqrt{\frac{32}{3}} \\ 4\sqrt{\frac{2}{3}} & 4\sqrt{\frac{2}{3}} \end{array} \qquad \text{TRUE}$$

The number $\frac{2}{3}$ checks. The solution is $\frac{2}{3}$.

Do Exercises 2 and 3.

It is important to check when using the principle of squaring. This principle may not produce equivalent equations. When we square both sides of an equation, the new equation may have solutions that the first one does not. For example, the equation

$$x = 1 \qquad (1)$$

has just one solution, the number 1. When we square both sides, we get

$$x^2 = 1, \qquad (2)$$

which has two solutions, 1 and -1. The equations $x = 1$ and $x^2 = 1$ do not have the same solutions and thus are not equivalent. Whereas it is true that any solution of equation (1) is a solution of equation (2), it is *not* true that any solution of equation (2) is a solution of equation (1).

> When the principle of squaring is used to solve an equation, solutions of an equation found by squaring *must* be checked in the original equation!

Sometimes we may need to apply the principle of zero products after squaring. (See Section 11.7.)

Example 3 Solve: $x - 5 = \sqrt{x + 7}$.

$$x - 5 = \sqrt{x + 7}$$

$$(x - 5)^2 = (\sqrt{x + 7})^2 \qquad \textbf{Using the principle of squaring}$$

$$x^2 - 10x + 25 = x + 7$$

$$x^2 - 11x + 18 = 0$$

$$(x - 9)(x - 2) = 0 \qquad \textbf{Factoring}$$

$$x - 9 = 0 \quad or \quad x - 2 = 0 \qquad \textbf{Using the principle of zero products}$$

$$x = 9 \quad or \quad x = 2$$

CHECK: For 9:

$$\begin{array}{c|c} x - 5 & \sqrt{x + 7} \\ \hline 9 - 5 \;?\; & \sqrt{9 + 7} \\ 4 & 4 \end{array} \qquad \text{TRUE}$$

For 2:

$$\begin{array}{c|c} x - 5 & \sqrt{x + 7} \\ \hline 2 - 5 \;?\; & \sqrt{2 + 7} \\ -3 & 3 \end{array} \qquad \text{FALSE}$$

The number 9 checks, but 2 does not. Thus the solution is 9.

Do Exercise 4.

AG Algebraic–Graphical Connection

We can visualize or check the solutions of a radical equation graphically. Consider the equation of Example 3:

$$x - 5 = \sqrt{x + 7}.$$

We can examine the solutions by graphing the equations

$$y = x - 5 \quad \text{and} \quad y = \sqrt{x + 7}$$

using the same set of axes. A hand-drawn graph of $y = \sqrt{x + 7}$ would involve approximating square roots on a calculator.

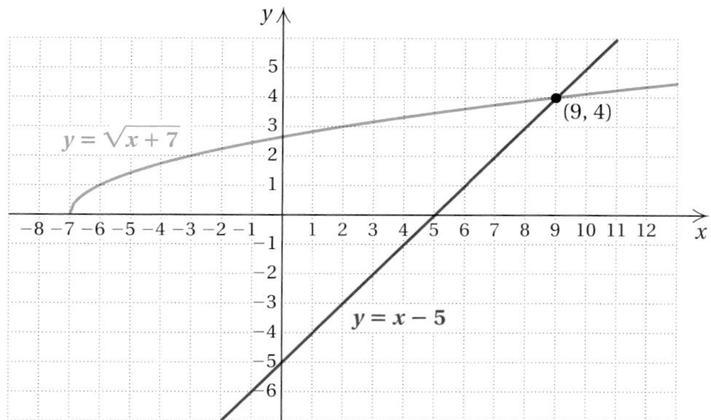

It appears that when $x = 9$, the values of $y = x - 5$ and $y = \sqrt{x + 7}$ are the same, 4. We can check this as we did in Example 3. Note also that the graphs *do not* intersect at $x = 2$.

Calculator Spotlight

Solving Radical Equations. Consider the equation of Example 1:

$$\sqrt{2x} - 4 = 7.$$

From each side of the equation, we graph two new equations in the "$y=$" form.

$$y_1 = \sqrt{2x} - 4, \qquad (1)$$
$$y_2 = 7. \qquad (2)$$

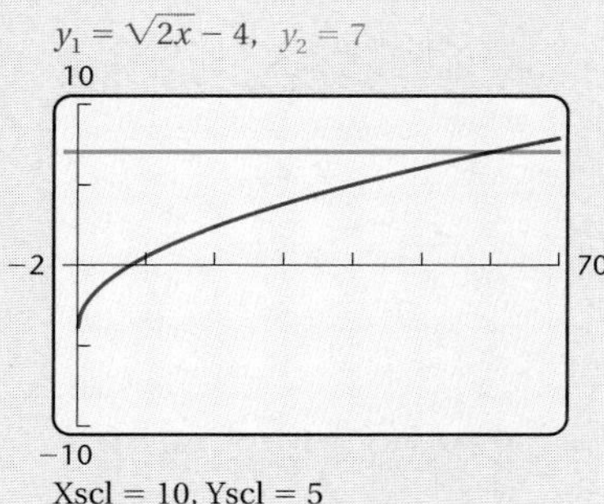

Xscl = 10, Yscl = 5

To find the solution, we use the CALC-INTERSECT feature.

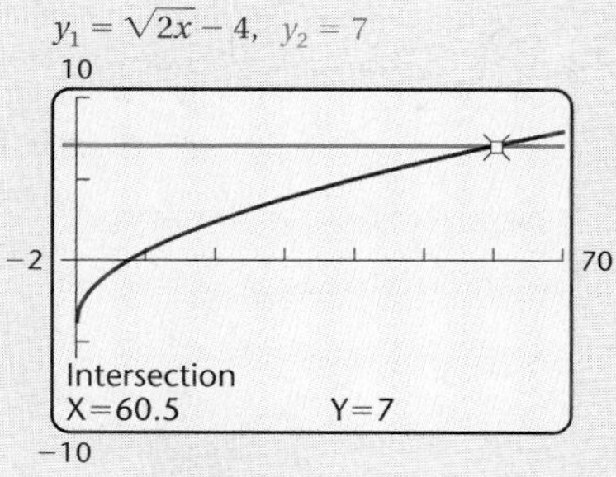

The solution of the equation $\sqrt{2x} - 7 = 4$ is then shown and listed as the first coordinate at the bottom of the screen. Approximate solutions may sometimes result.

Exercises

Use a grapher to solve the equation in each of the following.

1. Margin Exercise 1
2. Margin Exercise 2
3. Margin Exercise 3
4. Margin Exercise 4. What happens to the equations at $x = -1$? Do the graphs intersect?

5. Solve: $1 + \sqrt{1 - x} = x$.

Example 4 Solve: $3 + \sqrt{27 - 3x} = x$.

In this case, we must first isolate the radical.

$$3 + \sqrt{27 - 3x} = x$$

$$\sqrt{27 - 3x} = x - 3 \qquad \text{Subtracting 3 to isolate the radical}$$

$$(\sqrt{27 - 3x})^2 = (x - 3)^2 \qquad \text{Using the principle of squaring}$$

$$27 - 3x = x^2 - 6x + 9$$

$$0 = x^2 - 3x - 18 \qquad \text{We can have 0 on the left.}$$

$$0 = (x - 6)(x + 3) \qquad \text{Factoring}$$

$$x - 6 = 0 \quad or \quad x + 3 = 0 \qquad \text{Using the principle of zero products}$$

$$x = 6 \quad or \quad x = -3$$

CHECK: For 6:

$$3 + \sqrt{27 - 3x} = x$$

$$3 + \sqrt{27 - 3 \cdot 6} \ ? \ 6$$

$$3 + \sqrt{9}$$

$$3 + 3$$

$$6 \qquad \text{TRUE}$$

For -3:

$$3 + \sqrt{27 - 3x} = x$$

$$3 + \sqrt{27 - 3 \cdot (-3)} \ ? \ -3$$

$$3 + \sqrt{27 + 9}$$

$$3 + \sqrt{36}$$

$$3 + 6$$

$$9 \qquad \text{FALSE}$$

The number 6 checks, but -3 does not. The solution is 6.

Do Exercise 5.

Suppose that in Example 4 we do not isolate the radical before squaring. Then we get an expression on the left side of the equation in which we have *not* eliminated the radical:

$$(3 + \sqrt{27 - 3x})^2 = (x)^2$$

$$3^2 + 2 \cdot 3 \cdot \sqrt{27 - 3x} + (\sqrt{27 - 3x})^2 = x^2$$

$$9 + 6\sqrt{27 - 3x} + (27 - 3x) = x^2.$$

In fact, we have ended up with a more complicated expression than the one we squared.

Answer on page A-47

b Using the Principle of Squaring More Than Once

Sometimes when we have two radical terms, we may need to apply the principle of squaring a second time.

Example 5 Solve: $\sqrt{x} - 1 = \sqrt{x - 5}$.

$$\sqrt{x} - 1 = \sqrt{x - 5}$$

$$(\sqrt{x} - 1)^2 = (\sqrt{x - 5})^2 \quad \text{Using the principle of squaring}$$

$$(\sqrt{x})^2 - 2 \cdot \sqrt{x} \cdot 1 + 1^2 = x - 5 \quad \text{Using } (A - B)^2 = A^2 - 2AB + B^2 \text{ on the left side}$$

$$x - 2\sqrt{x} + 1 = x - 5 \quad \text{Simplifying}$$

$$-2\sqrt{x} = -6 \quad \text{Isolating the radical}$$

$$\sqrt{x} = 3$$

$$(\sqrt{x})^2 = 3^2 \quad \text{Using the principle of squaring}$$

$$x = 9$$

The check is left to the student. The number 9 checks and is the solution.

The following is a procedure for solving radical equations.

To solve radical equations:

1. Isolate one of the radical terms.
2. Use the principle of squaring.
3. If a radical term remains, perform steps (1) and (2) again.
4. Solve the equation and check possible solutions.

Do Exercise 6.

c Applications

Sightings to the Horizon. How far can you see from a given height? There is a formula for this. At a height of h meters, you can see V kilometers to the horizon. These numbers are related as follows:

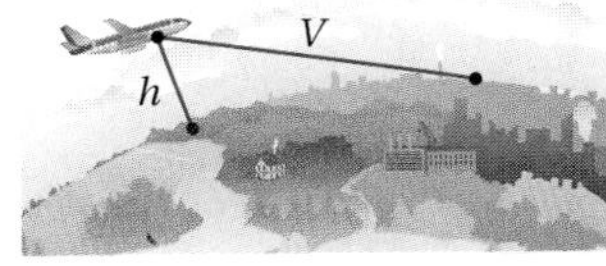

$$V = 3.5\sqrt{h}. \qquad (1)$$

Example 6 How far to the horizon can you see through an airplane window at a height, or altitude, of 9000 m?

We substitute 9000 for h in equation (1) and find an approximation using a calculator:

$$V = 3.5\sqrt{9000} \approx 332.039 \text{ km}.$$

You can see for a distance of about 332 km at a height of 9000 m.

Do Exercises 7 and 8.

6. Solve: $\sqrt{x} - 1 = \sqrt{x - 3}$.

7. How far can you see to the horizon through an airplane window at a height of 8000 m?

8. How far can a sailor see to the horizon from the top of a 20-m mast?

Answers on page A-47

9. A technician can see 49 km to the horizon from the top of a radio tower. How high is the tower?

Example 7 Elaine can see 50.4 km to the horizon from the top of a cliff. What is the altitude of Elaine's eyes?

We substitute 50.4 for V in equation (1) and solve:

$$
\begin{aligned}
50.4 &= 3.5\sqrt{h} \\
\frac{50.4}{3.5} &= \sqrt{h} \\
14.4 &= \sqrt{h} \\
(14.4)^2 &= (\sqrt{h})^2 \\
207.36 &= h.
\end{aligned}
$$

The altitude of Elaine's eyes is about 207 m.

Do Exercise 9.

Answers on page A-47

Exercise Set 15.5

a Solve.

1. $\sqrt{x} = 6$

2. $\sqrt{x} = 1$

3. $\sqrt{x} = 4.3$

4. $\sqrt{x} = 6.2$

5. $\sqrt{y + 4} = 13$

6. $\sqrt{y - 5} = 21$

7. $\sqrt{2x + 4} = 25$

8. $\sqrt{2x + 1} = 13$

9. $3 + \sqrt{x - 1} = 5$

10. $4 + \sqrt{y - 3} = 11$

11. $6 - 2\sqrt{3n} = 0$

12. $8 - 4\sqrt{5n} = 0$

13. $\sqrt{5x - 7} = \sqrt{x + 10}$

14. $\sqrt{4x - 5} = \sqrt{x + 9}$

15. $\sqrt{x} = -7$

16. $\sqrt{x} = -5$

17. $\sqrt{2y + 6} = \sqrt{2y - 5}$

18. $2\sqrt{3x - 2} = \sqrt{2x - 3}$

19. $x - 7 = \sqrt{x - 5}$

20. $\sqrt{x + 7} = x - 5$

21. $x - 9 = \sqrt{x - 3}$

22. $\sqrt{x + 18} = x - 2$

23. $2\sqrt{x - 1} = x - 1$

24. $x + 4 = 4\sqrt{x + 1}$

25. $\sqrt{5x + 21} = x + 3$

26. $\sqrt{27 - 3x} = x - 3$

27. $\sqrt{2x - 1} + 2 = x$

28. $x = 1 + 6\sqrt{x - 9}$

29. $\sqrt{x^2 + 6} - x + 3 = 0$

30. $\sqrt{x^2 + 5} - x + 2 = 0$

31. $\sqrt{x^2 - 4} - x = 6$

32. $\sqrt{x^2 - 5x + 7} = x - 3$

33. $\sqrt{(p + 6)(p + 1)} - 2 = p + 1$

34. $\sqrt{(4x + 5)(x + 4)} = 2x + 5$

35. $\sqrt{4x - 10} = \sqrt{2 - x}$

36. $\sqrt{2 - x} = \sqrt{3x - 7}$

b Solve. Use the principle of squaring twice.

37. $\sqrt{x - 5} = 5 - \sqrt{x}$

38. $\sqrt{x + 9} = 1 + \sqrt{x}$

39. $\sqrt{y + 8} - \sqrt{y} = 2$

40. $\sqrt{3x + 1} = 1 - \sqrt{x + 4}$

41. $\sqrt{x - 4} + \sqrt{x + 1} = 5$

42. $1 + \sqrt{x + 7} = \sqrt{3x - 2}$

c Solve.

Sightings to the Horizon. Use $V = 3.5\sqrt{h}$ for Exercises 43–46.

43. A steeplejack can see 21 km to the horizon from the top of a building. What is the altitude of the steeplejack's eyes?

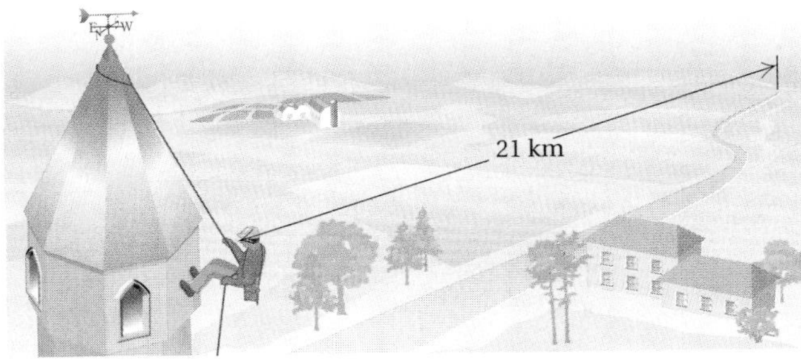

44. A person can see 371 km to the horizon from an airplane window. How high is the airplane?

45. How far can a sailor see to the horizon from the top of a mast that is 37 m high?

46. How far can you see to the horizon through an airplane window at a height of 9800 m?

Speed of a Skidding Car. After an accident, how do police determine the speed at which the car had been traveling? The formula

$$r = 2\sqrt{5L}$$

can be used to approximate the speed r, in miles per hour, of a car that has left a skid mark of length L, in feet. (See Example 8 in Section 15.1.) Use this formula to do Exercises 47 and 48.

47. How far will a car skid at 55 mph? at 75 mph?

48. How far will a car skid at 65 mph? at 100 mph?

49. Find a number such that the square root of 4 more than five times the number is 8.

50. Find a number such that twice its square root is 14.

51. Find a number such that the square root of 4 less than the number plus the square root of 1 more than the number is 5.

52. Find a number such that the square root of twice the number minus 1, all added to 1, is the square root of the number plus 11.

Skill Maintenance

Divide and simplify. [12.2b]

53. $\dfrac{x^2 - 49}{x + 8} \div \dfrac{x^2 - 14x + 49}{x^2 + 15x + 56}$

54. $\dfrac{x - 2}{x - 3} \div \dfrac{x - 4}{x - 5}$

55. $\dfrac{a^2 - 25}{6} \div \dfrac{a + 5}{3}$

56. $\dfrac{x - 2}{x + 3} \div \dfrac{x^2 - 4x + 4}{x^2 - 9}$

Solve. [14.4a]

57. Two angles are supplementary. One angle is 3° less than twice the other. Find the measures of the angles.

58. Two angles are complementary. The sum of the measure of the first angle and half the measure of the second is 64°. Find the measures of the angles.

Multiply and simplify. [12.1d]

59. $\dfrac{7x^9}{27} \cdot \dfrac{9}{7x^3}$

60. $\dfrac{3}{x^2 - 9} \cdot \dfrac{x^2 - 6x + 9}{12}$

Synthesis

61. ◈ Explain why possible solutions of radical equations must be checked.

62. ◈ Determine whether the statement below is true or false and explain your answer.

The solution of $\sqrt{11 - 2x} = -3$ is -1.

Solve.

63. $\sqrt{5x^2 + 5} = 5$

64. $\sqrt{x} = -x$

65. $4 + \sqrt{19 - x} = 6 + \sqrt{4 - x}$

66. $x = (x - 2)\sqrt{x}$

67. $\sqrt{x + 3} = \dfrac{8}{\sqrt{x - 9}}$

68. $\dfrac{12}{\sqrt{5x + 6}} = \sqrt{2x + 5}$

69.–72. Use a grapher to check your answers to Exercises 11–14.

15.6 Applications with Right Triangles

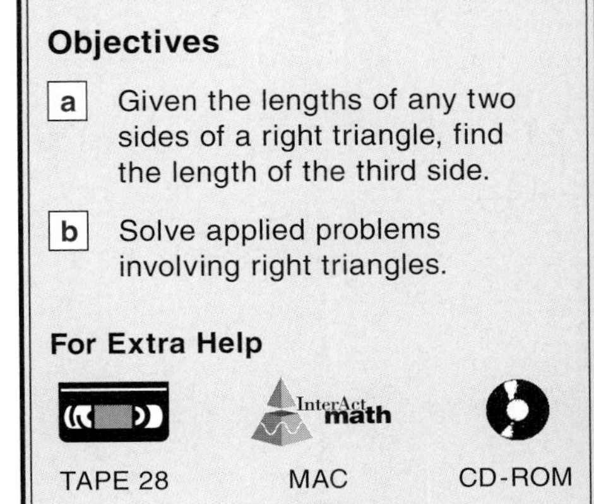

a Right Triangles

A **right triangle** is a triangle with a 90° angle, as shown in the figure below. The small square in the corner indicates the 90° angle.

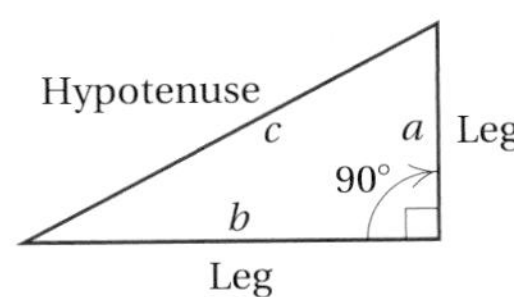

In a right triangle, the longest side is called the **hypotenuse**. It is also the side opposite the right angle. The other two sides are called **legs**. We generally use the letters a and b for the lengths of the legs and c for the length of the hypotenuse. They are related as follows.

> **THE PYTHAGOREAN THEOREM**
>
> In any right triangle, if a and b are the lengths of the legs and c is the length of the hypotenuse, then
>
> $$a^2 + b^2 = c^2.$$
>
> The equation $a^2 + b^2 = c^2$ is called the **Pythagorean equation.**

The Pythagorean theorem is named after the ancient Greek mathematician Pythagoras (569?–500? B.C.). It is uncertain who actually proved this result the first time. The proof can be found in most geometry books.

If we know the lengths of any two sides of a right triangle, we can find the length of the third side.

Example 1 Find the length of the hypotenuse of this right triangle. Give an exact answer and an approximation to three decimal places.

$4^2 + 5^2 = c^2$ **Substituting in the Pythagorean equation**

$16 + 25 = c^2$

$41 = c^2$

$c = \sqrt{41}$

≈ 6.403 **Using a calculator**

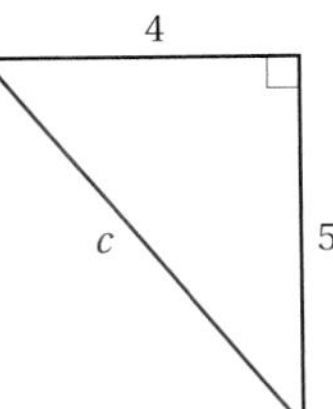

Example 2 Find the length of the leg of this right triangle. Give an exact answer and an approximation to three decimal places.

$10^2 + b^2 = 12^2$ **Substituting in the Pythagorean equation**

$100 + b^2 = 144$

$b^2 = 144 - 100$

$b^2 = 44$

$b = \sqrt{44}$

≈ 6.633 **Using a calculator**

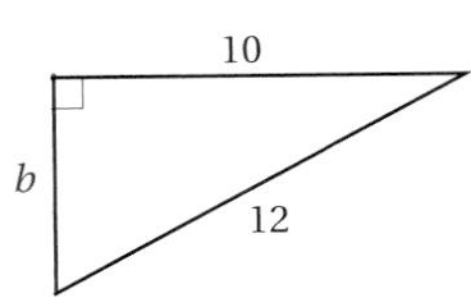

Do Exercises 1 and 2.

1. Find the length of the hypotenuse of this right triangle. Give an exact answer and an approximation to three decimal places.

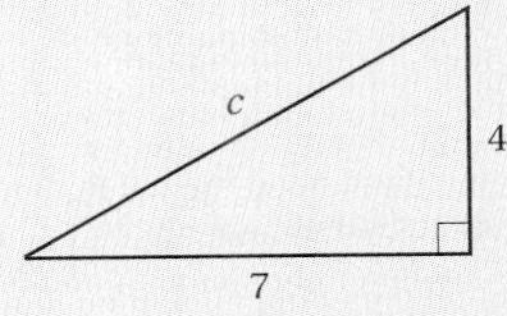

2. Find the length of the leg of this right triangle. Give an exact answer and an approximation to three decimal places.

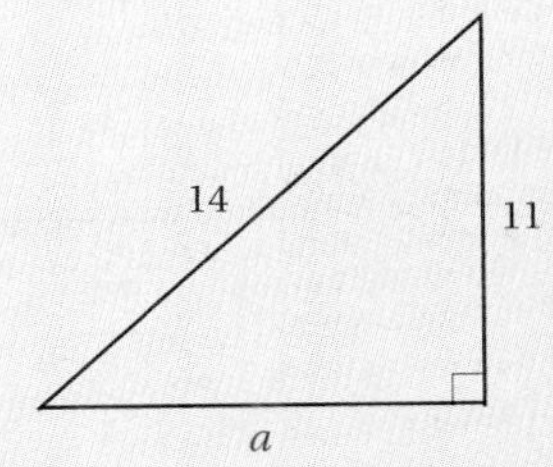

Answers on page A-47

Find the length of the leg of the right triangle. Give an exact answer and an approximation to three decimal places.

3.

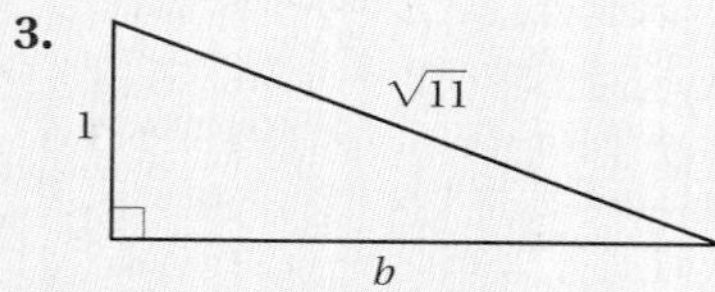

4.

5. *Guy Wire.* How long is a guy wire reaching from the top of a 15-ft pole to a point on the ground 10 ft from the pole? Give an exact answer and an approximation to three decimal places.

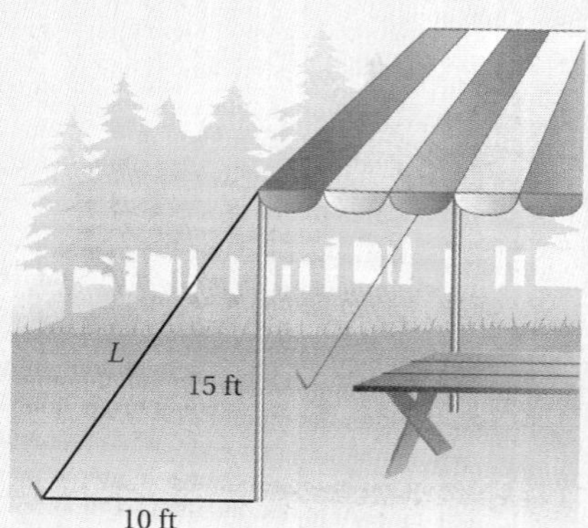

Answers on page A-47

Example 3 Find the length of the leg of this right triangle. Give an exact answer and an approximation to three decimal places.

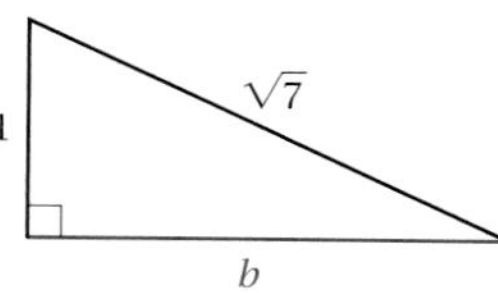

$$1^2 + b^2 = (\sqrt{7})^2 \quad \text{Substituting in the Pythagorean equation}$$
$$1 + b^2 = 7$$
$$b^2 = 7 - 1 = 6$$
$$b = \sqrt{6}$$
$$\approx 2.449 \quad \text{Using a calculator}$$

Example 4 Find the length of the leg of this right triangle. Give an exact answer and an approximation to three decimal places.

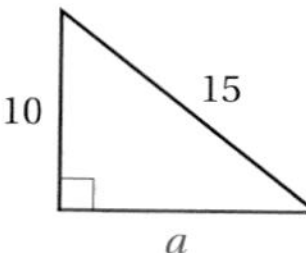

$$a^2 + 10^2 = 15^2$$
$$a^2 + 100 = 225$$
$$a^2 = 225 - 100$$
$$a^2 = 125$$
$$a = \sqrt{125}$$
$$\approx 11.180 \quad \text{Using a calculator}$$

Do Exercises 3 and 4.

b Applications

Example 5 *Dimensions of a Softball Diamond.* A slow-pitch softball diamond is actually a square 65 ft on a side. How far is it from home plate to second base? (This can be helpful information when lining up the bases.) Give an exact answer and an approximation to three decimal places.

a) We first make a drawing. We note that the first and second base lines, together with a line from home to second, form a right triangle. We label the unknown distance d.

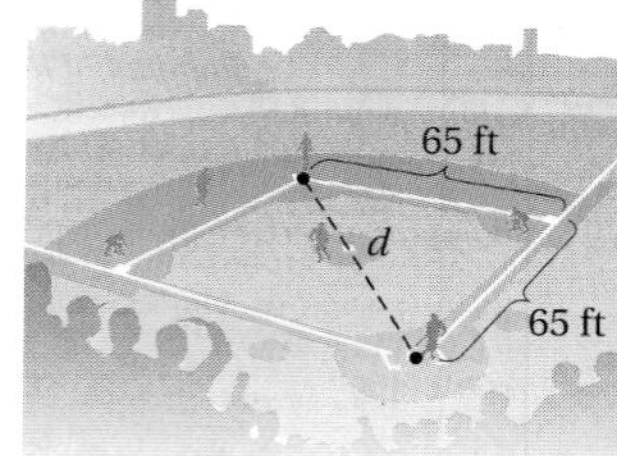

b) We know that $65^2 + 65^2 = d^2$. We solve this equation:

$$4225 + 4225 = d^2$$
$$8450 = d^2.$$

Exact answer: $\sqrt{8450}$ ft $= d$ *Approximation*: 91.924 ft $\approx d$

Do Exercise 5.

Exercise Set 15.6

a Find the length of the third side of the right triangle. Give an exact answer and an approximation to three decimal places.

1. 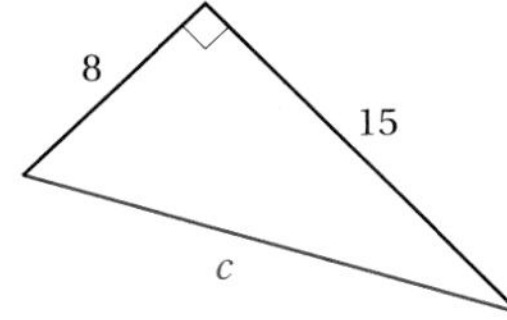

2. 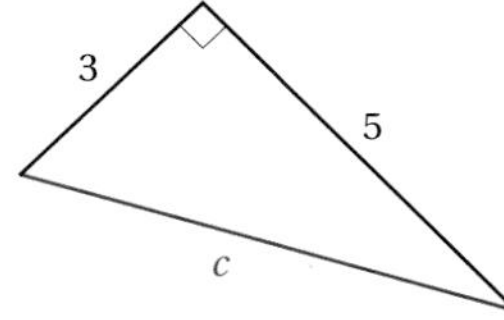

3. 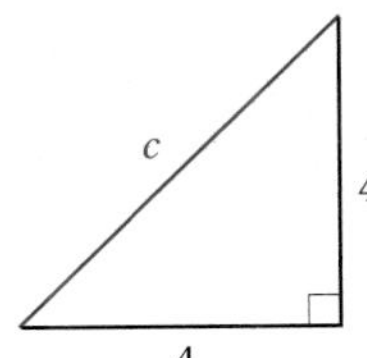

4. 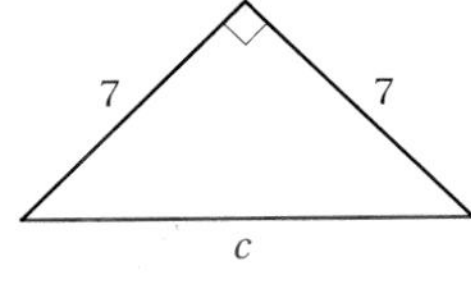

5. 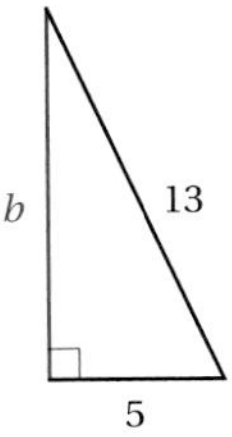

6. 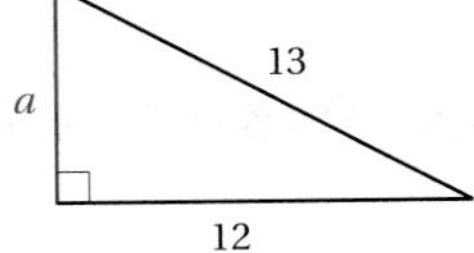

7. 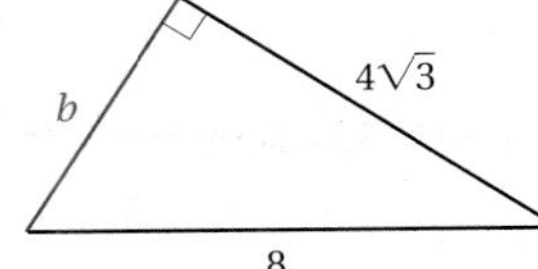

8. 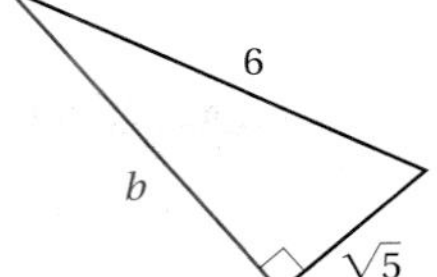

In a right triangle, find the length of the side not given. Give an exact answer and an approximation to three decimal places.

9. $a = 10,\quad b = 24$

10. $a = 5,\quad b = 12$

11. $a = 9,\quad c = 15$

12. $a = 18,\quad c = 30$

13. $b = 1,\quad c = \sqrt{5}$

14. $b = 1,\quad c = \sqrt{2}$

15. $a = 1,\quad c = \sqrt{3}$

16. $a = \sqrt{3},\quad b = \sqrt{5}$

17. $c = 10,\quad b = 5\sqrt{3}$

18. $a = 5,\quad b = 5$

19. $a = \sqrt{2},\quad b = \sqrt{7}$

20. $c = \sqrt{7},\quad a = \sqrt{2}$

b Solve. Don't forget to make a drawing. Give an exact answer and an approximation to three decimal places.

21. An airplane is flying at an altitude of 4100 ft. The slanted distance directly to the airport is 15,100 ft. How far is the airplane horizontally from the airport?

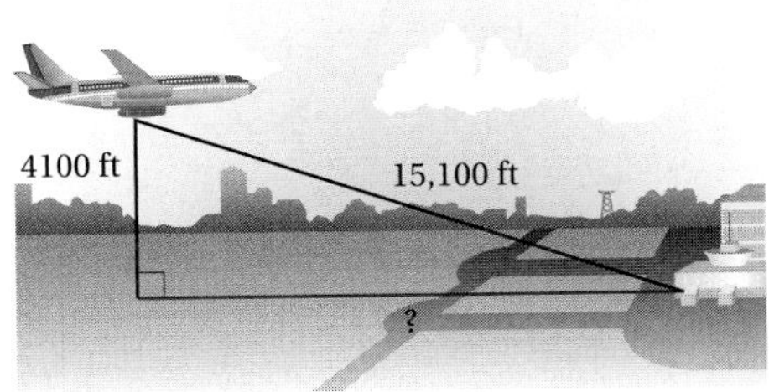

22. A surveyor had poles located at points P, Q, and R. The distances that the surveyor was able to measure are marked on the drawing. What is the approximate distance from P to R?

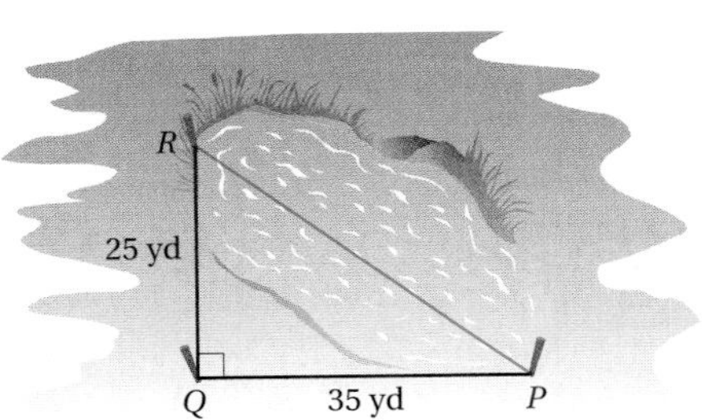

23. *Cordless Telephones.* Becky's new cordless telephone has clear reception up to 300 ft from its base. Her phone is located near a window in her apartment, 180 ft above ground level. How far into her backyard can Becky use her phone?

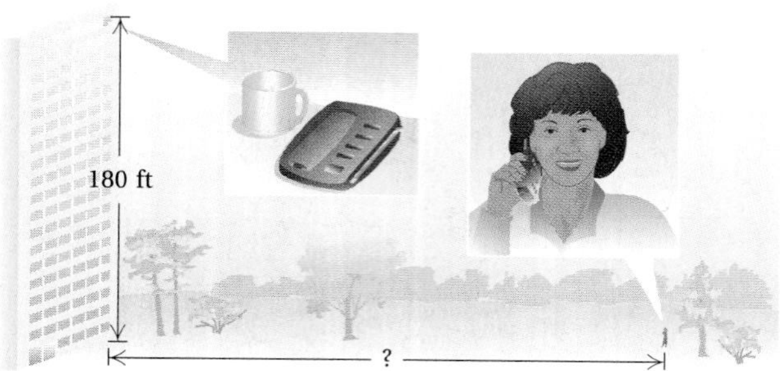

24. *Rope Course.* An outdoor rope course consists of a cable that slopes downward from a height of 37 ft to a resting place 30 ft above the ground. The trees that the cable connects are 24 ft apart. How long is the cable?

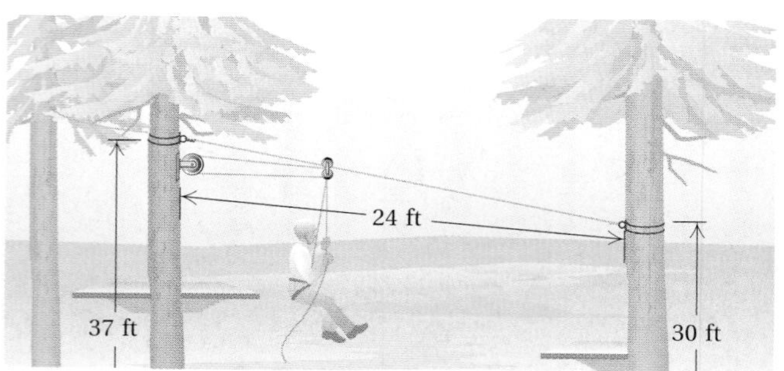

25. Find the length of a diagonal of a square whose sides are 3 cm long.

26. A 10-m ladder is leaning against a building. The bottom of the ladder is 5 m from the building. How high is the top of the ladder?

27. How long is a guy wire reaching from the top of a 12-ft pole to a point 8 ft from the pole?

28. The largest regulation soccer field is 100 yd wide and 130 yd long. Find the length of a diagonal of such a field.

Skill Maintenance

Solve. [14.3a, b]

29. $5x + 7 = 8y,$
$3x = 8y - 4$

30. $5x + y = 17,$
$-5x + 2y = 10$

31. $3x - 4y = -11,$
$5x + 6y = 12$

32. $x + y = -9,$
$x - y = -11$

33. Find the slope of the line $4 - x = 3y$. [13.1b]

34. Find the slope of the line containing the points $(8, -3)$ and $(0, -8)$. [13.1a]

Synthesis

35. ◆ Can a carpenter use a 28-ft ladder to repair clapboard that is 28 ft above ground level? Why or why not?

36. ◆ In an **equilateral triangle,** all sides have the same length. Can a right triangle ever be equilateral? Why or why not?

37. Two cars leave a service station at the same time. One car travels east at a speed of 50 mph, and the other travels south at a speed of 60 mph. After one-half hour, how far apart are they?

38. The length and the width of a rectangle are given by consecutive integers. The area of the rectangle is 90 cm^2. Find the length of a diagonal of the rectangle.

Find x.

39.

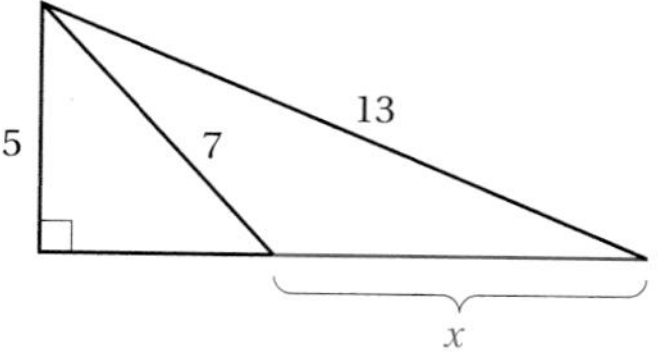

40.

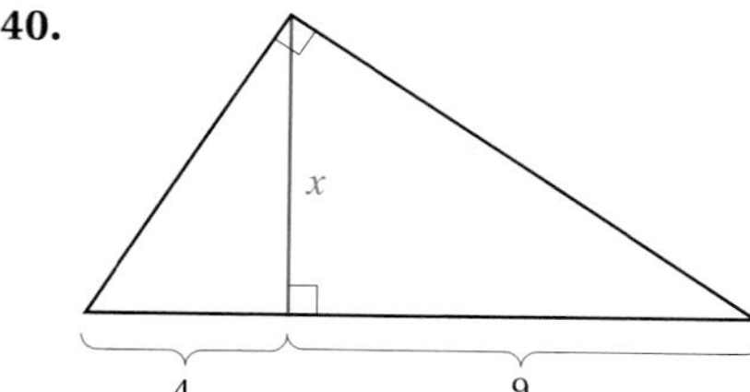

41.

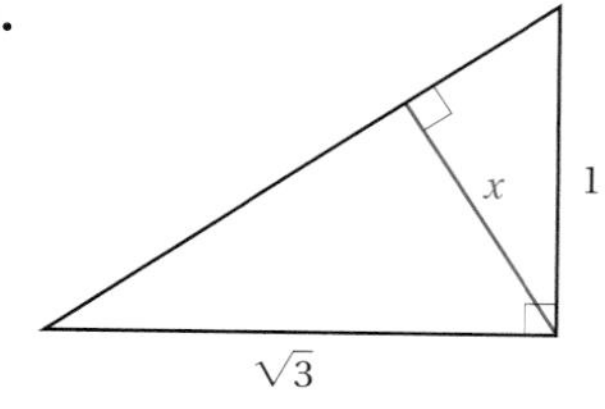

Collaborative Learning Manual
Use the Pythagorean theorem to draw a rectangle.

Summary and Review Exercises: Chapter 15

Important Properties and Formulas

Product Rule for Radicals: $\sqrt{A}\sqrt{B} = \sqrt{AB}$

Quotient Rule for Radicals: $\dfrac{\sqrt{A}}{\sqrt{B}} = \sqrt{\dfrac{A}{B}}$

Principle of Squaring: **If an equation $a = b$ is true, then the equation $a^2 = b^2$ is true.**

Pythagorean Equation: **$a^2 + b^2 = c^2$, where a and b are the lengths of the legs of a right triangle and c is the length of the hypotenuse.**

The objectives to be tested in addition to the material in this chapter are [12.2b], [13.5b], [14.3a, b], and [14.4a].

Find the square roots. [15.1a]

1. 64 **2.** 400

Simplify. [15.1a]

3. $\sqrt{36}$ **4.** $-\sqrt{169}$

Use a calculator to approximate each of the following square roots to three decimal places. [15.1b]

5. $\sqrt{3}$ **6.** $\sqrt{99}$

7. $-\sqrt{320.12}$ **8.** $\sqrt{\dfrac{11}{20}}$

9. $-\sqrt{\dfrac{47.3}{11.2}}$ **10.** $18\sqrt{11 \cdot 43.7}$

Identify the radicand. [15.1d]

11. $\sqrt{x^2 + 4}$ **12.** $\sqrt{5ab^3}$

Determine whether the expression is meaningful as a real number. Write "yes" or "no." [15.1e]

13. $\sqrt{-22}$ **14.** $-\sqrt{49}$

15. $\sqrt{-36}$ **16.** $\sqrt{-10.2}$

17. $-\sqrt{-4}$ **18.** $\sqrt{2(-3)}$

Simplify. [15.1f]

19. $\sqrt{m^2}$ **20.** $\sqrt{(x-4)^2}$

Multiply. [15.2c]

21. $\sqrt{3}\sqrt{7}$ **22.** $\sqrt{x-3}\sqrt{x+3}$

Simplify by factoring. [15.2a]

23. $-\sqrt{48}$ **24.** $\sqrt{32t^2}$

25. $\sqrt{t^2 - 49}$ **26.** $\sqrt{x^2 + 16x + 64}$

Simplify by factoring. [15.2b]

27. $\sqrt{x^8}$ **28.** $\sqrt{m^{15}}$

Multiply and simplify. [15.2c]

29. $\sqrt{6}\sqrt{10}$ **30.** $\sqrt{5x}\sqrt{8x}$

31. $\sqrt{5x}\sqrt{10xy^2}$ **32.** $\sqrt{20a^3b}\sqrt{5a^2b^2}$

Simplify. [15.3b]

33. $\sqrt{\dfrac{25}{64}}$ **34.** $\sqrt{\dfrac{20}{45}}$ **35.** $\sqrt{\dfrac{49}{t^2}}$

Rationalize the denominator. [15.3c]

36. $\sqrt{\dfrac{1}{2}}$ **37.** $\sqrt{\dfrac{1}{8}}$

38. $\sqrt{\dfrac{5}{y}}$ **39.** $\dfrac{2}{\sqrt{3}}$

Divide and simplify. [15.3a, c]

40. $\dfrac{\sqrt{27}}{\sqrt{45}}$ **41.** $\dfrac{\sqrt{45x^2y}}{\sqrt{54y}}$

42. Rationalize the denominator: [15.4c]

$$\frac{4}{2+\sqrt{3}}.$$

Simplify. [15.4a]

43. $10\sqrt{5} + 3\sqrt{5}$ **44.** $\sqrt{80} - \sqrt{45}$

45. $3\sqrt{2} - 5\sqrt{\dfrac{1}{2}}$

Simplify. [15.4b]

46. $(2 + \sqrt{3})^2$ **47.** $(2 + \sqrt{3})(2 - \sqrt{3})$

Solve. [15.5a]

48. $\sqrt{x-3} = 7$

49. $\sqrt{5x+3} = \sqrt{2x-1}$

50. $1 + x = \sqrt{1+5x}$

51. Solve: [15.5b]
$\sqrt{x} = \sqrt{x-5} + 1.$

In a right triangle, find the length of the side not given. [15.6a]

52. $a = 15, \quad c = 25$

53. $a = 1, \quad b = \sqrt{2}$

Solve. [15.6b]

54. *Airplane Descent.* A pilot is instructed to descend from 30,000 ft to 20,000 ft over a horizontal distance of 50,000 ft. What distance will the plane travel during this descent?

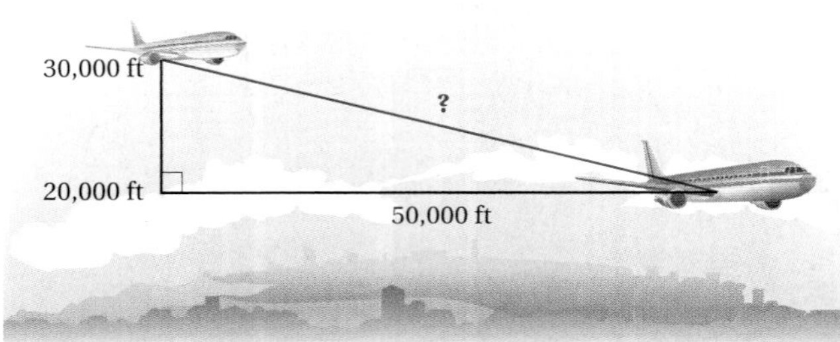

55. *Lookout Tower.* The diagonal braces in a lookout tower are 15 ft long and span a distance of 12 ft. How high does each brace reach vertically?

Solve. [15.1c], [15.5c]

56. *Speed of a Skidding Car.* The formula $r = 2\sqrt{5L}$ can be used to approximate the speed r, in miles per hour, of a car that has left a skid mark of length L, in feet.

a) What was the speed of a car that left skid marks of length 200 ft?

b) How far will a car skid at 90 mph?

Skill Maintenance

57. Solve: [14.3b]

$$2x - 3y = 4,$$
$$3x + 4y = 2.$$

58. Divide and simplify: [12.2b]

$$\frac{x^2 - 10x + 25}{x^2 + 14x + 49} \div \frac{x^2 - 25}{x^2 - 49}.$$

Solve.

59. A person's paycheck varies directly as the number of hours H worked. For 15 hr of work, the pay is \$168.75. Find the pay for 40 hr of work. [13.5b]

60. There were 14,000 people at an AIDS benefit rock concert. Tickets were \$12.00 at the door and \$10.00 if purchased in advance. Total receipts were \$159,400. How many people bought tickets in advance? [14.4a]

Synthesis

61. ◆ Explain why the following is incorrect: [15.3b]

$$\sqrt{\frac{9 + 100}{25}} = \frac{3 + 10}{5}.$$

62. ◆ Determine whether each of the following is correct for all real numbers. Explain why or why not. [15.2a]

a) $\sqrt{5x^2} = |x|\sqrt{5}$

b) $\sqrt{b^2 - 4} = b - 2$

c) $\sqrt{x^2 + 16} = x + 4$

63. Simplify: $\sqrt{\sqrt{\sqrt{256}}}.$ [15.2a]

64. Solve $A = \sqrt{a^2 + b^2}$ for b. [15.5a]

Test: Chapter 15

1. Find the square roots of 81.

Simplify.

2. $\sqrt{64}$

3. $-\sqrt{25}$

Approximate the expression involving square roots to three decimal places.

4. $\sqrt{116}$

5. $-\sqrt{87.4}$

6. $\sqrt{\frac{96 \cdot 38}{214.2}}$

7. Identify the radicand in $\sqrt{4 - y^3}$.

Determine whether the expression is meaningful as a real number. Write "yes" or "no."

8. $\sqrt{24}$

9. $\sqrt{-23}$

Simplify.

10. $\sqrt{a^2}$

11. $\sqrt{36y^2}$

Multiply.

12. $\sqrt{5}\sqrt{6}$

13. $\sqrt{x - 8}\sqrt{x + 8}$

Simplify by factoring.

14. $\sqrt{27}$

15. $\sqrt{25x - 25}$

16. $\sqrt{t^5}$

Multiply and simplify.

17. $\sqrt{5}\sqrt{10}$

18. $\sqrt{3ab}\sqrt{6ab^3}$

Answers

1. __________

2. __________

3. __________

4. __________

5. __________

6. __________

7. __________

8. __________

9. __________

10. __________

11. __________

12. __________

13. __________

14. __________

15. __________

16. __________

17. __________

18. __________

Answers

19. ______

20. ______

21. ______

22. ______

23. ______

24. ______

25. ______

26. ______

27. ______

28. ______

29. ______

30. ______

31. ______

32. ______

33. ______

34. ______

35. ______

36. ______

37. ______

38. ______

39. ______

40. ______

Simplify.

19. $\sqrt{\frac{27}{12}}$

20. $\sqrt{\frac{144}{a^2}}$

Rationalize the denominator.

21. $\sqrt{\frac{2}{5}}$

22. $\sqrt{\frac{2x}{y}}$

Divide and simplify.

23. $\frac{\sqrt{27}}{\sqrt{32}}$

24. $\frac{\sqrt{35x}}{\sqrt{80xy^2}}$

Add or subtract.

25. $3\sqrt{18} - 5\sqrt{18}$

26. $\sqrt{5} + \sqrt{\frac{1}{5}}$

Simplify.

27. $(4 - \sqrt{5})^2$

28. $(4 - \sqrt{5})(4 + \sqrt{5})$

29. Rationalize the denominator: $\frac{10}{4 - \sqrt{5}}$.

30. In a right triangle, $a = 8$ and $b = 4$. Find c.

Solve.

31. $\sqrt{3x} + 2 = 14$

32. $\sqrt{6x + 13} = x + 3$

33. $\sqrt{1 - x} + 1 = \sqrt{6 - x}$

34. A person can see 247.49 km to the horizon from an airplane window. How high is the airplane? Use the formula $V = 3.5\sqrt{h}$.

Skill Maintenance

35. The perimeter of a rectangle is 118 yd. The width is 18 yd less than the length. Find the area of the rectangle.

36. The number of switches N that a production line can make varies directly as the time it operates. It can make 7240 switches in 6 hr. How many can it make in 13 hr?

37. Solve:

$$-6x + 5y = 10,$$
$$5x + 6y = 12.$$

38. Divide and simplify:

$$\frac{x^2 - 11x + 30}{x^2 - 12x + 35} \div \frac{x^2 - 36}{x^2 - 14x + 49}.$$

Synthesis

Simplify.

39. $\sqrt{\sqrt{\sqrt{625}}}$

40. $\sqrt{y^{16n}}$

16

Quadratic Equations

An Application

In the not-too-distant future, a new kind of high-definition television (HDTV) with a larger screen and greater clarity will be available. An HDTV might have a 70-in. diagonal screen with the width 27 in. greater than the height. Find the width and the height of a 70-in. HDTV.

This problem appears as Exercise 4 in Exercise Set 16.5.

The Mathematics

We let w = the width and h = the height. Then the problem translates to

$$w^2 + h^2 = 70^2,$$

or

$$(h + 27)^2 + h^2 = 4900,$$

or

$$\underbrace{2h^2 + 54h - 4171 = 0.}$$

This is a quadratic equation.

World Wide Web For more information, visit us at www.mathmax.com

Introduction

A *quadratic equation* contains a polynomial of second degree. In this chapter, we first learn to solve quadratic equations by factoring. Because certain quadratic equations are difficult to solve by factoring, we also learn to use the *quadratic formula* to find solutions of quadratic equations. Next, we apply these equation-solving skills to applications and problem-solving. Then we graph quadratic equations.

Pretest: Chapter 16

Solve.

1. $x^2 + 9 = 6x$

2. $x^2 - 7 = 0$

3. $3x^2 + 3x - 1 = 0$

4. $5y^2 - 3y = 0$

5. $\frac{3}{3x + 2} - \frac{2}{3x + 4} = 1$

6. $(x + 4)^2 = 5$

7. Solve $x^2 - 2x - 5 = 0$ by completing the square. Show your work.

8. Solve $A = n^2 - pn$ for n.

9. The length of a rectangle is three times the width. The area is 48 cm^2. Find the length and the width.

10. Find the x-intercepts: $y = 2x^2 + x - 4$.

11. The current in a stream moves at a speed of 2 km/h. A boat travels 24 km upstream and 24 km downstream in a total time of 5 hr. What is the speed of the boat in still water?

12. Graph: $y = 4 - x^2$.

Objectives for Retesting

The objectives to be tested in addition to the material in this chapter are as follows.

[13.5c] Find an equation of inverse variation given a pair of values of the variables.
[15.2c] Multiply radical expressions and simplify, if possible.
[15.4a] Add or subtract with radical notation, using the distributive law to simplify.
[15.6b] Solve applied problems involving right triangles.

16.1 Introduction to Quadratic Equations

Algebraic–Graphical Connection

Before we begin this chapter, let's look back at some algebraic–graphical equation-solving concepts and their interrelationships. In Chapter 9, we considered the graph of a *linear equation* $y = mx + b$. For example, the graph of the equation $y = \frac{5}{2}x - 4$ and its x-intercept are shown below.

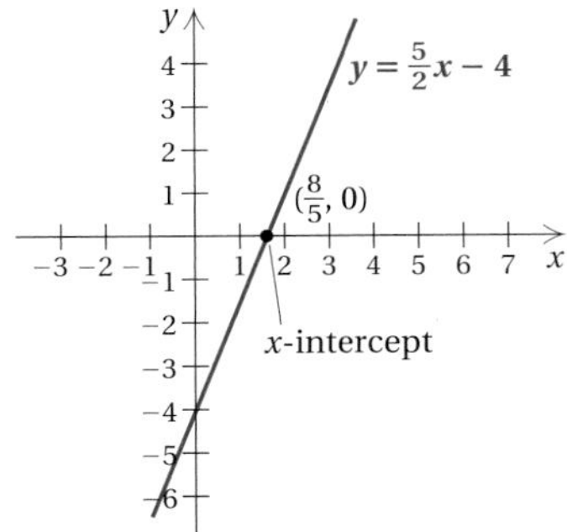

If $y = 0$, then $x = \frac{8}{5}$. Thus the x-intercept is $\left(\frac{8}{5}, 0\right)$. This point is also the intersection of the graphs of $y = \frac{5}{2}x - 4$ and $y = 0$.

In Chapter 8, we learned how to solve linear equations like $0 = \frac{5}{2}x - 4$ algebraically (using algebra). We proceeded as follows:

$$0 = \frac{5}{2}x - 4$$
$$4 = \frac{5}{2}x \quad \text{Adding 4}$$
$$8 = 5x \quad \text{Multiplying by 2}$$
$$\frac{8}{5} = x. \quad \text{Dividing by 5}$$

We see that $\frac{8}{5}$, the solution of $0 = \frac{5}{2}x - 4$, is the first coordinate of the x-intercept of the graph of $y = \frac{5}{2}x - 4$.

Do Exercises 1 and 2.

In this chapter, we build on these ideas by applying them to quadratic equations. In Section 11.7, we briefly considered the graph of a *quadratic equation*

$$y = ax^2 + bx + c, \quad a \neq 0.$$

For example, the graph of the equation $y = x^2 + 6x + 8$ and its x-intercepts are shown below.

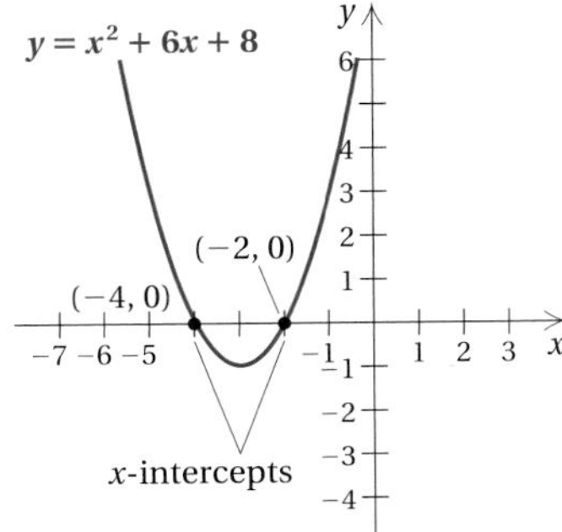

The x-intercepts are $(-4, 0)$ and $(-2, 0)$. We will develop in detail the creation of such graphs in Section 16.6. The points $(-4, 0)$ and $(-2, 0)$ are the intersections of the graphs of $y = x^2 + 6x + 8$ and $y = 0$.

Objectives

a Write a quadratic equation in standard form $ax^2 + bx + c = 0$, $a > 0$, and determine the coefficients a, b, and c.

b Solve quadratic equations of the type $ax^2 + bx = 0$, where $b \neq 0$, by factoring.

c Solve quadratic equations of the type $ax^2 + bx + c = 0$, where $b \neq 0$ and $c \neq 0$, by factoring.

d Solve applied problems involving quadratic equations.

For Extra Help

TAPE 29 — MAC WIN — CD-ROM

1. Consider $y = -\frac{2}{3}x - 3$. Find the intercepts and graph the equation.

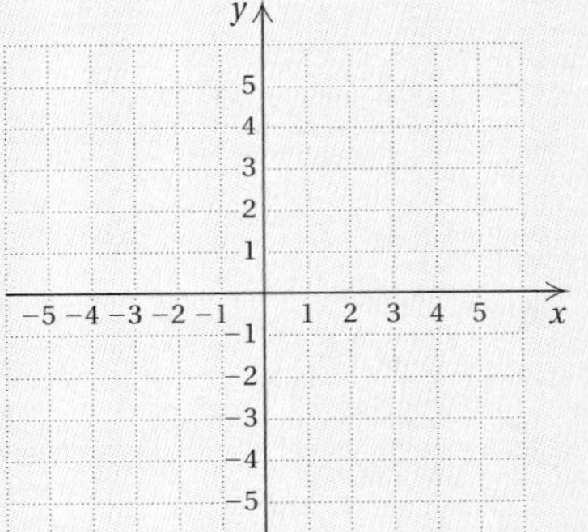

2. Solve the equation

$$0 = -\frac{2}{3}x - 3.$$

Answers on page A-48

Write in standard form and determine a, b, and c.

3. $y^2 = 8y$

4. $3 - x^2 = 9x$

5. $3x + 5x^2 = x^2 - 4 + 2x$

6. $5x^2 = 21$

Answers on page A-48

We began studying the solution of quadratic equations like $x^2 + 6x + 8 = 0$ in Section 11.7. There we used factoring for such solutions:

$$x^2 + 6x + 8 = 0$$

$$(x + 4)(x + 2) = 0 \quad \text{Factoring}$$

$$x + 4 = 0 \quad or \quad x + 2 = 0 \quad \text{Using the principle of zero products}$$

$$x = -4 \quad or \quad x = -2.$$

We see that the solutions of $x^2 + 6x + 8 = 0$, -4 and -2, are the first coordinates of the x-intercepts, $(-4, 0)$ and $(-2, 0)$, of the graph of $y = x^2 + 6x + 8$. We will enhance our ability to solve quadratic equations in Sections 16.1–16.3.

a Standard Form

The following are **quadratic equations.** They contain polynomials of second degree.

$$x^2 + 7x - 5 = 0, \qquad 3t^2 - \tfrac{1}{2}t = 9, \qquad 5y^2 = -6y, \qquad 5m^2 = 15$$

The quadratic equation $4x^2 + 7x - 5 = 0$ is said to be in **standard form.** Although the quadratic equation $4x^2 = 5 - 7x$ is equivalent to the preceding equation, it is *not* in standard form.

> A quadratic equation of the type $ax^2 + bx + c = 0$, where a, b, and c are real-number constants and $a > 0$, is called the **standard form of a quadratic equation.**

We define $a > 0$ to ease the proof of the quadratic formula, which we consider later, and to ease solving by factoring, which we review in this section. Suppose we are studying an equation like $-3x^2 + 8x - 2 = 0$. It is not in standard form. We can find an equivalent equation that is in standard form by multiplying by -1 on both sides:

$$-1(-3x^2 + 8x - 2) = -1(0)$$

$$3x^2 - 8x + 2 = 0.$$

Examples Write in standard form and determine a, b, and c.

1. $4x^2 + 7x - 5 = 0$ **The equation is already in standard form.**

$a = 4;\quad b = 7;\quad c = -5$

2. $3x^2 - 0.5x = 9$

$3x^2 - 0.5x - 9 = 0$ **Subtracting 9. This is standard form.**

$a = 3;\quad b = -0.5;\quad c = -9$

3. $-4y^2 = 5y$

$-4y^2 - 5y = 0$ **Subtracting $5y$**

Not positive!

$4y^2 + 5y = 0$ **Multiplying by -1. This is standard form.**

$a = 4;\quad b = 5;\quad c = 0$

Do Exercises 3–6.

b Solving Quadratic Equations of the Type $ax^2 + bx = 0$

Sometimes we can use factoring and the principle of zero products to solve quadratic equations. We are actually reviewing methods that we introduced in Section 11.7.

When $c = 0$ and $b \neq 0$, we can always factor and use the principle of zero products (see Section 11.7 for a review).

Example 4 Solve: $7x^2 + 2x = 0$.

$$7x^2 + 2x = 0$$

$$x(7x + 2) = 0 \quad \text{Factoring}$$

$$x = 0 \quad or \quad 7x + 2 = 0 \quad \text{Using the principle of zero products}$$

$$x = 0 \quad or \quad 7x = -2$$

$$x = 0 \quad or \quad x = -\tfrac{2}{7}$$

CHECK: For 0:

$$7x^2 + 2x = 0$$

$$7 \cdot 0^2 + 2 \cdot 0 \ ? \ 0$$

$$0 \quad \text{TRUE}$$

For $-\frac{2}{7}$:

$$7x^2 + 2x = 0$$

$$7\left(-\tfrac{2}{7}\right)^2 + 2\left(-\tfrac{2}{7}\right) \ ? \ 0$$

$$7\left(\tfrac{4}{49}\right) - \tfrac{4}{7}$$

$$\tfrac{4}{7} - \tfrac{4}{7}$$

$$0 \quad \text{TRUE}$$

The solutions are 0 and $-\frac{2}{7}$.

CAUTION! You may be tempted to divide each term in an equation like the one in Example 4 by x. This method would yield the equation

$$7x + 2 = 0,$$

whose only solution is $-\frac{2}{7}$. In effect, since 0 is also a solution of the original equation, we have divided by 0. The error of such division means the loss of one of the solutions.

Example 5 Solve: $20x^2 - 15x = 0$.

$$20x^2 - 15x = 0$$

$$5x(4x - 3) = 0 \quad \text{Factoring}$$

$$5x = 0 \quad or \quad 4x - 3 = 0 \quad \text{Using the principle of zero products}$$

$$x = 0 \quad or \quad 4x = 3$$

$$x = 0 \quad or \quad x = \tfrac{3}{4}$$

The solutions are 0 and $\frac{3}{4}$.

A quadratic equation of the type $ax^2 + bx = 0$, where $c = 0$ and $b \neq 0$, will always have 0 as one solution and a nonzero number as the other solution.

Do Exercises 7 and 8.

Solve.

7. $2x^2 + 9x = 0$

$0, -\frac{9}{2}$

8. $10x^2 - 6x = 0$

Answers on page A-48

Solve.

9. $4x^2 + 5x - 6 = 0$

10. $(x - 1)(x + 1) = 5(x - 1)$

Answers on page A-48

C Solving Quadratic Equations of the Type $ax^2 + bx + c = 0$

When neither b nor c is 0, we can sometimes solve by factoring.

Example 6 Solve: $5x^2 - 8x + 3 = 0$.

$$5x^2 - 8x + 3 = 0$$
$$(5x - 3)(x - 1) = 0 \quad \text{Factoring}$$
$$5x - 3 = 0 \quad or \quad x - 1 = 0 \quad \text{Using the principle of zero products}$$
$$5x = 3 \quad or \quad x = 1$$
$$x = \frac{3}{5} \quad or \quad x = 1$$

The solutions are $\frac{3}{5}$ and 1.

Example 7 Solve: $(y - 3)(y - 2) = 6(y - 3)$.

We write the equation in standard form and then try to factor:

$$y^2 - 5y + 6 = 6y - 18 \quad \text{Multiplying}$$
$$y^2 - 11y + 24 = 0 \quad \text{Standard form}$$
$$(y - 8)(y - 3) = 0 \quad \text{Factoring}$$
$$y - 8 = 0 \quad or \quad y - 3 = 0 \quad \text{Using the principle of zero products}$$
$$y = 8 \quad or \quad y = 3.$$

The solutions are 8 and 3.

Do Exercises 9 and 10.

Recall that to solve a rational equation, we multiply on both sides by the LCM of all the denominators. We may obtain a quadratic equation after a few steps. When that happens, we know how to finish solving, but we must remember to check possible solutions because a replacement may result in division by 0. See Section 12.6.

Example 8 Solve: $\frac{3}{x - 1} + \frac{5}{x + 1} = 2$.

We multiply by the LCM, which is $(x - 1)(x + 1)$:

$$(x - 1)(x + 1) \cdot \left(\frac{3}{x - 1} + \frac{5}{x + 1}\right) = 2 \cdot (x - 1)(x + 1).$$

We use the distributive law on the left:

$$(x - 1)(x + 1) \cdot \frac{3}{x - 1} + (x - 1)(x + 1) \cdot \frac{5}{x + 1} = 2(x - 1)(x + 1)$$
$$3(x + 1) + 5(x - 1) = 2(x - 1)(x + 1)$$
$$3x + 3 + 5x - 5 = 2(x^2 - 1)$$
$$8x - 2 = 2x^2 - 2$$
$$0 = 2x^2 - 8x$$
$$0 = 2x(x - 4) \quad \text{Factoring}$$
$$2x = 0 \quad or \quad x - 4 = 0$$
$$x = 0 \quad or \quad x = 4.$$

CHECK: For 0:

$$\frac{3}{x-1}+\frac{5}{x+1}=2$$

$$\begin{array}{r|l} \frac{3}{0-1}+\frac{5}{0+1} & ?\ 2 \\ \frac{3}{-1}+\frac{5}{1} & \\ -3+5 & \\ 2 & \end{array} \quad \text{TRUE}$$

For 4:

$$\frac{3}{x-1}+\frac{5}{x+1}=2$$

$$\begin{array}{r|l} \frac{3}{4-1}+\frac{5}{4+1} & ?\ 2 \\ \frac{3}{3}+\frac{5}{5} & \\ 1+1 & \\ 2 & \end{array} \quad \text{TRUE}$$

The solutions are 0 and 4.

Do Exercise 11.

d Solving Applied Problems

Example 9 *Diagonals of a Polygon.* The number of diagonals d of a polygon of n sides is given by the formula

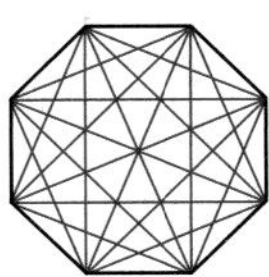

$$d=\frac{n^2-3n}{2}.$$

If a polygon has 27 diagonals, how many sides does it have?

1. **Familiarize.** We can make a drawing to familiarize ourselves with the problem. We draw an octagon (8 sides) and count the diagonals and see that there are 20. Let's check this in the formula. We evaluate the formula for $n = 8$:

$$d=\frac{8^2-3(8)}{2}=\frac{64-24}{2}=\frac{40}{2}=20.$$

2. **Translate.** We know that the number of diagonals is 27. We substitute 27 for d:

$$27=\frac{n^2-3n}{2}.$$

This gives us a translation.

3. **Solve.** We solve the equation for n, reversing the equation first for convenience:

$$\frac{n^2-3n}{2}=27$$

$$n^2-3n=54 \qquad \textbf{Multiplying by 2 to clear fractions}$$

$$n^2-3n-54=0$$

$$(n-9)(n+6)=0$$

$$n-9=0 \quad or \quad n+6=0$$

$$n=9 \quad or \quad n=-6.$$

4. **Check.** Since the number of sides cannot be negative, -6 cannot be a solution. We leave it to the student to show by substitution that 9 checks.
5. **State.** The polygon has 9 sides (it is a nonagon).

Do Exercise 12.

11. Solve:

$$\frac{20}{x+5}-\frac{1}{x-4}=1.$$

12. Use $d=\dfrac{n^2-3n}{2}$.

a) A heptagon has 7 sides. How many diagonals does it have?

b) A polygon has 44 diagonals. How many sides does it have?

Answers on page A-48

Calculator Spotlight

Solving Equations. We can check solutions to equations using the TABLE or CALC-VALUE features (see Section 11.7). We can also check solutions using the INTERSECT or ZERO feature. Let's consider the equation $(x - 1)(x + 1) = 5(x - 1)$ of Margin Exercise 10.

INTERSECT Feature. We set

$$y_1 = (x - 1)(x + 1) \quad \text{and} \quad y_2 = 5(x - 1)$$

and then use the INTERSECT feature (see the CALC menu) to determine points of intersection. The first coordinates of these points are the solutions of the equation $(x - 1)(x + 1) = 5(x - 1)$.

$y_1 = (x - 1)(x + 1),\ y_2 = 5(x - 1)$

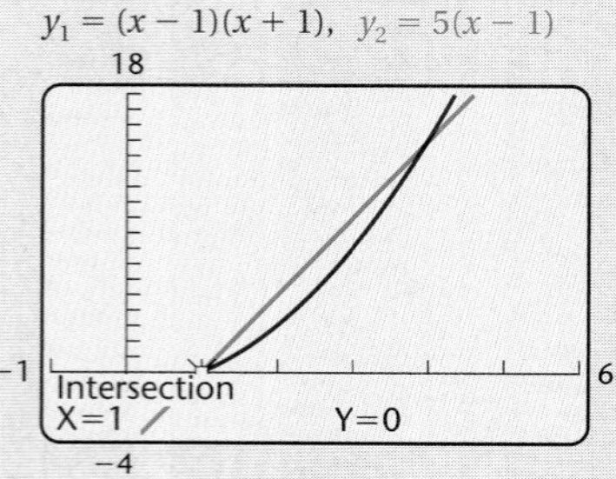

We first find the coordinates of the left-hand point of intersection. We see that 1 is a solution. We then find the coordinates of the right-hand point of intersection. We see that 4 is also a solution.

$y_1 = (x - 1)(x + 1),\ y_2 = 5(x - 1)$

18

−1 Intersection X=4 Y=15 6

−4

ZERO, or ROOT Feature. Some graphers have a ZERO, or ROOT, feature (see the CALC menu). Before we can use this feature, however, we must have a 0 on one side of the equation. So to solve

$$(x - 1)(x + 1) = 5(x - 1),$$

we consider

$$(x - 1)(x + 1) - 5(x - 1) = 0.$$

We then set

$$y_1 = (x - 1)(x + 1) - 5(x - 1)$$

and determine where the graph intersects the x-axis. We press CALC and ZERO, or ROOT, and continue. Performing the procedure twice, once for the left-hand value and once for the right-hand value, we see that both 1 and 4 are solutions. This also tells us that the x-intercepts of the equation $y = (x - 1)(x + 1) - 5(x - 1)$ are $(1, 0)$ and $(4, 0)$.

$y = (x - 1)(x + 1) - 5(x - 1)$

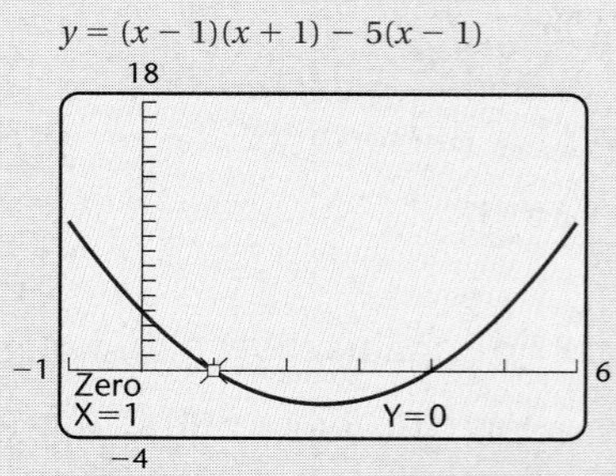

$y = (x - 1)(x + 1) - 5(x - 1)$

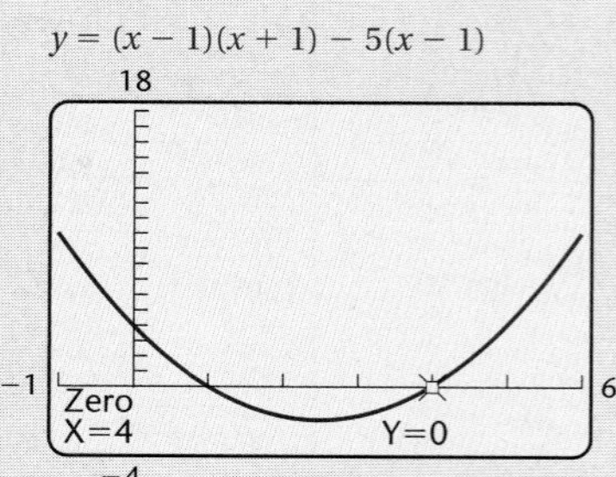

Exercises

Solve.

1. $5x^2 - 8x + 3 = 0$

2. $6(x - 3) = (x - 3)(x - 2)$

Exercise Set 16.1

a Write standard form and determine a, b, and c.

1. $x^2 - 3x + 2 = 0$

2. $x^2 - 8x - 5 = 0$

3. $7x^2 = 4x - 3$

4. $9x^2 = x + 5$

5. $5 = -2x^2 + 3x$

6. $3x - 1 = 5x^2 + 9$

b Solve.

7. $x^2 + 5x = 0$

8. $x^2 + 7x = 0$

9. $3x^2 + 6x = 0$

10. $4x^2 + 8x = 0$

11. $5x^2 = 2x$

12. $11x = 3x^2$

13. $4x^2 + 4x = 0$

14. $8x^2 - 8x = 0$

15. $0 = 10x^2 - 30x$

16. $0 = 10x^2 - 50x$

17. $11x = 55x^2$

18. $33x^2 = -11x$

19. $14t^2 = 3t$

20. $6m = 19m^2$

21. $5y^2 - 3y^2 = 72y + 9y$

22. $63p - 16p^2 = 17p + 58p^2$

c Solve.

23. $x^2 + 8x - 48 = 0$

24. $x^2 - 16x + 48 = 0$

25. $5 + 6x + x^2 = 0$

26. $x^2 + 10 + 11x = 0$

27. $18 = 7p + p^2$

28. $t^2 + 14t = -24$

29. $-15 = -8y + y^2$

30. $q^2 + 14 = 9q$

31. $x^2 + 10x + 25 = 0$

32. $x^2 + 6x + 9 = 0$

33. $r^2 = 8r - 16$

34. $x^2 + 1 = 2x$

35. $6x^2 + x - 2 = 0$

36. $2x^2 - 11x + 15 = 0$

37. $3a^2 = 10a + 8$

38. $15b - 9b^2 = 4$

39. $6x^2 - 4x = 10$

40. $3x^2 - 7x = 20$

41. $2t^2 + 12t = -10$

42. $12w^2 - 5w = 2$

43. $t(t - 5) = 14$

44. $6z^2 + z - 1 = 0$

45. $t(9 + t) = 4(2t + 5)$

46. $3y^2 + 8y = 12y + 15$

47. $16(p - 1) = p(p + 8)$

48. $(2x - 3)(x + 1) = 4(2x - 3)$

49. $(t - 1)(t + 3) = t - 1$

50. $(x - 2)(x + 2) = x + 2$

Solve.

51. $\frac{24}{x-2} + \frac{24}{x+2} = 5$

52. $\frac{8}{x+2} + \frac{8}{x-2} = 3$

53. $\frac{1}{x} + \frac{1}{x+6} = \frac{1}{4}$

54. $\frac{1}{x} + \frac{1}{x+9} = \frac{1}{20}$

55. $1 + \frac{12}{x^2-4} = \frac{3}{x-2}$

56. $\frac{5}{t-3} - \frac{30}{t^2-9} = 1$

57. $\frac{r}{r-1} + \frac{2}{r^2-1} = \frac{8}{r+1}$

58. $\frac{x+2}{x^2-2} = \frac{2}{3-x}$

59. $\frac{x-1}{1-x} = -\frac{x+8}{x-8}$

60. $\frac{4-x}{x-4} + \frac{x+3}{x-3} = 0$

61. $\frac{5}{y+4} - \frac{3}{y-2} = 4$

62. $\frac{2z+11}{2z+8} = \frac{3z-1}{z-1}$

d Solve.

63. A decagon is a figure with 10 sides. How many diagonals does a decagon have?

64. A hexagon is a figure with 6 sides. How many diagonals does a hexagon have?

65. A polygon has 14 diagonals. How many sides does it have?

66. A polygon has 9 diagonals. How many sides does it have?

Skill Maintenance

Simplify. [15.1a], [15.2a]

67. $\sqrt{64}$

68. $-\sqrt{169}$

69. $\sqrt{8}$

70. $\sqrt{12}$

71. $\sqrt{20}$

72. $\sqrt{88}$

73. $\sqrt{405}$

74. $\sqrt{1020}$

Use a calculator to approximate the square roots. Round to three decimal places. [15.1b]

75. $\sqrt{7}$

76. $\sqrt{23}$

77. $\sqrt{\frac{7}{3}}$

78. $\sqrt{524.77}$

Synthesis

79. ◆ Explain how the graph of $y = (x-2)(x+3)$ is related to the solutions of the equation $(x-2)(x+3) = 0$.

80. ◆ Explain how you might go about constructing a quadratic equation whose solutions are -5 and 7.

Solve.

81. $4m^2 - (m+1)^2 = 0$

82. $x^2 + \sqrt{22}\,x = 0$

83. $\sqrt{5}x^2 - x = 0$

84. $\sqrt{7}x^2 + \sqrt{3}x = 0$

Use a grapher to solve the equation.

85. $3x^2 - 7x = 20$

86. $x(x-5) = 14$

87. $3x^2 + 8x = 12x + 15$

88. $(x-2)(x+2) = x + 2$

89. $(x-2)^2 + 3(x-2) = 4$

90. $(x+3)^2 = 4$

91. $16(x-1) = x(x+8)$

92. $x^2 + 2.5x + 1.5625 = 9.61$

16.2 Solving Quadratic Equations by Completing the Square

a Solving Quadratic Equations of the Type $ax^2 = p$

For equations of the type $ax^2 = p$, we first solve for x^2 and then apply the *principle of square roots,* which states that a positive number has two square roots. The number 0 has one square root, 0.

THE PRINCIPLE OF SQUARE ROOTS

- The equation $x^2 = d$ has two real solutions when $d > 0$. The solutions are $\sqrt{d}$ and $-\sqrt{d}$.
- The equation $x^2 = 0$ has 0 as its only solution.
- The equation $x^2 = d$ has no real-number solution when $d < 0$.

Example 1 Solve: $x^2 = 3$.

$$x^2 = 3$$

$$x = \sqrt{3} \quad or \quad x = -\sqrt{3} \qquad \text{Using the principle of square roots}$$

CHECK: For $\sqrt{3}$:

$$x^2 = 3$$
$$(\sqrt{3})^2 \ ? \ 3$$
$$3 \quad \text{TRUE}$$

For $-\sqrt{3}$:

$$x^2 = 3$$
$$(-\sqrt{3})^2 \ ? \ 3$$
$$3 \quad \text{TRUE}$$

The solutions are $\sqrt{3}$ and $-\sqrt{3}$.

Do Exercise 1.

Example 2 Solve: $\frac{1}{8}x^2 = 0$.

$$\tfrac{1}{8}x^2 = 0$$
$$x^2 = 0 \qquad \text{Multiplying by 8}$$
$$x = 0 \qquad \text{Using the principle of square roots}$$

The solution is 0.

Do Exercise 2.

Example 3 Solve: $-3x^2 + 7 = 0$.

$$-3x^2 + 7 = 0$$
$$-3x^2 = -7 \qquad \text{Subtracting 7}$$
$$x^2 = \frac{-7}{-3} \qquad \text{Dividing by } -3$$
$$x^2 = \frac{7}{3}$$
$$x = \sqrt{\frac{7}{3}} \quad or \quad x = -\sqrt{\frac{7}{3}} \qquad \text{Using the principle of square roots}$$
$$x = \sqrt{\frac{7}{3} \cdot \frac{3}{3}} \quad or \quad x = -\sqrt{\frac{7}{3} \cdot \frac{3}{3}} \qquad \text{Rationalizing the denominators}$$
$$x = \frac{\sqrt{21}}{3} \quad or \quad x = -\frac{\sqrt{21}}{3}$$

Objectives

a Solve quadratic equations of the type $ax^2 = p$.

b Solve quadratic equations of the type $(x + c)^2 = d$.

c Solve quadratic equations by completing the square.

d Solve certain problems involving quadratic equations of the type $ax^2 = p$.

For Extra Help

TAPE 29

MAC
WIN

CD-ROM

1. Solve: $x^2 = 10$.

2. Solve: $6x^2 = 0$.

3. Solve: $2x^2 - 3 = 0$.

Answers on page A-48

CHECK: For $\frac{\sqrt{21}}{3}$:

$$\begin{array}{r|l} -3x^2 + 7 = 0 & \\ \hline -3\left(\frac{\sqrt{21}}{3}\right)^2 + 7 \; ? \; 0 & \\ -3 \cdot \frac{21}{9} + 7 & \\ -7 + 7 & \\ 0 & \quad \text{TRUE} \end{array}$$

For $-\frac{\sqrt{21}}{3}$:

$$\begin{array}{r|l} -3x^2 + 7 = 0 & \\ \hline -3\left(-\frac{\sqrt{21}}{3}\right)^2 + 7 \; ? \; 0 & \\ -3 \cdot \frac{21}{9} + 7 & \\ -7 + 7 & \\ 0 & \quad \text{TRUE} \end{array}$$

The solutions are $\frac{\sqrt{21}}{3}$ and $-\frac{\sqrt{21}}{3}$.

Do Exercise 3 on the preceding page.

b Solving Quadratic Equations of the Type $(x + c)^2 = d$

In an equation of the type $(x + c)^2 = d$, we have the square of a binomial equal to a constant. We can use the principle of square roots to solve such an equation.

Example 4 Solve: $(x - 5)^2 = 9$.

$$(x - 5)^2 = 9$$
$$x - 5 = 3 \quad \text{or} \quad x - 5 = -3 \qquad \text{Using the principle of square roots}$$
$$x = 8 \quad \text{or} \quad x = 2$$

The solutions are 8 and 2.

Example 5 Solve: $(x + 2)^2 = 7$.

$$(x + 2)^2 = 7$$
$$x + 2 = \sqrt{7} \quad \text{or} \quad x + 2 = -\sqrt{7} \qquad \text{Using the principle of square roots}$$
$$x = -2 + \sqrt{7} \quad \text{or} \quad x = -2 - \sqrt{7}$$

The solutions are $-2 + \sqrt{7}$ and $-2 - \sqrt{7}$, or simply $-2 \pm \sqrt{7}$ (read "-2 plus or minus $\sqrt{7}$").

Do Exercises 4 and 5.

In Examples 4 and 5, the left sides of the equations are squares of binomials. If we can express an equation in such a form, we can proceed as we did in those examples.

Example 6 Solve: $x^2 + 8x + 16 = 49$.

$$x^2 + 8x + 16 = 49 \qquad \text{The left side is the square of a binomial.}$$
$$(x + 4)^2 = 49$$
$$x + 4 = 7 \quad \text{or} \quad x + 4 = -7 \qquad \text{Using the principle of square roots}$$
$$x = 3 \quad \text{or} \quad x = -11$$

The solutions are 3 and -11.

Do Exercises 6 and 7 on the following page.

Calculator Spotlight

Solving Equations.

Example 1 can be visualized by letting

$$y_1 = x^2 \quad \text{and} \quad y_2 = 3$$

and finding the first coordinate of the points of intersection. We can use the INTERSECT feature to approximate the solutions.

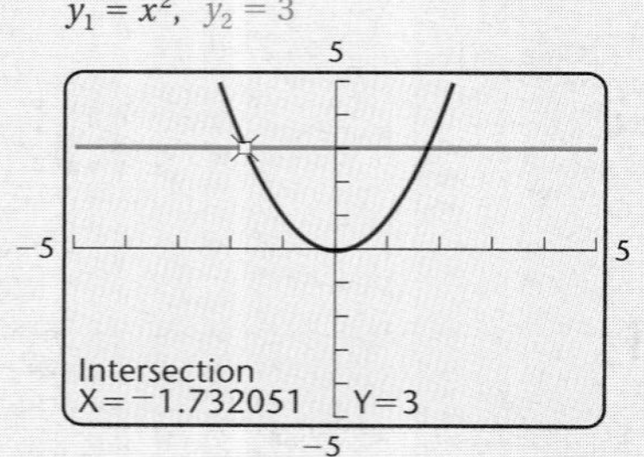

Note that

$$-\sqrt{3} \approx -1.732.$$

We would then repeat the procedure near the right-hand solution to approximate the other solution. The ZERO feature can also be used with the equation $x^2 - 3 = 0$.

Exercises

Use your grapher to approximate the solutions.

1. $x^2 = 10$

2. $-3x^2 + 7 = 0$

Solve.

4. $(x - 3)^2 = 16$

5. $(x + 4)^2 = 11$

Answers on page A-48

c Completing the Square

We have seen that a quadratic equation like $(x - 5)^2 = 9$ can be solved by using the principle of square roots. We also noted that an equation like $x^2 + 8x + 16 = 49$ can be solved in the same manner because the expression on the left side is the square of a binomial, $(x + 4)^2$. This second procedure is the basis for a method of solving quadratic equations called **completing the square.** *It can be used to solve any quadratic equation.*

Suppose we have the following quadratic equation:

$$x^2 + 10x = 4.$$

If we could add to both sides of the equation a constant that would make the expression on the left the square of a binomial, we could then solve the equation using the principle of square roots.

How can we determine what to add to $x^2 + 10x$ in order to construct the square of a binomial? We want to find a number a such that the following equation is satisfied:

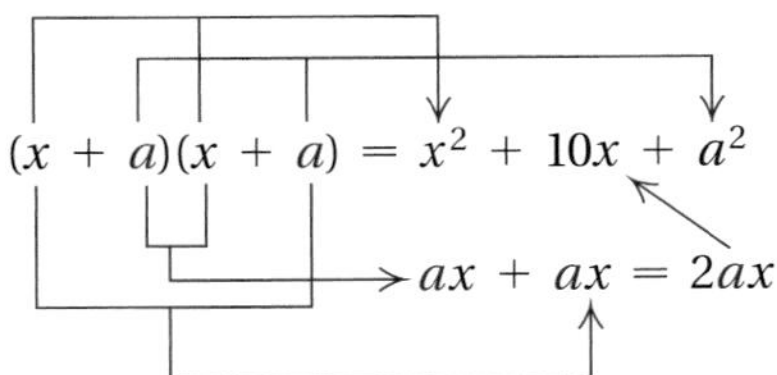

Thus, a is such that $2ax = 10x$. Solving for a, we get

$$a = \frac{10x}{2x} = \frac{10}{2} = 5;$$

that is, a is half of the coefficient of x in $x^2 + 10x$. Since $a^2 = \left(\frac{10}{2}\right)^2 = 5^2 = 25$, we add 25 to our original expression:

$x^2 + 10x + 25$ is the square of $x + 5$;

that is,

$$x^2 + 10x + 25 = (x + 5)^2.$$

> To **complete the square** of an expression like $x^2 + bx$, we take half of the coefficient of x and square. Then we add that number, which is $(b/2)^2$.

Returning to solve our original equation, we first add 25 on both sides to complete the square. Then we solve as follows:

$x^2 + 10x = 4$	**Original equation**
$x^2 + 10x + 25 = 4 + 25$	**Adding 25:** $\left(\frac{10}{2}\right)^2 = 5^2 = 25$
$(x + 5)^2 = 29$	
$x + 5 = \sqrt{29}$ *or* $x + 5 = -\sqrt{29}$	**Using the principle of square roots**
$x = -5 + \sqrt{29}$ *or* $x = -5 - \sqrt{29}$.	

The solutions are $-5 \pm \sqrt{29}$.

We have seen that a quadratic equation $(x + c)^2 = d$ can be solved by using the principle of square roots. Any quadratic equation can be put in this form by completing the square. Then we can solve as before.

Solve.

6. $x^2 - 6x + 9 = 64$

7. $x^2 - 2x + 1 = 5$

Answers on page A-48

Solve.

8. $x^2 - 6x + 8 = 0$

9. $x^2 + 8x - 20 = 0$

10. Solve: $x^2 - 12x + 23 = 0$.

11. Solve: $x^2 - 3x - 10 = 0$.

Answers on page A-48

Example 7 Solve: $x^2 + 6x + 8 = 0$.

We have

$$x^2 + 6x + 8 = 0$$
$$x^2 + 6x = -8. \quad \text{Subtracting 8}$$

We take half of 6 and square it, to get 9. Then we add 9 on *both* sides of the equation. This makes the left side the square of a binomial. We have now completed the square.

$$x^2 + 6x + 9 = -8 + 9 \quad \text{Adding 9}$$
$$(x + 3)^2 = 1$$
$$x + 3 = 1 \quad or \quad x + 3 = -1 \quad \text{Using the principle of square roots}$$
$$x = -2 \quad or \quad x = -4$$

The solutions are -2 and -4.

Do Exercises 8 and 9.

Example 8 Solve $x^2 - 4x - 7 = 0$ by completing the square.

We have

$$x^2 - 4x - 7 = 0$$
$$x^2 - 4x = 7 \quad \text{Adding 7}$$
$$x^2 - 4x + 4 = 7 + 4 \quad \text{Adding 4: } \left(\frac{-4}{2}\right)^2 = (-2)^2 = 4$$
$$(x - 2)^2 = 11$$
$$x - 2 = \sqrt{11} \quad or \quad x - 2 = -\sqrt{11} \quad \text{Using the principle of square roots}$$
$$x = 2 + \sqrt{11} \quad or \quad x = 2 - \sqrt{11}.$$

The solutions are $2 \pm \sqrt{11}$.

Do Exercise 10.

Example 7, as well as the following example, can be solved more easily by factoring. We solved it by completing the square only to illustrate that completing the square can be used to solve *any* quadratic equation.

Example 9 Solve $x^2 + 3x - 10 = 0$ by completing the square.

We have

$$x^2 + 3x - 10 = 0$$
$$x^2 + 3x = 10$$
$$x^2 + 3x + \frac{9}{4} = 10 + \frac{9}{4} \quad \text{Adding } \frac{9}{4}\text{: } \left(\frac{3}{2}\right)^2 = \frac{9}{4}$$
$$\left(x + \frac{3}{2}\right)^2 = \frac{40}{4} + \frac{9}{4} = \frac{49}{4}$$
$$x + \frac{3}{2} = \frac{7}{2} \quad or \quad x + \frac{3}{2} = -\frac{7}{2} \quad \text{Using the principle of square roots}$$
$$x = \frac{4}{2} \quad or \quad x = -\frac{10}{2}$$
$$x = 2 \quad or \quad x = -5.$$

The solutions are 2 and -5.

Do Exercise 11.

When the coefficient of x^2 is not 1, we can make it 1, as shown in the following example.

Example 10 Solve $2x^2 = 3x + 1$ by completing the square.

We first obtain standard form. Then we multiply on both sides by $\frac{1}{2}$ to make the x^2-coefficient 1.

$$2x^2 = 3x + 1$$

$$2x^2 - 3x - 1 = 0 \qquad \text{Finding standard form}$$

$$\frac{1}{2}(2x^2 - 3x - 1) = \frac{1}{2} \cdot 0 \qquad \text{Multiplying by } \frac{1}{2} \text{ to make the } x^2\text{-coefficient 1}$$

$$x^2 - \frac{3}{2}x - \frac{1}{2} = 0$$

$$x^2 - \frac{3}{2}x = \frac{1}{2} \qquad \text{Adding } \frac{1}{2}$$

$$x^2 - \frac{3}{2}x + \frac{9}{16} = \frac{1}{2} + \frac{9}{16} \qquad \text{Adding } \frac{9}{16}\text{: } \left[\frac{1}{2}\left(-\frac{3}{2}\right)\right]^2 = \left[-\frac{3}{4}\right]^2 = \frac{9}{16}$$

$$\left(x - \frac{3}{4}\right)^2 = \frac{8}{16} + \frac{9}{16} \qquad \text{Finding a common denominator}$$

$$\left(x - \frac{3}{4}\right)^2 = \frac{17}{16}$$

$$x - \frac{3}{4} = \frac{\sqrt{17}}{4} \quad or \quad x - \frac{3}{4} = -\frac{\sqrt{17}}{4} \qquad \text{Using the principle of square roots}$$

$$x = \frac{3}{4} + \frac{\sqrt{17}}{4} \quad or \quad x = \frac{3}{4} - \frac{\sqrt{17}}{4}$$

The solutions are $\frac{3 \pm \sqrt{17}}{4}$.

SOLVING BY COMPLETING THE SQUARE

To solve a quadratic equation $ax^2 + bx + c = 0$ by completing the square:

1. If $a \neq 1$, multiply by $1/a$ so that the x^2-coefficient is 1.
2. If the x^2-coefficient is 1, add so that the equation is in the form

$$x^2 + bx = -c, \quad \text{or} \quad x^2 + \frac{b}{a}x = -\frac{c}{a} \quad \text{if step (1) has been applied.}$$

3. Take half of the x-coefficient and square it. Add the result on both sides of the equation.
4. Express the side with the variables as the square of a binomial.
5. Use the principle of square roots and complete the solution.

Do Exercise 12.

12. Solve: $2x^2 + 3x - 3 = 0$.

Calculator Spotlight

Exercises

Use your grapher to find the solutions of the equation. Approximate to the nearest thousandth.

1. $x^2 + 10x = 4$
2. $x^2 - 4x - 7 = 0$
3. $2x^2 = 3x + 1$
4. $3 = 3x + 2x^2$

Answer on page A-48

13. The Transco Tower in Houston is 901 ft tall. How long would it take an object to fall to the ground from the top?

Answer on page A-48

d Applications

Example 11 *Falling Object.* The World Trade Center in New York is 1368 ft tall. How long would it take an object to fall to the ground from the top?

1. **Familiarize.** If we did not know anything about this problem, we might consider looking up a formula in a mathematics or physics book. A formula that fits this situation is

   $$s = 16t^2,$$

   where s is the distance, in feet, traveled by a body falling freely from rest in t seconds. This formula is actually an approximation in that it does not account for air resistance. In this problem, we know the distance s to be 1368 ft. We want to determine the time t for the object to reach the ground.

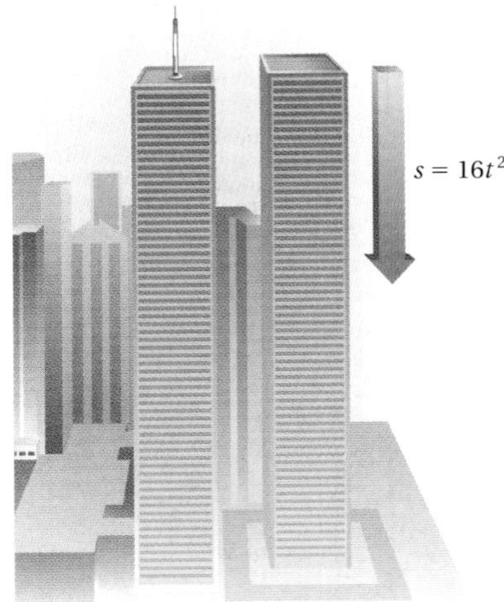

2. **Translate.** We know that the distance is 1368 and that we need to solve for t. We substitute 1368 for s:

   $$1368 = 16t^2.$$

   This gives us a translation.

3. **Solve.** We solve the equation:

   $$1368 = 16t^2$$

   $$\frac{1368}{16} = t^2 \qquad \text{Solving for } t^2$$

   $$85.5 = t^2 \qquad \text{Dividing}$$

   $$\sqrt{85.5} = t \quad or \quad -\sqrt{85.5} = t \qquad \text{Using the principle of square roots}$$

   $$9.2 \approx t \quad or \quad -9.2 \approx t. \qquad \text{Using a calculator to find the square root and rounding to the nearest tenth}$$

4. **Check.** The number -9.2 cannot be a solution because time cannot be negative in this situation. We substitute 9.2 in the original equation:

   $$s = 16(9.2)^2 = 16(84.64) = 1354.24.$$

   This is close; $1354.24 \approx 1368$. Remember that we approximated a solution, $t \approx 9.2$. Thus we have a check.

5. **State.** It takes about 9.2 sec for an object to fall to the ground from the top of the World Trade Center.

Do Exercise 13.

Exercise Set 16.2

a Solve.

1. $x^2 = 121$

2. $x^2 = 100$

3. $5x^2 = 35$

4. $5x^2 = 45$

5. $5x^2 = 3$

6. $2x^2 = 9$

7. $4x^2 - 25 = 0$

8. $9x^2 - 4 = 0$

9. $3x^2 - 49 = 0$

10. $5x^2 - 16 = 0$

11. $4y^2 - 3 = 9$

12. $36y^2 - 25 = 0$

13. $49y^2 - 64 = 0$

14. $8x^2 - 400 = 0$

b Solve.

15. $(x + 3)^2 = 16$

16. $(x - 4)^2 = 25$

17. $(x + 3)^2 = 21$

18. $(x - 3)^2 = 6$

19. $(x + 13)^2 = 8$

20. $(x - 13)^2 = 64$

21. $(x - 7)^2 = 12$

22. $(x + 1)^2 = 14$

23. $(x + 9)^2 = 34$

24. $(t + 5)^2 = 49$

25. $\left(x + \frac{3}{2}\right)^2 = \frac{7}{2}$

26. $\left(y - \frac{3}{4}\right)^2 = \frac{17}{16}$

27. $x^2 - 6x + 9 = 64$

28. $p^2 - 10p + 25 = 100$

29. $x^2 + 14x + 49 = 64$

30. $t^2 + 8t + 16 = 36$

c Solve by completing the square. Show your work.

31. $x^2 - 6x - 16 = 0$

32. $x^2 + 8x + 15 = 0$

33. $x^2 + 22x + 21 = 0$

34. $x^2 + 14x - 15 = 0$

35. $x^2 - 2x - 5 = 0$

36. $x^2 - 4x - 11 = 0$

37. $x^2 - 22x + 102 = 0$

38. $x^2 - 18x + 74 = 0$

39. $x^2 + 10x - 4 = 0$

40. $x^2 - 10x - 4 = 0$

41. $x^2 - 7x - 2 = 0$

42. $x^2 + 7x - 2 = 0$

43. $x^2 + 3x - 28 = 0$

44. $x^2 - 3x - 28 = 0$

45. $x^2 + \frac{3}{2}x - \frac{1}{2} = 0$

46. $x^2 - \frac{3}{2}x - 2 = 0$

47. $2x^2 + 3x - 17 = 0$

48. $2x^2 - 3x - 1 = 0$

49. $3x^2 + 4x - 1 = 0$

50. $3x^2 - 4x - 3 = 0$

51. $2x^2 = 9x + 5$

52. $2x^2 = 5x + 12$

53. $6x^2 + 11x = 10$

54. $4x^2 + 12x = 7$

d Solve.

55. *Sears Tower.* The height of the Sears Tower in Chicago is 1451 ft (excluding TV towers and antennas). How long would it take an object to fall to the ground from the top?

56. *Library Square Tower.* Library Square Tower in Los Angeles is 1012 ft tall. How long would it take an object to fall to the ground from the top?

57. *Free-Fall Record.* The world record for free-fall to the ground, by a man without a parachute, is 311 ft and is held by Dar Robinson. Approximately how long did the fall take?

58. *Free-Fall Record.* The world record for free-fall to the ground, by a woman without a parachute, into a cushioned landing area is 175 ft and is held by Kitty O'Neill. Approximately how long did the fall take?

Skill Maintenance

59. Find an equation of variation in which y varies inversely as x and $y = 235$ when $x = 0.6$. [13.5c]

60. The time T to do a certain job varies inversely as the number N of people working. It takes 5 hr for 24 people to wash and wax the floors in a building. How long would it take 36 people to do the job? [13.5d]

Multiply and simplify. [15.2c]

61. $\sqrt{3x} \cdot \sqrt{6x}$

62. $\sqrt{8x^2} \cdot \sqrt{24x^3}$

63. $3\sqrt{t} \cdot \sqrt{t}$

64. $\sqrt{x^2} \cdot \sqrt{x^5}$

Synthesis

65. ◆ Corey asserts that the solution of a quadratic equation is $3 \pm \sqrt{14}$ and states that there is only one solution. What mistake is being made?

66. ◆ If a quadratic equation can be solved by factoring, what type of number(s) will generally be solutions?

Find b such that the trinomial is a square.

67. $x^2 + bx + 36$

68. $x^2 + bx + 55$

69. $x^2 + bx + 128$

70. $4x^2 + bx + 16$

71. $x^2 + bx + c$

72. $ax^2 + bx + c$

Solve.

73. [calculator] $4.82x^2 = 12{,}000$

74. $\frac{x}{2} = \frac{32}{x}$

75. $\frac{x}{9} = \frac{36}{4x}$

76. $\frac{4}{m^2 - 7} = 1$

Visualize completion of the square using rectangles.

16.3 The Quadratic Formula

We learn to complete the square to prove a general formula that can be used to solve quadratic equations even when they cannot be solved by factoring.

Objectives

a Solve quadratic equations using the quadratic formula.

b Find approximate solutions of quadratic equations using a calculator.

For Extra Help

TAPE 29

MAC WIN

CD-ROM

a Solving Using the Quadratic Formula

Each time you solve by completing the square, you perform nearly the same steps. When we repeat the same kind of computation many times, we look for a formula so we can speed up our work. Consider

$$ax^2 + bx + c = 0, \quad a > 0.$$

Let's solve by completing the square. As we carry out the steps, compare them with Example 10 in the preceding section.

$$x^2 + \frac{b}{a}x + \frac{c}{a} = 0 \qquad \text{Multiplying by } \frac{1}{a}$$

$$x^2 + \frac{b}{a}x = -\frac{c}{a} \qquad \text{Adding } -\frac{c}{a}$$

Half of $\frac{b}{a}$ is $\frac{b}{2a}$. The square is $\frac{b^2}{4a^2}$. Thus we add $\frac{b^2}{4a^2}$ on both sides.

$$x^2 + \frac{b}{a}x + \frac{b^2}{4a^2} = -\frac{c}{a} + \frac{b^2}{4a^2} \qquad \text{Adding } \frac{b^2}{4a^2}$$

$$\left(x + \frac{b}{2a}\right)^2 = -\frac{4ac}{4a^2} + \frac{b^2}{4a^2} \qquad \text{Factoring the left side and finding a common denominator on the right}$$

$$\left(x + \frac{b}{2a}\right)^2 = \frac{b^2 - 4ac}{4a^2}$$

$$x + \frac{b}{2a} = \sqrt{\frac{b^2 - 4ac}{4a^2}} \quad or \quad x + \frac{b}{2a} = -\sqrt{\frac{b^2 - 4ac}{4a^2}} \qquad \text{Using the principle of square roots}$$

Since $a > 0$, $\sqrt{4a^2} = 2a$, so we can simplify as follows:

$$x + \frac{b}{2a} = \frac{\sqrt{b^2 - 4ac}}{2a} \quad \text{or} \quad x + \frac{b}{2a} = -\frac{\sqrt{b^2 - 4ac}}{2a}.$$

Thus,

$$x = -\frac{b}{2a} + \frac{\sqrt{b^2 - 4ac}}{2a} \quad \text{or} \quad x = -\frac{b}{2a} - \frac{\sqrt{b^2 - 4ac}}{2a},$$

so $$x = -\frac{b}{2a} \pm \frac{\sqrt{b^2 - 4ac}}{2a},$$

or $$x = \frac{-b \pm \sqrt{b^2 - 4ac}}{2a}.$$

We now have the following.

THE QUADRATIC FORMULA

The solutions of $ax^2 + bx + c = 0$ are given by

$$x = \frac{-b \pm \sqrt{b^2 - 4ac}}{2a}.$$

1. Solve using the quadratic formula:

$2x^2 = 4 - 7x.$

Note that the formula also holds when $a < 0$. A similar proof would show this, but we will not consider it here.

Example 1 Solve $5x^2 - 8x = -3$ using the quadratic formula.

We first find standard form and determine a, b, and c:

$$5x^2 - 8x + 3 = 0;$$

$$a = 5, \quad b = -8, \quad c = 3.$$

We then use the quadratic formula:

$$x = \frac{-b \pm \sqrt{b^2 - 4ac}}{2a}$$

$$x = \frac{-(-8) \pm \sqrt{(-8)^2 - 4 \cdot 5 \cdot 3}}{2 \cdot 5}$$ Substituting

CAUTION! Be sure to write the fraction bar all the way across.

$$x = \frac{8 \pm \sqrt{64 - 60}}{10}$$

$$x = \frac{8 \pm \sqrt{4}}{10}$$

$$x = \frac{8 \pm 2}{10}$$

$$x = \frac{8 + 2}{10} \quad or \quad x = \frac{8 - 2}{10}$$

$$x = \frac{10}{10} \quad or \quad x = \frac{6}{10}$$

$$x = 1 \quad or \quad x = \frac{3}{5}.$$

The solutions are 1 and $\frac{3}{5}$.

Do Exercise 1.

It would have been easier to solve the equation in Example 1 by factoring. We used the quadratic formula only to illustrate that it can be used to solve any quadratic equation. The following is a general procedure for solving a quadratic equation.

SOLVING USING THE QUADRATIC FORMULA

To solve a quadratic equation:

1. Check to see if it is in the form $ax^2 = p$ or $(x + c)^2 = d$. If it is, use the principle of square roots as in Section 16.2.
2. If it is not in the form of (1), write it in standard form, $ax^2 + bx + c = 0$ with a and b nonzero.
3. Then try factoring.
4. If it is not possible to factor or if factoring seems difficult, use the quadratic formula.

The solutions of a quadratic equation can always be found using the quadratic formula. They cannot always be found by factoring. (When $b^2 - 4ac \geq 0$, the equation has real-number solutions. When $b^2 - 4ac < 0$, the equation has no real-number solutions.)

Calculator Spotlight

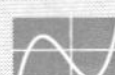

Exercises

Use your grapher to find the solutions of the equation. Approximate to the nearest hundredth.

1. $x^2 + 3x - 10 = 0$
2. $x^2 = 4x + 7$
3. $5x^2 - 8x = -3$
4. $2x^2 = 4 - 7x$

Answer on page A-48

Example 2 Solve $x^2 + 3x - 10 = 0$ using the quadratic formula.

The equation is in standard form. So we determine a, b, and c:

$$x^2 + 3x - 10 = 0;$$
$$a = 1, \quad b = 3, \quad c = -10.$$

We then use the quadratic formula:

$$x = \frac{-b \pm \sqrt{b^2 - 4ac}}{2a}$$
$$= \frac{-3 \pm \sqrt{3^2 - 4 \cdot 1 \cdot (-10)}}{2 \cdot 1} \qquad \text{Substituting}$$
$$= \frac{-3 \pm \sqrt{9 + 40}}{2}$$
$$= \frac{-3 \pm \sqrt{49}}{2} = \frac{-3 \pm 7}{2}.$$

Thus,

$$x = \frac{-3 + 7}{2} = \frac{4}{2} = 2 \quad \text{or} \quad x = \frac{-3 - 7}{2} = \frac{-10}{2} = -5.$$

The solutions are 2 and -5.

Note that the radicand ($b^2 - 4ac = 49$) in the quadratic formula is a perfect square, so we could have used factoring to solve.

Do Exercise 2.

2. Solve using the quadratic formula:

$$x^2 - 3x - 10 = 0.$$

Example 3 Solve $x^2 = 4x + 7$ using the quadratic formula. Compare with Example 8 in Section 16.2.

We first find standard form and determine a, b, and c:

$$x^2 - 4x - 7 = 0;$$
$$a = 1, \quad b = -4, \quad c = -7.$$

We then use the quadratic formula:

$$x = \frac{-b \pm \sqrt{b^2 - 4ac}}{2a} = \frac{-(-4) \pm \sqrt{(-4)^2 - 4 \cdot 1 \cdot (-7)}}{2 \cdot 1} \qquad \text{Substituting}$$
$$= \frac{4 \pm \sqrt{16 + 28}}{2} = \frac{4 \pm \sqrt{44}}{2}$$
$$= \frac{4 \pm \sqrt{4 \cdot 11}}{2} = \frac{4 \pm \sqrt{4}\sqrt{11}}{2}$$
$$= \frac{4 \pm 2\sqrt{11}}{2} = \frac{2 \cdot 2 \pm 2\sqrt{11}}{2 \cdot 1} \qquad \text{Factoring out 2 in the numerator and the denominator}$$
$$= \frac{2(2 \pm \sqrt{11})}{2 \cdot 1} = \frac{2}{2} \cdot \frac{2 \pm \sqrt{11}}{1}$$
$$= 2 \pm \sqrt{11}.$$

The solutions are $2 + \sqrt{11}$ and $2 - \sqrt{11}$, or $2 \pm \sqrt{11}$.

Do Exercise 3.

3. Solve using the quadratic formula:

$$x^2 + 4x = 7.$$

Answers on page A-48

4. Solve using the quadratic formula:

$$x^2 = x - 1.$$

Calculator Spotlight

To see that no real-number solutions exist for Example 4, we let

$y_1 = x^2 + x$ and $y_2 = -1$

and graph the equations in the same window.

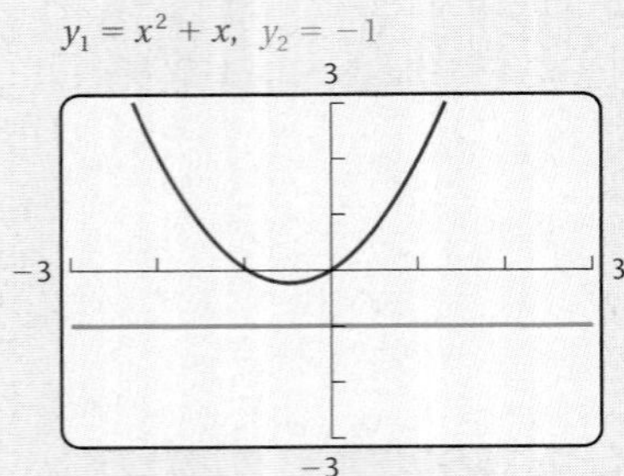

We see that the graphs do not intersect. This supports the conclusion that no real-number solution exists.

Exercises

1. What happens when the INTERSECT feature is used with these graphs?
2. How can the graph of $y = x^2 + x + 1$ tell us that no real-number solution exists?
3. Use a grapher to solve the equation $x^2 = x - 1$.

5. Solve using the quadratic formula:

$$5x^2 - 8x = 3.$$

6. Approximate the solutions to the equation in Margin Exercise 5. Round to the nearest tenth.

Answers on page A-48

Example 4 Solve $x^2 + x = -1$ using the quadratic formula.

We first find standard form and determine a, b, and c:

$$x^2 + x + 1 = 0;$$
$$a = 1, \quad b = 1, \quad c = 1.$$

We then use the quadratic formula:

$$x = \frac{-b \pm \sqrt{b^2 - 4ac}}{2a} = \frac{-1 \pm \sqrt{1^2 - 4 \cdot 1 \cdot 1}}{2 \cdot 1} = \frac{-1 \pm \sqrt{-3}}{2}.$$

Note that the radicand ($b^2 - 4ac = -3$) in the quadratic formula is negative. Thus there are no real-number solutions because square roots of negative numbers do not exist as real numbers.

Do Exercise 4.

Example 5 Solve $3x^2 = 7 - 2x$ using the quadratic formula.

We first find standard form and determine a, b, and c:

$$3x^2 + 2x - 7 = 0;$$
$$a = 3, \quad b = 2, \quad c = -7.$$

We then use the quadratic formula:

$$x = \frac{-b \pm \sqrt{b^2 - 4ac}}{2a} = \frac{-2 \pm \sqrt{2^2 - 4 \cdot 3 \cdot (-7)}}{2 \cdot 3} = \frac{-2 \pm \sqrt{4 + 84}}{2 \cdot 3}$$
$$= \frac{-2 \pm \sqrt{88}}{6} = \frac{-2 \pm \sqrt{4 \cdot 22}}{6} = \frac{-2 \pm \sqrt{4}\sqrt{22}}{6} = \frac{-2 \pm 2\sqrt{22}}{6}$$
$$= \frac{2(-1 \pm \sqrt{22})}{2 \cdot 3} = \frac{2}{2} \cdot \frac{-1 \pm \sqrt{22}}{3} = \frac{-1 \pm \sqrt{22}}{3}.$$

The solutions are $\frac{-1 + \sqrt{22}}{3}$ and $\frac{-1 - \sqrt{22}}{3}$, or $\frac{-1 \pm \sqrt{22}}{3}$.

Do Exercise 5.

b Approximate Solutions

A calculator can be used to approximate solutions.

Example 6 Use a calculator to approximate to the nearest tenth the solutions to the equation in Example 5.

Using a calculator , we have

$$\frac{-1 + \sqrt{22}}{3} \approx 1.230138587 \approx 1.2 \text{ to the nearest tenth, and}$$
$$\frac{-1 - \sqrt{22}}{3} \approx -1.896805253 \approx -1.9 \text{ to the nearest tenth.}$$

The approximate solutions are 1.2 and -1.9.

Do Exercise 6.

Exercise Set 16.3

a Solve. Try factoring first. If factoring is not possible or is difficult, use the quadratic formula.

1. $x^2 - 4x = 21$
2. $x^2 + 8x = 9$
3. $x^2 = 6x - 9$
4. $x^2 = 24x - 144$
5. $3y^2 - 2y - 8 = 0$
6. $3y^2 - 7y + 4 = 0$
7. $4x^2 + 4x = 15$
8. $4x^2 + 12x = 7$
9. $x^2 - 9 = 0$
10. $x^2 - 16 = 0$
11. $x^2 - 2x - 2 = 0$
12. $x^2 - 2x - 11 = 0$
13. $y^2 - 10y + 22 = 0$
14. $y^2 + 6y - 1 = 0$
15. $x^2 + 4x + 4 = 7$
16. $x^2 - 2x + 1 = 5$
17. $3x^2 + 8x + 2 = 0$
18. $3x^2 - 4x - 2 = 0$
19. $2x^2 - 5x = 1$
20. $4x^2 + 4x = 5$
21. $2y^2 - 2y - 1 = 0$
22. $4y^2 + 4y - 1 = 0$
23. $2t^2 + 6t + 5 = 0$
24. $4y^2 + 3y + 2 = 0$
25. $3x^2 = 5x + 4$
26. $2x^2 + 3x = 1$
27. $2y^2 - 6y = 10$
28. $5m^2 = 3 + 11m$

29. $\frac{x^2}{x+3} - \frac{5}{x+3} = 0$

30. $\frac{x^2}{x-4} - \frac{7}{x-4} = 0$

31. $x + 2 = \frac{3}{x+2}$

32. $x - 3 = \frac{5}{x-3}$

33. $\frac{1}{x} + \frac{1}{x+1} = \frac{1}{3}$

34. $\frac{1}{x} + \frac{1}{x+6} = \frac{1}{5}$

b Solve using the quadratic formula. Use a calculator to approximate the solutions to the nearest tenth.

35. $x^2 - 4x - 7 = 0$

36. $x^2 + 2x - 2 = 0$

37. $y^2 - 6y - 1 = 0$

38. $y^2 + 10y + 22 = 0$

39. $4x^2 + 4x = 1$

40. $4x^2 = 4x + 1$

41. $3x^2 - 8x + 2 = 0$

42. $3x^2 + 4x - 2 = 0$

Skill Maintenance

Add or subtract. [15.4a]

43. $\sqrt{40} - 2\sqrt{10} + \sqrt{90}$

44. $\sqrt{54} - \sqrt{24}$

45. $\sqrt{18} + \sqrt{50} - 3\sqrt{8}$

46. $\sqrt{81x^3} - \sqrt{4x}$

47. Simplify: $\sqrt{80}$. [15.2a]

48. Multiply and simplify: $\sqrt{3x^2}\sqrt{9x^3}$. [15.2c]

49. Simplify: $\sqrt{9000x^{10}}$. [15.2b]

50. Rationalize the denominator: $\sqrt{\frac{7}{3}}$. [15.3c]

Synthesis

51. ◆ List a quadratic equation with no real-number solutions. How can you use that equation to find an equation in the form $y = ax^2 + bx + c$ that does not cross the x-axis?

52. ◆ Under what condition(s) would using the quadratic formula *not* be the easiest way to solve a quadratic equation?

Solve.

53. $5x + x(x - 7) = 0$

54. $x(3x + 7) - 3x = 0$

55. $3 - x(x - 3) = 4$

56. $x(5x - 7) = 1$

57. $(y + 4)(y + 3) = 15$

58. $(y + 5)(y - 1) = 27$

59. $x^2 + (x + 2)^2 = 7$

60. $x^2 + (x + 1)^2 = 5$

61.–68. Use a grapher to approximate the solutions of the equations in Exercises 35–42. Compare your answers with those found using the quadratic formula.

16.4 Formulas

a To solve a formula for a given letter, we try to get the letter alone on one side.

Example 1 Solve for h: $V = 3.5\sqrt{h}$ (the distance to the horizon).

This is a radical equation. Recall that we first isolate the radical. Then we use the principle of squaring.

$$\frac{V}{3.5} = \sqrt{h} \qquad \text{Isolating the radical}$$

$$\left(\frac{V}{3.5}\right)^2 = (\sqrt{h})^2 \qquad \text{Using the principle of squaring (Section 9.5)}$$

$$\frac{V^2}{12.25} = h \qquad \text{Simplifying}$$

Example 2 Solve for g: $T = 2\pi\sqrt{\frac{L}{g}}$ (the period of a pendulum).

$$\frac{T}{2\pi} = \sqrt{\frac{L}{g}} \qquad \text{Dividing by } 2\pi \text{ to isolate the radical}$$

$$\left(\frac{T}{2\pi}\right)^2 = \left(\sqrt{\frac{L}{g}}\right)^2 \qquad \text{Using the principle of squaring}$$

$$\frac{T^2}{4\pi^2} = \frac{L}{g}$$

$$gT^2 = 4\pi^2 L \qquad \text{Multiplying by } 4\pi^2 g \text{ to clear fractions}$$

$$g = \frac{4\pi^2 L}{T^2} \qquad \text{Dividing by } T^2 \text{ to get } g \text{ alone}$$

Do Exercises 1–3.

In most formulas, the letters represent nonnegative numbers, so we need not use absolute values when taking square roots.

Example 3 *Torricelli's Theorem.* The speed v of a liquid leaving a bucket from an opening is related to the height h of the top of the liquid above the opening by the formula

$$h = \frac{v^2}{2g}.$$

Solve for v.

Since v^2 appears by itself and there is no expression involving v, we first solve for v^2. Then we use the principle of square roots, taking only the nonnegative square root because v is nonnegative.

$$2gh = v^2 \qquad \text{Multiplying by } 2g \text{ to clear fractions}$$

$$\sqrt{2gh} = v \qquad \text{Using the principle of square roots. Assume that } v \text{ is nonnegative.}$$

Do Exercise 4.

Objective

a Solve a formula for a given letter.

For Extra Help

TAPE 29

MAC
WIN

CD-ROM

1. Solve for L: $r = 2\sqrt{5L}$ (the speed of a skidding car).

2. Solve for L: $T = 2\pi\sqrt{\frac{L}{g}}$.

3. Solve for m: $c = \sqrt{\frac{E}{m}}$.

4. Solve for r: $A = \pi r^2$ (the area of a circle).

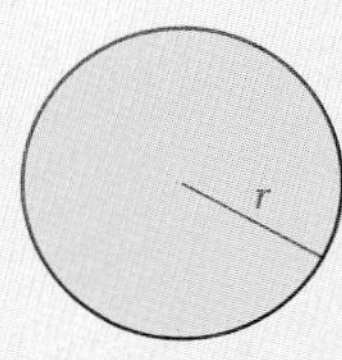

Answers on page A-49

5. Solve for d: $C = P(d - 1)^2$.

Example 4 Solve for r: $A = P(1 + r)^2$ (a compound-interest formula).

$$A = P(1 + r)^2$$

$$\frac{A}{P} = (1 + r)^2 \quad \text{Dividing by } P$$

$$\sqrt{\frac{A}{P}} = 1 + r \quad \text{Using the principle of square roots. Assume that } 1 + r \text{ is nonnegative.}$$

$$-1 + \sqrt{\frac{A}{P}} = r \quad \text{Subtracting 1 to get } r \text{ alone}$$

Do Exercise 5.

Sometimes we must use the quadratic formula to solve a formula for a certain letter.

6. Solve for n: $N = n^2 - n$.

Example 5 Solve for n: $d = \dfrac{n^2 - 3n}{2}$, where d is the number of diagonals of an n-sided polygon.

This time there is a term involving n as well as an n^2-term. Thus we must use the quadratic formula.

$$d = \frac{n^2 - 3n}{2}$$

$$n^2 - 3n = 2d \quad \text{Multiplying by 2 to clear fractions}$$

$$n^2 - 3n - 2d = 0 \quad \text{Finding standard form}$$

$$a = 1, \quad b = -3, \quad c = -2d \quad \text{The letter } d \text{ represents a constant.}$$

$$n = \frac{-b \pm \sqrt{b^2 - 4ac}}{2a} \quad \text{Quadratic formula}$$

$$= \frac{-(-3) \pm \sqrt{(-3)^2 - 4 \cdot 1 \cdot (-2d)}}{2 \cdot 1} \quad \text{Substituting into the quadratic formula}$$

$$= \frac{3 + \sqrt{9 + 8d}}{2} \quad \text{Using the positive root}$$

Do Exercise 6.

7. Solve for t: $h = vt + 8t^2$.

Example 6 Solve for t: $S = gt + 16t^2$.

$$S = gt + 16t^2$$

$$16t^2 + gt - S = 0 \quad \text{Finding standard form}$$

$$a = 16, \quad b = g, \quad c = -S$$

$$t = \frac{-b \pm \sqrt{b^2 - 4ac}}{2a}$$

$$= \frac{-g \pm \sqrt{g^2 - 4 \cdot 16 \cdot (-S)}}{2 \cdot 16} \quad \text{Substituting into the quadratic formula}$$

$$= \frac{-g + \sqrt{g^2 + 64S}}{32} \quad \text{Using the positive root}$$

Do Exercise 7.

Answers on page A-49

Exercise Set 16.4

a Solve for the indicated letter.

1. $P = 17\sqrt{Q}$, for Q

2. $A = 1.4\sqrt{t}$, for t

3. $v = \sqrt{\frac{2gE}{m}}$, for E

4. $Q = \sqrt{\frac{aT}{c}}$, for T

5. $S = 4\pi r^2$, for r

6. $E = mc^2$, for c

7. $P = kA^2 + mA$, for A

8. $Q = ad^2 - cd$, for d

9. $c^2 = a^2 + b^2$, for a

10. $c = \sqrt{a^2 + b^2}$, for b

11. $s = 16t^2$, for t

12. $V = \pi r^2 h$, for r

13. $A = \pi r^2 + 2\pi rh$, for r

14. $A = 2\pi r^2 + 2\pi rh$, for r

15. $F = \frac{Av^2}{400}$, for v

16. $A = \frac{\pi r^2 S}{360}$, for r

17. $c = \sqrt{a^2 + b^2}$, for a

18. $c^2 = a^2 + b^2$, for b

19. $h = \frac{a}{2}\sqrt{3}$, for a
(The height of an equilateral triangle with sides of length a)

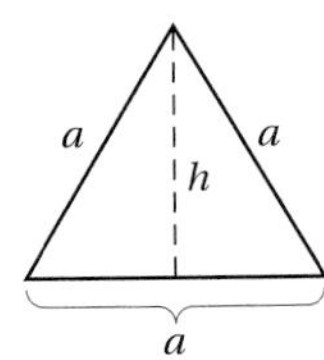

20. $d = s\sqrt{2}$, for s
(The hypotenuse of an isosceles right triangle with s the length of the legs)

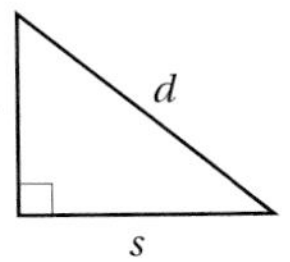

21. $n = aT^2 - 4T + m$, for T

22. $y = ax^2 + bx + c$, for x

23. $v = 2\sqrt{\frac{2kT}{\pi m}}$, for T

24. $E = \frac{1}{2}mv^2 + mgy$, for v

25. $3x^2 = d^2$, for x

26. $c = \sqrt{\frac{E}{m}}$, for E

27. $N = \frac{n^2 - n}{2}$, for n

28. $M = \frac{m}{\sqrt{1 - \left(\frac{v}{c}\right)^2}}$, for c

Skill Maintenance

In a right triangle, find the length of the side not given. Give an exact answer and an approximation to three decimal places. [15.6a]

29. $a = 4$, $b = 7$

30. $b = 11$, $c = 14$

31. $a = 4$, $b = 5$

32. $a = 10$, $c = 12$

33. $c = 8\sqrt{17}$, $a = 2$

34. $a = \sqrt{2}$, $b = \sqrt{3}$

Solve. [15.6b]

35. *Guy Wire.* How long is a guy wire reaching from the top of an 18-ft pole to a point on the ground 10 ft from the pole? Give an exact answer and an approximation to three decimal places.

36. *Street Width.* Elliott Street is 24 ft wide where it ends at Main Street in Brattleboro, Vermont. A 40-ft–long diagonal crosswalk allows pedestrians to cross Main Street to or from either corner of Elliott Street (see the figure). Determine the width of Main Street.

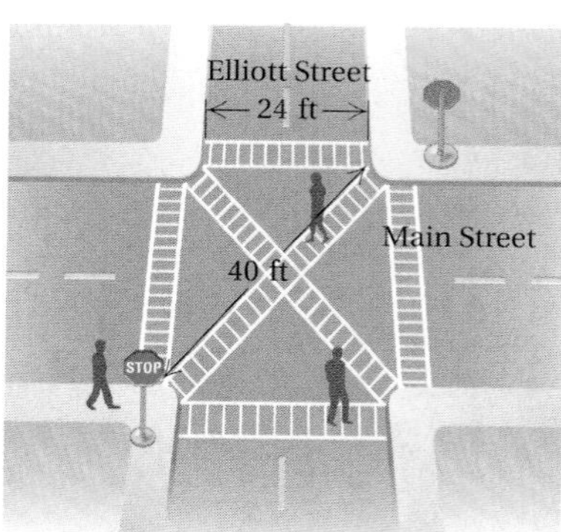

Synthesis

37. ◆ Explain how you would solve the equation $0 = ax^2 + bx + c$ for x.

38. ◆ Explain how you would solve the equation $y = ax^2 + bx + c$ for x.

39. The circumference C of a circle is given by $C = 2\pi r$.
a) Solve $C = 2\pi r$ for r.
b) The area is given by $A = \pi r^2$. Express the area in terms of the circumference C.

40. Referring to Exercise 39, express the circumference C in terms of the area A.

41. Solve $3ax^2 - x - 3ax + 1 = 0$ for x.

42. Solve $h = 16t^2 + vt + s$ for t.

16.5 Applications and Problem Solving

a Using Quadratic Equations to Solve Applied Problems

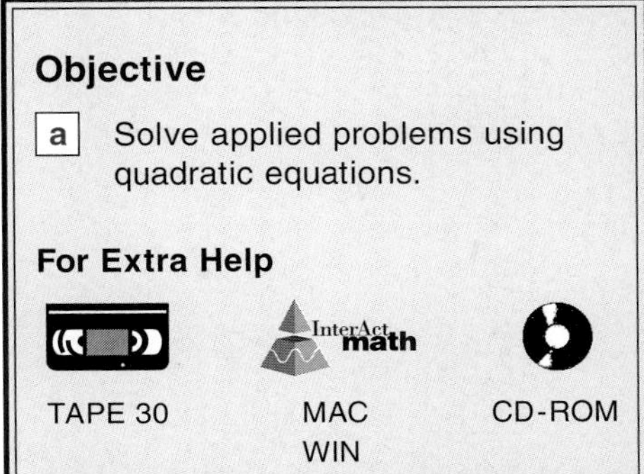

Objective
a Solve applied problems using quadratic equations.
For Extra Help
InterAct math
TAPE 30
MAC
WIN
CD-ROM

Example 1 *Red Raspberry Patch.* The area of a rectangular red raspberry patch is 76 ft^2. The length is 7 ft longer than three times the width. Find the dimensions of the raspberry patch.

1. **Familiarize.** We first make a drawing and label it with both known and unknown information. We let $w =$ the width of the rectangle. The length of the rectangle is 7 ft longer than three times the width. Thus the length is $3w + 7$.

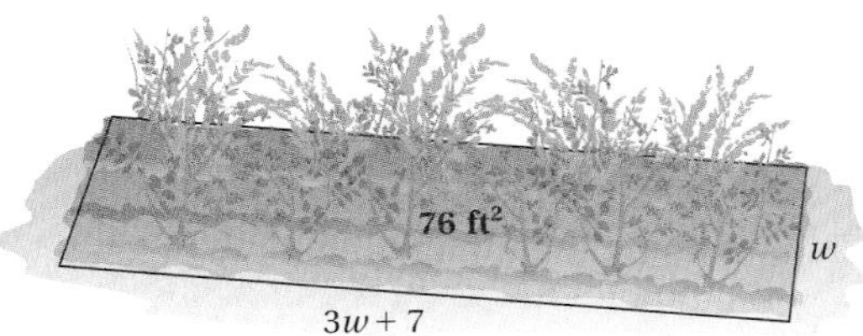

2. **Translate.** Recall that area is length × width. Thus we have two expressions for the area of the rectangle: $(3w + 7)(w)$ and 76. This gives us a translation:

$$(3w + 7)(w) = 76.$$

3. **Solve.** We solve the equation:

$$3w^2 + 7w = 76$$
$$3w^2 + 7w - 76 = 0$$
$$(3w + 19)(w - 4) = 0 \quad \text{Factoring (the quadratic formula could also be used)}$$
$$3w + 19 = 0 \quad or \quad w - 4 = 0 \quad \text{Using the principle of zero products}$$
$$3w = -19 \quad or \quad w = 4$$
$$w = -\tfrac{19}{3} \quad or \quad w = 4.$$

4. **Check.** We check in the original problem. We know that $-\frac{19}{3}$ is not a solution because width cannot be negative. When $w = 4$, $3w + 7 = 19$, and the area is $4(19)$, or 76. This checks.
5. **State.** The width of the rectangular raspberry patch is 4 ft, and the length is 19 ft.

Do Exercise 1.

1. *Pool Dimensions.* The area of a rectangular swimming pool is 68 yd^2. The length is 1 yd longer than three times the width. Find the dimensions of the rectangular swimming pool. Round to the nearest tenth.

Example 2 *Staircase.* A carpenter builds a staircase in such a way that the portion underneath the stairs forms a right triangle. The hypotenuse is 6 m long. The leg across the floor is 1 m longer than the leg next to the wall at the back. Find the lengths of the legs. Round to the nearest tenth.

1. **Familiarize.** We first make a drawing, letting $s =$ the length of one leg. Then $s + 1 =$ the length of the other leg.

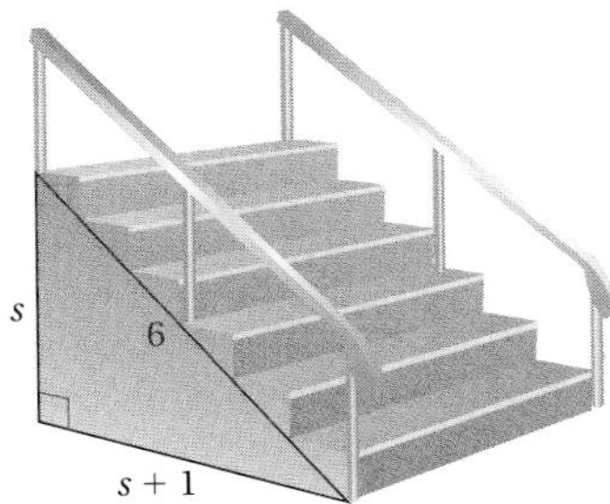

Answer on page A-49

2. The hypotenuse of a right triangular animal pen at the zoo is 4 yd long. One leg is 1 yd longer than the other. Find the lengths of the legs. Round to the nearest tenth.

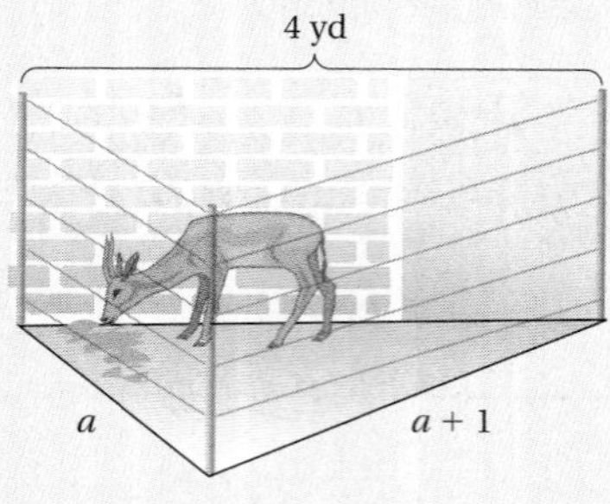

2.3 yd, 3.3 yd

Answer on page A-49

2. Translate. To translate, we use the Pythagorean equation:

$$s^2 + (s + 1)^2 = 6^2.$$

3. Solve. We solve the equation:

$$s^2 + (s + 1)^2 = 6^2$$
$$s^2 + s^2 + 2s + 1 = 36$$
$$2s^2 + 2s - 35 = 0.$$

Since we cannot factor, we use the quadratic formula:

$$a = 2, \quad b = 2, \quad c = -35$$

$$s = \frac{-b \pm \sqrt{b^2 - 4ac}}{2a}$$
$$= \frac{-2 \pm \sqrt{2^2 - 4 \cdot 2(-35)}}{2 \cdot 2}$$
$$= \frac{-2 \pm \sqrt{4 + 280}}{4}$$
$$= \frac{-2 \pm \sqrt{284}}{4}$$
$$= \frac{-2 \pm \sqrt{4 \cdot 71}}{4}$$
$$= \frac{-2 \pm 2 \cdot \sqrt{71}}{2 \cdot 2}$$
$$= \frac{2(-1 \pm \sqrt{71})}{2 \cdot 2} = \frac{2}{2} \cdot \frac{-1 \pm \sqrt{71}}{2}$$
$$= \frac{-1 \pm \sqrt{71}}{2}.$$

Using a calculator, we get approximations:

$$\frac{-1 + \sqrt{71}}{2} \approx 3.7 \quad \text{or} \quad \frac{-1 - \sqrt{71}}{2} \approx -4.7.$$

4. Check. Since the length of a leg cannot be negative, -4.7 does not check. But 3.7 does check. If the smaller leg s is 3.7, the other leg is $s + 1$, or 4.7. Then

$$(3.7)^2 + (4.7)^2 = 13.69 + 22.09 = 35.78.$$

Using a calculator, we get $\sqrt{35.78} \approx 5.98 \approx 6$. Note that our check is not exact because we are using an approximation for $\sqrt{71}$.

5. State. One leg is about 3.7 m long, and the other is about 4.7 m long.

Do Exercise 2.

Example 3 *Boat Speed.* The current in a stream moves at a speed of 2 km/h. A boat travels 24 km upstream and 24 km downstream in a total time of 5 hr. What is the speed of the boat in still water?

1. Familiarize. We first make a drawing. The distances are the same. We let r = the speed of the boat in still water. Then when the boat is traveling upstream, its speed is $r - 2$. When it is traveling downstream, its speed is $r + 2$. We let t_1 represent the time it takes the boat to go upstream and t_2 the time it takes to go downstream. We summarize in a table.

	d	r	t
Upstream	24	$r-2$	t_1
Downstream	24	$r+2$	t_2
Total Time			5

$\longrightarrow t_1 = \dfrac{24}{r-2}$

$\longrightarrow t_2 = \dfrac{24}{r+2}$

2. Translate. Recall the basic formula for motion: $d = rt$. From it we can obtain an equation for time: $t = d/r$. Total time consists of the time to go upstream, t_1, plus the time to go downstream, t_2. Using $t = d/r$ and the rows of the table, we have

$$t_1 = \frac{24}{r-2} \quad \text{and} \quad t_2 = \frac{24}{r+2}.$$

Since the total time is 5 hr, $t_1 + t_2 = 5$, and we have

$$\frac{24}{r-2} + \frac{24}{r+2} = 5.$$

3. Solve. We solve the equation. We multiply on both sides by the LCM, which is $(r-2)(r+2)$:

$$(r-2)(r+2)\cdot\left[\frac{24}{r-2} + \frac{24}{r+2}\right] = (r-2)(r+2)5 \qquad \textbf{Multiplying by the LCM}$$

$$(r-2)(r+2)\cdot\frac{24}{r-2} + (r-2)(r+2)\cdot\frac{24}{r+2} = (r^2-4)5$$

$$24(r+2) + 24(r-2) = 5r^2 - 20$$

$$24r + 48 + 24r - 48 = 5r^2 - 20$$

$$-5r^2 + 48r + 20 = 0$$

$$5r^2 - 48r - 20 = 0 \qquad \textbf{Multiplying by } -1$$

$$(5r+2)(r-10) = 0 \qquad \textbf{Factoring}$$

$$5r + 2 = 0 \quad \text{or} \quad r - 10 = 0 \qquad \textbf{Using the principle of zero products}$$

$$5r = -2 \quad \text{or} \quad r = 10$$

$$r = -\tfrac{2}{5} \quad \text{or} \quad r = 10.$$

4. Check. Since speed cannot be negative, $-\frac{2}{5}$ cannot be a solution. But suppose the speed of the boat in still water is 10 km/h. The speed upstream is then $10 - 2$, or 8 km/h. The speed downstream is $10 + 2$, or 12 km/h. The time upstream, using $t = d/r$, is 24/8, or 3 hr. The time downstream is 24/12, or 2 hr. The total time is 5 hr. This checks.

5. State. The speed of the boat in still water is 10 km/h.

Do Exercise 3.

3. *Speed of a Stream.* The speed of a boat in still water is 12 km/h. The boat travels 45 km upstream and 45 km downstream in a total time of 8 hr. What is the speed of the stream? (*Hint*: Let s = the speed of the stream. Then $12 - s$ is the speed upstream and $12 + s$ is the speed downstream. Note also that $12 - s$ cannot be negative, because the boat must be going faster than the current if it is moving forward.)

Answer on page A-49

Improving Your Math Study Skills

Preparing for a Final Exam

Best Scenario: Two Weeks of Study Time

The best scenario for preparing for a final exam is to do so over a period of at least two weeks. Work in a diligent, disciplined manner, doing some final-exam preparation *each* day. Here is a detailed plan that many find useful.

1. **Begin by browsing through each chapter, reviewing the highlighted or boxed information regarding important formulas in both the text and the Summary and Review.** There may be some formulas that you will need to memorize.
2. **Retake each chapter test that you took in class, assuming your instructor has returned it. Otherwise, use the chapter test in the book.** Restudy the objectives in the text that correspond to each question you missed.
3. **Then work the Cumulative Review that covers all chapters up to that point.** Be careful to avoid any questions corresponding to objectives not covered. Again, restudy the objectives in the text that correspond to each question you missed.
4. **If you are still missing questions, use supplements for extra review.** For example, you might check out the video- or audiotapes, the *Student's Solutions Manual,* or the Interact Math Tutorial Software.
5. **For remaining difficulties, see your instructor, go to a tutoring session, or participate in a study group.**
6. **Check for former final exams that may be on file in the math department or a study center, or with students who have already taken the course.** Use them for practice, being alert to trouble spots.
7. **Take the Final Examination in the text during the last couple of days before the final.** Set aside the same amount of time that you will have for the final. See how much of the final exam you can complete under test-like conditions.

Moderate Scenario: Three Days to Two Weeks of Study Time

1. **Begin by browsing through each chapter, reviewing the highlighted or boxed information regarding important formulas in both the text and the Summary and Review.** There may be some formulas that you will need to memorize.
2. **Retake each chapter test that you took in class, assuming your instructor has returned it. Otherwise, use the chapter test in the book.** Restudy the objectives in the text that correspond to each question you missed.
3. **Then work the last Cumulative Review in the text.** Be careful to avoid any questions corresponding to objectives not covered. Again, restudy the objectives in the text that correspond to each question you missed.
4. **For remaining difficulties, see your instructor, go to a tutoring session, or participate in a study group.**
5. **Take the Final Examination in the text during the last couple of days before the final.** Set aside the same amount of time that you will have for the final. See how much of the final exam you can complete under test-like conditions.

Worst Scenario: One or Two Days of Study Time

1. **Begin by browsing through each chapter, reviewing the highlighted or boxed information regarding important formulas in both the text and the Summary and Review.** There may be some formulas that you will need to memorize.
2. **Then work the last Cumulative Review in the text.** Be careful to avoid any questions corresponding to objectives not covered. Restudy the objectives in the text that correspond to each question you missed.
3. **Attend a final-exam review session if one is available.**
4. **Take the Final Examination in the text during the last couple of days before the final.** Set aside the same amount of time that you will have for the final. See how much of the final exam you can complete under test-like conditions.

   Promise yourself that next semester you will allow a more appropriate amount of time for final exam preparation.

Other "Improving Your Math Study Skills" concerning test preparation appear in Sections 6.7 and 7.5.

Exercise Set 16.5

a Solve.

1. The length of a rectangular area rug is 3 ft greater than the width. The area is 70 ft^2. Find the length and the width.

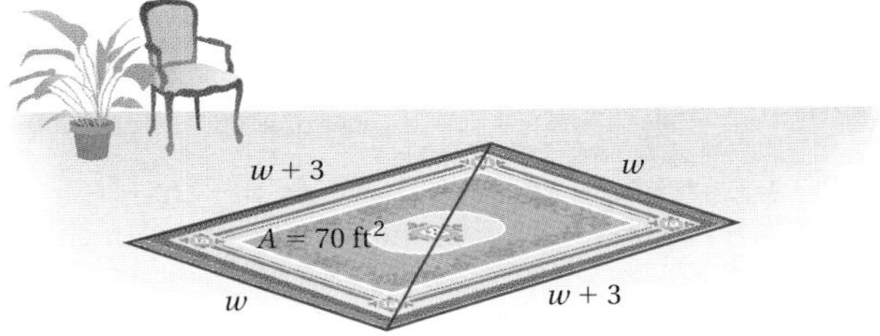

2. The length of a rectangular pine forest is 2 mi greater than the width. The area is 80 mi^2. Find the length and the width.

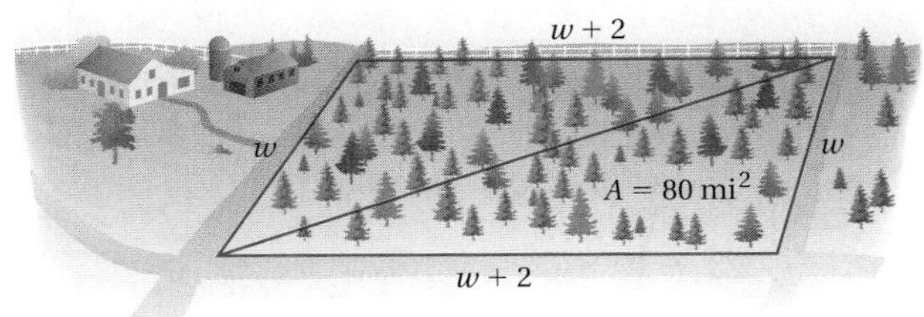

3. *Standard-Sized Television.* When we say that a television is 30 in., we mean that the diagonal is 30 in. For a standard-sized 30-in. television, the width is 6 in. more than the height. Find the dimensions of a standard-sized 30-in. television.

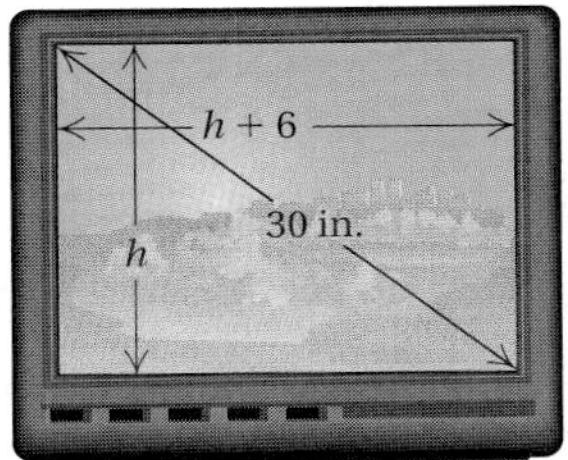

4. *HDTV Dimensions.* In the not-too-distant future, a new kind of high-definition television (HDTV) with larger screens and greater clarity will be available. An HDTV might have a 70-in. diagonal screen with the width 27 in. greater than the height. Find the width and the height of a 70-in. HDTV screen.

5. The width of a rectangle is 4 cm less than the length. The area is 320 cm^2. Find the length and the width.

6. The width of a rectangle is 3 cm less than the length. The area is 340 cm^2. Find the length and the width.

7. The length of a rectangle is twice the width. The area is 50 m^2. Find the length and the width.

8. *Carpenter's Square.* A ***square*** is a carpenter's tool in the shape of a right triangle. One side, or leg, of a square is 8 in. longer than the other. The length of the hypotenuse is $8\sqrt{13}$ in. Find the lengths of the legs of the square.

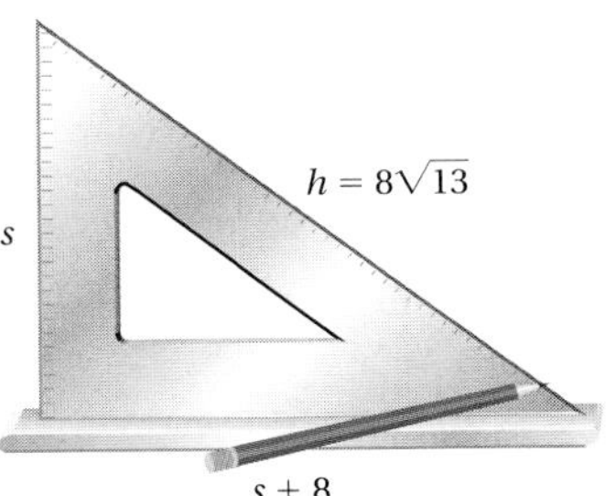

Find the approximate answers for Exercises 9–14. Round to the nearest tenth.

9. The hypotenuse of a right triangle is 8 m long. One leg is 2 m longer than the other. Find the lengths of the legs.

10. The hypotenuse of a right triangle is 5 cm long. One leg is 2 cm longer than the other. Find the lengths of the legs.

11. The length of a rectangle is 2 in. greater than the width. The area is 20 in^2. Find the length and the width.

12. The length of a rectangle is 3 ft greater than the width. The area is 15 ft^2. Find the length and the width.

13. The length of a rectangle is twice the width. The area is 20 cm^2. Find the length and the width.

14. The length of a rectangle is twice the width. The area is 10 m^2. Find the length and the width.

15. A picture frame measures 25 cm by 20 cm. There is 266 cm^2 of picture showing. The frame is of uniform thickness. Find the thickness of the frame.

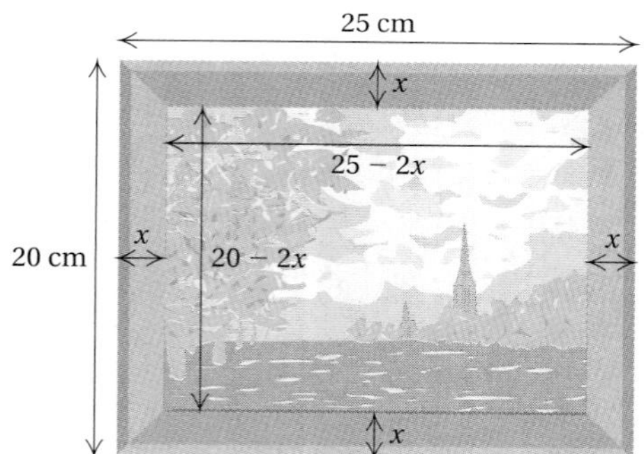

16. A tablecloth measures 96 in. by 72 in. It is laid on a tabletop with an area of 5040 in^2, and hangs over the edge by the same amount on all sides. By how many inches does the cloth hang over the edge?

For Exercises 17–22, complete the table to help with the familiarization.

17. The current in a stream moves at a speed of 3 km/h. A boat travels 40 km upstream and 40 km downstream in a total time of 14 hr. What is the speed of the boat in still water? Complete the following table to help with the familiarization.

	d	r	t
Upstream		$r-3$	t_1
Downstream	40		t_2
Total Time			

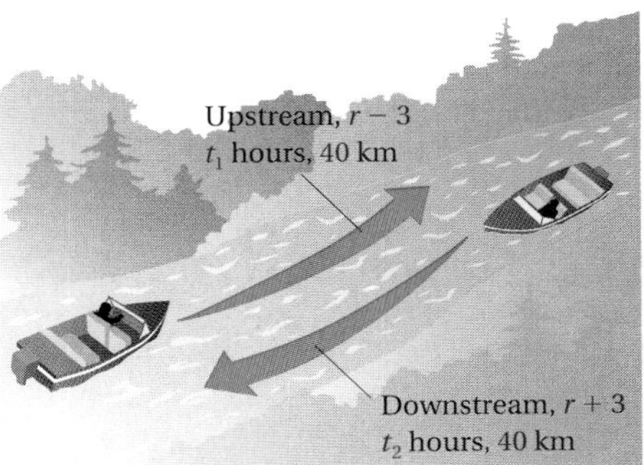

18. The current in a stream moves at a speed of 3 km/h. A boat travels 45 km upstream and 45 km downstream in a total time of 8 hr. What is the speed of the boat in still water?

	d	r	t
Upstream	45		
Downstream		$r+3$	
Total Time			

19. The current in a stream moves at a speed of 4 mph. A boat travels 4 mi upstream and 12 mi downstream in a total time of 2 hr. What is the speed of the boat in still water?

	d	r	t
Upstream		$r-4$	
Downstream	12		
Total Time			

20. The current in a stream moves at a speed of 4 mph. A boat travels 5 mi upstream and 13 mi downstream in a total time of 2 hr. What is the speed of the boat in still water?

	d	r	t
Upstream			
Downstream			
Total Time			

21. The speed of a boat in still water is 10 km/h. The boat travels 12 km upstream and 28 km downstream in a total time of 4 hr. What is the speed of the stream?

	d	r	t
Upstream			
Downstream			
Total Time			

22. The speed of a boat in still water is 8 km/h. The boat travels 60 km upstream and 60 km downstream in a total time of 16 hr. What is the speed of the stream?

	d	r	t
Upstream			
Downstream			
Total Time			

23. An airplane flies 738 mi against the wind and 1062 mi with the wind in a total time of 9 hr. The speed of the airplane in still air is 200 mph. What is the speed of the wind?

24. An airplane flies 520 km against the wind and 680 km with the wind in a total time of 4 hr. The speed of the airplane in still air is 300 km/h. What is the speed of the wind?

25. The speed of a boat in still water is 9 km/h. The boat travels 80 km upstream and 80 km downstream in a total time of 18 hr. What is the speed of the stream?

26. The speed of a boat in still water is 10 km/h. The boat travels 48 km upstream and 48 km downstream in a total time of 10 hr. What is the speed of the stream?

Skill Maintenance

Add or subtract. [15.4a]

27. $5\sqrt{2} + \sqrt{18}$

28. $7\sqrt{40} - 2\sqrt{10}$

29. $\sqrt{4x^3} - 7\sqrt{x}$

30. $\sqrt{24} - \sqrt{54}$

31. $\sqrt{2} + \sqrt{\frac{1}{2}}$

32. $\sqrt{3} - \sqrt{\frac{1}{3}}$

33. $\sqrt{24} + \sqrt{54} - \sqrt{48}$

34. $\sqrt{4x} + \sqrt{81x^3}$

Synthesis

Find and explain the error(s) in each of the following solutions of a quadratic equation.

35. ◆ $(x + 6)^2 = 16$
$x + 6 = \sqrt{16}$
$x + 6 = 4$
$x = -2$

36. ◆ $x^2 + 2x - 8 = 0$
$(x + 4)(x - 2) = 0$
$x = 4 \quad or \quad x = -2$

37. Find r in this figure. Round to the nearest hundredth.

r, r, r, 1 cm

38. The width of a dollar bill is 9 cm less than the length. The area is 102.96 cm^2. Find the length and the width.

39. What should the diameter d of a pizza be so that it has the same area as two 12-in. pizzas? Do you get more to eat with a 16-in. pizza or with two 12-in. pizzas?

16.6 Graphs of Quadratic Equations

In this section, we will graph equations of the form

$$y = ax^2 + bx + c, \quad a \neq 0.$$

The polynomial on the right side of the equation is of second degree, or **quadratic**. Examples of the types of equations we are going to graph are

$$y = x^2, \qquad y = x^2 + 2x - 3, \qquad y = -2x^2 + 3.$$

Objectives

a Graph quadratic equations.

b Find the x-intercepts of a quadratic equation.

For Extra Help

TAPE 30 | MAC WIN | CD-ROM

a Graphing Quadratic Equations of the Type $y = ax^2 + bx + c$

Graphs of quadratic equations of the type $y = ax^2 + bx + c$ (where $a \neq 0$) are always cup-shaped. They have a **line of symmetry** like the dashed lines shown in the figures below. If we fold on this line, the two halves will match exactly. The curve goes on forever. The top or bottom point where the curve changes is called the **vertex**. The second coordinate is either the largest value of y or the smallest value of y. The vertex is also thought of as a turning point. Graphs of quadratic equations are called **parabolas**.

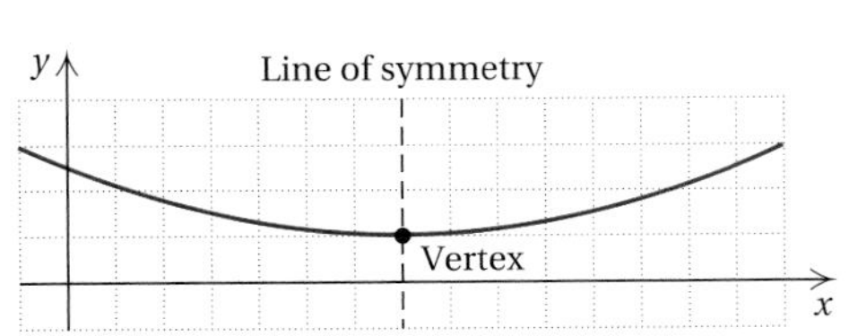

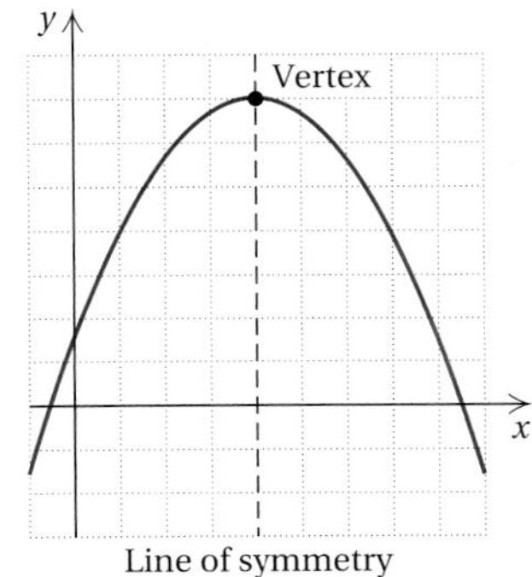

To graph a quadratic equation, we begin by choosing some numbers for x and computing the corresponding values of y.

Example 1 Graph: $y = x^2$.

We choose numbers for x and find the corresponding values for y. Then we plot the ordered pairs (x, y) resulting from the computations and connect them with a smooth curve.

For $x = -3$, $y = x^2 = (-3)^2 = 9$.
For $x = -2$, $y = x^2 = (-2)^2 = 4$.
For $x = -1$, $y = x^2 = (-1)^2 = 1$.
For $x = 0$, $y = x^2 = (0)^2 = 0$.
For $x = 1$, $y = x^2 = (1)^2 = 1$.
For $x = 2$, $y = x^2 = (2)^2 = 4$.
For $x = 3$, $y = x^2 = (3)^2 = 9$.

x	y	(x, y)
−3	9	(−3, 9)
−2	4	(−2, 4)
−1	1	(−1, 1)
0	0	(0, 0)
1	1	(1, 1)
2	4	(2, 4)
3	9	(3, 9)

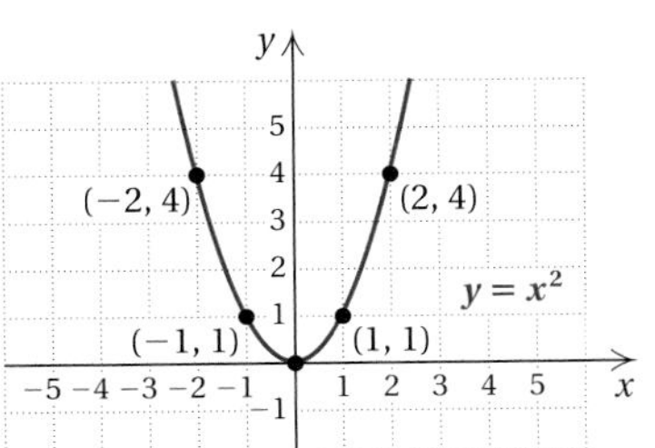

Calculator Spotlight

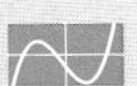
Graphing Quadratic Equations. The following is the graph of the quadratic equation $y_1 = x^2 + 3x - 4$.

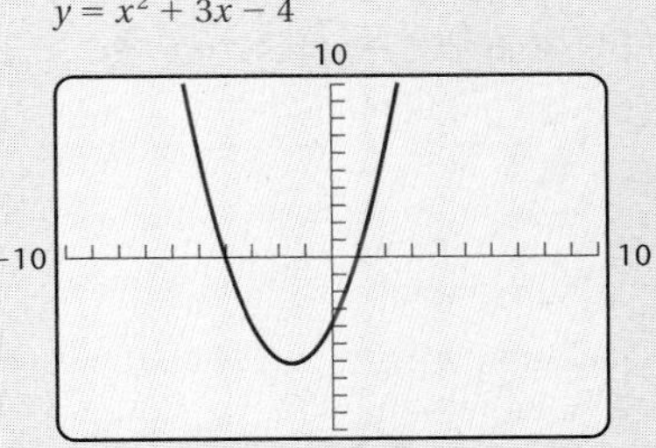

Exercises

Use your grapher to graph each of the following quadratic equations in the standard viewing window.

1. $y = x^2$
2. $y = x^2 + 2x - 3$
3. $y = 3.2 - x^2$
4. $y = 4 - 1.2x - 3.4x^2$
5. $y = -2.3x^2 + 4.1x + 1.8$
6. $y = -2.3x^2 + 4.1x - 3.8$

In Example 1, the vertex is the point (0, 0). The second coordinate of the vertex, 0, is the smallest y-value. The y-axis is the line of symmetry. Parabolas whose equations are $y = ax^2$ always have the origin (0, 0) as the vertex and the y-axis as the line of symmetry.

How do we graph a general equation? There are many methods, some of which you will study in your next mathematics course. Our goal here is to give you a basic graphing technique that is fairly easy to apply. A key in the graphing is knowing the vertex. By graphing it and then choosing x-values on both sides of the vertex, we can compute more points and complete the graph.

FINDING THE VERTEX

For a parabola given by the quadratic equation $y = ax^2 + bx + c$:

1. The x-coordinate of the vertex is $-\dfrac{b}{2a}$.
2. The second coordinate of the vertex is found by substituting the x-coordinate into the equation and computing y.

The proof that the vertex can be found in this way can be shown by completing the square in a manner similar to the proof of the quadratic formula, but it will not be considered here.

Example 2 Graph: $y = -2x^2 + 3$.

We first find the vertex. The x-coordinate of the vertex is

$$-\frac{b}{2a} = -\frac{0}{2(-2)} = 0.$$

We substitute 0 for x into the equation to find the second coordinate of the vertex:

$$y = -2x^2 + 3 = -2(0)^2 + 3 = 3.$$

The vertex is (0, 3). The line of symmetry is $x = 0$, which is the y-axis. We choose some x-values on both sides of the vertex and graph the parabola.

For $x = 1$, $y = -2x^2 + 3 = -2(1)^2 + 3 = -2 + 3 = 1$.

For $x = -1$, $y = -2x^2 + 3 = -2(-1)^2 + 3 = -2 + 3 = 1$.

For $x = 2$, $y = -2x^2 + 3 = -2(2)^2 + 3 = -8 + 3 = -5$.

For $x = -2$, $y = -2x^2 + 3 = -2(-2)^2 + 3 = -8 + 3 = -5$.

x	y	
0	3	← This is the vertex.
1	1	
−1	1	
2	−5	
−2	−5	

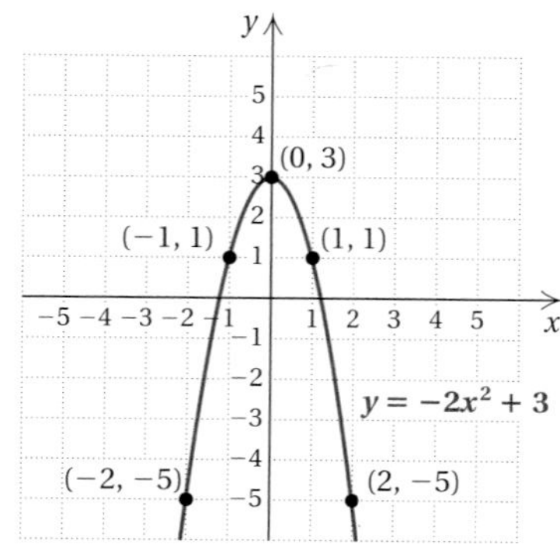

Calculator Spotlight

Use the GRAPH and TABLE features of your grapher to check the results of Example 2.

There are two other tips you might use when graphing quadratic equations. The first involves the coefficient of x^2. Note that a in $y = ax^2 + bx + c$ tells us whether the graph opens up or down. When a is positive, as in Example 1, the graph opens up; when a is negative, as in Example 2, the graph opens down. It is also helpful to plot the y-intercept. It occurs when $x = 0$.

Tips for Graphing Quadratic Equations

1. Graphs of quadratic equations $y = ax^2 + bx + c$ are all parabolas. They are *smooth* cup-shaped symmetric curves, with no sharp points or kinks in them.
2. The graph of $y = ax^2 + bx + c$ opens up if $a > 0$. It opens down if $a < 0$.
3. Find the y-intercept. It occurs when $x = 0$, and it is easy to compute.

Example 3 Graph: $y = x^2 + 2x - 3$.

We first find the vertex. The x-coordinate of the vertex is

$$-\frac{b}{2a} = -\frac{2}{2(1)} = -1.$$

We substitute -1 for x into the equation to find the second coordinate of the vertex:

$$\begin{aligned} y &= x^2 + 2x - 3 \\ &= (-1)^2 + 2(-1) - 3 \\ &= 1 - 2 - 3 \\ &= -4. \end{aligned}$$

The vertex is $(-1, -4)$. The line of symmetry is $x = -1$.

We choose some x-values on both sides of $x = -1$—say, -2, -3, -4 and 0, 1, 2—and graph the parabola. Since the coefficient of x^2 is 1, which is positive, we know that the graph opens up. Be sure to find y when $x = 0$. This gives the y-intercept.

x	y	
−1	−4	← Vertex
0	−3	← y-intercept
−2	−3	
1	0	
−3	0	
2	5	
−4	5	

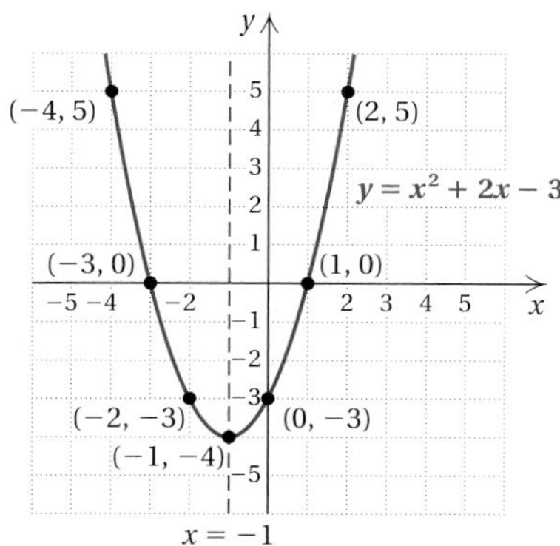

Do Exercises 1–3.

Graph. List the ordered pair for the vertex.

1. $y = x^2 - 3$

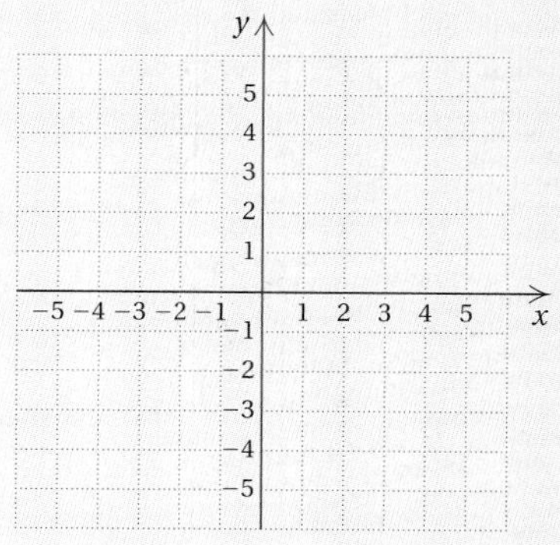

2. $y = -3x^2 + 6x$

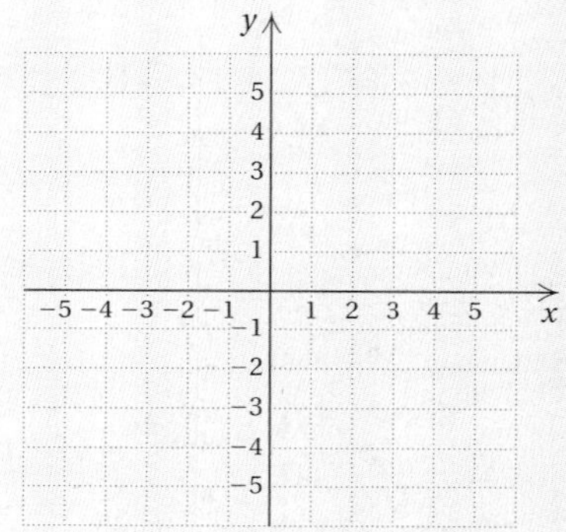

3. $y = x^2 - 4x + 4$

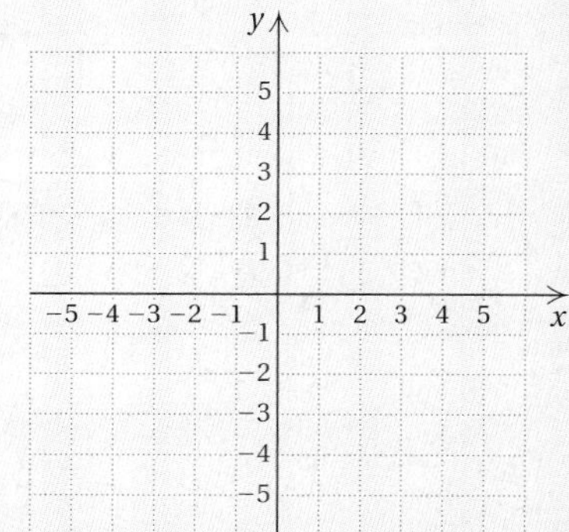

Answers on page A-49

Find the x-intercepts.

4. $y = x^2 - 3$

5. $y = x^2 + 6x + 8$

6. $y = -2x^2 - 4x + 1$

7. $y = x^2 + 3$

Answers on page A-49

b Finding the x-Intercepts of a Quadratic Equation

The x-intercepts of $y = ax^2 + bx + c$ occur at those values of x for which $y = 0$. Thus the first coordinates of the x-intercepts are solutions of the equation

$$0 = ax^2 + bx + c.$$

We have been studying how to find such numbers in Sections 16.1–16.3.

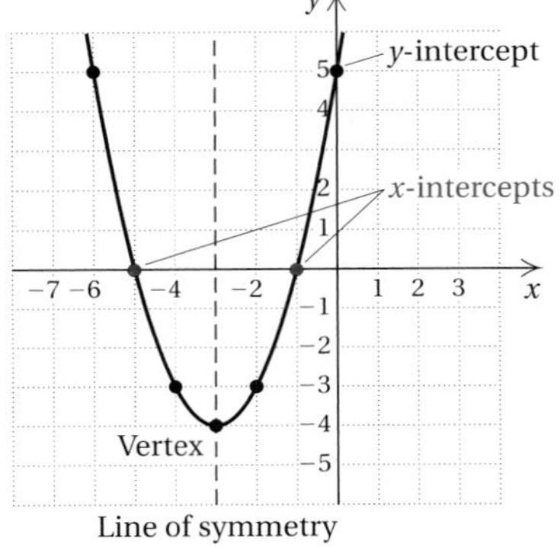

Example 4 Find the x-intercepts of $y = x^2 - 4x + 1$.

We solve the equation

$$x^2 - 4x + 1 = 0.$$

Factoring is not convenient, so we use the quadratic formula:

$$a = 1, \quad b = -4, \quad c = 1$$

$$\begin{aligned} x &= \frac{-b \pm \sqrt{b^2 - 4ac}}{2a} \\ &= \frac{-(-4) \pm \sqrt{(-4)^2 - 4(1)(1)}}{2(1)} \\ &= \frac{4 \pm \sqrt{16 - 4}}{2} \\ &= \frac{4 \pm \sqrt{12}}{2} = \frac{4 \pm \sqrt{4 \cdot 3}}{2} \\ &= \frac{4 \pm 2\sqrt{3}}{2} = \frac{2 \cdot 2 \pm 2\sqrt{3}}{2 \cdot 1} \\ &= \frac{2}{2} \cdot \frac{2 \pm \sqrt{3}}{1} = 2 \pm \sqrt{3}. \end{aligned}$$

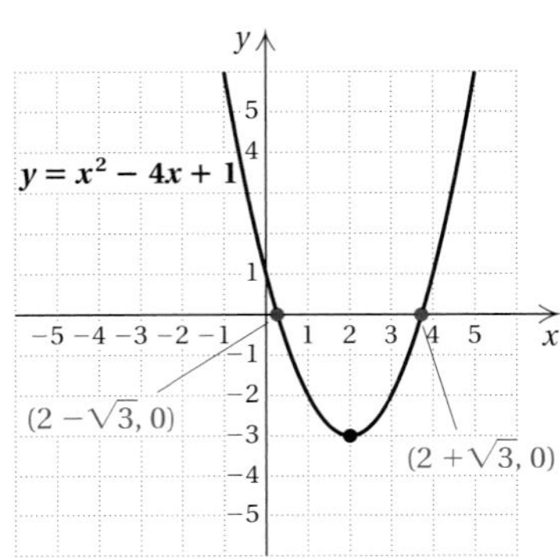

The x-intercepts are $(2 - \sqrt{3}, 0)$ and $(2 + \sqrt{3}, 0)$.

In the quadratic formula $x = \dfrac{-b \pm \sqrt{b^2 - 4ac}}{2a}$, the radicand $b^2 - 4ac$ is called the **discriminant**. The discriminant tells how many real-number solutions the equation $0 = ax^2 + bx + c$ has, so it also tells how many x-intercepts there are.

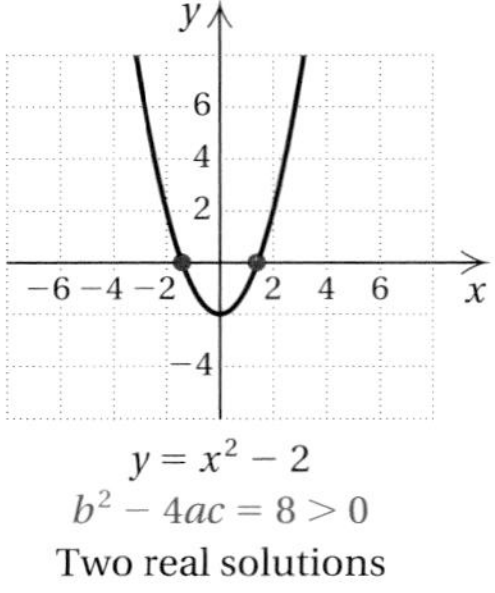

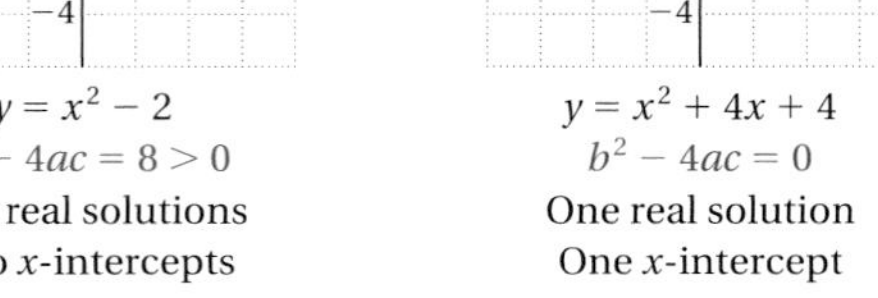

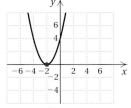

$y = x^2 - 2$
$b^2 - 4ac = 8 > 0$
Two real solutions
Two x-intercepts

$y = x^2 + 4x + 4$
$b^2 - 4ac = 0$
One real solution
One x-intercept

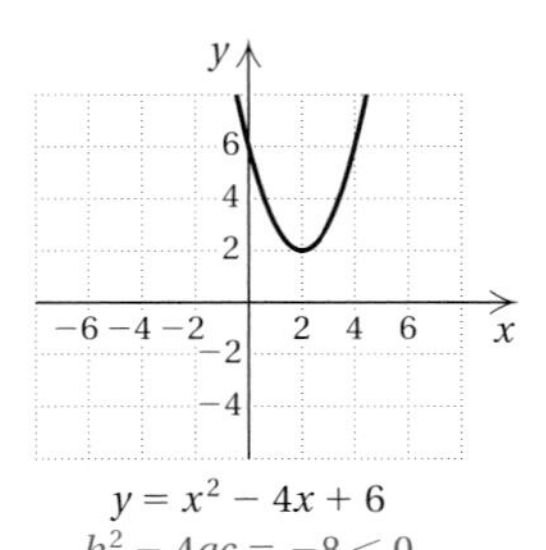

$y = x^2 - 4x + 6$
$b^2 - 4ac = -8 < 0$
No real solutions
No x-intercepts

Do Exercises 4–7.

Exercise Set 16.6

a Graph the quadratic equation. List the ordered pair for the vertex.

1. $y = x^2 + 1$

2. $y = 2x^2$

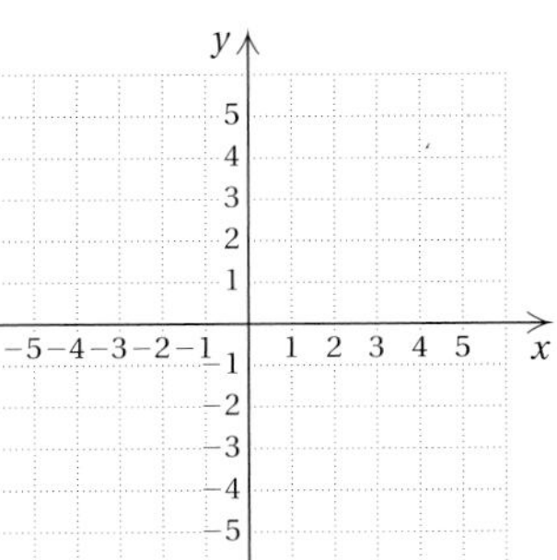

3. $y = -1 \cdot x^2$

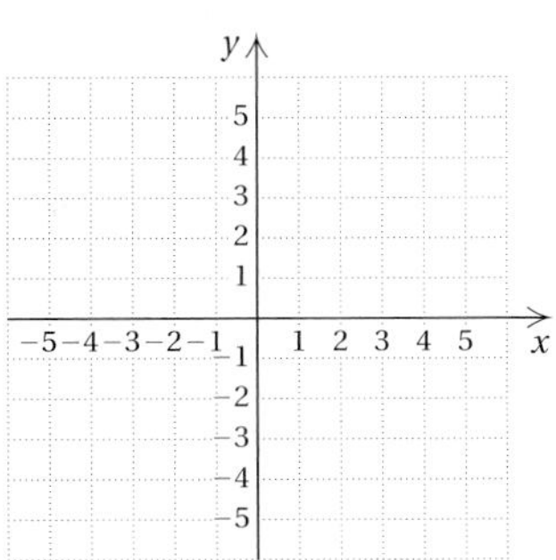

4. $y = x^2 - 1$

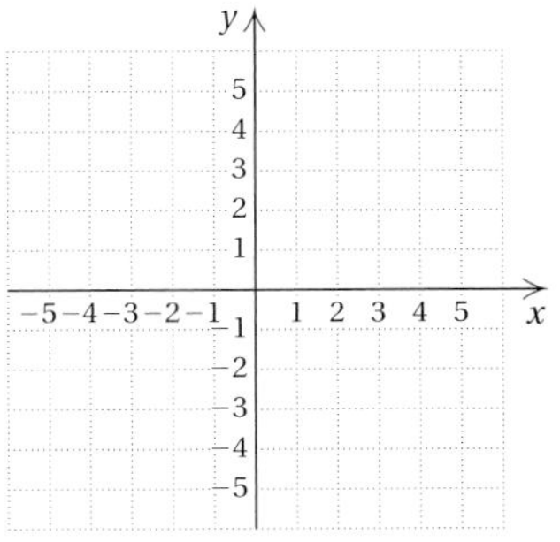

5. $y = -x^2 + 2x$

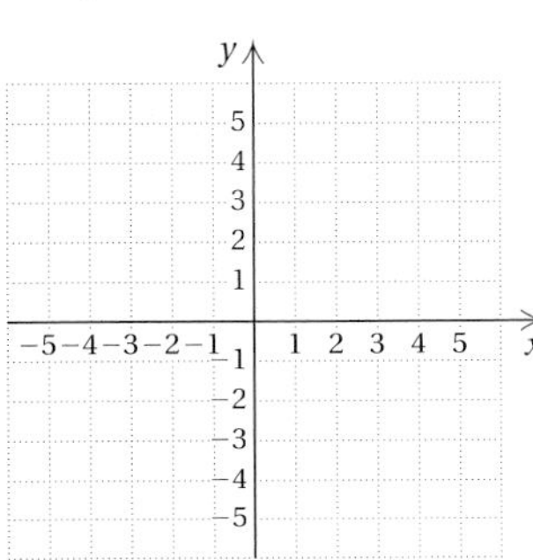

6. $y = x^2 + x - 2$

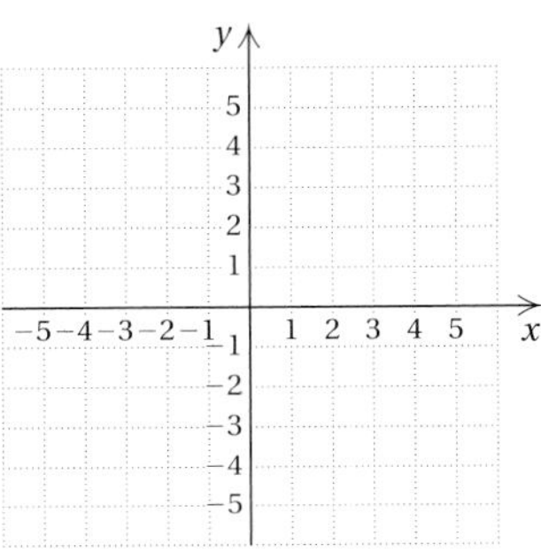

7. $y = 5 - x - x^2$

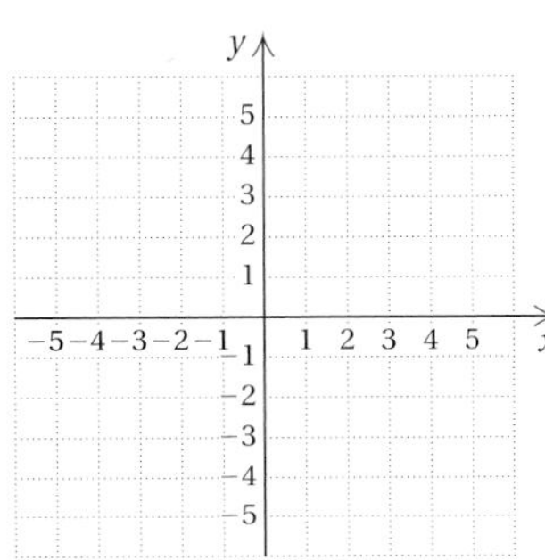

8. $y = x^2 + 2x + 1$

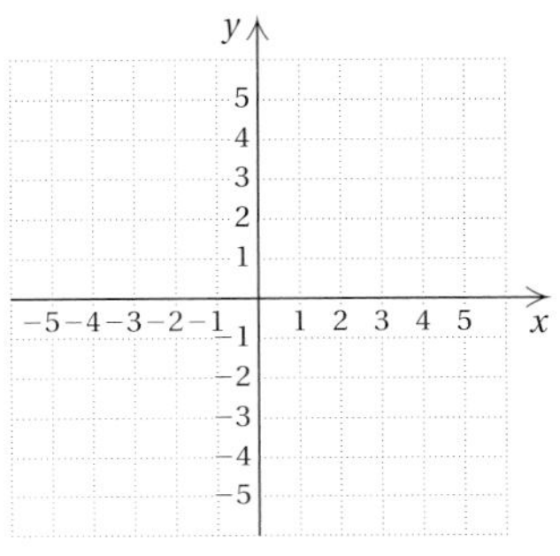

9. $y = x^2 - 2x + 1$

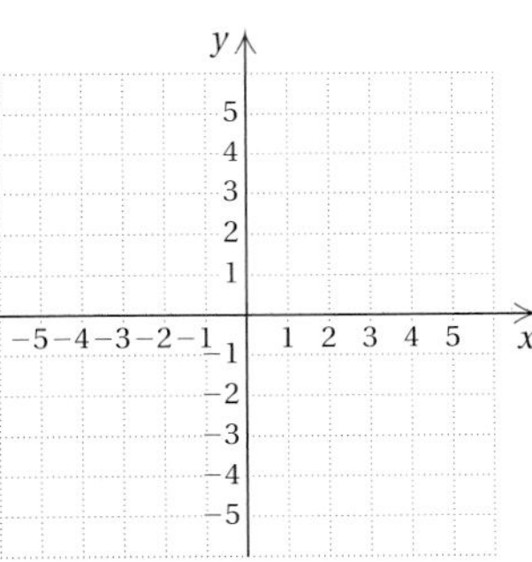

10. $y = -\frac{1}{2}x^2$

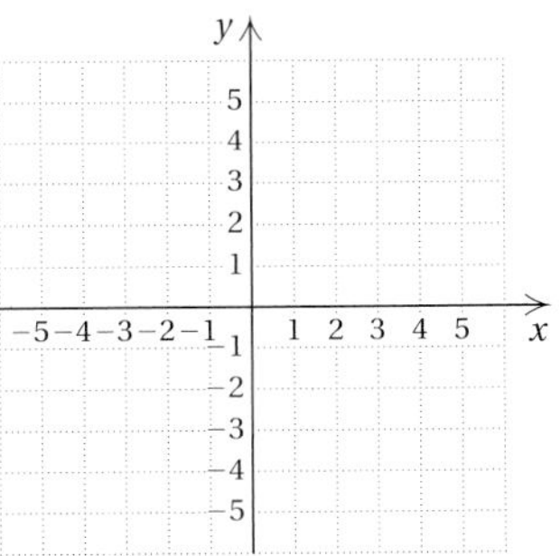

11. $y = -x^2 + 2x + 3$

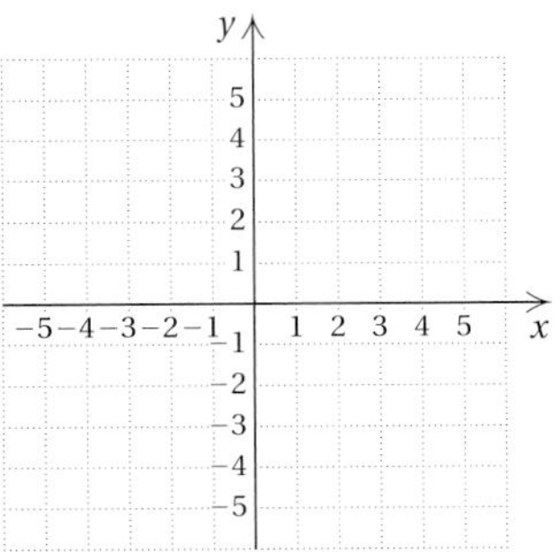

12. $y = -x^2 - 2x + 3$

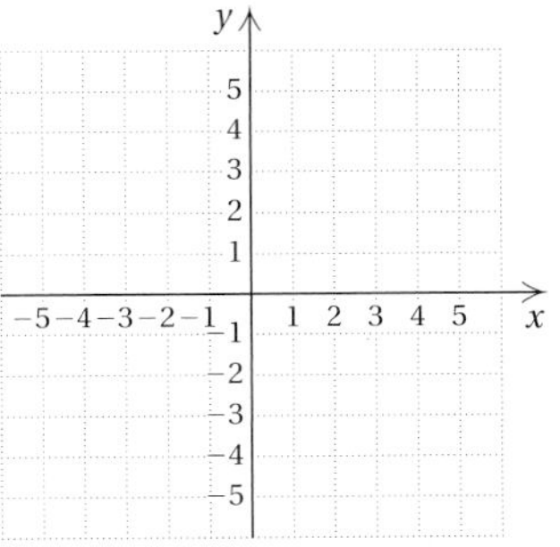

13. $y = -2x^2 - 4x + 1$

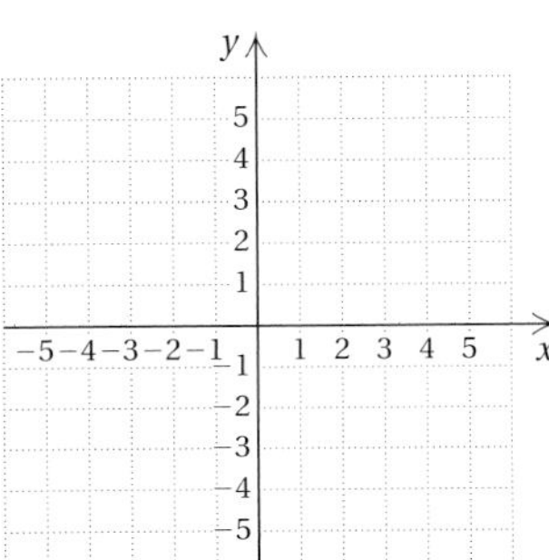

14. $y = 2x^2 + 4x - 1$

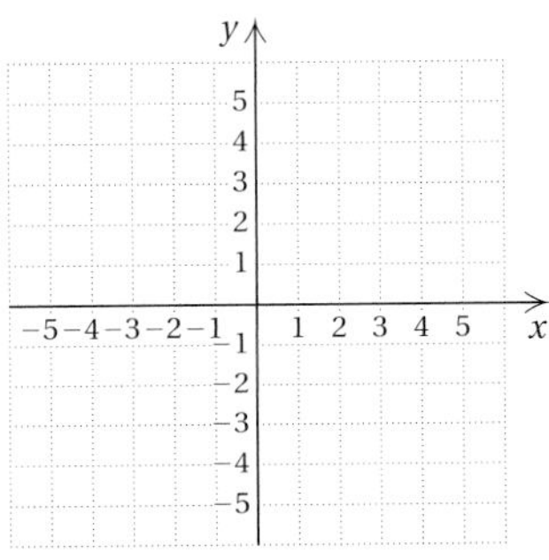

15. $y = 5 - x^2$

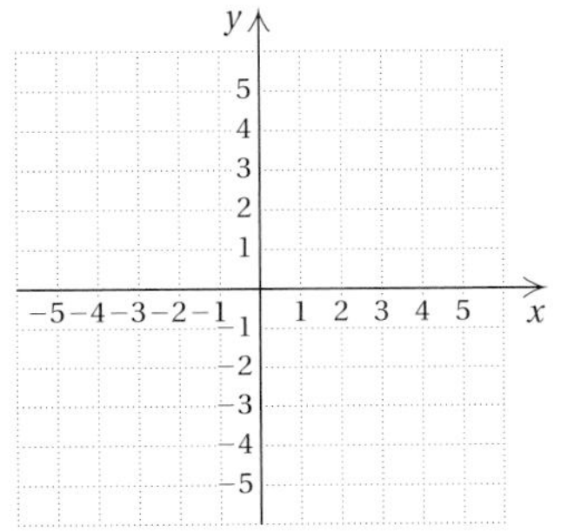

16. $y = 4 - x^2$

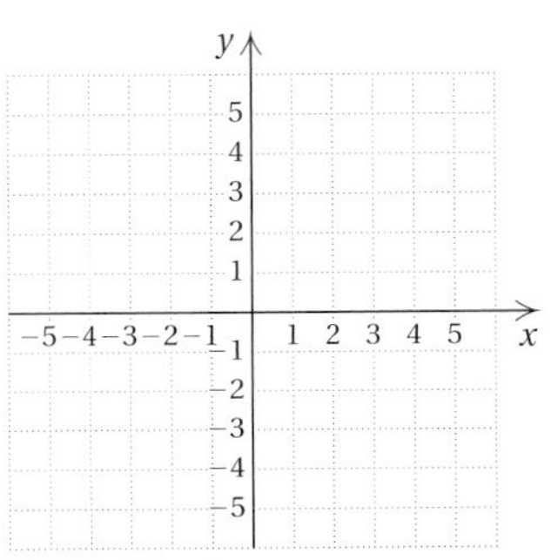

www.mathmax.com World Wide Web

Graph the quadratic equation. Use your own graph paper.

17. $y = \frac{1}{4}x^2$

18. $y = -0.1x^2$

19. $y = -x^2 + x - 1$

20. $y = x^2 + 2x$

21. $y = -2x^2$

22. $y = -x^2 - 1$

23. $y = x^2 - x - 6$

24. $y = 6 + x - x^2$

b Find the x-intercepts.

25. $y = x^2 - 2$

26. $y = x^2 - 7$

27. $y = x^2 + 5x$

28. $y = x^2 - 4x$

29. $y = 8 - x - x^2$

30. $y = 8 + x - x^2$

31. $y = x^2 - 6x + 9$

32. $y = x^2 + 10x + 25$

33. $y = -x^2 - 4x + 1$

34. $y = x^2 + 4x - 1$

35. $y = x^2 + 9$

36. $y = x^2 + 1$

Skill Maintenance

Add. [15.4a]

37. $\sqrt{x^3 - x^2} + \sqrt{4x - 4}$

38. $\sqrt{8} + \sqrt{50} + \sqrt{98} + \sqrt{128}$

Multiply and simplify. [15.2c]

39. $\sqrt{2}\sqrt{14}$

40. $\sqrt{5y^4}\sqrt{125y}$

41. Find an equation of variation in which y varies inversely as x and $y = 12.4$ when $x = 2.4$. [13.5c]

42. Find an equation of variation in which y varies inversely as x and $y = 264$ when $x = 18$. [13.5c]

43. Evaluate $5x^3 - 2x$ for $x = -1$. [10.3a]

44. Evaluate $3x^4 + 3x - 7$ for $x = -2$. [10.3a]

Synthesis

45. ◆ Suppose that the x-intercepts of a parabola are $(a_1, 0)$ and $(a_2, 0)$. What is the easiest way to find an equation for the line of symmetry? the coordinates of the vertex?

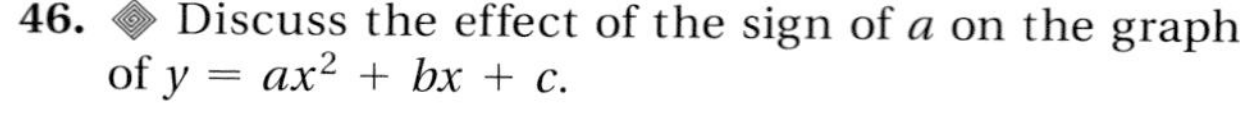

46. ◆ Discuss the effect of the sign of a on the graph of $y = ax^2 + bx + c$.

47. *Height of a Projectile.* The height H, in feet, of a projectile with an initial velocity of 96 ft/sec is given by the equation

$$H = -16t^2 + 96t,$$

where t = time, in seconds. Use the graph of this function, shown here, or any equation-solving technique to answer the following questions.

a) How many seconds after launch is the projectile 128 ft above ground?

b) When does the projectile reach its maximum height?

c) How many seconds after launch does the projectile return to the ground?

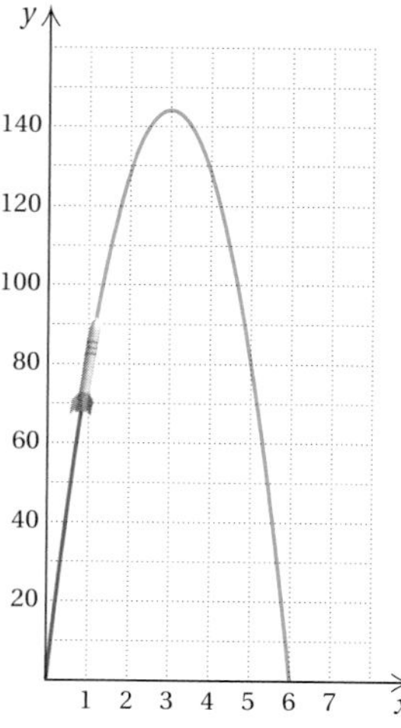

For each equation in Exercises 48–51, evaluate the discriminant $b^2 - 4ac$. Then use the answer to state how many real-number solutions exist for the equation.

48. $0 = x^2 + 8x + 16$

49. $0 = x^2 + 2x - 3$

50. $0 = -2x^2 + 4x - 3$

51. $0 = -0.02x^2 + 4.7x - 2300$

Collaborative Learning Manual — **Practice graphing and identifying the graphs of quadratic equations.**

16.7 Functions

a Identifying Functions

We now develop one of the most important concepts in mathematics, **functions**. We have actually been studying functions all through this text; we just haven't identified them as such. Ordered pairs form a correspondence between first and second coordinates. A function is a special correspondence from one set of numbers to another. For example:

> To each student in a college, there corresponds his or her student ID.
>
> To each item in a store, there corresponds its price.
>
> To each real number, there corresponds the cube of that number.

In each case, the first set is called the **domain** and the second set is called the **range**. Given a member of the domain, there is *just one* member of the range to which it corresponds. This kind of correspondence is called a **function**.

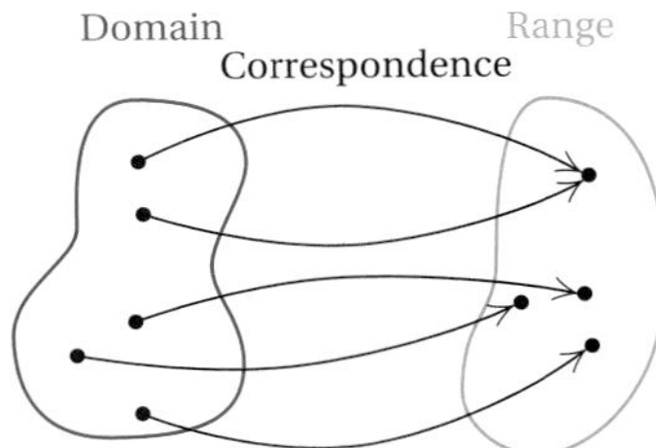

Example 1 Determine whether the correspondence is a function.

f:

Domain	*Range*
1 →	\$107.4
2 →	\$ 34.1
3 →	\$ 29.6
4 →	\$ 19.6

g:

Domain	*Range*
3 →	5
4 →	9
5 →	−7
6 →	−7

h:

Domain	*Range*
Chicago →	Cubs
Chicago →	White Sox
Baltimore →	Orioles
San Diego →	Padres

p:

Domain	*Range*
Cubs →	Chicago
White Sox →	Chicago
Orioles →	Baltimore
Padres →	San Diego

The correspondence *f* is a function because each member of the domain is matched to only one member of the range.

The correspondence *g* is also a function because each member of the domain is matched to only one member of the range.

The correspondence *h* is *not* a function because one member of the domain, Chicago, is matched to more than one member of the range.

The correspondence *p* is a function because each member of the domain is paired with only one member of the range.

> A **function** is a correspondence between a first set, called the **domain**, and a second set, called the **range**, such that each member of the domain corresponds to *exactly one* member of the range.

Do Exercises 1–4.

Objectives

a Determine whether a correspondence is a function.

b Given a function described by an equation, find function values (outputs) for specified values (inputs).

c Draw the graph of a function.

d Determine whether a graph is that of a function.

e Solve applied problems involving functions and their graphs.

For Extra Help

TAPE 30

MAC
WIN

CD-ROM

Determine whether the correspondence is a function.

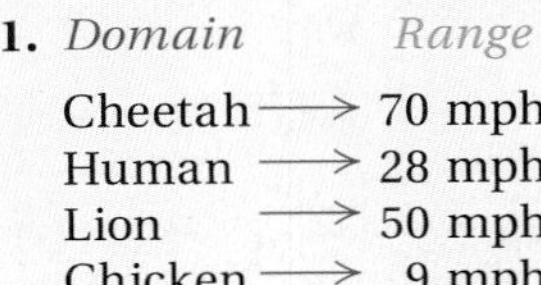

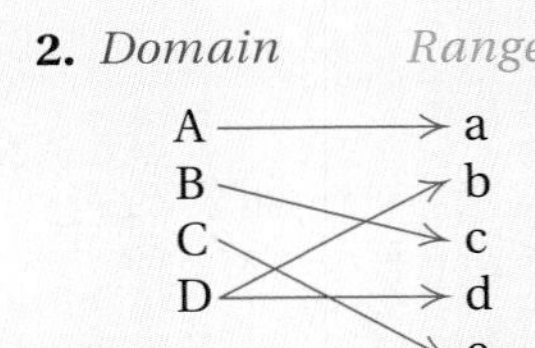

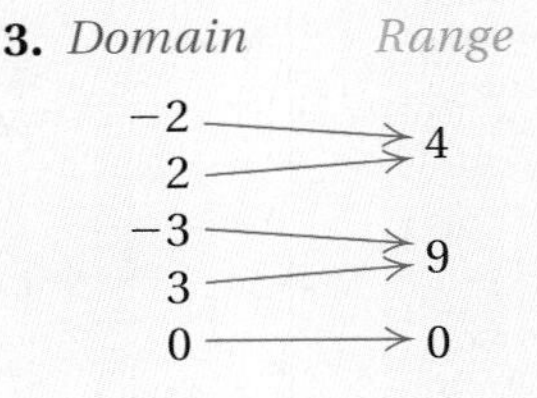

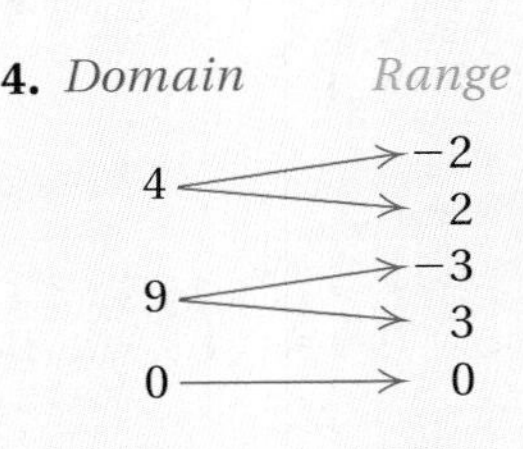

Answers on page A-50

Determine whether each of the following is a function.

5. *Domain*
A set of numbers
Correspondence
Square each number and subtract 10.
Range
A set of numbers

6. *Domain*
A set of polygons
Correspondence
Find the perimeter of each polygon.
Range
A set of numbers

Answers on page A-50

Example 2 Determine whether the correspondence is a function.

	Domain	*Correspondence*	*Range*
a)	A family	Each person's weight	A set of positive numbers
b)	The natural numbers	Each number's square	A set of natural numbers
c)	The set of all states	Each state's members of the U.S. Senate	A set of U.S. Senators

a) The correspondence *is* a function because each person has *only one* weight.

b) The correspondence *is* a function because each natural number has *only one* square.

c) The correspondence *is not* a function because each state has two U.S. Senators.

Do Exercises 5 and 6.

When a correspondence between two sets is not a function, it is still an example of a **relation**.

> A **relation** is a correspondence between a first set, called the **domain**, and a second set, called the **range**, such that each member of the domain corresponds to *at least one* member of the range.

Thus, although the correspondences of Examples 1 and 2 are not all functions, they *are* all relations. A function is a special type of relation—one in which each member of the domain is paired with *exactly one* member of the range.

b Finding Function Values

Most functions considered in mathematics are described by equations. A linear equation like $y = 2x + 3$, studied in Chapters 6 and 13, is called a **linear function.** A quadratic equation like $y = 4 - x^2$, studied in Chapter 16, is called a **quadratic function.**

Recall that when graphing $y = 2x + 3$, we chose x-values and then found corresponding y-values. For example, when $x = 4$, $y = 2x + 3 = 2 \cdot 4 + 3 = 11$. When thinking of functions, we call the number 4 an **input** and the number 11 an **output**.

It helps to think of a function as a machine; that is, think of putting a member of the domain (an input) into the machine. The machine knows the correspondence and gives out a member of the range (the output).

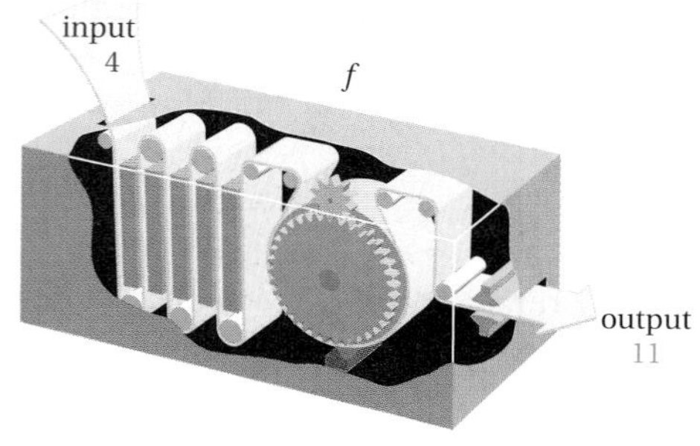

The function $y = 2x + 3$ has been named f and is described by the equation $f(x) = 2x + 3$. We call the input x and the output $f(x)$. This is read "f of x," or "f at x," or "the value of f at x."

CAUTION! The notation $f(x)$ *does not mean* "f times x" and should not be read that way.

The equation $f(x) = 2x + 3$ describes the function that takes an input x, multiplies it by 2, and then adds 3.

Input → $2x$

$$f(x) = 2x + 3$$

Double (for $2x$) **Add 3** (for $+3$)

To find the output $f(4)$, we take the input 4, double it, and add 3 to get 11. That is, we substitute 4 into the formula for $f(x)$:

$$f(4) = 2 \cdot 4 + 3 = 11.$$

Outputs of functions are also called **function values.** For $f(x) = 2x + 3$, we know that $f(4) = 11$. We can say that "the function value at 4 is 11."

Example 3 Find the indicated function value.

a) $f(5)$, for $f(x) = 3x + 2$

b) $g(3)$, for $g(z) = 5z^2 - 4$

c) $A(-2)$, for $A(r) = 3r^2 - 2r$

d) $f(-5)$, for $f(x) = x^2 + 3x - 4$

a) $f(5) = 3 \cdot 5 + 2 = 17$

b) $g(3) = 5(3)^2 - 4 = 41$

c) $A(-2) = 3(-2)^2 + 2(-2) = 8$

d) $f(-5) = (-5)^2 + 3(-5) - 4 = 25 - 15 - 4 = 6$

Do Exercises 7 and 8.

Find the function values.

7. $f(x) = 5x - 3$

a) $f(-6)$
b) $f(0)$
c) $f(1)$
d) $f(20)$
e) $f(-1.2)$

8. $g(x) = x^2 - 4x + 9$

a) $g(-2)$
b) $g(0)$
c) $g(5)$
d) $g(10)$

c Graphs of Functions

To graph a function, we find ordered pairs (x, y) or $(x, f(x))$, plot them, and connect the points. Note that y and $f(x)$ are used interchangeably when working with functions and their graphs.

Example 4 Graph: $f(x) = x + 2$.

A list of some function values is shown in this table. We plot the points and connect them. The graph is a straight line.

x	$f(x)$
−4	−2
−3	−1
−2	0
−1	1
0	2
1	3
2	4
3	5
4	6

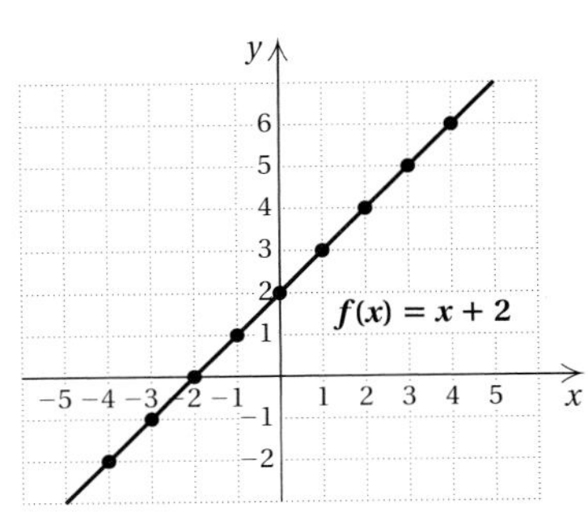

Answers on page A-50

Calculator Spotlight

 Graphs and Function Values. To graph the function $f(x) = 2x^2 + x$, we enter it into the grapher as $y_1 = 2x^2 + x$ and use the GRAPH feature. There are at least three ways to find function values. The first is to use the VALUE feature. We enter a value, say, $x = -2$. The y-value is given to the right and the point is indicated on the graph.

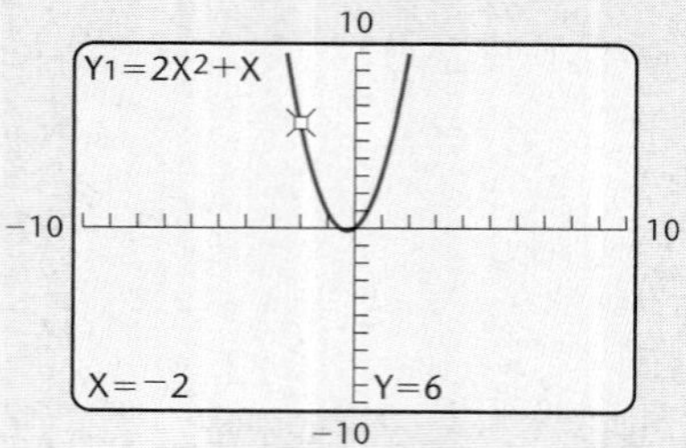

The second way to find a function value is to use the TABLE feature. The third is to bring Y1 to the screen using Y-VARS. We enter parentheses to complete the writing of Y1(−2). When ENTER is pressed, we obtain the output, 6.

Exercises

1. Graph $f(x) = x^2 + 3x - 4$. Then find $f(-5)$, $f(-4.7)$, $f(11)$, and $f(2/3)$.
2. Graph $f(x) = 3.7 - x^2$. Then find $f(-5)$, $f(-4.7)$, $f(11)$, and $f(2/3)$.
3. Graph $f(x) = 4 - 1.2x - 3.4x^2$. Then find $f(-5)$, $f(-4.7)$, $f(11)$, and $f(2/3)$.

Example 5 Graph: $g(x) = 4 - x^2$.

Recall from Section 16.6 that the graph is a parabola. We calculate some function values and draw the curve.

$$g(0) = 4 - 0^2 = 4 - 0 = 4,$$
$$g(-1) = 4 - (-1)^2 = 4 - 1 = 3,$$
$$g(2) = 4 - (2)^2 = 4 - 4 = 0,$$
$$g(-3) = 4 - (-3)^2 = 4 - 9 = -5$$

x	$g(x)$
−3	−5
−2	0
−1	3
0	4
1	3
2	0
3	−5

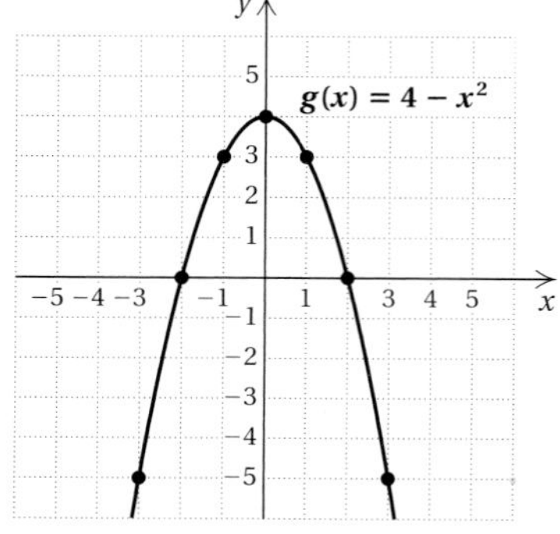

Example 6 Graph: $h(x) = |x|$.

A list of some function values is shown in the following table. We plot the points and connect them. The graph is a V-shaped "curve" that rises on either side of the vertical axis.

x	$h(x)$
−3	3
−2	2
−1	1
0	0
1	1
2	2
3	3

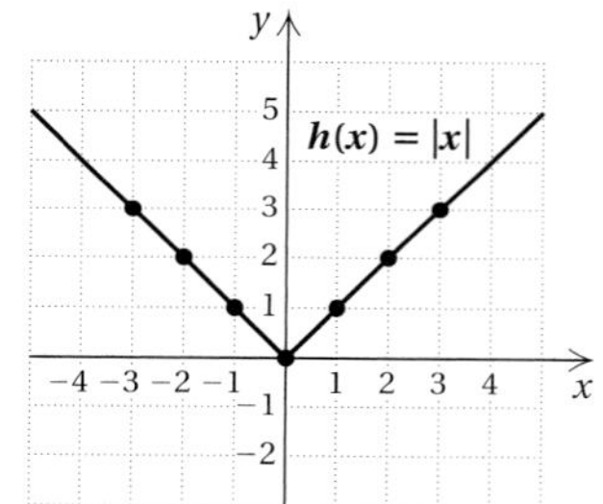

Do Exercises 9–11 on the following page.

d The Vertical-Line Test

Consider the function f described by $f(x) = x^2 - 5$. Its graph is shown at right. It is also the graph of the equation $y = x^2 - 5$.

To find a function value, like $f(3)$, from a graph, we locate the input on the horizontal axis, move vertically to the graph of the function, and then horizontally to find the output on the vertical axis, where members of the range can be found.

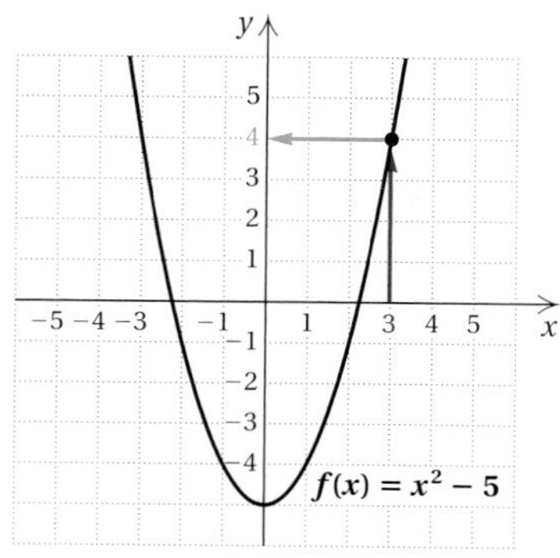

Recall that when one member of the domain is paired with two or more different members of the range, the correspondence is not a function. Thus, when a graph contains two or more different points with the same first coordinate, the graph cannot represent a function. Points sharing a common first coordinate are vertically above or below each other (see the following graph).

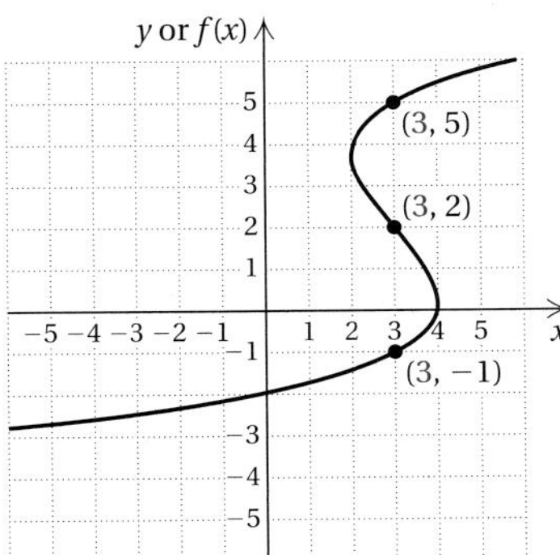

Since 3 is paired with more than one member of the range, the graph does not represent a function.

This observation leads to the *vertical-line test.*

> **THE VERTICAL-LINE TEST**
>
> A graph represents a function if it is impossible to draw a vertical line that intersects the graph more than once.

Example 7 Determine whether each of the following is the graph of a function.

a)

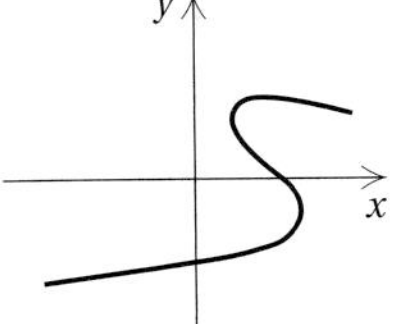

b)

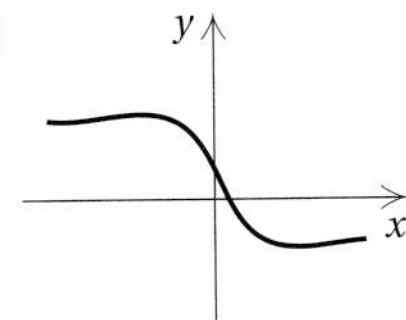

c)

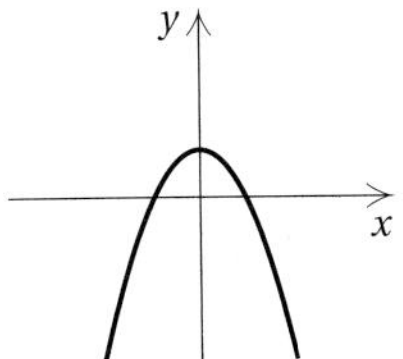

d)

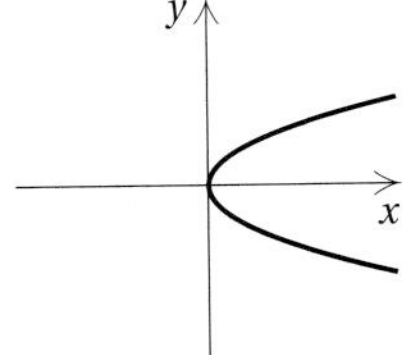

a) The graph *is not* that of a function because a vertical line crosses the graph at more than one point.

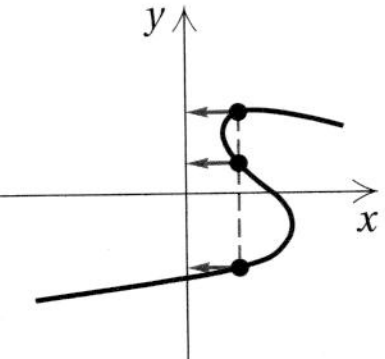

b) The graph *is* that of a function because no vertical line can cross the graph at more than one point. This can be confirmed with a ruler or straight edge.

Graph.

9. $f(x) = x - 4$

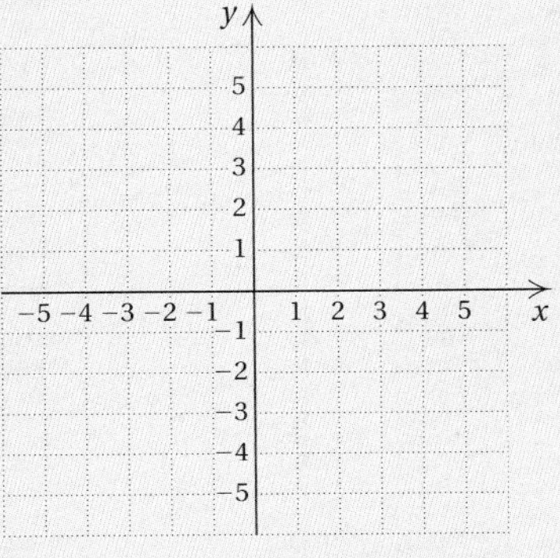

10. $g(x) = 5 - x^2$

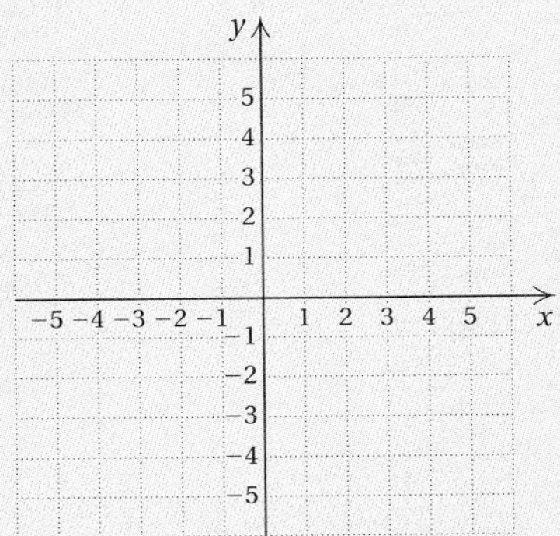

11. $t(x) = 3 - |x|$

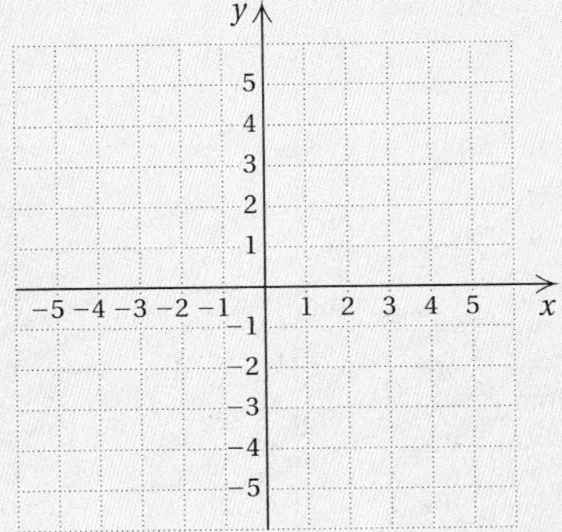

Answers on page A-50

Determine whether each of the following is the graph of a function.

12.

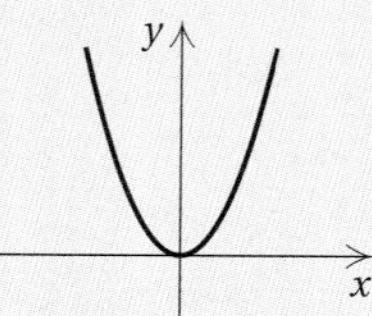

13.

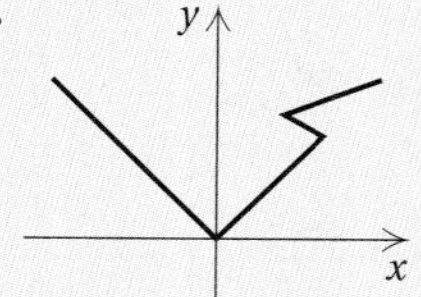

14.

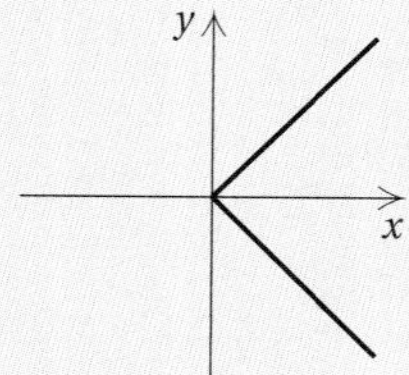

15.

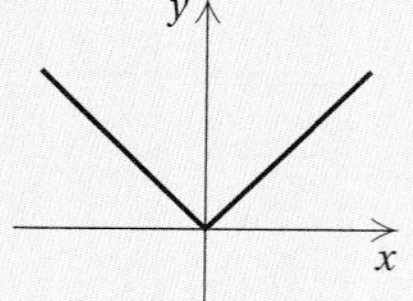

Referring to the graph in Example 8:

16. What was the movie revenue for week 2?

17. What was the movie revenue for week 6?

Answers on page A-50

c) The graph *is* that of a function.

d) The graph *is not* that of a function. There is a vertical line that crosses the graph more than once.

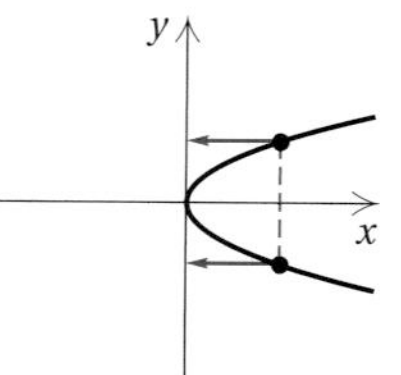

Do Exercises 12–15.

e Applications of Functions and Their Graphs

Functions are often described by graphs, whether or not an equation is given. To use a graph in an application, we note that each point on the graph represents a pair of values.

Example 8 *Movie Revenue.* The following graph approximates the weekly revenue, in millions of dollars, from the recent movie *Jurassic Park—The Lost World.* The revenue is a function of the week, and no equation is given for the function.

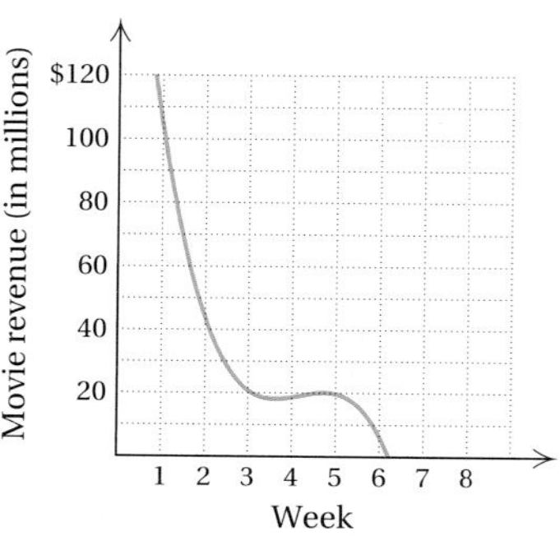

Source: Exhibitor Relations Co., Inc.

Use the graph to answer the following.

a) What was the movie revenue for week 1?

b) What was the movie revenue for week 5?

a) To estimate the revenue for week 1, we locate 1 on the horizontal axis and move directly up until we reach the graph. Then we move across to the vertical axis. We estimate that value to be about \$105 million.

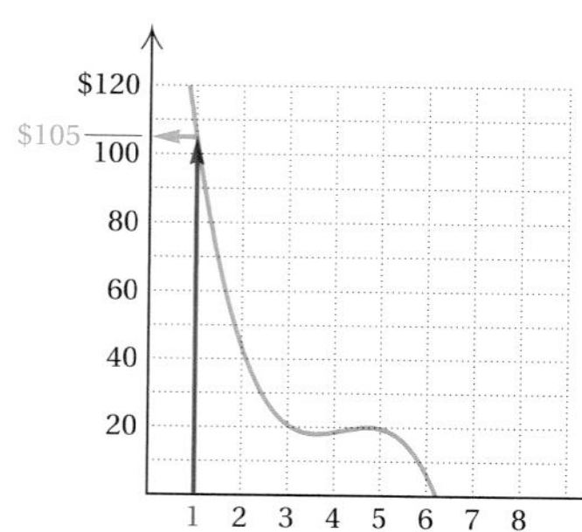

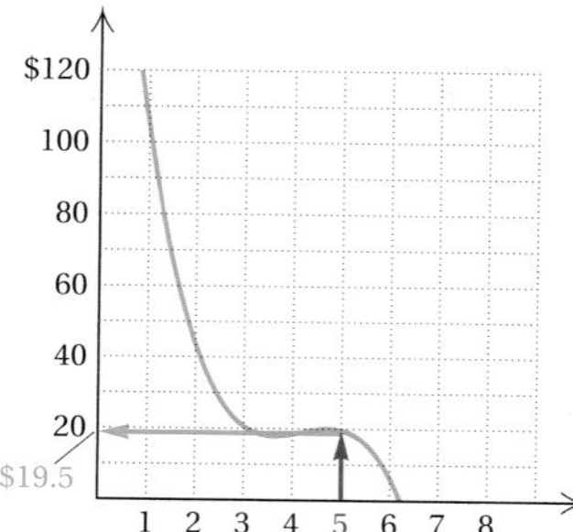

b) To estimate the revenue for week 5, we locate 5 on the horizontal axis and move directly up until we reach the graph. Then we move across to the vertical axis. We estimate that value to be about \$19.5 million.

Do Exercises 16 and 17.

Exercise Set 16.7

a Determine whether the correspondence is a function.

1. *Domain* *Range*

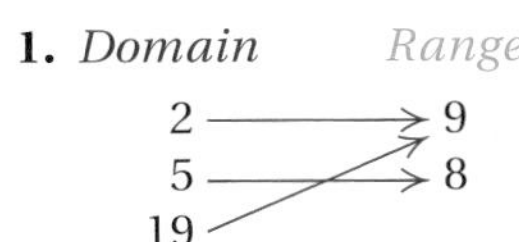

2. *Domain* *Range*

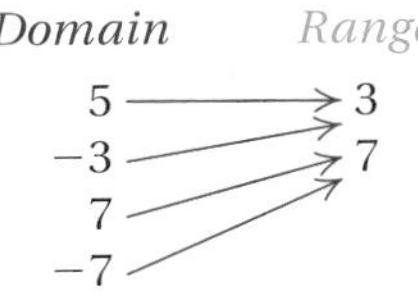

3. *Domain* *Range*

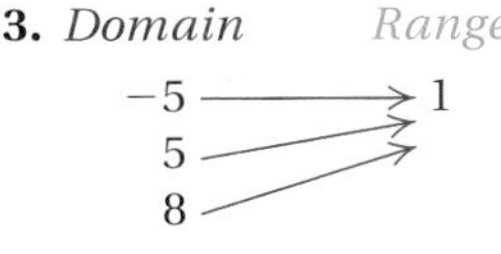

4. *Domain* *Range*

6 → −6
7 → −7
3 → −3

5. *Domain* *Range*

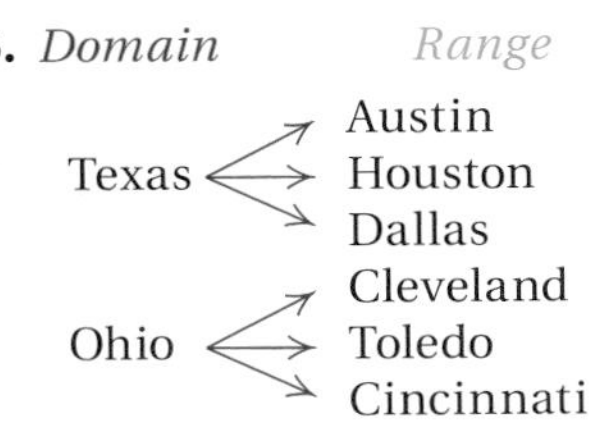

6. *Domain* *Range*

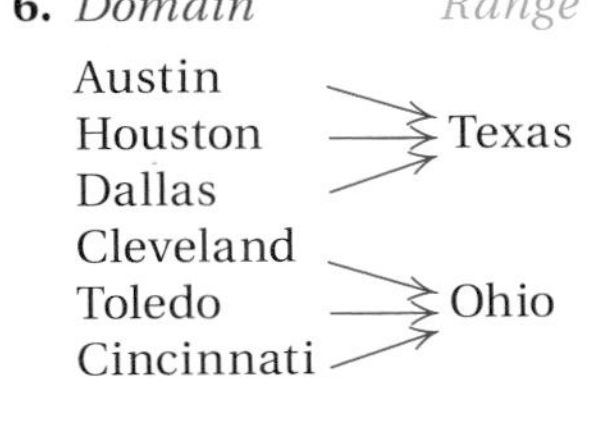

7.

Domain (Year)	*Range* (Consumption of Diet Cola, in gallons per person)
1991	11.7
1992	11.6
1993	11.7
1994	11.9

Source: U. S. Department of Agriculture Economic Research Service

8. *Domain* (Brand of Single-Serving Pizza) *Range* (Number of Calories)

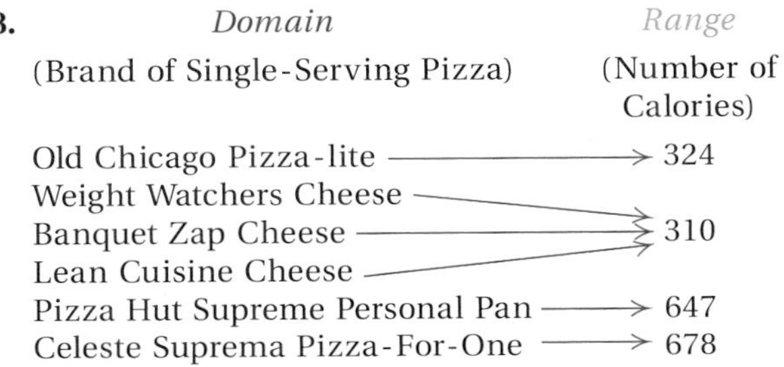

Determine whether each of the following is a function. Identify any relations that are not functions.

	Domain	*Correspondence*	*Range*
9.	A math class	Each person's seat number	A set of numbers
10.	A set of numbers	Square each number and then add 4.	A set of numbers
11.	A set of shapes	Find the area of each shape.	A set of numbers
12.	A family	Each person's eye color	A set of colors
13.	The people in a town	Each person's aunt	A set of females
14.	A set of avenues	Find an intersecting road.	A set of cross streets

b Find the function values.

15. $f(x) = x + 5$

a) $f(4)$ b) $f(7)$
c) $f(-3)$ d) $f(0)$
e) $f(2.4)$ f) $f\left(\frac{2}{3}\right)$

16. $g(t) = t - 6$

a) $g(0)$ b) $g(6)$
c) $g(13)$ d) $g(-1)$
e) $g(-1.08)$ f) $g\left(\frac{7}{8}\right)$

17. $h(p) = 3p$

a) $h(-7)$ b) $h(5)$
c) $h(14)$ d) $h(0)$
e) $h\left(\frac{2}{3}\right)$ f) $h(-54.2)$

18. $f(x) = -4x$

a) $f(6)$ b) $f\left(-\frac{1}{2}\right)$
c) $f(20)$ d) $f(11.8)$
e) $f(0)$ f) $f(-1)$

19. $g(s) = 3s + 4$

a) $g(1)$ b) $g(-7)$
c) $g(6.7)$ d) $g(0)$
e) $g(-10)$ f) $g\left(\frac{2}{3}\right)$

20. $h(x) = 19$, a constant function

a) $h(4)$ b) $h(-6)$
c) $h(12.5)$ d) $h(0)$
e) $h\left(\frac{2}{3}\right)$ f) $h(1234)$

21. $f(x) = 2x^2 - 3x$

a) $f(0)$ b) $f(-1)$
c) $f(2)$ d) $f(10)$
e) $f(-5)$ f) $f(-10)$

22. $f(x) = 3x^2 - 2x + 1$

a) $f(0)$ b) $f(1)$
c) $f(-1)$ d) $f(10)$
e) $f(2)$ f) $f(-3)$

23. $f(x) = |x| + 1$

a) $f(0)$ b) $f(-2)$
c) $f(2)$ d) $f(-3)$
e) $f(-10)$ f) $f(22)$

24. $g(t) = \sqrt{t}$

a) $g(4)$ b) $g(25)$
c) $g(16)$ d) $g(100)$
e) $g(50)$ f) $g(84)$

25. $f(x) = x^3$

a) $f(0)$ b) $f(-1)$
c) $f(2)$ d) $f(10)$
e) $f(-5)$ f) $f(-10)$

26. $f(x) = x^4 - 3$

a) $f(1)$ b) $f(-1)$
c) $f(0)$ d) $f(2)$
e) $f(-2)$ f) $f(10)$

27. *Estimating Heights.* An anthropologist can estimate the height of a male or a female, given the lengths of certain bones. A *humerus* is the bone from the elbow to the shoulder. The height, in centimeters, of a female with a humerus of x centimeters is given by the function

$$F(x) = 2.75x + 71.48.$$

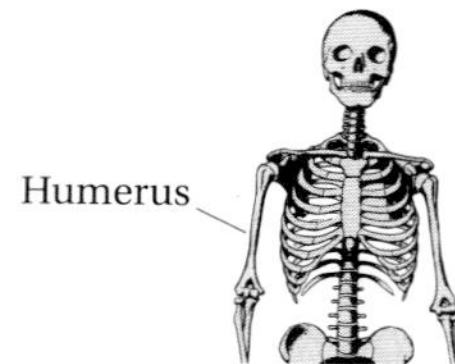

If a humerus is known to be from a female, how tall was she if the bone is **(a)** 32 cm long? **(b)** 30 cm long?

28. Refer to Exercise 27. When a humerus is from a male, the function

$$M(x) = 2.89x + 70.64$$

can be used to find the male's height, in centimeters. If a humerus is known to be from a male, how tall was he if the bone is **(a)** 30 cm long? **(b)** 35 cm long?

29. *Pressure at Sea Depth.* The function $P(d) = 1 + (d/33)$ gives the pressure, in *atmospheres* (atm), at a depth of d feet in the sea. Note that $P(0) = 1$ atm, $P(33) = 2$ atm, and so on. Find the pressure at 20 ft, 30 ft, and 100 ft.

30. *Temperature as a Function of Depth.* The function $T(d) = 10d + 20$ gives the temperature, in degrees Celsius, inside the earth as a function of the depth d, in kilometers. Find the temperature at 5 km, 20 km, and 1000 km.

31. *Melting Snow.* The function $W(d) = 0.112d$ approximates the amount, in centimeters, of water that results from d centimeters of snow melting. Find the amount of water that results from snow melting from depths of 16 cm, 25 cm, and 100 cm.

32. *Temperature Conversions.* The function $C(F) = \frac{5}{9}(F - 32)$ determines the Celsius temperature that corresponds to F degrees Fahrenheit. Find the Celsius temperature that corresponds to 62°F, 77°F, and 23°F.

c Graph the function.

33. $f(x) = 3x - 1$

34. $g(x) = 2x + 5$

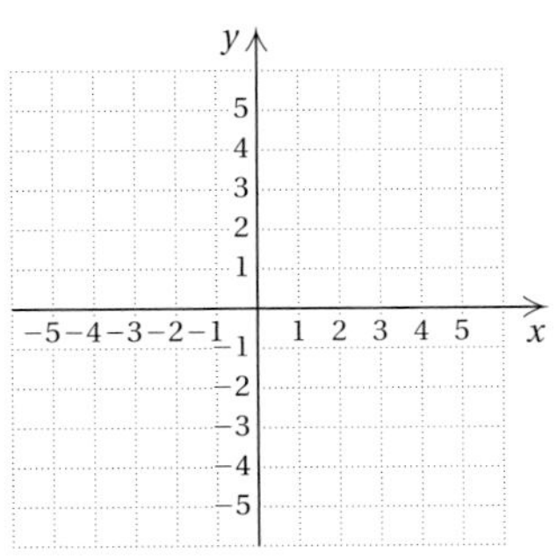

35. $g(x) = -2x + 3$

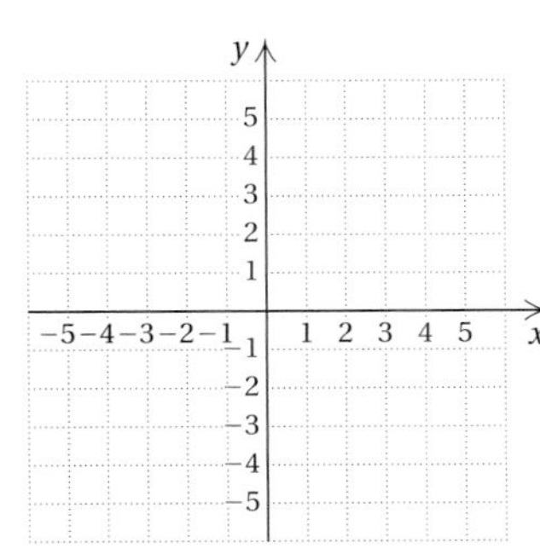

36. $f(x) = -\frac{1}{2}x + 2$

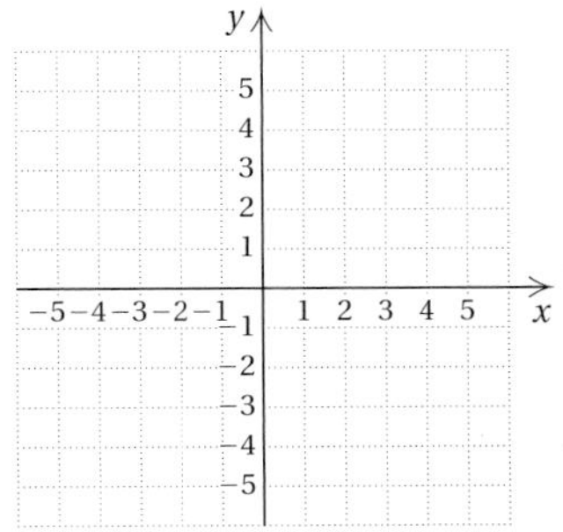

37. $f(x) = \frac{1}{2}x + 1$

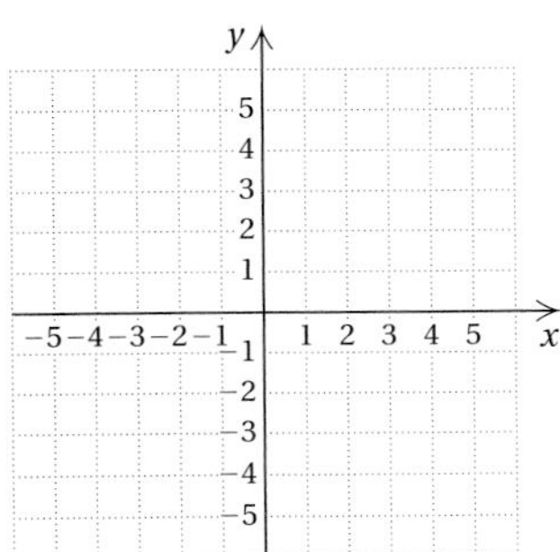

38. $f(x) = -\frac{3}{4}x - 2$

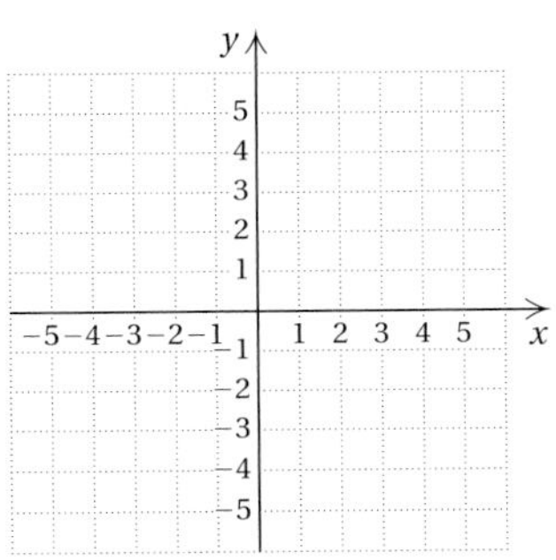

39. $f(x) = 2 - |x|$

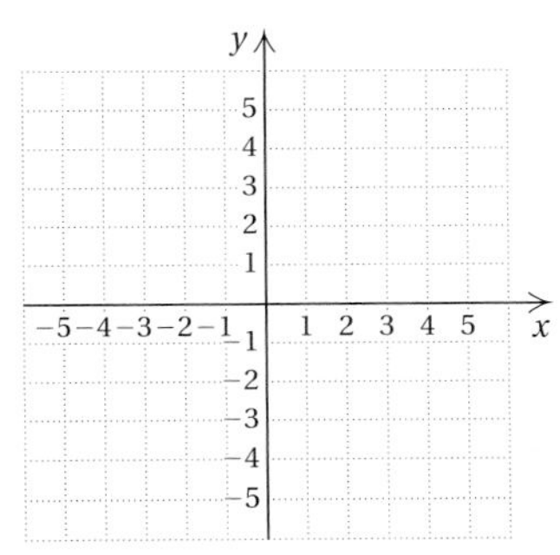

40. $f(x) = |x| - 4$

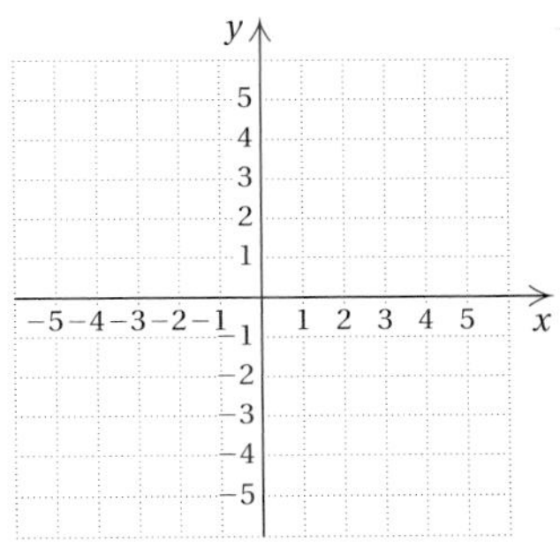

41. $f(x) = x^2$

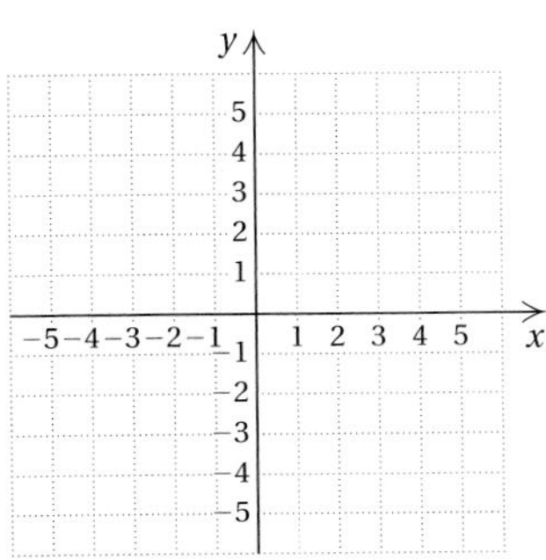

42. $f(x) = x^2 - 1$

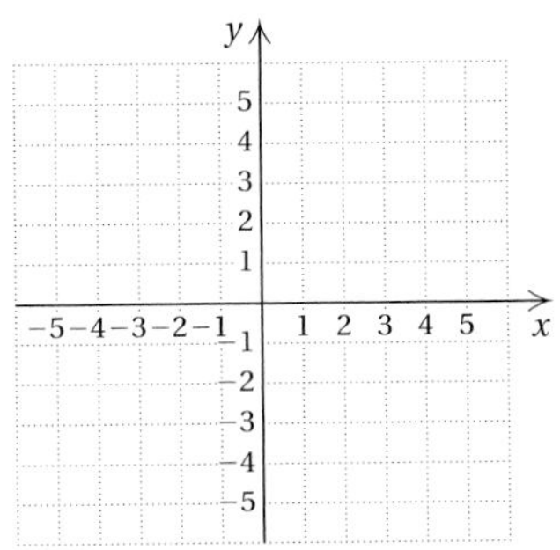

43. $f(x) = x^2 - x - 2$

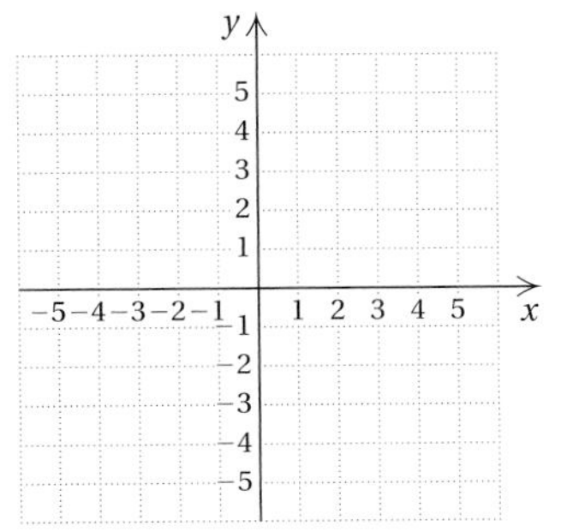

44. $f(x) = x^2 + 6x + 5$

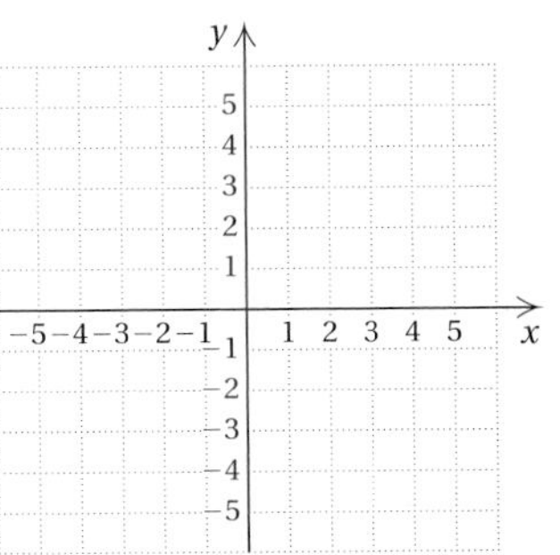

d Determine whether each of the following is the graph of a function.

45.

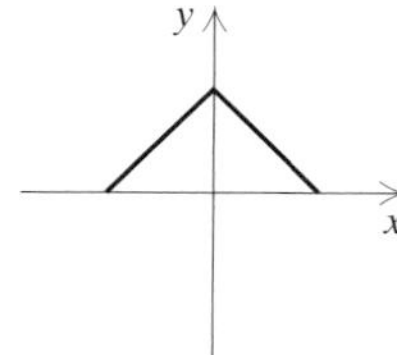

46.

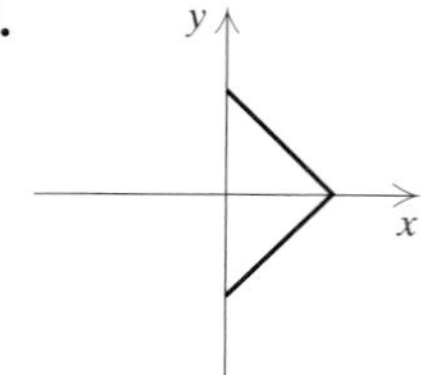

47.

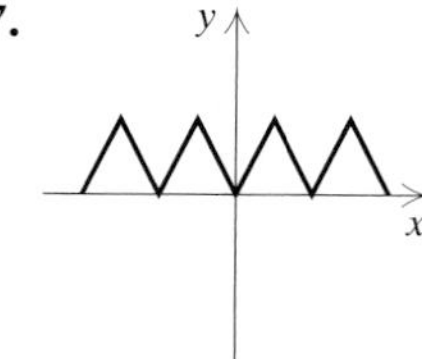

48.

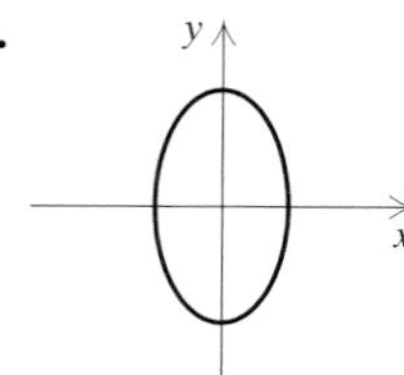

49.

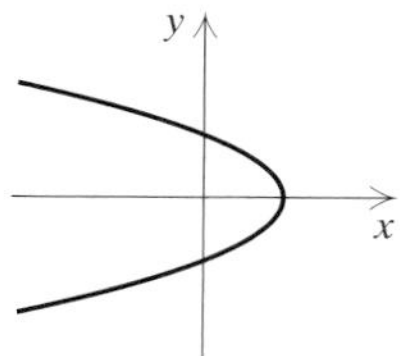

50.

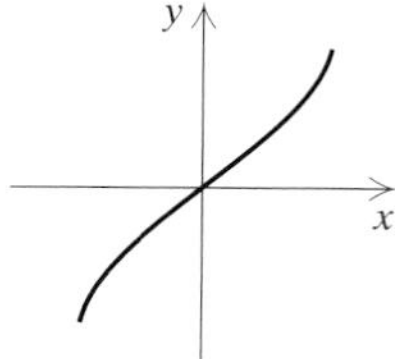

51.

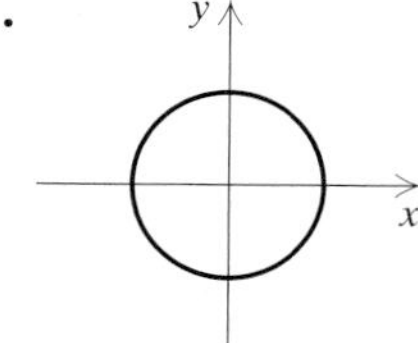

52.

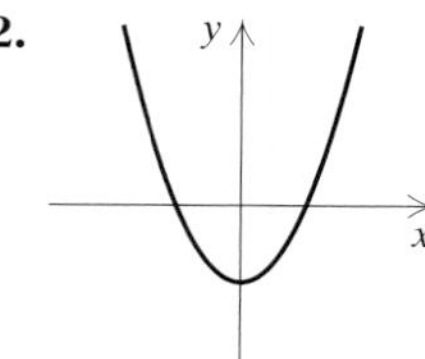

e *Cholesterol Level and Risk of a Heart Attack.* The graph below shows the annual heart attack rate per 10,000 men as a function of blood cholesterol level.

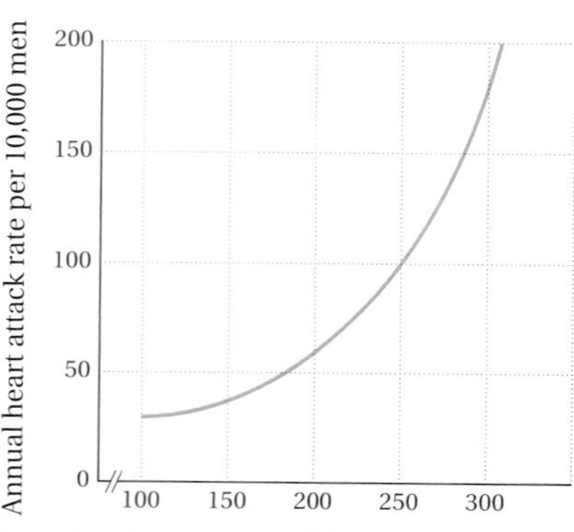

Source: Copyright 1989, CSPI. Adapted from *Nutrition Action Healthletter* (1875 Connecticut Avenue, N.W., Suite 300, Washington, DC 20009-5728)

53. Approximate the annual heart attack rate per 10,000 men for those whose blood cholesterol level is 225 mg/dl.

54. Approximate the annual heart attack rate per 10,000 men for those whose blood cholesterol level is 275 mg/dl.

Skill Maintenance

Determine whether the pair of equations represents parallel lines. [13.3a]

55. $y = \frac{3}{4}x - 7,$
$3x + 4y = 7$

56. $y = \frac{3}{5},$
$y = -\frac{5}{3}$

Solve the system using the substitution method. [14.2b]

57. $2x - y = 6,$
$4x - 2y = 5$

58. $x - 3y = 2,$
$3x - 9y = 6$

Synthesis

59. ◈ Is it possible for a function to have more numbers as outputs than as inputs? Why or why not?

60. ◈ Look up the word "function" in a dictionary. Explain how that definition might be related to the mathematical one given in this section.

Summary and Review Exercises: Chapter 16

Important Properties and Formulas

Standard Form: $ax^2 + bx + c = 0,\ a > 0$

Principle of Square Roots: The equation $x^2 = d$, where $d > 0$, has two solutions, $\sqrt{d}$ and $-\sqrt{d}$. The solution of $x^2 = 0$ is 0.

Quadratic Formula: $x = \dfrac{-b \pm \sqrt{b^2 - 4ac}}{2a}$

Discriminant: $b^2 - 4ac$

The x-coordinate of the vertex of a parabola $= -\dfrac{b}{2a}$.

The objectives to be tested in addition to the material in this chapter are [13.5c], [15.2c], [15.4a], and [15.6b].

Solve.

1. $8x^2 = 24$ [16.2a]

2. $40 = 5y^2$ [16.2a]

3. $5x^2 - 8x + 3 = 0$ [16.1c]

4. $3y^2 + 5y = 2$ [16.1c]

5. $(x + 8)^2 = 13$ [16.2b]

6. $9x^2 = 0$ [16.2a]

7. $5t^2 - 7t = 0$ [16.1b]

Solve. [16.3a]

8. $x^2 - 2x - 10 = 0$

9. $9x^2 - 6x - 9 = 0$

10. $x^2 + 6x = 9$

11. $1 + 4x^2 = 8x$

12. $6 + 3y = y^2$

13. $3m = 4 + 5m^2$

14. $3x^2 = 4x$

Solve. [16.1c]

15. $\dfrac{15}{x} - \dfrac{15}{x + 2} = 2$

16. $x + \dfrac{1}{x} = 2$

Solve by completing the square. Show your work. [16.2c]

17. $x^2 - 5x + 2 = 0$

18. $3x^2 - 2x - 5 = 0$

Approximate the solutions to the nearest tenth. [16.3b]

19. $x^2 - 5x + 2 = 0$

20. $4y^2 + 8y + 1 = 0$

21. Solve for T: $V = \dfrac{1}{2}\sqrt{1 + \dfrac{T}{L}}$. [16.4a]

Graph the quadratic equation. [16.6a]

22. $y = 2 - x^2$

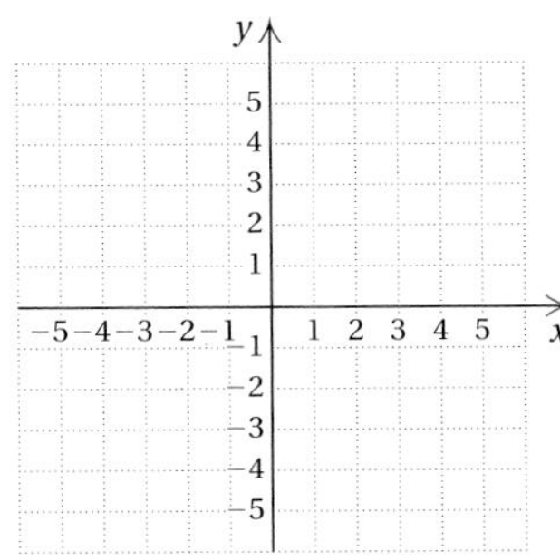

23. $y = x^2 - 4x - 2$

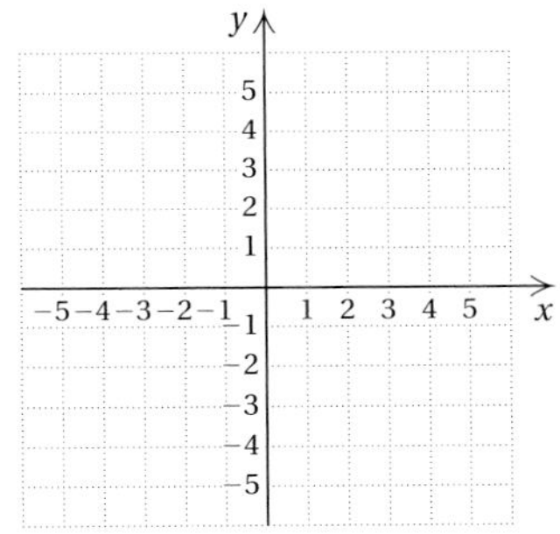

Find the x-intercepts. [16.6b]

24. $y = 2 - x^2$

25. $y = x^2 - 4x - 2$

Solve.

26. The hypotenuse of a right triangle is 5 cm long. One leg is 3 m longer than the other. Find the lengths of the legs. Round to the nearest tenth. [16.5a]

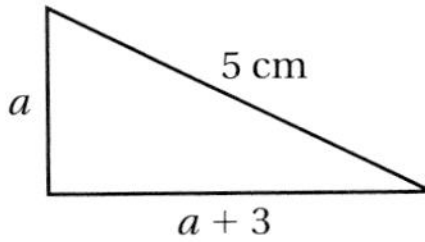

27. The hypotenuse of a right triangular freight ramp is 26 yd long. One leg is 14 yd longer than the other. Find the lengths of the legs. [16.5a]

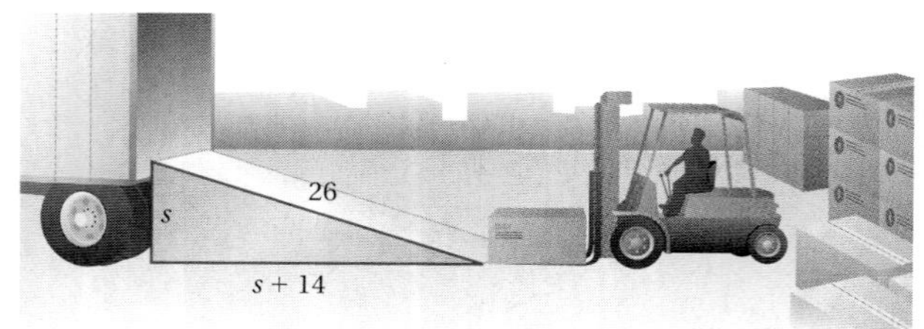

28. The height of Lake Point Towers in Chicago is 645 ft. How long would it take an object to fall to the ground from the top? [16.2d]

Find the function values. [16.7b]

29. If $f(x) = 2x - 5$, find $f(2)$, $f(-1)$, and $f(3.5)$.

30. If $g(x) = |x| - 1$, find $g(1)$, $g(-1)$, and $g(-20)$.

31. *Caloric Needs.* If you are moderately active, you need to consume each day about 15 calories per pound of body weight. The function $C(p) = 15p$ approximates the number of calories C that are needed to maintain body weight p, in pounds. How many calories are needed to maintain a body weight of 180 lb? [16.7e]

Graph the function. [16.7c]

32. $g(x) = 4 - x$

33. $f(x) = x^2 - 3$

34. $h(x) = |x| - 5$

Determine whether each of the following is the graph of a function. [16.7d]

35.

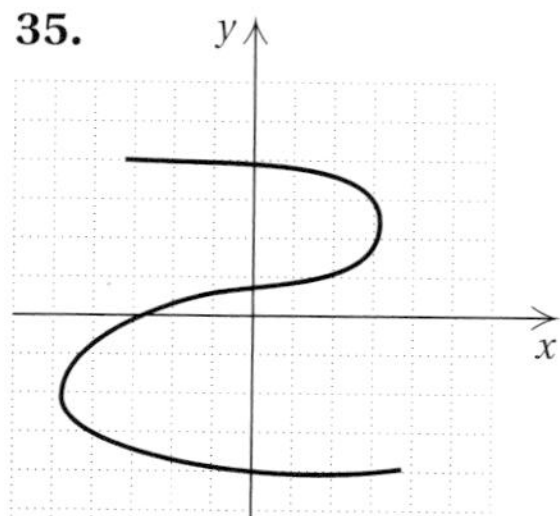

36.

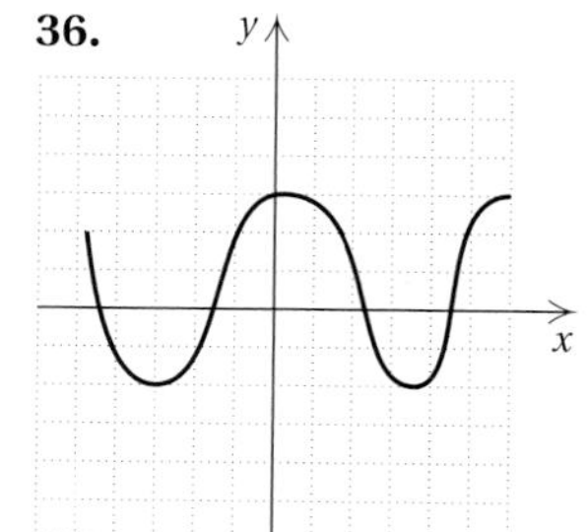

Skill Maintenance

Multiply and simplify. [15.2c]

37. $\sqrt{18a}\sqrt{2}$

38. $\sqrt{12xy^2}\sqrt{5xy}$

39. Find an equation of variation in which y varies inversely as x and $y = 10$ when $x = 0.0625$. [13.5c]

40. The sides of a rectangle are of lengths 1 and $\sqrt{2}$. Find the length of a diagonal. [15.6b]

Add or subtract. [15.4a]

41. $5\sqrt{11} + 7\sqrt{11}$

42. $2\sqrt{90} - \sqrt{40}$

Synthesis

43. ◆ List the names and give an example of as many types of equation as you can that you have learned to solve in this text. [8.1a], [12.6a], [14.1a], [15.5a], [16.1a]

44. ◆ Find the errors in each of the following solutions of equations. [16.1b]

a)
$$\begin{aligned} x^2 + 20x &= 0 \\ x(x + 20) &= 0 \\ x + 20 &= 0 \\ x &= 20 \end{aligned}$$

b)
$$\begin{aligned} x^2 + x &= 6 \\ x(x + 1) &= 6 \\ x = 6 \quad &or \quad x + 1 = 6 \\ x = 6 \quad &or \quad x = 5 \end{aligned}$$

45. Two consecutive integers have squares that differ by 63. Find the integers. [16.5a]

46. A square with sides of length s has the same area as a circle with a radius of 5 in. Find s. [16.5a]

47. Solve: $x - 4\sqrt{x} - 5 = 0$. [16.1c]

Test: Chapter 16

Solve.

1. $7x^2 = 35$

2. $7x^2 + 8x = 0$

3. $48 = t^2 + 2t$

4. $3y^2 - 5y = 2$

5. $(x - 8)^2 = 13$

6. $x^2 = x + 3$

7. $m^2 - 3m = 7$

8. $10 = 4x + x^2$

9. $3x^2 - 7x + 1 = 0$

10. $x - \frac{2}{x} = 1$

11. $\frac{4}{x} - \frac{4}{x + 2} = 1$

12. Solve $x^2 - 4x - 10 = 0$ by completing the square. Show your work.

13. Approximate the solutions to $x^2 - 4x - 10 = 0$ to the nearest tenth.

14. Solve for n: $d = an^2 + bn$.

15. Find the x-intercepts: $y = -x^2 + x + 5$.

Graph.

16. $y = 4 - x^2$

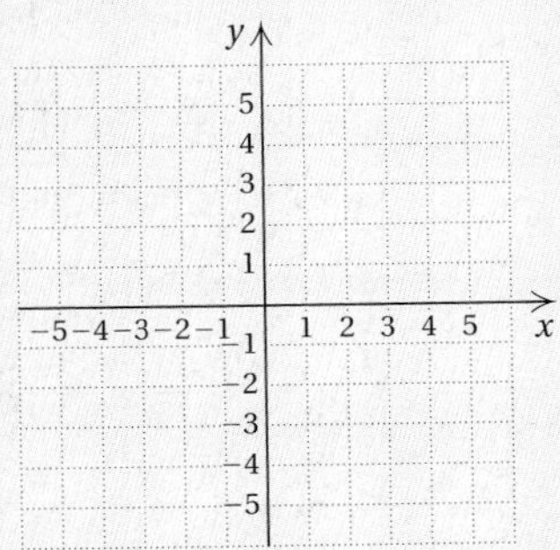

17. $y = -x^2 + x + 5$

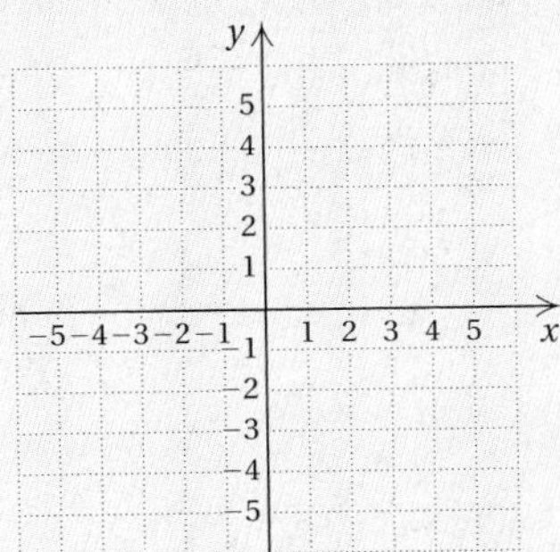

18. If $f(x) = \frac{1}{2}x + 1$, find $f(0)$, $f(1)$, and $f(2)$.

19. If $g(t) = -2|t| + 3$, find $g(-1)$, $g(0)$, and $g(3)$.

Answers

1. ______
2. ______
3. ______
4. ______
5. ______
6. ______
7. ______
8. ______
9. ______
10. ______
11. ______
12. ______
13. ______
14. ______
15. ______
16. ______
17. ______
18. ______
19. ______

Answers

20. ______

21. ______

22. ______

23. ______

24. ______

25. ______

26. ______

27. ______

28. ______

29. ______

30. ______

31. ______

32. ______

Solve.

20. The width of a rectangular area rug is 4 m less than the length. The area is 16.25 m^2. Find the length and the width.

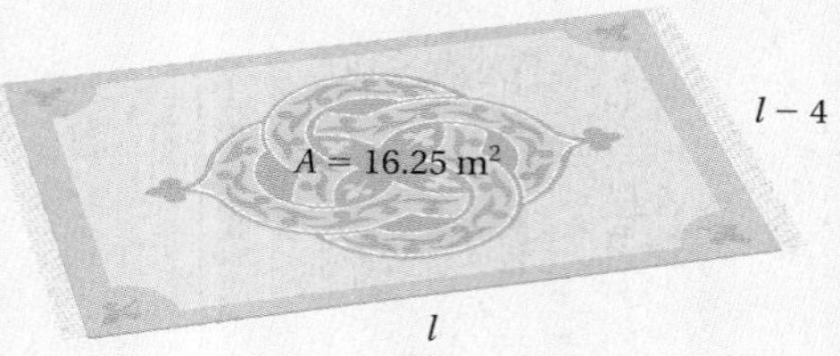

21. The current in a stream moves at a speed of 2 km/h. A boat travels 44 km upstream and 52 km downstream in a total of 4 hr. What is the speed of the boat in still water?

22. *World Record for 10,000-m Run.* The world record for the 10,000-m run has been decreasing steadily since 1940. The record is approximately 30.18 min minus 0.06 times the number of years since 1940. The function $R(t) = 30.18 - 0.06t$ estimates the record R, in minutes, as a function of t, the time in years since 1940. Predict what the record will be in 2000.

Graph.

23. $h(x) = x - 4$

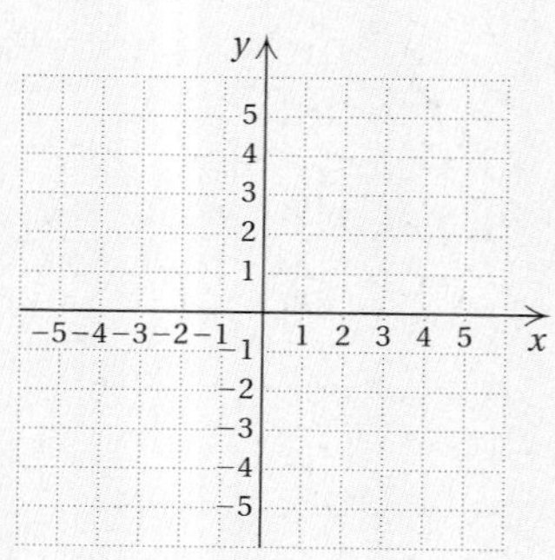

24. $g(x) = x^2 - 4$

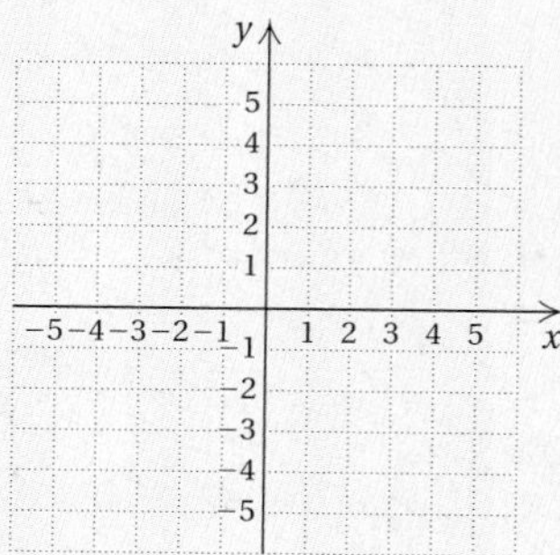

Determine whether each of the following is the graph of a function.

25.

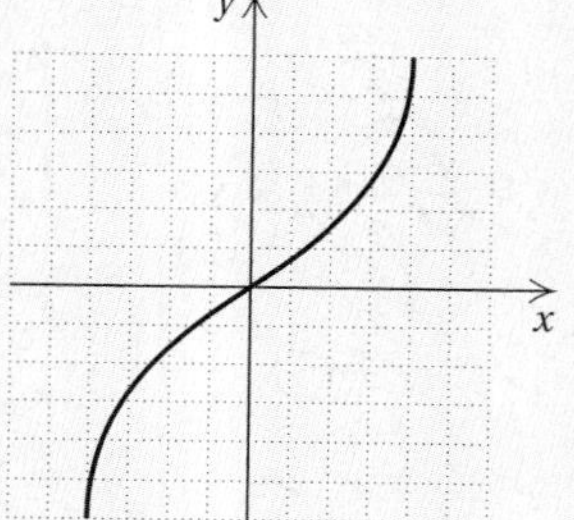

26.

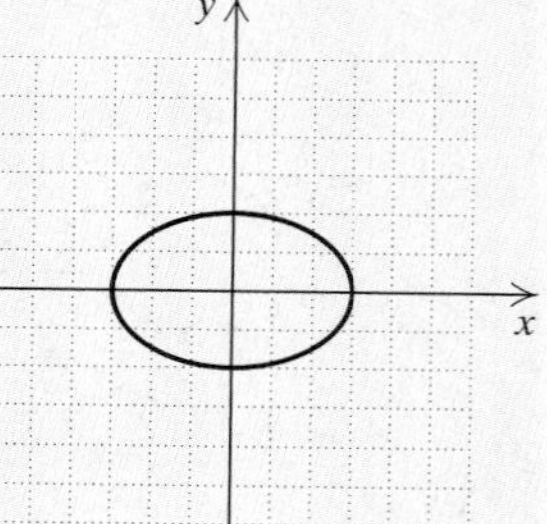

Skill Maintenance

27. Subtract: $\sqrt{240} - \sqrt{60}$.

28. Multiply and simplify: $\sqrt{7xy}\,\sqrt{14x^2y}$.

29. Find an equation of variation in which y varies inversely as x and $y = 32$ when $x = 0.125$.

30. The sides of a rectangle are of lengths $\sqrt{2}$ and $\sqrt{3}$. Find the length of a diagonal.

Synthesis

31. Find the side of a square whose diagonal is 5 ft longer than a side.

32. Solve this system for x. Use the substitution method.

$$x - y = 2,$$
$$xy = 4$$

Appendix A Linear Measures: American and Metric

Length, or distance, is one kind of measure. To find lengths, we start with some **unit segment** and assign to it a measure of 1. Suppose $\overline{AB}$ below is a unit segment.

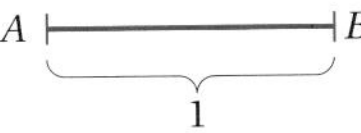

Let's measure segment $\overline{CD}$ below, using $\overline{AB}$ as our unit segment.

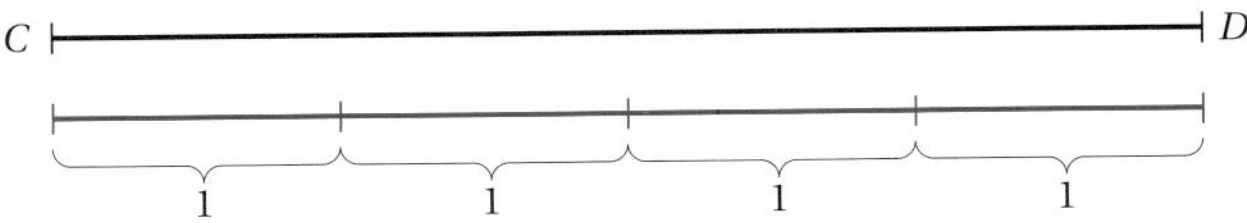

Since we can place 4 unit segments end to end along $\overline{CD}$, the measure of $\overline{CD}$ is 4.

Sometimes we have to use parts of units, called **subunits**. For example, the measure of the segment $\overline{MN}$ below is $1\frac{1}{2}$. We place one unit segment and one half-unit segment end to end.

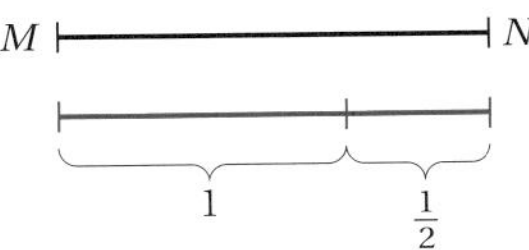

Do Exercises 1–4.

a American Measures

American units of length are related as follows.

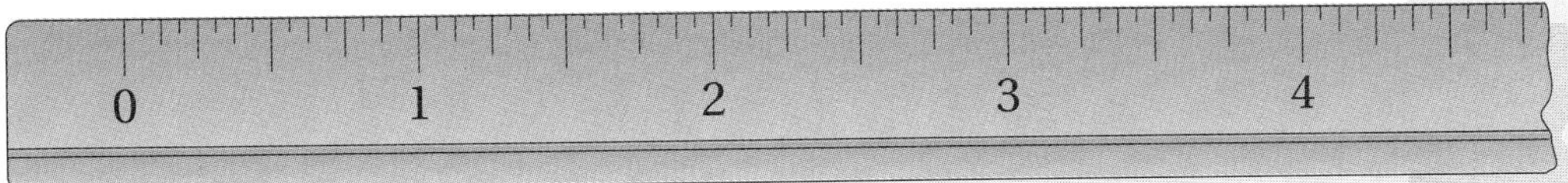

(Actual size, in inches)

American Units of Length

12 inches (in.) = 1 foot (ft)
36 inches = 1 yard
3 feet = 1 yard (yd)
5280 feet = 1 mile (mi)

The symbolism 13 in. = 13″ and 27 ft = 27′ is also used for inches and feet. American units have also been called "English," or "British–American," because at one time they were used by both countries. Today, both Canada and England have officially converted to the metric system. However, if you travel in England, you will still see units such as "miles" on road signs.

Objectives

a. Convert from one American unit of length to another.
b. Convert from one metric unit of length to another.
c. Convert between American and metric units of length.

Use the unit below to measure the length of each segment or object.

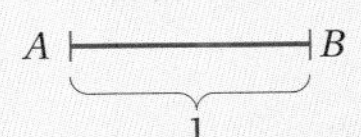

1.

2.

3.

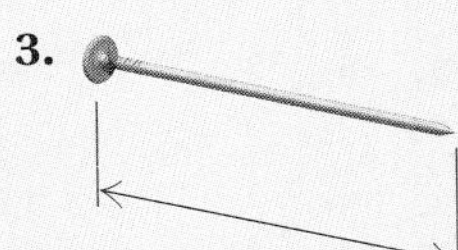

4.

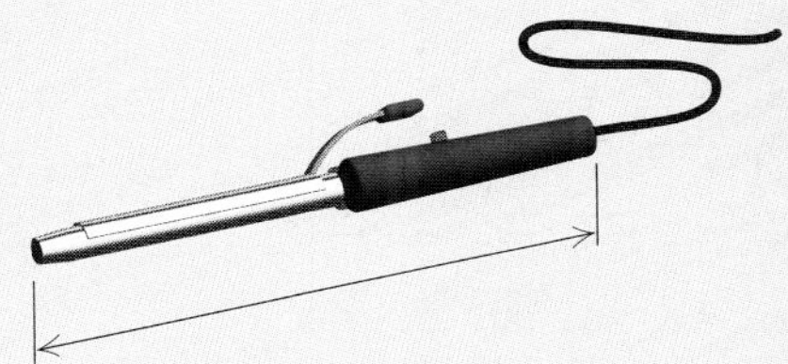

Answers on page A-52

Complete.

5. 8 yd = ________ in.

6. 14.5 yd = ________ ft

7. 3.8 mi = ________ in.

Complete.

8. 72 in. = ________ ft

9. 17 in. = ________ ft

Answers on page A-52

To change from certain American units to others, we make substitutions. Such a substitution is usually helpful when we are converting from a larger unit to a smaller one.

Example 1 Complete: 7 yd = ________ in.

$$
\begin{aligned}
7 \text{ yd} &= 7 \times 1 \text{ yd} \\
&= 7 \times 3 \text{ ft} && \textbf{Substituting 3 ft for 1 yd} \\
&= 7 \times 3 \times 1 \text{ ft} \\
&= 7 \times 3 \times 12 \text{ in.} && \textbf{Substituting 12 in. for 1 ft; } 7 \times 3 = 21;\ 21 \times 12 = 252 \\
&= 252 \text{ in.}
\end{aligned}
$$

Do Exercises 5–7.

Sometimes it helps to use multiplying by 1 in making conversions. For example, 12 in. = 1 ft, so

$$\frac{12 \text{ in.}}{1 \text{ ft}} = 1 \quad \text{and} \quad \frac{1 \text{ ft}}{12 \text{ in.}} = 1.$$

If we divide 12 in. by 1 ft or 1 ft by 12 in., we get 1 because the lengths are the same. Let's first convert from smaller to larger units.

Example 2 Complete: 48 in. = ________ ft.

We want to convert from "in." to "ft." We multiply by 1 using a symbol for 1 with "in." on the bottom and "ft" on the top to eliminate inches and to convert to feet:

$$
\begin{aligned}
48 \text{ in.} &= \frac{48 \text{ in.}}{1} \times \frac{1 \text{ ft}}{12 \text{ in.}} && \textbf{Multiplying by 1 using } \frac{1 \text{ ft}}{12 \text{ in.}} \textbf{ to eliminate in.} \\
&= \frac{48 \text{ in.}}{12 \text{ in.}} \times 1 \text{ ft} \\
&= \frac{48}{12} \times \frac{\text{in.}}{\text{in.}} \times 1 \text{ ft} \\
&= 4 \times 1 \text{ ft} && \textbf{The } \frac{\text{in.}}{\text{in.}} \textbf{ acts like 1, so we can omit it.} \\
&= 4 \text{ ft.}
\end{aligned}
$$

Do Exercises 8 and 9.

Example 3 Complete: 25 ft = ________ yd.

Since we are converting from "ft" to "yd," we choose a symbol for 1 with "yd" on the top and "ft" on the bottom:

$$
\begin{aligned}
25 \text{ ft} &= 25 \text{ ft} \times \frac{1 \text{ yd}}{3 \text{ ft}} && \textbf{3 ft = 1 yd, so } \frac{3 \text{ ft}}{1 \text{ yd}} = 1, \textbf{ and } \frac{1 \text{ yd}}{3 \text{ ft}} = 1. \textbf{ We use } \frac{1 \text{ yd}}{3 \text{ ft}} \textbf{ to eliminate ft.} \\
&= \frac{25}{3} \times \frac{\text{ft}}{\text{ft}} \times 1 \text{ yd} \\
&= 8\frac{1}{3} \times 1 \text{ yd} && \textbf{The } \frac{\text{ft}}{\text{ft}} \textbf{ acts like 1, so we can omit it.} \\
&= 8\frac{1}{3} \text{ yd, or } 8.\overline{3} \text{ yd.}
\end{aligned}
$$

We can also look at this conversion as "canceling" units:

$$25 \text{ ft} = 25 \cancel{\text{ft}} \times \frac{1 \text{ yd}}{3 \cancel{\text{ft}}} = \frac{25}{3} \times 1 \text{ yd} = 8\frac{1}{3} \text{ yd, or } 8.\overline{3} \text{ yd.}$$

Do Exercises 10 and 11.

Example 4 Complete: 23,760 ft = ________ mi.

We choose a symbol for 1 with "mi" on the top and "ft" on the bottom:

$$23{,}760 \text{ ft} = 23{,}760 \text{ ft} \times \frac{1 \text{ mi}}{5280 \text{ ft}} \qquad 5280 \text{ ft} = 1 \text{ mi, so } \frac{1 \text{ mi}}{5280 \text{ ft}} = 1.$$

$$= \frac{23{,}760}{5280} \times \frac{\text{ft}}{\text{ft}} \times 1 \text{ mi}$$

$$= 4.5 \times 1 \text{ mi} \qquad \text{Dividing}$$

$$= 4.5 \text{ mi.}$$

Let's also consider this example using canceling:

$$23{,}760 \text{ ft} = 23{,}760 \cancel{\text{ft}} \times \frac{1 \text{ mi}}{5280 \cancel{\text{ft}}}$$

$$= \frac{23{,}760}{5280} \times 1 \text{ mi} = 4.5 \times 1 \text{ mi} = 4.5 \text{ mi.}$$

Do Exercises 12 and 13.

b Metric Measures

The **metric system** is used in most countries of the world, and the United States is now making greater use of it as well. The metric system does not use inches, feet, pounds, and so on, although units for time and electricity are the same as those you use now.

An advantage of the metric system is that it is easier to convert from one unit to another. That is because the metric system is based on the number 10.

The basic unit of length is the **meter**. It is just over a yard. In fact, 1 meter ≈ 1.1 yd.

(Comparative sizes are shown.)

1 Meter

1 Yard

The other units of length are multiples of the length of a meter:

10 times a meter, 100 times a meter, 1000 times a meter, and so on,

or fractions of a meter:

$\frac{1}{10}$ of a meter, $\frac{1}{100}$ of a meter, $\frac{1}{1000}$ of a meter, and so on.

Complete.

10. 24 ft = ________ yd

11. 35 ft = ________ yd

Complete.

12. 26,400 ft = ________ mi

13. 2640 ft = ________ mi

Answers on page A-52

METRIC UNITS OF LENGTH

1 *kilo*meter (km) = 1000 meters (m)

1 *hecto*meter (hm) = 100 meters (m)

1 *deka*meter (dam) = 10 meters (m)

1 meter (m)

1 *deci*meter (dm) = $\frac{1}{10}$ meter (m)

1 *centi*meter (cm) = $\frac{1}{100}$ meter (m)

1 *milli*meter (mm) = $\frac{1}{1000}$ meter (m)

dam and *dm* are not used often.

You should memorize these names and abbreviations. Think of *kilo*- for 1000, *hecto*- for 100, *deka*- for 10, *deci*- for $\frac{1}{10}$, *centi*- for $\frac{1}{100}$, and *milli*- for $\frac{1}{1000}$. We will also use these prefixes when considering units of area, capacity, and mass.

Thinking Metric

To familiarize yourself with metric units, consider the following.

1 kilometer (1000 meters) is slightly more than $\frac{1}{2}$ mile (0.6 mi).
1 meter is just over a yard (1.1 yd).
1 centimeter (0.01 meter) is a little more than the width of a paperclip (about 0.3937 inch).

1 cm

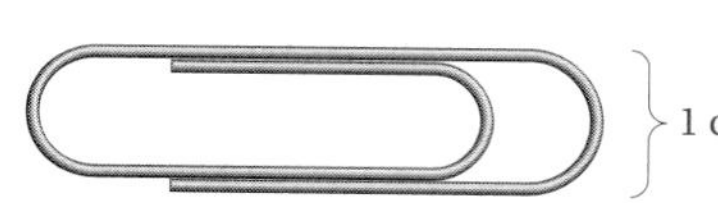

1 inch is about 2.54 centimeters.

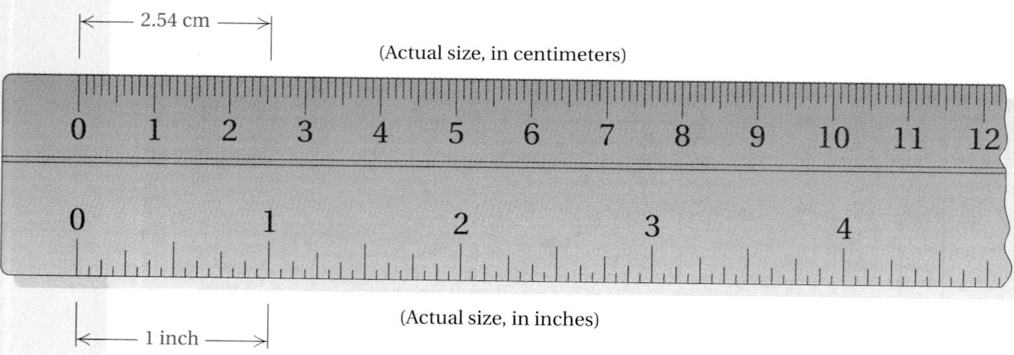

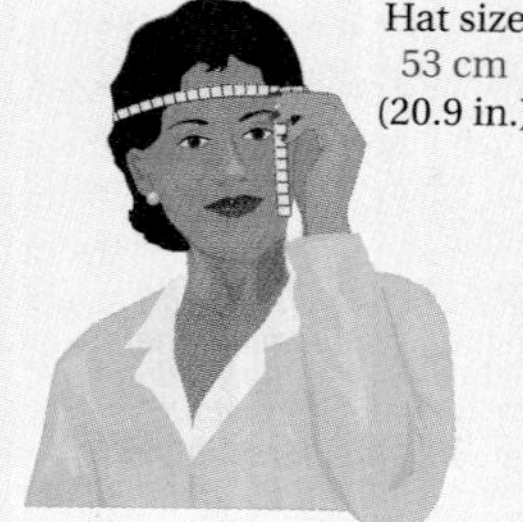

The millimeter (mm) is used to measure small distances, especially in industry. In many countries, the centimeter (cm) is used for body dimensions and clothing sizes.

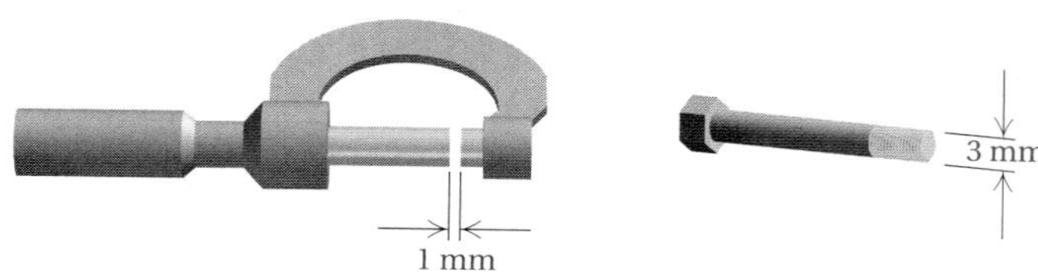

The meter (m) is used for expressing dimensions of larger objects—say, the length of a building—and for shorter distances, such as the length of a rug.

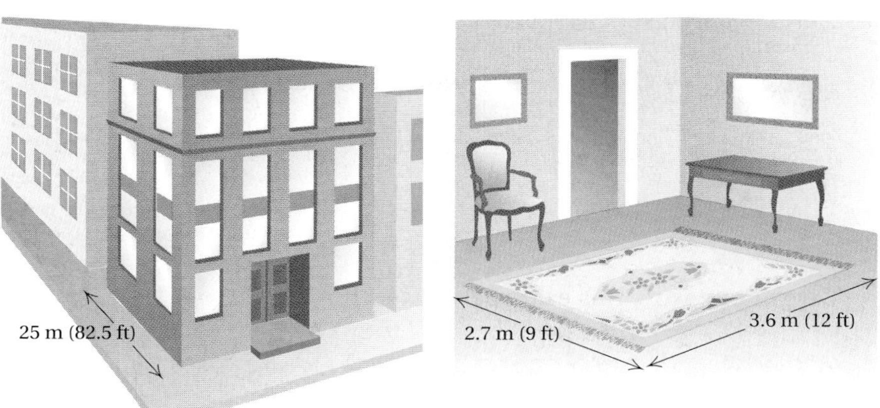

The kilometer (km) is used for longer distances, mostly in cases where miles are now being used.

Do Exercises 14–19.

Example 5 Complete: 4 km = _______ m.

$$\begin{aligned} 4\text{ km} &= 4 \times 1\text{ km} \\ &= 4 \times 1000\text{ m} \qquad \text{Substituting 1000 m for 1 km} \\ &= 4000\text{ m} \end{aligned}$$

Do Exercises 20 and 21.

Since

$$\frac{1}{10}\text{ m} = 1\text{ dm}, \quad \frac{1}{100}\text{ m} = 1\text{ cm}, \quad \text{and} \quad \frac{1}{1000}\text{ m} = 1\text{ mm},$$

it follows that

> 1 m = 10 dm, 1 m = 100 cm, and 1 m = 1000 mm.

Memorizing these will help you to write forms of 1 when making conversions.

Complete with mm, cm, m, or km.

14. A stick of gum is 7 _______ long.

15. Minneapolis is 3213 _______ from San Francisco.

16. A penny is 1 _______ thick.

17. The halfback ran 7 _______.

18. The book is 3 _______ thick.

19. The desk is 2 _______ long.

Complete.

20. 23 km = _______ m

21. 4 hm = _______ m

Answers on page A-52

Complete.

22. 1.78 m = ________ cm

23. 9.04 m = ________ mm

Complete.

24. 7814 m = ________ km

25. 7814 m = ________ dam

Answers on page A-52

Example 6 Complete: 93.4 m = ________ cm.

We want to convert from "m" to "cm." We multiply by 1 using a symbol for 1 with "m" on the bottom and "cm" on the top to eliminate meters and convert to centimeters:

$$93.4 \text{ m} = 93.4 \text{ m} \times \frac{100 \text{ cm}}{1 \text{ m}}$$ **Multiplying by 1 using $\frac{100 \text{ cm}}{1 \text{ m}}$**

$$= 93.4 \times 100 \times \frac{\text{m}}{\text{m}} \times 1 \text{ cm}$$ **The $\frac{\text{m}}{\text{m}}$ acts like 1, so we omit it.**

$$= 9340 \text{ cm.}$$ **Multiplying by 100 moves the decimal point two places to the right.**

We can also work this example by canceling:

$$93.4 \text{ m} = 93.4 \not{\text{m}} \times \frac{100 \text{ cm}}{1 \not{\text{m}}}$$

$$= 93.4 \times 100 \times 1 \text{ cm}$$

$$= 9340 \text{ cm.}$$

Example 7 Complete: 0.248 m = ________ mm.

We are converting from "m" to "mm," so we choose a symbol for 1 with "mm" on the top and "m" on the bottom:

$$0.248 \text{ m} = 0.248 \text{ m} \times \frac{1000 \text{ mm}}{1 \text{ m}}$$ **Multiplying by 1 using $\frac{1000 \text{ mm}}{1 \text{ m}}$**

$$= 0.248 \times 1000 \times \frac{\text{m}}{\text{m}} \times 1 \text{ mm}$$ **The $\frac{\text{m}}{\text{m}}$ acts like 1, so we omit it.**

$$= 248 \text{ mm.}$$ **Multiplying by 1000 moves the decimal point three places to the right.**

Using canceling, we can work this example as follows:

$$0.248 = 0.248 \not{\text{m}} \times \frac{1000 \text{ mm}}{1 \not{\text{m}}}$$

$$= 0.248 \times 1000 \times 1 \text{ mm} = 248 \text{ mm.}$$

Do Exercises 22 and 23.

Example 8 Complete: 2347 m = ________ km.

$$2347 \text{ m} = 2347 \text{ m} \times \frac{1 \text{ km}}{1000 \text{ m}}$$ **Multiplying by 1 using $\frac{1 \text{ km}}{1000 \text{ m}}$**

$$= \frac{2347}{1000} \times \frac{\text{m}}{\text{m}} \times 1 \text{ km}$$ **The $\frac{\text{m}}{\text{m}}$ acts like 1, so we omit it.**

$$= 2.347 \text{ km}$$ **Dividing by 1000 moves the decimal point three places to the left.**

Using canceling, we can work this example as follows:

$$2347 \text{ m} = 2347 \not{\text{m}} \times \frac{1 \text{ km}}{1000 \not{\text{m}}} = \frac{2347}{1000} \times 1 \text{ km} = 2.347 \text{ km.}$$

Do Exercises 24 and 25.

Sometimes we multiply by 1 more than once.

Example 9 Complete: 8.42 mm = ______ cm.

$$8.42 \text{ mm} = 8.42 \text{ mm} \times \frac{1 \text{ m}}{1000 \text{ mm}} \times \frac{100 \text{ cm}}{1 \text{ m}}$$

Multiplying by 1 using $\frac{1 \text{ m}}{1000 \text{ mm}}$ **and** $\frac{100 \text{ cm}}{1 \text{ m}}$

$$= \frac{8.42 \times 100}{1000} \times \frac{\text{mm}}{\text{mm}} \times \frac{\text{m}}{\text{m}} \times 1 \text{ cm}$$

$$= \frac{842}{1000} \text{ cm} = 0.842 \text{ cm}$$

Do Exercises 26 and 27.

Complete.

26. 9.67 mm = ______ cm

27. 89 km = ______ cm

Mental Conversion

Look back over the examples and exercises done so far and you will see that changing from one unit to another in the metric system amounts to only the movement of a decimal point. That is because the metric system is based on 10. Let's find a faster way to convert. Look at the following table.

1000 m	100 m	10 m	1 m	0.1 m	0.01 m	0.001 m
1 km	1 hm	1 dam	1 m	1 dm	1 cm	1 mm

Each place in the table has a value $\frac{1}{10}$ that to the left or 10 times that to the right. Thus moving one place in the table corresponds to one decimal place. Let's convert mentally.

Example 10 Complete: 8.42 mm = ______ cm.

Think: To go from mm to cm in the table is a move of one place to the left. Thus we move the decimal point one place to the left.

1000 m	100 m	10 m	1 m	0.1 m	0.01 m	0.001 m
1 km	1 hm	1 dam	1 m	1 dm	1 cm	1 mm

1 place to the left

8.42 0.8.42 8.42 mm = 0.842 cm

Example 11 Complete: 1.886 km = ______ cm.

Think: To go from km to cm is a move of five places to the right. Thus we move the decimal point five places to the right.

1000 m	100 m	10 m	1 m	0.1 m	0.01 m	0.001 m
1 km	1 hm	1 dam	1 m	1 dm	1 cm	1 mm

5 places to the right

1.886 1.88600. 1.886 km = 188,600 cm

Answers on page A-52

Complete. Try to do this mentally using the table.

28. 6780 m = ________ km

29. 9.74 cm = ________ mm

30. 1 mm = ________ cm

31. 845.1 mm = ________ dm

Complete.

32. 100 yd = ________ m
(The length of a football field)

33. 500 mi = ________ km
(The Indianapolis 500-mile race)

34. 2383 km = ________ mi
(The distance from St. Louis to Phoenix)

Answers on page A-52

Example 12 Complete: 3 m = ________ cm.

Think: To go from m to cm in the table is a move of two places to the right. Thus we move the decimal point two places to the right.

1000 m	100 m	10 m	1 m	0.1 m	0.01 m	0.001 m
1 km	1 hm	1 dam	1 m	1 dm	1 cm	1 mm

2 places to the right

3   3.00.   3 m = 300 cm

You should try to make metric conversions mentally as much as possible.

The fact that conversions can be done so easily is an important advantage of the metric system. The most commonly used metric units of length are km, m, cm, and mm. We have purposely used these more often than the others in the exercises.

Do Exercises 28–31.

c Converting Between American and Metric Units

We can make conversions between American and metric units by using the following table. Again, we either make a substitution or multiply by 1 appropriately.

Metric	American
1 m	39.37 in.
1 m	3.3 ft
0.303 m	1 ft
2.54 cm	1 in.
1 km	0.621 mi
1.609 km	1 mi

Example 13 Complete: 26.2 mi = ________ km. (This is the length of the Olympic marathon.)

$$\begin{aligned} 26.2 \text{ mi} &= 26.2 \times 1 \text{ mi} \\ &\approx 26.2 \times 1.609 \text{ km} \\ &\approx 42.1558 \text{ km} \end{aligned}$$

Example 14 Complete: 100 m = ________ yd. (This is the length of a dash in track.)

$$\begin{aligned} 100 \text{ m} &= 100 \times 1 \text{ m} \approx 100 \times 3.3 \text{ ft} \approx 330 \text{ ft} \\ &\approx 330 \cancel{\text{ft}} \times \frac{1 \text{ yd}}{3 \cancel{\text{ft}}} \approx \frac{330}{3} \text{ yd} \approx 110 \text{ yd} \end{aligned}$$

Do Exercises 32–34.

Exercise Set A

a Complete.

1. 1 ft = ______ in.
2. 1 yd = ______ ft
3. 1 in. = ______ ft
4. 1 mi = ______ yd
5. 1 mi = ______ ft
6. 1 ft = ______ yd
7. 13 yd = ______ in.
8. 10 yd = ______ ft
9. 84 in. = ______ ft
10. 29 ft = ______ yd
11. 3 mi = ______ ft
12. 3 mi = ______ yd
13. 3 in. = ______ ft
14. 4.6 yd = ______ ft
15. 10 ft = ______ yd
16. 15,840 ft = ______ mi
17. $4\frac{1}{2}$ ft = ______ yd
18. 10 yd = ______ in.
19. 36 in. = ______ yd
20. 240 in. = ______ ft
21. 330 ft = ______ yd
22. 1760 yd = ______ mi
23. 3520 yd = ______ mi
24. 25 mi = ______ ft
25. 100 yd = ______ ft
26. 360 in. = ______ ft
27. 1 in. = ______ yd
28. 13 in. = ______ ft
29. 2 mi = ______ in.
30. 63,360 in. = ______ mi

b Complete. Do as much as possible mentally.

31. a) 1 km = ______ m

    b) 1 m = ______ km

32. a) 1 hm = ______ m

    b) 1 m = ______ hm

33. a) 1 dam = ______ m

    b) 1 m = ______ dam

34. a) 1 dm = ______ m

    b) 1 m = ______ dm

35. a) 1 cm = ______ m

    b) 1 m = ______ cm

36. a) 1 mm = ______ m

    b) 1 m = ______ mm

37. 6.7 km = ______ m
38. 0.233 cm = ______ m
39. 98 cm = ______ m

40. 6770 m = ________ km

41. 8921 m = ________ km

42. 435 m = ________ cm

43. 56.66 m = ________ km

44. 5.666 m = ________ km

45. 5666 m = ________ cm

46. 3.45 mm = ________ m

47. 6.88 m = ________ cm

48. 6.88 m = ________ dm

49. 1 mm = ________ cm

50. 1 cm = ________ km

51. 1 km = ________ cm

52. 13.8 cm = ________ mm

53. 14.2 cm = ________ mm

54. 7.3 mm = ________ cm

55. 4500 mm = ________ cm

56. 6,000,000 m = ________ km

57. 0.024 mm = ________ m

58. 6.88 m = ________ hm

59. 6.88 m = ________ dam

60. 7 km = ________ hm

61. 392 dam = ________ km

62. 0.013 mm = ________ dm

C Complete.

63. 330 ft = ________ m
(The length of most baseball foul lines)

64. 12 in. = ________ cm

65. 1171.352 km = ________ mi
(The distance from Cleveland to Atlanta)

66. 2 m = ________ ft
(The length of a desk)

67. 65 mph = ________ km/h
(A common speed limit in the United States)

68. 100 km/h = ________ mph
(A common speed limit in Canada)

69. 180 mi = ________ km
(The distance from Indianapolis to Chicago)

70. 141,600,000 mi = ________ km
(The farthest distance of Mars from the sun)

71. 70 mph = ________ km/h
(An interstate speed limit in the United States)

72. 60 km/h = ________ mph
(A city speed limit in Canada)

73. 10 yd = ________ m
(The length needed for a first down in football)

74. 450 ft = ________ m
(The length of a long home run in baseball)

Appendix B Capacity, Weight, Mass, and Time

a Capacity

To answer a question like "How much soda is in the can?" we need measures of **capacity**. American units of capacity are ounces, or fluid ounces, cups, pints, quarts, and gallons. These units are related as follows.

AMERICAN UNITS OF CAPACITY

1 gallon (gal) = 4 quarts (qt) 1 pt = 2 cups = 16 ounces (oz)
1 qt = 2 pints (pt) 1 cup = 8 oz

Example 1 Complete: 9 gal = ______ oz.

We convert as follows:

$$\begin{aligned} 9 \text{ gal} &= 9 \cdot 1 \text{ gal} \\ &= 9 \cdot 4 \text{ qt} && \text{Substituting 4 qt for 1 gal} \\ &= 9 \cdot 4 \cdot 1 \text{ qt} \\ &= 9 \cdot 4 \cdot 2 \text{ pt} && \text{Substituting 2 pt for 1 qt} \\ &= 9 \cdot 4 \cdot 2 \cdot 1 \text{ pt} \\ &= 9 \cdot 4 \cdot 2 \cdot 16 \text{ oz} && \text{Substituting 16 oz for 1 pt} \\ &= 1152 \text{ oz.} \end{aligned}$$

Do Exercise 1.

Example 2 Complete: 24 qt = ______ gal.

In this case, we multiply by 1 using 1 gal in the numerator, since we are converting to gallons, and 4 qt in the denominator, since we are converting from quarts.

$$24 \text{ qt} = 24 \cancel{\text{qt}} \cdot \frac{1 \text{ gal}}{4 \cancel{\text{qt}}} = \frac{24}{4} \cdot 1 \text{ gal} = 6 \text{ gal}$$

After completing Example 2, we can check whether the answer is reasonable. We are converting from smaller to larger units, so our answer has fewer larger units.

Do Exercise 2.

Thinking Metric

One unit of capacity in the metric system is a **liter**. A liter is just a bit more than a quart. It is defined as follows.

Objectives

- a Convert from one unit of capacity to another.
- b Solve applied problems involving capacity.
- c Convert from one American unit of weight to another.
- d Convert from one metric unit of mass to another.
- e Convert from one unit of time to another.

1. Complete: 5 gal = ______ pt.

2. Complete:
80 qt = ______ gal

Answers on page A-52

Complete with mL or L.

3. The patient received an injection of 2 ________ of penicillin.

4. There are 250________ in a coffee cup.

5. The gas tank holds 80 ________.

6. Bring home 8 ________ of milk.

Complete.

7. 0.97 L = ________ mL

8. 8990 mL = ________ L

9. A physician ordered 4800 mL of 0.9% saline solution. How many liters were ordered?

Answers on page A-52

> **METRIC UNITS OF CAPACITY**
>
> 1 liter (L) = 1000 cubic centimeters (1000 cm^3)
>
> The script letter ℓ is also used for "liter."

The metric prefixes are also used with liters. The most common is **milli-**. The milliliter (mL) is, then, $\frac{1}{1000}$ liter. Thus,

> 1 L = 1000 mL = 1000 cm^3;
>
> 0.001 L = 1 mL = 1 cm^3.

A preferred unit for drug dosage is the milliliter (mL) or the cubic centimeter (cm^3). The notation "cc" is also used for cubic centimeter, especially in medicine. The milliliter and the cubic centimeter represent the same measure of capacity. A milliliter is about $\frac{1}{5}$ of a teaspoon.

> 1 mL = 1 cm^3 = 1 cc

Volumes for which quarts and gallons are used are expressed in liters. Large volumes in business and industry are expressed using measures of cubic meters (m^3).

Do Exercises 3–6.

Example 3 Complete: 4.5 L = ________ mL.

$$\begin{aligned} 4.5\text{ L} &= 4.5 \times 1\text{ L} \\ &= 4.5 \times 1000\text{ mL} \qquad \textbf{Substituting 1000 mL for 1 L} \\ &= 4500\text{ mL} \end{aligned}$$

Example 4 Complete: 280 mL = ________ L.

$$\begin{aligned} 280\text{ mL} &= 280 \times 1\text{ mL} \\ &= 280 \times 0.001\text{ L} \qquad \textbf{Substituting 0.001 L for 1 mL} \\ &= 0.28\text{ L} \end{aligned}$$

Do Exercises 7 and 8.

b Solving Applied Problems

The metric system has extensive usage in medicine.

Example 5 *Medical Dosage.* A physician ordered 3.5 L of 5% dextrose in water. How many milliliters were ordered?

We convert 3.5 L to milliliters:

$$3.5\text{ L} = 3.5 \times 1\text{ L} = 3.5 \times 1000\text{ mL} = 3500\text{ mL}.$$

The physician ordered 3500 mL.

Do Exercise 9.

Example 6 *Medical Dosage.* In pharmaceutical work, liquids at the drugstore are given in liters or milliliters, but a physician's prescription is given in ounces. For conversion, a druggist knows that 1 oz = 29.57 mL. A prescription calls for 3 oz of ephedrine. For how many milliliters is the prescription?

We convert as follows:

$$3 \text{ oz} = 3 \times 1 \text{ oz} = 3 \times 29.57 \text{ mL} = 88.71 \text{ mL}.$$

The prescription calls for 88.71 mL of ephedrine.

Do Exercise 10.

Example 7 *Gasoline Prices.* At a self-service gasoline station, regular lead-free gasoline sells for 29.3¢ a liter. Estimate the cost of 1 gal in dollars.

Since 1 L is about 1 qt and there are 4 qt in a gallon, the price of a gallon is about 4 times the price of a liter:

$$4 \times 29.3¢ = 117.2¢ = \$1.172.$$

Thus regular lead-free gasoline sells for about $1.17 a gallon.

Do Exercise 11.

c Weight: The American System

The American units of weight are as follows.

> **AMERICAN UNITS OF WEIGHT**
>
> 1 ton (T) = 2000 pounds (lb)
>
> 1 lb = 16 ounces (oz)

Example 8 A well-known hamburger is called a "quarter-pounder." Find its name in ounces: a "________ ouncer."

$$\frac{1}{4} \text{ lb} = \frac{1}{4} \cdot 1 \text{ lb}$$
$$= \frac{1}{4} \cdot 16 \text{ oz} \qquad \text{Substituting 16 oz for 1 lb}$$
$$= 4 \text{ oz}$$

A "quarter-pounder" can also be called a "four-ouncer."

Example 9 Complete: 15,360 lb = ________ T.

$$15{,}360 \text{ lb} = 15{,}360 \cancel{\text{lb}} \times \frac{1 \text{ T}}{2000 \cancel{\text{lb}}} \qquad \text{Multiplying by 1}$$
$$= \frac{15{,}360}{2000} \text{ T}$$
$$= 7.68 \text{ T}$$

Do Exercises 12–14.

10. A prescription calls for 4 oz of ephedrine.

a) For how many milliliters is the prescription?

b) For how many liters is the prescription?

11. At the same station, the price of premium lead-free gasoline is 43.9 cents a liter. Estimate the price of 1 gal in dollars.

Complete.

12. 5 lb = ________ oz

13. 8640 lb = ________ T

14. 1 T = ________ oz

Answers on page A-52

d Mass: The Metric System

There is a difference between **mass** and **weight**, but the terms are often used interchangeably. People sometimes use the word "weight" instead of "mass." Weight is related to the force of gravity. The farther you are from the center of the earth, the less you weigh. Your mass stays the same no matter where you are.

The basic unit of mass is the **gram** (g), which is the mass of 1 cubic centimeter (1 cm^3 or 1 mL) of water. Since a cubic centimeter is small, a gram is a small unit of mass.

$$1 \text{ g} = 1 \text{ gram} = \text{the mass of } 1 \text{ cm}^3 \text{ (1 mL) of water}$$

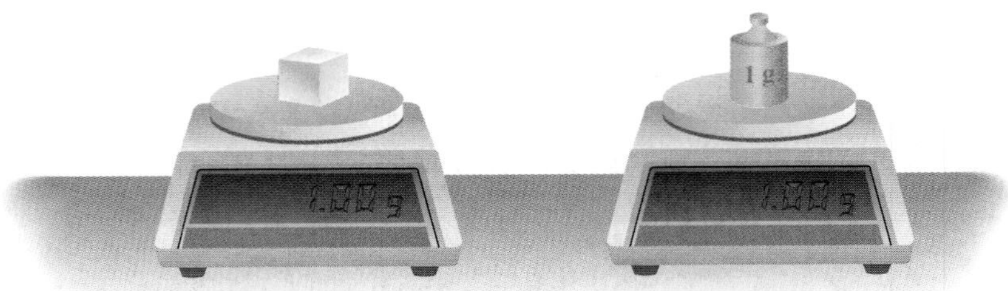

The following table shows the metric units of mass. The prefixes are the same as those for length.

METRIC UNITS OF MASS

1 metric ton (t) = 1000 kilograms (kg)
1 *kilo*gram (kg) = 1000 grams (g)
1 *hecto*gram (hg) = 100 grams (g)
1 *deka*gram (dag) = 10 grams (g)
1 gram (g)
1 *deci*gram (dg) = $\frac{1}{10}$ gram (g)
1 *centi*gram (cg) = $\frac{1}{100}$ gram (g)
1 *milli*gram (mg) = $\frac{1}{1000}$ gram (g)

Thinking Metric

One gram is about the mass of 1 raisin or 1 paperclip. Since 1 kg is about 2.2 lb, 1000 kg is about 2200 lb, or 1 metric ton (t), which is just a little more than 1 American ton (T).

1 gram

1 kilogram

1 pound

Small masses, such as dosages of medicine and vitamins, may be measured in milligrams (mg). The gram (g) is used for objects ordinarily measured in ounces, such as the mass of a letter, a piece of candy, a coin, or a small package of food.

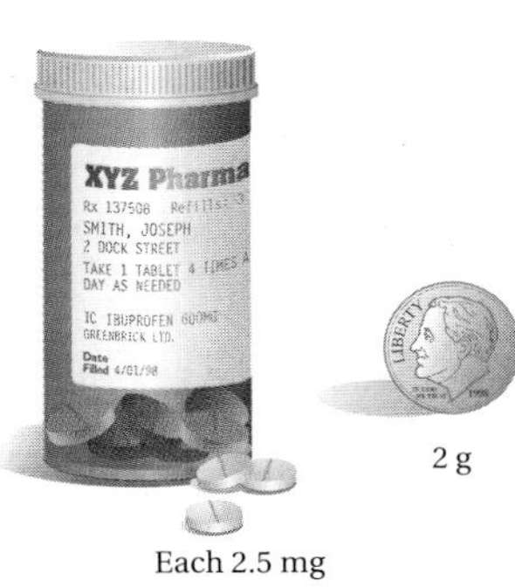

Each 2.5 mg

2 g

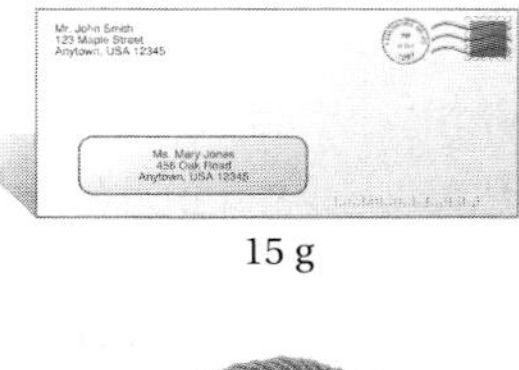
15 g

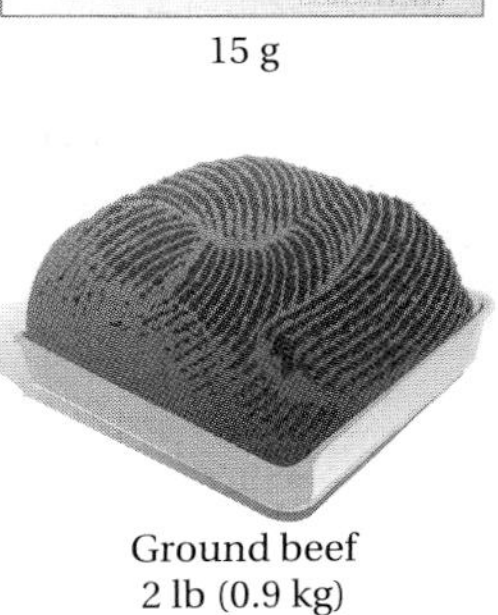
Ground beef
2 lb (0.9 kg)

90 kg

The kilogram (kg) is used for larger food packages, such as meat, or for human body mass. The metric ton (t) is used for very large masses, such as the mass of an automobile, a truckload of gravel, or an airplane.

Do Exercises 15–19.

Changing Units Mentally

As before, changing from one metric unit to another amounts to only the movement of a decimal point. We use this table.

1000 g	100 g	10 g	1 g	0.1 g	0.01 g	0.001 g
1 kg	1 hg	1 dag	1 g	1 dg	1 cg	1 mg

Example 10 Complete: 8 kg = ______ g.

Think: To go from kg to g in the table is a move of three places to the right. Thus we move the decimal point three places to the right.

8.0 8.000. 8 kg = 8000 g

Example 11 Complete: 4235 g = ______ kg.

Think: To go from g to kg in the table is a move of three places to the left. Thus we move the decimal point three places to the left.

4235.0 4.235.0 4235 g = 4.235 kg

Do Exercises 20 and 21.

Example 12 Complete: 6.98 cg = ______ mg.

Think: To go from cg to mg is a move of one place to the right. Thus we move the decimal point one place to the right.

6.98 6.9.8 6.98 cg = 69.8 mg

The most commonly used metric units of mass are kg, g, cg, and mg. We have purposely used those more often than the others in the exercises.

Complete with mg, g, kg, or t.

15. A laptop computer has a mass of 6 ______.

16. That person has a body mass of 85.4 ______.

17. This is a 3-______ vitamin.

18. A pen has a mass of 12 ______.

19. A minivan has a mass of 3 ______.

Complete.

20. 6.2 kg = ______ g

21. 304.8 cg = ______ g

Answers on page A-52

Complete.

22. 7.7 cg = ________ mg

23. 2344 mg = ________ cg

24. 67 dg = ________ mg

Complete.

25. 2 hr = ________ min

26. 4 yr = ________ days

27. 1 day = ________ min

28. 168 hr = ________ wk

Answers on page A-52

Example 13 Complete: 89.21 mg = ________ g.

Think: To go from mg to g is a move of three places to the left. Thus we move the decimal point three places to the left.

89.21 0.089.21 89.21 mg = 0.08921 g

Do Exercises 22–24.

e Time

A table of units of time is shown below. The metric system sometimes uses "h" for hour and "s" for second, but we will use the more familiar "hr" and "sec."

UNITS OF TIME

1 day = 24 hours (hr)	1 year (yr) = $365\frac{1}{4}$ days
1 hr = 60 minutes (min)	1 week (wk) = 7 days
1 min = 60 seconds (sec)	

Since we cannot have $\frac{1}{4}$ day on the calendar, we give each year 365 days and every fourth year 366 days (a leap year), unless it is a year at the beginning of a century not divisible by 400.

Example 14 Complete: 1 hr = ________ sec.

$$\begin{aligned} 1 \text{ hr} &= 60 \text{ min} \\ &= 60 \cdot 1 \text{ min} \\ &= 60 \cdot 60 \text{ sec} \qquad \text{Substituting 60 sec for 1 min} \\ &= 3600 \text{ sec} \end{aligned}$$

Example 15 Complete: 5 yr = ________ days.

$$\begin{aligned} 5 \text{ yr} &= 5 \cdot 1 \text{ yr} \\ &= 5 \cdot 365\frac{1}{4} \text{ days} \qquad \text{Substituting } 365\tfrac{1}{4} \text{ days for 1 yr} \\ &= 1826\frac{1}{4} \text{ days} \end{aligned}$$

Example 16 Complete: 4320 min = ________ days.

$$4320 \text{ min} = 4320 \cancel{\text{min}} \cdot \frac{1 \cancel{\text{hr}}}{60 \cancel{\text{min}}} \cdot \frac{1 \text{ day}}{24 \cancel{\text{hr}}} = \frac{4320}{60 \cdot 24} \text{ days} = 3 \text{ days}$$

Do Exercises 25–28.

Exercise Set B

a Complete.

1. 1 L = ______ mL = ______ cm^3

2. ______ L = 1 mL = ______ cm^3

3. 87 L = ______ mL

4. 806 L = ______ mL

5. 49 mL = ______ L

6. 19 mL = ______ L

7. 0.401 mL = ______ L

8. 0.816 mL = ______ L

9. 78.1 L = ______ cm^3

10. 99.6 L = ______ cm^3

11. 10 qt = ______ oz

12. 9.6 oz = ______ pt

13. 20 cups = ______ pt

14. 1 gal = ______ oz

15. 8 gal = ______ qt

16. 1 gal = ______ cups

b Solve.

17. *Medical Dosage.* A physician ordered 0.5 L of normal saline solution. How many milliliters were ordered?

18. *Medical Dosage.* A patient receives 84 mL per hour of normal saline solution. How many liters did the patient receive in a 24-hr period?

19. *Medical Dosage.* A doctor wants a patient to receive 3 L of a normal saline solution in a 24-hr period. How many milliliters per hour must the nurse administer?

20. *Medical Dosage.* A doctor tells a patient to purchase 0.5 L of hydrogen peroxide. Commercially, hydrogen peroxide is found on the shelf in bottles that hold 4 oz, 8 oz, and 16 oz. Which bottle comes closest to filling the prescription? (1 qt = 32 oz)

21. *Wasting Water.* Many people leave the water running while they are brushing their teeth. Suppose that 32 oz of water is wasted in such a way each day by one person. How much water, in gallons, is wasted in a week? in a month (30 days)? in a year? Assuming each of the 261 million people in this country wastes water in this way, estimate how much water is wasted in a day; in a year.

22. *World's Gold.* If all the gold in the world could be gathered together, it would form a cube 18 yd on a side. Find the volume of the world's gold.

c Complete.

23. 1 T = _______ lb

24. 1 lb = _______ oz

25. 6000 lb = _______ T

26. 8 T = _______ lb

27. 4 lb = _______ oz

28. 10 lb = _______ oz

29. 6.32 T = _______ lb

30. 8.07 T = _______ lb

31. 3200 oz = _______ T

32. 6400 oz = _______ T

33. 80 oz = _______ lb

34. 960 oz = _______ lb

d Complete.

35. 1 kg = _______ g

36. 1 hg = _______ g

37. 1 dag = _______ g

38. 1 dg = _______ g

39. 1 cg = _______ g

40. 1 mg = _______ g

41. 1 g = _______ mg

42. 1 g = _______ cg

43. 1 g = _______ dg

44. 25 kg = _______ g

45. 234 kg = _______ g

46. 9403 g = _______ kg

47. 5200 g = _______ kg

48. 1.506 kg = _______ g

49. 67 hg = _______ kg

50. 45 cg = _______ g

51. 0.502 dg = _______ g

52. 0.0025 cg = _______ mg

53. 8492 g = _______ kg

54. 9466 g = _______ kg

55. 585 mg = _______ cg

56. 96.1 mg = ________ cg

57. 8 kg = ________ cg

58. 0.06 kg = ________ mg

59. 1 t = ________ kg

60. 2 t = ________ kg

e Complete.

61. 1 day = ________ hr

62. 1 hr = ________ min

63. 1 min = ________ sec

64. 1 wk = ________ days

65. 1 yr = ________ days

66. 2 yr = ________ days

67. 180 sec = ________ hr

68. 60 sec = ________ hr

69. 492 sec = ________ min (the amount of time it takes for the rays of the sun to reach the earth)

70. 18,000 sec = ________ hr

71. 156 hr = ________ days

72. 444 hr = ________ days

73. 645 min = ________ hr

74. 375 min = ________ hr

75. 2 wk = ________ hr

76. 4 hr = ________ sec

77. 756 hr = ________ wk

78. 166,320 min = ________ wk

79. 2922 wk = ________ yr

80. 623 days = ________ wk

Synthesis

81. ◈ Give at least two reasons why someone might prefer the use of grams to the use of ounces.

82. ◈ Describe a situation in which one object weighs 70 kg, another object weighs 3 g, and a third object weighs 125 mg.

Complete. Use 1 kg = 2.205 lb and 453.5 g = 1 lb. Round to four decimal places.

83. 🖩 1 lb = ________ kg

84. 🖩 1 g = ________ lb

85. At \$0.90 a dozen, the cost of eggs is \$0.60 per pound. How much does an egg weigh?

86. Estimate the number of years in one million seconds.

87. Estimate the number of years in one billion seconds.

88. Estimate the number of years in one trillion seconds.

Medical Applications. Another metric unit used in medicine is the microgram (μg). It is defined as follows.

> 1 microgram $= 1\ \mu g = \dfrac{1}{1{,}000{,}000}\ g$; $1{,}000{,}000\ \mu g = 1\ g$

Thus a microgram is one millionth of a gram, and one million micrograms is one gram.

Complete.

89. 1 mg = ______ μg

90. 1 μg = ______ mg

91. A physician orders 125 μg of digoxin. For how many milligrams is the prescription?

92. A physician orders 0.25 mg of reserpine. For how many micrograms is the prescription?

93. A medicine called sulfisoxazole usually comes in tablets that are 500 mg each. A standard dosage is 2 g. How many tablets would have to be taken in order to achieve this dosage?

94. Quinidine is a liquid mixture, part medicine and part water. There is 80 mg of Quinidine for every milliliter of liquid. A standard dosage is 200 mg. How much of the liquid mixture would be required in order to achieve the dosage?

95. A medicine called cephalexin is obtainable in a liquid mixture, part medicine and part water. There is 250 mg of cephalexin in 5 mL of liquid. A standard dosage is 400 mg. How much of the liquid would be required in order to achieve the dosage?

96. A medicine called Albuterol is used for the treatment of asthma. It typically comes in an inhaler that contains 18 g. One actuation, or inhalation, is 90 mg.

a) How many actuations are in one inhaler?
b) A student is going away for 4 months of college and wants to take enough Albuterol to last for that time. Assuming that she will need 4 actuations per day, estimate about how many inhalers the student will need for the 4-month period.

Appendix C Factoring Sums or Differences of Cubes

Objective

[a] Factor sums and differences of two cubes.

[a] We can factor the sum or the difference of two expressions that are cubes.

Consider the following products:

$$(A + B)(A^2 - AB + B^2) = A(A^2 - AB + B^2) + B(A^2 - AB + B^2)$$
$$= A^3 - A^2B + AB^2 + A^2B - AB^2 + B^3$$
$$= A^3 + B^3$$

and
$$(A - B)(A^2 + AB + B^2) = A(A^2 + AB + B^2) - B(A^2 + AB + B^2)$$
$$= A^3 + A^2B + AB^2 - A^2B - AB^2 - B^3$$
$$= A^3 - B^3.$$

The above equations (reversed) show how we can factor a sum or a difference of two cubes.

$$A^3 + B^3 = (A + B)(A^2 - AB + B^2),$$
$$A^3 - B^3 = (A - B)(A^2 + AB + B^2)$$

N	N^3
0.2	0.008
0.1	0.001
0	0
1	1
2	8
3	27
4	64
5	125
6	216
7	343
8	512
9	729
10	1000

Note that what we are considering here is a sum or a difference of cubes. We are not cubing a binomial. For example, $(A + B)^3$ is *not* the same as $A^3 + B^3$. The table of cubes in the margin is helpful.

Example 1 Factor: $x^3 - 27$.

We have

$$x^3 - 27 = x^3 - 3^3.$$

In one set of parentheses, we write the cube root of the first term, x. Then we write the cube root of the second term, -3. This gives us the expression $x - 3$:

$$(x - 3)(\qquad\qquad).$$

To get the next factor, we think of $x - 3$ and do the following:

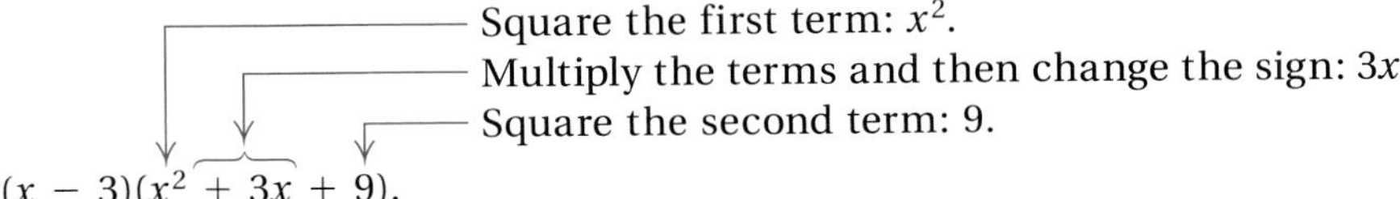

Square the first term: x^2.
Multiply the terms and then change the sign: $3x$.
Square the second term: 9.

$$(x - 3)(x^2 + 3x + 9).$$

Note that we cannot factor $x^2 + 3x + 9$. It is not a trinomial square nor can it be factored by trial and error.

Factor.

1. $x^3 - 8$

2. $64 - y^3$

Do Exercises 1 and 2.

Example 2 Factor: $125x^3 + y^3$.

We have

$$125x^3 + y^3 = (5x)^3 + y^3.$$

In one set of parentheses, we write the cube root of the first term, $5x$. Then we write a plus sign, and then the cube root of the second term, y:

$$(5x + y)(\qquad\qquad).$$

Answers on page A-52

Factor.

3. $27x^3 + y^3$

4. $8y^3 + z^3$

To get the next factor, we think of $5x + y$ and do the following:

Square the first term: $25x^2$.
Multiply the terms and then change the sign: $-5xy$.
Square the second term: y^2.

$(5x + y)(25x^2 - 5xy + y^2)$.

Do Exercises 3 and 4.

Example 3 Factor: $128y^7 - 250x^6y$.

We first look for a common factor:

$$128y^7 - 250x^6y = 2y(64y^6 - 125x^6) = 2y[(4y^2)^3 - (5x^2)^3]$$
$$= 2y(4y^2 - 5x^2)(16y^4 + 20x^2y^2 + 25x^4).$$

Example 4 Factor: $a^6 - b^6$.

We can express this polynomial as a difference of squares:

$$(a^3)^2 - (b^3)^2.$$

We factor as follows:

$$a^6 - b^6 = (a^3 + b^3)(a^3 - b^3).$$

One factor is a sum of two cubes, and the other factor is a difference of two cubes. We factor them:

$$(a + b)(a^2 - ab + b^2)(a - b)(a^2 + ab + b^2).$$

We have now factored completely.

In Example 4, had we thought of factoring first as a difference of two cubes, we would have had

$$(a^2)^3 - (b^2)^3 = (a^2 - b^2)(a^4 + a^2b^2 + b^4)$$
$$= (a + b)(a - b)(a^4 + a^2b^2 + b^4).$$

In this case, we might have missed some factors; $a^4 + a^2b^2 + b^4$ can be factored as $(a^2 - ab + b^2)(a^2 + ab + b^2)$, but we probably would not have known to do such factoring.

Example 5 Factor: $64a^6 - 729b^6$.

We have

$$64a^6 - 729b^6 = (8a^3 - 27b^3)(8a^3 + 27b^3) \quad \textbf{Factoring a difference of squares}$$
$$= [(2a)^3 - (3b)^3][(2a)^3 + (3b)^3].$$

Each factor is a sum or a difference of cubes. We factor each:

$$= (2a - 3b)(4a^2 + 6ab + 9b^2)(2a + 3b)(4a^2 - 6ab + 9b^2).$$

Factor.

5. $m^6 - n^6$

6. $16x^7y + 54xy^7$

7. $729x^6 - 64y^6$

8. $x^3 - 0.027$

Sum of cubes:	$A^3 + B^3 = (A + B)(A^2 - AB + B^2)$;
Difference of cubes:	$A^3 - B^3 = (A - B)(A^2 + AB + B^2)$;
Difference of squares:	$A^2 - B^2 = (A + B)(A - B)$;
Sum of squares:	$A^2 + B^2$ cannot be factored using real numbers if the largest common factor has been removed.

Do Exercises 5–8.

Answers on page A-52

Exercise Set C

a Factor.

1. $z^3 + 27$
2. $a^3 + 8$
3. $x^3 - 1$
4. $c^3 - 64$
5. $y^3 + 125$
6. $x^3 + 1$
7. $8a^3 + 1$
8. $27x^3 + 1$
9. $y^3 - 8$
10. $p^3 - 27$
11. $8 - 27b^3$
12. $64 - 125x^3$
13. $64y^3 + 1$
14. $125x^3 + 1$
15. $8x^3 + 27$
16. $27y^3 + 64$
17. $a^3 - b^3$
18. $x^3 - y^3$
19. $a^3 + \frac{1}{8}$
20. $b^3 + \frac{1}{27}$
21. $2y^3 - 128$
22. $3z^3 - 3$
23. $24a^3 + 3$
24. $54x^3 + 2$
25. $rs^3 + 64r$
26. $ab^3 + 125a$
27. $5x^3 - 40z^3$

28. $2y^3 - 54z^3$

29. $x^3 + 0.001$

30. $y^3 + 0.125$

31. $64x^6 - 8t^6$

32. $125c^6 - 8d^6$

33. $2y^4 - 128y$

34. $3z^5 - 3z^2$

35. $z^6 - 1$

36. $t^6 + 1$

37. $t^6 + 64y^6$

38. $p^6 - q^6$

Synthesis

Consider these polynomials:

$(a + b)^3;\quad a^3 + b^3;\quad (a + b)(a^2 - ab + b^2);$
$(a + b)(a^2 + ab + b^2);\quad (a + b)(a + b)(a + b).$

39. Evaluate each polynomial for $a = -2$ and $b = 3$.

40. Evaluate each polynomial for $a = 4$ and $b = -1$.

Factor. Assume that variables in exponents represent natural numbers.

41. $x^{6a} + y^{3b}$

42. $a^3x^3 - b^3y^3$

43. $3x^{3a} + 24y^{3b}$

44. $\frac{8}{27}x^3 + \frac{1}{64}y^3$

45. $\frac{1}{24}x^3y^3 + \frac{1}{3}z^3$

46. $7x^3 - \frac{7}{8}$

47. $(x + y)^3 - x^3$

48. $(1 - x)^3 + (x - 1)^6$

49. $(a + 2)^3 - (a - 2)^3$

50. $y^4 - 8y^3 - y + 8$

Appendix D Equations Involving Absolute Value

Objective

a Solve equations containing absolute-value symbols.

a There are equations that have more than one solution. Examples are equations with absolute value. Remember, the absolute value of a number is its distance from 0 on a number line.

Example 1 Solve: $|x| = 4$. Then graph using the number line.

Note that $|x| = |x - 0|$, so that $|x - 0|$ is the distance from x to 0. Thus solutions of the equation $|x| = 4$, or $|x - 0| = 4$, are those numbers x whose distance from 0 is 4. Those numbers are -4 and 4. The solution set is $\{-4, 4\}$. The graph consists of just two points, as shown.

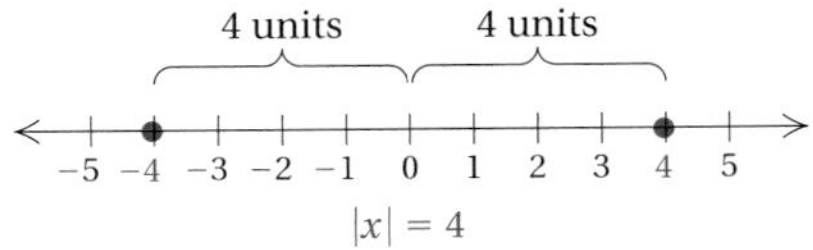

Example 2 Solve: $|x| = 0$.

The only number whose absolute value is 0 is 0 itself. Thus the solution is 0. The solution set is $\{0\}$.

Example 3 Solve: $|x| = -7$.

The absolute value of a number is always nonnegative. There is no number whose absolute value is -7. Thus there is no solution. The solution set is $\varnothing$.

Examples 1–3 lead us to the following principle for solving linear equations with absolute value.

> **THE ABSOLUTE-VALUE PRINCIPLE**
>
> For any positive number p and any algebraic expression X:
>
> **a)** The solutions of $|X| = p$ are those numbers that satisfy $X = -p$ *or* $X = p$.
>
> **b)** The equation $|X| = 0$ is equivalent to the equation $X = 0$.
>
> **c)** The equation $|X| = -p$ has no solution.

Do Exercises 1–3.

We can use the absolute-value principle with the addition and multiplication principles to solve equations with absolute value.

Example 4 Solve: $2|x| + 5 = 9$.

We first use the addition and multiplication principles to get $|x|$ by itself. Then we use the absolute-value principle.

1. Solve: $|x| = 6$. Then graph using a number line.

2. Solve: $|x| = -6$.

3. Solve: $|p| = 0$.

Answers on page A-53

Solve.

4. $|3x| = 6$

5. $4|x| + 10 = 27$

6. $3|x| - 2 = 10$

7. Solve: $|x - 4| = 1$. Use two methods as in Example 5.

Solve.

8. $|3x - 4| = 17$

9. $|6 + 2x| = -3$

Answers on page A-53

$$2|x| + 5 = 9$$

$$2|x| + 5 - 5 = 9 - 5 \quad \text{Subtracting 5}$$

$$2|x| = 4$$

$$|x| = 2 \quad \text{Dividing by 2}$$

$$x = -2 \quad \text{or} \quad x = 2 \quad \text{Using the absolute-value principle}$$

The solutions are -2 and 2. The solution set is $\{-2, 2\}$.

Do Exercises 4–6.

Example 5 Solve: $|x - 2| = 3$.

We can consider solving this equation in two different ways.

METHOD 1 This method allows us to see the meaning of the solutions graphically. The solution set consists of those numbers that are 3 units from 2 on the number line.

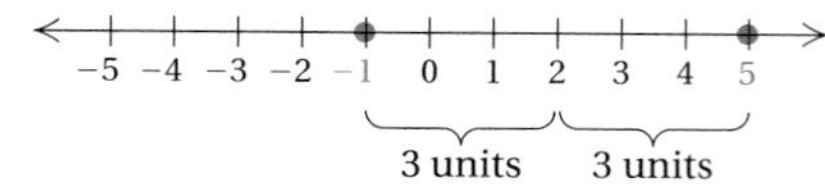

The solutions of $|x - 2| = 3$ are -1 and 5. The solution set is $\{-1, 5\}$.

METHOD 2 This method is more efficient. We use the absolute-value principle, replacing X with $x - 2$ and p with 3. Then we solve each equation separately.

$$|X| = p$$

$$|x - 2| = 3$$

$$x - 2 = -3 \quad or \quad x - 2 = 3 \quad \text{Absolute-value principle}$$

$$x = -1 \quad or \quad x = 5$$

The solutions are -1 and 5. The solution set is $\{-1, 5\}$.

Do Exercise 7.

Example 6 Solve: $|2x + 5| = 13$.

We use the absolute-value principle, replacing X with $2x + 5$ and p with 13:

$$|X| = p$$

$$|2x + 5| = 13$$

$$2x + 5 = -13 \quad or \quad 2x + 5 = 13 \quad \text{Absolute-value principle}$$

$$2x = -18 \quad or \quad 2x = 8$$

$$x = -9 \quad or \quad x = 4.$$

The solutions are -9 and 4. The solution set is $\{-9, 4\}$.

Example 7 Solve: $|4 - 7x| = -8$.

Since absolute value is always nonnegative, this equation has no solution. The solution set is $\varnothing$.

Do Exercises 8 and 9.

Exercise Set D

a Solve.

1. $|x| = 3$

2. $|x| = 5$

3. $|x| = -3$

4. $|x| = -5$

5. $|p| = 0$

6. $|y| = 8.6$

7. $|x - 3| = 12$

8. $|3x - 2| = 6$

9. $|2x - 3| = 4$

10. $|5x + 2| = 3$

11. $|4x - 9| = 14$

12. $|9y - 2| = 17$

13. $|x| + 7 = 18$

14. $|x| - 2 = 6.3$

15. $678 = 289 + |t|$

16. $-567 = -1000 + |x|$

17. $|5x| = 40$

18. $|2y| = 18$

19. $|3x| - 4 = 17$

20. $|6x| + 8 = 32$

21. $5|q| - 2 = 9$

22. $7|z| + 2 = 16$

23. $\left|\frac{2x - 1}{3}\right| = 5$

24. $\left|\frac{4 - 5x}{6}\right| = 7$

25. $|m + 5| + 9 = 16$

26. $|t - 7| - 5 = 4$

27. $10 - |2x - 1| = 4$

28. $2|2x - 7| + 11 = 25$

29. $|3x - 4| = -2$

30. $|x - 6| = -8$

31. $\left|\frac{5}{9} + 3x\right| = \frac{1}{6}$

32. $\left|\frac{2}{3} - 4x\right| = \frac{4}{5}$

Synthesis

33. From the definition of absolute value, $|x| = x$ only when $x \geq 0$. Thus, $|x + 3| = x + 3$ only when $x + 3 \geq 0$ or $x \geq -3$. Solve $|2x - 5| = 2x - 5$ using this same argument.

Solve.

34. $1 - \left|\frac{1}{4}x + 8\right| = \frac{3}{4}$

35. $|x + 5| = x + 5$

36. $|x - 1| = x - 1$

37. $|7x - 2| = x + 4$

Appendix E The Distance Formula and Midpoints

a The Distance Formula

We now develop a formula for finding the distance between any two points on a graph when we know their coordinates. First, we consider points on a vertical or a horizontal line.

If points are on a vertical line, they have the same first coordinate.

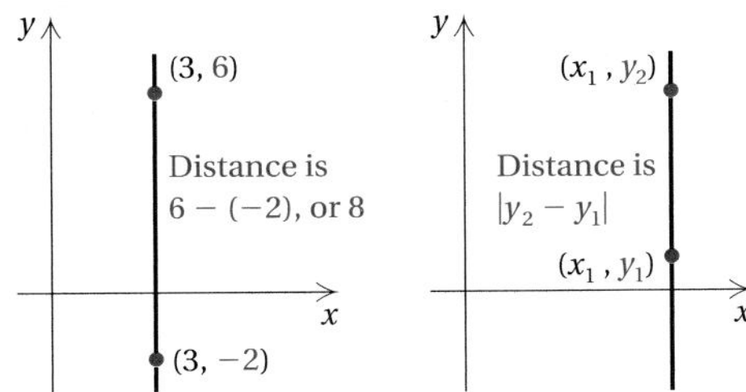

On the left above, we can find the distance between the points by taking the absolute value of the difference of their second coordinates:

$$|6 - (-2)| = |8| = 8. \quad \text{The distance is 8.}$$

If we subtract the opposite way, we get

$$|-2 - 6| = |-8| = 8.$$

If points are on a horizontal line, we take the absolute value of the difference of their first coordinates.

Examples Find the distance between these points.

1. $(-5, 13)$ and $(-5, 2)$

We take the absolute value of the difference of the second coordinates, since the first coordinates are the same. The distance is

$$|13 - 2| = |11| = 11.$$

2. $(7, -3)$ and $(-5, -3)$

Since the second coordinates are the same, we take the absolute value of the difference of the first coordinates. The distance is

$$|-5 - 7| = |-12| = 12.$$

Do Exercises 1 and 2.

Next, we consider two points that are not on either a vertical or a horizontal line, such as (x_1, y_1) and (x_2, y_2) in the figure on the following page. By drawing horizontal and vertical lines through these points, we form a right triangle. The vertex of the right angle of this triangle has coordinates (x_2, y_1). The legs have lengths $|x_2 - x_1|$ and $|y_2 - y_1|$. The distance we want is the length of the hypotenuse, d. By the Pythagorean theorem (Sections 11.8 and 15.6),

$$d^2 = |x_2 - x_1|^2 + |y_2 - y_1|^2.$$

Objectives

a Use the distance formula to find the distance between two points whose coordinates are known.

b Use the midpoint formula to find the midpoint of a segment when the coordinates of its endpoints are known.

Find the distance between the pair of points.

1. $(7, 12)$ and $(7, -2)$

2. $(6, 2)$ and $(-5, 2)$

Answers on page A-53

Find the distance between the pair of points. Where appropriate, find an approximation to three decimal places.

3. (2, 6) and (−4, −2)

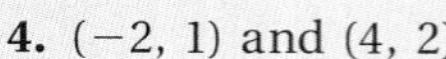

4. (−2, 1) and (4, 2)

Find the midpoint of the segment with the given endpoints.

5. (−3, 1) and (6, −7)

6. (10, −7) and (8, −3)

Answers on page A-53

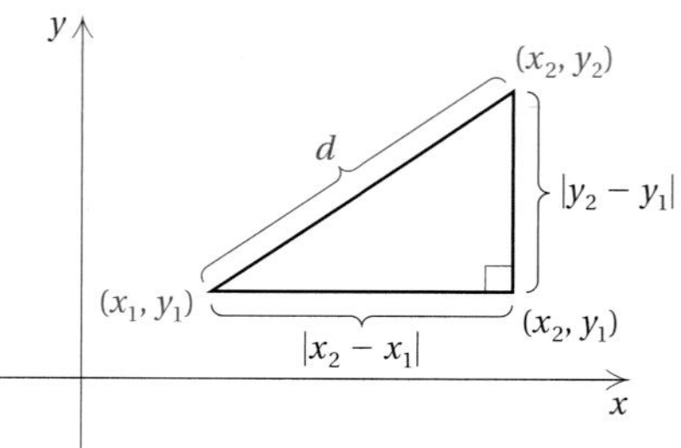

Since squares of numbers are never negative, we don't really need the absolute-value signs. Thus we have

$$d^2 = (x_2 - x_1)^2 + (y_2 - y_1)^2.$$

Taking the principal square root, we get the distance formula.

THE DISTANCE FORMULA

The distance between any two points (x_1, y_1) and (x_2, y_2) is given by

$$d = \sqrt{(x_2 - x_1)^2 + (y_2 - y_1)^2}.$$

This formula holds even when the two points *are* on a vertical or a horizontal line.

Example 3 Find the distance between (4, −3) and (−5, 4). Find an exact answer and an approximation to three decimal places.

We substitute into the distance formula:

$$d = \sqrt{(-5 - 4)^2 + [4 - (-3)]^2}$$
$$= \sqrt{(-9)^2 + 7^2} = \sqrt{130} \approx 11.402.$$

Do Exercises 3 and 4.

b Midpoints of Segments

The distance formula can be used to verify or derive a formula for finding the coordinates of the midpoint of a segment when the coordinates of the endpoints are known. We will not derive the formula but simply state it.

THE MIDPOINT FORMULA

If the endpoints of a segment are (x_1, y_1) and (x_2, y_2), then the coordinates of the midpoint are

$$\left(\frac{x_1 + x_2}{2}, \frac{y_1 + y_2}{2}\right).$$

Example 4 Find the midpoint of the segment with endpoints (−2, 3) and (4, −6).

Using the midpoint formula, we obtain

$$\left(\frac{-2 + 4}{2}, \frac{3 + (-6)}{2}\right), \text{ or } \left(\frac{2}{2}, \frac{-3}{2}\right), \text{ or } \left(1, -\frac{3}{2}\right).$$

Do Exercises 5 and 6.

Exercise Set E

a Find the distance between the pair of points. Where appropriate, find an approximation to three decimal places.

1. $(9, 5)$ and $(6, 1)$

2. $(1, 10)$ and $(7, 2)$

3. $(0, -7)$ and $(3, -4)$

4. $(6, 2)$ and $(6, -8)$

5. $(2, 2)$ and $(-2, -2)$

6. $(5, 21)$ and $(-3, 1)$

7. $(8.6, -3.4)$ and $(-9.2, -3.4)$

8. $(5.9, 2)$ and $(3.7, -7.7)$

9. $\left(\frac{5}{7}, \frac{1}{14}\right)$ and $\left(\frac{1}{7}, \frac{11}{14}\right)$

10. $(0, \sqrt{7})$ and $(\sqrt{6}, 0)$

11. $(-23, 10)$ and $(56, -17)$

12. $(34, -18)$ and $(-46, -38)$

b Find the midpoint of the segment with the given endpoints.

13. $(-3, 6)$ and $(2, -8)$

14. $(6, 7)$ and $(7, -9)$

15. $(8, 5)$ and $(-1, 2)$

16. $(-1, 2)$ and $(1, -3)$

17. $(-8, -5)$ and $(6, -1)$

18. $(8, -2)$ and $(-3, 4)$

19. $(-3.4, 8.1)$ and $(2.9, -8.7)$

20. $(4.1, 6.9)$ and $(5.2, -6.9)$

21. $\left(\frac{1}{6}, -\frac{3}{4}\right)$ and $\left(-\frac{1}{3}, \frac{5}{6}\right)$

22. $\left(-\frac{4}{5}, -\frac{2}{3}\right)$ and $\left(\frac{1}{8}, \frac{3}{4}\right)$

23. $(\sqrt{2}, -1)$ and $(\sqrt{3}, 4)$

24. $(9, 2\sqrt{3})$ and $(-4, 5\sqrt{3})$

Synthesis

Find the distance between the given points.

25. $(-1, 3k)$ and $(6, 2k)$

26. (a, b) and $(-a, -b)$

27. $(6m, -7n)$ and $(-2m, n)$

28. $(\sqrt{d}, -\sqrt{3c})$ and $(\sqrt{d}, \sqrt{3c})$

29. $(-3\sqrt{3}, 1 - \sqrt{6})$ and $(\sqrt{3}, 1 + \sqrt{6})$

If the sides of a triangle have lengths a, b, and c and $a^2 + b^2 = c^2$, then the triangle is a right triangle. Determine whether the given points are vertices of a right triangle.

30. $(9, 6)$, $(-1, 2)$, and $(1, -3)$

31. $(-8, -5)$, $(6, 1)$, and $(-4, 5)$

32. Find the point on the y-axis that is equidistant from $(2, 10)$ and $(6, 2)$.

33. Find the midpoint of the segments with the endpoints $(2 - \sqrt{3}, 5\sqrt{2})$ and $(2 + \sqrt{3}, 3\sqrt{2})$.

Appendix F Higher Roots, and Rational Numbers as Exponents

Objectives

a Find principal square roots and their opposites.

b Simplify radical expressions with perfect-square radicands.

c Find cube roots, simplifying certain expressions.

d Simplify expressions involving odd and even roots.

e Write expressions with or without rational exponents.

f Write expressions without negative exponents.

g Use the laws of exponents with rational exponents.

a Square Roots

When we raise a number to the second power, we say that we have **squared** the number. Sometimes we may need to find the number that was squared. We call this process **finding a square root** of a number.

> The number c is a **square root** of a if $c^2 = a$.

For example:

5 is a square root of 25 because $5^2 = 5 \cdot 5 = 25$;

-5 is a square root of 25 because $(-5)^2 = (-5)(-5) = 25$.

The number -4 does not have a real-number square root because there is no real number b such that $b^2 = -4$.

> **SQUARE ROOTS**
>
> Every positive real number has two real-number square roots.
>
> The number 0 has just one square root, 0 itself.
>
> Negative numbers do not have real-number square roots.

Example 1 Find the two square roots of 64.

The square roots of 64 are 8 and -8 because $8^2 = 64$ and $(-8)^2 = 64$.

Do Exercises 1–3.

> The **principal square root** of a nonnegative number is its nonnegative square root. The symbol $\sqrt{a}$ represents the principal square root of a. To name the negative square root of a, we can write $-\sqrt{a}$.

Examples Simplify.

2. $\sqrt{25} = 5$ *Remember*: $\sqrt{\ }$ indicates the principal (nonnegative) square root.

3. $\sqrt{\dfrac{81}{64}} = \dfrac{9}{8}$

4. $\sqrt{0.0049} = 0.07$

5. $\sqrt{0} = 0$

6. $-\sqrt{25} = -5$

7. $\sqrt{-25}$ Does not exist as a real number. Negative numbers do not have real-number square roots.

Do Exercises 4–10.

Find the square roots.

1. 9
2. 36
3. 121

Simplify.

4. $\sqrt{1}$
5. $\sqrt{36}$
6. $\sqrt{\dfrac{81}{100}}$
7. $\sqrt{0.0064}$

Find the following.

8. a) $\sqrt{16}$
   b) $-\sqrt{16}$
   c) $\sqrt{-16}$
9. a) $\sqrt{49}$
   b) $-\sqrt{49}$
   c) $\sqrt{-49}$
10. a) $\sqrt{144}$
    b) $-\sqrt{144}$
    c) $\sqrt{-144}$

Answers on page A-53

Identify the radicand.

11. $\sqrt{28 + x}$

12. $\sqrt{\dfrac{y}{y + 3}}$

Find the following. Assume that letters can represent any real number.

13. $\sqrt{y^2}$

14. $\sqrt{(-24)^2}$

15. $\sqrt{(5y)^2}$

16. $\sqrt{16y^2}$

17. $\sqrt{(x + 7)^2}$

18. $\sqrt{4(x - 2)^2}$

19. $\sqrt{49(y + 5)^2}$

20. $\sqrt{x^2 - 6x + 9}$

Answers on page A-53

> The symbol $\sqrt{}$ is called a **radical**.
> An expression written with a radical is called a **radical expression.**
> The expression written under the radical is called the **radicand.**

These are radical expressions:

$$\sqrt{5}, \quad \sqrt{a}, \quad -\sqrt{5x}, \quad \sqrt{y^2 + 7}.$$

The radicands in these expressions are 5, a, $5x$, and $y^2 + 7$.

Example 8 Identify the radicand in $\sqrt{x^2 - 9}$.

The radicand in $\sqrt{x^2 - 9}$ is $x^2 - 9$.

Do Exercises 11 and 12.

b Finding $\sqrt{a^2}$

In the expression $\sqrt{a^2}$, the radicand is a perfect square.

Suppose $a = 5$. Then we have $\sqrt{5^2}$, which is $\sqrt{25}$, or 5.
Suppose $a = -5$. Then we have $\sqrt{(-5)^2}$, which is $\sqrt{25}$, or 5.
Suppose $a = 0$. Then we have $\sqrt{0^2}$, which is $\sqrt{0}$, or 0.

The symbol $\sqrt{a^2}$ does not represent a negative number. It represents the principal square root of a^2. Note that if a represents a positive number or 0, then $\sqrt{a^2}$ represents a. If a is negative, then $\sqrt{a^2}$ represents the opposite of a. In all cases, the radical expression represents the absolute value of a.

> For any real number a, $\sqrt{a^2} = |a|$. The principal (nonnegative) square root of a^2 is the absolute value of a.

The absolute value is used to ensure that the principal square root is non-negative, which is as it is defined.

Examples Find the following. Assume that letters can represent any real number.

9. $\sqrt{(-16)^2} = |-16|$, or 16

10. $\sqrt{(3b)^2} = |3b| = |3| \cdot |b| = 3|b|$

> $|3b|$ can be simplified to $3|b|$ because the absolute value of any product is the product of the absolute values. That is, $|a \cdot b| = |a| \cdot |b|$.

11. $\sqrt{(x - 1)^2} = |x - 1|$

12. $\sqrt{x^2 + 8x + 16} = \sqrt{(x + 4)^2}$
$= |x + 4|$ ← CAUTION! $|x + 4|$ is *not* the same as $|x| + 4$.

Do Exercises 13–20.

c Cube Roots

> The number c is the **cube root** of a if its third power is a—that is, if $c^3 = a$.

For example:

2 is the cube root of 8 because $2^3 = 2 \cdot 2 \cdot 2 = 8$;

-4 is the cube root of -64 because $(-4)^3 = (-4)(-4)(-4) = -64$.

We talk about *the* cube root of a number because of the following.

> Every real number has exactly one cube root in the system of real numbers. The symbol $\sqrt[3]{a}$ represents the cube root of a.

Examples Find the following.

13. $\sqrt[3]{8} = 2$

14. $\sqrt[3]{-27} = -3$

15. $\sqrt[3]{-\frac{216}{125}} = -\frac{6}{5}$

16. $\sqrt[3]{0.001} = 0.1$

17. $\sqrt[3]{x^3} = x$

18. $\sqrt[3]{-8} = -2$

19. $\sqrt[3]{0} = 0$

20. $\sqrt[3]{-8y^3} = -2y$

When we are determining a cube root, no absolute-value signs are needed because a real number has just one cube root. The real-number cube root of a positive number is positive. The real-number cube root of a negative number is negative. The cube root of 0 is 0. That is, $\sqrt[3]{a^3} = a$ whether $a > 0$, $a < 0$, or $a = 0$.

Do Exercises 21–24.

d Odd and Even *k*th Roots

In the expression $\sqrt[k]{a}$, we call k the **index** and assume $k \geq 2$.

Odd Roots

The 5th root of a number a is the number c for which $c^5 = a$. There are also 7th roots, 9th roots, and so on. Whenever the number k in $\sqrt[k]{}$ is an odd number, we say that we are taking an **odd root.**

Every number has just one real-number odd root. If the number is positive, then the root is positive. If the number is negative, then the root is negative. If the number is 0, then the root is 0.

> If k is an *odd* natural number, then for any real number a,
> $$\sqrt[k]{a^k} = a.$$

Absolute-value signs are not needed when we are finding odd roots.

Find the following.

21. $\sqrt[3]{-64}$

22. $\sqrt[3]{27y^3}$

23. $\sqrt[3]{8(x+2)^3}$

24. $\sqrt[3]{-\frac{343}{64}}$

Answers on page A-53

Find the following.

25. $\sqrt[5]{243}$

26. $\sqrt[5]{-243}$

27. $\sqrt[5]{x^5}$

28. $\sqrt[7]{y^7}$

29. $\sqrt[5]{0}$

30. $\sqrt[5]{-32x^5}$

31. $\sqrt[7]{(3x + 2)^7}$

Find the following. Assume that letters can represent any real number.

32. $\sqrt[4]{81}$

33. $-\sqrt[4]{81}$

34. $\sqrt[4]{-81}$

35. $\sqrt[4]{0}$

36. $\sqrt[4]{16(x - 2)^4}$

37. $\sqrt[6]{x^6}$

38. $\sqrt[8]{(x + 3)^8}$

Answers on page A-53

Examples Find the following.

21. $\sqrt[5]{32} = 2$

22. $\sqrt[5]{-32} = -2$

23. $-\sqrt[5]{32} = -2$

24. $-\sqrt[5]{-32} = -(-2) = 2$

25. $\sqrt[7]{x^7} = x$

26. $\sqrt[7]{128} = 2$

27. $\sqrt[7]{-128} = -2$

28. $\sqrt[7]{0} = 0$

29. $\sqrt[5]{a^5} = a$

30. $\sqrt[9]{(x - 1)^9} = x - 1$

Do Exercises 25–31.

Even Roots

When the index k in $\sqrt[k]{}$ is an even number, we say that we are taking an **even root.** Every positive real number has two real-number kth roots when k is even. One of those roots is positive and one is negative. Negative real numbers do not have real-number kth roots when k is even. When we are finding even kth roots, absolute-value signs are sometimes necessary, as they are with square roots. When the index is 2, we do not write it. For example,

$$\sqrt{64} = 8, \qquad \sqrt[6]{64} = 2, \qquad -\sqrt[6]{64} = -2, \qquad \sqrt[6]{64x^6} = |2x| = 2|x|.$$

Note that in $\sqrt[6]{64x^6}$, we need absolute-value signs because a variable is involved.

Examples Find the following. Assume that letters can represent any real number.

31. $\sqrt[4]{16} = 2$

32. $-\sqrt[4]{16} = -2$

33. $\sqrt[4]{-16}$ Does not exist as a real number.

34. $\sqrt[4]{81x^4} = 3|x|$

35. $\sqrt[6]{(y + 7)^6} = |y + 7|$

36. $\sqrt{81y^2} = 9|y|$

The following is a summary of how absolute value is used when we are taking even or odd roots.

For any real number a:

a) $\sqrt[k]{a^k} = |a|$ when k is an *even* natural number. We use absolute value when k is even unless a is nonnegative.

b) $\sqrt[k]{a^k} = a$ when k is an *odd* natural number greater than 1. We do not use absolute value when k is odd.

Do Exercises 32–38.

e Rational Exponents

Expressions like $a^{1/2}$, $5^{-1/4}$, and $(2y)^{4/5}$ have not yet been defined. We will define such expressions so that the general properties of exponents hold.

Consider $a^{1/2} \cdot a^{1/2}$. If we want to multiply by adding exponents, it must follow that $a^{1/2} \cdot a^{1/2} = a^{1/2+1/2}$, or a^1. Thus we should define $a^{1/2}$ to be a square root of a. Similarly, $a^{1/3} \cdot a^{1/3} \cdot a^{1/3} = a^{1/3+1/3+1/3}$, or a^1, so $a^{1/3}$ should be defined to mean $\sqrt[3]{a}$.

> For any nonnegative real number a and any natural number index n ($n \neq 1$), $a^{1/n}$ means $\sqrt[n]{a}$ (the nonnegative nth root of a).

Whenever we use rational exponents, we assume that the bases are nonnegative.

Examples Rewrite without rational exponents, and simplify, if possible.

37. $x^{1/2} = \sqrt{x}$

38. $27^{1/3} = \sqrt[3]{27} = 3$

39. $(abc)^{1/5} = \sqrt[5]{abc}$

Do Exercises 39–43.

Examples Rewrite with rational exponents.

40. $\sqrt[5]{7xy} = (7xy)^{1/5}$ — We need parentheses around the radicand here.

41. $8\sqrt[3]{xy} = 8(xy)^{1/3}$

42. $\sqrt[7]{\dfrac{x^3y}{9}} = \left(\dfrac{x^3y}{9}\right)^{1/7}$

Do Exercises 44–47.

How should we define $a^{2/3}$? If the general properties of exponents are to hold, we have $a^{2/3} = (a^{1/3})^2$, or $(\sqrt[3]{a})^2$, or $\sqrt[3]{a^2}$. We define this accordingly.

> For any natural numbers m and n ($n \neq 1$) and any nonnegative real number a,
>
> $a^{m/n}$ means $\sqrt[n]{a^m}$, or $(\sqrt[n]{a})^m$.

Examples Rewrite without rational exponents, and simplify, if possible.

43. $(27)^{2/3} = \sqrt[3]{27^2} = (\sqrt[3]{27})^2 = 3^2 = 9$

44. $4^{3/2} = \sqrt[2]{4^3} = (\sqrt[2]{4})^3 = 2^3 = 8$

Do Exercises 48–50.

Rewrite without rational exponents, and simplify, if possible.

39. $y^{1/4}$

40. $(3a)^{1/2}$

41. $16^{1/4}$

42. $(125)^{1/3}$

43. $(a^3b^2c)^{1/5}$

Rewrite with rational exponents.

44. $\sqrt[3]{19ab}$

45. $19\sqrt[3]{ab}$

46. $\sqrt[5]{\dfrac{x^2y}{16}}$

47. $7\sqrt[4]{2ab}$

Rewrite without rational exponents, and simplify, if possible.

48. $x^{3/2}$

49. $8^{2/3}$

50. $4^{5/2}$

Answers on page A-53

Rewrite with rational exponents.

51. $(\sqrt[3]{7abc})^4$

52. $\sqrt[5]{6^7}$

Rewrite with positive exponents.

53. $16^{-1/4}$

54. $(3xy)^{-7/8}$

Use the laws of exponents to simplify.

55. $7^{1/3} \cdot 7^{3/5}$

56. $\dfrac{5^{7/6}}{5^{5/6}}$

57. $(9^{3/5})^{2/3}$

Answers on page A-53

Examples Rewrite with rational exponents.

45. $\sqrt[3]{9^4} = 9^{4/3}$

46. $(\sqrt[4]{7xy})^5 = (7xy)^{5/4}$

Do Exercises 51 and 52.

f Negative Rational Exponents

Negative rational exponents have a meaning similar to that of negative integer exponents.

> For any rational number m/n and any positive real number a,
>
> $$a^{-m/n} \quad \text{means} \quad \frac{1}{a^{m/n}},$$
>
> that is, $a^{m/n}$ and $a^{-m/n}$ are reciprocals.

Examples Rewrite with positive exponents.

47. $4^{-1/2} = \dfrac{1}{4^{1/2}}$

Since $4^{1/2} = \sqrt{4} = 2$, the answer simplifies to $\frac{1}{2}$.

48. $(5xy)^{-4/5} = \dfrac{1}{(5xy)^{4/5}}$

> *Caution!* Don't make the mistake of thinking that a negative exponent means that the entire expression is negative. That is, $9^{-1/2} \neq -\frac{1}{3}$, but $9^{-1/2} = \frac{1}{3}$.

Do Exercises 53 and 54.

g Laws of Exponents

The same laws hold for rational-number exponents as for integer exponents. We list them for review.

> For any real number a and any rational exponents m and n:
>
> **1.** $a^m \cdot a^n = a^{m+n}$ — In multiplying, we can add exponents if the bases are the same.
>
> **2.** $\dfrac{a^m}{a^n} = a^{m-n}$ — In dividing, we can subtract exponents if the bases are the same.
>
> **3.** $(a^m)^n = a^{m \cdot n}$ — To raise a power to a power, we can multiply the exponents.

Examples Use the laws of exponents to simplify.

49. $3^{1/5} \cdot 3^{3/5} = 3^{1/5+3/5} = 3^{4/5}$ — **Adding exponents**

50. $\dfrac{7^{1/4}}{7^{1/2}} = 7^{1/4-1/2} = 7^{1/4-2/4} = 7^{-1/4}$ — **Subtracting exponents**

51. $(7.2^{2/3})^{3/4} = 7.2^{2/3 \cdot 3/4} = 7.2^{6/12} = 7.2^{1/2}$ — **Multiplying exponents**

Do Exercises 55–57.

Exercise Set F

a Find the square roots.

1. 16 **2.** 225 **3.** 144 **4.** 9 **5.** 400 **6.** 81

Find the following.

7. $-\sqrt{\frac{49}{36}}$ **8.** $-\sqrt{\frac{361}{9}}$ **9.** $\sqrt{196}$ **10.** $\sqrt{441}$ **11.** $-\sqrt{\frac{16}{81}}$

12. $-\sqrt{\frac{81}{144}}$ **13.** $\sqrt{0.04}$ **14.** $\sqrt{0.81}$ **15.** $-\sqrt{0.0009}$ **16.** $\sqrt{0.0121}$

Identify the radicand.

17. $9\sqrt{y^2 + 16}$ **18.** $-3\sqrt{p^2 - 10}$ **19.** $x^4y^5\sqrt{\frac{x}{y - 1}}$ **20.** $a^2b^2\sqrt{\frac{a^2 - b}{b}}$

b Find the following. Assume that letters can represent any real number.

21. $\sqrt{16x^2}$ **22.** $\sqrt{25t^2}$ **23.** $\sqrt{(-12c)^2}$

24. $\sqrt{(-9d)^2}$ **25.** $\sqrt{(p + 3)^2}$ **26.** $\sqrt{(2 - x)^2}$

27. $\sqrt{x^2 - 4x + 4}$ **28.** $\sqrt{y^2 + 16y + 64}$

29. $\sqrt{4x^2 + 28x + 49}$ **30.** $\sqrt{9x^2 - 30x + 25}$

c Simplify.

31. $\sqrt[3]{27}$ **32.** $-\sqrt[3]{64}$ **33.** $\sqrt[3]{-64x^3}$

34. $\sqrt[3]{-125y^3}$ **35.** $\sqrt[3]{-216}$ **36.** $-\sqrt[3]{-1000}$

d Find the following. Assume that letters can represent any real number.

37. $\sqrt[4]{625}$
38. $-\sqrt[4]{256}$
39. $\sqrt[5]{-1}$
40. $\sqrt[5]{-32}$
41. $\sqrt[5]{-\dfrac{32}{243}}$
42. $\sqrt[5]{-\dfrac{1}{32}}$
43. $\sqrt[6]{x^6}$
44. $\sqrt[4]{(7b)^4}$
45. $\sqrt[10]{(-6)^{10}}$
46. $\sqrt[12]{(-10)^{12}}$
47. $\sqrt[7]{y^7}$
48. $\sqrt[3]{(-6)^3}$
49. $\sqrt[5]{(x-2)^5}$
50. $\sqrt[9]{(2xy)^9}$

e Rewrite without exponents.

51. $y^{1/7}$
52. $x^{1/6}$
53. $8^{1/3}$
54. $16^{1/2}$
55. $(a^3b^3)^{1/5}$
56. $(x^2y^2)^{1/3}$
57. $16^{3/4}$
58. $4^{7/2}$

Rewrite with rational exponents.

59. $\sqrt[3]{18}$
60. $\sqrt[3]{23}$
61. $\sqrt[5]{xy^2z}$
62. $\sqrt[7]{x^3y^2z^2}$
63. $(\sqrt{3mn})^3$
64. $(\sqrt[3]{7xy})^4$
65. $(\sqrt[7]{8x^2y})^5$
66. $(\sqrt[6]{2a^5b})^7$

f Rewrite with positive exponents.

67. $x^{-1/4}$
68. $y^{-1/5}$
69. $\dfrac{1}{x^{-2/3}}$
70. $\dfrac{1}{x^{-5/6}}$

g Use the laws of exponents to simplify.

71. $5^{3/4} \cdot 5^{1/8}$
72. $11^{2/3} \cdot 11^{1/2}$
73. $\dfrac{7^{5/8}}{7^{3/8}}$
74. $\dfrac{9^{9/11}}{9^{7/11}}$
75. $\dfrac{4.9^{3/5}}{4.9^{1/4}}$
76. $\dfrac{2.7^{3/4}}{2.7^{2/5}}$
77. $(6^{3/8})^{2/7}$
78. $(3^{2/9})^{3/5}$
79. $a^{2/3} \cdot a^{5/6}$
80. $x^{3/4} \cdot x^{5/12}$

Appendix G Nonlinear Inequalities

a Solving Polynomial Inequalities

Inequalities like the following are called **quadratic inequalities:**

$$x^2 + 3x - 10 < 0, \qquad 5x^2 - 3x + 2 \geq 0.$$

In each case, we have a polynomial of degree 2 on the left.

We will first consider solving a quadratic inequality, such as $ax^2 + bx + c > 0$, using the graph of a related equation, $y = ax^2 + bx + c$.

Example 1 Solve: $x^2 + 3x - 10 > 0$.

Consider the equation $y = x^2 + 3x - 10$ and its graph. Its graph opens up since the leading coefficient ($a = 1$) is positive.

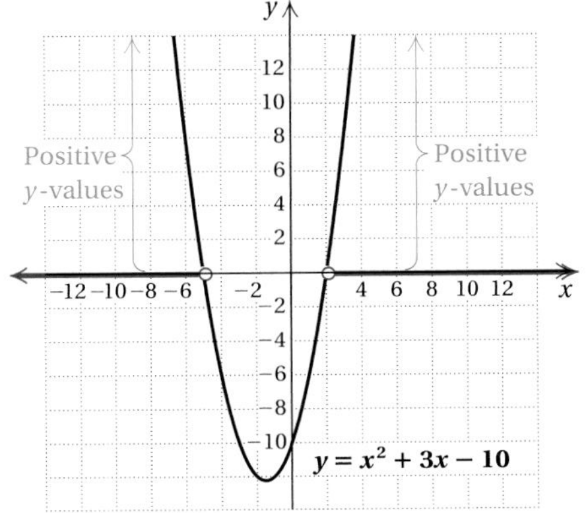

Values of y will be positive to the left and right of the intercepts, as shown. We find the intercepts by setting the polynomial equal to 0 and solving:

$$x^2 + 3x - 10 = 0$$
$$(x + 5)(x - 2) = 0$$
$$x + 5 = 0 \quad or \quad x - 2 = 0$$
$$x = -5 \quad or \quad x = 2.$$

Thus the solution set of the inequality is

$$\{x \mid x < -5 \text{ or } x > 2\}.$$

Do Exercise 1.

Objectives

a. Solve quadratic and other polynomial inequalities.

b. Graph quadratic inequalities.

1. Solve by graphing:

$$x^2 + 2x - 3 > 0.$$

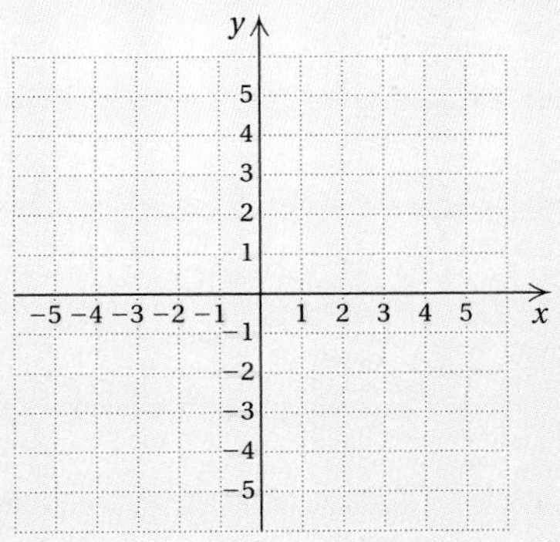

Answer on page A-53

2. Solve by graphing:

$x^2 + 2x - 3 < 0.$

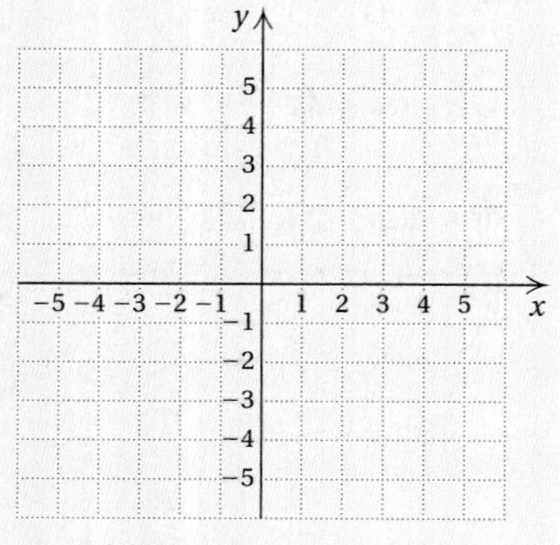

Example 2 Solve: $x^2 + 3x - 10 < 0$.

Looking again at the graph of $y = x^2 + 3x - 10$ or at least visualizing it tells us that y-values are negative for those x-values between -5 and 2.

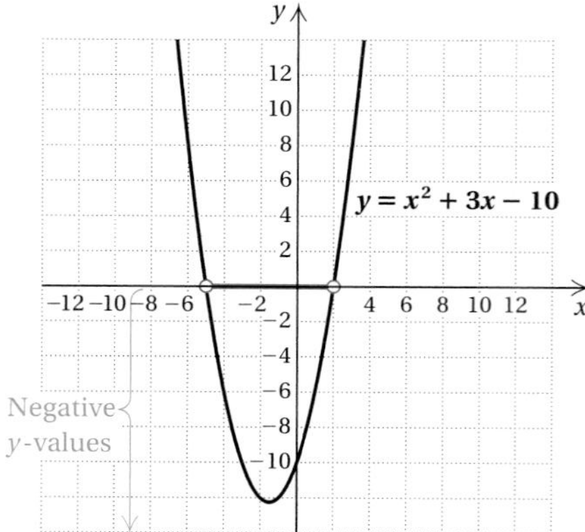

That is, the solution set is $\{x \mid -5 < x < 2\}$.

Do Exercise 2.

When an inequality contains $\leq$ or $\geq$, the x-values of the intercepts must be included. Thus the solution set of the inequality $x^2 + 3x - 10 \leq 0$ is $\{x \mid -5 \leq x \leq 2\}$.

Do Exercise 3.

3. Solve by graphing:

$x^2 + 2x - 3 \leq 0.$

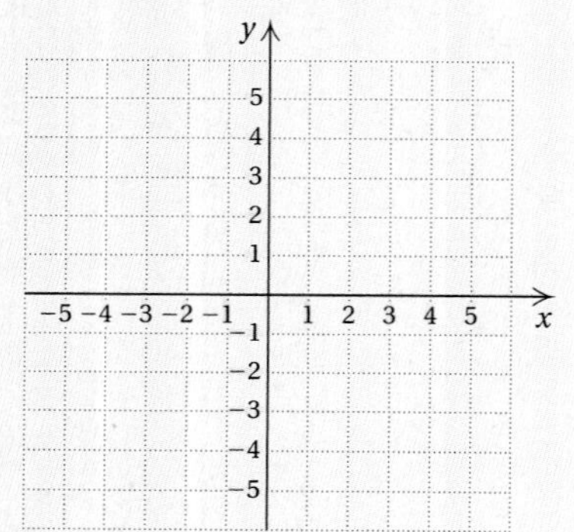

In Examples 1 and 2, we see that the intercepts divide the number line into intervals. If a particular equation has a positive output for one number in an interval, it will be positive for all the numbers in the interval. Thus we can merely make a test substitution in each interval to solve the inequality. This is very similar to our method of using test points to graph a linear inequality in a plane.

Example 3 Solve: $x^2 + 3x - 10 < 0$.

We set the polynomial equal to 0 and solve. The solutions of $x^2 + 3x - 10 = 0$, or $(x + 5)(x - 2) = 0$, are -5 and 2. We then locate them on a number line as follows. Note that the numbers divide the number line into three intervals A, B, and C.

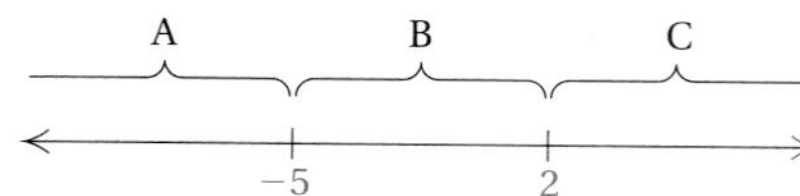

We choose a test number in interval A, say -7, and substitute -7 for x in the equation $y = x^2 + 3x - 10$:

$$y = (-7)^2 + 3(-7) - 10 = 49 - 21 - 10 = 18.$$

Note that $18 > 0$, so the y-values will be positive for any number in interval A.

Next, we try a test number in interval B, say 1, and find the corresponding y-value:

$$y = 1^2 + 3(1) - 10 = 1 + 3 - 10 = -6.$$

Note that $-6 < 0$, so the y-values will be negative for any number in interval B.

Answers on page A-53

Next, we try a test number in interval C, say 4, and find the corresponding y-value:

$$y = 4^2 + 3(4) - 10 = 16 + 12 - 10 = 18.$$

Note that $18 > 0$, so the y-values will be positive for any number in interval C.

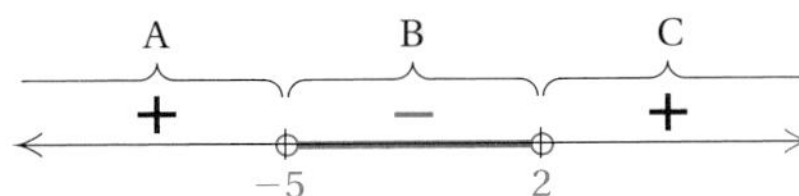

We are looking for numbers x for which $x^2 + 3x - 10 < 0$. Thus any number x in interval B is a solution. If the inequality had been $\leq$ or $\geq$, we would also need to include the intercepts -5 and 2 in the solution set. The solution set is $\{x \mid -5 < x < 2\}$.

Do Exercises 4 and 5.

Solve using the method of Example 3.

4. $x^2 + 3x > 4$

5. $x^2 + 3x \leq 4$

Example 4 Solve: $5x(x + 3)(x - 2) \geq 0$.

The solutions of $5x(x + 3)(x - 2) = 0$ are -3, 0, and 2. They divide the real-number line into four intervals, as follows.

A B C D

-3 0 2

We try test numbers in each interval:

A: Test -5, $y = 5(-5)(-5 + 3)(-5 - 2) = -350$.

B: Test -2, $y = 5(-2)(-2 + 3)(-2 - 2) = 40$.

C: Test 1, $y = 5(1)(1 + 3)(1 - 2) = -20$.

D: Test 3, $y = 5(3)(3 + 3)(3 - 2) = 90$.

A B C D

− + − +

-3 0 2

The expression is positive for values of x in intervals B and D. Since the inequality symbol is $\geq$, we will need to include the intercepts. The solution set of the inequality is

$$\{x \mid -3 \leq x \leq 0 \text{ or } 2 \leq x\}.$$

Do Exercise 6.

6. Solve: $6x(x + 1)(x - 1) < 0$.

b Graphing Quadratic Inequalities

Graphing quadratic inequalities involves the same procedure as graphing quadratic equations, but the graph will also include the interior region enclosed by the parabola or the exterior region outside the parabola.

Example 5 Graph: $y \leq x^2 + 2x - 3$.

We first replace the inequality symbol with an equals sign and graph the equation:

$$y = x^2 + 2x - 3.$$

Answers on page A-53

7. Graph: $y \geq x^2 - 2x - 3$.

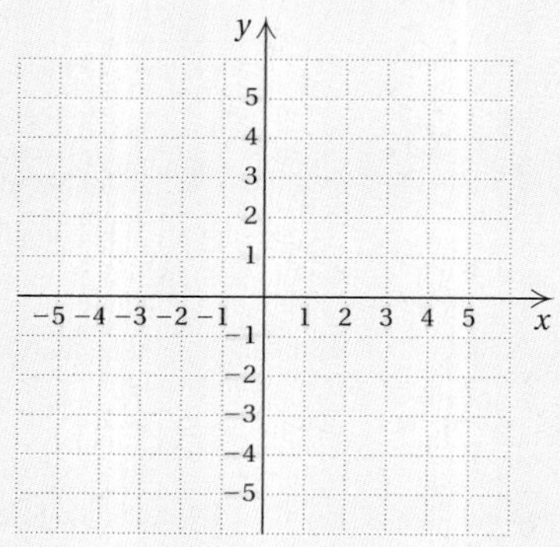

8. Graph: $y > -x^2 + 2x$.

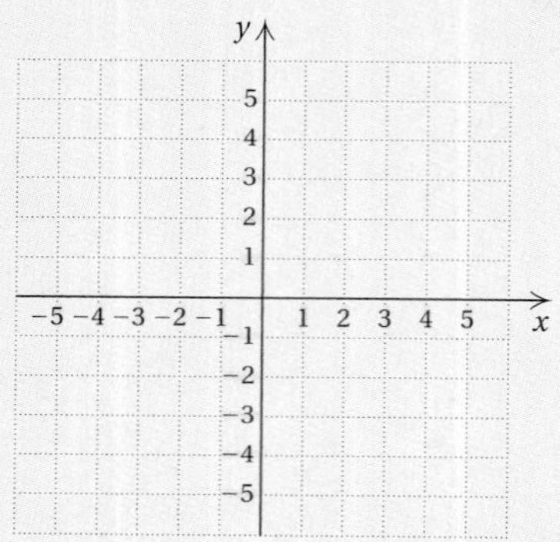

Answers on page A-53

The x-coordinate of the vertex is

$$-\frac{b}{2a} = -\frac{2}{2 \cdot 1} = -1.$$

We substitute -1 for x in the equation to find the second coordinate of the vertex:

$$\begin{aligned} y &= x^2 + 2x - 3 \\ &= (-1)^2 + 2(-1) - 3 \\ &= 1 - 2 - 3 \\ &= -4. \end{aligned}$$

The vertex is $(-1, -4)$. The line of symmetry is $x = -1$. We choose some x-values on both sides of the vertex and graph the parabola.

For $x = 0$, $\quad y = x^2 + 2x - 3 = 0^2 + 2 \cdot 0 - 3 = -3$.

For $x = -2$, $\quad y = x^2 + 2x - 3 = (-2)^2 + 2(-2) - 3 = -3$.

For $x = 1$, $\quad y = x^2 + 2x - 3 = 1^2 + 2 \cdot 1 - 3 = 0$.

For $x = -3$, $\quad y = x^2 + 2x - 3 = (-3)^2 + 2(-3) - 3 = 0$.

x	y
-1	-4 ← This is the vertex.
0	-3
-2	-3
1	0
-3	0

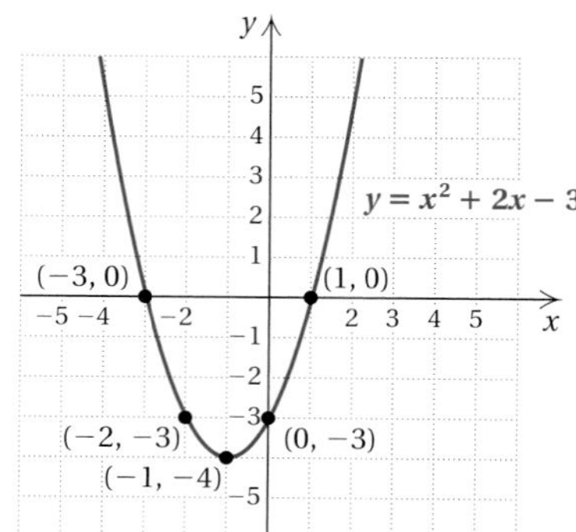

The inequality symbol is $\leq$, so we draw the curve solid.

To determine which region to shade, we choose a point not on the curve as a test point. The origin (0, 0) is usually an easy one to use.

$$\begin{array}{c|l} y \leq x^2 + 2x - 3 & \\ \hline 0 & 0^2 + 2 \cdot 0 - 3 \\ & 0 + 0 - 3 \\ & -3 \qquad \text{TRUE} \end{array}$$

We see that (0, 0) is *not* a solution, so we shade the exterior region. Had the substitution given us a true inequality, we would have shaded the interior.

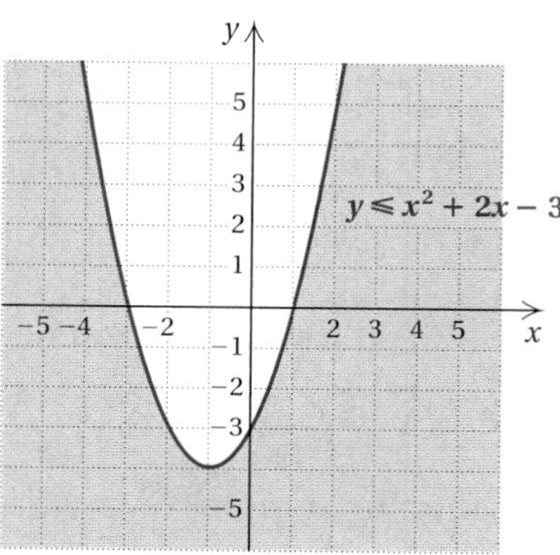

Do Exercises 7 and 8.

Exercise Set G

a Solve.

1. $(x - 6)(x + 2) > 0$

2. $(x - 5)(x + 1) > 0$

3. $3(x + 1)(x - 4) \leq 0$

4. $(x - 7)(x + 3) \leq 0$

5. $x^2 - x - 2 < 0$

6. $x^2 + x - 2 < 0$

7. $9 - x^2 \leq 0$

8. $4 - x^2 \geq 0$

9. $x^2 - 2x + 1 \geq 0$

10. $x^2 + 6x + 9 < 0$

11. $x^2 + 8 < 6x$

12. $x^2 - 12 > 4x$

13. $3x(x + 2)(x - 2) < 0$

14. $5x(x + 1)(x - 1) > 0$

15. $(x + 9)(x - 4)(x + 1) > 0$

16. $(x - 1)(x + 8)(x - 2) < 0$

17. $(x + 3)(x + 2)(x - 1) < 0$

18. $(x - 2)(x - 3)(x + 1) < 0$

b Graph.

19. $y \le 3x^2$

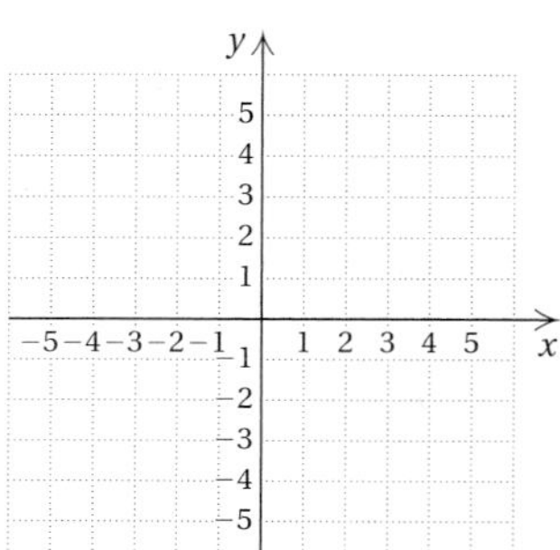

20. $y > -x^2$

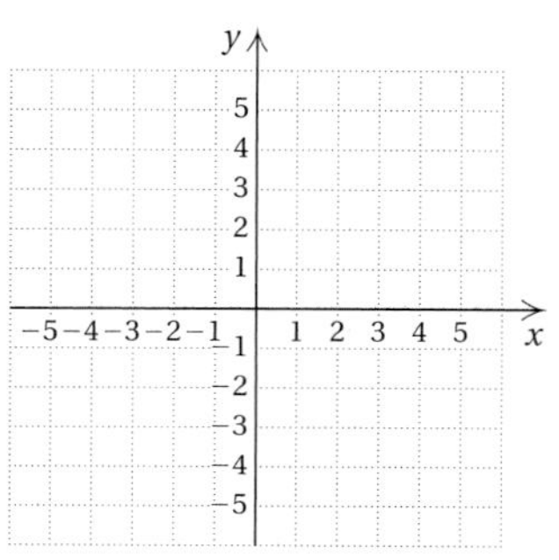

21. $y < 4 - x^2$

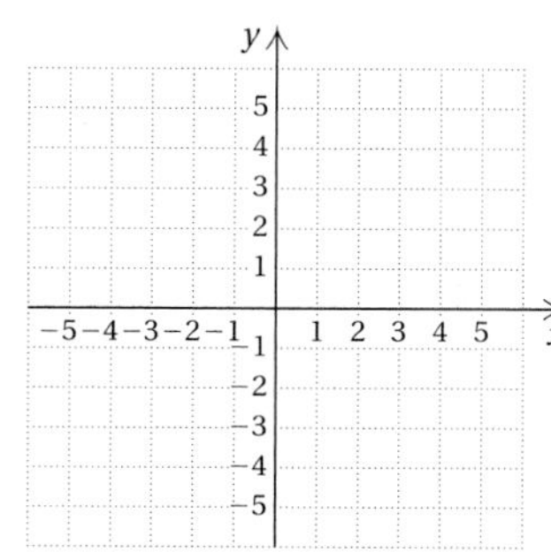

22. $y \le 2 - x^2$

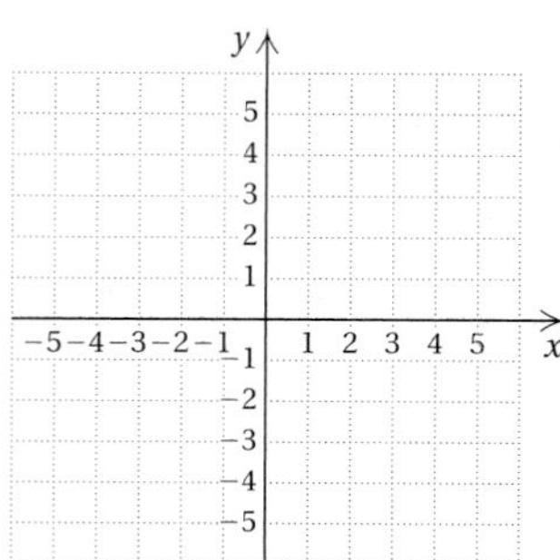

23. $y \le \frac{1}{2}x^2 - 1$

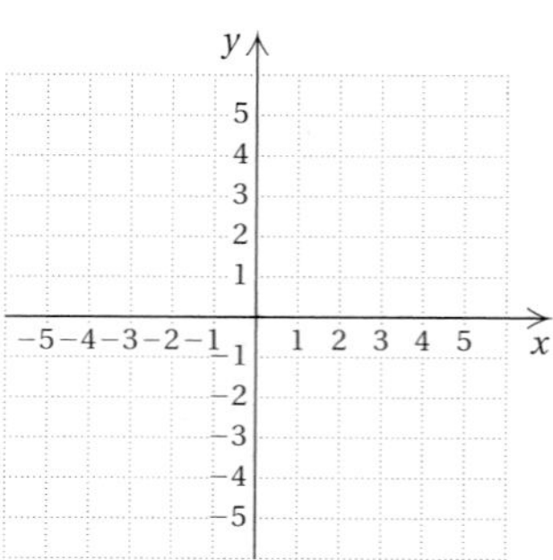

24. $y < -\frac{1}{2}x^2 + 2$

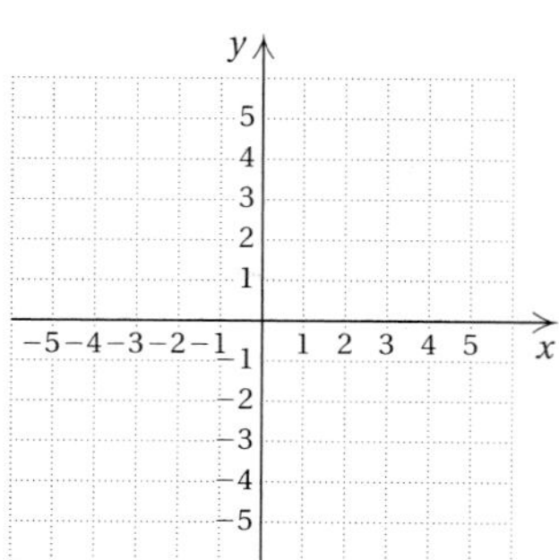

25. $y > x^2 + 4x - 1$

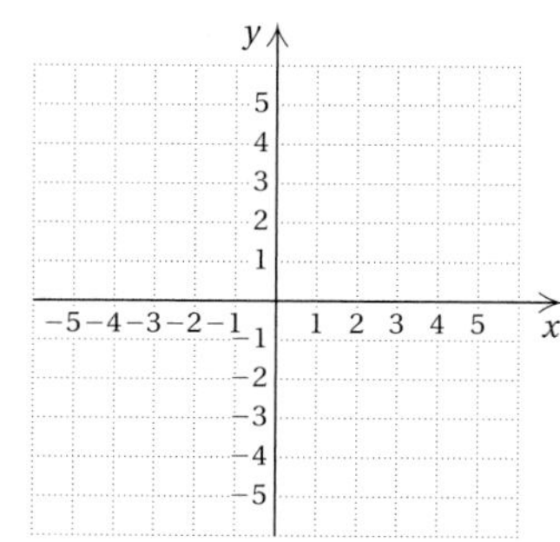

26. $y \le x^2 - 2x - 3$

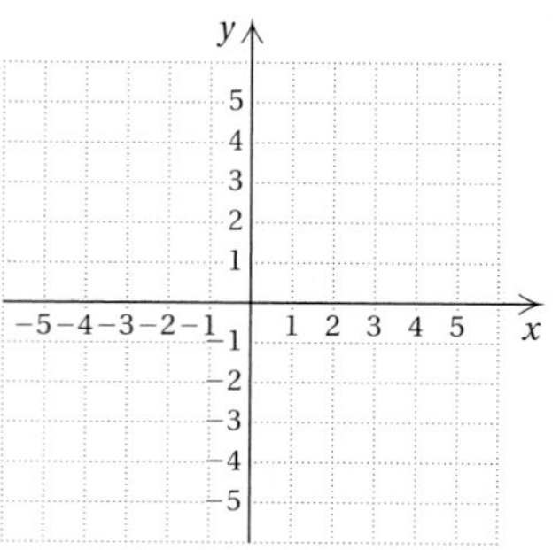

27. $y \ge x^2 + 2x + 1$

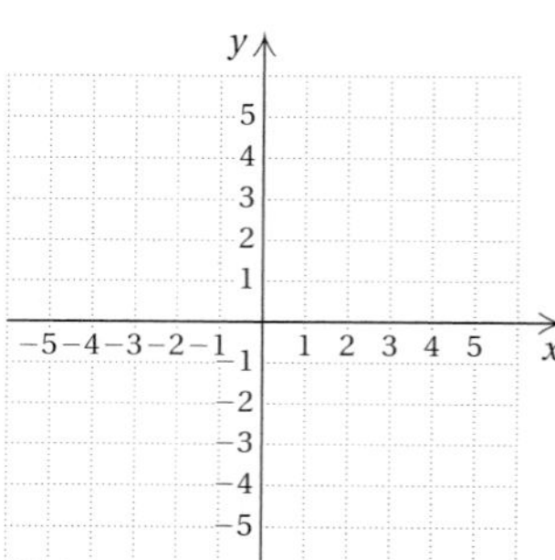

28. $y > x^2 + x - 6$

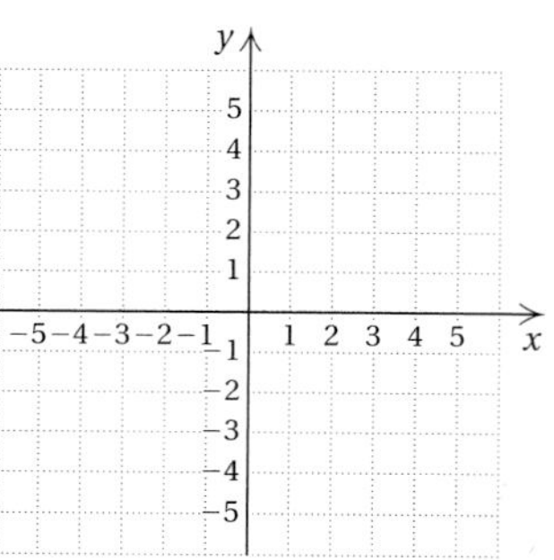

29. $y < 5 - x - x^2$

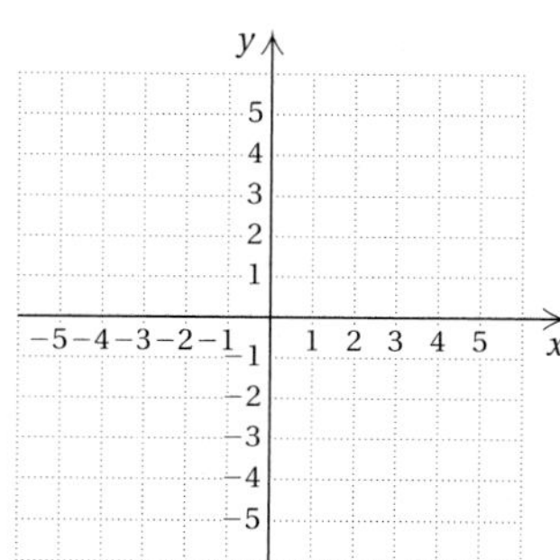

30. $y \ge -x^2 + 2x + 3$

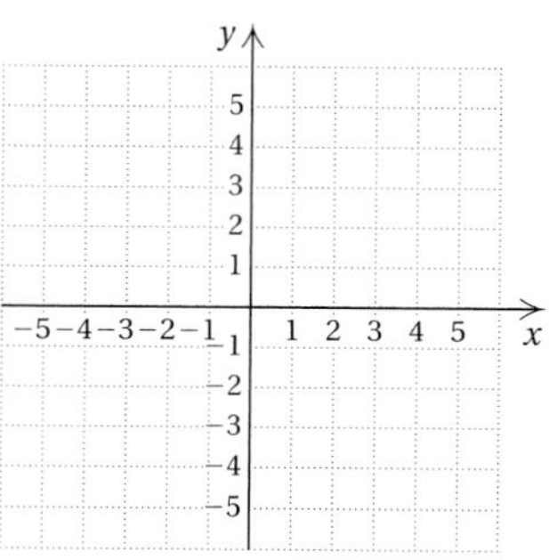

31. $y \le -x^2 - 2x + 3$

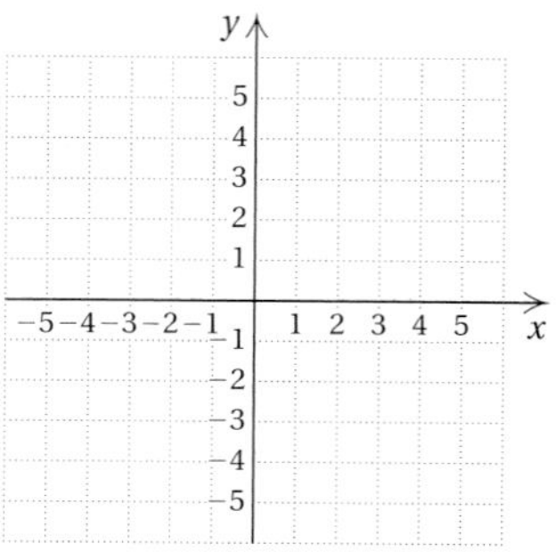

32. $y < -2x^2 - 4x + 1$

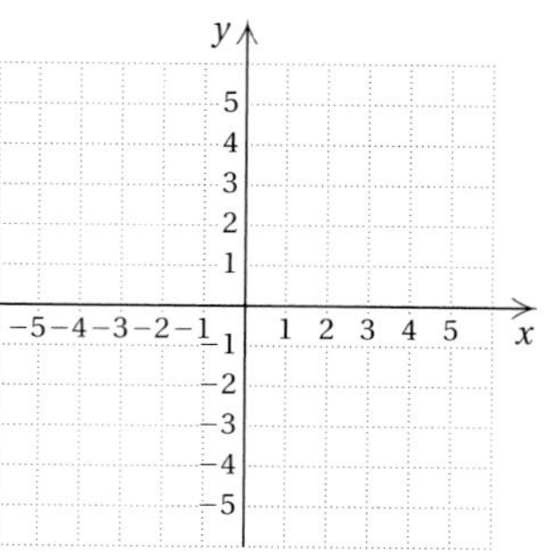

33. $y > 2x^2 + 4x - 1$

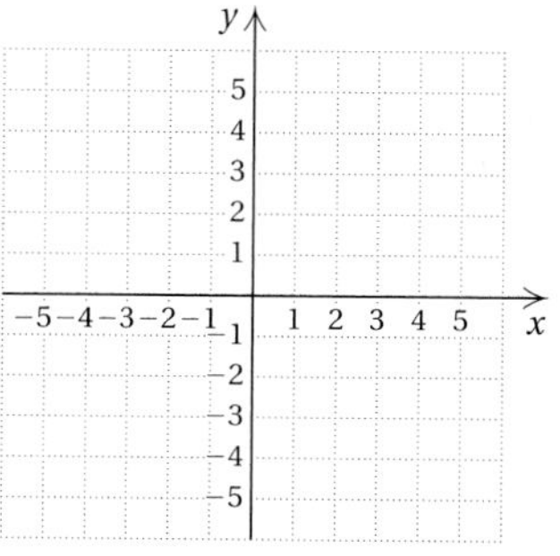

34. $y \le x^2 + 2x - 5$

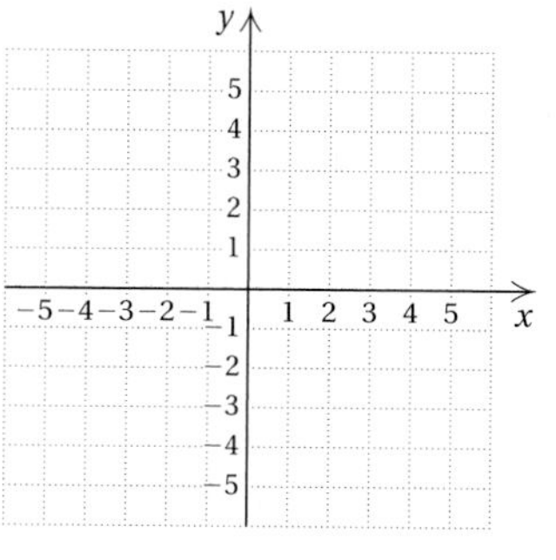

Appendix H Systems of Linear Inequalities

In Section 13.4, we studied the graphing of inequalities in two variables. Here we study *systems* of linear inequalities.

a Systems of Linear Inequalities

The following is an example of a system of two linear inequalities in two variables:

$$x + y \leq 4,$$
$$x - y < 4.$$

A **solution** of a system of linear inequalities is an ordered pair that is a solution of *both* inequalities. We now graph solutions of systems of linear inequalities. To do so, we graph each inequality and determine where the graphs overlap, or intersect. That will be a region in which the ordered pairs are solutions of both inequalities.

Example 1 Graph the solutions of the system

$$x + y \leq 4,$$
$$x - y < 4.$$

We graph the inequality $x + y \leq 4$ by first graphing the equation $x + y = 4$ using a solid red line. We consider (0, 0) as a test point and find that it is a solution, so we shade all points on that side of the line using red shading. The arrows at the ends of the line also indicate the half-plane, or region, that contains the solutions.

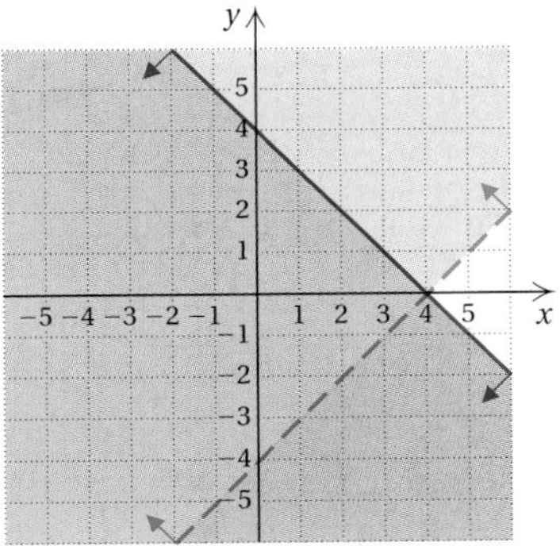

Next, we graph $x - y < 4$. We begin by graphing the equation $x - y = 4$ using a dashed blue line and consider (0, 0) as a test point. Again, (0, 0) is a solution so we shade that side of the line using blue shading. The solution set of the system is the region that is shaded both red and blue and part of the line $x + y = 4$.

Do Exercise 1.

Objective

a Graph systems of linear inequalities and find the coordinates of any vertices formed.

1. Graph:

$$x + y \geq 1,$$
$$y - x \geq 2.$$

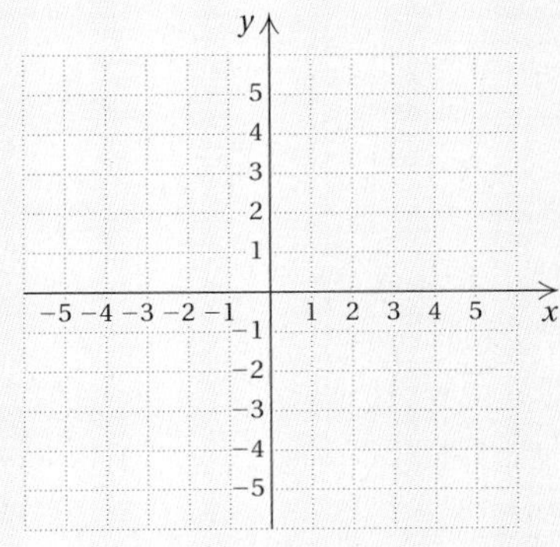

Answers on page A-54

2. Graph: $-3 \leq y < 4$.

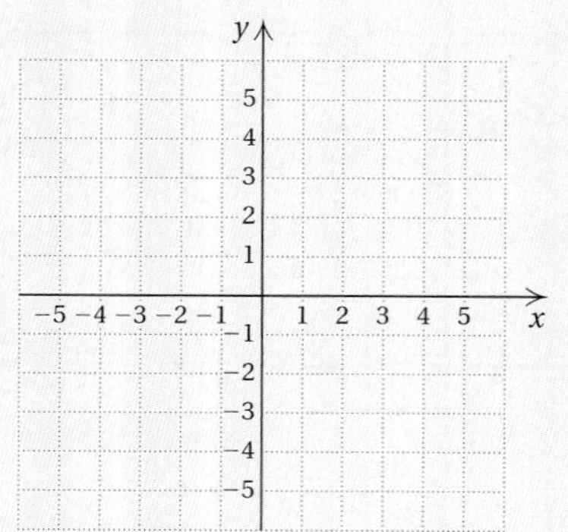

Answers on page A-54

Example 2 Graph: $-2 < x \leq 5$.

This is actually a system of inequalities:

$$-2 < x,$$
$$x \leq 5.$$

We graph the equation $-2 = x$ and see that the graph of the first inequality is the half-plane to the right of the line $-2 = x$ (see the graph on the left below).

Next, we graph the second inequality, starting with the line $x = 5$, and find that its graph is the line and also the half-plane to the left of it (see the graph on the right below).

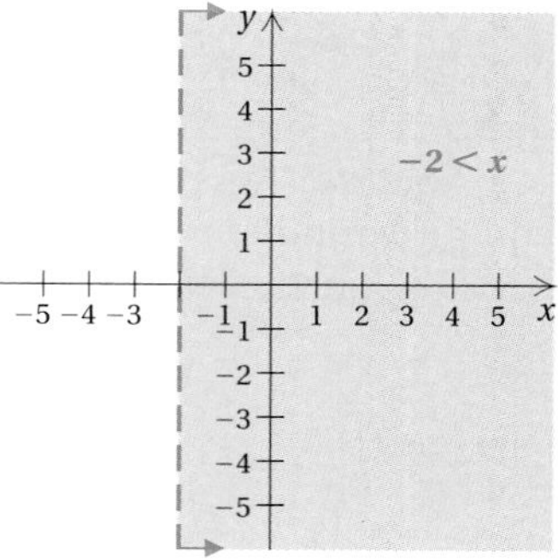

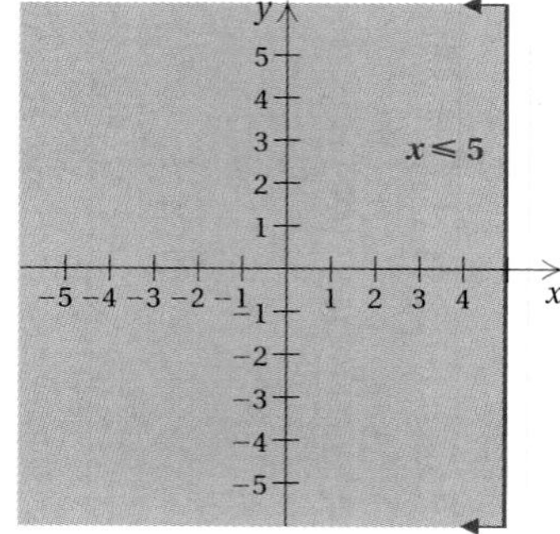

We shade the intersection of these graphs.

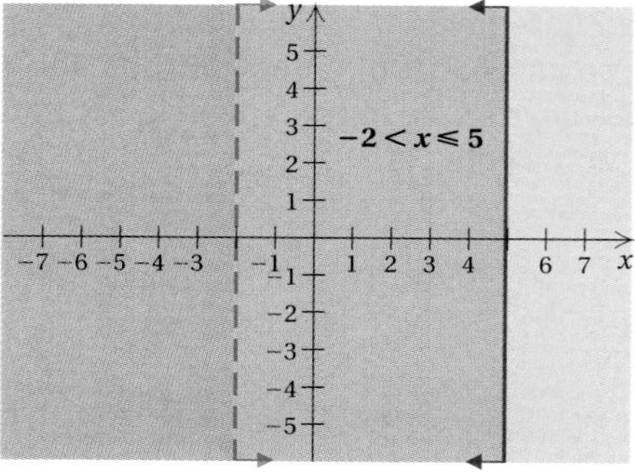

Do Exercise 2.

A system of inequalities may have a graph that consists of a polygon and its interior. In *linear programming,* which is a topic rich in application that you may study in a later course, it is important to be able to find the vertices of such a polygon.

Example 3 Graph the following system of inequalities. Find the coordinates of any vertices formed.

$$6x - 2y \le 12, \quad (1)$$
$$y - 3 \le 0, \quad (2)$$
$$x + y \ge 0 \quad (3)$$

We graph the lines $6x - 2y = 12$, $y - 3 = 0$, and $x + y = 0$ using solid lines. The regions for each inequality are indicated by the arrows at the ends of the lines. We then note where the regions overlap and shade the region of solutions using one color.

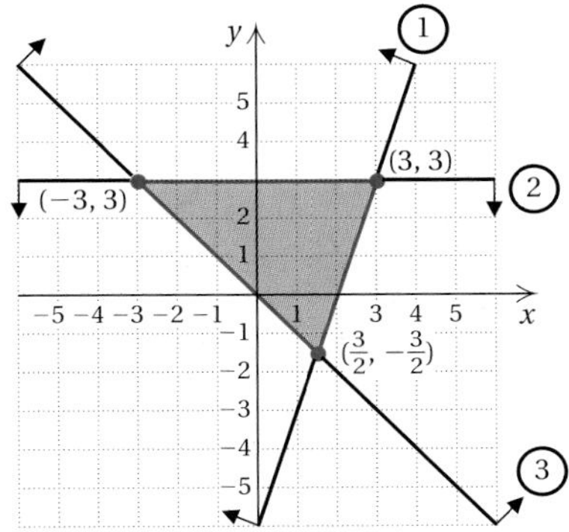

To find the vertices, we solve three different systems of equations. The system of equations from inequalities (1) and (2) is

$$6x - 2y = 12,$$
$$y - 3 = 0.$$

Solving, we obtain the vertex (3, 3).

The system of equations from inequalities (1) and (3) is

$$6x - 2y = 12,$$
$$x + y = 0.$$

Solving, we obtain the vertex $\left(\frac{3}{2}, -\frac{3}{2}\right)$.

The system of equations from inequalities (2) and (3) is

$$y - 3 = 0,$$
$$x + y = 0.$$

Solving, we obtain the vertex (−3, 3).

Do Exercise 3.

Example 4 Graph the following system of inequalities. Find the coordinates of any vertices formed.

$$x + y \le 16, \quad (1)$$
$$3x + 6y \le 60, \quad (2)$$
$$x \ge 0, \quad (3)$$
$$y \ge 0 \quad (4)$$

We graph each inequality using solid lines. The regions for each inequality are indicated by the arrows at the ends of the lines. We then note where the regions overlap and shade the region of solutions using one color.

3. Graph the system of inequalities. Find the coordinates of any vertices formed.

$$5x + 6y \le 30,$$
$$0 \le y \le 3,$$
$$0 \le x \le 4$$

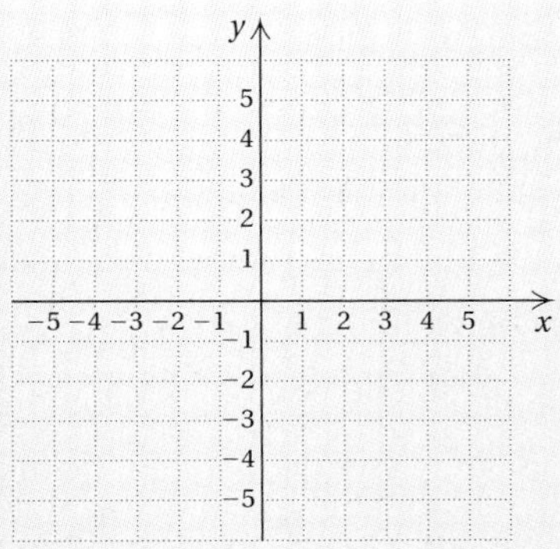

Answer on page A-54

4. Graph the system of inequalities. Find the coordinates of any vertices formed.

$$2x + 4y \leq 8,$$
$$x + y \leq 3,$$
$$x \geq 0,$$
$$y \geq 0$$

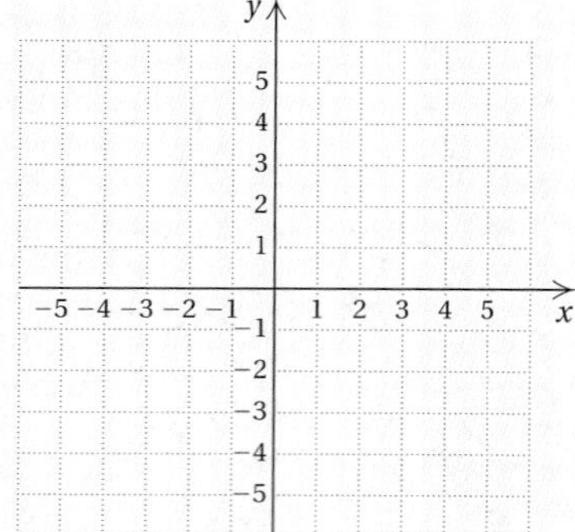

Answer on page A-54

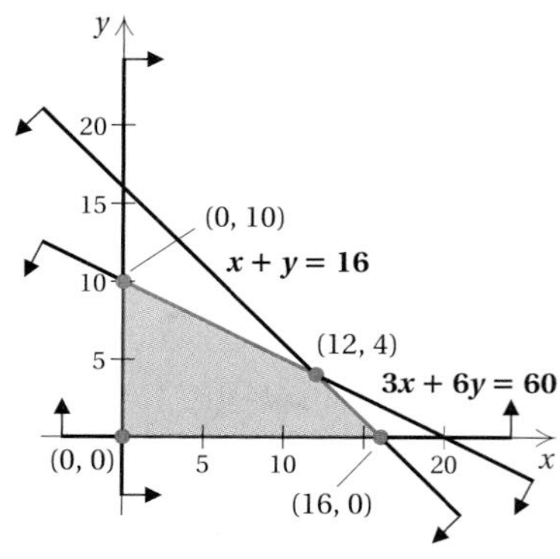

To find the vertices, we solve four different systems of equations. The system of equations from inequalities (1) and (2) is

$$x + y = 16,$$
$$3x + 6y = 60.$$

Solving, we obtain the vertex (12, 4).

The system of equations from inequalities (1) and (4) is

$$x + y = 16,$$
$$y = 0.$$

Solving, we obtain the vertex (16, 0).

The system of equations from inequalities (3) and (4) is

$$x = 0,$$
$$y = 0.$$

The vertex is (0, 0).

The system of equations from inequalities (2) and (3) is

$$3x + 6y = 60,$$
$$x = 0.$$

Solving, we obtain the vertex (0, 10).

Do Exercise 4.

Exercise Set H

a Graph the system of inequalities. Find the coordinates of any vertices formed.

1. $y \geq x,$
$y \leq -x + 2$

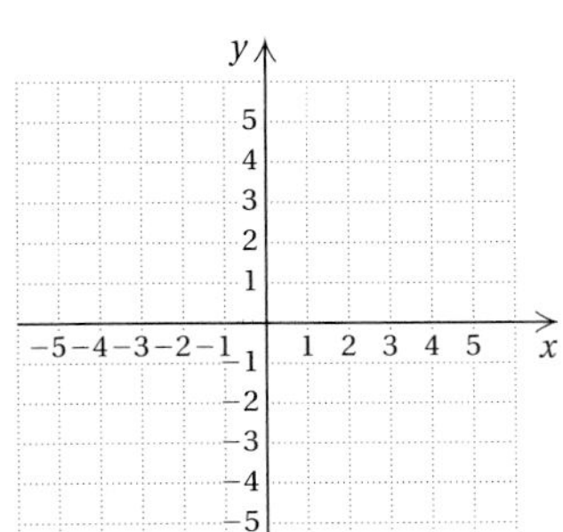

2. $y \geq x,$
$y \leq -x + 4$

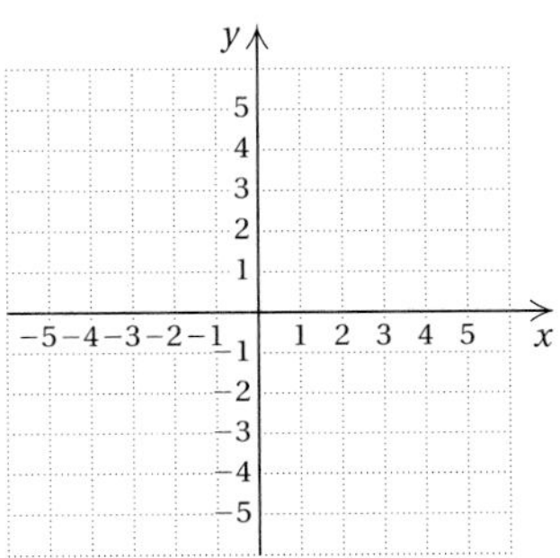

3. $y > x,$
$y < -x + 1$

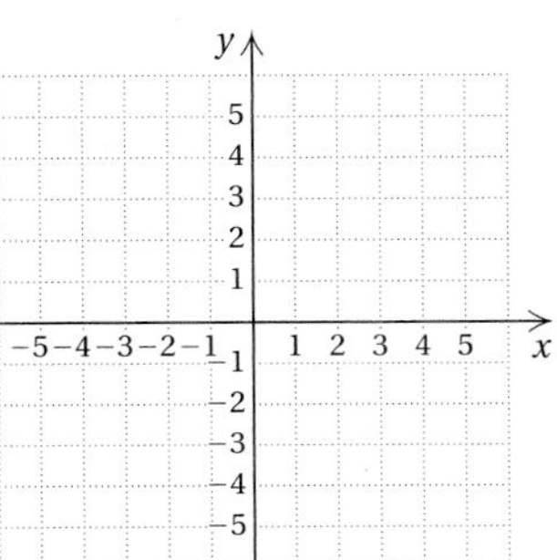

4. $y < x,$
$y > -x + 3$

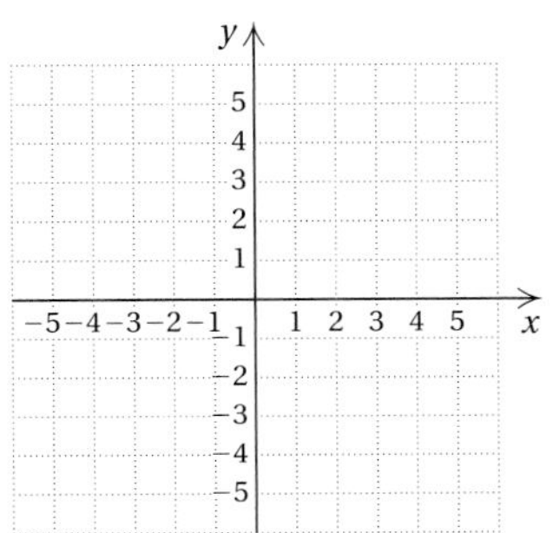

5. $y \geq -2,$
$x \geq 1$

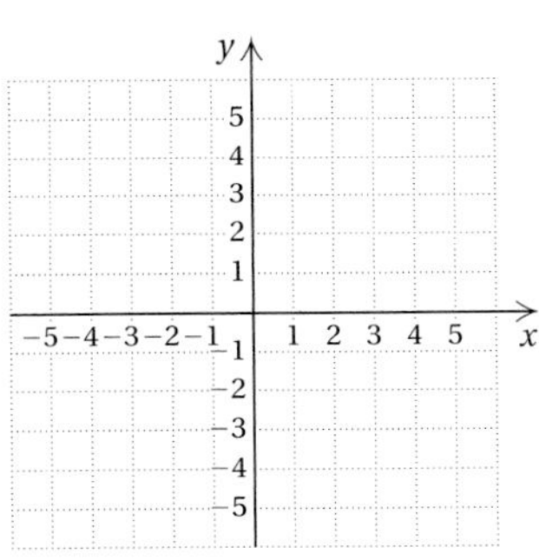

6. $y \leq -2,$
$x \geq 2$

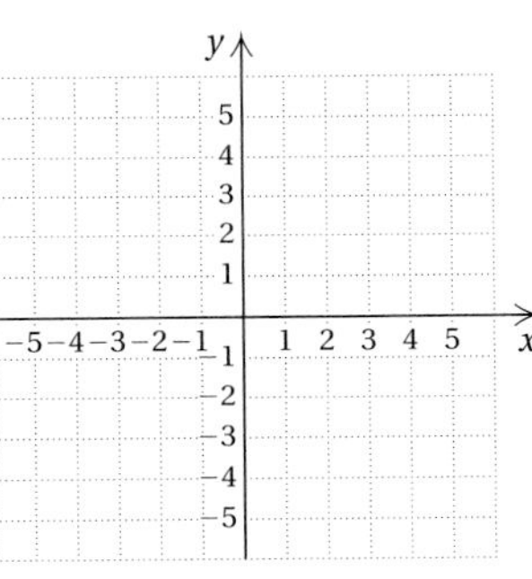

7. $x \leq 3,$
$y \geq -3x + 2$

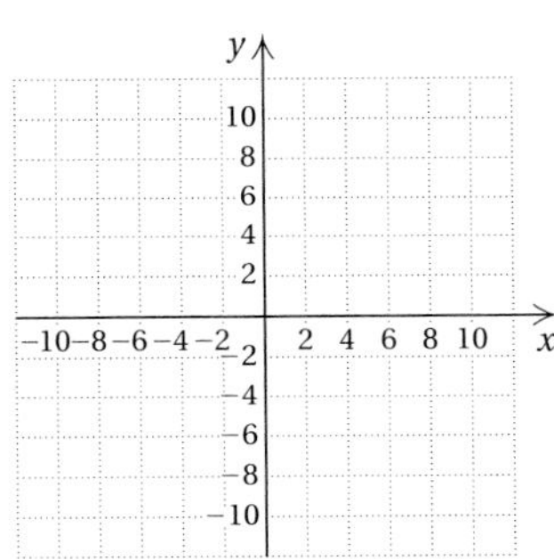

8. $x \geq -2,$
$y \leq -2x + 3$

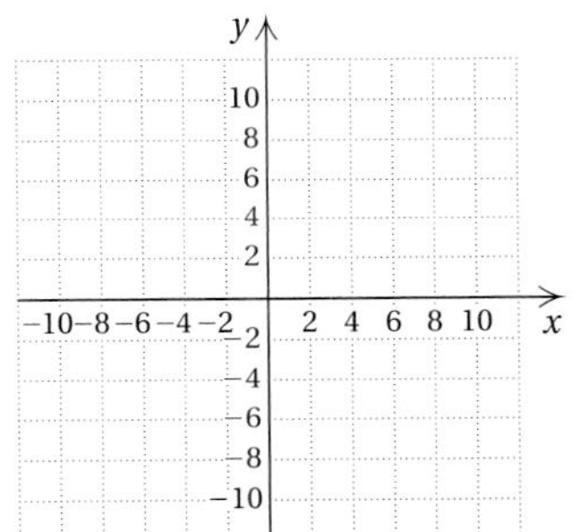

9. $y \geq -2,$
$y \geq x + 3$

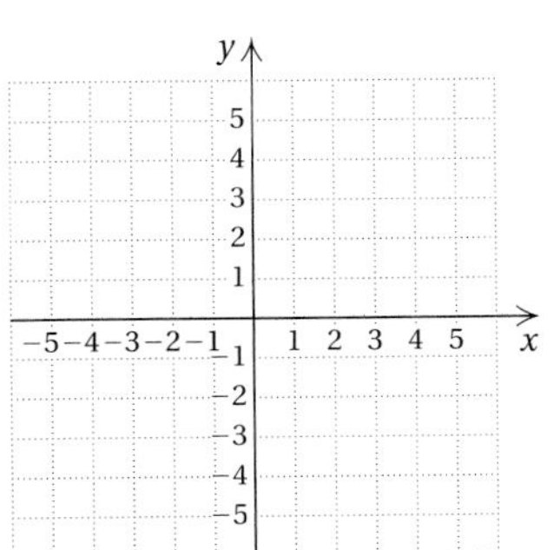

10. $y \leq 4,$
$y \geq -x + 2$

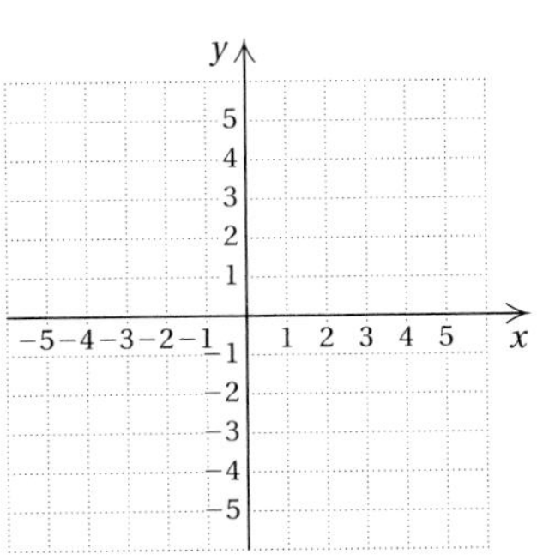

11. $x + y \leq 1,$
$x - y \leq 2$

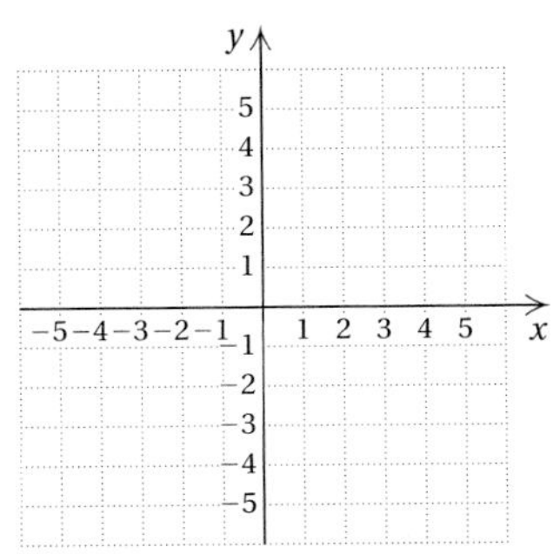

12. $x + y \leq 3,$
$x - y \leq 4$

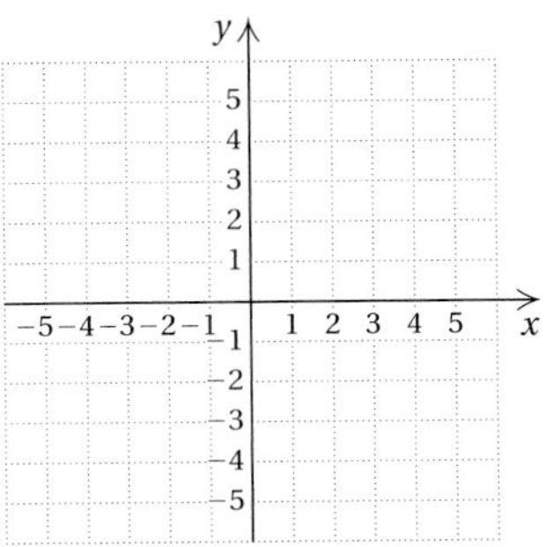

13. $y - 2x \geq 1,$
$y - 2x \leq 3$

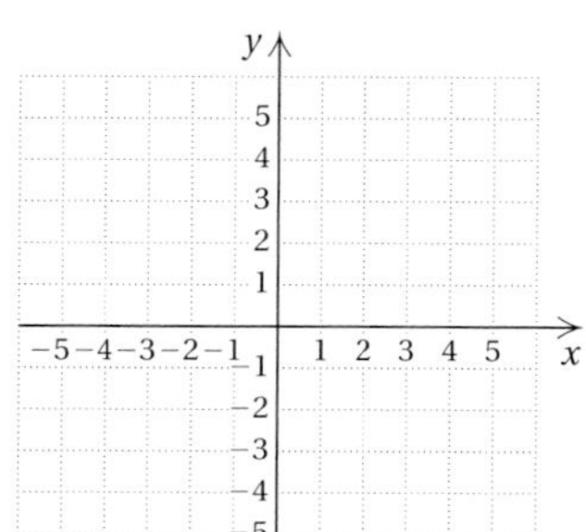

14. $y + 3x \geq 0,$
$y + 3x \leq 2$

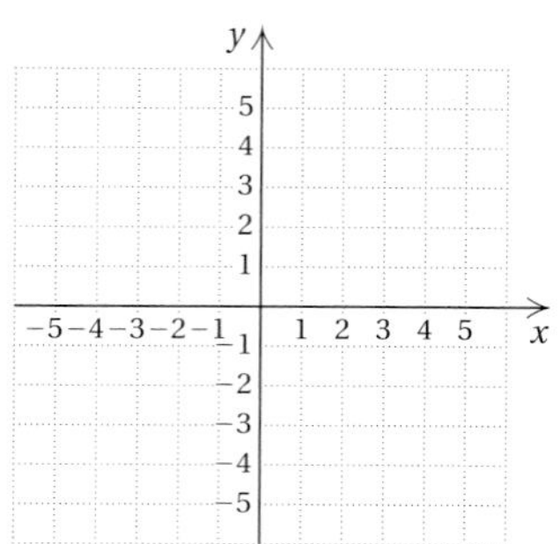

15. $y \leq 2x + 1,$
$y \geq -2x + 1,$
$x \leq 2$

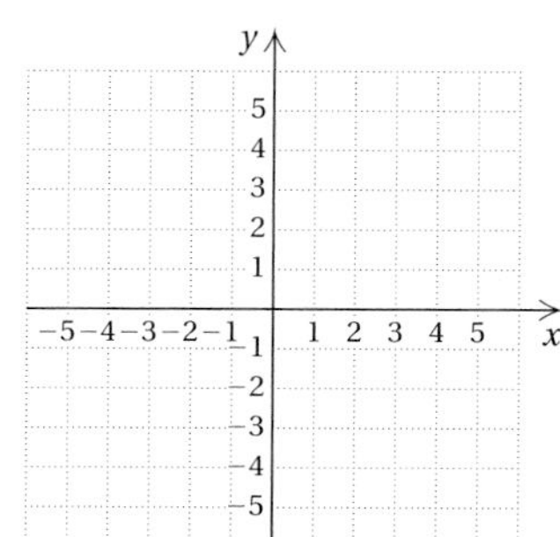

16. $x - y \leq 2,$
$x + 2y \geq 8,$
$y \leq 4$

17. $x + 2y \leq 12,$
$2x + y \leq 12,$
$x \geq 0,$
$y \geq 0$

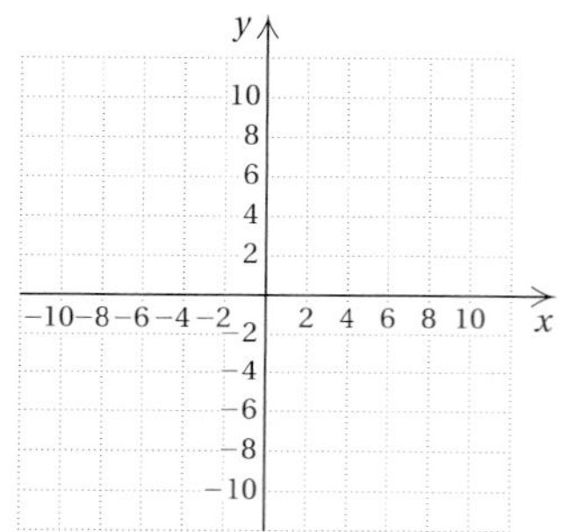

18. $4y - 3x \geq -12,$
$4y + 3x \geq -36,$
$y \leq 0,$
$x \leq 0$

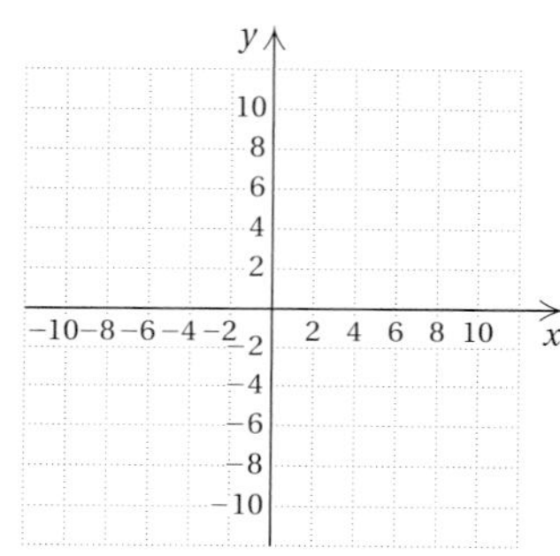

19. $8x + 5y \leq 40,$
$x + 2y \leq 8,$
$x \geq 0,$
$y \geq 0$

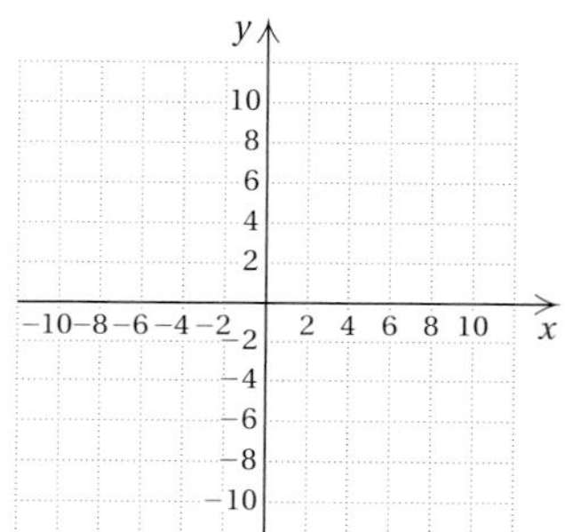

20. $y - x \geq 1,$
$y - x \leq 3,$
$2 \leq x \leq 5$

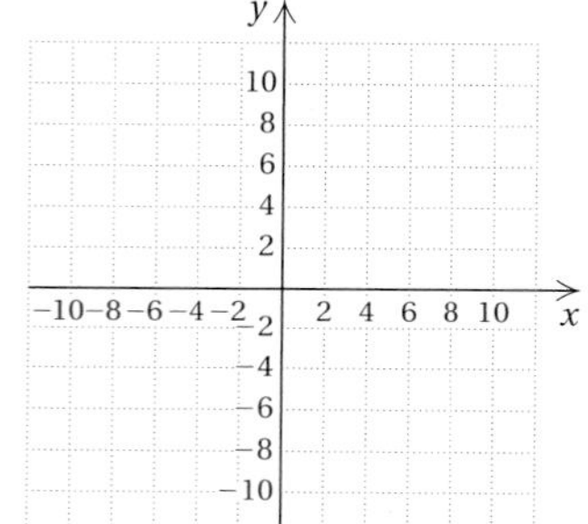

Synthesis

21. *Exercise Danger Zone.* It is dangerous to exercise when the weather is hot and humid. The solutions of the following system of inequalities give a "danger zone" for which it is dangerous to exercise intensely:

$$4H - 3F < 70,$$
$$F + H > 160,$$
$$2F + 3H > 390,$$

where F is the temperature, in degrees Fahrenheit, and H is the humidity.

a) Draw the danger zone by graphing the system of inequalities.
b) Is it dangerous to exercise when $F = 80°$ and $H = 80\%$?

Appendix I Probability

We say that when a coin is tossed, the chances that it will fall heads are 1 out of 2, or the **probability** that it will fall heads is $\frac{1}{2}$. Of course, this does not mean that if a coin is tossed ten times, it will necessarily fall heads exactly five times. If the coin is tossed a great number of times, however, it will fall heads very nearly half of them.

Objective

a Compute the probability of a simple event.

Experimental and Theoretical Probability

If we toss a coin a great number of times, say 1000, and count the number of times it falls heads, we can determine the probability of its falling heads. If it falls heads 503 times, we would calculate the probability of the coin falling heads to be

$$\frac{503}{1000}, \quad \text{or} \quad 0.503.$$

This is an **experimental** determination of probability. Such a determination of probability is quite common.

If we consider a coin and reason that it is just as likely to fall heads as tails, we would calculate the probability to be $\frac{1}{2}$. This is a **theoretical** determination of probability. Experimentally, we can determine probabilities within certain limits. These may or may not agree with what we obtain theoretically.

a Computing Probabilities

Experimental Probabilities

We first consider experimental determination of probability. The basic principle we use in computing such probabilities is as follows.

PRINCIPLE *P* (EXPERIMENTAL)

An experiment is performed in which n observations are made. If a situation E, or event, occurs m times out of the n observations, then we say that the **experimental probability** of that event is given by

$$P(E) = \frac{m}{n}.$$

Example 1 *Sociological survey.* An actual experiment was conducted to determine the number of people who are left-handed, right-handed, or both. The results are shown in the graph.

a) Determine the probability that a person is left-handed.

b) Determine the probability that a person is ambidextrous (uses both hands equally well).

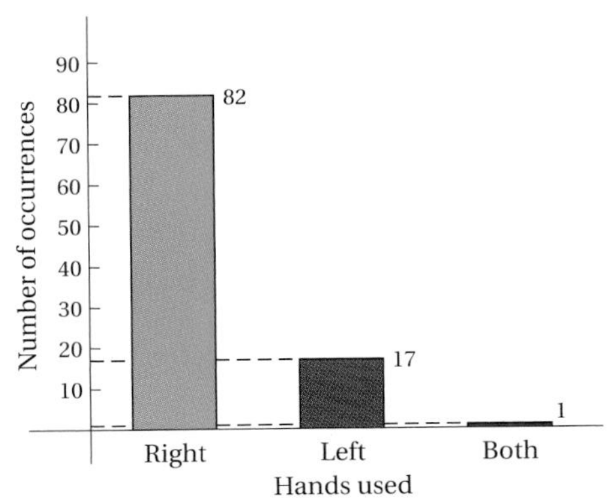

1. In reference to Example 1, what is the probability that a person is right-handed?

2. In Example 2, what is the probability that a home was tuned to NBC? What is the probability that a home was tuned to a network other than CBS, ABC, or NBC, or was not tuned in at all?

Answers on page A-55

a) The number of people who are right-handed was 82, the number who are left-handed was 17, and there was 1 person who is ambidextrous. The total number of observations was 82 + 17 + 1, or 100. Thus the probability that a person is left-handed is P, where

$$P = \frac{17}{100}.$$

b) The probability that a person is ambidextrous is P, where

$$P = \frac{1}{100}.$$

Do Exercise 1.

Example 2 *TV ratings.* The major television networks and others such as cable TV are always concerned about the percentages of homes that have TVs and are watching their programs. It is too costly and unmanageable to contact every home in the country so a sample, or portion, of the homes are contacted. This is done by an electronic device attached to the TVs of about 1400 homes across the country. Viewing information is then fed into a computer. The following are the results of a recent survey.

Network	CBS	ABC	NBC	Other or not watching
Number of Homes Watching	258	231	206	705

What is the probability that a home was tuned to CBS during the time period considered? to ABC?

The probability that a home was tuned to CBS is P, where

$$P = \frac{258}{1400} \approx 0.184 = 18.4\%.$$

The probability that a home was tuned to ABC is P, where

$$P = \frac{231}{1400} = 0.165 = 16.5\%.$$

Do Exercise 2.

The numbers that we found in Example 2 and in Margin Exercise 2 (18.4 for CBS, 16.5 for ABC, and 14.7 for NBC) are called the *ratings.*

Theoretical Probabilities

We need some terminology before we can continue. Suppose we perform an experiment such as flipping a coin, throwing a dart, drawing a card from a deck, or checking an item off an assembly line for quality. The results of an experiment are called **outcomes**. The set of all possible outcomes is called the **sample space.** An **event** is a set of outcomes, that is, the subset of the sample space. For example, for the experiment "throwing a dart," suppose the dartboard is as follows.

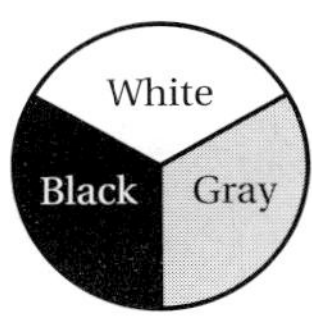

Then one event is

{black}, (the outcome is "hitting black")

which is a subset of the sample space

{black, white, gray}, (sample space)

assuming that the dart must hit the target somewhere.

We denote the probability that an event E occurs as $P(E)$. For example, "getting a head" may be denoted by H. Then $P(H)$ represents the probability of getting a head. When all the outcomes of an experiment have the same probability of occurring, we say that they are **equally likely.** A sample space that can be expressed as a union of equally likely events can allow us to calculate probabilities of other events.

> **PRINCIPLE *P* (THEORETICAL)**
>
> If an event E can occur m ways out of n possible equally likely outcomes of a sample space S, then the **theoretical probability** of that event is given by
>
> $$P(E) = \frac{m}{n}.$$

A die (pl., dice) is a cube, with six faces, each containing a number of dots from 1 to 6.

Example 3 What is the probability of rolling a 3 on a die?

On a fair die, there are 6 equally likely outcomes and there is 1 way to get a 3. By Principle P, $P(3) = \frac{1}{6}$.

Example 4 What is the probability of rolling an even number on a die?

The event is getting an *even* number. It can occur in 3 ways (getting 2, 4, or 6). The number of equally likely outcomes is 6. By Principle P, $P(\text{even}) = \frac{3}{6}$, or $\frac{1}{2}$.

Do Exercise 3.

3. What is the probability of rolling a prime number on a die?

Answer on page A-55

We now use a number of examples related to a standard bridge deck of 52 cards. Such a deck is made up as shown in the following figure.

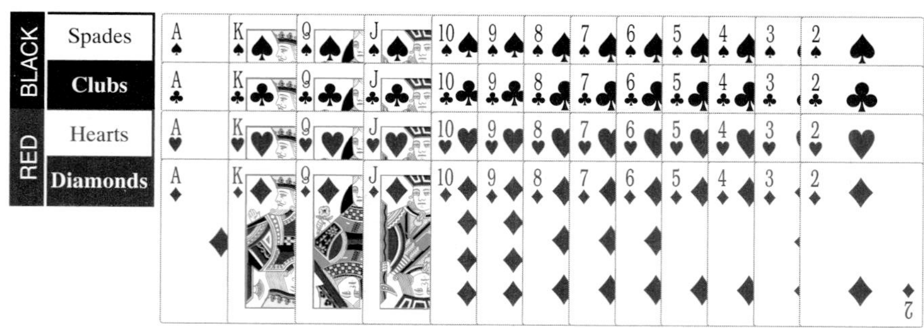

Example 5 What is the probability of drawing an ace from a well-shuffled deck of 52 cards?

Since there are 52 outcomes (cards in the deck) and they are equally likely (from a well-shuffled deck) and there are 4 ways to obtain an ace, by Principle *P* we have

$$P(\text{drawing an ace}) = \frac{4}{52}, \quad \text{or} \quad \frac{1}{13}.$$

Example 6 Suppose we select, without looking, one marble from a bag containing 3 red marbles and 4 green marbles. What is the probability of selecting a red marble?

There are 7 equally likely ways of selecting any marble, and since the number of ways of getting a red marble is 3,

$$P(\text{selecting a red marble}) = \frac{3}{7}.$$

Do Exercises 4 and 5.

If an event E cannot occur, then $P(E) = 0$. For example, in coin tossing, the event that a coin will land on its edge has probability 0. If an event E is certain to occur (that is, every trial is a success), then $P(E) = 1$. For example, in coin tossing, the event that a coin will fall either heads or tails has probability 1. In general, the probability that an event E will occur is a number from 0 to 1: $0 \leq P(E) \leq 1$.

Do Exercises 6 and 7.

4. Suppose we draw a card from a well-shuffled deck of 52 cards.

a) What is the probability of drawing a king?

b) What is the probability of drawing a spade?

c) What is the probability of drawing a black card?

d) What is the probability of drawing a jack or a queen?

5. Suppose we select, without looking, one marble from a bag containing 5 red marbles and 6 green marbles. What is the probability of selecting a green marble?

6. On a single roll of a die, what is the probability of getting a 7?

7. On a single roll of a die, what is the probability of getting a 1, 2, 3, 4, 5, or 6?

Answers on page A-55

Exercise Set I

1. In an actual survey, 100 people were polled to determine the probability of a person wearing either glasses or contact lenses. Of those polled, 57 wore either glasses or contacts. What is the probability that a person wears either glasses or contacts? What is the probability that a person wears neither?

2. In another survey, 100 people were polled and asked to select a number from 1 to 5. The results are shown in the following table.

Number Choices	1	2	3	4	5
Number of People Who Selected That Number	18	24	23	23	12

What is the probability that the number selected is 1? 2? 3? 4? 5? What general conclusion might a psychologist make from this experiment?

Linguistics. An experiment was conducted to determine the relative occurrence of various letters of the English alphabet. A paragraph from a newspaper, one from a textbook, and one from a magazine were considered. In all, there was a total of 1044 letters. The number of occurrences of each letter of the alphabet is listed in the following table.

Letter	A	B	C	D	E	F	G	H	I	J	K	L	M
Number of Occurrences	78	22	33	33	140	24	22	63	60	2	9	35	30

Letter	N	O	P	Q	R	S	T	U	V	W	X	Y	Z
Number of Occurrences	74	74	27	4	67	67	95	31	10	22	8	13	1

Round answers to Exercises 3–6 to three decimal places.

3. What is the probability of the occurrence of the letter A? E? I? O? U?

4. What is the probability of a vowel occurring?

5. What is the probability of a consonant occurring?

6. What letter has the least probability of occurring? What is the probability of this letter not occurring?

Suppose we draw a card from a well-shuffled deck of 52 cards.

7. How many equally likely outcomes are there?

8. What is the probability of drawing a queen?

9. What is the probability of drawing a heart?

10. What is the probability of drawing a 4?

11. What is the probability of drawing a red card?

12. What is the probability of drawing a black card?

13. What is the probability of drawing an ace or a deuce?

14. What is the probability of drawing a 9 or a king?

Suppose we select, without looking, one marble from a bag containing 4 red marbles and 10 green marbles.

15. What is the probability of selecting a red marble?

16. What is the probability of selecting a green marble?

17. What is the probability of selecting a purple marble?

18. What is the probability of selecting a white marble?

Synthesis

19. What is the probability of getting a total of 8 on a roll of a pair of dice? (Assume that the dice are different, say, one red and one black.)

20. What is the probability of getting a total of 7 on a roll of a pair of dice?

21. What is the probability of getting a total of 6 on a roll of a pair of dice?

22. What is the probability of getting a total of 3 on a roll of a pair of dice?

23. What is the probability of getting snake eyes (a total of 2) on a roll of a pair of dice?

24. What is the probability of getting box-cars (a total of 12) on a roll of a pair of dice?

Appendix J Applying Reasoning Skills

The two basic types of reasoning commonly used to solve problems are inductive reasoning and deductive reasoning.

Objectives

a Draw conclusions using inductive reasoning.

b Draw conclusions using deductive reasoning.

a Inductive Reasoning

Inductive reasoning is being used when conclusions are made on the basis of observations. This usually involves identifying a pattern exhibited by several items in a group and then drawing a general conclusion about the entire group. Four commonly occurring patterns are increasing, decreasing, alternating, and circular.

Example 1 Find the next number in the sequence

2, 5, 8, 11, ...

The numbers increase in value in this sequence, so we have an increasing pattern. We observe that 3 is added to each number to get the next number. Thus the next number in the sequence is 11 + 3, or 14.

Example 2 Find the next number in the sequence

50, 45, 40, 35, ...

The numbers decrease in value in this sequence, so we have a decreasing pattern. We observe that 5 is subtracted from each number to get the next number. Thus the next number in the sequence is 35 − 5, or 30.

Example 3 Find the next number in the sequence

3, 4, 6, 9, 13, ...

The numbers increase in value in this sequence, so we have an increasing pattern. We observe that, first, 1 is added to 3 to get 4, then 2 is added to 4 to get 6, then 3 is added to 6 to get 9, and then 4 is added to 9 to get 13. To find the next number in the sequence, we must add 5 to 13. Thus the next number in the sequence is 13 + 5, or 18.

Do Exercises 1–3.

Example 4 Find the next number in the sequence

4, 9, 6, 11, 8, ...

Note that the numbers go from smaller to larger to smaller to larger and so on. This is an alternating pattern. Observe that first 5 is added to 4 to get 9. Then 3 is subtracted from 9 to get 6. Next 5 is added to 6 to get 11, and then 3 is subtracted from 11 to get 8. The pattern is to add 5 and then subtract 3. To find the next number in the sequence, we add 5 to 8. Thus the next number in the sequence is 8 + 5, or 13.

Do Exercise 4.

Find the next number in each sequence.

1. 1, 6, 11, 16, ...

2. 35, 31, 27, 23, ...

3. 2, 4, 7, 11, 16, ...

4. Find the next number in the sequence

7, 13, 9, 15, 11, ...

Answers on page A-55

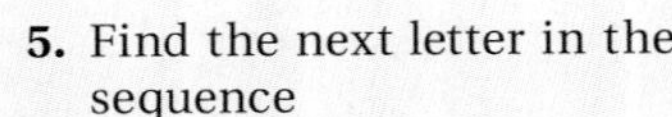

5. Find the next letter in the sequence

A, F, C, H, E, ...

6. Find the next number in the sequence

3, 13, 4, 12, 5, ...

Find the next figure in each sequence.

7.

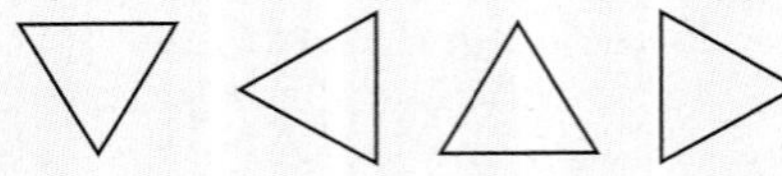

8.

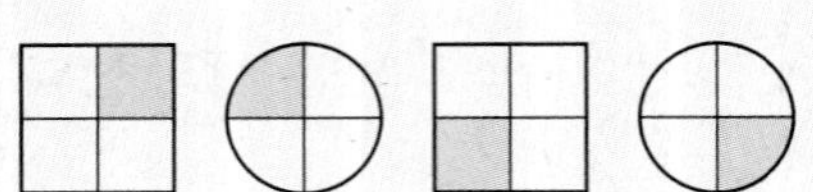

Answers on page A-55

Example 5 Find the next letter in the sequence

A, Z, B, Y, C, ...

First, we convert this sequence of letters to a sequence of numbers by assigning a numerical value to each letter. We let A = 1, B = 2, C = 3, and so on, ending with Z = 26. Rewriting the sequence in numerical form, we have

1, 26, 2, 25, 3, ...

The numbers alternate between small and large values, so we have an alternating sequence. We can use a diagram to find a pattern:

$$1 \xrightarrow{+25} 26 \xrightarrow{-24} 2 \xrightarrow{+23} 25 \xrightarrow{-22} 3$$

The diagram indicates that the next number in the sequence is 3 + 21, or 24. This corresponds to the letter X.

Do Exercises 5 and 6.

Example 6 Find the next figure in the sequence.

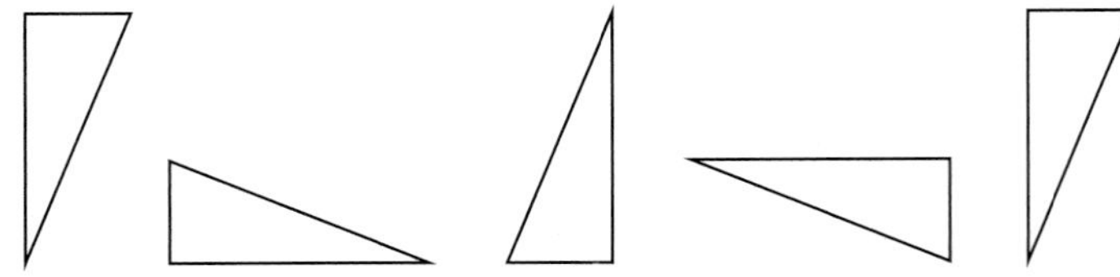

This pattern consists of a triangle that rotates one-fourth turn counterclockwise as the sequence moves from figure to figure. This is a circular pattern. To continue this pattern, we rotate the last triangle one-fourth turn counterclockwise to get the figure shown.

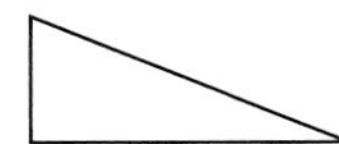

Example 7 Find the next figure in the sequence.

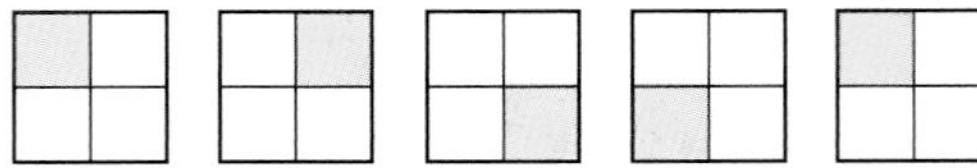

This sequence consists of a square divided into four regions, one of which is color. We observe that the color rotates clockwise as the sequence moves from figure to figure, so we have a circular pattern. To continue this pattern, the next figure in the sequence should be color in the upper righthand region, as shown here:

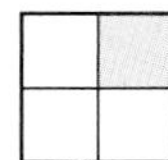

Do Exercises 7 and 8.

Example 8 Find the next figure in the sequence.

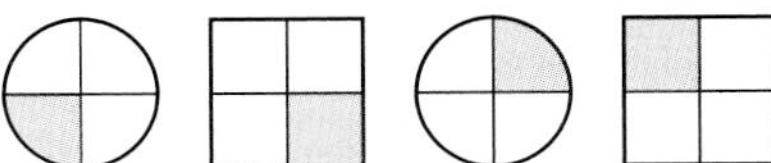

There are two patterns in this sequence. Circles and squares alternate to create one pattern. The second pattern is created by color that rotates counterclockwise as the sequence moves from figure to figure. The next figure in the shape pattern will be a circle. The next figure in the color sequence will be color in the lower lefthand region. Thus the next figure in the sequence is as shown below:

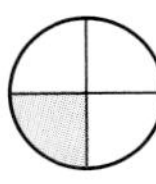

Do Exercises 9 and 10.

b Deductive Reasoning

Deductive reasoning is being used when conclusions are made on the basis of a set of facts. This usually involves deriving a series of new facts from those originally stated until a conclusion is reached.

Example 9 A survey of 100 people shows that 65 own a dog, 25 own a cat, and 18 own both a dog and a cat. How many of those surveyed own neither a dog nor a cat?

First, we find that 65 + 25, or 90 people, own a dog or a cat or both. From this number, we subtract the number of people who own both a dog and a cat, because they are counted twice—once as dog owners and once as cat owners. We have 90 − 18, or 72. Finally, we subtract this number from 100, the number of people surveyed, to find how many own neither a dog nor a cat. We have 100 − 72, or 28.

Do Exercises 11 and 12.

Find the next figure in each sequence.

9.

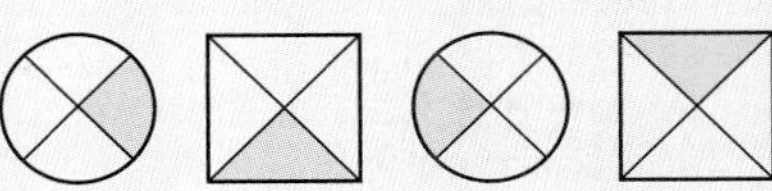

10.

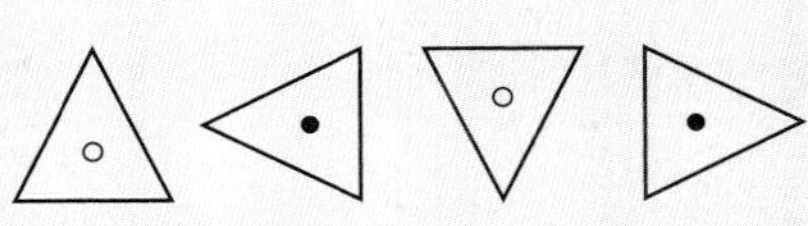

11. A poll of 125 people shows that 68 are on a low-fat diet, 39 are on a low-sodium diet, and 17 are on both a low-fat and a low-sodium diet. How many of those polled are on neither a low-fat nor a low-sodium diet?

12. A survey of 200 people shows that 125 of them subscribe to a daily newspaper, 70 subscribe to a weekly newsmagazine, and 65 subscribe to both. How many of those surveyed subscribe to neither a daily newspaper nor a weekly newsmagazine?

Answers on page A-55

Example 10 Amy, Chloe, Marc, and Pete all work in the same building. One is a doctor, one is a lawyer, one is a computer analyst, and one is an architect. Use the statements below to determine who is the architect.

1. Chloe and Marc carpool with the computer analyst.
2. Amy and Pete work on the same floor of the building as the architect.
3. Chloe eats lunch with the lawyer and the architect.

We set up a table as shown below. We use the given statements to determine which jobs each person *cannot* hold. In statement 1, we learn that neither Chloe nor Marc is the computer analyst, so we put an X under "Analyst" in the rows labeled "Chloe" and "Marc." In statement 2, we learn that neither Amy nor Pete is the architect, so we put an X under "Architect" in the rows labeled "Amy" and "Pete." Statement 3 tells us that Chloe is neither the lawyer nor the architect, so we put an X under both "Lawyer" and "Architect" in the row labeled "Chloe." Now we see that the only empty cell of the table under "Architect" corresponds to the row labeled "Marc," so Marc is the architect.

	Doctor	Lawyer	Analyst	Arhitect
Amy				X
Chloe		X	X	X
Marc			X	
Pete				X

Do Exercise 13.

13. Maria, Tonya, Carlos, and Frank all live in the same building. One is a student, one is a teacher, one is a salesperson, and one is a paramedic. Use the statements and the table below to determine who is the teacher.

1. Tonya lives on the same floor as the teacher and the paramedic.
2. Maria and Frank jog with the student.
3. Carlos and Frank eat dinner with the teacher.

	Sudent	Teacher	Salesperson	Paramedic
Maria				
Tonya				
Carlos				
Frank				

Answer on page A-55

Exercise Set J

a Find the next letter or number in each sequence.

1. 7, 10, 13, 16, ...

2. 25, 27, 21, 23, 17, ...

3. B, G, L, Q, ...

4. W, U, S, Q, ...

5. 40, 41, 39, 42, 38, ...

6. 1, 3, 4, 7, 11, ...

Find the next figure in each sequence.

7. 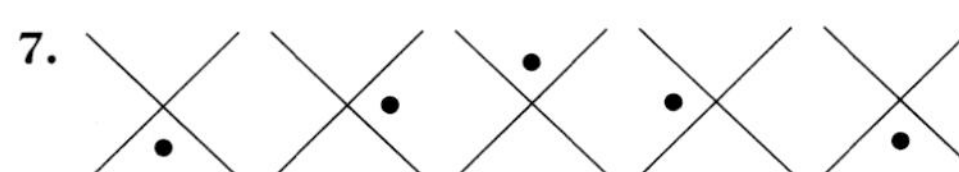

8. 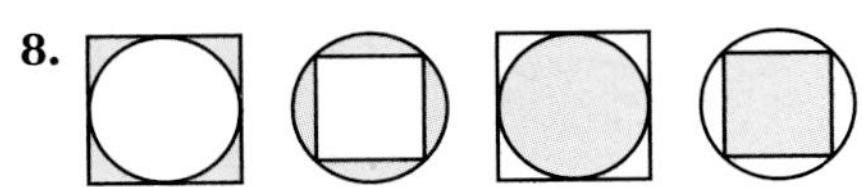

9. 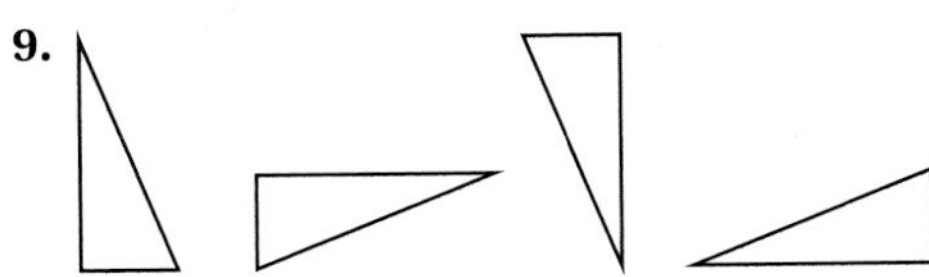

10. 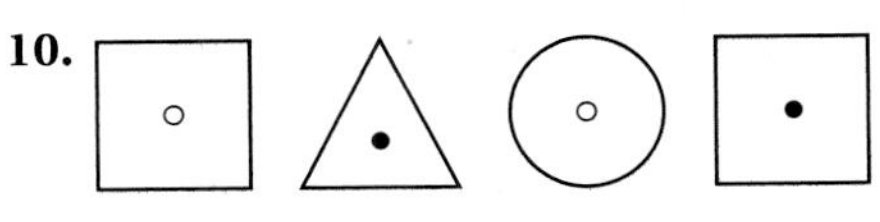

b

11. A survey of 100 households shows that 78 of them have a VCR, 42 have a CD player, and 37 have both a VCR and a CD player. How many of those surveyed have neither a VCR nor a CD player?

12. A poll of 80 library patrons shows that, in the past month, 60 of them checked out at least one novel, 25 checked out at least one audiotape, and 13 checked out both a novel and an audiotape. How many of those polled checked out neither a novel nor an audiotape in the past month?

13. Max, Jay, Bev, and Lea live on the same street. For exercise, one jogs, one walks, one swims, and one rides a bicycle. Use the statements and the table below to determine who is the swimmer.

1. Max and Lea carpool with the swimmer.
2. Jay and Bev live across the street from the walker.
3. Jay's next-door neighbors are the bike rider and the swimmer.

	Jogger	Walker	Swimmer	Biker
Max				
Jay				
Bev				
Lea				

14. Antonio, Dale, Janet, and Lynn attend the same college. One majors in English, one in business, one in math, and one in Spanish. Use the statements and the table below to determine Lynn's major.

1. Antonio and Lynn eat lunch with the Spanish major.
2. Dale and Janet study with the business major.
3. Lynn is in the same speech class as the English and business majors.

	English	Business	Math	Spanish
Antonio				
Dale				
Janet				
Lynn				

15. Pat, Chris, Casey, and Lee are in the same math class. One has brown hair, one has black hair, one has red hair, and one is a blond. Use the statements below to determine who has red hair.

1. Pat walks to class with the redhead and the blond.
2. Chris got a higher score on the last test than the students with black hair and blond hair.
3. Casey and Lee study with the redhead.

16. Joe, Sam, Fran, and Beth work in the same office. One is a supervisor, one is a typist, one is a receptionist, and one is a bookkeeper. Use the statements below to determine Joe's position.

1. Sam and Beth earn more than the typist.
2. Joe eats lunch with the receptionist and the bookkeeper.
3. The supervisor works later than Fran and Joe.

Appendix K Multistep Applied Problems

a Solving Multistep Applied Problems

Objective

a Solve multistep applied problems.

Often it is necessary to use a combination of mathematical skills in order to solve an applied problem. As you familiarize yourself with such a problem, you should break it down into small problems that can be solved in sequence in order to arrive at the desired result.

Example 1 Four students plan to drive 2440 mi during a one-week spring break. They rent an SUV that gets 25 miles per gallon (mpg) on the highway for a weekly rate of \$395 (with unlimited mileage). The gasoline costs average \$1.35 per gallon. If they plan to split the total cost of transportation equally, what is the transportation cost per student?

1. **Familiarize.** First, we find the number of gallons of gas required. Next, we find the cost of the number of gallons of gasoline. Finally, we determine the total cost of the transportation and divide by 4 to determine the cost per student. We let $n =$ the number of gallons required, $c =$ the cost of the gasoline, and $t =$ the total cost of transportation.

2., 3. **Translate** and **Solve.** First, we translate to an equation that will enable us to find n.

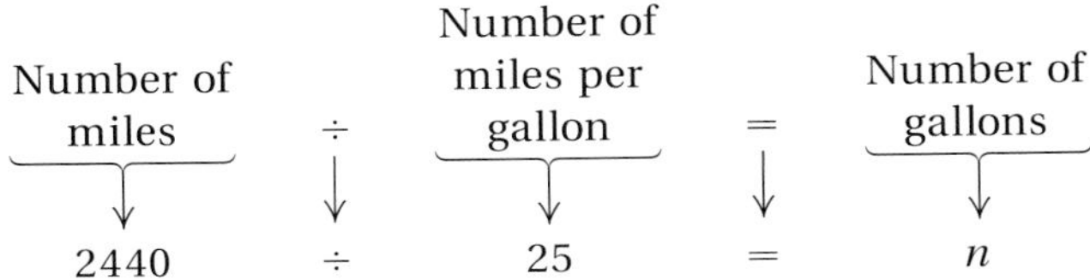

To solve the equation, we carry out the division:

$2440 \div 25 = 97.6.$

Thus, $n = 97.6$ gallons.

Next, we translate to an equation that will enable us to find c.

Price per gallon	×	Number of gallons	=	Cost of gasoline
↓	↓	↓	↓	↓
\$1.35	×	97.6	=	c

To solve the equation, we carry out the multiplication:

$1.35 \times 97.6 = 131.76.$

Thus, $c = \$131.76$.

We now find the total transportation cost, t.

Rental cost	+	Gasoline cost	=	Total cost of transportation
↓	↓	↓	↓	↓
\$395	+	\$131.76	=	t

To solve the equation, we add:

$395 + 131.76 = 526.76.$

The total cost of transportation is \$526.76. Since there are four students, we divide the total cost by 4:

$526.76 \div 4 = 131.69.$

Thus the cost per student is \$131.69.

1. Three professors are planning to drive 810 mi to a conference. They rent a minivan for a week at a weekly rate of \$285. The van gets 27 mpg and the average cost of gasoline is \$1.20 per gallon. If they plan to split the total cost of transportation equally, what is the transportation cost per professor?

2. A homeowner hires a painter to paint the exterior of her home. She buys 7 gal of paint at \$19.99 per gallon and two paintbrushes at \$10.95 each. The sales tax rate is 5%. If the painter charges \$20 an hour and it takes 32 hr to complete the job, what is the total cost of the job?

Answers on page A-55

4. **Check.** We multiply the cost per student by the number of students:

$$131.69 \times 4 = 526.76.$$

Then we subtract the rental cost of the SUV:

$$526.76 - 395 = 131.76.$$

Next, we divide the gasoline cost by the price per gallon to find the number of gallons used:

$$131.76 \div 1.35 = 97.6.$$

And finally, we multiply the number of gallons by the number of miles per gallon to obtain the total number of miles driven:

$$97.6 \times 25 = 2440.$$

The result checks.

5. **State.** The transportation cost per student is \$131.69.

Do Exercise 1.

Example 2 A law firm called A-1 Service to repair its air conditioner. It was charged \$55 for the service call, \$60 per hour for $1\frac{1}{2}$ hr of labor, and \$42.79 for parts. What was the total cost of the repair?

1. **Familiarize.** First, we find the cost of labor. Then we add the individual costs to find the total cost of the repair. We let x = the cost of the labor and c = the total cost of the repair.

2., 3. **Translate** and **Solve.** First, we translate to an equation that will enable us to find x.

$$\underbrace{\text{Cost per hour}}_{60} \cdot \underbrace{\text{Number of hours}}_{1\frac{1}{2}} = \underbrace{\text{Cost of labor}}_{x}$$

To solve the equation, we carry out the multiplication:

$$60 \cdot 1\frac{1}{2} = \frac{60}{1} \cdot \frac{3}{2} = \frac{180}{2} = 90.$$

Thus, $x = \$90$.

Now we add to find the total cost of the repair. We have

$$c = 55 + 90 + 42.79 = 187.79.$$

4. **Check.** We check by repeating the calculations. The result checks.

5. **State.** The total cost of the repair was \$187.79.

Do Exercise 2.

Exercise Set K

a Solve.

1. The Speedy Cab Company charges $3 for the first mile of a taxi ride and 75¢ for each additional mile. What is the cost of a 13-mi taxi ride?

2. A telephone company charges 25¢ for the first minute of a long-distance call and 10¢ for each additional minute. What is the cost of a 33-min long-distance call?

3. A driver plans to change her car's motor oil. She buys 4 qt of oil at $1.49 per quart and an oil filter for $4.95. What is the total cost of the oil change?

4. A consumer purchases 3 pairs of socks at $6.95 each and a belt for $36. What is the total cost of the purchases?

5. An employee at a sporting goods store is paid $8.50 per hour plus 2% commission on her sales. One week she worked 35 hr and had sales of $15,700. What was her total pay for the week?

6. A salesperson earns a commission of 5% on his first $20,000 of sales in a month and 8% on all sales over $20,000 for the month. One month he had sales of $45,000. How much did he earn in commissions for the month?

7. A family room measures 15 ft by 18 ft. Carpet costing $21 per square yard and pad costing $3.25 per square yard will be laid in the room. What is the total cost of the carpet and pad?

8. One pound of boneless honey-baked ham serves 6 people. A hostess is planning to serve this ham at a dinner party for 15 people. Ham costs $7.95 per pound. What is the cost of the ham for the dinner party?

9. A driver filled the gas tank when the odometer read 46,192.8. After the next fill-up, the odometer read 46,519.4. It took 11.5 gal to fill the tank. How many miles per gallon did the driver get?

10. A homeowner plans to paint the exterior of his home. He buys 9 gal of paint at $14.99 per gallon and a paint brush for $12.49. The sales tax rate is $4\frac{1}{2}$%. If he hires a painter who charges $15 per hour and it takes 45 hr to complete the job, what is the total cost of the job?

11. A cylindrical water tank has a height of 28 ft and a diameter of 10 ft. One cubic foot of water weighs 62.5 lb. How many pounds of water will the tank hold when it is full? (Use $\frac{22}{7}$ for π.)

12. A cylindrical tank with a height of 24 ft and a radius of 9 ft is filled with water. There is about 7.5 gal in a cubic foot of water. A pump can empty the tank at a rate of 120 gal per minute. How long will it take to empty the tank? (Use 3.14 for π.)

13. A lot measures 90 ft by 220 ft. A house measuring 30 ft by 85 ft sits on the lot. A bag of lawn fertilizer covers 5000 ft^2 and costs $7.95. How much will it cost to fertilize the lawn?

14. A caterer is preparing brunch for 32 people. She will serve 6 oz of orange juice to each person. A 64-oz carton of orange juice costs $3.49. What is the cost of the orange juice for the brunch?

15. A department store employee gets a 35% discount on shoes and a 30% discount on clothing. He buys a pair of shoes regularly priced at $69.95 and a shirt regularly priced at $54.95. What did the employee pay for these purchases?

16. A bedroom measures 4 m by 5 m. The ceiling is 3.5 m high. There are two windows in the room, each measuring 1 m by 4.5 m. The door to the room and the closet door each measure 1 m by 2.1 m. A liter of paint covers 20.2 m^2 and costs $3.98 per liter. How much will it cost to paint the walls of the room?

17. Sam, a paralegal who has just taken an entry-level job in state government, makes $26,300 a year. He can qualify for a home loan of up to $2\frac{1}{4}$ times his annual salary. With the current interest rate, his monthly payment on a mortgage is about 1% of the amount borrowed. If Sam borrows as much as he can, what will be his monthly payment?

18. John started a new job with an annual salary of $34,000. After three months, he received a cost-of-living adjustment that raised his salary by 2%. Then after one year, his salary was raised by 4%. What was his annual salary after both increases?

19. A decorative rug is placed in the center of a rectangular room. The distance between the carpet and the wall is the same all the way around the room. The room is 23 ft long and 18 ft wide. The rug is 150 ft^2. What is the distance between the carpet and the wall?

20. A wallpaper border is available in 15-ft rolls at $10.95 per roll. Sally is going to place this border around both the top and the bottom of a rectangular room that measures 22 ft by 12 ft. How many rolls will she need and what will be her total cost?

Appendix L TASP Practice Test

1. Green and White Landscaping employs 48 people during the summer. In August, $\frac{1}{3}$ of them return to school. Only $\frac{5}{8}$ of those remaining are employed for snow removal in the winter. How many people does Green and White employ during the winter?

   A. 10 **B.** 20 **C.** 46 **D.** 12

2. Debbie started a new job with an annual salary of $24,000. In January, she received a cost-of-living adjustment that raised her salary by 3%. In April, her salary was raised by 5%. What was her annual salary after both increases?

   A. $25,920 **B.** $25,956 **C.** $25,200 **D.** $24,800

3. Lindsay paid $4.60 for the use of 400 ft^3 of water. At that rate, how much would she have to pay for 700 ft^3 of water?

   A. $32.20 **B.** $2.63 **C.** $8.80 **D.** $8.05

4. Light travels at a speed of 3.0×10^5 km per second. Earth is, on average, 1.5×10^8 km from the sun. About how many minutes does it take for light to travel to Earth from the sun?

   A. 8.3 min **B.** 500 min **C.** 0.5 min **D.** 83 min

5. Jason began a hike at an elevation of 150 ft above sea level. He descended 265 ft, then climbed back up 75 ft. At what elevation was he at that point?

   A. 490 ft above sea level **B.** 340 ft above sea level **C.** 40 ft below sea level **D.** 190 ft below sea level

6. The following circle graph shows how college students, in general, use their spending money.

   College Spending Money

   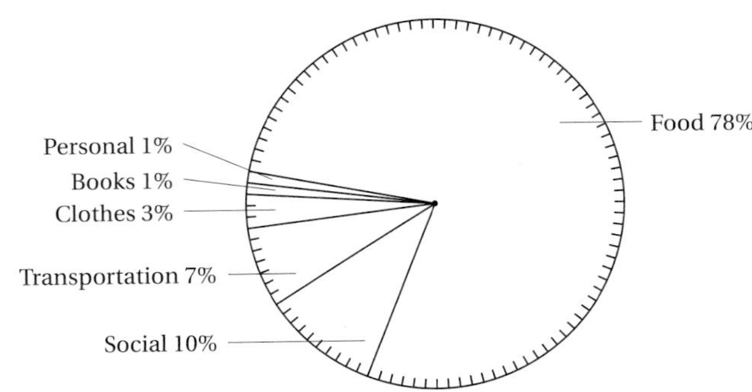

   After paying her fixed expenses, Carla has 25% of her income left for spending money. According to the circle graph, what percent of her income does she spend each month for social activities?

   A. 35%
   B. 2.5%
   C. 15%
   D. Not enough information given to determine the answer

7. Carol assembles 10 electronic parts per hour during the first half of her shift. She takes a half-hour lunch break. After lunch her rate drops to 8 parts per hour for the last half of her shift. Which of the following graphs represents the total number of parts assembled throughout the day?

A.

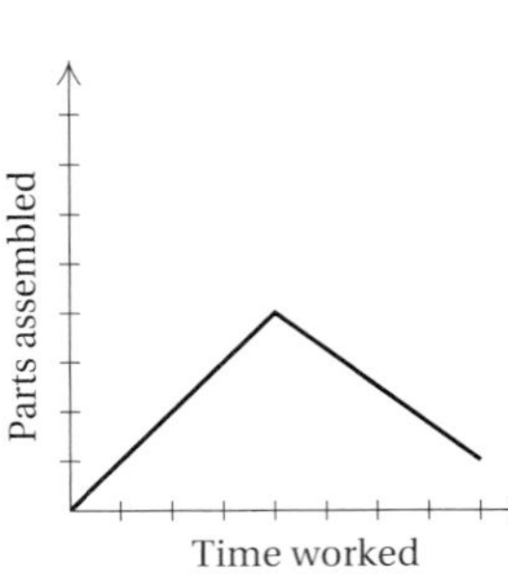

B.

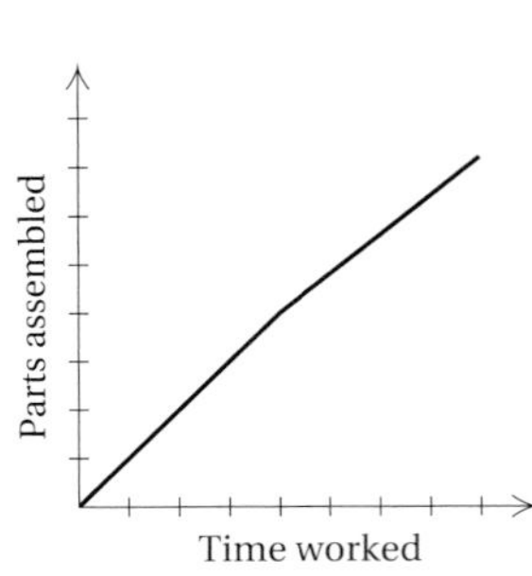

C.

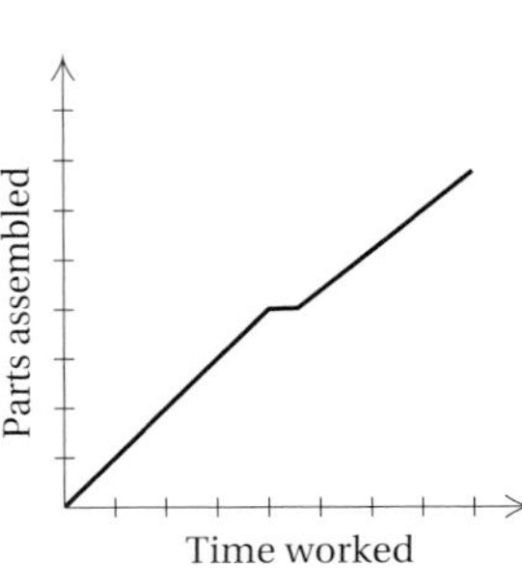

D.

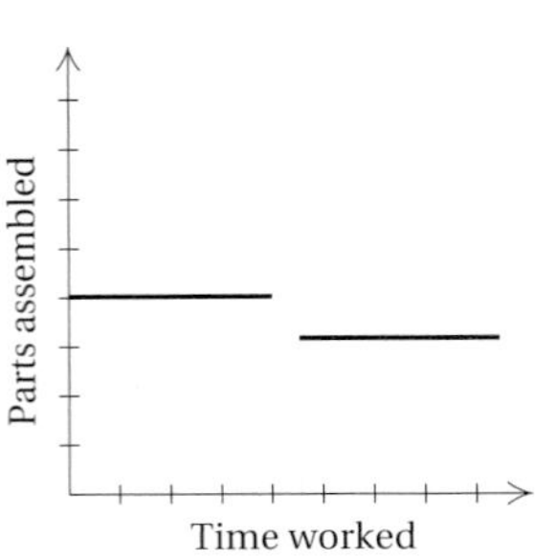

8. In order to estimate the market value of his home, Jordan needs to know the average selling price of other homes in his neighborhood. In the first half of the preceding year, three homes sold for an average selling price of $124,000. In the second half of the year, a home sold for $140,000 and another for $120,000. What was the average selling price of all the homes sold in the preceding year?

A. $124,000 **B.** $128,000 **C.** $130,200 **D.** $126,400

9. The curves below show the distribution of grades in two separate sections of a mathematics class.

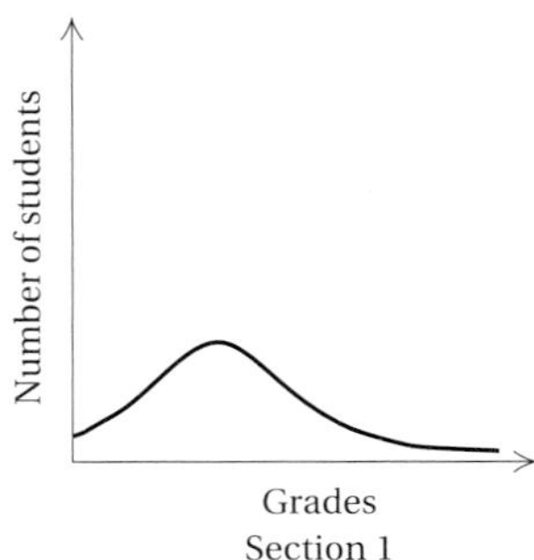

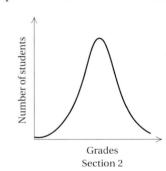

Which of the following statements correctly analyzes the information presented in these distributions?

A. The mean grade for both sections was the same.
B. There were more students in section 2 than in section 1.
C. There was more variability in the grades for section 1 than in those for section 2.
D. Students in section 2 received poorer grades than students in section 1.

10. Use the graph of the line AB at right to answer the question that follows.

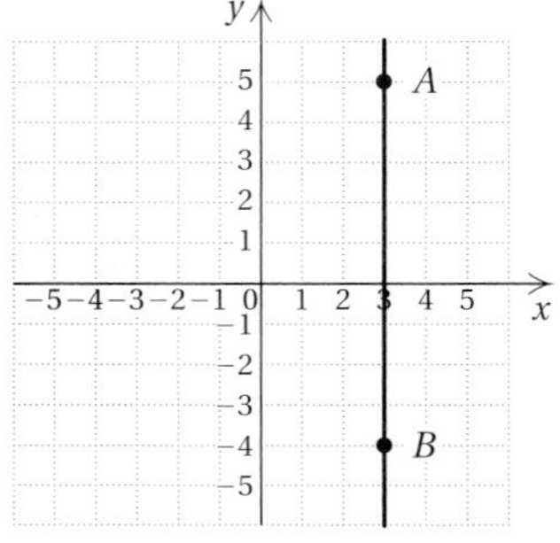

Which of the following equations represents the line AB?

A. $y = 3$ **B.** $x + y = 3$
C. $x = 3$ **D.** $y = x + 3$

11. Find the slope, if it exists, of the line containing the points (0, −4) and (−2, −1).

A. $-\frac{3}{2}$ **B.** $-\frac{2}{3}$ **C.** $\frac{3}{2}$ **D.** Undefined

12. What are the coordinates of the y-intercept of the line whose equation is $2x - 5y = 10$?

A. (0, −2) **B.** (0, 5) **C.** (5, 0) **D.** (−2, 0)

13. Which of the following is an equation of the line containing the points (−5, 6) and (−2, 4)?

A. $y = -\frac{3}{2}x - \frac{3}{2}$ **B.** $y = -\frac{2}{3}x + 6$ **C.** $y = \frac{3}{2}x + \frac{27}{2}$ **D.** $y = -\frac{2}{3}x + \frac{8}{3}$

14. The graph at right shows how the amount of time t that it takes to polish the rotor blade of a helicopter depends on the number of machinists n working on it. Which of the following statements about the relationship is true?

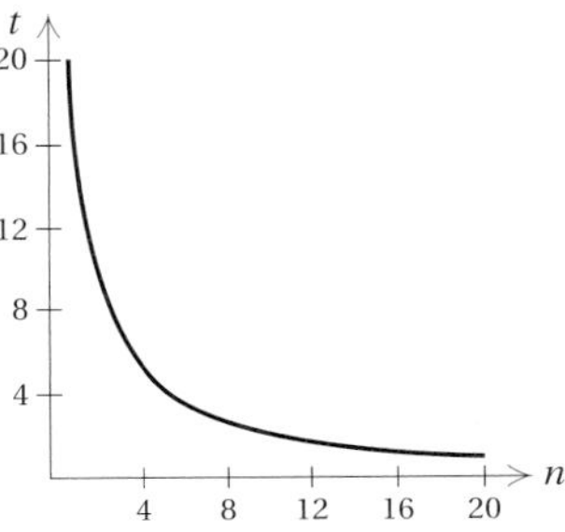

A. The more machinists working, the longer it takes to polish the blade.
B. It takes 4 machinists 10 hr to polish the blade.
C. If the blade must be polished in 2 hr, 5 machinists can get it done.
D. In 10 hr, 2 machinists can polish the blade.

15. If $\frac{2}{5}x + 7 = 1$, what is the value of $10 - 3x$?

A. −15 **B.** 55 **C.** 17 **D.** −35

16. Solve $t = \frac{1}{5}(r + 5)$ for r.

A. $r = 5t - 5$ **B.** $r = 5t - 25$ **C.** $r = \frac{t-1}{5}$ **D.** $r = \frac{1}{5}t - 1$

17. What is the solution of the system of equations $y = 2x^2 + 7x + 3$ and $4x + 3y = 9$?

A. $(-3, 7), \left(-\frac{1}{2}, \frac{11}{3}\right)$ **B.** $(0, 3), \left(-\frac{25}{6}, \frac{77}{9}\right)$
C. $(3\sqrt{3}, 3 - 4\sqrt{3}), (-3\sqrt{3}, 3 + 4\sqrt{3})$ **D.** No solution

18. Which of the following equations correctly translates this statement? The product of the length l of a fish and the square of the girth g of the fish is 280 times the weight w of the fish.

A. $(l \times g)^2 = 280w$ **B.** $280(l \times g^2) = w$ **C.** $l \times g^2 = 280w$ **D.** $280(l + g) = w$

19. Which of the following graphs shows the solution of the system of equations $y - 5 + x = 2x$ and $y = x^2 - 2$?

A.

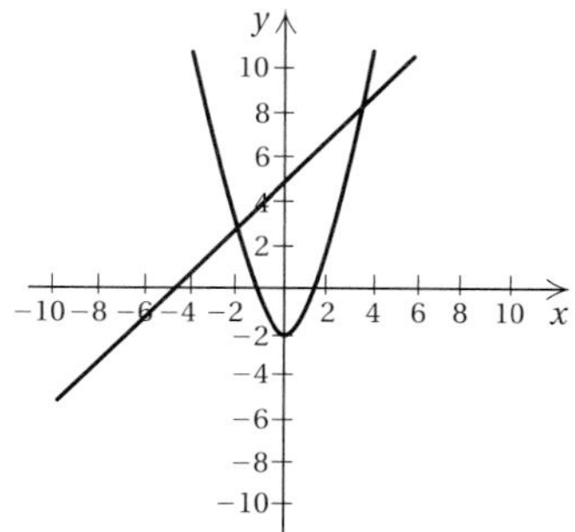

B.

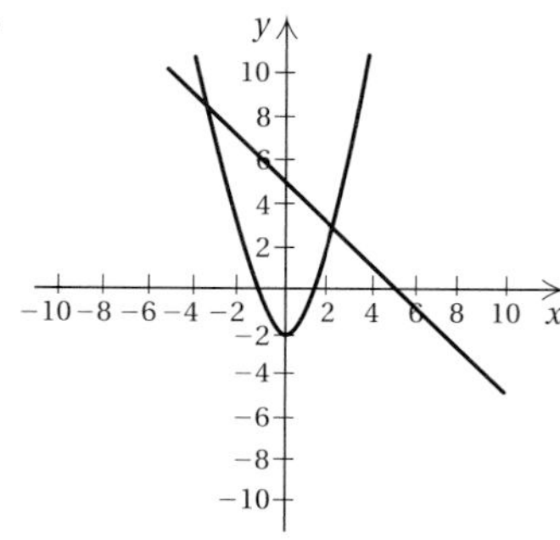

C.

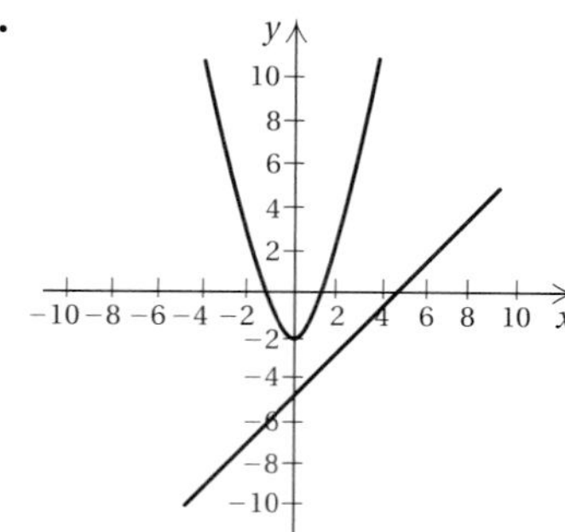

D.

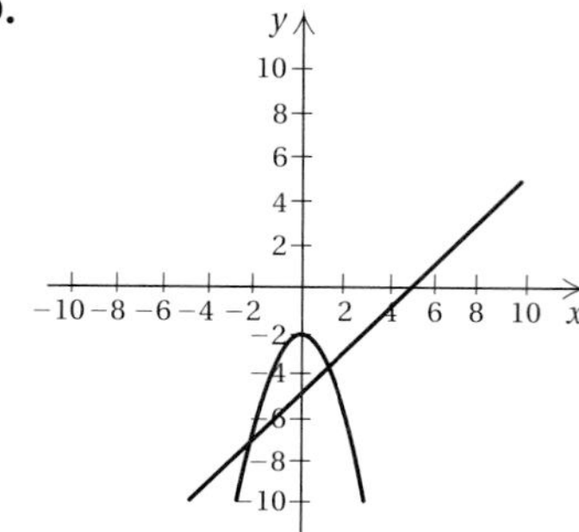

20. Karen earns 2% of the price of each home she sells as commission. Last week, she sold two houses and earned \$3500. The selling price of one house was $1\frac{1}{2}$ times the selling price of the other. What was the selling price of the more expensive house?

A. \$175,000 **B.** \$105,000 **C.** \$116,666 **D.** \$87,500

21. Party Time stores sell complete party packages. The Flyaway package contains paper products for 12 people plus 30 helium-filled balloons and sells for \$34.80. The Down-to-Earth package contains paper products for 15 people plus 1 balloon and sells for \$14.30. Party Time wants to begin selling a Party package that will contain paper products for 25 people plus 10 balloons. What should the selling price of the Party package be?

A. \$49.10 **B.** \$29.00 **C.** \$30.50 **D.** \$11.60

22. Which of the following is one factor of $3x^2 - 4x - 15$?

A. $(x - 5)$ **B.** $(3x - 5)$ **C.** $(x + 3)$ **D.** $(x - 3)$

23. Perform the multiplication: $(2x + 5)^2$.

A. $4x^2 + 25$ **B.** $4x^2 + 20x + 25$ **C.** $4x^2 + 10x + 25$ **D.** $2x^2 + 25$

24. Add and simplify, if possible: $\dfrac{x - 12}{x^2 + x - 6} + \dfrac{x}{x - 2}$.

A. $\dfrac{x + 6}{x + 3}$ **B.** $\dfrac{4x - 12}{x - 6}$ **C.** $\dfrac{2x - 12}{x^2 + 2x - 8}$ **D.** $\dfrac{2x - 12}{(x^2 + x - 6)(x - 2)}$

25. Add and simplify, if possible: $\sqrt{18x^3} - \sqrt{2x} + \sqrt{3x}$.

A. $\sqrt{18x^3 + x}$ B. $\sqrt{18x^3} + \sqrt{x}$ C. $(3x - 1)\sqrt{2x} + \sqrt{3x}$ D. $\sqrt{x}(2\sqrt{2} + \sqrt{3})$

26. If $f(x) = |2x - 7|$, find $f(2) \cdot f\left(\frac{1}{2}\right)$.

A. 18 B. 5 C. 25 D. 3

27. Shown at right is the graph of a quadratic function. Which of the following equations is represented by this graph?

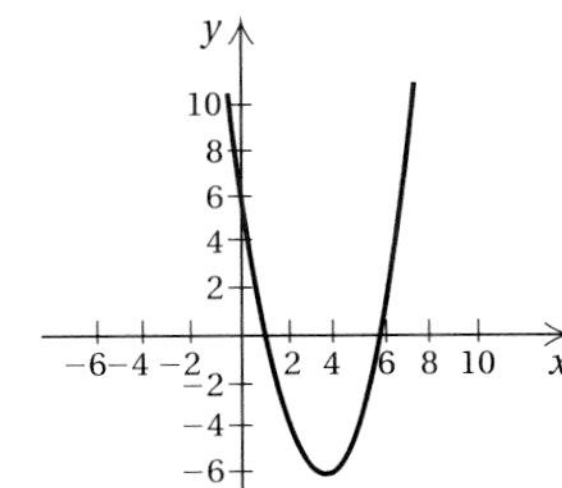

A. $y = x^2 - 5x - 6$ B. $y = x^2 + 5x - 6$
C. $y = x^2 + 7x + 6$ D. $y = x^2 - 7x + 6$

28. Shown at right is the graph of a quadratic inequality. Which of the following inequalities describes the shaded region?

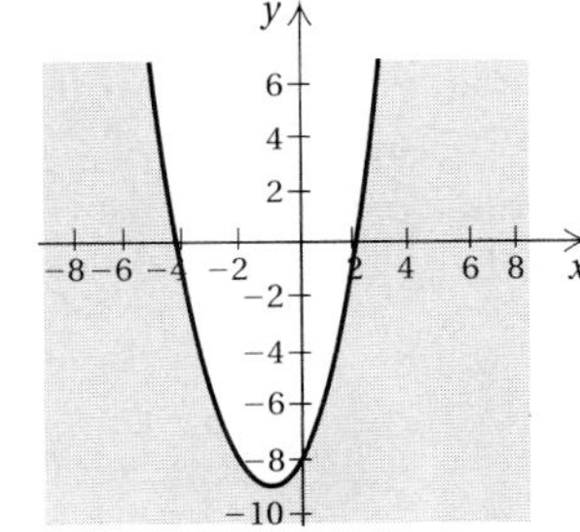

A. $y \leq x^2 + 2x - 8$ B. $y \geq x^2 + 2x - 8$
C. $y \leq (x - 1)^2 - 9$ D. $y \geq (x - 1)^2 - 9$

29. Which of the following numbers should be added to both sides of the equation $x^2 + 9x = 15$ in order to solve by completing the square?

A. 81 B. $\frac{81}{4}$ C. $\frac{9}{2}$ D. 125

30. Solve the equation $2x^2 - 5x = 8$ using the quadratic formula. In the solution, what is the number under the radical sign?

A. -39 B. 5 C. 41 D. 89

31. A rectangular piece of bound carpet lies in the center of a rectangular room. The distance between the carpet and the wall is the same all the way around the room. The room is 15 ft long and 12 ft wide. The carpet piece is 12 yd^2. What is the distance between the carpet and the wall?

A. 4 yd B. 4 ft C. $1\frac{1}{2}$ ft D. 3 ft

32. A window is square with a semicircular top section, as shown in the figure at right.

About how many square feet of glass will it take to make the window?

A. 16.1 ft^2 **B.** 12.5 ft^2
C. 23.1 ft^2 **D.** 28.3 ft^2

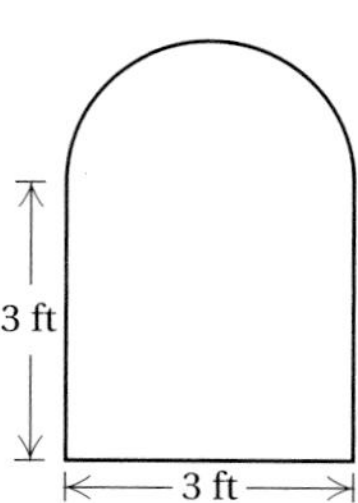

33. A decorative tin in the shape of a cylinder is full of candy. The tin is 8 in. high, and the diameter of the lid is 5 in. What is the volume of candy that the tin contains?

A. About 157 in^3 **B.** About 165 in^3 **C.** About 40 in^3 **D.** About 126 in^3

34. Metal Manufacturing is making a hand rest for a medical supply company in the shape shown at right.

How much metal is used to make the hand rest?

A. 1700 cm^2 **B.** 6500 cm^2
C. 2700 cm^2 **D.** 2550 cm^2

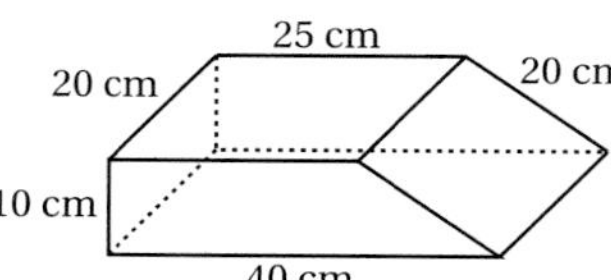

35. Buck Creek Township Fire Department has a fire truck with a ladder that extends straight to 60 ft. At a fire in an apartment building, the closest the truck could get to the burning building was 30 ft. About how high on the building did the ladder reach?

A. 52.0 ft **B.** 60.0 ft **C.** 30.0 ft **D.** 67.1 ft

36. Jessica is building a scale model of a house. The dining room is in the shape of a hexagon, as shown in the figure at right.

Polygon *ABCDEF*, the dining room of the house, is similar to polygon *PQRSTU*, the dining room of the model. *AB* measures 12 ft and *PQ* 4 in. How long should she make *UP* if *FA* is 5 ft?

A. 15 in. **B.** $\frac{3}{5}$ in.

C. $1\frac{2}{3}$ in. **D.** 3 in.

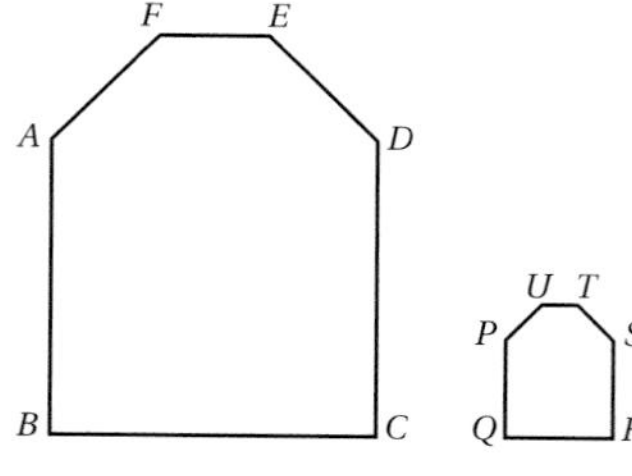

37. Shown at right are two triangles, $\triangle ABC$ and $\triangle DEF$. If you know that $\overline{AC} \cong \overline{DF}$ and $\angle ABE \cong \angle DEF$, which of the following is a valid conclusion about the triangles?

A. $\triangle ABC \cong \triangle DEF$ by SAS.
B. $\triangle ABC \cong \triangle DEF$ by ASA.
C. $\triangle ABC \cong \triangle DEF$ by SSS.
D. None of these is a valid conclusion.

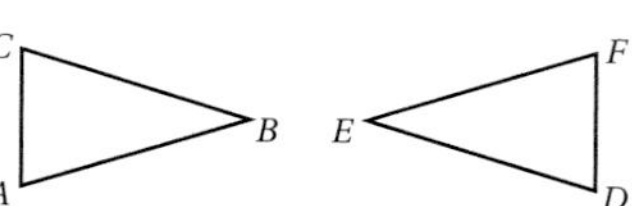

38. In the hexagon shown at right, $\overline{AB} \parallel \overline{DE}$. $\overline{BE}$ divides the figure into two quadrilaterals, $ABEF$ and $DEBC$.

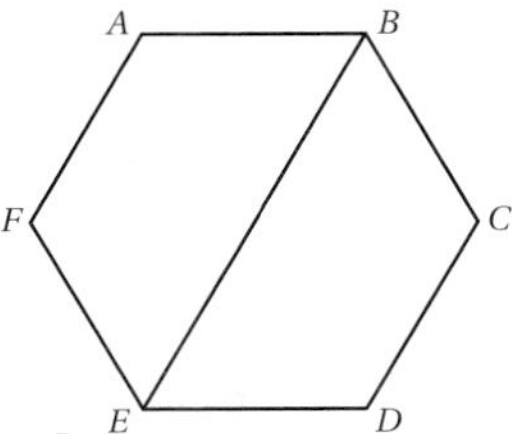

Which of the following is a valid conclusion about this figure?

A. $ABEF \cong DEBC$
B. $\angle CBE \cong \angle FEB$
C. $\angle ABE \cong \angle DEB$
D. None of these is a valid conclusion.

39. In the polygon shown at right, $\overline{BF}$ is perpendicular to both $\overline{BC}$ and $\overline{EG}$.

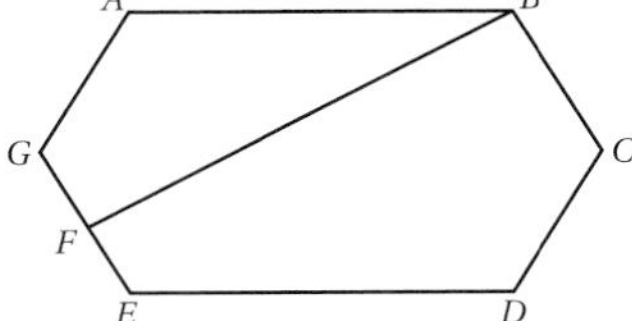

Which of the following is a valid conclusion?

A. $\overline{BC}$ is congruent to $\overline{EG}$.
B. $\overline{BC}$ is perpendicular to $\overline{EG}$.
C. $\overline{BC}$ is parallel to $\overline{EG}$.
D. Non of these is a valid conclusion.

40. Use the statements below to answer the question that follows.

1. All employees of Poster Prints will get a pay raise this year.
2. Some employees of Poster Prints will be transferred to another location.
3. No management personnel will be transferred.
4. Jackie is an employee of Poster Prints.

Which of the following statements must be true?

A. Jackie will be transferred.
B. Jackie is in management.
C. Jackie will get a pay raise.
D. None of these statements must be true.

41. Mike, Linda, Jeff, and Paula are brothers and sisters. One is an engineer, one is an author, one is a consultant, and one is an accountant. Use the statements below to determine which one is an author.

1. Mike lives near the author and the consultant.
2. Linda is older than the engineer and the accountant.
3. Paula is younger than the author and the engineer.
4. Jeff does not live near any of his brothers or sisters.

Who is the author?

A. Mike **B.** Linda **C.** Jeff **D.** Paula

42. Use the pattern sequence below to answer the question that follows.

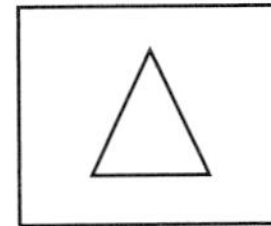 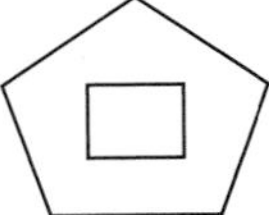 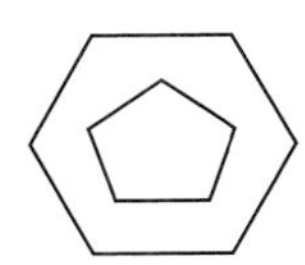 

What figure comes next in the sequence?

A. 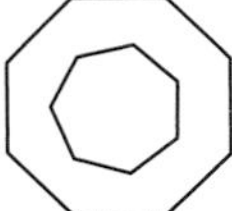

B. 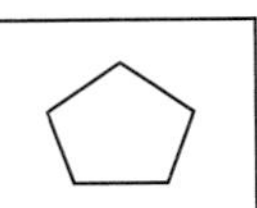

C. 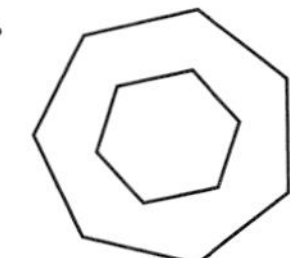

D. 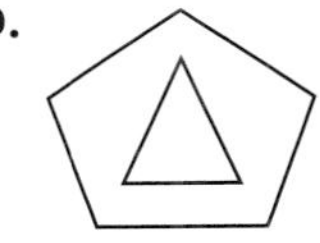

43. Use the pattern sequence below to answer the question that follows.

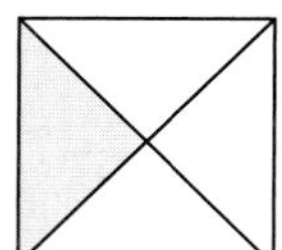 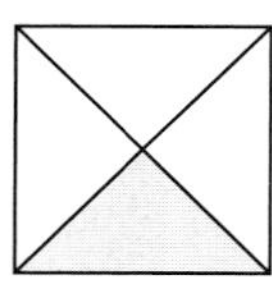 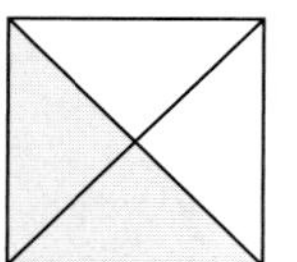 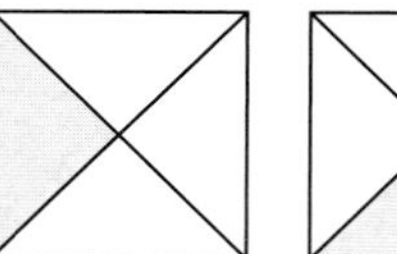 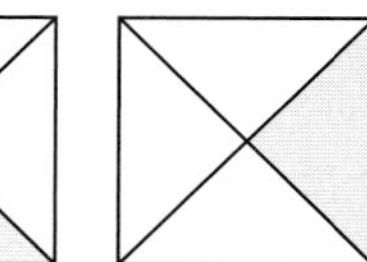  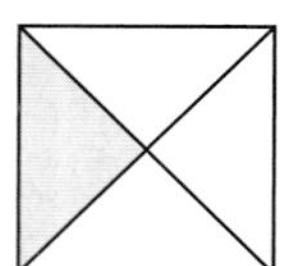 ?

Which figure comes next in the sequence?

A. 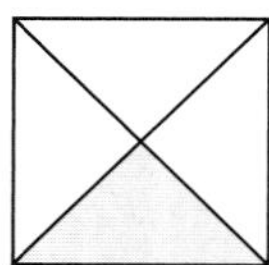

B. 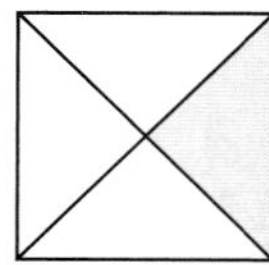

C. 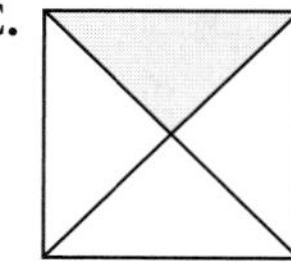

D. 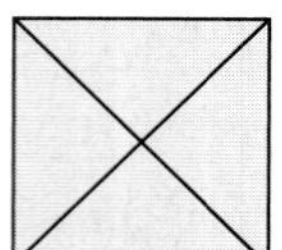

44. A flower garden is a square with a semicircle along each side, as shown in the figure at right.

A landscaper designer wants to plant flowers along the perimeter of the garden. If there should be 4 plants for every meter, and each plant costs \$0.70, how much will the flowers cost?

A. \$13.30 **B.** \$52.50
C. \$26.60 **D.** \$39.90

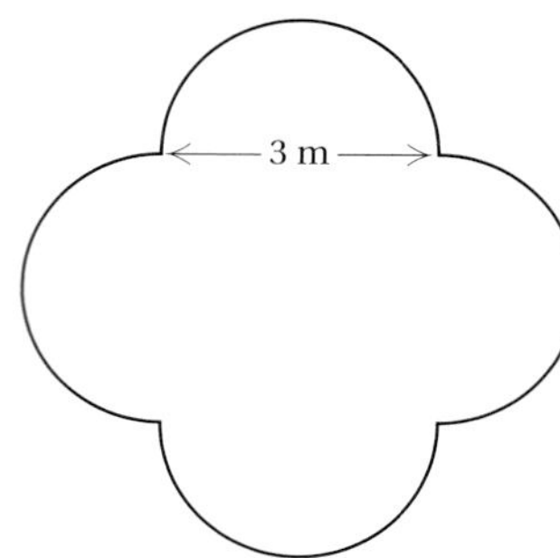

Use the information below for Questions 45 and 46.

Brenda is a graphics designer who has just accepted an entry-level job paying \$24,000 a year. She can qualify for a mortgage of up to $2\frac{1}{2}$ times her annual salary. With the current interest rate, the monthly payment on a mortgage is about 1% of the amount borrowed.

45. If Brenda borrows as much as she can, what will be her monthly mortgage payment?

A. \$600 **B.** \$240 **C.** \$480 **D.** \$120

46. Brenda has other fixed expenses. If she buys the house, she will have to pay \$100 a month for taxes and insurance. Income taxes take 20% of her earnings. Her car payment is \$175 a month, and auto insurance is \$100 a month. She spends 15% of her income on food. How much will she have left each month after paying her mortgage and other fixed expenses?

A. \$1675 **B.** \$325 **C.** \$925 **D.** \$1275

Answers

Chapter 1

Pretest: Chapter 1, p. 2

1. [1.1c] Three million, seventy-eight thousand, fifty-nine **2.** [1.1a] 6 thousands + 9 hundreds + 8 tens + 7 ones **3.** [1.1d] 2,047,398,589 **4.** [1.1e] 6 ten thousands **5.** [1.1f] $<$ **6.** [1.1f] $>$ **7.** [1.2b] 10,216 **8.** [1.2f] 4108 **9.** [1.3b] 22,976 **10.** [1.3e] 503 R 11 **11.** [1.4b] 5542 **12.** [1.4b] 22 **13.** [1.4b] 34 **14.** [1.4b] 25 **15.** [1.5a] 12 lb **16.** [1.5a] 126 **17.** [1.5a] 22,981,000 **18.** [1.5a] 2292 sq ft **19.** [1.6b] 64 **20.** [1.9a] 120 **21.** [1.6c] 0 **22.** [1.6d] 0 **23.** [1.7c] Prime **24.** [1.7d] $2 \cdot 2 \cdot 5 \cdot 7$ **25.** [1.8a] Yes **26.** [1.8a] No

Margin Exercises, Section 1.1, pp. 3–6

1. 1 thousand + 8 hundreds + 0 tens + 5 ones, or 1 thousand + 8 hundreds + 5 ones **2.** 3 ten thousands + 6 thousands + 2 hundreds + 2 tens + 3 ones **3.** 3 thousands + 2 hundreds + 1 ten **4.** 2 thousands + 9 ones **5.** 5 thousands + 7 hundreds **6.** 5689 **7.** 87,128 **8.** 9003 **9.** Fifty-seven **10.** Twenty-nine **11.** Eighty-eight **12.** Two hundred four **13.** Seventy-nine thousand, two hundred four **14.** One million, eight hundred seventy-nine thousand, two hundred four **15.** Twenty-two billion, three hundred one million, eight hundred seventy-nine thousand, two hundred four **16.** 213,105,329 **17.** 2 ten thousands **18.** 2 hundred thousands **19.** 2 millions **20.** 2 ten millions **21.** 6 **22.** 8 **23.** 5 **24.** 5 **25.** $<$ **26.** $>$ **27.** $>$ **28.** $<$ **29.** $<$ **30.** $>$

Exercise Set 1.1, p. 7

1. 5 thousands + 7 hundreds + 4 tens + 2 ones **3.** 2 ten thousands + 7 thousands + 3 hundreds + 4 tens + 2 ones **5.** 5 thousands + 6 hundreds + 9 ones **7.** 2 thousands + 3 hundreds **9.** 2475 **11.** 68,939 **13.** 7304 **15.** 1009 **17.** Eighty-five **19.** Eighty-eight thousand **21.** One hundred twenty-three thousand, seven hundred sixty-five **23.** Seven billion, seven hundred fifty-four million, two hundred eleven thousand, five hundred seventy-seven **25.** One million, eight hundred sixty-seven thousand **27.** One billion, five hundred eighty-three million, one hundred forty-one thousand **29.** 2,233,812 **31.** 8,000,000,000 **33.** 9,460,000,000,000 **35.** 2,974,600 **37.** 5 thousands **39.** 5 hundreds **41.** 3 **43.** 0 **45.** $<$ **47.** $>$ **49.** $<$ **51.** $>$ **53.** $>$ **55.** $>$ **57.** ◈ **59.** All 9's as digits. Answers may vary. For an 8-digit readout, it would be 99,999,999. This number has three periods.

Margin Exercises, Section 1.2, pp. 9–15

1. $8 + 2 = 10$ **2.** 100 mi + 93 mi = 193 mi **3.** 5 ft + 7 ft = 12 ft **4.** 4 in. + 5 in. + 9 in. + 6 in. + 5 in. = 29 in. **5.** 5 ft + 6 ft + 5 ft + 6 ft = 22 ft **6.** 30,000 sq ft + 40,000 sq ft = 70,000 sq ft **7.** 8 sq yd + 9 sq yd = 17 sq yd **8.** 6 cu yd + 8 cu yd = 14 cu yd **9.** 80 gal + 56 gal = 136 gal **10.** 9745 **11.** 13,465 **12.** 16,182 **13.** 27,474 **14.** 67 cu yd − 5 cu yd = 62 cu yd **15.** 20,000 sq ft − 12,000 sq ft = 8000 sq ft **16.** $7 = 2 + 5$, or $7 = 5 + 2$ **17.** $17 = 9 + 8$, or $17 = 8 + 9$ **18.** $5 = 13 - 8$; $8 = 13 - 5$ **19.** $11 = 14 - 3$; $3 = 14 - 11$ **20.** $67 + \square = 348$; $\square = 348 - 67$ **21.** $800 + \square = 1200$; $\square = 1200 - 800$ **22.** 3801 **23.** 6328 **24.** 4747 **25.** 56 **26.** 205 **27.** 658 **28.** 2851 **29.** 1546

Exercise Set 1.2, p. 17

1. $7 + 8 = 15$ **3.** 40 in. **5.** 1086 yd **7.** 387 **9.** 5198 **11.** 1010 **13.** 5266 **15.** 8310 **17.** 34,432 **19.** 2320 **21.** 22,654 **23.** $188 + \square = 564$; $564 - 188 = \square$ **25.** $7 = 3 + 4$, or $7 = 4 + 3$ **27.** $13 = 5 + 8$, or $13 = 8 + 5$ **29.** $6 = 15 - 9$; $9 = 15 - 6$ **31.** $8 = 15 - 7$; $7 = 15 - 8$ **33.** $17 + \square = 32$; $\square = 32 - 17$ **35.** 39 **37.** 298 **39.** 533 **41.** 1493 **43.** 7748 **45.** 84 **47.** 4206 **49.** 10,305 **51.** 7992 **52.** Nine hundred twenty-four million, six hundred thousand **53.** ◈

Margin Exercises, Section 1.3, pp. 20–31

1. 8 · 7 = 56 2. 10 · 75 = 750 mL 3. 8 · 8 = 64 4. 4 · 6 = 24 sq ft 5. 1035 6. 3024 7. 46,252 8. 205,065 9. 144,432 10. 287,232 11. 14,075,720 12. 391,760 13. 17,345,600 14. 56,200 15. 562,000 16. 112 ÷ 14 = □ 17. 112 ÷ 8 = □ 18. 15 = 5 · 3, or 15 = 3 · 5 19. 72 = 9 · 8, or 72 = 8 · 9 20. 6 = 12 ÷ 2; 2 = 12 ÷ 6 21. 6 = 42 ÷ 7; 7 = 42 ÷ 6 22. 6; 6 · 9 = 54 23. 6 R 7; 6 · 9 = 54, 54 + 7 = 61 24. 4 R 5; 4 · 12 = 48, 48 + 5 = 53 25. 6 R 13; 6 · 24 = 144, 144 + 13 = 157 26. 59 R 3 27. 1475 R 5 28. 1015 29. 807 R 4 30. 1088 31. 360 R 4 32. 800 R 47 33. 40 34. 50 35. 70 36. 100 37. 40 38. 80 39. 90 40. 140 41. 470 42. 240 43. 290 44. 600 45. 800 46. 800 47. 9300 48. 8000 49. 8000 50. 19,000 51. 69,000 52. 1800 53. 2600 54. 11,000 55. 210,000; 160,000

Exercise Set 1.3, p. 33

1. 21 · 21 = 441 3. 18 sq ft 5. 121 sq yd 7. 9600 9. 564 11. 1527 13. 64,603 15. 4770 17. 3995 19. 46,080 21. 14,652 23. 207,672 25. 798,408 27. 20,723,872 29. 362,128 31. 20,064,048 33. 25,236,000 35. 760 ÷ 4 = □ 37. 18 = 3 · 6, or 18 = 6 · 3 39. 22 = 22 · 1, or 22 = 1 · 22 41. 9 = 45 ÷ 5; 5 = 45 ÷ 9 43. 37 = 37 ÷ 1; 1 = 37 ÷ 37 45. 55 R 2 47. 108 49. 307 51. 92 R 2 53. 1703 55. 29 R 5 57. 90 R 22 59. 29 61. 370 63. 609 R 15 65. 50 67. 70 69. 730 71. 900 73. 100 75. 1000 77. 9100 79. 32,900 81. 6000 83. 8000 85. 45,000 87. 373,000 89. 180 91. 5720 93. 16,500 95. 5200 97. 31,000 99. 69,000 101. 50 · 70 = 3500 103. 30 · 30 = 900 105. 900 · 300 = 270,000 107. 400 · 200 = 80,000 109. 6000 · 5000 = 30,000,000 111. 8000 · 6000 = 48,000,000 113. 7 thousands + 8 hundreds + 8 tens + 2 ones 114. > 115. 21 = 16 + 5, or 21 = 5 + 16 116. 56 = 14 + 42, or 56 = 42 + 14 117. 47 = 56 − 9; 9 = 56 − 47 118. 350 = 414 − 64; 64 = 414 − 350 119. ◆ 121. 30 123. 247,464 sq ft

Margin Exercises, Section 1.4, pp. 37–40

1. 7 2. 5 3. No 4. Yes 5. 5 6. 10 7. 5 8. 22 9. 22,490 10. 9022 11. 570 12. 3661 13. 8 14. 45 15. 77 16. 3311 17. 6114 18. 8 19. 16 20. 644 21. 96 22. 94

Exercise Set 1.4, p. 41

1. 14 3. 0 5. 29 7. 0 9. 8 11. 14 13. 1035 15. 25 17. 450 19. 90,900 21. 32 23. 143 25. 79 27. 45 29. 324 31. 743 33. 37 35. 66 37. 15 39. 48 41. 175 43. 335 45. 104 47. 45 49. 4056 51. 17,603 53. 18,252 55. 205 57. 7 = 15 − 8; 8 = 15 − 7 58. 6 = 48 ÷ 8; 8 = 48 ÷ 6 59. < 60. > 61. 142 R 5 62. 334 R 11 63. ◆ 65. 347

Margin Exercises, Section 1.5, pp. 44–50

1. 1,424,000 2. $369 3. $18 4. $38,988 5. 9180 sq in. 6. 378 packages; 1 can left over 7. 37 gal 8. 70 min, or 1 hr, 10 min 9. 106

Exercise Set 1.5, p. 51

1. $12,276 3. $7,004,000,000 5. $64,000,000 7. 4007 mi 9. 384 in. 11. 4500 13. 7280 15. $247 17. 54 weeks; 1 episode left over 19. 168 hr 21. $400 23. **(a)** 4700 ft^2; **(b)** 288 ft 25. 35 27. 56 cartons; 11 books left over 29. 1600 mi; 27 in. 31. 18 33. 22 35. $704 37. 525 min, or 8 hr, 45 min 39. 3000 in^2 41. 234,600 42. 235,000 43. 22,000 44. 16,000 45. 320,000 46. 720,000 47. ◆ 49. 792,000 mi; 1,386,000 mi

Margin Exercises, Section 1.6, pp. 55–58

1. 5^4 2. 5^5 3. 10^2 4. 10^4 5. 10,000 6. 100 7. 512 8. 32 9. 51 10. 30 11. 584 12. 84 13. 4; 1 14. 52; 52 15. 29 16. 1880 17. 253 18. 93 19. 1880 20. 305 21. 93 22. 87 in. 23. 46 24. 4

Calculator Spotlight, p. 57

1. 1024 2. 40,353,607 3. 1,048,576 4. 49 5. 85 6. 135 7. 176

Exercise Set 1.6, p. 59

1. 3^4 3. 5^2 5. 7^5 7. 10^3 9. 49 11. 729 13. 20,736 15. 121 17. 22 19. 20 21. 100 23. 1 25. 49 27. 5 29. 434 31. 41 33. 88 35. 4 37. 303 39. 20 41. 70 43. 295 45. 32 47. 906 49. 62 51. 102 53. $94 55. 110 57. 7 59. 544 61. 708 63. 452 64. 13 65. 102,600 mi^2 66. 98 gal 67. ◆ 69. 675 71. 24; 1 + 5 · (4 + 3) = 36 73. 7; 12 ÷ (4 + 2) · 3 − 2 = 4

Margin Exercises, Section 1,7 pp. 61–64

1. 1, 2, 3, 6 2. 1, 2, 4, 8 3. 1, 2, 5, 10 4. 1, 2, 4, 8, 16, 32 5. 5 = 1 · 5, 45 = 9 · 5, 100 = 20 · 5 6. 10 = 1 · 10, 60 = 6 · 10, 110 = 11 · 10 7. 5, 10, 15, 20, 25, 30, 35, 40, 45, 50 8. Yes 9. Yes 10. No 11. 13, 19, 41 are prime; 4, 6, 8 are composite; 1 is neither 12. 2 · 3 13. 2 · 2 · 3 14. 3 · 3 · 5 15. 2 · 7 · 7 16. 2 · 3 · 3 · 7 17. 2 · 2 · 2 · 2 · 3 · 3

Calculator Spotlight, p. 62

1. Yes **2.** No **3.** No **4.** Yes

Exercise Set 1.7, p. 65

1. 1, 2, 3, 6, 9, 18 **3.** 1, 2, 3, 6, 9, 18, 27, 54 **5.** 1, 2, 4 **7.** 1, 7 **9.** 1 **11.** 1, 2, 7, 14, 49, 98 **13.** 4, 8, 12, 16, 20, 24, 28, 32, 36, 40 **15.** 20, 40, 60, 80, 100, 120, 140, 160, 180, 200 **17.** 3, 6, 9, 12, 15, 18, 21, 24, 27, 30 **19.** 12, 24, 36, 48, 60, 72, 84, 96, 108, 120 **21.** 10, 20, 30, 40, 50, 60, 70, 80, 90, 100 **23.** 9, 18, 27, 36, 45, 54, 63, 72, 81, 90 **25.** No **27.** Yes **29.** Yes **31.** No **33.** No **35.** Neither **37.** Composite **39.** Prime **41.** Prime **43.** $2 \cdot 2 \cdot 2$ **45.** $2 \cdot 7$ **47.** $2 \cdot 3 \cdot 7$ **49.** $5 \cdot 5$ **51.** $2 \cdot 5 \cdot 5$ **53.** $13 \cdot 13$ **55.** $2 \cdot 2 \cdot 5 \cdot 5$ **57.** $5 \cdot 7$ **59.** $2 \cdot 2 \cdot 2 \cdot 3 \cdot 3$ **61.** $7 \cdot 11$ **63.** $2 \cdot 2 \cdot 7 \cdot 103$ **65.** $3 \cdot 17$ **67.** 26 **68.** 256 **69.** 425 **70.** 4200 **71.** 0 **72.** 22 **73.** 1 **74.** 3 **75.** $612 **76.** 201 min, or 3 hr, 21 min **77.** ◆ **79.** Row 1: 48, 90, 432, 63; row 2: 7, 2, 2, 10, 8, 6, 21, 10; row 3: 9, 18, 36, 14, 12, 11, 21; row 4: 29, 19, 42

Margin Exercises, Section 1.8, pp. 67–70

1. Yes **2.** No **3.** Yes **4.** No **5.** Yes **6.** No **7.** Yes **8.** No **9.** Yes **10.** No **11.** No **12.** Yes **13.** No **14.** Yes **15.** No **16.** Yes **17.** No **18.** Yes **19.** No **20.** Yes **21.** Yes **22.** No **23.** No **24.** Yes **25.** Yes **26.** No **27.** No **28.** Yes **29.** No **30.** Yes **31.** Yes **32.** No

Exercise Set 1.8, p. 71

1. 46, 224, 300, 36, 45,270, 4444, 256, 8064, 21,568 **3.** 224, 300, 36, 4444, 256, 8064, 21,568 **5.** 300, 36, 45,270, 8064 **7.** 36, 45,270, 711, 8064 **9.** 324, 42, 501, 3009, 75, 2001, 402, 111,111, 1005 **11.** 55,555, 200, 75, 2345, 35, 1005 **13.** 324 **15.** 200 **17.** 138 **18.** 139 **19.** 56 **20.** 26 **21.** 234 **22.** 4003 **23.** 45 gal **24.** 4320 min **25.** ◆ **27.** $2 \cdot 2 \cdot 2 \cdot 3 \cdot 5 \cdot 5 \cdot 13$ **29.** $2 \cdot 2 \cdot 3 \cdot 3 \cdot 7 \cdot 11$ **31.** 7,652,341

Margin Exercises, Section 1.9, pp. 73–76

1. 45 **2.** 40 **3.** 30 **4.** 24 **5.** 10 **6.** 80 **7.** 40 **8.** 360 **9.** 864 **10.** 2520 **11.** 18 **12.** 24 **13.** 36 **14.** 210

Exercise Set 1.9, p. 77

1. 4 **3.** 50 **5.** 40 **7.** 54 **9.** 150 **11.** 120 **13.** 72 **15.** 420 **17.** 144 **19.** 288 **21.** 30 **23.** 105 **25.** 72 **27.** 60 **29.** 36 **31.** 900 **33.** 48 **35.** 50 **37.** 143 **39.** 420 **41.** 378 **43.** 810 **45.** 60 yr **47.** $<$ **48.** 250 **49.** 7935 **50.** 6939 **51.** 2 ten thousands + 4 thousands + 6 hundreds + 5 ones **53.** ◆ **55.** 2592 **57.** 18,900 **59.** 5 in. by 24 in.

Summary and Review: Chapter 1, p. 79

1. 2 thousands + 7 hundreds + 9 tens + 3 ones **2.** 5 ten thousands + 6 thousands + 7 tens + 8 ones **3.** 90,844 **4.** Sixty-seven thousand, eight hundred nineteen **5.** Two million, seven hundred eighty-one thousand, four hundred twenty-seven **6.** 36,260,064 **7.** 8 thousands **8.** $>$ **9.** $<$ **10.** $406 + $78 = $484, or $78 + $406 = $484 **11.** 986 yd **12.** 14,272 **13.** 66,024 **14.** 22,098 **15.** 98,921 **16.** $151 - 12 = 139$ **17.** $196 + $\square$ = $340, or $\square$ = $340 − $196 **18.** $10 = 6 + 4$, or $10 = 4 + 6$ **19.** $8 = 11 - 3$; $3 = 11 - 8$ **20.** 5148 **21.** 1153 **22.** 2274 **23.** 17,757 **24.** 345,800 **25.** 345,760 **26.** 346,000 **27.** 41,300 + 19,700 = 61,000 **28.** 38,700 − 24,500 = 14,200 **29.** $400 \cdot 700 = 280{,}000$ **30.** $32 \cdot 15 = 480$ **31.** $125 \cdot 368 = 46{,}000\ \text{yd}^2$ **32.** 420,000 **33.** 6,276,800 **34.** 506,748 **35.** 5,331,810 **36.** $176 \div 4 = \square$ **37.** $56 = 8 \cdot 7$, or $56 = 7 \cdot 8$ **38.** $4 = 52 \div 13$; $13 = 52 \div 4$ **39.** 12 R 3 **40.** 5 **41.** 4 R 46 **42.** 54 **43.** 452 **44.** 4389 **45.** 8 **46.** 45 **47.** 546 **48.** $2413 **49.** 1982 **50.** $19,748 **51.** 137 beakers filled; 13 mL of alcohol left over **52.** 4^3 **53.** 10,000 **54.** 36 **55.** 65 **56.** 56 **57.** 233 **58.** 260 **59.** 165 **60.** $2 \cdot 5 \cdot 7$ **61.** $2 \cdot 3 \cdot 5$ **62.** $3 \cdot 3 \cdot 5$ **63.** $2 \cdot 3 \cdot 5 \cdot 5$ **64.** No **65.** No **66.** No **67.** Yes **68.** Prime **69.** 36 **70.** 1404 **71.** ◆ Answers may vary. A vat contains 1152 oz of hot sauce. If 144 bottles are to be filled equally, how much will each bottle contain? **72.** $a = 8$, $b = 4$ **73.** 13, 11, 101

Test: Chapter 1, p. 81

1. [1.1a] 8 thousands + 8 hundreds + 4 tens + 3 ones **2.** [1.1c] Thirty-eight million, four hundred three thousand, two hundred seventy-seven **3.** [1.1e] 5 **4.** [1.2b] 9989 **5.** [1.2b] 63,791 **6.** [1.2b] 34 **7.** [1.2b] 10,515 **8.** [1.2f] 3630 **9.** [1.2f] 1039 **10.** [1.2f] 6848 **11.** [1.2f] 5175 **12.** [1.3b] 41,112 **13.** [1.3b] 5,325,600 **14.** [1.3b] 2405 **15.** [1.3b] 534,264 **16.** [1.3e] 3 R 3 **17.** [1.3e] 70 **18.** [1.3e] 97 **19.** [1.3e] 805 R 8 **20.** [1.5a] 1955 **21.** [1.5a] 92 packages, 3 cans left over **22.** [1.5a] 62,811 mi^2 **23.** [1.5a] 120,000 m^2; 1600 m **24.** [1.5a] 1808 lb **25.** [1.5a] 20 **26.** [1.4b] 46 **27.** [1.4b] 13 **28.** [1.4b] 14 **29.** [1.3f] 35,000 **30.** [1.3f] 34,580 **31.** [1.3f] 34,600 **32.** [1.3f] 23,600 + 54,700 = 78,300 **33.** [1.3f] 54,800 − 23,600 = 31,200 **34.** [1.3f] $800 \cdot 500 = 400{,}000$ **35.** [1.1f] $>$ **36.** [1.1f] $<$ **37.** [1.6a] 12^4 **38.** [1.6b] 343 **39.** [1.6b] 8 **40.** [1.6c] 64 **41.** [1.6c] 96 **42.** [1.6c] 2 **43.** [1.6d] 216 **44.** [1.6c] 18 **45.** [1.6c] 92 **46.** [1.7d] $2 \cdot 3 \cdot 3$ **47.** [1.7d] $2 \cdot 2 \cdot 3 \cdot 5$ **48.** [1.8a] Yes **49.** [1.8a] No **50.** [1.9a] 48 **51.** [1.5a] 336 in^2 **52.** [1.5a] 80 **53.** [1.6c] 83 **54.** [1.6c] 9

Chapter 2

Pretest: Chapter 2, p. 84

1. [2.1b] 1 **2.** [2.1b] 68 **3.** [2.1b] 0 **4.** [2.1e] $\frac{1}{4}$
5. [2.3d] $<$ **6.** [2.2b] $\frac{8}{7}$ **7.** [2.4a] $\frac{61}{8}$ **8.** [2.4b] $5\frac{1}{2}$
9. [2.4b] $11\frac{31}{60}$ **10.** [2.4e] $1\frac{2}{3}$ **11.** [2.2a] $\frac{6}{5}$
12. [2.3c] $\frac{1}{6}$ **13.** [2.3c] $\frac{2}{9}$ **14.** [2.2d] 30
15. [2.5a] $\frac{1}{24}$ m **16.** [2.5a] $21\frac{1}{4}$ lb **17.** [2.5a] $351\frac{1}{5}$ mi
18. [2.5a] $22\frac{1}{2}$ cups **19.** [2.6b] 0 **20.** [2.6b] $\frac{1}{2}$
21. [2.6b] 10 **22.** [2.6b] 2

Margin Exercises, Section 2.1, pp. 85–91

1. $\frac{1}{2}$ **2.** $\frac{1}{3}$ **3.** $\frac{1}{3}$ **4.** $\frac{1}{6}$ **5.** $\frac{5}{8}$ **6.** $\frac{2}{3}$ **7.** $\frac{3}{4}$ **8.** $\frac{4}{6}$
9. $\frac{5}{4}$ **10.** $\frac{7}{4}$ **11.** 1 **12.** 1 **13.** 1 **14.** 1 **15.** 1
16. 1 **17.** 0 **18.** 0 **19.** 0 **20.** 0 **21.** Not defined
22. Not defined **23.** 8 **24.** 10 **25.** 346 **26.** 23
27. $\frac{15}{56}$ **28.** $\frac{32}{15}$ **29.** $\frac{3}{100}$ **30.** $\frac{14}{3}$
31.

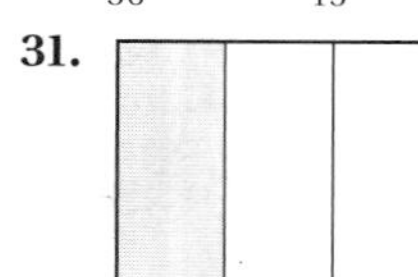

$\frac{1}{3}$

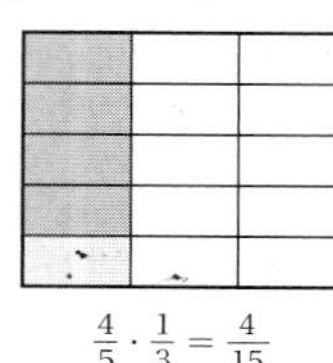

$\frac{4}{5} \cdot \frac{1}{3} = \frac{4}{15}$

32. $\frac{8}{16}$ **33.** $\frac{30}{50}$ **34.** $\frac{52}{100}$ **35.** $\frac{200}{75}$ **36.** $\frac{12}{9}$ **37.** $\frac{18}{24}$
38. $\frac{90}{100}$ **39.** $\frac{9}{45}$ **40.** $\frac{56}{49}$ **41.** $\frac{1}{4}$ **42.** $\frac{5}{6}$ **43.** 5
44. $\frac{4}{3}$ **45.** $\frac{7}{8}$ **46.** $\frac{89}{78}$ **47.** $\frac{8}{7}$ **48.** $\frac{1}{4}$
49. $\frac{2}{100} = \frac{1}{50}$; $\frac{4}{100} = \frac{1}{25}$; $\frac{32}{100} = \frac{8}{25}$; $\frac{44}{100} = \frac{11}{25}$; $\frac{18}{100} = \frac{9}{50}$

Calculator Spotlight, p. 91

1. $\frac{14}{15}$ **2.** $\frac{7}{8}$ **3.** $\frac{138}{167}$ **4.** $\frac{7}{25}$

Exercise Set 2.1, p. 93

1. $\frac{2}{4}$ **3.** $\frac{1}{8}$ **5.** $\frac{4}{3}$ **7.** $\frac{3}{4}$ **9.** $\frac{4}{8}$ **11.** $\frac{6}{12}$ **13.** 18
15. 0 **17.** 1 **19.** Not defined **21.** Not defined
23. 1 **25.** $\frac{1}{6}$ **27.** $\frac{5}{8}$ **29.** $\frac{2}{15}$ **31.** $\frac{4}{15}$ **33.** $\frac{9}{16}$
35. $\frac{14}{39}$ **37.** $\frac{21}{4}$ **39.** $\frac{49}{64}$ **41.** $\frac{5}{10}$ **43.** $\frac{20}{32}$ **45.** $\frac{75}{45}$
47. $\frac{42}{132}$ **49.** $\frac{3}{4}$ **51.** $\frac{1}{5}$ **53.** 3 **55.** $\frac{3}{4}$ **57.** $\frac{7}{8}$
59. $\frac{1}{3}$ **61.** 6 **63.** ◈

Margin Exercises, Section 2.2, p. 96–98

1. $\frac{7}{12}$ **2.** $\frac{1}{3}$ **3.** 6 **4.** $\frac{5}{2}$ **5.** $\frac{5}{2}$ **6.** $\frac{7}{10}$ **7.** $\frac{1}{9}$ **8.** 5
9. $\frac{8}{7}$ **10.** $\frac{8}{3}$ **11.** $\frac{1}{10}$ **12.** 100 **13.** 1 **14.** $\frac{14}{15}$
15. $\frac{4}{5}$ **16.** 32

Exercise Set 2.2, p. 99

1. $\frac{1}{3}$ **3.** $\frac{1}{6}$ **5.** $\frac{27}{10}$ **7.** $\frac{14}{9}$ **9.** 1 **11.** 1 **13.** 4
15. 9 **17.** $\frac{98}{5}$ **19.** 30 **21.** $\frac{1}{5}$ **23.** $\frac{9}{25}$ **25.** $\frac{11}{40}$
27. $\frac{5}{14}$ **29.** $\frac{6}{5}$ **31.** $\frac{1}{6}$ **33.** 6 **35.** $\frac{3}{40}$ **37.** $\frac{4}{5}$ **39.** $\frac{4}{15}$
41. 4 **43.** 2 **45.** $\frac{1}{8}$ **47.** $\frac{3}{7}$ **49.** 8 **51.** 35 **53.** 1
55. $\frac{2}{3}$ **57.** $\frac{9}{4}$ **59.** 144 **61.** 75 **63.** 2 **65.** $\frac{3}{5}$
67. 315 **69.** 67 **70.** 33 R 4 **71.** 285 R 2
72. 103 R 10 **73.** 67 **74.** 264 **75.** 8499 **76.** 4368
77. ◈ **79.** $\frac{3}{8}$

Margin Exercises, Section 2.3, pp. 101–106

1. $\frac{4}{5}$ **2.** 1 **3.** $\frac{1}{2}$ **4.** $\frac{3}{4}$ **5.** $\frac{5}{6}$ **6.** $\frac{29}{24}$ **7.** $\frac{5}{9}$ **8.** $\frac{413}{1000}$
9. $\frac{759}{1000}$ **10.** $\frac{197}{210}$ **11.** $\frac{1}{2}$ **12.** $\frac{3}{8}$ **13.** $\frac{1}{2}$ **14.** $\frac{13}{18}$ **15.** $\frac{1}{2}$
16. $\frac{9}{112}$ **17.** $<$ **18.** $>$ **19.** $>$ **20.** $>$ **21.** $<$
22. $\frac{1}{6}$ **23.** $\frac{11}{40}$

Exercise Set 2.3, p. 107

1. 1 **3.** $\frac{3}{4}$ **5.** $\frac{3}{2}$ **7.** $\frac{7}{24}$ **9.** $\frac{3}{2}$ **11.** $\frac{19}{24}$ **13.** $\frac{9}{10}$
15. $\frac{29}{18}$ **17.** $\frac{7}{8}$ **19.** $\frac{13}{24}$ **21.** $\frac{35}{24}$ **23.** $\frac{93}{100}$ **25.** $\frac{17}{24}$
27. $\frac{437}{500}$ **29.** $\frac{53}{40}$ **31.** $\frac{391}{144}$ **33.** $\frac{2}{3}$ **35.** $\frac{3}{4}$ **37.** $\frac{5}{8}$
39. $\frac{1}{24}$ **41.** $\frac{1}{2}$ **43.** $\frac{17}{60}$ **45.** $\frac{53}{100}$ **47.** $\frac{26}{75}$ **49.** $\frac{1}{24}$
51. $\frac{13}{16}$ **53.** $\frac{31}{75}$ **55.** $\frac{13}{75}$ **57.** $<$ **59.** $>$ **61.** $<$
63. $<$ **65.** $>$ **67.** $>$ **69.** $<$ **71.** $\frac{1}{15}$ **73.** $\frac{2}{15}$
75. $\frac{1}{15}$ **77.** 4673 **78.** 5338 **79.** 204 **80.** 1943 R 1
81. ◈ **83.** $\frac{227}{420}$ km

Margin Exercises, Section 2.4, pp. 109–114

1. $1\frac{2}{3}$ **2.** $8\frac{3}{4}$ **3.** $12\frac{2}{3}$ **4.** $\frac{22}{5}$ **5.** $\frac{61}{10}$ **6.** $\frac{29}{6}$ **7.** $\frac{37}{4}$
8. $\frac{62}{3}$ **9.** $2\frac{1}{3}$ **10.** $1\frac{1}{10}$ **11.** $18\frac{1}{3}$ **12.** $7\frac{2}{5}$ **13.** $12\frac{1}{10}$
14. $13\frac{7}{12}$ **15.** $1\frac{1}{2}$ **16.** $3\frac{1}{6}$ **17.** $3\frac{1}{3}$ **18.** $3\frac{2}{3}$ **19.** 20
20. $1\frac{7}{8}$ **21.** $12\frac{4}{5}$ **22.** $8\frac{1}{3}$ **23.** 16 **24.** $7\frac{3}{7}$ **25.** $1\frac{7}{8}$
26. $\frac{7}{10}$

Exercise Set 2.4, p. 115

1. $\frac{17}{3}$ **3.** $\frac{59}{6}$ **5.** $\frac{51}{4}$ **7.** $3\frac{3}{5}$ **9.** $5\frac{7}{10}$ **11.** $43\frac{1}{8}$
13. $6\frac{1}{2}$ **15.** $2\frac{11}{12}$ **17.** $14\frac{7}{12}$ **19.** $21\frac{1}{2}$ **21.** $27\frac{7}{8}$
23. $27\frac{13}{24}$ **25.** $1\frac{3}{5}$ **27.** $4\frac{1}{10}$ **29.** $15\frac{3}{8}$ **31.** $7\frac{5}{12}$
33. $13\frac{3}{8}$ **35.** $11\frac{5}{18}$ **37.** $22\frac{2}{3}$ **39.** $2\frac{5}{12}$ **41.** $8\frac{1}{6}$
43. $9\frac{31}{40}$ **45.** $24\frac{91}{100}$ **47.** $975\frac{4}{5}$ **49.** $6\frac{1}{4}$ **51.** $1\frac{1}{5}$
53. $3\frac{9}{16}$ **55.** $1\frac{1}{8}$ **57.** $1\frac{8}{43}$ **59.** $\frac{9}{40}$ **61.** 286 cartons; 2 oz left over **62.** 3728 lb **63.** ◈

Margin Exercises, Section 2.5, pp. 118–122

1. $\frac{3}{8}$ **2.** $\frac{63}{100}$ cm^2 **3.** 320 **4.** 200 gal **5.** $\frac{11}{20}$ cup
6. $\frac{3}{8}$ in. **7.** $23\frac{1}{4}$ gal **8.** $240\frac{3}{4}$ ft^2

Exercise Set 2.5, p. 123

1. $\frac{12}{25}$ m^2 **3.** $\$27$ **5.** 160 mi **7.** 32 **9.** 288 km; 108 km **11.** $\frac{23}{12}$ mi **13.** 690 kg; $\frac{14}{23}$ cement; $\frac{5}{23}$ stone; $\frac{4}{23}$ sand; 1 **15.** $\frac{5}{12}$ hr **17.** $\frac{1}{32}$ in. **19.** $6\frac{5}{12}$ in.
21. $\$17\frac{1}{8}$ **23.** $7\frac{5}{12}$ lb **25.** $28\frac{3}{4}$ yd **27.** $3\frac{4}{5}$ hr
29. $7\frac{3}{8}$ ft **31.** $82\frac{1}{2}$ in. **33.** 4 cu ft **35.** 16
37. 15 mpg **39.** 24 **41.** $35\frac{115}{256}$ in^2

43. $59{,}538\frac{1}{8}$ ft^2 **45.** 4992 ft^2 **46.** $928
47. 11 **48.** 32 **49.** 186 **50.** 2737 **51.** 5
52. 3520 **53.** 3 hr

Margin Exercises, Section 2.6, pp. 127–129

1. $\frac{1}{2}$ **2.** $\frac{3}{10}$ **3.** $20\frac{2}{3}$ **4.** $\frac{5}{9}$ **5.** $\frac{31}{40}$ **6.** $\frac{27}{56}$ **7.** 0
8. 1 **9.** $\frac{1}{2}$ **10.** 1 **11.** 12; answers may vary
12. 32; answers may vary **13.** $22\frac{1}{2}$ **14.** 132 **15.** 37

Calculator Spotlight, p. 130

1. $\frac{5}{8}$ **2.** $\frac{7}{10}$ **3.** $\frac{5}{7}$ **4.** $\frac{133}{68}$ **5.** $\frac{73}{150}$ **6.** $\frac{97}{116}$ **7.** $24\frac{3}{7}$
8. $\frac{115}{147}$ **9.** $13\frac{41}{63}$ **10.** $5\frac{59}{72}$ **11.** $3\frac{5}{8}$ **12.** $19\frac{5}{8}$

Exercise Set 2.6, p. 131

1. $\frac{1}{24}$ **3.** $\frac{59}{30}$, or $1\frac{29}{30}$ **5.** $\frac{211}{8}$, or $26\frac{3}{8}$ **7.** $\frac{7}{16}$ **9.** $\frac{1}{36}$
11. $\frac{3}{8}$ **13.** $\frac{25}{72}$ **15.** $\frac{103}{16}$, or $6\frac{7}{16}$ **17.** $\frac{8395}{84}$, or $99\frac{79}{84}$
19. 0 **21.** 0 **23.** $\frac{1}{2}$ **25.** 6 **27.** 12 **29.** 15
31. 6 **33.** 12 **35.** 16 **37.** 3 **39.** 2 **41.** 3
43. 100 **45.** $\frac{1}{2}$ **47.** $271\frac{1}{2}$ **49.** $29\frac{1}{2}$ **51.** $3077
52. $739 **53.** 848 **54.** 2203 **55.** 37,239 **57.** ◈
59. $\frac{3}{4}, \frac{7}{9}, \frac{17}{21}, \frac{19}{22}, \frac{13}{15}, \frac{15}{17}, \frac{13}{12}$

Summary and Review: Chapter 2, p. 133

1. $\frac{3}{5}$ **2.** $\frac{3}{100}; \frac{8}{100} = \frac{2}{25}; \frac{10}{100} = \frac{1}{10}; \frac{15}{100} = \frac{3}{20}; \frac{21}{100}; \frac{43}{100}$
3. 0 **4.** 1 **5.** 48 **6.** 6 **7.** $\frac{2}{5}$ **8.** Not defined
9. $\frac{1}{3}$ **10.** $\frac{1}{4}$ **11.** $\frac{2}{3}$ **12.** $\frac{3}{2}$ **13.** 24 **14.** $\frac{1}{14}$ **15.** $\frac{2}{3}$
16. $\frac{5}{4}$ **17.** $\frac{1}{3}$ **18.** $\frac{1}{4}$ **19.** $\frac{9}{4}$ **20.** 300 **21.** $\frac{4}{9}$ **22.** $\frac{63}{40}$
23. $\frac{19}{48}$ **24.** $\frac{29}{15}$ **25.** $\frac{7}{16}$ **26.** $\frac{1}{3}$ **27.** $\frac{1}{8}$ **28.** $\frac{5}{27}$
29. $\frac{11}{18}$ **30.** > **31.** > **32.** $\frac{15}{2}$ **33.** $\frac{67}{8}$ **34.** $2\frac{1}{3}$
35. $6\frac{3}{4}$ **36.** $10\frac{2}{5}$ **37.** $11\frac{11}{15}$ **38.** $4\frac{3}{20}$ **39.** $13\frac{3}{8}$
40. 16 **41.** $3\frac{1}{2}$ **42.** $1\frac{7}{17}$ **43.** $\frac{1}{8}$ **44.** $\frac{3}{10}$ **45.** $\frac{2}{5}$
46. 224 days **47.** $\frac{1}{3}$ cup **48.** $6 **49.** 15 **50.** $\$70\frac{3}{8}$
51. $\frac{173}{100}$ in., or $1\frac{73}{100}$ in. **52.** 1 **53.** $\frac{77}{240}$ **54.** $\frac{1}{2}$
55. 0 **56.** 1 **57.** 7 **58.** $2\frac{1}{2}$ **59.** 24 **60.** 469
61. $912 **62.** 774 mi **63.** 408 R 9 **64.** 3607
65. ◈ To simplify fractional notation, find the prime factorization of the numerator and of the denominator. Examine the factorizations for factors common to both the numerator and the denominator. Change the order of the factorizations, if necessary, so that pairs of like factors are above and below each other. Factor the fraction, with each pair of like factors forming a factor of 1. Remove the factors of 1, and multiply the remaining factors in the numerator and in the denominator, if necessary.
66. ◈ Taking $\frac{1}{2}$ of a number is equivalent to multiplying the number by $\frac{1}{2}$. Dividing by $\frac{1}{2}$ is equivalent to multiplying by the reciprocal of $\frac{1}{2}$, or 2. Thus taking $\frac{1}{2}$ of a number is not the same as dividing by $\frac{1}{2}$.
67. $a = 11{,}176$; $b = 9887$

Test: Chapter 2, p. 135

1. [2.1a] $\frac{3}{4}$ **2.** [2.3d] > **3.** [2.1b] 1 **4.** [2.1b] 0
5. [2.1e] $\frac{1}{14}$ **6.** [2.1b] Not defined **7.** [2.3a] 3
8. [2.3b] $\frac{37}{24}$ **9.** [2.3b] $\frac{79}{100}$ **10.** [2.3c] $\frac{1}{3}$ **11.** [2.3c] $\frac{1}{12}$
12. [2.3c] $\frac{1}{12}$ **13.** [2.2a] 32 **14.** [2.2a] $\frac{5}{2}$ **15.** [2.2a] $\frac{1}{10}$
16. [2.2b] $\frac{8}{5}$ **17.** [2.2b] $\frac{1}{18}$ **18.** [2.2c] $\frac{3}{10}$ **19.** [2.2c] $\frac{8}{5}$
20. [2.2c] 18 **21.** [2.2d] 64 **22.** [2.3e] $\frac{1}{4}$
23. [2.4a] $8\frac{2}{9}$ **24.** [2.4a] $\frac{7}{2}$ **25.** [2.4b] $14\frac{1}{5}$
26. [2.4c] $4\frac{7}{24}$ **27.** [2.4d] $4\frac{1}{2}$ **28.** [2.4e] 2
29. [2.5a] 28 lb **30.** [2.5a] $\frac{3}{40}$ m **31.** [2.5a] $17\frac{1}{2}$ cups
32. [2.5a] 80 **33.** [2.5a] $360\frac{5}{12}$ lb **34.** [2.5a] $2\frac{1}{2}$ in.
35. [2.6a] $3\frac{1}{2}$ **36.** [2.6a] $\frac{11}{20}$ **37.** [2.6b] 0 **38.** [2.6b] 1
39. [2.6b] 4 **40.** [2.6b] 16 **41.** [1.4b] 1805
42. [1.4b] 101 **43.** [1.5a] 3635 mi **44.** [1.3e] 380 R 7
45. [1.2f] 4434 **46.** [2.5a] $\frac{15}{8}$ tsp
47. [2.5a] Dolores runs $\frac{17}{56}$ mi farther.

Chapter 3

Pretest: Chapter 3, p. 138

1. [3.1a] Two and three hundred forty-seven thousandths **2.** [3.1a] Three thousand, two hundred sixty-four and $\frac{78}{100}$ dollars **3.** [3.1b] $\frac{21}{100}$
4. [3.1b] $\frac{5408}{1000}$ **5.** [3.1c] 0.379 **6.** [3.1c] 28.439
7. [3.1d] 3.2 **8.** [3.1d] 0.099 **9.** [3.1e] 21.0
10. [3.1e] 21.045 **11.** [3.2a] 607.219
12. [3.2b] 39.0901 **13.** [3.3a] 0.6179
14. [3.3a] 0.32456 **15.** [3.4a] 30.4
16. [3.4a] 0.57698 **17.** [3.4b] 84.26
18. [3.2c] 6345.157 **19.** [3.7a] 1081.6 mi
20. [3.7a] $285.95 **21.** [3.7a] $89.70
22. [3.7a] $3397.71 **23.** [3.6a] 224 **24.** [3.5a] 1.4
25. [3.5a] 0.925 **26.** [3.5a] 2.75 **27.** [3.5a] $4.\overline{142857}$
28. [3.5b] 4.1 **29.** [3.5b] 4.14 **30.** [3.5b] 4.143
31. [3.3b] $9.49 **32.** [3.3b] 490,000,000,000,000
33. [3.4c] 1548.8836 **34.** [3.5c] 58.17

Margin Exercises, Section 3.1, pp. 140–144

1. Twenty-one and one tenth **2.** Two and four thousand five hundred thirty-three ten-thousandths **3.** Two hundred forty-five and eighty-nine hundredths **4.** Thirty-one thousand, seventy-nine and seven hundred sixty-four thousandths **5.** Four thousand, two hundred seventeen and $\frac{56}{100}$ dollars **6.** Thirteen and $\frac{98}{100}$ dollars **7.** $\frac{896}{1000}$ **8.** $\frac{2378}{100}$ **9.** $\frac{56{,}789}{10{,}000}$ **10.** $\frac{19}{10}$
11. 7.43 **12.** 0.406 **13.** 6.7089 **14.** 0.9 **15.** 0.057
16. 0.083 **17.** 4.3 **18.** 283.71 **19.** 456.013
20. 2.04 **21.** 0.06 **22.** 0.58 **23.** 1 **24.** 0.8989
25. 21.05 **26.** 2.8 **27.** 13.9 **28.** 234.4 **29.** 7.0
30. 0.64 **31.** 7.83 **32.** 34.68 **33.** 0.03 **34.** 0.943
35. 8.004 **36.** 43.112 **37.** 37.401 **38.** 7459.355
39. 7459.35 **40.** 7459.4 **41.** 7459 **42.** 7460
43. 7500 **44.** 7000

Exercise Set 3.1, p. 145

1. Four hundred forty-nine and six hundredths **3.** One and five thousand five hundred ninety-nine ten-thousandths **5.** Thirty-four and eight hundred ninety-one thousandths **7.** Three hundred twenty-six and $\frac{48}{100}$ dollars **9.** Thirty-six and $\frac{72}{100}$ dollars **11.** $\frac{83}{10}$ **13.** $\frac{356}{100}$ **15.** $\frac{4603}{100}$ **17.** $\frac{13}{100,000}$ **19.** $\frac{10,008}{10,000}$ **21.** $\frac{20,003}{1000}$ **23.** 0.8 **25.** 8.89 **27.** 3.798 **29.** 0.0078 **31.** 0.00019 **33.** 0.376193 **35.** 99.44 **37.** 3.798 **39.** 2.1739 **41.** 8.953073 **43.** 0.58 **45.** 0.91 **47.** 0.001 **49.** 235.07 **51.** $\frac{4}{100}$ **53.** 0.4325 **55.** 0.1 **57.** 0.5 **59.** 2.7 **61.** 123.7 **63.** 0.89 **65.** 0.67 **67.** 1.00 **69.** 0.09 **71.** 0.325 **73.** 17.002 **75.** 10.101 **77.** 9.999 **79.** 800 **81.** 809.473 **83.** 809 **85.** 34.5439 **87.** 34.54 **89.** 35 **91.** 6170 **92.** 6200 **93.** 6000 **94.** 54 **95.** $6\frac{3}{5}$ **97.** ◈ **99.** 6.78346 **101.** 0.03030

Margin Exercises, Section 3.2, pp. 147–150

1. 10.917 **2.** 34.2079 **3.** 4.969 **3.** 3.5617 **5.** 9.40544 **6.** 912.67 **7.** 2514.773 **8.** 10.754 **9.** 0.339 **10.** 0.5345 **11.** 0.5172 **12.** 7.36992 **13.** 1194.22 **14.** 4.9911 **15.** 38.534 **16.** 14.164 **17.** 2133.5

Calculator Spotlight, p. 150

$8744.16 should be $8744.17; $8764.65 should be $8723.68; $8848.65 should be $8808.68; $8801.05 should be $8760.08; $8533.09 should be $8492.13

Exercise Set 3.2, p. 151

1. 334.37 **3.** 1576.215 **5.** 132.560 **7.** 84.417 **9.** 50.0248 **11.** 40.007 **13.** 771.967 **15.** 20.8649 **17.** 227.4680 **19.** 8754.8221 **21.** 1.3 **23.** 49.02 **25.** 45.61 **27.** 85.921 **29.** 2.4975 **31.** 3.397 **33.** 8.85 **35.** 3.37 **37.** 1.045 **39.** 3.703 **41.** 0.9902 **43.** 99.66 **45.** 4.88 **47.** 0.994 **49.** 17.802 **51.** 51.13 **53.** 2.491 **55.** 32.7386 **57.** 1.6666 **59.** 2344.90886 **61.** 11.65 **63.** 19.251 **65.** 384.68 **67.** 582.97 **69.** 15,335.3 **71.** 2720 **72.** $2 \cdot 2 \cdot 3 \cdot 19$ **73.** $\frac{1}{6}$ **73.** $\frac{34}{45}$ **75.** 6166 **76.** 5366 **77.** $16\frac{1}{2}$ **78.** $60\frac{1}{5}$ mi **79.** ◈ **81.** 345.8

Margin Exercises, Section 3.3, pp. 156–159

1. 529.48 **2.** 5.0594 **3.** 34.2906 **3.** 0.348 **5.** 0.0348 **6.** 0.00348 **7.** 0.000348 **8.** 34.8 **9.** 348 **10.** 3480 **11.** 34,800 **12.** $938,000,000 **13.** $44,100,000,000 **14.** 1569¢ **15.** 17¢ **16.** $0.35 **17.** $5.77

Exercise Set 3.3, p. 161

1. 60.2 **3.** 6.72 **5.** 0.252 **7.** 0.522 **9.** 237.6 **11.** 583,686.852 **13.** 780 **15.** 8.923 **17.** 0.09768 **19.** 0.782 **21.** 521.6 **23.** 3.2472 **25.** 897.6 **27.** 322.07 **29.** 55.68 **31.** 3487.5 **33.** 50.0004 **35.** 114.42902 **37.** 13.284 **39.** 90.72 **41.** 0.0028728 **43.** 0.72523 **45.** 1.872115 **47.** 45,678 **49.** 2888¢ **51.** 66¢ **53.** $0.34 **55.** $34.45 **57.** $3,600,000,000 **59.** 196,800,000 **61.** $11\frac{1}{5}$ **62.** $\frac{35}{72}$ **63.** 342 **64.** 87 **65.** 4566 **66.** 1257 **67.** ◈ **69.** 10^{21}

Margin Exercises, Section 3.4, pp. 163–168

1. 0.6 **2.** 1.5 **3.** 0.47 **4.** 0.32 **5.** 3.75 **6.** 0.25 **7.** (a) 375; (b) 15 **8.** 4.9 **9.** 12.8 **10.** 15.625 **11.** 12.78 **12.** 0.001278 **13.** 0.09847 **14.** 67.832 **15.** 0.78314 **16.** 1105.6 **17.** 0.2426 **18.** 593.44 **19.** 1.2825 billion

Exercise Set 3.4, p. 169

1. 2.99 **3.** 23.78 **5.** 7.48 **7.** 7.2 **9.** 1.143 **11.** 4.041 **13.** 0.07 **15.** 70 **17.** 20 **19.** 0.4 **21.** 0.41 **23.** 8.5 **25.** 9.3 **27.** 0.625 **29.** 0.26 **31.** 15.625 **33.** 2.34 **35.** 0.47 **37.** 0.2134567 **39.** 21.34567 **41.** 1023.7 **43.** 9.3 **45.** 0.0090678 **47.** 45.6 **49.** 2107 **51.** 303.003 **53.** 446.208 **55.** 24.14 **57.** 13.0072 **59.** 19.3204 **61.** 473.188278 **63.** 10.49 **65.** 911.13 **67.** 205 **69.** $1288.36 **71.** 59.49° **73.** $15\frac{1}{8}$ **74.** $5\frac{7}{8}$ **75.** $\frac{6}{7}$ **76.** $2 \cdot 3 \cdot 3 \cdot 3 \cdot 3$ **77.** $2 \cdot 2 \cdot 3 \cdot 3 \cdot 19$ **78.** $\frac{7}{8}$ **79.** ◈ **81.** 6.254194585 **83.** 1000 **85.** 100

Margin Exercises, Section 3.5, pp. 173–176

1. 0.8 **2.** 0.45 **3.** 0.275 **4.** 1.32 **5.** 0.4 **6.** 0.375 **7.** $0.1\overline{6}$ **8.** $0.\overline{6}$ **9.** $0.\overline{45}$ **10.** $1.\overline{09}$ **11.** $0.\overline{428571}$ **12.** 0.7; 0.67; 0.667 **13.** 0.8; 0.81; 0.808 **14.** 6.2; 6.25; 6.245 **15.** 0.72 **16.** 0.552 **17.** 9.6575

Exercise Set 3.5, p. 177

1. 0.6 **3.** 0.325 **5.** 0.2 **7.** 0.85 **9.** 0.475 **11.** 0.975 **13.** 0.52 **15.** 20.016 **17.** 0.25 **19.** 0.575 **21.** 0.72 **23.** 1.1875 **25.** $0.2\overline{6}$ **27.** $0.\overline{3}$ **29.** $1.\overline{3}$ **31.** $1.1\overline{6}$ **33.** $0.\overline{571428}$ **35.** $0.91\overline{6}$ **37.** 0.3; 0.27; 0.267 **39.** 0.3; 0.33; 0.333 **41.** 1.3; 1.33; 1.333 **43.** 1.2; 1.17; 1.167 **45.** 0.6; 0.57; 0.571 **47.** 0.9; 0.92; 0.917 **49.** 0.2; 0.18; 0.182 **51.** 0.3; 0.28; 0.278 **53.** 11.06 **55.** 8.4 **57.** $417.51\overline{6}$ **59.** 0 **61.** 2.8125 **63.** 0.20425 **65.** 317.14 **67.** 0.1825 **69.** 18 **71.** 2.736 **73.** 21 **74.** 10 **75.** $3\frac{2}{5}$ **76.** $30\frac{7}{10}$ **77.** $1\frac{1}{24}$ cups **78.** $1\frac{73}{100}$ in. **79.** ◈ **81.** $0.\overline{142857}$ **83.** $0.\overline{428571}$ **85.** $0.\overline{714285}$ **87.** $0.\overline{1}$ **89.** $0.\overline{001}$

Margin Exercises, Section 3.6, pp. 179–181

1. (b) **2.** (a) **3.** (d) **4.** (b) **5.** (a) **6.** (d) **7.** (b) **8.** (c) **9.** (b) **10.** (b) **11.** (c) **12.** (a) **13.** (c) **14.** (c)

Calculator Spotlight, p. 182

1. (a) [+] [×]; **(b)** [+] [×] [−] **2.** $a = 5$, $b = 9$ **3.** $66.70, $77.82, $88.94, $100.06, $111.18, $122.30, $133.42 **4.** $2029.66, $1950.88, $1872.10, $1793.32

5.

4.55	1.3	1.95
0	2.6	5.2
3.25	3.9	0.65

Magic sum = 7.8

6.

2.16	0.81	1.08
0.27	1.35	2.43
1.62	1.89	0.54

Magic sum = 4.05

7.

6.16	43.12	34.72	16.24
38.64	12.32	9.52	39.76
15.12	34.16	44.24	6.72
40.32	10.64	11.76	37.52

Magic sum = 100.24

Exercise Set 3.6, p. 183

1. (d) **3.** (c) **5.** (a) **7.** (c) **9.** 1.6 **11.** 6 **13.** 60 **15.** 2.3 **17.** 180 **19.** (a) **21.** (c) **23.** (b) **25.** (b) **27.** 7700 **29.** $2 \cdot 2 \cdot 3 \cdot 3 \cdot 3$ **30.** $2 \cdot 2 \cdot 2 \cdot 2 \cdot 5 \cdot 5$ **31.** $5 \cdot 5 \cdot 13$ **32.** $2 \cdot 3 \cdot 3 \cdot 37$ **33.** $\frac{5}{16}$ **34.** $\frac{129}{251}$ **35.** $\frac{8}{9}$ **36.** $\frac{13}{25}$ **37.** ◈ **39.** Yes **41.** No

Margin Exercises, Section 3.7, pp. 185–192

1. 8.4° **2.** 148.1 gal **3.** $55.92 **4.** $368.75 **5.** 96.52 cm^2 **6.** $0.89 **7.** 28.6 miles per gallon **8.** $221,519 **9.** $594,444

Exercise Set 3.7, p. 193

1. $39.60 **3.** $21.22 **5.** $3.01 **7.** 102.8° **9.** $21,219.17 **11.** 250,205.04 ft^2 **13.** 22,691.5 mi **15.** 8.9 billion **17.** mach 0.3 **19.** 20.2 mpg **21.** 11.9752 cu ft **23.** $10 **25.** 78.1 cm **27.** 2.31 cm **29.** 876 calories **31.** $1171.74 **33.** 227.75 ft^2 **35.** 0.305 **37.** $57.35 **39.** $349.44 **41.** 5.8¢, or $0.058 **43.** 2152.56 yd^2 **45.** $316,987.20; $196,987.20 **47.** 31 million **49.** 90.6 million **51.** 1.4°F **53.** No **55.** $435,976 **57.** $87,494 **59.** $67,972 **61.** 6335 **62.** $\frac{31}{24}$ **63.** $6\frac{5}{6}$ **64.** $\frac{23}{15}$ **65.** 13,766 **66.** 2803 **67.** $\frac{1}{24}$ **68.** $1\frac{5}{6}$ **69.** $\frac{2}{15}$ **70.** 2432 **71.** 28 min **72.** $7\frac{1}{5}$ min **73.** ◈ **75.** $1.44

Summary and Review: Chapter 3, p. 199

1. Three and forty-seven hundredths **2.** Thirty-one thousandths **3.** Five hundred ninety-seven and $\frac{25}{100}$ dollars **4.** Zero and $\frac{96}{100}$ dollars **5.** $\frac{9}{100}$ **6.** $\frac{4561}{1000}$ **7.** $\frac{89}{1000}$ **8.** $\frac{30,227}{10,000}$ **9.** 0.034 **10.** 4.2603 **11.** 27.91 **12.** 867.006 **13.** 0.034 **14.** 0.91 **15.** 0.741 **16.** 1.041 **17.** 17.4 **18.** 17.43 **19.** 17.429 **20.** 17 **21.** 574.519 **22.** 0.6838 **23.** 229.1 **24.** 45.551 **25.** 29.2092 **26.** 790.29 **27.** 29.148 **28.** 70.7891 **29.** 12.96 **30.** 0.14442 **31.** 4.3 **32.** 0.02468 **33.** 7.5 **34.** 0.45 **35.** 45.2 **36.** 1.022 **37.** 0.2763 **38.** 1389.2 **39.** 496.2795 **40.** 6.95 **41.** 42.54 **42.** 4.9911 **43.** $7.76 **44.** $5888.74 **45.** 24.36; 104.4 **46.** $239.80 **47.** 11.16 **48.** 6365.1 bu **49.** 272 **50.** 216 **51.** 4 **52.** $125 **53.** 2.6 **54.** 1.28 **55.** 2.75 **56.** 3.25 **57.** $1.1\overline{6}$ **58.** $1.\overline{54}$ **59.** 1.5 **60.** 1.55 **61.** 1.545 **62.** $82.73 **63.** $4.87 **64.** 2493¢ **65.** 986¢ **66.** 5,500,000,000,000 **67.** 1,200,000 **68.** 1.8045 **69.** 57.1449 **70.** 15.6375 **71.** $41.537\overline{3}$ **72.** $19\frac{4}{5}$ **73.** $6\frac{3}{5}$ **74.** $\frac{1}{2}$ **75.** $2 \cdot 2 \cdot 2 \cdot 2 \cdot 2 \cdot 2 \cdot 3$ **76.** 3300 **77.** ◈ Multiply by 1 to get a denominator that is a power of 10:

$$\frac{44}{125} = \frac{44}{125} \cdot \frac{8}{8} = \frac{352}{1000} = 0.352.$$

We can also divide to find that $\frac{44}{125} = 0.352$.

78. ◈ Each decimal place in the decimal notation corresponds to one zero in the power of ten in the fractional notation. When the fractions are multiplied, the number of zeros in the denominator of the product is the sum of the number of zeros in the denominators of the factors. So the number of decimal places in the product is the sum of the number of decimal places in the factors.

79. (a) $2.56 \cdot 6.4 \div 51.2 - 17.4 + 89.7 = 72.62$; **(b)** $(0.37 + 18.78) \cdot 2^{13} = 156,876.8$ **80.** $\frac{1}{3} + \frac{2}{3} = 0.33333333\ldots + 0.66666666 = 0.99999999\ldots$. Therefore, $1 = 0.99999999\ldots$ because $\frac{1}{3} + \frac{2}{3} = 1$. **81.** $2 = 1.\overline{9}$

Test: Chapter 3, p. 201

1. [3.1a] Two and thirty-four hundredths **2.** [3.1a] One thousand, two hundred thirty-four and $\frac{78}{100}$ dollars **3.** [3.1b] $\frac{91}{100}$ **4.** [3.1b] $\frac{2769}{1000}$ **5.** [3.1c] 0.074 **6.** [3.1c] 3.7047 **7.** [3.1c] 756.09 **8.** [3.1c] 91.703 **9.** [3.1d] 0.162 **10.** [3.1d] 0.078 **11.** [3.1d] 0.9 **12.** [3.1e] 6 **13.** [3.1e] 5.68 **14.** [3.1e] 5.678 **15.** [3.1e] 5.7 **16.** [3.2a] 405.219 **17.** [3.2a] 0.7902 **18.** [3.2a] 186.5 **19.** [3.2a] 1033.23 **20.** [3.2b] 48.357 **21.** [3.2b] 19.0901 **22.** [3.2b] 1.9946 **23.** [3.2b] 152.8934 **24.** [3.3a] 0.03 **25.** [3.3a] 8 **26.** [3.3a] 0.21345 **27.** [3.3a] 73,962 **28.** [3.4a] 4.75 **29.** [3.4a] 0.24 **30.** [3.4a] 30.4 **31.** [3.4a] 0.19

32. [3.4a] 0.34689 **33.** [3.4a] 34,689 **34.** [3.4b] 84.26
35. [3.2c] 8.982 **36.** [3.7a] 4.97 km
37. [3.7a] $6572.45 **38.** [3.7a] $1675.50
39. [3.7a] $479.70 **40.** [3.6a] 198 **41.** [3.6a] 4
42. [3.5a] 1.6 **43.** [3.5a] 0.88 **44.** [3.5a] 5.25
45. [3.5a] 0.75 **46.** [3.5a] $1.\overline{2}$ **47.** [3.5a] $2.\overline{142857}$
48. [3.5b] 2.1 **49.** [3.5b] 2.14 **50.** [3.5b] 2.143
51. [3.3b] $9.49 **52.** [3.3b] $2,800,000,000
53. [3.4c] 40.0065 **54.** [3.4c] 384.8464
55. [3.5c] 302.4 **56.** [3.5c] $52.339\overline{4}$ **57.** [2.4b] $2\frac{11}{16}$
58. [2.4c] $26\frac{1}{2}$ **59.** [2.1e] $\frac{11}{18}$ **60.** [1.9a] 360
61. [1.7d] $2 \cdot 2 \cdot 2 \cdot 3 \cdot 3 \cdot 5$ **62.** [3.7a] $35
63. [3.1c, d] $\frac{2}{3}, \frac{5}{7}, \frac{15}{19}, \frac{11}{13}, \frac{17}{20}, \frac{13}{15}$

Chapter 4

Pretest: Chapter 4, p. 204

1. [4.1a] $\frac{35}{43}$ **2.** [4.1a] $\frac{0.079}{1.043}$ **3.** [4.1d] 22.5
4. [4.1b] 25.5 mi/gal **5.** [4.2b] 0.87
6. [4.2c] 53.7% **7.** [4.3a] 75% **8.** [4.3b] $\frac{37}{100}$
9. [4.4b] $x = 60\% \times 75$; 45 **10.** [4.5b] $\frac{n}{100} = \frac{35}{50}$; 70%
11. [4.6a] 90 lb **12.** [4.6b] 20% **13.** [4.7a] $14.30; $300.30 **14.** [4.7b] $5152 **15.** [4.7c] $112.50 discount; $337.50 sale price **16.** [4.7d] $99.60 **17.** [4.7d] $20
18. [4.7e] $7128.60

Margin Exercises, Section 4.1, pp. 205–210

1. $\frac{5}{11}$, or 5:11 **2.** $\frac{57.3}{86.1}$, or 57.3:86.1 **3.** $\frac{6\frac{3}{4}}{7\frac{2}{5}}$, or $6\frac{3}{4}:7\frac{2}{5}$
4. $\frac{2.5}{0.8}$, or 2.5:0.8 **5.** $\frac{38.2}{56.1}$, or 38.2:56.1 **6.** 18 is to 27 as 2 is to 3 **7.** 3.6 is to 12 as 3 is to 10 **8.** $\frac{4}{3}$, or 4:3
9. 5 mi/hr **10.** 12 mi/hr **11.** 0.3 mi/hr
12. 1100 ft/sec **13.** 4 ft/sec **14.** 14.5 ft/sec
15. 250 ft/sec **16.** 2 gal/day **17.** Yes **18.** No
19. No **20.** 14 **21.** $11\frac{1}{4}$ **22.** 10.5 **23.** 2.64
24. 10.8 **25.** Approximately 14.1 gal **26.** 9.5 in.

Calculator Spotlight, p. 209

1. $11.\overline{6}$, or $\frac{35}{3}$; 1.1; 4.8; 14.5 **2.** 14; 11.25; 10.5; 2.64; 10.8

Exercise Set 4.1, p. 211

1. $\frac{4}{5}$ **3.** $\frac{56.78}{98.35}$ **5.** $\frac{4}{1}$ **7.** $\frac{3}{4}$ **9.** $\frac{3}{4}$ **11.** $\frac{32}{101}$
13. 40 km/h **15.** 11 m/sec **17.** 152 yd/day
19. 57.5 ¢/min **21.** 0.623 gal/ft^2 **23.** 124 km/h
25. 560 mi/hr **27.** No **29.** Yes **31.** Yes **33.** No
35. 45 **37.** 10 **39.** 20 **41.** 18 **43.** 0.06 **45.** 5
47. 1 **49.** 12.5725 **51.** 702 mi **53.** 309 **55.** 1980
57. 6 gal **59.** 8 ft **61.** $7\frac{1}{3}$ in. **63.** 2.72 **65.** 5.28
67. **(a)** 33.7 in.; **(b)** 85.598 cm **68.** **(a)** 2438 mi; **(b)** 3900.8 km **69.** ◈ **71.** 17 **73.** 7 gal

Margin Exercises, Section 4.2, pp. 216–218

1. $\frac{70}{100}$; $70 \times \frac{1}{100}$; 70×0.01 **2.** $\frac{23.4}{100}$; $23.4 \times \frac{1}{100}$; 23.4×0.01 **3.** $\frac{100}{100}$; $100 \times \frac{1}{100}$; 100×0.01 **4.** 0.34
5. 0.789 **6.** 0.83 **7.** 0.042 **8.** 24% **9.** 347%
10. 100% **11.** 90% **12.** 10.8%

Exercise Set 4.2, p. 219

1. $\frac{90}{100}$; $90 \times \frac{1}{100}$; 90×0.01 **3.** $\frac{12.5}{100}$; $12.5 \times \frac{1}{100}$; 12.5×0.01 **5.** 0.67 **7.** 0.456 **9.** 0.5901 **11.** 0.1
13. 0.01 **15.** 2 **17.** 0.001 **19.** 0.0009 **21.** 0.0018
23. 0.2319 **25.** 0.4 **27.** 0.025 **29.** 0.622 **31.** 47%
33. 3% **35.** 870% **37.** 33.4% **39.** 75% **41.** 40%
43. 0.6% **45.** 1.7% **47.** 27.18% **49.** 2.39%
51. 0.0104% **53.** 24% **55.** 58.1% **57.** $33\frac{1}{3}$
58. $37\frac{1}{2}$ **59.** $9\frac{3}{8}$ **60.** $18\frac{9}{16}$ **61.** $0.\overline{6}$ **62.** $0.\overline{3}$
63. $0.8\overline{3}$ **64.** $1.41\overline{6}$ **65.** ◈

Margin Exercises, Section 4.3, pp. 222–223

1. 25% **2.** 62.5%, or $62\frac{1}{2}\%$ **3.** $66.\overline{6}\%$, or $66\frac{2}{3}\%$
4. $83.\overline{3}\%$, or $83\frac{1}{3}\%$ **5.** 57% **6.** 76% **7.** $\frac{3}{5}$ **8.** $\frac{13}{400}$
9. $\frac{2}{3}$
10.

$\frac{1}{5}$	$\frac{5}{6}$	$\frac{3}{8}$
0.2	$0.83\overline{3}$	0.375
20%	$83.\overline{3}\%$, or $83\frac{1}{3}\%$	$37\frac{1}{2}\%$

Calculator Spotlight, p. 221

1. 52% **2.** 38.46% **3.** 107.69% **4.** 171.43%
5. 59.62% **6.** 28.31%

Calculator Spotlight, p. 224

1. 25.35; 1.68% **2.** 11.95; 0% **3.** 57.14; 2.5%
4. 39.41; 3.42%

Exercise Set 4.3, p. 225

1. 41% **3.** 5% **5.** 20% **7.** 30% **9.** 50%
11. 62.5%, or $62\frac{1}{2}\%$ **13.** 80% **15.** $66.\overline{6}\%$, or $66\frac{2}{3}\%$
17. $16.\overline{6}\%$, or $16\frac{2}{3}\%$ **19.** 16% **21.** 5% **23.** 34%
25. 36% **27.** 21% **29.** 24% **31.** $\frac{17}{20}$ **33.** $\frac{5}{8}$ **35.** $\frac{1}{3}$
37. $\frac{1}{6}$ **39.** $\frac{29}{400}$ **41.** $\frac{1}{125}$ **43.** $\frac{203}{800}$ **45.** $\frac{176}{225}$ **47.** $\frac{711}{1100}$
49. $\frac{3}{2}$ **51.** $\frac{13}{40,000}$ **53.** $\frac{1}{3}$ **55.** $\frac{11}{20}$ **57.** $\frac{19}{50}$ **59.** $\frac{11}{100}$
61. $\frac{1}{4}$ **63.** $\frac{569}{10,000}$

65.

Fractional Notation	Decimal Notation	Percent Notation
$\frac{1}{8}$	0.125	$12\frac{1}{2}\%$, or 12.5%
$\frac{1}{6}$	$0.1\overline{6}$	$16\frac{2}{3}\%$, or $16.\overline{6}\%$
$\frac{1}{5}$	0.2	20%
$\frac{1}{4}$	0.25	25%
$\frac{1}{3}$	$0.\overline{3}$	$33\frac{1}{3}\%$, or $33.\overline{3}\%$
$\frac{3}{8}$	0.375	$37\frac{1}{2}\%$, or 37.5%
$\frac{2}{5}$	0.4	40%
$\frac{1}{2}$	0.5	50%

67.

Fractional Notation	Decimal Notation	Percent Notation
$\frac{1}{2}$	0.5	50%
$\frac{1}{3}$	$0.\overline{3}$	$33\frac{1}{3}\%$, or $33.\overline{3}\%$
$\frac{1}{4}$	0.25	25%
$\frac{1}{6}$	$0.1\overline{6}$	$16\frac{2}{3}\%$, or $16.\overline{6}\%$
$\frac{1}{8}$	0.125	$12\frac{1}{2}\%$, or 12.5%
$\frac{3}{4}$	0.75	75%
$\frac{5}{6}$	$0.8\overline{3}$	$83\frac{1}{3}\%$, or $83.\overline{3}\%$
$\frac{3}{8}$	0.375	$37\frac{1}{2}\%$, or 37.5%

69. 70 **70.** 5 **71.** 400 **72.** 18.75 **73.** 4 **74.** $\frac{3}{44}$ **75.** $1\frac{9}{31}$ **76.** $5\frac{1}{4}$ **77.** $83\frac{1}{3}$ **78.** $20\frac{1}{2}$ **79.** $43\frac{1}{8}$ **80.** $62\frac{1}{6}$ **81.** $18\frac{3}{4}$ **82.** $7\frac{4}{9}$ **83.** ◈ **85.** $11.\overline{1}\%$ **87.** $0.01\overline{5}$

Margin Exercises, Section 4.4, pp. 229–232

1. $12\% \times 50 = a$ **2.** $a = 40\% \times 60$ **3.** $45 = 20\% \times t$ **4.** $120\% \times y = 60$ **5.** $16 = n \times 40$ **6.** $b \times 84 = 10.5$ **7.** 6 **8.** \$35.20 **9.** 225 **10.** \$50 **11.** 40% **12.** 12.5%

Exercise Set 4.4, p. 233

1. $y = 32\% \times 78$ **3.** $89 = a \times 99$ **5.** $13 = 25\% \times y$ **7.** 234.6 **9.** 45 **11.** \$18 **13.** 1.9 **15.** 78% **17.** 200% **19.** 50% **21.** 125% **23.** 40 **25.** \$40 **27.** 88 **29.** 20 **31.** 6.25 **33.** \$846.60 **35.** $\frac{9}{100}$ **36.** $\frac{179}{100}$ **37.** $\frac{875}{1000}$, or $\frac{7}{8}$ **38.** $\frac{9375}{10,000}$, or $\frac{15}{16}$ **39.** 0.89 **40.** 0.07 **41.** 0.3 **42.** 0.017 **43.** ◈ **45.** \$880 (can vary); \$843.20 **47.** 108 to 135 tons

Margin Exercises, Section 4.5, pp. 236–238

1. $\frac{12}{100} = \frac{a}{50}$ **2.** $\frac{40}{100} = \frac{a}{60}$ **3.** $\frac{130}{100} = \frac{a}{72}$ **4.** $\frac{20}{100} = \frac{45}{b}$ **5.** $\frac{120}{100} = \frac{60}{b}$ **6.** $\frac{N}{100} = \frac{16}{40}$ **7.** $\frac{N}{100} = \frac{10.5}{84}$ **8.** \$225 **9.** 35.2 **10.** 6 **11.** 50 **12.** 30% **13.** 12.5%

Exercise Set 4.5, p. 239

1. $\frac{37}{100} = \frac{a}{74}$ **3.** $\frac{N}{100} = \frac{4.3}{5.9}$ **5.** $\frac{25}{100} = \frac{14}{b}$ **7.** 68.4 **9.** 462 **11.** 40 **13.** 2.88 **15.** 25% **17.** 102% **19.** 25% **21.** 93.75% **23.** \$72 **25.** 90 **27.** 88 **29.** 20 **31.** 25 **33.** \$780.20 **35.** 8 **36.** 4000 **37.** 8 **38.** 2074 **39.** 100 **40.** 15 **41.** $8.0\overline{4}$ **42.** $\frac{3}{16}$, or 0.1875 **43.** $\frac{43}{48}$ qt **44.** $\frac{1}{8}$ T **45.** ◈ **47.** \$1134 (can vary); \$1118.64

Margin Exercises, Section 4.6, pp. 241–246

1. 16% **2.** 50 **3.** 9% **4.** 4% **5.** \$10,682

Calculator Spotlight, p. 242

1. 50 **2.** 7001.88 **3.** 83,931.456 **4.** 36,458.03724

Calculator Spotlight, p. 246

1. \$10,682 **2.** \$8918

Exercise Set 4.6, p. 247

1. 20; 100 **3.** 536; 264 **5.** 32.5%; 67.5% **7.** 20.4 mL; 659.6 mL **9.** 25% **11.** 166; 156; 146; 140; 122 **13.** 8% **15.** 20% **17.** \$30,030 **19.** \$12,600 **21.** 6.096 billion; 6.194 billion; 6.293 billion **23.** \$36,400 **25.** \$17.25; \$39.10; \$56.35 **27.** **(a)** 4.1%; **(b)** 31.0% **29.** 35.9% **31.** $2.\overline{27}$ **32.** 0.44 **33.** 3.375 **34.** $4.\overline{7}$ **35.** 0.92 **36.** $0.8\overline{3}$ **37.** 0.4375 **38.** 2.317 **39.** 3.4809 **40.** 0.675 **41.** ◈ **43.** 19% **45.** About 5 ft, 6 in. **47.** $83\frac{1}{3}\%$

Margin Exercises, Section 4.7, pp. 251–257

1. \$53.52; \$722.47 **2.** \$1.03; \$15.78 **3.** 6% **4.** \$5628 **5.** \$1675 **6.** \$33.60; \$106.40 **7.** \$602 **8.** \$451.50 **9.** \$27.62; \$4827.62 **10.** \$2464.20 **11.** \$8103.38

Calculator Spotlight, p. 257

1. \$16,357.18 **2.** \$12,764.72

Exercise Set 4.7, p. 259

1. \$16.56; \$281.56 **3.** 5% **5.** 4% **7.** \$2000 **9.** \$711.50 **11.** 5.6% **13.** \$2700 **15.** 5% **17.** \$980 **19.** \$5880 **21.** 12% **23.** \$420 **25.** \$460; 18.043% **27.** \$30; \$270 **29.** \$2.55; \$14.45 **31.** \$125; \$112.50 **33.** 40%; \$360 **35.** \$26 **37.** \$124 **39.** \$150.50 **41.** **(a)** \$128.22; **(b)** \$6628.22 **43.** \$484 **45.** \$236.75 **47.** \$4284.90 **49.** \$2604.52 **51.** \$4101.01 **53.** $\frac{93}{100}$ **54.** 37 **55.** $1.\overline{18}$ **56.** $2\frac{7}{11}$ **57.** ◈

Summary and Review: Chapter 4, p. 263

1. $\frac{47}{84}$ **2.** $\frac{46}{1.27}$ **3.** 23.54 mi/hr **4.** 0.638 gal/ft^2 **5.** 0.72 serving/lb **6.** No **7.** No **8.** 32 **9.** 24 **10.** \$4.45 **11.** 351 **12.** 832 mi **13.** 2,990,000 kg **14.** 48.3% **15.** 36% **16.** 37.5%, or $37\frac{1}{2}$% **17.** $33.\overline{3}$%, or $33\frac{1}{3}$% **18.** 0.735 **19.** 0.065 **20.** $\frac{6}{25}$ **21.** $\frac{63}{1000}$ **22.** $30.6 = x\% \times 90$; 34% **23.** $63 = 84\% \times n$; 75 **24.** $y = 38\frac{1}{2}\% \times 168$; 64.68 **25.** $\frac{24}{100} = \frac{16.8}{b}$; 70 **26.** $\frac{42}{30} = \frac{N}{100}$; 140% **27.** $\frac{10.5}{100} = \frac{a}{84}$; 8.82 **28.** \$598 **29.** 12% **30.** 20% **31.** 168 **32.** \$14.40 **33.** 5% **34.** 11% **35.** \$42; \$308 **36.** \$2940 **37.** **(a)** \$394.52; **(b)** \$24,394.52 **38.** \$121 **39.** \$7727.26 **40.** \$9504.80 **41.** $\frac{3107}{1000}$ **42.** $\frac{29}{100}$ **43.** 64 **44.** 7.6123 **45.** $3.\overline{6}$ **46.** $1.\overline{571428}$ **47.** $3\frac{2}{3}$ **48.** $17\frac{2}{7}$
49. ◈ No; the 10% discount was based on the original price rather than on the sale price.
50. ◈ A 40% discount is better. When successive discounts are taken, each is based on the previous discounted price rather than on the original price. A 20% discount followed by a 22% discount is the same as a 37.6% discount off the original price.
51. 105 min, or 1 hr, 45 min **52.** \$168

Test: Chapter 4, p. 265

1. [4.1a] $\frac{85}{97}$ **2.** [4.1a] $\frac{0.34}{124}$ **3.** [4.1b] 0.625 ft/sec **4.** [4.1b] $1\frac{1}{3}$ servings/lb **5.** [4.1c] Yes **6.** [4.1c] No **7.** [4.1d] 12 **8.** [4.1d] 360 **9.** [4.1e] 4.8 min **10.** [4.1e] 525 mi **11.** [4.2b] 0.89 **12.** [4.2c] 67.4% **13.** [4.3a] 137.5% **14.** [4.3b] $\frac{13}{20}$ **15.** [4.4a, b] $a = 40\% \cdot 55$; 22 **16.** [4.5a, b] $\frac{N}{100} = \frac{65}{80}$; 81.25% **17.** [4.6a] 50 lb **18.** [4.6b] 140% **19.** [4.7a] \$16.20; \$340.20 **20.** [4.7b] \$630 **21.** [4.7c] \$40; \$160 **22.** [4.7d] \$8.52 **23.** [4.7d] \$5356 **24.** [4.7e] \$1102.50 **25.** [4.7e] \$10,226.69 **26.** [4.7c] \$131.95; 52.8% **27.** [3.4b] 222 **28.** [3.1b] $\frac{447}{10}$ **29.** [3.5a] $1.41\overline{6}$ **30.** [2.4b] $3\frac{21}{44}$ **31.** [4.7b] \$117,800 **32.** [4.1e] 5888

Chapter 5

Pretest: Chapter 5, p. 268

1. [5.1a, b, c] **(a)** 51; **(b)** 51.5; **(c)** no mode exists
2. [5.1a, b, c] **(a)** 3; **(b)** 3; **(c)** no mode exists
3. [5.1a, b, c] **(a)** 12.75; **(b)** 17; **(c)** 4
4. [5.1a] 55 mi/hr **5.** [5.1a] 76
6. [5.4b]

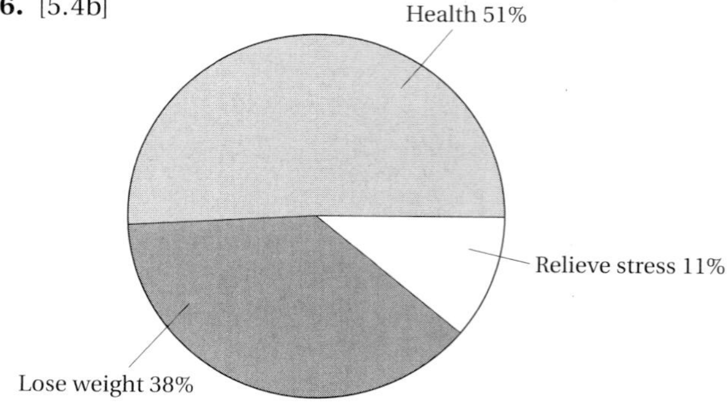

7. [5.2a] **(a)** \$208; **(b)** \$92
8. [5.3b]

\$320
310
300
290
280
270
260
Cost of insurance
31 32 33 34 35
Age

9. [5.3d]

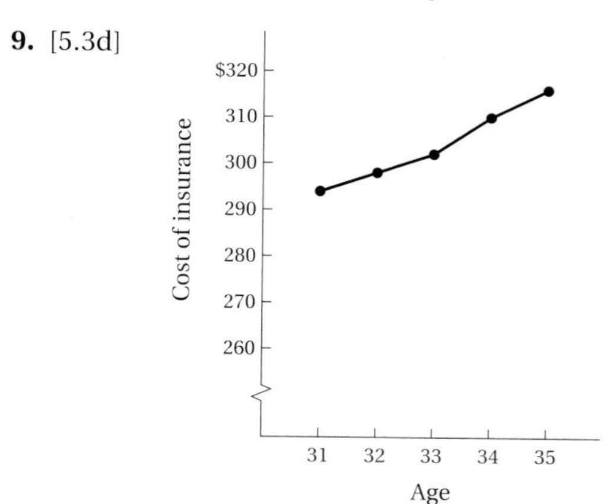

10. [5.3c] 260 **11.** [5.3c] 160

12. [5.3d], [5.5b]

(a)

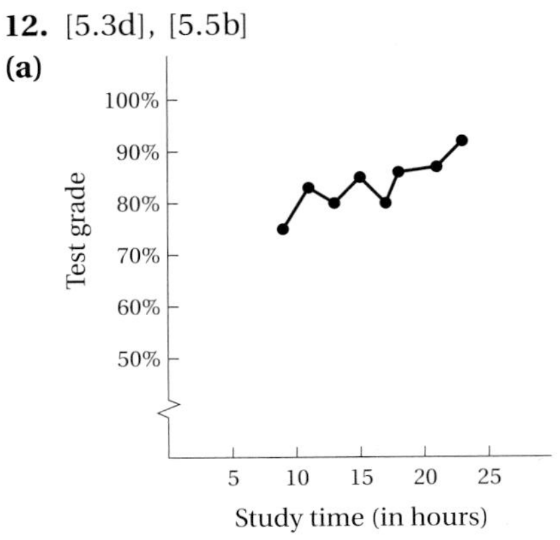

(b) Answers may vary; 94%

Margin Exercises, Section 5.1, pp. 271–274

1. 75 **2.** 54.9 **3.** 81 **4.** 19.4 **5.** 0.4 lb **6.** 19 mpg **7.** 2.5 **8.** 94 **9.** 17 **10.** 17 **11.** 91 **12.** $1700 **13.** 67.5 **14.** 45 **15.** 34, 67 **16.** No mode exists. **17.** **(a)** 17 g; **(b)** 18 g; **(c)** 19 g

Calculator Spotlight, p. 273

1. Only 79 would be divided by 3. The result would be $203\frac{1}{3}$. **2.** The answers are the same.

Exercise Set 5.1, p. 275

1. Average: 20; median: 18; mode: 29 **3.** Average: 20; median: 20; mode: 5, 20 **5.** Average: 5.2; median: 5.7; mode: 7.4 **7.** Average: 239.5; median: 234; mode: 234 **9.** Average: 40°; median: 40°; no mode exists **11.** 33 mpg **13.** 2.8 **15.** Average: $9.19; median: $9.49; mode: $7.99 **17.** 90 **19.** 263 days **21.** 196 **22.** $\frac{4}{9}$ **23.** 1.96 **24.** 1.999396 **25.** $1139.05 **26.** 3360 mi **27.** ◈ **29.** 182 **31.** 10

Margin Exercises, Section 5.2, pp. 277–281

1. Kellogg's Complete Bran Flakes **2.** Kellogg's Special K **3.** Kellogg's Special K, Wheaties **4.** Ralston Rice Chex, Kellogg's Complete Bran Flakes, Honey Nut Cheerios **5.** 510.66 mg **6.** 500 mg **7.** Mean: 232; median: 240; mode: 240 **8.** 20 **9.** Yes **10.** 440 mg **11.** 24% **12.** 6 g **13.** 60,000 **14.** Two and one half times as many in Zimbabwe as in Cameroon **15.** 55,000 **16.** 795; answers may vary **17.** 750; answers may vary **18.** 1830; answers may vary

19.

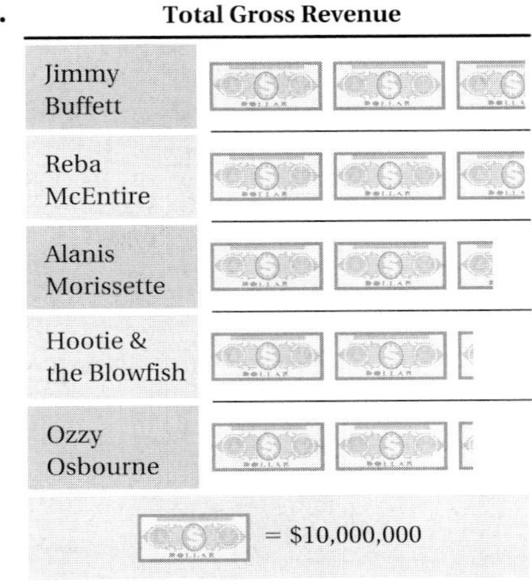

Exercise Set 5.2, p. 283

1. 483,612,200 mi **3.** Neptune **5.** All **7.** 11 **9.** Average: $27{,}884.\overline{1}$ mi; median: 7926 mi; no mode exists **11.** 92° **13.** 108° **15.** 3 **17.** 90° and higher **19.** 30% and higher **21.** 50% **23.** 59.29°, 59.58°; 0.5% **25.** 59.50°; 59.62°; 0.12° **27.** 1.0 billion **29.** 1999 **31.** 1650 and 1850 **33.** 2.0 billion; 50% **35.** 1998 **37.** 1994 and 1995 **39.** 7000 **41.** 1997

43.

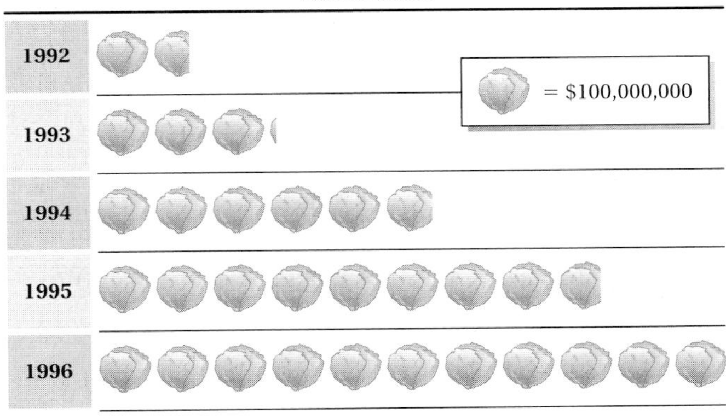

44. 12 **45.** 27,859.5 mi^2 **46.** $\frac{24}{100}$, or $\frac{6}{25}$ **47.** $\frac{4.8}{100}$, or $\frac{48}{1000}$, or $\frac{6}{125}$ **49.** ◈

50.

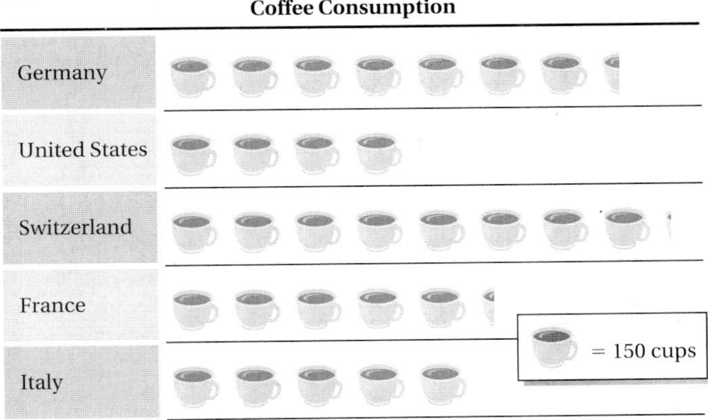

Margin Exercises, Section 5.3, pp. 289–294

1. 16 g **2.** Big Bacon Classic **3.** Single with Everything, chicken club, Big Bacon Classic **4.** 60 **5.** 85+ **6.** 60–64 **7.** Yes
8.

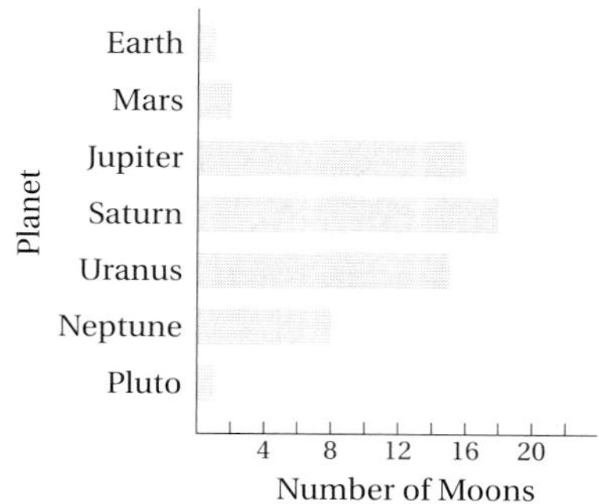

9. Month 7 **10.** Months 1 and 2, 4 and 5, 6 and 7, 11 and 12 **11.** Months 2, 5, 6, 7, 8, 9, 12 **12.** \$920 **13.** 40 yr **14.** \$1300
15.

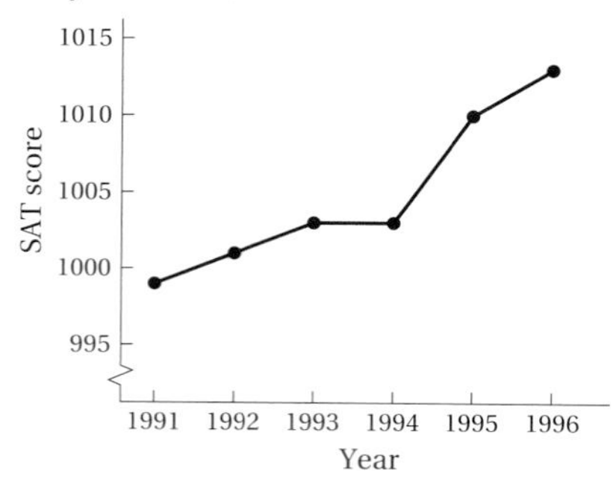

Exercise Set 5.3, p. 295

1. 190 **3.** 1 slice of chocolate cake with fudge frosting **5.** 1 cup premium chocolate ice cream **7.** 120 calories **9.** 920 calories **11.** 28 lb **13.** 920,000 hectares **15.** Latin America **17.** Africa **19.** 880,000 hectares
21.

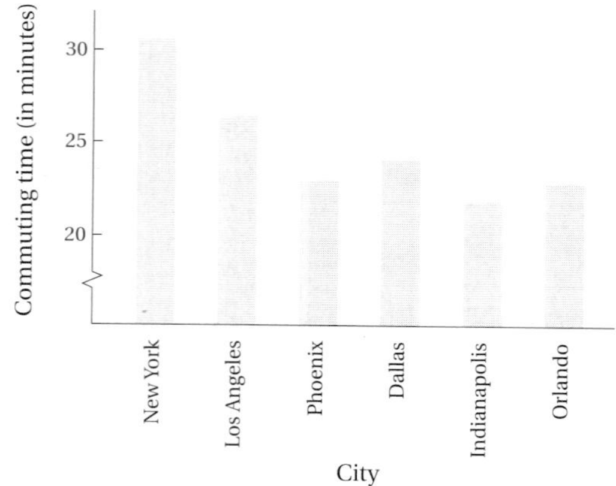

23. Indianapolis **25.** 23.55 min

26.

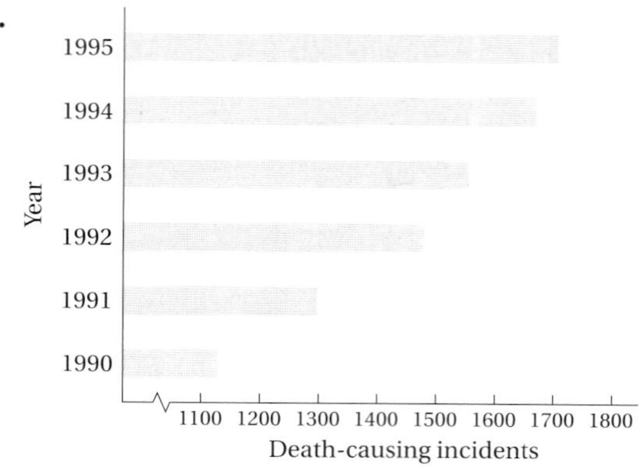

27. 1991 and 1992 **29.** $1472.\overline{6}$ **31.** 1997 **33.** About \$0.5 million **35.** 1994 and 1995
37.

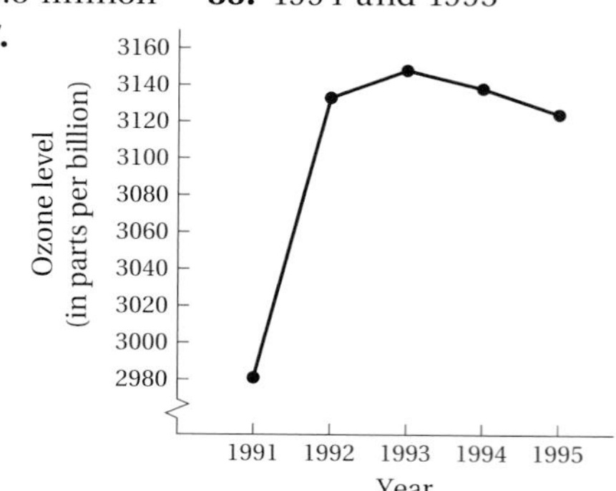

39. 1994 and 1995
41. 3133 parts per billion

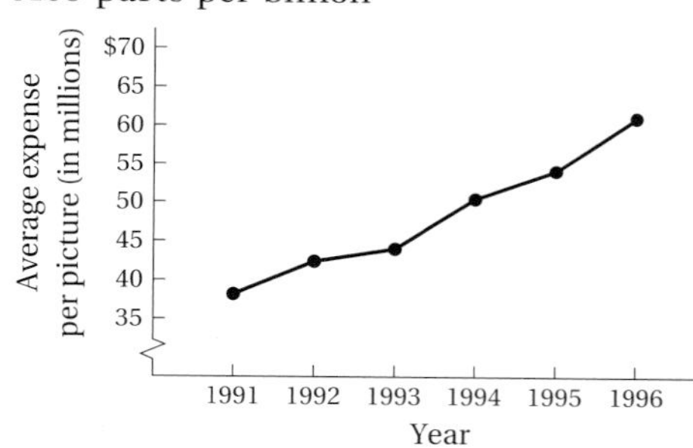

43. 1993 and 1994 **45.** \$48.35 million **47.** \$41.5 million **49.** 18 min **50.** 18% **51.** 82.5 **52.** $66\frac{2}{3}\%$ **53.** ◈ **55.** \$65.6

Margin Exercises, Section 5.4, pp. 301–302

1. Spaying **2.** 5% **3.** \$1122 **4.** 8% + 3%, or 11%
5. **Times of Engagement of Married Couples**

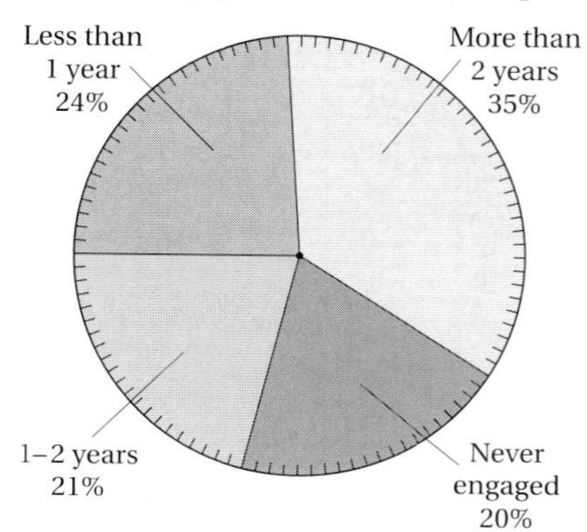

Exercise Set 5.4, p. 303

1. 3.7% **3.** 270 **5.** 6.8% **7.** Food **9.** 14%

11.

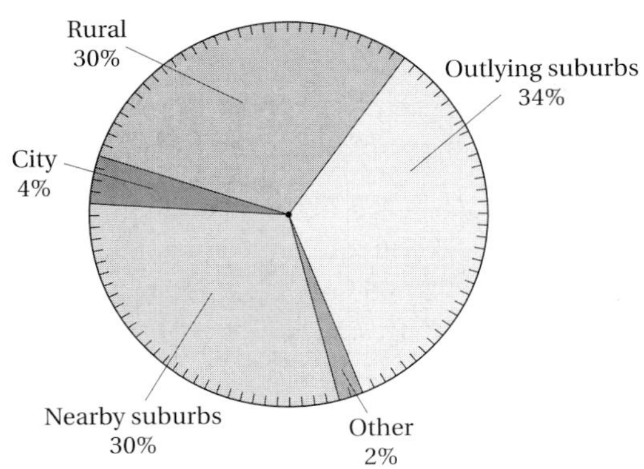

13.

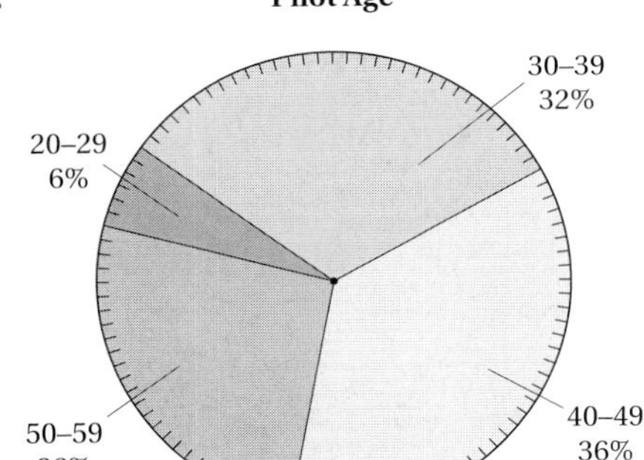

15.

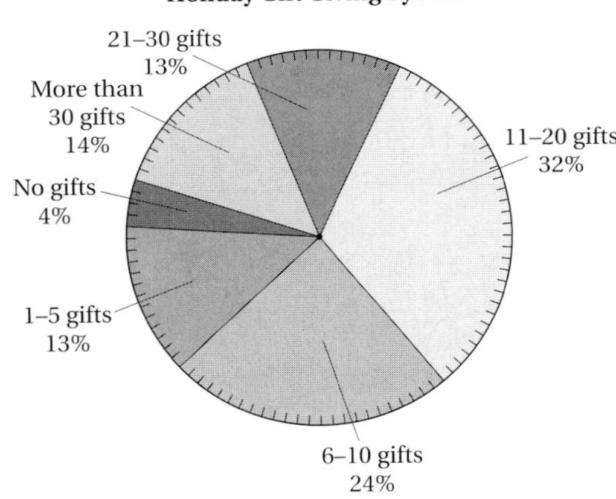

17. 300.6 **18.** 25% **19.** 115

Margin Exercises, Section 5.5, pp. 305–307

1. Wheat A: average stalk height ≈ 25.21 in.; wheat B: average stalk height ≈ 22.54 in.; wheat B is better. **2.** 111 million **3.** 94%

Exercise Set 5.5, p. 309

1. Bulb A: average time = 1171.25 hr; bulb B: average time ≈ 1251.58 hr; bulb B is better **3.** 83 **5.** 3112 parts per billion **7.** $148.8 billion **9.** $128,727 **10.** 0.625 cubic centimeter **11.** $1378.85 **12.** 5040 mi **13.** $\frac{15}{7}$, or $2\frac{1}{7}$ **14.** $\frac{1408}{3}$, or $469\frac{1}{3}$ **15.** $\frac{17}{25,000}$ **16.** $\frac{11}{12}$ **17.** ◈

Summary and Review: Chapter 5, p. 311

1. $21.00 **2.** $28.75 **3.** $10.00 **4.** $10.50 **5.** No **6.** $12.00 **7.** 5500 **8.** 1997 **9.** 1999 **10.** 4075 **11.** 38.5 **12.** 13.4 **13.** 1.55 **14.** 1840 **15.** $\$16.\overline{6}$ **16.** $321.\overline{6}$ **17.** 38.5 **18.** 14 **19.** 1.8 **20.** 1900 **21.** $17 **22.** 375 **23.** 26 **24.** 11; 17 **25.** 0.2 **26.** 700; 800 **27.** $17 **28.** 20 **29.** $110.50; $107 **30.** $66.1\overline{6}°$ **31.** 96 **32.** 420 **33.** 440 **34.** Big Bacon Classic **35.** Grilled chicken **36.** Plain single **37.** Chicken club **38.** 80 **39.** 250 **40.** Under 20 **41.** 12 **42.** 13 **43.** 45–74 **44.** 11 **45.** Under 20 **46.** 22% **47.** 11% **48.** 1600 **49.** 25%

50.

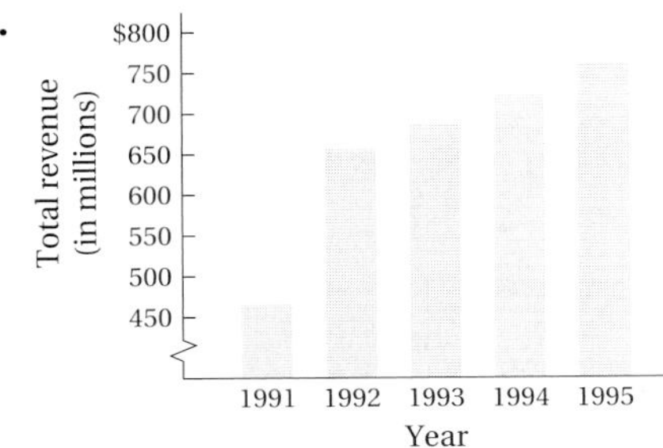

51.

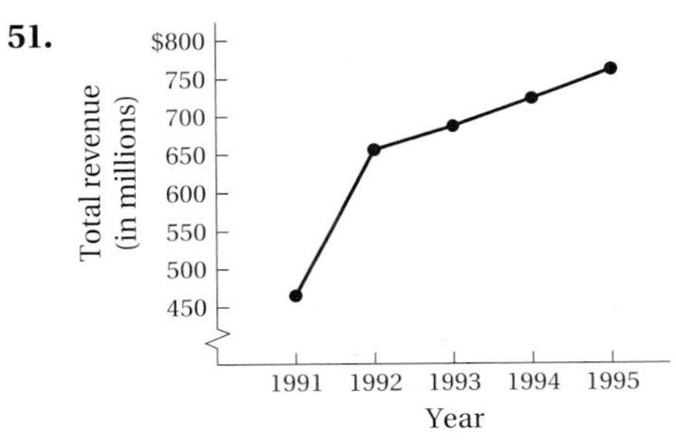

52. $800 million **53.** $806 million **54.** Battery A: average ≈ 43.04 hr; battery B: average = 41.55 hr; battery A is better. **55.** 12,600 mi **56.** 5.428 billion **57.** $222.\overline{2}\%$ **58.** 50% **59.** $\frac{9}{10}$ **60.** $\frac{5}{12}$ **61.** ◈ The average, the median, and the mode are "center points" that characterize a set of data. You might use the average to find a center point that is midway between the extreme values of the data. The median is a center point that is in the middle of all the data. That is, there are as many values less than the median than there are values greater than the median. The mode is a center point that represents the value or values that occur most frequently. **62.** ◈ The equation could represent a person's average income during a 4-yr period. Answers may vary. **63.** $a = 316$, $b = 349$

Test: Chapter 5, p. 315

1. [5.2a] $341,413 **2.** [5.2a] $2558 **3.** [5.2a] Couple, age 65; yearly income $100,000 **4.** [5.2a] Single female, age 45; yearly income $150,000 **5.** [5.2a] $78,169 **6.** [5.2a] $2264 **7.** [5.2b] 2003

8. [5.2b] 2002 and 2003 **9.** [5.2b] 7000 **10.** [5.2b] 2002 **11.** [5.1a] 50 **12.** [5.1a] 3 **13.** [5.1a] 15.5 **14.** [5.1b, c] Median: 50.5; no mode exists **15.** [5.1b, c] Median: 3; no mode exists **16.** [5.1b, c] Median: 17.5; mode: 17, 18 **17.** [5.1a] 58 km/h **18.** [5.1a] 76 **19.** [5.3c] \$6.5 billion **20.** [5.3c] \$4.4 billion **21.** [5.3c] \$3.1 billion **22.** [5.3c] 1996 **23.** [5.3c] \$3.9 billion **24.** [5.5b] \$7.9 billion **25.** [5.3b]

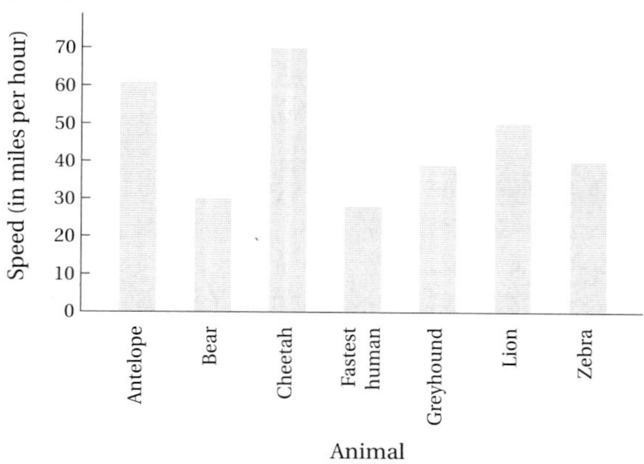

26. [5.3a] 42 mph **27.** [5.3a] No. The maximum speed of the fastest human is 22 mph slower than that of the lion. **28.** [5.3a] About 45.4 mph **29.** [5.3a] 40 mph **30.** [5.4b]

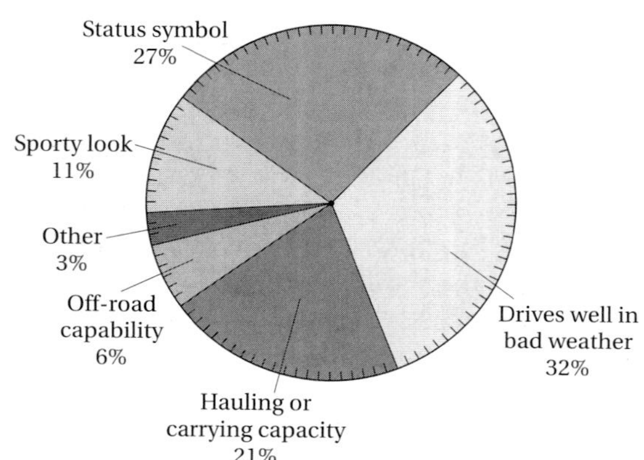

31. [5.3d]

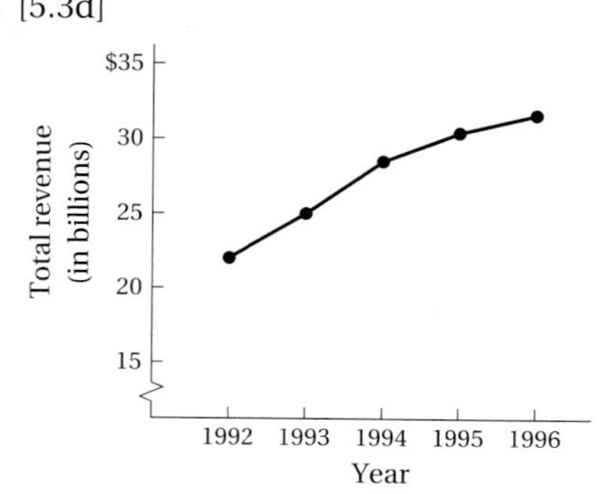

32. [5.5b] \$34.4 billion **33.** [5.5a] Bar A: average ≈ 8.417; bar B: average ≈ 8.417; equal quality.

34. [2.2c] $\frac{25}{4}$, or $6\frac{1}{4}$ **35.** [4.4b], [4.5b] 68 **36.** [4.6a] 15,600 **37.** [4.1e] 340 **38.** [5.1a, b] $a = 74$, $b = 111$

Chapter 6

Pretest: Chapter 6, p. 320

1. [6.2a] 131 mm **2.** [6.3b] 92 in^2 **3.** [6.4a] 22 cm^2 **4.** [6.4a] $32\frac{1}{2}$ ft^2 **5.** [6.4a] 4 m^2 **6.** [6.5a] 9.6 m **7.** [6.5b, c] 30.144 m; 72.3456 m^2 **8.** [6.6a] 160 cm^3; 256 cm^2 **9.** [6.6b] 1256 ft^3 **10.** [6.7c, d] $m\angle 1 = m\angle 7 = m\angle 3 = m\angle 5 = 151°$, $m\angle 2 = m\angle 6 = m\angle 4 = 29°$ **11.** [6.8a] $\overline{PQ} \cong \overline{ST}$, $\overline{QR} \cong \overline{TV}$, $\overline{RP} \cong \overline{VS}$; $\angle P \cong \angle S$, $\angle Q \cong \angle T$, $\angle R \cong \angle V$ **12.** [6.9b] $MA = 7$, $GT = 8$

Margin Exercises, Section 6.1, pp. 321–326

1. **(a)** and **(b)** E ——— F **(c)** $\overline{EF}$, $\overline{FE}$

2. • P • Q **3.** P ——— Q

4. P ——— Q ⟶, P

5. ⟵ P ——— Q, Q **6.** • R • S

7. R ——— S, R and S

8. R ——— S ⟶, R

9. ⟵ R ——— S, S

10. ⟵ R ——— S ⟶, no endpoints

11. $\overleftrightarrow{RS}$, $\overleftrightarrow{SR}$, $\overleftrightarrow{RT}$, $\overleftrightarrow{TR}$, $\overleftrightarrow{ST}$, $\overleftrightarrow{TS}$, n
12. Angle DEF, angle FED, $\angle DEF$, $\angle FED$, or $\angle E$
13. Angle PQR, angle RQP, $\angle PQR$, $\angle RQP$, or $\angle Q$
14. 127° **15.** 33° **16.** Right **17.** Acute **18.** Obtuse **19.** Straight **20.** No **21.** Yes **22.** **(a)** $\triangle ABC$; **(b)** $\triangle ABC$, $\triangle MPN$; **(c)** $\triangle DEF$; $\triangle GHI$, $\triangle JKL$, $\triangle QRS$ **23.** Yes **24.** No **25.** **(a)** $\triangle EDF$; **(b)** $\triangle GHI$, $\triangle QRS$; **(c)** $\triangle ABC$, $\triangle PMN$, $\triangle JKL$ **26.** Quadrilateral **27.** Hexagon **28.** Triangle **29.** Quadrilateral **30.** Dodecagon **31.** Octagon **32.** 64° **33.** **(a)** 3; **(b)** 180°; **(c)** 3, 540° **34.** 1080° **35.** 4140°

Exercise Set 6.1, p. 327

1. G ——— H, $\overline{GH}$, $\overline{HG}$

3. Q ——— D ⟶, $\overrightarrow{QD}$

5. $\overleftrightarrow{DE}$, $\overleftrightarrow{ED}$, $\overleftrightarrow{DF}$, $\overleftrightarrow{FD}$, $\overleftrightarrow{EF}$, $\overleftrightarrow{FE}$, l **7.** Angle GHI, angle IHG, $\angle GHI$, $\angle IHG$, or $\angle H$ **9.** 10° **11.** 180° **13.** 130°

15. Acute **17.** Straight **19.** Obtuse **21.** No **23.** Yes **25.** Scalene, obtuse **27.** Scalene, right **29.** Equilateral, acute **31.** Scalene, obtuse **33.** Quadrilateral **35.** Pentagon **37.** Triangle **39.** Pentagon **41.** Hexagon **43.** 1440° **45.** 900° **47.** 2160° **49.** 3240° **51.** 46° **53.** 43° **55.** 1.75 **56.** 2.34 **57.** 0.234 **58.** 0.0234 **59.** 13.85 **60.** $80\frac{1}{2}$ **61.** 4 **63.** ◈

Margin Exercises, Section 6.2, pp. 331–332

1. 26 cm **2.** 46 in. **3.** 12 cm **4.** 17.5 yd **5.** 32 km **6.** 40 km **7.** 21 yd **8.** 31.2 km **9.** 70 ft; $346.50

Exercise Set 6.2, p. 333

1. 17 mm **3.** 15.25 in. **5.** 13 m **7.** 30 ft **9.** 79.14 cm **11.** 88 ft **13.** 182 mm **15.** 826 m; $1197.70 **17.** 122 cm **19. (a)** 228 ft; **(b)** $1046.52 **21.** 0.561 **22.** 67.34% **23.** 112.5%, or $112\frac{1}{2}$% **24.** 25 **25.** 100 **26.** 961 **27.** 4,700,000 **28.** 4,300,000,000 **29.** ◈ **31.** 9 ft

Margin Exercises, Section 6.3, pp. 335–336

1. 8 cm^2 **2.** 56 km^2 **3.** $18\frac{3}{8}$ yd^2 **4.** 144 km^2 **5.** 118.81 m^2 **6.** $12\frac{1}{4}$ yd^2 **7.** 659.75 m^2

Exercise Set 6.3, p. 337

1. 15 km^2 **3.** 1.4 in^2 **5.** $6\frac{1}{4}$ yd^2 **7.** 8100 ft^2 **9.** 50 ft^2 **11.** 169.883 cm^2 **13.** $41\frac{2}{9}$ in^2 **15.** 484 ft^2 **17.** 3237.61 km^2 **19.** $28\frac{57}{64}$ yd^2 **21.** 1197 m^2 **23.** 630.36 m^2 **25. (a)** 819.75 ft^2; **(b)** 10 gal; **(c)** $179.50 **27.** 80 cm^2 **29.** 45.2% **30.** $33\frac{1}{3}$%, or $33.\overline{3}$% **31.** 55% **32.** 88% **33.** $\frac{27}{131}$; $\frac{131}{27}$ **34.** $\frac{1}{4}$; $\frac{3}{4}$ **35.** ◈ **37.** 16,914 in^2

Margin Exercises, Section 6.4, pp. 340–342

1. 43.8 cm^2 **2.** 12.375 km^2 **3.** 96 m^2 **4.** 18.7 cm^2 **5.** 100 m^2 **6.** 88 cm^2 **7.** 54 m^2

Exercise Set 6.4, p. 343

1. 32 cm^2 **3.** 60 in^2 **5.** 104 ft^2 **7.** 45.5 in^2 **9.** 8.05 cm^2 **11.** 297 cm^2 **13.** 7 m^2 **15.** 675 cm^2 **17.** 8944 in^2 **19.** 852.04 ft^2 **21.** $\frac{7}{20}$ **22.** $\frac{171}{200}$ **23.** $\frac{3}{8}$ **24.** $\frac{2}{3}$ **25.** $\frac{5}{6}$ **26.** $\frac{1}{6}$ **27.** 7500 **28.** 46; 2 cc left over **29.** ◈

Margin Exercises, Section 6.5, pp. 345–348

1. 9 in. **2.** 5 ft **3.** 62.8 m **4.** 88 m **5.** 34.296 yd **6.** $78\frac{4}{7}$ km^2 **7.** 339.62 cm^2 **8.** 12-ft diameter flower bed, by about 13.04 ft^2

Calculator Spotlight, p. 347

1. Answers may vary; 9 **2.** 1417.99 in.; 160,005.91 in^2 **3.** 1729.27 in^2 **4.** 125,663.71 ft^2

Exercise Set 6.5, p. 349

1. 14 cm; 44 cm; 154 cm^2 **3.** $1\frac{1}{2}$ in.; $4\frac{5}{7}$ in.; $1\frac{43}{56}$ in^2 **5.** 16 ft; 100.48 ft; 803.84 ft^2 **7.** 0.7 cm; 4.396 cm; 1.5386 cm^2 **9.** 3 cm; 18.84 cm; 28.26 cm^2 **11.** 151,976 mi^2 **13.** 2.5 cm; 1.25 cm; 4.90625 cm^2 **15.** 3.454 ft **17.** 65.94 yd^2 **19.** 45.68 ft **21.** 26.84 yd **23.** 45.7 yd **25.** 100.48 m^2 **27.** 6.9972 cm^2 **29.** 64.4214 in^2 **31.** 87.5% **32.** 58% **33.** $66.\overline{6}$% **34.** 43.61% **35.** 37.5% **36.** 62.5% **37.** $66.\overline{6}$% **38.** 20% **39.** 4 **40.** $8\frac{1}{2}$ **41.** 13 **42.** $39\frac{1}{2}$ **43.** 5 **44.** 0 **45.** 2 **46.** 3 **47.** $\frac{1}{2}$ **48.** 3 **49.** $275\frac{1}{2}$ **50.** $7\frac{1}{2}$ **51.** ◈ **53.** 3.142 **55.** $3d$; πd; circumference of one ball, since $\pi > 3$

Margin Exercises, Section 6.6, pp. 353–356

1. 12 cm^3 **2.** 20 ft^3 **3.** 128 ft^3 **4.** 38.4 m^3; 75.2 m^2 **5.** $1\frac{7}{8}$ ft^3; $10\frac{1}{4}$ ft^2 **6.** 785 ft^3 **7.** 67,914 m^3 **8.** $91,989\frac{1}{3}$ ft^3 **9.** 38.77272 cm^3 **10.** 1695.6 m^3 **11.** 528 in^3

Exercise Set 6.6, p. 357

1. 768 cm^3 **3.** 45 in^3 **5.** 75 m^3 **7.** $357\frac{1}{2}$ yd^3 **9.** 803.84 in^3 **11.** 353.25 cm^3 **13.** 41,580,000 yd^3 **15.** $4,186,666\frac{2}{3}$ in^3 **17.** 124.72 m^3 **19.** $1437\frac{1}{3}$ km^3 **21.** 113,982 ft^3 **23.** 24.64 cm^3 **25.** 33,880 yd^3 **27.** 367.38 m **29.** 143.72 cm^3 **31.** 32,993,441,150 mi^3 **33.** $19.20 **34.** $96 **35.** 1000 **36.** 225 **37.** 49 **38.** 64 **39.** 5% **40.** 11% **41.** ◈ **43.** About 57,480 in^3 **45.** 6.98 ft^3; 52.35 gal

Margin Exercises, Section 6.7, pp. 361–368

1. $\angle 1$ and $\angle 2$, $\angle 1$ and $\angle 4$, $\angle 2$ and $\angle 3$, $\angle 3$ and $\angle 4$ **2.** 45° **3.** 72° **4.** 5° **5.** $\angle 1$ and $\angle 2$, $\angle 1$ and $\angle 4$, $\angle 2$ and $\angle 3$, $\angle 3$ and $\angle 4$ **6.** 142° **7.** 23° **8.** 90° **9.** No **10.** Yes **11.** No **12.** Yes **13.** $m\angle 1 = 10°$, $m\angle 3 = 129°$, $m\angle 5 = 41°$, $m\angle 6 = 129°$ **14.** $\angle 1$ and $\angle 3$, $\angle 2$ and $\angle 4$, $\angle 5$ and $\angle 7$, $\angle 6$ and $\angle 8$ **15.** $\angle 2$, $\angle 3$, $\angle 6$, and $\angle 7$ **16.** $\angle 2$ and $\angle 7$, $\angle 6$ and $\angle 3$ **17.** $m\angle 7 = m\angle 1 = m\angle 5 = 51°$, $m\angle 8 = m\angle 2 = m\angle 6 = m\angle 4 = 129°$ **18.** $\angle CED \cong \angle BEA$, $\angle ECD \cong \angle EBA$, $\angle EDC \cong \angle EAB$, $\angle CEA \cong \angle BED$ **19.** $\angle TPQ \cong \angle TRS$, $\angle TQP \cong \angle TSR$

Exercise Set 6.7, p. 369

1. 79° **3.** 23° **5.** 177° **7.** 41° **9.** No **11.** Yes **13.** $m\angle 2 = 67°$, $m\angle 3 = 33°$, $m\angle 4 = 80°$, $m\angle 6 = 33°$ **15. (a)** $\angle 1$ and $\angle 3$, $\angle 2$ and $\angle 4$, $\angle 8$ and $\angle 6$, $\angle 7$ and $\angle 5$; **(b)** $\angle 2$, $\angle 3$, $\angle 6$, and $\angle 7$; **(c)** $\angle 2$ and $\angle 6$, $\angle 3$ and $\angle 7$ **17.** $m\angle 6 = m\angle 2 = m\angle 8 = 125°$, $m\angle 5 = m\angle 3 = m\angle 7 = m\angle 1 = 55°$ **19.** $\angle ABE \cong \angle DCE$, 95°; $\angle BAE \cong \angle CDE$; $\angle AEB \cong \angle DEC$; $\angle BED \cong \angle AEC$ **21.** $\angle AEC \cong \angle DCE$, 50°; $\angle BED \cong \angle EDC$, 41°

Margin Exercises, Section 6.8, pp. 372–376

1. $\angle A \cong \angle D$, $\angle B \cong \angle E$, $\angle C \cong \angle F$; $\overline{AB} \cong \overline{DE}$, $\overline{AC} \cong \overline{DF}$, $\overline{BC} \cong \overline{EF}$ **2.** $\angle N \cong \angle P$, $\angle M \cong \angle R$, $\angle O \cong \angle Q$; $\overline{NM} \cong \overline{PR}$, $\overline{NO} \cong \overline{PQ}$, $\overline{MO} \cong \overline{RQ}$ **3.** (a), (c) **4.** (a) **5.** (b) **6.** None **7.** SAS **8.** ASA **9.** SSS **10.** SAS **11.** $\triangle SRW \cong \triangle STV$ by ASA; $\angle RSW \cong \angle TSV$, $\overline{RS} \cong \overline{TS}$, $\overline{SW} \cong \overline{SV}$ **12.** $\triangle GKP \cong \triangle PTR$ by ASA. Thus, corresponding parts $\overline{GP}$ and $\overline{PR}$ are congruent, and P is the midpoint of $\overline{GR}$. **13.** $m\angle C = 27°$, $m\angle B = m\angle D = 153°$ **14.** $m\angle S = 114°$, $m\angle P = m\angle R = 66°$ **15.** $QR = 10$, $SR = 8$ **16.** $EF = 13.6$, $DE = GF = 20.4$

Exercise Set 6.8, p. 377

1. $\angle A \cong \angle R$, $\angle B \cong \angle S$, $\angle C \cong \angle T$; $\overline{AB} \cong \overline{RS}$, $\overline{AC} \cong \overline{RT}$, $\overline{BC} \cong \overline{ST}$ **3.** $\angle D \cong \angle G$, $\angle E \cong \angle H$, $\angle F \cong \angle K$; $\overline{DE} \cong \overline{GH}$, $\overline{DF} \cong \overline{GK}$, $\overline{EF} \cong \overline{HK}$ **5.** $\angle X \cong \angle U$, $\angle Y \cong \angle V$, $\angle Z \cong \angle W$; $\overline{XY} \cong \overline{UV}$, $\overline{XZ} \cong \overline{UW}$, $\overline{YZ} \cong \overline{VW}$ **7.** $\angle A \cong \angle F$, $\angle C \cong \angle D$, $\angle B \cong \angle E$; $\overline{AC} \cong \overline{FD}$, $\overline{AB} \cong \overline{FE}$, $\overline{CB} \cong \overline{DE}$ **9.** $\angle M \cong \angle Q$, $\angle N \cong \angle P$, $\angle O \cong \angle S$; $\overline{MN} \cong \overline{QP}$, $\overline{MO} \cong \overline{QS}$, $\overline{NO} \cong \overline{PS}$ **11.** No **13.** Yes **15.** Yes **17.** No **19.** Yes **21.** Yes **23.** Yes **25.** Yes **27.** Yes **29.** ASA **31.** SAS **33.** SSS or SAS **35.** $\overline{PR} \cong \overline{TR}$, $\overline{SR} \cong \overline{QR}$, $\angle PRQ \cong \angle TRS$ (vertical angles); $\triangle PRQ \cong \triangle TRS$ by SAS **37.** $m\angle GLK = m\angle GLM = 90°$, $\angle GLK \cong \angle GLM$, $\overline{GL} \cong \overline{GL}$, $\overline{KL} \cong \overline{ML}$; $\triangle KLG \cong \triangle MLG$ by SAS **39.** $\overline{AE} \cong \overline{CD}$, $\overline{AB} \cong \overline{CB}$, $\overline{EB} \cong \overline{DB}$; $\triangle AEB \cong \triangle CDB$ by SSS **41.** $\triangle LKH \cong \triangle GKJ$ by SAS; $\angle HLK \cong \angle JGK$, $\angle LHK \cong \angle GJK$, $\overline{LH} \cong \overline{GJ}$ **43.** $\triangle PED \cong \triangle PFG$ by ASA. As corresponding parts, $\overline{EP} \cong \overline{FP}$; thus, P is the midpoint of $\overline{EF}$ **45.** $m\angle A = 70°$, $m\angle D = m\angle B = 110°$ **47.** $m\angle M = 71°$, $m\angle J = m\angle L = 109°$ **49.** $TU = 9$, $NU = 15$ **51.** $KL = 3\frac{1}{2}$, $ML = JK = 7\frac{1}{2}$ **53.** $AC = 28$, $ED = 38$

Margin Exercises, Section 6.9, pp. 381–384

1. (a), (b), (d) **2.** $\overline{PQ} \leftrightarrow \overline{GH}$, $\overline{QR} \leftrightarrow \overline{HK}$, $\overline{PR} \leftrightarrow \overline{GK}$, $\angle P \leftrightarrow \angle G$, $\angle Q \leftrightarrow \angle H$, $\angle R \leftrightarrow \angle K$ **3.** $\angle J \cong \angle A$, $\angle K \cong \angle B$, $\angle L \cong \angle C$; $\frac{JK}{AB} = \frac{JL}{AC} = \frac{KL}{BC}$

4. $\frac{PN}{TS} = \frac{PM}{TR} = \frac{MN}{RS}$ **5.** $BT = 6\frac{3}{4}$, $CT = 9$ **6.** $QR = 10$ **7.** 24.75 ft **8.** 34.9 ft

Exercise Set 6.9, p. 385

1. $\angle R \leftrightarrow \angle A$, $\angle S \leftrightarrow \angle B$, $\angle T \leftrightarrow \angle C$, $\overline{RS} \leftrightarrow \overline{AB}$, $\overline{RT} \leftrightarrow \overline{AC}$, $\overline{ST} \leftrightarrow \overline{BC}$ **3.** $\angle C \leftrightarrow \angle W$, $\angle B \leftrightarrow \angle J$, $\angle S \leftrightarrow \angle Z$, $\overline{CB} \leftrightarrow \overline{WJ}$, $\overline{CS} \leftrightarrow \overline{WZ}$, $\overline{BS} \leftrightarrow \overline{JZ}$

5. $\angle A \cong \angle R$, $\angle B \cong \angle S$, $\angle C \cong \angle T$; $\frac{AB}{RS} = \frac{AC}{RT} = \frac{BC}{ST}$

7. $\angle M \cong \angle C$, $\angle E \cong \angle L$, $\angle S \cong \angle F$; $\frac{ME}{CL} = \frac{MS}{CF} = \frac{ES}{LF}$

9. $\frac{PS}{ND} = \frac{SQ}{DM} = \frac{PQ}{NM}$ **11.** $\frac{TA}{GF} = \frac{TW}{GC} = \frac{AW}{FC}$ **13.** $QR = 10$, $PR = 8$ **15.** $EC = 18$ **17.** 36 ft **19.** 100 ft **21.** $29\frac{2}{5}$ **22.** 0.244 **23.** 78 **24.** 61.1611 **25.** ◈

Summary and Review: Chapter 6, p. 387

1. 23 m **2.** 228 ft; 2808 ft^2 **3.** 81 ft^2 **4.** 12.6 cm^2 **5.** 60 cm^2 **6.** 35 mm^2 **7.** 22.5 m^2 **8.** 27.5 cm^2 **9.** 840 ft^2 **10.** 8 m **11.** $\frac{14}{11}$ in., or $1\frac{3}{11}$ in. **12.** 14 ft **13.** 20 cm **14.** 50.24 m **15.** 8 in. **16.** 200.96 m^2 **17.** $5\frac{1}{11}$ in^2 **18.** 93.6 m^3; 150 m^2 **19.** 193.2 cm^3; 240.4 cm^2 **20.** 31,400 ft^3 **21.** $33.49\overline{3}$ cm^3 **22.** 4.71 in^3 **23.** 60° **24.** Scalene **25.** Right **26.** 720° **27.** 8° **28.** 85° **29.** 147° **30.** 47° **31.** $m\angle 2 = 105°$, $m\angle 3 = 37°$, $m\angle 4 = 38°$, $m\angle 6 = 37°$ **32.** **(a)** $\angle 1$ and $\angle 5$, $\angle 4$ and $\angle 8$, $\angle 3$ and $\angle 7$, $\angle 2$ and $\angle 6$; **(b)** $\angle 4$, $\angle 5$, $\angle 2$, and $\angle 7$; **(c)** $\angle 4$ and $\angle 7$, $\angle 2$ and $\angle 5$ **33.** $m\angle 1 = m\angle 3 = m\angle 7 = m\angle 5 = 45°$, $m\angle 6 = m\angle 2 = m\angle 8 = 135°$ **34.** $\angle D \cong \angle R$, $\angle H \cong \angle Z$, $\angle J \cong \angle K$; $\overline{DH} \cong \overline{RZ}$, $\overline{DJ} \cong \overline{RK}$, $\overline{HJ} \cong \overline{ZK}$ **35.** $\angle A \cong \angle G$, $\angle B \cong \angle D$, $\angle C \cong \angle F$, $\overline{AB} \cong \overline{GD}$, $\overline{AC} \cong \overline{GF}$, $\overline{BC} \cong \overline{DF}$ **36.** ASA **37.** SSS **38.** None **39.** $\overline{IJ} \cong \overline{KJ}$, $\angle HJI \cong \angle LJK$, $\angle HIJ \cong \angle LKJ$; $\triangle JIH \cong \triangle JKL$ by ASA **40.** $m\angle C = 63°$, $m\angle B = m\angle D = 117°$; $BC = 23$, $CD = 13$

41. $\angle C \cong \angle F$, $\angle Q \cong \angle A$, $\angle W \cong \angle S$; $\frac{CQ}{FA} = \frac{CW}{FS} = \frac{QW}{AS}$

42. $MO = 14$ **43.** $54\frac{5}{8}$ **44.** 103.823 **45.** $\frac{1}{16}$ **46.** $\frac{73}{100}$ **47.** 47% **48.** 92% **49.** ◈ Linear measure is one-dimensional, area is two-dimensional, and volume is three-dimensional. **50.** ◈ See the area and volume formulas listed at the beginning of the Summary and Review Exercises for Chapter 6. **51.** 1038.555 ft^2 **52.** 100 ft^2 **53.** 7.83998704 m^2 **54.** 42.05915 cm^2

Test: Chapter 6, p. 391

1. [6.2a] 32.82 cm **2.** [6.2b], [6.3b] 14 ft; $11\frac{1}{4}$ ft^2 **3.** [6.4a] 25 cm^2 **4.** [6.4a] 12 m^2 **5.** [6.4a] 18 ft^2 **6.** [6.3a] 625 m^2 **7.** [6.3b], [6.5d] 103.815 km^2 **8.** [6.5a] $\frac{1}{4}$ in. **9.** [6.5a] 9 cm **10.** [6.5b] $\frac{11}{14}$ in. **11.** [6.5c] 254.34 cm^2 **12.** [6.6a] 84 cm^3; 142 cm^2 **13.** [6.6b] 1177.5 ft^3 **14.** [6.6c] $4186.\overline{6}$ yd^3 **15.** [6.6d] 113.04 cm^3 **16.** [6.1f] 35° **17.** [6.1e] Isosceles **18.** [6.1e] Obtuse **19.** [6.1f] 540° **20.** [6.7a] 149° **21.** [6.7a] 11° **22.** [6.7c] $m\angle 2 = 110°$, $m\angle 3 = 8°$, $m\angle 4 = 62°$, $m\angle 6 = 8°$ **23.** [6.7d] $m\angle 6 = m\angle 2 = m\angle 8 = 120°$, $m\angle 5 = m\angle 3 = m\angle 7 = m\angle 1 = 60°$ **24.** [6.8a] $\angle C \cong \angle A$, $\angle W \cong \angle T$, $\angle S \cong \angle Z$, $\overline{CW} \cong \overline{AT}$, $\overline{WS} \cong \overline{TZ}$, $\overline{SC} \cong \overline{ZA}$ **25.** [6.8a] SAS **26.** [6.8a] None **27.** [6.8a] ASA **28.** [6.8a] None **29.** [6.8b] $m\angle G = 105°$, $m\angle D = m\angle F = 75°$; $EF = 11$,

$DE = GF = 20$ **30.** [6.8b] $LJ = 6.4$, $KM = 6$
31. [6.9a] $\angle E \cong \angle T$, $\angle R \cong \angle G$, $\angle S \cong \angle F$;
$\frac{ER}{TG} = \frac{RS}{GF} = \frac{SE}{FT}$ **32.** [6.9b] $EK = 18$, $ZK = 27$
33. [3.3a] 10.626 **34.** [2.4d] 22 **35.** [1.6b] 1000
36. [1.6b] $\frac{1}{16}$ **37.** [4.3a] 81.25% **38.** [4.2b] 0.932
39. [4.3b] $\frac{1}{3}$ **40.** [6.4a] 1.875 ft^2 **41.** [6.6a] 0.65 ft^3

Chapter 7

Pretest: Chapter 7, p. 396

1. [7.1a] $\frac{5}{16}$ **2.** [7.1b] 78%x, or 0.78x **3.** [7.1a] 360 ft^2
4. [7.3b] 12 **5.** [7.2d] > **6.** [7.2d] > **7.** [7.2d] >
8. [7.2d] < **9.** [7.2e] 12 **10.** [7.2e] 2.3 **11.** [7.2e] 0
12. [7.3b] −5.4 **13.** [7.3b] $\frac{2}{3}$ **14.** [7.6b] $\frac{1}{10}$
15. [7.6b] $-\frac{3}{2}$ **16.** [7.3a] −17 **17.** [7.4a] 38.6
18. [7.4a] $-\frac{17}{15}$ **19.** [7.3a] −5 **20.** [7.5a] 63
21. [7.5a] $-\frac{5}{12}$ **22.** [7.6c] −98 **23.** [7.6a] 8
24. [7.4a] 24 **25.** [7.8d] 26 **26.** [7.7c] $9z - 18$
27. [7.7c] $-4a - 2b + 10c$ **28.** [7.7d] $4(x - 3)$
29. [7.7d] $3(2y - 3z - 6)$ **30.** [7.8b] $-y - 13$
31. [7.8c] $y + 18$ **32.** [7.2d] $12 < x$

Margin Exercises, Section 7.1, pp. 397–400

1. $2174 + x = 7521$ **2.** 64 **3.** 28 **4.** 60
5. 192 ft^2 **6.** 25 **7.** 16 **8.** 3.375 hr **9.** $x - 8$
10. $y + 8$, or $8 + y$ **11.** $m - 4$ **12.** $\frac{1}{2}p$ **13.** $6 + 8x$, or $8x + 6$ **14.** $a - b$ **15.** 59%x, or 0.59x
16. $xy - 200$ **17.** $p + q$

Calculator Spotlight, p. 398

1. 59.63768116 **2.** 11.9 **3.** 11.9 **4.** 34,427.16
5. 32 **6.** 27.5

Exercise Set 7.1, p. 401

1. \$20,400; \$46,800; \$150,000 **3.** 1935 m^2 **5.** 260 mi
7. 56 **9.** 8 **11.** 1 **13.** 6 **15.** 2 **17.** $b + 7$, or $7 + b$ **19.** $c - 12$ **21.** $4 + q$, or $q + 4$ **23.** $a + b$, or $b + a$ **25.** $y - x$ **27.** $w + x$, or $x + w$
29. $n - m$ **31.** $r + s$, or $s + r$ **33.** $2z$ **35.** $3m$
37. 89%x, or 0.89x **39.** 55t miles **41.** $2 \cdot 3 \cdot 3 \cdot 3$
42. $2 \cdot 2 \cdot 2 \cdot 2 \cdot 2$ **43.** $2 \cdot 2 \cdot 3 \cdot 3 \cdot 3$
44. $2 \cdot 2 \cdot 2 \cdot 2 \cdot 2 \cdot 2 \cdot 3$ **45.** 18 **46.** 96 **47.** 60
48. 48 **49.** ◈ **51.** $x + 3y$ **53.** $2x - 3$

Margin Exercises, Section 7.2, pp. 405–410

1. 8; −5 **2.** 134; −80 **3.** −10; 156
4. −120; 50; −80
5.

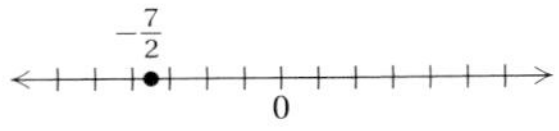

6.

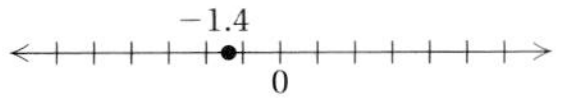

7.

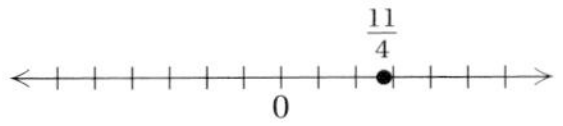

8. −0.375 **9.** $-0.\overline{54}$ **10.** $1.\overline{3}$ **11.** < **12.** <
13. > **14.** > **15.** > **16.** < **17.** < **18.** >
19. $7 > -5$ **20.** $4 < x$ **21.** False **22.** True
23. True **24.** 8 **25.** 0 **26.** 9 **27.** $\frac{2}{3}$ **28.** 5.6

Calculator Spotlight, p. 406

1. [(−)] [3] [ENTER]; −3
2. [(−)] [5] [0] [8] [ENTER]; −508
3. [(−)] [.] [1] [7] [ENTER]; −.17
4. [(−)] [5] [÷] [8] [ENTER]; −.625

Calculator Spotlight, p. 407

1. 8.717797887 **2.** 17.80449381 **3.** 67.08203932
4. 35.4807407 **5.** 3.141592654 **6.** 91.10618695
7. 530.9291585 **8.** 138.8663978

Exercise Set 7.2, p. 411

1. −1286; 13,804 **3.** 24; −2 **5.** −5,200,000,000,000
7.

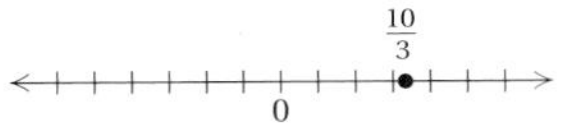

9.

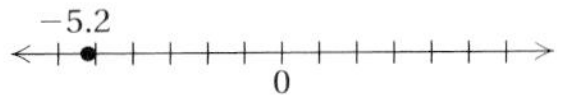

11. −0.875 **13.** $0.8\overline{3}$ **15.** $1.1\overline{6}$ **17.** $0.\overline{6}$ **19.** −0.5
21. 0.1 **23.** > **25.** < **27.** < **29.** < **31.** >
33. < **35.** > **37.** < **39.** < **41.** < **43.** True
45. False **47.** $x < -6$ **49.** $y \geq -10$ **51.** 3
53. 10 **55.** 0 **57.** 24 **59.** $\frac{2}{3}$ **61.** 0 **63.** 0.63
64. 0.083 **65.** 1.1 **66.** 0.2276 **67.** 75%
68. 62.5%, or $62\frac{1}{2}$% **69.** $83.\overline{3}$%, or $83\frac{1}{3}$%
70. 59.375%, or $59\frac{3}{8}$% **71.** ◈
73. $-\frac{5}{6}, -\frac{3}{4}, -\frac{2}{3}, \frac{1}{6}, \frac{3}{8}, \frac{1}{2}$ **75.** $\frac{1}{9}$ **77.** $5\frac{5}{9}$, or $\frac{50}{9}$

Margin Exercises, Section 7.3, pp. 413–416

1. −6 **2.** −3 **3.** −8 **4.** 4 **5.** 0 **6.** −2
7. −11 **8.** −12 **9.** 2 **10.** −4 **11.** −2 **12.** 0
13. −22 **14.** 3 **15.** 0.53 **16.** 2.3 **17.** −7.7
18. −6.2 **19.** $-\frac{2}{9}$ **20.** $-\frac{19}{20}$ **21.** −58 **22.** −56
23. −14 **24.** −12 **25.** 4 **26.** −8.7 **27.** 7.74
28. $\frac{8}{9}$ **29.** 0 **30.** −12 **31.** −14; 14 **32.** −1; 1
33. 19; −19 **34.** 1.6; −1.6 **35.** $-\frac{2}{3}; \frac{2}{3}$ **36.** $\frac{9}{8}; -\frac{9}{8}$
37. 4 **38.** 13.4 **39.** 0 **40.** $-\frac{1}{4}$

Exercise Set 7.3, p. 417

1. −7 **3.** −6 **5.** 0 **7.** −8 **9.** −7 **11.** −27
13. 0 **15.** −42 **17.** 0 **19.** 0 **21.** 3 **23.** −9
25. 7 **27.** 0 **29.** 35 **31.** −3.8 **33.** −8.1

35. $-\frac{1}{5}$ **37.** $-\frac{7}{9}$ **39.** $-\frac{3}{8}$ **41.** $-\frac{19}{24}$ **43.** $\frac{1}{24}$ **45.** 37 **47.** 50 **49.** −1409 **51.** −24 **53.** 26.9 **55.** −8 **57.** $\frac{13}{8}$ **59.** −43 **61.** $\frac{4}{3}$ **63.** 24 **65.** $\frac{3}{8}$ **67.** 0.57 **68.** 0.49 **69.** 0.529 **70.** 0.713 **71.** 125% **72.** 12.5% **73.** 52% **74.** 40.625% **75.** ◈ **77.** All positive **79.** −6483 **81.** Negative

Margin Exercises, Section 7.4, pp. 419–421

1. −10 **2.** 3 **3.** −5 **4.** −2 **5.** −11 **6.** 4 **7.** −2 **8.** −6 **9.** −16 **10.** 7.1 **11.** 3 **12.** 0 **13.** $\frac{3}{2}$ **14.** −8 **15.** 7 **16.** −3 **17.** −23.3 **18.** 0 **19.** −9 **20.** 17 **21.** 12.7 **22.** 77, 37, 25, 24, 23, 9, −4, −9, −12, −25, −28, −31, −41, −45; 53, 48, 45, 38, 29, 24, 21, 19, −10, −27, −35, −37, −53, −115 **23.** 50°C

Exercise Set 7.4, p. 423

1. −7 **3.** −4 **5.** −6 **7.** 0 **9.** −4 **11.** −7 **13.** −6 **15.** 0 **17.** 0 **19.** 14 **21.** 11 **23.** −14 **25.** 5 **27.** −7 **29.** −1 **31.** 18 **33.** −10 **35.** −3 **37.** −21 **39.** 5 **41.** −8 **43.** 12 **45.** −23 **47.** −68 **49.** −73 **51.** 116 **53.** 0 **55.** −1 **57.** $\frac{1}{12}$ **59.** $-\frac{17}{12}$ **61.** $\frac{1}{8}$ **63.** 19.9 **65.** −8.6 **67.** −0.01 **69.** −193 **71.** 500 **73.** −2.8 **75.** −3.53 **77.** $-\frac{1}{2}$ **79.** $\frac{6}{7}$ **81.** $-\frac{41}{30}$ **83.** $-\frac{2}{15}$ **85.** 37 **87.** −62 **89.** −139 **91.** 6 **93.** 107 **95.** 219 **97.** 2385 m **99.** $347.94 **101.** 100°F **103.** 125 **104.** $2 \cdot 2 \cdot 2 \cdot 2 \cdot 2 \cdot 3 \cdot 3 \cdot 3$ **105.** 100.5 **106.** 226 **107.** 0.583 **108.** $\frac{41}{64}$ **109.** ◈ **111.** −309,882 **113.** False; $3 - 0 \neq 0 - 3$ **115.** True **117.** True **119.** Up 15 points

Margin Exercises, Section 7.5, pp. 427–429

1. 20; 10; 0; −10; −20; −30 **2.** −18 **3.** −100 **4.** −80 **5.** $-\frac{5}{9}$ **6.** −30.033 **7.** $-\frac{7}{10}$ **8.** −10; 0; 10; 20; 30 **9.** 27 **10.** 32 **11.** 35 **12.** $\frac{20}{63}$ **13.** $\frac{2}{3}$ **14.** 13.455 **15.** −30 **16.** 30 **17.** 0 **18.** $-\frac{8}{3}$ **19.** −30 **20.** −30.75 **21.** $-\frac{5}{3}$ **22.** 120 **23.** −120 **24.** 6 **25.** 4; −4 **26.** 9; −9 **27.** 48; 48

Exercise Set 7.5, p. 431

1. −8 **3.** −48 **5.** −24 **7.** −72 **9.** 16 **11.** 42 **13.** −120 **15.** −238 **17.** 1200 **19.** 98 **21.** −72 **23.** −12.4 **25.** 30 **27.** 21.7 **29.** $-\frac{2}{5}$ **31.** $\frac{1}{12}$ **33.** −17.01 **35.** $-\frac{5}{12}$ **37.** 420 **39.** $\frac{2}{7}$ **41.** −60 **43.** 150 **45.** $-\frac{2}{45}$ **47.** 1911 **49.** 50.4 **51.** $\frac{10}{189}$ **53.** −960 **55.** 17.64 **57.** $-\frac{5}{784}$ **59.** 0 **61.** −720 **63.** −30,240 **65.** 441; −147 **67.** 20; 20 **69.** 180 **70.** $2 \cdot 2 \cdot 2 \cdot 2 \cdot 2 \cdot 2 \cdot 2 \cdot 2 \cdot 2 \cdot 3 \cdot 3$, or $2^9 \cdot 3^2$ **71.** $\frac{2}{3}$ **72.** $\frac{8}{9}$ **73.** $\frac{6}{11}$ **74.** $\frac{41}{265}$ **75.** ◈ **77.** 32 m below the surface **79.** **(a)** One must be negative, and one must be positive. **(b)** Either or both must be zero. **(c)** Both must be negative or both must be positive.

Margin Exercises, Section 7.6, pp. 433–436

1. −2 **2.** 5 **3.** −3 **4.** 8 **5.** −6 **6.** $-\frac{30}{7}$ **7.** Undefined **8.** 0 **9.** $\frac{3}{2}$ **10.** $-\frac{4}{5}$ **11.** $-\frac{1}{3}$ **12.** −5 **13.** $\frac{1}{1.6}$ **14.** $\frac{2}{3}$ **15.** First row: $-\frac{2}{3}, \frac{3}{2}$; second row: $\frac{5}{4}, -\frac{4}{5}$; third row: 0, undefined; fourth row: −1, 1; fifth row: 8, $-\frac{1}{8}$; sixth row: 4.5, $-\frac{1}{4.5}$ **16.** $\frac{4}{7} \cdot \left(-\frac{5}{3}\right)$ **17.** $5 \cdot \left(-\frac{1}{8}\right)$ **18.** $(a - b) \cdot \left(\frac{1}{7}\right)$ **19.** $-23 \cdot a$ **20.** $-5 \cdot \left(\frac{1}{7}\right)$ **21.** $-\frac{20}{21}$ **22.** $-\frac{12}{5}$ **23.** $\frac{16}{7}$ **24.** −7 **25.** $\frac{5}{-6}, -\frac{5}{6}$ **26.** $\frac{-8}{7}, \frac{8}{-7}$ **27.** $\frac{-10}{3}, -\frac{10}{3}$

Exercise Set 7.6, p. 437

1. −8 **3.** −14 **5.** −3 **7.** 3 **9.** −8 **11.** 2 **13.** −12 **15.** −8 **17.** Undefined **19.** $\frac{23}{2}$ **21.** $\frac{7}{15}$ **23.** $-\frac{13}{47}$ **25.** $\frac{1}{13}$ **27.** $\frac{1}{4.3}$ **29.** −7.1 **31.** $\frac{q}{p}$ **33.** $4y$ **35.** $\frac{3b}{2a}$ **37.** $4 \cdot \left(\frac{1}{17}\right)$ **39.** $8 \cdot \left(-\frac{1}{13}\right)$ **41.** $13.9 \cdot \left(-\frac{1}{1.5}\right)$ **43.** $x \cdot y$ **45.** $(3x + 4)\left(\frac{1}{5}\right)$ **47.** $(5a - b)\left(\frac{1}{5a + b}\right)$ **49.** $-\frac{9}{8}$ **51.** $\frac{5}{3}$ **53.** $\frac{9}{14}$ **55.** $\frac{9}{64}$ **57.** −2 **59.** $\frac{11}{13}$ **61.** −16.2 **63.** Undefined **65.** $\frac{22}{39}$ **66.** 0.477 **67.** 33 **68.** $\frac{3}{2}$ **69.** 87.5% **70.** $\frac{2}{3}$ **71.** $\frac{9}{8}$ **72.** $\frac{128}{625}$ **73.** ◈ **75.** ◈ **77.** Negative **79.** Positive **81.** Negative

Margin Exercises, Section 7.7, pp. 439–446

1.

	$x + x$	$2x$
$x = 3$	6	6
$x = -6$	−12	−12
$x = 4.8$	9.6	9.6

2.

	$x + 3x$	$5x$
$x = 2$	8	10
$x = -6$	−24	−30
$x = 4.8$	19.2	24

3. $\frac{6}{8}$ **4.** $\frac{3t}{4t}$ **5.** $\frac{3}{4}$ **6.** $-\frac{4}{3}$ **7.** 1; 1 **8.** −10; −10 **9.** $9 + x$ **10.** qp **11.** $t + xy$, or $yx + t$, or $t + yx$ **12.** 19; 19 **13.** 150; 150 **14.** $(r + s) + 7$ **15.** $(9a)b$ **16.** $(4t)u$, $(tu)4$, $t(4u)$; answers may vary **17.** $(2 + r) + s$, $(r + s) + 2$, $s + (r + 2)$; answers may vary **18.** **(a)** 63; **(b)** 63 **19.** **(a)** 80; **(b)** 80 **20.** **(a)** 28; **(b)** 28 **21.** **(a)** 8; **(b)** 8 **22.** **(a)** −4; **(b)** −4 **23.** **(a)** −25; **(b)** −25 **24.** $5x, -8y, 3$ **25.** $-4y, -2x, 3z$ **26.** $3x - 15$ **27.** $5x + 5$ **28.** $\frac{3}{5}p + \frac{3}{5}q - \frac{3}{5}t$ **29.** $-2x + 6$ **30.** $5x - 10y + 20z$ **31.** $-5x + 10y - 20z$ **32.** $6(x - 2)$

33. $3(x - 2y + 3)$ **34.** $b(x + y - z)$
35. $2(8a - 18b + 21)$ **36.** $\frac{1}{8}(3x - 5y + 7)$
37. $-4(3x - 8y + 4z)$ **38.** $3x$ **39.** $6x$ **40.** $-8x$
41. $0.59x$ **42.** $3x + 3y$ **43.** $-4x - 5y - 7$
44. $-\frac{2}{3} + \frac{1}{10}x + \frac{7}{9}y$

Exercise Set 7.7, p. 447

1. $\frac{3y}{5y}$ **3.** $\frac{10x}{15x}$ **5.** $-\frac{3}{2}$ **7.** $-\frac{7}{6}$ **9.** $8 + y$ **11.** nm
13. $xy + 9$, or $9 + yx$ **15.** $c + ab$, or $ba + c$
17. $(a + b) + 2$ **19.** $8(xy)$ **21.** $a + (b + 3)$
23. $(3a)b$ **25.** $2 + (b + a)$, $(2 + a) + b$, $(b + 2) + a$; answers may vary **27.** $(5 + w) + v$, $(v + 5) + w$, $(w + v) + 5$; answers may vary **29.** $(3x)y$, $y(x \cdot 3)$, $3(yx)$; answers may vary **31.** $a(7b)$, $b(7a)$, $(7b)a$; answers may vary **33.** $2b + 10$ **35.** $7 + 7t$
37. $30x + 12$ **39.** $7x + 28 + 42y$ **41.** $7x - 21$
43. $-3x + 21$ **45.** $\frac{2}{3}b - 4$ **47.** $7.3x - 14.6$
49. $-\frac{3}{5}x + \frac{3}{5}y - 6$ **51.** $45x + 54y - 72$
53. $-4x + 12y + 8z$ **55.** $-3.72x + 9.92y - 3.41$
57. $4x$, $3z$ **59.** $7x$, $8y$, $-9z$ **61.** $2(x + 2)$
63. $5(6 + y)$ **65.** $7(2x + 3y)$ **67.** $5(x + 2 + 3y)$
69. $8(x - 3)$ **71.** $4(8 - y)$ **73.** $2(4x + 5y - 11)$
75. $a(x - 1)$ **77.** $a(x - y - z)$ **79.** $6(3x - 2y + 1)$
81. $\frac{1}{3}(2x - 5y + 1)$ **83.** $19a$ **85.** $9a$ **87.** $8x + 9z$
89. $7x + 15y^2$ **91.** $-19a + 88$ **93.** $4t + 6y - 4$
95. b **97.** $\frac{13}{4}y$ **99.** $8x$ **101.** $5n$ **103.** $-16y$
105. $17a - 12b - 1$ **107.** $4x + 2y$ **109.** $7x + y$
111. $0.8x + 0.5y$ **113.** $\frac{35}{6}a + \frac{3}{2}b - 42$ **115.** $\frac{89}{48}$
116. $\frac{5}{24}$ **117.** 144 **118.** 30% **119.** $-\frac{5}{24}$ **120.** 60
121. **123.** Not equivalent; $3 \cdot 2 + 5 \neq 3 \cdot 5 + 2$
125. Equivalent; commutative law of addition
127. $q(1 + r + rs + rst)$

Margin Exercises, Section 7.8, pp. 451–454

1. $-x - 2$ **2.** $-5x - 2y - 8$ **3.** $-6 + t$ **4.** $-x + y$
5. $4a - 3t + 10$ **6.** $-18 + m + 2n - 4z$ **7.** $2x - 9$
8. $3y + 2$ **9.** $2x - 7$ **10.** $3y + 3$
11. $-2a + 8b - 3c$ **12.** $-9x - 8y$ **13.** $-16a + 18$
14. $-26a + 41b - 48c$ **15.** $3x - 7$ **16.** 2 **17.** 18
18. 6 **19.** 17 **20.** $5x - y - 8$ **21.** -1237 **22.** 8
23. 381 **24.** -12

Calculator Spotlight, p. 455

1. -11 **2.** 9 **3.** 114 **4.** 117,649 **5.** $-1{,}419{,}857$
6. $-1{,}124{,}864$ **7.** $-117{,}649$ **8.** $-1{,}419{,}857$
9. $-1{,}124{,}864$ **10.** -4 **11.** -2 **12.** 787
13. $-32 \times (88 - 29) = -1888$
14. $3^5 - 10^2 \times 5^2 = -2257$
15. $4 + 6 \cdot 8 - 2 = 4 + 8 \cdot 6 - 2 = 50$; the commutative law of multiplication
16. $5 + 9^2 \cdot 7 - 3 = 569$; because $a^2 \cdot b \neq b^2 \cdot a$, although students might phrase this verbally and not symbolically.

Exercise Set 7.8, p. 457

1. $-2x - 7$ **3.** $-5x + 8$ **5.** $-4a + 3b - 7c$
7. $-6x + 8y - 5$ **9.** $-3x + 5y + 6$ **11.** $8x + 6y + 43$
13. $5x - 3$ **15.** $-3a + 9$ **17.** $5x - 6$
19. $-19x + 2y$ **21.** $9y - 25z$ **23.** $-7x + 10y$
25. $37a - 23b + 35c$ **27.** 7 **29.** -40 **31.** 19
33. $12x + 30$ **35.** $3x + 30$ **37.** $9x - 18$
39. $-4x - 64$ **41.** -7 **43.** -7 **45.** -16
47. -334 **49.** 14 **51.** 1880 **53.** 12 **55.** 8
57. -86 **59.** 37 **61.** -1 **63.** -10 **65.** 25
67. -7988 **69.** -3000 **71.** 60 **73.** 1 **75.** 10
77. $-\frac{13}{45}$ **79.** $-\frac{23}{18}$ **81.** -118 **83.** $2 \cdot 2 \cdot 59$
84. 252 **85.** $\frac{8}{5}$ **86.** $\frac{5}{18}$ **87.** 81 **88.** 1000 **89.** 100
90. 225 **91.** **93.** $6y - (-2x + 3a - c)$
95. $6m - (-3n + 5m - 4b)$ **97.** $-2x - f$
99. **(a)** 52, 52, 28.130169; **(b)** -24, -24, -108.307025

Summary and Review: Chapter 7, p. 461

1. 4 **2.** 19%x, or $0.19x$ **3.** -45, 72 **4.** 38
5.
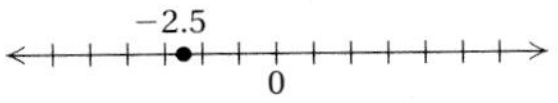

6.
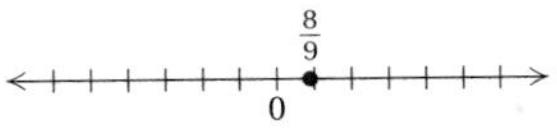

7. $<$ **8.** $>$ **9.** $>$ **10.** $<$ **11.** -3.8 **12.** $\frac{3}{4}$
13. $\frac{8}{3}$ **14.** $-\frac{1}{7}$ **15.** 34 **16.** 5 **17.** -3 **18.** -4
19. -5 **20.** 4 **21.** $-\frac{7}{5}$ **22.** -7.9 **23.** 54
24. -9.18 **25.** $-\frac{2}{7}$ **26.** -210 **27.** -7 **28.** -3
29. $\frac{3}{4}$ **30.** 40.4 **31.** -2 **32.** 8-yd gain
33. $-\$130$ **34.** $15x - 35$ **35.** $-8x + 10$
36. $4x + 15$ **37.** $-24 + 48x$ **38.** $2(x - 7)$
39. $6(x - 1)$ **40.** $5(x + 2)$ **41.** $3(4 - x)$
42. $7a - 3b$ **43.** $-2x + 5y$ **44.** $5x - y$
45. $-a + 8b$ **46.** $-3a + 9$ **47.** $-2b + 21$
48. 6 **49.** $12y - 34$ **50.** $5x + 24$ **51.** $-15x + 25$
52. True **53.** False **54.** $x > -3$ **55.** $\frac{55}{42}$
56. $\frac{109}{18}$ **57.** 6.25 **58.** 62.5% **59.** 0.0567 **60.** 270
61. $-\frac{5}{8}$ **62.** -2.1 **63.** 1000 **64.** $4a + 2b$

Test: Chapter 7, p. 463

1. [7.1a] 6 **2.** [7.1b] $x - 9$ **3.** [7.1a] 240 ft^2
4. [7.2d] $<$ **5.** [7.2d] $>$ **6.** [7.2d] $>$ **7.** [7.2d] $<$
8. [7.2e] 7 **9.** [7.2e] $\frac{9}{4}$ **10.** [7.2e] 2.7 **11.** [7.3b] $-\frac{2}{3}$
12. [7.3b] 1.4 **13.** [7.3b] 8 **14.** [7.6b] $-\frac{1}{2}$
15. [7.6b] $\frac{7}{4}$ **16.** [7.4a] 7.8 **17.** [7.3a] -8
18. [7.3a] $\frac{7}{40}$ **19.** [7.4a] 10 **20.** [7.4a] -2.5
21. [7.4a] $\frac{7}{8}$ **22.** [7.5a] -48 **23.** [7.5a] $\frac{3}{16}$
24. [7.6a] -9 **25.** [7.6c] $\frac{3}{4}$ **26.** [7.6c] -9.728
27. [7.8d] -173 **28.** [7.4b] 14°F **29.** [7.7c] $18 - 3x$
30. [7.7c] $-5y + 5$ **31.** [7.7d] $2(6 - 11x)$
32. [7.7d] $7(x + 3 + 2y)$ **33.** [7.4a] 12
34. [7.8b] $2x + 7$ **35.** [7.8b] $9a - 12b - 7$
36. [7.8c] $68y - 8$ **37.** [7.8d] -4 **38.** [7.8d] 448

39. [7.2d] $-2 \geq x$ **40.** [1.6b] 1.728 **41.** [4.3a] 12.5% **42.** [2.2c] 2 **43.** [1.9a] 240 **44.** [7.2e], [7.8d] 15 **45.** [7.8c] $4a$ **46.** [7.7e] $4x + 4y$

Chapter 8

Pretest: Chapter 8, p. 466

1. [8.2a] -7 **2.** [8.3b] -1 **3.** [8.3a] 2 **4.** [8.1b] 8 **5.** [8.3c] -5 **6.** [8.3a] $\frac{135}{32}$ **7.** [8.3c] 1 **8.** [8.7d] $\{x \mid x \geq -6\}$ **9.** [8.7c] $\{y \mid y > -4\}$ **10.** [8.7e] $\{a \mid a > -1\}$ **11.** [8.7c] $\{x \mid x \geq 3\}$ **12.** [8.7d] $\{y \mid y < -\frac{9}{4}\}$ **13.** [8.6a] $G = \dfrac{P}{3K}$ **14.** [8.6a] $a = \dfrac{Ab + b}{3}$ **15.** [8.4a] Width: 34 in.; length: 39 in. **16.** [8.5a] $460 **17.** [8.4a] 81, 82, 83 **18.** [8.8b] Numbers less than 17
19. [8.7b] **20.** [8.7b]

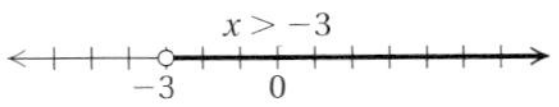

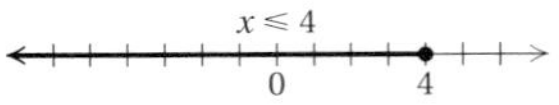

Margin Exercises, Section 8.1, pp. 467–470

1. False **2.** True **3.** Neither **4.** Yes **5.** No **6.** No **7.** 9 **8.** -5 **9.** 22 **10.** 13.2 **11.** -6.5 **12.** -2 **13.** $\frac{31}{8}$

Exercise Set 8.1, p. 471

1. Yes **3.** No **5.** No **7.** Yes **9.** No **11.** No **13.** 4 **15.** -20 **17.** -14 **19.** -18 **21.** 15 **23.** -14 **25.** 2 **27.** 20 **29.** -6 **31.** $6\frac{1}{2}$ **33.** 19.9 **35.** $\frac{7}{3}$ **37.** $-\frac{7}{4}$ **39.** $\frac{41}{24}$ **41.** $-\frac{1}{20}$ **43.** 5.1 **45.** 12.4 **47.** -5 **49.** $1\frac{5}{6}$ **51.** $-\frac{10}{21}$ **53.** -11 **54.** 5 **55.** $-\frac{5}{12}$ **56.** $\frac{1}{3}$ **57.** $-\frac{3}{2}$ **58.** -5.2 **59.** $50 - x$ **60.** $65t$ **61.** ◈ **63.** 342.246 **65.** $-\frac{26}{15}$ **67.** -10 **69.** All real numbers **71.** $-\frac{5}{17}$ **73.** 13, -13

Margin Exercises, Section 8.2, pp. 474–476

1. 15 **2.** $-\frac{7}{4}$ **3.** -18 **4.** 10 **5.** $-\frac{4}{5}$ **6.** 7800 **7.** -3 **8.** 28

Exercise Set 8.2, p. 477

1. 6 **3.** 9 **5.** 12 **7.** -40 **9.** 1 **11.** -7 **13.** -6 **15.** 6 **17.** -63 **19.** 36 **21.** -21 **23.** $-\frac{3}{5}$ **25.** $-\frac{3}{2}$ **27.** $\frac{9}{2}$ **29.** 7 **31.** -7 **33.** 8 **35.** 15.9 **37.** $7x$ **38.** $-x + 5$ **39.** $8x + 11$ **40.** $-32y$ **41.** $x - 4$ **42.** $-23 - 5x$ **43.** $-10y - 42$ **44.** $-22a + 4$ **45.** $8r$ **46.** $\frac{1}{2}b \cdot 10$, or $5b$ **47.** ◈ **49.** -8655 **51.** No solution **53.** No solution **55.** $\dfrac{b}{3a}$ **57.** $\dfrac{4b}{a}$

Margin Exercises, Section 8.3, pp. 479–484

1. 5 **2.** 4 **3.** 4 **4.** 39 **5.** $-\frac{3}{2}$ **6.** -4.3 **7.** -3 **8.** 800 **9.** 1 **10.** 2 **11.** 2 **12.** $\frac{17}{2}$ **13.** $\frac{8}{3}$ **14.** -4.3 **15.** 2 **16.** 3 **17.** -2 **18.** $-\frac{1}{2}$

Calculator Spotlight, p. 484

1. Both sides equal 9. **2.** Both sides equal -2. **3.** Both sides equal -8.18.

Exercise Set 8.3, p. 485

1. 5 **3.** 8 **5.** 10 **7.** 14 **9.** -8 **11.** -8 **13.** -7 **15.** 15 **17.** 6 **19.** 4 **21.** 6 **23.** -3 **25.** 1 **27.** 6 **29.** -20 **31.** 7 **33.** 2 **35.** 5 **37.** 2 **39.** 10 **41.** 4 **43.** 0 **45.** -1 **47.** $-\frac{4}{3}$ **49.** $\frac{2}{5}$ **51.** -2 **53.** -4 **55.** $\frac{4}{5}$ **57.** $-\frac{28}{27}$ **59.** 6 **61.** 2 **63.** 6 **65.** 8 **67.** 1 **69.** 17 **71.** $-\frac{5}{3}$ **73.** -3 **75.** 2 **77.** $\frac{4}{7}$ **79.** $-\frac{51}{31}$ **81.** 2 **83.** -6.5 **84.** $7(x - 3 - 2y)$ **85.** $<$ **86.** -14 **87.** -18.7 **88.** -25.5 **89.** $c \div 8$, or $\dfrac{c}{8}$ **90.** $13.4h$ **91.** ◈ **93.** 4.4233464 **95.** $-\frac{7}{2}$ **97.** -2 **99.** 0 **101.** 6 **103.** $\frac{11}{18}$ **105.** 10

Margin Exercises, Section 8.4, pp. 490–495

1. Top: 24 ft; middle: 72 ft; bottom: 144 ft **2.** 5 **3.** 313 and 314 **4.** 93,333 **5.** Length: 84 ft; width: 50 ft **6.** First: 30°; second: 90°; third: 60° **7.** **(a)** $10.03994 billion, $15.60842 billion; **(b)** 2001

Exercise Set 8.4, p. 497

1. 16 **3.** $-\frac{1}{2}$ **5.** 57 **7.** 180 in., 60 in. **9.** 305 ft **11.** $2.89 **13.** -12 **15.** $699\frac{1}{3}$ mi **17.** 286, 287 **19.** 41, 42, 43 **21.** 61, 63, 65 **23.** Length: 48 ft; width: 14 ft **25.** 11 **27.** 28°, 84°, 68° **29.** 33°, 38°, 109° **31.** **(a)** $1056 million, $1122.2 million, $1784.2 million; **(b)** 2002 **33.** $-\frac{47}{40}$ **34.** $-\frac{17}{40}$ **35.** $-\frac{3}{10}$ **36.** $-\frac{32}{15}$ **37.** 1.6 **38.** 409.6 **39.** -9.6 **40.** -41.6 **41.** ◈ **43.** 120 **45.** About 0.65 in.

Margin Exercises, Section 8.5, pp. 501–504

1. 32% **2.** 25% **3.** 225 **4.** 50 **5.** 11.04 **6.** About 33 **7.** 111,416 mi^2 **8.** $8400 **9.** $658

Exercise Set 8.5, p. 505

1. 20% **3.** 150 **5.** 546 **7.** 24% **9.** 2.5 **11.** 5% **13.** $16.77 billion **15.** **(a)** 8190; **(b)** 1.8% **17.** **(a)** 16%; **(b)** $29 **19.** **(a)** $3.75; **(b)** $28.75 **21.** **(a)** $28.80; **(b)** $33.12 **23.** $36 **25.** 200 **27.** $282.20 **29.** About 42.4 lb **31.** About 566 **33.** $15.38 **35.** $7800 **37.** $58 **39.** 181.52 **40.** 0.4538 **41.** 12.0879 **42.** 844.1407 **43.** 16 **44.** 189.6 **45.** 10 **46.** 18.4875 **47.** ◈ **49.** 6 ft, 7 in. **51.** $9.17, not $9.10

Margin Exercises, Section 8.6, pp. 509–510

1. 2.8 mi **2.** $I = \frac{E}{R}$ **3.** $D = \frac{C}{\pi}$
4. $c = 4A - a - b - d$ **5. (a)** About 306 lb;
(b) $L = \frac{800W}{g^2}$

Exercise Set 8.6, p. 511

1. $h = \frac{A}{b}$ **3.** $w = \frac{P - 2l}{2}$, or $\frac{1}{2}P - l$ **5.** $a = 2A - b$
7. $a = \frac{F}{m}$ **9.** $c^2 = \frac{E}{m}$ **11.** $x = \frac{c - By}{A}$ **13.** $t = \frac{3k}{v}$
15. (a) 57,000 Btu; **(b)** $a = \frac{b}{30}$ **17. (a)** 1423;
(b) $n = 15F$ **19. (a)** 1901 calories;
(b) $a = \frac{917 + 6w + 6h - K}{6}$; $h = \frac{K - 917 - 6w + 6a}{6}$;
$w = \frac{K - 917 - 6h + 6a}{6}$ **21.** 0.92 **22.** -90
23. -13.2 **24.** $-21a + 12b$ **25.** $\frac{1}{6}$ **26.** $-\frac{3}{2}$
27. ◈ **29.** $b = \frac{2A - ah}{h}$; $h = \frac{2A}{a + b}$
31. A quadruples. **33.** A increases by $2h$ units.

Margin Exercises, Section 8.7, pp. 513–520

1. (a) No; **(b)** no; **(c)** no; **(d)** yes; **(e)** no; **(f)** no
2. (a) Yes; **(b)** yes; **(c)** yes; **(d)** no; **(e)** yes; **(f)** yes
3. $x \le 4$
4.

5.

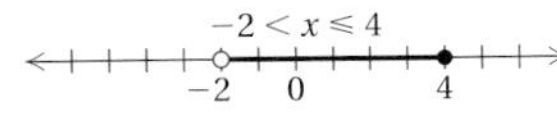

6. $\{x \mid x > 2\}$;
7. $\{x \mid x \le 3\}$; **8.** $\{x \mid x < -3\}$;
9. $\{x \mid x \ge \frac{2}{15}\}$ **10.** $\{y \mid y \le -3\}$
11. $\{x \mid x < 8\}$; **12.** $\{y \mid y \ge 32\}$;
13. $\{x \mid x \ge -6\}$ **14.** $\{y \mid y < -\frac{13}{5}\}$ **15.** $\{x \mid x > -\frac{1}{4}\}$
16. $\{y \mid y \ge \frac{19}{9}\}$ **17.** $\{y \mid y \ge \frac{19}{9}\}$ **18.** $\{x \mid x \ge -2\}$
19. $\{x \mid x \ge -4\}$ **20.** $\{x \mid x > \frac{8}{3}\}$

Exercise Set 8.7, p. 521

1. (a) Yes; **(b)** yes; **(c)** no; **(d)** yes; **(e)** yes
3. (a) No; **(b)** no; **(c)** no; **(d)** yes; **(e)** no
5. $x > 4$
7.

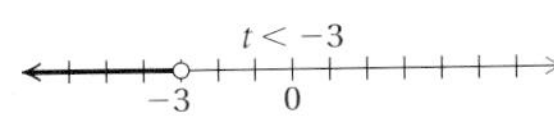

9. $m \ge -1$
11. $-3 < x \le 4$
13. $0 < x < 3$
15. $\{x \mid x > -5\}$;
17. $\{x \mid x \le -18\}$;
19. $\{y \mid y > -5\}$ **21.** $\{x \mid x > 2\}$ **23.** $\{x \mid x \le -3\}$
25. $\{x \mid x < 4\}$ **27.** $\{t \mid t > 14\}$ **29.** $\{y \mid y \le \frac{1}{4}\}$
31. $\{x \mid x > \frac{7}{12}\}$
33. $\{x \mid x < 7\}$; **35.** $\{x \mid x < 3\}$;
37. $\{y \mid y \ge -\frac{2}{5}\}$ **39.** $\{x \mid x \ge -6\}$ **41.** $\{y \mid y \le 4\}$
43. $\{x \mid x > \frac{17}{3}\}$ **45.** $\{y \mid y < -\frac{1}{14}\}$ **47.** $\{x \mid x \le \frac{3}{10}\}$
49. $\{x \mid x < 8\}$ **51.** $\{x \mid x \le 6\}$ **53.** $\{x \mid x < -3\}$
55. $\{x \mid x > -3\}$ **57.** $\{x \mid x \le 7\}$ **59.** $\{x \mid x > -10\}$
61. $\{y \mid y < 2\}$ **63.** $\{y \mid y \ge 3\}$ **65.** $\{y \mid y > -2\}$
67. $\{x \mid x > -4\}$ **69.** $\{x \mid x \le 9\}$ **71.** $\{y \mid y \le -3\}$
73. $\{y \mid y < 6\}$ **75.** $\{m \mid m \ge 6\}$ **77.** $\{t \mid t < -\frac{5}{3}\}$
79. $\{r \mid r > -3\}$ **81.** $\{x \mid x \ge -\frac{57}{34}\}$ **83.** $\{x \mid x > -2\}$
85. -74 **86.** 4.8 **87.** $-\frac{5}{8}$ **88.** -1.11 **89.** -38
90. $-\frac{7}{8}$ **91.** -9.4 **92.** 1.11 **93.** 140 **94.** 41
95. $-2x - 23$ **96.** $37x - 1$ **97.** ◈ **99. (a)** Yes;
(b) yes; **(c)** no; **(d)** no; **(e)** no; **(f)** yes; **(g)** yes
101. All real numbers

Margin Exercises, Section 8.8, pp. 525–526

1. $x \le 8$ **2.** $y > -2$ **3.** $s \le 180$ **4.** $p \ge \$5800$
5. $2x - 32 > 5$ **6.** $\{x \mid x \ge 84\}$ **7.** $\{C \mid C < 1063°\}$

Exercise Set 8.8, p. 527

1. $x > 8$ **3.** $y \le -4$ **5.** $n \ge 1300$ **7.** $a \le 500$
9. $2 + 3x < 13$ **11.** $\{x \mid x \ge 97\}$ **13.** $\{Y \mid Y \ge 1935\}$
15. $\{L \mid L \ge 5$ in.$\}$ **17.** $\{x \mid x > 5\}$ **19.** $\{d \mid d > 25\}$
21. $\{b \mid b > 6$ cm$\}$ **23.** $\{x \mid x \ge 21\}$
25. $\{t \mid t \le 0.75$ hr$\}$ **27.** $\{f \mid f \ge 16$ g$\}$ **29.** -160
30. $-17x + 18$ **31.** $91x - 242$ **32.** 0.25 **33.** ◈

Summary and Review: Chapter 8, p. 529

1. -22 **2.** 1 **3.** 25 **4.** 9.99 **5.** $\frac{1}{4}$ **6.** 7
7. -192 **8.** $-\frac{7}{3}$ **9.** $-\frac{15}{64}$ **10.** -8 **11.** 4 **12.** -5
13. $-\frac{1}{3}$ **14.** 3 **15.** 4 **16.** 16 **17.** 6 **18.** -3
19. 12 **20.** 4 **21.** Yes **22.** No **23.** Yes
24. $\{y \mid y \ge -\frac{1}{2}\}$ **25.** $\{x \mid x \ge 7\}$ **26.** $\{y \mid y > 2\}$
27. $\{y \mid y \le -4\}$ **28.** $\{x \mid x < -11\}$ **29.** $\{y \mid y > -7\}$
30. $\{x \mid x > -6\}$ **31.** $\{x \mid x > -\frac{9}{11}\}$ **32.** $\{y \mid y \le 7\}$
33. $\{x \mid x \ge -\frac{1}{12}\}$
34. $x < 3$ **35.** $-2 < x \le 5$

36. 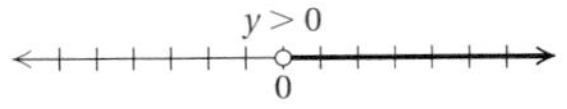

37. $d = \dfrac{C}{\pi}$ **38.** $B = \dfrac{3V}{h}$ **39.** $a = 2A - b$
40. Length: 365 mi; width: 275 mi **41.** 27
42. 345, 346 **43.** \$2117 **44.** 27 **45.** 35°, 85°, 60°
46. \$220 **47.** \$26,087 **48.** \$138.95 **49.** 86
50. $\{w \mid w > 17 \text{ cm}\}$ **51.** **(a)** 8.2 lb; **(b)** $L = \dfrac{800W}{g^2}$
52. $\frac{41}{4}$ **53.** $58t$ **54.** -45 **55.** $-43x + 8y$
56. ◈ The end result is the same either way. If s is the original salary, the new salary after a 5% raise followed by an 8% raise is $1.08(1.05s)$. If the raises occur the other way around, the new salary is $1.05(1.08s)$. By the commutative and associative laws of multiplication, we see that these are equal. However, it would be better to receive the 8% raise first, because this increase yields a higher salary the first year than a 5% raise.
57. ◈ The inequalities are equivalent by the multiplication principle for inequalities. If we multiply on both sides of one inequality by -1, the other inequality results. **58.** 23, -23 **59.** 20, -20
60. $a = \dfrac{y - 3}{2 - b}$

Test: Chapter 8, p. 531

1. [8.1b] 8 **2.** [8.1b] 26 **3.** [8.2a] -6 **4.** [8.2a] 49
5. [8.3b] -12 **6.** [8.3a] 2 **7.** [8.3a] -8
8. [8.1b] $-\frac{7}{20}$ **9.** [8.3c] 7 **10.** [8.3c] $\frac{5}{3}$ **11.** [8.3b] 2.5
12. [8.7c] $\{x \mid x \le -4\}$ **13.** [8.7c] $\{x \mid x > -13\}$
14. [8.7d] $\{x \mid x \le 5\}$ **15.** [8.7d] $\{y \mid y \le -13\}$
16. [8.7d] $\{y \mid y \ge 8\}$ **17.** [8.7d] $\left\{x \mid x \le -\frac{1}{20}\right\}$
18. [8.7e] $\{x \mid x < -6\}$ **19.** [8.7e] $\{x \mid x \le -1\}$
20. [8.7b] **21.** [8.7b, e]

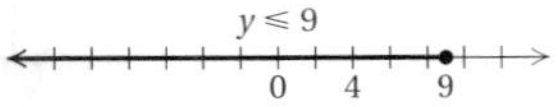

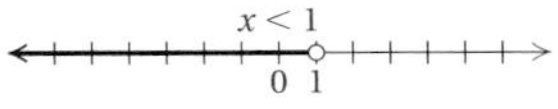

22. [8.7b] 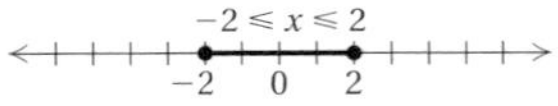

23. [8.4a] Width: 7 cm; length: 11 cm **24.** [8.4a] 6
25. [8.4a] 2509, 2510, 2511 **26.** [8.5a] \$880
27. [8.4a] 3 m, 5 m **28.** [8.6a] $r = \dfrac{A}{2\pi h}$
29. [8.6a] **(a)** 2650; **(b)** $w = \dfrac{K - 7h + 9.52a - 92.4}{19.18}$
30. [8.8b] $\{x \mid x > 6\}$ **31.** [8.8b] $\{l \mid l \ge 174 \text{ yd}\}$
32. [7.3a] $-\frac{2}{9}$ **33.** [7.1a] $\frac{8}{3}$ **34.** [7.1b] $73\%p$, or $0.73p$
35. [7.8b] $-18x + 37y$ **36.** [8.6a] $d = \dfrac{1 - ca}{-c}$, or $\dfrac{ca - 1}{c}$
37. [7.2e], [8.3a] 15, -15 **38.** [8.4a] 60

Chapter 9

Pretest: Chapter 9, p. 534

1. [9.2b] **2.** [9.3b]

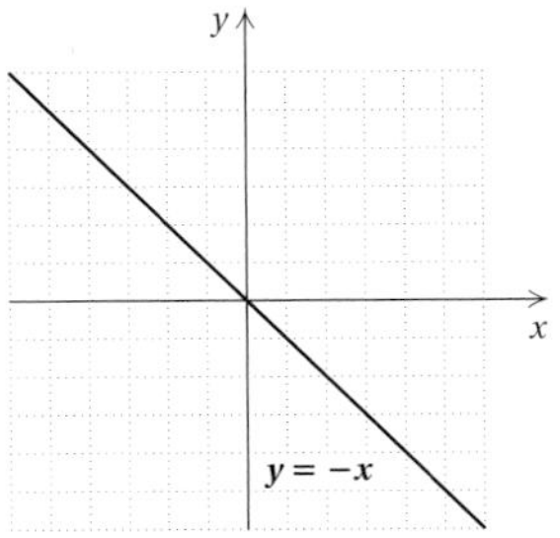

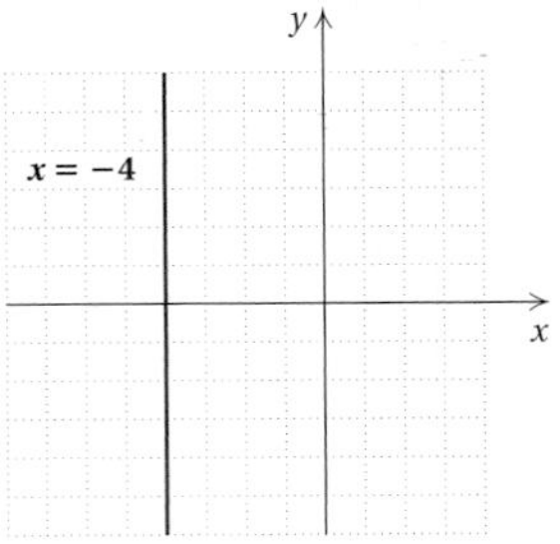

3. [9.3a] **4.** [9.2b]

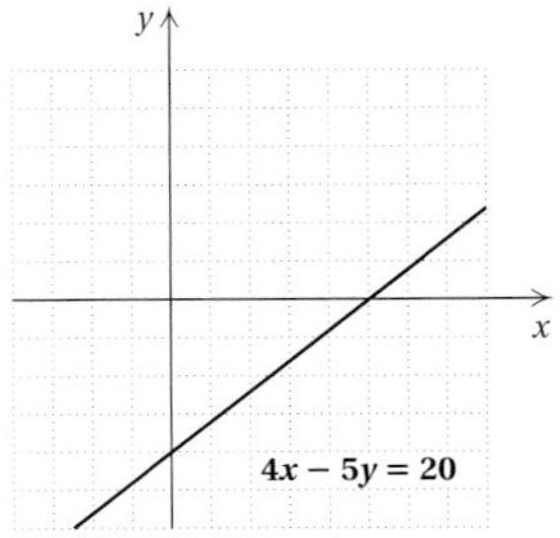

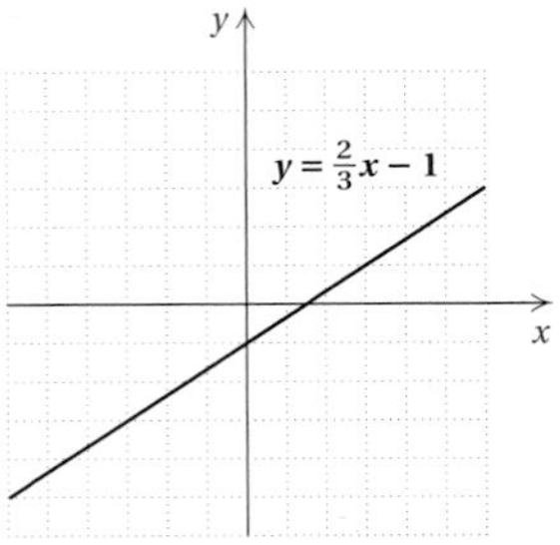

5. [9.1c] III **6.** [9.2a] No **7.** [9.3a] y-intercept: $(0, -4)$; x-intercept: $(5, 0)$ **8.** [9.2b] $(0, -8)$
9. [9.2c]

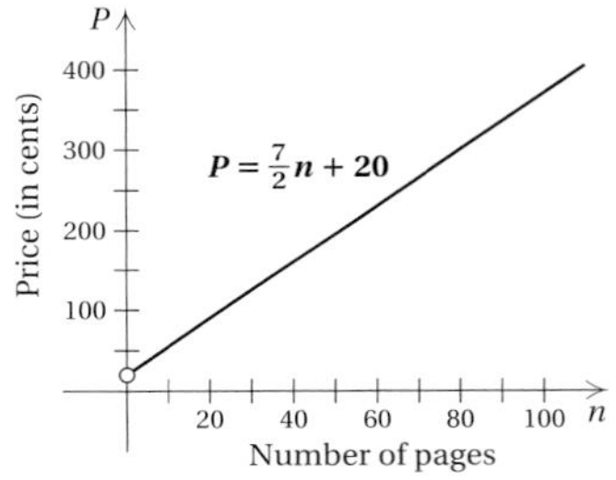

320¢, or \$3.20

10. [9.4a] Mean: 0.18; median: 0.18; mode: 0.18
11. [9.4c] 164 cm

Margin Exercises, Section 9.1, pp. 535–539

1. \$23.22 billion **2.** **(a)** 3 A.M.–6 A.M.; **(b)** midnight–3 A.M., 3 A.M.–6 A.M., 6 A.M.–9 A.M., 9 A.M.–noon **3.** **(a)** 2; **(b)** 60 beats per minute
4.–11.

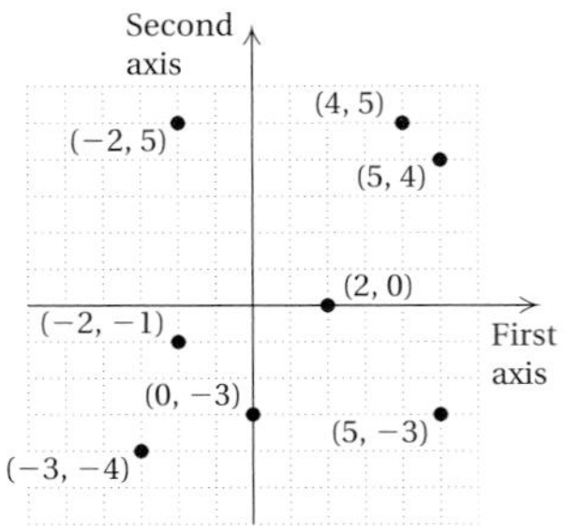

12. Both are negative numbers. **13.** First, positive; second, negative **14.** I **15.** III **16.** IV **17.** II **18.** Not in any quadrant **19.** A: $(-5, 1)$; B: $(-3, 2)$; C: $(0, 4)$; D: $(3, 3)$; E: $(1, 0)$; F: $(0, -3)$; G: $(-5, -4)$

Exercise Set 9.1, p. 541

1. 6 **3.** The weight is greater than 200 lb. **5.** The weight is greater than or equal to 120 lb. **7.** 32.4% **9.** $10,360.32 **11.** 20,000 **13.** 1994 **15.** About 500

17.

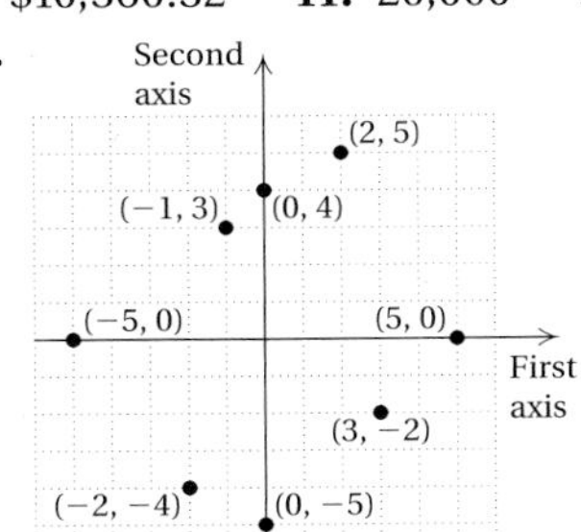

19. II **21.** IV **23.** III **25.** I **27.** II **29.** IV **31.** Negative; negative **33.** Second; first **35.** I, IV **37.** I, III **39.** A: $(3, 3)$; B: $(0, -4)$; C: $(-5, 0)$; D: $(-1, -1)$; E: $(2, 0)$ **41.** 12 **42.** 4.89 **43.** 0 **44.** $\frac{4}{5}$ **45.** $0.9 billion **46.** $18.40 **47.** ◈
49. $(-1, -5)$
51. Answers may vary. **53.** 26

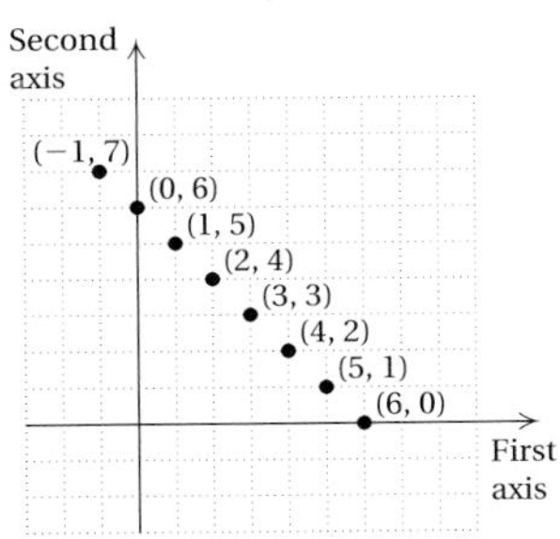

Margin Exercises, Section 9.2, pp. 545–552

1. No **2.** Yes
3. $(-2, -3)$, $(1, 3)$; answers may vary

4.

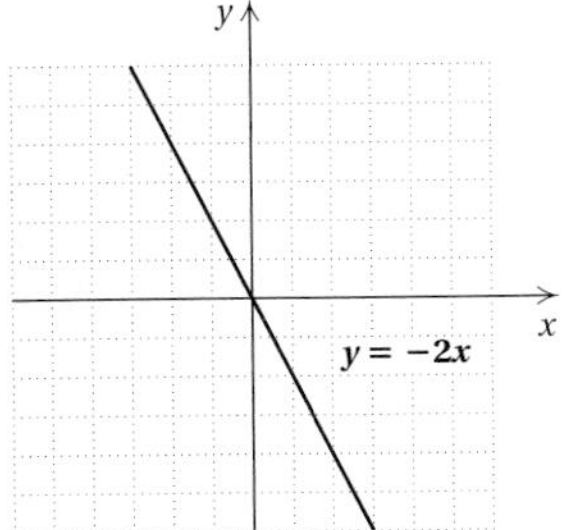

5.

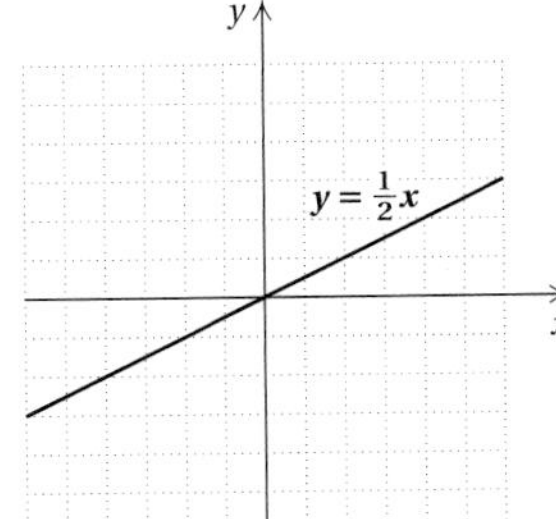

6.

y = 2x + 3

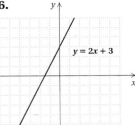

7.

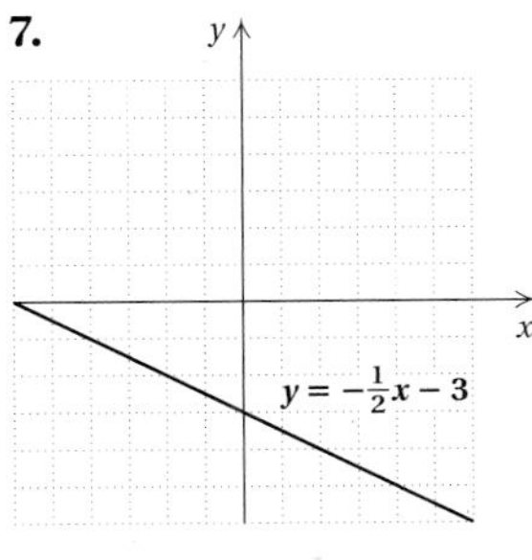

8.

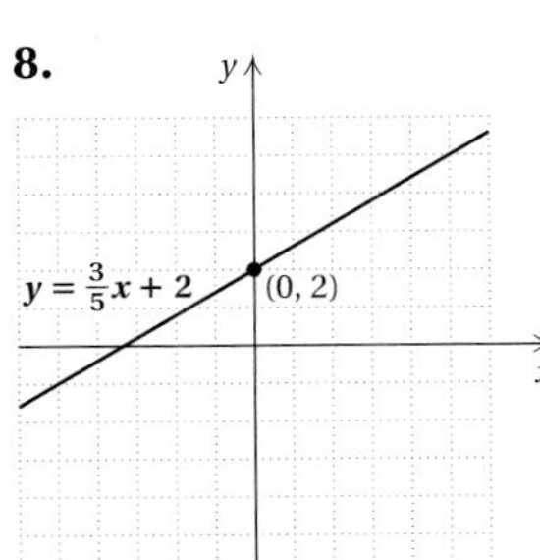

9.

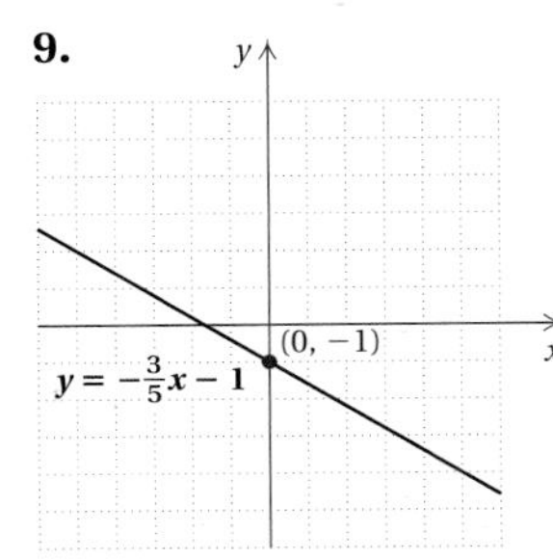

10.

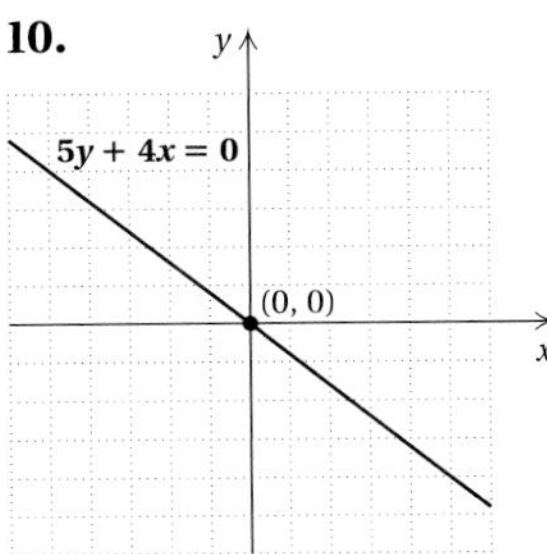

11.

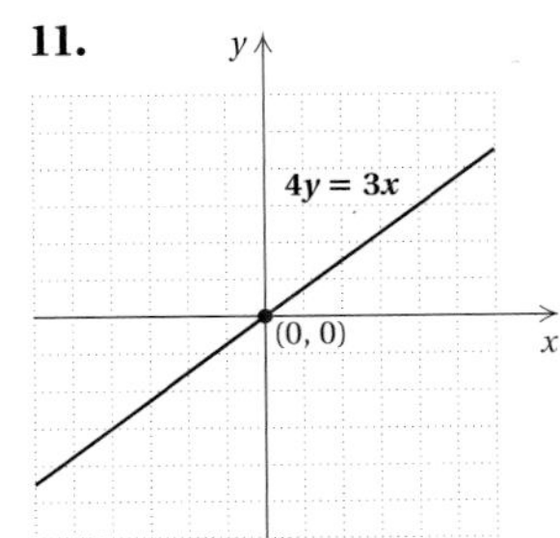

12.

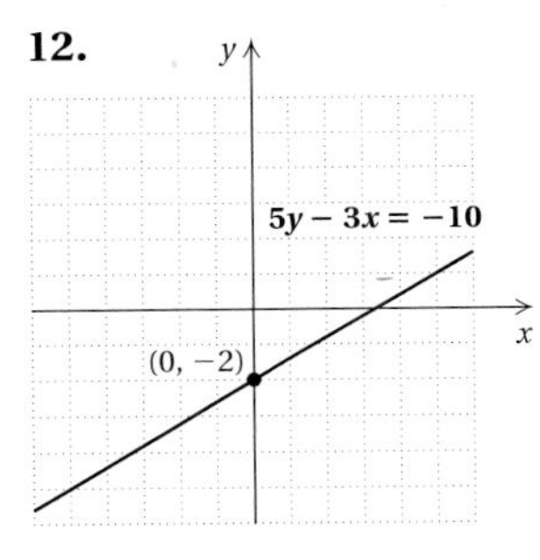

13.

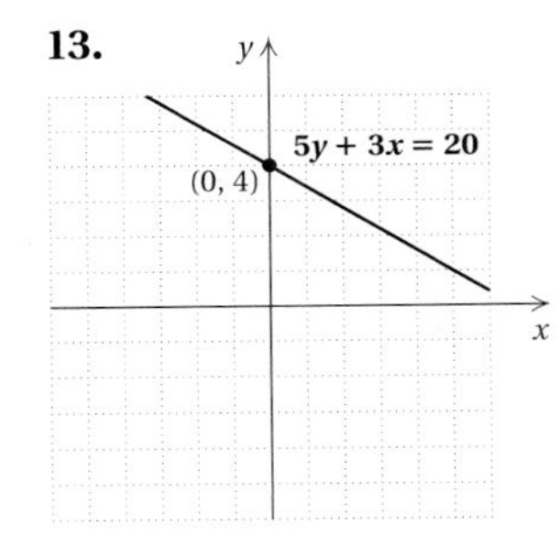

14. **(a)** $2720, $2040, $680, $0; **(b)** $1700 **(c)** About 2.8 yr

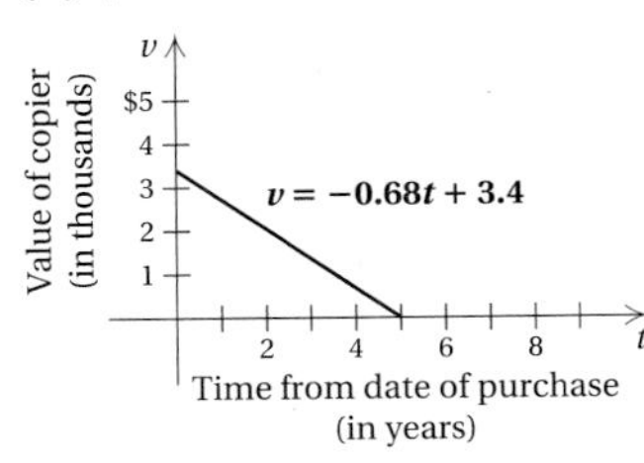

Calculator Spotlight, p. 554

1.

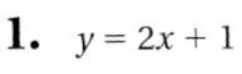

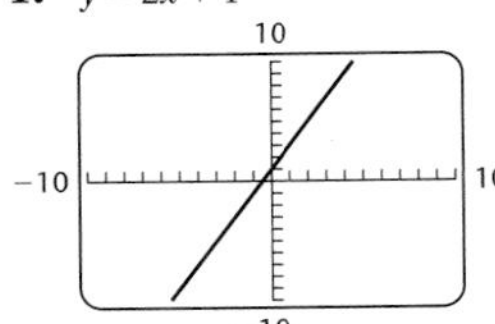

2. $y = -3x + 1$

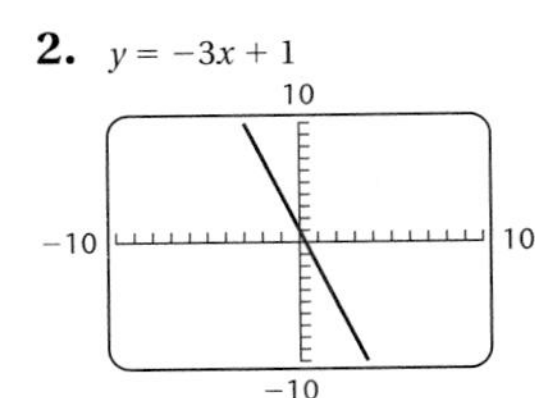

3.

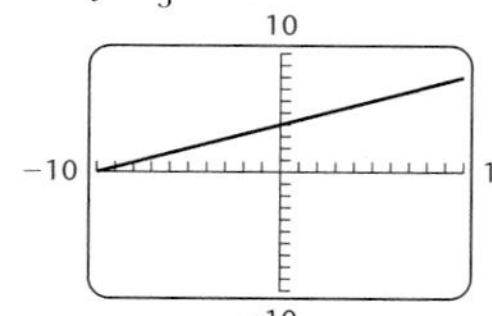

4.

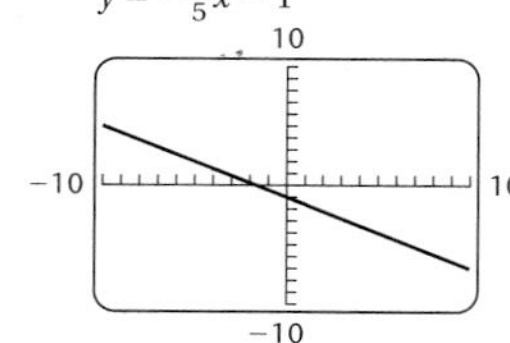

5.

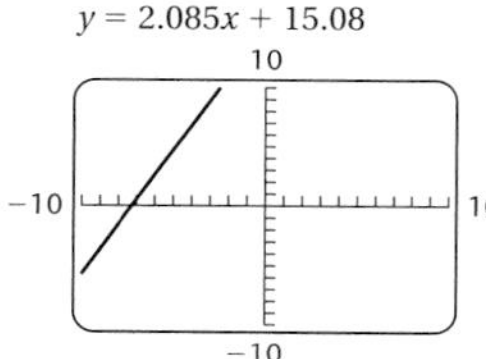

6.

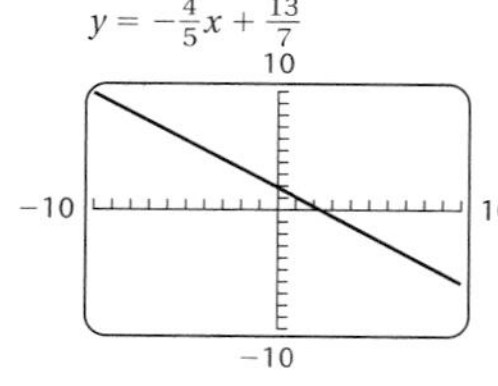

7.

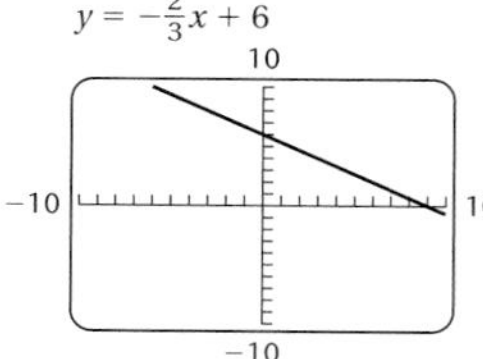

8.

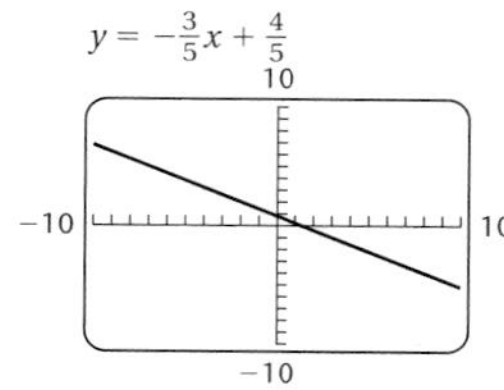

9.

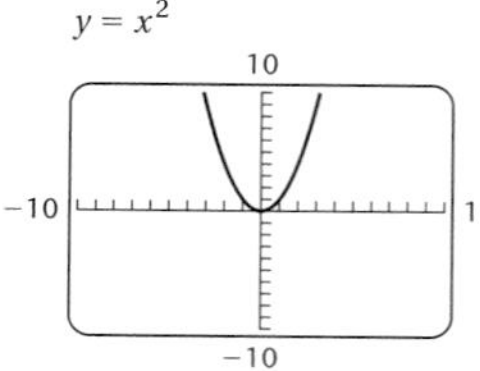

10.

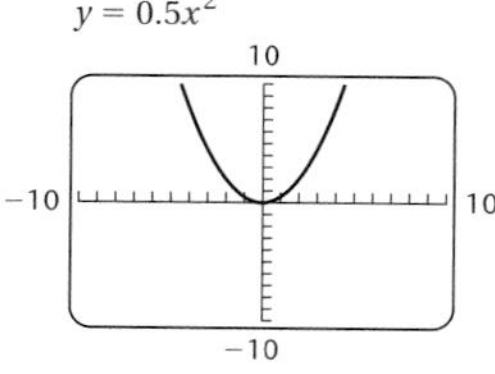

11.

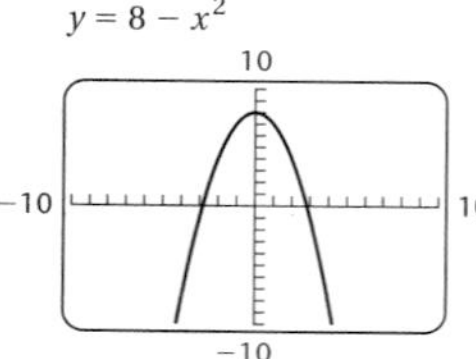

12.

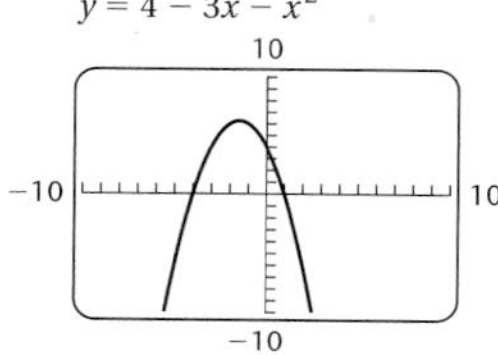

13.

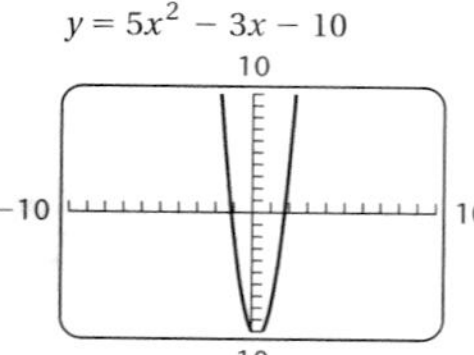

14.

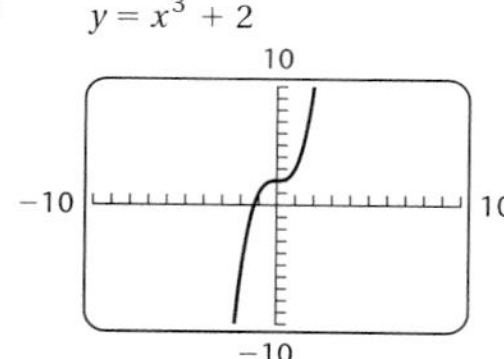

15.

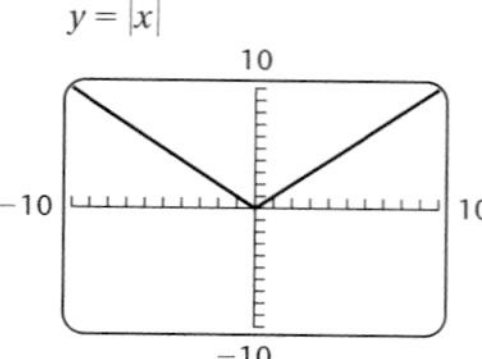

16.

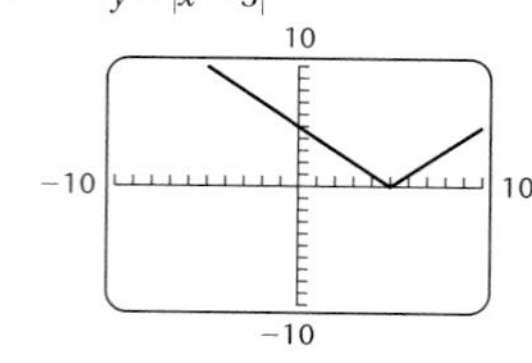

17.

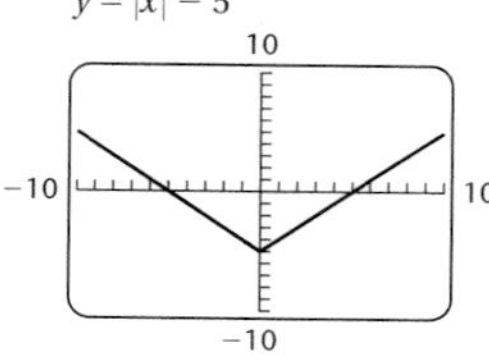

18.

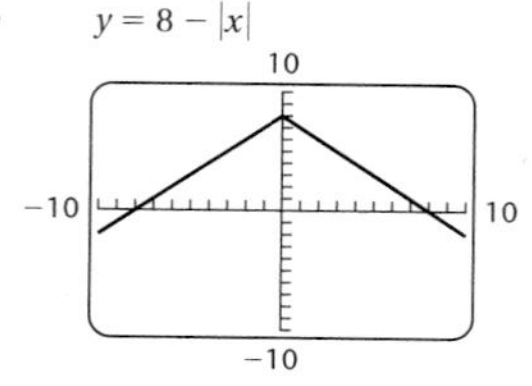

Exercise Set 9.2, p. 555

1. No **3.** No **5.** Yes

7.

$y = x - 5$
$-1 \;?\; 4 - 5$
$\quad -1$ TRUE

$y = x - 5$
$-4 \;?\; 1 - 5$
$\quad -4$ TRUE

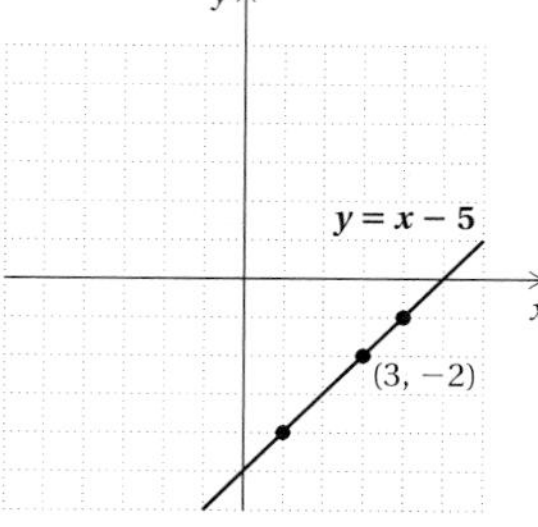

9.

$y = \frac{1}{2}x + 3$
$5 \;?\; \frac{1}{2} \cdot 4 + 3$
$\quad 2 + 3$
$\quad 5$ TRUE

$y = \frac{1}{2}x + 3$
$2 \;?\; \frac{1}{2}(-2) + 3$
$\quad -1 + 3$
$\quad 2$ TRUE

11.

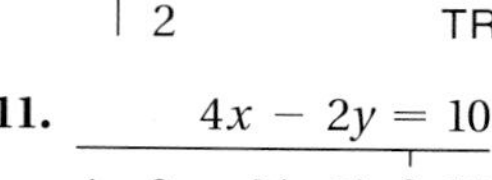

$4x - 2y = 10$
$4 \cdot 0 - 2(-5) \;?\; 10$
$0 + 10$
10 TRUE

$4x - 2y = 10$
$4 \cdot 4 - 2 \cdot 3 \;?\; 10$
$16 - 6$
10 TRUE

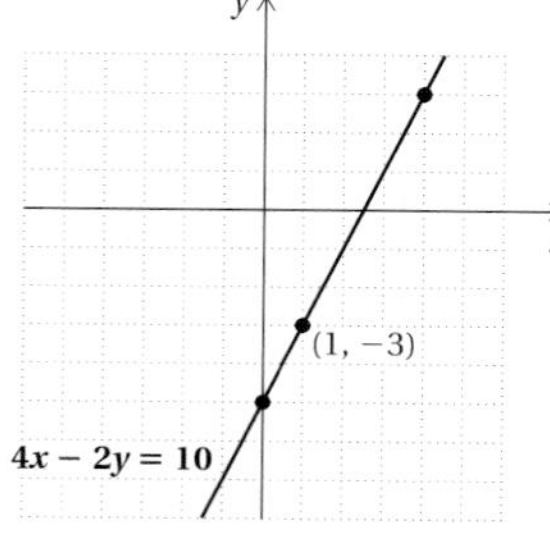

13.

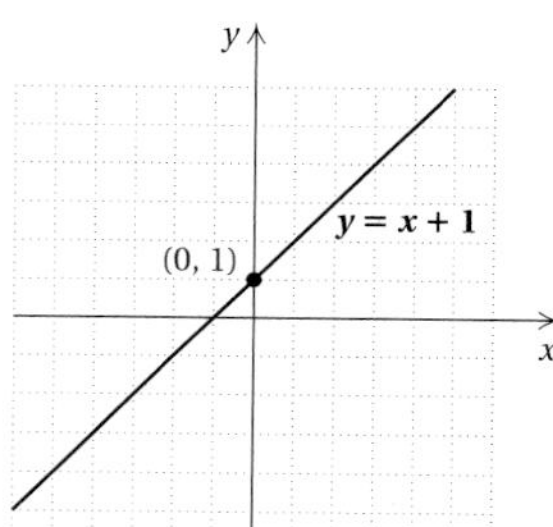

15.

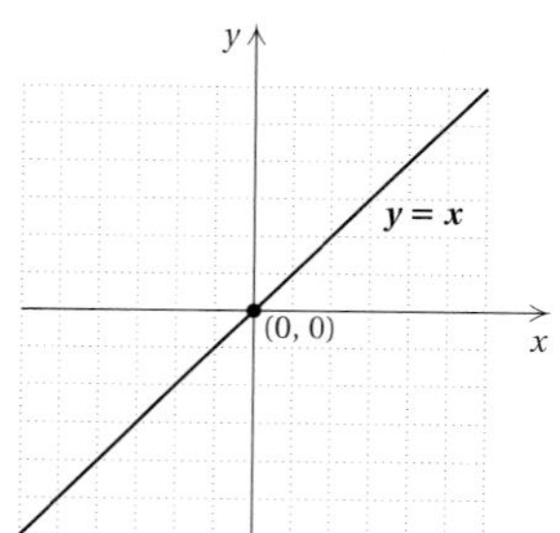

17.

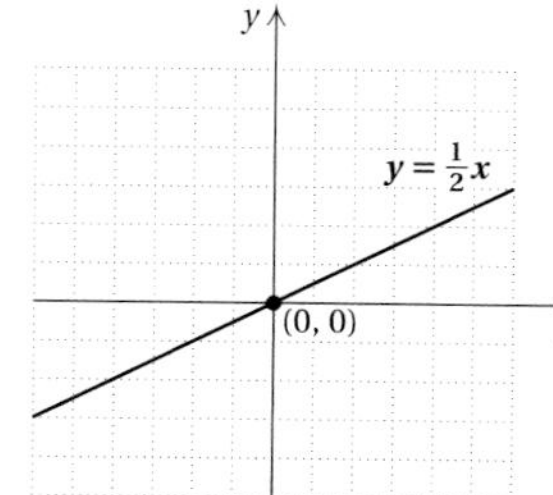

19.

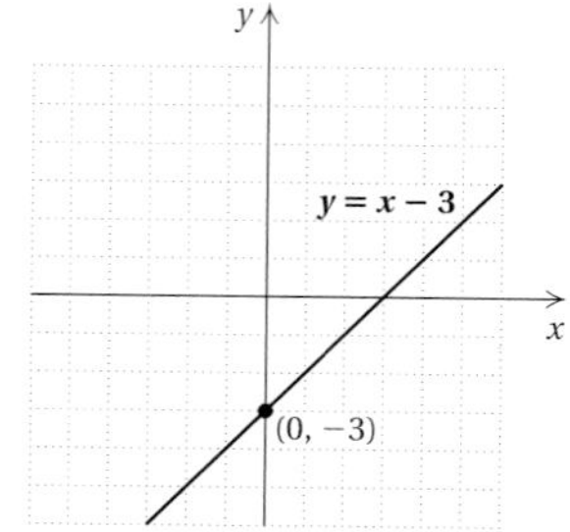

21.

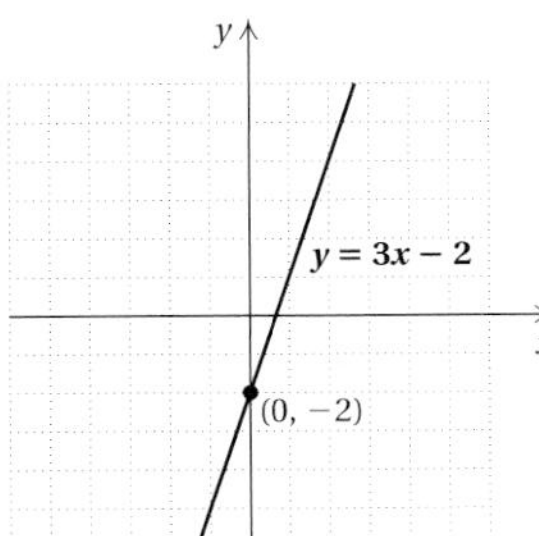

23.

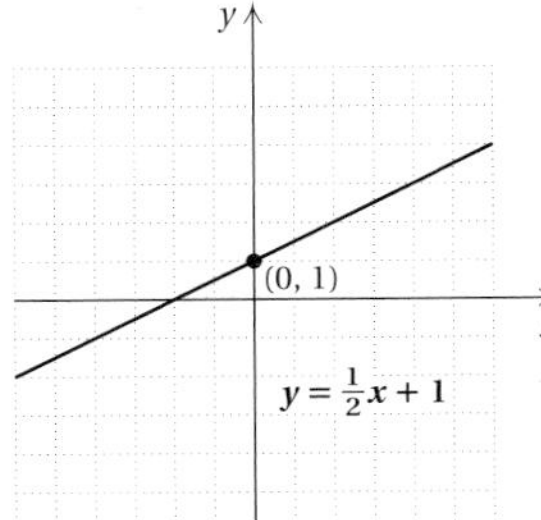

25.

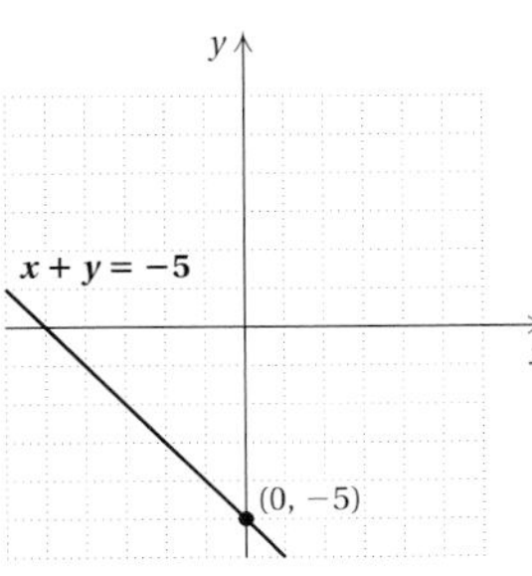

27.

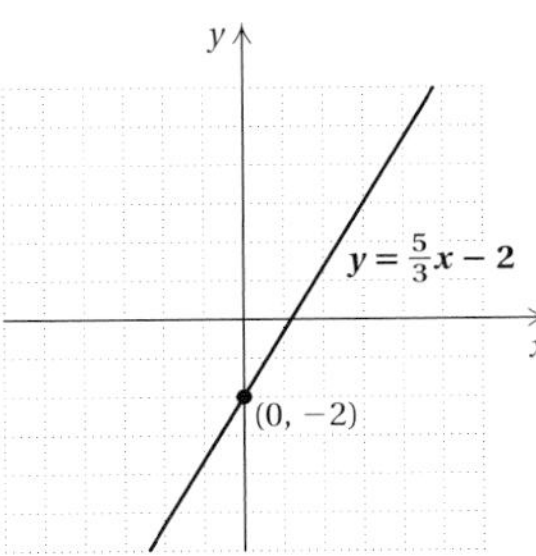

29.

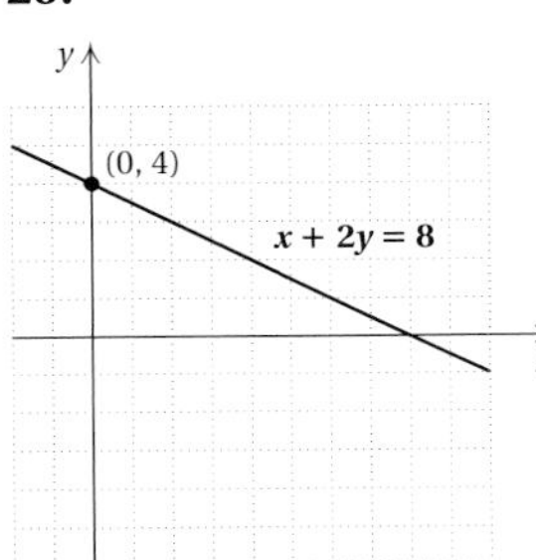

31.

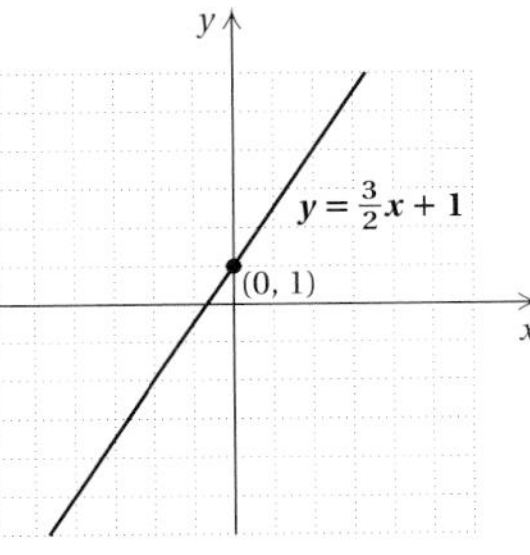

33.

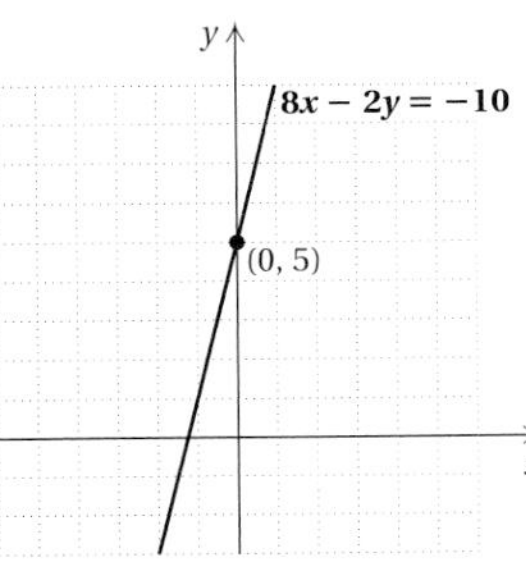

35.

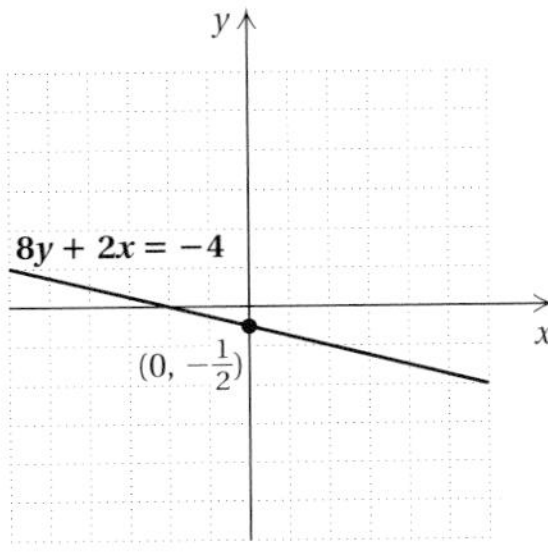

37. **(a)** \$300, \$100, \$0; **(b)**

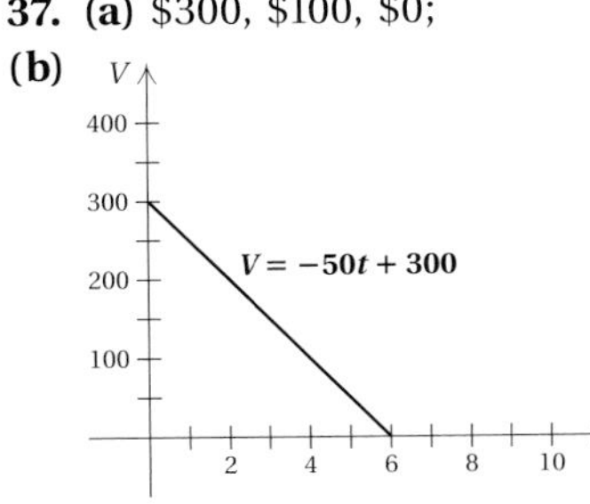

\$50; **(c)** 3

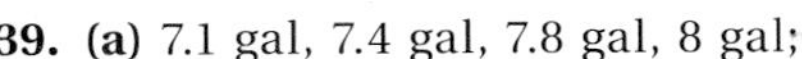

39. **(a)** 7.1 gal, 7.4 gal, 7.8 gal, 8 gal; **(b)**

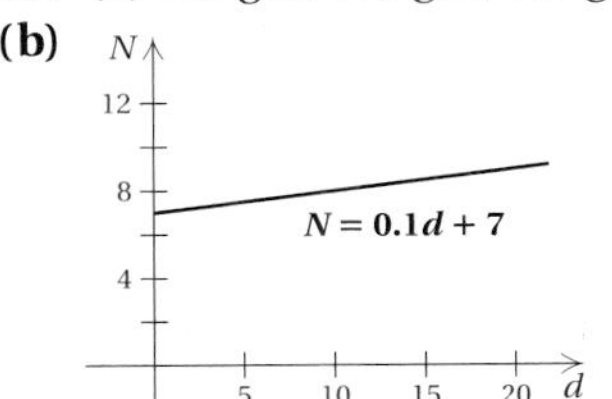

7.6 gal; **(c)** 2011

41. 3000 **42.** 125,000 **43.** 0 **44.** 6,078,000 **45.** −0.875 **46.** 0.71875 **47.** 1.828125 **48.** −2.25 **49.**  **51.** $y = -x + 5$ **53.** $y = x + 2$

Margin Exercises, Section 9.3, pp. 559–561

1. **(a)** (0, 3); **(b)** (4, 0)

2.

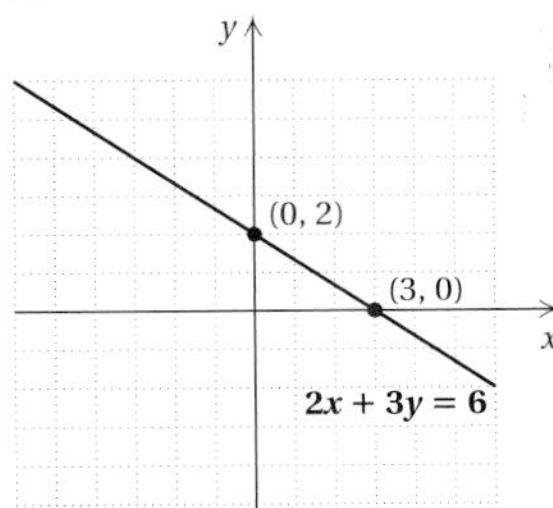

3.

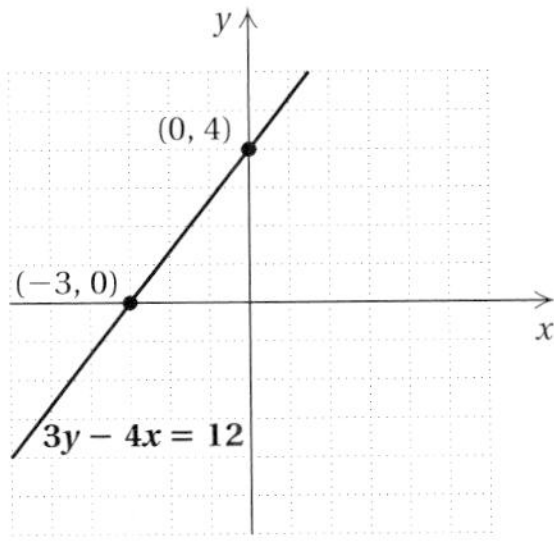

4.

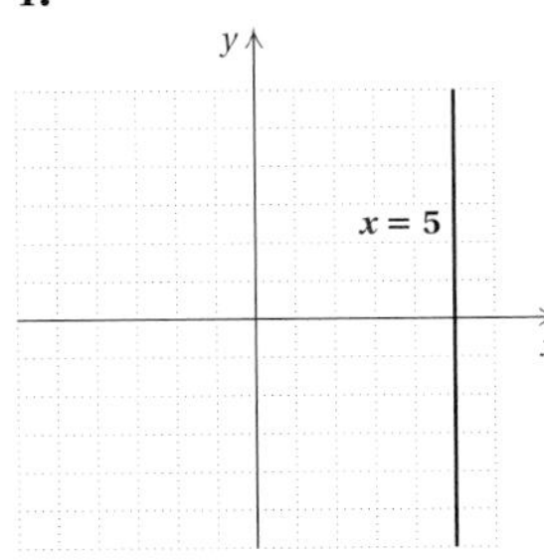

5.

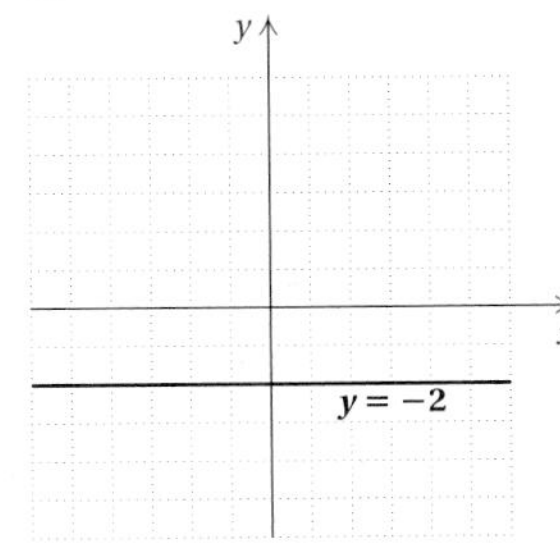

6.

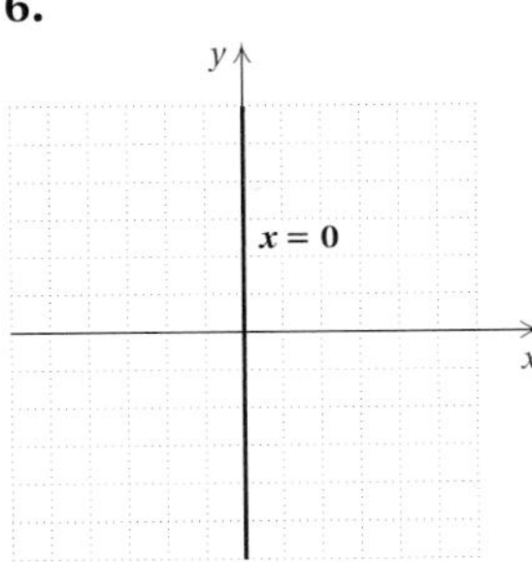

7.

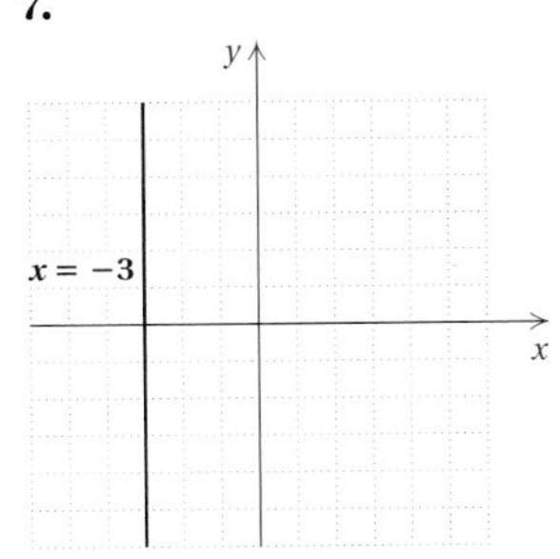

Calculator Spotlight, p. 562

1. y-intercept: (0, −15); x-intercept: (−2.08, 0)
2. y-intercept: (0, 27); x-intercept: (−12.68, 0)
3. y-intercept: (0, 14); x-intercept: (16.8, 0)
4. y-intercept: (0, −21.43); x-intercept: (75, 0)
5. y-intercept: (0, 25); x-intercept: (16.67, 0)
6. y-intercept: (0, −9); x-intercept: (45, 0)
7. y-intercept: (0, −15); x-intercept: (11.54, 0)
8. y-intercept: (0, −0.05); x-intercept: (0.04, 0)

Exercise Set 9.3, p. 563

1. **(a)** (0, 5); **(b)** (2, 0) **3.** **(a)** (0, −4); **(b)** (3, 0)
5. **(a)** (0, 3); **(b)** (5, 0) **7.** **(a)** (0, −14); **(b)** (4, 0)
9. **(a)** $\left(0, \frac{10}{3}\right)$; **(b)** $\left(-\frac{5}{2}, 0\right)$ **11.** **(a)** $\left(0, -\frac{1}{3}\right)$; **(b)** $\left(\frac{1}{2}, 0\right)$

13.

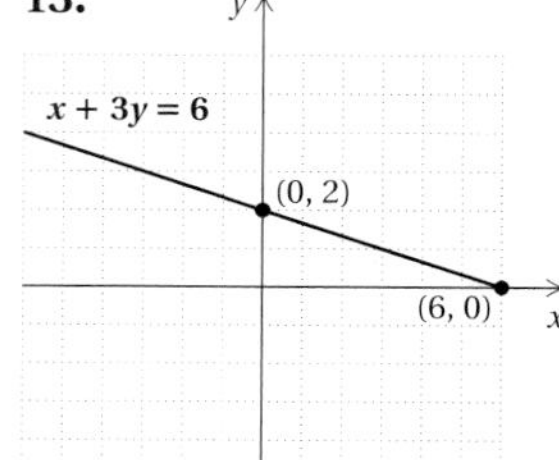

15.

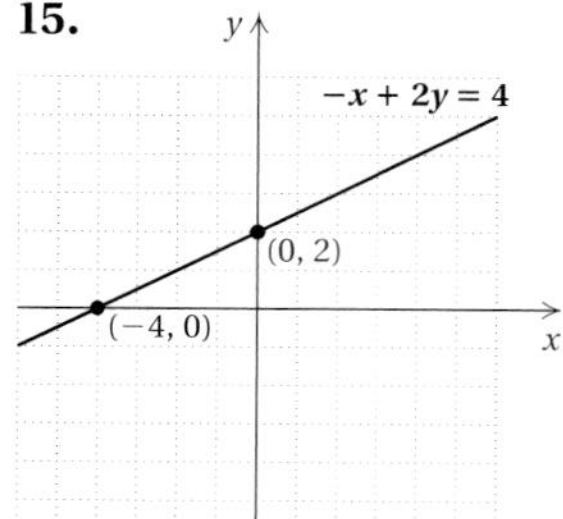

17.

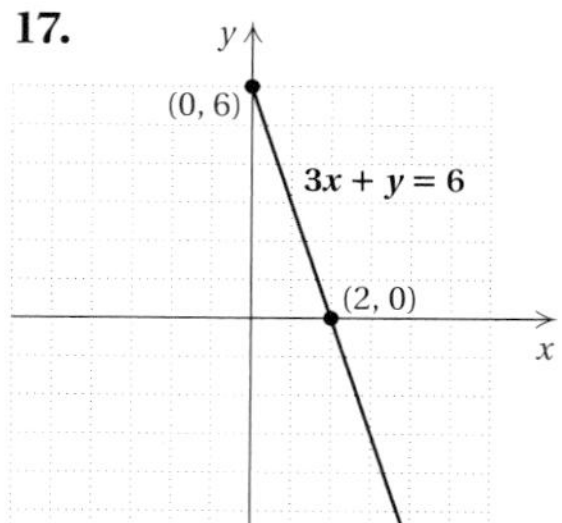

19.

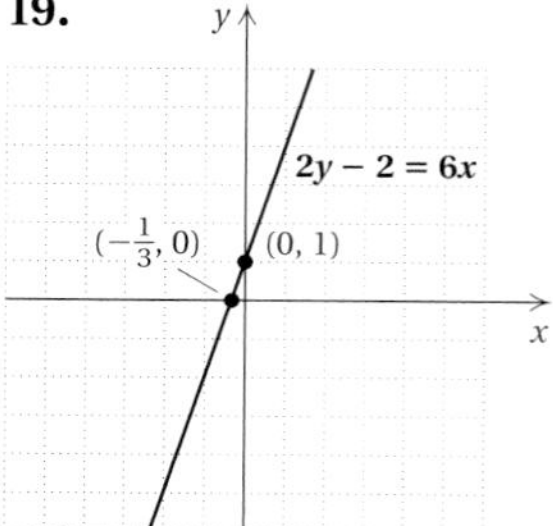

21.

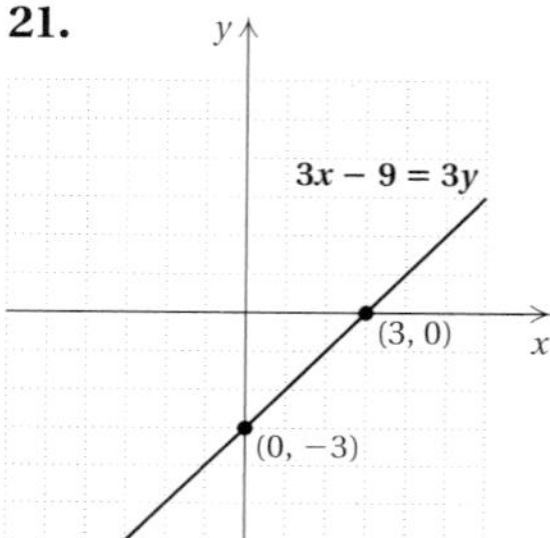

23.

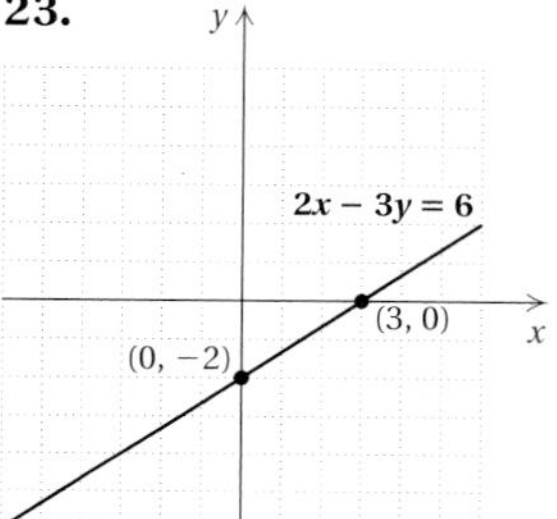

25.

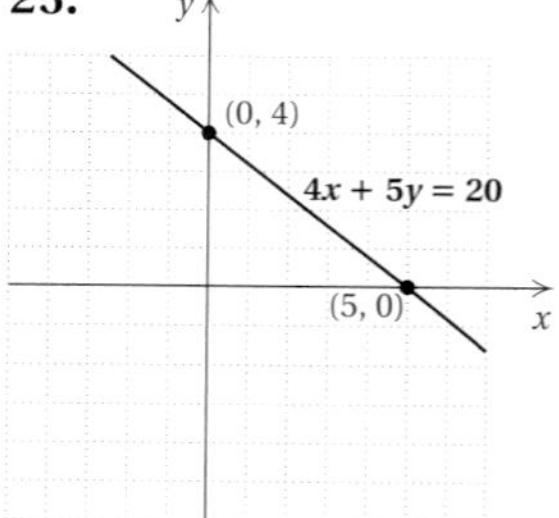

27.

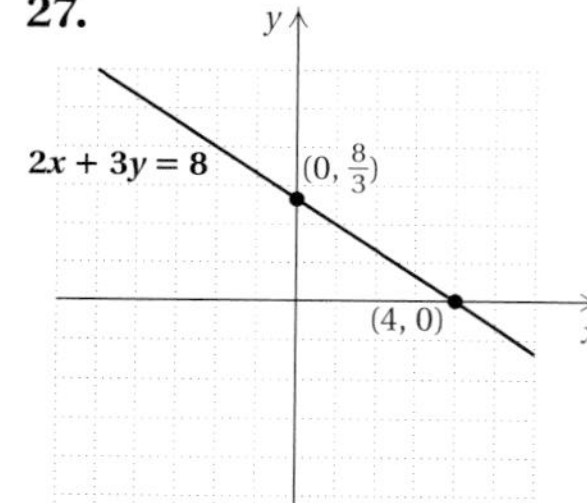

29.

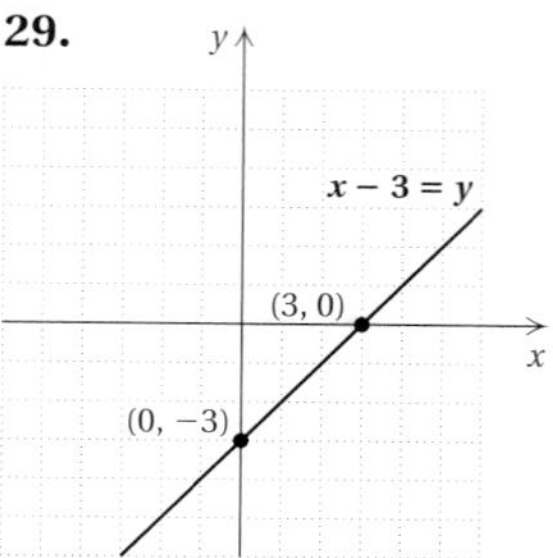

31.

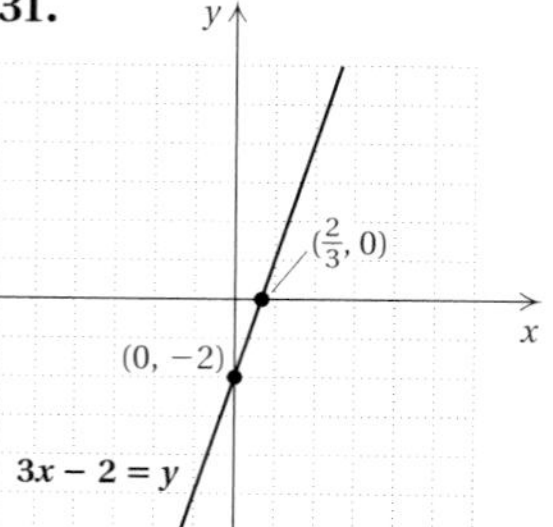

33.

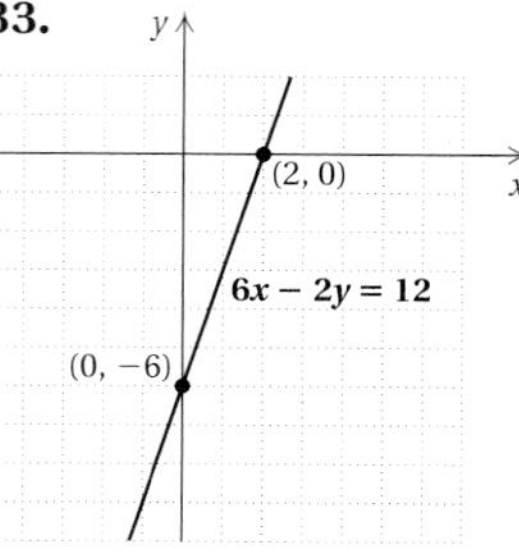

35.

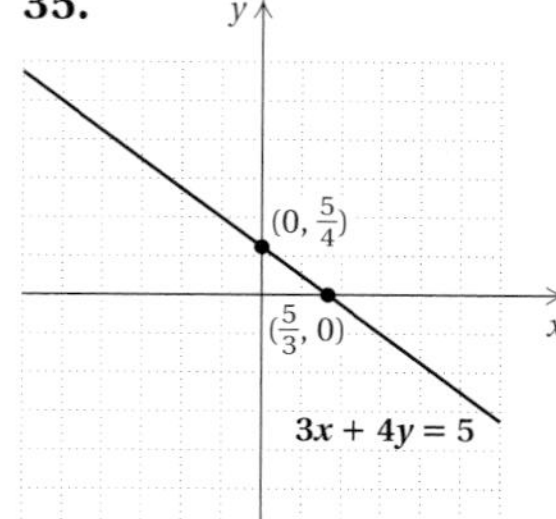

37.

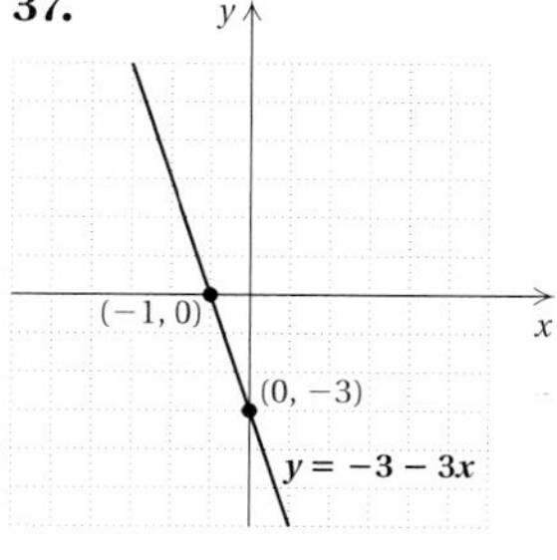

39.

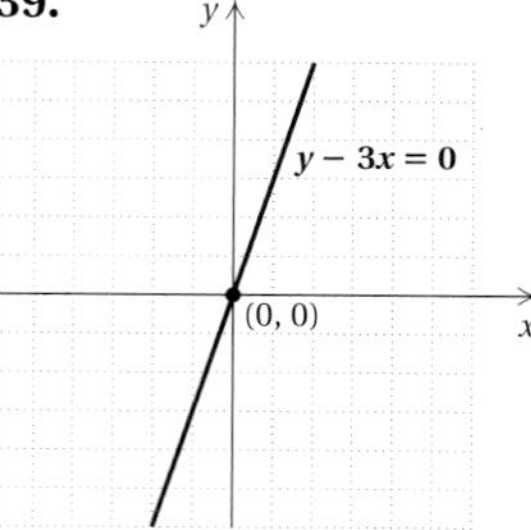

41.

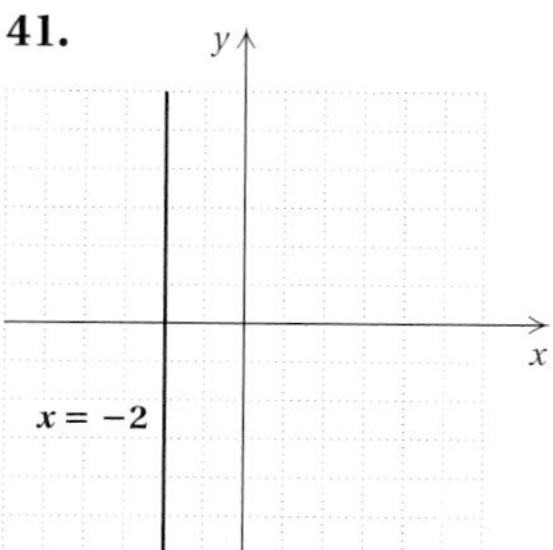

43.

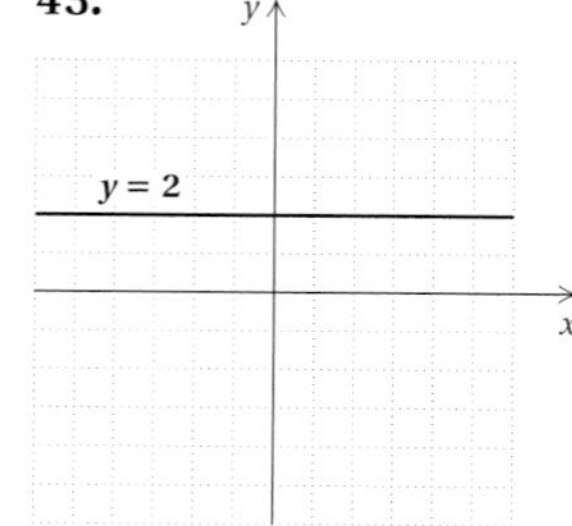

45.

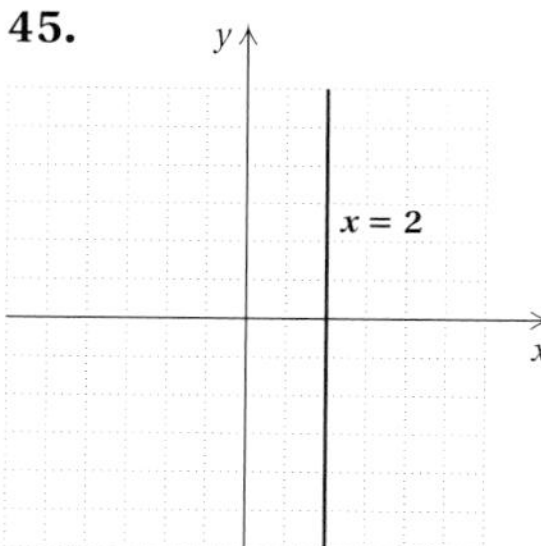

47.

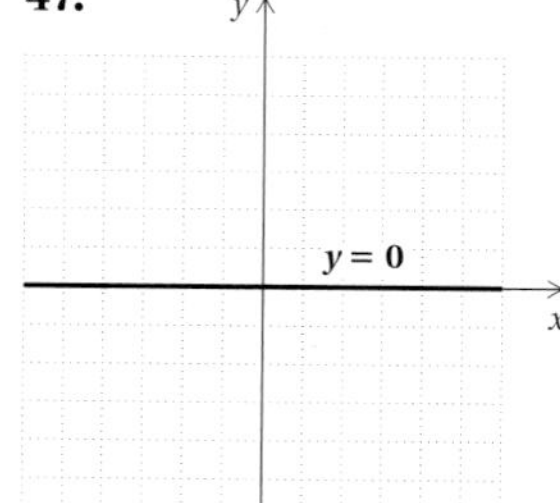

49.

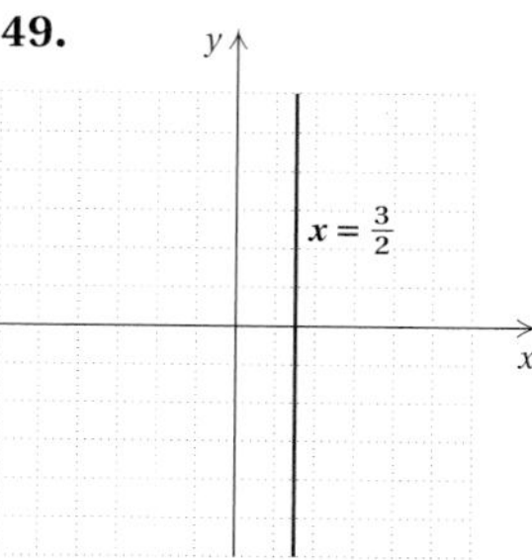

51.

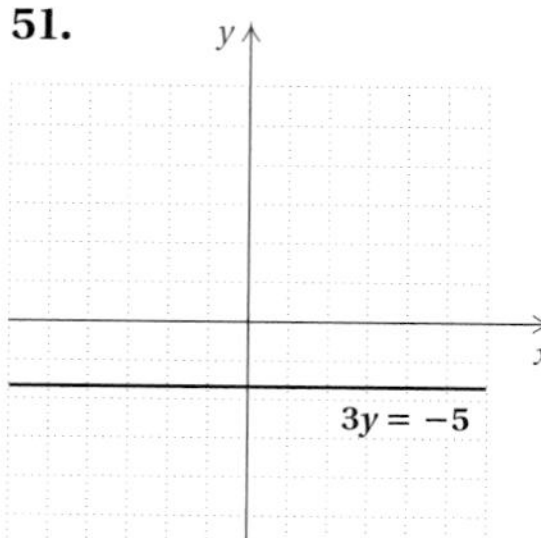

53.

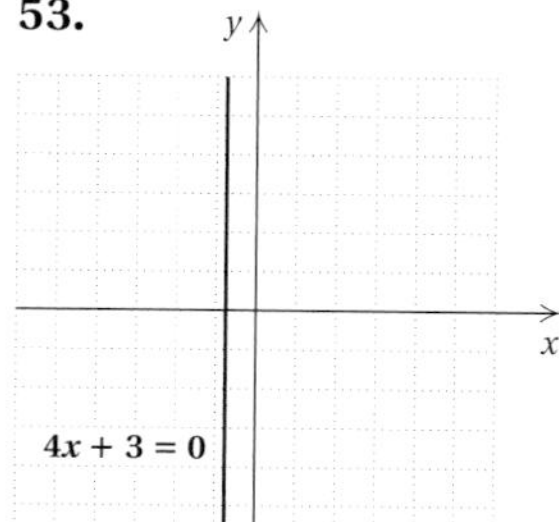

55.

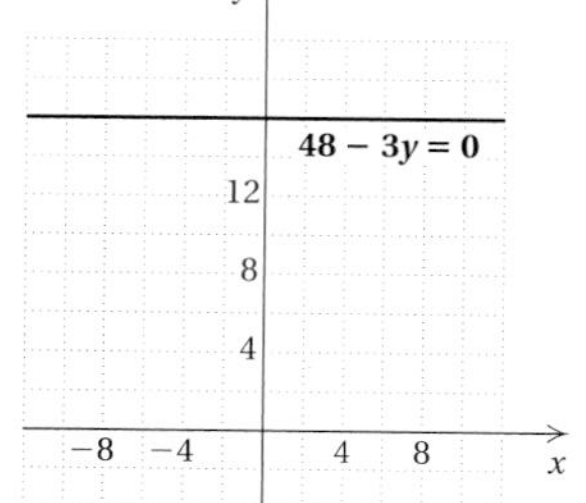

57. $y = -1$ **59.** $x = 4$ **61.** 16% **62.** 70 **63.** \$32.50 **64.** \$23.00 **65.** $\{x \mid x > -40\}$ **66.** $\{x \mid x \leq -7\}$ **67.** $\{x \mid x < 1\}$ **68.** $\{x \mid x \geq 2\}$ **69.** ◈ **71.** $x = 0$ **73.** $y = -4$ **75.** $k = 12$

Margin Exercises, Section 9.4, pp. 567–571

1. 56.7 **2.** 64.7 **3.** 87.8 **4.** 17 **5.** 16.5 **6.** 91 **7.** 55 **8.** 54, 87 **9.** No mode exists. **10. (a)** 25 mm^2; **(b)** 25 mm^2; **(c)** no mode exists. **11.** Ball A: 18.8 in.; Ball B: 19.6 in.; Ball B is better. **12.** \$866.79 **13.** 95

Calculator Spotlight, p. 572

1. (1.106383, 17.386809), (3.0425532, 21.423723), (5.6702128, 26.902394), (8.712766, 33.246117), (10.234043, 36.417979); answers may vary **2.** 223.58; 327.83 **3.** 404.98 **4. (a)**

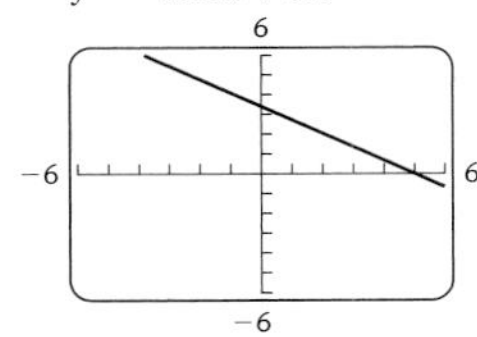

(b) no; **(c)** (5, 0); **(d)** $0 \leq x \leq 5$; **(e)** (0.76595745, 2.8791489), (1.787234, 2.1846809), (2.2978723, 1.8374468), (2.9361702, 1.4034043), (4.212766, 0.53531915); answers may vary; **(f)** 2.79 yr; **(g)** \$2244, \$1836, \$612, \$0

Exercise Set 9.4, p. 573

1. Mean: 28.6; median: 30; modes: 15, 30 **3.** Mean: 94; median: 95; no mode exists. **5.** Mean: 32; median: 35; mode: 23 **7.** Mean: 897.2; median: 798; no mode exists. **9.** Mean: 87.7; median: 88; modes: 85, 88 **11.** Battery B **13.** 162.4 **15.** 2008 **17.** 1997: 1.39; 2000: 1.54 **19.** 1997: \$1,500,000; 2000: \$1,700,000 **21.** 84 **23.** $\frac{4}{25}$ **24.** $\frac{1}{3}$ **25.** $\frac{3}{8}$ **26.** $\frac{3}{4}$ **27.** 20% **28.** \$18 **29.** \$45.15 **30.** \$55 **31.** ◈

33.

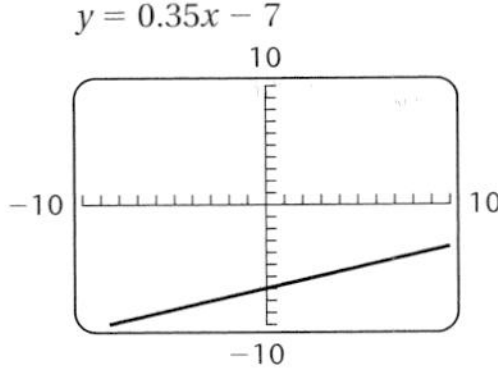

X	Y1
−10	−10.5
−9.9	−10.47
−9.8	−10.43
−9.7	−10.4
−9.6	−10.36
−9.5	−10.33
−9.4	−10.29

X = −10

35.

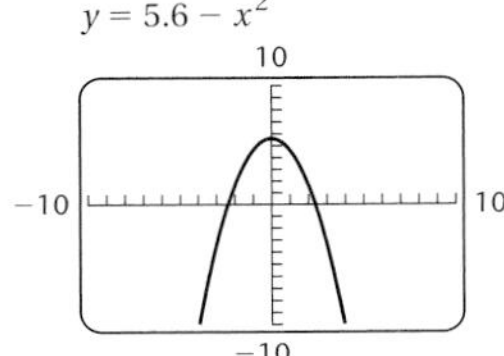

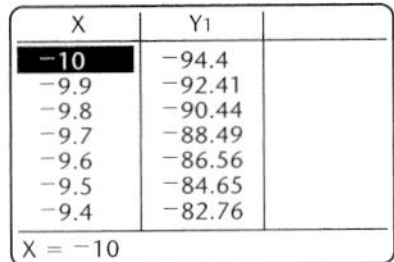

X	Y1
−10	−94.4
−9.9	−92.41
−9.8	−90.44
−9.7	−88.49
−9.6	−86.56
−9.5	−84.65
−9.4	−82.76

X = −10

Summary and Review: Chapter 9, p. 577

1. \$775.50; \$634.50 **2.** 47 lb **3.** 80 lb **4.** 33 lb **5.** 1995 **6.** 1990–1995 **7.** One shower **8.** One toilet flush **9.** One shave, wash dishes, one shower **10.** One toilet flush **11.** (−5, −1) **12.** (−2, 5) **13.** (3, 0) **14.–16.**

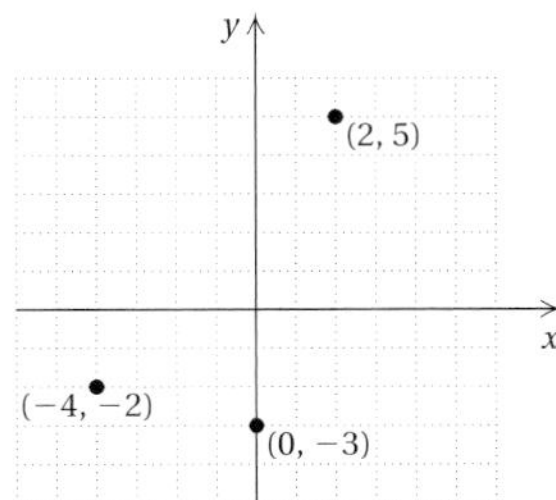

17. IV **18.** III **19.** I **20.** No **21.** Yes

22.

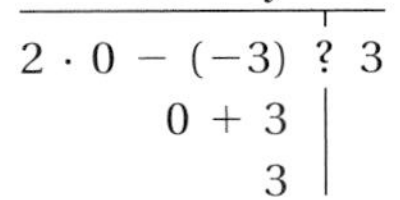

$$\begin{array}{c|c} 2x - y & 3 \\ \hline 2 \cdot 0 - (-3) & ?\ 3 \\ 0 + 3 & \\ 3 & \text{TRUE} \end{array}$$

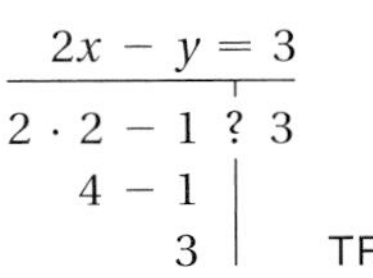

$$\begin{array}{c|c} 2x - y & 3 \\ \hline 2 \cdot 2 - 1 & ?\ 3 \\ 4 - 1 & \\ 3 & \text{TRUE} \end{array}$$

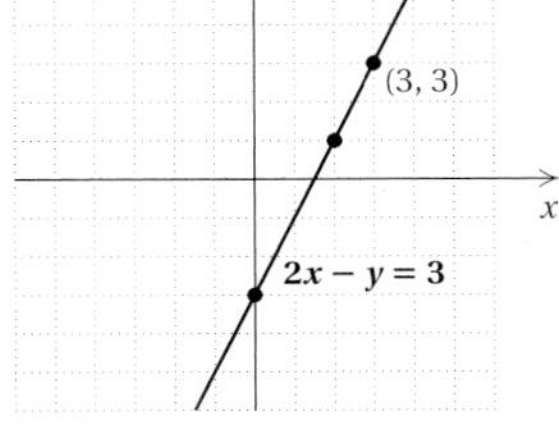

23.

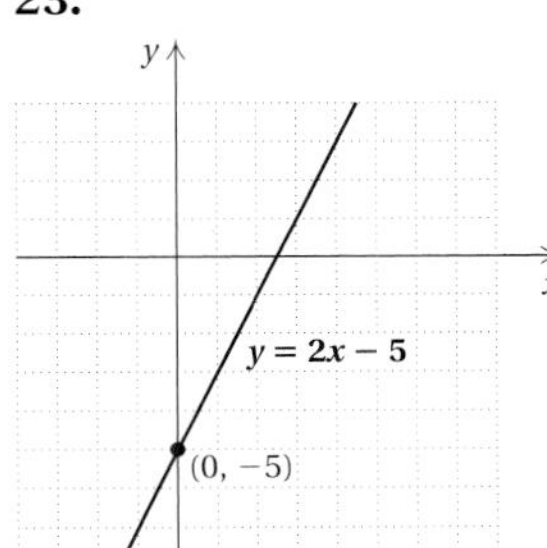

24.

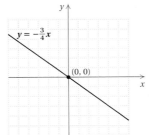

25.

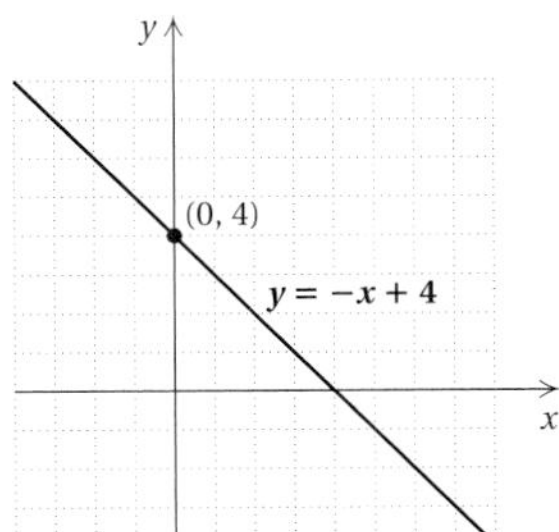

26.

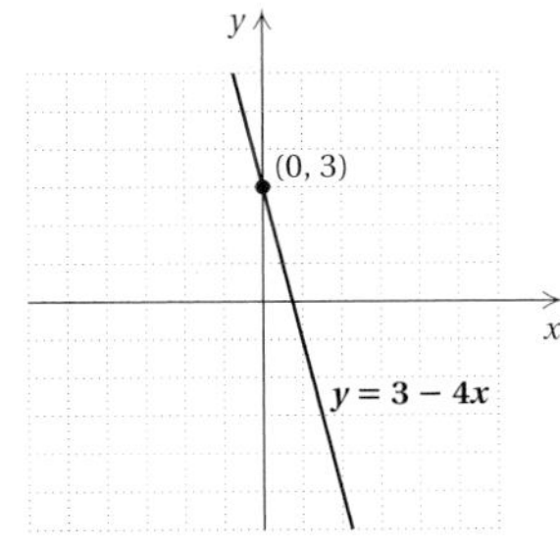

27.

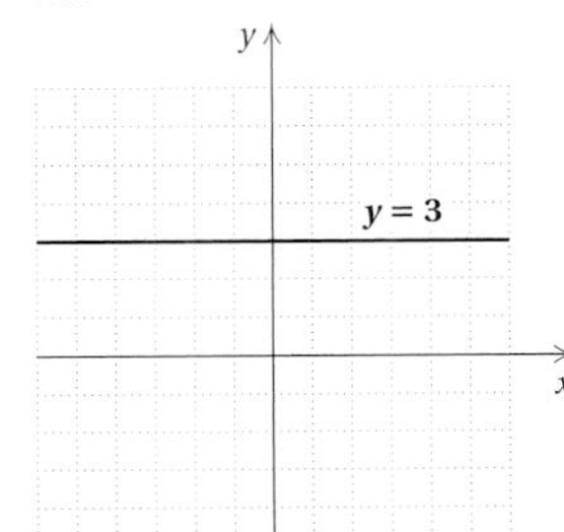

28.

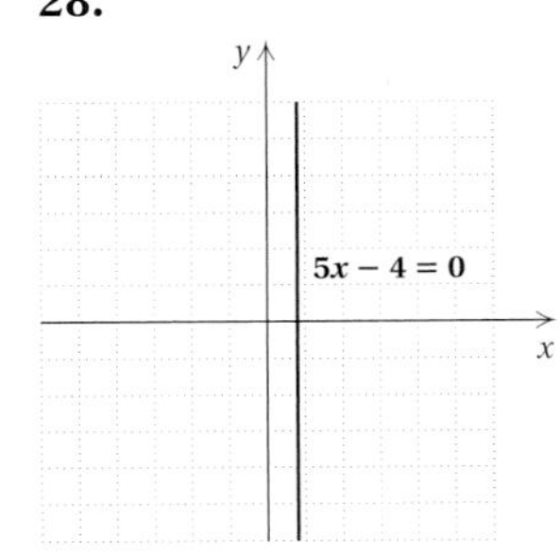

29.

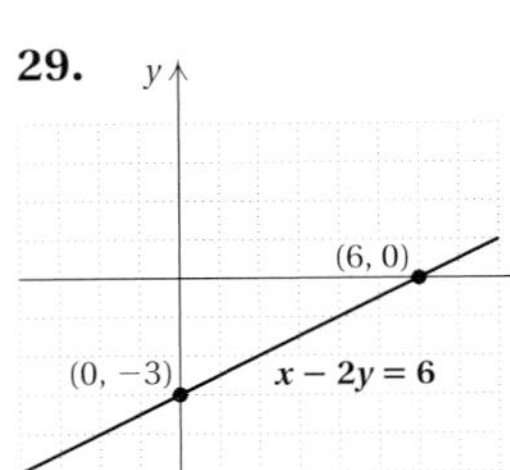

30.

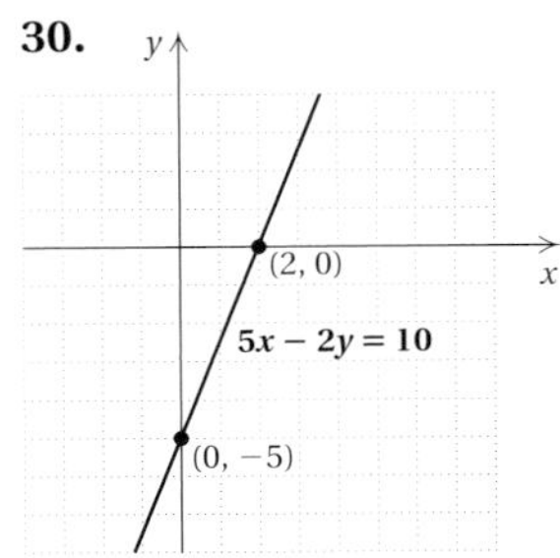

31. **(a)** $14\frac{1}{2}$ ft^3, 16 ft^3, $20\frac{1}{2}$ ft^3, 28 ft^3;
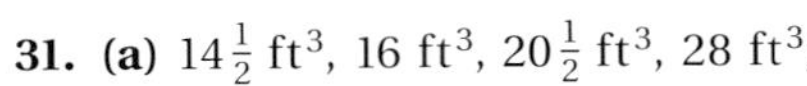
(b)

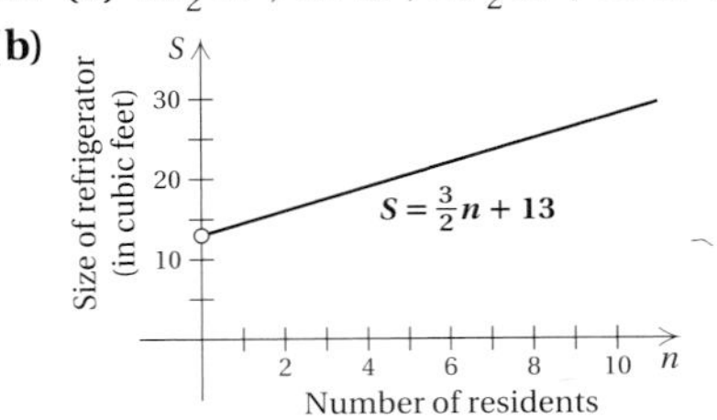

$17\frac{1}{2}$ ft^3; **(c)** 6

32. Mean: 39.5; median: 39.5; no mode exists.
33. Mean: 14.875; median: 16; modes: 12, 17, 19
34. Mean: 0.18; median: 0.175; no mode exists.
35. Mean: 229.8; median: 223; mode: 215
36. Mean: $50.1; median: $50.4; no mode exists.
37. Popcorn A **38.** 109.55 **39.** 1997: 64.4; 2000: 78.3
40. −0.34375 **41.** $0.\overline{8}$ **42.** 3.2 **43.** $\frac{17}{19}$ **44.** 42.71
45. 112.53 **46.** $9755.09 **47.** $79.95
48. ◈ A small business might use a graph to look up prices quickly (as in the FedEx mailing costs example) or to plot change in sales over a period of time. Many other applications exist. **49.** ◈ The y-intercept is the point at which the graph crosses the y-axis. Since a point on the y-axis is neither left nor right of the origin, the first or x-coordinate of the point is 0.
50. $m = -1$ **51.** 45 square units; 28 linear units

Test: Chapter 9, p. 581

1. [9.1a] $495,000,000 **2.** [9.1a] Crest and Colgate
3. [9.1a] Crest **4.** [9.1a] Arm & Hammer
5. [9.1a] June **6.** [9.1a] January **7.** [9.1a] March, April, May, June, July **8.** [9.1a] August
9. [9.1a] 1997 **10.** [9.1a] 1991 **11.** [9.1a] About $500,000 **12.** [9.1a] 1996–1997
13. [9.1a] 1994–1995 **14.** [9.1a] About $500,000
15. [9.1c] II **16.** [9.1c] III **17.** [9.1d] (3, 4)
18. [9.1d] (0, −4)
19. [9.2a]

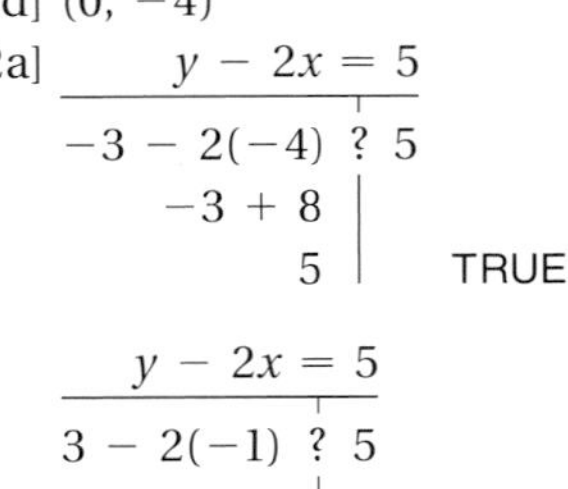

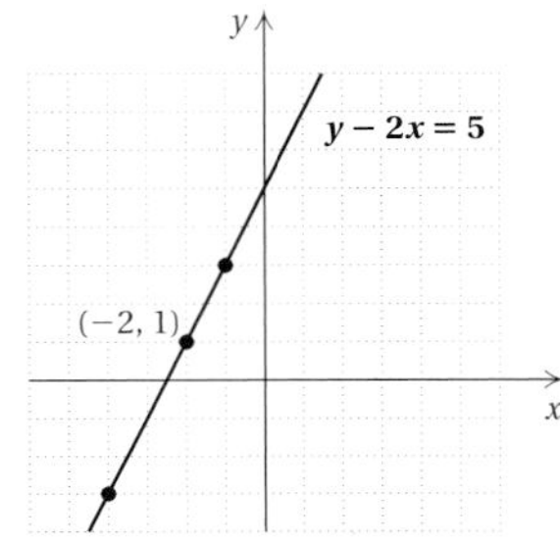

20. [9.2b]

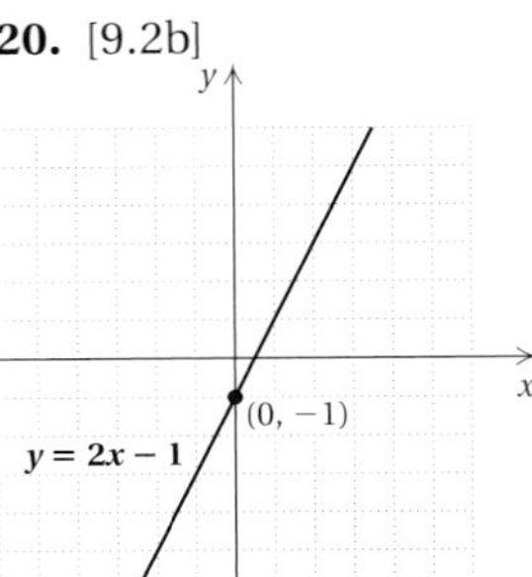

21. [9.2b]

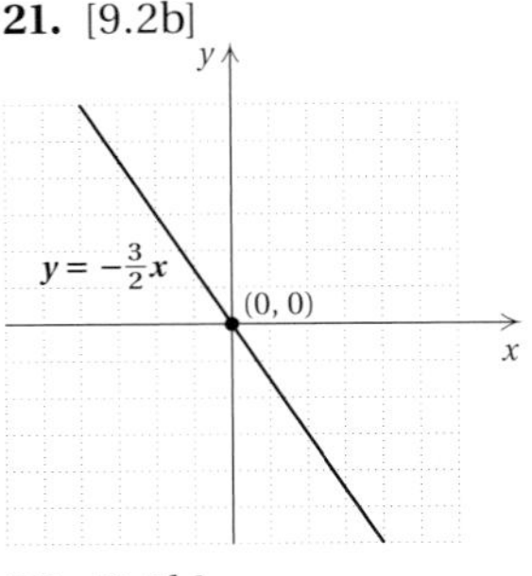

22. [9.3b]

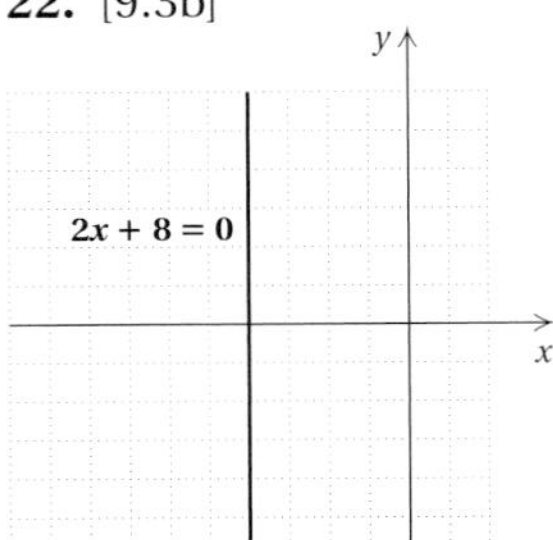

23. [9.3b]

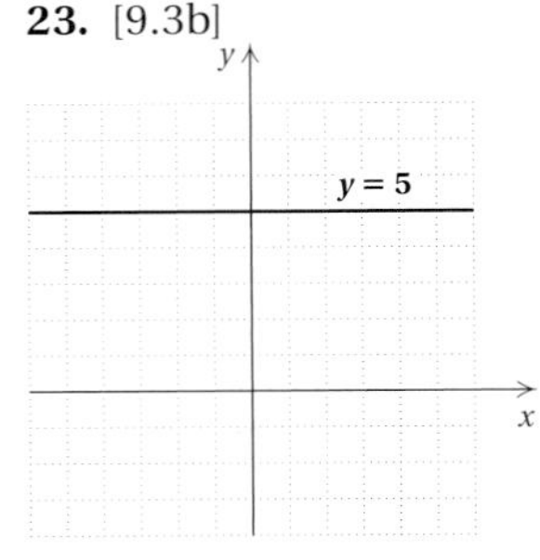

24. [9.3a]

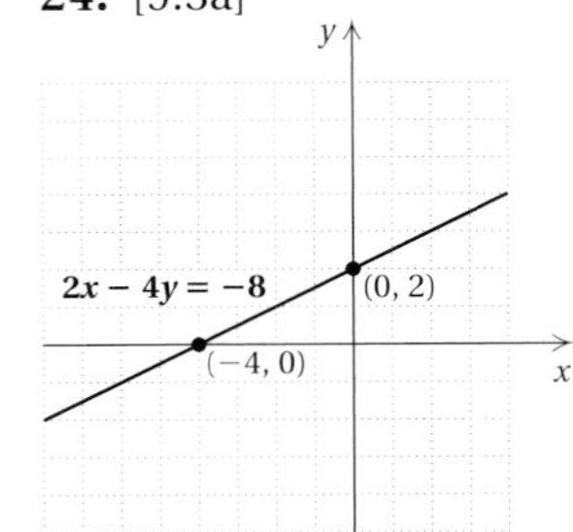

25. [9.3a]

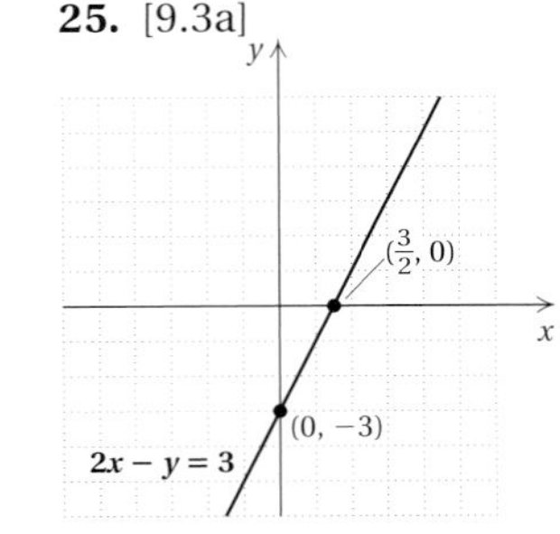

26. [9.2c] **(a)** $17,000; $19,400; $22,600; $24,200; $27,500; **(c)** 2002
(b)

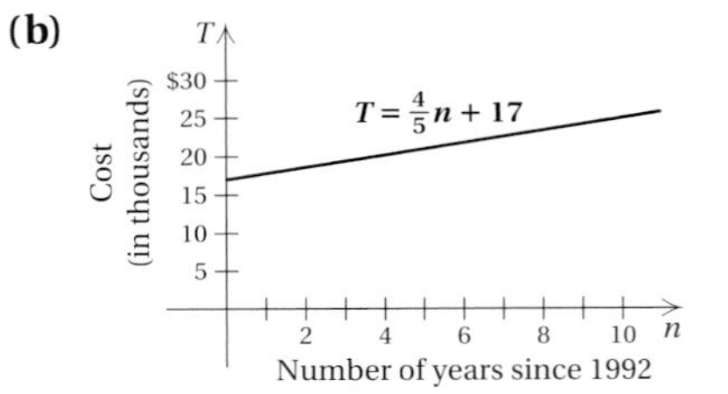

27. [9.4a] Mean: 51; median: 51.5; no mode exists.
28. [9.4a] Mean: 4.2; median: 4.5; mode: 5
29. [9.4a] Mean: 16.3; median: 18.5; modes: 18, 19
30. [9.4a] Mean: 45.4; median: 40; no mode exists.
31. [9.4b] Ball B **32.** [9.4c] 135.15 **33.** [9.4c] 1996: 1785; 2000: 2093 **34.** [7.2c] 0.975 **35.** [7.2c] $-1.08\overline{3}$
36. [7.2e] 71.2 **37.** [7.2e] $\frac{13}{47}$ **38.** [3.1e] 42.705
39. [3.1e] 112.527 **40.** [8.5a] $84.50
41. [8.5a] $36,400 **42.** [9.1b] 25 square units, 20 linear units **43.** [9.3b] $y = 3$

Chapter 10

Pretest: Chapter 10, p. 586

1. [10.1d, f] x^2 **2.** [10.1e, f] $\frac{1}{x^7}$ **3.** [10.2b] $\frac{16x^4}{y^6}$
4. [10.1f] $\frac{1}{p^3}$ **5.** [10.2c] 3.47×10^{-4}

6. [10.2c] 3,400,000 **7.** [10.3g] 3, 2, 1, 0; 3
8. [10.3e] $-3a^3b - 2a^2b^2 + ab^3 + 12b^3 + 9$
9. [10.4a] $11x^2 + 4x - 11$ **10.** [10.4c] $-x^2 - 18x + 27$
11. [10.5b] $15x^4 - 20x^3 + 5x^2$
12. [10.6c] $x^2 + 10x + 25$
13. [10.6b] $x^2 - 25$ **14.** [10.6a] $4x^6 + 19x^3 - 30$
15. [10.7f] $4x^2 - 12xy + 9y^2$
16. [10.8b] $x^2 + x + 3$, R 8; or $x^2 + x + 3 + \frac{8}{x-2}$

Margin Exercises, Section 10.1, pp. 587–592

1. $5 \cdot 5 \cdot 5 \cdot 5$ **2.** $x \cdot x \cdot x \cdot x \cdot x$ **3.** $3t \cdot 3t$
4. $3 \cdot t \cdot t$ **5.** 6 **6.** 1 **7.** 8.4 **8.** 1 **9.** 125
10. 3215.36 cm^2 **11.** 119 **12.** 3; −3 **13.** **(a)** 144; **(b)** 36; **(c)** no **14.** 3^{10} **15.** x^{10} **16.** p^{24} **17.** x^5
18. a^9b^8 **19.** 4^3 **20.** y^4 **21.** p^9 **22.** a^4b^2
23. $\frac{1}{4^3} = \frac{1}{64}$ **24.** $\frac{1}{5^2} = \frac{1}{25}$ **25.** $\frac{1}{2^4} = \frac{1}{16}$
26. $\frac{1}{(-2)^3} = -\frac{1}{8}$ **27.** $\frac{4}{p^3}$ **28.** x^2 **29.** 5^2 **30.** $\frac{1}{x^7}$
31. $\frac{1}{7^5}$ **32.** b **33.** t^6

Calculator Spotlight, p. 593

1. Yes **2.** No **3.** No **4.** Yes **5.** Yes **6.** Yes
7. No **8.** Yes **9.** Yes **10.** Yes **11.** No **12.** Yes
13. Yes **14.** No

Exercise Set 10.1, p. 595

1. $3 \cdot 3 \cdot 3 \cdot 3$ **3.** (1.1)(1.1)(1.1)(1.1)(1.1) **5.** $\left(\frac{2}{3}\right)\left(\frac{2}{3}\right)\left(\frac{2}{3}\right)\left(\frac{2}{3}\right)$
7. $(7p)(7p)$ **9.** $8 \cdot k \cdot k \cdot k$ **11.** 1 **13.** b **15.** 1
17. 1 **19.** ab **21.** ab **23.** 27 **25.** 19 **27.** 256
29. 93 **31.** 10; 4 **33.** 3629.84 ft^2 **35.** $\frac{1}{3^2} = \frac{1}{9}$
37. $\frac{1}{10^3} = \frac{1}{1000}$ **39.** $\frac{1}{7^3} = \frac{1}{343}$ **41.** $\frac{1}{a^3}$ **43.** $8^2 = 64$
45. y^4 **47.** z^n **49.** 4^{-3} **51.** x^{-3} **53.** a^{-5}
55. 2^7 **57.** 8^{14} **59.** x^7 **61.** 9^{38} **63.** $(3y)^{12}$
65. $(7y)^{17}$ **67.** 3^3 **69.** $\frac{1}{x}$ **71.** x^{17} **73.** $\frac{1}{x^{13}}$
75. $\frac{1}{a^{10}}$ **77.** 1 **79.** 7^3 **81.** 8^6 **83.** y^4 **85.** $\frac{1}{16^6}$
87. $\frac{1}{m^6}$ **89.** $\frac{1}{(8x)^4}$ **91.** 1 **93.** x^2 **95.** x^9 **97.** $\frac{1}{z^4}$
99. x^3 **101.** 1 **103.** 25, $\frac{1}{25}$, $\frac{1}{25}$, 25, −25, 25
105. 64%t, or $0.64t$ **106.** 1 **107.** 64 **108.** 1579.5
109. $\frac{4}{3}$ **110.** $8(x - 7)$ **111.** 8 in., 4 in. **112.** 228, 229
113. ◆ **115.** No **117.** No **119.** y^{5x} **121.** a^{4t}
123. 1 **125.** > **127.** < **129.** Let $x = 2$; then $3x^2 = 12$, but $(3x)^2 = 36$.

Margin Exercises, Section 10.2, pp. 599–604

1. 3^{20} **2.** $\frac{1}{x^{12}}$ **3.** y^{15} **4.** $\frac{1}{x^{32}}$ **5.** $\frac{16x^{20}}{y^{12}}$ **6.** $\frac{25x^{10}}{y^{12}z^6}$
7. x^{74} **8.** $\frac{27z^{24}}{y^6x^{15}}$ **9.** $\frac{x^{12}}{25}$ **10.** $\frac{8t^{15}}{w^{12}}$ **11.** $\frac{9}{x^8}$
12. 5.17×10^{-4} **13.** 5.23×10^8 **14.** 689,300,000,000
15. 0.0000567 **16.** 5.6×10^{-15} **17.** 7.462×10^{-13}
18. 2.0×10^3 **19.** 5.5×10^2 **20.** 1.884672×10^{11}
21. 9.5×10

Calculator Spotlight, p. 599

1. Yes **2.** No

Calculator Spotlight, p. 600

1. Yes **2.** Yes **3.** No **4.** No

Calculator Spotlight, p. 602

1. 2.6 E 8 **2.** 6.709 E −11

Exercise Set 10.2, p. 605

1. 2^6 **3.** $\frac{1}{5^6}$ **5.** x^{12} **7.** $16x^6$ **9.** $\frac{1}{x^{12}y^{15}}$ **11.** $x^{24}y^8$
13. $\frac{9x^6}{y^{16}z^6}$ **15.** $\frac{a^8}{b^{12}}$ **17.** $\frac{y^6}{4}$ **19.** $\frac{8}{y^6}$ **21.** $\frac{x^6y^3}{z^3}$
23. $\frac{c^2d^6}{a^4b^2}$ **25.** 2.8×10^{10} **27.** 9.07×10^{17}
29. 3.04×10^{-6} **31.** 1.8×10^{-8} **33.** 10^{11}
35. 1.135×10^7 **37.** 87,400,000 **39.** 0.00000005704
41. 10,000,000 **43.** 0.00001 **45.** 6×10^9
47. 3.38×10^4 **49.** 8.1477×10^{-13} **51.** 2.5×10^{13}
53. 5.0×10^{-4} **55.** 3.0×10^{-21} **57.** \$$1.32288 \times 10^{12}$
59. 1×10^{22} **61.** 3.3×10^5 **63.** 4.375×10^2 days
65. $9(x - 4)$ **66.** $2(2x - y + 8)$ **67.** $3(s + t + 8)$
68. $-7(x + 2)$ **69.** $\frac{7}{4}$ **70.** 2 **71.** $-\frac{12}{7}$ **72.** $-\frac{11}{2}$
73.

y = x − 5

74.

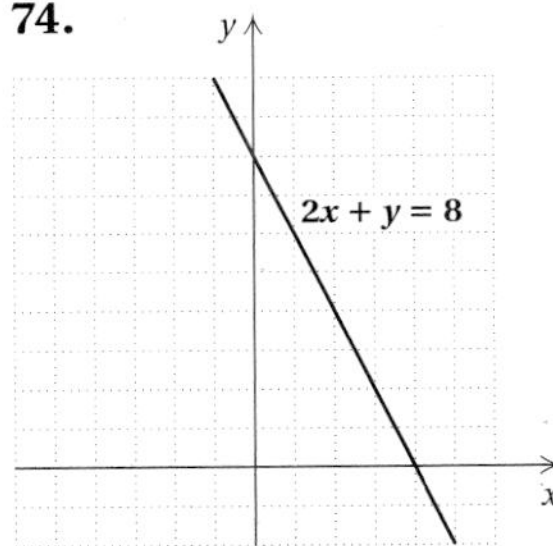

75. ◆ **77.** 2.478125×10^{-1} **79.** $\frac{1}{5}$ **81.** 3^{11}
83. a^n **85.** False **87.** False

Margin Exercises, Section 10.3, pp. 609–616

1. $4x^2 - 3x + \frac{5}{4}$; $15y^3$; $-7x^3 + 1.1$; answers may vary
2. −19 **3.** −104 **4.** −13 **5.** 8 **6.** 132 **7.** 360 ft
8. 6; −4 **9.** 7.55 parts per million **10.** 20
11. $-9x^3 + (-4x^5)$ **12.** $-2y^3 + 3y^7 + (-7y)$ **13.** $3x^2$, $6x$, $\frac{1}{2}$ **14.** $-4y^5$, $7y^2$, $-3y$, -2 **15.** $4x^3$ and $-x^3$
16. $4t^4$ and $-7t^4$; $-9t^3$ and $10t^3$ **17.** $5x^2$ and $7x^2$; $3x$ and $-8x$; −10 and 11 **18.** 2, −7, −8.5, 10, −4
19. $8x^2$ **20.** $2x^3 + 7$ **21.** $-\frac{1}{4}x^5 + 2x^2$ **22.** $-4x^3$
23. $5x^3$ **24.** $25 - 3x^5$ **25.** $6x$ **26.** $4x^3 + 4$

27. $-\frac{1}{4}x^3 + 4x^2 + 7$ **28.** $3x^2 + x^3 + 9$
29. $6x^7 + 3x^5 - 2x^4 + 4x^3 + 5x^2 + x$
30. $7x^5 - 5x^4 + 2x^3 + 4x^2 - 3$
31. $14t^7 - 10t^5 + 7t^2 - 14$ **32.** $-2x^2 - 3x + 2$
33. $10x^4 - 8x - \frac{1}{2}$ **34.** 4, 2, 1, 0; 4 **35.** x
36. x^3, x^2, x, x^0 **37.** x^2, x **38.** x^3 **39.** Monomial
40. None of these **41.** Binomial **42.** Trinomial

Calculator Spotlight, p. 612

1. 3, 1.81, −6.99 **2.** 2, 17, 21.55, 8.552
3. 13, 2, −8, −3.32, 7

Exercise Set 10.3, p. 617

1. −18, 7 **3.** 19, 14 **5.** −12, −7 **7.** −1, 5 **9.** 9, 1
11. 56, −2 **13.** 1112 ft **15.** $18,750; $24,000
17. −4, 4, 5, 2.75, 1 **19.** 66.6 ft, 86.6 ft, 66.6 ft, 41.6 ft
21. $2, -3x, x^2$ **23.** $6x^2$ and $-3x^2$ **25.** $2x^4$ and $-3x^4$; $5x$ and $-7x$ **27.** $3x^5$ and $14x^5$; $-7x$ and $-2x$; 8 and −9
29. −3, 6 **31.** 5, 3, 3 **33.** −5, 6, −3, 8, −2
35. $-3x$ **37.** $-8x$ **39.** $11x^3 + 4$ **41.** $x^3 - x$
43. $4b^5$ **45.** $\frac{3}{4}x^5 - 2x - 42$ **47.** x^4 **49.** $\frac{15}{16}x^3 - \frac{7}{6}x^2$
51. $x^5 + 6x^3 + 2x^2 + x + 1$
53. $15y^9 + 7y^8 + 5y^3 - y^2 + y$ **55.** $x^6 + x^4$
57. $13x^3 - 9x + 8$ **59.** $-5x^2 + 9x$ **61.** $12x^4 - 2x + \frac{1}{4}$
63. 1, 0; 1 **65.** 2, 1, 0; 2 **67.** 3, 2, 1, 0; 3
69. 2, 1, 6, 4; 6

71.

Term	Coefficient	Degree of Term	Degree of Polynomial
$-7x^4$	−7	4	4
$6x^3$	6	3	
$-3x^2$	−3	2	
$8x$	8	1	
−2	−2	0	

73. x^2, x **75.** x^3, x^2, x^0 **77.** None missing
79. Trinomial **81.** None of these **83.** Binomial
85. Monomial **87.** 27 **88.** −19 **89.** $-\frac{17}{24}$
90. $\frac{5}{8}$ **91.** −2.6 **92.** $\frac{15}{2}$ **93.** $b = \dfrac{cx + r}{a}$ **94.** 45%, 37.5%, 17.5% **95.** $3(x - 5y + 21)$ **97.** ◆ **99.** $3x^6$
101. 10 **103.** −4, 4, 5, 2.75, 1 **105.** 66.6 ft, 86.6 ft, 66.6 ft, 41.6 ft

Margin Exercises, Section 10.4, pp. 623–626

1. $x^2 + 7x + 3$ **2.** $-4x^5 + 7x^4 + x^3 + 2x^2 + 4$
3. $24x^4 + 5x^3 + x^2 + 1$ **4.** $2x^3 + \frac{10}{3}$ **5.** $2x^2 - 3x - 1$
6. $8x^3 - 2x^2 - 8x + \frac{5}{2}$
7. $-8x^4 + 4x^3 + 12x^2 + 5x - 8$
8. $-x^3 + x^2 + 3x + 3$ **9.** $-(12x^4 - 3x^2 + 4x)$; $-12x^4 + 3x^2 - 4x$ **10.** $-(-4x^4 + 3x^2 - 4x)$; $4x^4 - 3x^2 + 4x$ **11.** $-\left(-13x^6 + 2x^4 - 3x^2 + x - \frac{5}{13}\right)$; $13x^6 - 2x^4 + 3x^2 - x + \frac{5}{13}$
12. $-(-7y^3 + 2y^2 - y + 3)$; $7y^3 - 2y^2 + y - 3$
13. $-4x^3 + 6x - 3$ **14.** $-5x^4 - 3x^2 - 7x + 5$
15. $-14x^{10} + \frac{1}{2}x^5 - 5x^3 + x^2 - 3x$ **16.** $2x^3 + 2x + 8$
17. $x^2 - 6x - 2$ **18.** $-8x^4 - 5x^3 + 8x^2 - 1$
19. $x^3 - x^2 - \frac{4}{3}x - 0.9$ **20.** $2x^3 + 5x^2 - 2x - 5$
21. $-x^5 - 2x^3 + 3x^2 - 2x + 2$
22. Sum of perimeters: $13x$; sum of areas: $\frac{7}{2}x^2$
23. $\pi x^2 - 2x^2$, or $(\pi - 2)x^2$

Exercise Set 10.4, p. 627

1. $-x + 5$ **3.** $x^2 - 5x - 1$ **5.** $2x^2$
7. $5x^2 + 3x - 30$ **9.** $-2.2x^3 - 0.2x^2 - 3.8x + 23$
11. $12x^2 + 6$ **13.** $-\frac{1}{2}x^4 + \frac{2}{3}x^3 + x^2$
15. $0.01x^5 + x^4 - 0.2x^3 + 0.2x + 0.06$
17. $9x^8 + 8x^7 - 6x^4 + 8x^2 + 4$
19. $1.05x^4 + 0.36x^3 + 14.22x^2 + x + 0.97$
21. $-(-5x)$; $5x$ **23.** $-(-x^2 + 10x - 2)$; $x^2 - 10x + 2$
25. $-(12x^4 - 3x^3 + 3)$; $-12x^4 + 3x^3 - 3$ **27.** $-3x + 7$
29. $-4x^2 + 3x - 2$ **31.** $4x^4 - 6x^2 - \frac{3}{4}x + 8$
33. $7x - 1$ **35.** $-x^2 - 7x + 5$ **37.** −18
39. $6x^4 + 3x^3 - 4x^2 + 3x - 4$
41. $4.6x^3 + 9.2x^2 - 3.8x - 23$ **43.** $\frac{3}{4}x^3 - \frac{1}{2}x$
45. $0.06x^3 - 0.05x^2 + 0.01x + 1$ **47.** $3x + 6$
49. $11x^4 + 12x^3 - 9x^2 - 8x - 9$ **51.** $x^4 - x^3 + x^2 - x$
53. $5x^2 + 4x$ **55.** $\frac{23}{2}a + 10$ **57.** 6 **58.** −19
59. $-\frac{7}{22}$ **60.** 5 **61.** 5 **62.** 1 **63.** $\frac{39}{2}$ **64.** $\frac{37}{2}$
65. $\{x \mid x \geq -10\}$ **66.** $\{x \mid x < 0\}$ **67.** ◆
69. $20 + 5(m - 4) + 4(m - 5) + (m - 5)(m - 4)$; m^2
71. $z^2 - 27z + 72$ **73.** $y^2 - 4y + 4$
75. $5x^2 - 9x - 1$ **77.** $4x^3 + 2x^2 + x + 2$
79. Both columns are equal.

Margin Exercises, Section 10.5, pp. 631–634

1. $-15x$ **2.** $-x^2$ **3.** x^2 **4.** $-x^5$ **5.** $12x^7$
6. $-8y^{11}$ **7.** $7y^5$ **8.** 0 **9.** $8x^2 + 16x$
10. $-15t^3 + 6t^2$ **11.** $5x^6 + 25x^5 - 30x^4 + 40x^3$
12. $x^2 + 13x + 40$ **13.** $x^2 + x - 20$
14. $5x^2 - 17x - 12$ **15.** $6x^2 - 19x + 15$
16. $x^4 + 3x^3 + x^2 + 15x - 20$
17. $6y^5 - 20y^3 + 15y^2 + 14y - 35$
18. $3x^3 + 13x^2 - 6x + 20$
19. $20x^4 - 16x^3 + 32x^2 - 32x - 16$
20. $6x^4 - x^3 - 18x^2 - x + 10$

Calculator Spotlight, p. 634

1. Yes **2.** Yes **3.** No **4.** No **5.** No **6.** Yes

Exercise Set 10.5, p. 635

1. $40x^2$ **3.** x^3 **5.** $32x^8$ **7.** $0.03x^{11}$ **9.** $\frac{1}{15}x^4$
11. 0 **13.** $-24x^{11}$ **15.** $-2x^2 + 10x$ **17.** $-5x^2 + 5x$
19. $x^5 + x^2$ **21.** $6x^3 - 18x^2 + 3x$ **23.** $-6x^4 - 6x^3$
25. $18y^6 + 24y^5$ **27.** $x^2 + 9x + 18$ **29.** $x^2 + 3x - 10$
31. $x^2 - 7x + 12$ **33.** $x^2 - 9$ **35.** $25 - 15x + 2x^2$
37. $4x^2 + 20x + 25$ **39.** $x^2 - \frac{21}{10}x - 1$

41. $x^2 + 2.4x - 10.81$ **43.** $x^3 - 1$
45. $4x^3 + 14x^2 + 8x + 1$
47. $3y^4 - 6y^3 - 7y^2 + 18y - 6$ **49.** $x^6 + 2x^5 - x^3$
51. $-10x^5 - 9x^4 + 7x^3 + 2x^2 - x$
53. $x^4 - x^2 - 2x - 1$ **55.** $6t^4 + t^3 - 16t^2 - 7t + 4$
57. $x^9 - x^5 + 2x^3 - x$ **59.** $x^4 - 1$ **61.** $-\frac{3}{4}$ **62.** 6.4
63. 96 **64.** 32 **65.** $3(5x - 6y + 4)$
66. $4(4x - 6y + 9)$ **67.** $-3(3x + 15y - 5)$
68. $100(x - y + 10a)$
69.

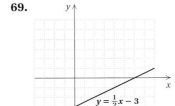

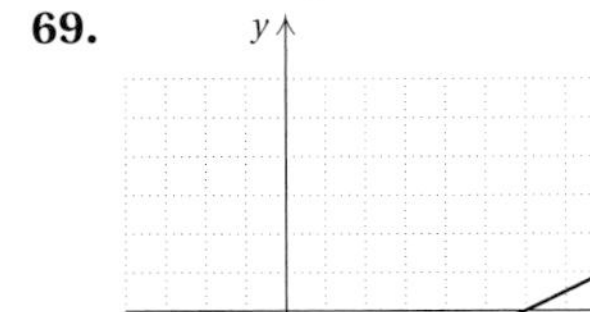

70. $\frac{23}{19}$ **71.** ◆
73. $78t^2 + 40t$
75. $A = \frac{1}{2}b^2 + 2b$
77. 0

Margin Exercises, Section 10.6, pp. 638–642

1. $x^2 + 7x + 12$ **2.** $x^2 - 2x - 15$ **3.** $2x^2 - 9x + 4$
4. $2x^3 - 4x^2 - 3x + 6$ **5.** $12x^5 + 10x^3 + 6x^2 + 5$
6. $y^6 - 49$ **7.** $t^2 + 8t + 15$ **8.** $-2x^7 + x^5 + x^3$
9. $x^2 - \frac{16}{25}$ **10.** $x^5 + 0.5x^3 - 0.5x^2 + 0.25$
11. $8 + 2x^2 - 15x^4$ **12.** $30x^5 - 27x^4 + 6x^3$
13. $x^2 - 25$ **14.** $4x^2 - 9$ **15.** $x^2 - 4$ **16.** $x^2 - 49$
17. $36 - 16y^2$ **18.** $4x^6 - 1$ **19.** $x^2 - \frac{4}{25}$
20. $x^2 + 16x + 64$ **21.** $x^2 - 10x + 25$
22. $x^2 + 4x + 4$ **23.** $a^2 - 8a + 16$
24. $4x^2 + 20x + 25$ **25.** $16x^4 - 24x^3 + 9x^2$
26. $60.84 + 18.72y + 1.44y^2$ **27.** $9x^4 - 30x^2 + 25$
28. $x^2 + 11x + 30$ **29.** $t^2 - 16$
30. $-8x^5 + 20x^4 + 40x^2$ **31.** $81x^4 + 18x^2 + 1$
32. $4a^2 + 6a - 40$ **33.** $25x^2 + 5x + \frac{1}{4}$
34. $4x^2 - 2x + \frac{1}{4}$ **35.** $x^3 - 3x^2 + 6x - 8$

Exercise Set 10.6, p. 643

1. $x^3 + x^2 + 3x + 3$ **3.** $x^4 + x^3 + 2x + 2$
5. $y^2 - y - 6$ **7.** $9x^2 + 12x + 4$ **9.** $5x^2 + 4x - 12$
11. $9t^2 - 1$ **13.** $4x^2 - 6x + 2$ **15.** $p^2 - \frac{1}{16}$
17. $x^2 - 0.01$ **19.** $2x^3 + 2x^2 + 6x + 6$
21. $-2x^2 - 11x + 6$ **23.** $a^2 + 14a + 49$
25. $1 - x - 6x^2$ **27.** $x^5 + 3x^3 - x^2 - 3$
29. $3x^6 - 2x^4 - 6x^2 + 4$ **31.** $13.16x^2 + 18.99x - 13.95$
33. $6x^7 + 18x^5 + 4x^2 + 12$ **35.** $8x^6 + 65x^3 + 8$
37. $4x^3 - 12x^2 + 3x - 9$ **39.** $4y^6 + 4y^5 + y^4 + y^3$
41. $x^2 - 16$ **43.** $4x^2 - 1$ **45.** $25m^2 - 4$
47. $4x^4 - 9$ **49.** $9x^8 - 16$ **51.** $x^{12} - x^4$
53. $x^8 - 9x^2$ **55.** $x^{24} - 9$ **57.** $4y^{16} - 9$
59. $\frac{25}{64}x^2 - 18.49$ **61.** $x^2 + 4x + 4$
63. $9x^4 + 6x^2 + 1$ **65.** $a^2 - a + \frac{1}{4}$ **67.** $9 + 6x + x^2$
69. $x^4 + 2x^2 + 1$ **71.** $4 - 12x^4 + 9x^8$
73. $25 + 60t^2 + 36t^4$ **75.** $x^2 - \frac{5}{4}x + \frac{25}{64}$
77. $9 - 12x^3 + 4x^6$ **79.** $4x^3 + 24x^2 - 12x$
81. $4x^4 - 2x^2 + \frac{1}{4}$ **83.** $9p^2 - 1$ **85.** $15t^5 - 3t^4 + 3t^3$
87. $36x^8 + 48x^4 + 16$ **89.** $12x^3 + 8x^2 + 15x + 10$
91. $64 - 96x^4 + 36x^8$ **93.** $t^3 - 1$ **95.** 25; 49
97. 56; 16 **99.** Lamps: 500 watts; air conditioner: 2000 watts; television: 50 watts **100.** $\frac{28}{27}$ **101.** $-\frac{41}{7}$
102. $\frac{27}{4}$ **103.** $y = \frac{3x - 12}{2}$, or $y = \frac{3}{2}x - 6$
104. $a = \frac{4 + cd}{b}$ **105.** ◆ **107.** $30x^3 + 35x^2 - 15x$
109. $a^4 - 50a^2 + 625$ **111.** $81t^{16} - 72t^8 + 16$
113. -7 **115.** First row: 90, -432, -63; second row: 7, -18, -36, -14, 12, -6, -21, -11; third row: 9, -2, -2, 10, -8, -8, -8, -10, 21; fourth row: -19, -6
117. $9x^2 + 24x + 16$ **119.** Yes **121.** No

Margin Exercises, Section 10.7, pp. 647–650

1. -7940 **2.** -176 **3.** 1889 **4.** $-3, 3, -2, 1, 2$
5. 3, 7, 1, 1, 0; 7 **6.** $2x^2y + 3xy$ **7.** $5pq - 8$
8. $-4x^3 + 2x^2 - 4y + 2$ **9.** $14x^3y + 7x^2y - 3xy - 2y$
10. $-5p^2q^4 + 2p^2q^2 + 3p^2q + 6pq^2 + 3q + 5$
11. $-8s^4t + 6s^3t^2 + 2s^2t^3 - s^2t^2$
12. $-9p^4q + 9p^3q^2 - 4p^2q^3 - 9q^4 + 5$
13. $x^5y^5 + 2x^4y^2 + 3x^3y^3 + 6x^2$
14. $p^5q - 4p^3q^3 + 3pq^3 + 6q^4$
15. $3x^3y + 6x^2y^3 + 2x^3 + 4x^2y^2$
16. $2x^2 - 11xy + 15y^2$ **17.** $16x^2 + 40xy + 25y^2$
18. $9x^4 - 12x^3y^2 + 4x^2y^4$ **19.** $4x^2y^4 - 9x^2$
20. $16y^2 - 9x^2y^4$ **21.** $9y^2 + 24y + 16 - 9x^2$
22. $4a^2 - 25b^2 - 10bc - c^2$

Exercise Set 10.7, p. 651

1. -1 **3.** -15 **5.** 240 **7.** -145 **9.** 3.715 liters
11. 110.4 m **13.** 56.52 in^2 **15.** Coefficients: 1, -2, 3, -5; degrees: 4, 2, 2, 0; 4 **17.** Coefficients: 17, -3, -7; degrees: 5, 5, 0; 5 **19.** $-a - 2b$
21. $3x^2y - 2xy^2 + x^2$ **23.** $20au + 10av$
25. $8u^2v - 5uv^2$ **27.** $x^2 - 4xy + 3y^2$ **29.** $3r + 7$
31. $-a^3b^2 - 3a^2b^3 + 5ab + 3$ **33.** $ab^2 - a^2b$
35. $2ab - 2$ **37.** $-2a + 10b - 5c + 8d$
39. $6z^2 + 7zu - 3u^2$ **41.** $a^4b^2 - 7a^2b + 10$
43. $a^6 - b^2c^2$ **45.** $y^6x + y^4x + y^4 + 2y^2 + 1$
47. $12x^2y^2 + 2xy - 2$ **49.** $12 - c^2d^2 - c^4d^4$
51. $m^3 + m^2n - mn^2 - n^3$
53. $x^9y^9 - x^6y^6 + x^5y^5 - x^2y^2$ **55.** $x^2 + 2xh + h^2$
57. $r^6t^4 - 8r^3t^2 + 16$ **59.** $p^8 + 2m^2n^2p^4 + m^4n^4$
61. $4a^6 - 2a^3b^3 + \frac{1}{4}b^6$ **63.** $3a^3 - 12a^2b + 12ab^2$
65. $4a^2 - b^2$ **67.** $c^4 - d^2$ **69.** $a^2b^2 - c^2d^4$
71. $x^2 + 2xy + y^2 - 9$ **73.** $x^2 - y^2 - 2yz - z^2$
75. $a^2 - b^2 - 2bc - c^2$ **77.** IV **78.** III **79.** I
80. II

81.

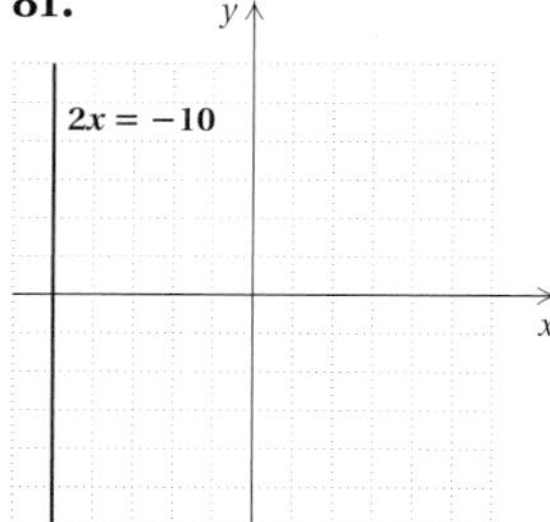

82.

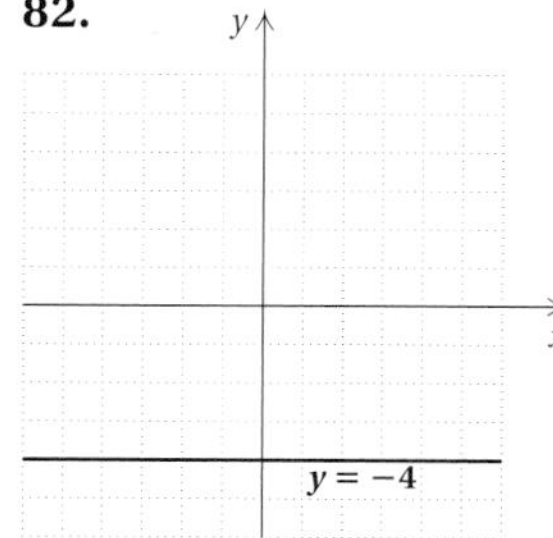

83.

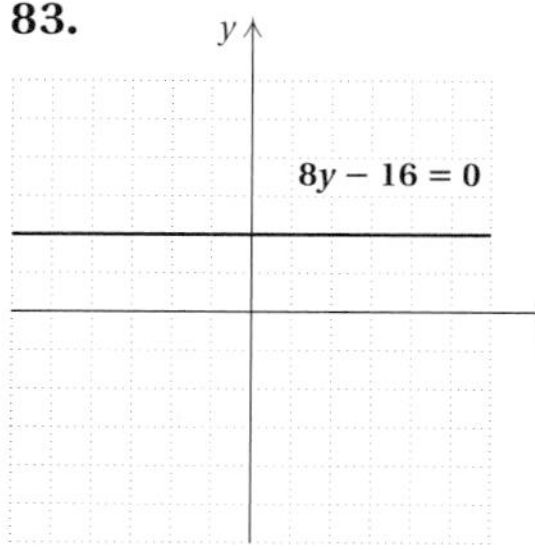

84.

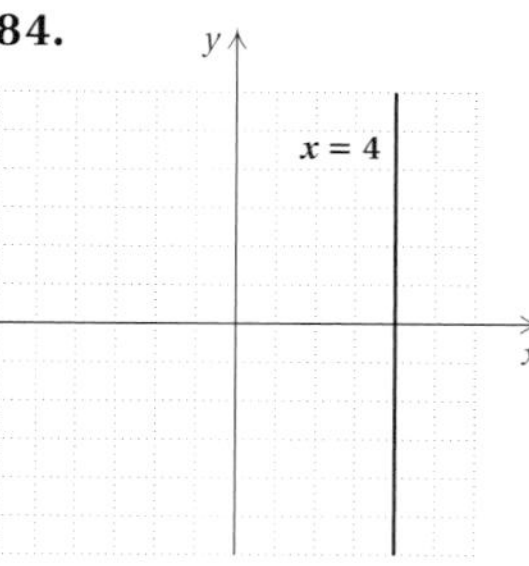

85. Mean: 27.57; median: 28; mode: 31
86. Mean: 5.69; median: 5.6; modes: 5.2, 5.6 **87.** ◈
89. $4xy - 4y^2$ **91.** $2xy + \pi x^2$
93. $2\pi nh + 2\pi mh + 2\pi n^2 - 2\pi m^2$

Margin Exercises, Section 10.8, pp. 655–658

1. $4x^2$ **2.** $-7x^{11}$ **3.** $-28p^3q$ **4.** $\frac{1}{4}x^4$ **5.** $7x^4 + 8x^2$
6. $x^2 + 3x + 2$ **7.** $2x^2 + x - \frac{2}{3}$ **8.** $4x^2 - \frac{3}{2}x + \frac{1}{2}$
9. $2x^2y^4 - 3xy^2 + 5y$ **10.** $x - 2$ **11.** $x + 4$
12. $x + 4$, R -2, or $x + 4 + \dfrac{-2}{x + 3}$ **13.** $x^2 + x + 1$

Exercise Set 10.8, p. 659

1. $3x^4$ **3.** $5x$ **5.** $18x^3$ **7.** $4a^3b$
9. $3x^4 - \frac{1}{2}x^3 + \frac{1}{8}x^2 - 2$ **11.** $1 - 2u - u^4$
13. $5t^2 + 8t - 2$ **15.** $-4x^4 + 4x^2 + 1$
17. $6x^2 - 10x + \frac{3}{2}$ **19.** $9x^2 - \frac{5}{2}x + 1$
21. $6x^2 + 13x + 4$ **23.** $3rs + r - 2s$ **25.** $x + 2$
27. $x - 5 + \dfrac{-50}{x - 5}$ **29.** $x - 2 + \dfrac{-2}{x + 6}$ **31.** $x - 3$
33. $x^4 - x^3 + x^2 - x + 1$ **35.** $2x^2 - 7x + 4$
37. $x^3 - 6$ **39.** $x^3 + 2x^2 + 4x + 8$ **41.** $t^2 + 1$
43. -28 **44.** -59 **45.** 6.8 **46.** $-\frac{11}{8}$
47. 25,543.75 ft^2 **48.** 51°, 27°, 102° **49.** $\frac{23}{14}$ **50.** $\frac{11}{10}$
51. $4(x - 3 + 6y)$ **52.** $2(128 - a - 2b)$ **53.** ◈
55. $x^2 + 5$ **57.** $a + 3 + \dfrac{5}{5a^2 - 7a - 2}$
59. $2x^2 + x - 3$
61. $a^5 + a^4b + a^3b^2 + a^2b^3 + ab^4 + b^5$
63. -5 **65.** 1

Summary and Review: Chapter 10, p. 661

1. $\dfrac{1}{7^2}$ **2.** y^{11} **3.** $(3x)^{14}$ **4.** t^8 **5.** 4^3 **6.** $\dfrac{1}{a^3}$ **7.** 1
8. $9t^8$ **9.** $36x^8$ **10.** $\dfrac{y^3}{8x^3}$ **11.** t^{-5} **12.** $\dfrac{1}{y^4}$
13. 3.28×10^{-5} **14.** 8,300,000 **15.** 2.09×10^4
16. 5.12×10^{-5} **17.** 4.2075×10^9 **18.** 10
19. $-4y^5, 7y^2, -3y, -2$ **20.** x^2, x^0 **21.** 3, 2, 1, 0; 3
22. Binomial **23.** None of these **24.** Monomial
25. $-2x^2 - 3x + 2$ **26.** $10x^4 - 7x^2 - x - \frac{1}{2}$
27. $x^5 - 2x^4 + 6x^3 + 3x^2 - 9$
28. $-2x^5 - 6x^4 - 2x^3 - 2x^2 + 2$ **29.** $2x^2 - 4x$
30. $x^5 - 3x^3 - x^2 + 8$ **31.** Perimeter: $4w + 6$; area: $w^2 + 3w$ **32.** $x^2 + \frac{7}{6}x + \frac{1}{3}$ **33.** $49x^2 + 14x + 1$
34. $12x^3 - 23x^2 + 13x - 2$ **35.** $9x^4 - 16$
36. $15x^7 - 40x^6 + 50x^5 + 10x^4$ **37.** $x^2 - 3x - 28$
38. $9y^4 - 12y^3 + 4y^2$ **39.** $2t^4 - 11t^2 - 21$ **40.** 49
41. Coefficients: 1, −7, 9, −8; degrees: 6, 2, 2, 0; 6
42. $-y + 9w - 5$
43. $m^6 - 2m^2n + 2m^2n^2 + 8n^2m - 6m^3$
44. $-9xy - 2y^2$ **45.** $11x^3y^2 - 8x^2y - 6x^2 - 6x + 6$
46. $p^3 - q^3$ **47.** $9a^8 - 2a^4b^3 + \frac{1}{9}b^6$
48. $5x^2 - \frac{1}{2}x + 3$ **49.** $3x^2 - 7x + 4 + \dfrac{1}{2x + 3}$
50. 0, 3.75, −3.75, 0, 2.25 **51.** $25(t - 2 + 4m)$ **52.** $\frac{9}{4}$
53. -12 **54.** -11.2 **55.** Width: 125.5 m; length: 144.5 m **56.** ◈ 578.6×10^{-7} is not in scientific notation because 578.6 is larger than 10.
57. ◈ A monomial is an expression of the type ax^n, where n is a whole number and a is a real number. A binomial is a sum of two monomials and has two terms. A trinomial is a sum of three monomials and has three terms. A general polynomial is a monomial or a sum of monomials and has one or more terms.
58. $\frac{1}{2}x^2 - \frac{1}{2}y^2$ **59.** $400 - 4a^2$ **60.** $-28x^8$
61. $\frac{94}{13}$ **62.** $x^4 + x^3 + x^2 + x + 1$

Test: Chapter 10, p. 663

1. [10.1d, f] $\dfrac{1}{6^5}$ **2.** [10.1d] x^9 **3.** [10.1d] $(4a)^{11}$
4. [10.1e] 3^3 **5.** [10.1e, f] $\dfrac{1}{x^5}$ **6.** [10.1b, e] 1
7. [10.2a] x^6 **8.** [10.2a, b] $-27y^6$
9. [10.2a, b] $16a^{12}b^4$ **10.** [10.2b] $\dfrac{a^3b^3}{c^3}$
11. [10.1d], [10.2a, b] $-216x^{21}$
12. [10.1d], [10.2a, b] $-24x^{21}$
13. [10.1d], [10.2a, b] $162x^{10}$
14. [10.1d], [10.2a, b] $324x^{10}$ **15.** [10.1f] $\dfrac{1}{5^3}$
16. [10.1f] y^{-8} **17.** [10.2c] 3.9×10^9
18. [10.2c] 0.00000005 **19.** [10.2d] 1.75×10^{17}
20. [10.2d] 1.296×10^{22} **21.** [10.2e] 1.5×10^4
22. [10.3a] -43 **23.** [10.3d] $\frac{1}{3}$, -1, 7
24. [10.3g] 3, 0, 1, 6; 6 **25.** [10.3i] Binomial
26. [10.3e] $5a^2 - 6$ **27.** [10.3e] $\frac{7}{4}y^2 - 4y$
28. [10.3f] $x^5 + 2x^3 + 4x^2 - 8x + 3$
29. [10.4a] $4x^5 + x^4 + 2x^3 - 8x^2 + 2x - 7$
30. [10.4a] $5x^4 + 5x^2 + x + 5$
31. [10.4c] $-4x^4 + x^3 - 8x - 3$

32. [10.4c] $-x^5 + 0.7x^3 - 0.8x^2 - 21$
33. [10.5b] $-12x^4 + 9x^3 + 15x^2$
34. [10.6c] $x^2 - \frac{2}{3}x + \frac{1}{9}$ **35.** [10.6b] $9x^2 - 100$
36. [10.6a] $3b^2 - 4b - 15$
37. [10.6a] $x^{14} - 4x^8 + 4x^6 - 16$
38. [10.6a] $48 + 34y - 5y^2$
39. [10.5d] $6x^3 - 7x^2 - 11x - 3$
40. [10.6c] $25t^2 + 20t + 4$
41. [10.7c] $-5x^3y - y^3 + xy^3 - x^2y^2 + 19$
42. [10.7e] $8a^2b^2 + 6ab - 4b^3 + 6ab^2 + ab^3$
43. [10.7f] $9x^{10} - 16y^{10}$ **44.** [10.8a] $4x^2 + 3x - 5$
45. [10.8b] $2x^2 - 4x - 2 + \dfrac{17}{3x + 2}$ **46.** [10.3a] 3, 1.5, −3.5, −5, −5.25 **47.** [8.3b] 13 **48.** [8.3c] −3
49. [7.7d] $16(4t - 2m + 1)$ **50.** [7.4a] $\frac{23}{20}$
51. [8.4a] 100°, 25°, 55°
52. [10.5b], [10.6a] $V = l^3 - 3l^2 + 2l$
53. [8.3b], [10.6b, c] $-\frac{61}{12}$

Chapter 11

Pretest: Chapter 11, p. 666

1. [11.1a] $4(-5x^6)$, $(-2x^3)(10x^3)$, $x^2(-20x^4)$; answers may vary **2.** [11.5b] $2(x + 1)^2$ **3.** [11.2a] $(x + 4)(x + 2)$
4. [11.1b] $4a(2a^4 + a^2 - 5)$
5. [11.3a], [11.4a] $(5x + 2)(x - 3)$
6. [11.5d] $(9 + z^2)(3 + z)(3 - z)$ **7.** [11.5b] $(y^3 - 2)^2$
8. [11.1c] $(x^2 + 4)(3x + 2)$ **9.** [11.2a] $(p - 6)(p + 5)$
10. [11.5d] $(x^2y + 8)(x^2y - 8)$
11. [11.3a], [11.4a] $(2p - q)(p + 4q)$ **12.** [11.7b] 0, 5
13. [11.7a] 4, $\frac{3}{5}$ **14.** [11.7b] $\frac{2}{3}$, −4 **15.** [11.8a] 6, −1
16. [11.8a] Base: 8 cm; height: 11 cm

Margin Exercises, Section 11.1, pp. 667–670

1. (a) $12x^2$; **(b)** $(3x)(4x)$, $(2x)(6x)$; answers may vary
2. (a) $16x^3$; **(b)** $(2x)(8x^2)$, $(4x)(4x^2)$; answers may vary
3. $(8x)(x^3)$, $(4x^2)(2x^2)$, $(2x^3)(4x)$; answers may vary
4. $(7x)(3x)$, $(-7x)(-3x)$, $(21x)(x)$; answers may vary
5. $(6x^4)(x)$, $(-2x^3)(-3x^2)$, $(3x^3)(2x^2)$; answers may vary
6. (a) $3x + 6$; **(b)** $3(x + 2)$ **7. (a)** $2x^3 + 10x^2 + 8x$; **(b)** $2x(x^2 + 5x + 4)$ **8.** $x(x + 3)$ **9.** $y^2(3y^4 - 5y + 2)$
10. $3x^2(3x^2 - 5x + 1)$ **11.** $\frac{1}{4}(3t^3 + 5t^2 + 7t + 1)$
12. $7x^3(5x^4 - 7x^3 + 2x^2 - 9)$ **13.** $2.8(3x^2 - 2x + 1)$
14. $(x^2 + 3)(x + 7)$ **15.** $(x^2 + 2)(a + b)$
16. $(x^2 + 3)(x + 7)$ **17.** $(2t^2 + 3)(4t + 1)$
18. $(3m^3 + 2)(m^2 - 5)$ **19.** $(3x^2 - 1)(x - 2)$
20. $(2x^2 - 3)(2x - 3)$ **21.** Not factorable using factoring by grouping

Exercise Set 11.1, p. 671

1. $(4x^2)(2x)$, $(-8)(-x^3)$, $(2x^2)(4x)$; answers may vary
3. $(-5a^5)(2a)$, $(10a^3)(-a^3)$, $(-2a^2)(5a^4)$; answers may vary **5.** $(8x^2)(3x^2)$, $(-8x^2)(-3x^2)$, $(4x^3)(6x)$; answers may vary **7.** $x(x - 6)$ **9.** $2x(x + 3)$ **11.** $x^2(x + 6)$
13. $8x^2(x^2 - 3)$ **15.** $2(x^2 + x - 4)$
17. $17xy(x^4y^2 + 2x^2y + 3)$ **19.** $x^2(6x^2 - 10x + 3)$
21. $x^2y^2(x^3y^3 + x^2y + xy - 1)$
23. $2x^3(x^4 - x^3 - 32x^2 + 2)$
25. $0.8x(2x^3 - 3x^2 + 4x + 8)$
27. $\frac{1}{3}x^3(5x^3 + 4x^2 + x + 1)$ **29.** $(x^2 + 2)(x + 3)$
31. $(5a^3 - 1)(2a - 7)$ **33.** $(x^2 + 2)(x + 3)$
35. $(2x^2 + 1)(x + 3)$ **37.** $(4x^2 + 3)(2x - 3)$
39. $(4x^2 + 1)(3x - 4)$ **41.** $(5x^2 - 1)(x - 1)$
43. $(x^2 - 3)(x + 8)$ **45.** $(2x^2 - 9)(x - 4)$
47. $\{x \mid x > -24\}$ **48.** $\left\{x \mid x \le \frac{14}{5}\right\}$ **49.** 27
50. $p = 2A - q$ **51.** $y^2 + 12y + 35$
52. $y^2 + 14y + 49$ **53.** $y^2 - 49$
54. $y^2 - 14y + 49$
55. **56.**

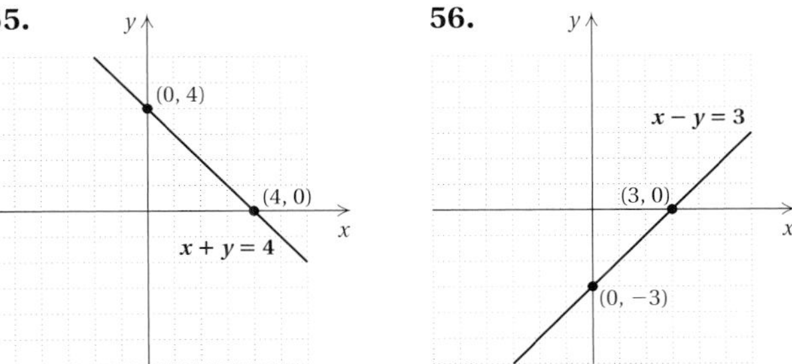

57. **58.**

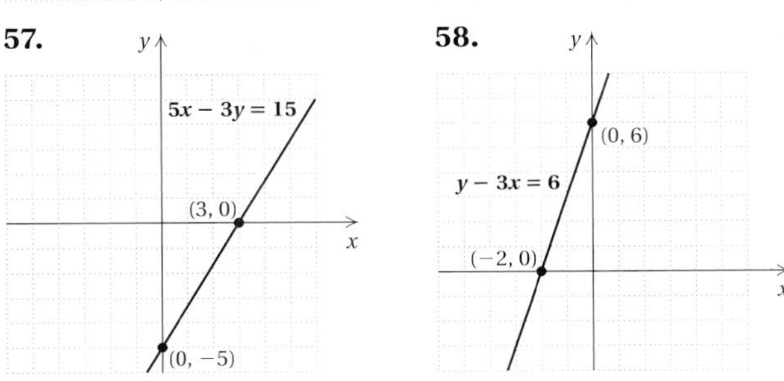

59. ◈ **61.** $(2x^3 + 3)(2x^2 + 3)$ **63.** $(x^7 + 1)(x^5 + 1)$
65. Not factorable by grouping

Margin Exercises, Section 11.2, pp. 673–676

1. (a) −13, 8, −8, 7, −7; **(b)** 13, 8, 7; both 7 and 12 are positive; **(c)** $(x + 3)(x + 4)$ **2.** $(x + 9)(x + 4)$
3. The coefficient of the middle term, −8, is negative.
4. $(x - 5)(x - 3)$ **5.** $(t - 5)(t - 4)$ **6.** 19, 8, 1; the positive factor has the larger absolute value. **7.** −23, −10, −5, −2; the negative factor has the larger absolute value. **8.** $x(x + 6)(x - 2)$ **9.** $(y - 6)(y + 2)$
10. $(t^2 + 7)(t^2 - 2)$ **11.** $p(p - q - 3q^2)$
12. Not factorable **13.** $(x + 4)^2$

Exercise Set 11.2, p. 677

1. $(x + 3)(x + 5)$ **3.** $(x + 3)(x + 4)$ **5.** $(x - 3)^2$
7. $(x + 2)(x + 7)$ **9.** $(b + 1)(b + 4)$ **11.** $\left(x + \frac{1}{3}\right)^2$
13. $(d - 2)(d - 5)$ **15.** $(y - 1)(y - 10)$
17. $(x - 6)(x + 7)$ **19.** $(x - 9)(x + 2)$
21. $x(x - 8)(x + 2)$ **23.** $y(y - 9)(y + 5)$
25. $(x - 11)(x + 9)$ **27.** $(c^2 + 8)(c^2 - 7)$
29. $(a^2 + 7)(a^2 - 5)$ **31.** Not factorable
33. Not factorable **35.** $(x + 10)^2$

37. $x^2(x-25)(x+4)$ **39.** $(x-24)(x+3)$
41. $(x-9)(x-16)$ **43.** $(a+12)(a-11)$
45. $(x-15)(x-8)$ **47.** $(12+x)(9-x)$, or $-(x+12)(x-9)$ **49.** $(y-0.4)(y+0.2)$
51. $(p+5q)(p-2q)$ **53.** $(m+4n)(m+n)$
55. $(s+3t)(s-5t)$ **57.** $16x^3-48x^2+8x$
58. $28w^2-53w-66$ **59.** $49w^2+84w+36$
60. $16w^2-88w+121$ **61.** $16w^2-121$ **62.** $27x^{12}$
63. $\frac{8}{3}$ **64.** $-\frac{7}{2}$ **65.** 29,555 **66.** 100°, 25°, 55°
67. ◆ **69.** 15, −15, 27, −27, 51, −51
71. $\left(x+\frac{1}{4}\right)\left(x-\frac{3}{4}\right)$ **73.** $(x+5)\left(x-\frac{5}{7}\right)$
75. $(b^n+5)(b^n+2)$ **77.** $2x^2(4-\pi)$

Margin Exercises, Section 11.3, pp. 680–683

1. $(2x+5)(x-3)$ **2.** $(4x+1)(3x-5)$
3. $(3x-4)(x-5)$ **4.** $2(5x-4)(2x-3)$
5. $(2x+1)(3x+2)$ **6.** $(2a-b)(3a-b)$
7. $3(2x+3y)(x+y)$

Calculator Spotlight, p. 683

1. Correct **2.** Correct **3.** Not correct **4.** Not correct **5.** Not correct **6.** Correct **7.** Not correct **8.** Correct

Exercise Set 11.3, p. 685

1. $(2x+1)(x-4)$ **3.** $(5x+9)(x-2)$
5. $(3x+1)(2x+7)$ **7.** $(3x+1)(x+1)$
9. $(2x-3)(2x+5)$ **11.** $(2x+1)(x-1)$
13. $(3x-2)(3x+8)$ **15.** $(3x+1)(x-2)$
17. $(3x+4)(4x+5)$ **19.** $(7x-1)(2x+3)$
21. $(3x+2)(3x+4)$ **23.** $(3x-7)^2$
25. $(24x-1)(x+2)$ **27.** $(5x-11)(7x+4)$
29. $2(5-x)(2+x)$ **31.** $4(3x-2)(x+3)$
33. $6(5x-9)(x+1)$ **35.** $2(3y+5)(y-1)$
37. $(3x-1)(x-1)$ **39.** $4(3x+2)(x-3)$
41. $(2x+1)(x-1)$ **43.** $(3x+2)(3x-8)$
45. $5(3x+1)(x-2)$ **47.** $p(3p+4)(4p+5)$
49. $x^2(7x-1)(2x+3)$ **51.** $3x(8x-1)(7x-1)$
53. $(5x^2-3)(3x^2-2)$ **55.** $(5t+8)^2$
57. $2x(3x+5)(x-1)$ **59.** Not factorable
61. Not factorable **63.** $(4m+5n)(3m-4n)$
65. $(2a+3b)(3a-5b)$ **67.** $(3a+2b)(3a+4b)$
69. $(5p+2q)(7p+4q)$ **71.** $6(3x-4y)(x+y)$
73. $q=\dfrac{A+7}{p}$ **74.** $x=\dfrac{y-b}{m}$ **75.** $y=\dfrac{6-3x}{2}$
76. $q=p+r-2$ **77.** $\{x\,|\,x>4\}$ **78.** $\left\{x\,|\,x\le\frac{8}{11}\right\}$
79.

y
$y=\frac{2}{5}x-1$
x

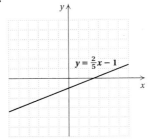

80. y^8 **81.** $9x^2-25$ **82.** $16a^2-24a+9$ **83.** ◆
85. $(2x^n+1)(10x^n+3)$ **87.** $(x^{3a}-1)(3x^{3a}+1)$

Margin Exercises, Section 11.4, p. 688

1. $(2x+1)(3x+2)$ **2.** $(4x+1)(3x-5)$
3. $3(2x+3)(x+1)$ **4.** $2(5x-4)(2x-3)$

Exercise Set 11.4, p. 689

1. $(x+7)(x+2)$ **3.** $(x-1)(x-4)$
5. $(2x+3)(3x+2)$ **7.** $(x-4)(3x-4)$
9. $(5x+3)(7x-8)$ **11.** $(2x-3)(2x+3)$
13. $(2x^2+5)(x^2+3)$ **15.** $(2x+1)(x-4)$
17. $(3x-5)(x+3)$ **19.** $(2x+7)(3x+1)$
21. $(3x+1)(x+1)$ **23.** $(2x-3)(2x+5)$
25. $(2x-1)(x+1)$ **27.** $(3x+2)(3x-8)$
29. $(3x-1)(x+2)$ **31.** $(3x-4)(4x-5)$
33. $(7x-1)(2x+3)$ **35.** $(3x+2)(3x+4)$
37. $(3x-7)^2$ **39.** $(24x-1)(x+2)$
41. $x^3(5x-11)(7x+4)$ **43.** $6x(5-x)(2+x)$
45. $3x^3(5+x)(1+2x)$ **47.** $\{x\,|\,x<-100\}$
48. $\{x\,|\,x\ge 217\}$ **49.** $\{x\,|\,x\le 8\}$ **50.** $\{x\,|\,x<2\}$
51. $\left\{x\,|\,x\ge\frac{20}{3}\right\}$ **52.** $\{x\,|\,x>17\}$ **53.** $\left\{x\,|\,x>\frac{26}{7}\right\}$
54. $\left\{x\,|\,x\ge\frac{77}{17}\right\}$ **55.** ◆ **57.** $(3x^5-2)^2$
59. $(4x^5+1)^2$ **61.–69.** Left to the student

Margin Exercises, Section 11.5, pp. 692–696

1. Yes **2.** No **3.** No **4.** Yes **5.** No **6.** Yes
7. No **8.** Yes **9.** $(x+1)^2$ **10.** $(x-1)^2$
11. $(t+2)^2$ **12.** $(5x-7)^2$ **13.** $(7-4y)^2$
14. $3(4m+5)^2$ **15.** $(p^2+9)^2$ **16.** $z^3(2z-5)^2$
17. $(3a+5b)^2$ **18.** Yes **19.** No **20.** No **21.** No
22. Yes **23.** Yes **24.** Yes **25.** $(x+3)(x-3)$
26. $4(4+t)(4-t)$ **27.** $(a+5b)(a-5b)$
28. $x^4(8+5x)(8-5x)$ **29.** $5(1+2t^3)(1-2t^3)$
30. $(9x^2+1)(3x+1)(3x-1)$
31. $(7p^2+5q^3)(7p^2-5q^3)$

Exercise Set 11.5, p. 697

1. Yes **3.** No **5.** No **7.** No **9.** $(x-7)^2$
11. $(x+8)^2$ **13.** $(x-1)^2$ **15.** $(x+2)^2$
17. $(q^2-3)^2$ **19.** $(4y+7)^2$ **21.** $2(x-1)^2$
23. $x(x-9)^2$ **25.** $3(2q-3)^2$ **27.** $(7-3x)^2$
29. $5(y^2+1)^2$ **31.** $(1+2x^2)^2$ **33.** $(2p+3q)^2$
35. $(a-3b)^2$ **37.** $(9a-b)^2$ **39.** $4(3a+4b)^2$
41. Yes **43.** No **45.** No **47.** Yes
49. $(y+2)(y-2)$ **51.** $(p+3)(p-3)$
53. $(t+7)(t-7)$ **55.** $(a+b)(a-b)$
57. $(5t+m)(5t-m)$ **59.** $(10+k)(10-k)$
61. $(4a+3)(4a-3)$ **63.** $(2x+5y)(2x-5y)$
65. $2(2x+7)(2x-7)$ **67.** $x(6+7x)(6-7x)$
69. $(7a^2+9)(7a^2-9)$ **71.** $(a^2+4)(a+2)(a-2)$
73. $5(x^2+9)(x+3)(x-3)$
75. $(1+y^4)(1+y^2)(1+y)(1-y)$
77. $(x^6+4)(x^3+2)(x^3-2)$ **79.** $\left(y+\frac{1}{4}\right)\left(y-\frac{1}{4}\right)$

81. $\left(5 + \frac{1}{7}x\right)\left(5 - \frac{1}{7}x\right)$ **83.** $(4m^2 + t^2)(2m + t)(2m - t)$
85. -11 **86.** 400 **87.** $-\frac{5}{6}$ **88.** -0.9 **89.** 2
90. -160 **91.** $x^2 - 4xy + 4y^2$ **92.** $\frac{1}{2}\pi x^2 + 2xy$
93. y^{12} **94.** $25a^4b^6$
95.

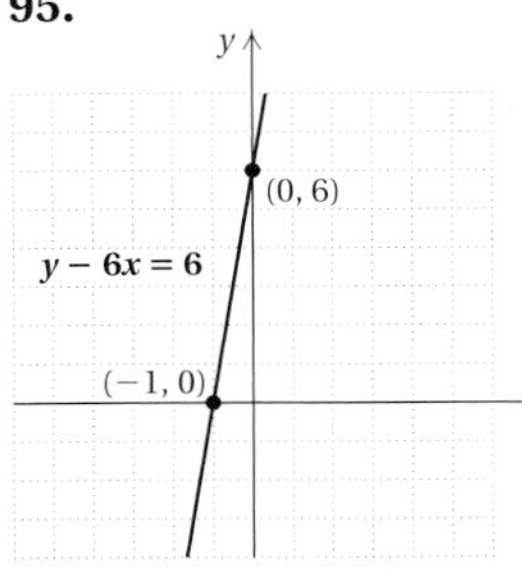

96.

y
3x − 5y = 15
(5, 0)
x
(0, −3)

97. ◆ **99.** Not factorable **101.** $(x + 11)^2$
103. $2x(3x + 1)^2$ **105.** $(x^4 + 2^4)(x^2 + 2^2)(x + 2)(x - 2)$
107. $3x^3(x + 2)(x - 2)$ **109.** $2x\left(3x + \frac{2}{5}\right)\left(3x - \frac{2}{5}\right)$
111. $p(0.7 + p)(0.7 - p)$ **113.** $(0.8x + 1.1)(0.8x - 1.1)$
115. $x(x + 6)$ **117.** $\left(x + \frac{1}{x}\right)\left(x - \frac{1}{x}\right)$
119. $(9 + b^{2k})(3 - b^k)(3 + b^k)$ **121.** $(3b^n + 2)^2$
123. $(y + 4)^2$ **125.** 9 **127.** Not correct
129. Not correct

Margin Exercises, Section 11.6, pp. 702–704

1. $3(m^2 + 1)(m + 1)(m - 1)$ **2.** $(x^3 + 4)^2$
3. $2x^2(x + 1)(x + 3)$ **4.** $(3x^2 - 2)(x + 4)$
5. $8x(x - 5)(x + 5)$ **6.** $x^2y(x^2y + 2x + 3)$
7. $2p^4q^2(5p^2 + 2pq + q^2)$ **8.** $(a - b)(2x + 5 + y^2)$
9. $(a + b)(x^2 + y)$ **10.** $(x^2 + y^2)^2$
11. $(xy + 1)(xy + 4)$ **12.** $(p^2 + 9q^2)(p + 3q)(p - 3q)$

Exercise Set 11.6, p. 705

1. $3(x + 8)(x - 8)$ **3.** $(a - 5)^2$ **5.** $(2x - 3)(x - 4)$
7. $x(x + 12)^2$ **9.** $(x + 2)(x - 2)(x + 3)$
11. $3(4x + 1)(4x - 1)$ **13.** $3x(3x - 5)(x + 3)$
15. Not factorable **17.** $x(x^2 + 7)(x - 3)$
19. $x^3(x - 7)^2$ **21.** $2(2 - x)(5 + x)$, or $-2(x - 2)(x + 5)$ **23.** Not factorable
25. $4(x^2 + 4)(x + 2)(x - 2)$
27. $(1 + y^4)(1 + y^2)(1 + y)(1 - y)$ **29.** $x^3(x - 3)(x - 1)$
31. $\frac{1}{9}\left(\frac{1}{3}x^3 - 4\right)^2$ **33.** $m(x^2 + y^2)$ **35.** $9xy(xy - 4)$
37. $2\pi r(h + r)$ **39.** $(a + b)(2x + 1)$
41. $(x + 1)(x - 1 - y)$ **43.** $(n + p)(n + 2)$
45. $(3q + p)(2q - 1)$ **47.** $(2b - a)^2$, or $(a - 2b)^2$
49. $(4x + 3y)^2$ **51.** $(7m^2 - 8n)^2$ **53.** $(y^2 + 5z^2)^2$
55. $\left(\frac{1}{2}a + \frac{1}{3}b\right)^2$ **57.** $(a + b)(a - 2b)$
59. $(m + 20n)(m - 18n)$ **61.** $(mn - 8)(mn + 4)$
63. $b^4(ab + 8)(ab - 4)$ **65.** $a^3(a - b)(a + 5b)$
67. $\left(a + \frac{1}{5}b\right)\left(a - \frac{1}{5}b\right)$ **69.** $(x - y)(x + y)$
71. $(4 + p^2q^2)(2 + pq)(2 - pq)$
73. $(1 + 4x^6y^6)(1 + 2x^3y^3)(1 - 2x^3y^3)$
75. $(q + 1)(q - 1)(q + 8)$ **77.** $(7x + 8y)^2$ **79.** 1990
80. 1995 **81.** 1992 and 1996 **82.** 75,000 **83.** 18,000
84. 59,000 **85.** $-\frac{14}{11}$ **86.** $25x^2 - 10xt + t^2$

87. $X = \dfrac{A + 7}{a + b}$ **88.** $\{x \mid x < 32\}$ **89.** ◆
91. $(a + 1)^2(a - 1)^2$ **93.** $(3.5x - 1)^2$
95. $(5x + 4)(x + 1.8)$ **97.** $(y + 3)(y - 3)(y - 2)$
99. $(a^2 + 1)(a + 4)$ **101.** $(x + 3)(x - 3)(x^2 + 2)$
103. $(x + 2)(x - 2)(x - 1)$ **105.** $(y - 1)^3$
107. $(y + 4 + x)^2$

Margin Exercises, Section 11.7, pp. 709–712

1. 3, -4 **2.** 7, 3 **3.** $-\frac{1}{4}$, $\frac{2}{3}$ **4.** 0, $\frac{17}{3}$ **5.** -2, 3
6. 7, -4 **7.** 3 **8.** 0, 4 **9.** $\frac{4}{3}$, $-\frac{4}{3}$ **10.** 3, -3
11. $(-5, 0)$, $(1, 0)$ **12.** 0, 3

Exercise Set 11.7, p. 713

1. -4, -9 **3.** -3, 8 **5.** -12, 11 **7.** 0, -3
9. 0, -18 **11.** $-\frac{5}{2}$, -4 **13.** $-\frac{1}{5}$, 3 **15.** 4, $\frac{1}{4}$
17. 0, $\frac{2}{3}$ **19.** $-\frac{1}{10}$, $\frac{1}{27}$ **21.** $\frac{1}{3}$, -20 **23.** 0, $\frac{2}{3}$, $\frac{1}{2}$
25. -1, -5 **27.** -9, 2 **29.** 3, 5 **31.** 0, 8
33. 0, -18 **35.** 4, -4 **37.** $-\frac{2}{3}$, $\frac{2}{3}$ **39.** -3 **41.** 4
43. 0, $\frac{6}{5}$ **45.** $\frac{5}{3}$, -1 **47.** $\frac{2}{3}$, $-\frac{1}{4}$ **49.** $\frac{2}{3}$, -1
51. $\frac{7}{10}$, $-\frac{7}{10}$ **53.** 9, -2 **55.** $\frac{4}{5}$, $\frac{3}{2}$ **57.** $(-4, 0)$, $(1, 0)$
59. $\left(-\frac{5}{2}, 0\right)$, $(2, 0)$ **61.** $(a + b)^2$ **62.** $a^2 + b^2$
63. -16 **64.** -4.5 **65.** $-\frac{10}{3}$ **66.** $\frac{3}{10}$ **67.** ◆
69. 4, -5 **71.** 9, -3 **73.** $\frac{1}{8}$, $-\frac{1}{8}$ **75.** 4, -4
77. **(a)** $x^2 - x - 12 = 0$; **(b)** $x^2 + 7x + 12 = 0$; **(c)** $4x^2 - 4x + 1 = 0$; **(d)** $x^2 - 25 = 0$; **(e)** $40x^3 - 14x^2 + x = 0$

Margin Exercises, Section 11.8, pp. 715–718

1. 5 and -5 **2.** 7 and 8 **3.** -4 and 5
4. Length: 5 cm; width: 3 cm **5.** **(a)** 342; **(b)** 9
6. 22 and 23 **7.** 3 m, 4 m

Exercise Set 11.8, p. 719

1. 5 and -5 **3.** 3 and 5 **5.** Length: 12 cm; width: 7 cm **7.** 14 and 15 **9.** 12 and 14; -12 and -14 **11.** 15 and 17; -15 and -17 **13.** 5
15. Height: 6 cm; base: 5 cm **17.** 6 km **19.** 4 sec
21. 5 and 7 **23.** 182 **25.** 12 **27.** 4950 **29.** 25
31. Hypotenuse: 17 ft; leg: 15 ft **33.** $9x^2 - 25y^2$
34. $9x^2 - 30xy + 25y^2$ **35.** $9x^2 + 30xy + 25y^2$
36. $6x^2 + 11xy - 35y^2$
37. y-intercept: $(0, -4)$; x-intercept: $(16, 0)$
38. y-intercept: $(0, 4)$; x-intercept: $(16, 0)$
39. y-intercept: $(0, -5)$; x-intercept: $(6.5, 0)$
40. y-intercept: $\left(0, \frac{2}{3}\right)$; x-intercept: $\left(\frac{5}{8}, 0\right)$ **41.** ◆
43. 5 ft **45.** 37 **47.** 30 cm by 15 cm

Summary and Review: Chapter 11, p. 723

1. $(-10x)(x)$; $(-5x)(2x)$; $(5x)(-2x)$; answers may vary
2. $(6x)(6x^4)$; $(4x^2)(9x^3)$; $(-2x^4)(-18x)$; answers may vary
3. $5(1 + 2x^3)(1 - 2x^3)$ **4.** $x(x - 3)$
5. $(3x + 2)(3x - 2)$ **6.** $(x + 6)(x - 2)$ **7.** $(x + 7)^2$

8. $3x(2x^2 + 4x + 1)$ **9.** $(x^2 + 3)(x + 1)$
10. $(3x - 1)(2x - 1)$ **11.** $(x^2 + 9)(x + 3)(x - 3)$
12. $3x(3x - 5)(x + 3)$ **13.** $2(x + 5)(x - 5)$
14. $(x^3 - 2)(x + 4)$ **15.** $(4x^2 + 1)(2x + 1)(2x - 1)$
16. $4x^4(2x^2 - 8x + 1)$ **17.** $3(2x + 5)^2$
18. Not factorable **19.** $x(x - 6)(x + 5)$
20. $(2x + 5)(2x - 5)$ **21.** $(3x - 5)^2$
22. $2(3x + 4)(x - 6)$ **23.** $(x - 3)^2$
24. $(2x + 1)(x - 4)$ **25.** $2(3x - 1)^2$
26. $3(x + 3)(x - 3)$ **27.** $(x - 5)(x - 3)$
28. $(5x - 2)^2$ **29.** $(7b^5 - 2a^4)^2$ **30.** $(xy + 4)(xy - 3)$
31. $3(2a + 7b)^2$ **32.** $(m + t)(m + 5)$
33. $32(x^2 - 2y^2z^2)(x^2 + 2y^2z^2)$ **34.** 1, −3 **35.** −7, 5
36. −4, 3 **37.** $\frac{2}{3}$, 1 **38.** $\frac{3}{2}$, −4 **39.** 8, −2
40. 3 and −2 **41.** −18 and −16; 16 and 18
42. $\frac{5}{2}$ and −2 **43.** −19 and −17; 17 and 19
44. Dining room: 12 ft by 12 ft; kitchen: 12 ft by 10 ft
45. 4 ft **46.** (−5, 0), (−4, 0) **47.** $\left(-\frac{3}{2}, 0\right)$, (5, 0)
48. $\frac{8}{35}$ **49.** $\{x \mid x \le \frac{4}{3}\}$ **50.** $4a^2 - 9$
51.

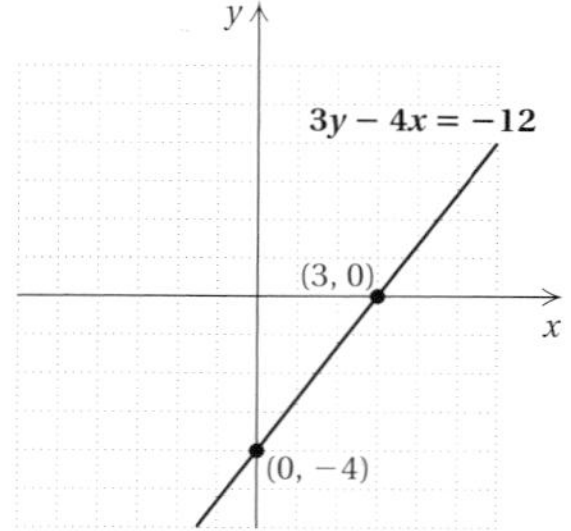

52. ◈ In this chapter, we learned to solve equations of the type $ax^2 + bx + c = 0$ (quadratic equations). Previously, we could solve only first-degree, or linear, equations (equations equivalent to those of the form $ax + b = 0$). The principle of zero products is used to solve quadratic equations, but it is not used to solve linear equations. **53.** ◈ Multiplying can be used to check factoring because factoring is the reverse of multiplying. The TABLE feature of a grapher can provide a partial check of factoring. When a polynomial and its factorization are entered as y_1 and y_2, the factorization is probably correct if corresponding values in the Y1 and Y2 columns are the same. The GRAPH feature of a grapher can also provide a partial check. If the graphs of y_1 and y_2 (entered as described above) coincide, then the factorization is probably correct. **54.** 2.5 cm
55. 0, 2 **56.** Length: 12; width: 6 **57.** No solution
58. 2, −3, $\frac{5}{2}$ **59.** a, i; b, k; c, g; d, h; e, j; f, l
60. 2^{100}; $2^{90} + 2^{90} = 2 \cdot 2^{90} = 2^{91} < 2^{100}$

Test: Chapter 11, p. 725

1. [11.1a] $(4x)(x^2)$; $(2x^2)(2x)$; $(-2x)(-2x^2)$; answers may vary **2.** [11.2a] $(x - 5)(x - 2)$ **3.** [11.5b] $(x - 5)^2$
4. [11.1b] $2y^2(2y^2 - 4y + 3)$ **5.** [11.1c] $(x^2 + 2)(x + 1)$
6. [11.1b] $x(x - 5)$ **7.** [11.2a] $x(x + 3)(x - 1)$
8. [11.3a], [11.4a] $2(5x - 6)(x + 4)$
9. [11.5d] $(2x + 3)(2x - 3)$ **10.** [11.2a] $(x - 4)(x + 3)$
11. [11.3a], [11.4a] $3m(2m + 1)(m + 1)$
12. [11.5d] $3(w + 5)(w - 5)$ **13.** [11.5b] $5(3x + 2)^2$
14. [11.5d] $3(x^2 + 4)(x + 2)(x - 2)$
15. [11.5b] $(7x - 6)^2$
16. [11.3a], [11.4a] $(5x - 1)(x - 5)$
17. [11.1c] $(x^3 - 3)(x + 2)$
18. [11.5d] $5(4 + x^2)(2 + x)(2 - x)$
19. [11.3a], [11.4a] $(2x + 3)(2x - 5)$
20. [11.3a], [11.4a] $3t(2t + 5)(t - 1)$
21. [11.2a] $3(m + 2n)(m - 5n)$ **22.** [11.7b] 5, −4
23. [11.7b] $\frac{3}{2}$, −5 **24.** [11.7b] 7, −4 **25.** [11.8a] 8, −3
26. [11.8a] Length of foot is 7 ft; height of sail is 12 ft
27. [11.7b] (7, 0), (−5, 0) **28.** [11.7b] $\left(\frac{2}{3}, 0\right)$, (1, 0)
29. [7.6c] $-\frac{10}{11}$ **30.** [8.7e] $\{x \mid x < \frac{19}{3}\}$
31. [9.3a]

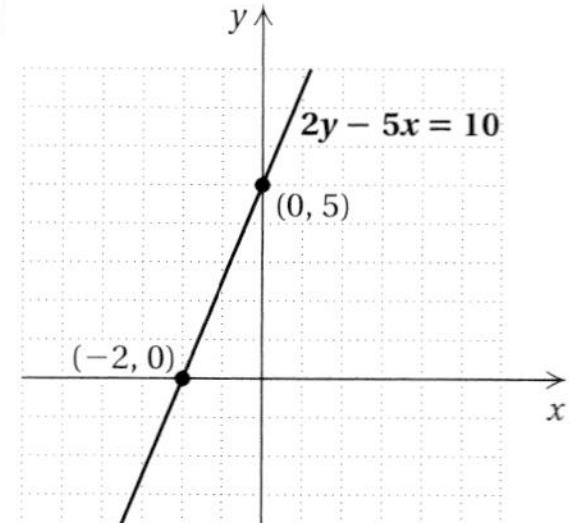

32. [10.6d] $25x^4 - 70x^2 + 49$
33. [11.8a] Length: 15; width: 3
34. [11.2a] $(a - 4)(a + 8)$ **35.** [11.5d], [11.7a] (c)
36. [10.6b], [11.5d] (d)

Chapter 12

Pretest: Chapter 12, p. 728

1. [12.3c] $(x + 2)(x + 3)^2$ **2.** [12.4a] $\frac{-b - 1}{b^2 - 4}$, or $\frac{b + 1}{4 - b^2}$
3. [12.5a] $\frac{1}{y - 2}$ **4.** [12.4a] $\frac{7a + 6}{a(a + 2)}$
5. [12.5b] $\frac{2x}{x + 1}$ **6.** [12.1d] $\frac{2(x - 3)}{x - 2}$
7. [12.2b] $\frac{x - 3}{x + 3}$ **8.** [12.9a] $\frac{y + x}{y - x}$
9. [12.6a] −5 **10.** [12.6a] 0 **11.** [12.8a] $M = \frac{3R}{a - b}$
12. [12.7b] 10.5 hr **13.** [12.7a] $\frac{30}{11}$ hr
14. [12.7a] 60 mph, 80 mph

Margin Exercises, Section 12.1, pp. 729–734

1. 3 **2.** −8, 3 **3.** None **4.** $\frac{x(2x + 1)}{x(3x - 2)}$
5. $\frac{(x + 1)(x + 2)}{(x - 2)(x + 2)}$ **6.** $\frac{-1(x - 8)}{-1(x - y)}$ **7.** 5 **8.** $\frac{x}{4}$
9. $\frac{2x + 1}{3x + 2}$ **10.** $\frac{x + 1}{2x + 1}$ **11.** $x + 2$ **12.** $\frac{y + 2}{4}$
13. −1 **14.** −1 **15.** −1 **16.** $\frac{a - 2}{a - 3}$ **17.** $\frac{x - 5}{2}$

Calculator Spotlight, p. 734

1. Correct **2.** Correct **3.** Not correct **4.** Not correct **5.** Not correct **6.** Correct **7.** Not correct

Exercise Set 12.1, p. 735

1. 0 **3.** 8 **5.** $-\frac{5}{2}$ **7.** 7, -4 **9.** 5, -5 **11.** None **13.** $\frac{(4x)(3x^2)}{(4x)(5y)}$ **15.** $\frac{2x(x-1)}{2x(x+4)}$ **17.** $\frac{-1(3-x)}{-1(4-x)}$ **19.** $\frac{(y+6)(y-7)}{(y+6)(y+2)}$ **21.** $\frac{x^2}{4}$ **23.** $\frac{8p^2q}{3}$ **25.** $\frac{x-3}{x}$ **27.** $\frac{m+1}{2m+3}$ **29.** $\frac{a-3}{a+2}$ **31.** $\frac{a-3}{a-4}$ **33.** $\frac{x+5}{x-5}$ **35.** $a+1$ **37.** $\frac{x^2+1}{x+1}$ **39.** $\frac{3}{2}$ **41.** $\frac{6}{t-3}$ **43.** $\frac{t+2}{2(t-4)}$ **45.** $\frac{t-2}{t+2}$ **47.** -1 **49.** -1 **51.** -6 **53.** $-x-1$ **55.** $\frac{56x}{3}$ **57.** $\frac{2}{dc^2}$ **59.** $\frac{x+2}{x-2}$ **61.** $\frac{(a+3)(a-3)}{a(a+4)}$ **63.** $\frac{2a}{a-2}$ **65.** $\frac{(t+2)(t-2)}{(t+1)(t-1)}$ **67.** $\frac{x+4}{x+2}$ **69.** $\frac{5(a+6)}{a-1}$ **71.** 18 and 20; -18 and -20 **72.** 3.125 L **73.** $(x-8)(x+7)$ **74.** $(a-8)^2$ **75.** $x^3(x-7)(x+5)$ **76.** $(2y^2+1)(y-5)$ **77.** $(2-t)(2+t)(4+t^2)$ **78.** $10(x+7)(x+1)$ **79.** $(x-7)(x-2)$ **80.** Not factorable **81.** $(4x-5y)^2$ **82.** $(a-7b)(a-2b)$ **83.** ◈ **85.** $x+2y$ **87.** $\frac{(t-9)^2(t-1)}{(t^2+9)(t+1)}$ **89.** $\frac{x-y}{x-5y}$

Margin Exercises, Section 12.2, pp. 739–740

1. $\frac{2}{7}$ **2.** $\frac{2x^3-1}{x^2+5}$ **3.** $\frac{1}{x-5}$ **4.** x^2-3 **5.** $\frac{6}{35}$ **6.** $\frac{x^2}{40}$ **7.** $\frac{(x-3)(x-2)}{(x+5)(x+5)}$ **8.** $\frac{x-3}{x+2}$ **9.** $\frac{(x-3)(x-2)}{x+2}$ **10.** $\frac{y+1}{y-1}$

Exercise Set 12.2, p. 741

1. $\frac{x}{4}$ **3.** $\frac{1}{x^2-y^2}$ **5.** $a+b$ **7.** $\frac{x^2-4x+7}{x^2+2x-5}$ **9.** $\frac{3}{10}$ **11.** $\frac{1}{4}$ **13.** $\frac{b}{a}$ **15.** $\frac{(a+2)(a+3)}{(a-3)(a-1)}$ **17.** $\frac{(x-1)^2}{x}$ **19.** $\frac{1}{2}$ **21.** $\frac{15}{8}$ **23.** $\frac{15}{4}$ **25.** $\frac{a-5}{3(a-1)}$ **27.** $\frac{(x+2)^2}{x}$ **29.** $\frac{3}{2}$ **31.** $\frac{c+1}{c-1}$ **33.** $\frac{y-3}{2y-1}$ **35.** $\frac{x+1}{x-1}$ **37.** $\{x \mid x \geq 77\}$ **38.** 4 **39.** $8x^3-11x^2-3x+12$ **40.** $-2p^2+4pq-4q^2$ **41.** $\frac{4y^8}{x^6}$ **42.** $\frac{125x^{18}}{y^{12}}$ **43.** $\frac{4x^6}{y^{10}}$ **44.** $\frac{1}{a^{15}b^{20}}$ **45.** ◈ **47.** $-\frac{1}{b^2}$ **49.** $\frac{(x-7)^2}{x+y}$

Margin Exercises, Section 12.3, pp. 743–744

1. 144 **2.** 12 **3.** 10 **4.** 120 **5.** $\frac{35}{144}$ **6.** $\frac{1}{4}$ **7.** $\frac{11}{10}$ **8.** $\frac{9}{40}$ **9.** $60x^3y^2$ **10.** $(y+1)^2(y+4)$ **11.** $7(t^2+16)(t-2)$ **12.** $3x(x+1)^2(x-1)$

Exercise Set 12.3, p. 745

1. 108 **3.** 72 **5.** 126 **7.** 360 **9.** 500 **11.** $\frac{65}{72}$ **13.** $\frac{29}{120}$ **15.** $\frac{23}{180}$ **17.** $12x^3$ **19.** $18x^2y^2$ **21.** $6(y-3)$ **23.** $t(t+2)(t-2)$ **25.** $(x+2)(x-2)(x+3)$ **27.** $t(t+2)^2(t-4)$ **29.** $(a+1)(a-1)^2$ **31.** $(m-3)(m-2)^2$ **33.** $(2+3x)(2-3x)$ **35.** $10v(v+4)(v+3)$ **37.** $18x^3(x-2)^2(x+1)$ **39.** $6x^3(x+2)^2(x-2)$ **41.** $(x-3)^2$ **42.** $2x(3x+2)$ **43.** $(x+3)(x-3)$ **44.** $(x+7)(x-3)$ **45.** $(x+3)^2$ **46.** $(x-7)(x+3)$ **47.** 54% **48.** 64% **49.** 74% **50.** 98% **51.** 1965 **52.** 1999 **53.** ◈

Margin Exercises, Section 12.4, pp. 747–750

1. $\frac{7}{9}$ **2.** $\frac{3+x}{x-2}$ **3.** $\frac{6x+4}{x-1}$ **4.** $\frac{x-5}{4}$ **5.** $\frac{x-1}{x-3}$ **6.** $\frac{10x^2+9x}{48}$ **7.** $\frac{9x+10}{48x^2}$ **8.** $\frac{4x^2-x+3}{x(x-1)(x+1)^2}$ **9.** $\frac{2x^2+16x+5}{(x+3)(x+8)}$ **10.** $\frac{8x+88}{(x+16)(x+1)(x+8)}$ **11.** $\frac{-2x-11}{3(x+4)(x-4)}$

Exercise Set 12.4, p. 751

1. 1 **3.** $\frac{6}{3+x}$ **5.** $\frac{2x+3}{x-5}$ **7.** $\frac{1}{4}$ **9.** $-\frac{1}{t}$ **11.** $\frac{-x+7}{x-6}$ **13.** $y+3$ **15.** $\frac{2b-14}{b^2-16}$ **17.** $a+b$ **19.** $\frac{5x+2}{x-5}$ **21.** -1 **23.** $\frac{-x^2+9x-14}{(x-3)(x+3)}$ **25.** $\frac{2x+5}{x^2}$ **27.** $\frac{41}{24r}$ **29.** $\frac{4x+6y}{x^2y^2}$ **31.** $\frac{4+3t}{18t^3}$ **33.** $\frac{x^2+4xy+y^2}{x^2y^2}$ **35.** $\frac{6x}{(x-2)(x+2)}$ **37.** $\frac{11x+2}{3x(x+1)}$ **39.** $\frac{x^2+6x}{(x+4)(x-4)}$ **41.** $\frac{6}{z+4}$ **43.** $\frac{3x-1}{(x-1)^2}$ **45.** $\frac{11a}{10(a-2)}$ **47.** $\frac{2x^2+8x+16}{x(x+4)}$ **49.** $\frac{7a+6}{(a-2)(a+1)(a+3)}$ **51.** $\frac{2x^2-4x+34}{(x-5)(x+3)}$ **53.** $\frac{3a+2}{(a+1)(a-1)}$ **55.** $\frac{2x+6y}{(x+y)(x-y)}$ **57.** $\frac{a^2+7a+1}{(a+5)(a-5)}$ **59.** $\frac{5t-12}{(t+3)(t-3)(t-2)}$

61. $x^2 - 1$ **62.** $13y^3 - 14y^2 + 12y - 73$ **63.** $\frac{1}{8x^{12}y^9}$

64. $\frac{x^6}{25y^2}$ **65.** $\frac{1}{x^{12}y^{21}}$ **66.** $\frac{25}{x^4y^6}$

67.

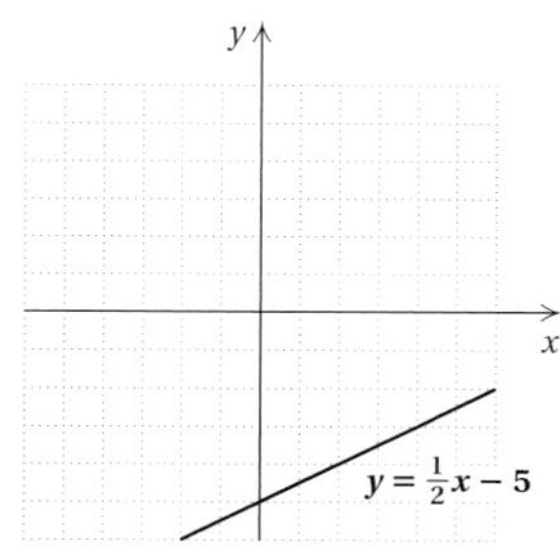

68.

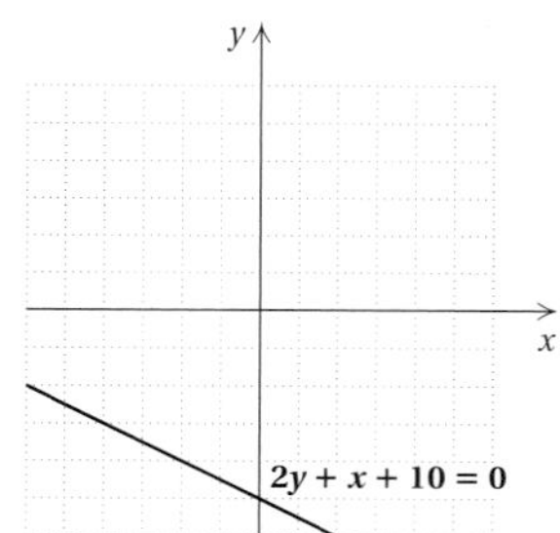

69.

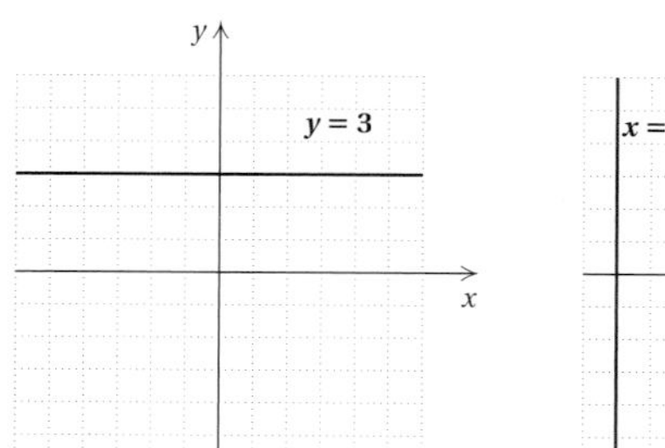

70.

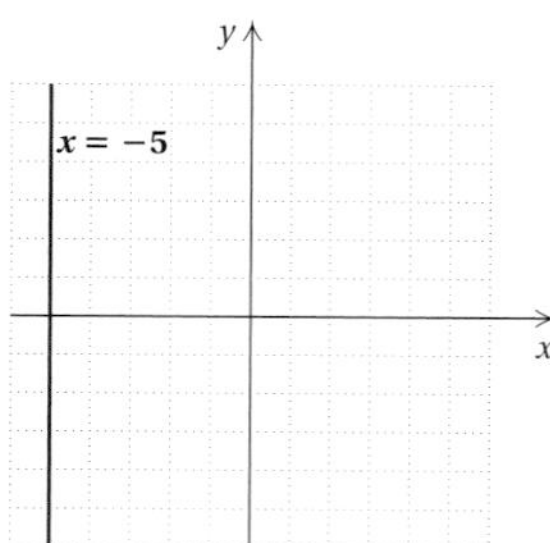

71. -8 **72.** $\frac{5}{6}$ **73.** 5, 3 **74.** 9, -2 **75.** ◈

77. Perimeter: $\frac{16y + 28}{15}$; area: $\frac{y^2 + 2y - 8}{15}$

79. $\frac{(z + 6)(2z - 3)}{(z + 2)(z - 2)}$ **81.** $\frac{11z^4 - 22z^2 + 6}{(z^2 + 2)(z^2 - 2)(2z^2 - 3)}$

Margin Exercises, Section 12.5, pp. 755–758

1. $\frac{4}{11}$ **2.** $\frac{5}{y}$ **3.** $\frac{x^2 + 2x + 1}{2x + 1}$ **4.** $\frac{3x - 1}{3}$

5. $\frac{4x - 3}{x - 2}$ **6.** $\frac{-x - 7}{15x}$ **7.** $\frac{x^2 - 48}{(x + 7)(x + 8)(x + 6)}$

8. $\frac{-8y - 28}{(y + 4)(y - 4)}$ **9.** $\frac{x - 13}{(x + 3)(x - 3)}$

10. $\frac{6x^2 - 2x - 2}{3x(x + 1)}$

Exercise Set 12.5, p. 759

1. $\frac{4}{x}$ **3.** 1 **5.** $\frac{1}{x - 1}$ **7.** $\frac{8}{3}$ **9.** $\frac{13}{a}$ **11.** $\frac{8}{y - 1}$

13. $\frac{x - 2}{x - 7}$ **15.** $\frac{4}{a^2 - 25}$ **17.** $\frac{2x - 4}{x - 9}$ **19.** $\frac{-9}{2x - 3}$

21. $\frac{-a - 4}{10}$ **23.** $\frac{7z - 12}{12z}$ **25.** $\frac{4x^2 - 13xt + 9t^2}{3x^2t^2}$

27. $\frac{2x - 40}{(x + 5)(x - 5)}$ **29.** $\frac{3 - 5t}{2t(t - 1)}$ **31.** $\frac{2s - st - s^2}{(t + s)(t - s)}$

33. $\frac{y - 19}{4y}$ **35.** $\frac{-2a^2}{(x + a)(x - a)}$ **37.** $\frac{9x + 12}{(x + 3)(x - 3)}$

39. $\frac{1}{2}$ **41.** $\frac{x - 3}{(x + 3)(x + 1)}$ **43.** $\frac{18x + 5}{x - 1}$ **45.** 0

47. $\frac{20}{2y - 1}$ **49.** $\frac{2a - 3}{2 - a}$ **51.** $\frac{z - 3}{2z - 1}$ **53.** $\frac{2}{x + y}$

55. x^5 **56.** $30x^{12}$ **57.** $\frac{b^{20}}{a^8}$ **58.** $18x^3$ **59.** $\frac{6}{x^3}$

60. $\frac{10}{x^3}$ **61.** $x^2 - 9x + 18$ **62.** $(4 - \pi)r^2$ **63.** ◈

65. $\frac{30}{(x - 3)(x + 4)}$

67. $\frac{x^2 + xy - x^3 + x^2y - xy^2 + y^3}{(x^2 + y^2)(x + y)^2(x - y)}$

69. $\frac{-2a - 15}{a - 6}$; area $= \frac{-2a^3 - 15a^2 + 12a + 90}{2(a - 6)^2}$

Margin Exercises, Section 12.6, pp. 763–767

1. $\frac{33}{2}$ **2.** 3 **3.** $\frac{3}{2}$ **4.** $-\frac{1}{8}$ **5.** 1 **6.** 2 **7.** 4

Improving Your Math Study Skills, p. 768

1. Rational expression **2.** Solutions **3.** Rational expression **4.** Rational expression **5.** Rational expression **6.** Solutions **7.** Rational expression **8.** Solutions **9.** Solutions **10.** Solutions **11.** Rational expression **12.** Solutions **13.** Rational expression

Exercise Set 12.6, p. 769

1. $\frac{6}{5}$ **3.** $\frac{40}{29}$ **5.** $\frac{47}{2}$ **7.** -6 **9.** $\frac{24}{7}$ **11.** $-4, -1$
13. 4, -4 **15.** 3 **17.** $\frac{14}{3}$ **19.** 5 **21.** 5 **23.** $\frac{5}{2}$
25. -2 **27.** $-\frac{13}{2}$ **29.** $\frac{17}{2}$ **31.** No solution **33.** -5
35. $\frac{5}{3}$ **37.** $\frac{1}{2}$ **39.** No solution **41.** No solution

43. 4 **45.** $\frac{1}{a^6b^{15}}$ **46.** x^8y^{12} **47.** $\frac{16x^4}{t^8}$ **48.** $\frac{w^4}{y^6}$

49. $32x^6$ **50.** $\frac{64x^{10}}{y^8}$

51.

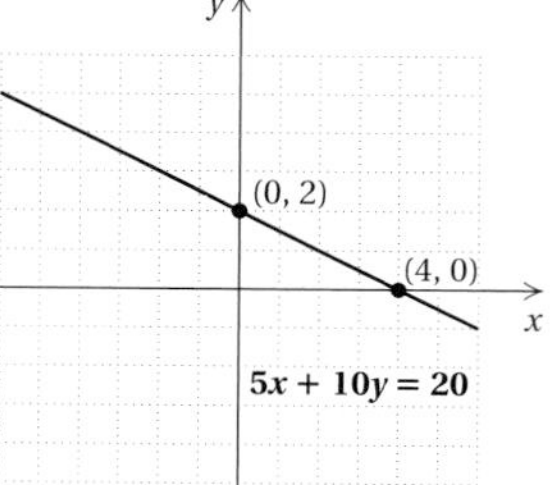

52.

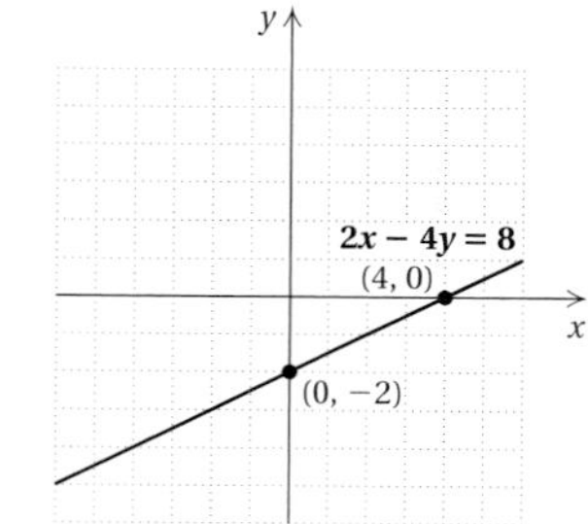

53. **54.**

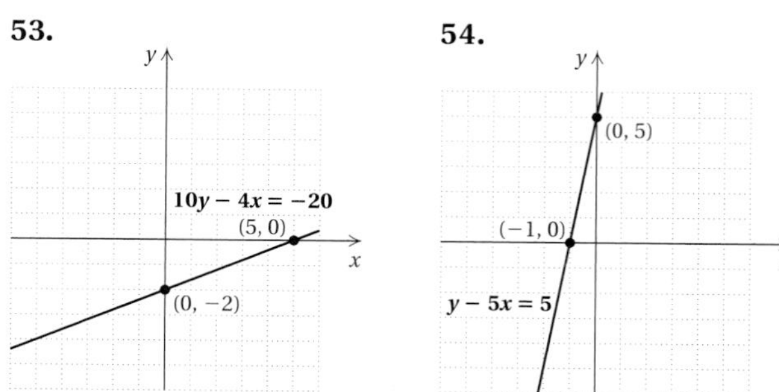

55. ◈ **57.** 7 **59.** No solution **61.** 2, -2 **63.** 4

Margin Exercises, Section 12.7, pp. 773–780

1. -3 **2.** 40 km/h, 60 km/h **3.** $\frac{24}{7}$, or $3\frac{3}{7}$ hr **4.** 58 km/L **5.** 0.280 **6.** 124 km/h **7.** 2.4 fish/yd^2 **8.** 81 gal **9.** **(a)** About 57; **(b)** Since $57 < 61$, it could not be predicted that McGwire would break Maris's record. **10.** 42 **11.** 2074 **12.** 24.75 ft **13.** 34.9 ft

Calculator Spotlight, p. 776

1. 20.6 min **2.** 0.54 hr

Exercise Set 12.7, p. 781

1. $\frac{24}{7}$ **3.** 20 and 15 **5.** 30 km/h, 70 km/h **7.** Passenger: 80 mph; freight: 66 mph **9.** 20 mph **11.** $4\frac{4}{9}$ min **13.** $5\frac{1}{7}$ hr **15.** 9 **17.** 2.3 km/h **19.** 66 g **21.** 702 km **23.** 200 **25.** **(a)** About 63; **(b)** yes **27.** 10,000 **29.** **(a)** 4.8 tons; **(b)** 48 lb **31.** 11 **33.** $\frac{21}{2}$ **35.** $\frac{8}{3}$ **37.** $\frac{35}{3}$ **39.** x^{11} **40.** x **41.** $\frac{1}{x^{11}}$ **42.** $\frac{1}{x}$ **43.** ◈ **45.** $9\frac{3}{13}$ days **47.** $\frac{3}{4}$ **49.** $27\frac{3}{11}$ min

Margin Exercises, Section 12.8, pp. 785–786

1. $M = \frac{fd^2}{km}$ **2.** $b_1 = \frac{2A - hb_2}{h}$ **3.** $f = \frac{pq}{p + q}$ **4.** $b = \frac{a}{2Q + 1}$

Exercise Set 12.8, p. 787

1. $r = \frac{S}{2\pi h}$ **3.** $b = \frac{2A}{h}$ **5.** $n = \frac{S + 360}{180}$ **7.** $b = \frac{3V - kB - 4kM}{k}$ **9.** $r = \frac{S - a}{S - l}$ **11.** $h = \frac{2A}{b_1 + b_2}$ **13.** $B = \frac{A}{AQ + 1}$ **15.** $p = \frac{qf}{q - f}$ **17.** $A = P(1 + r)$ **19.** $R = \frac{r_1 r_2}{r_1 + r_2}$ **21.** $D = \frac{BC}{A}$ **23.** $h_2 = \frac{p(h_1 - q)}{q}$ **25.** $a = \frac{b}{K - C}$ **27.** $-3x^3 - 5x^2 + 5$ **28.** $23x^4 + 50x^3 + 23x^2 - 163x + 41$ **29.** $(x + 2)(x - 2)$ **30.** $3(2y^2 - 1)(5y^2 + 4)$ **31.** $(7m - 8n)^2$ **32.** $(y + 7)(y - 5)$ **33.** $(y^2 + 1)(y + 1)(y - 1)$ **34.** $(a - 10b)(a + 10b)$ **35.** $x^2 + 2x + 8 + \frac{12}{x - 2}$ **36.** $x^2 - 3$ **37.** ◈ **39.** $T = \frac{FP}{u + EF}$ **41.** $v = \frac{Nbf_2 - bf_1 - df_1}{Nf_2 - 1}$

Margin Exercises, Section 12.9, pp. 790–792

1. $\frac{136}{5}$ **2.** $\frac{7x^2}{3(2 - x^2)}$ **3.** $\frac{x}{x - 1}$ **4.** $\frac{136}{5}$ **5.** $\frac{7x^2}{3(2 - x^2)}$ **6.** $\frac{x}{x - 1}$

Exercise Set 12.9, p. 793

1. $\frac{25}{4}$ **3.** $\frac{1}{3}$ **5.** -6 **7.** $\frac{1 + 3x}{1 - 5x}$ **9.** $\frac{2x + 1}{x}$ **11.** 8 **13.** $x - 8$ **15.** $\frac{y}{y - 1}$ **17.** $-\frac{1}{a}$ **19.** $\frac{ab}{b - a}$ **21.** $\frac{p^2 + q^2}{q + p}$ **23.** $4x^4 + 3x^3 + 2x - 7$ **24.** 0 **25.** $(p - 5)^2$ **26.** $(p + 5)^2$ **27.** $50(p^2 - 2)$ **28.** $5(p + 2)(p - 10)$ **29.** 14 yd **30.** 12 ft, 5 ft **31.** ◈ **33.** $\frac{(x - 1)(3x - 2)}{5x - 3}$ **35.** $-\frac{ac}{bd}$ **37.** $\frac{5x + 3}{3x + 2}$

Summary and Review: Chapter 12, p. 795

1. 0 **2.** 6 **3.** 6, -6 **4.** -6, 5 **5.** -2 **6.** 0, 3, 5 **7.** $\frac{x - 2}{x + 1}$ **8.** $\frac{7x + 3}{x - 3}$ **9.** $\frac{y - 5}{y + 5}$ **10.** $\frac{a - 6}{5}$ **11.** $\frac{6}{2t - 1}$ **12.** $-20t$ **13.** $\frac{2x^2 - 2x}{x + 1}$ **14.** $30x^2y^2$ **15.** $4(a - 2)$ **16.** $(y - 2)(y + 2)(y + 1)$ **17.** $\frac{-3x + 18}{x + 7}$ **18.** -1 **19.** $\frac{2a}{a - 1}$ **20.** $d + c$ **21.** $\frac{4}{x - 4}$ **22.** $\frac{x + 5}{2x}$ **23.** $\frac{2x + 3}{x - 2}$ **24.** $\frac{-x^2 + x + 26}{(x - 5)(x + 5)(x + 1)}$ **25.** $\frac{2(x - 2)}{x + 2}$ **26.** $\frac{z}{1 - z}$ **27.** $c - d$ **28.** 8 **29.** 3, -5 **30.** $5\frac{1}{7}$ hr **31.** 240 km/h, 280 km/h **32.** -2 **33.** 95 mph, 175 mph **34.** 160 **35.** 1.92 g **36.** 6 **37.** $s = \frac{rt}{r - t}$ **38.** $C = \frac{5}{9}(F - 32)$, or $C = \frac{5F - 160}{9}$ **39.** $r^3 = \frac{3V}{4\pi}$ **40.** $(5x^2 - 3)(x + 4)$ **41.** $\frac{1}{125x^9y^6}$ **42.** $-2x^3 + 3x^2 + 12x - 18$ **43.** Length: 5 cm; width: 3 cm; perimeter: 16 cm **44.** $\frac{5x + 6}{(x + 2)(x - 2)}$; used to find an equivalent expression for each rational expression with the LCM as the least common denominator **45.** $\frac{3x + 10}{(x - 2)(x + 2)}$; used to find an equivalent expression for each rational expression with the LCM as the least common denominator **46.** 4; used to clear fractions **47.** $\frac{4(x - 2)}{x(x + 4)}$; Method 1: used to multiply by 1 using LCM/LCM; Method 2: used the LCM of the denominators in the numerator to subtract in the numerator and used the LCM of the denominators in the denominator to add in the denominator. **48.** $\frac{5(a + 3)^2}{a}$ **49.** $\frac{10a}{(a - b)(b - c)}$ **50.** They are equivalent equations.

Test: Chapter 12, p. 797

1. [12.1a] 0 **2.** [12.1a] −8 **3.** [12.1a] 7, −7
4. [12.1a] 1, 2 **5.** [12.1a] 1 **6.** [12.1a] 0, −3, −5
7. [12.1c] $\frac{3x + 7}{x + 3}$ **8.** [12.1d] $\frac{a + 5}{2}$
9. [12.2b] $\frac{(5x + 1)(x + 1)}{3x(x + 2)}$
10. [12.3c] $(y - 3)(y + 3)(y + 7)$ **11.** [12.4a] $\frac{23 - 3x}{x^3}$
12. [12.5a] $\frac{8 - 2t}{t^2 + 1}$ **13.** [12.4a] $\frac{-3}{x - 3}$
14. [12.5a] $\frac{2x - 5}{x - 3}$ **15.** [12.4a] $\frac{8t - 3}{t(t - 1)}$
16. [12.5a] $\frac{-x^2 - 7x - 15}{(x + 4)(x - 4)(x + 1)}$
17. [12.5b] $\frac{x^2 + 2x - 7}{(x - 1)^2(x + 1)}$ **18.** [12.9a] $\frac{3y + 1}{y}$
19. [12.6a] 12 **20.** [12.6a] 5, −3 **21.** [12.7a] 4
22. [12.7b] 16 **23.** [12.7a] 45 mph, 65 mph
24. [12.8a] $t = \frac{g}{M - L}$ **25.** [12.7b] 15
26. [11.6a] $(4a + 7)(4a - 7)$ **27.** [10.2b] $\frac{y^{12}}{81x^8}$
28. [10.4c] $13x^2 - 29x + 76$
29. [11.8a] 21 and 22; −22 and −21
30. [12.7a] Team A: 4 hr; team B: 10 hr
31. [12.9a] $\frac{3a + 2}{2a + 1}$

Chapter 13

Pretest: Chapter 13, p. 800

1. [13.1b] 4 **2.** [13.1b] 0 **3.** [13.2a] Slope: $\frac{1}{3}$; y-intercept: $\left(0, -\frac{7}{3}\right)$ **4.** [13.1a] Undefined
5. [13.2c] $y = x - 4$ **6.** [13.2b] $y = 4x + 7$
7. [13.5a] $y = \frac{5}{2}x$ **8.** [13.5c] $y = \frac{40}{x}$
9. [13.4b] **10.** [13.4b]

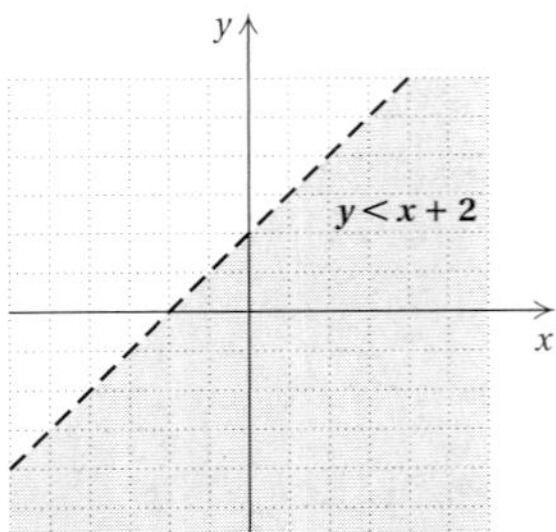

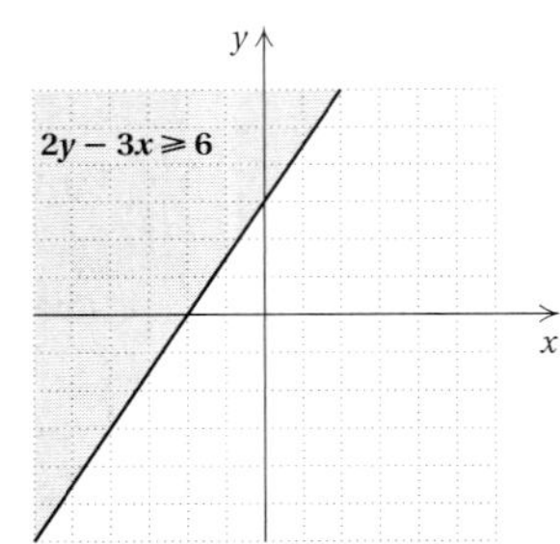

11. [13.3a] Parallel **12.** [13.3b] Perpendicular
13. [13.3b] Perpendicular **14.** [13.4a] Yes
15. [13.2a, c] **(a)** $y = 0.47x + 0.49$; **(b)** 0.47 billion dollars per year; **(c)** 5.19 billion dollars

Margin Exercises, Section 13.1, pp. 802–805

1. $\frac{2}{5}$ **2.** $-\frac{5}{3}$

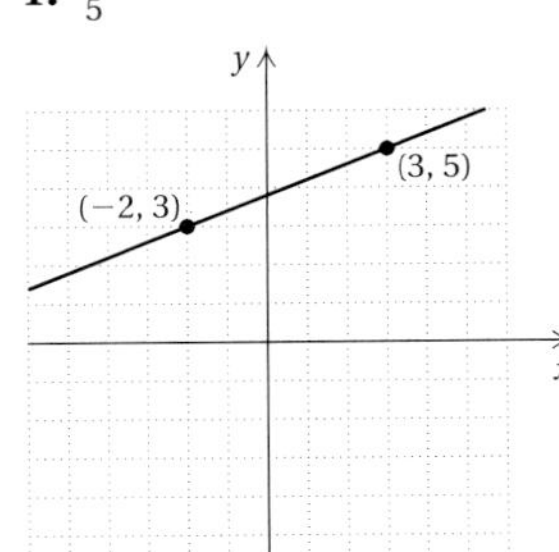

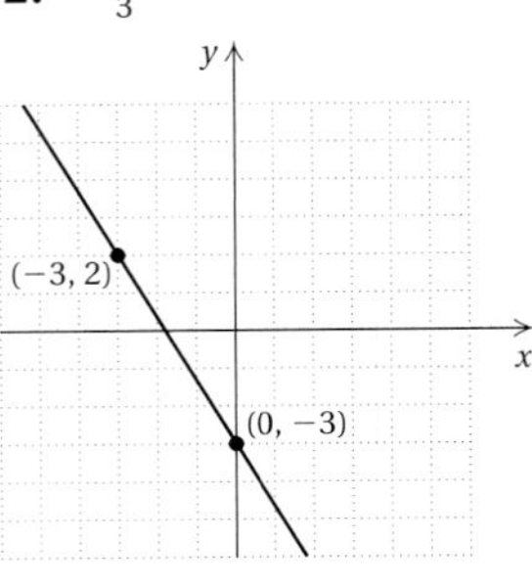

3. 4 **4.** −17 **5.** −1 **6.** $\frac{2}{3}$ **7.** −1 **8.** $\frac{5}{4}$
9. Undefined **10.** 0 **11.** $\frac{4}{5}$, or 80%

Calculator Spotlight, p. 803

1. This line will pass through the origin and slant up from left to right. This line will be steeper than $y = 10x$. **2.** This line will pass through the origin and slant up from left to right. This line will be less steep than $y = \frac{5}{32}x$.

Calculator Spotlight, p. 804

1. This line will pass through the origin and slant down from left to right. This line will be steeper than $y = -10x$. **2.** This line will pass through the origin and slant down from left to right. This line will be less steep than $y = -\frac{5}{32}x$.

Exercise Set 13.1, p. 807

1. $-\frac{3}{7}$ **3.** 0
5. $-\frac{4}{5}$ **7.** 3

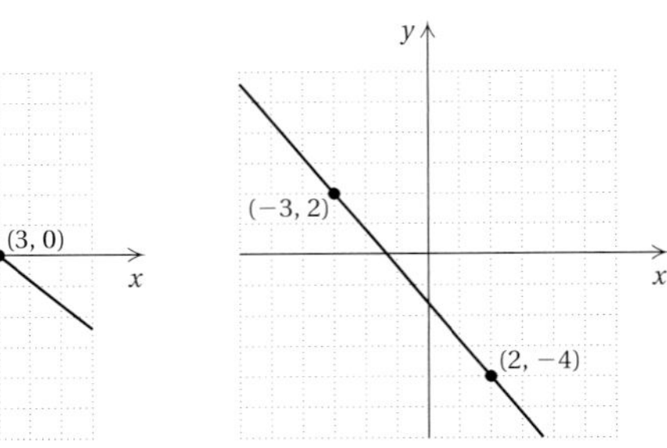

9. $\frac{2}{3}$ **11.** Undefined **13.** −10 **15.** 3.78 **17.** 3
19. $-\frac{1}{5}$ **21.** $-\frac{3}{2}$ **23.** $\frac{5}{7}$ **25.** −2.74 **27.** 3 **29.** $\frac{5}{4}$
31. 0 **33.** $\frac{12}{41}$ **35.** $\frac{28}{129}$ **37.** 15 cal per minute
39. 29.4% **41.** $\frac{44}{7}$ **42.** 10 **43.** −5
44. 391 **45.** $\frac{68}{109}$ **46.** 7412 **47.** $\frac{1}{2}$ **48.** 9 **49.** ◆

Margin Exercises, Section 13.2, pp. 809–810

1. Slope: 5; y-intercept: (0, 0) **2.** Slope: $-\frac{3}{2}$; y-intercept: (0, −6) **3.** Slope: $-\frac{3}{4}$; y-intercept: $\left(0, \frac{15}{4}\right)$
4. Slope: 2; y-intercept: $\left(0, -\frac{17}{2}\right)$ **5.** Slope: $-\frac{7}{5}$; y-intercept: $\left(0, -\frac{22}{5}\right)$ **6.** $y = 3.5x - 23$

7. $y = 5x - 18$ **8.** $y = -3x - 5$ **9.** $y = 6x - 13$
10. $y = -\frac{2}{3}x + \frac{14}{3}$ **11.** $y = x + 2$ **12.** $y = 2x + 4$

Exercise Set 13.2, p. 811

1. Slope: -4; y-intercept: $(0, -9)$ **3.** Slope: 1.8; y-intercept: $(0, 0)$ **5.** Slope: $-\frac{8}{7}$; y-intercept: $(0, -3)$ **7.** Slope: $\frac{4}{9}$; y-intercept: $\left(0, -\frac{7}{9}\right)$ **9.** Slope: $-\frac{3}{2}$; y-intercept: $\left(0, -\frac{1}{2}\right)$ **11.** Slope: 0; y-intercept: $(0, -17)$
13. $y = -7x - 13$ **15.** $y = 1.01x - 2.6$
17. $y = -2x - 6$ **19.** $y = \frac{3}{4}x + \frac{5}{2}$ **21.** $y = x - 8$
23. $y = -3x + 3$ **25.** $y = x + 4$ **27.** $y = -\frac{1}{2}x + 4$
29. $y = -\frac{3}{2}x + \frac{13}{2}$ **31.** $y = -4x - 11$
33. (a) $T = -0.75a + 165$; **(b)** -0.75 heart beats per minute per year; **(c)** 127.5 heart beats per minute
35. $0, -3$ **36.** $7, -7$ **37.** $3, -2$ **38.** $-5, 1$
39. $\frac{3}{2}, -7$ **40.** $-\frac{6}{5}, 4$ **41.** $-7, 2$ **42.** $\frac{2}{3}, -2$ **43.** $\frac{53}{7}$
44. $\frac{3}{8}$ **45.** $\frac{24}{19}$ **46.** $\frac{125}{7}$ **47.** **49.** $y = 3x - 9$
51. $y = \frac{3}{2}x - 2$

Margin Exercises, Section 13.3, pp. 813–814

1. No **2.** Yes **3.** Yes **4.** No

Exercise Set 13.3, p. 815

1. Yes **3.** No **5.** No **7.** No **9.** Yes **11.** Yes
13. No **15.** Yes **17.** Yes **19.** Yes **21.** No
23. Yes **25.** In 7 hr **26.** 130 km/h; 140 km/h
27. 4 **28.** -6 **29.** $\frac{30}{13}$ **30.** -11 **31.** $\frac{36}{11}$ **32.** 1
33. **35.–45.** Left to the student **47.** $y = 3x + 6$
49. $y = -3x + 2$ **51.** $y = \frac{1}{2}x + 1$ **53.** $k = 16$
55. A: $y = \frac{4}{3}x - \frac{7}{3}$; B: $y = -\frac{3}{4}x - \frac{1}{4}$

Margin Exercises, Section 13.4, pp. 817–820

1. No **2.** No

3.

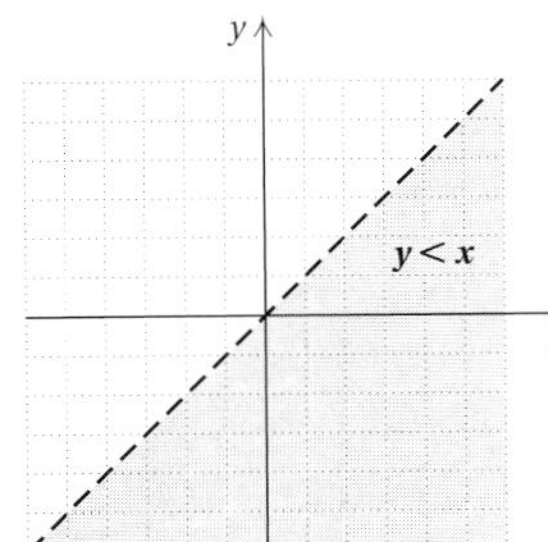

4.

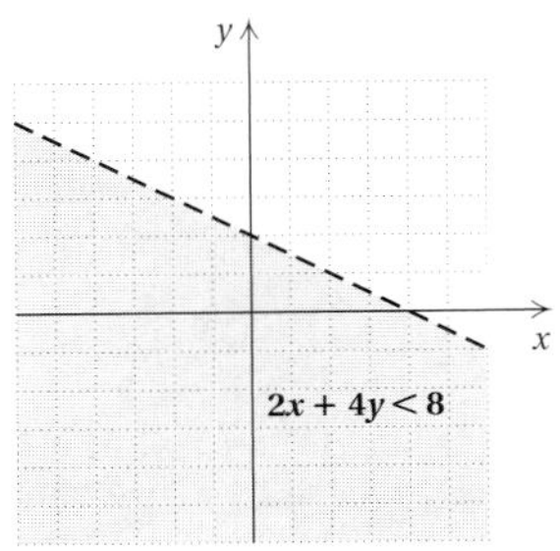

5.

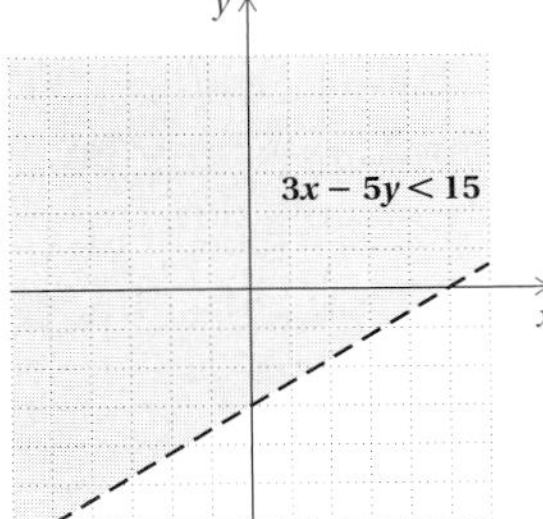

6.

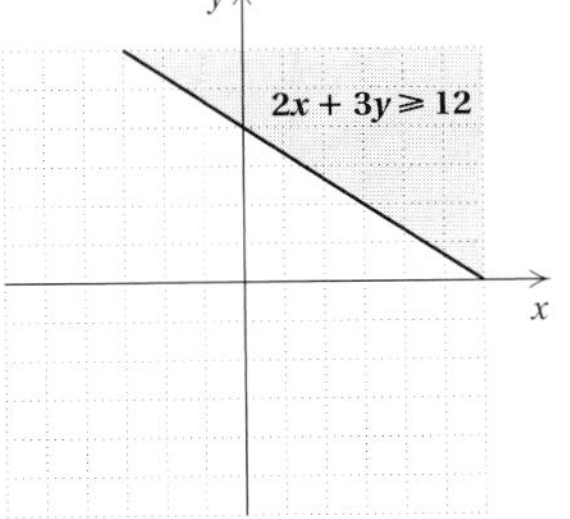

7.

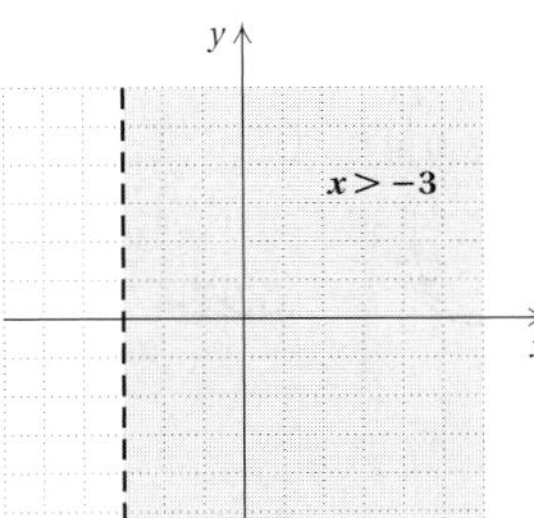

8.

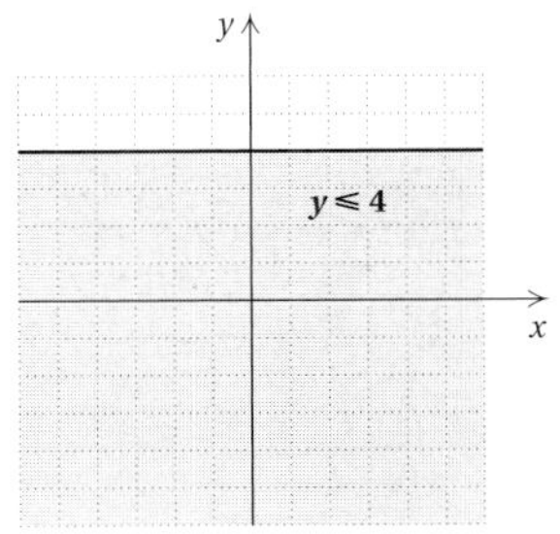

Exercise Set 13.4, p. 821

1. No **3.** Yes

5.

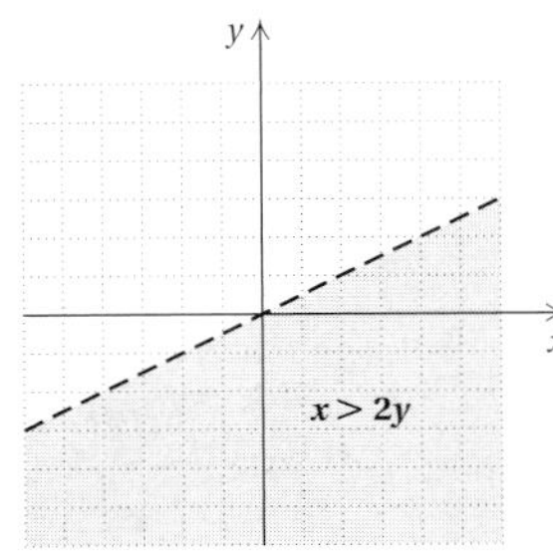

7.

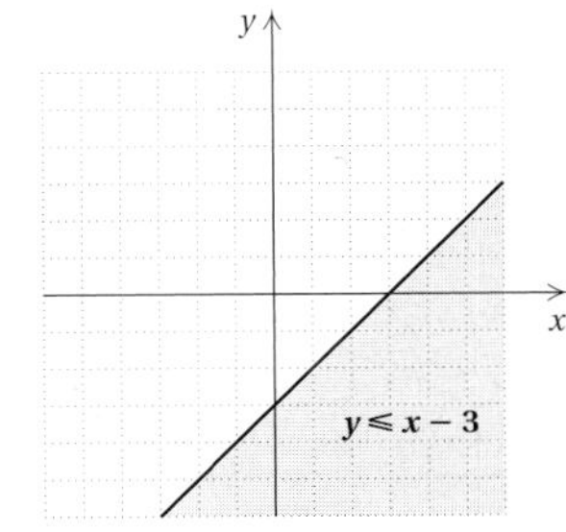

9.

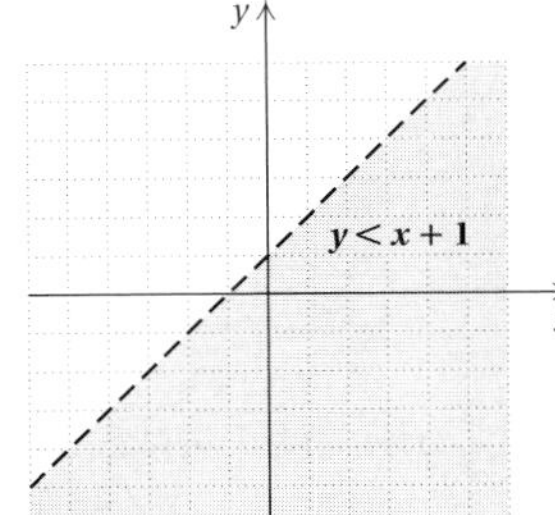

11.

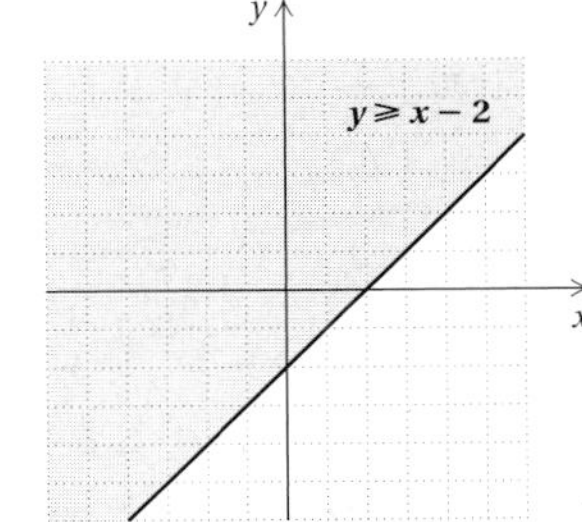

13.

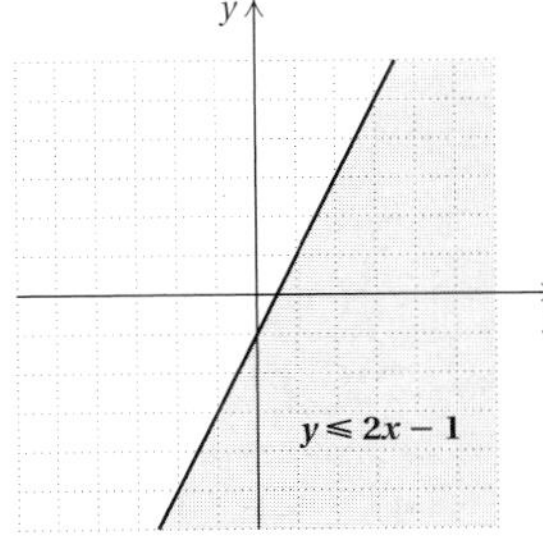

15.

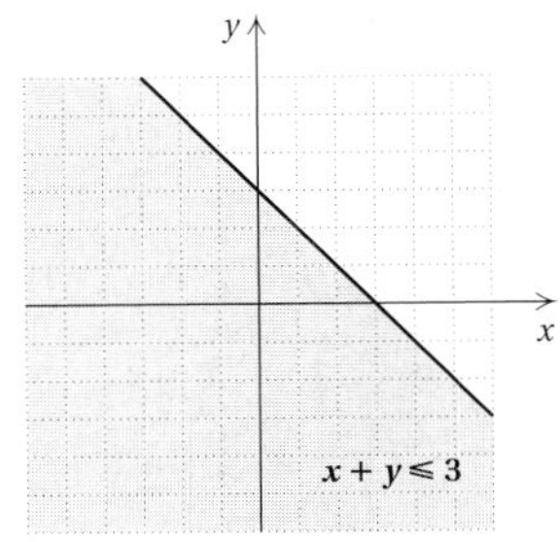

17.

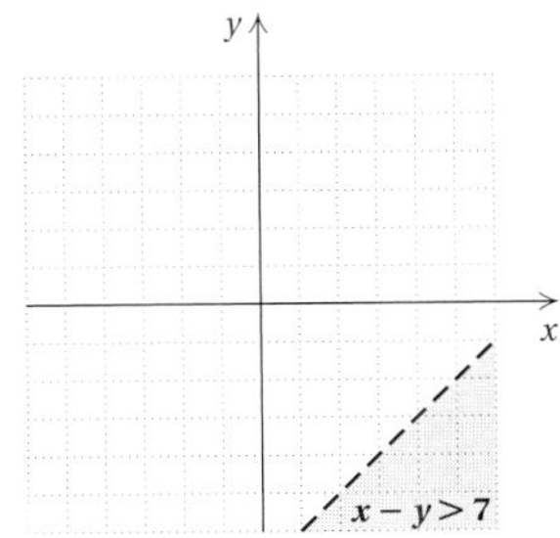

19.

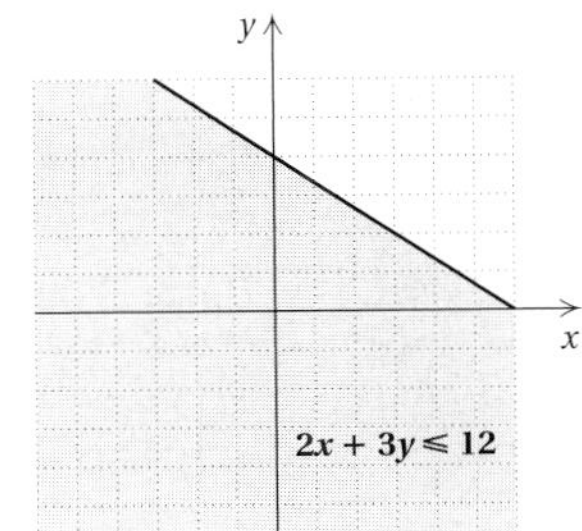

21.

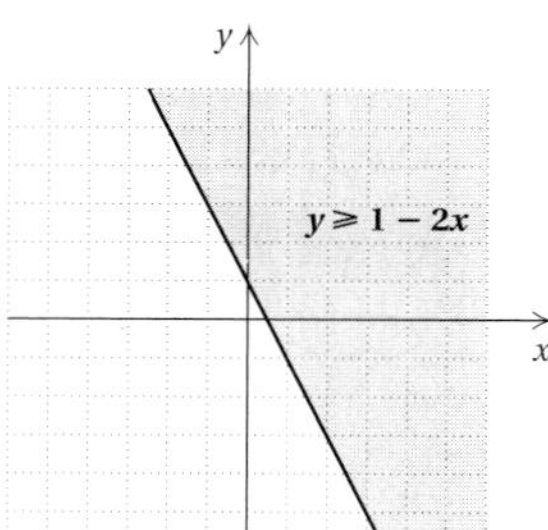

23.

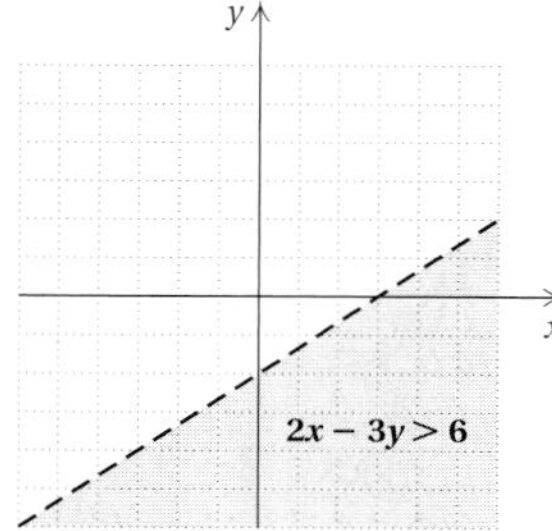

25.

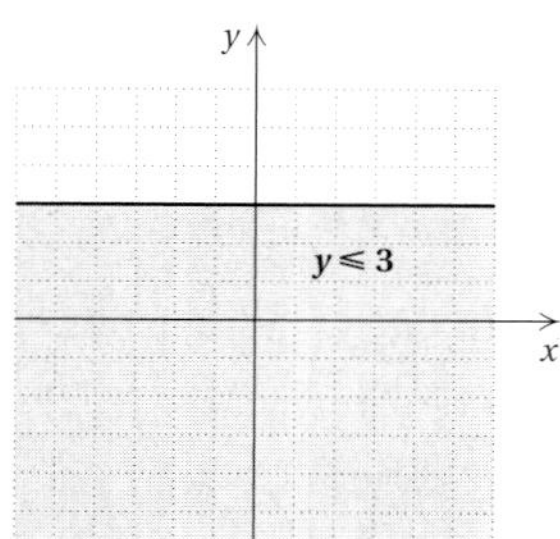

27.

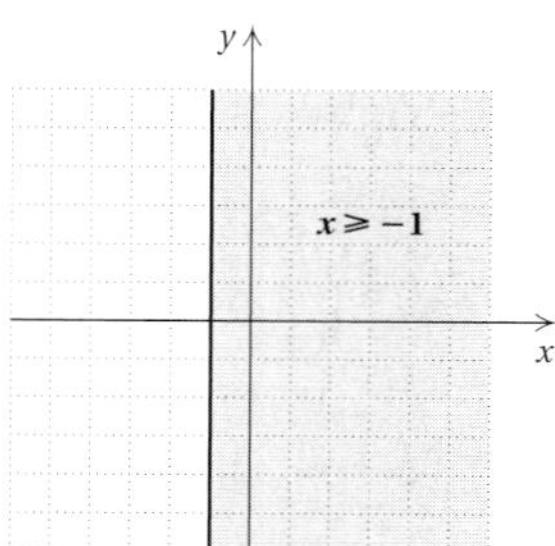

29. 3 **30.** $-2, -3$ **31.** 4 **32.** $\frac{1}{4}, -\frac{5}{3}$ **33.** ◆
35. $35c + 75a > 1000$

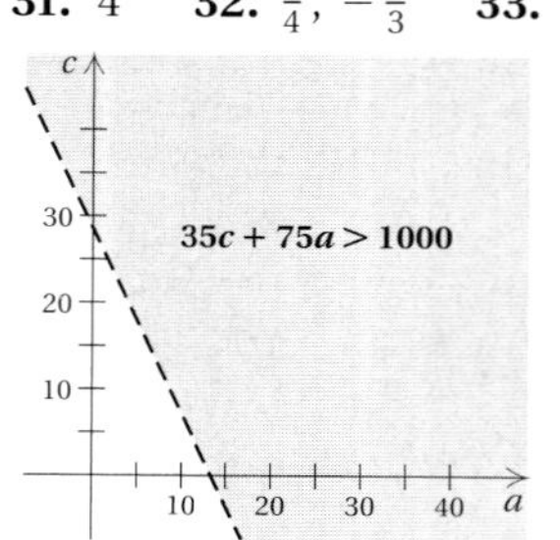

Margin Exercises, Section 13.5, pp. 824–826

1. $y = 7x$ **2.** $y = \frac{5}{8}x$ **3.** \$0.4667; \$0.0194
4. 174.24 lb **5.** $y = \dfrac{63}{x}$ **6.** $y = \dfrac{900}{x}$ **7.** 8 hr
8. $7\frac{1}{2}$ hr

Exercise Set 13.5, p. 827

1. $y = 4x$ **3.** $y = \frac{8}{5}x$ **5.** $y = 3.6x$ **7.** $y = \frac{25}{3}x$
9. \$196 **11.** \$100 **13.** $22\frac{6}{7}$ **15.** $36.\overline{6}$ lb
17. 16,000,000 **19.** $y = \dfrac{75}{x}$ **21.** $y = \dfrac{80}{x}$ **23.** $y = \dfrac{1}{x}$
25. $y = \dfrac{2100}{x}$ **27.** $y = \dfrac{0.06}{x}$ **29.** **(a)** Direct; **(b)** $69\frac{3}{8}$
31. **(a)** Inverse; **(b)** $4\frac{1}{2}$ hr **33.** 10 gal **35.** 32 amperes
37. 640 **39.** 8.25 ft **41.** $\frac{8}{5}$ **42.** 11 **43.** 9, 16
44. $-9, -12$ **45.** $\frac{4}{7}, \frac{2}{5}$ **46.** $-\frac{1}{7}, \frac{3}{2}$ **47.** $\frac{1}{3}$ **48.** $\frac{47}{20}$
49. ◆ **51.** ◆ **53.** The y-values become larger.
55. $P^2 = kt$ **57.** $P = kV^3$

Summary and Review: Chapter 13, p. 831

1. -1 **2.** Undefined
3. $-\frac{3}{4}$ **4.** 0

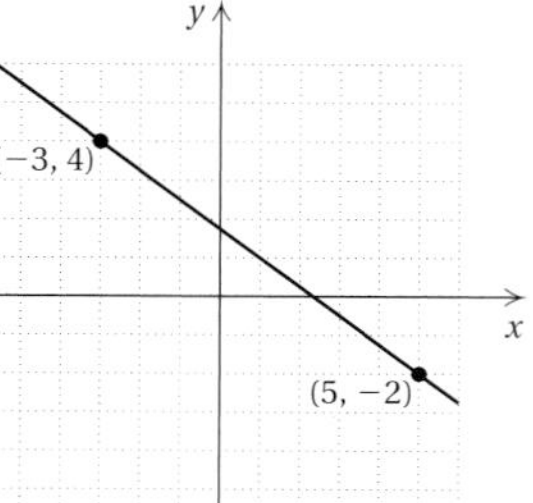

5. $\frac{3}{2}$ **6.** 2 **7.** 0 **8.** Undefined **9.** $-\frac{4}{3}$
10. Slope: -9; y-intercept: $(0, 46)$ **11.** Slope: -1; y-intercept: $(0, 9)$ **12.** Slope: $\frac{3}{5}$; y-intercept: $\left(0, -\frac{4}{5}\right)$
13. $y = -2.8x + 19$ **14.** $y = \frac{5}{8}x - \frac{7}{8}$ **15.** $y = 3x - 1$
16. $y = \frac{2}{3}x - \frac{11}{3}$ **17.** $y = -2x - 4$ **18.** $y = x + 2$
19. $y = \frac{1}{2}x - 1$ **20.** **(a)** $A = 0.233x + 5.87$; **(b)** 0.233; **(c)** 8.2 yr **21.** Parallel **22.** Perpendicular
23. Parallel **24.** Neither **25.** No **26.** No
27. Yes

28.

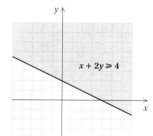

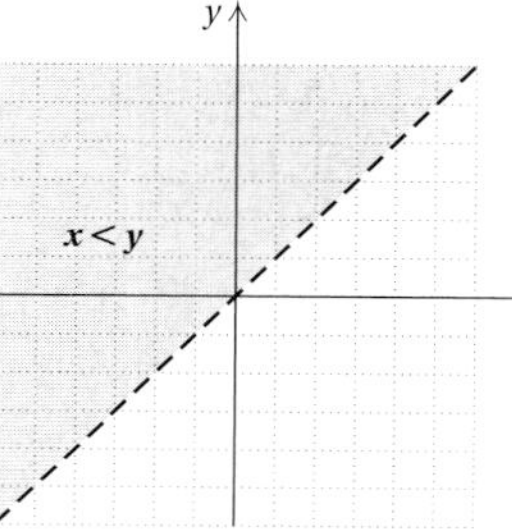

29.

30.

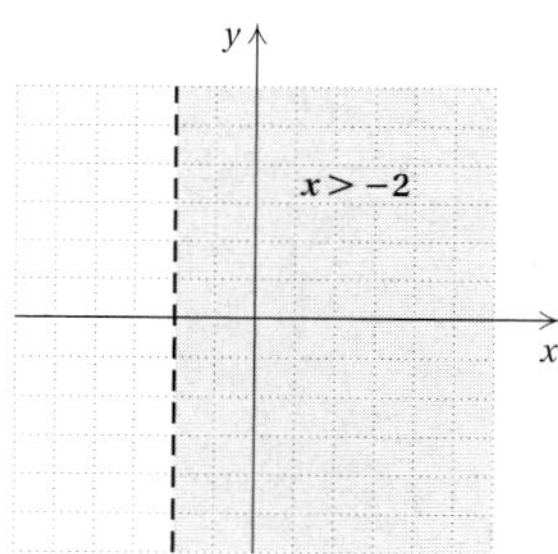

31. $y = 3x$ **32.** $y = \frac{1}{2}x$ **33.** $y = \frac{4}{5}x$ **34.** $y = \dfrac{30}{x}$
35. $y = \dfrac{1}{x}$ **36.** $y = \dfrac{0.65}{x}$ **37.** \$288.75 **38.** 1 hr
39. $3\frac{1}{3}$ hr **40.** 52 **41.** -4 **42.** $5, -11$
43. ◆ The concept of slope is useful in describing how a line slants. A line with positive slope slants up from left to right. A line with negative slope slants down from left to right. The larger the absolute value of the slope, the steeper the slant.

44. ◈ 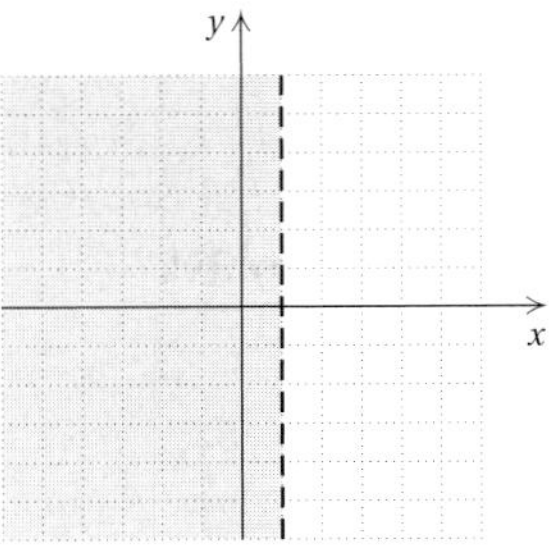

The graph of $x < 1$ on a number line consists of the points in the set $\{x \mid x < 1\}$. The graph of $x < 1$ on a plane consists of the points, or ordered pairs, in the set $\{(x, y) \mid x + 0 \cdot y < 1\}$. This is the set of ordered pairs with first coordinate less than 1.
45. $-\frac{1}{2}, \frac{1}{2}, 2, -2$

Test: Chapter 13, p. 833

1. [13.1a] -2 **2.** [13.1a] 0
3. [13.1a] $-\frac{2}{7}$,

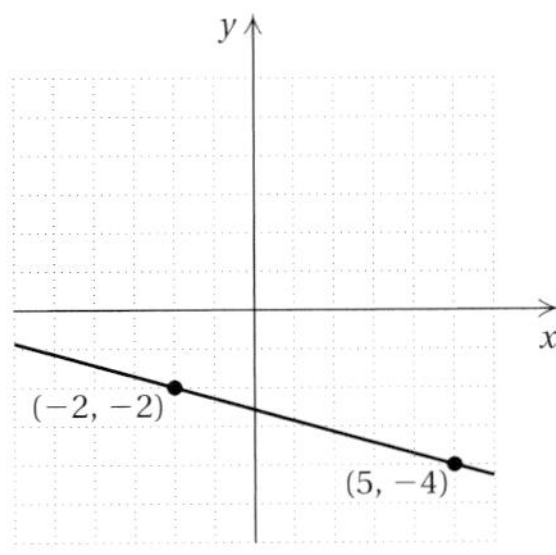

4. [13.1a] Undefined **5.** [13.1a] $\frac{7}{12}$ **6.** [13.1b] 0
7. [13.1b] Undefined **8.** [13.2a] Slope: 2; y-intercept: $\left(0, -\frac{1}{4}\right)$ **9.** [13.2a] Slope: $\frac{4}{3}$; y-intercept: $(0, -2)$ **10.** [13.2a] $y = 1.8x - 7$
11. [13.2a] $y = -\frac{3}{8}x - \frac{1}{8}$ **12.** [13.2b] $y = x + 2$
13. [13.2b] $y = -3x - 6$ **14.** [13.2c] $y = -3x + 4$
15. [13.2c] $y = \frac{1}{4}x - 2$ **16.** [13.2c] **(a)** $M = 102x + 2313$; **(b)** \$102 million per year; **(c)** \$3129 million **17.** [13.3a, b] Parallel
18. [13.3a, b] Neither **19.** [13.3a, b] Perpendicular
20. [13.4a] No **21.** [13.4a] Yes
22. [13.4b] **23.** [13.4b]

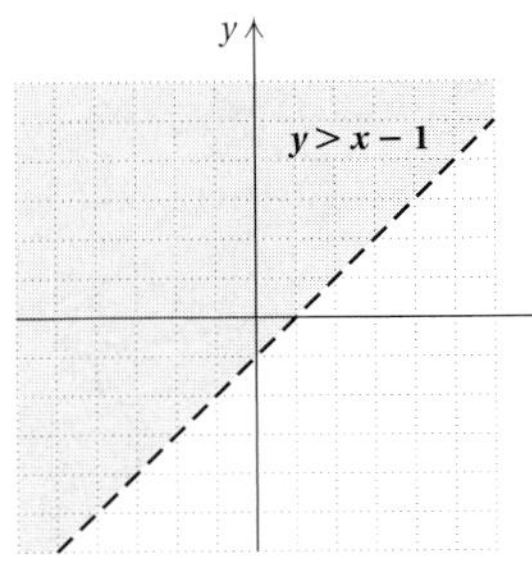

24. [13.5a] $y = 2x$ **25.** [13.5a] $y = 0.5x$
26. [13.5c] $y = \dfrac{18}{x}$ **27.** [13.5c] $y = \dfrac{22}{x}$
28. [13.5b] 240 km **29.** [13.5d] $1\frac{1}{5}$ hr
30. [12.7a] Freight: 90 mph; passenger: 105 mph
31. [7.8d] 1 **32.** [12.6a] 10 **33.** [11.7b] -7, 4
34. [13.3b] $k = 3$ **35.** [13.2b] $y = \frac{2}{3}x + \frac{11}{3}$

Chapter 14

Pretest: Chapter 14, p. 836

1. [14.1a] Yes **2.** [14.1b] No solution **3.** [14.2a] (5, 2)
4. [14.2b] (2, −1) **5.** [14.3a] $\left(\frac{3}{4}, \frac{1}{2}\right)$
6. [14.3b] (−5, −2) **7.** [14.3b] (10, 8) **8.** [14.3c] 50 and 24 **9.** [14.2c] 25° and 65° **10.** [14.5a] 8 hr after the second train leaves

Margin Exercises, Section 14.1, pp. 838–840

1. Yes **2.** No **3.** (2, −3) **4.** (−4, 3)
5. No solution **6.** Infinite number of solutions
7. **(a)** 3; **(b)** 3; **(c)** same **8.** **(a)** 3; **(b)** same

Exercise Set 14.1, p. 841

1. Yes **3.** No **5.** Yes **7.** Yes **9.** Yes **11.** (4, 2)
13. (4, 3) **15.** (−3, −3) **17.** No solution **19.** (2, 2)
21. $\left(\frac{1}{2}, 1\right)$ **23.** Infinite number of solutions
25. (5, −3) **27.** $\dfrac{108}{x^{13}}$ **28.** $3x^3$ **29.** $\dfrac{2x^2 - 1}{x^2(x + 1)}$
30. $\dfrac{-4}{x - 2}$ **31.** $\dfrac{9x + 12}{(x - 4)(x + 4)}$ **32.** $\dfrac{2x + 5}{x + 3}$
33. Trinomial **34.** Binomial **35.** Monomial
36. None of these **37.** **39.** $A = 2$, $B = 2$
41. $x + 2y = 2$, $x - y = 8$ **43.** (2, 1) **45.** (3, 2)
47. (−6, −2) **49.** Infinite number of solutions

Margin Exercises, Section 14.2, pp. 844–846

1. (3, 2) **2.** (3, −1) **3.** $\left(\frac{24}{5}, -\frac{8}{5}\right)$ **4.** Length: 84 ft; width: 50 ft

Calculator Spotlight, p. 845

1. (4.8, −1.6) **2.** (0.667, 0.429)

Exercise Set 14.2, p. 847

1. (1, 9) **3.** (2, −4) **5.** (4, 3) **7.** (−2, 1)
9. (2, −4) **11.** $\left(\frac{17}{3}, \frac{16}{3}\right)$ **13.** $\left(\frac{25}{8}, -\frac{11}{4}\right)$ **15.** (−3, 0)
17. (6, 3) **19.** 16 and 21 **21.** 12 and 40
23. 20 and 8 **25.** Length: 380 mi; width: 270 mi
27. Length: $3\frac{1}{2}$ in; width: $1\frac{3}{4}$ in.
29. **30.**

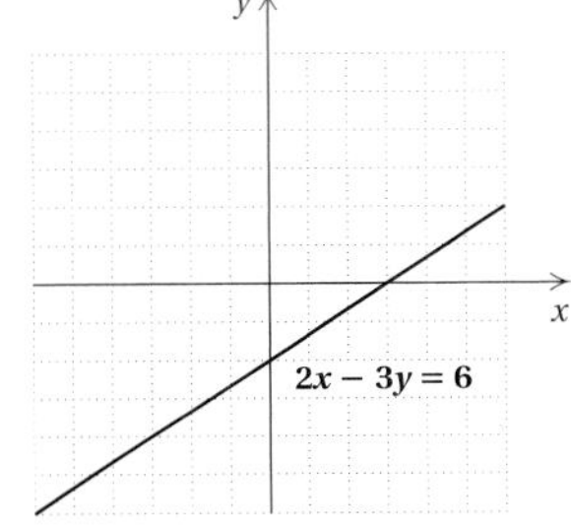

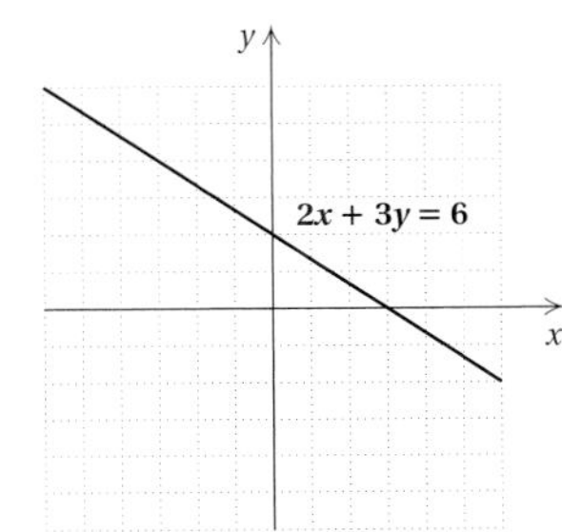

31. **32.**

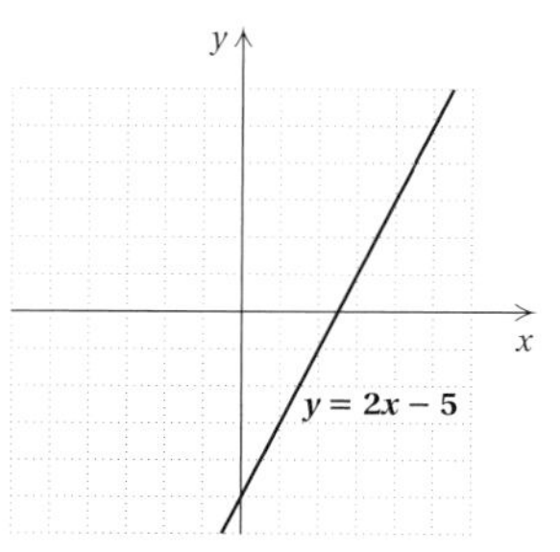

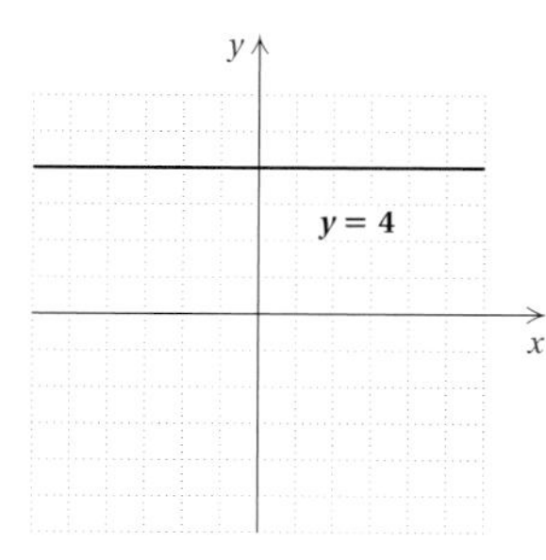

33. $(3x - 2)(2x - 3)$ **34.** $(4p + 3)(p - 1)$ **35.** Not factorable **36.** $(3a - 5)(3a + 5)$ **37.** **39.** $(5.\overline{6}, 0.\overline{6})$ **41.** $(4.38, 4.33)$

Margin Exercises, Section 14.3, pp. 849–854

1. $(3, 2)$ **2.** $(1, -2)$ **3.** $(1, 4)$ **4.** $(-8, 3)$ **5.** $(1, 1)$ **6.** $(-2, -1)$ **7.** $\left(\frac{17}{13}, -\frac{7}{13}\right)$ **8.** No solution **9.** Infinite number of solutions **10.** $(1, -1)$ **11.** $(2, -2)$ **12.** 75 mi

Calculator Spotlight, p. 852

1. You get equations of two lines with the same slope but different y-intercepts. The lines are parallel.
2. You get equivalent equations. The lines are the same.

Exercise Set 14.3, p. 855

1. $(6, -1)$ **3.** $(3, 5)$ **5.** $(2, 5)$ **7.** $\left(-\frac{1}{2}, 3\right)$ **9.** $\left(-1, \frac{1}{5}\right)$ **11.** No solution **13.** $(-1, -6)$ **15.** $(3, 1)$ **17.** $(8, 3)$ **19.** $(4, 3)$ **21.** $(1, -1)$ **23.** $(-3, -1)$ **25.** $(3, 2)$ **27.** $(50, 18)$ **29.** Infinite number of solutions **31.** $(2, -1)$ **33.** $\left(\frac{231}{202}, \frac{117}{202}\right)$ **35.** $(-38, -22)$ **37.** 200 mi **39.** 50° and 130° **41.** 62° and 28° **43.** Hay: 415 hectares; oats: 235 hectares **45.** $\frac{1}{x^7}$ **46.** x^3 **47.** $\frac{1}{x^3}$ **48.** x^7 **49.** x^3 **50.** x^7 **51.** $\frac{a^7}{b^9}$ **52.** $\frac{b^3}{a^3}$ **53.** $\frac{x - 3}{x + 2}$ **54.** $\frac{x + 5}{x - 5}$ **55.** $\frac{-x^2 - 7x + 23}{(x + 3)(x - 4)}$ **56.** $\frac{-2x + 10}{(x + 1)(x - 1)}$ **57.** **59.–67.** See answers for odd-numbered exercises 1–9. **69.–77.** See answers for odd-numbered exercises 21–30. **79.** Will is 6; his father is 30. **81.** $(5, 2)$ **83.** $(0, -1)$ **85.** 12 rabbits and 23 pheasants

Margin Exercises, Section 14.4, pp. 859–865

1. Hamburger: \$1.99; chicken: \$1.70 **2.** Sarah: 47; Malcolm: 21 **3.** 125 adults and 41 children **4.** 22.5 L of 50%; 7.5 L of 70% **5.** 30 lb of A; 20 lb of B **6.** 7 quarters, 13 dimes

Exercise Set 14.4, p. 867

1. Two-point shots: 14; one-point shots: 8 **3.** Kuyatts': 32 yr; Marconis': 16 yr **5.** Randy is 24; Mandy is 6. **7.** Brazilian: 200 lb; Turkish: 100 lb **9.** 70 dimes; 33 quarters **11.** One soda: \$1.45; one slice of pizza: \$2.25 **13.** 128 cardholders; 75 non-cardholders **15.** Upper Box: 17; Lower Reserved: 12 **17.** 40 L of A; 60 L of B **19.** Hay: 10 lb; grain: 5 lb **21.** 39 lb of A; 36 lb of B **23.** 12 of A, 4 of B; 180 **25.** Large type: $6\frac{17}{25}$; small type: $5\frac{8}{25}$ **27.** $(5x + 9)(5x - 9)$ **28.** $(6 - a)(6 + a)$ **29.** $4(x^2 + 25)$ **30.** $4(x + 5)(x - 5)$

31. **32.**

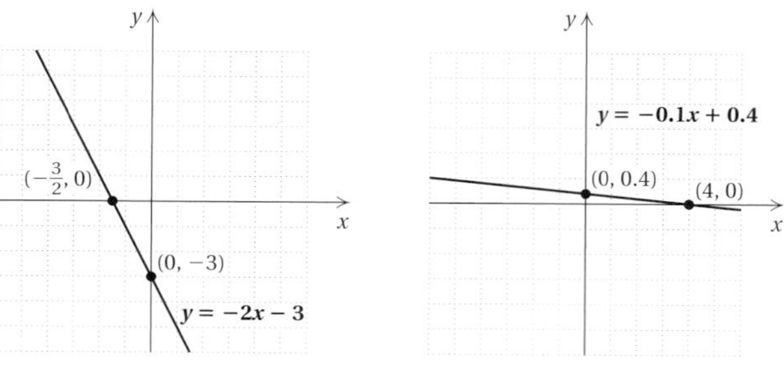

33. **34.**

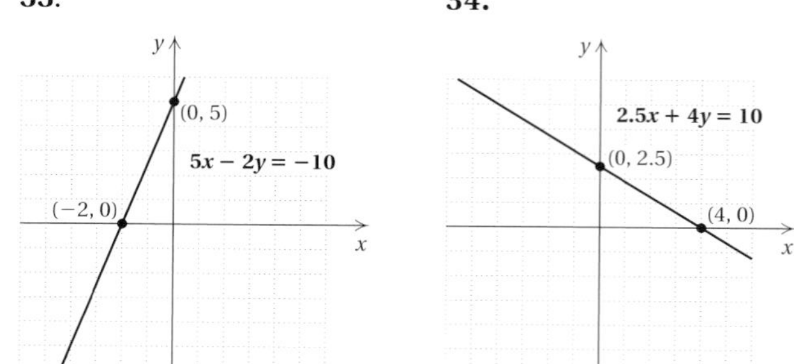

35. **37.** 12 gal of 87; 6 gal of 93 **39.** $4\frac{4}{7}$ L **41.** 43.75 L **43.** 11 in. by 14 in. **45.** \$25,000 at 6%; \$29,000 at 6.5%

Margin Exercises, Section 14.5, pp. 872–874

1. 324 mi **2.** 3 hr **3.** 168 km **4.** 275 km/h

Exercise Set 14.5, p. 875

1.

Speed	Time
30	t
46	t

4.5 hr

3.

Speed	Time	
72	$t + 3$	$\rightarrow d = 72(t + 3)$
120	t	$\rightarrow d = 120t$

$7\frac{1}{2}$ hr after the first train leaves, or $4\frac{1}{2}$ hr after the second train leaves

5.

Speed	Time	
$r + 6$	4	$\rightarrow d = (r + 6)4$
$r - 6$	10	$\rightarrow d = (r - 6)10$

14 km/h

7. 384 km **9.** **(a)** 24 mph; **(b)** 90 mi **11.** $1\frac{23}{43}$ min after the toddler starts running, or $\frac{23}{43}$ min after the mother starts running **13.** 15 mi **15.** $\frac{x}{3}$ **16.** $\frac{x^5y^3}{2}$

17. $\frac{a+3}{2}$ **18.** $\frac{x-2}{4}$ **19.** 2 **20.** $\frac{1}{x^2+1}$

21. $\frac{x+2}{x+3}$ **22.** $\frac{3(x+4)}{x-1}$ **23.** $\frac{x+3}{x+2}$

24. $\frac{x^2+25}{x^2-25}$ **25.** $\frac{x+2}{x-1}$ **26.** $\frac{x^2+2}{2x^2+1}$

27. ◈ **29.** Approximately 3603 mi **31.** $5\frac{1}{3}$ mi

Summary and Review: Chapter 14, p. 877

1. No **2.** Yes **3.** Yes **4.** No **5.** (6, −2) **6.** (6, 2) **7.** (0, 5) **8.** No solution **9.** (0, 5) **10.** (−3, 9) **11.** (3, −1) **12.** (1, 4) **13.** (−2, 4) **14.** (1, −2) **15.** (3, 1) **16.** (1, 4) **17.** (5, −3) **18.** (−2, 4) **19.** (−2, −6) **20.** (3, 2) **21.** (2, −4) **22.** Infinite number of solutions **23.** (−4, 1) **24.** 10 and −2 **25.** 12 and 15 **26.** Length: 37.5 cm; width: 10.5 cm **27.** Orchestra: 297; balcony: 211 **28.** 40 L of each **29.** Jeff is 27; his son is 9. **30.** Asian: 4800 kg; African: 7200 kg **31.** Peanuts: $6\frac{2}{3}$ lb; fancy nuts: $3\frac{1}{3}$ lb **32.** 135 km/h **33.** 412.5 mi **34.** t^8 **35.** $\frac{1}{t^{18}}$

36. $\frac{-4x+3}{(x-2)(x-3)(x+3)}$ **37.** $\frac{x+2}{x+10}$

38.

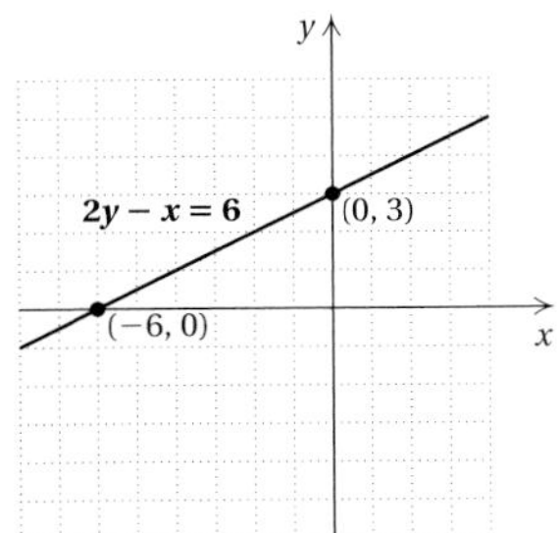

39. ◈ The methods are summarized in the table in Section 14.3.
40. ◈ The equations have the same slope but different y-intercepts, so they represent parallel lines. Thus the system of equations has no solution.
41. $960 **42.** $C = 1, D = 3$ **43.** (2, 0) **44.** $x + y = 5, -2x + 3y = 0$ **45.** $x + y = 4$, $x + y = -3$

Test: Chapter 14, p. 879

1. [14.1a] No **2.** [14.1b] (2, −1) **3.** [14.2a] (8, −2) **4.** [14.2b] (−1, 3) **5.** [14.2a] No solution **6.** [14.3a] (1, −5) **7.** [14.3b] (12, −6) **8.** [14.3b] (0, 1) **9.** [14.3b] (5, 1) **10.** [14.2c] Length: 2108.5 yd; width: 2024.5 yd **11.** [14.2c] 20 and 8 **12.** [14.5a] 40 km/h **13.** [14.4a] 40 L of A; 20 L of B

14. [12.5a] $\frac{-x^2+x+17}{(x-4)(x+4)(x+1)}$

15. [9.3a]

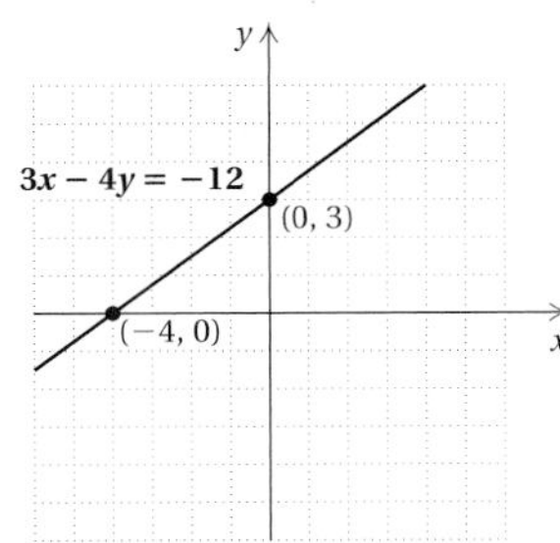

16. [10.1d, f] $\frac{10x^4}{y^2}$ **17.** [10.1e, f] $\frac{a^{10}}{b^6}$

18. [12.1c] $\frac{x+10}{2(x+2)}$ **19.** [14.1a] $C = -\frac{19}{2}; D = \frac{14}{3}$

20. [14.4a] 5 **21.** [13.2c], [14.1b] $x - 5y = -17$, $3x + 5y = 9$ **22.** [13.2c], [14.1b] $x = 3, y = -2$

Chapter 15

Pretest: Chapter 15, p. 882

1. [15.1a] 7, −7 **2.** [15.1d] $3t$ **3.** [15.1e] No **4.** [15.1e] Yes **5.** [15.1b] 6.856 **6.** [15.5a] 4 **7.** [15.1f] $2x$ **8.** [15.4a] $12\sqrt{2}$ **9.** [15.4b] $7 - 4\sqrt{3}$ **10.** [15.4b] 1 **11.** [15.2c] $2\sqrt{15}$ **12.** [15.4b] $25 - 4\sqrt{6}$ **13.** [15.3a] $\sqrt{5}$ **14.** [15.3b] $2a^2\sqrt{2}$ **15.** [15.6a] $\sqrt{89} \approx 9.434$ **16.** [15.6b] $\sqrt{193} \approx 13.892$ m **17.** [15.3c] $\frac{\sqrt{5x}}{x}$ **18.** [15.4c] $\frac{48 - 8\sqrt{5}}{31}$

Margin Exercises, Section 15.1, pp. 883–886

1. 6, −6 **2.** 8, −8 **3.** 11, −11 **4.** 12, −12 **5.** 4 **6.** 7 **7.** 10 **8.** 21 **9.** −7 **10.** −13 **11.** 3.873 **12.** 5.477 **13.** 31.305 **14.** −25.842 **15.** 0.816 **16.** −1.732 **17.** 28.3 mph, 49.6 mph **18.** 227 **19.** $45 + x$ **20.** $\frac{x}{x+2}$ **21.** $x^2 + 4$ **22.** Yes **23.** No **24.** No **25.** Yes **26.** 13 **27.** $|7w|$ **28.** $|xy|$ **29.** $|xy|$ **30.** $|x - 11|$ **31.** $|x + 4|$ **32.** xy **33.** xy **34.** $x - 11$ **35.** $x + 4$ **36.** $5y$ **37.** $\frac{1}{2}t$

Calculator Spotlight, p. 886

1.–8. Use your grapher. **9.** Not correct **10.** Correct

Exercise Set 15.1, p. 887

1. 2, −2 **3.** 3, −3 **5.** 10, −10 **7.** 13, −13 **9.** 16, −16 **11.** 2 **13.** −3 **15.** −6 **17.** −15 **19.** 19 **21.** 2.236 **23.** 20.785 **25.** −18.647 **27.** 2.779 **29.** 120 **31.** (a) 13; (b) 24 **33.** 200 **35.** $a - 4$ **37.** $t^2 + 1$ **39.** $\frac{3}{x + 2}$ **41.** No **43.** Yes **45.** c **47.** $3x$ **49.** $8p$ **51.** ab **53.** $34d$ **55.** $x + 3$ **57.** $a - 5$ **59.** $2a - 5$ **61.** $10,660 **62.** $\frac{1}{x + 3}$ **63.** 1 **64.** $\frac{(x + 2)(x - 2)}{(x + 1)(x - 1)}$ **65.** ◈ **67.** 1.7, 2.2, 2.6 **69.** −5 and −6 **71.** No solution **73.** $\sqrt{3}$

Margin Exercises, Section 15.2, pp. 889–892

1. (a) 20; (b) 20 **2.** $\sqrt{33}$ **3.** 5 **4.** $\sqrt{x^2 + x}$ **5.** $\sqrt{x^2 - 4}$ **6.** $4\sqrt{2}$ **7.** $x + 7$ **8.** $5x$ **9.** $6m$ **10.** $2\sqrt{23}$ **11.** $x - 10$ **12.** $8t$ **13.** $10a$ **14.** t^2 **15.** t^{10} **16.** h^{23} **17.** $x^3\sqrt{x}$ **18.** $2x^5\sqrt{6x}$ **19.** $3\sqrt{2}$ **20.** 10 **21.** $4x^3y^2$ **22.** $5xy^2\sqrt{2xy}$

Calculator Spotlight, p. 889

1. Correct **2.** Not correct

Calculator Spotlight, p. 892

1. Not correct **2.** Not correct

Exercise Set 15.2, p. 893

1. $2\sqrt{3}$ **3.** $5\sqrt{3}$ **5.** $2\sqrt{5}$ **7.** $10\sqrt{6}$ **9.** $9\sqrt{6}$ **11.** $3\sqrt{x}$ **13.** $4\sqrt{3x}$ **15.** $4\sqrt{a}$ **17.** $8y$ **19.** $x\sqrt{13}$ **21.** $2t\sqrt{2}$ **23.** $6\sqrt{5}$ **25.** $12\sqrt{2y}$ **27.** $2x\sqrt{7}$ **29.** $x - 3$ **31.** $\sqrt{2}(2x + 1)$ **33.** $\sqrt{y}(6 + y)$ **35.** x^3 **37.** x^6 **39.** $x^2\sqrt{x}$ **41.** $t^9\sqrt{t}$ **43.** $(y - 2)^4$ **45.** $2(x + 5)^5$ **47.** $6m\sqrt{m}$ **49.** $2a^2\sqrt{2a}$ **51.** $2p^8\sqrt{26p}$ **53.** $8x^3y\sqrt{7y}$ **55.** $3\sqrt{6}$ **57.** $3\sqrt{10}$ **59.** $6\sqrt{7x}$ **61.** $6\sqrt{xy}$ **63.** 13 **65.** $5b\sqrt{3}$ **67.** $2t$ **69.** $a\sqrt{bc}$ **71.** $2xy\sqrt{2xy}$ **73.** 18 **75.** $\sqrt{10x - 5}$ **77.** $x + 2$ **79.** $6xy^3\sqrt{3xy}$ **81.** $10x^2y^3\sqrt{5xy}$ **83.** (−2, 4) **84.** $\left(\frac{1}{8}, \frac{9}{8}\right)$ **85.** (2, 1) **86.** (10, 3) **87.** 360 ft^2 **88.** Two-point: 36; one-point: 28 **89.** 80 L of 30%, 120 L of 50% **90.** Adults: 350; children: 61 **91.** 10 mph **93.** ◈ **95.** $\sqrt{x - 2}\sqrt{x + 1}$ **97.** $\sqrt{2x + 3}\sqrt{x - 4}$ **99.** $\sqrt{a + b}\sqrt{a - b}$ **101.** 0.5 **103.** $3a^3$ **105.** $4y\sqrt{3}$ **107.** $18(x + 1)\sqrt{y(x + 1)}$ **109.** $2x^3\sqrt{5x}$

Margin Exercises, Section 15.3, pp. 897–900

1. 4 **2.** 5 **3.** $x\sqrt{6x}$ **4.** $\frac{4}{3}$ **5.** $\frac{1}{5}$ **6.** $\frac{6}{x}$ **7.** $\frac{3}{4}$ **8.** $\frac{15}{16}$ **9.** $\frac{7}{y^5}$ **10.** $\frac{\sqrt{15}}{5}$ **11.** $\frac{\sqrt{10}}{4}$ **12.** $\frac{10\sqrt{3}}{3}$ **13.** $\frac{\sqrt{21}}{7}$ **14.** $\frac{\sqrt{5r}}{r}$ **15.** $\frac{8y\sqrt{7}}{7}$

Calculator Spotlight, p. 897

1. Correct **2.** Correct **3.** Not correct

Exercise Set 15.3, p. 901

1. 3 **3.** 6 **5.** $\sqrt{5}$ **7.** $\frac{1}{5}$ **9.** $\frac{2}{5}$ **11.** 2 **13.** $3y$ **15.** $\frac{4}{7}$ **17.** $\frac{1}{6}$ **19.** $-\frac{4}{9}$ **21.** $\frac{8}{17}$ **23.** $\frac{13}{14}$ **25.** $\frac{5}{x}$ **27.** $\frac{3a}{25}$ **29.** $\frac{\sqrt{10}}{5}$ **31.** $\frac{\sqrt{14}}{4}$ **33.** $\frac{\sqrt{3}}{6}$ **35.** $\frac{\sqrt{10}}{6}$ **37.** $\frac{3\sqrt{5}}{5}$ **39.** $\frac{2\sqrt{6}}{3}$ **41.** $\frac{\sqrt{3x}}{x}$ **43.** $\frac{\sqrt{xy}}{y}$ **45.** $\frac{x\sqrt{5}}{10}$ **47.** $\frac{\sqrt{14}}{2}$ **49.** $\frac{3\sqrt{2}}{4}$ **51.** $\frac{\sqrt{6}}{2}$ **53.** $\sqrt{2}$ **55.** $\frac{\sqrt{55}}{11}$ **57.** $\frac{\sqrt{21}}{6}$ **59.** $\frac{\sqrt{6}}{2}$ **61.** 5 **63.** $\frac{\sqrt{3x}}{x}$ **65.** $\frac{4y\sqrt{5}}{5}$ **67.** $\frac{a\sqrt{2a}}{4}$ **69.** $\frac{\sqrt{42x}}{3x}$ **71.** $\frac{3\sqrt{6}}{8c}$ **73.** $\frac{y\sqrt{xy}}{x}$ **75.** $\frac{3n\sqrt{10}}{8}$ **77.** (4, 2) **78.** (10, 30) **79.** No solution **80.** Infinite number of solutions **81.** $\left(-\frac{5}{2}, -\frac{9}{2}\right)$ **82.** $\left(\frac{26}{23}, \frac{44}{23}\right)$ **83.** $9x^2 - 49$ **84.** $16a^2 - 25b^2$ **85.** $21x - 9y$ **86.** $14a - 6b$ **87.** ◈ **89.** 1.57 sec; 3.14 sec; 8.88 sec; 11.10 sec **91.** 1 sec **93.** $\frac{\sqrt{5}}{40}$ **95.** $\frac{\sqrt{5x}}{5x^2}$ **97.** $\frac{\sqrt{3ab}}{b}$ **99.** $\frac{3\sqrt{10}}{100}$ **101.** $\frac{y - x}{xy}$

Margin Exercises, Section 15.4, pp. 905–908

1. $12\sqrt{2}$ **2.** $5\sqrt{5}$ **3.** $-12\sqrt{10}$ **4.** $5\sqrt{6}$ **5.** $\sqrt{x + 1}$ **6.** $\frac{3}{2}\sqrt{2}$ **7.** $\frac{8\sqrt{15}}{15}$ **8.** $\sqrt{15} + \sqrt{6}$ **9.** $4 + 3\sqrt{5} - 4\sqrt{2} - 3\sqrt{10}$ **10.** $2 - a$ **11.** $25 + 10\sqrt{x} + x$ **12.** 2 **13.** $7 - \sqrt{5}$ **14.** $\sqrt{5} + \sqrt{2}$ **15.** $1 + \sqrt{x}$ **16.** $\frac{21 - 3\sqrt{5}}{22}$ **17.** $\frac{7 + 2\sqrt{10}}{3}$ **18.** $\frac{7 + 7\sqrt{x}}{1 - x}$

Exercise Set 15.4, p. 909

1. $16\sqrt{3}$ **3.** $4\sqrt{5}$ **5.** $13\sqrt{x}$ **7.** $-9\sqrt{d}$ **9.** $25\sqrt{2}$ **11.** $\sqrt{3}$ **13.** $\sqrt{5}$ **15.** $13\sqrt{2}$ **17.** $3\sqrt{3}$ **19.** $2\sqrt{2}$ **21.** 0 **23.** $(2 + 9x)\sqrt{x}$ **25.** $(3 - 2x)\sqrt{3}$ **27.** $3\sqrt{2x + 2}$ **29.** $(x + 3)\sqrt{x^3 - 1}$ **31.** $(4a^2 + a^2b - 5b)\sqrt{b}$ **33.** $\frac{2\sqrt{3}}{3}$ **35.** $\frac{13\sqrt{2}}{2}$ **37.** $\frac{\sqrt{6}}{6}$ **39.** $\sqrt{15} - \sqrt{3}$ **41.** $10 + 5\sqrt{3} - 2\sqrt{7} - \sqrt{21}$ **43.** $9 - 4\sqrt{5}$ **45.** −62 **47.** 1 **49.** $13 + \sqrt{5}$ **51.** $x - 2\sqrt{xy} + y$ **53.** $-\sqrt{3} - \sqrt{5}$ **55.** $5 - 2\sqrt{6}$ **57.** $\frac{4\sqrt{10} - 4}{9}$ **59.** $5 - 2\sqrt{7}$ **61.** $\frac{12 - 3\sqrt{x}}{16 - x}$ **63.** $\frac{24 + 3\sqrt{x} + 8\sqrt{2} + \sqrt{2x}}{64 - x}$ **65.** $\frac{5}{11}$ **66.** $-\frac{38}{13}$

67. 6, −1 **68.** 5, 2 **69.** Jolly Juice: 1.6 L; Real Squeeze: 6.4 L **70.** $\frac{1}{3}$ hr **71.** −9, −2, −5, −17, −0.678375 **73.** ◈ **75.** Not equivalent **77.** Not correct **79.** $11\sqrt{3} - 10\sqrt{2}$ **81.** True; $(3\sqrt{x+2})^2 = (3\sqrt{x+2})(3\sqrt{x+2}) = (3 \cdot 3)(\sqrt{x+2} \cdot \sqrt{x+2}) = 9(x+2)$

Margin Exercises, Section 15.5, pp. 913–917

1. $\frac{64}{3}$ **2.** 2 **3.** $\frac{3}{8}$ **4.** 4 **5.** 1 **6.** 4 **7.** Approximately 313 km **8.** Approximately 16 km **9.** 196 m

Calculator Spotlight, p. 915

1. 21.3 **2.** 2 **3.** 0.375 **4.** 4; for $x = -1$, $y_1 = x - 1 = -1 - 1 = -2$ and $y_2 = \sqrt{x+5} = \sqrt{-1+5} = 2$. The y-values are not the same; the graphs do not intersect.

Exercise Set 15.5, p. 919

1. 36 **3.** 18.49 **5.** 165 **7.** $\frac{621}{2}$ **9.** 5 **11.** 3 **13.** $\frac{17}{4}$ **15.** No solution **17.** No solution **19.** 9 **21.** 12 **23.** 1, 5 **25.** 3 **27.** 5 **29.** No solution **31.** $-\frac{10}{3}$ **33.** 3 **35.** No solution **37.** 9 **39.** 1 **41.** 8 **43.** 36 m **45.** Approximately 21.3 km **47.** 151 ft; 281 ft **49.** 12 **51.** 8 **53.** $\dfrac{(x+7)^2}{x-7}$ **54.** $\dfrac{(x-2)(x-5)}{(x-3)(x-4)}$ **55.** $\dfrac{a-5}{2}$ **56.** $\dfrac{x-3}{x-2}$ **57.** 61°, 119° **58.** 38°, 52° **59.** $\dfrac{x^6}{3}$ **60.** $\dfrac{x-3}{4(x+3)}$ **61.** ◈ **63.** 2, −2 **65.** $-\frac{57}{16}$ **67.** 13 **69.** 3 **71.** $4.25 = \frac{17}{4}$

Margin Exercises, Section 15.6, pp. 923–924

1. $\sqrt{65} \approx 8.062$ **2.** $\sqrt{75} \approx 8.660$ **3.** $\sqrt{10} \approx 3.162$ **4.** $\sqrt{175} \approx 13.229$ **5.** $\sqrt{325} \approx 18.028$ ft

Exercise Set 15.6, p. 925

1. 17 **3.** $\sqrt{32} \approx 5.657$ **5.** 12 **7.** 4 **9.** 26 **11.** 12 **13.** 2 **15.** $\sqrt{2} \approx 1.414$ **17.** 5 **19.** 3 **21.** $\sqrt{211{,}200{,}000} \approx 14{,}533$ ft **23.** 240 ft **25.** $\sqrt{18} \approx 4.243$ cm **27.** $\sqrt{208} \approx 14.422$ ft **29.** $\left(-\frac{3}{2}, -\frac{1}{16}\right)$ **30.** $\left(\frac{8}{5}, 9\right)$ **31.** $\left(-\frac{9}{19}, \frac{91}{38}\right)$ **32.** $(-10, 1)$ **33.** $-\frac{1}{3}$ **34.** $\frac{5}{8}$ **35.** ◈ **37.** $\sqrt{1525} \approx 39.1$ mi **39.** $12 - 2\sqrt{6} \approx 7.101$ **41.** $\dfrac{\sqrt{3}}{2} \approx 0.866$

Summary and Review: Chapter 15, p. 927

1. 8, −8 **2.** 20, −20 **3.** 6 **4.** −13 **5.** 1.732 **6.** 9.950 **7.** −17.892 **8.** 0.742 **9.** −2.055 **10.** 394.648 **11.** $x^2 + 4$ **12.** $5ab^3$ **13.** No **14.** Yes **15.** No **16.** No **17.** No **18.** No **19.** m **20.** $x - 4$ **21.** $\sqrt{21}$ **22.** $\sqrt{x^2 - 9}$ **23.** $-4\sqrt{3}$ **24.** $4t\sqrt{2}$ **25.** $\sqrt{t-7}\sqrt{t+7}$ **26.** $x + 8$ **27.** x^4 **28.** $m^7\sqrt{m}$ **29.** $2\sqrt{15}$ **30.** $2x\sqrt{10}$ **31.** $5xy\sqrt{2}$ **32.** $10a^2b\sqrt{ab}$ **33.** $\dfrac{5}{8}$ **34.** $\dfrac{2}{3}$ **35.** $\dfrac{7}{t}$ **36.** $\dfrac{\sqrt{2}}{2}$ **37.** $\dfrac{\sqrt{2}}{4}$ **38.** $\dfrac{\sqrt{5y}}{y}$ **39.** $\dfrac{2\sqrt{3}}{3}$ **40.** $\dfrac{\sqrt{15}}{5}$ **41.** $\dfrac{x\sqrt{30}}{6}$ **42.** $8 - 4\sqrt{3}$ **43.** $13\sqrt{5}$ **44.** $\sqrt{5}$ **45.** $\dfrac{\sqrt{2}}{2}$ **46.** $7 + 4\sqrt{3}$ **47.** 1 **48.** 52 **49.** No solution **50.** 0, 3 **51.** 9 **52.** 20 **53.** $\sqrt{3} \approx 1.732$ **54.** About 50,990 ft **55.** 9 ft **56.** (a) About 63 mph; (b) 405 ft **57.** $\left(\frac{22}{17}, -\frac{8}{17}\right)$ **58.** $\dfrac{(x-5)(x-7)}{(x+7)(x+5)}$ **59.** \$450 **60.** 4300 **61.** ◈ It is incorrect to take the square roots of the terms in the numerator individually. That is, $\sqrt{a+b}$ and $\sqrt{a} + \sqrt{b}$ are not equivalent. The following is correct:

$$\sqrt{\frac{9+100}{25}} = \frac{\sqrt{9+100}}{\sqrt{25}} = \frac{\sqrt{109}}{5}.$$

62. ◈ **a)** $\sqrt{5x^2} = \sqrt{5}\sqrt{x^2} = \sqrt{5} \cdot |x| = |x|\sqrt{5}$; the given statement is correct.
b) Let $b = 3$. Then $\sqrt{b^2 - 4} = \sqrt{3^2 - 4} = \sqrt{9-4} = \sqrt{5}$, but $b - 2 = 3 - 2 = 1$. The given statement is false.
c) Let $x = 3$. Then $\sqrt{x^2 + 16} = \sqrt{3^2 + 16} = \sqrt{9+16} = \sqrt{25} = 5$, but $x + 4 = 3 + 4 = 7$. The given statement is false.

63. 2 **64.** $b = \sqrt{A^2 - a^2}$

Test: Chapter 15, p. 929

1. [15.1a] 9, −9 **2.** [15.1a] 8 **3.** [15.1a] −5 **4.** [15.1b] 10.770 **5.** [15.1b] −9.349 **6.** [15.1b] 4.127 **7.** [15.1d] $4 - y^3$ **8.** [15.1e] Yes **9.** [15.1e] No **10.** [15.1f] a **11.** [15.1f] $6y$ **12.** [15.2c] $\sqrt{30}$ **13.** [15.2c] $\sqrt{x^2 - 64}$ **14.** [15.2a] $3\sqrt{3}$ **15.** [15.2a] $5\sqrt{x-1}$ **16.** [15.2b] $t^2\sqrt{t}$ **17.** [15.2c] $5\sqrt{2}$ **18.** [15.2c] $3ab^2\sqrt{2}$ **19.** [15.3b] $\dfrac{3}{2}$ **20.** [15.3b] $\dfrac{12}{a}$ **21.** [15.3c] $\dfrac{\sqrt{10}}{5}$ **22.** [15.3c] $\dfrac{\sqrt{2xy}}{y}$ **23.** [15.3a, c] $\dfrac{3\sqrt{6}}{8}$ **24.** [15.3a] $\dfrac{\sqrt{7}}{4y}$ **25.** [15.4a] $-6\sqrt{2}$ **26.** [15.4a] $\dfrac{6\sqrt{5}}{5}$ **27.** [15.4b] $21 - 8\sqrt{5}$ **28.** [15.4b] 11 **29.** [15.4c] $\dfrac{40 + 10\sqrt{5}}{11}$ **30.** [15.6a] $\sqrt{80} \approx 8.944$ **31.** [15.5a] 48 **32.** [15.5a] 2, −2 **33.** [15.5b] −3 **34.** [15.5c] About 5000 m **35.** [14.4a] 789.25 yd^2 **36.** [13.5b] $15{,}686\frac{2}{3}$ **37.** [14.3b] (0, 2) **38.** [12.2b] $\dfrac{x-7}{x+6}$ **39.** [15.1a] $\sqrt{5}$ **40.** [15.2b] y^{8n}

Chapter 16

Pretest: Chapter 16, p. 932

1. [16.1c] 3 **2.** [16.2a] $\sqrt{7}, -\sqrt{7}$ **3.** [16.3a] $\dfrac{-3 \pm \sqrt{21}}{6}$
4. [16.1b] $0, \frac{3}{5}$ **5.** [16.1b] $0, -\frac{5}{3}$ **6.** [16.2b] $-4 \pm \sqrt{5}$
7. [16.2c] $1 \pm \sqrt{6}$ **8.** [16.4a] $n = \dfrac{p \pm \sqrt{p^2 + 4A}}{2}$
9. [16.5a] Width: 4 cm; length: 12 cm
10. [16.6b] $\left(\dfrac{-1 + \sqrt{33}}{4}, 0\right), \left(\dfrac{-1 - \sqrt{33}}{4}, 0\right)$
11. [16.5a] 10 km/h
12. [16.6a]

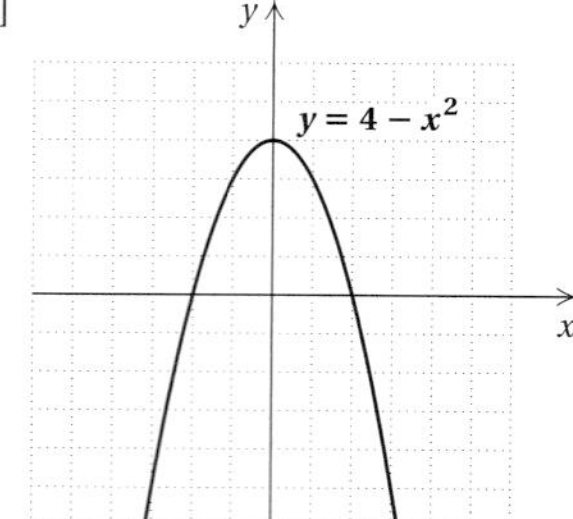

Margin Exercises, Section 16.1, pp. 933–937

1. y-intercept: $(0, -3)$; x-intercept: $\left(-\frac{9}{2}, 0\right)$;

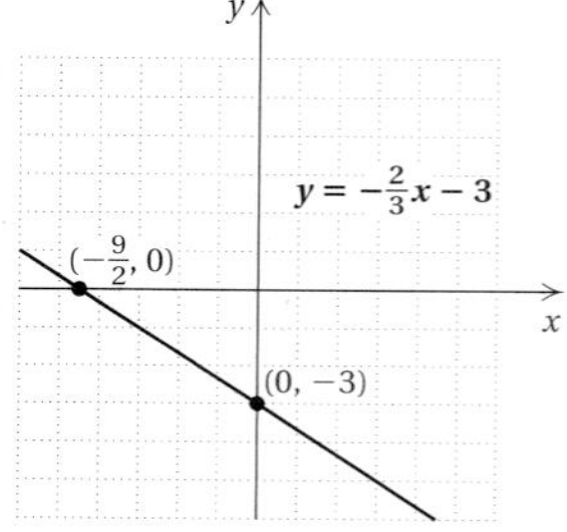

2. $-\frac{9}{2}$ **3.** $y^2 - 8y = 0$; $a = 1$, $b = -8$, $c = 0$
4. $x^2 + 9x - 3 = 0$; $a = 1$, $b = 9$, $c = -3$
5. $4x^2 + x + 4 = 0$; $a = 4$, $b = 1$, $c = 4$
6. $5x^2 - 21 = 0$; $a = 5$, $b = 0$, $c = -21$ **7.** $0, -\frac{9}{2}$
8. $0, \frac{3}{5}$ **9.** $\frac{3}{4}, -2$ **10.** 4, 1 **11.** 13, 5
12. **(a)** 14; **(c)** 11

Calculator Spotlight, p. 938

1. 0.6, 1 **2.** 3, 8

Exercise Set 16.1, p. 939

1. $a = 1$, $b = -3$, $c = 2$ **3.** $7x^2 - 4x + 3 = 0$; $a = 7$, $b = -4$, $c = 3$ **5.** $2x^2 - 3x + 5 = 0$; $a = 2$, $b = -3$, $c = 5$ **7.** 0, −5 **9.** 0, −2 **11.** $0, \frac{2}{5}$ **13.** 0, −1 **15.** 0, 3 **17.** $0, \frac{1}{5}$ **19.** $0, \frac{3}{14}$ **21.** $0, \frac{81}{2}$ **23.** −12, 4 **25.** −5, −1 **27.** −9, 2 **29.** 5, 3 **31.** −5 **33.** 4 **35.** $-\frac{2}{3}, \frac{1}{2}$ **37.** $4, -\frac{2}{3}$ **39.** $\frac{5}{3}, -1$ **41.** −1, −5 **43.** −2, 7 **45.** 4, −5 **47.** 4 **49.** 1, −2 **51.** $10, -\frac{2}{5}$ **53.** 6, −4 **55.** 1 **57.** 5, 2 **59.** No solution **61.** $-\frac{5}{2}, 1$ **63.** 35 **65.** 7 **67.** 8 **68.** −13 **69.** $2\sqrt{2}$ **70.** $2\sqrt{3}$ **71.** $2\sqrt{5}$ **72.** $2\sqrt{22}$ **73.** $9\sqrt{5}$ **74.** $2\sqrt{255}$ **75.** 2.646 **76.** 4.796 **77.** 1.528 **78.** 22.908 **79.** ◆ **81.** $-\frac{1}{3}, 1$ **83.** $0, \dfrac{\sqrt{5}}{5}$ **85.** −1.7, 4 **87.** −1.7, 3 **89.** −2, 3 **91.** 4

Margin Exercises, Section 16.2, pp. 941–946

1. $\sqrt{10}, -\sqrt{10}$ **2.** 0 **3.** $\dfrac{\sqrt{6}}{2}, -\dfrac{\sqrt{6}}{2}$ **4.** 7, −1 **5.** $-4 \pm \sqrt{11}$ **6.** −5, 11 **7.** $1 \pm \sqrt{5}$ **8.** 2, 4 **9.** 2, −10 **10.** $6 \pm \sqrt{13}$ **11.** 5, −2 **12.** $\dfrac{-3 \pm \sqrt{33}}{4}$ **13.** About 7.5 sec

Calculator Spotlight, p. 942

1. 3.162, −3.162 **2.** 1.528, −1.528

Calculator Spotlight, p. 945

1. −10.385, 0.385 **2.** −1.317, 5.317 **3.** −0.281, 1.781 **4.** −2.186, 0.686

Exercise Set 16.2, p. 947

1. 11, −11 **3.** $\sqrt{7}, -\sqrt{7}$ **5.** $\dfrac{\sqrt{15}}{5}, -\dfrac{\sqrt{15}}{5}$ **7.** $\dfrac{5}{2}, -\dfrac{5}{2}$ **9.** $\dfrac{7\sqrt{3}}{3}, -\dfrac{7\sqrt{3}}{3}$ **11.** $\sqrt{3}, -\sqrt{3}$ **13.** $\dfrac{8}{7}, -\dfrac{8}{7}$ **15.** −7, 1 **17.** $-3 \pm \sqrt{21}$ **19.** $-13 \pm 2\sqrt{2}$ **21.** $7 \pm 2\sqrt{3}$ **23.** $-9 \pm \sqrt{34}$ **25.** $\dfrac{-3 \pm \sqrt{14}}{2}$ **27.** 11, −5 **29.** 1, −15 **31.** −2, 8 **33.** −21, −1 **35.** $1 \pm \sqrt{6}$ **37.** $11 \pm \sqrt{19}$ **39.** $-5 \pm \sqrt{29}$ **41.** $\dfrac{7 \pm \sqrt{57}}{2}$ **43.** −7, 4 **45.** $\dfrac{-3 \pm \sqrt{17}}{4}$ **47.** $\dfrac{-3 \pm \sqrt{145}}{4}$ **49.** $\dfrac{-2 \pm \sqrt{7}}{3}$ **51.** $-\frac{1}{2}, 5$ **53.** $-\frac{5}{2}, \frac{2}{3}$ **55.** About 9.5 sec **57.** About 4.4 sec **59.** $y = \dfrac{141}{x}$ **60.** $3\dfrac{1}{3}$ hr **61.** $3x\sqrt{2}$ **62.** $8x^2\sqrt{3x}$ **63.** $3t$ **64.** $x^3\sqrt{x}$ **65.** ◆ **67.** 12, −12 **69.** $16\sqrt{2}, -16\sqrt{2}$ **71.** $2\sqrt{c}, -2\sqrt{c}$ **73.** 49.896, −49.896 **75.** 9, −9

Margin Exercises, Section 16.3, pp. 950–952

1. $\frac{1}{2}, -4$ **2.** 5, −2 **3.** $-2 \pm \sqrt{11}$ **4.** No real-number solutions **5.** $\dfrac{4 \pm \sqrt{31}}{5}$ **6.** −0.3, 1.9

Calculator Spotlight, p. 950

1. −5, 2 **2.** −1.317, 5.317 **3.** 0.6, 1 **4.** −4, 0.5

Calculator Spotlight, p. 952

1. The grapher indicates an error. **2.** The graph does not intersect the x-axis. **3.** No real-number solutions

Exercise Set 16.3, p. 953

1. $-3, 7$ **3.** 3 **5.** $-\frac{4}{3}, 2$ **7.** $\frac{3}{2}, -\frac{5}{2}$ **9.** $-3, 3$ **11.** $1 \pm \sqrt{3}$ **13.** $5 \pm \sqrt{3}$ **15.** $-2 \pm \sqrt{7}$ **17.** $\frac{-4 \pm \sqrt{10}}{3}$ **19.** $\frac{5 \pm \sqrt{33}}{4}$ **21.** $\frac{1 \pm \sqrt{3}}{2}$ **23.** No real-number solutions **25.** $\frac{5 \pm \sqrt{73}}{6}$ **27.** $\frac{3 \pm \sqrt{29}}{2}$ **29.** $\sqrt{5}, -\sqrt{5}$ **31.** $-2 \pm \sqrt{3}$ **33.** $\frac{5 \pm \sqrt{37}}{2}$ **35.** $-1.3, 5.3$ **37.** $-0.2, 6.2$ **39.** $-1.2, 0.2$ **41.** $2.4, 0.3$ **43.** $3\sqrt{10}$ **44.** $\sqrt{6}$ **45.** $2\sqrt{2}$ **46.** $(9x - 2)\sqrt{x}$ **47.** $4\sqrt{5}$ **48.** $3x^2\sqrt{3x}$ **49.** $30x^5\sqrt{10}$ **50.** $\frac{\sqrt{21}}{3}$ **51.** ◈ **53.** $0, 2$ **55.** $\frac{3 \pm \sqrt{5}}{2}$ **57.** $\frac{-7 \pm \sqrt{61}}{2}$ **59.** $\frac{-2 \pm \sqrt{10}}{2}$ **61.** $-1.3, 5.3$ **63.** $-0.2, 6.2$ **65.** $-1.2, 0.2$ **67.** $2.4, 0.3$

Margin Exercises, Section 16.4, pp. 955–956

1. $L = \frac{r^2}{20}$ **2.** $L = \frac{T^2 g}{4\pi^2}$ **3.** $m = \frac{E}{c^2}$ **4.** $r = \sqrt{\frac{A}{\pi}}$ **5.** $d = \sqrt{\frac{C}{P}} + 1$ **6.** $n = \frac{1 + \sqrt{1 + 4N}}{2}$ **7.** $t = \frac{-v + \sqrt{v^2 + 32h}}{16}$

Exercise Set 16.4, p. 957

1. $Q = \frac{P^2}{289}$ **3.** $E = \frac{mv^2}{2g}$ **5.** $r = \frac{1}{2}\sqrt{\frac{S}{\pi}}$ **7.** $A = \frac{-m + \sqrt{m^2 + 4kP}}{2k}$ **9.** $a = \sqrt{c^2 - b^2}$ **11.** $t = \frac{\sqrt{s}}{4}$ **13.** $r = \frac{-\pi h + \sqrt{\pi^2 h^2 + \pi A}}{\pi}$ **15.** $v = 20\sqrt{\frac{F}{A}}$ **17.** $a = \sqrt{c^2 - b^2}$ **19.** $a = \frac{2h\sqrt{3}}{3}$ **21.** $T = \frac{2 + \sqrt{4 - a(m - n)}}{a}$ **23.** $T = \frac{v^2 \pi m}{8k}$ **25.** $x = \frac{d\sqrt{3}}{3}$ **27.** $n = \frac{1 + \sqrt{1 + 8N}}{2}$ **29.** $\sqrt{65} \approx 8.062$ **30.** $\sqrt{75} \approx 8.660$ **31.** $\sqrt{41} \approx 6.403$ **32.** $\sqrt{44} \approx 6.633$ **33.** $\sqrt{1084} \approx 32.924$ **34.** $\sqrt{5} \approx 2.236$ **35.** $\sqrt{424} \approx 20.591$ ft **36.** 32 ft **37.** ◈ **39.** (a) $r = \frac{C}{2\pi}$; (b) $A = \frac{C^2}{4\pi}$ **41.** $\frac{1}{3a}, 1$

Margin Exercises, Section 16.5, pp. 959–961

1. Length: $\frac{1 + \sqrt{817}}{2} \approx 14.8$ yd; width: $\frac{-1 + \sqrt{817}}{6} \approx 4.6$ yd **2.** 2.3 yd; 3.3 yd **3.** 3 km/h

Exercise Set 16.5, p. 963

1. Length: 10 ft; width: 7 ft **3.** Width: 24 in.; height: 18 in. **5.** Length: 20 cm; width: 16 cm **7.** Length: 10 m; width: 5 m **9.** 4.6 m; 6.6 m **11.** Length: 5.6 in.; width: 3.6 in. **13.** Length: 6.4 cm; width: 3.2 cm **15.** 3 cm **17.** 7 km/h **19.** 8 mph **21.** 4 km/h **23.** 36 mph **25.** 1 km/h **27.** $8\sqrt{2}$ **28.** $12\sqrt{10}$ **29.** $(2x - 7)\sqrt{x}$ **30.** $-\sqrt{6}$ **31.** $\frac{3\sqrt{2}}{2}$ **32.** $\frac{2\sqrt{3}}{3}$ **33.** $5\sqrt{6} - 4\sqrt{3}$ **34.** $(9x + 2)\sqrt{x}$ **35.** ◈ **37.** $1 + \sqrt{2} \approx 2.41$ cm **39.** $12\sqrt{2} \approx 16.97$ in.; two 12-in. pizzas

Margin Exercises, Section 16.6, pp. 967–970

1. $(0, -3)$

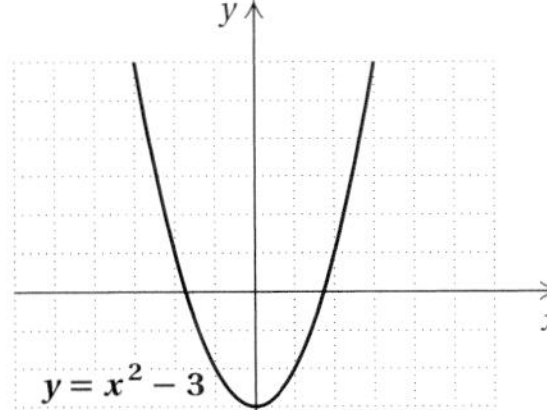

2. $(1, 3)$

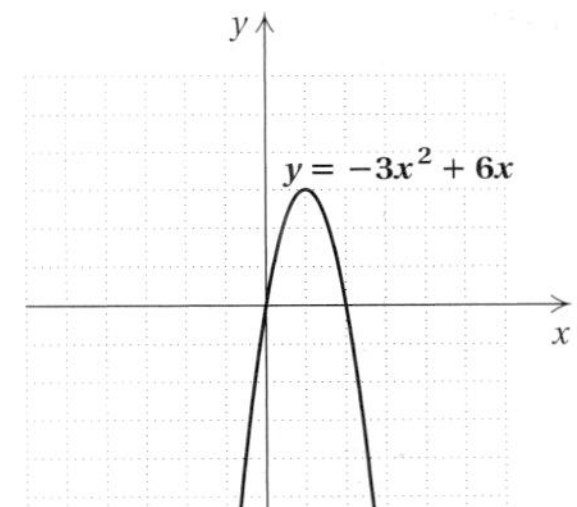

3. $(2, 0)$

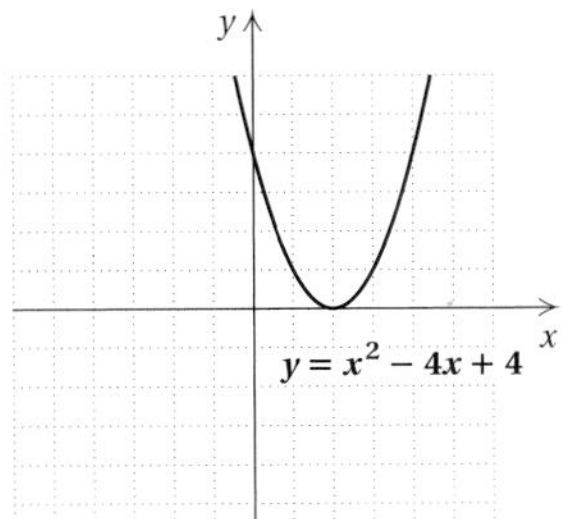

4. $(\sqrt{3}, 0)$; $(-\sqrt{3}, 0)$ **5.** $(-4, 0)$; $(-2, 0)$ **6.** $\left(\frac{-2 - \sqrt{6}}{2}, 0\right)$; $\left(\frac{-2 + \sqrt{6}}{2}, 0\right)$ **7.** None

Exercise Set 16.6, p. 971

1. (0, 1)

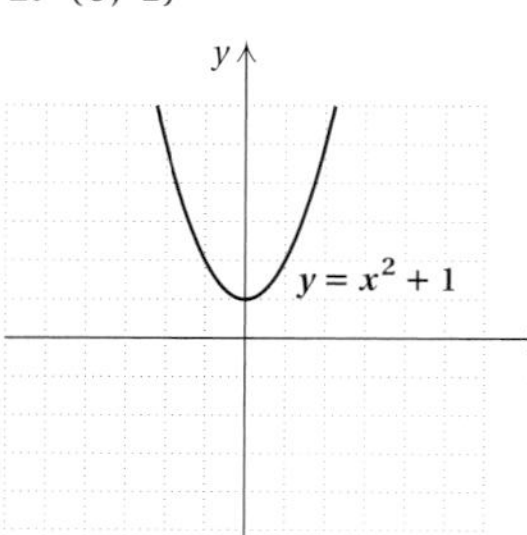

3. (0, 0)

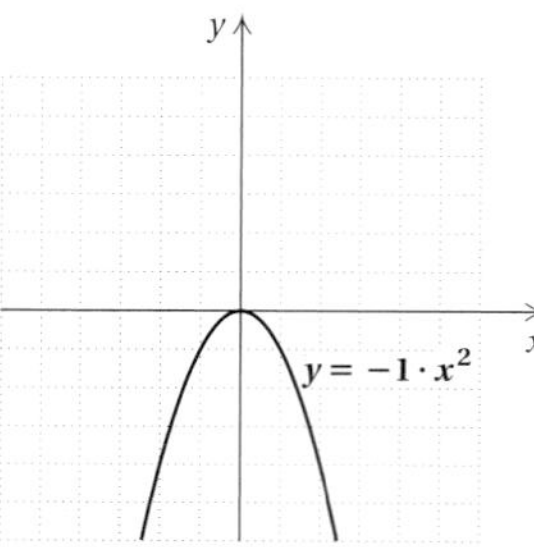

5. (1, 1)

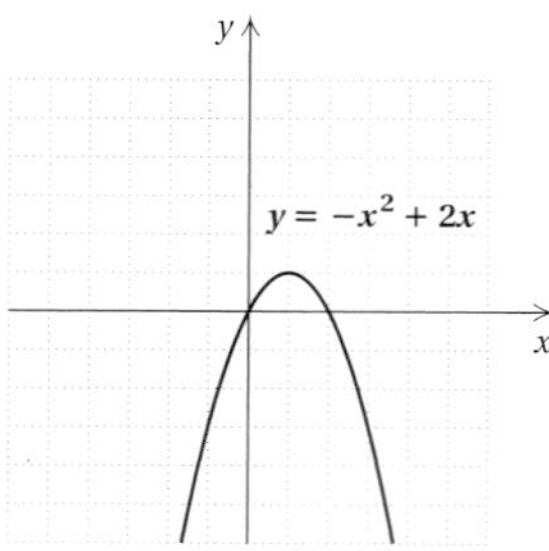

7. $\left(-\frac{1}{2}, \frac{21}{4}\right)$

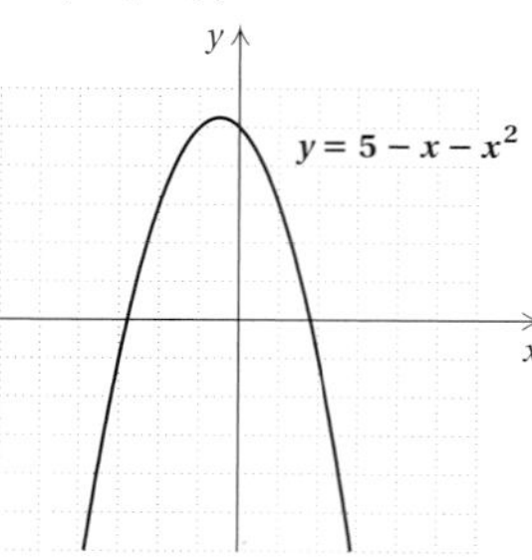

9. (1, 0)

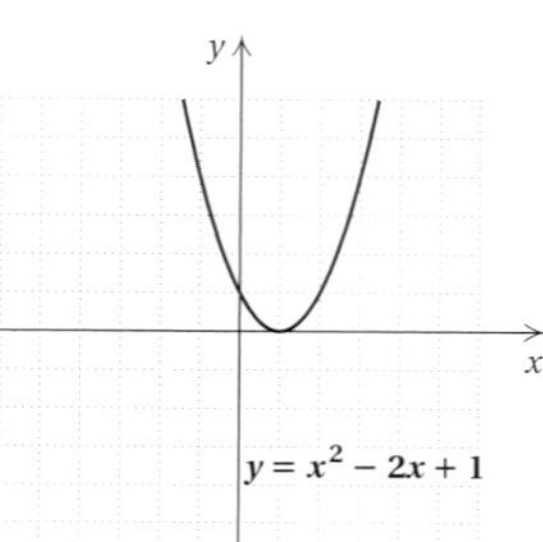

11. (1, 4)

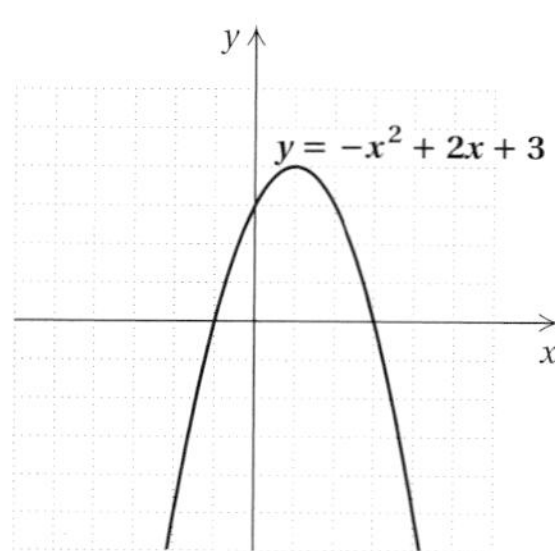

13. (−1, 3)

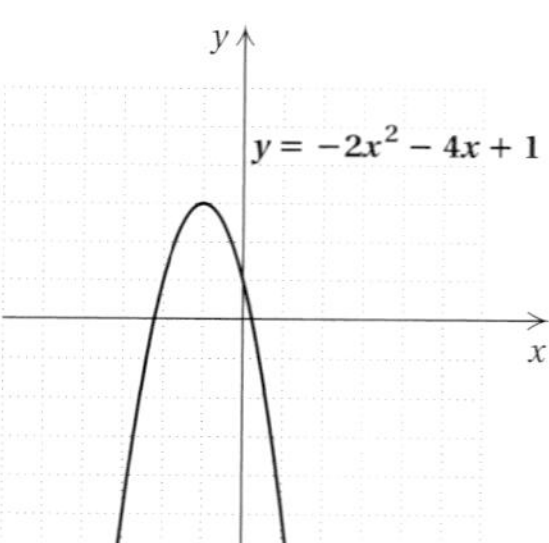

15. (0, 5)

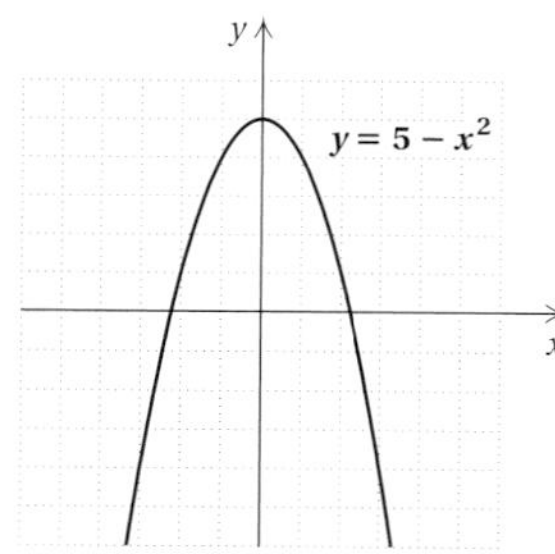

17.

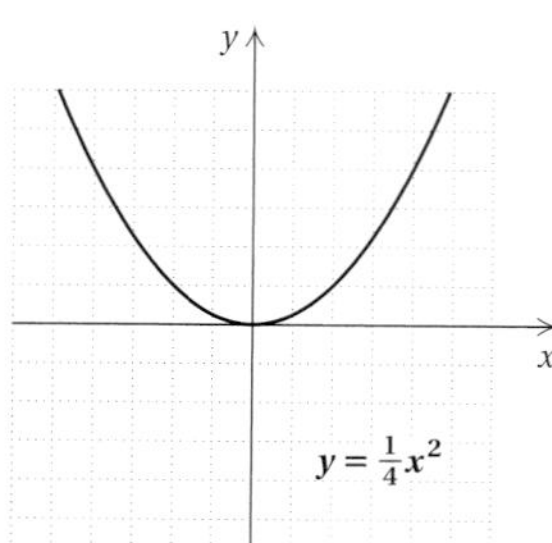

19.

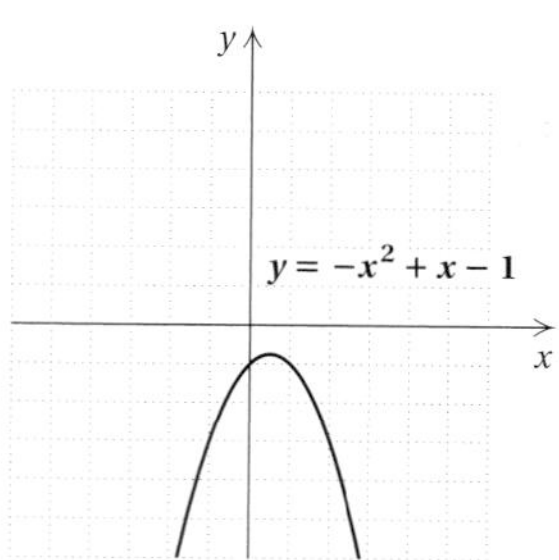

21.

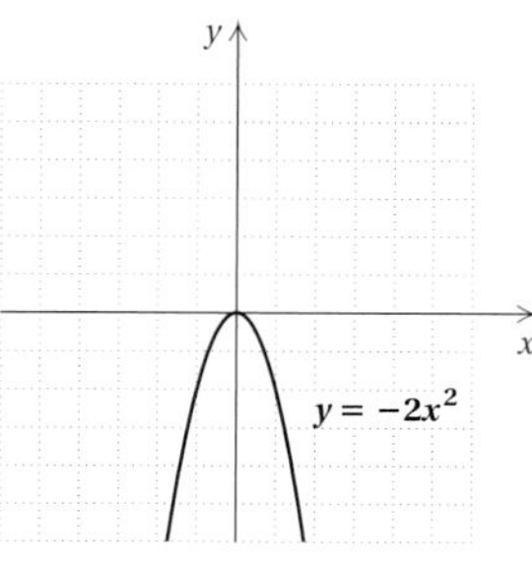

23.

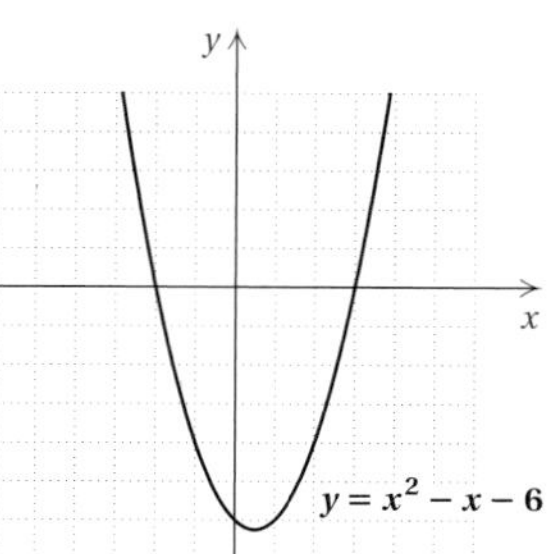

25. $(-\sqrt{2}, 0)$; $(\sqrt{2}, 0)$ **27.** (−5, 0); (0, 0)
29. $\left(\frac{-1 - \sqrt{33}}{2}, 0\right)$; $\left(\frac{-1 + \sqrt{33}}{2}, 0\right)$ **31.** (3, 0)
33. $(-2 - \sqrt{5}, 0)$; $(-2 + \sqrt{5}, 0)$ **35.** None
37. $(x + 2)\sqrt{x - 1}$ **38.** $22\sqrt{2}$ **39.** $2\sqrt{7}$
40. $25y^2\sqrt{y}$ **41.** $y = \frac{29.76}{x}$ **42.** $y = \frac{4752}{x}$
43. −3 **44.** 35 **45.** ◈ **47.** **(a)** After 2 sec; after 4 sec; **(b)** after 3 sec; **(c)** after 6 sec **49.** 16; two real solutions **51.** −161.91; no real solutions

Margin Exercises, Section 16.7, pp. 973–978

1. Yes **2.** No **3.** Yes **4.** No **5.** Yes **6.** Yes
7. **(a)** −33; **(b)** −3; **(c)** 2; **(d)** 97; **(e)** −9 **8.** **(a)** 21; **(b)** 9; **(c)** 14; **(d)** 69

9.

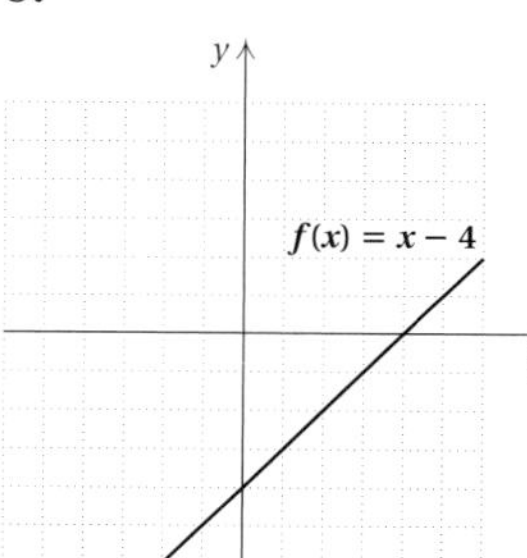

10.

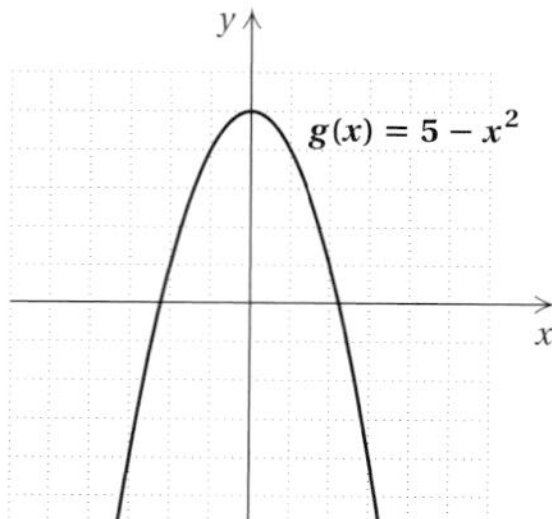

11.

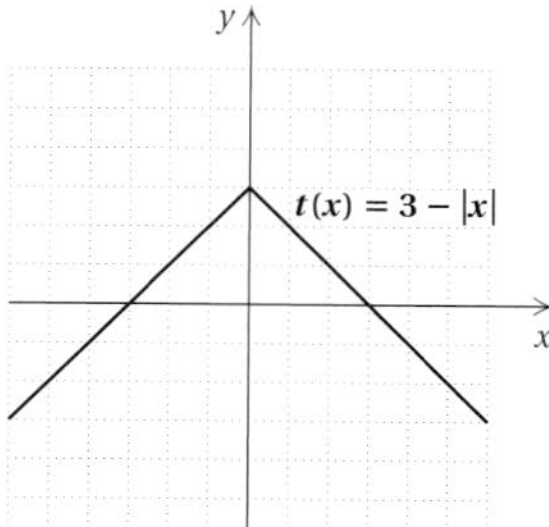

12. Yes **13.** No **14.** No **15.** Yes
16. About \$43 million **17.** About \$6 million

Calculator Spotlight, p. 976

1. 6, 3.99, 150, −1.6 **2.** −21.3, −18.39, −117.3, 3.26
3. −75, −65.47, −420.6, 1.69

Exercise Set 16.7, p. 979

1. Yes **3.** Yes **5.** No **7.** Yes **9.** Yes **11.** Yes **13.** A relation but not a function **15.** **(a)** 9; **(b)** 12; **(c)** 2; **(d)** 5; **(e)** 7.4; **(f)** $5\frac{2}{3}$ **17.** **(a)** −21; **(b)** 15; **(c)** 42; **(d)** 0; **(e)** 2; **(f)** −162.6 **19.** **(a)** 7; **(b)** −17; **(c)** 24.1; **(d)** 4; **(e)** −26; **(f)** 6 **21.** **(a)** 0; **(b)** 5; **(c)** 2; **(d)** 170; **(e)** 65; **(f)** 230 **23.** **(a)** 1; **(b)** 3; **(c)** 3; **(d)** 4; **(e)** 11; **(f)** 23 **25.** **(a)** 0; **(b)** −1; **(c)** 8; **(d)** 1000; **(e)** −125; **(f)** −1000 **27.** **(a)** 159.48 cm; **(b)** 153.98 cm **29.** $1\frac{20}{33}$ atm; $1\frac{10}{11}$ atm; $4\frac{1}{33}$ atm **31.** 1.792 cm; 2.8 cm; 11.2 cm

33.

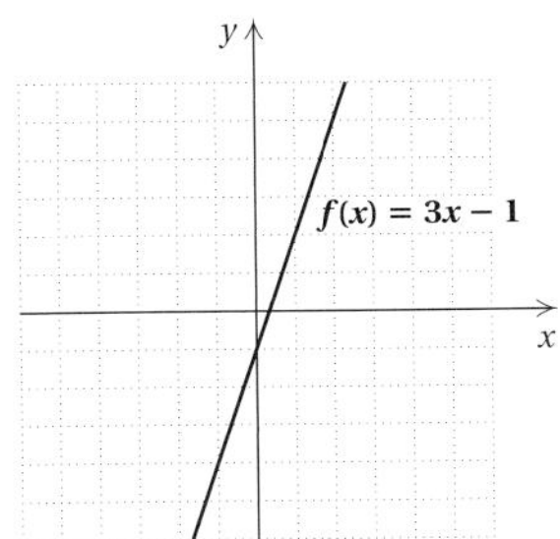

35.

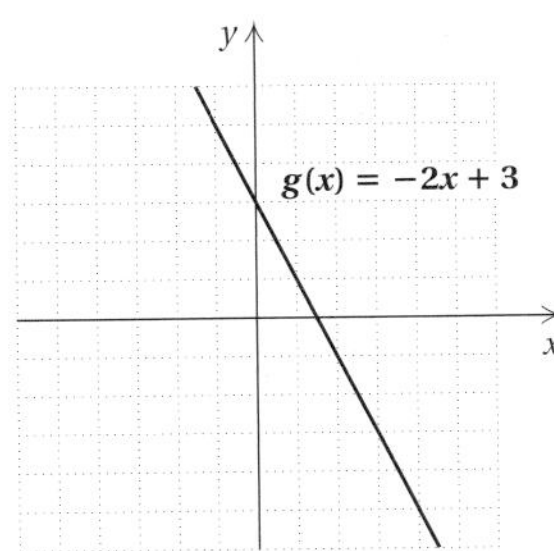

37.

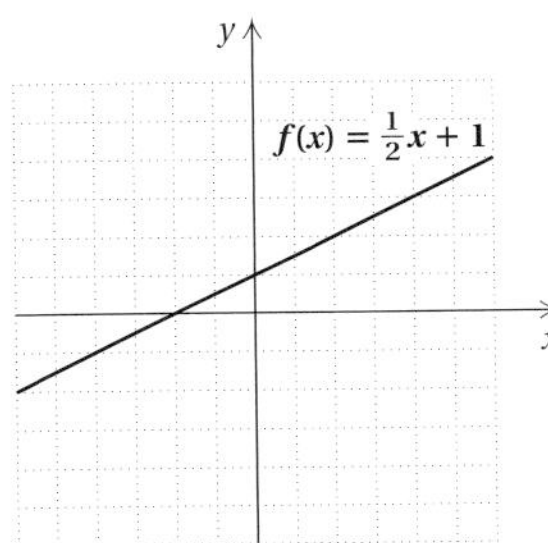

39.

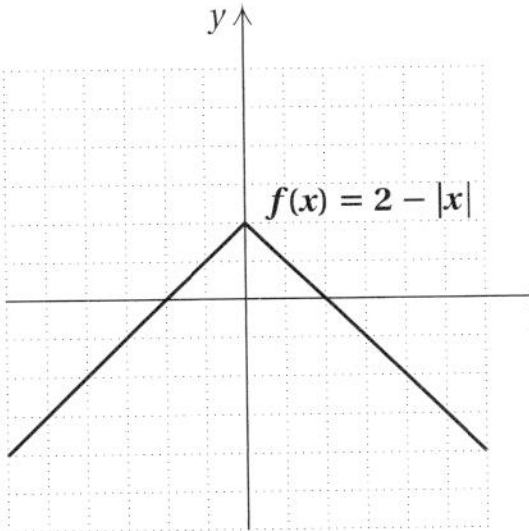

41.

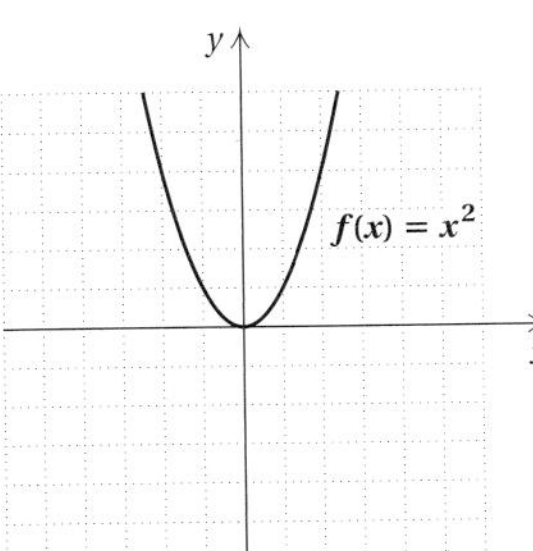

43.

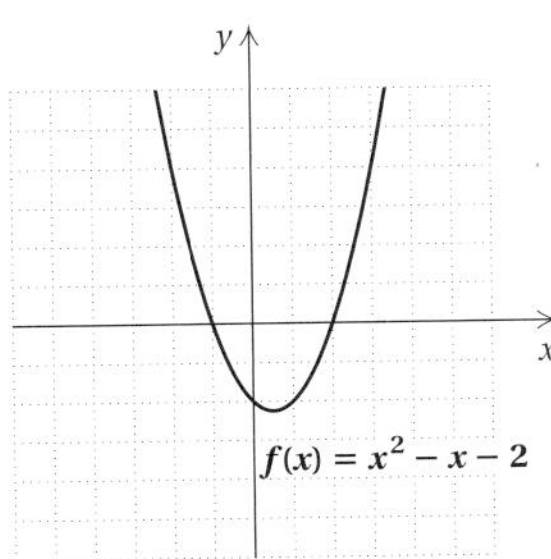

45. Yes **47.** Yes **49.** No **51.** No **53.** The rate is about 75 per 10,000 men. **55.** No **56.** Yes **57.** No solution **58.** Infinite number of solutions **59.** ◈

Summary and Review: Chapter 16, p. 983

1. $\sqrt{3}, -\sqrt{3}$ **2.** $-2\sqrt{2}, 2\sqrt{2}$ **3.** $\frac{3}{5}$, 1 **4.** $\frac{1}{3}$, −2 **5.** $-8 \pm \sqrt{13}$ **6.** 0 **7.** 0, $\frac{7}{5}$ **8.** $1 \pm \sqrt{11}$ **9.** $\frac{1 \pm \sqrt{10}}{3}$ **10.** $-3 \pm 3\sqrt{2}$ **11.** $\frac{2 \pm \sqrt{3}}{2}$ **12.** $\frac{3 \pm \sqrt{33}}{2}$ **13.** No real-number solutions **14.** 0, $\frac{4}{3}$ **15.** 3, −5 **16.** 1 **17.** $\frac{5 \pm \sqrt{17}}{2}$ **18.** $\frac{5}{3}$, −1 **19.** 4.6, 0.4 **20.** −1.9, −0.1 **21.** $T = L(4V^2 - 1)$

22.

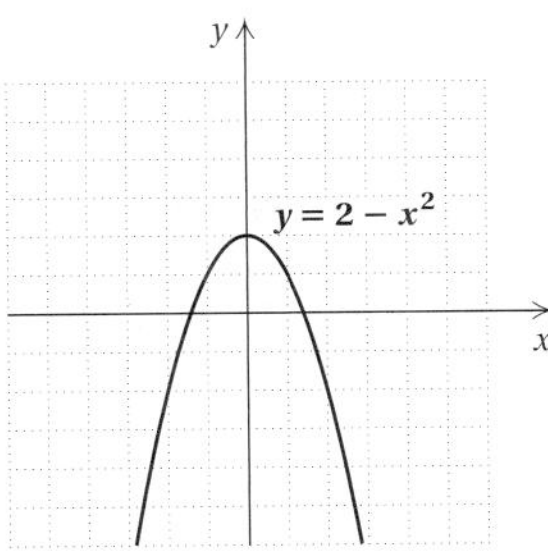

23.

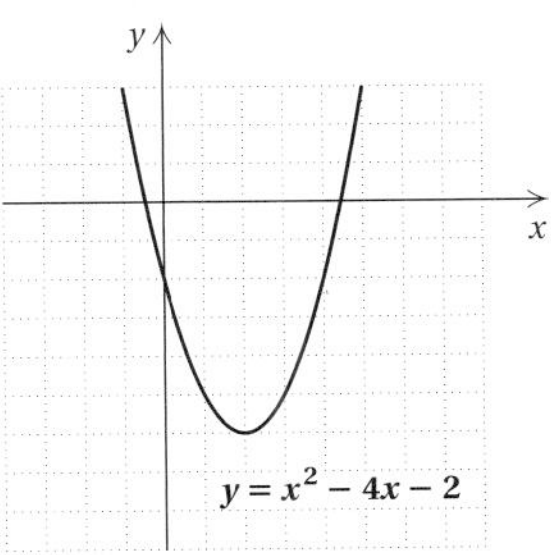

24. $(-\sqrt{2}, 0)$; $(\sqrt{2}, 0)$ **25.** $(2 - \sqrt{6}, 0)$; $(2 + \sqrt{6}, 0)$ **26.** 4.7 cm, 1.7 cm **27.** 10 yd, 24 yd **28.** About 6.3 sec **29.** −1, −7, 2 **30.** 0, 0, 19 **31.** 2700 calories

32.

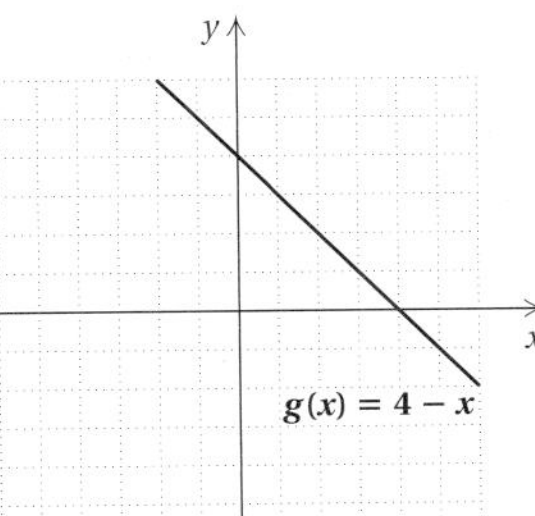

33.

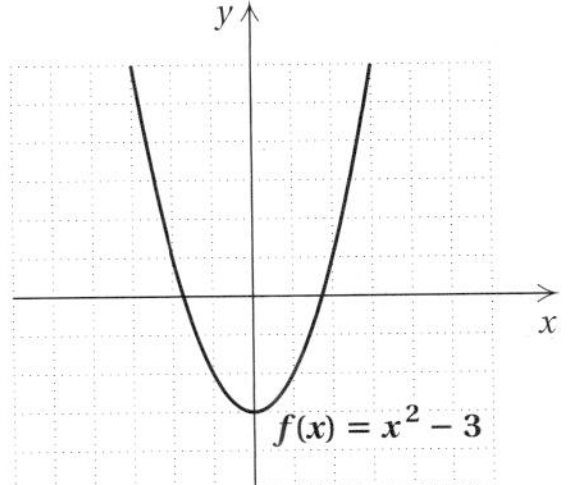

34.

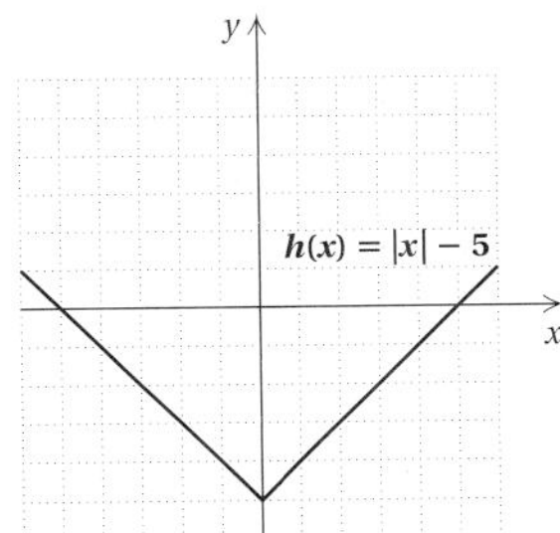

35. No **36.** Yes **37.** $6\sqrt{a}$ **38.** $2xy\sqrt{15y}$ **39.** $y = \frac{0.625}{x}$ **40.** $\sqrt{3}$ **41.** $12\sqrt{11}$ **42.** $4\sqrt{10}$ **43.** ◈

Equation	Form	Example
Linear	Reducible to $x = a$	$3x - 5 = 8$
Quadratic	$ax^2 + bx + c = 0$	$2x^2 - 3x + 1 = 0$
Rational	Contains one or more rational expressions	$\frac{x}{3} + \frac{4}{x - 1} = 1$
Radical	Contains one or more radical expressions	$\sqrt{3x - 1} = x - 7$
Systems of equations	$Ax + By = C$, $Dx + Ey = F$	$4x - 5y = 3$, $3x + 2y = 1$

44. ◈ **(a)** The third line should be $x = 0$ *or* $x + 20 = 0$; the solution 0 gets lost in the given procedure. **(b)** The addition principle should be used at the outset to get 0 on one side of the equation. Since this was not done in the given procedure, the principle of zero products was not applied correctly.
45. 31, 32; −32, −31 **46.** $5\sqrt{\pi}$, or about 8.9 in.
47. 25

Test: Chapter 16, p. 985

1. [16.2a] $\sqrt{5}, -\sqrt{5}$ **2.** [16.1b] $0, -\frac{8}{7}$
3. [16.1c] −8, 6 **4.** [16.1c] $-\frac{1}{3}, 2$ **5.** [16.2b] $8 \pm \sqrt{13}$
6. [16.3a] $\dfrac{1 \pm \sqrt{13}}{2}$ **7.** [16.3a] $\dfrac{3 \pm \sqrt{37}}{2}$
8. [16.3a] $-2 \pm \sqrt{14}$ **9.** [16.3a] $\dfrac{7 \pm \sqrt{37}}{6}$
10. [16.1c] 2, −1 **11.** [16.1c] 2, −4
12. [16.2c] $2 \pm \sqrt{14}$ **13.** [16.3b] 5.7, −1.7
14. [16.4a] $n = \dfrac{-b + \sqrt{b^2 + 4ad}}{2a}$
15. [16.6b] $\left(\dfrac{1 - \sqrt{21}}{2}, 0\right), \left(\dfrac{1 + \sqrt{21}}{2}, 0\right)$
16. [16.6a] **17.** [16.6a]

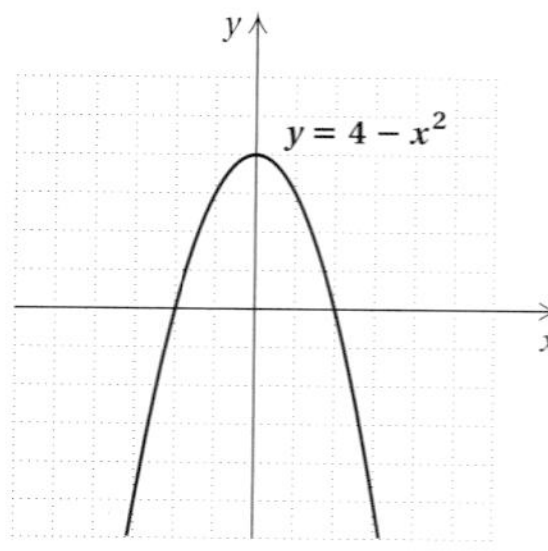

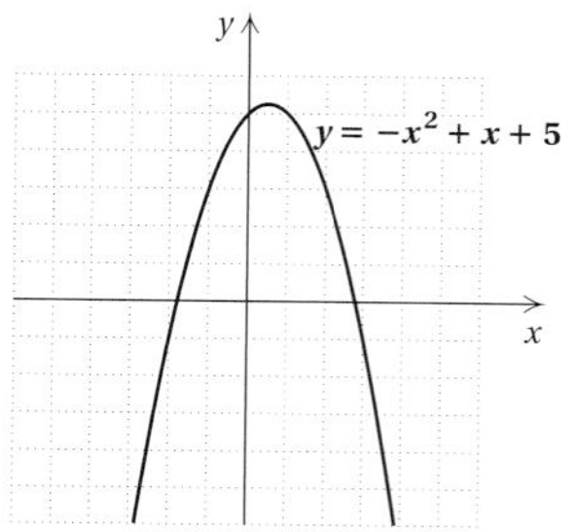

18. [16.7b] 1; $1\frac{1}{2}$; 2 **19.** [16.7b] 1; 3; −3
20. [16.5a] Length: 6.5 m; width: 2.5 m
21. [16.5a] 24 km/h **22.** [16.7e] 26.58 min
23. [16.7c] **24.** [16.7c]

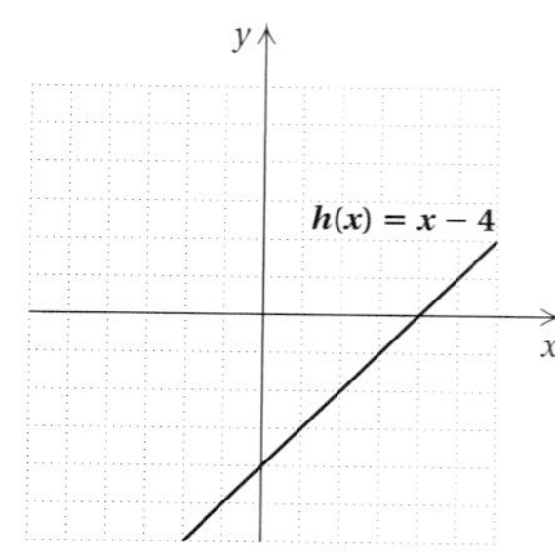

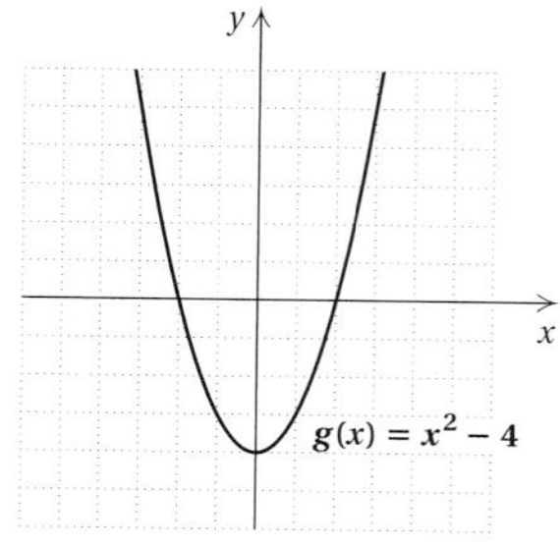

25. [16.7d] Yes **26.** [16.7d] No **27.** [15.4a] $2\sqrt{15}$
28. [15.2c] $7xy\sqrt{2x}$ **29.** [13.5c] $y = \dfrac{4}{x}$
30. [15.6b] $\sqrt{5}$ **31.** [16.5a] $5 + 5\sqrt{2}$
32. [14.2b], [16.3a] $1 \pm \sqrt{5}$

Appendixes

Margin Exercises, Appendix A, pp. 987–994

1. 2 **2.** 3 **3.** $1\frac{1}{2}$ **4.** $2\frac{1}{2}$ **5.** 288 **6.** 43.5
7. 240,768 **8.** 6 **9.** $1\frac{5}{12}$ **10.** 8 **11.** $11\frac{2}{3}$ **12.** 5
13. 0.5 **14.** cm **15.** km **16.** mm **17.** m
18. cm **19.** m **20.** 23,000 **21.** 400 **22.** 178
23. 9040 **24.** 7.814 **25.** 781.4 **26.** 0.967
27. 8,900,000 **28.** 6.78 **29.** 97.4 **30.** 0.1
31. 8.451 **32.** 90.909 **33.** 804.5 **34.** 1479.843

Exercise Set A, p. 995

1. 12 **3.** $\frac{1}{12}$ **5.** 5280 **7.** 468 **9.** 7 **11.** 15,840
13. $\frac{1}{4}$ **15.** $3\frac{1}{3}$ **17.** $1\frac{1}{2}$ **19.** 1 **21.** 110 **23.** 2
25. 300 **27.** $\frac{1}{36}$ **29.** 126,720 **31.** **(a)** 1000; **(b)** 0.001
33. **(a)** 10; **(b)** 0.1 **35.** **(a)** 0.01; **(b)** 100 **37.** 6700
39. 0.98 **41.** 8.921 **43.** 0.05666 **45.** 566,600
47. 688 **49.** 0.1 **51.** 100,000 **53.** 142 **55.** 450
57. 0.000024 **59.** 0.688 **61.** 3.92 **63.** 100
65. 727.409592 **67.** 104.585 **69.** 289.62
71. 112.63 **73.** 9.09

Margin Exercises, Appendix B, pp. 997–1002

1. 40 **2.** 20 **3.** mL **4.** mL **5.** L **6.** L **7.** 970
8. 8.99 **9.** 4.8 **10.** **(a)** 118.28 mL; **(b)** 0.11828 L
11. $1.76 **12.** 80 **13.** 4.32 **14.** 32,000 **15.** kg
16. kg **17.** mg **18.** g **19.** t **20.** 6200 **21.** 3.048
22. 77 **23.** 234.4 **24.** 6700 **25.** 120 **26.** 1461
27. 1440 **28.** 1

Exercise Set B, p. 1003

1. 1000; 1000 **3.** 87,000 **5.** 0.049 **7.** 0.000401
9. 78,100 **11.** 320 **13.** 10 **15.** 32 **17.** 500 mL
19. 125 mL **21.** 1.75 gal/week; 7.5 gal/month; 91.25 gal/year; 65,250,000 gal/day; 23,816,250,000 gal/year **23.** 2000 **25.** 3 **27.** 64
29. 12,640 **31.** 0.1 **33.** 5 **35.** 1000 **37.** 10
39. $\frac{1}{100}$, or 0.01 **41.** 1000 **43.** 10 **45.** 234,000
47. 5.2 **49.** 6.7 **51.** 0.0502 **53.** 8.492 **55.** 58.5
57. 800,000 **59.** 1000 **61.** 24 **63.** 60 **65.** $365\frac{1}{4}$
67. 0.05 **69.** 8.2 **71.** 6.5 **73.** 10.75 **75.** 336
77. 4.5 **79.** 56 **81.** ◈ **83.** 0.4535 **85.** 2 oz
87. 31.7 yr **89.** 1000 **91.** 0.125 mg **93.** 4
95. 8 mL

Margin Exercises, Appendix C, pp. 1007–1008

1. $(x - 2)(x^2 + 2x + 4)$ **2.** $(4 - y)(16 + 4y + y^2)$
3. $(3x + y)(9x^2 - 3xy + y^2)$
4. $(2y + z)(4y^2 - 2yz + z^2)$
5. $(m + n)(m^2 - mn + n^2)(m - n)(m^2 + mn + n^2)$
6. $2xy(2x^2 + 3y^2)(4x^4 - 6x^2y^2 + 9y^4)$
7. $(3x + 2y)(9x^2 - 6xy + 4y^2)(3x - 2y) \times (9x^2 + 6xy + 4y^2)$
8. $(x - 0.3)(x^2 + 0.3x + 0.09)$

Exercise Set C, p. 1009

1. $(z + 3)(z^2 - 3z + 9)$ **3.** $(x - 1)(x^2 + x + 1)$
5. $(y + 5)(y^2 - 5y + 25)$ **7.** $(2a + 1)(4a^2 - 2a + 1)$
9. $(y - 2)(y^2 + 2y + 4)$ **11.** $(2 - 3b)(4 + 6b + 9b^2)$
13. $(4y + 1)(16y^2 - 4y + 1)$
15. $(2x + 3)(4x^2 - 6x + 9)$ **17.** $(a - b)(a^2 + ab + b^2)$
19. $\left(a + \frac{1}{2}\right)\left(a^2 - \frac{1}{2}a + \frac{1}{4}\right)$ **21.** $2(y - 4)(y^2 + 4y + 16)$
23. $3(2a + 1)(4a^2 - 2a + 1)$
25. $r(s + 4)(s^2 - 4s + 16)$
27. $5(x - 2z)(x^2 + 2xz + 4z^2)$
29. $(x + 0.1)(x^2 - 0.1x + 0.01)$
31. $8(2x^2 - t^2)(4x^4 + 2x^2t^2 + t^4)$
33. $2y(y - 4)(y^2 + 4y + 16)$
35. $(z - 1)(z^2 + z + 1)(z + 1)(z^2 - z + 1)$
37. $(t^2 + 4y^2)(t^4 - 4t^2y^2 + 16y^4)$ **39.** 1; 19; 19; 7; 1
41. $(x^{2a} + y^b)(x^{4a} - x^{2a}y^b + y^{2b})$
43. $3(x^a + 2y^b)(x^{2a} - 2x^ay^b + 4y^{2b})$
45. $\frac{1}{3}\left(\frac{1}{2}xy + z\right)\left(\frac{1}{4}x^2y^2 - \frac{1}{2}xyz + z^2\right)$
47. $y(3x^2 + 3xy + y^2)$ **49.** $4(3a^2 + 4)$

Margin Exercises, Appendix D, pp. 1011–1012

1. $\{6, -6\}$; 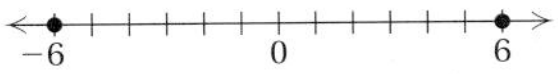

2. $\varnothing$ **3.** $\{0\}$ **4.** $\{2, -2\}$ **5.** $\left\{\frac{17}{4}, -\frac{17}{4}\right\}$ **6.** $\{4, -4\}$
7. $\{3, 5\}$ **8.** $\left\{-\frac{13}{3}, 7\right\}$ **9.** $\varnothing$

Exercise Set D, p. 1013

1. $\{3, -3\}$ **3.** $\varnothing$ **5.** $\{0\}$ **7.** $\{15, -9\}$ **9.** $\left\{\frac{7}{2}, -\frac{1}{2}\right\}$
11. $\left\{\frac{23}{4}, -\frac{5}{4}\right\}$ **13.** $\{11, -11\}$ **15.** $\{389, -389\}$
17. $\{8, -8\}$ **19.** $\{7, -7\}$ **21.** $\left\{\frac{11}{5}, -\frac{11}{5}\right\}$ **23.** $\{8, -7\}$
25. $\{2, -12\}$ **27.** $\left\{\frac{7}{2}, -\frac{5}{2}\right\}$ **29.** $\varnothing$ **31.** $\left\{-\frac{13}{54}, -\frac{7}{54}\right\}$
33. $\left\{x \mid x \ge \frac{5}{2}\right\}$ **35.** $\{x \mid x \ge -5\}$ **37.** $\left\{1, -\frac{1}{4}\right\}$

Margin Exercises, Appendix E, pp. 1015–1016

1. 14 **2.** 11 **3.** 10 **4.** $\sqrt{37} \approx 6.083$ **5.** $\left(\frac{3}{2}, -3\right)$
6. $(9, -5)$

Exercise Set E, p. 1017

1. 5 **3.** $\sqrt{18} \approx 4.243$ **5.** $\sqrt{32} \approx 5.657$ **7.** 17.8
9. $\dfrac{\sqrt{41}}{7} \approx 0.915$ **11.** $\sqrt{6970} \approx 83.487$ **13.** $\left(-\dfrac{1}{2}, -1\right)$
15. $\left(\frac{7}{2}, \frac{7}{2}\right)$ **17.** $(-1, -3)$ **19.** $(-0.25, -0.3)$
21. $\left(-\frac{1}{12}, \frac{1}{24}\right)$ **23.** $\left(\dfrac{\sqrt{2} + \sqrt{3}}{2}, \dfrac{3}{2}\right)$ **25.** $\sqrt{49 + k^2}$
27. $8\sqrt{m^2 + n^2}$ **29.** $6\sqrt{2}$ **31.** Yes **33.** $(2, 4\sqrt{2})$

Margin Exercises, Appendix F, pp. 1019–1024

1. 3, −3 **2.** 6, −6 **3.** 11, −11 **4.** 1 **5.** 6 **6.** $\frac{9}{10}$
7. 0.08 **8.** **(a)** 4; **(b)** −4; **(c)** does not exist **9.** **(a)** 7; **(b)** −7; **(c)** does not exist **10.** **(a)** 12; **(b)** −12;
(c) does not exist **11.** $28 + x$ **12.** $\dfrac{y}{y + 3}$ **13.** $|y|$
14. 24 **15.** $|5y|$, or $5|y|$ **16.** $|4y|$, or $4|y|$ **17.** $|x + 7|$
18. $|2(x - 2)|$, or $2|x - 2|$ **19.** $|7(y + 5)|$, or $7|y + 5|$
20. $|x - 3|$ **21.** −4 **22.** $3y$ **23.** $2(x + 2)$ **24.** $-\frac{7}{4}$
25. 3 **26.** −3 **27.** x **28.** y **29.** 0 **30.** $-2x$
31. $3x + 2$ **32.** 3 **33.** −3 **34.** Does not exist as a real number **35.** 0 **36.** $|2(x - 2)|$, or $2|x - 2|$
37. $|x|$ **38.** $|x + 3|$ **39.** $\sqrt[4]{y}$ **40.** $\sqrt{3a}$
41. $\sqrt[4]{16}$, or 2 **42.** $\sqrt[3]{125}$, or 5 **43.** $\sqrt[5]{a^3b^2c}$
44. $(19ab)^{1/3}$ **45.** $19(ab)^{1/3}$ **46.** $\left(\dfrac{x^2y}{16}\right)^{1/5}$
47. $7(2ab)^{1/4}$ **48.** $\sqrt{x^3}$, or $(\sqrt{x})^3$, which simplifies to $x\sqrt{x}$ **49.** $\sqrt[3]{8^2}$, or $(\sqrt[3]{8})^2$, which simplifies to 4
50. $\sqrt{4^5}$, or $(\sqrt{4})^5$, which simplifies to 32
51. $(7abc)^{4/3}$ **52.** $6^{7/5}$ **53.** $\dfrac{1}{16^{1/4}}$, or $\dfrac{1}{2}$ **54.** $\dfrac{1}{(3xy)^{7/8}}$
55. $7^{14/15}$ **56.** $5^{2/6}$, or $5^{1/3}$ **57.** $9^{2/5}$

Exercise Set F, p. 1025

1. 4, −4 **3.** 12, −12 **5.** 20, −20 **7.** $-\frac{7}{6}$ **9.** 14
11. $-\frac{4}{9}$ **13.** 0.2 **15.** −0.03 **17.** $y^2 + 16$ **19.** $\dfrac{x}{y - 1}$
21. $|4x|$, or $4|x|$ **23.** $|12c|$, or $12|c|$ **25.** $|p + 3|$
27. $|x - 2|$ **29.** $|2x + 7|$ **31.** 3 **33.** $-4x$ **35.** −6
37. 5 **39.** −1 **41.** $-\frac{2}{3}$ **43.** $|x|$ **45.** 6 **47.** y
49. $x - 2$ **51.** $\sqrt[7]{y}$ **53.** 2 **55.** $\sqrt[5]{a^3b^3}$ **57.** 8
59. $18^{1/3}$ **61.** $(xy^2z)^{1/5}$ **63.** $(3mn)^{3/2}$ **65.** $(8x^2y)^{5/7}$
67. $\dfrac{1}{x^{1/4}}$ **69.** $x^{2/3}$ **71.** $5^{7/8}$ **73.** $7^{1/4}$ **75.** $4.9^{7/20}$
77. $6^{3/28}$ **79.** $a^{3/2}$

Margin Exercises, Appendix G, pp. 1027–1030

1. $\{x \mid x < -3 \text{ or } x > 1\}$ **2.** $\{x \mid -3 < x < 1\}$
3. $\{x \mid -3 \le x \le 1\}$ **4.** $\{x \mid x < -4 \text{ or } x > 1\}$
5. $\{x \mid -4 \le x \le 1\}$ **6.** $\{x \mid x < -1 \text{ or } 0 < x < 1\}$
7.

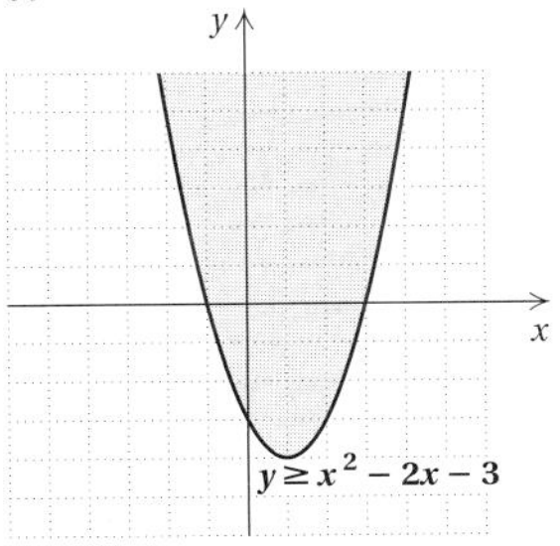

8.

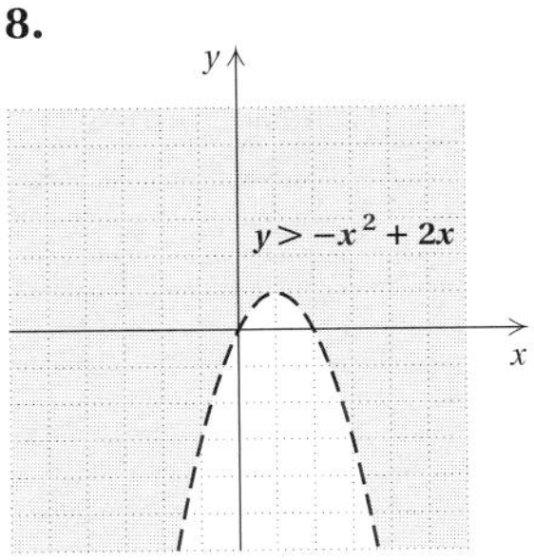

Exercise Set G, p. 1031

1. $\{x \mid x < -2 \text{ or } x > 6\}$ **3.** $\{x \mid -1 \le x \le 4\}$
5. $\{x \mid -1 < x < 2\}$ **7.** $\{x \mid x \le -3 \text{ or } x \ge 3\}$
9. All real numbers **11.** $\{x \mid 2 < x < 4\}$
13. $\{x \mid x < -2 \text{ or } 0 < x < 2\}$
15. $\{x \mid -9 < x < -1 \text{ or } x > 4\}$
17. $\{x \mid x < -3 \text{ or } -2 < x < 1\}$

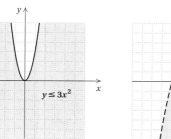

19.

$y \le 3x^2$

21.

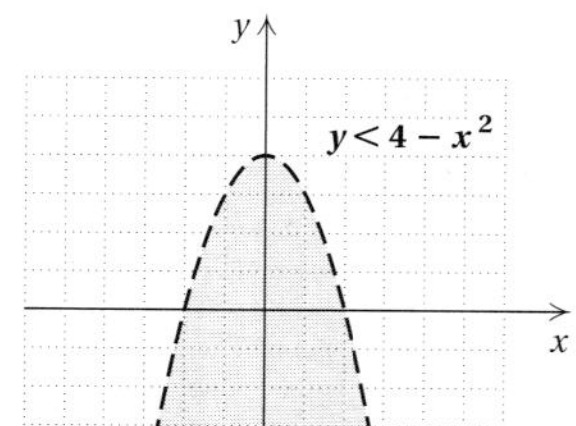

23.

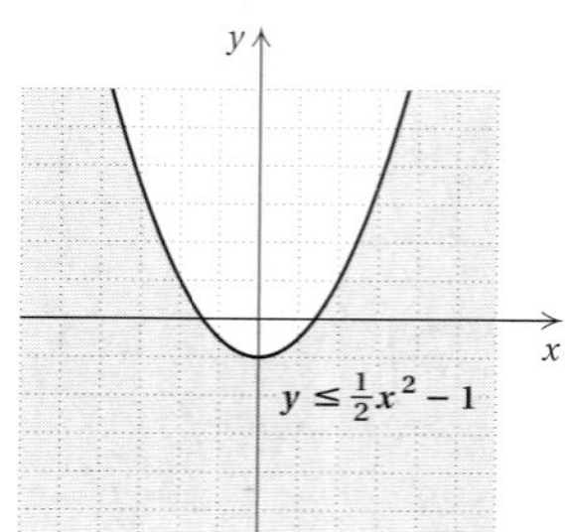

25.

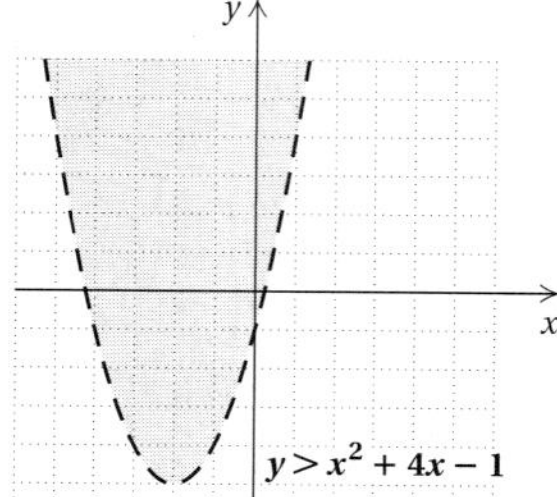

27.

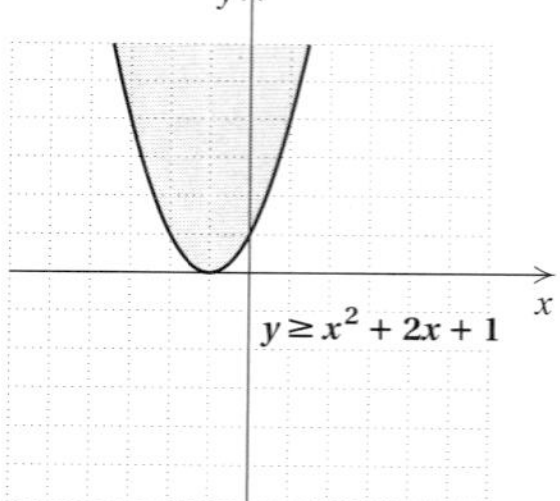

29.

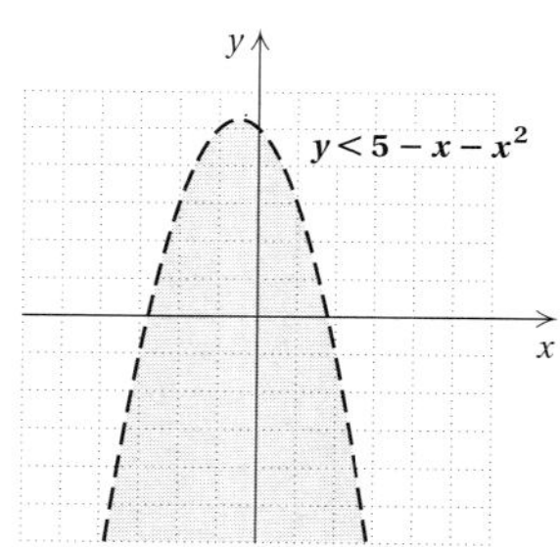

31.

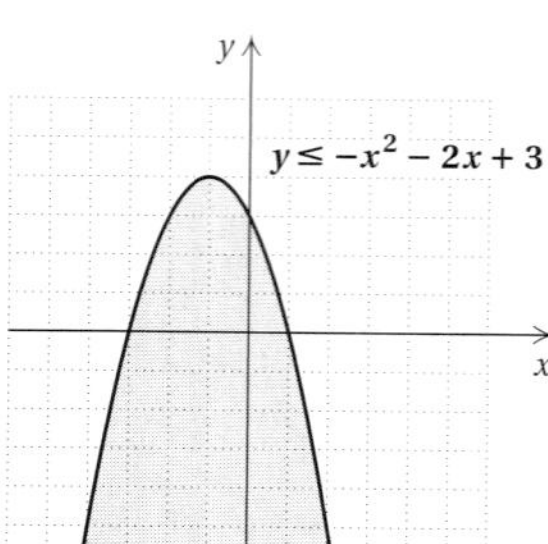

33.

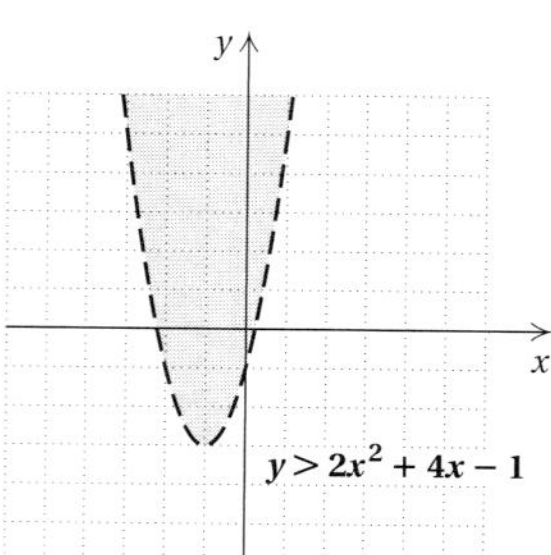

Margin Exercises, Appendix H, pp. 1033–1036

1.

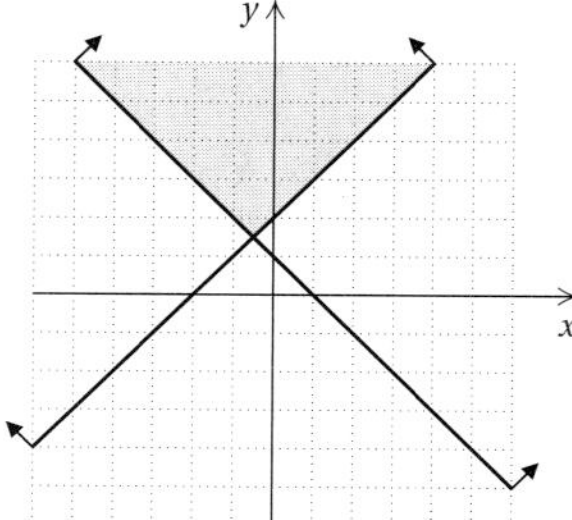

2.

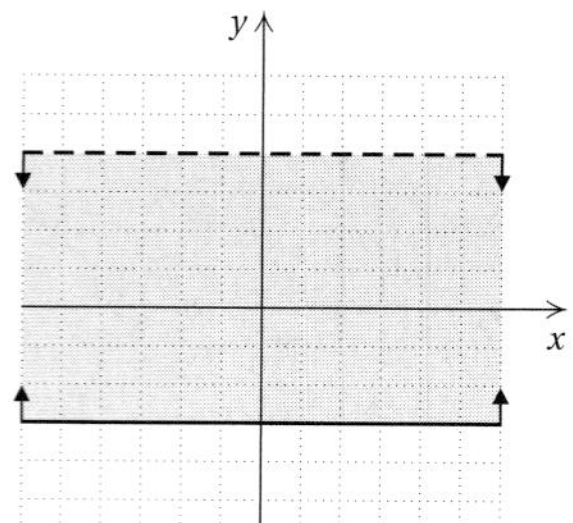

3.

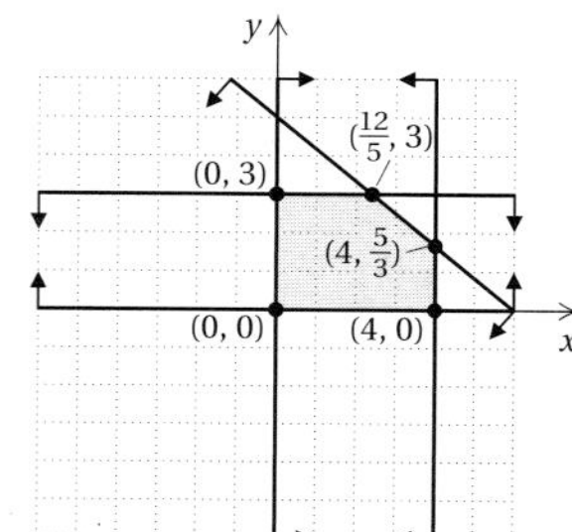

4.

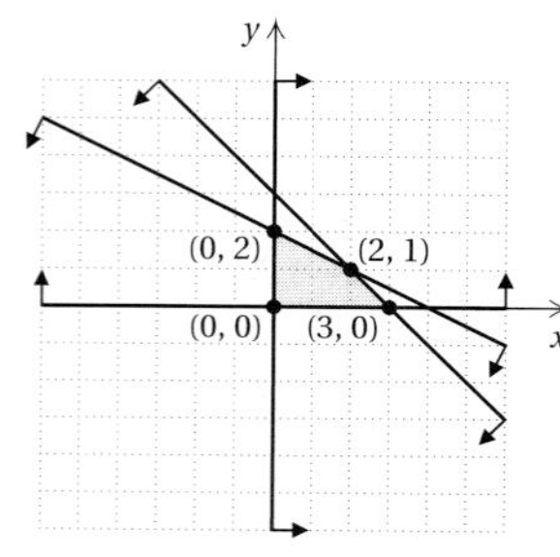

Exercise Set H, p. 1037

1.

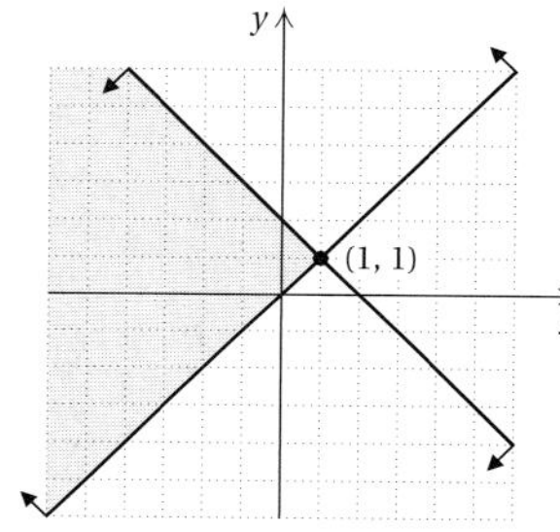

3.

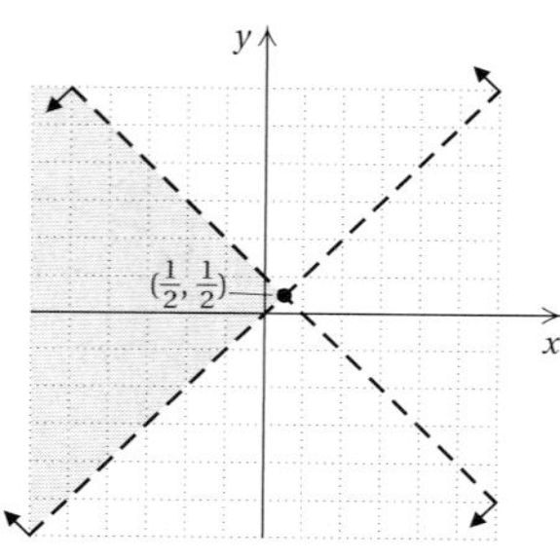

5.

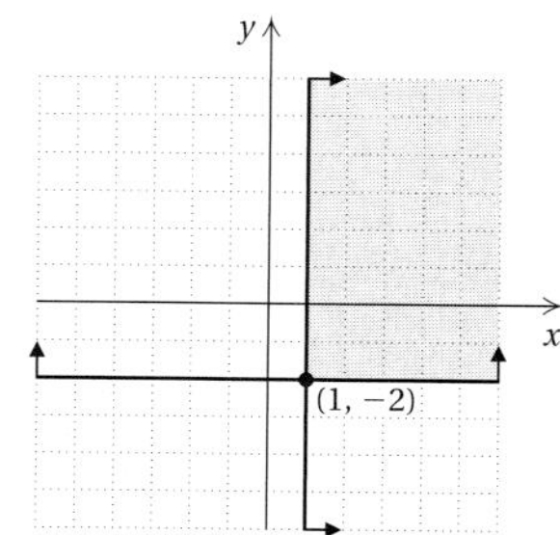

7.

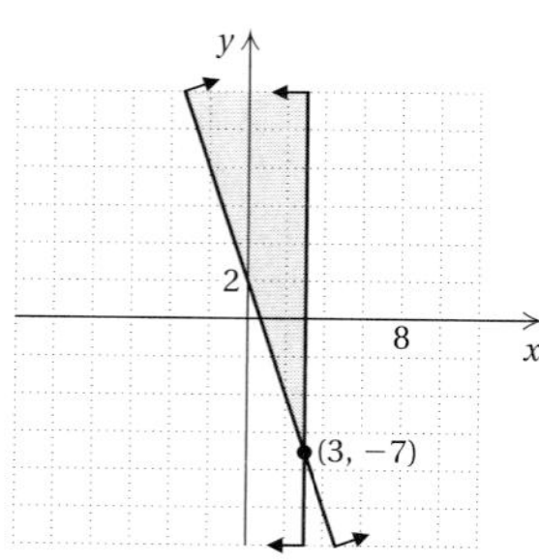

9.

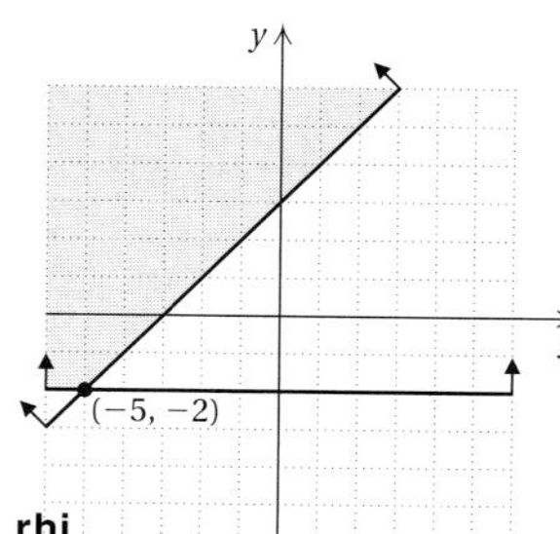

rhi

11.

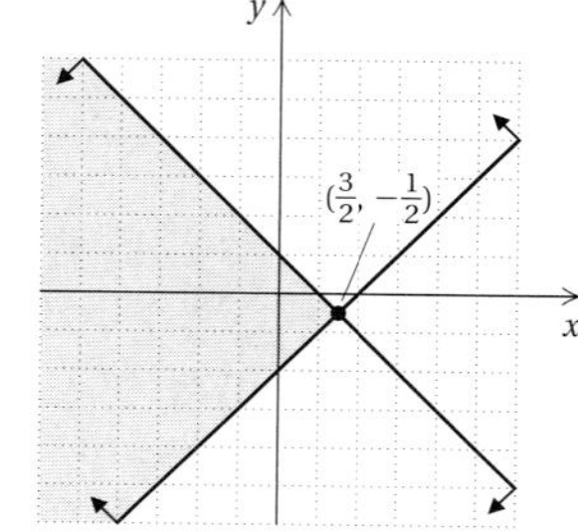

13.

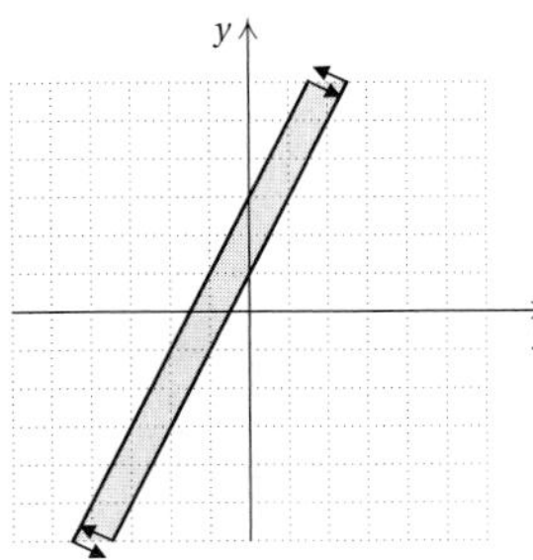

15.

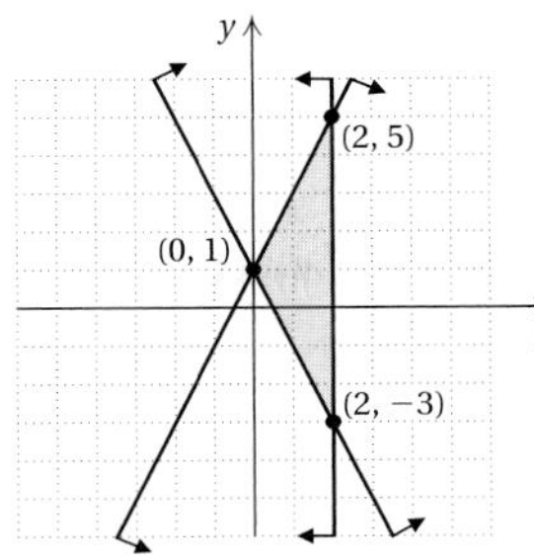

17.

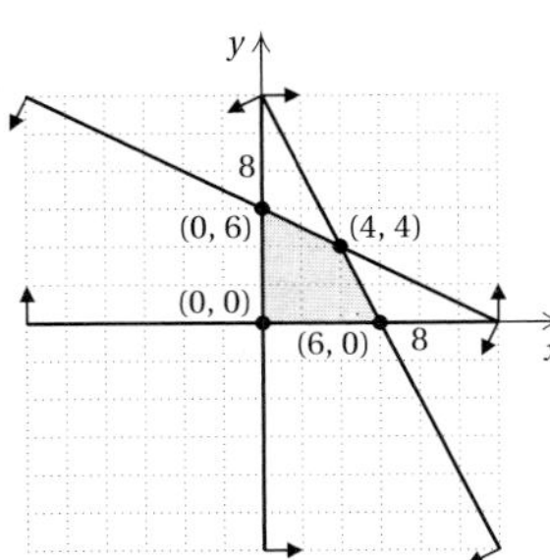

19.

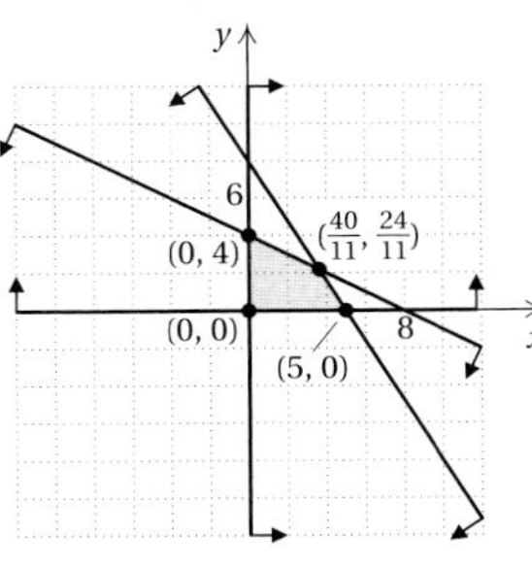

21. (a)

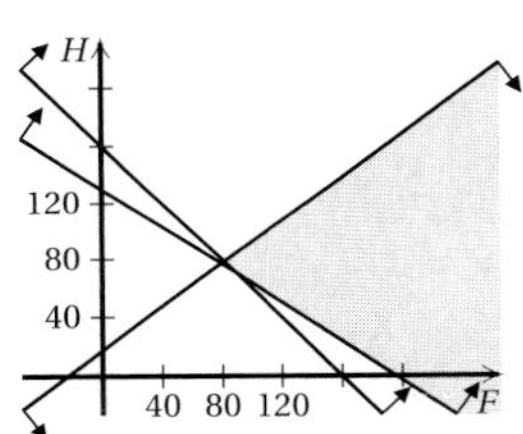

(b) no

Margin Exercises, Appendix I, pp. 1040–1042

1. $\frac{82}{100}$ **2.** 14.7%, 50.4% **3.** $\frac{1}{2}$ **4. (a)** $\frac{1}{13}$; **(b)** $\frac{1}{4}$; **(c)** $\frac{1}{2}$; **(d)** $\frac{2}{13}$ **5.** $\frac{6}{11}$ **6.** 0 **7.** 1

Exercise Set I, p. 1043

1. 0.57, 0.43 **3.** 0.075, 0.134, 0.057, 0.071, 0.030 **5.** 0.633 **7.** 52 **9.** $\frac{1}{4}$ **11.** $\frac{1}{2}$ **13.** $\frac{2}{13}$ **15.** $\frac{2}{7}$ **17.** 0 **19.** $\frac{5}{36}$ **21.** $\frac{5}{36}$ **23.** $\frac{1}{36}$

Margin Exercises, Appendix J, pp. 1045–1048

1. 21 **2.** 19 **3.** 22 **4.** 17 **5.** J **6.** 11

7.

8.

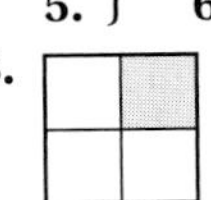

9.

10.

11. 35 **12.** 70 **13.** Maria

Exercise Set J, p. 1049

1. 19 **3.** V **5.** 43 **7.**

9.

11. 17 **13.** Bev **15.** Chris

Margin Exercises, Appendix K, pp. 1052

1. \$107 **2.** \$809.92

Exercise Set K, p. 1053

1. \$12 **3.** \$10.91 **5.** \$611.50 **7.** \$727.50 **9.** 28.4 mpg **11.** 137,500 lb **13.** \$31.80 **15.** \$83.93 **17.** \$591.75 **19.** 4 ft

Appendix L, p. 1055

1. B **2.** B **3.** D **4.** A **5.** C **6.** B **7.** C **8.** D **9.** C **10.** C **11.** A **12.** A **13.** D **14.** D **15.** B **16.** A **17.** B **18.** C **19.** A **20.** B **21.** C **22.** D **23.** B **24.** A **25.** C **26.** A **27.** D **28.** A **29.** B **30.** D **31.** C **32.** B **33.** A **34.** D **35.** A **36.** C **37.** D **38.** C **39.** C **40.** C **41.** B **42.** C **43.** A **44.** B **45.** A **46.** B

Index

A

E

F

G

H

I

K

L

M

S

T

U